AF526888

Question Bank on Agriculture for Competitive Exams

A Handbook for Pre PG (Ag) Test, JRF, SRF, ARS, NET, SET, PhD, ASRB, IRMA, UPSC, NABARD, Cooperative Bank, PO, BEd, Management and Allied Agricultural Examinations

Revised and Enlarged Third Edition

Neeraj Pratap Singh
Horticulture Executive
Planning and Architecture
Sahara India Commercial Corporation Limited
Sahara India Centre, 9th Floor, 2 Kapoorthala Complex
Aliganj, Lucknow 226 024 (UP) India

CBS Publishers & Distributors Pvt Ltd

New Delhi • Bengaluru • Chennai • Kochi • Kolkata • Lucknow • Mumbai
Gujarat • Hyderabad • Jharkhand • Nagpur • Patna • Pune • Uttarakhand

Disclaimer

Science and technology are constantly changing fields. New research and experience broaden the scope of information and knowledge. The author has tried his best in giving information available to him while preparing the material for this book. Although all efforts have been made to ensure optimum accuracy of the material, yet it is quite possible some errors might have been left uncorrected. The publisher, the printer and the author will not be held responsible for any inadvertent errors or inaccuracies.

Question Bank on Agriculture for Competitive Exams

ISBN: 978-81-239-2436-6

Copyright © Author and Publisher

Revised and Enlarged Third Edition: 2006

Reprint: 2007, 2008, 2010, 2012, 2014, 2015, 2016, 2017, 2019, 2020, 2021, 2023, 2024, **2026**

First Edition: 2004

Second Revised and Enlarged Edition: 2005

All rights reserved. No part of this book may be reproduced or transmitted in any form or by any means, electronic or mechanical, including photocopying, recording, or any information storage and retrieval system without permission, in written, from the author and the publisher.

Published by Satish Kumar Jain and produced by Varun Jain for

CBS Publishers & Distributors Pvt Ltd

4819/XI Prahlad Street, 24 Ansari Road, Daryaganj, New Delhi 110 002, India

Ph: 011-23289259, 23266838
Website: www.cbspd.com
e-mail: delhi@cbspd.com

Corporate Office: 204 FIE, Industrial Area, Patparganj, Delhi 110 092, India

Ph: 0114934 4934 Fax: 011-4934 4935 e-mail: publishing@cbspd.com; publicity@cbspd.com

Branches

- **Bengaluru:** Seema House 2975, 17th Cross, KR Road, Banasankari 2nd Stage, Bengaluru 560 070, Karnataka, India
  Ph: +91-80-26771678/79 Fax: +91-80-26771680 e-mail: bangalore@cbspd.com
- **Chennai:** 18/8B, Subbarayan Street, Shenoy Nagar, Chennai 600 030, Tamil Nadu, India
  Ph: +91-44-42032115, 45022115 e-mail: chennai@cbspd.com
- **Kochi:** 42/1325, 1326, Power House Road, Opposite KSEB, Kochi 682 018, Kerala, India
  Ph: +91-484-4059061-65 Fax: +91-484-4059065 e-mail: kochi@cbspd.com
- **Kolkata:** 147, Hind Ceramics Compound, 1st Floor, Nilgunj Road, Belghoria, Kolkata 700 056, West Bengal, India
  Ph: +91-33-25633055/56 e-mail: kolkata@cbspd.com
- **Lucknow:** Basement, Khushnuma Complex, 7-Meerabai Marg (behind Jawahar Bhawan), Lucknow 226 001, UP, India
  Ph: +91-522-4000032 e-mail: tiwari.lucknow@cbspd.com
- **Mumbai:** PWD Shed, Gala no. 25/26, Ramchandra Bhatt Marg, Next to JJ Hospital Gate no. 2, Opp. Union Bank of India, Noorbaug Mumbai 400 009, Maharashtra, India
  Ph: +91-22-66661880/89 e-mail: mumbai@cbspd.com

Representatives

- **Gujarat** • **Hyderabad** • **Jharkhand** • **Nagpur** • **Patna** • **Pune** • **Uttarakhand**

For trade terms please contact customercare@cbspd.com

For general enquiries please contact info@cbspd.com

Printed at: SRK Graphics, Delhi, India

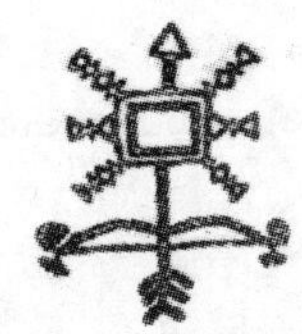

ॐ श्री हनुमते नमः

सब कुछ तेरा

क्या है मेरा

तेरा तुझको अर्पण

Preface to the Third Edition

It gives me immense pleasure to know that my readers warmly welcomed the second edition. These overwhelming responses encourage me to present the third edition of the book.

As the time is not static, information also keeps on changing, this requires you to be on your toes and update your knowledge and information from time to time. So it has the relative emphasis upon other segments of the Agriculture. For the purpose new chapters on Ecology & Environment and Seed Science have been incorporated.

The book has been extensively revised in the light of valued suggestions and proper guidance of University Teachers, Subject Matter Specialists, Publishers and Successful Candidates of Agricultural Competitive Exams.

I continue my efforts to strengthen your preparedness by inducting latest information and developments in the field of Agricultural Sciences. This package will make your knowledge more wider and serve well for the foundation as well as advanced studies required in the exams. I am sure, this edition will turn your vision into action and make a positive impact on your strategy towards the preparation for the forthcoming Agricultural Competitive Exams.

On the Eve Ram Navami **Neeraj Pratap Singh**

Preface to the First Edition

In the recent years competitive examinations have become a routine procedure of recruitment and admission for initial and higher placements. Most of the professional and technical education institutes, organizations and universities have adopted pre-admission tests. Government services agencies too have made it pre-requisite for appointments now-a-days. The questions in these tests are of objective type, large in number and of varied nature. These require quick answering within a specified period of time to secure high merit and assured success. A subject like agriculture encompasses a large number of disciplines and one has to scan through a number of text-books, journals, magazines and periodicals to come out triumphant in a competitive test. Thus, the examinee has to burn mid-night oil in true literary sense. A vacuum is felt regarding the availability of quality oriented agriculture competition books that quench the thirst of students. Keeping the aforesaid points in mind, an earnest effort has been made to craft 'Question Bank on Agriculture for Competitive Exams' on various aspects of general agriculture.

The book is framed holding in view the pattern of various competitive examinations viz. Pre PG (Ag.) Test, JRF, SRF, ARS, NET, SET, Ph.D., ASRB, IRMA, UPSC, NABARD, Co-operative Bank, PO, B.Ed., MANAGE and Allied Agricultural Exams. The approach of the book is to treat agriculture as a science and not merely as a mundane necessity of our day-to-day existence. In this book, several important facets of agriculture are dragged in a schematic way. After engulfing the book, students stride into the examination with clear concept and confidence and ultimately will become the winner.

I shall feel gratified if my sustained and specialized efforts give you just the best, which will be reflected through your enduring satisfaction. Constructive criticism and useful suggestions are cordially invited from the readers to improve upon the contents and scope of the publication in future. I am heartily grateful to publisher and the printer of this book whose sincere effort made it available to you in a short time.

KISAAN DAY **Neeraj Pratap Singh**

Think it over

Be like the bee...

Be like the bee which goes from one flower to another, savouring the nectar from each of them. A fly, on the other hand sits on a garbage, and then flies to a sweetmeat shop and again to the gutter and back to the rose garden. The fly is alternating between goodness and toxin. But the bee picks and chooses from what is available and is attracted to the good alone and hence flies from the flower to another. Being like a bee in life, will enable us to move from one form of goodness to another, and we will be attracted to all the sacredness of existence.

Positive thinking is vital to the processes that are guaranteed, to keep your spirit always high, so high indeed that discouragement cannot break it. So, try to think the best, not the worst, because in the long run, you tend to get what you think.

- Neeraj Pratap Singh

Contents

Keys

1. Boost-up your Agriculture Knowledge

Important Abbreviations and their Full Forms

Abbreviation	Full Form
AAR	Accumulated Assured Rainfall
ADB	Asian Development Bank
AGRIS	International Information System for Agricultural Sciences and Technology
AHRD	Agricultural Human Resource Development Project
AICRP	All India Co-ordinated Research Project
AIFPA	All India Food Preserver's Association, Delhi
APAARI	Asia-Pacific Association for Agricultural Research Institutions
APEDA	Agriculture and Processed foods product Export Development Authority
ARIC	Agricultural Research Information Centre
ASCII	American Standard Code for Information Interchange
ASI	Anthropological Survey of India, Kolkata
ASRIC	Asian Small Ruminant Information Centre
BARC	Bhaba Atomic Reseach Centre
BASIC	Beginners All Purpose Symbolic Instruction Code
BIS	Bureau of Indian Standards, New Delhi
BIOS	Basic Input Output System
BSI	Botanical Survey of India, Kolkata
Bit	Binary Digit (1 Byte = 8 Bits)
CABI	Centre for Agriculture & Biosciences International
CASAS	Commonwealth Association of Scientific Agricultural Societies
CEP	Commission on Environmental Planning
CEQ	Council of Environmental Quality
CFTRI	Central Food Technological Research Institute, Mysore
CGIAR	Consultive Group of International Agricultural Research, Washington, D.C. (U.S.A.)
CGPRT	Centre for Research and Development of Course Grains, Pulses, Roots and Tuber Crops
CIPET	Central Institute of Post harvest Engineering and Technology, Ludhiana (Punjab)
CPCB	Central Pollution Control Board (India)
CRIDA	Central Research Institute for Dryland Agriculture, Hyderabad (A.P.)
CRRI	Central Rice Research Institute, Cuttack
CSIR	Council of Scientific & Industrial Research
CTBT	Comprehensive Test Ban Treaty
DARE	Department of Agricultural Research & Education
DAVP	Directorate of Advertising & Visual Publicity
DOAC	Department of Agricultural and Co-operation
DRDO	Defence Research & Development Organisation, Hyderabad
DTH	Direct To Home
DWCRA	Development of Women & Children in Rural Areas
ESMA	Essential Service Maintenance Act
FAI	Fertilizer Assocation of India
FCI	Food Corporation of India
FICCI	Federation of India Chambers of Commerce & Industry
FORTRAN	Formula Translation
FRI	Forest Research Institute, Dehradun
FSI	Forest Survey of India, Dehradun
GATT	General Agreement on Tariffs & Trade (now it is WTO from Ist Jan, 95)
GIS	Geographical Information System
GPS	Global Positioning System

GS

The Khasis revolted in 1829-32 under Tirat Singh and Bar Manik.

GSI	Geological Survey of India, Kolkata
GSLV	Geo-Synchronous Satellite Launching Vehicle
html	Hyper Text Markup Language
http	Hyper Text Transfer Protocol
HYV	High Yielding Vareity
IBM	Internation Business Machine
ICICI	Industrial Credit and Investment Coorporation of India
IDBI	Industrial Development Bank of India
IFCI	Industrial Financial Corporation of India
IFFCO	Indian Farmers' Fertiliser Cooperative Ltd.
IGRMS	Indira Gandhi Rashtriya Manav Sangrahalaya, Bhopal
IMA	Indian Military Academy, Dehradun
IMF	International Monetary Fund
IPR	Intellectual Property Rights
IREDA	Indian Renewable Energy Development Agency
ISNAR	International Service for National Agricultural Research
IVLP	Institute Village Linkage Programme
JRSY	Jawahar Rojgar Smiridi Yojana
KRIBHCO	Krishak Bharathi Cooperative Ltd.
KVK	Krishi Vigyan Kendra
LAN	Local Area Network
MANAGE	National Institute of Agricultural Extension Management, Hyderabad)
MNC	Multi National Corporation
NAAS	National Academy of Agricultural Sciences (New Delhi)
NABARD	National Bank for Agriculture and Rural Development
NACA	Network of Aquaculture Centres of Asia
NAFED	National Agricultural Coperative & Marketing Federation of India Ltd.
NAL	National Agricultural Library (IARI, Pusa)
NARP	National Agricultural Research Project
NATP	National Agricultural Technology Project
NCPGR	National Centre for Plant Genomic Research, New Delhi
NDA	National Defence Academy, Khadakwasla near Pune
NDDB	National Dairy Development Board
NIAM	National Institute for Agricultural Marketing (Jaipur)
NIRD	National Institute of Rural Development (Hyderabad)
NNRMS	National Natural Resources Management Systems
NPA	National Policy on Agriculture
NRSA	National Remote Sensing Agency, Hyderabad
PBR	Plant Breeders Rights
Pixel	Picture Element
PRA	Participatory Rural Appraisal
PSLV	Polar Satellite Launching Vehicle
PUFA	Poly Unsaturated Fatty Acids
RAM	Random Access Memory
RAWEP	Rural Agricultural Work Experience Programme
RLEGP	Rural Landless Employment Guarantee Programme
ROM	Read Only Memory
RRA	Rapid Rural Appraisal
RVP	River Valley Project
SIDBI	Small Industrial Development Bank of India
SNF	Solids Not Fat

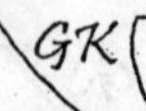

GK

The first Indian postal stamp was issued in 1852.

GS

STEP	Statellite Telecommunication Experiment Project
TCP/IP	Transmission Control Protocol / Internet Protocol
TPS	True Potato Seed
TRIMs	Treade Related Investment Measures
TRIPS	Trade Related Intellectual Property Rights
TRYSEM	Training of Rural Youth for Self Employment
TTC	Trainers Training Centre
UNDP	United National Development Programme
USAID	United States Agency for International Development
USDA	United States Department of Agriculture
WAN	Wide Area Network
WAP	Wireless Amplification Protocol
WMO	World Meterological Organization (of UNO)
www	World Wide Web
ZSI	Zoological Survey of India, Kolkata
WAN	Wide Area Network
WAP	Wireless Amplification Protocol
WMO	World Meterological Organization (of UNO)

Chronology of Agriculture & Related Fields

1875	Indian Meterological Department (IMD) at Pune
1881	First Department of Agriculture
1889	Imperial Bacteriological Laboratory (now IVRI) at Mukteshwar
1890	Foundation for Agricultural Research laid by Dr. J.A. Voelcker
1892	First appointment: Imperial Agricultural Chemist
1905	Imperial Agricultural Research Institute (now Indian Agricultural Research Institute, New Delhi) started at Pusa, Bihar
1928	Indian Council of Agricultural Research was established on the recommendation of Royal Commission on Agriculture
1929	Imperial Council of Agricultural Research on July 16
1943	Great Bengal Famine (caused by *Helminthosporium oryzae*)
1947	Grow More Food (GMF)
1948	Imperial Council of Agricultural Research to Indian Council of Agricultural Research, June 10
—	First Educational Commission headed by Dr. Radha Krishnan recommended establishment of Rural Universities
1952	Community Development Programme(CDP)
1953	National Extension Service (NES)
1956	Balwant Roy Mehta : Panchayati Raj (3-tier)
1960	Agricultural Universities in India are based on the recommendations of the Indo-American Team headed by Dr. M.S. Randhawa
1960	First Agricultural University at Pantnagar (G.B. Pant University of Agriculture & Technology) patterned on the Land-Grant system of USA
—	Dr. Ralph W. Cummings prepared blue print for establishment of Agricultural Universities in different States of India
—	Intensive Agricultural District Programme (IADP)
1964	Intensive Agricultural Area Programme (IAAP)
1965	First Scientist Vice President of ICAR, Dr. B.P. Pal
—	High Yielding Varieties Programme (HYVP) (Green revolution)
1970-71	Small Farmers Development Agency (SFDA)
—	Marginal Farmers Agricultural Labourers Programme (MFALP)
—	Intergrated Development of Dry Areas (IDDA)
1971	Drought Prone Area Programme (DPAP)

Sati was outlawed in 1829 by William Bentick.

1973	Agricultural Research Services (ARS) started by Dr. M.S. Swaminathan
1974	Command Area Development Authority (CADA)
—	Training & Visiting system (T & V system)
1975	Agricultural Scientist Recruitment Board (ASRB) based on Gajendra Gadkar Committee recommendations
1976	National Academy of Agricultural Research & Management, original name: Central Staff College for Agriculture.
1977	Food for Work Programme (FWP)
—	National Rural Extension Programme (NREP)
—	Antyodya Yojana
—	Desert Development Programme (DDP)
—	Ashok Mehta committee proposed two tier Panchayati Raj system
1978	Integrated Rural Development Programme (IRDP)
1979	Training of Rural Youth for Self Employment (TRYSEM)
1982	Development of Women & Children in Rural Areas (DWCRA)
—	National Rural Employment Programme (NREP)
1985	Crop Insurance Scheme (CIS)
1986	Technology Mission on Oil seeds (TWO)
—	Integrated Watershed Development Programmes
—	National Watershed Development Programmes for Rainfed Areas (NWDPRA)
1989	Jawahar Rojgar Yojana (JRY)
1993	Prime Minister Rojgar Yojana (new name of JRY) (PRMY)
1999	National Agricultural Insurance Scheme (new name of CIS) (NAIS)

About ICAR

Chairman of ICAR (General Body)	:	Director General
President of ICAR	:	Union Minister of Agriculture (Cabinet)
First President of ICAR	:	Habibulla
First Vice-President (now DG) of ICAR	:	Dr. B.P.Pal
First Director of IARI	:	Dr. Vishwanathan
First Agriculture Minister of independent India	:	Rafi Ahmed Kidwai
First All India Co-ordinated Research Project	:	on Maize
First National Research Centre (NRC)	:	for Groundnut
First KVK in India	:	Pondicherry by Mohan Singh
Number of Deemed Universities	:	4
Number of National Bureaux	:	5
Number of Project Directorates	:	10 (appx.)
Number of Central Research Centres	:	27 (appx.)
Number of National Research Centres	:	27 (appx.)
Number of AICRPs	:	84 (appx.)
Number of KVKs	:	376 (till 2004)
Number of Trainers Training Centres	:	8 (appx.)
Number of State Agricultural Universities in India	:	37
Number of Central Agricultural University in India	:	1

ICAR Technological Projects

National Demonstration Project (NDP): started in the year 1964

Operational Research Project (ORP): started in the year 1974-75

Krishi Vigyan Kendra (KVK) also known as Farm Science Centres: started in the year 1974

Lab to Land Programme (LLP): started in the year 1979 (ICAR Golden jubilee year)

National Agricultural Technology Project (NATP): started in the year 1998-99

Sind was annexed to the British empire in 1843.

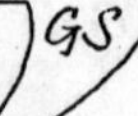

ICAR Publications

Journals	:	2 (Monthly, English) (i) The Indian Journal of Agricultural Sciences (ii) The Indian Journal of Animal Sciences
Semi-technical Journals	:	2 (Quarterly, English) (i) Indian Farming (ii) Indian Horticulture
Semi-technical Journals	:	1 (Monthly, Hindi) (i) Kheti
ICAR News	:	A science and technology news letter (every three months)
ICAR Reporter	:	House journal (every three months)

ICAR Awards

Award Name	Field	Periodicity
Hari Om Ashram Trust Award	Published research in Crop Science, Horticulture and Animal sciences	Biannual
Vasantrao Naik Award	Water Conservation & Dry land farming	Annual
Outstanding Women Agricultural Scientist Award	To encourage ICAR woman scientists	Annual
Jawaharlal Nehru Award	Best Ph.D. in the field of Agricultural & Allied Sciences	Annual
Rafi Ahmed Kidwai Memorial Award	Agriculture, Animal Husbandry and Allied Sciences	Two years
Fakruddin Ali Ahmed Award	Agricultural Research in Tribal area	Two years
ICAR Young Scientist	In the field of Agricultural & Allied Sciences	Two years
Outstanding Teacher Award (for 8 personalities)	Agricultural & Allied fields	Two years
ICAR Team Award	For disciplinary research work in Agricultural & Allied fields	Three years

- For Green Revolution, the ICAR got 'King Baudouin International Development Prize' in 1988.
- National Gene Bank at NBPGR, New Delhi holds more than 0.22 million accessions of plants.
- Number of patents by ICAR (till 31st July, 2002) : 50 (appx.)

Important ICAR Institutions and their Location

Institute	Full name	Place	State
Crop Science			
CICR	Central Institute for Cotton Research	Nagpur	Maharashtra
CRIJAF	Central Research Institute for Jute & Allied Fibres	Barrackpore	W. Bengal
CRRI	Central Rice Research Institute	Cuttack	Orissa
CTRI	Central Tobacco Research Institute	Rajahmundry	Andhra Pradesh
IARI	Indian Agricultural Research Institue	New Delhi	Delhi
IGFRI	Indian Grassland & Fodder Research Institute	Jhansi	Uttar Pradesh
IIPR	Indian Institute of Pulse Research	Kanpur	Uttar Pradesh
IISR	Indian Institute of Sugarcane Research	Lucknow	Uttar Pradesh
NBPGR	National Bureau of Plant Genetic Resources	New Delhi	Delhi
NCIPM	National Centre for Integrated Pest Management	New Delhi	Delhi
SBI	Sugarcane Breeding Institute	Coimbatore	Tamil Nadu
VPKAS	Vivekananda Parvatiya Krishi Anusandhan Sansthan	Almora	Uttaranchal
NRCG	National Research Centre for Groundnut	Junagarh	Gujarat

In 1856, the government passed the widow remarriage act.

GS

NRCPB	National Research Centre on Plant Bio-Technology	New Delhi	Delhi
NRCRM	National Research Centre on Rapeseed Mustard	Bharatpur	Rajasthan
NRCS	National Research Centre on Soybean	Indore	Madhya Pradesh
NRCS	National Research Centre on Sorghum	Hyderabad	Andhra Pradesh
NRCDF	National Research Centre for DNA Finger Printing	New Delhi	Delhi
DRR	Directorate of Rice Research	Hyderabad	Andhra Pradesh
DOR	Directorate of Oilseed Research	Hyderabad	Andhra Pradesh
DWR	Directorate of Wheat Research	Karnal	Haryana
DMR	Directorate of Maize Research	New Delhi	Delhi
DWMR	Directorate of Water Management Research	Patna	Bihar
PDBC	Project Directorate of Biological Control	Bangalore	Karnataka
ICARRCG	ICAR Research Complex for Goa	Goa	Goa
ICAR-RCNER	ICAR Research Complex for North-East Region	Barapani	Meghalaya
Horticulture			
CISTH	Central Institute of Subtropical Horticulture	Lucknow	Uttar Pradesh
CITH	Central Institute of Temperate Horticulture	Srinagar	J & K
CTCRI	Central Tuber Crops Research Institute	Trivendrum	Kerala
CPCRI	Central Plantation Crops Research Institute	Kasaragod	Kerala
CPRI	Central Potato Research Institute	Shimla	Himachal
IIHR	Indian Institute of Horticultural Research	Bangalore	Karnataka
IISR	Indian Institute of Spices Research	Calicut	Kerala
CIAH	Central Institute for Arid Horticulture	Bikaner	Rajasthan
CRIC	Central Research Institute for Chikoo	Muzaffarpur	Uttar Pradesh
NRCB	National Research Centre for Banana	Tiruchirapalli	Tamil Nadu
NRCC	National Research Centre for Cashew	Puttur	Karnataka
NRCC	National Research Centre for Citrus	Nagpur	Maharashtra
NRCG	National Research Centre for Grapes	Pune	Maharashtra
NRCMAP	National Research Centre for Medicinal & Aromatic Plants	Anand	Gujarat
NRCM	National Research Centre for Mushroom	Solan	Himachal
NRCO	National Research Centre for Onion & Garlic	Pune	Maharashtra
NRCO	National Research Centre for Orchids	Gangtok	Sikkim
IIVR	Indian Institute of Vegetable Research	Varanasi	Uttar Pradesh
Soil Science & Agronomy			
CARI	Central Agricultural Research Institute	Port Blair	Andaman
CAZRI	Central Arid Zone Research Institute	Jodhpur	Rajasthan
CRIDA	Central Research Institute for Dryland Agriculture	Hyderabad	Andhra Pradesh
CSWCRTI	Central Soil & Water Conservation Research and Training Institute	Dehradun	Uttaranchal
CSSRI	Central Soil Salinity Research Institute	Karnal	Haryana
IISS	Indian Institute of Soil Science	Bhopal	Madhya Pradesh
NBSS & LUP	National Bureal of Soil Survey & Land Use Planning	Nagpur	Maharashtra
NRCAF	National Research Centre for Agro-forestry	Jhansi	Uttar Pradesh
NRCWS	National Research Centre for Weed Sciences	Jabalpur	Madhya Pradesh
NRCWA	National Research Centre for Women in Agriculture	Bhubaneshwar	Orissa
PDCSR	Project Directorate for Cropping Systems Research	Modipuram	Uttar Pradesh
DWMR	Directorate of Water Management Research	Patna	Bihar

GK

The first jute mill was started in Rishra in 1855.

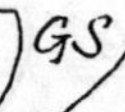

GS

Agricultural Engineering			
CIAE	Central Institute of Agricultural Engineering	Bhopal	Madhya Pradesh
CIPHET	Central Institute of Post Harvest Engineering and Technology	Ludhiana	Punjab
CIRCOT	Central Institute for Research on Cotton Technology	Mumbai	Maharashtra
ILRI	Indian Lac Research Institute	Ranchi	Jharkhand
NIRJAFT	National Institute for Research on Jute & Allied Fibres Technology	Kolkata	W. Bengal
Animal Sciences			
CARI	Central Avian Research Institute	Izatnagar	Uttar Pradesh
CIRB	Central Institute for Research on Buffaloes	Hisar	Haryana
CIRG	Central Institute for Research on Goats	Mathura	Uttar Pradesh
CSWRI	Central Sheep and Wool Research Institute	Avikanagar	Rajasthan
IVRI	Indian Veterinary Research Institute	Izatnagar	Uttar Pradesh
HSADL	High Security Animal Diseases Laboratory	Bhopal	Madhya Pradesh
NBAGR	National Bureau of Animal Genetic Resources	Karnal	Haryana
NDRI	National Dairy Research Institute	Karnal	Haryana
NIAGR	National Institute of Animal Genetic Research	Karnal	Haryana
NEBCAP	National Embryo Bio-Technology Centre for Animal Production	Karnal	Haryana
NIANP	National Institute of Animal Nutrition and Physiology	Bangalore	Karnataka
NRCC	National Research Centre on Camel	Bikaner	Rajasthan
NRCE	National Research Centre on Equine	Hissar	Haryana
NRCMP	National Research Centre for Meat & Poultry	Izatnagar	Uttar Pradesh
NRCM	National Research Centre on Mithun	Kohima	Nagaland
NRCY	National Research Centre on Yak	Dirang	Arunachal
PDC	Project Directorate on Cattle	Modipuram	Uttar Pradesh
PDP	Project Directorate on Poultry	Hyderabad	Andhra Pradesh
Fisheries			
CIBA	Central Institute of Brackish Water Aquaculture	Chennai	Tamil Nadu
CICFRI	Central Inland Capture Fisheries Research Institute	Barrackpore	W. Bengal
CIFA	Central Institute of Freshwater Aquaculture	Bhubaneshwar	Orissa
CIFE	Central Institute of Fisheries Education	Mumbai	Maharashtra
CIFT	Cental Institute of Fisheries Technology	Cochin	Kerala
NBFGR	National Research Centre on Coldwater Fisheries	Nainital	Uttaranchal
Others			
IASRI	Indian Agricultural Statistical Research Institute	New Delhi	Delhi
NCAP	National Central of Agricultural Economics and Policy Research	New Delhi	Delhi
NAARM	National Academy of Agricultural Research and Management	Hyderabad	Andhra Pradesh

State Agricultural Universities

University	Full name	Place	State
ANGRAU	Acharya N.G. Ranga Agricultural University	Hyderabad	Andhra Pradesh
AAU	Assam Agricultural University	Jorhat	Assam
BAU	Birsa Agricultural University	Ranchi	Jharkhand
RAU	Rajendra Agricultural University	Hissar	Haryana
HPKVV	Himachal Pradesh Krishi Vishwa Vidhyala	Palampur	Himachal
YSPUH&F	Dr.Yashwant Rao Parmar University of Horticulture and Forestry	Solan	Himachal Pradesh
UAS	University of Agricultural Sciences	Bangalore & Dharwad	Karnataka
GAU	Gujarat Agricultural University	S.K.Nagar	Gujarat

In Faizabad Maulvi Ahmadullah led the revolt of 1857.

JNKVV	Jawaharlal Nehru Krishi Vishwa Vidhyalaya	Jabalpur	Madhya Pradesh
IGKVV	Indira Gandhi Krishi Vishwa Vidhyalaya	Raipur	Madhya Pradesh
KAU	Kerala Agricultural University	Thrissur	Kerala
KKV	Konkan Krishi Vidhyapeeth	Dapoli	Maharastra
MPKV	Mahatma Phule Krishi Vidhyapeeth	Rahuri	Maharastra
MAU	Marathwada Agricultural University	Parbhani	Maharastra
PKV	Dr.Panjabrao Deshmukh Krishi Vidhyapeeth	Akola	Maharastra
OUA&T	Orissa University of Agriculture and Technology	Bhubaneshwar	Orissa
PAU	Punjab Agricultural University	Ludhiana	Punjab
RAU	Rajasthan Agricultural University	Bikaner	Rajasthan
TNAU	Tamil Nadu Agricultural University	Coimbatore	Tamil Nadu
TNV&ASU	Tamil Nadu Veterinary and Animal Science University	Chennai	Tamil Nadu
CSAUA&T	Chandra Shekhar Azad University of Agriculture and Technology	Kanpur	Uttar Pradesh
GBPUA&T	Govind Ballabh Pant University of Agriculture and Technology	Pantnagar	Uttaranchal
NDUA&T	Narendra Dev University of Agriculture and Technology	Faizabad	Uttar Pradesh
BCKVV	Bidhan Chandra Krishi Vishwa Vidhyalaya	Kalyani	W. Bengal
SKUA&T	Sher-e-Kashmir University of Agricultural Science & Technology	Srinagar	Jammu & Kashmir
MPUA&T	Maharana Pratap University of Agricultural Science & Technology	Udaipur	Rajasthan
CAU	Central Agricultural University	Imphal	Manipur

- Maximum number of campuses under a single Agricultural University : JNKVV, Jabalpur.
- Central Agricultural University, Imphal is the only multi-state, Agricultural University with following colleges.

College of Agriculture	:	Imphal
College of Veterinary Science	:	Aizwal
College of Horticulture & Forestry	:	Arunachal Pradesh
College of Fisheries	:	Tripura
College of Agricultural Engineering & Post Harvest Technology	:	Sikkim
College of PG Studies	:	Barapani
College of Home Science	:	Meghalaya

Consultative Group on International Agricultual Research (CGIAR)

Institute	Full name	Place	State
IRRI	International Rice Research Institute	Manila	Phillipines
CIMMYT	Centro International de Mejorimiento de Maizy Trigo (International Centre for the Improvement of Maize and Wheat)	Mexico	Mexico
WARDA	West Africa Rice Development Association	Monrovia	Liberia
IPGRI	International Plant Genetic Resource Institute	Rome	Italy
ICRISAT	International Crops Research Institute for Semi-Arid Tropics	Hyderabad	India
ICARDA	International Centre for Agricultural Research in the Dry Area	Aleppo	Syria
CIAT	Centro International de Agricultura Tropical (International Centre for Tropical Agriculture)	Coli	Columbia
IITA	International Institue for Tropical Agriculture	Ibadan	Nigeria
IIMI	International Irrigation Management Institute	Colombo	Srilanka
ICRAF	International Centre for Research in Agro-Forestry	Nairobi	Kenya
CIFOR	Centre for International Forestry Research	Bogor	Indonesia
ILRI	Interational Livestock Research Institute	Nairobi	Kenya
ICLARM	International Centre for Living Aquatic Resources Management	Manila	Phillipines
CIP	Centro International de la Papa International Potato Centre	Lima	Peru

Benjamin Disraeli described the revolt of 1857 as national rising.

GS

Important Publications Related to Agriculture & Allied Fields

Journal/Magazine	Published by	Place
Intensive Agriculture (English)	Ministry of Agriculture	New Delhi
Unnat Krishi (Hindi)	Ministry of Agriculture	New Delhi
Agriculture Extension Review (English)	Ministry of Agriculture	New Delhi
Krishi Vistar Samiksha (Hindi)	Ministry of Agriculture	New Delhi
Agricultural Situation in India	Ministry of Agriculture	New Delhi
Agricultural Statistics at a glance	Ministry of Agriculture	New Delhi
Kurukeshetra	Ministry of Rural Development	New Delhi
Yojana	Ministry of Information & Broadcasting	New Delhi
Indian Farmers Digest	G.B. Pant Agril. & Tech. Univ.	Pantnagar
Annals of Agricultral Research	Indian Society of Agril. Sciences	New Delhi
Fertilizer News	Fertilizer Associations of India	New Delhi
Indian Journal of Dairy Science	Indian Dairy Association	New Delhi
Current Science	Current Science Association & Indian Academy of Sciences	Bangalore
Science Reporter	C.S.I.R.	New Delhi
Resonance	Indian Academy of Sciences	Bangalore
Mausam	Indian Meteorological Department	New Delhi

Basic Information Related to Agriculture and Allied Sciences In India

Average Annual Rainfall	:	1190 mm
Total Geographical Area	:	329 mha
The Generated Rainfall Volume	:	400 mha-m
The Generated Runoff Volume	:	185 mha-m
Maximum area under Irrigation	:	Ganga basin
Maximum average annual runoff occurs	:	Brahmaputra basin
Gross Irrigated Area	:	76 mha
Gross Cropped Area	:	57 mha
Total Cropped Area	:	185.48 mha
Net Cropped Area	:	142.51 mha
Area sown more than once	:	42.97 mha
Cropping Intensity	:	135.1%
Total degraded land	:	175.0 mha
Severely degraded land	:	65.0 mha
Total area under forest	:	79.52 mha (23.03%)
Optimum forest area required	:	33% of Geographical area
Total % area under forage crops	:	6.15% of Cropped area
The average food grain production	:	1552 kg/ha.
Net availability of food grain per annum	:	167.2 kg
Net availability of cereals per day	:	391 g/capita/day
Net availability of pulses per day	:	26.4 g/capita/day
Total area under horticultural crops	:	12 m ha
Area under horticultural crops	:	7% of Total Cropped area
Area under pasture cum grazing lands	:	11.30 mha (4% of Total Geographical area)
Area under cultivable waste lands	:	14.63 mha (5% of Total Geographical area)
Under fruit crops-maximum production	:	Banana
Under fruit crops-maximum area	:	Mango

In Charter Act of 1813, one lakh rupee was earmarked for education.

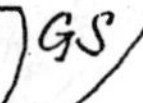

Total fruit production in India	:	47.5 mt
Total vegetable production in India	:	90.0 mt
Total Milk production in India	:	88 mt (15.2% of global production)
Milk availability in India	:	232 g / person / day
Average availability of milk in the world	:	300 g / person / day
Minimum requirement of milk	:	240 g / person / day
Area under minor irrigation	:	15.0 mha
Total area under drip irrigation	:	3,45,500 ha
Maximum area under drip irrigation	:	Maharashtra
Fertilizer consumption per unit of cropped area	:	92.1 kg/ha
Highest fertilizer consumption rate (State)	:	Punjab (181.0 kg/ha)
Lowest consumption rate of fertilizers	:	Andhra Pradesh (1.5 kg/ha)
Highest fertilizer consumption rate (U.T.)	:	Pondicherry 538 kg/ha
Total fertilizer consumption in India	:	17.0 mt
Total fertilizer consumption is maximum	:	Uttar Pradesh (includes Uttaranchal)
Irrigated area in India	:	40% of the total cultivated area.
Maximum area under irrigation	:	Punjab (92.7% of cultivated area)
Major source of irrigaiton in India	:	Canal
Number of electric pumps in India	:	1 million
Per capita availability of arable land	:	0.15 ha
Average size of operational holdings	:	1.57 ha
Maximum operational holding size	:	Nagaland (6.82 ha)
Minimum operational holding size	:	Kerala (0.33 ha)
Number of farm operational holdings	:	106.6 million
Annual soil loss due to erosion	:	5,334 million tonnes
Per capita water availability	:	2200 cu.m / person
Percentage of fresh water on the earth	:	2.8%
Agricultural & Livestock Census conducted	:	Every 5 years
Highest population below poverty line	:	Uttar Pradesh (529.89 lakh)
Total population below poverty line	:	2602.50 lakhs (26.10 %)
Population below poverty line is estimated	:	Every 5 years
Percentage of Agricultural exports to total exports	:	11.8 (in 2003-04)
Percentage of Agricultural imports to total imports	:	5.31 (in 2001)
Share of Agricultural & Allied sectors	:	25% of GDP at current prices

Agricultural Trade

Exports

Agricultural crops	Position
Rice	I
Cashew	II
Wheat and Spice	III
Tea	IV
Tobacco	V

Agricultural products	Position
Marine products	I
Rice	II
Oil meals	III
Cashew	IV
Wheat and Spice	V

GK GS

The army amalgamation scheme was passed in 1861.

Imports

Agricultural products	Position
Vegetable oil	I
Cashew	II
Pulses	III
Fruits and nuts	IV

Some Important Dates in Agriculture

11th March	Water Resource Day
21st March	World Forestry Day
1st May, 1960	Labour Day
5th June	World Environment Day
1-7th July	Van Mahotsava
17th July, 1976	Labour provident fund came into force
6th October	World Food Day
24th October, 1975	Bonded labour abolition law enforcement
4th December	Agriculture Women Day
23rd December	Kisan Day

Fathers of Different Concepts of Agricultural Sciences

Father of Sociology	Auguste Compte
Father of Botany	Theopharastus and Linnaeus
Father of Genetics	Gregeor John Mendel
Father of Pure Culture Technique	Oscar Brefeld
Father of Tissue Culture	Gottlieb Haberiandt
Father of Agricultural Chemistry	Justus von Liebig
Father of Indian Rust	K.C. Mehta
Father of Medical Bacteriology	Robert Koch
Father of Indian Plant Pathology	E.J. Butler
Father of Modern Plant Pathology	Anton de Bary
Father of Mycology	Pler A. Micheli
Father of Nematology	N.A. Cobb
Father of Bacteriology	Anton van Leewenhoeck
Father of Forest Pathology	Robert Haring
Father of Ornamental Gardening	M.S. Randhwa
Father of Statistics	R.A. Fisher
Father of Modern Tillage	Jethro Tull
Father of Microbiology	Louis Pasteur
Father of Plant Physiology	Stephen Hales

Revolutions

Rainbow Revolution	:	Agriculture
Black Revolution	:	Biofuel (Jatropha)
Green Revolution	:	Food grain production
Blue Revolution	:	Fisheries or Ocean lives production
White Revolution	:	Milk Production
Pink Revolution	:	Prawn production
Yellow Revolution	:	Oil seeds production
Brown Revolution	:	Non-Conventional energy sources

The first textile mill was started in Bombay in 1853.

GS

Red Revolution	:	Meat/tomato production
Golden Revolution	:	Fruit production
Silver Revolution	:	Egg/poultry production
Food-chain Revolution	:	Reduction in wastage of food grains, fruits & vegetables
Grey Revolution	:	Manures & Fertilizers
Round Revolution	:	Potato production
Prabhani Revolution	:	Okra production

Instruments and Their Use

Instrument	Use	Instrument	Use
Altimeter	Height	Lactometer	Milk Purity (Fat content) Density of milk
Aneroid barometer	Atmospheric pressure	Lysimeter	Evapotranspiration
Anemometer	Wind speed	Manometer	Pressure
Aerometer	Weight and density of air and gases	Planimeter	Area of an irregular figure
Current meter	Flow velocity in streams	Pyrometer	High temperature from distance
Crescograph	Growth of plant	Pyconometer	Specific gravity of soil
Cryometer	Measuring temperature below zero °C	Pyrheliometer	Amount of direct solar radiation
Dynamo	Converting Mechanical energy to Electrical	Pyranometer	Total incoming solar radiation
Cambel stokes recorder	Duration of sunshine	Sphegmomano-meter	Blood pressure
ECG (Electro Cardio Graph)	Movement of Heart	Tensiometer	Soil moisture tension
Fathometer	Depth of sea	Tachometer	Rotational speed
Galvanometer	Electric currents	Theodolite	Horizontal and vertical angles
Hydrometer	Measures relative density of liquids	Wind vane	Wind direction
Hygrometers (or) Psychrometer	Relative Humidity (Wet & Dry bulb thermometer)	Peizometer	Depth of water table

GENETICS & PLANT BREEDING

Quick Facts

- A single gene affecting more than one character is known as 'Pleiotropy'.
- Allogamy or Cross Pollination: Pollen grains from flowers of one plant transferred to receptive stigmas of flowers of another plant.
- Autogamy or Self Pollination: When pollen grains from an anther falls on receptive stigma of the same flower.
- Breeder's seed: Seed which is directly controlled by breeding institution or individual and which has the maximum genetic purity.
- C_3 plants have lower water use efficiency
- C_4 plants uses low concentration of CO_2.
- Carolus Linnaeus recognised the importance of sex in plants
- Chasmogamy: The flower opens but only after the pollination has taken place. e.g. Mung.
- Cleistogamy: Flower does not open at all, ensuring complete self pollination. e.g. Wheat
- Cytoplasmic genetic male sterility is exploited in hybrid seed production.
- Development of individual from an unfertilized egg is known as Parthenogenesis.
- Dioecious: Male and Female reproductive structures are found on separate plants.
- During metaphase the structure of chromosome can be best observed.
- Edible part of apple : Thalamus

Dastanbu was the name of Mirza Ghalib's account of the 1857 war in Delhi.

- Edible part of coconut : Endosperm
- Edible part of mango : Mesocarp
- Examples of C_3 type Plants : Rice, Wheat, Cotton, Soyabean, Potato etc.
- Examples of C_4 type Plants : Maize, Sorghum, Pearl millet, Sugarcane etc.
- Examples of CAM (Crassulaceae Acid Metabolism) type plants: Pineapple & Sisal.
- Geitonogamy: Pollen grains from flowers of one plant fall on the receptive stigmas of other flower of the same plant. eg. Maize.
- Ginger & Turmeric are examples of bulb
- Heterobeltosis: When F_1 is found superior to better parent.
- Heterosis: Occurrence of offspring more vigorous than either parent.
- Hugo de Vries defined the term Mutation.
- In CAM plants transpiration is negligible since stomata are closed during the day.
- Life was first originated in water
- Light saturation is high in C_4 plants compared to C_3 plants.
- Meiosis: Cell division in which one cell with two set of chromosomes gives rise to four cells with one set of chromosomes each.
- Mendel's laws of inheritance were rediscovered in 1900 by : de Varies & Tschermak
- Mitosis: Cell division in which one cell gives rise to two cells of genetically indentical to the cell which produced them.
- Monoecious: Male and Female reproductive structures are found on the same plant but in separate flowers.
- Pedigree record: Record that is used for maintaining relationship between parents and offspring.
- Robert Brown discovered nucleus of cell.
- Scientific name of bread wheat is *Triticum durum*.
- The inflorescence of Banana is known as spadix.
- The physical basis of life is protoplasm.
- The term Genetics was coined by : Bateson, W.
- The term heterosis was first used by : Mendel
- The theory of acquired characteristics was proposed by : Lamark
- The theory of genes in the chromosomes was introduced by : Morgan
- The theory of inheritance was proposed by : G. Mendal

Difference Between Prokaryotic Cell and Eukaryotic Cell

Prokaryotic cell	Eukaryotic cell
Nucleus is incipient. It lacks nuclear membrane, nucleolus and chromatin reticulum.	True nucleus. Nuclear membrane, nucleolus and chromatin network present.
DNA is circular and non-histone type, lacks protein association and lies in a tangled mass called nucleoid.	DNA is associated with protein to form chromosomes or chromatids.
All membrane bound structure *i.e.* chloroplast, ER, golgi complex, mitochondria, vacuoles are absent.	These structures are present.
Cell wall contains amino sugar and muramic acid.	These substances are absent in the cell wall.
Cytoplasmic streaming (cyclosis) is not observed.	It can be observed.
Endocytosis and Exocytosis are not found.	Both phenomenon are found.
Only 70S type of ribosome are found.	Both 70S and 80S ribosomes are found.
Mesosomes are found.	Mitochondria are found in place of mesosomes.
Flagella do not show 9+2 fibril arrangement.	Flagella comprise of 2 central and 9 peripheral fibrils.
Nucleosomes are microtubules are not present.	Both are present.
Cell wall and cell membrane do not fully separate during plasmolysis.	Cell membrane show complete shrinkage during plasmolysis.

A device used to measure height above sea level is known as Altimeter.

GS

Difference Between Plant and Animal Cell

Matter of difference	Plant cell	Animal cell
Cell wall	Present	Absent
Plastids	Present	Absent
Vacuoles	Present	Absent
Centrosomes	Absent	Present

HORTICULTURE

Area Production of Selected Fruit Corps

Crop	Area (mha)	Production (mt)
Mango	0.89	3.6
Guava	0.04	0.3
Banana	0.24	9.4
Citrus	0.28	2.3

Important Varieties of Selected Horticultural Crops

Variety	Remarks
Mango	
Arka Aruna	Banganapalli x Alphonso (High yields: 20-24 t/ha)
Mallika	Neelam x Dasheri
Amrapali	Dasheri x Neelam
Ratna	Neelam x Alphanso (Spongy tissue disease resistant)
Sindhu	Ratna x Alphanso (Seedless)
Grape	
Arkavati, Arka Neelmani, Dilkush	
Thompson	Seedless
Citrus	
Pramalini, Vikram	
Papaya	
Pusa Delicious, Pusa Majestic	
Guava	
Sardar (Lucknow-49), Safeda	
Rose	
Mrinalini, Dr.B.P.Pal, Jawahar, Priyadarshini, Dr. M.S. Randhawa, Chitra	Thornless variety
Chrysanthemum	
Indira, Rakhee, Shanti, Vasanti, Red gold	

Pest/Diseases in Major Horticultural Crops

Horticultural Crop	Major pests/diseases
Citrus	Citrus Decline, Cankar, Mealy Bug, Lemon Butterfly
Grapes	Powdery Mildew, Downy Mildew
Guava	Wilt
Mango	Malformation, Spongy Tissue, Stone Weevil, Leaf Spot
Potato	Black Heart (Physiological disorder), Late Blight, Wart
Coconut	Rhinoceros Beetle, Tanjore Wilt

Goa was given the status of state in 1987.

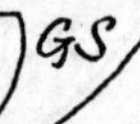

PLANT PATHOLOGY

Quick Facts

- Bacterial Technique was given by Dr. Robert Koch.
- Late Blight of potato caused by *Phytophthora infestans.*
- *Phytophthora infestans* caused Ireland femine in 1845.
- Brown spot of rice is caused by *Helminthosporium oryzae.*
- *Helminthosporium oryzae* caused Bengal femine in 1942.
- Coffee rust occurred in Sri Lanka in 1867.
- Maximum number of plant parasitic nematode belongs to *Tylenchida* order.
- Loose smut of wheat is internally seed-borne disease.
- Indian Phytopathological Society was established by E.J. Butler.
- Mycoplasma disease was first reported by Doi *et al.* and Ishiie *et al.*
- First plant parasitic bacteria was reported by T.J. Burrill.
- First plant parasitic bacteria was Fire blight of apple.
- First plant parasitic nematode was reported by Needham.
- First plant parasitic nematode was *Anguina tritici.*
- Stanley did crystallization of viruses.
- Antibiotic Penicillin was discovered by Alexander Flamming.
- Example of obligate parasite is *Albugo candida.*
- Green ear disease is soil and air-borne disease.
- Life cycle of wheat rust is given by K.C. Mehta.
- Systemic fungicide is discovered by Van Schleming and Kulka.
- Yellow mosaic virus was reported by Uppal *et al.*
- Tomato Mosaic Virus was reported by Adolfh Mayer.
- *Taphrina* causes peach leaf curl.
- Endoparasitic nematode is *Meloidogyne.*
- Black heart of potato is caused due to unfavourable O_2 conditions during storage.
- Black tip of mango is caused due to polluted gas (SO_2)
- Tip burn of paddy is due to reduced O_2 and N_2 under submerged conditions.
- Leaf roll of potato is caused by Luteo virus.
- Papaya mosaic is caused by Potex virus.
- Little leaf of egg plant (brinjal) is caused by MLOs.

AGRONOMY

Quick Facts

- Agro-forestry is a production technique that combines agriculture and forestry, horticulture or animal husbandry on the same piece of land in order to fully utilize the natural resources.
- Base period is the period for which water is supplied to a crop
- Cultivation of crops in areas receiving rainfall above 750 mm is known as dry land farming.
- Cultivation of crops in areas where rainfall is less than 750 mm per annum is called dry farming.

GK GS

Indian Aluminium Company was started in 1938.

- Delta represents total depth of water required by a crop.
- Dr. Varghese Kurien is the pioneer of Operation Flood (Milk revolution) in India.
- Duty is the volume of irrigation water required to mature a crop. It is the ratio of irrigated area to the quantity of water used (ha/l).
- Growing two or more crops on the same piece of land in one calender year is known as multiple cropping. It includes inter cropping, mixed cropping and sequence cropping.
- In low rainfall regions (<750 mm/annum) monocropping is practiced, where as in high rainfall regions (>750 mm/annum) inter cropping is practiced.
- Inter cropping is growing two or more crops simultaneously on the same piece of land with a define row pattern.
- Mixed cropping is growing of two or more crops simultaneously intermingled without any row pattern.
- Mixed farming is a system of farming on a particular farm which includes crop production, raising livestock, poultry, fisheries, bee keeping etc., to sustain and satisfy as many needs of the farmer as possible.
- Monocropping or Monoculture refers to growing of only one crop on a piece of land year after year.
- Nipping: Act of removal of ear heads from plant stalk.
- Number of days between irrigations during the crop period without rainfall is irrigation interval.
- Ratio of total cropped area in different seasons to the total land area is known as Cropping Intensity.
- Ratooning: The practice of taking the second crop from previous one is known as ratooning.
- Relay cropping can be defined as a cropping pattern in which crops are grown in the quick succession, second crop may be sown between the rows of first crop which is in the harvesting stage.
- Sequence cropping can be defined as growing of two or more crops in sequence on the same piece of land in a farming year.
- The common water distribution pattern in canal command areas in Northern India is Warabandhi
- The term Green revolution was cointed by William Gaude.

Special Purpose Crops

Truck crops	:	Potato, Onion
Cover crops	:	Groundnut, Methi
Nurse crops	:	Sannhemp in Sugarcane, Guar in Suran
Silage crops	:	Maize, Sorghum
Catch crops	:	Sunflower, Radish, Semi-rabi Sesame.
Soiling crop	:	Berseem
Pasture crops	:	Zinzvo
Row crop	:	Cotton
Support crops	:	Castor, Shevri
Mixed crops	:	Cowpea, Tur
Companion crop	:	Maize in Cowpea
Intercrop	:	Jowar in Fruit crop
Green manure crops	:	Sunnhemp, Cluster bean
Trap crops	:	French marigold around Tobacco, Castor around Sugarcane
Cash crops	:	Cotton, Sugarcane, Castor

Agro-Ecological Regions Classification by

ICAR	:	8
NSS & LUP	:	21 (now 20)

GK

The first industrial commission was appointed in 1875.

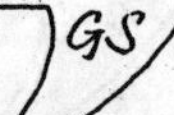

Agro-Climatic Zones Classification by

Planning commission : 15

NARP of ICAR : 131

Classification of Soils by ICAR : 8

Structure of Atmosphere

Troposphere: Extending upto a height of 8-18 km from the earth surface. Temperature decreases with altitude at the rate of 6 °C per km. Influences earth climate.

Stratosphere : From a height of 18 km up to 50 km from the earth surface. Uniform temperatures with altitude . Presence of Ozone layer.

Mesosphere : From a height of 50 km up to 80 km from the earth surface. Cold region.

Ionosphere: From a height of 80 km up to atmosphere. It includes thermosphere & exosphere. Satellite orbit. Temperature increases with altitude.

Composition of Atmosphere

Gases	Volume %
Nitrogen	78.03
Oxygen	21
Argon	0.94
Carbon dioxide	0.03

Iso's

Isobar	:	Imaginary lines connecting points of equal atmospheric pressures.
Isohytes	:	Imaginary lines connecting points of equal amounts of rainfall.
Isochrone	:	Imaginary lines connecting points of same times.
Isobath	:	Imaginary lines connecting points of equal water table elevations.
Isohels	:	Imaginary lines connecting points of equal sunshine hours.
Isopleuth	:	Imaginary lines connecting points of equal evaporation values.
Isostere	:	Imaginary lines connecting points of constant moisture contents.
Isogonals	:	Imaginary lines connecting points of same magnetic declination.
Isophene	:	Imaginary lines connecting points of similar seasonal phenomenon.
Isoseismals	:	Imaginary lines connecting points of equal seismic activities.
Isotherm	:	Imaginary lines connecting points of equal temperatures.
Isotach	:	Imaginary lines connecting points of equal wind velocities.

Forms of Precipitation

Drizzle	:	Fine drops of < 0.5 mm and intensity < 1mm/hr
Rain	:	Drops having size >0.5 mm to < 6 mm
Glaze	:	Freezing of rain droplets when they come in contact with cold objects
Snow	:	Ice crystals resulting from sublimation
Hail	:	Small lumps of ice (>5 mm dia)
Dew	:	Moisture condensed in small drops upon cool surfaces
Frost	:	A feathery deposit of ice
Fog	:	A thin cloud of varying size
Mist	:	A very thin fog

USSR became Russia in the year 1991.

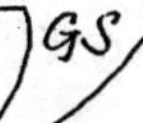

Terms & Associated Crops

Term		Crops
Curing	:	Tobacco, Tea
Stripping	:	Jute
Nipping	:	Cotton
Wrapping	:	Sugarcane
Propping	:	Banana
Trashing	:	Sugarcane

Important Varieties of Selected Crops

Variety	Remarks
Rice	
GEB 24, BCP 1, NP 130	Superior quality rice varieties
Basmati 370, Type 3	Aromatic
Pusa RH-10	First Super fine aromatic basmati hybrid
TKM, Pattambi	Diseases-Pest resistant
SR 26 B	Saline tolerant
FR 13A/43 B, Madhukar	Flood
Jalmagna, Jaisuria	Deep water ecology
Jaya	High yield
Vikramarya	Tungro-virus resistant
Ajaya	Bacterial blight resistant
Phalguna	Gall-midge resistant
Pusa Basmati-1	Worlds first high yield dwarf export quality aromatic rice
IR-8	First dwarf rice variety introduced in India
Lunishree	World's first super rice variety for saline/alkaline conditions
Yamini, Krishna, Hansa, Vasumati, Pusa Sugandh-2, Pusa Sugandh-3	
Wheat	
Lermoroja, Sonora-64	Direct introductions (imported)
Kalyan Sona, Sonalika	Selections
HD 2285, WH 542, UP 2338	North West plains
K 8804, K 9006, HP 1633	N-E plains
HI 1077, GW 190	Central zone
Sujata, C 306, K 8027	Rainfed region
Malviya, Wheat-510, HD 2733, HW 2045	
Pigeon pea (Redgram)	
ICPH-8	First Hybrid in the world
Mustard	
Pusa Jai Kisan	First Somaclonal Selection
Kalyani, Pusa Bold, Varuna	
Sugarcane	
CoC 671	Wonder Cane : Recovery = 10%
Gandak, Rajbhog	
Cotton	
Sujata	First Egyptian variety

First railway line was started from Mumbai to Thane in 1853.

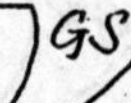

Three Largest Producing States of Important Crops during 2002-03

Production : Million Tonnes

Crops/Groups of Crops	States	Production	% Share of production to All India	Cumulative % share of Production
I. Foodgrains				
Rice	West Bengal	14.39	19.81	19.81
	Punjab	8.88	12.22	32.03
	Uttar Pradesh	8.11	11.16	43.19
Wheat	Uttar Pradesh	23.61	36.27	36.27
	Punjab	14.18	21.78	58.05
	Haryana	9.19	14.12	72.17
Maize	Madhya Pradesh	1.50	14.56	14.56
	Andhra Pradesh	1.49	14.47	29.03
	Karnataka	1.38	13.40	42.43
Total coarse cereals	Maharashtra	5.94	23.48	23.48
	Rajasthan	2.14	8.46	31.94
	Karnataka	3.63	14.35	46.28
Total pulses	Madhya Pradesh	2.21	19.84	19.84
	Uttar Pradesh	2.06	18.49	38.33
	Maharashtra	2.05	1840	56.73
Total food grains	Uttar Pradesh	36.30	20.84	20.84
	Punjab	23.49	13.49	34.32
	West Bengal	15.53	8.91	43.23
II. Oilseeds				
Groundnut	Gujarat	1.09	25.00	25.00
	Tamil Nadu	0.98	22.48	47.48
	Andhra Pradesh	0.82	18.81	66.28
Rapeseed & mustard	Rajasthan	1.32	33.67	33.67
	Uttar Pradesh	0.76	19.39	53.06
	Haryana	0.69	17.60	77.66
Soyabean	Madhya Pradesh	2.58	56.58	56.58
	Maharashtra	1.58	34.65	91.23
	Rajasthan	0.24	5.26	96.49
Sunflower	Karnataka	0.41	45.05	45.05
	Andhra Pradesh	0.28	30.77	75.82
	Maharashtra	0.15	16.48	92.31
Total oilseeds	Madhya Pradesh	3.00	19.92	19.92
	Maharashtra	2.33	15.47	35.39
	Rajasthan	1.75	11.62	47.01
III. Other cash Crops				
Sugarcane	Uttar Pradesh	116.32	41.31	41.31
	Maharashtra	37.02	13.15	54.46
	Karnataka	32.48	11.54	65.99
Cotton @	Maharashtra	2.60	29.82	29.82
	Gujarat	1.68	19.27	49.08
	Andhra Pradesh	1.09	12.50	61.58
Jute & mesta $	West Bengal	8.58	75.40	75.40
	Bihar	1.13	9.93	85.33
	Assam	0.72	6.33	91.65
Potato	Uttar Pradesh	10.16	43.88	43.88
	West Bengal	6.90	29.80	73.68
	Bihar	1.44	6.22	79.90
Onion	Mahashtra	1.22	29.08	29.08
	Gujarat	0.72	17.04	46.12
	Karnataka	0.45	10.57	56.69

@ : Production in million bales of 170 kgs. each
$: Production in million bales of 180 kgs. each

GK GS

In 1889 a British committee of INC was founded.

Normal Area, Production and Yield of Major Crops in India

Area - Million hectares; Production - million tonnes; Yield - Kg./hectare

Crops/Group of Crops	Season	Area	Production	Yield
I. Foodgrains				
Rice	Kharif	39.9	73.3	1835
	Rabi	4.0	11.9	2972
	Total	43.9	85.2	1938
Wheat	Rabi	26.4	71.0	2692
Jowar	Kharif	4.7	4.6	989
	Rabi	5.1	3.2	633
	Total	9.8	7.8	803
Bajra	Kharif	9.0	6.5	718
Maize	Kharif	5.8	10.0	1713
	Rabi	0.7	1.8	2749
	Total	6.5	11.8	1817
Total coarse cereals	Kharif	22.5	24.0	1064
	Rabi	6.5	6.5	999
	Total	29.0	30.5	1049
Tur	Kharif	3.4	2.4	703
Gram	Rabi	6.4	5.1	796
Total pulses	Kharif	10.2	4.7	455
	Rabi	11.2	8.1	723
	Total	21.4	12.8	595
Total foodgrains	Kharif	72.7	101.9	1401
	Rabi	48.1	97.5	2029
	Total	120.8	199.4	1651
II. Oilseeds				
Ground nut	Total	6.6	6.4	970
Rapeseed & mustard	Rabi	5.3	4.9	924
Soyabean	Kharif	6.3	6.0	958
Sunflower	Total	1.4	0.8	552
Nine Oilseeds	Total	23.4	19.9	850
III. Other cash crops				
Sugarcane	Total	4.3	298.5	69065
Cotton @	Total	8.7	10.4	204
Jute & mesta $	Total	1.0	10.8	1944
Potato	Total	1.3	23.6	18343
Onion	Total	0.5	4.8	10540

Note : 'Normal' is worked out as simple average of estimates for 5 years, i.e., 1998-99 to 2002-03.
@ : Production in million bales of 170 kgs. each
$: Production in million bales of 180 kgs. each.

Optimum Temperature and Water Requirement of Crops

Crop	Optimum temperature (°C)	Water requirement (mm)
Rice	30-32	900-2500
Wheat	25	450-650
Tobacco	28	500
Sugarcane	32-37	1500-2500
Sorghum	32-35	450-650
Maize	32-37	500-800
Cotton	20-30	500-700
Groundnut	20-25	750-850

The Congress Session of 1907 was presided by Rash Bihari Bose.

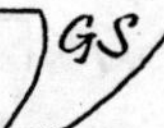

Comparative study of surface, drip and sprinkler irrigation system

Sl. No.	Surface irrigation	Drip irrigation	Sprinkler irrigation
1	More water is required e.g. for 1 ha of coconut field 10 m 10m x 10 m 505.23 cu m of water is required in flood irrigation	Less water e.g. for 10m x 10m coconut field 220 cu m of water is required	More water e.g. 1 ha coconut 10 m x 10 m spacing 280 cu m water is required
2	Salt water cannot be used. Moisture in the root zone becomes less and less till next irrigation by evaporation, transpiration and salt concentration is more and roots scorch, not good	Salt water can be used up to a certain limit.	Salt water cannot be used a layer of salt particles is depoisted on the leaves which will scorch the leaf and kill the plant
3	Suitable for all crops both widely spaced horticultural crops and closely spaced agricultural crops.	Suitable for widely spaced horticultural as well as row crops	Suitable for closely spaced field crops.
4	Medium pressure to irrigate 4 hectares; 10 H.P. pump set is required	Low pressure system to irrigate 4 hectares; one H.P. pumpset is required. 1/10 H.P. per acre is more than sufficient.	High pressure system to irrigate 4 hectares; 10 H.P. pump set is required
5	15-20% evaporation losses	No evaporation loss	20-30% evaporation losses
6	More labour to guide water and farmer has to walk a lot for guiding water	Less labour, only 10 minutes per day to open and close the gate value up to 10 ha. Once labour can irrigate.	More labour to lay and relay the pipelines in shushy soils
7	Even unwanted areas are irrigated	Only required area around the trees and bushes are irrigated.	Even unwanted areas are irrigated
8	More weed growth	Less weed growth	More weed growth
9	Insecticides and fungicides cannot be applied	Insecticides and fungicides can be applied	Insecticides and fungicides cannot be applied
10	Once in 7-10 days irrigated amount of moisture will not be constant. Hence maturity will be delayed and yield reduced	Water is applied daily or on alternate days, moisture remains constant and maturity is increased for more yield	Water applied on cyclic basis; moisture will not be a constant and decreases day by day till the next irrigation. Maturity is delayed. Yield will be less
11	Not possible to keep uniform moisture in the root zone	Soil moisture is maintained always at field capacity	Economically impossible to maintain moisture at field capacity
12	Annual electrical charges are less than sprinkler	Annual electrical charges are less	Electricity charges are high
13	Wind has no effect. Labourers feel tired to guide water from canal to field in windy condition	Wind has no effect	Water cannot be applied uniformly as most water is carried and sprinkled in different places other than the areas to be irrigated.
14	Land levelling is a must	Land levelling is not necessary as undulating lands also can be irrigated	Land levelling is not necessary
15	Irrigation Efficiency (IE) 30-35%	IE 90-95%	IE 70-5%

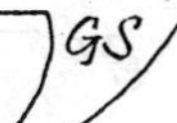

Jharkhand was formed in the year 2000.

Crop seasons and duration

Crop	Season	Duration of crop on land in months
Bajra	Kharif	4½
Barley	Rabi	5 - 5 ½
Castor	Kharif	Early 6,
		Others 8
Cotton	Kharif	Early 6-7
		Late 7-8
Gram	Rabi	6
Groundnut	Kharif	Early 4- 4 ½
		Late 4 ½ - 5
Jowar	Kharif	4 ½ - 5
	Rabi	4 ½ - 5
	Zaid Kharif	2 ½
Jute	Kharif	6-7
Linseed	Rabi	5 - 5 ½
Maize	Kharif	4 - 4 ½
Ragi	Kharif	3 ½
Rape & Mustard	Rabi	4-5
	Zaid Rabi	4
Rice	Winter	5 ½ - 6
	Autumn	4 - 4 ½
	Summer	2 - 3
Sugarcane	Perennial	10 - 12
Sesamum	Kharif	3 ½ - 4
	Rabi	5
Tobacco	Kharif	7
Wheat	Rabi	5 - 5 ½

Seed Rates of Important Crops

Crop	Seed rate (kg/ha)
Rice	
Transplanting	50
Broadcasting	80-100
Dribbling & drilling	80-90
Maize	20
Cotton	10-12
Hybrid cotton	2.5-3.0
Wheat	100-125
Bengal gram	60
Groundnut	120-140

GS

Jawaharlal Nehru was elected the chairman of the INC in 1929.

Oil Percentages in Selected Crops

Sesamum	46-52
Groundnut	44-50
Castor	35-58
Safflower	24-36
Niger	37-43
Rape Seed and Mustard	33
Soyabean	20
Coconut	60

Some important weeds and their association with various crops.

Sl. No.	Name of weed	Natural order	Field crops with which usually associated
1	*Justicia quinquangularis*	Acanthaceae	Kharif season week in field crops like rice, maize, sugarcane and cotton
2	*Amaranthus spinosus, A. viridi, A. polygamus, Celosia argentea, Digera arvensis, Achyranthus aspera*	Amaranthaceae	Reported to grow in sugarcane, maize, rice, onion and millet fields.
3	*Calotropis gigantea*	Asclepidiaceae	Mostly seen in sub-marginal lands
4	*Heliotopium eichwaldi, H. indicum*	Boraginaceae	Occurs in sugarcane, maize and cotton
5	*Cannabis sativa*	Canabinaceae	Occurs in wastelands and along water channels.
6	*Chenopodium album, C. murale*	Chenopodiaceae	Rabi season weeds dominant in fields of rabi cereals, peas, potato, linseed and carrot.
7	*Carthamus oxycantha,*	Compositae	Potato, peas, linseed
	Sonchus oleraceus, Ageratum conyzoides, Onicus arvensis		Reported to occur in all rabi crops like wheat, potato, linseed and peas
8	*Convolvulus arvensis*	Convolvulaceae	In non-cropped land and monocot crops
9	*Brassica sinensis, Sisymbrium iris, Nasturtium indicum*	Cruficerae	In mustard, peas and potato fields.
10	*Cyperus rotundus*	Cyperaceae	Perennial weeds in cropped and non-cropped land
11	*Euphorbia prostrata, E. hirta,*	Euphorbiaceae	Occurs in wheat, gram, millets and maize
	E. dracunculoides, Phyllanthus niruri		Occurs in rice fields
12	*Echinochloa crusgalli,*	Gramineae	Dominant in rice fields.
	E. colonum, Cyanodon dactylon		Perennial in all crops.
	Saccharum spontaneum,		Perennial in sugarcane, cotton and millets.
	Sorghum halepense,		Perennial in sugarcane, maize and cotton
	Avena fatua,		Annual in wheat, peas, potato, gram and linseed
	Phalaris spp.		Annual, occurs in rabi crops.
13	*Leucas aspera, Ocimum canum*	Labiatae	Occurs in cropped and uncropped lands
14	*Melilotus spp.*	Leguminosae	Annual weed in rabi crops
	M. indica		Annual in rabi crops
	Vica sativa		Annual in rabi crops
	Tephrosia purpurea		A perennial in arid tracts
	Prosopis juliflora		A perennial in non-cropped land
	Mimosa pudica		A perennial in cropped and uncropped land
15	*Orobanche cernua, O. indica*	Orobanchaceae	An annual root parasite on tobacco, mustard, etc.
16	*Argemopne mexicana*	Papavaraceae	Annual in all crops, especially wheat and mustard, under rainfed conditions
17	*Eichhornia crassipes*	Pontederiaceae	An aquatic, free-floating weed present in most parts of the world
18	*Striga asiatica, S. lutea, S. euphrasioides*	Scrophulariaceae	Annual root parasite in bajra, sorghum, sugarcane and rice.
19	*Solanum xanthocarpum,*	Solanacae	Perennial herb in crop lands
	S. elaeagnifolium		Perennial widely prevalent in Southern India
20	*Typha spp.*	Typhaceae	Aquatic emersed perennials
21	*Lantana camara*	Verbenaceae	Perennial shrub in border lands.

GK GS

Rashtriya Swayam Sevak Sangh was founded by Hedgewar in 1925.

Host-Parasite-Weed

Parasite	Weed	Host
Semi-parasite	*Striga* spp.	Jowar and Sugarcane
	Orobanche spp.	Tobacco
	Loranthus	Mango
Total Parasite	*Amarbel*	Lucerne

Weed Control in Field and Horticultural Crops

Crop	Weedicide	Rate Kg/ha (a.i.)	Time of Application
Wheat	2,4-D	1.2	Post-emergence
	Pendimethlin	0.5	Pre-emergence
Maize	Atrazine	2.0	Pre-emergence
Sugarcane	Metribuzin	1.5	Pre-emergence
	Simazine	4.5	Pre-emergence
	2,4-D	2.0	Post-emergence
Potato	Paraquat	2.5	Post-emergence
	Metribuzine	1.0	Pre-emergence
Gram	Pendimethalin	1.0	Pre-emergence
	Fluchloralin	1.0	Pre-planting
Mustard	Butachlor	1.5	Pre-emergence
Sunflower	Siramet	4.0	Pre-planting
Bajra	Atrazine	1.0	Pre-emergence
Sorghum	Simazine	1.0	Pre-emergence
Groundnut	Oxyflurofen	0.24	Pre-emergence
	Fluchloralin	1.5	Pre-planting
	Pendimethalin	0.75	Pre-emergence
Paddy	Benthiocarb	1.0	Pre-emergence
Pigeonpea	Pendimethalin	1.0	Pre-emergence
Cotton	Oxidizon, Diuron	1.0	Pre-emergence
Onion	Fluchloralin	2.0	Pre-planting
Garlic	Oxyflurofen	0.24	Pre-emergence
	Oxadiazon	0.75	Pre-emergence
	Fluchloralin	0.90	Pre-planting
Tomato	Metribuzine	0.7	Pre-emergence
Cauliflower	Aptec	4.5	Pre-planting
Cabbage	Fluchloralin	2.0	Pre-planting
Lucerne	Decathal	2.0	Pre-planting
	2,4-D	1.5	Post-emergence

On 20th July, 1905 Curron officially issued the order of partition.

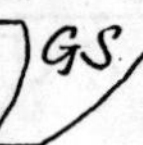

Pest/Diseases in Major Crops/Plants

Crop/Plant	Pest
Rice	Brown Plant Hopper, Stem Borer, Gall Midge, Khaira
Wheat	Cut Worm, Rust, Loose Smut, Karnal Bunt
Sugarcane	Shoot Borer, Pyrilla, Red Rot, Scale-Insect,
Cotton	Bolloworm, Jassid, White Fly, Wilt
Groundnut	White Grub, Tikka

Criteria for Harvesting and Storage of Important Crops

Crop	Criteria for harvesting	Moisture content for safe storage of grains
Rice	Moisture content of grains < 20%	14
Wheat	Moisture content of grains 15%	12
Sorghum	Moisture content of grains < 28%	12
Maize	Moisture content of grains 22-25%	12
Groundnut		6
Sugarcane	Brix 18-20%	
Cotton	Bolls fully opened	

Conversion Ratio Between Raw Material and Processed Products

Rice	Rice (cleaned production)	2/3 of paddy production
Cotton	Cotton lint production	1/3 of kapas production
	Cotton seed production	2/3 of kapas production
	One bale of cotton	180 kg.
Jute	100 yards of hession	2 times of cotton lint production
	One bale of jute	180 kg.
Groundnut	Kernel to nut in shell	70%
	Oil to nut in shell	28%
	Oil to kernel (crushed)	40%
	Cake to kernels	60%
Sesamum	Oil to seed (crushed)	40%
	Cake to seed (crushed)	40%
Rape and mustard	Oil to seed (crushed)	33%
	Cake to seed (crushed)	67%
Linseed	Oil to seed (crushed)	27%
	Cake to seed (crushed)	63%
Sugar	Gur from cane (crushed)	10%
	Crystal sugar from gur (refined)	62.4%
	Khandsari sugar from gur (refined)	37.5%
	Crystal sugar from cane (crushed)	9.97%
	Cane trash from cane (harvested)	10%

GK GS

All India Hindu Mahasabha was founded in 1915 under Maharaja of Kasim Bazar.

Agroforestry Systems

Agrisilviculture	Agricultural crops + Forest trees
Silvipastoral	Forest trees + Grasses
Agrisilvipastoral	Agricultural crops + Forest trees + Grasses
Agrihorti system	Agricultural crops + Fruit crops
Hortisilviculture	Fruit crops + Forest trees
Agrihortisilviculture	Agricultural crops + Horticultural crops + Forest crops
Aquasilviculture	Fish + Forest trees
Agrisilviaquaculture	Agricultural crops + Forest trees + Fish

SOIL SCIENCE

Quick Facts

- Bulk density is the mass of a unit volume of dry soil in its natural structure.
- Empty space between soil particles is called : Pore space
- Gneiss from Granite; Marble from Lime stone; Quartzite from Sand stone; Slate from Shales.
- Soil structure refers to the arrangement of soil particles viz., sand, silt and clay.
- Soil texture refers to the relative proportions of various types of soil particles.
- A mineral that forms the original component of a rock is known as primary mineral e.g. Feldspar, Hornblende, Mica etc.
- A mineral that has been formed as a result of subsequent changes in rocks is known as secondary mineral. e.g. Kaolinite, Montmorillonite, Illite, Gypsum ($CaSO_4.2H_2O$).
- A soil is said to be in good tilth when it is mellow, friable and adequately aerated and suitable for the germination of seeds.
- According to Marshall the most important phenomenon next to Photosynthesis is Cation Exchange Capacity (CEC).
- Alluvial soils are most recently formed soils
- Amendment for Acidic soils : Lime
- Amendments for Saline soils : Gypsum or Sulphur
- Average particle density : 2.67 g/cc (2.60 to 2.75)
- Azolla is widely used bio-fertilizer in Rice crop.
- Basic Nutrients: Carbon, Hydrogen, Oxygen provides basic structure to plant.
- Black cotton soils are deficient in Nitrogen
- Black soils are best suitable for dry land agriculture
- Bulk density of sub soils is higher as compared to surface soils
- Carbon : Nitrogen (C:N) ratio of normal soils 10-12 : 1
- Chisel ploughs are also used to break hard pans present even at 60-70 cm.
- Dapog system of rice nursery raising was adopted from Phillippines.
- Decomposition of organic matter is slowest at pH below 6 and fastest between 6 to 8.
- Detachment and transportation of soil particles by water/wind is Soil erosion.
- Dominant soil group in India : Alluvial; (The only Azonal soil)
- Electrical conductivity is generally as dS/m at 25°C and is used to express the salinity of the soil.
- For rice transplantation: Seedling bed preparation is about 1/10th of transplanting area.
- For waterlogged, saline, alkaline and dry conditions Sunhemp (Dhaincha) is suitable.
- Golden rice is genetically modified rice with yellow coloured seeds rich in Vitamin A.

- Green leaf manuring : e.g: *Glericidia acculata, Pongamia pinnetta, Ipomoea cornea.*
- Green manuring *in-situ* : e.g. *Crotolaria juncea* (Sunhemp) & *Sesbania acculata* (Dhaincha), Cowpea.
- Humus is a product of decomposition of plant and animal residues.
- In Black soils the dominant clay mineral is Montmorillonite
- In general plough are used for primary tillage and harrows for secondary tillage.
- In red soils the dominant clay mineral is Kaolinite
- In zero tillage, primary tillage is completely avoided and secondary tillage is restricted to seed bed preparation in the root zone only.
- Khaira disease of rice is caused due to the deficiency of Zinc.
- Loss of nutrients beyond root zone along with water is known as leaching.
- Minimum tillage is aimed at reducing tillage to a minimum necessary.
- Most commonly grown green leaf manuring crops is Sunhemp.
- Most preferable soil structure : Granular & Crumb
- Nitrate forms of nitrogenous fertilizers are subjected to higher leaching losses compared to other forms.
- Non Symbiotic Nitrogen fixation: Some of the free living bacteria capable of fixing atmospheric nitrogen. e.g. Azotobacter & Clostridium.
- On addition of organic matter bulk density decreases
- One cm of surface soil over one ha of land weights about 150 tonnes.
- Optimum bulk density : 1.2 to 1.6 g/cc
- Organic matter is responsible for the dark brown to very dark grey colour of soil.
- Permissible soil loss by water is 12 t/ha.
- Ploughing of undecomposed green matter into the soil to enrich the soil is called Green manuring.
- Pore space : Clay soils higher than sandy soils
- Preponderance of silica or lime results in Whitish or Greyish tinge to soil.
- Presence of Iron compounds like Hematite induces red colour to soil and Limonite: yellow colour.
- Puddling is ploughing the land with standing water (5-10 cm) so as to create an impervious layer below the surface to reduce deep percolation lossess of water and to provide soft seed bed for planting rice.
- Red soils are best suitable for irrigated agriculture
- Relationship between particle density & bulk density : Porosity = (1-Bulk density / Particle density) x 100
- Rice can tolerate very low level or even complete absence of oxygen in the soil.
- Stages of water erosion: Splash → Sheet → Rill → Gully → Ravine.
- Stages of wind erosion: Saltation → Suspension → Surface creep.
- Substances that influences the plant growth favourably by emulating the soil physical and chemical properties are called Soil amendments.
- Symbiotic nitrogen fixation: Bacteria belonging to genus Rhizobium infects roots of leguminous crop.
- The Red soils are red in colour due to presence of Iron Oxide
- The seedlings of tea, coffee, tobacco etc., are raised under shade to avoid damage by high soil temperature.
- Tillage is the mechanical manipulation of soil for loosening the surface crust and bringing about favourable conditions for the germination of seeds.
- Tilth is the physical condition of the soil in relation to plant growth.
- Urea ($CO(NH_2)_2$) in presence of urease forms into ammonium carbonate ($(NH_4)_2CO_3$).
- USA is the first country to introduce zero tillage.
- Water heats and cools slowly compared to soil.

GK GS

The first India-EU summit was held in Lisbon in 2000.

Spheres of Earth

Lithosphere = 93.2% by w.t.
Hydrosphere = 6.91 by w.t.
Atmosphere = 0.93% by w.t.

Earth's Chemical Composition

Oxygen	: 46.6%	Silicon	: 27.7%	Aluminum	: 8.1%%
Iron	: 5%	Calcium	: 3.6	Sodium	: 2.8%
Potassium	: 26%	Magnesium	: 2.1%	Other	: 1.5%

Soil Types

Alfisols : Relatively young soils, acid soils
Entisols : Alluvial soils (recently formed soils)
Histosols/Mollisols : Orgnaic soils
Vertisols : Black soils
Oxysols : Highly weathered soils

Rock Types

Igneous rocks : Granite, Basalt and Syenite
Sedimentary rocks : Lime stone, Sand stone and Dolomite
Metamorphic rocks : Gneiss, Marble, Quartzite and Slate

Classification of Soil Particle Size

Classification	Diameter (mm)	
	IIS	USDA
Gravel	> 2.0	> 2.0
Coarse Sand	0.2 - 2.0	0.5 - 1.0
Fine Sand	0.02 - 0.2	0.1 - 0.25
Silt	0.02 - 0.002	0.002 - 0.05
Clay	< 0.002	< 0.002

Chemical Properties of Different Soils

Parameter	Saline soil	Sodic soil	Alkali soil
pH	< 8.5	< 8.5	> 8.5
EC	> 4	> 4	< 4
ESP	< 15	> 15	>15

Soil Classification Based on pH

< 4.5	Extremely acidic	7.5-8.5	Slightly alkaline
4.5-5.5	Moderately acidic	8.5-9.5	Moderately alkaline
5.5-6.5	Slightly acidic	>9.5	Extremely alkaline
6.5-7.5	Neutral		

Density of Different Soil

Soil	Particle density (cm/cc)	Bulk density (cm/cc)
Normal soil	2.65	1-1.6
Sandy soil	2.6	1.6
Silt Soil	2.7	1.3
Clay soil	2.8	1.1

GK GS

Streptomycin was the first drug discovered in 1944, to be effective against tuberculosis (TB).

Total Pore Space of Soil

Soil	Pore space (%)
Sandy	20-30
Loamy	30-50
Clay	50-60

Component of Soil Colour

HUE: Denotes the dominant spectral colour (Yellow, red blue, brown etc.)

VALUE: Denotes the lightness and darkness of colour

CHROMA: Denotes purity of colour

Specific Heat (cal./gm)

Water	=	1
O.M	=	0.462
Clay	=	0.22
Air	=	24
Humus	=	0.4
Ice	=	0.5

Heat Capacity (cal. cm^{-3})

O.M.	=	0.462 x 1.3 = 0.60
Water	=	1x1 = 1
Air	=	24 x 0.000125=0.003

Cation Exchange Capacity (CEC) of Soils

Soil texture	CEC (meq/100 gm)
Sand (light colour)	3-5
Sand (dark colour)	10-20
Loam	10-15
Silt loam	15-25
Clay and clay loam	20-50
Organic soil	50-100

Composition of Soil Air

Gases	Volume %
Nitrogen	78.6
Oxygen	20
Argon	0.9
Carbon dioxide	0.5

Antagonistic Effect

Excess of nutrient	Causes deficiency
Ca	P
Ca, Mg	K
K, NH_4	Mg
Fe, SO_4	Mo
P	Zn
N, P, K	Cu
NO_3	Fe
Zn, Al	Cu
N, K, Ca	B

GK GS

Project Tiger was launched in 1973.

Plant Elements

Essential	C, H, O, N, P, K, Ca, Mg, S, Fe, Cu, Mn, Zn, Mo, Cl, B. = 16.
Primary	N, P, K
Secondary	Ca, Mg, S
Macro	N, P, K, Ca, Mg, S
Micro	Fe, Zn, Cu, B, Mo, Mn, Cl
Functional	Essential elements + Co, V, Si, Na = 20
Beneficial	Ru, Sr, Ni, Cr, As
Energy exchange	H, O
Energy storage	C, N, P, S
Translocation regulator	K, Ca, Mg, Na
Oxidation reduction	Fe, Mn, Zn, Cu, B, Co
Highly mobile	N, P, K, Mg, Mo
Moderately mobile	Zn
Less mobile	Fe, Cu, Mn, S
Immobile	Ca, B

Forms of Elements Absorbed by Plants

Elements	Form of availability
C	CO_3^{--}, HCO_3^{--}
H2O	H^+, OH^-
N	NH_4^+, NO_3^-
P	PO_4^{---}, $H_2PO_4^-$, HPO_4^{--}
K	K^+
Ca	Ca^{++}
Mg	Mg^{++}
Cu	Cu^{++}, Cu^+
Zn	Zn^+
Cl	Cl^-
Fe	Fe^{++}, Fe^{+++}
Mn	Mn^{++}, Mn^{+++}
S	SO_3^{---}, SO_4^{--}
B	$B_4O_7^{--}$, $H_2BO_3^-$
Mo	MoO_4^{--}

Soil Test Rating Chart for Available Nutrients

Available 'N'	Low	Medium	High
Available 'N'	Below 280 kg/ha	280-560 kg/ha	Above 560 kg/ha
Available 'P'	Below 10 kg/ha	10-25 kg/ha	Above 25 kg/ha
Available 'K'	Below 108 kg/ha	108-280 kg/ha	Above 280 kg/ha
Organic carbon	Below 0.5 per cent	0.5-0.75 per cent	Above 0.75 per cent

On 16th August, 1946 Muslim League observed the direct action day.

GS

Nutrient Contents of Some Products

Parameter	N	P	K
Farm Yard Manure	0.5	0.2	0.5
Blood meal	2.5	2	1
Fish meal	4-10	6-8	0.3-1.5
Groundnut cake	7.3	1.5	1.3
Safflower cake	7.9	2.2	1.9

Fertilizers

Nitrogenous fertilizer	N(%)
Nitrate fertilizer	
Sodium nitrate	16
Calcium nitrate	15.5
Ammonical fertilizer	
Ammonium phosphate	16-20
Ammonium chloride	24-26
Anhydrous ammonia	82
Ammonia solution	20-25
Nitrate and Ammonical fertilizer	
Ammonium nitrate	33-34
Calcium ammonium nitrate	26
Ammonium sulphate nitrate	26
Amide fertilizer	
Urea	46
Calcium cyanamide	21

Phosphatic fertilizer	P_2O_5 (%)
Fertilizer containing water soluble phosphoric acid or mono-calcium phosphate $Ca(H_2PO_4)_2$	
Single super phosphate	16
Double super phosphate	32
Triple super phosphate	46
Fertilizer containing citric acid soluble phosphoric acid or dicalcium phosphate $CaHPO_4$	
Basic slag	14-18
Dicalcium phosphate	34-39
Rhennia phosphate	23-26
Fertilizer containing insoluble phosphoric acid	
Rock phosphate	20-40
Raw bone meal	20-25
Steamed bone meal	22

Potassic fertilizer	K_2O (%)
Fertilizer containing potassium in chloride form	
Murate of potash	60
Fertilizer containing potassium in non-chloride form	
Sulphate of potash	48
Potassium nitrate	44
Sulphate of potash	25-30

India will host the Commonwealth games in the year 2010.

Fertilizer Production, Import, Consumption ('000 Tonnes)

Name of fertilizer	Production	Imports	Consumption	Max. consumption rate of fertilizer (State wise)	Total consumption (Maximum)
Nitrogen	10961	160	10920	Punjab	UP
Phosphorus	3743	409	4215	AP	UP
Potassium	No production	1627	1567	W.Bengal	Maharashtra
Total	14704	2196	16702	AP	UP

AGRICULTURAL ECONOMICS & FARM MANAGEMENT

Theories Developed by Different Economists

Theory	Economist
Theory of profit	A. Walker
Modern theory of interest	Hicks-Hansen
Wage fund theory	J.S. Mill
Population theory	Malthus
Theory of multiplier	Keynes
Theory of disguised unemployment	Nurkses
Theory of unlimited supplies of labour	Lewis
Low-level equilibrium trap	Nelson
Theory of undeveloped countries	Keynes
The stages of economic growth and Classical theory of economic development	W.W. Rostow
Theory of das capital	Karl Max
Theory of economic growth	John Fei & Gustav Ranis
Theory of big push	N. Rosenstein-Rodan
Theory of social dualism	J.H. Boeke
Theory of circular causation	Myrdal
Requirement of steady growth	Harrod-Damar
The model of distribution	Kaldor
The model of profit and growth	Psinette
The model of capital accumulation	Joan Robinson
The model of economic growth	R.E. Meade
The model of growth	Kaldor
The single sector model	Mahalanobis
Theory of rent	Ricardo
Theory of time preference	Fisher
Theory of employment	Keynes
Liquidity preference theory	Keynes
Monetory theory	R.G. Hawtrey
Theory of consumption	Keynes
Modern theory of wages and employment	Keynes
Sunspot theory	Stanley Jevon
Under consumption theory	J.A. Hobbson
Quantity theory of money	Milton Friedman
Theory of inflation	A.P. Lerner
Theory of absolute advantage	Adam Smith
Modern theory of international trade	Bertil Ohlin

GK GS

In 1960, Xerox company developed the small wheeled box attached to a computer, better known as a 'Mouse'.

Farm Efficiency Measures

Production efficiency $= \dfrac{\text{Yield per acre}}{\text{Yield of locality}} \times 100$

Intensity of cropping $= \dfrac{\text{Total cropped area}}{\text{Total cultivated area}} \times 100$

Productive man work unit per man $= \dfrac{\text{Total P.M.W.U.}}{\text{Man equivalent}}$

Machinery cost per crop acre $= \dfrac{\text{Total machinery cost}}{\text{Total cropped area}}$

Operating cost ratio $= \dfrac{\text{Total operation cost}}{\text{Total profit}}$

Overhead charges ratio $= \dfrac{\text{Total fixed cost per year}}{\text{Gross income}}$

Rate of capital turnover $= \dfrac{\text{Gross income}}{\text{Total farm income}} \times 100$

Net cash income = Total cash receipt - Total cash operating cost
Net farm income = Net cash income ± Change in inventory
Farm earnings = Net farm income + Value of farm product
Family labour earnings = Farm earning - Interest on farm capital
Percentage returns to capital = Ratio of farm earning - Imputed value of family labour to average capital investment.
Return to management = Family labour earning - Imputed value of family labour.

Types of Farmers

Farmers	Land Holding (ha)
Marginal	Less than 1
Small	1-4
Medium	4-10
Large	Above 10

Finance According to Period

Loan	Term (Yr)
Short term	1 to 1.5
Medium term	1 to 5
Long term	5 to 30

Relationship Between MP, TP & AP

1. What happens to MP, when TP is maximum ?	1. MP is equal to zero
2. When TP is zero, what happens to AP ?	2. AP is zero
3. When MP is increasing, how does AP behave ?	3. AP also increase
4. When MP is at its maximum, is TP at the point of inflation?	4. Yes
5. When AP is at its maximum, is the ratio of output to input maximum ?	5. Yes, because MP=AP
6. When AP is increasing, how does MP behave ?	6. MP is greater than AP
7. When AP is decreasing. What is the position of MP ?	7. MP is less than AP
8. How does TP behave, when	8.
a. MP is rising ?	a. TP increases at a increasing rate
b. AP is increasing ?	b. TP is increasing at a decreasing rate after MP at its maximum.

GS

John Pemberton founded the Coca-Cola company in 1986.

Difference Between

Production economics	Farm management
Macro approach	Micro approach
Interfarm science	Intrafarm science
Decisions are made for aggregate resources of nation	Decisions are taken for individual farm

Variable cost	Fixed cost
Known as cost of variable inputs	Known as sunk cost
It is always in cash	It may be cash and non-cash both
It vary with level of output	It does not change with level of output

Isoproduct curve	Indifference curve
It is curve of two input combinations which have same output	It is a curve which show combinations of two commodities which derive equal satisfaction to the consumer.

Agricultural Credit

Loans		Purposes
Short Term	:	For purchasing seeds, fertilizers and for meeting out family requirements.
Medium Term	:	To purchase agricultural equipments, animal and for land improvements.
Long Term	:	To purchase land and expensive agricultural equipments and for repayment of old loans.

EXTENSION EDUCATION

Stages of Extension in India

Stage I	- Pre-independence era	(1866-1947)
Stage II	- Post-independence era	(1947-1953)
Stage III	- Community development and extension service era	(1953-1960)
Stage IV	- Intensive agricultural development era	(1960-till date)

Important Projects and Associated Persons

Project Name		Associated Person
Gurgaon Project	:	F.L.Brain
Shantiniketan	:	Ravindra Nath Tagore
Rural Reconstruction	:	Daniel Hamilton
Marthandom	:	Spencer Hatch
Sewagram	:	M.K.Gandhi
Etawah Pilot	:	Albert Mayor
Nilokheri Experiment	:	S.K.Dey
Majdoor Manjil	:	S.K.Dey
Community Development	:	S.K.Dey
Indian Village Scheme	:	S.N.Gupta
National Demonstration Scheme (NDS)	:	Kalwar and Subramanyam
Young Farmer Association	:	Deshmukh
3 tier Panchayat Raj (First in Rajasthan)	:	B.R.Mehta
2 tier Panchayat Raj (First in Karnataka)	:	Ashok Mehta

GS

China will launch lunar orbiter Chang'e 1 by 2007.

STATISTICS

- AM is regarded as the best of all the averages.
- Arthimetic Mean (AM) > Geometric Mean (GM) > Harmonic Mean (HM)
- Correlation coefficient varies between -1 to + 1
- $GM^2 = AM \times HM$
- If any one of the observation is zero then the GM also becomes zero
- In Normal Distribution: Mean, Mode, Median are same
- Mode = 3 Median - 2 Mean
- Range is the difference between the value of largest and smallest.
- Regression coefficient varies between - a to + a
- The Normal Distribution Curve is Bell shaped curve.
- The value of probability varies between 0 to 1

REMOTE SENSING

- Human eye responds to wave length between 0.4 to 0.7 μm and which is also photosynthetically active radiation range.
- IRS-P4 launched in 1999 is used to measure physical and biological oceanographic parameters.
- LISS stands for Linear Image Self Scanner & WiFS stands for Wide Field Scanner.
- Near Infra Red wave length ranges between 0.7-1.3 μm and Middle Infra Red ranges between 1.3-3.0 μm.
- Simple Vegetation Index = Reflectance in Infra Red band / Reflectance in Red band.
- Vegetation Index is used in indentifing the crop stress that can be used in crop management.
- Velocity of Light in air: 3×10^8 m/s.
- In Remote Sensing optical wave lengths from 0.3 to 15 μm are used.
- Remote Sensing is the sensing of objects or a phenomena without being in contact.

ANIMAL HUSBANDRY

- 15% of world live stock population is in India.
- 25% of world cattle production is in India and milk production is 6%.
- Livestock population is increasing on an average about 20 lakh animals / year
- Total Livestock population in India : 470 millions

General Information About Livestock

Livestock	Period of heat	Average length of estrous cycle	Average heart beat
Cattle	18 hours	19-21 days	55 beats/min
Buffaloes	27 hours	21-23 days	55 beats/min
Sheep	1-3 days	18 days	75 beats/min
Goats	2-3 days	18-20 days	90 beats/min
Horses	4-5 days	21 days	44 beats/min
Swine	2-3 days	21 days	280 beats/min

Gestation Period and Body Temperatures of Animals/Birds

Animal / Bird	Group is known as	Meat	Gestation period / incubation, days	Body temp., °F
Cow	Herd	Calf: Veal, Cattle : beef	283-285	101
Buffalo	Herd	Bufen	300-310	101
Sheep (Ewe)	Flock	Mutton	150	103
Goat (Doe)	Flock	Chavon	148	101
Swine (Sow)	Herd	Pork	114 (3m 3w 3d)	101
Hen	Flock	Chicken	21	107

GK GS

India has 2,167 recorded natural wetlands.

Composition of Milk (% by weight)

Species	Water	Fat	Protein	Lactose	Ash
Cow	86.61	4.14	3.58	4.96	0.71
Buffalo	82.76	7.38	3.60	5.48	0.78
Goat	87.00	4.25	3.52	4.27	0.86
Sheep	79.50	8.50	4.00	4.30	0.97

Cattle

Quick Facts

- Bovine population of the country : 289 millions (Cattle + Buffalo)
- Buffalo population : 84 millions
- Calf < 1 yr old : Bull calf (m)
  Heifer (f)
- Cattle disease that is communicate to man is Anthrax.
- Cattle population : 205 millions (Cow : 165 + Bull: 40)
- Cow's Milk is light yellow or creamy in colour due to Carotene.
- Cow's Milk is white in colour due to presence of Casein.
- Female over 1 yr age and is yet to give birth to a calf is : Heifer
- For keeping good quality, cool the milk to 10°C as quickly as possible after milking.
- In India 1/3rd of the milk produced is consumed in fresh or whole stake, while the remaining 2/3 rd is converted into milk products like Curd (Dahi), Ghee etc.
- Indian buffalo belongs to the species : *Bubalus bubalis*
- Ox is used for both sexes
  Male Ox : Bull
  Female Ox : Cow
  Castrated Bull : Bullock
- Lactose is the carbohydrate or sugar constituent of the milk.
- Milk is deficient in Iron.
- Milk is rich in calcium and phosphorous and excellent source of Vitamin A.
- Milk Production is maximum in Uttar Pradesh
- Pasteurization is done to kill all the pathogenic and most of the non-pathogenic bacteria.
- Pasteurization refers to the processing of exposing milk to about (61-63 °C) for 30 min (LTLT) or to about 71-76.6 °C for 15 seconds (HTST), in Ultra High temperature milk is heated 148 °C for 1-5 seconds of 149 °C for 0.5 seconds followed by rapid cooling by 10 °C or below.
- Pratham, the world's first in vitro fertilized buffalo calf.
- Protein are built from simple compounds known as amino acids.
- Protein content of Colostrum : Cow milk : 12.8%, Buffalo milk : 21.4%.
- The average annual milk production per cows & buffaloes are 157 and 540 kg respectively.
- The fat of colostrum contains 9 times more vitamin D that the fat of normal milk.
- The Indian name of clarified butter is Ghee. It is pure butter fat.
- The mammary secretion for the first 3 or 4 days following calving is called Colostrum or Beestings.
- The targeted microbe in pasteurization is *Clostridum botulinum*.
- The three principal milk proteins are: Casein (80-83%); Lactoalbumin (8-10%); Lactoglobulin (3%).
- Used for Meat/Beef : Steer
- Young One : Calf

GK GS

The population density of India in 2001 was 324 per sq. km.

Classification of Cattle Breeds in India

S. No.	Breed name	Native of
Dairy Breeds		
1	Sahiwal	Montgomery, Pakistan
2	Sindhi (Cow)	Sindh, Pakistan
3	Gir (Cow)	Gir in Gujarat & Deccan
4	Deoni	North West, Andhra Pradesh
Draught Breeds		
1	Bargur, Kangayam	Tamil Nadu
2	Hallikar	Karnataka
3	Khillari, Dangi (Bull)	Maharashtra
4	Bachur	Bihar
5	Kankarh	Uttar Pradesh
6	Nagori	Rajasthan
7	Malvi	Madhya Pradesh
Dual Purpose Breeds		
1	Hariana (Cow)	Haryana
2	Kankrej	Gujarat
3	Rathi (Cow)	Rajasthan
4	Kosi	Uttar Pradesh
5	Gaolao	Madhya Pradesh
6	Krishna Valley	Karnataka
7	Tharparkar (Cow)	Pakisthan
8	Ongole	Guntur, Andhra Pradesh

Exotic Cattle Breeds

Breed				
Jersey				
Holstein Friesian				
Frieswal (Holstein x Sahiwal)	:	Highest milk yielding breed	:	14 litre/day
Brown Swiss				
Karan Fries: (Friesian x Tharparkar)	:	Milk yield	:	12.7 litre/day
Karan Swiss (Brown Swiss x Red Sindhi or Sahiwal)	:	Milk yield	:	11.3 litre/day

Buffalo Breeds

Western Type	:	Zaffrabadi	:	Native of Gujarat
		Surathi	:	Native of Kaira (Gujarat)
Northern Type	:	Nili Ravi	:	Pakistan
	:	Murrah	:	Southern Punjab
	:	Bhadawari	:	Native of Uttar Pradesh
Central & Southern Type	:	Toda	:	Neelgiri
		Parlakiddi	:	Orissa
Improved Breeds	:	Mehsana	:	Baroda
	:	Godavari	:	West Godavari (A.P.)

Milk Yield (lit) per Lactation Period & Fat Content of Milk of Indian Breeds

Breed	Milk Yield (lit)	Fat (%)
Nili Ravi	1600-1800	7.2
Murrah (Buffalo)	1700-1800	7.0
Malvi (Cow)	500-600	4.5
Ongole	600-700	5.0

Devaraya I kept 10,000 Muslims in his army.

Common Diseases of Cattle

Mastitis	:	Inflammation of udder
FMD	:	Foot & Mouth Disease: Viral diseases
Rinderpest	:	Viral disease (is also called Cattle Plague)
Anthrax	:	Bacterial disease (is also called Spleenic Fever).
Cow Pox	:	Viral disease

Sheep

Quick Facts

- Austalia stands first in Sheep population.
- India ranks first in production of carpet wool.
- Castrated male : Wether/Hogget
- Female aged about one year : Gimmer
- First clone of adult animal : Dolly
- In India wool production is maximum in : Rajasthan
- Male kept for breeding : Ram or Tup
  Female kept for breeding : Ewe
- Male Lamb : Ram
  Female lamb : Ewe
- Young one : Lamb
- Sheep population in India : 51 million

Sheep Breeds of India

Region	Breed name	Native	Remarks
Temporal Himalyan	Gureh, Karnah, Bhakarwal, Gaddi and Rampur	Jammu & Kashmir, Himachal Pradesh & Parts of Uttaranchal	Fine quality wool
Dry Western	Lohi	Pakistan & also seen in Punjab and Haryana	Largest Sheep
	Bikaneri	Rajasthan	Best Carpet wool
	Marwari, Sonadi, Jaisalmeri, Malpura, Kutchi	South Punjab, Rajasthan, Gujarat and Parts of Uttar Pradesh	
Southern	Deccani	Maharashtra, Karnataka & Andhra Pradesh	Good for mutton, Fit for weaving Kambal (Rug)
	Nellore	Andhra Pradesh	Tallest breed
	Mandya	Karnataka	White in colour
	Bellary	Andhra Pradesh	Black with white markings
	Nilgiri	Tamil Nadu	White coloured wool
Eastern	Shahabadi	Muzaffarnagar	Sheep do not thrive well in High rainfall regions

Exotic Breeds of Sheeps

Fine Wool Breeds

Merino	:	Native of Spain. About 80% of wool that is produced in the world comes from Merino and its crosses.
Rambouillet	:	Native of Spain. Dual purpose breed.

Medium Wool Breeds

Dorest horn	:	Native of England
Suffolk	:	Native of England

Long Wool Breeds

Lincoln & Licester	:	Native of England
Karkul	:	Native of USA. Wool is used for Felt marketing.

Diseases of Sheep

Anthrax	:	Caused by Bacteria
Circling diseases	:	Caused by *Listeria monocytogenes*
Stiff lamb disease	:	Caused by Vitamin E deficiency
Enterotoxemia	:	Caused by *Clostridium wechee* (fat lamb disease)
Foot-rot disease	:	Caused by Bacteria

Goat

Quick Facts

- Morocco leather a fine leather is prepared from Goat skin.
- Female kept for breeding : Doe
- Female of one year : Doeling
- Goat population in India : 115 millions
- Male kept for breeding : Buck
- Male of one : Buckling year
- Young one : Kid

Goat Breeds of India

Jamunapari	:	Near Ganga & Yamuna rivers, Dual purpose breed.
Barbari	:	Dairy breed
Beetul	:	Native of Punjab. Resembles Jamunapari
Surathi	:	Native of Gujarat State
Marwari	:	Native of Rajasthan
Kashmiri	:	White in colour, give Pashmina
Osmanabadi	:	Native of Maharastra, dual purpose
Malabari	:	Native of Kerala
Bengal	:	Excellent for Mutton

Exotic Breeds of Dairy Goat

Angora	:	Native of Turkey
Alpine	:	Native of Europe
Anglo-Nubian	:	Cross between Jamunapari (Indian breed) and Nubian (England)
Saanen	:	Presence of beard in both the sexes is the characteristic feature. This is breed is also known as Queen of Goats.
Toggenberg	:	Chocolate colour, Native of Switezerland

Diseases of Goat

Foot and Mouth, Ketosis, Bloat, Lumpy Jaw

A man jumping out of a moving train is thrown forward.

Swine (Pig)

Quick Facts

- Salted Smoked meat of pig is known as Bacon.
- Castrated Male : Beg
- Female kept for breeding : Sow
- Male kept for breeding : Bear
- Young one : Pigling
- Swine population in India : 13 million

Exotic Breeds of Swine

Lindrace	:	White in colour	:	Native of Denmark
Duroc	:	Red in colour	:	Native of U.S.A.
Yorkshire	:	White in colour	:	Native of U.K.

Poultry

Quick Facts

- Castrated Male : Capon
- Young One : Chick
- Mature male : Cock
- Mature female : Hen
- Poultry population in India : 308 million.
- Egg production in India : 31,320 million eggs.
- A medium sized egg contains 6.0 g protein, 6.5 $CaCO_3$ fat, 14 g vitamins.
- Andhra Pradesh has largest number of poultry and is the largest producer of eggs is India.
- Egg shell is made of $CaCO_3$.
- In our tropical climate it is always necessary to construct poultry houses East to West in order to avoid sun ray directly falling into the shed and heating up of the shed.
- There are four types of housing poultry: (i) Brooder house (ii) Grower house (iii) Layer house (iv) Broiler house. The yellow colour of egg yolk is due to the pigment Xanthophyll.
- Usually internal quality of eggs can be determined by Candling.

Important Disease of Poultry

Bacterial Diseases	:	Fowl Cholera, Greensh Diarrhoea
Viral Diseases	:	Ranikhet, New Castle or R.D., Fowl Pox, Marek's Disease
Protozoal Diseases	:	Coccidiosis
Nutritional Diseases		
Nutritional roup	:	Vitamin A deficiency
Rickets	:	Vitamin D deficiency
Curled toe paralysis	:	Vitamin B_2 deficiency

Fisheries

Quick Facts

- India is the second largest fish producer from inland sources after China.
- India ranks third in the world and first among Commonwealth countries in fish production.
- Maximum fish production is in West Bengal.
- Maximum marine fish production is in Gujarat.
- Out of 10.0 mt of fish production about 50% are from inland and 50% are from marine.

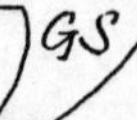

Sachin Tendulkar is the first cricketer to score 20,000 runs in international cricket (Test and One Day International included).

- The domestic consumption of fish per annum : 5.0 million tonnes.
- The total length of coastal line in India : 7516 km.
- Total fish production in India per annum : 10.0 million tonnes.
- Largest contributor in Agricultural Export - Marine products.

Vitamins, Enzymes

Quick Facts

- Calcium is essential component of all living cells, helps in body building.
- Enzymes are those organic substances which are capable of catalyzing chemical reactions in living systems. They speed up the reaction, they do not effect the equilibrium.
- Folic acid is essential for maturation of RBC.
- The largest concentration of minerals are found in bones and teeth.
- Vitamins are accessory food factors play vital role in growth, nutrition and metabolic processes.
- Fat Soluble Vitamins : A,D,E & K
- The term enzymes was coined by : W. Kuhne
- The term Vitamin is coined by : Funk
- Water Soluble Vitamins : B & C

Scientific Names of Vitamins

Vitamin	Scientific Name	Vitamin	Scientific Name
A	Retinol	B_{12}	Cobalamin
B_1	Thiamin	C	Ascorbic acid
B_2	Riboflavin	D	Calciferol
B_5	Niacin	E	Tocopherol
B_6	Pyridoxine	K	Phylloquinone

Deficiency/Excess of Vitamins/Effects and Sources of Availability

Deficiency of vitamin	Causes / Results in	Source of vitamin	Intake of Excessive quantities leads to
A	Growth failure, Night blindness	Liver, Rice source, Milk & its products, Egg yolk, Refined oils	Headache, Nausea, Enlargement of liver, Pain in the joints, Dryness of skin
B_1	Disturbance in digestive system and nervous system (Beri-Beri)	Rice polishings, Green leafy vegetables, Fish, Egg, Meat	
B_2	Photophobia, Inflamed lips, Sore tongue	Dried yeast, Milk powder, Fish, Pulses	
B_6	Degeneration of nerves	Dried yeast, Wheat germ, Legumes	
B_{12}	Pernicious anaemia	Only in animal foods, Liver, Meat, Fish, Egg	
C	Scurvy (soft tissues around the joints beca-me swollen and painful)	Aonla, Guava, Citrus, Green leafy vegetables	
D	Bone deformation (Rickets), Tooth decay	Sunlight, Fish & Liver oils (Rich), Egg yolk, Milk & its fat	Constipation, Nausea, Irritation
E	Reproductive failure & Liver damage in Animals, No definite proof in human beings	Outer layers of grain	
K	Prolong blood clotting time	Cauliflower, Liver, Soyabean	

Geostationary orbit is 36000 km away from the earth.

GS

Plant Growth Substances

Auxins	:	Stimulates elongation of cells of stems and coleoptiles. Cell division & root formation. As herbicide. Enhances fruit set and fruit ripening.
Cytokinins	:	To stimulate cell division. Promotes orderly development of embryos of seed. Break dormancy of seeds and buds. Delays senescence
Gibberellins	:	Increases cell division & cell elongation. Increase in size of leaf, flower, fruit. Dormancy is broken & flowering is induced
Abscisic Acid	:	Dormancy is induced. Facilitates stomata closure
Ethylene	:	Ripening of fruits. Iso-diametric growth of stems and roots

Plant Growth Substances used in Agriculture and Horticulture

Plant growth promoter	Use
2, 4 -D	Herbicide
Glyphosine, Ethephon	Sugarcane ripener
Abscisic acid	Cotton defoliant
NAA	Increasing number of flowers
Gibberllins	Production of seedless grape, increase in size of fruits
Ethylene	Regulating ripening of fruits

MISCELLANEOUS INFORMATION

- '*sui generis*' used in GATT proposals is a Latin word which means 'a system of its own'.
- A grain, vegetable and fodder crop, the amaranth has high protein (83%) with high lysine content, a good balance of essential amino acids.
- Black heart of potato is due to deficiency of Oxygen.
- Browning of caluliflower is caused by Boron deficiency.
- China is the largest producing and consuming country of tobacco in the world, India is the third highest producer of tobacco.
- Cotton contributes approximately 33% of the total forex earned by country.
- Crop having maximum area under irrigation : Wheat
- Fruit cracking of tomato is due to deficiency of Boron.
- Gas filled in fire extinguisher cylinders : CO_2.
- Green house effect main culprit : CO_2.
- Growth harmone helps in enlargement of grape fruit is 'Abscisic acid'.
- HD 2329 wheat variety contributed maximum producing during Green revolution in India.
- Highest number of hybrids in Rice are use in China followed by India.
- In sugarcane red rot is caused by 'fungus'.
- Incorporation of 'Norin 10' dwarfing gene from CIMMYT improved wheat yields during green revolution time.
- India contributes 15 per cent of the World's cotton production.
- India is the first country to develop and use hybrids in pearlmillet.
- India is the only country to have developed and grown hybrid cotton commercially.
- India is the Second country after China to have commercialized the hybrid rice technology.

In India a total area of 40494 sq. km is classified as wetlands.

- India is the second largest country after China in production of vegetables.
- India ranks first in prodcution and consumption of Sugar in the world.
- India ranks first in the production of tea.
- India ranks sixth in world coffee production.
- India stands first in Grapes productivity (40 t/ha).
- Indian variety HD 2172 has paved way for self sufficiency in food grain production in Sudan.
- Karanataka is largest producer of coffee in the country.
- Maharashtra is the largest cotton as well as cereals producing area in India.
- Mango is propogated through 'veneer grafting'.
- Minimum Support Prices of various crops are recommended by Agricultural Cost & Price Commission.
- Most extensively grown crop in the world : Mango
- National Agricultural Insurance Scheme (NAIS) or Rashtriya Krishi Bima Yojana is sponsored by General Insurance Corporation.
- Ozone depletion is mainly due to : CFC (Chloro Floro Carbon)
- Recently Germany has stopped import of textiles from India due to use of Azo Dyes in textile making.
- Recently Scientists separated germs in Maize corn for manufacturing contraceptives.
- Redness of apple is due to 'Anthocynin'.
- Rice is approximately 42% of total food grains produced in the country.
- Ricetec a company of America has got a marketing patent of 'Basmati Rice'.
- Sugar content in Sugarcane available in India is 10-11%.
- Sugar recovery in sugarcane in India is 6-8%.
- The tobacco contributes the highest excise revenue among agricultural commodities in India.
- White bud of maize is due to deficiency of Zinc.
- World's first genetically modified non-human primate, monkey, named ANDI.
- A cyclone is a roughly circular low pressure area whose diameter may be from hundreds to a thousand miles. Atmospheric pressure is always lowest in the centre of this region and increases rapidly outward.
- Albedo is the ratio between outgoing radiation to incoming radiation. Albedo of earth surface is equal to 0.4.
- Altitude: It is the height of a place from the mean sea level.
- Anticyclones are opposite to cyclones, wherein a region of high atmospheric pressure is located at the centre and spiral air movement is outward from the central area *i.e.*, pressure decreasing from the centre to the periphery.
- At equator, day and nights are always equal.
- Atmospheric pressure always decreases with altitude, it does not always decrease at the same rate.
- Climate is a generalized weather or summation of weather conditions over a given region during a comparatively longer period.
- Conduction is the process of heat transmission through a medium by contact of the minute particles of which the medium is composed.
- Contour : Lines joining equal elevations.
- Convection is the transfer of heat through air or liquid by means of currents.
- Cyclones are generally originate in water warmer than 27°C.
- Latitude : It is the distance of a place from equator expressed in degrees.
- North-East monsoon is limited mostly to Southern states (A.P. & T.N.)
- Of the total rainfall of India S-W monsoon contributes around 85%.
- Radiation is the process of heat transfer in wave form without the use of a transmitting medium.
- Silver iodide is used for seeding cold clouds.

Peninsular plateau is a part of ancient Gondwana land.

GS

- Sodium chloride is generally used for seeding warm clouds.
- Solar Constant is the energy falling in one minute on a surface of one square centimeter at the outer boundary of the atmosphere held normal to the sunlight, at the mean distance of the earth from the sun. Solar constant value is : 1.94 cal/ cm^2/min.
- South-West monsoon reaches India in First week of June in Kerala.
- Surface temperature of Sun : 6000 K, the earth has an average temperature of 288 K.
- The effect resulting from rotational movement of earth and the movement of air relative to the earth is called 'Corolis force'.
- The wind circulation in anticyclones is clockwise in Northern hemisphere and anticlock wise in Southern hemisphere.
- The wind circulation in cyclones is anti-clockwise in the Northern hemisphere and clockwise in Southern hemisphere.
- The word Monsoon means 'season' ; arabic word.
- Weather is a state or condition of atmosphere at a given place at a given instant of time.

GK GS

Village community emerged from Chalcolithic period.

2. The Recent Horizons of Agriculture

Biological Husbandry

It is a agricultural production system which avoids or largely excludes the use of synthetically compounded fertilizers, pesticides, growth regulators and livestock feed additives. To the maximum extent feasible organic farming systems rely upon crop rotations, crop residues, animal manures, legumes, green manures, mineral bearing rocks and aspects of biological pest control to maintain solid productivity and tilth to supply plant nutrients and to control insects, weeds and other pests.

Permaculture

Permatculture should involve successful management of resources for agriculture to satisfy changing human needs while maintaining or enhancing the quality of the environment and conserving natural resources.

Terra-forming

To develop the Mars (planet) according to the Earth's environment is called Terra-forming. At present bacteria and plants are being developed by creating the environmental condition of Mars on the Earth.

Mechanized Farming

Farming in which machine-drawn implements are used for the reduction of labour are used for the reduction of labour requirement or elimination of manual work, timeless of operations and improved quality of husbandry, resulting in higher output and better quality of produce for increased profit.

N-Serve

A trade name for 2-chloro-6-(trichloro-methyl) pyridine which is used as nitrification inhibitor. The rate is 1-2 % N-serve by weight of fertilizer (ammonical) nitrogen.

Sulphur-Coated Urea (SCU)

It has been developed as a slow release nitrogen fertilizer and is basically urea with a coating of elemental sulphur, including a binding agent, a sealant and microbicide.

You Don't Know!

Oldest slides

The rock-cut children's slide at Mamallapuram in Tamil Nadu is the oldest extant children's slide in the world.

The 'N' content of SCU ranges for 10 to 37% depending on the thickness of the sulphur coating.

Micropropagation

Production of plants from very small plant parts, tissue or cells aseptically in a test tube or other container where the environment, tube or other container where the environment, nutrition and hormonal balance are rigidly controlled. Enhancement of adventitious budding and then stimulation of root development form the base of the buds are accomplished in their propagation.

Kisan Khad

Kisan khad is most suitable for vegetables. It is commercially prepared from ammonium nitrate and ground limestone or dolomite containing 20% nitrogen. One half of the nitrogen is in nitrate form and the remaining half on the ammonical form. It is almost neutral in nature.

Multinutrient Fertilizer

The fertilizer containing more than one nutrient element. When fertilizers supply two major plant nutrients they are called binary fertilizers, e.g. ammonium phosphate and nitrophosphate. Fertilizers supplying three major plant nutrients are termed tertiary fertilizers, e.g. ammonium potassium phosphate.

Biofuel

Biofuel is also known as 'green fuel'. It will be prepared by extraction of *Jatropha*. It will cut vehicular emission by almost 50%. This plant requires low maintenance, does not need any water and grow on degraded land. The Government has agreed to spend nearly Rs. 1,500 crore to support awareness and cultivation of *Jatropha*. A "National Mission" has been set up for their implementation under the Ministry of Rural Development.

Kisan Credit Card (KCC)

KCC was introduced in 1998-99 for facilitating access to short-term credit to farmers from commercial banks, co-operative banks and RRBs. Its implementation has been taken up by 27 commercial banks, 378 district central co-operative banks/ state co-operative banks

The first meeting of Constituent Assembly was held on December 9, 1946.

and 196 RRBs throughout the country.

Dumping

Dumping is selling of goods in a foreign market at a price below the price realised in the home country, after allowing for all costs of transfer including transportation charges and duties. Thus it causes 'material injury' to an establishment industry or 'materially retards' the establishment of a domestic industry.

Rice-Fish Farming System

Many farmers in coastal and inland lowlaying areas are integrating fish culture with rich production both in irrigated and rainfed lowland rice to meet the problems of risk. This system appears to be economically sound. For successful rice-fish farming, the fields are made to hold water for longer periods. Rising of marketable fish and prawn along with rice has been accomplished successfully in W.B., Orissa and coastal A.P. Rice-Fish system in coastal saline areas involves growing salt tolerant rice varieties and fresh water fish/ prawn during monsoon season and sole cropping of salt water fish/ prawn in sequence during dry season.

Golden Rice

Golden Rice or GM rice is genetically engineered vitamin A rich rice. It was engineered to save million of children from vision impairment. Over 5,0000 children go blind every year in South Asia due to deficiency of this vitamin. The U.S. Food and Drug Administration has classified it as 'genetically acceptable and safe'.

Bt Cotton

Cotton with Bt gene is resistant against the pest, *Helicoverpa*. It is developed by U.S. based seed company Monsanto and registered under the name *'Bollgard'*. Bt variety obtained 25-27% more cotton also reduces the cost of pesticides and also protects the environment from the hazardous effect of pesticides.

Transgenic Crops

The term 'transgenic' or genetically modified organism (GMO) is used to describe new strains of organisms in which the DNA has been modified through *in vitro* insertion of genetic material from a foreign organism. Transgenic crops have multiple benefits which includes flexibility in term of crop management, decreased dependency on conventional insecticide and herbicides, higher yields and cleaner and higher quality grain.

The economic returns from transgenic crop have increased considerably along with consumer acceptability. Golden rice, transgenic banana and nutritionally improved potato would prove a great boon in eradicating blindness, immuness and malnutrition problem from developing countries respectively.

GRAMSAT

The GRAMSAT satellite network is tailored to meet the basic requirement for rural areas to disseminate culture and region-specific knowledge on health, hygiene, environment, family planning etc. The GRAMSAT network could also incorporate features for providing continuing education for special social groups, industrial workers and others.

Urja Gram

In an Urja Gram the renewable energy devices can find their applications to meet the just energy requirement. For example, a biogas plant working on locally available animal and agricultural waste supplies the cooking fuel and also fuel for lighting or for irrigation wherever required. Requirement of drying, cooking, hot water, etc. can be met by Solar Thermal System (STS).

Hydroponics

Solution culture is being used for raising flowers and vegetables at homes. This soilless production of plant is called hydroponics. Plants are raised in small tanks of concrete or metal. They are filled with a water solution containing appropriate quantities of all mineral elements. Hydroponics is useful in areas having thin, infertile soils. Out of season vegetables and flowers can also be obtained. However, the cost of setting up a hydroponics system is very high.

Artificial Rain

Injecting the clouds with a seeding agent like dry ice, sodium chloride and silver iodide from an aircraft or using a ground generator for producing artificial rain.

Biofertilizers

The use of fertilizers (chemical) has increased the productivity of agriculture crops in short term. These chemical fertilizers are expensive, pollutant and are lost in surface run-off. To combat the ill effects of synthetic agriculture chemicals, Biofertilizers have been introduced. More-over, Indian soils are usually poor in organic matters as well as in nitrogen. Biofertilizers are of biological origin. They are organisms which can bring about soil nutrient enrichment. The source of biofertilizers are:

Bacteria : *Rhizobium, Azospirillum*
Fungi : *Mycorrhiza*
Fern : *Azolla*
BGA : *Cynobacteria*

Acid Rain

Normal rain is slightly acidic because of carbon dioxide that dissolves in rain drops to form carbonic acid. Carbonic acid is very weak acid. So, even ordinary rain has a pH of little below 7. Air contains acidic gases such as sulphur dioxide (SO_2), nitrogen dioxide (NO_2) etc; as pollutants. These come in the atmosphere by the burning of fuels in homes, industries and automobiles. In the air, sulphur dioxide is oxidised to sulphuric acid (H_2SO_4) whereas nitrogen dioxide is oxidised to nitric acid (HNO_3). These acids are washed down to the earth with rain. The rain containing dissolved acids form the atmosphere is called *Acid Rain.*

Acid rain has adverse effects on vegetation. It results in suppression of growth, leaf drop and reduction in production. Acid rain damages the leaves of plants and trees. This is because the acid present in acid rain leaches extracts and nutrients from pant leaves.

Cloning

Cloning is the production of identical plants microorganisms or animals from a single individual. A clone is an organism that derived from a single parent through non-sexual activities. Natural cloning is common in plants, microorganisms and simple animals. Some plants such as carrot and tobacco can be cloned by taking a few cells from a living specimen and growing them on a nutrient jelly. The cells then grow and divide until a whole plant is formed. This form of tissue culture is not widespread.

Triticale

Triticale is the most successful synthetic allopolyploid produced by crossing wheat with rye. At present, triticales are being grown commercially in Canada. The desirable features of triticales are that they combine the yielding ability and grain qualities of wheat with the hardiness of rye. Extensive breeding work on triticales is going on at CIMMYT, Mexico. The results from such breeding programmes have been spectacular and have led to the release of several commercial varieties of triticale which yield as much as the best varieties of common wheat.

Vermicompost

Compost that is prepared with the help of earthworms is called vermicompost. The earthworms may be of local species or more vigorous exotic ones. Earthworms consume large quantities of organic matter and excrete soil as casts. The weight of the material passing through the body each day is almost equal to the weight of the earthworm. The casts of the earthworm have several enzymes and are rich in plant nutrients, beneficial bacteria and mycorrhizae. On an average, vermicompost contain 3% nitrogen, 1% phosphorous and 1.5 % potash.

Knowledge World!

Largest landing ship

INS *Magar*, built by Garden Reach Ship builders and Engineers Ltd., Calcutta, is meant for amphibious operations and can cary 15 tanks, 8 armoured vehicles, 4 assault-landing craft and 2 motorboats. In addition to its crew of 135, it has arrangements for transporting 500 troops for a duration of 24 hours. INS *Magar* was commissioned in July 1987.

Ferti-cum-Seed Drill

Fertilizers are placed at the depth of 5 cm and 5 cm away from seed row for effective utilization of fertilizers. Both operations *viz.* drilling seeds and fertilizers are done simultaneously by ferti-cum-seed drill. It is similar to seed drill, but with extra tyres and hopper for drilling fertilizers.

Star Weeder

It is a small implement pushed by manual labour. It consists of a long wooden or iron vertical rod with a small horizontal rod for holding the implements. To the other end, two star-like wheels and small blade of 10 cm are attached. The pointed teeth of rotating wheels loosen the soil and help in easy mobility of the implement while the blade helps in cutting the weeds. It is useful to control small weeds in close growing crops like groundnut, foxtail millet etc.

Cybrids

Somatic hydrids obtained by fusion of the complete protoplast of the one species with such protoplast of another species, which either have no nucleus or have an inactive nucleus; e.g., protoplast of species A + Cytoplast of species B = Cybrid.

GM Food

Genetically Modified Food is a food in which a foreign gene has been incorporated. Today most food stuffs in the USA contain GM ingredients as about 75% of all bioengineered crops are grown here. A large scale cultivation of GM plants has been approved in several other countries like Canada, Japan and most European nations. Food stuffs derived from GM plants bring health benefits to its consumers like those with enhanced nutritional status or having reduced fat and sugar content. But thoroughly testing the potential benefits or any likely drawbacks of the new life forms on human life before they are unleashed on society at large still

We include the money value of goods and services produced by nationals outside the country in calculating Gross National Product (GNP).

needed.

Molecular Farming

Genetic engineering is being employed to develop transgenic animal resistance to certain diseases, capable of faster growth rates and more efficient feed conversion, and with capacity to reduce certain valuable biochemicals and excreting them in milk, urine or blood from which they are isolated and purified is called Molecular Farming or Animal biotechnology.

Water Harvesting

Water harvesting is collection and storage of water on the surface of the soil for subsequent use. It is a method to induce, collect, store and conserve local surface for agriculture in arid and semiarid region. In humid climates, runoff usually occurs only when rain fall on a saturated soil. Under semi-arid conditions, the total annual precipitation may or may not be sufficient, but most of the rain falls in a short period causing humid climate humid conditions and resulting into runoff. In both conditions, the excess rainwater should be safely guided and collected in nearby pond for recycling in drier period for life saving irrigation.

Jalshakti

A chemical which when applied (mixed) in soil, improves the aeration, infiltration and water holding capacity of the soil.

Watershed Management

Watershed means an area of land whose runoff drains to common point *i.e.* in the river or waterpool. It may be a natural unit of land whose runoff collects and flows out of the area through a single common outlet into a river or other water body. Watershed management is the planned use of watershed lands in accordance with predetermined objectives, such as the control of erosion, stream flow, sedimentation, and the improvement of vegetative cover and other related resources. Such watershed management integrate plans for soil conservation, soil improvement, irrigation, drainage, flood prevention and development of water supplies for all purposes.

Forest Gardening

Forest Gardening or Agroforestry is primarily a system where agriculture and forestry are practised either simultaneously or separately on the same unit of land and has affinities with taungya system of regenerating forests which in Burmese means cultivation of trees and crops. Agriculture in forest is called agro-forestry.

Social Forestry

Social forestry means plantation of such trees which are useful for community dèvelopment. It is a forestry which aims at producing flow of protection and recreational benefits for the community. It devotes programmes for raising plants and trees for supply of fire wood, fodder and small timber of community.

No Tillage Concept

No tillage or zero tillage-concept is gaining momentum in developed countries replacing conventional tillage in crop production. Zero tillage reduces the pressure on labour and power in crop production. The system of no tillage reduced soil erosion and labour cost of farmers when it is tried in combination with herbicides. After the discovery of selective herbicides and their efficiency in controlling weeds and concept of 'no tillage' gained momentum. The results of the study indicated that there was not much yield difference between conventional and no tillage treatments but there was saving in the cost of cultivation of crops.

Trickle Irrigation

Surface irrigation gives 35 to 50 per cent irrigation efficiency depending on the nature of soil and crop raised. Low efficiency in irrigation requires large quantities of water to cover per unit of crop field. In recent times, an economic method of water application known as driptickle irrigation has come into practice. In this system, water is delivered to each plant at its root zone, through a network of tubing. It works under low or medium pressure and only the required quantity of water is given daily in order to avoid water stress to the plant. The field crops like cotton, maize, sugarcane, banana and tomato and plantation crops like coffee, orange, apple are quite suitable for this system of irrigation. The system saves water to the extent of 30 to 50 per cent, increase in yield reduce weed growth and labour cost and improve quality of produce.

Brackishwater Aquaculture

With the objective of utilizing the country's vast brackish water area for shrimpculture, an area of about 25,805 hectares has been developed for shrimpculture through 39 Brackishwater Fish Farmers Development Agencies (BFDAs). The BFDAs provide a package of technical, financial and extension support to shrimp farmers. Guidelines prepared for sustainable development and management of brackishwater aquaculture have been circulated to all maritime states and UTs and other agencies. The guidelines incorporate measures of mitigating the adverse impact, if any of shrimp farming on the coastal eco-system.

GK

The principle on which jet engine works is Newtonian Law of action and reaction.

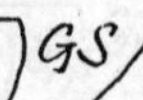

PURA

To tackle the problem of migration of people from rural to urban areas for employment, president APJ Abdul Kalam has proposed the concept of PURA (Providing Urban Amenities in Rural Areas) in the VISION 2020 project initiated by him. Its objective is to make rural areas as attractive to investor as cities. Then, rural areas too will generate urban style employment to halt, if not reverse, rural urban migration.

Global Warming

Since the beginning of the Industrial Revolution atmospheric concentration of CO_2 has increased by 30%, methane concentration by 100% and nitrous oxide concentration by about 15%. These increases have enhanced the heat-trapping capability of the earth's atmosphere.

Humanity's green house gas emissions have already disturbed the global balance by about 1% of the net incoming solar energy. The combustion of coal, oil and natural gas accounts for most of the heat-trapping green house gases emitted by human activities.

A global warming is likely to reduce the availability of agricultural land as river deltas may be in inundated due to rise in sea level. Agricultural pests may prosper to cause widespread damage to crop at increased temperature. As '*Kyoto protocol*', an initiative to tackle the problem of global warming at global level, the countries are to ensure that green house gases do not exceed their assigned amounts so that the overall emissions of such gases are limited to at least 5% below 1990 levels in the commitment period 2008-2012.

APEDA

Agricultural and Processed Food Products Exports Development Authority (APEDA) is a body engaged in the export promotion and development of markets for fruits, vegetables and their products, meat and meat products, rice, wheat, and other agricultural products. It is a promotional agency and does not undertake direct exports of any products on its own account.

Globalization

Globalisation means pursuing:

- reduction of trade barriers with a view to allowing free flow goods to (and from) the country;
- freer flow of foreign capital in terms of investment (direct and portfolio) by ensuring conducive atmosphere and easy approval of proposals;
- free flow of technology; and
- freer movement of labour and man power.

Quiz Kit!

Largest high-head spillway

The Hirakund high-head spillway on the Mahanadi river in Orissa completed in 1956 has a minimal discharge of 41,626 cusecs and a maximum discharge of 1.47 million cu ft.

GIS

Geographical Information System (GIS) are computer based means for assembling, analysis and storing varied forms of data corresponding to specific geographical areas, with the spatial location of the areas forming the basis of the system. These are not simple graphics or mapping but are system that interrelate, manipulate and analyse variety of geographical distributed data in addition to mapping. Thus it recreates new data. Its use in farm research is rather new. An effort has been made to elucidate its probable application in this sector.

Food for Work Programme

Launched in February 2001 with the aim of augmenting food security through wage employment in the drought affected rural areas. Wages are paid by state government partly in cash and partly in food grains, which are provided free of cost by the centre of these drought affected states.

Hariyali Scheme

Launched in January 2003 to harness rainwater to meet severe water problems of various areas, by involving the panchayats.

Maru Gochar Yojna

Launched in 2003-04 to be taken up for the desert districts of Rajasthan to provide for rehabilitation of the traditional pastures.

Golden Revolution

India, today is the second largest producer of fruits and vegetables in the world, contributing 10 and 13.38 per cent of the total world production of fruits and vegetables respectively. The availability of flower has increased significantly in all the cities, as indicated by the growing number of florists and sizable export of cut-and-dried flowers. India is a treasure house of medicinal and aromatic plants. It is the largest consumer, producer and exporter of spices and spice products in the world. It ranks first in the total production of coconut and arecanut, and the largest producer, processor, consumer and exporter of cashewnut in the world. There have

GK **Sedimentary rocks form the outer layer of Himalaya.** GS

been gluts in the production of several important crops. Thus, India is at the brink of a *Golden Revolution* in Horticulture.

Agri-clinic and Agri-business Scheme

The scheme was launched in 2001-02 in collaboration with NABARD to supplement the existing extension network and is the existing network and is an attempt to take technology to the doorstep of the farmer through the use of unemployed agriculture graduates.

I T in Agriculture

The long term vision on "Information and Communication Technology (ICT) in Agriculture Sector" is to bring farmers, researchers, scientists and administrators together by establishing "Agriculture-online" through exchange of ideas and information. The scheme was approved during the 9th Plan for strengthening the use of I T tools and network.

National Agriculture Insurance Scheme (NAIS)

NAIS was launched in 1999-2000 *rabi* season replacing Comprehensive Crop Insurance Scheme (CCIS) 1985. Its objective is to protect farmers against, losses suffered by them due to crop failure on account of natural calamities so as to restore credit worthiness for the ensuing season. It envisages coverage of all food crops.

Seed Bank Scheme

It is in operation since 1999-2000 for identified crops like paddy, wheat, maize, jowar, cotton etc, in the states of A.P., Gujarat, Haryana, Karnataka, M.P., Maharashtra, Orissa, Punjab, Rajasthan, U.P., to provide financial security to seed breeders/growers in the event of failure of seed crop.

Antodaya Anna Yojana

It was launched in December 2000 for the poorest of poor families and aims to provide them 35 kgs of food grains per family per month at a low price of Rs. 2/kg for wheat and Rs. 3/kg for rice. The scheme has been extended to over BPL families.

Gokul Gram Yojana

To improve the breed of animals and increase the production of milk an ambitious scheme was launched in Jharkhand on 14th October, 2004. In its first phase the scheme known as Gokul Gram Yojna, will be started in eight districts and one village in each district will be developed as 'Gokul Gram'. Under the scheme milking machine, milk freezing centres, pastures will be provided by the state government. In addition to these facilities, vaccination camps will also be organised in villages.

The Father of the Green Revolution in India

Dr. M.S. Swaminathan, eminent agricultural scientist, will head the National Commission on Farmers, which was set up to review the status of Indian agriculture and asses the condition of different categories of farmers in various regions. Dr. Swaminathan is known as 'The Father of the Green Revolution in India'. A graduate from Cambridge with a doctorate in Genetics, Dr. Swaminathan was awarded the World Food Prize in 1987. Currently he also heads the committee appointed by Agriculture Ministry to prepare a National Policy on Agricultural Biotechnology.

Agri-Export Zones (AEZ)

In the exim policy of 2001-02 the government announced the proposal to set Agri-export zones for the purpose of developing and sourcing raw materials and their processing and packaging leading to final exports. Till now, the Central Government has sanctioned and notified 45 AEZ which are being set up in 17 states.

National Commission for Farmers

In February, 2004 centre announced the setting up of the first ever National Commission on Farmers. The commission would recommend measures to enhance farm productivity, status of agriculture and farmers of different categories. The commission would focus on increasing the acreage under irrigation, analyse the effectiveness of the existing price and marketing policy and the legal regimes of improving income and general welfare of farmers.

World Trade Organization (W.T.O)

The WTO is the successor organisation to the multilateral General Agreement on Tariff and Trade (GATT). GATT came into force as an interim arrangement on January 1, 1948. The International Trade Organisation (I.T.O) never came into existence so this left the GATT as the framework for trade relations. There were three basic objectives behind the establishment of GATT:

- To provide a framework for the conduct of trade relations.
- To provide a framework for, and to promote, the progressive elimination of trade barriers.
- It was to provide a set of rules that would inhibit countries from taking unilateral action.

We get terracotta bulls from Malwa.

Since the start of the first round in 1947, there were eight rounds of GATT talks and important were : Dhillion Round (1960-61), Kennedy Round (1964-67), Tokyo Round (1973-79) and Uruguay Round (1986-93).
Uruguay Round gave birth to WTO which was officially constituted on January 1, 1995. Contrary to the temporary nature of GATT, WTO is a permanent organisation. It has completed the triad of organisations along with International Monetary Fund (IMF) and International Bank of Reconstruction and Development (IBRD). With the entry of China and Taiwan in 2001, its membership went upto 144. Uptil now five Ministerial Conferences of WTO have been held in Singapore (1996), Geneva (1998), Seattle (2000), Doha (2001), and Cancun (2003).

Intellectual Property Right (IPR)

Intellectual Property Right (IPR) is an idea, a design, an invention, a manuscript, etc. which can ultimately give rise to a useful product/application. The development of such a property, as a rule, requires intellectual inputs ingenuity and innovativeness; it also demands considerable monetary and other resources. Therefore, the inventor of an intellectual property would like to ensure at least a fair reward for his invention. But the major problem with intellectual properties is that they can be copied, imitated or reproduced, this minimizes the returns to the original inventor. The foregoing discussion recognizes the right of an inventor to derive economic benefits for his invention (*i.e.*, intellectual property); this right is called *intellectual property rights (IPR)*

Protection of IPR

The main form of IPR protection are as follows:

- **Patent:** A patent is the right granted by a government to an inventor to exclude others from imitating, manufacturing, using or selling the invention in question for commercial use during the specified period. In India it became effective on January 1, 2005.
- **Copyright:** Certain intellectual properties are not patentable; they are protected by copyright. Examples of such properties are authored and video cassettes etc. Recently, computer software has been included in the list of copyrightable properties.

Plant Variety Protection (PVP)

Plant varieties and animal breeds are developed through years of painstaking and scientifically planned work. These entities, therefore, should be regarded as intellectual properties of the breeders who have

Did You Know ?

Oldest High Court

The Calcutta High Court established in 1862 has 45 permanent and five additional judges as sanctioned strength.

developed them. It may be arranged that these entities are essentially derived from naturally occurring lines, but they usually represent a considerable reorganization of the existing gene combination and skillful selection work. Many countries recognize plant varieties as an intellectual property and grant a protection to them through a patent (in U.S.A) or a suitable form of plant breeders rights (PBR) (U.S.A., Japan, Australia, E.U., Israel, etc).

TRIPs

The TRIPs (Trade Related Intellectual Property Rights) agreements, which forms a part of the Uruguay Round of GATT (General Agreement on Tariff and Trade, signed by India and other countries), is to date the most comprehensive multilateral agreement on IPR, it became effective on January 1, 1995. The provisions of GATT are administered and enforced by World Trade Organization (WTO), Geneva. The member countries of WTO are obliged to meet all the articles of TRIPs. The provision of TRIPs cover a variety of Intellectual properties, including patents and protection of new varieties of plants. Each member country has the option to frame its patent laws within the broad framework defined in the GATT agreement.

Plant Breeder's Right (PBR)

Plant breeder's rights are the rights granted by the government to a plant breeder, originator or owner of a variety to exclude other form producing or commercializing the propagating material of that variety for a period of, minimally, 15-20 years. To qualify for PBR protection, a variety has to be novel, distinct from existing varieties and uniform and stable in its essential characteristics. A person holding PBR title to a variety can authorize other interested persons / organizations to produce and sell the propagation material of that variety. He should set reasonable terms of such transfers of PBR titles or for the sale of the propagating materials, otherwise the government can grant licenses of the title in public interest.

UPOV

The most significant event in the development of PBR system was the effort to harmonize PBR laws of different

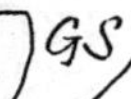

The objective Resolution adopted on January 22, 1947.

countries through UPOV (Union Internationale Pour la Protection des Obstentions Vegetales, International Union for Protection of New Plant Varieties). The first UPOV convention was signed in 1961 in Paris. India does not have any PBR system and, as such, is not a member of UPOV. The member states of UPOV adopt PBR systems conforming to the broad framework agreed upon in the convention.

Farmer's Right

Agriculture began some 10,000 years ago. During this vast period of time genetic resources have been selected, developed, used and conserved by farmer families and farming communities of, particularly, the gene-rich developing countries. These same materials have been and are being collected, conserved and used as raw materials to evolve the modern seed sales of these improved varieties earn huge profits for the seed corporation. It has been argued that the farmers should be allowed a share in this profit in recognition of their contribution by way of development of germplasms of the various crops. This has been recognized by FAO as farmer's rights, which arise from past, present and future contributions of farmers in conserving, improving and making available plant genetic resources, particularly in the centre of diversity. However, the farmer's rights are yet to be legalised in any country, and their implementation remains a far cry.

sui generis System

India is trying to evolve a *sui generis* system of PBR. 'A *sui generis* system' simply means ' a system of their own' *i.e*, being designed by them. The essential features of UPOV 1978 Act are being considered to be adopted. Some important features of the proposed Indian *sui generis* systems are as follows:

- Researcher's use right.
- Farmer's rights.
- Protection period, 15 years (annual crops)/ 18 years (fruit trees and vines).
- Compulsory deposit in national gene bank.
- National Authority for the protection of breeder's, farmer's and researcher's use rights.

National Agricultural Policy-2000

- An agricultural growth of over 4% per annum by 2005.
- Private sector participation would be promoted through contract farming and land leasing arrangements and also in areas like agricultural research and marketing.
- Emphasis on environmentally sustainable use of resources.
- Domestic agriculture would be liberalised and rural electrification will be high priority.
- National Livestock Breeding Strategy will be evolved and plant variety(ies) would be protected through a legislation to encourage research and breeding of new varieties.
- Consolidation of holdings in all states.
- Land reforms to provide land to poor farmers.

Agriculture and Tenth Five Year Plan

Although the Tenth Plan document categorically mentions that fall in GDP growth rate during the Ninth Plan is primarily due to decline in growth of agriculture and manufacturing sector, yet the allocation pattern of the Tenth Plan follows the beaten track. A careful perusal of the allocation pattern of the Tenth Plan reveals that it allocates 20% of total public sector outlay to agriculture, rural development and irrigation as against 21.4% during the Ninth Plan. In case of irrigation, relative share to allocation has been reduced from 7.4% in the Ninth Plan to 6.8% in the Tenth Plan and in case of rural development, a still smaller share of 8.0 % in the Tenth Plan as against 9.5% in the Ninth Plan. Whereas Tenth Plan recognises agriculture as a major employment generating sector, ironically it downplays in the allocation pattern. This is case of deviation and reality.

Agricultural Reforms under "Common Minimum Programme" of UPA Government

- Rural credit will be doubled in next three years.
- All existing projects will be completed within three-four years.
- All dues of farmers will be cleared at the earliest.
- Public investment in agriculture research, rural infrastructure and irrigation will be increased in a significant manner at the very earliest.
- Crop and livestock insurance schemes will be streamlined to make them more effective.
- Controls of the depress farm incomes will by systematically removed.
- All possible efforts will be made to maintain terms of trade in favour of agriculture.
- Government would also provide adequate protection to all farmers for imports, particularly when international prices fall sharply.

Eleventh Schedule of Constitution

The headquarter of SAARC is situated at Kathmandu.

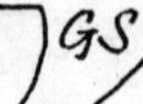

It mentions functional areas or subjects that are necessary for implementation of schemes for economic development and social justice in each Panchayat. To mention a few agriculture, social forestry, small scale industries, roads, rural housing, education, health and sanitation, poverty alleviation, non-conventional energy sources, etc.

Seventy-third Amendment (The Constitution Act 1993)

A new Part IX relating to Panchayats has been inserted in the Constitution to provide for among other things, *Gram sabha* in villages; constitution of Panchayats at village and other level or levels; direct elections to all seats in Panchayats at the village and intermediate level, if any and to the offices of chairpersons of Panchayats at such levels; reservation of seats for the Scheduled Castes an Scheduled Tribes in proportion to their population for membership of Panchayats and office of chairpersons in Panchayats at each level; reservation of not less than one-third of seats for women; fixing tenure of five years for Panchayats and holding elections within a period of six months in the event of supersession of any Panchayat.

Biopesticides

Biopesticides are pest management tools that are based on beneficial micro-organisms (bacteria, viruses, fungi and protozoa) and nematodes or other safe biologically active ingredients. Benefits of biopesticides include effective control of insects, plant diseases and weeds, as well as human and environmental safety. Biopesticides also play an important role in providing pest management tools in those area where pesticide resistance, nitch markets, and environmental concerns limit the use of chemical pesticide products.

Economic Poison

As defined under the Federal Insecticide, Fungicide, and Rodenticide Act, economic poison 'means any substance or mixture of substances intended for preventing, destroying, repelling or mitigating insects, rodents, nematode, fungi or weeds, or any other forms of life declared to be pests and any substance or mixtures of substances intended for use as a plant regulator, depoliant or desiccant'. As so defined, economic poisons are now known as generally as pesticides.

Integrated Fertility Management (IFM)

Technologies for management of fertilizing the crops with the objective to produce optimum crop yields with minimum fertilizer inputs taking into consideration ecological and socio-economic resources available for supply to the crop under a given agro-ecosystem, without deteriorating the soil fertility, such as application of organic manures, green manures, blue green algae and biofertilizers along with use of inorganic fertilizers for higher yields and improving the soil productivity.

You Don't Know!

First solary resturant

The solar restaurant at the *Dilli Haat,* Delhi, launched on Jan 1, 1996 serve solar cooked food between 12.30 to 3.30 p.m. The resturant also has a service counter for repairing solar cookers.

Integrated Pest Management (IPM)

It is a broad ecological approach that minimizes pest population below economic threshold level (ETL) by employing all available pest control techniques like mechanical, biological, chemical and crop management practices in a compatible manners.

Integrated Weed Management (IWM)

A weed management system that uses all suitable control methods in a compatible manner, to reduce weed population and maintain them at levels below those causing economic injury.

RESOURCESAT

In a remarkable launch, the PSLV-C5 put in orbit the RESOURCESAT, the world's most sophisticated remote-sensing spacecraft. Resourcesat has the most sophisticated cameras on board to produce imageries, which will be helpful in determining the health of crops, locating groundwater availability, surveying whether rivers, lakes and ponds are shrinking, assessing the severity of drought, real time monitoring of floods, mapping wasteland, studying the destruction of forests, delineating landslip-prone areas, and so on.

Bio-crop

A rapid composting technology has been developed by the Department of Environmental Sciences, Tamil Nadu Agricultural University, Coimbatore to produce an enriched organic manure term BIO-CROP. The manure is made from coir-dust which is a byproduct of coir industry. Application of the manure resulted in an increase in the available nitrogen, phosphorous and potassium level in soil. The high organic matter content of manure

GK **Net National Product (NNP) is obtained by subtracting depreciation value from Gross National Product.** GS

improves soil properties.

HS 375

The new wheat variety HS 375 (Himgiri) is suitable for its cultivation in very high altitude areas of northern hills zone in summer season. The average yield of the variety is 26.6 q/ha and possesses high degree of resistance against rust pathogens. Adoption of this variety by farmers of this hard core region would help for improving their economy and dietary requirements.

Terminator Technology

Technology meant for stopping gene flow from GM (Genetically Modified) Plants is the terminator technology. This technology is meant to 'Kill' the seeds of GM plants. This means that farmers will get a good crop in the first year of sowing but if they try to save the harvested seed for replanting, the crop would be sterile. In other words, farmers will be left with no choice but to buy seed afresh for every sowing.

International Year of Rice

The United Nations Food and Agriculture Organisation (FAO) on October 31, declared 2004 the "International Year of Rice". FAO Director General Jacques Diouf while launching the year, the slogan for which is "Rice is Life", said 'almost a billion households in Asia, Africa and the Americans depend on rice systems for their main source of employment and livelihood. The year of rice will act as a catalyst for country-driven programmes throughout the world.

Joint Forest Management (JFMA) Act

The National Forest Policy of 1988, Ministry of Environment and Forest, Government of India issued guidelines to all state forestry departments in 1990 to ensure participation of communities and NGO's in the management, regeneration and protection of degraded forest lands. Twenty seven states have so far issued resolutions for JFM. As on December 1, 2001, 14.25 million hectares of forest land in the country being managed and protected by 63,000 JFMs.

Biodiversity

Biodiversity is the variety and variability of all living forms on earth. In simple words, it is the sum total of all living plants, animals and microbes. Biodiversity is essential as mankind is dependent on other species for the sustenance of the biosphere and the supply of basic necessities, especially food. There is a growing concern about the biodiversity because a large number of species are extinct or facing extinction. There are about 5 to 30 million living species, of these only 1.4 million have been described. Whereas, the species extinction rates are approximately 10,000 to 20,000 per year. The warm tropical regions are the richest habitat of biodiversity. Conservation of biodiversity is of utmost importance because plants and animals provide us many known and unknown economic and essential services like cures for diseases and genes to increase the yield of our food crops.

Biogas

A mixture of gases containing methane, carbon dioxide, hydrogen and traces of few other gases produced by the anaerobic fermentation of easily decomposable cellulosic materials in presence of methane-forming bacteria.

Nuclear Agriculture

Radiation technology is a potent tool in the development of high yielding seed varieties by mutation increasing self life of food by way of delayed ripening and sprouting, and pest control by killing insects and pests - even those that hide within seeds. The Nuclear Agriculture Programme of the Department of Atomic Energy (DAE) focuses on the development of high yielding crop seeds, radiation processing of food items, fertilizer and pesticide related studies, and other areas. All these technologies have been benefiting Indian farmers and traders.

Women in Agriculture

Rural women are responsible for half of the world's food production and produce between 60-80% of the food in most developing countries. In developing countries in addition to food production women are responsible for multiple roles in household task such as care of family members, collection of water, fodder, fuel, forest products and tending the animals. The relationship of rural women have been associated with nature, agriculture, forests, flora and fauna dates back to human history. Women collecting and conserving plants contributed 50-70% of dietary requirements. Women continue to function as seed gatherers, seed conservers and possess the knowledge of plants that cure. In many regions of the world up to 90% of the planting material of the farming communities produce, select and save themselves. Women use their traditional knowledge in selecting, preserving, conserving seed and planting materials for subsequent use. In tribal culture where human and livestock ailments are cured with the medicinal plants, women play a vital role in collecting

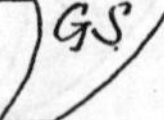

the medicinal plants. The rural households invariably have cattle, hens, goats, sheep, ducks and pigs according to their cultural requirements and available resources. Here women play a predominant role in selecting best breeds. It is clear that women have a central role in agriculture and they possess the precious indigenous knowledge passed down from generations.

Bioinformatics

The study of the application of computer and statistical techniques to the management of biological information. In genome projects, bioinformatics include the development of methods to search database quickly, to analyze DNA sequence information, and to predict protein sequence and structure from DNA sequence data.

World Bamboo Congress

The World Bamboo Congress, which is organised once in three years by the World Bamboo Organisation. National Mission on Bamboo Technology and Trade Development has plan to develop bamboo resources for industries and trade. Classifying bamboo as an agriculture product and terming it by 'golden green' would be an important step for promoting it. WBO is going to highlight India as a source for bamboo products and technology. The seventh World Bamboo Congress was held in New Delhi from February 27 to March 4, 2004.

Complete Chicken Genome Sequenced

The chicken has joined the growing group of animals whose genome has been sequenced. The finding reveal that, like us, the bird has between 20,000 and 23,000 genes. But it has only 1 billion DNA base pairs. The reduced number of base pairs in the fowl genome results in part from chickens possessing less so-called junk DNA. The results should help scientists better understand basic developmental biology, as well as improve vaccine production models.

Vertical Revolution in Agriculture

Maximizing production per unit land area per unit of time using intensive cropping systems, high production inputs and improved management practices.

Work Force in Agriculture

A majority of the work force still depends on agriculture for livelihood, the study said adding in Madhya Pradesh 71.6% of total work force is employed in the sector. The level of work force engaged in agriculture stood at 66% each in Rajasthan and Uttar Pradesh, 62% in Andhra Pradesh, 77% in Bihar and 65% in Orissa. However, the relatively prosperous state of Punjab has only 39% of its population depending on agriculture for its livelihood.

Knowledge World!

Youngest PM

Rajiv Gandhi assumed office when he was 40, on Oct. 31, 1984, following the assassination of his mother Indira Gandhi.

Mushroom Rubber

Japanese researchers have produced rubber from a substance extracted from a wild Chichitake Mushroom. It takes more than 10 kg of mushroom to make one kg of the rubber. Along with rubber produced, the end product also have the advantage of not containing a protein that can cause allergies among humans.

Soil Solarization

Soil solarization is a useful non-chemical and eco-friendly tool that can be integrated into the pest management programme. Various researchers have recommended soil solarization for weed control. The basic phenomenon helping weed control by soil solarization being the lethally high temperature build up in the soil. There are many locations in India experiencing extreme summers and in such places, soil solarization with transparent polythene sheet of 0.05 mm thickness for 30-45 days between March-May as an effective weed management technique can be followed. Hence, soil solarization seems to be the perfect answer to the age old problem of preventing weeds to certain extent as the dormant weed seeds required repeated manual interventions. The profitability of soil solarization should be viewed as an holistic approach in controlling weeds, diseases, insect-pests, plant nutrient availability and enhanced crop yields and it is economical if cost : benefit ratio is assessed based on break even cost of solarizing material with 2-3 years longevity.

Varsha-Bima Policy

The Agriculture Insurance Company of India Ltd (AIC) has launched a *Varsha-Bima* policy for farmers to provide cover against deficient rainfall during the Kharif season. The scheme was introduced on a pilot basis in 20 raingauge stations of the Indian Meteorological Department (IMD). These include Anantpur, Cuddapah, Kurnool, Khammam, Hanama-Konda and Nizamabad in Andhra Pradesh, Belgam, Gulbarga, Bellary and Karwar in Karnataka, Lucknow, Jhansi, Allahabad, Gorakhpur, Gonda and Bareilly in Uttar Pradesh and Alwar, Kota, Bikaner and Barmer in Rajasthan. The scheme would

In Siwaliks range of Himalayas, reminants of ancient human civilization have been found.

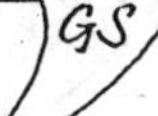

offer coverage against actual rainfall in a particular station during the June-September monsoon period falling 20% or more below the 'normal' for that period. The claim payout would be on a graded slab (i.e. Rs. per mm) basis and computed as per mm payout at given rainfall range times the deviation in actual rainfall.

Agri Vision 2020

- Conservation, planned enhancement and utilization of agro-biodiversity.
- Enhancing productivity through evolution of high-yielding hybrids and varieties.
- Research on diversification, quality improvement, post-harvest technology, value-addition and export-oriented commodities.
- Sustaining enhanced productivity of irrigated agriculture and judicious development and use of energy, especially renewable sources of energy
- Characterization and development of sustainable land-use models for rainfed agriculture in high-rainfall areas.
- Development of the Integrated Pest Management (IPM) and Integrated Nutrient Management System (INMS) approaches and systems for sustainable agriculture.
- Fostering excellence in relevant basic and strategic research.
- Generating research and technologies geared to promote equity among regions, sectors of society and gender.
- Strengthening social science, policy planning, agri-business and research monitoring mechanisms.
- Strengthening Agricultural Research Information System (ARIS).
- Promoting the Agricultural Human Resource Development (AHRD).
- Linking scientists with the farmers through institute- Village-Linkage Programme (IVLP) as an innovative technology transfer model.
- Institutionalization and strengthening linkages/ partnerships with CGIAR and under national and international agencies and research and development establishment, non-government organizations (NGOs), farmer organizations, private sector etc.
- Explore alternative or non-conventional plant resources for diversification of agriculture so as to enhance the food, fodder, fibre and fuelwood production.
- Provide solutions which are relevant, technically sound, economically viable, socio-culturally acceptable, eco-friendly and systems-responsive.
- Enhance the use of biofertilizers and biopesticides and biodegradable chemicals in agriculture.
- Strengthen the production base by harnessing modem scientific knowledge and translate it in the form of crop plants which are high-yielding, resistant to biotic and abiotic stresses, and efficient in input use.
- Improve skill and productivity of farm women with accent on social engineering and empowerment.
- State-of-art techology for post-harvest managment and agro-processing.

National Food Processing Policy

Agriculture often suffers from varietal constraints. Oranges in Nagpur and mangoes in Malihabad in UP are instances where though there are gluts at harvest time, processing has not taken off due to want of processable varieties, the policy document said. In context the policy document suggested various steps to augment the sector.

- Changes in state laws for regulated markets for allowing private sector participation.
- Contract farming
- Private sector participation in agricultural extension services
- Procurement of farmers' produces by large food retailers or markets
- Improving post-harvest management, cleaning, grading, packaging and storage.
- Existing cold storage facility caters to less than 10% of the produces. It is mainly for potatoes and is rudimentary in nature. Hence there is a need for expanding cold storage facility for other perishable produces.
- Most of the limited controlled atmosphere storage facilities, technologies, protocols and machineries are imported. There is, therefore a need to upgrade manufacturing capabilities.
- Upgradation of research labs.
- Removal of conditionalities and caps on financing.
- Warehousing receipts should be made negotiable instruments also there should be a system of accreditation of warehouses.
- Commodity markets should be stable and future markets more deepened. Banks, mutual funds and FIIs should be allowed in future trading and derivatives
- Liberal funding from RIDF for infrastructure development for food processing sector. State should participate in joint ventures for infrastructure development.

Telod is associated with Indus system.

GS

- R & D on ethnic foods.
- Apex research institute should be set up to address R & D problems for the sector.
- Norms relating to Codex, HACCP, GMP and GHP should be met.
- Integrated food law and liberal labour laws.

National Horticultural Mission

As per NCMP, the Government of India will pay focused attention on the overall development of horticulture in the country by launching a National Horticulture Mission. It was also highlighted in the budget speech of the Union Finance Minister that a National Horticulture Mission will be implemented for doubling horticulture production in the country by 2011-12. The mission will be implemented shortly. The Government has cleared a centrally-sponsored scheme on National Horticulture Mission (NHM)with an outlay of Rs. 2,300-crore in the Tenth Plan. This is likely to bring under horticultural crops about 33 lakh hectares additional area and to rejuvenate 16 lakh hectares of senile plantation by the end of the Eleventh Five Year Plan (2007-12). Besides this, about 21 lakh hectares would be covered under the integrated pest management and also about 2,040 nurseries would be established to make available good quality planting material.

The basic objective of the NHM is to double horticulture production to 300 million tonnes by 2011-12. The NHM will bring technological interventions through forward and backward linkages from production to marketing and processing of horticultural produce.

The importance of horticulture in improving the productivity of land, generating employment, improving economic conditions of the farmers and entrepreneurs, enhancing exports and, above all, providing nutritional security to the people, is widely acknowledged. Horticulture sector, which includes fruits, vegetables, spices, floriculture, and coconut, among others, covered 17.2 million hectares of land in 2003-04, accounting for 8.5 per cent of the gross cropped area of the country with production of 156.1 million tonnes in 2003- 04, the sector contributed 30 per cent of GDP from agriculture. It occupies 8.5 per cent area and 10 per cent of the total agricultural export earnings. The targeted growth rate during the Tenth Plan for the sector is 6 to 7 per cent.

With fruit production at 47.5 million tonnes in 2003-04, India accounted for about 10 per cent of the global production of fruits from an area of 4.0 million hectares and was the second largest producer of fruits in the world.

With 90 million tonnes of vegetable production in 2003-04, India ranked as the highest producer of vegetables in the world. In the world, India occupied first position in the production of cauliflower, second in onion and third in cabbage. A panel, appointed by the Government, reports that the country is facing recurring post-harvest losses of horticultural crops of about 5000 crore per annum.

Quiz Kit!

State contributing most Presidents

Andhra Pradesh has contributed three Presidents so far — Dr. Sarvepalli Radhakrishnan, Varahagiri Venkata Giri and Neelam Sanjiva Reddy.

The NHM will primarily focus on (i) horticultural research and development; (ii) to generate good quality seeds and planting material ; (iii) to have coverage of area with improved varieties; and (iv) to conduct productivity improvement programmes.

It will also address post-harvest management and marketing by fostering infrastructure facilities for cold chains, market yards and market intelligence and value addition. The Centre will also give suggestions to the States on what to cultivate with export also in mind. The mission will also be looking at marketing and export promotion.

The NHM will operate on three levels, the Mission at the Centre and State and District level missions.

Mission General Council is to be comprised of:

- Ministers and Secretaries of Ministries concerned.
- Representatives from FICCI.
- Exporters and Growers.
- Members of Indian Council of Agriculture Research.
- Council of Scientific and Industrial Research.
- National Bank for Agriculture and Rural Development.

The Development is planning to bring in 33 lakh hectares of additional area under cover of fruits and vegetables. 2,040 nurseries will be established so that good quality planting material is available for farmers. The NHM will also focus on post harvesting technology.

India has 39 Agri-Export Zones (AEZs). AEZs helps In exporting fresh and processed fruits and vegetables. The AEZs are going to be further strengthened by the investment of Rs. 1,018.85 crore from the private sector, the central and the state government.

National Bamboo Mission

Bamboo, a versatile grass, is yet another crop, whose potential has remained untapped. There is wide scope for growing bamboo both in the forest and non-forest areas, and thereby generating additional income to the farmers besides regenerating the soil health. A special

USA constitution influenced the 'Preamble' of Indian Constitution.

GS

scheme has been launched to cover an area of two million hectares during the Tenth Plan in a mission mode approach integrating the production. Post-felling management, value addition and marketing aspects. It is targeted to cover a total area of 6 million hectares under bamboo cultivation during the Tenth and the Eleventh Plans.

Impact of Advanced Agricultural Technology

Based on the case study conducted in Ludhiana district of Punjab. The following visible changes are noticed.

- A real agricultural has taken place which has integrated the rural revolution.
- Old traditions are disappearing, giving rise to new ways of living.
- New schools are springing up.
- Interior cultivation is practiced with higher level of inputs.
- Rural banks advances are availed and some are taking up additional professions to augment the income from agriculture.
- Huts and rural walled houses replaced by durable brick construction.
- Many houses are provided with furniture, vehicles, TV, radio etc.
- There is medical facility.
- Mechanization of good proportion of farm operation.
- There were many work opportunities available in agriculture.

System of Rice Intensification

System of rice intensification (SRI) concept is an alternate to the existing practises where fields are kept wet, not flooded with water. This system seems to be promising to overcome the shortage of water in irrigated rice. The SRI appear to be a viable alternate for rice cultivation that saves expensive inputs, improves soil health besides optimum water-use efficiency.

The system of rice intensification originated in Madagascar and was first synthesized in 1983 by Fr. Henri de Laulanie. Under the drought conditions of that year. There are three basic concepts constituting system of rice intensification. They are to be applied with adjustments in specific practices like spacing and timing as appropriate to local conditions. Transplanting the seedlings while still young, <15-days-old, i.e., prior to the start of the 4th phyllochron of growth. This preserves plants potential for tillering and root growth that is reduced by transplanting later. Direct seeding is an option, however, since what is important is that plant roots should not be traumatized after they start their growth trajectory, on or about the 15th day. Careful transplanting promotes rapid resumption of growth.

Wide spacing between plants is maintained with preferably just one plant per hill, and set them out in a square pattern, 25 cm x 25 cm or even wider if soil fertility is good. This gives room for profuse root and tiller growth, achieving the border effect throughout the whole field. Another important aspect in SRI is to keep the soil both moist and aerated, at least during the vegetative growth period, so that roots have access to both oxygen and water. Under continuously hypoxic conditions, rice roots degenerate, with as many as 75% become dysfunctional by panicle initiation.

Therefore, the aerobic conditions in SRI promote root growth substantially over submerged situations.

Specific practices include in addition to using young seedlings with still only two leaves, planted singly and widely spaced: frequent weeding, preferably with a rotating hoe that aerates the soil while it prevents weeds growth by churning them into the soil; application of small amounts of water daily, or alternate flooding and drying of fields for 3 to 6-day periods; and providing organic matter to the soil. Organic matter can be any decomposed biomass, including rice straw or weeds; manure or mulch can also be used. Green manures and cover crops are also useful. Correct implementation of the various, interdependent components (SRI practices) requires considerable field experience in which close observation and timeliness are crucial in minimizing labour requirements while maximizing total factor productivity.

Benfits associated with SRI

- Greater root growth
- Increased tillering
- Increased grain filling
- Improved grain quality and higher grain weight
- Water savings
- Less lodging
- Reduced pest and disease attacks
- Reduced or no need for chemical fertilizers
- Seed savings
- No need to purchase new seeds
- Low cost of production
- Increased factor productivity and profitability
- Reduced risks

Rice yields all over the world have leveled out under the present system of flooded cultivation. We need to be looking for alternatives to existing practices with an open mind. SRI is still evolving and it is hoped that the scientific community will collaborate in further refining

the SRI practices and working out the scientific reasons for the reported higher productivity. Planting young seedlings carefully and at wider spacing gives the plant more time and space for tillering and root growth. Careful water management keeping the field wet and not flooded gives better yield as it supports healthy root growth. This practice should be encouraged everywhere as the whole world is facing water shortages. Weeding rice fields with a rotary weeder helps by churning the soil and incorporating the weed biomass as it aerates the root zone. Therefore, seeing the overall benefits, the SRI needs to be evaluated further and refined to suit the local environments.

Hi-tech Horticulture

Hi-tech technologies are which can be defined as those, which are modern, less environment dependent, capital intensive and have capacity to improve productivity and quality.

Thematic areas

- Germplasm and crop improvement.
- Hi-tech crop production of horticultural crops.
- Protected cultivation.
- Hi-tech plant protection.
- Modern post harvest management.
- Logistics and polices for Hi-tech horticulture.
- Policy issues.

Overview of Hi-tech horticulture

Horticulture crops occupy 6.7% of gross cropped area, 18% of gross value of agriculture output and 50% of export earning. Horticulture today is not merely a means of diversification from an integeral part of food and nutritional security.

Genetic resource conservation of characteristics

- *In vitro* collection and conservation of fruit like mango, jackfruit.
- Tissue culture techniques.
- Cryo preservation
- Gene bank.

Hi-tech propagation

Micro propagation :

- Tissue culture units for propagation of banana, papaya, cardamom, bamboo etc.
- Hybrid seed and seedling production in vegetable and flower crops

Use of male sterile:

- Flower - Marigold
- Vegetable - Cabbage.

Did You Know ?

Shortest term as Speaker

Neelam Sanjiva Reddy was Speaker from March-July, 1977.

Hi-tech crop production

High density planting:

- *Mango*: Amarpali
- *Apple*: Malling (m) and malling marton (Mm) root stock.

Use of growth retardants

CCC, B-9

Integrated nutrient management :

- Fertilizer (slow release and coated)
- Organic manuring (FYM, poultry etc.)
- Green manuring
- Green leaf manuring
- Composting (farm-town compost)
- Bio fertilizer
- Vermiculture

Micro irrigation

- Drip irrigation – root zone application
- Sprinkler – for dew effect
- Fertigation (both water + fertilizer)
- Chemigation
- Herbigation

Organic farming

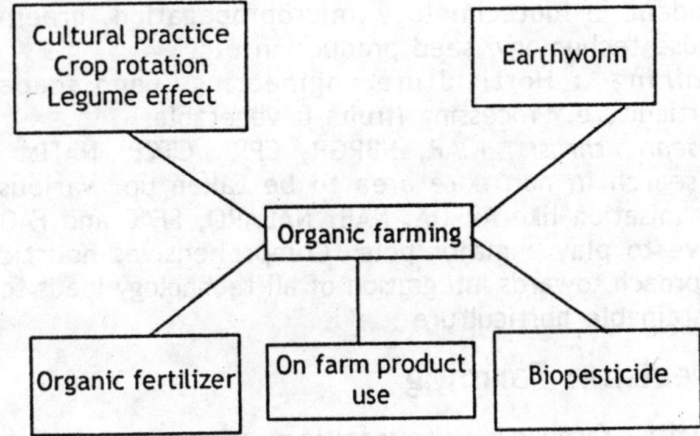

Protected cultivation

- Green house cultivation
- Low cost wire shade nets.
- Mist chamber for propagation

Crop protection

- Integrated pest management
- Pesticide application methods.

GK **Personal income is that income which is actually obtained by nationals.** GS

- Ultra low volume spraying
- Bio pesticide
  - NPV
  - GV
  - Bt formulation
- Pesticide residue management
- Molecular diagnostic
- Post harvest technology
  - Grading: maturity, colour, shape, texture, flavour
  - Packing and storage : modified atmosphere packaging, controlled atmosphere packaging, bio degradable film packaging

Cold chain : Cold chain is temperature management facility involving a number of equipment such as pre-cooling unit, cold storage, humidity controlled atmospheric storage, refrigerated and storage system.
Irradiation: It refers to physical means of exposing fresh or pre packed products to gamma rays. *Flowers:* rose, gladiolus, *Vegetables* : canned, preserved candy, *Fruits:* mango, citrus
Hi-tech processing : IQF –Individual quick freezing, Osmotic & Aseptic technologies.
MechanisationlAutomation
Quality standards : Sanitary and phytosanitary measure (SP), Technical Barriers to trade (TBT)

Logistics and policies
IPR issues
WTO concerns : Novelty, Distinctiveness, Imitating
Organisational support : National Horticultural mission, Budget allocation to NHC (National Horticulture Commission), NHB, APEDA
Human resource development : Advance training for student in biotechnology, micropropagation, green-house technology, seed production etc.
Training : Horticulture engineering, Land scape horticulture, Processing (fruits & vegetables)
Research thrust : ICAR, NBPGR, CPRI, CTCRI, NATP. Research in non core area to be taken up. Various organisation like APEDA, NAB, NABARO, SFAC and FAO have to play a major pole. Comprehensive, holistic approach towards integration of all technology leads to sustainable horticulture.

Precision Farming

Precision farming is a site specific management approach in modern crop production for applying various input to the field in an economical and sound fashion.

Aim
To identify and meet the real needs of variable parts of the field.

Used for
Precise application of agriculture inputs like 1. Seed, 2 Fertilizer, Pesticide, 3. Irigation etc.

Components
1. Yield monitoring, 2. Yield mapping, 3. Variable application of fertilizer, 4. Salinity mapping, 5. Guidance system, 7. Record and analysis.

Requirements
1. Field map, 2. Variable rate of input, 3. Differential Global Position System (DGPS).

Procedure
1. Selection of inputs and resource, 2. Proper study of GPS, GIS, range of sensor, 3. Monitoring and collection of input data, 4. Variable rate of application, 5. Achievements by harvesting.

Advantage
Less time, money, input loss.

Disadvantage
1. Input cost is high, 2. Fragment condition of the field.

India's scenario
1. Punjab, Haryana, it is possible because of 14 acre per capita, 2. Other states by cooperative or contact farming is possible, 3.It is in expanding stage.

Greenhouse Technology

It is a controlled environment (protected) cultivation. It has been evaluated to create favourable micro-climate in which crop production can be made possible throughout the year/part of the year as required. It can be called as "*food factories*". *It includes many of the followings :*

- Bench /beds
- Misters/fogger
- Drainage
- Light (fluorescent/PAR lamps)
- Heating & cooling system
- Irregation systems
- CO_2 enrichment
- Controlled system.

Benefits
- Raising any crop, any where, anytime.
- Yield is maximum
- High Quality Product (for exports)
- Raising of seedlings.
- Crop productivity several times higher than open conditions.
- Income generation for small farmers (for exports)
- Utilisation of wasteland

Types
- Thatch house: making dark area (structures)
- Shade nets for filtration of sun light (entire nets)

Example

- Vegetable; cauliflower, cabbage, tomato chillies, brinjal, cucumber
- Flowers; rose, orchids, chrysanthemum
- Fruits; strawberry

Disadvantages

- High labour requirement
- Higher cost of production
- Initial capital investment is high

Remote Sensing

Measurement or acquisition of information of some property of an object/phenomenon, by a recording device that is not in physical or inmate contact with the object/phenomenon under study.

Principle

Based on the reflected radiation energy from different areas on the land surface over a range of wavelength, the objects are distinguished.

Types of sensors

Active: Sensing device directs/creates EMR

Passive: Using other energy/natural energy sources (sun).

Types of sensing

Airborne remote sensing: It is the method of remotesensing in which cameras or other devices fixed in an aircraft flying at fixes altitude are used to take photographs of any sources on earth.

Types of aerial photographs.

Infrared aerial photograph

Black & white aerial photography: Particularly for mapping boundaries of water sources.

Colour aerial photography : Disease detection, Identification of tree species in temperature zones.

Space/Satellite remote sensing

It is the method of remote sensing in which cameras, sensors and other devices attached to a satellite, orbitting round the earth take photographs of earth and its resources.

Stages of remote sensing

- Origin of EMR
- Emission of energy from source to surface
- Interaction between energy in earth surface
- Transmission of reflected energy to the sensor.
- Detection of energy & converting into photographic image/electrical.
- Recording the sensor
- Data processing
- Collection of spacial information.

You Don't Know!

First hijack

A Fokker Friendship of the Indian Airlines was hijacked to Lahore on Jan. 30, 1971. The plane was later destroyed by the hijackers.

- Data processing & interpretation.

Basic concern

(i) Radiation from sun and other terestrial sources: EMR spans a large spectrum of length from 10^{-10} m (α-ray) to 10^6 m (long radiowaves).

In remote sensing useful regions are:

- Visible (0.4 - 0.7 μm)
- Infrared (0.7 - 3.0 μm)
- Thermal infrared (3 - 5 μm)
- Microwave regions (0.3 - 300 cm)

The sun is important source of EMR used in conventional optical remote sensing.

(ii) Effects of atmosphere: Electromagnetic radiation is scattered and absorbed by gases and particulate matters while passing through atmosphere.

Signatures concept

Any set of observable charateristic which is directly or indirectly related to the identification of an object and or its condition.

Spectral

Change in reflectance/emittance of objects as a function of wavelength. *e.g.* Forest.

A brief account of remote sensing applications in agriculture

A. Scientific applications of relevance in agriculture: All types of vegetation indices, Stress detection, Agro-climatology, Agro-ecology, Global change. Hydrology - surface run off ground water, Land surface climatology, Soil science, Terrestrial ecosystems, Vegetation sciences, Forest science

B. Management applications sensing in agriculture: Agriculture, Agro-ecology, Crop damage, Crop condition, Crop inventory - local, regional, national & global, Crop classification, Live stock inventory, Yield estimation, Irrigated lands inventory, Monitoring extent, Crop condition, Water utilisation, Regulation of water use, Land evaluation, Land use, Land degradation, Yield prediction, Degraded land studies, Environmental impact, Studies of agricultural development, urban development, GIS - contributions to regional agricultural information systems, Land reclamation, Agricultural & Urban Soil pollution, Vegetation, Range land management, Range land condition, Range land classification, Range production, Soil erosion, Soil salinity, Soil survey, Land drainage and Land classification

GK GS

Brahmaputra and Indus rivers make the eastern and western limit of Himalaya respectively.

Polarisation

Changes in polarisation of radiation reflected or emitted by an object.

Analysis & generated data

Visual analysis : Colour, texture, shape size etc.

Digital image processing techniques : Computer analyse the shapes signature so as to associate each pixel with a particular feature of imaging.

Application of RS in Agriculture in India

- Crop production forecasting.
- Land use
- Mapping of wastelands
- Soil mapping
- Drought monitoring & its assessment
- Monitoring of surface water bodies
- Ground water exploration
- Flood supply & damage assessment.

Mangrove Forests of India

Bordering the Indian ocean and its vicinity, mangrove vegetation is found in the Sunderbans in the estuary of the Ganges, Brahmputra system, the Andaman Islands and the Irrawady delta. Small remnants of mangroves are also found in the estuaries of the Mahanadi, the Godavari and the Krishna on the east coast. The west coast mangroves can be divided into two forms the open scrubby mangroves of the Kutch and Saurashtra coasts and the close forests extending from the mouth of the Narmada and Tapti southwards in the Karnataka and Kerala states.

(a) The coast line of Saurashtra and Kutch (Gujarat) :

On sea coast *Avicennia officinalis* associated with *Aegiceras mayus, Pandanus tectorius*; on salt marshes along the coast *Bruguiera gymnorhiza* together with *Aeluropus villosus* grass; in salt marshes and tidal creeks *Rhizophora mucronata* and *R. conjugata* are common.

(b) The coast line of Maharashtra from the mouth of the Narmada southwards:

Coastal sea water and semi-fluid mud area has *Avicennia officinalis, Acanthus ilicifolius, Rhizophora mucronata.* Salt marshes of tidal creeks have *Rhizophora mucronata, R. conjugata, Sonneratia apetala, S. acida, Excoecaria agallocha, Lumnitzera racemosa* etc.

(c) The coast tine of the Karnataka, Kerala and Tamilnadu:

The mangrove forests have been divided into three zones. The coastal sea water and semi-fluid mud zone contain *Rhizophoramucronata, R. conugata, Kandelia rheedii, Bruguiera gymnorhiza, B. caryophylloides, Acanthus ilicifolius* etc, Only the West coasts of Karnataka and Kerala contain *Ceriops candolleana* and *Bruguiera eriopetala.*

Salt marsh zone along the coast is dominated by *Lumnitzera racemosa, Excoecaria agallocha* and *Arthrocnemum indicum.* East coast of Tamil Nadu is more rich having species of *Avicennia, Salicornia, Scyphiphora* etc, In the salt marshes of tidal *creeks Aegiceras majus* and *Herritiera littoralis* occur dominantly in addion to *Sonneratia apetala, Avicennia ofjicinalis* (in west coast of Karnataka and Kerala only). Towards the interior dry marsh, lands of east coast of Tamil Nadu only, *Suaeda mudiflora, S. maritima, S. monoica* occur.

Weather forecasting

Within the year 2025 the world population would exceed 850 crores, but the production of Agricultural commodities remains uncertain. Agriculture has to meet this challenge, by increasing production in a sustainable way. Agriculture is still a gamble in a country like India as they lie near the equator.

The weather and climate near equator always remain dynamic and unstable. Hence the prediction of weather in these areas is highly difficult and not stable. Hence any strategy to increase agricultural production should take explicit amount of the complimentary of agrometeorology and thus development.

Already the area available under cultivation is getting reduced day by day due to urbanisation & other factors. Hardly there are lands to bring into cultivation. In this situation, increasing the production per unit area per unit time is essential for which the use of weather factors play a major role in our country. In India, crops on 67 percent of the 14.2 crores of arable land depends on rain. Therefore 44% of the food is from dryland and 40% of the Indian population depend on this dryland. Dryland cultivation is mainly dependent on the prevailing weather parameters. 40% of food grains, 80% of maize, 95% of pearl millet and sorghum, 95% of pulses, 75% of oil seeds are grown in dryland areas.

The following aspects of weather concepts are essential to decide the sustainability of production and productivity in our country.

- the date of onset and termination of rainy season and variability.
- the distribution of rainfall.
- micro climate of the cropped area to decide the occurrence of pest and diseases etc., i.e., air, temperature, relative humidity, dew fall, soil temperature wind speed of the cropped area.

By weather forecasting, decision can be taken in advance in agricultural operations and thus planning can be done.

The following table shows the emphasis and applicability of weather forcasting

Types	Validity & Accuracy	Forecast emphasis on	Applicability
1. Short range forecast	24-48 hours, 70-80% accuracy	Temperature, wind velocity & direcion, sunshine duration time & amount of time.	(i) Scheduling irrigation (ii) Adjusting time of agrlcultural operation.
2. Medium range forecast	up to 10 days, 60-70% accuracy	Types of weather, sequence of rainy days, normal weather hazards	(i) Decide the sowing time (ii) Depth of sowing (iii) Deciding harvesting (iv) Time of spraying (v) Irrigation (vi) Deci son of harvesting
3. Long range forecast	30 days to one season, 60% accuracy	Abnormality of temperature & precipitation	(i) To decide on soil moisture management. (ii) Selection of crop. (iii) Selection of suitable cropping pattern.

There are three types of weather forecasting. Among all these types of weather forecast, medium range weather forecast is given much importance as their applicability can be materialised easily, as their validity and accuracy is satisfactory. In case of short range forecast, the validity is for only one or two days which is a small period, for the farmers to collect the recommendation and then to follow. In case of long range weather forecast the accuracy is not as good as medium range weather forecast.

Contract farming

In this era of globalization, the concept of contract farming has been acquiring a greater momentum. Contract farming is a system for the production and supply of agricultural products under forward contracts between cultivators suppliers and buyers. Here to cultivator commits to provide an agricultural product of a specified type at a specified time and at a specified price that is required by the committed buyers.

The process of contract farming seems to attractive of implemented in its good sense. The main feature of contract farming is that the contractor supplies all the material inputs and technical advise required for cultivation to the cultivator. In turn the cultivator supplies the required land and labour.

In India the central government has drafted a model law on agriculture marketing to provide among other things, legal support to contract farming agreements. Some of the state government has entered into a contract farm for various crops.

Modernization of Agriculture

Agriculture is not an industry. It is an art and a way of life dealing with living beings. In the past years, it has undergone great changes all over the world. The important strategies of agriculture are:

1. Increased food production
2. Saving the environment and attaining sustainability There has been significant increase in the quantity of food produced and land conserved. This is the effect of modernization.

Drawbacks of modernization

- Cuts the number of jobs, people engaged.
- Workers replaced.
- Rural culture put under pressure.
- People migrate in search of work.
- Local institutions weakened.
- Resources once considered as precious, has now turned a waste to be disposed off.
- Water, soil contaminated.
- Agriculture has become fossil fuel intensive.
- Increased application of pesticide.
- Agriculture has begun contributing to global warming.

Reasons for the rejection of modern methods by the poor farmers

- High cost.
- Poor extension.
- Inappropriate incentives.
- Inappropriate innovation.
- General unawareness.
- Local practices are better than modern methods.
- Modern methods generate new problems.

Modernization of agriculture in Indian context

In India when the mechanization of bullock energy changed to power tiller use, about 8% increase in output required 43% increase in energy input. For tractor 13% increase in yield cost 74% extra energy. For wheat tractor 6% increase in yield cost 6% extra energy.

Biodiversity Loss due to Modernization of Agriculture

- In India, once 30,000 varieties of rice were grown.

From Canada constitution the expression 'Union of States' was borrowed.

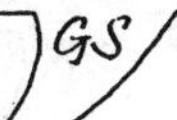

But now 10 varieties cover 75% of the whole rice area. In the U.S. 65 types of vegetables lose 80-100% of the varieties of each since the turn of the 20th century. Of the 8207 varieties listed in 1903, only 607 are stored in seed forms. The greater the uniformity the greater the risk and danger of pest, risk and disease attack.

- Loss of genetic diversity leads to future opportunity to raise adaptable crops and livestock.
- A farmer prefers multi-cropping, inter-cropping, mixed-cropping but does not standardize his practices.
- Farmers do not replace existing varieties of crops by a new variety. They add the new to the old adding to the variety.

Modernization of agriculture and breakdown of rural communities

- Number of jobs for local people gone.
- Standardization has reduced the management skills.
- Rural poverty increases.
- Food processing jobs are no more localized. Far away central factories do that work.
- Growing gap between the rich and the poor.
- Landscape homogenized; farming system simplified.
- Youth show less interest in labour intensive farming. Many farms have no successors.
- Farm size increases, family farms vanish; fewer people depend upon agriculture incomes.
- Frustration, financial crisis, suicide on the increase among farmers.

Self-Help Groups

It is a group of rural poor to have volunteer to organise themselves into a group for eradication of poverty of a members.

Stages

- Group formation.
- Capital formation through revolving fund & skill development.
- Income generation through economic activity.

Important features

- It consists of 10-20 persons
- All members of group belongs to families of BPL
- These can be one person from one family
- One person cannot be member of more than one group
- Broad alliance with foreign countries
- Both public & private investments should be encouraged
- An efficient marketing system available

Banking loans to SHG's

- Through NGO/Voluntary agency.
- NABARD promotes SHG's (improve rural poor).

Objectives

- Enabling member to become self dependent & self-reliant.
- Providing forum for members of discussing social & economic aspects.
- Enhancing social status of members.
- Providing patterns of members for exchange of idea;

Natural Farming

Concept

Natural farming requires no machines, no chemicals and very little weeding. In other words natural farming is an attempt to develop a method of farming which could help to reverse the degenerative momentum of modern agriculture.

Principles

The following steps are important to make farming in natural way.

- No cultivation-stirring of the soil by plough
- No chemical fertilizers-addition of crop residues and organic manures
- No tillage-mulching with plant materials
- No dependence on chemicals-plant extracts can be used for plant protection

The main purpose of farming is to add ecological sustainability and agricultural development. According to Dr. M.S. Swaminathan, the basic principles of sustainability include, land, water energy, nutrient supply, pest management, post harvest systems and systems approach. These factors are closely linked together for sustainability. Another important aspect in natural farmings is preventive crop protection. The farmers should know agricultural ecosystems, like mixed farming and diversification, crop rotation, optimum planting season for all crops, manuring and plant health, choice of varieties and field hygiene. Since insects very often attack crops and other pests, the knowledge of above items is important. Use of pesticides had led to disturbances in the natural eco-system and poisonous chemicals and up in ground water eventually having harmful environmental consequences. Hence, preventive crop protection principles are vital for natural farming. Some ecologically friendly plants, e.g., Neem, Pungam, Jatropha etc., help to promote ecological balance. The integrated farming systems can help in ensuring self-

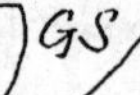

Bengal and Punjab, two provinces had non-Congress ministers after the 1937 elections.

sufficiency in farm holding, increase and help in ecological balance by recycling of farm residue and wastes.

Natural farming is a traditional form of agricultural practice popularly associated with a back-to-nature movement. Hence, it is to make maximum possible use of natural resources. thereby avoiding the use of chemicals liable to kill or reduce the activity of beneficial soil organisms and avoiding direct feeding of land with soluble minerals.

Phytoremediation

Phytoremediation is the direct use of living plants for remediation of contaminated soil, sludges, sediments, or ground water through contaminant removal, degradation, or containment of the contaminant(s). Growing and, in some cases, harvesting plants on a contaminated site as a remediation method is an aesthetically pleasing, solar-energy driven, passive technique that can be, used to remediate sites with shallow, low to moderate levels of contamination.

Significance of phytoremediation

Phytoremediation takes advantage of natural plant processes. It requires less equipment and labour than other methods since plants do most of the work. Trees and plants can make a site more attractive as well. The site can be cleaned up without removing polluted soil or pumping polluted ground water. This allows workers to avoid contact with harmful chemicals. Phytoremediation has been successfully tested in many locations, and is being used at several superfund sites. The time it takes to clean up a site using phytoremediation depends on :

- Type and number of plants being used
- Type and amounts of harmful chemicals present
- Size and depth of the polluted area
- Type of soil and conditions present

These factors vary from site to site. Plants may have to be replaced if bad weather or animals destroy them. This adds time to the cleanup. Often it takes many years to clean up a site with phytoremediation; tree roots take in water and pollution from the ground polluted soil water enters tree where pollution is cleaned up.

Principal

Phytoremediation works best at sites with low to medium amounts of pollutions. Plants remove harmful chemical from the ground when their roots take in water and nutrients from polluted soil, streams, and ground water. Plants can clean up chemicals as deep as their roots can grow. Tree roots grow deeper than smaller plants, so they are used to reach pollution deeper in the ground. Once inside the plant, chemicals can be :

Quiz Kit!

Longest epic

The *Mahabharata,* composed by Veda Vyasa around 1000 B.C. chronicles the battle between the children of two brother, Pandu and Dhritarashtra, in more than 1,00,000 slokas.

- Stored in the roots, stems, or leaves
- Changed into less harmful chemicals within the plant
- Changed into gases that are released into the air as the plant transpires (breathes).

Mechanisms

Phytoremediation uses one basic concept: the plant takes the pollutant through the roots. The pollutant can be stored in the plant (phytoextraction), volatized by the plant (phytovolatization), metabolized by the plant (phytodegradation), or any combination of the above. Phytoextraction is the uptake and storage of pollutants in the plant's stem or leaves. Some plants, called hyperaccumulators, draw pollutants through the roots. After the pollutants accumulate in the stem and leaves the plants are harvested. Then plants can be either burned or sold. Even if the plants cannot be used, incineration and disposal of the plants is still cheaper than traditional remediation methods.

In addition to transporting nutrients, certain plants are capable of transporting environmental pollutants such as metals, radionuclides, chlorinated solvents, petroleum hydrocarbons, and ammunition wastes through biological means.

Contaminants and contamination levels

Phytoremediation is effective in the remediation of all types of soil contaminants; howeyer it is more effective on lower concentration of contaminants.

Types of vegetation used

Some of the plants used in phytoremediation are :

- Alfalfa
- Hybrid Poplar Trees
- Blue-green Algae
- Duck Weed
- Arrowroot
- Sudan Grass
- Bermuda grass
- Alpine Bluegrass
- Yellow or White Water Lillies

Vetiver grass combined with applications of zeolite is being used to remove boron, cadmium and lead from

When personal direct taxes are subtracted from personal income; the obtained value is called disposable personal income (DPI).

indonesian soils contaminated by sludge from the textile industry.

Fate of contaminants

Once plants extract environmental contaminants, these contaminants are either broken down by the plant into non-hazardous substances or concentrated within plant tissues. Concentration of contaminants in plant material gives rise to the possibility of animals and insects consuming the contaminated material thereby introducing contaminants into the food chain.

Advantages

- Phytoremediation is cost effective.
- It is suited to remediation of large areas of soil.
- It is environmentally friendly.
- Phytoremediation sites are more aesthetically pleasing.
- Phytoremediation sites require low maintenance.
- It involves no noisy and expensive equipment.

Disadvantages

- Not as effective for sites with high contaminant concentration.
- Phytoremediation is slower than conventional methods.
- It does not work through the winter (seasonally effective).
- Risk to wildlife/food chain.

Ecotourism

It is the responsible travel to natural areas that conserves the environment and improves the welfare of the local people. Ecotourism is a growing component of the larger tourism industry.

Benefits of Ecotourism

- It is an environment friendly and potentially more sustainable alternative to extractive activities. Such as farming, logging and mining.
- It offers local people the chance to escape from a cycle of poverty and by sharing their knowledge of the local terrain ecology with visitors.
- Business initiated through ecotourism ventures can become self sufficient within a short span of time and can enhance the long term economic prospects of a community.
- It is an opportunity to preserve ecosystems and biological diversity.
- It can create jobs in remote regions.
- Protected areas and nature conservation provide many benefits to society including preservation of biodiversity.

Dangers of Ecotourism

- Tourism in a region that boast opportunities to view endangered species can cause the extinction of these species
- A primary value of ecotourism is the employment opportunities it provides to local people. Yet many tourism operations are owned and run by corporations, providing no benefit to the locals.
- If ecotourism operations fail to live up to conservation ideals can have very serious environmental consequences.

The guiding principles for ecotourism

- Tourist activities must not degrade the resources.
- Tourism must respect the intrinsic value of natural resources
- Tourist revenue must provide conservation, scientific or cultural benefits to the resources and local community-

Achieving the principle depends in large part upon the scale of the operation. Ecotourism can live up to the promises if it follows the principles, adequately monitor and protects its resources and ensure fair distribution of profits with in the community.

AGMARK

- "Agmark" is a "Quality Certification Mark" of the Government of India.
- Agmark products are free from adulteration; conform to the scientifically laid down quality standards.
- Each lot/batch of Agmark product is pretested for quality in well-equipped laboratory by qualified chemist.
- Agmark products are packed in food grade packing material and are never sold loose.
- Agmark products bear Agmark label or replica affixed on container/bag as identification mark, date of packing, expiry period, serial number, etc.
- Free of cost replacement is provided in rules, if Agmark products are found not up to the standard.
- Packing only in food grade packing material after lot/batch found conforming to the scientifically laid down Agmark specifications.
- Proper labelling of the lot with numbered Agmark label, Agmark replica, date of packing, date of expiry, net weight and name/address of manufacturer.
- Released for sale with accepted trade Brand label only.
- Products commonly certified under Agmark are:

Bengal and Punjab, two provinces had non-Congress ministers after the 1937 elections.

- Vegetable oils
- Groundnut oil, Mustard oil, Sunflower oil, Sesame (til) oil, Soyabean oil.
- Chillies, Turmeric, Curry powder, Garam masala, Cardamom (Elaichi), Cumin (Zira), Coriander (Dhania), etc.
- Cereals & Pulses
- Wheat Atta, Maida, Suji, Besan, Pulses
- Ghee, Butter
- Honey
- Asafetida (Hing) and many more items required by the consumers.

Methodology on Index Numbers

Base year

The Directorate of Economics and Statistics (DES) had adopted Triennium Ending (T.E) 1981-82 as base year for the purpose of Index Numbers of Area, Production and Yield in Agriculture until 1999-2000. In 2000-2001, it decided to adopt T.E. 1993-94 as a way of updating the base to a recent year and keep it in harmony with the other series of indices such as Index of Industrial Production, Whole Sale Price Index and the series of National Accounts Statistics. Since area and production in agriculture can fluctuate a great deal from year to year, the average of area/production over the TE 1993-94 is taken for determining the base level of area/ production against which the area/ production of the succeeding years is to be compared.

Weighting diagram

The weight of a commodity for the production index is taken as the average production of the commodity in the TE 1993 -94 and the national average price of the commodity during 1993-94 as obtained from the National Accounts Statistics. This has been done to fix the base production at its average level by eliminatng the cyclical variation and to evaluate the production with the same price for all states in view of wide variations observed in the state prices. A similar exercise was done for preparing weighting diagram for TE 1981-82.

Coverage

The index numbers of agriculture commodities cover 46 crops under two main groups and eight sub-groups. The crops and groups are listed below:

(a) Foodgrains

- Cereals: Rice, wheat, jowar, bajra, maize, ragi and barley and small millets. (crops except rice and wheat constitute the sub-group coarse cereals).
- Pulses : Gram, tur and other pulses.

(b) Non-foodgrains

- Oilseeds : Groundnut, sesamum, rapeseed & mustard, linseed, castor seed, safflower, nigerseed, soyabean, sunflower, coconut and cottonseed (the oilseed crops except coconut and cottonseed constitute 9 oilseeds).
- Fibres : Cotton, jute, mesta and sannhemp (jute and mesa constitute a sub-group).
- Plantation crops: Tea, coffee and rubber.
- Condiments and species : Pepper, ginger, garlic, chillies, turmeric, arecanut, coriander and cardamon.
- Fruits & vegetables : Potato, onion, banana, cashewnut, tapioca and sweet potato.
- Other crops : Sugarcane, tobacco and guarseed.

Did You Know ?

Introduction of court music

***Durbari* or court music was first introduced at the time of Akbar and developed side by side with the music of the temple and the drama.**

Methodology

Let

a_{ijk} : the area under i^{th} crop in the jth year in k^{th} state.

a_{iok} : the area under i^{th} crop in base year period in k^{th} state.

p_{ijk} : production of ith crop n the jth year in k^{th} state.

For the state k and for the year j, individual crop indices are calculated as below:

(a) Index number of area $= \frac{a_{ijk}}{a_{iok}} \times 100 = IA_{ijk}$

(b) Index number of production

$$= \frac{P_{ijk}}{P_{iok}} \times 100 = IP_{ijk}$$

(c) Index number of yield $= \frac{IP_{ijk}}{IA_{iok}} \times 100$

For any sub-group G of commodities, the indices for the year j and state k are calculated as below :

(a) Index number of area $= \frac{\sum a_{ijk}}{\sum a_{iok}} \times 100$

The state index is obtained by including all the items of the state in sub-group G where the summation is taken over items in G.

(b) Index number of production

GS

ISRO is located at Bangalore.

$$= \frac{\sum p_{ijk} P_{io}}{\sum p_{iok} P_{io}} \times 100$$

(c) Index number of yield

$$= \frac{\text{Index number of production}}{\text{Index number of area}} \times 100$$

Kisan Call Centres

(Farmers can dial a toll free number 1551 from any part of the country and get access to agricultural information)

The Department of Agriculture & Co-operations (DAC), Ministry of Agriculture, Government of India, launched Kisan Call Centres on January 21, 2004, across the country to deliver extension services to the farming community. The purpose of these call centres is to respond to issues raised by farmers, instantly, in the local language. These call centres are expected to handle traffic from any part of the country. Queries related to agriculture and allied sectors are being addressed through these centres. They would provide answers to questions from farmers in their local language.

To be with, Kisan Call Centres, have been initiated at 8 premier institutions of the Department of Agriculture and Co-operation on a pilot basis for meeting the requirements of different states. The eight centres are in Mumbai, Kanpur, Bangalore, Chennai, Hyderabad, Chandigarh, Delhi and Kolkata. These call centres are manned by agricultural graduates at the Ist level. They receive the telephone calls from the farmers and respond to a certain extent. At the second level 3-6 Subject Matter Specialists (SMSs) are positioned as a team who would be able to answer most of the queries directed here.

It was proposed to take advantage of the professionally managed call centers mechanism and dovetail it with the specialized Subject Matter Specialists knowledge of Agricultural Scientists knowledge of Agricultural Scientist and Extension Officers, so as to facilitate its reach to the farming community. The existing specialized infrastructure of Call centres is being utilized and this communication backbone made available to the Subject Matter Specialists of Agriculture, Horticulture, Animal Husbandry, Marketing and other related areas. The Kisan Call Centres consists of three levels:

Level-I: The call coming to the call centre is picked up by an operator (level-I functionary) who after a short welcome message takes down the basic information and the query of the caller, along with the question being asked by the farmer. These details are fed into a computer by the operator who is an agricultural graduate with a rural background and knows the local language.

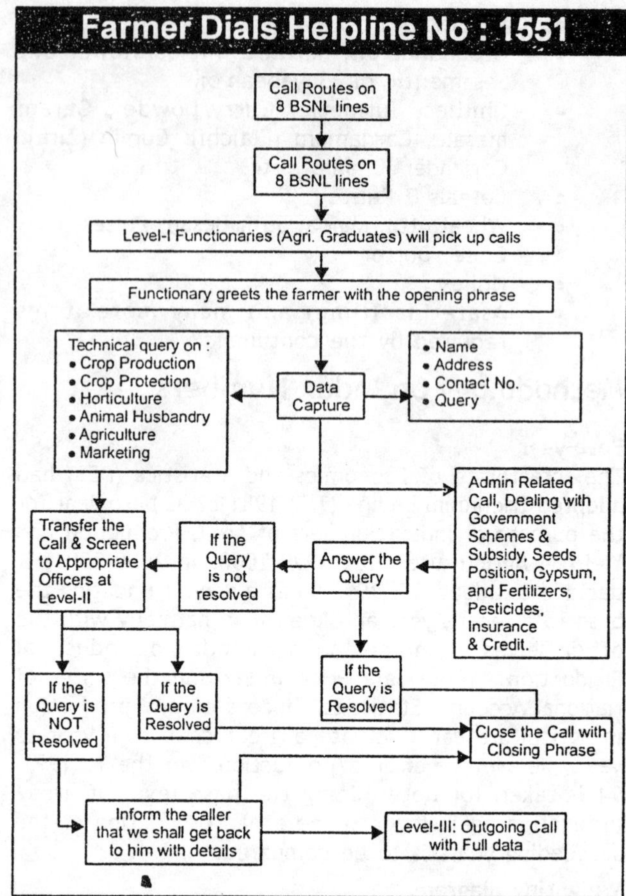

They are expected to answer a majority of the questions asked by the farmers.

Level-II : (Subject Matter Specialists on concerned important crops and enterprises, connected through good bandwidth telecom and computer connectivity) Level-II has Subject Matter Specialists (SMS) who are at their respective places (Research Stations; ATICs KVKs Agricultural Colleges) of work. In case the first level operator is not able to answer the question the operator forwards the call (in call sharing mode) to the concerned Subject Matter Specialist. Over 70% of the questions from the first level get answered at this level. In case, it is not possible to answer, there is a system to revert back to the caller by post/fax/e-mail or by telephone in 72 hours.

Level-III: (The Management Group to ensure ultimate resolution of all the farmer's queries which are not resolved at Level-II, connected on off line mode). Level - III consists of a dedicated cell located at the Nodal Office. This receives the questions that have not been answered at the first and the second levels as well as IVR (Interactive Voice Recording) calls. Appropriate replies to these questions are framed in consultation

Laddakh, Zaskar and Pir Panjal are important ranges of Kashmir Himalayas.
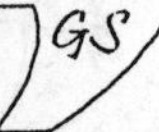

with the concerned specialists available within or outside the State by the nodal cell. The replies are sent to the farmers promptly by post/e-mail/fax/telephone etc. within 72 hours of receipt of the question.

LIST OF CALL CENTRES

Sl.	Location	States Covered
1.	Mumbai	Maharashtra, Gujarat, Goa & Diu & Daman
2.	Kanpur	Madhya Pradesh, Chhattisgarh, Uttar Pradesh, Uttranchal
3.	Bangalore	Karnataka, Kerala, Lakshadweep
4.	Chennai	Tamil Nadu, Andaman & Nicobar
5.	Hyderabad	Andhra Pradesh
6.	Chandigarh	Chandigarh, Jammu & Kashmir, Himachal Pradesh and Punjab
7.	Delhi	Rajasthan, Haryana, Delhi
8.	Kolkata	West Bengal, Bihar, Orissa, Jharkhand, Assam, Mizoram, Manipur, Meghalaya, Nagaland, Arunachal Pradesh, Sikkim and Tripura.

Agricultural Census

Agricultural Census is conducted in the country to meet the data need of planning for development and requirements of World Census of Agriculture. This census is conducted in the country at five-yearly intervals beginning from 1970-71 The census is followed by an Input Survey. So far, this Department has completed five Agricultural Censuses. The results of the Sixth Agricultural Census with reference year 1995-96 have been finalized and Report writing is in progress. The Seventh Agricultural Census with reference year 2000-01 (Agricultural year), which is being followed by an Input Survey with reference year 2001-02 is in progress in the States and Union Territories. Through the Agricultural Census, data are generated using "Operational Holding" as the unit of study. These data are compiled according to size-classes and size-groups of holdings (Marginal, Small, Semi-medium, Medium and Large), Social-group of the holder (SC, ST and Others), type Individual, Joint and Institutional) and gender of holder (Male & Female). The data collected through Agricultural Census throw light on the structural changes in Indian agriculture brought about by socio-economic factors and technological changes. The data include the information on number and area of operational holdings, land utilization, tenancy pattern, cropping pattern and irrigation status. Through the Input Survey, data are collected on use of vital agricultural inputs like use of fertilizers, manures, pesticides, agricultural machinery and equipment, livestock and agricultural credit.

Agricultural Census in the country is conducted using the administrative machinery of the States/UTs. However, for coordinating/overseeing all the census operations, this Ministry has provided skeleton staff in the States/UTs. The Department of Agriculture and Co-operation provides funds towards salary, travelling and office expenses etc. for these staff. Honorarium is paid to States UTs; for field functionaries whose services are utilized by them for collection of data from field.

You Don't Know!

Highest airport

Leh Airport in Ladakh at a height of 3,256 m (10,680 ft) is the highest non-military airport in the world.

Financing of the Scheme

The expenditure on Agricultural Census is met from a Centrally-Sponsored Plan Scheme under which full financial assistance is given to the States/UTs for payment of salaries, office expenses, travel expenses, honorarium and tabulation cost, printing of schedules and instructions, etc. The main responsibility of the Department of Agriculture and Co-operation is to give guidance to the States/UTs for conducting Agricultural Census and the Input Survey and to bring out all-India results based upon data received from States/UTs. The actual implementation of the Scheme is sought to be carried out according to a time schedule decided after/ necessary deliberations and consultations with States/ UTs.

Price Policy

The Commission for Agricultural Costs and Prices (CACP) recommends price and policy for important agricultural commodities. The recommendations aim to evolve a balanced and integrated price structure in the perspective of the overall needs of the economy with due regard to the interest of both the growers and consumers. The commodities under the purview of CACP number 25 and include rice, wheat, coarse grains, sugarcane, tobacco, cotton, jute, pulses and oilseeds. Five reports containing price and non-price recommendations are submitted by the Commission to the Ministry of Agriculture every year. They are :

- Report on price policy for kharif crops;
- Report on price policy for rabi crops;
- Report on price policy for sugarcane;
- Report on price policy for copra; and
- Report on price policy for raw jute.

It is recognized that the price policy followed by the Government has been largely instrumental in promoting technology adoption and investment by growers and has succeeded in making the country self-sufficient in

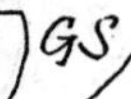

Foundation for Indian legislature were laid down in Indian Council Act 1861.

SUPPORT PRICING FOR AGRICULTURAL COMMODITIES

The main objectives of the Government's price policy for agricultural produce are aimed at ensuring remunerative prices to the growers for their produce with a view to encouraging higher investment and production and to safeguard the interests of consumers by making available supplies at reasonable prices. The price policy seeks to evolve a balanced and integrated price structure in the perspective of the overall needs of the economy. To achieve this end, the Government announces, each season Minimum Support Prices (MSP) for major agricultural commodities and organizes purchase operations through public and co-operative agencies such as * Food Corporation of India (paddy, wheat and coarse cereals), * Jute Corporation of India (jute) * Cotton Corporation of India (cotton), * National Agricultural Co-operative Marketing Federation of India (NAFED) (pulses and oilseeds) * and Tobacco Board (Tobacco) besides other agencies designated by the State Governments.

foodgrains. Recognizing the critical role played hitherto by the price policy, the National Agricultural Policy (July 2000) states that the Central Government would continue to discharge its responsibilities to ensure remunerative prices for agricultural produce through announcement of MSP for major agricultural produce. The major factor governing MSP recommendations is the cost of cultivation/production. The calculation of the cost of cultivation/production in major states is given in the Commission's reports.

Since the costing methodology is highly technical and cannot be understood by lay persons, some doubts had arisen in the minds of public/farmers. In order to remove any misgiving on this score, the commission, in collaboration with the Directorate of Economics and Statistics, had organized a National Seminar on Methodological Issues in fixing Minimum Support Prices. As a follow-up to the workshop, the Government has constituted an Expert Committee to examine the existing mandate of the Commission for Agricultural Costs and Prices and the need for repositioning it in the wake of liberalization, privatization and globalization.

The Commission has also expanded the scope of its regular interaction with officials/experts to include farmers to get a direct feedback from them on various issues having a bearing on agriculture.

Farm Income Insurance Scheme (FIIS)

In order to target the two critical components of a farmer's income, namely yield and price through a single policy instrument, the Department of Agriculture & Co-operation has formulated the Farm Income Insurance Scheme (FIIS). This scheme has been conceived to provide income protection to the farmers by integrating the mechanism of insuring production as well as market risks.

Farmer's income would be protected by ensuring minimum guaranteed income. The main features of the scheme are:

- If the actual income of the farmer falls short of the guaranteed income (product of average yield and MSP) of the farmers they would be eligible for compensation to the extent of indemnity from the Agriculture Insurance Company of India Ltd. (AIC)
- Area approach as in NAIS would be used for actual yield and price measurement of the insured crop.
- The scheme would be available for all the States and would be compulsory for farmers availing crop loans.
- NAIS will be withdrawn for the crops covered under FIIS but would continue to be applicable for other crops.

Pasteurised Milk

Milk heated to a temperature which kills the majority of the bacteria present, but without affecting its palatability. The treatment applied is that needed to destroy *Mycobacterium tuberculosis* (tuberculosis), the most heat resistant pathogen normally present in cow's milk. Most milk is pasteurised using the High Temperature. Short Time method (H.T.S.T.) by which it is heated to at least 72°C (161°F) for 15 seconds and then rapidly cooled to 10°C (50°F) or less. An alternative is the Holder Method by which milk is heated to between 63°C (140°C -150°F) for at least 30 min, and then cooled to 10°C (50°F) or less. The adequacy of the treatment and the keeping quality are determined, respectively, by two statutory tests, the phosphatase and methylene blue tests. Pasteurised milk should keep atleast 2-3 days. The term is derived from the French chemist, Louis Pasteur, who pioneered the process.

Space Age tractors for curing sick soil : Heavy modern farm equipment is killing the soil through repeated compaction and tilling, which in turn worsens salinity that kills crops. But a new world of automatically steered tractors, guided by satellites, will allow them to run on pre-determined tracks and confine their impact on the soil to 15 per cent of farmed land. Space-age tractors, first produced with Australian technology only a couple of years ago, are becoming increasingly popular and are leading to a revolution aimed at saving the world's soil - as well as boosting farm incomes.

Ankh was used as the symbol for eternal life in ancient Egypt.

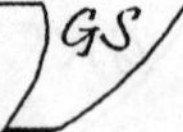

National Watershed Development Project for Rainfed Areas (NWDPRA)

The NWDPRA is the main scheme of the Department for watershed development of rain-fed area. The NWDPRA was launched in 1990-91 during the Eighth Five-Year Plan in 25 States and 2 Union Territories. The project was continued during the Ninth Five-Year Plan in 28 States (including 3 newly formed States of Uttranchal, Jharkhand & Chattisgarh and 2 UTs.

The scheme is being continued for implementation during the Tenth Plan under the revised guidelines with people's participation covering more than 6000 micro-water sheds. It is estimated to develop an area of about 2.0 million hectares during Tenth Plan at an estimated cost of Rs.100,00,000 thousand. The cumulative area developed since beginning of the Tenth Plan (April, 2002 to March, 2004) is 5.04 lakh hectares at a cost of Rs. 28,48,300 thousand.

Watershed Development Fund (WDF)

The Watershed Development Fund (WDF) with a corpus of Rs.20,00,000 thousand was set up in NABARD in the year 2000. The objective of the fund is to replicate and consolidate successful initiatives under different programmes in the Government, Semi-Government and Non-Government sectors, in the area of watershed development. Eight States are currently utilizing the support being provided by the WDF.

National Water Policy

The Water Policy 2002, adopted on April 1, 2002 the National Water Resources Council (NWRC), provides for the setting up of River Basin Organizations (RBO) by states and gives top-most priority to drinking water in water allocation, followed by irrigation, hydropower, ecology, agro-industries and non-agricultural industries and navigation. The policy says the priorities, however, can be modified depending on regional considerations. It says the concerns of the community should be taken into account for water resource development and management. The policy replaces with immediate effect the earlier one in force since 1987.

Plant Quarantine

The objectives of the scheme are to implement the provisions of Destructive Insects & Pests Act 1914 (DIP Act 1914) and the Plant Quarantine (Regulations of import into India) Order, 2003 so as to prevent introduction of exotic pests, diseases and weeds into India. Other allied responsibilities are issuance of Phyto-Sanitary Certificates (PSCs) as per the International Plant Protection Convention (IPPC), 1951, of the FAO and to undertake post-entry quarantine (PEQ) inspections in cases of imports of agricultural products.

Knowledge World!

First college

The Fort William College was established in Calcutta on August 18, 1800 by Lord Wellesley.

The new Plant Quarantine (Regulation of Import into India) Order, 2003, replacing the erstwhile Plants, Fruits and Seeds (Regulation of Import into India) Order, 1989, under the Destructive Insects and Pests Act, 1914, has been issued on 18.11.2003 with a view to making the Plant Quarantine Regulations in accordance with the International requirements under the WTO-SPS agreements. The order has come into force with effect from 1.1.2004.

There are 36 Plant Quarantine Stations at various international airports, seaports and land custom stations across the country. With a view to facilitating international trade in agricultural products, four new plant quarantine stations have already been established in the current financial year.

Five new Plant Quarantine Stations have been commissioned at Indo-Nepal border located at Rupadhia, Sanauli (Uttar Pradesh), Raxaul, Jogbani (Bihar) and Banbasa (Uttaranchal) to promote safe trade between India and Nepal.

Protocol on export of mangoes from India to China has been signed on 23.06.2003 which would boost the export of Indian mangoes to China.

Plant Food Ratio

A compound fertilizer containing 20 % N, 10% P and 10 % K has a ratio of 2:1:1. Such ratios are a guide to the type of compound. A number of fertilizers may have the same ratio and, therefore, be useable for the same purpose, but the quantity needed to be applied will vary in relation to the actual percentage nutrient content.

Plant Stand

It is the number of plants per unit area. Establishment of optimum plant stand is a prerequisite for successful crop production, especially in rainfed areas where seed germination is limited, by inadequate moisture and resultant hard crust formation. If the stand is too sparse, soil moisture will be rapidly lost, evaporation from the soil surface and available moisture cannot be used economically. If there are more plants than optimum, then there may be early mortality or under-nourishment. Punjab (including Delhi and Haryana) 630 mm; Rajasthan east 700 mm; Saurashtra and Kutch 480 mm; Rajasthan west 310 mm; Rayalaseema, interior Karnataka 560-700

Samudragupta granted permission to the Buddhist king of Ceylon Meghavarman to built monestry at Bodh Gaya.

mm; (including large pockets of Madhya Pradesh, west Tamil Nadu and Marathwada).

NCDEX

National Commodity & Derivatives Exchange Limited (NCDEX) is a professionally-managed online multi-commodity exchange promoted by ICICI Bank Limited (ICICI Bank), Life Insurance Corporation of India (LIC), National Bank for Agriculture and Rural Development (NABARD), and National Stock Exchange of India Limited (NSE). Punjab National Bank (PNB) and CRISIL Limited (formerly the Credit Rating Information Services of India Limited) by subscribing to the equity shares have joined the initial promoters as shareholders of the exchange. NCDEX is the only commodity exchange in the country promoted by national-level institutions. This unique parentage enables it to offer a bouquet of benefits, which are currently in short supply, in the commodity markets. The institutional promoters of NCDEX are prominent players in their respective fields and bring with them institutional building experience, trust, nation wide reach, technology and risk management skills.

NCDEX is a public limited company incorporated on April 23, 2003 under the Companies Act, 1956. It obtained its Certificate for Commencement of Business on May 9, 2003. It commenced its operations on December 15, 2003. NCDEX is a nation-level, technology driven de-mutualized on-line commodity exchange with an independent Board of Directors and professionals not having any vested interest in commodity markets. It is committed to provide a world class commodity exchange platform for market participants to trade in a wide spectrum of commodity derivatives driven by best global practices, professionalism and transparency.

NCDEX is regulated by Forward Market Commission in respect of futures trading in commodities. Besides, NCDEX is subjected to various laws of the land like the Companies Act, Stamp Act, Contracts Act, Forward Commission (Regulation) Act and various other legislations, which impinge on its working.

NCDEX is located in Mumbai and offers facilities to its members in more than 250 centres throughout India. The reach will gradually be expanded to more centres. NCDEX currently facilitates trading of twenty-three commodities –Gold, Silver , Soyabean, Refined Soyabean Oil, Rapeseed-Mustard Seed, Expeller Rapeseed-Mustard Seed Oil, RBD Palmolein, Crude Palm Oil, Cotton–medium arid long staple varieties, Pepper, Rubber, Jute Sacking bags, Chana, Guar Seeds, Castor Seed, Guar Gum, Raw Jute, Soyameal, Sugar, Turmeric, Urad, Wheat and Yellow peas. At subsequent phases trading in more commodities would be facilitated.

Plant Virus Indexing

Viruses are the most infectious group of disease-causing agents of plants. They are regarded as submicroscopic, infectious nucleoprotein particles that multiply only intercellularly and are potentially pathogenic. The appearance of chlorotic and distorted leaves and flowers in nature was usually attributed to some infections caused by micro-organisms. A disease of the tobacco plant noted in 1886 and termed Mosaik-Krankheit is now widely used for the description of the mottling type of the virus disease. It has been proved that the sap of the affected plant spreads the disease on a healthy plant and the term virus is used for the contagious, fluid. The plants infected with different viruses contain different antigens, the different strains of Tobacco Mosaic Virus (TMV) and potato virus diseases can be distinguished serologically. Viruses in a latent condition pass from generation to generation in the stock. More than 250 viruses have been identified and indexed and classified under six groups on the basis of shape, symmetry, type and relative molecular mass of nucleic acid involved.

Diversification

In view of the mounting demands of food, linking of enhanced food production with nutritional security, conservation of natural resources, enhancing farmers' income, employment generation etc., diversification of agriculture will be necessary. In this direction, technological improvements have been made in several non-traditional crops/regions/ seasons like soyabean, rajmash in plains and maize in rabi season. Hybrid breeding approach was extended to rice, safflower, castor, pigeon pea etc., and increased yields and stability were achieved through enhanced biotic and abiotic stress resistance/tolerance in plants and increased area under rabi season production. Diversification in terms of fruits, vegetables, mushrooms, flowers, livestock, poultry, fisheries and agro-forestry has contributed to a holistic development. Freshwater fish farming in terms of carps has become a major enterprise with an annual production of 2 million tonnes, valued at Rs. 5,000 crores. Prawn culture nets annual revenue of Rs. 3,500 crores, through export earnings. Improving cold storage, short period storage during transit of commodities on farm processing, and cool chain facilities are critical for insulating the farmer against risks and for the success of diversification.

Also, in view of the small and scattered farm holdings and resource-poor position of majority of our farmers, strengthening of the co-operatives, self-help groups, *apna mandi* type of institutions, and contract farming assumes significance. The markets have to play a decisive role for the survival and success of farming in the coming

years. Another area of importance is value addition, processing and product development. At present, value addition is only 79 per cent and agro-processing is only 2 per cent in case of perishables. There is a vast scope for value addition to an extent of 35 per cent and processing to an extent of 10 per cent in the next ten years. It is also estimated that about 80-85 million tonnes of food items can become available for consumption annually, through application of suitable post-harvest technologies, which otherwise is lost for want of proper storage and hand handling.

In view of considerable magnitude of agricultural produce being wasted at several stages, conversion of such wastes into wealth should receive our priority attention. Developing technologies for waste management and utilisation assumes considerable significance. Similarly, there is much scope to convert by-products into main products. Technologies and infrastructure, which promote primary processing, value addition and product diversification in and around the villages need to be promoted.

Resource Conservation

The effects of intensive agricultural practices, increasing population pressure, climatic changes, environmental pollution, loss of biodiversity, soil erosion, salinisation and water depletion are all threatening the sustainability of agriculture. We need to protect and conserve the natural resources not only for the present but also for posterity. The Council has established four National Bureaus, one each for plant, animal, fish and microbial genetic resources. The research efforts have successfully generated technologies that conserve resources and economize the resource use.

The impact of modern technologies in saving resources, increasing exports revenues, etc. is no less important. Modern varieties of rice saved nearly 39 million hectares and wheat varieties saved 37 million hectares. A single variety of wheat (PBW 343) in one season, covering less than 25 per cent acreage under wheat in India brought an estimated incremental revenue of over Rs. 1,000 crores during 2001-02. Similarly, the gain from exports of basmati rice during 2000-01 was over Rs. 2,000 crores. Resource conservation technologies like raised-bed planting, zero-till agriculture and laser leveling have 'the potential to be adopted over area particularly in the Indo-Gangetic plains, and save nearly Rs. 2,500 crores annually. Zero-till agriculture alone saved nearly Rs. 100 crores during 2002-03. These technologies led to a saving of irrigation water, fertilizer and seed besides significantly improving soil health.

3. General Agriculture

Multiple Choice Questions

1. % protein in fish is :
   a] Less than 30 b] 30-35
   c] 35-40 d] Above 40
2. (A x B) x C is :
   a] Single cross
   b] Double cross
   c] Three way cross
   d] Top cross
3. ___________ is gram-positive.
   a] *Xanthomonas* b] *Corynebacterium*
   c] *Pseudomonas* d] *Erwinia*
4. ___________ convert nitrite (NO_2) compounds into nitrates (NO_3).
   a] *Nitrobacter* b] *Nitrosomonas*
   c] *Rhizobium* d] *Azotobacter*
5. ___________ convert ammonia into nitrite.
   a] *Nitrobacter* b] *Nitrosomonas*
   c] *Rhizobium* d] *Azotobacter*
6. ___________ the physical methods used for controlling nematodes.
   a] Heat b] Radiation
   c] Osmotic pressure d] All of the above
7. 'Alley cropping' means :
   a] Growing of pastures in between two widely spaced rows of fast growing trees
   b] Growing of field crops in between two widely spaced rows of fast growing trees
   c] Growing of only short duration crops in between two widely spaced rows of fast growing trees
   d] Growing of only fodder crops in between two widely spaced rows of fast growing trees
8. 'Aruna' is a mutant variety of :
   a] Pea b] Cotton
   c] Castor d] Soyabean
9. 'Ascent of cell sap' from root to the tree top is exercised by :
   a] Phloem b] Xylem
   c] Cambium cell d] Lenticles
10. 'Awarodhi' variety of gram is resistant to :
    a] Wilt b] Blight
    c] Flooding d] Drought

Knowledge World!
First cotton mill
The first cotton mill in India was set up by Cowasjee Nanabhoy Davar in 1851, in Mumbai.

11. 'Layering' is useful for propagation of :
    a] Rose b] Garlic
    c] Lemon d] Sugarcane
12. 'Olour' is a polymbryonic variety of:
    a] Guava b] Citrus
    c] Mango d] Strawberry
13. 'Pusa purple round' is a___________variety of brinjal.
    a] Short duration b] Long duration
    c] Very low yielding d] High yielding
14. 'Queen' is the variety of :
    a] Grape b] Guava
    c] Pine apple d] Mango
15. 'Range' is the measurement between :
    a] Highest and lowest
    b] Medium and lowest
    c] Average and highest
    d] None of above
16. 'Ratna' variety of mango is a cross of :
    a] Neelam x Dashehari
    b] Dashehari x Neelam
    c] Neelam x Alphonso
    d] None of these
17. 'Record keeping' should be characterized by :
    a] Completeness b] Accuracy
    c] Simplicity d] All of the above
18. 'Tikka' disease is associated with :
    a] Wheat crop b] Tobacco crop
    c] Groundnut crop d] Sugarcane crop
19. "Fluid mosaic model" relates to the structure of :
    a] Cell Wall b] Protoplasm
    c] Biomembrane d] Nucleic acid
20. "Hollow-heart" disease of sugarbeet is caused due to :
    a] Salt toxicity b] Boron deficiency
    c] Moisture stress d] Air pollutant
21. "Kranz" anatomy of leaf is found in:
    a] Sugarcane b] Sunflower
    c] Soyabean d] Potato
22. 1 nm is equal to :
    a] 10^{-8} cm b] 10^{-9} m
    c] 10^{-10} d] None of the above
23. 2n + 1 state is referred to as :
    a] Monosomyb] Trisomy
    c] Tetrasomy d] Nullisomy

GK **Trachea is the proper name for the 'wind pipe'.** GS

24. Nitrogen is applied into transplanted rice in the proportion of :
a] 50% at basal + Nil at tillering + 50% at panicle emergence stage
b] Nil at basal + 50% at tillering + 50% at panicle emergence stage
c] 25% at basal + 50% at tillering + 25% at panicle emergence stage
d] 50% at basal + 25% at tillering + 25% at panicle emergence

25. 80Kg N, 40 kg P_2O_5 and 100 Kg K_2O; then the grade of this mixture will be :
a] 20-10-10 b] 8-4-10
c] 8-4-20 d] 10-4-10

26. Sugars present in DNA and RNA respectively are :
a] Glucose and Fructose
b] Deoxyribose and Ribose
c] Galactose and Raffinose
d] Erythrose and Starchyose

27. A 50 sq. m. plot is to be given an irrigation of 6 cm. depth. The total water requirement will be :
a] 300 liters
b] 3,000 liters
c] 30,000 liters
d] 3,00,000 liters

28. A balanced diet is :
a] The food that is rich in vitamins
b] The food that supplies all the essential nutrients in amounts that meet our nutritional needs
c] The food that provides the energy to maintain the body at its normal temperature
d] All the above

29. A character determined by a gene present on X-chromosome is called:
a] Sex-linked character
b] Sex-limited character
c] Sex-influenced character
d] Hollandric character

30. A chromosome which differs either in number or in morphology between male and female of a species is known as :
a] Autosome
b] Sex chromosome
c] Accessory chromosome
d] Homologous chromosome

31. A combination of horticultural crops, field crops and tree species is called:
a] Agro-forestry
b] Silvi-pastoral system
c] Multipurpose forest tree plantation system
d] None of the above

32. A combine cultivation of trees and grasses is referred to as :
a] Aqua silviculture
b] Agroforestry
c] Silvopastoral
d] Silviculture

33. A conference at which the views of several experts are given, one after the another, it seldom provides time for general discussion it refers to :
a] Symposium
b] Pannel Discussion
c] Seminar
d] Forums

34. A copping system where aerable crops are grown in the inter spaces between rows of planted trees :
a] Relay cropping
b] Mixed cropping
c] Inter cropping
d] Alley cropping

35. A cotton boll damaged by spotted boll worm (*Earias* spp) may be distinguished from that infested by pink boll worm. (*Pectinophora gossypiella*) in showing :
a] Internal rottening with no outer opening
b] A prominent hole with excreta at the opening
c] Presence of boll worm predators
d] None of the above

36. A demonstration which shows the value or worth of the new practice:
a] Result Demonstration
b] Method Demonstration
c] Whole Plot Demonstration
d] National Demonstration

37. A duck on average lays __________ than of a hen in a year.
a] 30-40 eggs lesser
b] 20-30 eggs more
c] 30-40 eggs more
d] At par

38. A form of low pruning to about 2 m up the stem :
a] Zero prunning
b] Brashing
c] Pollarding
d] None of the above

39. A perfect competative market is where :
a] Large firms b] One firms
c] Less firms d] Two firms

40. A photosynthetic unit consists :
a] 1000 light harvesting chlorophyll molecules
b] 800 light harvesting chlorophyll molecules and trapping centre
c] 200 light harvesting chlorophyll molecules
d] About 400 light harvesting chlorophyll molecules and trapping center

Hafiz is title given to a person who has learned the Quran by heart.

GS

41. A physical or chemical agent that frees plant, organ or tissue from infection is a :
a] Antisepticb] Deodorant
c] Disinfectant d] Germicide

42. A poccess of initiating a conscious and purposeful action is called :
a] Motivation b] Education
c] Coordination d] Action

43. A porous soil or geological formation which will yield sufficient quantities of water is called :
a] Well
b] Aquifer
c] Both a and b
d] None of the above

44. A protein produced by living cells of micro-organism, plants and animals that can catalyze specific organic reaction is known as :
a] Hormone
b] Enzyme
c] Catalyst
d] None of the above

45. A recessive gene for resistance to disease can be transferred from one parent to another using :
a] Pureline selection
b] Backcross breeding
c] Pedigree method
d] Bulk method

46. A short duration crop in between two main seasonal crops is termed as :
a] Cash crop
b] Inter crop
c] Companion crop
d] Catch crop

47. A single seed plant of tabacco may produce as many as :
a] 1,000 seeds
b] 10,000 seeds
c] 1,00,000 seeds
d] 2,00,000 seeds

48. A spherical particle embedded in the phospholipoprotein of thylakoid lamellae is known as :
a] Granule
b] Quantosome
c] Both a and b
d] None of these

49. A symbiotic association of fungus with subterranean part of a plant is said to be :
a] Ectophyte b] Endophyte
c] Mycorrhiza d] Parasite

50. A useful pruning practice to produce large quantities of foliage close to ground level for fodder species or for leaf extract product is called :
a] Coppicing b] Pollarding
c] Lopping d] Thinning

51. A variety produced by crossing in all combination a number of lines that combine well with each other is:
a] Composite b] Hybrid
c] Synthetic d] Elite variety

52. A variety produced by mixing seeds of several phenotypically outstanding lines and allowing all combinations is :
a] Synthetic
b] Composite
c] Hybrid
d] Germplasm complexes

53. Abscisic acid is a :
a] Gibbrellin b] Auxin
c] Retardant d] Inhibitor

54. According to projected estimate contribution of livestock to the total agricultural output in India is :
a] 5% b] 7%
c] 11% d] 14%

55. Aeciospores of *Puccinia recondita* infect to :
a] Barley crop b] Maize crop
c] Rice crop d] Wheat crop

56. Agricultural extension worker is :
a] A messenger boy
b] A passive person
c] An educator
d] Supplier of inputs for farmers

57. Agrosan G.N. is the name of :
a] Fungicide
b] Weedicide
c] Insecticide
d] None of the above

58. Aim of seed technology is :
a] Rapid multiplication of seeds
b] Assured high quality seeds
c] Timely supply of improved seeds
d] All of the above

59. All fruits are in general________ in nature.
a] Acidic b] Basic
c] Neutral d] All the above

60. The bacteroid *Rhizobium* is:
a] Macroaerobic
b] Anaerobic
c] Microaerobic
d] None of these

GK GS

January month in the Roman calendar is named after the two faced Greek God of Gates and Doors.

61. Alternaria blight disease can be effectively controlled by :
a] BHC
b] Aldrin
c] Treating seeds with hot water
d] All of the above

62. Ammonium sulphate contain approximately :
a] 10% nitrogen b] 20% nitrogen
c] 28% nitrogen d] 36% nitrogen

63. An accelerated hybrid sorghum breeding programme was initiated in India, in :
a] 1921 b] 1949
c] 1961 d] 1984

64. An acidic buffer solution can be prepared by mixing solution of :
a] Ammonium acetate and acetic acid
b] Ammonium chloride and ammonium hydroxide
c] Sulphuric acid and sodium sulphate
d] Sodium chloride and sodium hydroxide

65. An agent that causes gene mutations is known as :
a] Mutant b] Mutation
c] Mutagen d] Mutator

66. An area in which seedings are raised to be transplanted elsewhere is called :
a] Brashing
b] Shifting cultivation
c] Crop rotation
d] Nursery

67. An enzyme excreted by a micro-organisms into the environment is called :
a] Exoenzyme b] Endoenzyme
c] Coenzyme d] None of above

68. An enzyme formed within the cell and not excreted into the medium (intra cellular) is called :
a] Exoenzyme b] Apoenzyme
c] Endoenzyme d] Coenzyme

69. An example of parallel cropping is:
a] Sugarcane + Potato
b] Potato + Mustard
c] Potato + Radish
d] Wheat + Mustard

70. An example of the market classified on business basis:
a] Terminal b] Retail
c] Raw material d] Secondary

71. An extension method in which systematic display of models. Specimens, charts, graphs, information, posters, etc. is done in a sequence so as to be significant in teaching or creating interest among the farmers is called :
a] Group contact
b] Compaign
c] Exhibition
d] Tour and excursions

72. An extension programme should be carried out by :
a] Well trained personnel
b] B. D. O. and A. D. O.
c] Highly paid scientists
d] Local leaders

73. An indifference curve is alway :
a] Concave to the origin
b] A vertical straight line
c] A horizontal straight line
d] Convex to the origin

74. An indifference curve is defined as:
a] Crossing points of two curves of commodities of utility
b] Points of various combination of two commodities of same utility
c] Points of various combination of same price
d] Points of crossing points of various prices and same utility

75. An individual having the genotype AA is known as :
a] Heterozygote b] Hybrid
c] Homozygote d] F_1

76. An intensity of K-fixation is maximum in :
a] Kaolinite
b] Chlorite
c] Veumiculite
d] Micas

77. An outline of activities so arranged as to enable efficient execution of the entire programme and it answers the questions such as how, when, where and by whom is to be done is called :
a] A calendar of work
b] A plan of work
c] Extension programme plan
d] Project work

78. An unorganized mass of cells developed on an agar medium is known as :
a] Callus
b] Mass
c] Explant
d] Suspension

79. Anthesis is a phenomenon which occurs :
a] When the flower opens first
b] After pollination
c] When anthers are formed in the flower
d] When fruit drops due to low soil moisture

80. Apple crop thrives best within the________ temperature range.
a] 30-40^{0}F b] 40-45^{0}F
c] 45-60^{0}F d] Above 60^{0}F

81. Application of gypsum is required more for :
a] Paddy b] Berseem
c] Lucerne d] Groundnut

GK **Balram is also called Halajudha, from the plough which he wielded as a weapon.** GS

82. **Application of potash increases :**
a] Disease resistance in plants
b] Resistance for water logging
c] Frost resistance in plants
d] None of the above

83. **Approximately what percent of all light absorbed by a leaf is lost as heat?**
a] 50 to 60 b] 60 to 70
c] 80 to 85 d] 90 to 95

84. ***Arachis hypogea* L. is the botanical name of :**
a] Soyabean b] Rapeseed
c] Groundnut d] Sugarbeet

85. **Areas receiving heavy rainfall in India are :**
a] The great Himalayan barrier and hills of Assam
b] Plateaus of Deccan
c] East coast of south
d] Great Peninsular region

86. **Arka Chandan is a vriety of :**
a] Round melon
b] Pumpkin
c] Bottle gourd
d] Bitter gourd

87. **Arka Shyam is a cultivar of the following fruit crop :**
a] Jack fruit b] Sapota
c] Grape d] Pineapple

88. **Arrowing is known as :**
a] Tillering of sugarcane
b] Emergence of inflorescence in sugarcane
c] Arrow like shape of sugarcane leaf
d] None of the above

89. **Artifical application of water to the soil for the purpose of crop production is called :**
a] Drip irrigation
b] Flow irrigation
c] Irrigation
d] None of the above

90. **As an examples of a synthetic culture medium may be cited :**
a] Potato Dextrose Agar
b] Malt Agar
c] Heart Infusion Broth
d] Richard Agar

91. **As the oxygen concentration increases from zero, the rate of aerobic respiration :**
a] Decreases b] Stops
c] Increases d] No effect

92. **At DNA level, a gene mutation represents a change in :**
a] Base-pair sequence
b] Phosphate-sugar backbone
c] Hydrogen bonds
d] De-oxyribose

Quiz Kit!

First all-India union

All India Trade Union Congress was established on Oct. 31, 1920. Its first president was Lala Lajpat Rai.

93. **At higher pH, the availability of molybdenum is __________ in soils.**
a] Lower
b] Lowest
c] Higher
d] Not affected

94. **At molecular level, mutation represents a change in :**
a] Base-pair sequence
b] Phosphate-sugar backbone
c] Hydrogen bonds
d] Deoxy-ribose

95. **At what pH value, phosphate availability is the highest in the soil?**
a] 5.5 b] 6.5
c] 7.5 d] 8.5

96. **ATP and NADPH are produced on the stroma side surface of the :**
a] Thylakoid b] Granum
c] Membrane d] None of above

97. **Auxin is :**
a] An enzyme b] A vitamin
c] A hormone d] A protein

98. **Auxins :**
a] Induce enlongation in shoot cells
b] Resemble indole-3-acetic acid in physiological action
c] Are generally acids with an unsaturated cyclic nucleus
d] All the above

99. **Availability of phosphorous is ____ in acidic soils.**
a] Lesser b] More
c] Maximum d] Equal

100. **Available water is :**
a] Field capacity minus Wilting point
b] Wilting point minus Oven dry weight
c] Field capacity minus Hygroscopic coefficient
d] Per cent water present minus Field capacity

101. **Average gestation period of cow is :**
a] 261 days b] 282 days
c] 296 days d] 310 days

102. **Average lactation period of buffalo is of :**
a] 250 days b] 270 days
c] 300 days d] 350 days

GK GS

1 light year = 9.46×10^8 cm.

103. Average length of pregnancy (gestation) period in sheep is :
a] 100 days b] 118 days
c] 147 days d] 162 days

104. Average wool yield per live per clip is highest for the breed :
a] Gaddi b] Marwari
c] Jaisalmeri d] Billari

105. Avogadro Constant (N_0) is equal to :
a] 6.02217×10^{-23} mol^{-1}
b] $6.02217 \times 10^{-2.3}$ mol^{-1}
c] $6.02217 \times 10^{-0.23}$ mol^{-1}
d] 6.02217×10^{23} mol^{-1}

106. *Bacillus thuringesis* is :
a] Virus b] Bacteria
c] Fungi d] Insecticide

107. Bacterial diseases are controlled by use of chemicals :
a] Antibiotics b] Viricides
c] Fungicides d] Kelthane

108. Bacterial leaf blight of rice is caused by :
a] Xanthomonas
b] Erwinia
c] Pseudomonas
d] Bacillus

109. Balance sheet provides a picture of:
a] Only assets
b] Only liabilities
c] Net worth
d] Expenditure

110. Bavistin is a __________ fungicide.
a] Contact b] Systemic
c] Biological d] None of the above

111. Bending of oat coleoptile in response to unilateral illumination is termed as :
a] Photoperiodism
b] Photomorphogenesis
c] Phototropism
d] None of these

112. Ber in generally propogated by :
a] Layering b] Budding
c] Grafting d] Cutting

113. Best breed of goat having long ears originally found in U.P.
a] Barbari b] Gaddi
c] Sirohi d] Jamunapari

114. Bitter test in brinjal fruits is due to:
a] Glycoalkaloids
b] Allyl propyl disulphide
c] Carotenoids
d] Lycopene

115. Biuret content in urea should not exceed in foliar on wheat more than the following one :
a] 0.5% b] 1.0%
c] 1.5% d] 2.0%

116. Black heart disorder in potato is caused by :
a] Rapid growth b] Suboxidation
c] Frost d] Exposture to sun

117. Black soils are generally rich in :
a] N, P & K b] B, Na & Cu
c] Mg, Al & Fe d] Mn, Zn & B

118. Black soils of central high land and plateus have high content of :
a] Kaolinitic clay
b] Illitic clay
c] Montmorillonitic clay
d] None of the above

119. Blast disease of rice can be controlled by :
a] Urea b] Agrosan
c] B.H.C. d] DDT

120. Blue colour tag is used for __________ seed production.
a] Nuclear
b] Certified
c] Foundation
d] Registered seed

121. Botanical name of Ashok tree is :
a] *Lantana camera*
b] *Lawsonia alba*
c] *Saraca indica*
d] *Millingtonia hortensis*

122. Botanically pine-apple fruit is :
a] Sorosis b] Balusta
c] Berry d] Pome

123. Both chlorophyll 'a' and 'b' show absorption maxima in the :
a] Blue - violet region
b] Orange - red region
c] Both a & b
d] None of the above

124. Breeder seed is produced by :
a] Plant Breeder
b] Government Agencies
c] Registered Growers
d] N.S.C.

125. Brief visual messages on poster cards flashed before a group of audience to emphasise important points in a presentation is called :
a] Flannel Graph
b] Flannel Board
c] Bulletin Board
d] Flash Card

126. Brown rot of potato is caused by :
a] *Erwinia*
b] *Pseudomonas*
c] *Xanthomonas*
d] *Saccharomyces*

127. Brown rust of wheat is caused by :
a] *Puccinia recondita*
b] *Puccinia graminis tritici*
c] *Puccinia striformis*
d] *Puccinia hordei*

128. Bud necrosis of groundnut is transmitted by :
a] Thrips
b] White fly
c] Aphids
d] All the above

129. Budget which considers estimates of costs of returns from the farm as a whole is :
a] Partial budget
b] Enterprise budget
c] Complete budget
d] Family budget

130. Buff colour tag is used for :
a] Nuclear seed
b] Certified seed
c] Foundation seed
d] Registered seed

131. Bulk population breeding is suitable for :
a] Fruit crops
b] Vegetable crops
c] Small grain crops
d] Flower crops

132. Bunchy top in sugarcane is caused by :
a] Root borer b] Stock borer
c] Internode borer d] Top shoot borer

133. By means of 'Bulbils' which of the following crop is propagated?
a] Rose
b] Garlic
c] Lemon
d] Sugarcane

134. By which of the following method of raising seedlings of rice, we can get the ready seedling in early time?
a] Dapog method
b] Dry bed method
c] Wet bed method
d] None of the above

135. C.F.T.R.I. is located at :
a] New Delhi
b] Bangalore
c] Hyderabad
d] Mysore

136. C:N ratio for rapid decomposition is:
a] 30 b] 35
c] 40 d] 50

137. C:N ratio of orgaic matter is :
a] 1 to 10 b] 1 to 20
c] 10 to 1 d] 20 to 1

138. C-235 is a popular variety of :
a] Wheat b] Gram
c] Mustard d] Lentil

139. Cabbage is cross pollinated owing to:
a] Floral morphology
b] Protandry
c] Protogyny
d] Self incompatibility

140. *Cajanus cajan* belongs to family :
a] Compositae b] Euphorbiacea
c] Leguminoceae d] Cruciferae

141. Calvin selected the green algae particularly adopted to studies dealing with the assimilation of CO_2 are :
a] Chlorella and Tropacolum
b] Chlorella only
c] Chlorella and Scenedesmus
d] Elodea and Chlorella

142. Cane killer weed of sugarcane crop is the name of :
a] A weed b] An insect
c] A disease d] None of them

143. Carbon dioxide content in soil air is:
a] 0.03% b] 0.003%
c] 0.25% d] 0.50%

144. Carotenoids :
a] Are source of vitamin A
b] Protect against photo-oxidaton
c] Absorb and transfer light energy to chlorophyll 'a'
d] All the above

145. Cat face is a physiological disorder of :
a] Cauliflower b] Chillies
c] Tomato d] Potato

146. Cation exchange capacity of soil is expressed as :
a] Percentage
b] Parts per million
c] Millequivalent per 100 gm.
d] Milligram per 100 gm.

147. Cattle semen is frozen and stored for long periods in liquid nitrogen at temperature :
a] -5°C b] -79°C
c] -135°C d] -196°C

148. Cause of deterioration of variety is:
a] Mendelian variation only
b] Mutation only
c] Mechanical mixture only
d] All of the above

GK GS
Sir Edmund Hillary was born in Auckland.

149. Certified seed is produced from :
a] Foundation seed
b] Registered seed
c] Certified seed
d] All the above

150. Characteristic of informal gardening:
a] Uneven ground can be used
b] It is study of space relation
c] Least dominated by rules
d] All of the above

151. Characteristics of formal gardening is :
a] Suitable for city banglow
b] The land in forced to fit in the plants
c] Highly dominated by rules
d] All of the above

152. Characters not generally influenced by environment will have :
a] Low heritability
b] Medium heritability
c] High hertiability
d] None of the above

153. Charcol rot of sorghum pathogen have host range :
a] Very narrow b] Narrow
c] Wide d] None of above

154. Chemical formula of gypsum salt is :
a] $Ca\ SO_4.2H_2O$ b] $2\ CaSO_4.H_2O$
c] $Ca\ SO_4.5H_2O$ d] $CaSO_4.H_2O$

155. Chemical formula of single superphosphate is :
a] $Ca(H_2PO_4)_2$
b] $CaHPO_4$
c] $Ca_3(PO_4)_2$
d] $Ca(H_2PO_4)_2.CaSO_4.2H_2O$

156. Chemical which regulates opening and closing of stomata is :
a] Kaoline b] 2, 4-D
c] PMA d] Paraquat

157. Chicken-pox is related with :
a] Helminth b] Fungus
c] Bacteria d] Virus

158. Chlorophyl 'b' usually appears :
a] Blue green b] Yellow green
c] Red d] Red green

159. Chlorophyll 'a' usually appears :
a] Blue green b] Yellow green
c] Red green d] Brown green

160. Chlorophylls 'c' 'd' and 'e' are found only in :
a] Angiosperms b] Fungi
c] Gymnosperms d] Algae

161. Chloroplasts are absent in :
a] Fungi
b] Blue-green algae
c] Photosynthetic bacteria
d] All the above

162. *Chorchorus capsularis* belongs to family :
a] Tiliaceae b] Malvaceae
c] Euphorbiaceae d] Compositae

163. Citrus gummosis is spread through :
a] Air b] Insects
c] Irrigation water d] Man

164. Club root of crucifers is caused by :
a] Colletotrichum
b] Puccinia
c] Plasmodiophora
d] Sclerospora

165. Conidia are the sources of __________ inoculum in ergot of bajra.
a] Primary b] Secondary
c] Tertiary d] All the above

166. Conservation of rain water by collecting run-off precipitation for supplimental irrigation is called :
a] Moisture conservation
b] Water harvesting
c] Water collection
d] Watershed management

167. Contagious abortion in cow is caused by :
a] Clostridium
b] Streptococcus
c] Brucella
d] Entodinium

168. Yellow coloured fruits and vegetables are rich source of :
a] Vit. 'E' b] Vit. 'C'
c] Vit. 'A' d] Vit. 'B'

169. Conversion of nitrate nitrogen to molecular nitrogen is known as :
a] Denitrification
b] Nitrification
c] Ammonification
d] Nitrogen fixation

170. Conversion of proteins into ammonia is called :
a] Ammonification
b] Nitrification
c] Nitrogen fixation
d] None of above
a] Cooperative Extension Service : USA

171. Cooperative society is basic institution for :
a] Political growth of the villagers
b] Social growth of the villagers
c] Cultural growth of the villagers
d] Socio-economic growth of the villagers

172. Cordination is :
a] Subordination
b] Disordination
c] Integration
d] Between near equals

GK **Denmark administered the Nicobar Island before the British took over in 1961.** GS

173. *Corpora allata* is associated with :
a] Nervous system
b] Circulatory system
c] Digestive system
d] Excretory system

174. Covered floor space for each cow should be :
a] 10-15 Sq. ft. b] 16-19 Sq. ft.
c] 20-30 Sq. ft. d] 31-35 Sq. ft.

175. Crop which can be propagated by vegetative parts:
a] Pineapple b] Coconut
c] Papaya d] Phalsa

176. Cucurbits in early stage are mainly damaged by :
a] Aphids
b] Jassids
c] Borers
d] Red pumpkin beetle

177. Cultivation of durum wheat is primarily confined to:
a] Central India
b] Southern India
c] Central and Southern India
d] U.P. Hills

178. Cultivation of the apical meristems, particularly of shoot apical meristem, is known as :
a] Embryo culture b] Meristem culture
c] Pollen culture d] Tissue culture

179. Current cattle population (in millions) in India is :
a] 80 b] 160
c] 185 d] 255

180. Cut worm is a pest of :
a] Gram b] Patato
c] Wheat d] Maize

181. Cutting of braches of a tree is also called :
a] Coppicing b] Brashing
c] Lopping d] Prunning

182. Cytogenetic male sterility is utilized in :
a] Pure line selection
b] Hybrid seed production
c] Bulk method
d] Progeny test

183. Dahelia, Marigold and Gaillardia belong to the family:
a] Iridaceae b] Oleaceae
c] Malvaceae d] Compositeae

184. Dairy cows are kept in covered area the floor apace required per adult cow is :
a] 10-15 sq ft b] 20-30 sq ft
c] 30-40 sq ft d] 6-10 sq ft

185. Decomposition of organic matter in submerged soil is carried out by:
a] Bacteria
b] Actinomycetes
c] Fungi
d] Earthworm

Did You Know ?

First Party constitution

The first constitution of the INC was drawn up in 1908.

186. Demand for luxurious goods is :
a] Elastic
b] More Elastic
c] Inelastic
d] Less Elastic

187. Democratic Decentralisation is based largely on the recommendations of:
a] Grow More Food Enquiry Committee
b] Balwant Rai Mehta Committee
c] U.P. Panchayat Raj Act
d] Seventh Development Commissioner's Conference

188. Desuckering is done in :
a] Potato b] Tomato
c] Sugarcane d] Tobacco

189. Diethane M-45 is a local :
a] Insecticide b] Fungicide
c] Nematicide d] Weedicide

190. Difference in the energy of water in a solution and pure water at the same temperature and pressure is termed as :
a] Diffusion
b] Osmosis
c] Water potential
d] Matric potential

191. Demand for necessary commodities is :
a] Elastic
b] Inelastic
c] Zero elastic
d] Infinite elastic

192. Diplanetism in fungi refers to :
a] Presence of sexual and asexual phases
b] Presence of two types of sex hormones
c] Presence of two compatible mating types
d] Presence of two swimming periods with changes shape

193. Dissociation of phosphorus pentachloride is favoured by :
a] High temperature and high pressure
b] High temperature and low pressure
c] Low temperature and low pressure
d] Low temperature and high pressure

194. Dormancy due to condition within the dormant organ itself is called :
a] Innate dormancy
b] Imposed dormancy
c] Induced dormancy
d] None of above

GK **Durga Khote played the role of Jodhabai in 'Mughal-e-Azam'.** GS

195. **Downy mildew disease is caused by:**
a] Phytophthora b] Erysiphe
c] Peronospora d] Cercospora

196. **Downy mildew of bajra is caued by :**
a] *Erysiphe graminis*
b] *Claviceps purpurea*
c] *Claviceps microcephala*
d] *Sclerospora graminicola*

197. **Downy mildew of bajra pathogen is ________ parasite.**
a] Facultative b] Obligate
c] Not seroius d] None of above

198. **Dry and coarse soils are usually found at :**
a] High lands b] Low levels
c] Flooded area d] Coastal area

199. **During photorespiration**
a] No energy liberation occurs
b] O_2 is uptaken
c] CO_2 is liberated
d] All of the above

200. **During Prophase-I of meiosis crossing over occurs at :**
a] Zygotene b] Pachytene
c] Diplotene d] Diakinesis

201. **During Prophase-I of meiosis, chiasma formation occurs during:**
a] Leptotene b] Zygotene
c] Pachytene d] Diplotene

202. **EDCT mixture is :**
a] Attractant b] Fumigant
c] Repellant d] Antifeedant

203. **Ear cockle disease of wheat is caused by :**
a] Nematode b] Fungus
c] Protozoa d] All the above

204. **Economically Relevant Zone according to law of diminishing return is :**
a] Zone-I b] Zone-II
c] Zone-III d] None of these

205. **Economics is a science of "wealth". This definition is given by :**
a] J. K. Mehta b] Marshall
c] Robbins d] Adam Smith

206. **EDCT mixture is effective against:**
a] Storage pests
b] Fruit crop pests
c] Vegetable crop pests
d] Field crop pests

207. **Edible banana fruit is seedless because of :**
a] Embryo abortion
b] Absence of ovule
c] Vegetative parthenocarpy
d] Stimulated parthenocarpy

208. **Edible banana is :**
a] Diploid b] Triploid
c] Tetraploid d] None of the above

209. **Effect of one gene on the phenotypic expression of another non-allelic gene is called :**
a] Dominance b] Epistasis
c] Linkage d] Penetrance

210. **Embryo culture is helpful in**
a] Recovery of distant hybrids
b] Propagation of orchids
c] Shortening the breeding cycle
d] All of above

211. **Endosulfan belongs to :**
a] Chlorinated hydrocarbons
b] Organophosphates
c] Carbamates
d] Thiocynates

212. **Erisilk worms are commercially reared on :**
a] Mulberry leaves b] Castor leaves
c] Tobacco leaves d] None of above

213. **Establishments where disease free and quality seeds (fishes) are produed on scientific lines are called:**
a] Filatures b] Graingages
c] Seed farms d] All the above

214. **Ethylene dibromide is :**
a] An insecticide b] Nematicide
c] Weedicided] None of the above

215. **Etiolated plants contain proplastid is termed as :**
a] Chloroplast b] Etioplast
c] Carotene d] Xanthophyll

216. **Eubacteria belong to :**
a] Eukaryote b] Prokaryote
c] Cyanobacteria d] None of these

217. **Evaluation of an inbred includes :**
a] Single cross evaluation
b] Top-cross test
c] Phenotypic evaluation
d] All the above

218. **Evaluation of extension programme means :**
a] Measuring performance against pre-determined goal
b] Making progressive adjustments
c] Achieving a set of objectives
d] Knowing what is to be evaluated and how

219. **Evolution of 1 moles of O_2 is accompained by the reduction of ________ moles of NADP.**
a] 6 b] 4
c] 3 d] 2

220. **Examples of auxin-hormones are :**
a] IAA, IBA, NAA
b] MCPA, 2, 4-D, ATP
c] IPA, GA, PA
d] KCN, DNP, IPA

GK

R.K. Narayan was the first writer in English to be awarded the Sahitya Academy Award in 1961.

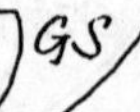

221. Examples of cytokinin are :
a] KCN and NAA
b] GA and 2, 4, 5-T
c] BA and Kn
d] AMP and GTP

222. Exanthema disease is caused due to deficiency of :
a] Copper
b] Zinc
c] Boron
d] None of the above

223. Excellent quality meat breed of goat found in India is :
a] Jamunapari b] Barbari
c] Gaddi d] Bengal

224. Exchange of homologous segment of chromatin between homologous chromosomes is known as :
a] Linkage
b] Segregation
c] Crossing over
d] Repulsion

225. Exhausting temperature of canned fruits and vegetables is :
a] 70 - 72°C b] 80 - 82°C
c] 90-92°C d] 100-102°C

226. World Environment Day is celebrated on :
a] May 5 b] June 5
c] July 5 d] August 5

227. Extension is :
a] Helping people to help themselves
b] Helping people who help other
c] Helping people who help himself
d] Helping people who are helpless

228. Factor which affects the dormancy of seed :
a] Dormant embryo
b] Rudimentary embryo
c] Germination inhibitor
d] All of the above factors

229. False smut of rice pathogen is placed in the family :
a] Clavicipitaceae b] Tilletiaceae
c] Ustilaginacae d] Dematiaceae

230. Farm management is an :
a] Intra-farm science
b] Inter-farm science
c] Inter-regional farm science
d] International farm science

231. Farm planning helps in increasing the :
a] Resourcesb] Farm income
c] Farm asset d] Farm cost

232. Farmers obtain short-term loans from :
a] Central Co-operative Bank.
b] Land Development Bank
c] National Bank for Agriculture and Rural Development
d] Primary Agriculture Co-operative Credit Societies

233. Filter chamber in insects is associated with :
a] Digestive system
b] Circulatory system
c] Respiratory system
d] Excretory system

234. Fine sand fraction in International Soil Science Society (ISSS) ranges between :
a] 0.06-0.20 b] 0.10-0.25
c] 0.02-0.20 d] 0.05-0.25

235. First discovery of bacteria as plant pathogen was made by :
a] E. J. Butler b] E. F. Smith
c] T. J. Burril d] A. S. Sussman

236. First products of photosynthesis in C_3 and C_4 plants respectively are :
a] Citric acid and Malic acid
b] Aspartic acid and Glutamic acid
c] Pyruvic acid and Glycolic acid
d] Phosphoglyceric acid and Oxaloacetic acid

237. First successful attempt at purification and crystalization of a virus was that of :
a] S. A. Waksman
b] W.M. Stanley
c] F.C. Bawden
d] F.O. Holmes

238. Fish contains ____________ per cent fat.
a] 5 b] 0.6 to 7.5
c] 15 d] 18-20

239. Foot and Mouth disease is caused by:
a] Clostridium b] Virus
c] Pasteurella d] Streptococcus

240. For a perfect incompressible liquid flowing in a continuous stream, the total energy of a particle remains the same, while the particle moves from one point of another is called. :
a] Equation of continuity
b] Bernoulli's equation
c] Rate of flow
d] Vena contracta

241. For biological control of tomato fruit borer, the idea trap crop is :
a] Marigold b] Mustard
c] Tobacco d] Garlic

242. For foundation seed production ____________ colour tag is used in the field is.
a] White b] Buff
c] Blue d] None of above

243. For improving self-pollinated crops, hybridization leads to :
a] Pedigree method
b] Bulk method
c] Back cross method
d] All of the above

GK **Jarawa and Sentinelese tribals are found in Andaman and Nicobar Islands.** GS

244. For most of the buffaloes breeding season is :
a] October - February
b] March-June
c] July-September
d] Throughout the year

245. For obtaining haploid plant ___________ is useful.
a] Multiplication stage of a variety
b] Level of genetic purity
c] Level of physical purity
d] All the above

246. For proper storage wheat grain should contain moisture per cent :
a] 0-5 b] 5-10
c] 10-14 d] 14-18

247. For the caliberation of othepressure gauges in laboratory we use :
a] Inverted differential manometer
b] Differential manometer
c] Dead weight pressure guage
d] All the above

248. For the incorporation of a disease resistance gene in an otherwise agronomically superior variety, the breeding method to be employed is:
a] Pedigree method
b] Backcross method
c] Mass selection method
d] Mass pedigree method

249. For trellis culture of tomato most suited variety belongs to :
a] Indeterminate group
b] Determinate group
c] Simple raceme group
d] All the above

250. Foundation seed is obtained from:
a] Certified seed
b] Registered seed
c] Farmer's seed
d] Breeder seed

251. Free living micro organisms which convert molecular nitrogen into its compounds in the soil are :
a] Azotobacter
b] Rhizobium
c] Nitromonas
d] Bacillus mycoides

252. Freezing point depression of a solution is measured with :
a] Viscometer
b] Osmometer
c] Calorimeter
d] Spectrophotometer

253. Fruits are blanched before canning at :
a] 80°C
b] 90°C
c] 100°C
d] None of the above

254. Garlic contain a colourless and odourless amino acid.
a] Metheonine b] Serine
c] Allicin d] All the above

255. Gene to gene host pathogen relationship was propounded by :
a] Stakman b] Walker
c] Flor d] Horsfall

256. Generally all kinds of nematodes can be destroyed at :
a] 30-45 °C b] 35-40 °C
c] 40-45 °C d] 50-55 °C

257. Generally seedless watermelon is:
a] Diploid b] Triploid
c] Tetraploid d] Hexaploid

258. Generally viruses are composed of:
a] Nucleic acid
b] Protein
c] Nucleic acid and protein
d] Nucleic acid and carbohydrates

259. Generally which acid is used to supplement the acidity of fruit for jam making :
a] Citric acid b] Tartaric acid
c] Maleic acid d] All the above

260. Generally, when carbohydrate is respired in the cell the RQ is :
a] 1 b] 0.5
c] 1.2 d] 1.5

261. Genetic basis of evolution is :
a] Availability of genetic variability
b] Selection from diversity
c] Isolation
d] All of the above

262. Genetically the purest seed is :
a] Breeder's seed
b] Foundation seed
c] Certified seed
d] Nucleus seed

263. Genotypes of zygotes of cross-pollinated cross are:
a] Homozygous
b] Heterozygous
c] Homogenous
d] Hetrogenous

264. Goal of Extension education is :
a] To promote income of the farmers
b] To promote production of the crops
c] To promote new crops
d] To promote scientific outlook

265. Goat consumes ___________ feed in proportion to its body weight.
a] Equal b] More
c] Less d] Any amount of

GK Royal Bengal Tiger was the national animal of India. GS

266. Grapes are good example of :
a] Tree b] Shrub
c] Vine d] Herb

267. Groundnut is a :
a] Fruit b] Modified stem
c] Modified leaf d] Stroage root

268. Growing agricultural crops with trees grown in and around wet land with other bodies of water in which fish etc can be grown, the combination is called :
a] Aquaforestry
b] Agro-silvi-aquaculture
c] Silvi-agriculture
d] Aqua-silviculture

269. Wind break are planted in orchards towards :
a] South-west b] West-north
c] North-east d] East-south

270. Growing of honey producing trees with bee keeping is called :
a] Sericulture
b] Silviculture
c] Apiculture
d] Aqua-silviculture

271. Guava plants bear fruits on :
a] Current seasonal growth
b] Spurs
c] Canes
d] Matured shoots

272. *Haemorrhiagic septicemia* disease is caused by :
a] Virus
b] *Streptococcus bovis*
c] *Clastridium septica*
d] *Pasteurella boviseptica*

273. Haploids are having :
a] Somatic chromosome number
b] Small chromosomes
c] Very long chromosomes
d] Gametic chromosome number

274. *Helithis armigera* is :
a] Monophagous b] Oligophagous
c] Polyphagous d] Saprophyte

275. Heterosis can be easily fixed in :
a] Self pollinated crops
b] Cross pollinated crops
c] Apomictic crops
d] Cleistogamaus crops

276. High salt concentration in the medium surrounding germinating seeds inhibits germination due to:
a] Poisioning of plasma membrane
b] Increased carbon-di-oxide availability
c] Lack of oxygen
d] Exosmosis

You Don't Know!

Oldest fort

The Kalinjar fort in Madhya Pradesh is the most ancient fort in all Hindustan. It was built by Kalinjara, the son of King Bharata, after whom India is said to be named.

277. Higher doses of nitrogenous fertilizers are known to accentuate the population growth in :
a] *Heliothis armigera*
b] Pyrilla spp
c] *Chilo partellus*
d] *Earias* spp

278. Highest milk producing dairy breed of India is :
a] Hariyana b] Sahiwal
c] Ongole d] Tharparkan

279. Hissardale breed has been developed by cross-breeding between :
a] Jamunapari and Barbari
b] Gaddi and Rambowllet
c] Bikaneri and Merino
d] Angora and Haryana

280. Hormone secreted by *corpus luteum* is :
a] Oestrogen b] Progesterone
c] Epinephrine d] Prostaglandin

281. Hormones that regulate growth are called :
a] Growth regulators
b] Growth hormones
c] Both a and b
d] None of above

282. How many agro-ecological zones have been identified in U.P. :
a] 20 b] 15
c] 25 d] 10

283. How many genes (recessive) for short stature have been identified in sorghum ?
a] 1 b] 2
c] 3 d] 4

284. How many villages generally constitute a Community Development block ?
a] About 100 b] About 300
c] About 400 d] About 50

285. How much is the porosity percentage if the particle density of soil is 2.6 and bulk density is 1.3 ?
a] 35% b] 40%
c] 50% d] 55%

286. How much seed of legume crop should be treated with one packet of *Rhizobium* culture?
a] 5 kg b] 10 kg
c] 15 kg d] 20 kg

GK **Europa, Gannymade and Calisto are moons of planet Jupiter.** GS

287. **Human body maintains constant temperture. This relates to :**
a] Physiologyb] Cytology
c] Ecology d] Psycology

288. **Hybrid varieties are the first generations (F_1) from crosses between ______ that are genetically dissimilar.**
a] Two pur lines
b] Two inbreds
c] Two open pollinated varieties
d] All the above

289. **Hybrid varieties were first commercially exploited in :**
a] Bajra b] Tomato
c] Cotton d] Maize

290. **Hybridization can be :**
a] Intervarietal b] Interspecific
c] Intergeneric d] All of the above

291. **I have observed a stained preparation of section of leaf under the microscope. The above statement is relate to :**
a] Ecology b] Histology
c] Physiologyd] Genetics

292. **IAA (Indole-3-Acetic Acid) belongs to which group of plant hormone :**
a] Auxin b] Gibberellin
c] Cytokinin d] None of these

293. **IBRD is synonyms to :**
a] World Bank
b] Asian Development Bank
c] International Monetary Fund
d] World Trade Organization

294. **Identify monoembryonic variety of mango :**
a] Olur b] Chandrakaran
c] Kurukkan d] Banganpalli

295. **Identify plant micro-nutrient :**
a] Hydrogen b] Sulphur
c] Boron d] Magnesium

296. **Identify rose variety evolved by Dr. B. P. Pal :**
a] Ganga b] Shola
c] Gulzar d] Homi Bhabha

297. **If n inbreds are to be tested in all possible single cross combinations, there would be __________ single crosses.**
a] n (n-1)/2 b] $n^2/2$
c] n(n-2)/2 d] n (n+1)/2

298. **If required water to be raised or lifted from its source to the field is called :**
a] Lift irrigation
b] Flow irrigation
c] Sprinkler irrigation
d] Drip irrigation

299. **If the Marginal Product is more than Marginal Cost, then further investment in production is :**
a] Profitable b] Harmful
c] No effect d] Equal

300. **If the soil contains 280 to 500 kg nitrogen per hectare than it is called __________ in available nitrogen.**
a] High b] Low
c] Medium d] Very high

301. **If the soils contain 110 to 280 kg K_2O/ha then soils are said __________ in this regard.**
a] High b] Low
c] Medium d] Very high

302. **If the Substitution Ratio is less than the Price Ratio, then the cost :**
a] Remains the same b] Increases
c] Decreases d] None of the above

303. **Immobilization of sulphur takes place when the 'S' content of orgnaic matter is less than :**
a] 0.60% b] 0.45%
c] 0.30% d] 0.15%

304. **Importance of fruits and vegetables in human diet is primarily because they are :**
a] Good source of carbohydrates
b] Good source of protiens
c] Good source of fats
d] Good source of vitamins and minerals

305. **Important achievements of plant breeding are :**
a] Production of dwarf cereal varieties
b] Production of hybrid varieties
c] Development of disease and pest resistance varieties
d] All of the above

306. **In AB blood group of human being, the genes A and B show :**
a] Dominance b] Epistasis
c] Linkage d] Co-dominance

307. **In an apomictic plant :**
a] Heterosis is increased
b] Heterosis is fixed
c] Heterosis is reduced
d] Heterosis is eliminating

308. **In C_3 plants the CO_2 compenstion point is usually __________ than it is in C_4 plants.**
a] Equal
b] More
c] Less
d] None of the above

309. **In C_4 plants the initial acceptor of CO_2 is :**
a] Phosphoenol pyruvic acid
b] Rubulose 1-5 biphosphate
c] Both a and b
d] None of above

GK **K_2 is the world's second highest peak.** GS

310. **In case of mungbean seed :**
a] The amount of carbohydrate is higher than that of protein
b] The amount of protein is higher than that of carbohydrate
c] Protein and carbohydrate are present in equal proportion
d] There is no carbohydrate

311. **In cows the first organ of the foetus that appears out of the reproductive tract of the mother under normal parturition is :**
a] Hind limbs b] Fore limbs
c] Head d] Tail

312. **In cucurbitaceous crops most serious pest at flowering and fruiting stage is :**
a] Red pumpkin beetle
b] Jassids
c] White grub
d] Fruit fly

313. **In dry lands which cropping system is becoming popular :**
a] Sequential
b] Inter cropping
c] Multiple cropping
d] Relay cropping

314. **In factor-factor relationship level of output is :**
a] Assumed b] Given
c] Forcastedd] Suggested

315. **In gastrointestinal tract of poultry gastric guice is secreted in :**
a] Crop
b] Proventriculus
c] Gizzard
d] Caeca

316. **In Gujarat ____________ is most common.**
a] Yellow rust b] Brown rust
c] Black rust d] None of above

317. **In honeybees, hind legs are modified into :**
a] Foragial b] Natatorial
c] Saltatoriald] Scansorial

318. **In India breeding season in buffaloes:**
a] Commences from February and continues upto April
b] Commences from May and continues upto August
c] Commences from September and continues upto February
d] Throughout the year

319. **In India maximum number of sheep are found in :**
a] Uttar Pradesh
b] Jammu and Kashmir
c] Rajasthan
d] Himachal Pradesh

320. **In India, almost all the recommended varieties of maize today are :**
a] Hybrids only
b] Synthetic varieties only
c] Either hybrid or synthetic varieties
d] Composites only

321. **In lactating cows if there is a case of milk fever it can be treated by injecting :**
a] Penicillin
b] Calcium borogluconate
c] Phosphorus
d] Streptomycin

322. **In mango high density (closer plantation) orcharding is possible by planting :**
a] Langra b] Dashehari
c] Chausa d] Amrapali

323. **In Munsell notation 10 YR 6/4-10 YR is :**
a] Hue b] Value
c] Chroma d] None

324. **In non-pregnant cows heat cycle (estrous cycle) is repeated every :**
a] 21 days b] 29 days
c] 34 days d] 30 days

325. **In normal production of hybrid maize seed which class of seed is produced?**
a] Inbred b] Single cross
c] Double cross d] All of the above

326. **In North India pruning in grape is done in the month of :**
a] October b] March
c] January d] July

327. **In personal contact most effective approach is :**
a] Personal letter
b] Telephone calls
c] Farm and home visit
d] Office call

328. **In sheep oestrus cycle is repeate every :**
a] 10 days b] 16 days
c] 21 days d] 28 days

329. **In some plants organic acids accumulate during the night and disappear during the day. Such plants are called :**
a] Hydrophytes b] Halophytes
c] CAM plants d] Indicator plants

330. **In some plants, particularly tropical plants the predominant compounds formed after very short period of photosynthesis in the presence of CO_2 are malic and aspartic acids these plants are called:**
a] C_3 plants
b] C_4 plants
c] CAM plants
d] None of above

GK **Nazi Germany's air force was called 'Luftwaffe'.** GS

331. In sugarcane breeding the initial selection after hybridisation is done in the generation :
a] F_0 b] F_1
c] F_2 d] F_6

332. In the digestive system of non-ruminant herbivores cellulase (present in plant material) is digested in :
a] Caecum-colon
b] Duodenum
c] Stomach
d] Jejunum

333. In the digestive tract of poultry food is crushed and ground in :
a] Caeca b] Gizzard
c] Proventriculus d] Crop

334. In the presence of O_2, fixation of CO_2 takes place in :
a] Chloroplasts only
b] Peroxisomes only
c] Mitochondria only
d] Chloroplasts, peroxisomes and mitochondria

335. In U.P. imposition of ceiling on land holding act was stormed in the year:
a] 1960 b] 1962
c] 1961 d] 1963

336. In which crop, the use of Blue Green Algae as a bio-fertilizer, will be most useful?
a] Maize b] Potato
c] Rice d] Sugarcane

337. In which of the following method of vegetative propagation plant part is usually detached before rooting :
a] Suckers b] Runners
c] Layers d] Corms

338. In which of the following plants, fruits are formed inside the soil?
a] *Solanium tuberosum*
b] *Raphanus sativus*
c] *Arachis hypogaea*
d] *Beta vulgaris*

339. In which of the following, fruits are botanically called as 'Hesperidium':
a] Jack fruit b] Orange
c] Litchi d] Apple

340. In which of the plot design the number of replications equals the number of varieties :
a] In systematic design
b] Randomized block design
c] Latin square design
d] None of the above

341. Incompatibility can be :
a] Heteromorphic b] Homomorphic
c] Both a & b d] None of the above type

342. Indian currency is :
a] Representative paper money
b] Token money
c] Convertable paper money
d] Full convertable paper money

343. Indian economy is concerned with:
a] Capitalistic
b] Socialistic
c] Mixed
d] Communalistic

344. Integrated Rural Development programme was first implemented in the year :
a] 1976-77 b] 1971-72
c] 1980-81 d] 1982-83

345. Inter cropping of mustard with potato is recommended in :
a] Replacement series
b] Additive series
c] Replacment cum Additive series
d] None of the above

346. Interculture in groundnut is avoided at :
a] Flowering stage
b] Seedling stage
c] Pegging stage
d] None of these

347. Intima is absent in :
a] Oesophagous
b] Proventriculus
c] Crop
d] Ventriculus

348. Introduced material can be utilized by :
a] Releasing directly as a variety
b] Using it in hybridization
c] Keeping it as a germplasm
d] All of above methods

349. Irrigations in gram are recommended at :
a] Early flowering and maturity stages
b] Late flowering and maturity stages
c] Early flowering and pod formation stages
d] Before flowering and pod formation stages

350. Isolation depends upon :
a] The nature of material to be protected by isolation
b] The nature of contaminant
c] The direction and speed of the prevailing winds
d] All of the above factors

351. Isolation distance for the production of double cross seed in maize should be at least :
a] 100 to 200 m
b] 200 to 300 m
c] 300 to 400 m
d] 400 to 500 m

352. Isoproturon is a :
a] Fungicide b] Insecticide
c] Nematicide d] Herbicide

GK **Kamal Amrohi directed the film 'Pakiza'.** GS

353. It occurs in presence of the sunlight due to deficiency of Fe, Zn and B or due to influence of low temperature, that is called :
a] Etiolation b] Gummosis
c] Bleeding d] Chlorosis

354. Jaffa and Hamlin are the varieties of :
a] Malta b] Papaya
c] Mango d] Banana

355. Jam can be prepared from :
a] Any fruit having pulp
b] Fruit which contain pectin
c] Citrus fruit only
d] All the above

356. Jassid of cotton can be controlled by :
a] DDT b] Aldrex
c] BHC d] All of them

357. Jelly can be pepared from :
a] Citrus fruits only
b] Fruits which contain pectin
c] Subtropical fruits only
d] Any fruit having pulp

358. Jelly is :
a] Non-transparent b] Semi transparent
c] Transparent d] None of the above

359. Karnal bunt is caused by :
a] Neovossia indica
b] *Neovossia horrida*
c] *Tilletia caries*
d] *Tilletia foetida*

360. Kashmiri is a breed of :
a] Cow b] Buffalo
c] Goat d] None of above

361. Who stated life is a photochemical phenomenon ?
a] Papov b] Virchov
c] Glass d] Rober brown

362. Kreb's cycle is also known as :
a] Citric acid cycle
b] Tricarboxylic acid cycle
c] Both a & b
d] None of the above

363. Krishanbhog is a variety of ?
a] Litchi b] Plum
c] Papaya d] Mango

364. Krishna and Kanchan are varieties of :
a] Grape b] Loquat
c] Aonla d] Apple

365. Lab of Land Programme was started by :
a] Govt. of India
b] Govt. of U.P.
c] Smt. Indira Gandhi
d] ICAR

Knowledge World!

Epic fort

The Vijayanagaram fort in the South is said to be on the site of ancient *Kiskinda* mentioned in the *Ramayana,* which was said to be the 'richest city in India and perhaps in the entire East'. The main temple within the fort — the temple of Pampapati — has a triple-headed Nandi which is extremely rare.

366. Larva of butterfly in known as :
a] Nymph b] Caterpillar
c] Grub d] Maggot

367. Late blight of potato can be controlled by :
a] HCl b] BHC
c] Bordeux mixture d] Aldrin

368. Lateral roots formation is the root hairs zone of :
a] Cortex b] Pericycle
c] Endodermis d] Epidermis

369. Laterite soils are poor in :
a] Nitrogen, Lime and Magnesium
b] Zinc & Iron
c] Copper & Cobalt
d] Phosphorous & Potash

370. Latest method for the improvement of cattle and buffaloes is :
a] Better feeding
b] Embryo transfer
c] Culling
d] Inbreeding

371. Law widely used for determining particle size distribution in soils is:
a] Stoke's law
b] Darcy's law
c] Schofield's law
d] Fick's law

372. Leaf rust of coffee is caused by :
a] *Hemileia vastatrix*
b] *Melampsora lini*
c] *Uromyces phaseoli*
d] *Puccinia rubigovera*

373. Lerma Rojo and Sonara 64 are wheat varieties introduced in India are from :
a] Australia
b] Britain
c] Mexico
d] Japan

374. Light-driven reactions in the chloroplasts are :
a] Phosphorylation and respiration
b] Fermentation and anaerobiosis
c] Ion-transport and CO_2 evolution
d] Protonation and transpiration

375. Lignin content (%) in soil organic matter is about :
a] 2-10 b] 28-35
c] 35-50 d] 60-80

GK **Bahmani Kingdom in South India was founded by Sultan Alauddin in 1347.** GS

376. **Line passing through the least cost points in the isoquants map is called:**
a] Isocline b] Ridge line
c] Expansion path d] Isoquant curve

377. **Linkage groups in maize are :**
a] 5 b] 10
c] 20 d] 40

378. **Little leaf of brinjal is caused by :**
a] Nematode b] Mycoplasma
c] Viroid d] Bacterium

379. **Live-stock insurance is included in:**
a] Variable cost b] Special cost
c] Fixed cost d] Marginal cost

380. **Llyod Botanic Garden is located at:**
a] Coimbatore b] Ootacamund
c] Mysore d] Darjeeling

381. **Lohi, Bikaneri, Marwari and Kutchi are the breeds of :**
a] Cow b] Buffalo
c] Goat d] Sheep

382. **Loose smut of sorghum is :**
a] Externally seed born disease
b] Internally seed born disease
c] Soil born disease
d] Infectious disease

383. **Loose smut of wheat is controlled by seed treatment with :**
a] Thiram b] Mancozeb
c] Sulphur d] Vitavax

384. **Lycopene pigment is present in :**
a] Beet root b] Chilli
c] Radish d] Tomato

385. ***Lycopersicumesculentum* is _____ flowering plant.**
a] Short-day
b] Long-day
c] Day-neutral
d] None of the above

386. **Maize (for green cobs) - Autumn Cane-Ratoon-Wheat will have cropping intensity of :**
a] 133% b] 233%
c] 333% d] 433%

387. **Maize was first introduced in India from:**
a] U.S.A.
b] U.S.S.R.
c] Mexico
d] Australia

388. **Mango hybrid Mallika is a cross between :**
a] Dasheri x Neelam
b] Neelam x Dasheri
c] Alphanso x Totapari
d] Langra x Totapari

389. **Marble is an example of :**
a] Plutonic rock b] Igneous rock
c] Metamorphic rock d] Sedimentary rock

390. **Marginal product is ratio of :**
a] Input-Output b] Output-Input
c] Cost-Income d] Price-Income

391. **Marketing surplus is :**
a] Total production - Total family requirement
b] Total consumption - Marginal consumption
c] Total cost - Marginal cost
d] Total production - Total cost

392. **Modern theory of wages and employment is given by :**
a] Keynes b] Karl Max
c] A. Walker d] Nelson

393. **Marmalades can be prepared from:**
a] Temperate region fruits only
b] Citrus fruits only
c] Fruits which contain pectin
d] Any fruit having pulp

394. **Maximum yield of mustard is obtained at a plant geometry of :**
a] 45 x 20 cm
b] 60 x 30 cm
c] 90 x 30 cm
d] 105 x 30 cm

395. **Meiosis leads to :**
a] Reduction in cell number
b] Reduction in chromosome number
c] Doubling of the chromosome number
d] Doubling of cell number

396. **Mendel's laws of inheritance operate during :**
a] Mitosis
b] Meiosis
c] Plant gametogenesis
d] Fertilization

397. **Meristem culture is useful for :**
a] Vegetative propagation
b] Recovery of virus-free stocks
c] Germplasm conservation
d] All the above

398. **Milk fever in cattle is caused due to:**
a] Decrease in the level of calcium in tissue fluid
b] Decrease phosphorous in blood
c] Inadequate exercise
d] High fever

399. **Milk production trait is a :**
a] Sex-linked trait
b] Sex-limited trait
c] Sex-influenced trait
d] Autosomal trait

GK **Scientific name of sweet flac is *Acorus Calamus.*** GS

400. Minimum wages for agricultural labourers were effective in U.P. from the date :
a] 1-3-1980 b] 1-11-1980
c] 9-1-1981 d] 11-2-1981

401. Mitochondria and Chloroplast both :
a] Are basically lipoprotein complexes
b] Contain respiratory enzymes
c] Produce ATP
d] All of the above

402. Mitochondria are found in :
a] Vacuole of eucaryotic cells
b] Cytoplasm of procaryotic cells
c] Plasmalemma of eucaryotic cells
d] Cytoplasm of eucayoic cells

403. Mohair is the long lustrous, soft hair covering produced by :
a] French Alpine b] Anglonubian
c] Angora d] Kashmiri

404. Moniliform antinnae is in :
a] Termite b] Grasshopper
c] Cockroach d] Beetle

405. Monocrotophos is :
a] Carbamate b] Pyrethroid
c] Hydrocarbon d] Organophosphate

406. Most common method of identification of pigs is :
a] Notching of ears
b] Branding on thighs
c] Tattooing
d] Nose-ring

407. Most commonly used nitrogen fertilizer these days :
a] Ammonium sulphate
b] Urea
c] Ammonium nitrate
d] Diammonium phosphate

408. Most critical growth stage for irrigation in wheat crop is :
a] Tillaring stage
b] Crown root initiation stage
c] Flage leaf stage
d] Jointing stage

409. Most effective growth regulator for rooting of most of the cuttings is :
a] Indole butyric acid
b] Indole acetic acid
c] Gibberellic acid
d] Naphthaline acetic acid

410. Most important and largest soils group in India is :
a] Deserts soils b] Alluvial soils
c] Black cotton d] Podsols

411. Most inhibitory effects of high auxin concentrations are due to formation of excessive amount of :
a] CO_2
b] Ethylene
c] Methane
d] HCN

412. Most serious disease of pea is :
a] Early blight
b] Powdery mildew
c] Late blight
d] Anthracnose

413. Motile zoospores are the chief characters of the subdivision :
a] Zygomycotina
b] Basidiomycotina
c] Mastigomycotina
d] Deuteromycotina

414. Movements of a plant organ in response to directional fluxes or gradients in environmental stimuli are called :
a] Nastic movement
b] Epinasty
c] Mutation
d] Tropism

415. Movements whose direction is determined by the morphology of the plant are called :
a] Nastic movement
b] Hyponasty
c] Epinasty
d] Mutation

416. Flooding of soil grown plant show :
a] Enhancement of phosphorous in shoots
b] Promotes ion accumulation by shoots
c] Depression in N & K in shoots
d] All of the above

417. MPP (Marginal Physical Product) of a variable input is negative in :
a] I stage b] II stage
c] III stage d] II & III stage

418. Mr. Arther. T. Mosher of New York and Shri B.N. Gupta established :
a] Indian Village Service
b] Agricultural Research Service
c] Indian Forest Service
d] Indian Administrative Service

419. Mulberry silk worm is :
a] *Bombyx mori* b] *Philosamia ricini*
c] *Antheraea* spp d] None of above

420. Mulching is an important agricultural practice. One of the chief aim of this practice is to :
a] Decrease the evaporation of moisture from the soil
b] Prevent insect pests from attacking plants
c] Provide support for the plant
d] Encourage the plant to grow taller

421. Multiple effects of a gene are called:
a] Pleiotropy
b] Multiple allelism
c] Crossing over
d] Polygenic effects

GK **Vacha is the Sanskrit version of sweet flac.** GS

422. **Muriate of potash is chemically known as :**
a] KCl b] K_2SO_4
c] KNO_3 d] K_2O

423. **Mustard sawfly belongs to family:**
a] Trypetidae b] Tenthridinidae
c] Curculionidae Sd] Termitidae

424. **Mycorrhiza makes _____________ available to plants.**
a] Nitrogen b] Potassium
c] Zinc d] Phosphorus

425. **Name the lowest tier under Three-Tier system of Democratic Decentralization :**
a] Village Schools
b] Village Cooperative
c] Village Punchayat
d] Chhetra Samiti

426. **Name the objective of training and pruning of plants:**
a] To control the direction of growth
b] To improve quality of product
c] To develop a strong frame work
d] All of the above

427. **Name the persons who discovered plant parasitic nematodes :**
a] R. Koch b] T. Needham
c] E.J. Butler d] T.S. Sadasivan

428. **Name the poultry bird used for broiler purposes :**
a] Australorp b] White Leghorn
c] Plymoth rock d] Minorca

429. **Neem cake is used to increase the efficiency of urea by :**
a] Reducing leaching and denitrification losses
b] Reducing leaching losses only
c] Reducing denitrification losses only
d] Reducing ammonia volatilization

430. **Neem seed kernel extract act as :**
a] Attractant b] Repellent
c] Deterant d] Antifeedant

431. **Nematode control by heat treatment can be done by :**
a] Steam sterilization of soil
b] Hot water treatment of plants
c] Both a & b
d] None of the above

432. **Nematodes can be controlled by :**
a] Physical Methods
b] Cultural Methods
c] Biological Methods
d] All the above

433. **Nematodes may be :**
a] Saprophyic
b] Predaceous
c] Plant parasites
d] All the above

434. **Nematods in citrus crop can be controlled by :**
a] Eldrine b] Thiodan
c] Sumithion d] Nemagon

435. ***Nicotiana tabacum* is____________ flowering plant.**
a] Short-day b] Long-day
c] Day-neutral d] None of the above

436. **Nitrogen (N) in NADP stands for :**
a] Nitrogenase b] Ammonia
c] Nicotinamide d] Nicotinic acid

437. **Nitrogen fixation is done by :**
a] Bacillus b] Striga
c] Agrobacterium d] Rhizobium

438. **Non-expanding 1 : 1 clay mineral is:**
a] Chlorite b] Illite
c] Vermiculite d] Kaolinite

439. **Normally average semen volume ejaculated at a time by a breeding bull is :**
a] 2-3 ml b] 3-5 ml
c] 5-8 ml d] 10-12 ml

440. **Normally mineral cantent (%) of soil volume is about:**
a] 70% b] 50%
c] 30% d] 20%

441. **North east monsoon brings maximum rainfall to :**
a] Mumbai b] Chennai
c] Punjab d] Kolkata

442. **Number of ATP molecules released from the oxidation of one molecule of glucose in:**
a] Two b] Thirty eight
c] Eight d] Sixteen

443. **Numerically the largest cattle population is found in:**
a] India b] Pakistan
c] U.S.A. d] Russia

444. **Nutrient availability to plant is maximum in pH range between :**
a] 4.5 to 5.5 b] 6.5 to 7.5
c] 5.5 to 6.5 d] 4.5 to 5.5

445. **Object of pruning is :**
a] Training of young trees
b] Maintenance of growth of trees
c] Bringing vigour on old trees
d] All of the above

446. **Objectives of hybridization are :**
a] To create genetical variability
b] To know the pattern of inheritance
c] To produce hybrid varieties
d] All the above

447. **Of the total milk produced in India, contribution (in percentage) from buffaloes is :**
a] 25% b] 40%
c] 60% d] 75%

Ninety three compounds have been detected in the oil of the tetraploid.

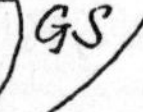

448. One gene may have no visible effect unless a second gene is present. This is :
a] Inhibiting action
b] Modifying action
c] Masking action
d] None of the above

449. One molecule of glucose on complete oxidation yields :
a] 30 ATP b] 32 ATP
c] 36 ATP d] 38 ATP

450. One nanometer is equal to :
a] 10^9 meter b] 10 Angstrom
c] 0.09 mole d] 0.001 meter

451. One of the best dual purpose breed of cow is :
a] Sahiwal b] Sindhi
c] Hissar d] Haryana

452. One of the edible species of mushrooms under cultivation is :
a] Volvaria volvacea
b] Amani a maesaria
c] Armillariella mellea
d] Amanita phalloides

453. One of the important fixed costs on a farm is :
a] Seed cost b] Land rent
c] Irrigation cost d] Feed cost

454. One of the most effective herbicide used in cultivation of C_4 crop plants is :
a] Butachlor b] Atrazin
c] Basalin d] 2, 4-D

455. Onion is used for medicinal value for:
a] Sunstroke b] Diabetes
c] Stomachache d] None of the above

456. Onion variety suitable for *kharif* crop is :
a] Pusa Red
b] Pusa Ratnar
c] Agrifound Dark red
d] Agrifound Light red

457. Opaque-2 maize composites are richer as compared to normal maize in :
a] Tryptophan
b] Lysine
c] Tryptophan and Lysine
d] Protein

458. Organic catalysts in the animates are called as :
a] Vitamins
b] Hormones
c] Enzymes
d] Regulators

459. P content in DAP is :
a] 20% b] 36%
c] 46% d] 50%

Quiz Kit!
First cellular phones
Modi Telstra Ltd. (MTL) was the first company to launch cellular phones in Aug, 23, 1995 in Kolkata. The company is vending Motorola, Nokia, Ericsson and Siemens handsets.

460. Parasitic flowering plant Mistletoe belongs to family:
a] Convolvulaceae
b] Loranthaceae
c] Orobanchaceae
d] Scroiphularaceae

461. Parthenogenesis in mustard aphid (*Lypaphis erysimi*) is combined with:
a] Psedogenesis
b] Polyploidy
c] Viviparity
d] Hermaphroditism

462. Patterns of evolution is :
a] Mendelian variation
b] Interspecific hybridization
c] Polyploidy
d] All of the above

463. Peat soil are generally characterised as :
a] Black, heavy and acidic soil
b] Red, light and acidic soil
c] Black, light and alkaline soil
d] Black, heavy and neutral soil

464. *Pectinophora gossipiella* is known as:
a] Spotted bollworm
b] American bollworm
c] Pink bollworm
d] Fruit borer

465. People's participation in an extension programme is significant when :
a] Literate section of villagers participate.
b] Gram Panchayat members participate
c] Majority of villagers participate
d] Local leaders participate

466. Percentage of total geographical area in India being cultivated is :
a] Below 30% b] 35-45%
c] 45-55% d] Above 75%

467. Periodic occurance over a large area, more population affected, the disease is :
a] Endemic b] Local
c] Epiphytotic d] Gummosis

468. pH value of freshly drawn milk is:
a] 6.2 to 6.4 b] 6.5 to 6.7
c] 6.7 to 7 d] 7.0 to 7.5

GK **Caterpillar are killed by Nuclear Polyhetosis Virus (NPV).** GS

469. *Phalaris minor* is an important weed of :
a] Rice crop b] Maize crop
c] Wheat crop d] Sugarcane crop

470. Photochemical exitation in chloroplasts, is directly responsible for :
a] Photo-oxidation
b] Photo-reduction
c] Photophosphorylation
d] All the above

471. Photoperiod required to induce flowering in plants is called :
a] Short day period b] Long day period
c] Critical day length d] Photoinductive cycle

472. Photoperiodically, high yielding triple dwarf wheat is a :
a] Photoinsensitive plant
b] Long day plant
c] Short day plant
d] Day neutral plant

473. Photorespiration is the characteristic of :
a] C_3 plants
b] C_4 plants
c] CAM plants
d] None of the above

474. Photorespiration takes place in :
a] Cristae b] Glyoxisomes
c] Peroxisomes d] Lysosomes

475. Photosynthetically, *Triticum aestivum* is a :
a] C_3 plant b] C_4 plant
c] CAM plant d] SD plant

476. Physical basis of life is :
a] Nucleus b] Cells
c] Protoplasm d] Food

477. Physical needs of a person is :
a] Physical assets
b] Food, shelter and clothing
c] Education
d] Entertainment

478. Physiological separation of plant parts is called as :
a] Abscissionb] Senescence
c] Cyclosis d] Pinocytosis

479. Piezometer is an instrument to measure :
a] Soil temperature
b] pH of the soil
c] Depth of water table
d] Soil moisture tension

480. Pigments found in photosynthetic bacteria are :
a] Bacterio-chlorophyll 'a'
b] Bacterio-chlorophyll 'b'
c] Chlorobium-chorophyll
d] All the above

481. Pinjore garden is located near :
a] Chennai b] Mysore
c] Chandigarh d] Hyderabad

482. Plant food manufactured by the process of photosynthesis in the presence of a nutrient in chlorophyll is :
a] Calcium b] Magnesium
c] Iron d] Boron

483. Plant Quarantine is included in which of the following nematode control methods :
a] Physical method
b] Cultural method
c] Regulatory method
d] Biological method

484. Planting of succeeding crop before the harvest of the standing crop is known as :
a] Alley cropping
b] Inter cropping
c] Relay cropping
d] Mixed cropping

485. Plants adopted to saline habitate are called :
a] Xerophytes
b] Hydrophytes
c] Mesophytes
d] Halophytes

486. Plants affected by seed born disease in certified seed production field must not be more than :
a] 1.5%
b] 1.0%
c] 0.5%
d] 2%

487. Plants that utilize primarily RuBP to fix CO_2, which results in the formation of the three carbon compound 3-PGA are called :
a] C_3 plants
b] C_4 plants
c] CAM plants
d] All the above

488. Plot design when varieties are arranged in the same order in every replication is known as :
a] Systematic arrangement
b] Randomized arrangement
c] Latin squares
d] None of the above

489. Plum moth is a pest of :
a] Black gram b] Potato
c] Tomato d] All of them

490. Plunger pump is used for__________ spraying.
a] Low volume
b] Medium volume
c] High volume
d] All of them

GK **Mediterranean type of vegetation occurs in Central California and Central Chile.** GS

491. **Polyembryonic variety of mango is :**
a] Chandrakaran b] Ratna
c] Neelam d] Mallika

492. **Porosity is highest in :**
a] Peat soil b] Sandy soils
c] Red soil d] Black soil

493. **Potassium content is maximum in:**
a] Feldspar b] Pyroxine
c] Amphibole d] Olivine

494. **Potassium metabisulphite is used as preservative in :**
a] Pomegranate juice
b] Phalsa juice
c] Coloured grape juice
d] Orange juice

495. **Potato leaf roll virus is transmitted by :**
a] Bacteria b] White fly
c] Aphids d] Rodents

496. **Potato provide vitamin :**
a] A b] B
c] C & E d] All of the above

497. **Potato tuber moth belongs to family:**
a] Gelechidae b] Arctidae
c] Noctuidaed] Sphingidae

498. **Off season variety of mango is**
a] Neeranjan b] Manjeera
c] Rumani d] Mulgoa

499. **Powdery mildew disease is a serious problem in :**
a] Capsicum b] Okra
c] Cucurbits d] Pea

500. **Powdery mildew of grapevine is caused by :**
a] *Uncinula necator*
b] *Uromyces phaseoli*
c] *Plasmopara viticola*
d] *Hemileia vestatrix*

501. **Price, which necessarily covers the variable cost and fixed price is the:**
a] Long-run price b] Short-run price
c] Market-price d] Equilibrium price

502. **Primary cooperatives provide :**
a] S. T. Loans
b] S. T. & M. T. Loans
c] All types
d] L. T. Loans

503. **Primary source of nitrogen for plants is :**
a] Soil b] Water
c] Air d] Manures and fertilizers

504. **Principle of Equi-Marginal Return is applied when :**
a] The resources are unlimited
b] The price of resources is high
c] The prices of resources vary
d] The resources are limited

505. **Principle of Extension education is:**
a] Learning by hearing
b] Learning by seeing
c] Learning by reading
d] Learning by doing

506. **Priyadarshani is the variety of :**
a] Dahlia b] Rose
c] Chrysanthemum d] Marigold

507. **Production of foundation seed is the responsibility of :**
a] Breeder b] NSC
c] Farmer d] Agricultural University

508. **Profit is due to :**
a] Not to risk taking
b] Uncertainly bearing
c] Higher prices
d] Large supply

509. **Protein portion of an enzyme responsible for specificity is known as :**
a] Enzyme b] Apoenzyme
c] Coenzymed] Exoenzyme

510. **Pruning in ber is done :**
a] During rainy season
b] During winter
c] During spring
d] During summer

511. ***Puccinia recondita* causes :**
a] Leaf rust of wheat
b] Stem rust of wheat
c] Yellow rust of wheat
d] Rust of pea

512. ***Puccinia triticina* is causal organism of :**
a] Leaf rust of wheat
b] Stem rust of wheat
c] Brown rust of wheat
d] None of them

513. **Pungency in radish is due to :**
a] Sinigrin
b] Allylpropyle disulphide
c] Capsaicin
d] Momordin

514. **Pure line may be defined as the progeny of :**
a] Any individual
b] A homozygous individual
c] A self-fertilized individual
d] A homozygous and self-fertilized individual

515. **Pure seed in foundation and certified seed must be:**
a] 90%
b] 95%
c] 100%
d] 98%

GK **Water Hyacinth an exotic plant species is known as 'Terror of Bengal'.** GS

516. **Purpose of Extension evaluation is:**
a] To identify the weak points
b] To identify the strong points
c] To identify the gaps and errors
d] To identify all the above

517. **Pusa Giant variety of papaya is :**
a] Dioecious b] Gynodioecious
c] Monoecious d] Andromonoecious

518. **Pusa Rasaraj is :**
a] A pureline variety of muskelon
b] A hybrid variety of muskmelon
c] A hybrid variety of watermelon
d] A seedless variety of watermelon

519. **Pusa Sawani is a variety of :**
a] Cucumber b] Water melon
c] Okra d] Pea

520. ***Pyricularia oryzae* is the causal organisms of :**
a] Blast of paddy
b] Stripe disease of barley
c] Ear cockle of wheat
d] Bacterial blight of paddy

521. ***Pyrilla* belongs to the family :**
a] Pyralididae b] Fulgoridae
c] Jassidae d] Pentatomidae

522. **Quality muskmelon fruits are produced under :**
a] Humid and dry climate
b] Dry and hot climate
c] Cool and humid climate
d] Temperate climate

523. **Quantasomes are present in the :**
a] Chloroplast b] Nucleus
c] Mitochondria d] Phytochromes

524. **Quartz is the principal component of :**
a] Quinine b] Granite
c] Felspar d] Hornblend

525. **Radio mass medium is characterised by :**
a] One way with immediate feedback
b] One way without instant audience-response
c] One way and colourful
d] One way and timeless

526. **Who is the author of training and visit system of Agricultural Extension?**
a] S.K. Dey b] V.T. Krishnamachari
c] Daniel Benore d] J. Paul Leagans

527. **Rancidity in sunflower oil is caused by :**
a] Reduction b] Oxidation
c] Esterification d] Nitrification

528. **Who in the chairman of block level panchayat samiti under Panchayati Raj system ?**
a] Block development officer
b] The M.L.A. of the area
c] An elected person from among the heads
d] A progressive farmer nominated by the Govt.

529. **Raspberry, chrysanthamum and garbera are vegetatively propagated by :**
a] Runners b] Suckers
c] Layers d] None of above

530. **Red hairy caterpiller is :**
a] Polyphagous b] Monophagous
c] Oligophagous d] Saprophyte

531. **Red rot in sugarcane is caused by :**
a] Fungus b] Bacteria
c] Virus d] Physiological disorder

532. **Refractometer is used to determine:**
a] Minerals b] T.S.S.
c] Vitamin d] None of the above

533. **Registered seed is produced from :**
a] Foundation seed b] Registered seed
c] Both a & b d] Certified seed

534. **Regular bearing variety of mango is:**
a] Desheri b] Langra
c] Mallika d] Chausa

535. **Relationship among micro organisms where one organism hinders the growth of other, is known as :**
a] Antibiosis b] Symbiosis
c] Synergism d] Commensalism

536. **Relationship between phenotypic variance (V_P) genotypic variance (V_g) and environmental variance V_e can be shown as :**
a] $V_p = V_g + V_e$
b] $V_p = V_g + V_e$
c] $V_g = V_p + V_e$
d] $V_e = V_p + V_g$

537. **Relationship in which one organism gets benefited without affecting the other is called :**
a] Symbiosis b] Commensalism
c] Synergism d] Antibiosis

538. **Who propounded the population theory first?**
a] Adam Smith b] Ricardo
c] Malthus d] Marshall

539. **Rental value of land is the part of:**
a] Cost C b] Cost A_1
c] Cost A_2 d] Cost B

540. **Repayment period of medium term loan is :**
a] 1-4 years b] 3-10 years
c] 3-5 years d] 5-10 years

541. **Representative minerals in early weathering stages (1 to 3) :**
a] Albite - Quartz-Muscovite
b] Gypsum-Calcite-Olivine
c] Muscovite-Vermiculite-Montmorillonite
d] Kaolinite-Gibbsite-Hematite

542. **Retail markets are considered to be:**
a] Perfect market
b] Imperfect market
c] More perfect market
d] Less perfect market

GK **In Asia the world's population is concentrated most.** 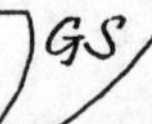

543. Retting of jute is a :
a] Physical process
b] Chemical process
c] Mirobiological process
d] Mechanical process

544. Reversion occurs due to :
a] Gene interaction
b] Mutation
c] Gene interaction and mutation
d] None of the above

545. Rhizobium bacteria is :
a] Symbiotic bacteria
b] Non-symbiotic bacteria
c] Anaerobic bacteria
d] Autotrophic bacteria

546. Rhizomes are referred to as__________because they grow horizontally.
a] Positive geotropic b] Diageotropic
c] Negative geotropic d] All the above

547. Richest source of vitamin A is :
a] Guava b] Aonla
c] Carrot d] Citrus

548. Richest source of vitamin C is :
a] Barbados cherry b] Grapes
c] Apple d] Citrus

549. Rima has the same colour of eyes as her mother. This is related to :
a] Histology b] Ecology
c] Genetics d] Physiology

550. Risk is minimised in :
a] Diversified farming
b] Specialised farming
c] Dry farming
d] Irrigation farming

551. Rock phosphate can be safely used in :
a] Saline soil b] Alkali soil
c] Calcarious soil d] Acid soil

552. Roller vane rotary pump is used for __________ spraying.
a] High volume b] Low volume
c] Medium volume d] All of them

553. Root-knot disease of tomato is caused by:
a] Fungus b] Protozoa
c] Nutritional disorder d] None of the above

554. Root-knot disease of tomato is caused by:
a] *Meloidogyne javanica*
b] *Meloidogyne incognita*
c] Both a & b
d] None of the above

555. Rural sociology is one of the specialized fields in :
a] Humanities
b] Biological sciences
c] Natural sciences
d] Social sciences

Did You Know ?

Smallest route train

In 1878, the Siliguri-Darjeeling mountain train, popularly known as Toy Train, ran a distance of 29 km between Siliguri and Tindharia. It was finally extended to Darjeeling on July 4, 1882.

The branch line of the N.E. Railway between Mankapur and Katra, built by the British in 1882, also covers a distance of 29 km.

556. Rural uplift movement on a mass scale was first started in the Gurgaon district of the Punjab, adjacent to Delhi. Under this programme a viallage guide is posted in each village to act as a channel through which the advise of experts could be passed on to the villagers. This rural uplift movement was started by:
a] Sir Daniel Hamilton
b] Shri Rabindranath Tagore
c] Mr. F.L. Brayne
d] Mr. Arther T. Moster

557. Rust disease of linseed is caused by:
a] Nectria b] Melampsora
c] Uromyces d] Ustilago

558. Sanitation, flooding, fallow and crop rotations are the _______ methods for controlling nematodes.
a] Biological b] Physical
c] Cultural d] Regulatory

559. Scarcity rent arises due to :
a] Surplus land
b] Low pressure on land
c] High pressure on land
d] Arise in the prices of other assets other than land

560. *Scirpophaga nivella* is :
a] Top shoot borer b] Root borer
c] Stalk borer d] Stem borer

561. Second most important food grain crop of India is :
a] Pulses b] Rice
c] Wheat d] Smaller millets

562. Secondary roots and shoots are __________that is, they grow to a position at an oblique angle to the gravitational force.
a] Plagiogeotropic
b] Diageotropic
c] Negative geotropic
d] Positive geotropic

563. Sedimentary rocks are :
a] Hard b] Porous
c] Brittle d] None of the above

564. Seed available for general distribution of farmers for commercial crop production is :
a] Breeder seed b] Foundation seed
c] Registered seed d] Certified seed

GK

Pygmies are found in Uruguay and Argentina.

GS

565. Seed developed by any cell except egg cell, is termed as :
a] Apospory b] Apomixis
c] Apogamy d] None of above

566. Seed production of late varieties of cauliflower can be taken in :
a] Tropical climate b] Sub-tropical climate
c] Arid climate d] Temperate climate

567. Seed testing is done for :
a] Purity b] Germination
c] Moisture content d] All of the above

568. Seedlessness in grapes is due to :
a] Parthenocarpy b] Embryo abortion
c] Gibberellins d] Excess nitrogen

569. Selective permeability is the property of the :
a] Leucopast b] Tonoplast
c] Cellwall d] Ectodesmata

570. Selfing :
a] Prevents the natural cross pollination
b] Increases the homozygosity
c] Maintains the purity of variety
d] All the above informations are correct

571. Self-pollination in a cross-pollinated crop plant leads to :
a] Increase in homozygosity
b] Increase in heterozygosity
c] Increase in heterosis
d] Increase in crossing over

572. Sexual spores of *Puccinia gramini tritici* are :
a] Basidiospores b] Urediospores
c] Teliospores d] Conidiospores

573. Share of agriculture in national income of India is around :
a] 25 per cent b] 30 per cent
c] 35 per cent d] 40 per cent

574. Short duration crop grown between two main crops is known as :
a] Companion crop b] Catch crop
c] Inter crop d] Relay crop

575. Short term credit facility is for the purpose :
a] Land development
b] Purchase of animals
c] Minor irrigation development
d] Crop production

576. Sieve plate is present in :
a] Plasmodesmata b] Mitochondria
c] Nucleus d] Phloem

577. Silk worm caterpillar require _________ gms of leaves during entire life.
a] 40 b] 70
c] 90 d] 110

578. Single plant progeny of self pollinated crop is :
a] Homozygous b] Heterozygous
c] Homogenous d] Heterogenous

579. Single superphosphate contains phosphorous mainly in the form of:
a] $CaHPO_4$
b] $Ca_2(PO_4)_2$
c] $Ca_2H_2(PO_4)_2 \cdot 2H_2O$
d] $Ca(H_2PO_4)_2$

580. Slate is made from sedimentary rock is an example of :
a] Metamorphic rock b] Sedimentary rock
c] Igneous rock d] Acidic rock

581. Socially prescribed forms of behaviour, transmitted by tradition and enforced by society is called :
a] Customs b] Folkways
c] Conventions d] Manes

582. Sodium benzoate is added for the preservation of fruit juice in the concentration of :
a] 0.01 - 0.5% b] 0.06 - 0.10%
c] 1.1 - 1.5% d] 1.6 - 2.0%

583. Soil having high concentration of soluble salts and exchangeable sodium percentage is greater 15 are:
a] Saline soils
b] Saline-sodic soils
c] Normal soils
d] Sodic soils

584. Soil is said to be medium in organic carbon if the percent organic carbon in soil is in between :
a] 0.5 to 0.75 b] 0.75 to 1
c] 1 to 1.5 d] 1.5 to 2

585. Soil strength is measured by :
a] Tensiometer b] Gypsum block
c] pH meter d] Penetrometer

586. Soils containing less than 280 kg nitrogen per hectare then these are called __________ in available nitrogen.
a] Low b] Medium
c] Very low d] High

587. Soils containing organic carbon less than 0.5% than soil is __________ in organic carbon.
a] Medium b] Poor
c] High d] None of above

588. Soils known as medium in available phosphorous contain :
a] 0-10 kg P_2O_5 b] 10-25 kg P_2O_5
c] 25-40 kg P_2O_5 d] Above 40 kg P_2O_5

589. Soils most retentive of moisture and cracks in dry season :
a] Black soils b] Sandy soils
c] Alluvial soils d] Red soils

GK **North-West part of Europe has high density of population.** GS

590. **Solubility of rock phosphate can be improved by :**
a] *Nitrosomonas* b] *Nitrobacter*
c] *Azotobacter* d] *Bacillus polymexa*

591. **Soluble sugars transported through phloem are :**
a] Glucose and Maltose
b] Raffinose and Galactose
c] Mannose and Raffinose
d] Sucrose and Starchyose

592. **Source of impurities in hybrid seed production is :**
a] Objectionable weed plants
b] Inseparable other crop plants
c] Pollen shedders
d] All of the above

593. **Soyabean originated in :**
a] China b] India
c] Ethiopia d] Mexico

594. **Specific example of a photonastic plant is :**
a] Oat b] Soyabean
c] Touch-me-not d] Sun-flower

595. **Specific surface (m^2/g) of Montmorillonite clay :**
a] 60-120 b] 120-200
c] 200-300 d] 500-800

596. **Splitting of pods is caused by :**
a] Hydrotropic movement
b] Hydration movement
c] Phototropic movement
d] Turgor movement

597. **Sponging type of mouth part present in :**
a] Mosquitoes b] Butterflies
c] House flies d] Honey bees

598. **Spot out family monocotyledony :**
a] Convolvulaceae b] Euphorbiaceae
c] Malvaceae d] Amaryllidaceae

599. **Starch and cellulose are also known as :**
a] Disaccharide b] Arabans
c] Glucosaus d] Invert sugar

600. **Steps in extension teaching are :**
a] Communicator-Encoder-Message-Channel-Decoder-Receiver
b] Attention-Interest-Desire - Conviction-Action-Satisfaction
c] Instructor-Learner-Subject matter-Teaching Equipment-Physical facilities
d] Awareness-Interest-Evaluation-Trial-Adoption

601. **Sterilization by steaming for 30 min on three successive days is known as :**
a] Pasturisation b] Solarisation
c] Tyndalisation d] Succerisation

602. **Striga is a partial ______________ parasite.**
a] Root b] Stem
c] Leaf d] Fruit

603. **Sub-Normal Price is :**
a] Very Short Period Price
b] Short Period Price
c] Long Period Price
d] Secular Price

604. **Sugar cane seed is supplied to the cultivators in the form of :**
a] Single cross
b] Double cross
c] Setts
d] All the above

605. **Sunflower originated in :**
a] India b] China
c] Mexico d] North America

606. **Surface area (cm^2) of 1 g of silt particles (0.02-0.002 mm) is about:**
a] 25 b] 90
c] 450 d] 1000

607. **Symptoms of stem gall of coriander development due to :**
a] Hypoplasia b] Hypertrophy
c] Necrosis d] Blight

608. **Symptoms of virus diseases are :**
a] Chlorosis b] Yellows
c] Necrosis d] All the above

609. **Synonym of Krebs cycle is :**
a] HMP shunt b] Calvin cycle
c] TCA cycle d] Glyoxalate cycle

610. **T Budding is also known as :**
a] I Budding b] L Budding
c] Shield Budding d] Layering

611. **Television broadcast for rural development, in India, started in :**
a] 1947 b] 1957
c] 1967 d] 1977

612. **Teliospores are ______________ spores.**
a] Asexual b] Sexual
c] Bud d] Vegetative

613. **Tensiometer can accurately measure soil moisture tension up to :**
a] 1.00 atm b] 0.85 atm
c] 0.95 atm d] 0.75 atm

614. **Test cross is :**
a] F_1 x dominant parent
b] F_2 x recessive parent
c] $F_1 \times F_1$
d] $F_1 \times F_2$

615. **That X-rays induce mutations was first established by :**
a] Gregor John Mendel
b] Charles Darwin
c] H.J. Muller
d] Hugo de Vries

616. **The 'hexagonal system' of orchard layout occupies ______________ trees than square system.**
a] 5% more b] 15% more
c] 15% less d] 25 more

GK GS

Twelve members to Rajya Sabha and two members to Lok Sabha are nominated by President.

617. The "carrier concept" in the transport of ions across the membrane :
a] Does not utilize metabolic energy
b] Utilizes metabolic energy
c] Utilizes water only
d] None of the above

618. The ability of an inbred to transmit desirable performance to its hybrid progenies is, its :
a] Specific combining ability
b] General combining ability
c] Combining ability
d] None of the above

619. The accepted ways of eating, meeting folks, wooing, training the young, supporting the aged etc. are called:
a] Mores b] Rituals
c] Beliefs d] Customs

620. The alternate host of black/stem rust of wheat is :
a] Mulberry b] Castor
c] Barberry d] Barely

621. The area that can be ploughed by a desi plough in a day is :
a] 1.0 hectare b] 0.85 hectare
c] 0.60 hectare d] 0.30 hectare

622. The arrangement of component fibrils in the zoospores of eukaryote organisms consists of :
a] 6 plus 5 b] 7 plus 4
c] 9 plus 2 d] 10 plus 1

623. The art and science of cultivation of forest tree species or management of forest tree species is called :
a] Agro-forestry b] Silviculture
c] Sericulure d] Apiculture

624. The asexual progeny of a plant is called :
a] Pure line b] Inbred line
c] Isogenic line d] Clone

625. The atmospheric concentration of CO_2 under which photosyntheis just compensates for respiration is referred to as the :
a] CO_2 compensation point
b] O_2 compensation point
c] Light compensation point
d] None of the above

626. The attack of pea stem fly is maximum in :
a] September b] November
c] January d] February

627. The basic haploid chromosome number in the genus *Arachis* is :
a] 8 b] 10
c] 14 d] 20

628. The best approach to study agricultural marketing is:
a] Institutional b] Functional
c] Decision making d] Commodity

629. The best suited area for poultry birds should have the temperature:
a] 37^0-40^0C
b] 36^0-37^0C
c] 13^0-26^0C
d] Less than 13^0C

630. The bitterness in colocasia corms is due to the presence of :
a] Calcium oxalate
b] Calcium carbonate
c] Calcium chloride
d] Potassium oxalate

631. The boiling point of milk is :
a] 100 to 102^0C b] 105^0C
c] 105.7^0C d] 107^0C

632. The botanical name of mandarin is :
a] *Citrus reticulata*
b] *Citrus sinensis*
c] *Citrus aurantifolia*
d] *Citrus maxima*

633. The breed of buffaloes whose milk has the maximum percentage of fat:
a] Bhadawari b] Murrah
c] Surti d] Nagpuri

634. The breed of sheep producing superior carpet wool is :
a] Magra b] Chokla
c] Marwari d] Jaisalmeri

635. The building in which cocoons are reared for the production of raw silk is called:
a] Filatures b] Grainage
c] Miniature d] None of the above

636. The capacity of a soil to resist appreciable change in pH value is known as :
a] Cation exchange capacity
b] Buffering capacity
c] Plasticity
d] Elasticity

637. The capacity to do work is known as:
a] Power b] Energy
c] Vigour d] Strength

638. The carbamate insecticide is :
a] Cryolite b] Cypermethrin
c] Sevin d] Chlordane

639. The carbohydrates produced in leaves cannot be translocated to different growing parts of plant in the absence of :
a] Manganese b] Boron
c] Zinc d] Iron

640. The causal agent of mango-malformation is :
a] Fungus b] Bacteria
c] Virus d] Viroid

GK **Oxygen found most abundantly in the Earth's crust.** GS

641. The causal agent of sandal spike is:
a] Virus b] Bacterium
c] Fungus d] MLO

642. The causal organism of black rust of wheat is :
a] *Pucinia graminis tritici*
b] *Ustilago nudo tritici*
c] *Puccinia triticina*
d] None of them

643. The causal organism of loose smut of wheat is :
a] *Puccinia triticina*
b] *Ustilago nudo tritici*
c] *Puccinia graminis*
d] None of them

644. The Central Agmark Lab is located at :
a] Mumbai b] Nagpur
c] Bangalore d] Kolkata

645. The centre of origin of soyabean is:
a] China b] India
c] Maxico d] North America

646. The characteristics of bioassay must include :
a] Specificity
b] Sensitivity
c] Relatively quick response
d] All the above

647. The chemical or mixture of chemicals released by an organisms to outside environment are :
a] Pheromones b] Hormones
c] Protines d] Carbohydrates

648. The choice of Combine Kafir 60, male sterile line was based on information that :
a] Male sterility in it is stable
b] It has high combining ability
c] It is photo-insensitive
d] All the above

649. The colour of egg yolk depends on absence or presence of :
a] Vitamin A b] Vitamin B
c] Chlorophyll d] Xanthophyll

650. The commercial utilization of synthetic varieties in maize was first suggested by :
a] Shull
b] Hugo de Vries
c] Hayes and Garber
d] Stadler

651. The common name of *Emmalocera dipressella* is :
a] Root borer b] Stem borer
c] Stalk borer d] Shoot borer

652. The common wheat is :
a] Diploid b] Tetraploid
c] Hexaploid d] Octoploid

You Don't Know!

First workable gene

Hargobind Khorana (b. 1922) and his team produced the first workable gene using of-the-shelf chemicals in 1976. The gene, *Escherichia coli,* is found in he intestines of human beings and animals.

653. The Community Development Programme was launched in India on:
a] 15th August b] 2nd October
c] 26th January d] 22nd October

654. The compounds that in the plant can be converted into auxins are called:
a] Auxins b] Auxin precursors
c] Antiauxins d] All the above

655. The concentration of CO_2 on high altitudes :
a] Does not changes b] Drops
c] Increases d] None of above

656. The concept of farming is best explained by :
a] Sectoral planning approach
b] Mixed farming approach
c] Holistic farming approach
d] Diversified farming approach

657. The concept of multipurpose village level workers (VLWs) originated from:
a] Marthandam project
b] Gurgaon project
c] Sriniketan project
d] Etawah pilot project

658. The crop grown in Uttar Pradesh during all the three seasons is :
a] Urd b] Sorghum
c] Moong d] Maize

659. The crop which can be propagated by hard wood cutting?
a] Pear b] Cherry
c] Grape d] Apple

660. The crown roots in wheat appear :
a] Above soil surface
b] Below soil surface but above seed
c] Below soil surface and below
d] Below seed

661. The degree to which the variability of a quantitative character may be transmitted to the progeny is referred to as :
a] Variation b] Heritability
c] G.C.A. d] S.C.A.

662. The depth of water stored in root zone between field capacity and permanent wilting point is known as:
a] Belt of soiled water b] Gravitational water
c] Available soil water d] Capillary water

GK **Indira Point touches the southern most tip of India.** GS

663. The downward movement of surface soil water is known as :
a] Infiltration b] Percolation
c] Leaching d] Wash out

664. The flood prone area of the country is ______ Mha.
a] 15 b] 19
c] 24 d] 28

665. The edible part of pomegranate is:
a] Thalamus b] Mesocarp
c] Endocarp d] Aril

666. The electrical conductivity (ds m^{-1}) limit of a saline soil is :
a] 2 b] 4
c] 8 d] 15

667. The electron transport carriers are located in the :
a] Mitochondrial membrane
b] Granum membrane
c] Stroma
d] Thylakoid

668. The element which is responsible for approximately fifty per cent of the dry matter in crop is :
a] Nitrogen b] Carbon
c] Phosphorus d] Potassium

669. The empirical formula of the chlorophyll molecule is:
a] $C_{40}H_{56}$
b] $C_{55}H_{72}H_5N_4Mg$
c] $C_{40}H_{56}O$
d] $C_{55}H_{70}O_6N_4Mg$

670. The end product of glycolysis is :
a] Ethanol b] Malic acid
c] Pyruvic acid d] CO_2

671. Biological precursor of ethylene is :
a] Leucine b] Methionine
c] Lycine d] Arginine

672. Root anaerobiosis in tomato resulted in :
a] Increase in glutamine
b] Decrease in alanine
c] Decrease in alpha-aminobutyric acid
d] Increase in glutamic acid

673. The energy possessed by a particle by virtue of its position is called :
a] Kinetic energy b] Positional energy
c] Potential energy d] Static energy

674. The family which one sets up after one's marriage is known as :
a] Family of origin
b] Family of procreation
c] Polygymous
d] Polyandrous

675. The farmer having land upto two hectares is called:
a] Big farmer b] Small farmer
c] Marginal farmer d] Medium farmer

676. The farming methods and cropping systems depend upon :
a] Climate and soil
b] Choice of the cultivator
c] Cost of inputs
d] Agricultural practices

677. The first commercial hybrid variety was released in:
a] 1910 b] 1921
c] 1935 d] 1957

678. The first Cooperative Credit Societies Act was passed in :
a] 1888 b] 1904
c] 1912 d] 1919

679. The first KVK was established in 1974 at :
a] Nagpur
b] Nilokheri
c] Ludhiana
d] Pondicherry

680. The food product preserved in common salt or in venegar is called:
a] Jam b] Squash
c] Nectar d] Prickle

681. The formation of most ATP in mitochondria takes place by means of a process known as :
a] Oxidative phosphorylation
b] Phosphorylation
c] Photosynthetic phosphorylation
d] None of above

682. The fungi are included among the eukaryotes because:
a] They are members of the Thallophyta
b] They have a nuclear membrane
c] They are not autotrophs
d] They can cause diseases

683. The genetic basis of pure line ws given by :
a] Charles Darwin
b] Gregor John Mendal
c] August Weismann
d] W.L. Johannsen

684. The genetically pure seed of an improved variety possess :
a] High yield
b] Wider adaptability
c] Synchronous maturity
d] Resistance to diseases and pests

685. The goat breed suitable for cities and towns is :
a] Pashmina b] Barbari
c] Kathiawari d] Marwari

686. The green lamellae of chloroplast from sac like structure, known as :
a] Granules b] Quantosome
c] Stroma Lamellae d] Thylakoid

GK

Aurora Australis occurs in Southern hemisphere.

687. The group of elements form part of micronutrients for plant growth:
a] Cu, Zn, Fe, Cl
b] B, S, Ca, Mg
c] C, H, O, Sl
d] Fe, Ca, Mg, Mo

688. The hard fore wings in an insect meeting together to form a straight mid dorsal line is a character of :
a] Coleoptera b] Hymenoptera
c] Heteroptera d] Orthoptera

689. The herbicides containing carbon and hydrogen in their molecule are called :
a] Arsenic b] Acid
c] Organic herbicide d] Salt

690. The hormones that initiate for formation of floral primordia or promote their development are called:
a] Flowering hormones
b] Flowering regulators
c] Phytohormones
d] Growth hormones

691. The huskless variety of barley is :
a] Manjula
b] K.K. 141
c] Dolma
d] R.S. 6

692. The hydrogen ion concentration at pH 6 is :
a] 10^{-6} b] 10^{-8}
c] 10^{-10} d] 10^{-12}

693. The initial acceptor of the carbon dioxide molecule:
a] Ribulose dinucletide
b] Ribulose 1, 5 biphosphate (RuBP)
c] Formaldehyde
d] Phosphoglyceric acid

694. The intensity of light in which the CO_2 utilized in photosynthesis is equal to that liberated by respiration is called the :
a] Light compensation point
b] Equillibrium point
c] Red drop
d] None of the above

695. The isolation distance for certified seed production of pea should be :
a] 30 m b] 10 m
c] 5 m d] 50 m

696. The isolation distance for producing certified seed of cole crops is maintained at :
a] 0.5 km b] 1.0 km
c] 1.5 km d] 2.0 km

697. The Law of Diminishing Return refers to an eventual fall in :
a] Marginal product of the variable factor
b] Average income of the farm
c] Total production of all factors
d] Total earning of the farm

698. The leaves of C_4 plants are characterized by :
a] A sheath of parenchyma cells that are radially arranged around each vascular bundle
b] The vascular bundle, is enclosed by loosely packed spongy mesophyll cells
c] Both a & b
d] None of above

699. The lowest temperature at which all bacteria of a given species are killed after an exposure of 10 minutes under standard conditions is called:
a] Thermal death time
b] Thermal death point
c] Minimum growth temperature
d] Lethal temperature

700. The lowest temperature at which growth and multiplication of an organism occur when other environmental factors are kept constant :
a] Minimum temperature
b] Survival temperature
c] Minimum growth temperature
d] Optimum growth temperature

701. The main objective of farm management is to obtain:
a] Maximum gross income
b] Maximum net income
c] Minimum gross income
d] Minimum net income

702. The main objective or preserving foods is :
a] To destroy micro-organisms
b] To prevent micro-organism
c] To destroy enzyme activities
d] All the above

703. The main states producing major amount of silk in our country are :
a] Karnataka, West Bengal and Jammu & Kashmir
b] Punjab, Haryana & Rajasthan
c] Gujarat, Maharastra and Tamil Nadu
d] Keral, M.P. & U.P.

704. The main theme of Indian's National Extension Service is :
a] To help people help others
b] To increase food production
c] To improve socio-economic standard
d] To help people to help themselves

705. The maximum adoption of F_1 hybrid seeds in India is in :
a] Tomato b] Cucumber
c] Chillies d] Cabbage

706. The maximum CO_2 concentration is found at high intensity of :
a] 2500-5000 foot candles
b] 6000-7500 foot candles
c] 1000-2000 foot cnadles
d] 2150-2450 foot candles

Financial Emergency has so far not been proclaimed in India.

GS

707. The maximum percentage of nitrogen is available in:
a] Jowar b] Dhaincha
c] Sunhemp d] Kidneybean

708. The modified F_2 ration for duplicate genes is :
a] 13 : 3 b] 15 : 1
c] 9 : 7 d] 9 : 3 : 4

709. The most common herbicide used for controlling weeds in pulse crops is :
a] 2, 4-D b] Batachlor
c] Atrazine d] Basalin

710. The most commonly used test to estimate the combining ability in species where pollination control is difficult is :
a] Open-pollinated progeny test
b] Top cross test
c] Single cross test
d] Polycross test

711. The most concentrated fertilizer used for nutrient supply is :
a] Urea b] DAP
c] Anhydrous Ammonia d] SSP

712. The most critical stage for irrigation in upland rice is:
a] Active tillering
b] Panicle initiation
c] Heading
d] Grain filling

713. The most desirable principle of insect control is :
a] Chemical control
b] Physical control
c] Cultural and mechanical control
d] Integrated control

714. The most difficult problem in the production of seed potato is the production of :
a] Good variety
b] Cold resistant variety
c] Virus free seed
d] Nematode resitance seed

715. The most important characteristic of prokaryotes is:
a] Absence of nuclear membrane
b] Presence of flagella
c] Absence of cell wall
d] Binary fission

716. The most important pulse crop of India is :
a] Pigeon pea b] Cow pea
c] Chick pea d] Black gram

717. The most influential type of demonstration is :
a] Method demonstration
b] Result demonstration
c] Composit demonstration
d] Half field demonstration

718. The most reliable method for pregnancy test is :
a] Physical b] Clinical
c] Biological d] Chemical

719. The most suitable stage of the pollen grain for pollen culture is :
a] After the second mitotic division of the pollen nucleus
b] The binucleate stage
c] After the first mitotic division of the pollen nucleus
d] Both b & c

720. The most suitable temperature for the germination of seeds of rabi crops is :
a] 5°C to 10°C b] 20°C to 30°C
c] 31°C to 40°C d] 41oC to 45oC

721. The most toxic isomer in BHC is :
a] Alpha b] Beta
c] Gamma d] Delta

722. The most virus prone crop is :
a] Pea b] Potato
c] Cabbage d] Onion

723. The most widely used method in hybrid seed production is :
a] Cytoplasmic-genetic male sterility
b] Cytoplasmic male sterility
c] Genetic male sterility
d] Manual emasculation

724. The nitrogen content (%) in Ammonium sulphate nitrate is :
a] 26 b] 33
c] 21 d] 46

725. The nucleus seed of potato is produced in the :
a] Plains b] Peninsular region
c] Ganga valley d] Hills

726. The object of agroforestry is :
a] To manage land efficiently
b] To generate employment apportunities to rural poor
c] To use available resources efficiently and econommically
d] All the above

727. The optimum plant population per hectare of sorghum is :
a] 50,000 plants b] 1,00,000 plants
c] 1,50,000 plants d] 2,00,000 plants

728. The optimum seed rate (kg/ha) for soyabean is :
a] 40 b] 60
c] 80 d] 100

729. The organic compounds other than nutrients that in small amounts promote, inhibit, or otherwise modify a physiological process in plants are called :
a] Plant regulators b] Plant hormones
c] Growth regultors d] Growth harmones

GK **Three categories of Civil Services are provided for the country by the Constitution.** GS

730. The origin home of rice is :
a] Tropical Asia b] Temperate Asia
c] South East Asia d] Central America

731. The pathogen of loose smut of wheat is :
a] Externally seed borne
b] Internally seed borne
c] Both externally and internally seed borne
d] Not seed borne

732. The per hectare seed rate for *Nicotiana rustica* is :
a] 20 to 30 gm. b] 35 to 50 gm.
c] 80 to 125 gm. d] 1.00 to 2.25 kg

733. The per hectare seed rate for *Saccharum officinarum* is :
a] 80 to 95 Quintals b] 20 to 30 Quintals
c] 35 to 45 Quintals d] 60 to 75 Quintals

734. The performance of a combination of two specific inbreds in a particular cross is :
a] General combining ability
b] Specific combining ability
c] Combining ability
d] None of the above

735. The pH is defined as :
a] pH = - log [H +] b] pH = log [1 - 1/H+]
c] pH = - log 1/[H +] d] pH = log 1/[H+]

736. The pH of a decinormal solution of HCl is :
a] 0-1 b] 1.0
c] 1.1 d] 10.0

737. The pH of a solution increases from 7 to 11, the solution becomes progressively more :
a] Concentrated b] Dilute
c] Acidic d] Alkaline

738. The pH of sodic soil is :
a] < 7.0 b] > 7.0
c] < 8.5 d] > 8.5

739. The phenomenon of heterothallism was first reported by :
a] Blakeslee b] Burgeff
c] Cutter d] Dodge

740. The phenomenon of uptake of water at the expense of energy by the cell and usually against osmotic phenomenon is known as:
a] Imbibition b] Osmosis
c] Active absorption d] Passive absorption

741. The photolysis of water takes place in the :
a] Thylakoid b] Granum
c] Stroma d] None of above

742. The physical characteristics of pure seed of an improved variety possess:
a] High purity
b] Optimum moisture content
c] Uniformity of seeds
d] All of the above

Knowledge World!
First anther culture
S.C. Maheswari and Sipra Mukerjee of Delhi discovered the anther culture technique in 1967. This technique, adopted on a large scale in China, enables agriculturists to obtain the same quality of crop from subsequent harvests.

743. The plant growth not sensitive or responsive to gravity is known as:
a] Negative geotropic
b] Diageotropic
c] Ageotropic
d] Positive geotropic

744. The plant part to be cultured *in vitro* is known as :
a] Miniature plant
b] Zygotic plant
c] Explant
d] Infant plant

745. The plants which flower after a critical day length is exceeded are called :
a] Short-day flowering plants
b] Long-day flowering plants
c] Day-neutral flowering plants
d] None of the above

746. The plants which flower when the day length is less than a certain critical length are called :
a] Short-day flowering plants
b] Long-day flowering plants
c] Day-neutral flowering plants
d] None of the above

747. The point where the total revenue line and total cost line intersect is known as :
a] Break even point b] Equilibrium point
c] Inflection point d] Crossing point

748. The pollen collecting basket in the honey bee lies on:
a] Fore leg b] Hind leg
c] Head d] Abdomen

749. The powerful inhibition of root and stem growth is :
a] Auxins b] Ethylene
c] CO_2 d] Cytokinins

750. The practice of rearing silkworms for obtaining raw silk is known as:
a] Agriculture b] Sericulture
c] Silviculture d] Worm culture

751. The preferred method of controlling loose smut of wheat is by :
a] Seed dressing with Thiram
b] Seed dressing with Agrosan
c] Seed treatment with Vitavax
d] Seed treatment with Agrimycin

GK **A Public Commission Member can be removed only on a report of the Supreme Court.** GS

752. The price which necessarily covers the variable cost and also the fixed price is :
a] The long run price b] The short run price
c] The market price d] The equilibrium price

753. The primary objective of green house cultivation of vegetables is :
a] To get early yield
b] To get off season yield
c] To protect the crop from diseases
d] To protect the crop from excessive rains

754. The principal method of breeding groundnut is :
a] Introduction b] Hybridization
c] Selection d] All the above

755. The principal way in which P and K ions move from the soil to the root of field crops :
a] Root interception b] Mass flow
c] Diffusion d] None

756. The process involved in the decomposition of organic matter and leading to the formation of humus is :
a] Nitrification b] Humification
c] Immobilization d] Intergrade

757. The process of removal of excess water from the land surface is called:
a] Sub surface drainage
b] Surface drainage
c] Slope drainage
d] None of the above

758. The process through which all existing organism in a medium are killed :
a] Fermentation b] Pasteurization
c] Sterilization d] Putrefication

759. The production of alcohol from hexoses is termed :
a] Glycolysis b] Fermentation
c] Reduction d] None of them

760. The production of which crop seed has received very little attention in India :
a] Cereal b] Fibre
c] Oil seed d] Forage

761. The progeny of a single self-pollinated homozygous plant is called :
a] Clone b] Inbred line
c] Pure line d] Isogenic line

762. The proportion of nitrogen in the air is :
a] 21% by weight b] 21% by volume
c] 10% by weight d] None of these

763. The pruning in hybrid-tea rose plants in Northern Indian Plains is done during :
a] 1st or 2nd week of May
b] 2nd or 3rd week of July
c] 2rd or 3rd week of October
d] 2nd on 3rd week of April

764. The pulse rate per minute of fowl is:
a] 72 b] 100
c] 272 d] 300

765. The pupae of silk worm are killed__________ days before the emergence of moth.
a] 1-2 b] 2-3
c] 4-5 d] 8-10

766. The quickest method for starting a lawn is by :
a] Seed sowing b] Dibbling
c] Turfing d] Turf-plastering

767. The rate of application of EDB and aluminium phosphide for control of stored grain insect pests is respectively :
a] 10 ml/tonne grain and 2 to 3 tablets/tonne of grain
b] 30 ml/room and 2 to 3 tablets/1000 cu ft. space
c] 3 ml/quintal grain and 2-3 tablets/tonne of grain
d] 10 ml/tonne of grain and 10 tablets/tonne of grain

768. The rate of ear parent to pollen parent rows, in production of maize does not exceed :
a] 1:1 b] 2:1
c] 3:1 d] 5:2

769. The recommended age of mating of rams is :
a] One year b] Two year
c] 2.5 year d] 36 months

770. The red colour in tomato is due to pigement :
a] Capsicin b] Lycopene
c] Xanthophyll d] None of the above

771. The regulators produced by plants, which in concentrations regulate plant physiological processes are called :
a] Growth regulators b] Plant regulators
c] Phytohormones d] Growth hormones

772. The regulators that effect flowering are called :
a] Flowering regulators
b] Flowering hormones
c] Both a and b
d] All the above

773. The requirement of fruit per day per capita is :
a] 50 g b] 60 g
c] 85 g d] 100 g

774. The rotation intensity of maize - wheat + gram - moong is :
a] 200% b] 250%
c] 300% d] 400%

775. The science which deals with the study of bacteria is called :
a] Virology
b] Bacteriology
c] Microbiology
d] Agricultural Microbiology

GK GS

Lord Amherst was the English Governor General in India at the time of the first Burmese war.

776. The science which deals with the study of virus is called :
a] Virology b] Mycology
c] Virulence d] Microbiology

777. The scientific name of Bihar hairy caterpiller is :
a] *Spilosomo obliqua*
b] *Euproctis fratorna*
c] *Prodenia literan*
d] *Agrotis psilon*

778. The scientific name of desert locust is :
a] *Schistocerca gregaria*
b] *Spilasoma obliqua*
c] *Anomala bengalensis*
d] *Agrotis ypsilon*

779. The second cephalic segment in insect contains the pailed appendages modified to form :
a] Antennae b] Ovipositors
c] Compound eyes d] Mandibles

780. The Second Stage of Production starts from the point where :
a] Marginal Physical Product is maximum
b] Average Physical Product is maximum
c] Marginal Physical Product is zero
d] Average Physical Product is zero

781. The seed produced by the breeder who developed the particular variety is :
a] Breeder seed
b] Foundation seed
c] Registered seed
d] Certified seed

782. The sequence of assimilation of CO_2 in photosynthesis was determined by:
a] Radioactive tracers
b] Radioautographs
c] Paper chromatography
d] All of the above

783. The shortest time required to kill all bacteria of a given species in a given substance of a prescribed temperature is called :
a] Thermal death time b] Thermal death point
c] Life span d] Sterilization

784. The silk consists of_____________ protein.
a] Fibroin b] Sericin
c] Both a & b d] None of the above

785. The soil structure in B-horizon of solonetz soils is :
a] Spheroidalb] Crumb
c] Prism like d] Columnar

786. The study of development and protection of forests for timber and other plant products used in industry:
a] Forestry b] Agro-forestry
c] Aqua-forestry d] None of the above

787. The study of plant pathology in India was initiated by:
a] B.B. Mundku b] E.J. Butler
c] K.C. Mehta d] J.F. Dastur

788. The sub-agroclimatic zones in the upper Gangetic plain are :
a] Three b] Two
c] Four d] Five

789. The substrate concentration at which the rate of enzyme catalysed reaction is half of its maximum velocity is termed as :
a] pH b] pK
c] K_m d] V_{max}

790. The substrate that are relatively reduced, such as fats will yield RQ values :
a] Less than one b] Equal to one
c] Greater than one d] All of them

791. The substrates that are highly oxidised such as the acids of kreb's cycle of R.Q value will be :
a] One b] More than one
c] Less than one d] None of above

792. The suppressed gene during epistasis is known as :
a] Hypostatic b] Suppreessed genome
c] Hyperstatic d] Static gene

793. The synthesis of messenger-RNA is known as :
a] Translation b] Transcription
c] Transmutation d] Transgenesis

794. The temperature which favours the most rapid growth of an enzyme is :
a] Maximum growth temperature
b] Suitable growth temperature
c] Minimum growth temperature
d] Optimum growth temperature

795. The term 'Market margin' implies:
a] Cost incurred on marketing goods and services
b] Profit of the middlemen
c] Profit of the producer
d] Price spread among various intermediaries

796. The term used to describe the use of living material to test the effect of known and putative biologically active substances such as phytochromes :
a] Biomass b] Biosynthesis
c] Bioassay d] All the above

797. The theory of germplasm was given by :
a] Charles Darwin b] Hugo de Vries
c] August Weismann d] Gregor John Mendel

798. The total depth of irrigation to crop:
a] Delta
b] Base
c] Intensity of irrigation
d] None of the above

GK GS

Lord Cornwallis had given proper shape to the police force.

799. The total nitrogen present in a fertilizer is calculated in term of :
a] NO_2 b] NO_3
c] Elemental form d] Ammonium form

800. The total number of elements essential for plant growth and development is :
a] 15 b] 16
c] 17 d] 18

801. The trade name of aluminium phosphide is :
a] Celphos b] Degesh
c] Rogor d] E D B

802. The type of male sterility exploited in the production of hybrids in pearl millet is :
a] Genetic
b] Cytoplasmic
c] Cytoplasmic-genetic
d] Gametocidal

803. The type of male-sterility exploited in the production of hybrids in sorghums is :
a] Genetic
b] Cytoplasmic
c] Cytoplasmic-genetic
d] Gametocidal

804. The types of gametes to be produce by an individual of genotype Aa BB Cc DD are :
a] 2 b] 4
c] 6 d] 8

805. The upper limit of available water in soil is determined at :
a] 1/3 bar b] 1/2 bar
c] 2/3 bar d] 1 bar

806. The variance contributed by alleles having linear quantitative effects is:
a] Dominance deviation
b] Additive genetic variance
c] Epistasis
d] None of a above

807. Term genetics was coined by:
a] Bateson b] Mendel
c] Johansen d] DeVries

808. The vertical distance between the water surface at the source and at the outlet is called :
a] Orifice
b] Cone of depression
c] Distance of influence
d] Static head

809. The volume of water discharged per unit of time from the well is called :
a] Capacity of well
b] Well yield
c] Well stock
d] None of the above

810. The wave length of light that are capable of affecting plant growth lie between :
a] 0.00003 -0.00009 cm
b] 0.003 - 0.009 cm
c] 0.003 - 0.009 cm
d] 0.03 - 0.09 cm

811. The wilt of gram crop can be controlled by :
a] Seed treatment with fungicide
b] BHC
c] Deep ploughing and late sowing
d] All the above

812. The word 'poulry strain' is synonym to :
a] Poultry feed b] Breed
c] Disease d] Poultry pest

813. The yellow colour in onion is due to the pigement :
a] Anthocyanin b] Quercetin
c] Lycopene d] Carotene

814. The zero-tillage concept was first postulated by :
a] Jethro Tullb] Wilcox
c] Garbar d] Russel

815. There is only one generation in a year in :
a] Mango mealy bug
b] Mango leaf hopper
c] Grass hopper
d] Leaf hopper

816. Thick white (firm white albumin) present in the eggs of poultry is secreted in :
a] Magnum region of the oviduct
b] Isthmus region of the oviduct
c] Uterus
d] Infundibulum

817. Tick the correct group of plant nutrients which act as energy exchangers :
a] K, Na, Ca and Mg
b] C, N, P and S
c] H and O
d] Fe, Mn, Mo, Cu, B and Zn

818. Tift 23 A, a cytoplasmic male sterile line is of :
a] Bajra b] Jowar
c] Maize d] Cotton

819. To measure moderate and high pressure as well as negative pressure, it is suitable to use :
a] Micromanometer
b] Simple manometer
c] Differential manometer
d] A dead weigh pressure gauge

820. To measure pressure above or below the atmospheric pressure :
a] Bourdon's tube pressure gauge
b] Micro manometer
c] Inverted diffential manometer
d] Differential manometer

Sisir Kumar Ghose was associated with newspaper *Amrita Bazar Patrika*.

GS

821. To obtain haploid plants __________ is useful.
a] Meristem culture
b] Anther culture/pollen culture
c] Tissue culture
d] None of the above

822. To prevent the horns from developing in calves dehorning (debudding) is done :
a] Within 2 days after birth
b] Within 10 days after birth
c] Within 21 days after birth
d] Within 40 days after birth

823. Tobacco is native to the :
a] America b] Australia
c] Africa d] Asia

824. Tomato crop is :
a] Self pollinated
b] Cross pollinated
c] Both a and b
d] None of the above

825. Total area which can be irrigated by the certain channel or project is called :
a] Gross command area
b] Culturable command area
c] Irrigated area
d] None of the above

826. Total revenue will remain unchanged if price elasticity of demand is :
a] More than one b] Less than one
c] Equal to one d] Infinite elastic

827. Total weed seed in foundation seed must not be more than ________ per kg of seed.
a] 10 b] 20
c] 5 d] 2

828. Transpiration in plants is most rapid when :
a] The atmosphere is saturated with water vapour
b] Excess of water is present in the soil
c] The air is still
d] The environmental conditions are dry

829. Treatment of fruits and vegetables with boiling water or steam for short periods, followed by cooling prior to canning is called :
a] Peeling b] Blanching
c] Washing d] Exhausting

830. Banana variety resistant to Panama wilt is:
a] Rasthali b] Basrai
c] Poovan d] Lal Kel

831. Tree crop which can be propagated by root cutting:
a] Apple b] Fig
c] Pomegranate d] Grape

832. *Trichogramma minutum* parasitises in the :
a] Egg stage b] Larval stage
c] Papal stage d] Adult stage

Quiz Kit!

Tiger fort

The forest for of Ranthambhor, a perfect example of the classical *vana durg,* is now a game sanctuary, popularly known as the Sawai Madhopur game reserve.

833. *Trifolium alexandrinum* is the botanical name of :
a] Soybean b] Chickpea
c] Sugarcane d] Berseem

834. Triticale is :
a] Interspecific cross b] Interaspecific cross
c] Intergeneric cross d] Intrageneric cross

835. Triticale is the most sucessful synthetic allopolyploid produced by crossing :
a] Wheat and Rice
b] Wheat and Barley
c] Wheat and Rye
d] None of the above

836. Tuber roots are used as a vegetable in the crop :
a] Potato b] Colocasia
c] Sweet potato d] Ginger

837. Two genes may produce the same effect but the effect will be intensified if both genes are present. This is called :
a] Duplicate
b] Additive effect
c] Complementary
d] Suplementary action

838. Umran is a variety of :
a] Ber
b] Loquat
c] Litchi
d] None of the above

839. Under deep-litter system of poultry farming space required per bind is :
a] 2-3 sq. ft. b] 4-5 sq. ft.
c] 6-7 sq. ft. d] 7-8 sq. ft.

840. Under dry land farming conditions, the most efficient crop from water use point of view is :
a] Linseed b] Lentil
c] Barley d] Gram

841. UPAS-120 is a variety of :
a] Wheat b] Arher
c] Barley d] Urd

842. Urea supergranule is efficient source of nitrogen for waterlogged paddy because :
a] Its leaching losses are minimum
b] Its ammonia volatilization losses are low
c] Its fixation in soil is low
d] Denitrification losses from it are minimum

GK **Nityanand Swami was the first Chief Minister of Uttaranchal.** GS

843. Using artificial insemination method by diluting bull semen, number of possible cows breed per ejaculate of bull semen is :
a] 200 b] 400
c] 600 d] 800

844. Utility is measured in terms of :
a] Satisfaction
b] Actual consumption
c] Price paid
d] Emotions of the consumer

845. V A micorrhiza promotes mainly the plant uptake of :
a] Nitrogen b] Phosphorus
c] Potassium d] Molybdenum

846. Variety of sorghum recommended of dry farming is :
a] CSH-1 b] CSH-2
c] CSH-5 d] CSH-3

847. Vegetable occupy __________ % of the total cultivated area.
a] 1.2 b] 2.2
c] 2.7 d] 3.3

848. Vegetables are canned in :
a] Brine b] Syrup
c] Distilled water d] None of the above

849. Vegetative propagation in fruits is adapted for :
a] True to type fruiting
b] Early fruiting
c] Uniform fruiting
d] All the above

850. Washing soda is the common name of :
a] Sodium carbonate b] Sodium bicarbonate
c] Calcium carbonate d] Calcium bicarbonate

851. Washington Navel orange belongs to species :
a] *Citrus sinensis*
b] *Citrus reticulata*
c] *Citrus aurantium*
d] *Citrus limon*

852. Water and inorganic nutrients are translocated through :
a] Xylem b] Phloem
c] Cambium d] Chlorenchyma

853. Wax coating treatment enhances the shelf life of fruits because it blocks:
a] Transpiration b] Respiration
c] Ripening process d] None of the above

854. What is full form of K G K :
a] Kisan Gyan Kendra
b] Kisan Gur Kendra
c] Krishi Gyan Kendra
d] Kisan Gossip Kendra

855. What is full form of KVK :
a] Krishi Vigyan Kranti
b] Krishi Vigyan Kendra
c] Kisan Vividh Kendra
d] Krishi Vivad Kendra

856. What is pork?
a] Meat obtained from goat
b] Meat obtained from buffaloes
c] Meat obtained from poultry
d] Meat obtained from swine

857. What is the cause of rinder pest diseases?
a] Virus b] Bacteria
c] Venom d] None of the above

858. What is the name of the first compartment of the stomach of ruminant?
a] Omasum b] Reticulum
c] Rumen d] Abomasum

859. What is the number of mammary glands in a buffalo?
a] One b] Two
c] Four d] None of above

860. What is the recommended composition of Bordeaux mixture for the control of late blight of potato at early stages?
a] 2 : 2 : 50 b] 4 : 4 : 50
c] 3 : 3 : 50 d] 6 : 6 : 50

861. What is the seat of milk synthesis in the body of a cow?
a] Milk veins
b] Milk well
c] Escutcheon
d] Alveolus of the mammary gland

862. What is White Yorkshire?
a] A breed of swine
b] A breed of buffaloes
c] A diseases of cattle
d] A breed of sheep

863. What will be the nutritive ratio of the feed which contains 70% T.D. N and 14% D. C. P.?
a] 1 : 4 b] 1 : 5
c] 1 : 0.2 d] None of the above

864. What will be the specific gravity of the liquid if its corrected lactometer reading was found to be zero?
a] Zero b] 1.0
c] 1.032 d] None of the above

865. When a growing plant is illuminated by a unilateral light, it responds by bending toward the light. This effect is called :
a] Positive geotropism
b] Negative geotropism
c] Phototropism
d] None of the above

GK **In all seven Peshwas ruled Maratha territory.** GS

866. When a maize variety is crossed with an inbred, it is said to a :
a] Top cross b] Test cross
c] Single cross d] Double cross

867. When a segment of a chromosome is missing, the condition is known as :
a] Translocation b] Inversion
c] Deletion d] Duplication

868. When all varieties appear in each replication of the experiment and are arranged in a random order within the replication, the design is called :
a] Systematic b] Randomized
c] Latin squares d] None of the above

869. When both sex mature at same time the condition is called :
a] Homogamous b] Heterogamous
c] Dichogamous d] Diclinous

870. When certain fungus are associated with roots of particular plants and both are benefited by this association, it in called :
a] Symbiosis b] Mycorrhiza
c] Synergism d] None of above

871. When gene hides the effect of a second gene when both are present. It is called :
a] Modifying action
b] Epistasis
c] Inhibiting
d] Additive effect

872. When marginal product increases at a increasing rate, then :
a] Total product increases at a decreasing rate
b] Total product increases at a increasing rate
c] Average product decreases
d] None of the above

873. When marginal utility is zero, total utility is :
a] Maximum b] Minimum
c] Zero d] Average

874. When MPP is zero, TPP is at its maximum, because :
a] Further addition to MPP is negative
b] Further addition to MPP is positive
c] MC = MR
d] Elasticity of production is more than one

875. When new variety of a crop is introduced, which farm management approach is appropriate?
a] Complete budgeting
b] Partial budgeting
c] No budgeting
d] Programme planning

876. When plant becomes elongated and yellow often loosing the strength, in absence of sunlight is kwown as :
a] Gummosis b] Etiolation
c] Chlorosis d] Necrosis

877. When seeds of two or more crops are mixed and sown without any set geometric relationship of arrangement, it is known as :
a] Intercropping
b] Mixed cropping
c] Sequence cropping
d] Relay cropping

878. When several commodities are demanded together for certain common purpose, it is called as :
a] Cross demand b] Derived demand
c] Joint demand d] Income demand

879. When the elasticity of production function is less then one it indicates:
a] Constant returns
b] Diminishing returns
c] Increasing returns
d] Negative returns

880. When the income from each of the different enterprises on a farm is less than 50 per cent of the total farm income, the farming is known as :
a] Mixed farming
b] Peasant farming
c] Diversified farming
d] Specialized farming

881. When the sex mature at different time condition is called :
a] Homogamous b] Heterogamous
c] Dichogamous d] Diclinous

882. When total product is maximum the marginal product is :
a] Maximum b] Negative
c] Zero d] Positive

883. When total utility is maximum, marginal utility is :
a] Negative b] Zero
c] Increasing d] Decreasing

884. When two non-allelic genes are required to produce a single effect then the gene action is :
a] Complementary b] Modifying
c] Additive d] None of above

885. When two single crosses say (A x B) and (C x D) are crossed the resulting hybrid population, (A x B) x (C x D) is known as :
a] Single cross b] Double cross
c] Three way cross d] Top cross

886. Which clay is most dominant in black soil :
a] Illite b] Kaolinite
c] Montmorillonite d] Vermiculite

887. Which clay mineral is rich in potash?
a] Montmorillonite b] Kaolinite
c] Illite d] Chlorite

888. Which country is the place of origin of potato ?
a] South America b] South Canada
c] South Africa d] South China

GK India's per capita requirement of foodgrains is 172 kg/annum. GS

889. Which element cause dispersion of soil particles ?
a] Na b] K
c] Ca d] Mg

890. Which element has greater significance in biological nitrogen fixation :
a] Nickel b] Sulphur
c] Potassium d] Molybdenum

891. Which element is the mineral constituent of chlorophyll :
a] Calcium b] Magnesium
c] Sulphur d] Zinc

892. Which form of nitrogen is available in urea?
a] Ammonicalb] Amide
c] Nitrite d] Nitrate

893. Which gas is released from paddy fields ?
a] CH_4 b] H_2S
c] CO_2 d] NH_3

894. Which is dwarfing rootstock for mango ?
a] Olour
b] Totapari Red Small
c] Alphonso
d] Dashehari

895. Which is the correct row ratio for intercropping of mustard with wheat?
a] 5 : 1 b] 9 : 1
c] 6 : 1 d] 10 : 1

896. Which molecule serves as the principal carbon acceptor in the Calvin-Benson pathway ?
a] Phosphoenol pyruvate
b] Ribulose 1-phosphate
c] Ribose 5-phosphate
d] Ribulose 1, 5-biphosphate

897. Which of part of the plant is used for rooting, of the *Bryophyllum, Glauxinia* and *Peperima* ?
a] Root cutting
b] Stem cutting
c] Leaf cutting
d] None of the above

898. Which of the disease does not occur on wheat crop in India?
a] Leaf blight b] Rusts
c] Smuts d] None of them

899. Which of the following agency is associated with agricultural development?
a] S.F.D.A. b] I.A.D.P.
c] A.N.P. d] I.C.D.P.

900. Which of the following approaches in extension has got the highest intensity if influence?
a] Group approach
b] Individual approach
c] Mass approach
d] Community approach

901. Which of the following chemicals is widely used for purifying drinking water?
a] Flourine
b] Patassium permagnate
c] Chlorine
d] Iodine

902. Which of the following compound provide the necessary energy and reducing power for CO_2 fixation and reduction?
a] ATP b] NADPH
c] Both a and b d] None of the above

903. Which of the following concentrates contains maximum amount of digestible nutrients ?
a] Jowar b] Barley
c] Maize d] Guar

904. Which of the following cost is substracted from Gross Income (G.I.) to get the value of Family Labour Income?
a] Cost A_1 b] Cost A_2
c] Cost B d] Cost C

905. Which of the following crops requires more amount of water to grow ?
a] Maize b] Paddy
c] Potato d] Banana

906. Which of the following disease is caused by mycoplasma-like-organisms?
a] Papaya ringspot b] Banana bunchy top
c] Tobacco leaf curl d] Brinjal little leaf

907. Which of the following factors affects the dormancy of seed ?
a] Seed coat impermeable to water
b] Mechanical resistance of seed coats
c] Seed coat impermeable to oxygen
d] All of the above factors

908. Which of the following fruit is not suitable for jam making ?
a] Mango b] Banana
c] Aonla d] Lemon

909. Which of the following fruits are indigenous to India?
a] Mango, Banana b] Bael, Jack fruit
c] Aonla, Phalsa d] All of these

910. Which of the following institutions do not finance for working capital to farmers ?
a] Nationalised banks
b] Primary co-operative credit societies
c] R.R.B's
d] Primary land development societies

911. Which of the following intercropping system has synergistic effect on component crop :
a] Autumn planted Sugarcane + Mustard
b] Autumn planted Sugarcane + Chickpea
c] Autumn planted Sugarane + Wheat
d] Autumn planted Sugarcane + Potato

912. Which of the following is a cross breed of sheep ?
a] Kutchi b] Nilgiri
c] Nellore d] Marwari

913. Which of the following is a ripening hormone ?
a] Ethylene b] Cytokinin
c] Auxin d] Gibberellin

914. Which of the following is also referred to as "Oportunity Cost"?
a] Money cost b] Alternative cost
c] Social cost d] Fixed cost

915. Which of the following is connected with Ascomycotina ?
a] Clamp connection
b] Cleistothecium
c] Zygospore
d] Sporangium

916. Which of the following is controlled by gibberellic acid ?
a] Fruit fall
b] Ripening of fruits
c] Vegetative growth
d] Prevention of the loss of flowers

917. Which of the following is ethylene absorbent ?
a] $KMnO_4$ b] KNO_3
c] K_2SO_4 d] KCl

918. Which of the following is example of photoperiodic responses exhibited by plants?
a] Flowering b] Seed germination
c] Leaf abscission d] All of the above

919. Which of the following is genetically heterogeneous?
a] Hybrid b] Composite
c] Clone d] Pure line

920. Which of the following is inflorescence vegetable ?
a] Potato b] Tomato
c] Cauliflower d] Spinach

921. Which of the following is low volume sprayer ?
a] Gear type rotary pump
b] Roller vane rotary pump
c] Plunger pump
d] All of the above

922. Which of the following is lowest in total digestible nutrients content?
a] Barley b] Bajra
c] Jowar d] Oats

923. Which of the following is more suitable for control of *Phalaris minor* in wheat field ?
a] Sulpho-sulphuran
b] Pendimethalin
c] Isoproturon
d] Methabenzthiazuron

924. Which of the following is not a method of group communication?
a] Demonstration b] Symposium
c] Flannel graph d] Circular letter

Did You Know ?

Establishment of agricultural banks

The establishment of agricultural banks was first suggested by Sir William Wedderburn after the Deccan riots of 1870s. The first scheme came about in 1881. An Agricultural Bank was proposed to be set up at Pune the same year. While the Governments of India and Bombay were in favour of the scheme, it fell through because the then Secretary of State opposed it.

925. Which of the following is not a organic mercury fungicides?
a] Ceresan b] $HgCl_2$
c] Agrosan G.N. d] Semesan

926. Which of the following is not *kharif* season weed ?
a] *Chenopodium album*
b] *Amaranthus viridis*
c] *Echynochloa colonum*
d] *Comelina benghalensis*

927. Which of the following is not the element of key step in the process of programme planning?
a] The procedure in programme planning
b] The product of programme planning
c] Demonstration
d] The annual plan of work

928. Which of the following is often cross pollinated ?
a] Wheat b] Rice
c] Soyabean d] Pigeonpea

929. Which of the following is root vegetable?
a] Celery b] Asparagus
c] Cabbage d] Tapioca

930. Which of the following is stem vegtable ?
a] Potato b] Tomato
c] Cauliflower d] Spinach

931. Which of the following is the rural institution helping extension work in the villages?
a] Community Development Block
b] Gram Panchayat
c] Village Community Centre
d] Rural Youths

932. Which of the following is the serious pest of bajra?
a] Gundi bug b] Stem borer
c] Gall midge d] All of them

933. Which of the following is the variety of cauliflower?
a] Patna White b] Pusa Himani
c] Patharnakh d] None of the above

934. Which of the following is tropical fruit crop?
a] Mango b] Grape
c] Citrus d] Guava

Both Supreme Court and High Court decides the reasonableness of restrictions imposed on the Fundamental Rights.

935. Which of the following is variety of lime ?
   a] Duncun   b] Honey
   c] Foster   d] Jaffa
936. Which of the following is visible through nacked eyes?
   a] Mould   b] Yeast
   c] Bacteria   d] None of the above
937. Which of the following media can be used for vegetative propagation?
   a] Soil   b] Sand
   c] Coconut coir   d] All the above
938. Which of the following micro organism is associated with ammonification ?
   a] *Rhizobium leguminosarum*
   b] *Bacillus mycoides*
   c] *Clostridium*
   d] *Nitrosomonas*
939. 'Hen and chicken' disease of grape is due to :
   a] Copper deficiency   b] Boron deficiency
   c] Iron deficiency   d] Zinc deficiency
940. Which of the following particle in soil is considered finest ?
   a] Gravel   b] Clay
   c] Sand   d] Silt
941. Which of the following rural development project was launched before independence :
   a] Etawah Pilot Project
   b] Cooperative Movement
   c] Community Development
   d] National Extension Service
942. Which of the following seed rate of TPS is recommended for planting one hectare area ?
   a] 150 g   b] 250 g
   c] 350 g   d] 450 g
943. Which of the following statements is correct regarding transpiration?
   a] It create suction force and help in the ascent of sap
   b] It affects the diffusion pressure deficit
   c] It maintains the temperature for the leaves
   d] All of the above
944. Which of the following statements represent true concept of Extension evaluation ?
   a] Extension evaluation begins before the start of an extension programme
   b] Extension evaluation begins at the mid step of the programme
   c] Extension evaluation begins at the last step of the programme
   d] Extension evaluation is inbuilt in each step of the programme
945. Which of the following variety of chickpea is recommended for late sowing ?
   a] KPG-59   b] Sadabahar
   c] Pragati   d] K-850
946. Which of the following you connect with Mycorales ?
   a] Phytopthora   b] Clamp connection
   c] Albugo   d] Zygospore
947. Which one is an essential element for plant growth?
   a] Aluminum   b] Barium
   c] Sodium   d] Chlorine
948. Which one is non-acid forming fertilizer ?
   a] Urea
   b] Ammonium sulphate
   c] Calcium ammonium nitrate
   d] Single super phosphate
949. Which one is not a tool of farm management?
   a] Farm Planning
   b] Farm Budgeting
   c] Production and Cost Functions
   d] Farm Book Keeping
950. Which one is not correctly matched?
   a] Ginger - Rhizome
   b] Garlic - Bulb
   c] Potato - Tuber
   d] Sweet potato - Stolon
951. Which one is the egg parasite ?
   a] *Trichogramma minutum*
   b] *Apanteles flavipes*
   c] *Pyrilla perpusilla*
   d] *Coccinella septumpunctata*
952. Which one is the recommended optimum plant populatiuon for hybrid maize ?
   a] 40-50 thousands per hectare
   b] 50-60 thousands per hectare
   c] 60-70 thousands per hectare
   d] 70-80 thousands per hectare
953. Which one is the rodenticide ?
   a] Zinc phosphide   b] Aldrin
   c] Lead arsenate   d] Endrin
954. Which one is the storage pest ?
   a] *Bemisia tabaci*
   b] *Dacus dorsalis*
   c] *Sitophilus oryzae*
   d] *Spodaptora litura*
955. Which one is used for primary tillage?
   a] Hand Hoe
   b] Deshi plough
   c] Mould board plough
   d] Blade harrow

The total wetland area of world is estimated to be around 85,58,000 sq.kms.

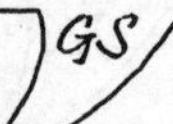

956. Which one of the following approach is not correctly matched with its technique ?
a] Mass = Exhibition
b] Individual = Farm visit
c] Group = Mela
d] Mass = Newspaper

957. Which one of the following crop is not correctly matched with its test weight (gm) ?
a] Pearl-millet = 5-7 b] Wheat = 20-22
c] Mungbean = 34-36 d] Sunflower = 60-62

958. Which one of the following factor is responsible for higher yield under protected cultivation ?
a] High photosynthetic efficiency
b] High O_2 availability
c] High temperature
d] All the above

959. Which one of the following fruit contains maximum vitamin C ?
a] Papaya b] Pineapple
c] Lime d] Aonla

960. Which one of the following fruit is monocot in nature?
a] Mango b] Pineapple
c] Litchi d] Papaya

961. Which one of the following fruits is a rich source of iron ?
a] Mango b] Datepalm
c] Apple d] Karonda

962. Which one of the following fruits is rich in vitamin A?
a] Aonla b] Litchi
c] Banana d] Mango

963. Which one of the following is a monocot vegetable ?
a] Radish b] Spinach beet
c] Garlic d] Pointed gourd

964. Which one of the following is a non-projected visual teaching aid?
a] Black board b] Film strips
c] Epidaiscope d] Overhead projector

965. Which one of the following is a regular bearing variety of mango?
a] Samarbahista Chausa
b] Langra
c] Amrapali
d] Alphanso

966. Who often referred to as the "father of plant physiology" ?
a] Papov b] Van Helmont
c] Hales d] Woodward

967. Which one of the following is a synthetic pyrethroid?
a] Pyrethrum b] Fenvalerate
c] Nicotene d] Rotenone

968. Which one of the following is a wam season vegetable?
a] Okra b] Pea
c] Cabbage d] Turnip

969. Which one of the following is an Audio-Visual Aid ?
a] Flannel Graph b] Photograph
c] Transparent slide d] Motion picture

970. Which one of the following is brinjal variety ?
a] Pusa Red b] K-2
c] Pant Samrat d] All Green

971. Which one of the following is complex fertilizer?
a] DAP b] CAN
c] SSP d] MoP

972. Which one of the following is dioecious plant ?
a] Peach b] Pomegranate
c] Pea d] Pointed gourd

973. Which one of the following is evergreen temperate plant ?
a] Apricot b] Chilgoza
c] Plum d] Almond

974. Which one of the following is mass-contact extension method ?
a] Holding farmers meeting in a village
b] Farmers training camp in a village
c] Farm broadcast from All-India Radio
d] Result demonstration on a crop

975. Which one of the following is non-selective herbicide?
a] Alachlor b] Butachlor
c] Paraquat d] Atrazine

976. Which one of the following is not a variety of arhar?
a] Type - 21 b] Type - 7
c] Type - 44 d] Type - 17

977. Which one of the following is not a variety of mango?
a] Arka Aruna b] Arka Puneet
c] Arka Anmol d] Arka Hans

978. Which one of the following is not correctly matched?
a] Unorganized market - organization
b] Regulated market - Control
c] Perfect market - Competition
d] Terminal market - Area

979. Which one of the following is not correctly matched?

	Crop	Variety
a]	Mustard	Basanti
b]	Groundnut	Chandra
c]	Lentil	Malika
d]	Arher	Neelum

980. Which one of the following is not narrow leaved weeds ?
a] *Cynodon dactylon* b] *Cyprus rotundus*
c] *Setaria glouca* d] *Melilotus indica*

GK **Article 226 of Constitution deals with the writ jurisdiction of High Court.** GS

981. Which one of the following is not the first-line extension system of the ICAR?
a] National Demonstration Project
b] Krishi Vigyan Kendra
c] Farmers Training Centre
d] Operational Research Project

982. Which one of the following is the gynodioecious cultivar of papaya?
a] Ranchi　b] Pusa drawf
c] Pusa Giant　d] Pusa Majesty

983. Which one of the following is the late ripening cultivar of ber?
a] Umran　b] Gola
c] Kaithli　d] Sindhura Narnaul

984. Which one of the following is the latest model of extension introduced in India?
a] Intensive Agricultural District Programme
b] Training and Vist System
c] Integrated Rural Development Programme
d] National Extension Service

985. Which one of the following is the structure concerned with photorespiration?
a] Mitochondria　b] Peroxisome
c] Ribosome　d] Lysosome

986. Which one of the following pairs is correctly matched?
a] Shantiniketan : F.L. Brayne
b] Gurgon Experiment : Daniel Hamilton
c] Etawah Pilot Project : Arthur T. Mosher
d] Marthandam Project : Spencer Hatch

987. Which one of the following seed rate (kg/ha) is not correctly matched with crop ?
a] Wheat = 100　b] Urd = 20
c] Maize = 18　d] Mustard = 20

988. Which one of the following statement is correct ?
a] Blue green algae fix 20-30 kg N/ha in a rice field
b] Blue green algae fix 40-50 kg N/ha in a rice field
c] Blue green algae fix 8-10 kg N/ha in a rice field
d] Blue green algae fix 60-70 kg N/ha in a rice field

GK **Article 243 of Constitution deals with Panchayat's election.** GS

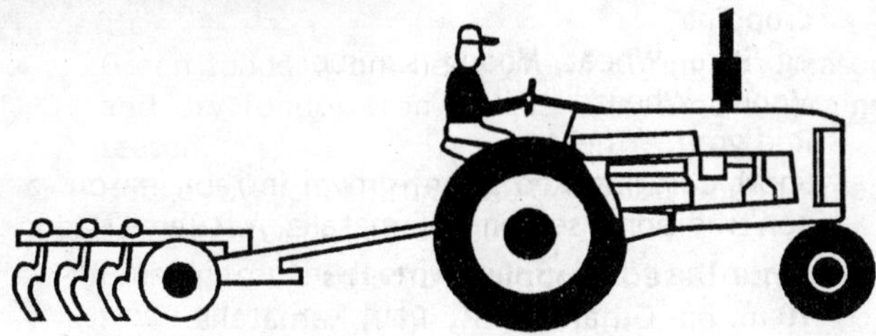

You Don't Know!

Smallest ship

The 122 dwt MV *Meenakshi* belongs to the Shipping Corporation of India.

Each Word Counts

FARMING SYSTEM

Efficient Cropping Systems for different Agro-climatic Zones of India

Agro-climatic region	*Soil type*	*Cropping systems*
1. Western Himalayas (cold-arid) Jammu and Kashmir, Ladakh plateau	Shallow, sandy to loamy	Rice-potato-wheat
2. Western Kachchh and part of Kathiawar	Desert and saline soils	Sorghum-wheat-millet, Maize- green manure Maize-wheat-sorghum
3. Deccan plateau (hot-arid) Andhra Pradesh, Karnataka	Red and black soils	Pigeon pea + sorghum/groundnut cotton - millet - sorghum
4. Northern plains and central highlands (semi-arid) Gujarat, Haryana, Madhya Pradesh, Punjab, Uttar Pradesh, Rajasthan	Alluvium derived soils	Maize-green manure Rice-wheat, Maize-wheat Cotton-wheat
5. Central highlands Gujarat plains Kathiawar peninsula (hot semi-arid)	Medium deep black soils	Groundnut-wheat Rice-sugarcane Rice-wheat
6. Deccan plateau (hot-semi-arid)	Shallow and Medium black soils	Pearl millet-potato-cotton Cotton-millet-sorghum
7. Deccan plateau & Eastern ghats (hot-semi-arid)	Red and black soils	Sorghum-safflower cotton-millet Pigeon pea-sunflower
8. Eastern ghats, Tamil Nadu uplands and Deccan plateau (hot semi-arid)	Red loamy soils	Rice-rice-pulses cotton-groundnut Groundnut-sorghum
9. Northern plain (hot sub-humid)	Alluvium derived soil	Rice-wheat, Maize-wheat, Maize-mustard-sugarcane
10. Central high lands (hot sub-humid)	Black and red soils	Sorghum-wheat, Maize-wheat
11. Eastern plateau (hot sub-humid) soils	Red and yellow soils	Rice-wheat-green manure Rice-pigeon pea
12. Eastern plateau and Eastern ghats (hot sub-humid)	Red and lateritic soils	Rice-Rice-Wheat Groundnut-sunflower
13. Eastern plain (hot sub-humid)	Alluvium derived soil	Rice-wheat, Rice-lentil Rice-chickpea, jute-rice
14. Western Himalayas (warm sub-humid)	Brown, forest and podzoilic soils	Maize-wheat, finger millet-rice-mustard
15. Bengal and Assam plains (hot sub-humid)	Alluvium derived soils	Rice-wheat-rice Rice-potato-sesame
16. Eastern Himalayas (warm humid)	Brown and red hill soils	Rice-rice, rice-wheat jute-rice
17. North-eastern Hills (warm humid)	Red and lateritic soils	Rice-rice, rice-wheat jute-rice
18. Eastern coastal plains (hot sub-humid to semiarid)	Coastal alluvium	Rice-rice, Rice-groundnut, Jute-rice
19. Western ghats and coastal plains (hot humid)	Red lateritic alluvium derived soils	Rice-rice, Rice-sorghum, Plantation crops
20. Islands (hot humid)	Red loamy and sandy soils	Rice-rice, Plantation crops

Cropping Patterns in India

Rice-Based Cropping Patterns

- Found in heavy rainfall areas (or) supplemental irrigation available areas.
- 46.4% areas in Monsoon belt – U.P., Bihar, M.P.
- 53.6% area in Multiseason belt – Assam, WB, Orissa, A.P., T.N, Kerala, Karnataka.
- 30- rice based cropping patterns in India.
- Bihar, WB, Orissa – Summer rice is emerging second crop competing jute.
- Introduction of Rice-non-traditional areas – Punjab Haryana, Western U.P. Rice - wheat due to

India has a total of 247 islands.

introduction of short duration of Rice has replaced maize in Punjab.

Wheat-Based Cropping Patterns

- 19 wheat-based cropping patterns in India.
- Introduction of photoinsensitive varieties grown wheat from Himalayas to Kanyakumari.
- Traditional areas -north west India Wheat -wheat rotation replaced by Moong -wheat; Red gram - wheat.

Reason : Availability of short duration varieties of moong and red gram.

Kharif sorghum-based Cropping Patterns

- Kharif sorghum -11 m.ha Maharashtra, M.P., Karnataka, Gujarat.
- 17 major cropping patterns.
- Maharashtra - Cotton, Pulses, Groundnut and Small millets-are alternatives.
- M.P. -Cotton, Pulses and Small millets are alternatives.
- Rajasthan -Wheat, Cotton, Bajra, Maize are alternatives.
- A.P. -Cotton, Pulses, Groundnut are alternatives.

Rabi sorghum-based Cropping Patterns

- 5 m.ha- Maharashtra, A.P, Karnataka
- 13 -cropping patterns -Maximum in Maharashtra with Bajra, Pulses, Oilseeds are alternatives.

Pearl millet-based Cropping Patterns

- Arid and semiarid tropics -l2 m. ha - Rajasthan and Gujarat.
- 20 major cropping patterns.
- Along with pulses, Groundnut, oilseeds, Kharif sorghum -Rajasthan.
- Cotton, tobacco -Gujarat.

Maize-based Cropping Patterns

- 6 m.ha - U.P., Bihar, Rajasthan, M.P. and Punjab.
- 12 cropping patterns.
- U.P, Bihar → Rice (Kharif) -Wheat (Rabi).

Ragi-based Cropping Patterns

- 2.5 m.ha-Karnataka,T.N.,A.P., Gujarat, Maharashtra, Hilly areas of U.P., H.P.
- Bajra, Sorghum, Maize, Pulses, and Oilseeds are alternative.

Cotton based Cropping Patterns

- 7.5 m.ha -Maharashtra, Gujarat, M.P.
- 16 cropping patterns.
- Cultivation extended -new command areas in Karnataka
- New short duration varieties -possible two crops of cotton in year.

Gram and other Pulses-based Cropping Patterns

- Short duration varieties of green gram, Bengal gram, red gram, cowpea made them suitable multiple cropping.
  eg. Rice -Wheat, Moong (summer)
  Moong -Wheat
  Red gram -Wheat
- Short duration Red gram -grown in rabi season -a non traditional season in Karnataka, A.P. Tamil Nadu.

Groundnut-based Cropping Patterns

- 70 m. ha- Gujarat, A.P., T.N., Karnataka
- Grown mainly rainfed conditions -9 cropping patterns.
- Kharif sorghum, Cotton, Pulses are alternative -A.P. and T.N.
- Kharif and rabi sorghum and small millets are alternative -Maharashtra.
- Sorghum, Cotton, Tobacco, Sugarcane - Karnataka.

Cropping System

It represents cropping patterns used on farm and their interaction with farm resources, farm enterprises and available technology to determine the farm makeup.

Cropping Pattern

Proportion of area under various crops at a point of time in a unit area.

Types of Cropping Systems

1. *Monocropping (or) Monoculture*

Growing of only one crop on a piece of land year after year.

Reasons :

(i) Climatological and socio-economic conditions.

(ii) Specialization of farmer growing particular crop.

2. *Multiple Cropping*

- Intensification of cropping for both time and space dimensions.
- Growing two (or) more crops on the same piece of land in one calendar year.

Types :

(a) *Intercropping*

- Growing two (or) more dissimilar crops simultaneously on the same piece of land in a definite row pattern.
- Cropping intensity in space dimension is achieved.

Intercropping is of two types–

(i) Additive Series

(ii) Replacement Series

(i) Additive Series

- One crop sown as a base crop.
- Another crop introduced into base crop by adjusting (or) changing crop geometry.

(ii) *Replacement Series :*

- Both the crops are component crops.

GK

There are 302 scheduled commercial banks in India.

GS

Requirements -Successful Intercropping

(i) The time of peak nutrients demands of component crops should not overlap. eg. Maize + green gram. Green gram -peak demand period- around 35 DAS. Maize - peak demand period -around 50 DAS.

(ii) Competition for light (solar radiation), nutrients, air space are minimum among component crops.

(iii) Complementary should exist between the component crops.

On the basis of above knowledge, suitable combinations are:

Principal Crop	*Intercrop*
Sugarcane	Wheat, Cowpea, Soyabean, Moong, Sunflower
Sorghum	Cowpea, Soyabean, Moong, Urd, Arhar
Maize	Cowpea, Soyabean, Urd, Arhar, Castor
Bajra	Cowpea, Soyabean, Urd
Cotton	Soyabean, Groundnut
Potato	Wheat, Radish

Advantages

(i) To raise productivity per unit area.

(ii) Better utilization of growth resources - nutrient, water, light, air.

(iii) Additional income from component crop.

(iv) Quick growing component crops, suppress harmful weeds by thriving in the interspace of base crop.

(v) Gainful utilization of labour -increasing more man days employment potential.

(vi) Less incidence of insect pests and disease attack

(vii) Less erosion losses.

Disadvantages

(i) Fertilizer management is difficult because nutrients requirements is different for different crops.

(ii) Difficult in harvest- different seeding time for different crops.

(iii) Certain circumstances (eg. allelopathy) - one suppress growth another crop.

(b) *Mixed Cropping*

- Growing two or more dissimilar crops simultaneously intermingled without any row pattern.

(c) *Sequential Cropping*

- Growing of two or more crops in sequence on the same piece of land in a farming year .
- Succeeding crop is planted after preceding crop harvested

(d) *Relay Cropping*

- Planting of succeeding crop before the harvest of preceding crop.

> **Knowledge World!**
> *First missile boat*
> **INS *Vinash*, a Soviet OSA 1, was the first Indian missile boat commissioned on Jan. 21, 1971.**

(e) *Ratoon Cropping (or) Ratooning*

- To raising a crop with regrowth coming out of roots or stalks after harvest of the crop.
  eg. - sugarcane, fodder, grasses

(f) *Multistoried Cropping/Multilevel Cropping*

- System of growing crops of different heights together at the same time on same piece of land.
- Using land, water, space most efficiently or economically.
  eg. Coconut + pepper + grasses, Coffee + banana + arhar.

3. *Crop Rotation*

- Recurrent succession of crops on the same piece of land either in a year (or) longer period of time, to get maximum profit with least investment and without impairing soil fertility.

Principles for Successful Rotation—

- Crop with tap roots and fibrous root system.
- Leguminous crops with non-leguminous crop.
- Selection of a crop based on demand of people of that area.
- Selection of crop should be problem-based:

(a) Sloping lands -alternate cropping of erosion promoting crops with erosion-resisting crops.

(b) Dryland (or) partially irrigated area → crops tolerate drought eg. oilseeds and pulses.

(c) Low lying (or) flood prone area -crops will tolerate water stagnation eg. Rice, jute.

Advantages

- Proper utilization of all the resources and inputs.
- Helps in controlling insect pests, diseases and weeds.
- Agricultural operation done timely for all crop.
- Soil fertility maintained.
- Get better price for produce.

Impact of Cropping Patterns

Factors responsible for Changes in Cropping Patterns

- Development of high-yielding and short duration varieties in main crops.
- Development of production technology to grow them.
- Incentives and innovations provided by central and state governments.
- Spread of irrigation.

GK GS

Indian Constitution originally consisted of 395 Articles.

- Development of fertilizer production and technology.
- Govt's special programmes -enhancing production of particular crop.
- Changes in the socio-economic conditions of the people and rapid industrialisation in the country.
- Transformation of subsistence agriculture into market-oriented economy.

Negative impact of -Changes in Cropping Patterns

- Introduction of HYV and short duration varieties. Consequently led to intensive cropping and resulted in decline in fertility and deficiency of Zn, Mn in Punjab due to introduction of paddy-wheat rotation.
- Cause more incidence of insect pests and weeds.
- More use of fertilizers -environmental pollution and salinity.
- Increased use of irrigation water- salinity and water logging.
- Indiscriminate use of pesticides -adverse effect on environment and microorganisms.
- Resistance of insects and pathogens towards.
- HYV -rice, wheat pushed pulses and oilseeds to marginal areas.

(a) Packages of Practices for Cereals and Millets

S.N.	*Crop*	*Varieties*	*Sowing Time*	*Spacing and Seed Rate*	*Manures and Fertilizers*	*Water Management*	*Weed Management*	*Yield*
1.	Rice	IR-8, IR-28, TKM-6, Jaya, Ratna, Basmati-370	Kh.-June-July, Rabi-Mid-Oct. Mid Nov.	(K) 20 x 15 cm, (R) 15 x 15 cm, Broad Casting - 100kg/ha, Nursery - 30-40 Kg/ha	FYM-10-15 t/ha, NPK-120:60:60 kg/ha	Critical stage (CS) Seedling, Tillering, Flowering, Grain, formation.	Pre-planting-Basalin @ 1 kg ai/ha, Pre-emr-2 kg ai/ha, Post-emr- 3 kg ai/ha	40-50 q/ha
2.	Wheat	Sonalika, Kalyansona, Arjuna	Rabi-Mid-Oct-Mid-Nov.	100 kg/ha	FYM-10-15 t\ha, NPK-80:40:40 kg/ha	CS-Crown Root initiation, Tillering, Flowering, Dough stage	Phalario-Isoproturon @ 7.5 kg ai/ha, 30 DAS	40-50 kg/ha (Irrigated), 20-25 q\ha (Rainfed)
3.	Barley	Kailash, Ratna, Jyoti	Rabi-Mid-Oct.-Mid-Nov.	20 x 100 cm, 100 kg/ha	FYM- 10-15t/ha, NPK-60:30:30 kg/ha	2-3 Irrigation	Phalario-Isoproturon @ 7.5 kg ai/ha 30 DAS	30-35 q/ha
4.	Maize	Ganga safed- 2 Vijay, Vikram	Kh.-Mid-June-July, Rabi-Mid-Oct.	50 x 20 cm, Hybrid - 20-25 kg/ha, Composite-20 kg/ha	FYM-10-15t/ha, NPK-100:40:40 kg/ha	CS-Seedling, Tasseling, Silking, Grain, formation	Atrazine @ 1 kg ai/ha (pre-emr)	40-50 q/ha (Hybrid), 30-40 q/ha (Composite)
5.	Sorghum	CSH-1-6, CSV-1-7	Kh.-June-July, Rabi-June-Oct.	50 x 15 cm, 15 kg/ha	FYM-10-15 t/ha, NPK-100:50:40 kg/ha	CS-Flowing, Grain filling	Atrazine @ 1 kg ai/ha (pre-emr)	50 q/ha (Irrigated), 25 q/ha (Rainfed)
6.	Pearl Millet	HB-5, PHB -10	June-July	50 x 15 cm, 5 kg/ha	FYM-15 cm, NPK-80:40:30 kg/ha	CS-Tillering, Flowering, Grain filling	Atrazine @ 1 kg ai/ha (pre-emr)	35 q/ha (Irrigated), 15 q/ha (Rainfed)

(b) Packages of Practices for Pulses

S.N.	Crop	Varieties	Sowing Time	Spacing and Seed Rate	Manures and Fertilizers	Water Management	Weed Management	Yield
1.	Gram	PUSA 209, Guasar, Ujjain 21	IInd fortnight of Oct.	80-100 kg/ha	N-P-K 15-20 : 50-60 : 0 kg/ha	Flowering, Grain development	Pre-planting Basalin @ 1 kg ai/ha Pre-emergence Tribunil @ 2.5 kg ai /ha	15-20 q/ha
2.	Red gram	UPAS 120, ICPL 151	June-July	60 x 15 cm, 10-15 kg/ha	N-P-K 20 : 50 : 30 kg/ha	Flowering, Grain development	Pre-planting Basalin @ 1 kg/ha Pre-emergence Tribunil @ 2.5 kg ai/ha	20-30 q/ha
3.	Green gram/Mung	T-44, Vamban-I	Kh.-June-July, Sum.-Mar-Apr.	30 x 45 cm, 12- 15 kg/ha	N-P-K 20 : 40 : 0 kg/ha	Flowering, Grain development	Pre-planting Basalin @ 1 kg ai/ha Pre-emergence Tribunil @ 2.5 kg ai/ha	12-15 q/ha
4.	Urd/Black gram	T-9, T-27, Pragati	Kh.-June-July, Sum.-Mar-Apr.	15 kg/ha	N-P-K 20 : 40 : 0 kg/ha	Flowering, Grain development	Pre-planting Basalin @ 1 kg ai/ha Pre-emergence Tribunil @ 2.5 kg/ha	12-15 q/ha
5.	Cowpea	Pusa Phalguni	Kh.-June-July, Sum.-Mar-Apr.		N-P-K 20 : 40 : 0 kg/ha	Flowering, Grain development	Pre-planting Basalin @ 1 kg ai/ha Pre-emergence Tribunil @ 2.5 kg ai/ha	10 q /ha
6.	Horse gram	CO-1, HPK-1 & 2	Kh.-June-July Sum.-Mar-Apr.	40 kg/ha	N-P-K 20 : 40 : 0 kg/ha	Flowering, Grain development	Pre-planting Basalin @ 1 kg ai/ha Pre-emergence Tribunil @ 2.5 kg ai/ha	7-8 q/ha
7.	Lentil	L9-12, Pusa 1		20 x 30 cm, 40 kg/ha	N-P-K 20 : 40 : 0 kg/ha	Flowering, Grain development	Pre-planting Basalin @ 1 kg ai/ha Pre-emergence Tribunil @ 2.5 kg ai/ha	18-20 q/ha

(c) Packages of Practices for Oilseeds

S.N.	Crop	Varieties	Sowing Time	Spacing and Seed Rate	Manures and Fertilizers	Water Management	Weed Management	Yield
1.	Groundnut	TMV-1, 3, 4, AK-12, 24	Kh.-Jun-July, Rabi-Nov., Sum.-Mar-Apr	Bunch-40 x 15 cm, 60 x 20 cm, 80-100 kg/ha	N-P-K 20 : 40 : 15 kg/ha	Branching, Flowering	Pre-planting Basalin @1 kg ai/ha, Pre-em. Pendimathalin @ 0.75 kg ai/ha.	30-35 q/ha
2.	Rapeseed/Mustard	Varuna, Prakash, Pusa Basari	Toria-Sept-Oct, Mustard-Oct	Toria - 30 x 10 cm, Mustard - 45 x 10 cm, 5-6 kg/ha	N-P-K 60 : 40 : 80 kg/ha	Pre-flowering, Pod filling	Pre. plt. Fluchloralin @ 1 kg ai/ha, Isoproturon @ 1 g ai /ha 30 DAS	Toria-20 q/ha, Mustard 25 q/ha

GS

Standard air pressure at sea-level is 1013.25 millibars.

S.N.	Crop	Varieties	Sowing Time	Spacing and Seed Rate	Manures and Fertilizers	Water Management	Weed Management	Yield
3.	Soyabean	Brag, Clask, Shilajeet	Mid June-I[st] week July	45 x 10 cm, 80 kg/ha	N-P-K 20 : 60 : 40 kg/ha	Seedling, Flowering, Pod formation	Pre. plt. Fluchloralin @ 1 kg ai/ha, Pre-em. Metribuzin @ 1 kg ai/ha	30-35 q/ha
4.	Sesamum	T-12, 13	Kh.-June-July, Rabi-Oct.-Nov	45 x 15 cm, 3-4 kg/ha	FYM- 15-20 t /ha N-P-K 30 : 60 : 30 kg/ha	Seedling, Flowering, Pod development	Pre plt. Basalin @ 1 kg ai/ha	8-10 q/ha
5.	Sunflower	CO1, BSH 1, KBSH 1		8-10 kg/ha	N-P-K 60 : 40 : 20 kg/ha	2 weeks before flowering to 2 weeks after flowering		20-25 q/ha
6.	Castor	HC-6, Aruna, T-3	June-July	90 x 40 cm, 8-10 kg/ha	N-P-K 40 : 40 : 20 kg/ha			400-950 kg seeds/ha

(d) Packages of Practices for Fibres

S.N.	Crop	Varieties	Sowing Time	Spacing and Seed Rate	Manures and Fertilizers	Water Management	Weed Managemer :	Yield
1.	Cotton	Hybrid- 4, Varalaxmi, Suria, Sujata	May-June	50 x 25 cm, American cotton - 20 kg/ha, Indian cotton-15 kg/ha	NPK 80 : 40 : 20 kg/ha	CS-Seedling, Branching, Flowering, Boll formation	Diuron @ 1 kg ai /ha (Pre-plant)	15-20 q/ha of seed cotton
2.	Jute	RC-212, 321, RO-632, 878	April-May	30 x 10 cm, 10 kg/ha	NPK 30 : 20 : 20 kg/ha	Pre-sowing Irrigation, Post-sowing Irrigation	2-3 weeding at 20 days interval	20-25 q/ha of fibre
3.	Sunhemp	Kanpur -12, ST 55, Bellary	Kh.-in North, Rabi- in South	30 cm (row) , 15 kg/ha				14 q\ha
4.	Flax	F895, F896	Rabi	20 x 2 cm, 40 kg/ha		1-2 irrigation		

(e) Packages of Practices for Fodder Crops

S.N.	Crop	Varieties	Sowing Time	Spacing and Seed Rate	Manures and Fertlizers	Water Management	Yield
1.	Berseem	Pusa Giant, T 526	Rabi - Oct.	25-35 kg/ha	15-20 t/ha organic Manure	10-12 irrigation	600-800 q/ha
2.	Lucene	Rambler	Rabi - Mid - Oct. Nov.	30 cm (row), 15 kg/ha		Irrigate the crop after every cut, Winter - 4 days interval, Summer- 10 days interval	500-600 q/ha 2000 q/ha
3.	Napier	Pusa Giant, Pusa Napier 1	July	90 x 30 cm, 12500 cuttings/ ha	2500-300 q Organic Manure	Sensitive to water logging	

GS

Highland climate is found in regions above 1500m.

WEED SCIENCE

Weed

Plants out of place. Weeds are unwanted and undesirable plants which interfere in the utilization of land and water resources and thus adversely affect human welfare.

- Relative weeds-crops plants in a crop
- Offtype plants-rogue-local wheat variety in Sonalika variety

Characteristics of Weeds

- Prolific seed producer
- Prolonged viability
- Dormancy
- Capable of reproducing by underground portions
- Dispersed through various agents
- Inherent hardiness against heat, cold, soil abnormalities
- Capable of escaping the destruction caused by man and animal
- Capacity to self-regenerate

Propagation of Weeds

1. Asexual - Algae, Ferns
2. Vegetative-Rhizomes, Runners, Stolons, Suckers, Offsets, Tubers, Bulbs, Bulbils, Stems and Roots

Dispersal/Dissemination of Weeds

- Reproductive unit may fall near the mother plant.
- A portion of mother plant may be carried away as contaminant with agricultural crop seeds
- Dispersed to short/long distances by dispersing agents

1. Wind :

- Weed seeds and fruits have specific organs to float in air
- Pappus hairs (dandelion)
- Whole plant may be detached and rolled-wild safflower
- Wind swings entire plant and cause mature capsule to disperse the seeds

2. Water :

- May drift either as whole plant or plant fragments.

3. Birds and animals :

- Fruits are eaten and seeds are dispersed
- Lantana seeds by Myna, Chinese turtle dove.
- Farm animals carry seeds and fruits on their skin, hair.

4. Man :

- Farm machinery, carts and automobiles

> **Quiz Kit!**
> *Longest canal*
> **The 649 km long *Indira Gandhi Nahar Pariyojana* (IGNP) in Rajasthan which took 35 years to build is the longest canal in the world.**

5. Manure :

- Farm yard manure.

Classification of Weeds

Based on Duration :

(a) Annuals: Weeds present along with annual crops. Duration is short.

- Multiply by seeds and produce large number of seeds/plants.
- Prevention of flowering or removal of weeds before flowering is important.
  e.g.*Amaranthus, Argemone, Cenchrus.*

(b) Biennials: *Alternanthera echinata, Eichorium intybus.*

- Life cycle two years
- I[st] year- vegetative; IInd year -produce flowers and seeds.

(c) Perennials: *Cyanodon dactylon and Cyperus rotundus*

- Grow more than two years.
- Propagate vegetatively in addition to seeds.

Based on cotyledons:

Monocotos- Dalapon and FLuchloralin

Dicots - 2, 4 D, MCPA.

3. **Morphological Characters:**

Grasses: *Echinochloa, Cyanodon.*

Sedges: *Cyperaceae - C. rotundus; C. deformis.*

Broad leaf weeds: All dicot weeds

Allelopathy

Definition

The phenomenon of one plant having detrimental effect on another through the production of chemical compounds.

- It depends upon a chemical compound being added to the environment by an allelopathic agent.
- Differs from competition in that competition leads to removal or reduction of some factors (water, nutrients and light) from the environment that is required by some other plant sharing the habitat.

Effect

- May be direct or indirect harmful effect.
- Inhibit seed germination or reduce the growth of other plant species.
- Phenolic acids, terpenoids, flavonoids.
- Released as vapour, as leachings from the foliage,

GK **Din-i-lahi was the new religion propagated by Akbar in 1582.** GS as exudates from the roots or decomposition of dead plant residues.

- It is the mechanism through which weeds affect crop growth.
- Significant role in maintaining the balance.

Types

True Allelopathy : The direct or indirect harmful effect on the other crops through the release of toxic substance as such from the plant.

Functional Allelopathy : When precursor is released which is converted into active substances by some microorganisms called functional allelopathy.

Based on Inhibition

Auto inhibitors: Toxic substances released from one species affect the other species.

Allo inhibitors: Toxic substances released from one species affect the same species.

I. Effect of weeds on crops: e.g. wheat. Seed exudates *Avena fatua* affect the germination and seedling growth of maize.
   Leaves and inflorescence, *Amaranthus spinosus* affect the vegetative growth.

II. Crop plants on weeds: Wheat, oats and peas *c .album* affect biomass production and leaf surface.

Activity and selectivity of Herbicides

Activity refers to the ability of herbicides to control weeds, whereas selectivity is the ability of herbicide to control weed without affecting the economic plant in a mixed population.

Herbicide activity is related to phytotoxic effects of a chemical on the plant growth and development.

Herbicide selectivity refers to the phenomenon wherein a chemical kills target plant species in a mixed population without harming or slightly affecting other plants.

Factors affecting herbicide activity and selectivity

(a) Formulation, Time, Method and Rate of application :

Formulation affect solubility, volatility and specific gravity and phytotoxicity of a herbicide.

Variation in the placement of a herbicide can affect its activity. A herbicide is selective at a lower rate may become non-selective when applied at higher rates.

(b) Plant Morphology :

Horizontally-arranged leaves intercept more of the spray than upright leaves. Presence of leaf hairs, thick cuticle prevents the absorption.

(c) Cultivation Practices :

Tillage practices bring out the dormant weeds to the top soil exposing them to sun resulting in germination, thereby increasing the susceptibility to herbicides.

(d) Absorption :

The differential absorption from one plant species to the other which determines herbicide activity and selectivity.

(e) Translocation :

Herbicide enters the plant must be transported to the site of action and disturb metabolic activities of plant.

(f) Metabolism :

Any herbicide which enters the plant undergoes metabolism. It is well known that tolerant plant inactivates a herbicide in one or many path ways.

(g) Environmental Factors :

The response of a plant to an applied herbicide depends upon the environmental stress occurs.

Crop-Weed Competition

Weed species absorb and accumulate nutrients faster than the crop plants. Application of high levels of fertilizers may increase the impact of the weed while weed growth is stimulated more than crop growth.

Competition for Light

Weeds grow faster and shade the crop plants if not checked weeds deplete photosynthetically active radiation resulting in reduction in photosynthesis and shortening the life of lower leaves. For eg. *Amaranthus hybridus* reduced the photon flux density reaching the cotton canopy by 90% at midday.

Competition for Water

The amount of water required to produce an unit amount of dry matter is known as to transpiration ratio. The transpiration ratio of weeds is high as compared to crops.

Critical Period of Crop-Weed Competition

The critical period of crop-weed competition is the period from the time of sowing upto which the crop is to be maintained in a weed-free environment to get higher yield.

Importance of critical period of crop-weed competition in weed management

Weeds generally germinate either before or at the same time as the crop, offer serious competition to the crop plants since they get opportunity to establish and accumulate dry matter faster than crop plant eg. In maize the first 2-3 weeks of its emergence, weeds put forth 15-18% of their growth while maize attained only 2-3%. Hence it is desirable to identify the critical period of crop-weed competition for all crops to get higher economic returns.

Integrated Weed Management (IWM)

"A weed population management system that uses all suitable techniques in a compatible manner to reduce weed population and maintain them at levels below those causing economic injury.

GK **The Danes formed their East India Company in 1616.** GS

Need for IWM

- No single control measure can be effective in all cropping situations.
- Usage of same control method leads to build-up of tolerant weeds.
- It is uneconomical to eradicate all weeds, that some may provide food and shelter for insect predators and predators that reduce other pests.

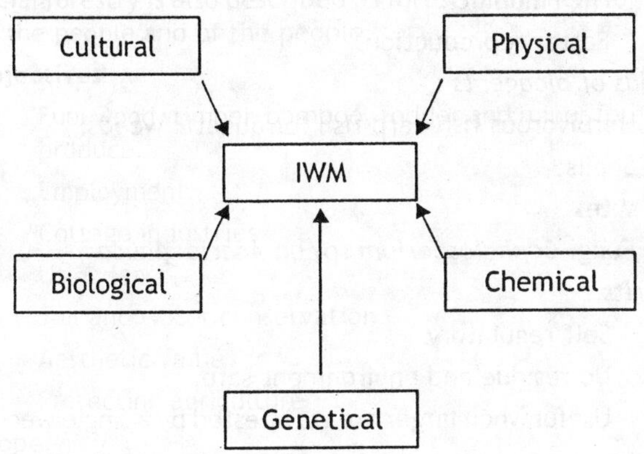

Cultural Methods–

(i) Field Sanitation - ploughing, burning

(ii) Sowing/Planting Method -weed seed free

(iii) Variety Selection -Selecting done

(iv) Planting Density/Spacing -closer

(v) Irrigation and Drainage -submergence alternate wet and dry

(vi) Cropping Systems -crop rotation, intercropping, mixed cropping

Physical Methods–

(i) Hand weeding -rice, tomato

(ii) Hand hoeing -onion, finger millet

(iii) Mowing -cutting at ground level

(iv) Dredging and chaining -aquatic weeds

(v) Burning

(vi) Mulching

(vii) Intercultivation

Biological Control Measures –

Insects - *Zygogramma (Parthenium sp.)*

Snails -*Mariaga sp.*

Fungi -*Water hyacinth.*

Chemical Methods –

Selective herbicides -Atrazine, 2,4 -D, Butachlor.

Non-selective herbicides -Paraquat, Diquat.

Did You Know ?

Strongest fort

The concentric Patiala fort was built in the 1760s by the Sikha on the plains. The people who protected it were strong enough to resist a fierce Maratha attack in 1794.

Genetical Methods –

Manipulation of herbicide resistant gene in cultivated crop that facilitate application of herbicides that are non-selective. e.g. tobacco, soyabean and maize etc.

Advantages of IWM

- All types of weeds can be controlled.
- Yield reduction is prevented.
- Evolution of tolerant weed is arrested.
- Ecologically sound and environment-friendly.
- Avoiding herbicides reduces the pesticide load in the environment.
- Prevent perennial weed shifts.
- Economically feasible.

(a) Principles of Weed Control

1. Prevention
2. Eradication
3. Control
   - Mechanical
   - Cultural
   - Biological
   - Chemical

1. Prevention

- All measures to deny the entry and establishment of new weeds in an area
- Use of clean crop seeds
- Care on the grains fed to cattle
- Prevent movement of cattle
- Clean farm implements
- Quarantine

2. Eradication

- Complete elimination of particular weed species from an area.
- Killing and destroying the viability of organs of reproduction.
- Eradication of perennial weeds cannot be accomplished in one season.

3. Control

(a) Mechanical or physical method :

(i) Hand pulling :

- In home gardens where hoe cannot reach.
- Best for annuals and biennials weeds.
- Useful in lawns after irrigating it.
- Time-consuming and costlier.

GK **In 1717 Farrukhsiyar granted to the English many trading privileges.** GS

(ii) Hoeing :

- Used in wide row crops.
- Complete destruction of all top growth. .
- Hand rotary weeder and wheel hoe.

(iii) Tillage :

- Weed control is one of the principles of the tillage.

(iv) Mowing :

- The object is to prevent seed production and remove unsightly weed growth.
- Used along roadsides, waste places and pastures.

(v) Flooding :

- In rice field.
- Water is stagnated 6-10 inches for 3-8 weeks.

(vi) Burning :

- To destroy top of weeds.
- Used in cotton to destroy weeds in rows.

(vii) Smothering with non-lining materials :

- Use of artificial mulch- straw, hay, plastic film, etc.
- Main purpose -to cut the light supply.
- 2-4 inch-thick mulch layer is required.

(b) Cultural method:

Not directly related to control weeds but helps to increase the yield.

(i) Crop stimulation :

- Modification of cropping conditions and soil leads to vigorous growth of crops which can compete better with weeds.
- Addition of sulphur, gypsum and altering the pH favourable to crops.
- Banding or seed dressing of fertilizers and manures.
- Foliar spray of fertilizers.

(ii) Crop rotation :

- Wild oat -using peas/gram as break crop for 2-3 years.
- Fodder -planting grain crops instead of lucerene.

(iii) Competitive or Smother cropping : (cowpea, millets, lucerene)

- They germinate quickly and form larger foliage and deep roots which exert their influence against weed seedlings.

(iv) Summer fallowing :

- Air temperature of 40-45°C.
- Not recommended for control of perennial weeds on light soils due to fear of erosion.

(c) Biological control :

- Control by living organisms called natural enemies which are encouraged or disseminated by man.
- Maintenance of balance between the target weed and its natural enemies.

Criteria for selection of biological agents :

- Host specific.
- Able to withstand new environment.
- Should be free of its parasites and predators.
- High mobility.
- Faster reproduction.

Kinds of bioagents :

(a) Herbivorous fish-carp fish for aquatic weeds.

(b) Snails

(c) Mites

(d) Fungi-*Cephalosporium* sp. on *Acacia glauca*

Merits :

- Self-regulatory.
- No residue and environment safe.
- Useful when larger area is infested by a single weed.

Demerits :

- Slow process and complete control is not possible.

(d) Chemical control :

Inorganic herbicide – Sodium chlorate, sodium arsenate, copper sulphate, copper nitrate.

Organic herbicide –Organic acids, diesel oil and salts.

Selectivity :

Selective herbicide –Simazine, Atrazine and 2,4-D.

Non selective herbicide - Paraquat and Diquat

Translocation :

Systemic – Simazine, 2,4 -D.

Contact – Paraquat and Diquat.

(b) Biological Weed Control

Use of living organisms viz. insects, disease organisms, herbivorous fish, snail or even competitive plants for the control of weed is called biological control.

Merits

1. Least harm to the environment
2. No residual effect.
3. Relatively cheaper and comparatively long-lasting effect
4. Will not affect non-targeted plants and safer in usage.

Demerits

1. Multiplication is costlier
2. Control is very slow
3. Success of controlling is very limited
4. Very few host specific bio-agents are available at present

GK GS
Bihar was attached to Bengal in 1733.

Outstanding and feasible examples of biological weed control

(A) *Specific bio-agents*

WEED CONTROL BY BIO-AGENTS

Sl. No.	*Weed species*	*Agent*
1.	*Salvinia molesta (water fern)*	Curculinoid weevil- *Crytobagus singularis*
2.	*Tribulus terrestris*	*Microlarinus lypriformis, M. lareynil*
3.	*Opuntia dilleni*	Cochineal scale insect - *Dactylopius tomentosus*
4.	*Lantana camara*	*Crocidosema lantana, Teleonemia scrupulosa*
5.	*Parthenium hysterophorus*	*Zygogramma bicolarata*

Bio-herbicies: The bio-herbicide approach involves spray of specific fungal spores or its fermentation products against the taget weed. These preparations are called 'Mycoherbicides'.

Non-specific agents: White amur feeds several weed species Lemna, Hydrilla, Potamogeton. Besides silver carp, common carp also feed large amount of algae.

Mycoherbicides and target weed

Sl. No.	*Product*	*Content*	*Target weed*
1.	Devine	*Phytophthora palmivora*	Strangle vine *Morrenia odorate* in citrus
3.	Bipolaris	*Bipolaris sorghicola*	Jhonson grass- *Sorghum halepens*
4.	Bilophos hygosco pius	*Steptomyces* general vegetation	Non-specific

Competitive plants: The weed parthenium is suppressed by growing *Cassia sericea*. The plant leachates of Cassia have 'Kolinei' which accumulate in the soil and interfere with the germination of parthenium.

Paragrass- *Brachiaria mutica* has been found highly competitive to *Typha sp.* in ditches.

Biological control of weeds in crop fields in India.

Eg. Control of *Ludwigia parviflora* in rice by steel blue beetle (*Haltica cynea*)

Control of *Cyperus rotundus by Bactra verutana.*

Recent Advances in Weed Science

Adoption of integrated weed management practices, appropriate, cultivation, technologies, sound crop rotation and crop diversification check the spread of weed seeds.

Weeds of Wetland, Garden land and Dry land

Wetland weeds	*Garden land weeds*	*Dry land Weeds*
I. Sedges	*I. Sedges*	*I. Sedges*
Cyperus diformis Cyperus iria Cyperus rotundus Fimbrishtylis miliacea	*Cyperus rotundus*	*Cyperus rotundus*
II. Grasses	*II. Grasses*	*II. Grasses*
Echinochloa colonum Echinochloa crusgalli Panicum sp.	*Cynodon dactylon Echinochloa colonum Dactyloctenium aegyptium Panicum repens Chloris barbata*	*Cynodon dactylon Dactyloctenium aegyptium Eleusine indica Panicum sp.*
III. Broad leaved weeds	*III. Broad leaved weeds*	*III. Broad leaved weeds*
Ammania baccifera	*Acalypha indica Amaranthus viridis Boerhaavia diffusa*	*Tridax procumbens Striga lutea Euphorbia hirta*

However, crop weed interactions are complex and needs application of recent advancements for the development of future weed management strategies.

A. Modelling crop-weed interactions

Competition between crop and weed for resources of growth (light, water and nutrients) is a critical process in agricultural ecosystems. The mechanisms of competition are not simple. Models of weed invasion, population growth and control will be useful for organising biological information on weeds and for developing weed control strategies.

B. Impact of herbicide use

The potential weed resistance to herbicides effect on non-target organisms and herbicide residues in the ecosystems needs to be monitored.

Herbicide resistance in weeds: The weeds may develop partial resistance, cross resistance and multiple resistance to applied herbicide.

Herbicides in agro-ecosystem : The persistence is very important factor because it determines the degree of weed control, the risk of carry over effect on the succeeding crop and the residue in soil and ground water. Various management techniques, viz., use of optimum dose of herbicides, application of FYM, crop rotation, and leaching the soil have been developed to minimise the residue

hazards in soils.

C. Biological control of weeds

The idea of biological control originated for dealing with noxious weeds on non-crop lands, and later it was extended to water bodies. But now, some success has been achieved in biological control of weeds in crop fields.

D. Biotechnological approaches

Bioherbicide approach Bioherbicide or Mycoherbicides viz. , **'Devine'** containing a formulation of soil borne fungus *Phytophthora palmivora* and **'Collego'** containing spores of *Colletotrichum gleosporoides* are commercialised *bioherbicides*

- Molecular markers in weed science
- Herbicide resistant crops

E. Allellopathy in weed management

Although the allelopathic suppression of crops by weeds has been well documented, the research on allelopathic effect of crops on weed has been very limited release of phytotoxic ferulic acid from wheat crop residues suppressed *Ipomea purpurea* and *Sida spinosa*.

DRY LAND

Dry Land Farming

Dry land farming is defined as farming dependent exclusively on rainfall which is often deficient, erratic and unreliable. There are no supplemental irrigation sources, annual rainfall is low, usually less than 750 mm.

Characteristic Features

- Farming dependent solely on rainfall
- Potential evapotranspiration is more than precipitation.
- Soil moisture storage is often inadequate for crop growth.
- Cropping is confined to short season only.
- Droughts are frequent.
- Crop yields are not stable and fluctuate widely from year to year.
- Farming is mostly of the subsistence type

Soil Moisture Conservation

It is using and managing land based on the capabilities of the land itself, involving the application of the best practices leading to profitable crop production.

Prerequisites for soil moisture conservation–

1. Physiographic
2. Rainfall
3. Flood, Drought

Need/Objective

1. Efficient collection of rain water
2. Reducing percolation losses and evaporation losses
3. Preventing soil erosion
4. Water harvesting at the time of availability
5. Efficient use of stored moisture
6. Absence of supplementary irrigation source

Types of Soil Moisture Conservation

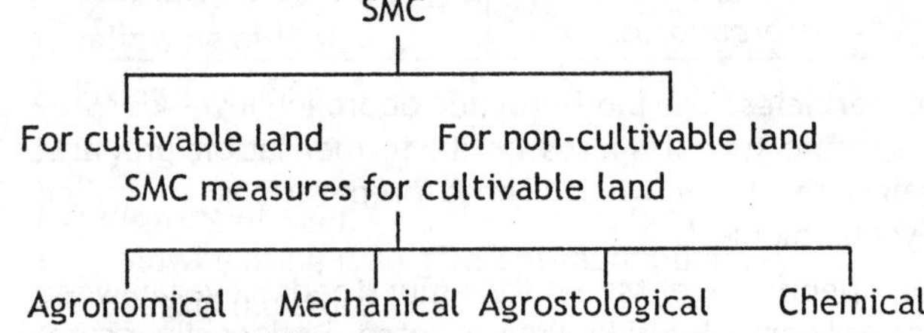

1. Agronomica/Cultural

- Mulching
- Crop management
- Conservation tillage

(I) Mulching

- Covering the soil surface with any material.
- Materials- crop stubble, straw, coir pith, groundnut shell, husk

Types of Mulching –

1. *Vertical*

trench - 40 x 15 cm at 2-4 m internal across slope -filled with organic matter .

2. *Live mulch*

Plant can grow by inter cropping.

e.g. - sorghum and cowpea

sorghums + Sword bean

3. Stover mulch/straw mulch

e.g. - cumbu, sorghum straw, sugar cane trash

Differences between Dry Farming and Rainfed Farming

Dry Farming	*Rainfed Farming*
Rainfall less than 750 mm	Rain fall more than 750 mm
Rainfall distribution is erratic	Rainfall is well distributed
PET (Potential Evapo Transpiration) exceeds rainfall	Rainfall is more than PET
Arid and semiarid climate	Humid and temperate climate
Crops are subjected to severe moisture stress	Crops don't face moisture stress
Soil moisture conservation are important	Runoff and erosion control are important

GS

First cotton textile mill was set up in 1818 at Fort Gloster.

4. *Stubble mulch*
- Stirring of soil -Crop residues protect from erosion by minimum disturbance.
- Even up to 3% slopes

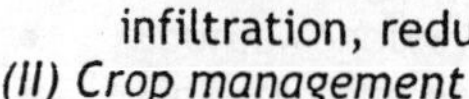

5. *Pebble mulch*
- Useful in dry land fruit tree culture on base of tree increase infiltration, reduce runoff loss

(II) Crop management
- Multiple cropping
- Agroforestry
- Crop alternation
- Strip cropping

1. Multiple crop :
- Avoiding fallow
- Crop rotation
- Smothering crop mixture
- Mix with shallow and deep-rooted crop (e.g.) Maize + Red gram

2. Agro forestry :
- Agri silvi culture
- Silvi pastural
- Agro silvi pastural

3. Crop alternation :
- Shallow-rooted crops with deep-rooted crops

4. Strip cropping :
- Crops are planted in narrow strips across the slope altered with strips of erosion-resistant crops and erosion-permissible crops
- 2 types

(i) Contour strip cropping :
- inter-tilled row crop + growing erosion resistant crop -right angle to flow of water

(ii) Field strip cropping :
- Uniform strip against slope but without contour.

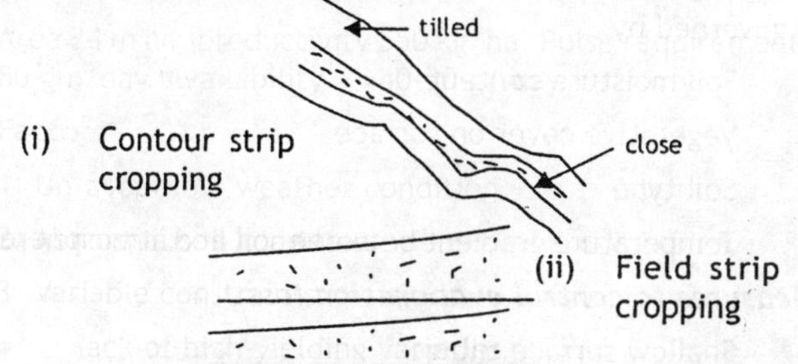

You Don't Know!
Longest breakwater
The 4.14 km Tuticorin north breakwater in Tamil Nadu is the longest breakwater in India.

(III) Conservation tillage
1. Off season/summer tillage
2. Contour ploughing/contour tillage
3. Minimum tillage

2. Mechanical Measures

(a) Basin listing
(b) Bunding
(c) Ridges and furrows
(d) Broad bed furrows
(e) Terracing

I. Basin Listing
- Small depressions of 10-15 cm depth at vegetative Internal by basin listing.

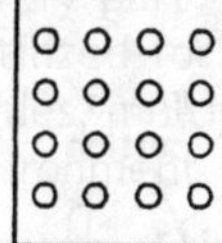

- Usually before sowing.

II. Bunding
- Across slope at suitable intervals

III. Ridges and Furrows
- 30-40 cm width, 15-20 cm ht -across the slope
- Suitable for medium to deep black and red soil
- Wide spaced crops grown i.e. maize, cotton, tomato

Types of bunding

Contour bunding	Graded bunding	Compartmental bunding
• 1 m basal width 0.5 m ht. along cotour • Suitable for red soil with 1% slope • Not for heavy black soil • Permanent	• 20-40 cm basal width, 15 cm ht. • Across + both direction • Deep red soil upto 1% slope • Not for black soil • Reshaped when required	• 15 cm width 15 cm ht. • Along + across both • Red, black soil upto 0.5 -1% slope • Suitable for broadcast crop

GK **Permanent settlement was introduced by Cornwallis in 1793.** GS

IV. Broad bed Furrow

- Beds with 1.5m width, 15 cm height
- Furrows of 30 cm ht.
- Suitable for heavy black soil
- Slope upto 0.6%

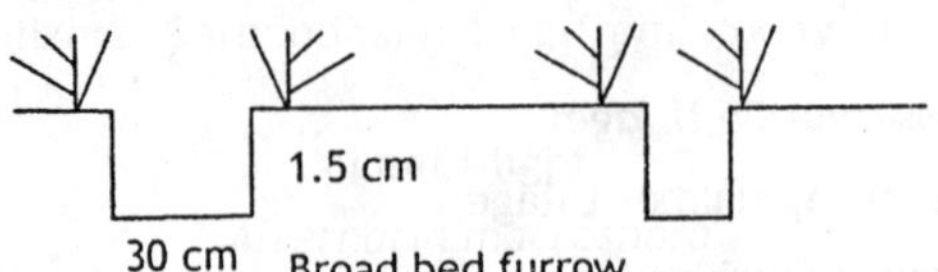

Broad bed furrow

Advantages–

- Accomodate wide range of crops
- Suitable for intercropping
- Runoff reduced
- Sowing can be done by seed drill

V. *Terracing*

(a) Bench terracing

- 16-33% slope

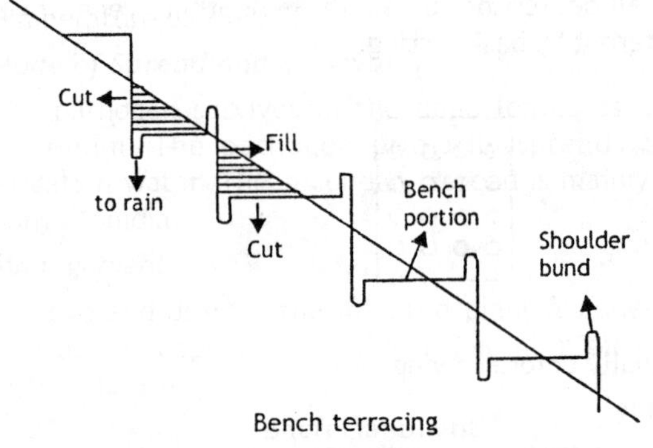

Bench terracing

- Series of level strips (or) platforms across slope
- Intensive cropping possible

(b) Zing terracing

- 3-10% slope
- Medium to deep soil of heavy rainfall areas
- Field divided -2 : l [donor; receiving area]

3. *Agrostological/Biological*

(i) Pasture/Grassland

(ii) Strip cropping with grasses

(iii) Ley farming

- Fodder grass + legume in rotation

4. *Chemical*

Soil Moisture Conservation measures for non-cultivable lands

1. Agroforestry
2. Pasture
3. Vegetative barriers
5. Mechanical measures

- Bunding
- Terracing

Advantages –

- Prevent direct 'splash' erosion by raindrop
- Increase infiltration
- Fit easily in any farming system
- Less expensive
- Less management

Disadvantages –

- Identification very difficult
- High initial cost

Reduction of Loss of Stored Soil Moisture

Rainfall infiltrates into the soil and permits downward and laterally gets stored in soil profile. Part of it percolates down to ground water. Stored water is absorbed by plants and weeds. It is lost from the soil surface as evaporation and from crop and weed canopy as transpiration.

Reduction of Evaporation Loss

Evaporation happens to maintain soil thermal regime and is governed by

- Soil moisture content
- Vegetative cover on surface
- Soil type
- Temperature gradient between soil and atmosphere

Measures to control evaporation loss

- Shallow surface tillage
- Mulching
- Use of anti-evaporant chemicals

The rate of transpiration is governed by:

- Soil moisture potential
- Atmospheric water demand
- Plant canopy characters such as leaf area, stomatal resistance

GK

GS

Fort William College was set up by Lord Wellesley in 1800 for the training of Civil Servants of the company.

Drought/Adoptic

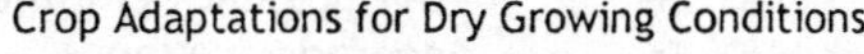

Crop Adaptations for Dry Growing Conditions

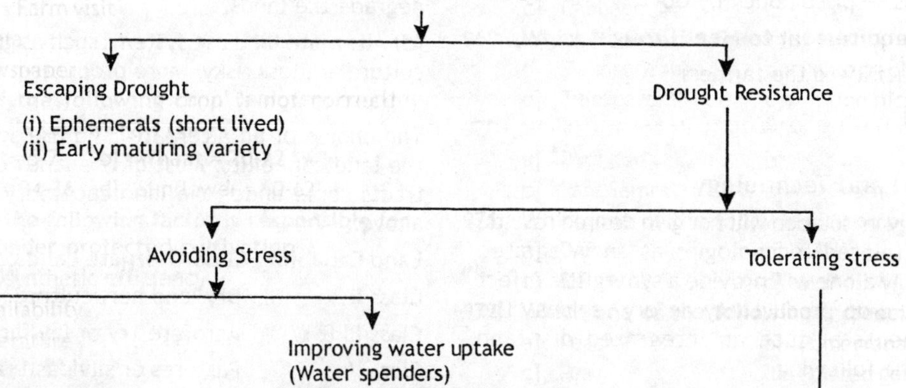

- **Escaping Drought**
  - (i) Ephemerals (short lived)
  - (ii) Early maturing variety
- **Drought Resistance**
  - **Avoiding Stress**
    - **Conserving water (Water savers)**
      - (a) Early closure of stomata
      - (b) Increased photosynthetic efficiency
      - (c) Low rates of cuticular transpiration
      - (d) Lipid desposites on leaves e.g. Soybean, sorghum
      - (e) Reduced leaf area
      - (f) Morphology of leaf surface (thick cuticle, waxy surface, spines, etc.
      - (g) Water storage in plants. The pineapple leaves contain substantial amounts of water in special non-chlorophyllous tissue which is utilised during drought
    - **Improving water uptake (Water spenders)**
      - (1) Efficient root system
      - (2) High root/top ratio
      - (3) Increaed plant osmotic potential
  - **Tolerating stress**
    - **Mitigating stress by maintaining high water potential**
      - (a) Resistance to dehydration
      - (b) Prevention of cell collapse (leaf with thick cuticle)
      - (c) By maintaining higher osmotic pressure by accumulating higher amounts of solutes.
    - **High degree tolerance to**
      - (a) Metabolic strain
      - (b) Plastic strain

Drought Management Technique for Dryland Development

Dry farming — Farming practiced in rainfall—areas of 400-700 mm/annum.

- Source of water only Rainfall

Rainfed farming — Farming practiced in rainfall-areas of > 700 mm Rainfall per annum

Impact of Drought

(i) shortage of food production-- failure of crop

(ii) shortage of fodder

(iii) shortage of drinking water leads to migration

(iv) shortage of fuel for cooking leads to deforestation

(v) death of cattle and humans

Management of Drought

1. Low monitory system

- Moisture conservation intercropping
- Choice of crops & varieties —Early maturity, deep root, dwarfshoot, thick leafs. More number of grain per pod
- Optimum plant population —Reduction in evaporation
- Pest management
- Mulching of plant material

2. High cost inputs

- Fertilizer use - to maximise the fertilizer use efficiency.
- Use of tractor power -improve infiltration & better root development.

3. *Water harvesting* —Farm ponds, percolation ponds, checkdams, retaining wall, etc

Benefits —

(i) Ground water available through percolation ponds

(ii) It creates droughtless condition

(iii) Less cost

(iv) Involvement of community participation

4. Transfer of Technology

Lab to land programmes through K.V.K.

5. Miscellaneous Efforts

- Financial assistance in the form of subsidy
- Low interest loans for the constructing of farm ponds
- Improving the socio-economic status of poor by Govt.

GK

The characteristic tools of mesolithic age is microlith.

GS

- Small scale & cottage industries in the rural areas.
- More emphasis on meterological research (computer & satellites)
- Comprehensive price policy by Govt.

Infrastructure Requirement to Meet Drought

(i) Seed banks -facility to the farmers

(ii) Fodder banks

(iii) Farm ponds

Integrated Dry Land Technology

A single technology in isolation will not give desired results. Adoption of all related technologies as an integrated dryland technology alone will provide a synergistic effect and improve the crop productivity on dry regions. The various components of such an integrated dry land technology are the following :

- *In situ* soil moisture conservation
- Choice of suitable crops and crop substitution
- Selection of high-yielding drought tolerant varieties
- Tillage to conserve moisture
- Establishment of optimum population
- Soil fertility management
- Crop protection against weeds, pests and diseases

Transpiration loss can be reduced by the use of antitranspirants and by some cultural methods.

Antitranspirants

Antitranspirants are substances or chemicals applied on plant foliage to control rate of transpiration. Based on their mechanism of action, antitranspirants are classified into various types.

- Stomatal closing type e.g. phenyl mercury acetate (PMA).
- Film forming type e.g. paraffin and wax emulsions
- Reflectant type e.g. kaolinite
- Growth retardant type e.g. cycocel (CCC)

Cultural methods of transpiration control

- Weed control
- Shelter belts
- Alley cropping

Alternative Land Use Systems for Dry Lands

Uncertain rainfall, poor soil conditions and low level of management has made annual cropping of field crops a non-remunerative enterprise in many pockets of dry lands. Sometime, cropping has been given up altogether and lands remain fallow and become wastelands overgrown with unwanted vegetation. To arrest this trend and to bring back the land under economically useful vegetation, alternate land use systems such as grasslands/pastures, agroforestry and horticulture are recommended. This has become necessary for the following reasons.

(a) Annual field crop production is non-viable and uneconomical in many ears.

(b) Yield of field crops is low and fluctuate.

(c) Continued use of the eroded and degraded lands under the present system of annual cropping may ecologically degrade the lands.

(d) Alternate land use systems such as grasslands and tree culture are less risky, more productive and remunerative in these marginal lands.

The choice of an alternate land use system depends on the land capability. Most of the lands under dry farming tracts come under the land capability classes of III and above.

Land Capability Class	Alternate Land Use Recommended
Class II	Dry land horticulture
Class III & IV	Agroforestry or lay farming
Class V	Pastures or silvipasture or tree farming
Class VI	Range lands or wood lots

Lay farming offers the following advantages

- Fodder for cattle
- Low risk system
- Soil moisture conservation
- Enrichment of soil fertility
- Control of perennial weeds

Choice of trees for dry lands

Trees suitable for dry lands have the following characters:

- Multipurpose tree species
- Adaptable to wide variations in soil and climate
- Withstand severe pruning
- Rapid growth e.g.

Black soil-*Azadirachta indica*

Red soil -*Albizzia lebbeck, Acacia nilotica*

Dry land Agriculture

The criteria for selection of fruit trees for drylands are :

- Drought tolerance
- Adaptability to varying soil conditions
- Quick regeneration after pruning
- Rapid recovery after stress is removed e.g.

Black soil- Ber, Sapota, Jamun, Amla

Red soil- Mango, Pomegranate, Amla

Silviculture System for Drylands

- Staggered planting -Trees are grown scattered in the field with annual crops
  e.g. neem + pulses/sorghum
- Border trees - Trees can be grown along farm boundaries. e.g. neem, tamarind
- Agrisilviculture

Alley Cropping

Annual field crops are grown in alleys formed by hedge rows of trees and shrubs. e.g. *Leucaena* as hedge with sorghum, maize as intercrop.

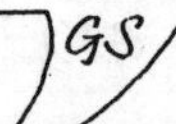

When beams of red, blue and green lights fall on the same spot, the colour of light becomes 'White'.

Benefits

- Green fodder from hedge rows during dry season and dry fodder from annual crops during rainy season.
- Hedge rows check runoff and erosion when formed along contour or across slope.
- Pruning from hedgerows can be used as fodder, fuel wood or for mulching.

Watershed

Watershed is a manageable hydrological unit that makes a harmonious use of the prevailing climate, soils, water, locally available material and human resources towards stepping up crop yields.

It is also referred to as catchment area or drainage basin.

Size of Watershed

- Micro
- Mini
- Macro

Principles of Watershed Management

1. Utilizing the land based on its capability
2. Minimizing silting up of tanks, reservoirs and lower fertile lands
3. Conservation of rainwater
4. Safe diversion of surface runoff to storage structures
5. Water harvesting for supplemented irrigation

Objectives

1. Production of food, fodder, fuel, fruit, fish combine on a sustained basis
2. Overexploitation of resources to be maintained
3. Water storage at convenient locations for different purposes
4. Erosion control, eco-system safety, economic stability
5. Recharge of ground water
6. Reduction of drought

Steps in Watershed Management

1. *Basic information* :

- Statistics of population and livestock
- Pattern of land ownership
- Topography, cropping system, yield and land capability for farming.
- Data on rainfall, erosion problems and ground water.
- Information on existing water sources like tanks, wells etc.
- Service facilities like schools, market, health and veterinary facilities.

2. *Developmental components :*

- Soil conservation measures
- Run off harvest in storage structures and its recycling for protective irrigation.
- Ground water recharge and development
- Efficient use of available water through proper field layouts, land shaping, leveling, lining of water courses and life-saving

Knowledge World!

Highest windmill

The highest windmill stands at Choglamsar in Leh at an altitude of 3,353 m (10,997 ft). It was set up by D. Ramachandra of Agro Aids, Bangalore, in 1985.

Main Components of Watershed

Management Programme :

1. Soil and water conservation
- Permanent measures-Bunds, Terrace
- Semi permanent measures-Inter bunds

Treatment

- Temporary measures- Vegetative barrier

2. Water harvesting
3. Crop management
- Optimum seeding time
- Fertilizers schedules and balanced use of plant nutrients
- Weed management and package of practices for aberrant weather
- Contingent crop planning

4. Alternate bund use system

Benefits of Water Shed Management

1. Increase soil health
2. Decrease silting
3. Increase natural resources

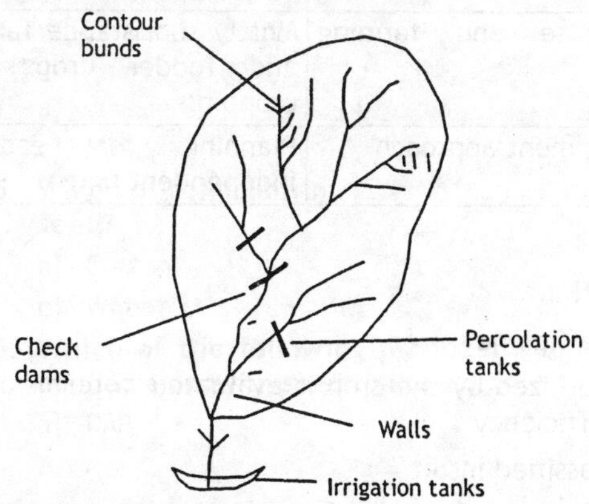

Watershed development for moisture conservation and water management

The term 'White Dwarf' is related to Astronomy.

Modern Concept In Dryland Farming

Aspect of Cultivation	*Traditional Practice*	*Modern Concept*
1. Choice of crop and variety	Local varieties, drought-tolerant, long duration, poor yield potential dependence on traditional cropping system	Improved varieties, drought tolerant, short or medium duration-high yield potential crop substitution. Sole cropping intercropping. Intercropping with suitable geometry sequential cropping
2. Tillage	Immediately before sowing summer tillage and off season tillage bullock power	Off season tillage deep tillage depending on soil mechanisation for early completion
3. *In situ* moisture conservation	No systematic efforts	Land shaping into various configurations depending on rainfall, soil, crop, slope
4. Erosion control	Absent	Mechanical cultural and biological methods of erosion control
5. Sowing		
(a) Time of sowing	After rains delay between rains and sowing	Early sowing-premonsoon sowing depending on soil, crop and rainfall.
(b) Seed treatment	Nill	Seed hardening-biofertiliser inoculation.
(c) Sowing method	Broadcasting sowing behind plough local seed drill	Mechanised seed drill bullock-drawn or tractor-drawn.
(d) Density	Low	Optimisation through sowing, seed treatment, proper depth and method of sowing and correct crop geometry.
6. Soil fertility managment	Limited or no use of organic manures and fertilisers	Integrated nutrient management through organic manures, optimal nutrient level time and method of application. Enriched FYM, biofertilizer.
7. Crop protection		
(a) Weed control	Manual weeding	Integrated weed management-use of efficient tools-interculture.
(b) Pest and disease control	Absent in food crops limited in commercial crops	Integrated pest management through resistant varieties early sowing, seed treatment, intercropping trap crops, summer tillage and dust formulations of chemicals if necessary.
8. Contingency planning and correction	Limited	Recommendations based on rainfall aberration.
9. Land use and farming system	Mostly subsistance farming with food and fodder crops mostly arable cropping	Land use system depending on land capability and suitability-integrated farming system.
10. Development approach	Planning for each farms as independent unit	Development based on watershed approach

IRRIGATION

Water use efficiency

The water utilized by the crop is evaluated in terms of water use efficiency.

It can be classified into :

1. Crop water use efficiency

It is the ratio of crop yield to the amount of water used by the crop for evapotranspiration.

2. Field water use efficiency

It is ratio of crop yield to the total amount of water used in the field.

3. Physiological water use efficiency

It is calculated in terms of the amount of fixed per unit of water transpired.

Types of Efficiency of water:

1. Application Efficiency:

$$= \frac{\text{Water directly available to the crop}}{\text{water received at field inlet}} \times 100$$

2. Conveyance Efficiency

$$= \frac{\text{Water received at inlet to block of fields}}{\text{water released at project head works}} \times 100$$

3. Project Efficiency

$$= \frac{\text{Water made directly available to the crop}}{\text{water released at head works}}$$

Scheduling of Irrigation

Definition

It's the process of determining when to irrigate and how much water to apply or frequency with which water is to be applied based on needs of the crop and nature of the soil.

Objectives

- Proper scheduling is essential for efficient use of irrigation water and production inputs, such as fertilizer etc.
- It leads to saving water and energy.
- Higher crop yield, efficient use of inputs and lower production costs.

Approaches to determine when to Irrigate –

1. *Soil approach*: Methods involve determination of water content of the soil and finding the deficit is available at which it's proposed to irrigate.

(a) Gravimetric approach

(b) Feel and appearance

(c) Tensiometers

(d) Electrical resistance

(e) Water budget tech (ET, Effective precipitation and soil water contents).

II. Plant approach: Monitoring plants is the most direct method of determining when to irrigate.

(a) Appearance and growth: Indicators which include leaf and shoot wilting, leaf colour, drooping of leaves, rolling of leaves etc., not often these are very effective and confusing.

(b) Leaf temperature: When partial or full stomatal closure occurs due to reduced transpiration, because of reduced availability of water and increase in heat temperature.

(c) Leaf water potential: Which indicates the plants need for water. Here the pressure used to measure the leaf moisture potential. This method is not extensively used since considerable time, care and training are required, to obtain reliable results.

(d) Stomatal resistance: Leaf resistance to vapour diffusion into the atmosphere is primarily governed by the degrees of stomatal closure. It's therefore an index to the need for water.

III. Meteorological Approach :

(a) IW/Epan ratio: Ratio between a fixed amount of irrigation water and cumulative pan evaporation (EPan) from an open pan evaporimeter minus effective rainfall since previous irrigation is used as a basis for scheduling irrigation to crops.

- Smaller IW /CPE ratio -Irrigation @ large intervals.
- Larger ratio -Irrigation @ shorter intervals.

IV. Rough Methods for Farmers :

(a) Soil-cum-sand miniplot technique: Based on plant as an indicator of water need. It's to reduce artificially available water-holding capacity, of the soil in root zone depth in the miniplot by mixing sand with it. The field is irrigated when the plants in the mini-plot show moisture deficiency symptoms by wilting.

(b) Can Evaporometer method: Evaporation from one litre can has a correlation with evopo transpiration from a crop. When the water level in the can has dropped to an amount equal to the amount of water to be applied in each irrigation, it's time to irrigate.

How much water to apply ?

The amount of water to be applied to the plants can be determined by the following methods:

Net Irrigation Requirement

It's the depth of irrigation water, required to bring the soil moisture in the effective root zone depth to field capacity.

$$NIR = \frac{(M_{FC} - M_{bi}U)}{100} \times D_b D$$

where,

M_{FC} - Moisture content @ field capacity (%)

M_{bi} - Moisture content before irrigation (%)

D_b - Bulk density of the soil (g/cc)

Quiz Kit!

Longest wall

The longest wall in recorded Indian history was the 38.5 km long fortification built by Rana Kumbha to prevent the invasion of Mewar.

GK **Harappan age forms part of proto-history.** GS

D - Depth of soil (cm)

Gross Irrigation Requirement

$$GIR = \frac{\text{Net irrigation requirement}}{\text{Field efficiency of system}}$$

Irrigation Interval

It refers to the number of days between two irrigations in the period of highest consumptive use of the crop growth without rainfall. It's the function of soil, crop and water.

$$\text{Design interval} = \frac{MF_c - M_{bi}}{P_{co}}$$

P_w = peak period moisture use rate of the crop (cm).

Irrigation Period

It's the number of days that can be allowed for applying the irrigation to a given design area during the peak consumptive use period of the crop being irrigated.

$$I_p = \frac{I_n}{ET(\text{crop})\text{Peak}}$$

I_n = net amount of moisture (cm)

ET = peak period water use rate of the crop (cm).

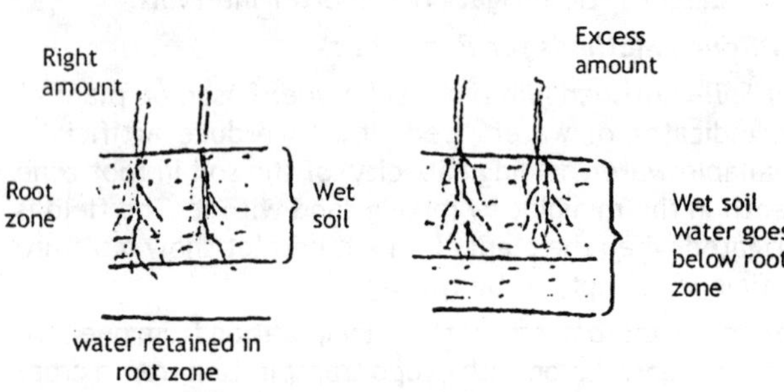

Irrigation requirement = Crop water requirement - Effective rainfall.

Can be expressed in mm/day or mm/month.

Methods of Irrigation

Definition

Application of irrigation water to cropped field by different types of layouts are called Methods of Irrigation.

Objective

- Necessary to save the water by proper methods to arrest run-off loss, percolation loss, evaporation loss etc.
- To optimise the crop water need.

Factors Influencing Methods

I. *Soil Type*: Physical properties such as texture, structure, porosity, infiltration rate influence the selection of method.

2. ***Soil Depth***: If the soil is shallow, holds less water, needs levelling and forming bunds to hold maximum water to increase irrigation interval.

If soil is deep, holds more water and needs longer irrigation interval.

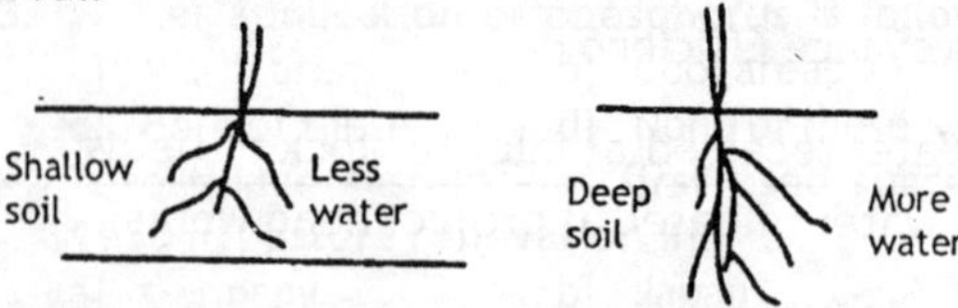

3. ***Topography of Land***: Topography decides the method of irrigation. If the land is more slopy, basin method can't be used. Instead strip method can be used. In undulating topography, sprinkler/drip irrigation method can be used.

4. ***Climate***: Rainfall temp, humidity, wind velocity, radiation etc., influence the irrigation methods. For e.g. heavy wind affects sprinkler irrigation and temperature affects surface method of irrigation by high evaporation loss.

5. ***Water source***: Flow velocity, quantity, quality of available water, decide the methods of irrigation methods.

6. ***Crops to be grown***: Value of the plant and geometry of the crop to be cultivated are main criteria.

For e.g., for high value/cash crop or wide spaced crop, sprinkler/drip can be used.

Classification of Irrigation Methods

The methods can be classified into :

1. Surface method or gravity method
2. Sub-surface/sub-irrigation
3. Pressurized or micro irrigation

(a) Drip-irrigation

(b) Sprinkler irrigation

I. ***Surface/Gravity Irrigation***: Common method of irrigation practised all over the world. Water is applied directly to the surface by providing some checks so the water flow.

1. ***Border strip method***: Field is divided into long parallel strips by providing small parallel earthen bunds or levees or dykes along both the side of the strips. The end of the strip may or may not be closed, based on the length of the strips. The application efficiency of this system is 75-85%.

- Suitable to soils with moderately low to moderately high infiltration rate, fields with 0-5%. Slope and not suited to very sandy soil and very clayey soil as they have too high and low infiltration rate.
- For closely spaced crops like pulses, wheat, barley, alfalfa, and ragi.

It can be divided into –

(i) Graded Borders, with slope 0.1 to 0.5% in the

longitudinal directions and these have no or very little slope across the strip.

(ii) Level border, where there is no slope.

2. *Check basin method* : Adopted for levelled land surfaces. Also known as basin method. Land is divided in to small basins. Area of basin surrounded by earthen bunds or levees or dykes.

- Suitable in fine textured soil.
- Suitable for crops like cereals, millets, pulses and oil seeds.

Disadvantages –

- Needs high degree of levelling.
- Not suitable for coarse textured soil.

It can be further classified into:

(i) Rectangular or square or irregular basin.

(ii) Ring basin -mostly for orchard crop. These are circular basins formed around individual trees.

3. *Furrow method of irrigation*: Small, evenly spaced shallow furrows or channels are formed in beds. Suitable for crops like, cotton, maize, sugar cane, potato, beet root, onion and vegetable crops.

It can be classified into –

(a) Corrugated furrow,

(b) Graded furrow,

(c) Level furrow and

(d) Contour furrow

(a) Corrugated furrow : Small furrow made by plough. 60 mm of bottom width and a depth of 100-150 mm.

(b) Graded furrow: Open ended which facilitate more time for infiltration and uniform distribution along the furrows. This is adopted in the land having more slope.

(c) Level furrow: Adopted for land having no or little slope or levelled lands.

(d) Contour furrow: Furrows formed along the contours.

4. *Surge Irrigation*: It's the application of water into the furrows intermittently in a series of relatively short on and off times of irrigation cycle. This method gives more uniform soil moisture distribution and storage in the crop root zone. The irrigation efficiency is 85-90%.

II. Sub-surface Irrigation: Water is applied below the ground surface through net work of pipes or some devices. Main aim is to reduce evaporation loss and to maintain an artificial water table near the root zone of the crop.

Mainly suitable for high temperature area, where ET losses are very high.

III. Pressurized Irrigation Methods: Under this method we have both sprinkler as well as drip irrigation methods, where water is applied through network of pipelines by means of pressure services.

Did You Know ?

Oldest railway institute

Vivekananda Institute, Asansol (earlier Durant Institute, named after, Sir Mortimer Durand), is the oldest railway institute in the country. It was established in 1878 in Asansol for the mechanical engineers of the East India Railway.

(A) Sprinkler Irrigation System: Irrigation water is sprayed into the air and allowed to fall on the ground surface more or less resembling rainfall.

It's made by pumping water under pressure through network of pipelines and allowing to eject out by means of small orifices or nozzles or holes.

It can be classified into –

1. Rotating head system: Special device to sprinkle the water called sprinkler head, which consists of small nozzles and metal ring or water ejected through the nozzle strike the metal ring which changes its direction by the help of the spring attached to it.

2. Perforated pipes system: Lateral pipes, small holes are made, based on the nature of the crops to distribute water uniformly.

Advantages :

- Highly suitable for sandy soil, where infiltration rate is more.
- Lands having undulating topography or steep slopes where levelling is economically not advisable.
- Stream size is very small, where surface flow is low.
- Controls canopy temperature.
- Wastage of land or basin, ridges and furrows and irrigation channels are reduced.
- Almost suitable for all crops except rice which needs stagnation of water.

Disadvantages :

- Distribution efficiency decrease in heavy windy areas.
- With saline water, causes clogging of pores.
- Continuous power supply required to operate system to maintain pressure.
- Costly to install and maintain.
- Uniformity of application is difficult.

(B) Drip or Trickle Irrigation: Water is applied through network of pipelines and allowed to fall drop by drop at crop root zone by a special device called emitters or drippers. The main principle is to apply the water at crop root zone based on the daily ET demand of the crop without any stress.

Advantages:

- Application of water in slow rate, facilitates the easy infiltration into the soil.

GK **Twenty seven per cent of Indian population is urban.** GS

- Root zone is always maintained with field capacity level.
- No seepage or percolation or evaporation losses.
- Restricted weed growth due to limited wetting zone.
- Fertigation, chemigation and herbigation possible.
- Salt content near the root zone is reduced.
- Saline water can also be used.
- Adopted for any type of topography.
- More area can be maintained with little quantity of water.
- Used for widely spaced crops like cotton, sugar cane, tomato, brinjal, coconut and orchard crops.

Disadvantages :

- Clogging in emitters due to salt content of water and other impurities like, dust, etc.
- Damage of pipelines by rodents.
- Not economical for closely spaced crops which require more number of pipes and drippers per unit area.
- Proper maintenance and periodical cleaning of drippers and pipelines (with 1% HCl) are very important to maintain the system efficiently.

Drainage

Definition

It's the process of removal of excess water or gravitational water from the surface and the sub-surface of farm lands with a view to avoid waterlogging and creates favourable soil conditions for optimum plant growth.

Need or Objective

- To have optimum soil water balance.
- It's of greater necessity due to heavy precipitation.
- Areas of salinity and alkalinity where annual evaporation exceeds rainfall and capillary rise of ground water occurs.
- Low lying flat areas surrounded by hills.

Characteristics of Good Drainage

- Should be permanent.
- Must have adequate capacity to drain the area completely.
- Should be minimum interference with cultural operations.
- Should be minimum loss of cultivable area
- Should intercept or collect water and; remove it quickly within shorter period.

Methods of Drainage

1. Surface method
2. Sub-surface method

1. Surface Method: It's to remove the excess water from the surface of the soil profile. It can be done by developing slope in the land.

It is suitable for, slowly permeable clay and shallow soil and for regions of high intensity rainfall. The land with less than 1.5% slope also needs a good drainage.

It can be made by land smoothing, making field ditches. The method can be further classified into:

(a) Lift drainage

(b) Gravity drainage

(c) Field surface drainage

(d) Ditch drainage

(a) Lift drainage: It's to drain the excess water from low lying area or areas having water due to embankment. It can be achieved by the use of scoops or pump.

(b) Gravity drainage: Water is allowed to drain from the areas of higher elevation to lower reaches through, the regulated gravity flow through, the outlet of various types.

(c) Field surface drainage: Excess water received from the rain or irrigation is drained through this method. Irrigated basins or furrows are connected with the drainage under lower elevation, which is connected to the main outlet to the farm pond used for water harvesting.

(d) Ditch drainage: Ditches of different dimension are constructed at distances inside the soil up to the depth of ditch. Such ditches are interceptors or relief drains. This method can be adopted for nurseries, seed beds and rainfed crops.

2. *Sub-surface Drainage System:* Sub-surface drains are underground artificial channels through which excess water may flow to a suitable outlet. The main purpose is to lower the ground water level below the root zone of the crop. It's mostly needed for medium textures soil, high value crop and high soil productivity.

Types of Sub-surface drainage –

(a) Tile drainage

(b) Vertical drainage

(c) Well drainage/Drainage wells

(a) Tile drainage: Continuous line of tiles laid at a specific depth and grade so that the excess water enters through the tiles and flow out by gravity. The tile drains are made up of clay and concrete.

(b) Vertical drainage : Disposal through well into porous layers of earth. Such a layer must be capable of taking large volume of water rapidly. These layers are found in river bed.

(c) Drainage wells: These wells are used for the drainage of agricultural lands especially in irrigated areas.

Systems of Drainage

1. Random
2. Herring bone
3. Grid iron

Major Harappan sites are in the Ghaghar-Hakra belt.

GS

4. Interceptor

1. *Random*: Lines are laid more or less at random to drain these wet areas. This is used where the wet areas are scattered and isolated from each other.

2. *Herring bone*: Here, the mains are in a narrow depression and the laterals enter the main from both sides at an angle of 45°, like the bones of a fish.

3. *Grid Iron*: Similar to herring bone, but the laterals enter the main only from one side at right angles. This is adopted in flat regularly shaped fields. This is an efficient drainage system.

4. *Interceptor*: Ditches of different dimensions are constructed at distances to drain the excess water accumulated on the surface and inside the soil upto the depth of ditch. Such ditches are interceptors or relief drains. Its an effective and efficient method, which is adopted for nurseries, seed beds and rain fed crops.

Quality of Irrigation Water

I. Suitability mainly depends on

- Amount of type of salts present in the water

II. Main constituents are

- Calcium, Magnesium, Sodium as cations.
- Chloride, Sulphate and Bicarbonate.
- Other ion → like Bo, Mo

III. Parameters which divide the quality of irrigation water

- Total salt concentration (Salinity)
- Sodium adsorption ratio (Sodicity)
- Bicarbonate and Boron content

1. Total concentration of Soluble Salts

- Total dissolved solids
- Electrical conductivity

2. Concentration of Na

- SAR

$$SAR = \frac{(Na^+)}{\sqrt{\frac{Ca^+ + Mg^+}{2}}}$$

- SSP

$$SSP = \frac{N_a^{2+}}{\sqrt{Ca^{2+} + Mg^{2+} + Na^+ + K^+}} \times 100 \quad \text{or}$$

$$\frac{\text{Soluble Sodium Conc.}}{\text{Total Soluble Cation Conc.}}$$

IV. Quality parameters for irrigation water

Class	*EC (dsm-1)*	*SAR*	*RSC(meq/l)*
Good	<2.0	<10	<2.5
Normal	2-4	<10	<2.5
Sodic	<4	>10	>2.5
Poor	>4	>10	>2.5

V. Management of saline water for irrigation

(a) Irrigation practices –

1. Salt tolerant crops -Barley, Sugarbeet, Cotton, Mustard.
2. Semi-tolerant- Sorghum, Rice, Castor.
3. FYM/green manure.
4. Application of fertilizer
5. Planting seed on the side of a ridge.

(b) Irrigation management –

1. Giving heavy - Pre-sowing irrigation
2. Frequent shallow irrigation
3. Drip irrigation
4. Furrow irrigation

(c) Improving water quality –

1. Adding chemical to precipitate harmful constituents
2. Good water + Saline water

(d) Soil management –

1. Irrigation and drainage
2. Mulching
3. Gypsum application

You Don't Know!
First/oldest highway
The 2,4000-km-long Grand Trunk road to Peshawar and beyond is the first and oldest highway started by Lord William Bentinck, then Governor General of Bengal (1824-34), who took the first steps to build a permanent metalled road.

Guide for Selecting Method of Irrigation

Irrigation Method	Topography	Crops	Remarks
Widely spaced borders	Land slopes capable of being graded to less than 1% slope and preferably 0.2%	Alfalfa and other deep rooted close growing crops like cotton, sunflower etc.	The most desirable surface method for irrigating close-growing crops where topographical conditions are favourable. Even grade in the direction of irrigation is required on flat land and is desirable but not essential on slopes of more than 0.5%. Cross slope is permissible when confined to differences in elevation between border strips of 6-9.

GK GS

Pluto was first discovered by P. Lowell.

Irrigation Method	Topography	Crops	Remarks
Closely spaced borders	Land slopes capable of being graded to 4% slope or less and preferably less than 1%	Pastures or grasses	Especially adapted to shallow soils underlain by claypan or soils that have a lower water intake rate. Even grade in the direction of irrigation is desirable but not essential. Sharp grade changes and reverse grades should be smoothed out. Cross slope is permissible when confined to differences is elevation between borders of 6-9.
Check basin and cross furrows	Land slopes capable of being graded to 0.2% slope or less	Groundnut, Soyabean, Rice, Wheat, etc.	This method is especially designed to obtain adequate distribution and penetration of moisture in soils with low water intake rates.
Corrugations	Land slopes capable of being graded slopes between 0.5% and 1.2%	Alfalfa pasture and grain Row crops and fruit trees	This method is especially adapted to steep land and small irrigation stream. An even grade in the direction of irrigation is desirable but not essential. Sharp grade changes and reverse grades should at least be smoothed out.
Graded contour furrows	Variable land slopes of 2-25% but preferably less.	Row crops and fruit trees	Especially adapted to row crops on steep land, though hazardous due to possible erosion from heavy rainfall. Unsuitable for rodent-infested fields or soils that crack excessively. Actual grade in the direction of irrigation 0-1.5%.
Contour ditches		Hay, pasture and grain	Especially adapted to foothill conditions. Requires little or no surface grading.
Rectangular checks (leaves)	Land slopes capable of being graded so single or multiple tree basins will be within 6 cm.	Orchards	Especially adapted to soils that have either a relatively high or low water intake rate.
Contour leaves	Slightly irregular land slopes or less than 1%	Fruit, rice, grain and forage crops.	Reduces the need to graded land. Frequently employed to avoid altogether the necessity of grading.
Sub-irrigation	Smooth-flat	Shallow rooted crops such as potatoes or grasses	Requires a water-table, very permeable subsoil conditions and precise levelling. Very few areas adapted to this method.
Contour bench terrace	Sloping land-best for slopes under 3% but useful to 6%	Particularly suited to cultivated crops	Considerable loss of productive land due to bunds. Requires expensive drop structures for water erosion control.
Localized (drip, trickle, etc)	Any topographic conditions for row crop farming.	Wide spaced row crops or fruit trees.	Perforated pipe on the soil surface drips water at base of individual vegetable plants or around fruit trees. Has been successfully used in Israel with saline irrigation water. Still it is in infant stage
Sprinkler	Undulating 1-35% slope	All crops	High operation and maintenance costs. Good for rough or very sandy land in areas of high production and good markets. Good method where power cost are low. May be the only practical methods in areas of steep or rough topography. Good for high rainfall areas where only a small supplemental water supply is needed.

FORESTRY

Social Forestry

Meaning

Practice of forestry on lands outside the conventional forest area for the benefit of rural and urban communities.

Since various social objectives are met out by such type of forestry that is why it is called social forestry.

Social forestry is also described as forestry for the people by the people and of the people.

Objectives

1. Fuel wood, timber, bamboo, fodder and minor forest produce.
2. Employment
3. Cottage industries
4. Using land
5. Soil and water conservation
6. Aesthetic value
7. Protecting agriculture

Scope

1. Forest area - 33% target
2. Wood requirement
3. Employment
4. Soil and water conservation
5. Farmers -own timber, fuel etc.

Components

I. Farm forestry (private land)

II. Extension forestry (public land)

(a) mixed forestry

(b) shelter belts

(c) linear strip plantation

III. Afforestation of degraded forests

IV. Recreational forestry

Trees

Jamun

Neem

Prosopis

Tamarindus

Pongamia

Agroforestry

Meaning

Suitable land management system which increases the yield of the land combines the crop and forests plants and/or animals simultaneously or sequentially on the same unit of land.

Knowledge World!

High Court with most judges

Allahabad High Court has the largest sanctioned strength of 60 judges.

Scope

1. Alternative to shifting cultivation
2. Sand dune stabilization
3. Employment opportunities
4. Rural development
5. Utilization of sub-marginal land

Benefits

1. Rational use of land
2. Complete use of land
3. Increase in production of agricultural crops
4. Increase fuel wood
5. Increase industrial wood
6. Pasture/fodder
7. Additional employment
8. Solving fuel, fodder and food problem

Components

A. Agri-silvi culture system

B. Silvi pastural system

C. Agro-silvi-pastural system

D. Multipurpose forest tree production system

Characteristics of Tree Species

1. Self-pruning
2. Artificial pruning
3. N-fixing
4. Permit-sunlight
5. High protein content
6. Different root zone
7. No allelopathy effect
8. Litter composition
9. Shade tolerant

Example –

I. *Albizzia, Eucalyptus* -Paddy -Millet
2. *Poplar, Acacia* -Wheat, Tapioca
3. *Cocos* -Maize, Mustard
4. *Leucaena, Gliricidia* -Potato, Pulses.

Propagation of various types of Forestry Plantations

Plants are raised by two ways –

GK **Aravalli is the oldest mountain range in India.** GS

(i) By seeds

(ii) By vegetative parts of the plant

Direct sowing :

Any artificial plantation can be raised either by sowing seeds directly at plantation site or in the nursery for raising seedlings, and when in nursery they obtain plantable size, are planted at permanent site.

Advantages

- No need of nursery
- Saving of time, man and material
- Plants grow freely without any disturbances.

Disadvantages

- More seed is needed
- At permanent site effective supervision is not possible for large area.
- At permanent site, it becomes very difficult to provide good soil conditions for good germination. Tree species raised by direct sowing. *e.g. Acacial nilotica. Albizia lebbeck, Prosopis alba.* Tree species raised by transplanting of seedlings.

Propagation by vegetative means:

Plant species which do not set seeds or their needs or not viable or by nature they have the capability of making union when they are layered or planted by their stem cuttings and a few by root cuttings. Tree species propagated by stem cuttings. *Acacia melanoxylon, Bombax ciba, Cestrum diurnum, Erythria indica, Punica* granatum.

OTHERS

Shifting Cultivation

Definition

Shifting cultivation is a primitive method of cultivation which is usually followed in hills by tribal people in which forest land is cleared by cutting down all the trees, shrubs, herbs and climbers near the ground level, felled materials are dried and set on fire, sowing is followed in cleared area by dibbling small holes .

Other names: Land rotation, Jhuming (North Eastern region) Podu (A.P. & Orissa)

Shifting Cultivation of India

1. N. E. hill region
2. Andhra Pradesh
3. Madhya Pradesh
4. Orissa
5. Bihar
6. Karnataka

Features

1. Cropping is done for few years on same unit of land, after clearing and burning.
2. Depletion of soil productivity causes cultivator move to other area for cultivating same crop.
3. Here crop is fixed burnt land is rotated
4. It causes soil erosion
5. Less productivity
6. Seldom use of manures and fertilizers
7. The size of plot from 1.0 to 2.5 ha per family having members 3-5

Demerits

1. Faulty land use (social evil)
2. Primitive method of cultivation (less productivity)
3. Change in soil microbial population
4. Loss nutrient through leaching, run off and percolation
5. C : N ratio reduced
6. Causes flood in adjoining area
7. Affect the flora and fauna

Ways for improving degraded areas

- Efficient land use management
- Preventing loss of soil fertility, by making contour bunds, levelling, terracing and good drainage system
- Soil management by growing cover crop, strip crops, mixed crop and erosion checking crops
- Adoption of production based crop management by growing HYV, weed and water management, fertilizer application and plant protection
- Problems created by this can easily be overcome by adoption of suitable agro forestry models

Taungya System

Taungya word derived from Burmese word, consis!ing of "Taung" means hill and "Ya" means cultivation i.e. cultivation in the hills.

Definition

Taungya is also a shifting cultivation in which cultivators are allowed to plant forest tree species and to raise agriculture crops in between rows of forest trees simultaneously and cleared for about 3-5 years. Thereafter shift to another patch of land.

Aim: Wood and food production.

History :

(1) First started in Burma in 1850 (teak plantation).

(2) Java in 1856.

(3) In India this was introduced by Brandis in1863.

(4) In India first plantation raised in North Bengal in 1863

Types of Taungya

(1) Departmental taungya

(2) Leased taungya

The Harappan writing is boustrophedon.

GS

(3) Village taungya

Crops grown in taungya system in India

State	*Tree crop*	*Associated agricultural crop*
Uttar Pradesh	Sal tree	Maize, Paddy
Andhra Pradesh	Cashew, Teak	Hill Paddy, Groundnut
Kerala	Teak, Eucalyptus	Paddy, Ginger
Assam	Sal tree	Paddy
Tamil Nadu	Teak, Bamboo	Millet, Pulses
West Bengal	Teak, Sal tree	Paddy, Maize

Merits

1. Suppression of weed and climbers.
2. Regeneration of forest species in cheap.
3. Employment opportunity to rural people.
4. Full fertilization of available land for production of food crops.
5. Increase forest wealth.

Demerits

1. Erosion and soil fertility loss
2. Exploitation of human labour
3. Danger of epidemics
4. Forest trees are not cared
5. Cultivators may claim legal right on land
6. Insecure land tenure

Puddling

Introduction

Rice growth and yield are higher when grown under submerged conditions. Maintaining standing water throughout the crop period is not possible without puddling.

Definition

Manipulation of soil under water with tools and implements to break down the soil segregates is called puddling.

Objective

- To provide optimum seed bed for planting rice.
- To create impervious layer below the surface to reduce deep percolation losses of water.
- Weed control.

Method of Operation

- 5-10 cm of water is applied and first ploughing is carried out.
- After 3-4 days, another 5 cm of water is applied and second ploughing is carried out.
- Within 3-4 days, another 4-5 cm of water is given and third ploughing is done in both direction.

Quiz Kit!

First bus ticket

The first bus ticket was issued to traffic manager Lucas on July 15, 1926.

Formation of Impervious Layer

- Unlike other tillage operations, puddling aims at destroying soil structure. The individual soil particles viz., sand, silt and clay are separated during this operation. The impervious layer are formed over compacted soil.
- The sand particle reach the bottom, over which silt particles settle and finally clay particles fill the pores.

Implements Used

Depending upon the availability of equipment and nature of the land, different implements are used.

1. Wet land puddler
2. Cage wheel
3. Helical blade puddler
4. Green manure trampler
5. Rotovator

Advantage

- Avoid percolation loss of water.
- Minimization of leaching loss of nutrients.
- Uniform and soft seed bed preparation.
- Incorporation of weeds, stubbles, green manure and green leaf manure is possible.
- To some extent, perennial weeds can be managed.

Ratooning

Cultivation of crop regrowth after harvest.

Advantage

- Reduced cost of production through saving in land preparation and seed material.
- Reduced crop cycle period for ratoon crop.
- Better utilization of growing season in monsoonal climates.
- Less input (irrigation, fertilizer) requirement than main crop.

Disadvantage

- Lower yield than plant crop.
- Insects, diseases problem.
- Greater cost per unit produced.

Ratooning in sugarcane

- Ratoon provide cheaper raw material.
- Better utilization of resources.
- Give fairly high sucrose in juice.

Good Management

- Harvesting at proper time.

GK GS

Procession is the conical motion of the spinning earth.

- Stubble shaving.
- Judicious fertilizer application. Varieties Co- 740, Co-6304

Cardinal Temperature

For each phase in the growth of cultivar, there is a temperature range within which growth and development is optimum. These temperature ranges are called cardinal temperature.

- Minimum -below which there is insufficient heat for biological activity.
- Optimum -at which the rate of metabolic processes are at their maximum 15-30°C
- Maximum -above which growth ceases.

Cool season crops

Wheat, barley, potato, oats

- Grow best in cool weather period and grown in winter season (November to February)
- Temperate crops
- Cardinal temperature:
  Maximum 30 -38°C
  Minimum 0 -5°C
  Optimum 21-30°C

Warm season crops

Rice, sorghum, maize, sugarcane

- Grown in monsoon season and summer season.
- Tropical crops
- Cardinal temperature:
  Maximum 45 -50°C
  Minimum 15- 20°C
  Optimum 30- 38°C

Flue Curing

Introduction

Curing is a process in tobacco by which the harvested leaves are made ready for marketing.The sole object of curing is dry the green leaf to get final product it has required colour, texture and aroma.

Curing

Curing is a process of drying, decomposition of chlorophyll till green colour disappear, hydrolysis of starch into sugar and fermentation of sugar.

Flue curing is an improved method of curing, in which drying of leaves under artificial condition without contact of green leaf with smoke.

Feature –

- Drying of leaf under controlled conditions.
- Starch get converted into sugar.
- Green tobacco becomes bright, aromatic and fine textured.

Variety Suitable for Flue Curing

- Virginia Gold
- Harison Special

Production Technique of Flue Cured Tobacco

1. Raised with low N_2 supply
2. Harvest at an advanced stage when rich in starch, low in nicotine content.
3. The leaves are harvested by priming 3-4 leaves at a time.

Method of Flue Curing

Harvested green leaves are strung on sticks

⇓

Stacked into a specially constructed flue cured pan

⇓

Artificial heating with gradual increase in temperature till leaves dry.

Stages in Flue Curing

1. Yellowing of leaf
2. Fixing of colour
3. Drying of leaf

1. Yellowing of Leaf

- Slow heating is to give optimum humidity.
- Retain moisture
- At end, leaf attain lescion yellow colour
- Starch converted to sugar
- Completed within 1 to 1½ day

2. Fixing of Colour

- Increase temperature gradually with decrease in humidity.
- Merely involves drying of the tissues to the point where oxidation of colour cannot take place.
- Completed within 1 to 1 1½ day.

3. Drying of Leaf

- Drying the midrib of leaf
- Temperature of pan increase @5°C for I hr. till reaches to 170-180°F for one day.
- After drying leave leaves in pan overnight with ventilation.

Harvest Index

Harvest Index defined as the ratio of economic yield to biological yield.

$$HI = \frac{\text{Ecological yield}}{\text{Biological yield}}$$

HI - can also be expressed as %.

Biological Yield: The total dry matter produced by a crop.

Economic Yield : Fraction of the biological yield which is useful for man.

B.Y x K = E.Y., K = constant

[Coeff. of effectiveness Harvest Index]

GK GS

Hormone adrenalin when secreted increases heart beat and produces a feeling of excitement.

Importance

A balance between the productive parts of the plant and reserves which form economic yield is essential. A considerable increase in the yield of the economic product is usually dependent on any increase in total dry matter produced.

Factors Affecting Economic Yield

- Source (photosynthetic organs)
- Sink (storage organs)
- Capacity to translocate assimilates from one part to other.
- Water
- Nutrient uptake/transport
- Plant population
- Radiation

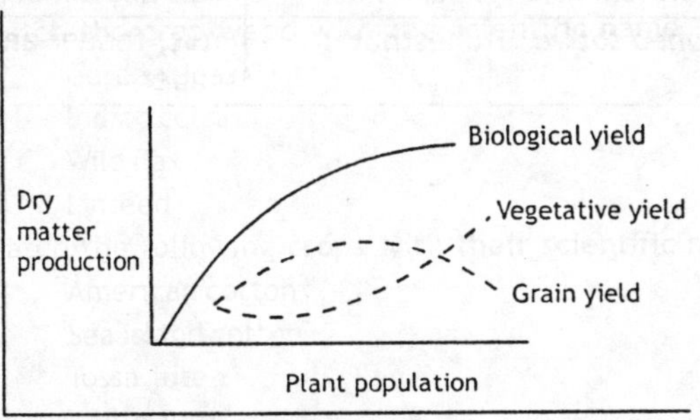

Recent Views/ Approach

Yield is limited by photosynthetic rate and attention is paid to selection of plants with more erect leaves (or) higher photosynthetic rate or Calvin cycle plants lacking photorespiration.

Constraints in Pulse Production

Area 24 m ha, productivity 550 kg/ha. Pulse requirement- 80 gm/day. Availability = >40-50 g/ day.

Factors

1. Unfavourable weather condition
2. Abnormal soil condition
3. Variable constraints:
   - lack of high-yielding varieties
   - high susceptibility to disease and Insect
   - flower drop
   - lack of short duration varieties
   - poor response to inputs
   - instability in performance
4. Seed constraints:
   - poor quality
   - high price
   - delay in supply
5. Agronomic constraints:
   - improper showing time
   - low seed rate
   - defective method of sowing
   - inadequate interculture
   - insufficient irrigation
6. Plant protective constraints
7. Inadequate extension facilities
8. Socio-economic constraints
9. Inadequate research infrastructure
10. Miscellaneous:
    - loss during storage
    - loss during processing

Did You Know ?

Tallest flagmast

The teakwood flagmast at Fort St. George, Madras, is 45.7 m (150 ft) high. Governor Yale first hoisted here in 1687.

Multi Purpose Trees and Shrubs (M.P.T.S)

In agroforestry systems, it is necessary and economical if selected trees and shrubs are having their various uses so that each and every part, produce and products of the trees are fully utilized for domestic purpose, industrial purpose etc. Besides, multipurpose trees and shrubs help in protecting environment.

Domestic and Industrial Uses

1. Fuel wood for cooking foods and warming houses.
2. Timber, posts and poles
3. Paper cardboards, and construction boards.
4. Vegetables, fruits and nuts, seeds and firewood.
5. Edible leaves and shoots for sauces, curries, salads and beverages.
6. Fodder for livestock
7. Host plant for silkworms and lac insects.
8. Green manure for incorporating into soil.
9. Medicines and pharmaceuticals.
10. Extractives such as gums, resins, rubber and dyes.
11. Oil (edible), and non-edible for illumination, lubrication and soap-making
12. Cake obtained after oil extraction for the use of animal feed, if edible and non-edible types are for the use as manure.
13. The fruit of anola, bahera are used as soap substitutes.
14. Essential oil, perfumes are for cosmetics.

GK GS **Hindi speaking Indians is about forty per cent.**

Environmental Protection and Other Uses

1. Shade and support for planations crops.
2. Aid to the soil improvement by controlling soil erosion and sand dunes stabilization.
3. Improvement in soil structure, chemical properties, water balance and weed control.
4. Useful in hedges, windbreaks, shelterbelts, and fire breaks.
5. Ornamental, aesthetic or recreational value.

A few multipurpose tree species for arid and semi-arid regions

Botanical name	*Common Name*	*Economic importance*
Acacia albida	African savana tree	Fodder, firewood, tannin, gum
Acacia aneura	Mulga acacia	Fodder, firewood
Acacia nilotica	Babul	Fodder, tannin, gum, timber
Azadirachta indica	Neem	Timber construction, farm implements, seed oil, cake for manure, tannin, firewood, gum, medicinal

A few multipurpose tree species for humid tropical regions

Botanical name	*Common Name*	*Economic importance*
Acacia auriculiformis		Small timber, firewood, charcoal, tannin,
Albizia lebbeck	Siris	Fodder, firewood, charcoal, tannin, gum, timber, farm implements
Anacardium occidentale	Cashew	Fruit, kernels, oil, timber, firewood
Anthocephalus chinensis	Kadam	Fodder, fruit, pulp-paper, firewood
Areca catechu	Supari	Chewing gum, tannin, firewood, manure, mulching, pulp.
Artocarpus heterophyllus	Jackfruit	Fruit, vegetable, fodder, packing, boxes, plywood, furniture, ladders, farm implements, firewood, medicinal
Cordia dichotoma	Indian cherry	Fruits, fire-wood, fodder, tannin medicinal

A few multipurpose tree species for temperate regions

Botanical name	*Common Name*	*Economic importance*
Acacia dealbata	Lilvere wattle	Gum, flower are used in perfumery, tannin and dye, ornamental
Acacia decurrens	Green wattle	Timber, firewood
Acer oblongum	Maple	Fodder, firewood
Aesculus india	Indian Horse	Timber
Bombax ceiba	Semal	Timber, firewood, fodder, silky floss, gum, medicinal
Cara illionensis	Pecan nut	Fruit and timber
Castanea sativa	Chestnut	Fruit, fodder and timber

Points to Ponder

- 'Penetration agents' improve absorption of a herbicide by solubilizing waxy cuticle (lipoidal part) of cell wall or the membrane of plant tissues.
- 'Woman', first observed that plant come up from seed.
- "Other seeds" shall include seed and seed-like structures of any plant species other than that of pure seed.
- 20-25 days old seedlings are ready for transplanting of short duration varieties of rice.
- A herbicide move faster in young plants than in old ones.
- A seed, strictly speaking, is an embryo, a living organism embedded in the supporting. or the food storage tissue.
- A significant quantity of plant available boron in soils is held in the organic matter fraction.
- A useful function of chlorine is the contributing to turgor of leaves and other plant parts.
- Absorption rates of $H_2PO_4^-$ by plant roots can be described by Michaelis-Menten kinetics.
- Acclimation is the process by which plants become adapted to an unfavourable set of climatic conditions.
- According to Bray's nutrient mobility concept crop yields obey the percentage sufficiency concept of Mitscherlich for such elements as phosphorus and potassium while are relatively immobile in the soil.
- According to Liebig's law of the minimum, "even if all but one of the essential elements be present, the absence of that one will render the crop barren".
- Acid forming nitrogenous fertilizers are expected to augment the uptake of both native and supplemental zinc.
- Acrolein is contact aquatic herbicide.
- Active ingradient ai is that part of the formulated product that is principally responsible for herbicidal effects.
- Active ion absorption is thought to take place by metabolically produced carriers which transfer the ions across otherwise impassable barriers.
- Additive effect refers to the total effect of a combination which is equal to the sum of the effects of the component independently.
- Aging refers to the accumulation of degenerative changes which ultimately results in death or the failure of the seed to germinate.
- Agricultural drought is a period of four consecutive weeks from May to mid-October or six consecutive weeks during rest of the year.

You Don't Know!
Oldest regional party
The Dravida Munnetra Mazhagam (DMK), founded by C.N. Annadurai in 1947, is the oldest regional party in India.

- Agriculture today is not merely production oriented, but has become sustainable agribussiness.
- Ammonium is the preferred nitrogen source since energy will be saved when it is used instead of nitrate in synthesis of protein.
- An excellent example of the use of plant analysis in crop production is the crop logging carried out for sugarcane in Hawaii.
- An extremely wide N/S ratio say 20:1 or more could restrict organic matter breakdown, if the soil was already deficient in this element.
- Any compact heavy texture soil that is calcareous and basic in reaction is potentially iron deficient.
- Any physical or chemical treatment that weakens the seed coat is called scarification.
- Application of phosphorus especially on phosphorus deficient soils considerably reduces the risk of winter damage.
- At sea, tornadoes are known as waterspouts.
- Awned varieties of wheat and rice are better suited under drought condition than awnless.
- Azospirillum is the nitrogen fixing bacterium on root surfaces of plants like maize, millet, rice and wheat etc..
- Azotobacter and Clostridium are asymbiotic nitrogen fixers.
- *Bacillius megatherium* var. *phosphaticum* (Phosphobacterin) increases the plant availability of native and applied soil phosphorus.
- Band placement reduces the surface of contact between the soil and fertilizer with a consequent reduction in phosphorus fixation.
- Because of immobility of copper in soil, it is supplied to plant roots mainly by root interception.
- Boron alters the phytohormonal activity with an accumulation of auxin in plants deficient in this nutrient.
- Boron is the only non-metal among the micronutrient elements having the constant valence of +3 coupled with a very small ionic radius.
- Breeder's seed of male sterile, restorer and inbred lines is highly homozygous, while that of synthetic and

Place on the same meridian have the same local time.

composite is highly heterozygous.

- Breeder's seed refers to the seed or the propagating material directly controlled by originating or sponsoring plant breeder.
- CDU (Crotonyledene Diurea) and IBDU (Isobutyledene Diurea) are slow nitrogen release forms of nitrogenous fertilizers.
- Central Seed Committee is the main source to advise Government (India) on administration of Seed Act.
- Certain species of Astragalus absorb many times as much selenium as do other plants growing in the same soil.
- Certified seed is the seed certified by the certification agency under Section 8 of the Seed act, 1966.
- Chloride containing fertilizers depress root-rot infection in winter wheat.
- Clean tillage means working of the soil of the entire field in a way that no living plant is left undisturbed.
- Coarse textured, well drained and sandy soils are low in boron.
- Coating, encapsulation and matrixes have been used to impede the dissolution of water soluble fertilizer materials containing nitrogen and other plant nutrients.
- Cobalt has been found to participate in leghaemoglobin metabolism and ribonucleotide reductase in Rhizobium.
- Cobalt in plants is essential for symbiotic nitrogen fixation.
- Coconut and oil palms are especially responsive to chlorine.
- Coleoptile in cereals is either colourless or contains a red colour of anthocynine.
- Complementary ion effect is the influence of one adsorbed ion on the release of the another from the surface of a colloid.
- Conductivity test technique of seed testing is based on the leakage of solutes that occurs from all seeds in water.
- Copper is capable of isomorphous substitution in octahedral positions of crystalline silicate clays.
- Crinkle leaf of cotton is caused by the toxicity of manganese on highly acid soils.
- Critical nutrient concentration (CNC) represents the level of nutrient below which crop yield, quality or performance is unsatisfactory.
- Crop rotation is an important approach for control of nematodes feeding on the roots of annual crops.
- Delayed transplanting rice leads to poor tillering, early flowering of main tillers and reduction in yield.
- Deprotonation or dissociation of H^+ from OH^- groups at the broken edges of clay particles is the prime source of negative charge in the 1:1 clay minerals.
- Depth of sowing is a very important aspect in successful cultivation of dwarf variety of wheat.
- Diagnostic and Recommendation Integrated System (DRIS), is a new comprehensive system to interpreting leaf or plant analysis which identified all the nutritional factors limiting crop production.
- Dibbling means putting a seed(s) or seed materials in a hole or pit or pocket made at a predetermined distance and depth.
- Diffusion becomes the dominant means for molybdenum transfer to plant roots when the concentration is less than 4 ppb.
- Diffusion is believed to the dominant mechanisms for transporting zinc to plant roots.
- Divalent cations enhance phosphorous adsorption relative to monovalent cations.
- Dropping of seeds in furrows of predetermined dimensions, at a predetermined distance and depth, covering with soil followed by compacting is known as drilling.
- Each unit increase in pH causes 1000-fold and 100-fold decrease in solubility of Fe^{3+} and Fe^{2+}, respectively.
- Electro-ultra-filtration (EUF) has been proposed for a satisfactory characterization of availability of potassium to crops.
- Emulsification is done primarily to disperse the small volume phase (concentrate) in the large volume phase (spray carrier).
- Ephimerals are drought escaping plants.
- Excessive concentration of soluble salts in contact with roots/germinating seeds causes injurious effects due to plasmolysis, restricted moisture availability or actual toxicity.
- Excessive nitrogen fertilization reduces the sugar content of sugar beets.
- Ferredoxin (stable Fe-S proteins) is the first stable redox compound of the electron transport chain.
- Flooding and submergence of soils can improve iron availability by enhancing the active Fe^{2+} ionic form.
- For hay making a crop should be harvested at flowering stage.
- Fumigants are volatile chemicals applied into confined spaces or into the soil to produce gas to destroy weed-seeds (an organism) or to act as soil sterilant.
- Generally, herbicides exist in ionic form in the soil between pH 4.3 and 7.5.
- Glutathione is a reducing agent which plays a very important role in the detoxification of some metabolites

GK

Intensity of an earthquake is measured on the Richter Scale.

GS

injurious to plant cells.

- Growth is defined as progressive development of an organisms.
- Herbicidal selectivity refers to the killing of the target plant species in a mixed population without harming or only slightly affecting other plants.
- Hidden hunger refers to a situation in which a crop needs more of given essential nutrient element, yet has show no deficiency symptoms.
- High availability of potassium enhances root development.
- High calcium saturation is indicative of a favourable pH for growth of most of the plants as well as for microbial activity.
- High levels of ammonium can retard plant growth and restrict the uptake of potassium.
- High levels of Cu, Fe or Zn in the growth medium will impede Mn uptake by plants.
- High levels of exchangeable potassium can interfere with magnesium uptake of a crop.
- Hot water is also an effective method of breaking hard-seededness in legumes.
- Immobilization of nitrogen is the reverse of mineralization and occurs when large quantities of low nitrogen crop residues begin decomposition.
- In 1878, Theodore Schloessing and Alfred Müntz, French bacteriologists concluded that nitrification was the result of bacterial action.
- In areas having problem of soil fixation of nutrients foliar application constitutes the most effective means of fertilizer placement.
- In case of nitrogen and phosphorus as much as 65-75 per cent of total uptake can be accounted for in just the harvest of grain.
- In cross-pollinated crops, field inspections are essential during flowering to verify freedom from genetic contamination.
- In groundnut the germination is hypogeal.
- In jute the ideal stage of harvest is when the plants are in small pods.
- In late sown wheat, deep sowing delays emergence of seedlings by 2-3 days and heading by 5-6 days.
- In maize a single cross in the hybrid progeny from a cross between two homozygous inbreds.
- In paddy the amount of variability observed in varieties is relatively higher than that of wheat.
- In pulses a minimum of 500 plants should be planted for the production of nucleus seed.
- In succulent plants the transpiration is maximum during nights.
- In temperate zone, dormancy comprises a survival mechanism that prevents the fall-germination of species that are not winter hardy.
- Intercropping refers to growing of two or more crops simultaneously in alternate rows in the same field during the same period either by replacement or addition to the recommended optimum plant density of the base crop.
- Iron comprises about 5 per cent of the earth's crust and is the fourth most abundant element in the lithosphere.
- Iron will form stable chelates with molecules containing oxygen, sulphur or nitrogen.
- It is generally agreed that turning of green manures results in better utilization of phosphorus by succeeding crops.
- It is possible to derive the maximum benefit from the applied phosphorus between 5.5 and 6.8 soil pH.
- It should be noted that SO_4^{2-} and SeO_4^{2-} ions can act antagonistically toward one another.
- Jean Baptiste Baussingault is called as the father of the field-pot method of experimentation.
- K^+ ion is held around negatively charged soil colloids by electrostatic attraction.
- Kaolinite minerals retain more sulphur than the montmorillonite group of clays.
- Land Equivalent Ratio (LER) is the relative land area under sole crop(s) that is required to produce the yield achieved in intercropping.
- Leghaemoglobin is confined to those nodule cells containing nitrogen fixing Rhizobia.
- Legumes and crucifers are very sensitive to insufficient molybdenum.
- Lime is seldom needed in those areas where rainfall is low and leaching is minimal.
- Magnesium with sulphur brings about significant increase in oil content of a number of oilseed crops.
- Maize, cotton and tabacco are highly responsive to magnesium.
- Male flower initiation stage in maize is also called tasseling stage.
- Manganese can substitute for magnesium in many of the phosphorylating and group transfer reactions.
- Many plants that embody the C-4 dicarboxylic photosynthetic pathway require sodium as an essential nutrient element.
- Marls represent the soft, unconsolidated deposits of calcium carbonate embodying a neutralizing value between 70-90%.
- Marth and Mitchell (1944) reported selective weed

control by 2, 4-D in a bluegrass lawn in U.S.A.

- Massflow could be a major factor in the potassium nourishment of crops grown in soil.
- Melanagromyza cucutae has been found useful against the weed, cuscuta (dodder).
- Micronutrients are applied in small amounts and placement methods are directed toward greatest efficiency.
- Minimum tillage or no-till system reduces the availability of potassium due to increased compaction, lessened aeration, lower temperature and positional availability.
- Mitscherlich's Growth Law states, "the increase in yield of a crop as a result of increasing a single growth factor is proportional to the decrement from the maximum yield obtainable in increasing the particular growth factor".
- Mitscherlich's Law of physiological relationship states that yield can be increased by each single growth factor even when it is not present in the minimum as long as it is not present in the optimum.
- Mixed cropping means growing of two or more crops simultaneously in the same field without keeping their identity with respect to field area.
- Molybdates depress sulphate adsorption.
- Molybdenum deficiency has been found more severe under dry land conditions.
- Molybdenum in lithosphere is about 2 ppm, while typically ranging 0.2-5.0 ppm in soils.
- Molybdenum is a structural component of nitrogenase, the enzyme actively involved in nitrogen fixation by root nodule bacteria of leguminous crops.
- Molybdenum is an essential component of the major enzyme nitrate reductase in plants.
- Most naturally occurring organic forms of phosphorus are esters orthophosphoric acid.
- Most of the ammonia in the world is produced synthetically by reacting nitrogen and hydrogen gas and only some of it as a by-product of coking coal.
- Most of the pH induced zinc deficiencies occur within the range 6 to 8 pH and calcareous soils are particularly prone to this nutritional problem.
- Most of the soil applied herbicide is usually concentrated in the upper 2-8 cm of top soil.
- Most plants are generally able to make good growth at light intensities of less than full daylight.
- Multi-storey cropping mean cultivation of different crops of different canopy heights simultaneously in a field.
- Nitrate is dominant form of nitrogen taken up by crops.
- Nitrogen fixation by a legume is at the maximum only when the level of available soil nitrogen is at the minimum.
- Nitrogen fixing organisms was first isolated by M.W. Beijerinck who called the organisms *Bacillus radicicola*.
- Nitrosomonas and Nitrobacter are usually referred to collectively as Nitrobacteria.
- No-tillage cropping system in the surface few centimeters makes the soil acid in a few years, primarily due to the surface application of nitrogen fertilizers.
- Nucleic acid and phytin are taken in by plants from sterile sand or solution culture.
- O.W. Wilcox employed the Mitscherlich equation to extend a branch of science, termed by him as Quantitative Agrobiology.
- On 29th December, 1966, the Seed Act (India) was passed which came into force on 2nd October, 1969.
- Once a herbicide is absorbed into the plant system, it moves either apoplastically or symplastically.
- Orange Seed Lot Certificates can be issued when sampling has been done according to the rules by an official body approved by the seed testing station.
- Ordinary superphosphate is manufactured by mixing gravimetrically parts of sulphuric acid and rock phosphate.
- Parasitic weeds need some stimuli from their host plants for germination.
- *Phalaris minor* came to India in 1960.
- Phosphate compounds act as "energy currency" within plants.
- Phosphatidylcholine (lecithin) and Phosphatidyethanolamine are the dominant phospholipids in soil.
- Phosphorous movement from the point of placement is generally limited, as its ion is only slightly mobile in soil.
- Phosphorus availability in soils induces zinc deficiency.
- Phosphorus concentration in earth's crust is only 0.11 per cent, whereas that potassium averages about 1.9 per cent.
- Phosphorus might be applied in a band at planting for maintenance of optimum levels of available soil phosphorus.
- Phosphorus should generally be placed in the zone of root development.
- Physiological drought plant roots fails to absorb water and nutrients from the soil due to increased concentration of soil solution.
- Pietro de Crescenzi (1230-1307) who published "Opus Ruralium Commodorum" (a collection of local agricultural practices) is referred to as the founder of Modern Agronomy.

GS

Mesopotamians called the Indus region Meluha.

- Plant deficiencies of magnesium can occur in soils with wide ratios of exchangeable Ca or Mg.
- Plants absorb molybdenum as MoO_4^{2-}.
- Plants normally contain between 1 and 5 per cent nitrogen by weight.
- Potassium can affect the rate of transpiration and water uptake by regulation of stomatal opening.
- Potassium diffusion to roots is limited to very short distances in soil, usually only 1-4 mm from the root surface during a growing season.
- Precipitation is associated with adiabatic cooling and condensation in rising air.
- Pulses are rich in vitamin A.
- Raj Kiran (RD-387) is nematode resistant variety of barley.
- Rape and alfalfa require relatively higher concentrations of sulphur.
- Rapeseed is tolerant to excessive amount to manganese, but very sensitive to its toxicity.
- Real value seed is equal to the product of purity (per cent) and germination (per cent) divided by 100.
- Relative Yield Index (RYI) refers to the yield of a crop in a district expressed as percentage of the whole country.
- Relatively older rice seedlings establish better than young seedlings in alkaline soils.
- Repeated tillage operations over a long period of time prove detrimental to surface soil granules.
- Rice seedling for transplanting under no circumstances should have more than 5 leaves.
- Robert Warrington demonstrated that nitrification was a two-step phenomenon, the ammonia first being conversion into nitrites and the nitrites subsequently to nitrates.
- Rotation can greatly improve the soil structure and tilth of many medium and fine-textured soils.
- Rotational cropping intensity (%) represent the total number of crops grown on a piece of land in one rotation-cycle and multiplied by 100.
- Rotational intensity of a crop refers to the number of times the crop recurs in a total rotational cycle.
- Rotational intensity of maize-potato-greengram and maize-mustard-summer vegetables respectively in first year will be 200.
- Seasonal drought occurs when the actual seasonal rainfall is defined by more than twice the mean deviation.
- Seed quality index refers to the vigour of germination or percentage of germinated seeds with normal seedlings.

Knowledge World!

First use of ballot paper

The general elections of 1962 introduced the ballot paper which had to be marked by the voter. The earlier two general elections followed a system, of balloting in which each candidate was given a ballot box.

- Seed vigour is the sum of those properties which determine the potential level of activity and performance of the seed or seed lot during germination and seedling performance.
- Sewage effluence is valuable nutrient resource for crops.
- Silica-gel primarily impregnates the walls of epidermal and vascular tissues, where it appears to strengthen the tissue, reduce water loss and retard fungal infection.
- Silking stage in maize is also known as cobbinitiation.
- Sodium in sugar beet influences water relations and enhances its resistance to drought.
- Sodium nitrate (Chilean saltpeter) and Rhenania phosphate contain 25 and 12 per cent of sodium respectively.
- Soil submergence and waterlogging lowers the redox potential and increase the amount of soluble manganese (Mn^{2+}) in soils.
- Soyabean and peas are inefficient nitrogen fixer in comparision to faba beans.
- Soyabean grows well in warm and moist climate.
- Sulphides and polysulphide do not exist in well-drained upland soils.
- Sulphur is a vital part of ferredoxin that participates in oxido-reduction process by transferring electrons.
- Sulphur, the thirteenth most abundant element has been estimated to vary between 0.06-0.10 per cent.
- Sulphuric acid is literally workhorse of the fertilizer industry, as more than 60 per cent of it is accounted for by the fertilizer industry alone.
- Sulphur-oxidizing organisms are the autotrophic bacteria belonging to the genus Thiobacillus.
- Surfactants are non-herbicidal compounds which enhance the absorption of aqueous solution of polar herbicides.
- Tetrazolium test is essentially an assessment of dehydrogenase enzyme activity.
- The "law of minimum" stated by Justus von Liebig in 1862 is a simple, but logical guide for predicting crop response to fertilization.

GK **Disease Alzheimer affect the brain.** GS

- The aim of all cropping systems should be sustained productivity and profits.
- The ammonifying microbial populations include both aerobic and anaerobic microorganisms.
- The application of fertilizer in irrigation water in either closed or open system is termed as fertigation.
- The average amount of solar energy reaching the outer limits of the atmosphere is called solar constant.
- The beneficial effects of sodium on plant growth tend to be the greatest when potassium nutrition is the poorest.
- The capacity of plants for uptake Ca^{2+} is limited because it can be absorbed only by young root tips in which the cell walls of the endodermis are still unsuberized.
- The cationic nature of ammonium (NH_4) permits its adsorption and retention by soil colloidal material.
- The cattle consuming grasses of low magnesium content suffer from hypomagne semia, commonly called Grass Tatany.
- The chemical stability of an active ingredient is often influenced by the reactions between active ingredient and inert material.
- The classification of crops in annual, biennial and perennial classes is known as ontogenic classification.
- The compound in rock phosphate deposits is apatite.
- The cropping pattern is most technically called as crop rotation.
- The discovery of 2, 4-D in early 1940's revolutionized the chemical weed control.
- The finer the texture of the soil, the less rapid will be the movement of soil moisture and the diffusion of ions through the water.
- The first cutting and inverting of soil during turn-around period is known as primary tillage.
- The foliage applied post-emergence herbicides effective against wild oat are barban, asulam, diclofop methyl, difenzoquat and benzoprop methyl.
- The growth of new roots throughout the soil mass and perhaps mycorrhizal infection enhances the significance of root absorption and contact exchange mechanisms of ion absorption.
- The impaired plant uptake of magnesium in many strongly acid soils is caused by high levels of exchangeable aluminium.
- The inclusion of an element in the list of essential elements based on its functioning in plant metabolism, whether or not its action is specific has been termed as functional or metabolic by D.J. Nicholas.
- The increase in absorption and translocation of herbicide at higher humidity is due to increased pholem transport.
- The indirect deficiencies of calcium are caused by it's shortage in fruit and storage organs which grow rapidly with restricted internal supplies of calcium.
- The inorganic phosphorus content of soil is frequently higher than that of organic phosphorus except organic soils.
- The kind of seeds which require specifically red (R) irradiance to germinate are known as phytochrome seed.
- The length of coleoptile of dwarf wheat varieties is dwarf as compared to tall varieties.
- The lower temperature tends to delay flowering in soyabean.
- The mineralization of organic nitrogen compounds takes place in essentially three step-by-step reactions: aminization, ammonification, and nitrification.
- The minimum genetic purity of certified seed should be 99 per cent as against 99.5 per cent for the foundation seed.
- The most commonly used instrument for seed sampling is sleeve-type trier or stick trier.
- The nitrate form of nitrogen is completely mobile and within limits moves largely with the water.
- The pH is defined as the logarithm of the reciprocal of the H^+ ion activity.
- The phosphorus is absorbed by plants largely as primary and secondary orthophosphate ions ($H_2PO_4^-$ and $H_2PO_4^{2-}$) which are present in the soil solution.
- The plants of canary grass (*Phalaris minor*) if pulled out during flowering, can set about 50 per cent of viable seed.
- The plants require different absolute amounts of nitrogen, phosphorus and potassium, but the amounts in pounds of each required to produce a yield, that is, 50 per cent of the maximum is termed as one baule unit.
- The point placement urea of supergranules will substantially increase its agronomic efficiency for rice production.
- The principle of uniformitarrianism govern a wide variety of geologic and climatic studies that involve organic residues.
- The quantities of chloride in excess of 30-40 kg/ha impairs the burning quality of tobacco.
- The quantity of nitrogen fixed by properly nodulated legumes averages about 75 per cent of the nitrogen used in the growth of the plant.
- The rain drops that reach the ground range from 1-5 mm in diameter.
- The rate of denitrification is very slow in acid soil (pH

3.6-4.8) and very rapid in soils of high pH (8.0-8.6).

- The residual heterozygosity in the variety becomes better visible through segregation within the progeny.
- The seed certification aims at ensuring genuineness and quality of seed and propagating material to a purchaser.
- The seeds of dwarf wheat varieties should not be sown deeper than 5 cm.
- The seeds of tobacco are positively photoblastic.
- The seeds that have a high germination after controlled deterioration (CD) are those which emerge and store well.
- The soil temperature of 15.5 °C or above favour rapid germination and vigorous seedling growth of soyabean.
- The subsistence form of agriculture was based on the principle grow it and eat it.
- The success of biological control of weeds through insects and pathogens depends upon host specific.
- The supply of nitrate and/or nitrite in soil is a prerequisite for denitrification.
- The temperature of 26.5 to 30 °C appears to be optimum for growth and development of most of the varieties of soyabean.
- The temperature range of growth for most of the agricultural plants are usually much lower, perhaps between 15 and 40°C, below and above these limits growth decreases rapidly.
- The translocation of sugar from leaves of sugarcane is greatly reduced in potassium-deficient plants.
- The transplanting of rice seedlings deeper than 2-3 cm delays and inhibits tillering.
- The ultimate objective of any farming enterprise is sustained maximum economic production.
- The usually recommended proportion of N:P:K per square meter of nursery areas is 8-14:8-17:10-18 grammes, respectively.
- The word Agronomy has been derived from Greek words-*agros* meaning field and *nomos* meaning to manage.
- The word cereal has originated from a Roman word *Ceres*, the name of a Roman Goddess who was the "giver of grain".
- Theodore de Saussure was able to demonstrate that the plants absorb oxygen and liberate carbon dioxide, the central theme of respiration.
- Though light is not usually essential for seed germination, it may be required for some kind of a seed like *Luctuca sativa.*
- Triazine herbicides also have growth stimulating effect on plants.
- Triazine, uracil and urea herbicides embodying low ability to enter symplast in leaves, move acropetally via. apoplast to the leaf mesophyll very rapidly with water.
- Trimethylene urea appears to be the most suitable single component as a slow release nitrogen material.
- Triticale may give more yield than wheat under dry condition.
- Two Germans, Hellrugel and Wilfarth (1886) gave the first specific information regarding nitrogen fixation by legumes.
- Urea on application to soil is acted upon by the enzyme urease which hydrolyzes it to unstable ammonium carbonate.
- Urea was first isolated in 1773 from urine.
- Virgil advocated the application of legume to boost the yield of yellow wheat.
- Water harvesting is a device to utilize the collected and preserved water for the purpose of crop production.
- Water logging of soil results in rapid denitrification due to impeded diffusion of oxygen to the sites of microbiological activity.
- Water use efficiency (WUE) represent the amount of dry matter that can be produced from a given quantity of water and it is also described as weight of dry matter derived from a hectare centimeter of water.
- Wettable powder (wp) contains finely divided solid particles measuring less than 3 microns that can be readily suspended in water to formulate suspension.
- When stubbles are left after harvesting to grow into a crop means ratooning.
- When the deficiency of the rainfall in an area in a year is above 50%, it is referred to as a severe drought.
- When two crops are grown together and their yields on area basis are higher than their pure cropping it is known as synergetic cropping.
- Wherever straw, stover or other organic materials are returned to the soil, adequate nitrogen and sulphur should be available for their rapid decomposition.
- With zero tillage, crop residues are allowed to remain on the soil surface rather than being worked into the soil.

In Harappa was found a bronze model of a cart with a seated driver.

Multiple Choice Questions

1. ________ has been recommended for the control of weeds in blackgram.
a] Atrazine 0.5 kg/ha
b] 2, 4-D 0.75 kg/ha
c] Fluchloralin 0.75 kg/ha
d] Simazine 0.5 kg/ha

2. ________ comprises an important mechanisms of herbicide detoxification in soil under field conditions especially surface incorporation subsequent to rainfall or irrigation.
a] Thermodecmoposition
b] Biodecomposition
c] Chemodecomposition
d] Photodecomposition

3. ________ is considered to be most important of the soil organisms bringing about the conversion of NH_4^+ to NO_2^-.
a] *Nitrosomonas* b] *Nitrobacter*
c] *Azotobacter* d] *Azospirillum*

4. ________ is probably the most important growth limiting factor in acid soils especially those having pH below 5.0-5.5.
a] Nitrogen b] Calcium
c] Potassium d] Aluminium

5. ________ of the NH_4^+ in a region of alkaline microsite helps to restore the conditions suitable for the normal conversion of NO_2^- to NO_3^-.
a] Diffusion b] Dilution
c] Both of these d] None of these

6. ________ degradation plays a major role in the persistence and behaviour of herbicides in the soil.
a] Microbial b] Physical
c] Chemical d] Thermal

7. ________ generally augment the utilization of phosphorus which is water soluble.
a] Broadcast
b] Broadcast and incorporation
c] Band placement
d] None of these

8. ________ in 1909 developed an equation that related growth to the supply of plant nutrients.
a] V.J. Kilmer
b] R.B. McKercher
c] J.W. Friedrich
d] E.A. Mitscherlich

9. ________ ion effect represents the effect of adsorbed ion on the release of another from the surface of a colloid.
a] Supplementary b] Complementary
c] Coordinated d] Integrated

10. ________ a.i./ha has been recommended for the control of weeds in mungbean.
a] Basalin @ 1 kg b] 2, 4-D @ 1 litre
c] Simazine @ 1 kg d] Atrazine @ 1 kg

11. *Aman* rice is sown in:
a] April-May b] June-July
c] Nov-Dec d] May-June

12. 'Krishi Pandit' is a title awarded every year by :
a] ICAR b] President of India
c] IARI d] FAO

13. 'Ratooning' is practised as a matter of routine in :
a] Sugarbeet and sugarcane
b] Sugarbeet and maize
c] Sugarcane and napier grass
d] Sugarcane only

14. 'V' notch is a :
a] Parshall flume b] Orifice
c] Weir d] Submerged orifice

15. Varalaxmi is a hybrid developed in :
a] Rice b] Cotton
c] Brinjal d] Brassica

16. Zineb is chemical used to control:
a] Bacterial diseases
b] Fungal diseases
c] Viral diseases
d] White fly

17. Arrowing refers to:
a] Flower primodia initiation in wheat
b] Flowering of cotton
c] Flowering of maize
d] Flowering of sugarcane

18. Boll shedding in cotton is due to:
a] Plant is over loaded with bolls
b] Inhibitory effect of ABA (Abscisic acid)
c] Heavy dressing of urea
d] Frequent irrigation

19. Hot water treatment of sugarcane setts is done because:
a] It increases germination
b] It controls the seed borne diseases
c] It breaks dormancy
d] It promotes the conversion of sucrose into glucose

20. Krishi Chayanika is published by :
a] ICAR
b] Association of Agricultural University
c] Food Corporation of India
d] Ministry of Agriculture

21. Reclamation disease due to copper deficiency occurs in plants grown on newly reclaimed acid soil:
a] Histosols b] Alfisols
c] Aridosols d] Ultisols

GK Continental drift theory was postulated by Alfred Wegner. GS

22. Trap crops refer to:
a] Crops grown around the main crop which are more susceptible to insect pest and diseases than the main crop
b] Crops which are resistant to pest and diseases
c] Crops efficient in trapping solar radiation
d] Crops efficient in trapping rainfall

23. 2, 4-D is a ________ herbicide.
a] Selective b] Non-selective
c] Contact d] Selective and systemic

24. Golden rice is rich in :
a] Vit. A b] Vit. B
c] Vit. C d] Vit. D

25. 2, 4-D is an important herbicide of:
a] Pulses b] Oilseed crops
c] Small grain crops d] Millets

26. 2, 4-D is the :
a] Translocated herbicide
b] Contact herbicide
c] Contact herbicide as well as translocated herbicide
d] All of above

27. The most sensitive stages of sunflower to water stress lie between:
a] Germination and seedling
b] Germination and flowering
c] Seedling and flowering
d] Flowering and grain filling

28. A basal dose of phosphorus at ________ g/m^2 to tobacco nursery area has been recommended.
a] 5.5-6.5 b] 7.5-8.0
c] 9.0-9.5 d] 10.0-10.5

29. A cloud type that occurs in low height in the atmosphere is :
a] Stratus
b] Cumulus
c] Nimbus
d] Cumulo-nimbus

30. A crop of wheat yielding 50 quintals of grain per hectare removes ________ from soil.
a] 80-100 g N + 50-60 kg P_2O_5 + 100-120 kg K_2O
b] 100-150 kg N + 70-80 kg P_2O_5 + 125-150 kg K_2O
c] 150-180 kg N + 100 kg P_2O_5 + 160-180 kg K_2O
d] 160-200 kg N + 80-100 kg P_2O_5 + 150-170 kg K_2O

31. A farmer is to select suitable crops for a drought prone unirrigated area. Which one of the following groups of crops should be chosen for mixed cropping ?
a] Sesame, maize, oats
b] Sorghum, wheat
c] Sesame, sorghum/pearl millet, niger
d] Maize, upland rice, berseem

Quiz Kit!
First singing of the national anthem

***Jana Gana Mana,* India's national anthem composed by Rahibdranath Tagore, was sung for the first time on Dec.27, 1911, at the Calcutta session of the Indian National Congress. It was adopted as India's national anthem on Jan. 24, 1950.**

32. A good crop of potato yielding about 400 q/ha on an average removes ________ N + P_2O_5 + K_2O kg/ha, respectively.
a] 170-180 + 100 + 250
b] 170-180 + 25 + 250
c] 170-180 + 50 + 125
d] 170-180 + 25 + 100

33. A high percentage of cropped area in India is covered by rice because:
a] There is adequate rainfall
b] Rice can be grown in different seasons
c] Rice is adaptable to most types of land and agriculture
d] A large part of the population of India is rice eating

34. A lentil crop grown with improved package of practices may record the yield upto t/ha:
a] 1.0-1.5 b] 3.0-3.5
c] 1.5-2.0 d] 2.0-2.5

35. A main shortcoming that was observed during green revolution:
a] Pulse crops were neglected
b] Paddy was neglected
c] Big farmer were neglected
d] Use offertilizers neglected

36. A millet crop is :
a] Sugarcane b] Bajra
c] Barley d] Jute

37. A minor portion of the phosphorus adsorption capacity of soils originates from:
a] Sodium bicarbonate
b] Sodium carbonate
c] Calcium oxide
d] Calcium carbonate

38. A screen which is used to measure the temperature of atmosphere is know as :
a] Clark screen b] Symon's screen
c] Thermocouple d] Stevenson's screen

39. A short duration crop in between two main crops is termed as:
a] Cash crop b] Catch crop
c] Companion crop d] Ephemeral

GK **The Sindhu is the river par excellence of Aryans.** GS

40. A tuber represents:
a] Englarged underground root
b] Enlarged underground stem
c] Enlarged underground nodules
d] None of these

41. A weed that spreads Asthma is :
a] Colotropisb] Parthenium
c] Kanuch weed d] Common grass weed

42. According to the modern concepts, the major function of inter-row tillage is :
a] Moisture conservation through soil mulching
b] Improved soil aeration
c] Improved soil granulation
d] Moisture conservation through weed destruction

43. Addition of ammonium sulphate to a herbicidal spray solution enhances the __________ against *Agropyron repens*.
a] Glyphosphate
b] Amylase
c] Reductase
d] Oxi-reductase

44. Adsali crop of sugarcane takes a period of maturity of :
a] 8 months b] 12 months
c] 18 monthsd] 24 months

45. Affinity of the enzyme RuDP carboxylase for carbon dioxide is enhanced by:
a] Sulphur b] Phosphorus
c] Nitrogen d] Magnesium

46. All India Co-ordinated Maize Improvement Project was started in:
a] 1957 b] 1955
c] 1965 d] 1967

47. American cotton should normally be planted at the distance of __________ cm.
a] 45 x 30 b] 60 x 30
c] 60 x 45 d] 90 x 45

48. Ammonium sulphate is preferred for nitrogen application in groundnut because it also provides:
a] Calcium b] Magnesium
c] Sulphur d] Boron

49. Among small grain crops _________ has an exceptional tolerance to low levels of soil copper supply.
a] Barley b] Wheat
c] Rye d] Oats

50. Among the oilseed crops, groundnuts ranks __________ in India.
a] First b] Second
c] Third d] Fourth

51. Among the states growing cotton as rainfed crop, the one which obtains the highest yield in the country is:
a] Maharashtra b] Haryana
c] Gujarat d] Punjab

52. Amorphous aluminosilicate minerals like allophane have ______ negative charge which is partly or entirely balanced by complex aluminium cation.
a] Small b] Zero
c] Large d] None of these

53. Anaerobic environment in paddy soil is responsible for gaseous loss of fertilizer nitrogen by:
a] Denitrification b] Volatilization
c] Nitrification d] Ammonification

54. Application of nitrogen in cowpea at the time of planting is known as:
a] Basic dose
b] Synergistic dose
c] Starter dose
d] Additional dose

55. Application of propanil after _________ in rice-paddies causes prohibitive vegetative injury.
a] DDT b] Endosulfan
c] Carbofuran d] Sevin

56. Application of urea, aqua-ammonia, anhydrous ammonia and diammonium phosphate form alkaline microsite at the point of application, and in rice __________ encourage the taking up of nitrite.
a] High NH_4^+ concentration
b] High pH
c] Both of these
d] None of these

57. Area required for raising seedlings for one hectare normal sown crop of paddy :
a] 250 sq meter b] 500 sq meter
c] 750 sq meter d] 1000 sq meter

58. *Arhar* (pigeonpea) is commonly known as:
a] Red gram b] Black gram
c] Bengal gram d] Green gram

59. As per principles of crop rotation pulses should be included in rotation because :
a] They increase total pulse production
b] They release land quickly for the second crop
c] They help in maintainance of soil fertility
d] They increase the yield of succeeding cereal crop

60. As per the area of cultivation, the most important crop of India is :
a] Rice b] Wheat
c] Cotton d] Sugarcane

61. At pH __________ H^+ ion toxicity is lessened by the removal of these ions.
a] 4.5 or lessb] 5.5 or less
c] 6.0 or lessd] 5.0 or less

62. Autumn sown crop of sugarcane requires _________fertilizer than that of winter sown ?
a] Equal b] More
c] Less d] Half

GK GS

Igneous rocks consist of silicate minerals.

63. Availability of phosphorus is maximum at pH :
a] Neutral b] Acidic
c] Alkaline d] All of the above

64. Azolla is a :
a] Water fern b] Water algae
c] Water bacteria d] Water protozoa

65. *Azotobacter* spp fix nitrogen :
a] As symbiouts on sugarcane
b] As symbiouts on rice
c] As symbiouts on legume
d] In free soil

66. Bacteria responsible for nitrification are :
a] Nitrosomonas b] Nitrobacter
c] Above both d] VAM

67. Balanced fertilizer is used :
a] To increase production
b] To improve quality of seed
c] To maintain soil productivity
d] All of these

68. Barley requires ________ climate.
a] Cool b] Dry
c] Humid d] Hot

69. Based on dark metabolism, glucose consumption for protein production is about ________ per cent when nitrogen is provided as nitrate rather than as ammonium.
a] 20 b] 30
c] 40 d] 50

70. Begasse, a by-product of sugar manufacturing industry, is used for the production of :
a] Paper b] Alcohol
c] Confectionaries d] Glass

71. Black soils are deficient in :
a] Potash b] Iron
c] Nitrogen d] Lime

72. Black soil is used for the cultivation of :
a] Jute b] Wheat
c] Cotton d] Sugarcane

73. Botanical name of sixed row barley:
a] *Hordeum irregular* b] *Hordeum distichon*
c] *Hordeum vulgare* d] None of above

74. *Brassica juncea* is the scientific name of :
a] Rai b] Toria
c] Soybean d] Wheat

75. Broadcast potassium is ________ than banded.
a] More efficient b] As efficient as
c] Less efficient d] None of these

76. Brown stripe downy mildew of maize can be controlled by use of:
a] Bordeaux mixture (4:4:50)
b] Bavistin
c] Karathane-45
d] Diathane M-45

77. Build up of soluble sugars and amino acids (due to inadequate potassium), low molecular weight compounds encourage the activity of:
a] Plant hormones b] Plant pathogens
c] None of these d] Both of these

78. Butylate stimulates germination in ________ seeds by enhancing gibberellic activity or by changing membrane or seed permeability.
a] Potato b] Pigeonpea
c] Safflower d] Barley

79. By which of the following method of raising seedlings of rice, we can get the ready seedling in early time:
a] Dapog method b] Dry bed method
c] Wet bed method d] None of the above

80. Indirect methods of measuring soil water content consist of resistance method and:
a] Gravimetric b] Volumetric
c] Neutron scattering d] Non of these

81. Capillary water will not be available to the plants it is held at a tension of:
a] Less than 15 atmospheres
b] 20 atmospheres
c] More than 15 atmospheres
d] Less than 5 atmospheres

82. Castor belongs to family:
a] Euphorbiaceae b] Leguminosae
c] Cruciferae d] Compositae

83. Central Dryland Research Institute is located at:
a] Jhansi b] Jodhpur
c] Hydrabad d] Jabalpur

84. Chickpea pod borer can be controlled by the spray of:
a] Monocrotophos @ 1 ml/litre of water
b] Aldrin (5%) @ 15 kg/ha
b] Seed treatment by Captan (0.25%) @4 g/kg of seed
c] BHC @ 10 kg/ha

85. Clays such as kaolinite will fix larger quantities of phosphorus with ________ SiO_2/R_2O_3 ratio.
a] Low
b] High
c] Equal
d] None of these

86. Common herbicide used in wheat is:
a] 2, 4-D b] Atrazine
c] Isoproturon d] Pendimethalin

87. Content of starch in maize is approximately :
a] 95% b] 70%
c] 55% d] 25%

88. Cotton belongs to the ________ family.
a] Malvaceae b] Compositeae
c] Crucifereae d] Lineaceae

89. Blackgram grown in summer should be seeded at ________ kg/ha.
a] 10-15 b] 20-25
c] 15-20 d] 25-30

GK GS
The branch of Physics that deals with the motion of a very small particle is called Quantum Mechanics.

90. Cowpea grown for grain purpose should be seeded at______kg/ha.
a] 10-15 b] 15-20
c] 20-25 d] 25-30

91. Crop lodging refers to surface:
a] Falling of the plants horizontally on the ground surface
b] Cutting of tree crops into loggs.
c] Record of the progress of the crop from sowing to harvesting
d] Determination of nutrient status of the plant

92. Crop mixtures have proved biologically most dynamic under:
a] Water stress conditions
b] Irrigated conditions
c] High soil fertility conditions
d] Large holding conditions

93. Crop production and animal husbandry collectively is know as:
a] Mixed cropping b] Relay farming
c] Mixed farming d] None of the above

94. Crop rotation is defined as a :
a] System of growing different kinds of crops on the same land
b] System of growing different crops in succession on the same land
c] A method of growing different crops
d] System of growing the same crop on different fields

95. Cropping intensity is:
a] The extent of the use of land for purposes other than cropping
b] The extent of the use of land for cropping purposes during a given year
c] Net cropped area
d] Gross cropped area

96. Cropping intensity means:
a] Percentage ratio of gross cropped area to net cropped area
b] Percentage ratio of net cropped area to gross cropped area
c] Percentage ratio of the number of crops in rotation to period of one rotation
d] Percentage ratio of gross cropped area to the net cropped area

97. Cropping pattern is defined as :
a] The proportion of area under different crops
b] Growing different crops in a recurrent succession
c] The proportion of area under different crops at a particular point of time
d] A method of growing different crops

98. Crops during any season are generally able to recover only _________ of the added fertilizer phosphorus.
a] Less than 25% b] Less than 30%
c] Less than 35% d] None of these

99. Dapog method of crop raising is associated with :
a] Interculture operation
b] Nursery raising
c] Seed-bed preparation
d] Intercultivation

1C0. Dapog method of raising seedlings is related to :
a] Sugarcane b] Tobacco
c] Paddy d] Jute

101. Dapog seedlings would be ready for transplanting within:
a] 8-10 days b] 18-20 days
c] 21-25 days d] 11-14 days

102. Deep ploughing of soil in a groundnut field should normally be avoided because it:
a] Makes harvesting difficult
b] Makes pegging difficult in upper layers
c] Delays germination
d] Delays pegging and pod development

103. Deficiency of nitrogen in maize causes:
a] Yellowing of leaves from tip to base in V shaped pattern
b] Chlorotic spots at base of leaves near the margins
c] Yellow streaking of leaf edges
d] Whitish-yellow spots on leaves

104. Deficiency symptoms on apical buds is due to :
a] Zn b] Mo
c] Bo d] Mn

105. Dehaulming in potato is done to obtain :
a] Higher yield
b] Quality seed tuber
c] Higher starch content
d] Higher protein content

106. Delinting of cotton help in:
a] Grading the seed
b] Killing of hibernating boll worm larvae and disease pathogen
c] Quicker germination
d] All of these

107. Delinting of cotton seed may be done with:
a] Sulphuric acid b] Citric acid
c] Sodium hydroxide d] Ammonium chloride

108. Denitrification losses of nitrogen in paddy field can be prevented by incorporating ammonical nitrogen in paddy field can be prevented by incorporating ammonical nitrogen into the:
a] Oxidized zone b] Reduced zone
c] Calcified zone d] Hydronized zone

109. Deshi variety of cotton requires _______ the seed rate than that of American variety.
a] Equal b] 3/5
c] Double d] 1/3

110. Dibbling (sowing) of wheat requires __________ of seed.
a] 15-20 kg/ha b] 25-30 kg/ha
c] 30-40 kg/ha d] 20-25 kg/ha

GK GS

A clone is produced asexually.

111. Dispersing agents reduce __________ between similar particles.
a] Adhesion b] Cohesion
c] Tension d] None of these

112. Dodder is a serious weed of :
a] Berseem b] Lucerne
c] Oats d] Kasni

113. Name of shifting agriculture in N-E India is ?
a] Taungya b] Milpa
c] Jhoom d] Logun

114. Dry land farming means :
a] Multiple cropping
b] Farming of high yielding varieties of maize, jowar and bajra
c] Checking of soil degradation and reclaiming problematic lands
d] Cultivation of drought resistant, short duration and high yielding varieties of crops, with package of soil and water management including water harvesting.

115. During *kharif* season sunflower takes __________ days to mature.
a] 80-90 b] 100-110
c] 120-130 d] 105-130

116. Dwarfing gene in rice is :
a] Norin - 10
b] Branchytic-2
c] Dee-gee-woo-gen
d] Opaque-2

117. Each unit increase in pH brings about a decrease in solubility of Fe^{2+} of __________ folds.
a] 50 b] 75
c] 100 d] 150

118. Embryo seed dormancy is mainly found in :
a] Tropical climate seeds
b] Temperate climate seeds
c] Sub-tropical seeds
d] None of the above

119. Enzyme responsible for hydrolysis of urea is :
a] Amylase b] Urease
c] Nitrate reductase d] Catalase

120. Excess of chlorides in tobacco:
a] Inhibits the leaf burn and reduces leaf storage capacity
b] Reduces the leaf size and aroma
c] Increases the leaf storage capacity and leaf burn
d] Increases the leaf size and aroma

121. Fertilizer dose recommended for a crop of groundnut is N + P_2O_5 + K_2O kg/ha, respectively:
a] 20-40 + 50-60 + 30-40
b] 40-50 + 50-60 + 30-40
c] 20-40 + 60-80 + 30-50
d] 40-50 + 60-80 + 30-50

Did You Know ?

President with least votes

In the elections held on Aug. 16, 1969, V.V. Giri polled a mere 50.2 per cent barely defeating his rival Sanjiva Reddy who polled 48.5 per cent.

122. Fertilizer requirement of castor has been recommended at __________ of N + P_2O_5 + K_2O kg/ha, respectively.
a] 60 + 40 + 40 b] 120 + 60 + 60
c] 120 + 40 + 40d] 60 + 60 + 30

123. Field water use efficiency is calculated by :
a] WUE = Y/WR b] WUE = Y/ET
c] WUE = Y/cu d] None of these

124. Fire curing is followed in :
a] Chewing type of tobacco
b] Bidi tobacco
c] Hookah tobacco
d] Cheroot tobacco

125. First Agriculture University of India is :
a] G.B. Pant Agriculture University, Pantnagar
b] Haryana Agriculture University, Hisar
c] Rajasthan Agriculture University, Bikanare
d] Punjab Agriculture University, Ludhiana

126. First herbicide produced in world is:
a] 2, 4-D b] Semazin
c] Atrazine d] Banavel-D

127. First irrigation to linseed is normally recommended __________ after sowing.
a] 30-40 b] 40-50
c] 20-30 d] 50-60

128. Fluchloralin can be used for weed control in soyabean as:
a] Pre-emergence
b] Pre-plant incorporation
c] Post-emergence
d] Through herbigation

129. Alachlor is used as :
a] Pre plant b] Pre energence
c] Post emergence d] Both b & c

130. For a fibre crop of sannhemp, recommended seed rate for one hectare is __________ kg/ha.
a] 5-10 b] 10-15
c] 15-20 d] 20-25

131. For common plants transpiration ratio is:
a] 1000 to 2000 b] 2000 to 3000
c] 2500 to 10,000 d] 250 to 1000

132. For each unit increase in pH, the concentration of Mn^{2-} decreases by __________ folds.
a] 50 b] 100
c] 150 d] 200

India is the largest producer of banana.

133. For fibre purposes, jute crop may be harvested at any time:
a] After pod formation
b] After flowering
c] Before flowering
d] Before pod formation

134. For harvest of sugarcane the brix ratio of the lower and above portion of the stem should be:
a] 0.2-0.3 b] 0.5-0.6
c] 0.7-0.8 d] 0.9-1.0

135. India Dry Land Agriculture contributes to :
a] 70 b] 50
c] 60 d] 42

136. For *kharif* season greengram, the recommended optimum seed rate is kg/ha:
a] 8-10 b] 12-15
c] 15-20 d] 20-25

137. For more effective weed control in soyabean fluchloralin should be used as:
a] Pre-emergence
b] Postemergence
c] Early postemergence
d] Preplant soil incorporation

138. For mungbean the period between ____________ days after sowing has been regarded as critical for weed control.
a] 10 and 30 b] 5 and 25
c] 15 and 40 d] 25 and 45

139. For nitrogen and phosphorus application in groundnut ________ fertilizers are normally preferred.
a] Urea and single superphosphate
b] Ammonium sulphate and single superphosphate
c] Urea and diammonium phosphate
d] Ammonium sulphate and diammonium phosphate

140. For raising a healthy nursery of tobacco on highly leached soils, besides other nutrients, borax are ________________ g/m^2 is also recommended.
a] 0.4-0.5
b] 0.6-0.7
c] 0.1-0.2
d] 0.3-0.4

141. For spring crop of soyabean seed should be __________ than that of *kharif* planting.
a] More b] Less
c] Similar d] Double

142. For urdbean, the suitable climate is:
a] Hot and humid b] Cool and humid
c] Cool and dry d] Hot

143. For wetland rice the best nitrogenous fertilizer is:
a] Urea
b] Calcium nitrate
c] Ammonium sulphate
d] Calcium ammonia nitrate

144. Glutathion-s-transferase is responsible for selectivity of :
a] Trizaines
b] Substituted ureas
c] Phenoxy herbicides
d] Carbamates

145. Grain to straw ratio in most of the maxican wheat varieties is:
a] 1:1 b] 2:1
c] 3:1 d] 1:2

146. Gravtational water is held in soil at a tension of :
a] 0.5 atmosphere
b] Less than 1/3 atmosphere
c] 1.25 atmosphere
d] 5.0 atmosphere

147. Green manuring of sannhemp adds approximately ____________ kg N/ha to the soil on decomposition.
a] 20-40 b] 40-60
c] 60-80 d] 80-100

148. Green revolution programme was introduced in the years :
a] 1964-65 b] 1966-67
c] 1967-68 d] 1969-70

149. Groundnut pegs when developed in the soil from :
a] Roots b] Stems
c] Tubers d] Fruits

150. Groundnut seed normally contains protein and oil ___________ per cent, respectively.
a] 25 and 45 b] 22 and 43
c] 45 and 26 d] 43 and 22

151. Growing more than two crops in a year in the same field is known as:
a] Mixed cropping b] Multiple cropping
c] Relay cropping d] None of the above

152. Growing of a crop after the failure of main crop is called :
a] Cash crop b] Cover crop
c] Catch crop d] None of these

153. Gujrat-67 and American Nectariless are parents of cotton hybrid:
a] Hybrid -4 b] Varalaxmi
c] Hybrid-6 d] Vikaram

154. Hand weeding in Arhar is necessary at the stage of :
a] 30 and 60 days
b] 90 days
c] 25 days
d] 90 and 120 days

155. Heavy rainfall or irrigation during fruiting periods of cotton may cause:
a] Lodging of crop
b] Shedding of flowers and young bolls
c] Yellowing of young bolls
d] Heavy incidence of boll worm

156. Herbicide is a chemical which used to:
a] Destroy weeds
b] Destroy crops
c] Destroy weeds and crops both
d] None of the above

157. Herbicide used for weed control in sorghum is ________ kg a.i./ha.
a] Atrazine at 0.5-1.0
b] Atrazine at 1.25-1.5
c] Basalin at 0.5-1.0
d] Basalin at 1.25-1.5

158. Heteromorphic seeds of the plant families asteraceae and chenopodiaceae do not permit ________ exchange.
a] Water b] Chemical
c] Gaseous d] Light

159. High humidity is harmful to field pea as it accentuates:
a] Incidence of insects
b] Excessive vegetative growth
c] Incidence of disease
d] Delays flowering

160. High temperature above 41°C flowering of castor results in:
a] Rotting of pollens and poor fertilization
b] Blasting of flowers and poor seed-set
b] Delayed pollination and poor-fertilization
c] Reduced pollination and rotting of seed

161. High yielding dwarf varieties of wheat were developed by:
a] Dr. B.P. Pal
b] Dr. Norman E. Borlaug
c] Dr. E.W. Burton
d] Dr. M.S. Swaminathan

162. High yielding varieties of which crop cover maximum area ?
a] Wheat b] Pady
c] Maize d] Sugarcane

163. High-yielding maize is obtained in regions where the atmospheric temperature during the growing season is :
a] High during the day and low during the night
b] High during the day and high during the night
c] Low during the day and high during the night
d] Low during the day and low during the night

164. How many irrigation is required in wheat crop :
a] 6 b] 7
c] 4 d] 3

165. How much area of nursery is needed for sowing of paddy ?
a] 1/5 of total area b] 1/10 of total area
c] 1/8 of total area d] 1/16 of total area

166. How much nitrogen will be supplied by 500 kg of Ammonium Sulphate ?
a] 50 kg b] 80 kg
c] 100 kg d] 150 kg

167. How much percent of *khandsari* can be obtained from gur refined?
a] 37.5% b] 20.0%
c] 42.5% d] 22.5%

168. Hybrid maize should be fertilized with ________ kg/ha of N + P_2O_5 + K_2O, respectively.
a] 120 + 30 + 15 b] 80 + 40 + 20
c] 120 + 60 + 40 d] 120 + 30 + 30

169. Hygroscopic water, which occurs as thin film held tenaciously around soil particles with a tension of 31 atmosphere or more, is:
a] Readily available to plants
b] Not available to plants
c] Harmful to plants
d] None of these

170. If a crop requires some amount of nitrogen and phosphorus which of the following fertilizers, farmer should apply in his field :
a] DAP
b] Potassium Nitro-phosphate
c] Urea and Ammonium phosphate
d] Ammonium phosphate

171. If an acid soil is limed to the pH level ________ it enhances nitrification.
a] 3.5-4.5 b] 4.5-5.5
c] 5.5-6.5 d] 6.5-7.5

172. If farmer practises 'sugarcane-cotton-wheat' the cropping intensity at his farm will be :
a] 100% b] 150%
c] 200% d] 300%

173. If temperature is high during maturation of sunflower, the linoleic acid:
a] Remains unaffected
b] Increases with slow rate
c] Decreases
d] Increases with rapid rate

174. If the average panicle density per square metre is 260, the average number of filled grains per panicle and test weight of rice are 136 and 20 g, respectively. The rice crop will yield:
a] 5.98 t/ha b] 7.08 t/ha
c] 9.12 t/ha d] 10.46 t/ha

175. If the germination and purity percentage; and test weight of weed are 95, 90 and 46 g, respectively and the wheat crop was planted at 20 x 3 cm. The seed rate of wheat per hectare will be:
a] 50.67 kg b] 60.12 kg
c] 79.67 kg d] 89.67 kg

176. If the land utilization index (LUI) of a crop is 0.5 then the duration will be :
a] 183 days b] 172 days
c] 200 days d] 135 days

GK Shale is the most abundant sedimentary rock GS

177. If the moisture contain of the seed is 8-10% than seed can be stored upto :
a] 5 yrs. b] 2 yrs.
c] 1yrs. d] 7 yrs.

178. If there is 500 kg seed than how much samples should be taken for testing ?
a] 8 b] 3
c] 5 d] 10

179. If urdbean is being grown for the first time in a field, the seed should be inoculated with:
a] *Azospirillum* b] *Rhizobium*
c] *Pseudomonas* d] *Clostridium*

180. IFFCO is a :
a] Farmer's Organisation
b] Private Fertilizer Company
c] Farmer's Fertilizer Co-Operative
d] Farmer's Cane Society

181. *Imperata cylindrica* is associated with :
a] Sugarbeetb] Tea
c] Lentil d] Cereals

182. In a cropping sequence of maize-wheat-moong (summer) the cropping intensity is:
a] 100% b] 200%
c] 300% d] 400%

183. In a predominantly biological system, mineralization of phosphorus from organic matter of crop residues is dependent on soil __________ activity.
a] Chemical b] Physical
c] Physico-chemical d] Biological

184. In a sub-tropical area if the average rainfall is 400-500 mm, which of the following crop is most suitable to cultivate:
a] Paddy b] Maize
c] Sorghum d] Cotton

185. In India amongst all the oilseed crops groundnut accounts for more than:
a] 40% acreage and 60 % production
b] 60% acreage and 40% production
c] 20% acreage and 40% production
d] 60% acreage and 80% production

186. In jute growing areas the usual alternate crop is :
a] Sugarcane b] Cotton
c] Rice d] Wheat

187. In most of the cases field capacity represents the water held at a tension of:
a] 5 atmospheres
b] 1/3 atmospheres
c] 2 atmospheres
d] 10 atmospheres

188. In north Indian conditions field peas should be sown in:
a] First fortnight of November
b] Second fortnight of September
c] First fortnight of October
d] Second fortnight of October

189. In Orissa summer or spring season rice is known as:
a] *Daula* b] *Boro*
c] *Aman* d] *Aus*

190. In parts of eastern region and peninsular India two or three crops of rice are taken in a year because of:
a] Favourable mean temperature
b] High rainfall
c] Soils having good water retention capacity
d] Prolonged sunshine

191. In phosphorus and potassium deficient soils, field pea crop should be fertilized with __________ kg/ha of P_2O_5 + K_2O, respectively.
a] 30-40 + 20-30 b] 50-60 + 20-30
c] 60-70 + 30-40 d] 50-60 + 30-40

192. In South India sesamum is grown in:
a] *Rabi* season only
b] *Kharif* season only
c] *Rabi* and *kharif* season
d] Summer season only

193. In terms of the area sown, which one of the following crops occupies the most important position in India?
a] Jute b] Cotton
c] Mustard d] Groundnut

194. In the fertilized soil with soil solution concentration of 1 ppm of phosphorus, mass flow contribution to phosphorus nutrition of crops could approach __________ per cent.
a] 0 b] 40
c] 30 d] 20

195. In the wet season , the initiation of panicle primordia in rice variety of 120 days', duration takes place in :
a] 60-65 days after sowing
b] 65-70 days after sowing
c] 70-75 days after sowing
d] 75-80 days after sowing

196. In wheat, seminal roots :
a] Induce extra strength againts lodging
b] Responsible for nourishing the plants during early stages of the crop
c] Take care of the plants during the later stages of the crop
d] Appear during the vegatative phase of the crop

197. India is largest soil group :
a] Cotton soils b] Desert soils
c] Saline soils d] Alluvial soils

198. Zinc deficiency in arhar (pigeonpea) can be rectified by spraying of:
a] 5 kg $ZnSO_4$ + 2.5 kg lime/ha
b] 20 kg $ZnSO_4$ + 2.5 lime/ha
c] 5 kg $ZnSO_4$ + 5 kg lime/ha
d] 10 kg $ZnSO_4$ + 5 kg lime/ha

199. Inherent capacity of soil to supply nutrients in balanced form is called:
a] Soil fertility b] Soil productivity
c] Fertility Index d] Productivity Index

200. Injury caused by relatively low temperatures, above the freezing points is called :
a] Freezing injury b] Mechanical injury
c] Dehydration injury d] Chilling injury

201. Intercropping is a system of cropping to make the maximum use of:
a] Solar energy b] Water resources
c] Soil resource d] All above

202. Iron deficiency usually occurs in soil in the pH range of :
a] 5.4-6.5 b] 7.4-8.5
c] 9.0-9.5 d] 9.6-10.1

203. Irrigated crop of castor should normally be planted at a row and plant distance of _________ cm.
a] 45 x 45 b] 60 x 45
c] 60 x 60 d] 90 x 45

204. Irrigated rapeseed and mustard is recommended to be fertilized with ___________ of N + P_2O_5 + K_2O kg/ha, respectively.
a] 60-90 + 60 + 40 b] 90-120 + 60 + 40
c] 60-90 + 90 + 40 d] 90-120 + 90 + 40

205. Irrigated safflower is recommended to be fertilized with _________ kg/ha of N + P_2O_5 + K_2O, respectively.
a] 80 + 60 + 20 b] 80 + 40 + 20
c] 40 + 20 + 20 d] 40 + 40 + 20

206. It has been found that _________ incorporation of fluchloralin provides relatively more effective control of weeds in groundnut.
a] Preemergence b] Preplant
c] Postemergence d] None of above

207. It has been found that ___________ incorporation of basalin provides relatively more effective control of weeds in mungbean.
a] Pre-flowering spray
b] Post-emergence (20 days after sowing)
c] Pre-emergence
d] Post-emergence (40 days after sowing)

208. It has been observed that the iron chelated by sorghum plant metabolites, on a moderately deficient soil, enhanced iron uptake by __________ per cent.
a] 25 b] 50
c] 75 d] 100

209. It has been suggested that the __________ of sunlight is generally most satisfactory for plant growth.
a] Full spectrum
b] Half spectrum
c] Little more than full spectrum
d] None of these

You Don't Know!
Largest Sivalingam
Located on the outskirts of Bhopal, the Bhojpur temple, built by Raja Bhoj (1010-53) houses the largest *Sivalingam,* 2.3 m (7.5 ft) high and 5.3 m (17.4 ft) in circumference.

210. It is generally accepted that the Gramineae members contain ______________ times the silicon concentration normally found in legumes and other dicotyledonous plants.
a] 5-10 b] 10-20
c] 20-30 d] 30-40

211. Jute belong to the family:
a] Malvaceae b] Compositeae
c] Lineaceaed] Tiliaceae

212. Kalyansona is a :
a] Single gene dwarf
b] Double gene dwarf
c] Triple gene dwari
d] Tall wheat

213. Karnal bunt or partial bunt was first observed in experimental seed plots at:
a] Karnal b] Pantnagar
c] New Delhi d] Hisar

214. Keeping in view the available technology for the *kharif* and *rabi* crops, which one of the following patterns would be more in demand for crop planting?
a] Multiple cropping
b] Monoculture
c] Location specific cropping
d] Mixed cropping

215. Late planting of autumn sugarcane reduces the germination due to:
a] Short day b] Frost
c] Low temperature d] High humidity

216. Legume crop should be included in rotation, because:
a] It increase the total pulse production
b] Emphasis has been given on pulses and oilseeds in the 20 point programme
c] It helps in maintenance of soil fertility
d] It increases cropping intensity

217. Lentil under late sowing situations should be seeded with:
a] Enhanced seed rate
b] Reduced seed rate
c] Widened row space
d] Widened plant space

218. Linseed can be used as ______crop.
a] Oilseed b] Fibre
c] Oilseed and fibre d] None of these

GK **Supreme Court upheld constitutionality of POTA.** GS

219. Linseed crop belongs to the family:
a] Compositeae b] Euphorbiaceae
c] Leguminoceae d] Lineaceae

220. Linseed has been recommended to be fertilized with ______ kg of P_2O_5 /ha to raise a good crop.
a] 10 b] 20
c] 30 d] 40

221. Linseed is a ______ season crop.
a] Warm b] Cool
c] Hot d] Dry

222. Low total iron content of soil is ______ the basis for plant deficiencies.
a] Commonly b] Often
c] Seldom d] Mostly

223. Maize can grow very well in the soils of pH ranging from :
a] 4.5 to 6.5 b] 6.0 to 8.5
c] 5.5 to 7.5 d] 7.5 to 8.5

224. Maize crop planted a spacing of 50 x 20 cm will have plant/ha
a] 50, 000 b] 1,00,000
c] 1,25,000 d] 1,50,000

225. Maize protein is deficient in:
a] Betadine
b] Tryptophane and lysine
c] Lysine
d] Tryptophane

226. Maize-potato-wheat-green gram is an example of :
a] Double cropping
b] Multistoried cropping
c] Relay cropping
d] Inter cropping

227. Major constituent of cotton fibre is:
a] Cellulose b] Lignin
c] Protein d] Fat

228. Major source of water used by the plants is:
a] Capillary water
b] Hygroscopic water
c] Gravitational water
d] Inter space water

229. Mull is :
a] Soil b] Mineral
c] Humas d] Manure

230. Maximum amount of fertilizers is used in :
a] Wheat b] Paddy
c] Potato d] Sugarcane

231. Maximum amount of nitrogen per unit wt. is in :
a] Groundnut cake b] Poultry manure
c] Compost d] Night soil

232. Maximum amount of P_2O_5 is in :
a] SSP b] DAP
c] DCP d] TSP

233. Maximum fixation of K occurs :
a] In 2:1 clay minerals b] In 1:1 clay minerals
c] In 2:2 clay minerals d] In 1:1 clay minerals

234. Measurement of irrigation water depends on :
a] Units of volume and water holding capacity of soil
b] Units of volume and units of soil
c] On tolerant crop thrive
d] None of the above

235. Mulch farming indicates :
a] Farming by traditional method
b] Farming by traditional and modern methods
c] Agriculture and dairy farming pursued together
d] None of the above

236. Mixed farming referred only to:
a] Growing of crop together by mixing the seeds
b] Growing of cash crops and food grains together
c] Crop production combined with dairy farming at farm
d] Growing of fruits and vegetable mixed

237. Molasses is a by-product of sugar industry contains about---% sugar.
a] 3 b] 6
c] 9 d] 11.5

238. Most of the annual rainfall in India occurs during the:
a] Pre-monsoon
b] South-east-monsoon
c] South-west monsoon
d] Month of December

239. Most of the herbicides embody ______ phytotoxicity at lower soil moisture content.
a] Higher b] Lower
c] Equal d] Unchanged

240. Most suitable temperature for growth of maize is:
a] 32^0 C b] 25^0 C
c] 37^0 C d] 15^0 C

241. Multistorey cropping system is practised in :
a] Jammu and Kashmir
b] Karnataka and Kerala
c] Bihar and MP
d] Rajasthan and Haryana

242. Mungbean cultured with the package of improved package of technological practices may yield______ q/ha.
a] 22-25 b] 12-15
c] 18-20 d] 8-10

243. Mustard crop planted at a spacing of 50x20 cm will have ______ plants/ha.
a] 75,000 b] 1,00,000
c] 1,25,00 d] 1,50,000

244. Mustard was planted at a plant to plant spacing of 12.5 cm with a population of 1,60,00 plants/ha, its row spacing will be:
a] 30 cm b] 40 cm
c] 50 cm d] 60 cm

GK GS
Permanent members in the United Nations Security Council are five.

245. Mycorrhiza is a symbiotic association between :
a] Bacteria and roots of higher plants
b] Algae and roots of orchids
c] Fungi and roots of higher plants
d] Protozoa and roots of higher plants

246. Name a cereal with maximum protein content :
a] Rice b] Sorghum
c] Maize d] Whole wheat

247. Name the acid forming fertilizer :
a] Urea
b] Sodium nitrate
c] Potassium-chloride
d] Calcium ammonium nitrate

248. Name the area where multiple crops are grown :
a] Crop catchment area
b] Inter crop area
c] Multiple plants area
d] All above

249. Name the crop which is used as pulse and oil purpose:
a] Blackgramb] Soyabean
c] Cowpea d] Arhar

250. Name the *kharif* season fodder crop:
a] Oats b] Lucerne
c] Sorghum d] Berseem

251. Name the mode of leaching :
a] Percolation b] Infiltration
c] Runoff water d] All the above

252. *Nicotiana rustica* has been derived as an amphidiploid from a cross between:
a] *Nicotiana tabacum* x *Nicotiana paniculata*
b] *Nicotiana paniculata* x *Nicotiana sylvestris*
c] *Nicotiana undulata* x *Nicotiana sylvestris*
d] *Nicotiana undulata* x *Nicotiana paniculata*

253. Nicotine content of the leaves of *Nicotiana rustica* ranges between ___________ per cent.
a] 2.0 and 5.5 b] 3.0 and 45
c] 3.5 and 8.0 d] 5.0 and 10.5

254. Nitrogen application of safflower influences:
a] Seed yield b] Seed composition
c] Both of these d] None of these

255. Nitrogen is taken by the plants in the form of :
a] Chloride b] Oxide
c] Nitrate d] None of the above

256. Normally recommended N + P_2O_5 + K_2O dose for soyabean is ___________ kg/ha, respectively.
a] 20-30 + 70-80 + 50-60
b] 20-30 + 100-120 + 50-60
c] 50-60 + 70-80 + 50-60
d] 50-60 + 70-80 + 80-100

257. Once the phenoxy herbicide penetrates the cuticle, it moves inside the plant:
a] Only apoplastically b] Only symplastically
c] By both mechanisms d] None of these

258. One acre is equivalent to :
a] 0.603 ha b] 0.040 ha
c] 0.404 ha d] 1.8 ha

259. One of the following herbicides is a selective herbicide :
a] Simazine b] Bisulphide
c] Pentachlorophenol d] Sodium chlorate

260. One of the following herbicides is a selective herbicide :
a] Atrazine b] Paraquat
c] Diquat d] Glyphosate

261. One of the following stages of rice is not critical in respect of irrigation:
a] Seedling b] Panicle initiation
c] Flowering d] Maturity

262. One quintal of sugarbeet contains ___________ of sugar.
a] 8 kg b] 10 kg
c] 12 kg d] 17 kg

263. One sq. km. consists of :
a] 10 ha b] 100 ha
c] 1000 ha d] 2000 ha

264. Origin of groundnut :
a] Africa b] India
c] China d] Brazil

265. Out of the following ___________ is commonly cultivated potato.
a] *Solanum demissum*
b] *Solanum stenotonum*
c] *Solanum tuberosum*
d] *Solanum andigenum*

266. Oxyfluorfen has been found to inhibit apical bud growth in young ___ plants to better bush formation.
a] Coffee b] Tea
c] Cacao d] Coconut

267. Percent protein in maize grain is about :
a] 2 b] 5
c] 10 d] 15

268. Percentage of kernal is groundnut with shell is :
a] 33% b] 50%
c] 70% d] 80%

269. Percentage of moisture held in soil at a tension of 15 atmospheres is referred to as :
a] Wilting point or wilting percentage
b] Wilting co-efficient of available water
c] Moisture capacity
d] None of the above

270. Percentage of nitrogen in castor cake is :
a] 2.0 b] 3.0
c] 4.0 d] 5.0

271. Permissible biuret level, in most of urea fertilizers is ___________ per cent.
a] 5 b] 4
c] 3 d] 2

GK GS
The Rig Vedic religion was primitive animism.

272. *Phalaris minor* weed can be controlled effectively by using _______ herbicide.
a] Isoproturon b] 2, 4-D
c] Fluchloralin d] Pendimethalin

273. *Phalaris minor* found with:
a] Dwarf wheat b] Tall wheat
c] Barley d] Oat

274. *Avena fatua* is a crop associated weed with :
a] Wheat b] Maize
c] Rice d] Cotton

275. Phosphatase enzymes play a major role in __________ of organic phosphates in soil.
a] Mineralization b] Immobilization
c] Mobilization d] None of these

276. Phosphorous content in Indian soil ranges from :
a] 0.03 - 0.06% b] 2.0 - 5.8%
c] 10-15% d] 8-10%

277. Photodecomposition plays an important role in the loss of urea herbicides from the soil surface in __________ areas.
a] Moist b] Wet
c] Dry d] None of these

278. Pigeonpea is commonly known as:
a] Redgram b] Greengram
c] Blackgramd] Bengalgram

279. Podsols are :
a] Acidic
b] Alkaline
c] Alkaline soil with oxidising agent
d] Marshy soils

280. Potassium is utilized more efficiently if applied:
a] Before/at planting b] Sidedressed
c] Top dressed d] None of these

281. Potassium salts are:
a] More mobile than nitrates
b] Less mobile than nitrates
c] Less mobile than phosphorus
d] As mobile as nitrates

282. Potato belongs to the family:
a] Solanaceae b] Compositeae
c] Leguminaceae d] Euphorbiaceae

283. Potato crop gives good response to liberal application of :
a] Nitrogen b] Phosphorus
c] Potassium d] Calcium

284. Potato tuber yield will be _______ q/ha, if crop was spaced at 60 x 25 cm, and average number of tubers/plant and average weight of tubers are 14 and 35 g, respectively.
a] 286.17 b] 317.29
c] 326.67 d] 356.87

285. Practicing agriculture without the use of chemicals and fertilizers so as to protect ecological balance is known as:
a] Inorganic farming
b] Natural farming
c] Extensive farming
d] Intensive farming

286. Projecting the current weather analysis into the future is know as:
a] Diagnosis b] Prognosis
c] Forecast d] Phenology

287. Propanil, a herbicide is readily absorbed by the:
a] Stem b] Root
c] Leaf d] None of these

288. Propanil is a _________ herbicide.
a] Pre-plant b] Pre-emergence
c] Post-emergence d] Soil-applicant

289. Proper _____________ nourishment for potato and sugar beet *etc.*, is essential since these are known to have its high requirement.
a] Sulphur b] Manganese
c] Zinc d] Calcium

290. Pulses fit well in cropping systems as they are :
a] Short duration crops
b] Long duration crops
c] Disease resistant crops
d] Moisture stress resistant crops

291. Quadricarpellatum capsule of sesamum contains _______________ carpels.
a] Two b] Three
c] Four d] Six

292. Rainfall at the time of maturity of tobacco is undesirable as it causes:
a] Lodging
b] Leaves to get rotten
c] Leaves to get shattered
d] Washing down of gums and resins from leaf

293. Raising a crop to supplement yield or monetary returns from the main crop is known as :
a] Alley cropping
b] Opportunity cropping
c] Augmentation cropping
d] Augmentation cropping

294. Raising of crop with least tillage operations is called:
a] Minimum tillage b] Zero tillage
c] No tillage d] All of above

295. Recommended approximate seed rate of bajra is :
a] 5-10 kg/ha b] 15-20 kg/ha
c] 40-50 kg/ha d] 125 kg/ha

296. Recommended dose of N + P_2O_5 + K_2O kg/ha is _______ respectively for sugarbeet.
a] 120 + 80 + 80 b] 90 + 60 + 40
c] 120 + 60 + 40 d] 90 + 80 + 60

GK 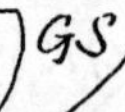

Cheese provide the nutrient for the growth of new tissues in the human body.

297. Recommended row spacing for summer season, urdbean is ____________ cm.
a] 10-15 b] 15-20
c] 20-25 d] 30-35

298. Recovery of sugar in sugarcane crop in India is :
a] 6-8% b] 8-10%
c] 10-15% d] 4-6%

299. Relay cropping is defined as:
a] A system of cropping where one crop hands over the land to the next crop in quick succession
b] A system of cropping where two or three crops are grown annually
c] A system of cropping where two or three crops are grown together by mixing the seeds
d] A system of cropping where an associate crop is grown in between the rows of principal crop.

300. Removal of a uniform thin layer of soil by the action of water is refered as :
a] Splash erosion b] Sheet erosion
c] Rill erosion d] Gully erosion

301. Repeated use of Bordeaux mixture, sewage sludge, pig and poultry manures and/or mine wastes may cause __________ toxicities.
a] Iron b] Aluminium
c] Zinc d] Copper

302. Retting is a process by which:
a] The poor fibres stalk is sorted-out
b] The fibres in the bark get loosened and separated from woody stalk
c] The good quality fibres are sorted out
d] Fibres are differentiated into various retted groups.

303. Rock phosphate is treated with _______ to break the apatite bond and to render the contained phosphate more soluble.
a] Acid alone b] Alkali alone
c] Heat alone d] Either acid or heat

304. Rothamsted Agricultural Experiment Station, England was found by:
a] J.B. Lawes
b] J.H. Gilbert
c] J.B. Lawes & J.H. Gilbert
d] Lord Hastings

305. Safflower as an oilseed crop was introduced into India in:
a] 1965 b] 1969
c] 1972 d] 1985

306. Safflower can be grown as an intercrop with:
a] Wheat b] Urdbean
c] Maize d] Sorghum

307. Safflower is a ________________ season crop.
a] Dry b] Summer
c] Arid d] Cool

Knowledge World!
Highest motorable road
The road at Khardungla in the Leh-Manali sector was constructed by the a Border Roads Organisation at an altitude of 5,682 m (18,383 ft).

308. Safflower rust can be controlled by:
a] Spray of Zineb
b] Treatment of seed with thiram
c] Destroying the infected material
d] Following the 3 to 4 years crop rotation.

309. Sannhemp belongs to the family:
a] Leguminoseae b] Compositeae
c] Liniaceae d] Tiliaceae

310. Sannhemp should be planted at a distance of ___________ cm to raise fibre crop.
a] 20 x 7-10 b] 25 x 5-7
c] 30 x 7-10 d] 30 x 5-7

311. Seed Plot Technique of potato is used to produce :
a] Insect-free seeds b] Nematode free seeds
c] Virus free seeds d] Big size tubers

312. Seed potato from hilly regions of India is preferred because :
a] It is cheaper
b] Good varieties are grown in those regions
c] It has better germination percentage
d] It is disease free

313. Seed rate for potato is :
a] 5-10 quintals/ha b] 10-15 quintals/ha
c] 20-25 quintals/ha d] 25-30 quintals/ha

314. Seed rate of lentil crop is :
a] 15-20 kg/ha b] 35-40 kg/ha
c] 40-50 kg/ha d] None of above

315. Seed rate of sorghum is ___________ kg/ha.
a] 12-15 b] 10-12
c] 8-10 d] 15-20

316. Seed treatment is done to :
a] Control the disease
b] To kill pests
c] To control Weeds
d] To enhance germination

317. Sesamum belongs to the family:
a] Leguminoseae b] Papilionaceae
c] Pedaliaceae d] Chinopodeaceae

318. Sesamum cake contains _______ per cent of N + P_2O_5 + K_2O respectively.
a] 6.0-6.2 + 2.0-2.2 + 1.0-1.2
b] 2.0-2.2 + 6.0-6.2 + 1.0-1.2
c] 6.0-6.2 + 1.0-1.2 + 2.0-2.2
d] 2.0-2.2 + 1.0-1.2 + 6.0-6.2

Fundamental Rights can be suspended in emergency.

319. Silicon is the ____________ most abundant element in the earth's crust.
a] Third b] Fourth
c] Sixth d] Second

320. Single super phosphate is a :
a] Citrate soluble fertilizer
b] Water soluble fertilizer
c] Citrate insoluble fertilizer
d] None of the above

321. Smoke raising prevents :
a] Radiation frost b] Advection frost
c] None of these d] Both of these

322. Soil borne diseases are found more in :
a] Cereal crops b] Fibre crops
c] Oil seed crops d] Pulse crops

323. Sorghum is considered as "Camel crop" because of:
a] Deep root system
b] Resistance to drought
c] Shallow root system
d] Nutrient exhaustiveness

324. Sorghum was planted at a plant spacing of 12.5 cm with a population of 1,60,000 plants/ha, its row spacing will be:
a] 30 cm b] 40 cm
c] 35 cm d] 50 cm

325. Sow summer mungbean in ________ for reasonably good productivity.
a] Middle of April
b] Middle of March to middle of April
c] Middle of March
d] Middle of February to middle of March

326. Sowing of a crop at optimum date is an example of :
a] Monetary input
b] Non-monetary input
c] Monetary and non-monetary input
d] None of the above

327. Soyabean besides good quantity of protein also contains about per cent of oil:
a] 40 b] 30
c] 20 d] 10

328. Soyabean crop should be free from weed competition during first ___ days for its good productivity.
a] 10-20 b] 11-12
c] 13-14 d] 14-15

329. Soyabean is a :
a] Short day plant
b] Long day plant
c] Day-nautral plant
d] All of the above

330. Soyabean is classified as :
a] Oilseed crop b] Legume
c] Pulse d] None of above

331. Soyabean planting in a field where wheat crop remains there is know as :
a] Mixed cropping b] Relay cropping
c] Multiple cropping d] All the above

332. Specialised farming is one in which:
a] A single crop contributes less than 50% to total production
b] A single crop contributes more than 50% to total production
c] More than three crops are taken on same piece of land per unit time
d] All the above

333. Split doses of nitrogen is used in the soil :
a] Saline b] Alkaline
c] Sandy d] Clay

334. Spreading varieties of groundnut are preferred in areas receiving late rains in the *kharif* season, because:
a] They are of longer duration
b] Their wate requierments are high
c] Their seed is dormant
d] Their seed is bold

335. Striga is crop associated weed with
a] Pearlmillet b] Mustard
c] Maize d] Gram

336. The per capita consumption of pulses in India is ______ g/day.
a] 63 b] 50
c] 17 d] 36

337. Sufficient soil moisture at ________ stage is essential for optimum productivity of blackgram.
a] Flowering and pod development
b] Branching and pod development
c] Branching and flowering
d] Pod development and maturity

338. Sugar recovery from sugarbeet is approximately :
a] 5-8% b] 15-18%
c] 25-28% d] 35-38%

339. Sugarbeet is a :
a] Short day plant b] Day neutral plant
c] Long day plant d] All the above

340. Sugarbeet is a ________________ plant.
a] Annual b] Biennial
c] Perennial d] None of these

341. Sugarbeet is a:
a] Vegetable crop b] Forage crop
c] Tuber crop d] Sugar crop

342. Sugarbeet is grown during ______________ season.
a] *Rabi* b] *Kharif*
c] *Zaid* d] *Rabi and Kharif*

343. Sugarbeet is produced maximum in:
a] Africa b] Europe
c] Asia d] None of these

GK **Fred Hale Sr. was world's oldest man (114 year old).** GS

344. Sugarcane belongs to the family:
a] Gramineae b] Cruciferеae
c] Leguminaceae d] Compositeae

345. Sugarcane cut for seed-setts should essentially have:
a] One bud b] Two buds
c] Three buds d] Four buds

346. Sugarcane is a ____________ plant.
a] Annual b] Perennial
c] Triennial d] None of these

347. Sugarcane setts are treated with a chemical before plantation :
a] Agallol b] Zineb
c] Thiram d] Aldrin

348. Sunflower is a ____________ crop.
a] Self-pollinated
b] Cross-pollinated
c] Self and cross-pollinated
d] Parthenocarpous

349. Superphosphate containing ______ selenium may provide enough of it to the plants in selenium deficient areas.
a] 10 ppm or more b] 20 ppm or more
c] 30 ppm or more d] 40 ppm or more

350. Temperature requirement for optimum growth of sorghum ranges from:
a] 16-20°C b] 20-25°C
c] 30-35°C d] 26-30°C

351. Ten square km. will have an area of how many hectare area ?
a] 100 b] 1,000
c] 10,000 d] 1,00,000

352. The _________ acids which undergo β-oxidation are MCPB, 2, 4-DB, and 2, 4, 5-TB.
a] Phenoxyacetic b] Phenoxyoxalic
c] Phenoxymaleic d] Phenoxybutyric

353. The _________ considered to be a progenitor of soyabean.
a] *Glycine max*
b] *Glycine sinensis*
c] *Glycine ussuriensis*
d] *Glycine usitattisimum*

354. The ___________ structure of the herbicide also has a great influence on its volatility through adsorption to soil.
a] Biological b] Chemical
c] Physical d] Biophysical

355. The 1000 grain wt. of Arhar is about:
a] 52 gm b] 29 gm
c] 60 gm d] 72 gm

356. The 2n chromosome number in Rugosa is:
a] 20 b] 24
c] 32 d] 36

357. The accessibility and utilization of soil applied copper can be enhanced by:
a] Reducing particle size
b] Reducing number of granules
c] Reducing droplets per unit soil volume
d] None of these

358. The active principle causing "Sorghum Poisoning" is:
a] Aspirin b] Caffein
c] Dhurrin d] Capsicin

359. The aim of topping and desuckering in tobacco is :
a] To reduce the plant height
b] To divert energy and nutrients from flower head to leaves
c] To increase the horizontal branching of plant
d] To protect the plants against lodging

360. The annual broad leaf and grassy weeds in sugarbeet can be controlled effectively by use of __________ kg a.i./ha.
a] Basalin @ 1.0 b] 2,4-D @ 1.5
c] Butachlor @ 2.0 d] Pyramin @ 3.0

361. The antiquality constituent in young seedings of forage sorghum is :
a] Coumarins b] Terpenes
c] Acetic acid d] Hydrocyanic acid

362. The application rates of nitrogen and phosphorus are 64 and 46 kg/ha to an experimental crop of wheat as basal, for that the quantity of urea and diammonium phosphate will be:
a] 50 kg urea, 100 kg DAP
b] 75 kg urea, 75 kg DAP
c] 75 kg urea, 100 kg DAP
d] 100 kg urea, 100 kg MoP

363. The area needed to raise rice seedlings by dapog method sufficient for transplanting one hectare area is:
a] 250-300 sqm b] 25-30 sqm
c] 40-50 sqm d] 400-500 sqm

364. The area where drought has occurred in _________ per cent of years during the period are considered drought areas.
a] 10 b] 20
c] 30 d] 40

365. The area where drought has occurred in more than _________ per cent of years are called chronic drought areas.
a] 30 b] 40
c] 50 d] 60

366. The *arhar* (pigeonpea) grows well in __________ climate.
a] Tropical and subtropical
b] Temperate
c] Humid and temperate
d] Sub-humid and temperate

GK **The language Hindi is spoken by the second largest number of the world.** GS

367. The autumn planting of sugarcane is most successful when planted in the month of:
a] September b] October
c] November d] February

368. The average rainfall of India is :
a] 112 mm b] 112 cm
c] 112 inches d] 112 ft.

369. The average yield of sugarcane in north India is low as compared to that in the southern parts mainly due to:
a] Climatic conditions b] Varietal status
c] Soil fertility status d] Level of management

370. The bacteria responsible for nitrogen fixation in soyabean is:
a] *Rhizobium japonicum*
b] *Rhizobium phaseoli*
c] *Rhizobium leguminosarum*
d] *Rhizobium glycicum*

371. The basic principal of taking crop rotation is:
a] To get higher crop production
b] To get higher returns per unit area of the soil
c] To maintain the fertility status of the soils
d] To keep the weeds under control

372. The best green manuring crop is :
a] Guar b] Sunhemp
c] Dhaincha d] None of these

373. The best observed pH for jute cultivation is:
a] 5.0-6.5 b] 6.0-7.5
c] 7.0-8.5 d] 8.0-9.5

374. The best soil for potato cultivation is :
a] Sandy loam b] Loamy sand
c] Clay loam d] Silty loam

375. The best soil for sugarbeet cultivation is :
a] Sandy loam to loam b] Silty clay loam
c] Clay loam d] Loamy sand

376. The best time for sowing of chickpea in northern India is:
a] Second fortnight of October
b] First fortnight of October
c] First fortnight of November
d] Second fortnight of September

377. The best time for sowing of toria is :
a] 15-30 September b] 15-20 September
c] 15-25 September d] All of the above

378. The black cotton soil in India are otherwise called :
a] Khadar b] Regur
c] Bangar d] Terai

379. The carbohydrate content in potato:
a] 5-6% b] 25-28%
c] 18-23% d] 28-30%

380. The causal agent of anthracnose disease in sorghum is:
a] *Puccinia purpurea*
b] *Helminthosporium turcicum*
c] *Colletotrichum graminicola*
d] *Gleourcospora sorghi*

381. The causal organism of brown rust in wheat is:
a] *Puccinia recondita tritici*
b] *Puccinia streiformis*
c] *Puccinia graminis tritici*
d] *Ustilago nuda tritici*

382. The Central Rice Research Institute is located at:
a] Cuttak b] Coimbtore
c] Hydrabad d] Bangalore

383. The Central Arid Zone Research Institute, Jodhpur (Rajasthan, India) was established in:
a] 1952 b] 1957
c] 1959 d] 1961

384. The centre of origin of wheat is:
a] South-western Asia b] South-eastern Asia
c] India d] China

385. The chickpea wilt is caused by:
a] *Fusarium orthocerus*
b] *Ascochyta rabi*
c] *Botrylis cinerea*
d] *Sclerotinia sclerotiorum*

386. The climate of India can be best described as :
a] Desert type
b] Mediterranean
c] Subtropical monsoon type
d] Tropical monsoon type

387. The concentrations of __________ ppm of SO_4^{2-} ion in the soil or solution cultures have proved adequate for the growth of many plant species.
a] 1-2 b] 3-5
c] 6-8 d] 9-11

388. The critical period of weed control in upland rice varies between ___________ days after sowing.
a] 15-20 b] 20-25
c] 25-30 d] 30-35

389. Antinutritional factor in Chickling vetch is :
a] Phytic acid b] Gossypol
c] Polyphenolics d] BOAA

390. The critical stages for irrigation in chickpea are:
a] Branching and pre-flowering
b] Branching and pod development
c] Pre-flowering and pod development
d] Branching, pre-flowering and pod development

391. The crop also sown for green manure is :
a] Maize b] Green gram
c] Black gram d] Bengal gram

392. The crop of _____________ is more liable to suffer from frost and cold.
a] *Toria* b] *Pahari rai*
c] *Sarson* d] *Banarsi rai*

393. The crop of sunflower can be effectively grown in ______season
a] Rabi and Zaid b] Rabi, Kharif and Zaid
c] Kharif and Zaid d] Kharif

394. The crop that can also be used as wind breaker :
a] Red gram b] Lentil
c] Soyabean d] Black gram

GK **66% earthquakes of world are recorded in Circum-pacific zone.** GS

395. The crop which requires maximum amount of nitrogen :
a] Wheat b] Barley
c] Paddy d] Sugarcane

396. The crops of group rapeseed and mustard (except taramira) belongs to the family:
a] Cruciferae b] Leguminoseae
c] Composite d] Euphorbiaceae

397. The crops which are grown during summer are known as:
a] Rabi b] Kharif
c] Zaid d] None of the above

398. The cultivation of which crop requires water logging?
a] Sugarcane b] Paddy
c] Cotton d] Maize

399. The dapog method of raising rice nursery was introduced in India from:
a] Philippinesb] Bangladesh
c] Burma d] Japan

400. The deficiency symptoms on lower leaves are due to:
a] N_2 b] P_2O_5
c] K_2O d] Zn

401. The dent corn is also known as:
a] *Zea mays indentata*
b] *Zea mays indurata*
c] *Zea mays tunicata*
d] *Zea mays everta*

402. The depth of sowing for sugarbeet should be :
a] 2-3 cm
b] 7-8 cm
c] 5-6 cm
d] None of the above

403. The dormancy of potato tuber cannot be broken with:
a] Thiourea
b] Ethylene chloropydrin
c] Potassium thiocynate
d] Ethylene thiocynate

404. The dwarf wheat varieties should be:
a] Sown at the same depth at which the tall wheat varieties are sown.
b] Sown at a depth deeper than for tall wheat varieties
c] Sown at a depth shallower than that of tall wheat varieties
d] Broadcas and incorporated in the top soil

405. The essentiality of copper for higher plants was first recognised in:
a] 1911 b] 1921
c] 1931 d] 1941

406. The family of sugarbeet is:
a] Chenopodiaceae b] Crucifereae
c] Euphorbiaceae d] Compositeae

Quiz Kit!
Longest platform
The platform at Kharagpur in West Bengal is 833 m (2,733 ft) long.

407. The fertilizer requirement in lentil is nearly ________ than that of 'arhar', 'moong' and 'urd'.
a] Equal b] Half
c] One of half d] Double

408. The fertilizer suitable for spraying to increase production of paddy :
a] Ammonium sulphate
b] Calcium ammonium nitrate
c] Urea
d] Ammonium phosphate

409. The fertilizer use recommendation for sunflower is ________ kg/ha of N + P_2O_5 + K_2O, respectively._
a] 30 + 60 + 40 b] 30 + 30 + 20
c] 70 + 40 + 30 d] 70 + 60 + 40

410. The fibre length of cotton is :
a] 1-5 cm b] 10-50 cm
c] 12-18 cm d] 20 cm

411. The first irrigation in wheat is normally recommended at :
a] Tillering stage
b] Crown root initiation stage
c] Flowering stage
d] Seedling stage

412. The fodder which has maximum protein content (on dry wt. basis) :
a] Lucerne b] Cowpea
c] Berseem d] Oat

413. The fruit of linseed is known as:
a] Pod b] Canum
c] Seedball d] Dutum

414. The fruit of rapeseed and mustard is known as :
a] Pod b] Siliqua
c] Grain d] Caryopsis

415. The grain legume which is also grown as an oilseed is:
a] Chickpea b] Urdbean
c] Lentil d] Soyabean

416. The gynophore of groundnut is commonly referred to as the:
a] Perianth b] Peg
c] Ovate d] Monadelph

417. The highest percentage content of nitrogen among the commonly used nitrogenous fertilizers is embodied by:
a] Urea b] Ammonium sulphate
c] Ammonium nitrate d] Anhydrous ammonia

GK **The later Vedic Aryans used painted grey ware pottery.** GS

418. The hulling percentage of rice is :
a] 20% b] 50%
c] 60% d] 80%

419. The humus myth was effectively deposed by ______ by presentation of a paper at a prominent scientific meeting.
a] Justus Senebier b] Jeans Young
c] Justus Von Liebig d] Justus Billright

420. The ideal moisture content in soyabean seed at the time of threshing has been estimated at ______ per cent.
a] 10-11 b] 11-12
c] 13-14 d] 14-15

421. The identified critical stage of irrigation in barley :
a] Tillering b] Flowering
c] Both of these d] None of these

422. The identified critical stage of irrigation in tobacco is:
a] Tillering b] Flowering
c] Topping d] Maturity

423. The identified critical stages for irrigation of safflower are:
a] Rosette and stem elongation
b] Rosette and flowering
c] Flowering and grain filling
d] Stem elongation and flowering

424. The inflorescence of sugarcane is called :
a] Spike b] Tassel
c] Panicle d] Arrow

425. The institute responsible for evolving new sugarcane varieties is:
a] Indian Institute for Sugarcane Research, Lucknow
b] Sugarcane Breeding Institute, Coimbatore
c] Sugarcane Improvement Research Station, Coimbatore
d] None of these

426. The intensity of crop rotation may be indicated by :
a] No. of crops x 100/No. of years
b] No. of crops x No. of years /100
c] No. of crops x 100
d] None of the above

427. The interculture operation in standing broadcast rice crop is commonly known as :
a] Thinning b] Puddling
c] Rotation d] Bushening

428. The International Centre for Agricultural Research in Dry Areas (ICARDA) was established in:
a] 1957 b] 1967
c] 1977 d] 1987

429. The isolation distance required to produce the certified seed of chickpea, lentil, mungbean and urdbean is ______ metres.
a] 25 b] 15
c] 10 d] 5

430. The latest botanical name of green gram is :
a] *Vigna radiata* b] *Vigna mungo*
c] *Phaseolus mungo* d] *Phaseolus vulgaris*

431. The leaf area per unit land area is known as:
a] LAR b] LAD
c] LAI d] LER

432. The losses of ammonia usually become greater with:
a] Increase in soil pH
b] Decrease in soil pH
c] Decrease in calcium
d] Decrease in temperature

433. The maize crop should be planted at the distance of:
a] 30 x 20-25 cm b] 60 x 20-25 cm
c] 30 x 10-15 cm d] 60 x 10-15 cm

434. The major part of agricultural land in India is under :
a] Cash crop b] Food crop
c] Oil seed d] Plantation

435. The major part of nitrogen uptake by the maize is over by the:
a] Knee high stage b] Tasselling stage
c] Silking stage d] Grain filling stage

436. The maximum carbon dioxide concentration at normal atmospheric levels occur in relatively low light intensities of ______ footcandles for some plants.
a] 2500-5000 b] 6000-7500
c] 1000-2000 d] 2150-2450

437. The maximum phosphorus availability in most of the soils is in the pH range:
a] 4.0-4.5 b] 5.0-5.5
c] 6.0-5.5 d] 7.0-7.5

438. The mineral calcite develops a negative charge due to ______ tendency for Ca^{2+} ions than CO_3^{2-} ions to go into solution.
a] Lower b] Greater
c] Equal d] None of these

439. The mineralization of organic phosphorus has been studied in relation to ______ ratio.
a] C/N/S b] C/N/K
c] C/N/P d] C/N/Ca

440. The moisture content at which the wilting is complete and the plant die is called :
a] Permanent wilting
b] Wilting co-efficient
c] Wilting range
d] Permanent wilting point

441. The moisture, which is present in the form of a continuous film outside hygroscopic water around soil particles and microspore spaces is called:
a] Available moisture b] Interspace water
c] Capillary water d] None of these

442. The most common cereal millet crop grown in arid climate is:
a] Sorghum b] Maize
c] Pearlmillet d] Paddy

GK GS

Molten rock materials are called magma.

443. The most common means of expressing soil water contentent is the _________ of water associated with their specified quantity of soil solids:
a] Volume b] Weight
c] Vapour d] None of these

444. The most critical stage for irrigation in potato is :
a] Stolon formation b] Tuberization
c] 25% tuber formation d] None of the above

445. The most critical stage of rice for moisture stress is:
a] Germination-tillering
b] Tillering-flowering
c] Panicle initiation-flowering
d] Flowering-ripening

446. The most destructive disease of sugarcane is:
a] Red rot b] Wilt
c] Smut d] Albino

447. The most effective cropping system for returning mineral elements to the soil is :
a] Monocropping b] Crop rotation
c] Double cropping d] Relay cropping

448. The most important crop grown on conserved moisture in our country is:
a] Gram b] Wheat
c] Mustard d] Barley

449. The most popular extra-early maturing variety of pigeonpea is :
a] UPAS-120 b] Gulbarga
c] Type 7 d] Type 17

450. The most suitable soil for groundnut is:
a] Sandy and sandy loam
b] Clayey loam
c] Loamy clay
d] Silty loam and clay loam

451. The most suitable soil for rice cultivation is :
a] Silty loam b] Clay or clay loam
c] Sandy loam d] Loam sand

452. The most suitable temperature for the growth of maize crop is :
a] 32°C b] 25°C
c] 21°C d] 38°C

453. The most suitable variety of wheat for late sown condition is:
a] Kalyansona b] Sonalika
c] Raj - 3077 d] WH- 147

454. The most suitable variety of wheat for rainfed conditions is:
a] Sonalika b] Kalyansona
c] Mukta d] Lok-1

455. The movement of nutrients, ions by mass flow is reduced at low temperatures because of:
a] Low respiration b] Low photosynthesis
c] Low transpiration d] None of these

456. The mustard crop is regarded ready for harvest when its pods turn _________ in colour.
a] Greenish-yellow b] Greenish-black
c] Yellowish-green d] Yellowish-brown

457. The nicotine content in cigarette tobacco is :
a] 1-2% b] 3-4%
c] 5-6% d] 6.5%

458. The number of known species in genus *Gossypium* have been estimated at:
a] 10 b] 15
c] 20 d] 25

459. The number of plants per hectare to be observed in taking one count in seed production of pearlmillet, sorghum, rice-paddies and wheat is _________ as recommended.
a] 500 b] 1000
c] 1500 d] 2000

460. The number of setts needed for planting one hectare of sugarcane is:
a] 25,000-30,000 b] 35,000-40,000
c] 45,000-50,000 d] 55,000-60,000

461. The observed best temperature for retting is _________ °C.
a] 14 b] 24
c] 34 d] 38

462. The optimum cardinal temperature points for germination of rice seeds are
a] 18-20 °C b] 20-22 °C
c] 30-32 °C d] 37-39 °C

463. The optimum period of sowing for *Capsularis* jute is:
a] February-March b] March-April
c] April-May d] May-June

464. The optimum period of sowing for *Olitorius* jute is:
a] February-March b] March-April
c] April-May d] May-June

465. The optimum pH range for sunflower is:
a] 5.5-7.5 b] 6.5-8.5
c] 7.5-9.5 d] 7.0-9.0

466. The optimum plant population for cotton has been estimated at _________ plants/ha.
a] 25,000-50,000 b] 30,000-60,000
c] 50,000-80,000 d] 80,000-1,00,000

467. The optimum plant population for mustard is around:
a] One lacs b] Two lacs
c] Three lacsd] Four lacs

468. The optimum population of hybrid cotton varieties is around _________ plants/ha.
a] 10,000 b] 20,000
c] 30,000 d] 40,000

469. Machete is effective against :
a] Annual grass b] Broad-leaved weed
c] Both a & b d] Narrow-leaved weed

470. The optimum seed rate of soyabean for normal planting is _________ kg/ha.
a] 50-60 b] 60-70
c] 70-80 d] 80-90

GK GS

Amount of nitrogen in air does not change in the process of respiration.

471. **The optimum sowing time for lentil in N.W. India is:**
a] Middle of October
b] Last week of October
c] First week of October
d] Last week of September

472. **The optimum sowing time of cotton in northern regions of India is:**
a] First fortnight of May
b] Second fortnight of May
c] First fortnight of June
d] Second fortnight of June

473. **The optimum stage of rice seedling for transplanting under irrigated coundition during *kharif* season is:**
a] 2 leaf stage b] 4 leaf stage
c] 6 leaf stage d] 8 leaf stage

474. **The optimum temperature for seed germination in *kharif* season is :**
a] 10-15 °C b] 20-25 °C
c] 35-40 °C d] 25-35 °C

475. **The optimum temperature for seed germination in *rabi* season is:**
a] 15-20 °C b] 5-10 °C
c] 20-25 °C d] None of the above

476. **The optimum temperature for the germination of soyabean is :**
a] 11°C b] 20°C
c] 15°C d] 22°C

477. **The optimum temperature range for sowing of wheat is:**
a] 10-15^0 C b] 15-20^0 C
c] 20-25^0 C d] 25-30^0 C

478. **The optimum time for *kharif* soybean planting is:**
a] First fortnight of June
b] First fortnight of July
c] Second fortnight of June
d] Second fortnight of July

479. **The optimum time for planting of castor varies between:**
a] 20th May and 5th June
b] 5th June and 20th June
c] 20th June and 5th July
d] 5th July and 20th June

480. **The optimum time for sowing nursery of tobacco is:**
a] First fortnight of August
b] Second fortnight of August
c] First fortnight of September
d] Second fortnight of September

481. **The optimum time of sowing of sunflower during *kharif* in India is:**
a] First fortnight of June
b] Second fortnight of June
c] First fortnight of July
d] Second fortnight of July

482. **The optimum time of sowing of *toria* in N. India varies between:**
a] First and second week of September
b] First and second week of October
c] Mid and last week of October
d] Mid and last week of September

483. **The organisms involved in herbicide detoxification include:**
a] Earthworms b] Nematodes
c] Algae d] None of these

484. **The origin place of potato is :**
a] North America b] Former U.S.S.R.
c] Afghanistan d] South America

485. **The pH range for successful cultivation of maize ranges from:**
a] 6.5-8.5 b] 7.5-8.5
c] 7.5-9.5 d] 5.5-7.5

486. **The pH range suitable for soyabean cultivation is:**
a] 5.0-6.0 b] 7.0-8.5
c] 5.5-6.5 d] 6.0-7.5

487. **The *Phalaris minor* can be controlled by using the weedicide called Isoproturon at __________ kg a.i./ ha depending upon soil type.**
a] 0.1-1.00 b] 0.75-1.25
c] 1-2 d] 2-3

488. **The place of origin of blackgram is:**
a] Burma b] India
c] Bangladesh d] Ceylon

489. **The planting of sugarcane by trench method:**
a] Improves germination
b] Reduces lodging
c] Improves tillering
d] Saves seeds setts

490. **The polish percentage in paddy is:**
a] 2% b] 4%
c] 6% d] 8%

491. **The practice of growing four crops in a year from same piece of land in such a way that one crop is sown immediately after or just before the harvest of the previous crop is called as:**
a] Double cropping
b] Relay cropping
c] Intensive cropping
d] Multiple cropping

492. **The presence of ectotrophic mycorrhiza enhances the uptake of nutrients in general, but particularly of :**
a] Molybdenum b] Calcium
c] Magnesium d] Phosphorus

493. **The protein content in linseed cake has been estimated at __________ per cent.**
a] 12 b] 24
c] 36 d] 48

GK GS
The northern kings were called Virat.

494. The protein content of soyabean is about __________ per cent.
a] 40 b] 20
c] 30 d] 10

495. The protein percentage in lentil has been estimated at:
a] 25 b] 20
c] 30 d] 35

496. The quantity of nutrients that can come in direct contact with the plant roots is the amount in a volume of soil __________ the volume of roots.
a] Equal to b] Less than
c] More thand] No relation with

497. The quantity of solid phase phosphorus that acts are reserve is often referred to as:
a] Intensity factor b] Quantity factor
c] Quality factor d] None of these

498. The quantity of urea needed for supplying 69 kg nitrogen to sorghum crop will be:
a] 100 kg b] 125 kg
c] 150 kg d] 175 kg

499. The rapid increase in denitrification at the elevated soil temperatures suggests that ____ microorganisms play a major role in denitrification.
a] Chemophilic b] Photophilic
c] Thermophilic d] Hydrophilic

500. The rate of application of atrazine in maize under heavy soils varies from __________ kg a.i./ha.
a] 0.75-1.00 b] 1.25-1.50
c] 1.7.5-2.00 d] 1.00-1.25

501. The ratio of cotton seed to lint is:
a] 1:2 b] 1:3
c] 2:1 d] 4:1

502. The ratio of grain and straw in paddy is :
a] 1:2 b] 2:2
c] 2:1 d] 3:1

503. The recommended dose of fertilizer for urdbean is __________ N + P_2O_5+ K_2O kg/ha, respectively.
a] 15-20 + 50-60 + 30-40
b] 15-20 + 20-30 + 8-10
c] 40-80 + 50-60 + 30-40
d] 15-20 + 0 + 30-40

504. The recommended dose of NPK in rice crop under flood condition is:
a] 120:60:40 b] 100:80:40
c] 80:40:40 d] 30:20:20

505. The recommended herbicide for weed control in chickpea is:
a] Basalin b] Butachlor
c] Alachlor d] 2, 4-D

506. The recommended isolation distance (metres) for foundation seed production of radish, turnip and cauliflower is :
a] 800 b] 1200
c] 1600 d] 2000

Did You Know ?

First electric train

Bombay was the first city to use electric trains. The first electric train ran from Victoria Terminus to Kurla on Feb. 3, 1925 via the harbour line.

507. The recommended optimum depth of sowing of castor is __________ cm.
a] 4 b] 6
c] 8 d] 10

508. The recommended optimum row spacing for planting of field pea is __________ cm.
a] 30 b] 20
c] 15 d] 25

509. The recommended optimum seed rate for lentil, in general, varies from __________ kg/ha.
a] 20-30 b] 40-50
c] 30-40 d] 10-20

510. The recommended optimum sowing time of *arhar* (pigeonpea) is:
a] Second fortnight of July
b] First fortnight of July
c] Second fortnight of July
d] First fortnight of June

511. The recommended planting geometry for sesame is __________ cm.
a] 25 x 75 b] 30 x 15
c] 30 x 10 d] 45 x 15

512. The recommended row and plant spacing for *kharif* cowpea is __________ cm.
a] 30-45 x 8-10 b] 30-45 x 4-6
c] 25-30 x 8-10 d] 25-30 x 4-6

513. The recommended row spacing for spreading type of groundnut is __________ cm.
a] 25-30 b] 35-50
c] 45-60 d] 55-70

514. The recommended row spacing normally maintained for safflower planting is __________ cm.
a] 25 b] 35
c] 45 d] 55

515. The recommended row to row spacing for bunch type of groundnut is __________ cm.
a] 20-30 b] 30-40
c] 40-50 d] 45-60

516. The recommended row to row spacing for linseed planting is __________ cm.
a] 20 b] 30
c] 40 d] 50

517. The recommended seed rate (kg/ha) of castor varies from:
a] 10 to12 b] 14 to 16
c] 16 to 18 d] 18 to 20

GK GS

Nilgiri mountain of India is an example of block mountain.

518. The recommended seed rate for pigeonpea is_______ kg/ha
a] 8-10 b] 12-15
c] 15-18 d] 20-25

519. The recommended seed rate of sunflower varies from ___________ kg/ha.
a] 6-8 b] 8-10
c] 10-12 d] 12-14

520. The recommended spacings for sugarbeet planting are _________ cm.
a] 30 x 15 b] 50 x 75
c] 30 x 25 d] 50 x 25

521. The recovery of 'Gur' from cane is about :
a] 15% b] 9-10%
c] 11-12% d] 18%

522. The roots emerging from a sugarcane sett are known as:
a] Sett roots b] Shoot roots
c] Primary roots d] Secondary roots

523. The sampling lot in agricultural seed of large size should not exceed __________ kg except that of maize (where 40,000 kg is allowed).
a] 5,000 b] 10,000
c] 15,000 d] 20,000

524. The scientific name of early maturing bold and black seeded urdbean varieties is:
a] *Vigna mungo* var. *viridis*
b] *Vigna mungo* var. *niger*
c] *Vigna mungo var. macropode*
d] *Vigna mungo* var. *micropode*

525. The scientific name of erect type of groundnut is:
a] *Arachis hypogaea subspecies fastigiata*
b] *Arachis hypogaea* subspecies *procumbens*
c] *Arachis hypogaea* subspecies oleifera
d] *Arachis hypogaea* subspecies *gigantea*

526. The scientific name of field pea is:
a] *Pisum sativum var. arvense*
b] *Pisum sativum var. hortense*
c] *Pisum sativum var. leguminosarum*
d] *Pisum sativum var. esculentum*

527. The second most important Indian pulse crop is:
a] Chickpea b] Pigeonpea
c] Cowpea d] Field pea

528. The seed ___________ kg/ha is required for sugarbeet planting.
a] 6-8 b] 8-10
c] 10-12 d] 12-14

529. The seed normally required for planting of sesamum is _________ kg/ha.
a] 2-3 b] 3-4
c] 4-5 d] 5-6

530. The seed of safflower should be treated with Agrosan G.N. @___________/kg of seed.
a] 30 g b] 30 mg
c] 30 decigram d] 100 decigram

531. The seed of wheat should be treated with ____________ against seed-born pathogens.
a] Vitavex b] Thiram
c] Both of these d] Any one of these

532. The seed rate (kg per ha) for sarson main crop is about :
a] 4-6 b] 2-3
c] 10-12 d] 12-15

533. The seed rate for composite of maize should be planted:
a] 18-20 b] 20-25
c] 28-30 d] 30-35

534. The seed rate of bold grain varieties of rice to raise seedlings enough for one hectare is:
a] 40-50 kg b] 20-25 kg
c] 30-35 kg d] 50-55 kg

535. The seed rate of bunching type groundnut is :
a] 85-100 kg/ha b] 60-80 kg/ha
c] All the above d] None of the above

536. The seed rate of cotton would be ___________ kg/ha, if the planting geometry is 75 x 50 cm, and test-weight, germination and purity percentages are 140 g, 90 and 95, respectively.
a] 2.17 b] 3.37
c] 4.37 d] 5.21

537. The seed rate of potato tubers would be _______ q/ha, if the crop spacing, average weight of tubers and rottage loss of tubers during storage are 60 x 25 cm, 35 kg and 5 per cent, respectively.
a] 2.17 b] 3.37
c] 4.37 d] 5.21

538. The seed rate of tobacco to raise seedlings for one hectare varies from_________ kg/ha.
a] 0.5-1.0 b] 1.0-2.0
c] 1.5-2.5 d] 2.0-3.0

539. The seed, the progeny of which comes from foundation seed or registered seed, is handled to maintain genetic purity, approved and certified by certifying agency is called:
a] Certified seed b] Foundation seed
c] Nucleus seed d] Registered seed

540. The seeds of ________________ are positively photoblastic.
a] Potato b] Tobacco
c] Jute d] Sugarcane

541. The seeds of oil palm require as high as ________________ °C temperature to break the dormancy.
a] 30-40 b] 40-50
c] 50-60 d] 60-70

GK GS
Vegetable becoming tender on cooking is an example of chemical change.

542. The sensitive stages to moisture stress in sorghum are:
a] Flowering b] Grain filling
c] Both of these d] None of these

543. The sequence in which amino acids are linked in proteins depends on:
a] Climatic control b] Agronomic control
c] Genetic control d] None of these

544. The soil moisture content at which the plants remain wilted day and night or in a humid chamber, the moisture content is called:
a] Permanent wilting point
b] Incipient wilting point
c] Temporary wilting point
d] None of these

545. The soil temperature for satisfactory growth of potato tuber should be __________ °C.
a] 7-9 b] 11-13
c] 17-19 d] 20-23

546. The soyabean crop can be intercropped with:
a] Wheat b] Chickpea
c] Maize d] Tobacco

547. The sprinkler system of irrigation is most suitable for :
a] Clay soils b] Undulated soil
c] Saline soild] Acidic soil

548. The starter dose of nitrogen for chickpea is recommended as __________ kg/ha.
a] 20 b] 10
c] 30 d] 40

549. The sucrose is found in __________ of sugarbeet.
a] Stem b] Leaves
c] Roots d] Inflorescence

550. The sufficiency range of iron is normally between __________ppm.
a] 10-40 b] 50-250
c] 300-400 d] None of these

551. The sunflower crop is recommended to be planted at __________ cm spacing.
a] 45 x 20 b] 60 x 20
c] 45 x 30 d] 60 x 30

552. The sunflower crop requires cool climate from:
a] Germination to seedling stage
b] Seedling to flowering stage
c] Flowering to maturity stage
d] Seedling to maturity stage

553. The surfaces on which phosphate ions enter into sorption reactions are:
a] Surfaces of constant charge
b] Surfaces of variable charge
c] None of these
d] Both of these

554. The technique, used for minimising the nitrogen losses in paddy field under Indian conditions comprises:
a] Use of pre-incubated urea
b] Use of neem seed cake
c] Use of mud ball
d] All of these

555. The temporary roots in maize root system are:
a] Seminal roots
b] Aerial roots
c] Crown roots
d] Coronal roots

556. The tendency of the young upper leaves to remain as the lower leaves yellow or die is an indication of nitrogen __________ in the plant.
a] Mobility b] Immobility
c] Stability d] Degradability

557. The term 'Cropping scheme' refers to:
a] Allotment of crops to the pieces of land on a farm
b] Allotment of crop rotations to the pieces of land in a farm
c] Planning and budgeting of the crop production in a farm
d] Planning of schedules for different farm operations

558. The time of planting of main crop of potato in northern plains is:
a] 10-15 October b] 15-25 October
c] 10-15 November d] 15-25 November

559. The time of planting of main crop of lucerne in northern plains is:
a] First fortnight of October
b] Second fortnight of October
c] First fortnight of November
d] Second fortnight of November

560. The total soluble iron concentration becomes sufficient to supply iron adequately to roots only at pH:
a] 2 b] 3
c] 4 d] 5

561. The twinning in sugarbeet crop should be done at the stage of :
a] 5 week age b] 2 week age
c] 8 week age d] 10 week age

562. The type of germination in field pea is called:
a] Hypogeal b] Epigeal
c] Hypo-epigeal d] Epi-hypogeal

563. The type of germination in greengram in known as:
a] Hypogeal b] Epigeal
c] Hypo-epigeal d] Epi-hypogeal

564. The type of germination in Sunflower is known as:
a] Epigeal b] Hypogeal
c] Hypo-epigeal d] Epi-hypogeal

The earliest mention of four ashramas is in Jabala Upnishad.

565. The usually recommended plant to plant distance for soybean is:
a] 2-3 cm b] 4-5 cm
c] 5-7 cm d] 8-10 cm

566. The varieties Harison special, Chetan and Gold of tobacco are grown for :
a] Cigarette b] Biri
c] Chewing d] Snuff

567. The variety of potato which possesses the resistance to frost conditions is:
a] Kufri Chandramukhi
b] Kufri Alankar
c] Kufri Sheetman
d] Kufri Chamatkar

568. The volume (in litres) of one cusec flow of water in one hour is :
a] 2.6×10^6 b] 3.6×10^6
c] 4.6×10^6 d] 5.6×10^6

569. The water requirement of sugarcane is the maximum during ________ phase.
a] Germination and formation
b] Formation phase and grand growth
c] Grand growth phase and maturity
d] Formation phase and maturity

570. The weeds can be controlled in linseed by the application of MCPB @________ kg a.i./ha.
a] 0.5 b] 1.5
c] 1.0 d] 2.0

571. The weight of potato tuber (for sowing) most economical for higher yields is:
a] 10-25 grams b] 30-50 grams
c] 50-80 grams d] 80-100 grams

572. The wheat crop should be harvested when the moisture content in the grain is :
a] 10-15% b] 25-30%
c] 15-17% d] 30-38%

573. The word 'Agros' means :
a] Soil b] Water
c] Field d] Crop

574. The World Environment Day is observed on:
a] May 17 b] June 5
c] September 8 d] March 8

575. The yield of wheat grain will be ________ t/ha from the following observation, spacing 20 x 3 cm, effective tillers/plant, 3, number of grains/ear, 32, 1000-grain weight, 40 g.
a] 5.8 t/ha b] 6.2 t/ha
c] 6.4 t/ha d] 6.8 t/ha

576. There are ________________ seed certification agencies in India at present.
a] 11 b] 15
c] 19 d] 23

577. There can actually be a build up of toxic concentrations of Fe^{2+} in the soil solution of :
a] Aridosol b] Histosol
c] Latosol d] Vertisol

578. Thinning of rapeseed and mustard is done to maintain a plant to plant distance of ________ cm.
a] 5-10 b] 10-15
c] 15-20 d] 20-25

579. Thiocarbamates *viz.*, EPTC, a soil-incorporated herbicide is readily absorbed by roots and coleoptile and translocated:
a] Only upward b] Only downward
c] Both of these d] None of these

580. Though tilth is dynamic in nature, can be measured by:
a] Aggregate analysis b] Chemical analysis
c] Biological d] None of these

581. Tillage of the soil ________ rate of disappearance of organic matter.
a] Reduces b] Increases
c] Maintains d] Has no effect

582. To assure adequate population in rice, the number of hills per square meter area should be maintained at:
a] 50 b] 80
c] 40 d] 60

583. To attain maximum yield of sorghum per hectare, plant population should be maintained at:
a] 1,50,000 b] 1,60,000
c] 1,75,000 d] 1,25,000

584. To avoid rotting in early planting of potato, seeds should be treated with ________.
a] 0.5 per cent Diathane M-45
b] 2.0 per cent Diathane M-45
c] 1.5 per cent Aretan
d] 2.0 per cent Tafasan

585. To encourage response of rice productivity in saline alkaline soil the age of seedling for transplanting should be :
a] 15-20 days b] 20-25 days
c] 25-30 days d] 30-40 days

586. To get maximum yield of maize, the number of plants required in one hectare land are :
a] 60-65 thousand b] 80-85 thousand
c] 25-35 thousand d] 90-95 thousand

587. To obtain maximum yield of wheat the number of tillers per square meter required is:
a] 100 b] 250
c] 500 d] 750

588. To quick up the sprouting of potato which of the following is used ?
a] Chlorine
b] Etherel
c] Carbon di-sulphide
d] Methyl Bromide

GK GS
Peneplain are end-product of normal cycle of erosion.

589. To supply 100 kg nitrogen and 50 kg potassium, the quantity of ammonium chloride and muriate of potash will be:
a] 299.12 kg AC, 73.3 kg MoP
b] 384.61 kg AC, 73.3 kg MoP
c] 2909.12 kg AC, 83.3 kg MoP
d] 384.61 kg AC, 83.3 kg MoP

590. Tobacco belongs to the family:
a] Compositeae b] Solanaceae
c] Crucifereae d] Leguminoseae

591. Tobacco has been classified as a __________ crop.
a] Cash b] Catch
c] Inter d] Relay

592. Tobacco requires __________ °C temperature for uniformly rapid germination.
a] 17-22 b] 19-25
c] 27-32 d] 29-42

593. Topping in tobacco means a process of :
a] Removal of the buds in the axil of leaves
b] Removal of leaves
c] Removal of terminal buds
d] Burning of leaves

594. Topping of safflower plants is beneficial:
a] To reduce the lodging
b] To reduce the water losses
c] To promote branching and flowering
d] To reduce losses of seed and fruit

595. *Tossa* jute is to be sown at a row spacing of ____ cm.
a] 15 b] 20
c] 25 d] 30

596. Transpiration is:
a] Evaporation
b] Wilting co-efficient
c] Water loss from plants
d] Changing status of water

597. Transplanting of seedlings is commonly practiced in:
a] Rice b] Maize
c] Sorghum d] Pearl millet

598. Treat the sugarcane seed setts with __________ against fungal diseases.
a] Diathane M-45 b] Zineb
c] Thiram d] Aretan

599. Treatment of cotton-seed with organo-mercurial compound(s) helps to minimise the __________ disease.
a] Soil borne b] Air borne
c] Water borne d] Seed borne

600. Triazine herbicides are said as notorious herbicides because of there:
a] Long residual toxicity b] Short residual toxicity
c] Non-selectiveness d] None of these

You Don't Know!

First car

The first automobile was imported by the Maharaja of Patiala. It was a 1902, French-made, De Dion Bonton car with the license plate number '0'.

601. Triazines are mostly used in :
a] Barley b] Wheat
c] Pulses d] Maize

602. Triazine-tolerant broadleaf weed biotypes include:
a] Lambsquarter (*Chenopodium album*)
b] Morning glory (*Convolvulus arvensis*)
c] Sweet clover (*Melilotus indica*)
d] None of these

603. Triticale may yield more than wheat under conditions:
a] Irrigated
b] Dry
c] Irrigated and dry
d] None of the above

604. Two hand weedings have been found adequate to control weeds in pigeonpea reasonably well __________ days after sowing.
a] 25-30 and 45-60 b] 40-45 and 60-65
c] 15-20 and 35-40 d] 30-35 and 60-65

605. Two hand weedings in safflower are required for its good growth at:
a] 15 and 30 days stages
b] 20 and 40 days stages
c] 25 and 50 days stages
d] 30 and 60 days stages

606. Under dry land conditions most preferred herbicide for maize is:
a] Atrazine b] Propanil
c] Fluchloralin d] Butachlor

607. Under high rainfall conditions maize should be planted:
a] In narrow furrows
b] On a flat bed
c] On the side of ridges
d] In widely spaced furrows

608. Under irrigated conditions, nitrogen should be splitted in castor as __________ days after sowing.
a] Half basal plus half at 60
b] Half plus half at 30
c] Three-fourth as basal and one-fourth at 60
d] Three-fourth as basal and one-fourth 30

609. Under late sown conditions area of nursery, sufficient for transplanting one hectare rice field is:
a] 800-1000 sqm b] 500-600 sqm
c] 1000-1200 sqm d] 600-700 sqm

GK GS

Carbon dioxide is the largest air pollutant.

610. Under moisture stress condition growing of ________ should be avoided.
a] Sugarcane b] Sunflower
c] Castor d] Green gram

611. Urea herbicides undergo conjugation with:
a] Lipids b] Carbohydrates
c] Proteins d] All of these

612. Urea is a :
a] Inorganic fertilizer
b] Organic fertilizer
c] Ammonical fertilizer
d] Nitrate fertilizer

613. Use of ________ provides effective control of weeds in groundnut.
a] 2, 4-D b] Simazine
c] Isoproturon d] Tok E-25

614. Use of linear measurements of plumule growth as a vigour test for cereals and sugarbeet was first suggested by :
a] S. Mathews in 1929
b] J.S. Burris in 1939
c] Germ in 1949
d] Hegarty in 1959

615. Use of________herbicide provides effective control of weeds in linseed.
a] 2, 4-D b] Atrazine
c] MCPB d] Pendimethalin

616. Variety of maize having high quantity of lysine :
a] Shakti b] Vikram
c] Jawahar d] Vijay

617. Various potassic fertilizers embody nearly an equal effectiveness in ____________ situations requiring its supplemental application.
a] Climatic b] Edaphic
c] Cropping d] Economic

618. Water available to plants to the maximum extent is the one held:
a] Soil moisture tension
b] Pan evaporation
c] Between field capacity and wilting point
d] At 20 atmospheres

619. Water enters into the plant by means of :
a] Diffusion b] Osmosis
c] Hydrolysisd] Symbiosis

620. Water requirement of a crop is related to its:
a] Evapotranspiration
b] Potential absorption coefficient
c] Dry matter content
d] Soil mineralogy

621. Weed competition in rice is more in the:
a] Transplanted crop b] Direct seeded crop
c] Flooded crop d] Crop sown in rows

622. What does 'jhuming' mean ?
a] A type of farm machine
b] A type of improved seed
c] Convential method of cultivation in hilly tracts
d] A type of fertilizer

623. What is crop rotation?
a] Growing more than one crop at a time
b] Growing of crops one after other to maintain soil fertility
c] Growing of an associate crop in between the rows of main crop
d] Growing of crops together in strips

624. What is CSH-I ?
a] Insecticide
b] Desi variety of ragi
c] Hybrid variety of jowar
d] Hybrid variety of maize

625. What is hulling percentage in rice?
a] 33 % b] 50%
c] 67% d] 75%

626. What is the effect of potassic fertilizers ?
a] It makes the plants more resistant to disease
b] It improve the grain quality
c] It makes the plants susceptible to disease
d] All of the above

627. What is the most important source of irrigation in India ?
a] Canal b] Pond
c] Charsa d] Tube well

628. What percentage of edible oils is contained in rice bran?
a] 5-8% b] 18-20%
c] 25-28% d] 30-32%

629. What should be the month of sowing of maize in maize-toria-potato-moong rotation ?
a] March b] June
c] July d] August

630. What will be the effect of potassic fertilizers ?
a] It makes the plants more resistant to disease
b] Improves the grain quality
c] Increases susceptibility to diseases
d] Reduces lodging

631. What will be the relative humidity, when the dry and wet bulb thermometer readings are 25°C ?
a] 10% b] 50%
c] 75% d] 100%

632. What will happen if the wheat applied with the more nitrogen in standing crop?
a] It will became green
b] The grain will be small in size
c] It will lodge
d] No effect

633. Wheat germination (90%) takes place after how many days ?
a] 3-4 b] 4-5
c] 5-6 d] 4-8

634. Wheat is of :
a] Mediterranean origin
b] Indian origin
c] Mexican origin
d] Chinese origin

635. Wheat protein is called as :
a] Zein b] Gluten
c] Dhurin d] Ricin

636. Wheat, maize, oat and jowar are the rich source of :
a] Carbohydrate b] Fat
c] Protein d] Vitamin

637. When the movement of water after rain or irrigation after 48-72 hours, the soil is said to be at:
a] Incipient wilting
b] Field capacity
c] Permanent wilting
d] None of these

638. When the phosphorus is either band placed or broadcasted, the relative efficiency is related to:
a] Its soil status
b] Its applied quantum
c] Both of these
d] None of these

639. Where is the headquarter of Central Soil and Water Conservation Research and Training Institute?
a] Dehradun b] Agra
c] Chandigarh d] Ootacamund

640. Wherever only one irrigation is available, it should be given at _____________ stage in safflower.
a] 20 days b] 30 days
c] 40 days d] 50 days

641. Wherever only two irrigations are available, the wheat crop should be irrigated:
a] CRI and tillering
b] CRI and flowering
c] CRI and milking
d] CRI and late jointing

642. Which crop is considered as king of the fodder crops?
a] Lucerne b] Berseem
c] Oat d] Sudan grass

643. Which disease occur most in groundnut ?
a] Blast b] Tikka
c] Blight d] Damping-off

644. Which fertilizer is cheaper?
a] Urea b] SSP
c] Groundnut cake d] Murrate of potash

645. Which grain is said to be poor man's food ?
a] Ragi b] Bajra
c] Maize d] Jowar

646. Which of following stages of a crop are more prone to weed competition?
a] Germination to seedling
b] Vegetative
c] Reproductive
d] Maturity

647. Which of one of the following fertilizer is more required for berseem crop?
a] Nitrogen b] Phosphorous
c] Potash d] Boron

648. Which of the agricultural operation requires maximum non-biological power ?
a] Ploughing b] Harrowing
c] Irrigation d] Harvesting

649. Which of the below mentioned crop is not a pulse ?
a] Cowpea b] Green gram
c] Cluster bean d] Kidney bean

650. Which of the crop is most sensitive to salinity ?
a] Mustard b] Potato
c] Gram d] Maize

651. Which of the following contains most of the plant nutrients ?
a] Bone meal b] Farm yard manure
c] Urea d] Rock phosphate

652. Which of the following criteria of irrigation scheduling is more acceptable to farmers?
a] Soil moisture tension
b] Pan evaporation
c] Water potential of plant
d] Critical growth stage

653. Which of the following crop is not used for making silage ?
a] Maize b] Jowar
c] Wheat straw d] Grass

654. Which of the following crop is used as fodder and grain?
a] Jowar b] Bajra
c] Maize d] Wheat

655. Which of the following crops does not have hybrid seed?
a] Cotton b] Castor
c] Wheat d] Jute

656. Which of the following fertilizer should not be applied alongwith seed to the soil?
a] DAP
b] Ammonium sulphate nitrate
c] Super phosphate
d] Ammonium sulphate

GK GS

The folk art form of Madhubani belong to Bihar state.

657. **Which of the following fertilizers should not be mixed together ?**
   a] Murate of Potash and Sulphate of Potash
   b] Sodium Nitrate and Super Phosphate
   c] Ammonium Sulphate and Sodium Nitrate
   d] Sodium Nitrate and F.Y.M.
658. **Which of the following herbicides is most suited for weed control in irrigated gram ?**
   a] 2, 4-D b] Simazine
   c] Diuron d] Fluchloralin
659. **Which of the following is a parenchymatus fibre?**
   a] Cotton b] Jute
   c] Silk d] All of the above
660. **Which of the following is a species of American cotton?**
   a] *Gossypium barbadense*
   b] *Gossypium hirsutum*
   c] *Gossypium herbaceum*
   d] Both a & b
661. **Which of the following is cash crop?**
   a] Rice b] Wheat
   c] Groundnut d] Gram
662. **Which of the following is cheaper fertilizer ?**
   a] SSP b] Urea
   c] DCP d] Ammonium sulphate
663. **Which of the following is not a primary tillage practice?**
   a] Ploughing b] Planking
   c] Harrowing d] Weeding
664. **Which of the following is not a short day plant ?**
   a] Maize b] Moong
   c] Jowar d] Wheat
665. **Which of the following is the cheapest method of irrigation?**
   a] Tube-wells
   b] Diversion from a flowing stream
   c] Canals
   d] Surface storage
666. **Which of the following is the correct statement?**
   a] Principle of agri-silviculture is similar to that of intercropping.
   b] Principle of agri-silviculture is similar to that of intensive cropping.
   c] Principle of agri-silviculture is similar to that of double cropping.
   d] Principle fo agri-silviculture is similar to that of mono-cropping.
667. **Which of the following is the leading state in *rabi* maize production?**
   a] Bihar b] Uttar Pradesh
   c] Punjab d] Gujarat
668. **Which of the following is the pre-requisite for multiple cropping?**
   a] Irrigation facility b] Intensive use of fertilizers
   c] Availability of HYV d] All of the above
669. **Which of the following management practices is more problematic in direct seeded rice cultivation?**
   a] Fertilizer management
   b] Weed management
   c] Water management
   d] Bushening
670. **Which of the following physiological growth stages of wheat crop is most susceptible to 2, 4-D?**
   a] Milk stage
   b] Panicle initiation stage
   c] Late tillering stage
   d] Dough stage
671. **Which of the following ratios between nursery area and transplantation is correct when rice seedling are raised by "Dapog method"?**
   a] 1:50 b] 1:25
   c] 1:100 d] 1:250
672. **Which of the following refers to cover crop?**
   a] Rapid growing crops grown to protect the soil from rain and wind action
   b] Tall growing crops which protects the shade loving crops
   c] Growing of a crop in strips all around the main crop
   d] Crops grown at right angle to the direction of prevailing winds
673. **Which of the following soils is most suitable for the cultivation of cereals?**
   a] Alluvial soils
   b] Red soils
   c] Lateritic soils
   d] Black soils
674. **Which of the following statement is correct ?**
   a] Blue-green algae causes plant disorders
   b] Blue-green algae fixes atmospheric nitrogen in soil
   c] Blue-green algae grows in wheat crop
   d] Dry conditions are the best for growth of blue-green algae
675. **Which of the following statement is true?**
   a] Fertile soil will always productive
   b] Productive soil will always fertile
   c] Both of above
   d] None
676. **Which of the following statement refers to the concept of critical stage of irrigation?**
   a] When irrigation is not given reduction of yield is maximum
   b] When water absorption rate is maximum
   c] When water loss by the plant is maximum
   d] All of the above
677. **Which of the following statements relates correctly to the green house effect ?**
   a] Protection of atmosphere by trees
   b] Rise in atmospheric temperatures
   c] Decrease in atmospheric temperature
   d] Maintenance of atmospheric temperature

In the Atharvaveda 'Gotra' appears with meaning class. GS

678. Which of the following supplies more plant nutrient?
a] Light soil b] Cotton seed cake
c] Castor cake d] Basic slag

679. Which of the following two crops are responsible for almost 75% of pulse production in India?
a] Gram and moong bean
b] Pigeonpea and moong bean
c] Moongbean and lentil
d] Gram and pigeonpea

680. Which of the following variety of maize has lysine content ?
a] Protina b] Sona
c] Jawahar d] Vijay

681. Which of the followings contains the highest protein content?
a] Pigeonpea b] Soyabean
c] Greengram d] Blackgram

682. Which of the mulch is best and cheap to minimise wind erosion ?
a] Polythene sheet mulch
b] Straw mulch
c] Stubble mulch
d] Saw dust mulch

683. Which of the species of sugarcane is referred as noble cane ?
a] *Saccharum spontaneum*
b] *Saccharum barberi*
c] *Saccharum officinarum*
d] *Saccharum sinense*

684. Which one is the most critical stage of irrigation in wheat?
a] Tillering b] CRI
c] Jointing d] Flowering

685. Which one of the following can be suitably grown as a mixed crop with wheat?
a] Lentil b] Cotton
c] Jowar d] Mustard

686. Which one of the following can be used for foliar spray?
a] Urea
b] Ammonium sulphate
c] Single super phosphate
d] Calcium ammonium nitrate

687. Which one of the following chemicals is used to treat seed potato to break its dormancy?
a] Ethrel b] Thiourea
c] IAA d] NAA

688. Which one of the following crop is not suitable for green manure ?
a] Sanai b] Dhaincha
c] Moong d] Berseem

689. Which one of the following crops has high transpiration ratio ?
a] Potato b] Cotton
c] Sorghum d] Millets

Knowledge World!
First missile

Indian successfully test-fired its first tactical surface-to-surface missile *Prithvi* on Feb. 25, 1988. *Prithvi* has a range capability of 250 km with different types of payload warheads. Prithvi has been inducted into the 11 Corps.

690. Which one of the following is composite variety of maize ?
a] Ankur
b] Arjun
c] Ganga-2
d] Kissan

691. Which one of the following is not a dicot weed ?
a] *Phyllanthus niruri*
b] *Chenopodium murale*
c] *Chenopodium album*
d] *Cyprus rotundus*

692. Which one of the following is the principal 'N' fixing algal community of blue-green algae?
a] *Rhizobium*
b] *Anabaena*
c] *Azotobactor*
d] *Azospirillum*

693. Which one of the following parts of the tobacco plant synthesises nicotine?
a] Root b] Stem
c] Branch d] Leaf

694. Which one of the following system needs irrigation throughout the year?
a] Diara land cropping system
b] Inter-cropping system
c] Mixed cropping system
d] Multiple cropping system

695. Which one of the following terms is used to denote the relative land area under sole crop required to produce the same yield as obtained under a mixed or an intercropping system at the same management level?
a] Land utilization index
b] Cropping intensity
c] Land equivalent ratio
d] Relay cropping

696. Which one of the following types of crop cultivations will be most appropriate and effective in forests?
a] Extensive cropping
b] Intensive cropping
c] Mixed cropping
d] Inter cropping

GK **Exfoliation refers to peeling off concentric shells of rock.** GS

697. Which one of the folowing pairs is not correctly matched?

a] Relay cropping	:	Sowing pulse in rice crop prior to harvest
b] Intercropping	:	Wheat after rice
c] Multiple cropping	:	Coconut-pepper-cacao-pineapple
d] Sole cropping	:	Potato

698. Which portion of sugarcane stalk is preferred for sowing purposes?

a] Top one-third to half portion
b] Top one-fourth protion
c] Bottom one-fourth portion
d] Bottom one-third to half portion

699. With too much application of nitrogenous fertilizer, sugar content in juice of sugarcane will :

a] Decrease
b] Increase
c] Constant
d] Crystallized

700. Which system of irrigation is not practiced usually ?

a] Canal irrigation
b] Sprinkler irrigation
c] Underground irrigation
d] Lift irrigation

701. Who introduce the tobacco cultivation in India?

a] Portuguese b] Persian
c] Chinese d] Japanese

702. Who is the founder of relay cropping?

a] M.S. Swaminathan
b] M.S. Benj Sardar
c] Dr. Borlaug
d] Dr. Balwant Raj

703. Winter application of phosphorus is followed in:

a] Warm areas b] Cool areas
c] Both of these d] None of these

704. Zero tillage system was first used successfully in 1950 in pasture renovation in:

a] Germany b] Japan
c] United Kingdom d] U.S.A.

705. An instrument that indicates the direction of the wind is :

a] Anemometers b] Hygrometer
c] Evaporimeter d] Wind Vane

706. Velocity of Gale is :

a] 62.4 to 73.6 km/hr b] 50.4 to 54.8 km/hr
c] 88.2 to 90.4 km/hr d] 104.2 to 112.8 km/hr

707. Scientific name of Cogon grass is :

a] *Avena fatua*
b] *Imperata cylindrica*
c] *Paspalum conjugatum*
d] *Portulaca oleracea*

708. Which of the following is perennial weed ?

a] Goat weed b] Pig weed
c] Black Jack d] Field bind weed

709. Which of the following soil is of ashy grey in colour:

a] Brown earth b] Chernozems
c] Tundra soil d] Podsols

GK — **In computer vocabulary CD stand for Compact Disc.** — GS

Matching

1. **Match the following crops with their class.**

a]	Kodo and Kakun	i]	Plantation crops
b]	Coffee and tea	ii]	Food crops
c]	Rice and wheat	iii]	Millet crops
d]	Berseem and lucerne	iv]	Forage crops

2. **Match the crops with their families.**

a]	Graminae	i]	Safflower and sunflower
b]	Cruciferae	ii]	Maize and sugarcane
c]	Solanaceae	iii]	Mustard and Rapeseed
d]	Compositae	iv]	Potato and tobacco

3. **Match the crops with their families.**

a]	Jute	i]	Chenopodiaceae
b]	Castor and tapioca	ii]	Papilionaceae
c]	Spinach and sugarbeet	iii]	Tiliaceae
d]	Groundnut and chickpea	iv]	Euphobiaceae

4. **Match the following crops with their agrarian or economic or agricultural class.**

a]	Wheat	i]	Field/forage crops
b]	Teosinte	ii]	Sugar crops
c]	Beet	iii]	Narcotic/fumitories/masticatories
d]	Tobacco	iv]	Cereal crops

5. **Match the following crops with the method of their harvesting.**

a]	Tobacco	i]	Tying
b]	Mustard	ii]	Reaping
c]	Potato	iii]	Priming
d]	Sugarcane	iv]	Earthing

6. **Match the purpose of growing or the special purpose served by the following crops.**

a]	Lathyrus	i]	Sod/turf crop
b]	Sorghum	ii]	Exhaustive crop
c]	Bermuda grass	iii]	Paira crop
d]	Linseed	iv]	Trap crop

7. **Match the classes with the crops given below.**

a]	Nurse crops	i]	Sugarcane and tapioca
b]	Riparian crops	ii]	*Tephrosia and Crotolaria*
c]	Energy crops	iii]	*Gliricidia* and pigeonpea
d]	Avenue crops	iv]	Pepperwort and water bind weed

8. **Match the nursery area required for transplantation of one hectare of the following crops.**

a]	Rice, dry bed	i]	1250-1500 m^2
b]	Rice, wet bed	ii]	50 m^2
c]	Rice, late sowing	iii]	1000 m^2
d]	Tobacco	iv]	75-80 m^2

9. **Match weedicide with the trade name.**

a]	Propanil	i]	Basalin
b]	Butachlor	ii]	Tok-E-25
c]	Fluchloralin	iii]	Machete
d]	Nitrofen	iv]	Stamp F-34

10. **Match type of tillage with the climate and/or agropractice(s).**

a]	Post-harvest	i]	Uncropped land
b]	Summer	ii]	Crop harvest
c]	Winter	iii]	Tropical zones
d]	Fallow	iv]	Temperate zone

11. **Match the ploughs with the working of soil and operation.**

a]	Country plough	i]	Walking type
b]	Disc plough	ii]	Soil turning type
c]	Bose plough	iii]	Riding type
d]	Gang plough	iv]	Soil stirring type

12. **Match the nitrogen content (%) on green matter basis with the green manuring crops.**

a]	Mungbean	i]	0.43
b]	Sesbania	ii]	0.49
c]	Clusterbean	iii]	0.53
d]	Cowpea	iv]	0.34

13. **Match the N + P_2O_5 + K_2O content (%) with the manures containing them.**

a]	0.5 + 0.2 + 0.5	i]	Castor cake
b]	4.3 + 1.8 + 1.3	ii]	Neem cake
c]	5.2 + 1.0 + 1.4	iii]	Blood meal
d]	10.5 + 2.5 + 0	iv]	Farmyard manure

14. **Match the cakes with the N + P_2O_5 + K_2O content (%).**

a]	7.3 + 1.5 + 1.3	i]	Rapeseed cake
b]	4.0 + 1.4 + 1.3	ii]	Groundnut cake
c]	5.2 + 1.8 + 1.2	iii]	Linseed cake
d]	2.5 + 0.8 + 1.8	iv]	Mahua cake

15. **Match the nitrogen fertilizers and their N-content (%).**

a]	Ammonium sulphate	i]	82
b]	Ammonium chloride	ii]	20-25
c]	Anhydrous ammonia	iii]	20-21
d]	Ammonia solution	iv]	24-26

16. **Match the phosphatic fertilizers with their P_2O_5 content (%).**

a]	Single superphosphate	i]	14-18
b]	Dicalcium phosphate	ii]	34-39
c]	Rock phosphate	iii]	16
d]	Basic slag	iv]	20-40

17. **Match the compound fertilizers with N + P_2O_5 content (%).**

a]	Nitrophosphate	i]	11 + 48
b]	Monoammonium phosphate	ii]	12.9 + 12.9
c]	Crude ammonium phosphate	iii]	21 + 94
d]	Diammonium phosphate	iv]	16 + 20

18. **Match the micronutrient fertilizers with their nutrient content (%).**

a]	Ferrous sulphate (Fe)	i]	23-35
b]	Manganese sulphate (Mn)	ii]	23
c]	Zinc sulphate (Zn)	iii]	37-39
d]	Sodium molybdate (Mo)	iv]	20

19. **Match the sulphur content (%) with the following micronutrient fertilizers.**

a]	18.8	i]	Copper sulphate
b]	12.4	ii]	Ferrous sulphate
c]	17.8	iii]	Manganese sulphate
d]	12.8	iv]	Zinc sulphate

20. **Match the crops with their critical stages for irrigation.**

a]	Tobacco	i]	Pre-flowering, capsule development
b]	Rapeseed and mustard	ii]	Topping
c]	Pearlmillet	iii]	Seedling, flowering
d]	Sorghum	iv]	Flowering

'Sare Jahan Se Achcha' written by poet Iqbal.

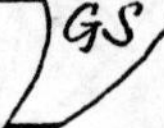

21. Match the crops with their critical stages for irrigation

a]	Rice-paddies	i]	Silking, cob development
b]	Maize	ii]	Flowering, grain formation
c]	Fingermillet	iii]	Emergence, tiller formation, elongation
d]	Sugarcane	iv]	Tiller initiation, flower primordia, flowering

22. Match the crop with the critical stages for irrigation.

a]	Chickpea	i]	Pegging to pod formation
b]	Pigeonpea	ii]	Preflowering, flowering
c]	Field pea	iii]	Flower initiation, pod filling
d]	Groundnut	iv]	Start of flowering, pod formation

23. Match the critical stage(s) irrigation with crops.

a]	Flowering, grain filling	i]	Safflower
b]	Branching, flowering	ii]	Linseed
c]	Preflowering, capsule development	iii]	Oats
d]	Ear emergence	iv]	Sunflower

24. Match the soil moisture storage capacity (mm) with broad soil groups.

a]	300/m	i]	Alluvial deep, loam
b]	140-180/m	ii]	Submontane, deep, loamy sand
c]	100-140/m	iii]	Black, deep
d]	90-100/m	iv]	Submontane, deep, sandy loam

25. Match the isolation distance (metres) from the fields of other prescribed crops for the production of foundation seed with the crops.

a]	Wheat	i]	1000
b]	Pearlmillet	ii]	400
c]	Cauliflower	iii]	50
d]	Maize	iv]	150

26. Match the crop with the diseases designated for seed certification.

a]	Rape and mustard	i]	Phyllody
b]	Sesame	ii]	Red rot
c]	Cowpea	iii]	Alternaria blight
d]	Sugarcane	iv]	Anthracnose

27. Match the weed species designated objectionable with the crops in seed production.

a]	*Convolvulus arvensis*	i]	Paddy
b]	*Chicoricum intybus*	ii]	Wheat
c]	*Cuscuta* spp.	iii]	Berseem
d]	*Echinochloa* spp.	iv]	Lucerne

28. Match the maximum permissible weight of a seed-lot the plant species given below.

a]	10,000	i]	*Vigna unguiculata*
b]	20,000	ii]	*Oryza sativa*
c]	25,000	iii]	*Zea mays*
d]	40,000	iv]	*Brassica juncea*

29. Match the minimum size of a working pelleted seed sample with the quality character to be determined.

a]	2500 pellets	i]	Size grading
b]	400 pellets	ii]	Determination of other seeds
c]	2000 pellets	iii]	Germination
d]	7000 pellets	iv]	Purity analysis

30. Match the prescribed number of days for final count for germination with the crop seeds.

a]	14 days	i]	Sorghum
b]	8 days	ii]	Wheat
c]	10 days	iii]	Okra
d]	21 days	iv]	Rice-paddies

Sulva-sutra marks the beginning of geometry.

31. **Match the following chemicals used to break the dormancy of seeds with their major chemical group.**

a]	Gibberellins	i]	Sulphydryl
b]	Indole acetate	ii]	Oxidant
c]	Hypochlorite	iii]	Growth regulators
d]	Dithiothreitol	iv]	Respiratory inhibitors

32. **Match the major chemical group with the chemicals used to break the dormancy in seeds.**

a]	Plant products	i]	Dinitrophenol
b]	Nitrogenous compound	ii]	Cotylenol
c]	Growth regulators	iii]	Thiourea
d]	Resipratory inhibitors	iv]	Ethylene

33. **Match the carbohydrate content (%) with seeds of the following.**

a]	65-76	i]	Sorghum
b]	56-69	ii]	Flax
c]	50-60	iii]	Wheat
d]	20-25	iv]	Chickpea

34. **Match the following seeds with their protein content (%).**

a]	Wheat	i]	20-25
b]	Chickpea	ii]	8-9
c]	Sorghum	iii]	26-41
d]	Sunflower	iv]	10-14

35. **Match the following seeds with their fat content (%).**

a]	Sunflower	i]	1.5-2.0
b]	Flax	ii]	40-50
c]	Chickpea	iii]	3-4
d]	Sorghum	iv]	35-45

36. **Match the quality component with the recommended standard percentage for hybrid cotton.**

a]	Germination	i]	98
b]	Physical purity	ii]	90
c]	Genetic purity	iii]	10
d]	Self-plants	iv]	60

37. **Match the following insect predators with the weed which they have been found to feed upon**

a]	*Acroclita* spp.	i]	*Cyperus rotundus*
b]	*Orseoliella javanica*	ii]	*Euphatorium adenophorum*
c]	*Bactra vermosana*	iii]	*Imperata cylindrica*
d]	*Procecidochares utilitis*	iv]	*Cuscuta* spp

38. **Match the organisms and weed controllable by it.**

a]	Cochinial mealy bug (*Dactylopius indicus*)	i]	*Striga* spp.
b]	Mexican gallfly (*Procecidochares utilitis*)	ii]	*Eupatorium adenophorum*
c]	Gall-forming weevil (*Smicronyx albovariegatus*)	iii]	*Parthenium hysterophorus*
d]	Larval of *Diacrisia obliqua*	iv]	*Opuntia vulgaris*

39. **Match the herbicide with their chemical group.**

a]	Dalapon	i]	Amides and acetamides
b]	Butachlor	ii]	Bipyridilium
c]	MSMA	iii]	Aliphatics
d]	Paraquat	iv]	Arsenicals

40. **Match chemical group with the herbicide.**

a]	Carbamates	i]	Nitrofen
b]	Benzoics and phenylacetates	ii]	Propham
c]	Dinitroanilines	iii]	TIBA
d]	Diphenyl esters	iv]	Pendimethalin

41. **Match the chemical group with the herbicide.**

a]	Nitriles	i]	PCP
b]	Phenols	ii]	Norflurazon
c]	Phenoxy acids	iii]	MCPA
d]	Pyridazinones	iv]	Ioxynil

Concept of Base Level of Erosion was forwarded by J.W. Powell.

GS

42. Match the herbicide with the chemical group.

a]	Benthiocarb	i]	Benthiazuron
b]	Triazines	ii]	Thiocarbamates
c]	Triazoles	iii]	Terbutryn
d]	Urea	iv]	Amitrole

43. Match the weeds with their botanical name.

a]	Star bur	i]	*Achyranthes aspora*
b]	Prickly chaff flower	ii]	*Aerva tomentosa*
c]	Javanese woolpant	iii]	*Acanthospermum hispidim*
d]	Camel thorn	iv]	*Alhagi camelorum*

44. Math the weeds with their botanical name.

a]	Khaki weed	i]	*Boerhaavia diffusa*
b]	Spreading hog weed	ii]	*Alternanthera echinata*
c]	Hog weed	iii]	*Brassica arvensis*
d]	Wild mustard	iv]	*Blumea wightiana*

45. Match the weeds with their botanical name.

a]	Chicory	i]	*Chichorium intybus*
b]	Cleome	ii]	*Coix lachryma*
c]	Conch flower creeper	iii]	*Cleome chelidonii*
d]	Sankru	iv]	*Clitoria ternatea*

46. Match the weeds with their botanical name.

a]	Jew's mallow jute	i]	*Crotolaria striata*
b]	Rattle box	ii]	*Corchorus olitorius*
c]	Bulb grass	iii]	*Cyperus iria*
d]	Yellow sedge	iv]	*Cyperus bulbosus*

47. Match the weeds with their botanical name.

a]	Crow foot grass	i]	*Datura fastuosa*
b]	Thorn apple	ii]	*Elephantopus scaber*
c]	False daisy	iii]	*Dactyloctenium aegyptium*
d]	Elephant's foot	iv]	*Eclipta alba*

48. Match the weeds with their botanical name.

a]	India lovegrass	i]	*Gynandropsis gynandra*
b]	Horseweed	ii]	*Hydrocotyle asiatica*
c]	Caravelleseed	iii]	*Erigecron canadensis*
d]	Asiatic pennywort	iv]	*Eragrostis pilosa*

49. Match the weeds with their botanical name.

a]	Indigo	i]	*Mimosa pudica*
b]	Ivy-leaved morning glory	ii]	*Ipomea hederacea*
c]	Leucas	iii]	*Indigofera glandulosa*
d]	Touch-me-not	iv]	*Leucas aspera*

50. Match the weeds with their botanical name.

a]	Carpet weed	i]	*Ocimum canum*
b]	Hoary basil	ii]	*Oxalis corniculata*
c]	Prickly pear	iii]	*Mollugo lotoides*
d]	Indian sorrel	iv]	*Opuntia dillenii*

51. Match the weeds with their botanical name.

a]	Knot grass	i]	*Polygonum convolvulus*
b]	Passion flower	ii]	*Paspalum sanguinale*
c]	Wild buck wheat	iii]	*Potamogeton pectinatus*
d]	Pond weed	iv]	*Passiflora foetida*

GK **Sunflower oil has high proportion of polyunsaturated fatty acid.** GS

52. Match the weeds with their botanical name.

a]	Yellow foxtail	i]	*Sida spinosa*
b]	Mallow	ii]	*Setaria glauca*
c]	Prickly sida	iii]	*Silene* sp.
d]	Sleepy catch fly champion	iv]	*Sida rhombifolia*

53. Match the weeds with their botanical name.

a]	White horse nettle	i]	*Spergula arvensis*
b]	Corn spurry	ii]	*Trianthema portulacastrum*
c]	Horse purslane	iii]	*Tribulus terrestris*
d]	Ground burnet	iv]	*Solanum elaegnifolium*

54. Match the weeds with their botanical name.

a]	Cattails	i]	*Xanthium strumarium*
b]	Ash-coloured flea bane	ii]	*Typha latifolia*
c]	Cocklebur	iii]	*Salvinia auriculata*
d]	Water fern	iv]	*Vernonia cinerea*

55. Match the crop with its scientific name.

a]	Rice	i]	*Zea mays*
b]	Wheat	ii]	*Sorghum bicolor*
c]	Maize	iii]	*Triticum aestivum*
d]	Sorghum	iv]	*Oryza sativa*

56. Match the crop with its scientific name.

a]	*Bajra* (pearlmillet)	i]	*Eleusine coracana*
b]	*Mandua* (fingermillet)	ii]	*Panicum miliaceum*
c]	*Cheena* (prosomillet)	iii]	*Echinochloa frumentacea*
d]	*Sawan* (barnyard millet)	iv]	*Pennisetum typhoides*

57. Match the crop with its scientific name.

a]	Chickpea	i]	*Cajanus cajan*
b]	Lentil	ii]	*Cicer arietinum*
c]	Pigeonpea	iii]	*Setaria italica*
d]	Foxtail millet	iv]	*Lens esculenta*

58. Match the crop with its scientific name.

a]	Greengram	i]	*Vigna mungo*
b]	Blackgram	ii]	*Arachis hypogaea*
c]	Cowpea	iii]	*Vigna radiata*
d]	Groundnut	iv]	*Vigna sinensis*

59. Match the crop with its scientific name.

a]	Barley	i]	*Pisum sativum*
b]	Kodomillet	ii]	*Hordeum vulgare*
c]	Soyabean	iii]	*Glycine max*
d]	Field pea	iv]	*Paspalum scrobiculatum*

60. Match the crop with its scientific name.

a]	Sesamum	i]	*Ricinus communis*
b]	Castor	ii]	*Linum usitatissimum*
c]	Linseed	iii]	*Carthamus tinctorius*
d]	Safflower	iv]	*Sesamum indicum*

61. Match the crop with its scientific name.

a]	Oats	i]	*Cyamopsis tetragonoloba*
b]	Egyptian clover	ii]	*Avena sativa*
c]	Lucerne	iii]	*Trifolium alexandrinum*
d]	Clusterbean	iv]	*Medicago sativa*

62. Match the crop with its scientific name.

a]	Sugarcane	i]	*Beta vulgaris*
b]	Sugarbeet	ii]	*Nicotiana tabacum*
c]	Potato	iii]	*Saccharum officinarum*
d]	Tobacco	iv]	*Solanum tuberosum*

63. Match the crop with its scientific name.

a]	Desi cotton	i]	*Corchorus olitorius*
b]	American cotton	ii]	*Gossypium herbaceum*
c]	White jute	iii]	*Corchorus capsularis*
d]	Tossa jute	iv]	*Gossypium hirsutum*

64. Match the crop with its scientific name.

a]	Sunflower	i]	*Crotolaria juncea*
b]	Napier grass	ii]	*Pennisetum purpureum*
c]	Sannhemp	iii]	*Nicotiana rustica*
d]	Tobacco	iv]	*Helianthus annus*

65. Match the crop with its scientific name.

a]	India rape	i]	*Brassica juncea*
b]	Mustard	ii]	*Gossypium barbadense*
c]	Black mustard	iii]	*Brassica campestris var. yellow toria*
d]	American cotton	iv]	*Brassica nigra*

66. Match the weed with its scientific name.

a]	Lantana	i]	*Echinochloa colonum*
b]	Kans grass	ii]	*Saccharum spontaneum*
c]	Jungli rice	iii]	*Eleusine indica.*
d]	Goose grass	iv]	*Lantana comera*

67. Match the weed with its scientific name.

a]	Saw thistle	i]	*Parthenium hysterophorus*
b]	Wild carrot weed	ii]	*Cuscuta chinensis*
c]	Common purslane	iii]	*Sonchus oleraceus*
d]	Dodder	iv]	*Portulace oleracea*

68. Match the weed with its scientific name.

a]	Canary grass	i]	*Azonpus compressus*
b]	Carpe grass	ii]	*Commelina beghalensis*
c]	Day flower	iii]	*Phalaris minor*
d]	Bermuda grass	iv]	*Cynodon dactylon*

69. Match the weed with its scientific name.

a]	Witch weed	i]	*Azolla caroliniana*
b]	Water velvet	ii]	*Sorghum halepense*
c]	Johnson grass	iii]	*Cyperus rotundus*
d]	Nut grass	iv]	*Striga spp.*

70. Match the weed with its scientific name.

a]	Sourgrass	i]	*Pennisetum cladestrium*
b]	Morning glory	ii]	*Paspalum conjugatum*
c]	Kikuyu grass	iii]	*Salvia auriculata.*
d]	Water fern	iv]	*Ipomea spp.*

71. Match the weed with its scientific name.

a]	Hairy beggar ticks	i]	*Ageratum conyzoides*
b]	Bran grass	ii]	*Bidens pilosa*
c]	Tropic ageratum	iii]	*Echinochloa crusgalli*
d]	Pimpernel	iv]	*Anagallis arvensis*

GK **The Pandayas were first mentioned by Megasthenese.** GS

72. Match the weed with its scientific name.

a]	Quack grass	i]	*Argemone maxicana*
b]	Prickly poppy	ii]	*Digitaria sanguinalis*
c]	Broom rape	iii]	*Agropyron repens*
d]	Crab grass	iv]	*Orobanche cernua*

73. Match the weed with its scientific name.

a]	Torepedograss	i]	*Borreria hispida*
b]	Black night shade	ii]	*Panicum repens*
c]	Wild oat	iii]	*Avena fatua*
d]	Button weed	iv]	*Solanum nigrum*

74. Match the weed with its scientific name.

a]	Lambsquarters	i]	*Mikania micrantha*
b]	Garden spurge	ii]	*Chenopodiun albun*
c]	Thistle	iii]	*Euphorbia hirta*
d]	Climbing hempvine	iv]	*Cirsium arvense*

75. Match the weed with its scientific name.

a]	Cox comb	i]	*Amaranthus viridia*
b]	Thatch grass	ii]	*Oxalis corymbosa*
c]	Wood sorrel	iii]	*Celosia argentina*
d]	Amaranth	iv]	*Imperata cylindrica*

76. Match the weed with its scientific name.

a]	Water lettuce	i]	*Lemma minor*
b]	Common duck weed	ii]	*Pistia stratiotis*
c]	Fragrant water lilly	iii]	*Eichhornia crassipes*
d]	Water hyacinth	iv]	*Nymphata odorata*

77. Match the weed with its scientific name.

a]	Water nut	i]	*Hydrilla vesticillata*
b]	Hydrilla	ii]	*Luphar advena*
c]	Yellow water lilly	iii]	*Pontederia sogittaria*
d]	Arrow head	iv]	*Nelumbo lutea*

78. Match the weedicide/fungicide with its method of application.

a'	Basalin	i]	Pre-emergence application
b]	Nitrofen	ii]	Seed treatment
c]	Ceresan	iii]	Post-emergence
d]	2, 4-D	iv]	Pre-plant incorporation

79. Match the following crops with the scientific practices

a]	Gram	i]	Retting
b]	Jute	ii]	Topping
c]	Maize	iii]	Silking
d]	Tobacco	iv]	Nipping

80. Match the following crops with the scientific practices.

a]	Maize	i]	Curing
b]	Groundnut	ii]	Buttoning
c]	Tobacco	iii]	Tasseling
d]	Sunflower	iv]	Pegging

81. Match the herbicides with its chemical class.

a]	Diallate	i]	Ureas
b]	Metoxuron	ii]	Phenoxy fatty acid
c]	Nitrofen	iii]	Ethers
d]	2, 4-D	iv]	Thiocarbamates

GS

Concept of cycle of erosion was given by W.M. Davis.

82. Match the herbicides with its chemical class.

a]	Propanil	i]	Dipyridyls
b]	MCPA	ii]	Amides
c]	Tricamba	iii]	Phenoxy fatty acid
d]	Paraquat	iv]	Acids

83. Match the herbicides with its chemical class.

a]	Linuron	i]	Anilines
b]	Trifluralin	ii]	Triazines
c]	Diquat	iii]	Urea
d]	Atrazine	iv]	Dipyridyls

84. Match the crop/weed with the scientific name.

a]	Wild oat	i]	*Zea mays indurata*
b]	Flint corn	ii]	*Avena fatua*
c]	Kabuli gram	iii]	*Cicer arietinum*
d]	Chickpea	iv]	*Cicer kabulium*

85. Match the crop/weed with the scientific name.

a]	Garden peas	i]	*Brassica compestris* var. yellow sarson
b]	Indian colza	ii]	*Linum angustifolium*
c]	Wild flax	iii]	*Linum usitatissimum*
d]	Linseed	iv]	*Pisum sativum* var. *hortense*

86. Match the following crops with their scientific names.

a]	American cotton	i]	*Corchorus capsularis*
b]	Sea island cotton	ii]	*Corchorus olitorius*
c]	Tossa jute	iii]	*Gossypium hirstums*
d]	White jute	iv]	*Gossypium barbadense*

87. Match the following crops with their product class.

a]	Sannhemp	i]	Forage crop
b]	Guar	ii]	Pulse crop
c]	Castor	iii]	Fibre crop
d]	Soyabean	iv]	Oilseed crop

88. Match the following with each other suitably.

a]	Maize kernel	i]	Stalk
b]	Castor inflorescence	ii]	Caryopsis
c]	Cotton square	iii]	Raceme
d]	Cotton fiber	iv]	Bracteoles

89. Match the following terms with the process.

a]	Desuckeing	i]	Removal of moisture from tobacco leaves
b]	Priming	ii]	Removal of fiber from jute stems
c]	Curing	iii]	Harvesting method of tobacco
d]	Extraction	iv]	Removal of buds in the axil

90. Match the following with each other.

a]	Steeping	i]	Fuel
b]	Retting	ii]	Sugarcane
c]	Bagasse	iii]	Sunnhemp
d]	Molasses	iv]	Jak

91. Match the following crops with their sowing period.

a]	Autumn sugarcane	i]	February-March
b]	Adsali sugarcane	ii]	July
c]	Spring sugarcane	iii]	October
d]	Sugarbeet	iv]	October-November

GK GS

Iodized common salts contain potassium iodate.

92. Match the following with each other.

a]	Lenticeles	i]	Red rusty spots
b]	Tuber	ii]	Respiratory structure
c]	Sugarbeet	iii]	Underground stem
d]	Spangles	iv]	Root

93. Match the following crop with the family.

a]	Canary weed	i]	Compositae
b]	Safflower	ii]	Leguminoceae
c]	Jute	iii]	Gramineae
d]	Pea	iv]	Tiliaceae

94. Match the following crop with the family.

a]	Wheat	i]	Gramineae
b]	Blackgram	ii]	Cruciferae
c]	Sesamum	iii]	Pedaliaceae
d]	Mustard	iv]	Leguminaceae

95. Match the crop with its critical period of weed control.

a]	Rice (Upland)	i]	25-30 days after sowing
b]	Rice (Transplanted)	ii]	15-25 days after sowing
c]	Groundnut	iii]	15-20 days sowing
d]	Urdbean	iv]	20-25 days after sowing

96. Match the following.

a]	Leaf area index	i]	Increase in plant dry weight/ unit dry weight/unit time.
b]	Crop growth rate	ii]	Increase in plant weight/unit leaf area/unit time
c]	Relative growth rate	iii]	Leaf area/unit land area
d]	Net assimilation rate	iv]	Increase in plant dry weight/unit time

97. Match the following soils with their chemical properties.

a]	Saline soil	i]	EC < 4 dsm^{-1}, ESP 15%, pH > 8.5
b]	Non-saline alkali soil	ii]	pH < 7.0
c]	Saline-alkali soil	iii]	EC > 4 dsm^{-1}, ESP < 15%, pH < 8.5
d]	Acid soil	iv]	EC > 4 dsm^{-1}, ESP > 15%, pH > 8.5

98. Match the threshold temperature with the crops given below.

a]	4°C	i]	Potato
b]	7°C	ii]	Cotton
c]	17-18°C	iii]	Oats
d]	6°C	iv]	Pea

99. Match the crop with the cultural practice

a]	Tobacco	i]	Earthing
b]	Cotton	ii]	Desuckering
c]	Peanut	iii]	Ginning
d]	Sugarcane	iv]	Pegging

100. Match the crop with their Important disease.

a]	Groundnut	i]	Blast
b]	Rice	ii]	Tikka disease
c]	Wheat	iii]	Late blight
d]	Potato	iv]	Black rust

101. Match the crop with their example.

a]	Silage crop	i]	Kidney bean
b]	Cover crop	ii]	Maize
c]	Row crop	iii]	Potato
d]	Cash crop	iv]	Cotton

102. Match the critical periods of weed with the field crops.

a]	Rice-lowland	i]	15
b]	Rice-upland	ii]	30
c]	Sorghum	iii]	60
d]	Fingermillet	iv]	35

103. Match phosphatic fertilizer with their origin/synthesis.

a]	Natural	i]	Bone char
b]	Treated	ii]	Nitrophosphate
c]	Industrial by-product	iii]	Bonemeal
d]	Chemical	iv]	Basic slag

Glossary of Technical Terms

Active Ingredient (*a.i.*) The chemical formulation which is directly responsible for the herbicides effect or the active toxic materials present in a formulation.

Agriculture Agriculture is defined as the art and science of producing crops and livestocks on a farm and marketing for use by the human race.

Agronomy Agronomy is a branch of agriculture that deals with the principles and practices of crop production and field management.

Agrostology It is the science which deals with the study of grasses, their classification, management and utilization.

Alien Weed Weed not native to our country.

Allelopathy Phenomenon of one plant having detrimental effect on another through the production of chemical compounds.

Antidote Safener Substance capable of antagonising specific herbicide phytotoxicity to plant.

Apophytes Facultative Weeds Weed species that grow primarily in wild communities but often escape to cultivated fields eg. cactus.

Arboricide A chemical used for killing shrubs or bushes, or trees (Silvicide).

Arid Climate Extremely dry climate with an annual average rainfall usually less than 250 mm and generally rainfall is well evapotranspiration demand of atmosphere.

Basal Treatment Herbicidal treatment applied to the stem of woody plants above the ground level.

Bioassay A testing method, using living organism to determine the presence of a chemical.

Biochemical Selectivity Is due to inheritence difference in the make up of plants with respect of their tolerance of certain toxic chemicals.

Bioherbicides Are usually native pathogens cultured artifically and spread just like pre-emergence application of herbicides on plants.

Blind Cultivation The cultivation after planting and before crop emerge.

Book Keeping It is an art and science of recording in books, business transactions in the regular and systematic manner so that their nature, extent and financial effects can be readily acertained at any time.

Border Effect The yield or any ancillary observations of plants nearer to the border of plot are affected due to different treatments of neighbouring plots, this effect is known as border effect.

Caryopsis The seed, grain or fruit of grass.

Cash Crops Crop grown for sale and brings money immediately e.g. cotton, tobacco, potato.

Catch Crops Crops which are grown as substitute for the main crop that has failed on account of unfavourable condition e.g. cowpea, sesamum.

Censer Mechanism When some weed plants are matured their pods are split and seeds are thrown away from their site upto some distance. This phenomenon of seed dispersion is know as censer mechanism eg., Maxican popy.

Clothes Line Effect Horozontal heat transfer from a zone of warm upwind over a relative cold irriaged crop field.

Companion Crops Two crops are taken together with the aim that they are benefitted to each other eg. maize and pea.

Contact Herbicide A herbicide that kills primarily by contact with plant rather than as a result of translocation.

Contingency Cropping It is growing of crops in aberrant situation like drought and flood. It aims at partial mitigation to misery by producing some food, feed and fodder to encounter emergency condition.

Cover Crops Cover crops are grown primarily to cover the soil and to reduce the loss of moisture and erosion by wind and wate eg. groundnut, kidney bean.

Crop Crop is defined as assembly or aggregates of useful plants which man can know how to use.

Crop Associated Weeds They may be associated with certain crops for one of the following causes viz. Need for specific microclimate, Mimicry and Ready contamination of weed seeds with crop seeds.

Crop Bound Weeds Species of weeds which usually parasitise the host crops eg., dodder on lucern, broom rape on tobacco.

Crop Rotation The repeatative cultivation in an ordered succession of crops on the same land eg. Jute-paddy-wheat.

Cropping Intensity It is the ratio between total cropped area and actual net cultivated area expressed in percentage.

Cropping Pattern The yearly sequence and special arrangement of crops and fellow on a given area.

Cropping Scheme Is a plan according to the crops are raised on individual plots of a farm or an object of getting the maximum return from each crop without impairing fetility of the soil.

Cropping System The cropping patterns used on the farm and their interaction with farm resources, other farm enterprises and available technology which determine their make up.

Cross Inoculation The inoculation by symbiotic bacteria of one legume species by those of another legume species.

Deep Tillage When soils tilled from 15 to 45 cm. It is carried out in root crops.

The chairman of the third Sangam was Nakkirar.

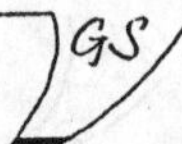

Defoliant A chemical which causes the leaf or foliage to drop from the plant.

Dessicant A chemicals which promotes the dehydration of plant tissue.

Detergent A chemical having the ability to remove oil.

Dibbling It is the method of sowing the crops with the help of manual labour or dibbler when specific spacing and number of plants are maintained between the rows and within the rows.

Dockage Sample Usually refers to a representative sample of seed used to determine the amount of inert material and other crop seeds etc.

Dry Farming Crops The crop which are grown under natural precipitation eg. sorghum.

Earthing Up To raise the soil at the base of plants eg. sugarcane, potato.

Emulsifying Agent A surface active material which facilitates the suspension of one liquid to another.

Endozoochory Mechanism The ingested weed seeds are passed in viable form from 0.2% to 9.0% with the animal excreta which is dropped wherever the animal moves eg., foxber.

Ephemerals The weed which completes its seed to seed cycle within two or four weeks e.g. niruri.

Facultative Weed Weed of wild community origin, escaping sometimes to crop yield e.g. cactus

Filler It is the inert inactive substance used to dilute the chemical for herbicide, insecticide or fungicide etc.

Forage Crop Crop grown primarily for livestock feed harvested either hay, silage, green feed grazing eg. lucern.

Gap Filling Resowing or replanting or retransplanting in sown, planted or transplanted field crops where gappings of plants found.

Grains The fruit of large seeded grasses. Botanically a grain or kernel is a fruit containing one seed.

Grass Botanically any plant that is in the family gramineae.

Green Manure Crops Any crops which are grown and burried into the soil for improving the soil condition by the addition of organic matter eg. sunhemp, dhaincha, gliricidia.

Green Manuring A practice of ploughing or turning into the soil the undecomposed green plant tissues for the purpose of improving physical structure as well as fertility of the soil.

Hectare Meter A volume of water necessary cover an area of one hectare to a depth of one meter. It is equivelent to 10, 000 m^3 ater or 10000 tornes of water.

Herbicidal Selectivity Is the foundation of phenomenal achievement in modern chemical vegetation management where the object is to some plants and retain others at the same place.

Quiz Kit!

First woman IAS officer

Anna Rajam George (b. 1927) of Calicut, Kerala was the first woman to qualify for the Indian Administrative Service in 1950. She was posted as a sub divisional officer in a small town in Tamil Nadu in 1951.

Herbicide A chemical used for killing or inhibiting the growth of plant.

Herbigation Application of herbicide along with irrigation eg. Benthiocarb in paddy.

Host Specificity Test The bioagent (insects, bacteria or fungi) choosen for biological control of weed should be tested for their basic dietory specific agent as there is possibility of any bioagent adopting new host.

Ideal Tillage Soil should be prepared in such a way that the tillage operations carried out to the normal sequence of tillage and upon moisture and type of soil.

Inter Cropping Growing two or more crops simultaneously in different rows in the field.

Inter Crops The crops raised in an orchard for increasing the income from land in short duration eg. vegetables, pulses, coconut.

Lay by Application It is the application of herbicide after the last cultivation of crops earthing up in sugarcane.

Lodging A state of standing crop in the field being flattend before harvest by heavy rain, wind or diseases.

Minimum Tillage The minimum manipulation necessary for crop production meeting tillage requirement under the existing condition.

Mixed Crops When two or more than two crops are grown together on the same piece of land at the same time eg. Bajra + Cowpea + Green gram.

Mixed Farming It is a combination of crop production with a significant amount of livestock raising eg. crops, dairying and poultry.

Monoculture The repeatative growing of the same crop on the same land eg. groundnut in Saurashtra.

Multiple Cropping Growing two or more crops on the same field in a year eg. maize-wheat, mung-maize, potato-wheat.

Nipping An act of removal of ear-heads from plant stalk.

Noxious Weeds (Objectionable Weeds) Those weeds being espe-cially undesirable, troublesome and difficult to control eg., *Cynodon dactylon*, *Cyperus rotundus*.

Nurse Crops Crops which are used to protect or nurse the other crops in their young stage eg. clusterbean in ginger, sunhemp in sugarcane.

Obligate Weeds Weeds that only in cultivated or distrub land

GK **Nile and Ganga are examples of arcuate delta.** GS

eg., field bind weed.

Pasture Crops Different types of vegetation found on pastures or grassland area which usually, grow eg. zinzvo, dharo.

Photolysis Decomposition of toxicant molecule structure under the influence of light from surface.

Planking A tillage operation carried out before sowing the crops for smothening and to cover the seeds after sowing.

Planting A method of sowing field crops in which plant parts are used eg. potato, sugarcane.

Polishing The removal of bran layer from brown rice by mechanical operation. It is also known as pearling or whitening.

Post-emergence Treatment Treatment made after the crop or weed plants have emerged above the soil surface.

Pre-emergence Treatment Treatment made after the crop is planted but before the emergence.

Preparatory Tillage Various tillage operations carried out on the farm before sowing the crops to create favourable soil condition for seed germination.

Puddling The ploughing or working of land in standing water with country or deshi plough or any other suitable implements by which soil looses granular structure and becomes deflocculated.

Rate Rate refers to the amount of active ingradient or acid equivalent of the herbicide applied per unit area.

Ratoon crop The practice of taking the second crop from previous one is known as ratooning.

Relay Cropping A cropping pattern in which crops are grown in the quick succession, second crop may be sown between the rows of first be sown between the rows of first crop which is in the harvesting stage.

Retting A process in which fibre plants like sunhemp, jute, etc. are kept in water for a few days to soften the fibres and fibre easily separated out from the stalk.

Row Crops Crops which are grown in rows with uniform spacing through out the field eg. cotton, sorghum, castor.

Rubbing A process of burning all farm residues of crops in the field after harvesting of crop to keep the soil free from pest, weeds etc.

Scarification The process of scratching the seed coat of hard or impervious seed in order to make germination easy.

Seed Seed is a dormant embryo with stored food, germinated under suitable conditions eg. bajra, wheat, rye.

Seedling Young germinated plantlets raised in the seed bed and then transplanted permanently in the field eg. paddy.

Shallow Tillage When soil is tilled upto depth of 2.5 cm to 15 cm. It is carried out for common field crops.

Silage Crops Crops which are preserved in a succulent condition by partial fementation in a tight silopit eg. maize, sorghum.

Smother Crop (Competitive crops) Crops which compete successfully with weeds by either excluding light or by utilizing large quantity of water, mineral nutrition or by limiting the quantities of all the three factors available to weeds eg., dhanicha, cowpea, kidney bean.

Soil Fumigant The materials used for killing all plants growth before planting or more desirable species eg., cynamide, methyl bromide.

Soil Sterilant Chemicals used where it is desire to prevent all plant growth for periods varying from months to years.

Sole Cropping One crop variety grown alone in pure stands at normal density.

Spot Treatment Application of spray to localised or restricted areas as differentiated from overall broadcast or complete coverage.

Stimulant Crop Crop or part of the plant stimulates or irritates the nervous system eg., tea, coffee, tobacco.

Stratification Means the practice of exposing imbibed seeds to cool (5 to 10 °C) and sometimes to hot (30 to 40 °C) condition for a few days prior to germination to break dormancy eg., grass seed.

Sugar Crops The crop grown for the purpose of producing sugar from them eg., sugarcane (10-13%), sugarbeet (15-18%).

Support Crops Certain fast growing crops work as supporter to vine crops eg. castor, shevri in betelvine.

Sward The grassy surface of pasture.

Sweath The strip of cut hurbage lying on the stubble.

Tame Pasture A pasture covered with cultivated plants and used for grazing.

Tannin A polyphenol compound found in tea leaves, is the second-most valuable constituent of tea, after caffeine.

Thinning Removal of excess plants from thickly populated crop field to give sufficient scope of growth to individual plant.

Threshing The process of removing grains from straw.

Till To plough or cultivate soil for seeding.

Tilth The physical condition of the soil with respect to its fitness for the planting or growth of a crop.

Topo-climatic Study of local climate as affected by slope and other surface characteristics.

Topping The removal of the terminal bud with or without some of the small top leaves just before or after the emergence of a flower bud.

Translcoated Herbicide Herbicide which is moved within the plant from the point of entry.

Transplanting It is the removal of an actively growing plant from one place and planting it in another place for further growth and producion.

GS

Stomach does not absorb indigestive food except alcohol.

Trap Crop Crops which are grown on boundary of the field for protection against insects, diseases, nematodes etc. eg. marigold.

Truck Crop Crop which yields in tonns and requires heavy transport vehicle eg., potato.

Tuber Crop Tubers enlarged under ground in the reproductive stem eg., potato.

Urease An universally occurring enzyme which activate urea hydrolysis to produce CO_2 and NH_3.

Vegetation Sum total of plants in a given area.

Volunteer Plants Unwanted plants growing from seed that remains on the field from the previous crop.

Water Regime It is the level or range of free water over the surface of soil especially paddy field.

Weed Competition Index The reduction in crop yields due to presence of weeds in comparison to weed free plot.

Weed A weed is an unsightly, useless injurious plant growing where it is not desired and something else should grow.

Wettable Powder (W.P.) Active ing-radiants carried on fine solid particles capable of forming suspension in a suitable carier.

Wood Ash A source of potash used as potassic manure containing about 2% P_2O_5 and 5% K_2O.

Xylophyte A woody plant

Yield Maximization Agronomic practices adopted to get highest possible crop production per unit area per unit time without considering either the cost of production as net return.

Zero Tillage A tillage system in mechanical soil manipulation is reduced to (soil compaction) and seed bed preperation only.

5. Horticulture

Each Word Counts

BASIC HORTICULTURE

Introduction

Horticulture is derived from the Latin words 'Hortus' meaning 'garden' and 'cultural' meaning cultivation.

Horticulture may be defined as the crop science which deals with the production, utilization and improvement of fruits, vegetables, ornamental plants, spices and plantation crops including medicinal and aromatic plants.

- Horticulture crops generally require intensive cultivation.
- Horticulture produces are highly perishable.
- Horticulture crops are the rich sources of vitamins and minerals.
- Cultural operations such as propagation, training, pruning, harvesting and marketing are skilled operations and are specific to horticulture crops.

Horticulture consists of the following divisions :

- **Pomology** - Study of fruit crops
- **Olericulture** - It is scientific study of vegetables.
- **Floriculture** -It is the art of growing, selling, designing and arranging flowers and foliage plants.
- **Spices** -Spices are those plants, the products of which are made use of as food adjuncts to add aroma and flavour *e.g.* pepper, cardamom, clove etc.
- **Condiments**- Condiments are also plants, products of which are used as food adjuncts to add taste only e.g. coriander, cumin.
- **Plantation crops** -Which are grown in extensive sale.

Importance of horticulture

- Horticulture crops contribute to rational income.
- Horticulture crops generate more employment opportunities.
- These are highly remunerative and profitable than cereals.
- Fruits and vegetables are regarded as 'protective foods' since they supply minerals and vitamins.
- Fruits yield more calorific value.

You Don't Know!

First double-decker bus

Taking into account the increasing rush and inability to cope with passenger traffic, the double-decker bus was introduced in Mumbai on Oct. 8, 1937. While a single-decker could carry 36 passengers, double-decker bus could take as many as 58.

- Many horticultural produces and their byproducts are the important raw materials for many industries.

Agro climatic zones for horticultural crops

1. Temperate Northern region :

Jammu and Kashmir, Himachal Pradesh, Uttaranchal, West Bengal. Temperate fruits and cool season vegetables are the important horticultural crops.

2. North Western arid region :

Rajastan, Gujarat, parts of Punjab and Haryana. Arid fruit crops like Ber, Pomegranate, Aonla, Date Palm, Wood apple and spices like coriander and cumin.

3. North Eastern subtropical humid region:

Arunachal Pradesh, Assam, Manipur, Meghalaya, Mizoram, Nagaland and Tripura. Fruits like Banana, Pineapple, Citrus, Jackfruit, Plantation crops like Tea etc.

4. North Central subtropical region :

U.P. Bihar, M.P. and parts of Maharashtra. Mango, Sapota, Sweet Oranges and Guava.

5. South Central tropical region :

Tamilnadu, Andhra Pradesh, Karnataka and parts of Maharashtra. Mango, Guava, Sapota, Pineapple and Turmeric.

6. Coastal tropical humid region :

The entire region stretching along the Bay of Bengal in the East and Arabic sea in the West. Banana, Mango, Cashew and Coconut.

7. Southern hilly region :

Comprising Western ghat and Eastern ghat. Coffee, Tea, Cardamom, Pepper, Oranges and Pineapple.

Establishment of orchard :

Establishment of orchard is a long-term investment and deserves a very critical planning. The selection of site, planting system and planting distance, choosing the varieties and the nursery plants have to be considered carefully to ensure maximum production.

Location and site

Criteria for selection :

1I. Location should be in a well-established fruit-growing region for getting benefits of others' experience and selling produce through co-operatives.

Jivaka Chintamani is written by Tiruttakadevar.

2. It should be closer to a market.
3. The climate should be suitable for the chosen fruit crops.
4. Adequate water supply should be available round the year.
5. Before a grower establish an orchard, he must have ensured the following factors.

(i) Suitability of soil, its fertility, the nature of sub-soil and soil debris.
(ii) Site must have proper drainage and no water stagnation during rainy season.
(iii) Irrigation water must be of good quality.
(iv) There must be proper transport facilities either by road or rail within the reach.
(v) Suitability of climate must be ensured for imminent dangers like cyclone, frost, hailstorms.
(vi) Whether there are seasonal gluts or overproduction in any particular period of the year.
(vii) Whether there is assured demand in the market for the fruits to be grown.
(viii) Availability of labour in the proximity.

Planning of an orchard :

A careful plan of the orchard is necessary for the most efficient economic management.

1. Optimum spacing to accommodate maximum number of trees.
2. Stores and office building in the orchard should be constructed at the centre for proper supervision.
3. Wells should be located at convenient places at the rate of one well for 2-4 ha.
4. Each kind of fruit should be assigned in a separate block.
5. Fruits ripening at the same time should be grouped together.
6. Pollinators should be provided in deciduous fruits. Some variety, require pollen from other variety, to set fruits in them, otherwise, they will be barren. Such pollen donors are pollinators. Each 3rd tree in every 3rd row should be planted with a pollinators.
7. Irrigation channels should be laid along the gradient.
8. Roads should occupy minimum space for the economy.
9. Short growing trees should be allotted at the front and tall at the back for easy watching and to improve appearance.
10. Fruits attracting birds and animals should be close to the watchman.
11. A good fence is essential. Live fence is economic, cheap, should be drought resistant, easy to propagate from seed, quick growing, dense foliage, standing severe pruning, thorny e.g. Agave, *Prosopsis juliflora, Pithecolobium dulse*.
12. Wind breaks, rows of tall trees planted close together around the orchard, are essential to resist velocity of wind.

Trees : Casuarina equisetifolia, Pterospermum, Polyalthia longifolia, Eucalyptus globlus, Grevillea robusta, Azadiracta indica.

Laying out of orchards :

Any method of layout should aim at providing maximum number of trees per hectare, adequate space for proper development of the trees and ensuring convenience in orchard cultural practices.

Broadly 2 types

(a) Vertical row planting pattern

(b) Alternate row planting pattern.

(a) Vertical row planting pattern

The trees set in a row is exactly perpendicular to those trees set in their adjacent rows (i.e. Square and Rectangular). In the latter planting pattern (i.e. Hexagonal, Quincunx and Triangular), the trees in the adjacent rows are not exactly vertical instead the trees in the even rows are midway between those in the odd rows.

The various layout systems used are :

1. *Square system*: Each corner of a square, whatever may be planting distance. This permits intercropping and cultivation in two directions.

2. *Rectangular system*: Any two rows is more than the distance between any two trees in a system. No equal distribution of space per tree.

(b) *Alternate row planting pattern* :

The various layout systems used are :

1. *Hexagonal system*: Planted in each corner of an equilateral triangle. 6 trees form a hexagon, 7th in the centre. This system provides equal spacing but difficult to lay out. Accommodates 15% more trees than the square system.

2. *Diagonal or quincunx system*: Square system with one tree at the centre, the central filler must be of short duration.

3. *Triangular system*: Square system, difference being that those in the even numbered rows are midway between those in the odd rows instead of opposite to them.

4. *Contour system*: On the hills, planted along the contour across the slope. It particularly suits land with undulated topography. The main purpose is to minimise land erosion and to conserve soil moisture. In South India tea is planted in contours either in single edge system or in double edge system.

Horticultural Soil

The soil act as a substratum and provides nutrients and water necessary for plant growth.

Tusks of elephants are modified incisor teeth.

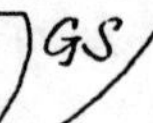

Kinds of Soil

Sandy soil is suitable for vegetables to obtain early crops and rapid root growth. It is the most common medium for propagating cuttings. Loam of tall types are highly suitable for horticultural production, it holds and releases nutrients for plant growth.

Soil fertility

The capacity of soil to nourish and sustain plant growth is known as soil fertility. Plantation crops like coffee, cardamom and species need fertile soils rich in organic matter and humus content.

Soil depth

Shallow soils with hard and compact subsoil layers like canker, heavy clay should be avoided for fruit crops as in such soils, the roots grow in the top layer of soil parallel to the surface. Such trees are easily uprooted by winds or premature death of trees may also occur.

Drainage

Drainage depends upon the nature of subsoil. In a good subsoil, trees can stand drought better because of deeper root penetration on wet, heavy soils with improvable sub-soil poor performance is noticed due to poor aeration and inadequate drainage.

Climate

Climate is one of the important complex factors which influences the production of horticultural crops.

Temperature

Optimum temperature range varies with crops and based on the optimum temperature requirement, horticultural plants are broadly classified into (a) temperate (b) subtropical (c) tropical plants. If a tropical plant is grown in subtropical zone their growth and performance may not be good.

Light

The intensity of light, quality and its duration affect the growth and development of horticultural plants.

Atmospheric humidity

Humidity affects the growth and development of horticultural crops. High humidity during flowering and fruiting promotes the incidence of pests and diseases.

Rainfall Total rainfall and its distribution decide the growth ahd development of horticultural crops.

Air pollutants The air pollutants like ozone, and photo chemical oxidants such as sulphur dioxide reduce the assimilation rate, growth and development of plants leading to reduction in yield and quality of the produces.

Hailstorms Occurrence of hail during pre-blooming or blooming period of apple, plum, peaches generally affect the fruit-set.

Knowledge World!

Most beautiful insect

The Common Peacock (*Papilio polyctor*) is thought to be perhaps the most beautiful butterfly. It occurs along the Himalayas from elevations of 1,200-1,800 m.

Harvesting and Marketing

Monetary returns of the fruit grower and the reputation of his fruit produce will depend on the following factors :

1. Stage of picking of fruits.
2. Packing methods.
3. Storing and transporting.
4. Processing and delivery to the consumer in the form and at the time the consumer desires to have the fruits.

Picking

Picking refers to the harvesting of fruits from the trees at the right time. The time of picking depends upon :

(i) Variety : Certain variety only ripen in trees and certain variety will ripen even when they are harvested at ¾th maturity.

(ii) Purpose: For which it is meant, e.g. tomatoes for table purposes can be harvested at red ripe stage, whereas for distant markets they may be harvested during breaker's (colour change) stage.

(iii) Distance to the market: The closer the market, ripe fruits may be picked, if distance is more mature green stage.

Methods to determine proper time

1. Maturity index

Following are the rough but ready maturity tests of fruits employed to pick the well matured fruits.

(a) Colour changes e.g. mango, papaya.
(b) Increase in size.
(c) Softening of the tissue of the fruits e.g. figs, grapes.
(d) Case of detachment from the stalk e.g. sapota, annona.
(e) Shrivelling of fruit stalk e.g. water melon.
(f) Time elapsed from the date of flowering to picking maturity.
(g) Sound by tapping jack and water melon produce hollow and dull sound at ripening stage.

2. Accurate tests

(a) *Colour charts*: Charts prepared for different stages of maturity.
(b) *Penetrometer*: Instrument measuring the softness of tissues as an index of maturity.
(c) *Sugar/acid or Brix/acid ratio*: Based on the principle

The largest tribal population of Rajasthan is Meena.

that acid content reduces and sugar increases on ripening.

The fruit growers should bestow more attention and considerable care during the picking season to reduce to a minimum level of careless handling of fruits by pickers.

1. Picking must commence from lower branches advancing towards the top to reduce dropping of fruits.
2. Dropping of the fruits from the tree should be avoided.
3. During picking, care must be taken to avoid any possible damages to the branches especially to the spurs as the subsequent cropping depends upon them.
4. Picking easily in the morning is best, fruits kept in shade. After picking fruits should be kept in coolest place available, which is well-ventilated to arrest respiration and breakdown as much as possible.
5. There should not be any bruises while picking, as it lowers the marketability of the fruits.
6. If picking is done in mid-day or hot weather, the fruits should be kept in a shed overnight to cool.

Name of fruits	*Storage temperature (C°)*	*R.H. (%)*	*Approximate storage*
Apple	0-2.0	85-90	4-8 weeks
Grapes	0-2.0	80-85	2-4 weeks
Peaches	0-2.0	80-85	2-4 weeks
Banana	12.5-14.5	85-95	3 weeks
Mango	9-10.0	85-95	4 weeks
Orange	6-6.5	85-90	18 weeks
Papaya	4.5-6.0	80-85	5 weeks
Guava	9-10.0	85-90	3 weeks
Tomato (ripe)	0-4.5	85-90	10 days
Tomato (mature green)	10-15.0	85-90	30 days
Peas	0-0.5	90-95	2 weeks

Measures to reduce respiration

Temperature regulation

High humidity

Ventilation i.e., outdoor air

Handling, Grading and Storage of fruits

Production of horticultural produce is limited by certain climatic condition and through proper storage of fruits, the availability, of produce is extended.

Factors influencing storage: Fruits and vegetables contain about 85 and 95% moisture content respectively. As these are living tissues, after harvesting respiration goes on. So any storage procedure should aim at slowing down, regulating or preventing the increased respiration.

Temperature: The rate of respiration can be controlled by temperature regulation, low temperature slows down.

Humidity : Maintaining high humidity during storage minimizes respiration.

Storage atmosphere: In addition to O_2 and CO_2 other gases may affect storage of fruits like ethylene by hastening ripening. These storage atmosphere can be improved by ventilation which mean introduction of outdoor air into the storage room.

Methods of storage

1. *Air cooled storage*: Fruits placed in racks in insulation box. Keeping the ventilator open during night to admit the cool night air and during the day shut down. Thus storage rooms can be kept cool.
2. *Air cooled storage refrigerated with ice* : A few storages are filled with ice and salt and equipped with furs for quick cooling.
3. *Treating with wax emulsions*: In the absence of cold storage, fruits are dipped for a minute in a wax emulsion and dried completely for 10-15 minutes.
4. *Refrigerated storage*: 1. Prompt cooling of produce 2. Maintenance of optimum R.H. 3. An even holding temperature with 1.0-2.0°C fluctuations 4. No need to depend upon the external conditions of air, their movements 5. Most efficient and scientific way of storing.
   Refrigerated storage involves the use of refrigerant (Freon NH_3-CH_3Cl).
5. *Controlled atmospheric storage (CAS)*: By reducing the O_2 supply available to the fruit and by increasing the amount of CO_2 around the fruit or by lowering the temperature, the respiration can be slowed down in fruits and vegetables. A combination of all these principles is involved in controlled atmospheric storage.
6. *Preservation by irradiation*: Potatoes, onion, garlic and other crops lose their commercial value by sprouting or rooting during storage. These processes can be delayed by irradiating the produce with gamma rays.
7. *By use of chemicals*: Certain chemicals increase the shelf-life of fruits by delaying the ripening and senescence $KMnO_4$ an ethylene absorbent absorbs ethylene, from banana-held sealed polythene bags.
8. *Evaporative cooling (cool chambers)* : Evaporation of water produces a considerable cooling effect and this evaporative cooling occurs when air that is not saturated with water vapour, is blown across the wet surface.

Cooling chamber works in the principle of evaporation cooling. This structure essentially contains a single layer of bricks as floor and a double layer of bricks as side wall with an interspace of 7.5 cm wide filled with river sand. The top of the storage space is covered with piloted vetiver roots or gunny cloth in a bamboo structure.

Once the cool chambers are saturated with water, sprinkling of water once in the morning and once in the

According to the World Bank, the world's population is 6.5 billion.

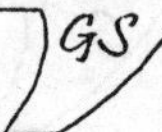

evening is enough to maintain temperature and humidity.

These storage chambers maintain a very high humidity of about 95% throughout the year. These chambers are ideal for storage of fresh fruits and vegetables for a short period.

Ripe mangoes can be stored upto 10 days as against 4 days in room temperature. Similarly citrus fruits like grape fruit and orange can be stored upto 60-90 days as compared to 8-15 days in ambient room temperature.

Training and Pruning

Certain fruit plants if left to themselves grow wild and do not bear abundantly unless trained or pruned to a specific form.

Training

When a plant is tied, fastened, staked or supported over a trellis or pergola in a certain fashion or some of its parts are pruned with a view to giving the plants a framework, the operation is called 'training'

Comparison between training and pruning

Training	*Pruning*
1. It concerns form primarily	1. It affects functions only
2. It determines plant's outline, its branching and framework	2. Assists more in what the tree does in respect to fruiting
3. Generally concerned with first few years of the tree growth to determine its annual framework	3. This operation is carried out annually

Objectives of training :

1. To admit more light and air to the centre of the tree and to expose maximum leaf surface to the sun.
2. To direct the growth of the tree so that various cultural operations, such as spraying and harvesting can be done at the lowest cost.
3. To protect the tree from sunburn and damage.
4. To secure a balanced distribution of fruit-bearing parts on the main limbs of the tree.

Method of leader training :

The method of leader training should be suited to the normal habit of the fruit tree the common system of training followed :

1. *Open centre*: The main stem is allowed, to grow only upto a certain height and the leader stem is pruned to encourage scaffold branches production. This system is also known as vase-shaped system.
2. *Central leader*: A tree is trained to form a trunk which extends from the surface of the soil to the top of the tree. This system of training is also known as closed centre.
3. *Modified leader*: Intermediate open centre and central leader. This is developed by first training the trees to the leader type, by allowing the central axis to grow unhampered for the first 4 or 5 years. The central stem is then headed back and lateral branches are allowed to grow. So it possesses the advantages of both the system.

> **Quiz Kit!**
> ***Computer to read handwritten scripts***
> **A computer system to read handwritten scripts was developed by the scientists at the Tata Institute of Fundamental Research (TIER) in 1994 which can handle a 2000-word vocabulary written in lower case English letters.**

Trees trained with the help of a support can be:

1. *Bower system* (pandal, pergola system) in grapes, cucurbitaceous vegetables, vines spread over a criss cross network of wires.
2. *Espalier system*: Trained to grow flat on trellis or on horizontal wires by training the branches perpendicularly to the main stem.
3. *Kniffin system*: Two trellis of wire are strung supported by vertical parts.
4. *Telephone system*: Overhead trellis (3-4 wires).
5. *Tatura trellis*: Multi-layered wire trellis (v-shaped).

Trees trained do not require support but will be trained to a particular shape:

(a) *Head system:* Grapevine is allowed to green single stem with the help of stakes.

(b) *Palmelle:* Central leader having 3-4 tiers of two opposite branches trained at an angle of 30 from the vertical.

(c) *Spindle brush*: Central stem with 15-20 semi-horizontal branches.

(d) *Dwarf pyramid*: Central leader tree and fully grown trees consist of central erect stem with 20-30 branches 45°-90°.

(e) *Head and spread system*: Variation of modified leader system. Primary branches are taken in different tiers, each tier being 60 cm above the lower tier.

Precaution should be taken to maintain the desired frame of the tree for long duration of time.

Pruning

On the other hand 'pruning' refers to the removal of plant's parts such as buds, developed shoots and roots to maintain a desirable form by controlling the direction and amount of growth.

Commonly, trees are pruned annually in two ways.

Shoots or branches considered undesirable are removed entirely without leaving any stub, called 'thinning out'.

Removal of terminal portion of the shoots, branches or limb, leaving its basal portion contact is called 'beading back' thinning out. Involving large limbs as in old and diseased trees: bulk pruning.

Objectives of pruning :

(i) To remove surplus branches.
(ii) To open the trees so that the fruits will colour more satisfactorily.
(iii) To train it to some desirable form.
(iv) To remove the dead and diseased limbs.
(v) To remove the water sprouts.
(vi) To improve fruiting wood and to regulated production of floral bud.

Season of pruning :

Winter pruning is preferred to reduce bleeding summer pruning may have a dwarfing effect or an invigorating influence.

A light summer pruning may aid in coloration of fruit in certain variety.

Special pruning techniques :

1. *Root pruning:* Culture of dwarf fruit trees.
2. *Ringing:* Removal of a complete ring of bark from a branch to increase fruit bud formation in certain fruit crops.
3. *Notching*: Partial ringing of a branch above a dormant lateral bud. The practice of notching below a bud (nicking) is effective in starting a floral shoot.
4. *Smudging*: Practice of smoking the trees like mango, to produce off-season crop in Philippines.
5. *Bending*: Bending of branches for increasing fruit production in guava.
6. *Coppicing*: Complete removal of the trunk in trees like, Eucalyptus and Cinchona leaving 30-35 cm stump alone.
7. *Pollarding* : Removing growing point e.g. silver oak.
8. *Pinching:* In herbaceous plants removal of growing point.
9. *Disbudding:* Removing unwanted flower bud.
10. Alternate differential pruning.

Propagation of Horticultural Crops

I. Sexual propagation

Seed: Most annuals, biennials and many perennial fruit plants; vegetable crops; plantation, aromatic and medicinal plants; ornamental, flowering and shade providing shrubs and trees.

II. Asexual propagation

(i) Propagation by apomicitic seedlings : Citrus, mango, apple etc.

(ii) Propagation by vegetative structures :

A. Propagation of division

Division of stem tubers: Potato, caladium etc.

Division of tuberous roots:Sweet potato, turmeric, dahlia, bigonia etc.

Division of rhizomes: Ginger, banana, blueberry, bamboo, iris, oxalis etc.

B. Propagation by separation

Bulbs: Onion, daffodil, Bird of Paradise, tulip, narcissus, hyacinth etc.

C. Propagation by sukers

Pineapple, banana, chrysanthemum, blackberry, raspberry etc.

D. Propagation by runners

Strawberry

E. Propagation by stolon

Doob grass.

F. Other vegetative structures

Bulbils : Lilies

Offsets: Pineapple, banana, cacti and succulent, tulip, iris, lily etc.

G. Propagation by cuttings

Stem cuttings: Grape, fig, pomegranate, mulberry, kiwi fruit, olive, quince, rose, honeysuckle wisteria, weeping willow, poplar etc.

Semi-hardwood cuttings: Mango, guava, jackfruit, camellia, lemon, rhododendron etc.

Softwood cutting: Juniper, hollyhock, azalea, lilac etc.

Leaf cuttings: African violet, bigonia, bryophyllum etc.

Leaf bud cuttings: Blackberry, lemon, cammelia, hydrangea, kalanchoe etc.

Root cuttings: Blackberry, fig, cherry, apple, pear etc.

H. Propagation by layering

(a) *Tip layering*: Trailing black berry, raspberry etc.
(b) *Simple layering*: Ornamental climbers, bougainvillea, grape, lemon etc.
(c) *Stooling/Mound layering*: Apple rootstocks, guava, mango, litchi, gooseberry etc.
(d) *Serpentine layering*: Clematis, American grape etc.
(e) *Trench layering*: Apple rootstocks, cherry, plum etc.

I. Propagation by grafting

(a) *Whip grafting*: Temperate fruit plants
(b) *Cleft grafting* : Avocado, apple, pear, plum, camelia etc.
(c) *Bridge grafting*: For different fruit plants
(d) *Side grafting*: Mango, avocado, junipers etc.
(e) *Inarching* : Mango, sapota, guava, litchi etc.
(f) *Veneer grafting*: Mango
(g) *Epicotyl grafting*: Mango

J. Propagation by budding

(a) *T-budding*: Citrus, plum, peach, cherry, ber, rose etc.
(b) *Patch budding*: Walnut, Indian rubber tree etc.
(c) *Chip budding*: Mango, grapes etc.
(d) *Ring budding*: Peach, plum, ber, mulberry etc.

K. Micropropagation

(a) *Shoot tip culture*: Citrus, papaya, grape, orchids, asparagus, lily, gladiolus, strawberry, carnation, chrysanthemum etc.
(b) *Embryo culture*: Citrus, rhododendron etc.
(c) *Single cell culture*: Carrot, sunflowers etc.
(d) *Embryo rescue*: Grape

Pulmonary veins carry de-oxygenated blood.

Botanical clasfication of fruits

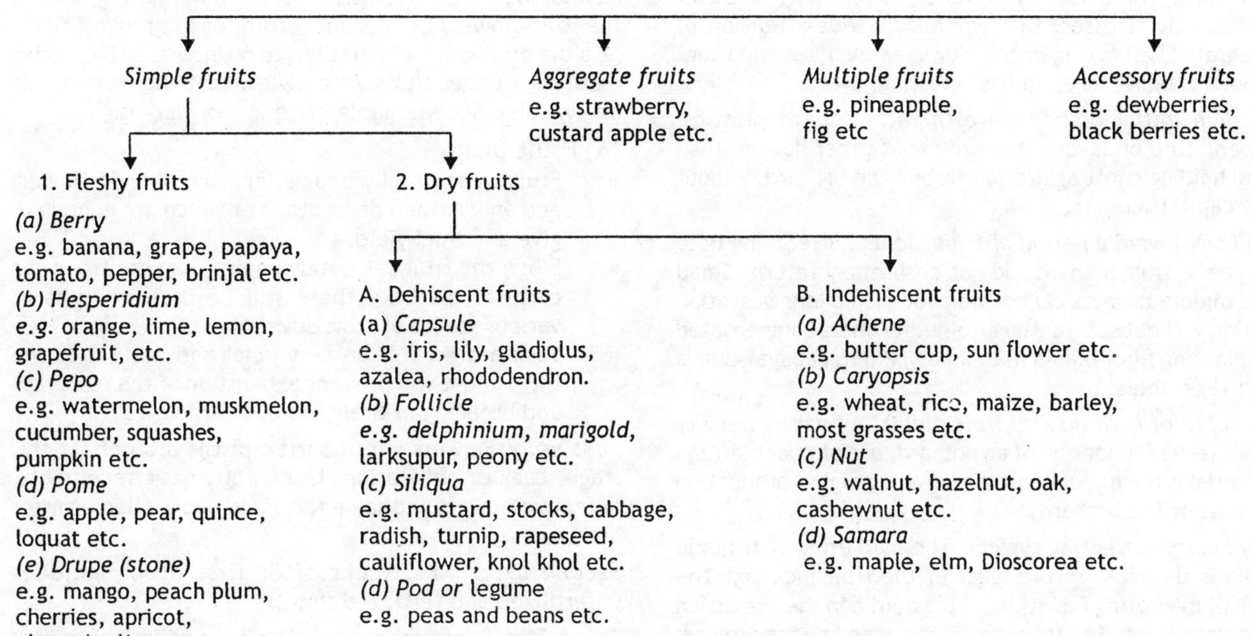

FRUIT SCIENCE

Scope of fruit growing

Cultivation of fruits contributes to the health, happiness and prosperity of the people.

In spite of the fact that India is climatically favourable for production of a variety of tropical, sub-tropical and temperate fruits, our per capita per day consumption of fruits is very low (80 kg) so in comparison to developed countries (250-450 gm) and developing countries like Hungary, Philippines, Pakistan (100-200 gm)

1. For that reason, there is a vast scope of growing fruit crops in our country because the total area under fruit crops is very less.
2. About majority of Indian population is vegetarian and production of fruits is far less.
3. With installation of Agro-industries and storage facilities, improvement in transport facilities, preservation and canning will have great demand. Govt's promotion of Agricultural exports through Agri export zones creation will ensure remunerative price to farmers.
4. The multiple cropping system in the orchards also generate more avenues for employment apart from giving additional income. The intensive horticultural practices may accommodate skilled and trained personnel where unemployment is a serious problem.
5. India processes only 2% of its produce as against 45% in Philippines. So there is a more distance to cover for catching up the world market with Govt. support in the value-added products.
6. The fruit production has a great possibility in trade and industrial development. A large number of ancillary industries and trade sectors are also developed.
7. With our rapid increasing population, the pressure on land is likely to continue and size of holding will remain small in the foreseeable future. Moreover fruit crops give much more production per unit of area and are certainly paying more than ordinary crops.
8. After achieving the food security when the Govt. is planning for agricultural diversification, at this moment fruits may well fit in crop diversification programmes

Problems of fruit cultivation

While fruit cultivation is having lots of promise but still some of the problems dog up the idea of a successful fruit

GK GS

Part XXI of Constitution deals with special provisions of J&K.

cultivation. Some of the national problems like malformation, alternate bearing and spongy tissue in mango still defining solution yet no headway has been achieved in production of superior quality planting materials. There is lack of varieties resistant to biotic and abiotic stresses. Due emphasis needs attention of Integrated Pest Management. However, the most important problems being faced in fruit growing, are:

(i) High initial orchard investment: Costs of plastics, expenditure on layout, the ordinary farmer having small land holding can't afford to establish an orchard without any capital support.

(ii) Long juvenile period of fruit plants : Stretching upto 10 years; fruit orchards do not ensure any return. Small and middle farmers cannot wait for such a long gestation period without substantial income. It can be compensated by planting filler plants like Kinnow, Plum, Papaya, Guava and Vegetables.

(iii) Lack of high quality fruit plants: A healthy nursery plant is the foundation of an orchard, use of poor planting material is the major constraint limiting higher production in most of the orchards.

(iv) Faulty marketing system: The marketing of fruits in India is the most sorry aspect of the fruit industry. The lack of marketing facilities is a big hurdle in the expansion of horticulture. The fruit crop is auctioned by the growers to the middle man who can get better profit from this produce.

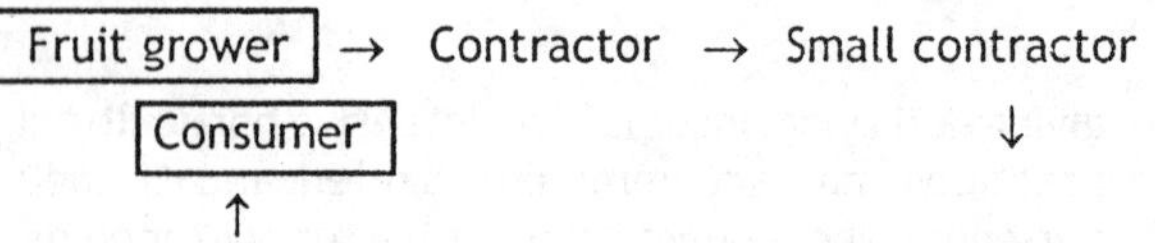

- Very lengthy procedure and cost more to the consumer .

(v) Lack of storage and transportation facilities: The post-harvest losses account for 25% of the fruits. After harvest, the fruits need to be transferred to cooling unit within half-an-hour. Due to poor cold storage and controlled atmospheric storage facilities the losses further increase. Inefficient transportation system aggravates the position.

(vi) Lack of processing units: When all over the world there is processing of fruits and vegetables. Processing units absorb the glut in the markets thus favouring the farmers income. But the processing industry may be located preferably established in the private sector with proper safeguards for the interests of the producer.

(vii) *Perishable nature of fruits* : Most of the horticultural produce is highly perishable because of high water content. In spite of very heavy losses, very little work has been done on post-harvest handling and storage systems.

(viii) Low purchasing power of people: The fruits are sold at very high cost in the markets, low income and middle income group has low purchasing power and do not purchase, the fruits regularly. There is a strong need for developing the co-operative system of marketing, so that the fruit growers get decent return for their investment. Also the production of fruits has to be increased by putting more area under fruits with a view that the fruit should be available at reasonable price in the market.

(A) Fruit Drop

- Fruit trees usually bear a large number of flowers and only a small percentage of which are enough to give a normal yield.
- When the fruit set is much more than the trees can carry to maturity there will be drop of fruits at various stages of fruit development.
- Such a drop will be beneficial and natural to the trees and it will prevent exhaustion of the resources and breaking of branches by overbearing.

First drop: Flowers with aborted pistils drop off at this stage. Lack of pollination, lack of stigmatic receptivity, defective flowers and some fertilized poor pollen transfer are causes.

Second drop : Fortnight after first drop. Includes unfertilized and fertilized flowers.

Third drop: Drop occurs when fruits are of *'minable size'* due to the formation of abscission layers on the young fruit stalks. The above drops which occur in 3 stages are beneficial to the fruit tree.

Pre-harvest drop: Loss to the grower. At this stage, half developed and 3/4th developed fruits are shed due to many causes. This is a loss to the growers and is a serious problem.

Causes of fruit drop :

1. *Mechanical*: Wind and hailstorms cause fruit drop.

2. *Climatic* : High temperature, low humidity, very low temperature bost on the formation of abscision layers, consequently fruits drop.

3. *Physiological* : Abnormal fluctuations of soil moisture favour heavy fruit drop.

4. *Nutritional*: Lack of available N and several other nutritional factors may cause fruit drop.

5. *Cultural* : Deep digging and deep ploughing during the fruit development phase will damage the roots cause the fruits to shed.

6. *Pathological* : Incidence of parts and diseases will cause more shedding of fruits.

7. *Varietal* : Within a kind of fruit, the varieties differ among themselves in the extent of fruit drop.

Prevention :

- The pre-harvest drop may be reduced by controlling

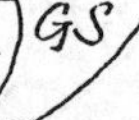

the causes to a certain extent. Proper and timely culture such as irrigation, plant protection, provision of polliniser and wind breaks will help to prevent or reduce the amount of fruit drop.

- In the final stage of fruit growth, a rapid decrease in auxin content is correlated to segmentation of endosperm causing pre-harvest fruit drop. In high concern of auxin supplied exogenously may inhibit fruit drop.
- NAA (10-20 PPM) reduces drop of fruits in pome, fruits such as apples and pears.
- 2,4-D (less than 20 PPM) checks drop in citrus fruits.
- GA at 60 PPM reduces the drop in Jamun.
- Great care, must be exercised on the use of hormones as it is possible to cause much damage to the foliage, and new growth.

(B) Alternate bearing

Alternate or biennial bearing is an age old horticultural problem and is also designated as irregular bearing.

A year of optimum or heavy fruiting is followed by a year of little or no fruiting. Essentially these trees, carry optimum load of crop in one year but fail to flower and produce a satisfactory crop in the following year. This phenomenon is observed in most of the perennial crops in the following year varying with degrees.

Any perennial plant, which carries a heavy crop load in the current year is likely to bear less in the following year. However this phenomenon is distinctly 'expressed in horticultural crops like mango, olive, coffee, tamarind, apple, coconuts.

The heavy crop year is 'on year' and the less bearing 'off year'.

If a 'on year' is regularly followed by an 'off year', then it is called "biennial or alternate bearing". Irregular bearing on the other hand refers to the rhythm of a heavy crop followed by more than one year of poor crops.

Disadvantages :

1. Compared to regular bearer, the average yield is less.
2. The quality of fruits may be inferior during 'on year' Cropping compared to regular bearer.
3. More breakage of heavy loaded branches is likely in 'on year'.

Causes :

1. *Genetic causes*: Certain trees which belong to the particular species or genus or family express this tendency always e.g., apple, pear, plum and apricot belong to Rosacea have a tendency. Within a species also cultivars differ on this tendency e.g. in mango, (Bangalora, Neelum, Mallika) are regular bearer but Mulgova, Dashari are alternate bearing.

Did You Know ?

Largest beetle

The largest Indian beetle is *Autocrates aenews* and is found in the Himalayas. Females measure up to 8.5 cm in length.

2. *Bearing habit*: Terminal flower bud formation is often regarded as an explanation for alternate bearing in apple. The formation of spurs (which is a terminal bearer) is strongly stimulated in the 'off year', hence in the on year, the crop bears heavily.

3. *Age of the tree and size of the shoots*: In biennial apple, this tendency is found to be more, in young trees than in older trees. In mango, it has been found that bearing shoots of 8-10 month-old branches produce flower buds.

4. *C/N ratio*: It plays important role in creating favourable conditions for the synthesis and action of the substances responsible for flowering.

5. *Hormonal balance*: Higher levels of auxin like substances and inhibitors (similar to ABA) and lower levels of GA_3 like substances are vital for a fruitfullness shoot In mango.

Remedial measures :

1. *Scientific management of orchards*: A proper schedule of numerus and fertilizer application, intercultural operations, irrigation at fruit set and fruit developmental stage, plant production measures etc. will help.

2. *Deblossoming*: Can be helpful only at full develop stage of panicles.

3. *Smudging and chemical regulation* : Practice of inducing flowering in mango by smudging is being followed in Java and Philippines.

4. *Pruning*: Only old trees may respond to pruning. It can be, helpful in reducing irregular bearing and not alternate bearing.

5. *Growing regular bearing variety* : Regular bearing varieties are to be preferred.

Nutritional value of fruit crops

Fruits being rich in vitamins and minerals, are known as protective food. Realizing the worth of fruits and vegetables in human health, Indian Council of Medical Reasearch (ICMR) recommends the use of 120 g fruits and 280 g vegetables per capita per day. Consumption of adequate amount of fruits helps in maintaining health and vigour of an individual.

Nutrient-wise richness of important fruits are as under :

Vitamin 'A' richness

Mango	4800 IU/100 g
Papaya	2020 IU/100 g

GK **Atmosphere absorbs 14% of incoming shortwave solar radiation.** GS

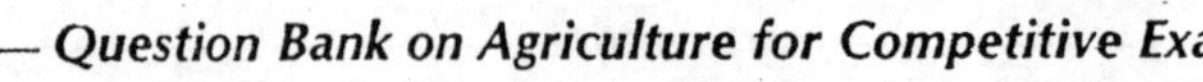

Vitamin 'B₁' (Thamine) richness

Cashewnut	630 mg/100g
Walnut	450 mg/100g
Apricot (dry)	120 mg/100g

Vitamin 'B₂' (Riboflavin) richness

Beal	1191 mg/100g
Papaya	250 mg/100g
Litchi	122.5 mg/100g

Vitamin 'C' richenss

Barbados cherry	1000-4000 mg/100g
Aonla	600 mg/100g
Guava	299 mg/100g
Lime	63 mg/100g
Sweet orange	50 mg/100g
Lemon	39 mg/100g

Carbohydrate richness

Apricot (dry)	72.81%
Date (pind)	67.30%
Karonda (dry)	67.10%
Banana	36.40%
Bael	30.60%
Custard apple	23.90%
Cashewnut	22.37%
Jamun	19.70%
Jackfruit	18.90%
Topioca	38.10%

Protein richness

Cashewnut	21.20%
Almond	20.80%
Walnut	15.60%

Fat richness

Walnut	64.50%
Almond	58.90%
Cashewnut	46.90%
Avocado	22.80%

Fibre richness

Guava	6.90%
Kaintha	5.20%
Pomegranate	5.10%
Aonla	3.40%
Grape	3.00%
Walnut	2.60%

Calcium richness

Litchi	0.21%
Karonda	0.16%
Kaintha	0.13%

Phosphorus richness

Cashenut	0.45%
Walnut	0.38%
Litchi	0.30%
Kaintha	0.11%

Iron richness

Dry Karonda	39.10%
Date (pind)	10.60%
Cashewnut	5.00%
Walnut	4.800%

Package of Practices of Fruits

	Mango	Banana
Soil	Alluvial to laterite soils. Slightly acidic sill with pH< 7.5 Black soils not suitable	Deep, well drained loamy soil Depth and drainage is very important.
Climate	Tropical fruit 24-27°C, 1400 mMSL. Low temperature and freezing.	Humid tropics Frost is the limiting factor.
Propagation	Veneer grafting (commercial) (season: March-September-North India)	Sword suckers (1.5-2.0 kg). Throughout the year, except heavy rains
Season	Monsoon season	February-March (hills of South India).
Spacing	(10x 10 m; HDP: 2.5 x 2.5m)	Tall var : (2.7 x 3.0m). Dwarf var: (1.8 x 1.8m).
Training pruning	Hardly needs any pruning Periodical removal of dead and diseased branches.	No training and pruning. Desuckering (of unwanted suckers).
Flowering	February-March (North India) January-February (East India) Duration 2-3 weeks	After 8-9 months of planting. 10-12 leaves with pseudostem (At the time of bunch disterentcation).

Gautam Buddha died in the Malla republic.

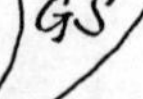

	Mango	Banana
Pollination & fruits growth & development	Highly cross-pollinated crop. Agent-housefly (Musca domestica), < 1.0 % fruit set. Sigmoid growth curve.	Vegetative parthenocarpic Double sigmoid growth curve. Parthenocarpic.
PGR used	Last week of March, 2nd week of June. NAA, GA, @ 25 PPM for fruit crop.	Ethrel-ripening.
Harvesting	Slight colour development of shoulders. Specific gravity 1.0 - 1.1%	Skin changes from angular to round. September-April.
Yield	10-15 t/ha (10th year).	Tall vars : 15-20t/ha. Dwarf var : 30-40t/ha.
Problems	Mango malformation less temperature). Spongy tissue (convective heat). Alternate bearing.	Panama wilt. Bunchy top of banana.
Varieties	Alphonso, Amarapali, Mallika, Dashari , Neelum.	Robusta, Rasthali, Poovan, Dwarf cavendish.
Manures and fertilizers	1000 : 500 : 1000 g NPK/tree 10 years and above age.	100-200 : 100-150 : 200-300 NPK / plant
	Citrus	**Grapes**
Soil	Loamy soil with pH 5.5-7.5. Sensitive to salts and presence of Kankar ($CaCO_3$).	Sandy loam soils. Moderately tolerance to salinity and alkalinity.
Climate	Subtropical fruit (Frost free). Light loving plant. 5-8 m depending on species. Monsoon season.	Long, warm to hot, dry summer and cool. Heat units for ripening winters. Early (1600 degree days), late (3000 degree days)
Propagation	Budding	Hardwood cutting (23-46 cm length with atleast 4 buds).
Season	February-March (major bloom). (North India)	January (1-year-old rooted plants).
Spacing	(8 x 8 m)	Bower system : (3x 3 m).
Training pruning	Pruning of young trees and training. Bearing trees- no pruning.	Training :Bower (S.I.) Head system (N.I.).
Flowering	Honey bees-cross pollination. Poor fruits set due to high drop.	N.I. (January) S.I. (March-April, September-October). Time of bud break : March-June. Full bloom : 6-8 weeks after bud break.
Pollination & fruits growth & development	Sigmoid growth pattern (Kinnow). Linear pattern (Sweet orange).	Parthenocarpy (Black Cornith). Stenospermocarpy (Petlette Pusa Seedless). Self pollination is rule. double sigmoid growth curve.
PGR used	2, 4-D is very effective for fruit drop. (M H-preharvest drop)	GA3 @ 40 PPM - flowering and bearing set stage improves size and quality of berry (Thompson seedless).
Harvesting	TSS/acid ratio-good indices (14.1). Generally green to pale yellow.	TSS/acid ratio is 15:1 Colour change and softening with taste.
Yield	Sweet orange: 1000-1500 fruits/tree.	15-20 tonnes/ha (Thompson seedless).
Problems	Citrus decline Granulation	Splitting of berries.
Varieties	Sweet orange: Mosambi, Sathugudi, Mandarins : Nagpur Cantia, Kinnow.	Thompson seedless, Pusa seedless, Bhokri, Beauty Seedless, Perltte Cornith.
Manures and fertilizers	Sweet orange: 25-30 kg. 1000 : 100 - 200 g NPK/tree	1000 : 500 : 1500 g NPK/vine fully grown vine (3rd year).

Requirements	*Sapota*	*Ber*	*Pomegranate*	*Apple*
Soil	Sandy loamy, red latarite	Marginal lands with high pH	Deep loamy	Slightly acidic soil
Climate	Tropical-warm and moist weather	Dry and hot climate	Subtropical fruit	Temperate
Propagation	Inarching	Budding	Rooting of Hard wood cuttings	Budding (shield/T budding)
Season	Early monsoon season	Aug-Sept	Monsoon season	Monsoon season
Spacing	(8-10m)	(8 x 8m) square system	(5x5 m)	(5-8m)
Training pruning	Central leader system	Modified leader system	Training-allowed to bush form	Modified leader system
PR used	GA_3 @ 100 ppm, NAA @ 300 ppm			NAA @ 100 ppm - better retention of fruit
Harvesting	March-May	Feb-April	Skin turns and light yellow (Index of maturity)	Soluble solids increase
Yield	100 fruits/tree	2 quintal/tree/year	100 fruits/tree/year	30-50 kg/tree/year
Problems	Fruit drop for its cultivation	Excess moisture is harmful	Fruit cracking internal browning	Apple woody aphid,
Varieties	Cricket ball	Illaichi,Gola	Ganesh, Khandari	Red Delicious, Red Gold
Requirements	**Guava**	**Papaya**	**Pineapple**	**Litchi**
Soil	Alluvial to laterite soil	Well drained sandy loam soil	Sandy and loamy soil	Acidic soils are good
Climate	Tropical and sub tropical	Warm, humid climate	Tropical and subtropical	Subtropical
Propagation	Stooling	Seeds	Slips	Air layering
Season	Onset of monsoon	Mid June-October	Early winter	August-September
Flowering	April-May, Aug-Sept	Starts flowering and fruiting 5 months after planting	Irregular flowering	Jan-Feb
PR Used	NAA @ 100 ppm, 2-4, D @ 30 ppm	Soaking seeds with GA_3 to increase germination	NAA @ 200-300 ppm	NAA, 2-4, D used to increase the fruit size
Harvesting	Fruits turns greenish yellow	Greenish to yellowish green	Orange yellow	Smoothness of epicarp
Yield	150-200 kg/tree	40-50 t/ha	20-25 t/ha	80-150 kg fruit/tree/year
Disease	Guava wilt	Viral disease	Irregular bearing	Fruit cracking
Manures and Fertilizers	350: 280 : 350 g NPK/tree	250: 200: 200 g NPK/plant	500: 40: 660 kg NPK/ha	60 kg FYM, CAN, Super
Varieties	L-49, Allahabad Safeda	Washington, Honey Dew Pusa Dwarf	Queen, Mauritius	Rose Scented

VEGETABLE SCIENCE

Types of vegetable garden

Vegetable gardens can be classified into 6 different types according to the purpose for which they have been developed.

These are *(i)* Home or kitchen garden *(ii)* Market gardens *(iii)* Truck gardens, *(iv)* Gardens for processing *(v)* Vegetable forcing *(vi)* gGardens for seed production.

Vegetable nutritional value

Natritional value: Vegetables, being rich in vitamins and minerals, are known as protective food. Realizing the

worth of fruits and vegetables in human health, the Indian Council of Medical Research (ICMR) recommends the use of 120 g fruits and 280 g vegetables per capita per day. Consumption of adequate amount of vegetables helps in maintaining health and vigour of an individual.

Nutrient-wise richness of important vegetables are as under :

Vitamin' A' richness

Beet leaf	9770 IU/100g
Spinach leaves	9300 IU /100 g
Colocasia leaves	10278 mg/100
Coriander leaves	6918 mg/100 g
Drumstick leaves	6780 mg/100 g
Fenugreek leaves	6450 IU /100 g
Mustard leaves	4370 IU /100 g

Vitamin' B_1' (Thamine) richness

Chillies	217mg/100g
Colocasia leaves	0.5 mg/100g
Tomato (red)	0.12 mg/100g

Vitamin 'B_2' (Riboflavin) richness

Fenugreek leaves	0.31 mg/lOOg
Amaranthus	0.30 mg/100g

Vitamin 'C' richness

Drumstick leaves	220 mg/100g
Coriander leaves	135 mg/100g
Chilli	111 mg/100g
Tomato	31 mg/100g

Carbohydrate richness

Sweet potato	28.2%
Potato	22.6%
Curry leaves	18.7%

Protein richness

Peas	7.2 mg/100g
Cowpea	4.3 mg/100g
Indian Bean	3.8 mg/100g

Fat richness

Mustard	0.8 mg/100g
Beet leaf	0.7 mg/100g
Spinach	0.6 mg/100g

Calcium richness

Agathi	1130 mg/100g
Curry leaves	813 mg/100g
Amaranthus	395 mg/100g
Fenugreek leaves	395 mg/100g
Radish leaves	265 mg/100g
Coriander leaves	184 mg/100g

You Don't Know!

Longest term as Speaker

Balram Jakhar served ten years as Speaker from 1980-89.

Iron richness

Agathi	83.9 mg/100g
Amaranthus (tender)	25.5 mg/100g
Coriander leaves	18.5 mg/100g
Beet leaf	9770 IU/100g

General Information about vegetables

Sex Forms

1. Cleistogamy : Lettuce
2. Chasmogamy : Tomato, capsicum
3. Protogyny : Cole crops
4. GMS : Tomato, Musk melon, Water melon, Cole crops, Beet, Onion
5. CMS : Carrot, Sweet Pepper, Cucumber
6. CGMS : Onion, Carrot, Radish, Beet, Tomato
7. Protandry : Onion, Carrot
8. Herkogamy : Lima Bean
9. Peas and Beans are self-pollinated crop due to presence of 'Keel'

Sex Forms in Cucurbits

1. Hermaphrodite : Satputia (Ridge Gourd), Cucumber - original form
2. Andromonoecious : Musk melon
3. Gynoecious : Cucumber
4. Gynoecious : Monoecious : Cucumber
5. Trimonoecious : Cucumber, Musk melon, Ridge gourd

- *Broccoli* : Highest Cross Pollination (95%)
- *Spinach* : 5 Sex forms present

Pollination

1. Anemophily : Spinach, Beet root, Palak, Amaranthus
2. Entmophily : Onion, Carrot, Radish, Cole Crops

Anthesis

1. Bottle gourd : evening 5-8 pm
2. Snake gourd : evening 6-9 pm
3. Ridge gourd : 5-8 pm
4. Musk Melon : Morning 5.3 -6.30 am
5. Cucumber : Morning
6. Sponge gourd : Morning

Fruit type

1. Peas and Beans: Pod/Legume
2. Orka: Capsule
3. Sweet potato, Yams: Capsule
4. Berry : Tomato, Potato, Taro
5. Carrot: Shizocarp
6. Seed ball : Turnip, Beetroot

Package of Practice of vegetables

Sl. No.	Name of the vegetables	Important varieties	Method of planting	Seed (rate/ha)	Spacing (cm)	Season of sowing	Manures and manuring (kg/ha)	Duration of the crops (days)	Yield (tons ha)	Other particulars
1.	Tomato	Co. 1, Co. 2, Pusa Ruby	Sown in nursery and transplanted 25 days after sowing	375 g	60 x 60	June-Nov, Nov-Apr, Feb-May	N-150 P-100 K-50, Hybrid Varieties 250 : 250 : 250	135-150	25-50	Harvest the half ripe fruits at 3 days interval
2.	Chillies	MDU-1, Pusa Jwala	Sown in nursery and tranplanted 40-45 days after sowing	1.0 to 1.5 kg	30 x 30 or 45 x 30	June-July to Dec-Jan, Sep.-Oct. to Mar-Apr, to Sep.-Oct.	N-70 P-35 K-35	210 -240	Dry pods: 2.25 Green chillies 7.5-10.3	-
3.	Brinjal	Pusa Purple Long, Arka Navneet (F_1), Palur-1	Sown in the nursery and tranplanted 45 days after sowing	375-500 g	75 x 60	May-Oct, Dec-May	N-100 P-75 K-50	150-165	20-25	-
4.	Bhendi	Pusa Sawani, Punjab-Padmini	Seeds dibbled in rows	7.5 kg	60 x 30	July-Nov, March-April	N-40 P-50 K-30	90-110	7-10	Immature fruits are edible and should be harvested at 3 days interval
5.	Peas	Bonneville, Telephone, Gloriosa	Seeds sown in raised beds (hills) and in ridges (plains)	80-100 kg	40 x 10 in hills, 60 x 10 in plains	Hills: Feb-Mar to May-June, Plains: Oct, Jan.	N-180 P-125 K-100	80 to 100	5.7.5	-
6.	French beans	Watex, Premier Contender	Sown in lines in the beds (hills), sown on the side of the ridges (plains)	Hills 125 kg., Plains 62.5 kg.	Hills: 50 x 20 Plains: 45 x 10	Hills: Mar-May, Plains: Oct-Jan	N-50 P-75 K-75	90-100	4-5	Also grown as rainfed crop with the half recommended dose of N and P. Picking should be done at correct maturity.
7.	Bellary onion	Bellary Red. N-53. Arka Niketan, Arka Pragati, Arka kalyan	Sown in nursery and transplanted in the main field after 45 days	8-12 kg	30-45 x 8 - 10, trans-planted on both the side of the ridges	May-June	N-75, P-150 K-75	145-150	15-18	Harvest the bulbs when 75% of the tops fall.
8.	Potato	Kufri Jothi, Kufri Muthu, Kufri Swarna, President	Healthy sprouted tubers of 40-45 g planted in furrows	3000 to 3750 kg	45 x 30	Summer: Mar to Apr, to July-Aug. Autumn: Aug. Dec. Irrigated: Feb-June	N-100 P-125 K-50	100 to 140	20-25	Fresh tubers exhibit dormancy, broken by treating carbon-di-sulphide @ 30g/100g of seeds. Harvest the crop when the foliage turns yellow and starts withering.
9.	Carrot	Puse Kesar, Half Long Danvers, Ooty-1	Seeds dibbled in ridges	3-4 kg	Hills: 25 x 5, Plains: 30 x 15	Throughout the year. Plains: Aug-Sep.	N-180 P-125 K-100	Hills: 120 Plains: 100	25 to 30	Thinning is essential
10.	Cabbage	September, Golden-Acre, Pusa, Drum Head, Hero	Sown in nursery and tranplanted 30-35 days after sowing	375 gm	Hills: 45 x 45, Plains: 60 x 45	Hills: Jan-Feb April, June-July-Plains: Aug-Sep.	N-180 P-125 K-100	105-120	30 to 40	Harvest when the head is firm but tender.

Floriculture and Landscaping

Current status of development in floriculture

India is endowed with various agroclimatic conditions suitable for growing large number of flowers. India is known for growing the traditional flowers such as Jasmine, Marigold, Chrysnthemum, Tuberose etc. Major cut flowers, which are grown in India, includes, rose, orchids, gladiolus, carnation, anthurium, gerbera etc. Important flower growing states are Tamil Nadu, Karnataka, Maharashtra, West Bengal, J & K, Sikkim and Andhra Pradesh. There has been tremendous appreciation of its economic value in addition to its aesthetic importance. During the last 10 years period, the flower trade in India has been increased many times in view of the significant increase in floriculture activities.

Major issues

Major issues confronting floriculture industry, particularly in the areas of production technology, post harvest, product quality and infrastructure and measures taken to initiate address the issues are as follows :

Acknowledging the potential floriculture holds, as a diversified economic activity in terms of production, employment, income and exports, the GOI have initiated several increases for supporting floriculture.

Developing a comprehensive database.

Developing specific programmes for traditional flowers and stepping up research and extension efforts.

Programme for assistance to small and large nurseries.

Strengthening of infrastructure for post harvest management and marketing.

Export promotion and market development.

Establishment of wholesale market-cum-auction centres for flowers.

Flower cultivation is a good source to earn foreign exchange considering India's geographical advantage, it is in a better climatic region unlike the severe winters prevailing in the west. But the trade in cut-flowers is negligible when compared to the Netherlands. This might be due to lack of enterprise, technical know-how, standard methods of growing and harvesting flowers of an internationally acceptable quality as well as the problems of packing and speedy and cheap transport.

Export potential

In view of the immense potential for export of ornamentals from the country, recently the GOI and its agencies like Ministry of Commerce, Processed foods APEDA, State Trading Corporation have given encouragement and support to it.

Future prospects

1. Emergence of new markets Japan, Europe, S-E Asian nations like Singapore, Hongkong and India may emerge as consumers of floricultural products in the year to come.
2. Many small units may exit from the market due to uneconomic returns and high overheads.
3. India has several advantages like: availability of skilled manpower, labour at competitive rates, presence of a reputed large tissue culture production facility which can propagate plant material at low cost etc.

Knowledge World!

Largest river basin

The Ganga basin, takes up 26 per cent of India's total geographical area.

Reasons for lukewarm response to floriculture

1. In India most of the farmers still use traditional methods of agriculture with poor infrastructure base.
2. Cost of establishment is quite high.
3. Cost of planting material is quite high.
4. Industry requires a technologically-advanced infrastructure to ensure quality.
5. Quality is the single best parameter for a good return.
6. The absolute necessity to ensure quality are specialized transport system, refrigerated, trucks, cold storage, regular international flights.
7. Flowers end up with a rather low priority in air cargo, as they are bulky, voluminous and perishable, the freight component is very high.

Strategy

1. Industry and Government should work in close co-ordination and Government support is necessary in foreign countries to certify credit worthy buyers and help Indian companies to penetrate new markets.
2. A high degree of transparency and sharing of market information and business experience is required to foster the growth of industry and face international market challenges collectively.
3. Increased frequency of domestic carriers.
4. Loan procedures should be simplified since industry is capital intensive (lower rate of interest may assist).
5. Policy initiatives are necessary to revamp the present insurance system for perishables.
6. Group marketing approach to exploit economics of scale needs to be developed.
7. Investing in marketing information on the present level of area, production and locations of flowers and ornamental plants in the country.
8. Product positioning-identifying niche markets and product differentiation. Flowers in demand be given preference in growing.

GK

Advocate General is the official legal adviser to the State Government.

GS

9. Invest in R & D for development of priority products through a breeding programme. Research on improvement of varieties which suits the demand of international markets.
10. Specialise in regional production of products that have a specific local advantage. In India also there is a budding market gradually catching up the flower trends.
11. Small growers of flower and live plants for export may require assistance in marketing their products. Export houses and big industrial organisations may be induced to enter into the export trade of floriculture items, like those in Kenya, Sri Lanka.
12. Visualising the acute shortage of qualified and trained staff in floriculture, the Government should depute floriculture staff for training abroad.

Style of Gardening

1. Formal Design

This design is very stiff and everything is done in a straight and narrow way in a symmetrical or a geometrical pattern. Everything is planted in straight lines. The hedges, edges and topiary are maintained in proper shape by regular training and pruning. Symmetry is always maintained in the garden by undertaking similar type of plantation and using similar garden technique and adornments.

2. Informal Design

In this design, the plants and features of the garden are arranged naturally without following any hard and fast rules. The plan first laid on is utilized as such for the purpose of beautification.

3. Picturesque Design

The idea of picturesque design of garden was expounded by William Robinson in the last decade of the nineteenth century. His idea was to naturalize plant in shrubbery. He also suggested that grass should not be mown and bulbous plants should be mown and bulbous plants should be grown scattered in the group to imitate wild scenery. He opined that passages should be opened in the woodland, trees, shrubs and bulbous plants should be planted in forest flora to have an effect of a wild garden. His idea was to allow the creeper to climb over the tree naturally to imitate a forest type of effect.

Types of Garden

1. Mughal Gardens

The gardens laid out during the regime of Mughal Emperor, Babar in India are known as Mughal Gardens, Babar (1494-1531 A.D.) was the first Mughal ruler to introduce this style in India. All other Mughal rulers like Akbar, Jahangir, Shah Jahan, Aurangzeb and Begum Nur Jahan shared their hand on developing Mughal gardens in India.

The main features of Mughal garden are :

(i) Site and Design
(ii) Walls and Gates
(iii) Terraces
(iv) Running water
(v) Baradari
(vi) Tomb or Mosque
(vii) Trees and Flowers

2. Persian Gardens

Persian garden style is one of the oldest garden styles. The garden styles were formal. Persians used crafted materials such as masonry, carved and pierced marble stones and highly polished stones. The gardens were laid out by cutting terraces on hill slopes. They tapped some natural spring to create straight water-course through the gardens and manipulated water course to undertake different movements along its run. If there was no natural source of water, it was created artificially by diverting some river or stream.

3. French Gardens

In accordance, the French style of garden were also very intricate and artificial. The new style of gardening now known as French style is largely due to contribution of Le-Notre who served in the Royal Garden in Louis-XIV from 1643 to 1700 A.D.

4. Italian Gardens

There is a similarity between Persian or Mughal styles and the Italian style of gardening. In all these styles of gardening, the similarity was the use of heavy masonry features. The Italian elite conceived that garden is just an extension of lavish palaces, as a glamorous outdoor hall for entertainment and showing off their wealth as well as status. Massive stairs, generally of marble, complete with balustrade to connect the different levels in the garden are the main features of Italian gardens. Decorative fountains generally in combination with stone sculpture are other important features of the Italian gardens

5. English Gardens

Due to favourable climatic conditions such as high annual rainfall, the natural ground cover in England is grass. In consonance to this the famous British Garden architects Repton and Brown advocated the concept that the British Garden should look like the countryside. The gardens were created with the idea that the garden should merge with the countryside without any artificial barriers such as fences or hedges etc. During the eighteenth century, the British architects Repon and Brown alongwith Kent brought through of a natural look in the garden by the middle of the sixteenth century. During the eighteenth century gardens were laid with more emphasis on architectural features. The main features of the garden during that

period were curved paths, informal group of trees, rivulet or streams, artificial waterfall and clipped hedges

6. Japanese Gardens

The plant materials were selected in such a manner that the garden would not undergo many major changes during the four seasons of the year. During the period of Muromachi (1392-1573 A.D) both the flat garden (Hira-niwa) and hill gardens (Tsukyama-sansui) developed. Japanese gardens were developed in close association with nature. This has contributed a lot in maintaining its popularity upto now whereas other styles of gardens have fallen into disfavour. The most important teaching of Japanese garden is not worth a place visiting. It should be a place where mind find rest and relaxation. The immutability is achieved also because rather rocks, stepping stones, streams, waterfalls, bridges and stone lanterns contribute to the structure of the Japanese gardens. The Japanese garden emphasizes to capture natural scenery. The three important elements for this purpose are water, stone and plants. Low sculptured bushes and trained dwarfed trees are attraction of Japanese gardens.

Lawn

A lawn can be defined as the green carpet for the landscape maintained by growing and mowing grasses. Thus lawn is an area maintained under thickly grown grass cover. It forms the heart of the garden and hold pivotal importance in adding beauty to the garden. No garden is considered complete without the establishment of the lawn. The lawn serves as a base for a flower bed, a border, a shrubbery or a specimen tree. Besides having monetary value, a lawn possesses aesthetic value of reducing tension of the mind after a day's hard work. It becomes very charming to enjoy the grandeur of the spacious lawn in the garden.

SUITABLE LAWN GRASSES

Doob or Bermuda grass: *Cynodon dactylon*, Manilla grass: *Zoysia matrella*

Korean grass: *Z. japonic, Korean velvet grass* : *Z. tenuifolia.*

PLANTING OF LAWN

Lawn grass is planted using one of the following methods suitably:

1. **By Seed.** When other planting material like cutting etc., are not available lawn is raised by seeds. The area over which grass planting needs to be taken is prepared well by tilling to a pulverized solid and adding sufficient quantity of FYM followed by proper levelling. About 500g seeds per 200-300 square metre area germinates sufficiently to prepare a thick lawn. The seeds of Doob being very light in weight are sown after mixing with double the quantity of finely sieved sand or ash. Mixing seed with sand or ash facilitates broadcasting of seed and, thus, ensures uniform germination of seed throughout the bed. After broadcasting the seed it is covered with a thin layer (1-2 cm thick) of finely sieved farmyard manure. The bed should be watered regularly using watering can fitted with fine nose of nozzles so as to keep the bed continuously moist. The seeds germinates in about 3-5 weeks.

Quiz Kit!

Largest delta

The Ganga and the Brahmaputra delta basin in West Bengal and Bangladesh covers 75,000 sq km.

2. **By Dibbling.** Dibbling implies planting in holes. The bits of the mown grasses are used for dibbling. The cutting should be taken from grass growing in shade. The grasses growing in sun having close internode are used for dibbling. Dibbling of grass is done at a distance of 10 cm in the moist bed. Continuous moistening of the bed is maintained to facilitate sprouting of grasses. The dibbled grasses start growth in about 5-6 weeks.

3. **By Turf.** This is the quickest method of developing lawn. The turf is a piece of earth with grass thickly grown over it. The roll of turf with a width of 1-2 ft. are used for setting a lawn. The pieces of turf are placed closely with each other. Before placing turf, the levelling of land is ensured. The joints of the turf are packed using sand. The turf is made firm by beating it with a wooden beater. The turf is watered immediately and irrigation is continued till it sets with soil.

4. **By Turf plastering.** Mixture of bits of grasses along with mud and water is termed as turf plastering. The paste of grass is spread evenly on prepared ground. It is then covered with 1-2 cm thick layer of sieved manure or soil. Watering is done at regular intervals so that the grass make constant with soil and grow successfully. This method is not suitable in dry areas where grasses dry after spreading.

MAINTENANCE OF LAWN

Weeding. Weed pose great problem in maintenance of lawn and infestation gives an ugly look to the lawn. Motha (*Cyperus rotundus*) and doob (*Euphorbia thymefolia*) are major weed in a lawn.

Weed is a problem in established lawn. Once established it becomes very hard to remove the weed, It is advisable to go for regular weeding in order to remove weeds. It is very difficult to eradicate motha from lawn completely. Deep hoeing removing stolon of the plant helps in minimizing infestation of the Doob. In established lawn the regular mowing can keep the growth of motha under check.

Liming. Frequent irrigation at close intervals and stagnation of water makes the soil of the lawn acidic. The growth of moss is indication of acidity of lawn. For ameliorating acidic situation lime @ 250 g per square metre is added in the soil.

Rolling. Rolling is practiced with the object of helping the grass in this anchorage to the soil. It is also worth in making the lawn level after weeding. The stone or pebbles also interfere with the mowing blade. To press the pebbles rolling is beneficial. Rolling is avoided in wet soil as it makes the soil compact.

Mowing. Mowing is done to prevent excessive growth of grass, higher the growth of grass, higher is the frequency of mowing.

Sweeping. The removal of cut-over grasses from lawn is mown, the sweeping using broom is practised. Sweeping is also done every morning by dragging the hosepipe of irrigation to allow draining and mixing of dew drop with soil.

Irrigation. Doob is a shallow-rooted crop and needs frequent and light irrigation. Nowadays provisions are being made to irrigate the lawn using sprinklers. The frequency of irrigation depends upon soil and climatic conditions.

Physiological Changes During Flower Senescence

The senescence of flowers is marked by loss of dry matter, primarily due to hydrolysis of macromolecules such as starch, sugars, protein and nucleic acids. The starch is broken down into sugars which are transported to the flowers. The breakdown of proteins release ammonia in the cells, which causes blueing of petals in red cultivars of proteins rose. Senescence is also marketed by changes in the property of membrane such as in microviscosity and loss of phospholipids and permeability which results in leakage of pigments, sugars, mineral ions and total electrolytes, thereby leading to death of the tissues. It has been observed that flowers show two peaks of respiration; first when the bud starts opening and the second when the senescence takes place. This is followed by decrease in respiration of flowers. This decrease in respiration has been ascribed to the depletion of sugars or inability of mitochondria to utilize available sugars.

From an investigation at IARI, New Delhi on the post harvest life of cut flower of ten different cultivars, it has been observed that the cultivars which lasted for longer duration recorded comparatively lower rate of respiration and less electrolyte leakage during the course of senescence over the cultivars which lasted for shorter duration. While studying the changes in vascular morphology of the cut surface of the rose stems of the above cultivars, the freshly cut rose stems of all the cultivars showed clear vascular system; in case of short lasting cultivars, on the third day itself, globular bodies were observed with partial breakdown of vessel cells on the cut surface.

There is also clear evidence that synthesis of new proteins is also associated with petal senescence. The newly synthesized enzymes during senescence thus, appear to play an important role in senescence.

Factors Affecting Shelf Life

Important factors which determine the storage life of flowers are quality of the flowers; stage of harvest; temperature; relative humidity; light; ethylene; and pathogens.

The flowers to be stored should be of good quality. Flowers should be devoid of any physical injury and free from attack of pests and diseases, The injured and diseased tissues are potential sources of ethylene and are highly prone to infection with pathogens. The flowers must be cut at an optimum harvesting stage. Some flowers like gladioli, roses, lilies, narcissi and irises are generally stored at bud stage as buds of these flowers open perfectly after storage.

Temperature management during storage is an important component which determine the storage life of flowers, temperature fluctuations in the cold room should be avoided. Relative humidity inside the cold room should preferably be maintained high (90-95). Low humidity causes excessive water loss from the flowers thereby stimulating their desiccation. Cold rooms should be cleaned at regular intervals to check the spread of pathogens.

Floral Preservatives

The term "Floral Preservative" is used for any chemical formulation which is used for to extend the life of flowers. Besides increasing life, floral preservatives help to improve flower opening, flower size, shape and colour.

Types of Floral Preservatives

Floral preservatives can be applied in three ways i.e. as (i) pulsing solution; (ii) bud opening solutions; and (iii) holding or vase solutions.

(i) *Pulsing solutions* : "Pulsing" refers to short duration (16-24 hr) pre-storage treatment. The effect of such a treatment lasts throughout the entire life of the flower. Sugar (sucrose) is one of the main components of the pulsing solution. Since pulsing involves short duration treatment, relatively higher levels of sucrose are used.

(ii) *Bud opening solution*: Immature buds of many flowers can be made to open in chemical solutions, referred to as "Bud opening solution". The components of bud opening solutions are essentially the same as those of pulsing solutions, but in case of bud opening solutions, lower concentration of sucrose are used.

(iii) *Holding or vase solutions*: Vase solutions are meant

to hold flowers continuously, till the termination of their vase life. The level of sucrose in vase solutions is therefore, also kept very low (0.5 to 2 per cent), due to long duration for which flowers are kept in the solution. Many commercial vase preservatives are also available for use as holding solutions and should be used as per recommendations of the manufacturers.

Constituents of Floral Preservatives

The main constituents of floral preservatives are (i) Water (ii) Sugar; and (iii) Biocides. Besides, mineral nutrients, acidifying agents, anti-ethylene compounds and growth regulators are also used in floral preservatives. The brief account of these constituents is given herewith :

(i) *Water*: Water is the most important component of floral preservatives. It affects vase life in terms of its quality as well as quality. Sometimes tap water can be harmful due to its high pH (alkalinity), presence of total dissolved solutes (TDS) and toxic ions. High pH reduces vase life. Therefore, acidifying water to low pH 3.0-3.5 is advantageous because it decreases microbial growth and considerably improves water absorption. The longevity of cut Super Star roses was improved by the use of tap water over that of single and double distilled water.

(ii) *Sugar*: Sucrose is the most widely used sugar in floral preservatives. Sugar acts as additional food source also improves water balance of cut flowers. The optimum concentration of sugar required however, varies from species to species, Since sugar promotes microbial growth, it is usually combined with biocides before use.

Did You Know ?

Longest beach

The Marina beach at Chennai stretches over 13 km and is the second longest beach in the world after Rio de Janeiro, Brazil.

(iii) *Biocides* : Biocides are chemical substances which are used to inhibit microbial growth in the vase water as well as on the stem surface. Important biocides used for treating cut flowers are 8-hydroxy quinone citrate, silver nitrate, aluminium sulphate, citric acid, slow release chlorine compounds and quarternay ammonium compounds.

8-hydroxy quinone citrate (8-HQC) is very effective broad spectrum biocide. It acidifies water and also indices partial closure of stomata. This compound is used at the concentration of 200-600 ppm. Silver nitrate ($AgNO_3$) is also very effective biocide; $AgNO_3$ at the concentration of 25 ppm completely inhibits the microbial growth.

Common house plants and their commercial propagation methods

Common name	*Scientific name*	*Commercial propagation method*
Agalonema	*Aglaonema commutatum*	Division of basal shoots, nodal cuttings
Anthurium	*Anthurium magnificum*	Seeds
Aphelandra	*Aphelandra squarrosa*	Half ripened wood cuttings
Aralia	*Aralia filiciplia*	Division of crown
Araucaria	*Araucaria heterophylla*	Seeds
Asparagus	*Asparagus africuns*	Division of crown
Aspidistra	*Aspidistra luridu*	Division
Bird of paradise	*Strelitzia reginae*	Seeds
Calathea	*Calthea insignis*	Root cuttings
Coleus	*Coleus hybridus*	Herbaceous cuttings of slips
Croton	*Codiaeum spp.*	Air layering
Dieffenbachia	*Dieffenbachia exotica*	Terminal and nodal cuttings
Draecena	*Dracaena deremensis*	Cane pieces
English ivy	*Cissus spp.*	Cuttings
Indian rubber tree	*Ficus elastica*	Air layering
Iron plant	*Aspidistra elator*	Division of crown
Monstera	*Monstera deliciosa*	Air layering
Palms	*Different species*	Offsets
Pandanus	*Pandanus utilis*	Offsets and division of suckers
Peacock plant	*Calathea spp.*	Division of crown
Philodendron	*Philodendron scandens*	Cuttings
Pitcher plant	*Nepenthes khasiana*	Hardwood cuttings
Podocarpus	*Podocarpus macrophyllus.*	Shoot tip grafting
Sago palm	*Cycas revoluta*	Offsets
Sansevieria	*Sansevieria spp.*	Offsets
Spider plant	*Chlorophytum comosum*	Division of crown
Tradescantia	*Tradescantia spp.*	Cuttings
Yucca	*Yucca gloriosa*	Seeds and offsets
Zebrina	*Zebrina pendula*	Herbaceous cuttings

Vidhan Sabha's term in J&K is of 6 years.

Common Name	*Scientific name*	*Commercial propagation technique*
Acalypha	*Acalypha spp.*	Hard-wood cuttings
Allamanda	*Allamanda cathartica*	Air layering
Azalia	*Azalia spp.*	Hard-wood cuttings and layering
Bottle brush	*Callistemon spp.*	Hard-wood cuttings and layering
Brya	*Brya ebenus*	Hard-wood cuttings and layering
Camelia	*Cammelia japonica*	Hard-wood cuttings and layering
Cestrum	*Cestrum nocturnum*	Hard-wood cuttings
Clerodendron	*Clerodendron spp.*	Hard-wood cuttings
Clematis	*Clematis spp.*	Leaf bud cuttings
Coral tree	*Erythrina indica*	Hard-wood cuttings
Hibiscus	*Hibiscus rosa-sinensis*	Hard-wood cuttings and layering
Hydrangea	*Hydrangea spp.*	Softwood and leaf-bud cuttings
Ipomea	*Ipomea carnea*	Hard-wood cuttings
Ixora	*Ixora spp.*	Hard-wood cuttings and layering
Jatropha	*Jatropha multifida*	Hard-wood cuttings
Juniper	*Juniperus spp.*	Hard-wood cuttings
Kachnar	*Bauhinia acuminata*	Hard-wood cuttings

Packages for export of cut flowers

Packaging for export of cut-flowers is as important as the quality of the flower. Only corrugated fibre board (CFB) packages are accepted in the trade. Different types of CFB for cut flowers.

Slotted box, telescopic boxes, rigid boxes, ready glazed boxes and folded type boxes.

An ideal package for export should be small in volume, air-tight non-water absorbent, of the required strength, should have minimal effect on quality and shelf/vase life of flowers. The size of the packages especially the length of boxes depend upon the stem length of cut flowers required in trade, e.g. Rose: 100 cm.

The strength of the containers should be enough to withstand a stack height of 3-5 times to protect the loads from damage due to shock and vibrations.

Ventilation holes are not necessary if the flowers are packed and transported in ideal conditions. The flowers are to be immobilized and cushioned to avoid abscission due to vibration. This is achieved by lining the inside of boxes with thin polyethylene film. Fine moist tissue paper spreading are also spread in each of the boxes to provide cushioning to the blooms.

Biotechnology for improvement of flower crops

The major constraints in flower breeding are, very limited gene pool for improving agronomic traits, which compels breeders to opt for sexual crossing, consuming lot of time and energy.

There is no possibility of directed plant breeding through mutation manipulation, because if we increase the flower size other traits get affected.

Biotechnology can provide many scientific pathways including cell and tissue culture techniques and molecular approaches for the flower breeders.

Plant biotechnology is applied to manipulate diverse traits including flower colour, shape and fragrance, flower and leaf longevity, plant form and habit, resistance to diseases and pests and induction and timing of flowering.

With the advent of *in vitro* culture techniques, plants can be cloned in a miniature form (microplants) from tiny pieces of leaves, scales axillary and apical buds. In some plants, cell suspension cultures can be used for plant regeneration.

It is estimated that there exists a global market of US $ 15 billion for tissue-cultured plants and products and the demand is still growing. In India, approximate 76 companies have plans for setting tissue cultured plants and more than 30 are already in the market with their products.

The various approaches used for improvement in flower crops are :

1. Micropropagation (Anthurium, Chrysanthemum, Carnation, Gerbera, Dieffenbachia, Gladiolus, Rose, Dahlia).
2. Meristem culture (Dahlia, Gladiolus, Lily, Rose, Antirrhinum, Gerbera, Kalanchoe, Dieffenbachia).
3. Embryo culture (Lilium, Chrysanthemum, Carnation).
4. Somaclonal variation (Carnation, Chrysanthemum, Lilium, Pelargonium).
5. Haploid culture (Pelargonium, Lilium, Gerbera).
6. Somatic embroyogenesis (Iris sp, Anthurium,

Geranium, Floribunda)

(1) Micropropagation in case of ornamental shrubs like, Jasmine can also be cloned through rapid proliferation of auxiliary buds. This technology is also used for production of disease-free motherstock cultures, which are then used as a source of conventional cuttings.

(2) The culture of shoot meristems has proven useful in eliminating virus from plants like Dahlia, Gladiolus, Rose, Lilium sp.

(3) Embryo rescue by culturing seeds devoid of endosperm is being used for obtaining new types of orchids and their multiplication. It is also used to obtain interspecific hybrids of Lilium, Chrysanthemum and Carnation

(4) Regeneration of plants through somatic embryogenesis has been obtained in Irish, Anthurium, Floribunda rose.

5) Somaclonal Variation where tissue culture results in plants that are physiologically and genetically different from their progenitors have been studied and utilized in number of important flowers crops Carnation, Chrysanthemum, Lilium, Pelargonium.

(6) The culture of anthers, ovules, rnicrospores is of considerable value to flower breeders to produce haploid plants

- *Bacillus thuringensis* (Bt) gene has been used in *Chrysanthemum* for resistance to insect pest armyworm.
- *In Petunia, transgenic plants* were obtained for resistance to *Lepidopteran* pest using potato protease inhibitor and bovine trypsin inhibitor.
- Increase in vaselife is an area of much concern to floriculturists, transgenic carnation having an antisense ACC oxidase, gene exhibited markedly low ethylene production and delayed petal senescence.
- *Many novel plants* - Blue anthurium, scented rose, chrysanthemum plants with altered flower morphology and flowers that do not fade will come into existence, flowers growing due to presence of bioluminescence genes (Lux Genes) from *Wibrio fischeri*.
- Biotechnology offers immense prospects to improve flower crops using the advanced tools, thus increasing the market value of flowers.

Cultivation practice of cut flowers

Flower name Family	*Varieties*	*Spacing (m)*	*Planting Time*	*Pruning Time*	*Propa-gation*	*Cultural practice*	*Harvesting stage*	*Stage temp.*	*Key feature*
1. Rose *Rosa x hybrida Rosaceae*	Dr. B.P. Pal, Raktima, First Red (GH), Nobleless (GH)	45 x 45	June	Oct.	T-Budding	Pruning	Tight bud	0.5 - 2°C	By products- Rose water, aktar, rose oil
2. Gladiolus *Gladiolus spp. Iridaceae*	Sachitra, Shagun, Dhiraj,	30 x 30	Sep. - Nov.	Sep.	Corms and Cormels	De-corming Thinning	Two florets open	4°C	No species is native to India
3.Chrysanthemum *Dendrothema spp. Compositae*	Diana, Neelambri	30 x 15	Aug. - Sep.	Sep.	Suckers, Stem cutting	De-suckering Thinning Disbudding	Halfly open	1°C	SDP plant
4. Carnation *Dianthus caryophylus Caryophylaceae*	Scarlet Elegance, Sam's pride	30 x 15	June - July	Aug.	Stem cutting	Pinching (- Half, full, half full)	Outer two whorls open	0 to 1°C	LD plants
5. Marigold *Tagetus erecta Compositae*	P. Narangi Gainda P. Basanti Gainda	30 x 30	July - Nov.	July	Seed	Pinching CCC, B-9, MH Growth retardants	Fully open flowers		Male sterility used for F_1 hybrid
6. Jasmine *J. sambac J. grandiflora J. auriculatum Oleaceae*	CO_1, CO_2, Surabhi	1 m x 1 m	June - July	Sep.	Stem cutting	Thinning Pruning	Bud stage		Used for essential oil extraction
7. Orchids Orchidaceae	Dendrobium Cymbidium Cyanthea Vanillila				Tissue culture Pseudo bulbs (kikus)	Velamen root for adsorptions of nutrition	Fully open flower	5-7°C	Lip and labelleum are ornamental

GK Smog is a special case of fog heavily laden with smoke. GS

POSTHARVEST TECHNOLOGY

Scope of fruit and vegetable preservation in India

Fruits and vegetables are an important supplement to the human diet as they provide the essential minerals, vitamins and fibre (roughage) required for maintaining health. India with a great diversity in climatic conditions and soil types, is ideal for growing a large variety of fruits and vegetables, both indigenous and introduced.

India - 2nd in vegetables (lst China), 3rd in fruit (Brazil, USA).

However for various reasons, this abundance of production is not fully utilized and wastage due to spoilage is about 20-30%.

Most fruits and vegetables are seasonal and perishable in nature. In a good season there may be a glut but because of insufficient transport facilities, lack of good roads and poor availability of packaging materials, the surplus cannot be taken quickly enough to markets in urban areas.

Moreover the surplus often cannot be stored for sale in off-season because of inadequate local cold storage facilities. Thus, the cultivators do not get a good price for their produce.

Although, the R and D efforts on the development of post-harvest handling has helped in reducing the spoilage, considerable losses continue to occur. Two approaches are possible for solving this problem.

(i) One is the creation/expansion of cold storage facilities in the fruit and vegetable-producing regions, as also in the major urban consumption centres, to ensure supply of fruit and vegetables throughout year.

(ii) Another approach is to process the fruit and vegetables into various products which could be preserved for a long time and add value to the product.
With increasing purchasing power of the middle class, there is increasing demand for factory made jams, jellies, pickles. In spite of all these, the fruit and vegetable preservation industry at present is able to utilise less than 2% of the total production for conversion into products like canned fruit, juices and their beverage squashes etc. as against 40% in developed countries. Thus there is considerable scope for expansion of the industry, which in turn would give a fillip to development of horticulture, especially in hilly areas and through export of value-added products, earn more foreign exchange.
India is one of the cheapest producers of horticultural produce. If the Government addresses some of the gaps in the processing industry, the Indian economy will get a boost up.

(i) Most of the varieties grown in India are not suitable for processing.

(ii) Seeds of processble varieties can be imported and multiplied using tissue culture technique.

(iii) Lowering sales tax on agriculture produce may help the industry.

(iv) In New Exim Policy of the Union Government, setting up of Agriculture Export Zone (AEZ) to identify each state's competitive advantage and to promote the cultivation and processing of fruit and vegetable accordingly.

Principles and methods of preservation

Food preservation can be defined as the science which deals with the methods of prevention of decay or spoilage of food, thus allowing it to be stored in a fit condition for future use. It is better if the following directions are kept in mind to control the spoilage.

1. Raw materials should be thoroughly examined and handled hygienically to avoid microbial spoilage.
2. Equipment must be cleaned every time before use.
3. Processing should take place as soon as possible after sealing ofcans or bottles.
4. Use of contaminated coater should be avoided.
5. The finished products after canning or bottling should be stored in well-ventilated rooms in a cool and dry place. Freshly prepared products are highly attractive in appearance and possess good taste and aroma, but deteriorate rapidly if kept for sometime. This is an account of several reasons.

(i) Fermentation caused by moulds, yeasts and bacteria.

(ii) Chemicals present in pulp/juice, may react with one another.

(iii) Air coming in contact with the product, may deteriorate.

(v) Traces of metal from the equipment may get into the product and spoil its taste and aroma.

In the preservation of foods by various methods, the following principles are involved :

1. Prevention or delay of microbial decomposition

(a) By keeping out micro-organisms (asepsis).

(b) By removal of micro-organisms e.g. by filtration.

(c) By hindering the growth and activity of micro-organisms.

(d) By killing the micro-organisms e.g. by heat or radiation.

2. Prevention or delay of self-decomposition of food

(a) By destruction or inactivation of enzymes e.g. by blanching.

(b) By prevention or delay of chemical reactions e.g. prevention of oxidation by means of an antioxidant.

3. Prevention of damage by insects, animals, mechanical causes.

To retain the natural taste and aroma of a product, it is necessary to preserve it soon after preparation, without

allowing it to stand for any length of time. The product must be prevented from the damage of insect and animals and proper transportation and packaging must be ensured to protect the food products from any mechanical problems.

Methods of Preservation

The spoliate swing to the different causes enumerated can be prevented by adopting different methods of preservation. These are :

1. *Asepsis* (Absence of infection) :

Asepsis means preventing the entry of microorganisms. Maintaining of general cleanliness while, picking, grading, packing and transporting of fruits and vegetables increases their keeping quality and the products prepared from them will be of superior quality.

2. *Drying* :

Sun-drying or mechanical drying of fruits and vegetables involves complete removal of the moisture from them. They are then treated with sulphur fumes to maintain colour and also to avoid spoilage by microorganisms. Dates, grapes, jack, peaches, beans, bitter gourd, onions, potatoes etc. are preserved by this method.

3. *Freezing* :

When fruits and vegetables are brought suddenly to a very low temperature at which all the chemical reactions stop. But it requires rather expensive equipment and stripping and storing under refrigeration.

(i) Cooler storage (about 15 °C) (ii) Refrigeration or chilling (0-5°C) (iii) Freezing (18-40°C)

4. *Preservation by heat* :

Consists basically of the application of heat in varying degrees to the food in close containers for a sufficient long time to sterilise the contents before they are hermetically sealed. This method of preservation by heat is also known as processing. Canned fruits and vegetables come under this category.

High temperature used for preservation: (i) Pasteurization temperature (below 100°C) (ii) Sterilization temperature (above 100°C).

5. *Preservation by sugar* :

No organisms can grow on a substance which contains sugar concentration above 65%, as they get desiccated by osmosis. Jams, jellies, marmalades, candies, crystallized fruits are preserved by this method.

6. *Preservation by salt* :

Products which contain about 15% salt makes it most unfavourable for the growth and multiplication of microorganisms. Salt acts both by osmosis and as a poison on microorganisms. Pickles of fruits and vegetables are so preserved.

7. *Preservation by chemicals* :

Chemicals like Benzoic acid, SO_2 at specific concentration help to preserve the fruit juices, squashes. Benzoic acid in the form of 0.06-0.10% sodium benzoate is sufficient and is effective against yeasts and its action is effective in the presence of CO_2. Potassium matabisulphite (KMS) when added liberates SO_2 which reacts with water to form sulphurous acid. The concentration of SO_2 recommended is 350 PPM. It is more active against moulds, spores and bacteria.

> **You Don't Know!**
> *Longest glacier*
> **Siachen Glacier on the Indo-Pak border is the world's longest, glacier. It is 75.6 km long and 2-8 km broad.**

8. *Preservation by fermentation* :

Fruit juices containing sugar are fermented to produce alcohol. These fermented substances keep off the spoilage organisms due to the presence of alcohol. 14% alcohol acts as preservative in wines. Wines, beers, vinegar, fermented drinks, fermented pickles etc. are prepared by these processes.

9. *Preservation by irradiation* :

Sterilization of food by ionising radiations is a recently developed method of preservation which is yet to gain acceptance. In the irradiation of food, the radiation dose must be carefully controlled. The WHO and IAEA have recommended that radiation dose of upto 1 kilo rad is not hazardous. Radiation energy must be provided in such a manner that it reaches every particle of food to ensure adequate killing of all microorganisms.

Ionizing radiations can be used for sterilisation of foods in hermetically sealed packs, reduction of the spoilage flora on perishable foods, control of infection in stored cereals, prevent of sprouting of potatoes, onions etc.

Preservation of Products

Jam

Jam is more or less a concentrated fruit pulp possessing a fruity heavy body form. It is rich in flavour. Pectin in the fruit gives it a good set.

It contains not less than 68% soluble solids and thus high concentration of sugar facilitate preservation.

Jam manufacture can chose fresh fruit, frozen, chilled or cold stored fruits, fruits or fruit pulp preserved by heat, sulphite, fruits or fruit pulp or dried fruits.

Sound fresh fruits free of any damage and for better felling effect, green fruits should be mixed with the ripe fruits overripe fruit should not be used.

The fruit must be washed thoroughly to remove any adhering dust.

Leaves, stalks and other undesirable portions of the fruits

Article 15 of Constitution available only to the Indian citizens.

must be removed.

Fruit is then subjected to preliminary treatment which varies from fruit to fruit, e.g. apples are peeled, cut into pieces and then pulped. Strawberries are crushed between rollers but in stone fruits parts, they are Lye peeled, stones removed and pulp is cut into pieces.

Addition of sugar: Properties of sugar to be added depends upon the degree of ripeness and acidity of the fruit. To get a minimum of 68.5% of sugar in the jam, generally 55 g of sugar is added to every 450g of fruit.

Addition of acid, colour, flavour: Generally citric acid, tartaric acid are used to supplement the acidity of fruits for jam-making because appropriate combination of pectin, sugar and acid is essential to give a set to the jam. A good result is obtained if the pH of mixture is about 3.0.

Ordinarily, jams do not require addition of flavours, if desired, may be added when jam boiling is nearing completion. Looking fruit and sugar mixture are boiled to concentrate the soluble solids to about 68.5% and also to bring about necessary degree of inversion of sugar. Determination of soluble solids is easily carried out by using hand refractometer while boiling is in progress.

Packing: Soon after the end point is reached the jam should be cooled in cooling pan to about 93°C and filled into jars at this temperature either mechanically or by hand. The surface of the jam in the jars should be covered with a thin disc of waxed tissue paper and allowed to cool. The jars should be stored in a dry place.

Jams can be prepared from fruits like apple, tomato, papaya, strawberry.

Jelly

In Jelly-making, pectin is the most essential constituent. This pectin is present in the cell walls of fruit. Precipitation of pectin while boiling causes jelly formation. Precipitation takes places, only when pectin, acid, sugar and water are in equilibrium.

Fruits for jelly: Many fruits are rich in pectin as well as acid and are thus well suited for jelly-making. Apple, grape, guava, lemon, orange, plum are rich in pectin and acid.

Selection of fruits : Sufficiently ripe, but not overripe and should have good flavour. Fruits are washed thoroughly with water, guava, apple, need not be peeled, while fruits like, orange, lemon require peeling before used.

Extraction of pectin : Only a minimum quantity of water should be used to the fruit for a simple extraction of pectin the amount of water required depend on the kind of fruit. To get the desired pectin requirements the fruit should be cooked. The cooking time also varies from fruit to fruit and the time of boiling of fruit Is required to extract pectin. The pectin extract can be clarified by allowing it to settle overnight and the supernatant pectin liquor separated as it may be filtered through a muslin cloth. Aluminium vessels used for boiling.

Pectin requirement: Usually about 0.5-1.0% : pectin of suitable quality in the extract is sufficient to produce a good jelly. If pectin is in excess of this a firm and tough jelly is formed and if it is less the jelly may fail to set.

Pectin, sugar, acid and water must be present approximately in the following proportion, pectin (1.0%), sugar (60-65%), fruit acid (1 %), water (33-38%).

The sugar should be sprinkled on the fruit extract while it is boiling and should be thoroughly mixed by stirring to ensure complete dissolution. During boiling, the scum which rises to the top is removed.

Cooking of Jelly: The mixture is boiled for about 20 minutes and the end point is when the final Brix reaches 65°. The end point can also be determined by flake test.

After training the end point, the jelly is then cooled slightly and poured into hot and dry containers.

Pickles

The preservation of food in common salt or vinegar is called pickling. Spices and oil may be added. Pickles are good appetizers and add to the partiality of a meal. They aid digestion by stimulating the flow of gastric juices.

Salt, vinegar and lactic acid are the three important ingredients used in pickling. These substances when used in sufficient quantities, act as preservatives either singly or collectively.

1. *Salt*: Vegetables don't foremost when covered with a large amount of salt.
2. *Vinegar*: This is a preservative for pickles, the final percentage should not be below 2%. To avoid dilution of vinegar by water from vegetable these can be kept in a strong vinegar of about (10% concentration).
3. *Lactic acid*: Sufficient quantity of lactic acid, prevents lactic acid bacteria, and any change in the composition of the material is prevented, provided it is air-tight.

Pickling process: Two stages

1. *Dry softening*: Vegetable is prepared, washed thoroughly and weighed. For every 100 kg of prepared vegetables 3 kg of salt is used. The vegetable is placed about a few cm and a small quantity of salt and another layer of vegetables thick layer is placed. This process of keeping the vegetables and sprinkling the salt is continued till three-quarter of the barrel full. The top of the vegetable is sealed with a choosen cloth and a clear stone placed over to give some weight to the vegetables placed in. The container is kept in a warm dry place and fermentation is allowed to proceed.

In a short time juice from the vegetable forms the brine covering the whole mases. Fermentation then starts and is usually complete in about 8-10 days. This may be confirmed by tapping the containers gently.

2. *Fermentation in brine*: Soaking of the vegetable in salt solution of predetermined concentration for a certain length of time is called brining. This treatment is given only to vegetables like cucumber which do not contain sufficient juice, to form brine with dry salt.

Problems in pickle-making

1. Bitter taste, (due to use of strong vinegar or excess spice).
2. Dull and faded product (insufficient curing, inferior quality).
3. Shivelling (when fruit and vegetable directly placed in strong sugar, vinegar or salt.
4. Swin formation: White swin always form on the surface, vegetable in brine.
5. Softness and stipperiness : Inadequate, covering with brine.
6. Blackening: Fe in the brine equipment used in pickling.

OTHERS

High density planting system (HDP)

Definition: Planting of fruit trees rather at a closer spacing than the recommended one using certain special techniques with the sole objective of obtaining maximum productivity per unit area without sacrificing quality is often referred to as HDP.

The introduction of HDP may well be one of the most important advances in fruit production all over the world.

First established in apple orchards in Europe in the sixties then system of otherwise orcharding has also been successfully adopted in peach, plum, sweet cherry, bananas, pineapple, papaya. In India HDP have been successfully demonstrated in banana, pineapple, papaya more recently mango.

Advantages of HDP :

1. Efficient land use 2 Reduced labour costs 3. Early cropping and bigger yield for a long time 4. Improved fruit quality with efficient harvesting.

In India, where large number of fruit crops are grown, there is a need to develop crop-specific strategies for developing HDP with adequate emphasis on :

(i) Availability of suitable sites, land and other physical resources (ii) Selection and identification of dwarf, compact variety of fruits (iii) Selection and identification of suitable dwarfing rootstock (iv) It gives direction of training and pruning practices to develop ideal tree structure and canopy architecture for efficient use of light.

Knowledge World!

Largest plateau

The deccan plateau covers a million sq km of volcanic rock with thickness varying from 1,200-1,500 m (3,936-4,920 ft) and up to 3,000 m (9,840 ft) near Mumbai.

(v) Use of plant growth retardants and mechanical devices.

Characteristics of HDP :

(a) Trees of HDP should have maximum number of fruiting branches and minimum number of structural branches.

(b) Trees are generally trained with a central leader surrounded by nearly horizontal fruiting branches.

(c) Pruning occur with canopy adjustment.

(d) The height should be according to the harvesting efficiency.

Planting system

Success of HDP depends upon planting system. Generally rectangular planting with single row (within the row spacing is less), double row (wider spacing after every two rows), triple row planting are followed.

In meadow orchards a bed of 10-15 rows are closely planted and separated by alleys of 2.5 m width between beds. This system is also called bed system.

Future strategies :

Fruit production in India is going through an evolutionary phase exemplified by changes in cultivar, root stocks, cultural practices handling and marketing. These changes are required to constrain in bringing into focus, the HDP to produce easily and sustained production of quality fruits.

In developing successful HDP of tropical and sub-tropical fruits, need-based, crop specific strategies are required to be formulated.

Latest technical skills in tree architecture and canopy management is required for which adequate manpower should be trained. The development and success of intensive orcharding need to be demonstrated for quicker adoption and practice by fruit growers.

Hot beds

The main use of hot beds is for starting plants to be grown in the main field and in certain cases to grow crops to maturity during non-growing season . Hot beds are usually located near the farm buildings and near a good water supply with south and south-east exposures for sunlight.

Generally farmers use hot bed made of wood and cement. These farmers may extend 30-45 cm above the surface of the ground on the backside and 15-30 cm in the front, thus giving a slope.

In Indian navy, Commander is equivalent of an army Lieutenant Colonel.

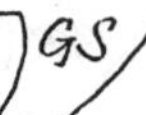

Hot beds have different heating systems:

(i) Manure heated: Fresh manure mixed with one post stress are placed in a pit where hot bed actually is needed with optimum moisture condition for fermentation and steaming.

(ii) Flue heated: Heated by hot air conducted through flues from a fire box located at one end of the bed.

(iii) Hot water heated: Most satisfactory method as the temperature can be controlled more easily, but is expensive.

(iv) Electrically heated: Electricity is passed through the specially constructed cable, consisting of electrically insulated resistance enclosed in a flexible metal sheath. Initial cost is high but beds are always ready for use and they are easily manipulated and readily removed.

Hydroponics

Intensive and profitable cropping in a greenhouse condition is possible after the advent of hydroponics. Hydro-water and pones labour literally meaning water working.

Definition: Science of growing plants without the use of soil, but use of an inert medium such as gravel, sand, peat, vermiculites, or sawdust to which is added a nutrient solution containing all the essential for normal growth and development.

Soilless culture: By W.F. Qreicke in USA in the year 1930.

Advantages

1. Much higher yields than those grown in the soil.
2. It can be used in places where ordinary agriculture is impossible.
3. Quality of the produce is always superior.
4. Freedom from soil-borne diseases and weeds.
5. Less cost of management.

Essentially there are three methods by means of which plants may be grown without soil.

(1) Water Culture (Plant root in solution)

(2) Drip Culture (Sand culture, watered with nutrient solution)

(3) Gravel Culture (Most widely used). It needs the following accessories:

(i) Nutrient tank : Made of concrete, wood or metal

(ii) Benches: Gently sloping benches made of concrete or wood.

(iii) Medium: Crushed graphite of irregular shape

(iv) Solutions : Different chemical solutions

Recent improvements in soilless cultures are Nutrient Film Technique (NFT), Tube culture and Aeroponics.

(NFT) is a water culture, plants grown with roots in plastic film through which nutrient solution is continuously circulated.

Tube culture is developed NFT where the nutrient solution is pumped through PVC dress pipes through which the seedlings are set into the solution flowing along the bottom of PVC.

Aeroponics is growing of plants in an opaque trough or supporting containers in which the roots are suspended and bathed in a nutrient bath mist rather than solution.

Orchard Cultivation

Refers to the careful management of the orchard soil in such a way that the soil is maintained in a good condition suitable to the needs of the tree with least expenses.

A good system of orchard cultivation should ensure

1. Weed control and saving in moisture.
2. Very little disturbance to soil and preventing soil erosion.
3. Reduced cost of cultivation.

Methods of soil management practices :

1. Clean culture : Regular ploughing and removal of weeds.
2. Clean culture with cover crops: Plant a green manure crop before the monsoon and plough it into the soil towards the end of monsoon season e.g. sunhemp, cowpea, dhaincha.
3. Mulching: One of the important soil management practices adopted. Crop residues like, strawdust, cotton stalks, leaves, sawdust, pine needles, air dust or polythene film spread in the tree basins.

The main objective is to conserve soil moisture end to control weed.

Advantages :

(i). Keeps soil cool in day, warm at night hours. (ii). Reduced surface run off. (iii). Adds humus to soil. (iv). Prevents soil erosion. (v). Fruits are protected and kept clean since they fall on mulches. (vi). It allows the absorption of more rain water. (vii). It reduces irrigation frequency.

4. Sod: Permanent cover of grass in orchard and no tillage. Useful in sloppy lands for preventing soil erosion. But they compete for soil moisture and available N.
5. Sod mulch: Similar to sod, but vegetation is cut frequently and the cut material is allowed to remain on the ground. Here moisture loss is not like sod. More of available N to the fruit trees should be applied in both the cases to compensate losses caused by vegetation.

Intercropping

Efficient utilisation of land upto the fruit trees for several years or in case of long duration host crops like tapioc, turmeric, ginger and banana, some area remain

nucleocapsid by the main crop for few months.

The practice of growing any economic crop in alley spaces of the fruit trees in the first few years is referred to as intercropping.

They also act as cover crop and the land benefits by the cultivation, irrigation, manuring given to crops.

Principles :

1. Intercrops should not occupy the area where the roots of the fruit trees are concentrated.
2. Soil fertility should be maintained or improved when intercrops are grown.
3. Water requirements of the intercrop should coincide with those of the main fruit trees. The intercrop may require an irrigation at a time when it is detrimental to the trees.
4. Intercrops should be selected with reference to their effect on soil moisture. Grain crops remove excessive moisture to the detriment of fruit trees.

Vegetables are the best intercrops compared to millets. The intercropping should be stopped when trees occupy the entire orchard space. Therefore green manuring or cover crop should be practised.

Some quick-growing fruit trees could be grown as intercrop as in some places in temperate regions peach and intercropped in apple.

In properly-spaced mango orchards guava tree can be planted to bear in 2-3 years and then removed. Such short-lived trees are known as fillers. Papayas, bananas or phalsa may be grown as fillers in orchards.

Crop	*Age*	*Intercrop*
Mango	upto 7 years	Legume vegetables, papaya (filler)
Grapes	Upto 8 months	Snake gourd, bitter gourd in pandal
Apple, Pear	upto 5 years	Potato, cabbage
Banana	upto 4 months	Sunhemp, onion
Arecanut	upto 10 years	Pineapple
Coconut	upto 3 years	Banana, tapioca, vegetables

Multi-tier system of cropping :

Certain plants like coconut, arecanut are grown for about 50 years in a particular land. It takes nearly 4- 7 years for the above trees to reach the bearing stage.

Adequate alley spaces (nearly 75%) are available in between these trees and being the palm trees, their root system will also not spread beyond one metre in diametre. Hence these vacant spaces can be utilised profitably for raising other crops, thereby increasing the employment opportunity and profit: This is the chief objective of multitier system of cropping.

Inter-cropping and mixed cropping involving jointly in it and is defined as a compatible comparison of crops having varying morphological frames and rooting habits, grown together in such a manner that their canopies intercept solar energy to varying heights and their roots forage, at different depth.

Quiz Kit!

Only active volcano

Barren Island, located 110 km north east of Port Blair, is India's only active volcano. The central core is about 300 m (984 ft) high. On April 10, 1991, the volcano erupted after a gap of two centuries.

The main principle here is land, water and sunlight should be effectively used.

An ideal combination of crops for multi-tier cropping in coconut and arecanut plantations are given :

Tier	*Crop*
First (Top)	Coconut or arecanut
Second	Pepper trained over the trunk of coconut trees
Third	Cocoa or cloves planted at the centre of four arecanut or coconut trees.
Fourth (ground)	Pineapple, ginger and dwarf coffee

Thinning in horticulture (fruit) :

Fruiting is an exhaustive process to the tree especially if the crop is heavy. Therefore the chief goal is to permit the tree to mature as large a crop as possible and yet conserve sufficient nutrients and carbohydrates for good shoot and spur growth, leaf development and flowers but formation for the next year crop.

If the tree is permitted to mature an excessive crop, it becomes devitalised to the point where it not only produces an inferior produce but it becomes increasingly susceptible to diseases. So in order to ensure proper balance, thinning is practised in certain fruit crops.

Thinning is the removal of a part of flower buds, flowers or fruits (before it natural) with the object of :

(i) To increase the annual yield and marketable, fruits.

(ii) To improve the fruit size and colour, TSS.

(iii) It reduces the chance of large breakage.

(iv) It promotes tree vigour and ensures more regular bearing.

(v) It promotes more thorough spraying and dusting of fruits.

(vi) It ensures uniform ripening.

Timing :

(i) Blossom thinning to prevent exhaustion.

GK GS

Pyruvic acid is the intermediate compound of respiration.

(ii) Marble stage thinning to prevent overcrowding.

(iii) Thinning after natural fruit drop of young fruits.

The amount of thinning of fruits depends upon 1. fruit set 2. response of the variety to thinning 3. nature of pruning gives 4. age of the tree 5. value of increased-sized fruits in the market.

Methods :

1. *Hand thinning*: Costly, advisable, if diseased 20% hand thinning of pea-sized berries is beneficial in grape.

Manual, thinning is commonly practised in date palm.

2. *Chemical thinning*: NAA when applied 2 or 4 weeks after petal fall stage in grapes at 100 ppm reduces fruit set in Anab-e-sahi var of grape.

In mandarin, NAA, 600 ppm on marble-sized stage is recommended to thin the overbearing fruits so as to increase the size and quality of fruit.

Seedlessness in horticultural fruits :

Craze for seedless fruits is increasing among the consumers. The seedlessness results in certain fruits due to "parthenocarpy".

Parthenocarpy' which refers to the development of fruits without fertilization or even without the stimulus that comes from pollination.

If a parthenocarpic fruit develops even without the stimulus from the pollination, then this phenomenon is vegetative parthenocarpy e.g. banana with the stimulus of pollination, stimulative guava, water melon.

Thompson seedless due to stimulative pollen- aborted embryo/ steno-par.

Need:

Greater preference among consumer owing to its quality and hence fetch higher prices than the seeded fruits e.g. seedless grapes, guava or oranges. Besides the problem of unfruitfulness may not arise if a fruit develops parthenocarpically and the grower is assured of his crop e.g. banana.

Drawback:

Fruits are usually small in size (grapes) and irregular in shape (guava).

Induction of seedlessness :

1. Use of growth regulators. Application of GA_3 7000-8000 PPM in lanolin paste on the cut end of the style of emasculated flowers of guava results in seedless.

2. By changing the ploidy level.

In Japan, it was domesticated that by developing a triploid water melon (3n = 33) by crossing tetraploid x diploid variety, seedlessness could be achieved.

The seedlessness in naturally available seedless guava var is due to *autopolyploidy* (triploid) and not due to parthenocarpic fruit development.

Hybrid seed production in flowers

During the early 1940s, the significant revolutionary change in floriculture was the exploitation of hybrid vigour in Petunia, when the technique for producing all double F_1 hybrids was developed for the first time.

Advantages of F_1 hybrids :

The main advantage in hybrid is the unique combination of appreciable vigour and uniformity. If selected judiciously, the F_1 hybrids have other advantages too.

Dwarf and compact plants, with basal branching, extremely free-blooming, early flowering, doubleness, large flowers, long duration of flowering, tolerance to heat and humidity and disease-resistance.

The triploid hybrid Nugget, F_1 hybrid of the African Marigold being sterile, the absence of fertilisation and seed set has the unique ability of holding the flowers fresh on the plants for a longer period.

Geranium hybrids raised from seeds do not carry disease unlike cutting or slips.

Hybrid breeding techniques :

The F_1 hybrid breeding process, though rewarding is long, costly and requires considerable skill.

The breeding cycle is accelerated by growing the breeding material out-of-season at any other suitable locations.

The process of testing inbred lines, making test crosses and testing the resulting F_1 hybrids are repeated several times to find a hybrid good enough to introduce.

Hundreds or thousands of last crosses are made to find one new hybrid.

The inbred lines come from the standard open pollinated varieties or occasionally from wild related species. The best plants from each inbred line are selected for test crosses. The promising inbreeds are crossed with a large number of plants from a different inbred line or lines.

The F_1 hybrids produced from these test crosses are then grown and observed in extensive greenhouse and field trials.

The year before a new hybrid is released, hundreds of samples are distributed to commercial plant growers. This provides additional information on performance and cultural method, under commercial conditions in each agro climatic regions.

There is a constant endeavour by flower breeders to produce new more attractive and brightly-coloured hybrids. With the advancements made in new bedding plant technology in USA such as automated systems, smaller containers, shorter crop time and energy stress,

Tata Institue of Fundamental Research is situated in Mumbai.

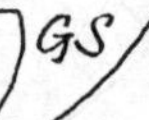

hybrid var are tailored to meet the requirements of commercial floriculture.

Maintaining and improving hybrids :

Once a new hybrid is produced, the work is not over. The parent lines must be maintained and retested to make sure some, of the desirable, attributes of the hybrid are not lost. The uniformity in hybrids depends upon the genetic purity of parental lines.

Production of F_1 hybrid seeds :

Most of the hybrid seed is produced by hand in greenhouse. Both emasculation and pollination are done by hand which enables the breeder to have a wider choice of parents for crossing.

In flowers of family compositae where emasculation by hand are difficult, like Marigold, Zinnia, male sterility is used for hybrid seed production.

Collection of pollen with the help of a vacuum pump, dehiscence of anthers for pollen under VU lamp, storage of pollen at cool temperature and use of pollen gun for pollination are some of the useful innovations technique for hybrid-seed production.

Harvesting of seeds at proper ripening stage, cleaning, and grading of seeds are essential processes before sale. At every stage of production care is taken to avoid selfing

Did You Know ?

Coldest desert

The cold desert region of Ladakh and Lahaul-Spiti is situated beyond the Himalayan peaks at a height between 3,500-5,000 m.

or natural cross pollination or both to ensure complete hybridity and genetic purity. Harvesting, processing and storage of seeds are done carefully so that there is no mechanical mixture.

Work done in India :

Pronounced hybrid vigour in Marigold, Antirrhinum, Hollyhock, Petunia.

Indo-American hybrid seeds:

Commercial hybrid seed production in Petunia, Antirrhiunum, Geranium, Marigold and Gloxinia.

The hybrid seed production being labour-intensive has a great potential for employment to youths in rural and semi-urban areas. It may also generate gainful income by setting up of ancillary industries dealing with equipment and other facilities needed like greenhouse, small seed processing machinery, seed packaging etc.

GK

All unisexual flowers are cross pollinated.

GS

Points to Ponder

- 2, 4, 5 trichlorophenoxy acetic acid is known to check fruit drop.
- A berry refers to a fruit derived from superior (sometimes inferior) ovary in which the entire ovary wall develops into a fleshy, often juicy and edible pericarp.
- A capsule (dry fruit) arises out of pistil with either superior or inferior ovary.
- A corm is an underground stem with a short, fleshy, vertical axis with dried leaf bases.
- A drupe is a fleshy one or more chambered and one or more seeded fruit developing from monocarpellary or syncarpous pistil.
- A natural mutant bougainvillea, Mary Palmer evolved spontaneously in a Kolkata garden.
- A planned removal of branches, twigs, shoots, limbs or roots is called prunning.
- A pollenizer must be a heavy pollen producer to ensure better changes of cross fertilization.
- A preharvest spray of potato with maleic hydrazide (MH) controls sprouting.
- A series of 16 vegetatively propagated apple rootstocks at East Malling comprises a classified example of effect of rootstock on plant-vigour.
- A tuber comprises a short and thickened part of a subterranean branch storing a large quantity of plant food, mainly starch.
- Acalypha plants are more suitable for ornamental internal hedge.
- According to DeCandolle citrus originated somewhere in the north-eastern parts of India.
- Acetic acid 1-4% preserves vinegar pickles permanently.
- Achene is a single seeded, small, dry fruit with seed attached to the wall of the ovary at one point.
- Actual pathogenecity tests are absolutely essential to claim actually disease pathogen-free plants through tissue culture.
- Addition of methanol to persimmon fruits aid ripening.
- Air layerage is also known as "Chinese layerage", "pot layerage", "circumportion", "marcotage" or "gootee".
- Air layering is most common in litchi.
- Akbar, established Nasim Bagh close to Dal lake in Kashmir (India).
- All the fruits do not have the same amount of increase in weight during the same period of growth.
- All vitamins are found in small or large quantities in common vegetable crops.
- Although a ripe tomato is 94 per cent water, it is a good source of vitamin A and B and an excellent source of vitamin C.
- American strain of *Prosopaltella perniciosi* has succeeded in the control of San Jose Scales (*Aspidiotus perniciosus*).
- An aggregate fruit is a collection of simple fruit and fruitlets developing from apocarpous pistil of a flower.
- Any product containing more than 65 per cent sugar can be permanently preserved.
- Apomixis, the development of seeds without complete sexual process is a form of non-sexual or vegetative propagation.
- Apple, pecan, quince and pear fruit buds are borne terminally that unfold to produce leafy shoots terminating into flower clusters.
- Arjun (*Terminalia arjuna*) is a large (20-25m), evergreen shade tree.
- Arka Nishant, Punjab Safed and Pusa Reshmi are Asiatic or tropical type of radish varieties.
- Arka Suryamukhi and Arka Chandan are the cultivars of pumpkin.
- Asexual method of propagation is used generally to increase the planting stock, while *sexual* for hybridization.
- Auxins also bring about the abortion of the developed or developing embryos leading to fruit thinning.
- Bacteria found in fruits and vegetables are destroyed by exposing to temperatures of 212 and 240°F for 15-30 and 30-60 minutes, respectively.
- *Baunhinia purpurea* syn. *B. triandra* (*gulabi kachnar*) flower is rosy purple or lilac or deep pink.
- Bitter gourd (*Memordica charantia*) is generally considered to be an old-world species with its native home in the tropical Africa and Asia.
- Bitter gourd has got heating and wormicidal effect on the body.
- Black-eyed susan (*Thumbergia alata*) and canary bird-flower (*Tropaeolum perigrinum*) are natives of south Africa and Peru respectively.
- *Bombax malabaricum* syn. *B. triandra* (*shimbal*) has egg-shaped fruit pods which yield silk cotton.
- Boron deficiency causes heart-rot in sugar beet.
- Bottle brush (*Callistemon anceolatus* syn. *C. citrinus*) bears crowded scarlet flowers.
- Bower system consists of climbing and spreading of vines in the trellised roof.

GS

Every sixth person in the world is an Indian.

- Bridge grafting is generally useful for the purpose of repairs.
- Broad leaved evergreen trees may not be successful in regions where temperature falls below 5°C.
- Brown heart in turnip is due to deficiency of boron.
- Bud opening is a method in which flowers are harvested at a stage earlier than the one considered as 'cutting stage'.
- Bulbils are tiny buds or small bulbs in axils of the leaves which on planting into ground develop into bulbs.
- Cabbage is heavy feeder of nitrogen and potassium.
- Cape forget-me-not (*Anchusa capensis*) of family borgginaceae has its origin in South Africa.
- Carrot is an excellent source of vitamin A and iron and is rich in sugar.
- Carrot, bitter gourd, onion and tomato comprise good source of iron.
- Caryopsis resembles the achene except that thin seed coat and the pericarp unite to form a single unit.
- Cauliflower has a high requirement of boron and molybdenum.
- Cauliflower is sensitive to high acidity.
- Champa, Rasthali, Dwarf Cavendish are the dessert banana varieties.
- Chekurmanis (*Sauronus androgynus*) is commonly known as multivitamin and multimineral packed leafy vegetable'.
- Chekurmanis is propagated by both hardwood and semi-hard wood cutting.
- Chiku (*Achras sapota*) is grown in the moist coastal tracts of peninsular India.
- The bean used for extraction of gum is cluster bean.
- Chrysanthemum, carnation, brachycome, marigold *etc.*, require pinching.
- Cleft grafting method is especially adopted for top worked trees.
- *Coccinella septempunctata* is very commonly used against cabbage aphid.
- Coccinia (*Coccinia indica*) , a semi-perennial cucurbit is propagated by stem cutting 25-30 cm long and 1.5-2.0 cm thick.
- Common salt at 15-20 per cent is required for permanent preservation.
- Cracking of fruits may also occur due to high absorption of water coupled with low transpiration exerting enhanced pressure on cell walls.
- Cracking of the skin of tomato-fruits has been associated with the deficiency of boron.

Did You Know ?

First Hindi version of the Ramayana

Tulsidas' 15th century version, *Ramcharit-manas,* was a path-breaking effort and is said to be more popular than the original *Ramayana* by Valmiki.

- Cucumber (*kheera*) is one of the oldest cultivated vegetables crops having its origin probably in India.
- Cultivated cowpeas consists of three groups *viz.*, asparagus or yard long bean, catjung and southern pea.
- Cuttings embodying a small portion of the woody branch attached to the base are known as mallet cutting.
- Cypsela is a dry one-chambered fruit developing from an inferior, bicarpellary ovary with the pericarp and the seed coat free as in composital.
- Dahelia is propagated by bulb.
- Damping off is most commonly occurring disease in nursery.
- Date-palm has been placed in high salt tolerant group.
- Date-plam is a single seeded berry.
- Defective graft unions are many times due to failure to maintain cambium continuity.
- Defoliation of tomatoes is caused by *Stemphylian solani* and *Alternaria solani*.
- Dieticians recommend inclusion of 285 grammes of vegetables in our daily diet.
- Double bean (*Phaseolus lunatus*) is a native of south America.
- During fruit ripening, starch is degraded to sugars adding sweetness to the fruits.
- Egg-plant (brinjal) contains vitamins A and B and is an appetizing vegetable.
- Elephant yam's botanical name is *Amorphophallus companulatus*.
- Espalier denotes training of fruit trees, to walls, trellises and fences.
- Ethylene could strongly promote ripening activities by stimulating the climacteric rise.
- Ethylene even in non-climacteric fruit like lemons and oranges stimulates respiration.
- European types of turnip are relatively sweeter and more palatable.
- Exclusion of light from stalks while plants are still growing makes them devoid of chlorophyll, is referred to as blanching.

GK **NAMA stands for Non Agricultural Market Access.** GS

- Faba bean is tolerant to water stress due to higher protein accumulation.
- Fibrous frame work of leaves, stems, bulbs, tubers and roots of vegetables yields fibrous mass which satisfies appetite and prevents constipation.
- *Ficus benghalensis* belonging to family moraceae is a native of India and tropical Africa.
- Filbert varieties of chestnut show both dichogamy and self incompatibility.
- Flame of the forest (*Butea monosperma* syn. *B. frondosa*) can withstand both frost and drought.
- Floriculture is the art and knowledge of growing flowers to perfection.
- Flower buds have higher maximum temperature requirement than leaf buds.
- Flowering panicle of mango with 100 to 4000 flowers may develop only one fruit, as large number of flowers are with male functional parts.
- Forest ecosystem is more stable than agro-ecosystem.
- French bean an ancient crop is probably a native of south America.
- Fruit crop enterprise suffers more from erosion than any thing else.
- Fruiting in mango can be augmented by mixed planting due to cross pollination.
- Fruits have waxy cuticles (apple), while tubers have corcky epidermis (potato) to cut down the water losses.
- Fruits of banana mature after 3-4 months of fruit setting.
- Ganga, Kaveri and Yamuna are hybrids of cabbage.
- Garden pansy (Heart's Ease), *Viola tricolor hortensis* originated in south France.
- Garlic is supposed to have originated from Central Asia and Mediterranean region.
- Generally, 11 per cent pollenizer trees are considered sufficient for deciduous fruit trees.
- Gibberellins overcome genetic and physiological dwarfism in plants.
- Gladiolus corms should be treated with 0.1 per cent of benlate/bavistin for about 15 minutes and then air-dried before planting.
- Graftage aims at cambial union.
- Grape phylloxera (*Phylloxera vitifoliae*) menace was combated by use of resistant rootstocks.
- Grape-fruits are closely related to pummelo.
- Growth substances have been used to increase size of the fruit.
- Growth substances or regulators accelerate the rooting of cuttings in vegetative plant propagation.
- Heavy irrigation to cabbage after development of heads causes bursting/splitting within 24 hours.
- Ibrahim Lodi (1526 AD) laid a garden at Agra, now known as Rambagh.
- Immature tomato blossoms drop rapidly during heat and drought due to increased transpiration.
- In a bouquet gladiolus in one plane and roses should be arranged in layers on different planes.
- In any squash 10-25% of fruit juice should be present.
- In April-May after first shower, when shoots come out form root-tubers is the best planting time for sweet gourd (*Memordica cochinchenesis*) in Assam.
- In budding it is essential to use vegetative and not flower buds.
- In bulbs, the stem is reduced extremely to a disc around which leaf scales remain attached.
- In carnation rooted cuttings have been found to give better results.
- In citrus, peach and plum fruit buds are borne laterally containing flower parts only and give rise to inflorescence without or with small leaves.
- In cutting 6-9 inches long part of one year age should be used.
- In early varieties of tomatoes concentric cracking netting or russeting is more common owing to their light, open foliage.
- In fruits developed from pistil of superior bior polycarpellary ovary where pericarp is hard and woody are known as nuts.
- In Head-to-seed method of seed production in cabbage, selected true to type fully matured plants are uprooted during November-December and are reset in new beds.
- In loquat and mango fruit buds are borne terminally containing flower parts and give rise to inflorescence without leaves.
- In mound layerage the young tip shoot should not be covered with soil at any stage, but may be covered half the way.
- In open centre system of tree training diverging branches of approximately equal diameter at the base are developed.
- In ornamental horticulture mostly 'T' or Shield budding is employed for propagation.
- In quick-dip method the basal portion is dipped from 5 seconds to 2 minutes depending upon the nature of the cuttings.
- In seed-to-seed method of seed production of cabbage, the plants are allowed to over-winter in their original

GK GS

Udayin founded the city of Patliputra.

position.

- In strawberry early maturing flowers are less fertile than those maturing late.
- In veneer grafting a shallow downward and inward cut of about 3-4 cm length is made above the crown of the stock plant.
- Inarching is a common method of propagation of mango and loquat.
- Incompatibility in grafting of *Cucumis melo* on *Cucurbita fictifolia* can be prevented by retaining few leaves of the stock (*C. fictifolia*).
- Indehiscent dry, 1-2 seeded fruit developing from a superior or tricarpellary ovary with flattened wing like out-growth are samara.
- Indian bean (*Dolichos lablab*), in west is known as Bonavist/hyacinth bean.
- Indian bean is grown throughout tropical regions of Asia, Africa and America.
- Intercrops preferably leguminous or vegetable crops can be successfully grown till the fruit trees attain the bearing age.
- Iodine substituted benzoic acids like, TIBA (2, 3, 5-triodobenzoic acid) is good for inducing profuse flowering in many crops.
- It has been observed that only 5-10% of blossoms develop into fruits on a full blossoming apple or pear tree.
- It is believed that small seeded peas originated from south Asia, while large seeded ones from Mediterranean.
- It is essential to gain sufficient knowledge about the trend in abnormal plant production before large scale plant production through tissue culture.
- It is estimated that potato in Nilgiri Hills of Tamil Nadu was introduced in 1822.
- Jack fruit is monoecious fruit tree.
- Jamun (*Eugenia jambolana*) fruits are purple-black oblong to roundish and edible.
- Jonathan apple in Australia, though mostly self-sterile on rich soils became self fruitful on poor soils.
- *Juniperous bermudiana* is a densely branched pyramidal tree.
- Karonda (*Carissa carandas*) is also known as Christ's Thorn.
- Kinnow is the first generation hybrid of the King sweet orange.
- Knolkhol (*Brassica oleracea* var. *gongylodes*) is little known in India.
- Lack of optimum soil moisture supply during bloom and fruit development stage is commonly highly detrimental.
- Lady's lace (*Pimpinella monoica*) has its origin in India.
- Large blocks of self-unfruitful trees improve in pollination when some of big trees are grafted over with pollenizer varieties.
- Lateral shoots pulled off from the stem containing a portion of it (stem) are called heal cutting.
- Leaf-cuttings or leaf-bud cuttings are raised in hot-humid and partially shaded places.
- Leaf-fall in potato occurs due to deficiency of boron.
- Litchi is heterostyle fruit tree.
- Litchi is indigenous to south east China.
- Long and short day plants accelerated flowering only in excess of carbohydrates and nitrogen, respectively.
- Long melon can stand cool climate better than muskmelon.
- Long melon variety Selection-3 is free from bitter principles.
- Longevity of brinjal seed is 4 years.
- Loquat varieties Golden Yellow and Tanaka are self-incompatible.
- Loquat varieties Golden Yellow, Pale Yellow and California Advance are self incompatible.
- Main crop of potato in Hills of Himalayan region (India) is planted in March-April.
- Maleic hydrazide spray, possibly due to its effect as check of respiration rate, lowers fruit drop.
- Malling II, a semi-dwarfing rootstock tends to have vigorous plants embodying good spreading habit and early bearing.
- Malling XII forms a very vigorous rootstock.
- Mango is a self-pollinated fruit crop.
- Mango malformation disease is caused by boron deficiency.
- Mango malformation is due to an imbalance between growth promoters and growth inhibitors.
- Mango variety Amrapali is a hybrid of the cross between Dashehri and Neelam.
- Mangoes blossom generally on the old spring and summer growths.
- Molybdenum has been shown essential for myrobalan seedlings of plum besides tomato and cabbage.
- Moonflower (*Calonyction aculeatum*) is of tropical origin and belongs to family convolvulaceae.
- Morning Glory (*Pharbitis purpurea*, *P. tricolor*) is a native of tropical America.
- Most commonly used shrubs for topiary are

Clerodendrom inerme and *Duranta plumeri*.

- Most ideal variety of radish grown in summer and rainy season only is Pusa Chetki.
- Most of the auxins do not only stimulate stem and coleoptile growth, but also affect differentiation and abscission.
- Most of the fruits in store are harmed when the CO_2 concentration exceeds 1 per cent.
- Muskmelon (*Cucumis melo*) is a native of north-west India.
- Muskmelon fruits on ripening when part completely and easily from the stem leaving a circular depression, it is said to be at full slip stage.
- Nitrogen deficiency causes yellowing of vein in monocot leaf, while entire leaf becomes yellowish green in dicots.
- Node cuttings are inserted horizontally in sand.
- Of the unisexual plants the pure staminate flowering ones are essentially barren.
- Onion bulbs and green onions are good source of vitamin B.
- Onion does not thrive well in areas receiving more than 750-1000 mm of annual average rainfall.
- Onion, okra, asparagus and summer squash supply iodine.
- Parthenocarpically developed fruits of tomato are seedless.
- Patch budding is commonly practised in guava.
- Peach can be successfully grown even with a single self-fertile variety except the self-barren varieties.
- Pepo is a type of berry in which the outer wall which is the receptacle becomes hard as in cucurbitaceae family.
- Per capita estimated consumption of potato in India is about 13 kg per annum.
- Perennial and annual types of Indian bean are known as *Typicus* and *Lignosus*, respectively.
- Pinching or stopping means removal of the growing point of a shoot along with a few leaves.
- Planting root materials are usually collected from old planting of pointed gourd during December and January.
- Plants like Saintpaulia, Peperomia, Bryophyllum and *Sedum spectabile* etc., are propagated by leaf cutting.
- Plants of litchi raised by seeds are not true-to-type and may take 8-20 years for bearing.
- Plants which bleed profusely should not be pruned when cell sap is at high pressure.
- Pointed gourd (*Trichosanthes diocia*) is a native of India and Assam is considered to be the primary centre of its origin.
- Pointed gourd is easily digestible, diuretic, laxative, invigorates the heart and brain, and is useful in disorders of the circulatory system.
- Pollination is not important for fruit development in persimmon.
- Polyembryonic stock induces more scion-vigour than monoembryonic ones.
- Pomes denote fleshy fruits embodying pericarp differentiated into epicarp, mesocarp and endocarp.
- *Pongamia glabra* (*karanj*) is a medium tall deciduous tree with trifoliate, bright-green, glossy leaves.
- Potassium metabisulphide (1:2000) contained in muslin bags when introduced in crates of grape wards off decay and augments storage life.
- Potato (*Solanum tuberosum*) is a native of South America.
- Potato crop removes large amount of potassium followed by nitrogen and phosphorus.
- Potato stored at -1-0°C suffers from internal break-down known as black heart.
- Premature producing of seed stocks in onion is known as bolting.
- Premature seeding and/or failure of the leaves to form a solid are common defects of early cabbage.
- Proper ripening of date-palm fruits needs a temperature of over 30°C for no less than 30 days.
- Proper setting of fruit jelly depends upon correct proportion of pectin, sugar and acid.
- Proteins are synthesized at increasing rates in apples during ripening.
- Pruning in effect improves the size and colour of fruits.
- Pruning is based on scientific principles to augment the overall performance of flowers fruits and trees.
- Pruning of nitrogen deficient trees induces more of unfruitfulness.
- Pseudostem is upper stem of banana.
- Pulsing / loading is a preshipment treatment employed by growers.
- Punjab Chhuhara is a determinate variety of tomato.
- Punjab Selection, Pusa Ratnar, and Arka Kalyan are the varieties of red onion.
- Pusa Early Bunching and Kasuri Selection are the cultivars of fenugreek.
- Pusa Kesar, a cultivar of carrot is a selection from cross between Local Red and Nantes Half Long.
- Pusa Sadabahar and Pusa Naubahar are the varieties of

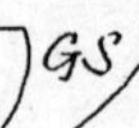

Baikal is the deepest lake of the world.

clusterbean.

- Pusa Sanyog is a F_1- hybrid of cucumber.
- *Putranjiva roxburghii*, Child-Life-Tree belongs to family euphorbiaceae.
- Quick-dip and Lanolin methods denote the application of growth substances to aid vegetative propagation.
- Rectangular system of planting is most common in grape, phalsa and peach.
- Red skinned tubers of sweet potato store better than the white-skinned ones.
- Removal of undesirable flowers from the plant is called deblossoming.
- Respiratory inhibitors prevent fruit ripening.
- Rhizocaline is mobilized to the morphological basal end of the cutting for root formation.
- Rhizome is a subterranean stem arising from a lateral bud from the main stem at ground level and extending underground near the surface horizontally.
- Ring budding is practised with jujuba and peach on commercial lines.
- Ringing leads to parthenocarpy only due to greater concentration of nutrition at the point of bearing on tips.
- Ripe watermelon fruit when thumped with finger(s) gives out a muffled dull/dead sound as against metallic and ringing sound by an immatured one.
- Ripening of muskmelon fruits begins first from the blossom end accompanied by a change in skin colour.
- Root cuttings should be taken from young roots of average thickness and the length may vary from 2.5 to 15.0 cm with one or a few buds.
- S-48, Arka Pragati and Hissar-2 are the varieties of white onion.
- SADH is very effective in controlling height of chrysanthemum.
- Sand-verbena (*Abronia umbellata* syn. *A. rosea*) of family *nictaginaceae* originated from California (USA).
- Savaji park, Baroda (Gujarat) was established by Maharaja Savajirao III in 1879.
- Scientific name of potato tuber moth is *Gnorimoschema operculelia*.
- Semi-hard wood cuttings are used for propagation in jasmine and hydrangeas.
- Short-period preservation of fruits and vegetables is achievable by retarding activity of enzymes and microbes by refrigeration, exclusion of moisture and addition of salt, sugar and chemicals.
- Shrubs trained to a single stem is called head system.
- Small grain crops need not be used as cover crops in orchards.

You Don't Know!

Earliest museum for forestry

The Gass Forest Museum in Coimbatore founded by H.A. Gass, is the earliest museum for forestry.

- Snake gourd (*Trichosanthes anguina*) is found growing wild in India and the Indian archipelago is thought to be its source of origin.
- Sod system is not favourable for foliage, tree growth and fruit yield.
- Soil sterilization can be done either with heat or with chemicals to eliminate soil borne diseases.
- Sour orange (*C. aurantium*) when budded on Eureka lemon suffered much more from winter injury than the unbudded seedlings.
- Spraying of tomato plants at flowering initiation using 25 and 50 ppm of GA_3, improves the quality of fruits.
- Staminant constants are those which bear pistillate flowers each year along with some staminate flowers.
- Successful orcharding consists of not only yield, but also economic production and efficient disposal.
- Sugarbeet is a long day plant.
- Summer squash (*Cucurbita pepo*) is of American origin.
- Supply of the required minimum of carbohydrate and nitrogen to plants brings the C:N ratio of the forefront to control flowering.
- Sweet Alison (*Alyssum maritimum* syn. *Lobularia maritima*) has its origin from Europe and West Asia and belongs to family cruciferae.
- Sweet orange is very susceptible to malda goma footrot.
- Sweet potato contains 16 and 4 per cent of starch and sugar, that is, 20 per cent alcohol producing material.
- Sweet potato weevil (*Cylas formicarius*) is the most important insect-pest of sweet potato.
- The amount of extracted pectin may be tested by sprit test and jel-meter test.
- The amount of RNA, DNA and proteins are always higher in healthy panicles than in malformed ones in mango.
- The area in the neighbourhood of Caspian sea is considered to be the place of origin of grapes.
- The best time for pruning of *ber* in northern India is mid May to mid June.
- The blue colour of the flowers of *Hydron macrophylla* is due to presence of more aluminium than that in the pink ones.
- The change of end-season fertility to self-sterility in

GK **Orissa ranks first in the poverty ratio.** GS

tomato, mango and grape plants or end-season sterility to self-fertility in strawberry are indicative of the effect of season.

- The citrus fruits are regarded as modified berries called hesperidium in which the rind is not edible.
- The combination of fruit and flower arrangement is called Morimono by Japanese.
- The commercial production of cabbage is done by seed-to-seed method.
- The control on the direction of growth of a plant is known as training.
- The cultivated carrot probably originated in the hills of the Punjab and Kashmir, with a secondary centre of distribution in Ethiopia and North Africa.
- The cultivated species of citrus are believed to be natives of tropical and subtropical regions of south-east Asia.
- The cuttings longer than 7 cm are placed vertically.
- The descending order of salt-tolerance of different fruit trees is grape, olives, figs, almonds, oranges, pears, peaches, apricot, lemons and mulberry.
- The dried fruit of *Vanilla planifolia* syn. *V. fragrans* yield the vanilla of commerce.
- The evergreen fruit trees suffer more from frost than the deciduous fruit trees.
- The first anti-auxin was possibly phenyl butyric acid.
- The first Japanese style garden was laid out in the Roshanara park Delhi.
- The flower buds of apple are formed mostly on the spurs.
- The flowers form a soul of garden and convey the message of nature to man.
- The NBRI popularly known as Sikandar Bagh was laid out by Nawab Saadat Ali Khan.
- The optimum offshoot size of date-palm for planting is 3 kg.
- The optimum temperature for most of the plants is 22-27°C.
- The original home of radish is probably China and India.
- The original races of pea came from Ethiopia.
- The pear are quite large and conspicuous forming a wider angle with the shoots.
- The plants bend towards lee-ward side under the stress of high winds.
- The potato variety Kufri Kuber has been developed at Potato Breeding Station (now CPRI) Shimla in 1943.
- The production of defective pistil is known to be a changing character associated with locality.
- The propagation of orchid clones through meristem culture (mericlones) was the first commercially successful venture in tissue culture technique.
- The pungency in onion is due to a volatile oil known as allyl-propyl-disulphide.
- The purpose of conditioning/hardening is to restore the turgidity of flowers.
- The seed rate of celery is about 125 g per hectare.
- The seed rate of garlic varies from 340 to 570 kg/ha of cloves/bulblets.
- The seed rate of knolkhol to raise seedlings for one hectare is 1250 grammes.
- The seed rate of lettuce for nursery and direct sowing is 500-750 and 1875-2000 g, respectively.
- The sexual embryos develop larger seedling than the asexual ones.
- The stakes can be interlinked with strings, forming an interlacing mesh, as is done in Sen-rin-tsukin method of chrysanthemum growing.
- The success of graftage depends on compatibility, cambium content and closeness between the scion and the rootstock.
- The success of *in vitro* propagation depends much on the chemical composition and physical qualities of the nutrient media.
- The temperature fluctuations of the storage chamber must not vary by more than 0.5°C at any time.
- The tomato scion produced poisonous fruits when grafted on *Datura strammonium*.
- The tomato varieties can be broadly classified into determinate and indeterminate categories.
- The tomatoes supplied with rich nutrition fail to produce normal flowers.
- The tuber production in potato totally stops at 36°C.
- The tubers embody buds in the axils of reduced leaves.
- The woolly apple aphid has been controlled in W. Himalayas by *Aphelinus mali*.
- Thinning out is the process of removal of twig or a part thereof according to need.
- Tip layerage is done commercially for black raspberries.
- Training controls the distribution of fruiting wood.
- Training has little effect on fruit bearing.
- Transferring of the young seedlings to another pan or tray is known as pricking.
- Turnip greens are a good source of minerals such as calcium and iron and vitamin A.
- Underground stem of lotus is a popular vegetable in N. India.

Saudi Arabia women do not even have the right to drive a car.

GS

- Unripe papaya fruit is rich in papain, an enzyme helpful in digestion of proteins.
- Unsaturated hydrocarbons (ethylene) can stimulate the process of fruit ripening.
- Vegetable growing is generally more lucrative than any other type of farming.
- Vegetable *viz.*, fenugreek, spinach and mountain spinach are rich in calcium.
- Vegetative propagation is preferred due to ease of maintenance of hereditary characters of the mother plant.
- Veneer grafting is the kind of detached grafting.
- Veni is a special kind of flower arrangement used in S. India to decorate long plant to hair.
- Vitamin A is abundantly found in leafy vegetables.
- When colour of ground spot (white spot) where the watermelon fruit rests on the ground assumes yellow tinge, is indicative of its maturity.
- When fruit juice is totally clear free from pulp or part the fruit and moderately sweetened, it is referred to as cordial.
- When the branches are alternatively covered and exposed over their entire length, the layerage is known as compound layerage.
- White Iscle, Scarlet Globe and Pusa Himani are European (temperate) type of varieties of radish.
- White Vienna, Purple Vienna and King of North are the cultivars of knol-khol.
- Wilted vegetables contain less of vitamin C than fresh ones.
- Wine is also made from faba bean beside grapes.
- Yeast ferments fruit-sugars and their products into alcohol and carbon-dioxide.
- 'Bird eye' is a disease of apple.
- Botanical name of loquat is *Eriobotrya japonica*.
- Botanically pineapple is sorosis.
- AVRDC is located in Taiwan.
- National Research Centre for Onion & Garlic is located at Pune.
- Amla Cultivar, Francis is highly susceptible to 'internal fruit necrosis'.
- 'Hen & Chicken' disease of grape is due to deficiency of boron.
- Predominant sugar of guava is fructose.
- Botanical name of bird pepper is *Capsicum fruitecens*.
- Seed rate of mid-season table pea is 65-75 kg/ha.
- Chilli variety, Punjab Lal is resistant to fruit rot, die-back and leaf-curl.
- Foot-yam is botanically *Amorpho-phallus companulatus*.
- CaC_3 is suitable for hastening ripening of fruits.
- Papaya and banana are climatric fruits.
- India's contribution in fruit production is 10%.
- Sylvia is a variety of pea.
- Gynodiocious varieties of papaya produce female and hermaphrodite plants.
- Neeranjan is off-season variety of mango.

Multiple Choice Questions

1. ______ is known as 'city of garden'.
   a] Bangalore b] Mumbai
   c] Chandigarh d] Delhi
2. __________ works for grape fruit setting.
   a] Alar b] IAA
   c] Cytokinin d] Abscissic acid
3. _________ is grown as a shade tree in coffee, areca and cardamom plantation.
   a] Jack-fruit b] Coconut
   c] Mango d] Guava
4. _________ ppm 2, 4-D spray can increase the yield of tomato three times.
   a] 25-35 b] 100-150
   c] 50-100 d] 1-5
5. ______________ is the cultivar of bottle gourd.
   a] Pusa Meghdoot b] CO-1
   c] Australian Green d] Straight Eight
6. __________ variety of *Brassica pekinensis* is a useful leaf vegetable for plains and hills of N. India.
   a] Chinese b] American
   c] Canadian d] Japanese
7. 'Heart of garden' is known as :
   a] Rose b] Border
   c] Hedge d] Lawn
8. 8-hydroxy quinaline sulphate is used in:
   a] Hormone preparation
   b] Cut flower solution
   c] Micro-nutrient spray
   d] Weed control
9. A cashewnut tree will start fruiting after ________ of planting.
   a] 4 years b] 9 years
   c] 7 years d] 2 years
10. A coconut tree will start fruiting after________ of planting.
    a] 2 years b] 6 years
    c] 1 year d] 4 years
11. A fruit that develops from a hollow pear-shaped fleshy receptacle (*hypanthodium*) which encloses a number of minute male and female flowers is known as :
    a] Aggregate fruit b] Hesperidium
    c] Pome fruit d] Syconus fruit
12. A perfect jelly have pectine of :
    a] 1.5% b] 1.0%
    c] 11.25% d] 2.0%
13. About 3 kg of seed or _________ root tubers of pointed gourd are enough to plant an area of one hectare.
    a] 20,000 b] 30,000
    c] 40,000 d] 50,000
14. Acacia is used for :
    a] Hedge b] Fruits
    c] Ornamentals d] None of these
15. According to All India Medical Science report per capita vegetable requirement in India is :
    a] 500 gm b] 400 gm
    c] 200 gm d] 300 gm
16. According to flowering habit datepalm is :
    a] Monoecious b] Dioecious
    c] Hermophrodite d] Andromonoecious
17. Age of plants at planting time should be_______ years.
    a] 2-3 b] 4-5
    c] 0.5-1.0 d] 1.0-1.5
18. Air layering in practised usually with:
    a] Grapes b] Persian lime
    c] Mango d] None of these
19. Air layering is most common in :
    a] Rose b] Litchi
    c] Grape d] Guava
20. Allahabad Safeda is the variety of :
    a] Mango b] Guava
    c] Loquat d] Grapes
21. Amimgola is a variety of :
    a] Guava b] Fig
    c] Mango d] Grape
22. Among the following which is the food preservator ?
    a] Celphos
    b] Calcium cynamide
    c] Sodium benzoate
    d] Potassium permagnate
23. Among vegetables___________has more of vitamin A and vitamin C.
    a] Potato b] Tomato
    c] Brinjal d] Table Chillies
24. An apple tree will give the fruits after____of planting
    a] 5 years b] 3 years
    c] 2 years d] 8 years
25. An isolation distance of ____________ between varieties of water melon is considered enough for pure seed production.
    a] 100-200 m b] 200-300 m
    c] 300-375 m d] 400-500 m
26. Any one type of flowers have :
    a] Plane disc b] Curled petals
    c] Doom shaped disc d] Ring around disc
27. Approximate number of sapota plants in one hectare is :
    a] 220 b] 120
    c] 320 d] 400
28. *Aristolochia*, family *Aristolochiaceae*, represents a group of :
    a] Herbaceous annuals
    b] Perennial shrubs
    c] Annual flowering plant
    d] Climbers
29. Artificial ripening is very difficult in :
    a] Lemon b] Papaya
    c] Persimon d] Plums

GK **Morphine is the insoluble solid waste product of plant.** GS

30. Asepsis means :
a] Exclusion of air
b] Exclusion of moisture/water
c] Exclusion of dust and spray
d] Exclusion of micro-organism
31. Banana planting time is :
a] June-July b] January
c] December d] October
32. Banana should be propagated by :
a] Sword suckers b] Water suckers
c] Rhizomes d] Stem cuttings
33. Banana suckers arise from :
a] Underground rhizomes
b] Underground corms
c] Stolons
d] Pseudostems
34. Bannana, papaya and jack fruits are used as :
a] Fruits b] Vegetables
c] Both d] Salad
35. Baradari is one of the main features of :
a] Italian garden
b] English garden
c] Mughal garden
d] Japanese garden
36. Botanical name of carnation is :
a] *Celosia cristata*
b] *Tagetis errecta*
c] *Dianthus ceryophyllus*
d] *Centuria cyanus*
37. Bougainvillea is propagated by :
a] Budding b] Layering
c] Cutting d] All these
38. Branch of science that deals with study of principles of propagation and production of fruits is called :
a] Olericulture b] Floriculture
c] Pomology d] Entomology
39. Branch of science that deals with study of principles of propagation and production of vegetables is called:
a] Olericulture b] Floriculture
c] Pomology d] None of these
40. Branch of science that deals with study of principles of propagation and production of flowers is called:
a] Olericulture b] Floriculture
c] Pomology d] None of these
41. Brinjal is considered to have had its origin in :
a] China b] Africa
c] India d] Japan
42. Browning of cauliflower is caused by the deficiency of :
a] Zinc b] Manganese
c] Iron d] Boron
43. *Caissa carandas* is used for :
a] Ornamental b] Hedge
c] Fruit d] Fence

Knowledge World!

Largest writing slate

A giant slate 3.14 m (10.3 ft) long and 2.13 m (7 ft) wide at the entrance of Atiya Kindergarten School, now known as Safa Model School, in Malkapet, Hyderabad, is the world's largest writing slate. It is made of cement concrete.

44. Canning is a method of
a] Heat processing b] Non heat processing
c] Sterilization d] Any other
45. Cardomom Hills are located in :
a] Nilgiries b] Kerala
c] Himalayas d] Satpura
46. Carrot seed will not germinate below ______ °C.
a] 7.9 b] 6.9
c] 5.9 d] 3.9
47. Celery seed takes nearly ________ to germinate.
a] 10 days b] 20 days
c] 30 days d] 40 days
48. Chenopodiaceae is the family of :
a] Potato b] Sweet potato
c] Okra d] Spinach
49. Chiku is propagated on root stock of :
a] *Mangifera indica*
b] *Melia azadirach*
c] *Jasmine sambac*
d] *Manikara hexandro*
50. Chrysanthemum flowers froms the emblem of:
a] Japan b] China
c] Russia d] Sri Lanka
51. Chrysanthemum is propagated by means of :
a] Suckers b] Cutting
c] Buddings d] Inarching
52. Citrus fruits can be preserved for a few weeks at room temperature by :
a] Submerging them in cold water and then drying
b] Keeping them in well ventilated room
c] Putting them in an atmosphere rich in CO_2
d] By wrapping individual fruit in HDP envelopes
53. Coffee plants are matured at the age of ______ years.
a] 15 b] 10
c] 20 d] 4
54. Coffee plants begin yielding a small crop from :
a] 4th year b] 8th year
c] 2nd year d] 10th year
55. Cordial is generally taken by mixing with :
a] Water and ice b] Tea
c] Coffee d] Gin
56. Crescent is well known for :
a] Flower arrangement
b] Pigment
c] Type of jasmine scent
d] English garden feature

GK **IUCN stands for International Union for Conservation of Nature.** GS

57. Culcutta, Early seedless and Rose scented are the varieties of :
a] Guava b] Papaya
c] Mango d] Litchi
58. Cutting from broad-leaved evergreen trees are usually taken in :
a] Early summer
b] Early winter
c] Late summer
d] None of the above
59. Dahalia is best propagated by :
a] Layering b] Rhizome
c] Sucker d] None of these
60. Datepalm is propagated by :
a] Off shoots b] Suckers
c] Both these d] Budding
61. Giant Taro should be propagated by :
a] Suckers b] Seeds
c] Offshoots d] Rhizomes
62. Delight Gold is a variety of :
a] Guava b] Papaya
c] Grape d] Mango
63. Die back of citrus can be controlled with the application of :
a] Bordeaux mixture b] Endosulfan
c] Zinc sulphate d] B.H.C.
64. Dry land scape is important feature of :
a] English garden b] Mughal garden
c] Desert garden d] Japanese garden
65. Edible part of arvi is :
a] Underground stem b] Underground root
c] Both these d] None of these
66. Edible part of onion is called :
a] Bulb b] Tuber
c] Corm d] Rhizome
67. Edible part of potato is scientifically termed as :
a] Tuber b] Rhizome
c] Corm d] Bulb
68. Elephant yam corms ("buds") weighting 100-200 g are used for planting of the ____________ years crops.
a] First b] Second
c] Third d] Fourth
69. Elephant yam corms at the end of ______________ year weight 6.8-9.0 kg.
a] Second b] Third
c] Fourth d] Fifth
70. Eschscholtzia flowers are :
a] Yellow b] Light blue
c] Pink d] Violet
71. Evergreen fruit like mango, guava and jack fruit should be planted in the month of :
a] February-March b] June-July
c] May d] December
72. Evidence for universal nature of flowering hormone has emerged from experiments involving :
a] Defoliation
b] Reciprocal grafting
c] Night interruption
d] Application of hormones
73. Expected average yield of rose per plant per year is :
a] 100 b] 300
c] 600 d] 1000
74. Fastening of vines to a upright support of a tree called panagara (*Erithrina indica*) is known as:
a] Kniffen system b] Head system
c] Single stake system d] Cordon system
75. Fig in India is grown in :
a] Madhya Pradesh b] Punjab
c] Kerala d] Gujrat
76. Flowering in pumpkin (*Cucurbita moschata*) starts ____________ days after sowing.
a] 30-40 b] 50-60
c] 60-80 d] 85-95
77. For arid climatic conditions which fruit should not be selected :
a] Date palm b] Ber
c] Pomegranate d] Banana
78. For hand pollination in date-palm 10 male trees are sufficient for ______________ female plants.
a] 50 b] 75
c] 100 d] 125
79. For jelly making the best fruit is :
a] Sour apple b] Ripe banana
c] Ripe mango d] Ber
80. For jelly preparation very important fruit is :
a] Guava b] Mango
c] Apple d] Grapes
81. For Malta blood red, the best root stock is :
a] Citrus jambhiri b] Karna khatta
c] Rangpur lime d] Cleoptera mandarin
82. For prickle making one can use maximum:
a] 40% salt b] 30% salt
c] 25% salt d] 35% salt
83. For processing tomato fruits are picked at :
a] Immature stage b] Pink stage
c] Hard ripe stage d] Over-ripe stage
84. For tomato ketchup, the tomato should have :
a] Green colour b] Yellow colour
c] High T.S.S. d] None of above
85. Free style gardens are nothing but :
a] Mughal gardens
b] Symmetrical gardens
c] Formal gardens
d] Informal gardens

GS

India accounts for 2.4% of world surface areas.

86. **Fruit candy is :**
a] Agra-ka-petha b] Aonla ka murabba
c] Murabba d] Sugar candy

87. **Fruit drop in mango can be controlled with the spray of :**
a] 2, 4-D b] B.H.C.
c] NaCl d] GA_3

88. **Fruit squash must have minimum of:**
a] 35% juice b] 25% juice
c] 20% juice d] 45% juice

89. **Fruit stored in a cold chamber exhibit longer storage life because:**
a] Exposure to sunlight is prevented
b] The concentration of carbon dioxide in the environment is increased
c] The rate of respiration is decreased
d] There is an increase in humidity

90. **Fruits and vegetables furnish ____________percent of vitamin C and _________ percent of vitamin A in the world.**
a] 30, 60 b] 60, 30
c] 90, 30 d] 90, 60

91. **Fruits of banana mature after _________ months of fruit setting.**
a] 1-2 b] 2-3
c] 9-10 d] 3-4

92. **Fruting in banana occurs after :**
a] 2 year b] 10-16 months
c] 4 months d] None of them

93. **GA_3____________ spray can increase the size of grape.**
a] 20 ppm b] 1000 ppm
c] 1 ppm d] 2 ppm

94. **Garlic freezes at the average temperature of ____________ °C.**
a] -5 b] -4
c] -3 d] -2

95. **Garlic is propagated by :**
a] Cloves b] Seeds
c] Scales d] Bulbs

96. **Golden Acre is a variety of :**
a] Cabbage b] Radish
c] Onion d] Tomato

97. **Golden shower (*Cassia fistula*) may take ____________ years to flower.**
a] 5 b] 10
c] 15 d] 18

98. **Good quality of pectin is found in guava fruits which are :**
a] Green in stage
b] Ripe but green stage
c] Yellow but hard stage
d] Over ripe stage

99. **Grafted mango trees becomes ready for transplanting :**
a] Early b] Late
c] Rarely d] Never

100. **Grape tree will start fruiting after_______of planting.**
a] 4 years b] 5 years
c] 1.5 years d] 2.5 years

101. **Grapes are generally dried in :**
a] Sun b] Dehydrator
c] Spray drier d] Vaccum drier

102. **Grapes are pruned only once in northern India while they are pruned twice in southern India. This difference is due to the face that grapes :**
a] Shed leaves in winter in northern India while they remain evergreen in southern India
b] Are early in bearing in northern India while they bear late in southern India
c] Bear lightly in northern India while they bear heavily in southern India
d] Are less prone to diseases in northern India while they are more prone to diseases in southern India

103. **Grapes contain which of the following acids ?**
a] Malic acid b] Tartaric acid
c] Hydrocyanic acid d] Glutonic acid

104. **Gros Michel is a variety of banana which is :**
a] Diploid b] Triploid
c] Tetraploid d] None of above

105. **Ground layering is most common in :**
a] Rose b] Litchi
c] Begonia d] Pomegranate

106. **Guava tree will give the fruits after________of planting.**
a] 4 years b] 2 years
c] 7 years d] 10 years

107. **Guava is a native of :**
a] Tropical south-east Asia
b] Tropical Africa
c] Tropical America
d] Sub-tropical India

108. **Hard-wood cutting is a propagating material used in:**
a] Date b] Fig
c] Grape d] None of above

109. **Hard-wood cuttings for propagation are used in :**
a] Dogwood b] Grape
c] Fig d] Begonia

110. **Herbaceous stem cuttings propogation are used in :**
a] Carnation b] Begonia
c] Grape d] Hollyhock

111. **Hillawi is very popular variety of ____________ through out the world due to adaptability to high humidity and rains.**
a] Coconut b] Date-palm
c] Cashewnut d] Arecanut

112. **Himrod and Perlette are the varieties of :**
a] Grape b] Mango
c] Papaya d] Guava

GK GS

Sohgaura inscription mentions the famine during Asoka's time.

113. Hogarth course is also known as :
a] Line of beauty
b] Circular curve
c] Tangent arrangement
d] None of these

114. Ideal pH for orchard crops is :
a] 6-8 b] 5-7
c] 4-6 d] 9-13

115. In apples small, narrow and mottled leaves are primarily because of the deficiency of :
a] Zinc b] Iron
c] Potassium d] Iodine

116. In budding the length of scion bud should be :
a] 1.0" to 2.0" b] 2.0" to 2.5"
c] 0.5" to 1.12" d] None of these

117. In *Cassia nodosa* _______________ flowers appear in bunches from the nodes all along the branches.
a] Yellow b] Red
c] Pink d] Purple

118. In cucurbits, the crop which is dioecious in nature is:
a] Bottle gourd b] Pointed gourd
c] Muskmelon d] Watermelon

119. In cutting___________ inches long part of one year age should be used.
a] 6-9 b] 12-16
c] 1-2 d] 15-20

120. In general, potato needs __________ mm of irrigation water for the optimum production.
a] 500 b] 600
c] 700 d] 800

121. In grafting lower part of plant is known as :
a] Stock b] T-Shield
c] Both of them d] None of them

122. In grafting upper part of stem is called :
a] Stock b] Scion
c] Bud d] Apical

123. In hexagonal method of tree planting, about_____________percent more trees can be planted than triangular method.
a] 25 b] 40
c] 10 d] 15

124. In hilly area which system of fruit planting should be used :
a] Contour
b] Strip cropping
c] Wind strip cropping
d] None of these

125. In onion bolting takes place due to:
a] High temperature
b] Low temperature
c] High nitrogen
d] Difficiency of nitrogen

126. In orchard layout smallest area should be allocated under :
a] Irrigation channels b] Roads and Path
c] Manures pit d] Farm building

127. In our diet staple vegetable is :
a] Brinjal b] Potato
c] Tomato d] Chillies

128. In potato 'Hollow heart' is due to :
a] Excessive watering and fertilization
b] Excessive watering and high temperature
c] Excessive watering and haulm cutting
d] Excessive watering and haulm cutting

129. In potato tree (*Solanum macranthum* syn. *S. maroniense*) flowers 2.5 cm across and resemble _____________ flowers.
a] Tomato b] Eggplant
c] Potato d] Sweet potato

130. In relation to crop rotation which one of the following statements is incorrect?
a] Legume may be followed by non-legume
b] Vegetable susceptible to a particular pest should be followed by a resistant one
c] Heavy feeders should be followed by low feeders
d] Deep rooted crop should be followed by same type of crop

131. In squash melon 100 g of seed accounts for about _______________ seeds.
a] 100 b] 200
c] 250 d] 350

132. In the case of potato, the seed rate per hectare is :
a] 1 quintal b] 5 quintals
c] 10 quintals d] 15 quintals

133. In tomato for fruit setting in adverse condition spray:
a] 2, 4 D
b] PCPA
c] Etheral
d] Cycocel

134. In watermelon the most important judging indication for maturity is :
a] Days after pollination
b] Dull sound when fruit is thumped
c] Metallic sound when fruit is thumped
d] None of the above

135. In which of the following crops, leaf and bud cutting are used for propogation ?
a] Tomato b] Juniper
c] Fig d] Sainpanba

136. In whip tongue grafting the length of the cut should be about _____________ time the diameter of the plants in question.
a] 3 b] 4
c] 5 d] 6

GK **Atolls are circular or horse-shoe shaped, enclosing a lagoon.** GS

137. It has been observed that the malta sweet orange budded on rough lemon produced trees ________ times larger than those budded on citrus (*C. medica*).
a] Two b] Three
c] Four d] Five

138. Jam comes under :
a] Concentrate
b] Sugar concentrate
c] Any other
d] Non heat processing

139. Jam is prepared with :
a] Fruit pectin and pulp
b] Fruit pulp alone
c] Fruit pectin alone
d] Pectin, pulp, seed and skin mix.

140. Jelly temperature at end point (at sea level) is :
a] 105.5 °C b] 100.5 °C
c] 105 °C d] 104 °C

141. Kinnow is a cross between :
a] King and Willow leaf
b] Mandarin and Sweet orange
c] Willow leaf and Wilking
d] King and Wilking

142. Kohir Safed and Safed jam are the varieties of :
a] Pomegranate b] Guava
c] Date palm d] Mango

143. *Lantana camara* syn. *L. aculeata* belong to family ________ and is a native of Jamaica.
a] *Lythraceae*
b] *Rosaceae*
c] *Acanthaceae*
d] *Verbenaceae*

144. Latest hybrid mango variety released from Maharashtra is :
a] Amrapali b] Mallika
c] Ratna d] Nileshwar dwarf

145. Lawn in Mughal garden are :
a] Slopy b] Star shape
c] Terraced d] Circular

146. Leaf and leaf bud cuttings for propagation are used in :
a] Juniper b] Begonia
c] Carnation d] Grape

147. Leafy vegetables are rich in :
a] Minerals b] Vitamin A
c] Vitamin C d] All these

148. Lemon squash is preserved for a longer period without discolouration by the use of :
a] Malic acid and potassium metabisulphite
b] Citric acid and potassium
c] Benzoic acid and citric acid
d] Potassium metabisulphite

Quiz Kit!
First State Electricity Board
On May 1, 1955, the West Bengal State Electricity Board went operational.

149. Litchi does not flower in south Indian plateau in winter months :
a] For want of sufficient low temperature in winter months
b] Because of the high temperature prevalent at the time of flower formation
c] Because of high relative humidity present at the time of flower formation
d] Because of the excessive light stimulus present at the time of flower formation

150. Litchi was introduced to India from :
a] Indonesia
b] Japan
c] China
d] Malaya

151. Lloyd Botanical Garden, Darjeeling (West Bengal) was laid out by ________ on a piece of land donated by William Lloyd.
a] Sir Robert Llyod b] Sir George King
c] Sir Peter Hunt d] Sir Holdgate Baird

152. Madonna lily is :
a] *Lilium candidum* b] *L. pardalimum*
c] *L. tigrinum* d] *L. auratum*

153. Mango beverages are prepared by using :
a] Colour and essence
b] Colour alone
c] Essence alone
d] None of the above

154. Mango malformation disease is caused by ________deficiency.
a] Bo b] Cu
c] Fe d] K

155. Marigold is native of :
a] Maxico b] S. America
c] Both of the above d] South Africa

156. Marmalade is prepared by :
a] Citrus b] Guava
c] Apple d] Mango

157. Marmalades must have :
a] 2% pectin b] 1.25% pectin
c] 0.75% pectin d] 1.0% pectin

158. Maturity time of muskmelon in north India is :
a] June-July
b] September - October
c] November - December
d] February - March

Bile is the metabolic waste product of liver.

159. Maximum number of fruit plants can be planted in the orchard by the system :
a] Digonal system
b] Square system
c] Hexagonal system
d] Rectangular system

160. Most popular method of propagation of mango is :
a] Veneer grafting b] Inarching
c] Forkert budding d] Epicotyl grafting

161. Most economic method of irrigation of orchard under water scarcity conditions is :
a] Flood system b] Ring system
c] Sprinkler system d] Drip system

162. Mucilage in cut flowers is reponsible for :
a] Improving flower life
b] Decreasing transpiration
c] Blackening of vessels
d] All of these

163. Mughal style of gardening was developed by :
a] Shah Jahan b] Akbar
c] Babar d] None of these

164. Multistorey cropping system is practised in :
a] Jammu and Kashmir
b] Karnataka and Kerala
c] Bihar and Madhya Pradesh
d] Rajasthan and Haryana

165. Muscat Red is a variety of :
a] Grapes b] Pomegranate
c] Plums d] Apple

166. Name a fruit plant which bears fruits only once in its life time.
a] Banana b] Plum
c] Arecanut d] Custard apple

167. Name of annual climber is :
a] *Quisqualis indica*
b] *Ipomea pentaphyla*
c] *Lathyrus odoratus*
d] *Tecoma grandis*

168. Name the fruit crop which is used as hedge :
a] *Durentta plumeri* b] Karonda
c] Kikar d] *Ticoma stans*

169. On weight basis ___________ kg of cuttings of sweet potato are required per hectare for planting.
a] 300 b] 400
c] 500 d] 600

170. One gramme carrot contains about ___________ seeds.
a] 550 b] 650
c] 750 d] 850

171. One gramme of celery seed holds about ___________ seeds.
a] 1570 b] 2070
c] 2270 d] 2470

172. One gramme of onion, as estimated, generally contains ___________ seeds.
a] 180 b] 240
c] 300 d] 360

173. One gramme of turnip seed holds ___________ seeds.
a] 460-500 b] 360-400
c] 560-600 d] 260-360

174. One gramme radish contains about ___________ seeds.
a] 700-1000 b] 500-600
c] 600-675 d] 1100-1150

175. Onion is propagated by :
a] Seeds b] Bulbs
c] Root cuttings d] Bud grafting

176. Orach or *chakwat* (*Atriplex hortensis*) or mountain spinach is mainly grown on small scale during winter in:
a] Maharashtra and Karnataka
b] Gujarat and Madhya Pradesh
c] Tamil Nadu and Kerala
d] Himachal Pradesh and J & K

177. Out of following which fruit plants need 'Bahar' treatment:
a] Litch and loquat
b] Guava and pomegranate
c] Ber and datepalm
d] Grape and anola

178. Paeony is a name for the class of :
a] Marigold b] Sunflower
c] Chrysanthemum d] Dahalia

179. Pagoda is a name of :
a] Japanese flowering tree
b] Japanese monument
c] Statue
d] None of these

180. Per capita minimum requirement of vegetables and fruits together is:
a] 37 grams per day
b] 87 grams per day
c] 137 grams per day
d] 187 grams per day

181. Pineapple is mainly propogated by planting :
a] Crowns b] Slips
c] Stumps d] Suckers

182. Planting distance of individual fruit species provides the following advantages :
a] Uniform growth
b] Allows easy orchard operation
c] Improper utilization of orchard space
d] Proper supervision

183. Plants not suitable for wind break in orchard is :
a] *Casurina equisetifolia*
b] *Prosopis juliflora*
c] *Dulbergia siso*
d] *Sizygium cuminii*

GK GS
West Bengal is the most densely populated states in India.

184. Pollination of the fig flower by the blastophage wasp takes place through :
a] Fasciation b] Parthenogenesis
c] Caprification d] Apomixis

185. Pomegranate, fig, olive and grape have been placed in __________ salt tolerant group of fruit trees.
a] High b] Low
c] Medium d] Very high

186. Popular method of orchard planting is :
a] Square system
b] Digonal system
c] Hexagonal system
d] Rectangular system

187. Potato tuber production is the maximum at __________ °C.
a] 15 b] 20
c] 25 d] 30

188. Pride of India is a variety of :
a] Apple b] Rose
c] Cardomom d] Cabbage

189. Production of fruits of commercial scale is known as:
a] Kitchen gardening
b] Truck gardening
c] Both a & b
d] None of these

190. Pusa Kesar is a variety of :
a] Carrot b] Radish
c] Turnip d] Chilli

191. Pusa Red, Ratnar, Early Grano, Patna White and Bellary Red are varieties of :
a] Onion b] Garlic
c] Ginger d] Leek

192. Pusa Sawani, Pusa Makhmali and Perkin's Long Green are varieties of:
a] Bhindi b] Squash
c] Bean d] Lobia

193. Qualitatively good fruit juice is one which is :
a] Extracted from healthy fruit
b] Extracted from healthy fruit and mixed with sugar, acid, colour and essence
c] Extracted from healthy fruit and added with sugar alone
d] Extracted from healthy fruit and added with essence alone

194. Quincunx in horticulture refers to :
a] Purning method of grapes
b] De suckering method of banana
c] Planting method of trees
d] Harvesting fruits from trees

195. Ratna variety of mango is a cross between :
a] Alphonso and Banganpalli
b] Neelum and Langra
c] Neelum and Alphonso
d} Dashehari and Neelum

196. Raw carrots are good source of :
a] Vitamin A b] Vitamin B
c] Vitamin K d] Vitamin E

197. Rectangular system of planting is most common in :
a] Peach b] Falsa
c] Grape d] All of them

198. Removal of undersirable leaves from the tree is known as :
a] Disbudding b] Defoliation
c] Defruitingd] Deblossoming

199. Removal of undesirable flowers from the plant is called :
a] Deblossoming b] Defruiting
c] Defoliation d] None of them

200. Removal of undesired branches by picking auxilliary buds is known as :
a] Clipping b] Nipping
c] Budding d] Topping

201. Richest source of vitamin C is :
a] Lemon b] Orange
c] Cucumber d] Anola

202. Rock garden is located at :
a] Bangalore b] Jaipur
c] New Delhi d] Chandigarh

203. Root cutting for propagation are used in :
a] Dog wood b] Blackberry
c] Fig d] Apple

204. Root cutting is a material used for propagation in :
a] Pomegranate b] Lime
c] Pummelo d] Grape

205. Rose cut flower must be cut at tight bud stage at least __________ cm stem length for export.
a] 20 b] 40
c] 60 d] 80

206. Rose-mix is having the NPK in the ratio of :
a] 2:5:5 b] 3:8:5
c] 5:8:18 d] 10:15:20

207. Roses are grown for :
a] Perfume or rose oil
b] Cut flower
c] Rose water, gulkand
d] All these

208. Salad crops are those crops used for:
a] Canning
b] Latex extraction
c] Raw consumption
d] Ornamental purpose

209. Scientific name of sapota or chiku is:
a] *Achras zapota*
b] *Phyllanthus amblica*
c] *Cynara scohemus*
d] *Pastinaca sativa*

GK **Seed Bank Scheme was introduced in 1999-2000.** GS

210. Seedlessness in grapes is due to :
a] Embryo abortion
b] Parthenocarpy
c] Aneuploidy
d] Lack of pollination

211. Semi-hard wood cutting is a propagating material used in :
a] Begonia b] Grape
c] Pomegranate d] Rose

212. Shade loving annual floweing plant is :
a] Salvia b] Forget-me-not
c] Dianthus d] Anchusa

213. Shalimar garden is situated in :
a] Bangalore b] Kashmir
c] Hydrabad d] Delhi

214. Sheet test is performed in :
a] Jam b] Jelly
c] Squash d] Murabba

215. Snow-ball - 16 is a variety of :
a] Cauliflower b] Cabbage
c] Potato d] Tomato

216. Sodium benzoate is added for preservation of most of fruit juice in the concentration of :
a] 0.01 - 0.05% b] 0.06 - 0.10%
c] 1.1 - 1.5% d] 1.6 - 2.0%

217. Soft-wood cutting is a propagation material used in :
a] Coleus b] Rose
c] Litchi d] Mango

218. Soft-wood cuttings for propagation are used in :
a] Dogwood b] Grape
c] Begonia d] None of these

219. Spacing between fruit plant is determined on the basis of :
a] Choice of farmer
b] Spreading behaviour of plants
c] Availability of land
d] All of these

220. Spinach produces seed stalks in about __________ and seed can be harvested in about 150 days.
a] 45 days b] 55 days
c] 65 days d] 75 days

221. Square method of planting is most common in :
a] Mango b] Litchi
c] Citrus fruits d] All of above

222. Squash melon (*Citrullus vulgaris*) seed should be sown when the temperature (air) warms upto __________ °C.
a] 15.6 b] 20.6
c] 24.6 d] 26.6

223. Stone lantarn was first designed by:
a] San chu keu b] Josef
c] Oribe d] Ashoka

224. Substituted adenine compounds which promote cell division in plant tissues would include :
a] Cytokinins b] Gibberellins
c] Morphactins d] Triacontanols

225. Sweet potato is propagated by :
a] Seeds
b] Buddings
c] Root cuttings
d] Vine cuttings

226. Temperate fruits like apple, peach, pear and plum should be planted in the month of :
a] March-April
b] December-January
c] July-August
d] September-October

227. The __________ is probably a native of South Asia and perhaps also of Africa, Australia and Polynesia.
a] Smooth sponge gourd
b] Bottle gourd
c] Pointed gourd
d] Bitter gourd

228. The approximate number of orange plants in one hectare would be :
a] 350 b] 375
c] 125 d] 275

229. The arbour system of training grape vine recommended for :
a] Citrus b] Thompson
c] Anab-e-Shahi d] Perlett

230. The best storage temperature for garlic is 0 to 2.2°C with __________ per cent of relative humidity.
a] 60 b] 50
c] 40 d] 30

231. The budded plants commence bearing in __________ year.
a] Second b] Third
c] Fourth d] Fifth

232. The citrus diseases namely 'reclamation disease' and 'dieback disease' are caused by the deficiency of :
a] Zinc b] Iron
c] Copper d] Molybdenum

233. The cold storage temperature for melons is 0°C and relative humidity __________ per cent.
a] 50-60 b] 60-70
c] 40-50 d] 80-90

234. The edible part of pomegranate fruit is :
a] Testa of fruit
b] Endosperm of seed
c] Fleshy ovary
d] Juicy embryo

235. The farming which takes place without season is known as :
a] Enforced farming b] Rabi farming
c] Kharif farming d] None of these

236. The first school of Ikebana called Ikenobo meaning "the arrangement of flowers of the hermitage along a pond" was started by Buddhist monk Semmu around __________ A.D.
a] 421 b] 521
c] 621 d] 721

GK GS
Sanstha and Sanchari were spies of the Mauryas.

237. The flat sour spoilage of canned vegetables is due to:
a] Development of sour odour without gas production by microorganisms
b] Gas formation by the microorganisms
c] Corrosion of the tin plate
d] A decrease in pH

238. The food crop which is not thoroughly propagated by vegetative means is :
a] Orange b] Banana
c] Grapes d] Pineapples

239. The ideal conditions for cold storage of mango are :
a] 8-10 °C and 85-90% RH
b] 7-8 °C and 70-80% RH
c] 6-7 °C and 75-80% RH
d] 10-12 °C and 90-85% RH

240. The ideal temperature for coffee crop is :
a] 15 °C - 30 °C b] 5 °C - 20 °C
c] 25 °C - 40 °C d] 30 °C - 50 °C

241. The ideal time for cutting evergreen trees is :
a] Rainy season b] Winter
c] Late spring d] None of these

242. The ill effects of submergence of roots of deciduous fruit plants in water, for very long time, is due primarily to :
a] Lack of nutrients
b] Lack of aeration
c] Excess of moisture
d] Poor absorption of moisture

243. The interfiller tree in mango garden is :
a] Papaya b] Litchi
c] Guava d] Loquat

244. The length of sweet potato cutting should normally be ____________ with 4-5 sound (healthy) buds.
a] 40 cm b] 30 cm
c] 20 cm d] 10 cm

245. The main insect reported to be responsible for affecting pollination in mango is :
a] Honey bee (*Apis indica*)
b] Common housefly (*Musca domestica*)
c] A dipterous fly (*Meliponta* spp.)
d] A hover fly (*Syrphidae* spp.)

246. The main objective of growing intercrops with the main fruit crop is to :
a] Help the fruit crop to grow better
b] Improve the soil fertility
c] Check soil erosion
d] None of the above

247. The Mandor Garden, Jodhpur (Rajasthan) was established by Raja __________ (1724-1749).
a] Abhai Singh
b] Ajay Singh
c] Hari Singh
d] Maan Singh

Did You Know ?

Restaurant Cars

Gap Railway introduced First Class restaurant cars on its mail trains. The new service was inaugurated on July 23, 1904 with standard British menu of that time. It listed: Bouile A Baise, Mullets, Mayonnaise, Risotto Milanese, Gateau St. Honore, etc. and also included Dessert and Cafe.

248. The Moribana meaning "Piled flowers" denoted the arrangement of flowers in:
a] Shallow containers
b] Deep containers
c] Paper wrappings
d] Papercones

249. The most common method of budding is :
a] T Budding b] L Budding
c] Eye budding d] All of the above

250. The most important quality of an onion for dehydration for export should have:
a] Red colour b] White colour
c] Brownish yellow d] None of the above

251. The most serious disease of okra is :
a] Powdery mildew b] Root rot
c] Yellow vein mosaic d] None of the above

252. The most widely used plant growth substance in grape culture for reducing compactness of bunches and improving berry size is :
a] N.A.A b] GA_3
c] I.A.A. d] 2, 4-D

253. The native place of rose is :
a] England b] U.S.A
c] India d] China

254. The number in an unit area becomes almost double of square system in :
a] Rectangular system
b] Triangular system
c] Quincunx system
d] Contour system

255. The number of pineapple plants to one hectare would be about :
a] 8,000 b] 6,000
c] 10,000 d] 12,500

256. The number of seeds of bitter gourd per 100 g are about:
a] 350 b] 450
c] 550 d] 650

257. The optimum temperature for brinjal germination is ____________ °C.
a] 5 b] 10
c] 15 d] 20

GK GS

Wheat is mainly a crop of temperate region.

258. The origin place of papaya is :
a] South America b] Pakistan
c] Russia d] Canada

259. The pea edible podded cultivar is:
a] Arkel
b] Bonneville
c] Perfection new line
d] Sylvia

260. The physiological disorder caused by boron deficiency in tomato is :
a] Pulpness b] Cracking
c] Blossom-end-rot d] Silvering

261. The plants bearing male and female flowers on the same plant are called:
a] Monoecious b] Monophrodite
c] Bisexual d] Hermophrodite

262. The recommended distance for banana planting is :
a] 5 x 5 meters b] 10 x 10 meters
c] 3 x 3 meters d] 8 x 8 meters

263. The seed of papaya for seeding should be sown in :
a] February-March b] April-May
c] June-July d] August-September

264. The seed rate (g/ha) of tomato is :
a] 400-500 b] 800-900
c] 1000-1100 d] 100-150

265. The seed rate of long melon is ____________ kg/ha.
a] 1-2 b] 2-3
c] 3-4 d] 4-5

266. The seed rate of onion is :
a] 8-10 kg/ha b] 15-20 kg/ha
c] 1-2 kg/ha d] None of them

267. The seed rate of radish is :
a] 12 kg/ha b] 5 kg/ha
c] 25 kg/ha d] None of them

268. The seed required for one hectare sowing of carrot is :
a] 1-2 kg b] 10-15 kg
c] 4-5 kg d] 15-20 kg

269. The seed requirement for one hectare planting of *palak* is :
a] 5-9 kg b] 15-20 kg
c] 20-25 kg d] 2-5 kg

270. The site for an orchard should be selected :
a] In a new locality
b] Away from market
c] In a remote place
d] In a predominantly fruit growing area

271. The stone, bark, leaves *etc.* of ____________ are used as an effective medicine against diabetes.
a] Litchi b] Mango
c] Indian blackberry d] Guava

272. The term scion and root-stock are used in :
a] Layering b] Grafting
c] Budding d] Cutting

273. The young potato plant grows the best at ____________ °C.
a] 16 b] 20
c] 24 d] 28

274. Tip, simple, trench and mound are the kinds of :
a] Layering b] Rootage
c] Budding d] Grafting

275. Tomato sauce is preserved by using
a] Acetic acid b] Sodium benzoate
c] Salt d] All the above

276. Tomato sauce must have not less than :
a] 16% T.S.S. b] 10% T.S.S.
c] 5% T.S.S. d] 12% T.S.S.

277. Upper stem of banana is called :
a] Pseudostem b] Rhizome
c] Sucker d] None of them

278. Vegetabels are subjected to drying after :
a] Blanching b] Without balancing
c] Sulphuring d] Any other

279. Vegetables are packed in cans and processed by heat in :
a] Retourt b] Water bath
c] Sun d] None of these

280. Vegetables occupy about ________% of the total cultivated area.
a] 2-2.5 b] 7-8
c] 12-13 d] 19-20

281. Very important constituent in fruits for preparing jelly is :
a] Sugar b] Pectin
c] Cellulose d] Vitamin

282. Vista vision is a theme for :
a] Japanese garden b] Mixed garden
c] Land scape d] None of these

283. Water requirement is a ratio between :
a] Water used and plant material produced
b] Water transpired and plant material produced
c] Water evaporated and plant material produced
d] Water percolated and plant material produced

284. What constitute bulb of onion ?
a] Fleshy petals
b] Fleshy sepals
c] Fleshy carola
d] Fleshy scales

285. What is the chemical formula of potassium metabisulphite, the most common preservative?
a] KMS b] K_2SO_2
c] KSO_2 d] $K_2S_2O_3$

286. What is TSS of jam?
a] 68.5 b] 60
c] 65 d] 75

GK GS

Tea and coffee increase the formation of urine.

287. When meristems of plants become dormant, there is:
a] Inhibition of auxins
b] Inhibition of cytokinins
c] Inhibition of gibberellins
d] Accumulation of abcisic acid

288. Where we have to contact to produce fresh seeds of potato?
a] New Delhi b] Shimla
c] Lucknow d] Darjiling

289. Which flower is grown in all seasons?
a] Chrysanthemum b] Candituft
c] Corn flower d] Marigold

290. Which fruit comes first in its calorific value?
a] Apple b] Pears
c] Dates d] Banana

291. Which fruit crop is grown in tropical region ?
a] Banana b] Datepalm
c] Coconut d] Apple

292. Which fruit crop occupies maximum production in India ?
a] Banana b] Grapes
c] Citrus fruits d] Guava

293. Which hormone is found in ripened fruits :
a] Ethylene b] Auxin
c] Gibberellins d] IAA

294. Which is the biggest formal garden in the India ?
a] Vrandhavan garden, Mysore
b] Rock garden, Chandigarh
c] Gardens of Bangalore
d] None of these

295. Which is the correct sequence of the floral parts of pumpkin?
a] Ovary, corolla and stigmatic lobes
b] Corolla, stigmatic lobes and ovary
c] Stigmatic lobes, ovary and corolla
d] Corolla, ovary and stigmatic lobes

296. Which is the most important factor considered while establishing an orchard?
a] Topography b] Approach road
c] Location d] Availability of water

297. Which is the popular and cheapest method of asexual propagation ?
a] Budding b] Layering
c] Cutting d] Grafting

298. Which of following is winter annuals?
a] Zinnia b] Kochia
c] Marigold d] Jasmine

299. Which of the following are salinity tolerant crops ?
a] Sugarbeet b] Date palm
c] Kale d] All these

300. Which of the following are the kinds of branch pruning?
a] Heading back b] Branch thinning
c] Branch tipping d] All of the above

301. Which of the following are the varieties of onion ?
a] Patna White b] Nasik White
c] Pusa Red d] All of them

302. Which of the following are the varieties of radish ?
a] Pusa Himani b] Rapid Red
c] Fresh Breakfast d] All of them

303. Which of the following are the variety of chilli ?
a] Pant C-1 b] California Wonder
c] Manjri Gota d] All of the above

304. Which of the following are the variety of tomato ?
a] Kalyanpur Angurlata
b] T-1
c] Punjab Chhuhara
d] All of the above

305. Which of the following crop is self-pollinated ?
a] Bottle gourd b] Cauliflower
c] Radish d] Tomato

306. Which of the following fruit crop is not belongs to temperate region ?
a] Apple b] Pea
c] Strawberry d] Lemon

307. Which of the following fruit tree is monoecious ?
a] Jackfruit b] Date
c] Fig d] Litchi

308. Which of the following fruits are drup ?
a] Ber b] Mango
c] Both these d] Raspberry

309. Which of the following growth promoter can develop the roots in mango ?
a] IBA b] IAA
c] 2,4, 5-T d] GA

310. Which of the following is a growth inhibitor ?
a] NAA b] ABA
c] IAA d] GA_2

311. Which of the following is a growth retardant ?
a] Alar b] Kinetine
c] GA_3 d] IBA

312. Which of the following is a temperate fruit crop ?
a] Walnut b] Lime
c] Strawberry d] Lemon

313. Which of the following is a tropical fruit crop ?
a] Sapota b] Walnut
c] Strawberry d] Mandarin

314. Which of the following is called the 'King of fruits'?
a] Pomegranate b] Grapes
c] Banana d] Mango

315. Which of the following is considered as 'Queen of spices'?
a] Turmeric b] Pepper
c] Cumin d] Cardomom

316. Which of the following is fruit and vegetable crop ?
a] Sweet potato b] Okra
c] Potato d] Spinach

GK GS

'Village on Wheels' a luxury train is scheduled to begin from Gaya, Bihar.

317. Which of the following heterostyly condition is found?
a] Litchi b] Date
c] Fig d] Walnut

318. Which of the following is not a self pollinated fruit crop ?
a] Guava b] Grape
c] Peach d] None of them

319. Which of the following is not a auxin?
a] GA_2 b] NAA
c] 2, 4-D d] IAA

320. Which of the following is not a cross pollinated fruit crop ?
a] Mango b] Litchi
c] Apple d] None of them

321. Which of the following is not a cross pollinated fruit crop ?
a] Papaya b] Jackfruit
c] Cherry d] None of them

322. Which of the following is not a growth promotor ?
a] Auxin b] Gibberellin
c] Abscissic acid d] Cytokinin

323. Which of the following is not a kind of detached grafting ?
a] Wedge, Splice b] Saddle, Tongue
c] Side, Crown d] None of them

324. Which of the following is not a method of tree planting?
a] Square method b] Triangular method
c] Hexagonal method d] None of them

325. Which of the following is not a self incompetible fruit crop ?
a] Pear b] Almoud
c] Apple d] Guava

326. Which of the following is not a sub-tropical fruit crop?
a] Litchi b] Peach
c] Loquat d] Papaya

327. Which of the following is not a sub-tropical fruit crop ?
a] Phalsa b] Grape
c] Loquat d] Papaya

328. Which of the following is not a tropical fruit crcp ?
a] Banana b] Cashewnut
c] Fig d] Mango

329. Which of the following is not a variety of apple ?
a] Beauty of bath b] Roma
c] Rome beauty d] Winter banana

330. Which of the following is not a variety of apple ?
a] Red delicious b] Blood red
c] Golden delicious d] None of the above

331. Which of the following is not a variety of brinjal ?
a] Pusa Purple Long b] Pusa Kranti
c] Black Beauty d] None of them

332. Which of the following is not a variety of radish ?
a] Pusa Desi b] Nasik Red
c] Pusa Chetaki d] Jaunpuri

333. Which of the following is not a variety of tomato ?
a] Pusa Rubyb] Best of all
c] Roma d] None of the above

334. Which of the following is not the kind of budding ?
a] Ring b] Flute
c] Chip d] None of above

335. Which of the following is self-pollinated crop ?
a] Reflesia b] Brinjal
c] Maize d] None of these

336. Which of the following is the early variety of cabbage?
a] Pride of India b] Early drum head
c] Chieftain d] None of the above

337. Which of the following is the kind of budding?
a] Patch b] Shield
c] Ring d] All of above

338. Which of the following is the kind of shield budding ?
a] Simple shield b] T-shield
c] Inverted T budding d] All of them

339. Which of the following is the variety of sweet orange?
a] Nagpur b] Foster
c] Blood red d] Marsh seedless

340. Which of the following is the variety of watermelon ?
a] Pusa Jyoti b] Arka Jyoti
c] Arka Nishant d] None of the above

341. Which of the following method is easiest and cheapest in lawn making
a] Turfing b] Turf plastering
c] Dibbling d] None of these

342. Which of the following plants are used for boundry tall hedge ?
a] Allysum b] Iresine
c] Tecoma d] Alocasia

343. Which of the following preservative is used in squash preparation ?
a] Sirka
b] Sodium benzoate
c] Sugar
d] Potasium metabisulphate

344. Which of the following variety of mango is developed by the result of a cross between Neelam and Desheri?
a] Alfanso b] Chusa
c] Amrapali d] None of them

345. Which of the folowing is the biennial vegetable?
a] Muskmelon b] Tomato
c] Onion d] Chilli

346. Which of the folowing is the kind of detached grafting?
a] Veneer b] Saddle
c] Tongue d] All of above

347. Which one of the following chemicals is used for the preservation of fruit products :
a] Potassium sulphate
b] Potassium managanate
c] Potassium metabisulphide
d] Potassium metabisulphite

348. Which one of the following crops can be grown successfully as an intercrop in mango orchard which has 10-year old trees?
a] Wheat b] Onion
c] Maize d] Turmeric

349. Which one of the following crops is best suited as intercrop in banana?
a] Papaya b] Persimon
c] Ginger d] Plum

350. Which one of the following fruits contains more fat than the rest?
a] Apple b] Guava
c] Avocado d] Jack fruit

351. Which one of the following fruits is cured in smoke for ripening?
a] Mango b] Orange
c] Jackfruit d] Banana

352. Which one of the following fruits will be most useful in human diet for avoiding scurvy disease?
a] Mandarin Orange b] Guava
c] Aonla d] Apple

353. Which one of the following genetic factors is responsible for poor fruit set in custard apple ?
a] Heterostyly
b] Dioecious nature
c] Dichogamy
d] Self-incompatibility

354. Which one of the following groups of crops is most suitable and recommended for inter-cropping in the young orchards?
a] Short duration fodders
b] Short duration fruit crops
c] Populars and Eucalyptus
d] Short duration legume vegetables

355. Which one of the following is the main cause for development of colour pigment in the flavedo of mature ripe orange fruit?
a] Hesperidin
b] Naringin
c] Carotene
d] Chlorophyll

356. Which one of the following pairs is a rich source of iron?
a] Sapota and ash gourd
b] Papaya and cabbage
c] Amaranthus and banana
d] Brinjal and citrus

Did You Know ?
First metalled road
Kolkatta had the first metalled road in 1825.

357. Which one of the following rose cultivars has striped flowers?
a] Amphisarica b] Bhim
c] Chitchore d] Mrinalini

358. Which one of the following soils is best for vegetable cultivation?
a] Sandy b] Sandy loam
c] Clay loam d] Clay

359. Which one of the following species of climbers is valued for the beauty of its foliage?
a] *Antigona leptopus*
b] *Clerodendron splendens*
c] *Monstera deliciosa*
d] *Tecoma grandiflora*

360. Which one of the following vegetables gives maximum calories per 100 gram of edible matter ?
a] Spinach b] Water melon
c] Sweet potato d] Tomato

361. Which one of the following vegetables is the richest source of protein?
a] Pea b] Fenugreek
c] Pointed gourd d] Cucumber

362. Which one of the following will be most suitable for growing under arid zone of India?
a] Jack fruit b] Papaya
c] Sweet Orange d] Ber

363. Which sugar is obtained from grapes?
a] Sucrose b] Lactose
c] Maltose d] Glucose

364. White flower plants are grown in which garden ?
a] Bottle garden b] Night garden
c] Public garden d] Water garden

365. Wilt disease takes place in :
a] Guava b] Papaya
c] Grapes d] Ber

366. CO_2 concentration in greenhouse for rose growing should be :
a] Upto 500 ppm b] 4000 - 6000 ppm
c] 1000 - 3000 ppm d] 500 - 1000 ppm

367. Which of following is not suitable greenhouse covering material in hills?
a] Glass b] Fiber Glass
c] Polythene d] Polycarbonate

368. Which of the following are dried by water drying method?
a] Hydrangea b] Cornflower
c] Baby's breath d] All of these

GK GS
Sugar is also obtained from Gardenbeet.

369. **Which of the following loose flower is not a shrub?**
   a] Tuberose b] Jasmine
   c] Crossandra d] Barleria

370. **Who of the following is associated with breeding of orchids in India?**
   a] S.S. Negi b] S.P.S. Raghava
   c] Foja Singh d] K.L. Chadha

371. **French marigold is :**
   a] Diploid b] Tetraploid
   c] Triploid d] Aneupliod

372. **Most of the cultivars of chrysanthemum are :**
   a] Short day b] Long day
   c] Day neutral d] All of the above

373. **International Bougainvillea Registration Centre is located at**
   a] IIHR b] NBRI
   c] IARI d] TNAU

Matching

1. **Match the common name with their scientific name.**

a]	Artichoke	i]	*Cichorium intybus*
b]	Head lettuce	ii]	*Cynara scolymus*
c]	Safflower	iii]	*Lactua sativa var. capitata*
d]	Chicory	iv]	*Carthamus tinctorius*

2. **Match the family with the vegetable which it belongs to.**

a]	Compositae	i]	Lady's finger
b]	Cucurbitaceae	ii]	Celery
c]	Malvaceae	iii]	Head lettuce
d]	Umbelliferae	iv]	Snap melon

3. **Match the family with the vegetable which it belongs to.**

a]	Amaryllidaceae	i]	Asparagus
b]	Liliaceae	ii]	Taro
c]	Araceae	iii]	Pigweed
d]	Chenopodiaceae	iv]	Garlic

4. **Match the family with the vegetable which it belongs to.**

a]	Cruciferae	i]	Pepper
b]	Leguminosae	ii]	Turnip
c]	Umbelliferae	iii]	Fenugreek
d]	Solanaceae	iv]	Parsley

5. **Match the mustard (*sarson*) species with the main growing states of India**

a]	*Brassica campestris* var. brown *sarson*	i]	U.P., Bihar, Bengal
b]	*B.campestris* var. yellow *sarson*	ii]	N. Bengal, Assam
c]	*B. juncea*	iii]	Punjab
d]	*B. campestris* var. *cuneifolia*	iv]	Delhi

6. **Match the common name of the vegetables with their scientific name.**

a]	Parsnip	i]	*Ipomea batata*
b]	Sweet potato	ii]	*Solanum melongena*
c]	Potato	iii]	*Pastinaca sativa*
d]	Brinjal	iv]	*Solanum tuberosum*

7. **Match the common name of the vegetables with their scientific name.**

a]	Tomato	i]	*Cucurbita pepo*
b]	Pepper	ii]	*Lycopersicon esculentum*
c]	Pungent pepper	iii]	*Capsicum annum*
d]	Summer squash	iv]	*Capsicum frutescens*

8. **Match the common name of the vegetables with their scientific name.**

a]	Wax gourd	i]	*Citrullus vulgaris*
b]	Water melon	ii]	*Cucumis melo*
c]	Cucumber	iii]	*Benincasa hispida*
d]	Muskmelon	iv]	*Cumunis sativus*

9. **Match the common name of the vegetables with their scientific name.**

a]	Snap melon	i]	*Citrullus vulgaris* var. *fistulosus*
b]	Long melon	ii]	*Cucumis melo* var. *memordica*
c]	Squash melon	iii]	*Luffa cylindrica*, var. *acutangula*
d]	Sponge gourd	iv]	*Cucumis melo* var. *utilissimus*

10. **Match the common name of the vegetables with their scientific name.**

a]	Bitter gourd	i]	*Trichosanthes diocia*
b]	Pointed gourd	ii]	*Lagenaria siceraria*
c]	Snake gourd	iii]	*Trichosanthes angunia*
d]	Bottle gourd	iv]	*Memordica charantia*

Article 21 of Constitution provides protection to life and liberty.

11. **Match the common name of the vegetables with their scientific name.**

a]	Taro	i]	*Allium cepa*
b]	Asparagus	ii]	*Colocasia esculentum*
c]	Onion	iii]	*Allium sativum*
d]	Garlic	iv]	*Asparagus officinalis*

12. **Match the common name of the vegetables with their scientific name.**

a]	Spinach beet	i]	*Spinacia oleracea*
b]	Spinach	ii]	*Brassica oleracea* var. *capitata*
c]	Pigweed	iii]	*Beta vulgaris*
d]	Cabbage	iv]	*Chenopodium album*

13. **Match the common name of the vegetables with their scientific name.**

a]	Cauliflower	i]	*Brassica campestris* var. *capitata*
b]	Turnip	ii]	*Raphanus sativus*
c]	Radish	iii]	*Brassica juncea*
d]	Leaf mustard	iv]	*Capsicum frutescens*

14. **Match the common name of the vegetables with their scientific name.**

a]	Garden pea	i]	*Vicia faba*
b]	Broad bean	ii]	*Cyamopsis psoraloides*
c]	Lima bean	iii]	*Psum sativum*
d]	Cluster bean	iv]	*Phaseolus lunatus*

15. **Match the common name of the vegetables with their scientific name.**

a]	Fenugreek	i]	*Daucus carota*
b]	Okra	ii]	*Trigonella* spp.
c]	Carrot	iii]	*Petroselinum crispum*
d]	Parsley	iv]	*Abelmoschus esculentus*

16. **Match the following varieties of *chaulai* with the species.**

a]	*Badi chaulai*	i]	*Amaranthus magnostanus*
b]	*Leotia chaulai*	ii]	*Amaranthus blitum*
c]	*Chhoti chaulai*	iii]	*Amaranthus dubius*
d]	*Khada sag*	iv]	*Amaranthus tricolor*

17. **Match type of potato crop with planting time.**

a]	Autumn	i]	July-August
b]	Winter	ii]	September-November
c]	Main	iii]	September-October
d]	Monsoon	iv]	December-January

18. **Match the following potato varieties with the duration (days of maturity).**

a]	Kufri Alankar	i]	85-95
b]	Kufri Chandramukhi	ii]	95-100
c]	Kufri Jyoti	iii]	110-120
d]	Kufri Chamatkar	iv]	75-80

19. **Match fungal diseases of potato with the causal fungus.**

a]	Dry rot	i]	*Corticium solani* and *Rhizoctonia solani*
b]	Black scurf	ii]	*Synchtrium endobioticum*
c]	Charcoal rot	iii]	*Fusarium oxysporum*
d]	Wart	iv]	*Corticium rolfsil* and *Sclerotium rolfsil*

20. **Match the bacterial diseases of potato with the causal bacterium.**

a]	Brown rot	i]	*Corneybacterium sepedonicum*
b]	Ring rot	ii]	*Streptomyces scabies*
c]	Black leg	iii]	*Erwinia carotobora*
d]	Scab	iv]	*Pseudomonas solanacearum*

Ramjet is supersonic aircraft.
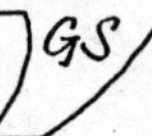

21. **Match the viral diseases of potato with the causal virus(es).**

a]	Leaf roll	i]	Virus X and Virus Y
b]	Mild mosaic	ii]	*Solanum* virus 14 (potato virus I)
c]	Black leg	iii]	Potato virus X and potato virus A
d]	Crinkle of potato	iv]	Potato mottle virus X, *Solanum* virus I

22. **Match the following pod vegetables with the botanical name.**

a]	Sword bean	i]	*Dolichos lablab*
b]	Velvet bean	ii]	*Phaseolus lunatus*
c]	Indian bean	iii]	*Mucuna cochinchinensis*
d]	Double bean	iv]	*Canavalia ensiformis*

23. **Match the soil characteristic with the weed usable as plant indicator.**

a]	Heavy clay soil	i]	*Polygonum aviculare*
b]	Damp soil	ii]	*Primula vulgaris*
c]	Light sandy soil	iii]	*Chrysanthemum segetum*
d]	Acid soil	iv]	*Juncus effusus*

24. **Match the soil characteristic with the weed usable as plant indicator.**

a]	Chalk soil	i]	*Spergula arvensis*
b]	Soil needling lime	ii]	*Papaver rheas*
c]	Damp soil	iii]	*Chichorium intybus*
d]	Light sandy soil	iv]	*Lychnis floscuculi*

25. **Match the pH range with fruit crop for which it has been found favourable.**

a]	Blue berry	i]	4.5-7.0
b]	Apple	ii]	5.2-6.7
c]	Citrus	iii]	5.8-8.0
d]	Grape	iv]	4.5-5.0

26. **Match the pH range with fruit crop for which it has been found favourable.**

a]	Pear	i]	5.0-5.5
b]	Peach	ii]	5.6-7.2
c]	Pineapple	iii]	5.2-6.7
d]	Red raspberry	iv]	5.1-6.7

27. **Match the pH range with fruit crop for which it has been found favourable.**

a]	Strawberry	i]	5.8-7.5
b]	Plum	ii]	4.5-7.0
c]	Cherry	iii]	5.0-6.0
d]	Citrus	iv]	5.8-7.0

28. **Match the following fruits with the soil found suitable for their cultivation.**

a]	Almond	i]	Medium/light loam with slightly heavier sub-soil
b]	Apple	ii]	Deep, rich, alluvial loamy soils
c]	Banana	iii]	Light loam/deep organic loam
d]	Citrus spp.	iv]	Light loam

29. **Match the following fruits with the soil found suitable for their cultivation.**

a]	Pear	i]	Light loam to heavy clay
b]	Pomegranate	ii]	Deep moist
c]	Date palm	iii]	Light sandy loam
d]	Guava	iv]	Deep loam/alluvial

30. **Match the following fruits with the soil found suitable for their cultivation.**

a]	Litchi	i]	Well drained, rich deep alluvial
b]	Jackfruit	ii]	Sandy loam to heavier soil, well drained light loam
c]	Grapes	iii]	Well drained clayey loam
d]	Loquat	iv]	Well drained, shallow medium black loamy

GK GS

Arjun Singh is the first Air Chief to receive the honour of 'Marshal'.

31. **Match the following fruits with the soil found suitable for their cultivation.**

a]	Pineapple	i]	Loamy, deep and well drained
b]	Papaya	ii]	New alluvial, lateritic hills soils, old alluvium
c]	Jamun	iii]	Red sandy and also rocky soils
d]	Custard apple	iv]	Rich loam, medium black and alluvial

32. **Match the choice of legume for sod in orchards with the soil or climatic conditions which these are suitable for.**

a]	Sesbania	i]	Well drained soils
b]	Sannhemp	ii]	Moderate conditions
c]	Vetch	iii]	Waterlogged conditions
d]	Clover	iv]	Cold region

33. **Match the tree(s) given below with their estimated longevity.**

a]	Sweet chestnut	i]	1000-3000 years
b]	Walnut, pecan	ii]	30-70 years
c]	Persimmon, avocado	iii]	1000 years
d]	Apricot, fig	iv]	50-100 years

34. **Match the trees(s) given below with their estimated longevity.**

a]	Strawberry	i]	20-40 years
b]	Peach, plum, pomegranate	ii]	25-30 years
c]	Currant, goose/raspberry	iii]	30-70 years
d]	Walnut, pecan	iv]	4-5 years

35. **Match the water content with the plant parts given below.**

a]	93%	i]	Dry beans
b]	40-50%	ii]	Barley roots
c]	10-15%	iii]	Sunflower stem
d]	80%	iv]	Woody tissues

36. **Match the fruits with its edible parts.**

a]	Apple, pear	i]	Endosperm
b]	Banana	ii]	Peduncles and cotyledons
c]	Cashewnut	iii]	Fleshy thalamus
d]	Coconut-palm	iv]	Mesocarp and endocarp

37. **Match the fruits with its edible parts.**

a]	Cucumber	i]	Pericarp
b]	Custard apple	ii]	Bracts, perianth and seed
c]	Date-palm	iii]	Flesh pericarp
d]	Fig	iv]	Fleshy receptacle

38. **Match the fruits with its edible parts.**

a]	Guava	i]	Pericarp and placentae
b]	Grape	ii]	Bracts, perianth and seed
c]	Jackfruit	iii]	Mesocarp including epicarp
d]	Indian plum	iv]	Thalamus and pericarp

39. **Match the fruits with its edible parts.**

a]	Litchi	i]	Outer portion of receptacle bracts and perianth
b]	Mango	ii]	Fleshy aril
c]	Citrus	iii]	Juicy placental hairs
d]	Pear (pome)	iv]	Mesocarp

40. **Match the fruits with its edible parts.**

a]	Tomato	i]	Succulent thalamus
b]	Strawberry	ii]	Fleshy thalamus
c]	Papaya	iii]	Pericarp and placentae
d]	Pineapple	iv]	Mesocarp

GK

Dasratha dedicated caves to Ajivikas.

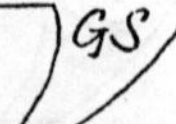

41. Match the fruits with its edible parts.

a] Pomegranate — i] Endosperm
b] Melon — ii] Fleshy receptacle
c] Coconut palm — iii] Mesocarp
d] Fig — iv] Juicy outer seed coat

42. Match the storage-life of fruits with the temperature.

a] Ripe banana, 3 weeks — i] 6-8ºC
b] Coorg orange, 10-12 weeks — ii] 2-5ºC
c] Nagpur santara, 10-14 weeks — iii] 12-13ºC
d] Punjab santara, 4-6 weeks — iv] 5-6ºC

43. Match the storage-life of fruits with the temperature.

a] Mosambi, 16 weeks — i] 2-5 ºC
b] Common malta, 16 weeks — ii] 8-9 ºC
c] Blood red malta, 16 weeks — iii] 5-6 ºC
d] Lemons, 8-12 weeks — iv] 7-8 ºC

44. Match the storage-life of fruits with the temperature.

a] Grapes, 6-8 weeks — i] 11-13 ºC
b] Guava, 3 weeks — ii] 8-9 ºC
c] Jackfruit, 6 weeks — iii] 0-2 ºC
d] Mango, 4-7 weeks — iv] 9-10 ºC

45. Match the storage-life of fruits with the temperature.

a] Papaya, 1-2 weeks — i] 0-2 ºC
b] Water melon, 2-4 weeks — ii] 9-10 ºC
c] Apple, 16-32 weeks — iii] 0-1 ºC
d] Pear 12-26 weeks — iv] 2-4 ºC

46. Match the plant part used for propagation with the growth substance and the concentration normally used.

a] Lemon, cuttings — i] IAA (week solution)
b] Sweet lime, hardwood cutting — ii] NAA (0.2%)
c] Trifoliate orange, cuttings — iii] Hortomone A
d] Guava, gootee — iv] IBA (200 mg/l with 2% sucrose)

47. Match the plant part used for propagation with the growth substance and the concentration normally used.

a] Tomato, seeds — i] 2, 4-D
b] Orange, seeds — ii] Heteroauxin aqueous solution and IBA
c] Peach, seeds — iii] IBA
d] Vine, seeds — iv] IAA

48. Match the plant part used for propagation with the growth substance and the concentration normally used.

a] Jackfruit, marcoting tests — i] IBA (3%)
b] Mango, ringed part of marcots — ii] IBA and NAA (3-5% in linolin)
c] Papaya, shoots — iii] Heteroauxin solution
d] Olive, cuttings — iv] IBA (1%)

49. Match the plant part used for propagation with the growth substance and the concentration normally used.

a] Grape cuttings — i] 2, 4-D
b] Vine cuttings — ii] IAA
c] Coconut seeds — iii] IBA
d] Fig cuttings — iv] NAA with honey (10 mg/l)

50. Match the common name with scientific name and family.

a] Aonla — i] *Persea americana, Lauraceae*
b] Avocado — ii] *Zizyphus mauritiana, Rhamnaceae*
c] Banana — iii] *Phyllanthus emblica, Euphorbiaceae*
d] Ber — iv] *Musa paradisiaca* var. *sampientum, Musaceae*

GK GS

Haryana has highest percentage of surface road.

51. Match the common name with scientific name and family.

a]	Bread-fruit	i]	*Citrus* spp., *Rutaceae*
b]	Chiku	ii]	*Annona squamosa, Annonaceae*
c]	Citrus	iii]	*Achras sapota, Sapotaceae*
d]	Custard apple	iv]	*Artocarpus altilis, Moraceae*

52. Match the common name with scientific name and family.

a]	Date-palm	i]	*Vitis vinifera, Vitaceae*
b]	Fig	ii]	*Phoenix dactylifera, Palmae*
c]	Grape	iii]	*Psidium guajava, Myrtaceae*
d]	Guava	iv]	*Ficus carica, Moraceae*

53. Match the common name with scientific name and family.

a]	Jack-fruit	i]	*Syzyguim cumuni, Myrtaceae*
b]	Jamun	ii]	*Eriobotrya japonica, Rosaceae*
c]	Litchi	iii]	*Artocarpus heterophyllus, Moraceae*
d]	Loquat	iv]	*Litchi chinensis, Sapindaceae*

54. Match the common name with scientific name and family.

a]	Mango	i]	*Morus* species, *Moraceae*
b]	Magosteen	ii]	*Carica papaya, Caricaceae*
c]	Mulberry	iii]	*Garcinia mangostana, Guttiferae*
d]	Papaya	iv]	*Mangifera indica, Anacardiceae*

55. Match the common name with scientific name and family.

a]	Passion fruit	i]	*Grewia asiatica, Tiliaceae*
b]	Phalsa	ii]	*Passiflora edulis, Passifloraceae*
c]	Pineapple	iii]	*Punica granatum, Punicaceae*
d]	Pomergranate	iv]	*Ananas comusus, Bromeliaceae*

56. Match the common name with scientific name and family.

a]	Apple	i]	*Malus pumila, Rosaceae*
b]	Apricot	ii]	*Prunus persica, Rosaceae*
c]	Cherry	iii]	*Prunus armeniaca, Rosaceae*
d]	Peach	iv]	*Prunus avium, Rosaceae*

57. Match the common name with scientific name and family.

a]	Japanese persimmon	i]	*Prunus* species, *Rosaceae*
b]	Pear	ii]	*Fragaria* species, *Rosaceae*
c]	Plum	iii]	*Diospyros kaki, Ebenaceae*
d]	Strawberry	iv]	Prunus domestica, *Rosaceae*

58. Match the common name with scientific name and family.

a]	Almond	i]	*Corylus avellana, Bitulaceae*
b]	Cashewnut	ii]	*Prunus amygdalus, Rosaceae*
c]	Hazlenut	iii]	*Carya illinoensis, Juglandaceae*
d]	Pecannut	iv]	*Anacardium occidentale, Anacardiaceae*

59. Match the common name with scientific name and family.

a]	Walnut	i]	*Citrus aurantium*, Rutaceae
b]	Sour orange	ii]	*Citrus limon*, Rutaceae
c]	Lemon	iii]	*Citrus aurantifolia*, Rutaceae
d]	Lime	iv]	*Jugulans regia*, Juglandaceae

60. Match the number of buds in each spur of grape varieties

a]	Seedless beauty	i]	2-3
b]	Thompson seedless	ii]	4-5
c]	Early-Muscut	iii]	5-6
d]	Anab-e-Shahi	iv]	8-10

Rubber is a crop of wet tropical climate.

61. **Match the fruit with the place of origin.**

a]	Loquat	i]	Brazil
b]	Mango	ii]	Malaya
c]	Mangosteen	iii]	South Asia
d]	Passion fruit	iv]	Japan and China

62. **Match the fruit with the place of origin.**

a]	Apple	i]	China
b]	Apricot	ii]	South-west Asia
c]	Peach	iii]	N. America and Japan
d]	Plum	iv]	Armenia

63. **Match the fruit with its protein content (per cent).**

a]	Litchi	i]	0.5
b]	Papaya	ii]	12.5
c]	Peach	iii]	1.15
d]	Plum	iv]	20.0

64. **Match the flower (plant) with place of origin.**

a]	Hollyhock (*Althaea rosea*)	i]	South Africa
b]	Sweet allyssum (*Alyssum*)	ii]	Orient, China
c]	Forget-me not (*Anchusa capensis*)	iii]	Europe
d]	Snapdragon (*Antirrhinum majus*)	iv]	Europe, W. Asia

65. **Match the flower (plant) with place of origin.**

a]	African daisy (*Aretotis*)	i]	South America
b]	Browallia (*Browallia demissa*)	ii]	Tropical Asia
c]	Pot marigold (*Calendula officinalis*)	iii]	South Africa
d]	Cockscomb (*Celosia argentea*)	iv]	South Europe

66. **Match the flower (plant) with place of origin.**

a]	Snapdragon (*Antirrhimum majus*)	i]	Calofornia
b]	Golden star (*Bartonia aurea*)	ii]	Tropical Asia
c]	Calendula (*Calendula officinalis*)	iii]	Europe
d]	Cockscomb (*Celosia argentea/plumosa*)	iv]	South Europe

67. **Match the flower (plant) with place of origin.**

a]	Sweet sultan (*Centaurea moschata*)	i]	Tropical America
b]	Wall flower (*Cheiranthus cheiri*)	ii]	Europe
c]	Giant spider flower (*Cleome spinosa*)	iii]	Australia
d]	Glory pea (*Clianthus formosus*)	iv]	Europe

68. **Match the flower (plant) with place of origin.**

a]	Mexican aster (*Cosomos bipinnatus*)	i]	Great Britain
b]	Fox glove (*Digitalis purpurea*)	ii]	South Europe
c]	Larkspur (*Delphinium ajacis*)	iii]	N. France
d]	Sweet William (*Dianthus barbatus*)	iv]	Mexico

69. **Match the flower (plant) with place of origin.**

a]	Blanket flower (*Gaillardia pulchella*)	i]	Caucasus
b]	Backelor's button (*Gamphrena globosa*)	ii]	Sicily and S. Italy
c]	Baby's breath (*Gypsophila elegans*)	iii]	America
d]	Sweet pea (*Lathyrus odoratus*)	iv]	India

70. **Match the flower (plant) with place of origin.**

a]	Scarlet flax (*Linum gradiflorum*)	i]	S. America
b]	Feverfew (*Matricaria eximia*)	ii]	N. Africa and rurope
c]	Nemesia (*Nemesia strumosa*)	iii]	N.Temperate Regions
d]	Petunia (*Petunia hybrida*)	iv]	S. Africa

GS

Coal is also known as Black Gold.

71. Match vegetable crop with the planting material.

a]	Yams	i]	Whole fruit
b]	Giant taro	ii]	Shoot tip
c]	Chow-chow	iii]	Rhizome
d]	Arrow roots	iv]	Minisett

72. Match vegetable crop with their nature of pollination.

a]	Amaranths	i]	Wind pollinated
b]	Radish	ii]	Insect pollinated
c]	Okra	iii]	Often cross pollinated
d]	Tomato	iv]	Highly self pollinated

73. Match vegetable crop with their class.

a]	Perennial crop	i]	Spine gourd
b]	Green	ii]	Okra
c]	Salad crop	iii]	Cress
d]	Fibre crop	iv]	Collards

74. Match fruit crop with the edible part.

a]	Chalta	i]	Thalamus
b]	Mangosteen	ii]	Endosperm
c]	Coconut	iii]	Fleshy calyx
d]	Guava	iv]	Aril

Glossary of Technical Terms

Annuals Are the annual plants that live for one year or less, that is, the plant makes its vegetative growth flowers, and produces seed within one year from the sowing date then plants die.

Asexual Propagation (Clonal Propagation) It does not involves, the gamate from parent in which vegetative parts such as leaf, stem or root are used, instead of seed.

Bark All the tissue of root stem from the cambium outward.

Biennials Those plants which require two years or parts of two growing seasons, with more or less of a dormant or resting season between, to complete their life period.

Blanching Heating of fruits and vegetables in water or in live steam before canning.

Bleeding Loss of water or sap from the plants as result of wound or pruning.

Bonsai It comprises a tree or shrub planted in a small container for developing as a miniature plant showing the general appearance of that plant species found in nature.

Bud A bud is undeveloped and elongated stems composed of a very short axis of meristem cells from which embryonic leaves, lateral buds, flower parts or all tree arises.

Budding Type of grafting in which a scion (vegetative bud) is placed in the stock/stock plant.

C.N. Ratio The relative proportion of carbon and nitrogen supply to a plant is very important in influencing growth and fruitfulness of trees.

Cacti Cacti are a group of plants with peculiar shape and size and mostly adapted for desert life.

Callus Wounded tissues which develops from cambium of other exposed meristem.

Cambium Between xylem and phloem there is a thin wall, undifferentiated tissues is known as cambium.

Canning (Appertizing) May be defined as the preservation of foods in harmetically sealed container by sterilization or by heat.

Cordial This is a sparkling, clear, sweetened juice from which all pulp and other suspend materials have been completely eliminated.

Cuttings A detached portion of a plant part placed in suitable medium like sand soil or saw dust in order that it may produces roots and shoots to form a new plant.

Deciduous Deciduous plants shed their leaves once in year when they go to rest.

Dehydration It is drying by artificially produced heat under carefully controlled condition of temperature humidity and air flow.

Deshooting Removal of newly formed shoots to prevent them from becoming branches. Also known as summer prunning.

Drying It is sun drying by exposing fruits and vegetables in sun.

Enzymes Chemical substances that are associated with the living cell. Its presence is necessary for working as a catalyst.

> **You Don't Know!**
> *Longest national highway*
> **NH-7 passing through six states, Uttar Pradesh, Madhya Pradesh, Maharashtra, Andhra Pradesh, Karnataka and Tamil Nadu is 2,369 km long.**

Evergreen Evergreen trees do not have definite resting season and they do not shed their leaves during any particular season. eg. mango, lime, sapota, lemon, orange, etc.

Filler Trees Trees placed in between permanent trees at the time of planting for getting some returns from them up to the stage, when permanent trees start production.

Floriculture Cultivation of flowers and ornamental plants for commercial purposes or merely for getting pleasure and as a hobby.

Flower Bud Differentiation The changing of vegetative bud into flower bud.

Formal Gardening It is the application of garden method and material geometrically balanced based on the bilateral symmetry.

Freezing Means reduction of temperature of the product to such an extent that the activities of micro organism stops.

Fruit Juice Concentrate This is a fruit which has been concentrated by removal of water either by heat or by freezing.

Garden A plot or land devoted to the growing of flower, shrubs, flowering and shade trees, creepers, herbs, other ornamental plants, fruit trees and vegetables in certain manners.

Graft Hybrid / Chimera A graft hybrid is a stem, branch or plant originating from a adventitious bud at the graft union.

Graftage It is process of joining a part of plant with another in such a way that both will unite to work as a unit and unit will continue to grow.

Green House Green house are man made devices treating condition favourable for the growth of plant.

Herbarium A collection and arrangements of plant specimens, according to taxonomy.

Horticulture Horticulture can be defined as culture of fruits, vegetables, flowers and ornamental plants and its preservation for long time.

Hydroponics The growth of plant in water without soil and supplied nutrient in water is known as hydroponics.

Indoor Gardening Defined as growing, arrangement and display of house plants for interior decoration of a room or house.

Informal Gardening Land Scaping Gardening It is defined as the application of garden method and materials for the improvement of any area, on which it is possible and desirable to developed new face.

Inter Cropping The programme should be fixed in such a manner that the main fruit crop does not suffer in any way on account of inter-cropping.

Frost is the biggest enemy of cotton crop.

GS

Jam It is produce prepared by boil in the fruit pulp with suffice quantity of sugar to a considerable thick consistency, firm enough to hold the fruit tissue in position.

Jelly It is a product prepared by boiling the fruit with or without water, straining, mixing the strained and clear juice extract with sugar and boiling the mixture to a stage at which it will be set to a clear gel.

Land Scaping Gardening Designing and laying out home gardens, public gardens, parks, road side plantation, avenues, etc. is called land scaping gardening.

Lawn Lawn is an open area with green grass of the garden e.g. *Cynodon dactylon*.

Layering A vegetative method of propagation in which a part of the plant is allowed to root being detaching it from the mother plant.

Matrix A place on the root stock that are prepared for joining the scion or bud.

Olericulture The cultivation of vegetable crops.

Orchard A group of fruit trees grown in a specified area.

Ornamental Garden An area established with valuable and pleasurable plants adjacent to a house or other building.

Over Walling This mode of healing is system mostly on prunning wounds.

Perennials Those plants which do not furnish their life cycle in one or two years, but lives for many years.

Pickling Preservation of fruits and vegetables in common salt or vinegar is called pickling.

Plantation Large scale Agricultural unit devoted to the production of particular fruit crop or any other commodity.

Polyembryony Some-seeds contains more than one embryo is known as polyembryony.

Pomology The science of growing fruit crops is called pomology.

Preservation of Fruits and Vegetables Canning of fruits and vegetables fruit juices, marmalades, jam, jellies, candies, dehydration etc.

Preservatives A chemical additive which preserve some desired characteristic of food making certain chemical structures more stable by retarding microbial growth.

Propagation Means production of new individuals.

Pruning It is art and science of cutting away a portion of a plant to improve its shape, to influence growth, flowering and fruitfulness, to improve the quality of the product.

Rejuvenlization Stimulation of new growth of old plants usually accomplished with prunning.

Scion The upper part of graft union .

Shrub Shrub may be defined as a perennial pant having many woody branches arising from the base of the plant.

Sod Culture Cultivation of grass in orchard is called sod culture.

Spoilage Physical and chemical changes brought about in food through the activities of enzymes, moulds, yeast and bacteria.

Squash This consists essentially of strained containing moderate quantities of fruit. In which cane sugar is added for sweetening.

Stock It is the part of a graft which has the roots and which support the growth made by the other component scion.

Succulents It is juicy plants are among the most specialized form and they are the expression of peculiar condition of soil and climate.

Suckers These are shoots growing from latent adventitious buds on roots.

Syrup This is a clear sugar syrup, which has been artificially flavoured.

Terrarium A closed type of glass container used to provide usual environment for the growth of plants.

Tomato Ketchup It is made by collecting tomato juice or pulp without seed and skin.

Training Means developing a desired shape of the tree with particular objectives by controlling the habit of growth.

Twig Stem one year old or less without leaves.

Vegetable Forcing This is a type of gardening which is concerned with the production of vegetables out of their normal season.

Vegetable Herbacious plant or part of plant which are use for culinary purpose.

Viability Capacity of seed to germinate is called viability.

Vitality strength or vigour of growth is called vitality.

Water Sprouts These are shoots growing form latent adventitious buds on stems or branches.

'Shyok Kaijan' is Indo-Japanese Coast guards joint antipiracy and safety exercsie in the Arabian Sea.

6. Soil Science

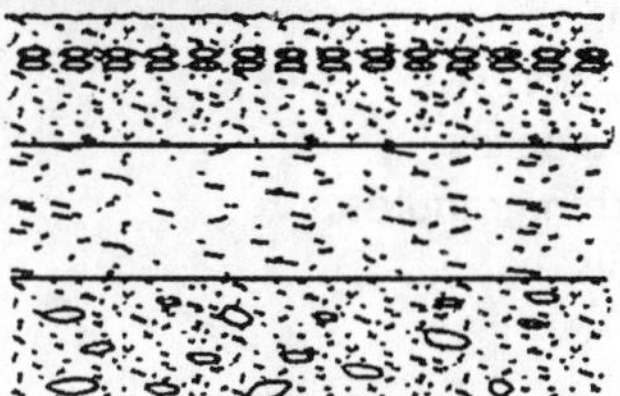

Did You Know ?

Highest highway

The 475-km-long Manali-Leh highway traverses seven passes ranging from 13,000-16,000 ft.

Each Word Counts

PROPERTIES OF SOIL

Physical	Chemical	Dynamic	Biological
1. Soil Texture	1. Soil Colloids		
2. Soil Structure	2. Soil Reaction	1. Soil Consistency	1. Macro-organisms
3. Density of Soil	3. Soil Organic Matters	2. Soil Compactions	2. Micro-organisms
4. Porosity of Soil	4. Soil Acidity	3. Soil Strength	
5. Soil Temperature	5. Soil Alkalinity	4. Soil Crusting	
6. Soil Colour	6. Neutrality		
7. Soil Water			
8. Soil Air			

Important Properties

Soil Texture

Can't be changed/altered.

1. Definition: It is the relative percentage of sand silt and clay in a soil.
2. Fundamental groups of textural classes
   - Sands: > 70% sand separate
   - Loams: Intermediate
   - Clays: > 35% clay separate
3. Determination of textural class
   - Feel Method
   - Laboratory Method (by USDA, based on the Mechanical Analysis)
4. Importance of soil texture
   - State of Hydration and Exchangeable cations
   - Deflocculation Na^+
   - Flocculation Ca^{2+}, $A1^{3+}$
   - Range of Workability
   - Heavy Soils
   - Light Soils

Soil Structure

Can be changed/altered.

1. Definition: The mutual arrangement, orientation and organisation of particles in the soil.
2. Types of structure
   - Plate like
   - Prism like
     a) prismatic
     b) columnar
   - Block like
     a) angular
     b) sub-angular
   - Spherical like
     a) granular
     b) crumb
3. Classes
   - Very fine
   - Fine
   - Medium
   - Coarse
   - Very coarse
4. Grades
   - Structureless
   - Break
   - Moderate
   - Strong

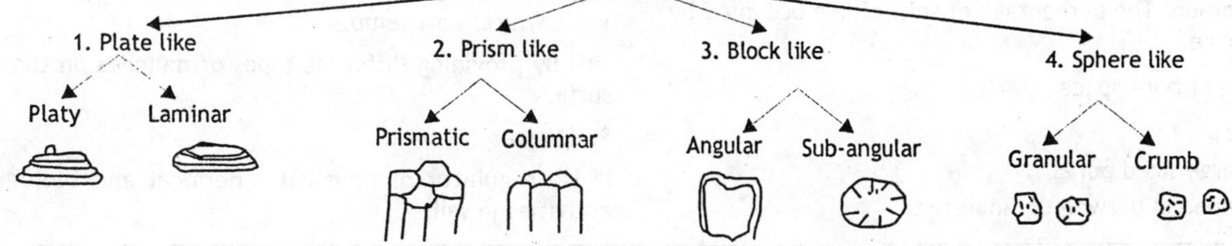

Taiga have coniferous forest vegetation.

5. Factors affecting soil structure

(a) Climate
(b) Organic matter
(c) Adsorbed cations
(d) Tillage
(e) Types of vegetation
(f) Plant roots
(g) Soil organisms
(h) Crop rotation
(i) Alternate wetting and drying

Density of Soil

1. Definition: Weight per unit volume of a substance

2. Types of Measurement

- Particle Density
- Bulk Density

(A) Particle density/True density

1. Definition: The weight per unit volume of the solid portion of soil.

2. $PD = \frac{\text{Mass of Solids}}{\text{Volume of Solids}}$

3. Unit: gm/cc.

4. Characteristics :

- Depends on density of individual inorganic and organic constituents of soil.
- Particle density decrease with increase in organic matter.
- For normal soil particle density is 2.65 g/cc.

(B) Bulk Density

1. Definition: The mass per unit volume of a dry soil.

$$BD = \frac{\text{Weight of Soil mass}}{\text{Soil Volume}}$$

3. Unit: gm/cc

4. Characteristics :

- Bulk density always smaller than particle density
- Low Bulk density favourable physical condition.
- Ideal Bulk density is 1.2 -1.4 g/cc.

5. Factors affecting: Amount of pore space. Compactness of soil, texture, structure, organic matter.

Porosity of Soil

1. Definition: The percentage of soil volume occupied by pore space.

2. Types of pore-space

Macro-pores :

- Large sized pores
- Found in between granules
- Sandy soils

Micro/Capillary pores:

- Small sized pores
- Found within granules
- Clayey soils.

Soil Colour

1. Kinds of soil colour

- *Lithochromic* : from Parent Material eg. Red Soil from Red Sandstone
- *Pedochromic*: during soil formation e.g. Red soils from granite.

2. Factors affecting soil colour

- Oxidation: Reduction
- Alternate wetting and drying
- Iron compounds
- Silica lime and salts

Soil Temperature

1. Sources of soil heat

- Solar radiation
- Bio-chemical reactions
- Conduction
- Precipitation, exposure, vegetation

2. Loss of soil heat

- Radiation
- Conduction
- Evaporation
- Precipitation

3. Importance on plant growth and nutrition

- Germination of seeds
- Soil structure
- Decomposition of organic matter in soil.
- Absorption of water
- Availability of nutrients

4. Management of soil temperature

(i) By regulating the amount of water

- Drain
- Add

(ii) The colour of surface soil

- Black: High temp.
- White: Low temp.

(iii) By providing different types of mulches on the soil surface

Soil Water

It is regulator of physical, chemical and biological activities in soil.

The Junagarh inscription is the first major inscription to be written in Sanskrit.

GS

Classification of Soil Water

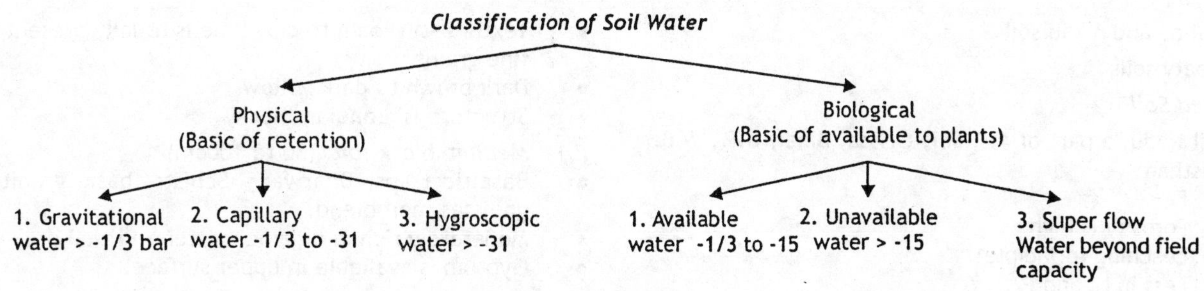

Movement of Soil Water

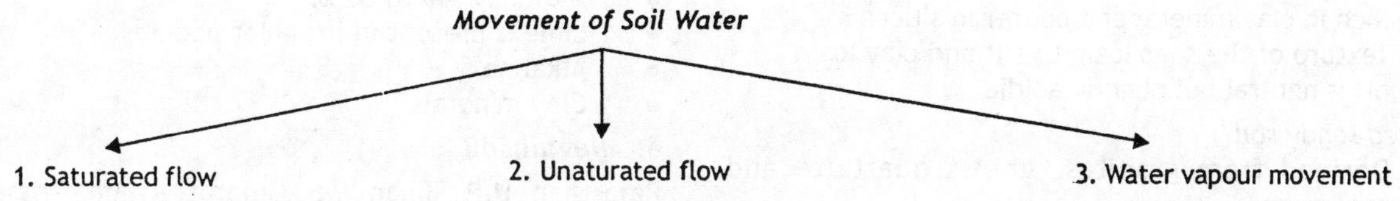

Factors Affecting Available Water

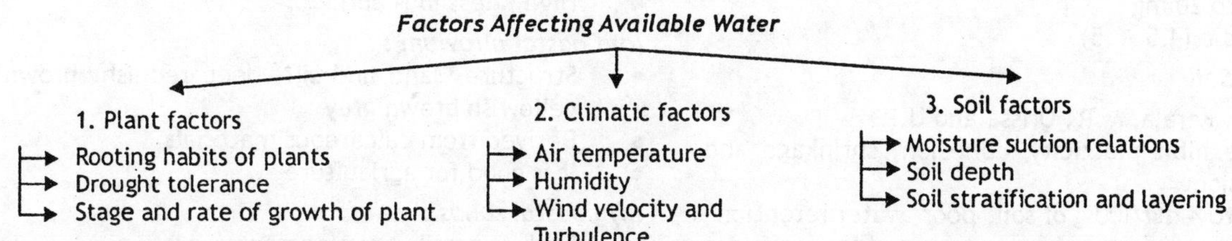

Soil Air

1. Definition: The exchange of CO_2 and O_2 gases between soil pore space and atmosphere.

2. Types of pores :

- Inter crumb pores -between crumbs
- Crumb pores -pores within crumbs

3. Reasons for poor aeration :

- Excess soil moisture
- Gaseous exchange
- Water logged conditions
- Saturated soil conditions

4. Mechanisms of exchange of gases

(i) Mass flow - due to pressure differences between atmosphere and soil air

(ii) Diffusion - molecular transfer of gases

5. Importance of soil air in plant growth and biological activity in soil

(i) Growth of plants

(ii) Development of plant

(iii) Adaptations of plant

(iv) Microbial population and activity

(v) Production of toxic substances

(vi) Absorption of water and nutrients

CLASSIFICATION AND GENESIS OF SOIL

Modern Classification

1. Red soil
   - Red loam
   - Red sandy soil (Fe++)
2. Laterite soil
   - Low level (leaching)
   - High level
3. Black soil
   - Shallow black
   - Medium black
   - Deep black soil
4. Alluvial soil
   - Coastal alluvium
   - Coastal sand
   - Deltaic alluvium
5. Desert soil
6. Tarai soil
7. Hill soil (brown)

8. Saline and Alkali soil

9. Peaty soils

1. Red Soil

Tamil Nadu, a part of A.P, M.P, Orissa, Bihar, U.P., W.B., Rajasthan

- Fe^{++}
- Porosity is high
- Leaching (principle)
- Less in N_2 and P_2O_5

(a) Red loam

- Derivative form weathering of granites, gneiss
- Rich in clay mineral and poorer in silicon
- Texture of the soils loam to silt and clay loam
- pH is neutral but slightly acidic.

(b) Red sandy soil

- Derived from granites, gneiss quartzite and sandstone
- Colour red hematite or yellow
- CEC 5 to 20 mg
- pH acidic (4.5 -6.5)

2. Laterite Soil

Maharashtra, Kerala, M.P., Orissa and U.P.

- They exhibit plasticity, cohesion, shrinkage and expansion.
- CEC 2 to 4 mg/100 g of soil, poor water retention
- High organic matter poor in lime and Mg, deficient in P and K.
- pH 5-6.

3. Black Soil

Maharashtra, Gujarat, M.P, Rajasthan, U.P, A.P., Tamil Nadu, Karnataka

- Clay % of 40 -60 %
- Clay -montmorilonite
- pH 7.5 to 8.5

(a) *Shallow black soil*: (depth- 30 cm)

- Texture -silt loam to clay lime is usually present in fine grains
- Dark brown to dark yellow
- Structure is granular, blocky

(b) *Medium black soil*: (30 to 100cm)

- Basaltic traps, Dharwar - Schists, basic granite, gneisses, hornblend.
- Moderately rich in organic matter and well drained
- Gypsum is available in upper surface.

(c) *Deep black soil* :

- Derived from basalt
- % of clay -40 to 60 %.
- Lime is present in irregular nodules
- Alkaline
- Clay minerals

4. ***Alluvial Soil***

Rajasthan, U.P., Bihar, W.B, Gujarat, Assam, Orissa

Mainly formed deposition of silt over ages.

- High K less in N and P_2O_5

(a) Coastal alluvium :

- Structure -sand and silt, deep reddish brown to yellowish brown grey
- Derived from calcareous materials
- It is good for agriculture

(b) Coastal sands :

- Lack of profile development.
- Salinity is high.

(c) Delatic alluvium

- It is formed by river deposition.

Soil Process and Formation

Soil Formation is a process of two distinct phases:

I. Weathering of Rocks and Minerals (Destructive Process)

II. Development of True Soil (Constructive Process)

1. **Weathering:** Process & transformation of solid rocks into soil

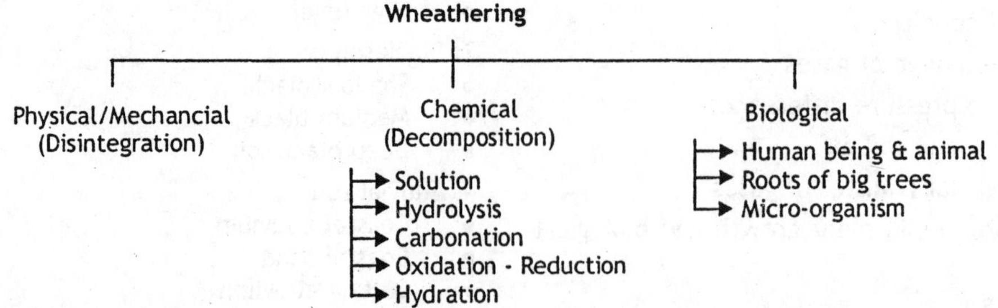

Iraq is the largest producer of Date-palm.

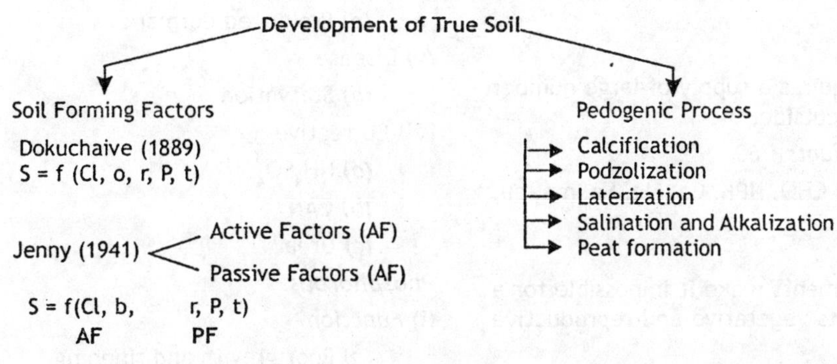

where, Cl - climate; o - organism ; b - biosphere ;
r - relief ; P - parent material ; t - time

Five Stages of Soil Development :

- Initial stage -Unweathered parent material
- Juvenile stage -Weathering just started
- Virile stage -Easily decomposing minerals
- Senile stage -Decomposition arrives final stage
- Final stage -Parent material completely weathered,

SOIL CONSTITUENTS

Organic Constituents

1. Influence the soil colour .
2. Improves the soil structure through aggregation.
3. Improves the rate of infiltration and percolation.
4. Increases the water-holding capacity.
5. Reduces the plasticity, cohesion in clay.
6. Increases the ability of soil to resist erosion.
7. Improves the porosity by changing bulk density.
8. Reduces the losses of surface soil by wind erosion.
9. Helps to reduce alkalinity of soil.
10. Acts as buffering agent.
11. Has solubilizing effect on some mineral compounds present in the soil.
12. Acts as the store house of various nutrients essential to plant growth.
13. Helps to clean contaminated water by adsorption of pollutants.
14. Acts as a chelate which helps to mobilize the plant nutrients.
15. Serves as a source of energy for both macro and micro-organisms.
16. Interacts with pesticides.

Harmful Effects of Organic Matter

1. Serves as source of energy and carbon for many disease organisms.
2. Excessive application creates problems in mixing with soil thoroughly and obstructs easy planting.
3. Most of the toxins identified are short chain organic acids as butyric, propionine etc.

C.N. Ratio (Carbon: Nitrogen Ratio)

1. Definition: The ratio of the weight of organic carbon to the weight of total nitrogen in a soil or organic material.
2. Ratio in soils :
   - Humus (10 : 1) is fairly constant in soils
   - Arable soils 5 : 1 to 15 : 1
   - Lower in warmer regions.
   - Narrow for sub-soils than surface soils.
3. Ratio in plant and microbes :
   - Plant material is 20 : 1 (legumes 30 : 1)
   - FYM is 100 : 1
   - Microbes 4 : 1 to 9 : I
4. Significance of C : N ratio :
   - The C : N ratio helps to conserve both the nitrogen and organic matter in the soil.
   - If organic residues with high C : N ratio is added, the N_2 is immobilized temporarily by micro-organic i.e. ratio becomes narrow.
   - The converted N_2 is reconverted to simple inorganic forms.
5. Constancy of C : N ratio :
   - Decomposition process continues
     → C - lost as CO_2
     → N - lost as leaching/absorbed by plant
   - The percentage of total N_2 being depleted from soil equal to the percentage of carbon being lost.
   - At this point, ratio in top soil range from 10: 1 to 12: 1 which is more or less constant.
6. C : N ratio and Organic matter:
   - Organic carbon: Humus (1: 1.7)

GK GS

West Bengal has the largest population of duck in the country.

- For most soils, the C:N ratio and organic matter is 20 : 1

Nutrients

1. *Nutrition*: Plant requires a supply of large number of substances from outside.
2. *Total elements in plants* : 60.
3. *Essential Elements* : CHO, NPK, Ca, Mg, Fe, mn, cu, Zn, mo, B, Cl

(a) Criteria :

(i) A deficit of the elements make it impossible for a plant to complete its vegetative and reproductive cycle.

(ii) It can't be replaced by another.

(iii) Elements directly involved in metabolism.

5. *Mineral content of plant*

(i) Water : 70%

(ii) Organic matter : 27%

(iii) Minerals : 3%.

6. *Factors determine nutrient availability:*

(i) Soil characters like texture, structure, moisture, pH. atmosphere and salinity.

(ii) Interception and contact exchange

(iii) Mass flow and diffusion

(iv) Soil solution

(v) Root growth and morphology

(vi) Root Exudation and rhizophore

(vii) mycorrhizae

Macronutrients

CHO

(i) Not mineral, structural compound

(ii) Source : Atmosphere, Water

Nitrogen

(i) Source

(a) Soil

(b) Biological N_2 fixation

(c) Industrial N_2 fixation

(ii) Function

Constituent of protoplasm of chlorophyll and coenzyme

(iii) Forms of absorbed

(a) NO_3^-

(b) NH_4^+

(iv) Deficiency symptoms

(a) Thin

(b) Pale yellow

(c) Protein decrease

(d) Root lengthening

(e) Prolonged dormancy

(v) Disease

(a) Starvation

(vi) Corrective

(a) NH_4SO_4

(b) CAN

(c) Urea

Phosphorous

(i) *Function*

(a) Root growth and ripening

(b) Help translocating of CH_2O.

(c) Electron Transport System

(ii) Forms of absorbed

(a) $H_2PO_4^-$

(b) HPO_2^{--}

(iii) Deficiency symptom

(a) Vein necrotic

(b) Tillering decrease

(c) Premature falling

(d) Stunting

(iv) Disease:

(a) Sickle leaf disease

Potassium

(i) Source

(a) Potash Feldspar

(b) Mica

(c) Glucomite

(ii) Function

(a) Cofactors for 40 Enzyme

(b) Stomatal regulation

(iii) Deficiency symptom

(a) Spot of dead tissue at tips.

(b) Margin of leaves (Interveinal chlorosis)

(iv) Disease

(a) Rosette

(b) Dieback

Calcium

(i) Source

(a) Chalk rock

(b) Ca CO_3

(ii) Function

(a) Constituents of lamella and cell wall

(b) Activator for many enzyme

GK **Cerebral cortex is the latest addition to the human brain.** GS

(iii) Deficiency

(a) Brown -scorching

(b) Terminal bud die

(c) Roots poorly developed.

(iv) *Disease*

(a) Tip hooking, blossom end rot of tomato

Magnesium

(i) Source

(a) Carbonate of Magnesium.

(ii) Functions

(a) Chlorophyll synthesis.

(iii) Deficiency

(a) Mottled Chlorosis

(b) Withering (defoliation)

(iv) Disease

(a) Sand-drown (Tobacco)

Sulphur

(i) Functions

(a) Odour in Onion

(c) Nodule formation

(ii) Deficiency

(a) Leaf fall is rapid.

(b) Tea leaf turning into yellow

(iii) Disease

(a) Akiochi (Rice)

Iron

(i) Source

(a) Soil oxide

(b) Fe^{3+}.

(ii) Functions

(a) N_2 Fixation

(b) Respiration

(iii) Deficiency

(a) Interveinal chlorosis

(b) Complete bleaching

(c) Scorching of leaf margin

Manganese

(i) Functions

(a) Enzyme cofactor

(b) Formation of Chlorophyll

(ii) Deficiency

(a) Dead spot scattered

(b) Grain formation decreased

(c) Root system become poor

(iii) Disease

(a) Pahla blight- sugarcane

You Don't Know!

Largest refinery

The IOC Koyali refinery in Gujarat has an annual capacity of 9.5 million tonnes.

(b) Marsh spot of pea

Copper

(i) Function

(a) Compounds of plastocyanin

(b) Essential for photosynthesis, respiration

(ii) Deficiency

(a) Vegetative growth is decreased

(b) Whitening of terminal bud

(c) Marginal chlorosis

Boron

(i) Function

(a) CH_2O transport.

(b) Pollen germination

(c) Water reaction

(d) N_2 Metabolism

(ii) Deficiency

(a) Plant dwarf

(b) Dieback

(iii) Disease

(a) Top sickness -tobacco

(b) Hard fruit -citrus

(c) Hen chick -grape

Molybdenum

(i) Function

(a) Nitrate reductase

(b) N_2 Fixation

(ii) Deficiency

(a) Light yellow and leaf blade fail to expand

(b) Mottling of leaves

(c) In oats, failure of grain formation

(iii) Disease

(a) Whiptail of Brassica

(b) Scald of legumes.

Chlorine

(i) Source

(a) Chlorides

(ii) Function

(a) O_2 evolution

(b) Photosynthesis

Pig rearing is common in the N-E states.

GS

(iii) Deficiency

(a) Basipetal chlorosis

(b) Necrosis

Zinc

(i) Functions

(a) Enzyme alcohol dehydrase

(b) CO_2 evolution

(c) Chlorophyll synthesis

(ii) Deficiency

(a) Leathery chlorosis

(b) Leaf margin sickle

(iii) Disease

(a) Khaira - Paddy

(b) Rosette - Fruits

(c) White bud - Maize

SOIL FERTILITY

Soil Fertility and Productivity

Soil Fertility : Soil fertility is the ability of the soil to provide all essential plant nutrients in available forms and in a suitable balance.

Soil Productivity : The capability of soil to produce specified crop yield under well defined and specified systems of management of inputs and environmental conditions.

Factors Governing Soil Fertility

Parent Material: Fertility of a soil depends on the chemical composition of parent material from which it is derived.

Topography: Soils on the upper slope are less fertile than the soils on lower slope because of high leaching and erosion on upper slope.

Climate: In tropical climate decomposition of organic matter is faster than temperate climate. Thus soils of tropical regions are less fertile when compared to temperate region.

Depth of Soil Profile:

Deep soils are more fertile than the shallow soils and the roots are spread well enough in deep soils than the shallow soils.

Physical Condition of Soil:

The soil texture and soil structure influence the soil fertility.

Artificial Factors:

- Waterlogging
- Cropping system
- Toxic chemicals and pesticides in the soil

Differences between Soil Fertility and Productivity

Soil Fertility	*Soil Productivity*
1. It is an index of available nutrient to plants	1. It is used to indicate crop yields
2. Influenced by the physical, chemical and biological factors of the soil	2. Depends upon fertility and location
3. It is the function of available nutrients of the soil	3. It is the function of soil fertility, management and climate
4. All fertile soils are not productive	4. All productive soils are fertile
5. It is an inherent property of the soil	5. It is not the inherent property of the soil

Soil Fertility Evaluation

It is the assessment of nutrient-supplying capacity of the soil.

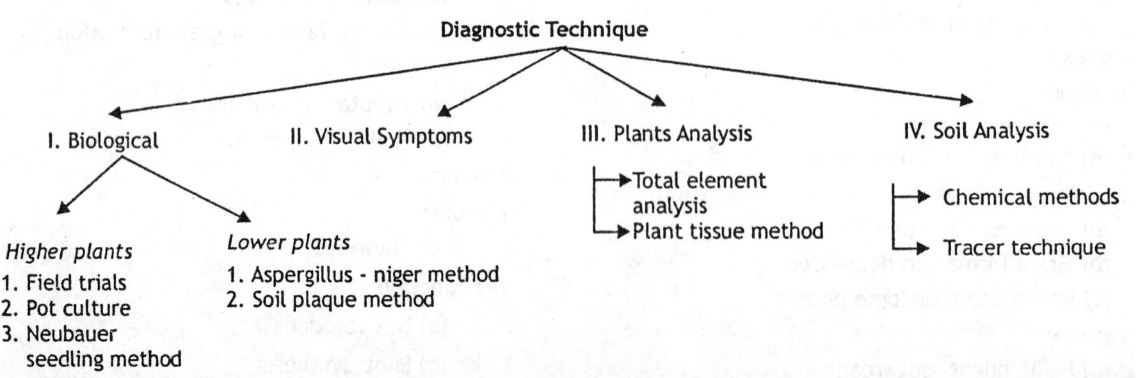

Soil Fertility Evaluation

It is assessment of nutrient supplying capacity of the soil.

I. A. Biological Methods (using higher plants as indicator)

1. *Field Trials*: Direct way to assess the nutrient status in soil.

- Trials with graded doses of nutrients will determine the exact requirement.
- The effect is very much location specific.
- Expensive and time-consuming and growth condition can't be fully controlled.

2. *Pot Culture Methods* :

- Same as field trials except that plants grown in pots with small volume of soil or under controlled field condition like greenhouse.
- Limitations:

(a) Preferential root growth

(b) Limited volume of soil

(c) Problems during packing of soil

3. *Neubauer Seedling Method* :

- Soil is exhausted of the available nutrients and the entire plant is then analysed for the uptake.

I. B. Biological Methods (using lower plants (micro-organism) as indicator)

1. *Aspergillus niger Method* :

- For P, K and Mg in soils.
- The amount of 'K' extracted in mycelium compares favourably with the content of exchangeable 'k' in soil under investigation.

2. *Soil Plaque Method*: by Sockett and Stewart for study of mineral deficiencies in soil.

- If soil is deficient in 'K' or 'P' or both, the calories of the Azotobacter will not develop.
- The intensity of growth of calories indicates the degree of deficiency.

II. Visual Symptoms

- Indicator plants
- Hidden hunger

III. Plant Analysis Method

1. *Total Elemental Analysis* :

- Leaf analysis widely used
- Approach:

(a) Visual Symptoms

(b) Deficiency

(c) Toxicity

(d) Critical nutrient level (CNL) - 'CNL' is a range of concentration at which growth of plants is restricted in comparison with that of plant at a higher nutrient level.

2. *Plant Tissue Tests/Rapid Chemical Analysis:*

- The conductive tissue consisting un assimilate nutrients in sap or extracts are analysed.
- N, P and K have been tested by this technique.

IV. Soil Analysis

1. *Chemical Methods* :

- Phosphorous -Olsen's method method
- Micro nutrient -Chelate DTPA
- Gypsum requirement -Schoonover's method

2. *Tracer Technique* :

- The plants of few supplied the phosphatic fertilizers with ^{32}p.
- The harvested plants tested for total 'P' as well as ^{32}P.

$$A = B\left(\frac{1-Y}{Y}\right)$$

where, A - available 'P' in soil

B - amount of fertilizer 'P'

Y- fraction of 'P' in plant desired from fertilizer 'P'

(1 -Y) - Fraction desired from soil.

Land Capability Classification

Evaluation of land for land use planning is a consequent step following the soil survey and mapping process. In recent years, it has been popularised in almost every land development programme. The system of land capability classification requires that every acre of land be used in accordance with its capability and limitations. The land capability classification is a broad grouping of soils based on their limitations and also serves as a guide to assess suitability of the land for cultivation, grazing and forest plantation.

The land capability classification scheme developed by soil conservation service, United States Department of Agriculture (USDA). The classification scheme has four categories namely land suitability class, land capability classes (comprising eight classes). Land capability subclass and land capability units.

Knowledge World!

Largest bug

The Giant Waterbug (*Lethocerus indica*) which measures over 10 cm in length is carnivorous and eats small fish and other insects.

GK **World's 90% population lives in Northern Hemisphere and only 10% lives in Southern Hemisphere.** GS

Land Capability Classification System

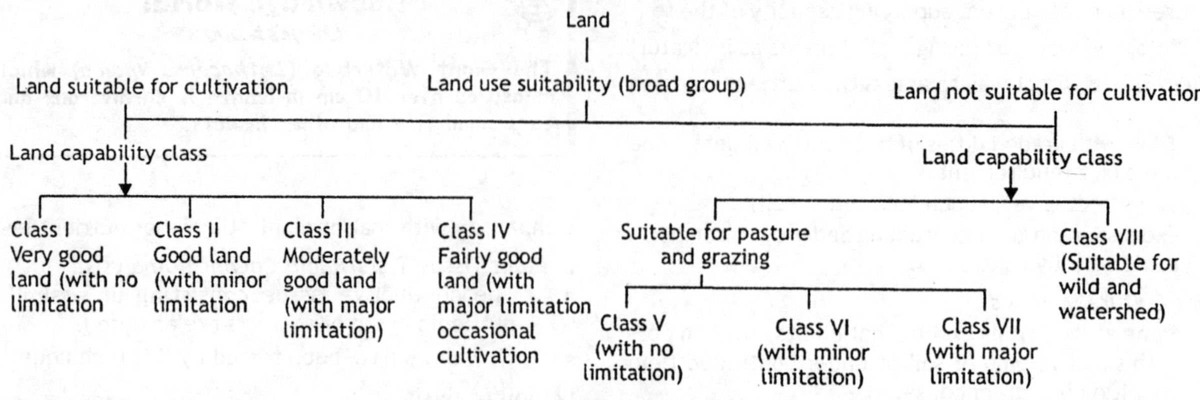

Land capability subclass
(Grouped according to kind of limitation in each subclass viz., erosion (e) climate (c), excess of wetness (w), soil limitation (s))

Land capability units

(Land management groups based on physical characteristics - land that is suited for essentially the same kind of management and same kind of soil and water conservation treatment).

PROBLEM SOILS

Acid Soil

Definition: The soil system proton (H^+ ions) donating capacity during its transition from a given state to a reference state.

Sources :

- Leaching due to heavy rainfall
- Acid fertilizer and soluble salts
- Humus and other organic acids
- Alumina silicate minerals
- Hydrous oxides - oxides of Fe and Al.

Classification of Acid Soils in India

- Laterite
- Lateritic red
- Mixed red, black and yellow
- Podsolic brown
- Foot hill soil
- Peat soil

Problems of Soil Acidity

***(i)* Toxic Effect**

(a) Acid Toxicity: The higher H^+ ion concentration is toxic to plants under strong acid conditions of soil.

(b) Toxicity of different nutrient elements: Fe and Mn → depends upon pH, organic matter and intensity of soil reduction increase organic matter in the soil the population of soil microbes increase the soil O_2 and results in reduction of soil: Toxicity of Al → problem in upland and lowland soils effect on crop by - root growth restrict, affect cell division, restrict the absorption and translocation.

***(ii)* Nutrient Availability**

(a) Non specific effect:

- Inhibition effect of root growth and thereby affects the nutrient availability.

(b) Specific Effect :

- Exchangeable bases - Ca, Mg

(i) Uptake *(ii)* Release

- Nutrient imbalances -

(i) Fe, Mn, Al → increase (ii) P, Mo, N, K, S → decrease

***(iii)* Microbial Activity**

- N -Fixation - *Azotobacter* sp.
- Symbiotic nitrogen fixation -*Rhizobium* sp.
- Fungi -grow well and cause disease

(i) Root rot of tobacco *(ii)* Blight of potato

Amelioration of Soil Acidity

Liming: Practicable feasible management practice.

Lime requirement: Amount of lime added to increase the pH.

Principle

$CaCO_3 + H_2O + CO_2 \rightarrow Ca(HCO_3)_2$

$Ca(HCO_3)_2 \rightarrow Ca^{2+} + 2HCO_3^-$

$H^+ + HCO_3^- \rightarrow H_2CO_3 \rightleftharpoons H_2O + CO_2$

U.S.A. is India's largest trading partner.

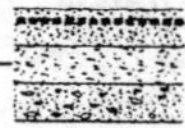

Change of pH with the application of lime

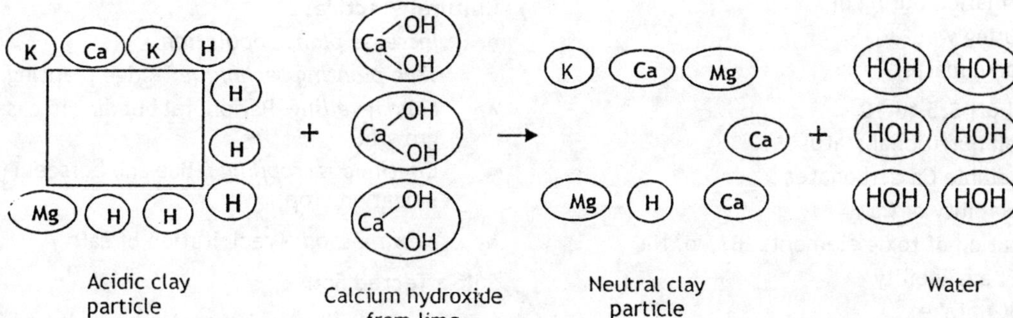

Factors Affecting Lime

- Moisture
- Temperature
- Amount of exchange acidity

Saline Soil

- Contain Sufficiently neutral soluble salts in chlorides and sulphate of Na, Ca, Mg in the root zone adversely affect the crop growth.
- Contain small to appreciable quantity of gypsum.
- Soluble carbonate always absent.
- Low permeability of soil cause poor drainage due to unfavourable soil structure and texture.
- Appearance of crop is patchy growth of crops with visible signs of salt injury (leaf tip burn/chlorosis).
- Presence of white salt crust on the surface so called White Alkali.

Problems of Salt-affected Soil

- Soils are usually barren but potentially productive
- Wilting coefficient of saline soil is very high.
- Amount of available soil moisture is low.
- Absorption of water and Nutrients: due to high salt concentration plants have to spend more energy to absorb water and to exclude salt from metabolically active sites. At the same time various nutrient elements become unavailable to plants.
- Salt toxicity: root injury, inhibition of seed germination.

Reclamation (or) Amelioration of Saline Soil

Pre-requisite:

- Quality of irrigation water should be determined.
- Quality of soil.
- Nature and distribution of salt in root zones.
- Level of subsoil water.

Technical requirements :

- Adequate drainage
- Availability of sufficient water to meet the demand of crop and also leach the salt below the root zone in the soil.
- Perfect land levelling, good bunding for irrigation and advanced agronomic practices.
- Good quality of irrigation water.

Methods :

(A) *Eradication*

Scraping: Scrap the surface salt by mechanical means, limited success and temporary cure.

Flushing: Wash away the surface soil by flushing water over salt crusts.

Leaching: Effective way. The fraction of the irrigation water that must be leached through the root zone. Amount of the salt leached depend on water.

Quantity depends on-

(i) Initial salt content

(ii) Final salt content desired

(iii) Extent of soil depth to be reclaimed under drainage.

(B) *Conversion*

In this conversion method different chemcial amendments are used for changing part of the caustic alkali carbonates into sulphates and ultimately lost by leaching.

(C) *Control*

Amendments	*Soil Conditions*
Gypsum	Saline and alklai soil haivng pH upto 9.
Sulphur Iron Sulphate Iron Pyrite	Alkaline and Saline-alkali soils having pH-8-9.
Limestone	Soil having pH < 6-8.

- Retardation of evaporation to control saline soil
- Use of salt-free irrigation water
- Avoid partial leaching of saline soil

Alkali/Sodic Soils

Exchangeable Sodium Cause

- Soil dispersion
- Affecting soil physical properties

- Poor permeability of air and water
- Crust formation during dry
- Sticky during wet
- Low organic matter

Increased Soil pH 8.5 to 10

- Affect soil nutritional status
- From insoluble Ca carbonates
- Non-availability of Ca
- Accumulation of toxic elements -B, Mo, Na
- Reduce N-availability
- Reduce K-uptake
- Fe deficiency

Appearance

- Dispersed soil structure
- Water stagnation problem
- Hard crust formation at surfaces
- Dispersion of organic matter in soil (due to↑pH) and evaporation brings this dispersed organic matter to soil surface give black coloured surface so called Black Alkali/Sodic.

Reclaiming Alkali Soil

Pre-requisites:

Aim: To replace most of the exchangeable Na with Ca.

For quick results: Proceed cropping by chemical amendment followed by leaching of salts.

For delayed Results : Include rice crop in the cropping sequence.

Along with this application of organic residues rice straw, farm yard manure etc.

Amendments and Practices

Direct Ca : Gypsum, $CaSO_4$

Indirect Ca : H_2SO_4 (or) Iron Pyrite, $FeSO_4$

Soluble Sodium carbonate often present - so irrigate before gypsum application.

- Gypsum broadcasting at lower soil depths (10-12 cm)
- Gypsum should be of 2 mm size.
- Organic matter with gypsum give fast reclamation.

Proper Crop Selection

Rice followed by other sensitive crops even grasses. This can be followed by other field crops.

Water Management

Drainage: Lowering of water-table from critical water-table depth.

Irrigation: Apply irrigation in small depths and more frequently. Sprinkler irrigation is beneficial.

Cultural Practices :

- *Increase plant population.*
- *Tree planting technique*: Ridge planting.
- *Land levelling:* Peripheral bunding in case of patchy growth.
- *Continuous cropping* : Rice and berseem (frequently irrigation crop).
- *New method*: Precipitation of salt.

Salt-affected Soil

Salt-affected soils are those soils, which have high amount of exchangeable soluble salt concentration or exchangeable sodium percentage variously called as saline soil/sodic soil. Salt affected intrazonal soils are halomorphic soil.

Distribution of Salt-affected Soils in India

- Arid, Semi-arid and Sub-humid regions.
- Coastal areas
- In India 8.4 mha salt-affected soils.

3 Main groups:

- Saline Soil
- Alkali-Soil/Sodic-Soil
- Saline-Alkali Soil

Sources of Salts

- *Rock/Mineral Weathering:* Find their way to sea.
- *Ocean/Sea water:* Upliftment of marine water, sea intrusion.
- *Primary minerals.*
- *Arid and Semi-arid climate:* Low rainfall and high evaporation with inadequate drainage measures.
- *Irrigated water:* Lack of drainage and leaching facilities increase water-table and surface salt content in soil.
- *Salt blown by wind:* Arid regions near sea.
- *Excessive use of basic fertilizer:* Sodium nitrate, Basic Slag may develop Soil Aklalinity.
- *Change in land use:* Deforestation, change in cropping/fallow pattern practices cause rise in water table

Characteristics of Salt-affected Soil

	EC	ESP	pH	SAR	TSS
Saline	> 4	< 15	< 8.5	< 13	> 0.1
Alkaline	< 4	> 15	8.5 -10	> 13	< 0.1
Saline-Alkali	> 4	> 15	> 8.5	> 13	> 0.1

National Centre for Antartic and Ocean Research is situated in Goa.

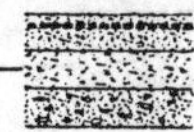

Characters of Salt and Acid Soil

Characters	*Saline*	*Alkaline*			*Acidic*
		Non - saline alkali	*Saline alkali*	*Degraded alkali*	
1. Content	Na of soluble salt	High exchangeable Na	Na soluble salt	H, Na	Exchangeable Hydrogen
2. Exchangeable Ca/Na	Ca	Na	-	-	H
3. Colour	White	Black	-	-	Red
4. Salt	SO_4, Cl, NO_3 of Na	Na_2CO_3	-	$NaCO_3^-$ (lower)	
5. SAR	< 13	> 13	> 13`	< 13	
6. ESP	< 15	> 15	>15	> 15	
7. pH	< 8.5	8.5-10	> 8.5	8.5	> 7
8. EC (dsm^{-1})	> 4	< 4	> 4		
9. Physical condition	Floculation	Defloculation	Floculation	Compact, Columnar structure	
10. Organic matter	Less than normal	Very low	Variable	Low	
11. Total soluble salt content (%)	0.1	< 0.1	> 0.1	< 0.1	

NUTRIENT MANAGEMENT

Integrated Plant Nutrient Management (IPNM)

It is the intelligent use of optimum combination of organic, inorganic and biological nutrient sources in a specific crop, cropping system and climatic situation so as to achieve and sustain the optimum yield and to improve or maintain the soil's physical, biological and chemical properties.

Objectives

- To maintain soil fertility
- Sustainable agriculture production
- Improve farmer's profit
- Increase availability of nutrient from all resources
- Match demand and supply
- Minimising loss of nutrients
- Key principle: Balanced fertilizer
- Aim of balanced fertilizer:

(a) Increase crop yield

(b) Increase farm Income

Components

1. *Chemicals* :

Urea, DAP to supply nitrogen, superphosphate and rock phosphate supply phosphorus, murate of potash and sulphate of potash supply potash. There are various complex fertilizers to supply combination of nutrients. Similarly there are many chemical compounds which supply various macro and micro nutrients in isolation or in combination.

2. *Organic matter* :

Farmyard manure, vermicompost, oil cakes of different oil seeds like castor, neem etc. Incorporation of organic residues of proceeding crop in the fields, if time and soil conditions permit also serves as a source of organic nutrients for the succeeding crop.

3. *Biological* :

(a) Green manures : cowpea, sunhemp, sesbania, etc. and green leaf manures like subabul, etc.

(b) Biofertilizers

Advantages

- Nutrients are easily and quickly available for the plants.
- Good for maintaining soil health
- Good for supplying micronutrient
- All sources can be generated on farm itself
- It has advantage of being diversified into small units to meet the demands of the specific problems of location
- Quality and shelf life of food products are increased

Disadvantages

- Excessive use of fertilizers leads to imbalance in the soil pH
- Most of the chemical fertilizers are high energy consuming
- Chemical fertilizers pollute the environment
- Organic sources are bulk in nature, low in nutrient content

GK GS

Rhine river of Europe is the busiest river of the world in term of cargo transport.

- Organic manures are having high C:N ratio and not properly composed can lead to temporary deficiency.

Nitrogen Losses

Nitrogen is an important macro nutrient for crop plants, which will be taken by the plants from soils. This nutrient released in soils either by weathering or by decomposition of organic matter or by addition of fertilizers, induce N_2 losses from soils.

Classes of N_2 Losses

1. Leaching loss/drainage loss of NO_3^-

Depends on

Mobility : N_2 Mobile for leaching loss

Region: Humid region N_2 loss is high than arid region.

Soil Type: High loss in sandy soil than loam/clay.

Land Nature: More loss in barren soil than crop covered land.

2. Run of loss like NO^-_3.

3. Gaseous losses : Takes place three ways.

(a) Volatilization/Non-biological loss of NH_3

Volatilization: Volatilization is a gaseous loss of N in which inorganic N ions can be converted to gas and lost to the atmosphere.

Causes for Volatilization -

- Poor drainage: Volatilization occurs in poorly drained soil. e.g. rice.
- Alkaline pH : Volatilization occurs in soil pH > 7.5
- Soil pH coupled with form of fertilizer applied :

Acidic soil + NH_4^+ Fertilizer → No Volatilization
(Amm. Ferti) (because soil pH don't increase)

Acid soil + Urea → Volatilization
(NH_4 Forming Ferti) (because soil pH ↑)

[Acid soil + $2H_2O^+$ + $CO(NH_2)^2$ → 2 NH_4^+ + HCO_3^-

NH_4 → ↑NH_3 (Volatialization)

(b) Chemical decomposition of NO_2 under acid condition:

(i) Decomposition of $NH_4\,NO_2$

$NH_4\,NO_2 \rightarrow 2\,H_2O + N_2\uparrow$

(ii) Van Slyke reaction

$R - NH_2 \rightarrow HNO_2 \rightarrow R - OH + H_2O + N_2\uparrow$

(iii) Spontaneous decomposition of Nitrous acid.

$3\,HNO_2 \rightarrow H_2O + 2\,NO\uparrow$

$NO + O \rightarrow NO_2\uparrow$

$2\,HNO_2 \rightarrow NO + NO_2\uparrow + H_2O$

(c) Microbial denitrification - Liberation of N_2 and N_2O:

- *Denitrification* : Nitrates are subjected to reduction in soils especially in those that are poorly drained or water-logged and low in aeration with the help of micro-organisms (or) biological reduction of NO_3 to NO_2.
- Micro-organism involved :

(a) *Pseudomonas denitrificans*
(b) *Thiobacillus denitrificans*
(c) *Bacillus* spp.

- Process of denitrification in soil :

$$2\,NO_3 \xrightarrow{-2(0)} 2NO_2 \xrightarrow{-2(0)} 2NO \longrightarrow N_2O \longrightarrow N_2\uparrow \text{ (Eliminated)}$$

Nitrate, (Nitrite), (Nitric oxide), (Nitrous oxide)

Microrganism removes O_2 act on e^- acceptor.

- Example: Rice ecosystem.

Causes of Denitrification –

- Waterlogged condition/anaerobic condition.

Significance of N_2 Losses

1. It maintains the N balances of N cycle.
2. N released during denitrification fixed by biological N fixation, makes available to plants.
3. In ecological point of view, leaching of NO_3 causes eutrophication, but denitrification prevents this eutrophication.

Way to Compensation

1. In case of leaching in alkali soil, add 25%, more N_2.
2. Denitrification can be reduced by addition of PO_4 residues.
3. Adequate drainage

Problems by N_2 Losses

- Eutrophication
- Atmospheric pollution

Fertilizers Use Efficiency (FUE)

What percentage of an applied fertilizer nutrient is utilized by the crops cropping system. This is termed as FUE.

Measures of FUE

It can be measured by the percentage of added fertilizer nutrient recovered by the harvested portion of a crop (or) kgms of economic produce per kg of nutrient applied.

Problems of 'N' fertilizer - Leaching and volatilization.

Problems of 'P' fertilizer -Fixation, immobility and transformation.

Nitrogen-Use Efficiency

- Mainly determined by various kinds of losses in the field, viz. volatilization in the form of ammonia, leaching and runoff, denitrification.
- N fertilizers are amide and ammoniacal forms.
- Converted into nitrate form, it becomes very much susceptible to loss by leaching with irrigation rain water
- Application of nitrogen fertilizer in split doses or as top dressing increases the efficiency of nitrogen use.

GS

The last Sunga King was Devabhumi.

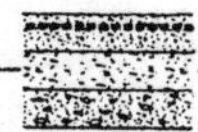

- The NH_4^+ ions in ammonical fertilizers are absorbed by the soil clay. They may be utilised directly by certain crops eg. rice, else they are transformed into nitrates by microbes and taken in this form.
- Higher the clay content of the soil and its exchange capacity the better is nitrogen use efficiency.

Urea Transformation in Soil

$$Co(NH_2) \xrightarrow{\text{Urease}} NH_4^+ \rightarrow NH_3 \xrightarrow{\text{Nitrosomonas}} NO_3^- \quad (\text{Volatilization loss from } NH_3)$$

$$\xrightarrow{\text{Nitrobacter}} NO_3^- \xrightarrow{\text{Denitification}} N_2O,\ NO,\ N_2 \quad (\text{Plant uptake/leaching loss from } NO_3^-)$$

Factors Determine Nutrient Uptake

- Crops, nature of root system
- Water
- Texture and soil pH
- Management practices (land preparation, choice of variety, timely sowing, optimum plant population, timely weed control, weed management, plant protection and balanced supply of essential plant nutrients)
- Agro-climatic conditions

Practices to increase Nitrogen Use Efficiency

- Split application ® apply 'N' in 2 or 3 instalments to coincide with the peak period of nitrogen requirement of the crop.
- Sub surface application
- Pelleting with soil
- Incubation of urea with moist soil (1: 6) for 2 or 3 days, resulting in the adsorption by soil clays of ammonia formed by the hydrolysis of urea.

Increase Efficiency through Slow Release

Nitrogenous Fertilizers

1. Urea aldehyde condensation products e.g. urea form, oxamide, isobutylidene diurea
2. Urea coated with sulphur, lac and neem. The coating forms a temporary barrier between urea granules and soil or soil water, thus reducing the rate of urea hydrolysis.
3. Blend Nitrogen fertilizer with a nitrification inhibitors:

- non toxic to plants, soil micro-organisms, animal and fish
- block the conversion of $NH_4^+ \rightarrow NO_3$ by inhibiting nitrosomonas activity.
- does not interfere with the transformation NO_2^- (nitrite) by nitrobactors.
- be able to move with the fertilizer so that it will be distributed uniformly throughout the soil zone contacted by nitrogen fertilizer.
- stable and long time inhibitory action
- relatively inexpensive, so that it can be used on a commercial basis. e.g. Nitrapyrin.

Quiz Kit!
Longest tunnel
The 6.5 km long tunnel in Karbude, Maharashtra, is the longest in the country.

4. Judicious mix of manures and fertilizers.
5. Placement of urea super granules, made up of ordinary granules below the soil surface has been found to increase, to varying extent the NUE with regard to rice crop.

Phosphorous Use Efficiency

- Fixation of phosphate is main problem.
- Water soluble phosphatic fertilizer after application to the soil react preferably with Fe and Al to form initial phosphate reaction products.
- Ca, Mg carbonates, pH and water status of the soif control the nature of the reaction products.
- Efficiency of phosphatic fertilizers depends primarily upon the release of 'P' from the products rather than the fertilizer.
- Reaction products vary in their ability to release 'P'.

Measures to Increase Efficiency

- Minimum contact of the fertilizer with the soil to restrict phosphate fixation.
- Rise pH of acid soil by liming.
- Liming, deep placement and combined use of super phosphate and organic manures to increase the PUE.
- Rice to increase PUE – surface broadcast, followed by mixing during puddling.
- Wheat - Phosphate placement applied in seed furrow or drilling by just below the seeds.
- Crop rotation -to utilise 'P' direct as well as residual effect. Gram + rice rotation.
- In India application of 'P' fertilizer with organic matter is beneficial. This increase the crop response and decrease fixation.
- Better utilisation of' P' in acid soil mixed with Farm Yard Manure.
- Sunhemp - Application of phosphate before ploughing best method to improve soil fertility.
- Crop may not use > 10% of' P' if applied broadcast upto 30% efficiency when applied as concentrated band along the plant row.
- Clay soil have greater phosphate fixing capacity than sandy soil.

GK
We are adding 80 million people each year to the world population.
GS

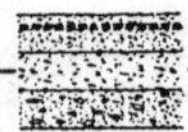

Potassium Use Efficiency

All potassium fertilizers are water soluble. Different 'K' fertilizer consist of 'K' in combination with chloride, sulphate, nitrate.

Both chloride and sulphate of K are soluble in water and on application to the soil then ionizes into K^{+}-, Cl^{-} and SO_4^{2-} ions. The released K^{+} ion from the fertilizer gets adsorbed on the soil colloids and also available to the plant through cation exchange reactions.

Nutrient Fixation

The process whereby readily soluble plant nutrients are changed into less soluble form by reaction with inorganic or organic compounds of the soil restricting their mobility in the soil and thereby suffer a decrease in their availability to the plants.

Two kinds of fixation:

1. Cation fixation: N, K, Fe, Mn, Cu, Zn
2. Anion fixation: P, B, Mo

Phosphorous Fixation in the Soil

Phosphorous was fixed in the soil by the 3 general types of reactions.

1. Adsorption
2. Isomorphous replacement
3. Double decomposition

(i) Adsorption

(a) Physical Adsorption

(b) Chemical Adsorption

In Physical Adsorption phosphate is held in the soil solid surface and Chemical Adsorption phosphate penetrate more or less uniformly into the soil surface.

Adsorption of 'P' takes place -

- on the surface of constant charge-crystalline clay minerals through cations.
- on the surface of variable charges like Fe, Al oxides, organic matter.
- on the surface of kaolinite and Allophanes which have pH dependent charges on their crystal edges and surfaces.
- on the surface of organic matter which have pH dependent cations on their surface.

(ii) Isomorphous Replacement Reaction

Phosphate is fixed by the Hydroxyl (OH) and silicate ions through attached to silicon and Al and are liable to either dissociate to give (Si, Al) OH (or), accept a proton (H ion) and give rise to positively.

Charged clay, which then take part in the anion exchange with PO^{-}_4

$$(Si, Al) - OH \rightarrow (Si, Al) + OH$$

$$(Si, Al) - OH + H + (Si, Al) - OH_2$$

Another important mechanism for the phosphate fixation is a certain amount of silicate released from the tetrahedron. But reaction of Fe, Al hydroxides with the phosphate ions are the most significant for phosphate fixation in soils.

(iii) Double Decomposition Reaction

This fixation (precipitation) is largely depend upon the pH of the system.

This phosphate fixation can be divided into two categories:

- Reaction involving Fe and Al - here P fixation can be reduced with increase in pH.
- Reaction involving $Ca(OH)_2$/Ca CO_3 - here P fixation can be increased by increasing pH.

Factors Affecting Phosphate Fixation

1. Nature and amount of Soil components

a. Hydrous oxides of Fe, Al

b. Type of clay

c. Amount of clay

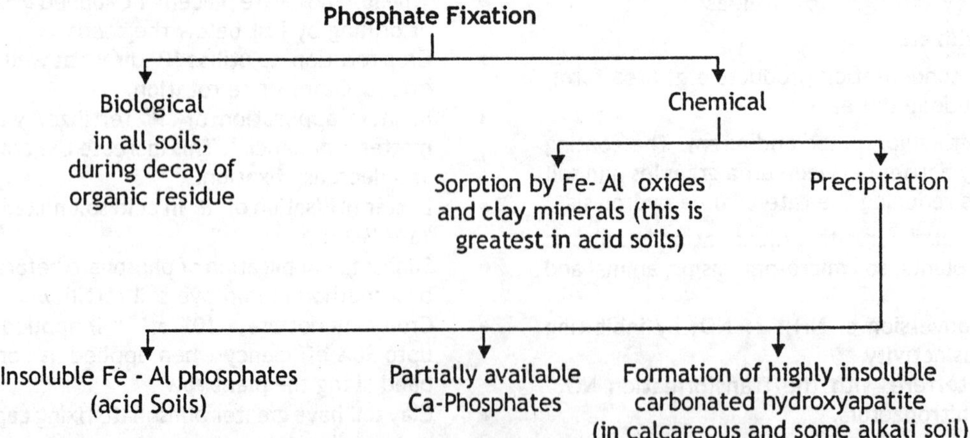

GK GS

U.S.A. is the largest producer of commercial timber.

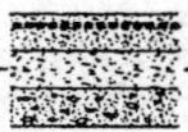

d. Calcium carbonates

2. pH
3. Other ions
4. Organic matter
5. Temperature
6. Over liming

Nitrogen Mineralisation and Immobilization

Nitrogen Mineralisation

It is a process of conversion of organic nitrogen to ammonia, which involves two reactions viz. aminization and ammonification.

Aminization

It is nothing but the process which converts the proteins into amimoacids and amines by the action of heterotrophic bacteria and fungi. Bacteria in alkaline and neutral soils. While fungi dominates in acid soils.

Proteins $\rightarrow$ Amino acids + Amines + Urea + Energy

Ammonification

Here amines and amino acids are decomposed by other heterotrophs, thereby releasing ammonia.

$$R - NH_2 + H_2O \rightarrow NH_3 + R - OH + Energy$$

$$NH_3 + H_2O \rightarrow NH_4^+ + OH^-$$

Did You Know ?

Youngest doctor

17-year-old Balamurali Krishna Amatio received his MD from Mount Sinai Medical School, New York, on May 19, 1995.

Fate of Ammonia

- It may be converted to NO_2^- and NO_3^- nitrification.
- Absorbed directly by higher plants.
- May be fixed in a biologically unavailable form.
- May be slowly released back to the atmosphere as N_2.

'N' Immobilization

It is the conversion of inorganic nitrogen to organic nitrogen and it is basically the reverse of nitrogen mineralisation. If decomposing organic matter contains low 'N' relative to 'C' micro-organisms will immobilize inorganic nitrogen in the soil. Soil microorganism compete very effectively with plants for inorganic nitrogen during immobilization and plants may become readily deficient in nitrogen. For this nitrogen fertilizers are applied to compensate for immobilization and crop requirements.

GK **Singapore is the most densely populated country of the world.** GS

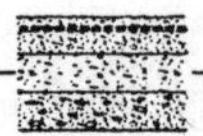

Points to Ponder

- "Aluminium can fit in the centre of the tetrahedron in place of the silicon without changing the basic structure" the process is known as isomorphous.
- "During summers (temperature > 35°C) covering of ground with transparent plastic sheet raises the temperature of upper few centimeters of soil to about 50°C resulting in control of wilt causing fungus and adversely effects the seeds of weeds and soil borne insects." This is called mulches.
- "Nitrogen fixation, a biochemical combination of atmospheric nitrogen with hydrogen to form ammonia which then incorporated into organic nitrogen compounds usable by higher plants" is largely dependent on bacteria.
- "The differential heating and cooling causes lateral stress which, in turn, peels surface layer away from the parent material" this has been termed as exfoliation.
- "The level of plant production can be no longer greater than that allowed by the most limiting of the essential plant growth factors" is called the "Law of minimum".
- A high gypsum-value indicates that only leaching is required, while low gypsum value indicates the need for amendments in an alkali soil.
- A natural body called as soil, characterized by three dimensions unit (pedon), the related groups of which are termed as soil series.
- A pH reading at saturation percentage in excess of 8.5 almost invariably indicates an ESP of 15 or more.
- A tolerable soil loss is the maximum combined water and wind erosion that can take place on a given soil without degrading the soil's long term productivity.
- A vertical exposure of horizon sequence is known as soil profile.
- Acid rain (acid precipitation) water contains nitric and sulphuric acids.
- Adsorbed Ca^{2+}, Mg^{2+}, or Al^{3+} may encourage soil aggregate formation starting with a process called flocculation.
- Alfisol soils are mostly high in respect of their fertility and are used either as cropland or forest or rangeland.
- Alley cropping is an agroforestry system that involves growing of food crops in alleys, the border of which are formed by fast growing trees or shrubs.
- Aluminium and iron hydroxy ions are partly responsible for the "pH-dependent" charge of soil colloids.
- Ammonium ions are subject to interlayer fixation (binding) by vermiculite and, to a certain extent, by fine-grained micas and organic matter.
- An average phosphorus content of soils is 100 kg P per HFS (hectare furrow slice).
- An average potassium content in soils is 42,000 kg K per HFS.
- Anion exchange together with cation exchange largely determines the ability of soils to provide nutrients to plants promptly.
- As minerals are translocated from one soil zone to another, soil horizon are formed.
- At hygroscopic co-efficient, water is so tightly held by soil particles (-31 bars) that it is considered non-liquid and can move only in vapour phase.
- Autotrophic bacteria have the ability to oxidize (or reduce) selected chemical elements in soil.
- Bench terraces are used where rather complete control of the water runoff must be achieved as in rice-paddies.
- Blue Green Algae, growing within the leaves of aquatic fern Azolla are also able to fix nitrogen in quantities significant for rice production.
- Boron concentrations below 0.7 ppm are safe for sensitive plants, 0.7 to 1.5 ppm is marginal limit and more than 1.5 ppm boron is unsafe.
- Both phosphorus and potassium content of peat are lower than those of a mineral soil.
- Calcium and magnesium cation in the soil solution are more strongly adsorbed by the exchange complex than sodium.
- Cation exchange capacity is expressed in centimoles of positive charge per kilogramme of soil.
- Clay and humus have a large surface area per unit weight due to their extremely small particle size.
- Club foot disease of cabbage may be controlled by raising the pH to 7.0 and above.
- Colloidal particles referred to as miscelles or microcells attract hundreds of thousands of cations (positively charged ions) giving rise to an ionic double layer.
- Contamination of soils with salts is one form of soil pollution primarily agricultural in origin.
- Copper, iron and molybdenum are capable of acting as "electron carriers" in enzyme systems that bring about oxidation-reduction reactions in plants.
- Doubling of the flow velocity enable water to move particles 64 times larger, allows it to carry 32 times more material in suspention with four times greater erosive power.

GK **Oxytosin hormone controls the milk secretion after birth.** GS

- Drainage in agriculture is the process of removal of excess water from soil.
- Drip or trickle method of irrigation is the most efficient of all the methods, but it is costly.
- Each unit change in pH make a ten fold change in the activity of H^+ and OH^- ions.
- Edaphology is the study of the soil properties in relation to higher plant.
- Effervescence upon application of acid can be used to detect as little as 0.5 per cent of alkaline earth carbonates in soils.
- $$\text{Exchangeable Sodium Percentage(ESP)} = \frac{\text{Exchangeable Sodium (Cmol/kg)}}{\text{Cation Exchange Capacity (Cmol/kg)}}$$
- Flooding methods of irrigation should be favoured if salinity constitutes a serious problem.
- Flourine and boron toxicities are generally localized problems.
- Frost penetration during the winter is considerably greater in bare non-insulated land.
- Fruits and seeds generally contain less radionucleotides than leaves hence they have a higher potential to contaminate forage crop.
- Furrow irrigation is well adapted to row crops and is suitable where topography is too rough or steep for other methods.
- Granular and crumb sub types of aggregation are commonly influenced by practical methods of soil management.
- Ground penetrating radar is used to improve the quality and to reduce the cost of soil surveys.
- $H_2PO_4^-$ ions is some what relatively more available to plants than is the HPO_4^{2-} ion.
- Heavy phosphate application reduces the availability of toxic cations, but has opposite effect on arsenic which is found in the anionic form.
- High levels of boron in soils can usually be reduced by leaching.
- Highly hydrated monovalent cations like sodium are not tightly held by the micelles, enhance soil dispersion.
- Histol are typically formed in a water-saturated environment such as that found in a peat bog.
- Histosols embody the maximum organic matter content (per cent) as against that the minimum embodied by Aridisols.
- Humic substances have been classified into three groups viz., fulvic acid, humic acid and humin.

Knowledge World!

Deepest oil producing well

Lakwa-351 owned by ONGC produces oil to a depth of 5113 m.

- Humus is a complex and rather resistant mixture of brown or dark brown amorphous and colloidal substances that result from microbial decomposition and synthesis, and has chemical as well as physical properties of great significance to soils and plants.
- If the entire ozone is brought down to the soil surface, it will make only 3 mm thick layer yet it is very important.
- In a strongly alkaline soil (pH > 8.5), the particles disperse and soil becomes unfavourable for entry and movement of water and for tillage.
- In case of saline-alkali soils the conductivity of the saturation extract is greater than 4 mmhos/cm at 25°C and the percentage of exchangeable sodium is more than 15%.
- In non-saline soils, there is usually a relatively abrupt transition from low to high moisture stress and wilting of plants indicates the need for irrigation.
- In physical classification of water, the terms gravitation, capillary and hygroscopic waters are used.
- Inclusion of a close growing forage crop in rotation with row crops helps to control both erosion and runoff.
- Indirect methods of measuring soil water content consist of resistance method and neutron scattering.
- It is expected that if the hydraulic conductivity of surface soil is as low as 0.1 cm/hr (0.04"/hr) leaching and irrigation may present serious difficulties.
- Justus von Liebig (1840) reported that crop yields were increased by adding minerals or inorganic elements to the soil.
- Large quantities of exchangeable and soluble potassium may occur in non-saline-alkali soils.
- Leaching requirement (LR) is the ratio of the equivalent depth of water to the depth of irrigation water expressed either as a fraction or per cent.
- Lead contamination of soils primarily occurs from the air borne source of it from automobile exhausts.
- Lime sulphur is an acid or acid-former type of chemical amendment applied in alkali soils.
- Loss of nitrous oxide (N_2O) from the soil into the atmosphere constitutes an environmental pollution.
- Maize and potato crops have lower soil protective ability than wheat and oats in respect of soil erosion.

GK **Kikuku tribes are found in Kenya.** GS

- Matric potentials are always negative.
- Mature residues, whether legume or non-legume generally have a higher C/N ratio than do younger succulent materials.
- Methyl mercury is quite soluble and available for plant.
- Most of the plant are likely to suffer if the soil volume occupied by air is less than 10-12 per cent.
- Moulds are more important than bacteria in humus formation and aggregate stabilization in soils.
- Mulches keep surface soils cooler than where no cover is used during periods of hot weather.
- Mushrooms are significance, especially in the break down of woody tissues.
- Neutral normal ammonium is the salt solution most commonly used for the saturation of the exchange complex in determination of cation exchange capacity.
- Nitrifying organisms are highly sensitive to pesticides.
- Nitrous oxide can be formed in small quantities by certain nitrosomonas bacteria during ammonia oxidation in aerobic soils.
- Non-saline-alkali soils having an exchangeable-sodium percentage considerably above 15 and yet the pH especially in the surface soil may be as low as 6, De Sigmond (1938) named such soils as degraded alkali soils.
- Non-saline-alkali soils show exchangeable-sodium percentage more than 15 and conductivity of the saturation extract less than 4 mmhos/cm at 25^{0}C, pH usually ranges between 8.5 and 10.
- On addition of acid forming chemicals to an alkaline soil in dryland area, the H^+ions would replace metal cations on soil colloids, and the pH would decrease.
- On an average sodium, calcium and magnesium each occur in earth's crust to the extent of 2-3 percent.
- On application of limestone to an acid soil, calcium replaces hydrogen and other cation, H^+ is neutralized by OH^- or CO_3^{2-} ions and the soil pH increases.
- Out of more than 13 billion hectare of global acreage, only 25 per cent has the physical potential for cultivation.
- Pedology refers to the study of the origin of the soils in their natural environment, its classification and description.
- Permeability of soil, in qualitative sense, refers to readiness with which the soil conducts or transmits fluids.
- Plant growth is a function of total soil moisture stress, that is, the sum of soil moisture tension and osmotic pressure of the soil solution.
- Planting of a desired crop among the native trees (plants) without their complete removal has been termed as mixed tree crop system.
- Platy type of structure of a soil is often inherited from the parent materials, especially those laid down by water or ice.
- Plutonic rocks are formed where solidification of magma occurs at the surface of the earth.
- Pore space found in clay soil are more than sandy soils.
- Potassium is essential for photosynthesis, protein synthesis, starch formation and the translocation of sugars.
- Potassium uptake by plants is limited by high levels of calcium in some soils due to antagonistic effect.
- Radon gas is formed from the radioactive decay of radium, a breakdown product of uranium found in minute quantities in most of the soils.
- Reduced tillage practices are collectively called conservation tillage in contrast to conventional tillage.
- Reduction of nitrates to nitrogen oxide compounds and to elemental nitrogen is known as denitrification.
- Resistance to a change in pH in soil solution is termed as buffering.
- Restricted drainage either owing to high ground watertable or to low soil permeability usually contributes to soil-salinization.
- Saline and alkali soils commonly contain alkaline-earth carbonates and a relatively high concentration of soluble salts.
- Saline soils occur for the most part in regions of arid or semi-arid climate.
- Saline-alkali soils are formed as a result of the combined processes of salinization and alkalization.
- Sandy soils have lower CEC than clay soils because the coarse textured soils are commonly lower in both humus and clay content.
- Sheet, rill and gully are three recognised types of water erosion.
- Shifting cultivation is primarily a system of nutrient conservation, accumulation and recycling.
- Sodium has been found partially to take place of potassium in the nutrition of certain plants.
- Sodium-Adsorption-Ratio (SAR) and Potassium-Adsorption-Ratio (PAR) are employed for discussing the equilibrium relationship between soluble and exchangeable cations.
- Soil algae are divided into the groups viz., blue, green, yellow green and diatoms.
- Soil consistence referred to the resistance of soil material to deformation or rupture.

Governor can sanction Minister's prosecution without the aid and advice of the council of ministers.

- Soil genesis is caused by weathering and breakdown of organic matter, translocation and accumulation.
- Soil refers to the upper biochemically weathered portion of the regolith.
- Soils are flocculated and permeable when saline may become deflocculated when leached.
- Some amino acids such as alanine and glycine can be directly absorbed by plants.
- Sodium Adsorption Ratio (SAR) = $\frac{Na^{+}}{\sqrt{\frac{Ca^{2+} + Mg^{2+}}{2}}}$
- Strip cropping help in the control of soil erosion.
- Sugar beets are very salt tolerant during later growth stages, but are extremely sensitive during germination.
- Sulphur is effective in the control of potato scab because the actinomycetes responsible are discouraged by acidity.
- Surface run-off waters from heavily fertilized lands may contain levels of nitrates toxic to livestock.
- Susceptibility to wind erosion is related to the moisture content of soil.
- Tetrahedral and octahedral sheets are the fundamental structural units of silicate clays.
- The activities of rotifers are mainly confined to peat bogs and wet areas of mineral soils.
- The aggregating can be an effective research tool for investigational work with saline-alkaline soils.
- The arrangement of sand, silt and clay particles within the soil is termed as Soil structure, while their relative amounts as soil texture.
- The availability of phosphorus in alkaline soils is largely determined by the solubility of the calcium compounds in which it is found.
- The average estimated chlorine and sulphur content of the earth's crust is 0.05 and 0.06 per cent respectively.
- The bare soils warm(-) up more quickly and cool (-) off more rapidly than those covered with vegetation or with plastic mulches.
- The best measurement of the aeration status of a soil is oxygen diffusion rate.
- The black colour of the vertisols may be due to the presence of vermiculite.
- The blocky and sub-angular blocky subtypes aggregate-fragment size varies from 1 to 10 cm.
- The colloidal matter of amorphous or crystalline structure not sufficiently ordered for detection of X-rays due to lack of three dimensional crystalline structure are sometimes referred to as short range order minerals.
- The colloidal particles are the seat of most of the chemical, physical and biological properties of soils.
- The colloidal surface area in the upper 15 cm of a hectare of a clay soil could be as high as 7,00,000 km^2.
- The conditions that encourage adsorption will discourage leaching.
- The conversion of oxygen to ozone call for the presence of oxygen in atomic form, the reaction gets energy from short-waves.
- The density of most of the mineral soils varies from 2.60 to 2.75 Mg/m^3.
- The earth worm casts (material passing through their bodies) are definitely higher in bacteria, organic matter and available nutrients than the soil.
- The erosion that exceeds 0.2-0.5 Mg/ha per annum becomes unusually destructive and it referred to as accelerated erosion.
- The excessive growth of algal and other aquatic species takes place in water overly enriched with nitrogen, phosphorus and other nutrients through the process known as entrophication.
- The higher the exchange capacity of a soil, the greater is its buffer capacity.
- The humic substances are characterized by aromatic, ring type structures that include polyphenol and polyquinones.
- The infiltration rate lower than 0.25 cm/hr (0.1"/hr) involves special water management problems.
- The interaction among the various forms of nitrogen in soils, plants and animals and nitrogen in atmosphere constitute the nitrogen cycle.
- The low pH tolerance of moulds is especially important in decomposing the organic residues in acid forest soils.
- The most abundant soil found in India is alluvial soil.
- The movement of dust particles of a fine sand size and small, parallel to the ground surface and upward is called suspention.
- The movement of soil by a short series of bounces along the surface of the ground is termed as saltation.
- The organic matter content of the grassland soils is generally higher than that of forest areas, especially in the surface horizons.
- The peat deposits are grouped in three general types viz., fibrous, woody and sedimentry.
- The percentage of the CEC that is satisfied by the base forming cations is known as percent base saturation.
- The pH is the negative logarithm of the concentration (activity) of H^{+} ions in the soil solution.

GK **The Kadamba dynasty was founded by Mayursarman.** GS

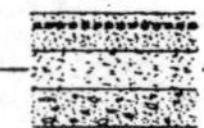

- The pH of gypiferous soil at saturation percentage is seldom in excess of 8.2 regardless of ESP.
- The presence of fine alkaline-earth carbonate particles is thought to improve the physical condition of soils.
- The process of tying up nitrogen in organic forms is called immobilization; whereas its slow release, specifically from organic to inorganic is termed as mineralisation.
- The rates of nitrification decline at temperature above 35 °C and essentially cease at 50°C.
- The ratio of the water flow velocity to the hydraulic gradient is called hydraulic conductivity.
- The reserved acidity is lower than active acidity.
- The salinity problem of principal economic importance arises when irrigation makes previously non-saline soils saline.
- The soil is a source of several gases involved in greenhouse phenomenon including carbon dioxide, nitrous oxide and methane.
- The soil is the primary source of radon gas that can cause lung cancer.
- The soil moisture content at which the plant remain wilted day and night (or in a humid chamber), that moisture content is called wilting co-efficient or permanent wilting point.
- The soil moisture retained in the soil between field capacity and wilting coefficient is known as available water.
- The soil texture is not changed by cultural management.
- The specific heat of pure water is 1.00 cal/g as against 0.2 cal/g that of a dry soil.
- The sum total of the exchangeable cations that a soil can adsorb is known as cation exchange capacity (CEC).
- The term chelate is applied to the compounds in which certain metallic ions are complexed or bounds to an organic molecule or anion.
- The tillage that provides a cloddy surface condition should be at right angle to the prevailing winds to control erosion.
- The water holding capacity of humus on mass basis is 4-5 times that of the silicate clays.
- The wind blown materials consisting primarily of silt with some fine sand and clay are called loess.
- The word saline refers to the soils for which the conductivity of saturation extract is more than 4 mimhos/cm at 25°C.
- There are three kinds of soil acidities viz., active acidity, easily exchangeable acidity and residual acidity.
- There is ordinarily no expansion of kaolinite on wetting as it's structure is fixed due to hydrogen bonding.
- Though boron toxicity is not limited only to saline/alkali soils, excess is common in saline soil.
- Three recognised types of soil water movement are saturated flow, unsaturated flow and vapour movement.
- Tilling of wet clay soils results in break down of aggregates, individual particles tend to act independently and disappearance of macropores, such a soil is said to be puddled.
- Tilth refers to physical condition of the soil in relation to plant growth, while the tillage involves mechanical manipulation to promote it.
- Transformation of rocks and minerals into parent material is known as Soil genesis.
- Unlike matric potential, the osmotic potential has little effect on the mass movement of water in soils.
- Usually more or less weathered rock material are known as bed rock.
- V.V. Dakuchaev investigated unique horizontal layer in soil that were associated with different combinations of climate, vegetation and underlying soil material.
- Variation in pH may result from local microbial action due to uneven distribution of organic residues in the soil.
- Very high salinity water is classed as S_3.
- Water flows, in both saturated and unsaturated soil in accordance with Darcy's law.
- Water molecules, along with magnesium and other ions, are strongly adsorbed in an interlayer space of vermiculite.
- When soil is ploughed and cultivated, the natural states of the upper 12-18 cm is modified and this manipulated portion is referred to as surface soil or top soil furrow slice.
- When the water movement after rain or irrigation becomes negligible after 48-72 hours, the soil is said to be at field capacity.
- Wind erosion is influenced by wind velocity, turbulence, soil surface conditions, soil characteristics and nature and orientation of vegetation.

Multiple Choice Question

1. 'A' horizon of the soil is also designated as :
a] Eluviated horizon b] Illuviated horizon
c] Both a & b d] None of above
2. 'A' value is given by :
a] Marshall b] Stevenson
c] Frid and Dean d] Larsen
3. 'Crop logging' is a method of :
a] Soil fertility evaluation
b] Plant analysis for assessing requirement of nutrients for crop roduction
c] Assessing crop damage
d] Testing suitability of fertilizers
4. 'Hue' is the term which is used for the determination of :
a] Relative thickness or whiteness
b] Purity of the colour
c] Dominant spectral or rainbow
d] All of the above
5. "Hidden hunger" means:
a] Deficiency symptoms are seen when the nutrient is deficient
b] Severe yield reduction may occur without appearance of deficiency symptoms
c] The nutrient is not deficient but apparently seems to be deficient.
d] Visual deficiency symptoms are suppressed by other elements.
6. 0.03 N, NH_4 F + 0.025 N, HCl, is called :
a] Bray No. 1 extractant
b] Olsen extractant
c] Bray No. 2 extractant
d] None of these
7. 0.5 N, NaH CO_3 (pH 8.5) is :
a] Bray No. 1 extractant
b] Bray No.2 extractant
c] Olsen extractant
d] None of these
8. 1.6 to 2.3 value of nutrient index indicates :
a] Average fertility of soil
b] Adequate fertility of soil
c] Both of the above
d] None of the above
9. The neutralizing values of the following liming materials are such that :
a] CaO > $Ca(OH)_2$ > $CaCO_3$ > $CaMg(CO_3)_2$
b] CaO > $Ca(OH)_2$ > $CaMg(CO_3)_2$ > $CaCO_3$
c] $Ca(OH)_2$ > CaO > $CaMg(CO_3)_2$ > $CaCO_3$
d] $CaCO_3$ > CaO > $Ca(OH)_2$ > $CaMg(CO_3)_2$
10. 95 to 99.5% portion of plant tissues are made up of :
a] N, P and K b] C, H and O
c] Ca, Mg and C d] None of these

Quiz Kit!

First newsprint factory

The factory at Nepanagar, Madhya Pradesh, started manufacture of newsprint in January 1955.

11. A combination of many taxonomic unit like soil series, phase or types is called :
a] Soil varient b] Soil series
c] Soil complex d] Soil association
12. A complex mixture of brown amorphous and colloidal substances synthesized by various soil organisms is referred to as:
a] Compost b] Humus
c] FYM d] Peat super compost
13. A man made epipedon which is similar to mollic epipedon in many characters except P_2O_5 content (more than 250 ppm) is :
a] Hisitic epipedon
b] Ochric epipedon
c] Anthropic epipedon
d] Umbric epipedon
14. A pH value of 6.0 indicates that the soil reaction is:
a] Acidic b] Alkaline
c] Neutral d] Highly alkaline
15. A property of fluids by virtue of which it reduces the relative motion among different layers of fluid, is termed as :
a] Soil consistency b] Soil specificity
c] Soil viscocity d] None of these
16. A sequence of soil of about same age, derived from same parent material and occuring under similar climatic conditions but have different characteristics due to variation in relief and drainage, is termed as :
a] Soil series b] Catina
c] Soil phase d] None of the above
17. A soil absorbs about ________ of incoming solar radiation.
a] 5% b] 10%
c] 15% d] 20%
18. A soil which have pH less than 8.5, ESP less than 15 and EC more than 4 mmhos/cm at 25 °C is called :
a] Saline soilb] Alkaline soil
c] Saline-alkali soil d] None of these
19. A soil which have pH more than 8.5, ESP more than 15 and EC less than 4 mmhos/cm at 25°C is called:
a] Saline soilb] Alkaline soil
c] Saline-alkaline soil d] All of the above

GK **Kimberley is the leading gem stones producer.** GS

20. A soil with 20 me ________ 100 g CEC and 5 me/100 gm exchangeable H will have a base saturation of :
a] 25% b] 75%
c] 80% d] 133.3%

21. A sub-surface horizon in which the accumulation of organic matter and clay occurs just below the plough:
a] A horizon b] B horizon
c] C horizon d] None of these

22. A sub-surface horizon with highly oxidized material and 1:1 type clay minerals is known as :
a] Cambic horizon b] Sodic horizon
c] Oxic horizon d] Agric horizon

23. A surface horizon which has very high organic matter is :
a] Histic epipedon b] Ochric epipedon
c] Umbric epipedon d] None of the above

24. A thick dark coloured horizon which have high base saturation (50% or more), high organic matter content and with strong soil structure, is called :
a] Umbric epipedon b] Mollic epipedon
c] Ochric epipedon d] Histic epipedon

25. A unit of soil classification as well as soil mapping which is similar to soil series in most of the soil properties except one or two but the area of such a unit is not enough to establish it as a separate series and show on a soil map, is known as :
a] Soil association b] Soil phase
c] Soil varient d] None of them

26. A vertical section of soil body which shows differnet layers is known as :
a] Soil horizons b] Soil profile
c] Regolith d] Parent material

27. A vertical section of the soil through all its horizons are extending into the parent material is known as :
a] Soil horizons b] Solum
c] Profile d] None of the above

28. A well developed soil have :
a] A horizon b] B horizon
c] C horizon d] All of the above

29. The quantity of single super phosphate needed for applying 80 kg P_2O_5/ha to sugarcane will be:
a]) 250 kg b] 500 kg
c] 750 kg d] 1000 kg

30. A=B (1-y) denotes A-value in which A is equal to :
a] Amount of available nutrients in soil
b] Amount of total nutrients in soil
c] Amount of unavailable nutrients in soil
d] None of these

31. A=RKLSCP denote universal soil loss equation in which K is equal to :
a] Soil erodibility factor
b] Soil productivity factor
c] Soil leaching factor
d] Crop management factor

32. Absorbed hydrogen and aluminium are largely responsible for :
a] Soil acidity b] Soil alkalinity
c] Neutralinity d] None of the above

33. Absorption of ions in plants occurring with the aid of metabolic energy is termed:
a] Passive absorption
b] Active absorption
c] Metabolic absorption
d] Mass flow absorption

34. According to "Universal Soils Loss Equation" the factors on which ratio of erosion depends are:
a] Rainfall, soil erodibility and slope
b] Rainfall, soil erodibility, slope length, slope gradient, crop management, erosion control practice
c] Rainfall, wind velocity, crop management
d] Soil erodibility, soil type, crop management

35. According to land capability classification, the soils which are not suitable for crop cultivation belong to:
a] Class I b] Class III
c] Class IV d] Class VIII

36. According to the carrier hypothesis the specific carriers are present on:
a] Plasmalemma
b] Tonoplast
c] Endoplasmic reticulum
d] Mitochondria

37. According to U.S.D.A. soil classification, the size of fine sand should be in between :
a] 0.01-0.1 mm
b] 0.1-0.25 mm
c] 0.25-0.5 mm
d] 0.001-0.0001 mm

38. Acid soils can be reclaimed by:
a] $CaCO_3$ b] H_2SO_4
c] Ca $SO_4.2H_2O$ d] HNO_3

39. Active soil forming factors would include :
a] Vegetation and topography
b] Time, topography and climate
c] Time and vegetation
d] Vegetation and climate

40. Addition of the following material makes it possible to take good crop in sodic soil:
a] FYM
b] Green manure
c] Gypsum
d] All of these

41. Adsorbed Ca^{++}, Mg^{++} or Al^{++} may encourage soil aggregate formation starting with a process called:
a] Flocculation b] Exfoliation
c] Soil solarization d] None of above

42. **Aeolian deposits are formed by the transportation of materials through:**
a] Running water b] Wind
c] Sea d] Ice

43. **Alfisol soils are mostly _________ in respect of their fertility and are used either as crop land or forest or range land:**
a] High b] Medium
c] Low d] Extremely low

44. **Alkali soils are generally found in:**
a] Sub-humid climate
c] Arid and semi-arid climate
d] Temperate climate

45. **All India Soil Survey and Land Use Organization was established in :**
a] 1952 b] 1956
c] 1960 d] 1962

46. **All potassium fertilizers are physiologically alkaline, there reaction with soils is :**
a] Acidic b] Alkaline
c] Neutral d] None of the above

47. **Alluvial soils are found mostly in:**
a] Kerala, Karnataka and Maharashtra
b] A.P., and M. P. and Assam
c] U.P., Punjab, Haryana, Bihar and West Bengal
d] None of these

48. **Alluvial soils are rich in :**
a] K_2O, P_2O_5 and Lime
b] N
c] Fe, Al
d] Si, Cu

49. **Ammonia lost through volatilization in significant quantities from:**
a] Alkaline soils b] Acidic soils
c] Saline soils d] None of these

50. **Ammonium and potassium ions are apparently just the right size to fit into the cavities between the crystal units of:**
a] Kaolinite b] Montmorillonite
c] Vermiculite d] None of these

51. **Ammonium sulphate contain:**
a] 20% sulphur b] 26% sulphur
c] 16% sulphur d] 24% sulphur

52. **Ammonium sulphate is preferred for nitrogen application in groundnut because it also provides:**
a] Calcium b] Magnesium
c] Sulphur d] Copper

53. **Among the fruits, lemon, apple, plum have been placed in the _________ salt tolerance group.**
a] High b] Low
c] Medium d] None of these

54. **Amount of heat required to raise temperature of 1 gm soil from 14 to 15 °C is known as :**
a] Heat capacity
b] Specific heat
c] Thermal conductivity
d] None of the above

55. **Amount of K in felspar and mica is about :**
a] 60% b] 20%
c] 90% d] 10%

56. **Amount of K in plants is :**
a] 8.0 to 15.0% b] 10.0 to 18.9%
c] 0.1 to 0.5% d] 0.3 to 6.0%

57. **Amount of K in soil plough layer is about :**
a] 1,000-10,000 kg/ha
b] 10,000-70,000 kg/ha
c] 5,000-6,000 kg/ha
d] 8.000-10,000 kg/ha

58. **An abrupt change in textural class or no continuity found between B and C horizon is termed as :**
a] Textural change
b] Lithological discontinuity
c] Regolith
d] None of them

59. **An argillic sub-surface horizon with high E.S.P. (more than 15) is called :**
a] Cambic horizon b] Natric horizon
c] Agric horizon d] Oxic horizon

60. **An epipedon which has dark colour low in base saturation is called :**
a] Umbric epipedon b] Histic epipedon
c] Ochric epipedon d] Anthropic epipedon

61. **Anaerobic environment of paddy soil is responsible for gaseous loss of fertilizer nitrogen by:**
a] Ammonification b] Nitrification
c] Denitrification d] Volatilization

62. **Anion exchange capacity is found maximum in the case of :**
a] Kaolinite b] Montmorillonite
c] Illite d] None of these .

63. **Apatite is a group of minerals which mainly contains**
a] Sulphur b] P_2O_5
c] N_2 d] Titanium

64. **Application of nitrogen in pulses at the time of planting is known as:**
a] Additional dose b] Starter dose
c] Synergistic dose d] Basic dose

65. **Aridisols soils are found in the region of**
a] Humid b] Arid
c] Submerged d] None of these

66. **Arrangement of soil particles is referred to as:**
a] Soil structure b] Soil texture
c] Soil organization d] None of the above

GK GS

Plant do not produce antibodies.

67. As plants grow to maturity, the concentration of plant nutrients changes. What kind of change occurs in respect of N, P, K, Ca and Mg ?
a] Concentration of all 5 elements decreases
b] N and P decrease K, Ca and Mg increase
c] N, P and K decrease K, Ca and Mg increase
d] N, P and K increase, Ca and Mg decrease.

68. *Aspergillus niger* test is used for :
a] Cu & Mg b] Cu & Zn
c] Mg & Zn d] N & S

69. At 1x 10^{-9} H^+ concentration, what will be the pH value ?
a] 5 b] 10
c] 9 d] 6

70. Available form of phorphorus ions is:
a] Phosphoric
b] Phosphorous chloride
c] Phosphorous
d] Phosphorous nitrate

71. Available nitrogen of less than 280 kg/hectare is considered:
a] Low b] Medium
c] High d] Trace

72. Available phosphorous of less than 10 kg/ha of soil is considered as:
a] Low b] Medium
c] High d] Trace

73. Available phosphorous of more than 25 kg/ha of soil is considered as:
a] Low b] Medium
c] High d] Trace

74. Available potassium of more than 280kg/ha in soil is considered:
a] Low b] Medium
c] High d] Trace

75. Available Zn in the soil, can be determined by using the solution of:
a] 1 N NH_4 OAC
b] DTPA+$CaCl_2$ (pH 7.3)
c] EDTA+$(NH_4)_2$ CO_7
d] All of these

76. Average annual rainfall of India is about :
a] 212 cm b] 112 cm
c] 52 cm d] 81 cm

77. Azolla can fix ________ kg N_2/ha/year.
a] 80-100 b] 30-40
c] 10-15 d] 100-120

78. *Azotobacter* and *Rhizobium* bacteria are:
a] Antibiotic producers
b] Nitrogen fixers
c] Plant pathogens
d] Animal pathogens

79. *Azotobacter* sp. fix atmospheric nitrogen:
a] As symbiounts on sugarcane
b] As symbiounts on rice
c] As symbiounts on legumes
d] In free soil

80. Barley, rape and cotton are ________ salt tolerant.
a] High b] Low
c] Medium d] None of these

81. Becteria present in the root nodules of legumes plants is :
a] Clostridium b] Azospirillum
c] Rhizobium d] Azotobacter

82. Beijerinckia may be present in the soils having pH :
a] 3-9 b] 1-5
c] 7-12 d] None of these

83. Beijerinckia may fixed the N_2 upto ________ mg N_2/ gm sugar.
a] 50-60 b] 10-12
c] 20-25 d] 25-50

84. Biofertilizer is a :
a] Mixture of organic matter and micro-organism
b] Mixture of inorganic fertlizer and micro-organis
c] Culture having the desired strain of micro-organisms
d] Decomposed compost enriched with mineral solubilising bacteria

85. Biological nitrogen fixation is :
a] Aerobic b] Anaerobic
c] Both a & b d] None of the above

86. Black and dark coloured soils absorb ______ heat from sunlight than light coloured soils.
a] Less b] More
c] Negligible d] Very high

87. Black soil is the ________ largest group of Indian soils.
a] First b] third
c] Second d] None of these

88. Black soils ________ in organic matter and high in bases.
a] Poor b] Medium
c] High d] None of these

89. Black soils are ________ when wet and very hard when dry.
a] Very hard b] Very soft
c] Soft d] Hard

90. Black soils have ________ amount of lime and P.
a] High b] Low
c] Very low d] Very high

91. *Azospirillum* fix nitrogen mainly in:
a] Sorghum b] Cereals
c] Sugarcane d] Legumes

GK GS

Cellphones emit signals at a frequency of around 800 mega hertz.

92. Blue-green algae grows best in:
a] Upland rice b] Jute
c] Lowland rice d] Sorghum

93. Bone meal is well suited for :
a] Alkaline soil b] Neutral soil
c] Acidic soils d] All of these

94. Bray suggested certain modification in the Mischerlich equation :
a] dy/dx = (A - Y) C
b] A = B (1 - Y)/Y
c] log (A - Y) = log A-CX
d] log (A - Y) = log A-C_1 b - CX

95. Buffering capacity of soil means resistance to a change in :
a] Soil microbial activity
b] Soil acidity
c] Soil pH
d] Base saturation

96. Bulk density is denoted by :
a] Weight of the given soil sample/Volume of the same soil sample (g/c.c.)
b] Volume of the soil sample/Weight of the soil sample (g/c.c.)
c] Weight of the soil sample x 100/Volume of the soil sample (g/c.c.)
d] None of these

97. By the formulae d = PW x As x D/100 we can calculate:
a] Quantity of irrigation water
b] Depth of irrigation water
c] Quality of irrigation water
d] None of the above

98. Zn, Fe, Cu, B, Mn and Mo are called micronutrient, because they are :
a] Nutrients of microscopic significance
b] Nutrients which produce microscopic effects
c] Nutrients that are required in minute qualities
d] Nutrients required by micro-organisms

99. $CaAl_2Si_3O_8$ is the formula of :
a] Anorthite b] Illite
c] Montmorillonite d] Albite

100. Calcite, Dolomite and Hornblends are :
a] Calcium minerals
b] Potassium minerals
c] Sulphur minerals
d] None of these

101. Calcium chloride is probably the most readily available source of soluble calcium, but seldom used due to:
a] Non-availability
b] High cost
c] Application problem
d] None of these

102. Calcium cynamide contain :
a] 16 to 18% nitrogen
b] 26 to 28% nitrogen
c] 21 to 26% nitrogen
d] None of the above

Did You Know ?

First shipping company

Freedom fighter V.O. Chidambaram Pillai established the Swadeshi Shipping Company in 1909.

103. Calcium is essential for:
a] Cell wall formation
b] Meristematic activity in roots
c] Both of these
d] None of these

104. Capillary movement of water is complemented by :
a] Stem elongation
b] Root extension
c] Leaf orientation
d] None of these

105. Carbon requirement of the plant is met by absorbing:
a] Carbon from soil minerals
b] Carbon from soil organic matter
c] Carbon monoxide from atmosphere
d] Carbon dioxide from atmosphere

106. Cation exchange capacity is expressed as :
a] Mole/litre b] Meq/100 g
c] Meq/lit d] Per cent

107. Cation exchange capacity is found minimum in the case of :
a] Montmorillonite b] Illite
c] Kaolinite d] Humus

108. Cation exchange capacity is highest in :
a] Kaolinite b] Montmorillonite
c] Illite d] Humus

109. Chemical weathering governs mostly by the process:
a] Solution b] Hydration
c] Hydrolysis d] All of these

110. Chlorite and hornblend are :
a] Mg minerals b] P minerals
c] K minerals d] Na minerals

111. Chlorosis is observed in upland rice due to deficiency of:
a] S b] Fe
c] Zn d] Mn

112. Chlorosis refers to the :
a] Yellowing of the leaf tissues
b] Burning of the leaf tissues
c] Death of the leaf tissues
d] Development of reddish brown colour on the leaves

113. Clay are classified as :
a] Primary minerals
b] Tertiary minerals
c] Hydroxy minerals
d] Secondary minerals

Dhanush is missile.

114. Commonly used method for enumeration of soil microbial population is :
a] Contact slide technique
b] Microscopic count
c] Dilution plate technique
d] Microbial respiration

115. Conservation tillage tends to encourage:
a] High microbial population
b] Lower number of earthworms
c] Reduced soil fauna
d] None of these

116. Continuous heavy rainfall leads to the formation of :
a] Saline soilb] Sodic soil
c] Acid soil d] Saline-sodic soil

117. Criteria of essentiality is given by:
a] Jean Baptist b] Liebig
c] Arnon d] Julius von Suches

118. C-spacing of kaolinite is :
a] 7.39 Å b] 8.39 Å
c] 10.0 Å d] 14.0 Å

119. DAP requirment for supplying 45 kg N and 115 kg P_2O_5 to one hectare peanut crops :
a] 100 kg b] 150 kg
c] 200 kg d] 250 kg

120. Deficiency of iron results in typical interveinal chlorosis in which :
a] Younger leaves are first affected
b] Interveinal area of the older leaves are first affected
c] Necrotic spots in the interveinal areas develops
d] None of the above

121. Deficiency of Mo in citrus plants results in :
a] Heart rot b] Yellow spots
c] Khaira disease d] All of above

122. Deficiency of several elements is exhibited first in the apical region of the growing part due to their :
a] Autolysis
b] Mobile nature
c] Translocation
d] Immobile nature

123. Degree of soil salinity is indicated by its:
a] Total soluble salt content
b] Organic matter content
c] Nitrogen content
d] Phosphorus content

124. Denitrification is more in :
a] Waterlogged soils
b] Well drained soils
c] Heavy soils
d] Light soils

125. Depth of cultivated soils is upto:
a] 5 cm b] 15 cm
c] 30 cm d] 45 cm

126. Desert soils are found in :
a] Arid and semi arid region
b] Humid and sub-humid region
c] Arid and humid region
d] All of the above regions

127. Diameter of clay particle is:
a] Less than 0.002 mm
b] 0.05-0.5 mm
c] 0.5-1.0 mm
d] 0.7-0.9 mm

128. Different soil separate may be determined with the help of :
a] International pipette method
b] Yet's chart
c] Munsell chart
d] None of the above

129. Diphenyl amine is an indicator which is used in determination of:
a] Organic carbon b] Phosphorus
c] Potassium d] Iron

130. Dominant organisms found in a compost pit are
a] Thermophilic
b] Psychrophilic
c] Mesophilic
d] Psychrophilic and mesophilic

131. Downward movement of food synthesized in leaves takes place through:
a] Xylme b] Phloem
c] Both these d] None of these

132. DQ =DA (dc/dx) dt is the :
a] Fick's law b] Darcy's law
c] Stoke's law d] None of these

133. Earth worms present in the soil:
a] Reduce soil fertility
b] Increase soil fertility
c] Harm plants
d] Causes plant diseases

134. Earth's total weight is contributed by different spheres as :
a] 6.91% lithosphere, 93.06% hydrosphere, 0.03% atmosphere
b] 93.06% lithosphere, 6.91% hydrosphere, 0.03% atmosphere
c] 92.03% lithosphere, 7.01% hydrosphere, 0.96% atmosphere
d] None of above

135. Efficiency of applied fertilizers in rice crop is :
a] 5-50% b] 10-75%
c] 50-75% d] 60—90%

136. Efficiency of applied fertilizers in wheat crop is :
a] 20-50% b] 35-50%
c] 60-80% d] 44-64%

GK GS

Godavari is the longest east flowering river.

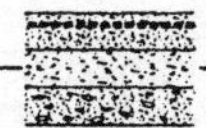

137. Efficiency of symbiotic biological nitrogen fixation depends on :
a] Host characteristics
b] Microbial strain characteristics
c] Ecological conditions
d] All the above

138. Elements which are absorbed from the soil are known as:
a] Plant nutrients b] Plant components
c] Plant feeders d] Plant supplements

139. Elluviation refers to :
a] Removal of clay from above horizons to lower horizons
b] Removal of clay and sesquioxide from above horizons to lower horizons
c] Removal of sesquioxide from above horizons to lower horozons
d] None of the above

140. Entry of K^+ ions through root hair having a low K^+ ion content in the soil is mediated through :
a] Ion exchange
b] Mass flow phenomenon
c] Donan equilibrium process
d] Utilization of metabolic energy

141. Erosion losses of nitrogen is ______ kg N/ha/year.
a] 10-40 b] 20-60
c] 40-80 d] 60-90

142. Etridiazol and nitrapyrin ______ oxidation of ammonia to nitrates.
a] Accelerates b] Maintains
c] Inhibits d] No effect

143. Examples of cumulose materials are
a] Gneiss b] Till plain
c] Colluvial d] Peat and Muck

144. Farm yard manure:
a] Balanced food for plants
b] Recommended for salinity reclamation
c] Tends to make soil acidic
d] Tends to make soil alkaline

145. Father of Soil Microbiology is :
a] S.N. Winogradsky b] Bejerinck
c] A. Waksman d] Liebig

146. Ferrous oxide can be obtained during chemical process weathering under the heads of :
a] Solution b] Oxidation
c] Hydration d] None of these

147. First 15 cm layer of land which is ploughed for cultivation of crop is called:
a] Furrow slice b] Rhizosphere soil
c] Soil d] Soil profile

148. For managing acid soils following chemical amendment is quite essential :
a] $ZnSO_4$ b] CuO
c] CaO d] $CaSO_4.2H_2O$

149. For preparing a mixed fertilizer, 100 kg of $(NH_4)_2SO_4$ and 125 k of SSP were mixed. 25 kg of filler was added to this mixture. In terms of N-P-K, the fertilizer grade of this mixture will be :
a] 5-5-0 b] 5-10-0
c] 8-8-0 d] 10-10-0

150. For reclamation of alkali soil, gypsum is mixed in the soil at the depth of :
a] 100 cm b] 30 cm
c] 60 cm d] 10 cm

151. For the determination of P by Bray method, the ratio of soil and solution should be :
a] 1:20 b] 1:100
c] 1:10 d] 1:25

152. Force (dynes) acting on the surface of water at right angle is one cm of length is called :
a] Cobersion
b] Surface tension
c] Adhesion
d] None of these

153. Formation of soil is the result of:
a] Action of wind
b] Chemical processes
c] Bio-chemical processes
d] Physical processes

154. Fraction of organic matter comparatively resistant to microbial degradaton is :
a] Cellulose b] Protein
c] Lignin d] Hemicellulose

155. Fungal population is comperatively high in a soil having:
a] Acidic pH
b] Neutral pH
c] Alkaline pH
d] Acidic to neutral pH

156. Geneal Soil Productivity Index (GSPI) can be calculated by :
a] Sum of PRI's of major crops /Number of crops
b] Sum of PRI's of major crops x 100/Number of crops
c] Number of crops/Sum of PRI's of major crops
d] None of these

157. Gravitational potential in soil water is potential created due to :
a] Soil solution
b] Mass of the soil solids.
c] Presence of organic matter in soil
d] Gravity

GK GS
Sambhar lake is the largest inland salt lake of India situated in Rajasthan.

158. Group of legume plant that develop nodules when exposed to bacteria nodules obtained from the nodules of any member of that plant group is known as :
a] Symbiotic prosmiscuity
b] Cross inoculation group
c] Symbiosis
d] None of the above

159. Gypsum requirement of the soil can be determined by 344 N (A-B) in which N is equal to :
a] Normality of EDTP
b] Molarity of EDTA
c] Amount of EDTP
d] None of the above

160. Gypsum, an amendment applied to alkali soils belongs to __________ type.
a] Acid former b] Low solubility
c] Soluble Ca-salt d] None of these

161. Heterotrophic organisms obtain energy from the biological oxidation of :
a] Organic carbon b] NH_4
c] SO_4^{-2} d] CO_2

162. Highly available form of phosphorus at pH 7.5 is :
a] H_2PO_4 b] PO_3
c] HPO_4 d] Inositol

163. Hilly soils are low in lime and :
a] Alkaline b] Acidic
c] Saline d] None of these

164. Horizontal layers in the soil profile are known as:
a] Soil columns b] Horizontal zones
c] Soil horizons d] Rhizospheres

165. How many orders are included in the 7th approximation system of soil classification:
a] 8 b] 15
c] 10 d] 12

166. Smectite groups includes:
a] Kaolinite b] Brucite
c] Saponite d] Illite

167. Humus colloids are composed basically of :
a] Carbon b] Nitrogen
c] Helium d] Ozone

168. Humus enhances the mineral breakdown and increases the availability of some:
a] Anions b] Cations
c] Neutral ions d] None of these

169. Hydraulic conductivity of soils varies directly with :
a] Capillary porosity b] Non-capillary porosity
c] Total porosity d] Water holding capacity

170. Hydrogen requirement of plants in met by:
a] Absorption of hydrogen from air
b] Absorptioin of hydrogen from water
c] Absorption of hydrogen from soil
d] Break-down of water within the plant

171. Hygroscopic moisture is held at :
a] 31 atm. b] 1000 atm.
c] 99 atm. d] 200 atm.

172. Identify which of the following is a passive soil forming factors :
a] Relief b] Biosphere
c] Climate d] None of these

173. If a soil has a bulk density of 1.50 mg/m^3 and a particle density of 2.65 ms/m^3 then the percentage pore space of the soil will be :
a] 35.0 b] 38.5
c] 43.4 d] 48.5

174. If crop residues have more than 2% nitrogen, then :
a] Net mineralization will take place
b] Net immobilization will take place
c] Partial mineralization will take place
d] None of these

175. If nutrient in India is less than 1.6, indicates that soils is________ in nutrients.
a] Deficient b] Average
c] Adequate d] None of these

176. If soyabean is being grown for the first time in a field, the seed should be treated with________________ culture.
a] Azospirillium b] Azotobacter
c] Rhizobium d] Phosphorbacterin

177. If the C:N ratio of crop residues is 20-30 then :
a] Mineralization may not take place
b] Immobilization may take place
c] Immobilization may not take place
d] Both immobilization and mineralization take place

178. If the C:N ratio of crop residues is less than 20 then :
a] Immobilization will take place
b] Mineralization will take place
c] Both of these
d] None of these

179. If the nitrogen content in crop residues is less than 2% nitrogen, then :
a] Net mineralization may take place
b] Net immobilization may take place
c] Partial mineralization may take place
d] All of these

180. If the particle size is 2 mm to 7.5 cm then it is known as :
a] Gravels b] Stones
c] Sand d] Silt

181. Igneous rocks are formed by :
a] Cooling of molten magma
b] Weathering of sedimentary rocks
c] Metamorphosis of sedimentary rocks
d] None of above

GS

Hastayurveda is written by Pulkapya.

182. Illuviation refers to :
a] Accumulation of clay and sesquioxide in lower horizons
b] Accumulation of clay and sesquioxide in upper horizons
c] Removal of clay and sesquioxide from lower horizons
d] None of the above

183. In 'Detailed Soil Survey' the base maps used having scale :
a] 1" - 1 mile b] 1" - 16 mile
c] 1" - 8 mile d] None of above

184. In a compost pit, the heat of combustion of organic matter raises the temperature when decay is occurring rapidly. During decay, the temperature rises by :
a] 75 to 90°C b] 60 to 75°C
c] 50 to 72°C d] 40 to 60°C

185. In a waterlogged soil the concentration of ________ is high.
a] Ethane b] Methane
c] Carbon dioxide d] Carbon monoxide

186. In acid soil rock phosphate should be applied:
a] At the time of sowing
b] 5 days before sowing
c] 30 days after sowing
d] 30 days before sowing

187. In acid soils, availability of Fe is :
a] High b] Low
c] Medium d] Not affected

188. In acid soils, phosphorous is fixed as :
a] Iron and Aluminium phosphate
b] Calcium phosphate
c] Fluroapatite
d] Hydroxyapatite

189. In CAN nitrogen is present in following forms :
a] Nitrate form b] Ammonical form
c] Both of the above d] None of the above

190. In case of ammonia synthesis the reaction is :
a] Endothermic b] Exothermic
c] Both the above d] None of the above

191. In chemical weathering the composition of minerals:
a] Changed
b] Unaffected
c] Remain the same as in original rocks
d] None of above

192. In degraded acid soils root extension into anaerobic and highly reduced layer is reduced due to injurious concentration of:
a] Calcium b] Iron
c] Aluminumd] Magnesium

You Don't Know!
Intelligence Bureau
The Intelligence Bureau was conceived and set up by the Secretary of State for India in London on Dec. 23, 1887. T.G. Sanjeev Pillai was the first Indian director of IB. He took over from Sir N.P.A. Smith in April 1947.

193. In general, waters with conductivity values below ________ mmhos/cm are satisfactory for irrigation in so far as salt content is concerned.
a] 1250 b] 1150
c] 950 d] 750

194. In high P-fixing soils, water soluble P_2O_5-fertilizers are applied to the crops essentially as :
a] Broadcast
b] Placement below the seed
c] Placement above the seed
d] Top dressing

195. In India, the area under red soil is:
a] 5.1 m. ha b] 7.5 m. ha
c] 8.0 m. ha d] 10.5 m. ha

196. In no-tillage systems, the surface soil layers have:
a] Higher bulk density
b] Lower bulk density
c] Higher total pore space
d] Lower water infiltration

197. In order to asses the immediately fertility level of a given soil for K, which one of the following forms of K present in soil would one test for?
a] Total K b] Exchangeable K
c] Water soluble K d] K held in the silt

198. In respect of a sugarcane crop to be grown in an acidic soil having high P-fixing capacity by the fertilizer of choice would be :
a] Tripple superphosphate
b] Finely ground rock phosphate
c] Coarse rock phosphate
d] Superphosphate

199. In soil where the negative charge is pH dependent, liming increase CEC which results in an increase potassium by the soil colloids and ________ in soil.
a] Increases b] Decreases
c] Little effect d] None of these

200. In the C:N : ratio of crop residues is more than 30, then:
a] Immobilization will take place
b] Mineralization will take place
c] Partial mineralization will take place
d] All of these

GK GS
Japan is the world leader in the shipping industry.

201. In the diffusion, the movement of ions takes place :
a] From the region of higher concentration to the lower concentration
b] From the region of lower concentration to the higher concentration
c] Both are correct
d] None of the above

202. In the P-determination by Olsen method, the ratio of soil and extractant should be :
a] 1:20 b] 1:100
c] 1:10 d] 1:50

203. Iron helps in chlorophyll synthesis because :
a] It is associated with chlorophyll
b] It is absent in chlorophyll
c] It activates enzyme nitrogen reductase
d] None of the above

204. Irrigation efficiency is very low in irrigated farming in India mainly because :
a] Large irrigated area is under canal irrigation system
b] A major part of the soil types is not suitable for irrigation
c] Of the poor quality of irrigation water
d] Of the adverse climatic conditions

205. Irrigation with sodic water increase the availability of :
a] Sodium b] Molybdenum
c] Zinc d] Chloride

206. Is called father of pedology :
a] V.V. Dokuchaev b] D.G. Vilensky
c] C.F. Marbut d] K.D. Glinka

207. Is potassium takes parts in metaboilic activity ?
a] Yes
b] No
c] May or may not be
d] None of the above

208. K in igneous rocks is about :
a] 8.9% b] 2.5%
c] 3.1% d] 18.5%

209. $K_2(Si_5\ Al_2)\ Al_4\ O_2\ (OH)_4$ is the formula of :
a] Muscovite b] Biotite
c] Montmorillonite d] Kaolinite

210. Kaolinite is a ________ type clay mineral.
a] 1:1 b] 2:1
c] 2:2 d] None of these

211. Khaira disease of rice is diagnosed by:
a] Necrotic lesions
b] Dark reddish brown pigmentation
c] Yellowish spots on leaves
d] Black pin head points on leaf

212. Land capability classes have :
a] 3 categories b] 4 categories
c] 5 categories d] 8 categories

213. Lignin like compounds in humus is about :
a] 45% b] 20%
c] 30% d] 70%

214. Lime is used for reclamation of:
a] Acidic soil b] Sodic soil
c] Acidic soil d] All of these

215. Limonite is formed by the process of :
a] Oxidation b] Carbonation
c] Hydrolysis d] None of these

216. Lithosphere contains about ________ sand stone and 0.25% lime stone.
a] 0.75% b] 0.25%
c] 2.25% d] 10.25%

217. Lithosphere contains about ________ shales.
a] 9% b] 4%
c] 1% d] 0.75%

218. Lithosphere has about ________ igneous rocks.
a] 75% b] 80%
c] 72% d] 95%

219. Mass flow of gases into the soil occurs due to the difference of :
a] Surface tension b] Pressure gradient
c] Both of the above d] None of the above

220. Number of master horizons are :
a] 4 b] 5
c] 3 d] 6

221. Maximum saline and alkaline soils are found in :
a] U.P. b] H.P.
c] M.P. d] A.P.

222. Maximum absorption of water by roots takes place through the :
a] Zone of elongation b] Zone of maturation
c] Root hair d] Root cap

223. Maximum population of microorganisms found in soils is of :
a] Bacteria b] Fungi
c] Actinomycetes d] Viruses

224. Mg and S are termed as :
a] Primary nutrients b] Micronutrients
c] Ulta micronutrients d] Secondary nutrients

225. Mica is :
a] 1:1 type clay mineral
b] 2:1 type clay mineral, non expandable
c] 2:2 type clay mineral, K is present inside the layer
d] None of the above

226. Microbial population of a soil in general
a] Decreases with depth
b] Increases with depth
c] Constant throughout the profile
d] No specific trend

GK GS

Viruses contain only one type of nucleic acid -DNA/RNA.

227. **Mineralization of N refers to the conversion of :**
a] Organic form into inorganic form
b] Inorganic form into organic form
c] Organic form to nitrogenous compound
d] All of the above

228. **Moisture content from lower upper plastic limit is know as :**
a] Sticky point b] Field capacity
c] Plastic number d] All of the above

229. **Molybdenum is a component of :**
a] Chlorophyll
b] Nitrogenase enzyme
c] Cell wall of plants
d] None of the above

230. **Montmorillonite is following type of clay mineral :**
a] 1:1 b] 2:2
c] 3:1 d] None of these

231. **Montmorillonite, illite and kaolinite are:**
a] Kinds of soil microbes
b] Clay minerals
c] Kinds of protozoa present in soil
d] Kinds of soil in U.S.A..

232. **More than 2.3 value of nutrient index indicates :**
a] Average productivity of soil
b] Adequate fertility of soil
c] Adequate productivity of soil
d] None of them

233. **Most of the plants are likely to suffer if the soil volume occupied by air is less than ______ per cent.**
a] 20-25 b] 15-20
c] 12-15 d] 10-12

234. **Most resistant mineral present in soil is :**
a] Mica b] Hornblende
c] Quartz d] None of above

235. **Murate of potash contain :**
a] 40% K_2O b] 36% K_2O
c] 50% K_2O d] None of the above

236. **Mycorrhiza-infected plants show better ability of absorption of the nutrients, especially:**
a] Iron b] Manganese
c] Potassium d] Phosphorus

237. **Mycorrihiza-infected plants show better growth than uninfected plants, especially on:**
a] Low fertility soils b] High fertility soils
c] Marshy soils d] Desert soils

238. **N, P, K per cent in cotton seed cake is :**
a] 4, 1.9, 2 b] 1.9, 2, 4
c] 4, 1.9, 1.7 d] 1.7, 4, 2

239. **Na Al Si_2 O_8 is the formula of :**
a] Anorthite b] Albite
c] Mica d] Illite

240. **National Bureau of Soil Survey and Land Use Planning's head quarter is at :**
a] Kolkata b] Nagpur
c] Dehradun d] Mumbai

241. **National Remote Sensing Agency was established in :**
a] 1973 b] 1977
c] 1968 d] 1970

242. **Nitrate levels in drinking water above about ______ mg per litre are considered a human health hazard.**
a] 5 b] 10
c] 15 d] 20

243. **Nitrification in the soil take place by:**
a] Reduction process
b] Oxidation process
c] Biological oxidation
d] Both a & b

244. **Nitrification is mainly carried out by :**
a] Hetero-trophic organism
b] Photoauto-trophic organism
c] Chemoauto-trophic organism
d] None

245. **Nitrogen is absorbed by the plants in the form :**
a] Only as NO_3^{-1}
b] Only as NH_4^+
c] Both NO_3^- and NH_4^+
d] None of the above

246. **Nitrogen is an essential nutrient element for the plants because :**
a] It imparts early vigorous vegetative growth and dark green colour
b] It provides extensive root system
c] It strengthens straw and prevent lodging
d] It stimulates seed formation

247. **Nitrogen is taken by the plants in the form of:**
a] Elemental nitrogen
b] Nitrate
c] Nitrite
d] None of these

248. **Nitrogen, phosphorous, potassium, calcium, magnesium and sulphur, which are used by plants in relatively large quantities, are referred to as:**
a] Macro nutrients b] Micro nutrients
c] Trace nutrients d] Bulk nutrients

249. **Nitrogenous fertilizer's are classified into :**
a] Two groups b] Three groups
c] Four groups d] None of the above

250. **Non-humic substances are less complex and are ____ to microbial attack than those belonging to humic group.**
a] More resistant b] Less resistant
c] More susceptible d] Less susceptible

GK GS

Kalapani Island is situated in Lakshwadweep.

251. Non-symbiotic bacteria fixes _____ amount of N_2 than symbiotic N_2 fixers.
a] Less b] More
c] Equal d] Not known

252. Nutrient Index is used to prepare the:
a] Soil productivity map
b] Soil fertility map
c] Both of the above
d] None of the above

253. Nutrients absorbed by plants are carried upward through the:
a] Xylem b] Phloem
c] Both these d] None of these

254. Occurrence of quartz, in high amounts makes a rock:
a] Hard b] Soft
c] Alkaline d] Acidic

255. Olsen method of P determination is more effective in :
a] Acidic soils b] Alkaline soils
c] Normal soils d] None of these

256. On the basis of pH measurements, the nearly neutral soils have pH value between:
a] 6.0 to 6.5 b] 6.6 to 7.5
c] 7.5 to 8.5 d] 5.0 to 6.0

257. On the basis of total mass, the relative abundance of the given elements in the earth's crust is such that :
a] Si > O_2 > Fe > Al b] O_2 > Al > Si > Fe
c] O_2 > Si > Fe > Al d] O_2 > Si > Al > Fe

258. Organic carbon content in the soil is a measure of:
a] Available potassium
b] Available phosphorus
c] Available nitrogen
d] Available magnesium

259. Organic matter contains about _____ percent organic carbon :
a] 32 b] 18
c] 68 d] 58

260. Organic matter content of 'histosols' and "hydromorphic" soils is :
a] 30-40% b] 50-57%
c] 60-70% d] 12-18%

261. Organism used as biofertilizer in wheat cultivation :
a] Azotobacter b] Rhizobium
c] Nostoc d] All of these

262. Oxidation of sulfur to sulphate is carried out by :
a] *Bacillus* spp.
b] *Aspergillus* spp.
c] *Azotobacter* spp.
d] *Thiobacillus* spp.

263. Oxygen requirement of the plant is met by absorbing:
a] Respiration b] Transpiration
c] Photosynthesis d] Chlorophyll formation

264. Peat soils are _____ in nature.
a] Alkaline b] Saline
c] Acidic d] Neutral

265. Peat soils have _____ organic matter.
a] 5-8% b] 10-40%
c] 50-55% d] 60-65%

266. Pedology is the study of:
a] Rocks
b] Soils
c] Nutrient management
d] Crop production

267. Percent nitrogen is equal to :
a] Percent organic matter x 5
b] Percent organic matter x 0.05
c] Percent organic matter x 5
d] None of the above

268. Percent of total N present in soil is in the form of organic N :
a] 95.99 b] 40.80
c] 100 d] 20

269. Percentage of soil solids in soil is :
a] 30% b] 40%
c] 50% d] 60%

270. pH value of 7.0 denotes that the soil reaction is:
a] Acidic b] Alkaline
c] Neutral d] Highly alkaline

271. Phosphates enzyme play a major role in _____ of organic phosphates in soil.
a] Adsorption b] Mineralization
c] Immobilization d] None of these

272. Phosphorous occurs in soil in :
a] Inorganic form only b] Organic form only
c] Both a & b d] None of these

273. Phosphorus fixation refers to the phenomenon of :
a] Absorption of P_2O_5 by plant
b] Leaching of P_2O_5 from soil by percolating waters
c] Conversion of soluble phosphorus into plant unavailable forms
d] Anion exchange with colloidal constituents of soil

274. Phosphorus is essential for:
a] Cell division
b] Development of meristematic tissues
c] Both of these
d] None of these

275. Physical weathering causes :
a] Alternation of rocks and minerals
b] Disintegration of rocks and minerals
c] Decomposition of rocks and minerals
d] Synthesis of rocks and minerals

GK The Satvahanas were the first to make land grants to Brahmins. GS

276. Plants meet their oxygen requirement by:
a] Breakdown of water in the plant body
b] Absorbing oxygen from air
c] Absorbing oxygen from soil
d] All the above methods

277. Poor quality water contains :
a] Higher amount of soluble salts
b] Lower amount of soluble salts
c] Higher amount of calcium
d] None of the above

278. Potassium fixation is higher in :
a] 1:1 type clays
b] 2:1 type clays
c] Oxide of Fe and Al
d] None of the above

279. Potassium is considered vital in:
a] Synthesis of protein and amino acids
b] Phtosynthetic activity of leaves
c] Movement of manufactured food from leaves to root
d] All of the above

280. Process of immobilization refers to:
a] Conversion of inorganic form to organic form
b] Conversion of organic form to in organic from
c] Conversion of N-compound to protein
d] None of the above

281. $Q=KA(P_1-P_1)/n_1$ denotes the :
a] Fick's law of diffusion
b] Stoke's law of velocity of particle falling
c] Darcy's equation of gas flow
d] None of these

282. Q=Ki is the ________ equation.
a] Darcy's equation b] Fick's law
c] Both d] None of these

283. Radioactive tracer technique is used for determining:
a] Translocation of minerals in plants
b] Cations exchange capacity of soil
c] Both these
d] None of these

284. Red soils are found maximum in :
a] U.P. b] M.P.
c] Assam d] Tamil Nadu

285. Refers to the study of the origin of soils in there natural environment, its classification and description:
a] Edaphology b] Pedology
c] Geology d] None of the above

286. Refers to the study of the soil properties in relation to crop production:
a] Soil taxonomy b] Geology
c] Pedology d] Edaphology

287. Refers to the upper most biochemically weathered portion of the regolith:
a] Soil b] Earth
c] Rock d] None of the above

Knowledge World!
First steam powered plant
The first steam powered plant was set up in Kolkata 1899 by the Calcutta Electric Supply Company. It has a power generation capacity of 1,000 KW.

288. Release of nutrients like nitrogen, phosphorus and sulphur from the added organic matter by M.O. at the initial stage of degradation will depend upon :
a] Soil pH
b] Soil temperature
c] Chemical composition of added organic matter
d] Method of incorporation

289. Rhizobia infects the host legumes through :
a] Epidermal cells b] Cortex cells
c] Root hairs d] All of the above

290. Rhizobium after infection converts to a form which if irregular in shape and is called :
a] Spore b] Cyst
c] Bacteriod d] None of the above

291. Rhizobium culture is used to increase biological nitrogen fixation in:
a] Pulse crops b] Cereals crops
c] Oilseeds d] Cotton

292. *Rhizobium japonicum* can fix the nitrogen in the roots of :
a] Soyabean b] Cowpea
c] Beans d] Lentil

293. *Rhizobium leguminoserum* can fix the nitrogen in the roots of :
a] Alfalfa, Sweet clover
b] Clovers
c] Pea, Sweetpea, Lentil
d] Beans

294. *Rhizobium phaseoli* can fix the nitrogen in the roots of :
a] Lentil b] Soyabean
c] Pea d] Bean

295. *Rhizobium* sp. fix atmospheric nitrogen:
a] As symbiouts on wheat
b] As symbiouts on cucumber
c] As symbiouts on pulses
d] In free soil

296. *Rhizobium trifolii* can fix the nitrogen in the roots of:
a] Sweet clover b] Clover
c] Lupines d] Soyabean

297. Rhizosphere refers to the small area:
a] Around the root
b] On the root
c] Near the primary root
d] All of the above

GK **Bacteria take their nutrition in solution-form only.** GS

298. Rhizospheres affect microbial population and nutrient availability by:
a] Exerting physical pressure
b] Secreating nutritive metabolites and organic solvents
c] Increasing pore space
d] Changing soil structure

299. Rock phosphate can be directly applied as fertilizer in soils:
a] Acidic b] Alkaline
c] Neutral d] All of these

300. Rock phosphate contain :
a] 20 to 22% P_2O_5 b] 16 to 18% P_2O_5
c] 12 to 16% P_2O_5 d] 20 to 40% P_2O_5

301. S.M.P. buffer is used in determination of :
a] Lime b] Potassium
c] Nitrogen d] Calcium

302. Saline soils are reclaimed/managed by :
a] Addition of gypsum
b] Addition of lime
c] Leaching the soluble salts
d] Addition of pyrite

303. Saline soils occur mostly in:
a] Himachal Pradesh
b] Haryana
c] Punjab
d] Madhya Pradesh

304. Saline-sodic soils has the value of:
a] EC more than 4 mmhos/cm, ESP more than 15, pH more than 8.5
b] EC less than 4 mmhos/cm, ESP more than 15, pH less than 8.5
c] EC more than 4 mmhos/cm, ESP less than 15, pH less than 8.5
d] EC less than 4 mmhos/cm, ESP more than 15, pH more than 8.5

305. Sand stone and lime stone are the examples of :
a] Metamorphic rock
b] Igneous rock
c] Sedimentary rock
d] All of these

306. Sedimentary rocks are formed by:
a] Solidification of lava
b] Increase in pressure
c] Increase in temperature
d] Cementation of broken parts

307. $Si_4 Al_4 O_{10} (OH)_8$ is a formula of :
a] Muscovite b] Mica
c] Kaolinite d] Biotite

308. $Si_4 O_{10} (OH)_8 n.H_2O$ is a formula of :
a] Kaolinite b] Halloysite
c] Muscovite d] Vermiculite

309. Silica content of an igneous rock is:
a] 80% b] 5-10%
c] 40-50% d] 60-75%

310. Silts have particle size range between:
a] 2.00 to 0.20 mm b] 0.02 to 0.002 mm
c] 0.20 to 0.02 mm d] 0.002 mm

311. Single super phosphate (SSP) contain:
a] 20% P_2O_5 b] 16 to 18% P_2O_5
c] 20 to 25% P_2O_5 d] 10 to 46% P_2O_5

312. Single super phosphate (SSP) is manufactured from :
a] Rock phosphate
b] Di-calcium phosphate
c] Bone meal
d] None of the above

313. Single super phosphate is preferred for phosphorus application in groundnut because it also provides:
a] Sulphur
b] Calcium
c] Sulphur and calcium both
d] None of these

314. Slow nitrification rate of non-edible mahua cake is due to :
a] The presence of oil
b] Wide C/N ratio
c] Low N_2 content
d] The presence of certain alkaloids

315. Sodic soils contain sufficient amount of ________ to interfere with plant growth:
a] Exchangeable calcium
b] Exchangeable sodium
c] Exchangeable potassium
d] Exchangeable magnesium

316. Sodium nitrate contain :
a] 18% nitrogen b] 24% nitrogen
c] 16% nitrogen d] 28% nitrogen

317. Soil air contains :
a] 0.25% CO_2 b] 0.02% CO_2
c] 2.5% CO_2 d] 3.0% CO_2

318. Soil colour determination is standardized and determined by the comparison of the soil colour to :
a] Tensiometer b] Gypsum block
c] Munsell chart d] None of the above

319. Soil formed under impeded drainage conditions, where movement of air and water is restructed is known as:
a] Gley soil b] Acid soil
c] Chernozem d] Podzol

320. Soil is biochemically weathered part of :
a] Bedrock b] Parent material
c] Regolith d] Organic matter

321. Soil organic matter has many remarkable effects and one of them is given below, identify the same:
a] Encouraged granulation
b] Increased plasticity
c] Enhanced cohesion
d] Low cation exchange

GK GS
Bangalore is also called as Hi-tech city.

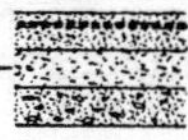

322. Soil samples are collected in:
a] Plastic bags b] Metal containers
c] Gunny bags d] Wooden boxes

323. Soil strength for engineering purposes is determinated by the characteristics:
a] Compressibility
b] Compactibility
c] Bearing resistance
d] All of these

324. Soil structural units having horizontal axis much longer than vertical axis. Such type of soil structure is known as :
a] Prismatic b] Blocky
c] Platy d] Columnar

325. Soil survey's are of different types:
a] Three types b] Four types
c] Two types d] None of the above

326. Soils are acidic for one of the reason:
a] Basic slag b] Acidic parent material
c] Low rainfall d] None of these

327. Soils are surveyed and classified on the basis of distinctive feature of:
a] Parent materials b] Soil horizons
c] Soil profiles d] Soil samples

328. Soils containing high proportions of organic matter have following colour:
a] Light colour b] Dark colour
c] Yellow colour d] Red colour

329. Soils with high pH are generally deficient in :
a] Zn and Mn b] B and Fe
c] Cu and Mo d] Ca and Mg

330. Soluble salt content of soil is determined by measuring its:
a] Electrical conductivity
b] pH value
c] Organic matter content carbon
d] None of these

331. Solum is made up of :
a] C horizon b] C+D horizon
c] A + B+C horizon d] None of these

332. Some pesticides with positively charged groups are also absorbed by:
a] Illite clays
b] Kaolinite clays
c] Montmorillonite clays
d] Silicate clays

333. Specific heat of the dry soil is equal to :
a] 1.0 cal./gm b] 0.5 cal./gm
c] 0.2 cal./gm d] None of the above

334. Specific heat of the mineral matter is :
a] 0.462 Cal gm^{-1} b] 0.18 Cal gm^{-1}
c] 1.0 Cal gm^{-1} d] 24.0 Cal gm^{-1}

335. Standard way of collecting a soil sample is to take:
a] One sample from centre of the field
b] One sample each from every corner of the field and mix
c] Several samples from all over the field and mix
d] None of these

336. Steamed bonemeal contain :
a] 18% P_2O_5 b] 16% P_2O_5
c] 22% P_2O_5 d] 25% P_2O_5

337. Sulphate of potash is preferable to MoP for tobacco because :
a] The chlorine in the MoP effects the burning quality of tobacco
b] The chlorine in the MoP increases the nicotine content of tobacco
c] The application of MoP results in the luxury consumption of potassium
d] The sulphate iron helps to increase the availability of other nutrients

338. Super-digested compost is called so because :
a] It is prepared by allowing it to decompose beyond the normal process of decomposition.
b] It contains plant nutrients more than present inordinary compost
c] Superhosphate is added to the compost during its preparation
d] Its P_2O_5 content is more than that present in the oridary compost

339. Symbiotic association of fungal hyphae with plant roots is known as:
a] Mycorrhiza b] Mutual infection
c] Lichen d] Co-parasitism

340. Tarai soils are generally deficient in:
a] Zinc b] Nitrogen
c] Phosphorus d] Potash

341. Tarai soils are high in :
a] Iron b] Sodium
c] Organic matter d] Calcium

342. Temperature has got significant effect(s) on :
a] Structure
b] Biological activity
c] Weathering of rocks and minerals
d] All of above

343. The acidic igneous rocks contains silica :
a] > 65% b] < 65%
c] 60 - 65% d] None of above

344. The acidity of H^+ absorbed on solid phase is known as :
a] Reserve acidity b] Active acidity
c] Both of the above d] None of the above

345. The acidity of H^+ in aquous phase of soils is called :
a] Reserve acidity b] Active acidity
c] Permanent charge d] None of these

GK **Natyashastra of Bharat is the earliest known work of Sanskrit.** GS

346. The addition of excess soluble phosphate fertilizers to soils may affect plant growth by :
a] Inhibiting the growth of micro-organisms
b] Precipitating Fe and Al
c] Reducing the availability of Zn
d] Reducing the availability of Ca

347. The adverse effect of poor quality water is/will be more on :
a] Clay soil b] Loamy soil
c] Sandy soil d] None of the above

348. The alkali soil can be identified by seeing :
a] White layer of salts on the surface
b] Black colour on the surface
c] Red colour on the surface
d] None of the above

349. The amount of N_2 fixed by Clostridium is ________ mg N_2 per gm sugar.
a] 15-25 b] 7-12
c] 2-7 d] 25-50

350. The amount of P in soil and soil solution that can exchange with orthophosphate in addition to soil as measured by plant growth in soil system is termed as:
a] A-value b] L-value
c] E-value d] None of these

351. The amount of work done by the water during its movement from initial point to reference point is known as :
a] Soil water potential
b] Solute water potential
c] Matric potential
d] Pressure potential

352. The area of alluvial soils in India is :
a] 150 m. ha b] 250 m. ha
c] 110 m. ha d] 75 m. ha

353. The area of black soils in India is :
a] 54.6 m. ha b] 5.46 m. ha
c] 28.5 m. ha d] 75.0 m. ha

354. The area of peaty soils in India is about :
a] 150 sq. km. b] 250 sq. km.
c] 275 sq. km. d] 25 sq. km.

355. The area of tarai soils in India is :
a] 5.7 m. ha b] 10.7 m. ha
c] 2.7 m. ha d] 3.7 m. ha

356. The area under alluvial soil affected by salinity and alkalinity in India is :
a] 6.9 m. ha. b] 8.9 m.ha.
c] 7.9 m. ha. d] 10.5 m. ha

357. The availability of phosphorous is maximum at soil pH :
a] 5.5 - 7.5 b] 7.0 - 8.0
c] 6.0 - 8.0 d] None of the above

358. The availability of phosphorous is optimum in the pH range of :
a] <7 b] >7-8
c] 8 d] 6-8

359. The available water present in soil for better plant growth in between:
a] Field capacity and Wilting percentage
b] Wilting percentage and Hygroscopic co-efficient
c] Field capacity and Air dry soil
d] None of the above

360. The average areable surface soil may be considered to have a particle density of about :
a] 2.70 gms/cc b] 2.45 gms/cc
c] 2.64 gms/cc d] 2.66 gms/cc

361. The average chemical composition of NPK of night soil is :
a] 3.7 - 4.0 - 2.0 b] 0.5 - 0.6 - 0.5
c] 2.0 - 2.6 - 0.5 d] 5.5 - 4.0 - 2.0

362. The average soil bulk density of cultivated loamy soil is approximately :
a] 0.5 - 1.0 gm cm^{-3}
b] 1.5 - 1.65 cm^{-3}
c] 1.1 - 1.4 gm cm^{-3}
d] None of the above

363. The bacteria considered most important to bring about the conversion of NH_4 to NH_2.
a] Nitrosomonas b] Nitrobacter
c] Both a & b d] Azotobacter

364. The bacteria responsible for fixation of nitrogen in soyabean is:
a] *Rhizobium leguminoserum*
b] *Rhizobium phaseoli*
c] *Rhizobium glycicum*
d] *Rhizobium japonicum*

365. The best estimate of the available P in phosphatic fertilizers is obtained from the :
a] Water soluble fraction
b] Citrate insoluble fraction
c] Citrate soluble fraction
d] Water soluble and citrate soluble fraction

366. The biological method for determining deficiency of lime, P and K is :
a] Neubauers method
b] Aspergillus niger test
c] Azotobacter plaque test
d] None of the above

367. The black cotton soils of central India have been derived from :
a] Granites b] Granodiorites
c] Basalt d] Rhyolite

GK GS

Country with world's lowest density of 1 person/km² is Mongolia.

368. The C.E.C. of illite is about :
a] 100 m.e./100 g
b] 140 m.e./100 g
c] 40 m.e./100 g
d] 10 m.e./100 g

369. The C.E.C. of mica is about :
a] 10 m.e./100 g
b] 100 m.e./100 g
c] 50 m.e./100 g
d] 140 m.e./100 g

370. The C/N ratio in the organic matter of the furrow slice (upper 15 cm) of arable soils commonly ranges from?
a] 3:1 to 4:1 b] 6:1 to 7:1
c] 18:1 to 19:1 d] 8:1 to 15:1

371. The capacity of soils to be moulded, that is to change shape in response to stress and peep that shape when stress is removed is termed as:
a] Plasticity b] Stickiness
c] Consistency d] Cementation

372. The carbon nitrogen ratio in cultivated soil is stabilized at:
a] 20:1 to 30:1 b] 4:1 to 9:1
c] 10:1 to 12:1 d] 20:1 to 30:1

373. The cation which competes with ammonium during fixation is :
a] Potassium b] Aluminium
c] Magnesium d] None of the above

374. The CEC of kaolinite clay varies between :
a] 3 and 15 b] 16 and 40
c] 41 and 80 d] 81 and 100

375. The cellulose content in humus is about :
a] 10% b] 4%
c] 15% d] 21%

376. The characteristics of surface soil which have important bearing in soil management and crop production, is known as :
a] Soil type b] Soil varient
c] Soil series d] Soil phase

377. The choice of a chemical amendment may be influenced by the ________ required for reaction in the soil.
a] Time b] Cost
c] Quantity d] Quality

378. The colloidal surface area in the upper 15 cm of a hectare of a clay soil could be as high as Km.
a] 700 b] 7,000
c] 70000 d] 7,00,000

379. The concentration of CO_2 in the soil increases after the rains, probably because of :
a] Increase nitrification and decomposition of organic matter
b] Increases denitrification
c] Increases leaching of nitrogen of CO_2
d] None of the above

Quiz Kit!
First bank
The Bank of Hindustan was established in 1770 by Alexander and Co. of Kolkata.

380. Gypsum is also called as :
a] Kyarite b] Dolomite
c] Salenite d] Baryte

381. The concept of soil as a natural body was first developed by:
a] C.F. Marbut b] K.D. Glinka
c] V.V. Dakuchaev d] E.W. Hillgard

382. The conditions responsible for the formation of salt affected soil is :
a] High rainfall
b] Low rainfall and high evaporation
c] Irrigation with good quality water
d] All of the above

383. The contribution of hill soils in India is about.
a] 4.5 m. ha b] 10.0 m. ha
c] 1.3 m. ha d] 2.8 m. ha

384. The conversion of amonia into nitrate by micro-organisms is known as :
a] Nitrogen fixation b] Nitrification
c] Denitrification d] Loss of nitrogen

385. The crop suitable for cultivation in saline sodic soil would include :
a] Soyabean and rice
b] Soyabean and sugarbeet
c] Sugarbeet and rice
d] Trice and maize

386. The C-spacing of halloysite is :
a] 10.2 Å b] 7.39 Å
c] 14.0 Å d] None of these

387. The cuticular membranes are generally more permeable to :
a] Anions rather than cations
b] Cations rather than anions
c] Cations and anions both
d] None of the above

388. The dark coloured soils are :
a] High in Fe and Mn
b] Medium in Fe and Mn
c] Low in Fe and Mn
d] Not known

389. The diagnostic surface horizons are called :
a] Epipedon b] Pedon
c] Polypedon d] All of them

390. The diameter of most erodible soils particle is:
a] 0.2 mm b] 0.1 mm
c] 0.3 mm d] 0.4 mm

391. The diameter of water molecule is approximately :
a] 2.64 Å b] 2.54 Å
c] 2.96 Å d] 1.33 Å

GK GS

Fifty per cent of world's population lives in less than five per cent of area.

392. The direct exchange of ions between the roots and clay colloids called :
a] Soil solution theory
b] Contact exchange theory
c] Diffusion
d] None of the above

393. The discovery of oxygen was done by ________ which proved the keystone of other discoveries that went far toward unlocking the mystery of plant life.
a] John Mayow
b] John Woodward
c] Benjamin Franklin
d] Priestly

394. The dominent colour of spectral (rainbow) in soil indicates :
a] Kalolinite type of clay mineral
b] Hue
c] Value
d] Chroma

395. The earth mainly consists of :
a] Four spheres b] Three spheres
c] Five spheres d] All of above

396. The earth's element crust is a part of :
a] Hydrosphere b] Lithosphere
c] Atmosphere d] All of above

397. The elements play direct role in formation of chlorophyll are :
a] Phosphorus and Potassium
b] Calcium and Zinc
c] Nitrogen and Magnesium
d] Iron and Sulphur

398. The example of amide group nitrogenous fertilizer :
a] Ammonium sulphate
b] Urea
c] Calcium ammonium Nitrate
d] None of the above

399. The factor responsible for the formation of acid soil is :
a] Low rainfall
b] High rainfall
c] High evaporation
d] Irrigation with saline water

400. The fertilizers which have two or more than two major plant nutrient in their chemical composition are called ?
a] Double fertilizer b] Complex fertilizer
c] Mixed fertilizer d] Organic fertilizer

401. The first passage through which nutrients enters into the leaves :
a] Intercuticular passage
b] Mesophyll cells
c] Cell wall plasmic membrane
d] None of the above

402. The foliated metamorphic rocks are formed due to:
a] Pressure - metamorphism
b] Heat - metamorphism
c] Hydro - metamorphism
d] None of above

403. The following factors other than gypsum solution may effect the extraction of calcium and sulphate from a gypsiferous soil:
a] Calcium solution of gypsum
b] Exchange reactions wherein calcium replaces no cations
c] Sulphate solution from a source other than gypsum
d] None of these

404. The gaseous exchange between soil and atmosphere is the :
a] Artificial process
b] Natural process
c] Change in soil structure
d] None of the above

405. The global average of solar radiation reaching the earth is :
a] 20% b] 30%
c] 40% d] 50%

406. The harmful effect of saline soil is due to :
a] High osmotic pressure of soil solution
b] Low osmotic pressure of soil solution
c] High pH
d] Low pH

407. The head office of All India Soil Survey and Land Use Organization is at :
a] Madras b] Mumbai
c] New Delhi d] Calcutta

408. The head quarter of NRSA is at :
a] New Delhi b] Mumbai
c] Secunderabad d] Banglore

409. The highest nitrogen content among the commonly used nitrogenous fertilizers is embodied by:
a] Anhydrous ammonia
b] Urea
c] Ammonium nitrate
d] Calcium ammonium nitrate

410. The horizons which make up the profile of a forest soil would include:
a] A, B and C
b] A, C and O
c] B and O
d] A, B, C and O

411. The illuviation zone of soil is :
a] Leaching b] Accumulation
c] Hard crust d] None of above

GK GS
Bacteria is used to produce vinegar industrially.

412. The important function of nitrogen is :
a] To increase growth and green colour in plants
b] To increase root development
c] To help in starch formation
d] To increase disease resistance

413. The inherent capacity of a soil to supply nutrients to plants in available and balanced proportion can be represented by:
a] Soil productivity b] Soil taxonomy
c] Edaphology d] Soil fertility

414. The light coloured soils are :
a] High in Fe and Mn
b] Low in Fe and Mn
c] Negligible in Fe and Mn
d] Not known

415. The lithosphere is mainly made up of :
a] Igneous rocks
b] Sedimentary rocks
c] Metamorphic rocks
d] All of rocks

416. The losses of ammonia usually become greater with____________ soil.
a] Increase in soil pH
b] Decrease in soil pH
c] Increase in soil temperature
d] Increase in soil calcium

417. The main characteristics which render the alkali soils unfavourable for crop growth is :
a] Excessive concentration of soluble salts
b] Low pH
c] Poor physical conditions
d] All of the above

418. The main source of heat energy for soil is :
a] Infra red radiations b] Ultra violet radiations
c] Solar radiations d] None of the above

419. The major nutrients are :
a] N, P, K b] Fe, Mn, Zn
c] Co, Cl, Na d] Cu, Mo, B

420. The manifesation of the physical forces of cohesion and adhesion acting within the soil at various moisture contents is called :
a] Soil plasticity b] Soil consistency
c] Soil elasticity d] All of these

421. The mass per unit volume of the solid mass of the soil excluding pore space is known as :
a] True density b] Bulk density
c] Both a & b d] None of the above

422. The maximum moisture is available to plants at :
a] Field capacity
b] Wilting point
c] Hygroscopic coefficient
d] Saturation

423. The maximum phosphorus availability in most of the soils in the pH range:
a] 4.0-4.5 b] 6.0-6.5
c] 7.0-7.5 d] None of these

424. The *maximum* size limit of clay fraction is considered to be
a] 0.02 mm b] 0.002 mm
c] 0.0002 mm d] None of the above

425. The method of survey adopted for intensive cultivation is :
a] Reconnaissance survey
b] Detailed survey
c] Detailed and reconnaissance survey
d] All of the above

426. The microbes which can survie at the temperature 25-30 °C is termed as :
a] Thermoduric b] Mesophylic
c] Psychrophylic d] Thermophytic

427. The mineral gypsum have chemical composition :
a] $CaSO_4$ b] $CaSO_4.2H_2O$
c] $CaSO_4.5H_2O$ d] None of above

428. The mineral limonite have chemical composition :
a] Fe_2O_3 b] $2Fe_2O_3.3H_2O$
c] Fe_3O_4 d] None of above

429. The mineralization of organic phosphorus has been studied in relation to:
a] C/N/P b] C/N/S
c] C/N/K d] C/N/Ca

430. The molecular transfer of gases due to concentration of partial pressure gardient in the soil is termed as :
a] Diffusion b] Mass flow
c] Both of the above d] None of the above

431. The most abundant mineral present in the earth's crust is :
a] Quartz b] Felspar
c] Mica d] Hornblende

432. The most important silicate clays are known as:
a] Photosilicates b] Phyllosilicates
c] Autosilicates d] Simply silicates

433. The N_2 fertilizer manufactured in India which occupies the first place in production today is :
a] $(NH_4)_2 SO_4$ b] $NH_4 H_2PO_4$
c] Urea d] CAN

434. The N_2 fixation by *Azqtobacter* and *Azospirillum* in the cereals is called:
a] Symbiotic N_2 fixation
b] Non-symbiotic N_2 fixation
c] Bothof these
d] None of these

435. The N_2 fixation by *Rhizobium* species in legume is called :
a] Symbotic N_2 fixation
b] Non-symbotic N_2 fixation
c] Both of these
d] None of these

436. The N_2 fixation in pea may be upto ________ kg N_2/ ha/year.
a] 80-150 b] 90-110
c] 15-75 d] 50-100

437. The N_2 fixation in soyabean may be upto ________ kg N_2/ha/year.
a] 100-175 b] 60-100
c] 65-115 d] 40-90

438. The N_2 fixation in the roots of Alfalfa is ________ kg N_2/ha/year:
a] 100-150 b] 125-300
c] 50-75 d] None of these

439. The N_2-fertilizer of choice for tea is:
a] CAN b] Urea
c] $(NH_4)_2SO_4$ d] NH_4NO_3

440. The natural soil aggregates are called as :
a] Concretion b] Mottling
c] Peds d] Soil separates

441. The negative logarithm of H^+ ion concentration (activity) in the soil-solution:
a] Acidity b] Salinity
c] Alkalinity d] pH

442. The Neubauer seedling method is used to evalutate the soil fertility. The test plants used in this method is :
a] Barley seedlings
b] Wheat seedlings
c] Rye or Oat seedlings
d] All of the above can be used

443. The new comprehensive soils classification is based on :
a] Genetic characters
b] Physical and chemical properties of soils
c] Both a & b
d] None of above

444. The number of essential elements required for the growth of most higher plant is :
a] 3 b] 6
c] 8 d] 16

445. The number of functional plant nutrient are :
a] 6 b] 20
c] 16 d] 10

446. The oblique type of soil structure have about :
a] 25-30% pore space b] 40-45% pore space
c] 5-10% pore space d] 15-20% pore space

447. The organic matter content of Indian peat and marshey soils is :
a] 10-40% b] > 40%
c] >10% d] None of the above

448. The organic matter of soils is ordinarily obtained by multiplying the organic carbon content by :
a] 1.52 b] 1.62
c] 1.72 d] 1.82

449. The organism which can survive at the temperature 10-15 °C, is called:
a] Psychrophylic b] Mesophylic
c] Thermoduric d] Thermophilic

450. The P_2O_5 content is P in the plough layer of 1 ha of soil containing 0.1%. P_2O_5 is :
a] 430 kg b] 860 kg
c] 1290 kg d] 1500 kg

451. The pattern of occurance of various soil taxonomic units on the detailed map is called :
a] Soil complex b] Soil association
c] Soil varient d] Soil phase

452. The percentage of soil water held with water potential - 15 bars is called as :
a] Hygrosopic coefficent
b] Wilting point
c] Field capacity
d] None of the above

453. The pH of a solution, containing $1x\ 10^{-7}$ gm ion/ml. will be :
a] 6 b] 7
c] 8 d] 9

454. The phosphate potential is denoted as
a] $1/2\ p\ Ca + pH_2PO_4$
b] $PH-1/2\ p\ Ca$
c] $RT\ in\ \frac{K^+}{\sqrt{Ca^{++}}}$
d] $p\ Ca + pH_2PO_4$

455. The plant nutrient when supplied in excess prolongs the growing period and delay crop maturity is:
a] Iron b] Phosphorus
c] Nitrogen d] Potassium

456. The porosity will be more in soils having particle size of :
a] 0.005 mm b] 0.02 mm
c] 2.0 mm d] 0.002 mm

457. The positive and high value of redox potential indicates:
a] Reducing conditions
b] Strong oxidizing conditions
c] Equally reducing and oxidizing conditions
d] None of these

458. The presence of autotrophic mycorrhiza enhances the uptake of nutrients in general, but particularly of:
a] Iron b] Molybdenum
c] Phosphorus d] Nitrogen

459. The prime objective of conservation tillage is:
a] Removal of plant residues
b] Complete retention of plant residues
c] Partial retention of plant residues
d] None of these

GK GS
Forced labour was known as Vishti.

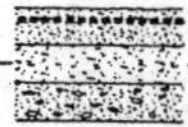

460. The principal cations that affect the availability/ fixation of phosphorus in alkali soils are :
a] Ca and Mg
b] Na and K
c] Fe and Al
d] None of the above

461. The process of ammonification is often ________ by the use of pesticides.
a] Harmed b] Benefited
c] Not affected d] None of these

462. The process responsible for the formation of sodic soil is :
a] Salinization b] Alkalinization
c] Podzolization d] Calcification

463. The pyrite is mostly found in :
a] Maharashtra b] Andhra Pradesh
c] Bihar d] Rajasthan

464. Zinc sulphate is generally take more production under:
a] Saline soilb] Acid soil
c] Water logged soils d] Alkali soils

465. The quantity of solid phase phosphorus that acts as reserve is often referred to as:
a] Intensity factor b] Quality factor
c] Quantity factor d] None of these

466. The quantity of urea needed for applying 115 kg nitrogen to maize will be (approx.):
a] 100 kg b] 250 kg
c] 200 kg d] 250 kg

467. The ratio of hthermal conductivity and heat capacity is known as :
a] Thermal diffusivity
b] Heat capacity
c] Thermal capacity
d] None of these

468. The relative amounts of air and water in the pore spaces :
a] Always remain constant
b] Fluctuate continuously
c] Half filled and half empty
d] All of the above

469. The relative proportions of sand, silt and clay in the soil is known as:
a] Soil structure b] Soil consistence
c] Soil texture d] None of the above

470. The representation of surface feature on map and their identification, is called :
a] Soil legand b] Soil map
c] Soil varient d] None of them

471. The residual effect of ammonical nitrogenous fertilizer in soil is :
a] Alkaline b] Neutral
c] Acidic d] None of the above

Did You Know ?

First nuclear power plant

The Tarapore nuclear power plant with an installed capacity of 380 MW was commissioned in 1969. When it went operational then plant had two enriched uranium-fuelled boiling water reactor units of 210 MW each.

472. The rice crop tolerates submergence because :
a] The plant does not need O_2
b] The plant has mechanism to transport O_2 through leaves
c] Bacteria keep the rhizosphere oxidised
d] O_2 penetrating to root zone through air movement is sufficient

473. The rock 'Quartzite' is a :
a] Sedimentary rock b] Igneous rock
c] Metamorphic rock d] None

474. The rock "Granite" is a :
a] Sedimentary rock b] Metamorphic rock
c] Igeneous rock d] None of above

475. The rock which is formed by sedimentation of primary rock, is called :
a] Igneous rock b] Metamorphic rock
c] Sedimentary rock d] None of these

476. The rock which is formed under high temperature and pressure and by sudden geological tectonic movement is called :
a] Igneous rock b] Sedimentary rock
c] Metamorphic rock d] None of these

477. The saline soil can be *managed* by :
a] Addition of gypsum
b] Addition of lime
c] Addition of pyrite
d] Leaching of salts

478. The sodicity of water can be expressed by :
a] ECE b] SAR
c] ESP d] All of the above

479. The soil layer of maximum leaching is indicated by the symbol
a] A_1 b] A_2
c] B_1 d] B_2

480. The soil solum includes :
a] A horizon b] B horizon
c] A and B horizon d] None of the above

481. The soil temperature under field conditions can be altered by suitable cultural practice such as :
a] Irrigation b] Mulching
c] Tillage d] All of the above

GK GS

Himalayas represent the youngest and highest folded mountains in the world.

482. The soils of ____________ zones can supply adequate potassium for many years even under irrigations.
a] Temperate b] Humid
c] Arid d] Sub-tropical

483. The species of Rhizobium present in the root nodules of soyabean is:
a] *R. japonicum*
b] *R. leguminosarum*
c] *R. trifolii*
d] *R. phaseoli*

484. The succulence and lodging in plants is due to 'excessive' application of :
a] Zinc b] Phosphorus
c] Nitrogen d] Potassium

485. The sufficiency limit of iron is normally between:
a] 10-45 ppm b] 50-250 ppm
c] 350-400 ppm d] 500-750 ppm

486. The systematic examination, description, classification and mapping of soil of an area is known as :
a] Soil survey
b] Soil mapping
c] Soil classification
d] All of the above

487. The technique used for minimizing the loss of nitrogen in paddy fields comprises:
a] Use of neem cake
b] Use of pre -incubated urea
c] Use of mud ball
d] All of these

488. The tendency of the young upper leaves to remain as the lower leaves yellow or die is an indication of nitrogen:
a] Immobility b] Mobility
c] Stability d] None of the above

489. The texture of surface soil or surface horizon is called:
a] Soil phase b] Soil type
c] Soil varient d] None of them

490. The tropical soils high in hydrous oxides of iron and aluminium embody ___________ to torrential rains.
a] Susceptibility b] Resistance
c] Both of these d] None of these

491. The unconsolidated material on the underlying rock is called :
a] Regolith b] Soil
c] Solum d] Earth

492. The unit of fertilizer or any other growth factor which is necessary to produce the yield which in 50% of maximum possible yield is called :
a] Optimum unit b] Standard unit
c] Baule unit d] Fertilizer unit

493. The value above ____________ ppm are associated with boron toxicity.
a] 150 b] 200
c] 250 d] 300

494. The water held with a water potential greater than -1/3 bar is known as :
a] Field capacity b] Gravitational water
c] Available water d] Unavailable water

495. The weathered material transported by means of waters and deposited in river basins are termed as:
a] Alluvial b] Marine
c] Lacustrine d] None of above

496. The weathering causes :
a] Disintegration and decomposition of rocks
b] Disintegration and decomposition of rocks
c] Disintegration and decomposition of rocks and minerals
d] All of above

497. The zinc deficiency in rice causes disease:
a] Khaira
b] Bacterial leaf streak
c] Bacterial leaf blight
d] Sheath blight

498. There is a net increase in the mineralization of orgnaic nitrogen in soil when :
a] C : N = 20-30 : 1
b] C : N = < 20 : 1
c] C : N = > 30 :1
d] Nitrogen is taken up by the crops

499. Thermal conductivity of different textural classes of the soils follows the following order :
a] Sand < Loam < Clay < Peat
b] Sand > Loam > Clay > Peat
c] Peat = Clay = Loam = Sand
d] None of the above

500. Til cake contains about :
a] 6.0% N b] 2.5% N
c] 10.2% N d] 9.8% N

501. To check nitrification, nitrification inhibitors are most useful in:
a] Paddy b] Cotton
c] Maize d] None of these

502. To prevent denitrification lossses of nitrogen in paddy field, nitrogen should be applied in the:
a] Reduced zone b] Oxidized zone
c] Hydronized zone d] Calcified zone

503. Tolerance to soil salinity of the given set of crops is such that
a] Barley > wheat > upland paddy > beans
b] Barley > wheat > beans > upland paddy
c] Wheat > barley > beans > upland paddy
d] Wheat > barley > upland paddy > beans

504. Total area of India under desert soils is :
a] 14.2 m. ha b] 18.2 m. ha
c] 24.2 m. ha d] 28.2 m. ha

GK GS

About 30 dB level of sound is produced when two persons in conversation.

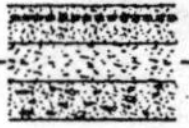

505. Transport of P_2O_5 from the cytosol (pH about 7) into vacuoles (pH nearly 5.5) involves
a] H_2PO_4 alone b] H_2PO_4 alone
c] PO_43 alone d] H_2PO_4 and HPO_4

506. Tricalcium phosphate is soluble in:
a] Water b] Weak acid
c] Strong mineral acid d] None of the above

507. Type of water vapour movement which occurs in soils is :
a] Internal b] External
c] Both of these d] None of these

508. Under tropical condition, the volatilization losses of nitrogen is:
a] 20-50% b] 10-20%
c] 40-80% d] None of above

509. Unit of measuring electrical conductivity is:
a] Micro-gram/sq.cm.
b] Angstran/cm.
c] Millimhos/cm
d] None of these

510. Upper plastic limit is also know as:
a] Liquid limit b] Lower limit
c] Index of plasticity d] None of these

511. Upward translocation of fluid in xylem takes place due to:
a] Pull of transpiration stream
b] Pumping action of guard cells
c] Suction by sun-rays
d] All above

512. Urea contain :
a] 24% nitrogen b] 60% nitrogen
c] 16.0 nitrogen d] 46.0 nitrogen

513. Urea is mineralized in the presence of enzyme :
a] Co-enzyme b] Peptase
c] Urease d] All of the above

514. Use of materials like S, polyform UF, isobutylidene diurea *etc.* improves fertilizer use efficiency, when such additions are used, the N-fertilizers are termed as :
a] Prilled urea
b] Controlled release N-fertilizers
c] Slow nitrifying N-fertilizers
d] Complex fertilizers

515. V= 2rg (PP-Pm)/gn is :
a] Darcy's law b] Stoke's law
c] Beer's law d] None of these

516. Vermicullite is :
a] 2:1 type mineral b] 1:1 type mineral
c] 1:1 type mineral d] None of the these

517. Volume of soil under the influence of roots of growing plants is known as:
a] Phyto-bio-sphere b] Phyto-bio-zone
c] Rhizosphere d] Phytozone

518. Volumetric heat capacity can be calculated by multiplying specific heat with the corresponding :
a] Volume b] Mass
c] Density d] Thermal conductivity

519. Water drains by gravitational force from pores larger than :
a] About 10 - 20 microns
b] About 20 - 30 microns
c] About 30 - 60 microns
d] None of the above

520. Weathered material transported by means of force of gravity the deposition is termed as :
a] Colluvial b] Alluvial
c] Cumulose d] Aeolion

521. What culture should be given priority in groundnut cultivation?
a] *Azosprililla* b] *Mycorrhiza*
c] Phosphobacteria d] *Rhizobia*

522. What is the effect of application of P_2O_5 fertilizer to a green manure leguminous crop on the availability of phosphorus to the succeeding crop ?
a] Greater amount of phosphorus is taken up and converted into organic form, which is released on decomposition of green matter
b] Green manure increases the availability in organic form to the next crop
c] It does not affect the availability of phosphorous to succeeding crop
d] It fixes the P_2O_5 through humus and reduces the avilability

523. What is the P_2O_5 content of various rockphosphates?
a] 20-35% b] 35%
c] 11-20% d] 18-20%

524. When a nutrient takes active part in growth and development of plant called :
a] Beneficial b] Essential
c] Required d] None of above

525. When a plant is under soil moisture stress, the roots:
a] Shrink in size
b] Expand in size
c] Remain unaffected
d] Both a + b

526. When emerging leaves of maize show yellow to white bleached bands in lower parts, it shows the deficiency of:
a] Zinc b] Iron
c] Copper d] Boron

527. When organic matter with wide C/N (70 : 1) is added to the soil, the net effect is the disappearance of mineral nitrogen due to :
a] Mineralization b] Immobilization
c] N_2 transformation d] N_2 reduction

528. Which nitrogenous fertilizer will you recommend for paddy crop ?
a] Nitrate fertilizer
b] Amide fertilizer

GK GS

Agar-agar is a red algae.

c] Ammonical nitrate ferlitizer
d] All the above

529. **Which nutrient is maximum given for raising the field crops :**
a] Nitrogen b] Potash
c] Phosphorus d] Calcium

530. **Which of the biochemically weathered constitutes soil?**
a] Upper
b] Middle portion
c] Lower portion
d] Portion adjacent to bedrock

531. **Which of the following are the slow release fertilizers?**
a] Surface coated urea
b] Neem cake and super grain urea
c] N-serve
d] All of the above

532. **Which of the following element is mobile in soil ?**
a] Ca b] Mg
c] K d] N

533. **Which of the following elements regulate oxidation-reduction reactions in plant?**
a] Hydrogen, oxygen and carbon
b] Nitrogen, phosphorus and sulphur
c] Potassium, calcium, sodium, and magnesium
d] Iron, manganese, molybdenum, copper, boron and zinc

534. **Which of the following fertilizer has maximum plant nutrient ?**
a] D.A.P. b] T.S.P.
c] Murate of Potash d] Ammonium Nitrate

535. **Which of the following is a high salt tolerant crop ?**
a] Lucerne b] Moong
c] Urd d] Arhar

536. **Which of the following is a igneous rock ?**
a] Quartz b] Marble
c] Mica schist d] Slate

537. **Which of the following is a nitrification inhibitor?**
a] A.M. (2, amino-4 6 methyl pyridine)
b] Thio-urea
c] Sulfur coated urea
d] CBC (β-chloroethyl, trimethyl ammonium chloride)

538. **Which of the following is considered as bio-fertilizer?**
a] Farm yard manure
b] Compost
c] Green manure
d] Blue-green algae

539. **Which of the following is correct explanation of ESP?**
a] Exchangeable Na/CEC
b] Exchangeable Na x 100/CEC
c] Exchangeable Na x 100
d] CEC x 100/Exchangeable Na

540. **Which of the following is harmful to the higher plants?**
a] Phenol carboxylic acid
b] Vanillic acid
c] Dihydroxystearic acid
d] All of these

541. **Which of the following is not a high salt tolerant crop?**
a] Sugarcane b] Barley
c] Sugarbeet d] Maize

542. **Which of the following is not a igneous rocks ?**
a] Granite b] Basalt
c] Dibase d] Quartzite

543. **Which of the following is not a kind of soil structure?**
a] Granular b] Columner
c] Compact d] None of these

544. **Which of the following is not a metamorphic rock ?**
a] Quartzite
b] Marble
c] Conglomerate schists
d] Sand stone

545. **Which of the following is not a method of placement of fertilizers?**
a] Plough sole b] Deep
c] Side d] None of these

546. **Which of the following is not a method of soil moisture measurement ?**
a] Gravitational method
b] Tensiometer
c] Gypsum block method
d] None of these

547. **Which of the following is not a primary major nutrient?**
a] N b] P
c] K d] S

548. **Which of the following is not a regional correlation centre of A.I.S.S.L.U.O ?**
a] Nagpur b] Bangalore
c] Hyderabad d] Jaipur

549. **Which of the following is not a secondary major nutrient ?**
a] S b] P
c] Ca d] Mg

550. **Which of the following is not a soil edaphic factor affect the plant growth ?**
a] Soil texture b] Soil structure
c] Soil temperature d] None of these

551. **Which of the following is wrong?**
a] Iron, manganese, copper, zinc, boron, molybdenum, chlorine and cobalt are called micronutrients.
b] Nitrogen, phosphorus and potassium are primary nutrients.
c] Calcium, magnesium and sulphur are secondary nutrients.
d] Nitrogen is converted into nitrate by earth worms.

552. **Which of the following methods is most useful to increase the availability of soil applied phosphorus?**
a] Broadcast
b] Band placement
c] Broadcast and soil incorporation
d] None of these

553. **Which of the following nutrient elements are regarded as energy exchangers?**
a] Hydrogen and oxygen
b] Carbon and nitrogen
c] Potassium and magnesium
d] Iron, manganese and molybdenum

554. **Which of the following organism is responsible for decomposition of organic matter?**
a] Algae b] Bacteria
c] Fungi d] Microorganism

555. **Which of the following primary minerals has no good water and nutrient holding capacity ?**
a] Quartz b] Feldspar
c] Mica d] All of these

556. **Which of the following rotations is likely to leave soil richer in organic matter?**
a] Continuous sorghum
b] Continuous maize
c] Continuous oats
d] Maize-oats-clovers

557. **Which of the following soil group covers largest area of India ?**
a] Saline and Alkaline soils
b] Alluvial soils
c] Red soils
d] Black cotton soils

558. **Which of the following soil texture is more heavier than others ?**
a] Loamy sand b] Silty clay loam
c] Loam d] Silty loam

559. **Which of the following soil texture is more lighter than others ?**
a] Silty clay loam b] Clay loam
c] Sandy loam d] Loamy sand

560. **Which of the following soils has higher buffering capacity ?**
a] Sandy soils b] Loamy soils
c] Loamy sand d] Clay soils

561. **Which of the following state has maximum black soils**
a] U.P. b] Maharashtra
c] Haryana d] Bihar

562. **Which of the following statement is correct?**
a] Blue-green algae causes plant diseases
b] Blue-green algae fixes atmospheric nitrogen in soil
c] Blue-green algae grows best in maize crop
d] Dry conditions are good for growth of blue-green algae

You Don't Know!
First encyclopedia
The country's first encyclopedia was compiled in the Marathi language by S.V. Ged. It was brought out in 28 volumes and was titled *Maharashtriya Gyana Kosu.*

563. **Which of the following statements is more correct as regards to "luxury consumption" of K found in minute?**
a] Maize has mechanism to increase yields by taking excess K above the optimum requirement.
b] Maize takes excess K above the optimum requirement even though growth and yield are not promoted.
c] Maize takes excess K above the optimum requirement the concen-tration which becomes toxic to plants:
d] None of the above

564. **Which one of the following concentrated organic manures has maximum P_2O_5 content?**
a] Fish meal b] Stera meal
c] Poultry manure d] Guano

565. **Which one of the following fertilizers is least hygroscopic?**
a] Urea b] $(NH_4)_2SO_4$
c] DAP d] CAN

566. **Which one of the following irrigation methods is best suited for an undulating topography?**
a] Flow irrigation b] Check basin method
c] Sprinkler irrigation d] Furrow irrigation

567. **Which one of the following is not a component of the moisture potential in soils?**
a] Matric potential
b] Turgor potential
c] Gravity potential
d] Osmotic potential

568. **Which one of the following is not a soil forming factor:**
a] Climate b] Organisms
c] Relief d] Moisture

569. **Which one of the following minerals is least weatherable in soil?**
a] Olivine b] Quartz
c] Calcite d] Biotite

570. **Which one of the following oil-cakes, in addition to its manurial value, acts also as a nitrification inhibitor?**
a] Mustard b] Groundnut
c] Neem d] Til

571. **Which one of the following organic matter has the narrowest C/N ratio?**
a] Biogas slurry b] FYM
c] Compost d] Groundnut cake

Apaddharma was the conduct to be followed in period of distress.

572. Which one of the following pairs is correct ?

a] Urea - 42%

b] DAP -20%

c] $(NH4)_2SO_4$ - 21.5%

d] MoP - 47.5%

573. Which one of the following pairs is not correctly matched?

a] Expanding type clay minerals – Higher water holding capacity

b] High calcium content in soil –Iron chloride

c] Water inundated soils – High nitrogen mineralization

d] Sandy soils – Split application of fertilizers

574. Which one of the following pairs of elements and the minerals from which they are obtained is not correctly matched?

a] Sulphur - Pyrite

b] Boron - Siderite

c] Zinc - Calamine

d] Copper - Malachite

575. Which one of the following sets of soil conditions is more suitable for plant growth?

a] Sandy soils with high aeration and low water holding capacity

b] Clay soils with higher water holding capacity and low aeration

c] Organic soils with high water holding capacity and low aeration

d] Loamy soils with medium aeration and medium water holding capacity

576. Which one of the following statements is correct ?

a] Rhizobia fix N_2 in soil non-symbiotically

b] Azotobacters fix N_2 symbiotically

c] BGA fix N_2 in paddy fields under submerged condition

d] Closterdia fix N_2 symbiotically

577. While classifying the soil according to textural class in equilateral triangle, suggested by the USDA, % of clay is plotted parallel to :

a] Silt b] Sand

c] Clay d] Loam

578. White, irregular spots between veins of upper leaves of maize plant shows deficiency of:

a] Copper b] Molybdenum

c] Manganese d] Boron

579. Who is the Father of the soil testing?

a] S. Larsen b] M.L. Jackson

c] M.L. Troug d] None of the above

580. Wind laden with soil particles embodies the abrasive action which is:

a] Greatly increased

b] Greatly decreased

c] Absolutely unaffected

d] None of these

581. Zinc deficiency occurs in crops growing in soils having high phosphorus:

a] Light soils

b] Calcareous soils

c] Soils low in phosphorous

d] Soil having pH 7.5

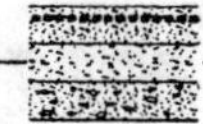

Matching

1. **Match nutrients with soil testing method.**

a]	Available N	i]	Morgan's method
b]	Available P	ii]	Chronic acid digestion method
c]	Available K	iii]	Olsen's method
d]	Organic matter	iv]	Alkaline permanganate method

2. **Match biofertilizers with its properties.**

a]	Azospirillum	i]	Helps in phosphorus uptake
b]	Azotobacter	ii]	Floating aquatic fern
c]	VAM	iii]	Grows and inside the roots of graminaceous crops
d]	Azolla	iv]	Free-living nitrogen fixing bacterium

3. **Match Crop with event.**

a]	Sunflower	i]	Gypsum application
b]	Groundnut	ii]	Potash
c]	Coconut	iii]	Poor seed setting
d]	Soyabean	iv]	Phosphate application

4. **Match host legume with *Rhizobium* sp.**

a]	Fenugreek	i]	*R. meliloti*
b]	Clovers	ii]	*R. phaseoli*
c]	Peas	iii]	*R. trifolii*
d]	Beans	iv]	*R. leguminosarum*

5. **Match microbial reactions with soil organisms.**

a]	Sulphate reduction	i]	Thiobacillus
b]	Sulphite oxidation	ii]	Desulfotomaculum
c]	Nitrite oxidation	iii]	Rhodospirillum
d]	Nitrogen fixation	iv]	Nitrobacter

6. **Match the following.**

a]	Gypsum	i]	Magnesium fertilizer
b]	17 : 17 : 17	ii]	Micronutrient fertilizer
c]	Serpentine	iii]	Calcium fertilizer
d]	Chelates	iv]	Compound/complex fertilizer

7. **Match the reactants used for the preparation of fertilizers with fertilizer produced.**

a]	Rock phosphate, phosphoric acid	i]	Urea
b]	Rick phosphate, phosphoric acid, nitric acid, ammonia	ii]	$(NH_4)_2SO_4$
c]	Ammonia, gypsum, CO_2	iii]	Nitrophosphate
d]	Ammonia CO_2	iv]	Concentrated superphos-phate

8. **Match principal soils of India with its taxonomic soils order.**

a]	Black soils of Maharashtra	i]	Alfisol
b]	Laterite soils of Kerala	ii]	Entisol
c]	Red soils of Chotanagpur	iii]	Ultisol
d]	Gangetic alluvial soils	iv]	Vertisol

9. **Match soil bacteria with its microbiological reactions.**

a]	Nitrobacter	i]	Symbiotic fixation of atmospheric N_2
b]	Azotobacter	ii]	Non-symbiotic fixation of atmospheric N_2
c]	Nitrosomonas	iii]	Conversion of ammonia into nitrite in soil
d]	Rhizobium	iv]	Oxidation of nitrite to nitrate

10. **Macth fertilizer with its equivalent acidity.**

a]	Urea	i]	148
b]	$(NH_4)_2SO_4$	ii]	128
c]	NH_4Cl	iii]	84
d]	Anhydrous NH_3	iv]	64

GK GS

Diatoms are used as mild abrasive in polishes.

11. Match the quantities of the following essential elements found in the soil solution in arid regions.

a]	N	i]	15-40 kg/ha
b]	P	ii]	0.1-0.2 kg/ha
c]	K	iii]	140-280 kg/ha
d]	Ca	iv]	5-20 kg/ha

12. Match the commonly dominant minerals in different types of sedimentary rocks.

a]	Limestone	i]	Clays
b]	Sandstone	ii]	$CaCO_3$ (Calcite)
c]	Shale	iii]	SiO_2 (Quartz)
d]	Conglomerate	iv]	Variable

13. Match the chemical formula with the primary mineral given below.

a]	Muscovite	i]	$K AlSi_3O_8$
b]	Biotite	ii]	$Ca Al_2Si_2O_8$
c]	Orthoclase	iii]	$K Al(Mg,Fe)_3Si_3O_{10}(OH)_2$
d]	Anorthite	iv]	$K Al_3Si_3O_{10}(OH)_2$

14. Match the horizon of a soil with the striking characteristics.

a]	A-horizon	i]	Eluviation of clay, iron and aluminium oxide
b]	E-horizon	ii]	Illuviation of minerals
c]	B-horizon	iii]	Unconsolidated material underlying the solum
d]	C-horizon	iv]	Strong mixture of humified organic matter

15. Match the nomenclature in relation to different categories employed in the soil classification system.

a]	Mollisols	i]	Great group
b]	Aquolls	ii]	Sub-group
c]	Argiaquolls	iii]	Order
d]	Typic Argiaquolls	iv]	Sub-order

16. Match the optimum temperature range with the growth stage of the crop given below.

a]	Maize seedlings	i]	16-21°C
b]	Maize root growth	ii]	24°C
c]	Potato-tuber development	iii]	7-10°C
d]	Best wheat growth	iv]	25°C

17. Match the ionic radius with the silicate clay.

a]	Si^{4+}	i]	0.074 nm
b]	Zn^{2+}	ii]	0.099 nm
c]	Ca^{2+}	iii]	0.155 nm
d]	OH^-	iv]	0.042 nm

18. Match cation exchange capacity with soil types.

a]	Sand	i]	18
b]	Sandy loams	ii]	25
c]	Loams and silt loams	iii]	5
d]	Silty clay loams	iv]	12

19. Match the biomass of soil flora and fauna based on the following estimates.

a]	Actinomycetes	i]	1120-11,200 kg HFS
b]	Fungi	ii]	11-110 kg HFS
c]	Nematodes	iii]	110-1100 kg HFS
d]	Earth worms	iv]	450-4500 kg HFS

20. Match the C/N ratios given below with the source.

a]	Legumes	i]	4:1 to 9:1
b]	Strawy residues	ii]	400:1
c]	Sawdust	iii]	100:1
d]	Microbe-bodies	iv]	20:1 to 30:1

21. Match the representative maximum CEC Cmol/kg of the following colloids.

a]	Organic humus	i]	120
b]	Vermiculite	ii]	3
c]	Kaolinite	iii]	300
d]	Hydrous oxide	iv]	5

Parandatta, the governor of Skandgupta got the Sudarshan lake repaired.

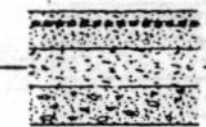

22. **Match the cross-inoculation groups of legumes and the associated *Rhizobia*.**

a]	*R. japonicum*	i]	Alfalfa
b]	*R. phaseoli*	ii]	Soyabean
c]	*R. leguminosarum*	iii]	Lentil
d]	*R. meliloti*	iv]	Runner bean

23. **Match the genus with family of the following actinomycetes-nodulated non-legume angiosperms.**

a]	Casuarina	i]	*Rhamnaceae*
b]	Ceanothus	ii]	*Betulaceae*
c]	Alnus	iii]	*Myricaceae*
d]	Myrica	iv]	*Casuarinaceae*

24. **Match the following antagonistic relationships between various essential elements.**

a]	Excess of Cu or SO_4	i]	Mn-uptake adversely affected
b]	Excess of Zn, Mn, Cu, Mo	ii]	Reduced B-uptake
c]	Excess of Na or K	iii]	Adverse effect on Mo
d]	Excess of lime	iv]	Fe-deficiency

25. **Match the pesticide-class with the target group.**

a]	Insecticides	i]	Weeds
b]	Fungicides	ii]	Diseases
c]	Herbicides	iii]	Insects
d]	Nematicides	iv]	Nematodes

26. **Match the following herbicides with the trade name.**

a]	Dalapon	i]	Karmex
b]	Chloramben	ii]	Lasso
c]	Alachlor	iii]	Dowpon
d]	Diuron	iv]	Amiben

27. **Match the common designations of the following herbicides with the common name.**

a]	Propachlor	i]	Treflan
b]	Trifluralin	ii]	Dacthal
c]	DCPA	iii]	Eptam
d]	EPTC	iv]	Ramrod

28. **Match the following observations with the observer.**

a]	Water is a principal nourishment of plants	i]	Jethro Tull (1674-1740)
b]	Plants embody salts, spirits, earth and oil, formed from water	ii]	Robert Boyle (1627-1691)
c]	Saltpeter (KNO_3) and not water was the principle of vegetation	iii]	Francis Bacon (1561-1624)
d]	Soil pulverization provides proper pabulum for the growing plants	iv]	J.R. Glauber (1604-1668)

29. **Match the conversion factor with the form to convert from one to other form.**

a]	Ca to CaO	i]	2.50
b]	Ca to $Ca(OH)_2$	ii]	1.40
c]	Ca to $CaCO_3$	iii]	1.67
d]	Mg to MgO	iv]	1.85

30. **Match the conversion factor with the form to convert from one to other form.**

a]	Mg to $Mg(OH)_2$	i]	3.50
b]	Mg to MgO_3	ii]	1.67
c]	Mg to Ca	iii]	4.17
d]	Mg to $CaCO_3$	iv]	2.42

31. **Match the conversion factor with the form to convert from one to other form.**

a]	MgO to $CaCO_3$	i]	1.19
b]	$MgCO_3$ to $CaCO_3$	ii]	3.50
c]	Ca to CaO	iii]	2.50
d]	Mg to $MgCO_3$	iv]	1.85

GK GS

Shivalik comprises the southern most range of Himalayas.

32. Match the following with the formula.

a]	Hydrated magnetite	i]	$Fe_3PO_4 . 8H_2O$
b]	Black hydroilite	ii]	$Fe_3O_4 . nH_2O$
c]	White siderite	iii]	$F_3 S. nH_2O$
d]	Blue vivianite	iv]	$Fe\ CO_3$

33. Match the following terms used to designate the status of essential nutrient elements in plants.

a]	Deficient	i]	Level lower than needed for growth, an imbalance with another nutrient, deficiency symptoms seldom seen
b]	Insufficient	ii]	Level sufficiently high to cause shortage in corresponding nutrient
c]	Toxic	iii]	Low concentration severely limiting yield, show more or less distinct deficiency symptoms, may lead to death
d]	Excessive	iv]	Concentration sufficiently high to reduce growth severely, may lead to plant death

34. Match the following zinc deficiency diseases as observed in various crops given below.

a]	White bud	i]	Rice
b]	Little leaf	ii]	Citrus crops
c]	Mottle leaf (frenching)	iii]	Maize and sorghum
d]	Khaira	iv]	Cotton

35. Chloride containing fertilizers have been found to reduce the incidence of some diseases in certain crops, match them suitably with the crop.

a]	Root-rot	i]	Maize
b]	*Fusarium* dryland rot	ii]	Spring barley
c]	Stalk rot	iii]	Wheat heads
d]	*Septoria*, leaf disease	iv]	Winter wheat

36. Match the generally observed cation exchange values (expressed in meq per 100 g) with the clay mineral.

a]	Mineral colloids 1:1	i]	20-40
b]	Mineral colloids 2:1	ii]	100-300
c]	Chlorites 2:1:1	iii]	1-10
d]	Organic colloids	iv]	80-150

37. Match the phosphate compound with the mineral name.

a]	$Al_5(NH_4)_3H_6(PO_4)_8 . 18H_2O$	i]	Leucophosphite
b]	$Al_5K_3H_6(PO_4)_8 . 18H_2O$	ii]	K-leucophosphite
c]	$Al_2K(PO_4)_2OH. 2H_2O$	iii]	K-taranakite
d]	$Fe_2K(PO_4)_2OH . 2H_2O$	iv]	NH_4-taranakita

38. Match the following phosphate compound with the mineral name.

a]	$Ca_8H_2(PO_4)_6 . 5H_2O$	i]	Flurapatite
b]	$Ca_{10}(PO_4)_6(OH)_2$	ii]	Octocalcium phosphate
c]	$Ca_{10}(PO_4)_6F_2$	iii]	Vivianite
d]	$Fe_3(PO_4)_2 . 8H_2O$	iv]	Hydroxyapatite

39. Match the formula of the following orthophosphate fertilizer (water soluble).

a]	Dicalcium phosphate dihydrate	i]	$Ca_4HPO_4 . 3H_2O$
b]	Anhydrous dicalcium phosphate	ii]	$CaHPO_4 . 2H_2O$
c]	Octocalcium phosphate	iii]	$CaHPO_4$
d]	Calcium ferric phosphate	iv]	$CaFe_2(HPO_4)_4 . 5H_2O$

40. Match the following formula with the phosphate compound.

a]	$KFe_3H_8(PO_4)_6 . 6H_2O$	i]	Basic potassium aluminium phosphate
b]	$(NH_4)Fe_3H_8(PO_4)_6 . 6H_2O$	ii]	Acid ammonium ferric phosphate
c]	$KAl_2(PO_4)_2 . OH\ 2H_2O$	iii]	Basic potassium ferric phosphate
d]	$KFe_2(PO_4)_2 . OH\ 2H_2O$	iv]	Acid potassium ferric phosphate

GS

Mushrooms are edible while toad stools are poisonous.

41. Match the following :

a]	Total	i]	Balance (remains) after water and citrate extraction
b]	Available	ii]	Soluble in water and soluble in citrate
c]	Citrate insoluble	iii]	Available + citrate insoluble
d]	Water soluble and citrate soluble	iv]	Water soluble + citrate soluble

42. Match the following P_2O_5 species with the formula.

a]	Ortho	i]	$H_4P_2O_7$
b]	Pyro	ii]	$H_6P_4O_{13}$
c]	Tetra	iii]	H_3PO_4
d]	Penta	iv]	$H_7P_5O_{16}$

43. Match the mineral with the formula.

a]	Microdine	i]	$[K(Mg, Fe)_3AlSi_3O_{10}(OH)_2]$
b]	Muscovite	ii]	$KAlSi_3O_8$
c]	Biotite	iii]	$[KMg_2Al_2Si_3O_{10}(OH)_2]$
d]	Phlogopite	iv]	$[KAl_3Si_3O_{10}(OH)_2]$

44. Match the following potassium fertilizer with the per cent K_2O content.

a]	Potassium sulphate	i]	60-62
b]	Potassium chloride	ii]	25-26
c]	Potassium orthophosphates	iii]	50-52
d]	Potassium calcium pyrophosphate	iv]	30-50

45. Match the per cent P_2O_5 with the fertilizer.

a]	Potassium orthophosphates	i]	39-54
b]	Potassium polyphosphates	ii]	30-60
c]	Potassium metaphosphate	iii]	40-60
d]	Potassium calcium pyrophosphate	iv]	55-57

46. Match the following four forms of the approximately estimated potassium fractions varying in their availability to plants in a soil system.

a]	Structural/mineral	i]	1-10 ppm
b]	Non-exchangeable/fixed	ii]	50-750 ppm
c]	Exchangeable	iii]	5,000-25,000 ppm
d]	Solution	iv]	40-600 ppm

47. Match the following sulphur bearing mineral found in rocks with the formula.

a]	Gypsum	i]	$MgSO_4 . 7H_2O$
b]	Epsomite	ii]	$Na_2SO_4 . 10H_2O$
c]	Mirabilite	iii]	$FeS_2 . FeAS_2$
d]	Arsenopyrite	iv]	$CaSO_4 . 2H_2O$

48. Match the following sulphur containing fertilizer material with the formula.

a]	Ammonium thiosulphate	i]	$Fe(NH_4)_2 . (SO_4)_2$
b]	Ferrous ammonium sulphate	ii]	$CaSO_4 . 4CO(NH_2)_2$
c]	Hydrated gypsum	iii]	$(NH_4)_2S_2O_3$
d]	Urea gypsum	iv]	$CaSO_4 . 2H_2O$

49. Match the formula given below with the material.

a]	$MgSO_4 . 7H_2O$	i]	Ammonium bisulphate
b]	$NaHSO_4$	ii]	Langbeinite (Sulphate of potash magnesia)
c]	$K_2SO_4 . 2MgSO_4$	iii]	Nitre cake (Sodium bisulphate)
d]	NH_4HSO_3	iv]	Epsom salt (Magnesium sulphate)

50. Match the magnesium content (Mg) with the material given below.

a]	55%	i]	Synthetic chelates
b]	16%	ii]	Magnesium silicate serpentine
c]	26%	iii]	Magnesia (MgO)
d]	2-4%	iv]	Magnesium nitrate $[Mg(NO_3)_2]$

GK GS

Baglihar Hydroelectric Dam is situated on river Chenab.

51. Match the boron fertilizer with the formula.

a]	Colemanite (Portabor)	i]	$Na_2B_{10}O_{16}.\ 10H_2O$
b]	Sodium pentaborate	ii]	$Na_2B_4O_7.\ 5H_2O + Na_2B_{10}O_{16}.10H_2O$
c]	Fertilizer borate-48	iii]	$Ca_2B_6O_{11}.\ 5H_2O$
d]	Solubor	iv]	$Na_2B_4O_7.\ 5H_2O$

52. Match the iron mineral with the formula from among the following.

a]	Olivine	i]	$[FeO(OH).\ nH_2O + Fe_2O_3.\ nH_2O]$
b]	Limonite	ii]	Fe_2O_3
c]	Hematite	iii]	$FeOOH$
d]	Geothite	iv]	$[Mg, Fe)_2\ SiO_4]$

53. Match the iron fertilizer with its formula as given below.

a]	Ferrous sulphate	i]	$(NH_4)_2SO_4.\ FeSO_4.\ 6H_2O$
b]	Ferrous ammonium phosphate	ii]	$Fe(NH_4)PO_4.\ H_2O$
c]	Ferrous ammonium sulphate	iii]	$Fe(NH_4)HP_2O_7$
d]	Iron ammonium polyphosphate	iv]	$FeSO_4.\ 7H_2O$

54. Match the iron chelate given below with the approximate content of Fe expressed in percentage.

a]	NaFe EDTA	i]	5-9
b]	NaFe HEDTA	ii]	10
c]	NaFe EDDHA	iii]	5-14
d]	NaFe DTPA	iv]	6

55. Match the manganese mineral with the formula as given below.

a]	Pyrolusite	i]	$MnSiO_3$
b]	Hausmannite	ii]	$MnCO_3$
c]	Rhodochrosite	iii]	MnO_2
d]	Rhodonite	iv]	Mn_3O_3

56. Match the following molybdenum fertilizer with the respective Mo content expressed in percentage.

a]	Ammonium molybdate	i]	39
b]	Sodium molybdate	ii]	1-30
c]	Molybdenum trioxide	iii]	54.
d]	Molybdenum frits	iv]	66

57. Match the source with the formula of zinc fertilizer.

a]	Zinc sulphate monohydrate	i]	$ZnSO_4.\ 7H_2O$
b]	Zinc sulphate heptahydrate	ii]	$ZnSO_4.H_2O$
c]	Basic zinc sulphate	iii]	$Zn_3(PO_4)_2$
d]	Zinc phosphate	iv]	$ZnSO_4.4Zn(OH)_2$

58. Match the following sodic soil with the exchangeable sodium percentage (ESP).

a]	Slightly sodic	i]	15-20
b]	Moderately sodic	ii]	20-30
c]	Very sodic	iii]	> 30
d]	Extremely sodic	iv]	7-15

59. Match the crop with the sodium uptake.

a]	Sugar beet	i]	Medium
b]	Potato	ii]	Very low
c]	Barley	iii]	High
d]	Maize	iv]	Low

60. Match the sodium uptake by the crop given below.

a]	High	i]	Buckwheat
b]	Medium	ii]	Millet
c]	Low	iii]	Cabbage
d]	Very low	iv]	Spinach

61. Match the per cent chlorine (Cl^-) content with the fertilizer source as given below.

a]	66	i]	Potassium chloride
b]	74	ii]	Ammonium chloride
c]	47	iii]	Calcium chloride
d]	65	iv]	Magnesium chloride

GK **Baglihar Hydroelectric Dam is situated on river Chenab.** GS

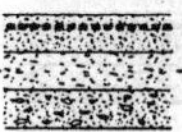

62. **Match the potassium mineral with the formula.**

a]	Carnallite	i]	$(K,Na)_2SO_4$
b]	Polyhalite	ii]	$K_2SO_4 . MgSO_4 . 2CaSO_4 . 2H_2O$
c]	Schoenite	iii]	$KCl_2 . MgCl_2 . 6H_2O$
d]	Aphthitalete	iv]	$K_2SO_4 . MgSO_4 . 6H_2O$

63. **Match the neutralizing value (CCE) of the pure forms of some commonly used liming materials.**

a]	$Ca(OH)_2$	i]	109
b]	$CaMg(CO_3)_2$	ii]	100
c]	$CaCO_3$	iii]	86
d]	$CaSiO_3$	iv]	136

64. **Match the plant part found representative for testing nitrate concentration in tissue testing technique.**

a]	Maize	i]	Petiole basal leaf
b]	Cotton	ii]	Base of stalk
c]	Maize	iii]	Lower stem
d]	Small grains	iv]	Mid-rib, basal leaf

65. **Match the soil test rating with fertility index.**

a]	Very low	i]	25-50
b]	Low	ii]	50-100
c]	Medium	iii]	0-10
d]	High	iv]	10-25

66. **Match the fertilizers with the salt index given below.**

a]	Ammonium sulphate	i]	2.990
b]	Diammonium phosphate	ii]	3.253
c]	Ammonium nitrate	iii]	6.060
d]	Sodium nitrate	iv]	1.614

67. **Match the salt index with the fertilizer as given below.**

a]	1.618	i]	Potassium sulphate
b]	0.390	ii]	Potassium chloride
c]	0.853	iii]	Urea
d]	2.189	iv]	Single superphosphate

68. **Match the following units of electrical conductivity when EC = 0.000694 mho/cm has been recorded for a soil saturation extract at 25°C.**

a]	EC	i]	69.4 ($=k \times 10^5$)
b]	$EC \times 10^3$	ii]	694 micromhos/cm
c]	$EC \times 10^5$	iii]	0.000694 mho/cm
d]	EC x 16	iv]	0.694 millimho/cm

69. **Match the ratio of SP (saturation percentage) of FAP (15-atmosphere percentage) in different soil textural groups.**

a]	Coarse (2.0-6.5)	i]	0.48
b]	Medium (6.6-15.0)	ii]	0.60
c]	Fine (> 15.1)	iii]	0.75
d]	Organic	iv]	1.15

70. **Match the pH value of $CaCO_3$ saturated solution with it's solubility (meq/litre).**

a]	6.21 pH	i]	1.1
b]	7.12 pH	ii]	0.82
c]	8.60 pH	iii]	19.3
d]	9.20 pH	iv]	7.1

71. **Match the amendments (in tonnes) equivalent to one tonne of sulphur.**

a]	Gypsum	i]	8.69
b]	Limestone	ii]	5.38
c]	Iron sulphate	iii]	3.06
d]	Sulphuric acid	iv]	3.13

72. **Match the leaching requirement with EC (mmhos/cm) of irrigation water if the EC (mmhos/cm) of the drainage water at the bottom of the root zone is 4 mmhos/cm.**

a]	2.5	i]	250
b]	6.2	ii]	750
c]	18.8	iii]	2250
d]	56.2	iv]	100

73. **Match the nutrients element with its function in the plants.**

a]	N_2	i]	Reduction of nitrate
b]	P_2O_5	ii]	Root growth
c]	K_2O	iii]	Acceleration of vegetative growth
d]	Mo	iv]	Translocation of photosynthetase

74. **Match classes of fertilizer with its examples.**

a]	Organic fertilizer	i]	Urea
b]	Inorganic fertilizer	ii]	Isobutylidene diurea
c]	Biofertilizer	iii]	$(NH_4)_2 SO_4$
d]	Slow release fertilizer	iv]	2-chloro 6-pyridine

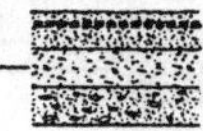

Glossary of Technical Terms

Absorption The process by which a substance is taken into and included within another substance i.e. intake of water by soil or intake of nutrients by plants.

Acidic Soil A soil that is acid in reaction throughout the root zone. They are reclaimed by adding liming material.

Additive A material added to fertilizers to improve its physical condition.

Adhesion Molecular attraction that holds the surfaces of two substances (e.g. water and sand particles) in content.

Adsorption The increased concentration of molecules or ions at a surface, including exchangeable cations and anions on soil particles.

Aeration The process by which air in the soil is replaced by air from the atmosphere. In a well aerated soil, the soil air is similar in composition to the atmosphere above the soil. Poorly aerated soils usually contain more carbon dioxide and correspondingly less oxygen than the atmosphere above the soil.

Air Porosity The proportion of the bulk volume of soil that is filled with air at any given time or under a given condition, such as a specified moisture potential, usually the larger pores.

Alkaline Soil A soil that is alkaline in reaction throughout the rootzone or for a major part of the root zone.

Alkalinization Refers to the process where by the exchangeable sodium content of a soil is increased.

Amendment Any material such as lime, gypsum saw-dust, sulphur or soil conditioner, that is added to soil to influence plant growth by changing the reaction of pH by improving the physical condition of a soil. Strictly a fertilizer is also an amendment, but the term is used most commonly for added a materials other than fertilizers.

Amminization Process of converting substance containing protein and organic nitrogen into amino group.

Ammonification Process in which amino (NH_2) group is converted first into (NH_3) ammonia and then into ammonium (NH_4) by *Bacillus, Clostridium and Psuedomonas.*

Anion Anion carrying negative charge of electricity.

Anion Exchange Capacity The sum total of exchangeable anions that a soil can absorb, expressed as a centimoles for kilogram (cmol/kg) of soil (or of other adsorbing material such as clay).

AP The surface layer of soil disturbed by cultivation or pasturing.

Available Nutrients in Soil A part of plant nutrient in the soil that can be taken up by growing plants immediately.

Band Placement The placement of fertilizers in the soil in continuous or narrow ribbons in between the plant rows or around each plant.

Basic Slag A by product of steel industry obtained from phosphate iron ores. Basic slag produced in India being of low grade is not used as a commercial fertilizers.

Buffering Capacity The ability of a soil to resist changes in pH. commonly determined by presence of clay, humus and other colloidal materials.

Knowledge World!

First Indian Buddha images

The Mathura school created an entirely Indian type of the Buddha for the first time, modeling them on the *yaksha* figure of the earlier centuries.

Bulk Density The mass of dry soil per unit of bulk volume, including the air space. The bulk volume is determined before drying to constant weight at 105°C.

Bulky Organic Manures These manures are bulky in nature and supply plant nutrients in small quantities and organic matter in large quantities.

Carbon Cycle The sequence of transformations where by carbon dioxide is fixed in living organisms by photosynthesis or by chemosynthesis, librated by respiration and by the death and decomposition of the fixing organism, used by heterotrophic species, and ultimately returned to its original state.

Carbonation It implies the action of CO_2 rocks and minerals with the formation of its carbonates and bicarbonates.

Cation Anion carrying positive charge of electricity. The common soil cations are calcium, magnesium, sodium, potassium and hydrogen.

Cation Exchange Capacity (CEC) Is the sum total of exchangeable cations that a soil can adsorb (meq/100 g soil at ph 7.0).

Cation Exchange is an interchange between a cation in solution and another cation on the surface of clay or organic matter.

Chelates An organic compound capable of holding the plant nutrient in a form which prevents it from getting tied with other elements in the soil, thus keeping it more or less in available form for the plant.

Chlorosis A condition in plant relating to the failure of chlorophyll (the green coloring matter) to develop chlorotic leaves range from light green through yellow to almost white.

Cohesion Force holding a solid or liquid together, owing to attraction between like molecules. Decrease with rise in temperature.

Complex Fertilizers The commercial fertilizers containing at least two or more of the primary essential nutrients.

Compost A mass of rotted organic manure obtained from waste.

Convection The transfer of heat through a gas or solution because of molecular movement.

Creep Slow mass movement of soil and soil material down relatively steep slopes primarily under the influence of gravity, but facilitated by saturation with water and by alternate freezing and thawing.

Denitrification Process is reverse of nitrification in which reduction of NO_3 into N_2 or its oxides or NH_3 by *Bacillus denitrificans* occur in the absence of O_2.

Diffusion The transport of matter as a result of the movement of the constituent particles. The intermingling of two

India's Chandrayan-1 will take pictures, map the moon and look for minerals.

gases or liquids in contact each other takes place by diffusion.

Edaphology The science that deals with the influence of soils on living things, particularly plants, including man's use of land for plant growth.

Eluviation The removel of soil material in suspension (or in solution) from a layer or layers of soil (usually, the loss of material in solution is described by the term leaching).

Erodibility The lack of resistance of a soil to soil erosion.

Erosion Detachment and movement of soil or rock by water, wind, ice, or gravity.

Eutrophic Soil solution (waters) containing nearly optimal concentrations of nutrients.

Facultative Organism An organism capable of both aerobic and anaerobic metabolism.

Fallow Crop land left idle in order to restore productivity, mainly through accumulation of water, nutrients, or both. Summer fallow is a common stage before cereal grain in regions of limited rainfall. The soil is kept free of weeds and other negotiation, thereby conserving nutrients and water for the next year's crop.

Fertilizer Any organic or inorganic material of natural or synthetic origin added to a soil to supply certain elements essential to the growth of plants.

Field Capacity The percentage of water remaining in a soil, two or three days after its having been saturate and after free drainage has practically ceased.

Field Placement This refers to applying fertilizer in bands or near the plants.

Flocculation It is a process by which individual tiny soil particles especially fine clay aggregate or clump together into small clumps or floccules. Adsorption of Ca^{++}, Mg^{++} and Al^{+} bring about flocculation.

Gel Condition When small amount of water is present in larger quantity of colloidal clay.

Geology It is a science which includes the study of earth from the beginning of its existence to the present time.

Gravitational Potential That portion of the total soil water potential due to differences in elevation of the reference pool of pure water and that of the soil water. Since the soil water elevation is usually chosen to be higher than that of the reference pool, gravitational potential is usually positive.

Gravitational Water Water that moves into, through, or out of the soil under the influence of gravity.

Green Leaf Manuring This refers to turning under the green leaves and tender green twigs collected from shrubs and trees grown on bunds, waste lands and nearby forest areas. Trees useful for this purpose are Glyricidia, Subanta, Karanj, etc.

Green Manuring *in-situ* A practice in which green manure crops are grown and burried in the same field which is to be green-manured.

Horizon It is a successive layer of soil, approximately parallel to the soil surface, whose characteristics are produced by soil forming process.

Humus That more or less stable fraction of the soil organic matter remaining after the major portions of added plant and animal residues have decomposed, usually dark in colour.

Hydration Chemical combination of water molecules with mineral.

Hydraulic Conductivity An expression of the readiness with which a liquid such as water flows through a solid such as soil in response to a given potential gradient.

Hydrolysis It is a decomposition reaction as H^+ ion of water tends to change places with the action of the mineral.

Hygroscopic Co-efficient The amount of moisture in the dry soil when it is equilibrium with some standard relative humidity near a saturated atmosphere (about 98%), expressed in terms of percentage on the basis of oven-dry soil.

Igneous Rock The rock formed by solidification of molten magma.

Immobilization Process in which simple inorganic compounds is changed to complex organic compound. This occur when C:N ratio is more than 30:1.

Immobilization The conversion of an element from the inorganic to organic form in microbial tissues or in plant tissues, thus rendering the element not readily available to other organisms or to plants.

Infiltration The downward entry of water into the soil.

Ion Exchange It is a reversible process in which cations and anions are exchanged between solid and liquid phases.

Irrigation Efficiency The ratio of the water actually consumed by crops on an irrigated area to the amount of water diverted from the source into the area.

Leaching Removal of plant nutrients of solution by the passage of water through soil.

Lime Requirement The amount of standard pound limestone required to bring a 6-inch layer of an acre of acid soil to some specific lesser degree of acidity.

Lithosphere It is a solid phase and about 93 percentage of earth mass by weight of earth crust.

Localised Placement of Fertilizers Application of fertilizer into the soil close to seed or plant.

Luster The reflection of light from the surface of mineral.

Mechanical Analysis It is method of determination of the percentage distribution of the various sizes of the primary particles of the inorganic mineral fraction.

Metamorphic Rock A rock that has been greatly altered from its previous condition through combined action of heat and pressure.

Mineral Minerals are the naturally occurring inorganic homogenous substance having a definite chemical composition, physical properties and characteristic geometrical forms.

Mineralization Process in which complex organic compound is changed to simple inorganic compound in the soil as a result of microbial decomposition. This occur when C:N ratio is less than 20:1 eg. organic nitrogen into inorganic nitrogen.

Mixed Fertilizers Two or more fertilizer materials mixed together may be as dry powders, granules, pellets, bulk blends or liquid.

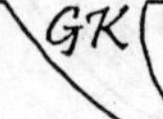

Indo-Gangetic-Brahmaputra plain is the largest alluvial tract.

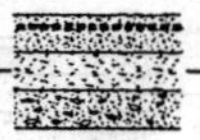

Mobilization It is a physiological process in which movement of elements, food or other materials occurs from one plant part to another part.

Muck Highly decomposed organic materials in which the original plant parts are not recognizable. Contains more mineral matter and is usually darker in colour thin peat.

Muck Soil Soil containing 20-50 per cent organic matter.

Nitrification Process in which NH_4 is converted into NO_2 and NO_3 by Nitrosomonas and Nitrobacter (autotrophic bacteria) occur in the presence of O_2.

Normal Solution When equivalent weight of a compound is dissolved in 1 litre of water the solution is one normal solution (n)

Optimum Dose Refers to the amount of fertilizer that produces the greatest profit per acre.

Parent Material It is the unconsolidated and more or less chemically weathered mineral from which the solum of soil is developed by pedogenic process.

Particle Density The mass per unit volume of the soil particles. In technical work, usually expressed as metric tons per cubic meter (mg/m^3) grams per cubic centimeter (gm/cm^3).

Peat Unconsolidated soil material consisting largely of un-decomposed, or only slightly decomposed, organic matter accumulated under conditions of excessive moisture.

pH The negative logarithm of the hydrogen ion activity (concentration) of a soil. The degree of acidity (or alkalinity) of a soil as determined by means of a glass, quinhydrone, or other suitable electrode or indicator at a specified moisture content or soil/water ratio, and expressed in terms of the pH scale.

Plastic Index The difference in per cent moisture content of a soil between the lower and upper plastic limit.

Plastic Limit The water content corresponding to an arbitrary limit between the plastic and semi-plastic state of consistency of a soil (At this state the soil can be rolled into a head of specified dimension without showing any sign of crumbing).

Plasticity It is a property of clay which enables clay to be pliable and capable of being molded into various shapes without rupturing and creaking.

Plough-sole Placement In this method, fertilization is placed in a continuous band on button of a furrow in the process of plough. Each furrow is covered as the next furrow turned.

Polynutrient Fertilizer A fertilizer contain more than one major plant nutrients.

Productivity The capacity of a soil for producing a specified plant or sequence of plants under a specified system of management. Productivity emphasizes the capacity of soil to produce crops should be expressed in terms of yields.

Profile A vertical cut of soil, form the surface down to the hard rock.

Rabbing The process of burning the farm waste on the soil where by the fertility of soil increases.

Rock Consolidated mass of one or more minerals, which have no definite chemical composition and geometry.

Salinization refers to the process of accumulation of soluble salts in the soil.

Sedimentary Rock A rock formed from materials deposited from suspension or precipitated from solution and usually being more or less consolidated. The principal sedimentary rocks are sandstones, shales, limestones, and conglomerates.

Shifting Cultivation A farming system in which land is cleared, the debries burned, and crops grown for 2-3 years when the farmer moves one to another plot, the land is then left idle for 5-15 years; then the burning and planting process is repeated.

Sludge Settled sewage solids combined with varying amounts of water and dissolved materials removed from sewage by screening, sedimentation, chemical precipitation or bacterial digestion.

Soil A dynamic natural body composed of mineral and organic materials and living forms in which plants grows.

Soil Air The soil atmosphere, the gaseous phase of the soil, being that volume not occupied by soil or liquid.

Soil Alkalinity The degree or intensity of alkalinity of soil, expressed by value greater than 7.0 on the pH scale.

Soil An organic soil containing more than 50 per cent organic matter.

Soil Classification The systematic arrangement of soil into groups or categories on the basis of their characteristics.

Soil Condition When small amount of colloidal clay is present in much larger quantity of water.

Soil Conditioner Any material added to a soil for the purpose of improving its physical condition.

Soil Conservation A combination of all management and land-use methods that safeguard the soil against depletion or deterioration caused by nature and/or humus.

Soil Genesis The mode of origin the soil, with special reference to the processes or soil-forming factors responsible for the development of the solum, or true soil, from the unconsolidated parent material.

Soil Microbiology A sub specialization of soil science concerned with soil inhabiting micro-organisms and with their relation to agriculture including both plant and animal growth.

Soil Monolith Permanent records of profile of important soil series.

Soil Morphology It is the constitutions of the soil including the texture, structure, consistence, colour and other physical, chemical and biological properties of a various soil horizons that makes up the soil profile.

Soil Physics It is a branch of soil science relating to physical properties of mineral soils applying the physics laws. It includes physical properties of soils such as texture, structure, density, pore space, consistence, tillage, soil water etc.

Soil Porosity The volume percentage of the total soil bulk not occupied by solid particles.

Soil Science The science dealing with soil as a natural resource on the earth surface including soil formation,

GK GS

Sweet potato is a form of modified root.

classification and mapping and the physical, chemical and biological and fertility properties of soils and these properties in relation to their management for crop production.

Soil Separates One of the individual sized groups of mineral soil particles (Viz. sand, silt or clay) called soil separate.

Soil Solution In a soil the water remains in the pore space, in which salts are dissolved and such solution is called soil solution.

Soil Structure Relates to the grouping and arrangement of soil particles into secondary grouping called aggregates or Peds.

Soil Survey The systematic examination, description, classification, and mapping of soils in an area.

Soil Texture Refers to the relative proportion of various soil separates (sand, silt and clay) in the soil.

Specific Gravity Ratio between the weight of mineral and weight of an equal volume of water.

Straight Fertilizer The fertilizers that have only one plant nutrient, e.g. Ammonium sulphate, Murate of potash etc.

Tillage The mechanical manipulation of soil for any purpose, but in agriculture it is usually restricted to the modifying of soil conditions for crop production.

Top Dressing An application of fertilizer to a soil after the crop stand has been established.

Water Use Efficiency Dry matter or harvested portion of crop produced per unit of water consumed.

Weathering It is the physical disintegration and chemical decomposition of rocks and minerals at or near the surface of the earth.

Wilting Point The moisture content of soil, on an oven-dry basis, at which plants wilt and fail to recover their turgidity when placed in a dark humid atmosphere.

7. Genetics, Plant Breeding and Biotechnology

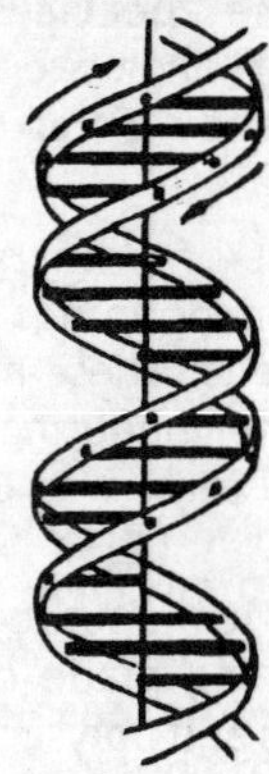

Each Word Counts

CELL

Cell Theory

The French scientist H. J. Dutrochet (1824), boiled the tissues in acid and separated the cells. On this basis he thought that all animals and plants are made of cells. This was the foundation of the *cell theory*. The German botanist *M.J. Schleiden* (1838) and the German zoologist *Theodor Schwann* (1839) found the essential similarity in the structure of plant and animal tissues. On this basis they gave an abstract generalization which is now known as cell theory.

The cell theory states that

1. All plants and animals are made of cells. The cell is a unit of structure and function.
2. Each cell behaves independently but also functions as an integral part of the complete organism.
3. The new cells are formed by a process similar to crystal formation.

Virchow (1858), a German scientist stated that *the new cells orginate from the pre-existing cells only (Omnis cellulae cellula)*. The main features of the cell theory as known today may be briefly summed up as follows :

1. All living organisms are composed of cells. Thus cell is the structural unit of living organisms.
2. All cells arise from the pre-existing cells of the similar kind. Thus continuity of life from one generation to another is through living cells.
3. The chemical composition and metabolism of all cells is basically alike.
4. The function of an organism as a whole is the outcome of the activities and interactions of the constituent cells.

Types of Cells

On the basis of presence or absence of organised nucleus, two types of cells have been recognised. These are *prokaryotic cells* and *eukaryotic cells*.

Did You Know ?

First bank for prisoners

The country's first prisoners' bank was started at Tihar Jail, New Delhi, on Sept. 30, 1993. Apart from the 8,000 prisoners the bank also caters to the needs of the 800 odd staff. The bank is an extension counter of the New Bank of India.

Differences between Prokaryotic and Eukaryotic cells

Prokaryotes	*Eukaryotes*
1. Protoplasm Relatively rigid, usually, non-vacuolate, relatively resistant to desiccation, osmotic shock and thermal denaturation.	Typical more fluid, generally vacuolate, more sensitive to shock and thermal denaturation.
2. Nucleus True nucleus absent; nuclear membrane and nucleolus not present, nucleoplasm undifferentiated from cytoplasm.	True nucleus present, nuclear membrane, nucleolus and nucleoplasm distinct.
3. DNA Scattered in the protoplasm without histone proteins	Organised with histone proteins to form distinct chromosomes.
4. Organelles Membrane bound organelles, like golgi bodies, plastids mitochondria, ER, etc., are absent.	Organelles like golgi bodies, plastids, mitochondria, ER, etc., are present.
5. Ribosomes 70 S type.	80 S type. (70S in Mit. Chl.)
6. Mitotic apparatus Absent.	Present
7. Cell division Mostly amitotic.	Typically mitotic.
8. Sexual process Typical sexual process lacking, but genetic material is genteic material is exchanged by parasexual processes.	Typical sexual process (with alternating nuclear fusion and reduction division) often present.
9. Examples Cells of mycoplasma, bacteria, blue-green algae.	Cells of all the higher organisms, except the three group mentioned as prokaryotes.

Cell Structure

The cell can be divided into three major structural parts:

(1) Cell wall, (2) Vacuoles and (3) Protoplasm

(1) *Cell wall* is characteristic of plant cells only, being

GK **The Mankharis are first mentioned in Patanjali's Mahabhashya.** GS

Differences between Plant and Animal cells

Plant Cell	Animal Cell
1. Cell wall	
A cell wall made of cellulose which is present outside the cell membrane.	Cell wall is absent. The outermost is the cell membrane or plasma membrane.
2. Chloroplasts	
The photosynthetic cells are green due to the presence of chloroplasts.	Chloroplasts are absent.
3. Centrosome	
Except for a few plant cells, centrosome (centriole) is absent.	In most of the animal cells centrosome (centriole) is generally present.
4. Vacuoles	
Plant cells generally have a large, centrally located vacuole. The protoplasm and the nucleus are present near the cell wall.	Animal cells mostly lack vacuoles. If present, these are small and remain scattered throughout the cell.
5. Cytokinesis	
This occurs due to the formation of the cell plate.	This is the result of formation of a furrow in the centre of the cell.

absent from animal cells. It forms the outermost limiting layer of the plant cell and is freely permeable.

(2) ***Vacuoles*** are spaces filled with cell sap or vacuolar sap. Vacuole is bound by vacuolar membrane or tonoplast.

(3) ***Protoplasm*** consists of cytoplasm and the nucleus, Cytoplasm includes all the structures outside the nucleus.

CELL WALL

The plants cells are bounded externally by a non-living, more or less firm covering called cell wall. Animal cells do not have a cell wall and thus their outermost covering is cell membrane. The cell wall does not act as a physiological boundary. Its main function is mechanical, *i.e.*, to provide protection and shape to the cell.

(I) Layers of the Cell Wall

Cell wall is derived from the living protoplast. It consists of *middle lamella, primary cell wall, secondary cell wall* and *tertiary cell wall.*

1. ***Middle lamella:*** It is the first formed layer, present between the two adjacent cells. Middle lamella is situated outside the primary cell wall. It is made of calcium and magnesium pectate. Middle lamella acts as a cement which holds the adjacent cells together..

2. ***Primary cell wall***: This layer occurs inner to the middle lamella. The major components of this layer include cellulose, hemicellulose, pectic substances, lipids, proteins, mineral elements and water. This thin and elastic cell wall becomes thick and rigid after the completion of cell enlargement.

During the development of primary cell wall and the middle lamella, certain openings are left at places between the adjacent cells. These maintain cytoplasmic continuity between the neighbouring cells and are known as *Plasmodesmata.* This term was used by *Strasburger* (1901) for the first time.

3. ***Secondary cell wall:*** It is situated inside the primary cell wall. It is thick being made of several layers of cellulose, hemicelluloses and polysaccharides. Secondary cell wall is deposited after the primary wall is fully formed, i.e., after the cell has undergone elongation. It is thick and non-elastic, being mostly made of lignin.

4. ***Tertiary cell wall:*** Sometimes tertiary cell wall is deposited in a few cells. It is considered to be a dried residue of the protoplasm.

(II) Cell Wall Depositions

Some other substances found in the cell wall are as follows:

1. ***Lignin:*** It is a complex chemical substance, found in woody and hard tissues. The cells become woody as a result of lignification. Woody tissues are, hence, generally dead. Most of the vegetable fibres are lignified. Lignin deposition is generally not uniform and may result in annular, spiral, scalariform, reticulate or pitted patterns. These thickenings provide mechanical support to the cell.

2. ***Cutin:*** This waxy layer forms a thin or a thick cuticle on the stem and leaf. It is impervious to water.

3. ***Suberin:*** This fatty substance is present in the walls of the cork and is totally impermeable to water and gases.

4. ***Mucilage:*** This slimy substance absorbs water and stores it. Mucilage becomes hard when dry and viscous when moist.

5. ***Mineral substances:*** Silica, calcium oxalate, calcium carbonate, etc., are present in the cell wall in the form of crystals.

VACUOLES

A major part of the plant cell is occupied by a vacuole. Animal cells, on the other hand, have either very small vacuoles or these may be absent. A young dividing cell is completely filled with cell sap and has no vacuoles. But as the cell grows, many small vacuoles appear which ultimately coalesce to form a large central vacuole. Thus in a mature plant cell, cytoplasm occurs as a thin layer near the periphery of the cell surrounding the large central vacuole. Vacuole is delimited by a membrane, known as *vacuolar membrane* or *tonoplast.* The vacuole is filled with *cell sap* or *vacuolar sap.*

Physiologically, vacuole is a very important part of the

GK Chhotanagpur is composed mainly of Gondwana rock. GS

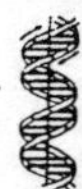

The Cell and its Parts

- CELL
  - 1. Cell wall
  - 2. Vacuoles
  - 3. Protoplasm
    - 1. Cytoplasm
      - 1. Protoplasmic membrane
      - 2. Mesoplasm
        - 1. Hyaloplasm
        - 2. Cytoplasmic organelles
          - Endoplasmic reticulum
          - Plastids
          - Mitochondria
          - Golgi complex
          - Ribosomes
          - Lysosomes
          - Microbodies
          - Microtubules
          - Cilia and flagella
        - 3. Metabolically inactive substances
          - **(a) Reserve materials**
            - Carbohydrates
            - Nitrogenous substances
            - Fats and oils
          - **(b) Excretory products**
            - Aklaloids
            - Essential oils
            - Gums
            - Glucosides
            - Latex
            - Organic acids
            - Resins
            - Tannins
            - Mineral crystals
          - **(c) Secretory products**
            - Enzymes
            - Colouring substances
            - Nectar
    - 2. Nucleus
      - 1. Nuclear membrane
      - 2. Nucleolus
      - 3. Nucleoplasm
      - 4. Chromosomes

cell. It serves as a storage place and a place for depositing the cellular wastes. Vacuole may contain a variety of dissolved substances like sugars, salts, acids, alkaloids and pigment anthocyanin. Anthocyanins give purple or red colour to different organs. The tonoplast is a differentially permeable membrane of lipo-proteins. It maintains the cell structure and also the osmotic gradient. It also checks the outwards flow of some ions, thus allowing their higher concentration in the vacuole than in the cytoplasm.

PROTOPLASM

Physical Characters

Protoplasm is the only substance that shows the properties of life. It is a complex, jelly-like, gelatinous, granular, viscous, elastic and colourless substance. The protoplasm responds to various stimuli like temperature, light, chemicals, etc. It also acts as a semi-permeable membrane.

Composition of the Protoplasm

I. ***Water:*** Water forms most of the protoplasm. In aquatic plants the amount of water may be as high as 95% while in dry seeds it may be as low as 10%.

2. ***Proteins:*** Proteins form about 60-65% of the total dry weight. These are made of carbon, hydrogen, oxygen and nitrogen. In some proteins, sulphur and phosphorus are also found. The fundamental units of proteins are amino acids. There are some 20 essential amino acids contributing to the protein structure. Each amino acid has an *amino group* -NH_2 and one or more acidic *carboxyl group* -COOH.

Thorn in Duranta is a modified stem.

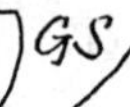

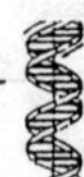

During the formation of proteins, amino group of one amino acid joins the carboxyl group of another amino acid, This is called *peptide bond*. A simple protein consists of about 100 amino acid molecules.

$$H_2N-\underset{H}{\overset{R}{C}}-COOH + H-\overset{H}{N}-\underset{R}{\overset{H}{C}}-COOH$$

Proteins are basic building blocks of protoplasm. Most of the enzymes found in plants and animals are proteins or they occur in conjugation with proteins.

3. *Lipids or fats:* Fats consist of carbon, hydrogen, oxygen, nitrogen and phosphorus. These are mostly insoluble in water. Lipids are of three types-true fats, waxes and phospholipids.

(a) True fats: These are made of glycerol and fatty acids.

(b) Waxes: These are lipids which are composed of fatty acids combined with long chain alcohols with 24 to 36 carbon atoms. These are found in combination with other substances as components of cell wall, *e.g.*, suberin and cutin.

(c) Phospholipids: They contain glycerol, fatty acids and atoms of phosphorus and nitrogen. In plants, these occur in least amounts among the three classes of lipids. These provide structural framework for various cell membranes and cellular organelles.

4. *Carbohydrates:* These are made of carbon, hydrogen and oxygen. Typically, the hydrogen and oxygen atoms are present in 2 : 1 ratio. The emperical formula of carbohydrates is CH_2O. On the basis of carbon atoms present, carbohydrates are divided into (i) monosaccharides, (ii) disaccharides and (iii) polysaccharides.

(a) Monosaccharides or simple sugars: They contain 3 to 7 carbon atoms and are accordingly named as *triose* (3 carbon), *tetrose* (4 carbon), *pentose* (5 carbon), *hexose* (6 carbon) and *heptose* (7 carbon) sugars. Common monosaccharides are sweet tasting, colourless solids and are soluble in water (*e.g.*, glucose).

(b) Disaccharides: Two monosaccharides join together to form disaccharides (double sugars). Sucrose, maltose and lactose are the common disaccharides occurring in protoplasm.

(c) Polysaccharides: They are composed of several monosaccharide units. The constituent monosaccharide units may be similar (homopolysaccharides; *e.g.*, starch, cellulose) or different (heteropolysaccharides; *e.g.*, pectin and chitins). Polysaccharides are tasteless and colourless amorphous powder which are little soluble in water, although some may form colloidal solution.

Carbohydrates constitute about 1% of the protoplasm. On oxidation they give energy. Cellulose, pectin, chitin, etc., are important parts of the cell structure.

5. *Inorganic substances:* The essential inorganic substances found in the protoplasm are in the form of solution. These occur in ionic form. Calcium, magnesium, potassium and iron are the main cations while nitrate (NO_3^-) sulphate (SO_4) phosphate (PO_4), chloride (Cl^-) and bicarbonate (HCO_3) are the major anions. Inorganic substances control pH of the cell and also influence numerous chemical reactions in the cell.

6. *Nucleic acids :* These are highly complex organic compounds, consisting of hydrogen, oxygen, carbon, nitrogen and phosphorus. These are of two types – *Ribose Nucleic Acid* (RNA) and *Deoxyribose Nucleic Acid* (DNA). Nucleic acid is made of many nucleotide units. Each nucleotide has three components– *nitrogenous bases* (purine or pyrimidine), *pentose sugar* (ribose or deoxyribose) and *phosphoric acid*.

7. *Other substances:* Besides the above major constituents, pigments, hormones, enzymes, alkaloides, latex, etc., are also found in the protoplasm.

Cell Membrane

Structure

Cell membrane is not clearly visible under the light microscope but can be observed with electron microscope. Two types of cell membranes are recognised – *cytoplasmic membrane* which surrounds the whole protoplasm and *internal membrane* which surrounds various cellular organelles.

The following structures of cell membranes have been proposed by different workers.

I. *Trilamellar structure:* It was given by *Danielli* and *Davson* (1935). They suggested that cell membrane consists of a middle bimolecular layer of phospholipids, which is bounded on both sides by a protein layer in figure A. The

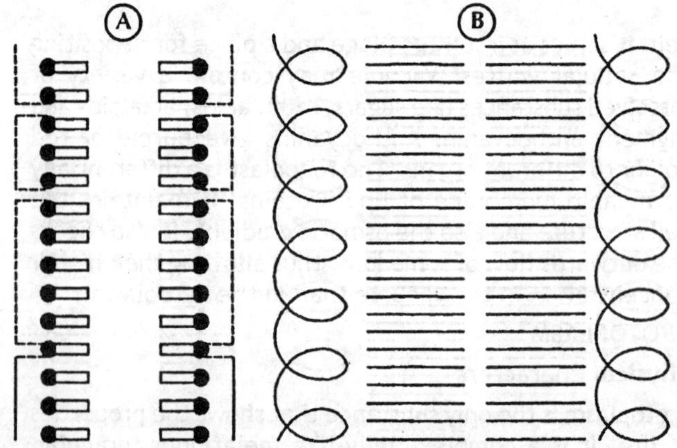

A-B Cell membrane as proposed by (A) Danielli and Davson, (B) Robertson

GK

Satyameva Jayate has been taken from the Mundokopnishad.

GS

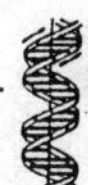

lipid molecule of each monomolecular layer has its hydrophilic and close to the protein layer while its hydrophobic end is away from it.

2. ***Unit membrane concept:*** It was proposed by *J.David Robertson* (1959). He found that not only the cell membrane is trilamellar but most of the membranes of cellular organelles are also trilamellar in figure B. Thus all the membranes in the cell are now called *unit membranes*. According to this concept, the unit membrane is made of three layers.

3. ***Fluid mosaic model:*** It was given by *Singer* and *Nicolson* (1974). This is one of the most accepted views. It differs from Robertson's model in the arrangement of proteins. According to this model, the proteins do not always cover the entire hydrophilic surface of the lipid bilayers but are of two types. These are: (i) *Peripheral* or *extrinsic* proteins: The peripheral proteins are present superficially on the surface and can be easily removed. (ii) *Integral* or *intrinsic* proteins of the integral proteins: Some are completely embedded in the lipid layer while others are partly embedded in the lipid layer and partly projected on the surface. These are tightly held and cannot be easily removed.

Lipids and integral proteins form a mosaic. The unit membrane is, hence, quasi-fluid (semi- fluid) in which lipids and integral proteins move within the bilamellar structure.

Chemical Composition

Cell membrane is mainly made of lipids and proteins. About 60-80% of the substances are proteins, some of which act as enzymes. Lipids constitute about 20-40% of the total substances. Besides, small amount of cellulose is also present. It is associated with proteins to give stability to the structure of cell membrane.

Functions

The primary function of the cell membrane is the regulation of passage of various substances in and out of the cell. Some of the important functions of the cell membrane are given below:

I. ***Permeability:*** Cell membrane allows the movement of small ions and molecules in and out of the cell. This membrane is *selectively permeable*.

2. ***Passive transport:*** Ions of different types move from their region of higher concentration to their region of lower concentration through cell membrane. It is called passive transport because it does not require chemical energy.

3. ***Active transport:*** During this process molecules move against the normal diffusion gradient. The process is called active because it requires chemical energy.

4. ***Exocytosis and Endocytosis:*** Macromolecules are known out of the cell through exocytosis and ingested through endocytosis. When the foreign substances ingested are fluids, the process is called pinacocytosis or *pinocytosis*, and if the substances ingested are solids, the process is known as *phagocytosis*.

You Don't Know!

First manufacture of wagons

Jessup and Co. Ltd. was the first to undertake assembly of good wagon with indigenous components in the late 1920s. The 4-seated wagons had a maximum load capacity of 12 tonnes.

5. ***Cellular recognition and adhesion:*** Two components of the cell membrane, glycoproteins and glycolipids, play an important role in selective adhesion of cells. It is due to this property of cell membrane that cells from different tissues like heart and liver do not adhere. Rejection or acceptance of transplanted tissues is determined by above mentioned components of the cell membrane. This occurs in animals only because the cell membranes of adjacent cells come in contact.

CELL ORGANELLES

Chloroplasts

The details of chloroplast are given below as they are of wide occurrence and great importance. These familiar type of plastids contain the photosynthetic pigment-the *chlorophyll*. Chloroplasts are present in all green coloured eukaryotic cells. Fungi lack chloroplasts and are, therefore, heterotrophic. In prokaryotic cells, like blue green algae and bacteria, organised chloroplasts are absent.

Structure

Chloroplast is surrounded by two unit membranes. There is a space between two membranes, called *periplastidial space*. The internal structure of the plastid shows two distinct parts- (i) colourless ground substance called *stroma* and (ii) closed, flat, stack-like membrane system called grana.

Stroma is watery and proteinaceous ground substance. It contains starch grains, lipid droplets, DNA and free ribosomes. The dark reaction of photosynthesis takes place in the stroma.

Grana are densely packed stacks of membrane layers called *thylakoids*. Each thylakoid is bounded by a single membrane but because of the flatness of these structures they appear as double membrane layers or lamellae. Large number of grana are formed at frequent intervals by packed stacks of thylakoids. This is the place where light reaction of photosynthesis takes place. Two adjacent grana are joined with one another by lamellae called *intergranal lamellae* or *stroma lamellae* or *fret channel*. Each

Tendril of pea is a modified leaf.

chloroplast contains about 40-60 grana.

Each lamella is made of a bimolecular layer of lipo-pigment sandwiched between two protein layers. The lipo-pigment layer contains chlorophyll and carotenoid pigments along with phospholipids. The chlorophyll molecules are so arranged in this layer that their hydrophilic porphyrin head is close to the outer layer of proteins. Amongst these lamellae, distinct groups, each consisting of about 230-300 chlorophyll molecules called *quantasomes*, are present. Quantasomes are the smallest morphologically distinct and photosynthetically functional units.

Chloroplasts also have 70S ribosomes which are either free or attached to intergranal lamellae. Pyrenoid is present in the chloroplasts of some algae and other plants. It consists of a protein core surrounded by peripheral starch plates.

Functions

The following are major functions of chloroplasts:-

1. Absorption of light energy.
2. Production of $NADPH_2$ and evolution of oxygen through the process of Hill reaction.
3. Transfer of carbon dioxide obtained from the air to 5 carbon sugar (ribulose I, 5 diphosphate).
4. Breaking of 6-carbon atom compound into two molecules of phosphoglyceric acid.
5. Hydrogenation of phosphoglyceric acid to form phosphoglyceraldehyde.

Mitochondria

Distribution

Mitochondria are present in the living eukaryotic cells and absent from prokaryotic cells like bacteria and blue green algae. Mitochondria are lost secondarily in some highly specialised eukaryotic cells like mammalian RBC.

Structure

Mitochondria can be stained *in vivo* with *Janus Green B*. It reveals that mitochondria has two parts – an outer *envelope* and a *central cavity* filled with matrix. The envelope that surrounds the mitochondria is made of two unit membranes. It is called *perimitochondrial space:* The outer membrane is smooth but the inner membrane has many infoldings which extend into the matrix. These infoldings are called *cristae*.

The cristae provide an increased surface area within the mitochondrion for enzymatic 'activity. The inner surface of the inner membrane (*i.e.*,the one facing the matrix) is covered by small tennis racket-like particles with a head and a stalk. These particles have been variously named as *inner membrane sub-units, elementary particles or oxysomes. Parsons* (1963) called them *electron transport particles* (ETP).

Each particle consists of a head called F_1 sub-unit, approximately 100 A° in diameter and is attached to a base piece called F_0 sub-unit. It is 35-50 A° in length F_1 sub-unit projects into the matrix. It is an integral protein of the inner membrane.

The inner membrane has all the enzymes required for electron transport. The F_1 -F_0 combination has special ATPase (ATP synthetase) for oxidative phosphorylation.

Functions

Mitochondria are known as '*power house of the cell*. This is because of the formation and storage of energy rich compound ATP (adenosine triphosphate) as a result of oxidation. ATP is a readily available source of energy for numerous cellular reactions.

Mitochondria are concerned with oxidation and phosphorylation reactions of Kreb's cycle during aerobic respiration. Mitochondria are absent in cells which do not respire aerobically. Besides oxidation of carbohydrates, mitochondria are also involved in the oxidation of proteins and fats.

Endoplasmic reticulum

Distribution

All cells do not possess ER. It is generally absent in egg and embryonic cells. It is well developed in differentiated cells. In spermatocytes, though present, it is poorly developed. Generally ER is fully developed in cells actively engaged in protein and hormone synthesis, The size and shape of the ER changes with the type of cells in which it occurs.

Structure

ER is composed of three different types of structures. These are –*cisternae, vesicles* and *tubules*.

1. *Cisternae:* These are long, flat and unbranched plates or lamellae arranged in parallel rows.
2. *Vesicles:* These are usually round or ovoid sacs. They often occur isolated in the cytoplasm.
3. *Tubules:* They are irregularly branched tube-like structures. These structures are surrounded by thin unit membrane of 50-60 A° thickness and their lumen is filled with the secretory products of the cell.

Types of ER

Two types of endoplasmic reticulum have been recognized. These may be present in the same or different types of cells.

1. ***Agranular or smooth endoplasmic reticulum (SER).*** The surface of this type of ER is smooth, ribosomes being not attached. Smooth ER is present in cells which are actively engaged in steroid synthesis (e.g., cholesterol, progesterone, testosteron, etc.), carbohydrate metabolism, pigment production, etc.
2. ***Granular or rough endoplasmic reticulum (RER).*** The rough endoplasmic reticulum have ribosomes

The gestration period of pig is 112-120 days.

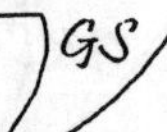

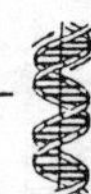

attached throughout the surface. This type of ER is present in cells which show active protein synthesis.

Functions

ER is involved in the following major functions:-

1. Functions common to both smooth and rough ER.

(i) Forms skeletal framework.
(ii) Active transport.
(iii) Metabolic activities due to enzymes.
(iv) Provides increased surface area.
(v) Formation of new nuclear membrane during cell division.

2. Functions of smooth ER.

(i) Lipid synthesis.
(ii) Glycogen synthesis.
(iii) Steroid synthesis like cholesterol, progesterone, testosterone, etc.

3. Functions of rough ER.

(i) It provides site for protein synthesis.
(ii) It helps in transport of proteins.

Golgi Complex

Distribution

Golgi apparatus is present in all the eukaryotic cells but it is absent in prokaryotic cells. Among eukaryotes, Golgi complex is not found in the cells of fungi, male gametes of bryophytes and pteridophytes, mature sieve tubes and mature sperms and red blood cells of animals.

Structure

It consists of three main parts—*cisternae, tubules* and *vesicles.* Cisternae, which form the central flattened or disc-shaped part of Golgi complex, are stacks of parallel double membranes or lamellae. In a plant cell 2 to 7 such cisternae constitute one Golgi apparatus.

Tubules arise from the periphery of the cisternae and form a highly branched anastomosing network. Vesicles are of two types—(i) *rough vesicles* present at the ends of the cisternae and (ii) *smooth vesicles* usually present within the network or sometimes near the centre of the stack of cisternae.

Golgi apparatus forms an extensive inter-communicating membrane system in association with cell membrane, endoplasmic reticulum, lyosome and nuclear membrane. It is also called *endomembrane system.*

Golgi apparatus is very rich in phospholipids, proteins, various enzymes, carotenoids, fatty acids and vitamin C.

Functions

The following are some of the major functions of Golgi apparatus.

1. Synthesis of polysaccharides.
2. Formation of glycoproteins by combining carbohydrates with proteins.

Knowledge World!

Highest per capita power consumption (U.T.)

Dadra and Nagar Haveli had a power consumption of 1,392 KW per head. Daman and Diu with 1,182 KW, ranked second.

3. Synthesis of pectins and other carbohydrates necessary for cell wall formation. Secretory vesicles of Golgi apparatus fuse and get incorporated into cell membrane. Similarly, in the plant cells, Golgi apparatus deposits pectic substances and cellulose microfibrils to form a cell plate during cell division.
4. It also secretes gum and mucilage.
5. It is associated with storage, condensation, packaging and transportation of various substances.
6. It is also active in the transformation of one type of membrane into another.
7. Secretory vesicles and lysosomes originate from Golgi apparatus.

Ribosomes

Distribution

Ribosomes are present in both prokaryotic and eukaryotic cells. In eukaryotic cells, they are free as well as attached to the endoplasmic reticulum and nuclear membrane. Ribosomes are also present in mitochondria and chloroplasts. There may be as many as 5,00,000 ribosomes in an ordinary plant cell. In prokaryotic cells ribosomes are free.

Structure

These dense and rounded granules are perhaps the *smallest cell organelles* and can be seen only with the electron microscope.

Ribosomes are isolated by differential centrifugation, and its sedimentation coefficient is measured in *Svedberg Units* or S units. On the basis of sedimentation coefficient two types of ribosomes have been recognised.

1. 70 S ribosome: This type of ribosome is smaller of the two. 70S type ribosomes are found in all prokaryotic cells like bacteria and blue green algae and also in chloroplasts and mitochondria of eukaryotic cells. These consist of two sub-units—50 S and 30 S,50 S sub-unit is larger and the smaller 30 Svub-unit is placed over it like a cap

2. 80 S ribosome: This type of ribosome is larger in size with a sedimentation coefficient of 80 S. It has a molecular weight of 40 x 10^6 daltons. These ribosomes are found in the cells of animals and plants (eukaryotic cells). 80 S ribosome consists of two sub-units –60 S and 40 S. 60 S sub-unit is larger and the smaller 30 S sub-unit is placed over it like a cap.

Sun is a hot sphere of gases.

Each sub-unit of ribosome is further composed of smaller sub-units. For example, 50 S sub-unit of 70 S ribosome consists of core particles 40 S and split proteins –SP 50. Similarly 80 S ribosomes are also made of smaller sub-units. Core particles are further composed of RNA and core particles. The amount of RNA present in the ribosomes is about 70-75% of the total amount of RNA present in the cell.

The association of ribosomal sub-units is dependent upon the concentration of Mg^{++} ions. This association requires very low concentration of Mg^{++} (0.001 M). If the concentration is lesser, both the ribosomal sub-units remain separate. If the concentration of Mg ++ ions is increased ten folds, two ribosomes get associated to form *dimer*. In bacterial cells, the two subunits occur freely in the cytoplasm.

During protein synthesis many ribosomes occur in a chain on common m-RNA strand and are called *polyribosomes or polysomes*

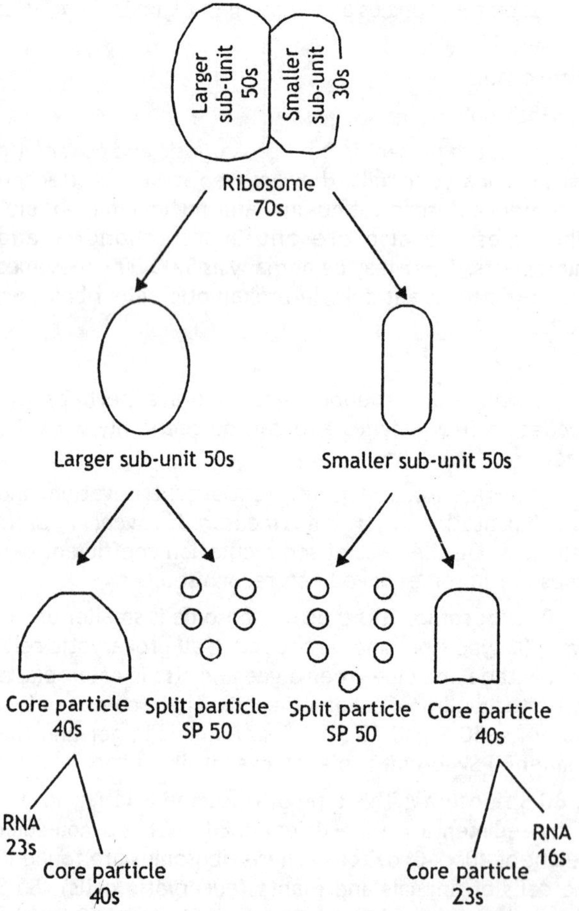

Ribosome: Components of 70 S ribosome.

Ribosomes (80S) remain attached to endoplasmic reticulum with their 60 S end by electrostatic reactions and also be the nascent polypeptide chain that grows from the ribosome and penetrates across the membrane. These could also be bound with ER by two transmembrane glycoproteins - *ribophorin* I and *ribophorin* II.

Functions

Ribosomes are the centres of protein synthesis.

Lysosome

Distribution

Lysosomes are common in animal cells but are also found in plant cells. These can be observed in secretory cells like pancreatic cells, spleen cells, leucocytes, liver cells, meristematic plant cells, etc.

Functions

The following are the main functions of lysosomes.

1. *Digestion of external particles:*Lysosomes, pinocytes and phagocytes digest the external food particles. Lysosomes of leucocytes digest proteins, bacteria and viruses.
2. *Digestion of intracellular substances:* Lysosomes digest stored materials like proteins, fats and glycogen to provide energy to the starved cells. Lysosomes in plant cells hydrolyse stored material in germinating seeds.
3. *Autolysis:* Under certain pathological conditions lysosomes digest cell organelles. This is also called *autophagy* and results in the death of the cell. Hence, lysosomes are also known as *suicidal bags*.
4. *Extracellular digestion*: Sometimes enzymes from lysosomes are liberated outside the cell, e.g., the lysosomes of sperm release enzymes which dissolve the protective coating of ovum.
5. *Mitotic division*. Lysosomes are said to initiate the process of mitosis.

Microbodies

In protozoans, fungi, some plants, liver and kidney, some small membrane-bound organelles are found associated with ER, mitochondria and chloroplasts. These are called microbodies. The following are some of the common types of microbodies:

Peroxisomes

The term peroxisome was introduced by *Beaufaytt* and *Berther* in 1963. These are found in many plant and animal cells. Peroxisomes mainly occur in photosynthesizing cells of higher plants, mature pear fruit, algae, fungi, liverworts, mosses and cells of ferns.

It consists of a granular matrix enclosed by a unit membrane. A nucleoid that consists of parallel tubules or twisted strands, occupies the centre, The enzymatic contents of animal and plant peroxisomes differ. However,

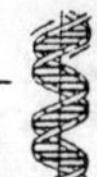

they contain some peroxodie producing enzymes like urate oxidase.

The exact function of peroxisomes is not known. However, glycolate metabolism is known to occur in them. Thus these are associated with photorespiration in plants (C_3 plants) and lipid metabolism in animal cells.

Sphaerosomes

They originate from endoplasmic reticulum by budding. Sphaerosomes are bound by a single membrane enclosing enzymatic proteins which synthesize fats and oils. Like lysosomes, sphaerosomes also contain phosphatases, esterases, ribonucleases and hydrolases. Thus, sphaerosomes do not appear to be different from lysosomes, except for their specific lipidic nature. The main function of spherosomes is to synthesize fats and their transportation.

Glyoxysomes

These occur in the cells of yeast, *Neurospora* and oil seeds of many higher plants. They have enzymes for fatty acid metabolism and gluconeogenesis (conversion of lipids of germinating seeds into carbohydrates).

Centrosome

Distribution

Centrioles are found in algae, fungi, bryophytes, ferns, gymnosperms, protozoans and most of the animals. However these are absent from the cells of prokaryotes, diatoms, yeast and most of the higher plants (e.g., conifiers and angiosperms).

Functions

Centrioles take part of the formation of basal bodies and mitotic spindle, and facilitate the movement of separating chromosomes during nuclear division.

Microtubules

Distribution

Microtubules are found in the cytoplasm of plant and animal cells, in cilia, flagella, centriole, basal body, nerve processes, mitotic apparatus, meristematic plant cells, etc.

Functions

The following are the main functions of mictotubules :

I. These are related with the movement of flagella and cilia.
2. During cell division they form spindle fibres which control the movement and alignment of chromosomes.

Nucleus

History

The credit for the discovery of nucleus goes to *Robert Brown* (1831). He observed the nucleus in the cells of an orchid. *J. Hammerling* (1953) conducted experiments on a greenalgae *Acetabularia* and proved hereditary role of nucleus.

Quiz Kit!

Most dangerous insect

The housefly has the dubious distinction of being the most dangerous to man since it is known to be a carrier of nearly 40 serious diseases.

Distribution

Nucleus is present in all the eukaryotic cells. Some cells contain nucleus when they are young but it degenerates when cells mature, e.g., sieve tubes in plants and RBC in mammals. In prokaryotic cells 'true nucleus' is absent.

Generally, there is one nucleus in each cell, but this number may vary in different types of cells. Depending upon the number of nuclei, cells are called uni–, bi–or multinucleate. Multinucleate condition may arise either by fusion of many cells (*e.g.*, syncytium of coconut endosperm) or due to repeated nuclear divisions (*e.g.*, coenocyte of *Vaucheria*).

Parts of the nucleus

The nucleus can be easily distinguished into following four parts :

(1) nuclear membrane,
(2) nucleolus,
(3) nucleoplasm and
(4) chromatin

1. ***Nuclear membrane***. The nuclear membrane or *karyotheca* is the outer envelope of the nucleus. It occurs in all the eukaryotic nuclei except during later part of the cell division. In prokaryotic cells, nuclear membrane is always absent.

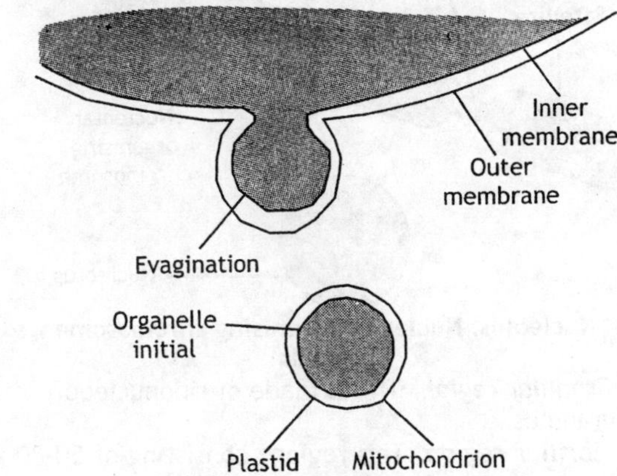

Nuclear membrane (envelope) : Formation of organelles

Nuclear membrane separates nucleus from the cytoplasm. It consists of two unit membranes. The space betweenthe two membranes, called perinuclear space, is about 75 Å. The outer unit membrane is continuous with endoplasmic reticulum. The nuclear membrane is interrupted by pores. These pores are usually octagonal and their diameter

GK GS

Tropical Grassland is found between 5° to 15° N & S.

varies from 300-1,000 Å. These are called nuclear openings or pores. They maintain continuity between nucleo-cytoplasmic or regions. Nuclear pores, however, do not freely communicate with the cytoplasm. They are plugged by a cylinder of protein material called annulus. The passage of ions and small molecules through the pores is regulated by annulus.

The nuclear membrane breaks into small vesicles during prophase of cell division and gets dispersed in the ground substance. Later, during telophase, these vesicles either fuse once again to form a nuclear membrane or new vesicles are formed to produce a nuclear membrane.

Nuclear membrane allows a free exchange of ions. Nuclear pores permit mRNA to move out of the nucleus into the cytoplasm. Evagination of nuclear membrane results in the formation of organelle initials which may later transform into organelles like mitochondria and chloroplasts.

2. ***Nucleolus:*** This organelle was first seen by Fontana (1781) and was termed nucleolus by Bowman (1840). Nucleolus is a spherical, dense, colloidal and acidophilic body that remains attached to a special type of chromosome—nucleolar organising chromosome at a specific place called nucleolar organising region or secondary constriction. Nucleolus disappears during prophase of mitosis and meiosis and reappears during telophase in mitosis and telophase 11 in meiosis.

A cell may have one or more nucleoli. The size of the nucleolus also varies depending upon its synthetic activity. Nucleolus consists of three parts as given :

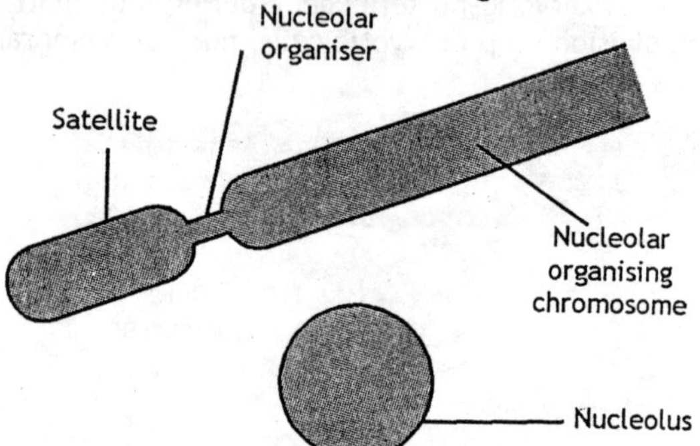

Fig. 4. Nucleolus: Nucleolar organising chromosome with nucleolus

(a) *Granular region.* This is made of ribonucleoprotein granules.

(b) *Fibrillar region.* This region, consisting of 50-80 Å long proteinaceous fibrils, is called nucleonema.

(c) *Amorphous matrix.* This, electrically less dense region, shows the presence of amorphous matrix and is called 'parsamorpha'.

3. ***Nucleoplasm.*** The nucleus is filled with a transparent, semi-solid, granular and acidophilic substance, known as nuclear sap, nucleoplasm or karyolymph. Nucleolus and chromatin fibres appear scattered in this sap.

Nucleoplasm is mainly made of nucleic acids, proteins, enzymes, lipids and minerals. The nucleic acids—DNA and RNA are found in the form of nucleoprotein. Proteins occur as basic proteins or histone proteins (*e.g.*, nucleoprotamines and nucleohistones) and acidic prote s or non-histone proteins (*e.g.*, phosphoprotein). Some of the important nuclear enzymes include DNA polymerase, RNA polymerase, nucleoside triphosphates, aldolases, enolases, etc. Lipids are found in small amounts. Minerals like phosphorus, potassium, sodium, calcium and magnesium are also present.

CELL PARTS AND THEIR DISCOVERERS

Part of the Cell	*Discoverer*
Cell theory, (1839)	Scheleiden & Schwann
Living Cell, (1723)	Antony von Leeuwenhoek
Primary cell wall, (1665)	Robert Hooke
Plasmodesmata, (1901)	E. Strasburger
Protoplasm, (1840)	J.E. Purkinje
Term Sarcode, (1835)	Dujardin
Hyaloplasm, (1846)	Hugo von Mohl
Protoplast, (1880)	Hanstein
Cell membrane, (1855)	Nageli & Cramer
Termed plasmalemma, (1931)	J.Q. Plowe
Unit membrane model, (1959)	Robertson
Fluid Mosaic model (1974)	Singer & Nicoloson
Endoplasmic reticulum, (1945)	K.R. Porter
Plastid, (1865)	E. Haechel
Term chloroplast, (1885)	A. F. W. Schimper
Mitochondrion, (1880)	A. Kolliker
Named fila, (1882)	W. Flemming
Named bioblast, (1894)	Altmann
Named mitchondrion, (1867)	Benda
Observed in plants, (1904)	F. Meves
Golgi body, (1898)	Camilo Golgi* (NL 1906)
Ribosome, (1943) observed & Termed microsome	A. Clande * (NL 1974)
Discovery of RNA in ribosome, (1955)	*Palade (NL 1974)
Term ribosome given, (1958)	Haguenau
Lysosomes, (1955)	Christian de Duve * (NL 1974)
Observed in plants, (1964)	P. Matile
Peroxisomes, Term given, (1963)	Beaufaytt & Berther
Microtubules, (1953)	De Robertis & Franchi
Centrosome, (1888)	T. Boveri
Nucleus, (1831)	Robert Brown
Nucleolus, (1781)	Fontana
Described, (1838)	Schleiden
Chromosome, observed, (1875)	E. Strasburger
Term chromosome given, (1888)	Waldeyer
Chromatin, (1879)	W. Flemming
Nucleosome, (1975)	Oudet, Gross-Bellard & Chambon
Nucleic acid,	
Term nuclein given, (1869)	F. Mlescher
DNA stained, (1924)	R. Feulgen & H. Rossenbeck
DNA, Structure proposed (1953)	Walson & Crick* (NL 1962)
DNA synthesized, (1959)	A. Kornberg (NL 1959)*
RNA synthesized, (1959)	S. Ochoa (NL 1959)*

* Nobel Prize winners, NL = Nobel Laureate or Nobel Prize winner

Pineapple is a composite fruit.
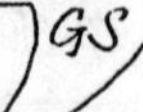

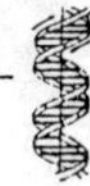

CELL ORGANELLES: STRUCTURE AND FUNCTION

Cell wall	A rigid cell wall surrounding the cell, consisting of cellulose microfibrils running through a matrix of other complex polysaccharides, namely hemicelluloses and pectic substances. May be secondarily thickened in some cells,	Provides mechanical support and protection. It presents osmotic bursting of the cell. It is totally permeable and hence pathway for movement of water and mineral salts. Various modifications, such as lignification, for specialised functions.
Middle lamella	Thin layer of pectic substances (calcium and magnesium pectates).	Cements neighbouring cell together.
Plasmodesmata	A fine cytoplasmic thread linking the cytoplasm of two neighbouring cells through a fine pore in the cell walls.	Enables a continuous system of protoplasm, to be formed between neighbouring cells for transport of substances between cells.
Plasma membrane	Two layers of lipid (bilayer) sandwiched between two protein layers.	A differentially permeable barrier controlling exchange of water and minerals between the cell and its outside.
Endoplasmic reticulum	A system of flattened, membrane-bounded sacs called cisternae, forming tubes and sheets. It is continuous with the outer membrane of the nuclear envelope.	If ribosomes are found on its surface it is called **rough ER** and transports proteins made by the ribosomes through the cisternae. **Smooth** ER (no ribosomes) is a site of lipid and steroid synthesis.
Chloroplast	Large plastid containing chlorophyll and carrying out photosynthesis. It is surrounded by an envelope of two membranes and contains a gel-like stroma through which runs a system of membranes that are stacked in places to form grana. It may store starch. The storma also contains ribosomes, a circular DNA molecule and lipid droplets.	It is the organelle in which photosynthesis takes place, producing sugars and other substances from carbon dioxide and water using light energy trapped by chlorophyll. Light energy is converted to chemical energy.
Mitochondria	Surrounded by two membranes, the inner being folded to form **cristae**. Contains a **matrix** with a few ribosomes and circular DNA molecule.	In aerobic respiration cristae are the sites of oxidative phosphorylation and electron transport, and the matrix is the site of Krebs cycle enzymes and fatty acid oxidation.
Golgi apparatus	A stack of flattened, membrane, bound sacs, called **cisternae**, continuously being formed at one end of the stack and budded off as vesicles at the other. Stacks may form discrete dictyosomes as in plant cells, or an extensive network as in many animals cells.	Processing in cisternae and transport in vesicles of many cell materials, such as enzymes from the ER. Often involved in secretion and lysosome formation.
Ribosomes	Smallest cell organelle with cell membrane, It consists of a large and a small subunit. They are made of roughly equal parts of protein and RNA. Ribosomes are found in mitochondria (and chloroplasts in plants).	Sites of protein synthesis. They are either bound to the ER or lie free in the cytoplasm. They may form **polysomes** (polyribosomes), chain of ribosomes—along messenger RNA.
Lysosome	A simple spherical sac bounded by a single membrane and containing digestive (hydrolytic) enzymes, contents appear homogeneous.	Many functions, all concerned with breakdown of structures of molecules like autophagy, autolysis, endocytosis and exocytosis.
Microbodies	A roughly spherical organelle bounded by a single membrane. Its contents appear finely granular except for occasional crystiloid or filamentous deposits.	All contain catalase, an enzyme that breaks down hydrogen peroxide. All are associated with oxidation reactions. In plants, are the site of the glyoxylate cycle.
Centriole	Cylindrical structure without limiting membrane. It consists of 9 triplet fibres, each consisting of three secondary fibres or tubules. These are enclosed in amorphous matrix. Centrosome separates into a pair of centrioles. Mostly found in animals and a few plants.	Centrioles take part in formation of basal bodies and mitotic spindle.These also facilitate the movement of chromosomes during cell division.

GS

Sun has three zones : photosphere, chromosphere and carona.

GENETICS

INTRODUCTION TO GENETICS

Genetics is a science which unravels the inheritance of various character from one generation to another. The term first used by Bateson in 1905.

Pre Mendelian Theory about Heredity

1. *Performation Theory*

Swammerdam and Bonnet (1770-1793). This theory states that a minute human called humanculus was already present in egg and sperm.

2. *Theory of Epigenesis*

Wolf (1738-94). Egg is differentiated into various organs takes place only after fertilization from the zygote resulting in various organs universally accepted. This concept view of origin of adult organ and tissue from zygote.

3. *Theory of Acquired Characters*

Lamarck (1744-1829). New characters once acquired an individual shall pass on to its progeny.

4. *Theory of Pangenesis*

Darwin (1809-1882). Small, invisible copies of each body organs (gemmules) are transported by the blood tissue.

5. *Germplasm Theory*

Weismann (1889). Body tissue in two parts germplasm and somatoplisms.

Contribution of some Geneticists

1. Mendel : Father of Genetics
2: Bateson : Linkage, Sex determination.
3. Johannsen W.L. : Pure line, Phenotype and Genotype
4. Morgan T .H. : Chromosomal theory of heredity
5. Muller H.J : X-ray - Physical mutant
6. Beadle and Tatum : One gene one enzyme hypothesis
7. Avery O.T ; MacLeod, C.M. and McCarthy : Bacterial transformation
8. Watson and Crick: Molecular Structure of DNA
9. Barbara McClintock : Jumping Gene Transposons
10. Benzer, S : Unit of functions cistron, Recon and muton
11. Jacob and Monod : Operon concept
12. Nirenberg M. W. and Khorana HG : Deciphers the genetic code

Organism Used in Genetics

I. Bacteria : *E. coli*
2. Bacterio Phage: X Phages
3. Insect : Drosophila
4. Fungi: *Neurospora crassa*
5. Plant :
(i) Corn/Maize
(ii) Sweet pea
(iii) Arabidopsis
6. Nematode

Cell Division

Cell is a basic unit of structure and function in all living systems. The process of formation of new cells from the pre-existing cells is referred to as cell division. The cell which undergoes division is known as mother cell and the new cells which are formed by the process of cell division are termed as daughter cells. In bacteria, cell division takes place by fission of pre-existing cell. In eukaryotes there are two types of cell division viz. mitosis and meiosis.

(a) Mitosis

It is the process by which a cell divides into two daughter cells, each having a nucleus with sets of chromosomes exactly identical to the parent cell. Cell cycle is divided into interphase, mitosis and cytokinesis.

I. *Interphase*

After the telophase stage of previous mitotic division and before the onset of prophase of the next one. Chromosomes are fully extended and uncoiled. Nuclei are surrounded by a nuclear envelope. It is the longest stage. Three substages -

(i) **G_1 Phase:** Between the beginning of interphase and the beginning of S phase. Protein and RNA Synthesis, Ribosome production in nucleoli, takes place during this phase.

(ii) **S Phase:** DNA Synthesis takes place as a result of which the chromosomes replicate. Production of two sisters chromatids from the single chromatid of each of the chromosomes in the nucleus. All the histones of the nucleus are synthesised. A DNA segment capable of undergoing replication is known as a *replicon.* Each chromosome has several replicons. Each replicon has a specific point the *initiation point* from where DNA replication is *initiated.* At the end of replication of all the replicons of an eucaryotic chromosome two sister chromatids are produced from the single chromatid of each chromosome. Since the two chromatids of a chromosome are produced by semi-conservative replication of a single parent chromatid they are known as sister chromatid.

(iii) **G_2 Phase:** Period between termination of DNA synthesis and beginning of prophase of the next division. RNA and protein, particularly components of microtubules synthesis. Some mRNA and Ribosomes are produced.

RNA and protein synthesis are abundant during G_1 and G_2 depressed during S, prophase and up to early to mid-metaphase and absent during late metaphase, anaphase and up to early telophase.

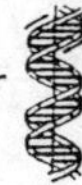

2. *Prophase*

The appearance of definite thread-like structures in nuclei. Chromosomes become increasingly shorter and thicker due to increased condensation. Nucleous and nuclear membrane remain present. Spindle apparatus is not yet organised.

Chromosome condensation due to coiling of chromosomes—

Relational Coiling :

- Plectonimic coiling (Mitotic prophase)
- Paranimic coiling (Meiotic prophase)

(i) *Plectonimic Coiling*: Sister chromatids are twisted round each other, cannot separate from each other.

(ii) *Paranimic Coiling*. The coils of one of the sister chromatids are simply slipped into those of the other can easily separate from each other.

3. *Metaphase*

(i) Disappearance of Nucleous

(ii) Breakdown of nuclear envelope and distribution of its components into the ER

(iii) Appearance of spindle fibre

(iv) Arrangement of the chromosomes on a single plane called equatorial plate.

(v) The shortest and the thickest form of chromosomes

(vi) Absence of relational coiling between sister chromatids.

4. *Anaphase*

Anaphase begins when the centromeres of chromosomes appear to divide longitudinally so that the sister chromatids separate from each other and ends with the reaching of the chromosomes to the opposite poles.

5. *Telophase*

(i) The chromosomes uncoil so that they become very long and thin

(ii) Nucleous reappears

(iii) Nuclear membrane is reorganised

(iv) Phragmoplast appears at the equatorial region of the cell.

Cytokinesis

It is completed by the end of telophase. At the equatorial plate, membranous elements of ER and products of golgi bodies organise into the phragmoplast gives rise to cell plate and subsequently to middle lamella. In plants the separation of cytoplasm begins in the centre of the cell and gradually extends outwards on each side in a plane perpendicular to the axis of the spindle. The two daughter cells produced by mitosis contain one nucleus each, each nucleus has the same number of chromosomes as the parent cell.

Significance of Mitosis—

- Mitosis is responsible for development of a zygote into adult organism.
  It is essential for normal growth and development of living organisms
- It is useful in maintaining the purity of types because it leads to production of identical daughter cells.

Did You Know ?

Largest post office

India's busiest and biggest post office is the Mumbai General Post Office (GPO) with 101 business counters. Around 40,000 people visit it every day.

(b) Meiosis

Meiotic division of one somatic (2n) cell gives rise to 4 haploid (n) daughter cells.

The redution in chromosome number of the daughter cells is mainly due to a single S phase during the pre-meiotic interphase followed by two successive nuclear divisions.

1. *Initial meiotic*: Meiosis occurs shortly after the zygotes are produced eg. fungi, yeast.
2. *Middle meiotic*: Meiosis produce spores which gives rise to the gametophyte generation. Gametes are produced through mitosis in the gametophyte eg. higher plants.
3. *Terminal meiotic*: Meiosis occurs at the end of the life cycle and it directly leads to the production of gametes eg. animals.

Stages of Meiosis

Premeiotic Interphase: The interphase stage just prior to the entry of a cell into meiosis.

During the S phase of premeiotic interphase chromosome replication takes place. But approximately 0.3% of the total DNA present in the nucleus does not replicate during the S phase. This DNA replicates during the zygotene substage of prophase I.

A special type of histone specific to cells preparing for meiosis is synthesized during S phase.

Thus during meiosis chromosomes replicate only once, while the nuclei divide twice.

First Meiotic Division:

Events—

(i) Pairing between homologous chromosomes.

(ii) Crossing over between them during prophase I.

(iii) Separation of homologous chromosome and their migration to the opposite poles of a cell during anaphase I.

First Prophase :

Longest in duration comparatively mitotic prophase.

Main events—

1. Intimate pairing between homologous
2. Condensation of chromosomes
3. Crossing over between homologous chromosome
4. Movement of homologous chromosome away from each other (without separation) so that chiasma become observable
5. Terminalization of chiasma
6. Movement of chromosomes during chromosome pairing and chiasma terminalization.

GK **Wular and Dal lakes are tectonic lakes.** GS

Substages :

(i) Laptotene :

(a) Increase in the nuclear volume

(b) Chromosome condensation

(c) RNA synthesis

(d) Proteins required for condensation are synthesised.

(ii) Zygotene

(a) Pairing between homologous chromosome.

(b) Completion of replication of the remaining 0.3% DNA of each nucleus. This DNA synthesis is referred to as Z-DNA synthesis. (or) zygotene DNA synthesis.

(c) Synthesis of a specific nuclear protein.

(d) Development of the synaptenemal complex.

(e) Progressive condensation of chromosomes.

Pairing between homologous chromosome is often referred to a synopsis.

(i) May begin at both the ends of a homologous pair and proceed toward its centre

(ii) It may begin at the centromere and progress toward the telomeres.

(iii) It may begin simultaneously at several places. Chromosome pairing occurs in two stages.

I[st] Homologous chromosome move close to each other and stay about 1000 Å away from each other.

2nd Synaptenemal complex develops in this 1000 Å region between the two homologous chromosome.

S. Complex is composed of protein. It consists of 3 distinct elements–

1. dense central element
2. two dense lateral elements
3. two less dense transverse elements.

(iii) Pachytene

(a) Further condensation of chromosomes.

(b) As a consequence of synapsis between homologous chromosome the number of chromosomes is only half of the somatic chromosome number, that is only n. The haploid number of chromosome pairs commonly referred to as bivalents. Since each of the two chromosomes of a bivalent consists of two sister chromatids, each bivalent has four chromatids as strands. So, bivalents are said to be in four strand (or) tetrad stage.

(c) Chromosomes are easily recognisable during this stage of the nucleous more crossing over between homologous chromosomes.

(iv) Diplotene

(a) Homologous chromosome of each bivalent begin to move away from each other

(b) The two homologous of each bivalent appear to be attached with each other; these attachments are known as chiasma.

(c) As diplotene progresses chiasmata slowly move round the ends of the homologous chromosome this movement is referred to as chiasma terminalization.

(v) Diakinesis

(a) Two homologous chromosome of each bivalent are now close to one or both the telomers only.

(b) Chromosomes become shorter and thicker.

(c) Nucleolus and nuclear envelope disappear

(d) Spindle apparatus is organised

Bivalents

(i) Closed Ring Bivalent : Metacentric and submetacentric chromosomes. Chiasma is formed in both the arms of the homologous chromosome.

(ii) Open Ring Bivalent : Chiasma is formed only one of the two arms of metacentric and submetacentric

(iii) Rod-shaped Bivalent : Acrocentric and Telocentric chromosomes chiasma in their long arms only.

Differences between Mitosis and Meiosis

MITOSIS	MEIOSIS
1. Mitosis occurs in almost all organisms	1. Occurs only in sexually reproducing organisms.
2. Mitosis takes place in all somatic cells whether haploid, diploid or polyploids.	2. Takes place in diploid cells such as zygote, spore mother cells, and gamete mother cells, also occurs in polyploid cells but never in haploid cells.
3. It completes in one sequence formation of two diploid nuclei from a diploid mother nucleus.	3. It completes in two successive nuclear divisions and results in the formation of four haploid nuclei.
4. Prophase -shorter duration no subphase.	4. Prophase-longer duration and five subphases.
5. Synopsis does not take place.	5. Synopsis in zygotene stage.
6. Crossing over, chiasmata formation absent.	6. Crossing over occurs during pachytene and chiasmata are formed.
7. Each chromosomes consists of two chromatids.	7. Each bivalent consists of four chromatids
8. Metaphase → the chromosome are arranged at the equatorial plane. The centromers lies on the equatorial plane and arms remain free.	8. Bivalents are arranged at the equatorial plane during metaphase I. Arms of chromatids lie on the equatorial plane and centromers are away from it.
9. During Anaphase the chromatids separate and move to opposite poles.	9. Anaphase I homologous chromosome of each bivalent separate and move to opposite poles (disjunction).
10. Telophase → daughter nuclei are organized.	10. Telophase → two daughter nuclei may be organized they are haploid.
11. The number of the chromosome in daughter nuclei is same as in the parent nucleus.	11. Daughter nuclei are haploid and contain half of the number of chromosome of the parent nucleus.
12. Chromosome of daughter nuclei are quantitatively and qualitatively similar.	12. Chromosome of daughter nuclei are qualitatively dissimilar due to crossing over.

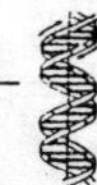

Life cycles of a typical haploid and diploids indicating :he stages when meiosis occur.

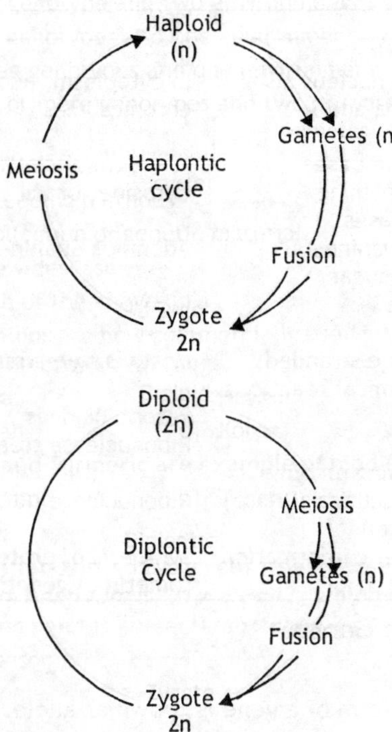

GENETIC MATERIALS

Nucleic Acid-Structure and Function

Chemical analysis of chromosomes shows two nucleic acids – DNA (Deoxyribose Nucleic Acid) and RNA (Ribose Nucleic Acid). They were first discovered by a Swiss biochemist *Fredrick Miescher* in 1869. He used the term *nuclein* to denote *nucleic* acids because of their acidic properties.

Nucleic acids are macromolecules, composed of many (polymers) small units called *nucleotides*.

Nucleic acid = many nucleotides

Each nucleotide consists of phosphoric acid (phosphate), a five carbon sugar and a nitrogenous base.

Nucleotide = phosphoric acid + sugar + nitrogenous base.

The sugar and base combination (without phosphoric acid) is called *nucleoside*.

phosphoric acid + sugar + nitrogenous base

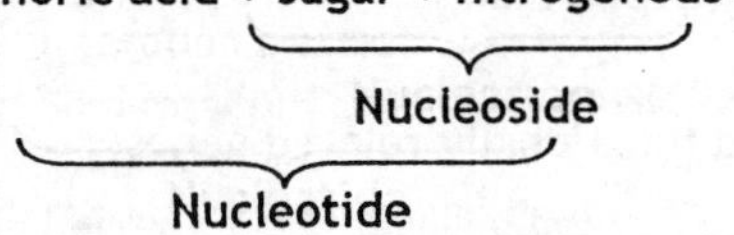

As such nucleotides are phosphoric esters of nucleosides.

I. Phosphoric acid

The acidic nature of nucleic acids is due to the presence

You Don't Know!

First at the Arctic

Sharat Kumar Roy (1897-1962), eminent paleontologist and renewed expert of meteorites, was the first Indian to visit the Arctic as a member of the Rawson-Macmillan Arctic expedition of the field museum in 1927.

of phosphoric acid. Sugar of the nucleoside combines with phosphoric acid by a phosphodiester bond.

II. Sugar

It is a five carbon (pentose) sugar. There are two types of sugars —*ribose* and *deoxyribose* sugars. The nucleic acid containing ribose sugar is called *ribose nucleic* acid (RNA) and the other with *deoxyribose sugar* is called *deoxyribose nucleic acid* (DNA).

III. Nitrogenous bases

Each nucleic acid has four nitrogenous bases —two purines and two pyrimidines. The *purine bases* are adenine and guanine and the *pyrimidine* bases are thymine and cytosine. In RNA, uracil (pyrimidine) is present in place of thymine.

DNA : Type and Structure

The DNA is a double stranded helix made of many nucleotides. The nucleotides consist of deoxyribose sugar, phosphoric acid, purine bases (adenine and guanine) and pyrimidine bases (cytosine and thymine). The arrangement of these substances in DNA molecule had long been a subject of curiosity. It was largely due to X-ray diffraction studies of *Wilkins* which provided basis for promising double helical structure of DNA. *Watson* and *Crick* (1953) finally described the structure of DNA and were awarded the Nobel Prize in 1962 along with Wilkins.

The following are some of the characteristic features of this model.

1. Each nucleotide consists of sugar phosphate and a nitrogenous base. Many such nucleotides are linked to form a polynucleotide chain or strand.
2. The adjacent nucleotides of the same strand are joined to each other by one phosphodiester bond between 5-carbon of sugar of one nucleotide and by another phosphodiester bond with 3-carbon of sugar of the next nucleotide.
3. Nitrogenous base is attached at 1-carbon of sugar. At this place purine is attached by its *9th-position* and pyrimidine by its *3rd-position*.
4. Thus a polynucleotide strand consists of sugar and phosphate forming its long axis.
5. The two polynucleotide strands are complementary to one another. If one strand has adenine, the other strand would have only thymine opposite it. Similarity guanine and cytosine form the other

GK

GS

The term 'socialist' has been inserted by 42nd Amendment Act.

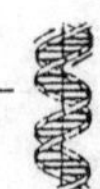

complementary base pair. Thus, if the base sequence of one strand is CAT TAG GAC, the base sequence of another strand would be GTA ATC CTG.

6. The two strands are joined to one another by hydrogen bonds between their complementary nitrogenous bases. There are two hydrogen bonds between adenine and thymine and three hydrogen bonds between cytosine and guanine.
7. Two polynucleotide strands are helically coiled around a common axis to form a DNA-molecule. The two strands are *antiparallel*, i.e., they run in opposite directions; the sugar molecule in one shows 5–P, 3–OH direction while the other shows 3-OH, 5-P direction.
8. The helical coiling of double strands is right handed. This DNA is called *B-DNA*.
9. Double stranded DNA molecule has a diameter of 20 Å, i.e., the distance between two polynucleotide strands is 20 Å.
10. The helix makes one complete turn every 34 Å along its length.

RNA : Type and Structure

RNA or Ribose Nucleic Acid is present in all the living cells. It is found in cytoplasm as well as nucleus.

I. Types of RNA

RNA is generally involved in protein synthesis but in some viruses it also serves as a genetic material. The following are the types of RNA.

I. **Genetic RNA: H. Fraenkel-Conrat** (1957) showed that RNA present in TMV (Tobacco Mosaic Virus) acts as a genetic material. Since then it is known to be the genetic material of most of the plant viruses and some bacteriophages.

2. **Non-genetic RNA:** This type of RNA is present in those cells where DNA is the genetic material. Non-genetic RNA is synthesized on DNA template. It is of the following three types.

(a) ***Messenger RNA (m RNA):*** It carries the genetic information present in the DNA. It forms about 5-10% of the total RNA present in the cell.

(b) ***Transfer RNA (t RNA):*** It is also known as **soluble** RNA (s RNA). These are the smallest molecules which carry amino acids to the site of protein synthesis. It forms about 10-15% of the total cell RNA.

(c) ***Ribosomal RNA (r RNA):*** It is the most stable type of RNA and is found associated with ribosomes. It forms about 80% of the total cell RNA.

RNA is generally single-stranded and made of a chain of polynucleotides. The single strand is folded in such a way that the chain formed by sugar and phosphate is external while the complementary nitrogenous bases are projected inside and joined by hydrogen bonds as in DNA.

The differences between DNA and RNA are given in the following table.

DNA and RNA: A Comparison of their Structure, Reactions and Role in the Cell

DNA	RNA
1. Localization	
Primarily in nucleus also in mitochondria and chloroplasts	In cytoplasm, nucleolus and chromosomes
2. Pyrmidine bases	
Cytosine, Thymine	Cytosine, Uracil
3. Purine bases	
Adenine, Guanine	Adenine, Guanine
4. Pentose sugar	
Deoxyribose	Ribose
5. Strands	
Mostly double stranded	Mostly single stranded
6. Cytochemical reactions	
Feulgen	Basophilic dyes with Ribonuclease treatment
7. Hydrolysing enzyme	
Deoxyribonuclease (DNase)	Ribonuclease (RNase)
8. Role in cell	
Always genetic information	Synthesis of proteins, sometimes genetic

Alleles and Genes

ALLELE

Alternative form of a gene is known as allele.

Main feature of Allele–

1. Govern same characters of an individual
2. A haploid cell has single copy of an allele, diploid two and polyploid more than two for a character
3. All individuals may have identical alleles
4. They may be dominant and recessive type

Test for Allelism

I. Recombination test
2. Complementation test

Multiple Allele

1. Always belong to the same locus
2. Control same characters
3. No crossing over in the multiple allelism
4. Wild type is always dominant
5. Don't show complementation

Example for multiple Allele–

1. Fur colour in rabbits
2. Wing type in drosophila
3. Eye colour in drosophila
4. ABO blood group in man

Pseudoalleles

Closely linked and functionally-related genes

Characters

1. Govern different expression of the same characters
2. Pseudo alleles occupy different position on the same locus
3. Low frequency of recombination by crossing over

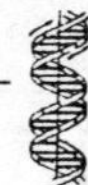

4. They exhibit cis-trans position effect

Isoalleles

An allele which is similar in its phenotypic expression to that of other independently occurring allele is known as isoallele.

Types of Isoallele

1. Mutant isoalleles
2. Normal isoalleles

GENES

Smallest and individually functional part of genetic material.

Properties of Gene

1. Form (alternative form - allele)
2. Location (chromosome, linear, locus)
3. Status (several unit)
4. Number (each diploid)
5. Sequence (specific sequence)
6. Expression incomplete, complete
7. Change in form mutation
8. Exchange of Gene translocation
9. Composition DNA bacteriophage

Function

I. Control the expression of a specific character in an organism.
2. Based on gene control over characters. Oligogenic traits, polygenic traits and pleiotropic.
3. Gene interaction (when two or more genes govern a character).
4. Linkage (two or more genes are inherited together).

Modern Concept of Gene

Genetical View

1. *Mendel Idea*

- Factor for gene → responsible for transmission of characters from parents to their offspring.
- Sutton and Boveri hypothesis → chromosomal theory of inheritance.
- Morgan → linkage studies given that idea of genes are located on the chromosome in linear fashion.

2. *Modern Concepts*

(a) Gene is divisible it was believed that gene is basic unit of structure which is indivisible by crossing over now observed that gene is divisible based on studies on Intragenic recombination like bar eye gene of drosophila.

(b) Part of a gene can function :
   1. Recon → Region between recombination.
   2. Muton → Elements of gene rise to mutant.
   3. Cistron → Unit of function of gene.

Biochemical View

(a) *Earlier Concept*: Gene is sequence of nucleotides in DNA which control a single polypeptide chain.

Knowledge World!

Longest antennae

Among the Longicorn beetles, perhaps the longest antennae in relation to body size is of the male *Chloridolum nymopha*, whose body is 3-4 cm long and antennae, 6-8 cm long.

(b) *Modern Concept*: (i) Fine structure of gene (Benzer) (ii) Split Genes.

The sequence of nucleotides were interrupted by intervening sequences. Such genes with interrupted sequence of nucleotides are referred to as split genes or interrupted genes.

It has two types of sequences. (i) Normal sequences (ii) Interrupted sequences

(c) *Jumping Genes*: A gene keeps on changing its position within the chromosome and also between the chromosomes of the same gene. Such genes are known as jumping genes of transposons or transposable elements.

The transposable elements are of two types -

(i) Insertion sequence

(ii) Transposons

(d) *Overlapping Genes*: Some nucleotide sequences (genes) can code for two or more proteins. The genes which code for more than one protein are known as overlapping genes.

(e) *Pseudo Genes*: There are some DNA sequences, especially in eukaryotes, which are non-functional and defective copies of normal genes. Such DNA sequences of genes are known as pseudogenes.

Application of Genetics

I. *Taxonomy* : To study evolution, classification and identification.
2. *Agriculture* : Improvement of crop plants yield, resistance to insects, disease, salinity, drought, frost, lodging and adaptability.
3. *In Medicine*: (i) Detection of Hereditary Disease (ii) Production of Antibiotics.
4. *Evolution* : Natural and Artifical selection have been responsible for evolution of various crop plants. (i) Polyploidy (ii) Mutation.

Genetic Material

I. Chemical Nature

Properties of the Genetic Material –

I. The Genetic material must be replicated with high fidelity.
2. Genetic material must be able to express itself.
3. Genetic material must be able to store the highly variable information.

GS

Vishnuvardhan founded the Chalukyas of Vengi.

4. The distribution of Genetic material must allow errors in a low frequency for the origin of new genetic variation.

Experiments of Griffith (1928)

His studies on Diploccocus pneumonia, different strains of Diploccocus from two types of colonies

1. Smooth 2. Rough.

Smooth colonies are enclosed in a polysaccharide capsule the strain are able to produce pneumonia, the virulent Rough colonies lack the polysaccharide capsule such strains are avirulent they can't produce pneumonia.

Virulent strains are classified into several types eg. II, III etc. on the basis of antigenic properties of the polysaccharides present in their capsule.

(a) Live IIIS (virulent) cells injected died (due to pneumonia).

(b) IIIS heat killed cells (virulent) → A live indicating that all the cells were killed by the heat treatment.

(c) Mix of heat killed IIIS and live IIR → died (pneumonia) since all the cells of heat killed IIIS culture were dead it was postulated that some of the cells IIIR had changed into the IIIS type due to influence of dead IIIS cells present in the mixture. This phenomenon was called transformation and the component of IIIS cells which induced the conversion of IIIR cells to IIIS was named transforming principle. Griffith demonstrated transformation but they did not hint at the identity of the transforming principle. It is now known that transformation is essentially a special type of recombination in which a segment from the transforming DNA replaces the homologous segment of the bacterial chromosome.

Experiments of Avery, Macleod and McCarthy–1944

Avery and associates carried out experiments of Griffith in vitro (in test tubes).

A culture of live IIR cells: Rough colonies heat killed IIIS cells or DNA isolated from IIIS cells no colony.

Live IIR + Head killed IIIS + Antibody IIR → IIIS colonies.

Live IIR + IIIS DNA + Anti IIR → IIIS colonies.

Anti IIR was used for inactivating IIR cells.

These finding show that DNA is the transforming principle.

In order to establish beyond any doubt that DNA is the transforming principle.

They treated IIIS DNA with RNA (the enzyme which digest RNA) or proteases (enzymes which degrade proteins) before it was mixed with IIR cells.

In both the experiments some IIIS type colonies were obtained. This showed RNA and proteins present as impurities in IIIS DNA preparations were not responsible for transformation.

Then next they treated IIS DNA with DNA (enzymes digests DNA) before it was mixed with IIR cells. There is no colony. They established beyond any doubt that DNA is the transforming principle.

Experiments of Hershey and Chase (1952)

This led to the universal acceptance of DNA as the genetic material.

They studied the life cycle of T_2 phage of *E. coli*. They showed that only the DNA component of T_2 particle is transmitted to the progency. T_2 has a hexagonal head and a contractile tail. Head coat and tail are made up of protein, while the DNA is packed inside the head coat. Infection begins when the tail plate of a T_2 particle comes in contact with the cell wall of an *E. coli* cell.

They labelled T_2 DNA and protein in two separate experiements, DNA contains P but no S. protein - S but no P. They labelled T_2 DNA→ ^{32}P or ^{35}S labelled phage particles E coli cells were grown for several generations on a culture medium containing ^{32}P or ^{35}S. The progeny phage particles thus obtained were labelled with either ^{32}P or ^{35}S.

In one experiment they mixed ^{32}P labelled T_2 particles with E-coli cells. The cells were than agitated in a blender to separate the empty phase particles (called ghosts) remaining outside the bacterial cells after infection. Most of the ^{32}P labelled form was present in the infected E-coli cells. The progency phage particles obtained after types of these E. coli cells also contained the ^{32}P labelled from this we come to know that DNA is to transmitted from one generation to the next.

In the other experiment they used ^{32}S labelled T_2 particles obtained through lysis of the infected E.coli cells was almost negligible. From this we come to know that proteins are not transmitted from one generation to the next.

RNA as Genetic Material (1957)

In TMV DNA is absent but these viruses are composed of RNA and protein.

The proteins and RNA of TMV can be separated when they are remixed they reassociate to produce TMV. In one experiment either RNA or proteins isolated from TMV were used for the infection of tobacco leaves. Mosaic symptoms developed only when RNA was used for infection and not the proteins were used.

They constructed two types of hybrid virus particles by mixing (1) RNA from strains A and proteins from strain B (2) Proteins from strain A and RNA from strain B.

When tobacco leaves were infected with hybrid TMV of the first type the disease symptoms of strain A developed the proteins also identical with those of strain A. Similarly when hybrid protein of the second type were used for infection of tobacco leaves symptoms of strain B. It is evident from these findings RNA (and not the protein) of TMV has the capacity to produce the disease and that the

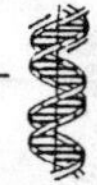

type of proteins present in the virus particle is determined by the RNA.

Chromosome Structure

The darkly stained, rod-shaped bodies visible under light microscope in a cell during metaphase stage of mitosis are referred to as chromosomes.

Main features of eukaryotic chromosomes:

- Chromosomes are not visible during interphase under light microscope.
- Transmission of characters from generation to generation.
- It varies in shape, size and number in different species of plants and animals.
- It has property of self duplication, segregation and mutation.
- It is composed of DNA, RNA and histones.

Chromosome Shape

It is usually observed during anaphase. It has three different shapes viz rod shape, S shape and V shape.

Chromosome Number

(i) *Haploid*: It refers to half of the somatic chromosome number of a species and is denoted by n.

(ii) *Diploid*: It refers to somatic chromosome number of a species and is denoted by 2n.

(iii) *Basic Number*: The genetic chromosome number of a true diploid species is called basic number.

Chromosome Morphology

- *Centromere*: The region of chromosome with which spindle fibres are attached during metaphase is known as centromere.
- *Chromatid*: One of the two distinct longitudinal subunits of a chromosome is called chromatid.
- *Secondary constriction*: The constricted or narrow region other than that of centromere is called secondary constriction.
- *Telomere*: The terminal region of chromosome on either side is known as telomere.
- *Chromomeres* : The linearly arranged bead-like structures found on the chromosomes are known as chromomeres.
- *Matrix*: A mass of acromatic material in which chromonemata are embedded is called matrix.

Karyotype

Karyotype is a phenotypic appearance of chromosomes of a particular species. It is represented by a diagram which is known as idiogram.

Karyotype is of two types viz. symmetrical and asymmetrical.

Special Types of Chromosomes

Lampbrush Chromosome: These are special types of chromosomes in which large number of loops are projected out from the chromatin axis giving a lampbrush appearance. They are formed in oocyte nuclei of both vertibrates and invertibrates and spermatocyte nuclei of drosophila during diplotene stage.

I. Extra-ordinary length
2. Large number of loops
3. Lamp-brush appearance.

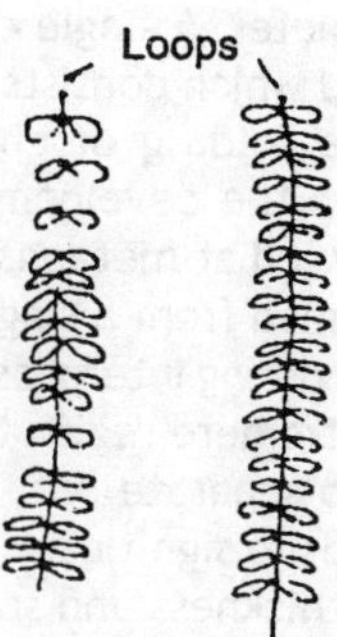

Two homologous chromosomes

Polytene or Giant Chromosomes: The multiple replicates of the same chromosome holding together in a parallel fashion resulting in very thick chromosome are known as polytene chromosomes.

These chromosomes have three main features:

1. Bands: The strips which are found in these chromosomes are known as bands.
2. Puffs: The smaller regions are known as chromosome puffs. These are the regions of genetic activity.
3. Giant Size .

B-Chromosomes:Some species possess extra chromosomes which are not members of normal chromosome complements. These are called as B chromosomes.

Classification

on the basis of stability :

- Stable
- Unstable

on the basis of size :

- Standard type
- Small type
- Very small type
- Large type

Behaviour at Mitosis and Meiosis

The meiotic behaviour of B chromosomes is studied during pachytene stage.

1. They do not pair with a chromosomes.
2. Lower degree of pairing is observed among B chromosomes.

The first ever Indian naval ship to circumnavigate the globe : INS Tarangini.

3. When single B chromosome is present, it remains univalent during pachytene.

Chromosome Models

Chromatin fibres are the basic units of chromosome structure.

1. *Folded Fibre Model*

Chromatin fibres are basic units of chromosome which are about 230 Å in diameter. A single chromatin fibre is found in each chromatid which consists of a single coiled double DNA helix. The folding of chromatin fibre in different ways results in the development of chromatin structure which is observed at metaphase. Two copies of chromatin fibre are formed from a single chromatin as a result of DNA replication during interphase. The replication of chromatin in the centromere region takes place where two chromatins have to separate out. Extensive folding of chromatin fibres leads to significant reduction in their length and increase in thickness and stainability.

2. *Nucleosome -Solenoid Model*

Chromatin is composed of DNA, RNA, histones and other proteins. Chromatin fibres are 300 Å in diameter. The nucleosomes are sub units of chromatin and have bead like appearance. Each nucleosome is composed of a histone octamer and 146 bp of DNA. Each nucleosome consists of (1) a core particle and (2) linker or spacer DNA. The core particle has two copies each of H2A, H2B, H3 and H4 histone molecules. The core particle is about 110 Å in diameter and 60 Å in height one molecule of histone H_1 is connected with linker DNA. The super coiled nucleosome fibre is known as solenoid.

Chromosomal Aberration - Structural Changes

Any change which alters the normal structure of a chromosome is known as structural chromosomal aberration.

Types

Alter gene number in chromosomes → (i) Deletion (ii) Duplication.

Alter the sequence of genes in the Chromosome → (i) Translocation (ii) Inversion.

Deletion (Deficiency)

Loss of a portion of segment from a chromosome.

Observed → Drosophila, maize, tomato, wheat. Depending upon location deletion are of two types:

I. *Terminal Deletion*: Loss of either terminal segment of a chromosome two types.

(i) Heterozygous Deletion → Deletion occurs only in one chromosome of a homologous pair.

(ii) Homozygous Deletion → Deletion occurs in both the chromosome of a pair.

2. Interstitial Deletion: Loss of a segment of chromosome from the intermediate portion or between telomere and centromere. The interstitial deletion generally does not involve centromere. In such deletion the break occurs at two places.

Detection

Cytological Method -- (i) Meiotic pairing (ii) Chromosome length.

In heterozygous Deletion: The pairing occurs between homologous segments.

In terminal Deletion: Normal chromosome remains unpaired at one end.

Interstitial Deletion: A loop is formed in the normal chromosome in the region of deletion. The loop confirms the presence of deletion.

Genetic Effects

1. *Fertility* : The pollen fertility is reduced in the presence of deletion. The pollens with deficient chromosomes are non-functional
2. *Viability* : A large deletion is lethal.
3. *Crossing over*: The crossing over is suppressed in the region of deficiency due to lack of corresponding segment in the area of deletion.
4. *Phenotype*: Deletion affects the phenotype. In the absence of dominant gene in the deletion region the recessive gene express. This results in the change in phenotype.
5. *Change in Karyotype*: The chromosomes with deletion can never revert to a normal condition. The gene number as well as the karyotype of the individual is changed.

Significance

It plays an important role in species formation and releasing variety through mutations.

Important cytological tools for mapping genes. Deletion mapping has been widely used in n drosophila to locate various genes in polytene chromosomes.

Translocation

One way or reciprocal transfer of segments between non-homologous chromosome is known as translocation.

Origin

Translocation originate through breakage and exchange of parts between non homologous chromosomes. When only one chromosome from each pair of two homologoues is involved, it gives rise to translocation hetrozygotes and when both chromosomes from each pair are involved it produces translocation homozygotes.

Detection

Translocation can be detected by cytological and genetic methods. Cytological methods includes study of pachytene configurations and metaphase configurations.

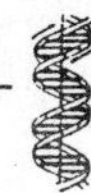

Difference between translocation and crossing over

Translocation	*Crossing over*
1. It involves in non-homologous chromosomes.	1. It involves in non-sister chromatids of homologous chromosome.
2. Change the linkage map.	2. Does not change the linkage map.
3. Breakage and reunion	3. Chiasma formation.
4. Pollen and ovule sterility.	4. No sterility.

At Anaphase the chromosome disjoin (segregate) in three different ways -

1. *Alternate Disfunction :*

When two normal chromosome (N_1 and N_2) move towards one pole and two translocated chromosome (T_1 and T_2) to another pole is known as alternate disfunction segregation. In such segregation all gametes receive full complement of genes and will give rise to viable individual.

2. *Adjacent one Segregation :*

The segregation of one normal chromosome with one translocated is called adjacent segregation. Such segregation occurs in open ring configuration. Here the chromosomes which go to one pole are non homologous ($T_1 + N_2$ and $T_2 + N_1$).

3. *Adjacent two Segregation :*

Sometimes in open ring configuration, two homologous chromosome (T_1 N_1) i. e. one normal one translocated move to one pole and other homologous (T_2 N_2) move to another pole. Such disjunction is known as adjacent 2 segregation. Both adjacent types of segregation will produce gametes with duplication and deficiencies which may cause some sterility. Translocation can be detected by genetic methods based on pollen sterility, gene segregation and linkage studies.

Genetic Effects

1. *Sterility*: Translocations lead to duplication and deletion of genes. Thus translocations result in pollen sterility and ovule sterility. If there is a ring of four chromosomes 50% sterility. If there is a ring of six chromosomes → 75% sterility
2. *Crossing over*: Crossing over is generally suppressed due to competition in pairing.
3. *Karyotype*: Changes in chromosome number and karyotype. They may alter the size of chromosome as well as position of centromere.

Phenotype

In human down syndrome (Mongolism) can arise in the progeny of an individual heterozygous for a translocation involving chromosome number 21.

Did You Know ?

First port of Independent India

Construction for the port at Kandla in Gujarat started in 1949 and the port was commissioned on April 8, 1955.

Significance

Translocation alter the chromosome size, chromosome number and karyotype and thus play an important role in the formation of species. Translocation are useful in locating the position of genes, centromere and other genetic markers on the chromosomes. They are useful tools in breeding programmes for transfer of desirable characters from one species to another.

Inversion

It refers to structural change in a chromosome in which a segment is oriented in a reverse order.

(i) *Paracentric Inversion*: The inversion in which centromere is not involved is called Paracentric Inversion. In this type both breaks occur in one arm of the chromosome.

1 2 3|4 5|6   1 2 3 5 4 6

Only one chromosome of a homologous pair has inversion is called inversion heterozygote. When both the members of a homologous pair have similar type of inversion it is called inversion homozygote. Meiosis is normal in inversion homozygotes. Crossing over within the inversion loop in a paracentric inversion heterozygote results in the formation of dicentric bridge and an accentric fragment after exchange. The other two chromosome remain normal. The dicentric chromosome leads to formation of bridge at anaphase. The bridge is later on broken due to pull from both the poles, thus a centric segment is lost due to lack of movement. Thus out of four two are normal and

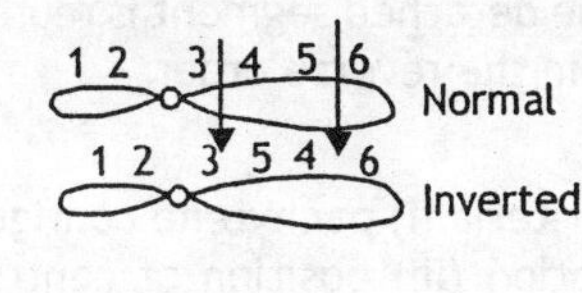

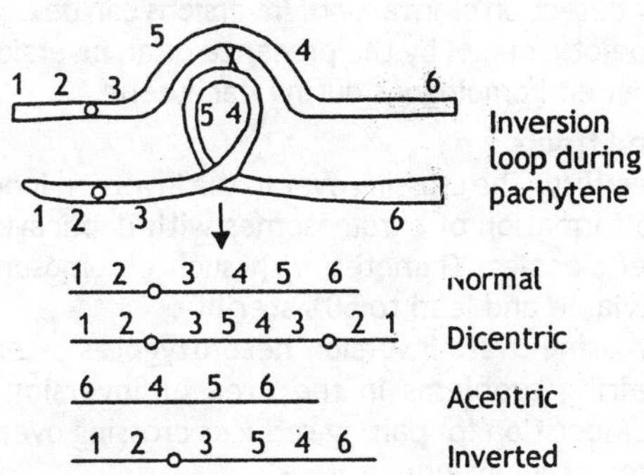

two are deficient for some genes.

(ii) *Pericentric Inversion*: When centromere is involved in the inversion it is known as pericentric inversion. When a break occurs in each of the two arms of a chromosome the centromere is included in the detached segment resulting in a pericentric inversion.

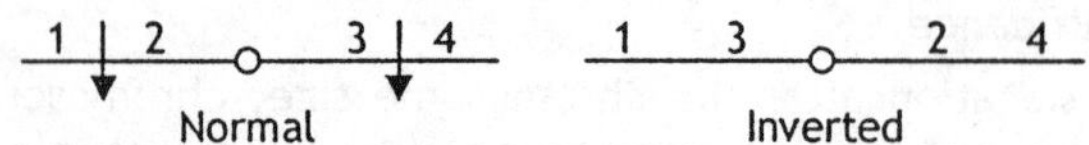

Crossing over within the inversion loop results in the formation of chromatids with duplication and deficiency. Out of four chromatids, two are cross over products and two are normal. One of the No. 6 cross over have original gene sequence and the other has inverted gene sequence.

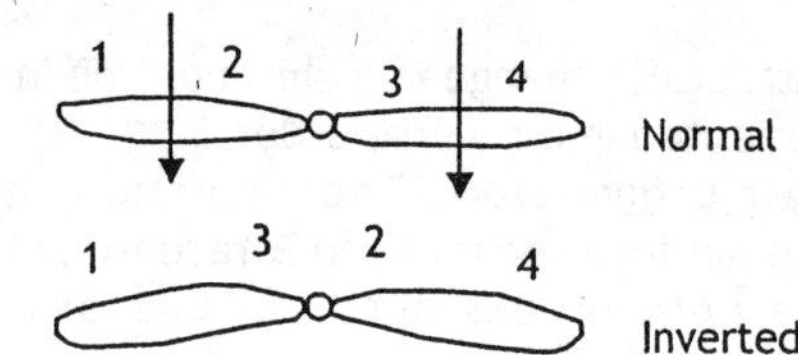

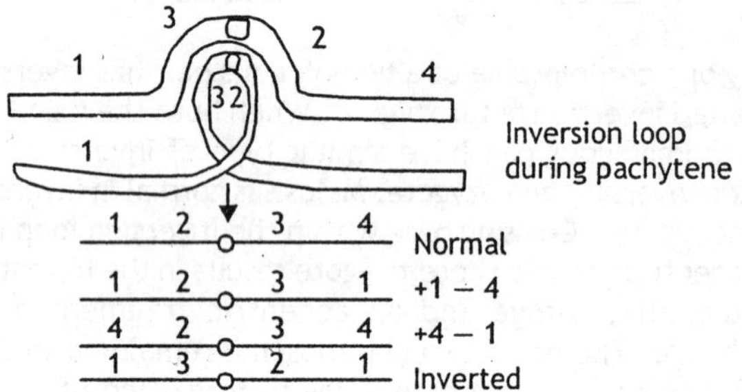

Origin

Inversion result when there are two breaks in a chromosome and the detached segment is reunited to the same chromosome in the reverse order.

Detection

Three cytological criteria (i) pachytene configuration (ii) anaphase configuration (iii) position of centromere are used for detection of inversion. Inversions can be detected in the meiotic nuclei by the presence of an inversion loop in the paired homologous during pachytene.

Genetie Effects

1. *Fertility*: The crossing over in the inversion loop leads to formation of chromosomes with duplication and deficiencies. Gametes with such chromosome are inviable and lead to 50% sterility.
2. *Crossing over*: Inversion heterozygotes often have pairing problems in the area of inversion. Thus competition for pairing reduces crossing over in the area of inversion.
3. *Gene order*: The Gene order is changed in the inverted segment of a chromosome. Inversion heterozygotes exhibit a linkage map with different gene order. In inverted chromosome there is no loss of genetic material provided crossing does not occur in the inversion loop.
4. *Karyotype*: Pericentric inversion sometimes results in change of karyotype by shifting the position of centromere in the inversion loop may lead to shift in the position of centromere.

Duplication

Occurrence of a segment twice in the same chromosome. It results in addition of one or more genes to a chromosome. First reported in drosophila by Bridges in 1919. Now it has been reported in maize, wheat. There are four types of duplication.

1. *Tandom Duplication*

Sequence of genes in the duplicated segment is similar to the sequence of genes in the original segment of a chromosome.

a b c [b c] d e f

2. *Reverse Tandom Duplication*

The sequence of genes in the duplicated is reverse to the sequence of genes in the original segment of a chromosome.

a b c [c b] d e f

3. *Displaced Duplication*

When the duplication is found away from the original segment but on the same arm of the chromosome.

a b c d e f g h i j k — Normal

a [d e] b c d e f g h i j k — Displaced

4. *Reverse Displaced Duplication*

Duplication is also away from the original segment but on the other arm of a chromosome.

These two types are known as non-adjacent duplication because they are away from the segment which shows duplication.

Origin

Duplication originate due to unequal crossing over during meiosis. The homologous chromosome usually pair in such a way that all the identical loci match with each other in their positions. This facilitates equal crossing over between non-sister chromatids. Sometimes homologous chromosome pair in such a mis-aligned manner that the corresponding identical loci do not fall opposite to each other. Such situation leads to unequal crossing over between non sister-chromatids. This gives rise to two types of chromatids viz one with duplication other with deletion. When a gamete with duplication unites with normal ovule,

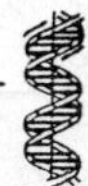

it leads to formation of zygote with duplicate genes in a particular segment of a chromosome.

Detection

b c

a b c d e f g h i j

Duplication loop can be observed during pachytene stage when homologous chromosome pair. Chromosome having duplicate segment are longer than normal chromosome. If a duplicate segment includes centromere it may be present as a small extra chromosome added to a normal chromosome complement duplication can also be detected by suppression of recessive characters.

A single dominant gene in the duplicate region is enough to suppress the expression of two recessive alleles.

Significance

Duplications are less harmful than deletions. They do not reduce the viability of an individual.

Duplications lead to addition of some genes in population.

Changes in Chromosome Number

Introduction

A basic or monoploid set of chromosomes of an individual is called genome. In a genome, each type of chromosome is represented only once. Most of the sexually reproducing plant species are diploid, i.e., have two sets of chromosomes. Any change in the chromosome number from the diploid state is referred to as heteroploidy and the individuals having chromosome number other than diploid are called heteroploids. The heteroploidy is of two types viz., I. euploidy and II. aneuploidy.

I. EUPLOIDY

The change in chromosome number which involves entire set is called euploidy. Euploidy includes, monoploids, diploids and polyploids.

Monoploids and Haploids

Monoploids contain a single chromosome set and are characteristically sterile. In a true diploid species, both monoploid and haploid chromosome number is the same *($n = x$)*. Thus a monoploid can be a haploid but all haploids cannot be monoploids.

Types of Haploids

Depending upon the origin, haploids are of two types viz., euhaploids and aneuhaploids. Euhaploids develop from a euploid species and have complete chromosome set. Euhaploids are of two types, viz., monohaploid- which develop from a normal diploid species, and polyhaploids which develop from autopolyploid species. When a haploid develops from a tetraploid species, it is called dihaploid.

Aneuhaploids develop from aneuploid species and have either one additional or missing chromosome. Aneuhaploids include disomic haploids (n + I), nullisomic haploids

Quiz Kit!

Twice on Mt. Everest

Santosh Yadav of the Indo-Tibetan Border Police is the first and only woman to have climbed Everest twice.

(n - 1), substitution haploids (n - I + 1), misdivision haploids etc. Mis-division haploids have an isochromosome which is produced by vertical division of centromere. Generally centromere divides longitudinally. Aneuhaploids are generally inviable.

Production

Now various methods are known by which haploids can be produced. These methods include (1) pollination with foreign pollen, (2) delayed pollination, (3) use of X-ray irradiated pollen for pollination, (4) temperature shock, (5) treatment is with chemicals like colchicine, (6) interspecific and intergeneric crosses, and (7) anther and pollen culture.

Uses of Haploids

Haploids have several applications in plant breeding. They are used for (I) development of pure lines, (2) disease resistance, (3) development of inbred lines, and (4) production of aneuploids. These aspects are briefly described below.

Diploids

Normal diploids are known as disomics. They have regular bivalent pairing during meiosis. Diploids also have disomic genetics with two alleles at each locus. Pure line or inbred lines of diploids have homozygosity at each locus. Polyploids which behave like diploids are known as disomic polyploids like wheat, cotton, etc.

Polyploids

An organism or individual having more than two basic or monoploid sets of chromosomes is called polyploid and such condition is known as polyploidy. Polyploidy is of two types, viz, (1) autopolyploidy (2) allopolyploidy.

1. *Autopolyploidy*

Polyploids which originate by multiplication of the chromosome of a single species are known as autopolyploids or autoploids and such situation is referred to as autopolyploidy. Autoploids include triploids (3x), tetraploids (4x), pentaploids (5x), hexaploids (6x), heptaploids (7x), octaploids (8x), and so on. Autoploids are also known as simple polyploids or single species polyploids.

Autotriploids

They have three sets of chromosomes of the same species. They can occur naturally or can be produced artificially by crossing between autotetraploid and diploid species.

GK **Mitochondria is called the power house of the cell.** GS

Triploids are generally highly sterile due to defective gamete formation. Triploids are useful only in those plant species which propagate asexually like banana, sugarcane, apple, etc.

Autoteraploids

They have four copies of the genome of same species. They may arise spontaneously or can be induced artificially by doubling the chromosomes of a diploid species with colchicine treatment. Tetraploids are usually very stable and fertile because pairing partners are available during meiosis. In such individuals diploid gametes (2n) are formed. Autotetraploids are usually larger and more vigorous than the diploid species. Rye, grapes, alfalfa, groundnut, potato and coffee are well known examples of autotetraploids.

2. *Allopolyploidy*

A polyploid organism which originates by combining complete chromosome sets from two or more species is known as alloploypioid or alloploid and such condition is referred to as allopolyploidy. Alloploids are also known as hybrid polypoids or bispecies or multispecies polyploids. An allopolyploid which arises by combining genomes of two diploid species is termed as allotetraploid or amphidiploid. Allopolyploidy can be developed by interspecific crosses and fertility is restored by chromosome doubling with colchicine treatment. Allopolyploids has played greater role in crop evolution than autopolyploidy, because allopolyploidy is found in about 50% of crop plants.

Natural Allopolyploids

Some important natural allopolyploid crops are wheat, cotton, tobacco, mustard, oats, etc. Interspecific crossing followed by chromosome doubling in nature have resulted in the origin of allopolyploids.

Induction of Polyploidy

Polyploidy is mainly induced by treatment with a chemical known as colchicine. This is an alkaloid which is obtained from the seeds of a plant known as ***Colchicum autumnale***, which belongs to the family **Liliaceae**.

The colchicine induced polyploidy is known as colchiploidy. In plants, colchicine is applied to growing tips, meristematic cells, seeds and axillary buds in aqueous solution.

Colchicine induces polyploidy by inhibiting formation of spindle fibres. The chromosomes do not line up on the equatorial plate and divide without moving to the poles due to lack of spindle fibres. The nuclear membrane is formed around them and the cell enters interphase. Thus nucleus has double the chromosome number.

Effects of Polyploidy

1. Stems are thicker and shorter.
2. Leaves are fleshy, thicker, larger and darker green in colour .
3. Roots are stronger and longer.
4. Flowers, pollens and seeds are larger than diploids.
5. Maturity duration is longer and growth rate is slower than diploids.
6. Water content is higher than diploids.

Applications in Crop Improvement

Polyploidy plays an important role in crop improvement. Both autopolyploidy and allopolyploidy are useful in several ways. However, allopolyploidy has wider applications than autopolyploidy. Applications of autopolyploidy and allopolyploidy in crop improvement are briefly presented below:

Autoploidy

Both triploids and tetraploids have been used in crop improvement. However, their applications have been limited to few species only. Autotriploids have been developed in sugar beets and water melon only. The triploid sugar beets have larger roots and higher sugar content than diploids. The triploid water melons are seedless, or have rudimentary and soft seeds like cucumber.

Alloploidy

Alloploidy is useful in four principle ways, *viz.*, (1) in tracing the origin of natural allopolyploids, (2) in creating new species, (3) in interspecific gene transfer, (4) as a bridging species.

Limitations of Polyloidy

Polyploidy has several limitations. Some important limitations of polyploidy in crop improvement are briefly presented below:

1. *Limited Use*: The single species polyploidy has limited applications. It is generally useful in those crop species which propagate asexually like banana, potato, sugarcane, grapes, etc.
2. *Difficulty in Maintenance*: The maintenance of monoploids and triploids is not possible in case of sexually, propagating crop species.
3. *Undesirable Characters*: In bispecies or multispecies polyploids characters are contributed by each of the parental species. These characters may be sometimes undesirable as in of **Raphanobrassica.**
4. *Some other Defects*: Induced polyploids have several defects such as low fertility, genetic instability, slow growth rate, late maturity, etc.
5. Chances of developing new species through allopolyploidy are extremely low.

II. ANEUPLOIDY

Aneuploids are of three types, *viz.*, (1) monosomics, (2) nullisomics, and (3) polysomics. These are described below:

(1) Monosomics

An individual lacking one chromosome from a diploid set

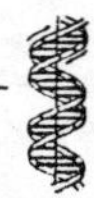

(2n-1) is called monosomic and such condition is known as monosomy. Monosomics may originate in three main ways, viz., (i) from diploids, (ii) from nullisomics, and (iii) from trisomics as described below.

(i) *From Diploids*: Monosomics may originate spontaneously from diploids. Sometimes nondisjunction during meiosis gives rise to n-1 gamete. If this gamete is fertilized by a normal (n) gamete, a monosomic zygote (2n-1) is produced.

(ii) *From Nullisomics*: Nullisomics produce n-1 gametes. Union of such gamete with normal gamete gives rise to monosomics as shown below:

Nullisomic	Gametes	Union with normal gamete	Result
2n-2	n-1	n-1 + n	2n-1

(iii) *From Trisomics*. Trisomics (2n+ 1) also give rise to monosomics. Sometimes non disjunction of three homologous chromosomes in a trisomic during meiosis gives rise to n-1 gametes. Union of such gametes with normal one results in the development of monosomic zygote.

Trisomics	Gametes	Union with normal gamete	Result
2n+1	$\frac{n-1}{n-1}$	n-1 + n	2n -1

(2) Nullisomics

An individual lacking one pair of chromosomes from a diploid set (2n-2) is called nullisomic and such situation is referred to as nullisomy.

(3) Polysomics

An individual having either single or one pair of extra chromosome in the diploid complement is known as polysomic and such condition is referred to as polysomy.

Applications in Crop Improvement

Aneuploids are useful in crop improvement in various ways. Some of the uses of aneuploids in plant breeding are briefly presented below:

1. Locating Genes: Aneuploids are useful tools for locating genes on a specific chromosome. Monosomics and nullisomics are used for this purpose.
2. Interspecific Gene Transfer: Monosomics are also used in transferring chromosomes with desirable genes from one species to another.
3. Aneuploids: Used for developing alien addition and alien substitution lines in various crops.
4. Primary trisomics: Useful in identification of chromosomes involved in translocations.

DNA Replication

The process by which a DNA molecule makes its identical copies is known as DNA replication. Three possible modes

Knowledge World!

Post office of the highest altitude

The Hikkim branch post office in Lahaul and Spiti district, Himachal Pradesh, is at an approximate height of 15,000 feet and was opened on Nov. 5, 1983.

of DNA replication.

1. Dispersive Replication
2. Conservative Replication
3. Semi-conservative Replication.

Semi-Conservative Replication

Main features—

1. Separation of two strands of parental DNA.
2. Complementary base pairing of the bases located in the single stranded regions.
3. Formation of phosphodiester linkages between the neighbouring deoxyribouncleotides.
4. This ensures that the base pairs of the new strands are complementary to the old strands.
5. The base sequence of a newly synthesized strand is dictated by the base sequence of old strand.

Evidence for Semi-Conservative Replication

1. *Meselson and Stahl 1958*

They grew *E. coli* in ^{15}N for 14 cell generations till nitrogen in the bacterial DNA was only ^{15}N. This heavy DNA had more density than DNA with ^{14}N.

Then the bacteria were transferred to culture medium ^{14}N. They observed that such DNA was half dense indicating the presence of DNA hybrid (one ^{15}N strand + one ^{14}N strand). After second round of replication there would be 4 DNA molecules of these, two molecules would be hybrid (^{14}N - ^{15}N).

Methods of DNA Replication

I. *Initiation of Replication*: Replication begins at a replication origin.

2. *Unwinding of Helix*: Unwinding is brought by enzyme 'helicase.' This unwinding may result in formation of supercoils which are removed by enzyme DNA Gyrase (topoisomerase). Formation of replication fork (y shaped structure). Single strands are stabilised by a single strand binding protein (SSB).

3. *Formation of Primer Strand*: An enzyme primase initiates transcription of 3' $\rightarrow$ 5' strands. This generates long primer RNA in 5' $\rightarrow$ 3' direction.

   The free 3-OH of this primer RNA provides the initiation point for DNA polymerase for the sequential addition of deoxyribo nucleotides.

   DNA polymerase Iil catalyse DNA replication.

GK **Indian sub-continent receives bulk of monsoon during S-W monsoon period.** GS

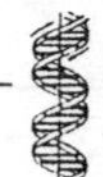

4. *Elongation of New Strand*: Synthesis of new strand occurs continuously along the upper strand known as leading daughter strand- Synthesis of another daughter strand along lower parental strand takes place in the form of short pieces lagging daughter strand. Short pieces of DNA is called Okasaki fragments. Discontinuous pieces of the lagging strand are joined together by the enzyme DNA ligase.

CLASSICAL GENETICS

The Laws of Mendel

Mendel's Predecessors

1. The hybrids obtained by mating of two different variety of a species. The hybrids of some characters are identical to those of one of the two parents while some others are intermediate between those of two parents. The concept of dominance is present in these observation.
2. Characters of F_1 and F_2 progeny produced from reciprocal crosses are identical.
3. F_1 progeny are uniform in their characters. But F_2 generation show a large variation for different characteristics.
4. The appearance of the parental forms in F_2 phenotypes were intermediate in appearance.
5. Some phenotype in F_2 have entirely new character forms.

Reasons for their Failure

1. They studied the plant as a whole.
2. As a result of the above, the plants could not be classified into few clear cut classes.
3. The data from different generations were not accurately.
4. Complete control on pollination in the F_1 was lacking.
5. In many studies the F_1 was an interspecific hybrid exhibiting partial to considerable sterility.
6. The number of plants studied in F_2 was relatively small.
7. Most of the characters studied by the earlier workers were quantitative in nature.

Pea-advantages as an Experimental Material—

1. Several characters had a contrasting form.
2. Pea-self pollination crop.
3. Flowers are relatively large. So, emasculation and pollination is very easy.
4. Duration single season every year one generation can be grown.
5. There is problem in germination.

Pairs of contrasting characters of pea

Characters	*Dominant*	*Recessive*
1. Length of stem	Tall	Dwarf
2. Position of flowers	Axial	Terminal
3. Pod shape	Full	Constricted
4. Pod colour	Green	Yellow
5. Cotyledon colour	Yellow	Green
6. Seed coat colour	Grey	White
7. Seed shape	Round	Wrinkled

Reason for Mendel's Success

- An accurate and incisive analysis of the reasons for failure of earlier workers.
- He accurately diagnosed the weakness of their experimental materials, techniques and approaches and avoided them in his own studies.
- Studied the inheritance of only one pair of contrasting characters at a time.
- His selected pea variety had clearly different forms of one or more characters.
- He classified the plants of a population on the basis of contrasting characters, and kept an accurate record of the number of plants in each category for every generation.
- He carried out his experiments with great care and elaborateness.
- His knowledge of mathematics which helped him during the interpretation of his findings.
- He was able to formulate appropriate hypothesis.

Mendel was Lucky

1. The characters chosen by him was qualitative inheritance.
2. The contrasting forms of each of seven characters were governed by a single gene.
3. The seven characters studied by him, the genes for two were located in one chromosome, while three others were present in another chromosome.

Reasons for the Neglect of Mendel's Finding

1. He used mathematical principles to explain a biological phenomenon. This was not acceptable to biologists.
2. He studied contrasting pairs of characters exhibiting discontinuous variation many of his contemporaries (Darwin) pre-occupied with characters exhibiting continuous variation.
3. The phenomenon of fertilization and the behaviour of chromosomes during cell divisions (mitosis and meiosis) were not known at that time.
4. Mendel did not publish his findings through further writing on the subject after his initial paper.
5. Mendel failed to domonstrate the validity of his conclusions in other species.

In Turner's Syndrome one X-chromosome is missing.

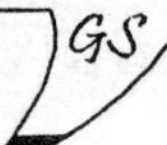

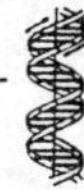

Highlights of the Mendel's Paper

1. Mendel concluded a complete fusion of male and female gametes, and equal contribution of the two parents to the development of characters of the hybrid (F_1).
2. He postulated the existence of genes, which are responsible for the development of various characters.
3. He clearly stated that genes were particulate.
4. He made clear distinction between the phenotype and genotype.
5. He gave the formula for determining the number of *(i)* different types of gametes produced by F_1 *(ii)* different gametes genotypes in F_2 *(iii)* homozygous genotypes.
6. He introduced the concept of dominance and recessiveness.
7. He described laws of segregation and independent assortment.
8. He concluded that a large number of progeny would increase the precision of observation.
9. He stated that his explantations were based on two important assumptions *(i)* equal proportion of the different gametes produced by F_1 *(ii)* equal chance for each gamete thus produced to effect fertilization.

The Law of Segregation

Inheritance of Seed Shape in Pea

He crossed round with wrinkled. The seeds resulting from hybridization (F_1 seeds) were all round. The F_2 seeds (seeds produced by selfing of F_1 plants) were on an average 3 out of every 4 seeds were round and 1 out of every 4 seeds wrinkled. He allowed the F_2 plants to self pollinate. He found that all the plants from wrinkled –F_2 seeds produced only wrinkled F_3 seeds. But F_2 round seeds were of two types an average 1/3 only round F_3 seeds, 2/3 both round and wrinkled F_3 seeds in the ratio 3: I. He noted that the result from reciprocal crosses were identical. The following conclusion :

1. Both male and female parents make equal contribution to the development of characters in the progeny results.
2. $F_1 \rightarrow$ character of only one of the two parents is expressed (dominant character).
3. F_2 generation both dominant and recessive characters appear in 3 : 1
4. The recessive character appears in F_2 unchanged and identical to that of the parent contributing this trait.
5. In F_2 1/3 of the individual with the dominant character are pure, 2/3 of them are hybrid.

The Law of Segregation

This law is explained by making the following assumptions.

> **Quiz Kit!**
> ***Pioneer agro technologist for medicinal plants***
> **Akhtar Hussain, (b. 1928) of Central Institute of Medicinal and Aromatic Plants, Lucknow, discovered that a group of carbohydrates called *Polysaccharides* produced by a special kind of bacteria caused various disease in plants, including wilting symptoms.**

1. A character is produced by a specific gene.
2. Each gene has two alternative forms (Alleles).
3. The two alleles, a gene govern the development of contrasting forms of the character governed by the gene.
4. Each somatic cell of an organism has two copies of each gene.

Definition

The two alleles of a gene present in the F_1 do not contaminate each other, they separate and pass into different gametes in their original form producing two different types of gametes in equal frequencies.

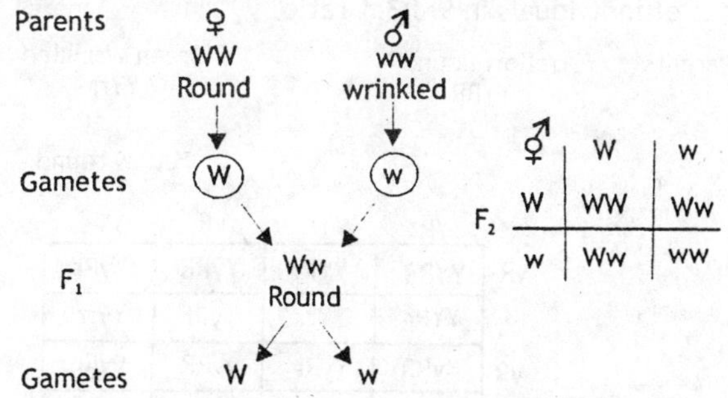

P' type Ratio → 3 Round : 1 wrinkled
G' type ratio → 1 WW : 2 Ww : 1 ww

Essential features of this explanation on the 3:1 ratio in F_2.

1. The presence of two copies of each gene in somatic cells.
2. Only one copy of each gene in gametes.
3. A lack of contamination or modification of each other by the two alleles of a gene during their stay together in the same cell of F_1 hybrids.
4. The separation of the two alleles of a gene and their transmission into separate gametes of F_1^1
5. The production of two types of gametes with respect to the heterozygote gene by the F_1 in equal frequencies.
6. Random union between male and female gametes.

Law of Independent Assortment

This law states that when two pairs of gene enter in F_1 combination, both of them have their independent

The Pala king Gopala founded Odantapuri.

dominant effect.

Main features:

- It explains simultaneous inheritance of two plant characters.
- In F_1 when two genes controlling two different characters, come together, each gene exhibits independent dominant behaviour without affecting other gene.
- These gene pairs segregate during gamate formation independently.
- The alleles of one gene can freely combine with the alleles of another gene.
- Each of two gene pairs when considered separately, exhibits typical 3:1 ratio in F_2 generation.

Example:

When plants of garden pea with yellow round seeds are crossed with plants having green wrinkled seeds, we get yellow round seeds in F_1. Thus yellow colour of seed exhibits dominance over green and round-seeds shape over wrinkled independently. The F_1 produces -Yellow round (YR) yellow wrinkled (YR), green round (YR), and green wrinkled (YR). Selfing of F_1 gives rise to all the above four types of individuals in 9:3:3:1 ratio

Parents — Yellow Round YYRR × Green Wrinkled YYrr

F_1 — YyRr Yellow round

F_2

	YR	Yr	yR	yr
yR	YYRR	YYRr	YyRR	YyRr
Yr	YYRr	YYrr	YyRr	Yyrr
yR	YyRR	YyRr	yyRR	yyRr
yr	YyRr	Yyrr	yyRr	yyrr

Independent assortment of two pairs of genes in garden pea.

Linkage

All the genes on a chromosome are said to be linked to one another and belong to the linkage group. The phenomenon of inheritance of linked genes in same linkage group is called linkage.

Features

- Two or more genes linked
- It may involve either dominant or recessive genes
- Linkage between closely located gene. Higher percentage of parent in F_1
- It leads to desirable or undesirable trait. Strength of linkage depends on the distance between the linked genes.
- Maximum number of linkage groups in an organism is equal to its haploid chromosome number.

Phases of Linkage

Coupling: Linkage between two or more either dominant (AB) or recessive (ab) allele is referred to as Coupling.

Repulsion: Linkage of dominant allele with recessive allele known as Repulsion.

Types of Linkage

1. *Based on crossing over*
   (a) Complete linkage: Linkage in which there is no crossing over
   (b) Incomplete linkage: Some frequency of crossing over between linked genes.
2. *Based on genes involved*
   (a) Coupling linkage
   (b) Repulsion linkage
3. *Based on Chromosome involved*
   (a) Autosomal linkage: Linkage of genes which are located in other than sex chromosomes.
   (b) X -chromosome linkage: Linkage of genes which are located in sex chromosomes.

Detection of Linkage

1. Test Cross
2. Individual heterozygous at two loci in self-pollinated linkage and pleiotropy.

The only way to distinguish between linkage and pleiotropy is to find out a crossover product between linked characters.

Significance of linkage in plant breeding

1. Effect on Selection.
2. Effect on Genetic variance.
3. Effect on Genetic Correlation.

Difference between crossing over and linkage

Crossing over	*Linkage*
1. Separation of linked genes	1. Keeps the genes together
2. Non-sister chromatids of homologous chromosomes involves.	2. Indiividual chromosome involves
3. Frequency of crossing over never exceed 50%	3. It can never be more than haploid chromosome number
4. It increases variability.	4. It reduces variability.
5. Provides equal frequency of parental and recombinant types in test cross progeny	5. Provides higher frequency of parental types than recombinant types in test cross progeny.

Crossing Over

Interchange of parts between non sister chromatids of homologus chromosomes during meiotic prophase.

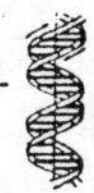

Main features

1. Take place during meiotic prophase
2. It occurs between non sister chromatids
3. It takes place at four strand stage
4. It yields two recombinant types or crossover types and two parental types or non crossover types
5. It leads to exchange of equal genes
6. The frequency of recombinants can be worked out from the test cross progeny

Chiasma and Crossing over

The point of exchange of segments between non sister chromatids of homologous chromosomes during meiotic prophase is called chiasma. Here only crossing over takes place.

Chiasma Terminalization

The movement of a chiasma away from the centromere and towards the end of tetrads is called terminalization. Chiasma terminalization occurs between diplotene and metaphase I.

Three theories to explain the mechanism of chiasma terminalization.

1. Electrostatic hypothesis
2. Coiling hypothesis
3. Elastic chromosome repulsion

Relationship between Crossing over and Chiasma formation

Two theories explain these relationship –

1. *Classical Theory* : This theory states that first chiasma is formed then crossing over takes place. The chiasma is formed at diplotene stage of meiosis and crossing over occurs between diplotene and anaphase.
2. *Chiasma type Theory*: Here first crossing over occurs and then chiasma is formed.

Molecular Mechanism of Crossing Over

1. *Copy Choice Theory* : The entire recombinant section arises from the newly synthesised section. The non sister chromatids when come in close contact they copy some section of each other resulting in recombination.
2. *Breakage and Reunion Theory*: The crossing over takes place due to breakage and reunion of non sister chromatids. The two segments parental chromosomes which are present in recombinants arise from physical breaks in the parental chromosomes with subsequent exchanges of broken segments.

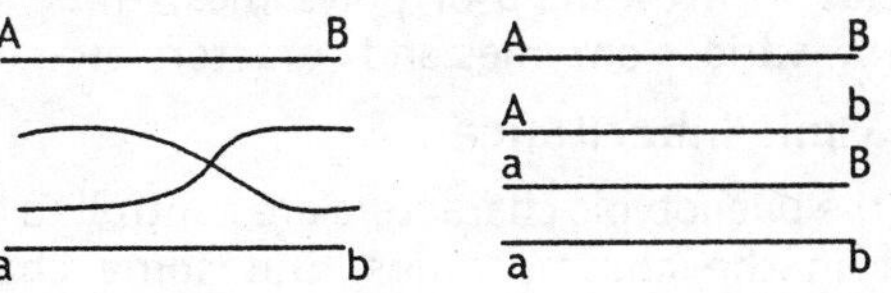

Interference (Muller)

Tendency of one crossover to reduce the chance of another crossover in its adjacent region.

Coefficient of interference (%) = 1 - Coefficient of Coincidence x 100.

Coincidence (Muller)

It explained the strength or degree of interference.

Coefficient of coincidence (%)

$$= \frac{\text{Observed double crossover}}{\text{Expected double crossover}}$$

Chromosome Mapping

A line diagram which depicts various genes present on a chromosome and recombination frequency between them. Such maps are also known as genetic maps or linkage maps. The mapping of chromosomes is done with the help of three point test cross. A three point test cross is a cross of trihybrid with its homozygous recessive parent.

Types of Crossing Over

Single Crossing Over: Formation of a single chiasma between non sister chromatids of homologous chromosomes.

Double Crossing Over: Formation of two chiasma between non sister chromatids of homologous chromosomes.

Multiple Crossing Over: Formation of more than two crossovers between non sister chromatids of homologous chromosomes.

Factors affecting crossing over

- Distance
- Temperature
- Sex
- Nutrition
- Chemicals
- Structural changes
- Centromere effect

Significance of crossing over

- Creation of variability
- Locating genes
- Linkage Maps

GENE ACTION

Protein Synthesis

Protein synthesis depends on the sequence of amino acids in its composition union of several amino acids leads to the formation of polypeptide chain. Protein synthesis involves two main process *(i)* Transcription *(ii)* Translation.

Transcription

The process of synthesis of messenger RNA from a DNA template is known as transcription.

$$\text{DNA} = \xrightarrow[\text{Transcription}]{\text{Transcriptase}} \text{mRNA}$$

Mechanism of Transcription

The process of transcription takes place during interphase and continues up to prophase of cell division. The process of transcription require (i) DNA tempelate (ii) Activated precursors (ATP, GTP, UTP and CTP) (iii) Divalent metal ions (Mg++ or Mn++) (iv) RNA Polymerase.

The process of transcription consists of three main steps:

1. Initiation of transcription
2. Elongation of RNA chain
3. Termination of RNA chain

1. *Initiation of Transcription*

- Start at specific site in a cistron called promotor site.
- The binding of RNA polymerase at initiation site results in separation of DNA double helix.
- After opening of DNA strands, synthesis of mRNA begins on one of the two strands of DNA. The RNA chain starts with triphosphate and the first base is always purine.
- After synthesis of terminal triphosphate, nucleotides are removed by endonucleases.

2. *Elongation of RNA Chain*

- The elongation of RNA chain takes place by addition of activated ribonucloside triphosphates like ATP, GTP, UTP, and CTP to one strand of DNA molecule. RNA synthesis takes place only on one strand of DNA.
- The elongation of mRNA chain takes place with the help of coenzyme which moves along the DNA template.

3. *Termination of RNA Chain*

- The chain is terminated either when DNA receives termination signal or an additional protein.
- The terminator gene acts as a signal to end transcription.
- RNA polymerase joins the nucleotides together to produce mRNA transcript.
- When mRNA transcript is detached, the DNA template strand reforms H-bands with its complementary strand resulting in the formation of double helical structure.

Translation

The process of protein synthesis from information in mRNA is known as translation.

$$\text{DNA} \xrightarrow{\text{Transcription}} \text{RNA} \xrightarrow{\text{Translation}} \text{Protein}$$

Mechanism of Translation

The translation process requires mRNA, rRNA, ribosomes, 20 kinds of amino acids and their specific transfer RNA's. It consists of five major steps :

(i) Activation of amino acids

(ii) Transfer of amino acids to tRNA

(iii) Chain initation

(iv) Chain elongation

(v) Chain termination

1. *Activation of Amino Acids*

There are 20 different amino acids which are used in protein synthesis. Each amino acid is activated in the presence of ATP by its own activating enzyme called aminoacyl synthetase to form aminoacyl adenylate.

2. *Transfer of Amino Acids to RNA*

The aminoacyl adenylate reacts with specific tRNA and gives rise to aminoacyl tRNA complex.

Thus synthetase has two active sites, one for recognizing the specific amino acid and the other for recognizing specific molecule.

3. *Chain Initiation*

- The 30S ribosome subunit attached to mRNA to form an mRNA 30S complex.
- The amino acid tRNA complex binds to the initiation codon AUG on mRNA through its anticodon UAC to form 30S initiation complex.
- The larger ribosomal subunit joins to the 30S initiation complex to form the complete initiation complex.
- The larger ribosomal subunit has two binding sites for tRNA i.e. acceptor site and peptide site.

4. *Chain Elongation*

- The second amino tRNA complex now occupies the acceptor site. There is enzymatic recognition of internal codons. This process requires EF- T_4, GTP and metallic ions Mg^{++}.
- Formation of peptide bonds takes place. This process requires two metallic ions Mg^{++} and K^{+}

5. *Chain Termination*

- The termination codon provides signal to the ribosome for the attachment of release factors.
- The release factors interact with peptydyl transferase causing hydrolysis of the bonds between tRNA and the polypeptide chain.
- Hydrolysis of GTP results in the dissolution of the release factors from the ribosome.
- The released polypetide chain undergoes processing. The formyl group of starting amino acid methionine is removed. Proteins are synthesized from amino acids with the help of polysomes, three types of RNA, various enzymes and cofactors etc.

Cytoplasmic Inheritance

Most of the phenotypic characters are controlled by genes present in the chromosomes. But some characters

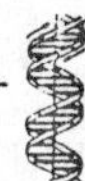

expressed by factors present in the cytoplasm. These factors lying in the cytoplasm are called plasmagenes. The transmission of characters controlled by plasmagenes are called cytoplasmic inheritance.

The first case of cytoplasmic inheritance was reported by Correns in 1909 in four '0' clock (*Mirabilis jalapa*) for leaf colour.

Important characteristics (or) features of cytoplasmic inheritance.

1. *Reciprocal Differences*

Characters which are governed by Cytoplasmic Inheritance invariably exhibit marked differences in reciprocal crosses in F_1

2. *Lack of Segregation*

In general F_2, F_3 and the subsequent generation do not show segregation for a Cytoplasmic Inheritance trait. This is because the F_1 individuals generally receive plasma genes from one parent only.

3. *Irregular Segregation in Biparental Inheritance*

In some cases, plasma genes from both the parents are transmitted to the progeny, this is known as Biparental inheritance. Biparental inheritance gives rise to irregular segregation ratio in the F_1 generation of higher plants.

4. *Somatic Segregation*

Plasma genes generally show somatic segregation during mitosis.

5. *Association with Organelle DNA*

Several plasma genes have been shown to be associated with CP-DNA not DNA.

6. *Mutagenesis*

Some Mutagens e.g. ethidium bromide, are highly specific Mutagens for plasma genes.

7. *Mappability*

It is very difficult to map cytoplasmic genes or prepare linkage map for such genes.

8. *Maternal Effects*

In case of cytoplasmic inheritance distinct maternal effects are observed. This is mainly due to more contribution of cytoplasm to the zygote by female parent than male parent.

Did You Know ?

Edible pesticides

Dr. C.N. Shastry of Bangalore has been making pesticides and insect repellents from herbs and medicinal plants which are lethal to insects but perfectly safe for humans. It took him 14 years of research to classify around 5,000 hill-oriented medicinal herbs which he used. Bugsac, an effective bug repellent, and Ullasini, a mosquito repellent, are two of his successful creations.

Significance in Plant Breeding

- Studies of cytoplasmic inheritance have played key role in mapping of chloroplast and mitochondrial genome in yeasts, chlamydomonas, maize, humans etc.
- Cytoplasmic inheritance has been useful in explaining the role of various cytoplasmic organelles in the transmission of characters in different organisms.
- Development of cytoplasmic male sterility (CMS).
- CMS line have been developed in several crops like maize, pearl millet, sorghum, cotton etc.
- Availability of CMS lines has facilitated the production of hybrid seed in these crops at a cheaper cost than with hand emasculated pollination method.
- Mitochondria can be used in the manifestation of heterosis.
- Mutation of chloroplast DNA and mitochondrial DNA leads to generation of new variants. Some of such variants are of special significance especially in ornamental plants.

Sex Determination

Sex determination is the process of sex differentiation, which utilises various genetical concepts to decide whether a particular individual will develop into male or female sex.

Mechanisms of Sex Determination

1. Chromosomal sex determination
2. Monogenic sex determination
3. Environmental sex determination

Difference between Mendelian Inheritance and Cytoplasmic Inheritance

Mendelian Inheritance	*Cytoplasmic Inheritance*
1. Governed by nuclear genes	1. Governed by plasma genes
2. Exhibits distinct segregation pattern	2. Does not exhibit distinct segregation
3. Reciprocal differences are not observed	3. Reciprocal differences are observed
4. Does not show maternal effects	4. Exhibits maternal effects
5. Genes can be easily mapped on chromosomes	5. Mapping of plasma genes is very difficult
6. Nuclear genes are associated with chromosomes	6. Plasma genes are associated with either chloroplast DNA or mitochandrial DNA

GS

Arid soils lack moisture and humas content.

1. *Chromosomal Sex Determination*

(a) Sex Determination by Allosomes

Sex with similar type of sex chromosomes (XX) is known as homogametic sex and with dissimilar type of sex chromosomes (XY) as heterogametic sex.

There are four different systems of allosomal sex determination

1. XX - XY female - male system
2. XX - XO female - male system
3. XO - XX female - male system
4. ZW - ZZ female - male system

(b) Diploid-Haploid System

In honey bees the females have diploid (2n = 32) chromosomes and males have haploid (n = 16) chromosomes unfertilized eggs develop parthenogenetically into haploid.

(c) *Genic Balance System*

Sex determining genes are present on both X-chromosomes as well as autosomes.

2. *Monogenic Sex Determination*

In some animals and plants, expression of sex influenced by a single gene.

3. *Environmental Sex Determination*

In horse tail plant development of sex depends on growing conditions. Plants grown under good conditions develop as females and those grown under stress conditions develop into males.

Sex-Linked Characters

The characters for which genes are located on sex chromosomes are known as sex-linked traits. Such genes are called sex linked genes, and linkage of such genes is referred to as sex linkage.

Main features –

1. *Location*: Sex-linked genes are located on X chromosomes only.
2. *Number*: In diploid organism, each homogametic sex (xx or zz) has two copies of linked alleles.
3. *Expression*: In homogametic sex, a recessive gene can express. In heterogamatic sex, a recessive allele can express.
4. *Transmission*: Sex linked genes are transmitted from female to male.
5. *Pattern of segregation*: Inheritance of sex linked characters does not follow normal segregation pattern.

Sex Limited Characters

The characters which are expressed in one sex only are referred to as sex limited characters.

Features -

- They express in one sex only.
- Sex limited genes may be located either in sex chromosome or autosomes:
- Sex limited genes control the expression of primary and secondary sex characters.

Sex Influenced Genes

The dominance expression of some genes depends on the sex of individual. Such genes are known as sex influenced genes.

Features –

- Such genes are located in the autosomes
- They express more frequently in one sex than other.
- Expression of such characters appears to be governed by sex hormones.

OTHERS

Mutation

Mutation	: Sudden heritable changes in the characters of organism.
Mutants	: Individual showing mutational changes.
Mutant allele	: The allele producing the changed pheno-type variant.

Mutation is the ultimate source of all the Genetic variation.

Historical

Seth Wright (1791)	: Male lamb with unusually short legs.
Hugo de vries (1900)	: The term mutation was introduced sudden heritable changes observed by him is another contribution.
Morgan (1910)	: White eye mutant of drosophila.
H.J. Muller (1927)	: Mutagenic action of X-rays in drosophila.

Characteristics

- Generally mutant alleles are recessive
- Random events
- Most have harmful effects
- Recurrent in nature
- Occurs at very low frequencies
- Mutable genes
- Mutator genes
- Highly mutable sites within a gene hot spots
- Rate of induced mutation varies from one gene to the other
- Most mutant alleles are pleiotropic
- Occurs in both forward and reverse directions

Classification

- The direction of Mutation

1. Forward 2. Reverse

GK GS

Abomasum is the true stomach of the cattle.

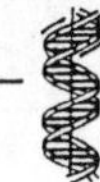

- The cause of Mutation

1. Spontaneous 2. Induced

- Dominance Relationship

1. Dominance Mutation
2. Recessive Mutation
3. Co dominant (blood group)
4. Partially dominant

- Tissue of Origin

1. Somatic Mutation (bud mutation)
2. Germinal Mutation

- Effect on survival

1. Lethal 2. Sublethal 3. Subvital 4. Vital 5. Super vital

In crop improvement → Vital and Super vital

- Type of Trait Affected

1. Visible Mutation 2. Character Mutation

- Intensity of Character Expression

1. Amorphic 2. Hypomorphic 3. Isoallelic
4. Hypermorphic

- Quantum of Morphological Effect Produced

1. Macro mutation 2. Micro Mutation

- Effect on the Expression of Neighbouring genes

1. Polar Mutation 2. Nonpolar Mutation

- Cytological Basis

1. Chromosomal Mutation
2. Gene Mutation
3. Cytoplasmic Mutation

- Molecular Basis

1. Base substitution
2. Deletion
3. Addition

- Type of Amino Acid Replacement in the Polypeptide

1. Missense 2. Non sense 3. Frame shift

Change in base sequence of DNA molecule occur in two ways :

1. Base substitution
2. Base addition and deletion

Base Substitution

When one base in DNA molecule is replaced by another one.

2 types: (i) Transition (ii) Transversion

(i) Transition:

Purine is replaced by another purine e.g.
Adenine is replaced by Guanine.

A ⟷ G

You Don't Know!

Caffeine benefits

Parthasarathy C. Kesavan (b. 1938) of School of Life Sciences, Jawaharlal Nehru University, New Delhi, demonstrated for the first time that caffeine which inhibits the repair of damage done to DNA by ultraviolet radiation also protects it (DNA) from the onslaught of the more harmful X-rays or gamma rays.

One base substitution affects the base sequence of only one codon as a result only amino acid is altered in the concerned protein.

(ii) Transversion

Pyrimidine (Thymine and Cytosine) is replaced by a purine (adenine and guanine) A ⟷ T are base substitution affects the base sequence of only one codon as a result, only one amino acid is altered in the concerned protein.

Base Addition and Deletion

Base Addition

Insertion of one or more bases in a DNA molecule

Base Deletion

Loss of one or more bases in a DNA molecule.

Mutator Gene

Some genes increase the frequency of spontaneous mutation of other genes located in the same cell.

Anti-Mutator Gene

Some genes decrease the frequency of spontaneous mutation of other genes located in the same cell.

Applications in Crop Improvement

1. *Development of Improved Varieties*

Wheat - Sarbati Sonara, Pusa Lerma
Rice - Jagannath
Tomato - Pusa lal, Meeruti
Cotton- MCU7, MCUI10
Sugarcane- Co8152, 8153

2. *Induction of Male Sterility*

Genetic male sterility has been induced in durum wheat using radiations.

3. *Production of haploids*

Use of X rays radiated pollen in production of haploids in many crops.

Polygenic Inheritance

Characters or traits are governed by one or few genes. Such traits are called as qualitative or oligogenic characters. Some characters are controlled by several genes. They are known as quantitative characters or polygenic characters.

Features of Polygenic Traits

- Each polygenic character is controlled by several independent genes.
- Polygenic character exhibit continuous variation.
- Highly sensitive to environmental changes.
- Effect of individual gene is not easily detectable.
- Expression is governed by additive gene action.
- Transmission is generally low because of high amount of environmental variation.

Analysis of Polygenic Traits

- Measurement based on degree of variation
- Observation recorded in population.
- Studies with the help of mean, variance and covariance.

Assumption of Polygenic Traits

- Each of the contributing genes involved in the expression of a character produces an equal effect.
- Each contributing alleles has either cumulative or additive effect in the expression of character
- There is no epistasis among genes at different loci.
- The linkage is in equilibrium.
- The environmental effects are absent.

Examples of Polygenic Traits

1. Kernel colour in wheat
2. Corolla length in tobacco
3. Skin colour inheritance in man

Transgressive Segregation

Appearance of transgressive segregants in F_2 is an important feature of polygenic inheritance.

The segregants which fall outside the limits of both the parents are known as transgressive segregants.

Example :

parents AAbb x aaBB

↓

F_1 AaBb

F_2 AABB and aabb

are transgressive segregants

Polygenic Variation

Three types –

1. *Phenotypic Variability*

It includes both genotypic and environmental variation.

2. *Genotypic Variability*

It is the inherent or genetic variability which remains unaltered by environmental conditions. It consists of additive, dominance and epistatic components.

3. *Environmental Variability*

It is non heritable variation. This uncontrolled variation is measured in terms of error mean variance.

Difference between Additive Variance and Dominace Variance

Additive Variance	*Dominance Variance*
• Difference between homozygotes	• Deviation of Aa from the mean of AA and aa
• Genes show lack of dominance	• Genes show incomplete, complete or over dominance.
• Associated with homozygosity	• Associated with heterozygosity
• It is fixable	• Non-fixable
• Selection is very effective	• Selection is ineffective.
• It is the cause of transgressive segregation	• Cause of heterosis.

Difference between Polygenic Traits and Oligogenic Traits

Polygenic Traits	*Oligogenic Traits*
• Governed by several genes	• Governed by few genes
• Effect of each gene is not detectable	• Effect of each gene is detectable
• Variation is continuous	• Variation is discontinuous
• Usually governed by additive genes	• Governed by non additive genes
• Highly influenced by environmental factors	• Not influenced by environmental factors.

Epistatic Variance

It is due to the deviations as a consequence of inter-allelic interaction, is interaction between alleles of two or more different genes or loci.

It is of three types –

1. Additive x Additive
2. Additive x Dominance
3. Dominance x Dominance

Significance of Polygenes

- It has great evolutionary significance
- Increases the crop yield
- Variation for speciation

PLANT BREEDING

A CONCEPT OF PLANT BREEDING

Objectives of Plant Breeding

- Non-shattering characteristics
- Higher yields
- Improved quality
- Disease and insect resistance

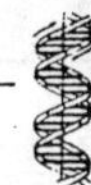

- Photoinsensitivity
- Synchronous Maturity
- Non-shattering characteristics
- Varieties for new seasons
- Moisture stress and salt tolerance

Activities in Plant Breeding

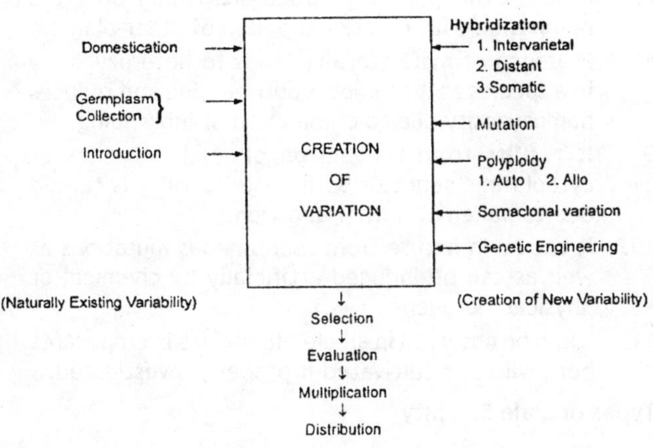

Activities in Plant Breeding

Mode of reproduction

(a) Sexual Reproduction

It means fusion of male and female gametes to form a zygote, which develops into an embryo.

In crop plants meiotic division of specific cells stamen and pistil yields microspores and megaspores respectively. This is followed by mitotic divisions of the spore nuclei to produce gametes like microspores and megaspores respectively.

Sporogenesis: Production of microspores and megaspores is known as sporogenesis.

Microspores are produced in anthers, while megaspores are produced in ovules.

1. *Microsporogenesis* : Each anther has four pollen sacs, which contain pollen mother cells. Each pollen mother cells undergoes meiosis to produce four haploid cells. This process is known as microsporogenesis.

2. *Megasporogenesis*: A single cell in each ovule differentiates into a megaspore mother cell. The megaspore mother cell undergoes meiosis to produce four haploid megaspores. Only one is functional.

Gametogenesis: The production of male and female gametes in the microspores and the megaspores respectively is known as gametogenesis.

1. *Microgametogenesis* : During the maturation of pollen, the microspore nucleus divides mitotically to produce a generative and a tube nucleus. When the pollen lands on to the stigma of a flower pollination occurs shortly after pollination, the pollen germinates. The pollen tube enters the stigma and grows through the style. The generative nucleus undergoes a mitotic division to produce two male gametes. The pollen tube enters the ovule through a small pore, micropyle and discharges the two sperms into the embryosac.

2. *Megagametogenesis* : The nucleus of a functional megaspore divides mitotically to produce four nuclei. Now megaspore nucleus undergoes three mitotic divisions to produce eight nuclei. Three of these nuclei move to one pole and produce a egg cell and two synergid cells. Another three nuclei migrate to the opposite pole to give rise to antipodal cells. The two nuclei remaining in the centre is called polar nuclei. The megaspore thus develop into a mature embryo sac. The development of embryo sac from a megaspore is known as megagametogenesis.

> **Knowledge World!**
>
> *Largest wild goat*
>
> **The sole mountain goat of the South, the Nilgiri Tahr (*Hemitragus hylocrius*), is marginally larger than the Himalayan Tahr, standing 100-110 cm at the shoulders, with horns measuring between 40 cm and 45 cm. Its habitat is from the Nilgiri to the Annamalai hills and along the Western Ghats between 1,220 m and 1,830 m.**

(b) Asexual reproduction

Asexual reproduction - Does not involve fusion of male and female gametes.

(a) Vegetative reproduction :

1. Underground Stems
   - Tubers
   - Bulb
   - Rhizomes
2. Subaerial Stems
   - Stolon
   - Suckers
   - Runners
3. Bulbils - garlic
4. Artificial Vegetative Reproduction
   - Grafting, layering
   - Tissue culture

(b) Apomixis:Seeds are formed but the embryos develop without fertilization.

1. Adventive embryony - embryo by nucellus, integeument and chalaza :
   - There is no formation of embryosac. *e.g.* Mango, Citrus.

India has monopoly in lac production.

GS

2. Apospory - non-reduced embryo by vegetative cells other than embryosac.

- Here embryosac is formed. *e.g.* Heiraceum, Ranunculus.

3. Diplospory - embryo formed by embryosac cells.

Mode of Pollination

Self Pollination

Pollen fall on to the stigma of the same flower.

Mechanism promoting self-pollination:

- *Cleistogamy*- Flowers do not open *e.g.* cereals.
- *Chasmogamy* -Flowers open, but only after pollination has taken place *e.g.* cereals.
- Position of anthers in relation to stigmas ensure self-pollination *e.g.* tomato, brinjal.
- Flowers open but the stamens and the stigma are hidden by other floral organs *e.g.* pea, soyabean.

Cross Pollination: Transfer of pollens from the anthers of a flower to the stigma of another flower of same species.

Mechanism promoting cross pollination :

1. Dicliny : The flowers are either staminate (male) or pistillate (female).

(a) Monoecy - Staminate and pistillate flowers occur in the same plant, either in the same inflorescence e.g. castor , mango.

(b) Dioecy -The male and female flowers are present on different plants *e.g.* papaya, datepalm.

2. Dichogamy : Stamens and pistils of hermaphrodite flowers mature at different times.

(a) Protogyny - Pistils mature before stamens *e.g.* bajra.

(b) Protandry - Stamens mature before pistil *e.g.* maize.

3. Stigmas are covered with a waxy film *e.g.* alfalfa.

4. Self-compatibility.

5. Male sterility

Male Sterility

Male sterility is a reproductive deficiency of some plants where male organs in hermaphrodite flowers are rendered defunctioning. It relates particularly to non-viable pollen grains which are formed through a chain of vital processes during microsporogenesis.

Characteristic Features

1. Better quality of hybrid seeds since human errors committed during detasselling are completely eliminated.
2. Larger quantity of hybrid seeds, since the production area no more remains a limiting factor, as also the manpower.
3. The possibilities of losses due to weather hazards at the time of detasselling are completely dissipated.
4. There is considerable wastage of plant energy in producing pollen grains in normal plants.
5. Male sterility can be utilized for developing mass reservoirs, particularly in outbreeders for the conservation of variability.
6. Male sterile lines can serve as tester genotypes for assessing the combining ability of a large number of stocks.
7. Male sterile plants produce seed only on cross pollination with functional pollen of other plants.
8. Presence of male sterility leads to heterozygocity in a species as it promotes outbreeding and reduces homozygosity due to elimination of inbreeding.
9. It results from the action of nuclear genes or cytoplasmic genes or both. Male sterility is caused due to pollen or anther abortion.
10. It occurs in nature from spontaneous mutations as well as can be induced artificially by chemical or physical mutagens.
11. It can be observed in all diploid species in crop plants both wild and cultivated if properly investigated.

Types of Male Sterility

It is of five types –

(i) Genetic male sterility

(ii) Cytoplasmic male sterility

(iii) Cytoplasmic genetic male sterility

(iv) Chemical-induced male sterility

(v) Transgenic male sterility.

Genetic Male Sterility (GMS)

The pollen sterility is caused by nuclear genes is termed as genic or genetic male sterility.

Reported in barley, wheat, maize, cotton, sorghum, lucerne, cucurbits, tomato and sugar beet.

The characteristics of GMS are :

1. Male sterility genes are usually recessive.
2. In majority of cases, sterility is caused by single gene. In few cases two or more genes control male sterility.

Merits

- GMS usually recessive and monogenic. Hence fertility restoration in the hybrid and crossing plants are relatively easy.
- Can be used for the production of hybrid seed both in seed as well as vegetively propagated crops.
- Generally doesn't have undesirable agronomic characters.
- Requires less area and labour, because the breeder has to maintain only A and B lines.

Demerits

- Less stable
- 50% plants are fertile, which have to be removed

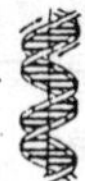

every year. This increases the cost of hybrid seed.

Cytoplasmic Male Sterility

Features –

1. Plants carrying particular type of cytoplasm are male sterile, but will produce seed if pollinators are present. The F_1 seeds (cross seeds) produce only male sterile plants, because their cytoplasm is derived entirely from the female gamete.
2. It consists of only A line and B line. A is male sterile and B is male fertile. For all other characters A and B lines are similar or isogenic.
3. CMS can't be utilised for hybrid seed production without the use of restorer line because F_1 seeds produce only male sterile F_1 plants.
4. CMS can be used for development of hybrids in vegetatively propagated crops and ornamental crops, where grain is not the economic part.
5. CMS is not influenced by environmental factors such as low or high temperature.

Merits

- Highly stable and is not influenced by environmental conditions such as temperature and day length.
- CMS requires less area because breeder has to maintain only A and B lines.

Demerits

- Can't be used for development of hybrids in those crops where seed is the economic product.
- Can be used only in asexually propagated crops viz., sugar cane, potato and forage crops.
- CMS is solely governed by plasma genes. Therefore, it is impossible to restore fertility in the hybrid.
- CMS line may have inferior agronomic performance.

Cytoplasmic Genetic Male Sterility

When pollen sterility is controlled by both cytoplasmic and nuclear genes it is known as cytoplasmic genetic male sterility.

Features –

- Male sterility is controlled by the cytoplasm and nuclear gene.
- System includes A, B and R lines. A is male sterile line, B is similar to A in all features but its male fertile and R restores fertility in F_1 hybrid, hence it's called restorer line (R). 'B' line is used to maintain the fertility and so it's called maintainer line 'B' line.

Development of new male sterile and restorer lines :

New male sterile lines are developed by following the same procedure as that CMS. But the nuclear genotype of the pollinator strain must be rr in this transfer, otherwise fertility would be restored.

Restorer lines are developed by following the procedure :

Quiz Kit!

Three in one

A female aphid has her children developing within her even before she emerges from her larval skin. These developing children have children developing within them, so that an expectant aphid mother is at the same time an aphid grandmother !

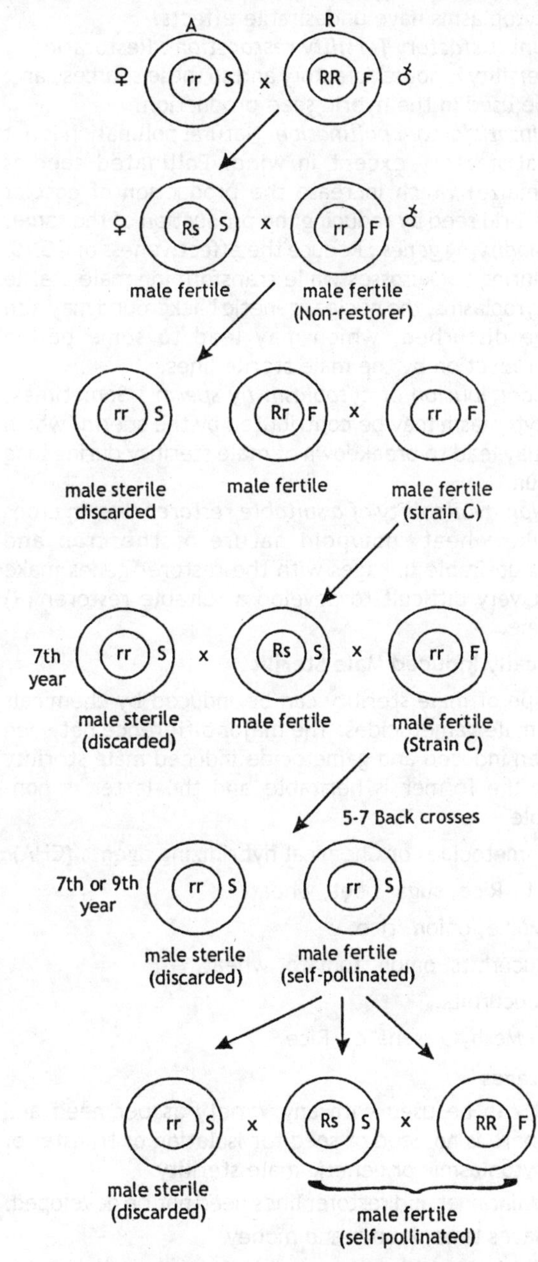

Male fertile progeny selfed. Individual pterogenesis grown in the next generation and non-segregating progenies are selected.

GK GS

The tails of a Comet always points away from the sun.

Utilization

- Commercially used to produce hybrid seed in maize, bajra, sunflower, cotton, jowar etc.
- Widely used for hybrid seed production in both seed as well as vegetatively propagated species.
- CGMS is highly stable and reliable as it's not affected by environmental factors.

Limitations

1. *Undesirable effects of cytoplasm*: Male sterile cytoplasms have undesirable effects.
2. *Unsatisfactory fertility restoration* : Restoration of fertility is not satisfactory and so these sources can't be used in the hybrid seed production.
3. *Unsatisfactory pollination*: Natural pollination is not satisfactory except in wind. Pollinated species (maize) which increase the production of cost of hybrid seed by reducing the production of the same.
4. *Modifying genes* : Reduce the effectiveness of CGMS. During backcrosses while transforming male sterile cytoplasms, the nuclear genetic background may also be disturbed, which may lead to some pollen production by the male sterile lines.
5. *Contribution of cytoplasm by sperm* : Sometimes, cytoplasm may be contributed by the sperm, which may lead to breakdown of male sterility during long run.
6. *Non-availability of a suitable restorer line*: In crops like wheat, polypoid nature of the crop and undesirable linkages with the restorer genes make it very difficult to develop a suitable restorer (R) line.

Chemically Induced Male Sterility

This type of male sterility can be induced by chemicals called male gametocides. The major difference between mutagen induced and gametocide induced male sterility is that the former is heritable and the latter is non-heritable.

Male gametocides or Chemical hybridizing agents (CHA)

Ethanol - Rice, sugar beet, wheat.

GA_3 - Maize, onion, rice.

MH - Cucurbits, onion, tomato, wheat.

NAA - Cucurbits.

Sodium Methyl Arsenate - Rice.

Advantages

1. It can be used with any variety as per need and there is no need of seed for isolation or transfer of cytoplasmic or genetic male sterility.
2. Maintainer and restorer lines need not be developed.
3. Saves time, labour and money.
4. This method is being used in China to develop rice hybrids.

Disadvantages

1. Pollen abortion is incomplete and variable.
2. Treatments are effective only at specific stage of crop growth and effect is short-lived.
3. Female fertility may also be reduced.
4. Production of undesirable side-effects.
5. High cost of CHAs and their repeated application increases the cost of hybrid seed.

Transgenic Male Sterility

The male sterility which is induced by the technique of genetic engineering is called TMS so the foreign gene (transgene) is used for the induction of male sterility. The gene responsible for male sterility is isolated from micro-organisms. It comes under genetic male sterility is heritable.

The TMS has been induced in tobacco by transferring a gene from B. amylolique faciens. Two types of transgenes are used, one gene (Barnase) causes male sterility, hence this gene is integrated in 'A' line. Another gene (Barstar) suppresses the male sterility gene (barnase) and leads to the restoration of fertility. This gene is transferred to restorer line. When a cross is made between A and B line the F_1 is fertile. But this is an expensive method of obtaining male sterility. Maintenance of male sterility is a problem.

Self-Incompatibility (SI)

SI was first reported by Koelreuter. Here, pollen grains fail to germinate on the sigma of the flower that produced them. If some pollen grains do germinate, pollen tubes fail to enter the stigma. In many species, the pollen tubes enters the style, but they grow too slow to effect fertilization before the flower drops. Sometimes fertilization is affected, but the embryos degenerate at a very early stage. All these conditions combinedly are referred to as self-incompatibility.

Types

SI can be classified on the basis of

(A) flower morphology.

I. Heteromorphic System

- Distyly
- Tristyly

II. Homomorphic System

- Diallelic
- Polyallelic

(B) Genes involved

(a) Mono-allelic

(b) Diallelic

(c) Polyallelic

(C) Site of expression

Rajendra II was crowned in the battlefield.

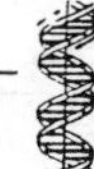

(a) Stigmatic
(b) Stylar
(c) Ovarian
(D) Pollen cytology
(a) Binucleate
(b) Trinucleate

Hetero-morphic system

Here, flowers of different incompatibility groups are different in morphology.

(SI) in primula two types of flowers Pin and Thrum and is called distyly.

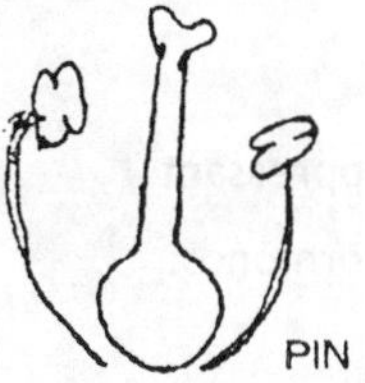

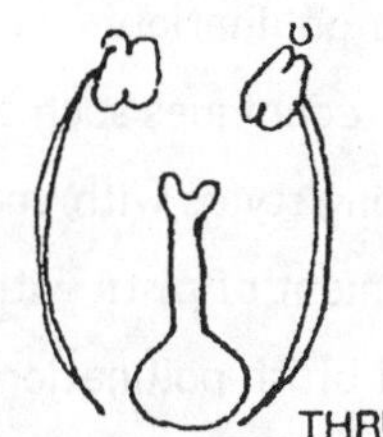

1. Pin and thrum flowers produced on different plants. The only compatible mating is between pin and thrum flowers, which is governed by a single locus S; Ss produces thrum, while ss produces pin flowers.
2. Here the *incompatibility reaction governed by the genotype of the plant producing them.*
3. *Allele 'S' is dominant over 's'* and *incompatibility system is hetero-morphic sporophytic.*
4. Results of four types of crosses in primula.

Pin x Pin	ss x ss	Incompatible
Thrum x Thrum	Ss x Ss	Incompatible
Thrum x Pin	Ss x ss	1 Thrum: 1 Pin
Pin x Thrum	sS x Ss	1 Thrum: 1 Pin

Tristyly

In tristyly anthers and style have three positions, short, medium and long. (Eg) members of lythraceae.

Three positions of style are genetically controlled by two genes (*Ss* and *Mm*). The 'S' gives rise to short style, *sM* to medium style and *sm* to long style. The compatible matings in primula are long x medlum, long x short, and medium x short.

Homomorphic System

The incompatibility reaction of pollen may be controlled by the genotype of the plant in which it's produced (sporophytic) by it own genetype (gametophytic control). This system is very much important in crop plants. It can operate in many ways like.

1. The pollen grains do not germinate on the stigma of same flower. If they germinate the pollen tube fails to penetrate the stigma as in a rye, cabbage and radish.
2. Pollen grains may germinate, but there is retardation of pollen tube growth.
3. There is slow rate of pollen tube growth and it rarely reaches the ovule.
4. Pollen tube growth may be normal, but it does not release the male gamete.

Did You Know ?

Colourful cocks

Cocks of the Manal Pheasant (*Lophophorus impejanus*) are reputed to have nine colours, black, white, blue, purple, green, red, yellow, orange and brown.

Gametophytic System (GS)

Described by *East* and *Mangelsdorf*

• Genotype of plant (Sporophyte)	$S_1 S_2$	$S_3 S_4$
Genotype of pollen and gametes	$S_1 S_2$ ↓ ↓	$S_3 S_4$ ↓ ↓
Incompatibility reaction of pollen grains	$S_1 S_2$	$S_3 S_4$
Incompatibility reaction of style	$S_1 S_2$	$S_3 S_4$

The incompatibility reaction of pollen is determined by own genotype and not by the genotype of the plant on which it is produced. This is because the biochemical substance involved in the S_1 reaction of the pollen is produced after meiosis.

Sporophytic system (SS)

Discovered by *Hugher* and *Babcock* (1950).

In sporophytic system SI reaction is governed by single genes with multiple alleles. In general, the number of 'S' alleles is considerably larger in the gametophytic system than in the sponophytic system.

The incompatibility reaction of pollen is governed by the genotype of the plant on which the pollen is produced and not by the genotype of the pollen.

In sporophytic system, the 'S' alleles may exhibit dominance individual action (codominance) or competition.

- Polygenes (modifying genes) are known to increase as well as decrease the activities of 'S ' alleles both in the GS and SS systems.

Mechanism of SI

1. *Stigmatic inhibition (pollen stigmq interaction)* : Inhibition of pollen germination or pollen tube growth occurs on the stigma. Normally it is confined to sporophytic system. This type of inhibition is found in Radish, Cabbage, Cauliflower, Sunflower etc.
2. *Stylar inhibition*: Inhibition of pollen tube growth

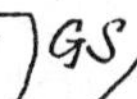

Darwin published his view in his book 'Origin of Species'.

in the stylar region is a common feature in SI plants with binucleate pollen. This inhibition is restricted to GS. Pollen tubes usually swell and burst at the apex in the upper region of the incompatible style. (*eg*) Beta, Helianths, Fagopyrum.

3. *Ovarian inhibition (pollen tube style in interaction)*: Self-incompatibility reaction occurs only when the pollen tubes reacts the ovary. Such inhibition occurs in the ovary in the species that have hollow styles. (Eg) Lilium, Anona etc.

Elimination of SI

1. Doubling the chromosome number in case of single locus gametophytic system.
2. Isolation of self fertile (S_f mutations.
3. Self-compatibility alleles may be transferred from related or same species through back cross.

Methods to overcome SI

Temporary self fertility can be achieved in many ways so then the SI is fully functional in the selfed progeny. Such fertility is called Pseudo fertility and is brought about by

1. Bud pollination.
2. End of season pollination.
3. Mentor pollen technique (recognition pollen).
4. High temperature exposure to pistils, tomato, Brassica.
5. Irradiation by x-rays γ-rays -Solanaceae.
6. Grafting.
7. Double-pollination.
8. Other techniques such as

(i) Injecting styles with immuno-suppressant.

(ii) Treatment of pistil with phytohormones.

(iii) Steel brush pollination.

Centres of Origin (Vavilov)

Name of the centre	*Areas covered*	*Plants*
1. Chinese centre	Mountaineous regions of central western China.	Buck wheat, Soyabean, Onion, Bamboo, Cherry, Citrus, Sugarcane Tea
2. Indian centre		
(a) Main centre	Assam and Myanmar	Rice, Sugarcane, Mango, Orange, Jute, Coconut, Pepper
(b) Indo Malayan	Indo-China and Malayan Archipelago	Banana, Sugarcane, Clove, Coconut, Manila hemp, Pepper.
3. Central Asiatic	Kashmir, NWFP of Punjab, Afghanistan, Uzbekistan, Tian-Shan	Wheat, Pea, Beans, Cotton, Carnet, Garlic, Apple, Almond, Apricot.
4. Near Eastern	Interior of Asia-minor, Transcaucasia, Iran highlands of Turkmenistan.	Cherry, Pomegranate, Fig, Almond, Clove.
5. Mediterranean	Borders of Mediterrranean sea	Cabbage, Turnip, Clove, Mustards, Duram and emmer wheat, Pepper etc.
6. Abyssinian	Ethiopia, Somalia, Eritrea	Wheat and Barley, Sesame, Castor, Coffee, Okra.
7. South Mexican and Central American	South Mexico, Guatamala, Honduras, Costa Rica	Maize, Bean, Sweet-potato, Papaya, Tobacco, Upland cotton.
8. South American	High mountainous areas	Many potato spp, Tomato, Pumpkin
(a) Peruvian-Bolivian		Egyptian cotton, Tobacco
(b) Chile		Common potato
(c) Brazilian-Paraguian		Groundnut, Cassava, Pineapple, Rubber and Cashew nut.

Primary gene centre: Native wild relatives, dominant genes, diversing due to ancient cultivation.

Secondary gene centre: Devoid of relatives recessive characters, ecological diversity, farming practice, human migration patterns and internal biological dynamics of hybridisation, segregation, selection.

GS

S.I. unit of Capacitance is Farad.

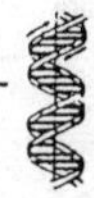

BREEDING METHODS

A generalized classification of selection procedure

I. Intra-population selection based on additive variance

(a) Based on sample size and degree of inbreeding

(i) Individual or pure-line or pedigree selection

(ii) Mass pedigree selection

(iii) Mass or bulk selection

(b) Based on family structure and relatives

(i) Individual selection

(ii) Family selection

(iii) Within-family selection

(iv) Combined selection

(c) Based on number of traits selected

(i) Tandem selection

(ii) Independent culling level

(iii) Index selection

2. *Inter-population selection based on non-additive variance*

(a) Reciprocal recurrent selection

(b) Recurrent selection for gca

(c) Recurrent selection for sca

II. Stabilizing selection

III. Disruptive selection

Plant Introduction

Taking a genotype or a *group* of genotypes of plants into a new environment where they are not being grown before.

The materials are introduced from other countries or continents.

Movement of crop variety from one environment into another within the same country is also introduction.

e.g. within the country introduction are the wheat in West Bengal and rice in Punjab.

Classification

(i) Primary introduction

When the introduced variety is well suited to the new environment it is released for commercial cultivation without any alteration in the original genotype.

e.g. dwarf wheat variety Sonar 64, Lerma Rojo dwarf rice variety -T.N. 1, IR8, IR28.

(ii) Secondary Introduction

The introduced variety may be subjected to selection to isolate a superior variety. Alternatively it may be hybridized with local variety to transfer one or few characters from this variety to the local ones.

You Don't Know!

Heaviest buffalo

The wild buffalo (*Bubalus bubalis*), which measures nearly 170 cm at the shoulders and weighing a ton, is heavier and much more aggressive than the domestic species. It is found in Nepal's terai region, the Brahmaputra plains of Assam, the Bastar district of Madhya Pradesh and Orissa's adjoining districts.

e.g. Kalyan Sona and Sonalika wheat variety selected from materials introduced from CIMMYT, Mexico.

Procedure for Plant Introduction

1. *Procurement of Germplasm*
   - Gift, Purchase or Collection
   - Propagules
2. *Quarantine*
   - Plant produce imported for consumption
   - Plant propagules imported for cultivation and Research
   - Quarantine procedure
3. *Cataloguing*
4. *Evaluation*
5. *Multiplication*
6. *Distribution*

Purpose or Objectives of Plant Introduction

- Provide an entirely new crop species *e.g.* Maize, Potato, Tobacco.

Directly released as New varieties *e.g.* Sonara 64, Larma Rojo

- Utilization in crop improvement programmes e.g. Pusa Ruby tomato was derived from Meeruty x Sioux.
- Saving a crop from a disease or pest coffee was introduced in South America from Africa to prevent losses from leaf rust.
- Utilisation in scientific studies.
- Used for Aesthetic value.

Merits :

- Provides entirely new crop plants.
- Provides superior variety.
- Introduction are the only feasible means of collecting germplasm and to protect variability from genetic erosion.
- Very quick and economical method of crop improvement.
- Crop species may be introduced in new disease from areas to protect them from damage *e.g.* coffee and rubber.

Demerits :

- Weeds entering through plant introduction *e.g.* Phalaris minor.

Forest contribute 2% in India's economy.

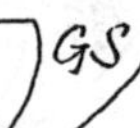

- Diseases: Late blight of potato from Europe, Coffee rust from Srilanka. Bunchy top of banana from Srilanka.
- Insect pest: Potato tuber moth from Italy, Woolly aphids of apple.
- Ornamental turned weeds: Water hyacinth, Lantana camara.
- Threat to ecological balance: Eucalyptus spp introduced from Australia causes a rapid depletion of the subsoil water reserves.

Selection

Selection operates only genetic variability occurs in plants or crops.

1. Mendelian variation :

All variations occur through gene mutation.

Macromutation : Maize, Brassica

Micromutation : Rice, barley, tomato

2. Interspecific hybridization :

Less contribution to evolution => sterile progenies

Introgressive hybridization => importance for evolution

Tripsacum x Primitive maize

↓

Maize

3. Polyploidy:

Autopolyploidy : more vigour, robust

1. auto triploids : banana, water melon (3x)
2. auto tetraploids : potato (4x)
3. auto hexaploid : sweet potato

Allopolyploidy : 50% of plants evolved

1. Allohexaploid : wheat, triticale
2. Allotetraploid : cotton, tobacco

Germplasm

(a) Land Races: Naturally selected strains/species
(b) Obsolete variety: Systematic breeding but not cultivated now.
(c) Varieties in cultivation
(d) Wild forms and relatives
(e) Breeding lines.

Pureline selection

A large number of plants are selected from a self-pollinated crop. The selected plants are harvested individually. The selected individual plants are grown in individual rows then evaluated and best progeny is selected.

Characters of pureline :

(1) All the individuals are homozygous.

(2) Variation in pureline is due to environment.

(3) Genetically become variable with time.

Procedure of Pureline Selection

Pureline selection consists of four major steps, *viz.* (1) selection of a heterogeneous population, from which purelines have to be isolated, (2) Isolation of purelines by individual plant selection, (3) testing of purelines in field trials, and (4) release of the best pureline as a variety. The year-wise procedure is given as follows :

First year: An old variety or land race is used for pureline selection. Single plants are selected from the heterogeneous population keeping in view the objective of selection. The number of individual plants to be selected may vary from 200 to 1000 in various crops.

Second year: The progeny of each selected plant is grown separately in few rows and evaluated for the character under consideration. The top 15-20 progenies are selected and seed of all plants in each progeny is bulked which constitutes strains.

Third year: The strains constituted in second year are evaluated in field trials for 2-3 years for yield performance. In India, the selected entries (strains) are evaluated in All India Coordinated Crop Improvement Project. The best genotype is identified on the basis of yield performance.

Eighth to Tenth year: The best performing strain is released and notified as a variety. Then the breeder, foundation and certified seeds are produced. The production of certified seed takes two years after release of a variety. Thus the seed of new variety reaches the farmers in tenth year.

Merits :

(1) Uniform
(2) Maximum improvement possible
(3) Variety is easily identified in seed certification programme

Demerits :

(1) Less adaptable
(2) More time, labour and money
(3) Upper limit and improvement is set by the genetic variation present in original population

Pedigree method

Individual plants are selected from F_2 and subsequent generations and their progenies are tested. During this process details about the plants selected in each generation is recorded in pedigree record.

Breeding Procedure

First parents are selected keeping in view the breeding objective. The cross is made between selected parents. The F_1 material is grown using wide spacings. The dominance behaviour for various characters is recorded.

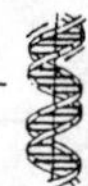

In F_2 also the material is grown using wide spacings. Individual plant selection is practised in F_2. The progeny of each selected plant is grown separately which constitutes F_3 generation. In F_3 and F_4 generations, selection is practised within and between families. From F_5 to F_8 between progeny selection is practised and superior progeny are identified and isolated in. These progeny constitute strains. These strains are evaluated in replicated multilocation trials for a period of 3-5 years. Based on superior performance the strains is released as a variety. Thus release of new variety by this method takes 14-15 years. The number of plants to be grown and selected in each generation is presented. This is a general procedure of pedigree method. The number of plants to be grown and selected in each generation is not fixed. It may slightly vary from crop to crop.

Merits :

- Gives maximum opportunity to the breeder to use his skill.
- Well-suited for characters which are simply inherited.
- Transgressive segregants can be easily identified through records.
- Information about inheritance is precisely obtained.

Demerits :

- Maintenance of pedigree record is time-consuming.
- The success in this method is largely dependent on skill of the breeder.
- Selection for yield in F_2 and F_3 is ineffective.

Achievements :

Rice	: Krishna, Ratna, Sabarmati, Padma, Jaya, Bala, Kaveri.
Wheat	: HD 2281, HD 2285, HD 2380, ND 2402, Janak.
Cotton	: LH 900, LH 1556, F 846, Vikas, Khandwa 3, Sharda, Abhadita, Sahana, Anjali, Surabhi, Suvin.
Pigeonpea	: T21, Prabhat
Greengram	: T2, T44, T51, Sheela.

Backcross method

Backcross refers to crossing of F_1 with either of its parents when the F_1 is crossed with homozygous recessive parent, it is known as test cross. A system of breeding in which repeated backcrosses are made to transfer a specific character to a well-adapted variety for which the variety is deficient is referred to as backcross breeding.

Main features :

Application - It is generally used to improve specific character of a well-adapted variety for which it is deficient such as resistance to a specific disease.

Knowledge World!

Longest enduring flower

Flower of the Lady's Slipper Orchid (*Paphiopedilum villosum*) last for six months, from mid-October to mid-March. Flowers of the closely related *P. fairieanum* remain fresh for six to eight weeks.

Parent material - In this method two types of parents like recipient parent and donor parent involves. The parent which receives a desirable character is known as recipient parent. The parent which donates desirable character is known as donor parent.

Genetic constitution - The new variety resembles the parent variety in all the characters except for the character under transfer.

Number of backcross - 5 to 6 backcross are sufficient to retain genotype of original variety with new character.

Procedure:

Transfer of dominant gene

Suppose wilt resistance in cotton is controlled by a dominant gene (RR). The donor parent is a strain (B) from the germplasm. The resistance has to be transferred to an adapted variety (A) which is susceptible to wilt. The adapted variety (A) will be used as recurrent parent and strain (B) as donor parent. The F_1 will be wilt resistant but heterozygous (Rr). Backcrossing of F_1 (Rr) with susceptible variety (rr) will produce resistant and susceptible plants in equal number in BC_1 F_1 (1Rr : 1 rr). The resistant cotton plant (Rr) can be identified by growing the material in wilt sick plot. The resistant plants (Rr) are then backcrossed to the adapted variety. Generally, 6-8 backcrosses are sufficient to obtain identical plants to adapted except for the added genes for wilt resistance. The wilt resistant plants are heterozygous (Rr). They are selfed for one generation to obtain homozygous (RR) resistant plants. All the resistant true breeding plants are bulked and new variety is released. The variety developed in this way is identical to the adapted variety (A) expect for wilt resistance.

Transfer of recessive gene

Suppose wilt resistance in cotton is governed by a recessive gene (rr). In such case, a progeny of each backcross will segregate into two genotypes (RR and Rr) which cannot be identified. Therefore, it is necessary to self the population after each backcross obtain resistant homozygous recessive plants (rr). The resistant plants are identified by owing the F_2 material in wilt sick plot. The resistant plants are backcrossed with adapted variety. Here each backcross is followed by one selfing, whereas with dominant one continuous backcrosses are made.

GK GS

Heliophytes are direct sunlight receiving plants.

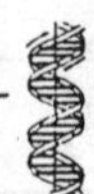

Merits :

- It retains all desirable characters of a popular adapted variety.
- It is a useful method for transfer of oligogenic character like disease resistance.
- The male sterility and fertility restorer genes are transferred to various agronomic bases.
- It is the only breeding method which is used for interspecific gene transfer.

Demerits :

- The backcrosses have to be made for 6-8 generations.
- Sometimes undesirable character is also transferred to the new variety.

Achievements :

Backcross method has been widely used for the development of disease resistant varieties in both self and cross pollinated species. It has also been used for interspecific gene transfer and development of multiline varieties in self-pollinated species. Several varieties resistant to various diseases have been developed by this method in wheat, cotton and several other crops. In cotton varieties V797, Digvijay, Vijalpa and Kalyan which belong to *Gossypium herboceus* have been developed by backcross method.

Mass selection

Definition: A number of plants are selected on the basis of their phenotype and the open-pollinated seed from them is bulked together to raise the next generation.

Procedure :

First year: (i) Several plants selected on the basis of phenotype (ii) open pollinated seed from the selected plants harvested and bulked.

Second year: (i) Bulk seed from the selected plants grown (ii) Mass Selection may be repeated.

Third to fifth year: Main yield trials are carried out to determine the performance of superior mass selected strains in comparison with standard varieties as checks.

Sixth to eighth year: Trials on the cultivator's holding.

Eighth year: The variety is produced and named.

Application :

- It cannot bring any new change in the phenotypes of the population but can only sort out the different types already existing in it.
- It can be applied to any crop which posesses some amount of heterozygosity *i.e.* variability, which is the basis of mass of selection.

In cross pollinated crops mass selection is more effective with following types of crop materials in which heterozygosity is found invariably:

- Wild varieties
- Local varieties growing on farmer's fields.
- Introduced varieties
- Mixed varieties of a crop

Uses :

- Production of new varieties
- Purification of unimproved varieties
- Maintaining the purity of a variety

Achievements :

- First commercial types of Hirsutum cotton in India *e.g.* Cambodias
- Variety of Groundnut such as TMV -1, TMV-2
- Maize - T13, Jaunpur, Darjeeling White Round
- Bajra - Pusa Moti
- Potato - K122

Demerits:

Varieties developed show variation and are not uniform as pureline varieties.

In the absence of progeny test, it is not possible to determine whether the selected plants are homozygous.

Improvement is less than that of pureline selection.

Single seed descent method

A breeding procedure used with segregating population of self-pollinated species in which plants are advanced by single seeds from one generation to the next is referred to as single seed descent method.

Procedure :

Modified from Bulk method (one seed is selected)

Merits :

- This is a simple, convenient, less expensive and time-saving method.
- Large number of crosses can be evaluated by this method, because less space and labour is required in each generation.
- This method is able to retain variability in a breeding population.

Demerits :

- It does not provide opportunity to select for superior plants till F_5 generation.
- The frequency of getting desirable genotypes in the advanced generation is reduced.
- The identity of superior F_2 plants cannot be maintained in this method.

Bulk method

In this method F_2 and subsequent generations are harvested as bulk to grow the next generation. The duration of bulking may be 6-7 generations. Selection can be made in each generation but harvesting is done as bulk. At the end of bulking period single plant selection is made and tested for yielding ability.

Procedure :

The bulk breeding method consists of four important steps, viz, (1) bulking period, (2) progeny selection and isolation

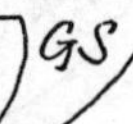

Weber is the S.I. unit of Magnetic flux.

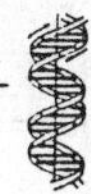

of superior progeny, (3) multilocation trials of superior progeny, and (4) release of the best progeny as a variety. These steps are briefly discussed below:

Bulk period: The F_1 plants are grown and their F_2 seeds are harvested in bulk. The F_2 plants are raised from a sample of F_2 seeds and F_3 seeds are harvested in bulk. This process is repeated until the desired level of homozygosity is achieved. In general, bulk period is allowed up to F_5 generation. The material is subjected to biotic and abiotic stresses during bulking period to eliminate undersirable genotypes.

Progeny selection: In F_6, the material is space planted and individual plant selection is practised. The progeny of each selected plant is grown separately in F_7 and superior progeny are selected and isolated in F_7 and F_8

Multilocation testing: The selected progeny constitute strains. In eighth year, preliminary yield trial is conducted. From 10th to 14th year multilocation testing is carried out and the best performing strain is identified on the basis of 3-4 years performance in the multilocation trials. The best strain is released and multiplied for seed distribution in the l5th year. Thus bulk method takes 15-16 years for release of new variety.

The bulk material should be tested under such environment which is expected to favour desirable genotypes. For example, screening for disease resistance should be carried out in the disease-prone area. Moreover, the screening should be done under natural field condition rather than in green houses.

Merits :

- No pedigree record is maintained.
- If bulking period is longer natural selection operates and desirable genotypes selected.
- Chances of obtaining more transgressive segregants.

Demerits :

- Takes much longer time to develop a new variety.
- A large number of progenies are to be selected in each generation which requires much labour, time and space.
- We cannot get information on inheritance.

Recurrent selection

Recurrent selection is reselection generation after generation with inter-breeding of selects to provide for genetic recombination. It is used to improve the frequency of desirable alleles for a character in a breeding population.

Main features :

Application - It is more commonly used in cross-pollinated species than in self-pollinated species.

Base population - A heterozygous base population is required to start recurrent selection.

Important steps - (i) Selection of superior plants from base population (ii) selfing of selected plants (iii) growing progeny of selected plants in the next season (iv) intermating among progeny (v) bulking of crossed seed.

Quiz Kit!

First Deputy Speaker

Ananthasayanam Ayangar was elected Deputy Speaker on May 30, 1952.

Use of end product - The population developed by recurrent selection can be used in producing homozygous inbreeds by selfing, in the production of hybrid varieties, in the production of synthetic varieties.

Impact - It is used to improve the frequency of desirable alleles for a character in a population.

Types of Recurrent Selection

1. Simple recurrent selection

A type of recurrent selection that does not include tester is referred to as simple recurrent selection. It is also known as phenotypic recurrent selection. This method is an extension of mass selection. The main features of simple recurrent selection are given below:

1. The tester is not used in this scheme.
2. It does not measure the combining ability.
3. The selection is based on phenotype or simple test.
4. This method is useful only for those characters which have high heritability.
5. This method requires only two seasons for the completion of one selection cycle.

2. Recurrent selection for GCA

First year: Superior plants for the character under improvement are selected from the base population. The selected plants are selfed and also crossed to a heterozygous tester having broad genetic base. The selfed seed is kept in cold storage.

Second year: The crossed seed is sown and the combining ability of the selected plants is evaluated and plants with good *gca* are identified.

Third year: The progeny of selected plants with good gca are grown from their selfed seed kept in cold store. These progeny are intermated in all possible combinations and their crossed seed is composited to form a new source population for further selection. This completes original selection cycle. In the same way another cycle can be completed in three years (4th to 6th year).

This is called first recurrent selection cycle. Many such cycles may be made to obtain desired results.

3. Reciprocal Recurrent selection of SCA

A form of recurrent selection that is used to improve the *sca* of a population for a character by using homozygous tester is referred to as recurrent selection for specific combining ability (RSSCA). It is also known as half sib recurrent selection with homozygous tester. This method was originally proposed by Hull in 1945.

93% of the total forests of India are tropical forests.

Comparison of recurrent selection for *gca* and recurrent selection for *sca*

Particulars	*Recurrent selection for gca*	*Recurrent selection for sca*
Application	Used to improve polygenic traits	Also used to improve polygenic traits
Basis of selection	Test cross performance	Test cross performance
Tester used	Heterozygous	Homozygous
Effectiveness	More effective with incomplete dominance	More effective with complete and overdominance
Condition of use	Used when additive gene action is important	Used when non-additive gene action is important
Impact	It improves *gca* of a character	It improves *sca* of a character

4. Reciprocal Recurrent Selection

A form of recurrent selection that is used to improve both *gca* and *sca* of a population for a character using two heterozygous testers is known as reciprocal recurrent selection (RRS). It is also termed as recurrent reciprocal half sib selection.

First year: Several phenotypically superior plants are selected from population *A* and *B*. The pollen of some selected plants of *A* population is used to cross large number of randomly selected plants of population *B*. Similarly, pollen of some selected plants of *B* population is used to cross large number of plants of population *A*. All the plants of population *A* and *B* used as pollen parents in the crosses are selfed.

Second year: The progeny of test crosses made with pollen parents of *A* and *B* population are evaluated in separate replicated trials. The superior progeny are identified.

Third year: The selfed seed of those *A* and *B* plants whose progeny were found among the progeny of *A* plants and also among the progeny of *B* plants. The crossed seeds of *A* block are composited in equal quantity to raise *A* generation. Similarly, crossed seeds of *B* block are bulked to raise B_1 generation. This completes original cycle of selection.

Fourth year: The A_1 and B_1 population are grown from the composite crossed seeds of respective population obtained in third year. Then operations of first year are repeated.

Fifth year: The operations of second year are repeated.

Sixth year: The operations of third year are repeated.

The last three years constitute first cycle of reciprocal recurrent selection. Such selection cycles may be continued till the desired improvement is achieved.

Merits :

It is an efficient breeding method for increasing the frequency of superior genes in a population.
- It helps in maintaining high genetic variability.
- Selection is made on the basis of test cross performance

Demerits :
- It is not used directly for the development of new varieties.
- It involves lot of selection, crossing and selfing work.
- It permits selfing which leads to loss of genetic variability.

OTHERS

Heterosis breeding

Heterosis hybrid vigour refers to the superiority of F_1 hybrids in one or more characters over its parents. Heterosis differs from luxuriance, as in the case of former, there is an increase of F_1 over parents in general vigour, yield and adaptation, while the latter shows only an increase in vegetative growth and not the yield.

Features of Heterosis

1. *Superiority over Parents*: Hybrid vigour/ Heterosis leads to superiority in adaptation, yield, quality, disease resistance, maturity and general vigour over its parents.
2. *Confined to F_1* : Heterosis is confined only to the F_1 generation of a cross. It declines and disappears in F_2 and subsequent generations of a cross as a consequence of segregation and recombination.
3. *Genetic control*: Expression of heterosis is governed by nuclear genes. It results due to interaction between nuclear genes and cytoplasm.
4. *Reproducible*: Heterosis once identified can be easily reproduced in a definite environment.
5. *Association with SCA* : It has positive association. SCA is a measure of dominant variance and a significant amount of dominant variance is essential for heterosis breeding.
6. *Effect of heterozygocity* : Magnitude of heterosis associated with heterozygocity because the dominance variance is associated with heterozygocity.
7. *Conceals recessive genes*: In case of heterosis, the deleterious effect of recessive genes are covered by the favourable effect of dominant genes.
8. *Low frequency*: The frequency of desirable heterotic combination is very low. All the F_1 crosses do not show desirable heterosis.

Fixation of heterosis :

1. *Vegetative propagation*: Which is based on mitosis. As a result genotype is transmitted intact to all its

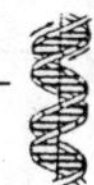

vegetative progeny e.g., in crops like, sugar cane, potato, sweet potato, banana, where vegetative propagation is a rule, hybrids can be easily developed.

2. *Apomixis*: Here the seed develops without fertilization. If F_1 and its progeny produce only apomictic seeds having 2n embryos, the F_2 and subsequent generations will be identical in genotype to the F_1 and thereby lead to the fixation of heterosis.
3. *Balanced lethal system*: In plants like, *Oenothera*, the homozygotes will die as they are lethal and only heterozygotes survive, which results in the fixation of heterosis.
4. *Polyploidy*: Heterosis can also be fixed by chromosome doubling or polyploidy especially in interspecific and intergeneric hybrids. For *e.g.* the heterosis in wheat-rye cross can be conserved in amphidiploid hybrids through chromosome doubling.

Factors affecting heterosis :

I. *Mode of pollination*: Level of heterosis higher in cross pollination than self-pollination crops.
II. *Genetic diversity of parents*: Higher heterosis associated with crosses of more distantly related parent and maximum heterosis occurs at optimum or intermediate level of parental diversity.
III. *Genetic base of parents*: If the genetic base is broad, then the heterosis will be higher.
IV. *Adaptability of parents* : If the parents are having wider adaptability, then there will be the occurrence of heterosis.

Advantages

- Higher yield potential
- More uniform and attractive
- Can be developed in cross pollination and self-pollination crops.
- Can be possible to reconstitute the hybrid with same genotype

Disadvantages

- Fresh seed to be produced every year
- Hybrid seed is costlier
- Cultivation required more input (fertilizer, water fungicides and insecticides)
- Requires more technical skill and areas

Hybrids

Hybrids are the first generation from crosses between two purelines, inbreeds, open pollinated varieties of other population that are genetically not similar.

Kinds of hybrids :

1. Single cross hybrids: Crossing two inbreeds or purelines, A x B.
2. Three way cross hybrid: A cross between a single cross hybrid and an inbreed (A x B) x C.

> **Did You Know ?**
> *First Atomic Power Plant*
> **The first atomic power plant for which every stage, from design to execution carried out by Indians was the Madras Atomic Power Station (MAPS) in Kalpakkam near Madras.**

3. Double cross hybrid: Double cross involving the four inbreeds A, B, C and D.
4. Double top cross hybrid: Double cross hybrid cross with open pollinated variety.

Development of inbreeds :

I. By inbreeding, selfing etc.
2. Development of inbreeds from haploids.

Evaluation of inbreeds

1. Phenotypic evaluation: Based on phenotypic performance. Highly suitable for characters with high heritability.
2. Top cross test: The selected inbreeds will be crossed to a tester parent with wide genetic base. The cross progenies will be evaluated in replicated progeny rows.
3. Single cross evaluation: The developed inbreeds can be crossed and the single crosses can be estimated in replicated trial standing hybrids tested over years in different locations, then released.

Production of hybrid seed

Methods

I. Hand emasculation and Dusting-cotton, tomato.
2. Use of male sterile lines
   (*a*) Cytoplasmic male sterility - ornamentals.
   (*b*) Genic male sterility - red gram, castor.
   (*c*) Cytoplasmic genic male sterility - jowar, bajra, rice.
3. Use of self incompatibility.

Merits :

- Hybrids from single cross varieties are more uniform than open pollinated varieties, synthetic and composite varieties.
- Hybrids produced in both self-pollinated and cross pollinated crops.
- It exploits both self-combining ability and general combining ability.

Demerits :

- Seeds are to be changed every year.
- Needs technical skill.
- In cross pollination crops require long isolation distance.

(a) ***Synthetic varieties***

A synthetic variety is produced by crossing in all

combinations a number of inbreeds that combine well with each other. Once synthesized, a synthetic is maintained by open pollination in isolation.

Steps involved in the production of synthetic varieties.

Procedure of developing synthetic cultivar from already available inbred lines

Year	*Main Breeding Activity*
First	Selection of inbred lines, crossing with a common tester, harvesting top crossed seed separately.
Second	Evaluation of top crosses in replicated trial using standard hybrid or open pollinated variety as check. Identification and selection of good general combining inbred lines based on top cross performance.
Third	Making all possible single crosses among inbred lines selected for good gca and harvesting crossed seed of single crosses separately.
Fourth and Fifth	Mixing seed of single crosses in equal quantity and seed multiplication by open pollination in isolation for one or two generations.
Sixth	Release as a new synthetic variety and distribution of seed to the farmers for commercial cultivation.

Merits :

- Farmer can maintain his synthetic variety for more seasons which is not possible in hybrids.
- Because of wider genetic base the synthetics are more stable over years and environments.
- Seed production is more skilled operation in hybrids and it is not so in synthetics.
- Synthetic varieties are good reservoir of genetic variability.
- Less costly compared to hybrids.

Demerits :

- Performance is little bit lower compared to hybrids because synthetics exploit only GCA while hybrids exploit both GCA and SCA.
- The performance may not be good when lines having low GCA are used.
- Synthetics can be produced and maintained only in cross pollinated crop species, while hybrid varieties can be produced both in self and cross pollinated crops.

Achievements

Name of crop	*Type of cross*	*Name of hybrids released*
(a) Cross pollinated crops		
Maize	Double and three way crosses	Ganga 2, Ganga 5, Ganga 11, Makka 1 and VL42
Pearlmillet	Single cross	HB 3, BJ 104, MBH 110, BK 560, MH 179
Sunflower	Single cross	BSH 1, HSFH 1, MSFH 10, KBSH 1 and KBSH 11
Castor	Single cross	GCH 2, GCH 3, Gauchi and GCH 4
Cucumber	Single cross	Pusa Sanyog
(b) Often cross pollinated crops		
Sorghum	Single cross	CSH1,CSH5,CSH6,CHS9,CSH10,CSNH
Cotton	Intraspecific	H 4, H 6, H 8, H 10, JKHY 1, JKHY 2, PKVHY 2, PKVHY 3, PKVHY 4
-	Insterspecific tetraploid hybrids	Varalaxmi, DCH 32, HB2 24, TCHB 213, DHB 105, Sruthi
-	Interspecific diploid hybrids	DH 7, DH 9, DDH 2
-	Intra-arboreum hybrid	LDH II
Pigeonpea	Single cross	ICPH 8 from ICRISAT
(c) Self pollinated crops		
Tomato	Single cross	Hybrid Tomato
Brinjal	Single cross	Vijay, Pusa Kranti, Arka Sheel, Arka Navneet and Azad Kranti
Jute	Single cross	Hybrid C

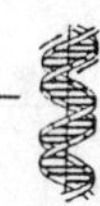

All the varieties of sugarcane and potato are hybrids. Several promising hybrids have been developed in mango. The important mango hybrids include, Mahmood, Bahar, Prabha Shankar, Neeluddin, Swarna Jehangir, Mallika, Amrapalli and Ratna. Hybrids have also been developed in chilli and bottle gourd.

In China, hybrid rice has been developed for commercial cultivation through the use of cytoplasmic genic male sterility. In India, efforts are being made to develop hybrids in rice. However, production of hybrid seeds is a major problem in wheat. In India, ICAR has launched a Hybrid Project entitled, "Promotion of research and development efforts on hybrids in selected crops" from 1989 to intensify hybrid research. The selected crops include rice, maize, sorghum, pearlmillet, pigeonpea, cotton, sunflower, rapeseed-mustard and castor. This project is aimed at developing high yielding superior quality hybrids in above nine crops.

Merits

1. Hybrid varieties have higher yield potential as compared to synthetics, composites and open pollinated varieties.
2. Hybrid varieties are more uniform and attractive than synthetics, composites and open pollinated varieties.
3. Hybrid varieties can be developed in both self and cross pollinated species, whereas synthetics and composites are relevant to cross pollinated crops.
4. It is possible to reconstitute the hybrid with same genotype, which is not possible in case of composite varieties and open pollinated varieties.

Demerits

1. Fresh seed has to be produced every year, because in F_2 the hybrid produces various types due to segregation and recombination. The farmer has to purchase fresh seed every year.
2. The seed of hybrid is costlier than synthetics, composites and open pollinated varieties.
3. Cultivation of hybrids requires more input (fertilizer, water, plant protection, etc,) to exploit their full potential.
4. Production of hybrid requires more technical skill and areas as compared to synthetics, composites and open pollinated varieties.

(b) Composite Varieties

In cross pollinated crops, the mixture of genotypes from several sources that is maintained in bulk from one generation to the next is referred to as composite variety. Composites are constituted by seed mixture of advanced generation material of intervarietal or inter racial cross.

Similarities and Dissimilarities between Synthetic and Composite cultivars

Particulars	*Synthetic cultivars*	*Composite cultivars*
Similarities		
Equilibrium	Polymorphic Stable	Polymorphic Stable
Relevant to	Allogamous Species	Allogamous Species
Genetic Constitution	Heterogeneous, mixture of several homozygotes and heterozygotes	Hetrogeneous, mixture of several homozygotes and heterozygotes
Adaptation	Wide, due to broad genetic base	Wide due to broad genetic base
Dissimilarities		
Base Material	Inbreds or clones	Open pollinated varieties or any other heterozygous source
Component Genotypes	Two to eight	Two to many
Evaluation for gca	Always done	Usually not done
Seed Replacement	After 4-5 years	After 3-4 years
Exact Reconstitution	Possible	Not Possible

Procedure of Developing Composite Variety

Year	*Main Breeding Activity*
First	Selection of base material, intermating in all possible combinations and harvesting crossed seed separately.
Second	Evaluation of F_1 crosses in replicated trial using hybrid or open pollinated variety as check.
Third to Fifth	Evaluation of F_2, F_3 and F_4 generations in replicated trials using standard check, identification of crosses exhibiting little or no inbreeding depression.
Sixth and Seventh	Mixing parental seed of superior crosses in equal quantity to constitute composite variety, seed multiplication by open pollination in isolation.
Eighth	Release of new composite variety, distribution of seed to the farmers for commercial cultivation.

Achievements

Several composite varieties have been developed in maize in India. The important varieties are Jawahar, Vijay, Kisan, Amber, Sona, Vikram, Rattan, Shakti, Vikas, Tarun, Renuka,

Kanchan, Co 1, etc. In *Brassica compestris* var. toria an early maturing variety known as Composite 1 has been released.

Mutation Breeding

The genetic improvement of crop plants for various economic characters through the use of induced mutations is referred to as mutation breeding. It is one of the special methods of crop improvement. Mutation breeding is commonly used in self-pollinated and asexually propagated species.

Significance

1. When the desired variability is not found in the cultivated varieties or in the germplasm of cultivated species.
2. When a high yielding variety has oligogenic defect such as susceptibility to a disease, mutation breeding is the best course of line for a breeder. Because induced mutagenesis leads to improvement of such variety without much alteration in the genetic background.
3. When there is a tight linkage between desirable and undersirable characters, mutation breeding is the best way of overcoming such problems.
4. When in fruit crops, the improvement has to be made without change in the taste and colour of the fruits, it can best be achieved through mutation breeding, because mutation breeding causes little changes in the genetic background of parental material.
5. When a particular reaction is to be blocked, mutation breeding is the only way. For example in synthesis of morphine taken place in step-wise manner. If the biochemical path way is blocked at bane level, it will block the synthesis of morphine without affecting the conversion process of the bane into useful pharmaceutical product.
6. In those crops where sexuality is absent and generation of variability through recombination process is not possible.
7. In those species where generation cycle is very long, such as plantation crops, fruit trees and forest trees, mutation breeding is the short cut way of genetic improvement.
8. When attactive flower and foliage colours have to be developed in ornamental plants which fetch high price in the market, mutation breeding is the only short cut method.

Mutagens and their mode of action

Mutagens refer to various physical or chemical agents which greatly enhance the frequency of mutations. Various radiations and chemicals are used as mutagens. Mutagens are of two types, viz. (1) physical mutagens, and (2) chemical mutagens.

Commonly used physical mutagens (radiations), their properties and mode of action

Type of radiation	*Main properties*	*Mode of action or changes caused*
1. X-rays	S.I., penetrating and non- particulate	Induce mutations by forming free radicals and ions. Cause addition, deletion, transitions and transversions.
2. Gamma Rays	S.I., penetrating and non- particulate	Induce mutations by ejecting atoms from the tissues. Cause all types of changes as above.
3. Alpha Particles	D.I., particulate, less penetrating and positively charged.	Act by ionization and excitation. Cause chromosomal and gene mutations.
4. Beta Rays Particles	S.I., particulate, more penetrating than alpha particles and negatively charged	Act by ionization and excitation. Cause chromosomal and gene mutations.
5. Fast and Thermal Neutrons	D.I., particulate, neutral particles, highly penetrating.	Cause chromosomal breakage and mutations.
6. Ultra Violet Rays	Non-ionizing, low rays mutations.	Cause chromosomal breakage and gene

Some commonly used chemical mutagens and their mode of action

Group of mutagen	*Name of chemical*	*Mode of action*
1. Alkylating agents	Ethyl methane Sulphonate	AT↔ GC Transitions
	Methyl methane Sulphonate	Transitions
	Ethyl ethane Sulphonate	GC ↔ AT Transitions
	Ethylene Imines	Transitions
2. Base analogues	5 Bromo Uracil	AT ↔ GC Transitions
	Amino Purine	AT↔ GC Transitions
3. Acridine dyes	Acriflavin, Proflavin	Deletion addition and Frameshifts
4. Others	Nitrous Acid	AT ↔ GC Transitions
	Hydroxylamine	GC ↔ AT Transitions
	Sodium Azide	Transitions

The mutation breeding process consists of four important steps. These are (I) choice of material, (2) choice of mutagen, (3) mutagen, treatment, and (4) handling of treated material.

GK GS

In electrolysis, oxidation takes place at anode.

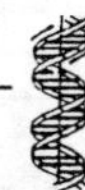

Handling of treated material

General procedure for handling of mutagen treated material consists of five important steps as discussed below:

1. **M_1 Generation:** A large M_1 population is raised from treated seeds. Plants are grown using wider spacing for easy identification. Generally, the mutants are recessive. If some dominant mutations are obtained, they are selected in M_1; otherwise observations are recorded on chlorophyll sectoring and fertility. All the M_1 plants are selfed to avoid contamination from cross pollination. Each M_1 plant is harvested separately.

2. **M_2 Generation:** The M_2 generation is raised from the seed obtained from M_1 using wider spacings. The oligogenic mutants with distinct features are identified and selected. The seed of such mutants is harvested separately.

3. **M_3 Generation:** The M_3 progeny are raised from selected M_2 and evaluated for homozygosity. The homologous M_3 progeny are bulked together to conduct yield trail in M_4 generation.

4. **M_4 Generation:** The M_4 progeny are raised in replicated trail using local check for comparison.

5. **M_5-M_9 Generation:** Selected lines are tested in coordinated multilocation trials. The best performing line is released as a variety.

Applications in crop improvement

Induced mutations play an important role in crop improvement. The main uses of induced mutations are: (1) development of improved varieties, (2) induction of male sterility, (3) production of haploids, (4) creation of genetic variability, (5) overcoming self-incompatibility, and (6) improvement in adaptation.

Achievements

Improved character	*Crops and mutant varieties*
Higher yield	High yielding mutation varieties have been developed in barley (DL 253), Chickpea (Pusa 408, Pusa 413, Pusa 417), Cowpea (Cowpea 88, V 16, V 37, V 38, V 240), Groundnut (Co 2, TG 17), Mungbean (ML 26-10-3). Oriental mustard (RLM 514). Okra (MDU 2), Pea (Hans). Ridge gourd (PKM 1), Tomato (PKM 1). Tossa jute (Mahadev TJ 40) and sorghum (Co 21).
Short stature Earliness	Barley (RDB1), Rice (Hybrid mutant), Sesame (Kalika), Rice (Prabhavati), Rice (AU I, IIT 48, IIT 60, Indira, K 84, Padmini, Sattari, CNM 20, CNM 25, CNM 6, CNM 31), Castor bean (Aruna) Cotton (MCU 7), Blackgram (Co 4), Mungbean (TAP 7), Pigeonpea (Co 5) and Sugarcane (Co 85035)
Stress resistance	Drought resistance in Cotton (MA 9, MCU 10), Salt tolerance in Rice (Mohan) and water-logging tolerance in Jute (Padma)

You Don't Know!

Most labour-intensive food

One kg of honey requires about 81,400 trips from the hive to the flowers. Honey is perhaps the most labour-intensive food made by insects.

Advantages

Mutation breeding has several advantages. Some important advantages of mutation breeding are briefly presented below:

1. Induced mutagenesis is used for the induction of cytoplasmic male sterility. Ethidium bromide (EB) has been used for induction of cytoplasmic male sterility in pearl millet and barley.

2. Mutation breeding is a cheap and rapid method of developing new varieties as compared to backcross, pedigree and bulk breeding methods.

3. Mutation breeding is more effective for the improvement of oligogenic characters such as disease resistance than polygenic traits.

4. Mutation breeding is the simple, quick and best way when a new character is to be induced in vegetatively propagated crops.

Limitations

Induced mutations have some limitations which are briefly presented below:

1. Most of the mutations are deleterious and undesirable.

2. Identification of micro-mutations, which are more useful to a plant breeder is usually very difficult.

3. Since useful mutations are produced at a very low frequency (0.1%), a very large plant population has to be screened to identify and isolate desirable mutants.

4. Mutation breeding has limited scope for the genetic improvement of quantitative or polygenic characters.

POLYPLOIDY

Introduction

An organism or individual having more than two basic or monoploid sets of chromosomes is called polyploid and such condition is known as polyploidy. It is estimated that about one third species of flowering plants are polyploids. In wild species of grass family polyploidy has been reported up to 70%. In animals, polyploidy is rare because of its lethal effects. It is found only in those species of animals which develop parthenogenetically like aphids. Polyploidy is of two types, *viz.* (I) autopolyploidy, and (2) allopolyploidy.

GK ... GS

The country's first ever tea museum has been setup in Munnar.

Types of polyploidy

Types of polyploid	*Definition/brief description*
1. Autopolyploidy	Multiplication of the same genome
(a) Autotriploidy	Three copies of the same genome (3x)
(b) Autotetraploidy	Four copies of the same genome (4x)
(c) Autopentaploidy	Five copies of the same genome (5x)
(d) Autohexaploidy	Six copies of the same genome (6x)
2. Allopolyploidy	A condition in which complete genomes of two or more species combine together
(a) Allotetraploidy	Two copies each of two distinct genomes ($2x_1 + 2x_2$)
(b) Allohexaploidy	Two copies each of three distinct genomes ($2x_1 + 2x_2 + 2x_3$)
(c) Allooctaploidy	Two copies each of four distinct genomes ($2x_1 + 2x_2 + 2x_3 + 2x_4$)

Induction of polyploidy

Polyploidy is mainly induced by treatment with a chemical known as colchicine. This is an alkaloid which is obtained from the seeds of a plant known as *Colchicum autumnale*, which belongs to the family Liliaceae. Colchicine does not affect Colchicum from which it is extracted, because this plant has an anticolchicine substance.

Colchicine is applied in a very low concentration, because high concentration is highly toxic to the cells. For effective induction of polyploidy, usually concentrations of 0.01 to 0.5% are used in different plant species. The colchicine induced polyploidy is known as colchiploidy. In plants colchicine is applied to growing tips, meristematic cells, seeds and axillary buds in aqueous solution or mixed with lanolin. The duration of treatment varies from 24 hours to 96 hours depending upon the species of plants.

Effects of polyploidy

Polyploidy has marked effects on the morphology of plants. The distinct features of autoploids are increase in general vigour and size of various plant parts. Such features are generally referred to as gigantism. Autoploids have the following important features :

1. Stems are thicker and stouter.
2. Leaves are fleshy, thicker, larger and deeper green in colour.
3. Roots are stronger and longer.
4. Flowers, pollens and seeds are larger than diploids.
5. Maturity duration is longer and growth rate is slower than diploids.
6. Water contents are higher than diploids etc.

Applications in crop improvement

Polyploidy plays an important role in crop improvement. The autopolyploidy and allopolyploidy are useful in several ways. However, allopolyploidy has wider applications than autopolyploidy. Applications of autopolyploidy and allopolypoloidy in crop improvement are briefly presented below:

Autotetraploids have been developed in forage crops like berseem, alfalfa and rye; vegetable like radish, turnip and cabbage; and fruits like grapes.

Alloploidy

Alloploidy is useful in four principal ways, *viz.*

(1) in tracing the origin of natural allopolyploids, (2) in creating new species, (3) in interspecific gene transfer, and (4) as a bridging species.

Limitations of Polyploidy

Polyploidy has several limitations. Some important limitations of polyploidy in crop improvement are briefly presented below:

1. **Limited use**: The single species polyploidy has limited applications. It is generally useful in those crop species which propagate asexually like banana, potato, sugarcane, grapes etc.

2. **Difficulty in maintenance**: The maintenance of monoploids and triploids is not possible in case of sexually propagating crop species.

3. **Undesirable characters**: In bispecies or multi species polyploids characters are contributed by each of the parental species. These characters may be sometimes undesirable as in case of *Raphanobrassica*.

4. **Some other defects**: Induced polyploids have several defects such as low fertility, genetic instability, low growth rate, late maturity, etc.

5. Chances of developing new species through allopolyploidy are extremely low.

Plant Breeders' Rights

Plant Breeders' Rights (PBR) refer to the legal rights provided to an originating plant breeder or owner of a variety to regulate production and marketing of his/her variety. In other words, the legal rights which will allow plant breeders to get benefits of their products are referred to as plant breeders' rights.

PBRs are specialized patent-like system for cultivated plants. PBRs are one of the most recent forms of Intellectual Property Law. The policy basis for PBR is the same as for patent, design and copyright laws. Features of PBR

1. PBRs are specialized plant-like system for cultivated plant species.
2. PBRs constitute one of the recent forms of

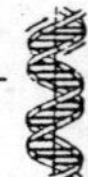

Intellectual Property Law.

3. PBRs provide legal rights to a plant breeder to get benefit of his/her innovation or variety.
4. PBRs differ from patents in the sense that the former allow farmers' privilege and research exemption. Farmers' privilege is the right to hold material as seed source for subsequent seasons. The research exemption refers to the right to use protected material as the basis to develop a new variety or other research use. Patents do not provide such exemptions. Research exemption is also called breeders' privilege.
5. PBRs protect the variety but not the standard breeding procedures that are used for development of a variety.
6. PBRs are generally considered to provide less protection than patents.
7. Laws related to PBRs were first framed in 1961 by the union for the protection of new plant varieties (UPOV) which were further revised in 1972, 1978 and 1991 UPOV conventions were held in Geneva.

Main Features of UPOV

The first UPOV convention was held and signed in Paris in 1961, but it came into force in 1968 with its headquarters at Geneva. Since then many changes have taken place in UPOV, because it was revised in 1972 , 1978 and 1991. The 1978 Act came into force in 1981 and the Act 1991 has not yet come into force. The main features of UPOV are briefly presented below:

1. Material to be Protected

A protection right can be granted for varieties of all botanical genera and species. In other words, all new varieties, of cultivated species are protected under: Plant Breeders' Rights (PBR). The new is variety must have a designation (name) as per the rules of International code of nomenclature.

2. Rights Provided

UPOV provides legal rights to the original plant breeder or owner of a variety for commercial production, marketing and export of his variety. Thus original breeder gets benefit of his intellectual property or his invention.

3. Basic Requirements

There are four basic requirements for protection of a variety under PBR. These are : (1) novelty, (2) distinctness, (3) uniformity and (4) stability. The last three criteria were given by UPOV 1978 convention and the first criteria *i.e.*, novelty was added in UPOV convention 1991. Novelty refers to newness of a variety. The variety should be a new one and it should not have been commercially cultivated for more than one year before granting protection under PBR. The second critetia is distinctness. The new variety must be distinguishable in one or more morphological, quality or other characters from previously available varieties. Thirdly, the variety should be uniform (homogeneous). In other words the variety should be pure and look uniform. Lastly, the variety must give stable performance in different generations. These four attributes are determined by grow out test.

Knowledge World!

First Antarctic expedition

The first Antarctic expedition, a 21-member team led by Dr. S.Z. Qasim, reached the frozen continent on Jan. 9, 1982. They set up a research station in Queen Maud land area named *Dakshin Gangotri*.

4. Power of Authorization

The holder of PBR has power of authorization. He can authorise any other person for commercial production, marketing, export and import of his variety. Prior authorization of a breeder (owner) of a variety is required for commercial production, marketing, export and import of his variety. The use of a protected variety for further research does not require authorization as per UPOV 1978 Act. However, the use in research requires written permission as per UPOV Act 1991.

5. Farmers' Rights

It refers to legal rights that are provided to farmers to save, use, exchange, share or sell his farm produce of a variety. Here the sale is restricted to non-commercial sale. The 1978 Act is beneficial, because it protects farmers' rights. According to the 1978 Act, farmers can use, reuse their produce as seed and have the right to dispose of their farm produces. However, the 1991 Act has curtailed farmers' rights. Farmers' rights is also referred to as farmers' privilege.

6. Breeders' Privilege

It refers to legal rights that are provided to plant breeders to use protected plant material for further research. Breeders' privilege is also known as Research Exemption or Breeders' Exemption. The UPOV Act 1978 provides Breeders' Privilege. However, the Act 1991 has curtailed Breeders Privilege to use the material of a protected variety for further research.

7. Period of Protection

The period of protection varies with plant species. For field crops, the minimum period of protection is 15 years as per 1978 Act and 20 years as per UPOV Act 1991. For forest trees, vines, fruit trees, ornamental trees and shrubs, the period of protection is 18 years as per UPOV Act 1978 and 25 years as per UPOV Act 1991. The UPOV Act 1991 is pro-Plant Breeders.

GK GS

Queen Didda belonged to the Utpala dynasty.

Advantages of PBR

Plant Breeders' Rights have several advantages. Some important advantages of PBRs are briefly presented below:

1. Incentive to breeders
2. Fast development of seed industry
3. Improvement in quality
4. Procurement of good material
5. Enrichment of genetic resources

Disadvantages of PBR

There are some disadvantages of PBR which are briefly presented below:

I. Exploitation
2. Encouragement of unhealthy practices
3. Increase in cost of seed
4. Reduction in genetic diversity
5. Ban on use of own seed

AGRICULTURAL BIOTECHNOLOGY

INTRODUCTION

Definition

It is defined as the application of scientific and engineering principles to the processing of material by biological agents to produce goods and service. Bio-technology consists of the controlled use of biological agents such as microbes cellular components for beneficial use.

History

1. *Pre Pastures Era*: Selection in plants and Fermentation
2. *Pasture Era*: Microbes are cause of Fermentation, Enzyme convert sugar into alcohol.
3. *Third Era*: Antibiotic (pencillin). Green revolution
4. *Modern Era*: DNA structure, Genetic engineering
5. *Recent Era*: Transgenic, marker, tissue culture etc.

Importance

1. To increase nutrient content
2. To increase resistance to biotic and abiotic stress
3. To introduce gene
4. To create variation
5. To minimize inputs and maximize output

Contribution of Biotechnology to Agriculture

	Technology	*Uses*
1.	Embryo culture to rescue otherwise inviable hybrids, to recover haploid plants from interspecific hybrids, micropropagation of orchids etc.	The first two applications are the most remarkable.
2.	Rapid clonal multiplication through meristem culture, e.g., many fruit and forest trees, such as, teak.	Very high rates of multiplication; conventional rates very low.
3.	Recovery of virus and other pathogen free stocks of clonal crops; meristem culture is generally combined with thermotherapy/cryotherapy.	Very useful in clonal crop; particularly for germplasm exchange.
4.	Germplasm conservation through storage in liquid nitrogen (at-196° C, cryo-preservation) or through slow growth.	Particularly useful in clonal crops, especially in those producing tubers, storage roots etc.
5.	Rapid isolation of homozygous lines by chromosome doubling of haploids produced through anther culture/interspecific hybridization/ovary culture.	Very successful in variety development in China, e.g., in rice and wheat.
6.	Isolation of stable somaclonal variants with improved yield/yield traits/disease resistance/resistance to cold, herbicides, metal toxicity, salt and other abiotic stresses.	Many examples of successful isolation; many variations are stable and heritable, often due to gene mutation, which may sometimes, be novel.
7.	Gene transfers (genetic engineering) for insect resistance, protection against viruses, herbicide resistance, storage protein improvement etc.	Mainly using the Tiplasmid of *Agrobacterium*; also through particle gun, free DNA uptake; revolutionary development in crop improvement.
8.	Molecular markers, e.g., RFLPs and RAPDs, for linkage mapping and mapping of quantitative trait loci.	A powerful tool for indirect selection for quantitative traits; several other important applications.

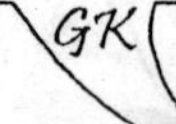

Lead storage cell are used in car.
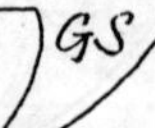

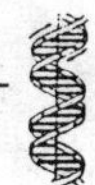

In agriculture, rapid and economic clonal multiplication of fruit and forest trees, production of virus- free stocks of clonal crop, creation of novel genetic variations through somaclonal variations, and transfer of novel and highly valuable genes *e.g.*, for disease and insect resistance through genetic engineering have opened up exciting possibilities in crop production, protection and improvement.

TISSUE CULTURE & MICRO PROPAGATION

It bypasses sexual process in the development of new crop varieties.

It overcomes the barriers of crops incompatibility in distant crossed (interspecific and intergeneric crosses).

It helps in development of transgenic (genetically engineered) plants with resistance to herbicides, biotic and abiotic stresses and improvement in quality.

It is a rapid method of crop improvement. e.g. tetraploid plants can be obtained in one step through protoplast fusion.

Plant Tissue Culture

Plant tissue culture refers to growth of living plant tissues in a suitable culture medium (*in vitro*). In a broad sense, it is referred to as cultivation of plant organs, tissues of cells in test tubes on an artificial media; often the techniques of plant cell and tissue culture are also called *in vitro* techniques.

"Culture medium" is a nutrient medium, which contains all essential micro and macronutrients, carbohydrates, vitamins and hormones. The p^H of the nutrient medium should be 5.5. However, the culture medium differs from species to species. In this technique, new plants are regenerated from tissues, cells and organs in test tubes in the nutrient medium. The organ includes any plant organ that has separate identity such as anther, ovule, embryo and bud.

The plant part, which is used for regeneration, is called '*explant*'. It may be a cell, a protoplast, a tissue or an organ. A mass of regenerated cells in culture medium is called '*callus*' and suspension of free cells of callus in a liquid medium is known as '*suspension culture*'. The regeneration capacity or ability of a plant cell to develop into a whole plant is known as '*totipotency*'.

Steps in Tissue Culture

Tissue culture technique consists of four important steps:

1. Isolation of tissues
2. Regeneration and callus formation in the culture medium.
3. Embryogenesis
4. Organogenesis

Quiz Kit!

Superbug

Ananda M. Chakrabarty (b. 1938, now a US citizen) invented in 1992 the superbug, the artificial microorganisms which can digest various petroleum components and can thus clean up oil spills on water.

Various steps involved in Tissue Culture Technique

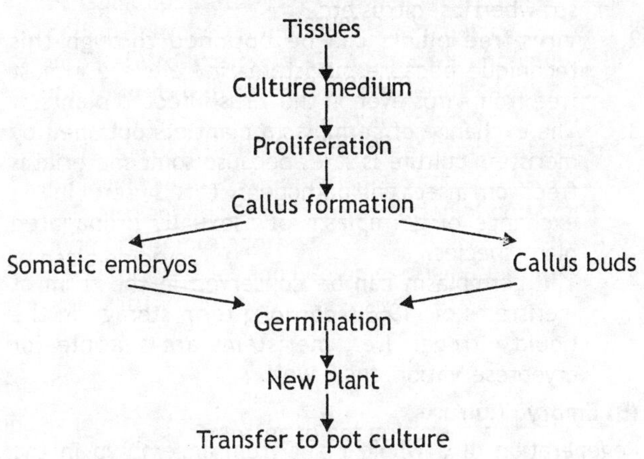

1. ***Isolation of tissues***: Disease-free tissues for regeneration can be isolated with the help of sterilized blade from any plant part *viz.* e.g. stem, apical bud, axillary bud etc.
2. ***Regeneration and callus formation*** : Tissues proliferate on the culture medium and give rise to a mass of cells called '*callus*'. The callus is generally of two types :

1. Friable callus
2. Compact callus

The friable callus can be easily manipulated for suspension culture.

However, compact callus is not suitable for suspension culture.

Techniques of Tissue Culture

Tissue culture techniques are also referred to as *in vitro* techniques. In vitro techniques give rise to new plants by regeneration of tissues and organs in the nutrient medium. Various plant parts *viz.* meristems, embryo, anther, cells and ovules are used for regeneration. Depending upon the plant parts used as explants, plant tissue culture techniques are of five main types *viz.* meristem culture, embryo culture, anther culture, cell culture and ovule culture.

GK GS

Daccan Gold Mines Ltd. is the country's first listed gold exploration company.

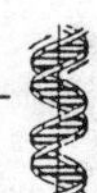

(A) Meristem Culture

Meristem culture refers to regeneration of whole plant from tissues of an actively dividing plant part such as stem tip, root tip or axillary bud. Generally, shoot apical meristem is used for regeneration. This technique is wisely used in vegetatively propagated plants such as sugarcane, potato, banana and several fruit trees and timber species.

The main applications of meristem culture in crop improvement are :

1. It is used for micro propagation (mass production of clonal progeny through tissue culture) in banana, strawberries, citrus etc.
2. Virus-free plants can be obtained through this technique because meristematic cells are almost free from virus even in the virus-infected plants.
3. The exchange of germplasm plantlets obtained by meristem culture is safe, because some material is free from insect and pathogens. This is useful in the exchange of germplasm of asexually propagated plant species.
4. The germplasm can be conserved in the form of meristems of -196° C for long-term storage in the liquid nitrogen *i.e.*, meristems are suitable for cryopreservation.

(B) Embryo Culture

Regeneration of a whole plant from an embryo in the culture medium is called '*embryo culture*'. Embryos of appropriate stage are removed from the seed and are transferred to the culture medium. This technique is used when there is disharmony between embryo and endosperm. This technique is used to make distant crosses successful.

e.g.

1. Interspecific hybridization in :

Trifolium, Lycopersicon

2. Intergeneric hybridization in :

Hordeum & *Secale*
Triticum & *Secale*
Triticum & *Aegilops* etc.

Main Advantages –

1. Useful in making interspecific and intergeneric (distant) crosses successfully
2. Useful in obtaining haploids *e.g.* barley, wheat.
3. In orchids, seeds are stored food and are unable to propagate. Hence embryo culture is useful
4. Long seed dormancy is noticed in *Prunus, Iris* and *Taxus* species. Embryo culture is useful in breaking the long dormancy and in obtaining seedlings.

(C) Anther Culture

Regeneration of whole plant from anther (or) pollen in the culture medium is called anther culture. The optimum stage differs from species to species. *e.g. Brassica, tobacco, rice, barley, wheat, potato, tomato, triticale etc.*

Advantages –

1. obtaining haploids
2. homozygous diploids are obtained simply by doubling chromosomes or obtained spontaneously from anther culture.
   (a) Mapping population
   (b) Breeding cycle is shortened

(D) Cell Culture

Regeneration of whole plant from callus and suspension cultures in the nutrient medium is known as *cell culture*. Protoplasts (naked cells or cells without cell wall) are also used for regeneration of whole plants. *e.g.*, wheat, rice, barley, maize, tobacco, fruit and timber trees.

Advantages –

1. Useful in mass clonal propagation of different species,
2. Protoplast culture is useful in somatic hybridization and overcoming the barriers of cross incompatibility.
3. Useful in the development of transgenic (genetically engineered) plants.
4. Useful/suitable for cryopreservation *i.e.* preservation in liquid Nitrogen at -196 degree Celsius.

(E) Ovule Culture

Regeneration of whole plant from the ovule in the nutrient medium is called *ovule culture*.

Two types of ovules :

Unfertilized → Haploids
Fertilized → Diploids

This technique is however used to a limited extent.

Applications In Crop Improvement

(A) Generation of Variability

Variation is created through *in vitro* techniques. Depending upon the explants, variation induced by tissue cultural techniques is of three types:

1. Gametoclonal variation: Variation observed among plants, which are regenerated from gametic culture (anther/ovule culture).
2. Somaclonal variation: The variation observed among plants obtained callus culture of somatic explants such as meristems.
3. Protoclonal variation: The variation observed among plants that are regenerated from callus cultures of protoplast.

Main features of the variation –

1. Variation caused is genetic in origin and hence heritable.
2. Somaclonal variation occurs both in sexually and asexually propagated species. But, the frequency is very high in a sexually propagated species.

India produces more than 50% of world jute.
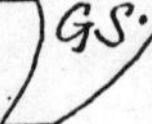

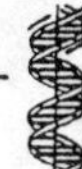

3. Occurs both in oligogenic and polygenic traits.
4. Somaclonal variation is heterozygous in origin (through mutation).
5. All three types of culture variations resulted due to chromosomal changes such as deletion, duplication, inversion and translocations.

Somaclonal variation is useful in isolation of diseases, herbicide tolerant and early maturing variants.

(B) Development of Haploids

Made possible through anther and pollen culture. In China, one variety in wheat (Jighua 1) and in rice (Guan 18) have been developed for commercial cultivation.

Protoplast Culture

Isolated plant protoplasts are cells from which the walls have been removed. Cell walls can be removed either by mechanical or enzymatic methods. Isolated protoplasts can be induced to divide in culture to form callus. From the callus whole plants can be regenerated, either through embryogenesis or organogensis.

Isolation of Plant Protoplasts

Plant protoplasts are generally isolated from leaves by enzymatic method.

The Need for an Osmoticum : During isolation and culture, plant protoplasts require a suitable osmotic stabiliser. Such a substance is called *osmoticum*. In the absence of an osmoticum the protoplasts may burst. Several ionic and non-ionic solutes have been tested for adjusting osmotic potential of various solutions used in protoplast isolation and culture. The important osmotica are *sorbitol, mannitol, glucose or sucrose* .

Culture of the Protoplasts

Isolated protoplasts are cultured *in vitro* to regenerate whole plants. The procedure is outlined below.

1. A culture medium (MS medium) with mannitol containing 1.5% agar in a petridish is used. To this medium the protoplasts are embedded and incubated at 25°C in dim light.
2. Within 48 hours the naked protoplasts form cell walls.
3. Two to seven days after culture, cell division occurs to form callus.
4. The callus is transferred to a medium without mannitol and auxin.
5. Within 3 to 4 weeks embryoids are formed.
6. These embryoids are dissected from the callus and transferred to filter-paper bridge in a culture tube containing liquid medium composed of :
   MS salts
   Kinetin- 0.2 mg/l
   Sucrose -2.0% w/v
7. After a few weeks plantlets are produced. These are transferred to sterile soil and grown to maturity.

Did You Know ?

Largest moving dome

The Kavalur Observatory in Tamil Nadu has the largest moving dome in India. It has a diameter of 21 m (68.8 ft) and weighs a staggering 260 tonnes.

Applications in Protoplast Culture

Protoplast culture has wide applications. Some of them are :

1. Cell Biology Studies

Isolated protoplast are ideal materials for the study of cell wall synthesis in plant cells:

2. Protoplast Fusion and Somatic Cell Hybridisation

Two protoplasts from the same source or different sources can fuse to form a hybrid protoplast. Whole plants can be regenerated *invitro* from this hybrid protoplast. This method is utilised to produce hybrids from two plants which show physical or chemical incompatibility in normal sexual crosses.

When protoplast isolated from two sources are allowed to fuse their cytoplasms mix. The nuclei of the fused protoplasts may fuse nuclei are called *heterokaryons or Heterokaryocytes*. Thus hybrid protoplasts are formed. These are now selected culture medium to regenerate hybrid plants. Thus isomatic cell fusion helps to obtain interspecific and intergeneric hybrids which are difficult to obtain by conventional plant breeding methods. This method of obtaining hybrids are called somatic hybrids.

Thus production of somatic hybrids is a multistep process:

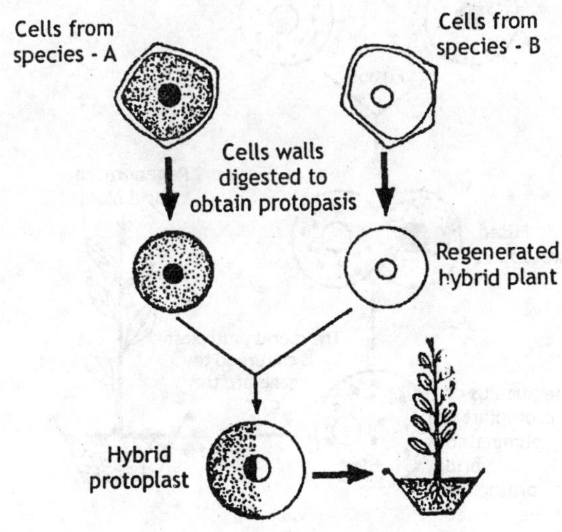

Scheme of somatic cell hybridisation

GK GS

Anantavarman built the Jagannath temple at Puri.

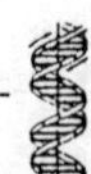

(i) Isolation of protoplasts from two different species

(ii) Fusion of protoplasts from two different species

(iii) Isolation of fused protoplasts, and

(iv) Regeneration of fertile hybrid plants from the fused protoplasts.

(v) Somatic hybrids have been produced in several crop plants like potato and tobacco.

3. Protoplast Fusion and Production of Cybrids

When protoplasts from two different species are allowed to fuse, their cytoplasms mix. Sometimes, the nucleus of one of the protoplasts completely disappears. The cytoplasms of the two parental protoplasts are still hybridized. This fusion product is called a *cybrid or cytoplasmic hybrid*. Some genetic factors are carried in the cytoplasm, *e.g.* factors for male sterility in some plants. Hence formation of cybrids has practical applications in plant breeding.

4. Production of Genetic Variants from Cell/ Protoplast Culture

Plants regenerated from cell cultures often show genetic variations. Since, such variation occures in cells of somatic origin, it is called *somaclonal variation*. Several kinds of mutations are formed inversion, duplication and deletion. Somaclonal variation in cell culture cannot be directed toward changing a particular kind of mutant. Beneficial genetic variations originating in cell cultures can be utilised in crop breeding for the development of improved varieties of crop plants. Using this method genetic variants with traits such as resistance to herbicides, salt tolerance and resistance to toxins such as resistance to hebicides, be evolved. Somaclonal genetic variants have been evolved in several crop plants like wheat and sugarcane.

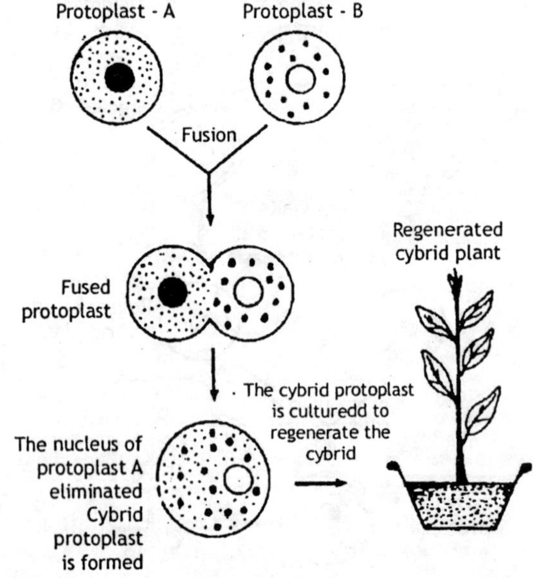

Production of cybrids

5. Genetransfer into Higher plants by Co-cultivation of Plant Protoplasts with *Agrobacterium*

Transgenic plants can be evolved by introducing foreign genes into isolated plant cells/protoplasts by co-cultivation with *Agrobacterium tumefaciens* that contains a co-integrated plasmid carrying the foreign gene. The transformed plant cells can be regenerated *in vitro* to evolve transgenic plants.

6. Introduction of Foreign Genes into Isolated Plant cells/ Protoplast by Microinjection, Electroporation and Microprojectile Bombardment etc.

Foreign DNA can be introduced into isolated plant cells by microinjection, electroporation and microprojectile bombardment. Liposomes containing foreign genes can be fused with isolated protoplasts. The transformed cells/ protoplasts are then selected by growing on suitable media and are then regenerated to produce transgenic plants.

Using these methods plants that are resistant to herbicides toxins, pests and pathogens have been evolved. Several transgenic plants that can produce compounds such as vaccines, interferons, plastics and thaumatin have been evolved.

7. Plant Cell Cultures and Biosynthesis of Secondary Products

Plant cells can be grown as micro-organisms in culture media and many secondary metabolites have been synthesised from different plant species

The yield of these substances is more when compared with the yield from whole plants.

A few secondary products produced by plant cells in culture

Plant Product	*Plant Species*
Codeine(alkaloid) - Analgesic	*Papaver somniferum*
Diosgenin (steroid) - Antifertility agent	*Dioscorea deltoidea*
Dogoxin (cardiac glycoside)	*Digitalis lanata*
Quinine (alkaloid) - Antimalarial agent	*Cinchona ledgeriana*
Pyrethrin (insecticide)	*Chrysanthemum cine rarialfolium*
Thaumatin (cholcone) Non-nutritive sweetener	*Thaumotcoccus danielli*
Ginseng (stimulant)	*Panax ginseng*
Jasmine (perfume)	*Jasminum sp.*

8. Plant Cell Culture in Biotransformation

Plant cell in culture can convert certain organic substances added to the cultures into useful products. This process is called *biotransformation*. Using this method useful

India is largest producer of milk in the world.

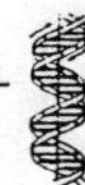

substances that have pharmacological properties can be synthesised.

Hardening or Acclimatisation of Plantlets Regenerated Through Culture

The plantlets regenerated through tissue culture methods are transferred to soil for futher growth and development. Different species of plants differ in their capacity to adust to the new environment. During this period the plantlet has to change from the heterotrophic to the autorophic state. Water loss from the regenerated plantlet is high. This is due to the following reasons:

1. The root system formed in the cultured plantlets is inadequate to maintain the plantlets in the soil.
2. The regenerated plantlets have thin leaves with reduced epicuticular wax on their leaves and stems.
3. The stomata do not respond efficiently to stress conditions. Due to these reasons the regenerated plants dehydrate more quickly than the normal plants. Hence the regenerated plantlets must be protected from dessication and hardened to attain some tolerance to moisture stress.

The regenerated plantlets have been grown under aseptic conditions. So, they will be easily attacked by soil pathogens. Such plants should be protected from soil borne pathogens so that they can grow and develop into healthy plants.

Hardening of regenerated plantlets is done by planting them in a sterile soil and covering the plants with a transparent hood. These plantlets are grown in shade for some time and then transferred to the field.

Artificial Seeds

Somatic embryo (embryoids), shoot buds or any other plant material obtained as a result of *in vitro* culture are covered (encapsulated) with a chemical membrane. Such encapsulated materials behave as seeds. These are called *artificial seeds* or *synthetic seeds*. The artificial covering acts as an artificial seed coat. Such seeds are bead-like and can "germinate" and plantlets are also formed. Several substances are used as artificial seed coats. Some of them are *agar, agarose, carrageenin, polyacrylamide, introcellulose, ethyl cellulose* and *sodium alginate*. Of sodium alginate is most commonly used.

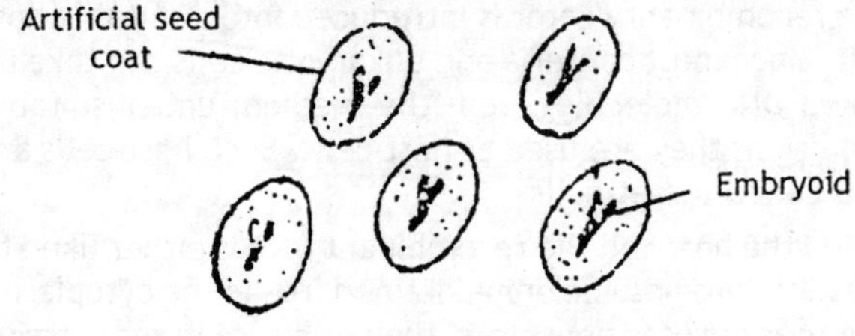

Artificial seeds

You Don't Know!

First scientist

Uddalaka Aruni, who lived at least 200 years. Thales of Mildetus (560 B.C.) is the first scientist in the world. He observed facts and used empirical methods to verify conclusions and understood the laws of nature. He is mentioned in the *Upanishads*.

Advantages of Artificial Seeds

1. The size of the artificial seeds is smaller when compared to the natural seeds of the plant.
2. Storage and transportation of such seeds is easier.
3. Viability of seeds is 100%.
4. Artificial seeds can be made to germinate uniformly on a suitable substratum.
5. Such seeds do not show dormant.
6. The plant grower can grow the desired plant any time and this is not season dependent.
7. Large scale production of seeds from any kind of plant part is possible.

Disadvantages

1. The artificial seeds cannot be stored for longer time and it is temperature dependent.
2. The initial cost for the production of artificial seed is more than that for the natural seeds.
3. Production and germination of artificial seeds require aseptic conditions. Any deviation will affect the quality of the seeds and their subsequent development.

Applications in Plant Tissue Culture

Plant tissue culture has applications in agriculture, horticulture, plant breeding and industry.

1. Rescue of hybrid embryos
2. Culture of immature embryos to propagate orchids
3. Culture of mature embryos to reduce seed dormancy.
4. Production homozygous lines of recessive genes through anther or pollen culture
5. Production of disease-free clones through nucellus culture and apical meristem culture
6. Production of somatic hybrids
7. Production-cybrids
8. Production of genetic variants from cell/ protoplast culture
9. Creation of transgenic plants
10. Biosynthesis of secondary products and biotransformation
11. Micropropagation

Tissue culture for Germplasm Storage

Invaluable plant species are stored to make them available for future breeding programmes. This is called *Germplasm storage*. Normally it is done by storing the seeds. It has

certain limitations: 1. the seeds may lose viability with the passage of time, 2. the seeds may be damaged by seed-borne pathogens, 3. this method cannot be effectively used for vegetatively propagated crops.

Tissue culture is used as an alternative method for the storage of germplasm. Culture tissues can be indefinitely stored at very low temperatures in a frozen state. This method is called *cryopreservation*. When there is a need whole plants can be regenerated from the tissues under *cryopreservation*. When there is a need whole plants can be regenerated from the tissues are brought to a state of non-dividing and zero- metabolism. Commonly tissues are stored in liquid nitrogen at -196°C. Certain substances are added to the culture media before freezing. These protect the tissues against ice damage. Such substances are called cryoprotectants, *e.g.* Glycerol, Proline, Mannitol, Sucrose, Glucose, Polyethylene glycol. Cryopreserved tissues can be recultured to produce, whole plants or they can be sub-cultured. The viability of cryopreserved tissues depends upon the plant species.

GENETIC ENGINEERING

Genetic engineering is the construction of a recombinant DNA molecule which contains DNA fragments (genes) from different and unrelated organisms. Hence a gene of interest or a desired gene has to be identified and isolated. It is then joined with the DNA of another organism. The final step will be to allow this " desired gene" to express itself in an appropriate host or new environment and to produce the gene product. In several cases the gene product will be isolated.

Major Steps of Genetic Engineering

1. Isolation of the Desired Gene

Specific DNA of fragments or desired genes are to be identified. They are then isolated and purified for being transferred to an appropriate host. This DNA of interest is also called *passenger DNA, target DNA or donor DNA.* Isolation of DNA is done by anyone with the following biochemical method.

(a) extraction of total cell DNA (b) synthesis of complementary DNA [cDNA]

2. Selection and Isolation of a Vector

If a desired gene is to be introduced into a host cell, a carrier molecule that can transport the gene into the host cell is required. Such a molecule is called a *cloning vehicle, carrier molecule* or a *vector*. A vector is a self-replicating molecule of DNA to which the donor gene can be linked. The vector should also be able to propagate in the host organism or cell. Commonly plasmids and the DNA viruses are used as vectors. A suitable vector is identified and then isolated. New techniques are now evolved to directly introduce the donor DNA into the host cell.

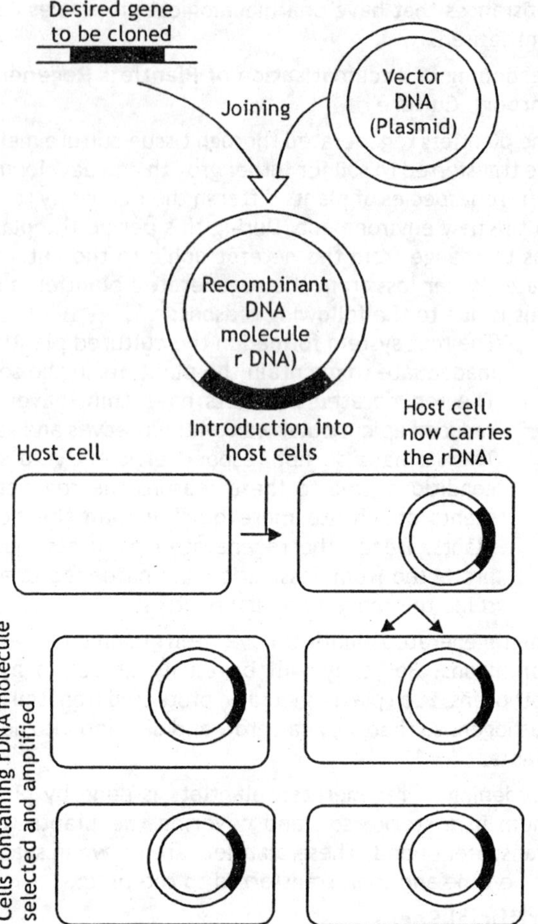

Outline of the synthesis and cloning of recombinant DNA molecules.

3. Construction of Recombinant Vectors

The donor DNA is inserted into the vector to produce a donor-vector hybrid DNA molecule. This hybrid is also called recombinant DNA a chimera or a recombinant vector as it contains DNA from unrelated sources. This recombinant DNA can now be introduced into a host cell.

4. Introduction of the Recombinant Vector into the Host Cell

The recombinant vector is introduced into a suitable host cell. Since most bacteria and eukaryotic cells can take up naked DNA molecules from the medium under suitable conditions they are used as host cells. Such host cells are also called target cells.

Inside the host cell the recombinant DNA is either liked to the host chromosome or maintained free in the cytoplasm. If viruses are used as vectors, the synthetic viruses carrying the recombinant DNA infect the target cells. The genome

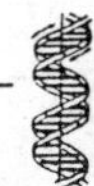

of such synthetic virus contains the gene of interest. This gene is present in the place of a segment of viral DNA that is not essential for its replication. Inside the target cell the recombinant vector DNA may multiply. Thus numerous copies of the recombinant DNA are formed. This means, the desired gene it carries is also multiplied.

5. Selection and Multiplication of the Host Cells Carrying the Recombinant DNA Molecule

The cells which have successfully taken up the recombinant DNA molecule are selected. These cells are allowed to divide. After a large number of cell division, a colony of identical host cells is produced. At the cellular level it is a clone of cells. Each cell in clone carries one or more recombinant DNA molecules. Thus the desired gene carried by the recombinant DNA is *cloned*.

6. Expression of the Desired Gene

The desired gene can now express itself in the body of the host organism and produces the desired trait. In many cases of genetic engineering the desired gene will produce the gene product in the new environment which can be isolated and purified, *e.g.*, synthesis of human insulin, synthesis of somatostatin etc.

Application in Genetic Engineering

1. R-DNA technology in nitrogen fixation.
- Introduce nif-gene into several other microbe
- To increase host range
- To transfer nif-genes from bacteria to plants
2. Transgenic plants Ti with Herbicide Resistance
- Atrazine resistant oil seed (*Brassicanapus*)
- Atrazine is detoxified by glutathione-s-transferase
3. Transgenic glyphosate tolerant plants glyphose is an active component of many herbicide.
4. Genetically-engineered plants that resists/repel pests and pathogens.
- Bacillus thuringensis crop: cotton, maize, Brassica
- NAPI-Nicogene natural insecticide effective against many pests from *Nicotiana alata*
5. Genetically-engineered protection against viruses in transgenic plants
- The presence of one virus in plants can interfere with infection by another virus or strain. This is called cross protection.
- Coat protein
6. Genetically engineered plants to produce novel compounds
- Transgenic Banana with heptitis virus
- GMO tobacco plants to produce interferon
7. Plants that produce plastic.
- Transgenic *Arabidopsis thaliana* to produce polyhydroxy butyrate (PHB).
8. Transgenic plants to improve Photosynthetic efficiency.

Knowledge World!

First steam boat

The first steam boat *Diana* was built in India in 1819 by Jesop and Company Ltd., Kolkata, and launched on July 12, 1823 at Kolkata. It has a splendid trial trip and was reported in local newspapers to have shown a 'velocity perfectly astounding'.

- Modification of RUBISCO to favour uptake CO_2 rather than O_2.
9. Transgenic plant to improve protein quality.
- Transgenic soyabean with sulphur containing amino acid
- Cereals with protein content.
10. To delay senescence.
- Tomato - "flavour savour" variety to delay ripening.
- Carnation - antisence RNA technology for delay flower senescence
11. Golden Rice
- β-Carotin gene is inserted in rice to improve vitamin - A content

Transformation Techniques

The uptake of foreign DNA or transgenes by plant cell is called transformation. A variety of techniques have been used to introduce transgenes into plant cells

1. Agro bacterium mediated

(a) Co-culture with tissue Explants

(b) In Planta transformation

2. Agro infection:

Introduction of a viral genome into plant by placing within the T -DNA of a Ti Plasmid and using the Agro bacterium containing this recombinant plasmid for co-culture with plant cell is called Agro infection.

3. Direct gene transfer (Biolisitic method)

(a) Chemical methods [(Poly ethylene glycol) PEG]

(b) Electroporation.

(c) Particle gun delivery

(d) Lipo fection

(e) Micro injection

(f) Macro injection

(g) Pollen transformation

(h) Delivery *via* growing pollen tubes

(i) Laser induced

(j) Fibre-mediated gene transfer

Transgenic Plants

The technology of Genetic engineering is the practice of altering or disturbing the genetic blueprints of living organism.

GK GS

India has largest area under irrigation.

Transgenic for transfer the desirable qualities from one organism to another.

Vectors

R-DNA technology is based on the insertion of a DNA fragments (gene) into a suitable cloning vector and then producing into a suitable host to propagate the recombinant DNA.

Plasmids :

Circular, double standard, extrachromosomal self-replicating DNA molecule.

Plasmids	*Size (nucleotide length in Kb)*
PBR 345	0.7
PBR 322	4.36
Col E_1	6.36

Other Vector

Vector	*Capability(Kbp)*
1. Cosmid	45-48
2. BAC	15
3. PAC	15
4. Phagemid	15-60

Identification of transgenic plants by reporter gene.

Certain marker genes on which no selection pressure is imposed are need to identify the transformed cells/plants. Such genes are called reporter genes.

Two types :

1. *Selectable marker*: Cell which persently survive under selective condition eg. Kanamycin, Neo mycine.
2. *Scorable marker*: Produce different phenotype which allow an easy identification of R-DNA cell eg. SVS, β-galactriodase.

Marker Genes used for Gene Transfers into Plants

Some of the marker genes used for gene transfers into plants.

Gene	*Scorable/ Selectable*	Characteristic
nos (Napoline synthase)	Scorable	Nopaline production
cat (Chlormaphemicol acetyl transfersase)	Scorable	Acetylation of chloramphenicol
lux (Luciferase)	Scorable	Phosphorescence
gusA (β-galactosidase)	Scorable	Blue colour
dhfr (dihydrofolate reductase)	Selectable	Methotrexate resistance
nptII (Neomycin phosphotranserase)	Selectable	Kanamycin, neomycin and G 418 resistance
hptIV (Hygromycin phosphotransferease)	Selectable	Hygromycin B resistance
bar (Phosphinothricin acetyl-transferase)	Selectable	Phosphinothricin resistance

Important Transgenes encoding valuable Polypeptides/Enzymes expressed in Plants

Polypeptide/ Enzyme	*Source*	*Applications*
POLYPEPTIDES		
Enkephalin	Human	Opiate activity
Hirudin	Synthetic	Thrombin inhibitor
α- Trichosanthin	Chinese medicinal plant	Inhibition of HIV replication
Epidermal growth factor	Human	Proliferation of specific cells
Erythropoietin	Human	Regulation of erythrocyte levels
Growth hormone	Trout	Growth stimulation
Human serum albumin	Human	Plasma expander
Interferon	Human	Anti-viral activity
ENZYMES		
α-Amylase	*Bacillus licheniformis*	Liquefaction of starch
(1-3,1-4) β-Glucanase	*Trichoderma reesei*; Hybrid of two *Bacillus sp.*	Brewing
Manganese-dependent lignin peroxidase	*Phanerochaete chrysosporium*	Bleaching and pulping of paper
Phytase	*Aspergillus niger*	Increased phosphate utilization from feed
Xylanase	*Clostridium thermocellum*; *Cryptococcus albidus*	Animal feed, paper and pulp, baking

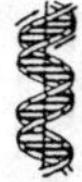

Transgenic Plants from Lab to Field

Transgenic lines obtained through transformation and regeneration

Laboratory analysis to confirm :
- stable integration of transgene(s)
- number of copies of the transgene(s)
- expression of transgene(s)

Monitoring and analysis of transgenics in a containment (glasshouse) facility for :
- stability of transgene expression
- agronomically desirable expression of transgenic trait(s)
- genetic behaviour of transgene(s)
- biosafety evaluation and risk assessment

Small-plot (<500 m²) field experiments to evaluate agronomic performance and to further analyze biosafety

Large-scale field testing at multiple sites

Commercialization of transgenic variety after risk assessment and evaluation of net benefit offered by the transgenic

Monitoring of transgenic variety during commercialization

Major Risk of Transgenic

1. Escape of engineered genes by gene flow or gene dispersal
2. Non-target effects or ecological effects
3. Invasiveness or weediness of transgenics
4. Creation of super weeds and super-virus
5. Toxicity and allergenicity to human beings and animals.
6. Expression of undesirable phenotypic traits
7. Erosion of Biological diversity

Risk Management

1. First to outline the risk assessment
2. Isolation zone
3. Trap crop
4. Refuge crop
5. Male sterility
6. Removal of flower from transgenic plants

Quiz Kit!

Largest planetarium

The Birla Planetarium in Kolkata, built in 1962, is the second largest planetarium in the world. It has an inner diameter of 33 m (108 ft) and can seat nearly 700 people.

MOLECULAR MARKERS

1. Morphological markers
2. Molecular markers (DNA)

Molecular Markers

Consisting of specific molecules which show easily detectable difference among different strains in the DNA sequences (polymorphism).

A linear map of different chromosomes of a species depicting the location of vaires molecular markers is known as molecular map.

Uses of Morphological Markers

1. Varietal and hybrid identification.
2. Identification of MS line of sunflower red leaf petiole (anthocyanin) pigment.
3. Identification of TGMS line in rice. PTGMS lines with receptive purple leaf colours.
4. Identification of cultivars resistant to pests and diseases.
   - Trichomeless trait in Cumbu -smut resistant.
   - Cotton-replant colour, hairiness of leaf glanded cotton offer resistance to various pests and disease.
5. Quality: Flax -varigated seed colour mutant used to identify fatty acid profile. Low linoleic acid varieties can be developed through this.

Groups of Molecular Markers

1. DNA hybridization-based assays
   - Restriction Fragment Length Polymorphism (RFLP)
   - Fluorescent *In situ* hybridization (FISH)
2. PCR-based techniques using arbitrary primers
   - Random Amplified Polymer DNA (RAPD)
   - DNA Amplification Fingerprinting (DAF)
   - Amplified Fragment Length Polymorphism (AFLP)

RFLP

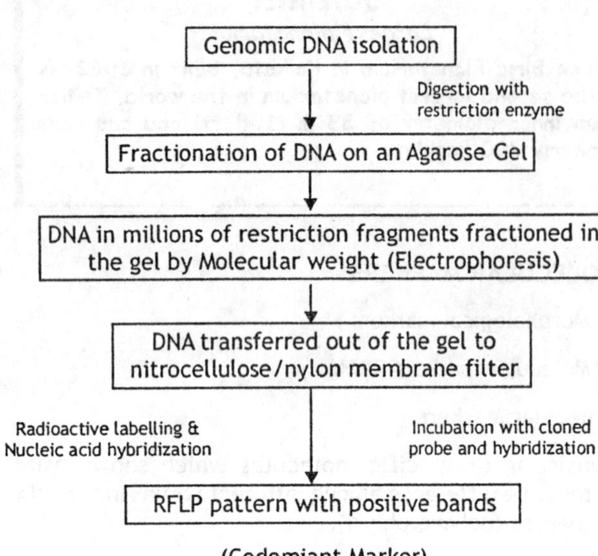

(Codomiant Marker)

RAPD

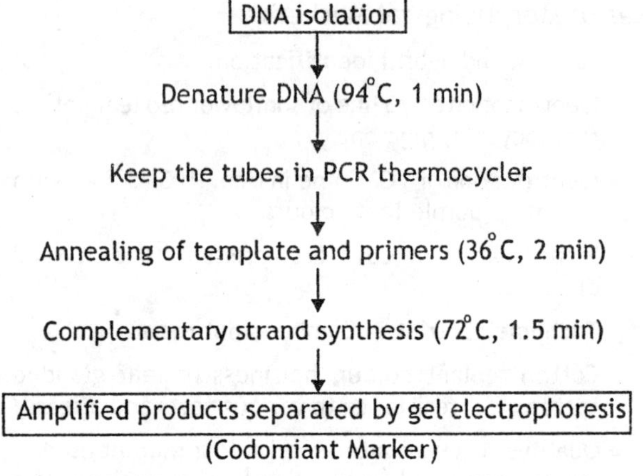

(Codomiant Marker)

Steps in RAPD

1. Isolation of DNA

- Powder the leaf in Liquid Nitrogen.
- Keep it in extraction buffer at 65°C for 1 hour.
- Add equal volume of chloroform : isoamyl-alcohol (24: 1)
- Take aqueous phase and precipitate with isopropanol.
- Wash with 70% alcohol.
- Dry it and dissolve in TE buffer.

2. Purification of DNA

- RNase treatment.
- Phenol: chloroform: Isoamylalcohol (25:24:1) treatment
- Chloroform: Isoamylalcohol (24:1) treatment
- Precipitate with isopropanol
- Dry and dissolve in TE buffer

3. PCR Amplification

- Ten random primers will be used
- PCR product will be resolved on 0.8% agarose gel

4. Polymerase Chain Reaction (PCR)

Principle: Selective amplification of required region of DNA molecule

Steps :

1. Denaturation of template DNA duplex by heating at 94°C.
2. Annealing of oligonucleotide primers to the target sequences of separated DNA strands at 35-65°C.
3. DNA synthesis from the 3'-OH end of each primer by DNA polymerase.

Merits of RAPD

- High degree of polymorphism
- Minimum quantity of DNA is required
- Rapid analysis
- Inexpensive compared to other methods

Applications of PCR

PCR can make use of very small amount of DNA. Even DNA in nanogram quantity can be easily handled. In view of this, PCR has many novel applications. Some of them are:-

1. Remnants of extinct organisms is sufficient to recover their DNA and amplify specific segments using PCR. Using this technique, from the remnants of woolly mammoth which lived about 40,000 years ago specific DNA segments were recovered and amplified. Now it is proved that this animal is related to present-day elephant.
2. PCR has revolutionised pre-natal detection of several genetic diseases. Only a very small sample of the foetus is sufficient to perform this work. Some of the diseases detected are phenylketonuria, cystic fibrosis, haemophillia A, beta thalassemia, sickle - cell anemia.
3. The sex of the *in vitro* fertilised egg can be determined using PCR. The presence of the γ - chromosome can be detected at the 6-10 cell cleavage embryo without affecting the life of the embryo.
4. The PCR has wide and significant use in forensic science also. Small pieces of tissue, even a single hair left at the scene of the crime or one sperm lingering after a rape is enough to fix the identity of the criminal.
5. One can pick out a single antibody-producing cell, amplify the genes encoding the antibody using PCR and clone them in bacteria to produce an anti- body

GS

Uttar Pradesh has the largest number of tubewells in India.

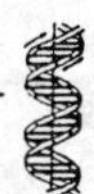

producing bacteria. Thereafter, any amount of antibody can be made by growing such bacterial cells in a fermenter.

OTHERS

Role of Biotechnology

In Agriculture

Biotech has revolutionised research activities in agriculture which includes

1. Tissue culture
2. Creating of new transgenic plants
3. Creating of transgenic animals which can be eventually used as bioreactors to produce drug from their milk, blood etc. This is known as Molecular Farming
4. Introduction of biopesticides and biofertilizers which will be the pillars of sustainable agriculture
5. Embryo transfer technologies

In Food

Can be effectively used to increase storage of food grains, increase nutrition value of food, enhance flavour etc. Fermented food through enzyme engineering (*beer biotech*).

In Health and Medicine

- Through human genetics, identification of defective genes, and gene therapy.
- In forensic medicine, identification of parents, criminals etc.
- Various vaccines are either produced or expected to produce eg: Rabies, Malaria, Hepatitis B, Cholera.
- Various vital ingredients like Insulin, Interferons can be produced. Also antibiotics like, penicillin, streptomycin etc.

Industry

Production of high fructose corn syrup - sweetening agent for soft drinks, enhance the production of alcohol, production of chemical from agri waste.

Environment

- Special Bacterial to eat away oil spilling by digesting hydro carbon.
- Bio-tech developed organisms to act as bio-indicators to indicate level of pollution in environment.

Biotech-in-India

- To disseminate biotech for the socio- economic development of the country.
- Department of Biotech -1986, entrusted with planning, promotion and co-ordination of major programmes undertaken by DBT.
- Development of vaccine and immunisation of especially vulnerable section.
- Increase oil seed production by tissue culture of coconut, oil palms etc.

> **Did You Know ?**
>
> *First universal adult franchise*
>
> **Travancore, which later became a part of Kerala, was the first region to have elections on the basis of universal adult franchise in 1948.**

- Cattle herd improvement by embryo technology.
- Development of bio-diagnostic kits for early detection of various diseases.
- Fertility control.
- In agri-development of crops-having high nutritional content, drought resistant, pest resistant etc.

Human Genome

- It refers to entire genetic makeup of human beings.
- Every human cell contains about 1 lakh of genes.
- Human genome project – an international effort to identify and locate all the genes and their variants, understand their functioning and write down the sequence of nucleotides that comprise the genes.

Uses

1. Better understanding of genome might help in diagnosing many diseases at an early stage or even before birth.
2. Knowledge may help in replacing defective genes.

Interferon

It is powerful anti-viral proteins made by cells against viral infection.

Functions

- It interferes with the ability of a virus to infect healthy cells.
- It affects the cellular membrane of the virus so that it cannot spread to other cells.
- It controls virus multiplication, by inhibiting their protein synthesis.
- Now interferon is produced outside the body with the help of genetic engineering.

Cloning

- Means making an identical copy physically as well as genetically.
- Clones are exact replica of plants or animals replicated asexually.
- Clones derive genetic matter from single parents.

Normally

Clones are not found in animals because offspring are sexually produced having genetic material from both parents.

With advent of genetic engineering an entire organism from a single somatic cell.

Acrolin is used as tear-gas.

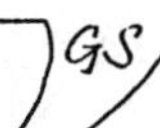

Recent Development

Dr. Ian Wilmut of Roslin Institute, UK.

- He used nuclear transfer technology.
- He arrested the cell-cycle by starving it of a vital nutrient – calf-serum.
- In this state – cell was alive but no multiplication.

Advantages

1. To clone animals that can produce pharmacologically useful proteins such as clotting factors.
2. To make animals with human disease and then test therapies.
3. To make superior animals of uniform quality in large numbers.
4. Childless couple-baby.
5. Organ transplant – no question of rejection.
6. Xenotransplantation. (Animals to Humans)

Disadvantages

1. Propagation of few species-less genetic diversity – a single disease may eliminate the entire spectrum.
2. In Human Cloning even if same genetic characteristics – social factor – man is a social animal.

Bioprospecting

- 30 million species of life on earth.
- Only 1.5 m have been described
- It is estimated that 50,000 species are disappearing every year.
- Genetic wealth of the planet holds the key to resolution of our problems of poverty and health

What is Bioprospecting?

It implies using the genetic diversity in the ecosystems of the earth for producing new drugs and new crops.

Somaclonal Variation

Variation arisen during tissue culture plants is known as somaclonal variation.

Causes

1. Change in chromosome number/structure
2. Gene mutation
3. Plasmagene mutation
4. Attention in gene expression
5. Gene amplification
6. Mitotic and crosing over
7. Rearrangements in cytoplsamic genes

Applications

1. Resistant to disease variants can be be isolated
2. Stress resistance
3. Low temperature mutant
4. Mutants for agronomic traits
5. Mutants for efficient nutrient utilization

Achievements

1. Scarlet variety in Sweet potato
2. Sigma variety in Alfalfa

Advantages

1. High frequency mutagensis
2. Variation is created
3. Novelty
4. Less time for variety release

Limitations

1. Selected cellines fail to regenerate.
2. Many of the selected clones are unstable

Flower that smells like rottenfish : Carcass Flower.

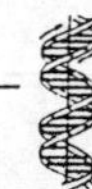

Points to Ponder

- "Evolution under domestication is a close counterpart of evolution in nature".
- "It is futile, indeed harmful, to expect that yields can continue to be raised simply by breeding superior new varieties".
- A breeder bases the hopes of improvement on evidence of genetic variations.
- A great deal of natural variability exists in plant tissues.
- A popular induced mutant variety of castor is Varuna.
- A synthetic variety is the one that has been evolved from all possible intercrosses among a number of selected genotypes to develop a population to increase by open-pollination.
- Actinomorphic flowers are found in family cruciferae.
- Adaptability refers to the capacity of a genotype to master macro-environmental factors in its favour to manifest a consistent performance over time and space.
- Adaptation refers to the capacity of a genotype to adjust to a specific environmental condition to attain a certain level of phenotypic expression.
- Aggressiveness is associated with the rapid spread of a disease.
- Albino plants tend to occur in patches.
- Alfalfa, besides an elaborate entophalous pollinating mechanisms, has some sort of compatibility system.
- Allopolyploidy is equally or even more unpredictable than autoploidy in morphological and physiological effects on the phenotype.
- Although free crossing is a danger on one side, while too close inbreeding is a hidden danger on the other.
- Androgamy denotes the development of embryo neither from egg-cell nor from synergids or antipodals, but from the gametes in or outside of embryo sac.
- Aneuploids are usually less vigorous than their diploid progenitors.
- Apogamy means the embryo-development of embryo neither from egg-cell nor from synergids or antipodals, but from the gametes in or outside of embryo sac.
- Asexual reproduction leads to perpetuation of the same genotype with precision.
- Asexual reproduction preserves the genotypes without variation.
- Auto polyploidy leads to sterility in plants.
- Autoploids have not been important in natural evolution due to aberrant chromosome behaviour and high degree of sterility.
- Auxine is responsible for the growth of plants.

> **Did You Know ?**
> *Tallest medieval minaret*
> **The tallest medieval minaret is the Qutub minor, started by Qutub-ud-din Aibak in 1199 and completed by his successors in 1239. The 239 ft tall minaret is also the oldest tower in India.**

- Back-cross is suited to transfer specific genes to a good variety deficient in one or a few characteristics.
- Backcross method can be carried out in any environment suitable for the development of the character being transferred.
- Backcrosses can serve a useful purpose in pedigree breeding programmes.
- Barley, rice, maize, flax, tomatoes and common beans have developed in their unique way, but the general pattern for each is that of Mendelian variation.
- Binary system of nomenclature was given by Carolus Linneus.
- Biotechnology consists of a cluster of techniques of genetic manipulation of living organisms at the cellular and the molecular level.
- Bulk-population handling often provides opportunity to make use of artificial aids to selection.
- Caryopsis fruit is present in wheat.
- Clonal propagation of heterozygous plants can be promoted by tissue culture of juvenile cells of apical meristem.
- Commercial tobacco (*Nicotiana tabacum*) does not occur as a natural species.
- Competition among homozygous genotypes indicate that adapted varieties are highly competitive against unadapted types.
- Consanguineous mating were favoured by ancient Egyptians, Hebrews, Greeks and Nordic tribes to maintain superior bloodliness and/or property within families.
- Crossing over and haploidization in fungi are not two parts of a single process, following in that order, as they do in meiosis.
- Crossing over is due to formation of chiasmata.
- Cucurbits are monoecious and cross-pollinating, on inbreeding some of their lines loose little vigour.
- Different levels of polypoids can be developed by haploid anther obtained from anther-culture derived from plants.
- Dioecy is not very common in plant species in general and occurs in only a few cultivated species.

GK **Malik Kafur was captured from Gujarat.** GS

- Disassortative mating best holds the population together.
- Disease resistance means the disease control by natural rather than by physical or chemical means and indicates hereditary struggle against some causal agents.
- DNA (nucleic acid) is the carrier of heredity in higher organisms.
- DNA are the pre-requisite for transformation.
- DNA can be demonstrated in the nucleus by the Feulgen strain.
- DNA contains phosphorus but no sulphur, where as proteins contain sulphur but no phosphorus.
- DNA duplicates by separation of two strands of the helix.
- Double cross hybrid gives higher yield than single cross.
- Eriksson in 1894 demonstrated that microorganisms can include strains differing in pathogenicity, even though morphologically undifferentiated.
- Euploidy refers to the situations in which the issue is number of repetitions of entire chromosome sets.
- Euploidy seems to arise most commonly through misadventures in meiosis.
- Fertilization is a synonym for syngamy.
- G.V. Yule (1906) suggested that there is needed no conflict between particulate and continuous inheritances if many genes having small and similar effects were postulated.
- Gene frequency also effects the efficiency of selection.
- Gene mutations are the building block of plant species evolution.
- Genes are the building blocks from which new varieties are fashioned.
- Genetic assimilation is an adaptive mechanisms of an organism in all natural selections.
- Genetic engineering is circumscribed by a non-sexual transfer of genetic information coded on the DNA, from one to the other.
- Guard cells take sodium ions during stomata opening.
- Hansen and Smith in 1932 demonstrated first that unlike nuclei occur together in the same hypha of *Botrytis cinerea*, an imperfect fungus.
- Haploid plants can be produced by anther culture, pollen culture makes it easier.
- Heat is liberated during germination of seeds.
- Heterocaryosis can lead to changes in pathogenicity in a plant-disease organisms.
- Heterocaryotic fungi can adjust themselves to circumstances as they develop.
- Heterosis can be regarded as converse of the deterioration that accompanies inbreeding hybrid.
- Heterozygotes show less variation than do homozygotes in an interbreeding population.
- High protein is associated with low yields in most of the cereals.
- Homozygosity within families increase with the continued self fertilization.
- Hunaan insulin, the first commodity product of biogenetic techniques was released for sale.
- Hybrid varieties (F_1) can be obtained by crossing clones, open pollinated varieties, inbred lines or other populations that are genetically dissimilar.
- Hybrid varieties depend for their superiority upon heterosis that characterizes the F_1 hybrids between certain genotypes.
- Identical twins result from single zygote.
- If the assortative mating is based on only two extreme phenotypes, fixation is achieved in one generation.
- If the heritability is 100 per cent, the phenotype provides a perfect measure of genotypic value.
- In 1939, Melchers coined the term Vernalin for the hypothetical activity factor that was thought to accumulate during vernalization.
- In angiosperm plants endosperm is formed as a result of triple fusion.
- In apomictic reproduction seeds are produced, but through processes other than normal meiosis and fertilization.
- In backcross breeding, a series of backcrosses are made to the variety to be improved and the desirable character(s) are maintained by selection.
- In data palm as male plants produce no fruits yet these are essential for the production of pollen.
- In embryo culture/rescue fertilized egg cells are usually excised one/two weeks after fertilization and are cultured on suitable media.
- In gametophytic incompatibility system, pollen tube growth is usually very slow in a style that contains the same allele of plants that are heterozygous at this locus.
- In gametophytic or haplo-diplo scheme, the incompatibility depends upon the genotype of the gemetophyte.
- In general, efficient inbreeding schemes are less difficult to secure than equally efficient outbreeding schemes.
- In genetic disassortative mating, individuals which are less closely related than they would be under random mating.
- In hops, the commercial product is dried pistillate inflorescence.

GK **India ranks seventh in the total fish production.** GS

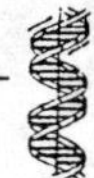

- In many fungi, mating type is governed by a single locus with two alleles, mt^+ and mt^-.
- In mass selection, the progeny of many pure lines are composed to form the new variety.
- In monoecism, protandry and protogyny all individuals of the population are genetically alike in respect of out-breeding mechanism.
- In protandry and protogyny, the anthers and stigma of a flower mature at different times.
- In pure-line breeding, the new variety is made up of the progeny of a single pure-line.
- In random mating, individuals are mated together at random or by chance.
- In sporophytic or in diplo-diplo schemes, the incompatibility is impressed upon by the gametophyte by its sporophytic parent.
- In the description of genetic variability, the significant hereditary parameters are defined precisely in terms of genetic models.
- In tomato, pollination follows opening of the flowers, but stamens form a cone enclosing the stigma almost to insure self-pollination.
- In wheat, anthers burst before extrusion so that the empty anthers hanging out are only a mockery.
- Inbreeders arise constantly from outbreeders.
- Inbreeding depression as observed is not a process of degeneration, but a consequence of Mendelian segregation.
- Introgressive hybridization is succeeded by recrossing with parental species in a way that some features of one transferred to the other without impairing taxonomic integrity.
- Isogenic lines can also be obtained by recurrent selection.
- It has been estimated that a normal maize tassel produces 25 million pollen grains.
- It has been observed that yield and bushel weight in wheat are influenced by the gene pair governing awned versus awnless lemmas.
- It has been possible to introduce N-fixing genes (nif-genes) into the bare protoplast during enzymatic digestion of the cell-wall.
- It is possible now to regenerate a mature plant or even a tree from an isolated callus.
- Leaves are eminently suitable for protoplast isolation.
- Linkage can be regarded as a conservative influence, tending to hold together existing combination of genes.
- Luxuriance is regarded as an accidental condition with different genetic antecedents and implications from true heterosis.
- Lysosomes are functioning as scavengers from removing unneeded or foreign material from the cell.
- Male sterile conditions is ordinarily recessive.
- Male sterility is a device to produce large quantities of seed of hybrid varieties.
- Male sterility unlike incompatibility is not a regular mechanism for controlling hybridity in natural populations.
- Mass selection is different than pure-line selection as a number of plants rather than one are selected to make up a new variety.
- Mechanism by which water is moved from roots to leaves is called Dixon theory of cohesion.
- Meiosis refers to double mitosis occurring in sexual reproduction which results in production of gametes with haploid (n) chromosome number.
- Minor millets are rich source of protein and lysine.
- Mitrochondria arises from de novo synthesis within the cell.
- Modifying genes means the genes that effect the expression of a nonallelic gene/genes.
- Mutants for genes affecting photorespiration in plants can be induced within a cell.
- Natural epiphytotics adequate to distinguish between disease resistant and susceptible plants may not occur regularly.
- Natural system of classification was produced by Betham and Hooker.
- *Nicotiana tabacum* (tobacco) arose as a hybrid polyploid between *N. sylvestris* and *N. tomentosa*.
- Occurrence of rust races in wheat gave an impetus to the back cross method.
- Osmotic pressure is higher due to salt concentration.
- Outbreeding species carry a substantial genetic load.
- Parasexual cycle is helpful in breeding improved strains of asexual fungi used in industry.
- Parthenogenesis refers to the development of embryo from egg-cell without fertilization.
- Pathogens manifest inherent variability.
- Penetrance refers to the ability of gene to be expressed in individuals which carry it.
- Perhaps the most successful case of induced polyploidy among crop plants is auto tetraploid rye with n=14 instead of n=7.
- Plants carrying particular types of cytoplasm are male sterile, but produce seed if pollinators are present.
- Plural segregating genes with similar effects explain

GK GS

Indigo is an example of vat-dyes.

the inheritance of the qualitative characters.

- Polyploidy refers to the variation arising through reduplication of chromosome sets.
- Post-resistance refers to collective heritable characters tending to reduce the probability of a plant to act as the host to a pest/insect.
- Production of 2 ATP in a chloroplast takes place during non-cyclic photophosphorylation.
- Protoplasts regardless of their origin, tend to fuse when brought into intimate contact with each other.
- Pure line selection has been widely used to select new varieties from the old land varieties.
- Pure-line represents the progeny of a single self-fertilized homozygous individuals.
- Pyruvic acid is the end product of glycolysis.
- Random mating has its place in plant breeding for example, in progeny testing.
- Rate of mutation varies with different genes.
- Reciprocal recurrent selection is useful in simultaneous for both general and specific combining ability.
- Recurrent selection involves selection of desirable genotypes, their selfed progeny and intercrossing in all combinations to produce population for reselection.
- Red fruited tomato belongs to sub-genus Eulycopersicon.
- Rhizobia can also function symbiotically with non-legume plant cells such as tobacco, wheat, carrot and rice etc.
- RNA is a genetic material of virus.
- Seedlessness in banana is due to triploid.
- Self-incompatibility in higher plants is far more important than dioecy.
- Self-pollination, excluding special genetic circumstances, rapidly reduces any population to the homozygous condition.
- Some fungi can be synthesized that carry diploid rather than haploid nuclei in their cells.
- Sometimes a single mutation with drastic or multiple effects may considerably augment the usefulness of a plant species.
- Stomata are closed due to higher concentration of water vapours around the leaf.
- Sunflower, a cross-pollinated species is fairly tolerant to inbreeding.
- Super hybrids or parasexual hybrids are developed by directly transferring regenes or genes guiding photo-respiration to plant cells through somatic fusion/ hybridization.
- Synthetic variety designates a variety maintained from open pollinated seed succeeding its synthesis by hybridization in all combinations among a number of selected genotypes.
- T.S. Venkatraman is the first person to start the sugarcane improvement in India.
- The smooth functioning of the reproduction mechanisms requires the simultaneous presence of two chromosomes of each kind.
- The 'Vilmorin Method' or 'Vilmorin Isolation Principle' was conceived by Louis de Vilmorin of France in 1843.
- The beneficial effects of crossing are more widely recognised that inbreeding depression as these are observable in nearly all F_1 hybrids.
- The best organised and most widespread use of Vilmorin Method was done by Swedish Seed Association established in Svalof in 1886.
- The cell tissue culture technique generates numerous progenies useful in mutation breeding.
- The chloroplast DNA (ct DNA) in plant codes for conversion of ammonia to glutamine.
- The chromosomal balance theory of sex determination applied to Drosophila.
- The coefficient of inbreeding devised by Wright is based primarily on the number and closeness of ancestral connections between male and female parents.
- The coefficient of inbreeding has a value of zero in random-mating populations and increases toward zero.
- The development of fruit without fertilization is known as parthenocarpy.
- The development of lima bean examplies a pattern of transformation characteristic of cultivated species of plants.
- The diastatic power of malting barley cannot be decided visually accurately.
- The direction and rate of water movement from cell to cell is based on DPD gradient.
- The domestication of micro organism has been most dramatic for the production of antibiotics.
- The double cross, suggested by D.F. Jone in 1918 made hybrid maize economically feasible.
- The enormous advantages of totipotency of plant cells could successfully be translated into standardization of bioculture techniques.
- The entry of pollen tube through micropyl is called programy.
- The essence of epidemiology is the disease cycle which has neither beginning nor end.
- The final product of back-crossing needs no agricultural

Sedna is the coldest and most distant object ever found in the solar system.

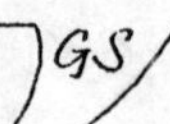

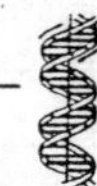

evaluation as the recurrent will be the same except for improved characteristics.

- The five phases of occurrence in host-parasite relationship are contact, preparation, penetration, establishment and development.
- The fungus extensively used for genetic study is Neurospora.
- The genes cannot cause a character to develop unless they have proper environment.
- The genetic effect of mating like to like on the basis of phenotypic characteristics depends on gene number and heritability.
- The genetic information is coded by the sequence of bases on the polynucleotide.
- The intensity of light in which the CO_2. utilized by respiration is called the light compensation point.
- The interallelic interaction refers to non-additivity of alleles.
- The isolated chloroplasts from cells belonging to a genotype carry the same karyotype.
- The main effect of apomixis is an increase in the proportion of maternal individuals through prohibition, or modification of genetic segregation and recombination.
- The major bottleneck of the bioculture techniques is the development of suitable method (s) of regeneration of plants especially in cereals and legumes.
- The mating systems of some crop species encourage, but do not enforce cross pollination.
- The maximum frequency of multiple crossing over is 50%.
- The Mendelian diversity arises basically from gene mutation.
- The nucleus contains a genetic material which is made up of chromatin material.
- The optimum rate of photosynthesis will take plase in visible light.
- The pedigree method of breeding derives its name from the records that are kept of ancestry of pedigree of each of the progenies.
- The pedigree record consists of a series of notes which give the relationships among the families that have been grown.
- The phenotypic ratio of inhibitory factor is 13:3.
- The polyploidy reduces gametophytic type of incompatibility.
- The populations suffering from inbreeding depression embody large number of recessive genes sheltered in heterozygotes.
- The process of meiosis or reduction division (of cells) occurs during the gametogenesis.

You Don't Know!

Shortest reign

Following Ala-ud-din's death on Jan . 16, 1316, Malik Kafur captured the throne of Delhi. His reign lasted precisely 35 days. He was outsted by Ala-ud-din's son, Sahibuddin Umar Shah.

- The progress in plant breeding depends on the nature, magnitude, and interrelations of genetic and non-heritable variation in agriculturally significant plant characteristics.
- The protoplasts of diverse karyotypes fuse together to produce heterokaryon in the process of somatic hybridization.
- The qualitative genes of classical genetics postulate more than a phenotypic manifestation.
- The regeneration of a mixed population of plants of various poloidy introduces further genetic variations in plants.
- The size of tobacco flower remains relatively constant under a range of environmental conditions.
- The smallest heritable extra chromosomal unit is called plasmagene.
- The specificity of legumes to establish a symbiotic relationship with Rhizobium species is due to their capacity to produce lectin protein directing bacterial invasion to root tissues.
- The stability of disease resistant varieties has been short lives.
- The stage between two successive mitotic divisions is called metabolic stage.
- The study of the survival of alternative alleles of a particular gene does not represent the effect of gene alone.
- The substitution of one allele for another produces a plus or minus shift on the scale of measurement, the shift is the same regardless of the presence of other genes.
- The synthesis of the RNA primers is catalyzed by enzymes called primerases.
- The term aneuploidy describes the organisms where chromosome number is not a whole-number multiple of the basis number of the group.
- The term heteorosis was coined by Shull.
- The two major areas of biotechnology are genetic engineering (gene splicing/recombinant DNA technology) and bioculture (tissue culture).
- The union of two pronuclear to form a common cleavage

The first Persian description of Jauhar was from Ranthambhor.

GS

nucleus is known as conjugation.

- The white eye mutant of *Drosophila melanogaster* affects pigmentation of eye, testicular membrane, spermatheca shape, length of life and general viability.
- Tissue regeneration in its culture is genetically controlled.
- Top cross is used to find out the combining ability of inbredlines.
- Turkey wheat is quite homogeneous for gross plant type.
- Two genes affecting same character in an individual are known as allelomorphs.
- Vertical disease resistance means resistance to some disease races more than to others.
- Water absorption takes place by root hairs.
- When the filament are united and anthers are free, this condition is known as syngenecious.
- When the resistance is evenly spread against all the races of a pathogen in a variety is known as horizontal resistance.
- Where there is embryonic degenera-tion especially after distant hybridization embryo culture might help in resurrecting viable hybrids.
- TTP (Test-tube pollination) was first used by Zenkteler (1990).
- *In-vitro* gynogenesis first observed in *Hordeum vulgare* by San Norum 1976.
- Melchers *et al.* (1978) produced somatic hybrid 'Pomato' by fusion between tomato and potato.
- RFLP's, RAPD and T-DNA and transposon taggings are proving helpful in gene mapping and identifying new genes.
- Pusa Jai Kisan a somaclone of Indian mustard released for cultivation.
- The plants containing foreign DNA are referred to as transgenic plant.
- The small segments of DNA with known base sequences, origin and function are known as DNA probes.
- The regeneration capacity of a plant cell to develop into a whole plant is called totipotency.
- Callus refers to a mass of unorganized regenerated cell in culture medium.
- In case of pollen-pistal incompatibility TTP (Test-tube pollination) is used.
- Harshey and Chase (1952) confirmed the role of DNA as the basic genetic material.
- Non-toxic somaclonal strain 'Moti' of *Lathyrus sativus* is developed by plant biotechnology.
- Centre for Cellular and Molecular Biology is situated in Hyderabad.

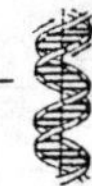

Multiple Choice Questions

1. ______________ estimated that incompatibility occurs in more than 3,000 species among 20 families of the flowering plants.
   a] D. De Winton
   b] M.B. Hughes
   c] W.H. Gabelman
   d] E.M. East
2. ______________ provides the best documented example of long time effects of selection.
   a] Sugarcane b] Sweet sorghum
   c] Sugarbeet d] Sugar maple
3. __________ deteriorates drastically upon inbreeding.
   a] Carrot b] Okra
   c] Brinjal d] Broccoli
4. ______________ in Ladino clover is governed by two dominant alleles, one produces cyanogenic glucoside, while the other releases the enzyme to catalyze its release from glucoside.
   a] Oxalic acid b] IAA
   c] HCN d] NAA
5. ______________ are examples of protandrous species.
   a] Avocados b] Walnuts
   c] Carrots d] None of these
6. 'Origin of Species by means of Natural Selection' was given by :
   a] Lamark b] Linnaeus
   c] Charles Darwin d] Erasmus
7. 'Radish' is an example of :
   a] Fusiform root b] Napiform root
   c] Conical root d] Tuberous root
8. "Expriments in Plant Hybridization" a paper of G. Mendal was first published in the year :
   a] 1905 b] 1787
   c] 1866 d] 1896
9. "If a gene is represented in an infinitely large random-mating population by adaptively neutral alleles, A and a, in the ratio of qA :(1-q)a, the genotypic proportions : $q^2AA + 2q(1-q)Aa + (1-q)^2 aa=1$" is known as :
   a] Stanford-Suneson law
   b] Weibe-Thomas law
   c] Briggs-Harlan law
   d] Hardy-Weinberg law
10. 2n - 1 is a:
    a] Monosomic b] Trisomic
    c] Tetrasomic d] Alloploid
11. 2n denotes :
    a] Number of homozygotes produce by selfing a F_1
    b] Number of types of gametes produced by F_1
    c] Number of types of heterozygotes produced by selling F_1
    d] None of the above
12. A cell or an organism having one chromosome pair in addition to the normal somatic compliment of the species is :
    a] Monosomic b] Nullisomic
    c] Tetrasomic d] Trisomic
13. A character dermined by many genes and does not show discrete variation is known as :
    a] Qualitative character
    b] Oligogenic character
    c] Quantitative character
    d] Multiple-allelic character
14. A complex of communities of plants and animals of a region existing under identical climatic conditions is called as :
    a] Biome
    b] Ecosystem
    c] Ecological complex
    d] Environment
15. A condition in which pollination and fertilization take place in an unopened flower is termed as :
    a] Homogamy b] Apogamy
    c] Polygamy d] Cleistogamy
16. A criss-cross inheritance is shown by:
    a] Sexinfluenced character
    b] Sexlimited character
    c] Sexlinked character
    d] Sexquantitative character
17. A cybrid is :
    a] A sexual hybrid
    b] A hybrid formed by cell fusion
    c] A plant produced by cell culture
    d] Synonym to hybrid
18. A dry, one chambered and one seeded fruit developing from a superior ovary with the hard and woody pericarp is :
    a] Achene b] Caryopsis
    c] Nut d] Cypsela
19. A fruit which develops from an inflorescence is known as :
    a] Simple fruit b] Aggregate fruit
    c] Composite fruit d] None of the above
20. A fruit which is collection of simple fruits developing from the apocarpous pistil of a flower is known as :
    a] Multiple fruit b] Aggregate fruit
    c] Complex fruit d] Simple fruit
21. A generative nucleus in the male gametophyte divide ____________ to form two male gametes.
    a] Mitotically
    b] Meiotically
    c] Mitotically and Meiotically
    d] None of the above
22. A head or capitulum is characteristic of :
    a] Compositeae b] Umbelliferae
    c] Solanaceae d] Gramineae

GK GS
Mohd. bin Tughlaq set up the Diwan-i-Amir-Kohi.

23. **A high yielding semi-dwarf variety of rice was first developed at :**
a] CRRI, Cuttack b] IARI, New Delhi
c] ICRISAT, Hyderabad d] IRRI, Manila

24. **A homozygous line developed by self fertlization in a cross-pollinated crop is termed as :**
a] Pureline b] Inbred
c] Clone d] Mutant

25. **A hybrid is always :**
a] Heterozygous b] Homozygous
c] Both of the above d] Often homozygous

26. **A large unit of natural vegetation in an area under identical condition (climatic) is called :**
a] Plant formation b] Association
c] Community d] Climax

27. **A maintainer line is :**
a] Male fertile and can restore fertility
b] Male sterile but can restore fertility
c] Male sterile but cann't restore fertility
d] Male fertile but cann't restore fertility

28. **A male child would be born to parents if :**
a] The father is healthier than the mother
b] The mother eats well during pregnancy
c] The genetic composition of the child includes XY combination of chromosomes
d] The genetic composition of the child includes XX combination of chromosomes

29. **A male gametophyte in angiosperms contain :**
a] One gametes only
b] Two male gametes only
c] Two male gametes and one tube nucleus
d] One male gamete and one tube nucleus

30. **A mendelian population means :**
a] Self-pollinated population
b] Inbred population
c] Random-mating population
d] Hybrid population

31. **A method of breeding most commonly used when the desired variation is required to be induced in a vegetatively propagated crop is:**
a] Polyploidy
b] Pedigree method of breeding
c] Mutation breeding
d] Backcross method of breeding

32. **A mixture of two immiscible liquids shaken up is :**
a] Saturated solution
b] A suspenson
c] An emulsion
d] Colloid

33. **A plant bearing both male and female flowers is said to be :**
a] Dioecious
b] Monoecious
c] Polygamous
d] None of above

34. **A plant breeder is interested to control pollination to:**
a] Prevent cross pollination
b] Control pollination
c] None of these
d] Both of these

35. **A plant cell when put into a hypotonic solution, it unlike animal cell does not brust because:**
a] It has no semi-permeable membrane
b] It has inelastic cell wall
c] All both a & b
d] None of the above

36. **A plant with 2n = 20 will have how many linkage group:**
a] 20 b] 10
c] 40 d] 5

37. **A plant with genotype Gg Ww will produce gametes of which of the following types ?**
a] GW, gw, gW, Gw b] GG, WW, gg, ww
c] Gw, GG, GW d] All of them

38. **A population at equilibrium developed from intermating a number of inbred line is called a :**
a] Composite variety
b] Synthetic variety
c] Hybrid variety
d] Pureline variety

39. **A restorer gene is present in :**
a] Nucleus b] Cytoplasm
c] Both d] None

40. **A state of meiotic cell division when homologous chromosomes separate and move to different poles is known as :**
a] Metaphase I b] Anaphase I
c] Telophase I d] Anaphase II

41. **A successful establishment of a vegetation in a new locality means:**
a] Commensalisnt b] Ecesis
c] Succession d] Psammosere

42. **A three way cross in maize is defined as a cross between a single cross and :**
a] An open pollinated variety
b] Synthetic variety
c] An inbred line
d] Another single cross

43. **A time honoured sugarcane variety (wonder cane) is:**
a] Co 975 b] Co 7508
c] Co 6907 d] Co 419

44. **Acclimatisation is :**
a] A process of adjustment by crop plants to environment
b] Related with the clay
c] Removal of female parts of a flower
d] Changing of climatical condition

GK GS

The average salinity in the oceans and seas is 35 per cent.

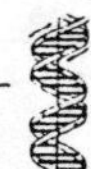

45. According to position of the ovule within the ovary. An ovule may be :
a] Ascending b] Pendulous
c] Suspended d] All the above

46. Actual petrified remains of ancient plants and animals or impressions left by them in rocks are called :
a] Fossils b] Models
c] Traces d] None of the above

47. Adjustment of individual to their environment by mean of special structures or of function is spoken of :
a] Mutation b] Variation
c] Heredity d] Adaptation

48. Adventitious roots grow from :
a] Radicle
b] Plumule
c] Hypophysis
d] Any part of the plant body other than the radicle

49. After the cell has undergone mitosis, the amount of DNA in daughter cells will be :
a] Equal to the parent cell
b] Half of the parent cell
c] Double of the parent cell
d] None of the above

50. Albinism in plants is associated with :
a] Epistatis b] Recessive lethal
c] Dominant lethal d] Chromosome duplication

51. Alternative forms of a gene is known as :
a] Mutation b] Gamete
c] Allele d] Genotype

52. An anticodon is a sequence of three nitrogenous bases found on :
a] DNA b] mRNA
c] rRNA d] tRNA

53. An association can be:
a] Hydrophytic b] Halophytic
c] Xerophytic d] All above

54. An egg or sperm cell in plants is called :
a] Gene b] Gamete
c] Zygote d] None of them

55. An embryo-sac of angiosperms generally contains :
a] Two nuclei b] Three nuclei
c] Four nuclei d] Eight nuclei

56. An endospermic cell of a seed after somatic chromosome number (2n=20) will have :
a] 30 Chromosomes b] 20 Chromosomes
c] 10 Chromosomes d] 40 Chromosomes

57. An individual having different alleles for any gene pair and producing two kinds of gametes is known as:
a] Hemizygous b] Homozygous
c] Heterozygous d] Heterogenous

Knowledge World!
Highest filling station
The world's highest filling station is situated at Kaza Valley at a height of 13,500 ft. IOC also operated the petrol pump at Leh, Ladakh, at 12,000 ft.

58. An organism having the gametic chromosome numbers is called as:
a] Genome b] Hybrid
c] Gamete d] Haploid

59. An organism with one extra chromosome (2n + 1), is insteed of the normal diploid (2n) is known as:
a] Monoploid b] Monosomic
c] Trisomic d] Tetrasonic

60. Ascending portion of the axis of the plant, developing directly from the plumule and bears leaves, braches and flower is known as :
a] Root b] Steam
c] Modified root d] None of above

61. At the molecular level point mutations are due to :
a] Destruction of the bases
b] Destruction of the double helix
c] Shifting of a portion of the helix and bases
d] Alterations of sequences of base in DNA

62. At the stage incipient plasmolysis the turgor pressure is :
a] Zero b] 2 atm.
c] 4 atm. d] 22.4 atm.

63. At the stage of fully turgid cell, the suction pressure will be :
a] Equal to wall pressure
b] Zero
c] Equal to osmotic pressure
d] Maximum

64. ATP synthesis occurs in :
a] Chloroplast
b] Chloroplast and mitochondria
c] Chloroplast, mitochomdria and nuclei
d] Mitochondria

65. Bacteria, protozoa and other living plants and animals are :
a] Climatic factors
b] Edaphic factors
c] Biotic factors
d] Topographic factors

66. Baldness in human being is a:
a] Sex-linked character
b] Sex-limited character
c] Sex-influenced character
d] None of the above

67. Barely is a _______ pollinated crop.
a] Often-cross b] Self and cross
c] Cross d] Self

68. Biffin in 1905 announced that resistance to yellow rust of wheat is governed by _____________ gene in crosses.
a] Single dominant b] Single recessive
c] Double dominant d] Double recessive

69. Binomial System of Nomenclature means that the name of plant is made up of two words which designate :
a] Family and genus b] Order and family
c] Species and family d] Genus and species

70. Bread wheat is :
a] Haploid b] Diploid
c] Tetraploid d] Auto-allopolyploid

71. Bulk population breeding is suitable for:
a] Fruit crops b] Vegetable crops
c] Small grains d] Flower crops

72. Calder in 1937 proposed phytogeographical region of India:
a] 7 b] 6
c] 8 d] 5

73. Callus is :
a] An undifferentiated mass of cells
b] A gamete
c] A tissue
d] A modification of leaf

74. Carolus Linnaeus is credited with introducing the :
a] Law of limiting factors
b] Law of independent assortment
c] Binomial system of nomenclature
d] Laws of inheritance

75. Cauliflower, cabbage, broccoli, kohlrabi and brussels sprouts have been derived from wild cabbage that still grows in coastal regions of :
a] Latin America
b] Mesopotamia
c] Europe and N. Africa
d] South Asia and South Africa

76. CAZRI does research on :
a] Arid plants and crops
b] Tropical crops
c] Sub-tropical crops
d] None of the above

77. Central Rice Research Institute is located (CRRI) at :
a] Madras b] Cuttack
c] Calcutta d] Dehradun

78. Characteristic feature of a xerophyte is :
a] Long taproot
b] Leaves and stem with thick cuticle
c] Sunken stomata
d] All above

79. Characteristic feature of the hydrophyte is :
a] Reduced protective tissues
b] Feebly developed root-system
c] Spongy stem
d] All the above

80. Chauncey Goodrich was the first _______ to publish a paper dealing with the resistance of potatoes to the blight diseases.
a] Britisher b] German
c] Dutch d] American

81. China rose is a :
a] Herb b] Shrub
c] Tree d] Climber

82. Chromosome numbers in somatic cells of a maize plant are :
a] 22 b] 26
c] 14 d] 20

83. Chromosome which pair at meiotic prophase and are similar size, shape, structure and function are known as:
a] Homologous b] Isocentric
c] Isochromosome d] Acentric

84. CIMMYT is located at :
a] Mexico b] India
c] Spain d] Nigeria

85. Cliestogamy encourages :
a] Self pollination b] Cross pollination
c] Apomixis d] Inbreeding depresion

86. Clone in general is :
a] Homozygous b] Heterogenous
c] Hetrozygous d] None of the above

87. Commercial bananas have ________ instead of 22 chromosomes of ordinary diploid varieties.
a] 13 b] 23
c] 33 d] 43

88. Complementary gene produces phenotypic ratio :
a] 3:1 b] 13:3
c] 15:1 d] 9:7

89. Composites are developed by using:
a] Advanced generation seed mixtures obtained from high yielding intervarietal or interracial crosses
b] Hybrid directly from intervarietal crosses
c] Recurrent selection for specific combining ability
d] The hybrid of an inbred with an open-pollinated variety

90. Conifers are abundant in :
a] Tropical zone b] Alpine zone
c] Temperate zone d] Subtropical zone

91. Cotton :
a] Is a self pollinated crop
b] Is a cross pollinated crop
c] Is an often-cross pollinated crop
d] Is an often-self pollinated crop

GK

The 'MKS' system of measurement originated in France.

GS
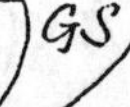

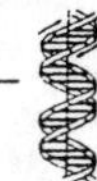

92. Creation of the world by God is basis of the :
a] Theory of eternity of present condition
b] Theoy of special creation
c] Theory of catastrophism
d] All the above

93. Cross between cytoplasmic male sterile x cytoplasmic male fertile :
a] Produce sterile F_1
b] Produce fertile F_1
c] Produce fertile, sterile in 1:1 ratio
d] None of the above

94. Cross of F_1 hybrid with either parents is called :
a] Test cross b] Single cross
c] Back cross d] None of them

95. Crossing over occurs at :
a] Two strand stage
b] Four strand stage
c] Three strand stage
d] None of the above

96. Cuscuta is :
a] Total stem parasite
b] Partial stem parasite
c] Partial root parasite
d] Total root parasite

97. Cytogeneticaly, the definitive nucleus is :
a] Haploid b] Diploid
c] Triploid d] Tetraploid

98. Date palm is :
a] Monoecious b] Dioecious
c] Polygamous d] None of the above

99. Degenerated parts of plants and animals which do not serve any useful purpose are known as :
a] Unuseful structures
b] Vestigial structures
c] Essential organs
d] Unwanted structures

100. Despite conspicuous deterioration of maize upon continuous selfing, it is more tolerant to inbreeding than:
a] Limabean b] Barley
c] Alfalfa d] Buckwheat

101. Details of the kreb's cycle were worked out by the use of :
a] Radioactive carbon compounds
b] Ultracentrifugation
c] Electronmicrosopy
d] X-ray crystallography

102. Development and formation of pollen grains in anther of the stamen is known as :
a] Pollination b] Fertilization
c] Microsporogenesis d] Megasporogenesis

103. Development of a fruit without fertilization is known as :
a] Parthenogenesis b] Polyembrohy
c] Budding d] Apomixis

104. Development of a successful variety by backcross breeding calls for:
a] Satisfactory recurrent parent
b] Retention of intensity of desired trait
c] Sufficient backcrosses to reconstitute recurrent parent
d] All of these

105. Development of seed without-sexual process is :
a] Apomixis b] Polyembryony
c] Both a & b d] None of the above

106. Development of the pollen tube, the seed and the 3n endosperm, double fertlization are characteristic features of :
a] Bryophytes b] Pteridophytes
c] Gymnosperms d] Angiosperms

107. Disease reaction, is by and large, a _________ process.
a] Physical b] Chemical
c] Mechanical d] Manual

108. Disease resistance in barley is :
a] Quantitative characteristics
b] Qualitative characteristics
c] Both a & b
d] None of the above

109. DNA exists as a double helix as a result of hydrogen bonding between:
a] Sugar molecules b] Phosphate groups
c] Nucleohistones d] Nucleosides

110. DNA polymerase :
a] Helps in DNA replication
b] Helps in RNA replication
c] Helps in protein synthesis
d] None of the above

111. DPD is the index of sucking power and is also known as :
a] Suction pressure b] Wall pressure
c] Osmotic d] Turgor pressue

112. During mitotic cell division the total chromosome number :
a] Remians the same b] Gets halved
c] Gets doubled d] None of them

113. Each cell of the female gametophyte is hapoid except:
a] Egg cell b] Synergid cells
c] Antipodals cells d] Definitive nucleus

114. Each gramme of water transpired absorbs - kj of latent heat from the leaf and its environment :
a] 2.0 - 2.1 b] 2.2 - 2.3
c] 3.4 - 3.5 d] 2.7 - 2.8

GK
Feroz Shah Tughlaq repaired the Qutub Minar as it was struck by lightening.
GS

115. Each mature anther has :
a] One pollen sac b] Two pollen sacs
c] Three pollen sacs d] Four pollen sacs

116. Each megaspore mother cell finally produces :
a] Four megaspores
b] Two megaspores
c] Only one megaspore
d] Several megaspores

117. Each pollen mother cell produces four haploid cells as a result of :
a] Mitosis
b] Meiosis
c] One mitosis and then meiosis
d] None of the above

118. Each stamen is differentiated into:
a] Filament b] Anther
c] Connective d] All above

119. Electron microscope was invented by :
a] Jansen b] Leeuwenhoek
c] Knoll and Ruska d] None of the above

120. Emasculation is :
a] Removal of male parts from a bisexual flower
b] Removal of female parts
c] Removal of male and female parts
d] None of the above

121. Embryosac is :
a] Male gametophyte b] Microgametophyte
c] Sporophyte d] Female gametophyte

122. Endosperm is store house for germinating embryo and it contains:
a] Hormones b] Vitamins
c] Enzymes d] All the above

123. Epicalyx is the characteristic of the family :
a] Gramineae b] Solanaceae
c] Malvaceae d] Compositeae

124. Epistasis should be carefully distinguished from dominance, which refers to non-additivity of alleles at :
a] The same locus of a homologus chromosome
b] Different loci of a homologous chromosome
c] The same locus of non-homologous chromosomes
d] Diferent loci of non-homologous chromosomes

125. Evaluation of inbreds in maize is carried out in four stages. Open pollinated variety is involved in the:
a] Phenotypic evaluation
b] Top cross evaluation
c] Single cross evaluation
d] Double cross evaluation

126. Exchange of genetic material takes place between non-sister chromatids of homologous chromosomes during the :
a] Four strand stage in prophase I
b] Two strand stage in prophase I
c] Metaphase I stage in meiosis
d] Two strand stage in prophase II

127. Female reproductive organ in the flower is :
a] Calyx b] Corolla
c] Stamens d] Carpels

128. Flower colour in plants is a :
a] Qualitative character
b] Quantitative character
c] Both of them
d] None of them

129. Flower is specialized :
a] Shoots b] Leaf
c] Stem d] Tendril

130. Flowers having both androecium and gynoecium are said to be :
a] Hermaphrodite b] Unisexual
c] Staminate d] Pistilate

131. For Hardy-Weinberg law to operate:
a] The population should be large
b] There should be open pollination
c] There should be no selection
d] All the above

132. Formation of new genes takes place due to :
a] Inversion b] Transduction
c] Transversion d] Mutation

133. Fruit of candytuft is known as :
a] Legume b] Siliqua
c] Silicula d] Capsule

134. Fruit of pea is known as :
a] Legume b] Follicle
c] Siliqua d] Capsule

135. Fruit of sun-flower is :
a] Caryopsis b] Samara
c] Cypsela d] Nut

136. Fruits of calotropis is :
a] Legume b] Siliqua
c] Capsule d] Follicle

137. Fusion of the one sperm (male gamete) and definitive nucleus results in the formation of :
a] Endosperm b] Embryo
c] Sporophyte d] Zygote

138. Ganga safed is a hybrid of :
a] Maize b] Sorghum
c] Cotton d] Pearl millet

139. Generally embryosac is :
a] Monosporic b] Bisporic
c] Tetrasporic d] Octosporic

140. Genes are made up of :
a] RNA only b] DNA only
c] RNA and DNA d] Proteins

141. Genetics is the study of :
a] Genes
b] Gene interaction
c] Heredity and variation
d] DNA

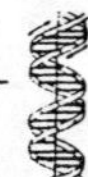

142. Genotype x Environment interaction would take place when there is/are
   a] One genotype and two environments
   b] Two genotypes and one environment
   c] Three genotypes and one environment
   d] Two or more genotypes and two or more environments
143. Germplasm means :
   a] A breeding method
   b] A collection of genetic material
   c] A new desease
   d] None of the above
144. Gibberellins are now commercially used into increase sugarcane growth and sugar yields :
   a] India b] Hawaii
   c] Japan d] Korea
145. Ginger and Turmeric are example of modified stem :
   a] Rhizome b] Tuber
   c] Bulb d] Corm
146. Given that the somatic chromosome number of *Triticum aestivum* is 2n = 6x =42, which one of the following pairs is correctly matched ?
   a] Monosome..... 2n = 40
   b] Trisome ..... 2n =42
   c] Tetrasome .....2n = 44
   d] Nullisome ..... 2n = 41
147. Mendel is famous for:
   a] Propounding the laws of inheritance
   b] Laws of limiting factors
   c] The theory of mutation
   d] The discovery of penicillin
148. Group of plants with similar characteristics of any rank is termed:
   a] Taxon b] Species
   c] Genus d] Order
149. Gynoecium or pistil in the female whorl differentiated into :
   a] Filament b] Style
   c] Stigma d] All above
150. Halophytes growing on muddy swamps form a special vegetation known as :
   a] Mangrove forests b] Coniferous forests
   c] Evergreen forests d] Riparian forests
151. Heritability may be defined as the:
   a] Interation product of genotype with environment
   b] Sum total of herditary material present in a species
   c] Degree of resemblance between the original and the selected plants
   d] Proportion of phenotypic variability which is due to heredity

Quiz Kit!

First woman cycle-rickshaw mechanic

Suhago Devi, a 57-year-old tribal from Bihar became a cycle rickshaw mechanic in the 1960's to supplement the family income.

152. Heterosis is commercially expolited in :
   a] Cross pollinated species
   b] Often cross-pollinated species
   c] Self-pollinated species
   d] Vegetatively propagated species
153. Heterosis over the better parent is called as :
   a] Relative heterosis b] Standard heterosis
   c] Pseudoheterosis d] Heterobeltiosis
154. How many cells will be produced from a cell if the cell divides meiotically ?
   a] 1 b] 2
   c] 3 d] 4
155. Hugo de Vries, who defined the 'mutation' was :
   a] Dutch Botanist
   b] French Bacteriologist
   c] English Botanist
   d] Russian Biologist
156. Hybrid seeds can be produced through open pollination with the use of :
   a] Inbreds b] Clones
   c] Pure lines d] Male sterile lines
157. Hydrilla is a :
   a] Hydrophyte b] Mesophyte
   c] Hygrophyte d] Xerophyte
158. ICRISAT deals with :
   a] Cotton b] Maize
   c] Millets d] Wheat
159. If a character is cytoplasmically controlled, then :
   a] The progeney resembles female parent
   b] The progeny resembles male parent
   c] It's segregation of 3:1 is observed in F_1
   d] 15:1 segregation in F_1
160. If a dihybrid is test crossed, the phenotypic ratio of progeny will be:
   a] 9:3:3:1 b] 9:3:4
   c] 1:1:1:1 d] 15:1
161. If a dihybrid without genic interaction is selfed the following ratio is obtained in F_2.
   a] 1:1:1:1 b] 9:3:3:1
   c] 9:7 d] 9:6:1
162. If a membrane allows the passage of both solvent and solute molecules it is said to be :
   a] Non-permeable b] Impermeable
   c] Semi-pemeable d] Permeable

GS

Tuberculosis (TB) kills approximately 2 million people each year.

163. If dissolved sugar in pure water the resulting solution will have osmotic potential than that of pure water :
a] Equal b] More positive
c] More negative d] Difficult to measure

164. If more solute is added to a saturated solution, it settles down but if heated the excess amount also gets dissolved. Such a solution said to be:
a] Unsaturated b] Saturated
c] Super saturated d] Crystalloid

165. If parents differ in ten gene pairs, no selection is done; six backcrosses will produce a population having __________ per cent homozygous and identical individuals to recurrent parent at all ten *loci*.
a] 65 b] 75
c] 85 d] 95

166. If the cells of a plant root tip contain 16 chromosomes, how many chromosoms will be found in its pollen grains ?
a] 4 b] 8
c] 16 d] 32

167. If the particles of dispersed phase have an affinity towards the particles of the dispersion medium the colloidal system is known as :
a] Lyophobic b] Hydrophobic
c] Lyophilic d] None of the above

168. In _______ pollen dispersal is poor and seed set is likely to be low on male sterile plants.
a] Tomato b] Cauliflower
c] Broccoli d] Carrot

169. In ____________, the first studies on artificial hybridization for which yields were reported were those of Beal during the period between 1877 and 1882.
a] Sorghum b] Pearlmillet
c] Maize d] Rye

170. In Angiosperms :
a] Sporophyte is complex and gametophyte incospicuous
b] Sporophyte simple and gametophyte complex
c] Sporophyte and gemetophyte complex
d] Sporophyte and gametophyte incospicuous

171. In asparagus one or more short, green cylindrical or sometimes flattened branches of limited growth develop from node of stem, such a branch is known as:
a] Cladode b] Phylloclade
c] Stem tendril d] Thorn

172. In bajra :
a] Cytoplasmic male sterility is found
b] Genetic male sterility is found
c] Cytoplasmic genetic male sterility is found
d] Self incompatibility is found

173. In C_4 pathway CO_2 combines with:
a] Ribulose monophosphate
b] Phosphoglyceric acid
c] Ribulose diphosphate
d] Phosphoenol pyruvate

174. In China rose flower is :
a] Hypogynous b] Perigynous
c] Epigynous d] None of the above

175. In compositeae the gynoecium is syncarpous bicarpellary with :
a] Marginal placentation
b] Axile placentation
c] Free central placentation
d] Basal placentation

176. In cotton, the popularly grown commercial hybrid 'varalaxmi's is the product of :
a] Intraspecific hybridization in *Gossypium hirsutum*
b] Intraspecific hybridization in *Gossypium barbadense*
c] Interspecific hybridization between *Gossypium hirsutum* and *Gossypium barbadense*
d] Interspecific hybridization between *Gossypium arboreum* and *Gossypium herbaceum*

177. In cruciferae the stamens are said to be :
a] Didynamous b] Tridynamous
c] Tetradidynamous d] Decadynamous

178. In family malvaceae the stamens are:
a] Diadelphous b] Gynandrous
c] Syngenesious d] Monoadelphous

179. In Ficus (banyan, pepal, fig) the inflorescence is
a] Cyathuim b] Verticillaster
c] Hypanthodium d] Spadix

180. In India for hybrid seed production in maize we use :
a] Cytoplasmic genetic male sterility
b] Genetic male sterility
c] Self incompatibility
d] Detasseling

181. In India male sterility is used in :
a] Maize b] Wheat
c] Pearl millet d] Chickpea

182. In mass selection, plants are selected on the basis of:
a] Phenotypes b] Genotypes
c] Homozygosity d] None of the above

183. In pedigree breeding single plants are selected in which one of the following generation :
a] F_5 b] F_3
c] F_2 d] F_4

184. In potato the inflorescence is :
a] Helicoid b] Scropioid cyme
c] Umbel d] Raceme

185. In *Primula* sp:
a] Sporophytic incompatibility found
b] Gametophytic incompatibility found
c] Heteromorphic incompatibility found
d] Homomorphic incompatibility found

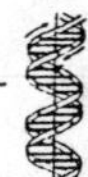

186. In RNA, nitrogen bases are same as in DNA except :
a] Uracil instead of Thymine
b] Cytosin instead of Thymine
c] Adenine instead of Guanine
d] Thymine instead of Adenine

187. In self-incompatibility system :
a] Pollen not produced
b] Self pollination takes place
c] Pollen does not germinate on self-pollination
d] All are correct

188. In solanaceae, the flowers are :
a] Actionomorphic, bisexual and epigynous
b] Zygomorphic, unisexual and perigynous
c] Actinomorphic, bisexual and hypogynous
d] Actinomorphic, bisexual and perigynous

189. In sorghum somatic chromosome number is 20. What would be its chromosome number in the endosperm?
a] 10 b] 20
c] 30 d] 40

190. In sugarcane the viable seed setting takes place only in :
a] North India b] Central India
c] East India d] South India

191. In sweet pea gene C or P alone produces white flower, where as both C and P are needed to produce purplecoloured flower. In a particular cross between two white flowered plants the offsprings were one half purple and one-half white. The possible genotypes of the parents is :
a] Cc Pp x ccPp b] CCpp x ccPP
c] ccPp x CCpp d] Ccpp x ccPp

192. In the cymose inflorescence the flowers develop in :
a] Basipetal succession
b] Acropetal succession
c] Centripetal succession
d] None of the above

193. In the flowering plants the process of fertilization is first discovered by:
a] Robert Brown b] Strasburger
c] Virchov d] Nawaschin

194. In the ovary of 'marigold' the placentation is :
a] Superficial b] Basal
c] Marginal d] Axile

195. In the pitcher plant, the pitcher is modification of :
a] Stem b] Root
c] Leaf d] Stem-branch

196. In the simple ovary (of one carpel) there is one common type of placentation known as :
a] Axile b] Parietal
c] Central d] Marginal

197. In wheat and oat there are ______ genes for red *versus* white kernels.
a] Two b] Three
c] Four d] Five

198. In wheat cross pollination is :
a] Less than 5% b] More than 5%
c] 15% d] 50%

199. In which one of the following crops maximum exploitation of hybrid vigour during the last three decades has been done ?
a] Wheat b] Rice
c] Maize d] Cotton

200. In which part of the cell, does protein synthesis take place ?
a] Mitochondria b] Nucleolus
c] Nucleus d] Ribosomes

201. Infloresence of wheat ear is :
a] Catkin b] Spike
c] Corymb d] Raceme

202. International Rice Research Institute (IRRI) is situated in :
a] Philippines b] India
c] China d] Burma

203. It has been estimated that a maize tassel produces at least __________ pollen grains for each kernel.
a] 5,000 b] 15,000
c] 25,000 d] 35,000

204. It is an established fact that higher and more complex forms of plants and animals have evolved from :
a] Recent and simpler forms
b] Earlier and simple forms
c] Recent and complex forms
d] Xerophytic forms

205. Jawahar, Vikram, Kisan and Sona, the important varieties of maize are :
a] Double-cross hybrids
b] Synthetic varieties
c] Composite varieties
d] Single-cross hybrids

206. Johansen proposed :
a] Theory of heterosis
b] Pure line theory
c] Male sterility
d] Hybridization

207. Kreb's cycle begins with :
a] Glycogen b] Lysine
c] Succinic acid d] Acetyl COA

208. Later formed rocks show fossils of:
a] Less complex and less advance types
b] More complex and more advance types
c] Simpler types
d] Simpler and more recent types

209. Leaf of the *Mimosa* is
a] Unipinnate
b] Bipinnate
c] Tripinnate
d] Decompound

Diwan-i-Insha was department of Correspondence.

210. Magnification of an object by electron microscope is:
a] 1,000 X b] 10,000 X
c] 1,00,000 X d] More than 2,00,000 X

211. Maize is :
a] Monoecious plant b] Dioecious plant
c] Polygamous d] None of the above

212. Most of the mutation have __ effects.
a] Deleterious b] Good
c] Bad d] Both b & c

213. Male gamete carries ____________ chromosome numbers.
a] n b] 2n
c] 3n d] n-1

214. Mango fruit is a :
a] Pepo b] Pome
c] Berry d] Drupe

215. Mangrove plants show adaptation like :
a] Stilt roots
b] Respiratory roots or pnematophores
c] Vivipary
d] All above

216. Mating between two individuals related by descent is :
a] Selfing
b] Inbreeding
c] Outbreeding
d] Maintenance of population

217. Megaspore mother cell produces four megaspores out of which usually:
a] Only one megaspore is functional and the remaining three megaspores abort
b] Two megaspores are functional and the remaining two megaspores abort
c] Three megaspores are functional and the fourth megaspore abort
d] All the four megaspores are functional

218. Meiosis is essential for the organism for:
a] Inducing polyploidy
b] Growth of the organism
c] Keeping the number of chromosomes constant from one generation to the other
d] Increasing the number of cells

219. Meiosis occurs in :
a] Generative cells b] New cells
c] Vegetative cells d] None of the above

220. Mendel has worked on the following crop :
a] *Pisum sativum* b] *Hordeum vulgare*
c] *Triticum aetivum* d] *Zea mays*

221. Mendel was born in the year :
a] 1821 b] 1820
c] 1828 d] 1822

222. Mendel's discovery contributed to agriculture through:
a] Increased production of fertilizer
b] Improved crop varieties obtained through hybridisation
c] New farm machines
d] The use of fungicides

223. Mendok is a gametocide used for inducing male sterility in
a] Onion b] Sunflower
c] Maize d] Cotton

224. Method of breeding can be used for studying the ancestral characteristic:
a] Pure line b] Pedigree method
c] Mass selection d] None of these

225. Microsporogenesis occurs in :
a] Roots b] Stems
c] Anthers d] Ovules

226. Microsporophylls are synonym to:
a] Carpels b] Gynoecium
c] Stamens d] Thalamus

227. Modern population genetics is found on a proposition deduced independently by Hardy in England in 1908 and __________ in Germany in 1909.
a] Weinberg b] Pugsley
c] Briggs d] Weibe

228. Monoploid is also a :
a] Haploid b] Diploid
c] Monosomic d] None of the above

229. Multiline varieties are mixtures of:
a] Several morphologically different plants having identical disease resistance genes
b] Bulked progenies of different inbred lines
c] Several similar pure lines having different genes for disease resistance
d] Morphologically identical plants with different genetic make-up

230. Mutation is known to be due to change in genes :
a] Loss b] Degeneration
c] Addition d] All the above

231. Neither beginning nor end to the universe is basis of:
a] Theory of special creation
b] Theory of catastrophism
c] Theory of eternity of present condition
d] None of the above

232. Nilsson-Ehle of ____________ was apparently the first to use the bulk population method to breed superior crop varieties.
a] Sweden b] Mexico
c] Germany d] France

GK GS

EU (European Union) comprises 25 member countries.

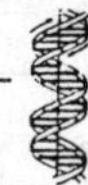

233. Normal female in human being is:
a] Homogametic sex
b] Heterogametic sex
c] Homizygous for a gene
d] Gynandromorph

234. Nucleus seed is :
a] 100% pure b] 80% pure
c] 50% pure d] 40% pure

235. Number of chromosome in endosperm is :
a] 2n b] 3n
c] n d] 4n

236. Nutritional improvement for the tryptophan content has been achieved in a __________ cell-line.
a] Tomato b] Cucurbits
c] Coffee d] Tobacco

237. Observation of Hugo de Vries were based on :
a] *Pisum sativum*
b] *Drosophila melanogastor*
c] *Oenothera lamarkiana*
d] *Solanum tuberosum*

238. Okra belongs to the family :
a] Solanaceae b] Malvaceae
c] Cucurbitaceae d] Liliaceae

239. Onion and Garlic are example of :
a] Rhizome b] Tuber
c] Bulb d] Corm

240. Operon concept was given by :
a] Watson and Crick
b] Beedle and Tatum
c] Jacob and Manod
d] Mosselson and Stahl

241. Orange or lemon fruit is known as:
a] Balusta b] Pepo
c] Pome d] Hesperidium

242. Osmotic pressure of cell sap is high among below given categories in :
a] Mesophytes b] Halophytes
c] Hygrophytes d] Hydrophytes

243. Oxisomes in plant cells are concerned with :
a] Kreb's cycle
b] Calvin cycle
c] Photorespiration
d] Photon-phosphate pathway

244. Paramutation in __________ is perhaps the one and only example of blending of alleles.
a] Sorghum b] Maize
c] Pearl millet d] Foxtail millet

245. Pea has stamens :
a] Diadelphous b] Monoadelphous
c] Polyadelphous d] Gynandrous

246. Persons with the following blood group is universal donor :
a] A b] B
c] AB d] O

Did You Know ?
Islamic principles
The Alai Darwaza built by Ala-ud-din Khilji in 1311 is the first building employing the Islamic principles of accurate construction and geometric ornamentation exclusively.

247. Phylloclade is :
a] Underground modification of stem
b] Sub-aerial modification of stem
c] Aerial modification of stem
d] None of all

248. Physical and chemical properties of soil its water and air contents are included in :
a] Climatic factors b] Edaphic factors
c] Biotic factors d] Topographic factors

249. Physiologically dry soil are :
a] Poor in moisture content
b] With adequate water
c] With saline water
d] Normal soils

250. Pigment which is responsible for skin colour is :
a] Melanin b] Glutin
c] Globulin d] Albumin

251. Pink (NN) flowered plants crossed with blue (nn) flowerd plants gave black (Nn) flowered plants in F_1 generation. If F_1 plants are selfed, which of the following will be the phenotype of F_2 progeny ?
a] All blue
b] 1/2 blue and 1/2 pink
c] All pink
d] 1/4 pink, 1/2 black, 1/4 blue

252. Plants grown from the seed of an open-pollinated variety of maize :
a] Vary in plant height as well as seed characters
b] Are uniform in plant height and seed characters
c] Vary in plant height but are uniform in seed characters
d] Are uniform in plant height but vary in seed characters

253. Plants that attach themselves to any neighbouring object often by means of some special devices and climb it are called :
a] Trailers b] Creepers
c] Climbers d] None of these

254. Plants that grown upon other plants but do not absorb food from them are called:
a] Parasites b] Epiphytes
c] Saprophytes d] Symbionts

255. Plants which feed upon insects and small animals absorbing only nitrogenous compounds are called:
a] Symbionts b] Epiphytes
c] Carnivorous d] Nocturnals

GK GS

'Sathiya Sodhanai' is the autobiography of Mahatma Gandhi.

256. **Plants which grow under average condition of temperature and moisture are :**
a] Hydrophytes b] Hygrophytes
c] Mesophytes d] Xerophytes

257. **Plasmid can :**
a] Self replicate
b] Integrate into host genome
c] Express the genes in it
d] Can do all the above

258. **Pneumatophores or respiratory roots are meant for :**
a] Absorption of salts
b] Absorption of water
c] Respiration
d] Photosynthesis

259. **Pointed gourd is a :**
a] Monoecious plant
b] Hermaphrodite plant
c] Dioecious plant
d] Polygamous plant

260. **Pollination taking place between the flowers borne by the same parent is called :**
a] Geitonogamy b] Alogamy
c] Xenogamy d] None of the above

261. **Pollination taking place between two flowers (bisexual or unisexual) borne by two separate plants of the same species is called :**
a] Self pollination b] Cross pollination
c] Xenogamy d] Geitonogamy

262. **Polygenes affecting the same trait, with each enhancing the phenotype are termed as :**
a] Amnions b] Alcaptonuria
c] Allosteric effect d] Additive factors

263. **Population improvement is used :**
a] To accumulate deleterious alleles
b] To maintain homozygosity
c] To accumulate favourable alleles
d] To increase heterozygosity

264. **Potato and Jerusalem artichoke are example of :**
a] Rhizome b] Tuber
c] Bulb d] Corm

265. **Probably the most useful measurable property in the soil -plant-air system is :**
a] Matric Potential b] Water Potential
c] Osmotic Potential d] Redox Potential

266. **Progeny of a hybrid plant :**
a] Breds true
b] Resembles maternal plant
c] Resembles paternal plant
d] Segregation

267. **Prop or still roots are found in :**
a] Acacia tree b] Tomato plant
c] Mango tree d] Banyan tree

268. **Protoplasm is a :**
a] True solution
b] Unsaturated solution
c] Suspension
d] Polyphasic colloidal system

269. **Pseudofertility may also occur late in the season as in:**
a] Tomato b] Tobacco
c] Tapioca d] Tea

270. **Pure line is :**
a] The progeny of selfed heterozygous plant
b] The progeny of selfed homozygous plant
c] The progeny of vegetatively progagated plant
d] The progeny of a nucleus seed

271. **Pureline selection is :**
a] A method of accumulating favourable alleles
b] Used to improve cross pollinated crops
c] Used to improve self pollinated crops
d] Used in vegetatively propagated crops

272. **Quarantine is :**
a] The name of a plant
b] The name of a character
c] The process of observation of alien material
d] None of the above

273. **Rainfall, temperature, light, wind and humidity are :**
a] Climatic factors b] Edaphic factors
c] Biotic factors d] Topographic factors

274. **Random assortment of genes will occur if the gene are located on :**
a] Same chromosome b] Different chromosomes
c] Sex chromosome d] Autosome

275. **Recessive lethal genes are carried in the :**
a] Heterogenous form b] Hetrozygous form
c] Homozygous form d] Homogenous form

276. **Recurrent selection is mostly used in :**
a] Cross pollinated crops
b] Self pollinated crops
c] Vegetatively propagated crops
d] Apomictic crop plants

277. **Reduction Division in microsporo-genesis occurs when:**
a] Pollen mother cells are produced from & porocytes
b] Sporocytes are produced
c] Ovules are formed
d] Four pollen grains are formed from a pollen mother cell

278. **Registered seed is directly produced form :**
a] Foundation seed b] Certified seed
c] Breeder's seed d] All the above

Jahandari was law made by the Sultans.

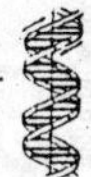

279. Replacement of purine by pyrimidine and *vice-versa* is known as :
a] Transversion b] Transduction
c] Translocation d] All the above

280. Restorer gene is needed
a] To make a plant correct
b] To bring fertility in male sterile plants
c] To make a plant resistant
d] To enhance the yield

281. Rice varieties IR 20, IR 26 and IR 36 have shown better resistance to:
a] Tungro b] Bacterial leaf blight
c] Bblast d] Brown plant hopper

282. Root bear :
a] Unicelluar hairs b] Multicellular hairs
c] Both d] None of the above

283. Root has :
a] Root cap
b] Region of cell division
c] Give proper anchorage to plant
d] All above

284. Secondary nucleus in the embryosac is :
a] Haploid b] Diploid
c] Triploid d] Hexaploid

285. Seeds of monocotyledons are :
a] Albuminous or non-endospermic
b] Ex-albuminous or non-endospermic
c] Both a & b
d] Neither a nor b

286. Seeds of pea, gram and beans are :
a] Albuminous b] Ex-albuminous
c] Both a & b d] None of the above

287. Selfing of a plant increases :
a] Heterozygosity b] Homozygosity
c] Homogenecity d] Heterogenecity

288. Sex in *Drosophila* is governed by the following mechanism.
a] XY methods b] OX method
c] ZZ method d] X autosomes ratio

289. Single gene affecting more than one character is :
a] Polymorphism b] Pleiotropy
c] Phenocopy d] Complete penetrance

290. Specific combining ability of the parents should be high in order to develop :
a] Hybrids b] Composites
c] Varieties d] Synthetics

291. Spikelets are characteristic of the family :
a] Umbelliferae b] Gramineae
c] Malvaceae d] Leguminoceae

292. Sporogenesis means :
a] Formation of zygote
b] Formation of embryo
c] Germination of seed
d] Development and formation of spores

293. Stems often bear :
a] Unicellular hairs b] Multicellular hairs
c] Both d] None of above

294. Sterility is most pronounced in the F_1 generation following :
a] Intervarietal hybridization
b] Intravarietal hybridization
c] Intraspecific hybridization
d] Interspecific hybridization

295. Sterility/ incompatibility of distant crosses can be overcome to produce hybrids, by :
a] Pollen culture b] Anther culture
c] Explant culture d] Embryo culture

296. Stilt roots are produced from :
a] Main root b] Secondary roots
c] Stem and branches d] None of above

297. Sudden and sharp change in individual and is directly inherited by the offspring is called :
a] Mutation b] Adaptation
c] Heredity d] Continuous variation

298. Sunflower is :
a] Annual herb b] Biennial herb
c] Perennial herb d] None of the above

299. Superiority of heterozygote leads to:
a] Fixation of recessive allele
b] Fixation of dominant allele
c] Maintenance of both the alleles
d] Depression of inbreeding

300. System of classification proposed by Hutchinson is :
a] Natural b] Artificial
c] Phylogenetic d] Numerical

301. Tetraploid potato has 48 chromosomes, whereas ordinary diploid species have __________ chromosomes.
a] 34 b] 24
c] 14 d] 10

302. The "Sonalika" variety of wheat has dwarfing genes derived from:
a] Rht 1-10 b] Stu 1
c] Norin -1 d] Norin -10

303. The 1st maize hybrid developed in India was :
a] Vikram b] Ganga-2
c] Vijay d] Kissan

304. The application of principles of genetics for the improvement of mankind is :
a] Herdity b] Linkage
c] Population genetics d] Eugenics

305. The attraction between two dissimilar molecules is known as :
a] Cohesion b] Adhesion
c] Surface tension d] None of the above

GK GS

Raindrops are spherical due to surface tensions.

306. The basic difference between a synthetic and composite variety lies in :
a] The difference in synthesis
b] Testing for picking ability
c] The mode of pollination
d] The number of components

307. The basis on which evolution works is :
a] Simpllcity b] Complexity
c] Variations d] Adaptation

308. The best known artificial system of classification is that promulgated by:
a] Sir Julian Huxley
b] Bentham and Hooker
c] Linnaeus
d] A.P. Candolle

309. The best method to break the unfavourable linkage is :
a] Single seed descent method
b] Diallel selective mating
c] Pedigree method
d] Bulk method

310. The best method to carry forward the maximum number of genotypes/genes to the advanced generations against the force of natural selection is :
a] Pedigree method
b] Bulk method
c] Single seed descent method
d] Back cross method

311. The botanical name of the onion is:
a] *Allium sativum* b] *Allum cepa*
c] *Pisum sativium* d] *Tagetes erecta*

312. The branch of genetics concerned with the study of frequencies of mating populations is :
a] Quantitative genetics
b] Behavioural genetics
c] Population genetics
d] Eugenetics

313. The cell sap of cell A has an osmotic potential of -14 bars and turgor pressure of 4 bars. Cell B has a water potential of - 16 bars and osmotic potential of -24 bars. The water movement will be :
a] From A to B b] From B to A
c] No movement d] None of the above

314. The chromosomal aberrations include:
a] Duplication or deficiency
b] Translocations or inversions
c] Multiplication or polyploidy
d] All of these

315. The chromosomes are attached to the spindle fibre by means of :
a] Pellicle b] Telomere
c] Satellite body d] Centromere

316. The commonest form of the ovule is :
a] Orthotropous b] Anatropous
c] Amphilotropous d] Campylotropous

317. The components of synthetic population would have already been tested for :
a] SCA b] Genetic advance
c] GCA d] Both GCA & SCA

318. The composite variety has an advantage over the hybrid variety in respect of :
a] Yield b] Synthesis
c] Homogeneity d] Seed production

319. The condition in which the anthers and stigmas of a bisexual flower mature at the same time is known as:
a] Cleistogamy b] Heterogamy
c] Syngamy d] Homogamy

320. The condition in which the pollen of a flower has no fertilizing effect on the stigma of the same flower is called :
a] Herkogamy b] Dichogamy
c] Dicliny d] Self-sterility

321. The definitive nucleus on fertilization gives :
a] Secondary
b] Zygote
c] Oospore
d] Primary endospermic nucleus

322. The diffusion of water through a differentially permeable membrane is called :
a] Osmosis b] Imbibition
c] Diffusion d] Water movement

323. The DNA is made up of the following four bases :
a] A.T.G.E. b] A.T.G.C.
c] A.T.U.C. d] A.G.U.C.

324. The edible part of the apple is :
a] Mesocarp and endocarp
b] Endosperm
c] Only mesocarp
d] Fleshy thalamus

325. The effect of the genotype of the pollen grain on the phenotype of the seed is termed as :
a] Apospory b] Pollinia
c] Endopolyploidy d] Xenia

326. The embroid (formed by vegetative cell) develops into a plant which, in fact, grows out of anther as in:
a] Tobacco b] Wheat
c] Barley c] Sorghum

327. The evolutionary history of a species is refered as :
a] Phylogeny
b] Progeny
c] Organic evolution
d] Natural selection

GK First 'baby-friendly state' in the world is Kerala. GS

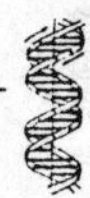

328. The factor that determines the movement of solutes through a membrane is :
a] Concentration gradient
b] Molecular size
c] Charge and active transport
d] All the above

329. The fibrous fruit of coconut is dispersed by :
a] Air b] Water
c] Animals d] None of the above

330. The first easily observed sign of a water deficit in a plant is decrease in the tugor of its :
a] Root hairs b] Stem cells
c] Xylem vessels d] Leaf cells

331. The first generation progeny of a hybrid are called:
a] Pure lines b] Segregates
c] Parents d] F_1 plants

332. The first inbreeding experiments with maize were those reported by:
a] Sprengel in 1866 b] Darwin in 1876
c] Sprague in 1886 d] Baird 1896

333. The first sunflower hybrid developed in India is :
a] MSFH-8 b] KBSH-1
c] APSH-11 d] BSH-1

334. The flattened, lateral, outgrowth of the stem or the branch, developing exogenously from a node and having a bud in its axil as known as :
a] Tendril b] Thorn
c] Leaf d] Flower

335. The fleshy edible part of the mango fruit is :
a] Epicarp b] Endocarp
c] Mesocarp d] Pericarp

336. The fleshy food storage tissue, lying immediately with in perisperm is called :
a] Endosperm b] Testa
c] Cruncle d] None of the above

337. The following character in human beings can explain the sex linked inheritance :
a] Baldness
b] Eye colour
c] Haemophilia
d] Pattern of the hair

338. The following organelles also contains DNA :
a] Endoplasmic reticulum
b] Centrosomes
c] Vacuoles
d] Chloroplast

339. The following part of the chromosome is responsible for the movement of chromosome/chromatid to different poles :
a] Chromatin b] Centromere
c] Chromomere d] Satellite

You Don't Know!

Earliest grain storage bins

Evidence of well-planned storage facilities for grain has been found at Harappa (2300-1750 B.C.). A typical bin had a floor area of 1,000 sq ft and was built on a raised podium. The bins were then systematically arranged in rows of six with a passage in between.

340. The formula for X^2 (Chi-square) is as follows :
a] $\sum\left(\frac{O-E}{E}\right)$
b] $\sum\left(\frac{O-E}{E}\right)^2$
c] (O-E)/E
d] (O-E)

341. The fruit of Acacia is :
a] Lomentum b] Double samara
c] Regma d] Cypsela

342. The fruit of cumin or coriander is known as :
a] Lomentum b] Cremocarp
c] Regma d] Cypsela

343. The fruits of cucurbitaceae is :
a] Pepo b] Pome
c] Berry d] Drupe

344. The fusion of a male gamete with two polar nuclei is often termed :
a] Fertilization b] Syngamy
c] Triple fusion d] Pollination

345. The genes that intensify or diminish the effect of major genes are known as __________ factors.
a] Modifying b] Nullifying
c] Falsifying d] None of these

346. The green expanded part of leaf is called as :
a] Leaf blade or lamina
b] Petiole
c] Pulvinous
d] Leaf base

347. The group of plants which bear flowers and seeds are known as :
a] Cryptogams b] Phanerogams
c] Pteridophyta d] None of the above

348. The haploid phase in plants is called:
a] Sporophyte b] Gametophyte
c] Zygote d] None of the above

349. The hydrophytes float on water because of the presence of :
a] Aerenchyma b] Parenchyma
c] Chlorenchyma d] Sclerenchyma

GK GS

India has 16.7 % of the world's total population.

350. The infloresence of banana is known as:
a] Catkin b] Raceme
c] Spike d] Spadix

351. The initial gene pool of a composite population is composed of :
a] Inbred lines
b] Isolines
c] Pure lines
d] Single plants derived from crosses and / or germplasm lines

352. The initial gene pool of a synthetic population is composed of :
a] Plants from an open-pollinated variety
b] Pure lines
c] Inbred lines
d] Isolines

353. The jointed stem with solid nodes and hollow internodes is called :
a] Caudes b] Culm
c] Scape d] None of above

354. The life first originated in :
a] Air b] Soil
c] Water d] All above

355. The life-cycle of an angiosperm is consist of
a] Sporophytic stage
b] Gametophytic stage
c] Both stages
d] None of above

356. The main body of ovule is called :
a] Hilum b] Chalaza
c] Nucellus d] Integuments

357. The main centre of sugarcane improvement in India is at :
a] Dehradun b] Lucknow
c] Coimbatore d] Bangalore

358. The main patterns concerning origin and domestication of cultivated plant species are :
a] Mendelian variation
b] Interspecific hybridization
c] Polyploidy
d] All of these

359. The main reason for the classification of plants is :
a] The difficulties arising in the study of plants because of their large number
b] Their mechanical separation
c] The nature of man
d] The environmental conditions

360. The main type of endosperm development is :
a] Nuclear type b] Cellular type
c] Both a & b d] None of above

361. The majority of halophytes show :
a] Mesomorphic character
b] Hydromorphic character
c] Xeromorphic character
d] None of above

362. The male sterile line is a cross to produce hybrid seed is known as :
a] A line
b] R line
c] C line
d] None of them

363. The mechanism of male sterility in self-pollinated crops can be successfully utilised to obtain :
a] Selfed seed
b] Mutated seed without radiation treatment
c] Hybrid seed without emasculation
d] Sterile seed

364. The Mendel's laws of inheritance were rediscovered in 1900 by :
a] Bateson and Punnet
b] Beadle, Tatum and Leaderberg
c] Watson and Crick
d] Correns, de Varies and Tschermak

365. The minute scaly outgrowths borne at the upper end of the leaf sheath, as in Gramineae is called :
a] Bract b] Bracteole
c] Stipule d] Ligule

366. The mode of arrangement of ovules in the cavity of ovary is known as :
a] Pollination b] Aestivation
c] Placenatation d] Fertilization

367. The mutation induced nullisomic variety of sugarcane is :
a] Co-8504 b] Co 8153
c] Co 527 d] Co 8152

368. The name of the botanist who recognised the importance of sex in plants is :
a] Birbal Sahni b] Carolus Linnaeus
c] T.N. Mitra d] Oswald Tippo

369. The net movement of a substance from an area of its own high concentration into another area of lesser concentration as result of kinetic energy of molecules is called:
a] Diffusion b] Osmosis
c] Imbibition d] Plasmolysis

370. The normal eye colour in *Drosphila* is red. But mutants occur having white eye and also different shades ranging between white and red which are all recessive to red, while white colour is recessive to all others. This phenomenon is due to :
a] Pseudodominance b] Pseudoallelism
c] Co-dominance d] Polygenes

371. The nucleus of the cell was first discovered by :
a] Robert Brown b] Robert Hooke
c] Strasburger d] Boveri

372. The nucleus of the functional megaspore divides :
a] Meiotically b] Asexually
c] Apomictically d] Mitotically

GK The city of Jaunpur was founded by Firoz Shah Tughlaq. GS

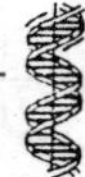

373. The number of chromosome in common wheat is :
a] 36 b] 38
c] 48 d] 42

374. The number of stamens is indefinite in :
a] Cruciferae b] Solanaceae
c] Papilionaceae d] Malvaceae

375. The nutritive tissue which is infact a remnant of the nucelus is called :
a] Endosperm b] Perisperm
c] Yolk d] None of the above

376. The one seeded fruit of rice or maize grain is called:
a] Berry b] Pepo
c] Siliqua d] Caryopsis

377. The only force responsible for the movement of water through a membrane is :
a] Molecular activity
b] Hydrostatic pressure
c] Osmotic pressure
d] All the above

378. The ovary of family liliaceae is :
a] Trilocular superior with axile placentation
b] Trilocular inferior with parietal placenation
c] Trilocular inferior with basal placentation
d] Trilocular superior with free central placentation

379. The particles of hemicolloides have size (diameter) round about:
a] Less than 1 mu
b] Between 1 mu and 100 mu
c] 1 mu
d] More than 100 mu

380. The phenomenon is called mesogamy, when the tube enters into the embryo sac through :
a] Micropyle b] Chalaza
c] Integuments d] None of the above

381. The physical basis of heredity is termed as :
a] Gene pool b] Gene frequency
c] Germplasm d] None of the above

382. The place on the stem or branch where one or more leaves arise is known as :
a] Internode b] Node
c] Apex d] Sheath

383. The placentation in mustard is
a] Marginal b] Parietal
c] Axile d] Basal

384. The plant that grew in very dry places and can withstand a prolonged period of drought uninjured are :
a] Hydrophytes b] Hygrophytes
c] Mesophytes d] Xerophytes

385. The plant which are growing in saline soils or saline water with prepondrance of salts in it are said:
a] Mesophytes b] Xerophyte
c] Hygrophytes d] Halophytes

386. The plants which grow in water or in very wet places are :
a] Hydrophytes b] Hygrophytes
c] Mesophytes d] Epiphytes

387. The plants which grow in water deserts are called :
a] Mesophytes b] Hydrophytes
c] Xerophytes d] Halophytes

388. The plants which occur in moist shady places in forests or in the moist soil near waterlogged localities are :
a] Hydrophyte b] Hygrophytes
c] Xerophytes d] Mesophytes

389. The plasma membrane is :
a] Impermeable membrane
b] Semi-permeable
c] Permeable
d] None of the above

390. The point of attachment of the body of the ovule is its funicle is known as :
a] Chalaza b] Raphe
c] Hilum d] Stalk

391. The pollen along with pollen tube is known as :
a] Microspore b] Pollengrain
c] Megaspore d] Male gametophyte

392. The popular variety, Ganga safed-2 of maize is a :
a] Single cross hybrid
b] Double cross hybrid
c] Three way cross hybrid
d] Double topcross hybrid

393. The pressure responsible for pushing the membrane against the cell wall is termed as :
a] Osmotic pressure
b] Suction pressure
c] Wall pressure
d] Turgor pressure

394. The process of formation of eggs in animals is called:
a] Oogenesis b] Telophase
c] Oogonium d] None of them

395. The process of formation of sperms in animals is known as :
a] Spermatogonium b] Spermatogenesis
c] Spematid d] None of them

396. The process of guttation is due to :
a] Leaf suction
b] Excess transpiration
c] Root Pressure
d] Wall Pressure

397. The process of making the expression of a non-allelic gene by another gene or gene pair is known as :
a] Epistasis b] Hypostasis
c] Nobilisation d] None of the above

398. The process of megasporogenesis occurs in :
a] Embryosac b] Egg cell
c] Pollensac d] Ovule

GK GS

Bihar has the lowest literacy rate.

399. The process, double fertlization was first discovered by :
a] Nawaschin b] Strasburger
c] Robert Brown d] Hugo de Vries

400. The production of double cross hybrid in maize was first suggested by :
a] Shull b] East
c] Jones d] Hull

401. The production of microspores and megaspores is known as :
a] Megasprogenesis
b] Microsporgensis
c] Gametogenesis
d] Sporogenesis

402. The production of sperms is known as :
a] Megagametogenesis
b] Microsporogenesis
c] Microgametogenesis
d] Megasporogenesis

403. The proportion of one allele as represented in a breeding population is known as :
a] Gene frequency b] Gene pool
c] Both of the above d] None of the above

404. The provision of light of shorter wavelength at the same time as longer red wavelengths ________ photosynthesis.
a] Maintains b] Enhances
c] Reduces d] Does not affect

405. The raw matrials required for photosynthesis in vascular plants are:
a] CO_2 and water
b] CO_2,water and mineral
c] CO_2, water, mineral salts and chlorophyll
d] CO_2, water and mineral salt

406. The removal of apical buds and young leaves ________ branching.
a] Decreases b] Increases
c] Maintains d] No effect

407. The replication of DNA is :
a] Conservative b] Non-conservative
c] Semi-conservative d] Direpersive

408. The rice variety containing "Dee-geo-woo-gen" gene is :
a] Indrasan b] Basmati
c] Tilak d] IR-8

409. The science dealing with variation and heredity is called :
a] Immunology b] Cytology
c] Genetics d] Radiology

410. The science which deals with the study of inter-relationship between the living organisms and the various factors of the environment surrounding them as:
a] Botany b] Biology
c] Ecology d] Zoology

411. The scientific name of bread wheat is :
a] *T. durum* b] *T. dicoccum*
c] *T. aestivum* d] *T. monecoccum*

412. The scientific name of maize is :
a] *Glycine max* b] *Hordeum vulgare*
c] *Zea mays* d] None of them

413. The segregation of individuals in the F_2 or in a later generation of a cross, which shows a more extreme development of a character than either parent is termed as :
a] Hybridisation
b] Heterosis
c] Linkage
d] Transgressive segregates

414. The series of changes in the vegetation of a pond lake, marsh or a stream are together known as:
a] Hydrosere b] Xerosere
c] Biosphere d] None of above

415. The series of changes in the vegetation of bare rocky beds, rocky hill slopes, sand beds into extreme scarcity of water are together known as :
a] Hydrosere b] Xerosere
c] Mesosere d] Xerophytes

416. The situation where an egg cell is developed into an embryo without fertilization, is decribed as :
a] Apomixis
b] Parthenocarpy
c] Sexual reproduction
d] Parthenogenesis

417. The slender stalk by which ovule is attached to the placenta is known as
a] Funicle b] Hilum
c] Raphe d] Chalaza

418. The stage of prophase occur in :
a] Mitosis b] Meiosis
c] Amitosis d] Free cell formation

419. The staminante ________ plants out yield pistillate plants by about 25 per cent.
a] Asparagus b] Lettuce
c] Spinach d] Hemp

420. The structure of chromosome can best be observed during :
a] Prophase b] Metaphase
c] Anaphase d] Telophase

421. The study of ecology from the standpoint of individual species is called :
a] Synecology b] Phytogeography
c] Autecology d] None of the above

422. The sugar found in DNA is :
a] Deoxyribose b] Ribose
c] Fructose d] Maltose

423. The sugarcane (*Saccharum officinarum* 2n = 80) variety are cytologically :
a] Autopolypoids
b] Octoploids
c] Segmental polyploids
d] Complex polyploids

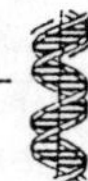

424. The surface tension in any liquid is ______ related to the temperature
a] Universally b] Directly
c] Indirectly d] Does not
425. The tap root system is normally meant to :
a] Absorb water and mineral salts from the soil
b] Conduct water and mineral salts to the stem
c] Give proper anchrage to plant
d] All above
426. The term epistasis was originally used by ______ in 1909 to describe genes whose effects mark or cover the effects of other genes.
a] Wagner b] Bateson
c] Hayman d] Atwood
427. The term genetics was coined by :
a] Gregor Mendel b] Bateson
c] T.H. Morgan d] H.J. Mujer
428. The term heterosis was first used by:
a] Mendel b] Shull
c] Bateson d] Punnet
429. The term used in sugarcane breeding to denote the crossing of *Sacharum officinarum* with related species followed by one or more back crosses is called:
a] Mass selection b] Crossing over
c] Cane improvement d] Nobilisation
430. The theory explaining the cause of variation and evolution, (Continuity of Gemplasm) was put forward by :
a] Charles Darwin b] Gregor John Mendel
c] Weismann d] Hugo de Vries
431. The theory of catastrophism was introduced by :
a] Christian chruch b] Cuvier
c] Lamark d] Darwin
432. The theory of gene s in the chromosomes was introduced by:
a] Mendel b] Morgan
c] Strasburger d] Hugo de Vries
433. The thoery of 'Natural Selection' was put forward in:
a] 1850 b] 1859
c] 1869 d] 1865
434. The topographic factors include :
a] Undulation b] Altitude
c] Slope d] All the above
435. The total volume of water on melting:
a] Increases b] Remains constant
c] Decreases d] None of these
436. The transference of pollengrain from the anther of a flower to the stigma of the same flower is known as:
a] Self pollination b] Cross pollination
c] Fertilization d] Mating

Knowledge World!
Only palace facade
The Hawa Mahal (the palace of winds) in Jaipur is a facade built onto a wall of the main palace compound. Royal ladies seated in to 953 niches and windows could look down at the main street below, specially the processions.

437. The transference of pollengrains from the anther to the stigma is called :
a] Fertilization
b] Double fertilization
c] Mating
d] Pollination
438. The turgor pressure is always equal and opposite in direction to:
a] Diffusion pressure b] Wall pressure
c] Osmotic pressure d] DPD
439. The two integuments of a ovule develop into :
a] Seed coats b] Disintegrate
c] Plumule d] Pericarp
440. The type of male sterility used in hybrid seed production in pearlmilet is :
a] Genetic male sterility
b] Cytoplasmic male sterility
c] Genetic cytoplasmic male sterility
d] None of the above
441. The usefulness of cortisome for medicinal purposes led to the identification of ______ a potentially rich source of it.
a] *Discorea* b] *Colocasia*
c] *Lantana* d] *Canabis*
442. The various modes in which the leaves are arranged on the stem or the branch, the term is called :
a] Phyllotaxy b] Phyllode
c] Phylloclade d] Cladode
443. The water potential of pure water is equal to :
a] Zero b] Always negative
c] Less than zero d] More than zero
444. Theory of 'Acquired characters' was proposed by :
a] Darwin b] Lamark
c] Bateson d] Weismann
445. Theory of 'Pangenesis' was given by:
a] Hugo de Vries
b] Charles Darwin
c] Lamark
d] Bateson
446. Theory of Natural Selection was based on :
a] Over-production of off springs and consequent struggle for existence
b] Variation and their inhertance
c] Elimination of unfavorable variations
d] All the above

GK GS
P.N. Bhagwati was the first Indian appointed Chairman of the New York based Human Rights Commission.

447. Theory of overdominance states :
a] Dominant homozygote is superior to recessive homozygote
b] Recessive homozygote is superior to dominant homozygote
c] Hetrozygote is superior to both the homozyote
d] None of the above

448. There are four stages of development of Genetics upto 20th century. The correct order of their occurrence is :
a] Darwinism, Lamarckism, Mendelism, Weismannism
b] Lamarckism, Darwinism, Weismannism, Mendelism
c] Lamarckism, Weismannism, Darwinism,Mendelism
d] Weismannism, Lamarckism, Darwinism, Mendelism

449. Thick and woody perennial climbers, met within forests, twine round tall trees and at their tops produce a canopy of foliage are :
a] Twiners b] Leaf climbers
c] Hook climbers d] Lianes

450. Thomas Andrew Knight first noted differences among wheat varieties in their resistance to:
a] Smut b] Bunt
c] Rust d] Wilt

451. Tissue culture in eucalyptus gave __________ seedlings per annum.
a] 4,000 b] 7,000
c] 10,000 d] 13,000

452. Tissue culture in teak produced ____________ seedlings per year.
a] 500 b] 1000
c] 1500 d] 2000

453. To produce hybrid seed we need :
a] A male sterile line, restorer, maintainer
b] A male fertile line, and maintainer
c] A maintainer and restorer
d] Only male sterile line

454. To which one of the following species does India's indigenous cotton belong ?
a] *Gossypium barbadense*
b] *Gossypium hirsutum*
c] *Gossypium arboreum*
d] *Gossypium herbacium*

455. Tomato fruit is a :
a] Drupe b] Berry
c] Siliqua d] Hesperidium

456. Transfer of resistance gene to high yielding self-pollinated varieties is achived by :
a] Hybridisation and pedigree selection
b] Recurrent backcrossing and selection
c] Hybridisation and bulk selection
d] Hybridisation and mass selection

457. Transmission of a gene in Bacterio-phase is known as :
a] Transversion b] Transduction
c] Inversion d] Fusion

458. Transmission of characteristics and qualities of parents of their offspring is known as :
a] Mutation b] Heredity
c] Adaptation d] Hybridization

459. Triticale is a cross between :
a] Wheat & barley b] Barley & rye
c] Wheat & rye d] Wheat & oat

460. Tropical Evergreen or Rain Forests occur in areas with annual rain fall:
a] Exceeding 2,000 mm
b] 1000 - 1,500 mm
c] 500 - 1,000 mm
d] Scanty

461. Umbel infloresence is characteristic of the family :
a] Mustard family b] Okra family
c] Pea family d] Coriander family

462. Unisexual flowers bear :
a] Only stamens
b] Only carpels
c] Either stamens or carpels
d] Both stamens and carpels

463. Varalaxmi' is a hybrid developed in:
a] Rice b] Cotton
c] Brinjal d] Brassica

464. Veticillaster inflorescence is characteristic of family:
a] Malvaceae b] Pedaliaceae
c] Liliaceae d] Labiateae

465. Vilmorin Principle of progeny testing stimulated ________ to devise Ear-to-Row Method of maize breeding in 1886.
a] H.L. Westover b] C.G. Hopkins
c] W.L. Brown d] F.D. Richey

466. Vivipary means :
a] Seed germinates inside the fruit while it is still on the parent tree
b] Seed germinates as usual
c] Seed dormancy is very long
d] None of above

467. Water potential concept was formulated by :
a] Peter Ray, 1960
b] M.H. Mathew, 1954
c] Otto Renner, 1915
d] W.C.Hoffer, 1920

468. Wheat crop is :
a] Cross pollinated b] Often cross pollinated
c] Self pollinated d] Anemophily

469. Oat is said to be :
a] Cross pollinated crop
b] Self pollinated crop
c] Mixed pollinatd crop
d] None of the above

GK GS
Rana Sanga defeated Ibrahim Lodhi at Ghatoli.

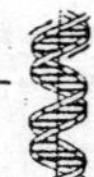

470. When a biological processes are made to occur outside the organism in test tube or vessel it is known as :
a] In vivo b] In vitro
c] In breeding d] Complete pentrance

471. When a cluster of leaves arises from the short underground stem as in pine apple, such leaves are said to be :
a] Radical b] Cauline
c] Pellate d] Centric

472. When a formation is dominated by one or more species which are abundant in it, is said to be :
a] Association b] Climax
c] Community d] None of the above

473. When a group of phenotypically similar appearing plant is selected and harvested and their seeds are bulked the process is known as :
a] Pedigree breeding
b] Mass selection
c] Bulk method of breeding
d] Pure line selection

474. When a plant heterozygous for two given characters Tt Rr, both of them are completely dominant, is self-pollinated, the resulting progeny will consist of :
a] Six phenotypes b] Four phenotypes
c] Three phenotypes d] Two phenotypes

475. When a Tall plant (Tt) in pea is crossed with another Tall plant (Tt), the proability of obtaining a dwarf plants is :
a] 1/2 b] 1/4
c] 1/8 d] 1/16

476. When an F_1 is crossed with the recessive homozygote parent the cross is known as :
a] Back cross b] Top cross
c] Test cross d] Single cross

477. When an individual is having both the alleles of contrasting characters, it is said to be :
a] Monoecious b] Homozygous
c] Dioecious d] Heterozygous

478. When breaks occur in two chromosomes simultaneously in a nucleus and the broken chromosomes rejoin in a new manner, it results in :
a] Dedeletion b] Duplication
c] Translocation d] Inversion

479. When homologous chromosomes fail to pair in prophase of meiosis, the phenomenon is known as :
a] Desynapsis b] Non-disjunction
c] Asynapsis d] Disjunction

480. When leaf fall of soon after it appears, then it is said to be :
a] Persistent b] Deciduous
c] Evergreen d] Caducous

481. When only one fruit develops from the single ovary of a flower it is said to be a :
a] Simple fruit b] Aggregate fruit
c] Multiple fruit d] Composite fruit

482. When stamens are united with the carpels, stamens are said to be :
a] Synandrous b] Syngenesious
c] Gynandrous d] None of above

483. When the anthers are united into a bundle or tube but the filaments are free, the stamens are said to be:
a] Syngenesious b] Synandrous
c] Monoadelphous d] Polyadelphous

484. When the flower can be divided in two equal halves by any plane, it is called :
a] Actinomorphic b] Zygomorphic
c] Hypogynous d] Epigynous

485. When the leaf is flat, with the blade placed horizontally showing a distinct upper surface and a lower surface, as in most dicotyledonous, it is said to be :
a] Peltate leaf b] Dorsoventral leaf
c] Isobilateral leaf d] Centric leaf

486. When the leaf is kidney shaped it is known as :
a] Acicular b] Linear
c] Lanceolate d] Reniform

487. When the leaf surface is smooth and free from hairs or outgrowth of any kind, it is said to be :
a] Rough b] Glutinous
c] Glabrous d] Glaucous

488. When the lobes of the base of the leaf partially enclose stem, the leaf is said to be :
a] Auriculate b] Amplexical
c] Semiamplexical d] None of above

489. When the male gametes are shed into embryo sac through micropyle, is called :
a] Porogamy b] Chalazogamy
c] Mesogamy d] Syngamy

490. When the ovary is one chambered and the placentae bearing the ovules develop on the inner wall of the ovary, then the placentation is known as :
a] Central b] Parietal
c] Axile d] Free central

491. When the ovule bends along the funicle that the micropyle lies close to the hilum, the ovule is said to be:
a] Orthotropous b] Anatropous
c] Amphilotropous d] Campylotropus

492. When the ovule is bent round like a horse-shoe it is said to be :
a] Orthotropous
b] Anatropous
c] Amphilotropous
d] Campylotropus

493. When the pistil has all the carpels united together it is said to be :
a] Apocarpous b] Syncarpous
c] Simple d] None of the above

GK GS

The length of national highway is longest in Madhya Pradesh.

494. When the stamens are united throughout their whole length by both the filaments and the anthers, they are said to be :

a] Syngenesious stamens
b] Polyadelphous stamens
c] Synadrous stamens
d] Diadelphous stamens

495. When two alleles of a gene are identical, the plant is said to be :

a] Recessive b] Dominant
c] Heterozygous d] Homozygous

496. When two genes show independent segregation, a genotype AaBb will produce ________ types of gametes.

a] 4 b] 3
c] 2 d] 1

497. When two organisms live together for mutual help to each other are called :

a] Saprophytes b] Symbionts
c] Parasites d] Epiphytes

498. When we place a living plant cell in a solution with an osmotic potential identical to that of its own cell sap the solution is :

a] Hypertonic b] Hypotonic
c] Isotonic d] None of the above

499. When we want to transfer a single gene from undersirable variety into a well adapted agronomically desirable variety the most suitable breeding method is :

a] Test cross b] Pedigree method
c] Back cross d] Recurrent selection

500. Which is a common method used in cross pollinated crops?

a] Mass selection b] Bulk method
c] Pure line method d] Pedigree method

501. Which is the dioecious plant ?

a] Maize b] Marigold
c] Vallisneria d] Rice

502. Which is the monoecious plant ?

a] Maize b] Datepalm
c] Rice d] Mango

503. Which of the following breeding procedures are not used for producing new varieties of self-pollined crop?

a] Recurrent selection
b] Mass selection
c] Pure line selection
d] Hybridization

504. Which of the following ions is involved in the opening and closing of stomata?

a] Mg^{2+} b] Fe^{2+}
c] K^{+} d] Na^{+}

505. Which of the following is not xerophytic plant?

a] Aloe b] Cacti
c] Gum tree d] Mango

506. Which of the following is ploidy level of common wheat ?

a] Tetraploid b] Hexaploid
c] Diploid d] Octaploid

507. Which of the following is sub-aerial modification of stem?

a] Corm b] Stolon
c] Thorn d] Bulb

508. Which of the following leads to the cross pollination?

a] Unisexuality b] Dichogamy
c] Heterostyly d] All the above

509. Which of the following material makes the protein chain?

a] Carbohydrates b] Sugars
c] Fatty acids d] Aminoacids

510. Which of the following scientists is called the father of genetics ?

a] Charles Darwin b] Robert Hooke
c] Gregor Mendel d] Gregor Morgon

511. Which of the following type(s) of male sterility is/are used in commercial seed production of double cross hybrid maize ?

a] Cytoplasm alone
b] Genetic alone
c] Cytoplasm and genetic
d] Cytoplasmic and genetic with restors

512. Which of the fruit is capsule ?

a] Beans b] Mustard
c] Cotton d] Wheat

513. Which one of the following categories of seeds is the farmer expected to buy afresh everytime?

a] Variety b] Hybrid
c] Synthetic d] Composite

514. Which one of the following chemicals is used to induce polyploidy ?

a] Ethyl methane sulphonate
b] Methyl methane sulphonate
c] Colchicine
d] Nitrous acid

515. Which one of the following improved crop varieties has been evolved through induced mutagenesis ?

a] IR-36 variety of rice
b] Aruna variety of castor
c] UP-262 variety of wheat
d] Varuna variety of mustard

516. Which one of the following is not dioecious plant ?

a] Datepalm b] Vallisnaria
c] Maize d] Papaya

517. Which one of the following is the site of protein synthesis ?

a] Ribosomes b] t-RNA
c] m-RNA d] r-RNA

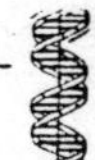

518. Which sterility is exploited in hybrid seed production:
a] Male genetic sterility
b] Cytoplasmic sterility
c] Cytoplasmic genetic male sterility
d] Genetic

519. Who established that character are transmitted from generation to generation through chromosomes?
a] Robert Brown
b] Robert Hooke
c] Nawaschin
d] Strasburger and Hertwig

520. Who founded the cell theory ?
a] Robert Hooke
b] Von Mohl and Nageli
c] Schwann and Schleiden
d] Grew and Malpigi

521. Who invented first compound microscope ?
a] Robert Hooke b] Jansen
c] Leeuwenhoek d] Von Mohl

522. Who is credited with the discovery of mitosis ?
a] Farmer and Moore b] Huxley
c] Flemming d] Robert Hooke

523. Who is regarded as the father of botany?
a] Theofrastus b] Gregor Mendel
c] Carolus Linnaeus d] Oswald Tippo

524. Who suggested the imbibition theory of ascent of sap?
a] Sir J.C. Bose b] Sachs
c] Strasburger d] Dixon and Jolly

525. Wind pollination is also known as:
a] Anemophyly b] Entomophyly
c] Zoophyly d] Hydrophyly

526. With four bases A, U, G and C the possible number of triplet codons are 64. If there is no uracil, the number of codons will be :
a] 3/64 b] 9/64
c] 27/64 d] None of above

527. With the increase in temperature, the solubility of solid in a liquid :
a] Increases b] Decreases
c] Both a & b d] None of the above

528. Zamikand and Taro (Colocasia) are example of :
a] Rhizome b] Tuber
c] Bulb d] Corm

GK GS

'My Days' is an autobiography of R.K. Narayan.

Matching

1. **Match the place of origin with the crops given below.**

a]	Tropical America	i]	Cotton
b]	Tropical N. America	ii]	Soyabeans
c]	Asia Minor	iii]	Wheat
d]	China	iv]	Maize

2. **Match the crops with their place of origin.**

a]	Tobacco	i]	Asia Minor or N. Africa
b]	Oats	ii]	Africa
c]	Potatoes	iii]	Tropical America
d]	Sorghum	iv]	South America

3. **Match the number of allelic pairs with kind of gametes possible in F_1.**

a]	2	i]	8
b]	3	ii]	1024
c]	4	iii]	4
d]	10	iv]	16

4. **Match the names of the aneuploids with the formula.**

a]	Nullisomic	i]	2x-1
b]	Monosomic	ii]	2x+1
c]	Trisomic	iii]	2x+2
d]	Tetrasomic	iv]	2x-2

5. **Match the crop with the chromosome number.**

a]	*Avene sativa*	i]	26
b]	*Gossypium arborium*	ii]	21
c]	*Gossypium barbadense*	iii]	7
d]	*Hordeum vulgare*	iv]	13

6. **Match the chromosome numbers with the crop.**

a]	16	i]	*Nicotiana sylvestris*
b]	12	ii]	*Trifolium alexandrinum*
c]	24	iii]	*Medicago sativa*
d]	8	iv]	*Nicotiana tabacum*

7. **Match the crop with chromosome number.**

a]	*Sorghum versicolor*	i]	8
b]	*Bromus carinatus*	ii]	5
c]	*Lotus corniculatus*	iii]	12
d]	*Trifolium hybridum*	iv]	28

8. **Match the gene/cytoplasm with the crop.**

a]	Norin gene and timophevi cytoplasm	i]	Rice
b]	Opaque-2 and T-cytoplasm	ii]	Pearlmillet
c]	Dee-gee-woo-gen	iii]	Wheat
d]	Tifton 23 A	iv]	Maize

9. **Match the following parts with the common term used.**

a]	Ovary	i]	Seed
b]	Ovule	ii]	Perisperm
c]	Definitive nucleus	iii]	Fruit
d]	Nucleus	iv]	Endosperm

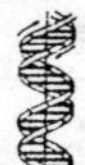

10. **Match the following plants with the type of apomixis.**

a]	Apple	i]	Adventive embryony
b]	Maize	ii]	Non-recurrent
c]	Citrus	iii]	Vegetative apomixis
d]	Grass species	iv]	Recurrent

11. **Match the manifestation of heterosis with the process it is associated with.**

a]	Molecular level	i]	Metabolic regulation and morphogenesis
b]	Metabolic path level	ii]	Growth and differentiation
c]	Cellular level	iii]	DNA duplication, transcription
d]	Organisms level and translation	iv]	Increased mitosis

12. **Match the following point mutation with acquired functions of the alleles.**

a]	Hypomorphs	i]	Impeding/antagonising the function of normal allele
b]	Hypermorphs	ii]	Allele that has acquired a new function
c]	Neomorphs	iii]	Normal functioning less efficiently
d]	Antimorphs	iv]	Allele functioning more efficiently than normal allele

13. **Match the symptoms with the disease which they are commonly associated with.**

a]	Necrosis	i]	Damping off
b]	Transformation	ii]	Foliage infection
c]	Hypertrophy and malformation	iii]	Green-ear disease
d]	Wilting	iv]	Ergot

14. **Match the following some of the most important events in plant breeding with the scientist(s) responsible for them.**

a]	Diverse germplasm as a source of variability	i]	G.F. Sprague and L.A. Tatum (1942)
b]	Conversion improvement and multiple convergent	ii]	M.T. Jenkins (1940)
c]	General and specific combining ability	iii]	S.K. Wu (1939)
d]	Simple recurrent selection techniques	iv]	F.D. Richey and G.F. Sprague (1931).

15. **Match the following crop varieties with their source of development/introduction in India.**

a]	Jaya and Padma (rice)	i]	CO_1 x Bonganhilo
b]	CO_2 (sorghum)	ii]	From Africa (introduced)
c]	Striga resistant sorghum	iii]	Sonora-64, isolated after irradiation
d]	Sharbati Sonora (wheat)	iv]	From a cross with TN-1 (from Taiwan)

16. **Match the following mutants (variety) with the crop.**

a]	Sharbati Sonora	i]	French bean
b]	TG-1	ii]	Wheat
c]	S-12	iii]	Peanut
d]	Pusa Parvathi	iv]	Tomato

17. **Match the following crops with their mutant varieties.**

a]	Rice	i]	Aruna
b]	Cotton	ii]	RDB-1
c]	Castor	iii]	MCU-7
d]	Wheat	iv]	IIT-60

18. **Match the range of concentration recommended of mutagens to treat cereal seeds.**

a]	0.05-0.30 M	i]	ENH and MNH
b]	0.015-0.02 M	ii]	EMS
c]	0.85-9.00 mM	iii]	dES
d]	1.20-14.00 mM	iv]	E1

19. **Match the spontaneous beneficial mutants with the crops wherein they occurred and later used in breeding.**

a]	GEB-24	i]	Sorghum
b]	CO2 and CO19	ii]	Potato
c]	ES593	iii]	Konamani Rice
d]	Red Triumph	iv]	Fingermillet

20. Match the stage of shoot differentiation in tissue culture of the following cereal crops.

a]	Root	i]	Maize
b]	Shoot just below first internode	ii]	Pearlmillet
c]	Mature/immature embryo	iii]	Sorghum
d]	Mesocotyl	iv]	Rice plants

21. Match the year of evolution of commercial hybrid with the field crops given below.

a]	Maize	i]	1969
b]	Grain sorghum	ii]	1965
c]	Pearlmillet	iii]	1921
d]	Wheat	iv]	1955

22. Match the crops with the commercial hybrid's year of release.

a]	1972	i]	Rice
b]	1968	ii]	Potato
c]	1975	iii]	Barley
d]	1980	iv]	Oats

23. Match the crop with the year of development of first commercial hybrid thereof.

a]	Soyabean	i]	1975
b]	Cotton	ii]	1972
c]	Sunflower	iii]	1933
d]	Sweet corn	iv]	1980

24. Match the following vegetable crops with the year of development of first commercial hybrid.

a]	Egg-plant	i]	1943
b]	Tomato	ii]	1939
c]	Onion	(iii)	1948
d]	Winter squash	(iv)	1950

25. Match the following crops with the year of release of its first commercial hybrid.

a]	Cabbage	i]	1961
b]	Broccoli	ii]	1964
c]	Carrot	iii]	1966
d]	Cauliflower	iv]	1954

26. Match the crop grains with their lysine content (mg/g N) as given below.

a]	Rice	i]	174
b]	Wheat	ii]	224
c]	Maize	iii]	237
d]	Pearlmillet	iv]	167

27. Match the crops with the tryptophan content (mg/g N).

a]	Rice	i]	65
b]	Wheat	ii]	78
c]	Pearlmillet	iii]	68
d]	Sorghum	iv]	122

28. Match the crop with varieties.

a]	Cotton	i]	Ratna
b]	Potato	ii]	Sujata
c]	Rice	iii]	Girija
d]	Wheat	iv]	Kufri Sinduri

29. Match crop varieties with different mutagenes.

a]	Rice CNM6	i]	Gamma rays
b]	Sugarcane Co997	ii]	X-rays
c]	Barley PL56	iii]	MMS
d]	Blackgram CO4	iv]	EMS

30. Match mutant varieties with its improved character.

a]	Jute (Padwa)	i]	Short stature
b]	Sesame (Kalika)	ii]	High yield
c]	Sorghum (Co-21)	iii]	Disease resistance
d]	Sugarcane (Co6608)	iv]	Stress resistance

31. Match the following.

a]	Certified seed	i]	Test of seed viability
b]	Growout test	ii]	Progeny of foundation seed
c]	Tetrazolium test	iii]	Initial seed of a variety
d]	Nucleus seed	iv]	Stress resistance

GK GS

National Bread Day : March 11.

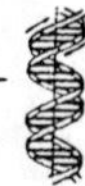

Glossary of Techincal Terms

'A' Line The fertile sterile parent in a cross being made to produce hybrid seed (male sterile line).

'B' Line The fertile counterpart parent of the 'A' line the maintainer line.

Acclimatization The adaptation or adjustment of an individual plant or a population of plants under the changed climate.

Allele or Allelomorph The alternative form of a gene.

Allopolypioidy A polyploid condition where all the sets of chromosomes are derived from the different species.

Amphiploid An individual originating by hydridization between species and possessing the total chromosome complement of the parent species. Generally produced by doubling the chromosome number of the F_1 hybrid plant.

Aneuploid An organism or a cell having one or few chromosomes more or less than the normal disomic number of the species, the chromosome number is not a multiple of the basic number (x).

Anthesis The process of dehiscence of the anthers, period of pollen distribution.

Antibody A substance that acts to neutralize a specific antigens in living organism.

Antigen Any substance usually a protein that causes antibody production when introduced into a living organisms.

Apomixis Seeds are formed and embryo develops without fertilization. The resulting plant will be identical in genotype to the parent plant.

Asexual Reproduction Reproduction without involvement of meosis and syngamy, resulting, in progeny which are gametically identical to the parents and each other. Also called vegetative reproduction.

Autopolyploidy An organism with more than two sets of chromosomes in body cells both sets derived from the same species.

Autosome Chromosome other than sex chromosomes, their kind and number is the same in males and females of a species.

Back-cross The cross of a hybrid to one of the parental types.

Biotype A population in which all individuals have an identical genotype.

Bivalent A pair of synapsed homologous chromosomes during first meiotic prophase.

Blending Inheritance Inheritance in which the offspring shows blending of the characters shown by the two parents.

Breeder's Seed Breeder's seed is directly controlled by the originating or sponsoring plant breeding institution, firm or individual, and is the source for the production of seed of the certified classes. It has the maximum genetic purity.

Centre of Origin Centre of origin of a species coincides with the areas where greatest diversity in that species.

Chasmogamy The flower opens but only after the pollination has taken place.

Chromosomal Aberrations Structural and numerical abnormalities in chromosomes.

Quiz Kit!

First Deputy PM

Sardar Vallabhbhai Patel was sworn in on Aug. 15, 1947 in the first cabinet of independent India which was headed by PM Jawaharlal Nehru.

Chromosomes These are thread like structures found in the nucleus and are carriers of genes.

Cistorn The functional (genetic) unit of DNA.

Cleistogamy The flower does not open at all. This ensures complete self pollination, since foreign pollen can not reach to stigma. Pollination and fertilization in an unopended flower bud.

Clone A group of plants originating by vegetative propagation from a single plant.

Codominance The condition in heterozygotes where both members of an allelic pair contributes to phenotype, which is then a mixture of phenotypic traits produced by individual alleles.

Codon A set of nucleotides that is specific for a particular amino acid in protein synthesis, generally consists of three nucleotides.

Co-incidence The observed frequency of double crossovers divided by their calculated or expected frequency.

Complete Flower The flower which possesses all the four floral whorls.

Complete Penetrance The ability of a gene to express itself in all the individuals which carry it in the appropriate genotype.

Cross Fertilization It is the union of an egg with a sperm from a plant of a different clone.

Cross pollination / Allogamy Pollen grains from flower of one plant are transfered to receptive stigmas of flowers of another plant.

Crossing Over A process where genes are exchanged between non-sister chromatids of a homologous pair and reunite with one another.

Cytogenetics A branch of biology dealing with chromosomes and their implications in genetics.

Cytokinesis Division of cytoplasm during cell division.

Cytoplasmic Inhertiance Inheritance dependent upon cytoplasmic structures and not nuclear genes.

Cytosine A pyrimidine base occurring in DNA and RNA pairs with guanine in DNA.

Deficiency or Deletion The loss of a part of chromosome involving one or more genes. If loss is terminal it is called deletion and if loss is interestetial then it called as deficiency.

Deoxyribose Nucleic Acid (DNA) A usually double stranded helically coiled nucleic acid molecule composed of phosphate deoxyribose backbones which are connected by paired bases attached to the paired bases to the deoxyribose sugar in the genetic material of all living organism.

Detassel Removed of the immature tassel as practiced in the production of hybrid seed corn.

GK GS

The First Indian-American to be elected to the U.S. Congress was Dalip Singh in 1956.

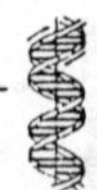

Diallel Cross Crossing of number of genotypes in all possible combinations.

Dioecious Condition in which the male and female reproductive structure are produced on separate plants.

Dominance In the herterozygote, one allele expresses itself at the expense of the other.

Double Cross A cross between two single crosses.

Emasculation Removal of anthers or killing of pollen grains of a flower without affecting the female reproductive organs in hermaphrodite flower.

Endosmosis The entrance of water of external environment inside the cell through osmosis is known as endosmosis.

Epistasis Interaction between non-allelic genes in which a gene or combination of genes exerts a dominant effect over another gene or combination of genes.

Eugenics The applications of principles of genetics for the improvement of mankind.

Euploid A cell or organism in which the chromosome number is an exact multiple of the basic number (x).

Evolution It is a process of change in organisms.

Fertilization Union of an egg with a sperm (gamete) to form zygote.

Foundation Seed The second link in the certified seed chain, produced from breeder seed and handled in such a way as to ensure genetic identity and varietal purity.

Fuzz The seeds and attached floral structures of cotton.

Gamete A cell which is capable of uniting with another gamete to produce a zygote or the one-celled stage of an embryo, therefore, the collective term for egg, sperm, and isogamete.

Geitonogamy Pollen grains from flowers of one plant fall on the receptive stigmas of the other flower of the same plant.

Gene A physiolological unit, carried on the chromosomes within the nucleus of each cell, which determines the hertiable characteristics of the cell.

Gene Sanctuary An area of diversity protected from interferences from man.

General Combining Ability (G.C.A.) Average performance of a strain in a series of crosses.

Genetic Erosion The gradual loss of variability by the cultivated forms and their wild relatives.

Genetics The science which deals with the study of heredity and variation.

Genome A complete set of chromosomes of a diploid species, all the members of a genome are district from each other in gene content and, often in morphology, members of a genome do not pair.

Genotype The genetic make up or constitution of an individual.

Germpalsm The sum total of hereditary material of genes present in a species.

Guanine A purine base occurring in DNA and RNA pairs normally with cytosine in DNA.

Gynandromorph An individual a part of whose body exhibits male sex characters and a part of female characters.

Haploid A cell or an organism having the genetic chromosome numbers.

Hemizygous A diploid organism having only one copy of a gene in place of the normal two.

Heredity The transmission of traits or genes from parents to offspring.

Heterobeltosis When F_1 is found superior to better parent.

Heterosis / Hybrid Vigour Occurrence of offspring more vigorous than either parent.

Heterozygote An individual having two different alleles for one or more genes.

Heterozygous Unlike genes paired together.

Homogenous Uniform, all members of a series being similar.

Homologous Chromosomes A pair of identical chromosomes, one paternal and one maternal.

Homozygote An individual having two copies of the same allele for one or more genes under consideration.

Hybrid The progeny obtained by crossing or mating two individuals having different genotypes.

Hybrid Variety First generation (F_1) from crosses between two purelines, inbreds, open pollinted varieties or other population, that are genetically dissimilar which are used for commercial planting.

Hybridization The method of producing new crop varieties in which two or more plants of unlike genetical constitution are together.

Idiogram Diagram representing the haploid set of an organism ordered in series of decreasing size of chromosomes.

Imperfect Flower The flower in which either androecium or gynoecium is absent.

In Vitro When biological processes are made to occur outside the organism in a vessel or test tube.

In Vivo Within the living organism.

Inbred Line A pure line usually originating by self pollination and selection.

Inbreeding Mating between two individuals related by descent i.e. having a common ancestor.

Incomplete Flower The flower in which one or more floral whorls are absent eg. maize.

Independent Assortment Segregation of two or more pairs of genes independently of each other.

Interferences A term used to explain that chiasma formation at a point reduces chances of chiasma formation in its vicinity.

Inversion Rearrangement of a chromosomal segment that reverse the order of genes on it.

Irradiation Breeding The use of irradiation to increase mutation rates for the purpose of obtaining mutual plants that may be useful in the development of improved variety.

Karyokinesis The division of nucleus into two daughter nuclei.

Karyotype The characteristics of entire set of chromosomes of an organism which describes chromosome number, size, shape and other features.

Lethal Gene An allele or gene which causes the death of all those individuals which carry it in the appropriate genotype, death occurs anytime before the individuals

GK GS

Paradeep on Orissa coast mainly handles iron ore and coal.

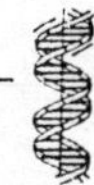

reach adulthood.

Linkage The tendency of genes to stay together during inheritance, due to the genes being located relatively closed to each other in the same chromosome.

Linkage Group A group of genes which are physically located on the same chromosome.

Locus The fixed position of a gene on a chromosome.

Male Sterility Inability to produce progeny usually due to nonfunctional male gametes.

Meiosis Cell division in which one cell with two sets of chromosomes gives rise to four cells with one set of chromosomes each. Reductional cell division.

Mericloning Vegetative multiplication through meristem culture.

Mitosis Cell division in which one cell gives rise to two cells both of them genetically identical to the cell which produced them. Equational cell division

Monoecious Condition in which the male and female reproductive structure are found on the same plant but in separate flowers.

Monogenic Governed by or involving a single gene.

Monohybrid The progeny derived by mating or crossing two individuals which differ for one gene or character

Monoploid An individual having one copy of a single genome, haploid from a diploid (2x) species.

Monosomic An individual having one chromosome less than the normal somatic complement of the species, denoted by 2n-1.

Multiple Cross Crossing of several pure lines together.

Multiple Genes Two or more different pair of alleles with a presumed cumulative effect growing quantitative traits.

Mutagen An agent capable of inducting mutation.

Mutation The sudden heritable changes in the characteristics of an organism other than those due to Mendelian segregation or recombination.

Muton A unit of DNA capable of undergoing mutation.

Nucleoside Portion of DNA or RNA molecule composed of one deoxyribose molecule (in DNA) a ribose (in RNA) plus a purine or pyrimidine.

Nucleotide Portion of DNA or RNA composed of one deoxyribose phosphate unit (ribose in RNA) plus a purine or pyrimidine.

Nullisomic An individual having one chromosome pair less than the normal somatic complement of the species.

Parthenogenesis The development of an individual from an unfertilized egg.

Pedigree Record Record in which we are maintaining the relationship between parents and offspring.

Perfect Flower The flower which possesses both androecium and gynoecium. It is also known as bisexual or hermaphrodite flower.

Phenocopy An individual whose phenotype has been modified by the environment to such an extent that is similar to that produced by another usually a mutant allele.

Phenotype External appearance of an organism as contrasted with its genetic constitution.

Phylogeny The evolutionary history of a species.

Phytogenetics The race history of a plant.

Pistillate Flower A flower bearing pistil only.

Plant Breeding It is the art as well as science of changing and improving the heredity of plant according to need and taste of consumer.

Plant Introduction The process of introducing plant from their growing locality to a new locality.

Pleiotropy A single gene affecting more than one character.

Polycross Open pollination in isolation among a number of selected genotypes.

Polymorphism Two or more phenotype maintained in a random mating population, due to the effect of environment.

Polyploid An organism with more than two sets (genomes) of chromosomes in its body cells.

Population All the individuals mating at random.

Population Genetics The branch of genetics concerned with the study of frequencies of genes and genotypes in random mating populations.

Pureline Progeny of a single homozygous self-pollinated plant.

Prepotency The capacity of a parent to impress characteristics on its offspring so they are more alike than usual.

Progeny Selection Selection based on the performance of the progeny.

Progeny Test A progeny or groups of progenies grown for the purpose of evaluating the genotype of the parent.

Propagule Plant part used for propagation.

Pyrimidines Nitrogenous bases occuring in DNA (thymine and cytosine) or RNA (uracil and cytosine).

Qualititative Character A character showing continuous variation, governed by polygenes one markedly affected by the environment.

Quantitative Character A character showing few distinct classes governed by one or few oligogenes, are relatively little affected by the environment.

Recessive Member of a pair of alleles which misses to express itself in the presence of its dominant allele and express itself only in the homozygous state.

Reciprocal Crosses The two crosses in which the same two parents are involved by the strain which serves as the male parent in one cross is used as the female parent in the other, and vice-versa.

Recombination New combination of genes or traits as a result of crossing over.

Recon A unit of DNA capable of recombination.

Recurrent Parent Parent to which hybrid material is crossed in a backcross.

Recurrent Selection A breeding system designed to increase the frequency of favourable genes for yield or other characteristics by repeated cycles of selection.

Ribonucleic Acid (RNA) A single stranded nucleic acid molecule synthesised in the nucleus from DNA composed of a phosphate ribose backbone with purines (adenine and guanine) and pyrimidines (uracil and

GK GS

Concave mirror is used as reflectors in Cinema projections.

cytosine) attached to the sugar ribose.

Seed Part of a plant used for raising a seed or commercial crop.

Segregation In genetics, the separation of two allelic genes into different meiospores or gametes during meiosis, also, the production by a hybrid offspring having different phenotypes.

Selection The process by which the individual plant or group of plants are sorted out from mixed population.

Self Fertilization Union of an egg with a sperm from the same flower or from another flower of the same plant or within a clone.

Self-Incompatibility The inability of fully functional pollen grains to effect fertilization and seed set on self-pollination, usually due to multiple alleles of a single gene S.

Self-pollinaton or Autogamy When pollen gains from an anther falls on receptive stigma of the same flower.

Sex Chromosome A chromosome which differs in number or kind between the male and female individuals of the same species and are involved in sex determination (Syn., allosome).

Sex-influenced Genes Genes that are expressed more frequently in one sex than in other e.g. pattern baldness in humans is dominant in males and recessive in females.

Sex-linked Gene Association of a gene or trait with sex. Genes for such traits are located in sex chromosome. In XY species gene are located only on the X chromsome.

Sib-mating Mating between individuals which are progeny of the same two parts.

Single Cross A cross between two lines.

Specific Combining Ability (S.C.A.) Deviation from performance of a cross combination from that predicted on the basis of general combining ability.

Staminate Flower A flower bearing stamens only.

Synapsis Pairing of homologous chromosomes.

Template A model or pattern DNA acts as a template that imparts its information to messenger RNA which if transcribes.

Test Cross The cross of an F_1 hybrid with an individual or strain having the recessive phenotype for the concerned trait.

Tetrasomic A cell or an organism having one chromosome pair in addition to the normal somatic complement of the species.

Three Way Cross A cross between a single cross (as female) and an inbred (as male).

Thymine A pyrimidine base occurring in RNA.

Totipotency It is the ability of a protoplast to regenerate into a whole organism.

Trait A morphological, anatomical or physiological feature of an organism; usually a product of the actions of both genotype and environment.

Transgenic Plants Plants in which a gene (or genes) from another organism has been introduced through genetic engineering and has become integrated in its genome.

Transgressive Variation Appearance in progeny of more extreme expression of a trait than occurs in the parents in the segragating generation.

Trisomic An individual having one extra chromosome of a set (2n + 1).

Uracil A pyrimidine base occuring in RNA.

Variation Occurrence of differences among various organisms.

Variety A population of plants having many characteristics in common, a variety may be a pure line, a mixture of pure lines, a Mendelian population, or a clone.

Wide Crossing Cross between two species of the same genus or of different genera.

Xenia Effect of the genotype of pollen grain on the phenotype of seed tissues.

Zygote The one-celled or beginning stage of an embryo, formed by fusion of two gametes and having, therefore, the diploid number of chromosomes.

GK **Athavana was land revenue department.** GS

8. Entomology

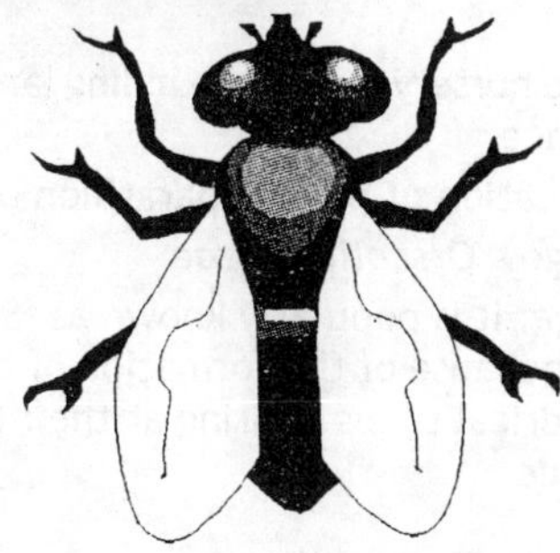

Each Word Counts

INSECT PESTS OF IMPORTANT CROPS

Sugarcane

1. Shoot Borer -*Chilo infuscatellus*

Nature of Damage: The Larva tunnels into the stem causing 'dead heart'. Its multiplication is favoured by high temperature and low humidity.

Life History :

The small pale grey brown moth with white hind wings. The egg, larval and pupal periods respectively 6, 25-30 and 6-8 days. The total life cycle is completed in 35 to 40 days.

Control: Infested tillers collected and destroyed. Avoid late planting.

2. Stem Borer (or) internode borer -*Chilo sacchariphagus indicus*

Nature of Damage: It occurs on sugarcane late in its growing phase. It affects juice quality.

Life History :

Egg (5-6 days) - Larva (37-53 days) - Bores (nodal region) -

Pupa (in leaf sheath) (8-10 days) - Adult (Pale brown)

Control: Release of egg parasite *Trichogramma australicum*.

3. Top Borer - *Scripophaga excerptalis*

Nature of Damage: It occurs in later stages of growth. Its infestation, is detected by presence of a reddish brown charred 'dead heart'.

Life History :

Egg (7-9 days) - Larva (35-50 days) - Pupa (inside stem) (8-10 days)

-Adult (Whitish in colour with buff coloured anal tuft) Life cycle 50- 70 days.

You Don't Know!

First Indian meteorologist

The first Indian, Lala Hem Raj, was appointed as imperial meteorologist on June 9, 1911.

Control :

- Egg masses collected and destroyed.
- Soil application of carbofuran at 2 kg a.i/ ha.

4. Sugarcane Leaf Hopper - *Pyrilla perpusilla*

Nature of Damage: It remains colonies on the lower surface of leaves and both the adults and nymphs suck up the sap causing yellowing and drying of leaves.

Life History :

Egg - Nymph (50-60 days) - Adult (brownish)

Control: Endosulfon 0.07% spray

Cotton

1. Cotton Aphid -*Aphis gossypii*

Nature of Damage: It infests tender shoots and the undersurface of leaves severe infestation results in curling of leaves, stunted growth and gradual drying and death of young plants.

Life History :

Nymphs - Adults (greenish brown) (7-9 days)

Control:

- Methylparathion 0.025 %
- Phosalone 0.05%

2. Cotton Leaf Hopper -*Amrasca devastans*

Nature of Damage: The nymphs and adults suck the plant sap from the undersurface of leaves. The leaves show symptoms of 'hopper burn' such as yellowing, curling, and bronzing.

Life History :

Eggs (4-10 days) - Nymphs (7-20 days) - Adult

Control :

- Methyl parathion 0.025 %
- Phosalone 0.05%

3. Cotton White Fly -*Bemisia tabaci*

Nature of Damage: The nymphs are found on the undersurface of leaves and drain of sap due to sucking in cases of severe infestation result in premature defoliation and development of sooty mould on honey dew excretion.

Life History :

Eggs (3-5 days) - Nymphs (9-14 days) Pupa (2-8 days) - Adult

GK **Black coloured object absorbs all seven colours.** GS

Control: Spray applications of phosalone 0.05% + neem oil 0.3%.

4. Spotted Boll Worm - *Earias insulana*

Nature of Damage: The larva causes drying of the shoot due to its feeding by boring into it. Heavy shedding of early formed flower buds due to this pest.

Life History :

Eggs (under surface of leaves) (3 days) - Larva (10-12 days) - Pupa (7-10 days) - Adult

Total life cycle -20-22 days.

Control :

- The infested portions of buds and bolls should be removed and destroyed.
- Syntheti.. pyrethroids like permethrin at 100-150 g ai/ha and cypermethrin at 80 g ai/ha.

5. Pink Boll Worm - *Platyedra gossypiella*

Nature of Damage: The larva enters the developing boll through the tip portion and the entrance hole gets closed up as the boll matures. The infested flower buds shed prematurely. A rosette shaped bloom ware found.

Life History :

Egg (4-25 days) - Pink larva (20-35 days) - Pupa (6-20 days) - Small dark brown moth

Control:

- Attacked parts collected and destroyed
- Fumigation of seeds with methyl bromide at 1.5 kg/ 100 cum for 24 hrs.

Rice

1. Rice Stem Borer - *Scirpophaga incertulas*

Nature of Damage: The larva feeds inside the stem causing drying of the central shoot or 'dead heart' in young plant and drying of the panicle or 'white ear' in older plant.

Life History :

Eggs (5-8 days) - Larva (dark brown head (35-40 days) - Pupa (inside the stem) (6-10 days) - Adult moth

Total life cycle 50- 70 days.

Control :

- Removal and destruction of rice stubbles from field.
- Collection and destruction of egg masses.
- Moths may collected by setting light traps.

2. Swarming Caterpillar - *Spodoptera mauritia*

Nature of Damage : The leaves are completely eaten away leaving behind the midribs of leaves. They feed mostly during night.

Life History :

Egg - Larva (20-25 days) - Pupa (10-15 days) - Adult (dark colour moth)

Control :

- Flooding the nursery makes the hiding larvae to come to the surface
- Spray application of methyl parathion 0.05%.

3. Rice Gall Midge - *Orseolia oryzae*

Nature of Damage: It is popularly known as 'Silver Shoot' or 'onion shoot' because of the formation of hollow pink, dirty white cylindrical tubes bearing at their tips a green reduced leaf blade.

Life History :

Eggs (3-4 days) - Maggots (15-20 days) - Pupa (2-8 days) - Adult (Yellowish brown fly)

Control:

- Spray application at 0.5/kg ai/ha are chlorfenvinphos and dicrotophos.

4. Rice Mealy Bug -*Heterococcus rehi*

Nature of damage: Its infestation causes round patches in a normal crop and plants have distorted panicles with chaffy grains.

Life History :

Eggs - Nymphs (17-35 days)

Control :

- The grass on the field bunds should be removed.
- The infested plants should be removed and destroyed.

5. Rice Leaf Roller - *Cnaphalocrocis medinalis*

Nature of Damage: The larva rolls the leaf blade by fastening its edges. In a severely infested field the whole crops give a sickly appearance with white patches.

Life History :

Eggs (4-7 days) - Larva (15-25 days) - Pupa (6-8 days) - Adult (Brownish orange colour moth)

Control :

- Removal of grass
- Periodical spray application of dicrotophos and chlorpyriphos at 0.5 kg ai/ha.

Wheat

Termites: *Odentetermes Obesus*

Nature of Damage: Attack young seedling, wither and die.

Control:

- Partially decomposed manure should be avoided apply Endosulfan 25 kg/ha.

Aphids: *Macrosiphum miscanihi*.

Nature of Damage: Nymphs and adults suck sap from leaves tender shoots, immature grain and result in poor yield

Control:

- Avoid heavy manure, and succulent crop, 0.02% dimethoate.

GK **The author of the book 'Success Vs Joy' is Geet Sethi.** GS

Stemborer - Sesamia inference

Nature of Damage: Dead heart

Control:

- Pullout and destroy dead heart's dust 0.05% sulfur.

Groundnut

1. Red Hairy Caterpillar -*Amascta albistriga*

Nature of Damage : The larvae feed the undersurface of tender leaflets and they also feed nature leaves leaving behind the petiole and mid-ribs of leaves.

Life History :

Eggs (2-3 days) - Larva (40-50 days) - Pupa - Adult

Control :

- Pupae collected and destroyed during summer ploughing.
- Putting light traps to attract moths.
- Larval parasites like *Apanteles flavipes* and *Apanteles bosei*.

2. Leaf minor - *Stomoperyx subrecivella*

Nature of Damage: The larva mines into tender leaf lots and feeds on the tissue. In a severely-infested crop the whole field presents a burnt-up appearance.

Life History :

Eggs (3 days) - Larva (14 days) - Pupa - Adults

Control :

- Putting light traps to attract adults.
- Spraying Fenitrothion 0.025% or Phosalone 0.05%.

3. Stem borer - *Sphenoptera perotetti*

Nature of Damage: The grubs of the dark brown shining beetle bore into the stem of groundnut plants and cause wilting and drying of plants.

Control:

- The infested plants should be collected and destroyed.

Coconut

1. Rhinoceros Beetle - *Oryctes rhinoceros*

Nature of Damage: The adult beetle cut across the leaf in the leaflets. Frequent infestation results in stunting of trees and death of growing point in young trees.

Life History :

Eggs (8-15 days) - Larval grub (90-180 days) - Pupa (10-25 days) - Adults

Life cycle -3 to 8 months.

Control :

- Decaying trunk of trees should be burnt
- Castor cake at 1 kg soaked in water in small mud pots attracts beetles.

2. Blackheaded Caterpillar - *Opisina arenosella*

Nature of Damage: Larvae live on the undersurface of leaf lots within galleries of silk material and feed by scraping the green matter.

Life History :

Eggs (1-5 days) - Larva (40 days) - Pupa (12 days) - Adult (Greyish white small moth)

Control :

- For young trees spray carbaryl 0.1 %.
- Releasing parasitoid such as *perisierola* and *Trichospilus pupivora*.

3. Red Palm Weevil - *Rhyncophocus ferrugineus*

Nature of Damage: It attack the central shoot shows sign of wilting. In the grown up trees the crown region is infested.

Life History :

Egg (2-5 days) Grub (35- 75 days) Pupa (12-35days) Adult (reddish brown weevil)

Control :

- Damaged palms should be destroyed.
- The infested portion should be scooped out.

PESTS OF VEGETABLES

Brinjal

1. Shoot and Fruit Borer - *Leucinodes orbonalis*

Nature of Damage: The larva bores into tender shoots in the early stage and causes 'dead hearts'. It also bores into flower buds and developing fruits.

Life History : The moth has brownish and red markings on the whitish forewings. The female lays eggs on tender shoots and developing fruits. The egg, larval and pupal periods respectively 5, 15 and 6-8 days.

Control:

- The damage portions should be removed.
- Periodical spray application of phosalone .07%.

2. Stem Borer - *Euzophera perticella*

Nature of Damage: The larva bores into the stem and covers the hole with excreta. The plants become stunted in growth and the infestation is generally noticeable in the later stage of crop growth life history.

The moth has greyish brown forewings and transverse lines in the middle wings. The yellowish larva with a red head. Life cycle ranges from 26 to 55 days.

Knowledge World!

Largest limestone cave

Siji cave in the Garo Hills, Meghalaya, contains caverns, the largest of which is 426.83 sq m (1,400 ft). It is 60.98 m (200 ft) long and 30.49 m (100 ft) broad.

GK **Babur won the battle of Panipat by using the Tulguma system.** GS

Control:

The attacked plants should be removed; periodical application of carbaryl 0.1 %.

3. Spotted Leaf Beetle - *Epilachana vigintioctopunctata*

The grubs and adults scrape the leaves and feed.

Life History : The female lays eggs in group of 10 to 20 on the undersurface of leaves. The egg period is 2-4 days. The yellowish spiny grubs become full grown in 10-20 days. The total life history takes 15-40 days.

Control:

Spray application of carbaryl 0.1 % malathion 0.1%.

4. The Grey Weevil - *Myllocerus subfasciatus*

Nature of Damage: The adult beetles feed on leaves of brinjal, the grubs feed on roots and cause wilting and death plants.

Life History : The weevil lays eggs in the soil. The incubation period is about a week. The grubs become full grown in 2 months. The adults emerge in 10-20 days.

Control:

The adult beetles should be collected and destroyed or killed by dust application of carbaryl.

Cucurbits

1. Snake Gourd Semilooper - *Plusia peponis*

Life History : The brownish moth has shiny brown fore-wings. The caterpillar is greenish with white longitudinal lines. The pupa is greenish but turns dark brown before the emergence of the adult moth. Life history occupies around 1 month.

Control:

- The larvae when found in small numbers may be hand-picked and destroyed. Spray application of parathion .025% controls the pest.

2. Snake Gourd Stem Weevil - *Baris trichosanthis*

The small black weevil feeds on the green leaves. It lays eggs in nodes. The incubation period is 5-6 days. The grubs bore into the stem and cause damage.

3. Fruit Flies - *Dacus cucurbitae*

Nature of Damage: It is slightly bigger with brown marking on wings. The maggots feed on the contents of fruits and cause premature dropping of fruits. The insects attack the fruits of snake gourd, bitter gourd etc.

Life History :

It lays eggs on flowers or tender fruits. The maggots that hatch out from the eggs in 2-9 days feed on the seeds of the fruit. The larval and pupal periods occupy 3 and 5 days respectively. The total life history occupy 12 to 35 days.

4. Pumpkin Beetles - *Raphidopalpa foveicollis*

Nature of Damage: The beetles bite holes on leaves and also feed on flowers. The roots as well as the stem and fruits that come in contact with the soil are damaged by the grubs.

Life History : The brownish eggs are laid in the soil. The egg period is 5 to 8 days. The grubs become full grown in 13-25 days and pupae in the soil. The pupal period from 7 to 17 days.

COLE CROPS

Cabbage

I. The Diamond Black Moth - *Plutella xylostela*

The larvae bite leaves and cause serious damage.

Life History : The moth is greyish brown with narrow wings having pale white markings anteriorly which form diamond-like white patches. It lays eggs on the undersurface of leaves. The larva is greenish. The period from egg to adult is 25 to 35 days.

2. The Cabbage Borer - *Hellula undalis*

Nature of Damage : The larvae web the leaves or bore into stem and cause damage.

Life History :

The moth lays eggs on the leaves and hatch in about 4 days. The brownish larva has a black head. It becomes full grown in about 9 days and pupates in the soil. Pupal period is about 6 days.

Radish

1. Leaf Webber - *Crocidolomia binotalis*

Nature of Damage: The larvae remain on undersurface of leaves in webs and feed on them. They also attack flower buds and pods.

Life History: The small moth with light brownish fore-wings lays eggs in masses. The incubation period ranges from 5-15 days. The larval becomes full grown in 20-25 days. Pupates emerges as adult in 14 to 40 days.

2. Painted Bug - *Bagrada cruciferarum*

Nature of Damage: The nymphs and adults suck the sap from leaves, shoots and pods.

Life History : The bug is small, black in colour with red and yellow lines. It lays barrel-shaped brownish eggs. The egg and nymphal periods take 1 and 3 weeks respectively.

STORAGE PESTS

Excess moisture and mould, insects and rodents are known to be responsible for damage and loss to stored products.

Internal Feeders

The larvae feed entirely within the kernels or grain or stored material.

1. Rice Weevil - *Sitophilus oryzae*

The weevil is dark brown. The female makes a small hole on the grain, deposits an egg. The grub feeds inside the grain, pupates there itself and emerges through an

irregular hole made on the grain.

2. The Lesser Grain Borer or Paddy Borer Beetle - *Rhizopertha dominica*

The dark brown beetle has its head bent under the thorax and the posterior abdominal end blunt. The grubs develop within kernels.

3. The Tobacco Beetle -*Lasioderma serricorne*

The brown round beetle has its thorax and head bent downward. The creamy white oval eggs are laid on the surface of stored material. The whitish hairy grubs feed on stored tobacco, ginger, turmeric and chillies.

4. Pulse Beetle - *Callosobruchus chinensis*

The beetle is small, dark brown. White elongated eggs are laid on the pulse. The grubs feed on the inner contents of pulses and pupate inside them. The egg, larval and pupal periods are 3, 12 and 4 days respectively.

5. Angoumois Grain Moth - *Sitotroga cerealella*

The caterpillars feed on the internal contents of grains and pupate inside the grains. It inflicts severe damage to unhusked paddy.

External Feeders

The larvae and adults feed on the grains from outside.

1. The Rice Moth - *Corcyra cephalonica*

The adult moth is greyish brown. The caterpillars web together the grains and feed within. It attacks broken grains and flour especially the milled products.

2. The Indian Meal Moth - *Plodia interpunctella*

The moth has brown fore wings with a white band. It lays about 300 eggs in clusters and the life cycle is completed in about 5-6 weeks.

3. Fig Moth - *Cadra cautella*

The small moth has greyish fore-wings. The caterpillars web together with the grains and feed on them.

4. The Khapra Beetle - *Trogoderma granarium*

The reddish brown beetle. The yellowish brown grubs are clothed with long hairs. It attacks stored wheat.

Secondary Pests and Scavengers

The larval or adults feed on broken grains which are out of condition and damp.

1. Saw-toothed Grain Beetle -*Oryzaephilus surinamensis*

The dark brown beetle has a row of saw-like sharp teeth on each side of the prothorax. It feeds by scarving of grain surface or burrowing holes in them.

2. Long-headed Flour Beetle - *Latheticus oryzae*

It occurs as a secondary infestation in stored wheat. The beetle is light reddish.

3. Flat Grain Beetle - *Cryptolestes minutus*

Measures for control of stored products

Preventive Methods :

- Before infestation grains should be cleaned and made free from insects.

Quiz Kit!

Earliest forest legislation

The Indian Forest Act was first enacted in 1865.

- Disinfestation is done by dusting BHC 10%.
- Material should be dried well in the sun.
- The grain bags should be stacked in such a way as to allow proper ventilation and sufficient moving space for periodical inspection.
- Malathion at 15 ml of 50% EC or 4.5 lit of water may be sprayed as a thin film on bags.

Curative Methods :

- The stocks of grains should be periodically examined and if insects are noticed to appear the stock should be removed; sun dried and cleaned.
- Heavily-infested material should be fumigated with hydrogen phosphide and methyl bromide.
- Seed material may be mixed with insecticides like DDT malathion.
- For tobacco beetle aluminium phosphide fumigation at 20-30 tablets/1000 cft for 100 hrs.

INSECT PEST CONTROL MEASURES

Biological Control

"The study and utilization of parasitoids, predators and pathogens for the regulation of pest population densities". In biocontrol, pest population is reduced to a level, that no longer of economic or health concern, yet leaves sufficient pest to allow survival of the control organisms. The biocontrol organism maintains its own population and prevents the pest from reaching to damaging level.

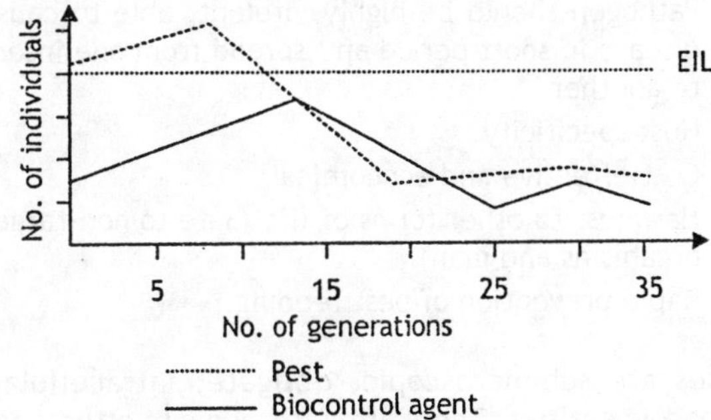

Factors Governed Biological Control

- Tolerance limit of injury: It is successful in crops with high tolerance limit.
- Crop value: It is successful in crops with high economic value.
- Crop duration: Long duration crops are highly suitable e.g. Sugarcane, fruit trees.
- Indigenous or exotic nature of pest: e.g. For cotton pests identify indigenous natural enemy by survey.

GK GS

Railways are an extremely efficient form of transportation.

- Level of control provided by the natural enemy on different stages of the pest (Egg-larval-pupal).
- Nature of pest complex (seasonal pest or perennial pest)
- Availability of selective insecticides.
- Insecticide resistant strains of natural enemies.
- Identical situation for successful control.

Qualities of an Effective Natural Enemy

- Host searching capacity
- Host specificity
- Adaptability
- Dispersal ability
- Amenability to culturing
- Ability to withstand competition
- Ability to outnumber the pest
- Survival capacity

Steps Involved in Classical Biological Control

- Conservation and encouragement of indigenous natural enemies
- Importation of exotic natural enemies
- Augmentation (mass rearing and release)

Microbial Control

Definition

"The control of pests by the use of micro-organisms such as viruses, bacteria, protozoa, fungi, rickettsia and nematodes or their by-products" is known as microbial control. These microbes can be grouped as **ingested microbes** (bacteria, viruses, rickettsia and protozoa) which enter the insect body with food (like stomach insecticide) and **penetrating microbes** (nematode and fungi) that enter by penetrating the integument (like contact insecticide).

Desirable Attributes

- Pathogen should be highly virulent, able to cause disease in short period and spread from one insect to another.
- Host specificity.
- Cost effective and economical
- Harmless to other forms of life (Safe to non-target organisms and man).
- Rapid prevention of pest feeding.

Viruses

Viruses are submicroscopic, obligate, intracellular, pathogenic entities. These are pathogenic to arthropods belongs to atleast eleven families.

Nuclear Polyhedrosis Virus (NPV)

Important features of NPV are -

- Occluded (rod shaped) singly or in groups in polyhedra (many sides) inclusion bodies.
- Site of multiplication is cell nucleus of epidermis, fat body, blood cells and trachea.
- Tree top disease: Diseased larvae not able to feed, move to the top of the tree or plant and hang inverterdly and die. (*e.g.*) NPV of *Spodoptera*, *Helicoverpa*.

Mode of Entry

The virus should be ingested to produce the disease. Due to alkaline gut juice, the virions are liberated from the polyhedral coat which attack nuclei of cells of tissues viz., fat body tracheal matrix, haemocytes, sarcolemma of muscles, neurilemma and nerve cells of anglion and brain.

Symptoms

Insects become dull in colour, feeding rate is reduced and larvae become pinkish white especially in the ventral side due to accumulation of polyhedra. In advanced stage larvae become flaccid, the skin becomes very fragile and eventually rupture. Diseased larvae hang upside down from the plants. This is called tree top disease.

Incubation period

4-6 days depending upon the stage of the infection, weather conditions and dose of virus.

Host Plant Resistant (HPR) in Pest Management

Methods of developing resistant varieties

Screening of available germplasm

Available germplasm collections are sown at a time in a single row in a location where there is a moderate to heavy incidence and evaluated by comparing with susceptible and resistant check varieties.

Selective Screening under Natural Infestation

Select promising lines from general screening and screen under natural condition in a single row or 2-3 rows in replicated trials.

Selecting Screening under Artificial Condition

To test true resistance, the selected varieties are screened under artificial condition in which insects are bombarded over the plant. Results are compared with a resistant and susceptible check. Breeders start screening of lines or cultivars from F_2 to F_6 stages for yield and resistance. If found suitable they will be forwarded to multilocation trial (MLT) and for Adaptive Research Trials (ART). If a line or cultivar is succeeding in all stages, it will be released as a variety.

Resistant Varieties

Some examples of resistant varieties are given below:

Rice –

Yellow stem borer : Vikas, Ratna,

Brown plant hopper : IR 64, IR 36

Sugarcane –

Early stem borer : CO 421, CO 661

Top shoot borer : CO 745, CO 6515

Padikaval was selling of policing rights.

Cotton –

Boll worm	:	Abhadita
Spotted boll worm	:	Hopi, Deltapine

Compatibility of HPR with other Pest Management Practices

HPR easily combines with other tactics of pest management, due to following reasons.

Resistant host will make the pests weak, or will suppress the population, so that less quantity of chemicals or botanicals or lesser rounds of chemicals or botanicals will control the pest. Resistant host will make the pests weak or will suppress the population, so that efficiency of predators, parasitoids or microbes (biological agents) on these population is increased.

e.g.,

- American root stocks of grape vine resistant to aphid *Phylloxera vitifolia.*
- Rice varieties IR 60, IR 36 are multiple resistance sources, PY3, IR 26 are BPH resistant and MDU3 gall midge resistant.

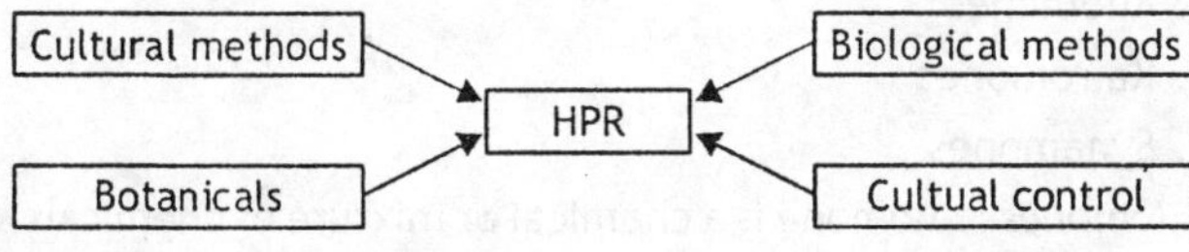

Merits

- Specificity
- Cumulative effect
- Ecofriendly
- Easily adoptable
- Effectiveness
- Compatibility

Demerits

- Time-consuming
- Genetic limitations
- High initial cost

Pheromones

Semi-chemicals are chemical substances that mediate communication between organisms. Semiochemicals may be classified into Pheromones (interspecific semiochemicals) and Allelochemicals (intraspecific semiochemicals).

Pheromones are chemicals secreted into the external environment by an animal which has a specific reaction in a receiving individual of the same species. Pheromones are volatile in nature and they aid in communication among insects.

Based on the responses elicited pheromone can be classified into 2 groups.

Primer Pheromones: They trigger off a chain of physiological changes in the recipient without any immediate change in the behaviour. They act through gustatory (taste) sensilla. Example, reproduction in social insects like ants, bees, wasps and termites are mediated by primer pheromones. These pheromones are not of much practical value in IPM.

Releaser Pheromones: These pheromones produce an immediate change in the behaviour of the recipient. Releaser phermones may be further subdivided based on their biological activity into :

1. Sex pheromones
2. Aggregation pheromones
3. Alarm pheromones

Releaser pheromones act through (smell) sensilla and directly act on the central nervous system of the recipient and modify their behaviour. They can be successfully used in pest-management programs.

1. Sex Pheromones

Sex pheromones are released by one sex only and trigger behaviour patterns in the other sex that facilitate in mating. They are most commonly released by females but may be released by males also. In over 150 species of insects, females have been found to release sex pheromones and about 50 species males produce. *Aphrodisiacs* are substances that aid in courtship of the insects after the two sexes are brought together. In many cases males produce aphrodisiacs.

Major differences between male and female produced pheromones are listed below:

Properties	*Female sex pheromone*	*Male sex pheromone*
Range	Acts at a long range Attracts males from long distance	Acts at a short distance
Action elicited the other sex	Attracts and excites males to copulate	Lowers females resistance to mating
Importance	More important	Less important

Insects orders producing sex pheromones : Lepidoptera, Orthoptera, Dictyoptera, Diptera, Coleoptera, Hymenoptera, Neuroptera and Mecoptera. In Lepidoptera, sex pheromonal system is highly evolved.

Pheromone producing glands: In Lepidoptera they are produced by reversible glands at the tip of the abdomen

> **Did You Know ?**
> ***Largest waterlogged area***
> **The largest waterlogged area in the country is Salt Lake in West Bengal (2,34,000 ha).**

of the females. The posture shown during pheromone release is called *'calling position'*. Aphrodisiac glands of male insects are present as scent brushes or hair pencils at the tip of the abdomen (e.g.. Male butterfly of *Danaus* sp.).

Pheromone reception: Female sex pheromones are usually received by olfactory sensilla on male antennae and males search upwind, following the odour corridor of the females. In pheromone perceiving insects, the antennae of male moths are larger and greatly branched than female moths to accommodate numerous olfactory sensilla.

Chemical nature of sex pheromones: In general pheromones have a large number of carbon atoms 10 to 20 and high molecular weight 180 to 300 Dalton. Narrow specificity and high potency are two attributes which depend on long chain carbon atoms and high molecular weight. But since pheromones are volatile their molecular weights cannot be very high as they cannot be carried by wind.

2. Aggregation Pheromones

Pheromones which induce aggregation or congregation of insects for protection, reproduction and feeding or combinations of these are called aggregation pheromones. They are released by either male or female which attract both male and female of the same species. Aggregation pheromones are known mostly in Coleoptera and Dictyoptera.

Females of bark beetle *Dendroctonus frontalis* produce aggregation pheromone which attract both sexes. Similarly males of Phloem beetle *Ips confusus* incorporate their pheromone in faecal matter which attract both male and female to the infested tree. Some common names of aggregation pheromones are :

Pheromone	***Insect***
Frontalin	- *Dendroctonus frontalis*
Ipsenol	- *Ips confusus*
Periplanone	- *Periplanata americana*

Aggregation pheromones of following insects are used for monitoring or mass trapping the pests in India.

Coconut rhinoceros beetle: *Oryctes rhinoceros*

Red palm weevil: *Rhynchophorus ferrugenius*

3. Alarm Pheromones

Alarm pheromones are chemical substances released by insects to warn members of the same species about the presence of or attack by an enemy. This warning elicits different behaviour in different insects as listed below:

- Dispersion or escape in aphids and bugs
- Aggression in ants and soldier termites
- Attraction in wasps and worker bees

Alarm pheromones have been reported in Hemiptera, Isoptera and Hymenoptera.

Organs producing alarm pheromones in various insects are listed below:-

Aphids	-	Cornicles
Termites	-	Cephalic glands
Worker bees	-	Sting glands
Ants	-	Anal, mandibular and Dufours's gland

The chemical nature of alarm pheromones are terpenese (aphids), aldehydes (hemipteran) and formic acid (ants). Alarm pheromones can be used in IPM for controlling aphids. When alarm pheromone of aphids is sprayed on plants, the aphids try to escape, fall down from the plants and get killed.

Allelochemicals

Allelochemicals are intraspecific semiochemicals. They mediate the communication between two different species of organisms or insects. Allelochemicals may be classified into :

1. Allomones
2. Kairomones
3. Synamones

Allomones : Allomone is a chemical or mixture of chemicals released by one organism that induces a response in another organism which is advantageous to the releaser. For example the defensive secretions of insects and plants that are poisonous or repugnant to attacking predators.

Ants release a defensive allomone called citral from its mandibular glands.

Kairomones : Kairomone is a chemical or mixture of chemicals released by one organism that induces a response in another organism which is advantageous to the recipient. The releaser turns out to be lower and the recipient (parasite or predator) a gainer.

Synamones : Synamone is a chemical or mixture of chemicals released by one organism that induces a response in another organism which is advantageous to both the releaser and the recipient. It encourages mutualistic relation between organisms. e.g., termites and protozoans.

Insect Growth Regulators

Insect Growth Regulators (IGRs) are compounds which interfere with the growth, development and metamorphosis of insects. IGRs include synthetic analogues of insect hormones such as acdysoids and juvenoids and non- hormonal compounds such as precocenes (Anti JH) and chitin synthesis inhibitors.

IGRs used in Pest Management

Ecdysoids : These compounds are synthetic analogues of

natural ecdysone. When applied on insects, kill them by formation of defective cuticle. The development processes are accelerated by passing several normal events resulting in integument lacking scales or wax layer.

Juvenoids : They are synthetic analogues of Juvenile Hormone (JH). They are most promising as hormonal insecticides.

Juvenoids have anti-metamorphic effect on immature stages of insect. They retain *status quo* in insects (larva remains larva) and extra mouldings take place producing super larva, larval- pupal and pupal-adult intermediates which cause death of insects. Juvenoids are larvicidal and ovicidal in action and they disrupt diapause and inhibit embryogenesis in insects.

Anti-JH: They act by destroying corpora allata and preventing JH synthesis. When treated on immature stages of insect, they skip one or two larval instars and turn into tiny precocious adults. They can neither mate, nor oviposit and die soon. e.g.. EMD, and PB (Piperonyl Butoxide).

Chitin synthesis inhibitors: Benzoyl phenyl ureas have the ability of inhibiting chitin synthesis *in vivo* by blocking the activity of the enzyme chitin synthetase. Two important compounds in this category are diflubenzuron (Dimilin) and Penfluron. The effects they produce on insects include

- Displacement of mandibles and labrum
- Adult fails to emerge from pupal skin leading to death
- Ovicidal effect.

IGRs from Neem : Leaf and seed extracts of neem which contains azadirachtin as the active ingredient, when applied topically causes growth, inhibition, malformation and mortality in insects.

Hormone mimics from other living organisms: Ecdysoids from plants have been reported from plants like mulberry; ferns and conifers. Juvenoids have been reported from yeast, fungi, bacteria, protozoans, higher animals and plants.

Advantages of using IGRs

- Effective in minute quantities and so are economical
- Target specific and so safe to natural enemies
- Bio-degradable, non-persistent and non- polluting
- Non-toxic to humans, animals and plants

Disadvantages of using IGRs

- Kills only certain stages of pest
- Slow mode of action
- Since they are chemicals, possibility of build-up of resistance
- Unstable in the environment

Antifeedants

Antifeedants are chemicals that inhibit feeding in insects when applied on the foliage without impairing their appetite and gustatory receptors or driving them away from the food. They are also called gustatory repellents, feeding deterrents and rejectants. Since insects do not feed on treated surface they die due to starvation.

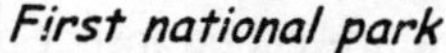

You Don't Know!

First national park

Corbett National Park established in 1935 was named Hailey National Park after the Governor of the United Provinces. In 1952, the name was changed to Ramganga National Park. The Park was renamed Corbett National Park in 1957 after the legendary Jim Corbett.

Groups of Antifeedants

Organotins : They are compounds containing tin. Triphenyl tin acetate is an important antifeedants effective against cotton leaf worm, Colarado potato beetle, caterpillars and grasshoppers.

Carbamates: At sublethal doses thiocarbamates and phenyl carbamates act as antifeedants of leaf-feeding insects like caterpillars and Colarado potato beetle. Baygon is a systemic antifeedant against cotton boll weevil.

Botanicals: Antifeedants from non-host plants of the pest can be used for their control. The following antifeedants are produced from plants.

(a) Pyrethrum: Extracted from non-host plants of the pest can be used for their control. The following antifeedants are produced from plants

(b) Neem : Extracted from leaves and fruits of neem *Azadiracta indica* is an antifeedant against many chewing pests and desert locust in particular

(c) Solanum alkaloids: Leptine, tomatine and solanine are alkaloids extracted from Solanum plants and are antifeedants to leaf hoppers.

Miscellaneous compounds: Compounds like copper stearate, copper resinate, mercuric chloride and phosphon are good antifeedants.

Mode of action: Antifeedants inhibit the gustatory (taste) receptors of the mouth region. Lacking the right gustatory stimulus the insect fails to recognize the treated leaf as food. The insect slowly dies due to starvation.

Advantages

- Affect plant feeders, but safe to natural enemies
- Pest not immediately killed, so natural enemies can feed on them
- No phytotoxicity or pollution

Disadvantages

- Only chewing insects killed and not sucking insects
- Not effective as sole control measure, can be included in IPM

GK **Babur defeated Medni Rai of Chanderi.** GS

Insect Attractants

Chemicals that cause insects to make oriented movements towards their source are called insect attractants. They influence both gustatory (taste) and olfactory (smell) receptors.

Types of Attractants

Pheromones: Pheromones are chemical secreted into the external environment by an animal which elicit a specific reaction in a receiving individual of the same species.

Food lures: Chemical present in plants that attract insect for feeding. They stimulate olfactory receptors.

Use of Attractants in IPM

Insect attractants are used in three ways in pest management.

(*a*) Sampling and monitoring pest population

(*b*) Luring pests to insecticide coated traps or poison baits

(*c*) In distracting insects from normal mating, aggregation, feeding or oviposition. The female insects if lured to wrong plants for egg laying, the emerging larva will starve to death.

Examples of Poison Baits

- *For biting insects*: Moistened Bran + Molasses + insecticides.
- *For sucking insects*: Sugar solution + insecticide
- *For fruitflies* : Trimedlure/Cuelure/Methyl eugenol + insecticides
- *For cockroaches*: Sweet syrup + white or yellow phosphorus
- *For sweet loving ants*: Thallous sulphate + sugar + honey + glycerine + water

Advantage: They are specific to target insects. Natural enemies not affected. But they cannot be relied as the sole method of control and can only be included in IPM as a component.

Insect Repellents

Chemicals that induce oriented movements in insects away from their source are called repellents. They prevent insect damage to plants or animals by rendering them unattractive, unpalatable or offensive.

Types of Repellents

1. ***Physical repellents*** :

Produce repellent by physical means.

Contact stimuli repellents: Substances like wax or oil when applied on leaf surface changes physical texture of leaf which are disagreeable to insects.

Auditory repellents: Amplified sound is helpful in repelling mosquitoes.

Barrier repellents: Tar bands on trees and mosquito nets are examples.

Visual repellents: Yellow light acts as visual repellents to some insects.

Feeding repellents: Antifeedants are feeding repellents. They inhibit feeding.

2. ***Chemical repellents*** :

Repellents of plant origin: Essentials oils of Citronella, Camphor and cedar wood act as repellents. Commercial mosquito repellent 'Odomos' uses citronella oil extracted from lemongrass.

Pyrethrum extracted from Chrysanthemum is a good repellent and has been used against tsetse-fly.

Synthetic repellents: Repellents synthetically produced.

Important synthetic repellents are listed below:

Insects	*Repellents*
Mosquito, bloodsuckers	Dimethyl pthalate
Mites (chigges)	Benzyl benzoate
Phytophagous insects	Bordeaux mixture
Wood feeders	Pentachlorophenol
Fabric eaters	Naphthalene or mothballs
Bees	Smoke

Uses of Repellents

- Applied on body to ward off insects
- Used as fumigants in enclosed area
- Used as sprays on domestic animals
- Used to drive away insects from their breeding place

Biotechnology in Pest Management

Molecular biology techniques are used for the management of insect pests. The following are some of the strategies:

Wide hybridization: This technique involves transfer of genes from one species to another by conventional breeding. The genes for resistance are transferred from a different species. e.g. WBPH resistant gene has been transferred to *Oryzae sativa* from *O. officinalis.*

Somaclonal variability: This variation is observed in tissue culture derived progeny. e.g. Somaclonal variants of sorghum resistant to *Spodoptera litura* has been evolved.

Transgenic plants: Transgenic plants are plants which possess one or more additional genes. This is achieved by cloning additional genes into the plant genome by genetic engineering techniques. Transgenic plants have been produced by addition of one or more following genes:-

- Bt endotoxin from *Bacillus thuringiensis*
- Protease inhibitors
- α-Amylase inhibitors

Cybernetics is study of communication between human and machine.

- Lectins
- Enzymes

Bt endotoxin Gene

The gram positive bacteria *Bacillus thuringiensis* produces a crystal toxin called δ (delta) endotoxin. The δ endotoxin is a stomach poison and kills the lepidopteran insects if consumed. The gene (DNA fragment) responsible for producing δ endotoxin is isolated from Bt and cloned into plants like cotton, potato, maize, etc. to produce Transgenic cotton, etc.

Transgenic Bt plants	*Target insect pests*
1. Cotton	Bollworms, *S. litura*
2. Maize	European corn borer
3. Rice	Leaf folder, stem borer
4. Tobacco, Tomato	Cut worms
5. Potato, Egg plant	Colarado potato beetle

Protease Inhibitors (PI) Gene

Insects have proteases in their gut which are enzymes helping in digestion of protein. Protease inhibitors are substances that inhibit the proteases and affect digestion in insects. The protease inhibitor gene are isolated from one plant and cloned into another to produce transgenic plants.

Transgenic apple, rice, tobacco contains PI gene.

Cowpea trypsin inhibitor (Cp TI) is a PI isolated from cowpea and cloned into tobacco. This transgenic tobacco is resistant to *Heliothis virescens*.

α-Amylase inhibitors gene

α-Amylase is a digestive enzyme present in insects for digestion of carbohydrate. α-Amylase inhibitor, affect digestion in insects. Transgenic tobacco and tomato expressing α-amylase inhibitor have been produced which are resistant to *Lepidopteran pests*.

Lectins genes

Lectins are proteins that bind to carbohydrates. When an insect feeds on lectins, it binds to chitin in peritrophic membrane of midgut and prevents uptake of nutrients.

e.g. Transgenic tobacco containing pea lectin gene is resistant to *H. virescens*.

Enzymes genes

Chitinase enzyme gene, and cholesterol oxidase gene have been cloned into plants and these show insecticidal properties.

Pyramiding genes

Engineering transgenic crops with more than one gene to get multi-mechanistic resistance is called pyramiding of genes. *e.g.* The Cp Ti gene and pea lectin gene were cloned to produce a transgenic tobacco.

Knowledge World!

Slowest growing flowering plant

Dioon (fam. *Cycadaceae*) has been known to grow only 2.01 m (6.6 ft) in a thousand years.

Potential Advantages of Biotechnology in IPM

- Slow Development of resistance against transgenic Bt, PI, lectins
- All plant parts express toxin and so no need for insecticide spray
- No need for continuous monitoring
- No environmental pollution, safe to natural enemies, non-target organism

INTEGRATED PEST MANAGEMENT (IPM)

Definition

"Integrated pest control is a pest management system that, in the context of associated environment and population dynamics of the pest species, utilizes all suitable techniques and methods in as compatible manner as possible and maintains pest population at levels below those causing economic injury".

Principles of strategies

Monitoring insect pests and natural enemies : It is a fundamental tool in IPM for taking management decisions.

Concepts of injury levels: Following the concepts of EIL and ETL to reduce the usage of insecticides and their impacts on environment.

Integration of tactics: Proper choice of compatible tactics and blending them so that each component complement the other.

Relationship between different components

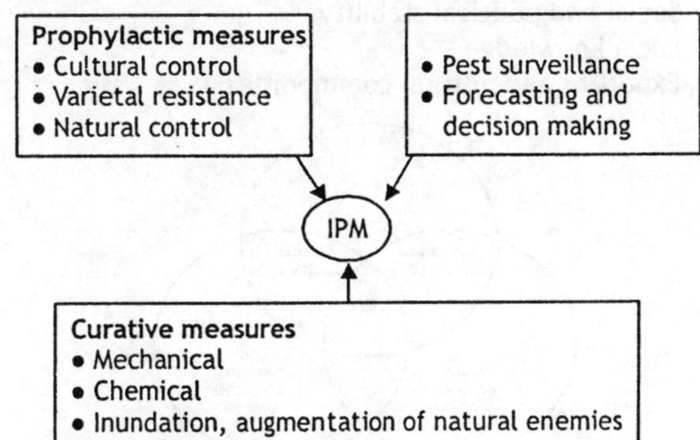

GK GS

Birbal died in the campaign against Yusuf Zahis.

Approaches in IPM when compared to normal method

Parameter	*Normal method*	*IPM*
Goal	Reduce losses due to pests	Reduce costs of production, ecological and social approach
Diversity	Low to medium	High
Ecosystem stability	Uncertain	Striving towards stability
Spatial scale	Single farm	Entire region
Time scale	Immediate	Long term/Single season
Target	Single pest or closely related groups of pests	Several pests around crops and natural enemies
Basis for control	Past experience or presence of pest	Economic Threshold Level (ETL)
Principal method	Cultural, mechanical, pesticides	All components
Effect on environmental quality	Highly detrimental	Moderately detrimental/ Negligible

Essential Requirements for IPM

- Proper identification of insect pests
- Life history and behaviour of the pest
- Natural regulating factors
- Need for control measures
- Timing of control measures
- Selection of suitable control measures
- Farmer's participation
- Government support

Advantages of IPM compared to Pesticide-based Plant Protection Programme

- Proper
- Sustainability
- Economics
- Health
- Environment quality
- Social and political stability
- Local knowledge
- Export of agricultural commodities

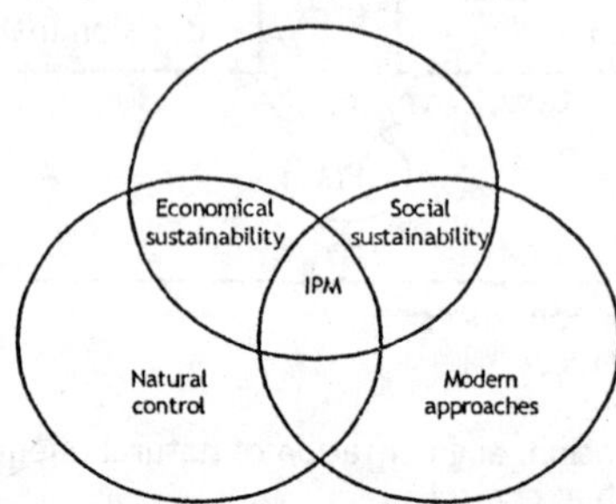

IPM for Rice, Cotton, Sugarcane

IPM for Rice

Observing ETL concept –

Insect	**ETL**
Stemborer	10% dead heart
Gall midge	10% silver shoot
Whorl maggot	25% damaged leaves
Leaf folder	10% leaf damage at vegetative stage
GLH	5 per hill at vegetative stage 10 per hill at flowering stage
BPH	1 per tiller 2 per tiller when spider is present at 1 per hill
Earhead bug	5 bugs per 100 panicle at flowering stage 10 bugs per 100 panicle at milky stage

- Avoiding use of excess nitrogen which induces BPH and leaf folder
- Alternate wetting and drying for BPH.
- Passing the rope on the crop and draining of water for case worm.
- Using light traps to monitor BPH, GLH and stem borer.
- Providing rogue spacing at every 2 m interval to take up the plant protection operation.
- Avoiding resurgence causing insecticides like synthetic pyrethroids, methyl parathion, and quinalphos for BPH.
- Avoiding shade for leaf folder.
- Use of Neem blended urea (I : 5) which helps in the slow release of nutrient.
- Use of egg parasitoid *Trichogramma japonicum* for stem borer .
- Use of egg parasitoid *T. chilonis* and bacterial agent *Bacillus thuringensis* for leaf folder .
- Use of Neem oil for Ear head bug.
- Use of resistant varieties. e.g. PY3, CO 42 for BPH

IPM for Cotton:

Observing ETL concept –

Pests	***ETL***
Leafhopper	1-2 per cent
Whitefly	5-10 per leaf
Boll worms	10% shoots, squares or bolls damaged
Stem weevil	10% infested plants
Spodoptera	8 egg masses per 100 m row

- Using light trap to monitor hoppers, *Spodoptera* and boll worms.
- Using pheromone traps to monitor and also for mass trapping of boll worms.
- Collection and destruction of infested plant parts,

squares and bolls for boll worms.

- Growing trap crop (e.g..) Castor for *Spodoptera*.
- Manual removal of egg masses (e.g.) *Spodoptera*.
- Hand-picking of larvae of boll worm.
- A voiding late sowing of crop to escape from whitefly damage.
- Following crop rotation with unrelated crops to check the life cycle of important pests like boll worms, stem weevil, leaf hopper, etc.
- Removal of alternate weed hosts.
- Judicious use of Nitrogen and water for hoppers and whiteflies management.
- Use of yellow sticky traps for whiteflies
- Avoiding early stage direct application of pesticides.
- Use of Neem-based formulations.
- Use of predators like *Chrysoperla camea*.
- Growing trap crop.
- Use of NPV virus.
- Use of egg parasitoid *Trichogramma* spp. for boll worms.

IPM for Sugarcane

- Trash mulching along the ridges to modify the moisture content of soil which checks the build up of early shoot borer.
- Detrashing on 105th day and 210th day for leaf hopper.
- Avoiding excess nitrogen for leafhopper.
- Draining excess water for scales and mealy bugs.
- Use of predators like Coccinellid beetles for scales and mealy bugs.
- Use of egg parasitoid *Trichogramma* spp. for early shoot borer and inter node borer.
- Use of Granulosis virus and egg parasitoid *Sturmiopsis inferens* for inter node borer.
- Use of egg parasitoid *Isotima javensis* for top borer.

OTHERS

Pesticide Formulations

Insecticides are available in different formulations after an insecticide has been commercially manufactured, formulators are added in various additives and prepare the insecticides in various forms such as dusts, granules, emulsifiable concentrates and solution etc.

3 types of formulations –

1. Solid 2. Liquid 3. Gaseous

1. Solid Formulations

(a) Dusts :

- Insecticides mixed with dust carriers called dusts.
- Toxicity of an insecticide increase as the particle size decrease.
- It should be applied in the morning hrs on wet foliage.

Quiz Kit!

Biggest tree trunk

The majestic deodar (*Cedrus deodara*) found in Jammu and Kashmir and Himachal Pradesh have trunks whose circumference measures 12 m (39.3 ft) or more.

(b) Granules :

- Dusts having 0.25 -2.4 mm particle size.
- Used as dressing on or in the soil.

(c) ***Insecticide -fertilizer Mixture***

- Granular insecticides added to fertilizers
- Applied as regular fertilization time to provide nutrition to plants as well as to control the soil insects.

2. Liquid Formulations

(a) Solutions :

- Pesticide is in a single liquid phase either in water or in an oil.
- Most organic insecticides are insoluble in water and need on organic solvent to dissolve.

(b) Emulsifiable Concentrates (EC)

- Commercially sold as EC to which a certain amount of water is added to make aqueous suspension suitable for spraying on plants.
- Clean solution and yields an emulsion when diluted witscreen. (e.g.) Alkaline Soaps, Organic amines, etc.

(c) Suspension: Active material is suspended as solid particles.

(d) Foam Spray: Chemical additive which when sprayed then special foam nozzles converses the pesticide.

3. Gaseous Formulations

Fumigants :

- Compound volatile. at ordinary temperature and sufficiently toxic.
- Advantage of using fumigant, places not easily accessible to other chemicals can easily be reacted due to the penetration and dispersal effect of the gas.
- Used to control pests in storage bins, buildings, and soil insects, nematodes.

Impact of Global Warming on Pests

What is global warming ?

Warmth from sun heats the surface of the earth. Earth absorbs most of the energy but reflects back some energy in the form of infra red radiation. Greenhouse gases e.g. CO_2, Methane, CFC (Chloro Fluoro Carbon). Nitrous oxide

GK **Abul Fazl was murdered by Bir Singh Bundela.** GS

present in atmosphere transmit the infra-red radiation and reflects back to earth. This reflected energy falls on earth and also lower atmosphere and keeps it warmer i.e., heats the earth's surface. This is called global warming or green house effect.

Effect of Global Warming on World and Agriculture

- Increases overall temperature on earth
- Causes changes in climate tremendously
- Results in melting of ice in polar region
- Increases sea level resulting in submerging of coastal areas
- Flooding and intense downpours
- Drought in warmer regions

Impact of Global Warming on Pest Status

- Due to change in climate, temperature and water availability, the farmers may change the type of crops grown.
- Due to increase in temperature, there can be outbreak of certain insect pests and diseases.
- In forest areas there will be a shift in tree species and also pest species.
- In agriculture lands since cropping pattern is changed, new crops are introduced to suit the climate and new pests are also introduced.
- When water availability is less, crops will be raised as rainfed. It will be difficult to take up control measures without water.

GK GS

Rakesh Sharma was the first Indian astronaut in space.

Points to Ponder

- "*Malus* yields a phenolic substance, phlorizin that stimulates probing in the apple feeding aphids and acts as feeding deterrent for non-apple (*Myzus persicae*)" is known as apple-factor.
- A Pest refers to an insect whose population increase to an extent that causes economic losses to crops or a nuisance or health hazard to human beings and other animals.
- A phytophagus insect, a leaf feeder nibbles the leaves either on margins or the surface that can be easily seen.
- A race producing single race a year is called univoltine.
- Absconding bees drink all the honey of their nest and leave behind empty combs, brood and sometimes even food.
- Adverse effect of the biology of an insect feeding on a resistance variety is called antibiosis.
- Aflatoxin causes serious change in metabolic activity of insects.
- Alarm pheromones are primarily an antipredator device.
- All honey bee eggs hatch in three days, but the period of development of larva and pupa differ among the castes.
- Allethrin, dimethrin, bathrin and resmethrin are synthetic analouges of pyrethrum.
- An emulsion is a two-phase dispersion involving water and an immiscible solvent.
- Any chemical that can inhibit the growth of gonads and interfere with reproductive capacity of an insect in any other way is called chemosterilant.
- Aphids are the main transmitters of viruses to the plants.
- *Apis dorsata* produce about 35 kg honey per colony per year.
- *Apis mellifera* originated in !taly and was introduced in different countries of the world.
- *Bacillus papillae* is pathogenic to white grubs and extensively used against *Papillio japoinica*.
- *Bacillus thuringinesis* has been found most effective against lepidopterous insects of the filed crops.
- *Bacillus thurningiensis* is crystalli-ferous, spore forming bacterium widely used as biopesticide; isolated by Berliner.
- Banding of mango tree tunk with alkathene sheet and applying grease on it prevent the infestation of mealy bug.
- Banding of stem is usually employed to prevent insect-pests from climbing to tree tops.
- Bees-wax is obtained from the combs of the wild hives, frame hives and cappings.
- BHC (Gemmexane, $C_6H_6C_{16}$) is benzene hexachloride.

Did You Know ?

First Speaker of Lok Sabha

G.V. Mavalankar was Speaker from May 15, 1952-April 1957.

- BHC was first sysnthesized by Faraday.
- Bioagent *Aphilinus mali* has been found effective in controlling wooly aphid infesting apple.
- Biological or biocontrol comprises a method of controlling pests by exposing them to their natural enemies.
- Brood-food is produced by the pharyngeal salivary glands of the workers.
- Carolus Linnaeus developed binomial method of classification.
- Chemicals present in the host plants act as attractant (phagostimulants) and those in the non-host plant as antifeedant.
- Clean culture method of pest control is specifically useful against vegetable and field crops.
- *Corpora allata* are paired organs and may fuse into one in some insects.
- *Corpora cardiaca* are small paired bodies immediately behind the brain and are believed to be modified ganglia (in insects).
- Crop rotation usually affects those insect which have a limited host-range and relatively immobile in some stage of their development.
- Cucurbitaceous weeds aids in survival and spread of the pathogens causing powdery mildew and mosaic in them.
- Cuticle of insect is differentiated into two main regions-procuticle and epicuticle.
- DDT "fall-out" is the second biggest after the nuclear fall-out.
- DDT was first described by a German chemist, Othmar Zeidler in 1874, but Paul Muller (of Switzerland) in 1939 discovered its insecticidal property.
- Decomposition of systemic insecticides into different compounds inside the plant body is known as detoxification.
- Destruxins causes paralysis and death in insect.
- Diaepam is used as antidote in the case of deltramethrin and fanvalerate poisoning.
- Diapause is a specialized condition of suspended activity with body metabolism at its minimum.
- Diflubenuron inhibits chitin synthesis.
- Ecdysone, juvenile, bursicon and eclosion are the hormones involved in cuticle production.
- Eggs of honey bee are fertilized to produce female.

Washing powder have 15-30% detergent.

- Entomophagous insect species may be either predators or parasites.
- Entomophagous pathogens produce fatal diseases in insects.
- Epicuticle is made up of cement layer, wax layer and cuticulin layer.
- ETL always represents a pest density lower than that of EIL.
- Eulphid (*Tetrastichus nyemitwus*) and chalcid (*Callitula bipartius*) are the larval parasites of Sorghum shoot fly (*Atherigona soccata*).
- Formation of species is called Speciation.
- Fumigation should be carried out during morning hours.
- General equilibrium of insect-pest population is maintained by the natural enemies.
- Helping to make imported species established it self in a new environment (area/country) is known as colonization.
- *Holotrichia consanguinea* is parasited by *Scolia aureipennis*.
- Honey bees employ both dance and sound in their language.
- Host-evation is the modification in plant growth pattern so as to bring asynchronies of insect-host phenologies.
- Hyper-sensitivity and hyper-activity followed by body tremors are symptoms of poisoning produced by allethrin and perinethrin.
- In adult Lepidoptera proboscis is formed from galeae.
- In BHC 10-20 per cent gamma isomer is present.
- In Cercopidae (Hemiptera) the spittle, in which the immature stage remains surrounded is produced by Malpighian tubules.
- In cryptonephridial condition the distal ends of Malpighian tubules are closely attached to the hind gut.
- In higher forms the coxa is articulated at two points which limits the movement of leg.
- In honey bee queen the mandibular gland produce pheromones which control the colony.
- In honey bee the hind legs are modified for brushing the pollens from the body.
- In honey bee, transformation of three castes depends on the amount of royal jelly (brood food).
- In honey-bee hive as the temperature falls in winter, they consume stored honey, become active, vibrate the wings to generate energy raising the inside temperature by many degrees.
- In hypognathous type of head the long axis of the head is vertical and mouth parts ventral.
- In India, Lefory was first to investigate silk worms and sericulture at IARI, New Delhi.
- In insects there are three types of cuticle rigid cuticle, membranous cuticle and elastic cuticle.
- In insects two types of eyes based on retinular cells can be distinguished-apposition eyes and superposition eyes.
- In light producing organs of the insects, the light is produced by the oxidation of luciferin.
- In mantids the pincers are formed by apposition of tibia and femur to catch hold the prey.
- In nymph / larva maximum selerotiation is found in mandibles.
- In Odonata the wings are not folded at the back when at rest.
- In the case of allethrin and permethin poisoning phenobarbital is useful.
- In the trichoid sensellium the potential is produced when the hair is bent in any direction.
- In woody plants absorption of an insecticide has been more effective through bark than through roots.
- Inflammable fumigants should be mixed with carbon tetrachloride (CCl_4).
- Insect is also called Hexapoda because they have three pairs of legs.
- Insect poisoning in organophosphorous insecticides occurs due to inhibition of acetylcholinesterase (AChE) enzyme causing accumulation of transmitter acetylcholine (ACh).
- Insecticide residues are referred to as the quantities of persistent insecticides that pollute environment and food stuffs.
- Insecticides can exist in three forms viz., pure, technical (commercial) forms and formulations.
- Insecticides used in gaseous form are called fumigants.
- Insects are arthropods with three pairs of legs belonging to invertebrate phylum Arthropoda.
- Insects living in fresh water or extremely moist environment excrete their nitrogen waste in the form of ammonia.
- Insects posses an exoskeleton, made up of chitin.
- Intercropping/mixed cropping helps in control of pests.
- Ionising radiations either kill the insect-pests outrightly or sterilize them.
- Karnataka is the biggest producer of raw silk in India followed by J & K.
- Lady birds has no special structural modifications as their preys are largely immobile.
- Larvae of *Callosobruchus chinensis* are internal feeder

Aurangzeb was called as Zindapir.

GS

of grain pulses.

- Life histories of insects make one important component of information about their biology that is crucial to their rational control.
- Light traps are used to attract nocturnul (night) insects which are strongly phototaxic.
- Lindane contains 99 per cent gamma isomer.
- Locusts do not eat *Calotropis procera* (aak), *Datura stramonium* (dhatura), *Melia azadirachta* (neem), *Psidium cuminii* (jamun), *Dalbergia sissoo* (sheesham) and *Ficus carica* (anjeer).
- Malphigian tubes are absent in parasitic insect.
- Malpighian tubules first discovered by Malpighi.
- Market male moths are known to have been attracted to a trap baited with the female sex pheromone from a distance of two miles.
- Masker (deodorant) are used to suppress the unpleasant odour of certain insecticides when used for house-hold purposes.
- Maximum sclerotization is found in the mesonotum of adult insect.
- Median residual life (RL_{50}) or half-life of an insecticide is the time required for half of its residues to react and kill 50 per cent of the insect-population.
- Melanization is a process by which the cuticle becomes dark.
- Meso and metathorax, the wing bearing segments are together called pterothorax.
- Methyl paration is a derivative of thio-phosphoric acid.
- Most of the insecticides are insoluble in water.
- Most of the insects excrete 80 to 90 per cent of their nitrogen waste in the form of uric acid.
- Most of the modern insecticides are a polar and, therefore, they easily penetrate the insect cuticle.
- Most of the organophosphate insecti-cides are esters if alcohols with phosphoric acid or anhydrides of phosphoric acid with some other acid.
- Moulting in insects is also considered to be a process of excertion particularly of nitrogenous waste products.
- Natural enemy of insects are host specific.
- Nereistoxin was isolated from a marine annelid.
- Neurosecretary cells (NSC) are neurons in the central nervous system that are specialized to secrete hormones.
- *Nilaparvata lugens* acts as vector of tungro disease in paddy.
- Only the grubs of *Trogoderma granarium* damage the stored grains.
- Organophosphorus insecticides produce two types of symptoms viz., muscarinic and nicotinic.
- *Perilissus cingulator* is the parasite of mustard sawfly *Athalia lugens proxima*.
- *Periplaneta americana* (a cockroach) is the largest and dark brown in colour.
- Pest resurgence occurs either due to development of resistance or due to destruction of parasites or both.
- Pink boll worm bores more in the new branches of cotton as compared to spotted worm.
- Pink bollworm (*Pectinophora gossipiella*) eggs are preyed upon by anthocorid bug *Orius tantilus*.
- Poisoning due to carbamate insecticides occurs due to AChE inhibition, by carbamylation unlike phosphorylation of organophorus group.
- Preventing the insects from entering the diapause is to force them to face unfavourable climatic conditions and perish.
- Pseudosynergist stabilize the droplet size of insecticides spray solution to ensure persistent contact with insect body for augmented kill.
- Pupation in fruit flies takes place in the soil.
- Relaxing is the process of re-sorfening the insect.
- Removal of bark of gape vines after pruning results in reducing the population of flea beetle.
- Repeated involuntary jerking movements in hands is the typical symptom of poisoning of deltamethrin and fenvalerate.
- *Rodonalia cardinalis* from Australia was introduced into California (U.S.A.) in 1888 to control cottony cushion sclae (*Iserya purchasi*) a serious pest of citrus.
- Root borer renders the moorings of the plant so weak that they can be easily pulled out or can be felled by gust or wind.
- Rotenone is obtained from the roots of *Derris elliptica*.
- Sclerotization is a process by which the cuticle becomes hard.
- Seeds and leaves (dried) of neem serve as locust repellant and protect stored grains against store-pests.
- Shedding or red tears due to the blood porphyrinuria in the case or organophosphate insecticide poisoning.
- Silk is the product of a pair of silk glands which are the slivary glands of the larva present in its thoracic cavity.
- Silver fish is primarily apterous insect.
- *Solanum* alkaloids extracted from some species of *Solanum* reduce feeding and survival time in potato leaf hopper, *Empoasca devastans*.
- Spinneret is a narrow pore situated on the hypopharynx

through which the sticky fluid secreted by silk glands comes out.

- Spore forming bacteria are more promising in insect control than the non-spore forming one.
- Stem borer will produce stunted growth, bunchy tops and dead hearts.
- Stone and Pedigo found it useful to describe the term economic damage as loss of marketable produce and called it gain threshold
- Synergist are chemicals embodying no insecticidal property of own, but an addition to an insecticide enhances, its toxicity manifold.
- Temporary development arrest largely of eggs in summer is reffered to as aestivation.
- Termites, commonly but wrongly called white ants, are neither white nor are related to ants.
- The ant (*Camponotus*), when attacked ejects formic acid in its own defence (as an allomone).
- The anterior tentorial pits lie in epi-stomal suture.
- The atennae are primarily the sense organs and most of the sensilla are found on the annuli in the middle of the flagellum.
- The biting and chewing type of mouth parts are considered to be primitive.
- The complete chemical name of DDT is Dichloro diphenyl trichloroethane.
- The concentration of fumigant per unit volume x duration of exposure is called Ct value.
- The correct name of BHC is Hexachlorocyelohexana.
- The damage threshhold refers to the level of pest population above which crop losses occur and below which the pest has no significance.
- The desirable substitutes for cyanide bottle (usually used for insect collection) are ethylacetate, ammonium hydroxide, benzene, chloroform, carbon tetrachloride, and trichloro-ethylene.
- The development of biotypes in the insect pest is due to the servers selection pressure exerted by the resistant vareity.
- The endoskeleton of insect head is called tentorium.
- The expansion of termite queen's abdomen to accommodate reproductive organs in known as physogastry.
- The female pyrilla lays eggs on the stem of sugarcane during summer.
- The filter chamber is a device for allowing excess water and soluble carbohydrates of the food to be eliminated by diffusing directly from the anterior part of the midgut into the anterior part of the hindgut.
- The first significant success in controlling a pest by predator was the control of *Icerya purchasi* by the *Rodlia cardinalis* action of fungal pathogens.
- The first transgenic plant resistant to *Manduce sexta* produced was of tobacco.
- The flight of insect is based on the Burnoullies principle of aerodynamics.
- The formic acid ejected by one ant is utilized by another ant (*Myrmecia gulosa*) to locate workers of this ant (*Camponotus*).
- The grubs of *Longitarsus nigripennis* bore and destroy the berries of black pepper.
- The host plant twig bearing lac alongwith eggs is called brood lac stick.
- The huscarinic effects of organo-phosphate insecticide poisoning are on the postganglionic parasympathetic cholinegic fibres.
- The infestation of most of the stored grain pests can be avoided if the moisture content of grain is kept below 9 per cent.
- The insect pathogenic viruses belong to the family Baculoviridae.
- The insect viruses can not be multiplied on artificial media.
- The insects are killed mainly by the histolytic action of fungal pathogens.
- The insects, response range to solar spectrum lies between below UV to red.
- The integument of insect is the out layer comprising epidermis and cuticle.
- The Johnston's organ is situated in the second segment of the antenna.
- The knowledge of life-history of insects helps in prediction of pest-outbreak and population build-up.
- The lady beetles (*Coccinella septumpunctata*) has been found predating on the larvae of stem borer (*Chilo zontus*).
- The larvae of Diptera are called maggot.
- The main function of taenidia is to prevent the collapse of the trachea when the pressure within is reduced.
- The male fleas use their antennae to clasp the female during mating.
- The male sterilization technique was successfully used in the island of Curaco for the control of *Cochliomya haminivorax*.
- The mosquito bug (*Helopeltis theivora*) nymphs are parasitised by *Agamermis paradecoudate*, nematode.
- The muscles of the insects are cross striated.

Akbar gave Ramdas a plot of land on which the golden temple was constructed.

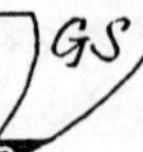

- The nest (comb) of a honey bee has three types of hexagonal cells, worker cells, drone cells and queen cells.
- The nicotinic effects of organo-phosphate poisoning are on sympathetic fibres.
- The ommitidia in which the Semper's cells produce hard in tracellular crystalline cone are called eucone type.
- The optical part of the compound eye gathers light and the sensory part perceives the radiations and transforms them into electrical energy.
- The Ostia are the openings in the wall of the heart.
- The Panel of FAO-Experts (1967) defined Integrated Pest Management (IPM) "as a system, that, in context of the associated environment and the population dynamics of the pest species, utilizes all suitable techniques and methods in as compatible a manner as it possible and maintains the pest populations at levels below those causing economic injury".
- The peculiarity of aldrin poisoning is that fully recovered patient may develop full-blown symptoms of poisoning as long as a month after the initial poisoning.
- The peritrophic membrane is absent in Hemiptera and adult Lepidoptera.
- The pharangeal glands of workers of honey bee produce brood feed which play an important role in caste determination.
- The pheromone produced by gypsy moth female has been estimated at 0.01 μg, it is enough to sexually stimulate one billion males of its species.
- The pheromones induce aggregation or congregation of insect for protection, reproduction, and feeding or combinations thereof.
- The pollen basket is formed by tibia and first tarsal segment of the hind leg of worker honey bee.
- The prolegs of caterpillars are hollow, cylindrical out growths of the body wall the lumen of which is continuous with the haemocoel.
- The rice variety "Mudgo" is resistant to brown plant hopper because it contains much smaller proportion of the amino acid asparagine as compared to the susceptible varieties.
- The silk (fiber) is composed of fibroin and sricin (proteins).
- The stubbles of maize should be burnt as they give shelter to hibernating larvae of *Chilo partellus*.
- The technical insecticides refers to a toxicant manufactured in bulk by a basic manufacturer after satisfactory lab tests.

You Don't Know!

First woman advocate

Corenelia Sorabji was the first woman graduate in law in 1894 although not allowed to practice till 1923 when she was finally enrolled to the Allahabad High Court.

- The term ecosystem was first proposed by Tansley.
- The termite king and queen of a colony are forced to live in royal chamber located in the centre of the nest, 0.5 m below the ground.
- The termite king develops from unfertilized egg and is smaller than queen in size.
- The termite workers develop from fertilized eggs, but remain stunted (smallest in size) because they are reared on ordinary food.
- Thimet (phorate) and disystox are used as seed systemics either directly or through soil treatment.
- Though cockroaches are cosmopolitan in distribution, they thrive in tropical and subtropical climates.
- Tillage tends to reduce soil insects.
- To reduce the spotted bollworm damage in cotton, bhindi is used as trap crop.
- Toxins produced by entomogenous fungi are known as Aflatoxin.
- Typical symptoms of stem borer are dead hearts, frass and weakening of the plants.
- USDA in 1963 demonstrated the efficiency of ULV spray on crops.
- Vaccum fumigation is done in case of cotton bales.
- Voltinism means broods or generations, but it is mainly used in reference to silk worm races.
- Well dried grains having moisture content between 8-10% escape insect damage.
- When an insecticide is absorbed on the particle surface of the insecticide of the diluent, the latter is called as carner, otherwise a filler.
- White ear heads in rice in caused by gundhibug larvae.
- White grub adults are parasitised by the fungus *Metarrhizium anisophiae*.

GK

The book 'Anand Math' contains the national song of India.

GS

Multiple Choice Questions

1. **11-Hexudecenal is sex pheromone of :**
   a] *Heliothis armigera* b] *Spodoptera litura*
   c] *Trichoplusia nigra* d] *None of the above*
2. **When a hectare of ground crop is completely covered by spraying less than 60 litres of liquid, is termed as:**
   a] Low volume spray
   b] Medium volume spray
   c] Very low volume spray
   d] High volume spray
3. **7-Dodecenyl acetate is sex pheromone of :**
   a] *Dacus cucurbitae* b] *Heliothis armigera*
   c] *Dacus dorsalis* d] None of the above
4. **A better of pest management depends upon :**
   a] The use of resistant variety
   b] Interlacing the known control methods
   c] Proper agronomic practices
   d] Use of pesticides
5. **A minimum period of ___________ years is to be given before assessing the effectiveness of an important biocontrol agent.**
   a] 2 b] 3
   c] 4 d] 5
6. **A notorious weed, *Lantana camara* is controlled by :**
   a] Lantana bug b] Shoot fly
   c] Cochnial insect d] None of them
7. **A typical 'Hopper burn' in rice is caused by feeding of :**
   a] Aphids
   b] Gundhi bug
   c] Brown plant hopper
   d] Mites
8. **Alternaria blight in mustard can be controlled by :**
   a] Treating seed with hot water
   b] Soil treatment with alderin
   c] Bordeaux mixture
   d] None of these
9. **Ammonium carbamate is mixed with aluminium phosphide because :**
   a] It releases ammonia gas alarming the operator
   b] It relases CO_2 which is fire extinguisher
   c] It gives cooling effect
   d] All of the above
10. **Amongst the following insects which one is monophagous?**
    a] *Spodoptera litura* b] *Spilosoma oblicua*
    c] *Bombyx mori* d] *Bemicia tabaci*
11. ***Aphelinus mali* has been a successful parasite in controlling :**
    a] Tobacco ceterpiller b] Apple wolly aphis
    c] Maize stem borer d] Sugarcane top borer
12. **Arachnids and insects contain malpighian tubules as organs of :**
    a] Respiration b] Excretion
    c] Digestion d] Secretion
13. **Axillary sclerites are absent in :**
    a] Lepidoptera b] Coleoptera
    c] Odonata d] Dictyoptera
14. ***Bacillus thurigiensis* was reported by Berliner from:**
    a] *Tribolium* b] *Heliothis*
    c] *Ephestia* d] *Spodoptera*
15. ***Bacillus thuringiensis* is most effective at pH :**
    a] 3 to 6 b] 9 to 10
    c] 11 to 14 d] All of the above
16. **Bagging is recommended for the control of either of these :**
    a] Anar butterfly b] Lemon butterfly
    c] Mango stone weevil d] Ber fruit fly
17. **Basic unit of classification is known:**
    a] Species b] Genus
    c] Family d] Order
18. **Berlese's theory was proposed to explain the origin of :**
    a] Egg stage b] Larval stage
    c] Pupal stage d] Adult stage
19. **Bio-types of brown plant hopper are of :**
    a] 2 types b] 3 types
    c] 4 types d] 5 types
20. **Black scale insect of citrus is controlled by :**
    a] *Metaphycus loundsbury*
    b] *Rodolia cardinalis*
    c] Both of them
    d] None of them
21. **Blood-porphyrinuria is due to accumulation of porphyrin in the:**
    a] Prothoracic glands b] Labial glands
    c] Salivary glands d] Lachrymal glands
22. **Bombay Natural History Society founded in**
    a] 1835 b] 1883
    c] 1903 d] 1913
23. **Burgundy mixture contians :**
    a] Copper sulphate and lime
    b] Copper sulphate, sodium carbonate and water
    c] Copper sulphate, gypsum and water
    d] None of these
24. **By radioactive carbon method the age of the fossil can be determined upto :**
    a] 10,000 Years b] 15,000 Years
    c] 20,000 Years d] 25,000 Years
25. **By the introduction of chlorine in 3 position of methyl parathion aromatic radical, the insecticide obtained is :**
    a] Ethyl parathion b] Malathion
    c] Chlorthion d] Fenitrothion

Muhtsib looked into censorship of public morals.

26. **Cane killing weed (*Striga* sp.) in sugarcane is :**
a] A weed b] A pest
c] A disease d] None of these

27. **Carbon disulphide control :**
a] Nematodes b] Fusarium
c] Virus d] All of these

28. **Classification of phylum mollusca is based on :**
a] Locomotory organs b] Respiratory organs
c] Digestive organs d] Reproductive organs

29. **Cotton jassid can be controlled by:**
a] DDT b] Aldrex
c] Copper sulphide d] None of these

30. **Cotton stainer is :**
a] *Pectinophora* b] *Ephestia*
c] *Dytiscus* d] *Dysdercus*

31. **Crushing of food in insects is done by :**
a] Rectal papila b] Green cells
c] Proventriculus d] Crop

32. **Crustacean posses the following distinguishing characters. Choose the odd one :**
a] One pair of antennae
b] Minimum five pairs of legs
c] Cephalothroax and abdomen
d] Metameric segmentation

33. **Cut worm is a pest of :**
a] Arhar b] Cowpea
c] Maize d] Gram

34. **DDT inhibits the enzyme which plays important role in the formation of egg shell in birds is :**
a] DDTase
b] Carbonic anhydrase
c] Carboxyesterase
d] None of the above

35. **DDT is toxic to insect pests by :**
a] Contact b] Ingestion
c] Both a and b d] Inhalation

36. **Destructive Insects and Pests Act was passed by Government of India in :**
a] 1914 b] 1974
c] 1977 d] 1968

37. ***Diacrisia obliqua* is the scientific name of :**
a] Termite
b] Top borer
c] Bihar hairy caterpillar
d] Leaf hopper

38. **Diflubenzuron is a derivative of :**
a] Difloro compounds b] Dithiophosphoric acid
c] Urea d] Phosphonic acid

39. **Directorate of Plant Protection, Quarantine and Storage started in**
a] 1940 b] 1946
c] 1950 d] 1552

40. **During the process of sclerotization which of the following is oxidised in the haemocytes :**
a] Arginine b] Glycene
c] Tyrosine d] Alanine

41. **Economic threshold level is always:**
a] Lower than economic injury level (EIL)
b] Equal to EIL
c] More than EIL
d] Reached when severe damage is done

42. **Enforcement of laws for the control of insect pests in India is governed by :**
a] Destructive Insects and Pest Act
b] Insecticide Act
c] Prevention of Food Adultration Act
d] East Punjab Pest Act

43. **Entomological Society of India started in :**
a] 1934 b] 1936
c] 1938 d] 1940

44. **Factor on which penetration of insecticide through insect cuticle does not depend :**
a] Thickness of the cuticle
b] Chemical nature of the cuticle
c] Components of the cuticle
d] Nature of the carriers and solvents

45. **Feeding in insects can be checked by the use of :**
a] Hormone b] Pheromone
c] Freeding deterent d] Antibiosis

46. **First insect- fossil was found in :**
a] England b] Scotland
c] Russia d] USA

47. **For the control of red locust in Mauritius the bird exported from India :**
a] Vulture b] Wild crow
c] House sparrow d] Mynah

48. **For the fumigation of rat burrows the fumigant required is :**
a] EDCT mixture
b] Methyl bromide
c] Aluminium phosphide
d] Chloropicrin

49. **Fore-gut is lined internally by :**
a] Peritrophic membrane
b] Epithelial membrane
c] Intima
d] Taenidia

50. **Fossil records indicate that insect evolved during :**
a] Cambrian period b] Devonian period
c] Silurian period d] Precambrian period

51. **Fully recovered patient may develop full- grown symptoms of poisoning as long as a month after the initial poisoning in the case of:**
a] DDT b] BHC
c] Aldrin d] Heptachlor

GK GS
Leila Seth was the first women Chief Justice of a High Court in India.

52. Give the name of a family of order Hymenoptera which includes crop pests :
a] Tenthredinidae b] Ichneumonidae
c] Braconidae d] Apidae

53. Giving complete coverage to one hectare crop more than 250 liters of liquid is called :
a] Low volume spray
b] Very low volume spray
c] Medium volume spray
d] High volume spray

54. Giving complete coverage to one hectare of crop by spraying 60 to 250 litres of liquid, is known as :
a] Low volume spray
b] Very low volume spray
c] Medium volume spray
d] High volume spray

55. Glycerophosphate shuttle is found in:
a] Sense organs b] Fat bodies
c] Flight muscles d] Pharyngeal muscles

56. Gram pod borer feeds or the following crop :
a] Gram b] Variety of crops
c] Tomato d] Cotton

57. Grassy cut worm of groundnut can be controlled by :
a] Aldrin b] Endrin
c] Lime sulphur d] None of these

58. Green Commandos is :
a] *Heterohabditis* b] *Steinemama*
c] *Neoplectana* d] *Howardula*

59. Green house, white fly can be controlled by :
a] *Apanteles erionotae*
b] *Encarsia formosa*
c] Both of them
d] None of them

60. Gughia weevil is a pest of :
a] Paddy b] Wheat
c] Jowar d] Maize

61. Hamulate type of wing couping appartus is found in :
a] Grasshopper b] Butterfly
c] Bug d] Honeybee

62. Heart, muscles and aorta are derived from :
a] Ectoderm b] Mesoderm
c] Endoderm d] All of the above

63. *Herioglyphus nigrorepletus* is the scientific name of:
a] Rice grass hopper. b] Gundhi bug
c] Pink stem borer d] Termite

64. Honey bee queen produces _______ thousand eggs per day.
a] 0.5-1.5 b] 2-3
c] 4.0-4.5 d] 5.0-5.5

65. Horizontal resistance is also known as :
a] Non-specific b] General
c] Polygenic d] All of the above

66. If equilibrium population level of a crop damaging insect is above the economic injury level, it is a :
a] Potential pest b] Not a pest
c] Sporadic pest d] Regular pest

67. If one gram of a pesticide formulation containing 50% active ingredient is mixed with 1 litre of water, what will be the concentration of the active ingredient in the spray fluid?
a] 0.5 ppm b] 5.0 ppm
c] 50 ppm d] 500 ppm

68. In aphids cornicals are projected from the dorsum of:
a] 5th abdominal segment
b] 6th abdominal segment
c] 7th abdominal segment
d] 8th abdominal segment

69. In binomial nomenclature different classes possessing common characters are included in a/an :
a] Sub class b] Order
c] Super class d] Phylum

70. In case of rotary type of duster the air current is developed by :
a] Bellows
b] Fans
c] Self-propelling jacket
d] Self-propelling blades

71. In grasshopper, the number of spiracles is :
a] 8 pairs b] 10 pairs
c] 6 pairs d] 12 pairs

72. In honey bees, when the larvae are fed on regular diet of pollen and honey, which cast is produced.
a] Queen b] Drones
c] Workers d] All of the above

73. In Hymenoptera the first abdominal segment which is fused with the metathorax is called :
a] Cornical b] Epiproct
c] Propodium d] Gaster

74. In India, lac is mainly produced in :
a] Tamil Nadu b] Bihar
c] Kerala d] Rajasthan

75. In leucerne resistance to spotted aphid is associated with :
a] Low sinigrin in leaves
b] High silica contents in leaves
c] High MBOA contents in leaves
d] High saponin in leaves

76. In order to be most effective control measures aimed at killing an insect-pest should be directed to attack:
a] Egg b] Pupae
c] Grub d] Adult

77. In organophosphate insecticide poisoning the muscarinic effects result in :
a] Anorexia b] Nausea
c] Bradycardia d] All of the above

GK GS
Malik Mohd. Jayasi wrote Padmavat during Sher Shah's reign.

78. In physiological considerations the resistance to insecticide may be developed due to :
a] Conversion of toxicant to non toxic metabolites
b] Excretion of toxicant
c] Storage of the toxicant
d] All of the above

79. In siphoning type of mouth parts proboscis is composed of :
a] Galeae b] Lacinia
c] Mentum d] Labial palp

80. In solid cone nozzle the solid cone of liquid is formed due to :
a] Bigger size of the cap
b] Bigger size of the orific plate
c] Central hole of swirl plate
d] Small size of the strainer

81. In the formation of cuticle which of the following layer is secreted first:
a] Exocuticle b] Endocuticle
c] Wax layer d] Cuticulin layer

82. In the reduction of a pest population, the balance of the natural enemies is maintained in :
a] Pest management
b] Pest control
c] Pest eradication
d] None of these

83. In which of the following insect the spiracles are absent ?
a] Cecidomyid larvae b] Chironomid larvae
c] Mycetophilid larvae d] Psycodid larvae

84. In which order females have no pupal stage but males have :
a] Mallophaga b] Strepsiptera
c] Epemeroptera d] Dictyoptera

85. India is exporting shellac to :
a] China b] Japan
c] USA d] Pakistan

86. India's place in the production of mulberry silk in the world is :
a] First b] Second
c] Third d] Fourth

87. Indicate the correct sequence through which the following insecticide came into use :
a] BHC, Rotenone, Malathion and Lead arsenate
b] Lead arsentate, Rotenone, BHC and Malathion
c] BHC, Rotenone, Lead arsenate and Malathion
d] Rotenone, BHC, Lead arsenate and Malathion

88. Indicate which represents the analogous group :
a] Wings of butterfly and birds.
b] Fore legs of horse and arms of a man.
c] Fore wings of moth and fins of fishes
d] Hind wings of butterfly and legs of birds.

Knowledge World!
Firs ICS officers
In 1869, Surendra Nath Banerji, Romesh Chandra Dutta and Bebarilal Gupta were the first three successful candidates in the ICS examinations then held in the United Kingdom.

89. Indiscriminate use of pesticides and cultivation of high yielding crop varieties often leads to :
a] A minor disease or insect pest problem becoming serious threat
b] Destruction of natural enemies of pests
c] Appearance of pesticides resistant pest strains
d] All of the above

90. Inorganic salts are toxic to insects by :
a] Contact b] Ingestion
c] Both of the above d] Inhalation

91. *Dasineura lini* is pest of
a] Castor b] Linseed
c] Seasamum d] Groundnut

92. Insect haemocoel consists of the following sinuses. Choose the odd one :
a] Pericardial sinus b] Peritarcheal sinus
c] Perivisceral sinus d] Perineural sinus

93. Insect haemolymph has no gas transport system except in-
a] Males of lac insect
b] Chironomid larvae
c] Nephrocytes
d] Coagulocytes

94. Insect having prolegs in larval stage:
a] *Thrips tabaci* b] *Athalia proxima*
c] *Dacus cucurbitae* d] *Holotrichia consanguinea*

95. Insecticide having antagonistic effect with NPV :
a] DDT b] Aldrin
c] Endosulfan d] Malathion

96. Insecticides approved for aerial spray :
a] Lindane 20 EC
b] Malathion 50 EC
c] Phosphamidon 85 WSC
d] All of the above

97. Insecticides banned by the Government of India for agricultural use :
a] Ethyl parathion b] Azinphos
c] BHC d] All of the above

98. Insects feeding on plants of several genera within a family are called :
a] Phytophagous b] Polyphagous
c] Oligophagous d] Monophagous

Sulphur dioxide (SO_2) gas is mainly responsible for Acid rain.

GS

99. **Integrated Pest Management (IPM) aims at :**
a] Keeping pest populations below injurious levels
b] Increasing natural enemies of the pest
c] Strengthening the host
d] Billing the pests

100. **International Congress of Zoology (1901) advised that scientific names should be printed in :**
a] Capital letters b] Small letters
c] Italics d] All the above

101. **Introduction of parasites, predators and/or pathogens of the pests into the environment to reduce the pest population constitutes :**
a] Biological control b] Autocidal control
c] Ecological control d] None of these

102. **Is a serious pest of jowar and bajra.**
a] Stem borer b] Midge
c] White borer d] All of these

103. **It has been estimated that insects had existed on this earth for at least ________ million years.**
a] 100 b] 150
c] 200 d] 250

104. **Johnston's organ is located in most of Pterygote insects in :**
a] Second antennal segment
b] Legs
c] Wings
d] Maxilla

105. **Juvenile hormone is secreted by :**
a] Thoracic glands
b] *Carpora allata*
c] *Carpora cardiaca*
d] Neurosecretary cells

106. **Ketone bodies are oxidised by :**
a] Fat bodies b] Flight muscles
c] Testes d] All of the above

107. **Most of the soil fumigants are applied with :**
a] Applicator b] Injector
c] Generator d] Dispenser

108. **Lac Reseach Institute is situated at:**
a] Bangalore b] Behrampur
c] Lucknow d] Namkum

109. **Light trapping of insects is done in the :**
a] Cultural methods
b] Legal methods
c] Mechanical methods
d] Preventive methods

110. **Light traps are used for the collection of :**
a] Positively phototropic
b] Negatively phototropic
c] Geotropic
d] Hydrotropic

111. **Locust eat almost all vegetation except :**
a] *Calotropis procera* b] *Melia azadirach*
c] *Datura stramonium* d] All of the above

112. **Maize stem borer hibernates in which stage ?**
a] Egg b] Larva
c] Pupa d] Adult

113. **Malathion is a ________ insecticide.**
a] Slow fumigant
b] Contact plus systemic
c] Contact plus stomach
d] Systemic

114. **Maximum loss to paddy is caused by *Tryporyza incertulas* in the state of:**
a] Punjab b] Bengal
c] Orissa d] Andhra Pradesh

115. **Metasystox is a :**
a] Contact poison b] Systemic poison
c] Both a and b d] Fumigant

116. **Metasystox-R is a ____________ insecticide.**
a] Systemic b] Fumigant
c] Contact d] Contact-plus stomach

117. **Monocondylic single mandible is found in :**
a] Hemiptera b] Thysanoptera
c] Hymenoptera d] Diptera

118. **Most of the insects excrete 80 to 90 per cent of their nitrogen waste in the form of :**
a] Urea b] Uric acid
c] Ammonia d] Ammonium nitrate

119. **Most of the caterpillars have legs or abdominal legs on segments :**
a] 2 to 5 and 10 b] 3 to 6 and 10
c] 6 to 10 d] 5 to 8 and 10

120. **Most of the insects cannot survive the temperatures above ____________ °C.**
a] 29.9 b] 39.9
c] 49.9 d] 59.9

121. **Most of the insects have abdominal segments :**
a] 6 to 7 b] 8 to 9
c] 10 to 11 d] 12 to 13

122. **Most of the insects possess :**
a] One pairs of wings b] Two pairs of wings
c] Three pairs of wings d] No wings

123. **Moulting fluid is able to digest :**
a] Cement layer b] Cuticulin layer
c] Endocuticle d] Exocuticle

124. **Name the fumigant which is used for killing the insects in storage as well as for plant fumigation ?**
a] Ethylene dibromide b] Methyl bromide
c] Aluminum phosphide d] EDCT mixture

125. **Name the insect pest which has migrated from India to other countries :**
a] *Sitophilus oryzae*
b] *Icerya purchasi*
c] *Trogoderma granarium*
d] *Sitotroga cereaella*

126. Name the insect which can survive at higher temperature and low humidity ?
a] Khapra beetle
b] Rice weevil
c] Rust red flour beetle
d] Angoumois grain moth

127. Name the insect which is known as oligophagous pest:
a] Pink boll worm of cotton
b] Brinjal fruit and shoot borer
c] Grass hopper
d] Cabbage butterfly

128. Name the insecticide which is harmless to the honey bees :
a] B.H.C. b] Carbaryl
c] Endosulfan d] Permethrin

129. Name the pest which has an international status ?
a] Pink boll worm of cotton
b] Locust
c] White grub
d] Army worm

130. Name the pest which has got thelytokous parthenogenesis in its life-cycle ?
a] Grass hopper b] Aphid
c] White flies d] Thrips

131. Nicotinic effects of organo-phosphatic insecticides results in :
a] Giddiness b] Stiffness of the neck
c] Ataxia d] None of the above

132. Number of abdominal legs in the larvae of mustard saw fly :
a] 4 pairs b] 6 pairs
c] 8 pairs d] 10 pairs

133. Number of insect species known :
a] About 50,000 b] About 1 million
ç] About 2 million d] About 3 million

134. *Odontotermes obesus* is the scientific name of :
a] Cutworm b] Termite
c] Shoot borer d] Top borer

135. One kilometer of locust swarm weighs as much as ________ tonnes of locusts.
a] 100 b] 200
c] 300 d] 400

136. Order Odanata includes :
a] May flies b] Dragon flies
c] Stone fliesd] Shad flies

137. Organisms which damage man's property, including plants and agricultural produce causing damage of significant economic importance are called :
a] Pests b] Pathogens
c] Parasites d] Aliens

138. Originally the phytophagous insects were :
a] Monophagous b] Oligophagous
c] Polyphagous d] All of the above

139. Out of the following poisons which works as an anti-coagulant for the control of rats ?
a] Zinc phosphide b] Strychmine
c] Warfarin d] Parathion

140. Out of the following which is considered to be of agricultural importance :
a] *Hydre* sp. b] *Pheritima* sp.
c] *Culex* sp. d] *Hirudinea* sp.

141. Palmitic and Oleic acids are prominent fatty acids in insects and comprise :
a] Over 40% of the total fatty acids
b] Over 50% of the total fatty acids
c] Over 60% of the total fatty acids
d] Over 70% of the total fatty acids

142. *Paratylenchus projectus* nematode has been found to cause consider-able damage in red clover yield at ________ nematodes/kg of soil.
a] 2-7 b] 8-13
c] 14-19 d] 20-25

143. Peritrophic membrane is absent in:
a] Orthoptera b] Isoptera
c] Coleoptera d] Hemiptera

144. Pest of maize belonging to family Crambidae :
a] *Chilo partellus* b] *Sesamia inferens*
c] *Tanyrhecus indicus* d] *Mythimna separata*

145. Pest of paddy belonging to family Fulgoridae :
a] *Tryporyza incertulus*
b] *Nilalparvata lugens*
c] *Nephotettix apicalis*
d] *Leptocorisa acuta*

146. Pest of sorghum belonging to order Diptera :
a] *Chilo partellus*
b] *Rhopalosiphum maidis*
c] *Bagrada cruciferarum*
d] *Atherigona soccata*

147. Pest population that produces incremental damage equal to the cost of preventing the damage is :
a] GEL b] PBL
c] EIL d] ETL

148. Pheromones are also known as :
a] Antibiotics b] Ectohomones
c] Hormones d] Attractants

149. Pheromones are synthesized by :
a] Thoracic glands
b] Glandular epidermal cells
c] Neuro-secretory cells
d] All of the above

150. Phorate is a ________ insecticide.
a] Systemic b] Contact plus systemic
c] Contact d] Contact plus stomach

151. **Phosdrin is a ________ insecticide.**
a] Slow fumigant
b] Systemic
c] Systemic plus contact
d] Contact

152. **Piercing and sucking type of mouth parts are found in :**
a] Grasshopper b] House fly
c] Aphids d] Beetles

153. **Plum moth and sphinx moth are the pests of :**
a] Bengal gram b] Green gram
c] Black gram d] All of the above

154. **Pneumatic sprayers have been used in pest control for many years. A pneumatic sprayer :**
a] Operates using wind or air
b] Contains poisonous pesticide
c] Is a knapsack type sprayer
d] Operates automatically

155. **Protein hydrolysate is used as an atractant for the control of :**
a] Fruit fly b] White fly
c] Butter fly d] Sawfly

156. **Respiratory system in which 1 mesothoracic and 7 abdominal spiracles are functional is called :**
a] Peripneustic b] Hemipneustic
c] Amphipneustic d] Metapneustic

157. **Rodentiside are used to kill :**
a] Rats and mites b] Insects
c] Weed d] None of these

158. ***Rodolia cardinalis* has been successful in controlling:**
a] Apple wolly aphis
b] Cottony cushion scale
c] Sugarcane shoot borer
d] Pink boll worm

159. **Sanjose scale is a pest of :**
a] Almond and Apple b] Pear and Plum
c] Mango and Guava d] Banana and Papaya

160. **Scale insects of citrus are controlled by :**
a] B.H.C. b] Boardaux mixture
c] Hydrocarbon oils d] D.D.T.

161. **Severe incidence of red hairy caterpillar occurs under which one of the following sets of conditions?**
a] High temperature and high soil moisture
b] High temperature and low soil moisture
c] Low temperature and high soil temperature
d] Low temperature and low soil moisture

162. **Small apterous insects, compound eyes reduced or absent ectopara-sites of birds and mammals are included in :**
a] Thysanoptera b] Hymenoptera
c] Mallophaga d] Strepsiptera

163. **Species of fossil-insects known :**
a] More than 10, 000 b] More than 20, 000
c] Less than 5, 000 d] Less than 1, 000

164. **Stem borer and pink borer can be controlled by :**
a] Thimet b] Dithane Z-78
c] Aldrin d] All of them

165. **Stick insects and leaf insects belong to the order :**
a] Orthoptera b] Phasmida
c] Embioptera d] Dictyoptera

166. **Stored grain insect-pests can easily be controlled by maintaining temperature at :**
a] 135^{0}F b] 150^{0}F
c] 165^{0}F d] 120^{0}F

167. **Stored grain pests can be killed by ________ hours exposure to 51.5-54.5^{o}C.**
a] 2 b] 3
c] 5 d] 7

168. **Sugarcane borer which does not belong to family pyralidae :**
a] *Emmalocera depressella*
b] *Chilo infuscatelus*
c] *Sesamia inferens*
d] *Chilo auricilia*

169. **Sugarcane leaf hopper can be effectively controlled by the use of:**
a] Malathion b] Lindane
c] Phosphamidon d] Toxaphene

170. **Termite control can laso be done by:**
a] Tilling b] Clean cultivation
c] Irrigation d] Trap cropping

171. **The gland which produces pheromone in queen honey bee :**
a] Mandibular gland
b] Maxillary gland
c] Labial gland
d] Pharyngeal gland

172. **The "Fauna of British India" started in :**
a] 1814 b] 1876
c] 1888 d] 1902

173. **The "Fauna of British India" was published by :**
a] Stebbing b] Hampson
c] Lefroy d] Fletcher

174. **The alimentary canal of insect is derived from :**
a] Ectoderm
b] Ectoderm and endoderm
c] Mesoderm and endoderm
d] Ectoderm and mesoderm

175. **The antennae are the appandages of which neuromere :**
a] Tritocerebrum b] Deutocerebrum
c] Labial ganglion d] Protocerebrum

GK GS
The India's largest fresh water lake is situated in Andhra Pradesh.

176. The author of "Indian Insect Life"-
a] Hussain b] Fletcher
c] Lefroy d] Mani

177. The average locust swarm spreads over :
a] 5 square kilometers
b] 10 square kilometers
c] 15 square kilometers
d] 20 square kilometers

178. The basis of antibiosis in corn against *Heliothis zea* is :
a] Nutritional imbalance
b] Presence of growth inhibitor
c] Lethal silk factor
d] Deficiency of nutrients

179. The basis of antibiosis in corn against European corn borer is :
a] Nutritional imbalance
b] Presence of growth inhibitors MBOA and DIMBOA.
c] Presence of toxins
d] Nutrition deficiency

180. The best method of control stored grain pests is :
a] Fumigation
b] Systemic insecticide
c] Proper drying and storage
d] Biological agents

181. The bio-type of BPH was first identified in Kerala on resistant variety :
a] IR-8 b] IR-10
c] IR-24 d] IR-26

182. The branch of zoology which deals with the naming and classification of animals is known as :
a] Taxonomy b] Binomics
c] Nomenclature d] Morphology

183. The caterpillars of *Indarbela qudarinotata* feed and damage :
a] Fruits b] Leaves
c] Flowers d] Bark

184. The Central Insecticide Act came into force from :
a] 1950 b] 1966
c] 1971 d] 1975

185. The centre for biological control in India is situated at :
a] Calcutta b] Lucknow
c] Bangalore d] Bombay

186. The chief excretory organ in insects is :
a] Malpighian tubules b] Green gland
c] Cuticle d] Fat body

187. The colour of grasshoppers blood is:
a] Red b] Blue
c] White d] Colourless

Quiz Kit!

First film star CM

M.G. Ramachandra, popularly known as MGR in the Tamil film world, was the first film star to become CM when he assumed office in Tamil Nadu on July 4, 1977.

188. The commodity totally banned for import :
a] Maize seeds b] Sunflower seeds
c] Groundnut seeds d] Sesame seeds

189. The components required for the study of ecosystem are/is :
a] Abiotic and biotic
b] Parasites and predators
c] Temperature and humidity
d] Soil

190. The control of caterpiller also done by :
a] *Bacillus thuriengiensis*
b] *Achromobacter* spp.
c] NPV
d] All of these

191. The damaging stage of cotton jassid is :
a] Caterpillar and Nymph
b] Nymph
c] Caterpillar
d] Nymph and Adult

192. The damaging stage of cut worm is :
a] Caterpillar b] Nymph
c] Adult d] All of these

193. The damaging stage of pink boll worm is :
a] Nymph b] Caterpillar
c] Adult d] Adult and Nymph

194. The damaging stage of top borer is :
a] Caterpillar b] Nymph
c] Adult d] None of these

195. The density of pest population at which control measures should be applied :
a] GEL b] PBL
c] EIL d] ETL

196. The device fitted at the end of the spray-lane and is used for atomisation is :
a] Hose b] Cut-off device
c] Boom d] Nozzle

197. The droplet size in normal high volume coarse spray is :
a] Less than 50 μm b] 50-100 μm
c] 100-300 μm d] 400-1000 μm

198. The duration of locust plague period lasts for ____________ years.
a] 2-3 b] 3-4
c] 5-10 d] 11-16

GK **Ustad Mansur specialised in animal painting.** GS

199. The entomogenous fungi tested under field conditions in India :
a] *Metarrhiizium anisopliae*
b] *Entomophthora virulenta*
c] *Beauveria bassiana*
d] All of the above

200. The enzyme that attacks the P-S bond in malathion metabolism is :
a] Carboxyesterase
b] Phosphates
c] Mixed function oxidase
d] P - 450

201. The exchange of gases in respiration in insects takes place through :
a] Spiracles b] Lungs
c] Book lungs d] Cuticle

202. The female mosquitoes possess :
a] Filiform antennae
b] Moniliform antennae
c] Plumose antennae
d] Serrate antennae

203. The fertilized female of *Laccifer lacca* lays eggs ranging form :
a] 100 to 200 b] 200 to 500
c] 500 to 700 d] Less than 100

204. The fertilized female of mulberry silkworm lays eggs within 24 hours :
a] 100 to 200 b] 200 to 300
c] 300 to 400 d] 400 to 500

205. The first Government Entomologist of India :
a] Niceville b] Stebbing
c] Pruthi d] Ramkrishna Ayyar

206. The first discovered protective insecticide was :
a] B.H.C. b] D.D.T.
c] Diazinon d] None of these

207. The first edition of "Systema Nature" was published in :
a] 1725 b] 1735
c] 1750 d] 1763

208. The first Indian Imperial Ento-mologist :
a] Hem Singh Pruthi
b] Afzal Hussain
c] Ram Chandra Rao
d] C.S. Mishra

209. The first known example of biological pest control was use of yellow ants to control the wormy fruits of orange in :
a] India b] U.S.A.
c] China d] U.S.S.R.

210. The first president of Entomological Society of India was:
a] Hem Singh Pruthi
b] Ramakrishan Ayyar
c] Afzal Hussain
d] Taskhir Ahmad

211. The first transformation with deltaendotoxin gene from Bt was done in Belglum in the year 1985 in:
a] Paddy b] Alfalfa
c] Tobacco d] Potato

212. The first volume of "Fauna of British India" which concerned Insects was on :
a] Beetles b] Bugs
c] Moths d] Wasps

213. The following chemosterilant is widely used against insects :
a] Apholate b] Altozar
c] Altocid d] Gossyplure

214. Coined the term 'Pest Management':
a] Geier b] Bartlett
c] Dr. Perry Adkisson d] Dr. Ray F. Smith

215. The function of peritrophic membrane is :
a] Absorption b] Assimilation
c] Protection d] All of above

216. The Government Bee of and Himachal Pradesh is located at :
a] Bhunter b] Bajaura
c] Manikarn d] Katrain

217. The grasshopper possess :
a] Natatorial legs b] Foragial legs
c] Saltatorial legs d] Ambulatorial legs

218. The grinding of food in grasshopper is done by :
a] Mandibles b] Crop
c] Gizzard d] Midgut

219. The groomingo of a new a queen in the honey bee colony depends upon the secretion known as :
a] Royal jelly b] Queen substance
c] Sucrose d] Honey

220. The incidence of "White grubs" is heavy in :
a] Heavily water-logged fields
b] Dry fields having sandy-loam soil
c] Dry fields of heavy clay soil
d] Desert land having little vegetation

221. The incidence of red hairy catepiller occurs under one of the following conditions :
a] High temperature and High soil moisture
b] High temperature and Low soil moisture
c] Low temperature and High soil moisture
d] Low temperature and Low soil moisture

222. The Indian Board of Wile Life was constititued in :
a] 1953 b] 1952
c] 1954 d] 1055

223. The Indian Lac Research Institute is located at :
a] Patna b] Cuttack
c] Gaya d] Ranchi

224. The infestation of which insect starts from the field?
a] Khapra beetle b] Angoumois moth
c] Rice weevil d] Tobacco beetle

Nagpur city has been proposed for setting up a plant converting plastic waste into oil.

225. The inorganic ions are absorbed into:
a] Fore-gut b] Fat bodies
c] Rectum d] None of above

226. The insect pest which introduced in our country before the enforcement of quarantine measures :
a] *Anthonomis grandis*
b] *Oxycaraenus laetus*
c] *Icerya purchasi*
d] *Gnorimochema heliopa*

227. The insecticide is considered highly selective if VSR ranges :
a] From 1 to 10
b] From 10 to 100
c] From 100 to 1000
d] From 1000 to 2000

228. The insects are able to maintain the balance while walking due to :
a] Hexapod b] Small size
c] Exoskeleton d] Wings

229. The largest locust swarm recorded so far :
a] 100 square kilometers
b] 200 square kilometers
c] 300 square kilometers
d] 400 square kilometers

230. The lateral movement theory of insecticidal penetration was proposed by :
a] Gerolt b] Matsmura
c] Ross d] Menton

231. The latest philosophy in insect pest control is :
a] Eradication b] Pest management
c] Control d] Integrated control

232. The legless young of a housefly is known as :
a] Caterpillar b] Maggot
c] Nymph d] Grub

233. The main characteristic feature of an insect is :
a] Presence of wings
b] Three pairs of legs
c] Three body regions
d] Pair of antennae

234. The main circulatory organ in grasshopper is :
a] Pericardial sinus
b] Haemocoel
c] Dorsal vessel
d] Perineural sinus

235. The main host tree for the cultivation of lac insect is:
a] Pipal b] Palas
c] Khejri d] Babul

236. The main host tree of eri-silk worm is :
a] Mulberry b] Sal
c] Castor d] Ber

237. The members of the phylum Echinodermata are exclusively :
a] Terrestrial b] Fresh water
c] Marin d] Amphibians

238. The minimum period to be given for the assessment of effectiveness of a biocontrol agent is :
a] 2 years b] 3 years
c] 4 years d] 5 years

239. The mouth parts of a plant bug are:
a] Bitting and chewing type
b] Coiled and suctorial type
c] Piercing and suctorial type
d] Suctorial proboscis type

240. The nematode *Paratylenchus* Microletzky (1992) members measure ______ mm in size.
a] 0.10-0.20 b] 0.25-0.50
c] 0.70-0.80 d] 0.9-0.10

241. The nervous system of insect is derived from :
a] Ectoderm b] Mesoderm
c] Endoderm d] Endoderm and mesoderm

242. The nozzle in which the spray fluid is fed into the air stream within the nozzle and spitted into spray droplets by the velocity of air is called :
a] Disc nozzle b] Annular nozzle
c] Shear nozzle d] Low volume nozzle

243. The nozzle used for producing mist:
a] Blast nozzle
b] Rotatory energy nozzle
c] Kinetic energy nozzle
d] Annular nozzle

244. The nozzle used to produce fog :
a] Gaseous energy nozzle
b] Thermal energy nozzle
c] Cermon energy nozzle
d] Kinetic energy nozzle

245. The number of segments which makes the insects head is :
a] 2 b] 3
c] 6 d] 8

246. The ommitidia in which the Samper cells produce hard crystalline cone are called :
a] Acone b] Eucone
c] Pseudocone d] Exocone

247. The origin of fore-gut is :
a] Ectodermal b] Mesodermal
c] Endodermal d] None of the above

248. The paddy variety 'Mudgo' is resistant to BPH because of :
a] The presence of growth inhibitors
b] Nutrition deficiency
c] Presence of toxins
d] Nutritional imbalance

GK GS
Raja Man Singh invented the Dhrupad.

249. The paralysis of honey bees is caused by :
a] Bacteria b] Virus
c] Fungus d] Mycoplasma

250. The population of fruit flies takes place in :
a] Soil
b] Rind of fruit
c] Stubbles in the field
d] Pulp of fruit

251. The potential produced in the dendrite of the sense organ by depolarisation is called :
a] Resting potential
b] Receptor potential
c] Generator potential
d] Action potential

252. The process of post-ecdysial hardening and darkening of the cuticle is regulated by :
a] Ecdysone
b] Juvenile hormone
c] PTTH
d] Bursicon

253. The prothoracicotropic hormone is secreted by :
a] Prothoracic glands
b] Neurosecretory cells
c] Corpora allata
d] Corpora cardiaca

254. The quantity of endosulfan 35% EC required for treating an area which needs 1000 litres of spray fluid at 0.05% (a.i.) strength is :
a] 35 x 0.05/1000 litres
b] 2.5 x 1000/1000 litres
c] 1000 x 0.05/35 litres
d] 35 x 1000/0.05 litres

255. The quantity of food grains in kg consumed by a single rat in a year is:
a] 5 b] 10
c] 15 d] 20

256. The queen in the honey colony feeds on :
a] Royal jelly b] Nectar
c] Queen substance d] Honey

257. The queen termite produces ___________ eggs per second.
a] 40 b] 50
c] 60 d] 70

258. The ratio of zinc phosphide in poison bait for rats is :
a] 1% b] 2%
c] 3% d] 4%

259. The reabsorption of water in insect takes place in :
a] Malpighian tubules
b] Hepatic caecea
c] Rectal papillae
d] Ventriculus

260. The recommended dose of NPV for the control of *Heliothis* and *Spodoptera* under field conditions is :
a] 125 to 250 LE b] 250 to 500 LE
c] 500 to 700 LE d] 750 to 1000 LE

261. The reproductive system of insect is derived form :
a] Ectoderm
b] Endoderm
c] Ectoderm and mesoderm
d] Ectoderm and endoderm

262. The safest method of disposing off the left-over pesticide is disposal by:
a] Pouring in the drain
b] Pouring in a pond
c] Pouring in a moving stream
d] Burying in the soil

263. The size (in micrometers in diameter) of droplets produced by high volume sprayers should most appropriately be :
a] 200 to 500 b] 100 to 150
c] 81 to 100 d] 30 to 40

264. The sperms after copulation in insects are stored in the female in :
a] Vagina b] Spermatheca
c] Accessory glands d] Seminal vesical

265. The spray is classified as aerosol when average droplet size (volume median diameter) is :
a] 50 μ b] 50-100 μ
c] 100-200 μ d] 200-400 μ

266. The stability of ecosystem is maintained by :
a] Homeostasis b] Diversity of species
c] Food web d] Trophic levels

267. The strain of the lac insect which survives on all hosts is known as :
a] Kusumi b] Rangini
c] Lakhini d] Namkumi

268. The striking difference in male and female cockroach is in :
a] Shape of the wings
b] Number of antennal segments
c] Presence or absence of anal style
d] Structure of the compound eyes

269. The synergist of malathion is :
a] Piperonyl butoxide
b] Sesamolin
c] N-isoibutylundecyleneamide
d] Sesamax

270. The terminal abdominal segments 6 to 9 are telescoped within the segment 5 when not in use. This condition is found in :
a] Butter fly b] House fly
c] Dragon flyd] White fly

271. The type of parthenogenesis where only males are produced is :
a] Arrhenotoky b] Thelytoky
c] Amphitoky d] None of the above

272. The ultimate control through biotic agencies is expected after :
a] One generation
b] Two generations
c] Three generations
d] More than three generations

GK GS

Gangotri glacier of Uttaranchal receding at a faster pace threatening the mighty river Ganga.

273. The virus used successfully for the control of caterpillars of gram pod borer is :
a] YMV b] TMV
c] Granulosis virus d] NPV

274. The wings of butterfly and birds are:
a] Homologous organs
b] Analogous organs
c] Vestigeal organs
d] Circulatory organs

275. Token stimuli theory was postulated by :
a] Throsteinson b] House
c] Painter d] Fraenkel

276. Tomato leaf curl virus is transmitted by :
a] Aphid b] Whitefly
c] Grasshopper d] Butterfly

277. Tree banding is useful for the control of :
a] Mango hopper
b] Bark eating caterpiller
c] Mango mealy bug
d] Mango stem borer

278. Trenching is recommended for the control of either of these :
a] White grub bettle
b] Maize stem borer
c] Red hairy caterpillar
d] Army worm

279. Tripple action nozzle provides one of the following spray patterns :
a] Holo cone, Flat fan, and Solid cone
b] Holo cone and Jet
c] Holo cone, Solid cone and Jet
d] Holo cone, Jet and Flat fan

280. Ultra-low volume (ULV) or low volume concentration (LVC) refers to the use of spray volume (liquid) less than ____________ litres.
a] 5 b] 7
c] 9 d] 11

281. Use of resistant varieties constitutes method of :
a] Genetic control b] Self control
c] Biotic control d] Natural control

282. Vegetables treated with B.H.C. are unfit for use for :
a] Two weeks b] One week
c] One month d] Two months

283. Vertical resistance is also called :
a] Oligogenic b] Specific
c] Monogenic d] All of the above

284. When 1 meso, metathoracic and 8 abdominal spiracles are functional, the respiratory system is called?
a] Holopneustic b] Peripneustic
c] Hemipneustic d] Metaneustic

Did You Know ?
First jute mill

Wellington Jute Mill in Rishra in West Bengal was established by George Auckland in 1855 as the Rishra Twine and Yarn Mills Ltd.

285. Where the Directorate of Plant Protection, Quarnatine and Stroage is situated ?
a] New Delhi b] Faridabad
c] Hissar d] Gurgaon

286. Where the nector is converted into honey ?
a] In the alimentary canal of queen
b] In the alimentary canal of worker
c] In royal chamber
d] In special cells of the hive

287. Which animal acts as natural tiller of soil ?
a] Termite b] Earthworm
c] Snake d] Millipede

288. Which group of insecticide is carcenogenic ?
a] Organophosphates
b] Carbamates
c] Synthetic pyrethroids
d] Fumigants

289. Which insecticide belongs to synthetic pyrethroid group ?
a] Pyrethrum b] Cypermethrin
c] Rotenone d] Sabadilla

290. Which insecticide has contact and fumigant action ?
a] Phosphamidon b] Diazinon
c] Lindane d] Endosulfan

291. Which insecticide has systemic action ?
a] Endosulfan b] Chlorpyiphos
c] Methyldemeton d] Diazinon

292. Which insecticide is more persistent in the environment ?
a] Parathion b] Aldrin
c] DDT d] BHC

293. Which insecticide is stored in the fat of milch animals?
a] Malathion b] DDT
c] Lindane d] Endosulfan

294. Which insecticide should be used on vegetables safely before harvesting?
a] Lindane b] Endosulfan
c] Malathion d] Carbaryl

295. Which is the best suiting species for bee culture ?
a] *Apis dorsata* b] *Apis florea*
c] *Apis indica* d] *Milipona* sp.

296. Which is the larvicidal fish ?
a] *Gambusia* sp. b] *Barbus* sp.
c] *Labeo* sp. d] *Catle* sp.

Afzal Khan belonged to Bijapur.

297. **Which is the international insect pest ?**
a] *Healiothis armigera*
b] *Amsacta moorei*
c] *Schistocerca gregaria*
d] *Holotrichia consanguinea*

298. **Which is the land frontier from the point of view of quarantine measures?**
a] Kalimpong b] Calcutta
c] Kandla d] New Delhi

299. **Which of *Entomophthora* species is pathogenic to aphids ?**
a] *E. grylli* b] *E. coronata*
c] *E. sphasperma* d] None of the above

300. **Which of following combination is synergistic ?**
a] Carbaryl +Piperonylbutoxide
b] Malathion + Piperonylbutoxide
c] Methyl parathion + Piperonylbutoxide
d] Diazinon + iperonylbutoxide

301. **Which of following insect infests the grains in the field and then comes to storage ?**
a] *Ephestia cautella*
b] *Tribolium castaneum*
c] *Sitotroga cerealella*
d] All of the above

302. **Which of the bee hive is more popular in India ?**
a] Thompson hive b] Newton hive
c] Dadant hive d] Langstroth hive

303. **Which of the fly is pest of cotton ?**
a] Fruit fly b] White fly
c] Stem fly d] Saw fly

304. **Which of the followiing fungi is predecious on nematodes ?**
a] *Didymella* sp. b] *Dactytaria* sp.
c] Both of them d] None of them

305. **Which of the following aphid feeds on pulses ?**
a] *Rhopalosiphum maidis*
b] *Aphis craccivora*
c] *Myzus persicae*
d] *Lipaphis erysimi*

306. **Which of the following are agricultural pests ?**
a] Sparrow, grasshopper, fungi, bacteria
b] Vultures, Cranes
c] Man
d] Mangoose

307. **Which of the following are not haemocytes ?**
a] Plasmocytes b] Cytocytes
c] Nephrocytes d] Coagulocytes

308. **Which of the following are systemeic fungicides ?**
a] Vitavax b] Benlate
c] Brestanol d] All of these

309. **Which of the following bug is a pest of brinjal ?**
a] Painted bug b] Plant bug
c] Lace bug d] Stink bug

310. **Which of the following chemical is used as fumigant against insect pests of stored food grains?**
a] D.D.T. b] Methyl bromide
c] Pyrethrin d] All of the above

311. **Which of the following combination is not antagonistic?**
a] Malathion + Parathion
b] Parathion + Azinphosmethyl
c] Demeton + Trichlorphon
d] Carbaryl + Seasmex

312. **Which of the following cotton pest belongs to family Pyrrhocoridae ?**
a] *Dysdercus singulatus*
b] *Pectinophora gossypiella*
c] *Earias fabia*
d] *Bamisia tabaci*

313. **Which of the following does not belong to order Lepidoptera ?**
a] Groundnut leaf miner
b] Citrus leaf miner
c] Pea leaf miner
d] None of the above

314. **Which of the following insecticide controls the white grubs effectively in the soil ?**
a] BHC 10% dust b] Phorate 10% G
c] Aldicarb 10% G d] Quinalphos 5% G

315. **Which of the following insects are migratory in nature**
a] *Danaus* b] *Aphids*
c] Locusts d] All of the above

316. **Which of the following is abdominal appendage?**
a] Gonopore b] Cerci
c] Waist d] Apophysis

317. **Which of the following is called "water scorpion" ?**
a] *Belostoma* b] *Nepa*
c] *Corixa* d] *Notonecta*

318. **Which of the following is cross vein?**
a] Costa b] Tegula
c] Scutum d] Phragmata

319. **Which of the following is larval parasitoid ?**
a] *Xanthopimpla*
b] *Tetrastichus*
c] *Apanteles*
d] None of the above

320. **Which of the following is larval-pupal parasitoid ?**
a] *Campoletis* b] *Apanteles*
c] *Exorista* d] *Carcelia*

321. **Which of the following is lateral sclerite?**
a] Basisternum b] Episternum
c] Pre-sternum d] Sternum

322. **Which of the following is neuropteran predator ?**
a] *Menochilus* b] *Chrysoperla*
c] *Brumus* d] *Crypotlaemus*

GK GS
Carbonmonoxide (CO) is main polluting gas in automobile emitted gases.

323. Which of the following is not a pest of stored cereals?
a] *Callosobrucus chinensis*
b] *Corcyra cephalonica*
c] *Sitotroga cerealella*
d] *Ephestia cautella*

324. Which of the following is not head sclerite ?
a] Gena b] Frons
c] Epimeron d] Vertex

325. Which of the following is not the pest of paddy ?
a] *Tryporyza incertulus*
b] *Lampides boeticus*
c] *Nilaparvata lugnes*
d] *Nephotettix nigropicta*

326. Which of the following is parasitic nematode ?
a] *Leptomonas pyrrhocoris*
b] *Nosema bombycis*
c] *Perezia meslini*
d] None of the above

327. Which of the following is parasitic protozoan attacking silkworm ?
a] *Leptomonas pyrrhocoris*
b] *Malpighamoeba mellifica*
c] *Adeline tipulae*
d] *Nosema bombycis*

328. Which of the following is pest :
a] *Cryptolaemus scymnus*
b] *Brumus saturalis*
c] *Coccinella septumpunctata*
d] *Epilachna dudocastigma*

329. Which of the following is pupal parasitoid ?
a] *Elasmus* b] *Tetrasticus*
c] *Xanthopimpla* d] All of the above

330. Which of the following is systemic insecticide ?
a] Parathion b] Malathion
c] Cabaryl d] Carbofuran

331. Which of the following is the European species ?
a] *Apis dorsata* b] *Apis indica*
c] *Apis mellifera* d] *Apis florea*

332. Which of the following is the pest of apple ?
a] *Sylepta lunalis*
b] *Nodostoma subcostalum*
c] *Cydia hemidoxa*
d] *Eriosoma lanigerum*

333. Which of the following is the pest of grapevine ?
a] *Phyllocnistris citella*
b] *Phytomyza atricomis*
c] *Cydia hemidoxa*
d] *Scelodonata strigicollis*

334. Which of the following muscles are directly attached to wing?
a] Dorsal muscles b] Axillary muscles
c] Basalar muscles d] Sub-alar muscles

335. Which of the following order comprises parasitic insects ?
a] Hemiptera b] Phasmida
c] Thysanoptera d] Lepidoptera

336. Which of the following order comprises phytophagous insects ?
a] Odonata b] Siphonoptera
c] Phasmida d] Mallophaga

337. Which of the following order comprises predaceous insects ?
a] Phasmida b] Neuroptera
c] Isoptera d] Lepidoptera

338. Which of the following organism is not storage pest ?
a] Rodents b] Nematodes
c] Bacteria d] Fungi

339. Which of the following organs are involved in excretion?
a] Malpighian tubules
b] Integument
c] Wall of the alimentary canal
d] All of the above

340. Which of the following parasites that attacks plants are useful to their hosts ?
a] Root nodule bacterium
b] Root knot nematode
c] Rice plant fungus
d] Rust fungi

341. Which of the following part of sprayer breaks the spray liquid into fine droplets ?
a] Nozzle b] Agitator
c] Pump d] None of these

342. Which of the following pest belongs to family Dermestidae ?
a] *Sitophilus oryzae*
b] *Trogoderma granarium*
c] *Tribolium casteneum*
d] *Latheticus oryzae*

343. Which of the following pesticides may act as contact poisons, stomach poisons and fumigants ?
a] B.H.C. b] D.D.T.
c] Malathion d] All of the above

344. Which of the following pests attack maize crop ?
a] Jassid b] Stem borer
c] Grassy cut worm d] None of these

345. Which of the following pests attack on brinjal ?
a] Shoot and stem borer
b] Stem fly
c] Cabbage borer
d] All of these

346. Which of the following predator does not belong to order Coleoptera ?
a] *Chilomenes* b] *Coccinella*
c] *Scymnus* d] Syrphus

347. Which of the following produces more lac and is important for commercial production ?
a] Male b] Female
c] Nymph d] All of above

GK GS

Kavi Kalas was a minister of Sambhaji.

348. Which of the following spray pump is used for high volume spraying ?
a] Plunger pump
b] Gear type pump
c] Roller vane rotary pump
d] None of the above

349. Which of the following spray pump is used for low volume spraying ?
a] Pneumatic pump
b] Piston pump
c] Gear type rotary pump
d] None of these

350. Which of the following spray pump is used for medium volume spraying?
a] Roller vane rotary pump
b] Gear type rotary pump
c] Plunger pump
d] All of the above

351. Which of the following statement is not correct ?
a] All pests are pathogens
b] All pathogens are pests
c] Some parasites are beneficial to plants
d] Some flowering plants are agricultural pests

352. Which of the following structure is associated with the articulation of the wings ?
a] Trochanter b] Tegula
c] Scutum d] Phragmata

353. Which of the following types of insecticides are not used against food grain pests in storage ?
a] Organic insecticides
b] Fumigants
c] Inorganic insecticides
d] All of the above

354. Which of the following types of pump is used in a hand compression sprayer ?
a] Air or pneumatic pump
b] Rotary pump
c] Piston pump
d] All of the above

355. Which of the toxicity creates widespread ecological problem?
a] Acute toxicity b] Dermal toxicity
c] Mammalian toxicity d] Chronic toxicity

356. Which oil acts as a repellent for mosquitoes?
a] Linseed oil b] Cedar oil
c] Citronella oil d] Sesamum oil

357. Which one is used for inoculation of host tree for lac cultivation ?
a] Seed lac b] Brood lac
c] Phunki lac d] Eri lac

358. Which one of the following can be classified as autocidal technique of insect control?
a] Introduction of bio-agents
b] Evolving resistant cultivars
c] Use of insect growth regulators
d] Release of sterile males

359. Which one of the following can be classified as hydraulic energy sprayer?
a] Knapsack mist blower
b] Controlled drip applicator sprayer
c] Knapsack sprayer
d] Wheel barrow sprayer

360. Which one of the following equipments can be effectively used for spraying an orchard crop?
a] Barrel pump b] Sikar pump
c] Foot sprayer d] Heli sprayer

361. Which one of the following insect is a gregareous pest?
a] Sorghum shoot fly b] Spotted bollworm
c] Migratory locust d] Rice caseworm

362. Which one of the following insecticides is capable of controlling mites as well?
a] Endosulfan b] Toxaphene
c] Cupermethrin d] Monocrotophos

363. Which one of the following insecticides is the most suitable for the control of gram pod borer?
a] BHC b] Aldrin
c] Endosulfan d] Dimethoate

364. Which one of the following insects has not much changed from it fossil records?
a] Butterfly b] Praying mantis
c] Cockroach d] Grasshopper

365. Which one of the following pairs is not correctly matched?
a] Biological control-Cottony cushion scale
b] Chemical control- Maize stem borer
c] Cultural control- Locust
d] Male sterlization technique -Screw wormfly

366. Which one of the following pests is controlled through the application of a grease band around the tree trunk?
a] Mango mealy bug
b] Trunk borer of mango
c] Codling moth of apple
d] Woolly aphis of apple.

367. Acricide is :
a] Pirmicarb b] Dicofol
c] Profenfos d] Sevidol

368. Which one of the following sprayers requires only one-third of the quantity of water while keeping the same quanity of active ingredient of the pesticide per hectare?
a] Foot/Pedal sprayer b] Mist blower
c] Knapsack sprayer d] Rocking sprayer

369. Which one of the following sprayers will cause more loss of spray droplets by drift?
a] Low volume sprayer
b] High volume sprayer
c] Ultra low volume sprayer
d] Duster

GK 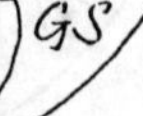

Trees of Mangroves forests are salt tolerant in nature.

370. Which one of the following sprayers would require the lowest volume of spray solution to cover a hectare of rice crop?
a] Power sprayer b] Foot sprayer
c] Knapsack sprayer d] Hand sprayer

371. Which one of the following truly diagnoses the infestation by the sorghum shootfly?
a] Formation of 'dead heart' in the early seedling stage of sorghum which can be easily pulled out
b] Formation of 'dead heart' and a bunchy top during the later seedling stage of sorghum
c] Formation of 'dead heart' which cannot be easily pulled out during the later seedling stage of sorghum
d] Formation of silvery shoot during the early seedling stage of sorghum

372. Which one of the following types of spraying is best done by manually operated knapsack sprayer having flat fan nozzle?
a] High volume b] Low volume
c] Ultra low volume d] Semi low volume

373. 'Hopper burn' appears in
a] Wheat b] Sugarcane
c] Sorghum d] Rice

374. Which state is famous for sericulture?
a] Kerala b] Karnataka
c] Andhra Pradesh d] Uttar Pradesh

375. Which term is used for the movement of coxa towards the body?
a] Protraction b] Promotion
c] Adduction d] Abduction

376. Which type of special reproduction is found in aphid?
a] Polyembryony b] Paedogenesis
c] Parthenogenesis d] Viviparity

You Don't Know!
First woman minister
Vijayalakshmi Pandit was Minister for Local Government and Health in 1937. A protagonist for India's Independence, she underwent more than two years of imprisonment by the British for her rebellious activities.

377. White ear-head in paddy is caused by :
a] Gall fly b] Green leaf hopper
c] Yellow stem borer d] Stink bug

378. White grub bettles emerge from soil in between :
a] 7.0 - 8.0 p.m. b] 8.0 - 9.0 p.m.
c] 4.0 - 5.0 a.m. d] 6.0 - 7.0 p.m.

379. Who for the first time studied Indian insect type?
a] Linnaeus b] Koenig
c] Fabricius d] Lefroy

380. Who is often referred to as the 'Father of Animal Classification'?
a] Aristotle b] Carolus Linnaeus
c] Wallace d] Hugo de Varies

381. Whorl application of granular insecticide provides effective control measure against :
a] Spotted bollworm b] Maize stem borer
c] Stem borer paddy d] Mango stem borer

382. Wingless hexapod with 6 abdominal segments :
a] Diplura b] Thysanura
c] Collembolad] Protura

383. *Xiphinema cobbi* nematode has been described after 1980 from:
a] Brazil b] South Africa
c] India d] Italy

GK GS
Amatya was finance minister.

Matching

1. **Match the pests with their families.**

a]	*Holotrichia consanguinea*	i]	Muscidae
b]	*Atherigonia varia soccata*	ii]	Scarabaeidae
c]	*Chilo partellus*	iii]	Arctiidae
d]	*Amsacta moorel*	iv]	Crambidae

2. **Match the pests with their families.**

a]	*Nilaparvata lugens*	i]	Coreidae
b]	*Niphotettix nigropicta*	ii]	Meloidae
c]	*Leptocorisa acuta*	iii]	Jassidae
d]	*Mylabris pustulata*	iv]	Fulgoridae

3. **Match the pests with their families.**

a]	*Lampides boeticus*	i]	Pyralidae
b]	*Etiella zinckenella*	ii]	*Lycaenidae*
c]	*Pyrilla purpusilla*	iii]	*Noctuidae*
d]	*Sesamia inferens*	iv]	*Fulgoridae*

4. **Match the pests with their families.**

a]	*Corcyra cephalonica*	i]	*Phycitidae*
b]	*Sitrotroga cerealella*	ii]	Gelechiidae
c]	*Ephestia cautella*	iii]	Dermestidae
d]	*Anthrenus flavipes*	iv]	Galleriidae

5. **Match the pests with their families.**

a]	*Eriosoma lanigerum*	i]	Coccidae
b]	*Sylapta lunalis*	ii]	Aleurodidae
c]	*Dialeurodes citri*	iii]	Aphididae
d]	*Drosicha mangiferae*	iv]	Pyralidae

6. **Match the pests with their families.**

a]	*Emmalocera depressella*	i]	Chrysomelidae
b]	*Longitarsus nigripennis*	ii]	Pyralidae
c]	*Spodoptera litura*	iii]	Anobiidae
d]	*Lasioderma serricorne*	iv]	Noctuidae

7. **Match the pests with their families.**

a]	*Agrotis ipsilon*	i]	Tenebrionidae
b]	*Triboloium castaneum*	ii]	Noctuidae
c]	*Myzus persicae*	iii]	Gelechiidae
d]	*Pectinophora gossypiella*	iv]	Aphididae

8. **Match the pests with their families.**

a]	*Earias fabia*	i]	Apionidae
b]	*Apion corchori*	ii]	Aleurodidae
c]	*Bemisia tabaci*	iii]	Pyralidae
d]	*Dysdercus singulatus*	iv]	Pyrrhocoridae

9. **Match the pests with their families.**

a]	*Phytomyza atricornis*	i]	Noctuidae
b]	*Plusia orichalcea*	ii]	Agromyzidae
c]	*Amrasca devastans*	iii]	Coccinellidae
d]	*Epilachna demurili*	iv]	Cicadellidae

10. **Match the pests with their families.**

a]	*Plutella xylostella*	i]	Noctuidae
b]	*Phyllotreta cruciferae*	ii]	Plutellidae
c]	*Leucinodes orbonalis*	iii]	Chrysomelidae
d]	*Heliothis arrnigera*	iv]	Pyraustidae

GK GS

'Aum Neemplus' is synthetic form of Neem can be used as a pesticide.

11. Match the pests with the crops they feed on.

a]	*Tanymecus indicus*	i]	Paddy
b]	*Calocoris angustatus*	ii]	Maize
c]	*Athalia proxima*	iii]	Sorghum
d]	*Tryporyza incertuhus*	iv]	Mustard

12. Match the pests with the crops they feed on.

a]	*Atherigona soccata*	i]	Pigeon pea
b]	*Nilaparvata lugens*	ii]	Brinjal
c]	*Exelastis erysimi*	iii]	Sorghum
d]	*Urentis serites*	iv]	Paddy

13. Match the pests with the crops they feed on.

a]	*Dacus dorsalis*	i]	Citrus
b]	*Othereis fullonica*	ii]	Mango
c]	*Lipaphis erysimi*	iii]	Sugarcane
d]	*Bissetia steniellus*	iv]	Mustard

14. Match the pests with the crops they feed on.

a]	*Longitarsus nigripennis*	i]	Cardamom
b]	*Dichocrosis punctiferallis*	ii]	Black pepper
c]	*Gnorimoschema heliopa*	iii]	Cotton
d]	*Platyedra gossypiella*	iv]	Tobacco

15. Match the pests with the crops they feed on.

a]	*Anomis sabulifera*	i]	Jute
b]	*Quadraspidiotus perniciosus*	ii]	Grapevine
c]	*Scelodonta strigicollis*	iii]	Apple
d]	*Etiella zinckenella*	iv]	Pea

16. Match the pests with the crops they feed on.

a]	*Gnorimoschema operculella*	i]	Cabbage
b]	*Plutella xylostella*	ii]	Potato
c]	*Aulacophora foveicollis*	iii]	Castor
d]	*Achaea janata*	iv]	Pumpkin

17. Match the pests with the crops they feed on.

a]	*Heliothis armigera*	i]	Lady's finger
b]	*Spodoptera litura*	ii]	Gram
c]	*Andraca bipunctata*	iii]	Cauliflower
d]	*Melanagromyza hibisci*	iv]	Tea

18. Match the pests with their common names.

a]	*Corcyra cephalonica*	i]	Almond moth
b]	*Sitotroga cerealella*	ii]	Indian meal moth
c]	*Polodia intepunctella*	iii]	Angoumois moth
d]	*Ephestia cautella*	iv]	Rice moth

19. Match the pests with their common names.

a]	*Myllocerus maculosus*	i]	Yellow stem borer
b]	*Mylabris pustulata*	ii]	Brown plant hopper
c]	*Tryporyza incertulus*	iii]	Blister beetle
d]	*Nilaparvata lugens*	iv]	Grey weevil

20. Match the pests with their common names.

a]	*Bissetia steniellus*	i]	Internode borer
b]	*Scirpophaga nivella*	ii]	Gurudaspur borer
c]	*Chilo sacchariphagus*	iii]	Root borer
d]	*Emmalocera depressella*	iv]	Top borer

GK **Saranjami system is related to Marathas.** GS

21. **Match the pests with their common names.**

a]	*Bagrada cruciferarum*	i]	Saw fly
b]	*Piens brassicae*	ii]	Diamond back moth
c]	*Plutella xylostella*	iii]	Cabbage butterfly
d]	*Athalia proxima*	iv]	Painted bug

22. **Match the pests with their common names.**

a]	*Trogoderma granarium*	i]	Carpet beetle
b]	*Tribolium castaneum*	ii]	Pulse beetle
c]	*Callosobruchus chinensis*	iii]	Confused flour beetle
d]	*Anthrenus flavipes*	iv]	Khapra beetle

23. **Match the common names of the honey bees with their scientific names.**

a]	European bee	i]	*Apis dorsata*
b]	Rock bee	ii]	*Apis indica*
c]	Indian bee	iii]	*Apis florea*
d]	Little bee	iv]	*Apis mallifera*

24. **Match the common names of locusts with their scientific names.**

a]	Desert locust	i]	*Locusta migratoria*
b]	Migratory locust	ii]	*Locustana parudalina*
c]	Red locust	iii]	*Schistocerca gregaria*
d]	Brown locust	iv]	*Nomadacris septumfasciata*

25. **Match the pests with the parts of the plants they feed on.**

a]	Italian locust	i]	*Patanga succinta*
b]	Australian locust	ii]	*Schistocerca paranensis*
c]	South Americal locust	iii]	*Calliptamus italicus*
d]	Bombay locust	iv]	*Chortocetes terminefera*

26. **Match the pests with the parts of the plants they feed on.**

a]	*Othereis fullonica*	i]	Inflorescence
b]	*Indarbela quadrinotata*	ii]	Stem
c]	*Tryporyza incertulas*	iii]	Bark
d]	*Mylabris pustulata*	iv]	Fruit

27. **Match the pests with their damage symptoms they cause.**

a]	*Sylepta derogata*	i]	White earhead
b]	*Tryporyza incertulas*	ii]	Dead-heart
c]	*Phyllocnistics citrella*	iii]	Rolling of leaf
d]	*Chilo partellus*	iv]	Mines in the leaf

28. **Match the sugarcane borers with their damage symptoms they cause.**

a]	*Emmalocera depressella*	i]	Bunchy top at later stage
b]	*Bissetia steniellus*	ii]	Dead-hearts at initial stage which can be pulled out easily
c]	*Scirpophaga nivella*	iii]	Spiral galleries
d]	*Chilo infuscatellus*	iv]	Dead-hearts which can not be pulled out easily

29. **Match the types of the silk with the silkworms producing them.**

(a)	Tasar silk	i]	*Attacus ricini*
b]	Mulberry silk	ii]	*Antheraea assama*
c]	Muga silk	iii]	*Bombax mori*
d]	Eri silk	iv]	*Antheraea paphia*

30. **Match the diseases of silkworms with their causal organisms.**

a]	Maggot disease	i]	Virus
b]	Pebrine disease	ii]	*Bacillus thuringiensis*
c]	Polydedrosis	iii]	*Nosema bombycis*
d]	Bacterial :ntoxication	iv]	*Tricholyga sorbillans*

31. Match the silkworms with their host-plants.

a]	Mulberry silkworm	i]	*Ricinus communis*
b]	Tasar silkworm	ii]	*Morus sp.*
c]	Muga silkworm	iii]	*Terminalia arjuna*
d]	Eri silkworm	iv]	*Machilus bombyciana*

32. Match the lac types with their inoculation periods.

a]	Kartiki	i]	January-February
b]	Baisakhi	ii]	July
c]	Aghani	iii]	October-November
d]	Jethi	iv]	June- July

33. Match the name of the scientists with the theory of resistance they postulated.

a]	House (1966)	i]	Non-preference and antibiosis theory
b]	Fraenkel (1959)	ii]	Dual discrimination theory
c]	Painter (1951)	iii]	Token stumull theory
d]	Kenedy and Booth (1951)	iv]	Nutrient imbalance theory

34. Match the parasitic fungi with their insect hosts.

a]	*Entomophthjora grylii*	i]	Lepidopterous caterpillars
b]	*Entomophthora coronata*	ii]	House fly
c]	*Entomophthora sphaesperma*	iii]	Grasshoppers
d]	*Entomophthora muscae*	iv]	Aphids

35. Match the pathogens with their groups they belong.

a]	*Serratia marcescens*	i]	Fungi
b]	*Metarhizium anisopliae*	ii]	Bacteria
c]	NPV	iii]	Virus
d]	*Neoplectana*	iv]	Nematode

36. Match the bacterial pathogens with their sporulation.

a]	Obligate	i]	Spore forming with phospholypase -C
b]	Crystalliferous	ii]	Spore forming, no toxin
c]	Facultative	iii]	Spore forming with protein crystal
d]	Non-crystalliferous	iv]	Spore forming without crystal.

37. Match the virus diseases with their aphid transmitter.

a]	Potato leaf roll virus	i]	*Bravicoryne brassicae*
b]	Sugarcane mosaic virus	ii]	*Myzus persicae*
c]	Crucifers mosaic virus	iii]	*Myzus persicae and Aphis gossypii*
d]	Bean mosaic virus	iv]	*Aphis maidis*

38. Match the predators with their families they belong.

(a)	*Menochilus*	i]	Coccinellidae
b]	*Chysoperla*	ii]	Pentatomidae
c]	*Andrallus*	iii]	Reduviidae
d]	*Rhinocoris*	iv]	Chrysopidao

39. Match the insects with their antennae.

(a)	Moth	i]	Aristate
b]	Pulse beetle	ii]	Pectinate
c]	Thrips	iii]	Serrate
d]	House fly	iv]	Moniliform

40. Match the antennae with the insects in which they are found.

a]	Bipectinate	i]	Mosquito
b]	Plumose	ii]	Silkworm
c]	Geniculate	iii]	Butterfly
d]	Clavate	iv]	Honey bee

Prithvi missile of India is capable of carrying nuclear weapons.

41. Match the insects with the type of mouth parts they have.

a]	*Thrips tabaci*	i]	Biting and chewing
b]	*Musca domestica*	ii]	Piercing and sucking
c]	*Dysdercus cingulatus*	iii]	Rasping and sucking
d]	*Tribolium cateneum*	iv]	Sponging

42. Match the insects with the type of legs they have.

a]	*Gryllotalpa*	i]	Jumping
b]	Mantid	ii]	Grooming
c]	Honey bee	iii]	Grasping
d]	Flea beetle	iv]	Digging

43. Match the receptors with which they are responsive.

a]	Progreoreceptors	i]	Light
b]	Hygoreceptors	ii]	Humidity
c]	Mechanoreceptors	iii]	Body movement
d]	Photoreceptors	iv]	Pressure

44. Match the terms with the movements they are used.

a]	Protraction	i]	Backward movement of the leg
b]	Retraction	ii]	Movement of coxa towards the body
c]	Adduction	iii]	Movement of coxa away from the body
d]	Abduction	iv]	Complete forward movement of the leg

45. Match the layers of the cuticle with the factors they are barries for.

a]	Cement layer	i]	Oxygen and other gas molecules
b]	Wax layer	ii]	Ions
c]	Cuticulin layer	iii]	Water
d]	Basement memberane	iv]	What enters from the blood

46. Match the enzyme with the substrate or site they act on.

a]	Exo - amylase	i]	Bonds within the starch molecule
b]	Endo - amylase	ii]	Sucrose
c]	α - glucosidease	iii]	Lactose
d]	β - glucosidase	iv]	Splits maltose from the end of starch molecule

47. Match the hormones with their producer.

a]	Prothoracicotropic	i]	Prothoracic glands
b]	Ecdysone	ii]	Perivisceral organs
c]	Juvenile	iii]	Corpora allata
d]	Bursicon	iv]	Neurosecretory cells

48. Match the Bt genes with their target order.

a]	Cry - I	i]	Coleoptera
b]	Cry - II	ii]	Diptera
c]	Cry - III	iii]	Lepidoptera
d]	Cry - IV	iv]	Lepidoptera and Diptera

49. Match the insecticides with the sources they are obtained.

a]	Pyrethrum	i]	*Lumbrinereis hetropoda*
b]	Nicotine	ii]	*Chrysanthemum cineriaefolium*
c]	Rotenone	iii]	*Nicotiana tabacum*
d]	Nereistoxin	iv]	*Derris elliptica*

50. Match the insecticides with the groups they belong.

a]	Isolan	i]	Oxime carbamate
b]	Carbaryl	ii]	Phenyl carbamate
c]	Carbofuran	iii]	Naphthyl carbamate
d]	Aldicarb	iv]	Heterocylic carbamate

Baji Rao I popularised the ideal of Hindu Pad Padshahi.

51. **Match the insecticides with their derivatives.**

a]	Phosphamidon	i]	Pyrophosphoric
b]	Parathion	ii]	Di-thiophosphoric acid
c]	Malathion	iii]	Phosphoric acid
d]	TEPP	iv]	Thiophosphoric acid

52. **Match the insecticides with the groups they belongs.**

a]	Aldiacarb	i]	Phenolic
b]	Endosulfan	ii]	Organophosphate
c]	Diazinon	iii]	Carbamate
d]	DNOC	iv]	Chlorinated hydrocarbon

53. **Match the systemic insecticide with their water solubility.**

a]	Dimethoate	i]	700 mg/litre
b]	Methyl - demeton	ii]	600 mg/litre
c]	Aldicarb	iii]	39 g/litre
d]	Carbofuran	iv]	330 mg/litre

54. **Match the insecticides with their nature of action.**

a]	Thiodemeton	i]	Chitin inhibitor
b]	Chlorthion	ii]	Fumigant
c]	Diflubenzuron	iii]	Contact
d]	Phostoxin	iv]	Systemic

55. **Match the insecticides with their site of action.**

a]	Pyrethroids	i]	Post-synaptic membrane
b]	Cylodines	ii]	Axonic membrane
c]	Nicotine	iii]	Synaptic juction
d]	Organophosphates	iv]	Pre-synaptic membrane

56. **Match the insecticides with the poisoning symptoms they produced in insects.**

a]	DDT	i]	Inactiveness
b]	BHC	ii]	Restlessness
c]	Rotenone	iii]	Fanning movements
d]	Parathion	iv]	Jitters

57. **Match the insceticides with their LD_{50} values.**

a]	Aldicarb	i]	8-14
b]	Bioresmethrin	ii]	850
c]	Carbaryl	iii]	7000-8000
d]	Carbofuran	iv]	0.93

58. **Match the insecticides with their LD_{50} values.**

a]	Acephate	i]	135-163
b]	Chlorfenvinphos	ii]	866-945
c]	Chlorpyrifos	iii]	1350
d]	Decamethrin	iv]	10-40

59. **Match the insecticides with their LD_{50} values.**

a]	Cypermethrin	i]	6-12
b]	DDT	ii]	57-106
c]	Demeton	iii]	113-118
d]	Demeton -S-methyl	iv]	303-4123

60. **Match the insecticides with their LD_{50} values.**

a]	Diazinon	i]	12-30
b]	Dichlorvos	ii]	46
c]	Dicrotophos	iii]	300-850
d]	Dieldrin	iv]	56-108

61. Match the insecticides with their LD_{50} values.

a] Diflubenzuron	i] > 4500
b] Dimethoate	ii] 2.6-8.6
c] Disulfoton	iii] 25-40
d] DNOC	iv] 320-380

62. Match the insecticides with their LD_{50} values.

a] Endosulfan	i] 17-24
b] Malathion	ii] 2800
c] Leptophos	iii] 50
d] Mythomyl	iv] 80-110

63. Match the insecticides with their LD_{50} values.

a] Monocrotophos	i] 30
b] Permethrin	ii] 1.6-3.7
c] Methyl parathion	iii] 430-4000
d] Phorate	iv] 14

64. Match the insecticides with their LD_{50} values.

a] Phosphamidon	i] 62-137
b] Quinalphos	ii] 17-30
c] TEPP	iii] > 4600
d] Tetramethrin	iv] 1.12

65. Match the insecticides with their ADI values.

a] Carbaryl	i] 0.003
b] Methyl parathion	ii] 0.0006
c] Monocrotophos	iii] 0.01
d] Carbofuran	iv] 0.001

66. Match the insecticides with their ADI values.

a] Diazinon	i] 0.0075
b] Dichlorvos	ii] 0.02
c] Dimethoate	iii] 0.002
d] Endosulfan	iv] 0.004

67. Match the insecticides with their ADI values.

a] Thiometon	i] 0.001
b] Lindane	ii] 0.01
c] Phosphamidon	iii] 0.006
d] Phosalone	iv] 0.005

68. Match equipments with insecticides/pesticides

a] High volume sprayer	i] Dust
b] Blowers	ii] Wettable powder
c] Back pack sprayer	iii] Liquids
d] Rocker sprayer	iv] Emulsions

69. Match equipment with the mode of use.

a] Rotary duster	i] Ultra low volume
b] Package duster	ii] Small scale use
c] Heli sprayer	iii] Feeding brush
d] Pedal pump	iv] High volume (HV) spray

70. Match the following pesticides/insecticides with the discoverer and his/her nationality.

a] DDT, 1939	i] Schrader, Germany
b] Parathion, 1944	ii] Geigy, Switzerland
c] Chlordane, 1945	iii] Muller, Switzerland
d] Carbamates, 1947	iv] Julius Hayman, USA/Riemschneider, Germany

GK

Bunda Bahadur was executed during Farrukhsiyar's reign.

GS
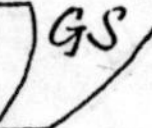

71. **Match the following insecticides pesticides with the country where these were developed and used first.**

a]	Paris Green, 1867	i]	Caucasus
b]	Chloropicrin, 1918	ii]	China
c]	Arsenite, 1900	iii]	France
d]	Pyrethroids, 1800	iv]	USA

72. **Match the following pesticides with their generation as has been grouped.**

a]	Petroleum products	i]	Third generation
b]	Dinitro compounds	ii]	First generation
c]	Juvenile hormone	iii]	Fourth generation
d]	Precocene	iv]	Second generation

73. **Match the pesticides with time interval (days) between application and consumption/harvesting of an edible crop or a part thereof.**

a]	All chlorinated hydrocarbons	i]	4
b]	Malathion	ii]	30
c]	Parathion, gusathion	iii]	14
d]	Phospdrin	iv]	7

74. **Match the LD_{50} values of the following insecticides, oral for rate given in mg/kg of body-weight.**

a]	Nicotine	i]	570
b]	Pyrethrins	ii]	750-1200
c]	Rotenone	iii]	70
d]	Ryania	iv]	25-132

75. **Match the LD_{50} values of the following insecticides, oral for rate given in mg/kg of body-weight.**

a]	Aldrin	i]	283
b]	Lindane	ii]	40-50
c]	Chlordane	iii]	250-300
d]	DDT	iv]	200

76. **Match the LD_{50} values of the following insecticides, oral for rate given in mg/kg of body-weight.**

a]	Rothane	i]	110
b]	Thiodan	ii]	3-6
c]	Endrin	iii]	40
d]	Heptachlor	iv]	400-300

77. **Match the LD_{50} values of the following insecticides, oral for rate given in mg/kg of body-weight.**

a]	Methoxychlor	i]	10-100
b]	Toxaphene	ii]	5,000
c]	Calcium arsenate	iii]	283
d]	Lead arsenate	iv]	35-100

78. **Match the LD_{50} values of the following insecticides, oral for rate given in mg/kg of body-weight.**

a]	Systox	i]	25-30
b]	Basudin	ii]	150-200
c]	Nuvan	iii]	137
d]	Rogor	iv]	3-5

79. **Match the LD_{50} values of the following insecticides, oral for rate given in mg/kg of body-weight.**

a]	Malathion	i]	3-5
b]	Phosdrin	ii]	20
c]	Nuvacron	iii]	57
d]	Metasystox-R	iv]	1400-1900

80. **Match the LD_{50} values of the following insecticides, oral for rate given in mg/kg of body-weight.**

a]	Parathion	i]	2-3
b]	Parathion-methyl	ii]	15
c]	Thimet	iii]	3-6
d]	Dimicron	iv]	12-16

81. **Match the LD_{50} values of the following insecticides, oral for rate given in mg/kg of body-weight.**
 a] Diptrex — i] 50-90
 b] Temic — ii] 650
 c] Sevin — iii] 6.2
 d] Pyrolan — iv] 400

82. **Match the following insecticides with their RL_{50} when used on different plants/plant products.**
 a] DDT on citrus — i] 150-190
 b] Malathion on stored grain — ii] 650
 c] Parathion on apple — iii] 6.2
 d] Pyrethrum on stored wheat — iv] 50-58

83. **Match the antidote with the following insecticides.**
 a] Atropine — i] PCP
 b] Potassium permanganate — ii] Nicotine
 c] Magnesium sulphate — iii] OP-insecticides
 d] Cold packs, tepid water — iv] Arsenals

84. **Match the following inorganic insecticides with their chemical formula.**
 a] Sodium fluosilicate — i] $Na_2B_4O_7$
 b] Cryolite — ii] Na_2SiF_6
 c] Sodium selenate — iii] Na_2AlF_6
 d] Borax — iv] Na_2SeO_4

85. **Match the following natural (organic) insecticides with their chemical composition.**
 a] Nicotine — i] $C_{21}H_{30}O_3$
 b] Pyrethroids — ii] $C_{23}H_{22}O_6$
 c] Rotenoids — iii] $C_{32}H_{49}N$
 d] Cevadine — iv] $C_{10}H_{14}N_2$

86. **Match the following insecticides with their chemical formula.**
 a] Endrin — i] $C_{10}H_6C_{18}$
 b] Endosulfan — ii] $C_{10}H_5C_{17}$
 c] Heptachlor — iii] $C_6H_8Cl_6O$
 d] Chlordane — iv] $C_9H_6Cl_6O_2S$

87. **Match the following insects with its food lure.**
 a] Onion fly — i] Sugar, molasses
 b] House fly — ii] Methyl eugenol
 c] Oriental fruitfly — iii] Heptyl butyrate
 d] Yellow jacket wasp — iv] Propylmercaptan

88. **Match the insects with the repellent.**
 a] Mosquito — i] Diacetyl pthalate
 b] Ticks, fleas — ii] Dimethyl pthalate
 c] Cattle fleas — iii] Benzyl benzoate
 d] Mites — iv] N-butylacetanilide

89. **Match the insect-pest with the scientific names.**
 a] Cucurbit stinking bugs — i] *Palpita indica*
 b] Pumpkin — ii] *Mylabris pustulata*
 c] Plume moth — iii] *Coridius janus*
 d] Blister beetle — iv] *Sphenarches caffer*

90. **Match the biological control stations with the country of their location.**
 a] CSIRO — i] India (Bangalore)
 b] CIBC — ii] Japan (Fukoda)
 c] FAKU — iii] Australia (Canberra, Sydney)
 d] IBS — iv] Germany (Dermstadt)

GK GS

Bourse is the common name for the Paris Stock Exchange.

91. Match the pest and crop with the natural enemy.

a]	Apple San Jose Scale	i]	*Capidosoma uruguayensis*
b]	Cotton bollworm	ii]	*Aphytis diaspidis*
c]	Paddy stem borer	iii]	*Trichogramma brasiliensis*
d]	Potato tuber moth	iv]	*Telenomous alecto*

92. Match the natural insect-enemy with the place where established in India.

a]	*Trichogramma fasciatum*, Sugarcane	i]	A.P. (Hyderabad)
b]	*Telenomus* sp., Castor	ii]	Orissa, Tamil Nadu
c]	*Rodolia cardinalis*, Citrus	iii]	West Bengal, U.P., Bihar
d]	*Spoggosia beniziana*, Coconut	iv]	Tamil Nadu, Kerala

93. Match the mouthpart type with the insects mentioned below.

a]	Biting and chewing	i]	Thrips and mites
b]	True piercing and sucking	ii]	Pea-moth and cutworms
c]	Superficial piercing and sucking	iii]	Larvae of diptera
d]	Rasping	iv]	Hemiptera of all kinds

94. Match the predatory stage of insects with their main prey.

a]	Lacewings, larvae and adults	i]	Plant feeding mites
b]	Robber flies, adult and larvae	ii]	Aphids
c]	Predatory mites, juvenile and adults	iii]	Various soft-bodied insect
d]	Predatory bugs, nymphs and adults	iv]	Various insects

95. Match the main group of insect pathogens with their examples cited below.

a]	Bacteria	i]	Nuclear polyhedrosis
b]	Virus	ii]	*Entomophthora* spp.
c]	Nematode	iii]	*Bacillus papilliae*
d]	Fungi	iv]	*Neoaplectana carpocapsae*

96. Match order of classification with the popular name of the insect.

a]	*Ephemeroptra*	i]	Stone flies
b]	*Odonata*	ii]	May flies
c]	*Plecophera*	iii]	Earwigs
d]	*Dermaptera*	iv]	Dragonflies

97. Match order of classification with the popular name of the insect.

a]	*Orthoptera*	i]	Leaf insects, stick insects
b]	*Phasmida*	ii]	Web spinners
c]	*Embioptera*	iii]	Cockroaches, mantids
d]	*Dictyoptera*	iv]	Grasshoppers, locusts, crickets

98. Match order of classification with the popular name of the insect.

a]	*Isoptera*	i]	Booklice
b]	*Psocoptera*	ii]	Thrips
c]	*Siphuculata*	iii]	Termites
d]	*Thysanoptera*	iv]	Sucking lice

99. Match order of classification with the popular name of the insect.

a]	*Neuroptera*	i]	Beetles
b]	*Coleoptera*	ii]	Stylopids
c]	*Strepsiptera*	iii]	Scorpion flies
d]	*Mecoptera*	iv]	Lacewings

100. Match order of classification with the popular name of the insect.

a]	*Lepidoptera*	i]	Ants, bees, wasps and saw-flies
b]	*Trichoptera*	ii]	Butter flies and moths
c]	*Hymenoptera*	iii]	Fleas
d]	*Siphonaptera*	iv]	Caddisflies

GK **Nijam-ul-mulk founded the state of Hyderabad.** GS

101. Match the following viruses with transmitter aphid.

a]	Potato leaf-roll Y virus	i]	*Brevicoryne brassicae*
b]	Sugarcane mosaic virus	ii]	*Myzus persicae*
c]	Mosaics of crucifers	iii]	*M. persicae* and *A. gosspyii*
d]	Bean mosaic virus	iv]	*Aphis maidis*

102. In India total loss due to pests, diseases and weeds is 18 per cent (Mehta, 1976), match the share of different agencies.

a]	Weeds	i]	26%
b]	Diseases	ii]	13%
c]	Insect-pests	iii]	33%
d]	Rates and storage losses	iv]	20%

103. Match Phylum with the example.

a]	Molusca	i]	Eels
b]	Nemathelminthes	ii]	Millepede
c]	Arthopoda	iii]	Cyst nematode
d]	Chordata	iv]	Slug

104. Match the following.

a]	Pheromones	i]	Intraspecific
b]	Allelochemicals	ii]	Interspecific
c]	Allomones	iii]	Favour emitter
d]	Kairomones	iv]	Favour receiver

105. Match the following.

a]	Brown plant hopper	i]	Delphacidae
b]	Zig-zag stem border	ii]	Pyralidae
c]	Yellow stem borer	iii]	Cicadellidae
d]	Paddy Gall fly	iv]	Cecidomylidae

106. Match transgenic plants carrying for insect resistance.

a]	Brinjal	i]	Cry 1A (b)
b]	Cotton	ii]	Bt
c]	Maize	iii]	Cry IA (c)
d]	Poplar	iv]	Cry III b

107. Match the following.

a]	Fumigants	i]	Aminocarb
b]	Acricides	ii]	MB
c]	Nematicides	iii]	Menazon
d]	Mollucides	iv]	Qamyl

108. Match the pattern of pesticide consumption.

a]	Fungicide	i]	21%
b]	Herbicide	ii]	60%
c]	Insecticide	iii]	14%
d]	Others	iv]	5%

109. Match the following.

a]	Egg parasitoid	i]	*Bracon kirkpatricki*
b]	Egg larval parasitoid	ii]	*Platygaster* sp.
c]	Larval parasitoid	iii]	*Trichogramma chilonis*
d]	Parasitoid of sucking pests	iv]	*Chelonus blackburi*

110. Match microbial pesticide with the targeted pest.

a]	Biolep	i]	*Helicoverpa armigera*
b]	Bioasp	ii]	*Lepidopterous caterpillars*
c]	Biopit	iii]	*Earias* spp.
d]	Halt	iv]	*Plutella xylostella*

111. Match household pests with its zoological name.

a]	Silver fish	i]	*Periplaneta americana*
b]	Black ant	ii]	*Cimex lectularius*
c]	Cockroaches	iii]	*Monomorium indicum*
d]	Bed bug	iv]	*Lepisma saccharina*

GS

Loktak lake is situated in Manipur.

Glossary of Technical Terms

Acaricide or Miticide The chemical substance that reduce the mite population of kill them are known as acaricide.

Antagonism It is the phenomenon where the toxicity of the two compounds is lesser than that expected from sum of their effect when applied separately. e.g. Malathion + Piperonyl butoxide.

Antibiosis Plants which have an adverse effect upon the insect due to the presence of a specific toxic chemical or due to black of specific nutritional requirements of the insects are said to exhibit antibiosis.

Antifeedants A chemical compound which prevents feeding of insets or animal on a treated material without known as antifeedants.

Artificial Control The control measures adopted by human agency are called as applied or artificial control, and they are entirely under his control and their success largely depends on human being.

Biocide A general posion or toxicant that will kill every from of life.

Bionomics The study of the habits, breeding and adaptations of living forms.

Botanical Insecticide Toxicant derived from plants and plant materials used to control insect pests of field crops. e.g. Nicotine, Rotenone, Pyrethrum.

Chemosterilants A chemical compound which deprive off insects of their ability to reproduce the progeny.

Clean Culture Clean culture means the removal of all undesirable plants, plants debris and other material from the fields and growing only healthy crops.

Cultural Method of Pest Control Cultural method of insect-pest control comprise the regular farm operation, so performed as to destroy the insect pest or to prevent them from causing injury.

Economic Damage The injury done to crop which will justify the cost of artificial control measures.

Economic Injury Level The lowest pest population density that cause economic damage.

Economic Threshold Level The pest population level at which control measure should be started to prevent the pest population from reaching the economic injury level.

Emulsion A suspension of fine droplets of the imisible liquids. e.g. Oil in water.

Entomology It is the branch of zoology that deals with the study of insects under class insecta of phylum Arthopoda.

Entomophagous An animal or plant which feeds upon insects is known as entomophagous.

Exuvae The old cuticle (skin) which is casted off is known exuvae.

Hormones The chemical compounds which are useful for controlling growth of insect pests. They are used to retard the development of pest species.e.g. Juvenile hormones : JH-I, JH-II, JH-III.

Knowledge World!

First refinery

The Assam Railway and Trading Co. erected a small refinery at Margherita in 1893 to refine oil from Makum and Digbhoi. The Digbhoi refinery, started in 1899, is the oldest existing refinery in India.

Hypermetamorphosis When a developing insects passes through two or more markedly different larval instars, it is said to undergo hypermetamorphosis e.g. Larvae of chrysopa.

Insect Attractant Chemical substance which is able to attract the insect pest of field crops by its specific smell or odour. eg. (i) Gossyplure - Male moth of pink boll worm. (ii) Methyl, euginol - *Dacus dorsalis* (iii) Sugar solution - Moths and Butterfly.

Insect-Pest When any insect cause economic damage is known as insect-pest.

Insects The tracheate invertebrates having segmented body divided into three regions head, throax and abdomen, with three pairs of leg, one pair of antenna, hard exoskeleton and development through metamorphosis and known as insects.

Instar The form of an insect, shape and size between two successive moultings is known as instar.

Integrated Pest Management (IPM) The FAO pannel of experts (1967) defined IPM as a pest management system that in the contact of the associated environment and the population dynamics of the pest species, utilizes all suitable techniques and methods in as compatible a manner as possible and maintains the pest population at levels below those causing economic injury.

LD_{50} Lethal dose required for killing 50% population of test animals is known as LD_{50} value. The unit of LD_{50} is mg/kg of body weight of test animals.

Legal Methods of Insect Control: It is defined as the legislative measures to prevent the introduction of new pest from one country of other and within country from one place to other is known as legal method of pest control

Major Pest Insect which normally cause loss more than 10 percent or more is called major pest.

Mechanical Control Mechanical control is one by which insect population is directly controlled by mechanical devices or manual operations.

Metamorphosis The changes which takes place is the shape and size of immature stage of insect before it becomes adult are collectively known as metamorphosis.

Minor Pest Insect which normally cause a loss ranging from 5 to 10 percent is known as minor pest.

GK GS

Surajmal was considered the Plato of Jat tribe.

Nature Control of Pest Control done by natural factors of environment which are not dependent and not influenced by man.

Nerve Poision A chemical associated with its solubility in tissue lipid and functions actively by blocking acetylocholinerase in insect and warm blooded animals is called nerve poision. e.g. Organophosphate etc.

Non-Preferred Plants The plants which do not possess the particular texture, colour, aroma, taste are not preferred by the insects are called non-preferred plants.

Pest Any organism detrimental to man or its properties.

Physical Control Physical control measures involve the use of certain physical forces for eradiction of insects. e.g. Heating, cooling, radiation etc.

Physical Poision The toxicant which brings about kill of an insect by exerting a physical effect is known as physical poision.

Phytosanitary Certificate Any consignment which is imported from foreign countries should be accompained with a certificate stating that the consignment is free from pests and diseases called as phytosanitary certificate.

Prophylactic Method of Pest Control Methods followed before the actual occurrance of the pest are termed as prophylactic method.

Protoplasmic Poision A toxicant reponsible for precipitation of protein especially destruction of cellular protoplasm of midgut epithelium is said to be a protoplasmic poision. e.g. Formaldehyde, Arsenic compounds etc.

Quarantine It is a process of checking material its healthyness before introduction and giving treatment if necessary at parts of entry.

Quiescence A state of inactivity, brought about by unfavourable conditions of temperature, humidity or other climatic conditions, by lack of some essential vitamins or other food substances, by failure to secrete particular hormone or by distance of the balance between opposing hormones. In this condition insect hardly move and may go without food for long period. This may occur at any growth stage and even in adult stage of insects.

Respiratory Poision A chemical which blocks cellular respiration as with the fumigants like hydrogen cyanide (HNC), carbon monoxide etc. is said to be respiratory poision.

Rodenticide The chemical substance that kills the rodents or reduce its population is known as rodenticide. eg. Barium carbonate, Zinc phosphide, Lead arsenate etc.

Sclerite The part of sclerotized, hard dark portion of the cuticle is known as sclerite.

Sclerotization The process of hardening of the exocuticle. It also becomes dark.

Stadium The time interval between two successive moultings is known as stadium.

Suture Unsclerotized, soft membrane left between two adjacent sclerites is known as suture.

Tolerant Plants The plants that withstand the insect attack and grow satisfactorily inspite of the attack by rapidly repairing the damage or by quick devlopment of new tillers, roots etc. In place of damaged ones are called tolerant plants.

Translaminar A pesticide which passes through from one surface of a leaf to the other (from lamina to lamina) through the leaf tissue.

Tropism A reflex response where by an insect turns in a particular direction in relation to a stimulus.

Uric Acid The chief nitrogenous waste products of insects.

Versons' gland The moulting glands of catterpillars. These epidermal glands secrets a fluid having a property liquifying the endocuticle and thus loosening the epicuticle prior to moulting.

Weismanns' Ring A gland surrounding the main dorsal blood vessel just above the cerebral ganglia in insects such as house fly and blow fly. It is believed to secrete a hormone which induces the formation of puparium.

Wing Venation The complete system of arrangement of veins in a wing is termed as wing venation.

Yellow Jacket A colloquial name sometimes used for any of the yellow and black wasps.

Zinc Phosphide It is a rodenticide now manufactured in India.

9. Plant Pathology

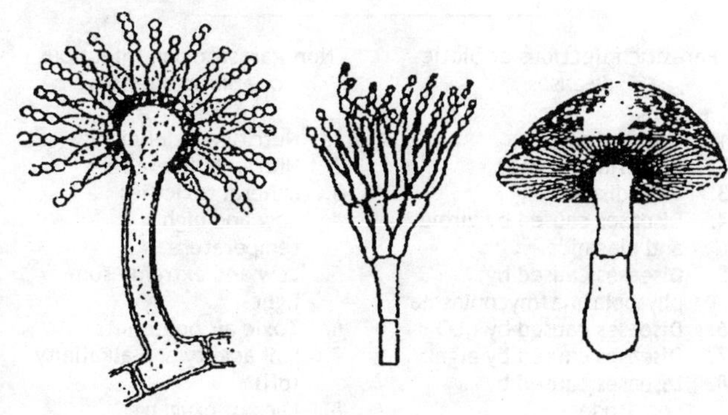

Did You Know ?

Largest ferns

Tree ferns of the family *Polypodiacceae* grow in the Eastern Himalayas and South India. A common species, *Cyathea spinulosa* grows to a height of 4 m and is found in the E. Himalayas.

Each Word Counts

INTRODUCTION

Definition

Plant pathology or phytopathology consists of three Greek words (1) *phyton* = plant, (2) *pathos* = ailments or diseases and (3) *logus* = discourse or knowledge. Thus, the term plant pathology or phytopathology means knowledge of plant disease.

Scope of Plant Pathology

Scope and responsibilities of plant pathology is unlimited. Its ultimate goal is to prevent and control plant diseases of economic importance. Responsibilities of the science of plant pathology may be summarized as under :

1. Survey of fields, orchards and areas in order to find out prevalence of diseases and their incidence.
2. Recording new diseases of economic importance if any, with their identification and extent of incidence.
3. Assessment of losses caused by different diseases of economic importance.
4. Study of etiology, symptoms, predisposing factors and recurrence of such diseases.
5. Find out suitable and economic methods of management of economically-important plant diseases.
6. Assisting in breeding of disease-resistant varieties.
7. To train the extension workers and subject matter specialists in order to bridge the gap between pathologists and farmers for better crop production.

Importance of Plant Pathology in Agriculture

1. Plant pathology has advanced the knowledge to protect the crop from losses due to diseases.
2. The science of plant pathology has contributed disease-free certified seed production.
3. Most of the diseases with known disease cycle can now be avoided by the modification of cultural practices and adoptation of crop rotations.
4. With the knowledge of mode of disease spread, many diseases of economic importance can now be checked, minimized or controlled.
5. Crop improvement and varietal resistance have been achieved against many diseases through the joint effort of breeder and plant pathologist.
6. Plant pathology has made possible to restrict the spread of plant diseases from one place to another and one country to another through suitable measures and quarantine legislation.
7. With the knowledge of plant pathology various prophylatic measures are adopted for successful management of diseases. These measures are seed treatment in seed-borne diseases, soil treatment and crop rotations.
8. Diseases can be avoided in cold storage by the application of plant pathological measures as per recommendations made for different diseases for protection of fruits and vegetables in storage.
9. Plant pathology has made possible to recognise, discard or utilize toxic substances by producing toxic substances or by competition or by parasitism. Organism exerting such lethal or damaging effect on the other is called antagonist.

Disease Triangle

Plant disease are caused by the interaction among susceptible host x abundance and virulence of pathogen x favourable environmental conditions for desirable time. These three factors constitute disease triangle, which are essential for disease incidence and its development. Each side of the triangle represents one of the three components. The size of each side is proportional to the sum total of the characteristics of each component favouring disease.

GK The first governor of Portuguese was Fransisco Almeida. GS

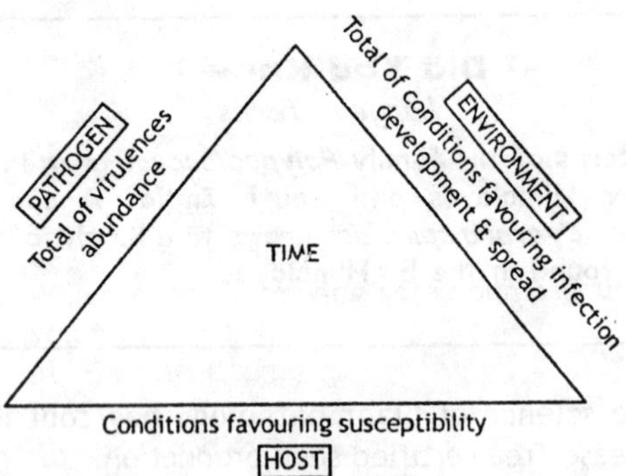

Plant Diseases

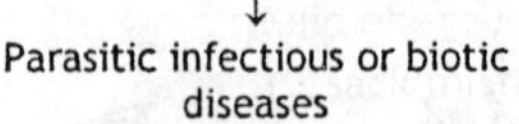

Parasitic infectious or biotic diseases	Non-Parasitic non-infectious or abiotic diseases
1. Fungal diseases	1. Nutritional deficiency
2. Bacterial diseases	2. Nutrients in excess
3. Viral diseases	3. Mineral toxicities
4. Diseases caused by viroid and plasmids	4. Low and high temperature
5. Diseases caused by phytoplasma/mycoplasma	5. Low and extreme sun-light
6. Diseases caused by RLO's	6. Toxic air pollutants
7. Diseases caused by algae	7. Soil acidity and alkalinity (pH)
8. Diseases caused by nematodes	8. Lack of oxygen
9. Diseases caused by insects	9. High moisture
10. Diseases caused by phenerogamic plant parasites	

Classification of Plant Diseases

Plant diseases have been classified into three groups on the basis of the occurrence of the pathogens.

***(i)* Endemic diseases:** Endemic diseases is one which is more or less constantly present in an area or locality. Such diseases are observed year to year in moderate to severe form. The causal agent of the diseases remains well established in the field or in the locality. By virtue of its ability of survival in weed, the disease is seen almost every year. Wilt of pigeon pea, red rot of sugar cane are observed as endemic diseases in Eastern U.P. Onion smut in Andhra Pradesh and flag smut of wheat in Punjab are endemic diseases observed over there every year in moderate or severe form depending upon presence or predisposing factors.

***(ii)* Epidemic or Epiphytotic diseases:** The word epidemic is popularly used for animal and human diseases occurring in severe form. Therefore, in case of plant diseases the word epiphytotic should be used which means the occurrence of a disease widely and severely but periodically. The casual agent of such diseases may or may not remain present constantly in the locality. It is because the environmental conditions which become favourable occasionally.

A disease may be endemic or sporadic but may take form of epiphytotic on occasions. Late blight of potato occurred in epiphytotic form in 1913 in the district of Rangpur now in Pakistan.

***(iii)* Sporadic diseases:** Sporadic diseases occur at very irregular intervals and locations. Green ear disease of bajra and black rust of wheat may be cited as examples of sporadic diseases which may appear in a locality or a district after one or more than one year or every year in sporadic form.

***(iv)* Pandemic diseases:** Diseases which occur all over the world and cause mass mortality, are called pandemic diseases. *Example* - late blight of potato.

***(v)* Classification according to causal agents:** According to causal agents the diseases may be classified into two main groups :

1. Parasitic or biotic diseases and
2. Non-Parasitic diseases

(a) Parasitic or infectious diseases: These diseases are caused by fungi, bacteria, viruses, nematodes and phanerogamic plant parasites. These diseases are usually contagious in nature and there is easy and efficient dispersal of the pathogens causing these diseases.

(b) Non-parasitic or non-infectious diseases: These diseases are not infected by biotic pathogens like fungi, bacteria, virus and nematodes etc. but are infected by the nutritional deficiencies or excess, by low and high temperature, by low and extreme light and by air pollutants etc. These diseases are not infectious in nature.

***(vi)* Classification according to nature of diseases:** Diseases classified according to the nature may be of the following types :

(a) Seed borne diseases: In such diseases the inoculum is found associated with the seed internally or externally and are called as internally seed borne or externally seed borne diseases.

(b) Soil borne diseases: These diseases are incited by the inoculum present in the soil in the form of oospores, sclerotia and mycelia etc.

(c) Air borne diseases: These diseases are incited by the inoculum present in the air in the form of conidia, chlamydospores and sporangia etc.

***(vii)* Classification according to Symptoms:** Symptoms of the diseases may also be considered as a base of classification and according to the symptoms, the diseases, may be grouped as rust diseases, leaf spot diseases, smut diseases, downy mildew diseases etc.

***(viii)* Classification according to the Host:** According to the host diseases may be grouped as cereal diseases and plantation diseases etc.

***(ix)* Classification according to the Nature of Infection:** According to the nature of infections, the diseases may be classified into two groups namely, (1) the systemic diseases such as green ear diseases of bajra and loose smut of wheat and, (2) Localized diseases such as leaf spot diseases.

GK

'Massacre of Innocents' is a famous painting by Peter Paul Rubenson.

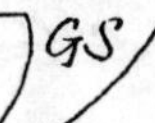

IMPORTANT DISEASES

Bacteria

Important diseases caused by bacteria

Name of Diseases	*Diagnostic symptoms*	*Management*
Yellow ear rot or Tundu disease of Wheat C.O. *Corynebacterium tritici*	1. Twisting of the leaves. 2. Exudation of blight yellow sticky slime covering the entire ear characteristic. 3. The slime envelops, the glume stems, and leaf sheaths. 4. Plant growth is checked and stem is distorted. 5. Distortion of' ear head and rotting of spikelets	1. Seeds free from nematode galls should be sown in nematode-free drained field. 2. Deep summer ploughing. 3. In standing crop disease plants should be located, uprooted carefully and destroyed. 4. Suitable crop rotation of 2-3 years should be adopted.
Bacterial blight or Angular leaf spot or Black arm of cotton *C.O. Xanthomonas exnopodis malvacearum*	1. Symptoms on all the parts of the plants above the ground level at different stages of growth. 2. The shape of the cotyledons are distorted causes them to dry and wither, 3. On the leaves similar spots appear on the undersurface.	1. Removal and destruction of the alternate host and diseased plant debris. 2. Crop rotation. 3. Early irrigation. 4. Seed treatment.· 5. Use of resistant varieties are recommended to control the diseases.
Bacterial brown rot of potato C.O. *Pseudomonas solanacearum*	1. Stunting. yellowing of foliage, wilting and finally collapse of palnt. 2. Brown ring is formed in the tubers due to discolouration of vascular bundles.	1. Disease could be minimized by selection of healthy seeds. 2. Crop rotation, proper drainage and tuber treatment with 0.02%. streptocycline for 30 minutes
Citrus canker C.O. *Xanthomonas compestris pv citri*	1. The disease affects all aerial parts of the plant. 2. Round watery raised, yellowish brown spots appear on the leaves which later become rough and corky and surrounded by yellow halo. 3. Cankers on fruits are the same as in leaves except that yellow hallow is absent.	1. Use of disease-free nursery stock for planting in new orchards. 2. In old orchard pruning of the affected twigs and spraying with 1 % copper oxychloride at periodical intervals.

Mycoplasma-like Organisms (MLOs)

Japanese workers in 1967. showed that mulberry dwarf disease was incited by a mycoplasma -like organism.

Plant diseases cuased by mycoplasma-like organisms (MLOs)

Name of Disease	*Vector*
Citrus greening	Psyllid
Clover phyllody	Leaf hopper
Legume little leaf	Leaf hopper
Papaya bunchy top	Leaf hopper
Potato witches broom	Leaf hopper
Rice yellow dwarf	Leaf hopper
Tomato big bud	Leaf hopper
Sugarcane white leaf	Leaf hopper

Rickettsia-like Organisms (RLOs)

These disease causing agents have been recognised in recent years. They have distinct cell walls and belong to the group of bacteria. They are readily seen in the electron microscope and are usually transmitted by leaf hopper. Unlike mycoplasma-like organism, they are susceptible to penicillin. Rickettsia like organisms usually occur in xylem and phloem vessels.

Transmission of Plant Viruses

The chief methods of transmission of plant viruses are:

1. ***Seeds and pollen grains transmission*** : Certain viruses like bean mosaic and curly top virus of beet sugar are transmitted by seed and pollen grains.
2. ***Transmission by vegetative propagation***: One of the chief methods of transmission of viral diseases especially of potato, raspberry, turnips, bulb plants, fruit trees and many ornamentals. The diseased vegetative parts such as tubers, bulbs, root, off shoots, buds and scions which are used for propagation will contain the virus present with parent plant. The new plants so raised by the above mentioned methods are usually infected.
3. ***Transmission by mechanical means*** : Many viral diseases especially mosaic viruses are mechanically transmitted by the following methods:

(a) By contact of infected leaves with healthy leaves brought about by wind.

(b) By rubbing the juice of the diseased plant over the surface of leaves of healthy plants.

(c) By grafting infected buds to healthy plants.

(d) Knife and other mechanical contaminated implements may spread the disease.

GS

The Dutch were defeated by the English in the battle of Bedara.

(e) Roots of diseased plant may come in contact of roots of healthy plants and thus spread the disease.

(f) Dodder (cuscutta) may serve as a bridge between the healthy and diseased plants resulting in viral infection.

4. *Soil transmission*: Soil-borne viruses are potato mosaic virus and oat mosaic etc. they are transmitted through the soil.

5. ***Insect transmission***: About 150 species of insects are known to transmit from one plant to other. Insects capable of virus transmission are known as insect (carries). Some of the insect vectors which play a major role in dissemination viruses are the aphids, leaf hoppers, beetles, scale insects, thrips and white flies; the insect vectors are sucking insects.

Some insect vectors of plant virus diseases

Crop	*Disease*	*Insect vectors*
Citrus	Veinal necrosis and twig decline	*Toxopetra citricidus*
Banana	Bunchy top	*Aphis gossypii*
Tomato	Leaf curl	*Bemisia tabaci*
Sugarcane	Grassy shoot	*Longuinuis sacchari*
Sesamum and sunhump	Phyllody	*Deltocephalus species*
Chinese-sarson	Mosaic	*Aphis gossypii, Myzus persicae*

6. ***Transmission by fungi:***

Gorgon (1985) found that the diseased lettuce was infected by Olpidium. He later discovered that the fungus act as a vector of the big vein mosaic. The virus acquired by the fungus remains in oospore. Similarly tobacco mosaic virus has been reported by Teakle (1960) to enter roots of its host by the zoospores of Olpidium brassicae.

7. ***Nematodes :***

Some soil borne viruses are transmitted by nematodes. Examples are tobacco rattle, pea early browning and colocasia mosaic etc.

Management of virus diseases

1. Elimination of source of viral infection:

(*a*) Perennial weed hosts, annual weed hosts, unrelated crops, volunteer crops and approach crops

(*b*) Plant residue

(*c*) Rogueing

2. Avoidance of vectors :

(*a*) Breaking the life cycle of hosts, vectors and viruses

(*b*) Creating physical barriers

(*c*) Early sowing

(*d*) Crop rotation

(*e*) Intercropping

3. Use of resistant varieties

4. Insecticidal sprays or neem oil sprays: In case of non-persistent viruses such as Potato Virus- Y, use of insecticides to control vectors can reduce the vector population and disease incidence but cannot check the entry of vectors and spread of the disease.

Important diseases caused by viruses

Name of the Diseases	Diagnostic Symptoms	Management
Yellow dwarf of rice *C.O. Mycoplasma like organism (Phytoplasma)*	1. Loss of chlorophyll of new leaves. 2. Affected plants may die immediately after seedling stage. 3. Severely infected plants may be devoid of grain formation.	1. Disease plant must be uprooted. 2. Grasses as alternate hosts nearby the field must be destroyed. 3. Leaf hoppers must be destroyed to check the spread of the disease.
Bunchy top of banana *caused by Banana virus*	I. Leaves are typically branched together forming a dense rossette at the apex, 2. Green steaks appear on the secondary veins underside of the lamina, 3. Older leaves show symptoms of and curling	1. Diseased plant material should be prohibited by law. 2. Diseased plant should be uprooted and burnt.
Mosaic of Papaya C.O. Papaya mosic virus	1. Profuse mottling and puckering of leaves especially the young ones. 2. Degeneration and marked reduction in growth. 3. First infection occurs on top leaves, 4. Leaves are often modified into tendril-like structures called shoe strings.	1. Spray 1% groundnut oil to help prevent infection.
Grossy shoot of sugarcane C.O. MLO now called as *Phytoplasma*	1. Production of numerous thin tillers from the base of affected shoots giving a dense or crowded bunch of tillers bearing pale yellow or chlorotic leaves. 2. Leaves remain thin, narrow, reduced. 3. Retarded elongation of internodes.	1. Use of disease free sets for sowing. 2. Precaution against transmission by mechanical means. 3. Vectors management by insecticides. 4. Use of resistant varieties.
Leaf roll of potato C.O. Potato leaf roll *virus*	1. Rolled leaf lets are rigid stiff thick and leathery, 2. Phloem necrosis also take place, 3. Tubers remain small in size.	1. Obtain certified seed, 2. Rogue out the sick plants, 3. Detop in the third or fourth week of December, 4. Spraying of crop with Metasystox (0.1%) to control the insect vectors

Cereals

RICE

Blast

Pyricularia oryzae

Symptoms

The typical symptoms appear on leaves, leaf sheath, nodes

and even the glumes are also attacked. On the leaves, the lesions start as small water soaked bluish green flecks, soon enlarge and form characteristic spindle shaped spots with grey centre and dark brown margin.

In infected nodes, irregular black areas encircle the nodes can be noticed.

Pathogen

Mycelium of the fungus, is hyaline to olivaceous, septate and highly branched. Conidia are produced in clusters on long septate, olivaceous slender conidiophores.

Favourable Conditions

Application of excessive doses of nitrogenous fertilizers, intermittent drizzles, cloudy weather, high relative humidity (93-99 per cent), low night temperature (between 15-20°C), more number of rainy days, longer duration of dew, cloudy weather, slow wind movement and availability of collateral hosts.

Mode of Spread and Survival

The disease spreads primarily through airborne conidia since spores of the fungus are present throughout the year. Mycelium and conidia in the infected straw and seeds are important sources of inoculum.

Management

Remove and destroy the weed hosts in the field bunds and channels. Treat the seeds with Captan of Thiram or Carbendazim or Carboxin or Tricyclazole at 2g/kg. Seed treatment with biocontrol agent *Trichoderma viride* @ 4g/kg.

Brown Spot or Sesame leaf spot

Helminthosporium oryzae

Symptoms

The fungus attacks the crop from seedling in nursery to milk stage in main field. Symptoms appear as lesions (spots) on the coleoptile, leaf blade, leaf sheath, and glume, being most prominent on the leaf blade and glumes. The disease appears first as minute brown dots, later becoming cylindrical or oval to circular.

Pathogen

H. oryzae produces greyish-brown to dark brown septate mycelium. Conidiophores may arise singly or in small groups.

Favourable Conditions

Temperature of 25-30°C with relative humidity above 80 per cent are highly favourable. Excess of nitrogen aggravates the disease incidence.

Mode of Spread and Survival

The infected seeds are the most common source of primary infection. The conidia present on infected grain and mycelium in the infected tissue may be viable for 2 to 3 years. The fungus may survive in the soil for 28 months at 30°C and 5 months at 35°C. Airborne conidia infect the plants both in nursery and in main field.

Management

Removal of collateral hosts and infected debris in the field. Crop rotation, adjustment of planting time and proper fertilization are suggested. Use of slow release nitrogenous fertilizers are suggested.

You Don't Know!

Tallest orchid

The tallest Himalayan orchid is the 5 m height *Galeola falconeri*, with its dark, yellow flowers. It is a spectacular sight, rising amongst wild banana in the Kameng forests.

Sheath rot

Sarocladium oryzae

Symptoms

Initial symptoms are noticed only on the upper most leaf sheath enclosing young panicles. The flag leaf sheath show oblong or irregular greyish brown spots. They enlarge and develop grey centre and brown margins covering major portions of the leaf sheath.

Pathogen

The fungus produces whitish, sparsely branched and septate mycelium. Conidiophore is slightly thicker than the vegetative hyphae. Conidia are hyaline, smooth, single celled and cylindrical in shape.

Favourable Conditions

Closer planting, high doses of nitrogen, high humidity and temperature around 25-30°C. Injuries made by leaf folder, brown plant hopper and mites increase infection.

Mode of Spread and Survival

Mainly through air-borne conidia and also seed-borne.

Management

Apply recommended doses of fertilizers. Adopt optimum spacing. Spray Carbendazim 250g or Edifenphos 1 lit or Mancozeb 1 kg or Chlorothalonil 1 kg/ha at boot leaf stage and 15 days later.

Sheath blight

Rhizoctonia solani

Symptoms

The fungus affects the crop from tillering to heading stage. Initial symptoms are noticed on leaf sheaths near water level. On the leaf sheath oval or elliptical or irregular greenish grey spots are formed. As the spots enlarge, the centre becomes greyish white with an irregular blackish

GK **The first Carnatic war ended with the treaty of Aixla Chapple.** GS

brown or purple brown border.

Pathogen

The fungus produces usually long cells of septate mycelium which are hyaline when young, yellowish brown when old. It produces large number of globose sclerotia, which are initially white, later turn to brown or purplish brown.

Favourable Conditions

High relative humidity (96-97 per cent), high temperature (30-32°C), closer planting and heavy doses of nitrogenous fertilizers.

Mode of Spread and Survival

The pathogen can survive as sclerotia or mycelium in dry soil for about 20 months but for 5-8 months in moist soil.

Management

Avoid excess doses of fertilizers. Adopt optimum spacing. Eliminate weed hosts. Apply organic amendments. Avoid flow of irrigation water from infected fields to healthy fields. Deep ploughing in summer and burning of stubbles.

Stem rot

Sclerotium oryzae

Symptoms

Small black lesions are formed on the outer leaf sheath and they enlarge and reach the inner leaf sheath also. The affected tissues rot and abundant sclerotia are seen in the rotting tissues. The culm collapses and plants lodge.

Pathogen

White to greyish hyphae produce smooth, spherical black and shiny sclerotia, visible to naked eyes as black masses.

Favourable Conditions

Infestation of leaf hoppers and stem borer and high doses of nitrogenous fertilizers.

Mode of Spread and Survival

The sclerotia survive in stubbles and straw are carried through irrigation water.

Management

Use of recommended doses of fertilizer. Deeply ploughing in summer and burning of stubbles. Avoid flow of irrigation water from infected fields to healthy fields.

Bacterial leaf blight

Xanthomonas oryzae p. v. oryzae

Symptoms

The disease is usually noticed at the time of heading but in severe cases occur earlier also. Seedlings in the nursery show circular, yellow spots in the margin, later enlarge, coalesce and cause drying of foliage.

In grown up plants water soaked, translucent lesions appear usually near the leaf margin. The lesions enlarge both in length and width with a wavy margin and turn straw yellow within a few days, covering the entire leaf.

Pathogen

The bacterium is strict aerobe, gram negative, non spore forming.

Favourable Conditions

Clipping of tip of the seedling at the time of transplanting, heavy rain, heavy dew, flooding, deep irrigation water, severe wind, temperature of 25-30°C and application of excessive nitrogen, especially late top dressing.

Mode of Spread and Survival

The infected seeds as a source of inoculum not be important since the bacteria decrease rapidly and die in the course of seed soaking.

Management

Burn the stubbles. Use optimum dose of fertilizers. Avoid clipping of tip of seedling at the time of transplanting. Avoid flooded conditions. Remove weed hosts. Grow resistant cultivars like IR 20 and TKM 6.

Maize

Downy mildew/Crazy top

Peronosclerospora sorghi

Symptoms

The most characteristic symptom is the development of chlorotic streaks appear on the leaves and the plants exhibit a stunted and bushy appearance due to the shortening of the internodes. White downy growth can be seen not only on the lower surface of leaf but also on the chlorotic streaks.

Pathogen

The fungus grows as white downy growth on both surface of the leaves, consist of sporangiophores and sporangia.

Favourable Conditions

Low temperature (21-23°C), high relative humidity (90 per cent), waterlogging condition and light drizzling. Young plants are highly susceptible.

Mode of Spread and Survival

The primary source of infection is through the oospores present in the soil and also the dormant mycelium present in the infected maize seeds. Secondary spread is through air-borne conidia. The oospores survive in the soil as well as in the infected plant debris.

Management

Deep ploughing. Crop rotation with pulses. Adjust the time of sowing. Rogue out infected plants. Treat the seeds with Metalaxyl compound (Apron 35SD) at 6 g/kg.

Leaf blight

Helminthosporium maydis

Symptoms

The fungus affects the maize plant at young stage. Small

yellowish round to oval spots are seen on the leaves. The spots gradually increase in area into bigger elliptical spots and it is greyish brown colour in the centre with dark brown margins.

Pathogen

Conidiophores are in group, geniculate, mid-dark brown, pale near the apex and smooth. Conidia are distinctly curved, fusiform; pale to mid-dark golden brown with 5-11 pseudosepta.

Favourable Conditions

Optimum temperature for the germination of conidia is 18 to 27°C. Infection takes place early in the wet season.

Mode of Spread and Survival

It is a seed-borne fungus.

Management

Treat the seeds with Captan or Thirarm at 4 g/ kg. Spray Mancozeb at 1.25 kg/ha.

Rust

Puccinia sorghi

Symptoms

Circular to oval, elongated, brown powdery pustules are scattered over both surface of the leaves. As the plant matures, the pustules become brown to black owing to the replacement of red uredospores by black teliospores.

Pathogen

Uredospores are globose or elliptical finely echinulate, yellowish brown with 4 germ pores. Teliospores are brownish black, or dark brown, oblong to ellipsoidal, rounded to flattened at the apex.

Favourable Conditions

Cool temperature and high relative humidity.

Management

Remove the alternate hosts. Spray Mancozeb at 1.25 kg/ha.

Head smut

Sphacelotheca reiliana

Symptoms

Symptoms are usually noticed on the cob and tassel. Large smut sori replace the tassel and the ear. Sometimes the tassel is partially or wholly converted into smut sorus. Under such conditions the individual spikelets on tassels may form shoot-like growths, or the entire tassel may develop into leaf-like structures.

Pathogen

Smut spores (chlamydospores) are produced in large numbers which are reddish brown to black, thick walled, finely spined, spherical or slightly irregular.

Knowledge World!

Only pitcher plant

***Nepenthes khasiana*, which captures and digests insects, is found in the Khasi and Jaintia Hills, Meghalaya, where there is a sanctuary for this rare plant.**

Favourable Conditions

Low temperature favours more infection and this fungus also infects the sorghum crop and vice versa.

Mode of Spread and Survival

The smut spores retains its viability for two years. The fungus is externally seed-borne and soil-borne. The major source of infection is through soil-borne chlamydospores.

Management

Field sanitation. Crop rotation with pulses. Treat the seeds with Captan or Thiram at 4 g/kg.

Charcoal rot

Rhizoctonia bataticola

Symptoms

The fungus attacks roots of seedlings and young plants. The affected plants first exhibit the wilting symptoms. The stalk of the infected plants can be recognized by grayish streaks and become weak.

Pathogen

The fungus produces large number of sclerotia which are round and black in colour. Sometimes, it produces pycnidia on the stems or stalks.

Favourable Conditions

High temperature and low soil moisture.

Mode of Spread and Survival

The fungus has a wide host range, attacking sorghum, cumbu, ragi and pulses. It survives for more than 16 years in the infected plant debris.

The primary source of infection is through soil-borne sclerotia.

Management

Long crop rotation with crops that are not natural host of the fungus. Irrigate the crops at the time of ear head emergence to maturity. Treat the seeds with Carbendazim or Captan at 2 g/kg.

GS

Modem is a device used to access internet.

Wheat Rust

Leaf or brown or orange rust	Yellow or stripe rust
Puccinia recondita	*Puccinia striiformis*
Mainly seen on leaves, rarely on leaf sheath. Rust pustules (Uredia) appear on leaves at early stage of crop and at maturity they turn to brown or orange (Telia).	Mainly occur on leaves than the leaf sheaths and stem. Bright yellow pustules (Uredia) appear on leaves at early stage of crop and at maturity, pustules are arranged in linear rows along with dull black lesions (Telia).
Brown, spherical, minutely echinulated with 7-10 germpores.	Yellow, spherical to oval, minutely echinulated with 6-16 germpores.
Brown, two celled, smooth, thick walled with rounded apex	Dark brown, two celled, thickwalled and flat at the top.
Alternate host: *Thalictrum* sp Uredospores and dormant mycelium persist on crop debris and weed host.	Alternate host: Unknown. Uredospores and dormant mycelium survive on stubbles and straws and also on weed hosts and self-sown wheat crops.
Wind-borne uredospores from hills	Wind borne uredospores from hills.
Warm weather (15°C) and heavy rain during January.	Very low temperature (11°C) and heavy rainfall during November-December.
Mixed cropping with suitable crops. Avoid excess dose of nitrogenous fertilizers. Dust Sulphur at 35-40 kg/ha. Spray Zineb at 2.5 kg/ ha. Grow resistant varieties like Lerma Rojo, Safed Lerma and Sonalika.	Mixed cropping with suitable crops. Avoid excess dose of nitrogenous fertilizers. Dust Sulphur at 35-40kg/ha. Spray Zineb at 2.5kg ha. Grow resistant varieties like Lerma Rojo Safed Lerma, Sonalika and Choti Lerma.

Smuts

Diseases	*Loose Smut*	*Flag Smut*
Causal organism	*Ustilago nuda*	*Urocystis tritici*
Symptoms	Symptom is visible only after emergence of ear. Diseases ear heads emerge out of the boot leaf earlier than healthy ones. The smutted spikelet is covered by a silvery membrance which contain black, dry powdery mass of spores.	The symptom can be seen on stem, culm and leaves from late seedling stage of maturity. The seedling infection leads to twisting and drooping of leaves followed by withering. Grey sori occur on leaf blade and sheath. The sorus contains black powdery mass of spores.
Spores	Spores germinate to produce promycelium with sporidia. Sporidia attack the ovary through stigma and remain dormant in embryo.	Aggregated sporeballs, consisting 1-6 lobose, brown, smooth walled spores surrounded by a layer of flat sterile cells.
Mode of Spread and Survival	Internally seedborne as dormant mycelium (Embryo infection)	Seed and soil borne. Smut sopres are viable for more than 10 years.
Favourable conditions.	Temperature of 18-24°C. High Humidity (60-85 per cent) during flowering.	Temperature of 18-24°C. Relative humidity 65 per cent and above.
Management	Hot water treatment: Soak the seeds in water at 26-30°C for 5 hrs to induce dormant mycelium to grow. Then immerse the seeds in hot water at 54°C for 10 min, to kill the mycelium.	Treat the seeds with Sulphur or Carboxin at 2g/kg. Grow resistant varieties like Pusa 44 and WG 377.

Bunt

Smooth spored bunt	Karnal bunt
Tilletia foetida	*Neovassia indica*
The fungus attacks seedling of 8-10 days old and become systemic and grows along the tip of shoot. At the time of flowering, hyphae concentrate in the inflorescence and spikeletes and transforming the ovary into smut sorus of dark green colour with masses of chlamydospores.	The infection occurs only at the flowering stage. Only a few spikelets are converted into dark coloured sori.
Reticulate, globose and smooth walled. No resting period. Produce primary sporidia upon germination.	Globose and smooth walled. Long resting period. Germinate to produce primary sporidia which are needle-shaped and then secondary sporidia which are sickle-shaped.
Externally seed-borne	Soil-borne and air-borne.
Temperature of 18-20°C. high soil moisture	Temperature of 15-20°C, High humidity, cloudy weather during flowering.
Treat the seeds with Carbendazin at 2g/kg. Grow the crop during high temperature periods. Adopt shallow sowing. Grow resistant varieties like Kalyan sona, HD2021, HD4513 and HD4519.	Deep ploughing during summer. Avoid continuous cropping of wheat in the same field. Grow resistant varieties like HD1907, H1358, HP753, and L176.

Cotton

Wilt

Fusarium oxysporum f. sp. vasinfectum

Symptoms

The disease affects the crop at all stages. The earliest symptoms appear on the seedlings in the cotyledons which

turn yellow and then brown. The base of petiole shows brown ring, followed by wilting and drying of the seedlings. In young and grown up plants, the first symptom is yellowing of edges of leaves and area around the veins i.e. discolouration starts from the margin and spreads towards the midrib.

Pathogen

Macroconidia are 1 to 5 septate, hyaline, thin- walled, falcate with tapering ends.

Favourable Conditions

Soil temperature of 20-30°C, hot and dry periods followed by rains, heavy black soils with an alkaline reaction.

Mode of Spread and Survival

The fungus can survive in soil as saprophyte for many years and chlamydospores act as resting spores. The pathogen is both externally and internally seed-borne.

Management

Remove and burn the infected plant debris in the soil after deep summer ploughing during June-July. Apply increased doses of potash with a balanced dose of nitrogenous and phosphatic fertilizers.

Root rot

Rhizoctonia bataticola

Symptoms

The fungus causes three types of symptoms *viz.*, seedling disease, sore-shin and root rot. Germinating seedling and seedlings of one to two weeks old are attacked by the fungus at the hypocotyl and cause black lesions, girdling of stem and death of the seedling, causing large gap in the field. In sore-shin stage (4 to 6 weeks old plants), dark reddish-brown cankers are formed on the stems near the soil surface, later turning dark black and plant breaks at the collar region leading to drying of the leaves and subsequently the entire plant.

Pathogen

The fungal hyphae are septate and fairly thick and produce black, irregular sclerotia.

Favourable Conditions

Dry weather following heavy rains, high soil temperature (35-39°C), low soil moisture (15-20 %).

Mode of Spread and Survival

The disease is mainly soil-borne and the pathogen can survive in the soil as sclerotia for several years. The spread is through sclerotia which are disseminated by irrigation water, implements, heavy winds and other cultural operations.

Management

Apply farm yard manure at 10t/ha or neem cake at 2.5 t/ha. Adjust the sowing time, early sowing (First week of April) or late sowing (Last week of June) so that crop escapes the high soil temperature conditions. Adopt intercropping with sorghum.

Quiz Kit!

Lightest spores

Fungus spores are so light that 20 billion spores make a gram. From an elevation of 5,500 m they would take 370 hours to reach the ground in perfectly still air.

Anthracnose

Colletotrichum capsici

Symptoms

The fungus infects the seedlings and produces small reddish circular spots on the cotyledons and primary leaves. The lesions develop on the collar region, stem may be girdled, causing seedling to wilt and die. In mature plants, the fungus attacks the stem, leading to stem splitting and shredding of bark. The most common symptom is boll spotting.

Pathogen

The pathogen forms large number of acervuli on the infected parts. The conidiophores are slightly curved, short, and club shaped.

Favourable Conditions

Prolonged rainfall at the time of boll formation and close planting.

Mode of Spread and Survival

The pathogen survives as dormant mycelium in the seed or as conidia on the surface of seed for about a year. The pathogen also perpetuates on the rotten bolls and other plant debris in the soil.

Management

Remove and burn the infected plant debris and bolls in the soil. Rogue out the reservoir weed hosts.

Boll rot

Symptoms

Initially, the disease appears as small brown or black dots which later enlarge to cover the entire bolls. Infection spreads to inner tissues and rotting of seeds and lint occur. The bolls never burst open and fall off prematurely.

Favourable Conditions

Heavy rainfall during the square and boll formation stage, punctures caused by the insects, especially red cotton bug *Dysdercus cingulatus*, close spacing and excessive nitrogen application.

GS

Elves helps Santa Claus to make his presence in the fairytales.

Mode of Spread and Survival

The fungi survive in the infected bolls in the soil. The insects mainly help in the spread of the disease. The fungi make their entry only through the insect punctures.

Management

Adopt optimum spacing. Apply the recommended doses of fertilizers. Carbendazim 1 kg from 45 day at 15 days interval.

Leaf blight

Alternaria macrospora

Symptoms

The disease may occur in all stages but more severe when plants are 45-60 days old. Small, pale to brown, irregular or round spots. Each spot has a central necrotic lesion surrounded by concentric rings.

Pathogen

The fungus produces dark brown, short, 1-8 septate, irregularly bent conidiophores with a single conidium at the apex.

Favourable Conditions

High humidity, intermittent rains and moderate temperature of 25-28°C.

Mode of Spread and Survival

The pathogen survives in the dead leaves as dormant mycelium. The pathogen primarily spreads through irrigation water. The secondary spread is mainly by air-borne conidia.

Management

Remove and destroy the infected plant residues. Spray Mancozeb or Copper oxychloride at 2 kg/ha at the initiation of the disease.

Bacterial blight

Xanthomonas campestris p. v. malvacearum

Symptoms

The bacterium attacks all stages from seed to harvest.

Pathogen

The bacterium is a short rod with a single polar flagellum. It is gram negative, non-spore forming.

Favourable Conditions

Optimum soil temperature of 28°C, high atmospheric temperature of 30-40°C, relative humidity of 85 per cent, early sowing, delayed thinning, poor tillage, late irrigation and potassium deficiency in soil.

Mode of Spread and Survival

The bacterium survives on infected, dried plant debris in soil for several years. The bacterium is also seed-borne and remains in the form of slimy mass on the fuzz of seed coat. The primary infection starts mainly from the seed-borne bacterium. The secondary spread of the bacteria may be through wind, wind blown rain splash, irrigation water, insects and other implements.

Management

Remove and destroy the infected plant debris. Rogue out the volunteer cotton plants and weed hosts. Follow crop rotation with non-host crops. Early thinning and early earthing up with potash.

Sugarcane

Red rot

Colletotrichum falcatum

Symptoms

Typical symptoms of red rot are observed in the internodes of a stalk by splitting it longitudinally. These include the reddening of the internal tissues which are usually elongated at right angles to the long axis of the stalk. The presence of cross-wise white patches is the important diagnostic character of the disease. The diseased cane also emits acidic-sour smell.

Pathogen

The fungus produces thin, hyaline, septate, profusely branched hyphae containing oil droplets. The fungus produces black, minute velvety acervuli with long, bristle-like, septate setae. Monoculturing of sugarcane, successive ratoon cropping, waterlogged conditions and injuries caused by insects.

Mode of Spread and Survival

The fungus is sett-borne. The fungus also persists in the soil on the diseased clumps and stubbles as chlamydospores and dormant mycelium. The primary infection is mainly from infected setts. Secondary spread in the field may be through irrigation water and cultivation tools.

Management

Adopt crop rotation by including rice and green manure crops. Select the setts from the disease-free fields or disease-free area. Avoid ratooning of the diseased crop.

Smut

Ustilago scitaminea

Symptoms

The affected plants are stunted and the central shoot is converted into a long whip-like, dusty black structure. The length of the whip varies from few inches to several feet. In early stages, this structure is covered by a thin, white papery membrane. The whip may be straight or slightly curved. On maturity it ruptures and millions of tiny black smut spores (teliospores) are liberated and disseminated by the wind. Affected plants are usually thin, stiff and remain at acute angle.

Nizamuddaulah signed the treaty of Allahabad with the British.

GS

Did You Know ?

Tallest trees

Fir (*Abies pindrow*) and deodar (*Cedrus deodara*) grow to an average height of 76.2 m (250 ft).

Pathogen

The fungal hyphae are primarily intercellular and collect as a dense mass between the vascular bundles of host cell and produce tiny black spores.

Favourable Conditions

Monoculturing of sugarcane, continuous ratooning and dry weather during tillering stage.

Mode of Spread and Survival

Teliospores may survive in the soil for long periods, upto 10 years. The spores and sporidia are also present in the infected plant materials in the soil. The smut spores and dormant mycelium are also present in or on the infected setts. The primary spread of the disease is through diseased. The secondary spread in the field is mainly through the smut spores developed in the whips, aided by air currents.

Management

Plant healthy setts taken from disease free area. Remove and destroy the smutted clump. Discourage ratooning of the diseased crops having more than 10 per cent infection. Follow crop rotation with green manure crops or dry fallowing.

Sett rot or Pineapple disease

Ceratocystis paradoxa

The disease primarily affects the setts usually two to three weeks after planting. The fungus is soil-borne and enters through cut ends and proliferates rapidly in the parenchymatous tissues. The affected tissues first develop a reddish colour which turns to brownish black in the later stages.

Pathogen

The fungus produces both macroconidia and microconidia. Conidiophores are linear, thinwalled with short cells at the base and a long terminal cell.

Favourable Conditions

Heavy clay soils, temperature of 25-30°C and prolonged rainfall after planting.

Mode of Spread and Survival

The fungus survives as conidia and chlamydospores in the soil and in the infected. buried cane tissues. The inoculum moves from field to field through wind-borne conidia or irrigation or rain water.

Management

Soak the setts in 0.05% Carbendazim or 0.25% Emisan for 15 minutes. Provide adequate drainage during rainy seasons.

Rust

Puccinia erianthi

Symptoms

Minute, elongated, yellow spots (uredia). The pustules turn to brown on maturity. Late in the season, dark brown to black telia appear on the lower surface of leaves.

Pathogen

The mycelium is hyaline, branched and septate.

Favourable Conditions

Temperature of 30°C, relative humidity between 70 and 90 per cent, high wind velocity and continuous cloudiness.

Mode of Spread and Survival

The fungus survives on collateral hosts like *Erianthus fulvus* and *Saccharum spontaneum*. The uredospores also survive in the infected stubbles in the soil. The disease is mainly spread through air-borne uredospores.

Management

Remove the collateral hosts. Spray Tridemorph 1 kg/ha.

Gummosis

Xanthomonas campestris p. v. vasculorum

Symptoms

The bacterium produces two distinct types of symptoms. On the mature leaves longitudinal stripes or streaks appear around the affected veins near the tip. Initially these stripes are pale yellow in colour, later turn to brown. The infected canes are stunted with short internodes, giving a bushy appearance.

Pathogen

The bacterium is a short rod, Gram negative, non spore forming with a single polar flagellum. It is a facultative anaerobe and it produces yellow slimy growth.

Mode of Spread and Survival

The bacterium remains viable in the soil as well as in infected canes. The primary transmission is through naturally-affected diseased setts or through soil-borne contamination. The secondary spread may be through wind splashed rain, harvesting implements, animals and insects.

Management

Remove and burn the affected clumps and the stubbles in the field. Select setts from disease-free areas. Avoid growing collateral hosts like maize, sorghum and cumbu near the sugarcane fields.

Grassy shoot

Mycoplasma Like Organism (MLO)

Symptoms

The disease symptoms are usually seen two months after

GK **The first Indian women to climb Mt. Everest was Bicchendri Pal.** GS planting. The disease is characterised by the production of numerous lanky tillers from the base of the affected shoots. Leaves become pale yellow to completely chlorotic, thin and narrow. The plants appear bushy and 'grass like' due to reduction in the length of internodes and premature and continuous tillering.

Pathogen

MLO is found to be present in the sieve cells of infected plants.

Mode of Spread and Survival

The pathogen is transmitted by aphids viz., *Rhopalosiphum maidis, Melanaphis sacchari*. In addition, leafhopper, *Proutista moesta* also involves in the transmission.

Management

Select setts from middle of the healthy fields or from the commercial seed nursery. Remove and burn the infected clumps periodically. Treat the setts before planting as follows: Hot Water Treatment (HWT) at 50°C for 2 ½ hr. Hot Air Treatment (HAT) at 54 °C for 8 hr or Aerated Steam Therapy (AST) at 50 °C for 1 hr or Moist Hot Air Treatment (MHAT) at 54 °C for 2 hr.

Oil Seeds

Groundnut

Rust

(*Puccinia arachidis*)

Symptoms

The disease attacks all aerial parts of the plant. The disease is usually found when the plants are about 6 weeks old. Uredosori appear on the lower surface of leaves. The epidermis ruptures and exposes a powdery mass of uredospores. Corresponding to the sori, small, necrotic, brown spots appear on the upper surface of leaves. The rust pustules may be seen on petioles and stem.

Favourable Conditions

High relative humidity (above 85 per cent), heavy rainfall and low temperature (20-25°C).

Mode of Spread and Survival

The pathogen survives as uredospores on groundnut plant. The fungus also survives in infected plant debris in soil. The spread is mainly through wind-borne inoculum of uredospores.

Management

Avoid monoculturing of groundnut. Spray Mancozeb 1 kg. or Wettable sulphur 2.5 kg. or Tridemorph 500 ml or Chlorothalonil 1 kg/ha.

Collar rot or seedling blight or crown rot

Aspergillus niger and A. pulverulentum

Symptoms

The fungus is both seed-borne and soil-borne. So the infection can be seen at any stage from sowing onwards.

Crown rot: The infection when occurs in adult plants show crown rot symptoms. Large lesions develop on the stem below the soil and wilting of plant.

Favourable Conditions

Deep sowing of seeds, high soil temperature (30-35°C) and low soil moisture.

Mode of Spread and Survival

The pathogens survive in plant debris in the soil, not necessarily from a groundnut crop. Soil-borne conidia cause disease carry over from season to season. The other primary source is the contaminated seeds. The fungi are carried on the seed surface or under the testa.

Management

Select good quality seeds. Treat the seeds with Carbendazim 2g or Thiram 4g/kg. Avoid deep sowing of seeds. Destroy the crop debris by burning.

Root rot

Macrophomina phaseolina

Symptoms

In the early stages of infection reddish brown lesion appears on the stem just above the soil level. The leaves and branches show drooping, leading to death of the whole plant. The decaying stems are covered by whitish mycelial growth. The rotten tissues contain large number of black or dark brown, thick walled sclerotia. When infection spreads to underground roots, the sclerotia are formed externally in the rotten tissue. Pod infection leads to blackening of the shells and sclerotia can be seen inside the shells.

Favourable Conditions

Prolonged rainy season at seedling stage and low-lying areas.

Mode of Spread and Survival

The fungus remains dormant as sclerotia for a long period in the soil and in infected plant debris. The primary infection is through soil-borne and seed-borne sclerotia. The secondary spread of sclerotia is aided by irrigation water, human agency, implements, cattle etc.

Management

Treat the seeds with Thiram 4 g or Carbondazim

Gingelly or Sesamum

Root rot or stem rot or charcoal rot

Macrophomina phaseolina

Symptoms

The disease symptom starts as yellowing of lower leaves, followed by drooping and defoliation The stem portion near the ground level shows dark brown lesions and bark at the collar region shows shredding. The sudden death of plants is seen in patches. In the grow-up plants, the

stems portion can be easily pulled out leaving the rotten root portion in the soil. The infection when spreads to pods, they open prematurely and immature seeds become shrivelled and black in colour.

Favourable Conditions

Day temperature of 30°C and above and prolonged drought followed by copius irrigation.

Mode of Spread and Survival

The fungus remains dormant as sclerotia in soil as well as in infected plant debris in soil. The infected plant debris also carry pycnidia. The fungus primarily spreads through infected seeds which carry sclerotia and pycnidia. The fungus also spreads through soil-borne sclerotia.

Management

Apply farm yard manure or green leaf manure at 10 t/ha or neem cake 250 kg/ha.

Powdery mildew

Erysiphe cichoracearum

Symptoms

Initially greyish-white powdery growth appears on the upper surface of leaves. When several spots coalesce, the entire leaf surface may be covered with powdery coating. In severe cases, the infection may be seen on the flowers and young capsules, leading to premature shedding. The severely affected leaves may be twisted and malformed.

Favourable Conditions

Dry humid weather and low relative humidity.

Mode of Spread and Survival

The fungus is an obligate parasite and disease penetrates through cleistothecia in the infected plant debris in soil. The ascospores from the cleistothecia cause primary infection.

Management

Remove the infected plant debris and destroy. Spray Wettable sulphur at 2.5 kg/ha or dust Sulphur at 25 kg/ha.

Leaf blight

Alternaria sesami

Symptoms

Initially small, circular, reddish brown spots appear on leaves which enlarge later and cover area with concentric rings. The lower surface of the spots are greyish brown in colour. Dark brown lesions can also be seen on petioles, stem and capsules.

Favourable Conditions

Low temperature (20-25°C), high relative humidity and cloudy weather.

You Don't Know!

Highest plants

Lichens have been found at an elevation of 7,000 m in the Himalayas and probably occur even higher. Lichens can survive - 200°C and can grow at - 24°C.

Mode of Spread and Survival

The fungus is seed-borne and also soil-borne as it remains dormant in the infected plant debris.

Management

Treat the seeds with Thiram or Carbendazim at 2 g/kg. Spray Mancozeb at 1 kg/ha.

Wilt

Fusarium oxysporum f. sp. sesami

Symptoms

The disease appears as yellowing, dropping and withering of leaves. The plants gradually wither, show wilting symptom leading to drying. They infected portions of fungus produces macroconida, microconidia and chlamydospores.

Mode of Spread and Survival

The fungus survives in the soil in the infected plant debris. It is also seed-borne and primary infection occurs through infected seeds or through chlamydospores in soil. The secondary infection may be caused by conidia disseminated by rain splash and irrigation water.

Management

Treat the seeds with Thiram or Carbendazim at 2g/kg. Apply heavy doses of green leaf manure or farm yard manure.

Bacterial leaf spot

Pseudomonas sesami

Symptoms

The disease appears as water-soaked yellow specks on the upper surface of the leaves. They enlarge and become angular as restricted by veins and veinlets. The colour of spot may be dark brown with shiny oozes of bacterial masses.

Mode of Spread and Survival

The bacterium remains viable in the infected plant tissues. It is internally seed-borne and secondary spread through rain-splash and storms.

Management

Keep the field free of infected plant debris. Spray with Streptomycin sulphate.

GS

Vijayalaxmi Pandit was the first woman president of the UN General Assembly.

Sunflower

Leaf blight

Alternaria helianthi

Symptoms

The fungus produces brown spots on the leaves, but the spots can also be seen on the stem, sepals and petals. The lesions on the leaves are dark brown with pale margin surrounded by a yellow halo. The spots later enlarge in size with concentric rings and become irregular in shape. The fungus produces cylindrical conidiophores, which are pale grey-yellow coloured, straight or curved, geniculate, simple or branched, septate and bear single conidium.

Favourable Conditions

Rainy weather, cool winter climate and late sown crops are highly susceptible.

Mode of Spread and Survival

The fungus survives in the infected host tissues and weed hosts. The fungus is also seed-borne. The spread is mainly through wind blown conidia.

Management

Remove and destroy infected plant debris. Rogue out weeds at periodical intervals. Sow the crop early in the season (June sowing). Treat the seeds with Thiram or Carbendazim at 2g/kg.

Rust

Puccinia helianthi

Symptoms

Small, reddish brown pustules (uredia) covered with rusty dust appear on the lower surface of bottom leaves. Infection later spreads to other leaves and even to the green parts of the head. In severe infection, when numerous pustules appear on leaves, they become yellow and dry. The black coloured telia are also seen among uredia on the lower surface.

Mode of Spread and Survival

The pathogen survives in the sunflower plants and in infected plant debris in the soil as teliospores. The disease spreads by wind-borne uredospores from infected crop.

Management

Remove and burn the infected plant debris in the field. Remove the volunteer sunflower plants. Spray Mancozeb at 1 kg/ha.

Root rot or charcoal rot

Rhizoctonia bataticola

Symptoms

The fungus is seed-borne and primarily causes seedling blight and collar rot in the initial stages. The grown-up plants also show symptoms after flowering stage. The infected plants show dropping of leaves and death occurs in patches. The bark of the lower stem and roots shreds and associated with a large number of sclerotia. Dark coloured, minute pycnidia also develop on the lower portion of the stem.

Mode of Spread and Survival

The pathogen survives in soil and in infected crop residues as sclerotia and pycnidia. The pathogen is seed-borne and it serves as primary source of infection. Wind-borne conidia cause secondary spread. The soil-borne sclerotia also spreads through rain splash, irrigation water and implements.

Management

Treat the seeds with Carbendazim at 2 g or Thiram at 4g/kg or treat the seeds with *Trichoderma viride* at 4g/kg.

Head rot

Rhizopus sp.

Symptoms

The affected heads show water-soaked lesions on the lower surface, which later turns brown. The discolouration may extend to stalk from head. The affected portions of the head become soft and pulpy and insects are also seen associated with the purified tissues. The larvae and insects which attack the head pave way for the entry of the fungus which attacks the inner part of the head and the developing seeds. The seeds are converted into a black powdery mass. The head finally withers and droops down with heavy fungal mycelial nets.

Favourable Conditions

Prolonged rainy weather at flowering and damages caused by insects and caterpillars.

Mode of Spread and Survival

The fungus survives as a saprophyte in host debris and other crop residues. The disease is spread by wind-blown spores.

Management

Treat the seeds with Thiram or Carbendazim at 2g/kg. Control the caterpillars feeding on the heads. Spray the head with Mancozeb at 1 kg/ha during intermittent rainy season and repeat after 10 days, if the humid weather persists.

Castor

Seedling blight

Phytophthora parasitica

Symptoms

The disease appears as circular, dull green patch on both the surface of the cotyledonary leaves. It later spreads gradually and cause rotting. The infection moves to stem and cause withering and death of seedling. In mature plants, the infection initially appears on the young leaves

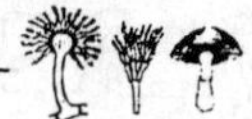

and spreads to petiole and stem causing black discolouration and severe defoliation.

Favourable Conditions

Continuously rainy weather, low temperature (20-25°C) and low-lying and ill-drained soils.

Mode of Spread and Survival

The fungus remains in the soil as chlamydospores and oospores which act as primary source of infection. The fungus also survives on other hosts like potato, tomato, brinjal, sesamum etc. The secondary spread takes place through wind-bome sporangia.

Management

Remove and destroy infected plant residues. Avoid low-lying and ill-drained fields for sowing. Treat with Thiram or Captan at 4g/kg.

Rust

Melampsora ricini

Symptoms

Minute, orange-yellow coloured, raised pustules appear with powdery masses on the lower surface of the leaves and the corresponding areas on the upper surface of the leaves are yellow. Often the pustules are grouped in concentric rings and coalesce together to form drying of leaves.

Pathogen

The fungus produces only uredosori in castor plants and other stages of the fungus are unknown. Uredospores are of two kinds, one is-thick walled and other is thin walled. They are elliptical to round, orange-yellow coloured and finely warty.

Mode of Spread and Survival

The fungus survives in the self-sown castor in crops in the off season. It can also survive on other species of *Ricinus*. The fungus also attacks *Euphorbia obtusifolia*, *E. geniculata*, and *E. marginata*. The infection spreads through air-borne uredospores.

Management

Rogue out the self-sown castor crops and other weed hosts. Spray Mancozeb at 1 kg/ha or dust Sulphur at 25 kg/ha.

Leaf blight

Alternaria ricini

Symptoms

All the aerial parts of plants viz., leaves, stems, inflorescences and capsules are liable to be attached by the fungus. Irregular brown spots with concentric rings form initially on the leaves and covered with fungal growth. When the spots coalesce to form big patches, premature defoliation occurs. The stem inflorescence and capsules are also shown dark brown lesions with concentric rings. On the capsules, initially brown sunken spots appear, enlarge rapidly and cover the whole pod. The capsules crack and seeds are also infected.

Knowledge World!

Highest growing moss

Members of the genus *Schistidium* have been found at 5,800 m. There are two species found in India, *S. alpicola* and *S. apocarpum*.

Favourable Conditions

High atmospheric humidity (85-90 per cent) and low temperature (16-20°C).

Mode of Spread and Survival

The fungus also survives on hosts like *Jatropha pandurifolia* and *Bridelia hamiltoniana*. The pathogen is externally and internally seed-borne and causes primary infection. The secondary infection is through air-borne conidia.

Management

Treat the seeds with Captan or Thiram at 4g/ kg. Remove the reservoir hosts periodically. Spray Mancozeb at 1 kg/ ha.

Horticultural Crops

BANANA

Panama disease/Fusarium wilt/Vascular wilt

Fusarium oxysporum f.sp. cubense

Symptoms

The infected plants show characteristic yellowing of leaf blades developing as a band along the margin and spreading towards midrib. The leaf wilts and the petiole buckles. The leaf hangs between the pseudostem while the middle of lamina is still green. All the leaves collapse whereas the petioles join the pseudostem and die.

Mode of Spread and Survival

The pathogen is soil-borne. It survives in soil as chlamydospores for longer periods. The primary spread of the disease is through infected rhizomes and secondary spread is through irrigation water.

Management

Infected plants should be removed and destroyed. After removal and destruction of affected plants with the rhizome, the pits have to be treated with lime at 1 to 2 kg/pit.

Anthracnose/fruit rot

Gloeosporium musarum

Symptoms

The fungus attacks the young banana fruits usually at the

GK **Neil Armstrong was the first man on the moon.** GS

distal end. The skin of the fruit turns black and shrivels and becomes covered with characteristic pink acervuli. Finally the whole finger is affected. Later the disease spreads and affects the whole bunch. The disease results in premature ripening and shrivelling of the fruits which are covered with pink spore masses.

Mode of Spread and Survival

The spread of the disease is by air-borne conidia and numerous insects which frequently visit banana flowers also spread the disease.

Management

In the field the distal bud should be removed when all the hands have opened to prevent infection. Infected materials should be burned.

Black spot or black tip

Deightoniella torulosa

Symptoms

The fungus causes three kinds of symptoms vis., black spot on leaves, black tip or black end on fruits and speckle (fruit spot). The earliest symptoms on leaves are round pin point, black spots on the main veins of the lamina, usually close to the leaf margin. These increase in size, become lenticular in the direction of the veins and are usually separated from the normal green leaf tissue by a narrow bright yellow peripheral band.

Mode of Spread and Survival

The conidia are spread by air.

Management

Affected leaves and fruits should not be left in the garden and they should be collected and destroyed along with dead and dried sheaths. Removal and destruction of trashes before heavy rain reduces the level of infection.

Bunchy top

Musa virus

Symptoms

The affected leaf shows green streaks on the secondary veins on the underside of lamina and on the midrib and petiole. The powdery bloom covers the midrib and petiole. If this is rubbed off the dark green streaks can be clearly seen. The streaks vary from a series of dark green dots to a continuous dark green line with a ragged edge.

Mode of Spread and Survival

It is not transmitted mechanically. The major vector is banana aphid viz., *Pentalonia nigronervosa.* It is transmitted in a persistent manner.

Management

The disease management can be attained by the following ways :

1. Selection of healthy suckers for planting.
2. Removal of affected trees and suckers at periodical intervals.
3. Control of aphid vector by suitable chemical methods.

Mango

Malformation

Fusarium moniliforme var. subglutinans

Symptoms

Three distinct types of symptoms are produced. They are bunchy top of seedlings (BT), vegetative malformation (MV) and floral malformation (MF). ***Bunchy top phase*** (BT) appears on young plants in the nursery beds when they are 4 to 5 months old. The characteristic symptoms of the disease is the formation of a bunch of thickened small shootlets bearing small rudimentary leaves or occasionally several bunches arising from a leaf axil at the top or lower down the main shoot. These shootlets are much thicker than main axis from which they arise. The shoot remains short and stunted. The growth of the plant is stopped and it gives an appearance of bunchy top. ***Vegetative malformation*** (MY) induces excessive vegetative branches of limited growth in young seedlings. They are swollen and are with very short internodes forming bunches of various sizes. *Malformation of inflorescence* (MF) shows variation in the panicle formation.

Mode of Spread and Survival

Diseased propagation materials help in the spread of the disease.

Epidemiology

The disease is serious in the northwest region where the temperature is between 10°C and 15°C during December-January before flowering.

Management

The disease incidence is reduced by spraying with NAA at 100 to 200 ppm during October. Pruning of diseased parts along with the basal 15 to 20 cm apparently healthy portions, followed by spraying of carbendazim 0.1 per cent or captafol 0.2 per cent effectively control the disease.

Powdery mildew

Oidium mangiferae

Symptoms

It usually occurs during December-March. A whitish bloom of fungal growth covers the stalks of the panicle, flowers and tender fruits. The affected flowers and fruits drop prematurely reducing the setting of fruits. At higher altitudes, the infection extends to the young leaves and twigs. Many of these are covered by the white powdery fungal growth and may exhibit distorted growth.

Mode of Spread and Survival

During off season, the pathogen has been reported in intact

green malformed panicles mostly hidden under dense foliage. During flowering period (January-March), the conducive environmental conditions activate the dormant mycelium in necrotic leaves. Abundant conidia are produced and blown over to the new flushes through wind on young panicles which provide sufficient spore load for initiating the disease.

Epidemiology

Rains or heavy mist in the morning accompanied by cool nights during the flowering period favour the disease.

Management

Three or four dustings with sulphur powder at fortnightly interval controls the disease.

Anthracnose

Colletotrichum gloeosporioides

Symptoms

The fungus attacks tender shoots and foliage. Brown or dark circular or irregular spots are formed on the leaves and such leaves are crinkled. The affected portions dry up and fall off and leave ragged margins. Often these leaves are shed leaving the twigs bare. The infection spreads to the green twigs and forms dark brown lesions on them. Young branches die-back. Older twigs and branches are infected through wounds and are killed.

Mode of Spread and Survival

Inoculum remains on dried leaves, defoliated branches, mummified flowers and flower brackets and they serve as primary inoculum. Secondary spread is through air-borne conidia. The fungus can enter the pores of green fruits.

Epidemiology

Under field conditions, the disease can be found on the fallen leaves and the blighted peduncles frequently remain *in situ* for many weeks. They produce spores under favourable moisture conditions and serve as foci of infection for the succeeding bloom. Even after they have fallen to the ground, they may continue to be the sources of infection for few weeks.

Management

Diseased twigs, leaves and fruits which fall on the ground in the orchard should be collected and all infected twigs should be pruned and burnt. Spraying of Bordeaux mixture 0.6 per cent in the young plants during February, April and September controls the disease.

Sooty mould

Capnodium ramosum

Symptoms

The fungi produce mycelium which is usually superficial and dark. They grow on the flowers, both tender and old leaves, stems and fruits. They grow and thrive on the sugary secretions of the plant hoppers. Black encrustations are formed on the surfaces of different parts of the plant.

Mode of Spread and Survival

Diseased leaves serve as primary inoculum.

Epidemiology

Disease is severe in old and dense orchards where light intensity is low. Trees exposed to eastern side has less incidence while the trees in centre of the orchard has more incidence. High infestation with plant hoppers and the sugary substances secreted by them favour development of sooty mould.

Management

Both the insects and the sooty moulds are to be simultaneously controlled in the eradication process. The insects are killed by spraying with carbaryl or phosphamidon 0.03 per cent.

Red rust

Cephaleuros virescens

Symptoms

The disease is characterised by red rusty spots on the leaves and young twigs. The spots are initially circular, slightly elevated and later coalesce to form irregular spots.

Epidemiology

The disease is more common on close plantations. The zoospores cause initial infection. High moist condition favours development of fruiting bodies of the alga.

Management

It is controlled by spraying with Bordeaux mixture 1.2 per cent or copper oxychloride 0.1 per cent or lime sulphur.

Anthracnose/Wither tip/Die-back

Colletotrichum gloeosporioides

Symptoms

The plant is affected at all stages. The weakened or injured twigs and branches are generally affected. The branches begin to wither from the tip downwards. The drying back gradually progresses downwards with the leaves turning yellow, withering and drooping and gum formation on the affected stem.

Mode of Spread and Survival

The fungus, survives in a dormant condition in the dead twigs and branches.

Quiz Kit!

Rose at highest elevation

The white flowered *Rosa sericea,* which is found east of Himachal Pradesh, grows at elevations up to 4,500 m.

GK **The sun is one but more than 100 billion stars in the Milkyway.** GS

Epidemiology

It causes decline of acid lime trees in North India. Deficiency of nitrogen and unfavourable soil conditions make the plant weak and susceptible to the disease.

Management

Dried twigs should be pruned. The cut ends should be protected by painting with Bordeaux paste or any other copper fungicide. Such trees may be sprayed thrice with carbendazim 0.1 per cent or captafol 0.2 per cent after pruning. The trees should be adequately manured with urea.

Powdery mildew

Acrosporium tingitaninum

Symptoms

All the aerial plant parts including young and actively growing leaves and twigs are affected. On the upper surface of the leaves white powdery patches of fungal growth are seen. The patches extend and cover the whole leaf. Petioles and stems are also covered by whitish fungal growth.

Mode of Spread and Survival

The fungus spreads through wind-borne conidia. The water shoots under the canopy are first infected.

Epidemiology

The disease appears usually from October to March and in higher elevations. Damp mornings with few hours of sunshine are the most favourable conditions for the onset of disease.

Management

The affected plant parts should be removed and destroyed carefully. Water shoots should be pruned to reduce infection. Dusting finely powdered sulphur gives effective control.

Scab/Verucosis

Elisnoe fawcetti

Symptoms

The disease attacks leaves, twigs and fruits. The lesions on the leaves in early stages consist of small, semi-translucent dots which become sharply defined pustular elevations usually on the underside, flat or somewhat depressed at the centre.

Mode of Spread and Survival

The pathogen perpetuates and survives in off- season as ascospores. The fungus spreads through conidia. It attacks sour orange, rough lemon and pummelo.

Epidemiology

The fungus infects tissues only when the surface is wet but prefers 16 to 20 °C temperature.

Management

The diseased leaves, twigs and fruits should be collected and destroyed. Spraying with Bordeaux mixture 1.0 per cent is quite effective.

Grapevine

Powdery mildew-*uncinula necator*

Symptoms

The fungus attacks all the stages of the crop growth. The characteristic symptom of this disease is the appearance of white powdery patches on affected parts. On the leaves the disease starts as small, white patches which later become larger in size and powdery in appearance. Sometimes the entire leaf is covered with dusty white fungal growth.

Mode of Spread and Survival

It survives as dormant mycelium and as cleistothecia on the shoots and buds from season to season. The disease spreads by the air-borne conidia.

Epidemiology

The disease occurs in severe form during October-November in North India and February-June in South India. Disease is favoured by warm weather and retarded by sunshine.

Management

Overcrowding and dense growth of the vines should be avoided by proper pruning. Effective control can be achieved by spraying wettable sulphur 0.2 per cent or dinocap 0.7 per cent or carbendazim 0.1 per cent or sulphur dusting at berry development stage with a prophylactic treatment.

Downy mildew

Plasmopara viticola

Symptoms

The disease is usually first observed as small translucent, pale yellow spots with indefinite borders on the upper surface of leaves. On the undersurface of leaves and directly under the spots, a downy growth of the fungus appears.

Mode of Spread and Survival

The pathogen survives on the infected leaves and vines as oospores and dormant mycelium. The secondary spread is through wind-borne sporangia and zoospores which are found on the new flush.

Epidemiology

Disease development is favoured during rainy season when there is a heavy dew, relative humidity is above 80 per cent and temperature is between 23°C and 27°C.

Management

Sanitation is very important in the management of this disease. Removing and burning of all diseased leaves,

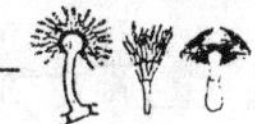

shoots, flowers and berries that may contain hibernating oospores help in preventing epidemics. Vines should be planted with proper spacing and should be trained in such a manner that leaves do not remain near the ground surface.

Anthracnose/Bird's eye spot

Elsinoe ampelina

Symptoms

The fungus attacks shoots, tendrils, petioles, leaves, veins and stems. Numerous spots occur on the young shoots. These spots may unite to girdle the stem and cause death of the tips. Spots on petioles and leaves cause them to curl or become distorted.

Mode of Spread and Survival

The pathogen survives as mycelium in the cankers on the stem and on the infected twigs. Secondary spread is through conidia which are carried by wind and rain water.

Epidemiology

The disease is severe during July-August and November-December months. Infection in new sprouts takes place during rainy season.

Management

The diseased leaves and twigs should be pruned and burnt. Spraying the pruned canes and leaves with a mixture of ferrous sulphate 2.5 kg + 0.5% sulphuric acid in 4.5 litres of water controls the disease.

Bacterial canker

Xanthomonas campestris pv. viticola

Symptoms

The disease starts as small water-soaked spots surrounded by yellowish halo at lower surface of the leaf. These spots enlarge in size and become dark brown and angular. Sometimes these spots coalesce to form larger patches. Leaves also show vein infection. Infected leaves after drying remain attached to the stem. Lesions are brown to black, elongated and cankerous on petioles and canes.

Mode of Spread and Survival

The alternate hosts are neem and mango. The bacterium survives in the infected, dry leaves upto 65 days. Secondary infection takes place through wind splashed rain. Disease spreads to distant places by diseased cuttings.

Epidemiology

Temperature range of 25 to 30°C is favourable for the disease development. Free water from dew, irrigation or rain on leaf are more important for pathogenesis.

Management

Regular inspection of vineyard, destruction of infected plant materials, use of disease-free cuttings and late-October pruning are recommended for its management.

Guava

Fusarium wilt

Fusarium oxysporum

> **Did You Know ?**
>
> ***Highest growing edible plant***
>
> ***Utica hyperborean*, a type of nettle, which is found across the Himalayas, Tibet and Mongolia, grows at elevations of 4,100 to 5,100 m. The young shoots are eaten as vegetables.**

Symptoms

The disease is characterised by yellowing and browning of leaves, discolouration of the stem and death of the branches along one side. Sometimes the infection girdles the stem and the whole plant may wilt. Leaves die and the twig barks split.

Mode of Spread and Survival

The fungus first colonises on the surface of the roots and enters the stem tissues at the basal portions near the ground level. It multiplies in vascular region and affects the cortical cells.

Epidemiology

Higher disease incidence is noticed during the monsoon period. The disease appears in August and increases sharply during September-October. It is severe in alkaline soils.

Management

Dry branches should be cut off and wilted plants uprooted. Soil should be treated with lime or gypsum to make the soil pH 6.0 to 6.5. Balanced nutrition of host reduces severity of the disease when organic nitrogen is supplied.

Fruit canker/Scab/Grey blight

Pestalotiopsis psidii

Symptoms

Infection generally occurs on green fruits. Minute, brown or rust-coloured, unbroken, circular, scabby lesions of 2 to 4 mm dia, appear on the fruit which later tear the epidermis open in a circinate manner. The margin of the affected area becomes raised.

Mode of Spread and Survival

Infected leaves and fruits serve as primary inoculum. Secondary spread is through the wind-borne conidia.

Epidemiology

The fungus is capable of growing at temperature between 20 and 25°C. Mycelial growth with intensive sporulation takes place at 5.5°C.

Management

Spread of the disease can be checked by three or four spraying with Bordeaux mixture 1.0 per cent or copper oxychloride 0.2 per cent.

Anthracnose/Die-back/Fruit spot/Twig blight

Gloeosporium psidii

Symptoms

Severity of the disease may show die-back of main

GK  GS

Sukumar Sen was India's first ever Chief Election Commissioner.

branches resulting in death of plants. The most characteristic symptoms appear during the rainy season as small pin-head sized spots on the unripe fruits. They gradually enlarge to form sunken and circular, dark brown to black spots.

Mode of Spread and Survival

The pathogen remains dormant for about three months in the young infected fruits. It becomes active and incites rot when the fruit begins to ripe. In moist weather, acervuli appear as black dots scattered throughout the dead parts of the twigs. From the twigs, the fungus penetrates the petioles and attacks the young leaves, which become distorted with dead areas at margins or tips.

Epidemiology

The cool season (January-March) and the hot, dry weather (April-June) prevent the spread of infection. In moist weather, acervuli are produced in abundance on dead twigs and pinkish spore masses are seen. Conidia initiate fresh infection.

Management

Spraying the trees with Bordeaux mixture 0.6 per cent or copper oxychloride 0.2 per cent before the onset of monsoon reduces the disease incidence.

Apple

Collar rot

Phytophthora cactorum

Symptoms

Important symptoms of collar rot are production of rough patches on the stem around the collar or crown region of the tree at soil level which develop into cankers. Rapid extension of the cankered area, both in lateral and vertical directions, results in girdling of the tree.

Mode of Spread and Survival

The pathogen survives as oospores in the soil.

Epidemiology

A soil temperature of 12°C to 20°C with pH of 5 to 6 is found to be the best for the survival of the fungus.

Management

Grafting of rootstocks 30 cm above the ground level reduces initial infection. The disease can be controlled by painting wounds with copper paints and drenching the area around the trees with fungicides like mancozeb or metalaxyl.

Stem cankers/ Black canker

Sphaeropsis malorum

Symptoms

It is also known as *black rot, die-back* or *smoky blight canker*. The disease appears in three phases, i.e., stem canker, leaf spot (frog eye leaf spot) and fruit rot (black rot). Among them the canker phase is the most destructive. In this, reddish brown sunken lesions develop on trunk and branches. The lesions turn smoky with alternate rings. The wood below is stained reddish brown.

Mode of Spread and Survival

The pathogen survives winter in the form of dormant mycelium and fruiting structure in cankered.

Epidemiology

High humidity and a temperature range of 20 to 22°C favour the disease development.

Scab

Venturia inequalis

Symptoms

The most striking symptoms of scab are commonly observed on leaves, fruits and rarely on 1 to 3-years-old shoots. In severe cases, petioles and blossoms also exhibit scab symptoms. On leaves, lesions first develop on the lower side. Most of these lesions do not have a definite margin in contrast to those on the upper side of the leaf. Lesions appear as olivaceous spots which turn dark brown to black and become velvety.

Mode of Spread and Survival

Pseudothecia formed in autumn and winter mature in spring to produce ascospores, the chief inoculum for primary infection. The secondary spread is through conidia.

Management

Clean cultivation reduces primary infection. Fallen leaves should be collected and destroyed along with pruned materials. Control programmes should have aim of elimination/reduction of pseudothecial formation and secondary infections.

Powdery mildew

Podosphaera leucotricha

Symptoms

The disease appears soon after the buds develop into new leaves and shoots. Sometimes buds are so heavily infected in the previous season that they are killed before developing into leaves. The disease may be seen on both surfaces of leaves in severe infections.

Mode of Spread and Survival

The fungus overwinters in the form of mycelium in diseased vegetative buds and fruits. Secondary spread is through wind-borne conidia.

Epidemiology

The conidial germination takes place when air is saturated and the temperature is between 10 and 25 °C.

Management

Pruning after shedding of leaves, thinning out and cutting back of laterals, removal and destruction of all diseased parts are important in the control of powdery mildew.

GK

Ram Mohan Roy was the founder of Brahma Samaj.

GS

Plant Disease Management

PDM

1. Prophylaxis

2. Curative measures

(a) Avoidance of pathogen	(i) Modification in cultural practices (ii) Disease escaping varieties (iii) Modification in cropping pattern, mixed cropping, catch crop and intercropping
(b) Exclusion of inoculum	(i) Seed certification and inspection (ii) Quarantine legislation (iii) Inhibition of disease spread
(c) Eradication of pathogen	(i) Deep summer ploughing and crop rotation (ii) Roguing (iii) Eradication of alternate and collateral host (iv) Sanitation (v) Heat treatment (vi) Soil treatment (vii) Destruction of crop residue
(d) Protective measures	(i) Seed treatment, systemic chemicals and antibiotics (ii) Control of insect-vectors (iii) Host resistance through proper nutrition (iv) Protective foliage spraying and dusting
(e) Development of host resistance	(i) Selection and hybridization (ii) Resistance through biotechnology and genetic engineering

Curative Measures
(Diseased Plant Therapy)

Chemotherapy	Tree Surgery	Biological Control
Use of fungicides bacterioides nematicides and other chemical means	Destruction of diseased plant part or entire plant	(i) Hyperparasitism (ii) Mycoparasitism (iii) Parasites of nematodes (iv) Cross protection and interference (v) Trap or catch crop (vi) Antagonistic plants (vii) Entomogenous fungi

Integrated Disease Management (IDM)

It is the integration of various disease management practices, to have a economic and long-lasting control over pathogens (Fungi, bacteria and virus) by manipulation of host, pathogen and environment interaction for a sustainable crop production and protection.

The general principles (or) components of IDM are :

1. *Avoidance of pathogen* : Planting at times when or in areas where inoculum is ineffective due to environmental condition.
2. *Exclusion inoculum (legislation)*: It is prevention of movement of diseased plant materials from infected area to healthy area through legislation.
3. *Eradication*: Eliminating or destroying inoculum at the source, either from a region or from an individual plant in which it is already established.
4. *Protection*: It means creating a barrier between host and pathogen.
5. *Disease resistance*: By genetic manipulation of the host, the disease resistance is developed.

The above principles are mainly preventing (prophylactic) and constitute the major procedures for plant disease management.

1. Avoidance of Pathogen

Avoid host contact with pathogen. Susceptible stages of the plant and favourable conditions for the pathogens should not coincide.

(i) *Choice geographic area*: Many bacteria diseases are severe in wet areas. Smut and ergot diseases of bajra are severe in wet regions. So avoid planting bajra in wet areas.

(ii) *Selection of the field*: Bacterial unit of potato, red rot of sugarcane, root knot nematodes. The causal organisms of above diseases persists in the soil. So avoid the same crop in the field for next time.

(iii) *Choice of time of sowing*: Avoid susceptible stage of growth of plant, when pathogen is in vigorous condition. The stem rust of wheat (*P. graminis tritici*) damages late sown crop than early sown because time of onset of disease and formation coincide.

2. Exclusion of Inoculum

It includes :

(i) *Plant quarantine*: It prevents the movement of diseased plant materials within the country (Domestic quarantine) and from other countries (International quarantine).

(ii) *Certification*: Only certified seeds and planting materials are allowed from one place to another.

((iii) *Notification of plant disease*: The grower himself should notify the occurrence of disease to the higher authority

(iv) *Preventing the sale of diseased plant materials* : It is done to prevent the further spread of disease. e.g. : Potato seed tubers are not allowed for sale from Nilgiri's to prevent the golden nematode spread.

3. Eradication

The measures are taken after establishment of the disease. The various measures are :

(i) *Eradication of alternate host*: Alternate host serves for the pathogens to complete its life cycle by eradicating alternate host, we can control several pathogens e.g. : To control wheat stem rust eradicate barbery plants. Brinjal serves as alternate host for cumbu rust

(ii) *Eradication over wintering host*: Several crops serve as over wintering host for pathogens during off-season. e.g. : Eradicate *Brachiaria mutica* and *Panicum ripens* to control rice blast pathogen (*Pyricularia oryzae*) Bean yellow mosaic over winters sweet clover. Little leaf of brinjal-pathogen over winters Datura and Vinca. Eradicate all over wintering hosts.

(iii) *Eradication of diseased plants*: By removing the diseased plant parts the inoculum potential of the pathogen is reduced

(iv) *Eradication of diseased plant parts* : Remove bleeding patches in the stem of coconut to control *Ceratocystis parodoxa* (stem bleeding of coconut) by knife and then sterilise it with fire and apply Bordeaux paste.

(v) *Sanitation*: Burning of crop residue – we can control RTV, sheath blight of paddy. In sugarcane smut (*Ustilago scitaminea*) collect the infected whips and burn it.

Avoid smoking in the tobacco field to prevent the spread of TMV. By rouging RTV in the rice field can be controlled.

(vi) *Sterilization*: Soil borne pathogens can be controlled by soil sterilization.

(vii) *Cultural practices:*

(a) *Crop rotation*: Monocropping is the main cause for the disease severity in the fields. The crop rotation is effective when the pathogen is soil invader (i.e. pathogen remain in the soil for short period as long as host is present) e.g. : Fusarial pathogens and Damping off caused by Pythium, etc.

(b) *Wider spacing*: Closer spacing enhances disease development due to higher humidity. By wider spacing we can control rice blast, sheath blight of rice and bacterial blight of rice.

(c) *Avoid growing of susceptible varieties*: By this we can reduce the inoculum potential. e.g. : IR-50 Susceptible to rice blast, Co- 43, IR-20 are susceptible to sheath blight of rice.

(d) *Raising raised nursery bed*: To avoid damping-off in chillies, tobacco caused by Pythium use raised nursery beds for seedling raising.

(e) *Mixed cropping*: Cotton root rot can be controlled by mixed cropping with *Phaseolus aconitifolius*.

(f) *Trap cropping*: Striga (partial root parasite) is controlled by trap crop like sudan grass. Orabanche (total root parasite) in tobacco field is reduced by chilli (trap crop). The alkaloid present in sudan grass and chilli reduce the germination of Striga and Orabanche

(g) *Adjusting the date of sowing*: Sugary disease of cholam and cumbu is severe during winter (Nov-Feb). So adjust date of sowing that flowering should not coincide with winter periods.

- Rice blast is severe in winter months. So do not plant IR-50, TKM 9 rice varieties in winter (Rabi-season).

(h) *Adjusting soil pH* : Alkaline condition favours *Streptomyces scabies* (potato scab). Apply gypsum to reduce the pH and to control this disease. Acid condition favours *Plasmodiophorum brassicae* (club root of cabbage). So apply lime to increase the pH to control this disease.

(i) *Adjusting irrigation*: High soil moisture favours damping-off, high soil temperature favours dry root caused by Macrophomina phaseolina (Dry root rot of sorghum, pulses) in summer months. Give frequent irrigations to control this disease. By reducing soil moisture we can control damphing-off.

(j) *Adjusting fertilizer application*: High 'N' favours blast, sheath blight and bacterial blight of rice, So minimise the disease incidence, apply more potash. Heavy application of FYM reduces soil-borne pathogens, FYM favours multiplication of antagonistic fungi.

(k) *Deep ploughing*: Summer ploughing expose soil-borne sclerotia of *Rhizoctonia solani, Sclerotium rolfsi*, under direct sunlight which is used to control the above pathogens.

(l) *Trenching*: Movement of disease is prevented from infected area to healthy area. e.g. For coconut wilt make trenches 3-4 feet, away from base of the trunk to prevent the further spread.

4. Protection

It includes physical, biological and chemical methods of control.

(i) *Physical method*: It includes -

(a) Hot water treatment
(b) Cold storage
(c) Sun drying
(d) Store in air tight container
(e) Irradiation.

(a) Hot water treatment: This was adopted for the control of internally seed-borne diseases. *Jensen* (1886) evolved it for the control of loose smut of wheat. Soak the seeds in hot water at 54°C for 10-15 min. or 4- 5 hrs in cold water.

- For *Helminthosporium* : Leaf spot and bacterial leaf blight of paddy, soak the seeds in hot water at so°C for 10 min.
- *Sugarcane mosaic and smut*: Sett treatment in 55°C for hours.

(b) Cold water treatment of cold storage : Fruits and vegetables are stored at 1-5°C in cold storage to prevent the storage diseases.

(c) Store in airtight containers: Soak the seeds in cold water for 4 hours and put it in airtight containers for 48 hours. It creates anaerobic condition to kill the spores.

(d) Irradiation: Fruits and vegetables are irradiated in UV rays or X-rays to prevent the post-harvest or storage diseases.

Fungicide

A good fungicide should not only be toxic to the parasite or inhibit the germination of its spore without causing phytotoxicity, but should also have the following qualities:

You Don't Know!

Commonest wild sheep

The Bharal or Blue Sheep (*Pseudovis nayaur*) is found at high elevation, between 3,000 m and 4,800 m in the Himalayas. At times herd size may top a hundred.

1. High field performance
2. Inherent fungitoxicity
3. Availability of the active constituents
4. Good coverage of host surface
5. Low phytotoxicity
6. Stability in storage
7. Stability after dilution to spray strength
8. Low toxicity to human beings and cattles
9. Should stick to the surface after drying
10. Good wetting property
11. Compatibility with pesticides, nematicides, herbicides, viricides and fertilizers.

Fungicides and their Methods of Application

According to their chemical nature and formulations the fungicides may be divided into the following categories:

1. *Spray materials*: Materials available in solutions the fungicides may be divided fungicides are used as spray which provide a protective covering over the surface.
2. *Dust treatment materials*: Insoluble powders are dusted over the surface.
3. *Seed treatments materials*: Seed treatment with fungicides are done with dry powder dips or slurries.
4. *Soil treatment fungicides*: Fungicides to control soil borne diseases in certain cases are used as soil treatment. The methods of soil treatments are following :
   - Drenching of soil with solution or suspension.
   - Soil fumigations such as chloropicrin, methyl bromide.
   - Furrow application.
   - Broadcasting of dusts, powders and granules.
5. *Treatment of tubers, bulbs and seed-materials*: Potato scab and black scurf disease are controlled by treating the seed materials in solutions of mercuric chloride, formaldehyde etc. Tubers are also dipped with mercury fungicides like aretan, emison etc.
6. *Treatment of tree wounds or tree surgery*: Cut-wounds are resulted from pruning or by cutting of diseased part of a plant. Such wounds may require wound dressing so as to avoid infection of cut parts. Therefore cut wounds are sterilized first by Swabbing of 0.5 to 1.0 p.c. of sodium hypochloride or 70 p.c. ethyl alcohol or with 1: 1000 solution of mercuric

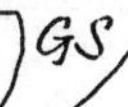

chloride. Wound parts so sterilized are dressed with mixture of lanolin 10 parts, rosin 2 parts and gum 2 parts. Bordeaux paste or Chaubattia paste are also used as wound dressing.

7. *Disinfection of warehouses*: Floors and walls of warehouses should be disinfected with copper sulphate solution.
8. *Control of insect-vectors* : Most of the virus diseases are transmitted by insects known as vectors. Such insect vectors should be controlled by use of suitable insecticides or through biological control in order to avoid transmission of viral diseases.

Systemic Fungicides

Systemic fungicides are those which are absorbed by the plants and remain present in protoplasms of different systems of the plant, making it artificially immune to parasitic infection. Some of the systemic fungicides are plantvax, vetavax and benlate etc.

Classification of Fungicides

According to chemical nature	Mode of action	According to general uses
1. Sulphur fungicide	1. Protectant	1. Seed protectant
2. Copper fungicide	2. Therapeuant	2. Soil fumigants
3. Mercurial fungicide	3. Eradicant	3. Foliage protectant
4. Systemic fungicide		4. Antibiotics
		5. Miscellaneous fungicides

Common Chemicals used as Fungicides

Sulphur fungicides:

1. *Inorganic sulphur*: Sulphur dust, wettable sulphur, lime sulphur etc.
2. *Organic sulphur*: Such as zinc dimethyl dithiocarbamate used as spray.

Copper Fungicides: Blitox-50, Gupramar, Fytolan and Copper Sandoz 53.

Organo-mercurial Fungicide: Ceresan and Emison etc., are commercial names and are used as seed dressers to control externally seed borne diseases of different crops.

Non-Mercurial Organic Fungicides : Dethane-Z- 78 (Zinc ethylene bisdithio-carbamate) is usually used as spray to control leaf spot and blight diseases of fruits, vegetables and flowers.

Biological Control

Garret (1965) explained biological control as methods of controlling a pathogen by any other organism and thus to reduce the infectivity of the pathogen is known as biological control. The following important methods practised as biological control are:

1. Biological control through hyperparasitism

(a) Bacteriophages or bacterial viruses

(b) Mycoparasitism

(c) Parasites of nematodes

2. Control through cross protection and interference: When any plant is infected by any mild strain of a virus the same plant escapes to be infected by the virulent stains of the same virus, such reaction is known as cross protection. If any plant is susceptible to two different strains of a virus – one being mild and the other virulent. The plant can be saved from being infected by virulent strain by vaccinating the plant with mild stream virus.

3. Disease control through trap or catch crop: There are some crop plants which are free from infection of some nematodes and secrete such biochemical substances which initiate hatching of eggs of nematodes and disrupt the life cycle of nematodes resulting in their death. Such crop is called trap or catch crop.

4. Disease control through antagonistic plants: Some plants like asparagus and marigolds are antagonistic to some nematodes as their roots secrete some toxic substances lethal to nematodes. In order to minimize the population of nematodes these antagonistic plants should be grown with the major crop.

5. Entomogenous fungi: Insecticides are not always effective to control some of the insect-pests. Besides, they also kill predators along with the pests. Keeping in view the demerits of insecticides, biological control of pests with entomogenous fungi may be suggested.

Control of Plant Diseases by Modification of Cultural Practices

Cultural practices are modified in such a way that the growth of the plant is not affected and the pathogen is directly or indirectly affected due to unsuitable environment conditions created by the modification of cultural practices. Some of the important practices may be summarized as under:

1.Crop rotation: The elimination of pathogen may be brought through crop rotation which may result due to the absence of suitable host to the pathogen till the period of its survival or due to toxic effects of root exudates of the intervening crops.

2. Mixed cropping: Growing a susceptible crop mixed with a non-susceptible crop have several advantages such as reduction of disease susceptible to host surface and also acts as barrier in disease spread in the fleld especially for vectors.

3. Spacing and method of sowing : Spacing between plants sometimes helps in reducing disease incidence due to reduced atmospheric humidity. Pathogens which flourish in high humidity are discouraged. Similarly regulation of depth of planting may help the young plants to escape infection.

4. Alternation of date of planting and irrigation: Some diseases are very destructive when susceptible stage of

the host and predisposing factor for pathogenesis coincide. Therefore, alternation of date of planting and irrigation frequencies may be helpful in some diseases to escape the disease.

5. **Drainage:** Accumulation of water in the field helps survival of inoculum, therefore proper drainage is essential to keep the field free from disease organisms as proper drainage makes the plants vigorous and resistant.

6. **Organic soil amendments and fertilizer application:** Decomposition of organic matter have a sanitary effect in the soil by inhibition and arrestation of pathogens by producing a number of toxic substances lethal to most of the pathogens. Similarly judicial application N, P and K, help in better and quicker development of roots which reduces the disease incidence.

7. **Destruction of crop residues and roguing:** Roguing of diseased plants or parts and eradication of alternate, collateral and weeds should be done so as to control most of the soil-borne pathogens.

8. **High ridging:** High ridging upto 10-15 cms. above the roots have been found to be beneficial against late blight of potato.

9. **Catch crop and trap cropping:** Catch crop and trap crop have been found effective to control some of the diseases incited by nematodes. Under such circumstances these crops should be cultivated in nematode manifested field.

10. **Adjustment of soil reaction** is essential to control some of the nematode and bacterial diseases.

LIST OF DISEASES

Diseases of Cereals

I. Rice

1. Blast - *Pyricularia oryzae*
2. Brown spot - *Helminthosporium oryzae*
3. Bakanae/foot rot/Fusarium blight - *Fusarium moniliforme*
4. Stem rot - *Sclerotium oryzae*
5. Sheath blight - *Rhizoctonia solani*
6. Sheath rot - *Sarocladium oryzae*
7. Udbatta disease - *Ephelis oryzae*
8. False/Green smut- *Ustilaginoidea virens*
9. Stackburn disease - *Trichoconis padwickii*
10. Bunt/Kernal smut/Black smut- *Tilletia barcylayana*
11. Bacterial blight - *Xanthomonas oryzae* pv. *oryzae*
12. Bacterial leaf streak - *Xanthomonas compestris* pv. *oryzicola*
13. Yellow dwarf - Mycoplasma-like organism (MLO)

Knowledge World!

Smallest blooms

***Pilea microphylla*, an artillery plant from India, has blooms whose diameter is only 1/72 of an inch.**

II. Wheat

1. Black/stem rust - *Puccinia graminis tritici*
2. Leaf, Brown/Orange rust - *Puccinia recondita*
3. Yellow/stripe rust - *Puccinia striiformis*
4. Loose smut - *Ustilago tritici*
5. Flag smut - *Urocystis tritici*
6. Stinking smut - *Tilletia caries*
7. Karnal bunt- *Neovassia indica*
8. Foot rot - *Pythium graminicolum*
9. Leaf blight - *Alternaria triticina*
10. Leaf blotch - *Septoria tritici*

III. Sorghum

1. Leaf spot - *Cercospora sorghi*
2. Anthracnose - *Colletotrichum graminicolum*
3. Leaf blight - *Helminthosporium turcicum*
4. Smut :
   (i) Covered/Kernal/Short grain - *Sphacelotheca sorghi.*
   (ii) Loose- *Sphacelotheca cruenta*
   (iii) Head- *Sphacelotheca reiliana*
   (iv) Long- *Tolyposporium ehrenbergii*
5. Downy mildew - *Pernosclerospora sorghii*
6. Charcoal-rot - *Macrophomina phaeseolina*
7. Ergot sugary disease - *Sphacelia sorghi*
8. Phenerogamics parasite - *Striga asiatica*

IV. Pearl Millet

1. Downy mildew - *Scelerospora graminicola*
2. Rust - *Puccinia penniseti*
3. Smut - *Tolyposporium pencillariae*
4. Ergot- *Claviceps fusiformis*

V. Maize

1. Smut- *Ustilago maydis*
2. Brown spot - *Physoderma maydis*
3. Rust- *Puccinia sorghi*
4. Leaf blight - *Helminthosporium maydis*
5. Downy mildew - *Peronosclerospora sorghi*

GK **Haryana has the lowest female per 100 male ratio.** GS

Sugarcane

Fungal Diseases:

1. Red Rot - *Colletotrichum falcatum*
2. Smut- *Ustilago Scitaminea*
3. Sett rot (or) Pineapple disease- *Ceratocystis paradoxa*
4. Wilt- *Cephalosporium sacchari*
5. Rust- *Puccinia erianthi*
6. Downy Mildew -*Peronosclerospora sacchari*
7. Damping off- *Pythium aphanidermatum*
8. Eye spot- *Helminthosporium sacchari*
9. Ringspot -*Leptosphaeria sacchari*

Bacterial Diseases:

1. Gummosis -*Xanthomonas campesris* pv. *vasculorum*
2. Red Stripe -*Pseudomonas rubrilineans*
3. Leaf Scald -*Xanthomonas albilineans*

Viral Diseases:

1. Mosaic - Sugarcane mosaic virus
2. Spike

M.L.O. Diseases :

1. Grassy shoot
2. White leaf

Oil Seeds

Groundnut

Fungal Diseases -

1. Early leaf spota : *Cercospora arachidicola*
2. Late leaf spot : *Phaeoisariopsis personata*
3. Rust- *Puccinia arachidis*
4. Collar rot/Seedling blight/Crown rot - *Aspergillus niger* and *Aspergillus pulverulentum*
5. Root rot - *Macrophomina phaseolina*
6. Stem rot- *Corticium rolfsii*
7. Wilt - *Fusarium oxysporum*
8. Anthracnose - *Colletotrichum dematium*
9. Grey mould - *Botrytis cinerea*
10. Yellow Mould - *Aspergillus flavus*

Bacterial Disease:

1. Bacterial wilt - *Pseudomonas solanacearum*

Viral Diseases:

1. Rosette virus
2. Peanut spotted wilt/Bud Necrosis/Groundnut ring mosaic/Tomato spotted wilt virus

Gingelly

1. Root rot/Stem rot/Charcoal rot- *Macrophomina phaseolina*
2. Leaf blight- *Alternaria sesami*
3. Leaf spot - *Cercospora sesami*
4. Wilt- *Fusarium oxysporum* f .sp. *sesami.*
5. Stem blight - *Phytophthora parasitica* var. *sesami*
6. Powdery mildew - *Erysiphe cichora-cearum.*
7. Anthracnose -*Colletotricum* sp.

Castor

1. Seedling blight - *Phytophthora parasitica*
2. Rust- *Melampsora ricini*
3. Leaf blight- *Alternaria ricini*
4. Brown leaf spot - *Cercospora ricinella*
5. Powdery mildew - *Leveillula taurica*
6. Stem rot- *Macrophomina phaseolina*
7. Bacterial leaf spot - *Xanthomonas campesteis* pv. *ricinicola*

Fibre Crops

Safflower

1. Rust - *Puccinia carthami*
2. Leaf spot - *Cercospora carthami*
3. Leaf blight - *Alternaria carthami*

Coconut

1. Bud rot- *Phytophthora palmivora*
2. Basal sterm rot/Thanjavur wilt/Bole rot - *Ganoderma lucidurm*
3. Stem bleeding - *Ceratocystis paradoxa*
4. Grey leaf blight - *Pestalotia palmarum*

M.L.O.:

1. Kerala wilt/Root wilt disease
2. Tatipaka disease

Cotton

1. Wilt - *Fusarium oxysporum* f.sp. *vasinfectum*
2. Verticillium wilt - *Verticillium dahaliae*
3. Root Rot - *Rhizoctonia bataticola*
4. Anthracnose - *Colletotrichum capsici*
5. Boll rot - *Fusarium moniliforme*
6. Leaf blight - *Alternaria macrospora*
7. Leaf spot - *Cercospora gossypina*
8. Myrothecium leaf spot - *Myrothecium roridum*
9. Rust- *Phakopsora desmium*
10. Sooty mould - *Capnodium* sp.
11. Bertical blight - *Xanthomomas campestris* pv. *malvacerum*

Diseases of Tobacco and Fruit Crops

Tobacco

1. Damping off- *Pythium aphanidermatum.*

Commonwealth is a multilateral organisation of 54 developed and developing nations around the world.

2. Black shank - *Phytophthora parasitica* var. *nicotianae*
3. Anthracnose - *Colletotrichum tabacum*
4. Powdery mildew - *Erysiphe cichoracearum* var. *nicotianae*
5. Frog eye spot - *Cercospora nicotianae*
6. Brown spot - *Alternaria longipes*
7. Tobacco mosaic - *Tobacco mosaic virus.*

Mango

1. Powdery mildew - *Oidium mangiferae*
2. Anthracnose - *Colletotrichum gloeosporioides*
3. Grey blight - *Pestalotiopsis mangiferae*
4. Stem-end rot - *Diplodia natalensis*
5. Black mould - *Aspergillus niger*
6. Sooty mould - *Capnodium ramosum*

Banana

1. Panama wilt/Fusarium wilt - *Fusarium oxysporum f. sp. cubense*
2. Anthracnose - *Gloeosporium musarum*
3. Sigatoka leaf spot - *Mycosphaerella musicola*
4. Cigar-end rot- *Verticillium theobromae*
5. Crown rot - *Botryodiplodia theobromae*
6. Bunchy top - *Musa virus*

Citrus

1. Gummosis - *Phytophthora palnivora*
2. Anthracnose - *Colletotrichum gloespo-rioides*
3. Bacterial canker - *Xanthomonas axonopodis pv. citri*
4. Powdery mildew - *Acrosporium tingitanium*
5. Scab - *Elsinoe fawcetti*
6. Tristeza or quick decline - *Citrus tristeza* virus
7. Citrus Greening - Rickettsia-*like organism*

Guava

1. Wilt - *Fusarium oxysporum f. sp. psidii*
2. Anthracnose- *Gloeosporium psidii*
3. Fruit canker - *Pestalotiopsis psidii*

Papaya

1. Damping off- *Pythium spp*
2. Stem rot - *Pythium aphanidermatum*
3. Powdery mildew - *Oidium caricae*
4. Anthracnose - *Colletotrichum papayae*

Quiz Kit!
Eggs galore

One termite queen may lay up to 15 million eggs in a lifetime.

5. Mosaic- *Papaya mosaic virus*

Grapes

1. Downy mildew - *Plasmopara viticola*
2. Powdery mildew - *Uncinula necator*
3. Anthracnose - *Elsinoe ampelina*

Ber

1. Powdrey mildew - *Oidium erysiphoides.* f.sp. *zizyphi*

Pineapple

1. Mealybug wilt - *Pineapple wilt virus*
2. Heart rot - *Phytophthora, Cinnamomi* and *P. parasitica*

Pomegranate

1. Leaf spot - *Cercospora punicae*

Sapota

1. Leaf spot - *Pestalotiopsis versicolor*
2. Sooty mould - *Capnodium sp*

Jack

1. Fruit rot - *Rhizopus artocarpi*
2. Leaf spot - *Collelotrichum gleosporioides*

Apple

1. Apple scab - *Venturia inaequalis*
2. Powdery mildew - *Podosphaera leucotricha*
3. Fire blight - *Erwinia amylovora*

Pear

1. Fire blight - *Erwinia anylovera*
2. Pseudomonas blight - *Pseudomnonas syringae*

Peach

1. Peach leaf curl - *Taphrina deformans*
2. Bacterial canker - *Pseudomonas syringae*

Plum

1. Wilt- *Verticillium alboatrum*
2. Leaf curl - *Taphrina deformans*

GK **Agra Fort was build by Akbar.** GS

Points to Ponder

- 'Bearding' or 'hairy root' and 'witch's broom' are caused by *Nacobbus* spp. of nematodes.
- "The tendency of treatments and conditions, acting before inoculation or introduction of incitant to affect susceptibility to biotic or abiotic pathogens" is referred to as Predisposition.
- A bacteria with a flagellum at each end of the cell is called amphitrichus.
- A host specific toxin is a metabolic product of a pathogenic microorganisms, selectively toxic only to susceptible host of the pathogen.
- A majority of phytophagous nematodes, survive through their dormant structures.
- A plant bacterium is prokaryotic organism.
- A Sclerotium is a compact, often hard mass of dormant fungus mycelium.
- A system is regarded as an interlocking complex of processes characterized by many reciprocal cause-effect pathways.
- Agar-agar is produced from Gelidium.
- Aggressiveness speaks of the capacity of a pathogen to invade and grow in its host plant and to reproduce on or in it.
- Algae posses chlorophyll whereas, fungi does not.
- Algal cells are eucaryotic.
- All organism living on dead organic matter is called saprophyte.
- All plant parasitic nematode feed with the help of a stylet.
- All the plant viruses contains RNA except cauliflower mosaic virus which contains DNA.
- Amphids and Phasmids are the sensory organs of nematodes.
- An epiphytotic disease is one which occurs widely but periodically.
- An organised nucleus is present in the fungi.
- *Anguina picridis* causes galls on the stem of *Aeroptilon picris* (orchid).
- *Anguina tritici* alone causes "tundu" disease of wheat.
- *Anguina tritici* larvae enter the flower primordia which instead of normal development into seed, change to rounded galls.
- *Anguina tritici* on wheat causes crinkling of leaves.
- *Anguina tumafaciens* causes leaf galls on *Cynodon transvalensis*.
- *Anguine tritici* was first plant parasitic nematode described by Needham in 1773.
- As a system, epidemic exists in a relatively very small compartment of a much bigger system, the agroecosystem.
- *Asparagus officinalis* produces asparaguisic acid (a toxic material) against the nematode *Tichodorus christici*.
- *Aspergillus flavus* produces aflatoxin.
- Bacteria and fungi are capable of producing poly-saccharides that can induce wilt symptoms in vitro.
- Bean bacterial blight (*Xanthomonas phaseoli*) spreads through irrigation and drainage water.
- Bordeaux mixture consist of $CuSO_4$ and $Ca(OH)_2$.
- Bordeaux paste, burgundy mixture and chestnut compounds are copper based home-made fungicides.
- Borellus (1956) first of all viewed a nematode under microscope.
- Carbon and nitrogen are major essential needs in fungus nutrition.
- Cereal rusts fungi over-summer in hills from where comes the primary inoculum through wind or insects.
- Certain infectious pathogens, such as viruses, are neither seen nor can be grown on artificial media.
- Chaubattia paste contains 800 g of copper carbonate, 800 g red lead and one litre of raw linseed oil or lanolin.
- Chemically different crops residues in rotation help in environmental modification in favour of host and against pathogen.
- Chemically viruses are composed by nucleic acid and protein.
- Chemically zined is called zinc ethylene disdithiocarbamate.
- Chlamydospores of smut fungi are inter-calary with regard their development.
- Chromosomal DNA is not degraded, but cytoplasmic DNA can be degraded by pathogens.
- Cleistothecia are born in genus *Erysiphe*.
- Cleistothecium possesses an ostiole.
- Clubrot of cabbage (*Plasmodiophora brassicae*) infestation is favoured by acid soils.
- Common scab of potato (*Streptomyces scabies*) incidence is favoured by water deficiency or dry soil.
- Compatible (susceptible) soybean - *Phytophthora* produce very little glyceollin, a pterocarpanoid phytolexin.
- Conidiophores of *Pernospora* are usually dichotomously branched.
- Continuous use of susceptible variety occupying large contiguous area help in build up of inoculum and augments chances of epiphytotics.
- Cowpea leaves resistant to *Cercospora* leaf spots possess toxic substances that inhibit the germination of Conidia.
- Crop sanitation is the best method for the control of

slow epidemics.

- Crystallization of viruses has done by Stanley.
- Cucurbits (bitter cucumber) produce a compound named cucurbitacin which is a repellent for nematode.
- Cyst forming nematodes belong to genus *Heterodera*.
- Deaminases enzyme separates the amino groups from nucleotides and makes them available.
- Disease severity is the product of inoculum potential and disease potential.
- Dissemination of red rot fungus and wilt organisms in active or dormant state may be transported by agency of water.
- During 1830-1845, late blight of potato spread fast in England, Ireland and continental Europe.
- Elicitors are cell wall components of the pathogen capable of inducing phytolexin synthesis.
- Elicitors are signals, presumably chemical signals, generated by an organisms to bring about metabolic shifts in another organisms.
- EPIDEM model could simulate, imitate or mimic an epidemic on a computer provided complete information about the epidemic is fed to the computer.
- Epidemiology deals with outbreaks and spread of disease in a population.
- Facultative parasites are basically saprophytic in their feeding mechanism.
- First reported bacterial plant disease is fire blight of apple and pear.
- Fractional sterilization to kill the spores was used by John Tyndall.
- Fungi and nematodes are the only group of plant pathogens employing force for direct penetration into the host.
- Fungi species like *Pythium, Phytophthora* etc., form zoospores provided with flagella for locomotion in films of soil water.
- Fungi, bacteria and nematodes embody pectic or pectionolytic enzymes which degrade the components of middle lamella and cell wall.
- Generally, the soil microflora reduce saprophytic activity in the process of saprophytic survival of fungi.
- Gum deposition in cells along the border of diseased tissues often serves as a protective mechanical barrier.
- Harmful effects of infection on host physiology comprise the chief causes of symptoms expression and loss.
- High birth rate coupled with low rate of mortality enhances the chances of occurrence of epiphytotics at a high level.
- *Hirschmanniella mucronata* nematode penetrates rice roots for feeding and ovisets inside the cortex.
- In 1675, the Dutch worker Leeuwenhoek developed the first microscope and in 1683 used it to describe bacteria seen with this.

Quiz Kit!

First tea company

The first tea company to commercialise tea was the Alubari Tea Company in 1856.

- In 1878, the downy mildew of grape vines was introduced into Europe from America.
- In 1898, Nocard and Roux discovered of disease incitants in animals similar to virus in size that was known as mycoplasma.
- In cereal rusts the infection-tube penetrates through the stoma.
- In citrus, canker is caused by *Xanthomonas citiri* and the bacterium survives in cankers on the standing trees.
- In Deutoromycotina, the fruiting body is known as synnema.
- In die-back or white tip disease plant organs especially stem or branches dry from the tip backward.
- In *Pythium* as a result of sexual reproduction oospore is formed.
- In red rot of sugarcane, causal fungus is *Colletotrichum falcatum* and its primary infection comes from diseased setts.
- In simple nature of disease-spread secondary infection rarely occurs during the same season.
- Indole acetic acid (IAA) is produced either by the pathogen in small quantities or by induce the plant to produce more of it.
- Inoculum and disease potentials vis-a-vis environment favourably or unfavourably influence the develop-ment of epidemics.
- It is believed that the cuticular enzymes mainly help in penetration of infection thread by pressure.
- Iwanowaski confirmed that virus could pass through filter with pores small enough to retain bacteria.
- K.R. Kirtikar was first Indian scientist who collected and identified fungi in India.
- Lichens are the association of algae and fungi.
- Longer the distance for the source of survival of the pathogen, the longer will be time required for build up of epiphytotic in a susceptible crop.
- Low night temperature favours blast development.
- Macroconidia and Microconidia are provuced by *Colletorichum*.
- Maneb is chemically known manganese ethylene disdithiocarba-mate.
- Marigold (*Tagetes erecta*) has been used in biological control of nematodes.

Hague is the headquarters of the International Court of Justice.

GS

- Molya disease of barley is caused by *Heterodera avenae*.
- Most of the nematodes grow best at pH values of 5-6 except *Heterodera avenae*.
- Most of the wilt causing species of *Fusarium* are host specific.
- Mycoplasma lack cell-wall but bounded by a triple layered membrane.
- Nematodes are bilaterally symmetrical.
- Nematodes cannot multiply outside the host as they depend for nutrition solely on the host.
- Nematodes have four larval stages before becoming adults.
- Nematodes possess well-developed reproductive and digestive system.
- Pathogens for direct penetration have to neutralize the natural barriers to establish in the host-plant system.
- Plant viruses have no resting stage and are transmitted through a continuous infection chain.
- *Pratylenchus penetrans*, *P. vulnis* and *P. coffeae* are bisexual nematodes with amphimictic reproduction.
- Pycnia is formed in members of Sphaeropsidales.
- Quinones are naturally present in plants and often exhibit antimicrobial activity.
- Root galls are characteristically produced by root-knot nematodes.
- Root inhabiting fungi have declining saprophytic phase limiting their survival in soil.
- Root knot of vegetable is caused by *Meloidogyne*.
- Rust pathogens are considered pathotrophs.
- Sporangiophore of *Sclerospora* are unbranched and clavat in shape.
- Stanley (1935) established crystalline nature of a virus particle.
- Stem rust of wheat (*Puccinia graminis tritici*) serves as an example of compound interest nature of disease spread.
- Stem segments of dodder (Cuscuta) are carried by birds to prepare nest and is thus transported to new sites.
- Stolons are absent in mucor.
- Sugarbeet sickness in Europe, yellows disease of pepper in Indonesia and decline of citrus in California (U.S.A.) are now established nematodes problems.
- Sulphur fungicide is effective against powdery mildew.
- Symptoms characterized by rapid killing of leaves, flowers and stems comes under blight.
- T.S. Sadasivan's school contributed to the concept of vivotoxin and worked out the mechanism of wilting in cotton due to *Fusarium vasinfectum*.
- Teliospore of Uredinales on germination produce sporidia.
- The ability of a pathogen to cause disease is known as pathogenicity.
- The basidium always bears the basidiospores exogenously.
- The best method to differentiate between Gram negative and Gram positive bacteria is Gram's Staining Techniques.
- The chain of events leading to disease development is referred to as patho/genesis disease cycle.
- The cross walls which divide the hypha into cells are called septa.
- The crown gall caused by *Agrobacterium tumefasciens* has served as classic model to study the role of auxins, IAA and related compounds in plant pathogenesis.
- The cytoplasmic defence reactions weaken, localize, and eliminate the endophytic nodule bacteria of the endomycorrhizal fungi.
- The energy of growth of a parasite available for infection of a host at the surface of the host organ to be infected" is known as inoculum potential.
- The epiphytotics develop in the same manner as the rise in human population.
- The facultative saprophytes live on dead organic matter.
- The first microscope was developed by Leeuwenhoek.
- The French scientist Prevost proved in 1807 that diseases are caused by microorganisms.
- The fungi causing late blight of potato, smut of maize, panama disease of banana, downy mildew of maize and pearlmillet and nematodes causing root knot not only induces to produce, but also produce IAA.
- The fungi or bacteria causing damping off, stalk rot, and canker can enter the xylem vessels.
- The gibberellins were discovered by Japanese, while investigating bakanae (foolish seedling) disease of rice caused by *Gibberella fujukuroi*.
- The glands in leaf hairs of chickpea contain maleic acid which is antifungal and provides resistance against rust fungus.
- The growth of pathogenic fungi in soil is either extremely slow or non-existent.
- The heteroecious rust requires two distinct host species to complete its life cycle.
- The infection by obligate parasites never kills the host.
- The infection occurring through wounds are known as trauma infection.
- The injury or wounds caused by pathogens to plant reduce their effective number and proportionately the water uptake.
- The juveniles of *Tylenchulus semipenetrans*, as estimated can last as long as nine years without host.
- The loss from disease is owing to reduction in

GK — **Arrah was suppressed by W. Taylor and V. Eyre.** — GS

reproductivity of the plant.

- The manifestation of resistant reaction of the host at a particular stage of growth is termed as latent infection.
- The most common insect vectors are aphids.
- The most common symptoms of plant diseases are those caused by disintegration tissues.
- The nematode of ear cockle of wheat may remain viable in the galls in quiescent state for 26 or more years.
- The nematodes can carry the fungal propagules on their body and help in entry through punctures for their own entry.
- The period of incubation and sporulation is short in epiphytotics.
- The primary inoculum of cucurbit wilt (*Erwinia tracheiphila*) by striped and spotted cucumber beetles in their body.
- The primary inoculum of rice leaf spot (*Helminthosporium oryzae*) and blast (*Pyricularia oryzae*) is believed to come from wind-borne spores.
- The species of *Pythium* invade only juvenile tissues in which secondary thickening has not yet developed.
- The spores are the main structures for dormant survival.
- The time lapses between penetration by pathogen and appearance of symptoms is known as incubation period.
- The toxin hypothesis with experimental evidence was first provided by metabolites of Cochliobolus (*Helminthosporium*) victoriae, a fungus which causes Victoria blight of oats.
- The uredospores of *Puccinia graminis* are disseminated by wind.
- The uredospores of *Puccinia graminis* var. *tritici* have been detected as high as 1400 feet above infected wheat fields.
- The vegetative body of the fungus is called thallus.
- There is no well defined nucleus in bacteria.
- Tomato leaves are known to excrete chemicals that provide resistance to attack of *Botrytis cinerea*.
- *Trichodorus* is a migratory ectoparasite.
- Tulasne brothers of France (R.L. Tulasne and C. Tulasne) produced illustrated description of rust and smut causing fungi.
- *Tylenchorhynchus agri*, nematode has been mainly reported as a parasite of leguminous and graminaceous plants.
- *Tylenchulus* and *Rotylenchus* are semi-endoparasitic nematodes.
- Tyloses are outgrowth of parenchyma adjacent to xylem and appear as peg-like structure.
- Virulence is a measure or degree of pathogenicity of an isolate or race of a pathogen.
- Virus transmitting nematodes belongs to order Dorylaimida.
- When a pathogen present in the host does not produce symptoms, the host is regarded as its carrier.
- When the dormant mycelium is present in the embryo, the disease is known as interally seed-borne.
- When the pathogen survives on only one genus or species of plants, it is called homogenous infection chain.
- When zinc is added to maneb it forms Dithane M-45 (Mancozeb).
- Wilt diseases are characterised by drying of the entire plant.
- Wilt of cotton (*Fusarium orysporium* f. sp. *vasinfectum*) and (*F. oxysporum* f. sp. *lycopersici*) is favoured by acidic soils.
- Wilt of pea (*Fusarium orysporium* f. sp. *pisi race 1*) and incidence is favoured by alkaline soils.
- *Xanthomonas malvacearum* is antigenically more similar to its host cotton than to other species of *Xanthomonas*.
- Phytophthora infestans do not produce oospore in India.
- Perfect stage of fruit-rot of brinjal is *Diphorthe vexans*.
- Soft rot of ginger is both soil and seed borne.
- Colocasia blight mostly occurs in the months of August and September.
- Downy mildew of grapevines is very common in Maharashtra.
- Mango malformation was first observed by a farmer Maries of Darbhanga in Bihar in 1891.
- Leaf curl of peach mostly affects the sub-Himalayan tracts of country.
- MLO's are bounded with triple layer membranes without true cell wall.
- The first symptom of powdery mildew of cucurbits appears on the leaves.
- *Fusarium oxysporium f.s. cubense* is fucultative parasite.
- Desi or Kagzi lime is more affected to citrus canker.
- Boron deficient fruit is susceptible to the disorder mango-necrosis.
- Large tubers are more affected by 'black-heart' than smaller ones.
- Kufri Naveen, Kufri Jeevan and Kufri Sinduri are resistant to early blight of potato.

Multiple Choice Questions

1. __________, before 1910 initiated exhaustive study of Indian fungi and diseases caused by them.
   a] E.J. Butler b] J.G. Moseman
   c] R.E. Nelson d] A. Livney
2. 'Didelphic' means nematodes with:
   a] One ovaryb] Two ovaries
   c] One testis d] Two testis
3. 'Spermatheca' is found in the nematode body of :
   a] Male b] Female
   c] Both d] None of the above
4. A bacterium is devoid of :
   a] Chlorophyll b] Well defined nucleus
   c] Cell wall d] Both a and b
5. A virus particle with both nucleic acid and portein coat is known as:
   a] Virion b] Viriod
   c] Capsid d] All above
6. Agallol is used for :
   a] Seed treatment b] Soil treatment
   c] Crop spray d] Aerial spray
7. Agar-agar is produced from :
   a] Nostoc b] Macrocrocystis
   c] Gelidium d] Oscillatoria
8. Albugo is a :
   a] Obligate parasite b] Facultative parasite
   c] Saprophyte d] Predator
9. Rust is a :
   a] Obligate parasite b] Saprophyte parasite
   c] Facultative parasited] Predator
10. Alexander Fleming (1929) discovered :
    a] Streptomycetease b] Penicillin
    c] Tyrocidine d] Giseofulvin
11. Alternate host of black stem rust of wheat :
    a] Mohania b] Thalictrum
    c] Barberry d] All above
12. Anterior tinsel flagellum is characteristics of :
    a] Chytridiomycetes
    b] Hypochytridiomycetes
    c] None of these
    d] Both a and b
13. Association of fungi with roots of higher plants is known as :
    a] Lichen b] Mycorrhiza
    c] Ascolichen d] Basidiolichen
14. B.B. Mundkur wrote :
    a] Fungi and Disease in plants
    b] Fungi and Plant diseases
    c] The Fungi
    d] All of the above
15. Bacilli are :
    a] Spiral shaped b] Spherical shaped
    c] Rod shaped d] None of the above
16. Bacterial cell surrounded by flagella is called :
    a] Monotrichous b] Lophotrichous
    c] Amphitrichous d] Peritrichous
17. Brassicol is ______ fungicide.
    a] Murcurious b] Sulphur
    c] Copper d] Miscellaneous
18. Bacterial leaf blight of rice caused by *Xanthomonaso* ryzae can be identified by :
    a] Wilting of the plant
    b] Yellowing of leaves
    c] 'Ooze test'
    d] Defoliation
19. Bengal famine was caused due to :
    a] *Phytophthora infestans*
    b] *Drechslera oryzae*
    c] *Puccinia graminis tritici*
    d] *Heterodera avenae*
20. Black heart of potato is caused due to deficiency of :
    a] Boron b] Zinc
    c] Iron d] Oxygen
21. Black stem rust of wheat a caused by a :
    a] *Puccinia graminis tritici*
    b] *Puccinia recondita*
    c] *Puccinia striformis*
    d] *Ustilagotritici*
22. Black tip of mango is caused by :
    a] Pythium b] Phytosynthetas
    c] Boron deficiency d] Zinc deficiency
23. Blast of rice is caused :
    a] Xanthomonas b] Drechslera
    c] Pyricularia d] Cercospora
24. Plant Pathology is _________ word.
    a] English b] Latin
    c] Greek d] Mandarin
25. Bordeaux mixture was discovered by P.A. Millardet of France during the year 1882 following his chance observation of farmer's practice for protection against:
    a] *Plasmopara viticola* on grapevine
    b] *Uncinula nectar* on grapevine
    c] *Podosphaera leucotricha* on apple
    d] *Venturia inequalis* on apple
26. Botulism is caused by species of :
    a] Bacillus b] Pseudomonas
    c] Agrobacterium d] Clostridium
27. Brown spot of maize is caused by :
    a] *Sclerospora sorghi*
    b] *Physoderus zeamaydis*
    c] *Alternaria solani*
    d] *Cercospora perfonats*

GK GS

Tipu was related to the pagalpanthi movement.

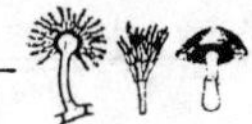

28. By 1857, Berkley and Schacht discovered root knot nematodes and cyst nematode in:
a] Sweet sorghum b] Sugarcane
c] Potato d] Sugarbeet

29. Carborundum powder is :
a] Silicon carbide
b] Aluminium oxide
c] Diatomaceous earth
d] Calcium carbonate

30. Cocci are :
a] Spiral shaped b] Spherical shaped
c] Rod shaped d] None of the above

31. Coffee rust is caused by :
a] *Uromyces hobsoni*
b] *Hemileia vastatrix*
c] *Cronartium ribicola*
d] *Ravenelia emblicae*

32. Conidia and sporangia are the spores produced by :
a] Bacteria b] Viruses
c] Fungi d] Phanerogams

33. Crystalization of viruses has done by:
a] Mayer b] Stanley
c] Iwanowski d] Bejerinck

34. Cysts are formed by :
a] Meloidogyne b] Aphelenchoides
c] Heterodera d] None of the above

35. Death of tissue in virus infected plant described as :
a] Mottling b] Mosaic
c] Necrosis d] Curling

36. Disease symptoms known as phyllody in sesamum is caused by:
a] Bacterium b] Fungus
c] MLO d] Virus

37. Dithane M-45 is a :
a] Bactericide b] Insecticide
c] Fungicide d] Nematicide

38. Domestic quarantine exists on :
a] Rice blast
b] Red rot of sugarcane
c] Wart of potato
d] Rust of wheat

39. Downy mildew of maize is caused by:
a] *Peronosclerspora sacchari*
b] *Peronospora triognella*
c] *Sclerospora graminicola*
d] *Sclerospora sacchari*

40. Dr. K.C. Mehta of Agra College made an outstanding contribution to plant pathology by discovering the disease cycle of cereal ____________ in India.
a] Smuts b] Rusts
c] Wilts d] Blights

41. Exclusion of plant disease by legislation is known as:
a] Disease resistance
b] Plant quarantine
c] Biological control of plant
d] Cultural control

Did You Know ?
First Governor
Sir George Oxdenden who was appointed President of East India Co. in 1662, took over as Governor of Mumbai (*c.* 1668). Oxenden died at Surat in July 1669.

42. First of all microscope discovered by:
a] Needham b] Spallanzani
c] Leeuwenhoeck d] Koch

43. First plant parasitic bacteria was reported by :
a] T.J. Burrill b] Needham
c] Louis Pasteur d] Leeuwenhoek

44. First plant parasitic nematode discovered was :
a] *Meloidogyne* b] *Heterodera*
c] *Anguina* d] *Globodera*

45. First plant parasitic nematode was reported by :
a] Needham b] Spallanzani
c] Cobb d] Mayer

46. Flag smut of wheat could be controlled by :
a] Seed treatment with vitavax
b] Spraying with vitavax
c] Soil application with vitavax
d] Above all

47. Flag smut of wheat is caused by :
a] *Ustilago tritici* b] *Urocystis tritici*
c] *Ustilago hordei* d] *Taphrina deformans*

48. Foot rot of papaya is caused by :
a] *Phytophthora palmivora*
b] *Pythium aphanidermatum*
c] *Peronospora parasitica*
d] *Albugo candida*

49. Fungi which can grow on living host plant are called:
a] Obligate saprophyte
b] Obligate parasite
c] Facultative parasite
d] Saprophytes

50. Fungi which can grow only on living host plant are called :
a] Obligate saprophytes
b] Obligate parasites
c] Facultative parasites
d] Saprophytes

51. Fusarium belong to order :
a] Melaconiales b] Moniliales
c] Sporobolomycetalesd] Cryptococcales

52. Gelatin plate technique was discovered by :
a] Burrill b] Fanny Hesse
c] Robert Koch d] Louis Pasteur

GK GS
Beighton Cup is associated with sport Hockey.

53. Genus *Radopholus* of nematode was first described in 1893 on the basis of a male specimen obtained from necrotic roots of ______ growing in Fiji Islands.
a] Sugarbeetb] Sugarcane
c] Banana d] Coconut
54. Heterocious nature of rust means:
a] A fungus require one host species to complete life cycle
b] A fungus require two different host species to complete life cycle
c] Both a and b
d] None of the above
55. Heterodera belongs to super family:
a] Atylenchoidea b] Neotylenchoidea
c] Tylenchoidea d] Heteroderoidea
56. Hot water treatment for control of loose smut of wheat was given by:
a] Luthra and Sattar b] Swindle
c] Jensen d] Butler
57. How much quantity of nematicide is used to control the nematodes in one hectare are ?
a] 10-15 kg b] 20-25 kg
c] 50-60 kg d] 100-200 kg
58. In bacterial brown rot and wilt of potato the amount of __________ produced by pathogens are proportional to severity of symptoms.
a] Gibberellins b] Lipids
c] Polysaccharides d] Proteins
59. In fermentation industry, Pasteuri-zation is done at temperature :
a] 90.7 °C b] 100.0 °C
c] 62.8 °C d] 80.0 °C
60. In *P. coffeae* (namatode), a _________ species, the optimum temperature for multiplication is reported to be 29.5°C on rough lemon.
a] Temperate b] Tropical
c] Humid d] Sub-humid
61. In plains of India which wheat rust appear first :
a] Black rust b] Orange rust
c] Yellow rust d] Orange and Yellow
62. In which rust pustules are arranged in end to end manner and forming strip ?
a] Yellow rust b] Orange rust
c] Black rust d] White rust
63. Indian Journal of Mycology and Plant Pathology is published from:
a] IARI, New Delhi
b] Botany Department, Lucknow University
c] R.C.A., Udaipur
d] Botany Department, Madras University
64. Indian Phytopathological Society was established by:
a] J.F. Dastur b] E.J. Butler
c] B.B. Mundkar d] K.C. Mehta
65. Irish famine was caused due to :
a] Wheat rust
b] Apple scab
c] Late blight of potato
d] Tomato wilt
66. Late blight of potato is caused by :
a] *Alternaria solani*
b] Xanthomonas campestris pv. *citrii*
c] *Phytophthora infestans*
d] *Sychytriu endobioticum*
67. Leaf roll of potato is caused by :
a] Fungus b] Bacterium
c] Virus d] Mycoplasm
68. Leaf roll of potato is transmitted by:
a] *M. persicae* b] *M. pseudosolani*
c] *M. circumflexum* d] All above
69. Levillula is a :
a] Ectoparasitic b] Endoparasitic
c] Both a and b d] None of the above
70. Lichen is an association between :
a] Algae and Fungus
b] Bacterium and Fungus
c] Fungus and Virus
d] Fungus and Nematode
71. Linseed rust is caused by :
a] *Uromyces* b] *Melampsora lini*
c] *Physaqmidium* d] *Ravenelia*
72. Acervuli is _______ shaped.
a] Round b] *Saucer*
c] Oblate d] Conical
73. Loose smut of wheat can be effectively controlled by :
a] Crop rotation
b] Soil treatment with non-systematic fungicides
c] Seed treatment with systemic fungicides
d] Spraying the crop with fungicides
74. Loose smut of wheat is a disease known as :
a] Internally seed borne
b] Externally seed borne
c] Air borne
d] Soil borne
75. Meloidogyne belongs to group of:
a] Ecotoparasite b] Endoparasite
c] Semi-ectoparasite d] Semi-endoparasite
76. Micro organisms were first observed by :
a] John Tyndall
b] L.J. Thernard
c] Antony von Leeuwenhock
d] Louis Pasteur
77. Microbiology is the science in which we study :
a] Viruses b] Bacteria
c] Fungi d] All above

Wahabi movement was started by Saiyad Ahmad of Rae Bareli.
GS

78. **Modification of floral parts into leafy structures by pathogenic infection called :**
a] Hyperplasia b] Hypertrophy
c] Phyllody d] Witch's broom

79. **Molya disease of wheat and barley is caused by :**
a] *Heterodera schachti*
b] *Heterodera rostochiensis*
c] *Heterodera avenae*
d] *Heterodera glycines*

80. **Most effective fungicide against powdery mildew disease :**
a] Dithane b] Captan
c] Karathane d] Bavistin

81. **Most of the plant parasitic nematodes belong to order:**
a] Tylenchida b] Dorylaimida
c] Aphelenchida d] Rhabditida

82. **Mycology is the study of :**
a] Bacteria b] Algae
c] Virus d] Fungi

83. **Mycoplasma disease of plant was first reported by :**
a] Butler
b] Brefeld
c] Anton de Barry
d] Doi *et al.* and Ishiie *et al.*

84. **Mycoplasma is sensitive to :**
a] Penicillin b] Steptomycin
c] Bavistin d] Tetracycline

85. **Plant overgrowth due to increase in size of cell of the tissue :**
a] Hypoplasia b] Hypertrophy
c] Hyperplasia d] All of above

86. **Nematode genus *Hoplolaimus* Daday was first recorded in ___________ from an Island in the Paraguay river.**
a] 1900 b] 1905
c] 1910 d] 1915

87. **Nematodes are also known as :**
a] Tapeworms
b] Trematodes
c] Earthworms
d] Round worms

88. **Nematodes penetrate in plant tissues with the aid of:**
a] Stylet b] Saliva
c] Labial papilla d] Cervicial

89. **Oil from American elm seed contains 55 per cent capric acid toxic to *Ceratocystis ulmi*, the pathogen causing the famous ___________ disease.**
a] English elm b] Dutch elm
c] American elm d] French elm

90. **Once the pathogen has established in an area, steps taken to remove or destroy is known as :**
a] Exclusion
b] Eradication
c] Protection
d] Management

91. **Pectinolytic enzymes of plant pathogens play a major role during pathogenesis in diseases like :**
a] Rusts
b] Downy mildew
c] Wilts, damping-off and root/foot rots
d] Powdery mildew

92. **Phycomycetes belong to :**
a] Fungi b] Bacteria
b] Yeast d] None of above

93. **Phyllactinis causes the disease :**
a] Powdery mildew b] Downy mildew
c] Leaf spot d] Blight

94. **Phytophthera belongs to family :**
a] Peronophythoraceae
b] Pythiaceae
c] Pareonosporaceae
d] Albuginaceae

95. **Plant disease in written by :**
a] B.B. Mundkur b] R.S. Singh
c] E.J. Butler d] R.S. Mehrotra

96. **Plant parasitic nematodes posses :**
a] Two-part oesophagus
b] Three-part oesophagus
c] Four-part oesophagus
d] None of these

97. **Plant pathogenic viroids contain :**
a] Externally seed-borne disease
b] Externally seed-borne and soil-borne disease
c] Air-borne disease
d] Internally seed-borne disease

98. **'Wilt Phase' is common in :**
a] Blast of Paddy
b] Bacterial blight of Paddy
c] White rust of Crucifers
d] Red rot of Sugarcane

99. **Powdery mildew disease is caused by :**
a] Bacterium b] Fungus
c] Virus d] Nematode

100. **Powdery mildew fungi belong to the family :**
a] Erysiphaceae
b] Hypocereaceae
c] Hypomycetaceae
d] Nectriaceae

101. **The symptom of red rot of sugarcane mostly found on :**
a] Stem b] Root
c] Leaf d] Arrow

102. **Powdery mildews are controlled by spray of :**
a] Copper fungicides
b] Mercury fungicides
c] Sulphur fungicides
d] Miscellaneous fungicides

GK GS

South Asian Games is the new name given to the South Asian Federation Games.

103. Primary source of infection to green ear disease of bajra is :
a] Insect b] Soil
c] Air d] Seed and soil

104. Pycnidia are formed in order :
a] Moniliales b] Melanoconiales
c] Sphaeropsidales d] All of the above

105. Pythium belongs to class :
a] Chytridiomycetes b] Oomycetes
c] Zygomycetes d] Hypochytridiomycetes

106. Radophyceae are also knwon as :
a] Red algae b] Green algae
c] Blue-green algae d] None of the above

107. Red rot of sugarcane is caused by :
a] Fusarium b] Alternaria
c] Dreschlera d] Colletotrichum

108. Red scales of _______ (red varieties) contain protocatechuic acid and catechol which may exude in drops to resist the attack of *Colletotrichum circinans*.
a] Garlic b] Onion
c] Leek d] Celery

109. Rhizopus belongs to class :
a] Oomycetes b] Zygomycetes
c] Trichomycetes d] Ascomycetes

110. Root knot of vegetables is caused by:
a] Meloidogyne b] Heterodera
c] Pratylenchus d] Xanthomonas

111. Root parasite of bajra is :
a] Cuscuta b] Striga
c] Orobanche d] Loranthus

112. Root rot of papaya could be controlled by application of :
a] Bordeaux mixture b] Dithane M-45
c] Benelate d] Calaxin

113. Rust teleutospore on germination produce :
a] Four basidiospores
b] Two basidiospores
c] Indefinite basidiospores
d] None of these

114. Rusts belongs to sub-division :
a] Zygomycotina
b] Ascomycotina
c] Basidiomycotina
d] Deuteromycotina

115. Secondary spread of ergot of bajra is through :
a] Seed b] Air
c] Insect d] Both b and c

116. Severity of disease club root of cabbage can be reduced by application of :
a] Lime b] Gypsum
c] FYM d] Fertilizers

117. Sexual dimorphism means :
a] Males and Females are separate
b] Males and Females are morphologically dissimilar
c] Females posses didelphic ovaries
d] None of the above

118. Shape of bacterial cells are :
a] Spheres b] Spiral
c] Rods d] All of the above

119. Smut gives the appearance of :
a] Compact hardness b] Sooty powder
c] Superficial growth d] Dusty layer

120. Smut of wheat is effectively controlled by :
a] Bavistin b] Vitavax
c] Benelate d] Thiram

121. Solar energy treatment for the control of loose smut of wheat was given :
a] Shrivdle b] Luthara and Sattar
c] Mitra d] Butler

122. Some __________ varieties of apple exude toxic wax on leaf surface to prevent germination of conidia of *Podosphaera leucotricha*.
a] Scab resistant
b] Leaf blight resistant
c] Powdery mildew resistant
d] Rust resistant

123. Some varieties of __________ are resistant to wilt due to the presence of hydrocyanidies (HCN) in root exudates.
a] Sunflower b] Safflower
c] Castor d] Linseed

124. Species Plantarum is the greatest contribution of :
a] Linnaeus b] P.A. Micheli
c] Saccardo d] Anton de Bary

125. Sporangium on germination produce vesicle in genera:
a] Phytophyhora
b] Pythium
c] Sclerospora
d] Personospora

126. Stipe disease of barley is caused by :
a] *Drechalera gramineum*
b] *Drechalera oryzae*
c] *Puccinia striiformis*
d] None of the above

127. Strawberry plants infected with *P. penetrans* (nematode) have been successfully treated with exposure in hot water at:
a] 49.4°C for 17.5 min.
b] 52.8°C for 6 min.
c] 54.5°C for 4 min.
d] 56°C for 2 min.

GK

Salim Ali Centre for Ornithology and Natural History is located in Coimbatore.

GS 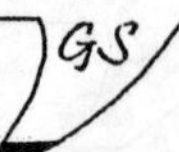

128. Striga is plant parasite included in :
a] Fungi
b] Mycoplasma like organisms
c] Phanerogamic plant parasites
d] Bacteria

129. Sugar beet cyst nematode was reported by :
a] Needham b] Cobb
c] Sachacht d] Smith

130. Systemic fungicides was discovered by :
a] Van Schleming and Kulka
b] Millardet
c] Tillet
d] Prevost

131. Taphrina causes the disease :
a] Chillies leaf curl b] Tomato leaf curl
c] Peach leaf curl d] Potato leaf curl

132. Term commensalism shows :
a] Living together of the two species, one of which is benefited by the association where as the other is neither beneficial nor harmful
b] Living together of two or more species of organisms for mutual benefit
c] The joint action of the two-organisms on a carbohydrate, resulting in the production of gas that is not formed by either organisms cultivated separately
d] Living two organism together which are harmful to each other

133. Tha anal and vulval openings in nematodes are located on :
a] Dorsal side b] Lateral side
c] Ventral side d] Dorso-lateral side

134. The ___________ botanist Micheli (1679-1737) was the first scientist who in 1729 studied fungi and saw their spores.
a] German b] Dutch
c] Italian d] Australian

135. The casual organism for earcockle disease of wheat is :
a] *Meloidogyne* spp. b] *Heterodera avenae*
c] *Anguina tritici* d] None of the above

136. The class ends with :
a] Mycotina b] Mycetes
c] Oomycetes d] Zygomycetes

137. The discovery of ___________ mosaic by Meyer in 1886 was the beginning of studies on viruses as incitants of diseases in plants.
a] Tomato b] Mungbean
c] Tobacco d] Soyabean

138. The epidermal cells of ___________ plant become filled with toxic terpenoid aldehyde after they are few days old and these cells are rarely penetrated by pathogens.
a] Jute b] Flax
c] Cotton d] Okra

You Don't Know!
First electrification
Kolkata was the first city to go electric when it received power in 1899. But the Maharaja of Bikaner was the first to switch on an electric light in 1896.

139. The founder of mycology was :
a] Linnaeus b] Anton de Bary
c] P.A. Micheli d] Saccardo

140. The genera *Claviceps*, *Alternaria* and *Sclerospora* are :
a] Bacteria b] Fungi
c] MLO d] Virus

141. The micro-organisms, used in brewing industry is :
a] Aspergillus b] Yeast
c] Rhizobium d] Virus

142. The nematode *Radopholus similes* has shown the maximum numbers at a soil depth of:
a] 5-15 m b] 20-30 m
c] 32-42 m d] 1-4 m

143. The organism included in order pernosporales cause the disease known as :
a] Downy mildew b] Smut
c] Rust d] Powdery mildew

144. The organism lacking a well defined necleus are termed :
a] Fungi b] Prokaryote
c] Eucaryote d] Protozoa

145. The organisms included in order Peronosporales cause the disease known as :
a] Rust b] Smut
c] Downy mildew d] Powdery mildew

146. The organisms lacking a well defined nucleus are termed :
a] Fungi b] Prokaryote
c] Eukaryote d] Protozoa

147. The plant disease caused by an algae is :
a] Blue mold of tobacco
b] Red rust of tea
c] Aster yellows
d] None of the above

148. The plant disease primarily responsive for the "Great Bengal Famine" of 1943 was :
a] Blast disease of rice
b] Bacterial blight of rice
c] Udbatta disease of rice
d] Brown leaf spot of rice

149. The plant mycoplasmal diseases are transmitted by :
a] Aphids b] Leaf hoppers
c] Mites d] Above all

GK GS
The Champaran Satyagraha was against the Tinkattia system.

150. The two important diseases of sugarcane would include :
a] Little leaf and bacterial wilt
b] Red rot and grassy shoot
c] Red rot and bund
d] Bakanae disease and stem canker

151. Tilletia causes the disease :
a] Rust b] Smut
c] Bunt d] Karnal Bunt

152. To which group Dithane Z-78 belongs:
a] Organomercurial
b] Dithiocarbamic acid
c] Heterocyclic nitrogen compound
d] None of these

153. Choose the odd one :
a] Leafspot = Paddy, Pigeonpea
b] Zonate leaf spot = Jowar
c] Red leaf = Sugarcane
d] Leaf blight = Wheat

154. Tobacco mosaic virus was first reported by :
a] Smith b] Martin
c] Mayer d] Beijerinck

155. Tobaco mosaic infections intity was isolated in pure cyrstalline form of protein by :
a] Stanley b] Beijerinck
c] Iwanowsky d] Bawden

156. *Tolyposorium penicillarie* cause :
a] Smut of sugarcane
b] Grain smut of sorghum
c] Bajra smut
d] Head smut of sorghum

157. Tomato leaf curl is caused by :
a] Bacteria b] Fungi
c] Mycoplasma d] Virus

158. Transduction is a process of multiplication in :
a] Fungi b] Bacteria
c] Bacteriophages d] Nematode

159. Transformation is a process of multiplication in :
a] Fungi b] Bacteria
c] Bactriophagus d] Nematode

160. Trichodorus belongs to family :
a] Trichodoridae b] Heteroderidae
c] Nacobbidae d] Longidoridae

161. Tundu disease of wheat is caused by:
a] Anguina b] Corynebacterium
c] Both a and b d] Heterodera

162. Two of the more important diseases of potato whose spread in India has been successfully restricted through plant quarantine measures include:
a] Brown rot and late blight
b] Wart and golden nematode
c] Ring rot and mosaic
d] Bacterial wilt and charcoal rot

163. Ufra disease of rice is caused by :
a] Pratylenchus b] Ditylenchus
c] Aphelenchoides d] Heterodera

164. Leaf curl of tobacco is caused by :
a] Nicotina virus Ca-1 b] Nicotina virus-10
c] MLO's d] Viriod

165. Uncinuls causes the disease :
a] Downy mildew b] Powdery mildew
c] Blight d] Wilt

166. Used lense for the first time to study micro-organism:
a] Robert Koch
b] Antony van Leeuwenhoeck
c] Louis Pasteur
d] Butler

167. Virus is transmitted by nematode:
a] Globodera b] Tylenchorhynchus
c] Anguina d] Xiphinema

168. Wheat rust life cycle in India was given:
a] E.J. Butler b] M.J. Thrimublachar
c] K.C. Mehta d] R. Prasad

169. When bacterial cells are subjected to unfavourable conditions they form :
a] Oospores b] Zoospores
c] Endospores d] Cyst

170. Which are the cellulose attacking fungi at the time of decomposition of organic compound?
a] *Trichoderma, Aspergillus*
b] *Ganoderma, Psalliotta*
c] *Albugo, Puccinia*
d] Pythium, Macrophomina

171. Which fungus produces aflatoxin?
a] Penicillium notatum
b] *Aspergillus flavus*
c] *Albugo candida*
d] *Trichoderma viride*

172. Which fungus supresses various pathogenic fungi present in soil ?
a] *Pythium aphanidermatum*
b] *Psalliotta* spp.
c] *Macrophomina* spp.
d] *Trichoderma viride*

173. Which is citrus namatode ?
a] Hoplolaimus b] Pratylenchus
c] Rotylenchulus d] Tylenchus

174. Which is sulphur bacteria ?
a] Nitrosomonas b] Cladethrix
c] Thiothrix d] Bacillus

175. Which is the root knot nematode ?
a] Angunia b] Heterodera
c] Meloidogyne d] Tylenchorhynchus

176. Which of the following nematicide is a carbamate ?
a] DBCP b] DD
c] Phorate d] Aldicarb

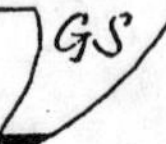

USA is the largest emitter of carbon dioxide in the world.

177. Which of the following nematodes attack above plant parts ?
   a] *Pratylenchus* b] *Anguina*
   c] *Meloidogyne* d] *Rotylenchulus*
178. Which of the nutrient is supplied by mycorrhiza to the plant ?
   a] Nitrogen b] Phosphorus
   c] Sulphur d] Iron
179. Which of them is a gram positive bacterium ?
   a] Xanthomonas b] Pseudomonas
   c] Erwinia d] Corynebacterium
180. Which of them is a poisonous mushroom ?
   a] Agaricus b] Volvariella
   c] Pleaurotus d] Amanita
181. Which one of the following correctly defines the period between inoculation and appearance of the first disease symptom?
   a] Penetration period b] Latent period
   c] Incubation period d] Infection period
182. Which one of the following groups consists of systemic fungicides?
   a] Bavistin, Topsin-M, Captan
   b] Plantvax, Emisan, Calixin
   c] Vitavax, Bavistin, Calixin
   d] Ziram, Plantvax, Blitox
183. Which one of the following helps in maintaining the shape of nematodes?
   a] Pseudocoelomic fluid
   b] Cuticle and muscles
   c] Pseudocoelome
   d] Annulation of cuticle
184. Which one of the following is known as downy mildew fungus?
   a] Erysiphe b] Sclerospora
   c] Taphrina d] Urocystis
185. Which one of the following is the best method of controlling ear cockle disease of wheat?
   a] Application of solar heat
   b] Mechanical separation of infected seed by floatation
   c] Application of balanced dose of fertilizer
   d] Treatment of seed with pesticide
186. Which one of the following is the cause of "spreading decline" of citrus?
   a] Citrus nematode
   b] Burrowing nematode
   c] Fungal diseases
   d] Nutritional deficiency
187. Which one of the following is the correct mode of infection in Ergot of bajra ?
   a] Shoot infection
   b] Local infection of blossom
   c] Seedling infection
   d] Systematic infection
188. Which one of the following is the correct order in which they appear during biocontrol operations?
   a] Parasite, Release, Mass rearing, Parasitism
   b] Mass rearing, Parasite, Release, Parasitism
   c] Release, Mass rearing, Parasite, Parasitism
   d] Parasite, Mass rearing, Release, Parasitism
189. Which one of the following pairs is not correctly matched?
   a] Karnal bunt - *Neovossia horrida* of wheat
   b] False smut - *Ustilaginoidea virens* of rice.
   c] Hill bunt - *Tilletia foetida* wheat.
   d] Ear cockle - *Anguina tritici* wheat.
190. Which one of the following pairs of crop and disease is correctly matched?
   a] Cabbage - Early blight
   b] Cucumber - Club root
   c] Potato - Black wart
   d] Tomato - White rust
191. Which one of the following remedies is to be suggested in case of failure of a manually operated sprayer to retain pressure?
   a] Tightening of loose nuts and clamps and replacement of gaskets
   b] Tightening of the lid of the tank, replacement of the gasket, if necessary
   c] Opening the nozzle and cleaning its various parts especially the orifice
   d] Straightening the plunger rod if found bent
192. Which one of the following sprayers would require the lowest volume of spray solution to cover a hectare of rice crop?
   a] Power sprayer b] Foot sprayer
   c] Knapsack sprayer d] Hand sprayer
193. Which one of the following statements correctly defines economic threshold level?
   a] The pest population level at which control measures should be taken to prevent the pest population from reaching economic injury level
   b] The lowest pest population density that will cause economic damage
   c] The optimum pest population density that will cause economic damage
   d] The pest population level at which application of pest control measures will be uneconomic
194. Which one of the following statements correctly defines the "alternate host" of a pathogen/parasite?
   a] A host on which a parasite attacks simultaneously with the main host
   b] A host which grows in alternate season or year
   c] One of the two kinds of hosts on which a parasitic fungus develops to complete its life cycle
   d] A host on which, the parasite does not attack in normal conditions

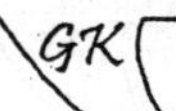

195. Which one of the following tissues is most suitable for tissue culture to produce virus-free plants of potato?
a] Leaf tissue b] Pollen
c] Embryo tissue d] Meristem tip tissue

196. Which shows the symbiotic relationship ?
a] Rhizobium b] Azotobacter
c] Clostridium d] Nitrosomonas

197. Which was the first plant bacterial disease reported?
a] Angular leaf spot of cotton
b] Citrus canker
c] Fire blight of apple
d] Bacterial blight of tomato

198. Whip smut of sugarcane is caused by:
a] *Ustilago nuda* b] *Ustilago hordei*
c] *Urocystis agrophri* d] Ustilago sciteminae

199. White rust of crucifer is caused by:
a] Puccinia b] Albugo
c] Sclerospora d] Peronospora

200. Who for the first time discovered antibiotics ?
a] Alexander Flemming
b] Stanley
c] Anton de Bary
d] Louis Pasteur

201. Who gave the term pasteurization ?
a] Anton de Bery b] Louis Pasteur
c] K.C. Mehta d] Leeuwenhoeck

202. Who is known as the father of nematology ?
a] N.A. Cobb b] G. Thorne
c] F.G.W. Jones d] De-man

203. Who proved that virus could pass through filters with pores small enough to retain bacteria ?
a] Stanley b] Ivanowski
c] Mayer d] Barkley

204. Wilt of *arhar* is caused by :
a] Verticillium b] Fusarium
c] Xanthomonas d] Meloidogyne

205. With regard to nutrition *Erysiphe* is:
a] Ectoparasite b] Endoparasite
c] None of the above d] Predator

206. Yellow ear rot of wheat is caused by:
a] Anguina b] Corynebacterum
c] Both a and b d] Heterodera

207. Yellow pigment in the culture is produced by :
a] Pseudomonas b] Xanthomononas
c] Erwinia d] Agrobacterium

208. Yellow vein mosaic of *bhindi* was reported by :
a] Jha and Mishra b] Uppal *et. al.*
c] Mayer d] Berckeley

209. Anthracnose of Grape is also known as:
a] Bird's-eye disease b] Buck eye rot
c] Black spot d] Both a & c

210. Choose the wrong combination :
a] Esso fungicide = Copper oxychloride
b] Ziram = Cuman
c] Thiram = Hexathir
d] Zineb = Kitazin

211. Dry root rot of legumes and beans is caused by :
a] *Erysiphe*
b] *Macrophomina*
c] *Xanthomonas*
d] *Phacidiopycnis*

212. Wrong information about Powdery mildew is :
a] Subdivision = Ascomycotina
b] Order = Erysiphales
c] Family = Melanconiaceae
d] All of the above

213. Whittaker (1969) proposed a ______ kingdom System of Classification of Microorganism.
a] 4 b] 5
c] 16 d] 18

214. _________ proposed a classification of fungi which is now widely accepted.
a] Walker (1957) b] Stakman & Harrar (1957)
c] Whittaker (1969) d] Ainsworth (1973)

215. Choose the odd one :
a] Lance nematode = Pratylenchulus
b] Lesion nematode = Pratylenchulus
c] Sting nematode= Belonolaimus
d] Awl nematode = Dolichodorous

GK GS

Lake Gatun is located in Panama Canal.

Matching

1. **Match crop species with its serious disease.**

a]	Rice	i]	Green ear
b]	Wheat	ii]	Bacterial leaf blight
c]	Maize	iii]	Stalk rot
d]	Bajra	iv]	Alternaria leaf blight

2. **Match crop disease with pathogen.**

a]	Tikka disease of groundnut	i]	*Erysiphe graminis f.s.p. hordei*
b]	Powdery mildew of barley	ii]	*Puccinia striiformis*
c]	Yellow rust of wheat	iii]	*Plasmopara viticola*
d]	Downy mildew	iv]	Cercospora personate

3. **Match diseases with disease incitants.**

a]	Karnal bunt of wheat	i]	*Sclerospora graminicola*
b]	Late blight of potato	ii]	*Tilletia indica*
c]	Flax rust	iii]	*Phytophothora infestans*
d]	Downy mildew of bajra	iv]	*Melampsora lini*

4. **Match pathogenic genera of fungi with its symptoms.**

a]	Cystopus	i]	Red rust
b]	Protomyces	ii]	Green ear
c]	Sclerospora	iii]	Leaf curl
d]	Taphrina	iv]	Stem gall

5. **Match diseases with its causal organisms.**

a]	Citrus canker	i]	*Pseudomonas solanacearum*
b]	Black rot	ii]	*Agrobacterium tumefaciens*
c]	Brown rot	iii]	*Xanthomonas campestris* pv. *campestris*
d]	Crown gall	iv]	*Xanthomonas campestris* pv. *citri*

6. **Match causal agent with disease caused.**

a]	Virus	i]	Sesamum phyllody
b]	Mycoplasma	ii]	Potato spindle tuber
c]	Fungus	iii]	Yellow vein mosaic of okra
d]	Viroid	iv]	Downy mildew of bajra

7. **Match cropos with smut diseases.**

a]	Barley	i]	Grain smut
b]	Paddy	ii]	Whip smut
c]	Sorghum	iii]	False smut
d]	Sugarcane	iv]	Covered smut

8. **Match the crop with nematode vector.**

a]	Grapevine	i]	*Trichodorus christiei*
b]	Strawberry	ii]	*Xiphinema index*
c]	Potato	iii]	*Longidorus elongatus*
d]	Red currant	iv]	*Xiphinema diversicaudatum*

9. **Match the viruses transmitted by nematodes with crop.**

a]	Raspberry ring spot	i]	Lucerne
b]	Tobacco rattle	ii]	Red currant
c]	Arabis mosaic	iii]	Potato
d]	Early browning	iv]	Strawberry

10. **Match the fungi with the virus disease transmitted by them.**

a]	*Olipidium brassicae*	i]	Oat mosaic
b]	*Polymixa graminis*	ii]	Potato mop top
c]	*Spongospora*	iii]	Pea false leaf roll
d]	*Pythium ultimum*	iv]	Lettuce big vein

'Coloris Basin' is found on the surface of planet Mercury.

GS

11. Match disease and host with the parasite.

a]	Small maize	i]	*Synchytrium andobioticum*
b]	Wart, potato	ii]	*Gymosporangium juneperivirginianae*
c]	Canker, apple	iii]	*Ustilago maydis*
d]	Rust, cedar apple	iv]	*Nectria galligena*

12. Match disease and host with the parasite.

a]	Downy mildew, sugarcane	i]	*Puccinia graminis*
b]	Bakanae disease, rice	ii]	*Sclerospora sacchari*
c]	Rust, wheat	iii]	*Gibberella fujukuroi*
d]	Leaf blight, cherry	iv]	*Gnomonia erythrostroma*

13. Match disease and host with the parasite.

a]	Rust, safflower	i]	*Omphalia flavida*
b]	Leaf spot, coffee	ii]	*Sclerospora sacchari*
c]	Downy mildew, maize	iii]	*Corynebacterium fasaiens*
d]	Witches broom, sweetpea	iv]	*Puccinia carthami*

14. Match the pathogen with the toxin produced by it.

a]	*Helminthosporium victoriae*	i]	T-toxin
b]	*H. maydis, race* T	ii]	HS-toxin
c]	*H. carbonum*	iii]	Victorin
d]	*H. sacchari*	iv]	HC-toxin

15. Match the pathogen with the toxin produced by it.

a]	*Alternaria kikuchiana*	i]	PM-toxin
b]	*Periconia circinata*	ii]	Phytoalternarin
c]	*Phyllosticta maydia*	iii]	Fumaric acid
d]	*Rhizopus* spp.	iv]	PC-toxin

16. Match the pathogen with the toxin produced by it.

a]	*Alternaria tenuis*	i]	Juglone
b]	*Fusarium* spp	ii]	Tentoxin
c]	*Pseudomonas tabaci*	iii]	Marticin
d]	*Juglans nigra*	iv]	Tabtoxin/wild fire toxin

17. Match the formation of appressoria with epidermal cells.

a]	53.0%	i]	Long cells
b]	16.2%	ii]	Hairs
c]	12.8%	iii]	Motor cells
d]	1.7%	iv]	Short cells

18. Match the antimicrobial compound with the crop(s) producing it.

a]	Allyl sulphoxides	i]	Pear and walnut
b]	Glucosinolate	ii]	Onion and garlic
c]	Cyanogenic glucosides	iii]	Cabbage
d]	Para-hydoquinone glucocides	iv]	Sorghum, limabean, peach.

19. Match the toxin with the crop(s) producing it.

a]	Benzoxazines	i]	Green skin of potato tubers
b]	Solanin	ii]	Potato leaves
c]	Chaconin	iii]	Tomato
d]	Tomatin	iv]	Maize, wheat, rye

20. Match the lignification agent imparting fungal resistance with the crop.

a]	*Alternaria japonica*	i]	Wheat
b]	*Septoria nodorum, S. tritici*	ii]	Radish
c]	*Cladosporium cucumerinum*	iii]	Sweet potato
d]	*Helicobasidim mompa*	iv]	Cucumber

21. Match the phytolexins with the plant (host) producing it.

a]	Rishitin	i]	Tobacco and pepper
b]	Phytuberin	ii]	Tobacco
c]	Capisdiol	iii]	Potato and tomato
d]	Glutinosone	iv]	Potato

22. Match the phytolexins with the plant (host) producing it.

a]	Hirinol	i]	Oats
b]	Ipomeamarone	ii]	Orchids
c]	6-methoxy mellein	iii]	Sweet potato
d]	Avenalumin, I, II & III	iv]	Carrot

23. Match the varieties of stem rust fungus *Puccinia graminis* with uredospores size (microns).

a]	*tritici*	i]	22 x 16
b]	*secalis*	ii]	32 x 20
c]	*avenae*	iii]	25 x 17
d]	*agrostis*	iv]	28 x 20

24. Match the host-parasite relationship given below.

a]	*Triticum*	i]	*Helminthosporium victoriae* (Victoria blight of oat)
b]	*Avena*	ii]	*Cladosporium fulvum* (Leaf mold of tomatoes)
c]	*Lycopersicon*	iii]	*Venturia inequalis* (Scab of apple)
d]	*Malus*	iv]	*Puccinia striiformis* (Yellow rust of wheat)

25. Match the host-parasite relationship given below.

a]	*Zea*	i]	*Hemileia vastatrix* (Coffee rust)
b]	*Solanum*	ii]	*Puccinia sorghi* (Rust of maize)
c]	*Gossypium*	iii]	*Globodera rostochiensis* (Cyst nematode)
d]	*Coffea*	iv]	*Xanthomonas malvacearum* (Black arm of cotton)

26. Match the predisposing soil temperature with the pathogen.

a]	30-35°C	i]	*Drechslera graminae* (Barley stripe)
b]	10°C	ii]	*Fusarium culmorum* (Wheat root rot)
c]	17-23°C	iii]	*Sclerotium rolfsii* (Apple root rot)
d]	24-28°C	iv]	*Thielaviopsis basicola* (Tobacco root rot)

27. Match the nematode infestations with the development of fungal and bacterial diseases which they enhance.

a]	*Criconematoides xenoplax*	i]	Cotton wilt
b]	*Meloidogyne* spp.	ii]	Bacterial canker in plum
c]	*Meloidogyne javanica*	iii]	Root rot of tomato
d]	*Meloidogyne incognita*	iv]	Tomato wilt

28. Match the nematode infestations with the development of fungal and bacterial diseases which they enhance.

a]	*Tylenchulus semipenetrans*	i]	Potato wilt
b]	*Pratylenchus thornei*	ii]	Root rot of tabacco
c]	*Meloidogyne incognita*	iii]	Soyabean root rot
d]	*Heterodera glycines*	iv]	Root rot of citrus

29. Match the disease with the pathogen exhibited by suppressive soils.

a]	Wilt of bean	i]	*Fusarium oxysporum* f. sp. *cucumerinum*
b]	Wilt of cucurbits	ii]	*Streptomyces scabies*
c]	Root rot of lettuce	iii]	*Fusarium solani* f. sp. *phaseoli*
d]	Common scab of potato	iv]	*Olpidium brassicae*

30. Match the disease with the pathogen exhibited by suppressive soils.

a]	Black root rot, cucurbits	i]	*Pseudomonas solanacearum*
b]	Damping off, vegetables	ii]	*Phomopsis sclerotioides*
c]	Wilt, vegetables	iii]	*Rhizoctonia solani*
d]	Bare patch, wheat	iv]	*Pythium ultimum*

Water vapour gas is the largest contributor to green-house effect.

31. Match the decoy crop(s) which reduce the pathogen populations.

a]	Ryegrass	i]	*Plasmodiophora brassicae*
b]	*Datura stromonium*	ii]	*Striga asiatica*
c]	Sudan grass	iii]	*Heterodera avenae*
d]	Maize	iv]	*Spongospora substerranea*

32. Match the decoy crop(s) which reduce the pathogen populations.

a]	*Crotolaria spectabilis*	i]	*Meloidogyne javanica*
b]	Groundnut	ii]	*Prtylenchus alleni*
c]	*Sesamum orientale*	iii]	*Pratylenchus* spp.
d]	Bean, chrysanthemum	iv]	*Meloidogyne incognita*

33. Match the effect of crop(s) grown in short term rotation on reduction or elimination of the pathogen.

a]	Rice-sesame/wheat	i]	*Gaeumannomyces graminis*
b]	Tobacco-groundnut	ii]	*Pratilenchus indicus*
c]	Wheat-pea	iii]	*Meloidogyne incognita*
d]	Cotton-paddy	iv]	*Verticillium dahliae*

34. Match the effect of crop(s) grown in short term rotation on reduction or elimination of the pathogen.

a]	Tomato-Sudangrass	i]	*Meloidogyne incognita*
b]	Cotton-barley	ii]	*Pratylenchus penetrans*
c]	Cereals-beet	iii]	*Pratylenchus indicus*
d]	Rice-sesame/wheat	iv]	*Pseudomonas solanacearum*

35. Match the nematodes with the crops(s) or disease which they are associated with.

a]	*Meloidogyne* spp.	i]	Citrus nematode
b]	*Tylenchulus semipenetrans*	ii]	Cereal root nematode
c]	*Heterodera avenae*	iii]	Root knot of tomato/okra/brinjal
d]	*Heterodera tobacum*	iv]	Tobacco cyst nematode

36. Match the nematodes with disease which they are associated with.

a]	*Globodera rostochiensis*	i]	Lesion nematode
b]	*Heterodera sachachtii*	ii]	Potato cyst nematode
c]	*Prtylenchus penetrans*	iii]	Sting nematode
d]	*Belonolaimum longicaudatus*	iv]	Sugarbeet cyst nematode

37. Match the antagonistic pathogens with each other.

a]	*Penicilium vermiculatum*	i]	*Ophiobolus graminis*
b]	*Didymella exitialis*	ii]	Conidia of many fungi
c]	*Gliocladium roseum*	iii]	Sclerotia of *Sclerotinia sclerotiorum*
d]	*Coniothyrium minitans*	iv]	*Rhizoctonia solani*

38. Match the pathogen with organic amendment of soil.

a]	*Fusarium oxysporum* f. sp. *cubense*	i]	Green manure with pea, *Melilotus* (Banana wilt)
b]	*Streptomyces scabies* (Potato scab)	ii]	Oat straw, maize stover, lucerne hay
c]	*Phymatotrichum omnivorum* (Cotton root rot)	iii]	Sugarcane residues
d]	*Thielavicopsis basicola* (Bean root rot)	iv]	Green manure with soyabean

39. Match the pathogen with organic amendment of soil.

a]	*Aphanomyces euteiches* (Pea root rot)	i]	Barley straw
b]	*Verticillium albo-atrum* (Potato wilt)	ii]	Wheat straw, saw dust, oilcakes
c]	*Rhizoctonia solani* (Potato black scarf)	iii]	Green manure with rape, pea mixed grass legume
d]	*Gaeumannomyces graminis* (Wheat lake-all)	iv]	Crucifer tissues

40. Match the crops with their variety/source of resistance against reniform nematode.

a]	Cotton	i]	Kalyanpur Selection I
b]	Cowpea	ii]	Lousiana 33
c]	Tomato	iii]	Flughum
d]	Oat	iv]	V-16

Kesab Chandra Sen founded the Brahma Samaj of India.

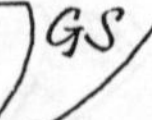

41. **Match the crops with their variety/source of resistance against reniform nematode.**

a]	Onion	i]	IPCL-95
b]	Pigeonpea	ii]	La rouge
c]	Potato	iii]	Gold rush
d]	Sweet potato	iv]	Evergreen

42. **Match *Hirschmanniella* species with the host(s).**

a]	*H. mucronata*	i]	*Colocasia esculenta*
b]	*H. miticausa*	ii]	Pearlmillet
c]	*H. areolata*	iii]	Rice
d]	*H. pomponiensis*	iv]	Lily, water chestnut

43. **Match the species of *Xiphinema* nematode described by a country.**

a]	*X. algeriense*	i]	Fiji
b]	*X. bacaniboia*	ii]	Italy
c]	*X. barbercheckae*	iii]	Algeria
d]	*X. barense*	iv]	South Africa

44. **Match the species of *Xiphinema* nematode described by a country.**

a]	*X. clavicaudatum*	i]	India
b]	*X. colombiense*	ii]	Nigeria
c]	*X. dolosum*	iii]	U.K.
d]	*X. hydrabadensis*	iv]	Brazil

45. **Match the species of *Xiphinema* nematode described by a country.**

a]	*X. incertum*	i]	Portugal
b]	*X. kosaigudensis*	ii]	Bulgaria
c]	*X. lusitanicum*	iii]	Malawi
d]	*X. malawiense*	iv]	India

46. **Match the species of *Xiphinema* nematode described by a country.**

a]	*X. manmillocaudatus*	i]	Canada
b]	*X. occiduum*	ii]	Germany
c]	*X. pseudocoxi*	iii]	Botswana
d]	*X. stockeri*	iv]	India

47. **Match the duration of life cycle of *Xiphinema* nematode with duration of life cycle.**

a]	*X. americanum*	i]	Two years
b]	*X. brevicolle*	ii]	One year
c]	*X. diversicaudatum*	iii]	4-7 months
d]	*X. vuitenezi*	iv]	Three years

48. **Match parasite with its nature**

a]	Striga	i]	Semi-parasiteon stem
b]	Cuscuta	ii]	Semi-parasite on root
c]	Loranthus	iii]	Total parasite on root
d]	Orobanche	iv]	Total parasite on stem

GS

Article 54 and 55 of Constitution deals with election of President.

Glossory of Technical Terms

Alternate Host Sometime a member of plant species serves as host of a pathogen. Each of these is known as an alternate host.

Antagonism It is a relationship between two organisms in which one inhibit or kill the growth other either fully or partly.

Antigen It is generally protein or carbohydrate which induces the formation of some other protein (antibody) in the blood when injected into he warm blooded receiver animal such that the produced material (antibody) can react with the antigen.

Autoecious When a plant parasite or pest completes its life cycle in one kind of host, it is called autoecious.

Autotrophic Organisms utilizing inorganic material as nutrition.

Bactericide Any substance killing bacteria.

Beaumont Period A period of two days (48 hours) during which the relative humidity does not exceeds 75% and temperature does not fall below 10°C. Presence of such period helps in the prediction of outbreak of potato scab, etc.

Biographic Organism that depends solely on another organisms for nutrient source.

Biotype It is a further division of species with groups of organisms genetically identical or near to identical.

Callus It is a mass of parenchymatous cambial tissues formed due to wound or infection on the plant.

Canker It is a dead area in the bark or cortex. The surface may be smooth, rough and some what shunken.

Chemotherapy Treating diseases by chemicals.

Colony Growth of hyphae mostly with spores or a collection of individual bacteria or yeast growing together.

Compound Interest Disease It refers to a disease occurring in a single season due to successive growth of previous generations.

Damping-off Death of seedlings due to lesion or stem at root level.

Dieback Shoot necrosis from top downward.

Differential Host Host giving different distinguishing reaction to isolates of a pathogen belonging to different physiologic races.

Diseases It refers to harmful deviation from normal functioning of physiological processes of an organism. This devotion may be caused by other organisms (pathogen, etc.) or due to mineral deficiency of the plant or due to other reason.

Disorder Diseases developed due to reason other than pathogens.

Enation It is an outgrowth from plant surface like stem, leaf.

Endemic Diseases It refers to a disease that is permanently occurring in a mild or severe form in a specific area.

Epidemic Disease It is a sudden outbreak of a disease in wider area during a season.

Epinasty Abnormal growth of a plant.

Etiolation Lack of green colour or excessive growth of internodes.

Fluorescent Antibody In this antibody the gamma-globulin protein are conjugated with the externally applied fluorescent dye to move it conveniently to observe the antigen distribution within cells.

Freeze-Drying It is a technique to move the water of plant tissues applying vaccume and the tissue is frozen for preservation purpose.

Fumigant A plant protection chemical volatile in nature.

Fungistat The chemical do not kill fungi but simply inhibit fungal growth temporarly. The phenomenon of temporary inhibiting the growth of the fungus is called fungistasis.

Gall Proliferation of plant tissues producing an outgrowth produced due to the action of a pathogen.

Heterotroph Organisms that utilize organic material as nutrient for their growth.

Host Organisms that harbour other organisms like pathogen.

Hyperparasite Parasite on another parasite.

Hypersensitive Reaction It refers to a reaction wherein the organism reacts with the pathogen so violently that infected tissues die of quickly and further spread is prevented.

Hypertrophy Enlargement of individual cell due to excessive growth.

In vitro Under artificial condition in the laboratory, etc.

In vivo Under natural condition.

Infection Entry of an organisms into a host and development of parasitic relationship subsequently.

Infection Peg When a hypha penetrates a host plant tissue, these form a structure around it. This is made of lignin, callose, cellulose, suberin, etc. Such structure is called infection peg.

Inoculum Mass of material containing virus or microorganisms for introduction or transfer to a host plant or medium for their proliferation.

Intergrated Control It refers to the complimentary use of biological, physical chemical or other methods of control to get an environment friendly process of disease and pest control.

Latent Infection Infection with no sign of symptom

Latent Period Period between infection and symptoms appearance.

Monocyclic Disease Here pathogen has one life cycle in each season on a host.

Necrosis Death of plant tissue, etc.

Parasite Organisms living on the host for nutrient and water for its own survival but does not benefit the host in return.

Passive Resistance Inherent resistance of the host plant without any defensive reaction with the pathogen.

Persistence Duration of time a virus remains within a vector after being acquired form an infected tissue.

Phototoxic Toxic to plants.

Phytoalexin Growth inhibiting substance for the micro-organisms. It is produced by higher plant in response to stimuli.

Balaghat hills are found between Bhima and Godavari valley.

GS

Phytopathogen Pathogen causing diseases to plant.

Plant Pathology Study of plant diseases.

Plasmid It is an extra nuclear organelle consisted on DNA and is found in a bacterial cell.

Polycyclic Disease Here the pathogen has many life cycles in each host season.

Powdery Mildew Powder like growth on plant surface due to some member of the Erysiphales.

Pre-emergence Before the emergence of the plant above the soil surface after sowing.

Propagule Part of an organisms that can spread and can reproduce itself.

Protactents Fungicides which is effective only when applied prior to fungal infection.

Pustule Blister-like spot on a plant leaf or stem from which fruiting structure of a fungus comes out on eruption.

Quarantine Governmental operations for preventing the dissemination of unwanted organisms between countries and between different areas within a country.

Resistance It is the ability of an organisms to oppose or, negative or lessen to overcome the effect of a pathogenic infection or injury.

Rot Tissue disintegration due to pathogen infection.

Rouging Removal of unwanted plants which are not true to type from a crop field.

Saprophyte Organism using dead organic material for its food.

Scab A surface lesion involving pilting, thickening and roughening of the plant surface on account of pathogenic infection.

Sequencing To determine the sequences of DNA bases for gene or amino acid for protein.

Simple Interest Disease A disease having single generation in a season developed from a constant inoculum source.

Smith Period A period of two days (48 hours) with temperature not below 10°C and relative humidity not below 90%. Such a period is favourable for potato blight development.

Smut Disease due to fungus of the Ustilaginales order.

Knowledge World!

First peasant organisation

The All-India Kisan Sabha was formed in 1936.

Sporadic Disease breakout occasionally without being constantly destructive is called as sporadic.

Suberisation Cork formation due to action of cork cambium, deposition of suberin on cells walls or in cell wall.

Substrat Medium (natural) for the microbial growth.

Syndrome It is a peculiar state of the affected host when the total effect produced by the pathogen is not known whether it happens simultaneously or successively or weather it can be detectable with naked eye or not.

Synergism Two or more organism acting on host making a charge together which cannot be brought out by either of the organisms singly.

Systemic Chemical Chemical absorbed through plant root, leaves and translocated subsequently throughout its body system.

Systemic Pathogen A pathogen present throughout the system of the plant(i.e. is root, shoot, stem, flower, etc).

Theraputants Fungicides capable of eradicating the fungus and so curing the disease after establishment of infection.

Vector Organisms that transports and transmits pathogen.

Vertical Resistance It refers to resistance to some of the races of a pathogen.

Virotoxin Toxin produced internally within the host by the pathogen.

Viruliferous Vector Vector that carries virus.

Wet Rot Rotting with complete degeneration of plant tissue.

White Rot Rotting of wood into white cellulose residue.

Wilt Loss of turgidity and collapse of leaves due to vascular infection by fungus, etc.

Zone Line A symptoms of narrow black or dark brown line on decaying wood due specially to fungal infection.

GK

Manas Wildlife Sanctuary is situated in Assam.

GS

10. Agricultural Economics and Farm Management

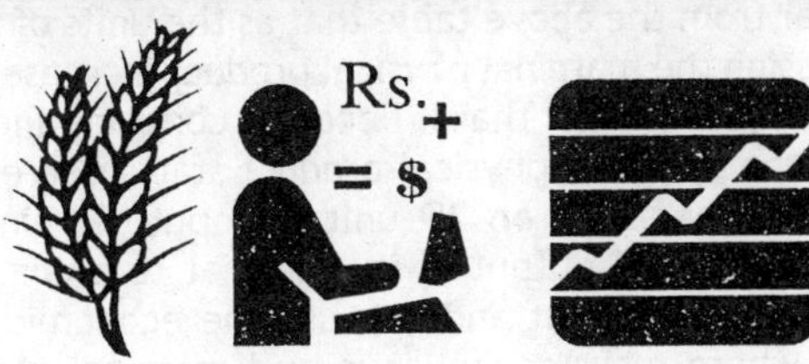

You Don't Know!

First budget

The first budget of free India was presented by the Finance Minister R.K. Shanmukham Chetty on November 26, 1947.

Each word counts

FARM MANAGEMENT

Farm management is a science which deals with the proper combination and operation of production factors including land, labour and capital and the choice of crop and livestock enterprises to bring about a maximum and continuous return to the most elementary operation units of farming.

Scope, Importance and Characters

The term farm Management consists of two words, 'Farm' and 'Management'. 'Farm' is a modified piece of land held or operated as a unit for the production of agricultural products. 'Management' means the art of managing the farm. It involves the process of determining how the farm unit shall be organised and operated.

Management performs three unique and vital functions –

(i) **Management makes decisions:** To make decision means to decide upon each course of action; to choose wisely among alternatives where such exist.

(ii) **Management acts on decisions**: After making decisions, the next step is to act, to work, the farm management selects the production level land technology and practices for products for producing and marketing the farm products. He must find and assemble all resources.

(iii) **Management takes financial responsibility** : The financial responsibility involves assembling, using and maintaining the farm resources. Farm management is concerned with decision-making. The main decisions of agriculture business which will maximise the returns to the resources used in the farm business are :

1. What to produce ? It decides which products to produce and shows the product-product relationship.

2. How to produce ? It relates to choosing the most effective method of producing a given quantity of a particular product. It shows the factor-factor relationship.

3. How much to produce ? It relates to the problems in converting the several farm resources into the final farm product. It shows the factor-product relationships.

4. Time relationship: It shows the relationship between the time that farmer makes investment of capital and that later date when this investment returns the physical production.

Scope of Farm Management

The study of farm management comes under microeconomics. The subject of farm management covers selection, size and appraisal of farm, appraisal of farm resources, investment decisions, enterprise relationships, planning the crops, farm labour and livestock, farm power machinery and equipment, costs returns on individual enterprises, complete budgeting, risk and uncertainty and marketing of farm produce.

Nature of Farm Management

Farm management is both pure and an applied pure science. It is pure science because it deals with the collection, analysis and explanation of facts and the discovery of principles, it is an applied science because the ascertainment and solution of farm problems are within its scope. The study of farming from the point of view of art would involve a careful examination of the thought, process and skill used by farmers in farming. Thus farm management is both a science and an art.

Limitations

The limitations of the farm management are :

(i) Small size of farm business: Due to fragmentation and subdivision of holding the average size of operational holdings is very small.

(ii) Inadequate capital: It is the most serious problem in agriculture. The farmers do not get adequate and timely financial aid from the institutional agencies.

(iii) Slow adoption of new technology: The farmers are not eager to adopt new technology, because it requires heavier inputs such as more fertilizers, timely irrigation, high yielding varieties of seeds and plant protection measures etc.

(iv) Under-employment: Under-employment reduces efficiency and productivity of rural people. Small size of operational holding, more family labour supply and seasonal nature of agricultural production are the source of the causes of underemployment in farming.

(v) Managerial problems: The illiterate farmers face difficult problems in managing the farm. Their managerial skill is very poor.

The Prarthana Samaj was founded by Atmaram Pandurang.

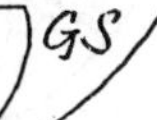

Economic Principles Applied to Farm Management

The economic principles applied to farm management have been developed directly or idirectly from the fundamental concepts and laws of economics. The economic science plays a very important part in all decisions of farm management. They are applicable to crop production, livestock production, machinery, farm buildings, labour and farm practices and production in general. The most important principles that help the farmers in making the choice and decisions to all phases of farming are :

1. *The Law of Diminishing Returns*

It is physical law of fundamental importance organising and operating the farm business. It is an important guiding factor in farming to decide the level at which a farmer can increase the output. It is practical experience of every farmer that after a certain stage in cultivation, the application of successive doses of labour and capital to a given area of land does not bring about much return as the previous dose. Additional dose of labour and capital applied to the same place gives Diminishing Returns. This tendency is known as the law of Diminishing Returns. Prof. Marshall has given three conditions which may delay the operation of the law in agriculture.

1. Improvement in the art (technique) of agriculture.
2. Increase in the efficiency or skill of the farmer.
3. Such previous doses of labour and capital applied to land that more than a proportional return is obtained from their increased application. The following table will illustrate the principle of Diminishing Returns.

Units of input applied	*Addition to total inputs*	*TTP*	*MPP*	*APP*	*Cost of additional inputs @ Rs. 20 per unit*	*Value of additional output @ Rs. 4 per unit*
0	1	0	0	0	-	-
1	1	8	8	8	20	32
2	1	20	12	10	20	48
3	1	31	11	10.3	20	44
4	1	40	9	10	20	36
5	1	45	5	9	20	20
6	1	48	-3	8	20	-24
7	1	42	-6	6	20	-24
8		32	-10	4	20	-40

TPP = Total Physical Product, MPP = Marginal Physical Product. APP = Average Physical Product.

It is clear from the above table that as the units of inputs is increasing the marginal physical product increases upto 3rd unit of input, after that it becomes constant and from 4th unit the marginal physical product starts decreasing, it becomes negative on 7th unit of input. To find out optimum level of output it is essential to change the physical units of input and output. The economic point will be where marginal revenue and marginal cost are equal. In the above table 5th unit is the optimum level from the economic point of view.

2. *Cost Principle*

This principle is of great importance in making decisions of farm business. Cost in general, refers to the expenses incurred on productive services and physical productivities are guided by the cost of various input factors. The cost are mainly of two types.

(i) Fixed Cost (FC): Fixed costs present farming expenses of an overhead nature and do not change with the output. e.g., interest on fixed capital, depreciation, revenue etc.

(ii) Variable Cost (VC): It varies with the level of output. It includes changes of hired, human labour, seed, fertilizer, irrigation, insecticides and pesticides etc.

(a) Total Variable Cost (TVC): It is obtained by multiplying the amount of variable input by the price per unit in input.

(b) Average Variable Cost (AVC): It is computed by dividing total variable costs by the amount of output. AVC varies with the level of production.

$$= \frac{TVC}{Output}$$

(c) Marginal Cost (MC): MC is the change in total cost as related to change in output.

$$= \frac{\Delta TC}{\Delta Y}$$

The marginal or added cost of each unit of output is important in determining how far we should push producing and how much of the various resources we should use.

(d) Total Fixed Cost (TFC): Total fixed cost shows the sum of expenditures which will be incurred irrespective of the output.

(e) Average Fixed Cost (AFC): It is the sum of total fixed cost and total variable cost.

TC = TFC + TVC

(g) Average Total Cost (ATC): It is worked out by dividing TC by output.

$$ATC = \frac{TC}{Output}$$

3. *The Law of Comparative Advantage*

The law of comparative advantage helps to explain regional specialization. Farmers tend to produce those items which give more income at a lowest relative cost. With the income, they buy the items needed for production and for living which are produced at lowest relative cost in other areas. In short, specialised and diversified farming are based on this principle.

4. *The Law of Equimarginal Returns*

Equi-means 'equal' and marginal means additional or incremental. The law states that profit will be greatest if each unit of labour, capital and is used where it adds the most to the return. The table below illustrates this principle.

Added or Marginal Returns to Capital on Three Enterprises

Amount of capital used (Rs.)	*Marginal returns*		
	Vegetable and crops	*Dairy*	*Poultry*
1000	1500	1400	1250
2000	1250	1250	1100
3000	1200	1000	1050
4000	1100	950	950
5000	1000	900	950
Total returns from Rs. 5000.00	6050	5500	5250
Net Profits	1050	500	250
Average Return/Rs.	1.21	1.10	1.05

In the above table, if we examine average return only, our advice to the farmer would be to produce only vegetable crops.

By investing Rs. 5000 he will get Rs. 1050 as net returns. But let us apply the principle of equimarginal returns for addition of resources. The farmer should apply Rs. 2000 on vegetable and crop, Rs. 2000 on dairy farming and Rs. 1000 on poultry keeping to get total returns of Rs. 6650 and net returns of Rs. 1650.00.

5. *The Law of Substitution*

The principle which helps selection from a number of alternatives is called law of Substitution. According to this law the least cost combination of inputs or practices is obtained when the inputs or practices replaced is equal to the value of the input or the practice added.

Procedure for finding out least cost combination -

1. Computation of Marginal Rate of Substitution (MRS) by

You Don't Know!

First High Courts set up in Independent India

Assam High Court with its seat at Guwahati, and Orissa High Court with its seat at Cuttack, were the first High Courts to be set up after independence, in 1948.

dividing the number of units of the replaced input (X_1) by the number of units of added input (X_2).

$$\text{Substitution Ratio} = \frac{\text{No. of units of replaced input}}{\text{No. of units of added input}}$$

Symbolically it may be expressed as

$$MRS = \frac{\Delta X_1}{\Delta X_2}$$

where is X_1 being replaced by X_2.

2. Computation of price ratio by dividing the price of added input (Px_2) by the price of the replaced input (Px_1).

$$\text{Thus price Ratio} = \frac{\text{Price of added input}}{\text{Price of replaced input}}$$

where x_1 is replaced by x_2. $\text{Price ratio} = \frac{Px_2}{Px_1}$

3. Calculation of least cost combination by

$$\text{equating (MRS)} = \frac{\Delta X_1}{\Delta X_2} = \frac{PX_2}{PX_1}$$

Farm Planning

Definition

It is a process of making decisions regarding the organisation and operation of a farm business.

Merits of Farm Planning

I. *Income Improvements*

Income maximizations could be achieved from a given resources by reorganising present type of production activities as well as introducing changes in technology.

2. *Educational Process*

Farm planning is an educational tool to bring about change in the outlook of the cultivators and the extension workers.

3. *Desirable Organizational Changes*

Farm planning introduces desirable changes in farm operations and makes a farm a viable unit.

4. *To Farmers*

- It helps him to examine carefully his existing resource situation and past experiences as a basis for decision.

GK **West Bengal of India has the largest area under mangrove forest.** GS

- To make rational decisions on what to do.
- To find out the credit needed.
- To give him an idea of the expected income after paying off his loans.

Characteristics of a Good Farm Plan

1. It should provide for efficient use of farm resources such as labour power and equipment.
2. Crop plan should have balanced combination of enterprises.
3. Avoid excessive risks.
4. Provides flexibility.
5. Utilize farmer's knowledge, training and experience and take account of the farmer's likes and dislikes.

Farm Budgeting

Definition

Farm budgeting is a method of analysing plans for the use of agricultural resources by the decision-maker.

Types of Farm Budgeting

(a) Partial Budgeting

- Partial Budgeting is an estimation of returns and expenses for a single enterprise or specific part of entire farm unit. e.g. Poultry unit (or) particular crop or one (or) few activities.
- It is used as an aid in full budgeting.
- It is used to estimate the effects or outcomes of possible adjustments in the farm business before such adjustments are actually made.

(b) Enterprise Budgeting

- It is used to estimate input required, costs involved, and expected returns from a particular enterprise.
- It is used to aid in selection of inputs and enterprises consistent with the resources available and to show combinations that will increase income from the farm business.

(c) Complete (or) Total Budgeting

- It involves estimation of returns and expenses for the entire farm as a single unit. Evaluates all enterprises of the farm i.e., crops, livestock, labour, machinery, buildings, managerial ability to obtain an overall financial picture of a farm.
- The process by which the farmer readjusts to his resources to his long-run objectives and opportunities.
- Farmers make both long-run and short-run or annual budget plans.

Merits of Complete Budgeting

- Brings about progressive changes in income.
- Takes care of all factors affecting farm income.

Differences between partial budgeting and complete budgeting

Partial Budgeting	*Complete Budgeting*
1. It considers a few alternatives.	1. In considers all alternatives.
2. Do not affect the organization vitally.	2. Affects the organization.
3. Easy and simple it requires less time and efforts.	3. It requires more time efforts and more data in accurate form.
4. It estimates only the changes in costs and returns associated with variations in organization and operation.	4. It involves estimating the results of a particular organisation and operation of a farm.

- Resources and enterprises are considered simultaneously.

Steps in Farm Planning and Farm Budgeting

1. *Inventory of limited resources*

Prepare a complete list of the farm resources which limit the size of the different farm enterprises.

2. *Examine the existing farm Planning*

Full information on how each resource is being utilized and what are the outputs obtained from various enterprises adopted on the farm.

3. *Locate the weakness of the present plan*

Careful analysis of the resource use in the existing plan will throw up the imbalances. The various weaknesses in the existing plan will act as guidelines for bringing about improvements in the alternative plan.

4. Prepare the alternative plans

Within the framework of resource restrictions and keeping in view the weakness of the existing plan the alternative plans are developed.

5. *Analysis of the alternative plans*

New plans are analysed for costs and returns and the optimum plan which promises the highest returns (to fixed resources) is selected.

6. *Implementing the plan*

- There may be certain difficulties in implementing the plan, unless all the problems are properly anticipated.
- Good plan must provide for flexibility.

GK **Mohd. Qasim Wanotavi and Rashid Ahmad Gangohi founded the Deoband movement.** GS

Technique of Farm Planning and Budgeting

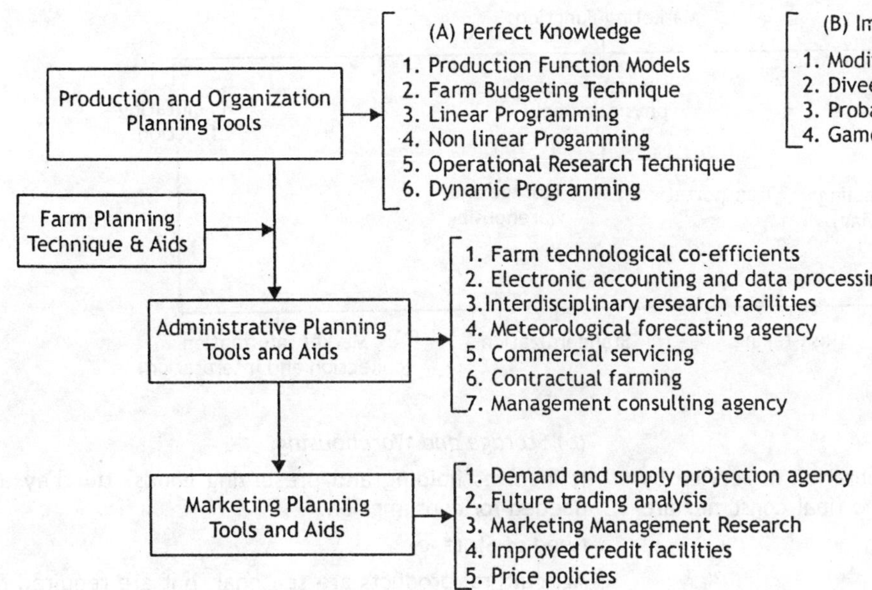

Types and Systems of Farming

Based on factors like crop and livestock raising and mode of social and economic planning, farming is classified into:

1. Types of Farming

When farms in a group are similar in the kinds and proportion of the crops and livestock that are produced and methods and practices followed in production.

II. Systems of Farming

Combination of products on a given farm and the methods or practices that are used in the production of the product.

I. Types of Farming

I. *Specialised Farming*: 50% or more of income derived from single source. In those areas where special market outlets and fairly uniform economic conditions for a long period are present.

2. *Diversified Farming*: No single product source income equals as much as 50% of the total receipts. Farmers depend on several sources of income.

3. *Mixed Farming*: Combination of crop production with particular amount of livestock raising.

4. *Ranching*: Livestock grazes on natural vegetation and multiply under natural surroundings. Ranch land not used for cultivation.

5. *Dry Farming*: In dry areas with less than 20 inch rainfall dry farming is practised. No irrigation is provided. Farming predominantly rainfed.

II. Systems of Farming

1. *Co-operative Farming*: All or part of agricultural operations carried out jointly by the farmers on a voluntary basis. Farmer retains the right of his own land. Land would be treated as one unit after pooling their land, labour and capital.

2. *Collective Farming*: Members surrender their land and livestock to the society. Members work together under management committee elected by themselves. Committee directs matters of allocation of work, distribution of income.

3. *Capital Farming*: Based on capitalistic methods. In India practised in commercialised agriculture areas where farm production is profit and market-oriented.

4. *State Farming*: Farms managed by government officials. Workers paid weekly or monthly based wages. In India all state farms governed by state farms corporation.

5. *Peasant Farming*: Agriculture practices followed in their own way. Entire farm family making and executing the decisions of farming programme.

GK GS

Kalika Mata temple is the oldest temple in Chittorgarh Fort.

MARKETING

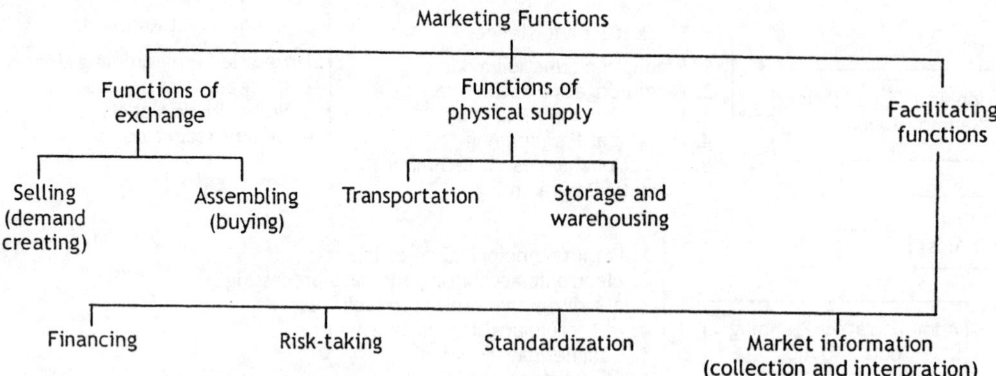

Marketing Function

A marketing function is an act, operation or service by which the original producer and the final consumer are linked together.

Broad Classification

1. Function of exchange
2. Function of physical supply
3. Facilitating functions

I. ***Functions of Exchange***

Functions of exchange is a process of passing goods into the consumer hands. It includes buying, assembling and selling.

(a) Buying

- First step in the process of marketing.

(i) What to buy ? (product)

(ii) When and how to buy ? (Time)

(iii) From whom and where to buy ? (Source)

(iv) On what terms and conditions and prices?

(b) Assembling

- Done after goods have been purchased.
- Involves creating and maintaining of stocks of goods

(c) Selling

- Right product available at right place
- In right quantity
- At right place
- At right time

II. ***Functions of Physical Supply***

(a) Transportation

Through which commodity is shifted from area of production to the markets.

Functions :

- Helps the growth of industries
- Increases demand for goods

(b) Storage and Warehousing

It involves holding and preserving goods, till they are needed for consumption.

Need of Storage-

Agricultural products are seasonal, but are required for whole year consumption.

Grain Storage Structure-

Small scale S.S. :

(a) PAU bin - Capacity 1.5- 15 quintal

(b) PUSA bin

(c) Hapur Tekka

Large Scale S.S. :

(a) CAP (Cover and Plinth) : by FCI.

(b) Ware house: by FCI, CWC, SWC

Warehousing-

Are scientific storage structure specially constructed for the protection of quantity and quality of stored products.

Functions :

- Scientific storage
- Financing
- Price stabilization
- Market intelligence

Types of Warehouses

Based on ownership	Based on commodity
1. Private	1. General
2. Public	2. Special commodity (cotton, wool, tobacco)
	3. Refrigerated warehouse Temp. at 30-50°F or less

GK In South India Veeraslingam Pantulu led the crusade for widow remarriage. GS

Warehouse in India -

(a) National Co-operative Development and Warehousing Board

(b) Central Warehousing Corporation

(c) State Warehousing Corporation

(d) Food Corporation of India

Reasons for Poor Utilization -

1. Lack of knowledge
2. Location
3. Complicated and time consuming procedure for deposit and withdrawing produce.

Processing

Activity which changes the form of a commodity

Farm Products $\xrightarrow{\text{Processing}}$ usable form

India's Position

- Primary processing
- 80% of value
- 12% food industry
- 33% in small-scale tiny and cottage sectors.

Standardization

Determination of the standards to be established for different commodities.

Grading

Sorting of the unlike lots of the produce into different lots according to the quality specifications.

III. Marketing Functions

Advantages of Marketing Function:

1. It can make inter functional comparison of marketing cost.
2. Inter-agency comparison of the cost of marketing function.
3. Inter commodity comparison of cost.

Disadvantages:

1. We don't know these functions are applied to specific business operations or not.
2. Numerous functions, so it is difficult to eliminate the unnecessary from necessary functions.

Marketing & Pricing of Agricultural Inputs and Outputs

Marketing of Agricultural Inputs

- No scope for extension of arable land, so production only through in crop productivity per unit of land.
- Judicious and efficient use of agro-inputs like seed, fertilizer, pesticide, irrigation, farm machines and credit play major role.
- These inputs are complementary through each other.
- Their interrelation is so strong that one of them can be viewed in isolation.

> **Knowledge World!**
> ***First STD***
> **Subscriber Trunk Dialling (STD) was first started between Lucknow and Kanpur.**

1. *Seed*

1. In India there are three stages, Breeder, Foundation, Nuclear.
2. 28 ICAR Projects, 24 Companies were producing seed.
3. 19 seed certification agencies and 86 seed-testing labs.
4. Private producing hybrid seed.
5. Planting material from biotech (tissue culture)
6. Critical Agricultural Inputs extra care should be needed.
7. State seed Co-operative marketing federation state Agricultural Industrial Co. providing seeds to the farmers.

Problem –

- Very cumbersome and difficult certification.
- Inadequate and untimely available seed.

Future strategies –

1. Create efficient marketing and delivery system.
2. Encourage greater participation of private, co-operative and NGOs.
3. Encouraging, public private sector to increase infrastructure.

2. *Fertilizer*

1. It contributes 50 to 60% of Incremental Agricultural Production.
2. Average fertilizer consumption -77 kg/hectare
3. 80% consumed in rice, wheat, sugarcane and cotton.

PRICING

Government Price Mechanism for Agriculture

Administered Prices

- It is the price fixed by the govt. with the objectives of protecting farmers against a decline in prices during years of bumper production.
- Protecting consumers from excessive price increases during lean production years.
- Ensuring procurement for the maintenance of the buffer stock or operation of PDS.

Minimum Support Price

- Govt. to protect the farmer against excessive fall in prices in the event of bumper production.

GK GS **Dr. Keshav Rao Hedegawar was the founder of R.S.S.**

- MSP for different crops are announced by the govt. before the start of the sowing season to provide a price guarantee to the farmers.
- For sugarcane in India, sugar mills are statutorily bound to pay the announced minimum price this is statutory minimum price.

Maximum or Ceiling Price

- It denotes upper price level of a commodity fixed by govt.
- This provision prohibits a trader from charging a price higher than the maximum price.
- It is fixed by govt. to protect the consumers from unwanted price increases.

Procurement Price

The procurement price of a commodity refers to the price at which the govt. procures it from producers to maintain a buffer stock and feed the PDS.

Issue Price

- This is the price at which the commodity is made available to consumers at fair price shops
- The issue price is always higher than the procurement price.
- The economic rationale of maintaining PDS lines in keeping issue prices lower than market retail price.

Fluctuations and Instability in Prices

Price Fluctuation

It refers to changes in prices of a commodity or group of commodities at a given market place or market areas.

Instability in Prices

It can be defined as the state in which prices continue to change with the passage of time at a given space.

Types of Instability

(i) *Short term instability* : The prices of a single commodity do not remains stable within a day, week or month.

(ii) Intra year instability (or) Inter Seasonal instability : Do not remain stable within a season.

(iii) *Inter year instability* : Prices of farm products never remain the same from one year to another.

(iv) *Long term instability* : Over the time, the economy may turn towards recession or depression and prices may start falling.

Factors Causing Price Fluctuations in Agricultural Commodities

Short Term Fluctuations –

(i) Unexpected changes in market arrivals.

(ii) Unexpected changes in offtake from the markets.

(iii) Bottlenecks in the movement of commodities.

(iv) Announcement of relevant govt. policy.

Intra Year Fluctuations –

(i) Seasonal nature of production

(ii) Perishability of products

(iii) Seasonal nature of demand for certain products

(iv) Change in govt. policy on procurement

(v) Change in credit availability.

Inter Year Fluctuations –

(i) Fluctuations in production due to variation in rainfall.

(ii) Fluctuations in suppły due to floods and droughts etc.

(iii) Change in import-export policy.

(iv) Change in public intervention policies like support or procurement prices.

Long Term Fluctuations –

(i) Change in technology like evolution of high-yielding seeds and control measures for pest diseases.

(ii) Change in money supply.

(iii) Change in credit and interest rate policy.

(iv) Change in pay scales, wage rates.

(v) Change in sales tax, excise tax.

(vi) Change in policy towards hoarding, black marketing.

(vii) Extent of processing facilities available for farm products.

Effect of High Instability in Prices

General Effect on the Economy:

(i) A balance between supply and demand cannot be brought.

(ii) Consumers are not able to rationally allocate their scarce income/resources

(iii) Planning and mobilisation of resources for longterm investment projects become difficult.

(iv) It affects the income and living standards of the population.

Special Effect on Farmers and Farm Sector :

(i) Instability in farm product prices leads to instability in farm incomes.

(ii) Uncertain prices of those products which form the raw material for industries like cotton, oil seeds and sugarcane affect the costs, profitability in such industries and this in turn affects the growers of these crops.

(iii) Uncertain prices of farm products lead to uncertainty about the repayment capacity of farmer.

(iv) Instability in farm product prices affects the employment prospects of labour in the agriculture sector.

(v) High fluctuations in farm product prices makes it difficult to plan for exports or imports of commodities.

Agricultural Price Stabilisation

Agriculture price stabilisation means reduction in price fluctuations and regulation of price movements within a certain range.

GK GS

The self-respect movement was started by E.V. Ramaswamy Naicker.

Need for Stabilisation of Agriculture Prices

(i) Farm products account for a considerable part of the consumption expenditure of a majority of the population. Uncertainty in farm product prices lead to uncertainty in real incomes.

(ii) Prices of agriculture commodities have a high weightage in the general price index.

(iii) In the agriculture sector, farmers cannot reduce production when prices fall due to rigidities on account of seasonality arid committed resources. So falling prices affect farmers adversely.

(iv) The efficiency of the farm sector and long-term adjustments in resource allocation in and away from the farm sector depend on a relatively stable cost-price relationship.

Measures for Stabilisation of Agriculture Prices

(A) *Non price Regulatory Measures* :

(i) Increasing and stabilising agriculture production.

(ii) Regulating imports and exports.

(iii) Building up buffer stocks.

(iv) Ensuring procurement and public distribution.

(v) Expanding infrastructure.

(vi) Improving the efficiency of the marketing system.

(B) *Direct price Control Measures* :

(i) Fixation of Ceiling (Max) price

(ii) Fixation of Floor (Min) price.

(iii) Fixation of an administered price.

OTHERS

Co-operatives

The co-operative movement was started in India largely with a view to providing agriculturists funds for agricultural operations at low rates of interest and protect them from the clutches of money lenders.

Quiz Kit!

First penal code

The Indian Penal Code and the Crimial Procedure Code were introduced in Bengal in 1862 by John Beames. They had been drafted by Macaulay's Committee 30 years earlier.

They meet all types of credit viz., short term, medium term and long-term credit requirements.

Co-operative Agricultural Credit Structure

Short term and Medium term credit (three tier):

1. State Co-operative Bank (SCB).
2. District Central Co-operative Bank (DCCB).
3. Primary Agricultural Credit Societies (PACS).

Primary Agricultural Co-operative Societies (PAC)

Functions

- To provide short term and medium term credit.
- To supply agricultural inputs like fertilizers and seeds.
- To provide marketing facilities for the sale of agricultural produce.
- To associate itself with economic and social welfare programmes of the village.

Sources of Funds

Share capital, deposits, reserve funds and loans borrowed from higher institutions or govt.

District Central Co-operative Banks

Functions

- To maintain close and continuous contact with PACS and provide leadership and guidance to them.

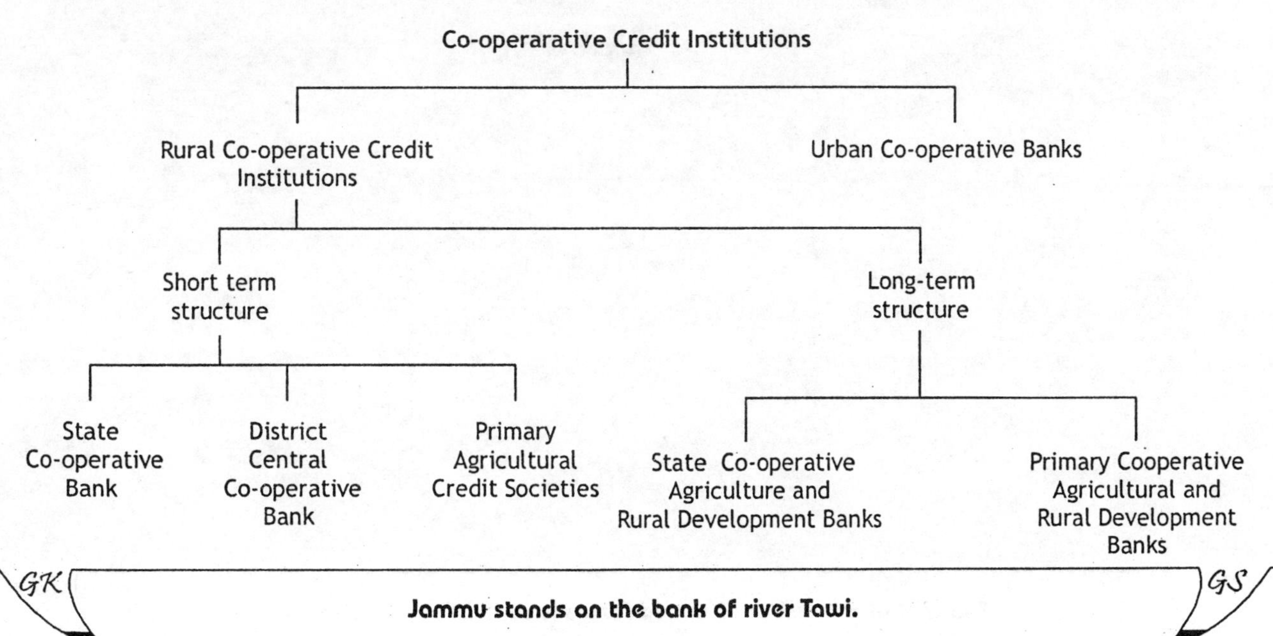

GK GS

Jammu stands on the bank of river Tawi.

- To undertake non-credit activities.
- To meet the credit requirement of member societies.

Sources of Funds

Share capital, deposit from public, borrowing from state cooperative banks, Govt. RBI, SBI and commercial banks.

State Co-operative Banks (SCB)

Functions

- Act as banker's bank to the district central co-operative banks.
- Uniform credit policies for the cooperative movement are formulated and executed by them.
- It also performs banking functions such as issuing drafts, cheques, collecting and discounting bills etc.

Sources of Funds

Share capital, reserve funds, deposit from members and non-members, borrowings from NABARD, SBI and State Govt.

Long-Term Rural Credit

The long-term requirements of the farmers were traditionally met by the money lenders but later by land mortage banks. These banks were later called land development banks. In recent years, they have been renamed as cooperative agricultural and rural development banks (CARDBs).

Functions

- To grant loans on the security of agricultural properties.
- They lend at fairly low rates of interest and enable the needy farmer to secure funds for long periods.

It provides credit for a variety of purposes such as redemption of old debts, improvement of land, purchase of costly agricultural equipment, construction of wells and erection of pumps.

Sources of Funds

Share capital, deposits and issue of bonds. They are subscribed by commercial banks, SBI and RBI.

GK The trial of a member of armed forces is called Court Martial. GS

Points to Ponder

- A farmer with limited funds may operate in a range of increasing returns to capital is however, not an irrational producer.
- A firm is a particular business firm.
- A higher rate of accumulation of capital relative to the rate of population growth adds to the real stock of capital of a nation.
- A production function which is homogenous of the first degree is known as linear homogenous production function.
- A production function with an elasticity of 1.0 throughout indicates constant returns.
- A production with two substitutable variable for one another within limits can be represented by a family of isoproduct curves/isoquants/production indifference curve.
- A profit maximising farmer will maximise the output for given level of outlay.
- A structural agricultural production involves acquisition of land and its preparation for agri-activities, seeds and agro-chemicals, tools and implements and organization of workers.
- Accumulation of capital refers to its ownership either by an individual or groups of individuals.
- Adam Smith is called as father of economics.
- Agricultural economics is both a social science as also a natural science.
- Agricultural economics is the science in which the principles and methods of economic are applied to the special conditions of agricultural industry.
- Agricultural output comes from many small units operation independently.
- Agricultural progress is normally a pre-requisite for industrial development.
- Agriculture is a unique industry in which the mode of life and business enterprise are combined together.
- Agro-industries and agro-based industries hold a great promise for employment in India.
- An isoquant curve is negatively inclined.
- As long as technology remains constant, the production function remains unchanged and indicates the greatest output from a given input mix.
- At equilibrium price both buyers and sellers are satisfied and no one can benefit from further recontracting.
- Average fixed cost is the total fixed cost divided by the number of units of output produced.
- Capacity to save is limited by risk taking competence and the level of personal income.
- Capital in agriculture is generally used to augment output per worker.
- Changing technologies and tastes of people have led to frequent changes in cropping patterns all over the world.
- Competitive products are those which compete for the use of farmer's resources.
- Dispersion is the process exactly the opposite to that of assembling.
- During a depression, the unpaid family labour serves as a cushion on many small farm units.
- Economic cost equals explicit cost plus implicit cost.
- Economic holding is the one which provides full employment to the farmers family and allows a reasonable standard of living.
- Economics is the body of knowledge which relates to wealth.
- Elasticity of production is less than one between the maximum average product and the maximum total product.
- Equalization of marginal gains on diverse farm activities is bidirection, that is, achieved either through reallocation of existing stocks or aquiring additional units of capital.
- F. Benham defined the law of variable proportions, as the proportion of one factor in a combination of factors is increased, after a point, first the marginal and then the average product of that factor will diminish.
- Factor productivity can be expressed as (i) average physical product = output/input (Y/X), (ii) marginal physical product = additional output/additional input (ÄY/X).
- Factor-product relationship explains the modus operandi of the various factors of production as converted into final farm products.
- Farm management is a study of the principles underlying the function of the farmer as a business proprietor.
- Farm management is an art of applying business and scientific principles to the organisation and operation of a farm.
- Farmers terms of trade is equal to the farm price divided by manufactured goods price.
- Farming in India is less a bussiness than a tradition.
- Farming is highly dynamic and receptive to technological innovations.

> **Quiz Kit!**
> *Largest power station*
> **The Korba thermal power station located on the west bank of the Hardeo river near Korba in Bilaspur District, Madhya Pradesh.**

The first real labour union was the printers union in Calcutta. GS

- Farming need to be scientifically managed to permit reasonable avoidance of shortage and redundancy of labour to minimise waste and overburden.
- Food production and the existence of mankind go together.
- For the world as a whole migrations are of no effect.
- If a farmer raises rice-paddies instead of maize, the social cost/opportunity cost of raising it will be the amount of maize sacrificed in the process.
- If the level of income does not rise, social values, attitudes and saving motivation may not come through.
- If the new capital restricts the freedom of the farmer to choose a crop or a variety or any other input, naturally an alternative with less or no restrictions is sought for.
- If two goods are not substitutes at all, the value of cross elasticity will be zero.
- Improved agri-technology enhances employment per unit of land and also productivity per unit of employment.
- In case of indirect tax, the tax payer is also the tax bearer.
- In first stage of production function, marginal product exceeds average product.
- In linear programming model, a farmer is interested in maximising the profits or minimising the cost of production with fixed and limited quantities of factors of production.
- In most profitable price, the rate of increase of marginal revenue must be less than the rate of increase of the marginal cost.
- In oligopoly, there is complete interdependence among the sellers with regard to their price output policies.
- In practice, not much use has been made of the linear programming technique in under-developed countries.
- In Ricardo's vision of economic society, there are three major groups of factors viz., landlords, capitalists and labours.
- In the amount produced and the amount consumed by the producer relationship, negative or zero values are not in tune with economic development.
- In the very long run, all costs are variable.
- Income elasticity of demand is equal to the percentage change in the quantity demanded divided by the percentage change in income.
- Increase in agricultural productivity raises agricultural incomes and rural purchasing power.
- Indifference curve is convex from below.
- Intensive system of farming generally need more labour per unit of land than that needed by extensive system.
- It is generally believed that the technique of linear programming is the product of World War II.
- It is the surplus generated by agricultural sector that would help Indian economy to golden stage.
- Japanese experience postulates that technological advancement in agriculture can labour-intensive in character.
- Labour is the primary factor in enhancing the production of traditional agriculture.
- Land is basic to agriculture.
- Land rent is an example of fixed cost.
- Large-sized farms are more economical in working and conducive to greater efficiency than small sized ones.
- Law of diminishing marginal utility does not apply on rare collected goods.
- Law of supply can be defined as the quantity of a commodity produced and offered for sale.
- Lenders are naturally guided by the rate of interest both at home and abroad.
- Less developed countries lack capital to develop agriculture at a desired rate of growth.
- Lewistype models visualized double-benefit from agriculture.
- Malthus failed to foresee the possibilities of technological revolution enabling man to produce more per unit of inputs including land.
- Malthus in his thinking faltered on man's inventive competence and acceptability of a cultural pattern of a controlled birth-rate.
- Malthusian thinking on population and food supply was greatly influenced by the scarcity of land.
- Marginal cost curve is equal to average variable cost curve at the lowest point of the letter curve.
- Mental quality of labour force shall depend upon the level of education, extension services and agricultural training programmes.
- More acquisition of land is not enough unless it is prepared to suit agricultural operations.
- Most of the underdeveloped countries have abundant of labour but, marginal productivity of labour being negligible or approaching zero.
- Non single production function can be used to characterize agricultural production under all environmental conditions.
- Non-redundancy ratio is directly proportional to labour utilization ratio and inversely proportional to the endowment ratio.
- Only agriculture turns out net product over and above its cost of production.
- Out-put of agriculture is largely a raw material.
- Over time labour has a special place in agriculture owing to its peculiar nature vis-a-vis industrial production.

Kynes gave the concept of 'Mixed Economy'.

- Perfectly substitutable factors replace each other at a constant rate regardless of level of output or proportion of combination.
- Physical quality of labour force refers to sound health, stamina and race characteristics.
- Physiocrats believed that agriculture alone is productive, for it yields the subsistence of its workers.
- Planning describes the courses of actions to be adopted and followed.
- Planning requires intellectual effort, reflective thinking, foresight and imagination.
- Price is one of the many factors affecting supply response of agricultural commodities.
- Price is the value received in exchange of goods and services and distinguished from money, merely a medium of exchange.
- Processing is often excluded from the list of marketing functions as it is essentially a form changing activity.
- Production denotes the process of conversion of certain inputs/resources into a consumable form.
- Production function is a mathematical relationship indicating the way in which the quantity of a product depends upon the qualities of particular input used.
- Product-product relationship is concerned with the determination of type of crops to grow or breed of livestock to rear and their suitable combination at a given amount of resources.
- Ranching refers to the practice of grazing animals on the public lands.
- Rate of wages, according to Ricardo is determined by the cost of subsistence.
- Rent is a form of direct tax.
- Resource - Resource Factor - Factor relationship refers to the most profitable combination of resources to produce a specific amount of a given output.
- Sale tax is an example of indirect tax.
- Saving decisions are conditioned by the desire and power to save.
- Sociological school of theories of population growth lays stress on institutional set-up devised by human beings and emphasizes social rather than environmental factors as influencing factors.
- Support price is the price at which the government would be under obligation to buy the entire stocks offered to it for sale.
- Terms of trade denote terms on which two sectors exchange their goods.
- The accumulation and use of capital coupled with suitable technologies does not contribute only to the growth of agriculture, but also to economy as a whole.
- The agricultural credit system in India is one of the largest in the world.
- The agricultural sector in most of the underdeveloped countries is particularly very weak to shed its traditional values.
- The allocation of capital is guided by the maximum return criterion.
- The amalgamation of farm enterprises effects labour efficiency.
- The application of game theory pertains to a situation in which two competitive farm firms are involved.
- The average total cost curve will fall as long as marginal cost curve is below it.
- The capital allocation decisions are more oftenly based on practical wisdom than on theoriticle postulates.
- The Cobb-Douglas function is based on the empirical study of the American manufacturing industry undertaken by Paul H. Douglas and C.W. Cobb.
- The constant marginal return function does not generally exist when inputs per unit (land/animal) are intensified.
- The demand for a commodity is a schedule of quantities that a buyer would be willing to purchase at different prices at any given time or period.
- The demand of agricultural labour depends upon the nature of operation(s) and seasonal sensitivity of crops.
- The desire for more land is also stimulated by its permanent value, industructivity and immovability.
- The elasticity of substitution between two factors of production is not in line.
- The factors substitute each other at a diminishing rate are the factors of production are incomplete substitute for each other.
- The failure of agriculture sector would spell disaster to the entire planning process.
- The farm size, system of farming and technological development influence the demand for agricultural labour.
- The Fei-Ranis model focuses attention on the transfer of labour from agricultural to non-agricultural sector as central to economic development.
- The Foodgrains Procurement Committee (1950) recommended the continuation and extension of the system of rationing in India.
- The income effect is associated with substitution effect.
- The introduction of new capital also calls for the study of nature and magnitude associated risks and uncertainties.
- The key to the process of economic expansion is the reinvestment of the capitalist system.
- The law of demand states that a commodity will be demanded more at a lower price and less when the

GK **Isaac Asimov coined the term 'Robotics'.** GS

price is higher.

- The level of saving in underdeveloped countries is low due to low level of income itself.
- The man's urge to own more land has been due to lack of alternative possibilities of employing the labour and capital.
- The marriages at younger age enlarge the reproductive age span and hence augment the rate of population growth.
- The monopolistic competition involves both the monopoly and competitive elements.
- The motive of specialized farming is profit, but that of diversified farming is self-sufficiency.
- The naturalistic school of theories on population growth emphasizes the factors like fecundity, fertility, sterility, food supply, physical resources, population density and standards of living as the growth influencing factors.
- The naturalistic school on population growth is founded on Malthusian theory on population growth.
- The nature of crops and the duration of the crop season are also the determinants of the use of capital in agriculture.
- The people in underdeveloped countries generally believe in the philosophy of contentment and do not have high expectations from life.
- The phenomenon of diminishing marginal returns has played a crucial role in shaping the classical theory of production.
- The policy of imposing a ceiling on land would be able to meet the desire to possess land and reducing the inqualities in ownership and use.
- The principle sources of the high productivity of modern agriculture are reproducible sources.
- The problems of planning form the core of modern economics.
- The productivity of pastures and grazing lands can be improved considerably by promoting permanent pasture land.
- The productivity per worker does not merely depend on the quantity of capital per worker, but also on quality of other inputs.
- The profit of the firm will be maximum when marginal revenue exceeds marginal cost.
- The real development in land development banking was made by Madras state in 1979.
- The relationship Log Y = Log a + b Log X has been estimated between yield per hectare/acre (Y) and the average of holding (X) for all crop production.
- The response of supply and demand to price changes in agriculture may not exactly match the law of demand and supply.
- The returns on capital in various sectors of economy act as a strong incentive in the competition for capital.
- The revenue of farm together with its costs determines the profits realized by the farmer.
- The reward for entrepreneurial functions is called profit.
- The share of capital is referred as interest.
- The simplex method is the most accepted procedure of solving a linear programming problem.
- The size of population can also be regulated by formulating public goals relative to population.
- The size of population depends upon the difference between births and corrected by in-and out-migration from a geographical territory.
- The suitability of otherwise of zero-rate of population growth is being debated and investigated.
- The surplus over a marginal land is rent.
- The term Cost generally refers to the outlay of funds for productive services.
- The total savings fund consists of savings form industrial sector and from agricultural sector.
- There is an advantage in using contract labour as it is self-organized and need no continuous supervision.
- Too much of diversification may lead to mismanagement.
- Total fixed cost represents the sum of expenditures which will be incurred.
- Total variable cost correspond to those factors of production whose supply is variable in the short period and change with the output.
- Transformation from traditional to dynamic agriculture does not depend only on farm labour force, but also on technological innovations.
- When an increase in output of one, with resources held constant in amount, also causes increase in the output of other are known as complementary enterprises.
- When average product attains the maximum value and is constant, the marginal product is equal to it.
- When average product is maximum, it is equal to marginal product.
- When average product is rising, marginal product rise more than average product.
- When competition is perfect the average revenue curve of the firm is a vertical straight line.
- When marginal cost in creases, it is more than the average cost.
- When output of one production increased without gain or sacrifice in another product at constant level of resources, the relationship is called supplementry.
- When the average product (AP) is maximum and constant, marginal product (MP) is equal to AP; when AP starts falling, MP falls faster than AP.

GK GS

Valentine Chirol wrote the Indian unrest.

Multiple Choice Questions

1. ___________ an application of the accounting principles to the business of farming.
   a] Farm planning b] Farm budgeting
   c] Farm accounting d] Marketing
2. 'Beef' and 'hides' are examples of :
   a] Independent enterprise
   b] Joint enterprise
   c] Supplementary enterprise
   d] Competitive enterprise
3. 'Indian Farming' is published by:
   a] Ministry of Agriculture Government of India
   b] ICAR
   c] Natioanl Seed Corporation of India
   d] None of these
4. "Rent is the price paid for the use of land" who said it ?
   a] Ricardo b] Carber
   c] Keynes d] None of above
5. "Under free and prefect competition there could not be any involuntary unemployment amongest wage earners in the labour market." This theory was established by :
   a] Prof. J.B. Say
   b] Prof. Pigou
   c] Prof. Keynes
   d] None of above
6. Which one of the following pairs is not correctly matched?
   a] Elasticity of production = Stage II of classical production function greater than unity
   b] Negative marginal product = Stage III of classical production function
   c] Negative elasticity of production = Stage III of classical production function
   d] Average product decreases but is greater than marginal product = Stage II of classical production function
7. Per capita land availability in India :
   a] 1/3 ha b] 1/6 ha
   c] 1/4 ha d] 1/7 ha
8. A consumer spends his income according to the law of :
   a] Diminishing marginal utility
   b] Increasing cost
   c] Equi-marginal utility
   d] Least cost
9. A crop loan over Rs. 1,00,000 will require :
   a] Equitable mortgage
   b] Registered mortgage of land
   c] Personal security
   d] None of these

> **Did You Know ?**
> ***Most prolific moths***
> **Female moths of the genus *Endoclita* (fam. *Hepialidae*) are estimated to be capable of laying upto 50,000 eggs each.**

10. A farmer having less than one hectare of irrigated the land, is krown as :
    a] Small farmers b] Marginal farmers
    c] Big farmers d] None of the above
11. A farmer which has land holding between 1 to 2 hectare known as :
    a] Marginal farmer b] Small farmer
    c] Optimal farmer d] None of the above
12. A farmer which has land holding between zero to one hectare is known as :
    a] Small farmer b] Big farmer
    c] Marginal farmer d] Agricultural labour
13. A group of technical units are known as :
    a] Farm firm b] Plant
    c] Economic unit d] All above
14. A holding which allow to a man a chance of producing sufficient to support himself and his family in resonable comfort after paying his necessary expenses is called :
    a] Basic holding b] Optimum holding
    c] Operational holding d] Economic holding
15. A holding which keeps in mind of the ceiling to the size, the marginal capacity and financial resources of an average cultivator, refers to the size not more than three times of the size of economic holding is called:
    a] Optimum holding b] Marginal holding
    c] Profitable holding d] None of the above
16. A market in which the purchase and sale is a commodity takes place at time 't' but the exchange of the commodity takes after t + 1 time, the market is known as :
    a] Spot b] Cash market
    c] Forward market d] General market
17. A market structure with large number of firms selling homogeneous product is known as:
    a] Monopoly market
    b] Monopolistic competition
    c] Perfect competition
    d] None of these
18. A normative farm plan is one which the farmer :
    a] Has already put into operation on his farm
    b] Should adopt during the current 5 year plan period
    c] Had actually adopted in the earlier 5 year plan periods
    d] Should not adopt

GS

New Celdonia is officially a colony of France.

19. A single convenient unit in production for which technical co-efficients are calculated is known as:
a] Units of accounting b] Technical units
c] Marginal units d] None of the above

20. A statement which shows the financial condition and stability of the business at a particular point of time is known as :
a] Net worth statement
b] Business statement
c] Physical efficiency measures
d] None of the above

21. A table is a :
a] Material good
b] Non-material good
c] Free good
d] None of the above

22. A tractor purchased with the bank assistance should be registered :
a] In the name of the bank
b] In the name of the borrower
c] In the joint names of the bank and borrower
d] In the name of tractor dealer

23. A.P.P. will be equal to (Y = output, X = input) :
a] X/Y b] X/Y x 100
c] Y/X d] None of above

24. About 15% farmers had land holding in between :
a] 2-4 ha b] 0-1 ha
c] 1-2 ha d] None of the above

25. About 70% farmers have less than ____________ hectare of land in India.
a] 5 b] 2
c] 4 d] 1

26. Absence of interaction between resources and activities in the process of production is called :
a] Linearity b] Non-linearity
c] Quasi-non-linearity d] Quasi-linearity

27. According to Hung Chao Tai land reforms aim at attaining just relationship among ____________ population.
a] Urban b] Industrial
c] Traders d] Agricultural

28. According to the Directorate of Marketing and Inspection ____________ per cent may be taken as the average marketing margin for all surplus food grains in India.
a] 15 b] 20
c] 25 d] 35

29. Adoption potato variety Kufri Sinduri-Wheat rotation in prefrence of Up-to-Date-Wheat rotation is example of :
a] Farm management practice
b] Farm planning
c] System of farming
d] Types of farming

30. Agricultural Development Banks are sponsored by :
a] Reserve Bank of India
b] State Bank of India
c] Nationalized Banks
d] Co-operative Banks

31. Agricultural price commission was estabished in :
a] 1964 b] 1966
c] 1965 d] 1967

32. Agricultural Refinance and Development Corporation (ARDC) was set up in :
a] 1963 b] 1975
c] 1952 d] 1979

33. Air is :
a] Economic goods b] Free goods
c] Transferable goods d] All above

34. As income increase the expenditure on light, fuel and rent generally :
a] Increases b] Decreases
c] Remains the same d] None of the above

35. As the price of a commodity goes up, the demand for it :
a] Contracts b] Increases
c] Expands d] Decreases

36. At a particular point on a Neo-classical Production Function graph, the magnitudes of TPP, APP and MPP are 18,6 and 7 respectively. The E_p (elasticity of production) at this point would be:
a] 3 b] 2.57
c] 1.16 d] 0.85

37. At equilibrium, the slope of the indifference curve is:
a] Equal to the slope of the budget line
b] Greater than the slope of the budget line
c] Smaller than the slope of the budget line

38. At present, the farming system of India has become :
a] Export oriented b] Technology oriented
c] Market oriented d] Domestic need oriented

39. Average fixed cost is equal to :
a] Total fixed cost/100
b] Tota fixed cost/total variable cost
c] Total fixed cost/output
d] Total fixed cost/input

40. Average product is equal to marginal product when :
a] Average Product is maximum
b] Average Product is minimum
c] Marginal Product is zero
d] None of the above

41. Basic fundamental law of agriculture is :
a] Law of diminishing return
b] Law of equimarginal return
c] Law of substitution
d] Law of demand and supply

GK GS
Lord Cornwallis laid the foundation of modern civil services.

42. Who was the first economist to have evolved the technique of Indifference Curve Analysis?
a] David Ricardo b] Bohm Bawerk
c] F.Y. Edgeworth d] A.C. Pigou

43. Break-even point is determined by:
a] MC = MR b] MC = AVC
c] MC = AC = AR d] TR = TC

44. Building, machinery and implements are examples of :
a] Variable resources b] Fixed resources
c] Flow resources d] None of the above

45. Buyer's market devotes the place where :
a] Supply receeds the demand
b] Supply exceeds the demand
c] Money value remains unchanged
d] None of these

46. By preparing chairs the wood creats:
a] Form utility b] Time utility
c] Place utility d] Knowledge utility

47. Capital market deals with :
a] Cotton b] Wheat
c] Raw material d] Bonds and shares

48. Central cooperative bank operates normally at :
a] District level b] Village level
c] National level d] None of these

49. Central government finance the RRB's the extent of:
a] 35% b] 17%
c] 50% d] 40%

50. Chairman of Planning Commission is:
a] Prime Minister
b] President
c] Appointed by Prime Minister
d] Appointed by President

51. Command Area Development Programme (CADP) were started in:
a] 1974 b] 1962
c] 1976 d] 1979

52. Commodity market deals with :
a] Bonds b] Shares
c] Wheat d] None of the above

53. Co-operative farming, Collective farming, Capitalistic farming, State farming and Peasant farming are :
a] Types of farming
b] Systems of farming
c] Both these
d] None of these

54. Co-operatives for tribes are called as :
a] Camp co-operative societies
b] Multi-purpose societies
c] Lead Bank
d] Savita

55. Creation of utility by way of marketing is known as :
a] Consumption b] Production
c] Distribution d] Exchange

56. Criteria for goods market information is :
a] Comprehensive b] Accuracy
c] Relevance d] All above

57. Crop farming and milk production is an example of :
a] Competitive enterprise
b] Complementary enterprise
c] Supplementary enterprise
d] None of them

58. Cropping intensity and gestation period of the crops grown are :
a] Inversely related
b] Directly related
c] Exponentially related
d] Not interrelated

59. Dairy farming, poultry farming, crop farming and pig farming are:
a] Types of farming
b] Systems of farming
c] Both a and b
d] None

60. Decisions are continuously made in the day to day operations of farm business and involve small investment are known as :
a] Strategic decisions
b] Marketing decisions
c] Operational decisions
d] None of the above

61. Decisions on construction of buildings are comes under :
a] Operational decisions
b] Administrative decisions
c] Strategic decisions
d] Selling decisions

62. Decisions regarding irrigation, conservation and reclamation programmes are comes under :
a] Buying decisions
b] Selling decisions
c] Marketing decisions
d] Strategic decisions

63. Decisions regarding selection of enterprise are known as :
a] Marketing decisions
b] Buying decisions
c] Administrative decisions
d] Operational decisions

64. Decrease in demand means :
a] Demand change due to change in price
b] Demand changes not due to the change in price but due to income
c] Both a and b
d] None of these

65. **Demand for necessary goods is :**
a] Elastic b] Inelastic
c] Perfectly elastic d] None of the above

66. **Diversified farming is a kind of :**
a] System of farming
b] Type of farming
c] Both of the above
d] None of the above

67. **The marginal productivity theory is based on the assumption of :**
a] Imperfect competition
b] Monopolistic competition
c] Perfect competition
d] Elastic competition

68. **Each SFDA is expected to cover how many small farmers ?**
a] 5,000 b] 10,000
c] 50,000 d] None of above

69. **Economic laws are not ascertain as:**
a] Physical laws b] Moral laws
c] Government laws d] All above

70. **Economics is a :**
a] Social science b] Political science
c] Natural science d] None of these

71. **Economics is a an :**
a] Art b] Science
c] Art and Science d] None of these

72. **Economics is primarily a study of :**
a] Man b] Wealth
c] Wants d] None of these

73. **Economy of an developing country is marked by :**
a] Predominance of agriculture
b] Predominance of industry
c] Predominance of transportation
d] None of them

74. **Enterprise budgeting deals with input-output relationship of :**
a] A set of different enterprises
b] A single enterprises of the farm
c] Both of the above
d] None of the above

75. **Expenditure on fertilizers is called:**
a] Fixed cost b] Variable cost
c] Marginal cost d] Total cost

76. **Export-Import Bank of India was set up on Ist January:**
a] 1980 b] 1982
c] 1981 d] 1983

77. **Extension is demand means :**
a] Change in demand due to change in price
b] Change in demand due to change in income
c] Both a and b
d] None of these

78. **Extent of land under forest in India is :**
a] 21% b] 20%
c] 27% d] 16%

79. **Facts and findings of other science are co-ordinated for the solution of various problems of individual farmers with view to achieving certain desired goals, hence, farm management is :**
a] Micro Approach Science
b] Practical Science
c] Integrating Science
d] Broader Field

80. **Farm budgets are used to :**
a] Compare economic profitability of different production activities within or among farms
b] Indicate wheather a proposed change will be profitable under a given set of factors
c] Explore conditions under which certain farm practices become profitable or unprofitable
d] Do all the above

81. **Farm credit may be defined as lending :**
a] For the development of land and labour
b] For more yield
c] For the development of agriculture
d] For the development of forestry, poultry, piggery etc.

82. **Farm management as a resource allocation proportion is related to:**
a] Inter-farm management
b] Intra-farm management
c] Both inter and intra farm management
d] Inter-regional management

83. **Farm management concern with obtaining the maximum yield per hetare and considers per unit cost therefore it is :**
a] Practical Science
b] Integrating Science
c] Profitability Oriented
d] Broader Field

84. **Farm management deals with :**
a] Judicious decisions
b] Use of scarce resources
c] Profit maximization and family satisfaction
d] All these

85. **Farm management is a practical science because of :**
a] It is interested in profitability
b] It is intergrating facts and findings of other sciences
c] It deals with facts of other physical sciences and testing the applicability of those facts and findings
d] All of these

GK GS
Satyendra Nath Tagore was the first Indian to qualify the Indian Civil Services.

86. Farm management treats every farm unit unique in available resources, problems and potentialities because it has :
a] Broader Field b] Micro Approach
c] Practical Field d] None of the above

87. Farm Planning involves :
a] Preparing farm budget in advance
b] Selecting and adopting best package of practices
c] Both of these
d] None of these

88. Who propounded the theory of optimum population first ?
a] Sidgwick b] Malthus
c] Carr-Saunders d] Kuczyansky

89. Farming planning means :
a] Farm budgeting
b] Cropping sequence
c] Types of enterprise
d] None of these

90. First Land Mortagage Bank was set up in Madras in the year :
a] 1929 b] 1951
c] 1934 d] 1921

91. Food Corporation of India (FCI) was established in January :
a] 1963 b] 1966
c] 1964 d] 1965

92. For pruchasing of seeds, fertilizers, etc., the loan distributed is called :
a] Short term loan
b] Long term loan
c] Medium term loan
d] All of the above

93. For purchasing the pumpsets, tractor, etc., the loan is distributed for :
a] Long term b] Short term
c] Medium term d] All of the above

94. Free goods have a great :
a] Value in use
b] Value in exchange
c] Transferability
d] Both b and c

95. Given that : $Y = 30 X_1^{0.25} X_2^{0.60}$ where Y = output, X_1 = input factor 1, and X_2 = input factor 2, the nature of return to scale will be :
a] Constant b] Increasing
c] Decreasing d] Fluctuating

96. Goodwill of a business is a :
a] Material good
b] Non-material good
c] Material and non-material
d] None of the above

97. Government of India financed Regional Rural Banks upto the extent of
a] 50% b] 35%
c] 15% d] 25%

You Don't Know!

Largest cockroach

The wingless *Sitophyga rhombifolia* is probably the largest cockroach. It is brown with yellow markings and is found in houses. It grown to over 6 cm in length.

98. Gross return plus charge in inventory minus total cost is equal to :
a] Farm income b] Family income
c] Tax d] None of the above

99. Oligopoly is a market situation which is characterized by the existence of:
a] One firm b] Two firms
c] Several firms d] Few firms

100. How many plots of one hectare area can be farmed from one square km area?
a] 10 b] 50
c] 100 d] 1000

101. How much farmers have land holding more than 10 hectares ?
a] 15% b] 8%
c] 15% d] 4%

102. How much per cent of total cost of production goes to labour investment ?
a] 25-30%
b] 46-50%
c] 40-45%
d] 15-20%

103. How the field of farm management is broader ?
a] It tests the applicability of facts and findings
b] It is interested in profitability
c] It gets information from more than one discipline
d] None of the above

104. Human wants are generally :
a] Static b] Recurring
c] Non-recurring d] Limited

105. Monopsony imply the existence of a :
a] Single buyer
b] Single seller
c] Single seller vs Single buyer
d] All of the above.

106. If a 2 per cent change in the price of a commodity is folowed by a 10 percent change in the quantity demanded, the elasticity of demand of that commodity is :
a] 0.2 b] 1.0
c] 0.5 d] 5.0

107. If by increasing the quantity of labour used by one unit the firm can substitute up to units of capital and still produce the same output, then the marginal rate of technical substitution of labour for capital is :
a] 1/2 b] 2
c] 1 d] 0

GK

Rahul Dravid, a famous cricketer is also known as 'The Wall'.

GS

108. If change in price is 1% and charge in demand is 1% demand will be called :
a] Elastic b] Proportionate elastic
c] Inelastic d] None of the above

109. If more and more capital and labour are applied to a fixed piece of land, the system of cultivation or farming is known as :
a] Intensive cultivation
b] Extensive cultivation
c] Co-operative farming
d] Large scale farming

110. If more is demanded at the same price, it is a case of:
a] Extension in demand
b] Contraction in demand
c] Increases in demand
d] Decreases in demand

111. If substitution ratio is equal to price ratio than cost will be :
a] Minimum b] Increasing
c] Maximum d] Decreasing

112. If substitution ratio is less than price ratio than cost will be :
a] Increasing b] Maximum
c] Decreasing d] None of the above

113. If technical progress leads to a rise in labour productivity of 5%, then an increase of 5% in the wage rate would imply that :
a] Workers will derive all of the benefits
b] There will be cost-push will remain unchanged
c] Unit labour costs will remain unchanged
d] Profit per unit will invariably fall

114. If the income elasticity of demand is greater than one, the commodity is :
a] A necessity b] A luxury
c] An inferior good d] A essentiality

115. If the members have not any right on land and can not take decisions regarding farming but guided by a general body and profit is given according to the labour and capital invested by the member is known as :
a] Co-operative joint farming
b] Co-operative collective farming
c] Co-operative better farming
d] Co-operative tenant farming

116. If the price of a commodity is decreased than the demand will :
a] Increase if other factors remain same
b] Minimum if other factors remain same
c] Maximum if other factors remain same
d] No effect if other factors remain same

117. If the production function is homogenous of degree one, then the scale elasticity of output will be :
a] Equal of unity b] Less than unity
c] Greater than unity d] Equal to zero

118. If the quantity of a commodity demanded does not change when its price, the demand curve is :
a] Negatively sloped b] Positively sloped
c] Verticaly sloped d] None of these

119. If the quantity of commodity demand remains unchanged as its price changes, the coefficient of price elasticity of demand is :
a] Greater than one b] Zero
c] Smaller than one d] Equal to one

120. If the small holders form a society for farming and get profit after deducting the expenses of his price of land and farmer also have right on his land the system is known as :
a] Co-operative better farming
b] Co-operative collective farming
c] Co-operative joint farming
d] None of these

121. If two items are complementary then demand of one item will increase for ____________ in price of another item.
a] Increase b] Decrease
c] No effect d] Not known

122. If two items are supplementry than demand of one item will decrease against the ______________ in price of another item :
a] Increase b] Maximum
c] Decrease d] Minimum

123. In a complete farm planning due consideration is given to :
a] Resource use and restrictives
b] Relationship among different enterprises
c] Goods managerial skill of the operator
d] All of above

124. In a manufacturing industry generally applies law of:
a] Constant return b] Increasing return
c] Decreasing return d] All of the above

125. In a market situation, when there is only one seller of a commodity, it is:
a] Duopoly Market b] Oligopoly Market
c] Monopoly Market d] None of the above

126. In capitalistic farming, the land is owned by :
a] Land lord b] Factory lord
c] Both a and b d] Government

127. In classical production function when average product is increasing:
a] Marginal product is less than average product
b] Marginal product is equal to average product
c] Marginal product is greater than average product
d] None of the above

128. In classical production function, rational zone is always :
a] First zone b] Second zone
c] Third zone d] None of the above

GK GS
Abhinav Bharat Scoiety was founded by Ganesh Savarkar.

129. In economics the term 'land' as a factor of production includes :
a] Mineral wealth
b] Sub-soil water and crops raised
c] Canals and dams
d] All above

130. In eonomics the term'labour' means:
a] Only manual labour
b] Only mental work
c] Manual and mental work
d] None of the above

131. In law of supply the relationship between price and quantity supplied is :
a] Direct b] Indirect
c] Both a and b d] None of the above

132. In mixed farming the contribution of livestock to gross farm income is :
a] At least 10% b] At least 30%
c] At least 40% d] More than 50%

133. Who wrote the book, 'The Theory of Monopolistic Competition?
a] Joan Robinson b] E.H. Chamberlin
c] A.C. Pigou d] Frederic Benham

134. In peasant farming, the owner of the land is :
a] Landlord b] Co-operative society
c] Government d] Farmer

135. In prefect competition, there are :
a] Large number of buyers and few number of seller in the market
b] Large number of sellers and few number of buyers in the market
c] Number of buyers and sellers are large and nearly equal
d] All of above

136. In terms of gross fertilizer consumption, India ranks:
a] First b] Second
c] Third d] Fourth

137. In the short run, Average Cost, Average Variable Cost and Marginal Cost Curves are 'U' shaped due to the operation of :
a] Law of equimarginal returns
b] Substitution between two factor inputs
c] Law of variable proportion
d] Time comparison principle

138. In unregulated markets the charges taken for impurity is known as :
a] Karda b] Dhelta
c] Muddat d] Darmada

139. In which method of sale, prices are fixed by mutual agreement ?
a] Dara sale
b] Under cover of.a cloth
c] Private negotiation
d] Close tender system

140. In which planning period profit rule says that gross return should cover variable cost :
a] Long run b] Short run
c] Both a and b d] None of the above

141. In which planning period wę can change the output by altering the size of form :
a] Long run b] Short run
c] Both a and b d] None of the above

142. In which planning period we can change the output without altering the size of the plant :
a] Long run b] Short run
c] Accounting period d] Base year period

143. In which relationship increase or decrease in the production of one product affects the production of the other commodity inversely :
a] Joint product b] Supplementary
c] Complementary d] Competitive

144. A boom is marked by:
a] Full employment
b] Under employment
c] Overfull employment
d] None of the above

145. In which system of sale commission agents takes the sample of the produce to the shopes of buyers ?
a] Quotations on samples
b] Open auction system
c] Private negotiation
d] None of the above

146. In zamidari abolition, the elimination of intermediaries led to excessive burden of :
a] Revenue b] Reservation
c] Supplementation d] Compensation

147. In zamidari system the land revenue rate was fixed :
a] Once for all b] For 30-40 years
c] For 20-25 years d] For 30 years

148. In zamidari system the land revenue was charged :
a] Through landlords
b] Directly from the farmers
c] Thorugh village panchayats
d] Through either of these

149. Income tax is a :
a] Direct tax b] Indirect tax
c] Both a and b d] None of these

150. India covers how many percentage area of the total world area :
a] 2.42 b] 3.42
c] 4.42 d] 2.00

151. Indifference curve are those curve:
a] Which are circular in shape
b] Which are concave in shape
c] Which does not intersect to each other
d] Which always cuts the axis

GK GS

Virender Sehwag is also known as Nazafgarh ka Sachin.

152. Individuals do not have physical control of the product but they render personal services are known as :
a] Speculated middle man
b] Commission agents
c] Brokers
d] Whole salers

153. Individuals which specialize in performing the various marketing functions and rendering their services are known as :
a] Producers b] Consumers
c] Middle man d] None of the above

154. Individuals who buy goods from whole salers and sell them to the consumers in small quantities are known as :
a] Whole saler b] Retailer
c] Consumer d] Producer

155. Initial utility is the utility of the :
a] First unit b] Last unit
c] Marginal unit d] Third unit

156. Keeping few birds, some fruit trees and a small garden on farm are the:
a] Complementry enterprise
b] Competitive enterprise
c] Supplementry enterprise
d] None of these

157. Land Development Banks are parts of :
a] Co-operative Banks
b] Nationalized Banks
c] Agricultural Development Bank
d] Regional rural banks

158. Land holdings of either 10 or more hectares have been classified as:
a] Semi-medium b] Medium
c] Semi-large d] Large

159. Land reforms have been equated with total ________ development.
a] Urban b] Agrarian
c] Industrial d] Educational

160. Level of output of a particular commodity depends upon the quantities of inputs used for its production. This relationship is known as :
a] Production process
b] Product mix
c] Production function
d] Cost function

161. Loans against tractor and implements are of :
a] Short term
b] Medium term
c] Long term
d] Both a and b

162. Long term loan is given for the period of :
a] More than 5 years b] 4 years
c] 2 to 5 years d] All of the above

163. M.P.P. is equal to (P = Production, X = input) :
a] rP/rX b] rX/rP
c] P/X d] rX/rP x 100

164. Main functions of a Central Bank are:
a] To serve as an ideal banking institution
b] To regulate entire banking institute of the state
c] To issue paper currency
d] To serve as a government bankers

165. Major 14 banks were nationalized in the year :
a] 1970 b] 1969
c] 1975 d] 1973

166. Man's wants are :
a] Limited b] Unlimited
c] Countable d] Scarce

167. Management decisions which involve lasting effects are known as :
a] Operational decisions
b] Strategic decisions
c] Marketing problem decisions
d] Administrative decisions

168. Marginal cost is equal to :
a] Change in total cost/change in output
b] Change in total cost + 100
c] Average fixed cost
d] None of the above

169. Marginal cost is equal to :
a] The average cost of the total output produced
b] The difference between average cost and the cost of the last unit produced
c] The cost of the last but one unit produced
d] The extra cost of an additional unit produced

170. Marginal cost is the :
a] Change in cost due to one unit change in input
b] Change in cost due to one unit change in output
c] Change in variable cost due to one unit change in input
d] None of the above

171. Marginal farmers are those farmers having land holding
a] Less than 5 acres b] Less than 2 acres
c] Less than 1 acre d] All of those

172. Marginal product is the :
a] Output per unit of input
b] Total output divided by total input
c] Change in output due to one unit change in input
d] Change in input due to one unit change in output

173. Maringal utility of a commodity is the utility derived from the purchase of :
a] The first unit
b] All the units
c] Any extra unit
d] The last but one unit

GK GS
In China there was a custom of binding little girl's feet in order to keep them small.

174. Market demand curve is faced by:
a] Monopolist b] Competitive firm
c] Joint firm d] Oligopolic firm

175. Marshall defines economics as :
a] The study of mankind in the ordinary business of life
b] The study of human behaviour as a relationship between ends and scarce means which have alternative uses
c] An enquiry into the nature and causes of the wealth of nations
d] Fundamentally a science of scarcity

176. Marshall's definition of economics has been criticised by :
a] Ricardo b] Robbins
c] Ruskin d] Carlyle

177. Maximum profit will be at the point where :
a] M.C. = M.R.
b] M.C. is more than M.R.
c] M.C. is less than M.R.
d] None of the above

178. Means to satisfy human wants are:
a] Unlimited b] Limited
c] Both a and b d] None of the above

179. Medium term loans are repayable in a period of :
a] 3 years b] 5 years
c] 7 years d] None of above

180. MFAL is expected to cover how many marginal farmers and agricultural labour ?
a] 10,000 b] 20,000
c] 30,000 d] 35,000

181. Mines and Fishes are subject to the law of :
a] Diminishing return
b] Increasing return
c] Constant return
d] None of the above

182. NABARD act was passed in ______________, 1979.
a] January b] February
c] March d] April

183. NABARD came into existence on July 12, :
a] 1980 b] 1981
c] 1982 d] 1983

184. Name the market in which parmanent or durable commodities are traded ?
a] Long period market
b] Short period market
c] Secular market
d] Bullion market

185. Net capital ratio is equal to :
a] Total assets/Total liabilities
b] Total liabilities/Total assets
c] Total assets/Current liabilities
d] None of the above

Knowledge World!
First woman mayor
Sulochana Modi became Mayor of Mumbai in 1956. However, she was mayor for only three months.

186. Net return per hectare is :
a] Over all efficiency measure
b] Partial efficiency measure
c] Specific efficiency measure
d] None out of the above

187. No advance is give for a farm situated beyond __________ miles from the branch.
a] 5 b] 10
c] 15 d] 20

188. Opportunity cost is :
a] Cost of supplementry enterprise
b] Cost of next best alternative foregone
c] Cost of cultivation
d] Cost of production

189. Optimizing the use of farm resources on an individual farm level. It is a :
a] Farm management
b] Production economics
c] Agricultural marketing
d] Macro economics

190. Who has contributed the modern theory of interest ?
a] Knut Wicksell b] Gunnar Myrdal
c] J.R. Hicks d] R.G. Hawtrey

191. Patents and copy rights are :
a] Material goods
b] Non-material goods
c] Transferable goods
d] Scare goods

192. Payment made to labour for its assistance to production, is called:
a] Rent b] Wage
c] Interest d] None of the above

193. Per hactare calculated cost is known as:
a] Cost of cultivation
b] Cost of production
c] Cost of farming
d] Total cost

194. Per Qunital calculated cost is known as:
a] Cost of cultivation
b] Cost of production
c] Cost of farming
d] None of the above

195. Percentage population of world lives in India is :
a] 16 b] 18
c] 20 d] 19

GK **Madanlal Dhingra murdered Curzon Wylie in London.** GS

196. Persons who expects that prices will go down in future are :
a] Bulls b] Bears
c] Retailers d] Whole salers

197. Persons who expects that prices will go up in future are known as :
a] Bulls b] Bears
c] Whole salers d] Dalal

198. Prices are settled by pressing the fingers under cover of a piece of cloth is :
a] Private negotiation b] Under cover of a cloth
c] Sale by sample d] Moghum sale

199. Primary wholesale market, secondary wholesale market and terminal market are the kind of :
a] Wholesale market b] Co-operative market
c] Both a & b d] None of the above

200. Profits are defined as p (pie) = R - (TFC + TVC), where TFC = total fixed cost and TVC = total variable cost and :
a] R = Revenue b] R = Reduction
c] R = Regression d] R = Co-efficient

201. Quantitated intrument of monetary policy includes :
a] Bank rate
b] Legal reserve requirements
c] Open market opertions
d] All above

202. Ratio of the areas cultivated more than once to the total cultivated area in India is?
a] 1:3 b] 1:5
c] 1:6 d] 1:6.5

203. Rational use of agricultural resources from national angle. It is a :
a] Farm management
b] Production economics
c] Micro economics
d] Agricultural finance

204. Regional Rural Banks (RRB) under the sponsorship of nationalized bank were established :
a] 1975 b] 1962
c] 1960 d] 1950

205. Regional rural banks are financed by:
a] Reserve Bank of India
b] Government of India
c] Nationalized banks
d] All of these

206. Regulated markets are those in which :
a] Business is conducted without any set rules and regulations
b] Business is conducted with the rules and regulations framed by statutory market
c] Bonds, shares and securities are bought and sold
d] None of the above

207. Rent arises under :
a] Extensive cultivation
b] Intensive cultivation
c] Both a and b
d] None of the above

208. Reserve Bank of India (RBI) was set up in :
a] 1934 b] 1949
c] 1904 d] 1935

209. Reserve Bank of India finances Regional Rural Banks upto the extent of :
a] 50% b] 35%
c] 15% d] 25%

210. Royal Commission of agriculture submitted its report in :
a] 1905 b] 1928
c] 1929 d] 1976

211. Selection and adoption of best packages of practices comes under:
a] Agricultural marketing
b] Systems of farming
c] Types of farming
d] Farm management

212. SFDA are formed on the recomendation of All India Rural Review Comittee headed by :
a] M.L. Danthawala
b] C. Swaminathan
c] K.N. Raj
d] Ashok Mehta

213. Share of agriculture in national income is :
a] 45% b] 30.5%
c] 25% d] 42%

214. Short term loans are mainly givern for the purchase of :
a] Seeds b] Pumpset
c] Bulocks d] Heavy machinery

215. Perfect Elasticity is shown by :
a] Vertical straight line
b] Horizontal straight line
c] Slopping curve right to left
d] Slopping curve left to right

216. Smaller than economic holding is known as :
a] Optimum holding b] Ownership holding
c] Basic holding d] None of the above

217. Some middle man do not buy and sell directly but assist in the marketing process, they are known as :
a] Speculated middle men
b] Commission agents
c] Facilitative middle man
d] Brokers

218. Specialised farming and Diversified farming are :
a] Types of farming
b] Systems of farming
c] Both a and b
d] None of these

Hemachandra was the last Hindu emperor to rule from Delhi.

GS

219. Sponsoring nationalised bank finances regional rural banks upto the extent of :
a] 50% b] 15%
c] 35% d] 25%

220. Star TV is controlled at :
a] Hongkong b] Tokyo
c] Singapore d] London

221. State having 100% rural electriciation is :
a] Haryana b] Andhra Pradesh
c] Punjab d] Madhya Pradesh

222. Storage function creates :
a] Form utility b] Possession utility
c] Time utility d] Place utility

223. Success of a farm plan depends to a great extent, on the accuracy of basic date regarding :
a] Estimates of produciton from crops/animals
b] Right production of crops and livestocks in relation to fodder and feed requirements
c] Availability of capital for investment in a particular enterprise
d] All these

224. Sufficient condition of farm efficiency is generally referred to as choice indicator because it :
a] Is a valid criterion of the assumption of certainty
b] Is consistent with the choices made under perfect competition
c] Is consistent with the welfare increase to the farming population
d] Helps to make allocation compatible with the given objectives

225. Sun light is :
a] Free good
b] Economic good
c] Both
d] None of these

226. Supply for perishables in very short run :
a] Can be increased b] Can be decreased
c] Remains constant d] None of these

227. Supporting prices of agricultural commodities are announced by :
a] ICAR
b] Agriculture Universities
c] Agriculture Price Commission
d] State Government

228. Suppose that cost A_2 on a farm is Rs. 26.450 and the rent of the leased in land is Rs.2,250. Cost A. on this farm will be :
a] Rs. 24,000 b] Rs. 24,200
c] Rs. 28,000 d] Rs.28,700

229. Taccavi loans were given by :
a] RBI b] State bank
c] Government d] World bank

230. The ____________ Year Plan sough to maintain a clear cut distinction between the minimum support price and the procurement price.
a] Second b] Third
c] Fifth d] Sixth

231. The 6 major banks is Andhra Bank, Corporation Bank, New Bank of India, Oriental Bank of Commerce, Punjab and Sind Bank and Vijaya Bank were nationalized on 15th April:
a] 1972 b] 1980
c] 1982 d] 1981

232. The amount added to total cost for producing additional unit of output is known as :
a] Marginal cost b] Average cost
c] Total cost d] None of them

233. The apex organisation for marketing co-operative societies at the national level is :
a] Marketing Band b] FCI
c] NAFED d] NCCF

234. The apple growers supplies apples from Kashmir to Lucknow and creates :
a] Form utility b] Place utility
c] Time utility d] Possession utility

235. The assets that are difficult to convert into cash to meet any current obligations are :
a] Working assets b] Current assets
c] Fixed assets d] None of the above

236. The author of the "wealth of nations" was :
a] Marshall b] Ricardo
c] Robbins d]. Adam Smith

237. The authorized share capital of NABARD is Rs. ____________ million.
a] 5,000 b] 4,000
c] 3,000 d] 2,000

238. The banks do not provide loan for which of the following in dairying?
a] Purchase of milch animals
b] Purchase of dairy equipments
c] Construction of dairy shed
d] Purchase of land for dairy business

239. The biggest nationalised bank is :
a] Union Bank of India
b] Bank of Baroda
c] Central Bank of India
d] Bank of India

240. The Canons of taxation were given by :
a] Adam Smith b] Robbins
c] Pigou d] Marshall

241. The classical theory of interest is also called as :
a] Real theory of interest
b] Monetary theory of interest
c] Lonable funds theory
d] Liquidity prefrence theory

The Congress session at Kolkata supported Gandhi's plan of non-cooperation.

242. The cost incurred by a firm in purchasing any factor of production is referred to as :
a] Explicit cost b] Implicit cost
c] Variable cost d] None of these

243. The cost of various operations from land preparation to threshing of the crop is known as :
a] Cost of cultivation
b] Cost of farming
c] None of the above
d] Both a and b

244. The cost which are not paid to others, are called :
a] Explicit cost b] Implicit cost
c] Both a and b d] None of these

245. The cost which do vary with the output is called :
a] Average variable cost
b] Average fixed cost
c] Variable cost
d] Fixed cost

246. The decisions about land resources and farming opeations, questions like what to produce, how to produce and how to equip the farm, are known as :
a] Administration problem decisions
b] Prodcution and organization problem decisions
c] Buying decisions
d] Selling decisions

247. The demand for common salt is :
a] Prefectly elastic b] Highly inelastic
c] Inelastic d] Both b and c

248. The demand for wheat is :
a] Perfectly elastic b] Perfectly inelastic
c] Very elastic d] Less elastic

249. The expenditure on __________ increases as income increases.
a] Necessaries b] Comforts
c] Luxuries d] All of these

250. The farmers which have land holding less than 2 hectares constitutes about ______________ of total land.
a] 15% b] 40%
c] 22% d] 50%

251. The farmers which have land holding more than 1 hectares constitues about ______________ of total land.
a] 15% b] 22%
c] 50% d] 5%

252. The first labour inquiry committee was formed in the year :
a] 1942-43 b] 1950-51
c] 1960-61 d] 1970-71

253. To whom do you attribute the 'Concept of Consumer's Surplus'?
a] A.C. Pigou b] Alfred Marshall
c] H.H. Gossen d] Barbura Wootton

254. The formula for rate of capital turn over is :
a] Gross Income/Total Farm Assests x 100
b] Net Income/Total Farm Asssests x 100
c] Total Farm Assets/Net Farm Income x 100
d] None of the above

255. The function of succesful farm manager is :
a] Formation of the goals
b] Recognition and definition of a problem
c] Decision making and taking action
d] All of these

256. The grain dealer stocks grain to supply it as the time of shortage and creates :
a] Place utility b] Time utility
c] Form utility d] Ownership utility

257. The industries directly or indirectly depend upon agricultural production by way of raw material to the extent of :
a] 40% b] 10%
c] 60% d] 20%

258. The inflection point on total product curve corresponds to the:
a] Maximum of marginal product curve
b] Maximum of average product curve
c] Maximum of total product curve
d] Minimum of marginal product curve

259. The large number of combination of two inputs which have some cost is known as :
a] Iso product b] Iso cost
c] Iso revenue d] Indifference curve

260. The law of dinimishing returns applies more generally to :
a] Cloth Industry
b] Chemical Industry
c] Agricultural Industry
d] Film Industry

261. The law which is very useful in determining the best use of limited resources is :
a] Law of opportunity cost
b] Law of diminishing return
c] Law of substitution
d] None of the above

262. The main cause of inflation is __________ in money supply and decrease in production.
a] Decrease b] Increase
c] Remains same d] None of them

263. The main function of Regional Rural Bank is :
a] To advance loan to weaker sections
b] To encourage deposit mobilization
c] To open branches in rural areas
d] To control inflation

264. The main objective of land reform measures undertaken after independence was removal of institutional and ____________ obstacles.
a] Emotional b] Motivational
c] Conventional d] National

GK GS
'Mein Kampf' is an autobiography of Adolf Hitler.

265. The meanings of the word 'Economics' is most closely associated with the word :
a] Scarce b] Free
c] Unlimited d] Expediture

266. The Ministry of Agriculture give subsidy of total cost for setting a bio-gas plant :
a] 15% b] 25%
c] 40% d] 75%

267. The modern theory of population is called is :
a] Classical theory of population
b] Optimum theory of population
c] Demand and supply theory of population
d] None of the above

268. The money paid for the use of money borrowed is nothing but :
a] Profit b] Rent
c] Premium d] Interest

269. The more we have of a things, the desire for additional units of it will be :
a] Less b] More
c] Remain same d] None of the above

270. The most common method exist in regulated markets is :
a] Close tender system b] Open auction system
c] Dara sale d] Moghum sale

271. The most influence factor of inflation is increase in money supply and __________ in production.
a] Increase b] No effect
c] Decrease d] None of the above

272. The most liquid assets are :
a] Fixed assets b] Working assets
c] Currents assets d] Liabilities

273. The object of subsistence farm is to produce :
a] Essential food crops for family and market
b] Essential food crops for family only
c] Food and fibre crops for family and market
d] Vegetable crops for family only

274. The oldest nationalised bank is :
a] Allahabad Bank
b] Punjab National Bank
c] Canera Bank
d] Central Bank of India

275. The owner of land in state farming is :
a] Farmer b] Capitalist
c] Government d] None of them

276. The paid up capital share of NABARD is Rs. __________ million.
a] 400 b] 600
c] 800 d] 1,000

277. The paid-out costs of a farmer is Rs. 500 and fixed costs is Rs. 250. If the gross returns from form business is Rs.3,700, the Capital Turnover would :
a] 17.42 b] 9.80
c] 4.93 d] 3.33

Quiz Kit!

First public sector credit card

Central Bank of India first entered the credit card market when it launched the Central Card in 1980.

278. The peculiarities of land as a factor of production is:
a] A free gift of nature
b] Permanent in nature
c] Limited in area
d] Completely immobile

279. The percentage of agricultural commodites comprising agriculture produce and commodities based on agriculture, contribute to the total exports of country to the extent of :
a] 20% b] 80%
c] 70% d] 45%

280. The percentage of cultivated waste land in India is :
a] 2% b] 5%
c] 10% d] 12%

281. The percentage of Indian population deriving livelihood from agriculture is about :
a] 66% b] 50%
c] 60% d] 90%

282. The physical property of a business along with their values at a specific date is enlisted in :
a] Farm book-keeping
b] Farm planning
c] Farm inventory
d] Farm management

283. The primary cooperative credit societies has :
a] Two-tier structure
b] Four-tier structure
c] Three-tier structure
d] None of the above

284. Who was the first economist to have propounded the theory of rent ?
a] Joan Robinson
b] R.G.D. Allen
c] Edward Chamberlin
d] Alfred Marshall

285. The principle of equi-marginal return states that the profits will be the greatest if unit resources are used where :
a] Marginal returns are the highest
b] Average returns are the highest
c] Total returns are the highest
d] Net returns are the highest

286. The principles of taxation were given by :
a] Marshall b] Adam Smith
c] Hicks d] None of the above

The Lahore Congress declared complete independence as the aim of the Congress.

287. The process of determination of wages, rent, interest and profit is called :
a] Consumption b] Production
c] Distribution d] None of the above

288. The process of transformation of certain resources or inputs is known as :
a] Product b] Production function
c] Production d] Consumption

289. The purchase or sale of a commodity at the present price with the object of sale or purchase at some future date at a favourable price is known as :
a] Hedging b] Speculation
c] Market intelligence d] None of the above

290. The reform of tenancy laid down that the rent should not exceed __________ per cent of gross produce.
a] 20-25 b] 10-15
c] 25-30 d] 35-40

291. The relationship between cost function and production function is:
a] Positive b] Negative
c] Increase d] None of above

292. The resources, if their services are not used, these cannot be stocked are known as :
a] Stock resources b] Flow resources
c] Fixed resources d] All above

293. The return from best alternate use of that input forgone by putting it to the present use is known as:
a] Average cost b] Total cost
c] Marginal cost d] Opportunity cost

294. The route through which agricultural products move from producers to consumers is known as :
a] Marketing information
b] Marketing channels
c] Marketing efficiency
d] Marketing institutions

295. The second labour inquiry committee was set up in :
a] 1947-48 b] 1954-55
c] 1949-50 d] 1955-56

296. The SFDA and MFAL was formed in :
a] 1961-62 b] 1980-81
c] 1971-72 d] 1976-77

297. The source of finance of State Co-operative Bank is :
a] Deposits from members and non-members
b] Borrow from R.B.I. and S.B.I.
c] Borrow from State Govternment etc.
d] All of the above

298. The spurt to Zero Marginal Productivity type of exercise was given by ______________ in 1954.
a] W. Arthur Lewis
b] W. David Hopper
c] W. Nelson Michael
d] W. Benzene Shoemaker

299. The statement "In the process of production and distribution of goods, income is generated and distributed amongest the factors of production" was given by :
a] J.B. Say b] Classical economist
c] Keynes d] Marshall

300. The technique which protect the traders from extreme falls in price is known as :
a] Speculation b] Future trading
c] Hedging d] For sale

301. The term 'operational holding' refers to the :
a] Land area owned by a farmer
b] Cropped area cultivated by a farmer
c] Net land area cultivated by a farmer
d] Land area leased in by a farmer

302. The term "Agricultural Economics" is used in preterence to the term "Rural Economics" for the following reason :
a] Agricultural economics is more widely under stood than the rural economics
b] Agricultural economics is far more popular term than the rural economics
c] Both a and b
d] None of the above

303. The term "land reform" should be used in broadest sense to serve as __________ for action.
a] Inscription b] Prescription
c] Description d] Formulation

304. The term monopolistic competition means :
a] A large number of sellers heterogenous commodity
b] A few number of sellers deal heterogenous commodity
c] A large number of buyers deal in heterogenous commodity
d] None of the above

305. The term monopoly means :
a] When there is only one seller of a commodity
b] When there is only two seller of a commodity
c] When there is only one buyer of a commodity
d] None of the above

306. Which of the following definitions of economics includes the economic concept of 'Scales of Preferences'?
a] 'Wealth' Definition
b] 'Welfare' Definition
c] 'Scarcity' Definition
d] 'Isolation' Definition

307. The term oligopoly means :
a] There are two or more buyer
b] There are two or more seller of a commodity
c] There is single seller of a commodity
d] None of the above

308. **The total fixed cost curve :**
   a] Increases with the level of output
   b] Remains constant at all the output levels
   c] Decreases as output increases
   d] Both b and c
309. **The want satisfying power of commodity is called :**
   a] Demand b] Utility
   c] Consumption d] Production
310. **Those individuals granted broad powers by those who purchase are known as :**
   a] Brokers b] Speculated middle man
   c] Commission agents d] Whole salers
311. **Time required for a resource to be completely transformed into a product is referred to :**
   a] Transformation period
   b] Production period
   c] Both a and b
   d] Planning period
312. **Total cost is equal to :**
   a] Fixed cost + Variable cost
   b] Fixed cost + Variable cost/100
   c] Fixed cost - Variable cost
   d] None of the above
313. **Total output is maximum at the point when :**
   a] Average product is maximum
   b] Averge product is zero
   c] Marginal product is zero
   d] Marginal product is maximum
314. **Tulsidas's works are :**
   a] Personal wealth
   b] Social wealth
   c] International weath
   d] None of the above
315. **Under the crop loan system, the farmer is given loan in the form of:**
   a] Cash
   b] Kind
   c] Partly case and partly kind
   d] None of these
316. **Uniform interest rate policy is not justified from the view point of economic soundness because it does not take into account :**
   a] Size of loan b] Risk
   c] Distance d] All above
317. **Utility in economics means :**
   a] Usefulness
   b] Pleasure
   c] Want-satisfying power of a commodity
   d] Satisfaction
318. **Utility of the last unit is called as :**
   a] Marginal utility b] Initial utility
   c] Total utility d] Zero utility
319. **Walchand Nagar Farm is an example of :**
   a] Peasant farming
   b] State farming
   c] Capitalistic farming
   d] Collective farming
320. **We get maximum satisfaction when:**
   a] Marginal utility is the same
   b] Total utility is the same
   c] Average utility is decreases
   d] Total utility is decreases
321. **Wealth is a :**
   a] Fund b] Flow
   c] Income d] Capital
322. **What is the average population of India's villages ?**
   a] Less than 500 b] More than 1000
   c] Less than 1000 d] More than 1500
323. **What is the basic consideration for advancing crop loan by commercial banks?**
   a] To access the repayment capacity of the farmers
   b] To advance loans at cheaper rates
   c] Cost of proudction per unit of area
   d] To advance loans to weaker sections
324. **What is the most important source of money lending to farmers in rural areas ?**
   a] Rural banks
   b] Nationalized banks
   c] Local money lender
   d] Co-operative agencies
325. **What per centage of total cultatived land is oppupied by small and marginal farmers?**
   a] 30 b] 50
   c] 70 d] 85
326. **What per centage of total geogrphical area in India is sown by crops?**
   a] 30% b] 44%
   c] 54% d] 81%
327. **When farmers pool their land, labour and capital and work jointly under the direction of an elected managing committe and divide the profit among themselves in proportion of land contributed and wages earned by each one of them it is called?**
   a] Co-operative farming
   b] Collective farming
   c] Capitalist farming
   d] Peasant farming
328. **When increase or decrease in the level of one product does not affect the production level of the another product, this relationship is known as ?**
   a] Competitive
   b] Complementary
   c] Supplementary
   d] Joint product

329. When land is owned by government and workers work as employees getting their wages regularly, it is called?
a] Peasant farming b] Collective farming
c] Capitalistic farming d] State farming

330. When marginal product is zero, the total product is ?
a] Maximum b] Minimum
c] Zero d] None of the above

331. When price rises, demand ?
a] Decreases b] Contracts
c] Increases d] Remain same

332. When quantity demanded of a commodity decreases with the increase in its price, it is called ?
a] Decrease in demand
b] Extension of demand
c] Contraction of demand
d] Increase in supply

333. When total cost is divided by total units of the production, we get ?
a] Marginal cost b] Average cost
c] Fixed cost d] Variable cost

334. When total product is highest, marginal product will be ?
a] Minimum b] Maximum
c] Zero d] Increasing

335. When total utility increases, marginal utility is ?
a] Zero b] Positive
c] Negative d] None of the above

336. When variable cost is zero, the toal cost will be ?
a] Equal to variable cost
b] Equal to fixed cost
c] Equal to average variable cost
d] None of the above

337. When we study a problem from particular to general, the method is known as ?
a] Deductive method b] Inductive method
c] Both a and b d] None of the above

338. Which of these statement are correct ?
a] Agriculture economic may be taken as the branch of both of agriculture and economics
b] Agriculture economic is more intimately related to economics then to agriculture
c] Agricultural economics is nothing but application of economic principles to agriculture
d] All of these

339. Which are of the following is administrative decision?
a] Financing the farm business
b] Accounting and book-keeping
c] Adjustments of government programmes and policies
d] All these

340. Which bank is lead bank in maximum number of districts ?
a] Bank of Baroda b] Syndicate Bank
c] Bank of India d] SBI

341. Which of the following banks is not nationalized ?
a] Union Bank of India
b] Indian Bank
c] Bank of India
d] Bank of Rajasthan

342. Which of the following contributes more national income and stand next to agriculture ?
a] Mining and manufactures
b] Animal husbandry
c] Transportation
d] Forestry

343. Which of the following factors does not effect the demand of bajra?
a] Income of the consumer
b] Price of wheat
c] Taste of the consumer
d] Price of groundnut

344. Which of the following is an economic factor affects the type or system of farming ?
a] Availability of capital
b] Water
c] Land
d] All above

345. Which of the following is eligible for getting the loan for tractor purchasing ?
a] A farmer having 5000 hours work per year
b] A farmer having 1000 hours work per year
c] A farmer having 100 hours work per year
d] A farmer having 400 hours work per year

346. Which of the following is not a component of farm business?
a] Land
b] Capital
c] Labour and Management
d] Water

347. Which of the following is not a factor affecting rate of market development ?
a] Nature of demand
b] Nature of product
c] Transportation of communication
d] Disturbance and unsecurity

348. Which of the following is not a farming system on the basis of ownership and management classification?
a] Tenant farming b] Co-operative farming
c] Mixed farming d] State farming

349. Which of the following is not a type of farming on the basis of land use and cultural practices classification?
a] Tenant farming
b] Specialized farming
c] Diversified farming
d] Mixed farming

350. Which of the following is not component of farm business ?
a] Capital b] Environment
c] Land d] Management

351. Which of the following is not merit of open auction system ?
a] It disposes of the market supply promptly
b] A wide variety of goods are available to buyers for selection
c] The buyers of small lots have disadvantage against the buyers of large lots
d] The payments of the price of the goods is made immediately

352. The Government established TRIFED in :
a] 2001 b] 1997
c] 1987 d] 2004

353. Which of the following loan is not advanced by land mortgage banks?
a] Crop loan
b] Medium term loan
c] Long term loan
d] Loan for farm mechanization

354. Which of the following statement is correct ?
a] When average cost is decreasing, it is more than marginal cost
b] When average cost is decreasing, it is less than marginal cost
c] When average cost is increasing, it is more than marginal cost
d] None of the above

355. Which of the following statement is wrong for perfect market ?
a] There is a large number of buyers and sellers
b] Buyers and sellers have imperfect knowledge of demand and supply
c] Price at any time are uniform
d] Prices are uniform over a geographical area at any one time

356. Which of the following system as followed for collecting land revenue during British period ?
a] Zamidari system b] Royattwari system
c] Mahalwari system d] All the systems

357. Which one of following statement is correct ?
a] Labour is inseparable from the labourer
b] The labour does not sells his services
c] Labour is not perishable than any other commodity
d] More mobile than capital and other goods

358. Which one of the following economic principles helps a farmer to allocate his limited resources to two different enterprises?
a] The law of variable proportions
b] The law of comparative advantage
c] The law of equimarginal returns
d] The principles of fixed and variable costs

Did You Know ?
First Bharat Ratna Award
Dr. Sarvepalli Radhakrishnan (1888-75) was awarded the Bharat Ratna in 1954 along with Dr. C.V. Raman (1888-70) and C. Rajgopalachari (1918-72).

359. Which one of the following formulae gives the correct measure of 'Rate of Capital Turnover'?
a] (Gross income/Total farm assets) x 100
b] (Net income/Total farm assets) x 100
c] (Farm business income/Total farm assets) x 100
d] (Capital investment/Total farm assets) x 100

360. Which one of the following has highest number in the country ?
a] Village level workers
b] Rural development banks
c] Branches of nationalized banks
d] Co-operative societies

361. Which one of the following has maximum share in total finance to agriculture ?
a] Rural Banks
b] Nationalized banks
c] District cooperative banks
d] Cooperative societies

362. Which one of the following is a farmers agency ?
a] IFFCO
b] FCI
c] Agro Service Centres
d] Shree Ram Fertilizers

363. Which one of the following is correct statement for market ?
a] A place where commodities are bought and sold
b] Potential buyers and sellers of the product
c] An organization which provides facilities for exchange of commodities
d] All of these

364. Which one of the following is not a physical factor affecting the type of farming ?
a] Climate b] Topography
c] Soil d] Labour supply

365. Which one of the following is not a system of farming?
a] Peasent farming
b] Co-operative farming
c] Mixed farming
d] Co-operative collective farming

366. Which one of the following is not capital ?
a] Goodwill of a business b] Skill of a doctor
c] Furniture in a house d] Cloth in a shop

367. Which one of the following is operational decision?
a] Enterprise mix b] Selection of least cost
c] Level of resource use d] All above

Singravellu Chettiyar was the first communist of South India.

GS

368. Which one of the following pairs is correctly matched?
a] Common cost = Cost associated with variable resources
b] Average cost = A study of returns and costs
c] Marginal cost = Additional cost to produce an additional unit of output
d] Prime cost = Cost related to fixed resources

369. Which one of the following principles is employed in production when resources are scarce?
a] Added cost is equal to added returns
b] Equal-marginal principle
c] Comparative advantage
d] None of these

370. Which one of the following principles will be applicable when a farmer is faced with the problem of purchase of heifers versus milch cows on his dairy farm ?
a] Cost principle
b] Opportunity cost principle
c] Comparative advantage principle
d] Time comparison principle

371. Which one of the following relationships will provide least cost combination of input use?
a] $rX_1/rX_2 = Px_2/Px_1$
b] $r X_1/r X_2 > Px_2/Px_1$
c] $rX_1/rX_2 < Px_2/Px_1$
d] $rX_1/rX_2 = Px_1/Px_2$

372. Which part of a pump-set should not be financed ?
a] Wirings and installation b] Pipes
c] Base plate d] Motor

373. Which state has maximum number of tube wells and pumpsets ?
a] Punjab b] Rajasthan
c] Haryana d] U.P.

374. Which state has maximum percentage of cultivated area to the total area ?
a] Bihar b] Haryana
c] U.P. d] M.P.

375. Which statement is used for classical economist to theory of income and employment ?
a] They thought that supply creates its own demand
b] Economy produces output and income corresponding to the level of full employment
c] Their theory was based on asumptions and full emloyment
d] Theory was not based on asumption and full employment

376. Who decides the procurement price of a particular commodity ?
a] Agricultural Production Commissioner
b] Agricultural Price Commission
c] Ministry of Food and Agriculture
d] Ministry of Finance

377. Who defined consumption as the use of economic goods and personal services in the satisfaction of human wants ?
a] Ely b] Keynes
c] Pigou d] Marshall

378. Who has defined Agriculture economics as the branch of economics dealing with the production and distribution of agricultural commodities and the institutions associated with agriculture ?
a] Philip Taylor b] L.C. Gay
c] E.O. Heady d] Marshall

379. Who said that "all those phenomenon which can be brought directly or indirectly into relation with measuring rod of money, fall under the subject matter of economics?
a] Marhsall b] Pigou
c] Chapman d] Keynes

380. Iso-revenue line is a :
a] Straight line b] Curve line
c] Horizontal line d] Vertical Line

381. Zamidari system was introduced by:
a] British b] Mughals
c] Marathas d] All of them

382. Zamidari-system was based on ________ to support British Imperialism in India.
a] Exploitation b] Elimination
c] Regulation d] Augmentation

383. Symbol used for purity of agricultural goods is :
a] ISI b] AGMARK
c] @ d] All of the above

384. The other name of overhead cost is:
a] Fixed cost b] Variable cost
c] Total cost d] Marginal cost

385. World Trade Organisation (W.T.O.) was established in :
a] 1992 b] 1998
c] 1995 d] 1997

386. Which of the following is not correctly matched ?
a] APP rises continuousely - Till APP and MPP are both equal.
b] MPP rises continuously - Till MPP is more than APP.
c] TPP rises continuously - Till MPP is not equal to Zero.
d] APP & MPP both fall but TPP rises - When MPP is less than APP but more than zero.

387. Marginal productivity of labour and capital on ridge lines is :
a] Zero b] Infinite
c] Positive d] Both a and b

388. Formula which indicates the least-cost combination point is?
a] MR = MC b] $\Delta Y/\Delta X_1$
c] $MRSY_1,Y_2$ d] $MRSX_1,X_2=PX_2/PX_1$

Soccer is the national sport of Greece. GS

Matching

1. **Match the enterprise relationship with the marginal rate of substitution.**

a]	Competitive	i]	More than zero
b]	Supplementary	ii]	More than zero
c]	Complementary	iii]	Equal to zero

2. **Match the following with their abbreviated expressions when *Y* = output and *X* = input**

a]	Output	i]	$(X/Y).(\triangle Y/\triangle X)$
b]	Elasticity of production	ii]	PX_2 / PX_1
c]	Marginal substitution ratio	iii]	$Y = A + BX$
d]	Price ratio	iv]	$\triangle X_1 / \triangle X_2$

3. **Match the class of land-holding as specified in the census reports with their actual hectarage in India.**

a]	Marginal holdings	i]	4-10 hectares
b]	Small holdings	ii]	Less than 1 hectare
c]	Semi-medium holdings	iii]	2-4 hectares
d]	Medium holdings	iv]	1-2 hectares

4. **Match the state with the prevalent rates of fair rent on agricultural lands in India.**

a]	Andhra Pradesh	i]	33%
b]	Jammu and Kashmir	ii]	30% on irrigated and 25% on drylands
c]	Jammu and Kashmir	iii]	25% on lands less than 5 hectares and wet
d]	Punjab	iv]	33% on drylands

5. **Match the marketable surplus of various commodities in India with the percentage of the total production.**

a]	Rice	i]	35
b]	Wheat	ii]	25
c]	Millet	iii]	90-95
d]	Cash crops	iv]	32

6. **Match decisions with techniqes.**

a]	Captial investment	i]	Least cost combination
b]	Economicallyoptimum level of resource use	ii]	Pay back period
c]	Economics of farm size	iii]	Linear programming
d]	Optimum enterprise combination under constraining resources	iv]	Average cost curve

7. **Match Nobel Prize Winners in Economics.**

a]	Input-output Model	i]	Tjalling Koopmans
b]	Optimum Resource Allocation	ii]	Moris Allies
c]	Decision Process in Organisation	iii]	W.W. Leontief
d]	Optimum Utilization of Resources	iv]	Herbert Simon

8. **Match the following.**

a]	Gini Multiplier	i]	Mobile in nature
b]	Fisical Drag	ii]	For promoting per capita income
c]	Footloose industry	iii]	Tax burden
d]	Technology Vision 2020	iv]	Coefficient of measuring the centralisation of a frequency distribution

9. **Match the following.**

a]	MODVAT	i]	1978
b]	CENVAT	ii]	2005
c]	VAT	iii]	2000
d]	Tobin Tax	iv]	1976

The book named 'We' has been written by M.S. Golwalkar. GS

10. **Match series of money stock measures in India.**

a]	M_1	i]	M_1 + Saving bank deposits with post offices.
b]	M_2	ii]	Most important component among all money stock measures.
c]	M_3	iii]	Most liquid form of money.
d]	M_4	iv]	Possesses the lowest liquidity.

11. **Match establishment years of major financial institutions in India.**

a]	ICICI	i]	1977
b]	HDFC	ii]	1982
c]	IDBI	iii]	1955
d]	EXIM	iv]	1964

12. **Match the following.**

a]	SENSEX	i]	Singapore
b]	DOW JONES	ii]	Germany
c]	MID DAX	iii]	New York
d]	SIMEX	iv]	Mumbai

GK GS

Leander Paes is the only Indian tennis player to win the Arjun Award, the Padma Shri and the Rajiv Khel Ratna Award.

Glossary of Technical Terms

Agricultural Economics It is defined as application of principles of economics for solving agricultural problems and to maximise agriculture production.

Agricultural Finance It is an economic study of financing the farm business.

Agricultural Prices It is the exchange value of agricultural commodity determined by buyers and sellers in the market and is expressed in terms of money.

Agriculture Marketing It comprises of all the operations involved in the movement of goods and raw materials from the farm to final consumer.

Average Product It is the total product divided by the number of units of variable input used to produce that product.

Barter Exchange of goods directly for other goods without the use of money.

Brokers Brokers render personal services to their clients in the market.

Budget It mentions in detail the expected income of the state from the various sources and proposed expenditure on the various needs for the next financial year.

Capital It is a factor of production which generates a flow of income spread over a certain period of time.

Cash Costs Cash costs are incurred when resources are purchased and used immediately in the production process.

Ceteris Paribus A latin phrase meaning "everything else remaining the same".

Commission Agent A commission agent is a person operating in the wholesale market who acts as the representative of either a seller or buyer.

Consolidation of Holding The amalgamation and re-distribution of fields constituting individual holding so as to reduce the number of fragments in the holdings.

Consumer Surplus Means what one prepare to pay minus one actually pays.

Consumption It is the use of goods and services for the satisfaction of human wants.

Co-operation It is a voluntary organisation of people meant for satisfying their economic interest.

Cost It refers to the expenses incurred or inputs.

Credit Credit means trust in borrower's ability to repay and willingness to repay.

Deflation Means a contraction of currency and credit leading to a fall in prices.

Demand Means desire backed by willingnent and ability to pay.

Demand Schedule A list showing the relationship between different demand prices and the quantity of the commodity demanded at each of these demand prices, at a particular place and time.

Depreciation It means decline the value of a given assest as a result of the use, wear and tear or due to any other cause.

Direct Tax It is one which is really paid by the person on whom it is imposed.

Distribution It deals with remuneration earned by four factors of production *viz.* land gets rent, labour earned wage, capital gains interest and organiser is rewarded with profits.

You Don't Know!

First central bank

The Reserve bank of India was established in 1935 to monitor all the other banks. Its main functions are establishment of a unified currency system and credit policies, maintenance of the money standard, issue of paper currency and maintenance of the external value of the rupee. The RBI was nationalised in 1948.

Division of Labour It may be defined as specialisation of persons in particular activties.

Duopoly A duopoly market is one which has only two sellers of a commodity.

Duopsony The market situation in which there are only two buyers of a commodity.

Economic Holding It is the size of land holding which provided a man the chance of production sufficient to support himself and his family in reasonable comfort after paying his necessary expenses.

Economic Surplus Those goods, services and resources available after minimum requirements for existence are provided for the people.

Economics It is a study of human behaviour as a relationship between ends and scare means which have alternative uses.

Elastic Demand A small change in price lead to a great change in demand.

Elasticity It is the rate at which demand changes in response to a change in price.

Expansion Path A line connecting the optimum combination of two products at level of inputs.

Farm Budgetting It is defined as a process of estimating costs, returns and the net profit of a farm or a particualar enterprise.

Farm Enterpreneur He is a person or group of persons who thinks of organising or operating the business and is responsible for results.

Farm Executive He is a person who carry out the orders of management.

Farm Firm Refers to aggregation of resources for which cost and returns are work out as a whole. It is production unit under one management.

Farm Management It is defined as a science which deals with judicious decision on the use of scarce farm resources having alternative uses to obtain the maximum profit and family satisfaction on a continuous basis from the farm as a whole and under sound farming programmes.

Farm Manager He is a person who manage or supervise the business according to the instructions of the farm enterpreneur.

Farm Means a piece of land where crops and livestock enterprises are taken up under a common management and has specific boundries.

The idea of Indian National Army was first conceived by Mohan Singh.

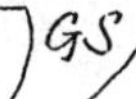

Farm Planning It is the process of deciding in the present, what to do in the future about the best combination of crops and livestocks to be raised through rational use of resources.

Farm Profit It is the net income in excess of cost of production.

Fiscal Policy Decisions of the federal government on the use of tax collection policy.

Fixed Costs Fixed costs are those which do not change in magnitude as the amount of output of the production process changes and are incurred even when production is not under taken.

Flow Resources There are some resources, if their services are not used they can be stocked.

Goods Anything that satisfies a human want.

Grading Means the sorting of the unlike lots of the produce into different lots according to the quality specifications laid down.

Gross Interest It is the total amount which the borrower pay to the lender for the use of capital.

Hedging Hedging is executing opposite sales or purchase in the future market to offset the purchases or sales of physical products made in the cash market.

Income Wealth refers to property or assets. The amount of money which these assests yield called income.

Indifference Curve Shows the various combinations of two commodities which yield equal satisfacation or utility to the consumer.

Indirect Tax It is one which is imposed on one person but paid by another.

In-elastic Demand A big change in price followed only by a small change in demand.

Inflation There is an increase in monetary units without any increase in physical units and characterised by a rise in the general level of prices.

Isocost Line It indicates all possible combination which can be purchased with a given outlay of funds.

Isoline A line or curve connecting the least cost combination of inputs for all output levels.

Isoquant Isoproduct curve Particular quantity of output can be produced by different combination of two inputs.

Isorevenue Line / Price Line It indicates the different combinations of two products which give the same amount of revenue or income.

Labour Labour is any effort of mind or body, undergoes partly or wholly with a view to get some reward other than the pleasure derived directly from the work.

Land That part of the surface of lithosphere which is not usually covered with water, forms the total natural and cultural environment with which production takes place.

Law of Demand When the price falls, demand expands and when the price rises, demand contracts.

Law of Diminishing Returns Beyond a certain level of output, when successive unit of variable inputs are added to a fixed quantity of water inputs, the addition to total product caused by each successive unit of variable input declines.

Law of Equi-Marginal Return (Opportunity Cost) The principle states that resources should be used where they bring not the greatest average retrun but the greatest marginal returns.

Law of Supply If price increases, supply will also increase, and if price falls, supply will also falls.

Lease The term used for an agreement in which one agent obtains the use of some property owned by another agent for a given period of time in return for an agreed fixed charge which is generally paid in periodic instalments.

Localisation It is defined as geographical specialisation or division of labour in particular industry.

Long Run Period It is a period which is sufficiently long for output level to be altered or changed by varying even the size of the plant.

Macro Economics Income Theory It explains the level at total production and why level rises and falls. It deals with aggregates and average of entire economy, e.g. National income.

Management Means act or act of managing the things.

Marginal Product It is defined as the quantity with which added unit of the variable input adds to the total product.

Markeing Cost It is the cost of performing the various marketing functions and operating various agencies.

Market Any body or persons who are in intimate business relations and carry on existensive transactions in any commodity.

Market Integration It is a process which refers to the expansion of firms by consolidating additional marketing functions and activities under a single management.

Marketing Efficiency It is defined as the effectiveness or competence with which a marketing structure performs its designed functions.

Marketing Includes those business activities that direct the flow of goods and services from producer to consumer.

Marketing Margins It refers to difference between prices prevalling at successive stages of marketing at a given point of time.

Micro Economics Price Theory It studies economic actions of individual firms and group individuals.

Money Anything which is widely accepted in payment for goods, or in discharge or other kind of obligations.

Monopolistic Competition When a large numbers of sellers deal in heterogenous and diferentiated form of a commodity.

Monopoly Market A market situation in which there is only one seller of a commodity.

Monopsony When there is only one buyer of a product.

Non Cash Costs Non fixed costs are consists of depreciation and payments to resources owned by the farmers.

Normative Science An attempt to prescribe a means to improve a situation.

Oligopoly Market A market in which there are more than two but still a few sellers of a commodity.

Oligopsony A market having more than two buyers.

Opportunity Cost Opportunity cost is the cost of next best alternative foregone.

Optimum Holding It is not exceeding three times than the economic holding except in the case of a charitable institutions.

GS

Mary, Queen of Scots is considered to be the first women golfer in the world.

Parity Price A parity price is one that will buy the same quantity of other products as it would during some specific base period.

Positive Science A discipline that describes and explains a situation as it is .

Price Elasticity of Demand Percent change in quantity

Price Index Number It device to show relative changes in prices or the value of money over period of time.

Price Stabilization It means prevention of voilent fluctuations in the prices at both lower and upper levels.

Price Means a service expressed in terms of money price is a value exchange or value of goods, which is expressed in terms of money.

Procurement Price Government purchase commodity at a specific price and put in a buffer stock and use at the time of scarcity.

Production Economics It is concerned with organizing resources in order to maximize the production of a single commodity i.e. to make choices from among various alternative ways of using resources and combinations of different commodities to produce.

Production Function Level of output of a particular commodity depends upon the quantities of input used for its production. This relationship between input and output is known as production-function.

Production Creation of utility is known as production. Process of transformation of certain resources of inputs (land, labour, seeds, fertilizer) into products or output (wheat, paddy).

Productivity Means output per unit of input employ. Increase in productiyity come about from increased efficiency on the part of capital or labour.

Quasi Rent It is the surplus earned by the instruments of production other than land.

Quick Assets Current assets which can be quickly converted into cash.

Regulated Market It is a wholesale market where dealings, buying and selling are regulated and controlled by the State Government through a maket committee.

Rent It refers to that part of the payment by a tenant which is made only for the use of land.

Resources The inputs which are consumed or transformed into products in the process of production are known as resources. e.g. seeds, manures.

Retailer Retailer buy goods from whole salers and sell them to the consumers in small quantities.

Reverse Dumping Refers to the practice of selling a good abroad at a price higher than the charged for the same good in the domestic market to take advantage of a monopoly position.

Ridge Line Represents the points of maximum output from each input give at a fixed amount of the other input.

Risk Bearing Ability It is a ability to withstand shocks of probable financial losses.

Short Run Period It is a period of time which is long enough to permit desired changes in output without altering or changing the size of the plant.

Social Science A science devoted to the study of people, especially their behaviour and their social institutions.

Speculation Means purchase or sale of a commodity at the present price with the object of sale or purchase at some future date of a favourable price.

Standard of Living Refers to our usual scale or expenditure, the goods we consume and services we enjoy our attitudes and values.

Standardization Means the determination of the standards to be established for different commodities. Standards are established on the basis of certain characteristics such as weight, size, colour, appearance, texture, moisture, content, staple length, amount of foreign matter, ripeness, sweetness, taste etc.

Stock Resources There are some resources, if their services are not used in one production period, they can be stored for a later period.

Strategic Decision These are the management decisions which involved heavy investment and have long lasting effect.

Supply Supply is the quantity of a commodity that a seller offers for sale at acertain price.

Supply Quantities of a commodity offered for sale.

System of Farming The combination of production on a given farm and the methods or practices that are used in the production of those products is known as the system of farming, that is followed on the farm.

Technical Unit Refers to a single convenient unit in production for which technical co-efficients are calculated.

Total Cost Are the sum of total variable costs and total fixed cost.

Total Product It is a amount of product produced by different quantities of the variable inputs.

Types of Farming : When farms in a group are quite similar in the kinds and proportions of the crops and live stocks that are produce and in the methods and practices followed in production, that group is described as a type of farming.

Up-Market Refers to that part of a market which caters the demands of the higher socio-economic groups.

Utility It is a capacity of goods to satisfy human wants.

Value Exchange power of commodity is known as value.

Variable Cost Variable costs are the costs of using the variable inputs. These costs vary with the level of production.

Victim Firm A firm which is the subject of a takeover bid.

Wage It is defined as the sum of money paid under contract by an employer to a worker for the services rendered by him.

Waste An inevitable by-product of economic activity.

Wealth Anything which has following three characteristics *viz*. Scarcity, utility and transferability is called wealth.

Whole Salers Whole salers are those merchant middlemen who buy and sell foodgrains in large quantities.

Working Capital Refers to the amount of current assets which has been financed from long term sources of finance.

Yield Gap The difference between an average yield on equities and a corresponding yield on long-dated fixed-interest securities.

Z-Chart A chart which relates to the output of a firm. It shows graphs of monthly totals, the cumulative monthly total and the moving annual these three graph together taking the shape of a letter 'Z'.

Subhash called Gandhiji as 'Father of the Nation'.

GS

11. Extension Education

Each Word Counts

Role of Extension Education in Agricultural Development

1. Making the adoption of new technology possible :
- To help the farmer to accept new agricultural technology for obtaining higher yields.
- Extension workers carry out field demonstrations about new technology.
2. Helps the farmers to visualize farming as a profitable occupation
- Helps farmers to understand their own situation, resources and critical analysis helps them to take rational decisions for the best use of the available resources.
3. Trains the local leaders
4. Helps getting necessary inputs
5. Helps in decision-making
6. Helps to change attitude
- To bring about this change through education.
7. It helps in achieving cooperation of people.
8. It develops, strengthens and organise groups, institutes and people to achieve their aims.
9. Extension helps the planners, policy-makers and administrators with local conditions and latest technologies suited to it.
10. Job of extension is to bridge the gap between the scientists on the one hand and the farmer's on the other. This is done by taking new farm technology and innovations to the farmers door and helping him to adopt them.

It also brings the farmer's problems to the research stations for further investigations and to find out the required solutions.

EXTENSION METHODS

The methods are the tools and techniques used to create situations in which communication can take place between the rural people and the extension workers.

Did You Know ?

Biggest bathing tank

The Brahma Sarovar at Kurukshetra, Haryana, which attracts as many as half a million pilgrims at the time of eclipses is 1,370 m long and 762 m wide (4,566 ft x 2,540 ft). It is said to be the largest tank in Asia.

New knowledge, skills and practices are extended to them by arousing their interest.

Classification of Extension-teaching Methods

(A) *According to use and nature of contact :*

1. *Individual contact*	2. *Group-contact*	3. *Mass community contact*
• Face to face (or) person to person contact between the rural and extension workers	• Group consists of 20-25 persons Groups are formed around common interest	• Large number of people are approached at shortest time to disseminate new information
• It is very effective in teaching new skills *Examples:* 1. Farm and Home visits 2. Office calls 3. Telephone calls 4. Personal letters	• It involves face-to-face exchanges and discussions *Examples :* 1. Method and Result demonstrations 2. National demonstration 3. Leader-training meetings 4. Conferences 5. Discussion meetings 6. Workshops 7. Field trips 8. Forum 9. Symposiums	• It is the quick method *Examples :* 1. Bulletins 2. Leaflets 3. Newspapers 4. Circulars 5. Radio 6. Televisions 7. Exhibitions 8. Fairs 9. Posters 10. Campaigns

(B) *According to form:*

1 Written	*2. Spoken*	*3. Objective or Visual*
1. Bulletins	1. General and Special meetings	1. Result demonstration
2. Leaflets	2. Farm and Home visits	2. Demonstration by posters
3. Folder	3. Official calls	3. Motion pictures (or) movies
4. News articles	4. Telephone calls	4. Charts
5. Personal letters	5. Radio	5. Slides
6. Circulars		6. Film strips
		7. Models

GK **World Earth Day is celebrated on April 22.** GS

		8. Exhibitions 9. Posters 10. Flannel graphs 11. Flash cards 12. Puppet shows 13. Photographs 14. Blackboards 15. Cultural programmes
Suitable for literates only	Suitable for both literates and illiterates	Suitable for both literates and illiterates

Observation Plot, Result Demonstration, Method demonstration (Comparison and Constrast)

Particulars	*Observation Plot*	*Result Demonstration*	*Method Demonstration*
1. Purpose	To try a new practice recommended by research workers with a view to observe its value, suitability or otherwise, in a given area under farmer's conditon	To show locally the worth or value of a recommended practice *i.e.* knowledge demonstration	To teach how to do job involving skill; (to teach 'doing' skills') *i.e.*, skill demonstration
2. Conducted by	Farmer (Co-operator) under close supervision of extension worker	Farmer (Demonstrator) under the guidance of extension worker	Extension worker himself or local leader specially trained for the purpose
3. For the benefit of	Extension worker to decide the suitabillity or otherwise of a new practice to a given locality	The demonstrator as well as other farmers	Person present at the demonstration
4.Comparison	Essential replications also necessary	Essential (not necessary to have replications in the same field)	Not essential
5. Mainte-nance of records	Absolutely necessary	Necessary	Not necessary
6. Time required	Substantial period	Substantial	Relatively very little
7. Cost	Costly	Costly	Relatively cheap
8. Inter-relation-ship	Usually precedes results demonstration	Usually follows observation plots; may involve one or more method demonstrations	Often paves the way for result demonstration

For effective use of extension, methods, the appropriate method/combination of methods should be selected. Farmers learn about new practices through several stages. These stages are (AIETA):

1. *Awareness stage*: Person comes to know about a new practice, but lacks the complete information.
2. *Interest stage*: He becomes interested in a new idea and wants to know more about it.
3. *Evaluation stage*: He mentally applies the new idea to his present situation and evalutes it.
4. *Trial stage*: He applies new ideas/practices on small scale to determine its utility under his condition.
5. *Adoption stage* : He decides to continue the full use of the practice.

Evaluation of Extension Work

In any organised system there is a need for stock-taking in respect of goals and objectives.

Purpose

- To facilitate effective decision-making
- To determine how well a desired behavioural changes have been taking place.

Types of Evaluation

1. *Casual everyday evaluation* :

Value judgements done everyday.

Limitations –

(i) Personal ideas used instead of standard measurements.

(ii) Institution and personal bias cannot be eliminated.

(iii) No systematic plan for arriving at conclusions.

2. *Self-checking evaluation* :

It makes conscious attempt to apply principles of evaluation e.g. checking the ordinary observations, talking with others, getting other people'a judgements.

3. *Do it yourself evaluation* :

Apply principles of evaluation and are more systematically done. These require surveys or score cards.

4. *Extension studies* :

More complex, use more scientific approach.

5. *Scientific approach* :

Experimental studies, scientifically carried out to determine cause and effect relationship. It must be -

Analytical : Analyse the relationship of various factors.

Objective : Free of bias.

Impartial : Approach with an open mind and spirit of enquiry.

Procedure for Conducting Evaluation

1. *Need for the evaluation*

(i) What extension project, problem, activity, job,

GK GS

Tom Matthews is the only player to get two hat tricks in a single Test match.

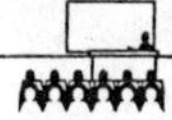

method or situation do you want to evaluate ?

(ii) Why evaluate it ?

(iii) How can the results of the evaluation be used in your extension work ?

2. *Purpose of the evaluation*

(i) An analysis of a situation.

(ii) What question should be answered by the evaluation.

3. *Questions to be answered by the evaluation*

(i) It analyses a situation, clarify the kind of information needed to answer the question.

(ii) If an evaluation of teaching, clarify the teaching objectives.

4. *Sources of information*

(i) People - farmers, local leaders, club members, extension workers etc.

(ii) Recorded information - reports, census etc.

5. *Collecting the information*

- How ? Interviews, mailed questionnaires observations etc.
- Whom ? person making the evaluation, extension personal, local leaders etc.

6. *Interpreting, reporting and applying the finding*

(i) Preparation of tables, charts and graphs

(ii) Summary of findings.

Advantages of Evaluation

1. *Results*

Evaluation helps us to determine the degree to which we are accomplishing what we set out to do.

(i) How effective extension work has been.

(ii) How much of the plan of work has been accomplished.

(iii) What improvements are desirable in plans for the year ahead.

2. *Programme Planning*

It gives a basis for adjusting the programme

(i) How to judge the effectiveness of programme planning procedure.

(ii) What modifications in the planning procedures are desirable.

(iii) How is the present programme contributing to the long-time programme.

3. *Extension Methods*

Evaluation serves as a check in teaching methods; which methods are most effective in what situations and with what people.

4. *Extension Organisation*

You Don't Know!

Earliest treaties on medicine

The *Atharva Veda* which belongs to the 12th century B.C.

(i) How can help the specialists be distributed more systematically and efficiently to the workers.

(ii) How can supervisors give more effective guidance to extension work.

5. *Reporting*

Evaluation provides a report to the public.

6. *Satisfaction*

Evaluation gives satisfaction to the local leaders and extension workers.

Training Programmes for Extension Workers

Definition

The process of training employees to gain effectiveness in their present or future work through the development of appropriate habits of thought and action, skills, knowledge and attitude knowledge

$$\text{Trainee} \xrightarrow[\text{belief and understanding}]{\text{knowledge skills, attitude, values}} \text{Farmers, Animal Breeders, Housewives}$$

Importance of Training

- Communication of the desired knowledge
- Knowledge of the techniques, job efficiency
- Syllabus of trainee should be according to the needs of Extension agents
- Teaching of manual skills

Types of Training

1. Pre-service training
2. In-service training

Pre-service Training

- It is the training the extension worker receives before joining the actual job.
- It includes education at school level and college level.
- The extension worker receive specific training at the village-level worker training centres.

In-service Training

It improves the ability of extension worker, keeps him abreast of the latest knowledge, gives him some special training in the new job he is required to do.

Steps

1. Oriented towards the organisational setup.

GK **The cabinet mission consisted of Lawrence, Cripps and Alexander.** GS

2. Induction training is given at the orientation centres where the extension workers are given the knowledge of working and organisation of community development and extension service.

3. Short-range courses organised by the state department of agriculture through state agriculture universities.

4. Periodical meetings and conferences are organized by state agriculture universities in collaboration with the state agriculture department for extension officers at district level or above.

Merits

- Training improves a person's skill and develops in him the desired attitude and values required for his work.
- Training helps to acquire occupational work skills and the latest knowledge.
- Makes him familiar with the objectives of the organization. It makes this contribution in promoting the goals of his organization.

Lab to Land Programme

The Lab-to-Land Programme (LLP) was launched by the ICAR in 1979 as a part of its Golden Jubilee celebration. The overall objective of the programme was to improve the economic condition of the small and marginal farmers and landless agricultural labourers, particularly scheduled castes and scheduled tribes, by transfer of improved technology developed by the agricultural universities, research institutes etc. The specific objectives of the lab-to-land programme.

1. Study and understand the background and resources of the selected farmers and landless agricultural labourers. To introduce low-cost relevant agricultural and allied technologies on their farms and homes for increasing their employment, production and income.

2. Assist the farmers to develop feasible farm plans keeping in view the availability of technologies, needs and resources of the farmers, and the resources which could be made available from external sources and agencies.

3. Guide and help the farmers in adopting improved technologies as per their farm plans, and demonstrate to them the economic viability of those technologies as well as methods of cultivation and farm management.

4. Organize training programmes and other extension activities, in relation to their adopted practices, and prepare them for active participation in agricultural development programmes of the State.

5. Make the farmers aware of the various opportunities and agencies which they could utilize to their economic advantage.

6. Develop functional relations and linkages with the scientists and institutions for future guidance, advisory services and help.

7. Utilize this project as a feedback and mechanism for the agricultural scientists and extension functionaries.

The programme was initiated with 75,000 farm families over the whole country. Major thrust of the programme was to introduce the most and appropriate technologies that would help in the diversification of labour use and introduction of supplementary sources of income. The programme had been in operation in a number of phases. Normally a phase ran for a period of two years with a particular set of farmers.

Training and Visit System

By the middle of 1970s it was felt that extension services in the developing countries were suffering from a number of weaknesses, including the dissipation of extension workers' energies on low priority tasks; the lack of a single, clear line of command; and a low level of agricultural knowledge and expertise among field level functionaries. As a means of reforming and strengthening the extension services, a reorganized system of agricultural extension, known as Training and Visit (T & V) system was introduced in India in 1974 with the World Bank assistance. It was presumed that transfer of technology through the 'contact farmers' shall benefit all farmers.

In two and a half decades, T & V became the dominant method of restructuring the extension services in over sixty countries in Asia, Africa and Latin America. The system tries to achieve changes in production technologies used by the majority of farmers through assistance from well-trained extension agents who have close links with agricultural research and supported by supply, service and marketing facilities.

T & V was regarded as an improved management system of agricultural extension and had the following *key features* -

1. *Professionalism*: Each extension agent is fully and continuously trained to handle one's particular responsibilities in a professional manner.

2. *Single line of command*: The extension service must be under a single line of technical and administrative command within the Ministry/Department of Agriculture.

3. *Concentration of effort*: All extension staff work only on agricultural extension. They are not responsible for any other activity not directly related to extension. In training sessions, attention is concentrated on important major points.

4. *Time-bound work*: Messages and skills be taught to farmers in a regular and timely fashion. The Village Extension Worker (VEW) must visit the farmers regularly on a fixed day, usually once each fortnight. All other extension staff must make timely and regular visits to the field. Recommendations for a specific area and for particular farming conditions for each two-fortnight

Abul Kalam Azad published the paper Al-Hilal.

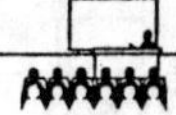

periods are discussed and learned by Subject Matter Specialists (SMSs) at regular monthly workshops; the recommendations are then presented to VEWs and Agricultural Extension Officers.

5. *Field and farmer orientation*: The contact with the farmers must be on a regular basis, on a schedule known to farmers, and with a large number of farmers representing all major farming and socio-economic types.

6. *Regular and continuous training*: Regular and continuous training of extension staff and discuss with them, and to prepare specific production recommendations required by farmers for the coming fortnights and to upgrade and update their professional skills.

Fortnightly training and monthly workshops, are the key means of bringing actual farmers' problems to the attention of research, of identifying research findings of immediate relevance to farmers.

7. *Linkages with research*: Problems faced by farmers that cannot be resolved by extension agents are passed on to researchers for either an immediate solution or investigation. Seasonal and monthly workshops, joint field visits, training of extension staff and formulation of production recommendations are some of the means by which linkages with research are maintained.

Organization of T & V

The T & V system has a hierarchical organization with one extension agent (VEW) for about 800 to 1000 farmers. The farmers are divided into eight groups of about equal size. Ten farmers in each group are selected as *'contact'* farmers. Extension message is to percolate to other farmers through these contact farmers. Eight VEWs are supervised by an AEO, and eight AEOs are supervised by a Sub-Divisional/District Extension Officer (SDEO/DEO), who are assisted by three to five Subject Matter Specialists (SMSs). Higher level supervision are provided by Zone Extension Officer (ZEO) and Director of Extension/Director of Agriculture of the State, with the assistance of SMSs.

OTHERS

Socio-economic survey and status of big, small and marginal farmers and landless agricultural labourers

Introduction

When planning rural development of a country, improving the socio-economic condition of the rural population is important.

In an agricultural-dominating economy, the landless labourers, marginal and small farmers are treated as target group, who should receive a high order of priority in the strategy of planning for rural development.

Agriculture is the dominant sector of the Indian economy.

- GDP contribution in 2.5%.
- Arable land is 143 in ha.

Knowledge World!

First panchayats

Panchayats were first established in villages under the Panchayat Act in 1912 but with little success until 1952 when a new Panchayat Act revived the system.

- Farm holding is about 89 million
- Average size of holding is 1.84 ha.
- The small and marginal farmers' holdings constitute around 75% of the total holdings in our country.
- Small farmers have increased due to subdivision of medium-sized holdings by their heirs.
- Marginal farmers have also grown in number when small farmers are shared by the heir.
- Large holdings have become medium-sized holdings due to the operation of land ceiling act under land reforms.
- According to National sample survey, 63.0% of the holdings are < 2 ha and they occupy nearly 19% of the total cultivated land.
- Agricultural work force of 299 million in India (1931) which has grown to nearly 74% of this work force is agricultural.

The landless labourer has no self-employment, they depend entirely on daily wages for their living. This category, constitutes, the core of the employment problem in the rural sector.

Characteristics :

Small farmers and marginal farmers

- Poor land use
- Surplus family labour
- Under nutrition
- Higher debt
- Distress sale of farm produce

Economic Condition :

COndition	*Small farmers*	*Landless labourers*
Work force	*65.31%*	*55.72%*
Household engaged in agriculture	*91.05%*	*98.90%*
Literacy	*37.45%*	*26.50%*
Bullock power	*13.43%*	*11.11%*
Own bullock carts	*2.2%*	*2.2%*
Average gross income	*Rs. 14,000*	*Rs. 3,400*
Per capita income	*Rs. 2,600*	*Rs. 765*
An average consumption	*Rs. 894*	*Rs. 623*

Subhash Chandra Bose wrote 'The Indian Struggle'.

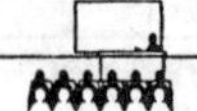

Social Status :

The status of small and marginal farmers varies only marginally and that of landless agricultural labourers is slightly less than the above 2 categories.

Extent of sub-division and fragmentation

Problems of small holdings :

1. Small size holdings are many.
2. Many owners have small pieces of plots, so there is difficulty to cultivate them together in an efficient manner.

Lack of capital surplus for investment :

Majority have less area (<1 ha), so they are cultivated and get only less yield, so this is insufficient for them to create capital surplus investment in land development.

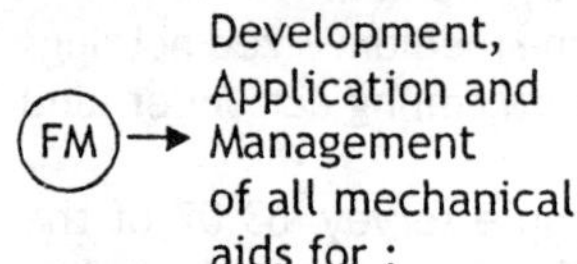

Status of Big farmers :

- They can mobilise surplus capital for investment in land development.
- More land under cultivation.
- Using modern equipment.
- Green revolution benefited the big farmers more than marginal and small farmers.

Farm Mechanisation

It is the application of engineering and technology in agricultural operations to do a job in a better way to increase productivity.

It is a need-based process which provides sufficient time gap for self-adjustment of various inputs without causing sudden impact of changes.

Scope :

1. Increased irrigation facility
2. Introduction of HYV and new crops
3. Introduction on high dose of fertilizers and pesticides.
4. Multi-cropping system and intensive cultivation.

Operational holdings in India

Size of holdings (ha)	*No. of holdings*		*Area of holdings*	
	1990-91	1980-81	1990-91	1980-81
Marginal (< 1 ha)	39.0	56.4	14.9	12.0
Small (1-2ha)	19.0	18.1	17.9	14.1
Small-medium (2-4ha)	13.2	14.0	23.2	21.2
Medium (4-10 ha)	7.2	9.1	27.2	29.6
Large (> 10ha)	1.6	2.4	17.4	23.0
	100.00	100.00	100.00	100.00

Benefits :

1. Timeliness of farm operation
2. Precision of operation
3. Enhancement of work and safety environment
4. Reduction of less crops and food products
5. Increased productivity of land
6. Increased economic return to farmer
7. Progress and prosperity in rural areas
9. To perform difficult operations which cannot be done by animal and manpower
10. Proper utilisation of inputs like water, seed, fertilizer etc

Limitations

1. Small land holdings (1.55 ha)
2. Low investing capacity of farmers
3. Agriculture labour is easily available
4. Adequate availability of draft animals
5. Lack of suitable farm machine for different operations
6. Lack of trained manpower
7. Lack of co-ordination between research organisation and manufacturer
8. Inadequate quality control.
9. Lack of repair and service facilities for machines.

Future Needs of Farm Mechanisation

1. To develop multifunctional machines.
2. Development of combination machinery like minimum tillage, zero tillage, till planter etc.
3. Better utilisation of solar, wind, energy and bio-fuels
4. Efficient fuel utilisation -through better design and matching machinery
5. Rain water harvesting and conservation
6. Low cost micro irrigation and sprinklers using surface water needs to be developed and commercialised

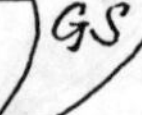

The author of 'The Clash of Civilizations' is Samuel Huntington.

Points to Ponder

- 'DIORAMA' is scenic representation of the original, with specimen, model, and painting.
- A decision is a choice among alternatives.
- A good extension programme should be flexible.
- A good extension service is a process of helping the people to help themselves.
- A new concept of budgeting is performance budgeting, the introduction of which has been recommended by the Administrative Reforms Commission.
- A prepared script should be divided into different parts depending upon the nature of the message to be conveyed.
- A screen is to provide the maximum possible visibility of the image with minimum glare or strain on the eyes.
- A single communication model consists of Source(S), Message(M), Channel (C), Receiver (R) and Effects of communication (E).
- A tape stored under normal room temperature and humidity is not erased, but holds sharp and clear quality for many years.
- Actions follows when desire, conviction and the prospect of satisfaction make it easier for the person to act rather than not to act.
- Administration refers to the guidance, leadership and control of efforts of a group of individuals towards some common goal.
- Aim of extension work is to bring about the changes towards betterment.
- All the rural listeners may or may not be educated in formal sense.
- An exhibition is a systematic display of models, specimens, charts, posters, information etc., in a sequence around a theme to create interest in the community.
- An extension worker is also learning, while giving the learning experience to the people.
- An extension worker should utilize the latent goodwill of the people in extension programmes.
- An innovation is an idea, practice or object perceived as new by an individual.
- Anthority in an organisation is the power in a position to exercise discretion in making decisions affecting others.
- Associations are organised for particular purposes.
- Attitude denotes the degree of positive or negative feelings associated with some psychological object towards which people differ in varying degrees.
- Basic education is designed to provide literacy, an elementary understanding of science and environment.
- Campaign denotes an intense educational activity for motivating and mobilizing a community to action, to solve a problem or satisfy a need urgently felt by it.

Knowledge World!

Oldest living language

Tamil is the oldest of all Indian languages and also the oldest living language in the world.

- Committee is a small group of person to who some matter is referred to mainly for detailed examination and decision.
- Contingency planning denotes planning for possible future environments which are not expected to occur, but which may occur; if this possible future is widely different from that premised, alternative premises and plans are required.
- Culture is defined as the socially standardized way of feeling.
- Drive is defined as a tendency initiated by shifts in the physiological balance in the body.
- Economic development is accorded priority in programmes of rural development.
- Education is the production of desirable changes is human behaviours.
- Education should be conceived as a life long process of learning.
- Efficiency-cum- performance audit is a technique of audit adopted to assess and evaluate the economy, efficiency and effectiveness of development projects in an organisation.
- Empathy is the capacity to see one self in other fellow's situation.
- Evolution should be in line with the overall purposes of the agency.
- Extension has to carry research findings to the rural people and establish their relevance in fulfilling their needs.
- Extension helps in adoption of innovations.
- Extension programme is a statement of situation, objectives, problems and solutions.
- Extension programmes should be people programmes with government aid.
- Extension work cannot thrive without trained specialists.
- Extension workers with superior knowledge and wisdom can help in solving problems of rural society and lead them on the voyage to modernization.
- Extention education is the education for rural people outside the regularly organised schools and class rooms to bring about social and cultural development.
- Facts are the foundation stones on which the community leaders and the problem committee build and carry

GK **Mountbatten's plan is also known as June 3rd Plan.** GS out their programmes.

- Films are one of the most effective means of arousing interest.
- Formal education is highly institutionalized, chronologically graded and hierarchically structured to being with primary school upto university education.
- Friends and neighbours are first in importance in the evaluation, trials and adoption stages of innovation(s).
- Heterophily is the degree to which pairs of individuals who interact are different in certain attributes.
- Home is the fundamental unit of civilization, family is the first training group of human race.
- Homophily is the degree to which pairs of individuals who interact are similar in certain attributes, such as beliefs, values, education, social status and the like.
- If a child of 8 can do all tests meant for the 10 years old then his I.Q. will be 125.
- Imagination will put new glitter on old things and will lead to seek new and better view points.
- In extension, educational psychology change agents are more concerned with the adult or grown up people.
- In presentation of a talk just talk to the people and do not speak naturally.
- In some cases use of progressive leader farmers is made to produce the multiplier effect.
- In the earlier days rural development was considered synonymous with community development.
- In the sound track, sound is photographed on the photographic film.
- Innovativeness is the degree to which an individual is relatively earlier in adoption of new ideas than others of his social system.
- Intelligence quotient is equal to mental age divided by chronological age and multiplied by 100.
- It is necessary to base the extension education programmes on the psychology of the rural people to make them effective.
- Judgment is an assignment of value to alternatives.
- Leading is the process of influencing people so that they will strive willingly and enthusiastically toward attainment of the organisational goals.
- Learning calls for effective communication.
- Learning is effected by physical and social environment.
- Literature is the basic of any teaching programme.
- Motivation is energy mobilization towards the attainment of goals.
- Need and interest based programmes yield better results than the imposed ones.
- Neglect of one member of a family by extension worker may lead to rejection of an innovation.
- Next to camera and film the most important ingredient for picture-taking is a good imagination.
- Non-formal education denotes any organised, systematic educational activity carried outside the framework of the formal system.
- Objectives present the directions of movement, while a goal is the distance in a given direction one expects to go during a given period of time.
- Personality is the dynamic organisation within the individual of those pshycological systems that determine the unique adjustments to his environments.
- Philosophy is a body of general principles/laws as a field of knowledge.
- Plan of work comprises a definite-outline of procedure for solving the different problems of an extension programme.
- POSDCORB denotes planning, organisation, staffing, directing, co-ordination, reporting and budgeting.
- Psychology motive refers to some internal activator within an individual.
- Radio belongs to the spoken word means of communication.
- Reference group is a group of persons whom an individual consults before taking an important decision.
- Results of programme evaluation/assessment greatly help in improving the quality of programme(s).
- Rural development is the result of many interacting forces and education is one of them.
- Rural development programmes should be form to meet short term change.
- Rural society is more homogeneous in nature, relatively independent and with a low degree of social differentiation.
- Rural sociology is the systematic study of people living in rural areas and who are living by or are immediately dependent on agriculture.
- Socialization is the process of inducting the individual into the social and cultural world to play his role as a member of society.
- Spontaneity and creativity are important parts of brain storming.
- Successful teaching contributes to effective learning.
- The ability and capability of an individual should be brought out and sharpened by an extention worker.
- The advantage of the over head projector is that a teacher can always face the class.
- The basic personal skills needed for effective administration are technical, human and conceptual.
- The basic philosophy of extension is directed towards changing the outlook of persons by educating them.
- The education designed to develop a specific knowledge

and the skills associated with economic activities beneficial to make respectable living is called occupational education.

- The extension work must be based on the need and interest of people.
- The extension worker should understand the basic wants and incentives of the people with he/she has to work.
- The extension workers should have faith in democratic values and should try to educate people.
- The family is the primary institution of society.
- The first principle of extension service rests on, "citizen is sovereign in the democracy".
- The fundamental objective of extension education is development of people.
- The leadership qualities are developed in rural people, if they participate in extension programmes.
- The main activities of Sevagram Attempt comprised organization of training centres for cottage industries, prohibition and removal of untouchability.
- The means by which sound is reproduced from a sound motion picture film is practically the reverse of the way it is recorded.
- The members of a secondary group has little personal affection and their relationship is governed by the fulfilment of some objective.
- The participation of the people in the programme planning process normally gives them a proprietary interest in seeing that an action programme
- The persons in the primary group have face-to-face relationship, such as a family and play group.
- The rural people should voluntarily participate in the extension work.
- The Sevagram Attempt was started under the guidance of Mahatma Gandhi in 1920.
- The success of the extension work lies in satisfaction/ participation of the people.
- The technical messages are carefully selected and are relayed during the visit to the farmers.
- The training and visit system comprises an excellent example of the training approach.
- The training approach emphasizes more systematic and deeper learning of specific basic skills and related knowledge.
- Working objectives are the specific items which are to be achieved.
- Farmers training centres imparts training to farmers, farmers leader, rural youths and farm women.
- I.V.L.P. is basically a linkage programme.
- N.A.T.P. stands for National Agriculture Technology Project.
- In extension programme people's participation is based on self help and self confidence.
- The need of training is to develop knowledge, skill and attitude.
- The principle of 'One village, One society' was suggested by Mclang Committee.
- DASP stands for Diversified Agriculture Support Project.
- 'Kisan Bharti' periodical is published from Pant Nagar.
- National Academy of Agriculture Research Management is located at Hyderabad.
- Khetihar Mazdoor Bima Yojna was started on 2001-02.
- Objective of Antyodaya Yojna (1977-78) is to make the poorest families of the village economically independent.
- Sampurna Gramin Rojgar Yojna (2001) provides employment and food security.
- NARD stands for National Fund for Rural Development.
- Method of mass communication are Newspaper, Television and Radio.
- Exhibition, Radio and Literature are 'Institutional' communication methods.

GK GS

Sugar test is used for the diagnosis of Diabetes.

Multiple Choice Questions

1. ________ concentrates on a rational structure with universal applications in social institutions.
   a] Bureacracy b] Autocracy
   c] Anarchy d] Monarchy
2. ________ is the degree of positive or negative effect associated with some psychological object.
   a] Attitude b] Skill
   c] Behaviour d] Knowledge
3. ________ answers four W's and one H.
   a] Calender of work b] Programme
   c] Plan of work d] Programme Planning
4. ________ are the devices in which communication take place between the instructior and the learner.
   a] Teaching aids
   b] Teaching methods
   c] Learning situation
   d] Learning theories
5. ________ are the expression of ends towards which our efforts are directed.
   a] Needs b] Principles
   c] Objectives d] Motivations
6. ________ is a statement of situation objective, problems and solution.
   a] Programme b] Plans of work
   c] Project d] Calender of work
7. ________ is the gap between what is and what ought to be.
   a] Knowledge b] Need
   c] Society d] Community
8. "Cone of Experience" devised by :
   a] Rogers
   b] A.A. Reddy
   c] O.P. Dahama and O.P. Bhatnagar
   d] Edger Dale
9. "Panchayati Raj" was started in the year :
   a] 1952 b] 1955
   c] 1953 d] 1959
10. A ________ mm camera is necessary for making slides that fit most of the projectors.
    a] 25 b] 35
    c] 45 d] 70
11. A block level rural development team consisting eight extension officers, 10-20 gram sevaks and 2 gram sevikas is headed by a :
    a] Sarpanch
    b] Panchayat Samiti President
    c] B.D.O.
    d] None of the above
12. A community development block passes through :
    a] 2 stages of development
    b] 1 stage of development
    c] 4 stages of development
    d] None of the above
13. A good extension programme should be :
    a] Flexible
    b] Rigid
    c] Both of the above
    d] None of the above
14. A good extension programme should be :
    a] Working for the people
    b] Helping the people to help themselves
    c] Giving a self made solution for progress
    d] All of the above
15. A, B, C's of journalism stands for :
    a] Attractive, brief and clear
    b] Accuracy, brief and clarity
    c] Accuracy, brevity and clarity
    d] None of the above
16. After a suitable extension teaching, the villagers attain desirable changes in his :
    a] Behaviour b] Health
    c] Habits d] All of them
17. Agricultural Extension deals with:
    a] Farmers
    b] Change agent
    c] Farmers and change agent
    d] None of the above
18. Aim of extension work is to bring about the changes:
    a] Towards betterment
    b] In the heart of the peoples
    c] In urban areas
    d] In the rural areas
19. Albert Mayer of U.S.A. started a project in September 1948 was :
    a] Grow more food campaign
    b] Marthandam Project
    c] Etawah Pilot Project
    d] Gurgaon experiment
20. Among the different types of motion picture projectors, ________ mm projector is most suitable for class room teaching.
    a] 16 b] 70
    c] 8 d] 20
21. An extension education :
    a] Should develop the programmes on the directions of gram pradhan
    b] Should develop the programmes on the directions of local leaders
    c] Develops programmes locally according to the need and desire of farmers
    d] All of the above
22. An extension educator's job is :
    a] To educate the people
    b] To advice the rural people
    c] To train the rural worker for agriculture
    d] All of the above

GK **Trial of Nanda Kumar took place during Warren Hasting's time.** GS

23. **An extension teacher helps farmers to attain desirabal change in his :**
a] Temperament b] Behaviour
c] Health d] All of the above

24. **An extension worker, while extending scientific knowledge to the farmers, starts with :**
a] Lectures
b] Practical demonstrations
c] Distribution of leaflets
d] All of the above

25. **At which of the following place, R.N. Tagore started his work for rural welfare?**
a] Shantiniketan b] Pune
c] Sevagram d] Bhatinda

26. **Balwant Rai Mehta Team stimulated an active consideration of ________ through democratic bodies.**
a] Centralization b] Mobilization
c] Stabilization d] Decentralization

27. **Based on the recommendations of the Grow More Food Enquiry Committee, ________ Community projects were taken up in 1952.**
a] 25 b] 35
c] 45 d] 55

28. **Basic operational unit for rural development in India is :**
a] Village b] District
c] Tahsil d] Block

29. **Block Samiti's :**
a] Guided by V.L.W.
b] Guided by Panchayat Samiti
c] Guided by Zila Parishad
d] Assisted by B.D.O. and his extension officers

30. **Bulletins, Leaflets, Circular letters, Posters, Fairs, Exhibitions, Radio and Television are the media of :**
a] Mass communication
b] Group communication
c] Public communication
d] Official communication

31. **Caste is a class gained by :**
a] Status b] Birth
c] Power d] Social relations

32. **Community Development Programmes were introduced in India for rural development :**
a] In early 40's b] In early 50's
c] In early 60's d] In early 70's

33. **Cottage industries play a significant role in the development of rural areas because :**
a] They are capable of providing full-time and part-time employment to the rural people
b] They help the rural people to supplement their income
c] They are specifically suitable for the rural areas
d] All the above statements are true

Quiz Kit!
First film shot in India
Coconut Fair was shot in India is 1897 by an unknown foreign camera man.

34. **Democratic Decentralization was recommended by a committee headed by :**
a] M.L. Mehta b] B.R. Metha
c] Spencer Hatch d] Mahatma Gandhi

35. **Desert Development Programme was started in :**
a] 1977-78 b] 1980-81
c] 1978-79 d] 1981-82

36. **Determination of Extension Programme consists of :**
a] Collection and analysis of data's determination of needs and objectives
b] Definition and finding solutions
c] Preparing plan of work and carry out the plan of work
d] Both a and b

37. **District Development Committee (D.D.C.) is presided over by :**
a] B.D.O.
b] District collector
c] Director of extension
d] None of these

38. **Efforts to revive panchayats in India were made in ________ by introducing Village Panchayat Act in several states.**
a] 1945 b] 1935
c] 1925 d] 1915

39. **Elements of communication are :**
a] Teacher, attention, learner and feeback
b] Message, channel and feed back
c] Source, message, channel, receiver and effect
d] None of above

40. **Execution of a plan assumed that a programme is planned, and a plan of work is:**
a] To be thought of b] To be developed
c] Already developed d] Being modified

41. **Extension Education is :**
a] An educational process
b] One way channel
c] Formal education
d] None of the above

42. **Extension education is ________ process.**
a] Only teaching
b] Only learning
c] Teaching and learning
d] None of these

GK **Lord Wellesley described himself as Bengal Tiger.** GS

43. **Extension of recent agricultural knowledge to rural peoples through trained persons of the same locality is called the principle of :**
a] Local leadership
b] Panchayati Raj
c] Centralised power
d] None of them

44. **Extension workers are :**
a] Lay leaders
b] Professsional leaders
c] Autocratic leaders
d] None of the above

45. **Family is included under :**
a] Primary group b] Secondary group
c] Pertaining group d] None of the above

46. **Farmers study problems, participation is purely voluntary, teaches and also learns from the farmers, teaching is also though local leaders, teaching if flexiable in comes under :**
a] Extension education
b] Formal education
c] Informal education
d] None of the above

47. **For demonstrating the value of a new extensions programme to rural peoples of the locality, the programme should be sponsored by a group of :**
a] Rural people in the locality
b] Urban people
c] Local leader
d] Gram pradhans and V.L.W.'s

48. **For providing inputs like quality seeds fertilizers and pesticides, the agency that is present at the Village Panchayat Samiti level is :**
a] NABARD
b] Nationalised banks
c] Co-operative society
d] Insurance companies

49. **For providing the precise and reliable scientific information in simple language the extension system used is :**
a] Demonstration b] Leaflets
c] Charts d] None of them

50. **For rural development, a conducive ecology and environment is essential because :**
a] Rural life styles have close links with nature and its resources
b] Rural houses are mostly an hygienic
c] Rural people do not care for ecology
d] None of the above

51. **Formations of clique ____________ an organisation.**
a] Damages b] Improves
c] Stabilizes d] None of these

52. **Fourth five year plan was started in the year :**
a] 1965 b] 1966
c] 1969 d] 1967

53. Functions of pre-extension stage block was :
a] To conduct agricultural demonstrations
b] To survey block area
c] To prepare ground for intensive stage
d] All of the above

54. **Gurgaon Attempt was organised by F.L. Brayne, Collector of Gurgaon district in Punjab in:**
a] 1915 b] 1920
c] 1925 d] 1930

55. **I.C.A.R. was established in :**
a] 1904 b] 1937
c] 1929 d] 1934

56. **IADP implementation should be so phased that in the first year 20 per cent and by the end of fifth year ____________ per cent area is covered.**
a] 85 b] 75
c] 65 d] 55

57. **If a new practice represents a real major improvement over the existing one, the content of such an innovation embodies:**
a] Compatibility b] Visibility
c] Utility d] Complexity

58. **If we want to impress the peoples about a particular recommended practice suitable for a specific area, then effective extension method is:**
a] Demonstration b] T.V.
c] Pamphlets d] Radio

59. **Implementation of Community Development (CD) and National Extension Service (NES) was brought to sharp focus by the report of the team for the Study of Community Projects and NES by ____________ in 1957.**
a] Balwant Rai G. Mehta
b] Bhalerao N. Khuspe
c] Dadajibhai R. Mehta
d] Chinnani M. Bathkal

60. **Important element of evaluation process is :**
a] Observations, collection of some information and final decision
b] Observations or collection and criteria to observations
c] Observations or collection of some information, applying some standards
d] Observations, criteria to observations and making some decisions

61. **In a normal community development block, the number of V.L.W's is :**
a] 10 b] 18
c] 25 d] 5

62. **In extension we prefer :**
a] Automatic approach
b] Democratic approach
c] Lessaiz faire approach
d] None of the above

GK GS
International Labour Organisation (ILO) was constituted in 1919.

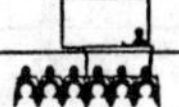

63. In extension we require :
a] Caste leader
b] Relegious leader
c] Lay leader
d] Local potential functional leader

64. In extension we want to bring about the desirable changes through :
a] Education b] Compulsion
c] Dictatorship d] None of the above

65. In India ________ of the population lives in villages.
a] Half b] One-fourth
c] One-fifth d] Two-thirds

66. In India there are about ________ village panchayats.
a] 1,15,000 b] 2,16,000
c] 5,00,000 d] 75,7500

67. In India, Community Development Projects were launched in :
a] 1955 b] 1952
c] 1967 d] 1947

68. In India, slightly less than half the villages are inhabited by less than ____________ persons.
a] 200 b] 300
c] 400 d] 500

69. In India, the development of poorers from all the sides is possible only by:
a] Extension education
b] More money supply
c] Making strong rules
d] Not known

70. In India, the use of television as a medium of communication for rural development was started in:
a] National interest
b] Local interest
c] All of the above
d] None of the above

71. In India, the use of television as a medium of communication or rural development was started in:
a] 1967 b] 1961
c] 1962 d] 1958

72. In later period, the name of National Extension Service is converted into:
a] Rural Development Block
b] Community Development Block
c] Village Development Block
d] Not changed

73. In NREP the share of centre and states will be :
a] 50:50 b] 40:60
c] 75:25 d] 80:20

74. In Panchayat Raj System, the responsibility of planning, co-ordination and consolidation of the development programme lies on :
a] Zila Parishad
b] Panchayat Samiti
c] State Development Committee
d] None of the above

75. In which of the following year, the planning commission came into existence in India ?
a] 1952 b] 1948
c] 1945 d] 1950

76. Incharge of an 'Intensive Agricultural District Programme' in district is :
a] Project Officer
b] Zila Parishad Chairman
c] District Collector
d] None of the above

77. Integrated Rural Development Programme (IRDP) has been conceived :
a] Essentially to develop adult education facilities in the rural areas
b] Essentially as an anti-poverty programme
c] Essentially to develop an anti-hoarding programme
d] Essentially to develop fisheries in the area

78. Intergrated Rural Energy System are significant :
a] For supplementing energies available in the rural areas
b] For meeting the energy needs of the dispersed rural population
c] For producing more and more gobar-gas for fuel purpose
d] All of the above

79. Krishi Vigyan Kendra's are functioning through the financial assistance of :
a] State Government b] UNICEF
c] ICAR d] IARI

80. Krishi Vigyan Kendra's are sponsored by :
a] IARI b] ICAR
c] UGC d] State Government

81. LLP was launched by the ICAR in 1979 as a part of its ____________ Jubilee Celebrations.
a] Silver b] Golden
c] Diamond d] Platinum

82. Natioanl Commission on Agriculture had gone through the aspects and submitted its report in:
a] 1974 b] 1976
c] 1981 d] 1983

83. Natioanl Extension Services in India was established in :
a] 1950 b] 1952
c] 1951 d] 1953

84. National Agricultural Extension Project was launched by :
a] State government b] Government of India
c] ICAR d] Agricultural universities

85. National demonstrations are 'first line' demonstrations conducted on the farmers field by :
a] Research scientists
b] Extension scientists
c] Development staff
d] Extension officers

GK GS

Prevention of Terrorism Act (POTA) first came into existence as an ordinance in 1995.

86. National Demonstrations were started in :
a] 1962 b] 1965
c] 1978 d] 1950

87. NERP was started in:
a] 1975 b] 1977
c] 1979 d] 1981

88. ORPs were initiated in:
a] 1970-71 b] 1972-73
c] 1974-75 d] 1975-76

89. Over-all responsibility of rural development programmes at district level goes on :
a] District Planning Officer
b] District Panchayat Raj Officer
c] District Collector
d] None of the above

90. Programme Evaluation Organization (PEO) was set up in 1952 as an independent organisation under general guidance and direction of:
a] Reserve Bank
b] Ministry of Agriculture
c] Finance Commission
d] Planning Commission

91. Puppet show in villages as a method for transfer of technology is an example of :
a] Mass media
b] Group media
c] Individualized media
d] Tribal media

92. Radio is a media for :
a] Official communication
b] Mass
c] Both of the above
d] It is not a media

93. Radio is :
a] Audio device b] Visual device
c] Both these d] None of above

94. Recenlty the government has changed the name of V.L.W. to :
a] V.L.O. b] C.D.O.
c] V.D.O. d] None of above

95. Rural community has :
a] Less dense population
b] More dense population
c] Moderately dense population
d] Nothing can be predicted

96. Rural development is directly linked with agricultural produciton, specially crop production. The main strategy for crop production during:
a] A steady growth of food-grains production
b] To aim at self-sufficiency in oil seed production to eliminate import of oil seeds.
c] To increase production of export oriented crops
d] All the above

97. Rural development programme should be formed to meet :
a] Short term changes
b] Emergent situation
c] Long term changes
d] All of the above

98. Rural Landless Employment Guarantee Scheme was started in the year :
a] 1978 b] 1983
c] 1984 d] 1985

99. Rural population is about ________ of the world population and lives mainly in developing countries.
a] One-third b] One-fourth
c] One-fifth d] Two-third

100. Rural Reconstruction Institute was established at Shriniketan in the year :
a] 1922 b] 1931
c] 1908 d] 1921

101. Rural sociology is expected to develop greater understanding of the behaviour of ________ people and society.
a] Urban b] Rural
c] Both of these d] None of these

102. Single line of command is one of the special features of :
a] Community development
b] National demonstration
c] T & V system
d] Krishi Gyan Kendra

103. Slightly less than three-fourth Indian villages have a human population of less than:
a] 1,000 b] 900
c] 800 d] 700

104. Special Livestock Production Programme was initiated in :
a] 1974-75 b] 1979-80
c] 1978-79 d] 1975-76

105. Staffing denotes ________________ maintaining and developing personnel in position to fulfill the organisational objectives.
a] Rejecting b] Eliminating
c] Selecting d] Upgrading

106. Steps of extension teaching are :
a] Attention, interest, desire, conviction, action and satisfaction
b] Awareness, interest, trial and evaluation
c] Source, message and channel
d] None of them

107. T&V system of agricultural extension was first introduced in the year :
a] 1964 b] 1974
c] 1977 d] 1954

GK **Lord Mayo was the governor general who was murdered.** GS

108. T&V system was proposed by :
a] Dr. Banerjee b] A.A. Reddy
c] Dr. Daniel Benor d] Kelsey and Hearney

109. T&V includes :
a] Block Samiti b] Zila Samiti
c] Panchayat Samiti d] Block Develop Advisory Committee

110. Tape recorder employes :
a] Audio device
b] Visual device
c] Audio and visual device
d] Not known, it is complicated item

111. Technical guidance for development programme at district level is provided by :
a] Head of technical development
b] District planning officer
c] District collector
d] All of the above

112. Television is a medium of mass communication employing :
a] Audio device b] Visual device
c] Both a and b d] None of the above

113. The _________ situation is the bench mark from where people should start the process of programme planning.
a] Regional b] Local
c] National d] None of these

114. The all-round development of Indian rural peoples can be achieved by means of :
a] Extension education
b] More money distribution
c] Adult education
d] None of the above

115. The authority responsbile for planning and implementation of rural development programme at state level is :
a] Planning commission
b] Agriculture Minister
c] State Development Committee
d] None of them

116. The basic guide, philosopher and friend in rural development is :
a] B.D.O.
b] V.D.O.
c] Pradhan
d] Pramukh

117. The basic unit of development under the Integrated Rural Development Programme is a :
a] District
b] Community development block
c] Village
d] Family

Did You Know ?
Largest bus station
Asia's largest bus station, the Imlibun bus terminus in Hyderabad is spread over 20 acres and has 49 platform for incoming and outgoing buses.

118. The basic village institutions are :
a] Family, Mahila Mandal and School
b] Panchayat, Hospital and School
c] Youth club, School and Co- operative
d] Panchayat, Co-operative and School

119. The benefit of first hand knowledge, good will and creating confidence can be developed by :
a] Organising village fairs
b] Visiting homes and farms
c] Telephone
d] None of the above

120. The C.D. Projects were extended to whole country in :
a] 1971 b] 1955
c] 1982 d] 1963

121. The Comilla Project approach was first used by the Academy of Rural Development at Comilla, in:
a] Pakistan b] Bangladesh
c] Mauritius d] Nepal

122. The development approach of National Extension Service Block was based mainly on:
a] Government help
b] Help from village panchayat
c] Self-help
d] None of the above

123. The development plans of a district and co-ordination of activities of panchayat samiti is the responsibility of :
a] Gram Pradhan b] Pramukh
c] Zila Parishad d] B.D.O.

124. The elements of behaviour are :
a] Knowledge and Education
b] Skill and Attitude
c] Attitude and Aspiration
d] Knowledge, Attitude and Skill

125. The elements of teaching-learning process are :
a] Teacher, subject matter and learner
b] Teaching equipments and physical facilities
c] All of the above
d] None of them

126. The execution of plan should be carried out :
a] Without change
b] With change
c] After a laps of time
d] In presence of leader

GK GS
The genome of chicken has been completely sequenced.

127. The extension practice varying in operation to varying extents and the results are showy is named as:
a] Complexity b] Visibility
c] Divisibility d] Utility

128. The extension work must be based on :
a] The need and interest of the peoples
b] The interest of gram pradhan
c] The interest of B.D.O.
d] None of them

129. The farmer, who adopts the new technology just after hearing about it, is called :
a] Early adopter b] Innovator
c] Awareness d] Laggards

130. The first Agricultural University of India is :
a] P.A.U. Ludhiana (Punjab)
b] H.A.U. Hissar (Haryana)
c] A.P.A.U. Hyderabad (A.P.)
d] G.B.P.U.A. and T. Pantnagar (Uttaranchal)

131. The first Agricultural University of India was established in :
a] 1960 b] 1995
c] 1962 d] 1965

132. The first five year plan was started in :
a] 1950 b] 1951
c] 1947 d] 1952

133. The first KVK was established in 1974 at:
a] Nagpur b] Nilokheri
c] Ludhiana d] Pondicherry

134. The first major efforts of ICAR to upgrade and strengthen the Regional Research Capability of the State Agricultural Universities is :
a] NAFP b] NREP
c] NARP d] None of above

135. The first step in Extension Programme Planning at local level is:
a] Determining the objectives
b] Analysis of the existing situation
c] Deciding the extension teaching methods to be used
d] Determining the targets

136. The fruits of research and development findings could reach the poorest of the poor through :
a] Research personnel themselves
b] NGO's
c] A strong and devoted extension agency
d] Agricultural universities and college

137. The fundamental objective of extension education is :
a] Modernization
b] Development of the people
c] Development of the pradhans
d] All of the above

138. The Government of India reorga-nised the ICAR and charged with the responsibility of coordina-ting, directing and promoting agricultural research and education in India in:
a] 1962 b] 1964
c] 1966 d] 1968

139. The Government of India set up Planning Commission in:
a] 1949 b] 1950
c] 1951 d] 1952

140. The government sponsored Firka Development Scheme of __________ was launched in 1966.
a] Bombay state
b] Madras state
c] Punjab state
d] U.P. state

141. The Gurgaon experiment was started by :
a] Shri Ravindra Nath Tagore
b] Mr. F.L. Brayne
c] Albert Mayer
d] Mahatma Gandhi

142. The head of National Development Council is :
a] Agriculture Minister
b] Prime Minister
c] Home Minister
d] None of them

143. The HYVP strteded in the year :
a] 1964 b] 1969
c] 1966 d] 1976

144. The idea of having a village guide in each village for introducing new skills among the rural people was introduced by :
a] R.N. Tagore b] B.P. Pant
c] Mahatma Gandhi d] F.L. Brayne

145. The immediate officer who supervises the VEWs working under T &V system is known as :
a] DEO b] SDEO
c] AEO d] ZEO

146. The Institute of Development Studies, University of Sussex, ________ developed the Farmer-First (FF) method in late 1980s.
a] U.K. b] U.N.
c] U.S.A. d] E. Germany

147. The Key Village Scheme was initiated in ______ 1952 comprising first systematic attempt to improve the quality and productivity of cattle and buffaloes in India.
a] June b] July
c] August d] January

148. The leaflet, is a single sheet of paper folded usually to make ____________ page piece of printed paper.
a] Two b] Four
c] Six d] Eight

GS

Abdul Gaffar Khan founded the Khudai Khidmatgars.

149. The main purpose of TV programmes is to :
   a] Identify research problems
   b] Bring about co-ordination in the administration of agricultural extension
   c] Update the technical knowledge and skills of agricultural extension workers
   d] Facilitiate farmers in the adoption of latest innovations in agriculture

150. The method of extension education for effectively impressing the people about a particular recommended practice suitable for a particular area is :
   a] Demonstration
   b] Radio
   c] Tours
   d] Charts

151. The methodology for accelerated rural development includes :
   a] Increasing the productivity and volume of production in agriculture
   b] Promoting marketing support for rural areas
   c] Skill formation and skill upgrading for the benefit of rural poor
   d] All of the above

152. The motion picture camera, actually a device for recording a series of still pictures at the rate of ____________ exposures per second.
   a] 12 b] 24
   c] 36 d] 48

153. The national level body for policy formulation of rural development programme in our country is :
   a] National Development Council
   b] Planning Commission
   c] National Development Committee
   d] Government of India

154. The National Rural Employment Programme (NREP) was replaced by Food and Work Programme during :
   a] 1977 b] 1982
   c] 1980 d] 1979

155. The new name of Community Development Programme is :
   a] Intensive Agriculture Programme
   b] Rural Development Programme
   c] Command Area Development Programme
   d] None of the above

156. The number of V.L.W's in an I.A.D.P. block is :
   a] 8 b] 5
   c] 20 d] 2

157. The objectives of education are decided by:
   a] Biological Science
   b] Physical Science
   c] Normatic Science
   d] Social Science

158. The period of pre-extension stage of a community development block was:
   a] 1 year b] 4 years
   c] 2 years d] None of above

159. The period of stage II of a community development block was :
   a] 10 years b] 5 years
   c] 7 years d] 2 years

160. The pilot project for rural development was started in :
   a] 1971 b] 1978
   c] 1952 d] 1948

161. The president of State Development Committee is :
   a] Agriculture Minister
   b] Governor
   c] Chief Minister
   d] None of them

162. The programme for transfer of technology started by ICAR at the golden jublee year occasion was :
   a] National Demonstration
   b] Operational Research Project
   c] Lab to Land
   d] National Agriculture Research Project

163. The programmes taken up in the village and small industries sector are :
   a] Project oriented
   b] Capital intensive
   c] Mainly promotional in nature
   d] None of the above

164. The programmes which have been merged under IRDP were :
   a] DRDA, MFAL and T&V
   b] SFDA, MFAL and Lab to Land
   c] SFDA, MFAL, DPAP and CADP
   d] MFAL, SFDA, IAAP and IADP

165. The purpose of introducing agricultural universities in India is to improve the standard of Agriculture :
   a] Extension
   b] Extension and Research
   c] Teaching
   d] Extension, Teaching and Research

166. The responsibility of development work at village level is of :
   a] V.L.W. b] Gram Pradhan
   c] B.D.O. d] Local Leaders

167. The role of an extension worker in extension education is to create an effective :
   a] Planning situation b] Learning situation
   c] Social atmosphere d] Awareness

168. The Rural Systems Research idea was motivated by ____________ in 1988.
   a] David E. Benor
   b] Ralph W. Cummings
   c] Guy B. Baird
   d] M.S. Swaminathan

GK GS

New species of monkey, *Munzla* has been discovered in Arunachal Pradesh recently.

169. The strategy for supporting the rapidly expanding agricultural programmes requires :
a] That cultivators are educated in favour of modern farming
b] That efforts are made to manufacture more fertilizers
c] That efforts are made to provide more irrigational facilities
d] All of the above

170. The term Community Development (CD) appears to have originated from Cambridge in England in:
a] 1950 b] 1948
c] 1946 d] 1944

171. The two important principles of extension education are :
a] Participation and leadership
b] Participation and classroom teaching
c] Leadership and classroom teaching
d] Leadership and subject-matter knowledge

172. The University Education Commission was headed by ______, recommen-ded the establishment of 'Rural Universities'.
a] Dr. Rajendra Prasad
b] Dr. S. Kothari
c] Dr. S. Radhakrishnan
d] Shri C. Rajgopalacharee

173. The Village Panchayat Act came into being in:
a] 1956 b] 1958
c] 1960 d] 1954

174. The villagers accept the new idea of development with least resistance by the local leaders. It is because of:
a] People are afraid of local leaders
b] People have more faith in local leaders
c] Local leaders gave money to the peoples
d] All of the above

175. There are __________ development block in India.
a] 2,000 b] 3,000
c] 4,000 d] 5,000

176. There are ______________ degree of evaluation.
a] Two b] Three
c] Four d] Five

177. To create interest in a wider range of people and to influence people to adopt better practices, the extension method generally used is:
a] Fair b] Field trails
c] Competition d] Exhibitions

178. To induce emotional participation of the local community and to create a conducive psycological climate for adoption of a new practice, the method of extension generally applied is :
a] Campaign b] Competition
c] Exhibition d] All of the above

179. Which of the following is not correctly matched ?
a] Gurgaon Project - 1928
b] Sewagram Project - 1920
c] Sriniketan Project - 1920.
d] Marthandom Project - 1928

180. To prove the worth of a new practice suited to a particular area, the best method of extension education is :
a] Demonstration b] Field trials
c] Farm visits d] Result demonstration

181. To which one of the following categories the discipline of extension education belong?
a] Pure science
b] Applied biological science
c] Applied behavioural science
d] Art

182. Training of Rural Youths for Self Employment (TRYSEM) was initiated in August :
a] 1979 b] 1977
c] 1978 d] 1980

183. Two Joint Indo-American Teams on Agricultural Research and Education recommended setting up of Agricultural Universities in the states on the pattern of Land Grant Colleges of:
a] Germany b] France
c] U.K. d] U.S.A.

184. Under NREP, the state government has its share :
a] 50% b] 20%
c] 75% d] 40%

185. Voluntary agencies are characterized by ______________ among the community.
a] Greater continuity
b] Less acceptability
c] Less commitment
d] Less intensity

186. What does the term DPAP denote ?
a] Development Project Area programme
b] Dry Project Area Programme
c] Drought Prone Area Programme
d] None of the above

187. What is the approximate total number of villages in India ?
a] 5.5 lakhs b] 8.1 lakhs
c] 7.0 lakhs d] None of the above

188. When an extension programme is sucessfully completed and the farmer and his family members are satisfied by solving the problems and getting a new idea it is called :
a] The principle of achievement
b] The principle of completion
c] The principle of maximum response
d] The principle of satisfaction

GK

Sandhurst Committee was appointed under Lord Reading.

GS

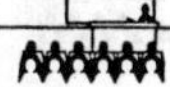

189. When an extension worker approaches large number of people, spread over large areas for disseminating new improved production technology it is called:
   a] Group communication
   b] Public communication
   c] Mass communication
   d] Unidirectional communication
190. When the villagers cooperate with the extension officers at different levels in implementing the new programmes it is known as the principle of :
   a] Satisfcation b] Implementation
   c] Co-operation d] None of the above
191. Which extension method is evolved in result demonstration, posters, movies, slides and film strips?
   a] Spoken b] Visual
   c] Written d] All of the above
192. Which method of extension teaching is used in meetings, farm and home visits and radio ?
   a] Written b] Spoken
   c] Visual d] All of the above
193. Which of he following statement is correct?
   a] Compatible innovations can more easily be adopted
   b] Complex innovations are easily adoptable
   c] Innovations are useless to the farmers
   d] Costly innovations are good
194. Which of the following are not commonly used for the purpose of extension?
   a] Black boards b] Flash cards
   c] Slides d] None of above
195. Which of the following are not methods of individual communi-cation ?
   a] Farm and home visit
   b] Office calls and telephone calls
   c] Personal letter
   d] Field trips
196. Which of the following body is responsible for development at base level ?
   a] Panchayat b] District
   c] Block d] All of them
197. Which of the following is not a method of group communication?
   a] Conference b] National demonstration
   c] Field trips d] Farm and home visits
198. Which of the following is not a method of individual communication?
   a] Office calls b] Personnel letter
   c] Farm visits d] None of the above
199. Which of the following is not true ?
   a] Programme planning is a spontaneous process
   b] Programme planning is an educational process
   c] Programme planning is a co-ordinating process
   d] Programme planning is a continuous process

You Don't Know!

Mughal-Persion architecture

The tomb of Humayun built by Haji Begam, his wife, in 1565 is the first distinct example of the proper Mughal style, inspired by Persian architecture.

200. Which of the following is responsible for village women education about rural development programme ?
   a] Gram Sevika b] Gram Sevak
   c] Both of the above d] B.D.O.
201. Which of the following is the group contact teaching method ?
   a] Farm and home visit
   b] Bulletins
   c] Result demonstration
   d] Method demonstration
202. Which of the following is the name of the non-official block level body in states not having Panchayat Raj System ?
   a] Block Development Advisory Committee
   b] Block Samiti
   c] Panchayat Samiti
   d] Zila Parishad
203. Which of the following is the operational unit of N.E.S.?
   a] Zila Parishad b] Village Panchayat
   c] N.E.S. Block d] None of them
204. Which of the following state has no Panchayati Raj?
   a] Karnataka b] Bihar
   c] Orissa d] Meghalaya
205. Which of the following statement is not correct ?
   a] V.L.W. is a connecting link between technical development department and the rural peoples
   b] Village panchayat is headed by V.L.W.
   c] Village panchayat is a non-official body
   d] The head of a block is B.D.O.
206. Which of the following statements about joint family system of rural areas is correct ?
   a] Joint families are the only factors of rural development
   b] There are no joint families in the rural areas
   c] Joint families are both an impediment and a factor of rural development
   d] Joint families are always a cause
207. Which of the following visual aids are commonly used by agricultural extension works in India ?
   a] Bulletin boards, cultural programmes and black boards
   b] Posters, photographs, flash cards and flannel graphs
   c] Slides, film-strips, models and puppets
   d] All the above

GK **Mount Kilimanjaro is located in Tanzania.** GS

208. Which one of the following extension programmes linked farm plan with credit for farmers?

a] IRDP
b] CD Project
c] IADP
d] Etawah Pilot Project

209. Who is mainly responsible for rural indebtedness ?

a] Rural people themselves
b] The village money lender
c] Socio-economic frame work of the rural areas
d] Absence of banks in rural areas

210. Working objective consist of :

a] Audience and knowledge
b] Audience, type of change and skill
c] Audience, type of change and content
d] Knowloedge, attitude and skill

211. Written form of extension teaching methods does not includes :

a] Telephone
b] News articles
c] Circular letters
d] Bulletins

GK GS

Kokkam is disease of banana.

Matching

1. **Match the following extension attempts with the founder and the year of organisation.**

a]	Etawah Attempt	i]	Hatch in 1921
b]	Martandom Attempt	ii]	F.L. Brayne in 1920
c]	Sriniketan Attempt	iii]	Albert Mayer in 1948
d]	Gurgaon Attempt	iv]	Rabindranath Tagore in 1921

2. **Match the audio-visual techniques with the acceptor's class.**

a]	Written material	i]	Who attend
b]	Radio programmes	ii]	Who see
c]	Meetings	iii]	Who can read
d]	Demonstrations	iv]	Who listen to radio

3. **Match the intelligence quotient (mental age/chronological age x 100) of child of 8 years old with tests performed by the child of the following ages.**

a]	12 years	i]	125
b]	10 years	ii]	100
c]	8 years	iii]	62.5
d]	6 years	iv]	150
e]	4 years	v]	75

4. **Match the elements of innovation diffusion with their components.**

a]	Receivers	i]	Origin
b]	Channels	ii]	New ideas
c]	Message	iii]	Members
d]	Source	iv]	Adoption/Rejection
e]	Effects	v]	Spreaders

5. **Match the category of an adoptor with the percentage of the total (100) adoption of an innovation.**

a]	Innovators	i]	34
b]	Early adopters	ii]	16
c]	Early majority	iii]	13.5
d]	Late majority	iv]	2.5
e]	Laggards	v]	34

6. **Match the social classes with the examples given below.**

a]	Defined class	i]	Farmers
b]	Economic class	ii]	Gathering of marriage function
c]	Self-identified class	iii]	Hindu
d]	Cultural class	iv]	Tribals
e]	Participation class	v]	Rotary club

7. **Match the social classes with the examples given below.**

a]	Defined class	i]	Mohammedan
b]	Economic class	ii]	Lion's club
c]	Self identified class	iii]	Non-tribal
d]	Cultural class	iv]	Congress I]
e]	Political class	v]	Business

8. **Match the training method with its *modus operandi*.**

a]	Lecture	i]	Discussion, learning and application under supervision
b]	Seminar	ii]	Modified panel discussion
c]	Colloquium	iii]	Formal group discussion
d]	Symposium	iv]	Verbal presentation on a topic
e]	Workshop	v]	Presentation of short papers by a small group

9. **Match the development agencies with the year of their initiation in India.**

a]	CDP	i]	1953
b]	NES	ii]	1957
c]	CDB	iii]	1952
d]	Panchayati Raj	iv]	1954

10. **Match the development agencies with the year of their initiation in India.**

a]	IADP	i]	1964-65
b]	IAAP	ii]	1960
c]	ICDP	iii]	1966
d]	HYVP	iv]	1964

11. **Match the development agencies with the year of their initiation in India.**

a]	SFDA, MFAL, DPAP	i]	1972-73
b]	PPTD	ii]	1974
c]	T&V	iii]	1978-79
d]	IRDP	iv]	1970-71

12. **Match the development agencies with the year of their initiation in India.**

a]	TRYSEM	i]	1980
b]	NREP	ii]	1983
c]	DWCRA	iii]	1979
d]	NAEP	iv]	1986
e]	TMO	v]	1982

13. **Match programme/plan with year of beginning.**

a]	SJSRY	i]	1993
b]	JGSY	ii]	2000
c]	PMGSY	iii]	1999
d]	EAS	iv]	1997

14. **Match programme/plan/institute with year of beginning.**

a]	CAPART	i]	1986
b]	ARDRS	ii]	1988
c]	SAA	iii]	1990
d]	NFRD	iv]	1984

15. **Match programme/plan/institution with their objective.**

a]	To provide loan, seeds, fertilizer tools to the farmers	i]	HYVP
b]	Overall development of rural areas with people's participation	ii]	IAAP
c]	To develop special harvest	iii]	IADP
d]	To increase productivity of foodgrains by adopting latest varieties of inputs for crops	iv]	CDP

16. **Match programme/plan/institution with their objective.**

a]	Food for work	i]	Providing foodgrains to labour for the work of development
b]	IRDP	ii]	Allround development of rural poor through a programme of asset endowment of self-employment
c]	NREP	iii]	For providing employment to landless farmers and labourers
d]	RLEGP	iv]	To provide profitable employment opportunity to the rural poor

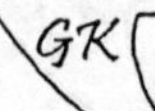

Soil salinity has become a serious problem where cotton is grown as a cash crop.

17. Match programme/plan/institution with year of beginning.

a]	FASC's	i]	April 1, 1985
b]	NFRD	ii]	1983-84
c]	CCIS	iii]	1984
d]	DRDA	iv]	1993

18. Match programme/plan/institution with year of beginning.

a]	Accelerated Rural Water Supply Programme	i]	1972-73
b]	Marginal Faermer and Agriculture Labour Agency	ii]	1973-74
c]	Twenty Point Programme	iii]	1975
d]	National Institution for Rural Development	iv]	1977

19. Match project name with associated person.

a]	2 tier Panchayat Raj	i]	Deshmukh
b]	Young Farmer Association	ii]	S.K. Dey
c]	National Demonstration Scheme	iii]	Ashok Mehta
d]	Majdoor Manjil	iv]	Kalwar and Subramanyam

Glossary of Technical Terms

Adopter Is a person who has continued full use of innovation.

Agricultural Extension Assistance to farmers to help them to identify and analyse their production problems and to become aware of the opportunities for improve-ment.

Aim It is a broad objective.

Andragogy Means art and science of helping adults to learn.

Association A group of persons organising themselves for the purpose of fulfilling certain interests.

Audience The intended receiver of messages.

Audio Visual Aid It is an Instructional device that can be heard as well as seen.

Calendar of Work It is plan of work arranged chronologically according to the time, when each step of work is to be done.

Caste When class is some what hereditory it is called as caste.

Change Agent A professional person who attempts to influence adoption decision in a direction that she feels is desirable.

Channel It is the physical bridge between the sender and the receiver of a message.

Communication It is a process by which two or more people exchange there ideas, facts feelings, or impressions in ways that each gains a common understanding of the meaning, intent and use of messages.

Community A sub group having many of the characteristics of society, but on a smaller scale, and with less intensive and co-ordinated common interest.

Contact Farmer A farmer who acts as intermediary between follower farmers and the extension agency.

Contact Man or Extension Agent The extension worker at the village level in direct day-to-day contact with local farmers.

Continuous Education or Life Long Education The Education which can be given through both formal & Informal types of education as and when need arises to educate for new technological development & for sociocultural development.

Culture It is the sum total of the ways in which human-beings lives transmitted from generation to generation.

Customs Uniformly accepted ways of acting about some social aspects of life.

Demonstration It is conducted by farmer under direct supervision of extension workers to prove the advantages of recommended practices.

Demonstration Plot A small area set aside for result demonstration by extension staff.

Distance Education It is any one of the various forms of the study which are not under the continuous, immediate supervision of tutors present to their students in lecture rooms.

Education It is the process of bringing desirable changes in human behaviour.

Esteem The evaluation of an individuals role or behaviour in a status which he occupies is called esteem.

Evaluation The study of the degree of success, we get in terms of our original objectives of the programme.

Extension Education It is an applied science consisting of content derived from research, accumulated field experiences and relevant principles drawn from behavioural sciences, synthesised with useful technology in a body of philosophy, principles, content and methods focused on the problem of out of school education for adults and youths.

Extension It is method of non formal education aimed at inducing behavioural changes in its clients for their increased knowledge, skills & income.

Extension Teaching Methods It is a method through which the results of research can be conveyed to the people.

Extension Worker Men and women employed by an extension agency or technical and professional staff employed at all levels.

Family It is a system of organised relationship involving workable and dependable ways of meeting basic social needs. It is a most multifunctional of all institutions in society.

Farming Need Are the gap between farming practices adopted by the farmers at present and the faming practices recommended by the research worker at present.

Field Trip A group of interested farmers and extension education worker goes on tour to see and obtain first hand knowledge of improved practices.

Filmstrips Series of still pictures in one roll.

Flannel Graph Piece of flannel or sand paper stick on flannel.

Folk Ways They are the customary ways of behaving in which society exerts some. They are not rigid.

Forum Two or more speakers who present talks on the same subject. The primary characteristics of the forum is that the subject is controversial and speakers present the opposite sides of question.

Functional Literacy The ability to read, write and calculate so that a person may engage in all.

Goal It is the distance on any given direction, proposed to be covered in a given period of time.

Illiteracy It means inability to read and write.

Illiterate Means person who has not acquired basic skills in reading, writing or arithmatic.

Innovation An idea perceived as new by the individual.

Innovativeness Degree to which an individual is relatively earlier in adopting new ideas than the other members of his social system.

Institution It is a system of relationships or a pattern for carrying out an idea or a desire which is regarded as necessary for the welfare of the group.

Integrated Rural Development It is defined as development which can help to increase the purchasing power of the rural poor through the generation of greater opportunities of gainful employment.

Interest It is defined as attention with a sense of concern focussed upon some object.

Leader An individual is a leader in any social situation in which his ideas and actions influence the thoughts and behaviour of others.

Leadership It is a process of influencing the behaviour of the individual in a real situation.

Learning Experience It is a mental and/or physical reaction, one makes through seeing, hearing or doing the things to be learned through which one gains meaning and understandings useful in solving new problems.

Learning It is the process by which an individual through his own activities becomes changed in his behaviour.

Lecture Note It is merely a store of information.

Lesson Plan It is the outline of activities that the teacher will follow in order to create effective learning situation for the learners. It is a plan for teaching.

Message It is the information, a communicator, wishes his audience to receive, understand, accept and act upon.

Models It is the replicas of real objects or scaled representation of things.

Motivation It is the process of creating a strong desire for applying to the fundamental wants of the people so that they are influenced to change the behaviour.

Need It is a difference between two situations that is "what is" and "what ought to be".

Norms Rules that govern action directed towards achieving values.

Objective It is the direction of movement.

Opinion Leaders Those individuals from whom others seek the information and advise.

Organisation It is a set of two or more parts which are so related that they function as a whole.

Over Time It is the time required to adopt from its origin.

Panchayati Raj (Democratic Decentrali-sation) 'Panch' means five. It refers to a member of an assembly of elders who settled disputes, according to custom in a village or within a caste.

Panel It is also called round table discussion. It is usually composed of 4-6 persons. The members of the panel carry on a discussion of the topic among themselves and the audience listens in, on the discussion.

Personality It is a pattern body of inherited traits.

Plan of Work It is an outline of activities, so arranged as to enable effective execution of the entire programme.

Planning It is a process of preparing systematic statement of the line of action for achieving certain overall and specific objectives in relation to needs and resources.

Poverty It is the social phenomena in which a section of the society is unstable to fulfill even its basic necessities of life.

Prestige Evaluation of the status

Programme It is a systematic statement of situation, objectives, problems and solution. It is relatively permanent but requires constant revision.

Programme Planning It is a process of working with rural people in an effort to recognise the problems and determine the possible solutions.

Project It is an outline of procedure and pertains to some phase of extension work.

Psychology It may use animals in behavioural research and collect information and insight relevant to their area of concern.

Reference Group It is a group of persons whom an individual consult before taking an important decision.

Role Function of the status.

Rural Development It is a process of a change from the traditional way of living to progressive way of living.

Rural Sociology It is the systematic study of the people living in rural areas and follow patterns of occupation and life somewhat different than those living in urban areas.

Social Centre The adult education classes will be organised through different adult education centres on the basis of needs and requirement of adult men and women.

Social Class It is an abstract categories persons arranged in levels according to the social status they possesses.

Social Control It is the process of constraining persons to conform to norms.

Social Group It is a unit of two or more persons in reciprocal interaction or communication with each other.

Social Problems It is a condition in society that is considered harmful or undesirable by society as a whole based on existing social values and in respect of which it is believed that improvement is possible, e.g. Dowry, Untouchability.

Social Sanction It is a reward or punishment applied to persons in order to encourage or discourage certain types of behaviour.

Social Stratification It has been defined as a pattern of super imposed categories of differential previlage.

Social System A population of individuals who are functionally differentiated and engaged in collective problems and solving the problems.

Social Values Defined as the attitude held by individuals, group or society as a whole as to whether material or non-material objects are good, bad, desirable or undesirable.

Social Welfare The term is very comprehensive in its meaning and includes a large number of activities connected with the lift up and welfare of the weak and exploited sections of the society.

Society It is that group within which men share.

Sociology It is the scientific study of men's behaviour in relation to other groups and individuals with whom the interacts.

Specimen It may be a part of an object or one of group or class to represent the whole group in other words a typical example.

Status It is a position or place in a social structure.

Subject Matter Specialist Professional staff who provide technical back up to extensionist.

Supervision It is the process of placing in to action standards of performance developing and utilizing human resources to achieve organization results.

Teaching It is the process of arranging situations that stimulate and guide learning activities. In order to bring desired changes in human behaviour.

Traditions The uniformly accepted ways of thinking.

Village Level Worker The multipurpose agent of India's community development programme.

Vrajrapathi is related to pasture.

GS

1. Animal Husbandry and Dairy Science

Quiz Kit!

First use of indelible ink

Indelible ink was used for the first in the 3rd Lok Sabha elections in 1962.

Each Word Counts

COMMON TERMS

Back fatter : A fat pig too heavy for bacon trade.

Bacon : Meat from back and sides of a pig.

Baconer : A bacon pig.

Bangers : Cows that have reacted to Brucellosis test.

Barren sow : Sterile mature female pig.

Barrow : A mature pig whose testes were removed before reaching breeding age.

Billy goat : Buck/Male goat

Bitch : A female dog.

Blue pig : Pig produced by crossing a white breed with black breed of pigs.

Boar : An uncastrated adult male pig used for breeding.

Bobby calves : Calves usually of only one week old or so.

Bow Wow : A small, stunted aged steer with no quality.

Brahmini bull : The bull which is left in the name of dead person, a practice prevalent in certain areas.

Breeding bull : A bull used for breeding purpose.

Breme : Female yak.

Brood mare : A mare kept for breeding.

Buck : Sexually matured male goat used for breeding.

Buckling : A male goat between 1 and 2 years age.

Bull : An uncastrated sexually matured male.

Bull calf : A male calf under one year age.

Buller : A cow apparently always in heat, oestrus.

Bullock or Steer : A castrated male ox of over 2 years age.

Cad/Critting/Runt : Smallest pig of litter farrowed in the last, gunner/rating runner-weaned pig.

Calf : A young animal of bovine species under one year age.

Calving : Parturition in cattle.

Cast ewe : An aged ewe culled from breeding flock.

Cattle : Members of the cow species—cow, bull, bullock, heifer calves etc.

Chevon : Meat of goats.

Closed gilt : A female pig after orariotomy or a young pregnant gilt.

Cob : A short age legged, stocky and small horse.

Cold slaughter : Carcass from dead of some cause, other than slaughter.

Colt : A young immature male horse usually above one year.

Counter feit : Description of cattle of good colour giving the impression of good breed that they do not possess.

Cow : A female of bovine species that has calved at least once.

Creep feeding : System of feeding young goats prior to meaning.

Crips/Downers : An animal that has been hurt or crippled.

Crock : An old ewe.

Crone : An old broken mouthed ewe retained in breeding flock beyond time due to excellent breeding performance. Ewes about 7 to 8 years old when teeth becomes loose.

Crossbreds : The animal produced by mating two different breeds of same species.

Cub : Young one of a tiger.

Dam : The cow that is served "Female Parent".

Deacon : A young calf not mature enough for veal.

Dinmont : Ram lambs after first shearing.

Doe : An adult female goat, rabbit, or deer

Doeling : A female kid from the first day of birth until first parturition.

Dog : Any member of canine, usually used for male dog.

Donkey : Synonymous to ass.

Double-rig : A horse of 1 year age or more whose both testes have been retained in the abdomen. (Cryptorchidism)

Downeres : Same as crips.

GK

The Government of India Act 1935 was the greatest influence on Indian constitution.

GS

Educated livestock : Animals consigned from agriculture college/university.

Entire : An adult uncastrated horse which is not used for breeding purpose.

Ewe : An adult female sheep.

Farrowing : Parturition in pigs.

Fattening cattle : Those nearly ready for butcher.

Fecund : Capable of producing many young ones.

Fill : Coarse feed and water given to animals to increase sale weights.

Filly : A young immature female horse of above one year age.

Filly foal : A female goat below one year age.

Flock : Collection of sheep/goats.

Foal : A young one of either sex upto one year age.

Fold : Enclosure of sheep.

Free Martin : When twin calves of different sexes are born, the female one is usually sterile which is called "Free Martin".

Gammen/Wam : The thigh of pig.

Gelding (Geld) : A castrated male horse of any age.

Gilt : Gilt not yet served.

Gilt/Milt/Yeltter: A young female pig kept for breeding purpose which has either not conceived as yet or going to farrow for first time.

Gimmer : A yearling ewe or ewe after first shearing till her first lamb is weaned.

Goatlings : A female goat between 1 and 2 years of age.

Growing pig : Between for first time.

Hard feeders : Heifers that have lost their figures having had one calf.

Heifer : A female animal of bovine species that has not yet calved. After calving she becomes a cow.

Heifer calf : Female calf under one year age.

Heiferettes : Dairy type steers that cannot be fattened economically.

Hinny : The cross of stallion and female donkey.

Hog : A male pig after being castrated, called Barrow.

Hogg/Hoggests/Tegs : A lamb after weaning but before first shearing (weaning to shearing) or whether lambs when ready for sale.

Hybrid vigour : A hybrid gains in comparison to either of the parents.

Hybridization : The process of mating animals of two species.

In-lamb : An ewe that is pregnant.

Jack pots : Mixed lot of cattle usually of common quality.

Kid : A young goat of either sex below 1 year age.

Kidding : Parturition in goats.

Kitten/Kit : Young one of a cat.

Lamb : A sheep upto 1 year age—ram lamb/ewe lamb.

Lamb hog : A male lamb from weaning time until it is shorn.

Lambing : Parturition in sheep.

Litter : A collective term used for all young members of each whelping.

Livestock : Domesticated animals kept on the farm—cattle, buffaloes, sheep, goats, pigs, horses, camels etc.

Maiden ewe : An ewe that has not been mated.

Maiden goat : An unbred goat.

Maiden heifer : An adult female that has not been allowed to breed.

Mare : An adult female horse.

Mithun : A semi-nocturnal animal reared at high altitude.

Mudders : Cattle fed in a muddy feed lot and always penalized in price.

Mule : A hybrid whose sire is a male donkey and dam a mare.

Mutton : Meat of sheep.

N.B. : Any cow, a dam and any bull a sire so long as both of them are mated together.

Naism : Dwarf growth.

Nanny : A young female goat.

Open gift : A young female pig yet not served.

Ovine : Animals of subfamily 'Ovidae'.

Ox : Same as cattle.

Pedigree bull : The bull whose ancestral record is known.

Pig : A domesticated animal "*Sus domesticus*" - applicable to every member.

Piggery : Place where pigs are housed.

Pigging/Piggy : Sow that has appearance of having been "suckled" pigs.

Piglets : Small pigs.

Plam shank : The hock end of ham.

Polipeds : Animals having clefted hoof.

Pony : A horse of smaller size.

Pork : Meat from pigs.

Porker : Pigs whose dead weight ranges between 27 and 36 kg.

Puppy or Pup : Young one of a dog species.

Queen : A female cat.

Jatin Das died after a fast in demand of better conditions in jail.

GS

Ram : A male sheep which is not castrated, Tnp.

Rannies : Small southern calves of poor quality.

Rig : A male sheep that has not been properly castrated.

Rig : Male pig with one testes undescended.

Rig and Rigling : A male horse of above one year age / Cryptorchid whose testes are retained in the abdomen.

Runt : Smallest piglet farrowed in the last.

Sausage : Product prepared from fresh minced pork.

Scrub bull : It is a nondescript type of stray village bull.

Seggy : A ram castrated after service.

Sheep : Used to denote any member of Genus "Ovis".

Shoat/Shote : A young pig of either sex between 27 and 72 kg.

Sire : The bull that serves a particular cow (male parent).

Skip : A worthless lamb.

Slink calves : Calves which have been aborted or those which are found in the uterus during slaughter.

Soliped : Animals having uncleftted hoof-horse.

Sow : An adult female used for breeding.

Springer cow : Cows due to freshen soon.

Stag : A male castrated sheep after it had developed full secondary sexual characteristics.

Stage, Steg or Seg : A male castrated late in life.

Stallion : An uncastrated male used for breeding.

Steer : Male bovine castrated before maturity.

Stickers : Animals difficult to sell.

Stillers : Cattle fed on distillery mash.

Stirk : Applied either to a heifer or a young bullock from about 15 months to 2 years of age.

Stock bull : One used for breeding purpose.

Stock cattle : Young steers and cows that are light in weight and usually immature.

Stocker cattle : Cattle that must be fed out to kill profitably.

Store : Young beast of either sex which are to be fed for butcher later on.

Stud bull : Same as breeding bull.

Sty : House of pig.

Swine : All type of domestic pigs.

Teaser ram : A vasectomised ram, sexually active but no service will be fertile.

Tnp : An uncastrated adult male sheep.

Tom cat : An uncastrated male cat.

Trutters : The feet of sheep, goat/pig.

Waler/Whaler : An Australian saddle horse imported into India and other Eastern countries.

Did You Know ?

First woman CM

Sucheta Kripalani (1908-74) was Chief Minister of Uttar Pradesh from 1963 to 1967.

Wallow : Water pool for pigs.

Wean : A condition when young calf is no longer fed milk from dam.

Weaner : Piglets separated from mother for tearing independently.

Wet cows : Milking cows.

Wether/Wedder : A castrated adult male sheep.

White faces : Hereford cattle or fine woollen sheep.

Wooden : Cattle, sheep or lambs improperly fed with low grade inferior meat yield.

Yearling : A horse between I and 2 years age (yearling colt/filly).

Yearlings : Uncastrated bovine between 1 and 2 years of age.

No separate specific term is used for buffaloes, commonly spoken of buffalo bull, buffalo bullock, buffalo cow, buffalo calves. He and she buffalo are also used to denote male/female respectively.

Significance of Farm-animal in Agriculture

1. It provides us with milk, draft power, ploughing, planking, irrigation, threshing, transport of produce, hide, skins, bones, hooves, manure, fuel, meat, egg, wool, honey, wax, fishes, etc.
2. Being 8th fishing nation of the world, India produces a meager 3-4% fish with only 1.7kg per capita consumption compared to as much as ten times in advanced countries.
3. Compared with other countries our per capita consumption of egg and chicken is much less.
4. Contributes 7% to national income.
5. Due to small-sized holding livestock fits well in mixed farming.
6. Honey production in India per bee-colony is only 7Kg compared to 30 to 50kg in many developed countries.
7. India, an agricultural country with 1/5th of the world's population of cattle and half of the world's buffaloes, the cows and bullocks are the backbone of the agriculture and play a major role in the rural economy.
8. Livestock contributes to about one-fourth of our GNP.

At present State Bank of India (SBI) is the largest commercial bank in the country.

9. Livestock supplies meat worth Rs.1200 million per year.
10. Makes better use of agricultural, industrial, dairy and animal by-products.
11. Milk and milk products from dairy animals are palatable and easy to digest, therefore an important human food.
12. Milk being the only source of animal protein in vegetarian diet; it is a protective and balanced food.
13. Milk contains fat, proteins, milk, sugar, minerals and vitamins in a much better proportion and quantity than any other food; so it is nearly perfect food.
14. Milk output accounts for 5.6% GNP.
15. Milk output accounts for more than 17% of India's agricultural production.
16. Offers opportunity of earning foreign exchange worth Rs. 800 million annually.
17. Our country produces only 8% of world milk.
18. Our pig population is one of the lowest in the world.
19. Provides organic manure for maintenance of soil fertility.
20. The investment in animal husbandry and dairy development is less than 10 per cent of the investment in the agricultural sector, but its contribution to the GNP from agriculture and animal husbandry is reported to be more than 31 per cent.

Thus, our livestock are poor producers. The reasons are many as follows:

1. About half of the cultivated area is under arid and semi-arid region where agriculture depends mostly on livestock.
2. Only 4-5% of our cultivated area is under forage crops unable to supply the bovine needs.
3. Lack of scientific management and prejudices of a large section of populace regarding rearing of poultry birds, consumption of egg and its meat are some of the causes of low poultry production.
4. Pigs are still reared by poor section of society under little care to the animals.
5. Apiculture is of recent development in India. Probably because of preference of a vegetarian diet and of regional habit, fish production is still at low ebb.

Breeds of Different Livestock

BREEDS OF CATTLE IN INDIA

There are 6 groups, 25 well-defined breeds of cattle and 6 breeds of buffaloes, in addition to a large number of non-descripts of low productivity in India.

Groups of cattle breed

	Group	Cattle breed
1	Lyre-horned grey cattle	Kankrej, Malvi, Kerigarh, Tharparkar
2	Short-horned white or light grey	Haryana, Ongole, Mewati, Nagauri, Rathi, Krishna valley, Goalao
3	Ponderous body with pendulous dewlaps and sheath, spotted or red or white	Gir, Sahiwal, Deoni, Red Sindhi, Nimari, Dangi
4	Medium sized powerful quarter and tight sheath. The horn emerges from the top of the forehead placed closed together in upward and backward direction (also known as Maysore breed).	Amritmahal, Hallikar, Kangayam, Khillari, Alambadi, Bargur
5	Heterogenous: These are small with black or red colour of the skin with large white patches, small sheath, short or slightly lyre-horned.	Siri, Ponwar
6	Cattles not falling on any of the above groups are grouped under this.	Dhanni breed of Pakistan.

Purpose of Breeds of Cows

Milch Breeds	Dual Purpose Breeds	Draught Breeds
1. Cows are high milk yielders.	Cows are medium milk yielders.	Cows are poor milk producers.
2. Bullocks are of poor quality.	Bullocks are good draught animals.	Bullocks are of excellent draught breeds.
3. Medium body size.	Average to large body size.	Heavy to medium size.

Key Characteristics of Cow Breeds

Breeds	Home	General features	Milk yield/lactation	During first calving Age (months)	During first calving Wt (kg)
I. Milch Purpose					
1. Deoni	Hyderabad (A.P.)	Pendulous ear. Red/black with white patches. Other characters are similar to Gir.	1650 kg.	46	355
2. Gir	Kathiawar (Gujarat)	Big horns curved turning backward. Hip bones prominent. Pendulous dewlaps. Ponderous body. Robust constitution. Sheath shades of red to almost black. Udder medium sized.	1900 kg.	47	350
3. Sahiwal	Montogmery (Pakistan), Delhi, M.P., U.P., W.B., Punjab & Haryana	Capacious udder. Heavy body. Heavy develop. Loose skin. Pendulous sheath. Red brown. Short legs. Stumpy horns. Other characters are similar to Gir.	1900 kg.	42	380
4. Sindhi	Karachi & Sindh (Pakistan)	Develop heavy hump. Sheath deep red. Horns thick & blunt. Medium sized. Medium size body. Udder larger.	1600 kg.	43	325
II. Dual Purpose					
1. Gaolao	Chindwara district (M.P.) Wardha, Nagpur in Maharashtra	Bulged head. Horn short thick. Pendulous sheath. Developed hump. Light built. Long face. Medium height. White colour.	1100 kg.	48	300
2. Haryana	Rohtak, Hissar, Delhi, Gurgaon, Karnal	Short horns curving upward and well formed. Narrow face. Small sheath. Well developed udder. White to light grey.	1600 kg.	47	330
3. Kantkrej	Rann of Kutch, Gujarat, Sindh	Broad chest. Heaviest of Indian cattle. Pendulous sheath. Silver grey colour. Strong lyre horns. *Swai chal*. Udder medium sized.	1300 kg.	43	420
4. Krishna Valley	Area of black cotton soil near Krishna valley	Large body. Pendulous ears. Short horn with prominent forehead. Neck thick and short. Short legs.	915 kg.	48	370

GK **In Kumaon range of Himalayas, Uttaranchal is situated.** GS

5.Nimari	Nimar district of M.P. Indore, Narmada Valley	Pendulous develop straight back. Prominent forehead. Red colour with white patches.	915 kg.	48	300
6.Ongole	Ongole, Guntur, Narsaraopet in Nellore of A.P.	Forehead broad with stumpy horn. Hump developed. Large muscular heavy body.	1250 to 2000 kg.	48	480
7.Tharparkar	Tharparkar district of Hyderabad; Jodhpur	Ears large & drooping. Medium sized udder. Lyre horned. Medium sized body. White/light grey.	1600 kg.	44	367
III. Draught Purpose					
1.Amritmahal	Mysore	Grey colour. Light skin. Longer sweeping horns. Forehead depressed in centre. Narrow face. Small sized body.	600 to 750 kg.	55	310
2.Dangi	Ahmednagar, Ghats of Nasik	Longer body. Medium sized. Pendulous udder. Red with large white patches in colour. Forehead bulging.	550 to 700 kg.	50	300
3. Gangatitri	Ghazipur, Ballia, Varanasi in U.P.	Long face. Medium sized. short blunt horns Small neck. Straight back. Udder medium sized. Wide protruded head in the centre.	1200	50	350
4.Hallikar	Hasan, Tumkur in Karnataka district	Massive body. Sharp upward horns. Develop prominent hump covered with tuft of hair. Tight sheath. Small head. Wide flat forehead.	400	50	380
5.Hissar	Hissar (Haryana)	Horn thick upward. White similar to Haryana except head is wider. Large pendulous ear. Medium sized.	800 kg.	38	310
6.Kangyam	Coimbatore (Tamil Nadu)	Medium sized. Wide forehead with depression, straight back. Horns thick and strong. Ears small and erect. Short neck. Navel tight. Small udder.	750 kg.	45	300
7.Kankatka	Bundelkhand in U.P.	Long developed sheath pendulous. Short length. Small head. Small sized. Wide forehead.	700	50	285

In Harappa, coffin burial is common.

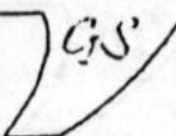

8. Kherigarh	Khiri in U.P.	Pointed horn backward. Small sized. Small udder. Straight back. Thick head. Tight sheath.	650	48	300
9.Mewati (Koshi)	Mathura (U.P.), Alwar & Bharatpur (Rajasthan)	Its bullocks are somewhat idle in nature. Medium sized body and mostly white. Resembles Haryana cattle. Sheath pendulous.	1000 to 1200 kg	48	320
Exotic Breeds of Cattle					
Ayreshire	Scotland	Large body. Leveled rump. Nervous temperament. Red/brown colour. Sharp upward curved horns. Short-thick head. Straight vertebra. Udder average sized.	4600	26	
Brown Swiss	Switzerland	Body larger wedge-shaped. Brown colour, thick loose skin. Head forehead broad and dished. Large udder. More heat tolerant than Jersey dual purpose (milk & beef).	5000	28	500
Holstein Friesian	Holland	Big chest. Body wedge shaped heavy voluminous. colour-black and white irregular. Head long and narrow. Large barrel. Long tail. Straight thigh. Large capacious udder.	5500 to 6000	26 months	520 to 550
Jersey	Island of Jersey in English Channel	Colour-fawn with/without marking. Compact body medium sized. Rump leveled. Straight back. Udder large and capacious. Wedge shaped.	4000	24	375

Taurus (European) Vs. Zebu (Indian) cattle

	Parameter	Taurus (European) (*Bos taurus*)	Zebu (Indian) (*Bos indicus*)
1.	Adaptability	Poor	Good
2.	Age at first calving (months)	30 months	36-42
3.	Age at maturity (months)	15 to 18	24 to 36
4.	Back	Slightly curved	Straight
5.	Body size	Large	Small
6.	Body weight	More	Less
7.	Calving interval (months)	13	24-28
8.	Colour	Black and white, brown	Shades of red, brown, white, grey and black
9.	Dewlap	Developed	More developed
10.	Dry period	Less than 60 days	More than 60 days
11.	Feed intake	More	Less
12.	Heat tolerance	Less	More
13.	Hump	Absent/humpless	Present
14.	Management	Difficult	Easy
15.	Milk fat and SNF	3.7% and 8.8%	4.5% and 8.9%
16.	Milk yield	More	Less
17.	Preference for shade	Relatively more	Yes
18.	Resistance to diseases	Less	More
19.	Water intake	More	Less
20.	Weaning	Easier	Easy
21.	Wedges	More prominent	Less prominent

Key Characteristics of Buffalo Breeds

Breed	Home	General features	Milk yield/ lactation	During first calving	
				Age (month)	Wt (kg.)
1. Bhadawri	Agra	Copper colour. Head bulged. Medium sized. Moderate udder. Wedge-shaped body.	1100 kg. (13% fat)	51	410
2. Jafrabadi	Kathiawar (Gujarat)	Black in colour. Developed udder. Horns inclined to droop on sides of neck. Longer body.	1500 kg. (7% fat)	47	404
3. Mehsarna	Mehsana distt. (Gujarat)	Black colour. Body-medium size. Hind quarters broad. Horns-sickle-shaped curved. Long neck. Udder-well developed. Wedge- shaped body. Wide forehead, with depression in the middle.	1400 to 1600 kg.	38	400-500
4. Murrah	Rohtak, Karnal (Haryana)	Body massive, Broad back. Less hair. Shiny jet scanty black colour. Short & tightly horns. Well developed large udder.	2000 kg. (7% fat)	44	462

5. Nagpuri, Ellichpuri	Marathwada, Aurangabad & Nagpur Wardha, Akola	Black colour but white marking on face, legs & switch of tail. Body medium. Horns very long reaching upto withers. Long and narrow face. Neck longer & brisket-heavy. Small sized udder.	1100 to 1200 kg.	41	400-500
6. Nili	Firozpur distt. (Punjab)	Black in colour with white patch on forehead. Developed udder. Head bulged and depressed between eyes. Small horns coiled.	1800 kg. (6% fat)	38	407
7. Surti	Anand, Baroda (Gujarat)	Black or brown colour. Medium sized. Sickle shaped horns.	1600 kg. (6% fat)	36	380

Key Characteristics of Goat Breeds

Breeds	Home	General features	Usefulness
EXOTIC BREEDS OF GOATS			
1. Alpine	France	White to black with spots, horn/hornless, pointed ears, adult weight 60 to 65 kg.	1.3 kg milk/doe/day
2. Nubian	Nubia (upper Egypt), Sudan	Black dark brown with white spots, body covered with white silky hair, developed udder.	1.2 kg milk/day
3. Saanan	Switzerland	Colour dull white, male horned but female hornless, upright erect ears, straight nose, large udder, mature weight 65 to 70 kg.	Milk yield 2-3 kg/day
4. Toggenberg	Toggenberg	Brown with white colour on legs, polled, mature body weight 55 kg large sized.	Milk yield 1 kg/day
5. Angora	Middle China	Short statured, small head, white coloured, long loped ears, flat pointed horns, small udder, body covered with Mohair.	Reared for meat and Angora fur, yield 1-1.5 kg/year.

Breeds	Home	General features	Usefulness
INDIGINOUS BREEDS OF GOATS			
A. Himalayan Region			
1. Chang Thang	Ladakh, Lahaul, Sipti valleys in Himachal Pradesh	Small sized. Dirty, white colour sometimes grey/brown. Leveled back long pendulous ears. Face covered with coarse hair. Longer fine Pashmina grows on sides and shoulders.	Pashmina yield 100-120 g/goat
2. Chegu	Sipti, Yaksar Kashmir mountainous range	Average body weight 50 kg in male and 30 kg in doe. Average height 55 to 65 cm at withers. Kids once a year and produce one kid/kidding. Medium sized.	Good meat. Small quantity of milk produces. Pashmina yield 110-150 g/goat/year
3. Gaddi	Himachal Pradesh (Kangra Kullu valley, Shimla) and Punjab	Long & coarse hair-20 to 25 cm. Strong & firm footed. Tough skin. Used for transport. White horns pointed at apex. Yield 1/2 to 1 kg.	Hair. Meat. Transport.

4. Kashmiri	Himachal, Kashmir, and Punjab	Large sized. Strong & hardy. Wt. 68 kg. Grey to white brown. White silky hair 10 to 12 cm beneath, fur coat. Pashmina grows in October & February.	60 g Pashmina/goat/combing. Draft purpose breed. Meat. Shawl, rugs making.
B. Dry Northern Region			
1. Barbari	Barbera (East Africa) Etah, Agra, Etawah, Mathura etc.	Legs short. Pendulous udder, white with tan spots. Short erect ears. Small sized. Straight facial line. Two kids per kidding, kids twice a year. Wedge-shaped.	Gestation period- 146 days. Milk 1 to 1.5 kg.
2. Beetal	Sialkot, Jhelum, Gurdaspur, Amritsar	Horns horizontally backward, spirally twisted. Long ear. Male bearded. Red or black, tan spotted with white. Resembles Jamunapari but smaller in size & wt. Roman nose.	Milk 1.5 to 2 kg. Twins/kidding.
3. Jammunapari	Between Jammu & Chambal; Etawah	Dual purpose. No standard colour, white with tan markings. Long pendulous ears of 25 cm. Roman nose. Short flat horn. Large udder.	Dual purpose. Good for meat and milk. Long hair on hocks. Milk 2.5-3 kg.
C. Central Region			
1. Berari	Nagpur, Wardha district, Nimar in MP	Tall breed with dark colour.	Milk yield/day 0.6 kg.
2. Kutchi	Kutch, Rajasthan, North Gujarat	Black colour. Good both for meat and milk. Long hair. Medium sized. Twisted horn.	Milk yield/doe per day 1.25 kg.
3. Marwari	Marwar area of Rajasthan	Jet black colour with white/speckled ears. Male have thick beard. Small ears. Long cork screw horns present in both sexes. Curved upward tail. Medium sized breed. Body weight 25 kg (doe) and 35 kg (buck). Height at withers 55-65 cm. Lustrous long hair (10-12 cm).	Triple purpose (hair, meat, milk). Daily milk yield 0.9 kg.
4. Mehsana	Mehsana district of Gujarat	Colour grayish black. Ears-white with black spots. Good milkers. Have Roman nose. Medium sized. Small beard. Udder well developed.	Milk yield 1.5-2.0 kg/day
5. Sirohi	Gujarat and Rajasthan	Average height 55-65 cm. Body densely covered with coarse, short hair. Body weight of doe 40 kg and buck-50 kg. Compact body. Goats-medium sized. White & brown in colour.	Suited for stall feeding. Good for meat and milk. Average milk yield is 1 kg per day.

6. Zalawadi	Zalawad in Kathiwar district of Gujarat	Body covered with black lustrous hair (15 cm long). Large sized goat. Straight screw shaped horns. Two lobular appendages of skin in throat region.	Good milker (1 kg). Kids once a year.
D. Southern Region			
1. Malabari	North Malabar, Tellicherry and Mangalore of Kerala	No uniform body colour. Black/brown/white or mixture of these colours. Medium sized. Flat head. Roman nose. Long ears. Average body weight of doe 40 kg.	Average milk yield per day 0.75 kg. Used for the meat purpose.
2. Osmanabadi	Osmanabad district of A.P	Black in colour, mixture of red white and black is also found, long horns, large sized goat, Two types breed-short and long haired.	Milk yield/dose/day 1 to 2 kg. Twinning is common.
3. Surti	Surat, Nasik (Maharashtra)	Short stature, unable to walk long distance, reared town, stall fed food waste.	1.2 kg milk/day
E. Eastern Region			
1. Assam Hill	Khasi, Naga, Lusai hills	White/grey, short legged, longer body and long hair. Mature weight 25-30 kg.	Kids twice a year. Twins common. Poor milker. Kept mainly for meat.
2. Bengal	W. Bengal	Chest wide. Ears upright. Excellent meat. Short and soft hair. Small legged. Twins multiples.	No milk. Meat excellent.
3. Ganjam	Puri (Orissa), coastal AP, South MP	Adult doe weight 28 kg. Black/white/grey colour. Compact goat. Horns parallel, curved backward and downward. Kids twice a year. Males bearded. Small sized. Twinning is common.	Good for meat. Milk yield 0.4 kg per day. Poor milker.

Key Characteristics of Sheep Breeds

Breed	Home	General features	Usefulness
EXOTIC BREEDS OF SHEEP			
1. Corriedale	New Zealand	Dual purpose for wool and mutton. Prolific.	Av. quality wool, yield 4 to 5 kg/year.
2. Leciester	England	Mature weight of stylish in features. Moderate sized.	Average quality wool. Yield 3 to 4 kg.
3. Lincoln	England	Large sized. Prolific breeder. Ram's weight 150 kg.	Coarse wool, yield 4.5 to 5 kg.
4. Merino	Spain	Hardy in nature. Rams with spirally-twisted horns, ewes- polled, folds in the skin. Small sized. White in colour.	Fine quality wool. Yield 2.5 kg/year.
5. Rambouillet	France	Av. weight of ram 120 kg. Large size head. Prolific breeder. White in colour.	Dual purpose for wool and mutton.
6. Southdown	England	Blocky in appearance. Grey in colour. Short round ears. Small-sized.	Fine quality wool. Yield 2 kg/year.

GK GS

J&K state has the Himalayan passes, Burzila and Zozila.

Breed	Home	General features	Usefulness
INDIGENOUS BREEDS OF SHEEP			
I.Temperature Himalayan breeds of sheep			
1.Bhadarwah	Kashmir, H.P., Garhwal (Uttranchal)	Body covered with fleece. Commonly white with brown face. Ewes hornless. Rams horned.	Fleece is fine, lustrous. Yield is 1.1.5 kg/year.
2.Bhakarwal	–do–	Generally white. Ewes hornless but rams horned. Long ears & drooping. Long sized. Straight superior back.	Wool yield 1.5 kg/year. Coarse wool. Mutton.
3.Gurez	–do–	Polled, white in colour coarse wool.	Wool fibre 6½" long. Wool yield 1.5 kg/year.
II. Dry Western Region breeds of sheep			
1.Chokla	Churn, Bikaner, Jaipur, Jhunjhunu, Nagaur (Rajasthan)	Body length 67 & height 67 cm. Both sexes polled. Coat is dense all over the body. Ears medium tubular. Face has no wool. Medium sized. Reddish brown. Tail thin of medium length. Wt. of male (M) 35 kg, female (F) 23 kg.	Relatively fine fur coat. Wool yield 2 to 2.8 kg per sheep annually.
2.Jalauni	Jalaun, Jhansi, Lalitpur in U.P.	Both sexes polled. Ewe wt. 26 kg. Large ears, drooping. Medium-sized. Tail-thin medium in length. White in colour.	Fleece coarse, short stapled. Yield/sheep, 0.5-1.2 kg.
3.Kathiawari	Parts of Kutch & Kathiwar district (Gujarat)	White in colour with tan/black face. Medium sized, well built.	Good quality carpet type wool. Yield 1.5 kg.
4.Lohi	Montgomery district (Pakistan)	Black/brown face. Long ears. Short & thick tail. Roman nose. White in colour polled.	Good quality mutton, carpet type coarse wool. Yield 2 kg.
5.Magra	Bikaner, Nagore, Jaisalmer, Churu district of Rajasthan	Adult body wt. M-27 kg, F-24 kg. Both sexes are polled, Ears are small and tubular. Face with light, brown patches, white in colour. Medium to large sized.	Fleece is of medium in carpet quality; white lustrous, not very dense. Yield, 1.7 to 2 kg / sheep/year.
6.Malpura	Jaipur, Tonk, Ajmer, Bhilawara & Bundi district of Rajasthan	Both sexes are polled. Light brown face. Long legs. Short tubular ears. Tail-thin-medium. Well built. White colour.	White fleece, extremely coarse. Yield of wool 2 kg/sheep.
7.Marwari	Southern Rajasthan	Black face. Body white. Short ears & tail. Small sized body.	Coarse white wool 1 to 1.5 kg/year.
8.Muzaffarnagari	Muzaffarnagar, Meerut, Bulandshahr, Saharanpur in U.P.	Body and face with black patches on white body colour. Body wt. M-50 kg and F-40 kg. Both sexes are polled. Face line slightly convex. Medium to large sized.	Fleece is white, coarse open with no wool on legs and belly. Wool yield 1 to 1.2 kg/sheep.

In Kalibangan, lower town was fortified.

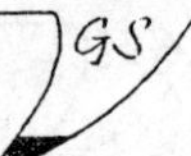

9.Nali	Ganganopar, Churu & Jhunghunu district of Rajasthan; Rohtak & Hissar in Haryana	Both sexes are polled. Ears are large and leafy. Fleece is white. Light brown face. Mature body weight of M-35 kg and F-24 kg. Medium sized. Tail is short and thin.	Fleece is white; coarse, dense and long stapled. Yield-2 kg/year.
10.Pugal	Bikaner & Jaisalmer district of Rajasthan	Black face. Body wt. M-32 and F-27 kg. Both sexes are polled. Ears short and tubular. Medium sized. Tail short medium and thin.	White fleece, medium carpet quality wool. Yield 0.8 kg/sheep.
11.Sonadi	Udaipur, Dungerpur, Chittorgarh	Adult wt. of M-36 kg and F-21 kg Both sexes are polled. Large ears flat and drooping. Light brown face. Medium-sized. Smaller than Malpura. Thin long tail. Udder, developed. Well-built.	White-fleece, extremely coarse and hairy. No wool on belly and legs.
III. Southern Region breeds of sheep			
1.Bellary	Bellary & Kurnool district (A.P.)	Colour black, grey and white patches. Rams-horned but ewes polled.	Coarse wool, yield is 1 kg.
2.Coimbatore	Madurai, Coimbatore district of Tamil Nadu	30 % males are polled. Body wt. of M-25 kg, ewe-25 kg. Medium-sized ear. Medium-sized. Tail small and thin. White body with black brown spots.	Fleece is white, coarse, hairy and open.
3.Deccan	Maharashtra and border area of A.P.	Black *grey colour. Rams-horned but ewes polled. Roman nose. Short tail. Top ears. Two long fleshy tassels grow under neck.	Meat purposes breed. Rough, hairy type coat for blankets.
4.Hasan	Hasan district of Karnataka	39 % males are horned. Ears-medium and drooping. Females are polled, adult body wt. M-26 kg, F-23 kg. Small-sized. White body with black spots.	Commonly for meat. Fleece white coarse, open. No wool on belly. Wool yield/sheep 0.7 kg.
5.Kenguri	Raichur district of Karnataka	Body wt. of M-32 kg, F-27 kg. Medium-sized, body colour-dark brown, sometimes white to black patches, Males are horned, female-polled.	Coarse hairy coat.
6.Macheri	Salem, Coimbatore district in Tamil Nadu	Both sexes are polled. Ears of medium-size. Hairy fleece. Light brown colour. Mature body, wt. of M-35 kg & F-22 kg. Medium sized. Tail short and thin.	Hairy type fleece short hairs, which are not shorn. Only for meat purpose.
7.Madras Red	Chingalpet & Chennai district of Tamil Nadu	Adult wt. of M-36 kg, F-23 kg. Dark brown colour. Ears medium-sized drooping. Medium-sized. Rams with twisted horns, ewes are polled. Short thin tail.	Body covered with short hair, not shorn.

GK GS

Yes! the preamble to the Indian Constitution can be amended.

8. Mandya	Mandya district of Karnataka	Body wt. M-35, F-24 kg. Both sexes are polled. Long ears, leafy drooping. Roman nose. Small sized. Tail short thin. White in colour with brown face.	Coat is extremely coarse and hairy. Only for meat purpose.
9.Nellore	Nellore (A.P.)	Deep red/fawn colour. Tallest breed of sheep in India with spirally twisted horns, short thin tail.	Hairy type wool yield 1 kg.
10.Ramnad	Ramnad district of Tamil Nadu	Body wt. of M-31 kg and F-23 kg. Males have twisted horns, females are polled. Medium-sized. Short thin tail. White body colour with black marking. Ears-medium sized.	Fleece very coarse, hairy and open.
11.Tirnchy	Tirnchy, Arcot, Salem, Dharmapuri district of Tamil Nadu	Adult wt. M-26, F-20 kg. Body colour-black. Males are horned; ewes are polled. Short tail and thin. Small-sized.	Fleece very coarse, hairy and open.

Key Characteristics of Poultry Breeds

Breed	Home	General features	Usefulness
EXOTIC BREEDS OF POULTRY			
1.Plymouth Rock	U.S.A.	Colour: white/buff/blue. Body: long, broad breast. Barred variety has a grey-white plumage. Males are lighter than female. Egg: big size.	Weight: cock 3.4 kg, hen 3.4 kg. pullet 2.7 kg, cockerel 3.6 kg/bird. Suitable for broiler.
2.Rhode Island Red	Rhode Island, U.S.A.	It has been developed by crossing number varieties. Colour: red buff/white and brown. Body: long and deep breast. Eyes and earlobe: red	Weight: cock 3.8 kg, hen 2.9 kg. pullet 2.5 kg and cockerel 3.4 kg/bird. It is good for meat purpose and some varieties are popular for brownish eggs.
3.New Hampshire	Developed from Rhode Island Red by the New Hampshire poultry farmers, U.S.A.	Colour: chestnut red plumage. Body: single comb, less rectangular in shape than Rhode Island Red.	Weight: cock 3.8 kg, hen 2.9 kg. pullet 2.5 kg and cockerel 3.4 kg/bird. It is a popular breed used as broiler. Egg brown shelled.
4.Leghorn	Italy It is popular breed in the Mediterranean area.	Colour: white/brown/black. Body: compact, small head, long shank, single comb.	Weight: cock 2.6 kg, hen 2 kg. pullet 2 kg and cockerel 2.2 kg/bird. Very popular for its early maturity, excellent laying capacity. It is highly demanded as broiler.
5.Minorca	Spain	Colour: red face, black shank.	Weight: cock 3.5 kg, hen 2.9 kg. pullet 2.5 kg and cockerel 2.9 kg/bird.
6.Cornish	England	Body: compact and close feather, heavy flesh very deep breast which is broad pea comb.	Weight: cock 4.5 kg, hen 3 kg. pullet 2.4 kg and cockerel 3.3 kg/bird.
7.Anstralorp	Australia Developed from the variety Black Orpington.	Colour: black beak, black or lead black shank and toes, feet bottom and toe bottom are pink-white, shiny green-black plumage. Body: long back, less massive.	Weight: cock 3.7 kg, hen 3 kg. pullet 2.5 kg and cockerel 3.4 kg/bird. Used also as a layer bird.

INDIGENOUS BREEDS OF POULTRY			
1. Desi	India	It is common indigenous breed seen in the country side. Body: small. Egg: small.	Good quality meat, highly priced. These are not good layers. But the eggs are tastier and good for stomach.
2. Bursa	Gujarat and Maharashtra of India. One of the four pure breeds of India.	Colour: a number of varieties of body color. Body: medium size, deep, lights feathered, resistant to parasitic and other common diseases.	Used both for meat and egg purpose.
3. Assel	Andhra Pradesh, U.P., Rajasthan. A pure Indian breed.	Colour: variety of colour like white, black, golden red, brown, light red. Body: long and slender face, long neck, round and short body and broad breast, small drooping tail, good fighter.	Weight: cock 4-5 kg, hen 3-4 kg. pullet 2.5-3.5 kg and cockerel 3.5-4.5 kg/bird. Good quality meat but not a good layer.
4. Chittagong	Eastern India A pure Indian breed.	Colour: long yellow beak, yellow featherless legs, plumage is buff, white, black or dark brown. Body: broad breast, fleshy, broad shoulder.	Dual purpose.
5. Karaknath	Jhabua and Dhar district of M.P. A pure Indian breed.	Colour: Silvery to bluish black plumage, slate like colour of skin, beak, shank, toe, sole, purple, comb, many internal portion are blackish, resistant to most of the diseases.	Weight: cock 1.5 kg, hen 1 kg.

DESIGN AND LAYOUT OF LIVESTOCK-POULTRY SHELTER

Livestock Housing

Main Aim

The main purpose of an adequate shelter is:

- Providing better housing at a cheaper cost.
- To give clean and comfortable accommodation.
- To guard animals from wild animals and theft.
- To save animal from sunburns, rain, hot and cold winds of the bad weather.

Benefits

The advantages of sufficient shelter are:

- Amplified production of milk.
- Appropriate disease control.
- Correct and controlled feeding of animals.
- Encouragement to other dairy farmers.
- Enhanced productive and reproductive efficiency of animals.
- Excellent fitness of animals.
- Improved care and management of animals.
- Production of top quality milk and milk products.
- Reduction in mortality rate of calves.
- Rising pride of dairy farmer.
- Superior employment of labour.

Site selection

A number of factors are to be considered for site selection of a dairy farm:

- Soil
- Sunlight
- Elevation
- Protection from incoming wind
- Availability of labour, market, electricity and water
- Durability
- Accessibility to the surroundings
- An overall view, etc.

Layout

It includes the following:

- Economic factors to be considered: capital, labour, nature of land, number of animals, forage land, etc.
- Distribution of land to different head: the land available in the entire farms should be distributed such that their proportion will be 2:2:2:1 for land under building+poddock+road, fodder, permanent pasture, cash crops, respectively.

A. Position and alignment of farm buildings.
1. Accommodation for cattle.
2. Buildings for dairy.
3. Additional essential buildings.

Housing requirement per animal

Age group	*Open Space (m²)*	*Manger Space (m²)*	*Covered Area (m²)*
Adult cows	11-12	0.8-1.0	1.8-2.0
Pregnant stock	15-20	1.0-1.2	8.5-10.0
Bulls	20-22	1.0-1.2	9.0-11.0
1-2 years	6-8	0.4-0.5	1.6.1.8
4-6 months	3-4	0.2-0.3	0.8-1.0
6-12 months	5-6	0.3-0.4	1.2.1.6

B. Separation of farmland into plots.

1. Land should be divided into sub-plots depending on the cropping system.
2. It should be near to the shed.
3. Well connected with roads and channels.
4. Plot should be easy to irrigate without much erosion.

C. Setting of pathways and roads.

1. Road should be straight, well-graveled, connected with all other houses and buildings.
2. Use necessary land with minimum wastage.

Heard Strength

The following would be strength of herd for 1000 milch animals

Animal class	*Number in the heard*	*Explanations*
1.Breeding heifers (1-2 yrs old)	250	-
2.Bullocks	80	-
3.Bulls	20	-
4.Cows in milk	750	(Nearly 3/4 of total cows)
5.Heifers below 1 yr	300	(Little more to replace and margin for culling)
6.Pregnant dry cows	250	(1/3 of the milch cows)
7.Pregnant heifers (2-3 yrs old)	200	(1/5th of cows to be culled every year)
8.Young calves below 6 months	400	

Housing Method

Basically, there are two types of housing systems for animal:

1. Loose Housing System

Merits and Demerits of Loose Housing & Barn System

Particulars	*Loose Housing*	*Barn System*
Comfort to animals	More	Less
Construction cost	Less	More
Cows remain cleaner	Less	More
Exercise due to movement	More	Less
Expansion	Easy and cheaper	Difficult and expensive
Heat detection in cows	Easier	Less, easy
Labour saving	More	Less
Profitability	More	Less
Quality of Milk	Better	Good
Sunlight benefit	More	Less

2. Cow House System or Barn System

a. Single storey

- Such barns are common in use.
- There is sufficient light and ventilation provision and bring saving in labour.
- It is standard stanchion barn.
  - i) Single row system (if cows number below 10)
  - ii) Double row system (for animals upto 50 in one)

- Tail to tail
- Head to head

} Arrangement by size, breed, health of udder or easiness of milking.

Good points and Bad Points of Tail to Tail & Head to Head Barns

Details	*Tail to Tail barn*	*Head to Head barn*
1.Back tracking of feed trolley	Yes	No
2.Chance of spread of disease	Less	More
3.Cleaning gutters	Easy	Difficult
4.Cost of construction	More	Less
5.Detection of injury on hind legs	Easy	Difficult
6.Feed distribution	Less Easy	Easier
7.Floor area of barn	More	Less
8.Gutter exposed to sunlight	Less	More
9.Look of animal's arrangement	Good	Better
10.Possibility of milk theft	Difficult	Easy
11.Supervision at milking	Easy	Difficult
12.Total floor area/cow	70 sq. ft.	65 sq. ft.
13.Walls remain cleaner	Yes	No

b. Double storey
- The storey on upstairs of "double storey barns" is used for storage of feed, concentrates and bedding materials.
- Ground floor is used for housing the dairy animals.

c. Basement storey

d. Round barns
- Such barns consist of silo tower in the centre and manger around it.
- Animals stand around the manger facing the silo tower.

e. Pen type barn
- It is one of the loose housing systems.
- Pen type housing system consists of 3 metres wide shed covered area for feeding purposes and protection from inclement weather like the sun, etc.
- Animals here move about freely.

Advantages and Disadvantages of Different Types of Barn

Specifics	*Single Storey*	*Double Storey*	*Round Type*	*Base Type*	*Pen Type*
1.Comfort to animal	Less	Less	Lesser	Least	More
2.Construction cost	More	Less	Lesser	Least	More
3.Contamination during clean milk production	Less	More	Less	More	More
4.Convenience in cleaning of mangers and gutters	More	More	Least	Less	Max.
5.Convenience in tying and untying of cows	More	More	Least	Less	Max.
6.Dampness	Less	More	Lesser	More	Least
7.Economy of floor space	Less	More	Max.	Lesser	Least
8.Floor space	Less	Less	Lesser	Least	More
9.Freedom to cows	Less	Less	Lesser	Least	More
10.Light and ventilation	More	Poor	Less	Poor	Max.
11.Protection from cold and heatstroke	Yes	Yes	Yes	Yes	No

Particulars of Barns Construction

1	Doors	Width-1.5 m in single row, 2.7 m in double row, Height-2.4 m.
2	Feed alloys	1.4 m wide enough for feed trolley.
3	Central path	1.8 m in tail to tail barn, smooth cemented with 1" slope from centre towards gutters.
4	Gutters	0.2 m depth, 0.6 m width and slope of 2.5 cm on every 3 metres length for easy cleaning.
5	Mangers	1.0 m width, Front wall-0.3 m, Rear wall-0.8 m high.
6	Floor	Stall per cow -Width 1.2 m and Length 1.5 m.
7	Walls	Height 1.52 m, Width-0.3m, Open space between wall and roof 0.9 m lengthwise only.
8	Roof-Height	4.6 m in the centre and 2.4 m on sides with the extension of 0.75 m beyond walls, asbestos or tile roof is good.

Indian Housing Standards for Livestock

Housing standards for livestock

Purpose		*Name of standard*	*Number given to the standard*
Housing of farm	1	Shed for 'Gaushala' and other organized milk producer.	IS: 8345 (Part III-1978)
	2	Shed for an average farmer for heavy rainfall and highly humid area.	IS: 5605 (Part I)
	3	Rural milk producer.	IS: 5605 (Part II-1970)
	4	Shed for 'Gaushala' and other other milk producer.	IS: 5605 (Part III-1970)
	5	Recommendation for shed for an average farmer of arid area.	IS: 8845 (Part II-1978)
	6	Shed for an average farmer recommendation for plain areas with medium rainfall.	IS: 4466 (Part II-1968)
	7	Shed for 'Gaushala' and other organized milk producer.	IS: 4466 (Part III-1968)
	8	Large dairy farms.	IS: 6027 (1970)
Milking shed	1	Recommendation for community.	IS: 5284 (1969)
Semen collection & transport	1	Recommendation for handling and transport of semen of breeding.	IS:8102 (1976)
Transport of livestock	1	Code for transport of cattle by rail.	IS: 4157 (Part II-1983). 1st revision
	2	Transport of sheep and goats by rail and road.	IS: 4157 (Part III-1983). 1st revision

Poultry Housing

Housing of poultry involves lot of cost only next to feeding. Adequate housing facility must be provided for better production of egg and meat.

Housing requirements

Light
- It should be distributed evenly.
- Young chicks require light all the time.

Ventilation
- It is required for fresh air circulation either naturally or artificially through electric fans.

Floor space
- Housing too many birds in limited space causes lot of problems. The following table shows the recommended space for different categories of bird:

GK — Through Jalep la pass generally, trade between India and Tibet takes place. — GS

For Chick	: Floor space requirement are 0.045, 0.09, 0.18, 0.18/chick for the age group of 0-4, 4-8, 8-12 and more than 12 weeks. The corresponding minimum feeder space requirements are 2.5, 6.5, 7.5 and more than 7.5 m^2/chick, respectively.
For Broiler	: Space requirements are 0.06-0.09 m^2/bird on floor for 0.12 weeks age group, 0.45-0.027 m^2/bird on plastic cage for 0-12 week age group.
For Breeders (adult)	: 0.09 m^2/bird in single cage, 0.045 m^2/bird in double cage and 0.045-0.066 m^2/bird in colony cage.
Space in Pen	: For leghorn with 25, 100, 200, 500 birds per pen the space requirements are 0.32, 0.30, 0.22 and 0.18 m^2/bird respectively. The corresponding figures for heavier birds are 0.36, 0.31, 0.30 and 0.22 m^2/bird, respectively.

Night shelter

- If some birds are reared on range system, i.e., they are set free during the day but comes back at night for shelter and feed.
- Shelter size per 12 birds- 1.2 x 0.9 x 0.9 m with 0.45 m high legs, weather-proof roof slope from 0.9 to 0.6 m above floor at the edges, water-pot, feed-vessel with feed.

Deep-litter shed

- It is a modern method to keep the poultry bird all the time inside the pen.
- The faeces of the birds make a good organic fertilizer with the litter material below.
- Those are made from straw, bamboo and mud.
- Floor space: 9 m^2/25 birds.
- The shed should be 2.7 m high.

Gable housing

- It is recommended for intensive system up to 500 layer birds accommodating in a number of pen each 2.5m x 1.8m for 12 to 16 birds.
- Suited to all weather conditions, low cost, collection of eggs is easy from the passage between the pens.
  Poultry house classification as per Bureau of Indian Standard.
- Brooder house-to accommodate chick of 1 to 10 day old.
- Layer house-to accommodate birds of age group 20 weeks and more.

ANIMAL NUTRITION

Nutrients and Their Functions

Nutrient	*Functions*
Carbohydrate	• Supplies energy for the different activities of the body like respiration, perspiration, digestion, blood flow, etc.
Fat	• Supplies energy to the body. • Makes the feed more palatable and tasty. • Essential fatty acids like linoleic and linolenic acids must be supplied for optimum performance of the animal. • Quality of milk depends on the quality of fat supplied in diet. • Energy is stored in the form of fat in animal.
Minerals	• Calcium and Phosphorus needed for bones and teeth formation in growth and general growth. • Calcium prevents milk fever in cows. • Cobalt is required for vitamin B_{12} synthesis. • Copper is essential for reproductive capacity, its deficiency may cause loss of heat of animals. • Iron is necessary for metabolism of energy, enzymes in the oxygen uptake, growth and reproduction. • Magnesium helps in prevention of tetany, slowing down of growth, etc. • Manganese in development skeleton, reproduction and general growth. • Minerals are needed for digestion, absorption and utilization of nutrients. • Phosphorus plays important role in carbohydrate metabolism, prevent Pica (loss of appetite) and coloured urination. • Selenium has been established as an essential element in enzyme system. It prevents muscular dystrophy. But sometimes it proves toxic causing peeling of skin. • Sodium and potassium maintain osmotic pressure of cells, prevention of loss of appetite, reduction of growth and milk production. • Sulphur is an essential component of protein required for proper growth of rumen microbes, a N : S of 10 : 1 in the feed of the animal is required for better utilization of non-protein nitrogen like urea. • Zinc is required in minute quantity for hair growth and skin.

Protein
- Building block of other proteins.
- Enzymes formation.
- Formation of blood protein required in the respiration.
- Formation of milk protein, egg protein.
- Primary need for making muscles and connective tissues.
- Ruminants can convert non-protein-nitrogen into protein. So they may be supplied with such material.
- Supply of amino acid.
- Supply of energy in case of energy-deficient ration fed to animals.
- To limited extent for bone formation.

Vitamins
- Nicotinic acid prevents pellagra type of disease in pig and poultry.
- Vitamin A and E prevent sterility.
- Vitamin A prevents loss of weight, loss of appetite increases resistance to disease and prevents night blindness.
- Vitamin B_1 prevents beriberi, loss of appetite and slow growth.
- Vitamin B_{12} prevents non-coordination of movements.
- Vitamin B_2 prevents anemia, general weakness, pellagra, stiff-leg and curled toe paralysis in poultry, skin thickening in swine, and loss of hair.
- Vitamin B_6 prevents skin lesions in poultry and pig.
- Vitamin C prevents scurvy.
- Vitamin D is the basic need for bones and teeth formation, metabolism of calcium and phosphorus. It prevents milk fever.
- Vitamin K helps in the coagulation of blood in poultry.
- Vitamins needed for nutrient utilization, health, growth and to prevent nutritional deficiency diseases.

Water
- Acts as solvent for absorption of nutrients and excretion of water.
- Approximately 80% part of growth tissues is water.
- Helps in chemical reaction and respiratory functions by gaseous exchange.
- Hydration of protoplasm of body cells.
- It provides rigidity to cells.
- Maintains osmotic pressures of body.
- Refractory medium for light in eyeball.
- Regulates body temperature.
- Softening and lubrication of food during digestion, transport of food to different parts of the body.

Nutrients in Feeds

Feeds

1. **Dry matter**

A. **Organic matter**

i. Nitrogenous matter

a. True protein
- Essential amino acids (10)
- Non-essential amino acids (14)

b. Non-protein nitrogenous matter (NPN)

e.g. amines, amides, nitrates, nucleic acid, nitrogenous glycosides (ammonium salts, ammonia, urea, creatinine).

ii. Non-nitrogenous matter

a. Carbohydrates
- Insoluble in water (crude fiber contains: cellulose and hemicelluloses).
- Soluble in water (sugars (nitrogen free extract), starch, pectins, organic acids, resins).

b. Fats or other extract

e.g. fat, oil, wax, organic acids, pigments, sterols, fat soluble vitamins.

B. **Inorganic matter**

i. Macro elements

e.g. Ca, Fe, P, K, S, Na, Mg, Ca

ii. Micro elements

e.g. Co, Cu, I, Mn, Zn, Mo, S, Bo

Mineral:
1. Acidic
2. Basic
3. Others

C. **Vitamins**

i. Water-soluble

e.g. Vitamin B, C

ii. Fat-soluble

e.g. Vitamin A, D, E, K

2. **Water**

Classification of Feed Stuffs

A. Concentrate (Contains less crude-fibre and more TDN)

a. Feed rich in carbohydrates

i. Grains (e.g. maize, oat, barley).

ii. Grains by-products (e.g. rice bran, wheat bran, gram and pulse chuni, rice polish).

b. Feed rich in protein

i. Oil seed and cake (e.g. cotton seed, mustard cake, linseed cake, groundnut cake).

ii. Animal and industrial by-products (fish meal, meat meal, blood meal, bone meal).

B. Roughage (Contains more crude-fibre and less TDN)

a. Non-leguminous fodder (Contains less DCP)

   i. Green fodder (e.g. sorghum, maize, napiar, paragrass).

   ii. Dry fodder (wheat bhusa, paddy straw, oat hay).

b. Leguminous fodder (Contains more DCP)

   i. Green fodder (e.g. berseem, cowpea, lucerne).

   ii. Dry fodder (e.g. pea bhusa, arhar bhusa, lucerne hay).

Digestible Nutrients in Different Feeds

Concentrates

Grains	DCP%	TDN%	Oilseeds and Cakes	DCP%	TDN%
Arhar	13.0	68.0	Seasame cake	38.0	78.0
Bajra	6.5	58.0	Linseeds	7.5	85.0
Barley	7.5	77.0	Cotton seed	8.0	87.0
Gram	12.8	75.0	Groundnut cake	41.0	74.0
Jowar	7.0	68.0	-	-	-
Maize	7.5	84.0	Linseed cake	30.0	65.0
Methi	18.0	62.0	Mustard cake	26.0	80.5
Oat	7.0	71.0	Cotton seed cake	18.0	72.0
Animal and Industrial byproducts			*By products of Cereals*		
Blood meal	89.0	90.0	Rice polish	12.0	67.0
Bone meal	21.0	69.0	Rice bran	7.0	60.0
Fish meal	45.0	64.0	Pulse chuni	12.0	68.0
Meat meal	50.0	65.0	Gram husk	0.3	54.0
Skim milk	3.5	10.0	Wheat bran	11.0	67.0

Roughages

Dry Roughages	DCP %	TDN %	Green Roughages	DCP %	TDN %
Arhar bhusa	4.0	45.7	Guar	1.5	10.0
Barley bhusa	0.8	44.0	Jowar silage	1.0	18.0
Berseem	2.8	12.0	-	-	-
Berseem hay	9.0	50.8	Pea	2.5	12.5
Cowpea hay	9.5	45.0	Sorghum	1.0	16.0
Dry doob	4.5	40.0	Para grass	1.5	11.4
Jowar karbi	1.0	48.0	Oat	2.6	16.7
Lucerne hay	11.0	50.0	Cowpea	3.2	10.8
Masoor bhusa	4.0	44.0	Maize	1.2	16.5
Oat-straw	0.8	41.0	Maize silage	1.1	10.0
Paddy straw	-	45.0	Napier	1.5	15.8
Pea straw	1.5	40.0	Lucerne	3.5	10.0
Wheat bhusa	-	40.0	Guinea Grass	1.1	12.5

DCP=Digestible Crude Protein Nutrients, TDN=Total Digestible Nutrients

Standards of Feeding for Livestock

These standards are the nutritional requirements of animals for their optimum growth, development and maintenance. There are both Indian and International feeding standards.

Feeding standards classification

A. Comparative type
- Thae's
- Scandinavian's

B. Digestive nutrient type
- Gruven's
- Kuhnis
- Wolff's

C. Productive type
- Hackeri's
- Kellner's
- Savage's
- Armsby's

A. Indian Feeding Standard (1957)

Based on Morrison's feeding standards, Dr. K.C. Sen collected the Feeding Standard. After 5 months of pregnancy, a cow requires 0.3 lbs DCP and 1.5 lb TDN. Requirement for bullocks were simplified into two categories, light and heavy work.

Requirement of Working Bullocks (kg)

Body wt.	For normal work			For heavy work		
(kg)	Dry feed	DCP	TDN	Dry feed	DCP	TDN
200	4.0	0.24	2.0	5.0	0.25	2.7
300	5.8	0.33	3.1	7.0	0.42	4.0
400	7.6	0.45	4.0	9.8	0.57	4.8
500	9.4	0.56	4.9	11.2	0.71	6.4
600	11.2	0.66	5.8	13.4	0.82	8.0

Maintenance requirement of mature breeding bulls (kg)

Body wt. (kg.)				
400	7.5	0.250	4.2	15.4
500	8.3	0.300	4.6	16.6
600	9.6	0.345	5.4	19.5
700	10.9	0.390	6.1	22.1
800	12.0	0.430	6.7	24.2

Daily Nutrient Requirements of Cattle

Body wt.(kg)	Dry feed (kg)	DCP (kg)	TDN (kg)	Mcal. (kg)
	Maintenance Requirement (kg)			
200	3.5	0.150	1.7	6.0
250	4.0	0.170	2.0	7.2
300	4.5	0.200	2.4	8.4
350	5.0	0.230	2.7	9.4
400	5.5	0.250	3.0	10.8
450	6.0	0.280	3.4	12.4
500	6.5	0.300	3.7	13.2
550	7.0	0.350	4.0	14:4
600	7.5	0.350	4.2	15.5

Maintenance Plus Pregnancy Requirement (kg)

	During last two months of gestation			
250	4.9	0.27	3.0	10.8
300	5.6	0.29	3.4	12.4
350	6.4	0.32	3.7	13.2
400	7.2	0.35	4.0	14.1
450	7.9	0.40	4.4	15.9
500	8.6	0.43	4.8	17.3
550	9.3	0.46	5.2	18.8
600	10.0	0.50	5.6	20.2

Production requirements per kg of milk (kg)

		Fat%		
3	-	0.04	0.270	0.97
4	-	0.045	0.315	1.13
5	-	0.051	0.370	1.28
6	-	0.057	0.410	1.36
7	-	0.063	0.460	1.54
8	-	0.069	0.510	1.80

B. International Feeding Standard

Thae's Feeding Standard (1810)

He introduced a feeding standard as early as knowledge of nutrient value was not developed that much. 100 lbs. of meadow hay = 91 lbs dry barseem = 200 lbs of potato. Meadow hay used as unit for comparison with no knowledge of chemical value of feed.

Scandinavian's Feeding Standards

Feeding Standard = 100 lbs body wt. requires 0.065 lbs digestible crude protein. For 1 lb milk production add 0.05 lb CP.

Gruven's Feeding Standards (1859)

For every 1000 lbs body wt. a cow requires dry matter 28.7 lbs, crude protein 2.67 lbs, crude fibre 0.86 lb, crude carbohydrate 14.55 lbs.

Wolff's Feeding Standards (1864)

For every 1000 lbs body wt. needs dry matter 24.5 lbs, digestible crude protein 2.5 lbs, digestible crude fibre 0.4 lb and digestible carbohydrate 125 lbs.

Kuhnis Feeding Standards (1867)

He worked out separately the nutrient requirement for maintenance, growth, work and production.

Wolff's Lehmann Feeding Standards (1896)

A cow of 1000 lbs body weight requires dry matter 18 lbs, digestible crude protein 0.7lb, crude fibre 0.1 lb, crude carbohydrate 8 lbs for maintenance and production need dm 25, digestible crude protein 1.6, crude fibre 0.3 and crude carbohydrate 10 lbs.

Hackeri's Feeding Standards (1903)

A cow of 900 lbs requires digestible crude protein. 63 lbs, digestible fat 0.09 lb, digestible carbohydrate 6.3

You Don't Know!

Youngest CM

Prafulla Kumar Mahanta (b. Dec. 23, 1952), became the youngest CM to be sworn in on Dec. 24, 1985 at the age of 33.

lbs and total digestible nutrients 7.132 lbs for maintenance and production of milk of 4% fat required 0.054 lb digestible crude protein, 0.021 lb digestible fat, 24 lbs digestible carbohydrate, 0.341 lb total digestible nutrients.

Kellner's Feeding Standards (1907)

He introduced a new feeding standard based on starch equivalent where 1 lb digestible crude protein = 0.94 lb starch, 1 lb fat from oil cake = 2.4 lbs starch, 1 lb digestible fat = 1 lb starch, 1 lb digestible fibre = 1 lb starch.

Savage's Feeding Standards (1912)

1000 lbs of body wt. needs 0.7 lb DCP, and 7.925 lbs TDN.

Armsby's Feeding Standards (1917)

He worked out a standard based on digestible true protein (DTP) and net energy value of the feed.

Morrison's Feeding Standards (1915)

He introduced a new feeding standard which he published in the book 'Feeds and Feeding Ist edition' which is subsequently revised in 1936 and 1956 respectively.

NRC Feeding Standards (1945)

On the line of Morrison the NRC introduced a somewhat modified feeding standards in terms of DCP and TDN which is revised in 1970.

PRINCIPLES OF RATIONING

Principles of Rationing of Dairy Cattle

1. Quality of an ideal ration:
    - Abundant succulent green feeds.
    - Ample amount of different nutrients.
    - Assorted feeds in it.
    - Cost-effective.
    - Makes desirable flavours to milk.
    - Non-toxic.
    - Possibly fresh.
    - Quite appetizing.
    - Simply palatable or digestible.
    - Well-balanced.
2. After milking, fodders like silage must be fed so that off flavour may not be imparted to milk.
3. Always either the dry fodder like wheat bhusa is

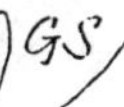

Head office of Reserve Bank of India is at Mumbai.

mixed or fed first and then greens.
4. Animals must be fed individually.
5. Animals should not be overfed with concentrate.
6. Avoid feeding hay just before and at milking time. It may create dusty atmosphere in the barn and adversely affect quality of raw milk.
7. Avoid sudden change in rations of dairy animals.
8. Bulky to keep normal tone of peristaltic movement of the alimentary canal.
9. Concentrate provided to dairy cows must contain 15 to 17.5 DCP and 7.5 TDN.
10. Dairy animals can consume total dry matter 2 to 3 % of their live body weight. Hence, total need of dry matter is accordingly determined.
11. Dairy animals must be fed liberally.
12. Dairy cows need large intake of water daily to meet their maintenance and production requirement.
13. Depending upon feed quality, palatability and demand of animal, maximum quantity of green fodder can be given to cow varies from 35 to 40 kg/day.
14. Feeding rations at regular scheduled intervals daily.
15. Green fodder must be sufficiently provided to ensure maximum consumption of dry matter.
16. Laxative to keep normal digestion.
17. Leguminous green fodder like berseem, lucerne, cowpea etc., must not be fed on empty stomach to dairy animals as this may upset the digestion and cause bloat.
18. Plentiful supply of clean water should be made available to dairy cows at least 3 times daily, if not at all times.
19. Prefer feeding grains at milking time as this becomes part of stimulus for letdown of milk.
20. Provide good effect on health.
21. Shortage of drinking water would cause serious depression of milk yield.
22. To replace concentrates @ 1 kg. concentrate for 12 kg. of greens, improved varieties of green fodders and legumes such as lucerne, berseem, cowpea could be fed to dairy animals.
23. Concentrate may be supplied to cows as follows:

	Maintenance	*Production*
Cows	1 kg	1 kg for every 3 kg milk
Buffaloes	1.5 kg	1 kg for every 2.5 kg milk

Actual Requirement of DM, DCP and TDN for a Cow

All the above standards are difficult to follow singularly. So actual requirement for cow could be arrived at (by using those standards) as follows:

a) For maintenance: 15 kg DM, 0.325 kg DCP, 4.2 kg TDN.

b) For production of 10 kg milk: 0.54 kg DCP, 4.1 kg TDN.

So, a) + b) = 15 kg DM, 0.865 kg DCP, 8.3 kg TDN for a milking cow. Considering 90% DM in concentrates. We can chalk out a balanced ration supplied through different feeds.

Balanced Ration for a Milking Cow/Day

The following ration will provide approximately the required feed for a 400 kg (body wt.) cow producing 10 kg milk (5% fat) per day: 25 kg green cereal supplying 0.30 kg DCP and 4 kg TDN + 15 kg green legume supplying 0.43 kg DCP and 1.9 kg TDN + 1 kg concentrate mixture supplying 0.14 kg DCP and 0.6 kg TDN + 2.1 kg cereal straw supplying o.8 kg TDN + 50 gm common salt + water (of free choice).

Supervision of Animals

Daily supervision with regard to:

- Culling.
- Dehorning.
- Disbudding.
- Drying-off of cows 6 to 8 weeks prior to calving cows.
- Exercise.
- Grooming.
- Handling and Disposal.
- Housing.
- Injuries.
- Kindness.
- Maintenance of tools.
- Milking.
- Observation of heat.
- Periodical trimming of hoofs.
- Permanent marking.
- Quantity and quality of food.
- Sanitation.
- Selection.
- Service.
- Training.
- Watering.

LIVESTOCK RECORDS

Purposes

1. Better pricing of animal.
2. Growth and health condition based on body weight.
3. Improvement of herd.
4. Income and expenses of farm and efficiency of labour.
5. Pedigree and history record.
6. Progeny testing of bulls.
7. Selection and culling of animals.
8. Supervision and management of herd.
9. Systematic programme of breeding.
10. To get help in evaluation.

Modern Periodic Law was started by Moseley.

GS

Methods

- Bound register.
- Envelopes.
- Loose leaf system.

Types

a) General records
  - Attendance.
  - Bill book.
  - Inventory record.
  - Ledger cash book.
  - Purchase book.
  - Record pay records.
  - Stock book.

b) Farm section records
  - Field register.
  - Fodder cultivation record.
  - Implements book.
  - Labour sheet.
  - Muster roll.

c) Cattle section records
  - Breeding and calving.
  - Bulls assignment register.
  - Calf register.
  - Cattle feed receipt.
  - Daily reports register.
  - Feed consumption record.
  - Feed order book.
  - Health registers.
  - History and pedigree sheet.
  - Livestock register.
  - Milk record register.
  - Milk summary.
  - Pregnancy record, weight register.
  - Progeny register.

MILKING SYSTEM

Benefits

- Helps in reducing cost of milk production.
- Producing high quality milk without detriment to health or udder.
- Proper labour utilization.

Milking methods

1. **Hand Milking**
   (i) Full hand method (Best method).
   (ii) Stripping.
   (iii) Knuckling.

Hold the diagonal teats and milking quickly, silently and completely—with dry hand.

Milking a cow should be completed within 6 to 7 minutes as the let-down effect of Oxytocin hormone remains up to 7 minutes only.

One milker under hand method can milk 12 cows at a stretch efficiently.

2. Machine milking

Milking machine is generally operated at 10-15" vacuum with a pulsation rate of 50 per min.

Steps

Step 1: preparing 1st cow for milking and then 2nd cow, placing the teat cups on the udder of Ist and then on the 2nd cow.

Step 2: machine on.

Step 3: now prepare 3rd cow for let-down.

Step 4: strip the Ist by machine and empty the milk pail.

Step 5: now rinse the teat cups in clean water and then in 200 ppm Chlorine solution and place the machine on 3rd cow.

Step 6: prepare 4th cow for let down and go back to 2nd cow for stripping and so on the routine is followed.

Number of units of milk machine to be handled/milker is as follows:

Type of Operator	*No. of units of machine/milker*	*Cows milked per hour*
1. Slow/beginner	One single unit	15-17
2. Good milker	Double single units	25-30
3. Very efficient and a helper	Three single units	40-55

PRODUCTION OF HYGIENIC MILK

Principle

- Free from harmful bacteria and other disease organisms.
- Handled under sanitary conditions.
- Have good flavour.
- Have longer keeping quality.
- Must be clean.

It is more important to produce good quality milk for fluid use than when milk is converted into dairy products.

Purpose

- To boost self-life of milk.
- To maintain milk free of disease organisms.
- To compose good quality dairy products.
- To check bad odour.
- To stop spread of milk-borne diseases.
- To make milk free of dirt.
- To safe milk of low bacterial count.

Hurdles

Milk production in India is characterized by—

1. Majority of non-descript 80% cows and 60% buffaloes.
2. Millions of small producers with little/no holdings.
3. Poor keeping quality of milk.
4. Scarce land for pastures and fodder production.
5. Seasonal and regional imbalances.

6. Use of recyclable crop residues and natural herbage.
7. Widely scattered milk producers.
8. With or without costly concentrates.

Limitations in improvement of milk production

A. Economic constraints
B. Infrastructural deficiencies
C. Managerial lackings
D. Situational problems
E. Technological problems

Advice

A. Genetic Improvement
B. Health Care
D. Reproductive Management for Improving Fertility
E. Method of Maintaining High Level of Fertility in Herd
F. Mineral Profile and Breeding Performance
G. Culling of Infertile Males
H. Inter-relationship of Semen Test Characteristics and Fertility
I. Factors Interfering with fertility
J. Artificial Induction of Lactation

SUSTAINABLE LIVESTOCK PRODUCTION

Enrichment of Milk Production

Nutritional Security

Rising Development

Sustainability

HEALTH CARE OF LIVESTOCK

Control Measures of Diseases

- Change in pastures
- Disinfection of calving boxes and sanitation
- Isolation/segregation of animal
- Personnel hygiene
- Prompt report to livestock authorities about outbreak
- Proper disposal of carcass and litter material
- Proper nutrition
- Quarantine
- Timely visit and constant watch
- Vaccination

Disease	Vaccine	Vaccination time	Immunity
Anthrax	Spore vaccine	May/June	One season
Black quarter (BQ)	Formal killed	May/June	One season
Brucellosis	Cotton strain-19	-do-	3 to 4 calvings (to be done in infected herd)
Foot & Mouth	Polyvalent tissue culture	6 months age and booster dose after 4 months	One season
Haemorrhagic Septicemia	Oil adjuvent	May/June	One season
Rinderpest	1. Caprinized for Zebu 2. Lapinised for crossbred	6 months age	Repeat after 3 years
T.B.	B.C.G.	-do-	2 years

Control of External & Internal Parasites

Control Measures of External Parasites

Insecticides	*On Animals*		*In Cow shed (in solution)*
	In solutions %	*Dusting powder %*	
Lindane	0.25	1.0	0.25
Malathion	0.5	4.5	0.5
Pestoban	2.5	–	2.0
Sevin	0.80	–	0.8
Sumithion	0.5	–	0.1

Control Measures of Internal Parasites (Worms)

Parasites	Therapeutic agents	Dose
(1) Flukes & Amphistomes		
Liver Fluke	Distodine tab. Carbontetrachloride Avlothane	10-15 single dose 1 ml/50 kg body wt. 200-250 mg
Stomach Fluke	Zanil	30 ml/100 kg body wt.
(2) Cestodes		
Tapeworm	Taenil (Indian Herbs) Wopell	Calf 10-12 gm and adult 60 gm. Calf 15 gm and adult 60 gm per kg body wt.
(3) Round Worms (Ascariasis)	Helmacid Piperazine adipate Vermex	4 ml 200-400 mg 0.5 ml
Strongyles (Gastro-intestinal Parasites lungworms) Oesophago-stomrum	Thiabendazole Nilverm Banminth II Phenovis	50-80 mg 15 mg 10-15 mg

Parasiticidal Therapy

ANTHELMINTICS

Drugs active against common helminths in domesticated animals.

Heart Worms: Arsenamide
Diethyl Carbamazine
Stibophen(Faudin)

Hook Worms:	Hexyl-resorcinal
	N. Butyl chloride
	Neo Bedermin & Bedermin
	Oil chenopodium
	Tetrachloroethylene
Liver flukes:	Carbon tetrachloride
Round worms:	Carbon tetrachloride (horse)
	Hexylresorcinol, Methyl Benzene (dog)
	Nicotine sulphate (poultry)
	Oil chenopodium (calves)
	Carbon disulphide (horse)
	Sodium fluoride (swine only)
	Tetrachoroethlene, N. Butyl chloride
	Hexachloroethane
Stomach worms:	Cu. Nicotine (cattle & goat)
	Phenothiazine (Cattle & goat)
	Carbon disulphide (swine & stomach Bots in horse)
Strongyles :	Phenothiazine (horse & cattle)
	N. Butyl chloride
Tapeworms:	Arecoline, Nemural, Cestrsol & Dicestal
	Cu-Nicotine
	Lead arsenate
	Oil of Turpentine
	Carbon tetrachloride
Whip worms:	Methyl Benzene
	N. Butyl chloride

Domestic Animals Diseases Spread via Arthropods

Arthropod species	Disease spread through	Animals affected by it	Control/ Treatment measures
1	2	3	4
1. Black and pitted tick	Gall sickness	Cattle	5 to 7 days dipping daily.
2. Bont Tick (*Amblyomma spp.*)	Heart water	Cattle, sheep	5 to 7 days dipping daily.
1. Brown Tick (*Rh. appendiculotus*)	Red wa ter & Gall sickness	Cattle, sheep	5 to 7 days dipping daily.
2. Common house fly (*Musca domestica*)	Various bacterial infections, including anthrax.	Cattle, sheep, goat, pigs, poultry	Good sanitation. Use of insecticides spray like Nuvon.
3. Dog Tick (*Rhipicephalus sanquineus*)	Gall sickness	Cattle	5 to 7 days dipping daily.
1. Hard tick (*Ixodes pilosus, Boophilus spp.*)	Tick paralysis Red water Gall sickness	Cattle, sheep Goats	Use of foot bath 14 days dipping.
2. Horse-fly (*Tamabides* spp.)	Surra Anaplamosis	Camels, horses, cattle, dogs and elephants.	Treatment of temporary protection of infected or exposed stock by chemotherapeutic methods.
3. Mideges (*Culicoides* spp.)	Blue-tongue	Sheep and Cattle	Measures designed to keep susceptible animals always from Culicoides-infected places.
4. Red Tick (*Rh. evertsi*)	Red water & Gall sickness	Cattle, sheep	5 to 7 days dipping daily.
5. Soft tick (*Argas persicus*)	Spirochatosis	Poultry	Disinfestations, use of benzene hydrochloride and 10 days quarantine.
6. Stable-fly (*Stomoxys* spp.)	Surra, Antrax, Anaplasmosis	All domesticated animals Cattle	Immunization and/or hygiene.
7. Tsetse-fly (*Glossina spp.*)	Trypanosomiasis (sleeping nagana, surra, etc.)	Cattle, horses, sheep, goats and pigs, etc.	Eradication of the tsetse fly may be attempted by : (i) Bush clearance. (ii) Destruction or removal of game & other hosts. (iii) Insecticides.

Knowledge World!

First woman CM

Sucheta Kripalani (1908-74) was Chief Minister of Uttar Pradesh from 1963 to 1967.

GK GS

States Reorganization Act was passed in 1956.

Control measures and treatments of common diseases of animals

Disease	Spread through	Incubative period signs	Symptoms	Animals affected by it	Treatment/Control measures
1	2	3	4	5	6
1. Anthrax (Bacillus anthracis)	Water and feed contaminated with blood and excretions. Wound infections.	1-3 days or longer	Bleeding from longer natural, orifices. Trembling, staggering. Convulsions. Sudden death. Very high fever.	Mostly cattle, buffaloes, sheep, goats and horses less commonly.	Animal vaccination by sores vaccine 1 ml s/c in May. Vector control. Antiserum 100 ml in diseased animals. Hygienic measures.
2. Black quarter (Clostridium chavuei)	Dystokia, Oversize calf, weakness. Infection, mishandling.	2-5 days	Sudden high fever. Gas gangrene swelling produces crepitating sound when pressed. Death before 48 hrs.	Cattle and sheep buffaloes, goats	Biannual vaccination Anti BQ serum 10 ml in diseased 2i m. oxysteclining 50 ml I/V.
3. Contagious abortion (Brucella abortus)	Feed, water etc., contaminated by discharge and aborted foetus.	21 days	Abortions. Fulltime still births. Retained fetal material afterbirths. Reduced milk joints.	Cattle and buffaloes	Avoid natural breeding. Sanitation in A.I. Vaccination before breeding.
4. Contagious bovine pleuro-pneumonia (Mycoplasma)	Direct contact (Diseases not found in North-East).	4 weeks	Deaths at frequent intervals. Frequent painful and subdued cough. Prolonged unthriftiness.	Cattle	Annual vaccination. Destruction of affected animal.
5. Enteroloxemia (Clostridium perfringens)	Contamination of food and water.	2-3 days	Abdominal pain. Blood tinged diarrhea. Convulsions and death. Lamb dysentery.	Goats	Multi-component vaccine 2.5 ml s/c inj. in goats. Use of hyper-immune serum.
6. Foot and Mouth Disease	Direct contact or material contaminated with discharge from lesions. Virus, A,O, C, S- I, II, II.	4 weeks	High fever. Lameness. Reduced feed intake. Salivation. Sores on feet and tongue.	All domestic animals	Sanitary measures. Segregation. Vaccination in February.
7. Fowl cholera	Contaminated with excretions (Pasteurella sp.) Direct contact by eating or inhaling material.	2-5 days	Discharge from eyes and nose. Gasping inability to stand. Sudden death.	Duck, geese, fowl	Antibiotics. Segregation. Sulphonamides (antíbacterial). Vaccination.
8. Fowl Pneumonia	Spores from infected litter, food etc.				
9. Fowl pox virus infection	Direct contact through scratches wounds, by Ectoparasites.	3-15 days	Discharge from nostrils. Rough, brown, wart-like sores on the head. Sore mouth and eyes.	Poultry, turkeys, sometimes guinea fowls, ducks and geese.	Egg adapted vaccine. Vaccination twice yearly by Fowl pox B.M. strain vaccine 1/m.

The Hathigumpha inscription is associated with Kharvela.

GS

Disease	Spread through	Incubative period signs	Symptoms	Animals affected by it	Treatment/Control measures
1	2	3	4	5	6
1. Fowl spiro-choetosis	Infective ticks (Blood parasite)	5-9 days	Diarrhoea. Fever profuse. Intense thirst.	Fowl, ducks and geese	1/m inj. Penicillin. Tick elimination.
2. Haemorrhagic septiceamia (Pasteurella-multocida)	Contaminated feed and water by blood-sucking insect. Direct contact. Fatigue.	2 days	Death in 25 hours. High fever (107°F). Swallowing and breathing difficult. Swelling in throat region.	Cows, Buffaloes	Antiserum 150 ml s/c. Dexona inj. 1/m. Diadin 34% inj.50 ml. Hygiene. Oil adjuvant vaccine in May. 2 ml s/c. in diseased. Segregation.
3. Johne's (Mycobacterium paratuberculosis)	Contaminated feed and water	6 months	Continuous emaciation. Anaemic exhaustion and death. Unthriftiness. Frequent diarrhoea. Swelling under jaw.	Cattle, Sheep, Goats	Proper disposal of excreta. Sanitation. Segregation. Vaccinations 0 5-10 ml of culture vac.
4. Mastitis (Str. Agalactiae, Corynebacteria, E. coli, Staphylococci, Fungal, Viral.	Infection of mammary gland by hand. Wet floor. Incomplete milking injury to teats.	3-4 days	Blood clots. Swelling of one or more quarter. Pain in udder. Clots of milk in the initial stage. Turbid thin milk.	Cattle buffaloes and goats	Antibiotic infusion in teats. CMT test. Hygienic milking- keep floor dry and clean use antiseptic wash for udder. Strip cup test.
5. Milk Fever	Acute calcium calving deficiency	72 hrs. after calving	Brief excitement. Muscular twitching animal staggers. No fever. Sits keeping head purined on broty. No rumination tympany.	Cows, buffaloes	Feeding bone meal 60 gm daily MIFEX-400 ml 1/v. Intravenous infusion of Calcium boron fluconate. Prenatal Vit. D.
6. Pleuropneumonia (Mycoplasma mycoides)	Direct contact. Throughout material contaminated with nasal discharge	2-6 days	Fever. Head kept down. Nasal discharge. Off the feed. Painful husky cough.	Goats and cattle	Sanitary. Segregation. Oxysteclin/Erythromycin 20 ml inj. 1/m daily.
7. Profuse vagina & uterus	Dystokia. Oversize calf. Weakness. Infection. Mishandling.	Before and after calving any time	Whole vagina or uterus come out and are visible.	All domestic animals	Formal killed. Homeo-Sepia 10m. Keep hind quarter on raised floor. Siquiling 5 ml 1/m. Use dystokia clamp. Wash organs with antiseptic wash and replace.
8. Rabies (virus)	Contamination of food and water. Direct contact through biting. Wound.	7 days	Biting other animal. Frothy saliva. Hydrophobia. Incardination.	All animals specially canine	Antirabies-Biomed (Hep.) vaccine 2 ml upto 7 days. BPL inactivated. Rabies cell culture 5% carbolised 20 ml in cows upto 154 days.
9. Ranikhet (virus)	Direct contact by eating or inhaling material contaminated with excretions.	6-9 days	Discharge from eyes and nose. Gasping inability to stand. Sudden death.	Duck, geese, fowl	Lasota F1-strain at 1 week. R_2B strain at 8 weeks age. Sanitary. Segregation. Vaccination.

Disease	Spread through	Incubative period signs	Symptoms	Animals affected by it	Treatment/Control measures
1	2	3	4	5	6
1. Retention placenta	Infection. Lack of tenacity of muscles. Weakness.	Just after calving	A portion of placenta hanging out of vulva, leads to enteritis. Neglected cases emit bad smell.	Cows, buffaloes goats	Give replenta 50 g. Inj. Oxytocin. Use cleaning drench. Ergor-30 g + Mag. sulph 200+Tr. ginger 30 valet 400 ml.
2. Rinderpest (Virus)	Direct contact. Material contaminated with discharge from lesions.	3-9 days	Blood-stained diarrhoea. High fever. Lesions on tongue and lips.	Cattle and buffaloes, sometimes, sheep and goats	Vaccination of tissue culture. Vac. 1 ml s/c in neck region.
3. Sheep pox (Virus)	Inhalation. Through broken skin. Material contaminated with discharge.	2-7 days	Abortion. Discharge from eyes and nose besides with pus on lips, faces, udder and teats. Higher fever.	Sheep	Vaccination.
4. Swine fever (Virus)	Direct contact by eating or inhaling material contaminated with excretions.	6-9 days	Blotched skin. Diarrhea. High fever. Increased respiration. No feed intake. Faeces mixed with blood sometimes. Vomiting.	Pigs	Annual vaccination of freeze dried tissue culture vac. 1 ml s/c. Sanitary. Segregation destruction of affected.
5. Trypanosomiasis (Surra, Naganna, etc.)	Infective flies. Tabanus fly. Stable fly.	Few days some weeks	Intermittent fever. Erratic behavior. Frequent urination. Staggering and death. Unthriftiness.	Cattle and buffaloes	Chemicals. Fly eradication Antrycide 3 g in 10 ml water.
6. Tuberculosis (M. tuberculi)	Contaminated feed and water.	6 months	General malaise. Remittent fever. Emaciation. Dry husky cough.	Cattle, Sheep, Goats	BCG vaccination at 6 months age. Follow weaning system. Segregation. Streptomycin inj. daily.

Dairy Science

Factors for Making Dairying More Cost-Effective

The entire edifice of dairying rests on four pillars—feeding, breeding, heeding and weeding.

Feeding

- It must be remembered that overfeeding is as harmful as underfeeding. Much progress is being made in the field of animal nutrition.
- Locally available feed and food need to be supplemented with minerals, vitamins and trace elements.
- The dairy farmer, besides making the best use of his experience and observation, should remain in close contact with the animal nutritionist who is in a better position to convey the latest on the scientific feeding.
- The feed and fodder should be fibrous, scientifically balanced, economical and palatable.
- Well-balanced ration plays an important role in improving the performance and health of the dairy animal.

Breeding

- The tool to improve livestock quality and production depends on artificial insemination (AI) of local cows and buffaloes with the semen of the bulls of high genetic potential.
- The evaluation requires a gynecologist to conduct a physical examination, a bacteriologist to conduct disease tests.
- When semen is of good quality and female to be inseminated is free from genital defects, the proficiency of inseminator matters and a semenologist to evaluate semen.
- Regular evaluation of each and every bull is

necessary.

Heeding

- Protection against contagious diseases and parasites (external and internal) will ensure health and efficiency of the dairy animals.
- It should be a routine with dairy farmers to vaccinate their animals against contagious diseases well in advance.
- The animals to be vaccinated should be free from worms for the optimum production of antibodies against the diseases.

Weeding:

- Timely disposal of animals suffering from incurable diseases like tuberculosis should be done to save time, labour and money spent on their management and feeding.

Causes of Loss in Dairying

- Deaths of animal.
- Infestation with worms.
- Outdated husbandry practices.
- Poor breeding.
- Self-medication of sick animals.
- Unvaccinated stock.

Livestock and Milk Productions Statistics

Annual average milk yield of cows in different countries

Countries	*Average milk yield of cow (Kg./year)*
World	1935
Canada	3552
Denmark	5164
India (1982)	917
Israel	6958
Japan	4633
New Zealand	3105
Russia	2052
Sweden	5485
Switzerland	4293
The Netherlands	5286
U.K.	5004
U.S.A.	5637

Availability of milk/capita/day

Country	*Milk (gm)*	*Year*	*India (g)*
India	236	1951	132
Switzerland	471	1961	127
New Zealand	637	1971-72	112
U.S.A.	623	1981-82	136
U.K.	509	1985-86	154
World (average)	265	1986-87	157
		1988-89	164
		2001-02	226
		2003-04	236
		2006-2007	246

Trend of Milk Production in India		*Total Livestock Population*	
Year	*Milk Production (Million Tons)*	*Livestock numbers*	*(000)*
1951	17.4	Cattle	185181
1961	20.4	Buffaloes	97922
1971	22.5	Sheep	61469
1981	34.3	Goats	124358
1991	57.5	Pigs	13519
1999	77.69	Camels	632
2000	80.8	Horses & Ponies	751
		Mules	176
		Donkeys	650
		Yalks	65
		Mithun	278
2001	84.0	Poultry	489012
2002	87.4	Others	2.8
2005-06	97.1		
2006-07	100.9		

Milk & Products of Milk

The body needs more than thirty distinct materials in its food supply. No single foodstuff supplies all of them but milk supplies nearly all. So we can say, milk is the most nearly perfect food available to us.

Meaning

"Milk is a whole, clean, lacteal secretion attained by the complete milking of healthy milch livestock, well-fed and kept, excluding that obtained within 15 days before and 5 days after calving."

Milk Composition

Average composition of milk of various mammals

Mammal	*Water %*	*Fat %*	*Sugar %*	*Protein %*	*Ash %*
Cow	86.54	4.92	4.58	3.21	0.75
Buffalo	83.50	7.16	4.81	3.77	0.76
Human	87.47	4.30	6.80	1.23	0.20
Goat	86.96	4.04	4.64	3.76	0.85
Sheep	81.58	8.63	4.82	4.60	0.67
Mare	90.18	1.67	5.77	2.02	0.36
Ass	91.61	0.62	6.10	1.39	0.28
Camel	86.57	3.07	5.59	4.00	0.77
Elephant	67.85	19.57	8.84	3.09	0.65
Dog	75.59	9.57	3.09	11.18	0.57
Cat	82.10	3.33	4.91	9.08	0.58

Reasons upsetting composition of milk

- The composition of the same animal is not always the same.
- The fat content of the milk is most variable. The other constituents vary in the following order: Protein, lactose, ash.

Karakoram range is the highest peak of India.

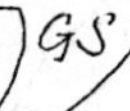

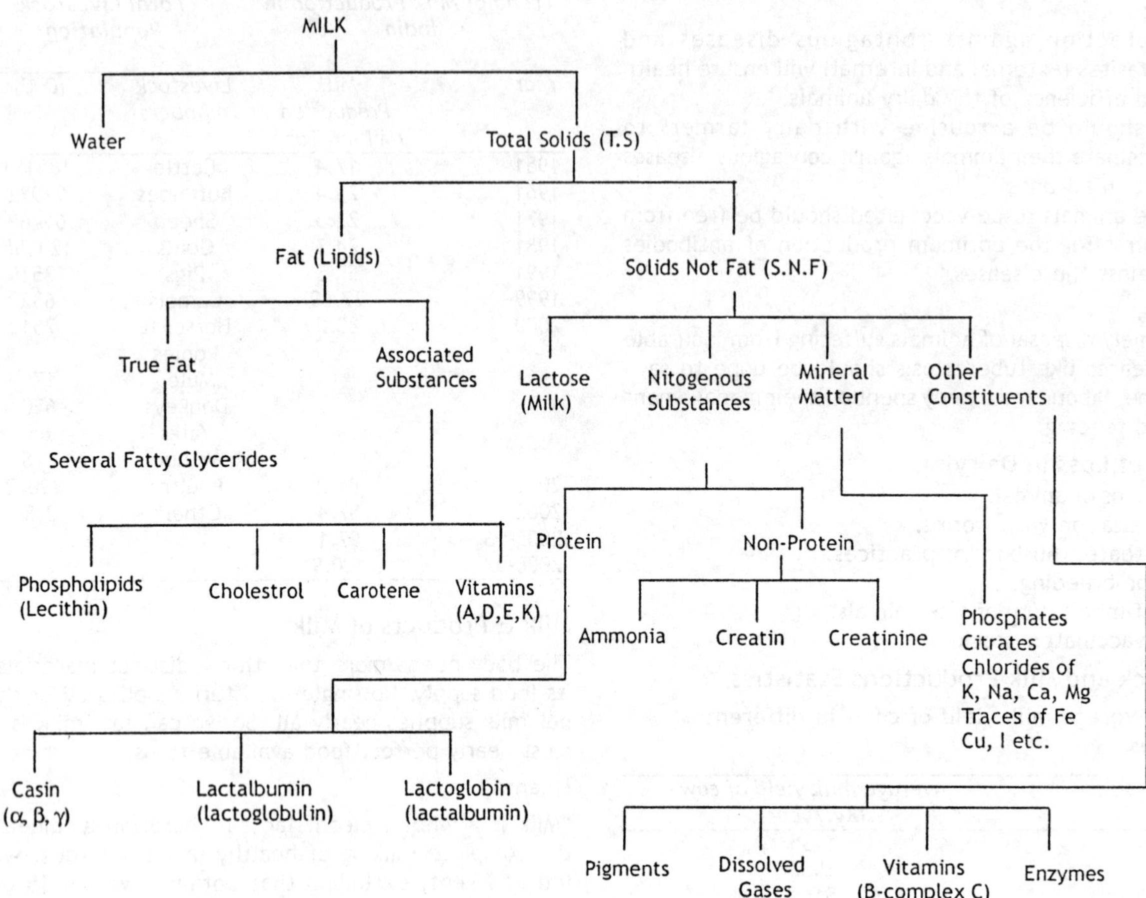

Milk compositions variation causes

Age effect

- There is a tendency for the fat content to rise from the 1st to 3rd lactation period, then remains fairly constant in subsequent lactation periods but later towards advancing age there is slight reduction in the fat content of the milk.

Fore and stripping milk

- The fore milk is very poor in fat and strippings are very rich in fat. This variation is more when the milk yield is high.

Animals feed

- When the feeding is changed there is some variation in the composition of milk but such variations are temporary.
- When the animals are given sufficient balanced ration, feed has no significant effect on composition.

Environment at the time of milking

- Anything which causes discontentment and uneasiness in the cow at the time of milking, causes the cow to be nervous and leads to the holding up of her milk. As the last portion of the milk which is held is rich in fat, hence it affects the composition.

Exercise undergone by the animal

- Light exercise often stimulates milk production without appreciable change in the percentage of fat.
- In the case of severe exercise, as when cows are used as work animals, the volume of milk is usually reduced but the fat content of milk is increased.

Breed influence

- The milk of some breeds of same species is comparatively higher in fat content than those of other breeds, e.g., milk of Holstein and Brown Swiss cow contains lesser fat than those of Red Sindhi.

Individual variations

It is partly due to unknown factors that operate in the case of individual animals and partly a matter of

inheritance.

Milking interval

- When the milking is done twice in 24 hours the milk drawn after longer interval contains less fat percentage than that drawn after shorter intervals.
- The morning milk which is drawn after longer interval contains lesser fat as compared to that of evening milk.

The milkman who milks

- If the milker is not an efficient one and not able to draw milk completely from the udder, the fat content of milk is reduced because the last milk is very rich in fat.

Animal health

- If the milk of cow or buffalo suffering from mastitis disease there is reduction in fat, protein and lactose content and marked increase in chlorides content.

Animal species

- Milk composition of different species of animals varies from each other.

Lactation period

- The period from the time the calf is born until the cow ceases to give milk is called the 'period of lactation'.
- The secretion of milk immediately after calving is known as colostrums. It may last from 3 to 6 days.
- Colostrum contains more protein and more total solids than those of normal milk.
- The composition of colostrum changes rapidly in successive milking.

Seasonal variation

- Generally fat content of milk tends to increase in cold weather and to decrease in warm weather.
- The solids-not-fat content also varies to some extent in similar manner.

Milking to milking differences

- The reasons for such variations are not yet known.
- Such variations are noted even when all other conditions are the same.

Weather

- During a long period of drought in summer, the yield and solids-not-fat content is reduced but the fat content is increased.

Colostrum

- The composition of colostrum is different from that of normal fresh milk.
- It changes rapidly in successive milkings and becomes normal milk in about 7 days after calving.
- 'Colostrum is the secretion of the mammary glands during the first few days of lactation after calving.'
- Colostrum is very rich in globulins which serves as the carrier of antibodies and protects the calves against disease producing organisms.
- For about three days it coagulates on heating.
- It has a strong odour and bitter flavour.

Composition of colostrum

Time after Calving	*Chlo-ride*	*Total protein*	*Casein*	*Albumin & Globulin*	*Lactose*	*Fat*	*Ash*	*T.S.*
	%	%	%	%	%	%	%	%
At once	0.153	17.57	5.08	11.34	2.19	5.10	1.00	26.99
6 hours	0.163	10.00	3.51	6.30	2.71	6.85	0.91	20.46
12 hours	0.156	6.05	3.00	2.96	3.71	3.80	0.89	14.53
24 hours	0.156	4.52	2.76	1.48	3.98	3.40	0.86	12.77
36 hours	0.160	3.98	2.77	1.03	3.97	3.55	0.84	12.22
72 hours	0.137	3.86	2.70	0.97	4.37	3.10	0.84	11.86
5 days	0.131	3.86	2.68	0.87	4.76	3.75	0.85	12.67

General Dairy Processing

A. Milk cooling

- Milk is not completely sterile, when drawn from the udder of animals.
- It contains some micro-organisms; their number increases during subsequent handling.
- The common milk micro-organisms grow best between 20° and 40°C.
- Bacterial growth is responsible for deterioration in market quality of milk due to development of off-flavours, acidity, etc.
- One way of preserving the quality of milk is prompt cooling to a low temperature.
- The principal object of cooling the milk is to check the growth of micro-organisms.

Bacterial growth factor in milk at different storage temperature:

Milk held for 18 hours at temperature (°C)	*Bacterial growth factor*	*Remark*
0	1 .00	Multiply initial count with this factor to get final count.
5	1.05	
10	1.80	
15	10.80	
20	200.00	
25	1,20,000.00	

B. Milk pasteurization

- Pasteurization derives its name from the eminent French scientist Louis Pasteur. During studies (1860-1864), he found out that heating of certain liquids especially wine at high temperatures improved their keeping quality.
- Abraham Jacob (1870) of New York was the first to recognize the value of pasteurization in protecting

the public health.

- Dr. Soxhlet of Germany in 1886 introduced the pasteurization of milk.

Definition

The process of heating every particle of milk to at least :-

63°C and holding it continuously at or above this temperature for at least 30 minutes.

or

72°C and holding it continuously at or above this temperature for at least 15 seconds in equipment properly operated and approved by the health authorities.

This is followed by the immediate cooling to 5°C or below.

Objectives

- To destroy cent per cent pathogenic (disease causing) micro-organisms.
- To improve the keeping quality of milk by destroying almost all spoilage micro-organisms (85 to 99%).

Pasteurization process

Pasteurization process involves the following three steps:

1. *Heating*: Milk must be heated to a temperature selected.
2. *Holding*: Milk must be held at that temperature long enough to ensure complete destruction of pathogens.
3. *Cooling*: Milk must be immediately cooled to such a low temperature that any bacteria surviving the process may not be able to grow at an appreciable rate.

Pasteurization methods

The most common methods adopted for pasteurizing milk in dairy industry are:

1. Low temperature longer time pasteurization (L. T. L. T.)
   - This method is also known as Batch or Holding method of pasteurization.
   - In this method milk is heated to at least 63°C and is held for at least 30 minutes followed by batch cooling to 5°C.
   - The process is carried out in a double jacketed vat, equipped with a mechanical agitator and a thermometer.
   - An outlet is provided in the bottom portion of the vat to drain out pasteurized milk.
   - Heating and cooling of milk is done by circulation of hot water or steam for heating and chilled water for cooling.
2. High temperature short time pasteurization (H. T. S. T.)
   - It is a continuous process of pasteurization in which milk attains a continuous flow in H. T. S. T. pasteurizer, where it is heated to 72°C for 15 seconds and promptly cooled to 5°C or below.
   - H. T. S. T. pasteurizer consists of four different sections viz.
     (a) Regeneration section,
     (b) Heating section,
     (c) Holding section, and
     (d) Cooling section.

Each section comprises stainless steel plates arranged in series. These plates are criss-crossed with knots, ripples or grooves to give a turbulent flow of milk. The heating and cooling of milk takes place through these plates.

- The flow of milk in H.T.S.T. pasteurizer is as follows:

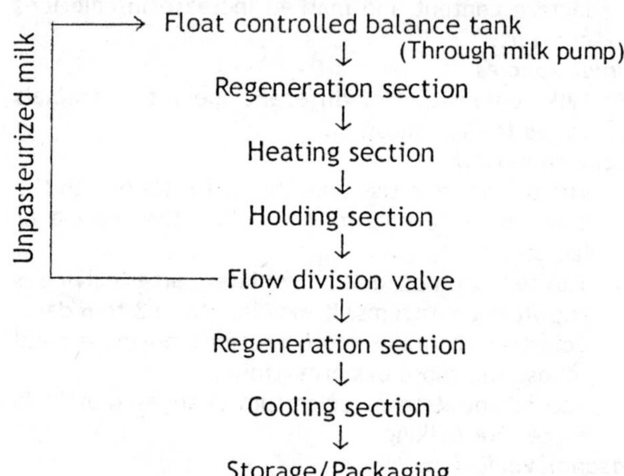

Basic considerations for temperature-time combination

1. The following considerations were involved in the formulation of the standards of pasteurization: Phosphates enzyme inactivation
   - Phosphate enzyme should be completely destroyed by pasteurization.
   - Pasteurized milk should show negative phosphate test.
   - Accuracy of pasteurization is checked by this test.
2. Killing of microbes/pathogens
   - Mycobacterium tuberculosis (causal organism of T.B.) is the most heat resistant bacteria.
   - Any temperature and time combination that kills T.B. germ, will also destroy all other pathogens in milk.
3. Least damage to cream volume
   - The cream line or volume or cream is reduced progressively with increase in the temperature-time of heating.

- Standards of pasteurization should be such that there should be minimum damage to the cream line.

The following table shows various temperature-time combinations which were considered while setting the standard of pasteurizations, i.e., 62°C for 30 minutes or 72°C for 15 seconds.

Temperature-time combination

Particulars	*Temperature for 30 minutes*	*Temperature for 15 seconds*
To kill T.B. bacteria	58.9°C	70°C
To get negative phosphate test	61.1°C	71.1°C
To obtain least damage to cream volume	62.2°C	72.2°C

C. Milk heat treatment (other than pasteurization)

Vacreation

This refers to pasteurization of milk/cream under reduced pressure by direct steam. The process was first developed in New Zealand by M/s Murray Deodorizers Ltd. The equipment designed for this purpose was called as 'Vacreator'. It was designed to remove feed and other volatile flavours from cream and to pasteurize it for butter making.

Vacreator

The vacreator consists of three stainless steel chambers connected to one another for steam heating and vacuum treatment with continuous product flow. The product, in the form of fine droplets, enters the first chamber of the vacreator where pasteurization occurs. Then the product and some free steam are moved from the bottom of the first chamber to the top of the second one. A multi-stage centrifugal pump removes the product from the third chamber. Altogether, it takes the product about 10 seconds to move continuously through the unit.

Stassanising

This device was invented by Dr. Henri Stassane of Italy and has undergone several changes and improvements after tests at the Danish Experimental Dairy in Silkeborg, Denmark.

The milk is heated to about 75°C under slight pressure in a thin layer between two heated surfaces in order that all carbonic acid may be retained. It is claimed that there is practically no milk stone formation, less destruction of vitamins, no evaporation of milk and more economy in steam utilization than in more recognized methods of pasteurization.

Ultra high temperature (URT) pasteurization

This method of pasteurization was developed in the 1950s. This usually encompasses temperature and time combinations of 135° to 150°C for no-hold (a fraction of seconds). The milk is packaged aseptically soon after the treatment.

Uperization

The term uperization has been derived from ultra-pasteurization. Uperization is the name applied to HTST process for sterilizing milk. The process was developed in Switzerland to preserve the nutritive value and vitamins after heat treatment. In this process milk is pumped through two pre-heaters in which its temperature is raised to 77.1°C. It is then passed to a high pressure unit where its temperature is raised to 148.8°C by direct injection of steam. The heated milk is cooled quickly to 78°C, as it is passed into a vacuum chamber where excess water derived from the injected steam as well as any foreign odours are removed from milk.

To avoid concentration or dilution of the product, the amount of live steam injected is balanced by the amount of vapour removed in the vacuum chamber.

Bactofugation

The process was developed by Professor Simonart of Belgium and the Alfa-Laval company in Sweden. It is a method for removing bacteria from milk by a centrifugal process. In the Bactofuge process, the centrifuge is operated at about 20,000 r.p.m. and is modified so that a small amount of skim milk, about 2.5% by volume, is discharged continuously. Milk is heated to 75°C and passed through two bactofuge one by one. In the first bactofuge about 90% of the bacteria present in milk are removed. Centrifugal milk when passed through the second bactofuge about 90% of the remaining 10% of bacteria is removed making a total of 99% removal. The 2.5% discharge from this operation is returned to the raw feed, keeping the loss of milk to about 2.5%.

D. Homogenization

It is a process of making stable emulsion of milk fat and milk serum by a mechanical treatment and the machines used for this purpose are known as homogenizers. This process is used mainly in the ice-cream mix and the evaporated milk trades, but there is also rapidly growing demand for homogenized fluid milk.

Homogenizers

Classification

(1) *Piston type high-pressure* (under 35 to 350 kg/cm^2 or 500-5000 p.s.i): Milk is forced through a very small orifice under high pressure, which results in the breaking down of larger fat globules into many smaller ones.

(2) *Rotary type low-pressure* (operation below35 kg/

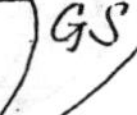

cm^2 or 500 p.s.i): They subject milk fat to a grinding and shearing action under low pressure.

(3) *Sonic oscillator/vibrator:* It involves tile subjection of milk to a high frequency vibration or oscillation.

Of these three the piston type high-pressure homogenizer is widely used.

Categorization of marketed milk

Pure milk

- Milk obtained from the animals (cow/buffalo) without altering the composition, is termed as pure or whole milk.
- The minimum requirements of fat and S.N.F. for these milk have been fixed.
- The cow milk should contain fat not less than 3.5% and S.N.F. not less than 8.5% whereas, buffalo milk should contain fat not less than 6% and S.N.F. not less than 9%.

Standardized milk

- This is milk whose fat and S.N.F. contents have been adjusted to a certain pre-determined level and can be done by partially skimming the fat in the milk or admixing skim milk in proper proportions.
- According to the PFA rules (1976), standardized milk contains a minimum of 4.5% fat and 8.5% SNF.

Toned milk

- Toned milk refers to milk obtained by the addition of water and skim 'milk' powder to whole milk and according to the PFA rules (1976), should contain a minimum of 3% fat and 8.5% SNF.

Double toned milk

- According to the PFA rules (1976), double toned milk should contain a minimum of 1.5% fat and 9% SNF.

Reconstituted milk

- This refers to milk prepared by dispersing whole milk powder in water approximately in proportion of 1 part powder to 7-8 parts water.

Recombined milk

- Product obtained when butter oil, skim milk powder and water are combined in the correct proportion to yield fluid milk.
- The milk fat may also be obtained from other sources like unsalted butter, fresh sweet cream or plastic cream.
- According to the PFA rules (1976), recombined milk should contain a minimum of 3% fat and 8.5% SNF.

Filled milk

- When milk fat in milk is substituted with some vegetable oil or fat, it is termed as filled milk.

CREAM

- It is fat-rich portion of milk.
- According to the PFA rules (1976), it should contain a minimum of 25% fat by weight.
- Cream may be separated from milk either by gravity or centrifugal method.
- The basic principle of cream separation, whether by gravity or centrifugal methods, is based on the difference in specific gravity of milk fat and milk serum.

Gravity separation

- When milk is allowed to stand for some time there is tendency for the fat to rise upward and being the heavier portion, serum is pulled downward.
- The velocity or rate at which the fat globules rise, which is known as Stokes Law.

Cream can be obtained by any of the following three methods under gravity separation:

1. Deep setting
2. Shallow method
3. Water-Dilution method

These methods were in use before the days of cream separators but now have become obsolete.

Centrifugal separation

- When milk enters the rapidly revolving bowl of the cream separator, it is immediately subjected to a tremendous amount of centrifugal force, which is 3000 to 6000 times greater than that of gravity.
- The skim milk is forced to the periphery while the fat portion moves towards the centre.
- Stokes Law applied to centrifugal separation.

Factors affecting the fat content of cream in centrifugally separation

- Amount of flushing
- Formation of separator slime
- Position of cream screw
- Rate of inflow of milk
- Richness of the milk
- Speed
- Temperature of milk

Categories of cream

Whipping cream

- A grade of market cream especially adapted for whipping in the open air.
- Whipping cream should contain 32 to 40% fat and should be aged for at least 24 hours at a temperature from 4.5°- 10°C to increase clumping of fat globules.
- Rapid agitation of whipping is important for best results.

Synthetic cream

- A product made of non-milk fats such as cotton seed oil, peanut oil, and certain ingredient of soyabean to which lecithin and other ingredients have been added.
- It is used as a substitute for cream.

GK **A mixture of T.N.T. and Ammonium nitrate is called Amatol.** GS

Artificial cream

- It is a food article made up of milk ingredients to which water has been added to make a product resembling naturally-produced cream.

Light cream

- Sweet cream containing 18-30% fat.
- The term sweet cream is used for the fresh cream which has an acidity of less than 0.2%.

Heavy cream

- Sweet cream containing not less than 36% rat.

Plastic cream

- A heavy concentrated cream containing about 80% fat.
- It is plastic in form, much the same as butter.
- In spite of its high fat concentration, the fat remains in substantially its original emulsion with the solids-not-fat present in the serum.
- Plastic cream is produced commercially by the use of a separator designed for skimming a very rich cream.
- Either whole sweet milk or sweet cream testing 40 % fat or less may be used.
- After separating, the cream is quickly cooled and packaged in better tubs.
- Plastic cream is used in the manufacture of butter, ice-cream, coffee cream, cream cheese and other dairy products.

Coffee cream

- Cream used in coffee.
- This cream is usually homogenized and tests about 20% butter fat.
- It is also called "Single Cream", "Table Cream" or "Homogenized Cream", if it has been homogenized.

The following features have been suggested for desirable table or coffee cream.

1. Fat content – 20-25%
2. Low bacterial count.
3. Maximum ability to colour coffee.
4. Maximum viscosity for a given fat percentage.
5. Minimum amount of serum of skim milk in bottom of bottle.
6. No "oiling off" in hot coffee, No "feathering" in hot coffee.
7. No cream plugs formation.
8. No developed acidity.
9. No visible sediments.
10. Pleasing flavor – no feed cooked or oxidized flavour.

Clotted cream

- Also known as evaporated cream and prepared by a process of scalding and cooling which results in a thick, heavy, clotted substance.
- It is used chiefly in place of butter and cultured cream.
- The product has an acidity of about 0.6% to 0.8% which is derived from a commercial starter culture.

Cereal cream

- A very light commercial sweet cream usually testing about 10 to 12% butter fat.
- It is more commonly known under the name of cereal milk but is sold in the market as "Half'-n-half".

Frozen cream

- It is usually stored in straight side tins or 50 to 60 lbs. capacity or in fibre containers and kept at temperature from -18° to -23°C. These containers should have no exposed iron or copper.
- Good quality fresh cream, which tests at least 40% fat and has been pasteurized, should be used.
- Use in ice-cream manufacture as and when required.

Sleepy cream

- In butter-making process, cream is churned into butter churn. Sometimes, cream does not break easily during the churning process and remains in semi-fluid state for long. Such condition of cream is known as sleepy cream.
- This condition is more pronounced when homogenized cream is used for churning.
- Other factors responsible for this are:
  a) excessive hardness of fat
  b) excessive loading of churn
  c) high protein content of cream
  d) low churning temperature
  e) small size of fat globules
  f) thinness of cream
- To overcome such churning difficulty un-homogenized cream is preferred for butter making. A reasonably rich cream (about 30-35% fat) ripened to approximately 0.25% acid, churned at a temperature 9°C to 10°C in summer and 11°C to 14°C in winter, in a churn from one-third to a little less than one-halful, will take care of this problem.

Common defects of milk and cream

Barny flavour
Bitter flavor
Bloody milk
Blue milk
Cooked flavor
Cream plug
Feathering
Feed flavor
Foreign flavor
High acid flavor
Malty flavor
Metallic flavor
Oiling-off
Rancid flavor

GK GS

Deccan Trap is covered by lava layers.

Ropy milk
Salty flavor
Sweet curdling

BUTTER

A fat concentrate which is attained by churning cream, milk or dahi is defined as butter. Butter milk is the watery substance left after taking out butter from these material.

Classification

Table butter

According to the PFA Rules (1976), table butter shall contain not less than 80% by weight of milk fat, not more than 1.5% by weight of curd and not more than 3% by weight of common salt. Diacetyl may be added as flavouring agent but, if used, the total diacetyl content must not exceed 4 ppm. It may or may not contain added annatto or carotene as colouring material.

White butter or Deshi butter

As per PFA Rules (1976), white butter or deshi butter shall contain not less than 76% by weight of milk fat, not more than 1.5% by weight of curd. It should not contain any added salt or colour.

Method of Manufacturing Table butter

STEPS

Cream separation and acceptance in the factory
↓
Cream neutralization
↓
Cream standardization
↓
Cream pasteurization
↓
Cream cooling
↓
Cream ripening
↓
Cream aging
↓
Cream churning
↓
Washing the butter
↓
Salt addition
↓
Working of butter
↓
Packaging and storing of butter

Cream separation and acceptance in the factory

The cream is either separated in the creamery or accepted from outside.

Cream neutralization

- Sometimes the cream received in the plant is sour, it is essential to neutralize it.
- Neutralization of cream for butter-making refers to a partial reduction in acidity of sour cream.
- Neutralization of sour cream improves the flavour, texture and keeping quality and helps to produce a butter of more uniform quality.
- In addition, it avoids excessive fat loss in butter milk that results from churning highly acidic pasteurized cream.

Neutralizers

- Among the commercial neutralizers, sodium bicarbonate or mixture of sodium bicarbonate and sodium carbonate is common.
- Lime or Soda neutralizers are commonly used for the neutralization of cream. Lime neutralizers are stronger alkalies; therefore, it should be used with great care.
- A mixture of hydrated lime and magnesium oxide or a mixture of calcium oxide and magnesium oxide is also used. The temperature of cream when adding the neutralizer should preferably be 29-32°C.
- The neutralizer should be added after diluting it with 10 to 15 times its weight of clean cold water. It should be distributed quickly and uniformly in the cream and mixed thoroughly by stirring vigorously.
- The cream acidity should be reduced to 0.06-0.08% before churning when it is used for manufacturing butter for long storage and to 0.25-0.03% when used for butter for early consumption.

The amount of neutralizer used can be calculated by using the following formula:

$$Q = A \times (I-D) \times NF$$

where Q = Amount of neutralizer in kg.

A = Amount of cream in kg.

I = % Initial acidity of cream.

D = % Desired acidity to which cream is neutralized.

NF = Neutralizing factor of the neutralizer used.

Cream standardization

- It is the adjustment of fat level in cream to the desired percentage.
- The fat percentage in cream is usually adjusted to the required level by addition of a calculated amount of skim milk or water.
- The fat content of cream for churning varies from 30-35% in winter and 35-40% in summer.

Cream pasteurization

This is done by any of the prescribed procedure of

pasteurization of milk.

Cream ripening

- Ripening means fermentation of cream with the use of culture and desirable starter.
- The butter starter culture contains two types of micro-organisms.
  1. Acid producing micro-organism (ferment lactose into lactic acids)
     a. *Streptococcus lactis*
     b. *S. eremaoris*
  2. Flavour producing micro-organism (ferment citric acid and produce a flavouring component known as diacetyl)
     a. *S. diacetilactis*
     b. *Leuconostoc citrovorum*
     c. *L. dextranicum*
- Ripening helps:
  1. to obtain exhaustive churning with low fat loss and butter milk.
  2. to generate butter with a pleasing pronounced characteristic flavour and aroma.
- The ripening of cream involves the addition of butter culture at a rate of 0.5-2% to the cooled cream (20-22°C) and incubating for 15-16 hours at 21°C.

Cream cooling and aging

- After ripening, the cream is cooled down to 5-10°C and kept for few hours ranging between 4-16 for allowing the fat to solidify for butter churning.
- During cooling and aging milk fat in fat globules gets partially solidified which makes churning easily possible.

Cream churning

- Churning of cream means:
  a. agitation at a suitable temperature until the fat globules adhere,
  b. forming larger and larger masses, and
  c. until a relatively complete separation of fat and serum occurs.
- The churn is properly sanitized and cooled.
- The cream is then poured into the churn.
- The amount of cream into the churn should be between 1/3 and 1/2 of the total air space of the churn.
- The butter colour (from annatto seeds in castor oil medium) @ 1-2 ml/kg butter fat is added.
- The lid of the churn is closed and churning is stored at 9-10°C during summer and between 11-13°C during winter.
- After 5 to 10 minutes of churning, the temperature rises by 1-3°C, foaming occurs, gas is formed. Gas is allowed to go out by the vent.
- Then breaking stage of cream fat comes.
- To reduce the temperature break water is added to the extent of 1/3rd the amount of cream.
- The churning action continued further till the size of butter granules becomes as large as a groundnut.
- After this the butter milk is removed and the cleaned butter is available as a lump.

Washing the butter

- To remove all loose butter milk adhered to butter grains.
- To correct defects in the firmness of butter by proper adjustment of wash water temperature.
- Chilled water having a temperature 1-2°C lower than the churning temperature is used for washing the butter.
- Equal to the quantity of butter milk, the amount of wash water is added into the churn.
- The churn is rotated for a while and all the water is removed to set clean butter.

Salt addition

- Common powdered salt is added either in the form of dry, wet or brine solution at a rate of 2 to 2.5 per cent of butter fat.
- Salting of butter is done in order to improve the keeping quality, enhance the taste and to increase the overrun in butter.

Working of butter

- After having added salt, butter is worked to completely dissolve, uniformly distribute and properly incorporate the salt.
- Working also controls the moisture content of butter.
- Working should be continued until the butter has compact body, a closely knit grain, a tough wax texture and an even distribution of salt and moisture.

Packing and storing of butter

- The butter is removed from the churn, packed in box or packets or in small packet of 25, 50, 100, 250, 500 gm for marketing.
- It is kept at -23°C to - 29°C for commercial purpose.

Overrun of butter

Overrun of butter is the percent increase in butter from given amount of fat because of presence of water, curd and salt in it.

% Overrun in butter

$$= \frac{\text{Amount of butter made - Amount of fat churned}}{\text{Amount of fat churned}} \times 100$$

Quality defects in butter, their causes and remedies

Name of defects	Causes	Remedies
(a) Colour defects		
Mottled	• Uneven distribution of water droplets in salted butter • Improper salting and working	• Adequate washing of the butter • Proper salting and working of the butter
Streaky/Wavy	• Improper working in two or more lots of butter together • Uneven working of butter	• Sufficient working of two or more lots of butter together • Proper working of the butter
(b) Flavour defects		
Acid/Sour	• Neutralization to a high acidity • Using sour cream	• Correct neutralization of cream • Using sweet cream
Neutralizer flavour	• Over-neutralization of cream	• Optimum neutralization of cream
Bitter flavour	• Feeding of bitter weeds to milch animals • Growth of proteolytic bacteria in cream • Lipase activity in raw cream	• Eradication of offending weeds • Storage of cream at 5°C or below • Checking lipase activity by treatment before separation
Cheesy	• Growth of proteolytic bacteria leading to casein break-down	• Storage of cream at 5°C or below
Cooked	• Overheating of cream during pasteurization	• Pasteurization at optimum temperature
Feed & weed	• Feeding of milk tainting feeds and weeds within three hours before milking	• Avoid feeding of milk tainting feeds and weeds before milking • Eradication of milk tainting weeds • Vacuum pasteurization
Fishy	• High acid salted butter in the presence of copper/iron contents under commercial cold storage conditions	• Unsalted sweet cream butter under commercial cold storage conditions
Flat	• Excessive washing of butter • Low diacetyl content • Low salt content in butter	• Optimum washing of butter • Adequate ripening of cream. • Correct salt content
Rancid	• Fat hydrolysis due to lipase action	• Inactivation of lipase enzymes by proper pasteurization
Oxidized/ Metallic/Tallowy	• Fat oxidation due to direct contact of milk/cream/butter with copper or iron, exposure of these products to sunlight etc.	• Storage in properly tinned or aluminum alloy or stainless steel vessels • Vacuum pasteurization of cream
Stale	• Butter stored at insufficiently low temperature for long • Cream of poor quality held too long before churning	• Correct temperature for storage of butter • Avoiding long storage of cream during churning
Yeasty	• Using old, yeasty cream for churning	• Using fresh sweet cream for churning
Body and texture defects		
Crumbly	• Churning at high temperature • Underworking of butter • High proportion of low melting glyceride in the butter fat	• Churning at optimum temperature • Adequate working
Sticky	• Overworking of butter	• Adequate working
Gritty	• Improper working • Undissolved coarse salt in butter	• Proper working of butter • Proper salting

CHEESE

It is made out of curd by coagulating the casein with the help of rennet or similar enzymes in the presence of lactic acid (added externally or produced by micro-organisms). The excess water is removed and the material is allowed to ripen to get the final product.

According to PFA Rules (1976), hard cheese shall contain not more than 43% moisture and not less than 42% milk fat in dry matter.

Classification

Cheddar cheese is the most popular variety among all varieties and was developed in the village of Cheddar in Somersetshine, England, in the 16th century.

GK

The Pallavas of the South Indian kingdom were involved in a bitter and continual struggle with the Chalukyas of Vatapi.

GS

Manufacturing method (of Bacterial ri ned Cheddar cheese)

Acceptance of milk

- Milk received at cheese factory should be carefully examined for its quality before accepting.
- Usually cow milk is preferred for cheese.
- Buffalo milk can also be used for the same.

Steps of Manufacturing Cheddar Cheese

Acceptance of milk
↓
Clarification/Filtration
↓
Milk standardization
↓
Pasteurization
↓
Calcium chloride ($CaCl_2$) addition
↓
Starter additions
↓
Colour addition
↓
Rennet addition
↓
Curd setting
↓
Curd cutting
↓
Curd cooking
↓
Whey removal
↓
Cheddaring
↓
Curd blocks milling
↓
Salt addition
↓
Hooping & pressing
↓
Blocks drying
↓
Dipping in paraffining
↓
Curing

CHEESE

- Soft
  - Unripened
    - Low fat e.g. Cottage cheese
    - High fat e.g. Cream cheese
  - Ripened
    - Bacterial ripened e.g. Pickled
    - Mould ripened e.g. Camembert
- Semi - soft/ Semi-hard
  - Bacterial Ripened e.g. Brick & Tilsit
  - Mould Ripened e.g. Silton & Roquefort
  - Surface slime e.g. Linverger
- Hard
  - Bacterial Ripened (Close texture) e.g. 1. Cheddar 2. Edam 3. Gouda
  - Bacterial Ripened (Open texture) e.g. 1. Swiss & 2. Herrgard
- Very hard e.g. Parmasam
- Whey Cheese e.g. Ricotia
- Processed Cheese

Clarification/filtration

- The milk is usually pre-heated to 35-40°C for efficient filtration/clarification.
- To remove any visible dirt in milk so as to improve the aesthetic quality of the cheese.

Milk standardization

- To maintain the fat and dry matter ratio.
- Fat and casein ratio of milk 1:0.7 (approximately) is generally adjusted in the milk.

Pasteurization

- Milk is pasteurized either by holding or H.T.S.T. method.

Advantage

- It helps to produce a more uniform product of high quality.
- To destroy pathogens and fault-producing micro-organisms.

Limitation

- It destroys the typical flavour and body of cheese.
- It encourages the use of low quality milk.
- It entails a longer ripening period.
- It increases the overall cost of cheese-making.

After pasteurization, cheese milk is put into the cheese vat where rest of the processing is carried out.

Calcium chloride addition

- After pasteurization, milk is put in the vat. Due to pasteurization, part of soluble calcium in milk becomes insoluble. This results in slower renneting action and a weaker curd. The gap is filled up by adding 0.01 to 0.3% calcium chloride to milk.

Starter additions

- Micro-organism inoculants of *Streptococcus lactis* or *S. cremoris* or both are added to pre-heated (30-31°C) milk in vat @ 0.5-1%.
- Before adding, the starter should be stirred until smooth and creamy in consistency, then strained and added in the required quantity and mixed thoroughly and uniformly into the milk.

Colour addition

- The colour is obtained from annatto seeds. It is soluble in water.
- It is added just before renneting: 30 to 200 ml colour per 1000 kg milk is added after diluting it with 20 parts of potable water for even distribution. It is vigorously agitated to ensure uniform mixing.

Rennet addition

- Rennet is a product obtained from the 4th rumen (stomach of a young calf). It contains an enzyme, which coagulates the casein with appreciable breaking down of protein.
- Rennet is added to the milk after the acidity has been increased by 0.02% over initial acidity.
- Rennet is added at a rate of 15 to 25 ml/100 liter of milk after diluting with 20 to 40 parts of water for uniform distribution.

Curd setting

- The temperature of the outer jacket of vat is maintained at 30-31°C and the vat is left undisturbed for settling of curd.

Curd cutting

- After the curd is properly settled, it is cut with horizontal and vertical knives to make into cubes of specific size.
- Acidity of the curd at the time of cutting is almost equal to acidity of initial milk.

Curd cooking

- The curd cubes are left for 2-3 minutes to allow them to be strong enough.
- Gentle stirring of the cubes started with the gradual increase in temperature of water in the outer jacket of the vat at the following rates: i.e., increase in the first 15 minutes, thereafter 1°C every 4 minutes up to 37-39°C temperature.
- The acidity of whey is raised to about 0.135%.

Whey removal

- After proper cooking, the whey is removed by draining it off the cheese vat.

Cheddaring

- After draining of whey, the curd cubes are allowed to mat for 5 to 10 minutes.
- Matted curd is cut into small slabs.
- These slabs are turned upside, one above the other to make a heap or pile of slab.
- After 2-3 hours, this process of cheddaring is completed.
- The curd slabs are dried to make them silky in touch and chicken meat to look like.
- The acidity of whey reaches 0.45 to 0.55%.

Curd blocks milling

- To remove the remaining part of whey (the watery part of the curd) from the curd, the blocks are milled in cheese mill.

Salt addition

- To improve the quality of cheese, to check the fermentation and to remove the whey further, powdered common table salt is spread over the green coloured cheese @ 1-2% by wt.

Hooping and pressing

- The salted green cheese is put in hoops (cloth lined) and kept under the press for half an hour at $0.5kg/cm^2$ pressure to remove the water.
- These are taken out of the hoops, dressed and again put under press at $2kg/cm^2$ for 10-12 hours.

In human body, the longest cell is Nerve cell.

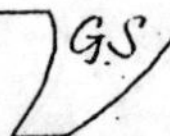

Blocks drying

- The cheese blocks are removed from the hoops and kept in drying room, for a few days to form rind all around the surface.
- In the drying room the temperature is maintained at 12 to 16°C and the relative humidity at 50%.

Dipping in paraffining

- To give a protective layer on the cheese blocks are inserted into molten paraffin wax for few seconds.
- Paraffining reduces the loss of moisture during ripening, prevents extensive mould growth and protects the cheese against insects.

Curing

- Curing or ripening is done for the development of desirable physical, chemical, bacteriological, textural properties and flavor.
- Store cheese for at least 2 to 3 months in a curing room where a temperature of 10-16°C and 75 to 85% relative humidity is maintained.

ICE-CREAM

- Ice-cream is made by suitable blending of cream, other milk products, sugar, flavour, with or without stabilizer, or colour and freezing it at proper temperature to get final product.
- According to PFA Rules (1976), ice-cream shall contain not less than 10% milk fat, 3.5% protein and 36% total solids. It may contain permitted stabilizers and emulsifiers not exceeding 0.5% by weight. In case when ice-cream contains fruits or nuts or both the content of milk fat may be proportionately reduced but may not be less than 8% by weight.

Ice-cream manufacturing method

Steps

Ingredients selection
↓
Amount of ingredients calculation
↓
Ingredients mixing
↓
Mixture pasteurization
(70°C for 30 min./80°C for 25 sec.)
↓
Mixture homogenization (I stage 2500 psi)
↓ (II stage 500 psi)
Mixture cooling and aging (0 to 5°C)
↓
Freezing (-4 to -5°C)
↓
Flavouring agent addition
↓
Packaging
↓
Hardening and storage (-23 to -29°C)

Ingredients selection

- Selection of ice-cream ingredients mainly depends upon:
  a. availability of milk products.
  b. convenience in handling.
  c. effect on flavour, body and texture of ice-cream.
  d. perishability of the products.

Amount of ingredients calculation

- Quantity of ingredients required depends upon:
  a. the composition of individual ingredients.
  b. the amount and the composition of mix.

Using each ingredient in proper proportions helps in maintaining uniform quality in day-to-day preparation.

Ingredients mixing

- The common formula of blending the ingredients is to:
  a. add liquid materials (cream, milk, other liquid milk materials) to a vat
  b. then add solids like dry milk, egg yolk, gelatine etc.
  c. start heating
  d. add sugar when mix is 50-55°C
  e. take precaution to ensure all materials are dissolved before pasteurization temperature is reached.

Mixture pasteurization

- Pasteurize the mixture at 70°C for 30 minutes or 80°C for 25 seconds.
- Pasteurization
  a. brings into solution and aids in blending the ingredients of the mix.
  b. improves flavor.
  c. improves keeping quality.
  d. produces a more uniform product.
  e. renders the mix free of pathogens.

Mixture homogenization

- It is done at pressure 2000 to 3000 psi on the first stage and 800 to 1500 psi on the second stage.
- Homogenization of mix
  a. aids in producing ice-cream with smooth texture.
  b. improves whipping properties.
  c. makes more uniform ice-cream.
  d. reduces fat globules to small uniform size

GK Kaimur hills are located along Son-valley. GS

which will prevent churning or separation of the fat in ice-cream.

Mixture cooling and aging

- The mixture is cooled down to 5°C immediately and kept at low temperature for some time for aging.
- It is advisable to age the mix for two to four hours but twenty-four hours is ample time to derive all the benefits of aging.
- Aging generally improves the body and texture and whipping properties.

Freezing

- Ice-cream is frozen in either the continuous or the batch freezers.
- Freezing is accomplished rapidly:
  a. a smooth texture.
  b. to ensure small ice crystals.
- The purpose of the freezing process is
  a. to incorporate air into the mix.
  b. to freeze a portion of the water of the mix.
- In continuous freezer the ice-cream is drawn at about -6°C and 50 to 55% of the water of the mix is frozen.
- The ice-cream is drawn from the batch freezer at approximately -4° C and 35 to 40% of water of the mix is frozen.

Overrun

- During the freezing process particular attention should be given to overrun and body and texture control.
- Overrun is an increase in volume of ice-cream due to incorporation of air during freezing.
- The overrun range in ice-cream is usually from 60-100%.
- Air incorporation (whipping) in ice-cream is necessary because it ensures smoother, more velvety product and improves palatability of ice-cream.

Overrun may be calculated either by volume or weight basis.

$$\%\text{overrun} = \frac{\text{Volume of ice-cream} - \text{Volume of mix}}{\text{Volume of mix}} \times 100$$

(By volume)

$$\%\text{overrun} = \frac{\text{Weight of unit volume of mix} - \text{Weight of unit volume of icecream}}{\text{Weight of unit volume of icecream}} \times 100$$

(By weight)

Flavouring agent addition

- Vanilla, chocolate, dry fruit, dry fruit powder, fruit pieces and nuts are common agents that are to be added at the time of freezing.
- Most common flavour in ice cream is vanilla.

Packaging

- Ice-cream is usually collected in containers which gives it the desired shape or size for convenient handling during the hardening and marketing.
- Most bulk ice-cream is packaged in fibre board cartons coated with wax or polythene-wax blends for protection against moisture and oxygen.
- Retail ice-cream may be packaged in cups, sticks or bars. Cups may be of paper or cardboard coated with wax.
- The most recent trend for such factory filled carry-home packages is the plastic cylindrical container with a re-closeable lid.

Hardening and storage

- Rapid hardening and uniform storage are essential to ensure good body and texture.
- The semi-frozen product after packaging is hardened in cold rooms or cabinets maintained at -23°C to -29°C and held at this temperature until delivered.
- Excessive storage must be avoided to protect the flavour and body texture of ice-cream.

CONDENSED AND EVAPORATED MILKS

- Condensed and evaporated milks are products obtained by evaporating part of the water of whole milk, or fully or partly skimmed milk, with or without the addition of sugar.
- The term condensed milk is commonly used for full cream sweetened condensed milk, while evaporated milk is commonly used for unsweetened condensed milk.
- According to PFA Rules (1976), condensed milk shall contain not less than 9% fat, not less than 31% total milk solids and not less than 40% cane sugar while evaporated milk shall contain not less than 8% fat and not less than 26% milk solids.

STEPS

Sweetened condensed milk

Milk reception
↓
Clarification/Filtration (38-40°C)
↓
Fore-warming (115-118°C for no holding)
↓
Standardization
↓

Sugar addition
↓
Condensing
↓
Cooling & Crystallization
↓
Packaging
↓
Storage

Evaporated milk

Milk reception
↓
Clarification/Filtration (38-40°C)
↓
Fore-warming (115-118°C for no holding)
↓
Standardization
↓
Condensing
↓
Homogenization
↓
Cooling (7°C)
↓
Pilot sterilization test
↓
Packaging
↓
Sterilization (116-118°C for 5 min)
↓
Cooling (27-32°C)
↓
Shaking
↓
Storage (5-15°C)

Manufacturing method

- The basic principle in the production of condensed and evaporated milk is that high quality milk is filtered or clarified, standardized fore-warmed and condensed to the desired level.
- The concentrated milk is preserved by the addition of sugar for condensed milk and by heat-sterilization for evaporated milk.
- Common in both, milk of high quality is received and clarified to remove visible sediments from milk.
- It is standardized to a desired fat and S.N.F. ratio so that desired ratio of the same can be obtained in the final product.
- The milk is pre-heated at 115°C to 118°C for no hold to ensure uninterrupted boiling in vacuum pan and to provide an effective means of controlling objectionable age thickening in the finished products.
- Fore-warning also ensures that the finished product is free from micro-organisms and enzymes.
- In sweetened condensed milk, manufacturing sugar is added for the purpose of preserving the condensed milk without sterilizing the products.
- The milk is condensed under vacuum in a vacuum pan. In some cases, two or more vacuum pans are attached in a series to economise the operation. Such evaporators are called as multiple-effect evaporators. The main advantages of condensing milk in vacuum are:
  a. can be readily reconstituted into the original milk.
  b. economy of operation.
  c. protection of milk against heat damage.
  d. rapidity of evaporation.
  e. the finished product is free from any cooked flavor.
- When required concentration has been obtained, the product is drawn from the pan.
- The condensed milk is promptly cooled to 30°C and sealed with 0.1 to 0.3% powdered lactose for forced crystallization. It is followed by vigorous stirring and slow cooling to 15°C in one hour. Stirring is continued until packaged.
- Skim milk powder and condensed milk from previous lot can also be used for seeding but the amount required is twice as much as of lactose.
- The evaporated milk after drawing from the vacuum pan is homogenized by applying a pressure of 2000 psi in first stage and 500 psi in second stage. It minimizes the chances of fat separation during storage. The product is then cooled to 7°C.
- Pilot sterilization test is carried out to determine the amount of chemical stabilizer to be added to any given batch of evaporated milk for the most satisfactory heat stability. The required amount of stabilizer is then added to the evaporated milk and packaged in cans. Cans or tins after sealing are sterilized at 116°C-118°C for 15 minutes.
- The evaporated milk is cooled within 15 minutes to 27-30°C after sterilization is over.
- Packed cans are shaken mechanically to break down any curd which has formed during sterilization to a smooth homogeneous consistency.
- The product is stored at below room temperature, generally between 5 and 16°C.

MILK POWDER

- Milk powder or dried milk is the product made

from whole milk, skimmed milk or partially skimmed milk by eliminating the water through heating or other suitable means such that the final product contains not more than 5% moisture.

- According to PFA Rules (1976), whole milk powder shall contain not more than 5% moisture and not less than 26% fat. Partially skim milk powder shall contain not more than 5% moisture and the fat content of the product shall be between 1.6 and 24%. Skim milk powder shall contain not more than 5% moisture and not more than 1.5% fat.

The following are the two systems of drying milk:-

1. Drum drying or roller drying.
2. Spray drying.

Drum drying

Continuously rotated metal drum of 8-12 ft length and 2-4 ft diameter is steam heated at 150°C. This moves at 12-20 rpm. Concentrated milk is poured in it as a thin film on its surface. The milk is in contact with the heated surface for less than 3 sec and is dried quickly to make into a film of dry milk which is continuously removed by a scraper, goes to the transporting station through conveyer belt.

Spray drying

Milk is first concentrated up to 35-40% total solid. It is sprayed through a atomizer in a chamber where this fog/mist like milk spray is mixed intimately with a hot air current (inlet air 150-230°C, outlet air 75-100°C). There is instant dehydration making the milk as a dry powder which is immediately collected in cyclone collector to avoid heat damage to the product, cooled, packed and marketed.

Infant milk food

- Normally, children below one year of age are called infant.
- Mother's breast milk is usually regarded as ideal food for the feeding of infants.
- Owing to various causes, an infant may not get sufficient breast milk due to number of reasons and it may become necessary to feed the infant either wholly or partly with milk from other sources.
- Infant milk food has been formulated to meet the feeding requirements of infants.
- Infant milk food refers to the material prepared either by spray drying or by roller drying of the milk of cow or buffalo or a mixture thereof.
- The milk may be modified by partial removal of fat and by the addition of different carbohydrates like sucrose, dextrose and dextrins, maltose and lactose; salts like phosphates and citrates; vitamins A, B group, C and D and minerals like calcium and iron.
- The infant milk food
  a. shall be free from starch, vegetable fat or any non-milk fat.
  b. should be free from extraneous matter which may be harmful to human health.
  c. should be free from lumps, reasonably free from brown or black specks.
  d. should be uniform in appearance.
  e. should be white or white with a greenish tinge to light cream in colour.
- The flavour of the infant milk food in the dry condition or of the reconstituted milk
  a. shall be fresh and sweet
  b. should not have any rancid taste or musty odour.
- The infant milk food is to be manufactured and packed under hygienic conditions in licensed premises.

Characteristics of Infant Milk Food

S.No.	*Features*	*Prerequisites*
1	Acid insoluble ash per cent by weight (Max)	0.1
2	Added Vitamin D.I.U. (100 gm)	400 to 800
3	Bacterial count, per gm (Max)	15,000
4	Coliform count, per gm (Max)	10.0
5	Iron mg /100 gm (Min)	4.0
6	Milk fat, per cent by weight	18 to 28
7	Moisture, per cent by weight (Max)	4.5
8	Solubility:	
	(a) Solubility index ml. (Max)	15 (If roller dried)
		2 (If spray dried)
	or (b) Solubility per cent by weight (Min)	85.0 (If roller dried)
		98.5 (If spray dried)
9	Total ash per cent by weight (Max)	8.5
10	Total carbohydrates including sucrose dextrose and dextrins maltose or lactose per cent by weight (Min)	35
11	Total milk protein, per cent by wt. (Min)	20.0
12	Vitamin A I.U./100 gm (Min)	1500

INDIGENOUS MILK PRODUCTS

Chhana

- It is prepared by coagulating the milk with some sort of acid or materials having acid action like citric acid, lactic acid, fruit juice from citrus fruits, calcium lactate, old whey of channa previously prepared.

- According to PFA Rules (1976), chhana shall contain not more than 70% moisture and the fat content shall not be less than 50% in dry matter.

Preparation method

Milk (preferably cow milk; buffalo milk or any other milk) is boiled thoroughly.

- The container is taken away from the fire and immediately the acidic material is added with gentle stirring till the coagulation occurs (the amount of acidic material depends upon the amount, type and nature of acidic material itself).
- This is strained through a clean cloth.
- The water is allowed to drain off by hanging the lot.

Composition of chhana

	% Moisture	% Fat	% Protein	% Lactose	% Ash
Cow milk chhana	53.4	24.7	17.6	2.2	2.1
Buffalo milk chhana	51.6	29.6	14.5	2.4	1.9

Paneer

- It is used as base material for preparation of large number of culinary dishes.
- Paneer is an acid coagulated milk product from which moisture is removed by pressing.
- Paneer contains about 54% moisture, 24% fat, 17% protein, 3% lactose and 2% ash.
- As per the PFA Rules (1976), paneer shall contain not more than 70% moisture and not less than 50% fat in dry matter.
- Like channa, it is also prepared by coagulating milk, but the acidic material used is 1% citric acid and the method of preparation is also somewhat different.

Preparation method

Steps

Milk standardization (5.5 to 6% fat)

↓

Milk heating (85-90°C)

↓

Holding (5 minutes)

↓

Cooling (70°C)

↓

Milk coagulation/coagulant addition (1% citric acid solution at 70°C)

↓

Curd settling in whey (5 minutes)

↓

Whey draining

↓

Pressing & hooping (10-15 minutes with 2 Kg/cm^2 pressure)

↓

Dipping in tap water (5 minutes)

↓

Immersion of paneer in chilled water (4-6°C for 2-3 hrs.)

↓

Loose water draining (10 minutes)

↓

Packaging

↓

Storing (under refrigeration condition)

Method of preparation

- Buffalo milk is generally used because of its high T.S. contents.
- Milk is standardized to a minimum fat of 5.5-6.0% to meet the legal standard.
- Milk heated to 85-90°C for 5 minutes followed by cooling to 70°C.
- 1% solution of citric acid heated to 70°C is slowly added to milk with continuous agitation till clear greenish whey is separated out.
- When coagulation is complete, stirring is stopped and the curd is allowed to settle down, whey is drained out and the curd is scooped into a hoop lined with cheese cloth and is pressed for about 10 minutes by applying a pressure of about 2 kg/cm^2.
- The pressed paneer is then removed and immersed in portable tap water for cooling and absorption of moisture till its temperature reaches that of cooling water.
- Paneer is then transferred to chilled water to bring its temperature to 5-10°C.
- Chilled paneer is then removed from the water and allowed to drain out till all the loose water is removed.
- The paneer is then cut into convenient sizes and packed in suitable packaging materials.
- It is stored at refrigeration temperature till disposal.

Ghee

Ghee is made by clarifying butter or cream obtained from milk of cow or buffalo.

Production method

1. Indigenous or desi method

- Normal cooking butter or cream is put in a pan placed over a slow and low fire with continuous stirring till the water is eliminated with a clear fat above and nice smell of ghee starts coming.
- It is filtered to remove the cerum solids or residue.

2. Pre-stratification method

- A modified technique for preparing ghee was developed at the Southern Regional Station, National Dairy Research Institute, Bangalore.
- Butter is put in a pan placed undisturbed at a uniform temperature of 80-85ºC for half an hour.
- Three layers are formed viz., bottom layer of butter milk containing 60-70% of solid not fat and 80% water present in butter initially, middle layer of fat and top layer of floating particles.
- The bottom watery layer is removed as such keeping the top two layers intact.
- Now the temperature is increased to 110-120ºC and after sometimes the ghee is obtained with its characteristics flavor.

4. Creamery-butter method

- This is the standard method adopted in almost all organized dairies, where the raw material used is unsalted creamery (white) butter.
- Butter is heated in a ghee boiler with careful stirring until the mass melts.
- Between 90 and 100°C the contents of the vessel vigorously boil with continuous evaporation of moisture.
- Sometimes a scum of foreign material that has entered the butter during collection and transport rises to the surface at his stage. This is removed by perforated ladle from time to time.
- When moisture has been completely driven off, the crackling sound, characteristic of evaporating moisture, is no longer heard.
- The temperature shoots up at the end of clarification and a few bubbles of escaping air form foam on the top of the ghee. This indicates the end point and simultaneously the characteristic ghee aroma also emanates.
- At this stage, the solids-not-fat in butter show a charred, light to dark brown colour and settle down at the bottom of the vessel.
- The temperature rises to about 120°C. Heating is then discontinued and the clear fat filtered while hot into sedimentation tanks or into retail containers.

3. Direct-cream method

- The cream usually obtained by normal separation of milk is used as raw material.
- The cream is heated in a ghee boiler, which consists of a stainless steel jacketed pan (vessel) provided with a manual stirrer.
- Heating is continued till practically all the moisture has been driven out.
- The final temperature of heating usually ranges from 110 to 120°C.
- After cooling and sedimentation, the ghee is filtered through a muslin cloth so as to separate it from the sediment known as ghee residue.

Khoa

- It is prepared by evaporating the pure milk under the fire such that its water content is reduced to a great extent (19-25% of moisture by wt. of Khoa).

Forms of Khoa

Danedar:

- It is used for the preparation of kalakand, milk cake and other granular milk sweets.
- Moisture content varies between 35 and 40%.
- This type of khoa is characterised by granular texture and uneven body.
- This typical body and texture of khoa is obtained by addition of about 0.1% citric acid (on product basis) during the heat desiccation of milk.

Dhap:

- It contains comparatively high moisture (37-44%).
- It is also brown as 'Kachha Mava'.
- It is most suitable for the preparation of Gulabjamun and Pantoa.
- Khoa has loose and sticky body and smooth texture.

Pindi:

- It is a circular ball of hemispherical pat with smooth and homogeneous body and texture.
- It contains 31 to 33% moisture and has characteristic cooked flavour.
- Sweets like burfi and peda are prepared from this khoa.

Preparation method

1. In general, buffalo milk is preferred for khoa making because of its high fat and T.S. content.
2. Milk is boiled in any iron container (*karahi*), placed over non-smoking fire, and is stirred continuously with circular motion by the flattened end of an iron stirrer (*khunti*).
3. During the operation all parts of the pan with which the milk comes in contact are lightly scraped.
4. Constant evaporation of moisture takes place and the milk thickens progressively.
5. Vigorous stirring and desiccation is continued until the viscous product reaches a pasty consistency and begins to dry up.
6. The final product is ready when it shows signs of leaving the sides of the pan and sticking together so as to make pat formation possible.
7. The khoa pat is invariably made after removing the pan off the fire and by working the contents up and down.
8. The finished product is in the form of balls or circular pats of uniform size.

Nucleus was first discovered by Robert Brown.

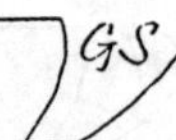

Composition of Khoa

	% Moisture	% Fat	% Protein	% Lactose	% Ash	% Iron
Cow milk Khoa	25.6	25.9	19.2	25.6	3.7	139
Buffalo milk Khoa	19.3	37.1	17.8	22.1	3.7	125

Dahi

- Dahi is a fermented milk product obtained from pasteurized or boiled milk by souring with the help of lactic acid producing bacteria.
- It should have the same percentage of fat and solids not fat as the milk from which it is prepared.

Preparation method

- The milk is standardized to a desired fat and SNF content to avoid variation in composition in day-to-day production.
- It is pasteurized at 80-90°C for 15-30 minutes and cooled to 22-25°C.
- Milk is then inoculated with 1-3% of specific starter culture and filled in suitable containers of required capacity followed by incubation at 22-25°C for 16-18 hours.
- During incubation, acidity reaches 0.6 to 0.7% and a firm curd is formed.
- The curd is cooled to less than 12°C for about an hour and then stored at about 5°C in a cold room.

GK Satpura range has seven bends. GS

Points to Ponder

- 1 kg concentrate per 1000 lbs of body weight is the thumb rule for calculating the required concentrate for cattle.
- 1 kg per every 2.5 lits of milk to calculate the concentrate requirement for a milch animals.
- 100 Ewes and 3 Rams produced 65 lambs.
- 25% of the total world cattle are presented in India and milk production is only 6%.
- 25-30 sq. ft. area is considered sufficient and economical for housing dairy cattle.
- 78 to 83% of egg laid before noon and 1-2% after 3 p.m.
- 80% income of dairy animals come from milk.
- A 60 gm egg is made up of 10% shell, 30% yolk and 60% white.
- A cattle disease communicate to man is Anthrax.
- A chronic buller or the cow having nymphomania will not usually breed until it passes one or two normal heat periods.
- A condition in cattle called teartness has been indentified as molybdenum poisoning.
- A cradle or bull ring is put in suckling cows.
- A good layer is among the first off the perch in the morning and among the last on it at night.
- A group of animals related by descent and similar in most characters like general appearance, features, size, configuration, *etc.*, are said to be a breed.
- A hormone of adrenal cortex called aldosterone regulates the sodium metabolism by promoting its reabsorption from kidney tubules.
- A hundred hens will drink 3-6 gallons of water, depending on size of birds, rate of production, salt contents of the ration, and weather conditions.
- A medium sized egg contains 6.0 gm protein, 6.5 gm fat, 14 vitamins and various minerals.
- A ram should be use for breeding purpose at 18 months of his age.
- A sheep is able to bite individual blades of grass unlike a cattle beast because it embodies split upper lip.
- A sheep require 1.7 sq meter of floor space under a shed.
- A well built up deep litter in operation for about one year has been found to contain 3, 2 and 2 per cent of nitrogen, phosphorous and potassium, respectively.
- Addition of aureomycin feed supplement to rations of fattening lambs has been found to reduce the incidence of enterotoxaemia or Over-eating disease.
- Among the bye-product of grains, rice bran contains the maximum (13.2 per cent) of fat content.
- Amrit Mahal breed of cattle have compact body frame with short straight back, well arched ribs, sloping quarters.
- Amrit Mahal breed of cattle have grey coloured body with dark head, neck, hump and quarters.
- Amrit Mahal breed of cattle have well developed dewlap and hump, very small sheath and close skin.
- Amrit Mahal is one of the best draught breeds of India.
- An early moulter under normal conditions, is a poor layer.
- Animal require 60-70 lit water for drinking.
- Annual laying capacity of an Indian hen is only about 60 eggs which is less than the half of average world is 130 eggs per year.
- Aseel (poultry) embodies noble qualities of fighting, is a superior sitter and has good parental qualities.
- At birth, pigs weight between 0.75 - 1.75 kilogrammes.
- At the junction of omasum and abomasum is an arrangement of folds of mucous membrane, the *vela terminalia*, derived from omasum in thecow, but from the abomasum in the sheep.
- Australia has the largest sheep population in world.
- Average milk production of exotic cow is 2000 to 3000 lit/lactation.
- Back crossing is not widely used by animal breeders.
- Beginning of early moult signals the lack of persistency.
- Between 300 and 500 gallons of blood must be passed through the udder for every gallon of milk.
- Bhadawari breed of buffaloes is found scattered in surroundings of Jamuna and Chambel rivers.
- Bhadawari buffaloes have medium size and wedge shaped body.
- Birds grow relatively faster and record higher yield, when kept in a 24 hours lighted house.
- Bovine veneral trichomoniasis is caused by a single celled protozoa, *Trichomones foetus* which brings about the destruction of embryo witnin 3-5 weeks after conception.
- Branding of animals should be done at an early age, preferably before the weaning.
- Broodiness in birds is the external evidence of the maternal instinct and is a dominant sex linked character.
- Brucellosis (Bang's disease) caused by *Bruclla abortus* is usually found in pregnant uterus, but may also localize in udder or testes.
- Brucellosis is a common genital disease in swine and is dangerous to man.
- Buffalo milk contain on an average 8.17% fat.
- Bullocks of Deoni breed are well suited for heavy work.
- Cannibalism in hens may be by inheritance, deficiency

in ration, and by close confinement.

- Capons are the male birds whose testicles have been removed either surgically of chemically.
- Castration ameliorates palatability of meat, accelerates the rate of increase in weight and improves the quality of skin is goats.
- Castration enhances the rate of gain, enables to produce more desirable type of meat and makes the animal more docile.
- Castration with the help of a Burdizzo's castrator is also known as bloodless castration.
- Cockerels from 6-8 weeks of age and weighing about 0.5-.1.0 kilogramme are considered the best subjects for the caponizing operation.
- Colostrum contains higher proportion of globulin than does normal milk, it contains 3-5 and 5-15 times more protein and vitamin A, respectively.
- Colostrum contains vitamin A.
- Colostrum is rich in protein (globulin), minerals and antibiotics.
- Comparatively small head bulging towards horns, short but stout legs, hooves of hind quarters more backward than those of fore quarters, short but well barrel, slightly broad forehead deep in the middle, are main characteristic of Bhadawari breed of buffalo.
- Cow milk contains on average 3.5 to 5% fat, 13 to 14% total solids and 3.5% protein.
- Cows can be re-bred in 9-12 weeks after parturition.
- Criss-cross breeding is proposed for utilizing the heterosis in both dam and progeny.
- Death rates in the inbred animals are higher than those in outbred ones.
- Deficiency of phosphorous causes pica in cattle.
- Dehorning of a calf should be done before it is 10 days old because horn button upto this age remains unattached to skull.
- Diesturm or dioestrum periods are transmitted by activities which norm,ally occur at the end of pregnancy such as nest making and the secretion of a milk and these periods are know as Pseudopregnant periods.
- Dietary protein may be measured either as crude protein or as CDP
- Dislike to feed, shaking of tail and restlessness *etc.*, are heat signs in doe (she goat).
- Double-yolked eggs may be due to ripening of two ova at the same time or one ovum being pushed back into the oviduct at the same time when another ovulation takes plaee.
- Ears of Deoni (Dongarpatti) breed of cattle have no notch near the tip.
- Egg shell is mode of $CaCO_3$

Knowledge World!

First Five Year Plan

The First Five Year Plan (1951-56) with an outlay of Rs. 2,378 crore laid emphasis on agriculture, irrigation, power and transport.

- Estradiol, estrone, estariol, equilin and equienin are called estrogen.
- Ewe's milk on an average has a much higher solid content than cow and goat milk.
- Ewes like cows need streaming-up to give good milk yield in early lactation.
- Exposure of ewes to artificial light brings them to heat, but darkness induces late maturity and reduces oestrus.
- Fatty sires are often indifferent and may even embody low fertility.
- Fecundity is the potential capacity of a female to produce function ova, regardless of what happens to them after they are produced.
- Feed the calf enough of colostrum for the first four days following its birth.
- First crosses of Sahiwal cow and Ayrshire bull were high milk yielders, but susceptible to F.M. disease.
- Flushing denotes low ground sheep farming practice of increasing fertility by pulling the ewes on to a higher plane of nutrition before introducing them to the rams.
- General colours in Sahiwal cattle are various shades of red, pale red and dark brown splashed with white.
- Grading is the practice of breeding sires of a given breed to non-descript females and their offpring for generation after generation.
- Guinea grass is a native of Tropical Africa.
- Gunn in Australia developed a method of semen collection from rams by electric stimulation of spinal cord.
- Hariana breed of cattle have proportionate body and compact graceful appearance.
- Hariana cattle breed animals have small and sharp ears, fine textured skin close to the body.
- Hariana, Ongole, Tharparkar and Kankrej are dual purpose breeds of Indian cattle.
- Hens with coarse, phlegmatic, masculine, or "beefy" heads are not likely to lay many eggs.
- Horse has the largest and most complex large intestine of any of the domestic animal.
- Hybrids of distant crosses like horse x ass = male and horse x zebra = zebroid are mostly sterile.
- Hypocalcaecia or milk fever in sheep is often associated with hypomagne-saemia.

Shilong plateau is a part of Chhotanagpur plateau.

GS

- If light provided is more than required it will be lead to cannibalism activity in birds.
- In cows and she buffaloes cessation of estrus, sluggish temperament, easier tractability and increase in body weight are some of the signs of pregnancy.
- In diary cattle the period of ovulation seems to be 14 hours after the end of heat.
- In guinea-pigs vaginal plug is formed due to semen coagulation after mating.
- In ruminants, the rumen is the largest, while the reticulum is the most cranial compartment in the stomach.
- In tail to tail system, of barn arrangement, any sort of minor disease or any change in the hind of the animals can be detected quickly and seen automatically.
- In the bull the spermatozoon is about 80 mu in length (head is 9mu long and 5.5mu wide and about 65mu in length).
- In the milk of ruminants, fatty acids are also synthesized in the mammary cells from acetate and about 25 per cent of them are derived from fatty acids coming from dietary fat.
- Increased deficiency of essential fatty acids markedly enhances, the water consumption in animals.
- Indian goats commonly kid twice in one year and usually thrice in two years.
- It has been observed that Nili and Ravi breeds of buffalo are distinct without difference and are officially treated as one breed.
- It has been recommended that sows should be mated on the second of heat for the optimum fertlity.
- It is the face and forehead of Nili and Ravi buffaloes that distinguish this breed mainly from the Murrah.
- Jaffarabadi breed of buffaloes in their purest form are seen in Gir forest of Kathiawar where they are bred in large numbers, almost solely for *ghee* (butter-oil) production.
- Kangayam breed of bullocks/bulls are excellent type for hard work.
- Kangayam breed of cattle are also known as Kongu or Konganad.
- Kankrej breed of cattle are also known as Banniai or Vagadia or Wadhiar.
- Kankrej cattle have broad chest, their forehead is dished in the centre, have strong curved horns accompanied by skin upto some length, powerful body and straight back.
- Karakul breed of sheep is famous throughout the world for pelt production.
- Kellner's idea of feeding animals on the basis of net energy rather than by TDN was first recognised in Germany in 1905 and in England in 1909.
- Lactose content in different animals milk is 3 to 5.5% and women milk is 6%
- Lambs that are weaned from their mothers and go for further feeding prior to slaughter are known as store lamb.
- Light in a poultry house does not only destroy disease germs and supplies Vitamin D, but also makes the birds happy.
- Long sweeping horns of Amrit Mahal breed are typical of Mysore type cattle.
- Loose horns are common in females of Sahiwal breed of cattle.
- Lucerne (alfalfa) besides protein is also rich in calcium and phosphorus.
- Mastitis, in sheep is most commonly caused by *Staphylococcus aureus*.
- Meat or blood spots may result due to degenerated blood in general or of clots due to haemorrhages of small blood vessel in the ovary or oviduct.
- Medium and dark brown poultry eggs hatch better than light brown eggs.
- Milk "let down" mechanism is greatly influenced by sensory and efferent motor sympathetic nerve fibres.
- Milk production enhancement programme is known as Operation Flood.
- Milk production increase rapidly following parturition, reaches a peak 2-4 weeks where it remains for a short period and then gradually declines.
- Milk yield and egg production are sex-linked traits.
- More than 40% of the wool produced in the world comes from the merino and their derivatives.
- Most of the Indian animal breeds are well known for their draught resisting qualities and to withstand diseases and parasites.
- Most of the mares are bred to foal in spring, the length of estrus cycle is 15-24 days and the heat period is 3-7 days.
- Most of the milk is stored in alveoli.
- Murrah buffaloes have short, characteristic tightly curled horns, well developed udder and a long tail with a white switch reaching the fetlock.
- Murrah buffaloes have deep massive frame with short, broad back and a compratively light neck and head.
- Nellore is the tallest breed of sheep in India.
- Nili and Ravi breeds of cattle, are medium sized, have deep frame with an elongated, coarse and heavy head bulging at the top and depressed between the eyes.
- Normally ovulation occurs about 30-35 minutes after laying each egg.
- One of the first diluent of semen in veterinary science for the insemi-nation of cattle in Russia was Milovanov's

GK **The crops cultivated by the Harappan included : wheat, rice, sugarcane.** GS

"SGE-2 dilutor".

- Ongole breed animals are greatly alert and docile with good gait.
- Ongole breed of cattle is also known as Nellore.
- Oxytocin and Vasopressin are produced by hypothalmic neurosecretory cells.
- Parathormone provide optimal level of calcium, while insulin regulates carbohydrate metabolism during lactation period.
- Parturition is presumably caused by an increase in the level of estrogen, a decrease in the level of progesterone, increase in the sizse of foetus, and inadequacy of nutrients.
- Pashmina is hairy under coat obtained from Pushmina goat.
- Penicillin is more effective than other antibiotics as feed supplement to poultry ration especially for growing chicks and turkey poults.
- Penninsular region has the highest sheep population in India followed by North Western region their wool production is 28 and 64 percent respectively.
- Perosis disease of chickens is due to the deficiency of manganese.
- Phillips and Lardy (1939-40) introduced an advance technique of semen storage known as egg yolk phosphate (EYP) dilutent.
- Pituitary has frequently been described as, "the conductor of the endocrine orchestra."
- Poultry egg contains 12.4% protein, 10.7% fat.
- Prolificacy is more or less restricted in its application to the female or groups such as breed, strain or heard.
- Ram produce more wool than the ewes but quality of wool is inferior.
- Reversion of vagina its protrusion out through the vulva is known as ballooned vagina.
- River buffaloes originate from the Indian subcontinent, have chromo-some number of 2n = 50 and are improved domesticated animals with high milk and meat production.
- Sahiwal and Sindhi are muilch breeds of Indian cattle.
- Sahiwal is also called as Lola.
- Sahiwal is one of the best dairy breeds in India.
- Semen is a suspension of sprmatozoa in seminal fluid, it is opaque, white to light cream-coloured fluid.
- Severe riboflavin deficiency in chicks causes swelling and softening of the sciatic and brachial nerves producing a contraction of toes and typical symptoms known as curled-toe paralysis.
- Shantagurtadia breed is well known for meat purpose.
- Sharp horns, relatively small ears, well placed hump covered with a tuft of hair at the top, strong legs and feet, and tight sheath are some of the characters of Siri breed of animals.
- Sheep is a golden hoofed animals.
- Sheep is seasonally poly-gastrous animals.
- Sheep require 3 to 4 kg of dry matter per 100 kg of their body weight.
- Shell membrane is formed in the Isthmus.
- Sindhi breed of cattle has medium size and is compact, it has well proportioned body and is extremely docile.
- Station (male horse) x Janned (female ass) = Hinny which is generally inferior to mule.
- Tharparkar breed of cattle are also known as Thari.
- Tharparkar breed of cattle are medium sized, deep built and embody short, straight and strong limb.
- The active principle of the thyoid gland was first isolated by Kendall who named it thyroxine, found in a protein material called thyroglobulin.
- The age of bull at maturity will be about 3 years of village condition.
- The animal specially cattle diseases *viz.*, "salt sick" and "lechsucht" are caused by copper deficiency.
- The anterior pituiary secretes F.S.H. in increasing quantities bringing about the multiplication of puberty.
- The antibiotic vitamin B12 feed supplement is largly included in the commercial formula feed for chicks, pigs and dairy calves for their growth and fattening.
- The average annual milk production per cows and buffalo are 157 kg and 504 kg respectively.
- The average percentage increase in homozygosity or decrease in hetrozygosity is an inbred animal in relation to an average animal of the same breed of the foundation stock is known as the co-efficient of inbreeding.
- The average specific gravity of whole semen is 1.028, 1.011 and 1.03 in man, dog and bull, respectively.
- The best known and commonly used method of semen collection is Artificial Vagina (AV) method.
- The best known breeds of Indian buffaloes are Murrah, Jaffarabadi, Nili and Bhadawari.
- The best time for cutting a crop for hay making is when it is 30-50% in blossom.
- The boiling point of milk is 100.17^{0}C.
- The cells which supply nutrition to the maturing spermatide are known as sertoli's cells.
- The colostrum contains greater percentage of protein, a large part of which is due to the presence of gamma globulins the antibody proteins.
- The colour of the Kangayam bull is of grey with dark grey to black markings, while the cow is white with black marking just in front of the fetlocks on all the four legs sometimes on the knees.

GK **'Rule of Law' is the meaning of the term 'Equality before Law'.** GS

- The deficiency of F.S.H. and L.H. hormone may limit or retard reproduction in either sex.
- The deficiency of fat leads to a number of troubles in animals namely, lack of libedo in males, reabsorption of the foetus, abortion or dead birth etc.
- The Digestibility co-efficient may be defined as the percentage of the total amount consumed which is digested and absorbed.
- The ears of Gir breed of cattle are markedly long, pendulous resembling a tiny curled leaf.
- The eggs having a proportion of white to yolk of about 2 : 1 usually hatch better than those having either wider or narrower ratio.
- The embryonic mortality may be due to hormone deficiency or imbalance, failure of implanation of fertilized ova, death due to lethal genes, accidents in development, insufficient nutrition, overcrowding or infection in the uterus.
- The female organ extending from the cervix posteriorly upto the urogential sinus or vestibule from which it is separated by the hymen or the hymenal constriction is the organ of copulation in females.
- The first breed of poultry named Plymouth rock was exhibited at America's first poultry show, held at Boston in 1849.
- The general colour of the Malvi breed of cattle is grey to iron grey black in the neck and quarters.
- The Gir (also called Kathiawarhi Surti or Desan) breed of Indian cattle probably originated in the Gir forests of South Kathiawar..
- The Government of India started artificial insemination centres at Calcutta, Hissar, Madras, Bangalore and Nagpur in 1945.
- The hens kept in battery system produce vegetative eggs.
- The hens start laying eggs at about 15 days age. At this age 14 hours light is necessary.
- The home of Sindhi breed of Indian cattle is Karachi and Hyderabad (Sindh) in Pakistan.
- The hormone called testosteron has been considered to be the primary sex hormone.
- The Indian Veterinary Research Institute, Izatnagar, started experimental work on artificial insemination under the direction of Dr. P. Bhattachareya, in 1942.
- The Kangayam breed of animals have moderate sized hump, wide muzzle strong limbs, small dewlap, fire skin, very small sheath, well developed quarters and a fine tail.
- The Kankrej breed of cattle is one of the heaviest of Indian breeds.
- The length of estrous cycle for heifers is 20 days and 21 to 22 days for cows.
- The main problem in horse breeding is the usually long heat period and the variation in ovulation time.
- The Malvi breed bullocks are good for road and field work, economical feeders and have good adaptability.
- The Malvi breed of cattle have deep, short and compact body, their ears are alert and short in size, horn short but thick tapering to a blunt point.
- The Malvi breed of cattle is also known as Mahadespuri or Manthai.
- The maternal dystokia may be due to uterine abnormalities like torsion of uterus, uterine intertia, and uterine hernia.
- The mixture of grasses in a good pasture should have high percentage of leguminous grasses (forage crops).
- The most popular fine-wool breed in the world is Merino.
- The most reliable method of pregnancy diagnosis in bovine animals is palpation of the uterus per rectum.
- The Murrah breed of buffalo is the most efficient producer of milk, not only in India but probably in the world.
- The noticeable feature of Jaffarabadi buffalo-breed is the very prominent forehead and heavy horns, inclined to droop on each side of the neck and then turn up at the points.
- The organisms *Fusarium nodosus* attacks only sheep and goats.
- The ovaries of a sow are rounded and the surface is lobulated with follicles and corpora lutea.
- The oviduct (fallopian tube or ovarian tube) has a diameter of about 0.1 cm in case of the cow and it widens into a funnel shaped tube at the ovarian end which is called the as Infundibulum.
- The ovualation in cows takes place by the action of leutenizing hormone.
- The percentage of thick white in egg tends to decrease from about April to July in most of the flocks.
- The popular colour of murrah buffaloes is jet black with white markings on the tail, face and extremities.
- The portion termed as germspot on the yolk of the egg gives rise to an blastoderm which ultimately develops into an embryo.
- The poultry breed Rhode Island Red is one of the hardiest of all the breeds.
- The purity of the breed is maintained by confining the mating of animals to within the breed.
- The ram is in full vigour for breeding during his age of 2.5 to 5 yrs.
- The recommended calcium-phosphorus ratio for reproduction in pigs is approximately 1.5:1.
- The retractor muscle in bulls which is attached to the anterior arms of the flexure helps in pulling the erected

and exposed penis back into the sheath for protection against injury and infection.

- The silent or shy breeder cows and buffaloes go through the normal ovarian changes in the estrous cycle but show no signs of heat and receptivity to males.
- The sire is half of the heard.
- The Siri breed cattle have massive body, small head, squarecut, wide and flat forehead, presenting no convexity.
- The Siri breed of cattle are found in Darjeeling hill tracts, Sikkim and Bhutan.
- The spermatozoa are generatedin the testes and stored in epididymis, whereas seminal plasma is contributed by the secretory fluids produced in accessory organs.
- The sperms, in case of a hen, remain in the oviduct for 2-3 weeks after mating, but the newest sperm is the one most likely to fertilize the egg.
- The stages of parturition (a continuous act) comprises the stages *viz*, preliminary dilation foetus expulsion and after-birth stage.
- The suckling stimulus and intra-mammary pressure are two important factors influencing the level at which lactation is maintained.
- The swam water buffalo (*Bubalus bubalis*) is indigenous to southeast Asia, southeast China and Assam in India.
- The systematic cross breeding of sheep on a regional scale is called stratification.
- The tail of Tharparkar breed of cattle is fine with black witch.
- The term class is used to designate groups of poultry breeds developed in certain areas.
- The term concentrate is applied to that group of feeds which are relatively high in total digestible nutrients and low in crude fibre.
- The thick walled portion of the reproductive tract lying between the uterus and the vagina is known as cervix.
- The udder secretion, immediately before parturition changes from honey like secretion to an yellowish one, known as colostrum
- The unborn calf makes most of its growth during the last 2 to 3 months before birth.
- The yellow colour of egg yolk is due to the pigment xanthophyll.

Quiz Kit!

First Indian in the ICS

The Indian Civil Service or ICS competitive examination was devised in 1853 by a committee set up by Macaulay. Satyendranath Tagore, elder brother of Nobel Laureate Rabindranath Tagore, was the first Indian to join the service.

- Two medium sized eggs supply about 26% of iron 20% of phosphorus, 10% of iodine, 8% of calcium and 20% of protein of the daily requirement of an adult male.
- Two types of pigments in milk are carotene and riboflavin.
- Two-toothed females (sheep) are referred to as gimmers or shearling eves, while the males as shearling rams.
- U.D. Sharma (1957) used the semen dilutor known as IVT (Illine Variable Temperature) to preserve it at room temperature.
- When a female calf is born as a twin to a male calf, a sterile female is known as Freemaretin or neuter.
- White leghorn is the most popular breed for egg purpose in the world originated in Italy.
- Wool of Patanwadi is suitable for manufacturing of carpet and blankets.
- Yellow colour in milk is due to carotene and yellowish green due to riboflavin.
- Zinc deficiency in cattle causes parakeratosis.
- Rinderpest is also called as Cattle Plague.
- The mammary secretion for the first three or four days following calving is called beestings.
- The yellow colour of egg yolk is due to the pigment Xanthophyll.
- National Research Centre for Meat & Meat Products is situated in Hyderabad (A.P.).
- Project Directorate on Cattle is situated in Modipuram (U.P.)
- Anthrax is also called spleenic fever.
- Milk is deficient in iron.
- The carbohydrate of sugar constituent of the milk is lactose.
- Stiff lamb disease caused by vitamin-E deficiency.

GK GS

The Reserve Bank occupies the apex position in the Indian money market.

Multiple Choice Questions

1. ________ is a breed of buffalo.
   a] Berkshire b] Beetal
   c] Jaffarabadi d] Sindhi
2. ________ is a breed of cattle.
   a] Bhadawari b] Tharparkar
   c] Kashmiri d] Landrace
3. ________ is a breed of goat.
   a] Berari b] Large White Yorkshire
   c] Sahiwal d] Nili
4. ________ is a breed of swine.
   a] Malvi b] Ravi
   c] Gaddi d] Chester White
5. ________ kg of feed is required per kg gain weight of chicken.
   a] 2.5 kg b] 1.5 kg
   c] 5.0 kg d] 4.0 kg
6. ____________ per cent of animals contracting bang's disease become either temporarily or permanently sterile.
   a] 10-15 b] 15-20
   c] 18-23 d] 25-30
7. __________________ are excellent feed convertors.
   a] Goats b] Sheep
   c] Pigs d] Hens
8. 'Balanced ration' is the amount of feed :
   a] Which is given during pregnancy
   b] Which contains all the nutrients
   c] Which is supplied during 24 hours
   d] Which has correct proportion of all the feed nutrients supplied in 24 hours
9. 'Merino' is the most popular breed which produces one of the following types of wool :
   a] Medium wool b] Long wool
   c] Carpet wool d] Fine wool
10. 'Operation flood I' was taken up in :
    a] 1972 b] 1970
    c] 1974 d] 1975
11. 'Operation Flood II' was launched in:
    a] 1977 b] 1982
    c] 1979 d] 1978
12. 'Zebu' is a term which refers to :
    a] Indian cattle as known in Europe
    b] European cattle
    c] Breed of buffalo in China
    d] Flock of sheep
13. 103°F is the average normal body temperature of :
    a] Cow b] Goat
    c] Fowl d] Horse
14. 12-14 month old sheep is known as:
    a] Hogget b] Shaearer
    c] Gimmer d] Feeder
15. 9-10 month old sheep is called :
    a] Gimmer b] Hogget
    c] Shaearer d] Feeder
16. 9-10 months old male and female weaned lambs are called as :
    a] Gimmer b] Hogget
    c] Gummer d] Yearling
17. In most of the cows ovulation occurs ________ hours after the end of heat.
    a] 5-9 b] 10-14
    c] 15-17 d] 18-20
18. A 100 kg live weight pig requires about ________ kg of daily feed.
    a] 3.2 b] 5.3
    c] 7.5 d] 4.2
19. A calving box should have an area of about ________ square feet with ample of soft bedding.
    a] 100-150 b] 75-95
    c] 60-75 d] 45-55
20. A cow during her productive life must remain pregnant only for ______ months in every year.
    a] 6 b] 7
    c] 8 d] 9
21. A cow must not be allowed to eat its placenta becuase it result in reduced milk production due to :
    a] It causes infection in the digestive tract
    b] It contains poisonous substance
    c] It is high in protein and appetite is lost
    d] It is high in fat and disturbes digestion
22. A cow which came in heat and conceived on January Ist is expected to parturate during :
    a] First week of October
    b] Second week of October
    c] First week of November
    d] Second week of November
23. A farmer has established two separate farms, one for layer strain and other for broiler purpose. Which one of them will be more profitable:
    a] Both are equally profitable
    b] Layers are more profitable
    c] Broilers are more profitable
    d] Both are not profitable
24. A farmer should purchase layer chicks in the month of :
    a] February to April
    b] May to July
    c] September to October
    d] November to December
25. A fertile herd takes less than _______ months between calving.
    a] 13 b] 14
    c] 15 d] 16

GK
Hargobind Khurana's work relates to understanding of genetic code.
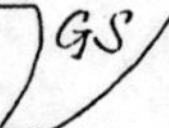
GS

26. A goat breed which has superior quality skin and is in great demand in foot wear industry is :
a] Marwari b] Jamunapuri
c] Kathiawari d] Bengali
27. A hormone released from the posterior pitutary required for milk let-down is:
a] Vassopressin b] Epinepherin
c] Oxytocin d] A.D.H.
28. A Jersey or Holstein half bred heifer when will maintained will attain puberty at the age of :
a] 6-9 months b] 1.0-1.5 year
c] 2.0-2.5 year d] More than 3 years
29. A loss of 10 per cent of water causes restlessness, trembling and weakness, and loss of _________ per cent of water results in death through thirst.
a] 14-16 b] 17-19
c] 20-22 d] 23-25
30. A normal dry period of cow should be :
a] 60-75 days b] 90-100 days
c] 100-120 days d] 120-150 days
31. A pregnant cow does not come in heat because :
a] Corpus luteum is present
b] Corpus luteum is absent
c] Grafian follicle is present
d] Grafian follicle secretes progesterone
32. A system of grazing known as Henbenhein system was developed in :
a] Holland b] Poland
c] England d] Germany
33. A well fed buck keeps fit for breeding till _______ years of age.
a] 5-7 b] 8-10
c] 11-13 d] 14-16
34. About __________million tonnes of cattle manure is produced annually in India.
a] 1200 b] 900
c] 1500 d] 424
35. According to P.F.A. act chhana should contain fat not less than :
a] 25% b] 30%
c] 50% d] 60%
36. According to standard classifiaction chickens have been classified into :
a] 8 classes b] 11 classes
c] 15 clases d] 20 classes
37. Acidity in milk is not attributed to:
a] Carbon dioxide b] Citric acid
c] Casein d] Acetic acid
38. Acute rise in temperature, lameness, skin become dry and dark and hot, tense and painful swelling in hind quarter shoulders and neck, are the symptoms of :
a] B.Q. b] F.M.D.
c] H.S. d] Anthrax
39. After milking milk should be cooled to a temperature of _________ as soon as it is possible.
a] 0^0C b] 4^0C
c] 30^0C d] 37^0C
40. Age of castration of a bull is about:
a] 1.5 year b] 2.5 year
c] 1 year d] 3.0 years
41. Among the following, which instrument is used to test the purity of milk :
a] Tube b] Chronometer
c] Sphynomonometer d] Lactometer
42. Amongest Indian breeds of sheep which one produces good quantity of milk :
a] Malpura b] Sonadi
c] Lohi d] Marwari
43. Amongst the exotic breeds of cattle maximum milk is produced by :
a] Holstein b] Jersey
c] Red Dane d] Brown Swiss
44. An area of 100 square feet per head for an young stock of __________ calves and an increase of 50 square feet for every additional calf makes a good paddock.
a] 12 b] 10
c] 8 d] 6
45. An examination of the rainfall map of India will show thatn it is invriably in the higher elevation, and in humid areas that the cattle are :
a] Large and non-descript
b] Large and well-descript
c] Small and well-descript
d] Small and non-descript
46. Animal protein factor is the name given to :
a] Riboflavin b] Biotin
c] Pyridoxine d] Cobalamine
47. Annual wool yield from a sheep varies from :
a] 0.1 to 0.3 kg b] 0.3 to 1.8 kg
c] 5.2 to 6.5 kg d] 2.2 to 2.3 kg
48. Approximately 43-53 is the dressing percentage obtained from one of the following species :
a] Goat b] Sheep
c] Swine d] Poultry
49. Argine, methionine, lysine and tryptophan are :
a] Carbohydrates b] Fats
c] Protein d] Minerals
50. Artificial hatching was practised some _______ years ago in China and Egypt.
a] 2,500 b] 3,000
c] 3,500 d] 4,000
51. At one time a full mature ram ejaculates semen volume of about:
a] 0.5 ml b] 1 ml
c] 2-3 ml d] 4-6 ml

GK GS

The average elevation of Western Ghat is 1200 m.

52. At present the most economic animal producing milk at least cost in India is :
a] Deshi cow b] Exotic cow
c] Cross-bred cow d] Buffalo

53. Aureomycin hydrochloride is fed to young calves 80 mg/calf/day because:
a] It treats calf scours
b] Increases growth from 10 to 30 per cent
c] Develops rumen
d] Helps in digestion of milk

54. Average amount of blood per goat is about :
a] 0.75 litre b] 2.5 litre
c] 3.0 litre d] 1.5 litre

55. Average fat in Cow's milk is about:
a] 5.2% b] 4.2%
c] 4.9% d] 3.8%

56. Average gestation period of doe is:
a] 114 days b] 146-150 days
c] 62 days d] 281 days

57. Average lactation period of buffalo:
a] 300 days b] 281 days
c] 310 days d] 200 days

58. Average yield of Indian cow per year is :
a] 1500 kg b] 500 kg
c] 173 kg d] 100 kg

59. Bacteria responsible for making curd are :
a] Streptococcus lactis
b] Mycobacterium tuberculosis
c] Streptococcus agalactic
d] Cornybacterium pyogens

60. Bacterial fermentation of food in the digestive tract of horse occurs in :
a] Rumen b] Crop
c] Gizzard d] Caecum

61. Barned variety of Plymoth Rock was developed in about _______ from the cross of the Dominique and Black Cochin breeds.
a] 1835 b] 1845
c] 1855 d] 1865

62. Best broilers are produced by crossing the following breeds :
a] White Leghorn and White Rock
b] White Leghorn and Rhode Island Red
c] White Cornish and White Rock
d] White Cornish and Minorca

63. Black-quarter is an acute infectious and highly fatal disease of cattle which affects cattle of the age group of :
a] 0-3 days
b] More than 4 yrs.
c] 6 months-2 yrs.
d] 2-4 yrs.

64. Blisters on tongue, lips and skin around clave of the feet; blisters my cause a heavy flow of saliva and affected animal shake legs frequently, are the symptoms of :
a] F.M.D. b] B.Q.
c] H.S. d] Anthrax

65. Boars reach sexual maturity between _________ days.
a] 60-90 b] 100-180
c] 190-200 d] 210-220

66. Bonneman, a French physician, in ________ hatched chicks by heating a hatching oven with circulated hot water.
a] 1770 b] 1777
c] 1779 d] 1787

67. Breeding of goats at the age from _________ months yields good results.
a] 8-12 b] 14-18
c] 10-13 d] 19-22

68. Breeding season in goat is :
a] April-May b] March-April
c] June-September d] September-February

69. *Brucella abortus* is a causitive agent of :
a] Anthrax b] Tentanus
c] John's disease d] Abortion

70. Buffalo milk is not yellowish because:
a] It contains more fat
b] It contains no lactose
c] It contains no carotene
d] It contains more calcium

71. Buffaloes are 30% of cow's population and they produce _________ per cent of the total milk yield of India.
a] 20 b] 50
c] 40 d] 63

72. Bullock poplation in India is about:
a] 40 million b] 60 lakhs
c] 1.2 million d] 7.0 million

73. By homogenization of milk the size of fat glouble is reduced below:
a] 4 micron b] 5 micron
c] 2 micron d] 1 micron

74. By water-glass and lime-water methods eggs can be preserved for about _________ months.
a] 2 b] 4
c] 6 d] 8

75. Carotene is precursor of vitamin :
a] A b] D
c] E d] K

76. Casein constitutes about _________ per cent of total milk protein present in milk.
a] 80 b] 95
c] 60 d] 22

77. Casein is precipitated from milk at:
a] pH 4.4 b] pH 4.6
c] pH 4.8 d] pH 6.6

78. Casting means :
a] Making an animal sterile
b] Throwing the animal on the ground
c] Securing the animal in travis
d] Restraining a bullock with rein ropes

79. Castrated male bird is known as :
a] Cockerel b] Broiler
c] Capon d] Cock

80. Castration in male calf is done to produce :
a] Capon b] Stage
c] Bullock d] Barrow

81. Castration of male pig should be done in the age of :
a] 3-4 weeks b] 1-2 months
c] 1-2 weeks d] 2-4 weeks

82. Cattle plague is the common name given to the disease of cattle known as :
a] Anthrax b] Black quarter
c] Mastitis d] Rinderpest

83. Central Sheep and Wool Research Institute is located at :
a] Hissar b] Karnal
c] New Delhi d] Malpura

84. Chedder and Cottage are the names of :
a] Ice-Cream b] Cheese
c] Ghee d] Milk powder

85. Cheese contains about ________ percent of water.
a] 30.3 b] 35.5
c] 40.3 d] 50.9

86. Chevon is the name of ________ meat.
a] Sheep b] Cattle
c] Buffalo d] Goat

87. Cleaning and sterilization of dairy utencils and equipment means :
a] Removal of milk residue and visible dirt
b] Destruction of all bacteria
c] Removal of milk residue, visible dirt and destruction of all bacteria including spores
d] Removal of milk residue, visible dirt

88. Coccidiosis in cattle is caused due to:
a] Bacteria b] Coccidia
c] Virus d] Fungi

89. Coccidiosis is a ________ disease of poultry.
a] Bacterial b] Protozoan
c] Viral d] Bacteial and viral

90. Cockerels are caponzied when they are usually :
a] 3 weeks old b] 5 weeks old
c] 8 weeks old d] 10 weeks old

Did You Know ?

Most Presidential ordinances

President Fakhruddin Ali Ahmed promulgated 29 ordinances in a single year in 1976.

91. Colostrum coagulates on heating because :
a] It contains more fat
b] It contains less lactose
c] It contains more albumin and globulin
d] It contains less fat

92. Colostrum is fed to new bron calf as it provides immunity against diseases due to the presence of :
a] Gama globulins b] Serum albumins
c] Casein d] Lacto-globulins

93. Milk production is maximum in :
a] U.P. b] A.P.
c] M.P. d] Punjab

94. Colstrum is :
a] Oil extracted from coconut
b] The blood corpuscles
c] The fat derived from hog
d] Secretion of udder just after claving

95. Cow and Buffalow milk contain on an average of fat :
a] 2 and 4.5 b] 4.5 and 7.1
c] 6.00 and 8.5 d] 3 and 9.5

96. Cow milk is generally richer than buffalo milk in :
a] Fat b] Carotene
c] Minerals d] Proteins

97. Cow milk is not complete food because :
a] It is deficient in fat content
b] It is deficient in iron content
c] It is deficient in protein content
d] It is deficient in vitamin content

98. Cow milk is very different from human milk because:
a] It has more carotene
b] It contains less protein
c] It contains less lactose
d] It contains less fat

99. Cow should ordinarily be bred during the ________ month after calving.
a] 2nd to 3rd b] 1st to 2nd
c] 3rd to 4th d] All of the above

100. Cows come in heat regularly at intervals of 16-24 days and heat lasts ________ hours.
a] 8-12 b] 13-15
c] 16-20 d] 21-24

101. Dahi is a milk product in the category of :
a] Concentrated product
b] Fermented product
c] Coagulated product
d] Frozen product

GK **Offices of Governor and President are exempted under Article 14.** GS

102. Dairy cows and buffaloes require _______ gallons of water per diem and an additional amount of about one kilogramme need be allowed for each pound (nearly half a kilogramme) of milk produced.
a] 4-5 b] 5-6
c] 7-8 d] 9-10

103. Deep litter system of management is used for housing of :
a] Pigs b] Sheep
c] Poultry d] Ducks

104. Dehorning of female goat should be done at the age of :
a] 12 days b] 18 days
c] 2 days d] 25 days

105. Dipping of eggs in hot water at a temperature of 135-145 °F for about _________ minutes kills the germ of the egg.
a] 10 b] 15
c] 20 d] 25

106. Dr Sampath Kumaron employed artificial insemination method for the first time in India in _____ at the Palace Dairy Farm, Mysore.
a] August, 1939 b] September, 1939
c] August, 1938 d] September, 1938

107. Duration of heat in sheep is :
a] 10-12 hours b] 11-30 hours
c] 36 hours d] 3-4 days

108. During first week the most comfortable brooding temperature is :
a] 28.5 to 30.5°C b] 32.5 to 35.0 °C
c] 36.5 to 40.0°C d] 40.5 to 44.0°C

109. Durock is a breed of :
a] Swine b] Cattle
c] Buffalo d] Goat

110. Dystokia, among the domesticate species is mostly met with in :
a] Dogs b] Cats
c] Monkeys d] Diary cattle

111. Each broiler should be provided with floor space of :
a] 700-900 sq. cms
b] 460-500 sq. cms
c] 1000-1500 sq. cms
d] 2300-2800 sq. cms

112. Egg production records per bird can best be kept, if the birds are housed using :
a] Free range system
b] Semi-intensive system
c] Deep litter system
d] Cage system

113. Enzyme to break cellulose is present in :
a] Cow b] Hen
c] Pig d] Dog

114. Enzyme used in digestion of milk casein is called :
a] Pepsin b] Trypsin
c] Renin d] Oxytocin

115. Estrous cycle of pig is :
a] 19 days b] 23 days
c] 11 days d] 27 days

116. Ewes are matured with _________ months.
a] 2-3 b] 9-14
c] 5-6 d] 18-19

117. Ewes remain in heat for about _______ hours and ovulate shortly before then end of heat.
a] 20 b] 25
c] 30 d] 33

118. Exposure of hens to brilliant light say _________ watt-bulb flashed on for 15 seconds effectively stimulates winter egg production.
a] 750 b] 1,000
c] 1,250 d] 1,500

119. Farrier is a person :
a] Who grazes sheep
b] Who does milking
c] Who fixes shoes in cattle

120. Fat in milk is percent in :
a] Colloidal form
b] Solution form
c] Emulsion form
d] None of the above form

121. Feed requirement of a hen in total economic life is a:
a] 55 kg b] 20 kg
c] 65 kg d] 25 kg

122. Feeding cow with cottonseed :
a] Increase the fat content of milk
b] Decrease the fat globule size
c] Increases sugar content
d] Induces no change in milk

123. Feeding extra concentrates to pregnant sheep is known as :
a] Flushing b] Steaming-up
c] Holding-up d] Grazing

124. Feeding of cottonseed of buffaloes:
a] Decreases fat content in milk
b] Has no effect on fat content in milk
c] Increases butter fat content in milk temporarily
d] Increases butter fat content in milk permanently

125. Final temperature of ghee finishing should not be above :
a] 105°C b] 110°C
c] 120°C d] 125°C

126. First vaccination for Ranikhet disease for poultry should be done in the chicks at the age of :
a] 1 week b] 15 day
c] 1 month d] 1 -7 day

127. Fish meal contains __________ D.C.P.
a] 20% b] 40%
c] 50% d] 75%

128. Fish usually contains __________ protein on wet weight basis.
a] 5-10% b] 13-20%
c] 20-30% d] 30-40%

129. Floor space requirement per adult bird of light breed in deep litter system should be :
a] 1900 sq. cms b] 1230-2800 sq. cms
c] 2800-3700 sq. cms d] 1000-1500 sq. cms

130. Foot and mouth disease does not occur in :
a] Cattle b] Buffaloes
c] Sheep d] Poultry

131. For breeding sheep a ram should not be used until it attains the age of __________ months.
a] 12 b] 18
c] 24 d] 30

132. For early development of rumen cud can be fed to young calf, The curd is:
a] An antibiotic
b] Regurgitated feed from mother's mouth
c] Colostrum
d] Bacteiral culture developed in laboratory

133. For hay making fodder should be rich in :
a] Carbohydrates b] Proteins
c] Water d] Fat

134. For most of the farm animals the ratio is of the order of _______ parts by weight of water to every one part of dry matter eaten.
a] 1-2 b] 3-4
c] 5-6 d] 7-8

135. For thermo-stabilization to preserve the quality of eggs the temperature of water used should be :
a] 38.3°C b] 45.4°C
c] 54.5°C d] 62.5°C

136. Freshly drawn milk has a pH value.
a] 6.3-6.5 b] 6.5-6.7
c] 6.7-7.0 d] 7.0-7.2

137. Gestation period in mare is :
a] 340 days b] 280 days
c] 310 days d] 150 days

138. Gestation period in pregnant ewes last _______ days.
a] 135 b] 140
c] 147 d] 170

139. Gestation period in sow is of the duration of :
a] 60-62 days b] 112-115 days
c] 275-285 days d] 300-310 days

140. Gestation period of camel is :
a] 340 days b] 370 days
c] 390 days d] 450 days

141. Glucose, fructose, cellulose, hemicellulose etc. are
a] Carbohydrates b] Fats
c] Proteins d] Minerals

142. Goat meat is called :
a] Pork b] Beaf
c] Mutton d] Chevon

143. Goat milk has 7 to10 times as much minerals as human and milk.
a] Cow b] Sheep
c] Buffalo d] Pig

144. Goat milk have ________ percent fat.
a] 4.5 b] 2.9
c] 5.3 d] 6.0

145. Goat produce about __________ percent of total meat in India.
a] 40.5 b] 52
c] 41 d] 10

146. Grains of maize, sorghum barely and oats are rich sources of :
a] Carbohydrates b] Fat
c] Protien d] Minerals

147. Gaolao is :
a] Dual purpose breed
b] Draught breed
c] Dairy breed
d] Breed name of buffalo

148. Grasstetany is caused due to the deficiency of :
a] Ca b] Mg
c] Bo d] P

149. Haemacytometer contains in the centre 16 big squares divided into ________ small squares.
a] 128 b] 256
c] 384 d] 512

150. Haemorrhagic sapticaemia, Black quarter, Anthrax, T.B. and Titanus are the __________ disease.
a] Viral b] Bacterial
c] Protozoan d] None

151. Heat cycle comes in ewes after __________ days.
a] 17-19 b] 15-16
c] 21-23 d] 20-22

152. Heat period of buffalo is :
a] 12-16 hours b] 24-25 hours
c] 18-36 hours d] None of them

153. Heavy milkers will produce more milk if they are milked :
a] Once a day b] Twice a day
c] Three times a day d] Four times a day

154. Heifer should be bred when it has attained :
a] ½ of adult body wt.
b] ¾ of adult wt.
c] Full adult wt.
d] $^{7}/_{8}$ of adult wt.

GK GS

Money market is that organization which provides short term credit.

155. Hereford is a breed of :
a] Cattle b] Goat
c] Swine d] Buffalo

156. High fat product of milk is :
a] Milk powder b] Ghee
c] Cheese d] Anenner

157. High fever, hot swelling in thorat and sometime upto fore limbs are the symptoms of :
a] Anthrax b] B.Q.
c] Cutaneous H.S. d] F.M.D.

158. High fever, ulcers on the mucus membrane of lips and tongue and foetid diarrhoea with offensive order are the symptoms of :
a] Rinderpest b] H.S.
c] Cowpox d] Anthrax

159. Holding of milk in cow is due to :
a] Secretion of epinephrine
b] Over-eating
c] Secretion of oxytocin
d] Secretion of thyroxine

160. Home tract of Gir cattle is :
a] U.P. b] M.P.
c] Gujrat d] Punjab

161. Hormone produced by thyroid gland is :
a] Thyroxine b] Parahormone
c] Insulin d] ACTH

162. How many kilograms of feed will be consumed by a layer of 25 weeks upto the age of 1 year 6 months?
a] 20-25 kg b] 25-30 kg
c] 30-35 kg d] 40-45 kg

163. How many kilograms of feed will be consumed by a layer upto the age of 24 weeks ?
a] 5 kg b] 10 kg
c] 15 kg d] 20 kg

164. How much manure is produced by one sheep in a year?
a] 2 tonnes b] 1.2 tonnes
c] 0.5 - 0.7 tonnes d] Less than 0.2 tonnes

165. If a goat has six permanent teeth, its age will be :
a] 3-4 years b] 2-2.5 years
c] 5-6 years d] More than 6 years

166. If birds make groups in the room, it indicates that they are feeling :
a] Hot b] Cold
c] Both of the above d] None of the above

167. In 1770, an Englishman named ______ , hatched chicks by passing hot air through the room where the eggs were located.
a] John Watchman
b] John William
c] John Champion
d] John Laggard

168. In a new born suckling calf the functional compartment of stomach is :
a] Rumen b] Reticulum
c] Omasum d] Abomasum

169. In a poultry bird if the eye rings and ear lobes are bleached; it indicates that the bird has been laying eggs since :
a] 1-2 weeks b] 2-4 weeks
c] 6-8 weeks d] 10 weeks

170. In bovine ________ per cent of the total rumen bacteria have a proteolytic activity.
a] 32 b] 34
c] 36 d] 38

171. In calculation of T.D.N. which nutrient is multiplied by 2.25?
a] Digetible fat-extract
b] Digestible crude protein
c] Digestible crude fibre
d] Digestible N.F.E.

172. In case of ________, the ovaries are enclosed by membranous covering which is continuous with the wall of the tube to ensure that the discharged ova passes into the tubes and is not lost into the body.
a] Dog b] Cow
c] Goat d] Rabbit

173. In case of cow, the vagina is about _______ cm long.
a] 7.5-10.0 b] 12.5-15.0
c] 16.5-19.0 d] 20.0-25.0

174. In certain physiological conditions excess water is retained in the body tissues. The condition is known as :
a] Edema b] Dehydration
c] Flue d] Anaemia

175. In Denmark, Edward Sorensen and Jens Gylling Ham organized the first Cooperative Artificial Breeding Association in :
a] 1916 b] 1926
c] 1936 d] 1946

176. In each ovary of farm animals there may be as many ______________ folliciles.
a] 25,000 b] 50,000
c] 75,000 d] 100,000

177. In electrical dehorning the rod embodies an automatic control to maintain the temperature at about _______ °F and its application to horn button for 10 seconds destroys the horn cells.
a] 1,000 b] 900
c] 800 d] 700

178. In Holstein cow, between two months and the time of puberty (7-10 months), the mammary gland grows three and half times faster than the body weight and reaches the peak pubertal development at ______________ months of age.
a] 6-8 b] 10-12
c] 11-13 d] 14-16

GK GS

The genetic constitution of a child is exactly the same as that of identical twin.

179. In India the maximum goat is found in :
a] Haryana b] Rajasthan
c] Karnataka d] Assam

180. In lamb four pairs of milk incisors appears at the age of :
a] 3 weeks b] 4 weeks
c] 5 weeks d] 6 weeks

181. In poultry proventriculus is the organ that secretes digestive juices. From here the food goes to :
a] Small intestine b] Large intestine
c] Gizzard d] Crop

182. In poultry, the part of digestive tract which acts as a reserve to hold the moistened feed is known as:
a] Gizzard b] Proventriculus
c] Rumen d] Crop

183. In ram semen, sperm concentration per ml. is about:
a] 1 billion b] 2-3 billion
c] 0.5 billiond] 4-5 billion

184. In sheep all permanent teeth will be first present at the age of :
a] 20-24 months b] 24-30 months
c] 30-40 months d] 4.5-50 months

185. In sow fertilization occurs only ________ hours after mating.
a] 2-5 b] 6-10
c] 11-13 d] 14-15

186. In species *viz.*, ________ semen after ejaculation remains liquid unlike in others wherein it tends to gelate or coagulate shortly after ejaculation.
a] Horse and Ass b] Sheep and Goat
c] Bull and Dog d] Cat and Camel

187. In tail to tail system of barn arrangement ________ per cent of time is spent behind the cows.
a] 40 b] 50
c] 60 d] 70

188. In the case of bulls the seminal vesicles are about ________ cm long 5 cm wide.
a] 16.0-13.5 b] 10.0-12.5
c] 14.0-15.5 d] 13.0-14.5

189. In the world Merino sheep contributes ________ per cent of total wool production :
a] 40-45 b] 25-30
c] 60-65 d] 70-80

190. In unipara as the cow, mare *etc.*, usually single ovum is released, in sheep one or two and occasionally three, whereas in swine several ova upto ________ are released in a single ovulation.
a] 15 b] 20
c] 25 d] 30

191. Incubation period of fowl egg is :
a] 15-16 days b] 22-25 days
c] 20-21 days d] 26-28 days

You Don't Know!
First modern bank owned by Indians
The Oudh Commercial Bank, established in 1881, lasted till 1958.

192. Indicate one of the following which is not breeding season for sheep :
a] March-April b] July-August
c] October-November d] January-February

193. Intercalving period in normal cows should be :
a] 13-15 months b] 15-18 months
c] 2 years d] 2-2.5 years

194. It is estimated that ________ volumes of blood flow through the mammary gland for each volume of milk synthesized.
a] 600 b] 500
c] 400 d] 300

195. It takes about 2.5 hours for feed to go from mouth to cloaca in the laying hen, and ______ hours in a non-laying hen.
a] 4 b] 6
c] 8 d] 10

196. Jamunapari and Barabari are breeds of :
a] Cattle b] Buffalo
c] Swine d] Goat

197. Jersey calves suffering from ________ have broad skull, large eye-sockets, short and broad nasal bones and forehead broader than normal.
a] Aguatha
b] Cerebral hernia
c] Umblical hernia
d] Bulldog head (prognathism)

198. Ketosis is caused due to the faulty utilization of :
a] Sugar b] Ca
c] Protein d] Mg

199. Lactometer is the instrument used for :
a] Measuring purity of milk
b] Measuring density of milk
c] Measuring refrective index of milk
d] Measuring fat percentage of milk

200. Land requirement for 1000 layers farm is :
a] 1.00 acre b] 1.50 acre
c] 0.25 acre d] 0.50 acre

201. Land requirement for 2500 broiler farm is :
a] 1.00 acre b] 0.50 acre
c] 0.25 acre d] 2.00 acre

202. Late lactation milk contain :
a] More chloride b] Less fat content
c] More lactose d] More water

203. Lecithin is a component of :
a] Milk proteins b] Milk vitamins
c] Milk lipids d] Milk pigments

204. Leghorn matures usually at the stage of _______ months.
a] 5-6 b] 2-3
c] 9-10 d] 1-2

205. Let down of milk in cow in due to:
a] Secreton of oxytocin
b] Secretion of thyroxine
c] Ingesion
d] Over eating

206. Main carbohydrate of cow milk is:
a] Glucose b] Fructose
c] Lactose d] Sucrose

207. Malathion is used livestock as :
a] Puragative astringent
b] Lactose
c] Laxative
d] Anti-ectoparasitic

208. Marek's disease in poultry is caused by :
a] Bacteia b] Virus
c] Protozoa d] Fungi

209. Maximum amount of nitrogen is in :
a] Ground nut cake b] Compost
c] Night soil d] Poultry manure

210. Maximum lactose is present in the milk of :
a] Tharparkar b] Sahiwal
c] Jersey d] Holstein

211. Merino breed of sheep originated from :
a] U.S.A. b] Australia
c] Denmark d] Spain

212. Micro organism responsible for flavour production in butter is :
a] Streptococcus lactis
b] Streptococcus cremoris
c] Bacillus subtillis
d] Leuconostoc citrovorum

213. Mild fever, eruptions over skin occuring most commonly on the udder and teats are the symptoms of :
a] F.M.D. b] Cow pox
c] H.S. d] Anthrax

214. Milk contains about _______ percent of phospholipids.
a] 2.5-3.0 b] 3.0-3.5
c] 0.2-1.0 d] 1.0-1.5

215. Milk fat is easily separated from milk because :
a] It is lighter than other constituents
b] It is not uniformly mixed in milk
c] It has more specific gravity
d] It has less surface tension

216. Milk fat is rich in :
a] Fat soluble vitamins
b] Water soluble vitamins
c] Soluble minerals
d] Soluble proteins

217. Milk fever in cattle is caused due to:
a] Decrease in the level of calcium in tissue fluid
b] Decrease in the phosphorus in the blood
c] Inadequate feeding
d] High fever

218. Karan Fries is cross between:
a] Brown Swiss x Red Sindhi
b] Friesian x Sahiwal
c] Friesian x Tharaparkar
d] Tharparkar x Brown Swiss

219. Milk from cows suffering from the following diseases is unfit for human consumption as the diseases are communicable to man:
a] Tubercullosis and Rinderpest
b] Tubercullosis and Brucellosis
c] Brucellosis and Black quarter
d] Brucellosis and Foot and mouth disease

220. Milk production enhancement programme is known as :
a] ICDP b] White revolution
c] I.D.C. d] Operation flood

221. Milk vein of a cow contains :
a] Milk b] Water
c] Pure blood d] Impure blood

222. Minimum fat present in table butter is :
a] Haryana b] Nagori
c] Jersey d] Holstein

223. Mohair is obtained from :
a] Angora b] Surti
c] Marwari d] Beetal

224. Mohair is produced from one of the following breeds of :
a] Sheep b] Horse
c] Camel d] Goat

225. Most of the gilts reach puberty between _______ months.
a] 4-5 b] 6-8
c] 9-10 d] 11-12

226. Most of the growth of the unborn young pig is during last _______ days of pregnancy.
a] 20 b] 30
c] 40 d] 50

227. Name the poultry bird used for egg production:
a] Cornish b] Rhode Island Red
c] White Leghorn d] Australorp

228. National Dairy Research Institute is located at :
a] Anand b] Ludhiana
c] Karnal d] Hissar

GK GS

Protection of Civil Right Act 1955 is the changed name of untouchability (Offence Act 1955).

229. Numbering of young calves is usually done by one of the following sets of methods:
a] Branding and Notching
b] Tagging and Tattooing
c] Tagging and Branding
d] Notching and Tattooing

230. On attainment of maturity, between first and second year a bull can serve ________ cows if properly spaced in repsect of time interval.
a] 18 b] 20
c] 23 d] 25

231. One kg. of pork can be produced on as little as ________ kg of feed.
a] 5 to 5.5 b] 4. to 4.5
c] 3 to 3.5 d] 2 to 2.5

232. One kilogramme of work can be produced on ________ kilogrammes of feed.
a] 1.5-2.0 b] 2.5-3.5
c] 4.0-4.5 d] 5.0-5.5

233. One of 100 kg of mixed milk how much kilograms of ghee can be obtained :
a] 5.75 kg b] 5.00 kg
c] 4.50 kg d] 4.00 kg

234. One of the best dual purpose breed of cows is :
a] Hissar b] Sahiwal
c] Sindhi d] Haryana

235. One ram may be sufficient for ________ ewes.
a] 30-40 b] 80-90
c] 1-2 d] 70-80

236. Optimum age of ram for mating is ________ years.
a] 4.0 to 5.0 b] 5.0 to 9.0
c] 2.5 to 7.0 d] 1.0 to 5.0

237. Optimum brooder temperature at a hieght of 2" above floor level should be :
a] 95-100^{0}F b] 65-70^{0}F
c] 102-104^{0}F d] 75-90^{0}F

238. Ordinarily, lambs are not used for servicing until ______ months of age.
a] 20 b] 18
c] 16 d] 14

239. Over feeding of cow causes :
a] Increase in fat % of milk
b] Increase in protein % of milk
c] Increase in milk yield
d] No change in milk composition

240. *Parturient pareis* is also known as:
a] Grass tetany b] Cyanide poisoning
c] Pica d] Milk fever

241. Pashmina is produced from one of the following breeds of goat :
a] Jamunapari b] Beetal
c] Kashmiri d] Angora

242. *Pasteurella multicida* is a causative agent of :
a] T.B. b] Cowpox
c] Rinderpest d] H.S.

243. Per capita annual consumption of eggs in India is :
a] 24 eggs b] 10 eggs
c] 48 eggs d] 250 eggs

244. Per capita chicken availability per year in India is about :
a] 1 b] 0.5
c] 2.5 d] 0.75

245. Per day light requirement to layer is:
a] 8-9 hours b] 14-16 hours
c] 10-12 hours d] 24 hours

246. Percentage of fat is highest in the milk of :
a] Red Sindhi b] Gir
c] Jersey d] Holstein

247. Percentage of fat is minimum in the milk of :
a] Haryana b] Nagori
c] Jersey d] Holstein

248. Percentage of total lactose is maximum in the milk of cattle breed know as :
a] Tharparkar b] Sahiwal
c] Jersey d] Holstein

249. Percentage of total solids is maximum in the milk of:
a] Tharparkar b] Sahiwal
c] Jersey d] Holstein

250. Pica in cattle is caused due to the deficiency of :
a] Calcium b] Phosphorus
c] Copper d] Cobalt

251. Pig milk contains ________ per cent fat, 5 per cent milk sugar, and 6 per cent protein
a] 10 b] 8
c] 6 d] 5

252. Pork is the name of ________ meat.
a] Goat b] Cow
c] Sheep d] Pig

253. Provision of ________ hours of light artifical or natural maximizes egg production.
a] 10 b] 12
c] 14 d] 16

254. Ranikhet disease of fowl is caused by :
a] Bacteria b] Protozoa
c] Virus d] Fungi

255. Removal of ovaries in female is called :
a] Caponization b] Castration
c] Spaying d] Immaculation

256. Rinderpest is a disease of :
a] Poultry b] Man
c] Cat d] None of the above

257. Rinderpest is caused by :
a] Bacteria b] Protozoa
c] Virus d] Fungi

GK GS

'Pashmina' is a breed of sheep.

258. Foot and mouth disease, cowpox and Rabies are the _________ disease in cattle.
a] Viral b] Protozoan
c] Bacterial d] All of the above

259. Roughages are feed embodying more than _______ per cent of crude fibre.
a] 24 b] 22
c] 20 d] 18

260. Rumen is found in :
a] Hog b] Poultry
c] Cattle d] Dog

261. Ruminants can convert low quality protein feed into usable amino acids as they have _________ in the rumen.
a] Pancreatic juice b] Enzymes
c] Ptylin d] Micro-organisms

262. Rumination is a long process and occupies about ________ hours out of 24 hours.
a] 6 b] 7
c] 8 d] 9

263. Scientific castration is done by using an instrument known as :
a] Jamboo b] Burdizzo
c] Clippers d] Scissors

264. Scurvy is a disease due to difficiency of :
a] Mn b] Vitamin C
c] Protein d] Peptone

265. Nutritional Roup of poultry is caused due to the deficiency of :
a] Magnese b] Vitamin C
c] Protein d] Vitamin A

266. Second vaccination for Ranikhet disease of poultry should be done in the birds at the age of :
a] 2-4 weeks b] 12-16 weeks
c] 9-12 weeks d] 4-5 weeks

267. Sharp teeth of piglet are known as:
a] Milk teeth b] Wolf teeth
c] Suckling teeth d] Weaning teeth

268. Sheep and goat together constitute nearly _________ per cent of total meat suppliy in India.
a] 20 b] 40
c] 60 d] 80

269. Sheep waste product contains twice quantity of P and K than in __________ waste material.
a] Pig b] Cattle
c] Poultry d] Goat

270. Shell, yolk and white in egg are in proportion of :
a] 11:30:59 b] 11:15:74
c] 20:30:50 d] 12:36:52

271. Sindhi milk contains about _________ fat.
a] 4.5% b] 4.2%
c] 4.9% d] 5.8%

272. Skim milk contain approximately % of water :
a] 70 b] 80
c] 95 d] 90

273. Space required in cage system for layer is :
a] 1 sq. feet b] 2.5 sq. feet
c] 0.75 sq. feet d] 4.0 sq. feet

274. Space requirement for a layer in deep litter system is :
a] 2.0 sq. ft. b] 2.5 sq. ft.
c] 3.0 sq. ft. d] 3.5 sq. ft.

275. Castrated male of swine (pig) :
a] Beg b] Bear
c] Bacon d] Boar

276. Sudden death without apparent cause, high temperature, blood discharge from natural body openings and may cause death with 10-36 hours, are the symptoms of :
a] Anthrax b] T.B.
c] H.S. d] Tetanus

277. Sugar content in sweetened condensed milk should not be less than :
a] 40% b] 30%
c] 50% d] 35%

278. Sugar content of milk in cow and buffalo ranges :
a] 2-3% b] 6-7%
c] 4-5% d] 3-4.5%

279. Sugar of milk is called :
a] Glucose b] Lactose
c] Sucrose d] Fructose

280. Surti and Beetal are breeds of :
a] Swine b] Goat
c] Cattle d] Buffalo

281. Swiss baby's foaster mother is the name given to :
a] Cow b] Sheep
c] Goat d] Buffalo

282. Tamworth and Middle White Yorkshire are breeds of:
a] Buffalo b] Goat
c] Cattle d] Swine

283. Th lactometer in milk without cream when compared to whole milk :
a] Sink more b] Sink less
c] Sink completely d] Float horizontally

284. Exotic breed of swine 'Duroc' is :
a] Red in clour b] White in colour
c] Brown in colour d] Black in colour

285. The age of maturity of a bull heifer is about :
a] 4.5-5.5 yrs b] 2.5-3.0 yrs
c] 1.5-2.0 yrs d] None of these

286. The American class of exotic breeds of poultry are specialy bred for the production of :
a] Meat b] Meat and eggs
c] Eggs d] Broilers

GK

Capital market is the organization which provides long term credit.

GS 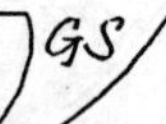

287. The amount of ______ needed by an adult cow for efficient reproduction is about 10-12 g per day.
a] Potassium b] Phosphorus
c] Calcium d] Sodium

288. The amount of feeds given to the animal during a period of 24 hrs is defined as :
a] Balanced ration b] Optimum ration
c] Ration d] Ideal ration

289. The amount of manure produced per year by a flock of 40 birds is about :
a] 400 kg b] 1 tonnes
c] 2 tonnes d] 3 tonnes

290. The animal teeth contains approximately ______ percent water.
a] 5 b] 10
c] 15 d] 20

291. The animals suspected as carries of any contagious disease must immediately be :
a] Quarantined b] Segregated
c] Vaccinated d] Innoculated

292. The average annual capacity of Desi hen for eggs production is only :
a] 20 eggs b] 60 eggs
c] 100 eggs d] 200 eggs

293. The average fat content in buffalo milk is about :
a] 6.9% b] 2.3%
c] 7.3% d] 8.5%

294. The average fat content in camel milk is :
a] 5.4% b] 8.5%
c] 5.8% d] 5.0%

295. The average fat content in mare milk is :
a] 3.8% b] 1.6%
c] 4.5% d] 2.8%

296. The average heat period in ewe is about :
a] 14-18 hrs b] 34-36 hrs
c] 2 days d] 304 days

297. The average life of a goat is usually ______ years and should give milk for about ten lactations.
a] 8 b] 10
c] 12 d] 14

298. The average yield of Haryana breed of cow per lactation is :
a] 1560 litre b] 900 litre
c] 1350 litre d] 2600 litre

299. The average yield of Murrah breed of buffalo per lactation is :
a] 2500 litre b] 1200 litre
c] 3800 litre d] 4500 litre

300. The average yield of Sahiwal breed of cow in good management conditions per lactation is :
a] 2500 litre b] 2000 litre
c] 4200 litre d] 3000 litre

Knowledge World!

First woman bus driver

Vasanthakumari of Kanyakumari, Tamil Nadu steers a State-run Nesamoni Transport Corporation city bus on the Parvathipuram circular route in Nagarkoil.

301. The bacterial content in milk is maximum in :
a] Fore milk b] Middle milk
c] Strippings d] Whole milk

302. National Research Centre on Equine is situated in :
a] Hisar b] Bikaner
c] Kohima d] Dirang

303. The best age for castration of male goat (buck) is :
a] 2-4 week b] 1 year
c] 6 month d] 9 month

304. The best breed if sheep for pelt production is :
a] Malpura b] Sonadi
c] Karakul d] Lincoln

305. The best breed of fowl for egg production is :
a] Rhode Island Red b] Cornish
c] White Leghorn d] White Rock

306. The best milk production record which should be kept on every farm is :
a] Two times milk record
b] Daily milk record
c] History sheet
d] Milk summary register

307. The best single measures of the ability of a sow to produce is the weight of her litter at ______ day, the usual weaning time.
a] 60 b] 58
c] 56 d] 54

308. The best time for castration of calves, lambs and pigs is between ______, 2 and 1 weeks respectively.
a] 2-4 b] 5-7
c] 8-10 d] 11-13

309. The biggest compartment of ruminant stomach is :
a] Rumen b] Reticulum
c] Omasum d] Abomasum

310. The biological value of buffalo meat:
a] 82% b] 79%
c] 72% d] 65%

311. The biological value of cow meat is:
a] 75% b] 55%
c] 65% d] 45%

312. The biological value of goat meat is:
a] 60% b] 90%
c] 40% d] 85%

313. The birth weight of a lamb produced by an exotic sheep is about :
a] 1-2 kg b] 2.5-3.5 kg
c] 4.5-5 kg d] 6-8 kg

GK GS

Chromosomes are principally composed of DNA and proteins.

314. The birth weight of lamb produced by an Indian sheep is about :
a] 1-2 kg b] 2.5-3.5 kg
c] 4.0-0.5 kg d] 6-8 kg

315. The body temperature of cattle is :
a] 100-102^{0}F b] 103-104^{0}F
c] 99-100^{0}F d] 105-107^{0}F

316. The body temperature of fowl is :
a] 100-102^{0}F b] 103-104^{0}F
c] 99-100^{0}F d] 105-107^{0}F

317. The boiling of milk is about :
a] 99.17^{0}C b] 100.9^{0}C
c] 100.17^{0}C d] 102.5^{0}C

318. The bovine population of India is approximately of the world population :
a] 1/5th b] 1/3rd
c] 1/11th d] 1/8th

319. The bread of goat known as "Jersey cow of goat" of world is :
a] Angora b] Nubian
c] Toggenburg d] Alpine

320. The breed of buffalo which has great capacity to consume coarse feed is :
a] Jaffarabadi b] Bhadawari
c] Nagpuri d] Murrah

321. The breed of buffalo which produces milk containing highest amount of fat is :
a] Jaffarabadi b] Bhadawari
c] Nagpuri d] Murrah

322. The breed of poultry Australorp develped from in Australia :
a] Black Orpington b] Sussex
c] Plymouth Rock d] Aseel

323. The breed producing highest wool per clip amongst following Indian breeds is :
a] Gaddi b] Jaisalmeri
c] Marwari d] Bellari

324. The breeding between indegneous cows with exotic bulls should be more scientifically called as :
a] Line breeding
b] Inter mating
c] Species hybridization
d] Cross-breeding

325. The breeding in which pure bred sires of a given breed are used grade females of the same bred is called :
a] Line-breeding b] Pure-breeding
c] Cross-breeding d] Up-grading

326. The breeding in which the mating of two animals both of which are pure breed but represent different breeds is known as :
a] Out-crossing b] Line breeding
c] Up-grading d] Close-breeding

327. The breeding of closely related animals such as son to dam, sire to daughter or brother to sister is known as :
a] System of breeding
b] Pure breeding
c] Inbreeding
d] Line breeding

328. The breeds Deccani, Nellore, Bellary Mandhya and Bandur of sheep are of :
a] Temperate himalayan region
b] Dry western region
c] Southern region
d] Eastern region

329. The buffalo meat is known as :
a] Beef b] Veal
c] Chevon d] Buffen

330. The bull shed should have covered and open space respectively:
a] 4 sq.m. x 8 sq.m.
b] 6 sq. m. x 12 sq.m.
c] 9 sq.m. x 50 sq.m.
d] 12 sq.m. x 120 sq.m.

331. The bulls of __________ breed of buffalo are extensively used for upgrading inferior stock.
a] Ravi b] Bhadawari
c] Murrah d] Jaffarabadi

332. The cake which contains highest D.C.P. is :
a] Cotton seed cake b] Mustard cake
c] Mahuwa cake d] Groundnut cake

333. The calcium to phosphorus ratio for the maximum reproduction efficiency in cattle should between:
a] 1:2 and 2:1 b] 1:1 and 3:1
c] 1:3 and 3:1 d] 1:4 and 4:1

334. The carbohydrate content in Khoa is :
a] 27.5% b] 18.5%
c] 24.8% d] 42.5%

335. The carbohydrate content in skimmed milk powder (cow) is :
a] 51% b] 47%
c] 59% d] 81%

336. The carbohydrate content of cheese is about :
a] 75% b] 6.3%
c] 51% d] 18.5%

337. The carbohydrate content of goat milk is about :
a] 2.9% b] 6.8%
c] 5.8% d] 7.8%

338. The castrated male lambs are called:
a] Wethers b] Shaearer
c] Feeder d] Hogget

339. The casual organism of Anthrax in cattle :
a] *Bacillus anthracis*
b] *Mycobacterium tuberculosis*
c] *Staphylococcus aureus*
d] *Streptococcus pyogenes*

340. The cattle will produce more milk if they are watered:
a] Once a day b] Twice a day
c] Adlibitum d] Thrice a day

341. The causal organism of Circling disease of sheep is:
a] *Mycobacterium tuberculosis*
b] *Listeria monocytogenes*
c] *Escherichia coli*
d] *Brucella abortus*

342. The causal organism of Bangis diseases in cattle is :
a] *Mycobacterium paratuberculosis*
b] *Brucella abortus*
c] *Vibrio foetus*
d] *Streptococcus pyogenes*

343. The causal organism of *Haemorr-hagic septicaemia* in cattle is :
a] *Baccillus anthracis*
b] *Pasteurella boviseptica*
c] *Pasteurella bubaliseptica*
d] *Pasteruella hemolytica*

344. High Security Animal Disease Laboratory is situated in :
a] Bhopal b] Izatnagar
c] Mathura d] Hisar

345. The central poultry breeding farm, Hasserghatta has developed a poultry strain which lays 260 eggs per year. The name of the strain is:
a] HH 260 b] CP 260
c] HC 260 d] HG 260

346. The central walk (of a cow stall) should have a width of 5-6 feet exclusive of gutters when cows face out, and ______ feet when they face in.
a] 2-3 b] 3-4
c] 4-5 d] 1.5-2

347. The chicks hatched in an incubator are called as :
a] Sexed chicks b] Straight run chicks
c] Artificial chicks d] Broilers

348. The chicks of about eight weeks old specially reared for meat production and attain a body weight of about 1 kg are called as :
a] Culled chicks b] Broilers
c] Pullets d] Cockerels

349. The chief function of gizzard is to:
a] Mix juice in food
b] Absorb digestible portion of feed
c] Grind and crush the coarse feed
d] None of the above

350. The colour of fat in milk is depend on :
a] Carotene b] Albumin
c] Water d] Lactose

351. The colour of fat in milk is dependent on :
a] Albumin b] Lactose
c] Carotene d] Water

352. The colour of the egg yolk depends on the absence or presence of :
a] Vitamin A b] Riboflavin
c] Xanthophyll d] Chlorophyll

353. The commercial egg farms in India maintain Austro White, a cross between :
a] Australorp male x White Leghorn female
b] Australorp male x Langshan female
c] Australorp male x New Hampshire female
d] Australorp male x Wyandotte female

354. The compartment of ruminant stomach in which digestive juices are secreted is :
a] Rumen b] Reticulum
c] Omasum d] Abomasum

355. The concentrate mixture fed to milch cows should contain a protein percentage of :
a] 16-18% b] 18-20%
c] 12-14% d] 10-12%

356. The cotton seed cake is not considered suitable for feeding to calves as it contains :
a] Aflatoxins b] Gossypol
c] Ricin d] Hydrocyanic acid

357. The country from which Saanen breed of goat originated is :
a] Switzerland b] U.S.A.
c] Australia d] U.S.S.R.

358. The daily per capita availability of milk is :
a] 210 gms b] 50 gms
c] 191 gms d] 200 gms

359. The density of milk at 60^{0}F temperature is :
a] 1.230 b] 1.032
c] 1.320 d] 1.003

360. The dental formula of cattle and bufalloes at the age of 6 years :
a] 0033/4033 b] 0033/3033
c] 0033/3033 d] 0033/3034

361. The diameter of mature follicles is about ______ mm.
a] 4-5 b] 7-8
c] 9-10 d] 11-12

362. The diameter of sow-corpora lutea is about 15-18 cm, while the body of the uterus is about ______ cm long.
a] 3 b] 4
c] 5 d] 6

363. The diameter of wool fibre of Marino wool ranges in between.
a] 15-25 microns b] 30-40 microns
c] 40-50 microns d] 50-60 microns

364. The digestibility coefficient of buffalo meat is :
a] 71% b] 82%
c] 74% d] 89%

365. The digestibility coefficient of cow meat is :
a] 82% b] 84%
c] 88% d] 91%

GK **The most important functionary during the Rig Vedic period was Purohit.** GS

366. The digestibility coefficient of goat meat is :
a] 95% b] 82%
c] 99% d] 65%

367. The digestibility of colostrum increases when it is given at a temperature between :
a] 89-92°F b] 93-98°F
c] 97-98°F d] 99-102°F

368. The disease in which animal aborts or gives birth to a dead or weak calf is known as :
a] Rinderpest b] Calf scours
c] Brucellosis d] Anaplasmosis

369. The disease ricket in animals ocurs due to the deficiency of :
a] Vitamin K b] Vitamin A
c] Vitamin C d] Vitamin D

370. The disease which can not occur in epidemic form is:
a] Rinderpest b] Black quarter
c] Anthrax d] Ketosis

371. The disease which is caused by shortage or faulty utilization of sugar in the body is called :
a] Ketosis b] Tympanitis
c] Anthrax d] Rinderpest

372. The disease which is not caused by bacteria is :
a] Anthrax b] Tuberculosis
c] H.S. d] Rabies

373. The dressing percentage in pigs is:
a] 40-45 percent
b] 55-60 percent
c] 65-80 percent
d] More than 80 percent

374. The duration of oestrous cycle in ewe is about :
a] 10 days b] 17 days
c] 21 days d] 28 days

375. The egg, after the formation of fertilization membrane decreases in diameter by _______ per cent due to the secretion of some fluid.
a] 4 b] 8
c] 11 d] 14

376. The English class of exotic poultry breeds have good producing capacity of :
a] Eggs b] Meat
c] Meat and eggs d] Chicks

377. The entry from where Toggeenburg breed of goat originated is :
a] Spain b] Switzerland
c] Denmark d] U.S.A.

378. The estrous cycle of goat comes after:
a] 12 days b] 19 days
c] 16 days d] 21 days

379. The fat content of Holstein Frisien milk is :
a] 3.4% b] 5.1%
c] 2.8% d] 4.5%

380. The fat content of Jersey milk is :
a] 5.9% b] 6.2%
c] 5.4% d] 6.5%

381. The fat content of Sahiwal milk is :
a] 4.2% b] 4.5%
c] 4.3% d] 4.9%

382. The fat percent of Chhana (cow) is about
a] 28.8% b] 18.5%
c] 24.8% d] 20.8%

383. The fat percentage is maximum in the portion of milk:
a] Fore milk b] Middle milk
c] Strippings d] Separated milk

384. The first scientific research on artificial insemination of domestic animals was conducted with _________ by Lazarro Spallanzani in 1780.
a] Dogs b] Sheep
c] Goats d] Cows

385. The foot-rot disease in sheep is caused by :
a] Virus b] Protozoa
c] Fungi d] Bacteria

386. The freezing point of milk is about:
a] -29°C b] -0.10°C
c] -4.5°C d] -0.55°C

387. National Embryo Bio-technology Centre for Animal Production is situated in :
a] Karnal b] Bhopal
c] Avikanagar d] Izatnagar

388. The gastration period of buffalo is:
a] 310 days b] 285 days
c] 340 days d] 390 days

389. The gastration period of cow is :
a] 260 days b] 310 days
c] 280 days d] 390 days

390. The gestation period in ewes is about:
a] 145-155 days b] 281 days
c] 310 days d] 62 days

391. The gestation period of goat is -
a] 144 days b] 120 days
c] 146 days d] 150 days

392. The gestation period of pig is :
a] 112-115 days b] 128-140 days
c] 142-145 days d] 190-195 days

393. The goat breed not considered good for milk production is :
a] Jamunapari b] Beetal
c] Barbari d] Berari

394. The goat breed suitable for the cities and towns is :
a] Marawari b] Kathaiwari
c] Pashmim d] Barbari

395. The highest D.C.P. percentage is found in the following oil cakes :
a] Ground nut cake
b] Cotton seed cake
c] Toria cake
d] Mustard cake

396. The highest total solids are found in _________ milk.
a] Buffalo b] Sheep
c] Mare d] Camel

397. The interval between one heat period to the next is known as :
a] Gestation period
b] Oestrus cycle
c] Service period
d] Intercalving period

398. The intestine represents a long tube approximately _________ feet long in average dairy cattle.
a] 180 b] 160
c] 200 d] 220

399. The lactating pig can produce _________ g of protein in her milk every day.
a] 50-100 b] 110-220
c] 350-375 d] 400-450

400. The lambs maturing early and producing mere mutton are called as :
a] Prime lambs b] Mutton lambs
c] Productive lambs d] Prepared lambs

401. The large intestine of _________ is the shortest and simplest of all the domestic animals.
a] Cat b] Dog
c] Cow d] Horse

402. The largest goat skin producing country is :
a] U.S.A. b] Denmark
c] Switzerland d] India

403. The mammary glands in sow are _________ number, are arranged along the belly in two rows.
a] 14-16 b] 10-12
c] 6-8 d] 17-19

404. The manure gutter should have a gradient of one inch for every _________ feet length.
a] 15 b] 14
c] 12 d] 10

405. The mating of animals of the same breed having no close relationship in 4 to 6 generations is know as :
a] Line breeding b] Out crossing
c] Cross-breeding d] Inbreeding

406. The maximum amount of nutrient present in the ration of cows and buffaloes is :
a] Protein b] N.F.E.
c] Crude fibre d] Fat

407. The maximum amount of water present in table butter under is :
a] 10 per cent b] 16 per cent
c] 20 per cent d] 24 per cent

408. The maximum bacterial count of pasteurized milk per ml is :
a] 1,000,00 b] 10,000
c] 50,000 d] 1,50,000

Quiz Kit!
Youngest Governor
Swaraj Kaushal was appointed Governor of Mizoram in February 1990 when he was 37 years and 7 months. His wife Sushma was appointed Minister in the Devi Lal cabinet in Haryana on July 16, 1977 when she was 25 years and 5 months.

409. The maximum dry matter require-ment per 100 kg body weight is of :
a] Sheep b] Goat
c] Cow d] Buffalo

410. The maximum protein content is present in the milk of :
a] Sheep b] Buffalo
c] Goat d] Cow

411. The meat availability per person per year in India is:
a] 149 gm b] 120 gm
c] 135 gm d] 157 gm

412. The meat of calf is called as :
a] Pork b] Veal
c] Chevon d] Mutton

413. The meat of sheep is called :
a] Beef b] Pork
c] Mutton d] Chevon

414. The Medaterranean class of exotic breeds of polutry are generally considered most profitable for the production of :
a] Eggs b] Meat
c] Eggs and meat d] Broilers

415. The method not used for hay making is :
a] Ground method
b] Artificial drying
c] Tripod method
d] Fence method

416. The milch type breed of sheep is :
a] Marino b] Dorset
c] South Down d] Corsican

417. The milk exudes from _________ separate gland tubes that open at the base of the "mammary hairs".
a] 50-95 b] 100-150
c] 40-80 d] 155-195

418. The milk sugar is :
a] Monosaccharide b] Disaccharide
c] Oligosaccharide d] Polysaccharide

419. The milk yield of a cow rises to a maximum after calving in about :
a] 4 weeks b] 6 weeks
c] 8-10 weeks d] 12-16 weeks

420. The minimum amount of water present in table butter is :
a] 75 b] 80
c] 85 d] 90

GK GS
Akbar was the first Indian ruler who organised Haj pilgrimage at the expense of the state.

421. The most convenient folding unit to house poultry is the one made for _______ hens.
a] 15 b] 20
c] 25 d] 30

422. The most potent source of bacterial contamination in milk is :
a] Udder of cow b] Milker
c] Atomsphere d] Utencils

423. The most potent source of off-flavours in goat milk is:
a] Weeds and other plants eaten by goat
b] Dirty goat shed and surroundings
c] Presence of buck with females
d] Natural characteristics of goal milk

424. The most variable constituent of milk is :
a] Casein b] Lactose
c] Fat d] Minerals

425. The normal content of protein in cow milk is :
a] 3.5% b] 4.5%
c] 6.5% d] 2.5%

426. The normal life span of sheep is :
a] 8-9 years b] 15-16 years
c] 10-12 years d] 12-15 years

427. The normal respiration rate per minute of cattle is :
a] 8-16 b] 20-25
c] 12-20 d] 5-7

428. The number of chromosomes in a somatic cell of buffalo is :
a] 22 pairs b] 24 pairs
c] 30 pairs d] 42 pairs

429. The number of lambs born per ewe, on an average rise untile the age of _______ years.
a] 6 b] 5
c] 4 d] 3

430. The number of pairs of chromosome present in the somatic cell of fowl are :
a] 23 b] 33
c] 20 d] 39

431. The oestrus cycle in sow is of the duration of :
a] 18-24 days b] 27 days
c] 120 days d] 140 days

432. The oestrus period in sow is of the duration of :
a] 12-18 hrs b] 20-25 hrs
c] 25-35 hrs d] 40-65 hrs

433. The only triple purpose Indian breed of sheep is :
a] Lohi b] Deccani
c] Hissardale d] Magra

434. The onset of puberty in a female (swine) normally occurs at around _______ days of age and 90 kg of live weight.
a] 160 b] 170
c] 180 d] 190

435. The optimum body weight of pig for pork production:
a] 45-90 kg b] 90-135 kg
c] 135-180 kg d] 180-225 kg

436. The optimum time of insemination for a cow in heat is :
a] Early heat b] Mid heat
c] Late heat d] After heat

437. The origin of Merino breed of sheep is from :
a] U.S.S.R. b] Denmark
c] Australia d] Spain

438. The ovary can be examined upto _______ months of pregnancy after which it sinks deeper into abdominal cavity.
a] One b] Two
c] Three d] Four

439. The parathyroid glands were discovered by Sondstorm in :
a] 1840 b] 1860
c] 1880 d] 1900

440. The percent of total supply of meat in India derived from goat and sheep respectively is :
a] 20 and 30 b] 30 and 30
c] 35 and 52 d] 25 and 35

441. The percentage of abnormal semen in a normal bull should not exceed:
a] 21-56 b] 5-12
c] 15-20 d] 26-30

442. The percentage of goats of the world's goat population found in India :
a] 35-45% b] 50%
c] 25% d] 15%

443. The pH of the duodenal juice from pigs ranges from :
a] 4.5-5.6 b] 6.2-7.9
c] 8.4-8.9 d] 8.0-8.2

444. The pH value of milk is :
a] 6.6 b] 7.5
c] 6.9 d] 7.9

445. The phase estrous cycle not observed in continuous breeders :
a] Estrus b] Metaestrus
c] Diestrus d] Anestrus

446. The phase of estrous cycle in which level of estrogen is maximum :
a] Proestrus b] Estrus
c] Metaestrus d] Diestrus

447. The pork producers prefer to produce pork of the type :
a] More lean and more fat
b] More lean and less fat
c] More fat and less
d] None of the above

GK GS

Those banks whose names are included in the second schedule of the RBI Act of 1934 are called Scheduled Commercial Bank.

448. The poultry breed New Hampshire developed in _________ about 2000 years ago, primarily as a table bird.
a] Poland b] Holland
c] New Zealand d] England

449. The poultry breed New Hampshire was developed both for meat and egg production in New Hampshire from single comb Rhode Island Red and Red Stock during :
a] 1910 b] 1920
c] 1930 d] 1940

450. The poultry specially raised for meat production is known as :
a] Pullet b] Grower
c] Broiler d] Layer

451. The process of spermatogenesis in the male and oogenesis in the female finally gives rise to ______ daughter cells.
a] 2 b] 4
c] 6 d] 8

452. The production of atleast five to ten daughters is compared with the production of their dams. If the daughter show a significant milk increase over that of the respective dams, then the sire is designated as:
a] Improved dam
b] Improved sire
c] Proven sire
d] Hybird sire

453. The protein content in goat milk is about :
a] 2.9% b] 4.3%
c] 3.1% d] 5.8%

454. Sheep breed fit for weaving rug (*Kambal*):
a] Deccani b] Bikaneri
c] Lohi d] Gaddi

455. The protein content of cow milk is :
a] 4.5% b] 5.5%
c] 3.6% d] 6.3%

456. The protein content of Khoa is about:
a] 20.1% b] 28.5%
c] 25.5% d] 32.5%

457. The protein percentage in poultry egg is about :
a] 5-6 per cent b] 6-8 per cent
c] 8-10 per cent d] 11-13 per cent

458. The protein percentage in sheep milk is :
a] 5.2 b] 6.9
c] 4.8 d] 2.8

459. The protein percentage in the buffaloes milk is :
a] 2.6% b] 4.6%
c] 1.6% d] 3.6%

460. The pulse rate per minute of sheep and goat is about:
a] 70-80 b] 80-100
c] 120-140 d] 140-160

461. The rate at which cows require dry matter per 100 kg bodyweight is :
a] 2-2.5 kg b] 3-3.5 kg
c] 1.0-2.0 kg d] 4-5.0 kg

462. The rate of gas production in the rumen is most rapid immediately after meal and in the cow may exceed _________ per hour.
a] 15 b] 20
c] 25 d] 30

463. The recommended method of breeding for F_1 cross-bred animals is :
a] Line breeding b] Criss-crossing
c] Interse mating d] Back croosing

464. The Refractive Index at 20°C of milk is :
a] 1.345-1.348 b] 1.112-1.215
c] 1.54-1.548 d] 1.211-1.222

465. The replacement of central milk teeth by two permanent teeth occurs at 12-14 months of age in sheep. At this stage the animal is called as :
a] Gimmer b] Hogget
c] Infant d] Giglet

466. The reproductive organ of female in which fertilization of ovum occurs is :
a] Ovary b] Fallopian tube
c] Uterus d] Vagina

467. The reproductive organ of female in which foetus develops is :
a] Ovary b] Fallopian tube
c] Uterus d] Vagina

468. The sac of the scrotum in bulk is thermoregulatory in function, that is, it provides protection to the testes and maintains them at a temperature lower by _____ °F than the abdomen which is necessary for active spermatogenesis.
a] 2-4 b] 5-7
c] 8-10 d] 11-13

469. The signs of heat are shown by a cow during one of the phases of estrous cycle. The phase is :
a] Pro-estrusb] Estrous
c] Metaestrus d] Diestrus

470. The space requirement for broiler is :
a] 1 sq. feet b] 2.5 sq. feet
c] 2 sq. feet d] 1.25 sq. feet

471. The space requirement for layer is:
a] 1.5 sq. feet b] 3.5 sq. feet
c] 2.5 sq. feet d] 4.5 sq. feet

472. The specific gravity of cow milk is :
a] 1.2-1.345 b] 0.95-1.25
c] 1.028 d] 1.312-1.416

473. The standard weight of an egg in ounces in international market is :
a] 1 b] 2
c] 3 d] 4

GK GS

A lactometer is used to test the purity of milk.

474. The state in India where per capita availability of milk is highest is :
a] Rajasthan b] U.P.
c] Punjab d] Himachal Pradesh

475. The sterol of the milk is :
a] Phytosterol b] Cholesterol
c] Ergosterol d] None of the above

476. The system of housing cattle, which costs less, saves labour and provides more comfort to the animals is :
a] Tail to tail system
b] Face to face system
c] Single row arrangement
d] Loose housing system

477. The table butter must not contain more than ________ water.
a] 20% b] 10%
c] 16% d] 18%

478. The table butter should contain minimum fat of :
a] 75% b] 80%
c] 85% d] 90%

479. The term 'Flushing' means :
a] To clean the animals using a sprayer
b] To drain the waste material
c] Lambing in sheep
d] Feeding liberally before breeding of sheep

480. The term 'wether' refers to the :
a] Removal of feather from a fowl
b] Male lamb that has been castrated after one month of birth
c] Male lamb that has been castrated before reaching sexual maturity
d] Climate suitable for rearing the poultry birds

481. The term "let down" of milk refers to :
a] Release of milk by contraction of alveoli
b] Collection of milk in milk bucket
c] Milk milked before calving
d] Colostrum

482. The term "Tandem" refers to :
a] Selection of one character at a time until it is improved by grading
b] Selection of good cows
c] Random
d] Control pregnancy

483. The term poultry includes which one of the followings:
a] Chicken
b] Chicken, duck and geese
c] Chicken and duck
d] Chicken, duck, geese, turkey and Guinea fowl

484. The thickness of wool fibre of Merino is usually 20 u. It can also be expresed as :
a] 0.2 cm b] 0.02 cm
c] 0.002 cm d] 0.0002 cm

485. The time taken in hatching fowl eggs into chicks in the incubator is:
a] 15-16 days b] 20-21 days
c] 22-25 days d] More than 25 days

486. The total amount of cattle manure produced annually in the country is approximately :
a] 1200 tonnes
b] 1200000 tonnes
c] 1200000000 tonnes
d] 12000000 tonnes

487. The udder usually increases about ______ in size between milking, depending upon the quality of the udder and the amount of milk secreted.
a] One-fourth b] Half
c] One-third d] Two-third

488. The vitamin in which colostrum is richest is :
a] Vitamin A b] Vitamin D
c] Vitamin E d] Vitamin K

489. The volume of semen ejaculated by mature ram at one time is :
a] 0.5 ml b] 0.72 ml
c] 2-3 ml d] 4-6 ml

490. The water content in Chhanan is about :
a] 57% b] 92%
c] 475 d] 54%

491. The water content of milk is :
a] 30-40% b] 87%
c] 50-60% d] 70-80%

492. The water requirement per day for 100 adult birds for drinking is about:
a] 10-15 litres b] 20-25 litres
c] 30-35 litres d] 40-45 litres

493. The weight of 8 weeks of chicken will be about :
a] 1.0 kg b] 1.75 kg
c] 1.5 kg d] 2.2 kg

494. The whole milk containing 5.0 per cent fat is to be standardized to 4.0 per cent fat by the addition of skim milk containing 0.1 per cent fat. The whole milk and skim milk are to be mixed in the proportion of:
a] 20:40 b] 3.9:1
c] 1:40 d] 2.9:1

495. The word "Poultry strain" is assocaiated with its :
a] Feed b] Breed
c] Disease d] Class

496. The young boar weight of large white yorkshire breed of pig is about :
a] 200-250 kg b] 100-150 kg
c] 250-260 kg d] 300-400 kg

497. The young one weight of middle white yorkshire is about :
a] 180-270 kg b] 300-400 kg
c] 275-300 kg d] 275-300 kg

GK GS

Anaimundi is the highest peak of peninsular India.

498. The__________ is the most economical animal to keep.
a] Cow b] Buffalo
c] Swine d] Goat

499. There are ______ breds of buffaloes in India, in addition to a large number of non-descripts.
a] 4 b] 6
c] 10 d] 12

500. There are about ______ breeds of cattle in India besides hybrids.
a] 14 b] 20
c] 26 d] 32

501. Thiamine deficiency of fowl results in the deficiency symptom is known as :
a] Curled to paralysis b] Star-gazing positio..
c] Dermatitis d] Hamorrhages

502. To gain one kg of weight a broiler bird consumes :
a] 1.9 kg feed b] 2.5 to 3.0 kg feed
c] 3.5-4.0 kg feed d] 4.0-5.0 kg feed

503. Total digestible nutrients required for production of one kg. of milk by a cow depends upon the :
a] Weight of the animal
b] Age of the animal
c] Percentage of fat in milk
d] Percentage of protein in milk

504. Total solid percentage in milk can be calcuated by the formulae :
a] CLR/4 + 1.2 F + 0.14
b] CLR/4 + 2.2 F + 1.04
c] CLR/4 + 1.2 F + 0.14
d] CLR/4 + 2.2 F + 2.24

505. Total solids in buffalo milk is ________ percent.
a] 17.3 b] 13.1
c] 15.2 d] 19.2

506. Two pairs of permanent incisors are present in the lower jaw of cattle at the age of :
a] 1 year b] 2 year
c] 3 year d] 4 year

507. Udder infection generally reduces in milk :
a] The serum protein percent
b] Lactose and casein percent
c] Chloride percent
d] pH of milk

508. Under average conditions the separated milk should not contain more fat than :
a] 0.2% b] 0.1%
c] 0.5% d] 1.0%

509. Under feeding of cow results in lowering of :
a] Lactose and Protein in milk
b] Minerals in milk
c] Lactose in milk
d] Milk yield and increase in milk fat

Did You Know ?

First science park

The first science park was established by the Birla Institute of Technology at Ranchi, Bihar, in 1972 to promote entrepreneurship in science and technology, create a technocrat industrial society provide facilities for R&D and to use untapped facilities and expertise available with technological institutions.

510. Under natural system of mating one bull is sufficient for :
a] 10-20 cows b] 20-40 cows
c] 50-60 cows d] 60-80 cows

511. Vaccination for Fowl pox disease of poultry should be done in the birds at the age of :
a] 6-10 weeks b] 1-2 weeks
c] 12-18 weeks d] 4-5 weeks

512. Vitamins ________ are synthesized by the ruminants in their body.
a] A and D b] B and K
c] E and D d] C and D

513. Well developed gilts should be used for breeding at the age of :
a] 6 months b] 12-14 months
c] 2 yrs. d] 4 yrs.

514. Well maintained bird can lay ________eggs per year.
a] 180 b] 240
c] 295 d] 120

515. Wessex Saddleback is a breed of :
a] Goat b] Sheep
c] Pig d] Cow

516. Which of the folloiwng breed of goat is not found in north-region?
a] Jamanapari b] Barabri
c] Beetal d] Surti

517. Which of the following animal's milk is recommended in peptic ulcer, liver disfunction, jaundice and insomania?
a] Goat b] Sheep
c] Cow d] Buffalo

518. Which of the following breed of cow has highest fat content in milk?
a] Jersey b] Sindhi
c] Gir d] Tharparkar

519. Which of the following breed of cow is heaviest and most powerful?
a] Haryana b] Kankrej
c] Rathi d] Sahiwal

520. Which of the following breed of goat is found in central region ?
a] Marwari b] Bengal
c] Malabari d] Assami

GK GS
Long description of plough share ritual is mentioned in Satpath Brahman.

521. Which of the following breed of goat is found in hilly tract ?
a] Pashmina b] Barbari
c] Beetal d] Marwari

522. Which of the following breed of goat is not found in southern region ?
a] Surti b] Beetal
c] Osmanabadi d] Malabari

523. Which of the following breed of poultry is American ?
a] New Hampshire b] Dorking
c] Brahma d] White Leghorn

524. Which of the following breed of poultry is Asiatic?
a] Sussex b] Rhode Island Red
c] Cochin d] None of these

525. Which of the following breed of poultry is excellent for broiler purpose?
a] Wyandotte b] Cochin
c] Plymouth rock d] Australorp

526. Which of the following breed of poultry is not English breed?
a] Orphington b] Dorking
c] Sussex d] Cochin

527. Which of the following breed of sheep is considered as a farmer's sheep?
a] Gaddi b] Hassan
c] Nili d] Lohi

528. Which of the following breed of sheep is known for superior carpet wool?
a] Merino b] Lohi
c] Gaddi d] Chokla

529. Which of the following breeds of cattle is not a pure breed ?
a] Kherigarh b] Rathi
c] Malvi d] Amritmahal

530. Which of the following breeds of cattle is not found in Rajasthan ?
a] Rathi b] Ongole
c] Mewati d] Nagori

531. Which of the following dry-roughage has maximum digestible crude protein ?
a] Berseem hay b] Anjan hay
c] Spear grass d] Lucerne hay

532. Which of the following is a breed of goat found in eastern region ?
a] Pasmina b] Himalayan
c] Assami d] Malabari

533. Which of the following is a draft breed of cow?
a] Sahiwal b] Haryana
c] Hallikar d] Gir

534. Which of the following is a dual purpose bred of cow?
a] Gir b] Nagore
c] Thaparkar d] Amrit Mahal

535. Which of the following is a Himalayan region breed of sheep?
a] Nili b] Deccani
c] Gaddi d] Hassan

536. Which of the following is a Rajasthan breed of sheep?
a] Lohi b] Shahabadi
c] Nellore d] Truchi

537. Which of the following is excellent layer?
a] White Leghorn b] Brahma
c] Wyandotte d] Cochin

538. Which of the following is not a breed of buffalo?
a] Murrah b] Jafarabadi
c] Red dane d] Nagpuri

539. Which of the following is not a dairy breed of cow ?
a] Sahiwal b] Haryana
c] Sindhi d] Gir

540. Which of the following is not a draft breed of cow ?
a] Nagore b] Kherigarh
c] Amritmahal d] Kankrej

541. Which of the following is not a dual purpose breed of cow?
a] Kherigarh b] Ongole
c] Haryana d] Mewati

542. Which of the following is not a exotic bred of goat ?
a] Beetal b] Alpine
c] Toggenberg d] Angora

543. Which of the following is not a exotic breed of cow?
a] Jersey b] Hallikar
c] Brown Swiss d] Red Dane

544. Which of the following is not a exotic breed of sheep?
a] Merino b] Rambouillet
c] Cheviot d] Truchi

545. Which of the following is not a factor affect the milk secretion ?
a] Sanitationb] Health of animal
c] Species d] Stage of lactation

546. Which one is not a ruminant animal?
a] Cattle b] Sheep
c] Goat d] Pig

547. Which one of the breeds of cattle originates from Pakistan ?
a] Kankrej b] Mewati
c] Malvi d] Sahiwal

548. Which one of the following disease of animals is caused by virus ?
a] Brucellosis b] Tuberculosis
c] Pox d] Anthrax

549. Which one of the following dry roughage has highest total digestible nutrients :
a] Berseem hay b] Lucerne hay
c] Anjan hay d] Doob hay

GK **The colour of a star is an indication of its temperature.** GS

550. Which one of the following fodder is generally not used for silage making ?
a] Maize b] Sorghum
c] Wheat straw d] Bajra

551. Which one of the following has high total digestible nutrient content ?
a] Cotton seed b] Gram
c] Maize d] Soyabean

552. Which one of the following is a breed of sheep ?
a] Gaddi b] Marwari
c] Osmanabadi d] Karakul

553. Which one of the following is an infectious diseases ?
a] Ketosis b] Tympanitis
c] Milk fever d] Warts

554. Which one of the following is not a breed of goat :
a] Kashmiri b] Gaddi
c] Chamba d] Cheviot

555. Which one of the following is not a character of good layer ?
a] Bright eyes
b] Red and full comb
c] Two finger distance between pin bones
d] Light yellow beak

556. Which one of the following is not a characteristics of Marawari sheep?
a] Black face
b] Coarse to medium wool
c] Produces good quantity of milk
d] Sturdy body and long legs

557. Which one of the following is not a determining factor in selecting a dairy cattle ?
a] Pedigree record
b] Production record
c] Body capacity
d] Physical appearance

558. Which one of the following is not a method of dehorning ?
a] Chemical method
b] Suppressor method
c] Hot iron
d] Clippers

559. Which one of the following is not an American poultry breed ?
a] Plymouth Rock b] Rhode Island Red
c] New Hampshire d] Orpington

560. Which one of the following is the cross-bred sheep ?
a] Kutchi b] Nilgiri
c] Marwari d] Nellore

561. Which one of the following poultry breeds does not lay brown shelled eggs ?
a] Orpington b] Sussex
c] Corrish d] Dorking

562. Which one of the following species does to have 30 pairs of chromo-somes in the somatic cell ?
a] Cattle b] Buffalo
c] Goat d] Horse

563. Which one of the followings is the breed of duck ?
a] Wyandotte b] Darking
c] Khaki Campbell d] Minorca

564. Which one of the symptoms given below are not the symptoms of rinder pest ?
a] Ulcers on the mucus membrane
b] Foetid diarrhoea
c] Hot swelling in the region of throat
d] Lachrymation and dryness of the muzzle

565. Which part of sheep's stomach constitue about 80% stomach?
(a) Reticulum (b) Abomassum
(c) Omassum (d) Rumen

566. White colour of milk is because of:
a] Casein b] Globulin
c] Carotene d] Albumin

567. White Leghorn pullets start laying eggs at the age of:
a] 4 months b] 5 to 6 months
c] 7 to 8 months d] After 8 months

568. Whole milk containing 6 per cent fat is to be standardized to 3 percent fat by addition of skim milk of 0.1 per cent fat proportions of whole and skim milk would be:
a] 50:50 b] 25:75
c] 2.9:3 d] 3:2.9

569. Wool is rich in cystine and contains ______ per cent of sulphur.
a] 1.5 b] 2.0
c] 3.0 d] 4.0

570. Wool of sheep can be clipped for __________ times in a year.
a] Twice or thrice b] 4 times
c] 6 times d] Each month

571. Young chicken about 8-12 weeks of age is called :
a] Layer b] Broiler
c] Chick d] None of these

572. Which of the following resembles as Jamunapari goat breed ?
a] Surati b] Beetul
c] Marwari d] Barbari

573. Lumpy Jaw is the disease of :
a] Cattle b] Sheep
c] Goat d] Poultry

574. Choose the dairy breed of goat :
a] Jamunapari b] Beetul
c] Bengal d] Barbari

575. Yorkshire, an exotic breed of swine is native of :
a] Denmark b] U.S.A.
c] U.K. d] China

GK GS

The Industrial Development Bank of India (IDBI) is an apex institution in the field of industrial development banking in the country.

576. In tropical climate, it is always necessary to conduct poultry houses:
a] North to South b] South to North
c] East to West d] West to East

577. New Castle, a poultry disease caused due to :
a] Virus b] Protozoa
c] Bacteria d] Nutritional disorder

578. Maximum fish production is in :
a] West Bengal b] Goa
c] Gujarat d] Kerala

579. Maximum marine fish production is in :
a] West Bengal b] Goa
c] Gujarat d] Tamil Nadu

580. India is the _____ largest fish producer.
a] Second b] Third
c] Fourth d] Sixth

GK GS
Malarial parasites is brought to human body by female Anopheles mosquito.

Matching

1. **Match the somatic chromosome number with species.**

a]	78	i]	Sheep
b]	54	ii]	Cattle
c]	60	iii]	Pig
d]	38	iv]	Hens

2. **Match the sex chromosome number with the animals embodying it.**

a]	39 pairs	i]	Cattle
b]	30 pairs	ii]	Sheep
c]	27 pairs	iii]	Horse
d]	32 pairs	iv]	Dog

3. **Match the heat-period with the animal.**

a]	12-24 hours	i]	Ewe
b]	1-2 days	ii]	Mare
c]	4-5 days	iii]	Sow
d]	2-3 days	iv]	Cow

4. **Match the term used in parturition with the animal.**

a]	Foaling	i]	Ewe
b]	Lambing	ii]	Sow
c]	Farrowing	iii]	Bitch
d]	Whelping	iv]	Mare

5. **Match the disease with the symptom.**

a]	Salpingitis	i]	Uterus inflamnation
b]	Piosalpingitis	ii]	Pus accumulation in uterus
c]	Metritis	iii]	Fallopian tube-inflammation
d]	Pyometra	iv]	Pus-filled fallopian tube

6. **Match the percentage with the component as estimated in colostrum.**

a]	14.0	i]	Lactose
b]	2.7	ii]	Fat
c]	6.7	iii]	Total solids
d]	23.9	iv]	Protein

7. **Match the reproductive rate with the animals.**

a]	One offspring/year	i]	Sheep and goat
b]	One to three progeny/year	ii]	Swine
c]	Ten progeny twice/year	iii]	Poulry
d]	100 or more porgeny/year	iv]	Cattle, buffalo, horse

8. **Match the animals breed with the parents.**

a]	Hisardale, sheep	i]	Lincoln rams x Rambouillet ewes
b]	Columbia, sheep	ii]	Tamworth x Landrace
c]	Panama, sheep	iii]	Bikaneri ewes x Merino rams
d]	Minnesota No. 1, swine	iv]	Rambouillet rams x Lincoln ewes

9. **Match the year with the event in the artificial insemination.**

a]	1939	i]	Experimental work started on A.I.
b]	1942	ii]	Establishment of A.I. centers by GOI
c]	1945	iii]	Sahiwal cows impregnated by A.I. successfully
d]	1940	iv]	First A.I. in cows

10. **Match the animal with the average volume of ejaculate per cc.**

a]	Bull	i]	5-100
b]	Ram	ii]	200-500
c]	Stallion	iii]	4.0
d]	Boar	iv]	0.1-1.5
e]	Cock	v]	1.0

GK GS

The culmination point of Eastern Ghat exist in Vishakhapatnam district.

11. Match the animal with average concentration of ejaculated semen per cc in million.

a]	Bull	i]	800-4000
b]	Ram	ii]	25-1000
c]	Stallion	iii]	600-1200
d]	Boar	iv]	0.5-0.6
e]	Cock	v]	50-200

12. Match the animal with ejaculated semen storage survival time with retention of fertilizing capacity expressed in days.

a]	Bull	i]	40-100
b]	Ram	ii]	4-8
c]	Stallion	iii]	6-10
d]	Boar	iv]	8-12
e]	Cock	v]	100-600

13. Match the shape of the head of spermatozoa with the animal which pertains to.

a]	Ovoid	i]	Fowl
b]	Like elongated cylinder	ii]	Human species
c]	Like a hook	iii]	Bull, ram, boar and rabbit
d]	Fattened oval body	iv]	Mouse and rat

14. Match the average semen-volume per ejaculation with the animal.

a]	Bull	i]	0.6 ml
b]	Ram	ii]	100 ml
c]	Stallion	iii]	1 ml
d]	Boar	iv]	200 ml
e]	Cock	v]	5 ml

15. Match the frequency of occurrence or its effect on the bacterium in the semen of a bull.

a]	*Diptheroids*	i]	Frequently
b]	*Staphylococci*	ii]	Detrimental
c]	*Psedomonas aeriginosa*	iii]	Predominast
d]	*E. Coli*	iv]	Occasionally

16. Match the source of development with the semen diluent.

a]	Egg-yolk phosphate	i]	Salisbury *et al.* (1941)
b]	Egg-yolk citrate	ii]	Almquist (1952)
c]	Homogenized cow's milk	iii]	Roy and Bishop (1954)
d]	Glycine plus egg-yolk	iv]	Phillips and Lardy (1939-40)
e]	Colloidon-filtered sea-water	v]	In France, (1958)

17. Match the protein contents (%) with concentrate.

a]	Wheat bran	i]	20.0
b]	Rice bran	ii]	25.0
c]	*Chuni*	iii]	12.0
d]	Linseed cake	iv]	46.0
e]	Groundnut cake	v]	10-15

18. Match the feeding stuff with digestibility coefficient.

a]	Egyptian clover	i]	61
b]	Gucnea grass	ii]	45
c]	Maize (feed)	iii]	81
d]	Mixed grass hay	iv]	0
e]	Rice/wheat straw	v]	58

19. Match the one gramme of substance oxidized with the metabolic water produced (gramme).

a]	Glucose	i]	0.77
b]	Starch	ii]	1.11
c]	Tributrin	iii]	0.56
d]	Tristearin	iv]	0.42
e]	Protein	v]	0.60

GK **Kural is known as Bible of Tamil Nadu.** GS

20. **Match the saturated fatty acids with its common source.**

a]	Caproic acid ($C_5H_{11}COOH$)	i]	Palm oil
b]	Caprylic acid ($C_7H_{15}COOH$)	ii]	Not known
c]	Palmitic acid ($C_{15}H_{31}COOH$)	iii]	Mutton tallow
d]	Stearic acid ($C_{17}H_{35}COOH$)	iv]	Butter fat
e]	Arachidic acid ($C_{19}H_{39}COOH$)	v]	Laurel kernel oil

21. **Match the unsaturated fatty acids with its common source.**

a]	Palmetoleic acid ($C_{15}H_{29}COOH$)	i]	Linseed oil
b]	Oleic acid ($C_{17}H_{33}COOH$)	ii]	Glandular organs
c]	Linoleic acid ($C_{17}H_{31}COOH$)	iii]	Milk fat
d]	Linolenic acid ($C_{17}C_{29}COOH$)	iv]	Soyabean oil
e]	Arachidonic acid ($C_{19}H_{31}COOH$)	v]	Pork fat

22. **Match the protein with the example/occurrence.**

a]	Histones	i]	Serum alpha and beta
b]	Collagens	ii]	Casein of milk
c]	Glycoprotein	iii]	Tendons
d]	Chromoproteins	iv]	Nucleoprotein
e]	Phosphoproteins	v]	Hemoglobulin

23. **Match the food source with the coversion factor for converting nitrogen into crude protein.**

a]	Soybean	i]	5.30
b]	Milk	ii]	6.25
c]	Cotton seed	iii]	5.71
d]	Maize	iv]	5.83
e]	Wheat	v]	6.38

24. **Match the element with the approximate cocentration (%) found in animal body.**

a]	Calcium	i]	0.2
b]	Phosphorus	ii]	0.16
c]	Potassium	iii]	1.0
d]	Sodium	iv]	0.15
e]	Sulphur	v]	1.5

25. **Match the calcium content (%) with the source.**

a]	0.12	i]	Whole rice
b]	0.05	ii]	Milk
c]	0.02	iii]	Fish meal
d]	5.50	iv]	Wheat or rice bran
e]	0.10	v]	Whole egg

26. **Match the source with the estimated phosphorus content (%).**

a]	Milk	i]	0.21
b]	Whole egg	ii]	1.40
c]	Wheat bran	iii]	0.14
d]	Rice	iv]	0.09
e]	Whole rice	v]	1.20

27. **Match the name of the disease caused by cobalt deficiency with the country where in it has been called as.**

a]	Bush sickness	i]	South Australia
b]	Coast disease	ii]	Kenya
c]	Wasting disease	iii]	Great Britain
d]	Nakuruitis	iv]	Western Australia
e]	Pinning	v]	New Zealand

28. **Match the molybdenum content (ppm) determined by xanthine oxidase method with the food/feed stuff.**

a]	1.00	i]	Whole soyabeans
b]	1.80	ii]	Dehydrated alfalfa meal
c]	2.50	iii]	Cabbage
d]	1.40	iv]	Liver and glandular meal
e]	0.35	v]	Peas

29. Match the animal with the requirement of vitamin D expressed in I.U/day (0.025 microorgamme of pure crystalline vitamin D_3 is equal to one International Unit).

a]	Man	i]	200
b]	Dairy cattle	ii]	400
c]	Sheep	iii]	200
d]	Growing poultry	iv]	500
e]	Laying/breeding poultry	v]	350

30. Match the pathological condition caused by the deficiency of vitamin E with the animal (s) affected.

a]	Reproductive failure	i]	Ewe
b]	Embryonic degeneration	ii]	Rat, pig
c]	Sterility	iii]	Chick, turkey
d]	Liver necrosis	iv]	Female rat, hen and turkey
e]	Blood protein loss	v]	Male rat, dog. cock, hamsterand guinea pig

31. Match the food or feed stuff with α tocopherol content (mg.100g)

a]	Wheat germ oil	i]	13
b]	Brown rice	ii]	145
c]	Whole sunflower seed	iii]	0.8-2.0
d]	Dehydrated clover	iv]	1-2
e]	Fish meal	v]	10

32. Match the vitamin K_1 contents (mg) per 100g of edible portion with its source.

a]	Glucose	i]	0.7
b]	Body and food fats	ii]	0.97
c]	Tripalmitin	iii]	1.00
d]	Triolein	iv]	0.704
e]	Protein	v]	0.712

33. Match the form of sugar with the absorption value taking glucose absorption value as 100.

a]	Galactose	i]	19
b]	Fructose	ii]	110
c]	Mannose	iii]	9
d]	Xylose	iv]	43
e]	Arabinose	v]	15

34. Match one kilogramme of digestible nutrient with starch equivalent.

a]	True portein	i]	1.00
b]	Crude fibre	ii]	2.10
c]	Fat (roughages)	iii]	2.40
d]	Fat (cereals)	iv]	0.94
e]	Fat (oil cakes)	v]	1.90

35. Match the animal protein food with biological value.

a]	Whole milk	i]	67
b]	Whole egg	ii]	94
c]	Potato	iii]	83
d]	Oats	iv]	65
e]	Egg white	v]	85

36. Match the breed (s) of cow (s) or buffalo(es) with the year of beginning of herd book by the Indian Council of Agricultural Research, New Delhi.

a]	Sindhi and Sahiwal	i]	1942
b]	Hariana and Murrah	ii]	1947
c]	Gir and Kankrej	iii]	1941
d]	Tharparkar, Kangayam and Ongole	iv]	1945

The Dinesh Goswami Committee was concerned with electoral reforms.
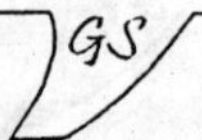

37. Match the animal with its daily water requirement expressed in gallons.

a]	Cows and buffaloes	i]	20-25
b]	Camel (watered daily)	ii]	1-5
c]	Pig	iii]	1
d]	Elephant	iv]	7-8
e]	Dog	v]	4-8

38. Match the class with the breed of poultry.

a]	American	i]	Orpington
b]	Asiatic	ii]	Wyandotte
c]	English	iii]	Ghagus
d]	Indian	iv]	Langshan
e]	Mediterranean	v]	Minorca

39. Match the description with the body part (s) of a hen (poultry).

a]	Largest bone of the limb	i]	Backbone
b]	Scapulae, coracoids, clavicle make	ii]	Drumstick (tibia)
c]	Shows fusion of vertebrates	iii]	Pelvic girdle
d]	Is large, project back beyond ribs	iv]	Pectoral girdle
e]	Consists of ilia, ischia and pubes	v]	Sternum

40. Match the gland with the hormone produced in hens.

a]	Anterior pituitary	i]	Cortical hormones
b]	Posterior	ii]	Androgen
c]	Adrenal	iii]	Insulin
d]	Islets of langerhans	iv]	Thyrotrophic hormone
e]	Jests	v]	Oxytocin

41. Match the age of a chick (in weeks) with brooding temperature °F (measured about 5 cm) above the floor at the age of the brooder.

a]	0-1	i]	78-80
b]	1-2	ii]	70
c]	2-3	iii]	93-95
d]	3-4	iv]	83-85
e]	5-6	v]	68-90

42. Match the ration ingredient with protein contents (%).

a]	Sesame cake	i]	5.2
b]	Rice polish	ii]	1.8
c]	Maize	iii]	1.5
d]	Maize gluten meal	iv]	4.5
e]	Wheat bran	v]	1.6

43. Match the popular terms used in swine husbandry (SH) with the meaning thereof.

a]	Swine	i]	A domesticated animal with long broad snout, thick fat body and coarse bristles
b]	Pig	ii]	Female swine having produced piglets
c]	Hog	iii]	Young female intended for breeding
d]	Sow	iv]	A pig or hog usually used collectively
e]	Gilt	v]	Denotes entire swine family or weighs more than 120 lb.

44. Match the popular term used in swine husbandary with the meaning therof.

a]	Litter	i]	Castrated male swine before attaining sexual maturity
b]	Ratling	ii]	To bring forth a number of young at one time
c]	Boar	iii]	Male swine castrated on becoming a mature boar
d]	Stag	iv]	An uncastrated male swine
e]	Barrows	v]	The smallest in a litter, the last born, called rit, erit or cad pig

Industrial Development Bank of India (IDBI) was set up in February 1976.

45. Match the popular term used in swine husbandary with the meaning therof.

a]	Bacon	i]	The part of the leg behind knee (back of thigh)
b]	Ham	ii]	Fat melted out from fresh, fatty pork tissue
c]	Sausage	iii]	Feeding concentrates to young in enclosures away from dam
d]	Lard	iv]	Chopped fine highly seasoned, stuffed ordinarily in prepared intestine tube
e]	Creep feeding	v]	Salted and smoked meat from hog's back or sides.

46. Match the breed of the pig with its place or origin.

a]	Land race	i]	Central England
b]	Bershire	ii]	Northeastern U.S.A.
c]	Tamworth	iii]	Denmark
d]	Duroc	iv]	South Central England
e]	Hereford	v]	Missouri, U.S.A.

47. Match the breed of the pig with its place or origin.

a]	Chester White	i]	Northern England
b]	Large White Yorkshire	ii]	Southeastern Pennsylvania, U.S.A.
c]	Middle White Yorkshire	iii]	South Central England
d]	Berkshire	iv]	Denmark
e]	Landrace	v]	England

48. Match the amino acid requirement expressed as percentage of diet of weaning pigs.

a]	Isoleucine	i]	0.60
b]	Leucine	ii]	0.40
c]	Lysince	iii]	0.20
d]	Valine/threonine	iv]	1.00
e]	Arginine/tryptophan	v]	0.70

49. Match the vitamin with the deficiency symptoms in pigs.

a]	Vitamin B_{12}	i]	Diarrhea necrotic enteritis
b]	Pantothenic acid	ii]	Dermatitis
c]	Niacin	iii]	Microcytic anaemia
d]	Riboflavin	iv]	Liver necrosis
e]	Vitamin E	v]	Loss of hair follicles

50. Match the diet ingredient with the commonly considered dry matter percentage thereof.

a]	Root crops	i]	15-25
b]	Fish meal	ii]	85-90
c]	Liquid milk by products	iii]	87
d]	Cereal/soyabean mixes	iv]	10-25
e]	Drymatter of commonly used diet	v]	6-13

51. Match the lower limits for energy density range with the type of pig.

a]	14 MJ DE/kg	i]	Grower (20-60 kg)
b]	13 MJ DE/kg	ii]	Pregnant breeders
c]	12 MJ DE/kg	iii]	Lactating breeders
d]	< 11 MJ DE/kg	iv]	Grower (60-100 kg)
e]	12.5 MJ DE/kg	v]	Baby pig (5-20 kg)

52. Match the type of pig with the nutrient concentration of diet.

a]	Young grower	i]	13
b]	Pregnant breeder	ii]	13
c]	Lactating breeder	iii]	14
d]	Improved entire male grown	iv]	14
e]	Unimproved entire male grown	v]	12.5

53. Match the food with degradability of protein (%).

a]	Oats and barley	i]	60-70
b]	Maize	ii]	65-85
c]	Soyabean meal (extracted)	iii]	85-95
d]	Groundnut meal (decorticated)	iv]	25-45
e]	Fish meal	v]	45-65

54. **Match the diseases with the causal organism as commonly observed in sheep.**

a]	Coccidiosis	i]	*Clostridium welchii type D*
b]	Johne's disease	ii]	*Fusiformis necrophorus*
c]	Pulpy Kidney (in lambs)	iii]	*Clostridium welchii type B*
d]	Lamb dysentery	iv]	*Protozoan parasite*
e]	Lameness	v]	*Mycobacterium paratuberculosis*

55. **Match the poisonous plant(s) found to cause poisoning in sheep with the country where such observation have been made.**

a]	Yew, rhododendron, laurel	i]	New Zealand
b]	Ragwort	ii]	Scotland
c]	Laurel	iii]	Britain
d]	Bog asphodel	iv]	Eastern Canada
e]	Some strain of subterranean clover	v]	Australia

56. **Match the temperature in respect of housing standards for pigs with the body weight or sex.**

a]	26-30°C	i]	10-20 kg
b]	22-26°C	ii]	50-100 kg
c]	18-22°C	iii]	< 10 kg
d]	16-20°C	iv]	20-50 kg

57. **Match the basic physical and physiological properties of sheep with each other.**

a]	Average temperature	i]	16 days
b]	Average pulse rate	ii]	75-80/minutes
c]	Oestrus cycle	iii]	40°C
d]	Natural life time	iv]	8-10 years

58. **Match the age with the name given according to sex of the sheep.**

a]	Birth to weaning (0-14/16 weeks)	i]	Wether hogget (castrated male)
b]	Weaning to shearing (14/16 weeks to 5 months)	ii]	Gimmer (female)
c]	First to second shearing	iii]	Full mouthed tup (entire male)
d]	Second to third shearing	iv]	Ewe lamb (female)
e]	After fourth shearing	v]	Two shear ram (entire male)

59. **Match the following.**

a]	Castrated male of sheep	i]	Beg
b]	Castrated male of swine	ii]	Bullock
c]	Castrated male of poultry	iii]	Hogget
d]	Castrated bull	iv]	Capon

60. **Match buffalo breeds.**

a]	Western type	i]	Toda
b]	Northern type	ii]	Godavari
c]	Central & Southern type	iii]	Murrah
d]	Improved type	iv]	Zaffrabadi

61. **Match common diseases of cattle with their causal organism.**

a]	FMD	i]	Inflammation of udder
b]	Mastitis	ii]	Viral
c]	Cowpox	iii]	Bacterial
d]	Anthrax	iv]	Viral

62. **Match composition of milk fat% (by wt.) of species.**

a]	Cow	i]	8.5
b]	Buffalo	ii]	7.3
c]	Goat	iii]	4.1
d]	Sheep	iv]	4.2

63. **Match composition of milk lactose% (by wt.) of species.**

a]	Cow	i]	4.30
b]	Buffalo	ii]	4.27
c]	Goat	iii]	5.48
d]	Sheep	iv]	4.96

64. **Match livestock with their average heart beat.**

a]	Swine	i]	280 beats/min
b]	Horse	ii]	90 beats/min
c]	Goat	iii]	44 beats/min
d]	Sheep	iv]	75 beats/min

GK GS

Mahandragiri is the second highest peak of Eastern Ghat.

Glossary of Technical Terms

Alergic Diseases Allergic reactions of animai body also cause some diseases e.g. photosterilization, serum shock, allergic termatitis, hay fever etc.

Animal Science The science which deals with the feeding, breeding and management practices of the livestock.

Artificial Insemination It is a technique in which semen with live sperms collected from the male and deposited in the female genetical tract by mechanical means.

Autosexing Sex differentiation at day old age some visual characters such as colour of down, early feathering etc.

Balanced Ration A ration which provide essential nutrients to the animal in such proportion and amounts that are required for the nourishment of the particular animal for 24 hrs.

Boar An uncastrated adult male pig used for breeding purpose.

Breed A breed is a group of individuals of common ancestors which are genetically identifiable because all members of the genes present in one breed are also present at low frequency in all other breeds of the same species.

Broiler A chicken of either sex, usually of 8-10 weeks age that is tender meated flexible breast bone, cartilage, suitable for flying .

Brooder / Hover An apparatus used for keeping chickens warm.

Brooding Rearing of chicks after hatching till are upto which warmth has to be provided

Buck-Billy An uncastrated male used for breeding purpose.

Bull An uncastrated male ox used for breeding purpose.

Bull-Calf A male calf under one year of age.

Bullock / Steer A castrated male ox of over two years.

Calf A young one of cow class (ox) of either sex.

Calving An act of giving birth to young one by a female of cow class.

Calving Interval It is the period between the two successive calving.

Candling Examination of eggs for finding egg objects such as cracks on shell, blood spots, loose cell etc. by holding the eggs between the eye and light source.

Capon A castrated male chicken.

Castration The process of removal of testicles.

Chicken Domestic fowls including chicks, hens, pullets, cockerals, cocks.

Cock A mature male chicken.

Cockeral A male fowl below one year of age.

Colostrum Milk of cow for first 3-4 days after calving is known as colostrum. It contains 3 to 4 time more protein, minerals and vitamins also work as laxative.

Colt A young male horse usually over one year age.

Colt-Foal A male foal under one year age.

Concentrates They are feeds which contain less than 18% of crude fibre and having more than 60% TDN.

Congenital Defects They may be hereditory or non-hereditory but are found at the time of birth e.g. free martin, cleff palate, white heifa disease, atrecla in animal etc.

Cow A female ox (cow class) which has produced at calf at least once.

Deficiency Diseases Nutritional deficiences lead to some animal diseases e.g. rickets, night blindess, ostomatesia etc.

Digestibility Co-efficient It is the percentage of total amount consumed, which is digested and absorbed.

Digestible Energy Represented by that portion of energy consumed which is not excreted in faeces. DE = GE - FE. Where, FE = Faeceal Energy.

Doe / Nanny A female goat which has produced young one atleast once.

Domestication Process of adapting of life in intimate association with and to the advantages of man.

Down-Calve / Springer A female ox nearly ready to give birth to its calf.

Dystokia Difficulty at the time of parturition

Estrous (Heat) It is a period during which animal strongly desired matting with a male partner.

Ewe An.adult female sheep which has produced a kid atleast once.

Farrowing An act of delivery in swine.

Fecundity It is the potential capacity of the female to produce functional ova, regardless of what happens to them after they are produced.

Feeding Standards These are tables standard amount of various nutrients that should be present in the daily ration of different classes of live stock for optimum result in growth, work production.

Fertility It is the ability of an animal to produce large number of living young ones.

Filly A young female horse over one year age.

Filly-Foal A female foal under one year age.

Flavoured Milk Milk to which some flavours have been addded when the milk is used, the product contain milk fat percentage at least equal to the minimum legal requirement for market. But when the fat level is lower (1-2%) the term drink is used.

Flock A group of goat.

Foaling An act of delivering a young female horse.

Follicular Phase Both the proestrous and estrous phase of estrous cycle are the part of which known as follicular phase because there is predominance of the hormone estrogen during these two phases.

Food Edible substance which satisfy the instinct of hunger is food. Food for animal is know as feed.

Free Martin A heifer calf born as a twin with male most of these heifers are found to be service when they become adult.

Geld A castrated male horse of any age.

Gestation Period It is the period between the day of conception to the day of calving.

Gilt A young female pig not farrowed atleast once. She becomes a sow after farrowing.

Gonads The sex-cells are produced by a specilised glands known as gonads.

Gross Energy Amount of heat/energy released through the complete combination feed ingredient.

Hatching To bring forth young from the eggs by natural or artificial incubation.

Heifer A female ox of over one year, which has not calved. After calving she becomes a cow.

Heifer-Calf A female calf under one year of age.

Hen A mature female chicken.

Herd Group of animals of cow class.

Hinny A hybrid whose sire is a stallion dam is a female donkey. It is also sterile.

Hog / Stag A castrated male pig.

Hogget A sheep of either sex between and first shearing.

Homogenization The homogenization consists in the sub division of the fat globules into cause the milk to loose its creming property.

Hygiene The science of preserving health by removing or reducing the hazzards, associated with the domestic life in more or less artficial environment to enhance animal well being.

Incubation Hatching of eggs by means of natural or artificial heat.

Injuries Damages done to body due to cuts, burms, blows, falls, accidents etc.

Kid A young one of goat of either sex exceeding one year.

Lactation Period The period between the date of calving to the date of drying.

Lamb Young one of sheep of either sex male lamb known as ram lamb while female lamb known as ewe lamb.

Live Stock Those come in the class of mammalian *viz.* cow, buffalow, sheep, goat swine, horse etc. which are domesticated or pet are considered as live stock.

Maintenance Ration It is the food required by an animal to keep it in good bodily without gain or loss in weight and maintains body temperature.

Mare An adult female horse, which produced a young one atleast once.

Metabolic Diseases Some times metabolic processes in animal body get disturbed and disease condition e.g. milk fever, acetonemia/ketosis (due to upset in CHO metabolism).

Metabolic Energy When energy losses in urine and combustible gases substracted from digestible the remaining energy is called metabolic energy.

Miscellaneous Diseases Can not be classified under heads, e.g. Prolapse of vagina bloat, horn cancer etc.

Mule A hybrid whose size is a donkey and is a mare it is usually sterile.

Net Energy This is the net reminder of useful energy after all the losses accounted for faeces, urine, and heat increment are substracted from gross energy.

Nutrient An element or compound that is required in the diet of a given animal to permit normal functioning of the life process.

Ornithology The study of birds which are not as poultry is known as ornithology.

Ovulation During the metestrous phase of estrous cycle mature follicle of the ovary ruptures and ovum is released. This process is known as ovulation.

Parturition Parturition is the expulsion of the foetus and its membranes from the uterus through the birth canal by natural forces and in such a state of development that foetus is capable of the independent life.

Poultry Domesticated species of birds reared for egg, meat or feathers includes chickens, ducks, turkey, etc.

Production Ration It is the food requirement by an animal in addition to the maintenance ration supply the nutrients needed for some forms of production viz. growth, fattening, milk etc.

Prolificacy It is used to denote wheather many or few offsprings result from a given mating or from a certain individual during its lifetime.

Puberty A stage of physical growth of the body at which the functional gametes are produced for the first time by the animal after its birth is known as puberty. But generally, all related processess necessary for successful reproduction are not started right at puberty.

Pullets Female birds of 18-26 weeks age.

Ram An adult uncastrated male sheep used for breeding purpose.

Ration It is often used interchangeable with diet but it may also mean a daily supply of food or feed.

Reproduction A process of producting or creating a new individual similar to the parents.

Residual Milk The residual milk is the amount of milk left in the udder after a normal milking. It can obtained only after the injection of oxytocin and remilking the animal.

Roughages They are bulky feeds containing relatively large amount of less digestible material i.e. crude fibre more than 18% and low in the TDN (about 60%).

Sand Check Eggs with a crack in the shell so small that it can be detected only at by candling.

Semen Semen is a suspension of spermatozoa in seminal fluid, it is opaque, white to light cream coloured fluid.

Seminal Plasma It is a composite mixture of fluids secreated by organs which in the higher species comprise the epididymis, the vasdeferens, ampullae, prostate, seminal vesicles, cowpers glands and contain other glands located in the wall of the urethral canal.

Service Period It is the period between the date of calving to the date of next successive service.

Serving Act of mating/copulation of male and female ox used as the cow has been served by bull. Also used as the service has been given by a bull.

Sexual Maturity The stage of growth at which the animal can reproduce successfully is called sexual maturity.

Sow An adult female pig used for breeding purpose.

Stallion A mature uncastrated house used for breeding purpose.

Tonned Milk It refers to milk obtained by the addition of water and skim milk. It should contain minimum 3% fat.

Total Digestible Nutrient (TDN) % digestible protein + % digestible nitrogen free extract + % digestible crude fibre + % digestible other extract x 2.25.

Toxicity Poisoning due to eating certain poisonous substances.

Tupping An act of matting male and female sheep.

Vasectomy It is an act of cutting down the portion of vas-deferens and ligating it reaving the blood and nerves supply to testicles infect.

Wedder A castrated male sheep.

GK GS

Irish Constitution influenced the makers of Indian Constitution to incorporate Directive Principle of State Policy.

13. Plant Physiology

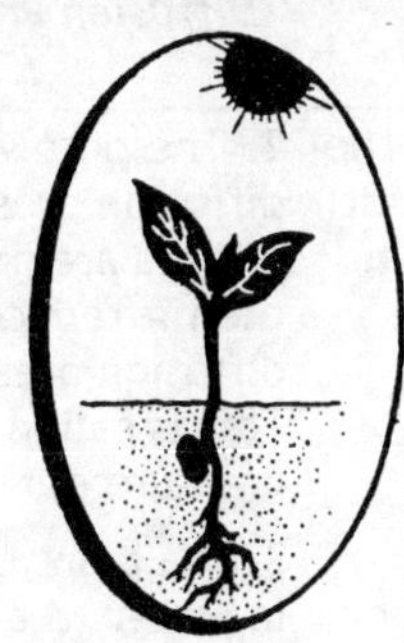

Did You Know ?

Smallest wild pig

The nocturnal Pygmy Hong (*Sus salvanius*) measures 25 cm at shoulder height. It usually moves in herd of 5 to 20 animals. Believed to be extinct, it has now been sight in the Himalayan foothills of Bhutan and north-west Assam.

Each Word Counts

WATER RELATION

Diffusion

Movement of particles or molecules from a region of higher concentration to a region of lower concentration is called as diffusion.

Significance

- It is an essential step in the exchange of gasses during respiration and photosynthesis.
- During passive salt uptake, the ions are absorbed by simple process of diffusion.
- In stomatal transpiration the diffusion of water vapours from the intercellular spaces into the outer atmosphere through open stomata.

Osmosis

If a solution and its pure solvent are separated by a semipermeable membrane the solvent molecules diffuse into the solution. This diffusion of solvent molecules into the solution through a semipermeable membrane is called as osmosis.

Significance

- Large quantities of water are absorbed by roots from the soil by osmosis.
- Cell to cell movement of water and other substances dissolved in it is involved in this process.
- Opening and closing of stomata depend upon the turgor pressure of the guard cells.

Plasmolysis

If a plant cell is placed in a highly concentrated sugar or salt solution, water from the cell sap flows out due to exosmosis through the plasma membrane outside the cell. The loss of water from the cell sap causes contraction or shrinkage of the protoplasm. Since the cell wall is firm and less elastic, it cannot keep pace with the contraction of plasma membranes. Ultimately the protoplasm separates from the cell wall and assumes spherical shape. This condition is called plasmolysis.

Significance

- It indicates the semi-permeable nature of the plasma membrane.
- This phenomenon is utilized in addition of concentrated sugar solution to jams and jellies.
- It is also used in determining the osmotic pressure of the cell sap.

Imbibition

Certain substances like dry wood or dry seeds when placed in water, absorb the water quickly and swell up considerably so that their volume is increased. These substances are called imbibants and the phenomenon as imbibition.

Significance

- The first step in the absorption of water by the roots of higher plants is the imbibition of water by the cell walls of the root hairs.
- Imbibition of water is very essential for dry seeds before they start germination.

Distinction amongs osmotic pressure turgor pressure, wall pressure and diffusion pressure deficit

Osmotic Pressure (OP)	*Turgor Pressure (TP)*	*Wall Pressure (WP)*	*Diffusion Pressure Deficit (DPD)*
1. It is a colligative property of the molecules particularly in a solution. The OP of a concentrated solution is more than that of a dilute solution.	1. It is the hydrostatic pressure generated within the cell against cellwall as a result of entry of water into it, due to osmosis.	1. It is equal and opposite pressure of the cell wall against the cell content to the turgor pressure.	1. When a cell is placed in pure water, the water enters into the cell as a result of the diffusion pressure deficit of cell sap.

GK GS

Copper was the first metal to be used by man.

Osmotic Pressure (OP)	*Turgor Pressure (TP)*	*Wall Pressure (WP)*	*Diffusion Pressure Deficit (DPD)*
2. If the two solutions are separated by a semi permeable membrane, the water moves from the solution having less OP to the solution of more OP.	2. Turgor pressure of a cell is maximum when it is fully turgid and it is reduced as the cell loses water.	2. It is the resistance of the cell wall generated against the expanding protoplasm due to entry of water.	2. The pure water has more diffusion pressure. If some solutes are mixed into it then it reduces the diffusion pressure of water. This is called diffusion pressure deficit.
3. The OP is responsible for the movement of water across a semi-permeable membrane.	3. The actual growth of a cell occurs when the TP exceeds wall pressure.	3. At a given time wall pressure equals the turgor pressure WP = TP.	3. If a plant cell is placed in water, the water enters into it due to OP. But this entry is restricted by the turgor pressure. Therefore, the actual pressure due to which water enters will be equal to OP minus TP. DPD=OP-TP.

Absorption of Water

Soil Water Relations

Plants absorb water mainly from the soil.

Five stages of water in the soil –

1. Gravitational water
2. Capillary water
3. Imbibed water
4. Runaway water
5. Chemically combined water.

Mechanism of water absorption

1. Active absorption of water
2. Passive absorption of water

1. *Active absorption of water*

When the water is absorbed due to the activity of root itself, particularly root hairs, it is termed active absorbtion.

(a) Osmotic Absorption

The first step in the absorption of water is its imbibition on the cell wall surface, The water enters through the cell wall and comes in contact with differentially permeable plasma membrane and cytoplasm. The cell sap, filled inside the vacuole is now separated from external soil water by a differentially permeable membrane.

(b) Non-osmotic Absorption

Sometimes, it has been observed that absorption of water takes place even when the osmotic pressure of cell sap. This type of absorption which is non osmotic and against the osmotic gradient requires the expenditure of metabolic energy probably through respiration.

2. Passive Absorption of water

Passive absorption of water takes place when rate of transpiration is usually high. Rapid evaporation of water from the leaves during transpiration creates a tension in water in the xylem of leaves. This tension is transmitted to water in xylem of roots through the xylem of stem and the water rises upward to reach the transpiring surfaces.

Pathway of water across root cells – There are three possible pathways for the movement of water across the cortex :

(i) Through a system of inter connecting cell walls and intercellular spaces.

(ii) Through the living cytoplasm via plasmodesmata.

(iii) Through the living cytoplasmic membranes, cytoplasm and vacuoles of the critical cells.

Difference between active and water absorption

Active water absorption	*Passive water absorption*
It occurs mainly due to the activity of root hairs. Needs energy.	It occurs mainly due to the activity of upper part of plant such as shoot and leaves.
The water is absorbed by the osmotic and non-osmotic mechanisms.	The water is absorbed due to the process of active transpiration in the upper part.
The osmotic process involves diffusion pressure deficit of the cells.	The passive absorption occurs due to the tension created in xylem sap by transpiration pull.
The water first enters into the cell sap and then passes from one cell to another (symplast movement).	The water in passive absorption moves probably through the free spaces (apoplast movement).

Factors affecting the rate of water absorption

1. Available soil water: If water in the soil is present below the permanent wilting percentage the absorption is affected.
2. Temperature of the soil: The low temperature of the soil reduces absorption of water.
3. Concentration of soil solution.
4. Aeration of soil : Poorly aerated soil retards absorption of water.

Transpiration and Guttation

The loss of water in the form of vapours from the living tissues of aerial parts of the plant is termed as transpiration.

Types of Transpiration —

1. Stomatal transpiration
2. Cuticular transpiration
3. Lentricular transpiration

Factors affecting the rate of transpiration

External Factors —

1. *Atmospheric Humidity*: The rate of transpiration decreases with the increase in relative humidity of atmosphere.

2. *Temperature*: The rate of transpiration increases with the increase in atmospheric temperature.

3. *Light*: The rate of transpiration increases in light and decreases in dark.

4. *Atmospheric Pressure*: Lowering of atmospheric pressure permits more rapid diffusion of water vapours.

5. *Available Soil Water*: If the available water in the soil is not sufficient the rate of transpiration is decreased.

Internal Factors —

1. *Leaf Area*: By reducing the size of leaves, plants decrease their rate of transpiration.

2. *Leaf Structure*: Thick cuticle on the surface of plant parts, compact mesophyll cells in the leaves and reduction in number of stomata reduce the rate of transpiration.

3. *Age of Plants*: Rate of transpiration is slow at the seedling stage and maximum at maturity state.

Advantages of transpiration —

- It controls the rate of absorption of water.
- It helps in the mineral absorption and ascent of sap.
- It regulates the plant temperature.

Disadvantages of transpiration —

- Nearly 97% of the total water absorbed by the plants is lost in the form of vapours.
- Transpiration often results in water deficits which causes injury to the plants by desiccation.
- Rapid transpiration causes temporary wilting.
- Deciduous trees have to shed their leaves to check transpiration.

You Don't Know!

Oldest rock

The greatest reported age for any scientifically dated rock is 3775 + 89 Ma (million years) found in Champua, Orissa.

Guttation

The process of exudation of liquid drops from the edges of leaves is called Guttation.

Difference between Transpiration & Guttation

Transpiration	*Guttation*
Occurs in day	Night
Water is given out in the form of vapour	In the form of liquid
Occurs through stomata, lenticle or cuticle	Occurs through hydathodes
It is a controlled phenomenon	It is a uncontrolled phenomenon

Ascent of Sap

The water is absorbed mainly by the roots and is moved upward to all the parts of plants via stem. Thus the upward movement of water through stem is called ascent of sap.

Path of Ascent of Sap

After absorption by root hair cells, the water moves through several root tissues such as cortex, endodermis, pericycle and finally enters the xylem tissue. Once the water, it reaches the mesophyll tissues of the leaves. The bulk of water enters the mesophyll cells and finally evaporates and transpires through the stomata, only small amount of water is used in metabolism and growth.

Many theories have been put forward to explain the upward movement of water.

1. Root pressure theory
2. Vital theories
3. Physical theories:

(a) Root pressure theory - Stocking

(b) Vital theories of ascent of sap - Godlewski, Bose and Westermeir

3. Physical theories :

(a) Imbibition theory -Unger and Sachs

(b) Capillary theory -Boehm

(c) Transpiration pull and cohesion of water theory - This theory was first proposed by Dixon and Joly in 1894.

The theory is based on the following features :

(i) Cohesive and Adhesive properties of water molecules

to form a continuous water column in the xylem.

(ii) Transpiration pull exerted on this water column.

Water molecules remain joined to each other due to the presence of H-bonds between them.

Although H-bond is very weak when they are present in enormous numbers as in case of water, a very strong mutual force of attraction or cohesive force develops between water molecules and hence they remain in the form of a continuous water column in the xylem.

The adhesive properties of water further ensure the continuity of water column in xylem.

When transpiration takes place in leaves at the upper parts of the plant, water evaporates from the intercellular spaces of the leaves to the outer atmosphere through stomata. More water is into the intercellular spaces from the mesophyll cells. Due to all this tension is created in water in the xylem elements of the leaves. This tension is transmitted downward to water in xylem elements of the roots through the xylem of petiole and stem and the water is pulled upward in the form of continuous unbroken water column to reach the transpiring surfaces up to to the top of the plants.

Factors affecting Ascent of Sap

High temperature, low atmospheric humidity, high atmospheric pressure and wind velocity permit more rapid transpiration and thus these factors also promote upward transport of water. Deficiency of water in the soil also cause decrease in the ascent of sap indirectly by influencing the absorption of water. Stomatal opening and closing also regulates movement of water in plants.

METABOLISM

Photosynthesis

It is an anabolic process, where the carbon containing compounds are formed from CO_2 and water by illuminated green cells, water and oxygen being used to produce photo synthesis.

Blueprint

1. Light Reaction (Grana)

   $H_2O \rightarrow ATP + NADPH_2 + O_2 \uparrow$

   1.a → Non cyclic Phosphorylation

   1.b → Cyclic Phosphorylation

   1.c → Pseudocyclic Phosphorylation

2. Dark Reaction (Stroma)

   $ATP + NADH_2 + CO_2 \rightarrow C_6H_{12}O_6$

   2.a → C_3 Cycle

   2.b → C_4 Cycle

   2.c → CAM

Photosynthetic Apparatus

The chloroplasts in green plants constitute the photosynthetic apparatus. The chloroplast is bounded by two membranes consisting of lipid bilayer and proteins. Internally the chloroplast is filled with a hydrophilic matrix called stroma in which are embedded grana. Each granum consists of disk shaped grana lamella. In cross section these lamella are paired to form sac-like structures and have been called as thylakoids.

Chlorophylls and other photosynthetic pigments are found in the form of protein pigment complexes mainly in thylakoid membranes of grana. The grana are sites of primary photochemical reaction. Dark reaction of photosynthesis occurs in stroma.

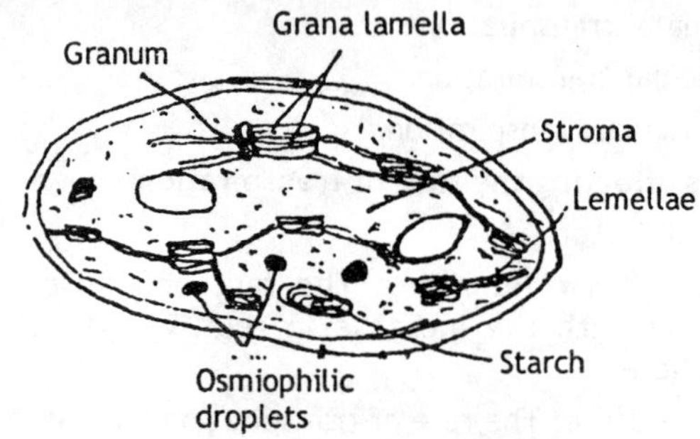

Internal structure of chloroplast

Photosynthetic Pigments are of three types :

1. Chlorophylls 2. Carotenoids 3. Phycobillins.

Different pigments absorb light of different wavelengths.

Red Drop and Emersons Effect

Proposed by Robert Emerson, (1943) while working on quantum yield of Photosynthesis in monochromatic light of different wavelengths in chlorella and noticed a sharp decrease in quantum yield at wavelengths greater than 680 nm. Because this decrease in the quantum yield took place in red part of the spectrum and this phenomenon is called red drop.

Emerson and Lewis later found that, inefficient far-red light in chlorella, beyond 680 nm could be made fully efficient if supplemented with light of shorter wavelengths. The quantum yield from the two combined beams of light was found to be greater than the sum effect of both beams used separately. This enhancement of photosynthesis is called as Emerson's enhancement effect.

Two Pigment System

Photosynthesis is driven by two pigment systems called PSI and PSII. Wavelength of light shorter than 650 nm affect both the pigment systems and greater than 680 nm affect only PSI.

(1) PSI Relatively Weak Fluorescent –

- Consists of chlorophylls, carotenoids, cytochrome

7, molecule of P700, Plastocyanin, cyt b, FRS (Ferredoxin reducing substance) and rich in chlorophyll-a, iron and copper.

- Controls, the process of producing a strong reductant to reduce NADP to NADPH + H+
- Mainly found in stroma Lamellae

(II) PSII Strongly Fluorescent –

- In addition to chlorophylls, carotenoids, moecule of P 680, Plasto quinone, four Mn molecules bound to one or more proteins. Cyt b, cyt b_6, and chloride.
- Concerned with generation of strong oxidant and weak reductant coupled with the release of oxygen.

Production of Assimilatory Powers in Photosynthesis

ATP and NADPH are the assimilatory Powers. The process of NADP into NADPH + H^+ may be denoted as ETS in Photosynthesis or Reduction of NADP, while the process of formation ATP from ADP and inorganic phosphate using light energy called photo phosphorylation.

Light Reaction

The light reaction and its concerned activities are found in thylakoids and grana.

$$6\ CO_2 \rightleftharpoons C_6H_{12}O_6 + 6O_2$$

The light reaction phase is a complicated process with several important events.

Events in Light Reaction –

(i) *Absorption of light energy by Chloroplast Pigments*: Different chloroplast pigments absorb light in different regions on the visible part of the spectrum.

(ii) *Transfer of light energy from Accessory Pigments to Chlorophyll-a* : All the Photosynthetic Pigments except Chlorophyll-a, are called accessory pigments. Light energy absorbed by them is transferred by resonance to Chl-a which alone take part in the primary photo chemical reaction. Chl-a alone can absorb the light energy directly. In PSII, reaction centre is P680 and in PSI it is P700.

Knowledge World!

Largest school

South Point High School, Kolkata, has a record number of 11,683 regular students. The school was founded on April 1, 1954 with 17 students.

(iii) *Activation of Chl-a molecule by Photons of Light*: When P680 or P700 forms chlorophyll -a molecule in two pigment systems receive a photon of light, it becomes an excited molecule having more energy than ground state energy and finally expels its energy along with an e^- and a+ charge comes on the Chl-a which now becomes oxidised.

$$\text{Chl-a} \xrightarrow{\text{Light}} \text{Excited triplet state of chl-a} \longrightarrow (\text{Chl-a})^+ + e^-$$

(iv) *Photolysis of water and O_2 Evolution (Oxidation of water)* : Associated with PSII and catalysed by presence of Mn^{++} and Cl^- ions. PSII receives light and water molecules split into OH^- and H^+ ions (Photolysis of water). OH^- ions unite and form some more water molecules again and release O_2 and electrons.

$$2H_2O \longrightarrow 4H^+ + O_2 + 4\ e^-$$

(v) Transfer e- from PS II to PS I, (Pc) Plasto cyanin is blue, Cu containing protein art as Mobile carrier.

(vi) *Electron transport and the production of Assimilatory power (NADPH + H^+ + ATP)* : When the expelled e- travelling through number of electron carriers either cycled back or is consumed in reducing $NADP^+$ to NADPH + H^+. The extra light energy liberated along with the e^- is utilised in the formation of ATP molecules at certain places during its transport, the process of which is called photo phosphorylation.

There are 3 types of photophosphorylation, viz.

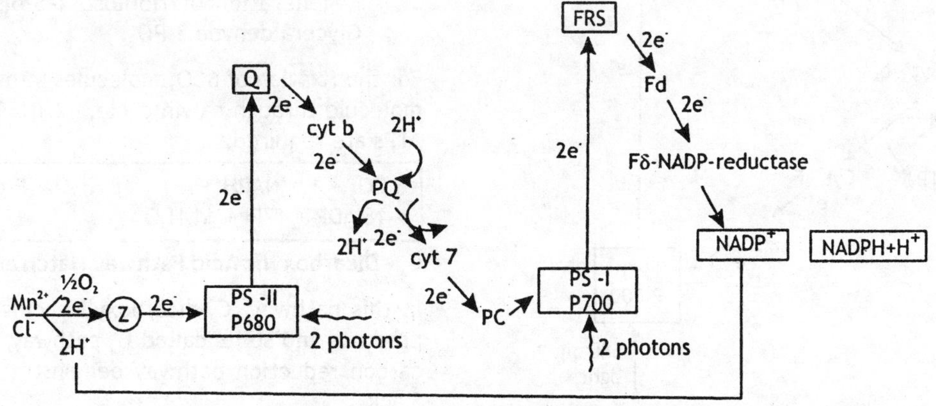

Non cyclic Photo Phosphorylation

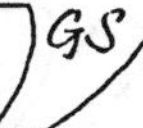

GK GS

The Gold coin was the *nishka*.

1. Non-cyclic photo phosphorylation/e⁻transport (z-scheme).

In non-cyclic photophosphorylation, the flow of electron is unidirectional, that is, electron denoted by PSII after passing through plasto quinones, Cyt b_6, Cyt F, plastocyanin and PSI eventually reaches ferredoxin which in turn denotes to reduced NADP. The reduced NADP is utilised for the reduction of CO_2. The electron does not complete the cycle. It starts from PSII reduction. So the ATP synthesis resulting from this type of non-cyclic electron transport chain is known as non cyclic photophosphorylation. Water molecule is utilised as a source of electron in this system. In this process, two molecules of ATP are formed per two molecules of NADP+ reduced.

During light reaction, the protons accumulate inside the thylakoid membrane resulting in a proton gradient. The energy released by the protons when they diffuse across the thylakoid membrane into the stroma is used to produce ATP.

2. Cyclic Photophosphorylation

Here ATP formed from PSI and wavelength of light greater than 680 nm is used. Here initially energy is provided from absorbed light Quantum instead of glucose. Electrons used here do not come from water. In the cyclic photophosphorylation light lifts the electron from P700 or feredoxin. The excited electron returns to P700 through two to three transfer steps to decreasing redox potentials. It is during such a downhill migration of the electron that enough energy is released for ATP Synthesis. In this process ATP can be formed between cytochrome b_6 and cytochrome and between ferredoxin and cytochrome b6.

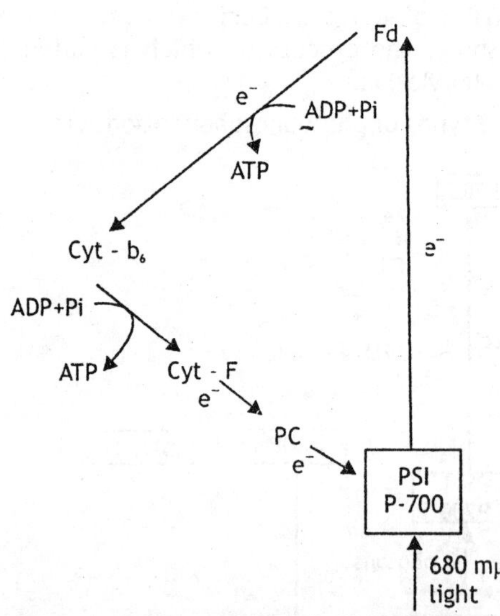

Cyclic Phosphorylation

3. Pseudocyclic Photophosphorylation

Arnon and co workers demonstrated, even in the absence of CO_2 and NADP, if chlorophyll molecules are illuminated it can produce ATP from ADP and Pi in presence of FMN and oxygen. The process otherwise is called oxygen dependent FMN catalylsed photophosphroylation, which involves reduction of FMN with the production of oxygen.

FMN is reoxidisable Hydrogen acceptor with the effect that the reduced FMN is to be oxidised by oxygen. Thus the process can continue repeatedly to produce ATP.

Here the source of both hydrogen for the reduction of FMN and of oxygen for its oxidation in the water.

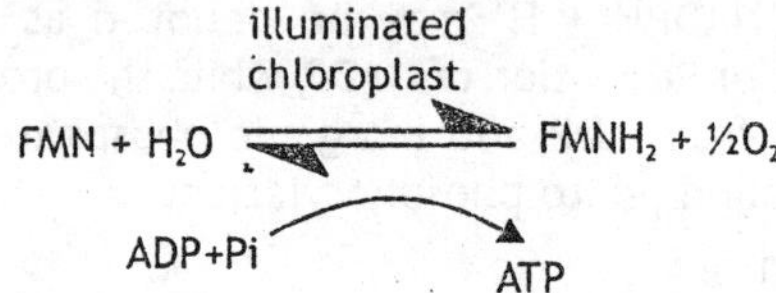

Dark Reaction

It occurs in Stroma –

The dark reaction process of photosynthesis has been named variously such as Calvin cycle, Blackman reaction, Carbon assimilation, Reductive pentose phosphate cycle etc. It consists of –

(a) Synthesis of carbohydrate

(b) Regeneration of ribulose diphosphate

It can otherwise be called as Reductive Pentose Phosphate Pathway. (RPP) the cycle includes 3 phases.

1. Carboxylation of the ribulose 1-5 - bisphosphate forming 3-phosphoglycerate.
2. Reduction of 3-phosphoglycerate forming Glyceraldehyde-3-PO_4.
3. Regeneration of ribulose 1-5-bisphosphate from Glyceraldehyde 3-PO_4.

For the fixation of $6CO_2$ molecules into one hexose sugar molecule through calvin cycle, 12 (NADPH + H+) and 18 ATPs are required.

$6\ CO_2 + 12\ (NAPH + H^+) + 18\ ATP \rightarrow F\text{-}6\text{-}P + 12\ NADPH^+ + 18\ ADP + 17Pi + 11\ H_2O$

C_4 - Dicarboxylic Acid Pathway Hatch and Slack Pathway

In this pathway, C_4-dicarboxylic acids are the earliest products and so its called C_4 pathway, which is another carbon reduction pathway demonstrated by Hatch and Slack.

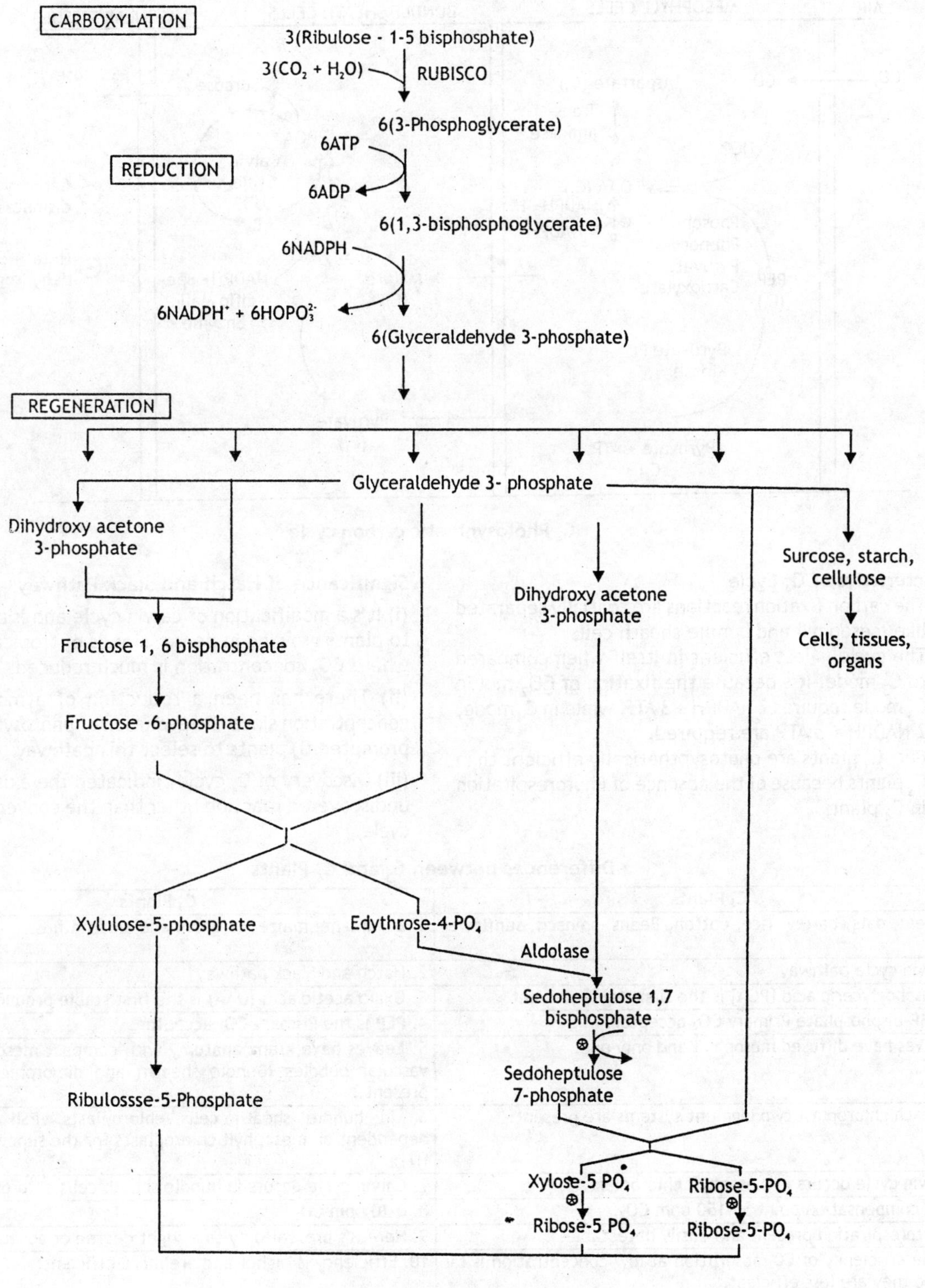

CARBOXYLATION

3(Ribulose - 1-5 bisphosphate)

$3(CO_2 + H_2O)$ → RUBISCO

6(3-Phosphoglycerate)

REDUCTION

6ATP → 6ADP

6(1,3-bisphosphoglycerate)

6NADPH → $6NADPH^+ + 6HOPO_3^{2-}$

6(Glyceraldehyde 3-phosphate)

REGENERATION

Glyceraldehyde 3- phosphate

Dihydroxy acetone 3-phosphate

Fructose 1, 6 bisphosphate

Fructose -6-phosphate

Dihydroxy acetone 3-phosphate

Surcose, starch, cellulose

Cells, tissues, organs

Xylulose-5-phosphate

Edythrose-4-PO_4

Aldolase

Sedoheptulose 1,7 bisphosphate

Sedoheptulose 7-phosphate

Ribulosse-5-Phosphate

Xylose-5 PO_4 → Ribose-5 PO_4

Ribose-5-PO_4 → Ribose-5-PO_4

Calvin Cycle

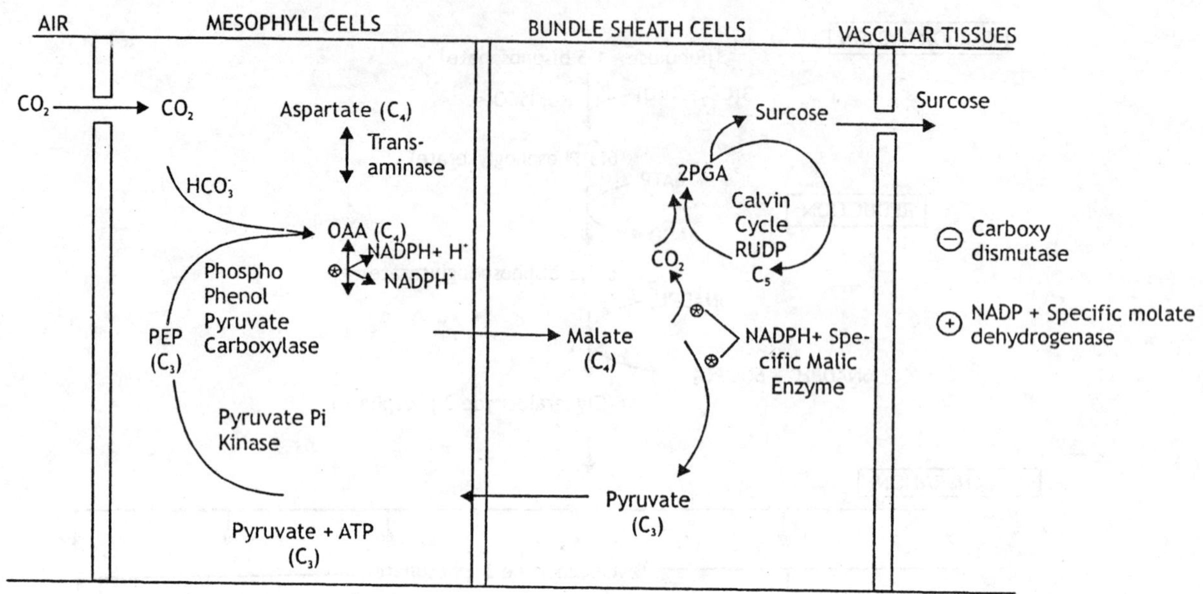

C_4 Photosynthetic carbon cycle

Characteristics of C_4 Cycle

- The carbon fixation reactions are spatially separated like mesophyll and bundle sheath cells.
- The cycle is less efficient in itself when compared to C_3 mode. It's because the fixation of CO_2 mol in C_3 mode requires 2 NADPH + 3 ATP, while in C_4 mode, 2 NADPH + 5 ATP are required.
- But, C_4 plants are photosynthetically efficient than C_3 plants because of the absence of Photorespiration in C_4 plants.

Significance of Hatch and Slack Pathway

(i) It's a modification of calvin cycle and is advantageous to plants growing in dense, stands of tropical vegetation where CO_2 concentration is much reduced.

(ii) There has been a reduction of atmospheric CO_2 concentration since the evolution of photosynthesis which prompted C_4 plants to select this pathway.

(iii) Discovery of C_4 cycle, indicated the existence of yet undiscovered reaction other than the conventional calvin cycle.

Differences between C_3 and C_4 Plants

C_3 Plants	C_4 Plants
1. Wheat, oats, barley, rice, cotton, Beans, spinach, sunflower etc.	1. Sugarcane, maize, sorghum and amarnathus.
2. Calvin cycle pathway	2. Hatch and Slack pathway
3. Phosphoglyceric acid [PGA] is the first stable product.	3. Oxalo acetic acid (OAA) is the first stable product.
4. RUBP-bisphosphate Primary CO_2 acceptor.	4. PEP is the Primary CO_2 acceptor.
5. Leaves have diffused mesophyll and only one.	5. Leaves have kranz anatomy with compact mesophyll around vascular bundles (Bundle sheath) and dimorphic chloroplasts present.
6. In each chloroplast two pigment systems are present.	6. In bundle sheath cells chloroplasts, PSII absent. So, dependent on mesophyll chloroplasts for the supply of NADPH + H+.
7. Calvin cycle occurs in mesophyll chloroplast.	7. Calvin cycle occurs in bundle sheath cells chloroplasts.
8. CO_2 compensation point 50-150 ppm CO_2.	8. 0-10 ppm CO_2.
9. Photorespiration present and easily detectable.	9. Here it's present only to a slight degree or absent.
10. The efficiency of CO_2 absorption at low concentration is far less. So they are less efficient.	10. Efficiency is higher and are more efficient.
11. Only C_3 cycle is found.	11. Both C_3 and C_4 cycles found.
12. Optimum temperature 10-25°C	12. Optimum temeprature 30-45°C.
13. 18 ATP for one glucose molecule	13. 30 ATP for one glucose molecule.

Western narrow hilly area of Mysore plateau is called Mainad.

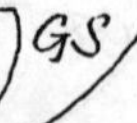

Crassulacean Acid Metabolism (CAM Cycle)

Here the CO_2 is concentrated at the site of Rubisco, which is found in *Crassulaceae*, Euphorbiaceae, Orchidaceae and agavaceae. This mechanism enables plants to improve water use efficiency and so, these plants have a competitive advantage in dry environment.

Unlike C_4 plants, the formation of C_4 acids, is both temporarily and spatially separated.

In CAM plants, stomata open at night and closed during most of the day. The distinctive diurnal fluctuation in acidity in plants showing CAM due to changes in amounts of vacuolar malic acid in mesophyll cells. During night, malic acid synthesied using CO_2 which accumulates in the vacuole. During day this malic acid consumed resulting in decrease of acidity. Citric and isocitric acid also contribute to acidity, but their amount is negligible and these do not show consistent diurnal patterns of fluctuation as shown by malic acid.

Synthesis of Malate during night or Dark CO_2 Fixation

Large amounts of starch are consumed during acidification which indicates that carbohydrates are the source of malate synthesis

$$C_6H_{12}O_6 + 2CO_2 \rightarrow 2\ C_4\ H_6{+}O_5$$
(malate)

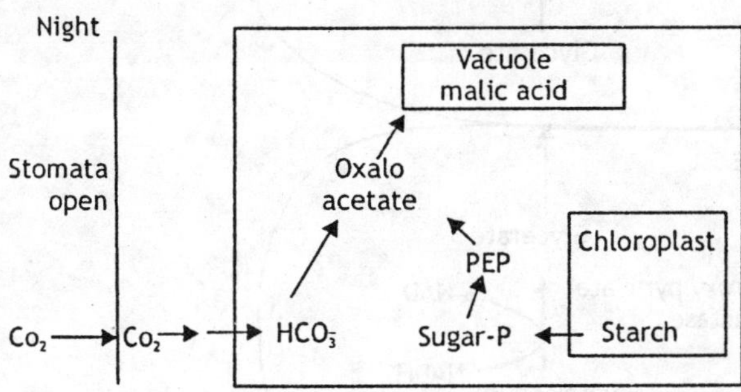

MESOPHYLL CELLS

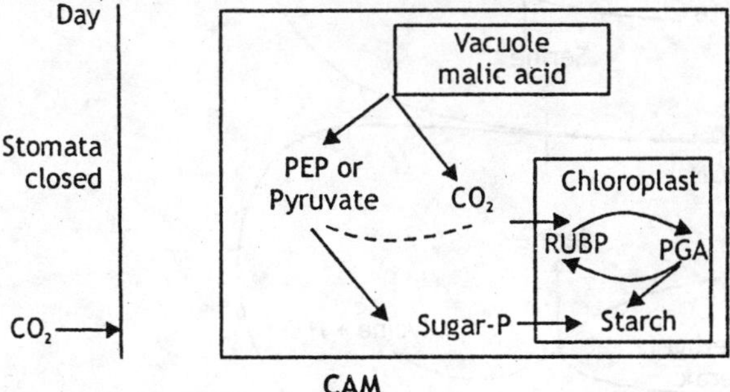

CAM

Malate is synthesized during night in reaction in which some malate product derived from carbohydrate reserves e.g. pyruvate or PEP is carboxylated to produce malate either directly or first forming oxalo acetic acid which is then reduced to malate.

Quiz Kit!

Longest-lived wild cattle

The Asiatic Buffalo lives to a maximum age of over 29 years in captivity.

Consumption of malate in Light Deacidification

During the day, when acidified organs exposed to light, rapid consumption of malate occurs resulting in deacidification (due to releases CO_2 from malate). The malate may be decarboxylated in two ways.

(i) $\text{Malate} + NADP^+ \xrightleftharpoons{\text{NADP-malic enzyme}} \text{Pyruvate} + CO_2 + NADPH + H^+$

(ii) (a) $\text{Malate} + NAD^+ \xrightleftharpoons[\text{dehydrogenase}]{\text{Malate}} \text{Oxaloacetate} + NADH + H^+$

(b) $\text{Oxalo acetate} + ATP \xrightleftharpoons[\text{Kinase}]{\text{PEP-carboxy}} \text{PEP} + CO_2 + ADP$

Pyruvate and PEP utilised for carbohydrates synthesis during the day. Pyruvate first converted to PEP in the presence of Pyruvate Po_4 dikinase.

$$\text{Pyruvate} + ATP + Pi \rightarrow PEP + Amp + PPi$$

PEP in both types of CAM plants is converted into 3 PGA by release reactions of glycolyis. Then, 3-PGA is utilised in calvin cycle.

Photo Respiration

Otherwise known as photorespiratory carbon oxidation cycle (PCO). Rubisco, the most abundant protein in the world is having a distinct property, that it can catalyse both the carboxylation and oxygenation of RuBp. Oxygenation is the primary reaction in a process known as photorespiration.

It's the process, in which, CO_2 is lost from cells that are simultaneously, fixing CO_2 by the calvin cycle. The cycle acts as a scavenger operation to recover the fixed CO_2 lost by the oxygenase reaction of rubisco. Three organelles are involved in the process viz., chloroplast peroxisomes and mitochondria.

- Photorespiration occurs only in chlorophyllous tissues of plants. Glycolate is the chief metabolite and also its substrate. Other important metabolites are the amino acids glycine and serine.
- Like usual mitochondrial respiration, its also an oxidative process where oxidation of glycolate occurs with subsequent.

GK

The Magadha society was anti-traditional.

GS

- Glycolate cycle/C_2-cycle.

Factors Affecting

1. CO_2 compensation point
2. CO_2 concentration when high, photo respiration is also high and vice versa.
3. Inhibitors of Glycolic acid oxidase such as α-hydroxy-sulphonates inhibit the process of photo respiration.

Significance

1. It's a double wasteful process
2. Resulting in considerable decrease in photosynthetic productivity.
3. Unlike usual mitochondrial respiration, neither reduced co-enzymes are generated nor the oxdiation of glycolate coupled with the formation of ATP molecules.
4. Moreover, there is consumption of reduced coenzyme and ATP.
5. However, the knowledge on photo respiration is important. Therefore by manipulating different atmospheric conditions, use of inhibitors of glycolic acid oxidase such as α-hydroxy sulphonates and through genetic control, the process can be reoulated and photosynthetic productivity can be increased.

Factors Affecting Photosynthesis

External Factors –

1. *Light:* It affects the rate of photosynthesis in three ways:

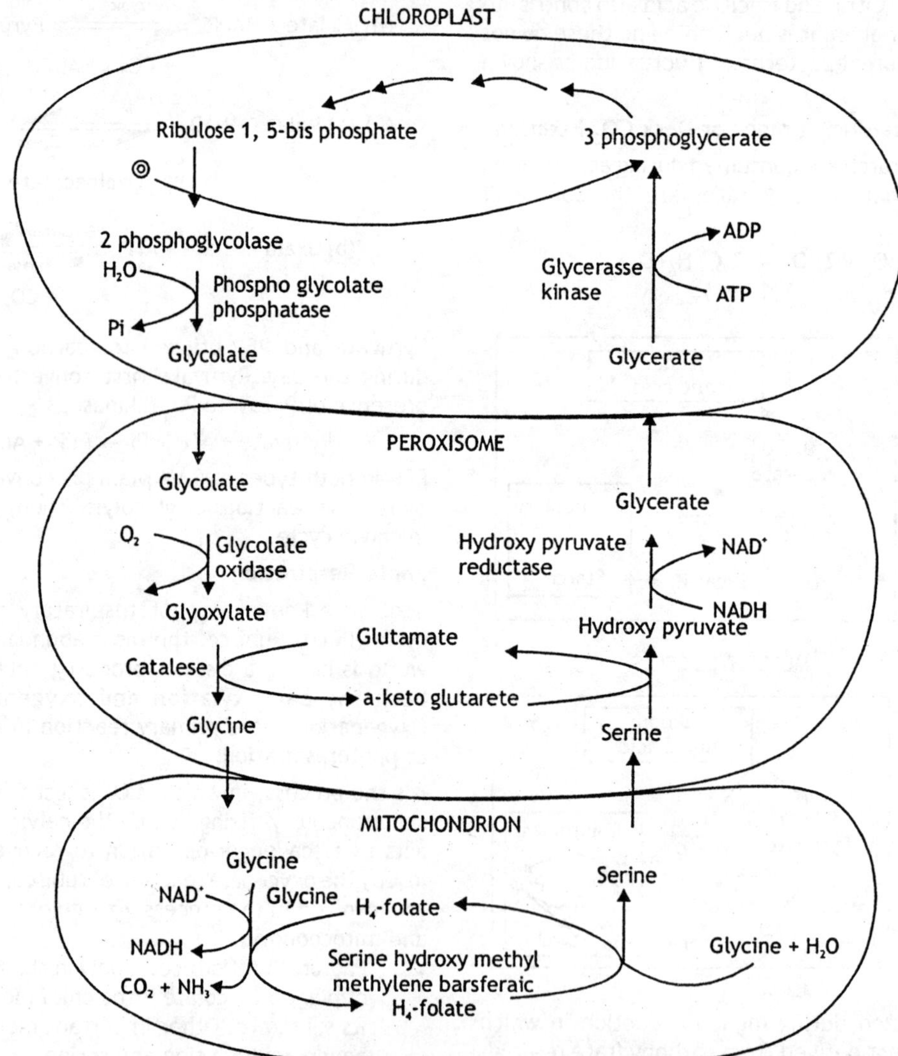

Operation of the photorespiratory carbon oxidation cycle involves cooperative interaction among three organelles.

GK GS

Article 45 directs the State to provide free and compulsory education for children upto 14 years of age.

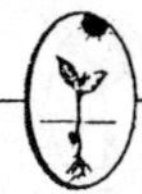

(a) *Light Quality*: Due to heavy absorption of light rays in the red part of the spectrum of chlorophylls, maximum photosynthesis takes place in red light.

(b) *Light Intensity*: The rate of photosynthesis is greater in intense light than in the diffused light.

(c) *Duration of Light*: The rate of photosynthesis is greater in intermittent light than in continuous light.

2. *Carbon dioxide*: An increase in CO_2 concentration up to about 1% increases the rate of photosynthesis.
3. *Temperature*: Photosynthesis will stop in some plants at about freezing point and in some plants beyond 40-50°C.
4. *Water*: The rate of photosynthesis may go down if the plants are inadequately supplied with water.

Internal Factors –

1. *Chlorophyll content*: The rate of photosynthesis should increase with an increase in the chlorophyll content.
2. *Protoplasmic factors*: Proper hydration of the protoplasm is essential for photosynthesis.
3. *Accumulation of the end products of photosynthesis*: Accumulation of carbohydrates in the photosynthesizing cells retards the rate of photosynthesis.
4. *Anatomy of leaf*: The thickness of the cuticle and epidermis, structure and distribution of stomata, distribution and relative proportion of chlorophyllus and non-chlorophyllus mesophyll tissue all influence the rate of photosynthesis.

CATABOLISM

Respiration

- It is a catabolic process.
- Plant respiration (aerobic) is the biological process by which reduced organic compounds are mobilized and subsequently oxidised in the controlled manner.
- During respiration, free energy is released and incorporated into a form ATP that can be readily utilized for the maintenance and development of the plant.

$$C_6H_{12}O_6 \rightarrow CO_2 \uparrow + H_2O + ATP$$

- Production of ATP is called phosphorylation (3 types):

1. Photo phosphorylation: Light $\rightarrow$ ATP (light reaction)
2. Substrate level phosphorylation: Fructose 1, bisphosphate $\rightarrow$ ATP $\rightarrow$ Glycolysis
3. Oxidative phosphorylation : ADP $\rightarrow$ ATP $\rightarrow$ ETS.

Respiration Process

1. Glycolysis $\Rightarrow$ $C_6H_{12}O_6 \rightarrow$ Pyruvate (Cytosol/cytoplasm)
2. Acetyl CoA $\Rightarrow$ Pyruvate $\rightarrow$ Acetyl COA (Mitochondrion Matrix inner wall)
3. TCA/Kreb cycle $\Rightarrow$ Acetyl COA
   ATP
   $\rightarrow$ $FADH_2$ (Matrix of Mitochandrion)
   NADH
4. Electron transport system $\Rightarrow$
   FADH2
   NADH (Inner Mitochandrion Membrane)

Did You Know?

Smallest plants

***Wolfe arrbiza*, a duckweed with leaves as small as one mm, is found floating on still waters in Asia and Africa. The flowers are almost invisible to the naked eye.**

Glycolysis

- Greek word: Glykos -Sugar
  Lysis -Splitting
- The first stage of respiration, six carbon molecule of glucose split into two 3 carbon molecules of sugar, are then oxidised and rearranged to yield 2 molecules of Pyruvate.
- Glycolysis occurs in all living organisms (Prokaryotes and Eukaryotes).
- No oxygen required to convert glucose to pyruvate and glycolytic mechanisms can become the primary mode of energy production in plant tissue when oxygen levels are low.
- In animal cells the enzymes of glycolytic pathway do not exist independently in the cytosol but associated in a supermolocular complex that may be loosely bound to the surface of outer mitochondrion membrane.
- This complex is thought to facilitate the conversion of substrate to products in the multistep glycolytic process.
- In this process phosphofructokinase is one of the control points in both plants and animals.
- This type of ATP synthesis is referred to as substrate level phosphorylation because it involves the direct transfer of phosphat moiety from the substrate molecule to ADP to form ATP.

Alternative Glycolytic Pathway

- The synthesis of glucose through reversal of the glycolytic pathway known as gluconogenesis.
- It is not common in plants, but it does operate in seed of some plants (esp. oil seeds sunflower).
- Carbon stored in the form of oil (Triglycerols). After seed germinate, the oil, are converted into sugar by gluconogenesis.

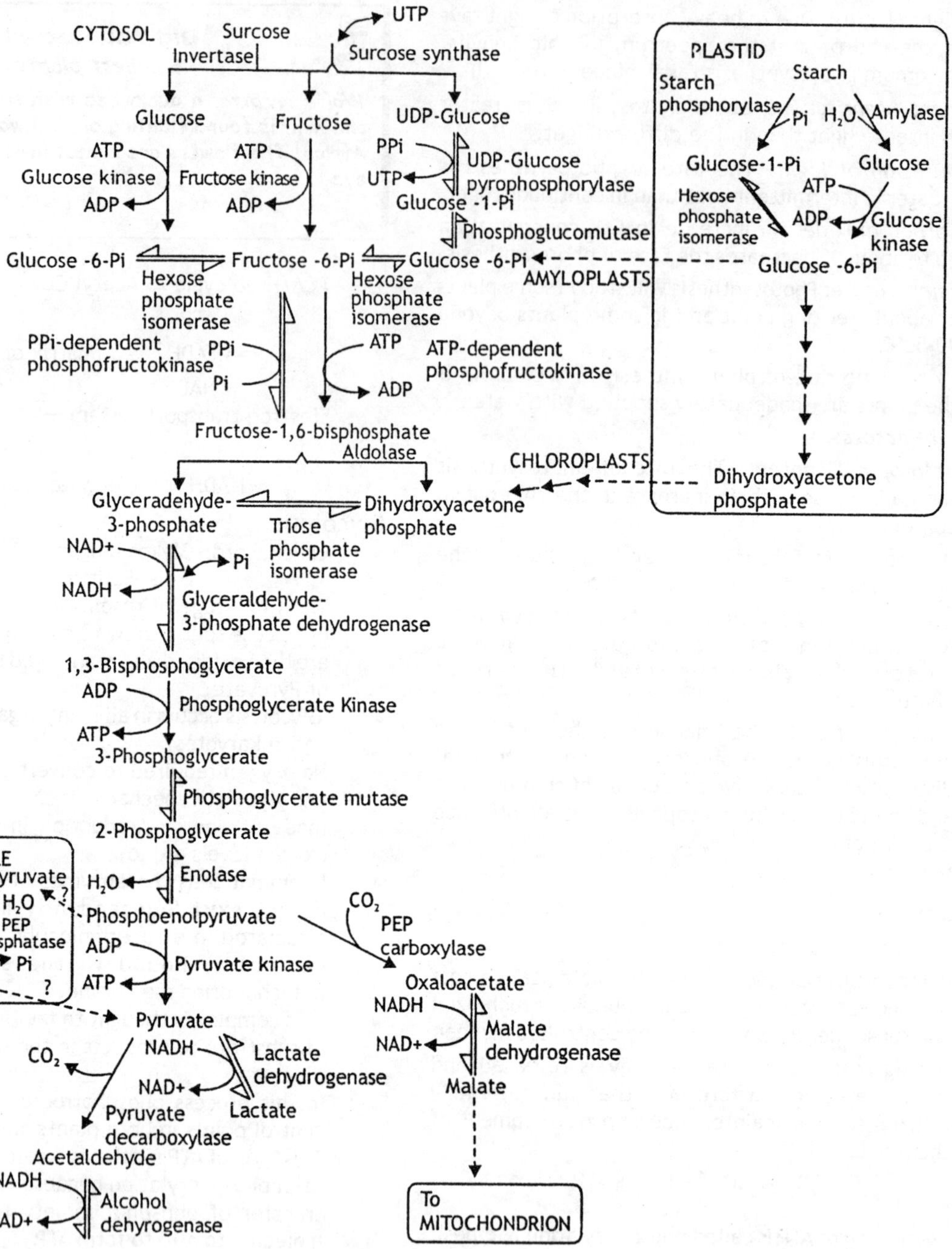

CYTOSOL
Surcose
Invertase
UTP
Surcose synthase
Glucose
Fructose
UDP-Glucose
ATP
Glucose kinase
ADP
ATP
Fructose kinase
ADP
PPi
UTP
UDP-Glucose pyrophosphorylase
Glucose -1-Pi
Phosphoglucomutase
Glucose -6-Pi
Fructose -6-Pi
Glucose -6-Pi
Hexose phosphate isomerase
Hexose phosphate isomerase
AMYLOPLASTS
PLASTID
Starch
Starch phosphorylase
Pi
H_2O
Amylase
Glucose-1-Pi
Glucose
Hexose phosphate isomerase
ATP
ADP
Glucose kinase
Glucose -6-Pi
PPi-dependent phosphofructokinase
PPi
Pi
ATP
ATP-dependent phosphofructokinase
ADP
Fructose-1,6-bisphosphate
Aldolase
CHLOROPLASTS
Dihydroxyacetone phosphate
Glyceradehyde-3-phosphate
Dihydroxyacetone phosphate
Triose phosphate isomerase
NAD+
Pi
NADH
Glyceraldehyde-3-phosphate dehydrogenase
1,3-Bisphosphoglycerate
ADP
Phosphoglycerate Kinase
ATP
3-Phosphoglycerate
Phosphoglycerate mutase
2-Phosphoglycerate
VACUOLE
Phosphoenolpyruvate
H_2O
PEP phosphatase
Pi
Pyruvate
?
?
H_2O
Enolase
Phosphoenolpyruvate
CO_2
PEP carboxylase
ADP
Pyruvate kinase
ATP
Oxaloacetate
NADH
NAD+
Malate dehydrogenase
Malate
Pyruvate
NADH
Lactate dehydrogenase
NAD+
Lactate
CO_2
Pyruvate decarboxylase
Acetaldehyde
NADH
NAD+
Alcohol dehyrogenase
To MITOCHONDRION

Glycolysis

Acetyle CoA Formations

- Inside the mitochandrial matrix, pyruvate is oxidatively decarboxylated by the enzyme pyruvate dehydrogenate to produce NADH, CO_2 and acetic acid.
- The acetic acid is linked via thioester bond to a sulphur-containing cofacor, coenzyme.

TCA/Kreb/Citric Acid Cycle

- It occurs in mitochondrial Matrix.
- 36 ATP released in Eukaryotes.

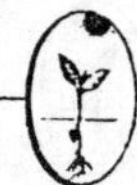

Pyruvate
CoA
NAD^+
Pyrivate dehydrogenase
NADH
CO_2
Acetyl CoA
Citrate synthase
CoA
Citrate
Aconitase
Isocitrate
NAD^+
Isocitrate dehydrogenase
NADH
CO_2
α-Oxoglutarate
CoA
NAD^+
α-Ketoglutarate dehydrogenase
NADH
CO_2
Succinyl CoA
CoA
Succinyl-CoA synthetase
ATP
ADP
H_2O
CoA
Succinate
FAD
$FADH_2$
Succinate dehydroenase
Fumarate
Fumarase
H_2O
Malate
Malate dehydrogenase
NAD^-
NADH
Oxaloacetate
TCA Cycle

TCA Cycle

Electron Transport System (ETS)

Oxidative phosphorylation

- Here the high energy electron capture during the TCA cycle (In the form of $FADH_2$ and NADH) can be converted into ATP.
- This O_2 dependent process occurs on the inner mitochondrial membrane and invites series of electron carriers known as electron transport chain.
- Two molecules of NADH from cytosol, 8 molecules of NADH and two molecules of $FADH_2$ from TCA - appear in mitochondrial matrix.
- The individual electron transport proteins are organised into a series of four multiprotein complexes and localised in the inner membrane of mitochondrial matrix.
- Complex I: TCA cycle NADH dehydrogenase, Flavin Mono Nucleotide (FMN) and several sulphur proteins for the transfer of electrons.
- Complex II:TCA cycle enzyme succinyl dehydrogenase, electron derived from the oxidation of succinate are transfered via $FADH_2$.
- Complex III: Act as a ubiquinol : Cytochrome C Oxidoredutase, oxidising reduced ubiquinol and transfering electron through an iron-sulfur centre, two b-type cytochrome and a membrane bound Cyt. C. Cytochrom C - is the only protein in ETC that is not an integral membrane properties, and it serves as a mobile carrier to transfer the electron from complex III to IV.

GK

GS

Barometer was invented by Benjamin Franklin.

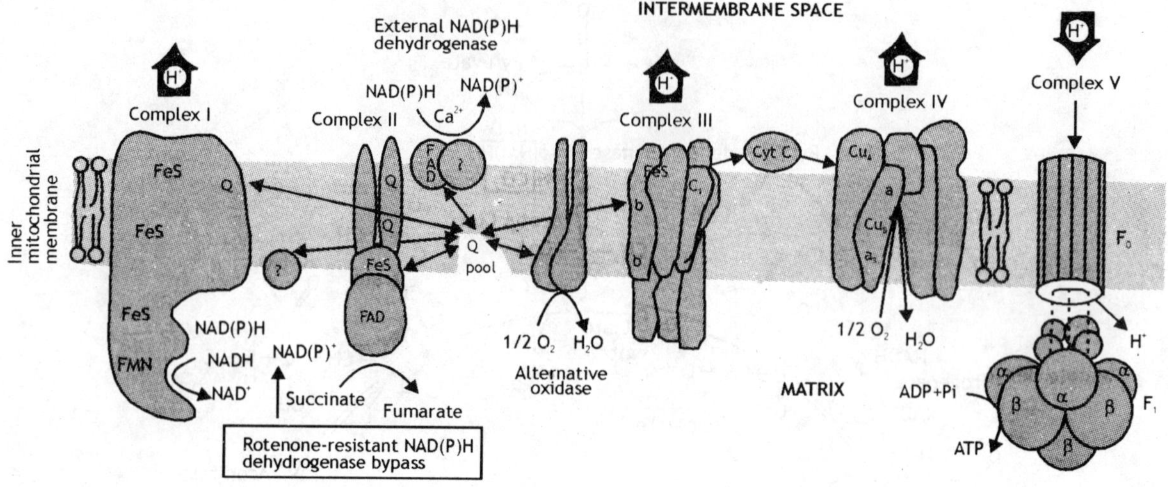

Electron Transport System

- Complex IV: Represents the cytochrom C oxidase, which contain two copper centre (CUA and CUB) and Cyt a.
- Electron carriers within the mitochondrial inner-membrane allows for transfer of proton. The electron transport is associated with a transfer of protons from the mitochondrial matrix to the inner membrane space.
- Complex V: This complex consist of two major components

F_1 : Peripheral membrane protein complex contain at least five different subunits and contains a catalytic centre for converting ADP + Pi to ATP.

F_2 : Integral membrane protein complex that consists of at least three different polypeptides that form the channel through which protons are able to cross the inner membrane.

Uncouplers :e.g. Dinitrophenol. PMA break the membrane leads to free flow H+ atoms.

So, no energy production.

Factors affecting Respiration

External Factors –

1. *Oxygen*

- O_2 is a important factor because of its role as substrate in the overall process.
- The diffusive movement of O_2 represents a limitation on plant respiration.

2. *Temperature*

- In the physiological temperature range, the respiration is temperature dependent.

3. *Injury*

- Physical injury to plant tissues often stimulate O_2 uptake because of increases in respiration and in mitochondria oxygen consuming enzyme activity as catalysed by lipoxygenase, polyphenol oxidase and peroxidase.

4. *Water*

- Proper hydration of respirating cell is essential for respiration.

5. *Mechanical Effects*

- Gentle rubbing or bending of leaf blade shows increased rate of respiration.

6. *Light*

- Indirect effect through synthesis of organic matter.

Internal Factors –

1. *Protoplasmic Factor*

- The amount of protoplasm in the cells and its state of activity influence rate of respiration.
- High rate of young meristamatic cells which divide actively require more energy.
- Rate of respiration is undoubtedly affected by internal structural relationship of various cell organelles and occurance and distribution of enzymes.

2. *Concentration of Respiratory Factor or Substrate*

- Favourable and increased concentrate respirable food material brings about an increase in the rate of respiration. Under starvation condition, leaves are etiolated and slow downs the respiration.

Bhabar is the name given to a piedmont fan-shaped plain.

GROWTH AND DEVELOPMENT

Carbohydrate Metabolism

Carbohydrates are the storage materials that are stored in plants which upon its breakdown yield energy required for the vital processes in plants. The prominent carbohydrates in plants are, sucrose and starch.

Synthesis and Breakdown of Sucrose

Synthesis: It may take place by three different ways :

1. From *Glucose-1 phosphate* and fructose in presence of enzyme *sucrose phosphorylase*. (eg.) in bacteria.

$$\text{Glucose-1 phosphate + Fructose} \underset{\text{Phosphorylase}}{\overset{\text{Sucrose}}{\rightleftharpoons}} \text{Sucrose + inorganic phosphate.}$$

2. From UDPG (*Uridine Diphosphate Glucose*) and *Fructose* in the presence of *sucrose synthetase*. (eg.) in higher plants.

$$\text{UDPG + Fructose} \underset{\text{Synthetase}}{\overset{\text{Sucrose}}{\rightleftharpoons}} \text{UDP + Sucrose}$$

3. From UDPG and *Fructose-6-phosphate* in the presence of the enzyme *sucrose phosphate synthetase* (eg) Higher Plants.

$$\text{UDPG + Fructose} \underset{\text{Synthetase}}{\overset{\text{Sucrose Phosphate}}{\rightleftharpoons}} \text{UDP + sucrose phosphate.}$$

Sucrose phosphate thus produced is *hydrolysed* in the presence of the enzyme *phosphatase* to yield sucrose.

$$\text{Sucrose Phosphatase} \xrightarrow[+ H_2O]{\text{Phosphatase}} \text{Sucrose + Phosphate}$$

Breakdown: Sucrose is broken down or hydrolysed to yield glucose and fructose in the presence of enzyme invertase which is a reversible reaction.

$$\text{Sucrose} + H_2O \xrightarrow{\text{Invertase}} \text{Glucose + Fructose}$$

Synthesis and Breakdown of Starch

Synthesis of Amylose or α-1, 4 Glycosidic Linkages.

As the starch is made up of *Amylose* and *Amylopectin*, both the components are synthesised simultaneously with α-*1, 4 Glycosidic linkages* and α-1, 6 linkages respectively.

Synthesis of amylose may take place by the following ways:

1. Amylose can be synthesised in the presence of the enzyme starch *phosphorylase* from *glucose-1-phosphate* and an acceptor molecule consists of 3-20 glucose units joined together by α-1, *4 glycosidic linkages*.

You Don't Know!

Largest seed pod

The climber *Entada phaseoloides,* found in the E. Himalayas and tropical Asia, bears pods that are up to 120 cm (4 ft) long and 10 cm broad, containing upto 30 seeds.

$$\text{n(Glucose-1-phosphate) + Acceptor} \underset{\text{Phosphorylase}}{\overset{\text{Starch}}{\rightleftharpoons}} \text{Amylose + n (ip)}$$

2. Formation of α-*1, 4 glycosidic linkages* may also take place in the presence of the enzyme UDPG-transglycosylase (*amylose synthetase)* by the transfer of *glucose* from UDPG to an *acceptor* molecule with 2-4 or more *glucose* -units joined together to the α-l, 4 glycosidic linkages or even a starch molecule.

$$\text{UDPG + Acceptor} \underset{}{\overset{\text{UDPG trans glycosylase}}{\rightleftharpoons}} \text{UDP} + \alpha\text{-}1, 4 \text{ Glucose acceptor.}$$

Synthesis of AmyloPectin or α- 1, 6 Glycosidic Linkages

Transfer of small chains of glucose units joined together by *α.-1, 4* linkages to an acceptor molecule consisting of at least four *α.-1, 4* linked glucose units in the presence of 'Q' enzyme. The *α.-1, 6* Glycosidic bond established between C-1 of the terminal glucose units of donor molecule and C-6 of one of the glucose units of the acceptor molecule.

Breakdown by Means of Two Ways

1. By the enzyme Diastase

$$\text{Starch} \xrightarrow[+ H_2O]{\text{Diastase}} \text{Glucose}$$

Diastase is a complex of many enzymes viz., α-*amylase*, *β-amylase*, *R-enzyme* and *maltase*.

2. By starch phosphorylase Starch

$$\text{Starch + Phosphate} \underset{}{\overset{\text{Starch Phosphorylase}}{\rightleftharpoons}} \text{Glucose-1-phosphate.}$$

$$\text{Glucose-1-phosphate} \xrightarrow{\text{Phosphate}} \text{Glucose + Phosphate}$$

GK

Vijarayana flourished under Pala dynasty.

GS

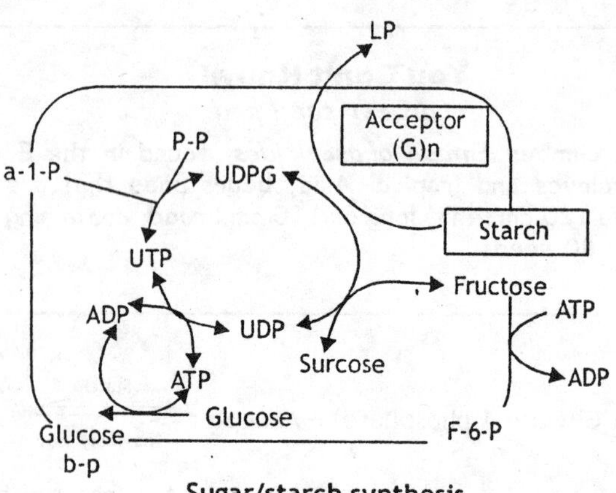

Sugar/starch synthesis

Fat Metabolism

Lipids may be simple, compound or derived. Simple lipids are esters of fatty acids alcohol. If alcohol in the constitution is glycerol they are called fats.

Fat metabolism consists of two parts:

(a) Anabolism :

(a) Synthesis of fatty acid.

(b) Synthesis of glycerol.

(c) Condensation of fatty acids and glycerol into fats.

$$\text{Glycerol} + 3 \text{ Fatty acid CoA} \xrightarrow[\text{Mg}^{++}]{\text{ATP}} \text{lipid} + 3\text{CoA}$$

(b) Catabolism :

3 Pathway

(a) β-oxidation

(b) α-oxidation

(c) Peroxidation

β-oxidation is the *chief process* of fatty acid degradation in plants which is *well established* in *saturated fatty acids* not in unsaturated fatty acids. The place of occurrence of this reaction is mitochondria and also in glyoxysomes which involves sequential removal of 2-C in the form of acetyl-CoA from the carboxyl and of the fatty acid. This is called *β-oxidation*, because β-c of the fatty acid is oxidised during this process.

Reactions Involved

Fatty acid

Thio kinase ⇅ ATP, CoAStd

Fatty acyl - CoA

Acyl-CoA Dehydrogenase ⇅ FAD → $FADH_2$

Trans-α, β unsaturated fatty acyl-CoA

Enoyl Hydrase ⇅ + H_2O

β - hydroxy acyl - CoA

β - Hydroxyacyl-CoA dehydrogenese ⇅ NAD^+ → $NADH + H^+$

β - keto fatty acyl - CoA

β - keto acyl thiolase ⇅ + CoASH

Fatty acyl-CoA + Acetyl CoA

Here in this process, huge amount of energy is generated in the form of ATP molecules by the mitochandrial oxidation of fatty acids through β-oxidation spiral and TCA cycle. Acetyl-CoA may enter into TCA cycle or in case of germination of fatty seeds, they are converted to soluble sucrose through glyoxylic acid cycle.

Glyoxylate Cycle (or) Glyoxylic acid Cycle

In this cycle, the fats could be converted into sucrose during the germination of fatty seeds in plants. This cycle occurs in many other bacteria, yeasts, moulds, and higher plants or is completed in glyoxysomes, mitochondria and cytosol.

Significance

- During germination of fatty seeds, unsoluble fats are converted into soluble sucrose, which is then available to the different growing regions by Glyoxylate cycle.
- Micro organisms that can grow in C_2H_5OH/ acetate as a sole source use this cycle to produce longer carbon chains.
- It's an example of Gluconeogeneris process.

Nitrogen Metabolism and Nitrogen Fixation

Role of Nitrogen in Plants

- Proteins are the building blocks of protoplasm which are made up of nitrogenous substances called amino acids, which are synthesised when inorganic 'N' in the environment is converted to organic 'N' inside the plants.
- Proteins are the major substances in cell next only to water and mineral salts.
- 'N' is the constituent element of important organic compounds like chlorophylls, cytochromes, alkaloids, vitamins and nucleic acids.
- Thus 'N' plays an important and fundamental role in metabolism, growth, reproduction and heredity.

Sources of 'N'

1. Atmospheric/Molecular Nitrogen: Only some bacteria, some BGA, leguminous plants having root nodules can fix atmospheric 'N'.

2. Nitrates, Nitrites, Ammonia in soil (In organic 'N') : Nitrates is the chief form of 'N' taken up by the plants in the soil.

3. Amino acids (Organic Nitrogen) in Soil.

4. Ammonium Compounds of which NH_4^+ ions are absorbed directly by the roots of plants along with nitrate ions.

5. Organic Nitrogenous Compounds in the bodies of Insects.

'N' fixation in Plants: It can be achieved by the conversion of Nitrate in the soil to Ammonia by the plants which occurs as a two-step process.

(i) Reduction of Nitrate to Nitrite

(ii) Reduction of Nitrite to Ammonia

The process of conversion is basically a reductive process, as the Nitrate (NO^-_3) which is a highly oxidised form gets converted into the reduced form *i.e.* Ammonia (NH_4) under the two major steps, there are many intermediate steps involved in the conversion, which are mediated by specific enzymes.

$$\underset{\text{Nitrate}}{NO^-_3} \xrightarrow{2e^-} \underset{\text{Nitrite}}{NO^-_2} \xrightarrow{2e^-} \underset{\text{Hypo Nitrite}}{H_2N_2O_2} \xrightarrow{2e^-} \underset{\text{Hydroxyl Amine}}{NH_2OH} \xrightarrow{2e^-} \underset{\text{Amonia}}{NH_3}$$

At each step R electrons are added and ultimately NO_3 in which 'N' has 5 positive charges converted to Ammonia (NH_3) with 3 negative charges. The electrons are supplied by NADH and NADPH.

1. *Reduction of Nitrate to Nitrite:*

Takes place in green leaves and roots and the enzyme is found in cytoplasm.

$$NO_3 + NADH + H^+ \xrightarrow[\text{reductase}]{\text{Nitrate}} NO_2 + NAD+ + H_2O$$

The enzyme is the molybdoflavoprotein which contains FAD (Flavin Adenine Dinucleotide) as its Prosthetic Group associated with molybdenium.

NADH₂ or NADPH₂ | FAD (Oxidation) | MO (Reduction) | NO_3^-

NAD (or) NADP | FAD H_2 (Reduction) | Mo (Oxidation) | $NO_2^- + H_2O$

Nitrogen reduction

Electrons are transferred from reduced conenzyme to FAD which becomes reduced $FADH_2$. From $FADH_2$, electrons are finally tranferred to NO_3^- through molybdenum so that NO^-_2 and H_2O are formed.

II. *Reduction of Nitrite to Ammonia:*

$$NO^-_2 \xrightarrow[OH^-]{2e^- + 2H^+} NOH \xrightarrow{2e^- + 2H^+} NH_2OH \xrightarrow[H_2O]{2e^- + 2H^+} NH_4^+$$

The reduction takes place in the presence of Nitrite Reductase which catalyses by-reduction from NO_2^- to NH^+_4 with the formation of Nitroxyl/Hypo Nitrite and Hydroxyl amine as intermediates.

Electron donors for this reduction are, reduced ferredoxin and also reduced pyridine nucleotides.

As against the presence of Nitrate reductase in cytoplasm, this enzyme is found in chloroplasts. Sirohaem, an iron-porphyrin is known to be associated with this enzyme and probably mediates the transfer.

The involvement of two intermediate compounds in the process is doubtful because.

1. Hyponitrite is unstable

2. Hydroxyl amine is toxic

3. These two, they never occur in free state in the cells. It's now generally believed that hyponitrite and hydroxyalamine may be formed at the surface of the enzyme and leave the surface only when they are completely reduced to further intermediate or ammonia.

There is yet another type of formation of Ammonia, through the Reduction of hydroxylamine.

The reaction is catalysed by hydroxylamine reductase which requires manganese for its activity.

$$NH_2OH + NADH + H^+ \xrightarrow{Mn} NH_3 + NAD + H_2O$$

The ammonia combines with organic acids to produce amino acids.

Biological Nitrogen Fixation

It can be defined as the conversion of molecular nitrogen of the atmosphere into inorganic nitrogenous compounds through the activity of some living organic acid.

The process of Biological Nitrogen Fixation can be divided into two types.

Non Symbiotic Nitrogen Fixation

In case of non-symbiotic fixation, some of the components are necessary in the cell system of 'N' fixing bacteria.

1. Presence of hydrogenase and nitrogenase enzyme systems. Hydrogenase reversibly catalyses the reduction of hydrogen ion to molecular hydrogen while, nitrogenase reduces molecular nitrogen to NH_3 in the presence of molecular hydrogen.

2. Presence of Ferredoxin as an electron carrier.

3. Presence of pyruvate which acts as e^- donor and energy

Knowledge World!

Water fort

The Vellore fort with its crocodile-infested moat and blue granite bastions housed the family and retinue of Tipu Sultan after the British conquered Seringapatnam in 1799.

GK

GS

Article 44 directs the state to make efforts to secure a uniform civil code throughout the territory of India.

source.

4. Presence of co-factors like TPP, CoA, ip and Mg^{2+} required in the metabolic utilization of pyruvate, that forms the energy (ATP).

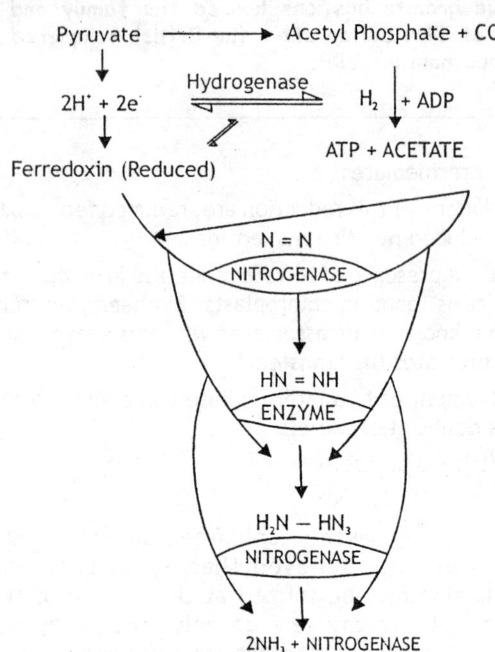

NH_3 formation

Symbiotic Nitrogen Fixation

It is carried out by bacteria which is frequently found in the root nodules of leguminous plants. The most common bacterium is Rhizobium which has two types of strains.

1. Fast grower: e.g. *R. leguminosarum/Bacillus radicicola* and *R. trifolii*.
2. Slow growers : e.g. *R. japanicum* and *R. lupinii*.

Nodule Formation

It is the principal event in this process, which is a complicated process. In the initial stage, host cells have to recognise the correct strain of symbiotic bacteria where a critical specificity is involved in the association of various species of root nodule bacteria and leguminous plants and this step is controlled by specific protein found in the host cell called lectin. Lectins are the key proteins involved in the Rhizobium-legme symbiotic association and in the process calcium plays a critical role in modifying the ability of root to absorb rhizobia.

Rhizobia accumulate in the soil near the roots of the legumes probably due to the secretion of some growth factors by their roots. The bacteria now penetrate the root through soft or injured root hairs whose tips become curved.

They now enter the cells of the inner layers of the cortex through infection threads and cause cortical cells to multiply which ultimately result in the formation of nodules on the surface of the roots. The cells of the bacteria enlarge, assume pleomorphic shape and are called bacteriods.

The nodule contains a pink coloured leghaemoglobin, like true haemoglobin combine with O_2 and CO_2 gets readily oxidised to brown form with trivalent ion. The development of leghaemoglobin and Nitrogen fixing capacity are outcome of symbiotic interaction. The leghaemoglobin is located in the space between bacteriods and membrane enclosing them or its present outside the membrane and in the host cytoplasm.

Nitrogenase

The process of 'N' fixation, whether its symbiotic or non-symbiotic involves the reduction of atmospheric 'N' to ammonia (NH_3) by the enzyme Nitrogenease. The enzyme is made up of two protein components one containing iron and molybdenum and the other containing only iron called Fe-protein or azoferredoxin. The enzyme is extremely sensitive to oxygen. The development of leghaemoglobin pigment in nodule cells and heterocysts in BGA for protecting nitrogenase from oxygen inactivation. Thus the enzyme remains active under anaerobic conditions. The pigment leghaeomoglobin regulates the course of oxygen and provides suitable atmosphere to the enzyme. It combines rapidly with oxygen to avoid any inhibitory effect of it on enzyme, at the same time it's able to make oxygen available to bacteriods for ATP production, required for 'N' fixation. Thus the process of 'N' fixation requires anaerobic conditions even though Rhizobium is aerobic.

Mechanism of Nitrogen Fixation

It's to be noted here, during the process of 'N' fixation, the free atmospheric nitrogen first bound to the enzyme surface and is not released until it's completely reduced to ammonia. 'N' bound to the enzyme surface is reduced in step-wise reactions before N-N bond is ruptured. Pathway of e-transport in nodule bacteriods.

Enzymes involved –

1. Glucose-6-PO_4 dehydrogenase
2. NADPH-Ferredoxin oxido reductase
3. 'Fe' protein compound of Nitrogenase.
4. Fe-Mo protein compound of Nitrogenase

The reduction of 'N' into NH_3 by nitrogenase in bacteroids depend upon the availability of ATP and reduced substrate capable of donating 'H^+ atoms to 'N' Glucose-6-PO_4 is the reduced substrate for the process and reduced NADP together with ferrodoxin function as electron carriers.

ATP interact with non-haem iron (NHI) protein component

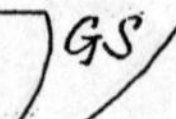

Maulvi Ahmadullah was the leader of the 1857 Revolt at Faizabad.

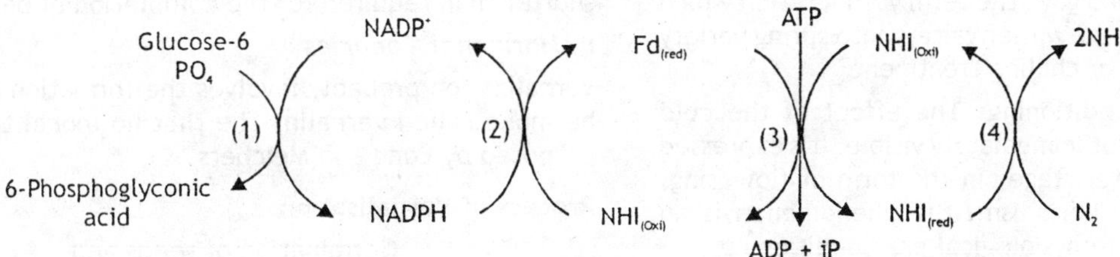

Nitrogen fixation

of Nitrogenase and brings about confirmational change to convert it to a powerful reductant, which becomes capable of transferring electrons to reduce N_2 and NH_3.

Thus the reduction of one molecule of 'N' into two molecules of ammonia requires twelve molecules of ATP because six electrons are required per molecule of 'N' reduced.

$N_2 + 6H^+ + 6e^- + 12ATP \longrightarrow 2NH3 + 12\ ADP + 2\ Pi$

Photoperiodism

The plants in order to flower require a certain day length (*i.e.*) relative day length and night is called photoperiod. The response of plants to the photoperiod expressed in the form of flowering is called photoperiodism.

The observation of Garner and Allard (1920) that a mutant variety of tobacco, Maryland mammoth flowers when the relative length of the day was shorter than the length of the dark period, which led to the discovery of photoperiodism. Depending upon the duration of photo period, they classified plants into three categories.

I. Short Day Plants (SDP):

- Qualitative (flowering occurs only in short days).
- Quantitative (Flowering accelerated by short days).
- The plants which require a relatively short daylight period (8-10 hrs) and a continuous dark period about 14-16 hrs for subsequent flowering.
- Examples are, Maryland mammoth variety of tobacco, rice, datura, onion, etc.

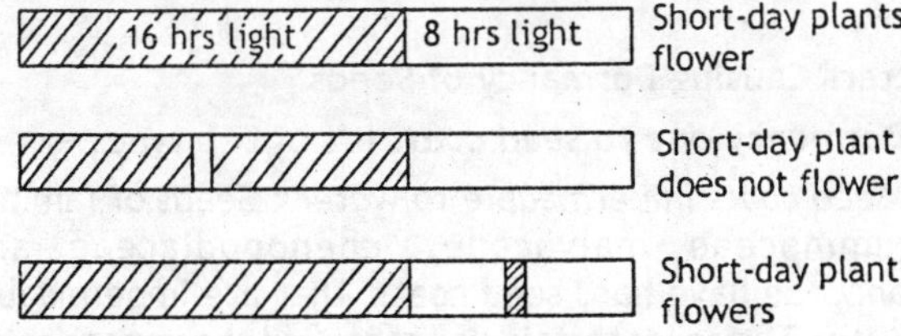

Effect of night on short-day plants

- In short day plants, dark period is critical and must be continuous. If this dark period is interrupted even with a brief exposure of red light, short-day plant will not flower .
- Maximum inhibition of flowering with red light occurs at about the middle of the critical dark period.
- The inhibitory effect of red light can be overcome by a subsequent exposure with far-red light (730-735 nm).
- Interruption of the light period with red light does not have inhibitory effect on flowering in short day plant.
- Prolongation of the continuous dark period initiates early flowering in short-day plants.

II. Long Day Plants (LDP):

- Qualitative -flowers only in long days.
- Quantitative -Flowering accelerated by long days.
- These plants require a longer daylight for a period of 14-16 hrs. in a 24 hrs cycle for subsequent flowering.
- Examples are spinach, lettuce, radish alfalfa, sugar beet, opium, poppy, oats and wheat.
- In the long day plants, the light period is critical.
- The best flowering of long day plant usually occurs in continuous light.
- For flowering they require either no dark period or a very short dark period.
- A flash of light in long dark periods can induce flowering in them even during short-day periods.
- Long day plants can flower even in short-day periods, if these short-day periods are accompanied with still shorter dark periods.
- Too long dark periods inhibit flowering and so these are called short-night plants.
- A brief exposure in the dark period the prolongation of the light period stimulates flowering in long day plants.

III. Day Neutral Plants (Ambiphotoperiodic species):

- The plants that flower in all photo periods ranging from 5 hrs to 24 hrs continuous exposure.
- Examples are tomato, cotton, cucumber, pea, sunflower, maize and dandelion etc.
- Their flowering is not affected by the length of the day and they can flower even if the light period provided ranges from few hrs to continuous illumination.

Vernalisation

It can be defined as the method of inducing early flowering in plants by pretreatment of their seeds at very low temperatures. Lysonko gave the term vernalisation which means conversion of the winter variety into spring variety by low temperature or chilling treatment.

Physiological Pre-conditioning: The effect of the cold stimulus on plant is not immediately visible. It's expressed only at a certain later stage in the form of flowering. Thus like the photo periodism, the phenomenon is an excellent example of physiological pre-conditioning.

Perception of the Cold Stimulus

The cold stimulus is perceived by the apical meristerms. And also all dividing cells including those in roots or leaves may be the potential sites of vernalization.

The perception of the cold stimulus results in the formation of a floral hormone, which is transmitted to other parts of the plant. It can also be possible, the cold stimulus may even be transmitted to another plant across a graft union and the floral hormone has been named as vernal by Melchers.

Conditions Necessary for Vernalization

1. Age of the Plant: In cereals, effective only at germinating seed.

2. Appropriate low temperature and duration of exposure: 1 to 6°C, one and a half months or more.

3. Oxygen: Vernalisation an aerobic process and requires metabolic energy provided by oxygen.

4. Water: In dry seeds, vernalisation is not possible.

Mechanism of Vernalization

Two theories are prevailing to deal about the mechanism of vernalization.

I. Phasic Development Theory

According to this theory, the process of development of an annual seed plant consists of a series of phases which must occur in some predetermined sequence. An important condition is, commencement of any of these phases will take place only when the preceding phase has been completed.

Thermo Stage: First stage conditions required are –

1. Temperature (0-20°C)
2. Moisture
3. Aeration
4. Varied time

Photo Stage: When thermo stage is completed, the plant passes through the next stage. There is pronounced effect of the relative length of day and night on the production of flowers, and this phenomenon is called photoperiodism.

Third Stage: It's quite necessary for seed germination as it has been connected with formation of sexual elements and is associated with garmetogeneris. The photoperiod requirement for the commencement of this stage is little shorter than required for the completion of photo stage.

II. Hormonal Theories

Vernalization probably involves the formation of a floral hormone called vernalin. The first hormonal theory was proposed by Long and Melchers.

Process of Vernalisation

Germination of seeds and
cold treatment at 0-5°C

↓

Drying and after that
sowing of seeds

After the cold treatment, seedlings are allowed to dry for some time and then down. The drying period also should not be so long as the vernalize response decreases and it continues till seedlings are completely devernalised.

Significance

1. Shortens the vegetative period of the plants.
2. Increases the cold resistance of the plants.
3. Increases resistance of plant to Fungal diseases.
4. Crops can be produced earlier *i.e.* crop can be harvested much earlier than the control crop.
5. Crops can be grown in the regions where they do not naturally reproduce.
6. Plant breeding work can be accelerated.
7. In colder countries like Russia, the harmful effects of the cold can be overcome and thus improving crop production.

Dormancy of Seeds

It may be defined as any phase in the lifecycle of plant in which active growth is temporarily suspended.

All the viable seeds have capacity to germinate if placed under suitable conditions necessary for germination. When seeds fail to germinate for some time even if placed under suitable conditions, that are ordinarily favourable for germination. During this period the growth of the seeds, remain suspended and they are said to be in dorminant stage.

Factors Causing Dormancy of Seeds

1. ***Dormancy due to seed coat***: It's of two types –

(a) Seed coats impermeable to water : Seeds of families leguminoceae, malvaceae, chenopodiaceae, and solanaceae have hard seed coats, that are impermeable to water. The seeds remain dormant until the impermeable layer of testas decay by soil micro organisms.

(b) Seed coats impermeable to oxygen : In cocklebur (Xanthium), many grasses, compositae, the seeds coat is impermeable to oxygen. During the period of dormancy of seed coats gradually become more permeable to oxygen so that they may germinate afterwards.

2. *Dormancy due to immature embryos* : Some seeds are shed, before the embryo is matured and they will not grow, which require an "after ripening period" during which certain changes occur within the seed. The changes include acidity, enzyme activity and respiratory rate etc. and once the embryo development is complete, the seed then germinates without any special treatment.

3. *Mechanically resistant seed coats* : Seeds of certain weeds such as pigweed (Amaranthus), Shepherd's Purse (capsella), water plantain (Alismas) etc. remain dormant because their hard seed coat prevent any appreciable expansion of the embryo.

4. ***Dormancy due to chemical inhibitors*** : Seeds present in the juicy fruits such as orange, tomato do not germinate immediately. Para-ascorbic acid, Phenolic acids, Dehydro acetic acid in the embryo, endosperm or other tissues of the seed fruit is also the cause of seed dormancy.

5. ***Need for after-ripening in dry storage*** : In many plants e.g. Barley, oats, wheat, seeds though containing fully developed embryos are dormant, when they are harvested they require no special, treatment to overcome this dormancy and germinate if kept under dry storage conditions at normal temperature for few weeks to several months. This type of dormancy can be overcome by removing the seed coats or temperature treatments. (35-40°C storage).

6. ***Chilling or low temperature treatment***: Seeds of temperature species show dormancy which is overcome by chilling e.g. Apple, Rose, Peach, Grapes, etc., which will not germinate, if planted under moist conditions at 20°C, but will germinate if kept at 0-5°C for several weeks and then transferred to warmer conditions.

7. ***Dormancy due to light sensitivity***: The germination of many seeds affected by light. Such seeds are called *photoblastic*. The seeds in which germination is stimulated by light are called positively photoblastic whereas those in which germination is inhibited by light are negatively photoblastic.

Positive Photoblastic: Amaranthus, Shepherd's Purse, tobacco, and tomato etc.

Negative Photoblastic : *Helleborus niger, Nigella damascena* and *Silene arnemia*.

Mechanism to Seed Dormancy

I. Regulation of dormancy is primarily hormonal.

II. Importance of interference with gaseous exchange by seed coats and other structure.

The hormonal theory visualises that the prevention of growth may be either (1) because of *deficiency* of some essential growth promoting substance (endogenous gibberelins). (2) because of the presence of some inhibitory substances. (dormins/abscisins).

Quiz Kit!

Longest coastline

Gujarat has 1,600 km (26 per cent) on the 6,100 km long coastline of peninsular India.

Methods of Breaking Dormancy

1. *Scarification*: The process of rupturing or weakening the seed coats by mechanical or other means and is employed, where the dormancy is due to impermeable seed coat. Scarification can be done either mechanically or chemically.

2. *Pressures*: When the dormant seeds are subjected to 2000 atm at 18°C for about 5-20 mins. and this works in cases, where the seed coats are impermeable to water.

3. *Low Temperature*: When the dormancy is due to chilling requirement, it can be overcome, if the seeds are treated in moist medium at low temperature (5-10°C) for sufficient period of time, which is called stratification.

4. *Alternating Temperatures*: An alternation of low and high temperatures improves the germination of seeds.

5. *Light Dormancy of Positively Photoblastic*: Seeds can be broken by exposing them to red light or white light. This promotion of germination by red light and inhibition by far-red light probably involves phytochrome.

6. *Germination Stimulating Compounds* : Such as gibberellins induce germination in tobacco, lettuce etc., and chemicals such as KNO_3, thiourea and ethylene also involved.

Advantages of Seed Dormancy

1. In temperate zones, to overcome the severe colds.

2. In tropical zones, seed dormancy due to hard seed coat improves the chances of survival.

3. Dormancy of cereal seeds, increase the availability of them as food and otherwise, they may not be used as food grains.

Seed Germination

The process of seed germination starts with the imbibition of water by seed coats and emergence of growing root tip of embryo. It ends when the embryo has developed into a seedling which is out of bounds of seed coats and has its own photosynthetic system.

Conditions for Germination

(i) Water (ii) Aeration (iii) Temperature (iv) Light.

(i) Water: It is absorbed initially by imbibition followed by absorption.

- Period required for soaking of seeds in water varies depending upon the nature of seed coat.

The Small Industries Development Bank of India (SIDBI) was established on April 2, 1990.

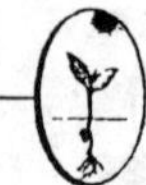

- Seeds with impermeable seed coats are to be soaked longer than those with permeable seed coats.
- Prolonged soaking affects, as the metabolites leached out which are essential for germination e.g. Barley when soaked for 2-3 days.

(ii) Aeration: Oxygen requirement of different seeds vary considerably, but most seeds germinate at a lower oxygen level.

- High requirement of oxygen may be due to the presence of permeabilty barrier i.e. seed coat.
- Most seeds require active aerobic oxidation system for germination.

(iii) Temperature : The seeds of temperate plants require relatively low temperature as compared to tropical plants.

(iv) Light: Photoblastic seeds.

Physiological, Biochemical Changes During Germination

Water Uptake: Seed coat imbibes water through seed coat, which is purely a physical process various hydrophilic groups such as $-NH_2$, -OH, -COOH Proteins and CHO attract dipolar water molecules and swelling occurs. Due to imbibition of water, seed coats become (i) more permeable to oxygen and water and (ii) less resistant to outward growth of the embryo.

Respiration: Uptake is accompanied by rapid increase in respiration rate of embryo. Initially its anaerobic, which is soon replaced by aerobic due to the availability of oxygen.

Mobilization of Reserve Materials: Occurs to provide:

(i) Building blocks for the development of embryo.

(ii) Energy for the biosynthetic processes.

(iii) Nucleic acids for control of protein synthesis and overall embryonic development changes in these components are as follows.

Nucleic Acids: In monocots, during imbibition stage, there is a rapid increase of DNA and RNA contant in the endosperm. Appreciable amount of RNA appears in the aleurone layer after 16 hrs. Due to more cell division the DNA content increased.

Carbohydrates: Insoluble ones like starch are important reserve materials of cereals. During germination starch is hydrolysed first into maltose in the presence of α-amylase and β-amylase and then maltose is converted into glucose by maltase. Glucose is absorbed by scutellum, converted into soluble sucrose and transported to the growing axis.

Gibberellin is secreted during germination by embryonic axis into aleurone layer, causing: denovo synthesis of α-amylase. This enzyme is not found in ungerminated seeds. GA induced synthesis of α-amylase is countered by ABA.

In contrast to α-amylase, β-amylase is already present in the seed in inactive form, which gets activated during germination. The activity of maltase is also regulated by GA.

Lipids: During germination, mobilization of lipids is brought about by hydrolysis of fats to fatty acids and glycerol by lipases and β-oxidation of fatty acids to acetyl-CoA. Some of the acetyl CoA is converted into sucrose via glyoxylate cycle and is transported to the growing embryonic axis.

Proteins: Some plants store proteins as reserve food in their seeds in the form of aleurone grains. Mobilization of these proteins involves the hydrolytic cleavage into amino acids by peptidases.

Inorganic Materials: Number of inorganic materials such as Phosphate, Ca, Mg, and Potassium are also stored in seeds in the form of Phytin.

Enzyme

Enzyme has been called the "agents of life" since they control almost all life process. It is catalysts of life.

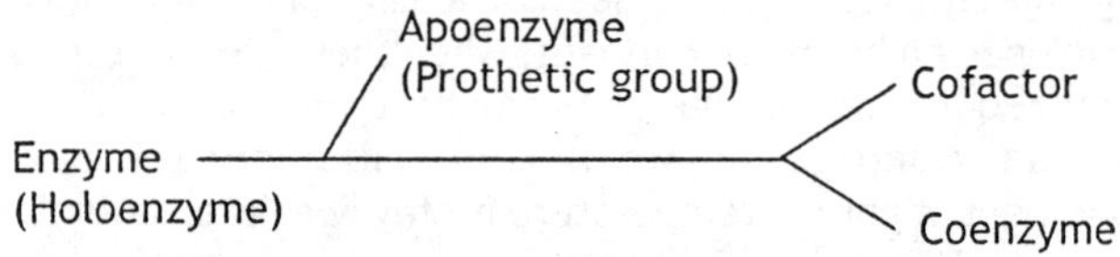

Cofactors : (Inorganic in Nature)

1. Cu -Cytochrome oxidase
2. Fe -Cytochrome oxidase, catalase, peroxidase
3. K-Pyruvate kinase
4. Mg-Hexokinase, Pyruvate kinase, Glucose-6-phosphate
5. Mn- Arginase, Ribonuclease reductase.
6. Mo -Dinitrogenase
7. Ni -Urease
8. Se -Glutathione Peroxidase
9. Zn -Carboxianhydrase, Alcohol dehydrogenase.

Coenzyme: (Organic Group Transfer)

1. Biotin	CO_2
2. Coenzyme A	Acyl
3. Deoxy adenosolyl	H^+
4. FAD	e^-
5. NAD	H^+
6. Pyridoxy phosphate	amino
7. TPP (Thymine pyrophosphate)	

Basic equation:

For enzyme kinetic $V_0 = \frac{V_{max}(s)}{km+[s]}$

- Small value of km -maximal catalytic efficiency at low substrate concentration.

The attraction between similar molecules is called cohesion.

Enzyme substrate Interaction :

-Weak interaction between enzyme and substrate

Turnover Number :

It is equivalent to the number of substrate molecules converted to product is a given unit of time on a single enzyme molecule when the enzyme in saturated with substrats.

Characters of enzyme

1. Colloidal in nature
2. React with both acidic and alkaline substance
3. Show sensitivity
4. Inhibited
5. Activity can be accelerated
6. Size is greater than substrate
7. Specific in nature
8. pH sensitive

Effect of inhibitors on Michalis-Menton reaction

Inhibitors	Equation	Curve	Effect
1. Normal	$V_0 = \frac{V_{max}(s)}{km + [s]}$		
2. Competitive (action active site)	$V_0 = \frac{V_{max}(s)}{\alpha\, km + [s]}$	Interception $1/V_0$ axis at $1/V_{max}$	↑ km
3. Un-competitive (conformation in enzyme)	$V_0 = \frac{V_{max}(c)}{km + \alpha'(s)}$	Identical shape	km-un-affected V_{max}↓
4. Mixed (both)	$V_0 = \frac{V_{max}(s)}{\alpha\, km + \alpha'(s)}$	Lift of $1/V_0$ axis	V_{max} ↑ km ↑

Mode of enzyme action

Michaelis and Menton (1913)

- Enzyme have active sites for the attachment of substrate molecule whereas enzyme can form an intimate rehabitable with the substrate.

Enzyme + substrate ⇌ enzyme substrate complete
⇌ End products + Enzyme

Factors affecting enzyme activity

1. Temperature
2. pH
3. Anaesthetics
4. Substate + Enzyme concentration.

PLANT PHYSIOLOGY IN AGRICULTURE

1. Role of Plant Physiology in the Improvement of Varieties

Plant breeders use the method of selection for crop improvement. Selecting on the basis of only external characters is not fruitful. Therefore, its protein content, oil content, carbohydrate content and various other substances of nutritional importance must be tested before proper selection. Plant physiology thus plays an essential role in specifying the features for which the plant breeder must select.

Did You Know?

First air crash

On July 17, 1950, a DC-3 crashed near Pathankot, Punjab, killing all 22 passengers on board.

2. Role in Soil Fertility

Our knowledge of mineral nutrition of different crop plants provided extensive revisions in the application of fertilizers. It is now possible to recommend specific organic fertilizers or inorganic salts using knowledge of plant physiology.

3. Role of Auxins

(a) Auxins are used to prevent premature fruit drop, when applied close to the time of harvest.

(b) Auxins (particularly IBA) induce formation of roots on cuttings of many ornamental, horticultural and forest trees.

(c) Many auxins are used to promote sprouting, induce flowering and enhance yield in potatoes, peas, beans, etc.

4. Role of Gibberellins

(a) Gibberellins are used to break dormancy and induce uniform crop emergence, e.g., seed potatoes.

(b) It is used to enhance the growth of vegetative parts of several floricultural species and used to increase flower size.

(c) Gibberellins are also used to produce parthenocarpic (seedless) fruits.

(d) Cytokinins have less importance in agriculture. It is used in extending self-life of lettuce, tomatoes, apples, etc. It is also used to induce efficient root system in maize and rice.

5. Role of Ethylene

(a) Ethylene is a hormone which is well known to cause fruit ripening.

(b) Ethylene causes abscission.

(c) It is used to induce bulbing in onion.

(d) Ethylene increases tillering in rye, peas, etc.

6. Role of Inhibitors

(a) A well known inhibitor CCC (2-chloroethyl-trimethyl

ammonium chloride) is used to induce stiffening of plants, increase number of roots and litters, prevent lodging and increase yield in wheat, oat and several strains of Secale (rye).

(b) CCC is also used to increase bolting, flowering and increase yield in cotton.

(c) Many inhibitors, such as, CCC, Phosphon, B-9, etc. are used in ornamental horiculture to retard the height of plants like chrysanthemum.

7. Role in Seed Production

Knowledge of plant physiology is used in production of high quality, viable and pure seeds from weed seeds.

8. Role in Food Technology

Plant physiology provides many techniques required for food preservation, food processing, such as drying, pickling and food storage.

Plant Growth Regulators

Contents	Auxin	Abscissic acid (ABA)
1. Definition	A compound that has a spectrum of biological activity similar to but not identical with IAA.	-
2. History	Kogl-named as Auxin.	Weering : • First identified in *Acer pseudoplatanus* (Abscisin-II) • From mature cotton fruit (Abscissin-I)
3. Site of synthesis	Shoot, Root tip	Matured leaves
4. Transport	Polar, Active	Bipolar, Mainly through phloem
5. Mode of action at physiological level	1. Cell Elongation 2. Increase the extentibility of cell wall 3. Promote growth in stem 4. loosening of cell wall	1. Inhibit seed germination
6. Natural hormone	IAA, NAA,	Cis-ABA
7. Morphological effect	Tallness of Plant	Yellowing of leaves
8. Physiological effect	1. Apical dominance 2.Promote lateral and adventitious roots 3. Delay abscission 4. Vascular differentiation	1. Induce seed dormany 2. Accumulated in dormant bud 3. Inhibit α.-amylase 4. Close stomata
9. Molecular effect	Stimulating cyclin dependent protein kinase (CDK) which is regualating cell cycle transitions from G_1 to S and from G_2 to mitosis	(ABA-response elements) ABA → DNA —Transcription→ mRNA → Protein → Stress Protein / adaptation to low temperature / Salt tolerance
11. Commercial use	1. Initiate rooting in cutting (IBA) 2. Induce flowering in dry material 3. Herbicide (2,4-D) 4. Parthenocarpy eg. Brinjal	1. Used as stress harmone 2. Initiation of flower in short-day plants. -

Contents	GA	Cytokinine (CDK)
1. Definition	Stimulate cell elongation and cell division or both.	Promote regulatory function
2. History	"Hori" observed, kurosawa -discovered.	Habenlandt -I[st] observed
3. Transport	All shoot, leaves	Acropetal, through xylem and phloem.
4. Mode of action	Increase the number of cells, reduces the cell cycle	Increase cell number reduces the cell cycle.
5. Morphology action	1. Substitute for low temperature 2. Flowering in long day plants 3. Parthenocarpy 4. Fruit elongation 5. Breaking seed dormancy	1. Delay senescence 2. Weakens the apical dominance 3. Promote lateral bud development

Contents	GA	Cytokinine (CDK)
6. Physiological action	1. Induce the activity of α-amylase 2. Break the genetic dwarf	1. Mainly cell division 2. Shoot and root differentiation
7. Molecular action	1. Extensionability of cell wall 2. Stress relaxation	1. Increase the mitosis division 2. Increase in m-RNA synthesis in cytokinin
8. Commercial application	1. Parthenocarpy in grapes 2. Breaking seed dormancy 3. Substitute for kernalization -	1. Mainly in tissue culture 2. Uniform production of lateral branches 3. Delay senescence 4. Shoot elongation
9. Regulation	1. Carboxyl group => gibberellin glycoside 2. Hydroxyl group => gibberellin glycosyl ether	1. Zeatine 7-glucoside 2. Zeatine 9-glucoside

Points to Ponder

- Prokaryotic cells contain no membrane-bound organelles.
- Living organisms are self-generating structures.
- In living organisms, as in other machines structure and function are intimately wedded.
- Microfilaments are smaller solid structures (5-7 nm in diameter) that act alone or in coordination with microtubules to produce cellular movement.
- Microbodies are spherical organelles bound by a single membrane.
- Plastids are special structures occurring in plants and some protistants, bound by a double membrane system.
- The nucleus is the control centre of eukaryotic cell, and often is its most conspicuous organelles.
- Golgi bodies have been named after the name of an Italian Camillo Golgi who discovered them in 1898.
- Most of the unique properties of water can be assigned to line segments connecting the centre of the oxygen atom forming an angle of about, 105° closer to the right angle than to straight line.
- In water molecule, the hydrogen atoms approximate naked protons on the surface of the oxygen atom.
- In a water molecule, the protons distributed 105° apart on the surface of the oxygen atom provide a slight posi‹ ›e charge balanced by an equal negative charge on the side, renders it neutral.
- Hydrogen bonds are weaker than ionic and covalent bonds.
- Van der waals attractive forces are even weaker than hydrogen bonds.
- Water has a molecular weight of only 18, is a liquid at room temperature because hydrogen bonds provide a disproportionately high attractive forces among its molecules, inhibiting their separation and escape as vapour.
- For practical purposes liquid water is incompressible.
- 586 calories or 2452 joules are required to convert one gramme of water to one gramme water vapour at 20°C.
- Unusually high latent heat of vaporization can be assigned to the tenacity of the hydrogen-bond.
- Tetrahedrons are so arranged that the ice crystal is basically hexagonal demonstrated in the pattern of snowflakes.
- The total volume of water decreases during melting.
- The density of ice is lower than the density of water.
- The number of bonds per oxygen atom are Fewer in ice than in water molecules.
- Viscosity of water decreases markedly with increasing temperature.
- A falling water drop becomes spherical due to surface tension.
- Absolutely pure water (extremely difficult to obtain) is a poor conductor of electricity.
- Near room temperature K = 10^{-14} $[H^+]$ $[OH^-]$ = K], so in pure water $[OH^-]$ and $[H^+]$ = 10- (10^{-7} x 10^{-7} = 10^{-14}) molar (M).
- Diffusion often occurs in response to differences in the concentration of substances between one point and another.
- Kinetic theory state that the elementary particles (atoms, ions and molecules) are in constant motion at temperatures above absolute zero.
- The average hydrongen molecule, near room temperature moves at 6,433 km per hour.
- Energy occupies no space and has no mass, but can transform or act on matter.
- Everything not in the system is known as surroundings (environ-ment) and is separated by the boundary.
- Photosynthesis will never be 100 per cent efficient because some of light energy driving the process will be converted into heat energy.
- Gibb's free energy is a measure of the maximum energy, available for conversion to work (at constant, temperature and pressure).
- J. Williard Gibbs in 1870 developed thermodynamic measure of energy passing across the boundary between a system and its surroundings available for work interaction.
- Chemical bond energy in gasoline is Potential energy.
- The basic equation for the Gibb's free-energy has an enthalpy factor and entropy factor.
- The enthalpy factor of a system consists of the internal energy plus the absolute pressure multiplied by volume.
- The entropy factor (entropy) multiplied by absolute temperature, is subtracted enthalpy to determine Gibb's free energy.
- Chemical potential like solute concentration and temperature is independent of the quantity of the substance.
- Activity (effective) concentration for solute particles is the most common and important factor to establish chemical potential gradients that drive diffusion.
- Particles diffuse from points of high to points of low activity.
- Liquid water or water vapour often diffuses from deep into the soil to the surface when it is cooled at night.
- The cooler gas diffuses into the warmer gas.
- An increase in the pressure increases free-energy and hence chemical potential of a system.
- Mole fraction of a solvent is equal to moles of solvent divided by moles of solute plus moles of solvent.
- The concentration of water molecules in vapour (gas)

Head office of SIDBI is at Lucknow.

GS

phase is expressed as gm^{-3} or grammes per cubic metre.

- The relative humidity above a solution (at equilibrium) is equal to the mole fraction of solvent expressed as per cent.
- The rate of diffusion is given by a simple, but highly significant equation cakked Fick's First Law formulated in 1855.
- Permeability is simply inverse of the resistance.
- Q_{10} for diffusion of many gases is about 1.03 times faster when temperature is raised by 10°C.
- The solutes in water have Q_{10} values of 1.2-1.4 for diffusion.
- The cell wall and the cell membrane are two different structures.
- Animal cells do not have walls and hence they often burst when pressure(s) is (are) built.
- Turgid cells provide much of the rigidity to non-woody plant parts, while many animals are supported by a skeleton.
- A cell never comprises a perfect osmotic system.
- Equilibrium is achieved when water Potentials are equal both side (the water potential difference between the two sides of the membrane is zero).
- As equilibrium water potential will be equal in all parts of the system.
- Osmotic potential has also been termed as solute potential caused by the presence of solute particles.
- Pressure is usually positive in living cells, but is often negative in dead xylem elements.
- In an osmometer, an ideal 1.0 molar sugar (say, glucose) solution at 28°C has an osmotic potential of -2.5 KJ kg^{-1} or -10.75 $Cal.mol^{-1}$ or 2.5 x 10^7 ergs g^{-1}.
- The osmotic potential of a 1.0 molar glucose solution (or water potential of the same solution at atmospheric pressure) can be expressed in pressure terms as -2.5 MPa.
- In the SI system kj kg^{-1} have the numerical value as Mega pascals.
- In 1887, J.H. Von't Hoff discovered an emperical relationship to determine an approximate osmotic potential from the molal concentration of a solution.
- The most meaningful property that can be measured in soil-plant-air is the water potential.
- In 1948, V.S. Chardakov who devised a simple and efficient method of determining the test solution in which no change in concentration occurs was Russian scientist.
- The saturation vapour density approximately doubles for each 10⁰C or 20⁰F increase in temperature.
- A thermocouple operates on the principal of the thermoelectric effect.
- The Peltier effect is the opposite of the thermoelectric effect.

You Don't Know!

First national gene bank

R.S. Paroda (1942) Director General, Indian Council of Agricultural Research, Delhi, established the first National Gene Bank at Delhi in 1985.

- In 1877, Wilhelm F.P. Pfeffer made nearly perfect, rigid and semi-permeable membrane to measure osmotic potential.
- Vapour pressure method applies equally well to the measurement of osmotic potential of a fret liquid.
- The properties of solutions that are functions of the mole fraction are called Colligative properties.
- It has been assumed that incipient plasmolysis occurs in tissues in which about half the cells are just beginning to plasmolyze.
- Osmotic potentials of about -0.1 MPa in aquatic plants and -20 MPa in halophytes have been observed.
- The matric potential is a measure at atmospheric pressure of the tendency for a matrix to adsorb additional water molecules.
- When colloidal particles are observed in a light microscope, they appear as points of light due to Tyndall effect.
- Brownian movement was discovered by the Scottish botanist Robert Brown in 1827.
- Brownian movement is the spectacular confirmation of Kinetic theory.
- A single solid cube 1 cm on each of its edges, when cut to parts each measuring 1 mm long on its edges, will have total surface of 600 m^2 or 6.000,000 cm^2.
- The Meniscus (plural menisci) is the basis of the phenomenon of capillarity.
- At equilibrium, the matric potentials of all the hydrophilic surfaces in a system are equal.
- It has been estimated that 225 kg of water were transpired to produce one kg of dried plant material.
- When soil moisture becomes limiting, transpiration and carbon dioxide uptake are inhibited because stomates close. .
- If the leaf is warmed above air temperature by sunlight, wind lowers its temperature causing a decrease in transpiration.
- Water molecules diffuse 1.56 times as fast as carbon dioxide because of their lower molecular weight.
- The atmosphere at 35°C normally contains 0.3-1.4 mol m^{-3} of water and of 0.0156 mol m^{-3} carbon dioxide.
- The amount of water used in plant growth is less than one per cent of the final dry weight of the plant.
- Lysimeter method was used to measure water loss through transpiration by Stephen Hales about 200 years

GK

The most abundant constituent in air is Nitrogen.

GS

ago.

- Lysimeters provide the most reliable field method to study evapotranspiration.
- Cuvettes and Photometers are appropriate in laboratory, where purposes to study transpiration are special.
- Stomatal openings on leaf make up only one per cent of surface area.
- Greater the distance of diffusion through atmosphere, higher is the resistance.
- Plants are ideally adapted for carbon dioxide from the atmosphere, but also for loss of water by transpiration.
- Water evaporates inside the leaf from the palisade and spongy parenchyma cell walls to intercellular spaces which is continuous with outside air when stomates are open.
- Carbon-dioxide follows the reverse diffusional path into the leaf.
- Sunken stomates are apparently an adaptation to reduce transpiration.
- Typical stomates of dicots consist of two kidney shaped guard cells whereas; in grass, guard cells tend to be more elongated (dumbbell shaped).
- The guard cells of some species as noticed in 1856 were slightly thickened along the concave wall adjacent to the stomatal opening.
- Radial micellation is as important in grass as in dicot guard cells. .
- Some succulent plants native to hot and dry conditions open their stomates at night and fix carbon dioxide into organic acids in the dark and close their stomates during the day.
- Opening of stomates generally need about an hour and closing occurs gradually throughout the afternoon.
- Removal of carbondioxide by mesophyll cells during photosynthesis comprises the main reason that stomates of most species open in light.
- The stomates close with the decrease in water potential.
- Potassium ions (K^+) move from the surrounding cells in the guard cells when stomates open.
- Stomates also close in response to the application of abscisic acid as it causes loss of potassium from guard cells.
- A leaf, at a temperature different from that of the surrounding air exchanges heat directly first by conduction and then by convection.
- Leaves at night often radiate more energy than they absorb.
- Much of the green wavelength is reflected and transmitted, that is why leaves appear green.
- Most of the ultraviolet radiation is absorbed by ozone in the upper atmosphere.
- The velocity of all forms of radiant energy in vacuum is the same and it is equal to 3×10^{10} cms^{-1} or 300000.
- Green light has a frequency of about (vibrations or waves) 6×10^{14} pulsations S^{-1}.
- The energy of a photon or quantum is directly proportion to the frequency of radiation.
- The wavelengths 400 to 600 nm are most active in photosynthesis.
- The boundary/unstirred layer is thinnest and offers the least resistance to convective heat transfer for small leaves and high wind velocities.
- The driving force for transpirations is the difference in water vapour density within the leaf and in the atmosphere beyond the boundary layer.
- Even if the leaf is exactly at the same temperature as the atmosphere beyond the boundary layer, there is under most conditions a vapour density gradient with higher values inside the leaf.
- Temperature, can greatly accentuate the vapour density gradient.
- Many desert plants, perhaps most have small leaves, providing thin boundary layers with consequent efficient convective heat transfer.
- A suction pump can lift water only to the barometric height, which is the height that is supported by atmospheric pressure (1.0 atm.) from below.
- According to the Guinness Book of World Records (1983), the tallest tree in the world is the "Tallest Tree" a *Sequoia sempervirens* in Redwood Creek Grove, Redwood National Park, California (U.S.A.).
- At zero pressure, water will normally vacuum boil even at 0°C (actually it boils at 0.610 kP), its vapour pressure at 0°C.
- Droplets of water exuded from openings in the tips or edges of grass or strawberry leaves is due to guttation.
- Root pressures are not found in conifer trees.
- The cohesion theory for the ascent of sap consists of three basic elements namely, driving force, hydration in the pathway, and the cohesion of water.
- Special cells called Tyloses grow into vapour locked tracheids and vessels in some plant species act as seals against water loss and pathogen invasion.
- The absorption of ammonium (NH_4^+) and other cations occurs simultaneously with transfer of hydrogen (H^+) from the root to the surrounding solution.
- When sheep and cattle eat grasses abundant in silica, it is largely excreted in the urine, but sometimes it forms kidney stones.
- Cobalt is required for nitrogen fixation by bacteria in root nodules on legumes.
- Perennial range grasses conserve minerals by translocation to roots and to lower stem tissues making up the crown in late summer.

GK The Bhur is gravellous soil. GS

- Significant amounts of N, P, K and Mg move out of tree leaves into twigs and branches before leaf fall.
- Siderophores form extremely stable chelates with iron (Fe^{3+}) and are transported into microbial cells where Fe^{3+} is reduced to Fe^{2+} and released.
- Plants respond to an inadequate supply of an essential element by forming characteristic deficiency symptoms.
- Inadequate supply of boron causes several disorders related to disintegration of internal tissues like "heart rot" of beet, "stem crack" of celery, "water core" of turnip and "drought spot" of apples.
- Zinc deficiency causes disorders namely, "little leaf' and "rosette" of apples, peaches and pecans; "*khaira*" of rice and "white bud" of maize.
- Citrus orchards are occasionally deficient in copper, in which case the dying young leaves led to the name "die back" disease.
- "Whip tail' disease of cauliflower and broccolli is caused by deficiency of molybdenum.
- In acid soils addition of lime augments the availability of molybdenum.
- A well-documented function of molybdenum in plants is as a part of the enzyme nitrate reductase.
- In soyabean and tomato, it is difficult to differentiate tap rot and branch roots.
- Plants adapted to dry regions are not necessarily deep rooted as shallow roots take better advantage of brief rains.
- R.F. Sutton (1980) reported that a 100-year old-scot pine (*Pinus-sylveslris*) tree 500 root tips and line tips, had million root tips, and a mature red oak (*Quercus rubrum*) had respectively million live tips.
- Cylindrical and filamentous-long and narrow form of roots prove unexpectedly important for absorption of water and solutes from soils.
- Exploration of large soil volume is important for roots to contact water and ions.
- Two main groups of mycorrhizae are ectomycorrhizae and endomycorr-hizae.
- In ectomycorrhizae, there is no intercellular penetration into epidermal or cortical cells, but an extensive net-work called heating net is formed between these cells.
- VAM fungi seem to penetrate directly into the cytosole of cortical cells (to form vesiscles and arbuscules), that part of the fungus is surrounded by an invaginated plasma-membrane of the cortex cell.
- Plants grown in shade are deficient in sugars and hence have poor mycorrhizal development.
- The most well documented advantage of mycorrhizae to plants is increased phosphate absorption.
- The surfaces of clay minerals become negatively charged because some atoms of Si^{4+} are replaced by Al^{3+}, and some atoms of Al^{3+} are replaced by Mg^{2+} or Fe^{2+}.
- The availability of negative sites on surfaces of clay minerals results in attraction and adsorption of dissolved cations namely H+ > Ca^{2+}> Mg^{2+} > K^+ = NH4$^+$ > Na^+.
- The organic matter fraction of soils has an excess of negative charges due to ionized carboxyl groups (–COOH → COO+H^+) and also ionized OH groups of phenolic compounds.
- H^+ ions can teplace any cation adsorbed to a soil colloid and make it available to the roots.
- Nitrate (NO_3^-), sulphate (SO_4^{2-}) and chloride (Cl^-) are soluble, are repelled by negatively charged colloids and consequently are leached with water.
- Most of the crops (except legumes) require relatively large amounts of nitrogen fertilizers because of high nitrogen need and loss of nitrate or NO_3^- by leaching.
- Most soils contain little ammonium (NH_4^+) as it is rapidly oxidised to nitrate (NO_3) form by bacteria.
- The anions and cations most readily available to roots are those freely dissolved in the soil solution.
- The symplastic and the apoplastic pathways contribute substantially to the transport of most ions.
- Calcium (Ca^{2+}) is mainly transported by apoplastic pathway in plants.
- The ions absorbed into root cell vaculoes usually move out slowly.
- Both plant and animal cells keep concentration of calcium (Ca^{2+}) in their cytosoles at low concentrations partly because it precipitates both inorganic and organic phosphate compounds.
- The loss of epidermis and cortex in herbaceous dicots and grasses in the older region leaves endodermis or pericycle as the outermost layer of cells.
- Electron micrographs show that most biological membranes are similar regardless of their location.
- The principal lipids of most plant membranes are phospholipids glycolipids and sterols.
- Plant membranes largely consist of proteins and lipids.
- Proteins, in composition of plant membranes represent one-half to two-thirds of the membrane dry weight.
- Fatty acids are hydrophobic whereas; glycerol is more hydrophilic.
- Integral proteins are bound tightly within the plant membrane but peripheral proteins are loosely bound to one or the other side of the membrane surface.
- In some plasma-membranes of animal cells. some peripheral proteins called glycoproteins contain short polysaccharides attached at the outer membrane surface.
- If plant cells are not alive and metabolizing, their membranes become much more permeable to solutes.
- Water molecules and dissolved gases (N_2, O_2, CO_2)

Maitreynath is the first propounder of Yogachara School.

penetrate all membranes rapidly.

- Hydrophobic solutes penetrate at rates positively related to their lipid solubility.
- Leaves absorb 2,4-D more effectively at pH 5 than at pH 8.
- Hydrophilic molecules and ions with similar lipid solubility penetrate at rates inversely related to their size.
- The ion size relevant to penetration is that attained after hydration is attached.
- Transportation of accumulated sucrose and other carbohydrates (in high concentration) is generally essential for photosynthetic leaf cells.
- The diffusion to roots is a limiting factor in most plants in nature for the nutrients (nitrate, ammonium, phosphate and potassium) used in relatively larger amounts.
- Absorption of ions largely through channels in protein has been termed as facilitated diffusion.
- Each ionophore is especially selective for a certain cation, consistent with selective properties of membranes.
- Water with its dissolved minerals moves primarily upward in the plant through xylem tissues.
- Plant assimilates more from source to sink.
- PCMBS does not penetrate the plasmalema of the cells and thus does not enter the symplast.
- Increase in agricultural yields is associated with the increase in harvest index.
- The phloem sap change significantly in its composition, while underway from leaf to sink such as developing fruits or tubers.
- The formation of large molecules from small molecules is called anabolism (from Greek *ana*, upwards) while the breakdown of large molecules to small molecules is termed as catabolism (from Greek *kata*, downwards).
- Anabolism requires energy whereas, catabolism releases energy.
- Enzymes generally speed up reaction by factors between 10^8 and 10^{20}.
- Proteins consist of polypeptide chains tens to thousands of amino acids.
- The $-NH_2$ is the amino group and the $-COOH$ is the carboxyl group.
- Electrophoresis is essentially the separation of dissolved proteins or other charged molecules in an electric field.
- Opaque-2 and Fluory-2 cultivars of maize are relatively richer in both lysine and tryptophane amino acids.
- Water was implicated by N.T. de Saussure when in 1804, he made the first quantitative measurements of photosynthesis.
- The thylakoid membranes in plants consist of two kinds of green chlorophyll a and chlorophyll b.
- In most of the plants including green algae, β-carotein and lutein are the most abundant carotenoids in thylakoids.
- One mole (6.02 x 1023) of photons has been called an Einstein.
- According to Stark Einstein Law any molecules can absorb only one photon at a time and this photon causes the exitation of only one electron.
- In brown algae and dinoflagellates xanthophylls called fucoxanthin and periodinin are highly effective in transfer of energy.
- During last 15 years, many weed species have become resistant to triazines because a chloroplast mutation has caused them to form a modified thylakoid protein.
- The Calvin cycle occurs in the stomata of chloroplasts and mainly consists of carboxylation, reduction and regeneration.
- Calvin cycle is synonymous to C-3 photosynthetic pathway (as first product 3PGA contains three carbon atoms) and photosynthetic carbon reduction cycle.
- The species that produce 4 carbon acids as the primary initial $C0_2$ fixation products are commonly referred to as C-4 species, those fixing $C0_2$ initially in 3 PGA are called C-3 species.
- Most of the C-4 species are monocots, especially grasses and sedges.
- All *gymnosperms*, *pteridophytes*, *bryophytes*, and algae that have been studied are C-3 plants.
- C-4 plant species under high light and warm temperatures can photosynthesize more rapidly and produce substantially more biomass than can C-3 plants.
- The most striking feature of CAM (crassulacean acid metabolism) plants is the formation of malic acid at night and its disappearance during day light.
- Worldwide increases in atmospheric carbon dioxide comprise a cause of concern because they could lead to warming of the earth' s surface.
- The more rapid photosynthesis of C-4 species under high irradiance results in lower water requirement per gramme of dry matter produced, where CAM plants have much lower than either C-4 or C-3 species.
- Oxidized forms of nitrogen in the atmosphere are important ecologically because when converted to NO_3^- they contribute to acidic rainfall or "acid rain"
- Important non-legumes that fix N_2 are primarily trees and shrubs.
- Conversion of carbohydrates to fats requires production of fatty acids of glycerol backbone to which the fatty acids become esterified.
- The entire shoot system of a herbaceous plant is covered by cuticle that shows water loss from its parts.
- Most of the cuticle is composed of a heterogeneous mixture of compounds collectively called cutin.
- Allelophathy (Greem *allelon*, of one other, *pathos*,

disease) is special case of allelochemy involving a negative chemical interaction between different plant species.

- All sterols (steriod alcohols) triterpenoids are built from six isoprene units.
- Milkweeds (*Aesclepias* spp.) produce a number of bitter-tasting cardiac glycosides that protect them against berbivory by most insects and even cattle.
- Saponins (sterols attached to a short chain of one or more sugar) cause foaming in the intestinal tract leading to bloat in cattle that eat young alfalfa plants.
- β-carotein is the most abundant carotenoid found in higher plants and is the compound imparting the orange colour to carrot root.
- Rubber, an isoprenoid compound contains some 3,000-6,000 isoprene units linked together in very long unbranched chains.
- Resins represent a complex mixture of terpenes containing 10-30 carbon atoms.
- Phenylalamine, tyrosine, and tryptophane are aromatic amino acids.
- Lignin is a strengthening material that occurs with cellulose and other polysaccharides in the cell walls of all higher plants.
- Most fruits and many flowers owe their colour to anthocyanin (Greek, *anthos*, a flower; *kyanos*, dark blue).
- The red pigment of beets is a β-cyanin, from red beet root, a betalain pigment coloured, but different from anthocyanins.
- Those interested in crop productivity measure the plant growth by increase in dry weight (freshly harvested plant material dried for 24-48 hours at 70-80^{0}C) of a plant or a part thereof.
- Determinate structure grows to a certain size and then stops, eventually undergoing senescence and death.
- In a great majority of species, seed germination begins with radical (embryonic root) rather than epicotyl (shoot) protrusion through the seed coat.
- Anthesis is the opening of flowers with parts available for pollination, is sometimes a spectacular phenomenon: usually associated with full development of colour and scent.
- Normal production of fruits lacking seeds is called parthenocarpic fruit development.
- The term auxin was first used by Frits Went, who as a graduate student in Holland in 1926, discovered that some unidentified compound probably caused curvature of oats coleoptyle.
- Bioassays take advantage of the extreme sensitivity and specificity of certain plant parts to a particular hormone.
- In stems of most of the plant species, the apical bud exerts an inhibitory influence (apical dominance) upon the lateral (axillary buds) preventing or slowing their development.

Knowledge World!

First Constituent Assembly meeting

The Indian Constituent Assembly held its first meeting on Dec. 4, 1946, at 11 a.m. at the Constitution Hall, New Delhi.

- Acid growth hypothesis states that auxins cause receptive cells to secrete H^+ into their surrounding primary walls, H^+ lowers pH and growth occurs.
- In the 1930s, T. Yabuta and T. Hayashi isolated an active compound from the fungus (*Gibberella fujikuroi*), which they named as Gibberellin.
- Ethylene stimulates ripening of fruits, causes senescence and abscission of leaves.
- The highest known rate of ethylene release is from fading flowers of *Vanda* orchids; 3.4 μl per 9 of fresh weight per hour.
- In many fruits, little ethylene is produced until just before respiratory climacteric signalling the onset of ripening.
- Non-climacteric fruits synthesize little ethylene and are not induced to ripening by it.
- ABA (dormin), a bud dormancy hormone if applied directly to non-dormant buds caused dormancy.
- Batasins are compounds in yam plants (*Discorea batatus*) that seem to cause dormancy in bulbils arising from swelling of aerial lateral buds.
- An upward bending of a leaf (or any organ) is called hyponasty; whereas downward bending is epinasty.
- Nastic movement of leaves occurs due to water movement in and out of motor cell at the base of the petiole blade or leaflet.
- Nyclinastic movements in plants are rhythmic processes controlled by biological clock.
- At night, water moves out of subepidermal cortical cells in the upper (ventral) side of the pulvinus and into cortical cells on the lower (dorsal) side.
- Nastic movement results from touch (*thigmonasty*; Greek *thigma* = touch) in plants belonging to subfamily Mimosoideae in the Fabaceae (Leguminosae).
- A number of plants are capable of solar tracking in which the flat blade of the leaf remains at right angles to the sun throughout the day maximising the light-harvest by the leaf.
- Many plant cells are totipotent, that is, a nonembryonic cell has the potential to differentiate into an embryonic cell and then into a new plant under a suitable environment.
- Phytochrome provides seeds with a clue whether they are covered by a canopy of other plants or exist in an open area.

'Poverty & Un-British Rule in India' was written by Dadabhai Naoroji.

GS

- Quiescence in seeds is due to external factors, while dormancy is due to its internal conditions.
- Some plants grow better when the environment fluctuates on a 24-hour cycle, presumably to coincide with the phases of their circadian clock.
- Photoperiodic response/photoperio-dism responses of organisms better the understanding of natural ecosystems, but out of about 300,000 plant species, only a few hundred have been grown with different artificial photoperiods.
- Sex expression in many plant species is often strongly influenced by photoperiod.
- The operational environment of a plant represents the complex of climatic, edaphic and biotic factors.
- The rate of cooling is an expression of the adiabatic gradient or adiabatic lapse rate for air in the atmosphere.
- Desert ephemerals are annual plants that escape the drought by existing only as dormant seeds during the dry season.
- The gramme molecular weight (1 mol) of sucrose is 342.3 grammes which is dissolved in 1000 grammes (55.508 mol) of water, its mole fraction will be 0,9823 (98.23%).

GK **In July 1988 National Housing Bank (NHB) was established.** GS

Multiple Choice Questions

1. Resolving power of a light micro-scope is :
   a] 0.1 micro metres
   b] 0.2 micro metres
   c] 0.3 micro metres
   d] 0.4 micro metres
2. Mitochondria were seen first in :
   a] 1850 b] 1900
   c] 1940 d] 1975
3. Each gramme of water transpired absorbs _______kJ of latent heat from the leaf and its environment.
   a] 2.0-2.1 b] 2.2-2.3
   c] 3.4-3.5 d] 2.7-2.8
4. Quantasomes contain :
   a] 200 chlorophyll molecules
   b] 230 chlorophyll molecules
   c] 250 chlorophyll molecules
   d] 300 chlorophyll molecules
5. Rice plants exposed to temperatures below_______ °C at the time of pollen-mother cell division will not produce a crop.
   a] 22 b] 20
   c] 19 d] 16
6. Soil water that is available to plants seldom has a water potential more negative than :
   a] 3.0 MPa b] −1.5 MPa
   c] −2.0 MPa d] −2.5 MPa
7. Considering the supplies of total PAR to land area on growing crops, overall biomass product in efficiencies are always much below _______per cent.
   a] 30 b] 25
   c] 18 d] 22
8. The amount of carbon dioxide has increased slowly since about 1850 from 28 μl/l to slightly more than _______μl/l in 1962.
   a] 310 b] 320
   c] 330 d] 340
9. It has been estimated that carbon dioxide content within the maize plant-canopies was 260 or less during day, whereas it rose to _______ μl/l during darkness.
   a] 400 b] 500
   c] 600 d] 700
10. Direct sunlight at noon in late June in the northern hemisphere corresponds to _______ footcandles.
    a] 5,000 b] 7,500
    c] 10,000 d] 12,500
11. The atmosphere has less carbon than other reservoirs but, it still contains about 700 billion tonnes and roughly per cent is annually used in photosynthesis.
    a] 5.0 b] 7.5
    c] 10.0 d] 12.5
12. H^+ ions in thylaoid channels arise from oxidation of H_2O and PQH_2 cause H^+ ions concentration in channels become; about _______times as great as in the stroma when photosynthesis occurs.
    a] 500 b] 750
    c] 1000 d] 1250
13. The total thickness of unit membrane fall in the range of :
    a] 7-10 Å b] 70-100 Å
    c] 700-1000 Å d] 7,000-10,000 Å
14. The following percentage of water absorbed by herbaceous plants is lost in transpiration :
    a] 80 b] 60
    c] 99 d] 40
15. At equilibrium, the vapour density or pressure above pure water in a closed container is :
    a] 85% b] 98%
    c] 95% d] 100%
16. In water molecule, line-segments connecting the centres of the two hydrogen atoms with the centre of oxygen atoms form an angle of about :
    a] 85° b] 95°
    c] 105° d] 115°
17. A maize plant to produce one kg of grain (dry) transpires _______kg of water.
    a] 900 b] 800
    c] 700 d] 600
18. Prokaryota means :
    a] A cell with many nuclei
    b] A cell without chloroplast
    c] A cell with diffused nucleus
    d] A cell with one nucleus
19. Which one of the following factors causes an increase in transpiration rate?
    a] A hot dry windy day
    b] A cool damp windy day
    c] A hot damp windy day
    d] None of the above
20. Wilting of plant results from excessive :
    a] Absorption b] Photosynthesis
    c] Respiration d] Transpiration
21. Process of transpiration in plants helps in :
    a] Absorption of CO_2
    b] Upward conduction
    c] Opening of stomata
    d] None of these
22. Plasmolysis will occur when the cell is placed in a :
    a] Acidic solution b] Isotonic solution
    c] Hypotonic solution d] Hypertonic solution
23. Blue green algae are :
    a] Actinomycetes b] Acellular
    c] Prokaryotes d] Eukaryotes

Ammonia gas is extremely soluble in water.

24. The solute transport into the cells is strongly dependent on_______and ability of cells to respire aerobically and to produce it.
    a] ADP b] ATP
    c] RNA d] DNA
25. Which one of the following is a prokaryote?
    a] *Agaricus* b] *Salmonella*
    c] Green algae d] Bacteriophage
26. Enzymes, vitamins and hormones can be classified into a single category of biological chemicals because all of them :
    a] Aid in regulating metabolism
    b] Are proteins
    c] Are synthesized within the body
    d] Enhance the oxidative metabolism
27. ABA (abscic acid) appears to be universal among :
    a] Algae b] Liveworts
    c] Vascular plants d] All mosses
28. Simplest amino acid is :
    a] Alanine b] Glycine
    c] Leucine d] Aspartic acid
29. Phytochrome is found in :
    a] Algae b] Fungi
    c] Gymnosperms d] Angiosperms
30. Which of the following can fix atmospheric nitrogen?
    a] Alkaloids b] Fats
    c] Alcohols d] Anabaena
31. Fermentation is conducted by :
    a] All fungi
    b] All bacteria
    c] Some fungi and some bacteria
    d] All micro-organisms
32. Peptide linkages are formed among:
    a] Amino acids b] Glucose molecules
    c] Sucrose molecules d] Glycerol molecules
33. Proteins are made up of :
    a] Amino acids b] Nucleic acids
    c] Sugars d] Fatty acids
34. Which of the following denotes the correct sequence of appearance of compounds in a cell?
    a] Amino acids→Proteins→enzyme
    b] Glucose→starch→sucrose
    c] Protein→enzyme→amino acids
    d] Disacchardies→monosaccharies→plysaccharides
35. If a plant cell is immersed in water, the water continues to enter the cell until the :
    a] Amount of water is the same inside the cell as outside
    b] Diffusion pressure is the same inside the cell as outside
    c] Concentration of the salt is the same inside the cell as outside
    d] Cell bursts
36. A characteristic feature of ripening of some fruits (like banana) is a sudden increase in respiration which is known as :
    a] Anthesis b] Photorespiration
    c] Climacteric d] Climatic
37. High energy compounds :
    a] Are produced in respiration only
    b] Are compounds which link exergonic to endergonic process
    c] Refer to sugar, the oxidation of which releases large amounts of energy during respiration
    d] Are produced when ATP looses two of its phosphate groups
38. The Singer and Nicholson model of plasma membrane differs from Robertson's model in the :
    a] Arrangement of lipid layers
    b] Arrangement of proteins
    c] Absence of proteins in the Singer model
    d] Number of lipid layers
39. Carbon dioxide is liberated during:
    a] Ascent of sap b] Transpiration
    c] Photosynthesis d] Respiration
40. A final component of thylakoid necessary for photophos-phorylation is a complex protein called :
    a] ATPase/coupling factor
    b] Detaching factor
    c] Reductase factor
    d] None of these
41. In 1948, V.S. Chardakov who devised a simple and efficient method to determine the test solution in which no change in concentration occurs was :
    a] Austrian Scientist
    b] American Scientist
    c] Bulgarian Scientist
    d] Russian Scientist
42. The ultrastructure of cell can be best studied by :
    a] Autoradiography
    b] X-ray diffraction method
    c] Phase contrast microscope
    d] None of these
43. An apparatus for measuring transpiration in plants is called :
    a] Auxanometer b] Potometer
    c] Hygrometer d] Lactometer
44. In unisexual plants sex can be changed by application of :
    a] Auxins b] ABA
    c] Ethanol d] Cytokinins
45. A free living anaerobic bacterium capable of N_2 fixation in soil is :
    a] *Azotobacter* b] *Streptococcus*
    c] *Clostridium* d] *Rhizobium*

46. Except which one, all other processes help in adding nitrogen to soil :
a] Bacterial decay of animal remains
b] Electric lightning
c] Addition of lime in soil
d] Nitrogen fixing bacteria

47. Which of the following plants can not fix atmosphere nitrogen directly?
a] Bean b] Castor
c] Pea d] Gram

48. Whose technique is employed for the extraction and elimination of hormones?
a] Beck b] Beer
c] Garner d] Allard

49. A natural growth regulator (hormone) is :
a] Benzaldehyde b] 2, 4-D
c] N.A.A d] Ethylene

50. The action of ethylene is effectively inhibited by _______ ion.
a] Boron b] Manganese
c] Zinc d] Silver

51. In both transpirations and evaporation water is lost in the form of vapour yet they differ because :
a] Both transpiration and evaporation are similar but the rate of water loss differs
b] Frequency of water loss is different in both of them
c] Transpiration is a physical process and evaporation is physiological process
d] Transpiration is a physiological process and evaporation is physical process

52. In land plants the gaseous exchange takes place through the open stomata, in submerged hydrophytes it takes place/through:
a] By general surface of the cells by diffusion
b] Lenticels
c] Stomata
d] Hydathodes

53. Which is a micronutrient?
a] Ca b] Zn
c] P d] Mg

54. Which of the following is a micronutrient?
a] Calcium b] Phosphorous
c] Copper d] Magnesium

55. A trace element essential for plant growth and radio-isotope, which is used in cancer therapy is known as:
a] Calcium b] Cobalt
c] Sodium d] Iron

56. CO_2 is necessary for photo-synthesis; the chemical used to remove this gas most effectively from entering a control apparatus is :
a] Calcium oxide
b] Distilled water
c] Sodium carbonate
d] Potassium hydroxide solution

Quiz Kit!

First Jnanpith award

G. Sankara Kurup, won the Jnanpith Award - the country's highest literary award in 1965, the year of its inception, for his *Odakuzhal* (The Flute), an anthology of poems in Malayalam.

57. Position and frequency of stomata can be determined by :
a] Calculating the loss of water
b] Cobalt chloride paper method
c] Potometer
d] Porometer

58. Which of the following effects of auxins on plants is the basis for commercial application?
a] Callus formation
b] Curvature of stem
c] Induction of root formation in stem cuttings
d] All of the above

59. Chemical phase in photosynthesis was discovered for the first time by :
a] Calvin b] Blackman
c] Hill d] Arnon

60. Carbon becomes available to crop plants in the form of :
a] Carbon dioxide
b] Element carbon (coal)
c] Amino acids
d] Carbonates

61. If naked pea seeds are kept in four respiratory flasks, they would germinate best if flask contains :
a] Carbon dioxide b] Oxygen
c] Hydrogen d] Nitrogen

62. During the photosynthesis the oxygen in glucose comes from :
a] Carbon dioxide
b] Water
c] Both from carbon dioxide and water
d] Oxygen in the air

63. When yeast and certain bacteria carry on anaerobic respiration they produce :
a] Carbon dioxide b] Nitrogen
c] Water d] Oxygen

64. One of the following is a water soluble pigment :
a] Carotene b] Chlorophyll
c] Anthocyanin d] Xanthophyll

65. Photosynthesis is :
a] Catabolic process b] Anabolic process
c] Exothermic process d] Exergonic process

66. Gibberellins differ from auxins since they produce:
a] Cell division b] Stem elongation
c] Root initiation d] Shortening of internodes

67. Sialic acid is a constituent of :
a] Cell membrane b] Cell wall
c] Chromosomes d] Dictyosomes

68. Which pair of structures are usually found in both plant and animal cells?
a] Cell membrane and nucleolus
b] Cell membrane and cell wall
c] Nucleolus and chloroplast
d] Nucleus and cell wall

69. Plant cells do not burst in distilled water because :
a] Cell wall is permeable
b] Cell wall is living
c] Cell wall is elastic, rigid and get stretched
d] Cell wall is outer most layer of plant cell

70. Transpiration is minimized by the deposition of :
a] Cellulose b] Pectin
c] Cutin d] Mucilage

71. Unit membrane or (Plasma membrane) is made of :
a] Cellulose b] Lipoproteins
c] Phosphorproteins d] Phospholipids

72. Which of the following parts of a cell is non-living?
a] Centriole b] Vacuole
c] Ribosome d] Mitochondira

73. Which of the following structures controls the transport of the material into and out of living cells?
a] Centrosome b] Cell membrane
c] Cell wall d] Ribosome

74. Atmospheric nitrogen in the soil is fixed by :
a] Cereals b] Pulses
c] Sugarcane d] Cotton

75. The pH gradient across the cell membrane provides a powerful form of _______ potential energy largely responsible for driving phosphorylation.
a] Chemical b] Physical
c] Electrical d] Biological

76. _______is obviously useful to a plant as the excitation by an insect of one or more sensory hairs of the venus's-flytrap (*Dionea muscipula*).
a] Chemical potential b] Action potential
c] Physical potential d] Biological potential

77. Both respiration and photo-synthesis require :
a] Chlorophyll b] Cytochromes
c] Sunlight d] Glucose

78. The role of molybdenum is :
a] Chromosome contraction
b] Carbon assimilation
c] Nitrogen fixation
d] Flowering induction

79. If a cell is placed in a solution and the cell shrinks, the solution is :
a] Clear b] Isotonic
c] Hypotonic d] Hypertonic

80. What determines the diffusion of water accumulation of sugars?
a] CO_2 accumulates, reduces pH, stimulates enzymes resulting in accumulation of sugars
b] Increase in CO_2 concentration, conversion of organic acids into starch resulting in the increased uptake of potassium ions and water
c] Low CO_2 concentration, accumulation of roganic acids reulting in the increased concentration of cell sap
d] None of the above

81. During aerobic respiration oxidation of glucose results in the formation of :
a] CO2 and water
b] Hydrogen and water
c] Hydrogen and oxygen
d] Oxygen and CO_2

82. If CO_2 is given off in respiration, why does the amount of CO_2 in the atmosphere remains relatively constant?
a] CO_2 forms carbonate rocks
b] CO_2 is a buffer
c] CO2 is converted in photo-synthesis to carbohydrates
d] CO_2 is split up during photosynthesis

83. Non-essential element for the proper growth of plant is :
a] Cobalt b] Magnesium
c] Chlorine d] Calcium

84. Fermentation is :
a] Complete oxidation of carbohydrates
b] Incomplete oxidation of carbohydrate
c] Anaerobic respiration
d] None of the above

85. Exogenous application of Gibberellins induces male flowers formation on genetically female plants in:
a] *Cucumis* b] *Cucurbita*
c] *Carica* d] *Coccinia*

86. Basic structure of all chlorophylls comprises :
a] Cytochrome system
b] Flavoproteins
c] Poryhyrin system
d] Plastocyanin

87. Who proposed the concept of unit membrane for tripartite structure of lipoproteins?
a] Davson and Danielli
b] Robertson
c] Seifriz
d] Altmann

88. Deficiency of iron causes :
a] Decrease in protein synthesis
b] Reduced leaves and stunted growth
c] Interveinal chlorosis first on young leaves
d] Bending of leaf tip

GK GS

Article 40 directs the state to take steps to organize village panchayats.

89. The removal of apical buds and young leaves _______ branching.
a] Decreases b] Increases
c] Maintains d] No effect

90. If a few cells of an alga *Spirogyra* are kept in a strong solution, the cell contents contract away from the cell wall, this is due to :
a] Dehydration b] Plasmolysis
c] Hydrolysis d] Contraction

91. Non-living substances of the protoplasm are collectively known as :
a] Deutoplasm b] Plasmagel
c] Mesoplasm d] Cytoplasm

92. D.P.D. stands for :
a] Diffusion potential deficit
b] Diffusion pressure deficit
c] Daily phosphate demand
d] Daily photosynthetic depression

93. The real force causing entry of water into a cell from other cells is:
a] Diffusion pressure deficit
b] Turgor pressure
c] Osmotic pressure
d] Wall pressure

94. Turgidity of a cell is maintained by:
a] Diffusion pressure b] Wall pressure
c] Osmotic pressure d] Turgor pressure

95. Mineral salts which are absorbed by the roots for the soil are in the form of :
a] Dilute solution
b] Very concentrated solution
c] Concentrated solution
d] Very dilute solution

96. The correct sequence in protein synthesis is
a] DNA→Amino acids→ RNA→Proteins
b] DNA→RNA→ Amino acids→ Proteins
c] Amino acid→DNA→ RNA→ Proteins
d] RNA→DNA→Amino acids→Proteins

97. Aerobic respiration is more advantageous to a large organism than anaerobic respiration because aerobic respiration :
a] Does not require sunlight
b] Produces oxygen as a waste product
c] Does not require molecular oxygen and hydrogen
d] Releases more energy from an equal amount of nutrients

98. In a fully turgid cell the values (arbitrary values are shown) of DPD, OP and TP will show the tendency :
a] DPD 02 atm; OP 07 atm; TP 05 atm
b] DPD 00 atm; OP 15; TP 05 atm
c] DPD 10 atm; OP 15 atm; TP 05 atm
d] DPD 05 atm; OP 12 atm; TP 07 atm

99. Light harvesting complexes surrounded by both photosystems (I and II) and _____ light energy to them.
a] Draw b] Accumulate
c] Transfer d] None of these

100. Functions and vital activities of the plants are studied under :
a] Ecology b] Cytology
c] Physiology d] Morphology

101. Plasmagel is the name of
a] Ectoplasm b] Endoplasm
c] Protoplasm d] None of these

102. The isolation of individual organelle from cell homogenates is achieved through :
a] Employment of different solvents
b] X-ray diffraction
c] Chromatography
d] Differential centrifugation

103. Excessive use of fertilizer causes death of plants due to :
a] Endosmosis b] Exosmosis
c] Imbibition d] Turgidity

104. Swelling of wooden door during rainy season is due to :
a] Endosmosis b] Imbibition
c] Capillary d] Deplasmolysis

105. In fermentation :
a] Energy is liberated
b] Energy is consumed
c] Energy is neither liberated nor consumed
d] Energy is stored

106. Ethylene especially in tomato plants has the following effects:
a] Enhances stem elongation
b] Reduces radial expansion
c] Retards formation of adventitious roots
d] Causes leaf senescence

107. The condition under which transpiration would be lowest :
a] Environmental conditions are very dry
b] When there is high humidity in the atmosphere
c] High wind velocity
d] There is excess of water in the cells

108. The nucleoplasm and the cytoplasm are continuous with one another through the nuclear pores in :
a] Eukaryotic cells
b] Prokaryotic cells
c] Cells of blue green algae
d] Bacteria

109. The conditions under which transpiration would be most rapid is :
a] Excess of water in soil
b] Low humidity, high temperature, guard cells are turgid and moist soil
c] Low velocity of wind
d] High humidity

Indian Councils Act, 1909 recognised the principles of election, through in special cases.

110. Nitrogen is essential component of:
a] Fat b] Protein
c] Carbohydrates d] Mg

111. Stomata are bound by a pair of guard cells and open guard cells are :
a] Flaccid b] Large
c] Turgid d] Green

112. Osmosis means :
a] Flow of solute from low concentration to higher concentration
b] Flow of solute from high concentration to low concentration
c] Flow of solvent from low concentration to high concentration
d] Flow of solvent from high concentration to low concentration

113. When the dark period of short day plant is interrupted by a brief exposure of light then the plant :
a] Flowers immediately
b] Turns into a long day plant
c] Will not flower at all
d] Gives more flowers

114. Out of the following, which one is the most common type of transpiration?
a] Foliar b] Stomatal
c] Lenticular d] Cuticular

115. The rate of transpiration will greatly depend upon :
a] Frequency of stomata
b] Position of stomata
c] State of stomata
d] None of the above

116. Photochemical reactions of photosynthesis take place in :
a] Frets b] Stroma
c] Thylakoids d] All of these

117. If fermentation is allowed to proceed in a closed vessel :
a] Gas pressure will develop because of excess of oxygen
b] Gas pressure will develop because of excess of carbon dioxide
c] Vacuum will be created
d] No change will take place

118. In submerged leaves respiration takes place through:
a] General surface of the leaf
b] Stomata present on dorsal surface of the leaf
c] Stomata present on ventral surface of leaf
d] Stomata present on both surfaces of the leaf

119. Cytokinins were first discovered in 1913 by Gottlieb Haberlandt in :
a] Germany b] Holland
c] Poland d] Austria

120. Which of the following is indispensable in cell culture?
a] Gibberellin b] Kinetin
c] Ethylene d] Auxin

121. Plants hormone which is translocated to other parts for growth of the plant as :
a] Gibberellin b] Cytokinin
c] Indole acetic acid d] None of the above

122. Which of the following is capable of delaying yellowing of leaves?
a] Gibberellins b] Cytokinins
c] ABA d] Auxins

123. Which of the following is an essential part of photosynthesis?
a] Glycolysis b] Photorespiration
c] Plasmolysis d] Photolysis

124. Hydroponics are :
a] Growing of plants in water culture
b] Growing of aquatic plants
c] Growing of floating aquatic plants
d] Soil less cultivation of plants

125. The end product of fermentation is :
a] H_2O and CO_2
b] C2H5OH and CO2
c] CO_2
d] Pyruvic acid and H_2O

126. Most plants obtain their nitrogen from the soil in the form of :
a] HNO_3 b] Nitrate
c] Free nitrogen d] Nitrogen oxide

127. The optically hyaline portion of the protoplasm is called as :
a] Hyaloplasm b] Cytoplasm
c] Mesoplasm d] Endoplasm

128. Which of the following elements can land plants normally obtain directly from the air?
a] Hydrogen b] Phosphorus
c] Nitrogen d] Carbon

129. Which of the following is not naturally occurring plant hormone?
a] IAA b] GA_2
c] 2,4-D d] Gibberellin

130. One of the following causes rootings in cuttings :
a] IBA b] 2,4-D
c] GA_3 d] ABA

131. When a seed is placed in a suitable medium for germination, by which of the following first process water enter into the seed coat?
a] Imbibition b] Absorption
c] Osmosis d] Root pressure

132. Plasma membrane is :
a] Impermeable
b] A selective barrier
c] A non-selective barrier
d] Made of up cellulose

GK GS
NHB is an apex institution for providing housing loan on easy conditions.

133. The efforts to enhance the partition of assimilates to storage organs of interest to farmers have quite successful due to :
a] Improved component technology
b] Improved control of insects
c] Better disease control
d] Better breeding techniques

134. Gibberellic acid has been successfully employed to induce flowering :
a] In long day plants under short day conditions
b] In short day plants under long day conditions
c] For some plants
d] None of the above

135. The prokaryota are a group of organisms :
a] In which cells are a group of organisms
b] In which nuclear membrane, nucleolus and chromosomal matrix are lacking
c] In which chlorophyll is lacking
d] With a well defined nucleus

136. Which one of the following will reduce the rate of transpiration?
a] Increase in windy velocity
b] Rise in temperature
c] Increase in water uptake by plants
d] Decrease in light intensity

137. The total volume of water on melting :
a] Increases
b] Remains constant
c] Decreases
d] None of these

138. Gibberellins are now commercially used ______into increase sugarcane growth and sugar yields.
a] India b] Hawaii
c] Japan d] Korea

139. A hypothetical chemical involved in the flowering of plants is :
a] Indole acetic acid
b] Florigen
c] Gibberellin
d] Kinetin

140. Anaerobic respiration is also known as :
a] Intra-molecular respiration
b] Inter-molecular respiration
c] Extra-molecular respiration
d] Molecular respiration

141. Which of the following is not a part of chlorophyll?
a] Iron b] Nitrogen
c] Carbon d] Hydrogen

142. Wilting of leaves in hot weather is due to :
a] Iron and calcium
b] Sodium and copper
c] Calcium and potassium
d] Iron and magnesium Fe, Mg

Did You Know ?

Oldest gharana

The Gwalior *gharana,* of the Hindustani stream of music was founded in the 16th century by Nathu Peerbash, a musician at the court of Raja Mansingh Tomar and is the oldest school of Hindustani music.

143. A micro-essential element is one which :
a] Is less important than the major essential element
b] Is more important than the major essential element
c] Is present in small quantities in the soil
d] Is needed in very small amounts but is as important as a major essential element

144. Rotation of crops is practised in plants because :
a] It decreases the nitrogen content of the soil
b] It increases the nitrogen content of the soil
c] It kills harmful bacteria of the soil
d] It increases the water content of the soil

145. It is advantageous to the *Hydrilla* to demonstrate photosynthesis than land plant because :
a] It respires slowly
b] It does not transpire
c] It photosynthesizes rapidly
d] Oxygen bubbles can be collected over water

146. If you place a piece of tomato in concentrated solution:
a] It would become limp due to loss of water from its cells
b] It would become turgid due to absorption of water from salt solution
c] If would die
d] The plant is rapidly respiring

147. When a piece of potato tuber is placed in concentrated salt solution :
a] It would become turgid by absorbing water from salt solution
b] It would become limp (shrink) due to loss of water from its solution
c] It would die
d] Nothing would happen

148. According to the fluid mosaic model the correct sequence of plasmalemma is :
a] L,-P,L,-P b] L,-P,-P,-L
c] P,-L,-L,-P d] P,-P,-L,-L

149. In the process of respiration in plants 180 gms. of sugar plus 192 gms. of oxygen produce :
a] Large amount of CO_2, no water and no energy
b] 132 gm. of CO_2, 54 gms. of water and 337 kilo cals. of energy
c] 264 gms. of CO2,108 gms. of water and 674 kilo cals. of energy
d] 528 gms. of CO_2, 216 gms. of water and 1348 kilo cals.of energy

Soft water gives lather easily with soap. GS

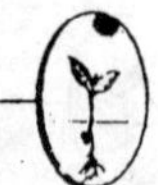

150. Leaves with yellow or colourless patches are termed as :
a] Lecorine b] Decolourized
c] Etiolated d] Variegated

151. Oxygen evolved in photosynthesis is :
a] Less than CO_2 consumed
b] More than CO_2 consumed
c] Equal to CO2 consumed
d] None of these

152. The dry weight of a green leaf will be :
a] Lesser in evening
b] Greater in the afternoon
c] Greater in morning
d] Almost same throughout the whole day

153. Etiolated plants are formed due to lack of :
a] Light b] Hg
c] Fe d] Mg

154. For photosynthesis (*i.e.* for the synthesis of organic matter) the green plants need only :
a] Light b] Chlorophyll
c] CO_2 and water d] All of these

155. Growth is primarily affected by two climatic factors which are :
a] Light and temperature
b] Temperature and relative humidity
c] Light and wind
d] Rainfall and temperature

156. For the purpose of photosynthesis all except one of the following items are essential, point out the exception :
a] Light, chlorophyll
b] Oxygen, sucrose
c] CO_2, optimum temperature
d] Water, minerals

157. One of the following is not found in intact cells :
a] Lomasomes b] Dictyosomes
c] Microsomes d] Ribosomes

158. A plant which needs not less than 10 hours of light to flower is :
a] Long day plant b] Short day plant
c] Day neutral plant d] None of these

159. If cells are broken up and sedimented by centrifugation, the new structure formed in one of the fractions is :
a] Lysosome b] Ribosome
c] Microsome d] Centrosome

160. Which of the following element is concerned with the photolysis of water?
a] Magnesium b] Manganese
c] Calcium d] Zinc

161. Which one of the following elements is not required by plants for their normal healthy growth?
a] Magnesium b] Lead
c] Iron d] Calcium

162. The framework elements of plants are :
a] Magnesium, Copper and Iron
b] Copper, Carbon and Oxygen
c] Manganese, Calcium and Nitrogen
d] Carbon, Hydrogen and Oxygen

163. Fertilizers have formula consisting of three figures such as 8-10-10. These figures stand for percentages of :
a] Magnesium, iron, potassium
b] Calcium, iron, boron
c] Nitrogen, phosphorous, potassium
d] Magnesium, manganese, sulphur

164. Anaerobic respiration was first of all reported by :
a] Maguenne b] Kostychev
c] Klein d] Pfeffer

165. The provision of light of shorter wavelength at the same time; as longer red wavelengths _______photosynthesis.
a] Maintains b] Enhances
c] Reduces d] No effect

166. Probably the most useful measurable property in the soil-plant-air system is :
a] Matric potential b] Water potential
c] Osmotic potential d] Redox potential

167. Every living cell has a :
a] Membrane b] Food vacuole
c] Chloroplast d] Cell wall

168. Cytochrome oxidase contains :
a] Mg b] Fe
c] Hg d] Co

169. The figures of cork cells as seen by Robert Hooke were published in this book :
a] Micrographia b] Origin of species
c] Plant kingdom d] Genera plantarum

170. Proteins are synthesized on :
a] Mitochondria b] Centrosome
c] Golgi bodies d] Ribosomes

171. Tonoplast is a differentially permeable membrane surrounding the :
a] Mitochondria b] Cytoplasm
c] Vacuole d] Nucleus

172. Cell organelle found only in plants is :
a] Mitochondria b] Golgi complex
c] Ribosomes d] Plastids

173. Chlorophyll contains :
a] Mn b] Fe
c] K d] Mg

174. Organisms which fix atmospheric nitrogen in the soil are found among :
a] Mosses b] Green algae
c] Soil fungi d] Bacteria

175. Which of the following is a weed killer?
a] N.A.A b] 2, 4-D
c] ABA d] GA

GK GS
Khol is the name given to the difference of relief in the Doab of Ganga and Yamuna.

176. In the nodules of roots in leguminous plants we find:
a] N_2 producer bacteria
b] Denitrifying bacteria
c] N2 fixing bacteria
d] Nitrifying bacteria

177. When chlorophyll is burnt which one is obtained?
a] Na b] Mn
c] Mg d] Ca

178. The following is a naturally occurring growth inhibitor:
a] NAA b] GA_3
c] ABA d] IAA

179. Which is not essential micronutrient?
a] Ni b] Co
c] Mn d] B

180. During 24 hours there is a time when plants neither give O_2 nor CO_2. This is the time of :
a] Night b] Day light
c] Twilight d] None of these

181. Plants absorb nitrogen in the form of :
a] Nitrate b] Nitrogen gas
c] Nitrogen dioxide d] All of the above

182. The production of nitrates from Ammonia through *Nitrosomonas* is called :
a] Nitrification
b] Ammonification
c] Nitrogen fixation
d] Denitrification

183. The element with is required in largest quantities by plant of water?
a] Nitrogen b] Calcium
c] Sulphur d] Phosphorous

184. One the following is micronutrient for plants :
a] Nitrogen b] Manganese
c] Phosphorous d] Calcium

185. Bacteria that change proteins to ammonia in nitrogen cycle are :
a] Nitrogen fixing bacteria
b] Nitrate bacteria
c] Decay bacteria (ammonifying bacteria)
d] Denitrifying bacteria

186. Bacterioids are :
a] *Nitrosomonars* bacteria in soil
b] A mobile bacterium
c] A bacterial cell infected with viruses
d] Enlarged non-mobile cellular bacteria-*Rhizobium legumino-sarum* in root nodules of legumes

187. Double membrane is absent in :
a] Nucleus b] Lysosomes
c] Mitochondria d] Chloroplast

188. Chloroplasts fix :
a] O_2 b] H
c] CO2 d] N_2

189. Increase in temperature and velocity of wind cause an increase in transpiration initially but later it slows down because :
a] Of closure of stomata
b] Water is not made available
c] The air around the plant becomes humid
d] Of mechanical disturbances

190. Ethrel (trade name) of ethephone (common name) is used in various aspects of :
a] Olericulture b] Aquaculture
c] Horticulture d] Apiculture

191. ABA has the following effects depending upon the tissue involved:
a] On plasma membrane of roots
b] Inhibition of RNA-synthesis (transcription)
c] Inhibition of protein synthesis (translocation)
d] All of these

192. Chlorophyll is present :
a] On the surface of chloroplasts
b] In the stroma of chloroplasts
c] In the grana of chloroplasts
d] Dispersed throughout the chloroplasts

193. When a cell is fully turgid its :
a] OP = Zero b] DPD = OP
c] DPD = Zero d] SP = TP

194. DPD is equal to :
a] OP X TP b] OP-TP
c] TP-OP d] OP + TP

195. Seeds swell when placed in water due to :
a] Osmosis b] Imbibition
c] Hydrolysis d] Plasmolysis

196. The loss of water in the form of vapour from aerial plant parts is known as :
a] Osmosis b] Respiration
c] Photosynthesis d] Transpiration

197. Seeds when soaked in water imbibe it because :
a] Osmotic pressure inside the seed is low
b] Seed coat contains lot of salts
c] The process of absorption works
d] There are many vacuoles in the endosperm

198. When a cell is fully turgid, which of the following will be zero?
a] Osmotic pressure b] Turgor pressure
c] Wall pressure d] Suction pressure (D.P.D)

199. Stomata open because of :
a] Oxygen in the air
b] Increased turgidity of the guard cells brought about by exposure to light
c] Vacuoles in guard cells
d] All the above

200. In anaerobic respiration in plants:
a] Oxygen is taken in
b] Oxygen is given out
c] Carbon dioxide is given out
d] Carbon dioxide is taken in

GK GS

Atvika were forest tribes.

201. Before opening of stomata, accumulation of one of the ion is seen in guard cells :
a] P b] K
c] Mg d] Na

202. In nature, the main organisms which take part in nitrogen cycle are :
a] Parasitic fungi b] Legume plants
c] Bacteria d] Saprophytic flowering plants

203. In terms of permeability the cell wall and plasmalemma are :
a] Permeable and differentially permeable
b] Both differentially permeable
c] Both semi-permeable
d] Semi-permeable and differentially permeable

204. Water potential concept was formulated by
a] Peter Ray, 1960
b] M.H. Mathew, 1954
c] Otto Renner, 1915
d] W.C. Hoffer, 1920 ,

205. One of the following is a precursor of auxin :
a] Phenylalanine b] Methionine
c] Tryptophan d] Cystosine

206. Gibberellins were first discovered in _______ in studies with diseased rice plants that grew excessively tall.
a] Philippines b] Thailand
c] China d] Japan

207. Minerals absorbed by roots move to the leaf through:
a] Phloem
b] Sieve tube
c] Xylem
d] None of the above

208. According to the mosaic model (proposed by Singer and Nicholson) plasma membrane is composed of:
a] Phospholipids and integrate proteins
b] Phospholipids, extrinsic proteins, intrinsic proteins
c] Phospholipids, hemicellulose
d] Cellulose, hemicellulose

209. Commonest free ion in the cell is of :
a] Phosphorous b] Potassium
c] Iron d] Sulphur

210. Two chief functions of leaves are :
a] Photosynthesis and respiration
b] Photosynthesis and transpiration
c] Transpiration and respiration
d] Respiration and digestion

211. Alcohol is produced during the process of :
a] Photosynthesis
b] Aerobic respiration
c] Anaerobic respiration
d] None of the above

212. Which pigment is essential for nitrogen fixation by leguminous plants?
a] Phycocyanin b] Leghaemoglobin
c] Phycoerythrin d] Anthocyanin

213. A plant is kept in 30 ppm. of CO_2 concentration what will happen to it?
a] Plant will not grow but will not die
b] Plant will show normal photosynthesis
c] Plant will die soon
d] Plant will grow faster

214. Semi-permeability is the property of :
a] Plasma membrane b] Cell wall
c] Golgi body d] Endoplasmic reticulum

215. If a thoroughly cleaned goat bladder if filled with a syrup, tightly tied immersed in a vessel of water :
a] Plasmolysis will occur
b] The volume of the liquid within the bladder will remain constant
c] The volume of liquid within the bladder will increase
d] The volume of liquid within the bladder will decrease

216. Protoplasm is
a] Polyphasic colloidal system
b] Collodial solution
c] Suspension
d] True solution

217. Which of the following elements has not been proved to be essential for plants?
a] Potassium b] Sodium
c] Zinc d] Iron

218. The apparatus generally used for studying the rate of transpiration stream is :
a] Potometer b] Evaporimeter
c] Atmometer d] Perimeter

219. High concentration of synthetic auxins would :
a] Prevent lateral buds to grow
b] Kills weeds
c] Cause root initiation
d] Control cell enlargement

220. The process of photophos-phorylation was discovered by
a] Priestley b] Warburg
c] Arnon d] Calvin

221. Which level of protein structure is affected by DNA (or for which level of protein structure DNA carries information) ?
a] Primary structure
b] Secondary structure
c] Tertiary structure
d] Quarternary structure

222. Chlorophyll 'b' is present in :
a] Prokaryotes
b] Eukaryotes
c] Both prokaryotes and eukaryotes
d] Bacteria

223. Supra-optimal concentration of auxins :
a] Promotes flowering
b] Kills the plants
c] Prevents shortening of the internodes
d] Promotes growth in both stem and root apex

224. Cell membrane consist of :
a] Proteins
b] Phospholipids
c] Both phospholipids and proteins
d] Cellulose

225. In anaerobic respiration the end product is :
a] Pyruvic acid
b] Ethyl alcohol
c] Starch
d] Sugar

226. Chloroplast contains maximum quantity of :
a] Pyruvic carboxylase
b] Hexokinase
c] RUDP carboxylase
d] None of the above

227. Which of the following are photosynthetic units?
a] Quantasomes b] Oxysomes
c] Phycobilinosome d] F_1 particles

228. Which of the following are short day plants?
a] Radish b] Maize
c] Soyabean d] Sugar beet

229. Auxins promote growth :
a] Rarely b] Always
c] Mostly d] Never

230. Potometer is used to measure :
a] Rate of respiration
b] Rate of transpiration
c] Growth
d] Osmosis

231. The most effective wavelength of visible light in photosynthesis is in the region of which of the following?
a] Red b] Yellow
c] Violet d] Green

232. Photosynthesis is a :
a] Reductive, endergonic, catabolic process
b] Reductive, endergonic, anabolic process
c] Reductive, exergonic, anabolic process
d] Oxidative, exergonic, catabolic process

233. In an *E.coli* cell according to operon theory an operator gene combines with :
a] Regulator gene to "switch off" structural gene transcription
b] Regulator protein to "switch off" structural gene transcription
c] Regulator protein to "switch on" structural gene transcription
d] Inducer gene to "switch on" structural gene transcription

You Don't Know!
First octagonal tomb

The tomb of Khan-i-Jahan, the prime minister of Firuz Shah Tugluq, which has a domed octagonal chamber enclosed by a verandah, is the first octagonal tomb in Delhi.

234. More energy is released by the process of aerobic respiration than by fermentation because fermentation :
a] Requires oxygen both as a hydrogen acceptor and donor
b] Is a process in which only some kind of organic matter can be oxidized
c] Occurs in micro-organisms which have very few enzymes
d] Causes incomplete oxidation of organic food

235. In nitrogen cycle, which of the following plays an important role?
a] *Rhizopus* b] *Mucor*
c] *Nitrobacter* d] *Spirogyra*

236. If we separate the cell organelles of a living cell what part should be alive?
a] Ribosome b] Endoplasmic reticulum
c] Chloroplast d] Cell wall

237. Which of the following is the site of respiration within the cell?
a] Ribosomes b] Nucleus
c] Golgi body d] Mitochondrion

238. Foliar application of colloidal suspensions of triacontanol (30-carbon saturated primary alcohol isolated first from alfalfa shoots) enhanced the growth of seedlings of________ and also maize.
a] Rice b] Soybean
c] Castorbean d] Clusterbean

239. The oxygen produced during photosynthesis comes from water was proved by :
a] Robert Mayer b] Blackman
c] Ruben d] M. Calvin

240. Cytology and genetics were merged into "Cytogenetics" by one of the following :
a] Robertson b] Muller
c] T.H. Morgan d] Altmann

241. Distribution of growth in a root by marking it at equal intervals with Indian ink was originally studied by:
a] Sachs b] Strasburger
c] Went d] Nitsch

242. Phytochrome is used in :
a] Seed germination only
b] Transpiration only
c] Flowering only
d] All physiological processes exhibited by the plants such as seed germination, flowering, stem elongation and transpiration

Export and Import Bank (EXIM) was established on January 1, 1982.

243. Chlorosis occurs in plants growth in :
a] Shade b] Iron free medium
c] Dark d] Strong light

244. In the mechanism of opening of stomata, the important factor is the :
a] Shape of the guard cells
b] Chlorophyll content of the cells
c] Hormone content of the cells
d] Protein content of the cells

245. Under what conditions the transpiration rate will be maximum?
a] Soil is wet and atmosphere is dry
b] Soil is dry and atmosphere is humid
c] Soil is wet and atmosphere is humid
d] Soil is dry and atmosphere is dry

246. Ethylene is a :
a] Solid hormone
b] Gaseous enzyme
c] Gaseous hormone
d] Liquid gas mixture

247. One of the following is present outside the plasma membrane but inside the cell wall :
a] Spherosome b] Peroxisome
c] Lomasome d] Golgi body

248. All are membrane bounded cell organelles except :
a] Spherosomes b] Mitochondria
c] Ribosomes d] Lysosomes

249. The first visible product of photosynthesis is
a] Starch b] Glucose
c] Fructose d] Sucrose

250. Plants synthesis protein from :
a] Starch b] Sugar
c] Amino acids d] Fatty acids

251. Carbohydrates are present in the plasmalemma in the form of :
a] Starch
b] Cellulose
c] Hemicellulose
d] Phospholipids (glycolipids) and phosphoproteins (glycoproteins)

252. The end products of respiration in plants are :
a] Starch and O_2
b] H_2O and energy
c] CO2, H2O and energy
d] Sugar and O_2

253. Guard cells are found in :
a] Stomata b] Root tips
c] Ovary d] Lenticels

254. Transpiration occurs through :
a] Stomata b] Cuticle
c] Lenticles d] All the above

255. The protoplasm theory was established in plant cell by :
a] Strasburger b] Robert Brown
c] De Bary d] Max Schultze

256. The transpiration is regulated by the movements of :
a] Subsidiary cells of the leaves
b] Guard cells of the stomata
c] Mesophyll tissue cells
d] Epidermal cells of the leaves

257. Cytokinins have the ability to :
a] Substitute cold treatment
b] Cause root formation
c] Cause morphogenesis
d] Convert dwarf plants into tall plants

258. Ripening of banana is accompanied with :
a] Sudden rise in cytokinin
b] Sudden rise in auxin
c] Sudden rise in ethylene
d] Sudden rise in gibberellin

259. A bulk of nitrogen in nature is fixed by :
a] Symbiotic bacteria
b] Chemical industries
c] Lightning
d] Denitrifying bacteria

260. The first event in photosynthesis is :
a] Synthesis of ATP
b] Photoexcitation of chlorophyll and ejection of electron
c] Photolysis
d] Release of oxygen

261. Mg for Fe are needed by plants for
a] Synthesis of chlorophyll :
b] Energy transfer
c] Translocation of carbohydrates
d] Stomatal opening

262. Brassins, recently discovered steroid growth-promoters were first isolated from pollen grains of rape plants, but now known to be present in ______ also.
a] Tea b] Bean
c] Rice d] In all of these

263. Which one of the following is most limiting factor for nitrification in the soil?
a] Temperature b] Tillage
c] Soil reaction (pH) d] Moisture

264. The attraction of like molecules for each other (because of hydrogen bonds) is called :
a] Tension
b] Adhesion
c] Cohesion
d] None of these

GK GS
Pyrene is commercial name of carbontetrachloride.

265. A trace element is :
   a] That which draws other elements out of protoplasm
   b] Radoactive can be traced by Geiger counter
   c] Required in very minute amounts
   d] That which was first found in protoplasm
266. Tick-out the correct statement :
   a] The cell wall is permeable
   b] The cell wall is semi-permeable
   c] The cell wall is an impermeable membrane
   d] The cell wall does not react to the permeability
267. In an experiment a plant is dried, placed in a crucible and strongly heated in air for a long time. The residue is composed of :
   a] The oxides or carbonates of about ten elements
   b] Carbon
   c] Starch
   d] The oxides and carbonates of three elements
268. If CO_2 content of the atmosphere is as high as 300 parts per million:
   a] The plants would not grow properly
   b] The plants would thrive well
   c] All plants would be killed
   d] Plants would grow for some time and then die
269. Which among the following is the most appropriate reason for storing green coloured apples at low temperatures?
   a] The rate of photosynthesis is reduced
   b] The rate of respiration is reduced
   c] The rates of photosynthesis and respiration are reduced
   d] Respiration and photosynthesis are completely inhibited
270. In many plants the changes over from vegetative to reproductive phase takes place in response to :
   a] The severity of temperature
   b] The length of the day
   c] The oxygen present in the air
   d] Mainly the food material available in the soil
271. When a plant cell is placed in a solation which is hypotonic to the cell sap, which of the following conditions will not apply?
   a] The wall pressure of the cell will fall
   b] The cell will become turgid
   c] The suction pressure of the cell sap will fall
   d] The water potential of the cell sap will rise
272. Without adding nitrogenous manures good yield or rice crop can be obtained because :
   a] Their roots have nitrogen fixing bacteria
   b] There are nitrogen-fixing algae in rice fields
   c] Rice plants do not require nitrogen
   d] They require minute quantities of nitrogen
273. One of the alkaloids listed below is found in cocoabeans:
   a] Theobromina   b] Caffeine
   c] Quinine   d] Colchicine
274. Members of bean family are of particular importance in crop rotation programmes primarily because :
   a] They add nitrates to the soil
   b] They manufacture nitrogenous compounds
   c] They have nitrogen fixing bacteria associated in their roots
   d] They provide green manure
275. Algae float in water during day time and sink during night time because :
   a] They become light due to consumption of food materials in respiration
   b] They become buoyant in light due to accumulation of oxygen bubbles released in photosynthesis
   c] They come up to enjoy sunshine
   d] They loose weight at night
276. It is not advisable to sleep under the trees at night because :
   a] They release carbon dioxide during night
   b] They release oxygen during night
277. How do most of the plants obtain air?
   a] Through a system of branches
   b] Through the tips of branches
   c] Through opening in the leaves
   d] From water that comes up from soil
278. By hormone application it is possible to obtain seedless fruits. In which of the following it is desirable to obtain seedless fruits?
   a] Tomato   b] Orange
   c] Watermelon   d] Pomegranate
279. The process by which proteins are synthesized in a cell is called :
   a] Translation   b] Transduction
   c] Translocation   d] Transcription
280. Root pressure is maximum when :
   a] Transpiration is very high and absorption is also high
   b] Transpiration and absorption both are low
   c] Transpiration is high and absorption is very low
   d] Transpiration is very low and absorption is high
281. Which one is the precursor of chlorophyll?
   a] Tryptophan
   b] Protochlorophyll or chlorophyllides
   c] Bacteriochlorophyll
   d] Bacterioviridin
282. Water which is an essential part of protoplasm may constitute :
   a] Up to a quarter of it
   b] Up to a half of it
   c] Up to three quarters of it
   d] More than three quarters of it

We find the fan-shaped alluvial cones in Ganga plain of Bihar.

283. Which of the following leaves would dry up last?
a] Upper surface greased
b] Lower surface greased
c] Both surfaces greased
d] Both surfaces ungreased

284. One of the following is sulphur containing amino acid:
a] Valine b] Methionine
c] Tryptophan d] Phenylalanine

285. In which of the following plants would metabolism be hindered if the leaves are coated with wax on their upper surface?
a] Vallisneria b] Pista
c] Lotus d] Hydrilla

286. Enzymes are basically :
a] Vitamins b] Fats
c] Proteins d] Carbon

287. The actual pressure with which water enters into the cell is called:
a] W.P b] O.P
c] D.P.D d] Diffusion

288. The normal form of a plant is maintained by the presence of :
a] Water b] Gases
c] Solids d] None of these

289. The incomplete breakdown of sugar in anaerobic respiration results in the formation of :
a] Water and carbon dioxide
b] Alcohol and carbon dioxide
c] Fructose and water
d] Glucose and carbon dioxide

290. If cell A with OP =5 and TP = 4 is surrounded by the cells with OP =3 and TP = 1 what will be direction of water movement?
a] Water will not move up
b] Water will move up
c] From cell A to other cells
d] From other cells to cell A

291. Which one of the following is a long day plant?
a] Wheat b] Jowar
c] Bajara d] Soyabean

292. Zeatin has been isolated and identified from the milky endosperm of :
a] Wheat b] Sorghum
c] Maize d] Barley

293. Process of endosmosis stops :
a] When water concentration becomes equal
b] When solution become isotonic
c] When the leaves fall
d] When there is no light

294. Most of the _______ flowers contain carotenoid especially xanthophyll type and it is believed that they benefit certain plants by attracting pollinating insects.
a] White b] Red
c] Pink d] Yellow

295. Symbiotic nitrogen fixing bacteria occur in the genus:
a] *Xanthomonas* b] *Pseudomonas*
c] *Rhizobium* d] *Azotobacter*

296. Which pigment is absent in chloroplast?
a] Xanthophyll b] Anthocyanin
c] Chlorophyll-a d] Carotene

297. Deficiency of Mo causes :
a] Yellowing of leaves
b] Mottling and necrosis of leaves
c] Poor development of vasculature
d] Bending of leaf tip

298. On the basis of symptoms of chlorosis in leaves a student inferred that this was due to deficiency of nitrogen. This inference could be correct only if we assume that yellowing of leaves appeared first in :
a] Young leaves
b] Old leaves
c] Matured leaves followed by young leaves
d] Young leaves followed by mature leaves

299. Pineapple can be made to flower in off season by :
a] Zeatin b] Ethylene
c] Temperature d] Short days

300. 'Die-back' of citrus is caused due to the deficiency of :
a] Zinc b] Copper
c] Manganese d] Iron

301. Which of the following element is involved in nitrogen metabolism in reduction of nitrates?
a] Zinc b] Molybdenum
c] Boron d] Manganese

302. Which of the following elements is involved in nitrogen metabolism in reduction of nitrates?
a] Zinc b] Molybdenum
c] Boron d] Manganese

303. Cellulosic cell wall can be specifically stained by :
a] Zinc chloride b] Phloroglucinol
c] Methylene blue d] Sudan IV

304. One of the following is needed in the reduction of nitrate to nitrite :
a] Zn b] Mo
c] Fe d] B

305. Pigment system is concerned with:
a] Photolysis of water b] Reduction of CO_2
c] Flowering d] None

306. The researches on photosynthesis started :
a] In the 17th century b] In the 18th century
c] In the 19th century d] In the early 20th century

307. Respiration is :
a] Exothermic process
b] Endothermic process
c] Endergonic process
d] Anabolic process

308. Which of the following is the source of respiration ?
a] Stored food b] RNA
c] DNA d] ATP

309. Instantaneous source of energy is:
a] Amino acid b] Fats
c] Glucose d] Proteins

GK GS

No mention of Malecha is in Rig Veda.

Matching

1. The coefficient of viscosity (centipoises) at 20°C is.

a]	Water	i]	1.60
b]	Ethyl alcohol	ii]	830.0
c]	Mercury	iii]	1.002
d]	Glycerine	iv]	1.200

2. Empirically determined osmotic potentials of leaves of the following plants were.

a]	Sunflower	i]	-8.9
b]	Chickweed	ii]	-4.8
c]	Pickleweed	iii]	-1.9
d]	Mandarin orange	iv]	-0.74

3. Stomates are variously arranged in various plants.

a]	Waterlily	i]	None at all
b]	Submerged plants	ii]	On both sides equally
c]	Grasses	iii]	On top
d]	Pine	iv]	Sub-stomatal crypt

4. Quantity and units (with symbols) are given below, correspond to each other, match them.

a]	Frequency	i]	Pascal (Pa)
b]	Pressure	ii]	Newton (N)
c]	Energy/work/heat	iii]	Joules (J)
d]	Force	iv]	Hertz (Hz)

5. At 20°C, match the atmospheric water potential with relative humidity (RH) per cent as given below.

a]	98% of RH	i]	-14.2 MPa
b]	90% of RH	ii]	-311 MPa
c]	50% of RH	iii]	-2.72 MPa
d]	10% of RH	iv]	-93.5 MPa

6. Match the texture class with the diameter range as specified in World System of Soil Classification.

a]	Coarse sand	i]	0.02-0.002 mm
b]	Fine sand	ii]	0.002 mm and less
c]	Silt	iii]	2.0-0.2 mm
d]	Clay	iv]	0.2-0.02 mm

7. Match the salts and their molarity as in Hoagland's solution.

a]	KNO_3	i]	0.003
b]	$Ca (NO_3)_2$	ii]	0.230
c]	H_2PO_4	iii)	0.490
d]	$MgSO_47H_2O$	iv]	0.010

8. The following are the concentrations of some essential elements considered adequate for most of the higher plants (in dry tissues), match them suitably.

a]	Molybdenum	i]	20 ppm
b]	Copper	ii]	0.10 ppm
c]	Zinc	iii]	100 ppm
d]	Iron	iv]	6.0 ppm

9. Match the following elements with their available form.

a]	Copper	i]	MoO_4^-
b]	Molybdenum	ii]	Cu^+, Cu^{2+}
c]	Zinc	iii]	H_3BO_3
d]	Boron	iv]	Zn^{2+}

10. Match the following concentrations with the essential (considered adequate on dry tissue basis for most of the plants).

a]	Sulphur	i]	5,000 ppm
b]	Magnesium	ii]	2,000 ppm
c]	Calcium	iii]	1,000 ppm
d]	Potassium	iv]	10,000 ppm

Proportional representation by means of single transferable vote is the method of presidential election in India.

11. **Nutrient** — **Concentration**
    - a] Phosphorous — i] 50 ppm
    - b] Nitrogen — ii] 2,000 ppm
    - c] Chlorine — iii] 15,000 ppm
    - d] Manganese — iv] 100 ppm
12. **Nutrient** — **Concentration**
    - a] Hydrogen — i] 450,000 ppm
    - b] Carbon — ii] 50 ppm
    - c] Boron — iii] 60,000 ppm
    - d] Manganese — iv] 20 ppm
13. **The form of essential nutrients available to plants and the nutrients are as follows, match them.**
    - a] Nitrogen — i] Fe^{3+}, Fe^{2+}
    - b] Iron — ii] $H_2PO_4^-$, $HPO_4^=$
    - c] Phosphorous — iii] $SO_4^=$
    - d] Sulphur — iv] NO_3^+
14. **Enzymes are not uniformly mixed throughout cells, but are located at different sites to catalyse different processes in plants as.**
    - a] For photosynthesis — i] Nuclei
    - b] For aerobic respiration — ii] In mitochondria
    - c] For respiration — i] In chloroplasts
    - d] For mitosis/synthesis of DNA/RNA — iv] In cytosol
15. **Match the following enzymes with their general reactions.**
    - a] Proteinases — i] Transfers phosphate group especially from ATP
    - b] Transferases — ii] Hydrolyzes lipids (esters)
    - c] Lyases — iii] From double bonds by elimination of a chemical group
    - d] Lipases — iv] Hydrolyze proteins
16. **Match the enzyme with their general reactions.**
    - a] Ribonucleases — i] Rearrange atom/molecule tolower form
    - b] Deoxyribonucleases — ii] Hydrolyze RNA
    - c] Isomerases — iii] Breakdown chemical bonds
    - d] Hydrolases — iv] Hydrolyse DNA
17. **Match the carbohydrate content of the following seeds.**
    - a] Maize — i] 2 per cent
    - b] Wheat — ii] 14 per cent
    - c] Soyabean — iii] 51.74 per cent
    - d] Sunflower — iv] 60-75 per cent
18. **Match the plants with the temperature at which they can photosynthesize.**
    - a] Bacteria, BOA — i] −60°C
    - b] Conifers (extremely slowly) — ii] 35°C
    - c] Exposed to bright sunshine — iii] -18°C
    - d] Antarctic lichens — iv] 70°C
19. **Match the following plants with their maximum photosynthetic rate under natural conditions in µmol CO_2 per square metre per second.**
    - a] Maize (C-4 pathway) — i] 0.6-2.4
    - b] Soyabean (C-3 pathway) — ii] 3-12
    - c] Century plant (CAM) — iii] 20-40
    - d] European beech — iv] 10-20
20. **Match the scientists with the observation which they made.**
    - a] Stephen Hales, 1927 — i] Implicated oxygen (dephlogisticated air)
    - b] Joseph Priestley, 1771 — ii] Light, necessary for air purification
    - c] Jan Ingenhousz — iii] Noxious gas CO_2 stimulated O_2 production by plants
    - d] Jean Senebier, 1782 — iv] Partial nourishment from atmosphere and light, participation

21. Match the following sterols:

a]	Sitosterol	i]	Widespread in trace amounts in plants
b]	Cholestrol	ii]	Sex attraction in aquatic fungus *Achlya bisexualis*
c]	Ergosterol	iii]	Most abundant in green algae and higher plants
d]	Antheridiol	iv]	Rare in plants, common in some fungi

22. Match the phytoalexins (Greek, phyton, plants; alexin, a warding off substance} with their plant parts wherein they occur.

a]	Pisatin	i]	Carrot roots
b]	Phaseolin	ii]	Bean pods
c]	Isocoumarin	iii]	Red clover roots
d]	Trifolirhizin	iv]	Pea pods

23. Match the alkaloids with the plants, they are associated with.

a]	Morphine	i]	Flowering heads of cactus *Lophophora williamsii*
b]	Nicotine	ii]	Cultivated tobacco varieties
d]	Meascaline	iv]	Poisonous black night shade

24. Match the following plant responses to the environment.

a]	Gravitroplsm	i]	Growth towards light
b]	Phototropism	ii]	Nastic movement due to touch
c]	Thigmotropism	iii]	Growth of several organs on intermediate angles
d]	Plagiotropis	iv]	Response to gravity

25. Match the longevity of the seeds of the plants given below.

a]	Cocklebur (*Xanthium strumariumy*)	i]	221 years
b]	Canada thistle (*Circium arvense*)	ii]	21 years
c]	Kentucky blue grass (*Poa pratensis*)	iii]	16 years
d]	Sensitive plant (*Mimosa glonlerata*)	iv]	30 years

26. Match the following.

a]	Short day plant	i]	Kidney bean (*Phaseolus vulgaris*)
b]	Long day plant	ii]	Sugarbeet (*Beta saccharifera*)
c]	Intermediate day plant	iii]	Soybean (*Glycine max*)
d]	Day neutral plant	iv]	Tarweed (*Madia elegans*)

27. Match the organisms with compatible solute and increase in their cells during osmotic adjustment.

a]	Bacteria	i]	Sucrose
b]	Fungi	ii]	Sodium chlolride
c]	Microalgal	iii]	Proline
d]	Angiosperms	iv]	Arabitol

28. Match the following.

a]	R.Q. for carbohydrates	i]	< 1
b]	R.Q. for succulents	ii]	1
c]	R.Q. for protein	iii]	> 1
d]	R.Q. for organic acids	iv]	0

29. Match the following.

a]	Term photo-respiration	i]	Krotkov
b]	Glycolysis also called	ii]	Respiratory organelle
c]	First observed anaerobic respiration	iii]	W. Cruickshank
d]	Mitochondria pathway	iv]	Embden Meyerhof Paranas

30. Match the following.

a]	Named chlorophyll	i]	Schleiden & Schwann
b]	Term protoplasm	ii]	F. Mischer
c]	Isolated nucleic acid	iii]	Pelletier Caventou
d]	Introduced the cell theory	iv]	J.E. Purkinje

EXIM is the apex institution for financing exports and imports.

Glossary of Technical Terms

Accessory or Subsidiary Cells One or two modified epidermal cells adjacent to each guard cell are known as Accessory or subsidiary cells.

Accumulation Ratio The extent to which the concentration of essential solutes becomes greater within the cell than that in external solution is known as Accumulation ratio.

Active Absorption If the actual ratio measured experimentally is more than the predicted ratio between outside and inside the membrane concentrations of ions at equilibrium the cell uses energy to absorb the ion, and this has been defined as Active absorption originally by H.H. Ussing (1949).

Adenine A purine base CSHSNs that is a constituent of nucleic acid and various enzyme systems 11 is white and crystalline.

Adenosine A nucleoside $C_{10}H_{13}NsO_4$, that is a constituent of ribo-nucleic acid yielding adenine and ribose on hydrolysis. It is white, odourless and crystalline powder.

Adenosine Diphosphate (ADP) A compound made up of adenine, five carbon sugar ribose, and two phosphate groups. It is involved in the mobilization of energy in cellular metabolism.

Adenosine Triphosphate (ATP) ADP with an additional phosphate group attached by a high energy bond; its decomposition into ADP and phosphate results in the release of energy needed for cellular metabolism.

Adhesion The attraction between unlike molecules is called adhesion.

Adsorption The adhesion in an extremely thin layer of molecules (as of gases, solutes, or liquids) to the surfaces of solid bodies or liquids which they are in contact.

Alkaloids Aromatic nitrogenous compounds are called alkaloids.

Assimilation Number It is the amount of CO_2 absorbed in grams per hour for each gram chlorophyll during photosynthesis photosynthesis.

Autonomic Movement (Spontaneous Movement) Movement due to internal causes or influences.

Bioassay (Biological Assay) Determination of the relative strength of a substance (hormone, drug etc.) by comparing its effect on a test organism with that of a standard preparation.

Black Body A surface that absorbs all the radiation falling upon it is a black body.

Bulk Flow The flow occurring in response to difference in pressure and involving groups of atoms or molecules together is termed as bulk flow.

Capillarity The interaction between contacting surfaces of a solid and a liquid that distorts the liquid surface from a planar shape refers to capillarity.

Carrier A substance by whose agency some elements or group is transferred from one compound to another.

Chelate A chelate is the soluble product formed when certain atoms in an organic molecule called chelating agent or ligand donate electrons to the metal cation.

Chemical Potential The free-energy per unit quantity of substance, specifically per gramme molecular weight (the free-energy mol^{-1}) is called Chemical potential.

Chemiosmotic Theory Peter Mitchell in England in 1961 first proposed that pH gradients could provide energy for ATP formation in chloroplast, mitochondria and bacteria, it was later known as Chemiosmotic theory.

Chloroplasts Chlorophyll containing plastids are known as Chloroplasts.

Chromoplasts The plastids that contain the red pigments of tomato skin have been termed as chromoplasts.

Circumnutation "A stem appears to trace a more or less regular ellipse as the stem grows" Darwin (1881) called this phenomenon as circumnutation.

Coagulation Becomes soft, semisolid mass of albuminoid substance.

Coenzyme An organic substance of low molecular weight (prosthetic group) that can unite with a given protein (apoenzyme) to form an enzyme system.

Colloids Larger particles when suspended in water, sometimes form a glue and hence they have been termed as Colloids = *Kolla* (Greek word) or glue.

Compensation Point It is the light intensity, at which the photosynthetic intake of carbon dioxide is equal to the respiratory output of carbon dioxide.

Conjugated Protein A complete enzyme with a protein and a non-protein (prosthetic group) part is called holoenzyme or conjugated protein.

Denaturation Change in the structure of protein by heat, acids, alkalis, etc. so that the original properties are greatly changed or eliminated.

Desmotubule In the central region of plasmodesmata lies a tubular structure, called desmotubule.

Diffusion The net movement from one point to another due to random kinetic activities of ions and molecules is called diffusion.

Electrophoresis (Cataphoresis) The movement of colloidal particles suspended in a fluid toward the electrodes in the fluid, through which an electric current is passed, with the resulting collection of the particles at the electrodes.

Emersion Enhancement Effect The enhancement in photosynthesis by supplying shorter wavelengths and longer red wavelengths at the same time is known as Emersion Enhancement Effect.

Enzymes Any of various protein like substances, formed in plant and animal cells, that act as organic catalysts in initiating or speeding up specific chemical reactions and that usually become inactive or unstable at high temperatures.

Etiolation Blanching or bleaching due to lack of sunlight; pale and unhealthy growth of plant.

Fermentation Breakdown of complex organic compounds into simple compounds, caused by a ferment like bacteria, yeast etc.

Human body contains about 70% water.

GS

First law of Thermodynamics "You cannot get something for nothing" is the First law of Thermodynamics.

Flurescence Flurescence refers to light production accompanying rapid decay of excited electrons.

Freezing point or cryoscopic Measuring freezing points to work out osmotic potential is known as Freezing point or cryoscopic.

Gel A jelly like substance formed by the coagulation of a colloidal solution into a solid phase.

Gelation The coagulation of a sol to form a gel.

Glycolation Glycolation refers to the attachment of one or more sugars to the enzyme.

Golgi bodies The stacks of flattened hollow discs with convoluted margins surrounded by spherical bodies are called Golgi bodies.

Granum Stack of green thylakoids layered on top of each other within the chloroplast.

Growth regulators The substances, not synthesized by plants but by chemists, nor they are hormones, have been classified as Growth regulators.

Guttation Exudation of water usually through structures called *hydathodes* present at the tips of veins of leaves.

Heat interaction If the pushing, pulling or striking at the interaction boundary is random, chaotic and on the scale of atoms and molecules, it is termed as Heat interaction.

Hexosane Polysaccharides (carbohydrates) formed by the condensation of hexose molecules.

Hydathodes Structures present at the ends of veins lets of leaves; specially meant for exudation of water through a pore.

Hydration or inhibition The adsorption of water by hydrophillic surface is called Hydration or inhibition.

Hydrogen bond The positive side of one water molecule is attracted to the negative side of another forming a bond known as Hydrogen bond.

Hydroponics The plants with roots grown in aqueous nutrient solution are known as Hydroponics.

Hypertonic Having an osmotic pressure lower than that of an isotomic solution.

Hypotonic Having an osmotic pressure lower than that of an isotomic solution.

Intussusception The process of taking in food or other foreign matter and interposing tiny particles among those already present in the cell walls.

Leaf Area Index (LAI) Leaf Area Index (LAI) refers to the ratio of leaf area of a crop to the ground area upon which that crop grows.

Lecithin Any of several phosphatides found in blood, milk, egg-yolk, soyabeans, etc.

Light compensation point The irradiance at which photosynthesis just balances respiration, that is, net CO_2 exchange is zero, is known as Light compensation point.

Light-Harvesting Complex Green bands separate from chloroplast by electrophoresis contain chlorophylls a and b, xanthophyll, carotenoid but, little of β-carotene, all bound to proteins, are known as light-harvesting complex.

Knowledge World!

Only interim President

Sachidanand Sinha, the oldest member of the Constituent Assembly, was action President before President Rajendra Prasad in 1946.

Matrix A material with surfaces that bind water is called Matrix.

Mesophyll Cells Palisade and spongy parenchyma cells are collectively called mesophyll cells.

Net Radiations The difference between the radiation absorbed by an object and that emitted is the net radiations.

Night Interruption Phenomenon The night interruption by light inhibited flowering in short day plants and promoted flowering in short day plants and promoted flowering in long day plants, this phenomenon has been termed as night interruption phenomenon.

Nitrogen Fixation The process by which N_2 is reduced to NH^{4+} is known as Nitrogen fixation.

Nuclear Envelope Two parallel surface membranes surrounding the nucleus are together termed as nuclear envelope.

Nucleic Acid Complex organic acids found specially in the nucleus of all living cells and essential to life. They consist of a combination of phosphoric acid, a carbohydrate (Pentose-ribose or deoxyribose), and a base derived from purine or pyrimidine, usually bound in helical chains by hydrogen bonds between bases.

Nucleoside Any of various compounds consisting of a purine or pyrimidine base linked to a carbohydrate (pentose).

Nucleotide Any of several phosphate esters of nucleosides.

Nucleus The Nucleus contains one or more roughly spherical bodies called Nucleoli (singular), nuclelus each having a diameter of about 3-5 mm (up to 10 mm).

Ohm's law "The amount of a current flowing through a wire is proportional to the driving force and inversely proportional to the resistance encountered by the current in the wire" is the Ohm's law.

Osmometer A device that measures osmosis is termed as osmometer.

Passive Absorption If the actual ratio is either equal or less, between ion concentration inside and outside the membrane, the ion absorption is known as Passive absorption.

Periplasm A peripheral layer of protoplasm surrounding the functional gametic nuclei.

Perisperm Storage tissue in seed similar to endosperm but derived from nucellus.

Phosphorylation Phosphorylation means esterification of phosphate to the hydroxyl group of one or more serine or threonine.

Phosphorylation or Photo-phosphory-lation Synthesis of ATP in isolated chloroplasts only during light is known as phosphorylation or photo-phosphorylation photo-

Barind is called the layer of laterite, found in the plain between peninsular plateau and river Hoogly.

synthetic.

Photodormant Seeds that require light for germination are said to be photodormant.

Photon or quantum A quantum of the electromagnetic field is known as Photon or quantum.

Photosynthetically Active Radiation (PAR) The wavelengths between 400 and 700 nm are called photosynthe-tically active radiation (PAR).

Phytochrome A bluish green plant protein that in response to variations in red light regulates the growth and flowering of plants or plant structures.

Phytoferritin One abundant and stable form of iron in leaves is stored in chloroplasts as an iron-protein complex called phytoferritin.

Pigment Pigment refers to any substance that absorbs light.

Planck's Constant A universal constant of nature relating the energy of a photon to the frequency of the oscillator that is emitted is known as Planck's constant.

Plasmodesmata Protoplasmic connections between adjoining cells through pits in the cell wall.

Polar molecule A molecule having equal positive and negative charges on opposite side is called a Polar molecule.

Polyribosomes/Polysomes Ribosomes forming a chain like, a string of beads are called Polyribosomes/Polysomes.

Primary Growth The growth in plant body from the time of the initiation by the apical meristem until their expansion is completed. It starts in apical meristems and continues in their derivative meristems, protoderm, ground meristem and procambium, and even in still older tissue.

Prosthetic Group Non-protein inorganic or organic part of an enzyme.

Purine A colourless, crystalline, organic compound, $C_5H_4N_4$, the parent substance of uric-acid group of compounds.

Pyrimidine A colourless, liquid, crystalline organic compound, $C_4H_4N_2$, the fundamental group of bases, some of which are constituents of nucleic acid.

Q_{10} The factor by which a reaction rate decreases with a 10°C increase in temperature is called Q_{10}.

Quantasomes Photosynthetic units present in the unit membrane of the granum disc.

Quantum Yield Quantum yield refers to an expression when the absorption of a photon by a molecule results in photochemical reaction.

Quiescent Centre A group of cells, with a low DNA content and few mitosis, at the apical meristem.

Ramification Branching.

Relative Humidity The amount of water vapour in air at a particular temperature compared with the amount of water vapour that the air could hold at that temperature is known as Relative Humidity.

Rudiment An initial or primordial group of cells which gives rise to a structure.

Second Law of Thermodynamics "No real process can be 100 per cent efficient" is the second law of Thermodynamics.

Secondary Growth A type of growth which results in increase in thickness of stems and roots in gymnosperms, most dicotyledons and some monocotyledons. By this growth secondary vascular tissues are added by the vascular cambium and phellogen to the primary body.

Secretory Cavity A space lysigenous in origin having secretion or excretion of one or more usually organic materials.

Secretory Duct A schizogenously originating canal with living cells called as epithelial cells which produce secretion.

Seed A mature ovule with its normal coverings. A seed consists of the seed coat, embryo, and in certain plants, an endosperm.

Sigmoid Double curved like the letter S.

Stomate Technically, the term stomate (stoma is singular, while stomata is plural) refers only to the openings between the guard cells.

Suberin The waxy substance develop in a thicked cell wall which is characteristic of work tissues. It is closely related to cutin.

Symplast A continuous system of protoplast comprising protoplasts of adjoining cells connected with each other by means of plasmodesmata.

System Region of space or quantity of matter on which attention is focussed is called system.

Thermonasty Thermonasty refers to temperature induced movement of the tepals caused by growth.

Transamination The formation of aspartate from malate occurring in cytosol requires another amino acid such as alanine as a source of amino acid group is known as transamination

Transfer cells Plant cells in some species with wall ingrowths and expanded membrane surfaces are called transfer cells.

Tubules In many cells endoplasmic recticulum is less like a collapsed sack than it is like an assemblage of thousands of tiny tubes called tubules.

Vernalisation Vernalisation means exposure of plants to low temperatures for a few weeks to induce flowering on their return to normal temperatures.

Warberg Effect It has been recorded in 1920 that photosynthesis in algae is inhibited by O_2 and this inhibition occurs in all C-3 species studied, this effect is termed as Warberg effect often the name Otto Warburg a famous German biochemist.

Water Vesicle Pores on surface of leaves for excretion of water.

Work Interaction If the pulling or pushing across the boundary is highly organised, it is called Work interaction.

Zymogens An inactive antecedent form of an active enzyme that becomes functional by the action of an appropriate kinase or other activator.

Asoka took the title of Devanampiyadasi.

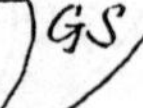

14. Agricultural Engineering

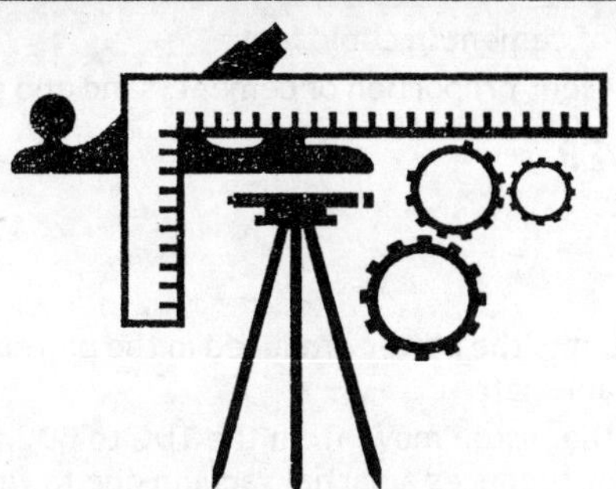

Each Word Counts

FARM STRUCTURE AND BUILDINGS

Dairy Barn

Principles for planning a stall dairy barn:

1. Maximum labour efficiency is obtained when cows face out.
2. Sweep-in mangers make feeding and cleaning easier.
3. Provision should be made for expansion of herd and for increased feed storage.
4. The milk house should be conveniently located.
5. Use mechanical equipment for handling materials.
6. Store material in a convenient location with chutes properly placed.
7. A minimum width of 32 ft. is required for two rows of cows.
8. Provide adequate cross alley to save steps.
9. Larger stalls aid in reducing injuries. By using cow trainers the cows can be kept cleaner in a larger stall.
10. Provide suitable pains for sick, injured and maternity cows.

Loose Housing Barn

Principles and practices adopted for maximum efficiency in loose housing barn:

1. Located bedded resting area away from traffic lanes to reduce bedding requirements and to keep cows clean.
2. Avoid concentrated traffic lanes.
3. Arrange building and board fences to act as wind breaks for the lot.
4. Have resting area open on the south or east side for good ventilation, sunlight and easy access.
5. Provide not more than 60 sq. ft. of bedded area per cow for efficient use of bedding and a warm farm manure pack.
6. Apply bedding evenly and only when needed.
7. Feed hay outside or from a self-feeding ground level hay storage building.
8. Keep young stock and dry cows separate from the lactating herd.
9. Protect calves under 10 days old from drafts by individual box pens.

Knowledge World!

Most dances in a film

Uday Shankar's *Kalpana* was the first film structured in choreographic terms with the maximum number of dance sequences.

10. Locate milking rooms so that cows enter from the lot or cover holding area but not from bedded area.

Fencing

Principles and practices for fence erection :

1. Do not drop the roll of fence on its end.
2. Tighten the fence until the bends in the horizontal wires are half straightened. These tension bends prevent damage from contraction in cold weather and sagging in hot weather.
3. Stretch from corner post, not around the corner.
4. Do not drive head of staple tightly against the wire.
5. Drive staples diagonally across the grain in wooden posts.
6. Alternate wood and steel line posts on opposite sides of the fence if both are used.

Electric Fencing

Purpose:

1. For temporary enclosing of pastures or any others field to be grazed by animals.
2. For confining cows and heifers in a lofting barn.
3. For placing inside the bull pens to decrease the damage to the main fencing.
4. For the fencing of marginal land which does not need costly permanent fencing.
5. For keeping wild-life fields from the forest nursery.

Farm Machinery Housing

The location of building site should have following characteristics:

1. Good drainage.
2. Space around the building: 30 feet from other buildings, 50 feet in front of building for turning and space for enlargement.
3. Conveniently located with respect to other buildings and field.
4. Openings protected from weather.
5. Utilities readily accessible.

Septic Tank

Practices for successful operation of septic tank:

1. Clean tank periodically every 3 to 5 years when 2 ft sludge has accumulated in the tank.
2. Starters are not needed to promote growth of bacteria.
3. Inspect septic tank annually.

Chief Justice of India administers the oath of office to the President.

4. Dispose of sludge in a burry pit and cover 2 ft of soil.
5. Do not light a match to inspect the tank because an explosion may occur.
6. Normal household soaps and detergents do not harm or hinder the operation of the tank.

Poultry Production

Principles and practices:

1. Keep all hens in one building, if possible.
2. Keep hens in large pens to reduce work.
3. Use portable brooders and range shelters to raise healthy chicks.
4. Use self-feeders in hen house and on rage. Henhouse feeders should hold several day's supply and have a chute through the wall of filling from the outside.
5. Store feed in the house if self-feeders are not used.
6. Pipe water to poultry house, and heat water in cold weather to prevent freezing.
7. Use deep built-up litter, and clean once or twice a year.
8. Use low roots over boarded-off dropping pits which require cleaning only three or four times a year.
9. If home grown feeds are used, balance with ratio pasture in summer and bright legume hay in racks in winter.
10. Provide doors and layout which will permit driving a manure spreader through the house for cleaning or provide a mechanical cleaner.
11. Have trap doors for chutes for manure removal from multiple story poultry houses.

Mortar

Purpose

1. It binds several units into one stable unit.
2. It fills the space between stones or bricks, thereby making the construction wind and water proof.
3. It distributes the vertical load of masonry evenly throughout the mass.
4. It is also used as a plaster on the walls and ceiling to give a pleasing appearance, as well as to protect it from rain.

FORMULAE

Foundation

$$w = \frac{W}{P}$$

Where,

w = width of the foundation, m

W = dead load coming on foundation per unit length, m

P = safe bearing pressure of soil, kg/m^2

Quantity of aggregated (Fuller formula)

$$C = \frac{42}{c+s+r}$$

Where,

C = quantity of cement in cubic feet

c, s, r = represent proportion of cement, sand and gravel.

FARM POWER

Engine Parts

Piston

Functions:

1. It transmits the power produced in the power stroke to the crankshaft
2. When the piston move from the TDC to BDC in suction stroke it produces a partial vacuum due to which fresh charge is sucked in i.e. it works as pump for taking in the cylinder.
3. It works as a guide for upper part of the connecting rod.
4. It carries rings, which seal the combustion chamber from the sump.
5. Some of the pistons have combustion chamber made in head for proper swirling action of air charge.
6. In exhaust stroke it pushes out the exhaust gases.
7. It has to withstand high temperature and pressure in power stroke.

Piston Rings

Functions:

1. They must seal the pressure in the cylinder and not allow the combustion gases and its product to leak to the crankcase.
2. They must carry away as much heat away from the piston to the cylinder wall.
3. They must properly lubricate the cylinder wall, supplying sufficient amount of lubricating oil and scraping out extra oil back to the crankcase.
4. They must be able to withstand pressure up to 1000 psi i.e between 2000 and 3000 times per minute in an engine.
5. They must withstand without losing their tension the temperature upto 300°C.
6. They must be quiet in operation.
7. They must absorb minimum amount of power in order to reduce friction losses.
8. They must be durable.

Flywheel

Functions:

1. It stores the energy received in power stroke and keeps the crankshaft rating in idle stroke.
2. It maintains the speed of crankshaft.
3. It transmits the power to gearbox through clutch.
4. For starting the engine starter ring is fitted on its periphery.

Carburettor

Parts of Carburettor:

1. Float chamber
2 Strainer
3. Throttle
4. Choke

5. Metering system
6. Idling system
7. Accelerating system
8. Economizer system

Qualities of Carburettor:

1. To supply rich mixture when engine is cold at the time of starting.
2. To correct the supply of fuel when engine gets heated-up.
3. To supply correct quantity of fuel at idle speed.
4. To supply enough fuel when required
5. Response to sudden acceleration and supply rich mixture.
6. To go on atomize fuel with air at all load and road conditions.

Fundamental Consideration in Proper functioning of Any Carburettor:

1. A constant and specific fuel level must be maintained in the nozzle.
2. There must be atleast partial if not complete vaporization of the liquid fuel, regardless of surrounding temperatures.
3. The correct mixture of vaporized fuel and air must be maintained at all times, regardless of engine load, speed, temperature, atmospheric pressure and other operating conditions.

Spark Plug

Functions:

1. To give a hot spark in the combustion chamber for igniting the charge.
2. There should be a high voltage so that it could jump from one electrode to another especially in high compressed air which is non-conductor of electricity.
3. Spark plug should have good insulation so that high tension current cannot leak.
4. Spark plug should work efficiently both in hot and cold working conditions.
5. Spark plug should give spark at an exact time. There should not be more time lag.

Fuel Injection Pump

Functions:

1. It should pump metered quantity of fuel.
2. It should pump equal quantity of fuel to all cylinders.
3. It should commence supplying the metered quantity of fuel at the same degree before TDC to all cylinders.
4. It should build up enough pressure so that diesel when injected is in the form of atomized spray.
5. The injection of diesel should be such that combustion process is kept under control.

Injector

Functions:

1. It should inject the fuel in fine atomized form.
2. The atomized fuel should mix up with hot air to form homogeneous mixture for proper combustion.
3. Fuel supply and cut-off should be rapid.

Quiz Kit!

Oldest post office still in use

Calcutta GPO is the oldest post office still in use.

4. It should not dribble.

Injection Chamber

Direct Injection Combustion Chamber:

Advantages:

1. Minimum heat loss during compression because of lower surface area to volume ratio and hence better efficiency.
2. No cold starting problems.
3. Fine atomization because of multi-hole nozzle.

Disadvantages:

1. High fuel injection pressure required and hence complex design of fuel injection pump.
2. Necessity of accurate metering of fuel by the injection system, particularly for small engines.

Indirect Injection Combustion Chamber:

Advantages:

1. Injection pressure required is low.
2. Direction of spraying is not very important.

Disadvantages:

1. Poor cold starting performance requiring heated plugs.
2. Specific fuel consumption is high because there is loss of pressure due to air motion through the duct and heat loss due to large heat transfer area.

Tractor Parts

Chassis

Functions :

1. To support the load of assemblies such as engine, transmission, front axle etc.
2. To provide connecting link for front and rear wheel.
3. To carry load of implements.
4. To withstand load caused due to working in uneven field.
5. To withstand load when tractor is working in field with implement.
6. To withstand load caused due to sudden braking or acceleration.
7. To withstand load / force caused due to cornering.

Front axle

Functions:

1. It carries the weight of the front end of the tractor.
2. It carries kingpin, steering arm through which the tractor can be steered.
3. It keeps the tractor in level position by its swinging action.
4. It absorbs the shocks in case fitted with spring.
5. In case of 4WD it also transmits the power to road wheels.

Muffler

Purposes:

1. To reduce the temperatures of exhaust gases.

Andhra Pradesh is the first state which estimated all controls on co-operatives.

2. To damp down the sound of exhaust gases.
3. To trap the flame of hot gases or unburnt gases.
4. To reduce the speed of outgoing gases.

Alternator

Advantages:

1. It is light in weight.
2. It is a simple machine.
3. It has fewer parts and as such cheap to manufacture.
4. It produces enough electricity even at slow speed.
5. It requires less maintenance.
6. It can be made to work faster at high speed.

Tractor Systems

Steering System:

Qualities of good steering system:

1. It should be light in operating.
2. It should be foolproof.
3. It should be self-adjusting and self-aligning.
4. It should not transmit field shocks to the steering wheel.
5. It should not disturb steering geometry on bad fields or on overloading.
6. It should have provision for front wheel spacing.

Clutch System:

Quality of a good clutch:

1. It should pick up load smoothly without grab or chatter.
2. It should have driven disc of low moment of inertia to permit easy shifting of gear.
3. Damp out vibrations of crankshaft, if any.
4. It should require little pedal pressure to operate.
5. It should be easy to adjust.
6. It should be easy to overhaul.
7. Friction material for lining must have high coefficient of friction.
8. Complete clutch assembly should be as light as possible to minimize inertia effect which otherwise causes gear-shifting problem.
9. The friction member must not slip under high torque where on the other hand it should slip when clutch is pressed.
10. The clutch component namely clutch plate should have long life.
11. Provision should be made for quick dissipation of heat generated due to slippage and friction.
12. Clutch assembly should be well balanced to avoid production of vibration at high speed.

Brake System:

Requirement of a good braking system :

1. The brake should stop the vehicle in shortest possible distance and in less time.
2. The brakes should work equally well on fair to bad roads.
3. Pedal efforts applied by the driver should not be much so as not to strain the driver.
4. Brakes should work equally well in all weather.
5. When brakes are applied the vehicle should not pull to one side.
6. It should have less wearing parts.
7. It should require little maintenance.
8. There should be minimum sound when brake is applied.
9. An independent parking brake system should be available to put to use in case of emergency.
10. It should have provision for independent wheel braking for negotiationg sharp turns.

Hydraulic System:

Merits:

1. It is easy and simple in design.
2. The hydraulic working units can be fitted anywhere and in any position.
3. Speed can be easily controlled.
4. Working capacity of any individual unit can be controlled.
5. There is practically no wear and tear as these parts always run in hydraulic oil and get lubricated.
6. It is cheaper to maintain.
7. Heavier jobs can be done quickly with less force.

Cooling System:

Distribution pattern of fuel energy:

1. Fuel energy is used for useful work through which we get brake horsepowers is 20 per cent.
2. Amount of fuel energy required for overcoming friction 5 per cent.
3. Amount of fuel energy, which escapes through exhaust is 45 per cent.
4. Amount of fuel energy that is absorbed by cooling system is 28 percent.
5. Amount of fuel energy required for overcoming friction by 2 percent in transmission

Efficiency of Air Cooling

Depending factors:

1. Area of fins.
2. The speed and quantity of air touching the fins.
3. Temperature of the engine.
4. Fitting position of the engine.
5. The conductivity of metal used for cylinder head and block.

Merits:

1. Engine weight is less.
2. Engine required less space.
3. It is simple in design and manufacturing.
4. Engine attains operating temperature quickly resulting in less wear of cylinder liner .
5. Due to higher working temperature, fuel consumption is less.
6. Due to higher working temperature, deposit, of carbon is comparatively less.
7. Air-cooled engine can work efficiently if one or two fins are broken.

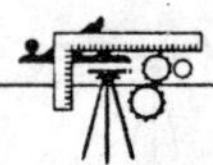

8. Engine performance is good due to higher mean cylinder temperature.
9. No topping of wear.
10. No water freezing hazards in cold countries.

Demerits:

1. Air-cooled engines produce more sound.
2. Volumetric efficiency is less due to higher working temperature of cylinder.
3. Air-cooling is not suitable for multi-cylinder engines.

Water Cooling System

Merits:

1. As water is in direct touch with cylinder wall and cylinder head, it takes away heat quickly.
2. The water circulates freely between freezing and boiling points.
3. The availability of water in water-jackets damp down sound of engine to some extent.
4. Engine working temperature can be controlled easily.
5. Engine with water-cooled system can be fitted in any position in the vehicle.

Demerits:

1. In water-cooled engine weight of radiator with water pump increases dead weight of engine.
2. Radiator is usually fitted in front of vehicle. Due to its presence in front slope on bonnet cannot be given to avoid wind resistance.
3. Water in water-cooled system freezes at zero degree temperature.
4. Water boils and evaporates easily at 100°C.
5. Water corrodes the metal part in the system.
6. In case of careless handling of water-cooled engine, it may overheat soon and seize.
7. As the water-cooled engines warm-up slowly during start, cold running accelerates sludge formation and more wear parts.

Essential Characteristics of fuel:

1. They must have reasonably calorific value.
2. They must vaporize at least partially at comparatively low temperatures.
3. The fuel vapours must ignite and burns readily when mixed in proper proportions with oxygen.
4. They must be of such a nature that they can be handled and transported with comparative ease and safety .

Lubrication System

Merits:

1. Reduce wear of parts.
2. It absorbs heat and brings it along with it to sump.
3. It seals the piston ring with cylinder wall.
4. It cleanses and flushes the moving parts.
5. Lubrication oil film works as cushion for moving parts.
6. Lubrication system damps down the sound of the engine.

Parts Lubricated by Lubrication System:

1. Cylinder valves, pistons and piston rings

Did You Know ?

First oil well

The first regular oil well was discovered at Borbil (later known as Digbhoi) in 1890.

2. Piston pin
3. Crankshaft and connecting rod bearing
4. Camshaft bearing
5. Valves and valve operating mechanism

Ignition System

Requirements:

1. It should provide a good spark between the electrodes of the plugs at the correct timing.
2. It should function efficiently over the entire range of engine speed.
3. It should be light, effective and reliable in service.
4. It should be compact and easy to maintain.
5. It should be cheap and convenient to handle.

Transmission System

Purpose:

1. It must provide a means of self-propulsion and proper speed reduction between the engine crankshaft and traction members to give the required travel speed.
2. It must provide for an equalization of the power transmission to the traction members on both gradual and short, quick turns.
3. It must provide means of reversing direction of travel.
4. It must provide brakes for controlling the movement of the tractor in specific situations or holding it stationary for certain operations.

Hitching System

General rules for tractor hitches and stability:

1. A long, low-mounted drawbar is preferable to a short, high one from a safety stand point and allows for better draft and traction as well.
2. A low centre of gravity and a wide tread consistent with the desired crop clearance is preferred to a high centre of gravity and a narrow tread.
3. The location of centre of gravity for a two-wheel drive tractor along a longitudianal axis should be such as to provide only sufficient weight on the front wheel to give satisfactory steering and control under normal hitch and drawbar pull conditions
4. For four-wheel drive tractors, the centre of gravity along the longitudinal axis should be located so that when the tractor is under load, part of the front weight plus implement weight is transferred to rear axle, dividing by weight equally between front and rear axle.
5. The longitudinal stability of tractor is reduced when the drive wheel dig in or drop into a ditch. Placing a timber cross wise in front of the wheels is particularly hazardous and may cause overturning.

6. Sudden engagement of the clutch of tractor pulling a heavy load under good traction conditions may lift the front end or cause overturning.

Engines

Supercharged Engine:

Merits:

1. Due to greater volume of air finding its way to the cylinder a higher thermal efficiency is achieved.
2. Supercharger achieves high-pressure rate.
3. More power for given size of engine approximately 20 to 30 percent is achieved.
4. Better fuel consumption for given power, approximately.
5. Cleaner exhaust.

Demerits:

1. To withstand higher-pressure, engine components are made sturdy as such costly materials are used.
2. As the part of made sturdy, engine is slightly heavier than conventional engine.
3. Some part of the power is used for driving supercharger unless turbo supercharger.
4. Because of higher temperature, bigger size radiator is used.

Operator Seating

Principles for high level of operator comfort:

1. The seat should be adjustable, preferably for forward and backward-movement and height, in order to accommodate as wide a range of operators as possible.
2. The seat back should provide adequate back support.
3. The seat should be adjusted so that legs can be comfortably extended, reach the floor or other support, and can readily reach all-foot activated controls.
4. The seat should be positioned so that there is a minimum amount of pressure on the back of the thighs.
5. The seat should be adjustable so the operator can have a good, unobstructed vision of the area around the tractor.

FORMULAE

Efficiency of Otto cycle

$$\eta = 1 - \frac{1}{r^{r}-1}$$

Efficiency of Diesel cycle

$$\eta_{Diesel} = 1 - \frac{1}{r^{(r-1)}}\left[\frac{r^{r}_{c}-1}{\gamma(r_c-1)}\right]$$

Where, r = compression ratio

r_c = Cut-off ratio

γ = Constant (1.4)

Thermal efficiency

$$\text{Thermal efficiency} = \frac{\text{Available useful work}}{\text{Heating value of supplied fuel}} \times 100$$

$$\text{Power} = \frac{\text{Work done (m)}}{\text{Time (S)}}$$

Indicated thermal efficiency

$$\eta\, i_{th} = \frac{\text{ip (kJ/s)}}{\text{Energy in fuel per second (kJ/s)}}$$

$$= \frac{\text{ip}}{\text{mass of fuel per second x calorific value of fuel}}$$

Brake thermal efficiency

$$\eta_{bth} = \frac{\text{bp}}{\text{mass of per second x calorific value of fuel}}$$

Volumetric efficiency

$$\eta_v = \frac{\text{mass of charge actually inducted}}{\text{mass of charge represented by swept volume at ambient temperature and pressure}}$$

$$= \frac{\text{Volume of charge aspirated per stroke at ambient conditions}}{\text{swept volume}}$$

Firing interval

$$F.I = \frac{720^\circ}{\text{No. of cylinder}} \text{ For four-stroke cycle engine}$$

$$F.I = \frac{360^\circ}{\text{No. of cylinder}} \text{ For two-stroke cycle engine}$$

Indicated horse power

$$I.H.P = \frac{P.L.A.N.}{4500} \times \frac{n}{2} \rightarrow \text{Four - stroke cycle engine}$$

$$I.H.P = \frac{P.L.A.N.}{4500} \times n \rightarrow \text{Two - stroke cycle engine}$$

Where,

P = Mean effective pressure kg/cm^2

L = Length of stroke in meter

A = Area of cylinder in cm^2

N = Number of revolution of crankshaft per minute

n = Number of cylinders

Indicated mean effective pressure (N/m^2)

$$P_{im} = \frac{\text{Area of indicated diagram}}{\text{length of the indicator diagram}}$$

Brake horse power

$$B.H.P = \frac{\text{I.H.P x Mechanical efficiency}}{100}$$

Drawbar horse power

$$D.B.H.P. = \frac{\text{Pull (kg) x Speed (meter/min)}}{4500}$$

Mechanical efficiency

$$\eta = \frac{B.H.P}{I.H.P} \times 100$$

GK Chandragupta took the title of Priyadarshan. GS

Compression Ratio

$$\text{Compression Ratio} = \frac{\text{Total cylinder volume}}{\text{Clearance volume}} = \frac{V_1}{V_2}$$

$$= \frac{\text{Swept volume + Clearance volume}}{\text{Clearance volume}} = \frac{V_s + V_c}{V_c}$$

Piston speed/mean piston speed

$$\text{Piston speed} = \frac{2 \times L \times N}{100} \text{ meter/minute}$$

Where, L= Length of stroke in cm

N = Number of revolution per minute

Displacement volume

$$\text{Displacement volume} = A \times L \times N \times \frac{n}{2} \text{ For four-stroke engine}$$

$$= A \times L \times N \times n \text{ Four-stroke engine}$$

Break mean effectve pressure

$$\text{B.M.E.P.}(kg/cm^2) = \frac{BHP \times 75 \times 60}{L \times A \times N \times \frac{n}{2}} \text{ For four-stroke engine}$$

Where, L = Length of stroke in cm

A = Area of the piston in m^2

N = No. of revolution per minute

Break thermal efficiency

$$\text{Break thermal efficiency} = \frac{641 \times 10^2}{\text{S.F.C (Kg/bhp/hr)} \times \text{Calorific value (calories per kg)}}$$

Efficiency of air cleaner

$$\text{Efficiency} = \left(\frac{\text{Weight of dust collected in the cleaner}}{\text{Dust in the cleaner and dust in the absolute cleaner}}\right) \times 100$$

Specific fuel consumption

$$\text{sfc} = \frac{\text{Fuel consumption per unit time}}{\text{Power}}$$

API degree of oil

$$\text{API} = \frac{141.5}{\text{Specific gravity of oil at } 60^\circ \text{ F}} - 131.5$$

Heating value of fuel

Heat value = Quantity of fuel x Calorific value fuel

Specific gravity of fuel

$$\text{Sp. gravity} = \frac{\text{Density of fuel}}{\text{Density of water}}$$

Fuel weight = Density of fuel x Volume of fuel

Relative efficiency

$$\text{Relative efficiency} = \frac{\text{Indicated thermal efficiency}}{\text{Air standard efficiency}}$$

You Don't Know!

Earliest history of raga

The concept of *raga* had not developed in Bharata's time. The full history of *ragas* appears in the *Brihaddesi* in the 6th century.

Governor regulation (R)

$$R = \frac{N_1 - N_2}{\left(\frac{N_1 + N_2}{2}\right)} \times 100$$

Where, R = Regulation in %

N_1 = Speed of engine at no load

N_2 = Speed of engine at load

Weight transfer

$$\text{Weight transfer} = \frac{\text{Pull in Kg} \times \text{Hitch height in meter}}{\text{Wheelbase in meter}}$$

Wheel slip

$$\text{Wheel slip (\%)} = \frac{N_1 - N_0}{N_1} 100$$

Where, N_1 = Number of revolution at load

N_0 = Number of revolution at no load

Rim pull

$$\text{Rim pull } (lbs) = \frac{375 \times HP \times \text{efficiency}}{\text{speed (Mile/hour)}} \text{ OR}$$

$$\text{Rim pull (kg)} = \frac{170 \times HP \times \text{efficiency}}{\text{Speed (km/hr)} \times 100}$$

Rolling resistance (R)

$$R = \frac{2}{(n+1)(K_c + BK\Phi)^{1/n}} \left(\frac{W}{2L}\right)^{\frac{n+1}{n}}$$

Where, n = Coefficient of wheel sinkage

K_c = Cohesive modulus of soil deformation

$K\Phi$ = Friction modulus of soil deformation

B = Width of each track

W = Vertical load over shear area

L= Length of track in contact

Frequency (f)

$$f = \frac{1}{T}$$

Where, T= time period in seconds

Amplitude

$$A_{RMS} = \frac{1}{\sqrt{2}} A_{Peak}$$

Where, A_{RMS} = Amplitude, N/m^2

A_{Peak} = Peak amplitude, N/m^2

GK **The term of office of the President is 5 years.** GS

Sound pressure level (SPL)

$$SPL = 20 \log \frac{P}{P_0} dB$$

Where, *SPL* = Sound pressure level, decibel

P = RMS sound pressure, N/m^2

P_0 = Reference sound pressure, N/m^2 (0.00002)

Log = Logarithm to the base 10

Transmissibility

$$\text{Transmissibility} = \frac{\text{Output vibration intensity}}{\text{input vibration intensity}}$$

$$T = \left[\frac{1+4\xi^2\left(\frac{\omega t}{\omega s}\right)^2}{\left[1-\left(\frac{\omega t}{\omega s}\right)^2\right]^2+4\xi^2\left(\frac{\omega t}{\omega s}\right)^2}\right]^{0.5}$$

Where,

T = Transmissibility

ξ = Damping ratio

ωt = Tractor chassis frequency

ωs = Undamped natural frequency

Damping ratio

$$x = \frac{C}{Cc}$$

Where, *C* = seat suspension damping rate, N/m.s

Cc = critical damping rate, N/m.s

Critical damping rate (Cc)

$Cc = 2\ M\ \omega s$

Where, M = Mass of seat and operator, kg

ωs = Undamped natural frequency of operator seat, rad/second

$$\omega s = \sqrt{\frac{K}{M}}$$

Where, *K*= spring rate of seating system, N/m

M = mass of seat and operator, kg

Torque (T)

$T = \mu . F . r_m . n$

Where,

μ = Coefficient of friction

F = Effective clamping force

r_m = Mean radius of the clutch facing

n = Number of torque transmitting surfaces

FARM MACHINERY

Tillage

Objectives:

1. To obtain deep seed bed suitable for different type of crops.
2. To add more humus and fertility to soil by covering the vegetation.
3. To destroy and prevent weeds.
4. To aerate the soil for proper growth of crop.
5. To increase water absorbing capacity of soil.
6. To destroy the insects, pests and their breeding place.
7. To reduce the soil erosion.

Functions :

1. To obtain a seedbed of good tilt.
2. To add humus and fertility to the soil by covering vegetation and manure.
3. To destroy the weeds and to prevent their growth.
4. To leave the soil in such a condition that air will circulate freely.
5. To leave the soil in a condition to retain moisture from the rain.
6. To destroy the insects and their eggs, larvae and their breeding place.
7. To leave the surface in a condition to prevent erosion by wind.
8. Precision levelling of land for irrigation and other operations.

Ploughing

Purposes:

1. To obtain a deep seed bed of good texture.
2. To increase water-holding capacity of soil.
3. To improve soil aeration.
4. To destroy weeds and grasses.
5. To destroy insects and pests.
6. To prevent the soil erosion.

Rolling Coulter (Part of MB plough)

Functions:

1. It cuts furrow slice edge levers, cleans furrow wall.
2. It cuts the trash over the surface of the field for better coverage by the furrow slice.
3. It reduces the plough draft.

Secondary Tillage

Objectives:

1. To pulverize the soil of the seedbeds in the field.
2. To destroy grasses and seeds in the field.
3. To cut crop residues and mix them with top soil of the field.
4. To break the big clods and make the field surface uniform and levelled.

Harrowing

Purpose of harrowing

1. Tillage tool kills weeds by cutting them, covering and dragging them out of the surface of soil to die.
2. Tillage tool loosens the soil so that moisture can get into the soil easily.
3. Tillage tool smoothens and level the ground.
4. Tillage tool tear apart the trash and mixes it with the soil.
5. Tillage tool break-up.

GK The Planning Commission of India was set up in 1950. GS

Knowledge World!

First film shot by an Indian

Harishchandra S. Bhatavadekar shot two short films - *The Wrestlers* and *Man & Monkey* - both released in November 1899.

Secondary Tillage Implements

Disc Harrow

Advantages:

1. A disc plough can be forced to penetrate into soil, which is too hard and dry for working with a mould board.
2. It works well in sticky soil.
3. It is more useful for deep ploughing.
4. It can be used safely in stony and stumpy soil without much danger of breakage.
5. A disc plough works well even after a considerable part of the disc is worn off in abrasive soil.
6. It works in loose soil also without much clogging.

Disadvantages:

1. It is not suitable for covering surface trash and weeds as effectively, as mould board plough does.
2. Comparatively the disc plough leaves the soil in rough and cloddier condition than mould board plough.
3. Disc plough is much heavier than mould board plough.

Cultivation

Objectives:

1. To control weeds so that they do not compete with crops for water and nutrients.
2. To prevent the surface evaporation losses.
3. To maintain the seedbed in a good tilt during the growth of the crop.
4. To achieve rapid infiltration of rainfall and adequate aeration.

Cultivator

Functions:

1. Interculture of the fields.
2. Destroys the weeds in the field.
3. Aerates the soil for proper growth of crops.
4. Conserves moisture by preparing mulch on the surface.
5. To sow seed when it is provided with sowing attachments.
6. To prevent surface evaporation and encourage rapid infiltration of rain water into the soil.

Seed Drill

Functions of seed drill:

1. To carry the seeds.
2. To open furrow at uniform depth.
3. To meter the seeds.
4. To place them furrow in an acceptable pattern.
5. To cover the seeds and compact the soil around the seed.

Mechanical functions:

1. It opens furrows to a uniform depth.
2. It meters the seed and drops uniformly at a desired depth without injury.
3. It meters the fertilizer and drops at a desired depth.
4. It covers the seeds and compact the soil around them.

Planter

Functions:

1. To open the furrow.
2. To meter the seeds.
3. To deposit the seeds in the furrow.
4. To cover the seeds and compact the soil.

Sprayers

Purposes of spraying/sprayers:

1. Application of herbicides to remove the weeds.
2. Application of fungicides to minimize the fungus diseases.
3. Application of insecticides to control the insects and pests.
4. Application of micronutrients on the plants.

Functions of sprayer:

1. To break the liquid into droplets of effective size.
2. To distribute them uniformly over the plants.
3. To regulate the amount of liquid to avoid excessive application.

Function and work:

1. It should distribute the chemical uniformly on the plant surface.
2. It should have provision to increase or decrease the quantity of insecticide application.
3. It should break the liquid into fine droplets.
4. It should have the provision to control or regulate the size of droplets.
5. If the droplets are very fine or small these may be carried away with the wind and if they are too big, the spray may not cover the entire plant.

Harvesting

Actions involved in harvesting operation:

1. Slicing action with sharp tool.
2. Tearing action with rough serrated edge.
3. High velocity single element.
4. Two-element scissors action.

Function of power thresher:

1. To feed the harvest to the threshing cylinder.
2. To thresh the grain out of the head.
3. To separate the grain from straw.
4. To clean grain.
5. To make bhusa suitable for animal feeding.

Main function of combine:

1. Cutting the standing crop.
2. Feeding the cut crops to threshing unit.

3. Threshing the crops.
4. Cleaning the grain from straw.
5. Collecting the grain in a container.

Advantages of combine harvester:

1. It saves the cost of harvesting and threshing the crops.
2. It reduces labour requirement of the field.
3. The grains can be obtained easily from the crops and thus the risk of fire, rain and other hazards are minimised.
4. It cleans the field earlier, which permits easy dry weather farming.
5. As the grains come in hand earlier, there is a very possiblity to get good profit in the market.
6. The straw has to be spreading the straw at the time of harvesting itself.

Disadvantages:

1. Higher initial cost.
2. Inadaptable for mixed farming.
3. It cannot be used in fragmented piece of land
4. Operating and maintenance is costly.

FORMULAE

The effective field capacity of a machine is expressed as

$$\text{Effective field capacity} = \frac{1000 \times S \times W \times Ef}{10000 \times 100} = \frac{SW \times Ef}{1000}$$

Where, C = Effective field capacity in ha/h
S = Speed of travel in kmph
W = Width of implement in meter
Ef = Field effieciency in per cent

Effective field efficiency of the machine

$$Ef = \frac{To}{Te \times Th \times Ta} \times 100$$

Where, Te = Time required to cover one hectare considering the overlap
K = Percentage of acutal width of machine utilized
Th = Time lost per hectare due to interruptions which is not proportional to area like time of filling
Ta = Time lost per hectare due to interruptions which is proportional to area like time taken for turning.

Draft

$$D = \cos\theta$$

Where, D = draft in kgf
P = pull in kg
θ = angle between line of pull and direction of motion.

Side draft

$$= P \cos\theta \sin\theta$$

Unit draft

$$\text{Unit draft} = \frac{\text{Draft (kgf)}}{\text{Crosssection area of furrow (cm}^2)}$$

Vertical Component of pull

$$= P \sin\theta \cos\theta$$

Power

Power (watt) = Draft (N) x Speed (m/s)

Horsepower

$$HP = \frac{\text{Draft (kg)} \times \text{Speed (m/s)}}{75}$$

$$HP = \frac{\text{Pull (kg)} \times \text{Speed (m/min)}}{4500}$$

Internal friction

$$S = C + N \tan\theta$$

Where, S = shear stress at soil failure surface, kg/cm^3
C = cohesion of soil
N = normal stress on the plance of shear failure
θ = angle of internal friction

Friction

$$\mu = \frac{F}{N} = \tan\theta$$

Where, μ = coefficient of friction
F = frictional force
N = normal force perpendicular to the surface
θ = angle of friction

Speed and draft relationship

$$Ds = Do + K S^2$$

Where, Ds = draft at speed S
Do = Static component of draft, independent of speed
S = forward speed
K = constant, depending on types of implement and soil conditions

Velocity ratio

$$VR = \frac{N_1}{N_2} = \frac{D_2}{D_1}$$

Where, N_1 = number of revolution of driving pulley
N_2 = number of revolution of driven pulley
D_1 = diameter of driving pulley
D_2 = diameter of driven pulley

Seed rate

$$\text{Seed rate (kg/ha)} = \frac{\text{Seed Collected (kg)}}{\text{Area (ha)}}$$

Spraying and dusting equipments
Field capacity

$$\text{Area covered (ha/h)} = \frac{N \times S \times W}{1000} \times \eta$$

Where, N = number of nozzles
W = spacing between nozzles, m
S = speed of travel, kmph
h = spraying efficiency, %

The compounds produced by plants in the process of photosynthesis are carbohydrates.

GS

Application rate

Spray volume

$$Q = Cd \times a\ \sqrt{2gh}$$

Where, $\sqrt{2gh}$ = pressure developing in the nozzle, m

a = projected area if the nozzle, m^2

Cd = discharge coefficient

Time taken for spraying/dusting

$$= \frac{\text{Total dischared volume}}{\text{Discharge rate}}$$

Discharge

Q = Area covered x Application rate

Calculation for testing of threshing equipments

(1) Total grain input (A)

$$A = B + C + D$$

Where, B = Quantity of grains collected at main outlets per unit times

C = Quantity of clean grains from sieve overflow, sieve underflow and thus at outlet per unit time

D = Unthreshed grains collected from all outlets per unit time

(2) Percentage of broken grains

$$\frac{E}{F} \times 100$$

Where,

E = quantity of broken grains in the sample at grain outlet

F = total quantity of sample taken at main grain outlet

(3) Percentage of blown grains

$$\frac{G}{A} \times 100$$

Where, G = quantity of clean grains obtained at straw outlet per unit time

A = total grain input

(4) Percentage of unthreshed grain

$$= \frac{H}{A} \times 100$$

Where, H = quantity of unthreshed grains obtained from all outlets per unit time

(5) Cleaning efficiency

$$\eta_c = \frac{M}{F} \times 100$$

Where, M = quantity of clean grains obtained from sample taken at main grain outlet

F = total quantity of sample taken at main grain outlet

Chaff cutter

Capacity of chaff cutter:

$$C = V \times d$$

Where, V = volume chopped fodder per unit time, m^3

d = density of chaff, kg/m^3

> **Quiz Kit!**
> ***First rails***
> **First rails were imported from England and consisted of 65 lb. wrough iron double-headed rails laid on stone sleepers and the first track was laid by *GIP* in 1853.**

Capacity of chaff cutter

$$C = W \times H \times L \times N \times R \times K$$

Where, C = weight of cut fodder tonne/hour

W = throat width, cm

H = throat height, cm

L = length of chaff, cm

R = revolution per minute

K = constant (2×10^{-6} for power chaff cutter and 1 for hand operated chaff cutter)

N = number of knives

Length of chaff

$$\chi = \frac{\Pi D}{n} \times \frac{N'}{N}$$

Where, χ = length of cut, mm

D = diameter of feed roller, cm

n = number of knives

N' = revolution of feed roller

N = revolution of fly wheel

Belt drive

Relation between tensions

$$\frac{T_1}{T_2} = e^{\mu\theta}$$

Where, T_1 = tensions on slack side

T_2 = tensions on tight side

μ = coefficient of friction

θ = angle of lap of belt over pulley

Length of open belt

$$L = \frac{\Pi}{2}(D_1 + D_2) + \frac{(D_1 - D_2)^2}{4C} + 2C$$

Length of cross belt

$$L = \frac{\Pi}{2}(D_1 + D_2) + \frac{(D_1 - D_2)^2}{4C} + 2C$$

Where, L = length of belt (open /cross), m

D_1 = diameter of driver pulley, cm

D_2 = diameter of driven pulley, cm

C = distance between centre of both pulley, m

Length of V belt

$$L = 2C + 1.57\,(D + d) + \frac{(D - d)^2}{4C}$$

GK GS

Golakhnath is the highest place of Kathiawar.

$$C = \frac{1}{2}(D + d) + 3t$$

Where, L = length of belt, m

D = diameter of driver pulley, cm

d = diameter of driven pulley, cm

C = distance between centre of both pulley, m

t = belt thickness, mm

Belt life in years

$$\frac{\text{Belt length in mm}}{\text{Total fatique rate}} \times 100$$

Theoretical field capacity (ha/h)

$$C_{th} = \frac{\text{Width (cm)} \times \text{Speed (m/s)} \times 36}{10000}$$

Field efficiency

$$\eta f = \frac{\text{Effective field capacity}}{\text{Theoretical field capacity}} \times 100$$

Effective field capacity

$$\text{Ceff} = \frac{S \times W}{10} \times \frac{E}{100}$$

Where, S = speed of implement/machine, kmph

W = swath, m

E = efficiency, %

Soil inversion

$$= \frac{\text{No. of weeds before ploughing} - \text{No. of weeds after ploughing}}{\text{No. of weeds before ploughing}}$$

Performance Index of weeder

$$PI = \frac{Q_1 \times Q_2}{P}$$

Where, PI = Performance index

Q_1 = quantity of weeding work done

Q_2 = quantity of weeding work done per unit time

P = average power used in the weeding operation

$$PI_{water} = \frac{\text{fieldcapacity (ha/h)} \times \text{soil inversion index} \times \text{weeding efficiency}}{\text{Appliedhorsepower}}$$

Puddling Index

$$PI = \frac{Vs}{V} \times 100$$

Where, Vs = volume of settled soil, cm^3

V = total volume of soil, cm^3

Diameter of disc (cm)

= 3.5 x depth of ploughing (cm)

Utility index (implement)

$$K = \frac{\text{Number of working hours}}{\text{Years}} \times 1000$$

Horsepower

$$HP = \frac{\text{draft (kg)} \times \text{speed (m/s)}}{75}$$

$$HP = \frac{\text{draft (kg)} \times \text{speed (m/min)}}{4500}$$

Where, HP = horse power

Area of coverage (harrow)

$$= \frac{\text{Width (m)} \times \text{Speed (kmph)}}{10} \times \text{efficiency}$$

RENEWABLE SOURCES OF ENERGY

Solar Energy

Application of Solar Energy:

- Solar water heating
- Thermal electric conversion
- Solar distillation
- Agricultural and industrial process heat
- Solar cooking
- Solar green house
- Space heating and cooling
- Photovoltaic electric conversion
- Solar pumping
- Solar furnace
- Solar production of hydrogen

Application of Solar Air Heater:

- Heating of buildings
- Drying of agricultural products
- Heating greenhouse
- Air-conditioning buildings utilize desiccant beds or an absorption refrigeration process
- Using air heaters as the heat sources for a heat engine such as Brayton or Stirling cycle

Solar Energy Collectors

Concentrating Collector

Advantages:

1. Reflecting surfaces require less material and are structurally simpler than flat plate collector.
2. For a concentrator system the cost per unit area of solar collection surface is less.
3. The absorber area of a concentrator system is smaller than flat plate for same solar energy collection.
4. Concentrating system can be used for electric power generation when not used for heating and cooling.
5. The amount of heat which can be stored per unit volume is larger and consequently heat stored costs are less.
6. Efficiency of concentrator system is high.

Disadvantages:

1. Additional requirements of maintenance particularly to retain the quality of reflecting surface against dirt, weather oxidation etc.

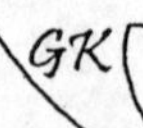

GS

Changez Khan died in 1227.

2. Non-uniform flux on the absorber.
3. Additional optical losses such as reflectance loss and the intercept loss, so that they introduce additional factor in energy balance.
4. Initial cost is high.

Water Heating System

Advantages:

1. In case of water heating, common heat transfer and storage medium water is used; this avoids temperature drop during transfer of energy into and out of storage.
2. It requires relatively smaller storage volume.
3. It can be easily adopted to supply of energy absorption air conditioners.
4. Relatively low energy requirements for pumping of the heat transfer fluid.

Disadvantages:

1. Water heaters may also operate at excessively high temperatures and means must be provided to remove energy and avoid boiling and pressure build-up.
2. Collector storage has to be designed for overheating during the period of no energy level.
3. Care has to be taken to avoid corrosion problems.

Solar Air Heating System

Advantages:

1. There is no problem with freezing in collectors.
2. Corrosion problems are minimized.
3. Conventional control requirement for air heating is already available and can be readily used.
4. Problems of designing for overheating during periods of no energy removal are minimized.
5. Working fluids in air and the warm air heating system is in common use.

Disadvantages :

1. Relatively higher power cost for pumping air through the storage medium.
2. Relatively large volume of storage units.

Solar Photovoltaic System

Applications

1. Water pumping sets for micro-irrigation and drinking water supply.
2. Radio beacons for ship navigation at ports.
3. Community radio and television sets.
4. Cathodic protection of oil pipe line.
5. Weather monitoring.
6. Railway signalling requirement.
7. Battery charging.
8. Street lighting.

Advantages:

1. Direct room, temperature conversion of light electricity through a simple solid state device.
2. There are no moving parts.
3. Ability to function unattended for long periods as evidence in space programme.

Did You Know ?

First sperm bank

Dr. Aniruddha Malpani and Dr. Anjali Malpani of Mumbai opened a sperm bank in 1990 to help infertile patients.

4. Maintenance cost is low as they are easy to operate
5. They do not create pollution.
6. They have a long effective life and are highly reliable.
7. They consume no fuel to operate, as the sun's energy is free.
8. They have rapid response in ouput to input radiatlon changes.
9. They have wide power handling capabilities.
10. They are easy to fabricate, being one of the simplest of semi-conductor devices.
11. They have high power to weight ratio.
12. They can be used with or without sun tracking.

Disadvantages:

1 They have high cost
2. The energy storage is required because of no isolation at night.

Solar Furnace

Advantages:

1. In a solar furnace heating is carried out without any contamination and changing the position of the material in focus easily controls temperature.
2. It gives an extremely high temperature.
3. It provides very rapid heating and cooling.
4. Various property measurements are possible on an open specimen.
5. Proper desirable atmosphere can be provided to the specimen.

Disadvantages:

1. Its use is limited to sunny days and to 4-5 hours only.
2. Cost of solar furnace is high.

Wind Energy Conversion System

Advantages:

1. It is a renewable source of energy.
2. Wind energy system avoids fuel provision and transport.
3. On small scale upto a few kilowatt system it is less costly.
4. On large scale cost can be competitive wlth conventional electricity and lower cost could be achieved by mass production.
5. Like all forms of solar energy, wind power systems are non-polluting. So it has no adverse influence on the environment.

Disadvantages:

1. Wind energy available in dilute and fluctuating in nature.
2. Unlike water energy wind energy needs storage capacity because of its irregularity.

GK **President addresses his resignation to the Vice-President.** GS

3. Wind energy system is noisy in operation.
4. Wind energy system has a relatively high overall weight.

One Blade Rotor

Advantages:

1. Simple blade controls lower blade weight and lower gear box cost.
2. Counter weight costs less than second blade.
3. Counter weight can be inclined to reduce blade coning.
4. Pitch bearing do not carry centrifugal force.
5. Blade root spar can be a large diameter.

Disadvantages:

1. Vibration produced, due to aerodynamic torque.
2. Unconventional appearance.
3. Large blade root cause of bending moment.
4. Starting torque reduced by ground boundary layer.
5. One-per-rev coriolis torque produced due to flapping.

Savonius Rotor

Advantages:

1. The machine performs at low velocity ranges.
2. It is useful more of the time and less dependent on storage of supplementary power.
3. The weight of electric generator may be carried at ground level without use of level gears.
4. The cost of vertical axis wind turbine may be significantly lower than that of standard wind turbines.
5. It has simple structure, hence easy to manufacture.

Disadvantages:

1. It is not useful for a very tall installation.
2. This not only leads to excessive weight for a large installation but also leaves the machine at the mercy of severe storm, since there is no way to reduce the effective area.

Energy From Biomass

Floating Drum Type Plant

Advantages:

1. It has less scum troubles because solids are constantly submerged.
2. No separate pressure-equalizing device needed when fresh waste is added to the tank or digested slurry is withdrawn.
3. It is danger of mixing oxygen with the gas to form an explosive mixture minimised.
4. Higher gas production per cubic metre of digester volume is achieved.
5. Floating drum has welded braces, which helps in bleaking the scum by rotation.
6. No problems of gas leakage.
7. Constant gas pressure.

Disadvantages:

1. It has higher cost.
2. Heat is lost through metal gasholder.
3. Gasholder requires painting once or twice a year.
4. Flexible pipe joining the gas holder to the main gas pipe requires maintenance.

Fixed Dome Type Plant

Advantages:

1. It has low cost.
2. It has no corrosion trouble.
3. In this type plant heat insulation is better as construction is beneath the ground. Temperature will be constant.
4. Cattle and human excreta and long fibrous stalks can be fed.
5. No maintenance.

Disadvantages:

1. This type of plants needs the service of skilled masons.
2. Gas production per cubic metre of digester volume is less.
3. Scum formation is a problem as it has no stirring arrangement.
4. It has variable gas pressure.

Biological Conversion of Solar Energy

Advantages:

1. The initial investment is comparatively low.
2. Processing of biomass for fuel from sewage and waste reduces the environmental hazards.
3. Small unit can be located near the accumulation of waste in the forming region.
4. The by-product of such a system can be fully recycled.
5. The method of power production is relatively low polluting.
6. Biogas can be distributed through GI pipes for domestic use.
7. It can be stored in container, which can be transported to consumers
8. In rural areas hygienic and sanitation conditions are improved.

Disadvantages:

1. The land area required is relatively large.
2. Relatively low concentration of biomass per unit area of land.
3. Collection and transportation of biomass becomes expensive due to high moisture content.
4. Photosynthesis efficiency or a small percentage of solar radiations are converted into biomass by plants.

Geothermal Energy

Applications:

1. Generation of electric power.
2. Industrial process heat.
3. Space heating for various kinds of buildings.

Advantages:

1. Geothermal energy is versatile in its use.
2. It is cheaper, compared to the energies obtained from other sources.
3. Geothermal energy delivers greater amount of net energy from its system than other alternatives or conventional systems.

Durgapur, Bhilai and Rourkela plants were set up during Second Plan.

GS

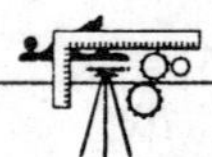

4. Geothermal power plants have the highest annual load factors of 85 to 90 per cent compared to 45 to 50 per cent for fossil fuel plants.
5. Geothermal energy is the least polluting compared to the other conventional sources.
6. The greatest attraction of geothermal energy is its amenability for multiple uses from a single resource.

Disadvantages:

1. Overall efficiency of power production is low.
2. Drilling operation is noisy.
3. Large area is needed for exploitation of geothermal energy as much of it is defused.
4. Withdrawal of large amount of steam or water from a hydrothermal reservoir may result in surface subsidence.

Energy from Oceans

Tidal power

Advantages:

1. It is inexhaustible source of energy.
2. Tidal power generation is free from pollution.
3. These power plants do not demand large area of the valuable land.
4. Peak power demand can be effectively met when it works in combination with thermal or hydroelectrical system.

Disadvantages:

1. The fundamental drawback to all methods of generating tidal power is the variability in output caused by the variations in tidal range.
2. Tidal ranges is highly variable and thus the turbines have to work on a wide range of head variation.
3. The tidal range is limited to a few metres.
4. Seawater is corrosive and it was feared that machinery may get corroded.
5. Construction in a sea is difficult.
6. Cost is not favourable.

Wave Energy

Advantages:

1. The energy is naturally concentrated by accumulation.
2. It is free and renewable energy sources.
3. Wave power devices do not use up large landmasses unlike solar or wind.
4. These devices are relatively pollution-free.

Disadvantages:

1. The extraction equipment must be operated in marine environment with all that employs in terms of maintenance, construction cost, lifetime and reliability.
2. Wave energy converts must be capable of withstanding very severe peak stress in storms.
3. There is relatively scarcity of accessible sites of large wave activity.
4. Wave energy conversion devices that have been proposed are relatively complicated. .

Small Scale Hydroelectrical Electric System

Advantages:

You Don't Know!

Largest river basin

The Ganga basin, takes up 26 per cent of India's total geographical area.

1. Smaller hydro projects take shortest time for developing a unit.
2. Once it is built the running expenditure is almost negligible.
3. The operation and maintenance of such power station is simplest.
4. The energy source is free from hazards of pollution.

Disadvantages:

1. General lack of awareness of benefits from development.
2. Prohibitive economics compared to that of convention hydropower which is still abundantly available in the country.
3. Remoteness of site especially in the hilly areas and adverse geological conditions in Himalaya regions.

Chemical Energy Sources

Fuel Cell

Applications:

1. Domestic use
2. Central power stations
3. Automobile vehicles
4. Special applications

Advantages:

1. It has very high convergent efficiency.
2. Fuel cell can be installed near the use of point.
3. They have few mechanical components hence they operate fairly quietly and require little attention and less maintenance.
4. Atmospheric pollution is small.
5. There are requirement for large volume of cooling water.
6. As fuel cells do not make noise, they can easily be accepted in residential areas.
7. The fuel cell takes little time to go into operation.
8. Space requirement is less.

Disadvantages:

1. It has high initial cost.
2. It has low service life.

Magneto Hydro Dynamic System

Advantages:

1. Large amount of power is generated.
2. It has no moving part so is more reliable.
3. The closed cycle system produces power free of pollution.
4. The size of plant is smaller than conventional.
5. Direct conversion of heat into electricity.
6. Better fuel utilization

Security and Exchange Board (SEBI) of India was established in 1988 to safeguard the interests of investors in share market.

FORMULAE

Coefficient of performance

$$Cp = \frac{P}{P_{max}} = \frac{P}{\frac{1}{2}\rho AV^3}$$

Where, ρ = density of air
A= swept area
V = velocity of wind

Wind power

$$hp = 0.00000525\ D^2W^3$$

where, D = maximum diameter of wind wheel or circle, ft
W = wind velocity, mph
Available wind power

$$Pa = \frac{1}{2}\rho\frac{\Pi}{4}D^2V^3$$

$$= \frac{1}{8}\rho\,\Pi D^2V^3$$

Where, D = diameter of rotor

Power coefficient

$$\text{Power coefficient} = \frac{\text{Power of wind rotar}}{\text{Power available in the wind}}$$

Power coefficient

$$Cp = \frac{\text{Power output from wind machine}}{\text{Power available in the wind}}$$

Torque

$$T = \frac{P}{\varpi} = \frac{P}{\Pi DN}$$

Where, *T* = torque in kgf or N
ω = angular velocity of turbine wheel, m/sec
D = diameter of turbine wheel
N = Number of blades

Tip speed ratio

$$TSR = \frac{Vtip}{V}$$

where, Vtip = speed of the rotor tip
V = free wind speed

Torque coefficient

$$CT = \frac{T}{Tmax}$$

Where, T = Shaft torque
T max = torque at maximum efficiency

Performance of wind machines

Overall conversion efficiency

$$\eta o = \frac{\text{Useful output power}}{\text{Windpower input}} = \eta_A \times \eta_G \times \eta_C \times \eta_{Gen}$$

Where, c_A = efficiency of aero turbine
c_G = efficiency of gearing
c_C = efficiency of mechanical coupling
c_{Gen} = efficiency of generator

$$\eta A = \frac{\text{Useful shaft power output}}{\text{Windpower input}} = \text{coefficient of performance}$$

Capacity of digestion tank

$$\text{Capacity} = \frac{V_1 + V_2}{2} \times t$$

Where, V_1 = volume of the raw waste added daily
V_2 = volume of waste after digestion
t = period of digestion in days

AGRICULTURAL PROCESSING

Drying

Natural Air Drying

Advantages:

1. Lowest Inltlal Investment and maintenance cost.
2. No fuel cost and fire hazards
3. Required least supervision and least mould growth compared to supplemental heat.

Disadvantages:

1. Very slow drying rate.
2. Dependent on the weather.
3. Require more drying space.
4. Useful for short-term storage of product in farm.
5. Not successful for humid tropics.

Low Temperature Drying

Advantages:

1. Maintain high quality of grain.
2. Avoid overdrying or overheating of grain.
3. Problems of labour and management are reduced.
4. Lower temperature drying usually controlled by turning the fan on and off during drying of grain.
5. Low capital investment.

Disadvantages:

1. Independent process on favourable weather conditions.
2. Require longer drying time as compared to other systems.
3. High risks of grain spoilage during conditions of high ambient temperature and humidity.

Supplemental Heat Drying

Advantages:

1. Involve low cost equipments and maintenance.
2. Independent of weather.
3. Requires less supervision.
4. Most efficient use of bin capacity.

Disadvantages:

1. Fire hazards to a certain extent.

GK **The unit of energy is Joule.** GS

2. Danger of accelerated mould growth.
3. Low rate of drying.
4. Use for short-term storage in the farm.

Advantages:

1. Faster drying rate independent of weather.
2. High drying capacity per fan horsepower.
3. Useful for both long and short-term storage of grains.

Disadvantages:

1. Higher initial investment and maintenance cost.
2. Considerable fuel expenditure.
3. Danger of fire hazards.
4. Requires skilled manpower for control of condition
5. Due to direct firing with liquid fuel, the products may be contaminated with fuel gas.

Sun Drying (Radiation Drying)

Advantages:

1. No need of fuel or mechanical energy.
2. Simple in operation.
3. Preserved viability, germination and baking qualities of product.
4. Reduces microbial activity and insects / pests infestation.
5. No pollution and labour oriented.

Disadvantages:

1. Completely weather dependent.
2. Not possible to carry out throughout the year.
3. Excessive losses occur due to shattering, birds and rodents etc.
4. Requires specially constructed large floor area, restricting the capacity of mill to a certain limit.
5. The centre process is unhygienic.
6. Unsuitable for handling of large quantity of grain within a short period of harvest.

Infrared Drying

Advantages:

1. Small thermal inertia.
2. Simplicity and safety in operation.

Disadvantages:

1. Requires high expenditure of electric power.
2. Lower utilization.

Conduction Drying

Advantages:

1. Economical in favourable conditions only.
2. Used mainly in preparation of grain for processing.
3. Achieved more substantial reduction of moisture.

Disadvantages:

1. Inefficient drying.
2. Requires considerable expenditure of fuel.
3. Does not ensure, the conservation of quality of the grain as it is heated and dried very unevenly.
4. Requires heavy manual labours.

Chemical Drying

Advantages:

> **Knowledge World!**
> *Tallest MP*
> **Ratilal Kalidas Verma, a BJP MP elected from Dhandhulka (Gujarat) in the 9th Lok Sabha, was 1.98 m (6ft 5 in) tall.**

1. Dries paddy and also reduces the damage due to fungal, microbial and enzymatic activities and heat of respiration.
2. Viability of the grain is not affected.
3. Milling quality of paddy is satisfactory.
4. Dry matter loss is negligible.
5. Quality of rice bran is not affected.

Disadvantages:

1. Total removal of moisture is not possible due to the presence of sodium chloride on the husk.
2. Reduces the life of gunny.
3. The colour of husk changes to dark yellow.
4. The common salt treated paddy requires an additional drying subsequently.
5. Economy of process has yet to be established.

Dryers

Flat Bed Type Batch Dryer (Spray Dryer)

Advantages:

1. Fairly reasonable price.
2. Intermittent drying can also be used.
3. Operation is very simple.
4. Can manufacture locally using various types of materials like steel sheet and wood pieces etc.
5. Used as seed dryer and also as storage after drying.

Disadvantages

1. Low rate of drying.
2. Uneven drying results in higher percentage of broken grains.
3. Low holding capacity.

Recirculatory Batch Dryer

Advantages:

1. Price is reasonable.
2. Simplest design amongst all flow type dryers.
3. Easy to operate.
4. Can be used on the farm and rice mill.
5. Low operating cost with husk fired furnace.

Disadvantages:

1. Uneven drying as compared to mixing type.
2. After a long time use of the perforation of the cylinders may be clogged with the parboiled paddy.

LSU -Dryer

Advantages:

1. Uniformly drying of product if dryer designed properly.
2. Can be for drying different type of grains.

Disadvantages:

1. High capital investment.

GK **Northern Circars is name given to the east coastal plain near Golkunda.** GS

2. High cost of drying.

Baffle Dryer

Advantages:

1. Uniform drying of product.

Disadvantages:

1. Ratio of the volume of plenum to the total volume of the dryer is relatively high.
2. Grains on the baffle plates move slowly than that other sections

Rotary Dryer (Horizontal)

Advantages

1. Fast drying rate
2. Uniform drying of all grains.
3. Milling quality of parboiled paddy is high if it is dried in two phases under optimum drying conditions.

Disadvantages:

1. Complicated design.
2. Needs careful attention.
3. Higher capital investment.
4. High power requirement.
5. High operating cost may be high due to higher consumption of electricity and steam.
6. Dryer requires horizontal larger floor space.
7. Generally only 30 per cent of the dryer volume is utilized.
8. Not used for all types of freshly harvested grains.

Parboiling

Parboiling of Paddy

Advantages:

1. The milling in field is higher and quality is improved, as there are fewer broken grain.
2. The grain structure becomes compact and vitreous, even if some kernels were entirely or partially chalky.
3. The milled rice becomes translucent and shiny.
4. The grain remains firmer during cooking, and less likely to be become sticky.
5. A greater amount of water is absorbed during cooking causing the rice to swell.
6. After cooking the rice absorbs less fat from added conditions.
7. When cooked, the rice keeps larger and it will not go rancid so easily.
8. Parboiled rice retains more proteins, vitamins, and minerals which have spread during parboiling in the endosperm.
9. After cooking parboiled rice is digestible.
10. Fewer solids are kept behind in the cooking water.
11. Shelling of parboiled paddy is easier.
12. Rice bran obtained from parboiled rice gives higher percentage of oil.

Disadvantages:

1. It develops a relatively darker colour compared to raw rice.
2. The traditional parboiling process produces an undesirable smell.
3. Parboiled rice takes more time to cook than raw rice to the same degree of softness.
4. Because of long soaking in traditional process, mycotoxins may develop in parboiled rice and cause health hazard.
5. Heat treatment during parboiling destroys some natural antitoxidants hence parboiled rice develops more rancidity during storage.
6. Shelled parboiled rice requires more power for polishing.
7. Parboiled paddy may choke the polisher because of the higher oil content of bran.
8. Parboiling process requires an additional investment of capital.

Parboiling of wheat

Advantages:

1. The light coloured product gives an appearance resembling rice.
2. The product is suitable for quick cooking cereals, breakfast cereals and as an ingredient of body food.

Disadvantages:

1. Production cost is comparatively high.

Sheller/Husker

Under Runner Disk Sheller

Advantages:

1. Life of machine is long.
2. Low-running cost.
3. Simple in construction and operation.
4. Satisfactory performance of the machine for raw and parboiled paddy of uniform size.

Disadvantages:

1. Gives less head and total yield compared to rubber roll husker.
2. Damage to rice kernel is done due to the formation of scratches on it by the hard abrasive coating.

Blade Type Husker

Advantages:

1. Easy to operate.
2. Does not require much attention.

Disadvantage:

1. Yield of cracked and crushed grain is high.

Single Huller

Advantages:

1. Low initial investment and operating cost.
2. Easy to manufacture by local artesian and operated with unskilled labour .
3. The huller can be utilized for whitening of parboiled rice as it produces uniformly white rice.

Disadvantages:

1. Both total and head yield are low, as the degree of whitening cannot be adjusted to a low value.

2. The huller bran contaminated with large amount of husk cannot be utilized for oil extraction and for feed purpose.

Sheller Mill

Advantages:

1. Sheller bran can be used for oil extraction.
2. Low-operating cost.
3. Life of machinery is long.
4. The emery cone can be recoated at the mill sight.
5. Works well with parboiled paddy.
6. Sheller mill can be manufactured locally.

Disadvantages:

1. Head and total yield is less than modern rice mill.
2. Some scratches may be performed on the rice kernels which are undesirable for long storage.

Open Double Sieve Precleaner

Advantages:

1. Price is low
2. Locally manufactured.

Disadvantages:

1. It is an open sieve; dust formation in the mill premises is considerable.
2. Unavailability of self -cleaning device the bottom sieves with small perforation is often clogged, resulting in lower separation efficiency.
3. Impossible to separate same size impurities as the paddy grains.

Paddy Cleaner with Stoner (Japan)

Advantages:

1. The roto scalper and aspirator is capable of removing about 30 per cent impurities content.
2. The same machine separates large impurities, light impurities, stones, sands and metallic parts.

Disadvantages:

1. The separating capacity of the stones-separating tray is limited up to 2 per cent stone content only.
2. Careful operation and maintenance are necessary.
3. The separating tray is to be imported from the manufactuer only.

Impact Type Paddy Husker (Japan)

Advantages:

1. Price is reasonably low.
2. Low operating cost due to small power requirement and long life of the rubber rim.
3. Total and head yield are higher.
4. Husking efficiency and head yield are quite satisfactory for parboiled rice.
5. Machine is compact and portable.

Disadvantages:

1. Critical adjustment is required for different verities and moisture content of paddy.

> **Quiz Kit!**
> *First garden tomb*
> **The tomb of Sikandar Lodi (1451) is the first garden tomb in the country.**

2. Breakage percentage increases with the increase of husking ratio.

Rubber Roll Husker

Advantages:

1. Highest percentage of sound and whole husked rice is produced as the risk of breaking the kernel is small and chance of scratching on kernel is also nil.
2. The mixture of different size and varities of paddy can be used without any significant increase of broken in husked rice.
3. Husking ratio can be increased to 0.9 without reduction in head yield.
4. It does not remove germ.

Disadvantages:

1. High operating cost due to wear of rubber rolls.
2. It requires skilled labour to operate the machine.
3. Storage life of rubber rolls is limited.

Indented Tray Type Separator

Advantages:

1. Perfect separation can be achieved if different varities are not mixed up.
2. Separating condition of the mixture on the dimpled tray is always visible and can be adjusted accordingly.
3. Simple in operation.
4. A compact unit and requires minimum space.
5. Power requirement is also small.

Disadvantages:

1. It is difficult to manufacture the dimpled tray locally.
2. Difficulties in perfect separation are encountered when the size and shape of dimples do not correspond to the length of paddy and brown rice.

Compartment Separator

Advantages:

1. Minimum running cost as there is part to be changed or subject to service wear.
2. Very long life of machine due to its sturdy construction.
3. If it is operated properly the separation is perfect.

Disadvantages:

1. Expensive machine to buy and transport.
2. Needs strong foundation at ground level.
3. Requires skilled labour.
4. The wooden part of the machine is subjected to termite attacks.

Vertical Whitener or Pearler

Advantages:

1. The degree of whitening is adjustable.

GK

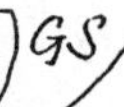

GS

Narshimha Committee was appointed by the Government on August 14, 1991 to review the financial structure of the country.

2. Long life of machine.
3. Low operational cost.
4. It can deal with any variety of rice.
5. Excellent performance of machine with parboiled rice.

Disadvantages:

1. The machine is big and heavy.
2. Requires skilled operator and attention.
3. High power requirement.

Horizontal Whitener

Advantages:

1. Total and head yield is high.
2. Bran removal is easier and uniform due to jet air system.
3. The degree of whitening can be adjusted.

Disadvantages:

1. The machine needs careful attention.
2. Clearance between screen and abrasive roll cannot be adjusted.
3. Once the diameter of abrasive roll is reduced by about 1.5 cm it has to be replaced.

Friction Type Whitening Machine

Advantages:

1. Production of uniformly whitened rice with shining appearance is the main feature of the machine.
2. Strong air current helps in lowering rice temperature, generated by friction and removing from whitened rice.
3. Pressure is adjustable.
4. It can produce uniformly whitened parboiled rice without increase of broken.

Disadvantages:

1. Broken percentage may be increased if the machine is not operated properly.
2. In case of parboiled paddy, the screen may be chocked with the oily bran if higher percentage of polish is given particularly in the humid atmosphere.
3. The slotted screen and steel cylinder are subjected to wear and are replaceable.

FORMULAE

MOISTURE CONTENT

Moisture content on wet basis

$$m = \frac{Wm}{Wm + Wd} \times 100$$

Where, m = moisture content on wet basis, per cent

Wm = weight of moisture, g

Wd = weight of bone dried material, g

Mositure content on dry basis

$$M = \frac{Wm}{Wd} \times 100 = \frac{m}{m-1} \times 100$$

Where, M = moisture content on dry basis, per cent

m = moisture content on wet basis, per cent

Wm = weight of moisture, g

Wd = weight of bone dried material, g

$$\frac{W'_m}{W_1} = \frac{m_1 - m_2}{100 - m_2} = \frac{M_1 - M_2}{100 + M_1}$$

$$\frac{W'_m}{W_2} = \frac{m_1 - m_2}{100 - m_1} = \frac{M_1 - M_2}{100 + M_2}$$

Where, W_1 = initial weight of wet material = $(Wm + Wd)$ kg

W_2 = final weight of dried product, kg

$W'm$ = weight of moisture evaporated, kg

$m_1\ m_2$ = initial and final moisture contents respectively, per cent, wet basis

$M_1 M_2$ = initial and final moisture contents respectively, per cent, dry basis

Henderson equation for EMC

$$1 - RH = \exp[-cTMe^n]$$

Where, RH = equilibrium relative humidity, decimals

Me = EMC dry basis, per cent

T = Temperature, °K

c and n = product constants varying with materials

Constant rate drying

Rate of drying in constant rate drying

$$R = \frac{Wd\ dX}{A d\theta}$$

Where, R = constant drying rate period, kg of moisture evaporated/hr-m²

Wd = weight of dry solid, kg

A = wet surface, m²

dX = change in moisture content, kg

dθ = change in drying time, h

Drying time

$$\theta c = \frac{Wd}{A}\left(\frac{X_1 - X_2}{Rc}\right)$$

Where, θc = drying time in h

Wd = weight of dry solid, kg

A = wet surface, m²

X_1 = initial moisture content, kg moisture/kg dry solid

X_2 = Final moisture content, kg/kg

Rc = Rate of drying in the constant rate period, kg moisture evaporated/hr m²

The constant drying rate by use of wet bulb temperature theory

$$\frac{dw}{d\theta} = 18\ kA\ (ps - pa) = hfaA\ (ta - ts)/\lambda s$$

Where, $\frac{dw}{d\theta}$ = constant rate of drying, kg/hr

pa and ps = water vapour pressure at ta and ts respectively, atm

ta, ts = air and water temperature respectively, °C

h_{fa} = film heat transfer coefficient of air water interface,

GK GS

Pyrometers are instruments used for measuring high temperature.

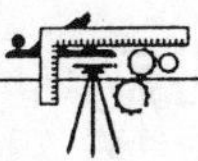

kcal/hr m² °C

λs = latent heat of water at ts, kcals/kg

A = water surface area, m²

k = water vapour transfer coefficient at the water air interface, kgmole (hr m² atm)

Deep bed drying

Time of advance of drying front

$$\theta_1 = \frac{Wd(M_1 - M_x)}{AG(Hs - H_1) \times 100}$$

Where, M_1 = initial moisture content of grain (db), %

Mx = average moisture content, (db), % at the end of drying front advance at the top

θ_1 = time of advance, h

A = cross sectional area of the dryer through which air passes

G = mass flow rate of dry air, kg/h m²

Hs = humidity of the saturated air leaving the dryer kg/kg

H_1 = humidity of the air entering into the dryer, kg/kg

Wd = weight of dry grain in the bin, kg

Decreasing rate period drying

$$\theta_2 = (1/K)1n\left(\frac{Mx - Me}{M - Me}\right)$$

Where, Θ = time of drying decreasing rate period, h

Me, = equilibrium moisture content of the grain (db)

K = drying constant, 1\h

M = average moisture content (db) at the end of decreasing rate period

Mx = average initial content (db) at the beginning of decreasing period

Thermal efficiency of grain dryer

$$\eta_{thermal} = \frac{\left(\frac{dM}{d\theta}\right) wd\lambda}{q}$$

$$\theta = \frac{60\ VA}{v}(h_1 - h_0)$$

$\frac{dM}{d\theta}$ = drying rate, kg/hkg

Wd = weight of dry material, kg

λ = latent heat of evaporation, kcal/kg

θ = rate of heat flow, kcal/h

V = air rate, m³/min m²

A = area, m²

v = humid volume of air (at the point of rate measurement) m³/kg

h_1 and h_0 = enthalpy of drying and ambient air, kcal/kg

Heat Utilization Factor

$$HUF = \frac{\text{Heat utilized}}{\text{Heat supplied}} = \frac{t_1 - t_2}{t_1 - t_0}$$

Coefficient of performance

$$C.O.P. = \frac{t_1 - t_0}{t_1 - t_0}$$

Where, t_2 = dry bulb temperature of exhaust air, °C

t_0 = dry bulb temperature of ambient air, °C

t_1 = dry bulb temperature of drying air, °C

Effective heat efficiency (EHE)

$$EHE = \frac{t_1 - t_2}{t_1 - tw_1}$$

Where, tw_1 = wet temperature of drying air, °C

Separation

Velocity of particle at any point of inclined plane

$$v_1 = \sqrt{2gK_1x_1}$$

Where, $K_1 = \sin\alpha - \mu 1 \cos\alpha$

x_1 = distance travelled by a body which has initial velocity zero

α = angle of inclination

μ_1 = coefficient of friction

Effectiveness of separation

$$\varepsilon A = \frac{(X_F - X_D)X_C(X_F - X_C)(1 - X_D)}{(X_D - X_C)^2 X_F(1 - X_F)}$$

Where, εA = Separating effectiveness for a mixture of two components A and B

F = rate of feeding to the separator, kg/h

C = rate of separating the product, kg/h

D = rate of reject, kg/h

X_F, X_C, X_D = fractions of components A in the feed, separated product and reject respectively.

Storage Structure

Grain pressure theories

$$P\gamma = \rho Y$$

Where, $P\gamma$ = lateral pressure exerted by the grain on the bin wall

ρ = density of grain, kg/m³

Y = depth of grain from the top of the bin surface

Rankin's earth pressure coefficient

$$Py = K\gamma Y$$

> **Did You Know ?**
> *First Court*
> **Court of Judicature was set up as directed by the East India Company and English Law was introduced in India on Aug. 1, 1672 in Kolkata for the first time.**

GS

Andaman-Nicobar islands are submerged uplands.

$$K = \frac{1-\sin\phi}{1+\sin\phi}$$

Φ = angle of internal friction of grain

SOIL AND WATER CONSERVATION ENGINEERING

Principle characteristics of centrifugal pump

1. Smooth, even flow.
2. Adapted to high speed operation.
3. Capacity and head depend upon rpm, impeller and width.
4. Horsepower is a function of capacity, head and pump efficiency.
5. When the speed is constant, the capacity decreases as the head increases and the power is reduced.
6. When the operating speed is changed, the capacity will change in direct proportion to the variation in speed.
7. The capacity varies directly with the diameter of the impeller, the head varies as the square of the diameter and horse power wanes as the cube of the diameter.

Well Irrigation

Advantages:

1. In well irrigation, water is available all the time and crops can be irrigated as per their needs.
2. The wells are located in the areas which they irrigate and the conveyance losses are generally less and irrigation efficiencies are higher.
3. Intensity of irrigation can be higher as compared to canal irrigation as water is available throughout the year.
4. In areas where water table is high and affects the crop growth, wells to a certain extent, the lower is the water table.

Disadvantages:

1. The discharge available from the wells is smaller and as such the areas which they can irrigate is limited,
2. Well irrigation is costlier than canal irrigation. In wells some water lifting devices has to be invariably used to lift the water. Failure of such device will interrupt the supply of water.

Sprinkler Irrigation

Advantages:

1. Seepage losses, which occur in earthen channels of surface irrigation methods, are completely eliminated. Moreover, only optimum quantity of water is used in this method.
2. Land levelling is not required and thus avoiding removal of top fertile soil, as happens in other surface irrigation methods.
3. No cultivation area is lost for making ditches, as happens in surface irrigation methods. It, thus, results in increasing about 16% of the cropped area.
4. In sprinkler system, the water is to be applied at a rate lesser than the infiltration capacity of the soil, and thus avoiding surface runoff, and its bad effects, such as loss of water, washing of top soil, etc.
5. Fertilizers can be uniformly applied, because they are mixed with irrigation water itself.
6. This method leaches down salts and prevents water logging or salinity.
7. It is less labour-oriented and hence useful where labour is costly and scarce.
8. Upto 80% efficiency can be achieved, it up to 80% of applied water can be stored in the root zone of plants.

Limitations:

1. High winds may distort sprinkler pattern, causing non-uniform spreading of water on the crops. In areas of high temperature and high wind velocity considerable evaporation losses of water may take place
2. They are not suited to crops requiring frequent and larger depths of irrigation, such as paddy.
3. Initial cost of the system is very high, and the system requires a high technical skill. Only sand and silt-free water can be used, as otherwise pump impellers lifting such waters will get damaged.
4. It requires larger electrical power.
5. Heavy soil with poor intake cannot be irrigated efficiently.
6. A constant water supply is needed for commercial use of equipment.

Drip irrigation system

Advantages:

1. Water distribution in close proximity to plant roots along plant row and therefore water distribution is uniform and controlled.
2. Estimation of land levelling and irrigation on steeper slopes.
3. No surface flow as such no tail-water loss or soil erosion.
4. Concurrent water and fertilizer application.
5. Concurrent cultural operation during irrigation on such crops as trees and vines.
6. Restriction of weed growth to wetted areas.
7. Water saving and generally increased yield.

Disadvantages:

1. Higher initial cost.
2. The problem of blockage of outlet does not appear (but to have been completely solved).

Drainage

You Don't Know!

First air-conditioned train

The first fully air-conditioned and air braked passenger train was the Rajdhani Express introduced on the Mumbai-Delhi route. It covers the distance of 1,384 km in 16 hr 15 min.

Factors influencing the drainage:

1. Topography of the affected area and the adjoining watershed which contribute flow into the problem.
2. Soil characteristics.
3. Rainfall and other climatic conditions.
4. Crop factors.

Type of soil requiring drainage:

1. Flat lands with fine textured soil.
2. Humid regions with continuous or heavy rainfall.
3. Excessive soil moisture content.
4. The water-table is within or near to the root zone.
5. When water stands on the land surface for long periods.
6. In regions where annual evaporation exceeds annual rainfall.

Benefits of drainage:

1. Drainage provides a better environment for plant growth.
2. Free water in the soil fills the pore space and air is excluded.
3. Drainage improves the soil structure and the infiltration characteristic of the soil is improved.
4. Optimum conditions for tillage are provided over a longer range of time.
5. Drainage hastens the warming of the soils and maintains desirable soil temperature.
6. Drainage promotes to increase leaching of salts and prevent their accumulation in soil.
7. Larger crop growing season can be achieved with good drainage due to earlier possible planting dates.
8. Certain toxic substances and disease organism are removed from the soil due to better drainage and soil aeration.
9. The less chance of uprooting of plants due to high winds.
10. A good variety of crops can be grown successfully on a well-drained farm land.
11. Good drainage reduces disease that thrives on wet land.

Benefits of subsurface drainage:

1. Aeration of root zone for maximum development of plant roots.
2. Opportunity for desirable soil micro organisms to develop through aeration and higher soil temperatures.
3. Availability of soil for early cultivation and thus increased crop growth period.
4. Improvement of soil moisture conditions for operation of farm machinery.
5. Removal of undesirable salts from root zone.
6. Greater storage of rain water in the root zone because of a low initial water table before the rains.

Seepage losses depend upon:

1. Type of seepage, it whether 'percolation' or 'absorption'.
2. Soil permeability.
3. The condition of the canal; the seepage through a silted canal is less than that from a new canal.
4. Amount of silt carried by the canal; the more the silt, lesser are the losses.
5. Velocity of canal water; the more the velocity , the lesser will be the losses.
6. Cross-section of the canal and its wetted perimeter.

> **Knowledge World!**
> *First woman engine driver*
> **Surekha Bhonsle of Mumbai joined the Indian Railways as trainee assistant driver in 1990, when she was 24.**

Tile Drains

Advantages:

Tile drainage helps in increasing crop yields by draining the water or by lowering the water-table in the following manner:

1. Removes the free gravity water that is not directly available to the plants.
2. Increases the volume of soil from which roots can obtain food.
3. Increases air circulation.
4. Increases bacterial activity in the soil, thus improving soil structure and making the plant food more readily available.
5. Reduces soil erosion. A well-drained soil has more capacity to hold rainfall, resulting in less runoff and hence, reduced erosion.
6. Reduces and removes toxic substances such as sodium and other soluble salts, which when present in large concentrations may retard plant growth.
7. Lesser time and labour is required for tilling and harvesting the soils, as tile drains do not obstruct fanning operations. With a crop such as com, a delay in planting may decrease the yields. Planting in wet soils is also likely to decrease yields. All such troubles are removed in tile-drained soils.
8. Tile drains permit deep roots development by lowering the water-table, especially during spring months.

Comparison of cable tool and hydraulic rotary methods of drilling

Advantages of Cable Tool Method:

1. A more accurate sample of the formation can be obtained
2. Lesser amount of water is required during operations.
3. Cable tool rig is lighter and easy to transport.
4. Very useful for consolidated rocks and less useful for loose formations.
5. For shallow tubewells in unconsolidated materials too, it comes out to be cheaper.

Advantages of Hydraulic Rotary Method:

1. Can be used for larger holes up to 1.5 m diameter .
2. Can be best used for drilling test holes, because the hole can be abandoned with minimum cost.

Supreme Court of India decides disputes regarding presidential election.

3. Rotary drilled hole can be gravel packed, which, increases its specific capacity, and keeps the fine particles away, thus causing less sand trouble.
4. Casing is to be driven only after the hole completion, and hence, can be set at any desired depth.
5. It is the fastest method of drilling and especially useful in unconsolidated formations.
6. It can handle alternate hard and soft formations with ease and the danger of accidents is lesser. In quicksands, clays, etc, cable tool method is likely to give trouble, as there is a danger of freezing.

Advantages and disadvantages of tubewell irrigation over canal irrigation

Advantages:

1. Isolated lands which cannot be served by canals can be easily irrigated by tube wells.
2. Cultivators can have their own private tube wells and thus have not to depend on the government-owned canal waters.
3. Canal irrigation projects require huge funds and considerable time, while wells can be constructed wherever required in a small time and with lesser funds.
4. The supply of water from a well can be started as soon as required and can be stopped at any moment, thus taking advantage of momentary rainfall.
5. Tubewell channels are of shorter length and generally, lined, thus resulting in lesser percolation losses.
6. Tubewell water is generally sold on a volumetric basis, which results In optimum utilization of water at the correct time.
7. Pumping from ground water by means of tube-wells helps in lowering the ground water-table, and thus helps in reducing waterlogging, which on the other hand is generally caused by canal irrigation.
8. With the help of irrigation, more than one crop may be grown in an year.
9. The well water which is warmer in cold weather and colder in hot weather is more suitable for crops.
10. The well water is more pure than the canal water, and thus the irrigation scheme may be combined with rural water supply scheme.
11. The land acquisition is less for tubewell irrigation.
12. Unless drought continues for many years, well irrigation does not fail in droughts, while a canal supply may fail in a single drought or at the most in two or three consecutive drought years.

Disadvantages:

1. If the electric supply fails (which generally does in drought years), the pumps of the tubewells cannot be operated, and hence, the well water cannot be made available to the crops, unless diesel power is available.
2. Canal irrigation projects are generally combined with flood control and hydropower projects, thus giving added benefits.
3. The well water which is generally free from silt is not so good from manauring point of view as is the silted canal water.
1. The tubewell water proves much costlier than the canal water since the well water has to be lifted by pumps.
2. The life of a tubewell is limited.
3. Frequent breakdown of power and motor parts cause large-scale interruptions in the working of tubewells, so much so that the usual running hours have been estimated to be 3,760 out of annual of 8,760.

Selection of a suitable site for a reservoir

It is almost impossible to select a perfect ideal reservoir site. But its selection is guided by the following factors:

1. A suitable dam site is available. The cost of the dam is generally a controlling factor in the selectlon of a reservoir site.
2. The geological formationsior the reservoir banks, walls.etc should be such as to entail minimum leakage.
3. The geology of the catchments area should be such as to entail minimum water losses through absorption and percolation.
4. The site should be such that a deep reservoir is formed. A deep reservoir is preferred to a shallow one, because of lower land cost per unit of capacity, less evaporation loss, and less possibility of weed growth.
5. The reservoir site must have adequate capacity.
6. Too much silt-laden tributaries should be avoided as far as possible.
7. The reservoir basin should have a deep narrow opening in the valley, so that the length of the dam is minimum.

FORMULAE

Darcy Law

$$V = \frac{k(h_1 - h_2)}{l}$$

Where, V = Velocity of flow, m/day

k = hydraulic conductivity, depends upon the properties of the sand and liquid, m/day

$h_1 - h_2$ - difference in hydraulic head, m

l = distance along the flow between the points h_1 and h_2 m

Discharge of steady state flow to well in confined aquifer

$$Q = \frac{\pi k\,(H+hw)\,(H-hw)}{\ln \frac{R}{r_w}}$$

Where, Q = constant discharge rate, m^3/sec

H = original elevation of the water surface measured from the impervious base, m

hw = depth of water in the well measured from the impervious base, m

R = radius of island, m

r_w = radius of well, m

Transmissibility of confined aquifer

$$T = \frac{Q \ln(r_2/r_1)}{2\pi\,(S'_1 - S'_2)}$$

Where, T = Transmissibility of the aquifier, m^2/sec

Q = Constant discharge rate, m^3/sec

r_1 and r_2 = distance of the observation wells from the centre of the last well respectively

S_1' and S_2' = steady state drawdown at point 1 and 2 respectively

$$S_1' = \frac{S_1 - S_1^2}{2H} \qquad S_2' = \frac{S_2 - S_2^2}{2H}$$

Steady state flow to wells in confined aquifer

$$Q = \frac{2\pi T (S_1 - S_2)}{\ln(r_2/r_1)}$$

Where, Q = discharge rate, m^3/sec

T = transmissibility

S_1 and S_2 = values of drawdown of r_1 and r_2 distance

Unsteady state drawdown in confined aquifer

$$S = \frac{Q}{4\pi T} W_{(u)}$$

Where, S = unsteady state drawdown in confined aquifer, m

Q = constant discharge rate , m^3/sec

$W_{(u)}$ = Theis well function

T = Transmissibility of the aquifier, m^2/sec

Steady state drawdown in a semi-confined aquifer

$$S = \frac{Q}{2\pi Kb} Ko(r/B)$$

Where, S = steady state drawndown in an obervation well at a distance r from the pumped well, m

Q = steady state constant discharge, m^3/sec

$Ko\,(r/B)$ = modified Bessel function of the second kind and or order zero

$B = \sqrt{kbc}$, leakage factor of the semipervious layer, m

Irrigation pumps

Efficiency of ram (D' Aubuission's efficiency)

$$\pi_d = \frac{qh_d}{(Q+q)H}$$

Where, q = volume of water delivered, m^3/sec

Q = volume of water escape through waster valve, m^3/sec

H = effective supply head, m

h_d = effective delivery head, m

Specific speed

$$n_s = \frac{nQ^{1/2}}{H^{3/2}}$$

Where, n_s = specific speed, rpm

> **Quiz Kit!**
> *Oldest mosque*
> **The Cheraman Jama Masjid in Kodungallur, Kerala, was built in 629 A.D.**

n = pump speed, rpm

Q = pump discharge, US gallons per minute

H = total head, feet

Specific speed in dimensionless form

$$n_s = \frac{nQ^{1/2}}{(gH)^{3/4}}$$

Where, n_s = specific speed, rpm

n = pump speed, rpm

Q = pump discharge, US gallons per minute

H = total head, feet

Maximum practical suction lifts of pumps

$$Hs = Ha - Hf - e_s - NPSH - Fs$$

Where, Hs = maximum practical suction lift, m

Ha = atmospheric pressure at water surface, m

Hf = friction losses in stream, pipe, fittings and valve on the suction line, m

e_s = saturated vapour pressure of water, m

$NPSH$ = net positive suction head of the pump including losses at the impeller and velocity head, m

Water horsepower

$$WHP = \frac{Q \times H}{75}$$

Where, Q = discharge in litres per second

H = total head, m

Shaft horsepower

$$SHP = \frac{WHP}{\eta_p}$$

Where, WHP = water horse, hp

η_p = pump efficiency

Brake horsepower

(i) with direct drive pump

Brake horsepower = shaft horsepower

(ii) with belt or other horsepower

$$BHP = \frac{WHP}{\eta_p \times \eta_d}$$

Where, BHP = brake horsepower

WHP = water horsepower

η_p = pump efficiency

η_d = drive efficiency

Effect of impeller diameter on pump performance

GK GS

Four measures of money supply are being used in India. These are M1, M2, M3, M4.

$$Q = Q_1 \left(\frac{n}{n_1}\right)$$

$$H = H_1 \left(\frac{n}{n_1}\right)^2$$

$$P = P1 \left(\frac{n}{n_1}\right)^3$$

Where, n = new speed desired, rpm

n_1 = speed at which characteristics are known, rpm

Q = capacity at desired speed n, lit/sec

Q_1 = capacity at desired speed n_1, lit/sec

H = Head at desired speed n, metres

H_1 = Head at desired speed n_1, metres

P = BHP at desired speed n, H and Q

P_1 = BHP at speed n_1, H_1 and Q_1

Effect of impeller speed on pump performance

$$Q = Q_1 \left(\frac{D}{D_1}\right)$$

$$H = H_1 \left(\frac{D}{D_1}\right)^2$$

$$P = P_1 \left(\frac{D}{D_1}\right)^3$$

D and D_1 are change diameter and original diameter of impeller in mm

Water horse power

$$WHP = \frac{\text{Dischage (litre/second) x Total head (metre)}}{7}$$

OR

$$= \frac{\text{Dischage (cubic metre hr) x Total (hr)}}{273}$$

Shaft horse power

$$\text{Shaft horsepower} = \frac{\text{Water horse power}}{\text{Pump efficiency}}$$

Input horse power

$$\text{Input horse power} = \frac{\text{Brake horse power}}{\text{Motor efficiency}}$$

$$\frac{\text{Water horse power}}{\text{Pump efficiency x Drive efficiency x Motor efficiency}}$$

Discharge through an orifice

$$Q = 0.61 \times 10^{-3}\, a\, \sqrt{2gh}$$

Where, Q = Discharge, litre per seconds

A = Area of cross section of the orifice, cm^2

g = Acceleration due to gravity, cm/s^2

h = Depth of water over the centre of orifice, cm

Discharge through a weir (Basic formula)

$$Q = cLH^m$$

Where, Q = Discharge

c = coefficient, depending on the nature of the crest and approach conditions

L = Length of crest

H = Head on crest

m = Exponent depending upon weir

Discharge through rectangular weir (suppressed)

$$Q = 0.0184\, LH^{3/2}$$

Where, Q = Discharge, litre per second

L = Length of crest, cm

H = Head on the crest, cm

Discharge through rectangular weir (contracted)

$$Q = 0.0184\, (L - 0.1nH)^{3/2}$$

Where, n = Number of end contractions

Discharge over a Cipolletti weir

$$Q = 0.0186\, LH^{5/2}$$

Discharge through 90° V-notch

$$Q = 0.0138\, H^{5/2}$$

Manning's formula

$$V = \frac{R^{2/3}\, S^{1/2}}{n}$$

Where, V = velocity in metre per second

R = Hydraulic radius, metre

S = Hydraulic slope

n = Rough coefficient of the channel lining

Capactiy of channel

$$Q = av$$

Where, Q = Discharge, litre per second

a = Cross-sectional area of channel, m^2

v = Velocity, meter per second

Discharge from a siphon tube

$$Q = 0.65 \times 10^{-3}\, a\, \sqrt{2gh}$$

Where, Q = Discharge, litre per second

a = Area

g = Acceleration due to gravity, cm/s^2

h = Effective head causing flowing, cm

Discharge through pipeline

$$V = \sqrt{\frac{Hdg}{2fl}}$$

Where, V = Velocity flow of water through the pipe, cm/sec

H = Available head causing flow, cm

d = Diameter of pipe, cm

g = Acceleration due to gravity, cm/s^2

l = length of pipe, cm

f = Darcy's roughness coefficient

Gross irrigation requirement

Heat is transmitted from one end of a metal rod to its other end by the method of conduction.

Gross irrigation requirement = $\frac{\text{Net irrigation requirement}}{\text{Field efficiency of the system}}$

Duty of water

$$D = 8.6\ \frac{b}{d}$$

Where, D = Duty of canal water, h/cm/s

b = Number of base days for duty

d = Water depth, m

Irrigation period

Irrigation period =

$$\frac{\text{Net amount of moisture in soil between start of irrigation and the lower lim it of moisture depletion}}{\text{Peak period moisture use rate of the crop}}$$

Water conveyance efficiency

$$Ec = \frac{Wf}{Wr} \times 100$$

Where, Ec = Water conveyance efficiency

W_f = Water delivered to the irrigated plot

W_r = Water supplied at the source

Water application efficiency

$$Ea = \frac{Ws}{Wf} \times 100$$

Where, E_a = Water application efficiency

W_s = Water stored in the root zone during irrigation

W_n = Water needed in the root zone prior to irrigation

Water storage efficiency

$$Es = \frac{Ws}{Wn} \times 100$$

Where, E_s = Water storage efficiency

W_s = Water stored in the root zone during irrigation

W_n = Water needed in the root zone prior to irrigation.

Water distribution efficiency

$$Ed = 100\left(1 - \frac{y}{d}\right)$$

Where, E_d = Water distribution efficiency

y = Average numerical deviation in depth of water stored from average depth stored during irrigation

d = Average depth of water stored during irrigation

Cristiansen formula for uniformity coefficient

$$C_u = \left(1.0 - \frac{\sum x}{mn}\right)$$

Where, C_u = Uniformity coefficient

m = Average application rate, mm/hr

n = Total number of observations

> **Did You Know ?**
> *First Chief Justice*
> **Justice Harilal J. Kania was Chief Justice (CJ) of India from J. 26, 1950 to Nov. 6, 1951.**

x = Deviation of individual observation from average application, mm/hr

mn = Total of all observations

Pump capacity

$$Q = 278\ \frac{A \times D}{F \times H \times E}$$

Where, Q = Pump discharge, litre per second

A = Area, ha

D = Depth of application, cm

E = Field application efficiency (Fraction)

F = Irrigation interval, days

H = Operation period per day, hrs

Sprinkler irrigation system

Application rate

Application rate (cm/hr) =

$$\frac{\text{Discharge of sprinkler (li/sec)} \times 360}{\text{sprinkler spacing (m)} \times \text{lateral move (m)}}$$

Basic dyraulics of drip system

Blassius equation

$$f = \frac{0.316}{Rn^{1/4}}$$

Where, f = friction coefficient,

Rn = Reynolds number

William and Hazen equation

$$\Delta H = 15.27\left(\frac{Q^{1.852}}{D^{4.811}}\right)L$$

Where, ΔH = Energy drop by friction, m

Q = Total discharge in the pipe, lps

L = length of pipe, m

D = inside diameter of pipe, cm

Drainable porosity

Drainable porosity =

$$\frac{\text{Volume of water drained}}{\text{total volume of saturated soil}} \times 100$$

Leaching requirement

$$LR = (E\text{-}P)\ C\ \frac{Ci}{F(Cfc - Ci)}$$

Where, LR = leaching requirement, mm

E = Evapotranspiration, mm

P = Effective rainfall, mm

Ci = Salt concentration in irrigation water, me/l

GK **Arkan mountain ranges are parts of Andaman-Nicobar islands.** GS

Cfc = Salt concentration of soil at field capacity, me/l

F = Leaching efficiency in fraction

Universal soil loss equation

A = RKLSCP

Where, A = Average annual soil loss, mg/ha (meteric tonne/ha or mega gram/ha)

R = Rainfall-runoff erosivity factor for specific geographic location

K = Soil erodibility factor

L = Length of slope factor

S = Slope steepness factor

C = Plant cover management factor

P = Conservation practice factor

Ration method of prediction of runoff rates

$Q = CIA$

Where, Q = Design peak runoff rate, cubic ft. per seconds

C = Runoff coefficient

I = Maximum average rate of rainfall, inch per hour

A = Watershed area, acres

Estimation of soil erosion by wind

$E = f\ (IKCLV)$

Where, E = Estimated average annual soil loss, mg/ha/year

I = Soil erodibility index, mg/ha/year

K = Ridge roughness factor

C = Climate factor

L = Unsheltered length of eroding field, m

V = Vegetative cover factor

Points to Ponder

- The residential building on a farm is the heart of the farmstead.
- To take maximum advantage of the sun the residence should face south or north.
- A typical bedroom of 3.6 x 3 m will accommodate two single beds of 1 x 2 m.
- The minimum size of a bedroom should be 4.5 x 3.6 m.
- Floor space of kitchens usually varies from 9.3 to 14 square metres.
- The common size of a guest room is 3.6 x 3 m.
- A single storey building will have its ceiling at a height of about 3.3 m.
- Two storey buildings usually have first floor of 3.0m height and the second floor of 2.7 m height.
- A 37.5 cm wall can support the ceiling as well as its own load.
- The location of septic tank with respect to residence should be within 1.5 m.
- For an average size family of five members including two adults and three children, the tank capacity may be kept about 2.8 cu. m.
- For additional member the capacity should be increased by 0.42 cu. m.
- The capacity of tank should not be smaller than 2.1 cu. m.
- The shape of space tank may be rectangular, cylindrical or square.
- The tank should be inspected at least once in two years and cleaned before the sludge level reaches the bottom of the outlet.
- All subsurface adsorption systems should be kept 30 m away from any water supply well, 15 m from any stream and 6 m from dwelling.
- A slope of 0.25 to 0.40 per cent gives quite reasonable adsorption.
- For a tank size of 2.80 cu. m, a soakage pit having a total surface area over 15 square metres will be sufficient.
- The soakage pit should be filled with coarse and fine aggregates.
- Woven wire fences are most popular for general farms.
- In woven wire fences the space between vertical wires is kept between 15 and 30 cm whereas space between horizontal wires varies between 7.5 and 22.5 cm.
- The spacing between barbs varies between 7.5 and 15 cm.
- Barbed wire is also used for supplementing the woven or plain wire fencing.
- Barbed wire fencing is not very effective against the goats and rabbits.
- Barbed wire fencing cannot be used for poultry.
- Plain wire fencing is mostly used for large cattle fencing.
- The electric fence is one of the low cost fences that is widely used for confining dairy cattle.
- Continuous voltage is not applied to the electric fences.
- In electric fencing, the wooden or concrete posts are placed about 10 metre apart.
- In the stanchion barn, the cows are housed and milked in the same building.
- In loose housing barn, cows are housed in covered or partially covered yard and milked in milking parlour.
- Open air barn does not have any cover.
- Individual feeding and its management of animals are possible in stanchion barn.
- The face out barn are preferred for milch animals.
- The face in type barn gives a saving of 20 per cent in floor area of the barn as compared to the face out type barn.
- The maximum length of stanchion barn should be restricted to 40 metre.
- The width of manger should have ranges as 75 to 90 cm.
- The passage between the outer wall and the manger is called feed alley.
- If gable roof is used, the maximum height of the ceiling may vary between 5 and 5.5 metres and for a barn 10 and 11 metres wide.
- In a double range stall barn, the door way should be 2.4 to 2.7 metres wide on both sides.
- For adequate lighting in the barn, 0.37 square metre of window area per cow is desirable.
- The water bowls are placed in the manger at about 60 cm above the floor level.
- The floor space both for loafing as well as for feeding may be required in the order of 9 to 11 square meters per cow.
- The feed manger space may be between 70 and 75 cm wide for each cow, where all of them eat at the same time.
- The space requirement is increased by 0.046 square metres per bird in hot climate.
- On an average a man develops nearly 0.1 horse power
- The average force a bullock exerts is nearly equal to one-tenth of their body weight.
- Medium-size bullock can develop between 0.50 and 0.75 hp.
- A small size bullock is not fit for heavy work and cannot develop even 0.50 hp.
- The efficiency of a diesel engine varies between 32 and 38 per cent.

GK **Yes! the Indian President has the power to promulgate ordinance.** GS

- The efficiency of a petrol engine varies between 25 and 32 per cent.
- The extra high speed engines used in knapsack sprayers are powered by petrol.
- On an average about 1/10th of the total electric power generated in India is consumed by farm work.
- The velocity required to operate a windmill is more than 5 kmph.
- The average capacity of windmill would be 0.50 hp.
- The power produced by a windmill varies from 0.1 to 0.9 hp with velocity varying from 6.4 to 37 kmph.
- Ten humans equivalent to 1.0 hp.
- Two bullocks (pair of bullock) is equivalent to 1.0 hp.
- Two windmills is equivalent to 1.0 hp.
- The small farmers hold less than 4 ha land while marginal farmers hold between 4 and 10 ha land and large farmers hole more than 10 ha land.
- In carburettor type engine the liquid fuel is mixed with air.
- In diesel engine, liquid fuel injected alone is under high pressure of 105 kg per sq.
- In dual fuel engine, both diesel and gases like methane, biogas etc is used as fuel.
- Vertical single cylinder, high speed engines are generally used as portable engine for pumping and electric generation.
- Vertical multiple cylinder, high speed engines are mostly used as automotive engine on farm tractors.
- A tractor is equipped with multi-cylinder engine for providing frequent power impulse per revolution of crankshaft.
- A single-cylinder engine requires large flywheel and counter balance to carry load through idle stroke and to overcome the friction between moving parts.
- The power stroke of six cylinder four-stroke engine occurs after every 120° of flywheel travel.
- The power stroke of four-cylinder four-stroke engine occurs after every 180° of flywheel travel.
- The power stroke of single cylinder four stroke engines occurs after every 720° of flywheel travel.
- The power stroke of single cylinder two stroke engines occurs after every 360° of flywheel travel.
- The low speed engine has below 350 rpm, medium engine have between 350 and 1000 rpm while high speed engine have over 1000 rpm.
- Tractors are mostly equipped with high speed engines running at 1500 rpm.
- The stroke bore ratio of tractor engine varies between 1.00 and 1.45 and is about 125. The compression ratio of carburettor type engine ranges between 4:1 and 8: 1.
- The compression ratio of diesel engine ranges between 14:1 and 20:1.
- The piston displacement is also known as swept volume.
- The piston speed of high-speed engine varies between 300 to 500 metres per minute.
- The rate of doing work is equivalent to 75 kg metres per second.
- The indicated horsepower is developed by all cylinders and measured at top of piston.
- The brake horsepower is available at the crankshaft.
- The frictional horse, power is power loss between piston ends to crankshaft end.
- The two-stroke engine has an airtight crankcase.
- Specific fuel consumption of diesel engine is about 0.2 kg per bhp hour.
- Specific fuel consumption of petrol engine is about 0.29 kg per bhp per hour.
- The heart of diesel engine is fuel injection system, injector and fuel pump.
- The two-stroke engine develops power after every revolution of crankshaft (two stroke of piston).
- The four-stroke engine develops power after every two revolution of crankshaft (four stroke of piston).
- The valves are provided in four-stroke engine while ports are provided in twostroke engines.
- The thermal efficiency of petrol engine is low as compared to diesel engine.
- The four-stroke engine gives half power that of two-stroke engine.
- The two stroke engines are high-speed engine while four stroke engines are low, medium
- and high-speed engine.
- The two-stroke engine have comparatively smaller flywheel than four stroke engine.
- In four-stroke engine, camshaft speed is exactly half the speed of the crankshaft.
- The clearance between rocker and and valve stem is tappet clearance provided with 0.38 mm.
- The piston ring gap is about 1 mm per 200 mm diameter of the piston.
- The connecting rod of a four-stroke engine is subjected to compression as well as tension, while crank shaft subjected to bending as well as twisting from the connecting rod thrust. The inlet valveopeils 10° after TDC and remains open up to 30° after BDC.
- The exhaust valve opens 30 to 45° before BDC and remains open 9° after TDC.
- The exhaust valve opens about 225° arid 1/21 seconds.
- Both the valves (inlet and exaust) remain closed for about 285° during compression and power stroke.

- Oil soaked cleaners used on stationary engine is not expected to work in dusty condition.
- Oil bath cleaners are used on tractor works in very dusty conditions.
- Pre-cleaners are used part of air cleaning and placed 45 cm above the hood where the intensity of dust is not severe.
- The oil of oil bath cleaner is replaced after every 10 hours of work under normal dusty condition.
- Fuel with Cetane rating of between 30 and 60 are available as commercial diesel fuel.
- The air fuel ratio of carburettor engine is 15: 1.
- The fuel supply pump of diesel engine forces fuel at a pressure of 2 kg/cm^2.
- The fuel is injected through a nozzle of 126 to 175 kg/cm^2 or more.
- The single hole nozzle has advantage of not being clogged.
- The pintle nozzle is preferred over other nozzles.
- Primary filter removes particles down to 5 microns (5/25000 in).
- Secondary filter removes particles down to 2 microns (2/25000 in).
- The variation of 100 rpm is found between no load and full load conditions.
- The governor used on tractor engine is variable speed governor.
- The governor used on stationary engine is constant speed governor.
- The percentage regulation at no load and full load in 7 per cent.
- The battery on tractor is six cell and developing 2.2 volts.
- Positive plate of battery is made of lead and antimony whereas spongy lead is used for negative plates.
- The fully charged battery will have electrolyte of specific gravity 1.280.
- The electrolyte is a mixture of sulphuric acid and distilled water.
- A completely discharged battery will have electrolyte of specific gravity of 1.150 or less.
- The battery should recharge when the specific gravity falls below 1.225.
- Hydrometer is used to check specific gravity of electrolyte.
- The level of electrolyte above the plate should be about 6 mm.
- Hot plug are larger insulator and used in powering engine or engine operated under light loads.
- Cold plug are shorter insulator and used on petrol engine.
- The spark plug gap settings are kept between 0.50 and 0.85 mm.

Quiz Kit!

First hydro power station

India's first small hydro power station was set up in Ganderbal, near Srinagar, on the river Sindh in Jammu and Kashmir. Commissioned in 1955 by the Jammu and Kashmir State Electricity Board, it had a power generating capacity of 15 MW.

- Magneto is used to supply the spark for ignition.
- The points of magneto should wear and regapped after 200 to 300 hours of operation.
- A 500 amp electric current is required by motor to crack the engine of compression ignition engine.
- The cooling system of an internal combustion engine, maintains the engine temperature at 71° to 82° for petrol engine.
- The cooling system of an internal combustion engine, maintains the engine temperature at 88° to 90° for heavier fuels.
- The air-cooled engine runs hotter than water cooled engine and requires heavier lubricating oil.
- Thermostat valve begins to open at about 74°C and opens widely at about 80°C.
- In the forced feed cooling system pressure of about 0.34 to 0.4 kg/cm^2 is created and thereby the boiling temperature of water is raised to about 110°C.
- Bellows type thermostat valve is used on tractors. Flexible bellows are filled under vacuum with alcohol, which has narrow boiling range.
- Oil pump of lubricating system is run by camshaft.
- A pressure of 3 kg/cm^2 is needed to transfer the oil in the system
- The oil filter of lubricating system may change after 120 hours of operation.
- The crankcase oil is usually an engine or motor oil and is changed after 120 hours of operation.
- Crankcase breathers (of oil wetted screen) are provided to remove the vapours when the engine is operating.
- It is a device to disconnect the tractor engine from its power trains for changing gear ratios known as master clutch.
- Clutch transmits power by means of friction, between driving and driven elements.
- The single disc or single plate clutch is a dry type clutch.
- Multiple disc clutches are either dry type or wet type.
- Steering clutches are operated by hand lever.
- Driving plate is attached to flywheel and drive plate to clutch drum.
- Transmission systems on modern tractors are selective sliding gear type.

CRISIL-500 is share price Index.

- In constant mesh type gear, the gears are always meshed and usually helical gear is used.
- Lubricants used for tractor transmission are termed as gear box or rear axle lubricant.
- SAE 90 oil is recommended as lubricating oil for gear box and recommended for use in temperature of above 2°C.
- Reduction of the engine speed through the transmission gearbox, range between 1 and 1/ 5th at the highest and lowest speeds respectively.
- The tractor speed in top gear varies between 18 and 24 kmph.
- The forward speed of crawler tractor in top gear is restricted to about 10 kmph.
- Differential is the second speed reduction unit in tractor.
- Differential lock helps in developing full traction on both the drive wheels.
- Final drive is additional gear reduction between engine and rear axle.
- The final drive provides 3 to 5 : 1 gear reduction.
- The PTO speed operating under load is 540 ± 10 rpm or 1000 ± 10 rpm.
- The vertical distance of PTO shaft of a drawbar is 20 cm and within the limit of 7.5 cm to the right or left of the centerline of the tractor.
- Wheels or track used on a tractor are primarily meant for converting engine torque to the drawbar pull.
- Tractor is self -propelled machine, used either for pulling or pushing loads or for stationary belt work.
- Swinging drawbar advantageous for taking short turn with machines wider than tractor.
- Swinging drawbar reduces side draft and leaves small area uncovered.
- Traction refers to the pulling force a tractor can develop on ground.
- The size of tractor's rear tyre are 12.00 inch (sectional diameter of tyre) -38 inch (rim diameter).
- The size of tractor's front tyre are 8.00 inch (sectional diameter of tyre) -16 inch (rim diameter to 7.50-16).
- The inflation pressure of rear wheel of the tractor varies between 0.8 and 1.2 kg/cm^2.
- The inflation pressure of front wheel of the tractor varies between 2 and 2.5 kg/cm^2.
- Toe-in of the tractor's front wheel should not be more than 6 mm for a 1.22 m width.
- The useful life of pneumatic tyre under normal operating condition would be about 6000 hours.
- The extra width row crop tractor having tread spacing of -1.22 to 2.15 m.
- The normal width row crop tractor having tread spacing of 1.22 to 2 m.
- The extra narrow width row crop tractor having tread spacing of less than 1.22 m.
- The clutch and brake type steering mechanism used on crawler tractors.
- The tractor tyre is provided with camber angle to improve steering.
- The angle made by the vertical line passing through the centre of the tyre and the vertical line is called camber angle.
- The angle made by the line passing through the centre of king pin and the vertical line is called caster angle.
- The worm gears are provided to transmit power of stearing to the front wheel.
- The fluid used in hydraulic system is the mixer of alcohol and glycerine.
- Hydraulic pump of hydraulic system develop a pressure of about 150 kg/cm^2.
- In point hitch of tractor, the upper link is compression link and lower two links are tension links.
- A few drops of castor oil help to keep belt soft and clean and helps to prevent slipping.
- Speed cones are pulley with several steps of varying diameters on which common belt may be used.
- Slip is usually expressed as the percentage speed reduction.
- A V -belt is advantageous of reduced space requirement and having greater contact area per unit of belt width than flat belt.
- V- belts are used for relatively short transmission distance.
- Chains are providing positive drive, which is not possible with frictional drives as belt.
- Gears are used for changing speed, shaft direction and power.
- Spur gears have teeth cut parallel to the axis of the shaft and used to transmit motion between two shafts, which are parallel.
- Helical gear which are having teeth inclined to the axis of wheel and used on parallel shaft.
- Rack and pinion gears are specially used on fuel injection pump.
- Bevel gears are, commonly employed for two shafts with axes, which are right angle and can intersect.
- Bevel gears transmit motion slightly greater or less than 90°.
- Worm gears are used for two shafts with axes which are usually at right angles but cannot intersect.
- Worm gears are used where high velocity ratio is required with limited space.
- Clutches are used to transmit power between two shafts where drive has to be disconnected frequently.
- Coupling used to connect two shafts, which may slightly be out of order.

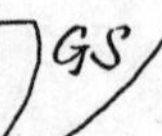

GK GS

Heat and smoke in a kitchen are transferred through the chimney by convection.

- Bearing reduces friction between the moving parts and loss of available power.
- Bearing has to resist radial and axial load.
- Thrust bearing designed to resist axial load while journal bearing resists radial load.
- Bush bearing is made of soft material with low coefficient of friction.
- Ball and roller bearing are used on various agricultural machines.
- The thermal efficiency of petrol engine varies between 25- 32 % while the thermal efficiency of diesel engine varies from 32-35 %.
- Otto cycle engine (CVC) are those engines in which heat is taken at a constant volume and rejected at another constant volume.
- Diesel Engine (CPC) are those engines in which heat is taken at one constant pressure and rejected at another constant volume.
- Two-stroke engine is the engine which cycle is completed in one revolution of crankshaft.
- Four-stroke engine is that engine which cycle is completed in two revolution of crankshaft.
- The camshaft runs at half the speed of crankshaft.
- Scavenging is the process of removal of burnt gases from the engine cylinder.
- The compression ratio of diesel engine varies between 14 and 22:1 and pressure varies from 30-45 kg/cm^2 with a high temperature of about 500°C.
- The compression ratio of petrol engine varies between 4 and 8:1 and pressure varies from 10 kg/cm^2 and a temperature of 260°C.
- Cylinders are usually made of high-grade cast iron.
- Dry cylinder liners are not come in contact with water.
- Piston is made of aluminum and its alloys due to lightness.
- Piston pin is also called wrist or gudgeon pin.
- Connecting rod is made of drop forged steel.
- Crankshaft converts reciprocating motion of piston into rotary motion of flywheel.
- Flywheel is made of cast iron and provides energy during idle stroke and provides uniform rotary motion by virtue of its inertia.
- Camshaft gear is also called halftime gear.
- The common face and seat angle of valve is 45° but 30° angle is also used for intake valve.
- L-head arrangements of valve is quite extensively used on tractor's engine and automobile.
- Tappet is called valve lifter.
- Valve time is function of engine speed.
- Inlet valve opens at 5° before TDC and closes 30° after BDC while exhaust valve opens 40° before BDC and closes 5° after TDC.
- The firing order of four-cylinder engine is 1-3-4-2.
- The firing order of six-cylinder engine is 1-4-2-6-3-5 or 1-5-3–6-2-4.
- Four-stroke cycle engine completes all strokes in 720° revolution of crankshaft.
- Two-stroke cycle engine completes all stroke in 360° of crankshaft.
- The stroke bore ratio of tractor engine is 1.25 and varies between 1 and 1.45.
- One horsepower equals 75 kg m/sec.
- IHP is generated in engine cylinder and received by piston.
- BHP is delivered by engine and available at the end of crankshaft.
- Displacement volume is total swept volume of all pistons during power strokes occurring in one minute.
- Mechanical efficiency is the ratio of brake horsepower to indicated horsepower.
- Brake horsepower is measured by prony brake, electric and hydraulic dynamometer.
- Drawbar horsepower is measured by spring, strain gauge and hydraulic dynamometer.
- The specific gravity of LSD is 0.920 while calorific value is 10300 Kcal/kg.
- The specific gravity of HSD is 0.820 while calorific value is 10550 Kcal/kg.
- The specific gravity of kerosene is 0.827 while calorific value is 10850 Kcal/kg.
- The specific gravity of petrol is 0.730 while calorific value is 11000 Kcal/kg.
- Cetane number indicates knocking characteristics of diesel fuel.
- Cetane is mixture of cetane and alphamethyle nepthelene.
- Cetane number for commercial diesel fuel is varying from 30 to 60.
- Higher the Cetane numbers better the ignition quality of fuel.
- Octane number is indicating knocking characteristics of petrol fuel.
- Octane is the mixture of iso-octane and normal heptanes.
- For excellent knocking quality, the rating of octane number is 100.
- Pre-ignition is due to spark advance or excessive heat in cylinder.
- The specific gravity of liquid is ratio of its weight to the weight of an equal volume of water at 60°C.
- Pure water has got API (American Petroleum Institute) is 10.
- Reid vapour pressure test is commonly used to measure the vapour pressure of a fuel and pressure read on the

bourdon gauge at 100°F temperature.

- Immersing a polished strip of copper in the fuel for 3 hours at 122° F indicates cohosive sulphur.
- Best result was obtained when carburettor are so adjusted that air and fuel mixture is about 15:1.
- Firing order of six-cylinder engine is 1-5-3-6-2-4, 1-5-4-6-2-3, 1-2-4-6-5-3 and 1-2-3-6-5-4.
- Firing order of eight-cylinder V-shape engines is 1-5-4-8-6-3-7-2, 1-8-4-3-6-5-7-2, 1-6-2-5-8-3-7-4, 1-8-7-3-6-5-4-2 and 1-5-4-2-6-3-7-8.
- Firing order of a 12 cylinder engine is 1-4-9-8-5-2-11-10-3-6-7-12.
- Firing order of a 16-cylinder engine is 1-4-9-12-3-16-11-8-15-14-7-6-13-12-5-10.
- SAE stands for Society of Automotive Engineers and Society of Agricultural Engineers also.
- ASTM stands for American Society of Testing for Materials.
- About 15-20% slip is permissible for wheel type tractors
- 7 to 10% wheel slip is permissible in crawler tractors.
- Tillage system for dry land agriculture is called till plant system.
- Till plant equipment includes sweeps to open the seed bed.
- Minimum tillage is called plough plant method.
- Surface planting with conservation feature of contour listing is called ridge planting.
- Ridge planting is most suitable for high rainfall areas.
- No tillage system reduce runoff by about 90 per cent and soil loss nil as compared to conventional tillage.
- Chisel ploughs are used for reduced tillage or conservation tillage system.
- With two type walking type of implement, the operator has to walk about 64 Km to plough 1 ha land.
- Trailed type implement is attached to the tractor drawbar by a pin joint.
- A semi-mounted is rigidly attached to tractor drawbar and wheels to support part of its load.
- A mounted implement is one that is attached to the tractor as a integral part.
- Indigenous plough cuts trapezoidal furrow cross section.
- The energy and time requirement in using indigenous plough are greater compared to other type of ploughs.
- The inside angle between shoe and body varies from 116° to 160°.
- Mould board-plough used in areas where sufficent rainfall to produce good crop because it does not conserve moisture.
- Two-way plough suitable for terraced land of hilly tracts and do not upset the slope of land.
- Two-way plough does not leave any back furrow and dead furrow.
- Two-way ploughs also used for ploughing irrigated land and left land level without depression.
- Two way ploughs also called turn wrest plough because its plough bottom turns through 180° angle.
- Mould board plough works on principle of suction (horizontal and vertical).
- Stubble mould board adapted for ploughing an old ground.
- Sod or breaker mould board used in tough sod (grass land).
- Slat M. B. preferred for sticky soils and high speed MB used on tractor ploughs for general farm use.
- Landslide provides stability against tilting furrow wall side ways, due to soil pressure acting on the mould board.
- The width of the landslide of animal-drawn ploughs varies between 5 and 10 cm
- Gunnel: It is vertical face of the share, which slides along the furrow wall. It takes side thrust of the soil and supports the plough bottom against the furrow wall.
- Share made of chilled cast iron consists of about 0.70 to 0.80% carbon and about 0.50 to 0.80% manganese besides other minor element.
- Slip share with curved cutting edge, having no additional part used mostly by farmers.
- Slip nose share: Having a small detachable piece.
- Shin share: The share having shin as an additional part.
- Bar point share: It is provided with an adjustable and replaceable bar.
- General purpose MB: It turns the furrow slice and pulverizes the soil thoroughly.
- Stubble type MB: Has short but broader MB used in stubble soil.
- Sod or breaker type MB: Have long MB with gentle curvature design used in sod soils.
- Slat type MB: Surface is made of slats placed along the length of mould board used in a sticky soil
- Jointer: It is a miniature plough, purpose to turn over a small ribbon-like furrow slice directly in front of main plough.
- Coulter: It is used to cut furrow slice vertically from land ahead of plough bottom.
- Rolling type coulter: It is a round steel disc, which is sharpened on the edge and suspended on shank and yoke from the beam. It may be either plain or notch.
- Sliding type coulter: It is stationary knife fixed downward in the vertical position on the beam.
- Gauge wheel: An auxiliary wheel for maintaining uniform depth of working.
- Land wheel is front wheel of plough that runs on unploughed land.
- The draft requirement of disc plough is less than the mould board plough, turning same volume of soil in similar condition due rolling action of disc.

- Front and rear furrow wheel: Mounted in front and rear of plough runs in furrow.
- Rolling coulters are used to cut trashes and keep plough from clogging and set 2 cm outside the landside of the plough.
- Notch coulters used to cut heavy trashes.
- Jointer is a miniature plough set to cut 4 to 5 cm deep.
- Gauge wheel being used as a transporting aid and helps in keeping uniform depth of ploughing.
- Size of mould board is expressed by width of furrow that is designed to cut and measure the perpendicular distance from the share to the landside.
- Walking type plough is medium sized (150 to 200 mm).
- The light plough have sizes above 100 mm and 150 mm width of cut, while heavy plough having size of above 200 mm.
- The horizontal clearance in mould board plough is provided with 5 mm while vertical clearance is 3 to 5 mm.
- Disc ploughs are used where climate is dry and where soil is rough and stony.
- Disc plough is preferred for deep ploughing and also for conservation tillage.
- Standard disk plough has single disc on a stub, while the diameter and curvature of disc is larger than vertical disc plough.
- The tilt angle of standard disc plough varies from 15 to 25°, while disc angle varies from 42 to 45°.
- The vertical disc plough also called harrow plough, because of number of discs mounted on a single gang.
- In vertical disc plough, tilt angle of disc is zero.
- The vertical disc plough used for shallow ploughing and standard disc plough used for deep ploughing.
- The vertical disc plough used is preferred for use in wheat growing hence also called wheat plough or one-way plough.
- Disc plough is advantageous for deep ploughing and penetrates any one of soil like hard, stice and loose.
- Disc plough is disadvantageous of not covering trashes, weeds and leaves surface in rough condition.
- Standard disc plough difference with vertical disc plough in respect of diameter and curvature of disc.
- Standard disc plough is used for deep ploughing, while vertical disc plough is used for shallow phoughing.
- Standard disc plough has disc angle of 42-45° and tilt angle of 15-25°
- Vertical disc plough has disc angle of 42-45° and no tilt angle
- Vertical disc plough is also called as harrow plough or one-way plough.
- Increasing disc and tilt angle penetration of disc increases while be increasing disc angle width of cut reduced.
- The width of cut of standard and vertical disc plough changed by adjusting the angle between the frame and the land wheel axle to rotate the entire frame in their horizontal frame.
- Penetration improves by increasing disc angle and by decreasing tilt angle of standard disc plough.
- Chisel plough: Used to cut through hard soils by means of number of narrow types.
- Subsoiler: Penetrates to a depth of 40 cm or more in hardpan to drain heavy soil and deep root growth.
- Rotary plough is used for rotary operation of the rotor is 200 to 300 rev/min.
- The mulch tillage is suitable for low rainfall areas where soil is subjected to erosion.
- Deep tillage and subsoiling are usually beneficial in conserving moisture in monsoon season.
- Deep ploughing to the depth of 15 to 20 cm is beneficial in unirrigated areas.
- Chisel plough penetrates to the depth of 25 to 35 cm. while sub-soiler operated to the depth of 60 to 70 cm.
- Sub-soiling is a desirable tillage practice in areas having hard or compacted layers below the plough depth.
- Forces acting on implements are force of gravity, soil reaction on the implement and pull of the power unit. These forces are a must be in equilibrium for satisfactory operation.
- The vertical component V has effect of adding load to the tractor rear wheels, this removes load from the front wheels of the tractor
- The casting is an important method used in tractor ploughing
- Lister preferred in dry land areas and where soil is crumbly.
- Sweeps are subsurface tillage tool that works at a shallow depth and not turn the soil.
- Ridgers are required for row crop planting.
- The main objective of puddler is to reduce leaching of water.
- The indigenous plough is the most prevalent implement used as puddler in the paddy growmg areas.
- The centre of plough load on a single bottom is a point about 7.5 cm while 5 cm on a small size animal drawn ploughs to the right of shin.

Did You Know ?

First heavy water plant

India's first heavy water plant was set up at Nangal in Punjab in 1962. Heavy water or deuterium oxide (D_2O) is the basic fuel used by first generation nuclear power plants.

- Draft of plough increases with increasing rate of ploughing, blunt surface of disc and any other attachments.
- The hitching of plough is done by placing plough a few centimetres below the ground level.
- The useful life of walking type ploughs may take as much as 1500 hrs. and 2000 hrs, for tractor-drawn ploughs.
- The repair and maintenance cost for indigenous plough is 20 %, for walking type it is 10% and 5 % for tractor-drawn ploughs.
- Housing cost of ploughs is taken as zero.
- The concavity of disc harrow is less than disc plough.
- In trailed type tractor-drawn harrows, the depth is usually controlled by changing the angle between the gangs and by addition of weight.
- Drag harrows are used to cut branches from trees and leveling the soil.
- Spring type harrow penetrate much deeper than spike tooth harrow and used in soil having obstacles like stones, roots and weeds.
- Spike tooth harrow and spring type harrow does not require lubrication.
- The blade harrow called bakhar used in clayey soil for preparation of seed bed of both seasons.
- The depth of cultivation limited to 4 to 5 cm.
- Hoe is the versatile form of implement used for seedbed preparation, ridge making, and channel shaping and weeding.
- A man can weed out about 0.025 hectare in a day under normal working condition.
- The wheel hoe is used for cultivating land between rows.
- Harrows cuts soil to shallow depth for smoothening and pulverizing the soil as well as to cut weeds.
- Single action disc harrow: Have two gangs placed end to end which throws the soil in opposite direction.
- Double disc harrow: A set of two gangs followed by another set of one or two gangs arranged in such a way that front and back gang throws the furrow slice in opposite direction.
- Tandom disc harrow comprising four gangs in which each gang can be angled in opposite direction.
- Offset disc harrow: Comprising four gang in tandem and soil is thrown in both the direction.
- Spike tooth harrow: Is used to smoothen and level soil after ploughing.
- The animal spike tooth harrows are rigid type.
- Spring type harrow: Are suitable for working in hard and stony field.
- The guardrails are provided to give rigidity and support to the harrow.
- Acem harrow is special harrow with curved knife. The good pulverization as well as good mulching occurs with this harrow
- Patela is used for smoothening the soil and crushing the clods.
- Bakhar: Used for shallow working soil with minimum soil inversion.
- Bodela: It is a twin blade harrow.
- Rolling or rotary harrow: It is a harrow consisting of rigid or flexible tines, mounted on one or more horizontal or vertical shaft. It is also called as Gyro harrow.
- Cultivators: Used for seedbed separation, sowing with seeding attachment and inter cultivation. They are tine, disc and rotary cultivator.
- Cultivator with spring loaded tine: Are swung back when an obstacle is encountered.
- Cultivator with rigid tines: are those, which are not deflecting during work in field.
- Duck foot cultivator: It is rigid tine cultivator used for shallow ploughing.
- Puddler: Used for churning soil in standing water of 5 to 10 cm depth.
- Bund former: Forms ridges or bunds by collecting soil.
- Ridger: It cuts soil and turns in opposite direction simultaneously for forming ridge and it is also called a furrower.
- Ridger: Are used for formation of ridges, field furrow or channel and earthing up.
- Green manure trample is used to trample and press the green manure crop.
- Dibbler is a planter.
- Dibbling is suitable for small plots and vegetable crop.
- By dibbling method seed rate is reduced to 1/5th or more.
- Seeding behind ploughing is used for wheat and barley.
- Check row planting is method makeing possible inter-culture from both directions.
- Hoe opener with spring released are especially suitable for stony or root-infected fields.
- Disc type furrow openers are favoured where considerable plant debris or trash mulches are used.
- The seed drill drop seed in a continuous stream therefore the distance between plants is not maintained.
- The planter drop seeds after successive interval, therefore distance between plant is maintained.
- Seed dropping behind plough used for seeds like maize, gram, peas, wheat and barley drop in furrow behind plough.
- In drilling, the row-to-row spacing maintained but plant-to-plant is not.
- Seed dropping behind plough: Seeds like maize, gram. peas, wheat and barley drop in furrow behind plough.

USA generates more municipal waste per head than any other country.

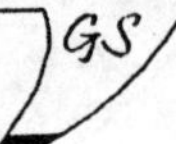

- Sowing behind plough can be done by device called as Malobansa.
- Seeding behind plough is slow and laborious method.
- Check row planting: Plant-to-plant and row-to-row, both the distance is maintained and facilitates weeding in both the direction.
- The fluted feed mechanism is more positive in its metering action than the internal double run method.
- Seedling behind plough is slow and laborious method.
- The internal double which has double-faced-wheel, one face for larger opening for larger seeds and other face has smaller opening for smaller seeds.
- Cup seed mechanisms common is British seed drill.
- A number of bullock drawn planters have brush feed mechanism.
- Shovel type furrow opener are best suitable for stone or root-infected fields.
- Shoe type furrow openers are used in trashy soil where seedbed is not smoothly prepared.
- Shoe type furrow openers are especially suitable for black cotton soil.
- Disc type furrow openers are found suitable where plants debris or trash mulches are used and also in sticky soil.
- Shoe made of carbon steel having minimum carbon content of 0.5 per cent with minimum thickness of 4 mm.
- Shovel made of carbon steel having carbon content of 0.5% of and minimum thickness 4 mm.
- Boot device that convey seeds from delivery tube to the furrow.
- Sticky belt and sand bed are the two methods used for testing uniformity of seeds.
- Edge drop carries seeds on edge in the cell of plate.
- Flat drop carries the seed on a flat in the cell of the plate.
- Hill drop, the cell round the edge of the plant is large enough to admit several seeds are a time.
- The field capacity of semi-automatic 2-4 potato planter is 0.15-0.25 ha/hr.
- Manual rice transplanter 0.2-25 ha/day of 8 hrs.
- Sugarcane planter has capacity of 0.6 ha/hr.
- Seed dresser is a machine to apply coating of protective chemicals to seeds.
- Fumigator is a machine to generate and distribute gases or smoke.
- Flame gun is a appartus to kill weeds by flame.
- Swath is a material left by harvesting machine.
- Windrower is a machine to cut crops and deliver them in a uniform manner in a row.
- The cutter bar is set at about 88° to the direction of motion i.e., 2° given to it in order to overcome the back pushing action of crops.
- When cutter is properly aligned, the knife and pitman run in a straight line.
- Generally 2 cm lead per metre length of cutter bar is recommended.
- The reaper harvest is 5 to 8 cm above ground.
- The reaper binder cuts crop 10 cm from the ground and capacity is 0.25 - 0.35 ha/hr.
- Olpad thresher is mainly a wheat thresher and pulled by a pair of bullocks.
- Olpad thresher origin is at a small place, Olpad in Gujarat state.
- Threshing efficiency depends on peripheral speed ofthe cylinder, cylinder concave clearance, type of crop and moisture content of crop, weather condition and feed rate.
- Concave clearance is clearance between beater and concave.
- Blower loss is the percent of grains blown by blower along with blusa.
- The fluted feed mechanism is more positive in its metering action than the internal double run method.
- The internal double wheel has double faced wheel, one face for larger opening for larger seeds and other face has smaller opening for smaller seeds.
- Windrower, is machine to cut crops and deliver them in uniform manner in a row.
- Primary tillage is an operation performed on any cultivable land to prepare seedbed for growing crops.
- Secondary tillage is a lighter and finer operation performed on the soil after primary tillage.
- Minimum tillage is a minimum soil operation to meet tillage requirement for crop production, also called harrow plough.
- Strip tillage is a tillage operation in which only isolated bands of soil are tilled.
- Rotary tillage is a tillage operation employed by rotary operation to cut, break and mix the soil.
- Mulch tillage is the preparation of soil in such a way that plant residues or mulching materials are specially left on or near the surface.
- In combined tillage operation simultaneously utilized in two or more different types of tillage tools.
- Tool is an individual working element such as disc and shovel.
- Implements are the equipment which generally does not have driven moving part. As harrow plough.
- Machine is a combination of rigid body having definite motion.
- Normal ploughing: Ploughing about the depth of 15 cm.
- Contour Ploughing: Ploughing in which soil is broken and turned along the contours.

The Planning Commission was constituted in India as a non-constitutional and advisory corporation.

- Vertical suction is the maximum clearance under the landside and the horizontal surface when the plough is resting on the horizontal surface in the working position.
- Horizontal suction is maximum clearance between the landside and horizontal plane touching point of share at its gunnel's side and rear heal of landside.
- Throat clearance is the perpendicular distance between point of share and lower point of the beam of the plough.
- Vertical clevis is to control depth of operation and line of pull.
- Horizontal clevis is a device to made lateral adjustment of the plough relative to the line of pull.
- Plough size is the perpendicualar distance from wing of the share to the line joining the point of the share and heel of landside is called the size of plough.
- Trailed type implements are pulled and guided from single hitch point but weight is not supported by the tractor.
- Semi-mounted implements are one, which attached to tractor along a hinge axis are not at a single hitch point but weight, is partially supported by tractor.
- Mounted implement is one which is attached to the tractor and controlled directly by tractor steering unit. The implement is fully carried by the tractor when out of work.
- Centre of power is true point of the tractor.
- Centre of resistance is point at which the resultant of all the horizontal and vertical forces acts. The centre lies at a distance equal to 3/4 th size of the plough from share wing.
- Line of pull is imaginary line passing from the centre of resistance through the clevis to the centre of pull (power).
- Draft is horizontal component of the pull, parallel to the line of motion.
- Side draft is horizontal component of pull, perpendicular to the line of motion. This is developed when centre of resistance is not directly behind the centre of pull.
- Effective field capacity is the actual area covered by implement based on its total time consumed with its width.
- Theoretical field capacity is rate of field coverage by implement, based on 100 per cent of time at the rated speed and coving 100 per cent of the rated width.
- Field efficiency is ratio of effective field capacity to theoretical field capacity.
- Soil inversion is quality of work in terms of soil aggregates and clod size. It is measured by penetrometer.
- One-way plough throws furrow slice on the one side of the direction of motion.
- Two-way plough throws furrow slices to left and right side of direction of travel. The plough bottom is changed from right hand to left hand by rotating it through approximately 180° about longitudinal axix.
- Disc angle is angle made by plane of cutting edge of disc with direction of travel and it varies between 42°-45° for standard disc plough.
- Tilt angle is angle made plane of cutting edge with the vertical line and it varies from 15°-25°.
- Concavity is the depth measured at center by disc by placing concave side of disc on flat surface. The concavity of standard disc plough is 8 cm for 60 cm diameter of disc and 16 cm for 95 cm diameter of disc.
- Crown is top portion of turned furrow slice.
- Back furrow is a raised left at the centre of the strip of land when ploughing starts from centre to side.
- Dead furrow is an open trench left in between two adjacent strips of land after finishing the ploughing.
- Head land is a strip of unploughed land left at each end of field for tractor turning.
- Gathering in which plough works carried round the strip of ploughed land.
- Casting in which plough works carried out round the strip of unploughed land.
- Gang angle is the angle between the axis of the gang and the lone perpendicular to the direction of travel.
- Dibbling is method of placing seeds in hole at predetermined depth and placing.
- Drilling is dropping of seeds in a line in a continuous flow and covering them with soil.
- Transplanting is method of placing plants in a soil raised in nursery.
- Hill dropping in which seeds are dropped at fixed spacing and not in a continous stream.
- Check row planting in which plant to plant and row to row; both the distance are maintained and facilitate weeding in both the directions.
- Cleaning efficiency is the clean grain received at main grain outlet with respect to the total grain mixture received at main outlet expressed in per cent.
- Threshing efficiency is the threshed grain received from all outlets with respect to the total grain input expressed as per cent by mass.
- Sieve loss is the percentage of healthy grain dropped from sieve with respect to healthy grains received from main outlet.
- Cylinder loss is the percentage of unthreshed grains from all outlet with respect to total input in the thresher by weight.
- Visible loss: The percentage of broken or cracked grain from all outlets with respect to total grain input in the thresher by weight.
- Solar energy reaching the top of earth 's atmosphere consists of about 8 per cent ultraviolet radiation, 46 percent visible lights and 46 per cent infrared radiation.
- Energy radiates by the sun as electromagnetic waves of which 99 per cent have wavelengths in the range of 0.2

to 4.0 micrometer.

- Ultraviolet radiation has short wavelength, less than 0.39 micrometre.
- Wavelength of visible light ranges from 0.39 to 0.78 micrometre.
- Infrared radiations have long wavelength more than 0.78 micrometer.
- The rate at which solar energy arrives at the top of the atmosphere is called solar constant.
- Solar radiation, which reaches the ground directly from the sun, is called direct radiation or beam radiation.
- Solar radiation received from the sun after its direction has been changed by reflection and scattering by the atmosphere is called diffuse radiation.
- India lies between latitude 7° and 37° N, and receives an annual average intensity of solar radiation between 16700 and 29260 kJ/ m^2/ day.
- A device used for collecting solar radiation is called solar collector.
- Solar energy is a time dependent and intermittent energy resource.
- Solar energy can be stored by heating, melting or vaporization of material and the energy becomes available as heat, when the process is reversed.
- Thermal energy storage is used for both domestic water and space heating.
- Solar energy also may be stored by battery for a long period.
- Photosynthesis is method of solar-energy conversion.
- Solar pond is used for solar energy collection and sensible heat storage.
- Wind energy is one of America's greatest natural resources.
- Conversion of kinetic energy of the wind into mechanical energy that can be utilized to perform useful work or to generate electricity.
- Wind speed increases with height because of reduction in drag effect of the earth's surface.
- A 100 % efficient aero generator would be only able to convert up to a maximum of around 60 % of the available energy in wind into mechanical energy.
- The windmill works on the principle of converting kinetic energy of the wind into mechanical energy .
- Geothermal energy from the earth's interior is almost as inexhaustible as solar or wind energy.
- The hot molten rock is called magma, is commonly present at depths greater than 24 to 40 km.
- Geothermal steam is of two kinds that originates from the magma, are called magnetic steam and that from ground water heated by the magma called magnetic steam.
- The total amount of energy in the outer 10 km of the earth' s crust exceeds greatly that obtainable by the combustion of coal, oil and natrural gas.

You Don't Know!

Neem coated urea cake as fertiliser

Rajendra Prasad of IARI, Delhi, created in the early 70s a neem cake coated with urea, now used as a fertiliser by farmers. Its efficiency is 20% higher than urea.

- There are about 340 known thermal areas in India.
- The conversion of solar energy stored as heat in the ocean into electrical energy by making use of the temperature difference between the warm surface of water and colder deep water.
- Tide is periodic rise and fall of the water level of the sea, which are carried by the action of sun and moon on the water of the earth.
- The tide range varies from few centimetres to about 8 to 10 metres.
- Solar energy absorption by the water takes place according to Lambert's law of absorption.
- For the OTEC the best sites are in the tropical belt between 20ºN and 20°S latitude.
- It has been estimated that about a billion kW of tidal power is dissipated by friction and eddies alone.
- Fuel cells are the source of chemical energy.
- The main use of fuel cells are in power production, automobile vehicles and in special military use.
- The pH value for digestion of sewage solids is reported to be in range of 7 to 7 .5.
- Methane bacterial work best at temperature between 35°C and 38°C.
- The gas production starts falling at 20°C and stops at 10°C.
- When temperature was raised to 28.3°C the gas production increased by 50% to 3.75 cu.m per day.
- Raw cowdung contains 80-82% of moisture.
- Normal value of the retention period is between 30 and 45 days.
- pH value slightly higher than 8.2 has been reported to be optimum for digestion of raw animal or plant waste.
- The optimum thermophilic temperature is around 55°C.
- The optimum mesophilic temperature lies at about 35°C.
- The cowdung mix usually in proportion of 1: 1 (by weight) in order to bring useful solid content to 8 to 10%.
- Most municipal sewage treatment plants operate at a loading rate of 0.5 to 1.6 kg of volatile solids per m^3 per day.
- The optimum C: N ratio that best suits for maximum microbiological activity is 30: 1.

- Gas production per unit volume of digested capacity is maximum when the diameter to depth ratio is in the range of 0.66 to 1.00.
- The most activity digesting sludge is in the lower half of digester.
- Regulating the feed volume daily can control the retention time.
- Methanogenic bacteria multiply at a slower rate.
- The retention time should not be less than 24 days for gas production.
- Biogas is the mixture of methane and carbon dioxide.
- Biogas production may be 0.037 m^3/kg for wet cow dung.
- Calorific value of biogas is 4500 Kcal/kg.
- Percentage of methane in the biogas is varied from 45 to 70%.
- Biogas is produced by anaerobic fermentation of organic matter.
- In biogas plant proportion of cattle dung mixed with water is 4:5.
- Position of the sun directly overhead is zenith.
- In aerobic digestion, the biogas production by bacterial decomposition is done in presence of oxygen.
- The rate of respiration of paddy increases sharply (at 25°) at 14 to 15 % moisture content, which is called critical point.
- Rate of respiration increases with increase of temperature to 40° C.
- The removal of relatively small amount of moisture from a solid material by evaporation is drying.
- Drying involves both heat and mass transfers operations.
- The bed thickness in thin layer drying is 20 cm.
- The bed thickness in deep bed drying is above 20 cm.
- The grain drying process in which all grains are fully exposed to the drying as under constant drying condition i.e. at constant air temperature and humidity in thin layer drying.
- Usually the moisture content of a substance is exposed in percentage by weight on wet basis.
- Direct method of moisture determination measures directly with moisture content by the moisture evaporation.
- The air-drying method can be accomplished in a single stage or double stage in accordance with grain containing less than 13% or than 13% moisture content.
- Electrical resistance of grain varies with moisture temperature and degree of compaction.
- Universal moisture meter, tag-happenstall moisture meter Kett moisture meter are resistance type moisture meter.
- The dielectric properties of grains depend on grain moisture content.
- EMC is useful to determine whether a product will gain or lose moisture under a given set of temperature and relative humidity.
- Grain EMC curve or isotherms are generally S-shaped and attributed to multi-molecular adsorption.
- The EMC is determined by static and dynamic method.
- EMC is determined under constant RH and temperature.
- The phenomenon of reaching equilibrium by absorption, desorption of the moisture is hystresis.
- Wheat is dried under constant rate period when its moisture content exceed 72%.
- In the constant rate period of drying takes place by surface evaporation and moisture moves by vapour pressure.
- The moisture content at which the drying rate ceases to be constant is critical moisture of solid.
- The average critical moisture content for given type of material depends, upon the surface moisture concentration, bed thickness of the material and rate of drying.
- In the constant rate period the rate of evaporation under any given set of aircondition is independent of the solid.
- Cereals are dried entirely under falling rate period.
- The falling rate period is due to ability of the moisture to be conveyed from the centre of the body to the surface at a rate comparable with the moisture evaporation from its surface to the surrounding.
- The falling rate period is characterised by increasing temperature both at the surface and within the solid.
- The drying rate is dependent upon air temperature, air velocity, air humidity and air exposure time.
- The rate of air flow per unit mass of grain in deep bed drying is small compared to the thin layer drying.
- The contact drying of the heat for drying is transferred to wet solid mainly by conduction through solid surface.
- The value of moisture constant of grain at dry basis is more than the moisture content of grain at wet basis.
- The rate of heat transfer in conductor drying depends upon the thermal conductivity of the heated surface and also depends on the heat transfer coefficient from heating medium to the surface.
- Thermal efficiency of constant drying is higher.
- The sensible heat of heated air is transferred to the wet product by convection.
- In freeze drying the heat is supplied by radiation or conduction from heated trays and the temperature of the product is not raised above 0°C.
- In radiation drying heat energy can be supplied to the products by electromagnetic waves.
- Osmotic drying is successfully applied for dewatering of cellular products such as fruits, vegetable and meat.

- Fluidized bed drying used for drying material having high initial moisture content and lighter in weight such as vegetable seeds.
- Airflow rate of deep bed dryer is 2.94-3.92 m^3/min per tonne.
- For deep bed drying, the bed depth of grain should be limited to 3 m for grain having moisture content upto 18%.
- For deep bed drying, the bed thickness should be limited to 3 m for grain of moisture content above 18 %.
- Paddy with 2.5 m bed depth takes 20 days for drying during favourable weather and 40 days during bad weather.
- Flat bed dryer are usually 1-2 tonne capacity.
- In mixing dryers, baffles are provided to cause the grain mixing.
- In non-mixing type dryer, baffles are not provided.
- The airflow rate of mixing type dryers is 50-95 m^3/min-tonne and high drying temperature as 65°C.
- The airflow rate of non-mixing type dryer is high airflow rate 125-250 m^3/min-tonne.
- Drying air temperature used in non mixing dryer is 45°C.
- LSU dryer was developed at Louisiana State University.
- LSU dryer was developed specially for rice.
- LSU dryer is mixing type of continuous flow dryer with inverted V- shaped channel-
- In commercial rotary type dryer has been developed by TNAU.
- Try dryer are used for drying vegetables.
- The flow of heated air in a tunnel dryer may be concurrent.
- Solar drying provides higher air temperature and lower RH than simple sun drying.
- Sun drying is an example of radiation drying.
- Various chemicals like sodium chloride; sodium propionate, copper sulphate, ferrous sulphate, urea etc. have been tried for preservation of wet paddy.
- In sack drying an air temperature of 45°C with an airflow rate of 4 m3/min at 3-4 cm static pressure per bag of 60 kg is dried for fastest drying.
- Natural air-drying cannot be used as the ambient relative humidity exceeds 70%.
- Grain containing moisture content is higher than 20% and should not be dried with natural air.
- Rotary dryer can employ for rapid drying parboiled paddy using temperature as high as 100 to 110°.
- Hydrology is the science, which deals with the occurrence, circulation and distribution of the water and earth's atmosphere.
- Sun is the source of energy for the hydrologic cycle.
- The area of land draining into a stream or a watercourse at a given location is called catchments area.
- All forms of water that reach the Earth from the atmosphere is known as precipitation.
- The magnitude of precipitation varies with time and space.
- The net precipitation at a place and its form depends upon a number of metrological factors.
- A tropical cyclone, also called cyclone in India, hurricane in USA and typhoon in South-East Asia.
- The centre of storm is called eye.
- The precipitation is collected and measured in a rain gauge.
- Mass curve of rainfall are very useful in extracting the information on the duration and magnitude of a storm.
- Intensity of storm decreases with increase in storm duration.
- The parts of precipitation, which infiltrates and moves lateral through upper layer of soil and meets the stream is called inter flow.
- The rate of evaporation is proportional to the difference between the saturation vapour pressure at the water temperature and actual vapour pressure in the air.
- Rate of evaporation increases with wind speed upto a critical speed.
- Energy-budget method is an application of the law of conservation of energy.
- The maximum rate at which a given soil at a given time can absorb water is defined as the infiltration capacity.
- Average infiltration rate is called infiltration index.
- Wire gauge is used to measure the water-surface elevation from above the surface.
- Dilution method of flow measurement depends upon continuity principle.
- A perennial stream is one which always carries some flow.
- An intermittent stream has limited contribution to the groundwater.
- An ephemeral stream is one, which does not have any base flow contribution.
- The flow-duration curve is also known as discharge frequency curve.
- The total catchments area of all the rivers in India is approximately 30.5 million square kilometres.
- Minor catchments have an area less than 2000 square kilometres.
- Medium catchments have an area between 2000 and 20000 square kilometres.
- Large catchments have an area greater than 20000 square kilometres.
- The missing data of rainfall can be computed by hydrology

budget method.

- Flood hydrograph is useful for analyzing the stream characteristics.
- The time interval between centre of mass of rainfall and the centre mass of hydrograph is called as lag time.
- S-curve is also known as S-hydrograph.
- S-hydrograph is a hydrograph produced by a continuous effective rainfall at a rate for an infinite period.
- The limiting case of a unit hydrograph of zero duration is known as instantaneous unit hydrograph.
- An imaginary surface representing the artesian pressure in artesian aquifer is called piezometric surface.
- The hydrostatic pressure within artesian aquifer is same times greater enough to cause the water to rise in a well.
- The piezometric level in semi-confined aquifer is some as that of pheratic water table before pumping.
- The piezometric head in semi-confined aquifer is always more than the pheratic head.
- The ratio of head is depending on the hydraulic conductivity and thickness of semi-pervious layer.
- A perched water-table is a special case of an unconfined aquifer.
- A perched water is a local zone of saturation which exists at some level above the main water-table.
- The perched water table occurs when an impervious strata within the zone of aeration interrupts percolation and case water accumulation in limited area above that stratum.
- Well-logging consists of recording the characteristics properties of the sample of subsurface materials obtained during process of drilling.
- Cable tool method is slower but provides more accurate samples from the bails.
- The rotary method is faster but difficult to determine the exact characters of formation.
- The properties of the aquifer may express in terms of its hydraulic conductivity, transmissibility, storage coefficient and specific yield.
- In semi-confined aquifer additional properties are leakage factor and hydraulic resistance.
- The rate of flow of groundwater is response to a given hydraulic gradient which is dependent upon the hydraulic conductivity.
- Hydraulic conductivity- the rate of flow of water in litres per day through a horizontal cross-sectional area of one square metre of the aquifer under hydraulic gradient is of one metre per metre at the prevailing temperature of well.
- Capability of entire thickness of confined and semi-confined aquifer is termed as transmissibility.
- Transmissibility ranges from less than 12000 litres per day per metres to nearly 1200000 litres per day per metre.
- Transmissibility value of tubewell is 120000 litres per day per metre or more considered satisfactory for irrigation.
- Coefficient of storage is properties of confined and semi-confined aquifer.
- The coefficient of storage ranges from 0.00001 to 0.001 for confined aquifer.
- Specific yield is the property of unconfined aquifer.
- Specific yield- volume of water released or stored per unit surface area of aquifer per unit change in component of head normal to that surface.
- The quantity of water that unit volume of aquifer retains when subjected to gravity drainage is called specific retention.
- The hydraulic resistance is also called reciprocal leakage factor.
- The volume leakage factor of semi-pervious layer generally ranges between 100 and 10,000000 min.
- High value of leakage factor indicates a greater resistance of the semi-pervious strata to flow as compared to the resistance of strata itself.
- Low head water lifts are suitable when the depth of water surface does not exceed 1.2 metre.
- Medium head lifts are suitable when the height of lift is within the range of 1.2 to 10 metres.
- The height of high head lift is above 10 metres.
- Specific speed is often used as an index to the operating characteristics of pump.
- The part of rainfall, which enters the stream immediately after the rainfall, is called surface runoff.
- That portion of rainfall which first leaches into the soil and moves laterally without joining the water table to the stream is known as subsurface runoff.
- Ratio of total channel length in the watershed to total watershed area is called drainage density.
- The unit hydrograph method used to compute the direct runoff, is limited to the watershed area of about 5000 square kilometre.
- The top 30 cm soil depth is very useful for the human being and wild life.
- Accelerated erosion is in excess of geologic erosion.
- Raindrop erosion is also known as splash erosion.
- The soil may splash into the air up to a height of 50 to 60 cm depending upon size of raindrop.
- Splash erosion is caused by impacting the raindrop.
- Uniform removal of soil in the form of a thin layer by the flowing of water from a given width of sloping land is called sheet erosion.

Beaufort Scale is used for measuring velocities of winds.

GS

- Gully erosion is the advance and last stage of water erosion.
- In other words gully erosion is the advance stage of rill erosion.
- Glacial erosion is found where mean temperature is below zero degree celsius.
- Photogenic erosion is the destruction of soil, caused by plants root system.
- Zoogenic erosion is caused by animals.
- Anthropogenic erosion is associated with the activities of human beings.
- The annual soil loss from 0.05 to 0.5 mm depth or 0.5 to 5 cubic metres per hectare is counted as week erosion.
- In medium erosion the annual soil loss is ranging from 0.5 to 1.5 mm of soil depth or 5 to 15 cubic metres per hectare per year.
- The range of soil loss in serious erosion varies from 1.5 to 5 mm of soil depth or 15 to 50 cubic metres per hectare per year.
- The rate of extent of gully development are closely related to the amount and velocity of runoff water.
- U-shape gullies are generally found in the alluvial plains.
- V -shape gullies are often developed where the subsoil are tough to resist the rapid cutting of soil by runoff flow.
- Drop spillways are used where head drop is 1.2 to 2.4 metres.
- Chute spillways are used at the locations where head drop is fairly large ranging from 5 to 6 metres.
- Wind erosion is the process of detachment, transportation and deposition of soil particles by the action of wind.
- The desert areas are severely affected by wind erosion problem.
- Soil erosion due to wind is mainly the result of creation of wind turbulence over the ground surface.
- Suspension is the movement of very fine soil particles generally less than 0.1 mm in diameter.
- In saltation process, the size range of particles may be from 0.05 to 0.5 mm in diameter.
- Among all types of soil movement the saltation is responsible for transporting the maximum amount of soil particles along ground surface by rolling and pushing action which ranges from 55 to 72 % of total soil erosion.
- The size of soil particles associated with surface creep ranges from 0.5 to 1 or 2 mm in diameter.

Did You Know ?

Earliest folk instruments

The drum is the oldest musical instrument used in folk music and dancing. The typical ones include the *poong, dholak, kohl* and *chimta*, among others.

- The extent of total desert in India is about 20.7 Mha.
- Wind breaks and shelter belts are used for controlling wind erosion.
- Erosivity is defined as the potential ability of rain to cause the erosion.
- Erosivity depends upon the physical characteristics of rainfall include raindrop size, drop size distribution, kinetic energy and terminal velocity etc.
- Erodibility is the susceptibility of soil to get erosion.
- In low rainfall regions the primary purpose of contour cultivation is to conserve the rain water into the soil as much as possible.
- In humid regions the basic purpose of contour cultivation is to reduce the soil erosion by retarding the overland flow.
- Terracing is the engineering soil conservation practice, used to control the soil erosion in highly sloped areas.
- Bench type terraces are generally constructed on the land of 6 to 33 % slope.
- Graded terrace is also known as channel type terrace.
- Level terraces are constructed for conserving the moisture and controlling soil erosion.
- Level terraces are also known as conservation terraces.
- The conservation terraces are normally, found most suitable where rainfall is low and land slope is less than 6 per cent.
- Bunding is used for retaining the water, creating obstruction and thus to control erosion.
- Bunding practices are found most suitable for the land having the slope ranges from 2 to 10 %.
- Contour bunding is not technically feasible on the land slopes greater than 6 per cent.
- Graded bunding is used in the area having rainfall greater than 700 mm per year.
- Pipe culverts are used for very small drainage works.
- Box culverts are preferred for the construction, especially in loose soil condition and for larger spans.
- Box culvert can be used for a single span of 3 metres for a double span of 6 metres width.

GK

Rice is main food crop of India.

GS

Multiple Choice Questions

1. A man can develop
   a] 0.1 hp b] 0.5 hp
   c] 0.75 hp d] 1.0 hp
2. The average force that a bullock can exert
   a] 1/5 th of their body weight
   b] 1/10 th of their body weight
   c] 1/20 th of their body weight
   d] 1/25 th of their body weight
3. Medium size bullock can develop
   a] 0.50 to 0.75 hp b] 0.75 to 1.0 hp
   c] 0.75 hp to 1.25 hp d] 1.0 to 1.5 hp
4. The thermal efficiency of diesel engine varies between
   a] 25 and 32 per cent b] 32 and 38 per cent
   c] 35 and 40 per cent d] 40 and 45 per cent
5. The thermal efficiency of petrol engine varies between
   a] 25 and 32 per cent b] 30 and 35 percent
   c] 32 and 38 per cent d] 30 and 40 percent
6. The extra high speed engines used in knapsack sprayers are powered by
   a] Kerosene b] Diesel
   c] Petrol d] Dual fuel
7. On an average about 1/10th total electric power generated in India is consumed for
   a] Domestic work b] Farm work
   c] Industrial work d] None of these
8. The velocity required to operate wind mill is more than
   a] 5 km ph b] 10 km ph
   c] 5 miles per hour d] 10 miles per hour
9. The average capacity of wind mill would be
   a] 0.50 hp b] 1.0 hp
   c] 1.5 hp d] 2.0 hp
10. The power produced from a wind speed 6.4 to 37 km ph by windmill varies from
    a] 0.1 to 0.5 hp b] 0.1 to 0.9 hp
    c] 0.5 to 1 hp d] 1 to 9 hp
11. A ten human power equals to
    a] 0.1 hp b] 0.5 hp
    c] 1.0 hp d] 10.0 hp
12. A pair of bullock develop power equivalent to
    a] 0.5 h b] 1.0 hp
    c] 1.5 hp d] 2.0 hp
13. Two wind mills equivalent to
    a] 1.0 hp b] 1.25 hp
    c] 1.5 hp d] 2.0 hp
14. The land holding of small farmer is
    a] Less than 4 ha land b] Less than 10 ha land
    c] More than 4 ha land d] More than 10 ha land
15. A marginal farmer has land holding
    a] Less than 4 ha b] Between 4 and 10 ha
    c] Less than 10 ha d] More than 10 ha
16. The land holding of large farmer is
    a] Less than 4 ha b] Between 4 and 10 ha
    c] Less than 10 ha d] More than 10 ha
17. The power availability on Indian farm is
    a] 1 hp per ha b] 1.5 hp per ha
    c] 2.0 hp per ha d] 2.5 hp per ha
18. The main source of power on Indian farm is
    a] Bullock b] Tractor
    c] Power tiller d] Diesel engine
19. The cheapest source of power on Indian farm is
    a] Animal power b] Human power
    c] Electrical power d] Mechanical power
20. The low speed engine runs at speed of
    a] 350 rpm b] Less than 350 rpm
    c] More than 350 rpm d] More than 1000 rpm
21. The speed range of medium speed engine is
    a] Less than 350 rpm
    b] Between 350 rpm and 1000 rpm
    c] Less than 1000 rpm
    d] More than 1000 rpm
22. The speed of high-speed engine
    a] Less than 350 rpm
    b] Between 350 rpm and 1000 rpm
    c] Less than 1000 rpm
    d] More than 1000 rpm
23. The single cylinder engine is generally used in
    a] Tractor b] Stationary engine
    c] Motor cars d] Power tiller engine
24. The steam engine is
    a] Single stroke engine
    b] Two-stroke engine
    c] Four-stroke engine
    d] None of these
25. The two-stroke engine is
    a] Diesel engine b] Steam engine
    c] Petrol engine d] None of these
26. The diesel engine used on tractors are
    a] One-stroke engine b] Two-stroke engine
    c] Four-stroke engine d] None of these
27. The carburettor is main part of
    a] Diesel engine b] Steam engine
    c] Petrol engine d] Gas engine
28. Diesel engine works on principle of
    a] Diesel cycle b] Otto cycle
    c] Carnot cycle d] None of these

GS

Solid carbon dioxide is also known as Dry ice.

29. The injector and fuel pump are the heart of
a] Diesel engine b] Steam engine
c] Petrol engine d] Gas engine

30. In two-stroke cycle engine, one power stroke is obtained after every
a] Half revolution of crank shaft
b] One revolution of crank shaft
c] Two revolution of crank shaft
d] Three revolution of crank shaft

31. In four-stroke cycle engine, one power stroke is obtained after every
a] Half revolution of crank shaft
b] One revolution of crank shaft
c] Two revolution of crank shaft
d] Three revolution of crank shaft

32. The carburettor of an engine is used to mix fuel with
a] Air b] Water
c] Oil d] None of these

33. The firing order of 4-stroke 4 cylinder engine is
a] 1-3-2-4 b] 1-3-4-2
c] 1-4-3-2 d] 1-2-3-4

34. The firing order of 4 strokes 6 cylinder engine is
a] 1-5-3-6-2-4 b] 1-3-4-2-6-5
c] 1-3-2-5-4-6 d] 1-3-5-6-4-2

35. The firing interval of 2-stroke two-cylinder engine is
a] 180° b] 360°
c] 120° d] 720°

36. The firing interval of 4-stroke four-cylinder engine is
a] 90° b] 180°
c] 360° d] 720°

37. The compression ratio of diesel engine is
a] 4 to 8: 1 b] 14 to 20: 1
c] 4 to 15: 1 d] 14 to 22: 1

38. The compression ratio of petrol engine is
a] 4 to 8:1 b] 4 to 10:1
c] 8 to 15:1 d] 14 to 20:1

39. The air fuel ratio of diesel engine is
a] 8:1 b] 15:1
c] 20:1 d] None of these

40. The air fuel ratio of petrol engine is
a] 10:1 b] 15:1
c] 20:1 d] None of these

41. The IHP of an engine indicates
a] Power on piston b] Power on flywheel
c] Power on drawbar d] Loss of power

42. The BHP of an engine indicates
a] Power on piston b] Power on flywheel
c] Power on drawbar d] Power on PTO

43. In diesel cycle engine, heat is added at
a] Constant pressure b] Constant temperature
c] Constant volume d] None of these

44. In Otto cycle, heat is added at
a] Constant pressure b] Constant temperature
c] Constant volume d] None of these

45. The square engine has
a] Cylinder of square section
b] Combustion chamber of square section
c] Equal bore and stroke
d] Piston of square section

46. The cross sectional area of one cylinder of an engine multiplied by its stroke is
a] Clearance volume b] Combustion volume
c] Swept volume d] Piston volume

47. A power stroke in single cylinder 4- stroke engine occurs at
a] 720° of crankshaft rotation
b] 360° of crankshaft rotation
c] 180° of crankshaft rotation
d] 120° of crankshaft rotation

48. In three cylinder in-line engine, the power impulse occurs after every
a] 720° of crankshaft rotation
b] 360° of crankshaft rotation
c] 240° of crankshaft rotation
d] 120° of crankshaft rotation

49. In six cylinder in-line engine, the power impulse occurs after every
a] 720° of crankshaft rotation
b] 360° of crankshaft rotation
c] 240° of crankshaft rotation
d] 120° of crankshaft rotation

50. The fluctuation of engine speed during a cycle depends upon
a] Mass of flywheel b] Mass of crankshaft
c] Speed of flywheel d] Governor speed

51. The material used for construction of cylinder block is
a] Cast iron and steel
b] Cast iron and aluminum alloy
c] Brass and steel
d] Steel and aluminum alloy

52. The compression rings mounted on piston are generally made of
a] Low carbon steel b] Aluminum
c] High carbon steel d] Chromium

53. In an engine, the angle between the cylinder axes and the crankshaft centreline is
a] 60° b] 90°
c] 180° d] 120°

GK GS

Near Ponta Saheb Yamuna comes down to plains.

54. **The gasket sealing is the most difficult problem at**
a] Head b] Oil pan
c] Intake manifold d] Timing cover

55. **The camshaft in an engine is always mounted**
a] Perpendicular to the crankshaft
b] Inclined to the crankshaft
c] Parallel to the crankshaft
d] Inline to crankshaft

56. **The piston skirt clearance at room temperature is**
a] 0.004 mm b] 0.04 mm
c] 0.50 mm d] 1.00 mm

57. **The piston pin of some engine is offset to**
a] Major thrust side b] Right side
c] Minor thrust side d] Left side

58. **The main purpose of piston rings is**
a] To control combustion pressure
b] To control cylinder wall lubrication
c] To drain out excessive oil
d] All are correct

59. **The primary purpose of ring expander is to**
a] Make up for cylinder's wear
b] Increase static ring tension
c] Decrease dynamic ring tension
d] Reduce ring vibration

60. **The connecting rod of an engine connects the crankshaft with**
a] Cylinder head b] Piston
c] Piston pin d] Cylinder liner

61. **The oil pan of an engine is made of**
a] Steel or aluminum b] Cast iron or brass
c] Steel or cast iron d] Cast iron or zinc

62. **The counter weights on a crankshaft are located opposite to**
a] Small end bearing b] Big end bearing
c] Main bearing d] None of these

63. **The camshaft controls**
a] Valve closing b] Valve opening
c] Valve timing d] All are correct

64. **The pushrod is located between**
a] Valve and rocker arm
b] Tappet and rocker arm
c] Tappet and camshaft
d] Valve and camshaft

65. **A six-cylinder engine with possible number of firing orders are**
a] 6 b] 4
c] 3 d] 2

66. **Valve timing diagram is a function of**
a] Engine speed b] Engine compression
c] Engine torque d] I.M.E.P.

67. **The specific fuel consumption of diesel engine is**
a] 180 gm/ bhp/hr b] 200 gm/ bhp/ hr
c] 150 gm/ bhp/hr d] 290 gm/ bhp/ hr

68. **The specific fuel consumption of petrol engine is**
a] 180 gm/ bhp/ hr b] 200 gm/ bhp/ hr
c] 150 gm/ bhp/hr d] 290 gm/ bhp/ hr

69. **When the stroke: bore ratio of an engine increases, the volumetric efficiency of the engine**
a] Decreases b] Increases
c] Remains unchanged d] None of these

70. **The piston speed of an engine is equal to**
a] 2 LN b] ALN
c] 4 LN d] None of these

71. **Oil pump is driven by**
a] Camshaft b] Timing gear
c] Crankshaft d] All are correct

72. **The compression pressure of diesel engine varies from**
a] 8-10 kg/sq. cm b] 10-40 kg/sq.cm
c] 35-50 kg/sq.cm d] 45-50 kg/sq.cm

73. **Flywheel is made of**
a] Cast iron b] Aluminium alloy
c] High carbon steel d] All are correct

74. **The stroke: bore ratio of tractor engines varies from**
a] 1.0 to 1.5 b] 1.0 to 1.25
c] 1.0 to 1.35 d] 1.0 to 1.45

75. **The inlet valve of four-stroke engine opens at**
a] 5° before TDC b] 10° before TDC
c] 15° before TDC d] 20° before TDC

76. **Working temperature of diesel engine ranges from**
a] 71-82°C b] 82-88°C
c] 75-80° C d] None of these

77. **Working temperature of petrol engine ranges from**
a] 71-82°C b] 82-88°C
c] 75-80°C d] None of these

78. **White smoke indicates**
a] Presence of water in fuel
b] Burning of lubricant in cylinder
c] Presence of water in lubricant
d] Rich air and fuel mixture

79. **Blue smoke indicates**
a] Presence of water in fuel
b] Burning of lubricant in cylinder
c] Engine is overloaded
d] Rich air and fuel mixture

80. **The specific fuel consumption of diesel engine is**
a] More than petrol engine
b] Less than petrol engine
c] Equal to petrol engine
d] None of these

GK **Gulbadan Begam wrote Humayunama.** GS

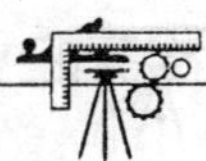

81. The purpose of governor in the engine
a] To increase the engine speed
b] To regulate the engine speed
c] To decrease the engine speed
d] None of these

82. The first IC engine was invented in the year
a] 1876 b] 1890
c] 1900 d] 1905

83. The IHP of an engine is always
a] Less than BHP b] More than BHP
c] Equal to the BHP d] Equal to the FHP

84. The clearance volume is the volume of air or mixture left in the cylinder when piston is at
a] BDC b] TDC
c] Between TDC and BDC d] All are correct

85. The compression pressure of diesel engine inside the cylinder varies from
a] 30-40 kg/cm^2 b] 35-45 kg/cm^2
c] 40-50 kg/cm^2 d] 50-60 kg/cm^2

86. The compression pressure of petrol engine inside the cylinder varies from
a] 6-10 kg/cm^2 b] 10-15 kg/cm^2
c] 8-12 kg/cm^2 d] 12-15 km/cm^2

87. The firing interval of 2-stroke engine is determined by
a] 180° / No. of cylinder
b] 270° /No. of cylinder
c] 360° / No. of cylinder
d] 720° / No. of cylinder

88. The firing interval of 4-stroke engine is determined by
a] 180° / No, of cylinder b] 270° / No. of cylinder
c] 360° / No. of cylinder d] 720° / No. of cylinder

89. The rate of doing work is equal to
a] 75 kg m/sec b] 4500 kg m/min
c] 27000 kg m/ hr d] All are correct

90. The efficiency of IC engines varies from
a] 25 to 32 per cent b] 30 to 40 per cent
c] 35 to 40 per cent d] 32 to 38 per cent

91. The efficiency of external combustion engine ranges from
a] 10 to 15 per cent b] 10 to 20 per cent
c] 15 to 30 per cent d] 30 to 40 per cent

92. The weight of diesel engine per horsepower is
a] Lighter than petrol engine
b] Equal to petrol engine
c] Heavier than petrol engine
d] None of these

93. The thermal efficiency of CI engine is
a] Higher than SI engine b] Lower than SI engine
c] Equal to SI engine d] None of these

You Don't Know!
First stationmaster
Rinku Sinha Roy joined as the first woman stationmaster of the Eastern Railway in August 1994 in the Dum Dum Cantonment Station of the Seldah-Bangaon section of the Eastern Railway.

94. Piston of an engine is made of
a] Cast iron b] Aluminium alloy
c] Nickel iron alloy d] All are correct

95. Gudgeon pin of engine piston is made of
a] Case hardened steel
b] Cast iron
c] Aluminium alloy
d] All are correct

96. Piston pin connected piston with
a] Crankshaft b] Connecting rod
c] Camshaft d] Flywheel

97. Connecting rod convert
a] Reciprocating motion piston into rotary motion crankshaft
b] Rotary motion of piston into reciprocating motion of crankshaft
c] Rotary motion of piston into oscillatory motion of crankshaft
d] Oscillatory motion of piston into rotary motion of crankshaft

98. Oil ring on the piston are located
a] Below the compression ring
b] Above the piston skirt
c] Above the compression ring
d] Both A & B

99. The work timing gear is
a] Control the ignition timing
b] Control the fuel injection timing
c] Control the timing of opening and closing of valves
d] All are correct

100. The fuel used in EC engines for generation of power is
a] Petrol b] Diesel
c] Kerosene d] Steam

101. Internal combustion engine was invented by
a] Hudson b] Rudolf Diesel
c] James Watt d] Nicolaus Otto

102. The work of crankshaft is to
a] Power the piston b] Turn the flywheel
c] Operate valves d] All are correct

103. In diesel cycle engines, air enters in the cylinder through the
a] Inlet port b] Intake valve
c] Injector d] Nozzles

GS

Sometimes which substitutes a hard punishment with a soft one is known as Commutation.

104. The common face and seat angle of valve is

a] 45° b] 60°
c] 90° d] 120°

105. The four-stroke engines are

a] High speed engine
b] Low speed engine
(c) Medium speed engine
d] All are correct

106. Two-stroke engines are

a] High speed engine
b] Low speed engine
c] Medium speed engine
d] All are correct

107. The power develops by four-stroke engine for every one revolution of crankshaft is

a] Half of two-stroke engine
b] Equal to two-stroke engine
c] Twice two stroke engine
d] Thrice two stroke engine

108. In four-stroke diesel engines, the air enters in the cylinder during

a] Suction stroke b] Power stroke
c] Compression stroke d] Exhaust stroke

109. F- type valves are used in engines has

a] Pre-combustion chamber
b] External combustion chamber
c] Don't have combustion chamber
d] Both a] & b]

110. I- head type valves are generally used in

a] Tractor engines b] Stationary engines
c] Steam engines d] Both a] & b]

111. In T- head type engine, the valves are

a] Both valves on one side
b] Both valves are opposite to each other
c] Both valves are opposite but inline
d] Both b] & c]

112. The cam lobes of camshaft helps in

a] Opening and closing the valves
b] Driving the timing gear
c] Driving flywheel
d] Operated crankshaft

113. The teeth on the crankshaft gear is

a] Half to camshaft gear
b] Equal to camshaft gear
c] Double of camshaft gear
d] Four times camshaft gear

114. When cam gear rotates one time the crankshaft gear will rotate

a] Half time
b] Two times
c] Same rotation of cam gear
d] Three times

115. Firing order of an engine indicates

a] Sequence of power stroke
b] Sequence of all strokes
c] Interval of between two successive power strokes
d] Interval between all power strokes

116. Engine is a device, which converts

a] Chemical energy into mechanical
b] Mechanical energy into chemical
c] Chemical energy into electrical energy
d] Mechanical energy into electrical

117. To and fro motion of piston is converted to

a] Reciprocating motion
b] Rotary motion
c] Oscillatory motion
d] Vibrating motion

118. External combustion engine is more powerful than

a] Diesel engine
b] Petrol engine
c] Internal combustion engine
d] All are correct

119. The volume displaced by one-stroke of piston is

a] Cylinder volume b] Swept volume
c] Clearance volume d] Total volume

120. Pre-combustion chamber is found in

a] Petrol engine b] Diesel engine
c] Steam engine d] All are correct

121. Distance travelled by piston from TDC to BDC is

a] Stroke b] Bore
c] Piston displacement d] Piston speed

122. The diesel engine does not have

a] Carburettor b] Spark plug
c] Ignition coil d] All are correct

123. As the 'cc' of engine increases, the power of engine will

a] Decrease b] Increase
c] Remains same d] None of these

124. Ratio of brake horsepower to indicated horsepower is

a] Thermal efficiency b] Volumetric efficiency
c] Mechanical efficiency d] Brake efficiency

125. A gear used in timing gear is

a] Spur gear b] Bevel gear
c] Helical gear d] Warm gear

126. Gaskets are made of

a] Copper b] Asbestos
c] Fibre d] Both a] & b]

127. The gaskets are used to

a] Prevent leakage b] Prevent dust
c] Prevent wear d] All of above

Pyrolysis, chemical process is used to receive valuable materials from solid waste.

128. The total number of rings on piston may vary from
a] 2- 4 b] 3 -5
c] 3-7 d] 2-7

129. A connecting rod of four-stroke engines is subjected to
a] Compression and tension
b] Compression and side thrust
c] Side thrust only
d] Compression and twisting.

130. The arrangements of valves used in the engine may be
a] I-headtype b] T-headtype
c] L- head type d] All are correct

131. The firing order of eight-cylinder engine is
a] 1-5-4-2-6-3-7-8 b] 1-2-6-3-4-7-5-8
c] 1-5-3-4-2-7-8 d] 1-2-4-6-7-3-5-8

132. The useful life of stationary engine is
a] 10 years b] 15 years
c] 12 years d] 20 years

133. The expected life of stationary engine is
a] 10,000 hrs b] 12,000 hrs
c] 15,000 hrs d] 20,000 hrs

134. The flywheel is mounted on
a] Front end of crankshaft
b] Rear end of crankshaft
c] Front end of camshaft
d] Rear end of camshaft

135. Push rod is operated between
a] Tappet and rocker arm
b] Camshaft and tappet
c] Tappet and rocker arm
d] Valve and piston

136. Constant speed engine is used in
a] Tractor b] Power tiller
c] Motor car d] Electric generator

137. Variable speed engines are used in
a] Tractors b] Four wheel vehicles
c] Two wheel vehicles d] All are correct

138. Constant speed governor is used in
a] Petrol engine b] Stationary engine
c] Tractor engine d] All are correct

139. If stroke: bore ratio of an engine increase, the thermal efficiency
a] Increase b] Decrease
c] Remains same d] None of these

140. Black smoke indicates
a] Burning of lubricant in the cylinder
b] Presence of water in fuel
c] Engine is overloaded
d] None of these

141. The violent noise head in engine during process of combustion is
a] Scavenging b] Detonation
c] Wearing d] All are correct

142. The process of removal of burnt gases from the engine cylinder is
a] Scavenging b] Detonation
c] Knocking d] Burning

143. The indicated horsepower of an engine was measured by
a] Dynamometer b] Tachometer
c] Indicator d] None of these

144. The location of carburettor in tractor from front is in
a] Center b] Left side
c] Right side d] None of these

145. The best hydrocarbons from detonation point of view are
a] Paraffin b] Naphthenes
c] Aromatics d] Olefins

146. The calorific value of gasoline is
a] 45 MJ/kg b] 45 KJ/kg
c] 45 J/kg d] None of these

147. The temperature at which fuel flow ceases is
a] Pour pointb] Flash point
c] Cloud point d] Smoke point

148. The temperature at which fuel catches fire is
a] Pour pointb] Flash point
c] Cloud point d] Smoke point

149. The lowest temperature at which fuel begins to crystallize is
a] Pour pointb] Flash point
c] Cloud point d] Smoke point

150. The temperature at which fuel begins to vaporize is
a] Pour pointb] Flash point
c] Cloud point d] Smoke point

151. The specific gravity of petrol at 15° is
a] 0.73 b] 0.82
c] 0.90 d] 0.85

152. The specific gravity of powering is
a] 0.82 b] 0.88
c] 0.91 d] None of these

153. The specific gravity of high-speed diesel is
a] 0.73 b] 0.87
c] 0.92 d] 0.95

154. The specific gravity of light diesel oil is
a] 0.82 b] 0.85
c] 0.92 d] 0.95

155. The calorific value of high-speed diesel is
a] 10,000 Kcal/ kg b] 12,000 Kcal/ kg
c] 10,550 Kcal/ kg d] 81,00 Kcal/ kg

GK GS

Green Revolution in the country was actually started in the mid sixties.

156. The calorific value of petrol is
a] 10,000 Kcal/ litre b] 11,000 Kcal/ litre
c] 15,500 Kcal/ litre d] 81,50 Kcal/litre

157. The calorific value of diesel oil is
a] 9400 Kcal/ litre b] 10,000 Kcal/litre
c] 11,000 Kcal/ litre d] 12,000 Kcal/ litre

158. The anti-knock quality of petrol fuel is determined by
a] Cetane number b] Octane number
c] Both a] & b] d] None of these

159. The anti-knock quality of diesel is determined by
a] Cetane number b] Octane number
c] SAE number d] All are correct

160. The viscosity of oil denoted by a number is
a] Cetane number b] Octane number
c] SAE number d] None of these

161. The temperature of cloud point is
a] 5° above the pour point
b] 10° above the pour point
c] 15° above the pour point
d] All are correct

162. Bomb calorimeter is used for determining
a] Calorific value of fuel
b] Gravity of fuel
c] Cetane number of fuel
d] SAE number of fuel

163. Junker's calorimeter is used for determining the calorific value of
a] Liquid fuels b] Solid fuels
c] Gaseous fuels d] None of these

164. The boiling range of petrol is
a] 30° to 180°C b] 30° to 200°C
c] 30° to 230°C d] 30° to 250°C

165. General formula of paraffin is
a] Cn G2n+2 b] Cn Gn+2
c] Cn G3n+ 1 d] None of these

166. General formula of Naphthenes is
a] Cn H2n b] Cn H3n
c] Cn H2n+1 d] Cn H2n-1

167. General formula of aromatics is
a] Cn H2n-2 b] Cn H2n-6
c] Cn H2n+2 d] Cn H2n-4

168. General formula of olefins is
a] Cn H2n b] Cn Hn
c] Cn H2n+2 d] Cn H2n-2

169. The boiling range of kerosene is
a] 100° to 200°C b] 200° to 300°C
c] 300° to 400°C d] 300° to 500°C

170. The specific gravity of gasoline ranges from
a] 0.70 to 0.72 b] 0.70 to 0.78
c] 0.72 to 0.82 d] 0.72 to 0.92

171. For Indian high-speed diesel, minimum Cetane number is
a] 42 b] 45
c] 50 d] 55

172. The API gravity of light diesel oil is
a] 22 b] 25
c] 40 d] 50

173. The API gravity of petrol is
a] 61 b] 65
c] 70 d] 80

174. The API of high-speed diesel is
a] 25 b] 31
c] 40 d] 50

175. The quality of fuel is judged by
a] Colour b] Volatility
c] Both a] & b] d] None of these

176. The volatility of fuel effects
a] Efficiency of engine b] Power of engine
c] Speed of engine d] All are correct

177. The vaporizing temperature of petrol fuel is
a] Higher than diesel fuel b] Lower than diesel fuel
c] Equal to diesel fuel d] None of these

178. Molecular formula of diesel is
a] $C_{11}H_{15}$ b] $C_{11}H_{16}$
c] $C_{12}H_{15}$ d] $C_{11}H_{14}$

179. High speed diesel is used in
a] High speed diesel engine
b] Low speed diesel engine
c] Petrol engine
d] Aircraft engine

180. The specific gravity of fuel is measured by
a] Hygrometer b] Thermometer
c] Hydrometer d] All are correct

181. The API gravity of water is
a] 8 b] 10
c] 15 d] 20

182. Lighter fuels have
a] High API gravity b] Low API gravity
c] High density d] All are correct

183. If the volatility of fuel is high, it ignites
a] Shortly b] Late
c] Will not ignite d] All are correct

184. A fuel has higher octane number, its ignition quality of fuel will be
a] Poor b] Better
c] Not affected d] All are correct

185. Detonation refers to the
a] Engine knocking b] Engine speed
c] Engine efficiency d] All are correct

186. The detonation of fuel takes place during
a] Combustion b] Injection
c] Exhaust d] None of these

GK Lead pencil contains graphite. GS

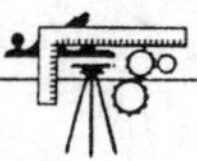

187. Distillation test measures the
a] Calorific value b] SAE number
c] Volatility d] Power

188. The volatility quality of fuel indicates
a] Specific gravity of fuel
b] Density of fuel
c] Vaporizing ability of fuel
d] All are correct

189. Cetane rating of commercial diesel fuel is
a] 30-60 b] 30- 70
c] 30-65 d] 30-80

190. The specific heat of petrol is nearly
a] 2.1 KJ/kg°C b] 2.5 KJ/kg°C
c] 2.8 KJ/kg°C d] 3.0KJ/kg°C

191. The specific heat of diesel is nearly
a] 1.5 KJ/kg°C b] 1.9 KJ/kg°C
c] 2.0 KJ/kg°C (D 2.5 KJ/kg°C

192. Effect of detonation is
a] Loss of power b] Poor combustion
c] Overheating d] All are correct

193. Vapour lock is associated with
a] Ignition system b] Cooling system
c] Fuel supply system d] Ignition system

194. The work of choke is
a] Supply the air into cylinder
b] Supply of electric spark
c] Supply of rich mixture
d] All are correct

195. A pump which supplies the fuel to different cylinder of multi cylinder engine is
a] Centrifugal pump
b] Oil pump
c] Fuel injection pump
d] None of these

196. Air injection pumps are mostly used in
a] Tractor engine
b] Railway engine
c] Aircraft engine
d] Heavy duty stationary engine

197. The pipeline used to supply the fuel to the injector is
a] Low pressure line b] High pressure line
c] Zero pressure line d] All are correct

198. In diesel engines, fuel injected in the cylinder is through
a] Carburettor b] Fuel injection pump
c] Fuel injector d] All are correct

199. In Otto cycle engines, air fuel mixture comes in cylinder through
a] Injector b] Carburettor
c] Both a] & b] d] None of these

Knowledge World!
Only mosque in marble
Moti Masjid built by Aurangzeb within the Lal Qila is made entirely of marble.

200. The first butterfly valve of carburetor controls the flow of
a] Air fuel mixture b] Fuel supply
c] Oil supply d] All are correct

201. The second butterfly valve of carburettor is
a] Carburettor b] Choke
c] Plug d] All are correct

202. Choke controls the amount of
a] Air b] Fuel
c] Water d] All are correct

203. Function of carburettor is
a] To mix the air and fuel
b] To regulate air fuel ratio at different speed
c] To supply correct amount of mixture
d] All are correct

204. Venturi tube is provided in carburetor for
a] Producing low pressure
b] Producing high pressure
c] Both a] & b]
d] None of these

205. A device for regulating the fuel supply in main jet of carburetor is
a] Choke b] Throttle
c] Economizer d] All are correct

206. Fuel pump used in carburetor engines is
a] Fuel injection pump b] Centrifugal pump
c] Gasoline pump d] None of these

207. A pump that transfers fuel from the fuel line to the fuel injection pump is
a] Fuel lift pump b] Feed pump
c] Transfer pump d] All are correct

208. A pump that delivers metered quantity of fuel to each cylinder at appropriate time under high pressure is
a] Transfer pump b] Fuel injection pump
c] Centrifugal pump d] Oil pump

209. The high-pressure pump used in diesel engine has pressure range of
a] 120-300 kg/cm^2 b] 120-350 kg/cm^2
c] 120-380 kg/cm^2 d] 120-400 kg/ cm^2

210. For petrol engine, the maximum power obtained from one kg of fuel is
a] 10.5-13.8 kg b] 11-15 kg
c] 12-16 kg d] 15-20 kg

GK GS

Ken is the main tributary of Yamuna.

211. Chemically correct air mixture ratio for Otto cycle engine is
a] 12: 1 b] 14: 1
c] 15: 1 d] 20: 1

212. In leanest mixture, the air fuel ratio is about
a] 12: 1 b] 14: 1
c] 15: 1 d] 20: 1

213. In richest mixture, the air fuel ratio is about
a] 4: 1 b] 8: 1
c] 15: 1 d] 16: 1

214. A primary fuel filter is made of
a] Cloth b] Paper
c] Both a] & b] d] None of these

215. Secondary fuel filter is made of
a] Paper b] Thin wire
c] Wooden d] All are correct

216. Primary fuel filter is used to remove particles of size
a] 5 Micron b] 8 Micron
c] 10 Micron d] 12 Micron

217. Secondary fuel filter is used to remove the particles of size
a] 1 Micron b] 2 Micron
c] 3 Micron d] 4 Micron

218. The fuel injection pressure of CI engine ranges from
a] 10 to 45 kg/cm^2 b] 50 to 120 kg/ cm^2
c] 120 to 175 kg/ cm^2 d] 100 to 300 kg/ cm^2

219. The fuel injected through a nozzle of carburetor is at a pressure of
a] 10 to 45 kg/cm^2 b] 50 to 120-kg/ cm^2
c] 126 to 175 kg/ cm^2 d] 100 to 300 kg/cm^2

220. The specific fuel consumption of diesel engine is
a] Less than in the petrol engine
b] More than in the petrol engine
c] Same as a petrol engine
d] Incomparable with petrol engine

221. The carburettor employed on most of tractor engines are
a] Down draft b] Middle draft
c] Updraft d] None of these

222. Purpose of lubrication is
a] Reducing frictional effect
b] Cooling effect
c] Sealing effect
d] All are correct

223. The lubricants are obtained from
a] Animal fat b] Vegetables
c] Minerals d] All are correct

224. Animal fat lubricants are
a] Suitable for machine
b] Less viscous
c] It is gummy and waxy type
d] All are correct

225. Lubricants obtained from vegetables is suitable for
a] Small and simple implements
b] Tractor and stationary engines
c] All heavy implements
d] None of these

226. Mineral lubricants are most popular for
a] Simple implements
b] Engines and Machines
c] Both a] and b]
d] None of these

227. Petroleum lubricants are less expensive and suitable for
a] Steam engines b] IC engines
c] Both a] & b] d] None of these

228. A good lubricant should have
a] Sufficient viscosity
b] Remains stable under changing temperature
c] It should not corrode metallic surface
d] All are correct

229. The method used for testing of lubricating oil is
a] Viscosity test b] Flash fire test
c] Pour point test d] All are correct

230. Say bolt viscometer is used for
a] Flash point test b] Viscosity test
c] Pour point test d] Colour and carbon residue

231. The temperature at which vapour flashes momentarily is
a] Pour pointb] Flash point
c] Fire point d] All are correct

232. The temperature at which flame persists approximately for 5 seconds is
a] Pour pointb] Flash point
c] Fire point d] All are correct

233. Gravity test is done by using the
a] Hydrometer b] Hygrometer
c] Thermometer d] All are correct

234. Splash system of lubrication is generally used in
a] Single cylinder & stationary engine
b] Multi-cylinder engines
c] Tractor engines
d] All are correct

235. A pump used in force feed lubricating system is
a] Positive displacement type
b] Gear type
c] Vane type
d] All are correct

236. Force-feed lubrication system is commonly used in
a] High speed multi-cylinder engine
b] Tractor engines
c] Truck and automobile engine
d] All are correct

GS

N.D. Ranga established the Indian Peasant Institute to train peasant workers.

237. In force feed lubrication system pump is driven by
a] Camshaft b] Crankshaft
c] Both a] & b] d] None of these

238. Lubrication system of tractor engine develops pressure of about
a] 3 kg/ cm^2 b] 5 kg/ cm^2
c] 8 kg/ cm^2 d] 12 kg/cm^2

239. Oil filters are made of
a] Cloth b] Paper
c] Felt or wire screen d] All are correct

240. A type of oil filter is used in tractor engine is
a] Full flow type b] Bi-pass type
c] Both a] & b] d] None of these

241. Oil pressure gauge is used
a] To check the quality of oil
b] To indicate oil temperature in pipeline
c] To warn the operator of any irregularity
d] All are correct

242. Grease is semi-fluid lubricant used for
a] Slow moving parts
b] Not easily accessible parts
c] Both a] & b]
d] None of these

243. Lubrication reduces
a] Wear between two rubbing surfaces
b] Friction effect in moving parts
c] Heat from moving parts
d] All are correct

244. Vegetable lubricants are made from
a] Castor oil b] Cotton seed oil
c] Olive oil d] All are correct

245. Mineral lubricants are obtained from
a] Castor and cotton seed oil
b] Crude petroleum
c] Both a] & b]
d] None of these

246. Suitable lubricant for 2-stroke engines is
a] Crude petroleum lubricant
b] Vegetable lubricant
c] Mineral lubricant
d] None of these

247. Suitable lubricant for 4-stroke engines is
a] Mineral lubricant
b] Vegetable lubricant
c] Animal fat lubricant
d] All are correct

248. A pump used in force-feed lubrication system is
a] Centrifugal pump
b] Reciprocating pump
c] Positive displacement pump
d] None of these

249. Excessive oil consumption in engines is due to
a] High speed of engines
b] Excess going of oil into combustion chamber
c] More pressure of line
d] All are correct

250. Lubricants are available in the form of
a] Fluid b] Semi-fluid
c] Semi-solid d] All are correct

251. Semi-solid lubricant is
a] Gear oil b] Engine oil
c] Grease d] None of these

252. If the viscosity of fluid is high then flow will be
a] Rapidly b] Slowly
c] Irregular d] All are correct

253. Lubrication oils are rated by
a] Gravity b] Viscosity
c] Density d] All are correct

254. Grease is used as lubricant in the
a] Engine liner b] Universal coupling
c] Bearings d] Both b] & c]

255. Type of pump used in lubrication system
a] Diaphragm type b] Gear type
c] Jet pump d] Plunger type

256. Crankcase breather is used for
a] Cleaning the dust particles
b] See the level of oil
c] Removing water vapour from crankcase
d] All are correct

257. In force-feed lubrication system, the gear used is
a] Helical gear b] Spur gear
c] Both a] & b] d] None of these

258. Oil filter of tractor engine is generally changed after
a] 50 hours of work b] 100 hours of work
c] 120 hours of work d] 150 hours of work

259. The elements of oil filter is
a] Cloth & felt b] Paper
c] Thin metal plate d] All are correct

260. The crankcase breather filter is mostly made of
a] Paper b] Wire
c] Oil wetted screen d] All are correct

261. American Petroleum Institute adopted the classification of engine oils in the year
a] 1920 b] 1950
c] 1952 d] 1960

262. ML- light service, MM- medium service and MS- severe service are service classes for the
a] SI engines b] IC engines
c] Steam engines d] All are correct

263. DG-general service, DM -moderately service and DS-very severe services are service classes for the
a] Petrol engines b] Steam engines
c] Kerosene engines d] Diesel engines

264. DG and MS oils have high additive contents, which help in keeping
a] Engine cool b] Engine clean
c] Both a] & b] d] None of these

265. If diesel fuel contains high sulphur content the best class of lubricating oil for use is
a] DS lubricating oil b] DM lubricating oil
c] DG lubricating oil d] MS lubricating oil

266. Lubrication reduces
a] Wear and tear b] Abrasive elements
c] Noise d] All are correct

267. Viscosity of oil is measured in terms of
a] Its resistance to flow b] Its cooling effect
c] Both a] & b] d] None of these

268. 'Say bolt Universal Seconds' is related with the
a] Viscosity of oil b] Density of oil
c] Gravity of oil d] All are correct

269. Viscosity of oil expressed at two temperatures are
a] -18°C and 99°C b] -20°C and 100°C
c] -20° C and 120° C d] -18°C and 150°C

270. Viscosity measured in
a] Centistokes b] Centipoises
c] Redwood seconds d] All are correct

271. Smaller change in viscosity of oil with temperature indicates
a] High viscosity index number
b] Low viscosity index number
c] Both a] & b]
d] None of these

272. The variation in viscosity of oil with change in temperature is measured by
a] High viscosity number
b] Viscosity index
c] Both a] & b] d] None of these

273. The density and specific gravity of lubricating oil is determined with the help of
a] Hydrometer b] Specific gravity bottle
c] Both a] & b] d] None of these

274. Viscosity of lubrication oil is measured by
a] Hygrometer b] Thermometer
c] Viscometer d] All are correct

275 For engine bearings, a type of lubrication system adopted is
a] Gravity feed lubrication
b] Wick lubrication
c] Ring lubrication
d] All are correct

276. For engine parts, a type of lubrication system adopted is
a] Splash lubrication b] Pressure lubrication
c] Petroil lubrication d] All are correct

277. Wick lubrication is also called
a] Siphon lubricator
b] Hydrostatic lubricator
c] Mechanical lubricator
d] None of these

278. Ring lubrication system is generally used for lubrication of
a] Vertical bearings b] Horizontal bearings
c] Inclined bearings d] None of these

279. Petroil lubrication system generally is used in
a] Two-stroke engines b] Four-stroke engines
c] Tractor and vehicles d] Aircraft

280. A type lubricator is used in steam engines is
a] Siphon type
b] Hydrostatic type
c] Mechanically operator lubricator
d] Both b] & c]

281. Purpose of cooling
a] To maintain the optimum temperature of engine
b] To dissipate surplus heat for protection
c] To maintain the lubricating property of oil
d] All are correct

282. Air cooled engine has
a] Radiator b] Water jacket
c] Fins d] None of these

283. Air cooling system is generally found in
a] Single-cylinder engines
b] Multi-cylinder engines
c] Both a] & b]
d] None of these

284. In the air cooled engines, cylinder normally enclosed in a sheet metal casing is called
a] Cowling b] Pre-combustion chamber
c] Fins d] None of these

285. A disadvantage of air cooling system is
a] Uneven cooling of engine parts
b] Engine temperature is high
c] Both a] & b]
d] None of these

286. The fan of radiator is driven with the help of
a] V- belt b] Flat belt
c] Chain sprocket d] Gear

287. The fan of radiator is driven by
a] Camshaft b] Crankshaft
c] Flywheel d] All are correct

288. In water-cooled engines the water pump used
a] Centrifugal type b] Submersible type
c] Both a] & b] d] None of these

289. In water cooled engines, water pump is connected with
a] Radiator b] Flywheel
c] Fins d] Combustion chamber

The term 'Nuclear Winter' is associated with aftermath of a nuclear holocaust.

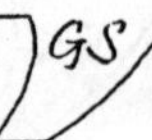

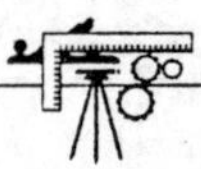

290. In radiator, hot water flows at the
a] Top b] Bottom
c] Middle d] None of these

291. Radiator is generally fabricated as
a] Tabular type b] Cellular type
c] Both a] & b] d] None of these

292. Honeycomb type radiator is also known as
a] Cellular type b] Tabular type
c] Both a] & b] d] None of these

293. Thermostat valve is used in
a] Air-cooled engines b] Water-cooled engines
c] Steam engines d] All are correct

294. Thermostat valve begins open at
a] 70-75°C b] 70-80°C
c] 65-75°C d] 75-80°C

295. In petrol engines, thermostat valve fully opens at
A) 72°C b] 78°C
c] 82°C d] 85°C

296. In diesel engines, thermostat valve fully opens at
a] 88-90°C b] 90-95°C
c] 90-100°C d] 75-110°C

297. The liquid used in bellows type thermostat is known as
a] Alcohol b] Ether
c] Mercury d] Both a] & b]

298. In force-feed cooling system, thermostat valve begins to open at
a] 70°C b] 74°C
c] 78°C d] 80°C

299. Thermostat valve widely opens at about
a] 78°C b] 80°C
c] 85°C d] 90°C

300. In most tractor engines, a cooling system used is
a] Splash cooling b] Pressure cooling
c] Gravity feed type cooling d] All are correct

301. The pressure of radiator cap is about
a] 0.34 to 0.4 kg/ cm^2 b] 0.4 to 0.5 kg/ cm^2
c] 0.4 to 0.6 kg/ cm^2 (0) 0.4 to 1.0 kg/ cm^2

302. The boiling temperature of water in the radiator is raised to about
a] 90°C b] 100°C
c] 110°C (0) 120°C

303. The energy removed by cooling is about
a] 10 per cent b] 20 per cent
c] 30 per cent d] 40 per cent

304. Heat losses by exhaust gases is about
a] 25 per cent b] 30 per cent
c] 33 per cent d] 38 per cent

305. Heat obtained to IHP of total heat generated is
a] 32 per cent b] 36 per cent
c] 40 per cent d] 46 per cent

Quiz Kit!

Largest climber

The Camel foot climber, *Bauhinia vablii,* is found up to 1,500 m in the W. Himalayas. The stems reach a girth of 1.6 m (over 5 ft).

306. Heat obtained to BHP of total heat is about
a] 20 per cent b] 26 per cent
c] 32 per cent d] 38 per cent

307. Heat losses by friction and compression etc is about
a] 5 per cent b] 10 per cent
c] 12 per cent (0) 15 per cent

308. The main drawback of thermo syphon cooling is that
a] Cooling depends on only temperature and independent of engine
b] Cooling depends on engine speed and independent of temperature
c] Both a] & b]
d] None of these

309. Cooling independent of temperature is main drawback of
a] Thermo Syphon cooling
b] Forced pump system
c] Thermostat
d] Air cooling system

310. In which cooling system, the engine warms up quickly after starting
a] Thermo Syphon cooling
b] Forced pump system
c] Thermostat
d] Air cooling system

311. In pressurized water cooling, the boiling point of coolant can be increased by increasing its
a] Temperature
b] Pressure
c] Both a] & b]
d] None of these

312. In pressurized water cooling the boiling point of coolant can be increased by increasing its pressure
a] For 2 bar it is 121°C
b] For 5 bar it is 153°C
c] For 10 bar it is 1180°C
d] All are correct

313. In pressurized water cooling, the water pressure is kept between
a] 1.5 and 2 bars b] 1.5 and 3 bars
c] 1.5 and 5 bars d] 1.5 and 10 bars

314. Evaporating cooling system is used for cooling of
a] Steam engines b] Aircraft engines
c] Railway engines d] Industrial engines

315. Evaporating cooling is also called
a] Steam of vapour cooling
b] Water cooling
c] Air cooling
d] None of these

316. In evaporating cooling the temperature of cooling water is allowed to reach a temperature of
a] 50°C b] 100°C
c] 110°C d] 130°C

317. In case of air-cooled engines, the reduction of weight of engines is about
a] 10 per cent b] 20 per cent
c] 25 per cent d] 30 per cent

318. Volumetric efficiency of air-cooled engines in comparison to water cooled engines is
a] Higher b] Lower
c] Equal d] None of these

319. A cooling system that requires more maintenance is
a] Steam of vapour cooling
b] Water cooling
c] Aircooling
d] None of these

320. The engine performance becomes more sensitive to climatic conditions in case of
a] Evaporation cooled engine
b] Water cooled engine
c] Air cooled engine
d] All are correct

321. In tractor engines, a commonly used thermostat valve is
a] Bimetallic type b] Bellows type
c] Both a] & b] type d] None of these

322. In forced circulation system, a thermostat valve is placed at
a] Forward end of cylinder head
b] Backward end of cylinder head
c] Between cylinder and radiator
d] None of these

323. In thermo siphon system, fan is operated by
a] Crankshaft b] Camshaft
c] Flywheel d] All are correct

324. In IC engines, a commonly used cooling system is
a] Splash b] Thermo Siphon
c] Force feed system d] Hopper system

325. Melting point of cast iron is about
a] 100°C b] 200°C
c] 500°C d] 1100°C

326. Melting point of aluminium is about
a] 100°C b] 500°C
c] 660°C d] 800°C

327. A metal having highest melting point is
a] Aluminium b] Cast-iron
c] Copper d] Steel

328. A metal with lowest melting point is
a] Aluminium b] Copper
c] High carbon steel d] Nickel

329. In air cooled engine, purpose of fans is to
a] Increase contact area
b] Cooling the engine
c] Both a] & b]
d] None of these

330. In water cooling system, the cold water flows at the radiator from the
a] Top b] Bottom
c] Middle d] All are correct

331. In IC engines, the method used for ignition of fuel is
a] Electric sparks
b] Heat of compression
c] Open flame and hot tube
d] All are correct

332. In spark ignition engines, fuel is ignited by
a] Battery ignition system
b] Magneto ignition system
c] Both a] & b]
d] None of these

333. In battery ignition system primary circuit includes
a] Primary coil b] Secondary coil
c] Both a] & b] d] None of these

334. Primary coil has
a] Small number of turns of heavy wire
b] More number of turns of thin wire
c] Both a] & b]
d] None of these

335. Ignition circuit gets current from the
a] Spark plugb] Battery
c] Alternator d] All are correct

336. Battery consists of
a] Positive plate b] Negative plate
c] Both a] & b] d] None of these

337. Positive plate of battery is made of
a] Lead b] Antimony
c] Both a] & b] d] None of these

338. Negative plate of battery is made of
a] Spongy Lead b] Antimony
c] Silicon d] All are correct

339. A chemical used in the battery for reaction is
a] Electrolyte b] Distilled water
c] Both a] & b] d] None of these

340. Electrolyte is consists of
a] Sulphuric acid b] Distilled water
c] Both a] & b] d] None of these

341. Specific gravity of a fully charged battery is
a] 1.270 b] 1.280
c] 1.150 d] None of these

GK **Solid CO_2 is used as a refrigerant.** GS

342. Specific gravity of a fully discharged battery is
a] 1.150 or less b] 1.20 or more
c] Both a] & b] d] None of these

343. Specific gravity measured by an instrument is
a] Hydrometer b] Hygrometer
c] Thermometer d] All are correct

344. A container of battery is made of
a] Steel b] Cast iron
c] Rubber d] Plastic

345. The battery used on tractor consists of
a] Three or six cells b] Two or four cells
c] Both a] & b] d] None of these

346. The voltage developed by each cell of battery is about
a] 1.2 volts b] 2.2 volts
c] 1.5 volts d] 3.0 volts

347. Battery should be recharged when specific gravity goes below
a] 1.225 b] 1.25
c] 1.27 d] 1.28

356. Most of diesel starting motors are equipped with an
a] Overrunning clutch
b] Dog clutch
c] Both a] & b] d] None of these

357. Hot plugs are used in
a] Diesel tractors b] Petrol tractors
c] Powerine tractors d] All are correct

358. Hot plugs used in tractor is operated by
a] Light load b] Heavy load
c] No load d] None of these

359. Cold plugs is used on
a] Powerine engine b] Petrol engine
c] Diesel engine d] All are correct

360. Spark plug gap setting should be kept between
a] 0.50 and 0.85 mm b] 0.5 and 0.90 mm
c] 0.50 and 1.00 mm d] 0.50 and 1,20 mm

361. In ignition system, magneto is used for
a] Supply of fuel
b] Supply of spark
c] Supply of air fuel mixture
d] Supply of air

362. Diesel engine is
a] Compression ignition engine
b] Spark ignition engine
c] External combustion engine
d] None of these

363. Glow plug is used for
a] To ignite the fuel
b] To heat the air charge of engine
c] To ignite the diesel
d] All are correct

364. Primary winding of ignition coil consists of
a] Thin wire b] Thick wire
c] Steel wire d] All are correct

365. Primary winding of ignition coil consists of
a] 200-300 turns b] 200-400 turns
c] 200-450 turns d] 200-500 turns

366. A primary winding wire has resistance of
a] One ohm b] Two ohms
c] Three ohms d] Four ohms

367. Secondary winding consist of
a] 10000-15000 turns b] 15000-20000 turns
c] 15000-25000 turns d] None of these

368. The wire (standard wire gauge) used in primary coil is of
a] 10 SWG b] 15 SWG
c] 20 SWG d] 25 SWG

369. A wire (standard wire gauge) used in secondary coil is about
a] 25 SWG b] 30 SWG
c] 40 SWG d] 50 SWG

370. A modern spark plug has an economical life of about
a] 8000-10000 Km b] 8000-12000 Km
c] 10000-16000 Km d] 10000-20000 Km

371. The gap of spark plug generally should be
a] 0.1 mm b] 0.3 mm
c] 0.5 mm d] 0.6 mm

372. Magneto system is
a] Rotating armature type
b] Rotating magnet type
c] Both a] & b]
d] None of these

373. Contact breaker points is generally made of
a] Steel b] Copper
c] Nickel d] Tungsten

374. The material commonly used for insulator of spark plug is
a] Alumina b] Asbestos
c] Copper d] Steel

375. The life of spark plug used in 2-stroke engine as compared to 4-stroke engines is
a] Same b] Twice
c] Thrice d] One-half

376. Spark plug having white centre insulator indicates
a] Retarded ignition
b] Leaded fuel
c] Over advanced ignition
d] All are correct

377. A spark plug with a black center insulator indicates
a] Over advanced ignition
b] Retarded ignition
c] Stuck-up choke valve in carburettor
d] None of these

GK GS

Badlands is the name given to the heavily eroded land by Chambal.

378. Cold spark plug has
a] Long insulator b] Short insulator
c] Very long insulator d] None of these

379. Hot plugh has
a] Long insulator
b] Small thread
c] Both a] & b]
d] None of these

380. Spark advance is associated to
a] Low speed engine b] Medium speed engine
c] High speed engine d] Stem engine

381. Ignition temperature of IC engines varies from
a] 300-500°C b] 500- 700°C
c] 500-900°C d] 700-1000°C

382. In battery ignition system, the ignition fails due to
a] High speed of engine
b] Heavy use
c] Not proper setting of breaker point
d] None of these

383. The common type of battery used in IC engines is
a] Dry battery b] Lead acid battery
c] Both a] & b] d] None of these

384. Lead acid battery is generally used in
a] Tractors b] Aircraft
c] Light motors d] None of these

385. Charge of battery is measured in terms of
a] Density of electrolyte
b] Weight of electrolyte
c] Specific gravity of electrolyte
d] All are correct

386. In lead acid battery, percentage of sulphuric acid is about
a] 35% b] 50%
c] 65 % d] 75 %

387. Capacity of battery is expressed in tenns of
a] Amp power b] Voltage
c] Amp- hour d] None of these

388. Ignition timing depends on
a] Engine speed b] Compression ratio
c] Valve timing d] All are correct

389. The period between start of injection and ignition is called
a] Ignition period b] Injection time
c] Ignition delay d] All are correct

390. Ignition delay of high-speed engine should be about
a] 0.01 second b] 0.001 second
c] 0.0001 second d] None of these

391. Ignition delay of low-speed engine
a] 0.01 second b] 0.002 second
c] 0.003 second d] 0.004 second

392. Starting motor consists of
a] Alternate current b] Direct current
c] Both a] & b] d] None of these

393. Starting motor is engaged with ring gear of engine with the help of
a] Bevel gear b] Helical gear
c] Pinion gear d] Sliding pinion gear

394. Alternator of a tractor engine is driven by
a] Gear system b] V- belt system
c] Flat belt d] Chain sprocket

395. Speed of alternator with respect to engine speed is
a] One-half b] Equal
c] Twice d] Thrice

396. Main function of fuel injection system is
a] To measure the correct amount of fuel as required by engine speed and load
b] To maintain correct timing for beginning and end of injection
c] To atomize the fuel in the cylinder
d] All are correct

397. Air injection system is mostly used in
a] Tractor's englnes
b] Petrol engines
c] Heavy dusty stationary
d] None of these

398. In solid injection system, fuel pump is used for
a] Measure the quantity of fuel
b] Forcing fuel into combustion chamber
c] Metered the fuel
d] All are correct

399. In common rail system, fuel is supplied by
a] One pressure line to all injectors
b] Separate pressure line to all injectors
c] Both a] & b]
d] None of these

400. In line system of injection, the pump used is
a] Constant stroke type
b] Lapped- plunger type
c] Both a] & b]
d] None of these

401. Pump used in inline system of injection is driven by
a] Camshaft b] Crankshaft
c] Belt-pulleyd] None of these

402. In individual pump system, fuel is delivered to injector at a high pressure of
a] 500-1000 KPa b] 1000-5000 KPa
c] 5000-10000 KPa d] 10000-20000 Kpa

403. Unit injectors are driven by
a] Separate crankshaft
b] Separate cams
c] Both a] and b]
d] None of these

GK **Mahadev Desai was the secretary of Gandhiji.** GS

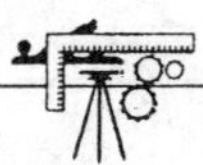

404. Fuel injector is also known as
a] Nozzle b] Atomizer
c] Fuel valve d] All are correct

405. Hole type nozzle is generally used in
a] Open type combustion chamber
b] Swirl combustion chamber
c] Pre-combustion chamber
d] None of these

406. Pintle type nozzle is commonly used in
a] Pre combustion chamber
b] Swirl combustion chamber
c] Both a] & b]
d] None of these

407. Pressure in hole type nozzle varies from
a] 15-20 MPa b] 15-25 MPa
c] 17-34 MPa d] None of these

408. Pressure in pintle type nozzle varies as
a] 5-10 MPa b] 7-15 MPa
c] 7-20 MPa d] 10-25 Mpa

409. In air injection system, fuel pump develops
a] High pressure b] Low pressure
c] Very high pressure d] None of these

410. The advantage of air injection system is
a] Good atomization
b] Good mixing of air fuel
c] Injection of heavy and viscous fuels
d] All are correct

411. Solid injection system has
a] Pressurizing unit b] Atomizing unit
c] Both a] & b] d] None of these

412. Solid injection system is also known as
a] Airless mechanical injection
b] Air injection
c] Jerk pump system
d] All are correct

413. Accumulator is used in
a] Jerk pump system
b] Unit injector system
c] Common rail system
d] Distributor system

414. The work of accumulator is
a] To develop high pressure
b] To keep constant pressure
c] To develop low pressure
d] To metered fuel

415. The size of hole of single hole nozzle is usually larger than
a] 0.1 mm b] 0.2 mm
c] 0.3 mm d] 0.4 mm

416. In multi hole nozzle, the number of holes varies as
a] 2-5 b] 3-8
c] 4-10 d] 4-18

Did You Know ?
Smallest wild sheep
The Urial or Shapu (*Ovis orientalis*) measures a little less than a meter at the shoulder as weight about 85 kg. The horns measure up to 99 cm.

417. Governor is a device used in
a] Tractor engine b] Stationary engine
c] Both a] & b] d] None of these

418. The purpose of governor is
a] To maintain the constant speed of engine at different loads
b] To protect the engine and attached equipment against high speed
c] Both a] & b]
d] None of these

419. A governor automatically regulates
a] Flywheel speed
b] Engine speed
c] Speed of Rear wheel of tractor
d] All are correct

420. Hit and miss system of governor is to regulate the
a] Engine load b] Frequency of explosions
c] Power strokes d] Both b] & c]

421. Throttle system controls the
a] Engine load b] Amount of air fuel mixture
c] Both a] & b] d] None of these

422. Pneumatic governor is used on
a] Diesel engine b] Petrol engine
c] Both a] & b] d] None of these

423. A governor based on the principle that the change of pressure in one part of system produces an opposite change of much greater magnitude in other part is called
a] Centrifugal governor
b] Pneumatic governer
c] Hydraulic governor
d] None of these

424. Centrifugal governor is common in
a] Tractor engine b] Stationary engine
c] Both a] & b] d] None of these

425. The weight of centrifugal governor is to regulate the
a] Fuel control of diesel injection pump
b] Load of engine
c] Both a] & b] d] None of these

426. The variation in engine speed is observed at full load and no load is approximate
a] 100 rpm b] 150 rpm
c] 200 rpm d] 250 rpm

GS

Article 65 provides for discharge of duties by the Vice-President in a presidential capacity.

427. The governor is used on tractor engine is called
a] Constant speed governor
b] Variable speed governor
c] Both a] & b]
d] None of these

428. The governor used on stationary engine is
a] Constant speed governor
b] Variable speed governor
c] Both a] & b]
d] None of these

429. Of the following which type of governing system is used in the gas engines?
a] Hit and miss system
b] Throttle system
c] Variable speed governor
d] Constant speed governor

430. Throttle type governor is kind of
a] Constant speed governor
b] Variable speed governor
c] Hit and miss governor
d] Hydraulic governor

431. Variation in engine speed at full load and no load is called
a] Pressure drop b] Speed drop
c] Pressure hike d] Speed hike

432. Centrifugal governor is directly connected with
a] Accelerator b] Engine speed
c] fuel injection pump d] None of these

433. Centrifugal governor is also termed as
a] Maximum speed governor
b] Minimum speed governor
c] Both a] & b]
d] None of these

434. All speed governors is directly connected with
a] Accelerator pedal b] Pump control rack
c] Both a] & b] d] None of these

435. The main part of pneumatic governor is
a] Venturi unit b] Diaphragm unit
c] Both a] & b] d] None of these

436. Venturi unit is connected to the
a] Fuel injection pump
b] Engine inlet manifold
c] Engine exhaust manifold
d] None of these

437. Diaphragm unit is fitted on
a] Fuel injection pump
b] Engine inlet manifold
c] Exhaust manifold
d] None of these

438. Venturi unit and diaphragm unit are connected by
a] Accelerator b] Vacuum pipe
c] Fuel pipeline d] All are correct

439. An angular moment of front wheels is governed by operator through a system is known as
a] Steering wheel
b] Steering gearbox
c] Steering mechanism
d] All are correct

440. Of the following, which is the essential part of steering mechanism?
a] Tie rod b] Steering wheel
c] Drag link d] All are correct

441. A gear used in steering system is
a] Helical gear b] Spiral gear
c] Spur gear d] Warm gear

442. A pump used in power steering system is
a] Gear pump b] Vane pump
c] Both a] & b] d] None of these

443. The sequence of gears and shafts through which the engine power transmited to tractor wheel is called
a] Gear system b] Transmission
c] Both a] & b] d] None of these

444. The complete path of power from engine to wheel is cal1ed
a] Power system b] Power train
c] Gear unit d] All are correct

445. Function of transmission system is
a] To transmit the power from engine to rear wheel
b] To reduce the crankshaft speed to give desired speed
c] To transmit power through right angle
d] All are correct

446. Clutch is a device used to
a] Connect or disconnect the engine power from trasnmission system
b] Reduce the engine power
c] Brake the engine power
d] All are correct

447. Essential feature of good clutch is
a] Ability to taking load withou⁺ dragging and chattering
b] Capacity to transmit maximum load without slipping
c] High resistant to friction and easy control by hand or pedal lever
d] All are correct

448. Pressure plate is an essential part of
a] Transmission system
b] Clutch
c] Gear reduction
d] None of these

449. Friction plate is part of
a] Clutch b] Final drive
c] Differential d] None of these

Trishul is surface to air missile.

GS

450. The power of engine transmitted to rear wheel of tractor is connected and disconnected by
a] Flywheel b] Clutch
c] Gear d] None of these

451. The main shaft of an engine is also called
a] Output shaft b] Input shaft
c] Lay shaft d] Counter shaft

452. Lay shaft is also termed as
a] Output shaft b] Input shaft
c] Counter shaft d] All are correct

453. The gear used in constant mesh type system is
a] Spur gear b] Bevel gear
c] Warm gear d] Helical gear

454. The device that permits one of the rear wheels of tractor to rotate faster than the other at the turning is called
a] Power train b] Clutch
c] Differential lock d] None of these

455. Crown wheel is a part of
a] Clutch system b] Steering system
c] Differential system d] Cooling system

456. The main function of crown wheel is
a] To transmit power at right angle
b] To reduce the speed of rotation
c] Both a] & b] d] None of these

457. The device used for reduction of speed and multiplication of torque immediately before rear wheel of tractor is called
a] Differential b] Final drive
c] Power train d] All are correct

458. A splined shaft used for taking off power from engine is
a] Final drive b] Drawbar
c] Power take off d] Differential

459. Gear drive is recommended when two shafts are
a] Close b] At 90 degree
c] At 45 degree d] All are correct

460. Single disc or single plate type clutch is
a] Dry type b] Wet type
c] Both a] & b] d] None of these

461. Multiple disc clutch is
a] Wet type b] Dry type
c] Both a] & b] d] None of these

462. Multiple disc type clutches are mostly used in
a] General purpose tractor
b] Crawler tractor
c] Power tillers
d] None of these

463. Gears are generally made of
a] Cast iron b] High carbon steel
c] Aluminium d] Alloy steel

464. Lubricant used in gear box is
a] SAE 30 b] SAE 40
c] SAE 90 d] All are correct

465. Transmission oil should be changed after
a] 750 hours b] 800 hours
c] 1000 hours. d] 1200 hours

466. Reduction of engine speed through transmission gearbox ranges between
a] 1 and 1/5th highest and lowest
b] 1 and 1/6th
c] l and 1/8th
d] None of these

467. Generally tractor speed in the top gear varies from
a] 12-18 kmph b] 12-25 kmph
c] 18-24 kmph d] All are correct

468. Forward speed of crawler tractor is restricted to about
a] 5 kmph b] 6 mph
c] 10 kmph d] Both b] & c]

469. Differential lock helps to
a] Developing full traction in both drive & wheel
b] Developing full traction in one drive wheel
c] Both a] & b]
d] None of these

470. Speed reduction ratio of final drive ranges from
a] 3 to 5:1 b] 2 to 4:1
c] 3 to 6:1 d] 2 to 5:1

471. PTO shaft gets drive from
a] Engine b] Transmission
c] Output shaft d] All are correct

472. PTO drive taken from output shaft, the speed of PTO is
a] Proportional to speed of tractor
b] Directly proportional to engine speed
c] Inversely proportional to engine speed
d] None of these

473. PTO shaft drive taken from engine transmission the speed of PTO will be
a] Directly proportional to engine speed
b] Inversely proportional to engine speed
c] Both a] & b]
d] None of these

474. A PTO shaft with diameter 3.6 cm and 6 splined, standard PTO rpm will be
a] 540 ± 10 rpm b] 540 ± 20 rpm
(C} 1000 ± 10 d] None of these

475. A PTO with 21 splined has speed of
a] 540 ± 10 rpm b] 1000 rpm
c] 1500 rpm d] None of these

GK GS

The base year for Index of Industrial Production (IIP) has been changed to 1993-94.

476. The rear mounted pulley of tractor gets speed of a full throttle is
a] 950 metre per minute
b] 950 rpm
c] 950 kmph
d] All are correct

477. Tractor pulley is made of
a] Cast Iron b] Cast steel
c] Compressed paper d] All are correct

478. Advantages of swinging drawbar is
a] Taking short turns
b] Reduces side draft
c] Leaves only small area uncovered on corners of field
d] All are correct

479. Spline in PTO are used for
a] Joining two shafts in line
b] Joining of two different kind of shaft and pulley
c] Transmit power at 120 degree
d] All are correct

480. In sliding gear transmission system, the gear used is
a] Spiral b] Spur
c] Bevel d] Helical

481. In constant mesh transmission system, the gear used to control the noise is
a] Spur gear b] Bevel gear
c] Helical gear d] Spiral gear

482. The hydrostatic transmission system is popular in
a] Tractors b] Combines
c] Power tillers d] Crawlers

483. Spur and helical gears transmit power between two
a] Parallel shafts
b] Parallel shafts with intersecting axis
c] Both a] & b] d] None of these

484. Bevel gears transmit power between two shafts are
a] Intersecting axis b] Parallel shaft
c] Inline shaft d] All are correct

485. Spline shafts are used when the torque transmission is in
a] Small amount b] Large amount
c] Both a] & b] d] None of these

486. Hydraulic pump develops a pressure of about
a] 150 kg/cm^2 b] 180 kg/cm^2
c] 200 kg/cm^2 d] 250 kg/cm^2

487. Hydraulic system in tractor is mainly used for
a] Operating three point hitch
b] PTO drive
c] Both a] & b] d] None of these

488. A gear used in the tractor hydraulic system normally is
a] Spiral type b] Helical type
c] Bevel type d] Spur type

489. A pump used in the tractor hydraulic is
a] Internal gear pump b] Gyrator gear pump
c] Vane type pump d] All are correct

490. An accumulator used in hydraulic system is for
a] Storing energy b] Increasing pressure
c] Operating pump d] All are correct

491. An accumulator also serves as
a] Shock absorber b] Pump
c] Both a] & b] d] None of these

492. A type of valve used in hydraulic system is
a] Control valve b] Relief valve
c] Priority valve d] All are correct

493. Mass per unit volume of substance is called
a] Specific gravity b] Density
c] Mass density d] None of these

494. Ratio of mass density of substance at certain temperature to the mass density of water at same temperature is called
a] Specific gravity b] Density
c] Specific weight d] None of these

495. Specific gravity of petroleum-based hydraulic-fluid is
a] 0.5 b] 0.7
c] 0.8 d] None of these

496. Mass density increases as
a] Pressure increase
b] Pressure decrease
c] Temperature increase
d] Temperature decreases

497. Mass density decreases with
a] Temperature increase
b] Temperature decrease
c] Both a] & b]
d] None of these

498. Amount of heat required to raise the temperature of a unit mass by one degree is called
a] Specific gravity b] Thermal conductivity
c] Specific heat d] Viscosity

499. Nudging system is used for
a] Rising implement b] Lower the implement
c] Both a] & b] d] None of these

500. An automatic position control system is associated with
a] Two point hitch b] Three point hitch
c] Both a] & b] d] None of these

501. Average body surface area of a man will be
a] 122 m^2 b] 1.5 m^2
c] 2.0 m^2 d] 25 m^2

502. The frequency range of human ear extends from as low as
a] 10 Hz b] 16Hz
c] 20 Hz. d] None of these

GK The first organic compound prepared in laboratory was Urea. GS

503. The frequency range of human ear extends from as high as
a] 10000 Hz b] 15000 Hz
c] 20000 Hz. d] None of these

504. Unit of noise is
a] Kg per sq.cm b] Decible
c] Both a] & b] d] None of these

505. For air bone sounds the reference sound pressure is
a] 0.001 N/m^2 b] 0.00002 N/m^2
c] 0.0003 N/m^2 d] 0.00004 N/m^2

506. Maximum noise level from a tractor near the operator's ear should not exceed
a] 85 dB b] 90 dB
c] 95 dB d] 100 dB

507. The upper safe noise level is
a] 80 dB b] 85 dB
c] 90 dB d] 92 dB

508. A noise level for eight hours of working is
a] 85 dB b] 90 dB
c] 95 dB d] 100 dB

509. A noise level for six hours of working is
a] 80 dB b] 85 dB
c] 90 dB d] 92 dB

510. A noise level for four hours of working is
a] 90 dB b] 95 dB
c] 100 dB d] 105 dB

511. An oscillatory motion of mechanical system is called
a] Noise b] Vibration
c] Both a] & b] d] None of these

512, Vibration is measured by
a] Accelerometer b] Tachometer
c] Tensiometer d] None of these

513. A six-cylinder engine rated at 2200 rpm will generate a fundamental sound frequency of
a] 150 Hz b] 220 Hz
c] 250 Hz d] None of these

514. A spring mass damper system vibrates with
a] 5 degree of freedom
b] 6 degree of freedom
c] 7 degree of freedom
d] 8 degree of freedom

515. Tractor seat suspension should have its natural frequency of
a] 0.5 to 1.5 cycle per second
b] 0.5 to 2.0 cycle per second
c] 1 to 2.5 cycles per second
d] All are correct

516. Ratio of output vibration to input vibration is called
a] Vibration intensity
b] Transmissibility
c] Both a] & b]
d] None of these

You Don't Know!

Smallest hibiscus

Hibiscus trionum, **which attains a height of 60 cm, with 4 cm wide flowers which are pale yellow with a purple centre, is an annual weed found in fields in Kashmir and in the warmer parts of Eurasia.**

517. Range of transmissibility for satisfactory performance is
a] 0.5 to 0.65 b] 0.5 to 0.75
c] 0.5 to 1.0 d] None of these

518. In proper seat suspension design, the transmissibility should be
a] Less than unity b] More than unity
c] Less than 2 d] More than 2

519. In general tractor, frequency should be
a] 3-10 Hz b] 3-12 Hz
c] 5-15 Hz d] 5-20 Hz

520, The undamped natural frequency of seat suspension system with maximum travel of 76 mm should be
a] 1.0 Hz b] 1.5 Hz
c] 2.0 Hz d] 1.5 Hz

521. A cylinder block is normally made of
a] Aluminium b] Cast iron
c] High carbon steel d] Mild steel

522. A connecting rod of tractor is made of
a] High carbon steel b] Cast iron
c] Mild steel d] Forged steel

523. The standard speed of a 21 spline PTO shaft is
a] 540 ± 10rpm b] 540 rpm
c] 1000 rpm d] 10000 rpm

524. The type of starting aid generally used in the diesel power tiller is called
a] Decompression lever
b] Hot plug
c] Cold plug
d] choke

525. Toe-in provide in tractor is approximately
a] 7-10 mm b] 10-15 mm
c] 12-16 mm d] 15-20 mm

526. The first tractor company which introduced air-cooled tractor
a] Eicher b] Mahindra
c] Escort d] Ford

527. In CI engine of tractor injection pressure is in range of
a] 5-20 kg/cm^2 b] 20-100 kg/cm^2
c] 120-200 kg/cm^2 d] None of these

528. Traction coefficient is maximum in a field when
a] It is dry b] It is cultivated
c] It is irrigated d] All are correct

GK **Koshi river is called Sorrow of Bihar.** GS

529. **Farm tractor is**
   a] Prime mover b] Traction engine
   c] Both a] & b] d] None of these
530. **The first tractor demonstration was held at**
   a] Nebraska b] Germany
   c] Russia d] India
531. **The first tractor manufactured in India was**
   a] Eicher b] Mahindra
   c] Escort d] Ford
532. **Pressure due to radiator caps**
   a] Lower the boiling point b] Raise the boiling point
   c] Cools water d] All are correct
533. **A power tiller operates most satisfactory with**
   a] Reaper b] M B plough
   c] Disc harrow d] Rotary tillage tool
534. **The maximum torque in a tractor is generated at**
   a] Rated rpm b] Rated horsepower
   c] Less than rated rpm d] More than rated rpm
535. **Weight transfer in a tractor implement system is caused by**
   a] Application of pull b] Tractive force
   c] Traction coefficient d] All are correct
536. **Coefficient of traction is defined as**
   a] Ratio of BHP to IHP
   b] Ratio of dbhp to PTO hp
   c] Ratio of drawbar pull to dynamic loads
   d] All are correct
537. **Turning radius is affected by**
   a] Centre of gravity b] Wheel base
   c] Both a] & b] d] None of these
538. **Crawler tractor is also known as**
   a] Chain type tractor b] Track type tractor
   c] Botha] & b] d] None of these
539. **For hot climate the suitable type of tractor is**
   a] Water cooled b] Air cooled
   c] Both a] & b] d] None of these
540. **Power train indicates**
   a] PTO drive
   b] Injection system
   c] Complete power transmission path
   d] Reduction unit
541. **Clutch and differential are part of**
   a] Injection system
   b] Power transmission system
   c] PTO drive
   d] None of these
542. **Final drive is a**
   a] Speed increasing unit
   b] Speed reduction unit
   c] Torque increasing unit
   d] Torque reduction unit
543. **Clutch works on principle of**
   a] Friction b] Cohesion
   c] Adhesion d] All are correct
544. **Highest tractive efficiency occurs in a**
   a] Irrigated soil b] Compact and firm soil
   c] Ploughed soil d] All are correct
545. **Pain in eardrum starts at a noise level of**
   a] 85 dB b] 95 dB
   c] 110 dB d] 115 dB
546. **Tractor brake test should be carried out at speed of**
   a] 15 kmph b] 20 kmph
   c] 25 kmph d] All are correct
547. **Range of pressure angle in spur gear is**
   a] 5 -8° b] 5 -12°
   c] 14 -20° d] 20 -25°
548. **Cold spark plug is used in**
   a] IC engine b] CI engine
   c] Light duty engine. d] Heavy duty engine
549. **Valve timing diagram is a function of**
   a] Torque b] Engine speed
   c] Air fuel ratio d] All are correct
550. **In a tractor centre of gravity is located at**
   a] 113 of wheel base ahead of rear axle
   b] 2/3 of wheel base ahead of rear axle
   c] 114 of wheel base ahead of rear axle
   d] All are correct
551. **Friction clutch is most commonly used in**
   a] Two wheel drive tractor
   b] Four wheel drive tractor
   c] Both a] & b] d] None of these
552. **A clutch used in power tiller is**
   a] Single disk clutch b] Double disk clutch
   c] Dog clutch d] Dual plate clutch
553. **Single plate clutch is also known as**
   a] Single disk clutch b] Double disk clutch
   c] Dog clutch d] All are correct
554. **Pitman arm is part of**
   a] Gear system b] Hitch system
   c] Brake system d] Steering system
555. **Drag link is component of**
   a] Clutch system b] Fuel supply system
   c] Transmission system d] Steering system
556. **Hydraulic brake works on the principle of**
   a] Newton's Law b] Pascal's Law
   c] Boyles Law' d] None of these
557. **In tractor three-point hitch is operated by**
   a] Clutch system b] Gear system
   c] Hydraulic system d] All are correct
558. **Top link of tractor is also known as**
   a] Tension link b] Mast
   c] Pressure link d] All are correct

GK Kudappa was associated with Forest Satyagrah. GS

559. Lower links of tractor reduces
a] Side draft b] Unit draft
c] Total draft d] All are correct

560. If stroke bore ratio increases then thermal efficiency
a] Decreases b] Increases
c] Remains same d] None of these

561. Tractor operation in lower gear causes
a] Less torque b] More pull
c] Less pull d] All are correct

562. Muffler is provided on tractor to control
a] Noise b] Fuel
c] Fuel consumption d] Vibration

563. Optimum engine temperature is about
a] 50°C b] 62°C
c] 72°C d] 82°C

564. Supercharger of an engine is driven by
a] Camshaft b] Crankshaft
c] Exhaust gases d] Flywheel

565. Turbocharger of an engine is driven by
a] Dynamo b] PTO shaft
c] Exhaust gas d] Air

566. One bar is equal to
a] 1 KPa b] 5 KPa
c] 10 KPa d] 100 Kpa

567. Bar is the unit of
a] Work. b] Pressure
c] Viscosity d] Energy

568. Fluctuation of engine speed during a cycle depends upon
a] Crankshaft b] Governor
c] Mass of flywheel d] All are correct

569. Camshaft controls
a] Valve opening b] Valve closing
c] Both a] & b] d] All are correct

570. Engine requires overhauling in case of
a] Poor compression b] Mechanical failure
c] Both a] & b] d] All are correct

571. Coolant pump is
a] Vane type b] Gear pump
c] Centrifugal pump d] All are correct

572. Radiator core is made of
a] Cast iron b] Steel
c] Plastic d] Brass

573. The grade of oil used during summer season is
a] SAE 90 b] SAE 110
c] SAE 30 d] SAE 40

574. The grade of oil used in winter season is
a] SAE 30 b] SAE 90
c] SAE 110 d] All are correct

575. A function of drawbar in tractors is to
a] Lift the load b] Hitch the implement
c] Both pull load d] All are correct

576. In tractor weight transfer is greater when
a] Speed is greater b] Speed is less
c] Pull is higher d] All are correct

577. Weight transfer in tractor is affected by
a] Hitch height b] Wheel base
c] Centre of gravity d] All are correct

578. Dbhp of tractor indicates
a] Power on flywheel b] Power on drawbar
c] Power on PTO d] None of these

579. PTO hp indicates
a] Power on piston b] Power on crankshaft
c] Power on PTO shaft d] All are correct

580. Ballasting is done by
a] Water b] Oil
c] Calcium chloride d] All are correct

581. Ballasting in front tyre of tractor is done for
a] Increasing traction
b] decreasing load
c] Increasing stability
d] All are correct

582. Ballasting in rear tyre of tractor is done for
a] Increasing traction
b] Reducing tractor vibration
c] Increasing weight of tractor
d] Prevent overturning of tractor

583. Inflation pressure of rear tyre of tractor is
a] 0.8 to 1.2 kg/ cm^2 b] 0.8 to 1.5 kg/cm^2
c] 1.2 to 2.0 kg/ cm^2 d] 2.0 to 2.5 kg/ cm^2

584. Inflation pressure of front tyre of tractor is
a] 1.2 to 2.0 kg/ cm^2 b] 2.0 to 2.5 kg/ cm^2
c] 2.0 to 2.8 kg/ cm^2 d] All are correct

585. Front wheel of tractor is smaller than rear wheel that helps
a] More pull of load b] Increasing traction
c] Sharp turning d] All are correct

586. Ply rating of tyre indicates
a] Load-bearing capacity
b] Load-pulling capacity
c] Both a] & b]
d] None of these

587. Toe-in and toe-out are related to
a] Front axle of tractor
b] Rear axle of tractor
c] Tractor steering
d] Transmission system

588. Angle between centreline of kingpin of tractor and vertical line is called
a] Tilt angle b] Camber angle
c] Caster angle d] All are correct

GK GS

Radhakrishnan was the first Vice-President of India.

589. Angle between centerline of tyre and vertical line is called
a] Camber angle b] Disc angle
c] Helix angle d] Relief angle

590. The oil from oil pump delivers directly to
a] Oil straine b] Oil filter
c] Mall bearing d] All are correct

591 Detonation results in
a] Interruption in lubrication
b] Delay in ignition
c] Loss of power
d] All are correct

592. An indication of ignition quality of diesel fuel is given by
a] Octane number b] Cetane number
c] SAE number d] All are correct

593. Removal of burnt gases from engine cylinder is called
a] Pre-ignition b] Detonation
c] Scavenging d] Supercharging

594. Lean air fuel mixture is required for
a] Starting b] Cruising
c] Idling d] All are correct

595. Cage wheel is used in tractor for
a] Increasing the weight
b] Improving traction
c] Reducing load
d] Weight transfer

596. Depreciation of machine value depends on
a] Capital investment
b] Nos. of working hours/ year
c] Salvage value
d] All are correct

597. Depreciation of machine is computed on its
a] Fixed cost
b] Operating cost
c] Repair and maintenance cost
d] All are correct

598. The most accurate dynamometer is
a] Prony brake b] Hydraulic type
c] Swinging field type d] Eddy current type

599. Brake mean effective pressure of an engine depends upon its
(A} Speed b] Power
c] Torque d] All are correct

600. Spark occurs when
a] Points open b] Points close
c] Both a] & b] d] None of these

601. The colour of positive plate of a lead-acid battery is
a] White b] Grey
c] Brown d] Black

602. The colour of negative plate of lead acid battery is
a] Black b] Grey
c] White d] Brown

603. Positive plate of lead-acid battery is made of
a] PbO_2 b] Pb
c] $PbSO_4$ d] H_2SO_4

604. Negative plate of lead-acid battery is made of
a] $PbSO_4$ b] H_2SO_4
c] PbO_2 d] Pb

605. In an alternator the magnetic field is produced in the
a] Frame b] Stator
c] Regulator d] Rotar

606. The number of winding in the stator of alternator is
a] 1 b] 2
c] 3 d] 4

607. The output of an alternator is controlled by
a] Current regulator b] Voltage regulator
c] Cutout device d] All are correct

608. The ratio of output power to input power for a traction device is called
a] Mechanical efficiency b] Thermal efficiency
c] Tractive efficiency d] None of these

609. Ratio of net pull produced for the dynamic normal load on traction device is called
a] Tractive force
b] Traction constant
c] Net traction coefficient
d] All are correct

610. In 4WD tractors power occurs in
a] Both front wheels
b] Both rear wheels
c] One front and one rear wheel
d] All four wheels

611. The power required to run the engine at any given speed without production of useful work is called
a] Drawbar power b] Indicated horsepower
c] PTO horsepower d] Friction power

612. Brake horsepower of an engine can be measured by
a] Spring dynamometer
b] Brake dynamometer
c] Strain gauge dynamometer
d] All are correct

613. Revolution of an engine (engine speed) is measured by
a] Tachometer
b] Spring Dynamometer
c] Eddy current dynamometer
d] All are correct

614. Drawbar pull is measured by the
a] Prony brake dynamometer
b] Brake dynamometer
c] Hydraulic dynamometer
d] All are correct

615. Useful life of stationary engine is
a] 8 years b] 10 years
c] 12 years d] 15 years

616. Engine oil should be changed at every
a] 50 hours b] 100 hours
c] 120 hours d] 150 hours

617. Tractor transmission oil should be changed at every
a] 100 hours b] 300 hours
c] 500 hours d] 750 hours

618. Slogging or lugging ability is associated with
a] Engine speed b] Engine torque
c] Engine load d] All are correct

619. A device to control the engine speed within specifiec limit is called
a] Carburettor b] Governor
c] Radiator d] Differential

620. The useful life of tractor is
a] 8,000 hours b] 10,000 hours
c] 12,000 hours d] 15,000 hours

621. A part that connects piston with connection rod is
a] Piston ring b] Gudgen pin
c] Safety pin d] Camshaft

622. Power tiller operates most satisfactorily with
a] Rotary tools
b] Ploughs
c] Transport wheel
d] Harrows

623. In an adiabatic process system
a] Temperature is constant
b] Pressure is constant
c] Heat is constant
d] Volume is constant

624. Weight transfer of tractor is directly affected by
a] Pull b] Hitch height
c] Both a] & b] d] Wheel base

625. Compression ratio of diesel engine is
a] 4:1 b] 6:1
c] 15:1 d] 25:1

626. Maximum torque in a tractor is generated at
a] Rated rpmb] Less than rated rpm
c] Rated horsepower d] Maximum idling

627. Brakes works on principle of
a] Friction b] Vibration
c] Sliding d] Suction

628. Present tractor population in India is around
a] 1 million b] 1.5 million
c] 2.5 million d] More than 3.5 million

Knowledge World!

Largest duck

The White winged Wood duck (*Cairina scutulata*) measures up to 81 cm. It was once found in Bangladesh, Assam and Manipur but is now restricted to Arunachal Pradesh.

629. Brake fluid used in hydraulic brake is a mixture of
a] Glycerine and alcohol
b] Alcohol and methane
c] Glycerine and methane
d] None of these

630. When brake is applied the kinetic energy of body is converted into
a] Electrical energy b] Heat energy
c] Chemical energy d] Mechanical energy

631. The optimum engine temperature is about
a] 60 deg. b] 82 deg.
c] 100 deg. d] 120 deg.

632. Cetane number for commercial diesel fuel varies from
a] 20-40 b] 30-60
c] 60-90 d] 90-100

633. Loss of value of machine with passage of time is
a] Depreciation b] Interest
c] Hiring cost d] None of these

634. Pressure due to radiator cap
a] Raises boiling temperature
b] Reduces boiling temperature
c] Cools the water
d] None of these

635. The voltage in spark plug at the time of spark is
a] 300-400 V b] 100-200 V
c] 1000-10000 V d] 20000 V

636. Heavy smoking in an engine during operation may be due to
a] Rich mixture b] Overloading
c] Late injection d] All are correct

637. Relative movement of the wheel in direction of travel for given distance under load and no load condition is
a] Wheel slip
b] Coefficient of traction
c] Coefficient of rolling resistance
d] Tractive efficiency

638. BHP of the engine indicates
a] Power in cylinder b] Power on flywheel
c] Power drawbar d] Power at PTO pulley

Small Industrial units are those units which have investment of rupees one crore in plants and machinery.

639. **Which of the following is transmission dynamometer?**
a] Torsion dynamometer
b] Froude's hydraulic dynamometer
c] Belt dynamometer
d] Prony brake dynamometer

640 **Turbo chargers are driven by**
a] Engine b] Dynamo
c] Exhaust gas d] PTO shaft

641. **Centre of gravity of tractor is determined by**
a] Suspension method b] Balancing method
c] Weighing method d] All are correct

642. **The PTO HP of a tractor is**
a] Less than drawbar
b] More than engine HP
c] More than drawbar HP
d] Less than friction HP

643. **The most commonly used method for determination of centre of gravity of tractor is**
a] Suspension method b] Balancing method
c] Weighing method d] All are correct

644. **Main constitute of biogas is**
a] Methane b] Ethane
c] Butane d] Carbon dioxide

645. **In SI engine intake opens about**
a] 10 deg before TDC b] At TDC
c] 10 deg after TDC d] 20 deg after TDC

646. **The basic function of governor in an engine is to regulate the**
a] Fuel consumption b] Power development
c] Engine speed d] None of these

647. **The ratio of rolling resistance to dynamic load is**
a] Coefficient of traction
b] Tractive efficiency
c] Coefficient of rolling resistance
d] Rolling resistance

648. **The power developed by an average pair of bullocks varies from**
a] 0-0.5 hp b] 0.5-1.0 hp
c] 1.0-1.5 hpd] 1.5-2.0 hp

649. **Ratio of total force output of the traction device in direction of travel to dynamic weight on traction device is**
a] Tractive efficiency
b] Coefficient of traction
c] Coefficient of rolling resistance
d] Wheel slip

650. **The rear part of a tractor is heavier than the front part of tractor to get**
a] Higher tractive efficiency
b] Lower tractive efficiency
c] Less wheel slip
d] All are correct

651. **An engine is called I head engine when**
a] Valves are immediately above the cylinder
b] Both valves are on one side
c] Valves are in opposite direction
d] None of these

652. **A steam engine is an**
a] Spark ignition engine
b] Compression ignition engine
c] External combustion engine
d] IC engine

653. **In hydraulically control hitching system, the equipment vibrates due to**
a] Lower oil level
b] Clogging of oil filter
c] Trapping of air in oil filter
d] All are correct

654. **Engine in which heat is taken at one constant pressure and rejected at another constant pressure is**
a] Otto cycle b] Auto cycle
c] Diesel cycle d] Petrol cycle

655. **The total swept volume of all pistons during power strokes occurring in one minute is**
a] Piston volume
b] Clearance volume
c] Displacement volume
d] None of these

656. **When crankshaft of the engine completes one cycle, the camshaft completes**
a] Two revolution
b] Half revolution
c] Equal revolution of crankshaft
d] None of these

657. **The camshaft gear is called**
a] Two time gear b] One time gear
c] Halftime gear d] None of these

658. **Tractive efficiency is affected by**
a] Slope of land b] Wheel size
c] Height of hitch d] All are correct

659. **In hydraulically controlled hitching system, the implement does not remain in lifted position due to**
a] Defectlve valve b] Oil leakage
c] Both a] & b] d] Lower oil lever

660. **The stroke bore of tractor engine varies from**
a] 1 to 1.45 b] 1 to 2
c] 1.45 to 2 d] 2 to 2.45

661. **The permissible slip for wheel type tyre during slippage testing is**
a] 3 to 5% b] 7 to 10%
c] 15 to 20% d] 20 to 30%

662. **The permissible slip for crawler tractor during slippage testing is**
a] 3 to 5% b] 7 to 10%
c] 15 to 20 %d] 20 to 30%

GK **Soda water contains carbonic acid.** GS

663. Traction can be increased by
    a] Rubber tyre with grooves
    b] Placing chains
    c] Putting lugs
    d] All are correct
664. The operation to perform to open any cultivable land is
    a] Primary tillage b] Secondary tillage
    c] Minimum tillage d] Reduced tillage
665. Finer operation performed for seedbed preparation is
    a] Primary tillage b] Secondary tillage
    c] Strip tillage d] Rotary tillage
666. Indigenous plough is
    a] Primary tillage tool
    b] Secondary tillage tool
    c] Multi-purpose tool
    d] All are correct
667. Of the following, which are the primary tillage implements?
    a] Chisel plough and disc harrow
    b] Chisel plough and subsoiler
    c] Disc plough and disc harrow
    d] Leveler and clod crusher
668. A cultivar whose tynes swings back when obstacle encounter is
    a] Spike tooth cultivator
    b] Spring type cultivator
    c] Rigid type cultivator
    d] All are correct
669. A bullock drawn implement have
    a] Higher speed and higher working capacity
    b] Lower speed and lower working capacity
    c] Higher speed and lower working capacity
    d] Lower speed and higher working capacity
670. Tractor drawn implements have
    a] Higher speed and higher working capacity
    b] Lower speed and lower working capacity
    c] Higher speed and lower working capacity
    d] Lower speed and higher working capacity
671. An implement that pulled and guided by single hitch point of a tractor is
    a] Trailed implement
    b] Mounted implement
    c] Semi mounted implements
    d] All are correct
672. An implement that attached to a tractor along hinge axis and controlled directly by tractor steering unit is
    a] Trailed implements
    b] Mounted implements
    c] Semi mounted implements
    d] All are correct
673. An implement that fully supported by the tractor is
    a] Trailed implement
    b] Mounted implement
    c] Semi-mounted implement
    d] All are correct
674. Ploughing is done to
    a] Improve soil aeration
    b] Destroy weeds
    c] Increase water-holding capacity
    d] All are correct
675. The plough bottom as combined unit consists of
    a] Share, landside, frog and MB
    b] Coulter, jointer and frog
    c] Beam, handle and MB
    d] None of these
676. Jointer and coulter are the parts of
    a] Disc plough b] Harrow plough
    c] Indigenous plough d] MB plough
677. The part of plough that penetrate into the soil and cuts the soil in horizontal direction is
    a] Share b] Coulter
    c] Mould board d] Land side
678. The forward end of cutting edge that actually penetrate into the soil and cuts the soil is
    a] Share point b] Cutting edge
    c] Wing of share d] Gunnel of share
679. The front edge of the share which makes horizontal cut in the soil is
    a] Gunnel of share b] Cutting edge
    c] Wing of share d] Shin of share
680. The shar of MB plough is made of
    a] Cast iron b] High carbon steel
    c] Chilled cast iron d] White metal
681. The MB plough works on the principle of
    a] Friction b] Rolling
    c] Suction d] Scouring
682. The pwer consumed in secondary tillage operation per unit area is
    a] Less compared to primary tillage operation
    b] More compared to primary tillage operation
    c] Same as the primary tillage operation
    d] None of these
683. A single piece share with curved cutting edge is
    a] Slip nose share b] Bar point share
    c] Shin shared] Slip share
684. A type of share provided with adjustable bar is
    a] Slip nose share b] Bar point share
    c] Shin shared] Slip share
685. In the bar point share, the bar serves as
    a] Share onlyb] Landside only
    c] Share and landside d] None of these

GK GS

Soda water contains carbonic acid.

686. The stubble type of mould board is mostly suitble for
a] Grassy land b] Tough soil
c] Sticky soil d] None of these

687. The slat type of mould board is used in
a] Grassy land b] Sod soil
c] Sticky soil d] None of these

688. A long mould board with gentle curvature used in tough soil full of grasses is
a] Stubble type MB b] Breaker type MB
c] Slat type MB d] General purpose

689. The part of MB plough to which all other parts of plough bottom is attached is
a] Mould board b] Land side
c] Frog d] Coulter

690. In MB plough, a small irregular piece of metal has similar shape to ordinary plough bottom is
a] Jointer b] Coulter
c] Gunnel d] None of these

691. A device used to cut the furrow slice vertically from a land ahead of bottom is
a] Jointer b] Coulter
c] Gunnel d] None of these

692. A part of mould board plough which lifts, turns and break furrow slice is
a] Share b] Mould board
c] Landside d] Jointer

693. A wheel used for maintaining uniform depth of ploughing in different soils is
a] Transport wheel b] Land wheel
c] Furrow wheel d] Gauge wheel

694. The part of MB plough that resists the side pressure exerted by the furrow slice on the mould board is
a] Mould board b] Gunnel
c] Landside d] Frog

695. In MB plough, the perpendicular distance from wing of share to the line joining the point of the share and heel of landside is
a] Width of furrow b] Size of plough
c] Plough width d] All are correct

696. The size of mould board plough is expressed by
a] Width of cut b] Depth of cut
c] Size of furrow d] Plough width

697. A point at which the resultant of all horizontal and vertical forces acts is
a] Centre of gravity b] Centre of resistance
c] Centre of pull d] Line of pull

698. A straight line passing through the centre of the pull, the point of hitch and the centre of resistance is
a] Centre of gravity b] Centre of resistance
c] Centre of pull d] Line of pull

699. According to Colling, the draft of mould board plough on the ground is
a] 34% b] 18%
c] 48% d] Above 48%

700. According to colling, the draft of MB plough due to furrow slice is
a] 34% b] 48%
c] 68% d] Above 68%

701. According to colling, the draft of MB plough due to cutting slice is
a] 34% b] 48%
c] 68% d] Above 68%

702. The part of indigenous plough that helps in stabilizing and balancing the plough while operation is
a] Share b] Shoe
c] Body d] Handle

703. The part of MB plough to which the plough bottom and handle attached is
a] Beam b] Standard
c] Frog d] All are correct

704. The maximum clearance under the land side and horizontal surface when the MB plough is resting on ground surface is
a] Horizontal suction b] Vertical suction
c] Throat clearance d] All are correct

705. In MB plough, the maximum horizontal clearance between the land side and furrow wall is
a] Horizontal suction b] Vertical suction
c] Throat clearance d] All are correct

706. The vertical clearance of the MB plough ranges from
a] 2 to 5 mm b] 2 to 5 cm
c] 3 to 5 mm d] 3 to 5 cm

707. Vertical clearance of MB plough influences
a] Depth of cut
b] Width of cut
c] Both depth and width of cut
d] None of these

708. The horizontal suction of MB plough is
a] 5mm b] 3 cm
c] 5 cm d] above 5 cm

709. The horizontal suction of MB plough influences
a] Depth of cut
b] Width of cut
c] Both depth and width of cut
d] None of these

710. The horizontal clevis of MB plough helps in
a] Adjusting line of draft
b] Adjusting line of pull
c] Both a] & b]
d] None of above

GK **Maxwell-Broomfield committee is associated with Baradoli Peasant Revolt.** GS

711. Centre of resistance of plough lie at a distance of
a] 1/3rd size of plough from share wing
b] ¾th size of plough from share wing
c] ½ size of plough from share wing
d] All are correct

712. A MB plough that throws furrow slice on one side to the direction of travel is
a] One-way plough b] Two-way plough
c] Turn wrest plough d] Irreversible plough

713. The part plough that transmit the animal power to the plough is
a] Standard b] Beam
c] Handle d] Body

714. A MB plough that throws furrows slice to both the direction of travel is
a] One-way plough b] Irreversible plough
c] Reversible plough d] All are correct

715. A MB plough whose plough bottom turns through 180° about a longitudial axis is
a] One-way plough b] Irreversible plough
c] Turn wrest plough d] All are correct

716. The size of light MB plough is
a] Below 100 mm width of cut
b] Between 100 mm and 150 mm width of cut
c] Above 100 mm width of cut
d] Above 150 mm width of cut

717. The size of medium MB plough is
a] Below 150 mm width of cut
b] Above 150 mm width of cut
c] Between 150 mm and 200 mm width of cut
d] Above 200 mm width of cut

718. The size of heavy MB plough is
a] Below 150 mm width of cut
b] Above 150 mm width of cut
c] Between 150 mm and 200 mm width of cut
d] Above 200 mm width of cut

719. In standard disc plough, all the discs are
a] Fitted on a single shaft
b] Fitted on separate standard
c] Both a] & b]
d] None of these

720. In disc plough, the angle at which the plane of cutting edge of disc is inclined to the direction of travel
a] Disc angle b] Tilt angle
c] Gang angle d] None of above

721. In disc plough, the angle at which the plane of cutting edge of disc is inclined to the vertical line is
a] Disc angle b] Tilt angle
c] Gang angle d] None of above

722. Disc angle of standard disc plough varies from
a] 15-25° b] 25-35°
c] 35-42° d] 4-45°

Quiz Kit!
Smallest hibiscus
***Hibiscus trionum*, which attains a height of 60 cm, with 4 cm wide flowers which are pale yellow with a purple centre, is an annual weed found in fields in Kashmir and in the warmer parts of Eurasia.**

723. Tilt angle of standard disc plough varies from
a] 15-25° b] 25-35°
c] 35-42° d] 2-45°

724. In disc plough, a device used to pervent disc from clogging is
a] Spacer b] Scraper
c] Standard d] Spool

725. Back furrow and dead furrows are left while ploughing with
a] One-way plough b] Two way plough
c] Both a] & b] d] None of these

726. A MB plough which is suitable for ploughing along hillsides is
a] One-way plough b] Two way plough
c] Both a] & b] d] None of these

727. The tilt angle of wheat plough is
a] 0° b] 15-25°
c] 35-42° d] 42-45°

728. Tillage system which is suitable for dry land agriculture is
a] Till plant system b] Zero tillage
c] Minimum tillage d] All are correct

729. A tillage system in which the field is ploughed and followed by row crop planter directly is
a] Zero tillage system
b] Plough plant
c] Rotary tillage system
d] Till plant system

730. The depth measured at the centre of the disc by placing its concave side down on a flat surface is
a] Convexityb] Concavity
c] Both a] & b] d] None of these

731. In disc harrow, a device used to provide a distance between two discs mounted on same gang axle is
a] Scraper b] Gang
c] Spool d] Arbor bolt

732. A shaft of disc harrow on which the number of discs are mounted is
a] Standard b] Arbor bolt
c] Spacer d] Scraper

733. The increase in tilt angle of disc plough influences
a] Penetration of disc
b] Width of cut
c] Reduce penetration of disc
d] Reduce width of cut

Prime Minister has the real executive authority, under the Indian Constitution.

734. The tilt angle of disc plough influences
a] Depth of cut b] Width of cut
c] Both a] & b] d] None of these

735. The disc angle of disc plough influences
a] Depth of cut b] Width of cut
c] Both a] & b] d] None of these

736. In disc plough, the penetration of disc improves but width of cut reduces due to
a] Increasing tilt angle
b] Reducing tilt angle
c] Increasing disc angle
d] Reducing disc angle

737. The method of ploughing in which the plough works round a strip of unploughed land is
a] Gathering b] Casting
c] Head land d] None of these

738. The method of ploughing in which the plough works round the strip of ploughed land is
a] Gathering b] Casting
c] Head land d] None of these

739. An open trench left between two adjacent strips of land after the plouging is
a] Furrow wall b] Dead furrow
c] Back furrow d] Furrow slice

740. An undisturbed soil surface of a furrow is
a] Furrow slice b] Furrow wall
c] Furrow crown d] All are correct

741. A strip of unploughed land that is left at each end of the field for tractor turning is
a] Gathering b] Casting
c] Head land d] None of these

742. The top portion of a turned furrow slice is
a] Furrow slice b] Crown
c] Ridge d] None of these

743. A harrow with two gangs placed end to end is
a] Offset disc harrow
b] Tandem disc harrow
c] Single action disc harrow
d] Double action disc harrow

744. A disc harrow with four gangs in which a set of two gangs follow behind the set of other two gang is
a] Offset disc harrow
b] Tandem disc harrow
c] Single action disc harrow
d] Triangular harrow

745. A raised left at centre of the strip of land when ploughing is started from center to side is
a] Furrow wall b] Dead furrow
c] Back furrow d] Furrow slice

746. A harrow which is based on a principle that side thrust against the front gang is opposed by side thrust of rear gang
a] Offset disc harrow b] Tandem disc harrow
c] Spike tooth harrow d] Spring tyne harrow

747. A harrow whose tynes swing back when obstacle encountered is
a] Spike tooth harrow b] Spring tyne harrow
c] Patela harrow d] Triangualr harrow

748. A special type of harrow with curved knives is
a] Spike tooth harrow b] Spring type harrow
c] Patela harrow d] Acme harrow

749. The internal angle between the gangs of single action disc harrow is
a] Less than 90° b] Exactly 90°
c] More than 90° d] None of above

750. In disc harrow the angle made by the axis of the arbor bolt with the direction of travel is
a] Disc angle b] Tilt angle
c] Gang angle d] None of these

751. The offset disc harrow are mostly suitable for working in
a] Garden b] Orchard
c] Field d] All are correct

752. The spacing between disc in light duty harrow is
a] 5-6 inch b] 6-9 inch
c] 10-12 inch d] Above 12 inch

753. The bearing used on disc harrows are subjected to
a] Heavy radial load b] Thrust load
c] Both a] & b] d] None of these

754. The bearing used on disc harrows are
a] Ball bearing b] Taper roller bearing
c] Wooden bearing d] All are correct

755. The size of disc harrow is determined by
a] Maximum width of cut
b] Maximum depth of cut
c] Number of disc
d] Number of gangs

756. A bullock drawn harrows mostly have
a] Rigid tynes
b] Flexible tynes
c] Both rigid and flexible tynes
d] None of these

757. Acme harrows are mostly used for
a] Clod crushing b] Mulching
c] Levelling d] All are correct

758. Blade harrows are used for seed bed preparation in
a] Sandy soil b] Hard soil
c] Sticky soil d] Clayey soil

759. The spacing between disc in heavy duty harrow is
a] 5-6 inch b] 6-9 inch
c] 10-12 inch d] Above 12 inch

760. A tractor-drawn harrow mostly has
a] Rigid tynes
b] Flexible tynes
c] Both rigid and flexible tynes
d] None of these

In Hindu mythology, god Agni is said to have seven tongues.

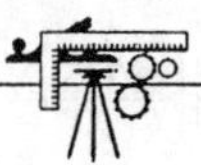

761. **A ridger is used for**
a] Clod crushing b] Making channel
c] Mulching d] Seed bed preparation

762. **Heavy draft of a disc harrow is due to**
a] Blunt disc b] Furrow too wide
c] Loose bearing d] All are correct

763. **Macermic cultivator is a**
a] Bullock drawn b] Tractor drawn
c] Manually operated d] None of these

764. **Animal drawn patela is**
a] Clod crusher b] Leveller
c] Clod crusher d] None of these

765. **A stubble type mould board of MB plough gives**
a] Less pulverization
b] Medium pulverization
c] Tharow pulverization
d] All are correct

766. **Sweeps are used for**
a] Seedped preparation
b] Ridging
c] Earthing plants
d] Mulching

767. **A mould board of MB plough is made of**
a] Cast iron b] Chilled cast iron
c] Mild steel d] Soft centre steel

768. **A mormal ploughing depth is**
a] 15 cm b] 30 cm
c] 40 cm d] 100 cm

769. **A subsoiler is used to break a hard pan of soil upto a depth of**
a] 15 cm b] 20 cm
c] 30 cm d] 100 cm

770. **Rotary tillers are used for**
a] Shallow cultivation and weed control
b] Deep cultivation and weed control
c] Both a] & b]
d] None of these

771. **The lower penetration of disc plough is due to**
a] Blunt disc b] Plough too light
c] More tilt angle d] All are correct

772. **The uneven furrow formation in disc ploughing is due to**
a] Disc angle is not uniform
b] Loose bearing of disc
c] Hitching defects of disc plough
d] All are correct

773. **The excessive side draft in disc plough is due to**
a] Disc angle is not uniform
b] Loose bearing
c] Hitching defect
d] All are correct

774. **In disc harrow, the penetration of disc improves by**
a] Increasing disc angle
b] Lowering hitch point
c] Regulating optimum speed
d] All are correct

775. **The gang angle of disc harrow is adjusted in the range of**
a] 0-30° b] 30-60°
c] 60-90° d] Above 90°

776. **The width and depth of cut of disc harrow is controlled by**
a] Weight of the implement
b] Gang angle mechanism
c] Diameter of discs
d] All are correct

777. **The condition of soil that determines the optimum condition for tillage is**
a] Plastic b] Hard
c] Crumbly d] Cemented

778. **The type furrow opener used in zero tillage implement is**
a] Reversible type b] Inverted T type
c] Shoe type d] Hoe type

779. **In vertical disc plough, the numbers of disc varies from**
a] 2-12 b] 2-22
c] 2-32 d] 2-42

780. **The shape of furrow cut by indigenous plough is**
a] V-shape b] Rectangular shape
c] Trapezoidal shape d] None of these

781. **The shape of furrow cut by MB plough is**
a] V-shape b] L - shape
c] W-shape d] M-shape

782. **The indigenous plough is used for**
a] Ploughing
b] Puddling
c] Harvesing tuber crop
d] All are correct

783. **The draft per unit cross-section area of furrow is called**
a] Draft b] Unit draft
c] Side draft d] None of these

784. **The horizontal component of pull parallel to the direction of travel is**
a] Draft b] Unit draft
c] Side draft d] None of these

785. **The horizontal component of pull perpendicular to the direction of travel is**
a] Draft b] Unit draft
c] Side draft d] None of these

786. **An instrument used to measure soil compaction is**
a] Hydrometer b] Cone penetrometer
c] Tachometer d] Hygrometer

The small units having investment upto Rs. 25 lakh in plants and machinery are called tiny units.

787. **A tractor mounted on rotary tillers are used for seedbed preparation in**
   a] Seed bed preparation
   b] Puddling
   c] Both a] & b]
   d] None of these

788. **The L-shaped blade of rotary tillers are used for seed bed preparation in**
   a] Hard and sticky soil
   b] Stony and trashy soil
   c] Trahsy and sandy soil
   d] None of these

789. **The C-shape blade of rotary tillers are used for seedbed preparation in**
   a] Sticky soil b] Stony soil
   c] Trashy soil d] Sandy soil

790. **The true point of hitch of the implement is called**
   a] Line of pull
   b] Centre of pull
   c] Centre of resistance
   d] All are correct

791. **An implement that is used in till plant system is**
   a] Ridger b] Sweeps
   c] Furrower d] All are correct

792. **The purpose of ploughing is**
   a] To improve soil aeration
   b] To prevent soil erosion
   c] To add fertility to soil
   d] All are correct

793. **No till system reduces runoff by**
   a] 25% b] 50%
   c] 75% d] 90%

794. **The conservation of moisture is not possible when working with**
   a] Chisel plough b] Sub surface
   c] MB plough d] None of these

795. **The uneven fields are left when working with**
   a] Chisel plough b] Rotary plough
   c] MB plough d] None of these

796. **Area covered per day of 8 hours by 4 bottoms, 35 cm tractor plough if the speed of ploughing is 6 Km/h is**
   a] 626 m^3 b] 6.26 ha
   c] 62.6 m^3 d] 62.6 ha

797. **The purpose of harrowing is to**
   a] Pulverise soil b] Destroy weeds
   c] Break clods d] All are correct

798. **A vertical plate that is used to control depth and adjust line of pull in plough is**
   a] Vertical suction b] Vertical clevis
   c] Horizontal suction d] Horizontal clevis

799. **A device that is used to adjust the lateral movement of plough is**
   a] Vertical suction b] Vertical clevis
   c] Horizontal suction d] Horizontal clevis

800. **A draft required to pull 4-bottom 30 cm tractor plough working to a depth of 15 cm operated at speed of 6 Km/h. The soil resistance is 0.7 kg/ cm^2**
   a] 12.6 kg b] 126 kg
   c] 1260 kg d] None of these

801. **The group of number of discs that is mounted on a gang axle is**
   a] Gang b] Gang axle
   c] Arbor bolt d] Gang bolt

802. **The forces acting on MB plough are**
   a] Plastic b] Parasitic
   c] Elastic d] None of these

803. **A two-way plough eliminates**
   a] Back Furrow b] Dead furrow
   c] Both a] & b] d] None of these

804. **One-way plough requires laying out a field land, starting with**
   a] Back Furrow b] Dead furrow
   c] Both a] & b] d] None of these

805. **Two-way plough are advantageously working in**
   a] Levelled field b] Contour field
   c] Undulated field d] None of these

806. **A two-way plough left field nearly level for**
   a] Drainage and Irrigation b] Runoff and Erosion
   c] Both a] & b] d] None of these

807. **The two-way plough are designed to turn furrow slices to**
   a] Right and left b] Only right
   c] Only left d] All are correct

808. **Pain coulter are used in relatively**
   a] Trashy field b] Heavy trashy field
   c] Clean fieldd] All are correct

809. **Notched coulter work well in**
   a] Trashy field b] Heavy trashy field
   c] Clean fieldd] All are correct

810. **Share of mould board plough is made of**
   a] Cast iron b] Mild steel
   c] C-1095 solid steel d] All are correct

811. **The coulter of MB plough are generally mounted**
   a] Above share point b] Little ahead of share point
   c] Both a] & b] d] None of these

812. **One-way plough requires laying out a field in lands, ending with**
   a] Dead furrow b] Back Furrow
   c] Both a] & b] d] None of these

Diamonds are weighed in carrots.

813. A plough that works effectively in hard, dry and sticky soil is
a] Mould board plough b] Disc plough
c] Harrow plough d] Indigenous plough

814. The mould board plough absorb side forces mainly through the
a] Share b] Mould board
c] Land side d] Frog

815. The bearings used in most of disc harrow is about
a] Sealed ball bearing
b] Unsealed ball bearing
c] Sealed taper roller bearing
d] Unsealed taper roller bearing

816. The maximum operating depth for disc harrow is about
a] One-half of the disc diameter
b] One-third of the disc diameter
c] One-third of the disc diameter
d] One-fourth of disc diameter

817. Puddling is done mainly for
a] Reducing seepage of water
b] Reducing deep percolation of water
c] Smoothing seed bed
d] None of these

818. Puddling is done in standing water of
a] 2-5 cm b] 5 - 10 cm
c] 25-30 cm d] Above 40 cm

819. Land shaping becomes necessary where often the following implement is used
a] MB plough b] Disc harrow
c] Cultivators d] Rotary plough

820. Land plane is ideally suited for
a] Land levelling b] Earth moving
c] Final grading d] Top finishing

821. Two bullocks weighing 750 Kg each are pulling an implement with speed of 2 km/h. The hp that required pulling an implement is
a] 1.11 hp b] 11.1 hp
c] 6.66 hp d] 66.66 hp

822. A coulter attachment is used with
a] Disc harrow b] Seed drill
c] MB plough d] Subsoiler

823. Plough share is made up of
a] Mild steel b] Cast iron
c] Gun metal d] None of these

824. The size of MB plough is expressed in
a] Width of cut b] Depth of cut
c] Length of share d] None of these

825. The spike tooth harrows are used mainly for
a] Primary tillage b] Intercultivation
c] Surface finishing d] None of these

Did You Know ?

Largest duck

The White winged Wood duck (*Cairina scutulata*) measures up to 81 cm. It was once found in Bangladesh, Assam and Manipur but is now restricted to Arunachal Pradesh.

826. Puddling is done to achieve
a] Soft soil bed
b] Non-leaching of water
c] Size reduction of clods
d] Conservation of moisture

827. Dead furrow is made by
a] One-way MB plough b] Two-way MB plough
c] Rotary plough d] Disc harrow

828. A low mould board with gradual twist comparatively inverts the furrow slice with a minimum of break-up, thus covering vegetative thoroughly is called
a] Sod bottom b] Stubble bottom
c] Slated mould board d] General purpose bottom

829. Till plant equipment which are used for seed bed are
a] MB ploughs b] Sweeps
c] Cultivators d] Subsoilers

830. Surface planting with conservation feature of contour listing is
a] Flat planting b] Furrow planting
c] Ridge planting d] None of these

831. Ridge planting is most suitable for
a] Low rainfall areas b] Average rainfall areas
c] High rainfall areas d] None of these

832. The minimum tillage method is termed as
a] Furrow plant method
b] Plough plant method
c] Ridge plant method
d] None of these

833. The disc force into the ground by
a] Suction b] Slicing
c] Scouring d] Weight

834. The disc plough is advantageous over MB plough in respect of
a] Suitable tor deep ploughing
b] Complete covering of trashes and weeds
c] Leaves field levelled'
d] All are correct

835. The implement suitable for terraced hilly land and do not upset the slope of land is
a] One-way plough b] Two-way plough
c] Ridger d] None of these

GK **Brahmni river is called Vridh Ganga, Dakshin Koyal in Jharkhand.** GS

836. The plough that does not leave any back as well as dead furrow is
a] One-way plough b] Disc plough
c] Two-way plough d] Shabas plough

837. The plough bottom of wrest plough turn through
a] 90° b] 120°
c] 180° d] 360°

838. Mould board plough works on the principle of
a] Suction b] Scouring
c] Sliding d] None of these

839. The mould board used on tractor ploughs for general farm use is
a] Stubble MB
b] Breaker MB
c] General purpose MB
d] High speed MB

840. The coulters used tor cutting heavy trashes are
a] Plain coulter b] Rolling plain coulter
c] Notch coulter d] None of these

841. The draft requirement in MB plough compared to disc plough for same depth of ploughing is
a] Less b] More
c] Equal d] None of these

842. The plough used for deep ploughing is
a] MB plough b] Standard disc plough
c] Vertical disc plough d] None of these

843. The vertical disc plough is
a] Harrow plough b] Wheat plough
c] Both a] and b] d] None of above

844. The centre of plough load in a single bottom plough is
a] 7.5 cm to left of shin
b] At the shin
c] 7.5 cm to right of shin
d] None of these

845. The hitching of plough is done by placing the plough
a] Few centimetres below ground level
b] On the ground level
c] Few centimetres above ground level
d] None of these

846. The useful life of walking type ploughs may take about
a] 1000 hrs b] 1500 hrs
c] 2000 hrs d] 2500 hrs

847. The tillage practice is desirable in the areas of having hard or compacted layers below the plough depth is
a] Minimum tillage b] Conservation tillage
c] Sub soilingd] None of these

848. The versatile implement used for seed bed preparation, ridge-making and channel-shaping is
a] Indigenous plough b] MB plough
c] Hoe d] Cultivator

849. The harrow used for smoothening and levelling the soil after ploughing is
a] Spring tyne harrow b] Spike tooth harrow
c] Offset disc harrow d] Tandem disc harrow

850. The random scattering of seed on the surface of seedbed is
a] Dibbling b] Broadcasting
c] Hill dropping d] Check row planting

851. The seed rate required in broadcasting method in comparison to drilling is
a] More b] Less
c] Equal d] All are correct

852. Dibbler is a
a] Seed drill b] Planter
c] Transplanter d] None of these

853. The dibbling is mostly used for sowing
a] Cereal grains b] Vegetables
c] Plantation crop d] All are correct

854. The dropping of seeds in furrow lines in continuous flow is
a] Drilling b] Planting
c] Dibbling d] Hill dropping

855. The equipment used for dropping seeds in a continuous stream and the spacing between plant to plant in a row is not constant is
a] Seed drill b] Planter
b] Transplanter d] All are correct

856. The method of planting in which row to row as well as plant to plant distance is uniform is
a] Drilling b] Hill dropping
c] Check row planting (D} All are correct

857. The most positive seed metering mechanism used in seed drill is
a] Fluted roller b] Internal double run
c] Cup feed type d] All are correct

858. Of the following, the precision planter is
a] Seed drill b] Broadcaster
c] Dibbler d] Dofaan

859. A seed metering mechanism has double faced wheel, one face has a larger opening for larger seeds and other face has smaller opening for smaller seeds is
a] Cell feed b] Internal double run
c] Cup feed type d] All are correct

860. A metering mechanism commonly used in bullock drawn planter in country is
a] Cup feed mechanism
b] Cell fed type mechanism
c] Brush feed mechanism
d] None of these

861. The furrow opener which works well in trashy soils are
a] Inverted T type b] Shoe type
c] Hoe type d] All are correct

During Akbar's period, the lowest mansabdari for the nobles was : 10.

GS

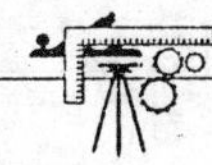

862. The hoe type furrow openers work weli in
a] Trashy soils b] Sticky soils
(G) Root infested soil d] All are correct

863. The part of seed drill that conveys seed or fertilizer from delivery tube to furrow is
a] Shoe b] Boot
c] Furrow opener d] Seed tube

864. The procedure of testing seed drill for correct seed rate is
a] Metering b] Calibration
c] Calculation d] None of these

865. The device used to sow seeds, which are larger in size and cannot be sown by usual seed drill is
a] Planter b] Transplanted
c] Seed drill d] All are correct

866. In hill drop metering mechanism, the cells round the edge of the plate admits
a] One seed at a time
b] Two seeds at a time
c] Several seed at a time
d] None of these

867. In a seed metering mechanism, only one seed is allowed at a time is
a] Flat drop b] Hill drop
c] Edge drop d] None of these

868. The seeding behind plough is done by a device known as
a] Seed drill b] Malobansa
c] Planter d] Dibbler

869. The function of seed drill is
a] To carry seeds
b] To meter seeds
c] To deposit seeds in furrow
d] All are correct

870. A seed drill of size 10 x 20 cm has seed drop in one revolution is 40 gram. The area covered by seed drill in one revolution is 9.42 m^2. The seed rate (kg/h) is
a] 2.2 Kg b] 22 Kg
c] 1.36 Kg d] 136 Kg

871. The most common method in irrigated areas and high rainfall areas to improve surface drainage is
a] Flat planting b] Ridge planting
c] Furrow planting d] All are correct

872. The common method of planting in semi-arid areas for row crop is
a] Flat planting b] Ridge planting
c] Furrow planting d] All are correct

873. The hopper design to plant accurately meter of all varieties of maize and all type of seeds from tomato to beans is
a] Duplex hopper b] Richmond hopper
c] Both a] and b] d] None of these

874. The furrow opener used for planting cotton seeds in areas where the uncertainty of favourable weather for germination is
a] Variable depth furrow opener
b] Constant depth furrow opener
c] Disc type furrow opener
d] Stub runner type furrow opener

875. The constant depth furrow opener is
a] Stub runner b] Disc type
c] Hoe type d] All are correct

876. The press wheel used for compaction of soil around the seeds in low moisture conditions are
a] Opelt center wheel
b] Solid steel narrow wheel
c] Hallow steel narrow wheel
d] None of these

877. The press wheel used for compaction of soil around the seeds in high moisture conditions are
a] Open centre wheel
b] Solid steel narrow wheel
c] Hallow steel narrow wheel
d] None of these

878. Seed drill are used for sowing
a] Small seeds b] Bolder seeds
c] Seedlings d] Plants

879. Planters are used for sowing
a] Small seeds b] Bolder seeds
c] Seedlings d] Plants

880. A nursery required for transplanting paddy seedling in one hectare of land is
a] Half of area to be sown
b] 1/10th of area to be sown
c] 1/3rd of the area to be sown
d] 1/5th of the area to be sown

881. The size of seed drill expressed by
a] Length and wIdth of machine
b] Area covered per unit time
c] Number of furrow opener x spacing between two furrow openers
d] All are correct

882. Planter is different from seed drill in respect to
a] Power transmission b] Metering mechanism
c] Furrow opener d] None of these

883. The metering mechanism suitable for metering small and large seeds is
a] Fluted roller b] Cup feed type
c] Cell feed type d] Internal double run type

884. In cell feed type-metering mechanism, cell fill is affected by
a] Speed of cell b] Size of cell
c] Size of seeds d] All are correct

885. A double disc type furrow opener is suitable for
a] Stony soil b] Sticky soil
c] Trashy soil d] All are correct

886. The seed tubes are made of
a] Rubber b] Plastic
c] Mild steel d] All are correct

887. The diameter of seed tube for different seeds should be
a] 10 mm b] 20 mm
c] 25 mm d] 30 mm

888. Seed drills are suitable for maintaining distance between
a] Seeds b] Rows
c] Both seeds and rows d] None of these

889. The planters are suitable for maintaining distance between
a] Seeds/plant, b] Rows
c] Both a] and b] d] None of these

890. The most common power transmission system used in seed drill is
a] Belt and pulley b] Chain and sprocket
c] Gears d] All are correct

891. A metering mechanism that does not required cut-off device is
a] Inclined plate b] Vertical plate
c] Horizontal plate d] All are correct

892. 229, The metering mechanism used in potato planter is
a] Fluted roller b] Brush feed type
c] Picker wheel type d] Cell feed type

893. In cup feed metering device, the seed rate is controlled by
a] Shaft rotation b] Speed of machine
c] Size of cups d] All are correct

894. The furrow opener used in black cotton soil is
a] Disc type
b] Shoe type
c] Reversible shovel type
d] Hoe type

895. The shovel of seed drill is made up of
a] Cast iron b] Mild steel
c] White metal d] Carbon steel

896. Of the following, blower is part of
a] Planter
b] Zero till drill
c] Pneumatic seed drill
d] Till plant machine

897. Pneumatic seed drills are suitable for sowing
a] Small seeds b] Bolder seeds
c] Both a] and b] d] None of these

898. The fluted roller of seed drill is made of
a] Aluminum b] Cast iron
c] Plastic d] All are correct

899. For sowing of wheat seed, a suitable metering mechanism is
a] Cup feed b] Cell feed
c] Fluted roller d] Brush feed

900. A zero till seed-cum-fertilizer drill is designed for sowing
a] Paddy b] Wheat
c] Potato d] Vegetable

901. A potato planter is
a] Automatic planter
b] Semi-automatic planter
c] Self-propelled planter
d] All are correct

902. For sowing mustard seeds, the most suitable seed drill is
a] Plain seed drill b] Pneumatic seed drill
c] Zero till drill d] All are correct

903. The metering mechanism used for mustard seed is
a] Fluted roller b] Vertical plate type
c] Cell fill type d] Stud roller type

904. The furrow planting of crops are widely used in
a] Heavy rainfall conditions
b] Arid conditions
c] Semi-arid conditions
d] None of these

905. In seed drill, the depth of seed placement for better germination should be
a] 3 cm b] 5 cm
c] 7 cm d] 10 cm

906. Edge cell and edge drop plates of metering mechanism are well suited for planting
a] Large seeds and flat seed
b] Large and round seed
c] Small seeds and flat seeds
d] Small and round seed

907. Vertical rotor metering device is used for
a] Bed planting b] Precision planting
c] Furrow planting d] Ridge planting

908. The blower of pneumatic seed drill is operated by
a] Tractor PTO b] Tractor hydraulic
c] Chain and sprocket d] Gear operated

909. In pneumatic seed drill, seeds are released in the furrow under
a] Gravity b] Pressure
c] Both a] and b] d] None of these

910. The commonly used power transmission system in fertilizer seed drills are
a] Chain sprocket and gear
b] Gears and belt pulley
c] Belt pulleys and cam
d] Belt and chain sprocket

GK GS

At Rome Olympic Milkha Singh broke the world record in the 400 metres event.

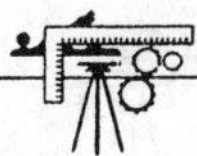

911. The dibbling method of sowing reduces seed rate by
a] 1/2 nd b] 1/3 rd
c] 1/5 th d] 2/3 rd

912. Equipment for placing seedlings in the soil is called
a] Seed drill b] Seed-cum-fertilizer drill
c] Planter d] Transplanter

913. ASPEE is famous for manufacturing
a] Tractor b] Power tillers
c] Ploughs d] Sprayers

914. Hand operated sprayers are suitable for
a] Small holdings b] Medium holdings
c] Large holdings d] Very large holdings

915. Pressure of hand operated sprayers varies from
a] 1 -5 Kg/cm^2 b] 1 -6 kg/cm^2
c] 1-7 kg/cm^2 d] 1-8 kg/cm^2

916. Power operated sprayers are suitable for
a] Small holdings b] Large holdings
c] Medium holdings d] None of these

917. Pressure of power operated sprayers ranges from
a] 10-25kg/cm^2 b] 10-35 kg/cm^2
c] 10-45 kg/ cm^2 d] 10-55 kg/ cm^2

918. Airplane sprayers and dusters are suitable for
a] Small scale work b] Medium scale work
c] Large scale work d] All are correct

919. The volume of liquid required for ground spraying of field crop is
a] 100 litres/ha b] 200 litres/ha
c] 300 litres/ ha d] 400 litres/ ha

920. The volume of liquid required for orchard spraying is about
a] 1000 litres/ ha b] 2000 litres/ ha
c] 4000 litres/ ha d] 5000 litres/ ha

921. Bucket type sprayer consist of
a] Single and double acting pump
b] Centrifugal pump
c] Plunger type pump
d] All are correct

922. The pump is mostly made of
a] Aluminium b] Copper
c] Brass d] Plastic

923. A tank capacity of knapsack sprayer is about
a] 5-10 litresb] 8-16 litres
c] 9-22.5 litres d] 10-25.5 litres

924. Area that one man can spray in a day is
a] 0.1 ha b] 0.2ha
c] 0.3 ha d] 9.4 ha

925. Amount of liquid that a man can spray in a day is
a] 60 litres liquid b] 70 litres liquid
c] 80 litres liquid d] 90 litres liqutd

You Don't Know!
First at Antarctica
Lt. Ram Charan of the Indian Navy reached Antarctica in 1960.

926. Tank capacity of compression sprayer is
a] 10 litres b] 12 litres
c] 14 litres d] 20 litres

927. Hand atomizer is used for spraying in
a] Nursery b] Orchard
c] Field crop d] None of these

928. The pump is used in power-operated sprayer is
a] Plunger type b] Diaphragm type
c] Geartype d] Piston type

929. The pump used in airplane sprayers
a] Centrifugal and gear pump
b] Gear and Diaphragm pump
c] Diaphragm and centrifugal pump
d] Plunger and Centrifugal pump

930. In power-operated sprayer, the pump works at a pressure of
a] 2-5 kg/cm^2 b] 3-8.5 kg/cm^2
c] 4-12 kg/cm^2 d] 5-15 kg/cm^2

931. In airplane sprayer, the tank capacity ranges between
a] 450 -1000 litres b] 450-1500 litres
c] 450-1800 litres d] 450-2200 litres

932. A plunger type hand duster is suitable for dusting
a] Small area b] Medium area
c] Large area d] All are correct

933. A tank capacity of rotary type hand duster is about
a] 2.6-4.5 kgb] 3.6-4.5 kg
c] 4.6-6.5 kgd] 5.0-8.5 kg

934. A 3.6 to 4.5 kg dust is enough for dusting of
a] 0.2-0.4 hab] 0.4-0.6 ha
c] 0.5-0.7 had] 0.6-0.8 ha

935. The height of spray for rotary type hand sprayer for effective spraying is
a] 1 metre b] 2 metres
c] 3 metres d] 4 metres

936. In rotary type power duster, the speed of blower is
a] 1000 rpm b] 2000 rpm
c] 3000 rpm d] 4000 rpm

937. In rotary type power duster, the dust can be blown at a height of
a] 3 metres b] 4 metres
c] 5metres d] 6 metres

938. The tank capacity of airplane duster varies from
a] 100-500 kg b] 200-800 kg
c] 220-900 kg d] 220-120 kg

GK **Cloth industry is the largest industry in country.** GS

939. **The hand operated compressed air sprayer utilizing the pressure from**
a] 1.4-4.0 kg/cm^2 b] 1.75-5.0 kg/cm^2
c] 2.0 -6.0 kg/cm^2 d] 2.5 -7.5 kg/ cm^2

940. **277 Per hectare application rate of low volume sprayer is**
a] 2 litres b] 3 liters
c] 4 litres d] 5 litres

941. **Per hectare application rate of low volume sprayer is**
a] 5-100 litres b] 5 - 2001itres
c] 5-300 litres d] 5 - 400 litres

942. **Per hectare application rate of high volume sprayer is**
a] Less than 400 litres b] 400 litres
c] More than 400 litres d] None of these

943. **Per hectare application rate of smogging is**
a] 1 liter b] Less than liter
c] 2 liter d] Less than 2 liter

944. **A nozzle used in foam spraying is**
a] Flat fan nozzle b] Solid cone nozzle
c] Flooding nozzle d] Air aspirating nozzle

945. **Wet dusting is beneficial for**
a] Drought area
b] Semi-arid area
c] Low rainfall areas
d] Heavy rainfall areas

946. **Insecticide capsules are used at**
a] 2.5 cm below the soil surface
b] About 2.5 cm from each hill
c] From 3 to 5 days after transplanting
d] All are correct

947. **ONK-B sprayer duster is used as**
a] Spraying of liquid
b] Dusting of dry pesticides
c] Dusting of wet pesticides
d] All are correct

948. **The main function of sprayer is**
a] To break the liquid into droplet of effective size
b] To distribute the droplet in uniformly over the plants
c] To regulate the amount of liquid to avoid excessive application
d] All are correct

949. **Swirl plate is a part of**
a] Flat fan nozzle
b] Cone nozzle
c] Spinning disc nozzle
d] None of these

950. **Spray gun is known as**
a] Nozzle b] Lance
c] Nozzle disc d] Trigger

951. **The motor of ultra low volume sprayer is powered by battery of**
a] 3-6 volt b] 6-12
c] 12-18 volt d] 12-24 volt

952. **Spinning disc is used in**
a] Low volume spray
b] High volume spray
c] Ultra low volume spray
d] Foam spraying

953. **Speed of spinning disc of sprayer varies from**
a] 2000-4000 rpm b] 2000-6000 rpm
c] 2000-8000 rpm d] 2000-9000 rpm

954. **Droplet size of ultra low volume sprayer varies from**
a] 25-50 micron b] 35- 70 micron
c] 35-100 micron d] 40-120 micron

955. **Ultra low volume sprayer is used for application of**
a] Insecticides b] Pesticides
c] Weedicides d] All are correct

956. **Rocker sprayer is used for spraying**
a] Small crops
b] Tall crops
c] Godown and factories
d] All are correct

957. **Another name of positive displacement pump is**
a] Piston pump b] Plunger pump
c] Centrifugal pump d] All are correct

958. **Volumetric efficiency of piston pump is about**
a] 60 per cent b] 70 per cent
c] 80 per cent d] 90 per cent

959. **Rotary pump is used as**
a] Low pressure sprayer
b] High pressure spray
c] Ultra low volume sprayer
d] All are correct

960. **Discharge rate of centrifugal pump varies at**
a] 14 th of speed b] Square of speed
c] Cube of speed d] None of these

961. **Power of centrifugal pump varies as**
a] Square of speed b] Cube of speed
c] Half of speed d] None of these

962. **A machine used to generate and distribute gases or smoke is known as**
a] Sprayer b] Duster
c] Fumigatord] None of these

963. **Flame gun is used for killing the**
a] Insects b] Pests
c] Weeds d] All are correct

964. **In hydraulic sprayers, the degree of atomization is primarily a function of**
a] Air velocity
b] Size of nozzle
c] Size and shape of atomizer
d] Liquid pressure and the nozzle characteristics

GK GS
The charing of sugar on heating is due to escaping of water molecules.

965. Hollow cone nozzle employed on boom type sprayer have spray angle of
a] 30-50 degree b] 40-60 degree
c] 60-95 degree d] 90-120 degree

966. Nozzle spacing on boom is governed by
a] Height of crop b] Density of crop
c] Type of spray d] All are correct

967. Nozzle spacing in boom is governed by
a] Inclined b] Axial horizontal
c] Axial vertical d] All are correct

968. Flooding nozzles have spray angle of
a] 30-45 degree b] 60-90 degree
c] 90-120 degree d] 100-150 degree

969. In electrostatic spraying, the volume of liquid per hectare is
a] Less than 1 litre b] Less than 51itres
c] Less than 50 litres d] Less than 100 litres

970. The amount of liquid required for spraying plantation crop is
a] 400- 60 l/ha b] 600- 800 l/ha
c] 800-1000 l/ha d] 1000-1200 l/ha

971. The application rate of dusters is
a] Less than 10 kg/ha b] 10- 15 kg/ha
c] 20- 25 kg/ha d] 30- 40 kg/ha

972. The immersion method is used for identifying
a] Nozzle discharge b] Droplet size
c] Spraying pressure d] Spray pattern

973. More spraying pressure results in
a] Reduction in droplet size
b] Increase in droplet size
c] No change in droplet size
d] None of these

974. The mechanical efficiency of spray pumps ranges from
a] 50- 60% b] 50- 70%
c] 50- 80 % d] 50- 90 %

975. Air blast sprayer is used for
a] Gardens b] Field crops
c] Orchards d] Nursery

976. Air blast sprayer was developed in
a] 1920 b] 1940
c] 1960 d] 1980

977. The maximum application rate of air blast sprayer for orchards may be
a] 700- 1700 l/ha b] 1700- 2700 l/ha
c] 2700- 3100 l/ha d] 3700- 4700 l/ha

978. The minimum application rate with air blast sprayer for orchards may be
a] 10- 20-1/ha b] 25- 50 l/ha
c] 45- 90 l/ha d] 150-2501/ha

979. The pump used in most of the hydraulic sprayers is
a] Plunger pump
b] Positive displacement pump
c] Reciprocating pump
d] Vane pump

980. Tank of hydraulic pump is made of
a] Fibre glass b] PVC
c] Plastic d] Metal

981. The forward operating speed of power sprayer varies from
a] 2- 5 kmph b] 4.8- 8 kmph
c] 6.8- 12 kmph d] 10- 25 kmph

982. Knapsack sprayer is suitable for
a] Brushes b] Small trees
c] Row crop d] All are correct

983. The nozzle used for band treatments of herbicide along row crop is
a] Hallow cone b] Solid cone
c] Flat fan d] Jet stream

984. The solid cone nozzle is used for
a] Weedicides application
b] Herbicide application
c] Fungicide application
d] Insecticide application

985. The nozzle used for herbicide and fungicide application is
a] Hallow cone b] Solid cone
c] Flat fan d] Jet stream

986. The term harvesting is associated with
a] Cutting of crop b] Picking of pods
c] Plucking & digging d] All are correct

987. A device used for manual harvesting of crops
a] Sickle b] Reaper
c] Mower d] Combine

988. An action involving manual harvesting of crop is
a] Slicing and tearing action
b] Tearing and beating action
c] Scissors and Slicing action
d] Scissors and tearing action

989. A machine used to cut herbage crops and leave them in swath is
a] Reaper b] Mower
c] Combine d]. Baler

990. A machine used for cutting of cereal crop
a] Reaper b] Combine
c] Mower d] Windrower

991. A machine that cuts the crops and ties them into a knot and uniform sheaves is called
a] Combine b] Reaper
c] Reaper binder d] All are correct

GS

Krishna river has the tributaries like, Koyna, Bhima etc.

992. **The material that is left by the harvesting machine in row is**
a] Swath b] Stubble
c] Straw d] All are correct

993. **A machine used to cut crops and deliver them in a uniform manner in a row is**
a] Combine b] Mower
c] Reaper binder d] Windrower

994. **A mower with rotating helical blades used for cutting is**
a] Cylinder mower b] Reciprocating mower
c] Horizontal mower d] Flail mower

995. **Helical blades of mower are arranged in**
a] Horizontal cylindrical form
b] Vertical cylindrical form
c] Axial cylindrical form.
d] None of these

996. **A mower with high speed knife rotating in horizontal plane is known as**
a] Reciprocating mower
b] Gang mower
c] Horizontal rotary mower
d] Flail mower

997. **A mower in which swinging knives are used for cutting is**
a] Flail mower b] Reciprocating mower
c] Cylinder mower d] All are correct

998. **In flail mower, knives operating is in**
a] Horizontal plane and in cylinder
b] Vertical plane and in cylinder
c] Horizontal plane and around cylinder
d] Vertical plane and around the horizontal

999. **In bullock-drawn mower, the power transmitted to the axle is through**
a] Transport wheel b] Starwheel
c] Gauge wheel d] All are correct

1000. **A tractor drawn in semi-mounted mower is operated by**
a] Belt pulley drive
b] Tractor hydraulic
c] Tractor power take off
d] Tractor hitch

1001. **Cutter bar of reaper is made of**
a] Mild steel b] Cast iron
c] High grade steel
d] Medium carbon steel

1002. **The knife section of a reaper makes a**
a] Rotating motion
b] Reciprocating motion
c] Both a] and b]
d] None of these

1003. **In proper registered reaper, after each stroke a knife shall stop**
a] At the centre of guard
b] Ahead of guard
c] Behind the guard
d] None of these

1004. **In power tiller mounted reaper, the rotating power of power tiller in prime mover is converted into**
a] Rotary power of cutter bar
b] Reciprocating power of cutter bar
c] Vibrating power of cutter bar
d] None of these

1005. **The knife moves on the**
a] Guard b] Wearing plate
c] Ledger plate d] All are correct

1006. **In reaper, knife clips are placed with**
a] Ledger plate b] Wearing plate
c] Guard d] Cutter bar

1007. **In the reaper, a space between knife clips and wearing plate is about**
a] 5-10 cm b] 10- 20 cm
c] 20- 30 cm d] 30- 40 cm

1008. **In the reaper, shoe is provided with**
a] One end of cutter bar
b] Each end of cutter bar
c] Middle of cutter bar
d] Each knife of cutter bar

1009. **The work of shoe in the reaper is**
a] To regulate the blade
b] To regulate the height of cut
c] To regulate the width of cut
d] To regulate the engine speed

1010. **The wearing plate of cutter bar of reaper is made of**
a] Soft centred steel b] Mild steel
c] Harden steel d] Cast iron

1011. **In reaper, a flat plate with two cutting edges is known as**
a] Cutter bar b] Knife section
c] Knife d] All are correct

1012. **In reaper, a knife head of cutter bar is connected to**
a] Cutter bar b] Guard
c] Pitman d] Knife section

1013. **Pitman that is used in reaper is a type of**
(A) Crankshaft b] Connecting rod
c] Camshaft d] Flywheel

1014. **The pitman of reaper is made of**
a] Metal b] Wood
c] Plastic d] None of these

1015. **A wooden pitman is generally used in**
a] Reaper b] Mower
c] Combine d] All are correct

Akbar introduced the system of Zat and Sawar in mansabdari system.

GS

1016. Breaking of knives of cutter bar in reaper is due to
a] Play in bearing b] Improper alignment
c] Both a] and b] d] None of these

1017. When knife and pitman run in a straight line, then cutter bar is in
a] Proper registration b] Improper registration
c] Proper alignment d] Improper alignment

1018. For better alignment, a cutter bar should be set at
a] 80° b] 82°
c] 86° d] 88°

1019. When the knife section of cutter bar stops in the center of its guard after every stroke is
a] Alignment b] Registration
c] Reel index d] Stroke length

1020. In reaper, lead generally recommended for alignment purpose is of
a] I cm per metre length
b] 1.25 cm per metre length
c] 2 cm per metre length
d] 2.5 cm per metre length

1021. An improper registration of reaper results in
a] Uneven harvesting b] Excessive clogging of crop
c] Unbalance load d] All are correct

1022. An animal drawn reaper is designed to harvest the crop up to a height of
a] 3-5 cm b] 5-8 cm
c] 8-10 cm d] 10-15 cm

1023. The knife of cutter bar of animal-drawn reaper is made of
a] Mild steel b] Cast iron
c] High carbon steel d] Low carbon steel

1024. The vertical conveyer reaper is mostly used for harvesting
a] Cereal crop b] Pulse crop
c] Vegetables d] Oilseed crop

1025. In vertical conveyer reaper, the crop to be harvested are guided by
a] Star wheel b] Reel
c] Row divider d] Cutter bar

1026. In vertical conveyer reaper, crop is held in vertical position with the help of
a] Star wheel b] Springs
c] Row divider d] Reel

1027. The capacity of tractor-operated vertical conveyer reaper is about
a] 0.2- 0.4 ha/h b] 0.4-0.6 ha/h
c] 0.6-0.8 ha/h d] 0.8-1.0 ha/h

1028. The length of cutter bar of power tiller operated reaper is about
a] 100-120 cm b] 100-140 cm
c] 100-160 cm d] 100-180 cm

Knowledge World!
Highest gateway
The 53.5 m (175 ft) high Buland Durwaza at Fatehpur Sikri is the highest gateway in India. It was built by Akbar to commemorate his conquest of the Ranthambhor fort and also his conquest of Khandesh in Gujarat.

1029. The capacity of power tiller operated vertical conveyer reaper is
a] 0.15 ha/h b] 0.25-0.35 ha/h
c] 0.35-0.45 ha/h d] 0.45-0.60 ha/h

1030. The cutting height of reaper binder from ground is about
a] 5 cm b] 10 cm
c] 15 cm d] 20 cm

1031. In manual harvesting, a man-hour needed to harvest one hectare of paddy is about
a] 120-150 b] 150-170
c] 170-200 d] 200-225

1032. The harvesting of wheat, barley and grams are mostly done under
a] Immature condition
b] Mature condition
c] Over mature condition
d] Dry condition

1033. The action involved in harvesting of crops is
a] Scissors action b] Tearing action
c] Slicing action d] All are correct

1034. For impact cutting, a single element sharp edge requires a velocity of
a] 5 m/sec b] 10 m/sec
c] 15m/sec d] 20m/sec

1035. For impact cutting, dual edge single element blade requires a velocity of
a] 10 m/sec b] 20 m/sec
c] 30 m/sec d] 40 m/sec

1036. For harvesting of agricultural crops most widely used element action is
a] Single element scissors action
b] Two element scissors action
c] Single element impact action
d] Two element impact action

1037. The inclined angle between cutting edge of knives of cutter bar is
a] 18° b] 28°
c] 38° d] 48°

1038. Manually operated push type harvester is commonly called
a] Reaper b] Hand dropper
c] Sickle d] None of these

GK **Article 73(3) talks about collective responsibility.** GS

1039. The hand dropper was developed in
a] India b] America
c] Japan d] Israel

1040. Hand dropper consists of
a] A serrated blade b] A pair of serrated blade
c] A plane blade d] A pair of plane blade

1041. In hand dropper the serrated blade are mounted at an angle of
a] 10° to each other b] 20° to each other
c] 30° to each other d] 40° to each other

1042. The capacity of hand dropper is about
a] 0. 1 ha/day b] 0.2 ha/day
c] 0.3 ha/day d] 0.4 ha/day

1043. An engine used in rotary power scythe is
a] High speed diesel b] High speed petrol
c] Gasoline engine d] Kerosene engine

1044. The blade used for harvesting in rotary power scythe is
a] A plane blade b] Aserrated disc
c] Both a] and b] d] None of these

1045. The output capacity of rotary power scythe is about
a] 0.1-0.2 ha/day b] 0.2-0.4 ha/day
c] 0.4-0.6 ha/day d] 0.4 -0.8 ha/day

1046. The output capacity of rotary power scythe depends on
a] Field condition
b] Physique of operator
c] Both a] and b]
d] None of these

1047. A rotary power scythe is operated by
a] Power tiller b] A pair of bullock
c] Tractor d] Human

1048. The cutter bar of conventional combine is operated at
a] 350-800 strokes/min
b] 500-800 strokes/min
c] 800-900 strokes/min
d] 800-1200 strokes/min

1049. The conventional animal drawn mower has a stroke length of
a] 5 cm b] 7.5 cm
c] 10 cm d] 12 cm

1050. The star wheel of reaper lifts up the crop and turns at
a] 45° b] 60°
c] 90° d] 180°

1051. The length of cutter bar of tractor front mounted reaper is
a] 1.8-2.0 m b] 1.9-2.1 m
c] 2-2.5 m d] 2.25-2.5 m

1052. Tractor front mounted reaper is operated at a forward speed of
a] 1.5 km/h b] 2.0 km/h
c] 2.5 km/h d] 3.0 km/h

1053. The width of cut of animal drawn reaper is
a] 500 mm b] 600 mm
c] 800 mm d] 900 mm

1054. The mowers are designed to cut
a] Wheat b] Paddy
c] Mustard d] Grasses

1055. The cutter bar used in cutting unit of combine is
a] Mower type b] Reaper type
c] Both a] and b] d] None of these

1056. The cutting platform of combine operated at height above ground is
a] 6.5 -8 cm b] 7:5- 9 cm
c] 8.5-11 cm d] 9.5-12 cm

1057. The suitable clearance between reel and cutter bar for all purposes is
a] 10.5-15 cm b] 12.5 -25 cm
c] 12.5-30 cm d] 15-27.5

1058. The size of combine is determined by
a] Overall dimensions b] Width of harvest
c] Weight of machine d] None of these

1059. The cutting width of small size propelled combine ranges from
a] 1.5-2.1 m b] 1.5-2.25 m
c] 1.5-2.5 m d] 1.5-3.0 m

1060. Tractor mounted groundnut digger shaker is driven by
a] Tractor hydraulic, b] Belt pulley
c] Tractor PTO d] None of these

1061. The average output capacity of tractor mounted groundnut digger shaker is
a] 0.15 ha/h b] 0.20 ha/h
c] 0.25 ha/h d] 0.27 ha/h

1062. The horsepower required for tractor-operated potato digger is
a] 30 hp b] 35 hp
c] 45 hp d] 50 hp

1063. Of the following, the impact cutting principle is applied in
a] Rotary cutter b] Flail shredder
c] Both a] and b] d] None of these

1064. A rotary cutter has knives to rotate in
a] Vertical plane b] Horizontal plane
c] Both a] and b] d] None of these

1065. The knives of shredders are rotated in
a] Horizontal plane with direction of travel
b] Vertical plane with direction of travel
c] Both a] and b]
d] None of these

GK GS

Admiral Karl Doenitz succeeded Hitler as the leader of Germany in April 1945.

1066. A cutter bar of mower is able to cut grasses from ground satisfactory upto a height of
a] 50 mm b] 75 mm
c] 100 mm d] All are correct

1067. A conventional pitman drive mower has crank speed of
a] 750-900 rpm b] 850-1100 rpm
c] 850-120 rpm d] 900-1200 rpm

1068. A dynamically balanced mower has crank speed of
a] 900-1200 rpm b] 1000-1200 rpm
c] 1000-1250 rpm d] 1000-1300 rpm

1069. The knives used in dynamically balanced mower are
a] Single b Double
c] Many d] All are correct

1070. A gang mower is a group of
a] Two cylindrical mower
b] Three cylindrical mower
c] Four cylindrical mower
d] None of these

1071. In flail type mower, cutting section has
a] Fixed knives
b] Swinging knives
c] Reciprocating knives
d] Rotating knives

1072. At proper alignment of mower the cutter bar is set at
a] 20° inward b] 5° inward
c] 10° inward d] 15° inward

1073. The main function of potato digger elevator is
a] Digging of potatoes
b] Windrowing of potatoes
c] Both a] and b
d] None of these

1074. The operating speed of combine in field varies from
a] 2-6.4 kmph b] 2-8 kmph
c] 4-10 kmphd] 4-12 kmph

1075. The combine harvester is a machine designed for
a] Harvesting crop
b] Threshing of harvested crop
c] Separation and cleaning of harvested crop
d] All are correct

1076. The size of cutter bar of self-propelled combine varies from
a] 1-2 m b] 1.5-2.5 m
c] 2-4 m d] None of these

1077. Self-propelled combine harvester is provided with
a] Powering engine b] Petrol engine
c] Diesel engine d] Kerosene engine

1078. The operating speed of combine harvester varies within a range of
a] 600-800 rpm b] 900-1000 rpm
c] 600-1400 rpm d] Above 1500 rpm

1079. The size of cutter bar of tractor PTO operated combine is
a] 1-2m b] 1.5-2.5 m
c] 2-4 m d] 4-6 m

1080. The blade assembly of combine is operated at a speed of
a] 1.2-1.5 m/sec b] 1.2-1.5 m/min
c] 1.5-3 m/sec d] 1.5-3 m/min

1081. A reel of combine helps in
a] Preventing shattering of the crop
b] Allow crop to move uniformly to platform
c] Handle matted crop
d] All are correct

1082. The reel most commonly used in combine now a days is
a] Slat type b] Pick up type
c] Wooden d] None of these

1083. An assembly, which diverts the straw coming from cylinder towards straw walker is
a] Front beater b] Rear beater
c] Straw walker d] None of these

1084. The rack loss should remain between
a] 0.2-0.4 % b] 0.6-1.0%
c] Above 1.0 d] None of these

1085. The grain that is carried over the rear of the sieve with chaff and blown out of the combine is
a] Cylinder loss b] Rack loss
c] Sieve loss d] None of these

1086. The optimum value of reel index for minimum cutter bar loss is
a] 1 to 2 b] 1.25 to 1.5
c] 2 to 3 d] 2.25 to 3.5

1087. A thresher causes more seed damage if
a] Speed is increased b] Clearance increased
c] Speed is decreased d] Feed rate is reduced

1088. A type of cylinders used on power wheat threshers are
a] Serrated disc and rasp bar
b] Spike tooth and wire loop
c] Spike tooth and hammer mill
d] Hammer mill and syndicate

1089. In thresher, the percentage of healthy grain dropped from sieve with respect to healthy grain received from main outlet is
a] Cylinder loss b] Blower loss
c] Sieve loss d] Visible loss

1090. The threshing of crop by stripping is done with the help of
a] Hoves of animal b] Peg/hoop of threshing
c] Both a] and b] d] None of these

1091. An action involve in threshing of crops are
a] Rubbing action b] Impact action
c] Stripping action d] All are correct

GK GS

In December 1999 Foreign Exchange Regulation Act (FERA) had been replaced by Foreign Exchange Management Act (FEMA).

1092. The hold type thresher originated in
a] India b] Japan
c] America d] Canada

1093. The output capacity of hand beating method is about
a] 5-10 kg/h b] 10-15 kg/h
c] 17-20 kg/h d] 20-25 kg/h

1094. The output capacity of treading with bullock is
a] 100 kg/h b] 120 kg/h
c] 140 kg/h d] 160 kg/h

1095. The stripping method of threshing is used for
a] Wheat b] Paddy
c] Gram d] Corn

1096. The Olpad thresher is mostly used for threshing
a] Paddy b] Wheat
c] Oil seeds d] All are correct

1097. The disc used in Olpad thresher for threshing purpose is
a] Plane b] Notched
c] Serrated d] All are correct

1098. An Olpad thresher is
a] Bullock drawn
b] Tractor driven
c] Power tiller-operated
d] Electric motor driven

1099. The size (diameter) of disc used in Olpad thresher for threshing is
a] 30 cm b] 35 cm
c] 40 cm d] 45 cm

1100. The discs mounted on olpad thresher for threshing are
a] 10 b] 15
c] 20 d] 25

1101. The spacing between two adjacent discs in Olpad thresher is
a] 5 cm b] 10 cm
c] 15 cm d] 20 cm

1102. The cylinder speed of rotary type thresher is
a] 200 rpm b] 300 rpm
c] 400 rpm d] 500 rpm

1103. The Japanese type rotary thresher used for threshing
a] Paddy b] Wheat
c] Sunflower d] Safflower

1104. The drum diameter of rotary type thresher is
a] 23 cm b] 33 cm
c] 43 cm d] 55 cm

1105. The capacity of manual operated rotary paddy thresher is
a] 1 to 2 q/day b] 3 q/day
c] 4 q/day d] 5 q/day

1106. The output capacity of tractor operated peg type thresher is
a] 3.5 q/h b] 4.5 /h
c] 5.2 q/h d] 6.5 q/h

1107. The size of power thresher is determined by
a] Width of cylinder and straw rack
b] Diameter of cylinder and straw rack
c] Rpm of cylinder and straw rack
d] All are correct

1108. A man can thresh average grain in a day is
a] 11 kg b] 14 kg
c] 19 kg d] 22 kg

1109. In power thresher, a semi-circular disc is inserted between cylinder and straw thrower to achieve
a] Higher threshing efficiency
b] Higher winnowing efficiency
c] Minimize breakage loss
d] Minimize blown losses

1110. The average threshing of capacity bullock-drawn Olpad thresher per day is
a] 150 kg b] 353 kg
c] 562 kg d] 600 kg

1111. Threshing capacity of a large stationary thresher is
a] 2.5 tonnes b] 3.55 tonnes
c] 4.17 tonnes d] 5.0 tonnes

1112. Main part of thresher that removes seeds from grain head is
a] Cylinder b] Concave
c] Pegs on cylinder d] Sieve

1113. Threshing in rotating drum cylinder is due to
a] Stripping b] Tearing
c] Impact d] Rubbing

1114. A hammer mill type thresher has
a] Hammers
b] Beaters
c] Close cylinder casing
d] All are correct

1115. Hammer mill thresher is
a] Bullock-drawn
b] Power tiller-operated
c] Prime mover-operated
d] Self-propelled

1116. The drum type thresher is
a] Rasp bar type b] Hammer mill type
c] Syndicate type d] All are correct

1117. A Multicrop thresher is used for threshing
a] Wheat b] Paddy
c] Soyabean d] All are correct

1118. The loop type threshing drum is used for threshing
a] Wheat b] Paddy
b] Gram d] Soyabean

1119. Of the following the main parts of syndicator type thresher are
a] Corrugated flywheel
b] Serrated chopping knives
c] Closed cylinder casing
d] All are correct

1120. The syndicator type thresher is termed as
a] Hammer mill type thresher
b] Rasp bar type thresher
c] Chaff cylinder type thresher
d] All are correct

1121. The drummy type thresher is also termed as
a] Rasp bar type thresher
b] Hammer mill type thresher
c] Spike tooth type thresher
d] Syndicator type thresher

1122. A multi crop threshers is equipped with
a] Spike tooth cylinder b] Rasp bar cylinder
c] Hammer mill cylinder d] All of above

1123. In angle bar cylinder type thresher, the clearance between cylinder and concave unit at entrance is
a] 7-13 mm b] 13-19mm
c] 15-20 mm d] 20-25 mm

1124. In angle bar cylinder type thresher, the clearance between cylinder and concave unit at the end
a] 1-3 mm b] 3-6 mm
c] 5-8 mm d] 6-9 mm

1125. In raspbar type cylinder, the bars varies from
a] 4-6 b] 6-8
c] 8-10 d] 10-12

1126. The threshing cylinder used in groundnut thresher is
a] Spike tooth type b] Rasp bar type
c] Hammer mill type d] Angle bar type

1127. The threshing efficiency of thresher depends on
a] Cylinder peripheral speed
b] Cylinder concave clearance
c] Feed rate, moisture content and type of crop
d] All are correct

1128. The clean grain received at main grain outlet with respect to total grain mixture received at main gram outlet is
a] Threshing efficiency
b] Cleaning efficiency
c] Winnowing efficiency
d] None of these

1129. The threshed gram received at main grain outlets with respect to total grain input expressed as percentage by mass is
a] Cleaning efficiency
b] Winnowing efficiency
c] Threshing efficiency
d] None of these

1130. The percentage of grain blown by blower along with bhusa in respect to total grain input in the thresher by weight is
a] Cylinder loss b] Sieve loss
c] Blower loss d] Visible

Quiz Kit!

Longest Constitution

The Constitution of India is the longest written constitution in the world, with 395 Articles and 12 Schedules.

1131. The percentage of healthy grain obtained from all outlets with respect to healthy grains received from main outlet by weight is
a] Blower loss b] Sieve loss
c] Cylinder d] Visible loss

1132. For wheat, the recommended speed of threshing drum is
a] 10-15 m/s b] 20-30 m/s
c] 10-15 m/min d] 20-30 m/min

1133. For paddy, the recommended speed of threshing drum is
a] 10-15m/s b] 16-25 m/s
c] 22-32 m/s d] 25-32 m/s

1134. The broken gram received from main outlet is due to
a] Less concave clearance
b] Low speed of drum
c] More concave clearance
d] All are correct

1135. The process of separation of grain from a mixture of grain and chaff is
a] Cleaning b] Winnowing
c] Separation d] Grading

1136. The material used for construction of biade of winnower is
a] Aluminium sheet b] Low grade carbon steel
c] Mild steel d] High grade carbon steel

1137. A system used to operate hand-operated winnowing fan is
a] Chain sprocket system
b] V-belt pulley system
c] Gear system
d] All are correct

1138. The speed of hand-operated winnowing fan varies from
a] 100-200 rpm b] 200-350 rpm
c] 300-450 rpm d] Above 450 rpm

1139. In precision planting, the height of seed drop should be
a] Close to ground
b] At least 50 cm from ground
c] More than 50 cm from ground
d] None of these

1140. The main objective of tillage is
a] Loosing soil
b] Breaking compact earth surface
c] Improve aeration of soil
d] All are correct

GK **Kaveri river rises in Brahmgiri range of Western Ghat.** GS

1141. The power tillage is most suitable for
a] Stationary operation
b] Deep ploughing
c] Rotary operation
d] All are correct

1142. The minimum soil manipulation is essential for crop production in
a] No tillage b] Minimum tillage
c] Strip tillage d] Reduced tillage

1143. The equipment that has no driven/moving component is
a] Tool b] Machine
c] Implementd] Equipment

1144. 481 The area covered by dry land weeder is
a] 0.l ha/dayb] 0.2 ha/day
c] 0.3 ha/day d] 0.4 ha/day

1145. The seed drill, which is used for direct sowing of seeds in unploughed land after paddy harvesting is
a] Strip drill b] Zero till drill
c] Conventional drill d] None of these

1146. A vertical disc plough is also termded as
a] Wheat plough b] Harrow plough
c] Both a] and b] d] None of these,

1147. The normal depth of operation (ploughing) in subsoiler is
a] 15-35 cm b] 35-65 cm
c] 45- 75 cm d] 60-90 cm

1148. The shovel of cultivator is made up of
a] Mild steel b] Cast iron
c] Chilled cast iron d] Soft centre steel

1149. The per cent saving of water in bed sowing of wheat as compared to flat sowing is
a] 10 % b] 20 %
c] 30 % d] 40 %

1150. The carbon percent in mild steel varies from
a] 0-0.25 % b] 0.1-0.25 %
c] 0.25-0.3 % d] 0.25-0.5 %

1151. The percentage of carbon in medium carbon varies from
a] 0.1to 0.25 % b] 0.12 to 0.3 %
c] 0.25 to 0.5 % d] 0.5 to 0.75 %

1152. The percentage of carbon III high carbon steel varies from
a] 0.25 to 0.75 % b] 0.5 to 0.75%
c] 0.5 to 1.0 % d] 0.5 to 1.5 %

1153. The percentage of carbon in cast iron is
a] 0.5 to 1.0 % b] 1.5 to 2.0 %
c] 2.0 to 2.5 % d] 2.5 to 4.5 %

1154. The speed restriction of light belt drive is up to
a] 5 m/s b] 7.5 m/s
c] 10 m/s d] 12 m/s

1155. The speed variation of medium belt drive is
a] 5-15 m/s b] 10-22 m/s
c] 12-24 m/s d] 15-30 m/s

1156. The speed variation of heavy belt drive is about
a] 15-20 m/s b] 20-22 m/s
c] < 22 m/s d] > 22 m/s

1157. To plough a one hectare land with 15 cm width country plough, a man has to walk
a] 52 km b] 60 km
c] 66 km d] 72 km

1158. The cutting blade used in subsoiler is made from
a] Mild steel b] Cast iron
c] Low carbon steel d] High carbon steel

1159. The thickness of disc used in Olpad thresher is
a] 2 mm b] 3 mm
c] 4 mm d] 5 mm

1160. The cotton stripper is used for
a] Pod picking b] Ginning
c] Harvesting d] Threshing

1161. The term turrel is associated with
a] Sickle b] Reaper
c] Mower d] Combine

1162. The seeding behind the plough method of sowing is generally used for
a] Maize b] Wheat
c] Paddy d] Soyabean

1163. The mounted type seed drill attached to tractor by
a] Single hitch point b] Two hitch point
c] Three point hitch d] Shank

1164. The working capacity of two-row potato planter is about
a] 0.2 to 0.3 ha/h b] 0.5 to 1.5 ha/h
c] 1.0 to 1.5 ha/h d] 1.5 to 2.0 ha/h

1165. The useful life of self-propelled combine is
a] 5 years b] 6 years
c] 8 years d] 10 years

1166. The useful working hours of self-propelled combine is
a] 1000 hours b] 2000 hours
c] 3000 hours d] 4000 hours

1167. The useful working hours of mounted type combine is
a] 1000 b] 1200
c] 1800 d] 2000

1168. The expected annual use of plough is
a] 1200 hours b] 150 hours
c] 2000 hours d] 2400 hours

1169. The expected annual use of seed-cum-fertilizer drill is
a] 600 hours b] 800 hours
c] 1200 hours d] 1000 hours

GS

Adil Shah appointed Hemu the wazir with the title of Vikramjit.

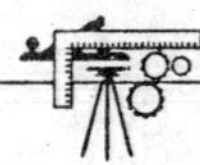

1170. The expected annual use of planter is
a] 800 hours b] 1000 hours
c]1200 hours d] 1600 hours

1171. The expected life of tractor mounted combine is
a] 5 years b] 7 years
c] 8 years d] 10 years

1172. The expected annual use of seed drill is
a] 800 hours b] 1000 hours
c] 1200 hours d] 1500 hours

1173. The expected annual life of harrow is
a] 8 years b] 10 years
c] 12 years d] 14 years

1174. The expected life and useful hours of cultivator is
a] 8 years, 2400 hours
b] 12 years, 3000 hours
c] 16 years, 3500 hours
d] 20 years, 3800 hours

1175. The expected annnul use and life of puddler is
a] 1500 hours. 8 years
b] 1000 hours, 12 years
c] 2000 hours, 12 years
d] 2500 hours, 10 years

1176. The coupling joining two shafts out of alignment is
a] Flanged coupling b] Universal coupling
c] Oldham coupling d] Flexible coupling

1177. The coupling used for joining two shafts appreciably out of line is
a] Oldham coupling b] Flanged coupling
c] Flexible coupling d] Universal coupling

1178. The coupling used for joining two parallel shaft whose centre and short distance apart is
a] Flexible coupling b] Oldham coupling
c] Flanged coupling d] Universal coupling

1179. The coupling used to join to shaft in line is
a] Flexible coupling b] Oldham coupling
c] Flanged coupling d] Universal coupling

1180. The corrugated seed tube used in seed drill is made up of
a] Plastic b] Rubber
c] Metal d] PVC

1181. In horizontal plate planter, the percentage is
a] Less than 100 b] 100
c] More than 100 d] None of these

1182. The ratio of reel peripheral speed to forwards speed of combine is termed as
a] Reel velocity b] Reel index
c] Reel diameter d] Reel speed

1183. Renewable source of energy is
a] Exhaustible b] Inexhaustible
c] Nuclear based d] Biogas

1184. Renewable source of energy can be renewed by
a] Chemical process
b] Mechanical process
c] Reproduction process
d] All are correct

1185. Renewable energy is mainly obtained by
a] Sun b] Biomass and Biogas
c] Wind d] All are correct

1186. Tidal energy is the
a] Renewable energy b] Non-renewable energy
c] Exhaustible d] None of these

1187. Tidal energy is associated with
a] Wind b] Sun
c] Sea d] None of these

1188. Wave energy is associated with
a] Sun b] Ocean
c] Wind d] All are correct

1189. The energy obtained from the sun is
a] Atomic energy b] Tidal energy
c] Solar energy d] Wave energy

1190. A reflector used in solar power plant to reflect sunlight is
a] Convex reflector b] Concave reflector
c] Plane reflector d] All are correct

1191. Solar cells are made of
a] Silicon
b] Germanium
c] Semi-conductor elements
d] All are correct

1192. In solar power plant, the turbine is run by
a] Hot waterb] Steam
c] Sunlight d] Photovoltaic cell

1193. Solar pond is used to
a] Collect and store the solar energy
b] Reflect the solar energy
c] Divert solar energy
d] None of these

1194. The solar cells used in solar panel are used to convert solar energy into
a] Chemical energy b] Mechancal energy
c] Electrical energy d] None of these

1195. A type of solar reflector used in solar furnace
a] Concave and spherical
b] Convex and spherical
c] Plane and spherical
d] None of these

1196. The sun's rays has higher intensity
a] On the ground surface
b] Near the polar region
c] Near the equator
d] All are correct

GK GS

Article 72(2) says that the Ministers shall be individually responsible to the President.

1197. The solar radiation receives by every square meter area of the earth surface is
a] 0.64 KJ/sec b] 1.36 KJ/sec
c] 1.50 KJ/sec d] 1.80 KJ/sec

1198. In sunlight, the heating rays are
a] Violet rays b] Ultraviolet rays
c] Infra-red rays d] All are correct

1199. Wind is a source of
a] Inexhaustible energy
b] Exhaustible energy
c] Non-renewable energy
d] None of these

1200. For working a windmill the minimum speed of wind should be
a] 5 kmph b] 10 kmph
c] 20 kmph d] 30 kmph

1201. In case of windmill, the ratio of power delivered by rotor to the maximum power available in the wind is
a] Rotor coefficient
b] Coefficient of performence
c] Coefficient of gyration
d] None of these

1202. High wind energy region of India is
a] Assam b] Andhra Pradesh
c] Gujarat d] West Bengal

1203. Wind due to its motion possesses the
a] Potential energy b] Kinetic energy
c] Chemical energy d] Electrical energy

1204. A sailboat utilizes the energy of
a] Wind b] Sun
c] Biogas d] Water

1205. A windmill suitable for water lifting is
a] Single blade type b] Double blade type
c] Triple blade type d] Multiple blade type

1206. The ratio of the speed at the tip of rotor to the wind speed is called
a] Wind speed ratio b] Rotor speed ratio
c] Tip speed ratio d] All are correct

1207. In multi-blade rotor, the number of blades varies from
a] 8-12 b] 10-16
c] 12-20 d] 14-26

1208. The speed of multi-blade rotor is
a] 20- 40 rpm b] 40-60 rpm
c] 60- 80 rpm d] 80-100 rpm

1209. A sail type unit has
a] 2 blades b] 3 blades
c] 6 blades d] 8 blades

1210. Blades of sail unit are made of
a] Wood b] Canvas cloth
c] Metal d] All are correct

1211. Propeller type unit runs at a speed of
a] 100 -200 rpm b] 200- 300 rpm
c] 300- 400 rpm d] 400- 500 rpm

1212. Pump used in the windmill may be
a] Reciprocating type b] Rotor type
c] Diaphragm type d] All are correct

1213. The major constitute of biogas is
a] Methane b] Ethane
c] Carbon dioxide d] Carbon monoxide

1214. The percentage of methane in biogas is
a] 30 per cent b] 45 per cent
c] 65 per cent d] 80 per cent

1215. Digester diameter of biogas plants ranges from
a] 1 -3 metres b] 1.2- 4 metres
c] 1.2- 6 metres d] 1.6- 8 metres

1216. In biogas plants, the depth of digester varies from
a] 2 -4 metres b] 3 -6 metres
c] 4- 8 metres d] 5- 10 metres

1217. The percentage of carbon dioxide in biogas is about
a] 12-28 b] 32-45
c] 40- 60 d] 65- 85

1218. Major constituent of biogas is
a] Ethane b] Methane
c] Butane d] All are correct

1219. Gobar gas plant is a
a] Solar plant
b] Biogas plant
c] Nuclear power plant
d] None of these

1220. The dung requirement for a small size of gobar gas plant per day is
a] 10 kg b] 20 kg
c] 30 kg d] 40 kg

1221. The economically smallest size of gobar gas plant is
a] 30 cubic ft b] 50 cubic ft
c] 60 cubic ft d] 70 cubic ft

1222. The capacity of biogas plant varies trom
a] 2.0 -2.75 m^3 b] 10- 50 m^3
c] 2 -150 m^3 d] 5 -200 m^3

1223. The capacity of smallest biogas plant is about
a] 1 m^3 b] 2 m^3
c] 6 m^3 d] 12 m^3

1224. The percentage of nitrogen in gobar gas manure is
a] 1.5% b] 2.0 %
c] 4.5% d] 8 %

1225. Biogas is used for
a] Cooking food b] Operating engine
c] Burning gas stove d] All are correct

1226. A quantity of gobar gas required to cook one-person meal is
a] 3 kg b] 5 kg
c] 8 kg d] 10 kg

Dilip Bose became the first Indian player to be seeded at Wimbledon in 1950.

GS

1227. The production of gas from one kilogram of wet dung in biogas plant is
a] 0.037 m^3 b] 0.085 m^3
c] 1.0 m^3 d] 1.5 m^3

1228. The pH value of slurry for better gas production should be between
a] 3 -5 b] 5 -6
c] 7- 8 d] 8- 9

1229. Calorific value of biogas is
a] 450 Kcal/ m^3 b] 1000 Kcal/ m^3
c] 3000 Kcal/ m^3 d] 4500 Kcal/ m^3

1230. Biogas is obtained by the anaerobic fermentation of
a] Animal dung b] Plant waste
c] Animal waste d] All are correct

1231. The gasholder used in floating gas type biogas plant is
a] Conical b] Spherical
c] Inverted drum d] All are correct

1232. The KVIC biogas is
a] Fixed type b] Floating
c] Both a] & b] d] None of these

1233. In fixed dome type biogas plant, the gas holder is made of
a] Metal b] Cement and bricks
c] Wooden d] Stone

1234. In floating type biogas plant, gas holder is made of
a] Metal b] Wood
c] Cement and bricks d] All are correct

1235. Which of the biogas plant has digester tank and gasholder as combined unit?
a] Floating dome type b] Fixed dome type
c] Both a] & b] d] None of these

1236. A high gas production achieved from biogas plant when inside temperature of the chamber is
a] 15° b] 20°C
c] 35°C d] 50°C

1237. Insulated metal of solar cooker is
a] Black from outside
b] Black from inside
c] Black from out and inside
d] White from out and inside

1238. The temperature inside the solar cooker ranges from
a] 40 - 80°C b] 50-100°C
c] 100-140°C d] 120- 180°C

1239. A mirror used to reflect sunlight in box type solar cooker is
a] Convex b] Concave
(C) Plane d] All are correct

1240. In solar furnace, temperature can be raised upto
a] 1000°C b] 2000° C.
c] 3000° C d] 4000° C

Did You Know ?

First demand for election

A demand for representation on the basis of elections was made by K.T. Toland, a member of the Indian National Congress at its very first session in 1885.

1241. Which of the following items utilizes direct energy of sun
a] Solar heater b] Solar cooker
c] Solar cell d] All are correct

1242. The use of reflector in solar heating devices is to
a] Increase efficiency
b] Decrease efficiency
c] Create greenhouse effect
d] None of these

1243. A group of solar cell is
a] Solar cap b] Solar panel
c] Solar noded] All are correct

1244. In India, thermal power contribution in total energy production is
a] 25 % b] 40 %
c] 56 % d] 70 %

1245. In India, hydro electrical power contribution in total energy production is
a] 20 % b] 36 %
c] 45 % d] 55 %

1246. In India, nuclear power contribution in total energy production is
a] 10 %. b] 12 %
c] 16 % d] 20 %

1247. An instrument used for measuring the solar radiation is
a] Thermometer b] Pyrheliometer
c] Thermocouple d] All are correct

1248. An instrument, which measures the total or global radiation over a hemispherical field of view is
a] Pyranometer b] Thermister
c] Thermocouple d] Thermometer

1249. In India, the average intensity of solar radiation receives about
a] 200- 500 cal/cm^2/day
b] 400- 700 cal/cm^2/day
c] 500 -1000 cal/cm^2/day
d] 600 -1200 cal/cm^2/day

1250. A vertical angle of the projection of sun's rays on the horizontal plane and direction of sun's rays is called
a] Zenith angle b] Altitude angle
c] Hour angle d] Incident angle

1251. Flat plate collector is generally used for
a] Room heating b] Water heating
c] Lighting d] All are correct

GK GS

Biogas is obtained during fermentation of animal and plant waste.

1252. Of the following solar air heater is applicable in
a] Heating buildings
b] Drying of agricultural products
c] Air-conditioning of buildings
d] All are correct

1253. Wind energy is a form of
a] Solar energy b] Wave energy
c] Biomass energy d] Ocean energy

1254. The energy wind is converted into
a] Chemical energy b] Mechanical energy
c] Electrical energy d] Biomass energy

1255. In a windmill, tip speed ratio increases with
a] Increase of number of blades
b] Decrease of number of blades
c] Independent of blade number
d] None of these

1256. An instrument used to measure wind speed is
a] Dynamometer b] Anemometer
c] Hygrometer d] Thermometer

1257. Hot magma is a source of
a] Solar energy b] Wind energy
c] Geothermal energy d] Tidal energy

1258. In India which of the following state has geothermal energy plants?
a] Himachal Pradesh b] Karnataka
c] Jammu & Kashmir d] All are correct

1259. Heat flow in the ground is measured in bore holes to a depth of about
a] 25 metres b] 50 metres
c] 75 metres d] 100 metres

1260. One heat flow unit is equal to
a] 10^{-6} cal/cm^2/sec b] 10^{-8} cal/cm^2/sec
c] 10^{-6} cal/cm^2/sec d] 10^{-8} cal/cm^2/sec

1261. According to US geological survey, geothermal energy is all heat stored in the earth's crust above 15°C to a depth of
a] At 10 km b] Below 10 km
c] Above 10 km d] All are correct

1262. Geothermal energy from the earth's interior is
a] Exhaustible b] Inexhaustible
c] Non-renewable d] None of these

1263. Tidal power is dissipated by
a] Friction b] Eddies current
c] Both a] & b] d] None of these

1264. The tides in a period of 24 hours 50 minutes are
a] One high and one low
b] One high and two low
c] Two high and two low
d] Two high and one low

1265. The neap tides occurs when
a] Sun is at right angle with respect to the earth
b] Moon is at right angle with respect to the earth
c] Sun and moon are at right angle with respect to the earth
d] None of these

1266. The gradient for generation of power by heat engine in OTEC plant utilizes
a] Pressure gradient b] Temperature gradient
c] Moon gradient d] None of these

1267. The fossil fuel is
a] Coal b] Natural gas
c] Oil d] All are correct

1268. Calorific value of coal is about
a] 2000 kcal/ kg b] 4600 kcal/ kg
c] 6300 kcal/ kg d] 8300 kcal/kg

1269. Calorific value of natural gas ranges between
a] 10 -15 KJ/ kg b] 20- 30 KJ/ kg
c] 33-50 KJ/kg d] 50-75 KJ/g

1270. Caloritlc va]ue of hydrogen gas is
a] 50 KJ/ kg b] 100 KJ/ kg
c] 150 KJ/ kg d] 180 KJ/kg

1271. Solar energy can be directly converted into e1ectrical energy, by a method is
a] Thermoelectric b] Thermionic
c] Magneto hydrodynamic d] All are correct

1272. Paraboloidal type solar cooker was first developed in
a] USA b] India
c] China d] England

1273. The first nuclear power station came into operation (1957) at
a] USA b] England
c] Germany d] China

1274. In biogas plant digestion (biological process) occurs in absence of
a] Oxygen b] Carbon dioxide
c] Hydrogen d] Methane

1275. Biogas is a mixture of
a] Methane and carbon dioxide
b] Ethane and carbon dioxide
c] Ethane and methane
d] Ethane and carbon mono oxide

1276. The methane bacteria work best at a temperature range of
a] 30- 32°C b] 35 -38°C
c] 40- 45°C d] 48 -52°C

1277. A biogas production starts falling at an atmospheric temperature of
a] 15°C b] 20°C
c] 25°C d] 30°C

1278. The biogas production stops at an atmospheric temperature of
a] 5°C b] 10°C
c] 14°C d] 18°C

1279. When the temperature is raised to 38.3°C, the gas production increased by
a] 25 % b] 40 %
c] 50 % d] 75 %

GK GS
India can use 20% water of Indus river system by Indus Agreement.

1280. The optimum C: N ratio for maximum microbiological activities is
a] 10:1 b] 15:1
c] 25:1 d] 30:1

1281. The gas production per unit volume of digester capacity will be maximum when the diameter to depth ratio of digester will be in the range of
a] 0.66 to 1.00 b] 0.75 to 1.20
c] 0.85 to 1.30 d] 0.90 to 1.40

1282. Solar radiation reaches the earth in the form of
a] Beam radiation
b] Difuse radiation
c] Beam and defuse radiation
d] None of these

1283. Temperature attainable with concentrating collector as compared to flat plate collector is
a] Lower b] Higher
c] Equal d] None of these

1284. According to NASA, a standard value for solar constant is
a] 1.353 kW/m^2 b] 1.533 kW/m^2
c] 1.755 kW/m^2 d] 1.850 kW/m^2

1285. Thermal efficiency of Magneto hydrodynamic system is about
a] 40- 45.0 b] 50 -55
c] 60 -65 % d] 70 -75

1286. For every 100 metre below ground level the temperature of rock increases about
(A) 2°C b] 3°C
c] 4°C d] 5°C

1287. Ocean waves contain
a] Kinetic energy b] Potential energy
c] Static energy d] All are correct

1288. A latitude for wave energy in both Northern and Southern Hemispheres is between
a] 10 and 20° b] 20 and 30°
c] 30 and 50° d] 40 and 60°

1289. A contribution of hydropower in total world energy power is about
a] 6 % b] 8 %
c] 10 % d] 12 %

1290. The amount of solar energy reaching a specific location on the surface of the earth at a specific time is
a] Solar constant b] Insolation
c] Solar radiation d] None of these

1291. If the pH range of slurry is between 4 and 6 the slurry is
a] Acidic b] Alkaline
c] Neutral d] None of these

1292. The proportion of wet cow dung and water in slurry feed to the digester is
a] 1:1 b] 1:2
c] 1 :3 d] 1 :4

1293. The quantity of gas produced by one kg of dry dung is about
a] 0.156 m^3 b] 0.160 m^3
c] 0.175 m^3 d] 0.186 m^3

1294. In a 30 days of retention period, the production of gas/kg of fresh dung is
a] 0.04 m^3 b] 0.05 m^3
c] 0.06 m^3 d] 0.08 m^3

1295. Biogas can be used to operate
a] SI engine
b] CI engine
c] Both SI and CI engine
d] Not possi'Lle to operate engine

1296. Heat value of the gas produced in gasfire is about
a] 950- 1200 kcal/ m^3 b] 1000- 1300 kcal/ m^3
c] 1200- 1500 kcal/ m^3 d] 1500- 1800 kcal/ m^3

1297. The boiling temperature of methanol is
a] 45°C b] 65°C
c] 470°C d] 85°C

1298. In anaerobic digestion, biogas production by bacterial decomposition is
a] Presence of oxygen b] Absence of oxygen
c] Presence of CO^2 d] Absence of CO^2

1299. The wet process of biomass digestion involves
a] Pyrolysis and Gasification
b] Hydrogenation and liquefication
c] Liquefication and pyrolysis
d] Anaerobic digestion and fermentation

1300. The dry process of biomass digestion involves
a] Pyrolysis and liquefication
b] Anaerobic digestion and fermentation
c] Fermentation and pyrolysis
d] All are correct

1301. Solar energy reaches top of the earth's surface consist of
a] Ultraviolet radiation
b] Infrared radiation
c] Visible light
d] All are correct

1302. The solar radiation receives from the sun after its direction has been changed is
a] Beam radiation b] Diffuse radiation
c] Direct radiation d] All are correct

1303. The solar radiation received directly from the sun is
a] Beam radjation b] Scattered radiation
c] Diffuse radiation, d] Reflected radiation

1304. The air mass will be one
a] At Zenith
b] At 60° to Zenith
c] Above earth's atmosphere
d] None of these

GK GS

Zabti system of land revenue was called Todar Mal's bandobast.

1305. An angular distance of the sun rays north (or south) of the equator is
a] Latitude angle b] Hour angle
c] Declination d] Inclination

1306. A vertical between the projection of the sun's rays on the horizontal plane and direction of sun's rays is
a] Zenith angle b] Solar altitude
c] Solar angle d] Latitude angle

1307. Equinox means
a] Equal days b] Equal nights
c] Equal day and night d] Equator

1308. A angle through which the Earth must turn to bring the meridian of a point directly in time with sunrays is
a] Altitude angle b] Zenith angle
c] Hour angle d] Solar azimuth angle

1309. The process of removing husk and bran of the paddy in one operation is called
a] Hulling b] Shelling
c] Milling d] All are correct

1310. The process of removing husk and hull from the paddy grain is known as
a] Shelling b] Milling
c] Husking d] All are correct

1311. In rubber roll huller breakage is
a] Less b] More
c] No breakage d] None of these

1312. In rubber roll huller the rubber rolls turn at
a] Same speed
b] Different speed and different directions
c] Same speed and same direction
d] All are correct

1313. In forced air-dring method the moisture moves from.
a] Grain to air b] Centre to grain surface
c] Air to grain d] All are correct

1314. The drying process involves
a] Heat transfer
b] Mass transfer
c] Heat and mass transfer
d] None of these

1315. In thin layer drying, the depth of grain layer should be
a] 20 cm b] Less than 20 cm
c] More than 20 cm d] None of these

1316. Thin layer drying includes
a] Batch type dryer b] LSU dryer
c] Flat bed dryer d] All are correct

1317. Deep bed drying includes
a] Baffle dryer b] Batch dryer
c] LSU dryer d] None of these

1318. In deep bed drying, the depth of grain layer is
a] 10 cm b] 20 cm
c] Less than 20 cm d] More than 20 cm

1319. A sack dryer is suitable for drying
a] Small quantity of grain
b] Large quantity of grain
c] Any quantity of grain
d] None of these

1320. For air-drying, the recommended normal air temperature is
a] 30- 40°C b] 40- 50°C
c] 50-60°C d] 60-65°C

1321. In batch type dryer, the maximum depth of grain is about
a] 200 cm b] 225 cm
c] 250 cm d] 300 cm

1322. In batch type dryer the minimum airflow rate required for drying per cubic metre of paddy is
a] 2 - 3 cu m/min b] 3 -4 cu m/min
c] 4- 5 cu m/min d] 4 -6 cu m/min

1323. In rotary dryer, the diameter of drum ranges between
a] 2 -4 metres b] 3 -5 metres
c] 3 -6 metres d] 4 -8 metres

1324. Recirculatory batch dryer is developed at
a] Pantnagarb] Mysore
c] Kharagpur d] Ludhiana

1325. The capacity of RPEC dryer is about
a] Half tonne per hour
b] One tonne per hour
c] Two tonne per hour
d] All are correct

1326. LSU dryer was developed in/at
a] 1949 (USA) b] 1950 (England)
c] 1960 (India) d] 1965 (Japan)

1327. The main advantage of LSU dryer is
a] Shorter drying period
b] Maximum drying capacity
c] Minimum drying losses
d] All are correct

1328. For LSU dryer, the recommended airflow rate is
a] 50 cu m/min/tonne b] 60 cu m/min/tonne
c] 70 cu m/min/tonne d] 80 cu m/min/tonne

1329. In continuous flow dryer, drying air temperature may be
a] 65°C b] 70°C
c] 75°C d] 80°C

1330. In batch type or the maximum drying air temperature is
a] 30°C b] 35°C
c] 40°C d] 45°C

Zabti system of land revenue was called Todar Mal's bandobast.

GS

1331. LSU dryer is
a] Batch type dryer
b] Continuous flow type dryer
c] Continuous flow mixing type dryer
d] Continuous flow non -mixing type dryer

1332. For LSU dryer, the recommended drying air temperature is
a] 60°C b] 70°C
c] 75 °C d] 80 °C

1333. The relation of equilibrium relative humidity with moisture content of grain at a particular temperature is expressed by a curve known as
a] Grain isotherm
b] Relative humidity
c] Equilibrium moisture content
d] All are correct

1334. In drying process, the heat required to convert one kilogram of grain moisture to water vapour is
a] 450 kcal b] 500 kcal
c] 550 kcal d] 650 kcal

1335. The optimum moisture content (wb) for paddy harvesting is
a] 12 -15 % b] 15 -18 %
c] 18-20 % d] 20-22 %

1336. The optimum moisture content (wb) for wheat harvesting is
a] 12-14 % b] 14-16 %
c] 16 -18 % d] 18- 20 %

1337. The percentage of hull in paddy grain is about
a] 15 % by weight b] 16 % by weight
c] 18 % by weight d] 20 % by weight

1338. The percentage of protein by weight in pulse is
a] 15- 20 % b] 17- 5 %
c] 18- 27 % d] 20 -30 %

1339. The density and specific gravity of grain is essential for
a] Calculating the thermal diffusivity
b] Determining the Reynolds's number
c] Pneumatic and hydraulic handling
d] All are correct

1340. The yardstick of sharpness of solid materials is
a] Roundness b] Sphericity
c] Terminal velocity d] Drag coefficient

1341. The ratio of the diameter of a sphere of the same volume as that of particle and diameter of smallest circumscribing of sphere is called
a] Drag coefficient b] Terminal velocity
c] Sphericity d] None of these

1342. Geometrical mean diameter of particles are known as
a] Specific diameter
b] Equivalent diameter
c] Geometric diameter
d] None of these

You Don't Know!
First demand for Pakistan
The All-India Muslim League at its meeting in Lahore in 1940 approved the demand for a separate Muslim state and declared the formation of Pakistan as its goal.

1343. Specific gravity of grains is determined by
a] Pycnometer
b] Specific gravity bottle
c] Both a] and b] d] None of these

1344. The density of wheat is about
a] 850- 1100 kg/ m^3 b] 998- 1238 kg/ m^3
c] 1150-1300 kg/m^3 d] 1200-1450 kg/m^3

1345. Terminal velocity of wheat grain is
a] 8-10 m/s b] 9-11.5 m/s
c] 10 -12.5 m/s d] 10 -13 m/s

1346. The frictional force acting between the surface of contact at rest with respect to each other is called
a] Dynamic friction b] Static friction
c] Kinetic friction d] None of these

1347. The friction forces existing between the surfaces in relative motion is called
a] Static friction b] Coefficient of friction
c] Kinetic friction d] Angle of internal friction

1348. Angle of repose for paddy ranges between
a] 25- 30 degree b] 30 -40 degree
c] 30-45 degree d] 35-50 degree

1349. The value of angle of repose for wheat is between
a] 18- 23 degree b] 23-28 degree
c] 28- 33 degree d] 33 -38 degree

1350. The value of angle of repose for millets is about
a] 15 -20 degree b] 20 -25 degree
c] 25-30 degree d] 30-35 degree

1351. The amount of heat in kilocalorie that must be added to or removed from one kg of substances to change its temperature by one degree is
a] Latent heat b] Specific heat
c] Enthalpy d] Entropy

1352. Thermal conductivity of single grain ranges from, 1-5 is
a] 0.3- 0.6 Kcal /mhr/°C
b] 0.4- 0.8 Kcal /mhr
c] 0.5 -0.9 Kcal /m
d] 0.6- 1.0 Kcal /mhr

1353. Thermal conductivity of bulk grain varies from
a] 0.10- 0.15 Kcal/mhr.°C
b] 0.15- 0.20 Kcal/mhr °C
c] 0.20- 0.25 Kcal /mhr. °C
d] 0.25- 0.30 Kcal /mhr.°C

GS

Part VI of Constitution deals with Union Legislature or Parliament.

1354. Thermal conductivity of air is
a] 0.01 Kcal/mhr.°C b] 0.02 Kcal /mhr.°C
c] 0.03 Kcal /mhr. °C d] 0.04 Kcal /mhr.°C

1355. Cleaning of a material is done by
a] Washing b] Screening
c] Handpicking d] All are correct

1356. The removal of few large particles in an initial process is
a] Scalping b] Cleaning
c] Grading d] Sorting

1357. Cleaning of material is done on the basis of
a] Shape b] Size
c] Weight d] Density

1358. In the cylinder sorter, the inclination of cylinder for dry granular material is kept up to
a] 75 mm / mb] 100 mm /m
c] 125 mm / m d] 150 mm / m

1359. Disc separator separates materials on the basis of
a] Length b] Width
c] Weight d] Shape

1360. Indented cylinder separator separates the grains on the basis of
a] Weight b] Relative length
c] Length d] All are correct

1361. Spiral separator separates the grains on the basis of
a] Weight b] Sphericity
c] Roundness d] Size

1362. Specific gravity separator makes the separation as per difference in
a] Density b] Shape
c] Roughness d] Relative length

1363. Specific gravity separator works on the principle of
a] Characteristics of grain to flow down over an inclined surface
b] Flotation of the particles due to upward movement of air
c] Both a] and b]
d] None of these

1364. The shape of perforated deck used in specific gravity separator is
a] Triangular b] Rectangular
c] Circular d] Hexagonal

1365. In inclined draper, the major factor causing the separation of grain are
a] Shape and size b] Shape and surface texture
c] Shape and weight d] Weight and size

1366. The major controlling factor in separation of grain by fluidized bed separator is
a] Air velocity and weight of grain
b] Shape and size of grain
c] Shape and texture of grain
d] Magnetic and aerodynamic properties of grain

1367. Fluidized bed separator used to separate the seed are
a] Lighter seeds b] Medium seeds
c] Heavier seeds d] All kinds of seeds

1368. The separation of grain in magnetic separator is carried out on the basis of
a] Specific gravity b] Surface texture
c] Roundness d] Relative length

1369. Cyclone separator works on
a] Centrifugal force b] Gravity force
c] Inertia force d] None of these

1370. In India, morai type structure used in the
a] Eastern region b] Western region
c] Northern region d] Southern region

1371. The capacity of morai type storage structure is about
a] 2.5 to 15 tonnes b] 3.5 to 18 tonnes
c] 4.5 to 20 tonnes d] 5 to 25 tonnes

1372. The storage capacity of mud kothi is
a] 1 to 20 tonnes b] 1 to 30 tonnes
c] 1 to 40 tonnes d] I to 50 tonnes

1373. Safe storage moisture level of grain crop is generally below the range of
a] 4 to 6 % b] 8 to 9 %
c] 10 to 11 %d] 12 to 13 %

1374. In rice, angle of internal friction is
a] 15 degrees b] 18 degrees
c] 20 degrees d] 24 degrees

1375. Cyclone separator is used for separating the
a] Fine particles b] Medium particles
c] Heavy particles d] All are correct

1376. Safe moisture content of paddy (wb) for storage over one year is
a] 8 per cent b] 10 per cent
c] 12 per cent d] 14 per cent

1377. In batch type dryer, the air temperature seldom exceeds
a] 35°C b] 40°C
c] 45°C d] 50°C

1378. The relative humidity will be reduced to half its original value if the temperature of air is increased by
a] 8 degrees b] 11 degrees
c] 14 degrees d] 17 degrees

1379. The pressure of water vapour for each 10° C rise in temperature is nearly
a] Half b] Double
c] Triple d] None of these

1380. The capacity of kothar type storage structure varies from
a] 5 to 15 tonnes b] 7 to 25 tonnes
c] 9 to 35 tonnes d] 10 to 40 tonnes

1381. The nature of crop grain is
a] Aerodynamic b] Hygroscopic
c] Thermodynamic d] None of these

1382. The law applicable in size reduction of fine powder is
a] Bond's lawb] Kick's law
c] Rittinger's law d] None of these

1383. Stroke law is applied when Reynolds's number is
a] Less than 1 b] More than
c] Less than 2 d] Less than or equal to

1384. The moisture content for safe storage corn is about
a] 10 per cent b] 13 per cent
c] 14 per cent d] 16 per cent

1385. For seed production, maximum grain temperature usually recommended is
a] 38°C b] 40°C
c] 43°C d] 50°C

1386. Heat required to evaporate one kg of water from grain is
a] 450 Kcal b] 500 Kcal
c] 600 Kcal d] 650 Kcal

1387. Natural drying needs dehumidification if atmospheric humidity is
a] 30 % b] 60 %
c] 70 % d] Exceed 70 %

1388. For biological materials, the relationship between EMC and RH has been given by
a] Janssen b] Rankine
c] Henderson d] Chung fast

1389. The heating in infrared drying in comparison to convection drying is higher by
a] 5 times b] 8 times
c] 10 times d] 12 times

1390. The cost of mechanical drying per unit of grain as compared to sun drying is
a] High
b] Low
c] No significant difference
d] Exactly same

1391. The capacity of thermal dryer depends on the
a] Rate of mass transfer
b] Rate of heat transfer
c] Rate of heat and mass transfer
d] None of these

1392. The recommended high temperature limit for seed grains is
a] 65°C b] 70°C
c] 80°C d] 85°C

1393. A critical moisture content varies with
a] Rate of drying b] Thickness of material
c] Both a] and b] d] None of these

1394. Area required to dry one tonne of grain in the sun is about
a] 8 sq.m b] 10 sq.m.
c] 15 sq.m d] 20 sq.m.

1395. In sun drying, the relative humidity (per day) of grain under favourable conditions is reduced by
a] 1-2 % b] 3-4 %
c] 5- 6 % d] 7- 8 %

1396. Sun radiation temperature corresponding to black body temperature is at about
a] 3000°K more b] 5000°K more
c] 6000°K more d] 7000dK more

1397. In freeze-drying, the liquid phase comes at the pressure above
a] 1.7mm b] 2.7mm
c] 3.7mm d] 4.7mm

1398. In conventional freeze drying, dehydration - temperature are kept between
a] 10°C -30°C b] 30.2°C -65.5°C
c] 35.5°C-75°C d] 37.8°C-93.3°C

1399. In attrition mill, the size of food grains is reduced due to
a] Impact and shear
b] Shear and crushing
c] Impact and crushing
d] Impact only

1400. In ball mill, the size of food grains is reduced due to
a] Crushing b] Shearing
c] Impact d] All are correct

1401. Standard screens are used and measurement of the particle size range between
a] 70 mm and 30 micro metre
b] 76 mm and 38 micrometre
c] 78 mm and 38 micro metre
d] 80 mm and 40 micro metre

1402. The opening of test sieve is
a] Rectangular b] Circular
c] Square d] Triangular

1403. Tyler sieves originated in/at
a] 1910 (USA) b] 1920 (Japan)
c] 1925 (India) d] 1930 (England)

1404. Crushing law was proposed by
a] Rittinger b] Kick
c] Bond d] None of these

1405. A law, which is based on stress analysis of plastic defoliation within the elastic limit was proposed by
a] Bond b] Kick
c] Chung fast d] Henderson

1406. The method for estimating power required for crushing and grinding operation is given by
a] Chung fast b] Kick
c] Henderson d] Bond

Consumer price index for agriculture labourers have been accepted as basis for estimating rural poverty in Lakdawala formula.

1407. Hammer mill and Burr grinder are used for grinding the
a] Grain b] Oil seed
c] Pulse d] All are correct

1408. In hammer mill, the force applied on grains is cut by
a] Shear b] Impact
c] Crushing d] Crushing and shear

1409. For grinding operation, operating speed of hammer mill is
a] 700 -1000 rpm b] 1000 -1400 rpm
c] 1400-1500 rpm d] 1600-1800 rpm

1410. The hammer of hammer mill is made of
a] Cast iron b] Stone
c] Trough steel d] Wood

1411. In hammer mill, the speed of hammer tip ranges from
a] 700- 1000 rpm b] 1000- 1800 rpm
c] 1500- 4000 rpm b] 200- 6000 rpm

1412. The capacity of hammer mill depends upon the
a] Feed rate b] Hammer speed
c] Size of sieve d] All are correct

1413. Plate of burr grinder is made of
a] Stone b] Trough steel
c] Mild sheet d] Chilled iron

1414. In burr grinder, the plates for grinding are
a] 1 b] 2
c] 3 d] 4

1415. The size of most common type burr grinder is
a] 20 cm diameter b] 30 cm diameter
c] 40 cm diameter d] 55 cm diameter

1416. The operating speed of burr grinder varies from
a] 500- 600 rpm b] 650- 750 rpm
c] 750- 1000 rpm d] 1000 -1400 rpm

1417. The ratio of the overall surface area of the ground product to overall surface area of the feed is called
a] Grinding efficiency b] Coefficient of grinding
c] Degree of grinding d] Milling efficiency

1418. Crushers are reducing the size of grains by
a] Squeezing and pressing
b] Impact and crushing
c] Impact and shearing
d] Shearing and crushing

1419. In a gyratory crusher, the speed of crushing head ranges from
a] 100- 200 gyration /min
b] 125- 425 gyration /min
c] 150- 500 gyration /min
d] 175 -650 gyration /min

1420. Size reduction of fruits and vegetables are mostly performed by
a] Crushing b] Shearing
c] Cutting d] Squeezing

1421. In rotary knife cutter, a horizontal rotor rotates
a] 100- 500 rpm b] 200- 900 rpm
c] 300-1000 rpm d] 400-1400 rpm

1422. Rietz mill is applicable for milling
a] Dry material b] Semi-dry material
c] Wet material d] All are correct

1423. Rietz mill suitable for grinding material below is of
a] 15 micron size b] 20 micron size
c] 25 micron size d] 30 micron size

1424. Rotor diameter of Rietz mill ranges from
a] 5- 30 cm b] 10-60 cm
c] 15- 70 cm d] 20- 80 cm

1425. Centrifugal sheller is used for milling
a] Paddy b] Wheat
c] Pulses d] Oilseeds

1426. The speed of hammer tip of a Rietz mill varies from
a] 4.5- 80 m/s b] 5.2 -111 m/s
c] 10- 150 m/s d] 12- 160 m/s

1427. Colloid mill is used for
a] Puree making b] Food pastes formation
c] Pulp formation d] All are correct

1428. In oil extraction unit, the most commonly used solvent is
a] n- hexane b] n- heptane
c] Nepthene d] All are correct

1429. The boiling point of n-hexane is about
a] 60°C b] 65.5°C
c] 75.5°C d] 90°C

1430. Percentage of husk in paddy is about
a] 15 -18 per cent b] 18 -22 per cent
c] 20- 24 per cent d] 25 -30 per cent

1431. Percentage of bran in therice is
a] 2 -4 % b] 4- 6 %
c] 6- 8% d] 8- 10 %

1432. Percentage of endosperm in the paddy
a] 68-70 % b] 70-72 %
c] 72-74 % d] 74-78 %

1433. The Englberg huller removes
a] Husk from paddy
b] Bran from paddy
c] Husk and bran from paddy
d] Germ from rice

1434. Clean and good quality of paddy gives a yield of
a] 65 % b] 70 %
c] 74 % d] 78 %

1435. Parboiling of paddy is done to
a] Achieve maximum recovery of head rice
b] Minimise the broken percentage
c] Reduce the milling losses
d] All are correct

1436. The breakage office may be reduced by
a] Drying the paddy at optimum moisture content
b] Gelatinizing the starch
c] Both a] and b]
d] None of these

1437. The temperature for gelatinisaation is about
a] 65°C b] 70°C
c] 75°C d] 80°C

1438. Rice recovery in parboiling as compared to raw rice milling is
a] 1 -2 % b] 2 -3 %
c] 3- 4 % d] 4- 5 %

1439. In single boiling method, paddy is soaked in water for
a] 12- 24 hours
b] 24- 48 hours
c] 24- 72 hours
d] 48 -72 hours

1440. In CFTRI method, the water is heated to a temperature of
a] 70°C b] 75°C
c] 80°C d] 85°C

1441. In steaming process, a soaked paddy exposed to steam at a pressure of about
a] 2 kg/ cm^2 b] 3 kg/ cm^2
c] 4 kg/ cm^2 d] 5 kg/ cm^2

1442. In Jadavpur University Method, temperature of soaking water ranges from
a] 60- 70°C b] 65 -75°C
c] 70- 80°C d] 80- 90°C

1443. In Jadavpur University Method, steaming time is about
a] 3.5 minutes b] 10 minutes
c] 30 minutes d] 45 minutes

1444. The pressurised parboiling method was developed at
a] Punjab b] Tamil Nadu
c] West Bengal d] Andhra Pradesh

1445. In pressurised parboiling, the soaking time of method is about
a] 20 minutes b] 30 minutes
c] 40 minutes d] 60 minutes

1446. In pressurfised parboiling method, the temperature of soaking water is about
a] 75°C -80°C b] 80°C -85°C
c] 85°C- 90°C d] 90- 95°C

1447. In comparison to raw rice bran, parboiled rice bran contains
a] Less starch and more oil
b] More starch and less oil
c] More starch and more oil
d] Less starch and less oil

Knowledge World!

Only domesticated insect

The only domesticated insect is the silkworm or the silkmoth *Bombax mori*. There is no wild population of this moth, which has been bred for over 4,000 years. Closely related wild species occur in China and the Himalayas.

1448. Which of the method, the separation of bran without loss of endosperm is possible?
a] Raw rice milling b] Parboiling rice method
c] Both a] and b] d] None of these

1449. In comparison to raw rice is parboiled rice rich in
a] Protein b] Vitamin
c] Minerals d] All are correct

1450. Compared to raw rice, parboiled rice needs
a] Less polishing
b] More polishing
c] Doesn't need polishing
d] None of these

1451. Parboiled rice is inferior in
a] Starch b] Protein
c] Fat d] Mineral

1452. Cooking quality of rice is represented by
a] Time of cooking
b] Swelling capacity and expansion ratio
c] Gruel quality and pastiness
d] All are correct

1453. A time required by parboiled rice to attain same level of softness in cooking needs for as compared to raw rice is
a] Same b] Double
c] Triple d] None of these

1454. Which of the following rice cooks more flaky?
a] Parboiled rice b] Raw rice
c] Both a] & b] d] None of these

1455. In Jadavpur University Method, the soaking of paddy is done at high temperature of water for a period of
a] 1 -3 hours b] 3 -5 hours
c] 5- 8 hours d] 10 -12 hours

1456. The rice kernel which are lesser than 6/8 of the actual size are called
a] Head yieldb] Head rice
c] Broken rice d] All are correct

1457. In pounding method, rice recovery ranges from
a] 55 -60 % b] 60- 65 %
c] 65 -68 % d] 68- 70 %

1458. A pounding method of rice milling requires
a] Mechanical power b] Electrical power
c] Human power d] All are correct

GK **Orsan is the main tributary of Narmada .** GS

1459. The most common paddy huller used in India is
a] Englberg huller b] Under runner disc huller
c] Rubber roll Sheller d] All are correct

1460. The speed of roller in huller mill is
a] 500- 700 rpm b] 600- 900 rpm
c] 600- 1200 rpm d] 1000- 1500 rpm

1461. The capacity of paddy huller varies from
a] 200 -500 kg/hr b] 250- 750 kg/hr
c] 300- 800 kg /hr d] 250- 1000 kg/hr

1462. In case of parboiled rice, the average yield obtained from huller is
a] 58- 60 % b] 62- 64 %
c] 64-68 % d] 68-70 %

1463. In case of raw rice, the average yield obtained from huller is
a] 52 % b] 56 %
c] 60 % d] 65 %

1464. The shelling of paddy by centrifugal dehusker is due to
a] Shear b] Crushing
c] Rubbing d] Impact

1465. 157, A speed of centrifugal dehusker ranges between
a] 1500-2000 rpm b] 2000-2500 rpm
c] 2500- 3000 rpm d] 3000-3500 rpm

1466. A capacity of centrifugal dehusker is about
a] 450 kg/h b] 500 kg/h
c] 550 kg/h d] 600 kg/hr

1467. The shelling of paddy in rubber roll sheller is due to
a] Rubbing b] Crushing
c] Shearing d] All are correct

1468. In rubber roller sheller, the clearance between rubber roll is
a] Equal to paddy thickness
b] Smaller than paddy thickness
c] Greater than paddy thickness
d] None of these

1469. Dehusking capacity of rubber roll sheller in different climatic condition ranges between
a] 25- 50 tonnes/hour
b] 50- 100 tonnes/hour
c] 100- 200 tonnes/hour
d] 200- 300 tonnes/hour

1470. A white shiny talc powder and glucose solution is added to the rice during
a] Polishing b] Glazing
c] Whiteningd] Hulling

1471. Glucose is added to the rice with water solution in the ratio of
a] 1:1 b] 1: 2
c] 1:3 d] 1:4

1472. Dry talc powder required to add to rice by weight of
a] 1-1.2 % b] 1-1.5 %
c] 1-1.75 % d] 1-2 %

1473. One kilogram glucose solution is sufficient for
a] 500 kg rice b] 1000 kg rice
c] 1500 kg rice d] 2000 kg rice

1474. Percentage of protein in pulse by weight is
a] 10- 15 % b] 15 -20 %
c] 17-25 % d] 20-28 %

1475. A bearing used in belt conveyor is
a] Friction b] Antifriction
c] Bush d] Ball

1476. The belt speed for transportation of grams should not increase by
a] 3.5 m/s b] 4.5 m/s
c] 5.0 m/s d] 6.5 m/s

1477. The recommended belt speed for grain conveying ranges is
a] 2.3- 2.5 m/s b] 2.5- 2.8 m/s
c] 30-3.5 m/s d] 3.5-4.5 m/s

1478. For belt conveying of paddy, the most suitable trough angle is
a] 10°C b] 15°C
c] 20°C d] 25°C

1479. The flat belt idlers are used for granular materials having an angle of repose not less than
a] 20°C b] 25°C
c] 30°C d] 35°C

1480. The capacity of bucket elevator varies as
a] 1-500 tonne/h b] 2-1000 tonnes/h
c] 3- 1500 tonnes/h d] 4- 2000 tonnes/h

1481. The capacity of bucket elevator mainly depends on
a] Bucket design, size and spacing
b] Number of buckets and characteristics of material
c] Way of loading and unloading of material
d] All are correct

1482. The speed of bucket elevator ranges from
a] 1.5 to 3.0 m/s b] 2.5 to 4.0 m/s
c] 3.0 to 4.5 m/s d] 4.0 to 6.0 m/s

1483. In bucket elevators, the total stretch of belt under maximum load should not exceed more than
a] 1 -2 % of belt length b] 1 -2.5 % of belt length
c] 1- 3 % of belt length d] 1- 3.5 % of belt length

1484. The type of belt used in bucket elevator is
a] Duck type
b] Solid woven cotton
c] Striched canvas type
d] All are correct

1485. A screw conveyor has
a] Triangular trough b] U-shape trough
c] V-Shape trough d] Trapezoidal shape trough

GK **Akbar raised the age of marriage to 14 for girls and 16 for boys.** GS

1486. The screw conveyors are generally used to move grains
a] Vertically b] Horizontally
c] At 45° inclination d] All are correct

1487. The specific heat (Kcal / kg °C) of grains ranges from
a] 0.15-0.25 b] 0.25-0.35
c] 0.35 -0.45d] 0.40 -0.60

1488. The percentage of shelling outturn of groundnut decorticator is
a] 50 b] 60
c] 75 d] 80

1489. The per cent of edible oil in rice bran is about
a] 10-12 b] 14-16
c] 18-20 d] 22-24

1490. ' R ' pocket type disc separators are used for separation of
a] Broken rice from whole rice
b] Stones from grains
c] Mustard from wheat
d] Cowpeas from rice

1491. 'V' pocket (vetch) type disc separator is used for separation
a] Oblong shape material
b] Irregular shape material
c] Round shape material
d] All are correct

1492. After pre-treatment (pulse with oil) the pulses are kept on floors for oil diffusion for about
a] 6 hours b] 8 hours
c] 12 hours d] 16 hours

1493. In dry milling process, the following operation are done prior to oil treatment is
a] Conditioning b] Pitting
c] Polishing b] Grading

1494. The amount of heat required to raise the temperature of a given mass substance by unit is
a] Thermal conductivity
b] Specific heat
c] Heat capacity
d] None of these

1495. The number of holes per square inch in a 20-mesh screen will be
a] 20 b] 400
c] 800 d] 1600

1496. One tone of refrigeration implies the cooling effect equal to
a] 144 BTU/sec b] 200 BTU/sec
c] 60 kcal /min d] 120 kcal/min

1497. Percentage of germ in rice, is about
a] 1-2 b] 2-3
c] 3-4.5 d] 4-5

1498. Percentage of bran in paddy
a] 2-4 % b] 4-6 %
c] 6-8 % d] 8-10 %

1499. Shedding and shattering losses of grain in the field can be considerably reduced by harvesting paddy at moisture content of
a] 14-16 % b] 16-18 %
c] 18-23 % d] 23-25 %

1500. Rice yield can be increased with improved mechanical drying method by
a] 1 % b] 1.5 %
c] 2 % d] 3 %

1501. Rice yield can increased due to improved mining equipments by
a] 1-2 % b] 2-3 %
c] 3-4 % d] 4-5 %

1502. Centrifugal and rubber roll sheller may be used in place of
a] Hullers b] Dehusker
c] Polishers d] Graders

1503. Losses in fruits and vegetables generally varies from
a] 1-10 % b] 10-30 %
c] 30-40 % d] 40-50 %

1504. In case of seed production, a maximum grain temperature usually recommended is
a] 35°C b] 43°C
c] 48°C d] 55°C

1505. The high temperature limit for the drying of seed grains is considered as
a] 50°C b] 65°C
c] 70°C d] 80°C

1506. Natural air-drying needs dehumidification if the atmospheric humidity exceeds
a] 30 % b] 50 %
c] 70 % d] 80 %

1507. The temperature in sun drying corresponding to black body is about
a] 3000°K b] 4000°K
c] 5000°K d] 6000°K

1508. In sack drying, a maximum air temperature should be
a] 41.3°C b] 42.3°C
c] 43.3°C d] 45.3°C

1509. Natural airflow takes place in the dryer due to variation in
a] Air moisture b] Air temperature
c] Air-pressure d] Volume of air

1510. Air temperature in solar drying in comparison to sun drying is
a] Equal b] Low
c] High d] None of these

1511. Solar energy reaching per square metre of the earth's atmosphere is called
a] Solar coefficient b] Solar constant
c] Solar radiation d] All are correct

1512. Drying temperature for using batch-bin dyers is
a] 55°C b] 60°C
c] 65°C d] 70ºC

1513. A moisture content of threshed paddy which is sun dried on floor is upto
a] 8-10% b] 10-12%
c] 13-14% d] 15-16%

1514. The operational life of a pair of rubber rolls is estimated at
a] 80 tonnes b] 90 tonnes
c] 100 tonnes d] 120 tonnes

1515. Hardness of rice endosperm ranges from
a] 5 to 7 kg/mm^2
b] 8.5 to 10 kg/mm^2
c] 11.5 to 14.5 kg/mm
d] 12.5 to 16.5 kg/mm

1516. Hot pulses are tempered in the tempering bin after each pass for
a] 4 hours b] 6 hours
c] 48 hours d] 10 hours

1517. Recovery of dhal in each pass is reduced by
a] 0.5-0.7 % b] 0.8-1.2 %
c] 11.5 -2.0 % d] 2.0- 3.5 %

1518. In dry milling of pulses with seed oil treatment, oil is used at a rate of
a] 0.5 to 1.5 kg /tonne
b] 1.5 to 2.5 kg /tonne
c] 2.5 to 3.5 kg /tonne
d] 3.5 to 4.5 kg /tonne

1519. In dry milling, the percentage of pulse dehusked in one pass or single operation is
a] 40 % b] 50 %
c] 60 % d] 70 %

1520. In bullock driven wooden ghani, the percent oil loss (left in cake) is
a] 7-8 % b] 10-12 %
c] 13- 15% 1 d] 16-18%

1521. In power driven iron ghani, the percentage oil loss (left in cake) is
a] 4- 6% b] 7.5- 8.5 %
c] 10.5-11.5 % d] 12.5-13.5 %

1522. In hydraulic press ghani, the percent oil loss (left in cake) is
a] 3- 6 % b] 7- 8 %
c] 8- 10 % d] 10- 12 %

1523. The storage life of non-perishable fruits exceed by
a] 8 weeks b] 10 weeks
c] 12 weeks d] 14 weeks

1524. During processing of fruits and vegetables, transportation losses occurs in as
a] 15 % b] 20 %
c] 25 % d] 30 %

1525. During fruit juice canning, pasteurization is done at the temperature of
a] 71°C b] 74º C
c] 78°C d] 80°C

1526. The total depth of water application to the crop is called
a] Duty b] Delta
c] Base c] None of these

1527. A period in days for which water applied to a crop is
a] Duty b] Delta
c] Base d] None of these

1528. The actual area irrigated in a year from outlet is
a] Irrigation period b] Intensity of irrigation
c] Irrigation frequency d] None of these

1529. The upper strata of the earth where the openings are partially filled with water is
a] Zone of fracture b] Zone of aeration
b] Zone of saturation d] None of these

1530. An aquifer is a permeable bed only partly filled with water and overlying a relatively impervious layer is
a] Pheratic aquifer
b] Water-table aquifer
c] Non arteslan aquifer
d] All of these

1531. The upper strata of the earth where the openings are completely filled with water are
a] Zone of fracture b] Zone of aeration
c] Zone of saturation d] None of these

1532. When a well is dug in water-table aquifer the static water level in the well stands at
a] Upper elevation of water table
b] Same elevation of term
c] Lower elevation of water table
d] None of these

1533. A aquifer found between impermeable layer is
a] Non-artesian aquifer
b] Artesian aquifier
c] Unconfined aquifer
d] Free aquifer

1534. The pressure of water in confined aquifier is
a] Above atmospheric pressure
b] At atmospheric pressure
c] Below atmospheric pressure
d] All are correct

1535. In artesian aquifer, the elevation to which water level rises in wall is
a] Water-table level b] Piezometric level
c] Piezometric head d] Piezometric surface

Seventy per cent of Switzerland is covered by mountain ranges.

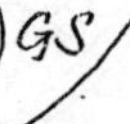

1536. A completely saturated aquifer is
   a] Free aquifer b] Artesian aquifer
   c] Leaky aquifer d] Non-artesian aquifer

1537. In artesian aquifer, the imaginary surface representing artesian pressure is
   a] Water table pressure
   c] Piezometric level
   c] Piezometric head
   d] Piezometric surface

1538. In semi-confined aquifer, when there is time lag between lowering of two water levels is
   a] Semi confined aquifer with prompt yield
   b] Semi confined aquifer with delayed yield
   c] Both a] and b]
   d] None of these

1539. A perched water-table is a special case of .
   a] Artesian aquifer b] Non-artesian aquifer
   c] Confined aquifer d] Non-confined aquifer

1540. The properties of aquifer may be expressed in terms of
   a] Hydraulic conductivity
   b] Transmissibility
   c] Specific yield
   d] All are correct

1541. In semi-confined aquifer with prompt yield, the pheratic level is lowered when
   a] Simultaneously lowered with piezometric level
   b] Simultaneously increased with piezometric level
   c] No change in piezometric level
   d] None of these

1542. Leakage factor is the property of
   a] Confined aquifer b] Semi-confined aquifer
   c] Unconfined aquifer d] All are correct to

1543. Specific yield is the property of
   a] Confined aquifer b] Semi confined aquifer
   c] Unconfined aquifer d] All are correct

1544. The value of leakage factor for semi pervious layer ranges between
   a] 100 and 1000 min b] 100 and 10000 min
   c] 100 and 100000 min d] loo and 1000000 min

1545. The water level stands in the well before pumping is
   a] Static water level
   b] Pumping water level
   c] Piezometric level
   d] Draw down

1546. The difference between static water level and pumping water level is known as
   a] Drawdown b] Well yield
   c] Specific capacity d] All are correct

1547. The volume of water discharged per unit time is
   a] Drawdown b] Well yield
   c] Specific capacity d] All are correct

Quiz Kit!

Longest horns

The Indian Water Buffalo has the longest horns in the animal kingdom. One specific bull had horns which measured 4.24 m (14 ft) from tip to tip along the outside curve of its forehead.

1548. Arch median screw is
   a] Low head water lift
   b] Medium head water lift
   c] High head water lift
   d] All are correct

1549. Specific speed of pump expresses relationship between
   a] Pump speed b] Pump discharge
   c] Total headd] All are correct

1550. A 24-litre capacity bucket is filled in 12 seconds by the discharge of Persian wheel, then the rate of flow is
   a] 2 lit/min b] 20 lit/ min
   c] 120 lit/ min d] None of these

1551. A Persian wheel with an average discharge of 230 lit/ min irrigates one hectare wheat crop in 50 hours. The average depth of irrigation is
   a] 6.9 cm b] 7.2 cm
   c] 9.0 cm d] 9.6 cm

1552. The device used for measuring discharge of irrigation channel, well or canal outlet is called
   a] Weir b] Notch
   c] Meter gate d] All are correct

1553. A sheet of water, which overflows a weir, is
   a] Head b] Nape
   c] Both a] & b] d] None of these

1554. A basic formula to calculate the discharge through weir is
   a] Q = CLH b] $Q = CLH^m$
   c] $Q = LH^m$ d] None of these

1555. A contracted trapezoidal weir in which each side of the notch has a slope of 1 horizontal to 4 vertical is
   a] Trapezoidal weir b] Cipoletti weir
   c] Rectangular weir d] V-notch

1556. A cipoletti weir has a breadth of 60 cm at its crest and head of water flowing over the crest is 30 cm the discharge of weir will be
   a] 180 lit/ sec b] 180 lit/ min
   c] 186 lit/ sec d] 183 lit/min

1557. An open channel flow in which flow characteristic at any point do not change with
   a] Steady flow b] Unsteady flow
   c] Uniform flow d] Non-uniform flow

GK GS

Under Jawahar Rozgar Yojna, 30% employment opportunities have been reserved for women.

1558. An open channel flow in which velocity changes from crossection to crosssection is
a] Uniform b] Non uniform
c] Steady d] Non steady

1559. The velocity of flow in open channel is determined by
a] Darcy's formula b] Manning's formula
c] Both a] & b] d] None of these

1560. A brick or stone masonry linings are recommended for lining water courses from channels where discharge is
a] Less than 30 lit/ sec
b] More than lit/ min
c] Only 30 lit/ sec
d] None of these

1561. A structure used to carry out irrigation water across the stream, canals, gullies and other natural depression is
a] Weir b] Notche
C) Flume d] Gate

1562. A recommended safe limit of land slope for efficient irrigation in heavy soil is
a] 0.05 to 0.25 % b] 0.20 to 0.40 %
c] 0.25 to 0.65 % d] All are correct

1563. The relative proportion of sand, clay and silt determines
a] Soil structure b] Soil texture
c] Soil mass d] Soil volume

1564. The particle size of clay according to USWA is
a] 0.002 mm b] <0.002 mm
c] 0.002 to 0.05 mm d] 0.5 to 1.0 mm

1565. The particle size of silt according to USWA is
a] >0.002 mm b] <0.002 mm
c] 0.002 to 0.05 mm d] 0.25 to 0.5 mm

1566. The particle size of gravel according to USWA is
a] 0.5 to 1 mm b] 1 to 2 mm
c] <2 mm d] >2 mm

1567. The porosity index of relative volume of pores is influenced by
a] Texture b] Structure
c] Both a] & b] d] None of these

1568. In case of porosity the volume of pores may change
a] With volume change of voids
b] With volume change of solids
c] Without volume change of solids
d] All are correct

1569. The porosity of sandy soil usually ranges from
a] 30 to 35 %b] 35 to 50 %
c] 40 to 60 %d] 50 to 60 %

1570. The porosity of clayey soil usually ranges from
a] 30 to 35 %b] 35 to 50 %
c] 40 to 60 %d] 45 to 65 %

1571. The ratio of volume of pores to volume of solids is
a] Porosity b] Void ratio
c] Particle density d] None of these

1572. The true specific gravity of dominant material is
a] 1.65 b] 2.65
c] 3.65 d] 4.65

1573. The density of soil organic matter is usually about
a] 1.15 to 1.29 g/cc
b] 1.3 to 1.5 g/cc
c] 1.5 to 1.65 g/cc
d] 1.65 to 2.65 g/cc

1574. When the bulk density of medium to fine textured subsoil exceeds about 1.7 g/cc the value of hydraulic conductivity of soil is
a] Low b] Medium
c] High d] None of these

1575. A ratio of mass of water to mass of soil is referred to as
a] Volume wetness
b] Gravimetric water
c] Degree of saturation
d] All are correct

1576. A soil consistency in the characteristics of the soil at various moisture content when influenced by physical force is
a] Cohesion b] Adhesion
c] Both a] & b] d] None of these

1577. Water that holds highly to the surface of soil particles by adsorption force is
a] Hygroscopic water b] Capillary water
c] Gravitational water d] None of these

1578. Water held by force of surface tension is
a] Gravitational water b] Hygroscopic water
c] Capillary water d] None of these

1579. The capillary water is held between tensions of about
a] 15 atm b] 31 atm
c] More than 31 atm d] None of these

1580. The capillary adjustment are very sluggish between
a] 15 and 30 atm b] 31 and 15 atm
c] 31 and 41 atm c d] 41 and 15 atm

1581. Infiltration rate is generally lower in soils of
a] Heavy texture b] Light texture
c] Medium texture d] None of these

1582. The viscosity is measured in
a] Dynes b] Newton
c] Poise d] Joule

1583. The viscosity of water at 20 degree centigrade is almost
a] 1 dyne b] 1 dyne / sq.cm
c] 1 poise d] 1 centistokes

GK GS

Air pressure is usually high when the air is cool and dry.

1584. **The capillary rise of the dry range of soil moisture content is**
a] Much faster than wet range
b] Much slower than wet range
c] Uniform in wet range
d] None of these

1585. **The tension of water at saturation capacity is almost**
a] 0 b] 1
c] 10 d] 20

1586. **The moisture tension of a soil at permanent wilting point ranges from**
a] 7 to 15 atm b] 15 to 32 atm
c] 7 to 32 atm d] 15 to 45 atm

1587. **The moisture content at which the wilting is complete and the plant dies**
a] Permanent wilting point
b] Ultimate wilting
c] Wilting range
d] Temporary wilting

1588. **The soil moisture at permanent wilting point may be**
a] 32 atm b] 60 atm
c] 70 atm d] 150 atm

1589. **The indicator plant used for determining the permanent wilting percentage is**
a] Wheat b] Safflower
c] Sunflower d] Paddy

1590. **The soil moisture between field capacity and permanent wilting point is**
a] Free moisture b] Bound moisture
c] Available moisture d] None of these

1591. **The downward movement of water through saturated or nearly saturated soil is response to force of gravity is**
a] Infiltration b] Percolation
c] Seepage d] Inter flow

1592. **The lateral seepage of water in a relatively pervious soil above a less pervious layer is**
a] Percolation b] Infiltration
c] Interflow d] Seepage

1593. **The movement of water in seepage is**
a] Downward b] Lateral
c] Both a] & b] d] None of these

1594. **The saturated flow in soil occurs when**
a] Water is under tension
b] Water is not under tension
c] Both a] & b]
d] None of these

1595. **The application of Darcy's Law is limited by the condition that flow through the porous medium must be**
a] Laminar b] Turbulent
c] Intermediate d] None of these

1596. **Darcy's law is applicable when Reynolds number is**
a] Equal to 1 b] Less than
c] More than 1 d] Less than 2

1597. **The quantity of water going through various individual path of hydrological cycle can be described by**
a] Continuity equation
b] Water budget equation
c] Hydrologic equation
d] All are correct

1598. **The area of land draining into a water course at a given location is known as**
a] Catchments area b] Drainage area
c] Drainage basin d] All are correct

1599. **Total quantity of water in the world is about**
a] 1386 Mkm^3 b] 1386 Mkm^3
c] 1580 Mkm^3 d] None of these

1600. **The quantity of fresh water available is**
a] 25.0 Mkm3 b] 35.0 Mkm3
c] 45.0 Mkm3 d] 55.0 Mkm3

1601. **The percentage of total quantity of water in the world that is saline is**
a] 67 b] 75
c] 88 d] 97

1602. **The annual runoff from the catchments in cm is**
a] 42.0 b] 45.4
c] 47.3 d] 110

1603. **Of the following, which is the form of precipitation?**
a] Rainfall b] Snowfall
c] Hail d] All are correct

1604. **The magnitude of precipitation varies with**
a] Time b] Space
c] Both a] & b] d] None of these

1605. **The maximum size of rain drops about**
a] 4 mm b] 5 mm
c] 6 mm d] 6.5 mm

1606. **The initial density of snow varies from**
a] 0:06 - 0.15 g/cm^3 b] 0.06- 0.25 g/cm^3
c] 0.08- 0.35 g/cm^3 d] 0.08- 0.45 g/cm^3

1607. **In drizzle the size of droplet is**
a] Less than 0.5 mm b] More than 0.5 mm
c] Equal to 0.5 mm d] None of these

1608. **The intensity of drizzle is**
a] 1.0 mm/h b] 1.5 mm/h
c] 2.0 mm/h d] 2.5 mm/h

1609. **The average annual rainfall in India is estimated at**
a] 115 cm b] 119 cm
c] 125 cm d] 165 cm

1610. **One cm of rainfall over a catchment area of one km^2 represents the volume of water equal to**
a] $10^3 m^3$ b] $10^4 m^3$
c] $10^6 m^3$ d] $10^8 m^3$

GK **Ganga delta is the largest delta of the world.** GS

1611. Symons's gauge is which type of rain gauge
a] Recording type
b] Non-recording type
c] Both a] & b]
d] None of these

1612. Which is the recording type of rain gauge
a] Tipping bucket type
b] Weighing bucket
c] Natural siphon type
d] Ail are correct

1613. The standard recording rain gauge adopted in India is
a] Telemetry type
b] Tipping bucket type
c] Weighing bucket type
d] Natural siphon type

1614. Double mass curve technique is used for checking the
a] Consistency of a record
b] Inconsistency of a record
c] Both a] & b]
d] None of these

1615. Mass curve of a rainfall of a storm is plotted between
a] Accumulated precipitation and time in chronological order
b] Accumulated rainfall Intensoty and time
c] Rainfall intensity and time in chronological order
d] None of these

1616. A hyetograph is plot of
a] Rainfall intensity versus time
b] Rainfall depth versus duration
c] Cumulative rainfall versus time
d] Discharge versus time

1617. To convert the point of rainfall value at various stations into an average value over a catchment area is possible by
a] Arithmetical-mean method
b] Thiessen-polygon method
c] Isohyetal method
d] All are correct

1618. An isohyetal is a line joining point having
a] Equal barometric pressure
b] Equal evaporation value
c] Equal rainfall depth in given duration
d] All are correct

1619. Evaporation from water bodies and soil masses together with the transpiration from vegetation is termed as
a] Evapotranspiration
b] Water loss
c] Total loss or total evaporation
d] All are correct

1620. The rate of evaporation is dependent on
a] Vapour pressure at the water surface and air above
b] Air and water temperature and wind speed
c] Atmospheric pressure and quality of water
d] All are correct

1621. The ISI standard pan evaporimeter is the
a] Same as US class A pan
b] Has average pan coefficient 0.7
c] Both a] & b]
d] None of these

1622. Chemical that is found to be most suitable as water evaporation inhibitor is to
a] Methyl alcohol b] Ethyl alcohol
c] Cetyl alcohol d] All are correct

1623. Evapotranspiration is confined
a] Day time b] Night time
c] Both a] & b] d] None of these

1624. Which method is used for estimation of evaporation
a] Water budget method
b] Energy balance method
c] Mass transfer method
d] All are correct

1625. Lysimeter is used to measure
a] Evaporation
b] Evapotranspiration
c] Atmospheric pressure
d] None of these

1626. The lines on a map through places having equal depth of evapotranspiration is called
a] Isohyets b] Isobar
c] Isopleths d] None of these

1627. In which state of India evapotranspiration is highest
a] Rajasthan b] Gujarat
c] Madhya Pradesh d] Andhra Pradesh

1628. Interception losses includes
a] Evaporation loss only
b] Evaporation through flow and stream
c] Evaporation and transpiration losses
d] Stream flow only

1629. Infiltration is measured by the
a] Infiltrometer b] Evaporimeter
c] Hydrometer d] None of these

1630. The average infiltration rate is called
a] Infiltration capacity
b] Infiltration index
c] Both a] & b]
d] None of these

1631. The science which deal with the measurement of flow is known as
a] Hydro-meteorology b] Hydrometry
c] Hypsometry d] None of these

GK

Nana Patil was associated with Satyasodhak Movement.

GS

1632. **Which of the following instruments is connected with stream flow measurement?**
a] Electro-magnetic flow meter
b] Sounding weight
c] Eco-depth recorder
d] All are correct

1633. **The slope area method is extensively used in**
a] Case where shifting control exist
b] Case where back water effect is present
c] Development of rating curve
d] Estimation of flood discharge based on high-water mark

1634. **A hydrograph is a plot of**
a] Stream discharge against time
b] Rainfall intensity against time
c] Cumulative runoff against time
d] Cumulative rainfall vs time

1635. **Direct runoff is made up of**
a] Overland flow
b] Surface runoff and prompt inter flow
c] Surface runoff and infiltration
d] Rainfall and evapotranspiration

1636. **The flow-duration curve is plot of**
a] Discharge versus time in chronological order
b] Stream discharge against per cent of time the flow is equaled or exceeded
c] Accumulated flow versus time
d] None of these

1637. **Flow mass curve is a plot of**
a] Cumulative flow versus time
b] Cumulative discharge volume against time in chronological order
c] Both a] & b]   d] None of these

1638. **Flow mass curve is an integral curve of**
a] Flow-duration curve
b] S-curve
c] Hyetograph
d] Hydrograph

1639. **A situation where more than 25 % decrease in precipitation from the normal over an area is called**
a] Hydrological drought
b] Meteorological drought
c] Agricultural drought
d] None of these

1640. **When the soil moisture and rainfall are inadequate during the growing season to support healthy crop growth to maturity is called**
a] Metrological drought
b] Hydrological drought
c] Agricultural drought
d] None of these

Did You Know ?

Earliest food preservation

The concentration of sugar in honey is so high that bacteria cannot service in it. Honey found in the 4,000-year-old Egyptian pyramids was not spoilt.

1641. **Factor, which affects the flood hydrograph, is**
a] Shape of basin   b] Size of basin
c] Slope of basin   d] All are correct

1642. **The rising limb of a hydrograph is also known as**
a] S-curve   b] Concentration curve
c] Flow mass curve   d] Double mass curve

1643. **A unit hydrograph has**
a] One unit of direct runoff
b] One unit of rainfall
c] One unit of peak discharge
d] All are correct

1644. **A hydrograph produced by a continuous effective rainfall at a constant rate for an infinite period is known as**
a] S-hydrograph   b] Direct run off hydrograph
c] Hyetograph   d] Double mass curve

1645. **Rational formula is only applicable to small size catchments which has an area**
a] Less than 50 km^2 b] More than 50 km^2
c] Less than 5000 km^2 d] More than 5000 km^2

1646. **The use of unit hydrograph for estimating floods is limited to catchments of size less than**
a] 500 km^2 b] 1000 km^2
c] 5000 km^2 d] 6000 km^2

1647. **The hydrological flood routing method used is**
a] Energy equation
b] Equation of motion
c] Equation of continuity
d] All are correct

1648. **The hydraulic flood routing method used is**
a] Equation of motion
b] Energy equation
c] Continuity equation
d] Both a] & c]

1649. **A geological formation which is essentially impermeable for flow of water even though it may contain water in its pores, is called**
a] Aquifuge   b] Aquiclude
c] Aquitard   d] Aquifer

1650. **An aquifer confined at the bottom but not at the top is called**
a] Confined aquifer   b] Semi-confined aquifer
c] Unconfined aquifer   d] Perches aquifer

GK **Article 80 details the composition of Rajya Sabha.** GS

1651. A geological formation through which only seepage is possible is called
a] Aquifuge b] Aquitard
c] Aquiclude d] Perches aquifer

1652. A geological formation which is neither porous nor permeable and there are no interconnected openings hence it cannot transmit water is called
a] Perch aquifer b] Aquiclude
c] Aquifuge d] Aquitard

1653. The unit of intrinsic permeability is
a] Cm/day b] m/ day
c] Cm^2 d] Cm^3

1654. The dimension of the coefficient of transmissibility is
a] L^3/T b] L^3T^2
c] LT^2 d] L/T^3

1655. The specific storage is
a] Specific capacity per unit depth of aquifer
b] Same as specific yield
c] Storage coefficient
d] None of these

1656. The discharge per unit draw down at a well is known as
a] Specific yield b] Specific storage
c] Specific capacity d] None of these

1657. Spattering of small soil particles, caused by raindrop impact on wet soil is called as
a] Inter rill erosion b] Rill erosion
c] Chemical erosion d] All are correct

1658. In splash erosion the raindrop energy acts in form of
a] Kinetic energy b] Potential energy
c] Chemical energy d] None of these

1659. Kinetic energy of raindrop depends on
a] Terminal velocity of raindrop
b] Drop diameter
c] Rainfall intensity
d] All are correct

1660. For attaining a raindrop to terminal velocity, the fall height should be
a] At least 5 m b] At least 7 m
c] Less than 7 m d] 15 m

1661. Detached particles are transported by the mechanism of
a] Displacement caused by physical impact
b] Entrainment caused by overland flow
c] Both a] & b]
d] None of these

1662. Soil loss in geologic erosion is
a] Less than accelerated erosion
b] Greater than accelerated erosion
c] Medium
d] None of these

1663. Geologic erosion is also called as
a] Normal erosion b] Natural erosion
c] Chemical erosion d] Both a] & b]

1664. Accelerated erosion is caused by
a] Water b] Wind
c] Both a] & b] d] None of these

1665. Soil erosion is a
a] Fluvial work b] Geomorphologic work
c] Both a] & b] d] None of these

1666. Soil erosion involves the detachment and transportation of
a] Insoluble soil materials b] Soluble soil materials
c] Both a] & b] d] None of these

1667. Soil erosion refers to the gross amount of
a] Soil loss b] Sediment yield
c] Both a] & b] d] None of these

1668. Denudation rate is expressed in the unit of
a] gram/m^2 b] gm/see
c] m^3/m^2/sec d] Both b] &c]

1669. Erosivity is the
a] Climatic aggressiveness
b] Soil aggressiveness
c] Both a] & b]
d] None of these

1670. Soil erosion is more when
a] Panly impact of raindrop
b] Raindrop impact along with overland flow
c] Only overland flow
d] All are correct

1671. Slope steepness affects more to the
a] Rill erosion b] Splash erosion
c] Sheet erosion d] None of these

1672. Soil Erodibility is a
a] Dynamic process b] Static process
c] Both a] & b] d] None of these

1673. Kinetic energy of raindrop depends on
a] Drop size b] Terminal velocity
c] Particle size d] Both a] & b]

1674. The erosion caused by concentration flow in small channels or rills is called as
a] Gully erosion b] Rill erosion
c] Torrent erosion d] All are correct

1675. The most relevant flow to cause soil erosion is
a] Laminar flow b] Uniform flow
c] Turbulent flow d] All are correct

1676. V -shaped gullies are developed when
a] Now velocity is high but runoff volume per unit time is less
b] Now velocity is less but runoff volume is more
c] Flow velocity is less
d] Runoff volume is more

Article 80 details the composition of Rajya Sabha.

1677. **Mass movement of soil occurs at**
a] Critical slope b] Mild slope
c] Steep slope d] None of these

1678. **Wind erosion is common in**
a] Arid zones b] Arid and semi-arid zones
c] Humid zones d] All are correct

1679. **Detachability of soil particle increases with**
a] Decreasing median grain size
b] Increasing medium grain size
c] Both a] & b]
d] None of these

1680. **Reliability is lower in**
a] Unstable soil b] Stable soil
c] Both a] & b] d] None of these

1681. **V-shaped gullies are found in**
a] Hilly areasb] Flat areas
c] Ravine areas d] All are correct

1682. **The intensity of wind erosion is affected by**
a] Wind speed b] Soil particle size
c] Soil structure d] All are correct

1683. **Wind erosion is influenced by**
a] Clod size b] Wind velocity
c] Soil moisture d] All are correct

1684. **Wind velocity varies with**
a] Temperature b] Density of air
c] Viscosity of air d] All are correct

1685. **In wind erosion movement of particle is carried out by**
a] Saltation and surface creep
b] Suspension
c] Saltation, suspension and surface creep
d] None of these

1686. **Surface creep is carried out by the action of**
a] Rolling and jumping
b] Rolling or sliding
c] Both a] & b]
d] None of these

1687. **Particles less than 0.1 mm diameter is moved by the action of**
a] Suspension b] Saltation
c] Surface creep d] None of these

1688. **A 0.2 mm diameter particles are counted as**
a] Saltation particles
b] Particles of surface creep
c] Suspension particles
d] None of these

1689. **Particles size for saltation process ranges from**
a] 0.1 to 0.5 mm diameter
b] 0.5 to 1.0 mm diameter
c] 0.1 to 0.01 mm diameter
d] None of these

1690. **Wind erosion is more in**
a] Cohesive soils b] Non-cohesive soils
c] Rocky soils d] None of these

1691. **Wind erosion is influenced by**
a] Clod size surface roughness and soil moisture
b] Wind velocity, clod size, surface roughness and soil moisture
c] Wind velocity only
d] None of these

1692. **Critical diameter of clod takes place in the wind erosion is**
a] 0.50 mm b] 0.84 mm
c] 0.10 mm d] 100 mm

1693. **At the wind velocity greater than threshold value, the vibration of particles will be**
a] Zero b] More
c] Slight d] None of these

1694. **Threshold wind velocity is directly proportional to**
a] Soil particles diameter
b] Square root of particles diameter
c] Height from ground surface
d] None of these

1695. **'Sand dune' is associated with**
a] Wind erosion b] Chemical erosion
c] Water erosion d] None of these

1696. **Wind erosion can be controlled by use of**
a] Mulch over soil surface
b] No tillage
c] Both a] & b] d] None of these

1697. **Shelterbelts are constructed by**
a] One row b] Two rows
c] More than two rows d] None of these

1698. **Stubble mulching is associated with**
a] Water erosion b] Wind erosion
c] Glacial erosion d] None of these

1699. **Splash erosion takes place when rainfall intensity exceeds the**
a] Infiltration rate b] Evapotranspiration rate
c] Both a] & b] d] None of these

1700. **Gully erosion is the advance stage of**
a] Splash erosion b] Sheet erosion
c] Rill erosion d] All are correct

1701. **Universal Soil Loss Equation (USLE) is named because**
a] Its validity is more
b] It accounts all parameters affecting the soil loss
c] It is universally adopted
d] All are correct

1702. **USLE was developed for**
a] African condition b] U.K. conditions
c] American condition d] None of these

GK **The World Literacy Day is celebrated on 8th September.** GS

1703. USLE is used for predicting
a] Cropping system b] Farm planning
c] Erosion potential d] All are correct

1704. Which would be more effective for erosion
a] Combination of crops trees and livestock
b] Combination of crop and livestock
c] Combination of trees and livestock
d] None of these

1705. Erosion is the interaction
a] Erosivity and rainfall intensity
b] Erodibility and land use
c] Erosivity and erodibility
d] None of these

1706. Sheet erosion is also called
a] Surface flow erosionb] Direct flow erosion
c] Channel flow erosiond] None of these

1707. Fine clay particles have low detachability but
a] High transportability
b] High dispersive
c] High soil loss
d] None of these

1708. Watershed is a
a] Drainage basin contributing the runoff to a single outlet
b] Hydrological entity
c] Unit of area extent in which hydrologic cycle completes
d] All are correct

1709. Tile position of seepage line is influenced by
a] Permeability of fill materials
b] Permeability of foundation materials
c] Both a] & b]
d] None of these

1710. Settlement of an embankment depends on
a] Void ratio b] Density of soil
c] Water viscosity d] All are correct

1711. Control of gully erosion can be achieved by use of
a] Spurs b] Wooden plank dams .
c] Retaining wall d] None of these

1712. The side slope of bund is function of
a] Bund's height.
b] Angle of repose off ill material
c] Both a] & b]
d] None of these

1713. 188 The common conservation measures for agricultural lands are
a] Bunding and broad based terracing
b] Bench terracing
c] Contour trenching
d] All are correct

1714. Contour blinds are recommended for the areas of
a] Low rainfall b] Medium rainfall
c] High rainfall d] None of these

1715. Graded terraces are used in the areas of
a] Medium to high rainfall
b] High rainfall, only
c] Medium rainfall, only
d] None of these

1716. A nominal size of shoulder bund is provided on bench
a] Terrace sloping inward
b] Terrace sloping outward
c] Level top bench terrace
d] All are correct

1717. The area lost in bench terracing depends on
a] Better slope b] Depth of cut
c] Depth of fill d] Both b & c]

1718. Hill-type bench terraces are the
a] Terraces with reverse grade towards hill
b] Terraces with reverse grade outside the hill
c] Level bench terrace
d] None of these

1719. Contour trenching is done mainly in
a] Low rainfall areas for moisture conservation
b] High rainfall areas for moisture conservation
c] Arid zones for wind erosion control
d] All are correct

1720. Vertical interval of bench terrace depends on
a] Land slope b] Soil depth
c] Both a] & b] d] None of these

1721. Shoulder bunds are provided to
a] Retain the runoff over benches
b] Provide stability to the outer edge of bench
c] Both a] & b]
d] None of these

1722. Contour trenches are constructed
a] On contour of the land
b] Not on the contour of the land
c] Along the land slope
d] All are correct

1723. Contour trenching is performed to
a] Control the landslide from the hill face
b] Conserve the moisture
c] Both a] & b]
d] None of these

1724. Downstream section of straight drop spillway is called as
a] Apron b] Stilling basin
c] Both a] & b] d] None of these

1725. Chute spillways are constructed at
a] Gully head b] Along the gully bed
c] Both a] & b] d] None of these

Most of the disguised unemployment persons in India are found in agriculture.

GS

1726. **Chute spillways are used at**
a] Mild slope b] Steep slope
c] Both a] & b] d] None of these

1727. **Drop-inlet spillways are constructed for**
a] Water detention along the gully bed
b] Controlling water velocity
c] Creating silt deposition
d] Both a] & b]

1728. **Soil texture represents the**
a] Size and proportions of particles making a soil
b] Arrangement of soil particles
c] Both a] & b]
d] None of these

1729. **Which soil is more resistant to erosion?**
a] Sandy soil b] Cay soil
c] Loamy soild] All are correct

1730. **A granular structured soil is**
a] Less erodible b] Least erodible
c] More erodible d] None of these

1731. **Runoff coefficient is the ratio of**
a] Annual rainfall to annual runoff
b] Runoff to rainfall
c] Runoff to rainfall intensity
d] All are correct

1732. **Rainfall intensity is the ratio of**
a] Time to rainfall amount
b] Rainfall amount to time
c] Rainfall amount to falling velocity
d] Both b] & c]

1733. **Curve number method estimates the**
a] Runoff rate
b] Direct runoff
c] Sub-surface runoff
d] Both a] & b]

1734. **Bench terraces are constructed at slope gradient of**
a] 5 to 10 % b] 10 to 20 %
c] Less than 5% d] All are correct

1735. **At steep slope the depth of flow is**
a] Less than the flat surface
b] More than the flat surface
c] Same flow depth
d] None of these

You Don't Know!
First at Antarctica
Giriraj Singh Sirohi, a plant physiologist, was the first Indian to set foot on Antarctica as part of an American team in 1960 to conduct studies on biological clocks.

1736. **At greater temperature the infiltration rate is**
a] More b] Less
c] No effect d] None of these

1737. **Surface sealing problem is less on**
a] Steep slope b] Flat land
c] Vegetated surface d] Both a] & c]

1738. **Primary objective of graded terrace is to**
a] Reduce the field slope length
b] Remove runoff with non-erosive way
c] Both a] & b]
d] None of these

1739. **The primary objective of level bench terrace is to**
a] Conserve water and control erosion
b] Reduce runoff
c] Make steeply land cultivable
d] All are correct

1740. **Crop management system reduces the**
a] Detachment of soil particles
b] Transport of soil particles
c] Both a] & b]
d] None of these

1741. **Grassed waterways are used as**
a] Diversion channels
b] Outlets
c] Inlet to terrace system
d] All are correct

1742. **Bench terracing is also called as**
a] Hill farming
b] Staircase farming
c] Erosion control farming
d] None of these

1743. **Broad base terrace is also known as**
a] Bench terrace b] Channel terrace
c] Drainage type terrace d] None of these

GK **Glacial types of lakes are in Kumaon Himalayas.** GS

Matching

1. Match the following

a]	Large woven wire fencing	i]	6000 to 15000 volts
b]	Close mesh woven wire fencing	ii]	Support entire fence
c]	Barbed wire fencing	iii]	Septic tank
d]	Electric fencing	iv]	Cattle's
e]	Plain wire fencing	v]	Poultry
f]	Farm gate	vi]	Large animals
g]	Farm post	vii]	Small animals and birds
h]	Sanitation	viii]	Goat and rabbits
i]	Woven wire fencing	ix]	14 gauge wire
j]	Close mesh type fencing	x]	Saving inconvenience

2. Match the following

a]	Wire floored house	i]	Milking parlour
b]	Brooder house	ii]	Lofing barn system
c]	Pen barn	iii]	12-14 gauge wire
d]	Milk house	iv]	Dairy barn
e]	Stanchion barn	v]	General purpose barn
f]	Loose housing barn	vi]	Govt or cooperative organization
g]	Open air barn	vii]	Rice husk
h]	Cage house	viii]	Warm region
i]	Deep litter poultry	ix]	Dropping pits
j]	Community barn	x]	Chicks

3. Match the following

a]	Gutter	i]	Space for feeding animals
b]	Stanchions	ii]	Space for collection of egg
c]	Manger	iii]	Space for standing animals
d]	Perches	iv]	Act as perch
e]	Egg retainer	v]	Egg counting
f]	Feeder	vi]	Tough for feeding animals
g]	Trap nest	vii]	Tough for feeding birds
h]	Roost	viii]	Sitting birds
i]	Cow stall	ix]	Tying of animals
j]	Feed alley	x]	Sewage disposal

4. Match the following

a]	Pug mill	i]	Cement
b]	Slump cone	ii]	Strengthening of concrete
c]	Portland	iii]	Compacting concrete
d]	Slaking	iv]	Measurement of strength of concrete
e]	Fuller's formula	v]	Plasticity
f]	Seasoning	vi]	Bricks
g]	Slump test	vii]	Lime
h]	Curing	viii]	Quantity of sand, cement and gravel
i]	Ramming	ix]	Heat flow
j]	Fire bricks	x]	Timber

GK GS

Singarvelu was the president of Indian Communist Conference held at Kanpur in 1925.

5. **Match the following**

a]	Piston	i]	Cloth
b]	Compression ring	ii]	Energy store
c]	Oil ring	iii]	I-section
d]	Piston pin	iv]	Lubrication
e]	Connecting rod	v]	Wrist pin
f]	Flywheel	vi]	SI engine
g]	Primary filter	vii]	Aluminium
h]	Secondary filter	viii]	Leakage
i]	Carburettor	ix]	Diesel engine
j]	Injector	x]	Paper

6. **Match the following**

a]	BHP	i]	Petrol rating
b]	IHP	ii]	Swirl
c]	Inlet valve	iii]	Diesel rating
d]	Intake manifold	iv]	Power on flywheel
e]	Decompression lever	v]	Four stroke engine
f]	Gasket	vi]	Diesel engine
g]	Oil bath	vii]	Asbestos
h]	Octane number	viii]	Two stroke engine
i]	Cetane number	ix]	Power on piston
j]	Combustion chamber	x]	Air cleaner

7. **Match the following**

a]	Venturi	i]	Battery
b]	Governor	ii]	SI engine
c]	Magneto system	iii]	Ignition
d]	Electrolyte	iv]	Regulation
e]	Spark plug	v]	Carburetor
f]	Thermo siphon	vi]	Lubrication
g]	Thermostat valve	vii]	Cooling
h]	Radiator	viii]	Alcohol
i]	Splash	ix]	Spur gear
j]	Oil pump	x]	Fins

8. **Match the following**

a]	Bypass type	i]	Brake
b]	Pressure plate	ii]	Mast
c]	Constant mesh	iii]	PTO
d]	Spline	iv]	Tyre
e]	Swinging	v]	Drawbar
f]	Radial	vi]	Steering
g]	King pin	vii]	Oil filter
h]	Drag link	viii]	Helical
i]	Disc	ix]	Clutch
j]	Upper link	x]	Front axle

9. **Match the following**

a]	Seat	i]	Noise
b]	Super charger	ii]	Vibration
c]	Turbocharger	iii]	Exhaust gases
d]	Ballasting	iv]	Hitching
e]	Weight transfer	v]	Traction
f]	API degree	vi]	Crankshaft
g]	Firing order	vii]	Sequence of power stroke in a multi-cylinder engine
h]	Firing interval	viii]	Gravity measurement
i]	Volatility	ix]	Vaporizing ability
j]	Decibel	x]	Interval between two power strokes

10. **Match the following**

a]	Additive properties	i]	Governor
b]	Viscosity	ii]	Noise and spark
c]	Valve timing	iii]	Tachometer
d]	Lobes	iv]	Camshaft
e]	FIP	v]	Engine speed
f]	Pintle	vi]	MICO
g]	RPM	vii]	Viscometer
h]	Lead-acid	viii]	Injector
i]	Centrifugal	ix]	Fuels
j]	Muffler	x]	Battery

11. **Match the following**

a]	Damper	i]	Piston
b]	Frequency	ii]	Seat
c]	Slogging ability	iii]	Diesel engine
d]	Tread	iv]	Spur gear
e]	Hydraulic pump	v]	Hertz
f]	Crown gear	vi]	Differential
g]	Free play	vii]	Tyre
h]	Taper roller bearing	viii]	Brake pedal
i]	Skirt	ix]	Fuel supply
j]	Sediment bowl	x]	Front wheel

12. **Match the following**

a]	Hunting	i]	Hydraulic control
b]	Dipstick rod	ii]	Removal of burnt gases from cylinder
c]	Bellows type	iii]	Cost estimation
d]	Friction plate	iv]	Noise of machines
e]	Inflation pressure	v]	Brakes horsepower
f]	Prony brake dynamometer	vi]	Tyre
g]	Sound level recorder	vii]	Clutch
h]	Break even point	viii]	Governor
i]	Scavenging	ix]	Oil level
j]	Nudging system	x]	Thermostat valve

13. **Match the following**

a]	Constant speed engine	i]	Tractor engine
b]	Variable speed engine	ii]	Ballasting
c]	Poppet valve	iii]	Steering
d]	Star pinion	iv]	Half time gear
e]	Swinging drawbar	v]	Valve opening and closing
f]	Timing gear	vi]	Differential
g]	Push rod	vii]	Electric generator
h]	Warm gear	viii]	Modern tractor
i]	Inflation pressure	ix]	Side draft
j]	Calcium chloride	x]	Front and rear tyre

14. **Match the following**

a]	Centre of gravity	i]	Oils
b]	High pressure line	ii]	One-third of wheel base ahead from rear axle
c]	Calorific value	iii]	Fuels
d]	Bevel gear	iv]	Fuel injection pump
e]	Otto cycle	v]	Temperature at which fuel catches fire
f]	Flash point	vi]	Petrol engine
g]	Cushion	vii]	Right angle
h]	Balancing method	viii]	Seat
i]	Boiling point	ix]	Centre of gravity
j]	SAE numbers	x]	Thermostat

15. Match the following

a]	Fuel supply system	i]	Thermostat valve
b]	Ignition system	ii]	Centrifugal weight
c]	Cooling system	iii]	Synchronizer
d]	Lubrication system	iv]	Disc
e]	Governing system	v]	Filter
f]	Hydraulic system	vi]	Friction plate
g]	Brake system	vii]	Drag link
h]	Clutch system	viii]	Dipper
i]	Steering system	ix]	Battery
j]	Transmission system	x]	Master cylinder

16. Match the following

a]	HP	i]	IHP-FHP
b]	IHP	ii]	AL
c]	BHP	iii]	ALn
d]	FHP	iv]	2LN
e]	Mechanical efficiency	v]	BHP x 4500/LAN
f]	Piston displacement	vi]	PLAN/4500
g]	Swept volume	vii]	IHP-BHP
h]	Piston speed	viii]	BHP/IHP x 100
i]	Comperession ratio	ix]	V_1/V_2
j]	BMEP	x]	75 kg-m/sec

17. Match the following

a]	Secondary tillage	i]	Direct sowing of seed without land preparation
b]	Ridge planting	ii]	Narrow strip along the row
c]	Combined tillage	iii]	Conservation of soil and moisture
d]	No tillage	iv]	Reduce water runoff
e]	Primary tillage	v]	Mould board or disc plough
f]	Minimum tillage	vi]	High rainfall area
g]	Till plant	vii]	Operation for opening the line of soil
h]	Conservation tillage	viii]	Finer tillage operation
i]	Rotary strip tillage system	ix]	Plough plant
j]	Conventional tillage	x]	CTBFF

18. Match the following

a]	Combine	i]	Harvest cereal crops
b]	MB plough	ii]	Cutting fodder
c]	Subsoiler	iii]	Multi-purpose tool
d]	Chaff cutter	iv]	Weeding in paddy crop
e]	Disc harrow	v]	Cutting grasses
f]	Indigenous plough	vi]	Harvesting, threshing, separation and cleaning
g]	Winnower	vii]	Shallow ploughing
h]	Reaper	viii]	Cut, turn and invert the furrow slice
i]	Mower	ix]	Deep ploughing
j]	Cono weeder	x]	Cleaning of threshed crop

19. Match the following

a]	Slipnose share	i]	Trashy soil
b]	Stubble type MB	ii]	MB with gradual curvature
c]	Shin share	iii]	Slot place along the length of MB
d]	Bar point share	iv]	Stony and root infected soil
e]	Sod type MB	v]	Curved cutting edge with small detachable piece
f]	General purpose MB	vi]	Shin as additional part
g]	High speed MB	vii]	MB with abrupt curvature
h]	Slat type MB	viii]	MB with gentle curvature
i]	Slip share	ix]	Zero till drill
j]	Shoe type furrow opener	x]	Share provided with adjustable bar
k]	Hoe type furrow opener	xi]	Plant debris and sticky soil
l]	Disc type furrow opener	xii]	Used on tractors
m]	Inverted 'T' type furrow opener	xiii]	Curved cuttng edge having no additional part

Seasonal unemployment is a feature of agricultural sector.

20. Match the following

a]	Beam	i]	Share, MB and landside attached rigidly
b]	Landside	ii]	Plough bottom and standard attached
c]	Plough bottom	iii]	Penetrate into the soil
d]	Frog	iv]	Transmit power of animal to implement
e]	Mould board	v]	MB, share frog and land side
f]	Share	vi]	Invert furrow slice
g]	Standard	vii]	Control depth of plough
h]	Coutler	viii]	Lateral stability in the plough
i]	Jointer	ix]	Circular disc
j]	Gauge wheel	x]	Miniature plough

21. Match the following

a]	Vertical disc plough	i]	Group of disc
b]	Concavity	ii]	Influence width of cut
c]	Standard disc plough	iii]	Depth of disc from center
d]	Gang angle	iv]	Number of disc on single shaft
e]	Arbor bolt	v]	Shallow ploughing
f]	Disc angle	vi]	Prevent disc from clogging
g]	Tilt angle	vii]	Maintain space between two adjacent disc
h]	Spacer	viii]	Deep ploughing
i]	Gang	ix]	Influence depth of cut
j]	Scraper	x]	Influence depth and width of cut in disc harrow

22. Match the following

a]	Offset disc harrow	i]	Having four gang one behind other
b]	Harrow plough	ii]	Having fixed mould board
c]	Tandem disc harrow	iii]	Having two gang place end to end
d]	Double action disc harrow	iv]	A mould board turns through 180°
e]	Single action disc harrow	v]	Shallow ploughing
f]	Spring tyne harrow	vi]	One left and one right gang
g]	One-way plough	vii]	Curved knives
h]	Two-way plough	viii]	Tynes swings back when obstacle encountered
i]	Indigenous plough	ix]	Front and back gangs throw the soil in opposite directions
j]	Acme harrow	x]	Multipurpose tool

23. Match the following

a]	Back furrow	i]	Mass of soil cut
b]	Furrow wall	ii]	Ploughing round strip of ploughed land
c]	Crown	iii]	Ploughing round strip of unploughed land
d]	Head land	iv]	Raised ridge left at centre of the strip of land
e]	Dead furrow	v]	Undisturbed face of furrow
f]	Furrow slice	vi]	Start ploughing between first ridge and the side land
g]	Casting	vii]	Top portion of toped furrow slice
h]	Gathering	viii]	Ridge and furrow interface with cultivation
i]	Continuous ploughing method	ix]	Unploughed land for tractor turning
j]	Round to round ploughing	x]	Open trench between two adjacent strips of land

24. Match the following

a]	High volume sprayer	i]	Less than 2 kg per hectare
b]	Low volume sprayer	ii]	Weedicides
c]	Ultra low volume sprayer	iii]	More than 4000 lit per hectare
d]	Fumigator	iv]	5 to 400 lit per hectare
e]	Aero blast sprayer	v]	Lance
f]	Flame gun	vi]	4000 to 9000 rpm
g]	Spinning disc	vii]	Herbicides
h]	Solid cone nozzle	viii]	Less than 5 lit per hectare
i]	Hallow cone nozzle	ix]	Orchard
j]	Spray gun	x]	Weed killing

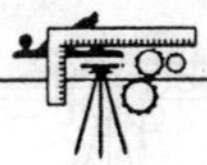

25. Match the following

a]	Draft	i]	Width of cut
b]	Side draft	ii]	MWD
c]	Horizontal suction	iii]	Perpendicular distance between point of share and lower position of the beam of the plough
d]	Unit draft	iv]	Horizontal component of pull parallel to direction of travel
e]	Vertical suction	v]	Resultant of all horizontal and vertical forces
f]	Pulverization	vi]	Straight line passing through the centre of pull, point of hitch and centre of resistance
g]	Plough size	vii]	Maximum clearance under landside and the horizontal surface
h]	Throat clearance	viii]	Horizontal component of pull perpendicualr to direction of travel
i]	Centre of resistance	ix]	Clearance between landside and furrow wall
j]	Line of pull	x]	Draft per unit section area of furrow

26. Match the following

a]	Light plough	i]	15-25°
b]	Disc angle	ii]	150-200 mm
c]	Heavy plough	iii]	3 mm
d]	Tilt angle	iv]	42-45°
e]	Medium plough	v]	< 150 mm
f]	Horizontal suction	vi]	88°
g]	Vertical suction	vii]	>200 mm
h]	Reel index	viii]	3-5 mm
i]	Alignment	ix]	8 cm for 60 cm diameter
j]	Concavity	x]	1.25 to 1.50

27. Match the following

a]	Disc	i]	Wood
b]	Mould Board	ii]	Plastic
c]	Share	iii]	Forged steel
d]	Reel	iv]	Soft centered steel
e]	Flutted roller	v]	Polythene
f]	Frog	vi]	High grade harden steel
g]	Landside	vii]	Rubber
h]	Spool	viii]	Cast iron
i]	Seed tube	ix]	Chilled cast iron
j]	V-belt	x]	Mild steel

28. Match the following

a]	Broadcasting	i]	Planter
b]	Seed drill	ii]	Dropping seed in a continous stream
c]	Check row planting	iii]	An equipment for placement of seedling in the field
d]	Hill dropping	iv]	Placing seeds in hole
e]	Transplanting	v]	Seeds are dropped at fixed spacing not in continuous stream
f]	Drilling	vi]	Seed to seed distance is not maintained
g]	Dibbling	vii]	Planting of seedlings
h]	Transplanter	viii]	Malobansa
i]	Dibbler	ix]	Scattering of seed on surface of soil
j]	Seeding behind plough	x]	Row to row as well as plant to plant distance is uniform

GK **Charles was not tried under Meerut conspiracy case of 1931.** GS

29. Match the following

a]	Star wheel	i]	Transplanter
b]	Reel	ii]	Duster
c]	Flutted roller	iii]	Weeder
d]	Float	iv]	Reaper
e]	Blades	v]	Sprayer
f]	Rasp bar	vi]	Combine
g]	L-shaped blade	vii]	Chaff cutter
h]	Cone	viii]	Rotary tiller
i]	Spinning disc	ix]	Thresher
j]	Blower	x]	Seed drill

30. Match the following

a]	Cast iron	i]	60 to 70 % copper
b]	Grey cast iron	ii]	0.25 to 0.5 % C
c]	Mild steel	iii]	3 % copper
d]	Medium carbon steel	iv]	0.5 to 1.5 % C
e]	High carbon steel	v]	88 % copper
f]	Brass	vi]	0 to 0.75 % C
g]	Bronze	vii]	0.15 to 0.25 % C
h]	White metal	viii]	2.5 to 4.5 % C

31. Match the following

a]	Coal	i]	Up to 10 kW
b]	Wood	ii]	3000 KJ/kg
c]	Biogas	iii]	35-40000 KJ/kg
d]	LPG	iv]	10-50 kW
e]	Small size gasfire	v]	17000 KJ/kg
f]	Medium size gasfire	vi]	50 kW and above
g]	Large size gasfire	vii]	55000 KJ/kg

32. Match the following

a]	Solar cell	i]	Direct radiation
b]	Beam radiation	ii]	Ultraviolet
c]	Scattering	iii]	Beam radiation
d]	Spherical reflector	iv]	Silicicon
e]	Ozone	v]	Diffuse radiation
f]	Pyrheliometer	vi]	Solar furnace
g]	Pyranometer	vii]	Greenhouse
h]	Sunshine recorder	viii]	Global radiation
i]	Flat plate collector	ix]	Duration of bright sunshine
j]	Solar air heater	x]	Home heating

33. Match the following

a]	Box type solar cooker	i]	Wind energy
b]	Solar pond	ii]	Plane mirror
c]	Thermo siphon	iii]	Animal and human waste
d]	Convective loop	iv]	Thermo chemical conversion
e]	Solar energy	v]	Collection and storage
f]	Rotor	vi]	Pumping
g]	Sailboat	vii]	Wind mill
h]	Aero-generator	viii]	Solar water heater
i]	Biomass	ix]	Solar space heating
j]	Gasification	x]	Wind

Rajya Sabha do not have any power in matters relating to Money Bill.

GS

34. Match the following

a]	Aerobic and anaerobic	i]	Wave energy conversion device
b]	Methane	ii]	Nuclear fusion energy
c]	Fixed dome	iii]	Open and close cycle
d]	Digester	iv]	Fermentation
e]	Hot magma	v]	Ocean
f]	Seismic monitoring	vi]	Geothermal energy
g]	Tides	vii]	Biogas
h]	Dolphin	viii]	Gobar gas
i]	Fuel cell	ix]	Slurry
j]	Magneto hydro dynamic system	x]	Heat engine
k]	Thermionic conversion	xi]	Chemical energy
l]	Plasma	xii]	Geothermal reservoirs

35. Match the following

a]	LSU dryer	i]	Below 0ºC
b]	Sack dryer	ii]	Fruit and vegetables
c]	Rotary dryer	iii]	Milk powder
d]	Recirculatory dryer	iv]	Continuous flow dryer
e]	RPEC dryer	v]	Small quantity of grain
f]	Baffle dryer	vi]	Rotating drum (batch dryer)
g]	Cabinet dryer	vii]	Continuous flow non-mixing type
h]	Spray dryer	viii]	Kharagpur
i]	Mechanical dryer	ix]	Zigzag fashion
j]	Freeze dryer	x]	Air flow rate

36. Match the following

a]	Thin layer drying	i]	Solid foods
b]	Deep bed drying	ii]	Sublimation
c]	Tray drying	iii]	Liquid foods
d]	Freeze drying	iv]	Up to 15 cm
e]	Drum drying	v]	More than 15 cm
f]	Unheated air-drying	vi]	Direct radiation drying
g]	Sun drying	vii]	Rate of moisture migration from centre to surface and evaporation of moisture from surface is constant
h]	Solar drying	viii]	Rate of moisture migration from centre to surface and evaporation of moisture from surface differs
i]	Constant rate period drying	ix]	Batch or bin drying
j]	Falling rate period drying	x]	Indirect radiation drying

37. Match the following

a]	Disk separator	i]	Centrifugal force
b]	Indented cylinder separator	ii]	Shape and surface texture
c]	Spiral separator	iii]	Stickiness
d]	Specific gravity separator	iv]	Roundness
e]	Inclined draper	v]	Density
f]	Pneumatic separator	vi]	Length
g]	Fluidised bed separator	vii]	Relative length
h]	Magnetic separator	viii]	Electrostatic force
i]	Cyclone separator	ix]	Photo electric cell
j]	Electrostatic separator	x]	Difference in size and density
k]	Colour separator	xi]	Aerodynamic property

38. Match the following

a]	Morai	i]	1.5 to 60 tonnes
b]	Bukhari	ii]	GI Sheet
c]	Kothar	iii]	2 to 4 tonnes
d]	Mud kothi	iv]	Bihar and Uttar Pradesh
e]	Muda	v]	1 to 20 tonnes
f]	Kanaj	vi]	1 to 50 tonnes
g]	Kuthla	vii]	9 to 35 tonnes
h]	Pusa bin	viii]	1 to 3 tonnes
i]	Bricks and cemented bin	ix]	Eastern and Southern India
j]	Metal bin	x]	3.5 to 18 tonnes

39. Match the following

a]	Henderson equation	i]	Freezing time
b]	Hakins jura equation	ii]	Stress
c]	Plank's equation	iii]	Moisture adsorption by solid material
d]	Ergun's equation	iv]	Grain pressure
e]	Jansen equation	v]	EMC curve
f]	Klevin equation	vi]	EMC model

40. Match the following

a]	Kick's law	i]	Lateral pressure of shallow bin
b]	Fick's law	ii]	Loading pressure on fruits
c]	Bond's law	iii]	Energy requirement in size reduction
d]	Rittinger's law	iv]	Deep bed drying
e]	Fourier law	v]	Molecular diffusion
f]	Hertz contact theory	vi]	Relationship between energy consumed and size attained
g]	Aeiry theory	vii]	Molecular transport of heat
h]	Hukill's analysis	viii]	Size reduction of fine powder

41. Match the following

a]	Black body	i]	Reflect the entire incident energy
b]	White body	ii]	Viscosity
c]	Grey body	iii]	Plasticity
d]	Hookean body	iv]	Absorb a definite fraction of incident radiation
e]	St. Venant body	v]	Visco-elasticity
f]	Newtonian body	vi]	Absorb all incident radiation
g]	Real bio-material	vii]	Elasticity

42. Match the following

a]	Shearing	i]	Classification of cleaned product into quality fraction
b]	Crushing	ii]	Removal of silver skin and bran layer of brown rice
c]	Cutting	iii]	Removal of few large particles in an initial process
d]	Soaking	iv]	Loosen the husk to facilitate its separation from kernal in legumes
e]	Steaming	v]	Removal of foreign and undesirable matters
f]	Canning	vi]	Size reduction having multi-directional rupture
g]	Blanching	vii]	Size reduction by sharp knife
h]	Scalping	viii]	Separation of cleaned product into quality fraction on the basis of size, surface texture and colour
i]	Grading	ix]	Edible oil and water mixed with dal
j]	Sorting	x]	Combination of cutting and crushing
k]	Whitening	xi]	Heat for gelatinisation of a starch
l]	Conditioning	xii]	Process of water absorption
m]	Polishing	xiii]	Vegetable processing
n]	Pitting	xiv]	Fruit processing
o]	Cleaning	xv]	Oil penetration process for loosening of husk

43. Match the following

a]	Universal-moistre meter	i]	Thermal conductivity
b]	Oven	ii]	Specific gravity
c]	Thermometer	iii]	Indirect method of determination of moisture content
d]	Pilot tube	iv]	Direct method of determination of moisture content
e]	Pycnometer	v]	Moisture dishes
f]	Hypodermic needle	vi]	Static pressure of an air stream
g]	Desicator	vii]	Temperature measurement
h]	Hygrometer	viii]	Pressure
i]	Manometer	ix]	Wet bulb temperature
j]	Sling Psychrometer	x]	Relative humidity

44. Match the following

a]	NTU	i]	Packaging
b]	LSU	ii]	Refrigeration
c]	CAP	iii]	Thermal unit
d]	COP	iv]	Dryer
e]	MAP	v]	Storage
f]	HTST	vi]	Unit of refrigeration
g]	Ton	vii]	Pasteurisation

45. Match the following

a]	Hammer mill	i]	Huller
b]	Attrition mill	ii]	Spray drying of liquid foods
c]	Rubber roll Sheller	iii]	Impact
d]	Paddy separator	iv]	Paddy milling
e]	Englberg	v]	Food texture
f]	Rubber rolls	vi]	Feed grinding
g]	Rheology	vii]	Bag storage
h]	Dunage	viii]	Gravity type
i]	Twin fluid nozzle	ix]	Shelling

46. Match the following

a]	Area velocity method	i]	Unsteady state flow confined aquifer
b]	Slip circle method	ii]	Stream measurement
c]	Dupit-Theim's method	iii]	Evaporation
d]	Theis method	iv]	Mean precipitation
e]	Electro-magnetic method	v]	Mean velocity
f]	Cook's method	vi]	Soil moisture monitoring
g]	Panman method	vii]	Pumping test
h]	Time-domain refectory method	viii]	Stability of slope
i]	Orthographically weight isohytel method	ix]	Recharge area
j]	Chow's method	x]	Drainage

47. Match the following

a]	Rational method	i]	Prediction of flood peak
b]	Curve number method	ii]	Measurement of precipitation
c]	Gumble's method	iii]	Estimation storage coefficient and transmissibility of aquifer
d]	Muskingum method	iv]	Single auger hole
e]	Dip-stick method	v]	Equal rainfall magnitude
f]	Isohytel method	vi]	Peak runoff rate
g]	Slope-area method	vii]	Synthetic unit hydrograph
h]	Jacob's method	viii]	Estimation of flood discharge based on high water marks
i]	Hoogdout's method	ix]	Flood routing
j]	Snynder's method	x]	Direct runoff

GK — **Mausynram in India experiences the maximum rainfall in the world.** — GS

48. **Match the following**

a]	Rational method	i]	Uniform coefficient
b]	Manning's formula	ii]	Critical height
c]	Blaney-Criddle formula	iii]	Lake evaporation
d]	Christensen formula	iv]	Flow rate
e]	Cypress Creak formula	v]	Discharge of rectangular weir
f]	Dicken's formula	vi]	Depth of flow
g]	Francis formula	vii]	Flow in open channel
h]	Darcy's-Weisbach formula	viii]	Unconfined aquifer
i]	Meyer's formula	ix]	Potential evapo-transpiration
j]	Khosla's formula	x]	Relationship between monthly runoff and rainfall
k]	Kennedy's formula	xi]	Peak runoff
l]	Dupuit's formula	xii]	Flood peak area relationship

49. **Match the following**

a]	Glover-Duman equation	i]	Drainage
b]	Laccy's equation	ii]	Conservation of energy
c]	Bernoulli's equation	iii]	Catchments area
d]	Laplace equation	iv]	Lake evaporation
e]	Universal soil loss equation	v]	Drain spacing
f]	Water budget equation	vi]	Estimation of aquifer parameter
g]	Dalton's equation	vii]	Average energy soil loss
h]	St. Vanant equation	viii]	Velocity in channel
i]	Theis equation	ix]	Unsteady open channel flow

50. **Match the following**

a]	Pitot-static tube	i]	Static pressure of flowing fluid
b]	Lysimeter	ii]	Velocity of flow
c]	Piezometer	iii]	Rainfall
d]	Tensiometer	iv]	Non-recording type
e]	Current meter	v]	Wind speed
f]	Pressure plate apparatus	vi]	Storm rainfall
g]	Rain gauge	vii]	Recording type
h]	Symons' gauge	viii]	Evapotranspiration
i]	Tipping-bucket type gauge	ix]	Capillary potential
j]	Colorado sunken pan	x]	Soil moisture tension
k]	Infiltrometer	xi]	Difference in total and static pressure
l]	Anemometer	xii]	Small change in water elevation
m]	Bubble gauge	xiii]	Infiltration
n]	Planimeter	xiv]	Evaporation
o]	Radar measurement	xv]	Leaf surface area

51. **Match the following**

a]	Darcy's law	i]	Pressure
b]	Pascal's law	ii]	Construction of venturi meter
c]	Blight creep theory	iii]	Inertia force to gravitation force
d]	Bernaulli's theorem	iv]	Flow through porous medium
e]	Reynolds number	v]	Seepage flow
f]	Froud number	vi]	Inertia force to viscous force

52. **Match the following**

a]	Envelop curve	i]	Stream discharge versus per cent time the flow in equaled or preceded
b]	S-curve	ii]	Unit hydrograph
c]	Mass curve of rainfall	iii]	Concave side
d]	Double mass curve	iv]	Flood peak area relationship
e]	Flow mass curve	v]	Convex side
f]	Flow-duration curve	vi]	Change of region of rain gauge station
g]	Scoring due to curve	vii]	Accumulated precipitation versus time in chronological order
h]	Sitting due to curve	viii]	Stream flow versus time in chronological order

Shyamji Krishna Verma founded Home Rule Society in London.

53. Match the following

a]	Light rain	i]	One station for every 100000 sq. km zone
b]	Moderate rain	ii]	Trace to 2.5 mm per hour
c]	Heavy rain	iii]	One station for every 50000 sq. km zone.
d]	Arid zone	iv]	Stream flow versus time chorological order
e]	Humid temperate climate	v]	2.5 to 7.5 mm per hour
f]	Cold region	vi]	More than 7.5 mm per hour
g]	Hydrograph	vii]	Peak runoff rate
h]	Hyetograph	viii]	One station for every 30000 sq. km zone
i]	Unit hydrograph	ix]	Rainfall intensity versus time
j]	Time of concentration	x]	One cm of rain off

54. Match the following

a]	Ephermeral stream	i]	Sewage water/rain like irrigation
b]	Effluent stream	ii]	Bubbler/twin flood nozzle
c]	Sprinkler irrigation	iii]	Perennial stream
d]	Ring basin irrigation	iv]	Tubewell
e]	Drip irrigation	v]	Parashall flume
f]	Lift Irrigation	vi]	Peak flood
g]	Border irrigation	vii]	Trickle irrigation

55. Match the following

a]	Tubewell	i]	Tubewell
b]	Positive displacement	ii]	Tubewell spacing
c]	Semi-open pump	iii]	Propeller
d]	Diffuser	iv]	Systematic record of different strata
e]	Radial flow	v]	Total head
f]	Cone depression	vi]	Water harvesting
g]	Well log	vii]	Takes water from shallow depth
h]	Open well	viii]	Centrifugal pump
i]	Water pumping	ix]	Submersible pump
j]	Rainfed farming	x]	Booster pump

56. Match the following

a]	Sub-surface drainage	i]	Rainfall
b]	Tile drain	ii]	Suitable slope less than 3 m
c]	Hoogoute	iii]	Suitable slope more than 3 m
d]	Wind erosion	iv]	Mole plough
e]	Water erosion	v]	Drainage
f]	Erosivity	vi]	Saltation
g]	Erodibility	vii]	Earth dam
h]	Chute spillway	viii]	Hilly area
i]	Drop spillway	ix]	Envelop material
j]	Water erosion control	x]	Soil
k]	Bench terrace	xi]	Stream erosion
l]	Shoulder bund	xii]	Strip cropping

Two Anglo-Indians can be nominated by the President in Lok Sabha.

Glossary of Technical Terms

'V' Notch It is commonly used to measure the small and medium size of flow streams.

Agricultural Drainage Removal of free/gravitational water both from the surface and below the surface of a farm land in order to create the conditions favourable for plant growth is known as agricultural drainage.

Agricultural Surveying Agricultural surveying comprises laying out of contour terrace, drainage, profile lines, establishment and preparation of property-maps, computation of farm and fields areas as well a laying out farm buildings and roads.

Agriculture Engineering The application of knowledge and techniques of engineering in the field of agriculture to raise the crop production and standard of living of rural mass is defined as agricultural engineering.

Aquifer The formation or strata within the saturated zone below the ground surface from which ground water can be obtained are called aquifers.

Artesian Well The water rises to a height equal to the piezometric level and this level (piezometric) is higher than the gound level, water flows out due to its own pressure and such a well originating from a broken confining stratum is known as artesian well.

Basic Infiltration Rate Either the constant or nearly constant rate at which the soil takes in water after an extended period of time denotes the Basic infiltration rate.

Bearing It is the direction of a line with reference to the given meridian (directions).

Bench Mark (BM) Bench mark is a relatively permanent point of reference whose elevation is known with reference to some assured datum.

Bore It is the diameter of the cylinder expressed in centimetre.

Bunding In the slopy land, the bunding is the measure to check the runoff water either by storing the water in the field itself or the drainout the water with safe flow velocity without eroding the soil.

Capacity of Pump It is the volume of water pumped per unit time. It is expressed in litres per second, cubic metre per second.

Centre or Resistance All the forces acting upon a plough at work may be considered as balanced about a theoretical point located between the mouldboard and the landslide is called Centre or resistance.

Chaff Cutter The machine which chops the fodder is known as chaff cutter.

Chisel Plough Any type of cultivator that is capable fo working compacted soil at moderate depth is indiscriminately called a chisel plough.

Cone of Depression The piezometeric surface around a well existing in the form of an inverted cone which is devoid of water has been named as cone of depression.

Confined Aquifer The aquifer laying below an impervious layer and wherein water is under pressure is known as confined aquifer.

Contour Its an arbitrary line joining the points of same elevations.

Contour Interval The contours and the contour lines differing by a fixed vertical distance are known as contour interval.

Datum Datum is any surface to which the elevations are referred.

Delivery Type Spinning disc A broadcasting machine embodying a hopper, often of truncated cone shape, situated above a disc which rotates at high speed is delivery type spinning disc.

Delta It is the total depth of irrigation given to a crop in centimetres.

Detention Storage/Surface Detention The water availabe for infiltration or overland runoff is known as detention storage/surface detention.

Diesel Engine In this type of engine the addition of heat takes place at constant pressure but the rejection of heat takes place at a constant volume.

Discer Usually very wide, embodying moderately small discs in gangs of six and capable of independent vertical movement on the main frame is a popular mounted implement called discer.

Draft It can be defined as the force per unit area of soil cut. It is expressed in kg. Sometimes it is also called the 'pull'.

Draw Bar Horse Power (Dbhp): The actual power available on the draw bar of tractor to pull the implement attached with it is called draw bar horse power.

Duty It is the ratio between the irrigated areas and the quantity of water used.

Engine Cycle It consists of the number of event taking place in each cylinder of engine between two successive explosions.

Engine Size It is expressed as Diameter (D) x Stroke length (L) in centimetre.

Engineering It can be defined as the art and science by which the natural forces and materials are utilized in proper ways to find out its utility by changing its forms. In real sense it is the applied science.

Feed Type Agitator One of the simplest in operation and best for sowing baic slag or any other free-flowing material is the delivery roller called Feed Type Agitator.

Field Book The measured position of the objects on field is recorded in a recording book called field book.

Field Capacity Per hour area covered by a machine is termed as field capacity of machine.

Field Efficiency It is ratio between the actual field capacity and the theoretical field capacity.

Flail Mowers The term flail mowers is normally applied to mowers employing a horizontally arranged shelf with winging cutting blades rotating at high speed in a direction opposite to that of the tractor wheels.

Fly Levels Fly levels, taken at certain representative points, provide an approximate idea of the nature of ground enablig the engineer to select the line along which to take the detailed levels to prepare project designs.

Jhumpa Lahri is the author of *Namesake*.

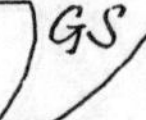

Fogging In this method of application of pesticides the mist is formed in surrounding of the crops.

Fore or Forward Bearing The bearing of a line in the direction of progress of survey is known as fore or forward bearing.

Formation Level Formation level is the level to which the earth work will have to be carried out.

French Drains/Blind Inlets When the quantity of surface water is small or when the sediment load is too large, the drains likely to be installed are called French drains/ blind inlets.

Froude Number The ratio of the inertia force to gravity forces is known as Froude Number and can be written as

$$Fn = \frac{V_1}{\sqrt{gD_1}}$$

Geodetic Surveying Geodetic surveying is deonted as trigonometrical surveying which considers the curvature of the earth because large distances and areas are surveyed.

Gross Irrigation Requirement Net irrigation requirement divided by field efficiency of the system under field conditions equals gross irrigation requirement.

Heat Engine The machine is known as heat engine when it receives heat energy and transforms it into mechanical energy.

Horse Power It is unit of measurement of power.

Hydrokinetic When torque converter is found in automobile transmission system, the energy is transmitted by the oil impelled at high speed to impinge on the blades of turbine, such a transmission system is known as hydrokinetic.

Infiltration Opportunity Time The vertical distance along the border (furrow) between the advance and recession curves at the point denotes Infiltration opportunity time.

Intensity of Irrigation The actual area irrigated from outlet denotes Intensity of irrigation.

Internal Combustion Engine Heat is generated by burning of fuel in enclosed chamber.

Irrigation Efficiency It refers to that how efficiently the available supply of water is being used.

Irrigation Frequency It refers to the number of days between irrigation during crop period.

Levelling It is a branch of surveying consists of the determination of the relative elevation of the points on the surface of the earth.

Local Attraction It is the term used to denote any influence due to any magnetic field which prevents the needle from pointing to the magnetic 'North' during the survey.

Local Attraction The deviation of compass needle due to various factors from the magnetic meridian is called Local attraction.

Magnetic Bearing Magnetic Bearing of the line represents the angle which a line makes with the magnetic meridian.

Magnetic Declination Magnetic Declination denotes the angle between the true meridian and the magnetic meridian.

Mangum Terraces An American farmer in 1885 introduced the broad based terraces in order to carry out the tillage operation over the entire structure, these are also refered to as mangum terraces.

> **You Don't Know!**
> ***First at Antarctica***
> ***Giriraj Singh Sirohi, a plant physiologist, was the first Indian to set foot on Antarctica as part of an American team in 1960 to conduct studies on biological clocks.***

Mole A mole plough has a long blade like shank to which is attached a cylindrical bullet-nosed plug called mole.

Mouldboard The part of plough which inverts furrow slice in a general purpose plough and pulverises it in the digging type is the mouldboard.

Mower This machine is used to harvest the fodder. It is manual, animal and power operated.

Non-consumptive Use of Water The water used for hydro power generation navigation, pollution control, recreation, fish culture etc., has been classified as non-consumptive use of water.

Offsets Offsets denotes the perpendicular distance taken from any point of a chain line to any object of feature.

Open Channel Conduit in which the water flowing with a free water surface is termed as open channel.

Oppel One of the most widely used beet lifting devices, used mainly on, complete harvesters consists of a pair of inclinded lifting wheels called 'Oppel', which pull the roots and can result in low dirt tares.

Optical Square The optical square is a compact hand instrument used to set out right angles with greater accuracy than a cross staff.

Orifice Usually circular, rectangular opening in a vertical buck head (wall) through which water flowing is cut named as orifice.

Pacing It is the rough measurement of distance on ground.

Permanent Filter Strip Diversion ditch located immediately below the inadequately protected fields from sheet erosion need a closely growing vegetation embodying a minimum width of 15 metres and it is called permanent filter strip.

Potato Combiner The machine is used to dig the potato and separate it from the soil. The potato can also be graded and separated from each other.

Pulverization This term is referred to break the clodes after plowing into fine aggregates.

Rain Gauge An instrument consisting of a galvanised iron cylinder 225 cm in diameter and 60 cm high with a funnel top equipped with a brass knife edge periphery commonly is known as rain gauge.

Ranging Ranging means establishment of imtermediate points on a straight line between the terminal points.

Reaper It has a cutting bar which works on the principles of shearing action. It is commonly used for harvesting of cereal crops.

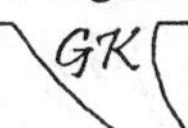

Fiscal policy is related to public revenue and expenditure.

Representative Fraction (RF) The ratio between any length on the drawing and the correspnding length on the ground which is independent of units of measurements is denoted as representative fraction (RF).

Rotary Weeder/Rolling Cutivator An implement with ground-driven spiked wheels, with curved teeth embodying slicing action is a typical rotary weeder/rolling cutivator.

Scale of a Map The scale of a map is the fixed relation that every distance on the map bears to the corresponding distance on the ground.

Seed Drill It is a machine which drops the seed randomly in the furrow at proper depth.

Skidding When the ground wheel does not revolve but the machine proceeds forward the phenomenon is termed as skidding.

Slip It is referred to the free rotation of wheel without covering any forward distance in the ground. It is due to improper grip of wheel with ground. It is determined in percentage.

Soil Resistance It is the resisting force which is imposed on the implement by soil during the field operation. It is expressed in kg/cm^2.

Square Ploughing A system of ploughing popular especially in Fens (U.K.) for deep work, where the fields are well shaped and must be left level is known as Square ploughing.

Statia Haris Levels and theodolites are usually equipped with two horizontal cross hairs known as Statia Haris.

Stroke It is the distance travelled by a piston from the bottom dead centre to top dead centre. It is expressed in cm.

Stroke-Bore Ratio It is the ratio between the piston stroke and diametre of cylinder. In case of engine it varies from 1 to 1.5.

Surveying It is the art of determining of relative positions of various objects on the surface of the earth and location of these object on the map.

Technology It differs from the engineering which is directly related to process of the product. It is the art for processing of a product by combining the engineering and other activities to get the final shape of a produce.

Terrace A terrace is an embankment or ridge of earth constructed across a slope to control the runoff water and minimise the soil erosion.

Threshing In this process the grain is separated from the plants either by an impact action or by rubbing and stripping action.

Tillage Manipulation of soil to create a proper tilth for crop planting is termed as tillage.

Transverse The area between two or more than two survey lines is termed as transverse.

Tube Well Tube well, a long pipe or tube is bored or drilled deep into the ground intercepting one or more water bearing stratum to obtain a larger discharge mechanically.

Vertical Suction In adjustment of plough, the straight edge is placed under the point of the share and the land side, a clearance of 0.3 to 0.5 cm should be measured and this is known as Vertical suction.

Water Course According to northern India canal and drainage act, any channel receiving water from a canal but which is not maintained at the cost of the state including all the subsidiary works of such a channel is known as a Water course.

Weir The weirs are used to measure the flow in an irrigation channel. A weir is a notch of regular form through which the irrigation stream is flowing.

Fiscal policy is related to public revenue and expenditure.

GS

Each word counts

BASICS

Aims of Statistics

There are three aims of statistics

1. To study the population.
2. To study the variation and its causes.
3. To study the methods of reduction of data.

Definition of Statistics

Statistics are the numerical statements of facts capable of being analyzed and interpreted and the science of statistics is the study of principles and methods used in the collection analysis and interpretation of numerical data in any sphere of enquiry.

Limitations of Statistics

The important limitations of statistics are as given below:

1. It does not study qualitative phenomena.
2. Statistical studies are true only on an average.
3. It does not study individuals.
4. It does not reveal the entire story.
5. It is liable to be misused.

MEASURES OF CENTRAL TENDENCY

Central tendency may be defined as a value of the variate which is thoroughly representative of the series or the distribution as a whole. The measure of this character are known as *averages*.

Averages

Types of Averages

The following are the different types of averages which are in common use are :

1. Arithmetic Mean
2. Median
3. Mode
4. Geometric Mean
5. Harmonic Mean
6. Weighted Mean

Arithmetic Mean

This is the only form of average which is of practical importance. Its value depends upon all the observations. If $x_1, x_2, x_3, \ldots\ldots\ldots x_n$ are the N observations. The mean

Quiz Kit!

First visa card

Andhra Bank was the first public sector bank to launch the visa card in India in 1981.

$\bar{x}$ will be given by the following expression :

$$\bar{X} = \frac{1}{N}(x_1 + x_2 + x_3 + x_n) = \frac{\Sigma x}{N}$$

Method of Computation –

(A) Computation from the raw data.

Example 1 : Find out the arithmetic mean of the following fifteen numerical quantities.

62, 85, 103, 107, 109, 110, 114, 115, 126, 151, 154, 161, 165, 170, 175.

Here $A.M = \frac{\Sigma x}{N}$

$$= \frac{62+85+\ldots\ldots\ldots+175}{15}$$

$$= \frac{1907}{15} = 127.133$$

(B) Computation from the frequency distribution

(i) Discrete series :

Example 2 : The arithmetic mean of the following discrete distribution of wages of 100 workers has been computed by calculating the value of N and Σfx.

Daily wages (Rupees) (x)	*Frequency (f)*	*fx*	*Calculations*
2	2	4	Here N = 100
4	5	20	and $\Sigma fx = 1212$
6	8	48	Arithmetic Mean
8	10	80	$= \frac{\Sigma fx}{N}$
10	12	120	$= \frac{1212}{100}$
12	20	240	= 12.12
14	13	182	
16	15	240	
18	11	198	
20	4	80	
	N = 100	$\Sigma fx = 1212$	

(ii) Continuous series (grouped data):

Example : By short cut method

$$\text{Arithmetic Mean} = A + \frac{\Sigma fd}{N}$$

Let A = 25

The minimum age requirement for a person to become member of Lok Sabha is 25 years.

Class interval	Frequency (f)	Mid value (x)	A = 25(d)	Fd	Calculations
0-10	2	5	-20	-40	Here, A = 25
10-20	4	15	-10	-40	Arithmetic Mean = A+ $\frac{\Sigma fd}{N}$
20-30	6	25	0	0	
30-40	8	35	10	80	$= 25 + \frac{200}{30}$
40-50	10	45	20	200	= 25 + 6.666 = 31.67
	N = 30			Σfd = 200	

Merits of Arithmetic Mean

1. It is rigidly defined.
2. It is based on all the observations.
3. It can be easily calculated.
4. Its algebraic treatment is specially easy and definitely possible.

Demerits

1. It is very much affected by the values at extremes and hence, often abnormal results are obtained due to some abnormal value.
2. If the items of the beginning and end of the series are not known, arithmetic mean can not be calculated.

Uses of A.M.

1. It is calculated to compare two or more series with respect to certain character.
2. It is used in calculating the standard deviation of the data.
3. It is calculated for calculating the regression coefficient and the correlation coefficient.

Median

When all the observations are arranged in ascending or descending order of magnitude, the middle one is known as the median.

If N is the total number of observations, the value of the $\left(\frac{N+1}{2}\right)^{th}$ th item will be called median $\left(\frac{N+1}{2}\right)$ which stands for the serial order of the median is known as *median number*.

Method of Computation –

(A) Computation from the raw data.

Example 1 : (When the number of item is odd).

Find the median of the following figures :

10, II, 13, II, 13, 10, 13, II, 12, 15, 12

Here, median number = $\frac{11+1}{2} = 6$

Arranging the figures in the ascending order of magnitude, we get

10, 10, 11, 11, 11, 12, 12, 12, 13, 13, 15

Here 6th figure is 12.

Therefore, the median is 12.

Example 2 : (When the number of item is even).

Find the median of the following figures :

10,7,11,9,8,10,11,7,9,12

Here, median number = $\frac{10+1}{2} = 5\frac{1}{2}$

But, serial order of a term can never be fractional. In such a case there are two middle terms 5th and 6th. The mean of these two terms will be median.

Arranging the terms, we get

7, 7, 8, 9, 9, 10, 11, 11, 12

Here, 5th term = 9

and 6th term = 10

$$\therefore \; Median = \frac{9+10}{2} = 9.5$$

(B) Median for grouped data

We first prepare the cumulative frequency table. Let the total number of observations be *n*. We determine the median class *i.e.*, the class in which $(n/2)^{th}$ observation lies, then we use the formula

$$\text{median} = l + \frac{h}{f_1}\left(\frac{n}{2} - f_0\right)$$

where,

l - is the lower limit of the median class,

n - is the total frequency,

GK Sachin Tendulkar made his cricket debut in age of sixteen. GS

h - is the range (upper limit -lower limit) of the median class;

f_o - is the cumulative frequency of the class just before the median class and f_1 is the frequency of the median class.

Merits of Median

1. It is easy to calculate.
2. It is not affected by the extreme values.
3. It can be located graphically.
4. It is best suited for open-end classes.

Demerits of Median

1. It can be located precisely, when the items are grouped.
2. It is not based on all the observations in the given data.
3. It may not be a true representative of a given data, when the items vary greatly in magnitude.
4. For finding out the median of a data, it is necessary to make an array in ascending or descending order. So when the number of item is quite large, this operation becomes tedious.

Mode

This is the value of the variable which occurs most frequently or whose frequency is maximum. Also, if several samples are drawn from a population, the important value which appears repeatedly in all the samples is called the *mode*.

Mode for Individual Data

Example: The marks obtained in a test by 12 students in a class are given below:

37, 23, 16, 19,34, 23, 5,27,36,23,20,38

Find the modal mark.

Solution: Arranging the data in ascending order.

5,16,19,20,23,23,23,27,34,36,37,38

Obviously, 23 occurs the maximum number of times

∴ Mode = 23

Mode for Grouped Data.

$$\text{Mode} = l_1 + \frac{f_2}{f_1 + f_2} \times i$$

where,

l_1 = lower limit of the modal class

f_1 = Frequency of the next lower class.

f_2 = Frequency of the next higher class

i = width of the class interval

Merits of the Mode

1. It is easily comprehensible.

Did You Know ?

Largest gurudwara

The Golden Temple at Amritsar was built in 1577 by Guru Ram Das, the fourth Guru and completed by Guru Arjun Dev, his successor.

2. It can be obtained simply by inspection. Even if it is to be computed, it can be computed very easily.
3. This is the value whose expectation is the greatest in whole of the statistical series.
4. Neither the extremes are needed in its computation nor is it affected by them.

Demerits

1. In various cases, there is no single and well defined mode.
2. Its computation is not based on all the observations.
3. It is unsuitable for any algebraic treatments.
4. When there are more than one mode in the series, it becomes difficult and takes much time to compute it.
5. In any set of observations, where, there are very small differences among the items of the same size, mode does not prove to be a good representative.

Use of mode

Its use is increasing everyday especially in the field of business.

Measures of dispersion: The measures that are constructed to indicate the spread of the data with respect to the average in a set of observations are called the measures of dispersion.

Range: Range is the simplest measure or dispersion. It is the difference between the highest and the lowest terms of a series of observations.

Mean Deviation

If the deviations of all the observations from their mean are calculated, their algebraic sum will be zero. When this sum is always zero, it is impossible to get the average of these deviations. To overcome this difficulty these deviations are added irrespective of plus or minus sign and then the average is calculated.

The deviations without any plus or minus sign are known as absolute deviations. The mean of these absolute deviations is called the mean deviation about the mean.

It is defined by the following expression :

Mean deviation about the mean

$$= \frac{1}{N}\Sigma|x| = \frac{1}{N}\Sigma|X - \overline{X}|$$

GK GS

Chelliah Committee recommended Income Tax for agriculture income of more than 25,000 p.a.

where, x = Deviation from the mean = $x - \bar{x}$

$|x|$ = Absolute deviation

N = Number of observations.

Standard Deviation

Its calculation is based on the deviations from the arithmetic mean.

In case of mean deviation the difficulty that the sum of the deviations from the arithmetic mean is always zero is solved by taking these deviations irrespective of plus or minus signs. But here, that difficulty is solved by squaring them and taking the square root of their average.

It is defined by the following formula

$$S.D. = \sqrt{\frac{\Sigma(x-\mu)^2}{N}}$$

where, x = An observation or variate value

μ = Arithmetic mean of the population

N = Number of given observations.

Variance

Variance is the square of the standard deviation variance $=(S.D.)^2$

Characteristics

1. If all the variate values are the same, S.D. is equal to zero.
2. S.D. is least affected by fluctuations of sampling.
3. Its computation is based on all the observations.
4. It is affected by the change of scale, but not affected by the change of origin.

Uses

S.D. is used in computing statistical quantities like, regression coefficients, correlation coefficient, etc.

Coefficient of Variance

This is also a relative measure of dispersion. It is important on account of the widely used measure of central tendency and dispersion *i.e.*, Arithmetic Mean and SD.

It is given by –

$$C.V. = \frac{S.D.}{A.M.} \times 100$$

PROBABILITY AND DISTRIBUTION

Probability

The probability of a given event is an expression of likelihood or chance of occurrence of an event. A probability is a number which ranges from zero to one, zero for an event which can not occur and one for an event certain to occur.

Definitions

1. *Experiment and outcomes :*

Any operation on certain object or object which gives different results is called an 'Experiment' and the different possible results is called its 'outcomes'.

Example: Tossing of a coin is an experiment and appearance of head or tail are its outcomes.

2. *Event :*

One outcome or a group of outcomes associated with certain conditions. It is called , '*Event*'.

Example: If one card is drawn from a pack of 52 playing cards, then drawing of: *(i)* queen of heart; *(ii)* a card of spade; *(iii)* king of diamond etc., will be the examples of an event.

3. *Equally likely cases :*

The cases are said to be equally likely when we have no reason to expect any one more rather than the other.

Example: In a throw of an unbiased dice any one out of the six faces may turn up so that all the six faces are equally likely.

4. *Favourable cases :*

Out of all equally likely cases, those cases which fulfil the conditions of the 'Event' or the cases which entail the happening of the "Event" are known as favourable cases for that event.

Example: If a dice is rolled and the event is $x > 4$, then the favourable cases are two viz. , $x = 5$ and $x = 6$, the remaining cases are unfavourable.

5. *Exhaustive cases :*

The total number of possible outcomes of a random experiment in a trial is known as the exhaustive number of cases.

Example: In throwing of a dice the exhaustive number of cases is 6, since anyone of the six faces marked with 1,2,3,4,5,6 may come uppermost

6. *Mutually exclusive events :*

Events are said to be mutually exclusive or incompatible if the occurrence of any one of them prevents the occurrence of all the others *i. e.*, if no two or more of them can occur simultaneously in the same trial.

7. *Simple events :*

The occurrence of a single event is known as the simple event.

8. *Compound event :*

Events obtained by combining together two or more elementary events are known as the compound events.

Example: In a throw of a dice the event getting a multiple of 2, is a compound event because this event occurs if

Loo is the name of very hot winds which blow in north-central part of India in summer.

GS

anyone of the elementary events 2, 4 or 6 occurs.

9. ***Independent event :***

Events are said to be independent if the happening of one event is not affected by the happening of others.

Example: If two cards are drawn from a well shuffled pack of 52 cards one after the other with replacement, then getting an ace in the first draw is independent to getting a king in the second draw but if the first card drawn in the first draw is not replaced, then second draw is dependent on the first draw.

Correlation

If the change in one variable is accompanied by a change in the other variable in such a way that an increase in one variable results in an increase or decrease in the other and also greater change in one variable results in a corresponding greater change in other, then the two variables are said to be correlated.

Positive Correlation

- If the increase or decrease in one variable results in a corresponding increase or decrease in the other correlation is said to be positive.
- If the increase or decrease in one variable results in a corresponding decrease or 'increase, in other correlation is said to be negative.
- If the change in one variable is followed by corresponding and proportional change in other variable, correlation is as said to be perfect.

The method to determine whether two variables are correlated or not

Scatter Diagram Method –

In this method we use the rectangular coordinate axes to mark a dot corresponding to each pair of x and y values and thus obtain as many points as the number of ordered pairs in the given bivariate distribution. This diagram of dots is called the scatter diagram.

Regression

If the variables in a bivariate distribution are the correlated, then points in the scatter diagram generally condense around a curve, which we call the curve of regression and we say that there is a curve of linear regression between variables.

The Properties of Regression Coefficients –

1. The correlation coefficient and the two regression coefficients are of the same sign.

2. The correlation coefficient is the geometric mean between the regression coefficients.

3. Both the regression coefficients cannot be numerically greater than unity.

4. Regression coefficients are independent of change of origin but not of scale.

You Don't Know!

Longest coastline

Gujarat has 1,600 km (26 per cent) on the 6,100 km long coastline of peninsular India.

Distributions

(a) Binomial Distribution

It is a very useful distribution for dealing with discontinuous variates. It is the distribution in which the different frequencies form a binomial series. Binomial series is obtained by expanding a binomial on the basis of the binomial theorem.

$(q+p)^n = q^n + {}^nC_1\, q^{n-1}\, p + {}^nC_2 q^{n-2} p^2 + \ldots\ldots\ p^n$

(b) Normal Distribution

Normal distribution represented by following equation:

$$Y = \frac{N}{\sigma\sqrt{2\pi}} e^{-\frac{1}{2}\left(\frac{x-\mu}{\sigma}\right)^2}$$

(c) Poisson Distribution

$$e^{-m}\left(1+m+\frac{m^2}{2!}+\frac{m^3}{3!}+\ldots..\infty\right)$$

where, $m = np$

This expression is known as Poisson distribution.

FIELD DESIGN AND TEST OF SIGNIFICANCE

Experimental Designs

1. *Treatments*: The objects of comparison, which an experimenter has to try in the field for assessing their values are known as treatments e.g. the varieties, manures, cultivation practices etc.

2. *Experimental Units*: Experimental material is divided into a number of ultimate units to which the treatments are applied and are known as experimental units.

3. *Experimental Material*: This is the name given to the material on which the experiment is performed *e.g.* a field, soil, patients in a hospital etc.

Basic Principles of Field Experimentation

There are three basic principles of field experimentation which have been discussed below:

1. *Replication*: The allocation of the treatments under investigation is known as replication.

2. *Randomization*: The allocation of the treatments to the different experimental units by a random process is known as randomization of treatments.

GK **The normal term of a member of Lok Sabha is 5 years.** GS

3. *Local control*: The principle of making use of greater homogeneity in groups of experimental (e.g. blocks of a number of plots homogeneous within themselves) for reducing the experimental error as explained in the foregoing paragraph is known as local control. This is done by dividing the experimental material (*e.g.*, field) into homogeneous groups of experimental units (*e.g.*, blocks).

Condition for Application of Designs

(i) C.R.D. is applied when the experimental material is limited and homogeneous, such as the soil in the pot experiments.

(ii) R.B.D. is appropriate when the fertility gradient of the field is in one direction. It may be adopted upto 20 treatments without an appropriate loss of efficiency.

(iii) L.S.D. can replace the R.B.D., when the fertility gradient is in two directions instead of one. It may be adopted for number of treatments ranging from 5 to 8 or from 5 to 12 at the most.

(a) Completely Randomized Design

This is the simplest type of design in which the whole experimental material is divided into a number of experimental units depending upon the number of treatments and the number of replications for each. After that the treatments are allotted to the units entirely by chance.

Advantages

1. The method of analysis remains simple when the results from some units are rejected.

2. This design is specially useful in small experiments where, the supply of experimental material is scarce and homogeneous.

3. In this design any number of treatments and replicates may be used.

4. The statistical analysis of the data is very easy.

5. The relative loss of information due to missing data is smaller than that with any other design.

Application Conditions of C.R.D.

1. C.R.D. is used when the experimental material is limited in quantity and homogeneous.

2. Where it is expected that some of the units will be destroyed.

3. In small experiments where the increased accuracy from the alternative design is not sufficient to exceed in importance the loss of error degree of freedom.

(b) Randomized Block Design

In this design the whole experimental material is divided into homogeneous groups, each of which constitutes a single replication. Each of these groups is further divided into the number of experimental units which are equal in all respects. The treatments are applied to these units by any of random process.

Some important points are given below about this design:

1. In R.B.D., the number of blocks must be equal to the number of replications fixed for each treatment.

2. The number of plots in each block should be equal to the number of treatments.

3. Randomization of treatments in each block should be afresh.

4. The experimental error within each block are to be kept as small as practically possible and the variation from block to block as great as possible. In this way all the treatments which are assigned to one block experience the same type of environmental effects and therefore comparable.

Advantages

1. When the material is heterogeneous the residual variance can be reduced by choosing blocks or plots such that the plots within any block are fairly similar.

2. This design allows of any number of treatments and any number of replications but when the number of treatments is very large the efficiency of error control decrease.

3. If a part of the experimental material is damaged by some agricultural disaster like insect, floods, waterlogging etc.

4. By means of grouping more accurate results are usually obtained than with the C.R.D.

5. Although a reduction in the number of replications leads to a larger standard error, yet it furnishes a result of some value at least.

(c) Latin Square Design

In this design it is essential that each row and each column should contain the same application of treatments and hence, the variation between the row means and between the column means can be assessed and eliminated from the error increasing the precision of the estimates.

- The experimental design which simultaneously control the variation in two directions is known as Latin Square Design.
- The L.S.D. is very reliable to give precised results when the number of treatment is from 5 to 8 or at most 12. This design should not be used for less than 5 treatments.

The important points about L.S.D.

1. No. of replications = No. of treatments

2. No. of rows = No. of column = No. of treatments,

3. Randomization of treatments is done in such a way that each treatments occurs once and only once in each

row and each column.

Advantages

1. This design gives precision high enough to reduce the standard error to less than 1 %.
2. L.S.D. control the variations due to two sources of variation.

Disadvantages of L.S.D.

1. The analysis becomes very complicated when several plot yield are missing.
2. An L.S.D. with two treatments is not possible.
3. L.S.D. is less flexible in nature.

Test of Significance

The statistical procedure for deciding whether the difference under study is significant or non- significant, is called the test of significance.

Null hypothesis is the hypothesis which is a test for possible rejection under assumption that it is true.

Alternate hypothesis: Any hypothesis which is complementary to the null hypothesis is called alternate hypothesis. This is denoted by H_1.

Two Type of Error

1. *Type Ist error*: Type 1st error is committed when we reject the hypothesis when in reality it is true. The probability of committing a type Ist error is denoted by α.

2. *Type IInd error*: Type IInd error is committed by not rejecting the null hypothesis when it is false. The probability of committing a type IInd error is denoted by β.

Level of Significance

The maximum probability of rejecting the hypothesis when it is true or in other words, the maximum probability of type Ist error is known as level of significance.

χ^2 (Chi-square)- test

The value of the quantity of χ^2 is calculated by the help of the following expression :

$$\chi^2 = \Sigma\frac{(O-E)^2}{E}$$

where, *O* = Observed frequency in a cell

Knowledge World!

Largest Hanuman statue

The steel city of Rourkela, Orissa, houses a 75 ft concrete idol of Lord Hanuman perched on a 10 ft high pedestal. It is part of a five-acre temple complex dedicated to various god and goddesses known as Hanuman Vatika. The sculptor is Togu Laxmana Swamy.

E = Expected frequency of the same cell

S = Summation taken over all the cells

Expected value calculated by the following equation :

$$E = \frac{RT \times CT}{N}$$

where, *E* = Expected frequency

RT = Rows total

CT = The column total for the column containing the cell.

N = The total number of observation.

Properties of χ^2 Distribution

Some properties are given below:

1. χ^2 ranges from 0 to ∞ and is always positive.

2. Mean of the chi-square distribution is equal to V, and variance = 2 V, where V stands for degree of freedom.

3. When there is a perfect agreement of the observed with the hypothetical distribution of frequencies, the value of χ^2 is zero.

Applications of χ^2-Test

1. Testing the homogeneity of correlation coefficients.
2. Testing the independence of attributes.
3. Testing the expectation of a ratio.
4. Comparison with expectation of normal Binomial and Poisson distributions.
5. Application in genetic problem and detection of linkage in genetics.
6. Comparison of a sample variance with population variance.

The most important source of public revenue is taxes.

Points to Ponder

- Chi-square is an index of dispersion
- A student tossed a coin 800 times, getting 440 heads. The value of chi-square in relation to hypothesis that heads and tails will be equally likely 8.
- The use of 5 per cent and 1 per cent levels of significance is simply a working convention.
- Statistical inference is concerned with attempts to make quantitative statement about the properties of a population form a knowledge of the results given by a sample.
- A point estimate is stated single number as an estimate of a some quantitative property of the population.
- An interval estimate is a statement that a population parameter has a value lying between two specified limits.
- The target population denotes the aggregate about which an investigator tries to make inference from the sample.
- In a frequency distribution, the values in the sample are grouped into a limited number of classes.
- A test of significance is, in general terms, a calculation by which the sample results are used to throw light on the truth or falsity of a null hypothesis.
- A null hypothesis is a specific hypothesis about a population that is being tested by means of the same results.
- Chi-square= Σ (observed-expected)/(expected) is a test criterion for the null hypothesis that the population with some attribute in the population has a specified value.
- Normal distribution is completely determined by two constants (parameters) *viz.*, mean (m), which locates the centre of distribution and standard deviation (σ), which measures the spread or variation of the individual measurements.
- The rescaled measurement is given by the relation Z = X - μ/σ
- The quantity Z in rescaled measurement is known as Standard normal variety/deviate.
- The sample standard deviation denoted by s, is most widely used in statistical work and $s = \sqrt{\frac{\Sigma(x-\bar{x})^2}{n-1}} = \sqrt{\frac{\Sigma x^2}{n-1}}$
- The standard deviation $\bar{x}$,σ is often called alternatively, the SE (Standard Error) of $\bar{x}$.
- Standard error and standard deviation are synonymous.
- In estimating the mean height of large group of boys with a standard deviation of 1.5 inches, a sample of 56 boys must be taken if the standard error of mean height is to be 0.2 inch.
- Compute the mean (μ) and standard deviation (σ) when perfect dice are thrown, the probability is 1/6 that each of the faces 1, 2, 3, 4, 5, 6 turns up = μ =3.5, σ = 1.71
- If the boys and girls are equally likely, the probabilities that a family of size two contains 0, 1, 2, boys respectively 1/4 , 1/2 and 1/4. The mean and standard deviation for this population will be μ =1, σ = 0.71 respectively.
- The heights of a random sample of 16 persons from a population with s = 2.6 inches, the confidence probability (p), the x does not differ from m by more than 1 inch will be 0.876
- If you wish to estimate the monthly house rent in a city correct to within ± Rs. 20, apart from a 1 in 20 chance, and standard deviation (σ) is about Rs. 60; n = 36 houses must be included in the sample.
- The "students" t-distribution was discovered by W.S. Gosset in 1908 and perfected by R.A Fisher in 1926.
- An investigative survey on growth of school children revealed the sample mean height of 265 boys of 13.5-14.5 years was 63.84 inches, with standard deviation σ = 3.08 inches, the 95 per cent confidence interval for μ will be 63.5 to 64.2 inches.
- In describing the amount of variation in a population, a measure often used is Coefficient of variation (C = σ/m).
- The coefficient of variation is often between 5 and 15 per cent.
- Coefficient of variation denotes the ratio of averages having the same unit of measurement and hence is independent of unit.
- In an experiment on 20 rats gave ΣX^2 =23,000, σ = 35 g as suspected, as a check test of null hypothesis: σ =26 g, against the alternative; σ >26 g, X_2 = 23,00/ $(26)^2$ =34.02, d.f = 19
- In a sample of 61 patients, the amount of an anaesthetic to produce anaesthesia required for surgery was found to have a standard deviation of σ =10.2 mg (from patient to patient), the 90 per cent confidence limit will be 8.9 and 12.0 mg by using X^2 0.950 and X^2 0.50.
- If 3 + 4 + 5 + 8 + 2 + 3 = 25 samples have |t| >2.262; X^2 value to test the hypothesis that 5 per cent of the population values are more than 2.262 will be 0.0124.
- In X^2 Goodness of fit test, the data are grouped into classes to form a frequency distribution and the sample mean $\bar{X}$ and standard deviation s are computed.
- In a population, a measure of kurtois is the average values of $(X-\mu)^4$ divided by σ.
- K. Pearson produced a family of theoretical non-normals curves intended to simulate the shapes of frequency distribution having any special values of λ_1 and λ_2

Nineteen per cent of Indian land is under forest cover.

GS

provided, the non-normality was not too extreme.

- The investigations are often designed to discover and evaluate differences between effects rather than the effects themselves.
- In an investigation when two groups or two procedures are compared, these procedures are called treatments.
- The objective of pairing is to enhance the precision of the comparison of the two procedures.
- The variables on which we pair, should predict accurately the performance of the subjects on the measurement by which the effects of treatments are to be judged.
- Environmental variations often calls for pairing.
- A population variance is defined as the average, over the population of the squared deviations for the population mean.
- The pH readings of a soil with a modified quinhydrone electrode were 5.78, 5.74, 5.84, and 5.80, while with modified $Ag/AgCl^-$ electrode, the readings were 5.82, 5.87, 5.96, and 5.89; with the hypothesis that μ_1-μ_2 = 0 and t = 2.66
- Two sets of maize grains (kernels) harvested at full maturity (A) and dough stages (B) were tested for crushing resistance, the resistance (pounds) for set A were 50, 36, 34, 45, 56, 42, 53, 25, 65, 33, 40, 42, 39, 43, 42, and for set B were 43, 44, 51, 40, 29, 49, 39, 59, 43, 48, 67, 44, 46, 54, 64; the test of significance of the difference between the two means t = 1.38.
- Randomization does not guarantee to balance out the natural difference between the members of a pair exactly.
- Randomization gives each treatment an equal chance of being allotted to any subject giving unusually good or poor response exactly as assumed in the theory of probability.
- Randomization serves primarily to protect against sources of bias that are unsuspected
- If σ_D = 5, 17 pairs are needed to make the half width of the 90 per cent confidence interval for the difference between the two population means = 2.
- Distribution free methods are needed when sampling from population far from normal is done.
- Range-test like t-test assumes a normal distribution and has become popular, particularly in industrial work.
- Range-test is more sensitive than t-test to skewness in the population and to the appearance of gross errors.
- The median of a population has the property that half the values in population exceed it and half fall short of it.
- If the distribution is symmetrical about it's mean, the mean and the median coincide.
- In reporting on frequency distribution from large samples, the investigators often quote percentiles of the distribution.

Knowledge World!

First credit card

Diner's credit card entered the Indian market in 1969.

- For the 179 cows, the 90th percentile is estimated by order statistic whose number i =(180) (90)/100 = 162.
- The t-test is sometimes called a parametric test while the sign-test is called non-parametric.
- In sampling from normal distribution the efficiency of the sign-test relative to the t-test is about 65 per cent.
- The sign-test save time at the expenses of a loss of sensitivity in the test.
- The signed rank-test due to Wilcoxon, is another substitute for the t-test in paired samples.
- The Mann-Whitney test applies to unequal group sizes as well as to equal group sizes.
- The practice of quoting the test that agrees the ones predilections vitiates the whole technique.
- In a paired two-sample, the ten values of differences D were 3, 3, 2, 1, 1, 1, 1, 0, 0, 1. The randomization test gives P =3/64 = 0.047, while the value of t for continuity is 2.457 corresponding to a P value of about 0.036.
- In his studies of inheritance Galton developed the idea of regression.
- The correlation coefficient (r) is a measure of the degree of Closeness between two variables.
- Correlation coefficient (r) is a pure number without units or dimensions and always lies between -1 and +1.
- The population correlation coefficient P and its estimate r are intimately connected with a bivariate population known as the bivariate normal Distribution.
- If $\overline{X}$ =83 g, $\overline{Y}$= 60 mg, åC^2 = 1000, åY^2 = 6854, XY = 2302, then the regession of body weight X on comb weight Y of chicks will be X = 83 + 0.336 (Y-60)g.
- A convenient way to draw samples from a normal bivariate population is by use of an old device called common elements.
- Assuming that μ_1, μ_2, μ_3 independent draws from the same population, and the $X_1 = 3\mu_1 + \mu_2 + X_2 = 3\mu_1 + \mu_3$; the correlation P between X_1 and X_2 will be 0.9.
- The study of Probability began around three hundred years ago.
- If an event is satisfied by any of a group of mutually exclusive outcomes, the probability of the event is the sum of the probabilities of the outcomes of the group.
- In second of probability, the condition of mutual exclusiveness is essential.

GK GS

A member of Lok Sabha addresses his resignation to the Speaker.

- In a series of independent trials, the probability that each of a specified series of events happen is the product of the probabilities of the individual events.
- A bag contains the letter *A, B, c, D, f, e, G, h, I*; if each letter is equally likely to be drawn, the probability drawing a capital letter, a vowel and either a capital or a vowel will be 4/9, 1/3 and 5/9 respectively.
- Ordered classifications are common in the study of human behaviour and preferences, and more generally whenever different degrees of some phenomenon can be recognized.
- The variance ratio *F* was first tabulated by Fisher in whose honour the criterion was named *F* by Snedecor.
- Split-plot/nested design is often desirable to get precise information on one factor and on the interaction of this factor with a second, but to forego such precision on the second factor.
- The essential feature of the split-plot experiment is that the sub-plot treatment are not randomised over the whole large block, but only over the main plots.
- With more than two variables, an extension of the bivariate normal distribution called multivariate normal distribution forms the basic model in correlation studies.
- The coefficient is the correlation between variables 1 and 2 in a cross section of individuals all having the same value of variable 3; the third variable is held constant so that only 1 and 2 are involved in the correlation partial correlation.
- The multiple correlation coefficient applies to the situation in which one variable, say *Y*, has been singled out to examine its joint relation with the other variables.
- The analysis of covariance is a technique that combines the features on analysis of variance and regression.
- The discovery of a precise description of the relation between two or more quantities is one of the problems of curve fitting, known as curvilinear regression.
- The law that the rate of increase at any stage is proportional to the size already attained is described mathematically as $dW/dX = cW$ where c is the constant.
- The probability of selection need not be equal for all members of the population : it is sufficient that these probabilities be known.
- In order to apply probability, have some way of sub-dividing the population into units, called sampling units, which form the basis, for the selection of the sample.
- If a sample of 4 from the 16 townships of a count has a standard deviation 45, show that the standard error of the mean is 19.5.
- Systematic sampling is easier to draw, since only one random number is required, and it distributes the sample more evenly over the listed population.
- In stratified sampling, the size of the sample chosen to be taken from any stratum.

GK **Armin Meiwes was the name of Germany's self-confessed cannibal.** GS

Multiple Choice Questions

1. Arithmetic mean is most commonly used because:
a] Based on all the observations
b] Largest and smallest observations
c] Located in the middle of the series.
d] None of the above

2. As a general rule, the number of classes in the frequency distribution should be about:
a] 30 b] 15
c] 6 d] None of the above

3. Biometry deals with:
a] Observation with living things
b] Collection of observations
c] Interpretation of data
d] None of the above

4. Coefficient of variation is calculated by the formula:
a] $\frac{\text{Mean}}{\text{S.D}}$x100 b] $\frac{\text{S.D.}}{\text{Mean}}$x100
c] $\frac{\text{S.D.}}{\text{Mean}}$ d] $\frac{\text{Mean}}{\text{S.D.}}$

5. Coefficient of variation is useful for comparing:
a] Between different populations
b] Same population
c] Both a] and b]
d] None of the above

6. Correlation coefficient is:
a] More than +1
b] Less than -1
c] Lies between -1 and +1
d] None of the above

7. Correlation of continuity in 2x2 contingency table should be used when expected frequency of any cell is :
a] Greater than 5 b] Less than 5
c] Less than 10 d] None of the above

8. For any two values a and b, the following relationship among Airthmatic mean (A) Gemetric mean (G) and Harmonic mean (H) exists:
a] A = HG^2 b] G = A.H.
c] G2 = A.H. d] A = G.H.

9. For compairing 4 treatments with R.B design, the 15 d.f. for error can be obtained when the number of replications are:
a] 5 b] 6
c] 11 d] 12

10. For testing the hypothesis about mean on one population we use t-test when:
a] Sample size is large and population S.D. not known
b] Sample size is small and population S.D. not known
c] Sample size is small and population S.D. known
d] Sample size is large and population s.d. known

11. For testing the independence of two attribute the test used is:
a] t-test b] S.D.-test
c] X2 - test d] F-test

12. For testing the significance of corelation coefficient, we use:
a] Chi-square test b] F-test
c] S.N.D.-test d] t-test

13. For testing the significance of difference between two means, we use the following test:
a] X^2-test b] F-test
c] t-test d] None of above

14. F-test can be used for testing the significance of:
a] Serval differences
b] A single difference
c] Not used for testing the difference
d] None of the above

15. If all variate values are negative, the S.D. will be:
a] Positive b] Negative
c] No definte sign d] Impossible to calculate

16. If fertility variation in the field is in two directions at right angles, which of the experimental design is suitable?
a] CRBD b] RBD
c] Latin Square d] None of the above

17. In a fertilizer trail, 40 kg N/ha has to be applied to the plot of 10.0 sq.m. area. What quantity of urea is needed?
a] 95 b] 89
c] 115 d] None of the above

18. In a field experiment the net plot size was 2.5 x 4.0 m and the yields were recorded in kg. What is the conversion factor for q/ha?
a] 10 b] 100
c] 50 d] None of the above

19. In a normal distribution
a] Mean and median are equal
b] Median and Mode are equal
c] All the three are equal
d] None of the above

20. In an experimental plot, 7 rows were sown with 50 cm row spacing and 5 m length, while harvesting one row on either side of the plots were discarded. The net plot size will be:
a] 2.0 x 5.0 m
b] 2.5 x 5.0 m
c] 3.5 x 5.0 m
d] None of the above

21. In field experiments the commonly used design is:
a] C.R.design b] R.B.design
c] L.S.design d] None of these

GK **Interest payment is the most important component of public expenditure.** GS

22. In industries for quality control, the most important measure of dispersion is:
a] S.D b] M.D.
c] Range d] C.V.

23. Latin square design is suitable for compairing:
a] 5 to 12 treatments
b] Three treatments
c] 15 treatments
d] 2 to 4 treatments

24. Local control helps in
a] Reducing the experimental error
b] Homogeneity of experimental error
c] Both these
d] None of the above

25. Range of distribution is:
a] The difference between largest and smallest observations
b] Averages of all observations
c] Middle of the series
d] None of the above

26. Standard deviation is:
a] Expressed in the same unit of measurements
b] Free from unit of measurements
c] It is in percentage
d] None of the above

27. The A.M. of age of 20 students in a class is 15 years. If we include the age of class teacher also, then the mean the increase by one year. The age of class teacher:
a] 16 years b] 21 years
c] 36 years d] 25 years

28. The analysis of variance technique is used for:
a] Compairing the means of two population
b] Compairing the means of more than two population
c] Compairing the variances of more than two population
d] Compairing the variances of two populations.

29. The correlatin coefficient lies between:
a] 0 to + 1 b] -1 to 0
c] -1 to +1 d] 0 to 1.5

30. The degree of freedom for error in R.B. design with 10 treatments and 4 replicatin will be:
a] 30 b] 27
c] 36 d] 40

31. The degree of freedom to test the significance of difference between two means based on n_1 and n_2 observations is:
a] $n_1 + n_2$ b] $n_1 + n_2 - 1$
c] n1 + n2 -2 d] None of above

32. The first step of summarising the data is:
a] Tabulation
b] Graphical representation
c] Classification
d] Collection

33. The following relationship among the correlation coefficients(r) and the two regression coefficients b_{yx} and b_{xy} exists:
a] r1 = bxy bxy b] $r = \sqrt{b_{xy} \quad b_{xy}}$
c] $r = \frac{b_{yx}}{b_{xy}}$ d] None of the above

34. The formula for G.M. on n values $X_1, X_2........X_n$ is :
a] $(X_1, X_2........X_n)^n$
b] $(X_1, X_2........X_n)^{4/9}$
c] (X1, X2........Xn)1/n
d] $(X_1, X_2........X_n)^N$

35. The measure of cetral tendency to be used to study average ratio of change in the population is:
a] A.M. b] Mode
c] G.M. d] Median

36. The minimum sample size for using chi-square test should be:
a] 100 b] 30
c] 50 d] 5

37. The most commonly used measure of central tendency is:
a] Harmonic mean b] Mode
c] Geometric mean d] Arithmatic mean

38. The randomization of treatments in the field experiments is essential:
a] The statistical procedures hold good only when randomization is done
b] To average out the effects of environmental differences
c] For comparison of treatment effects:
d] None of the above

39. The range of the values -6, -8, -2, -12, -20 will be:
a] +18 b] -18
c] -26 d] -14

40. The regression coefficient lies between:
a] -1 to + 1 b] -0 to ∞
c] -1 to +1 d] None of above

41. The relation between A.M., G.M. and H.M. is:
a] AM>HM>GM b] AM>HM>GM
c] AM>HM<GM d] AM>HM<GM

42. The relation between mean deviation and standard deviation is:
a] M.D. is greater than S.D.
b] M.D. is less than S.D.
c] M.D. is =2/3 S.D.
d] M.D. = S.D.

43. The standard deviation for th values 4,5, 6, 7, 8 will be:
a] $(5)^{1/2}$ b] (2)1/2
c] (2) d] $(2.5)^{1/2}$

GK GS

Teak tree are found abundantly in Chhatisgarh.

44. The student's t-test was discovered by:
a] R.A. Fisher b] F. Yates
c] P.C. Mahalanobis d] W.S. Gosset

45. The suitable measure to find average speed when time for each speed is fixed would be:
a] A.M. b] G.M.
c] H.M. d] None of these

46. The test of significance of differences between two means in small samples is known as:
a] *t*-test
b] F-test
c] On the basis of normal deviate
d] None of these

47. The value of X_2 always lies between:
a] 0 to ∞ b] .00 to 00
c] 0 to +1 d] -1 to 0

48. To find the average heights of plants, we should use:
a] A.M. b] Median
c] Mode d] H.M.

49. To find the average of quantity prices, the measure of cetral tendency to be used is:
a] H.M. b] G.M.
c] A.M. d] None of these

50. To find the average size of shoes sold in the market, we should use:
a] A.M. b] G.M.
c] Median d] Mode

Quiz Kit!

Extinct duck

The pink headed Duck (*Rhodonessa caryophillacea*) which used to be found from Bihar and Orissa to Assam and Manipur was last recorded in 1935.

51. To find the averge speed of a vehicle when same distance is covered with different speeds, the suitable measure is-
a] A.M. b] H.M.
c] G.M. d] None of these

52. To reduce the experimental error with heterogenous material we need:
a] More replications
b] Use of randomization
c] Use of local control technique
d] More replication & use of local control technique

53. To test the agreement between observed frequencies and expected frequencies we use:
a] t-test b] S.N.D-test
c] Chi-square test d] F-test

54. Under the assumption for analysis of variance, the parent population should be:
a] Binormal b] Normal
c] Poisson d] Any type

Matching

1. Match

a]	Correlation coefficient	i]	0 to 1
b]	Regression coefficient	ii]	- a to+g
c]	Probability	iii]	3 Median-2 Mean
d]	Mode	iv]	- to +1

2. Match

a]	AM	i]	Difference between the value of largest and smallest
b]	Normal Distribution	ii]	AMXHM
c]	Range	iii]	Best of all averages
d]	GM2	iv]	Bell shaped

3. Match source of variation and Degress of freedom in CRD-ANOVA.

a]	Treatment	i]	1(n-1)
b]	Error	ii]	r (t-1)
c]	Total	iii]	(1+1)

4. Match sources of avariation and Degree of freedom in RBD-ANOVA.

a]	Replication	i]	(r+-1)
b]	Treatment	ii]	(r-1) (t-1)
c]	Error	iii]	(t-1)

5. Match sources of variation and Degree of freedom in LSD-ANOVA.

a]	Column	i]	(t-1)
b]	Row	ii]	(t-1)
c]	Treatments	iii]	(t-1)
d]	Error	iv]	(t-1) (t-2)
e]	Total	v]	(t^2-1)

6. Match the following.

a]	Null hypthesis	i]	(H_1)
b]	Alternate hypothesis	ii]	(H_0)
c]	Level of Signifiance	iii]	Conclusion
d]	Interpretation	iv]	(5% to 1%)

7. Match the following.

a]	A.M.	i]	$\sqrt{\frac{\Sigma d^2}{N}}$
b]	S.D.	ii]	$\frac{\Sigma x}{N}$
c]	C.V.	iii]	$\frac{\sigma}{\sqrt{N}}$
d]	S.E.	iv]	$\frac{\sigma}{\mu} \times 100$

8. Match the following.

a]	Population	i]	Aggregate of all units
b]	Median	ii]	Positional average
c]	Mean	iii]	Arithmetic mean
d]	Central values	iv]	Average

GK GS

Hindu god Kamadeva is also known as Kusumesa, as his arrows are decorated with flowers.

9. Match the following.

a]	Replication	i]	Homogenous
b]	Randomization	ii]	Repeatation
c]	Local control	iii]	Equal chance

10. Match the following.

a]	CRD	i]	Widely adopted in Agricultural Research
b]	RBD	ii]	Limited use for Agricultural Research
c]	LSD	iii]	Flexible design
d]	Factorial Design	iv]	Interaction effect can be precisely determined

11. Match the following.

a]	Empirical model	i]	Attempt to explicitly represent casuality between variables
b]	Mechanistic model	ii]	Describe relationship between variables without referring to any underlying biological or physical structure that may exist between variables.
c]	Phenological model	iii]	A broad class of models that predict crop development from one growth stage to another.
d]	Stochastic model	iv]	Based on the probability of occurance of some event or exogenous variable.

12. Match the following.

a]	Population	i]	The entire bulk of the observational material from which samples are drawn.
b]	Sample	ii]	Any character, quantitative or qualitative, which changes from observation to observation.
c]	Variable	iii]	A relatively small portion of a population.
d]	Frequency	iv]	The number of observations that fall within a particular class interval.

GK GS

From tax revenue, excise duty is the largest single source of the government's earning.

Glossary of Technical Terms

Absolute Mean Devaluation The ratio of the sum of the absolute deviation measured for the arithmetic mean of a given series to the number of observations present in a given series is absolute mean deviation.

Arthmatic Mean The arthmatic mean can be defined as the ratio of the sum of the observational values to its number of observations presented in a given series or distribution.

Attribute A characteristics of population which can not be measured quantitatively but described qualitatively e.g., honesty, morality, skill, sex.

Biometry When the principles of statistics are applied to living organism or living thing, the science is called biometery eg., age of cow, height of plant.

Bivariate Population A population in individual exhibits paired value of two variable known as bivariate population.

Census Survey In census survey each and every item of population is studied so that nothing is left being inquired e.g., population census.

Central Tendency The centrally located value which represents the group of values.

Chi-square Test The Chi-square test is defined as the sum of the ratio of square of difference between observed frequency and expected frequency to the expected frequency of the respective.

Class The different groups which are to be formulated on the basis of similarity of individual are termed as classes.

Class Interval The difference between upper limit and lower limit of a given class.

Classification Defined as process by which the data are arranged in groups or classes according to resembleness or similarities.

Co-relation It is the method of studying association or relationship between two variables.

Co-relation Coefficient It is the measure of degree of association relationship between variables.

Coefficient of Skewness The measure of the degree of the departure from the symmetry (asymmetry) is known as coefficient of skewness.

Coefficient of Variance (C.V.) It is defined as the ratio of standard deviation to the mean of a given series expressed into percentage.

Confidence Interval or Limit The ratio between the lower limit and upper limit is confidence limit.

Contingency Table When individual in a same have two characters and a frequency distributes to be made by classifying them on the basis both the characters so as to show the relation between the two characters the resulted termed as contingency table.

Continuous Variable It is the quantity which can take any numerical value in the given range e.g. weight of student, milk yield, height of plant.

Data Defined as mass or aggregates of variable.

Degree of Freedom (d.f) Degree of freedom is defined as the number of observation minus independent restriction which are to be imposed.

Discontinuous/Discrete Variable It is the quantity which can take only specific value *i.e.* whole figures without fraction and there is definite gap between two values e.g., number of boys in class, number of plants, per hectare.

Error In statistic, error refers to a difference between population parameters and sample values.

F-test It is a ratio of bigger variance to smaller variance.

Frequency The number of observation included in a given class or number of times an observation is repeated.

Frequency Distribution The manner in which frequencies are distributed over the different classes.

Frequency Table The table indicating the frequency distribution.

Geometric Mean (G.M.) The G.M of 'n' positive non zero observation say X_1 X_2,.........Xn is the nth root of the product of the numbers.

Harmonic Mean (H.M.) The ratio of number of observations to the sum of the reciprocal of the value of different observations.

Hypothesis Any statement is to be made in respect to the parameters is termed as hypothesis.

Hypothetical Population Hypothetical population is one which does not exist in reality but it is only in our imagination e.g., tossing of a coin. **Kurtosis** It refers to the degree of flatness or peakness in the region about the mode of a frequency curve.

Level of Significance The maximum probability of rejection the hypothesis when it is true. Maximum probability of type-I error is known as level of significance.

Median If the observation are arranged in ascending or descending order of magnitude array the value of centrally located obervation is considered as median.

Mid Value/Class Mid Point It is the value laying half way between upper and lower limit of a given class.

Mode That value of the variety which occurs most frequently.

Mode Class A class of a given distribution having a maximum frequency.

Moment The arithmatic mean of various powers of the deviation in any distribution is called moment.

Mutually Exclusive Events 1. When the events are such that occurrence of any one is incompatible with the occurrence of any of the others, they are known as mutually exclusive events. 2. Two events A and B are said to be mutually exclusive if the occurrence of one prevent the occurrence of the other e.g. sunset and sun rise.

Null Hypothesis (H_o) It is the statement of equality (H_o: A = B) or it is hypothesis which is tested under the assumption.

Parameters Any quantity which is to be worked out by using the information of entire population is called parameters.

Population or Universe Population is an aggregate or individual having some common observable

GK **Maharashtra have the maximum extent of black soil.** GS

characteristics.

Probability It is the ratio between the numbers of favourable cases to the total number of equally likely cases.

Qualitative Variable It is distinguished by some quality of attributes e.g., flower colour.

Quantitative Variable It is distinguished by the magnitude of attributes e.g., yield of crop.

Random Sample Every individual of the population has an equal and independent chance of being selected or chosen for sample is called random sample.

Range It is defined as the difference between maximum value and the minimum value of the given distribution.

Regression It is the measure of the relationship between two or more variables in of the original units of the data.

Regression Coefficient It is defined as average change in the dependent variable unit change in the independent one.

Sample A part of population under study.

Sample Size (n) The number of individuals in a sample.

Sampling It is process or method of selecting the individuals or sample from the population is known as sampling.

Skewness It is a departure from symmetry.

Standard Deviation (S.D.) It is defined as the positive square root of the variance.

Standard Error of Mean (SEM) It is the s.d. of sampling distribution of means.

Statistics In singular form it means subject or scientific methods for collecting, organising, summerising, presenting and analysing the data as well as drawing valid conclusion and making reasonable decision for population on the basis of analysing data of the sample of the population.

t-test It is defined as the ratio of the difference the sample mean and hypothetical or population mean to the standard error of which is to be worked out from the information.

Tabulation Defined as process by which the classified data are presented in an ordinary manner by being placed in proper rows and columns of a table in order to bring out its chief characteristics and essential features.

Tally Mark Each observation of a given set of data is to be placed against the different classes according to magnitude by horizontal or vertical or cross line which is known as tally mark.

Test of Significance It is a statistical method of process which is to be utilized or testing the difference under study is significant or not is called test of significance.

Type-I Error It is committed by rejecting hypothesis of the basis of samples evidence it is true in reality. It is devoted by α.

Type-II Error It is committed by accepting null hypothesis on the basis of sample when it is false in reality. It is denoted by β.

Variance (S^2) The variance can be defined as a ratio of the sum of the square of deviation measured from the arithmetic mean of a given series to the number of observation minus one or degree of freedom.

Weighted Mean (W.M.) If all the items are not of equal importance in the calculation of mean, at that time they are given proper weights according to their relative importance and the average (mean) which is calculated on the basis of these weights is called weighted mean.

Yate's Correction In order to avoid irregular caused by smaller frequency in 2x2 counting table, a correction for continuity is called correction.

Z-test It can be defined as the ratio of between the estimated value the parameter value of the population to the standard error mean of the population.

16. Seed Science

Quiz Kit!

Earliest observation of Pole Star

Ashwalayana and Patanjali observed the Pole Star, *Dhruva Tara* in 1300 B.C.

Each Word Counts

SEED

A ripened ovule containing an embryo in arrested state of development usually with a food reserve and a protective coat.

Different Kinds of Seed

Multiplication of quality seed under supervision of breeder or seed certification agency to distribute quality seed of notified varieties for sowing purpose.

(a) Nucleus Seed: The true seed or vegetative propagules used for crop production and propagation are produced by the breeders. 100% gentically pure.

(b) Breeder's Seed: Progeny of Nucleus Seed multiplied in larger area under the supervision of breeder .

(c) Foundation Seed: Progeny of Breeder Seed.

It is handled by seed producing agencies in public and private sector under the supervision of Seed Certification Agencies (SCA).

SCA issues white colour certificate.

(d) Certified Seed: Progeny of Foundation Seed produced by registered seed growers under, supervision of Seed Certification Agencies. SCA gives Blue colour certificate.

(e) Truthfuly labelled seed (TLS) minimum standard prescribed for the variety

(i) Production of Nucleus Seed

Basic Seed

↓

Sown in field with row to row and plant to plant spacing

↓

Plants are marked and observed

↓

Harvested

Breeding of Nucleus Seed—

Notification of variety

↓

Nucleus Seed (Bulk the seeds of selected rows)

↓

Breeders seed production plot

↓

Breeders Seed

(ii) Production of Breeder Seed

- Nucleus seed is sown with less seed rate in line with all requirements.
- Roguing is complete before flowering
- Harvesting, threshing and processing
- Monitored by a team of breeder, representative of crop co-ordinator, representative of National Seed Corporation Agency.
- Quality of Processed Breeder seed is submitted to ICAR. Golden yellow certification is given

(iii) Production of Foundation and Certified Seed

- Foundation and certified seed produced under supervision of seed certificationAgency.)
- The grower get breeder seed from seed corporation.
- Grower harvests, threshes and dries the seeds, lastly seeds transported to seed processing plant.
- Seed is processed under the observation of seed inspector .
- Seed inspector submits a sample from the lot to seed testing lab.

Seed Dormancy

Condition in a viable seed by which it prevents germination even under optimal conditions provided for germination.

Types of Seed dormancy

Structural component of Seed:

- Seeds cannot germinate until maturity of embryo is complete.

Physiological - pH and osmotic pressure in the immediate environment of seed may act as inhibitors for germination.

Chemical - Inhibitors like phenols, benzoic acids etc.

Breaking of Dormancy

Scarification

- Physical or chemical treatment which weakens or softens seed coat.

Suez Canal connects the Red Sea with the Mediterranean Sea.

GS

Bitter gourd -Piercing with needle

Rubber -Removing the seed coat

Cotton -Soaking in conc. H_2SO_4 for 20-30min.

Stratification (Temperature treatment) :
Pre chilling (0-5°C) for 3-10 days *e.g.* mustard
Pre heating (40-50°C) eg. Rice
Hot water treatment -80°C for 1 to 5 minutes.

Treatment with Growth Regulators and other chemicals

GA -For endogenous inhibitors

KNO_3 -For light requiring seeds - *e.g.* Tomato

Thiourea -For prechilling *e.g.* lettuce.

SEED PRODUCTION

Package of Practices

Agronomic Principles of Seed Production

- Selection of suitable agro climatic region
- Selection of seed plot
- Isolation distance
- Selection of variety
- Seed treatment
- Seed rate
- Sowing method
- Roguing
- Weed control
- Pest and disease control
- Harvesting
  (i) Time of harvesting
  (ii) Method of harvesting
- Drying of seeds
- Storage

SEED PROCESSING AND MARKETING

Processing Technique

1. Seed Drying

Lowering of seed moisture for maintaining seed viability and vigour.

- Sun drying
- Air drying

2. Pre conditioning

Includes all operations of shelling, debearding etc.

3. Pre cleaning

Removal of extraneous matter like pieces of trash and stones etc.

4. Basic cleaning

Removal of larger, smaller, lighter and thicker, adulterants as compared to crop seed. It was done on the basis of weight, size and density using cleaner with air screen.

- Grader : Based on density and size
- Scalper : Inert matter of larger size than seeds.
- Aspirator : Lighter inert matter

5. Grading

Removal of smaller and shrivelled seeds from the healthy seeds. Specific gravity separators on the basis of resistance to air flow.

6. Seed treatment

Methods of seed treatment -

- Physical : Cotton delinting, scarification
- Chemical : Fungicide, insecticide
- Biological : Rhizobium

7. Packing

- Filling of seed bag to the exact weight
- Attaching labels regarding seed quality
- Storage

8. Seed Storage

Maintenance of high seed germination and vigour after harvest until planting is important.

- Stored in dry and cool place
- Only high quality seed should be stored.
- Proper sanitation should be there.
- Diseases, insects and rodents should be controlled.

Seed Marketing

It is a vital activity controlling the supply and price of seeds for the progress in crop production and productivity.

Marketing organisation —

Central Marketing Cell

↓

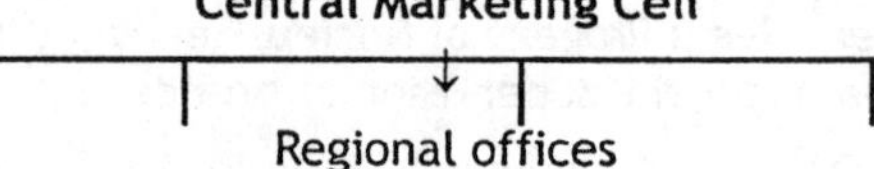

Regional offices

Private dealers | Dealers (Co-operative) | Dealers (Agro sales services) | Company

Marketing Chain

Production
↓
Procurement
↓
Processing
↓
Quality Control
↓
Distribution
↓
Sales

Problems with Marketing

- Irregular supply
- Adulteration

GK Largest single source of government revenue is indirect taxes. GS

- Storage and warehouse inadequate
- Lack of financial facilities and market facilities
- Absence of proper grading and standardisation

OTHERS

Role of seed Industry

Seed sector is heavily dependent on research which has to provide a continuous supply of improvement seeds to keep pace with growing requirement. So far, more than 3,000 varieties have been notified under the Seeds Act, 1966 in different crops at national level. Today, the Indian seed programme boasts one of the biggest seed markets in the world, with annual sales at around US $920 million. Of this, domestic accounts for US $900 million and sales in the global market accounts for the remaining US $ 20 million.

The National Seeds Project (NSP) launched in 1975 with the assistance of World Bank had an objective of developing a broad-based decentralised network of seed production agencies throughout the country capable of meeting the requirements of different quantities of certified seed. That is, with adequate infrastructure for production, processing and storage of seeds. The National Seeds Projects I and II provided the basis for today's seed industry. Under NSP I and II, State Seed Corporations were set up in Andhra Pradesh, Bihar, Haryana, Karnataka, Maharashtra, Orissa, Punjab, Rajasthan and Uttar Pradesh.

The implementation of the New Policy for Seed Development in 1988, greatly encouraged foreign participation in the seed industry. The policy permitted private companies to import seeds of vegetables, flowers, and ornamental plants, subject to prior evaluation by the Indian Council of Agricultural Research (ICAR). Imports regarding seeds of coarse cereals, pulses, and oilseeds were also permitted but only under the condition that the Indian company importing the seed negotiate a technical cooperation agreement with the foreign supplier stipulating that parent lines, breeder seed, and seed production technology would be ceded to the Indian company within two years. Import duties on seed and seed processing equipment were lowered significantly.

NSP III was launched in March 1990 with the assistance of the World Bank. The main objectives of the project were:

(i) To assist the farmers by ensuring timely and adequate availability of certified/quality seeds of suitable varieties at reasonable prices.

(ii) To improve the working efficiency of the National and State level public sector seed companies so as to make them economically viable.

(iii) To provide facilities for the growth of private seed industry through adequate institutional finance.

Public sector

Like many agriculturally developed Asian nations, India too has a sizeable amount. of public and private sector seed businesses. Giant public sector players include the National Seeds Corporation (NSC), the State Farms Corporation of India (SFCI) and the thirteen State Seed Corporations (SSCs). NSC was the first public sector organisation, established in 1963, and remained virtually the only agency in seed production for about 13 years. Its role extended to several development programmes including training, quality control and extension activities in seeds. This was followed by the setting up of the SSCs under two consecutive plan periods, supported by the World Bank, and these largely adopted to role of the NSC in the Indian states. These corporations mainly engage in production and marketing of seed of high yielding and hybrid varieties developed by the public sector.

Did You Know ?

Oldest planetarium

The Birla Planetarium at Kolkata was established on September 29, 1962 by the Birla Education Trust.

Private Sector

Although private seed companies such as Poacha and Sutton have been established since the pre-independence era, accelerated growth of the private sector began only after the introduction of the new seed policy in 1988 which ushered in a liberal business climate. Currently there are over 200 private seed companies, together with a few multinationals, and these tend to focus on low volume, high value crops with the principal effort being placed on creating hybrids for oilseeds, maize, cotton and vegetable crops. The private sector accounts for 70 per cent of the market turnover whereas the public sector has the greater share in volume sales.

SEED AGENCIES

State Seed Corporation (SSC)

- Production and supply of certified and foundation seeds of national and state varieties.
- Distribution and marketing of certified seeds within the state.

State Seed Certification Agencies

- Responsible for seed certification in the concerned state.
- Make field inspections and conduct seed tests.

State Farms Corporation of India (SFCI)

- Production of foundation and certified seed on behalf of NSC and SSC, respectively.
- Taken up large multiplication programme of oilseeds and pulses.

Private Seed Companies

- 60% of seed in our country is produced by private seed companies.

GK GS

Regur is the name given to the black soil in the area of Daccan Trap.

Points to Ponder

- A genotype released for commercial cultivation either by state variety release committee or central variety release committee is referred to as Variety.
- A legal system which ensures production of high quality seed in terms of purity and germination is called Seed certification.
- Absence of seeds of other variety of the same crop and also of other crops in a seed lot refers to Genetic purity.
- Accelerated ageing test was developed by Delouche and Baskin (1942).
- Breeder seed in the progeny of nucleus seed.
- Breeder seed is not meant for general distribution to the farmers for commercial cultivation.
- Breeder seed is used for the production of foundation seed.
- Breeder seed production is taken up from the nucleus seed. B and A line are planted in an isolation of 1 km in a ratio of 1:4.
- Burries et al (1969) showed a positive correlation between tetrazolium staining with seed quality in soyabean seeds.
- Certification is not required for breeding seed.
- Certified seed is the progeny of either foundation or registered or certified seed.
- Certified seed produced from certified seed shall not be eligible for further seed increase under certification.
- Emergence of normal seedlings from the seeds under ideal condition of light, temperature, moisture, oxygen and nutrients is referred to as Germination.
- For growing breeder seed every six rows one row is not sown.
- Foundation seed has physical purity of 98% in majority of crops.
- Foundation seed is the progeny of breeder seed.
- Freedom from inert matter and defective seeds of a seed sample indicates its Physical purity.
- Genetic purity is determined by Grow out test.
- Germ and Kietreiber (1953, 1954) showed a close relationship between the viability of the aleunone layer and the seedling emergence from the maize seed.
- Hybrid programme has been successfully utilized in India in crops like sorghum, maize and bajra.
- Improved seed refers to the seed of a released and popular variety produced by scientific method.
- In a seed sample, non living materials such as sand, pebbles, soil particles, straw, etc are combinedly known as Inert matter.
- In crop plants in which cotyledons come above the soil surface and generally because green and photosynthetic and called Epigeal germination.
- In cross pollinated species are highly heterozygous and heterogeneous.
- In foundation and certified seeds of carrot, physical purity of 85% is allowed.
- In foundation seed of groundnut and spinach, physical purity of 96% is permissible.
- In India national seed project was launched in 1976.
- In India, breeder seed is produced by NSC.
- In India, foundation seed is produced by the National Seeds Corporation.
- In paddy the amount of variability observed in varieties is relatively higher than that of wheat.
- In pulses, a minimum of 500 plants progenies should be planted for the production of nucleus seed.
- In soyabean, physical purity of 97 is permissible for foundation and registered seeds.
- Initial seed of an improved variety, limited in quantity and produced by originating breeder is called nucleus seed.
- Kitock and Law (1963) described a method of estimating seed vigour on the basis of colour intensity of the stained seeds.
- Kotowski (1926) gave the formula for coefficient of velocity of germination.
- Maguire (1962) gave the formula for speed of germination.
- Methods to identify cultivars of cotton and pearl millet were standardized using polyacrylamide get electrophoresis of total soluble proteins and isoenzymes.
- Moore (1973) described the use of tetrazolium staining for assessing seed quality and the basis of topographical patterns.
- National Seeds Corporation of India was established in 1960.
- National seeds corporation was established in 1963 to take up the work of quality seed production in India.
- Nicolls and Heydecker (1962) gave the formula for mean germination time.
- Notification of released varieties is done by the Government of India.
- Nucleus seed is planted in large plots to rise breeder seed.
- Oka (1975) has suggested that the population size suitable for maintenance and seed production of a commercial crop of a self-pollinated species should be 300.
- Permanent reduction either in genetic or agronomic value of a released variety is known as varietal deterioration.

GK | GS

A new committee system was launched in Parliament in April 1993.

- Plant breeders' rights enable the breeder to multiply and market seeds of his verity and also allow him to authorise other interested persons for its commercial production and marketing.
- Plants which cotyledons do not come above the soil surface are called Hypogeal germination.
- Process of evaluation of seeds in terms of purity and germination is called seed testing.
- Process of removal of off plants from the field of an improved variety is known as roguging.
- Product of a fertilized ovule which consists of embryo, seed coat and cotyledon(s) is called Seed.
- Progeny of breeder seed is referred to as Foundation seed.
- Progeny of foundation seed produced under the supervision of seed certification agency is known as Certified seed.
- Progeny of nucleus seed is called Breeder seed.
- Registered seed is the progeny of either foundation seed or registered seed.
- Seed is a product of fertilized ovule which consists of embryo, seed coat and cotyledon(s).
- Seed production is different from crop production.
- Seedlings which are unable to develop into normal plants are called Abnormal seedlings.
- Seeds of wheat, paddy and maize showed positive coordination between germination (%) and viability (%) as determined by TZ test.

Did You Know ?

Smallest duck

The Cotton Teal or Quacky Duck (*Nettapus corfomandelianus*) measures 33 cm. It is found over most of India.

- Seeds other than hard seeds, which remain firm and apparently viable at the end of the test period are called Fresh ungerminated seeds.
- Seeds which at the end of the test period are neither hard not fresh and have not produced any part of a seedling are called Dead seeds.
- Seeds which do not absorb water are referred to as Hard seeds.
- Separation of the field of a variety to the prescribed distance from that of another variety to avoid contamination is called Isolation.
- The minimum isolation distance in foundation seed of berseem is 400 metres.
- Two inspections are necessary at different stages for producing foundation seed of berseem.
- Use of linear measurements of plumula growth as a vigour test was first suggested by Germ (1949).
- Variety refers to genotype which has been released for commercial cultivation either by state variety release committee or central variety release committee.
- Woodstock (1976) suggested that tetrazolium test of maize and soyabean embryos was more reliable for evaluating seed vigour.

GS

India's first experimental geo-stationary communication satellite : Apple.

Multiple Choice

1. Argemone mexicana is an objectional weed in
a] Wheat b] Barley
c] Rapeseed/Mustard d] Chick pea
2. Breeder seed is the progeny of
a] Foundation seed b] Registered seed
c] Nucleus seed d] Certified seed
3. Certification is not required for
a] Nucleus seed b] Breeder seed
c] Foundation seed d] Certified seed
4. Cuscuta is an objectional weed of
a] Cowpea b] Berseem
c] Sorghum d] Lucerne
5. Headquarters of the Union for the Protection of New Plant varieties is in
a] Thailand b] USA
c] Denmark d] Switzerland
6. Improved seed includes
a] Nucleus seed b] Breeder seed
c] Foundation seed d] All of the above
7. In Bhindi, production of foundation seed needs an isolation distance of
a] 100 metres b] 50 metres
c] 200 metres d] 3 metres
8. In India, foundation seed is produced by
a] Registered growers b] SSCA
c] NSC d] all the above
9. In sunflower, production of foundation seed requires an isolation distance of (metres)
a] 400 b] 800
c] 200 d] 100
10. In wheat, production of foundation seed needs an isolation distance of
a] Three metres
b] Five metres
c] Ten metres
d] Twenty metres
11. Initial seed of an improved variety is called
a] Nucleus seed b] Breeder seed
c] Foundation seed d] Certified seed
12. Physical purity of 95% is permissible for the foundation and certified seed of
a] Soyabean b] Groundnut
c] Spinach d] Carrot
13. Physical purity, is permissible (98%) in the crop
a] Groundnut b] Rice
c] Soyabean d] Spinach
14. Plant Breeders' Rights are operating in
a] Germany b] Denmark
c] Netherlands d] All of the above
15. Production of breeder seed in cotton requires an isolation distance of
a] 20 metres b] 30 metres
c] 50 metres d] 75 metres
16. Progeny of a nucleus seed is referred to as
a] Certified seed b] Foundation seed
c] Registered seed d] Breeder seed
17. Seed certification requires
a] An improved variety
b] Genetic purity
c] Physical purity
d] All of the above
18. Seed meant for generation distribution to the farmers for commercial crop production refers to
a] Foundation seed b] Breeder seed
c] Certified seed d] Nucleus seed
19. The recommended ratio of male rows to female rows is in hybrid bajra production
a] 2 to 4 b] 1 to 2
c] 4 to 8 d] 3 to 6
20. Freedom from inert matter and defective seeds
a] Genetic purity b] Physical purity
c] Defective Purity d] Normal purity
21. International Crop Improvement Association (ICIA) in _____ classified seed into different categories
a] 1964 b] 1946
c] 1963 d] 1972
22. Seed is a
a] Immature embryo b] Mature embryo
c] Developed embryo d] Undeveloped embryo
23. Presently ICAR has _____ breeder seed production units
a] 45 b] 54
c] 92 d] 107
24. Cotyledons in gymnosperms are called
a] Embryo b] Integuments
c] Megagametophyte d] Endosperm
25. Cotyledons in monocots are called
a] Endosperm b] Megagametophyte
c] Embryo d] Integuments
26. The first private seed came into existence in the year
a] 1918 b] 1987
c] 1959 d] 1912
27. First private seed company came into existence in
a] Mumbai b] Kolkata
c] Lucknow d] New Delhi
28. Seed moisture varies from crop to crop in ranges from:
a] 15-20% b] 30-40%
c] 1-2% d] 9-12%

Tax on income and wealth are important direct taxes.

GS

29. The hybrids developed by Government agencies or Government Institutions and Agricultural Universities are called
   a] Private hybrids b] Institutional hybrids
   c] Public hybrids d] Government hybrids
30. The coarse rice generally has a _____ aleurone layer
   a] Larger b] Smaller
   c] Bigger d] Smoother
31. Coloured varieties of rice have _____ aleurone layer
   a] Thinner b] Thicker
   c] Coarse d] Smooth
32. TDC came into existence in
   a] 1996 b] 1962
   c] 1926 d] 1969
33. Contamination permitted in maize is
   a] 1% b] 2%
   c] 0.1% d] 0.2%
34. Breeder seed production units of ICAR producing ____ tonnes of breeder seed
   a] 7300 b] 37
   c] 3700 d] 73
35. The first sybolic importance of seed started after the report of Famine Commission in
   a] 1881 b] 1890
   c] 1891 d] 1892
36. Pure Live Seed (PLS) is related to
   a] Physical purity
   b] Genetic purity
   c] Germination percentage
   d] Contamination
37. Breeder seed is _____%pure.
   a] 90 b] 100
   c] 70 d] 99.99
38. In flowering plants a second seed coat is known as
   a] Integument b] Aleurone layer
   c] Tegamen d] Inner ventral scale
39. Standards of germination for seed certification in chillies
   a] 70% b] 90%
   c] 80% d] 60%
40. Indian Society of Seed Technology founded in the year
   a] 1969 b] 1971
   c] 1924 d] 2006
41. There are _____ seed testing laboratories in India
   a] 301 b] 103
   c] 205 d] 502
42. First seed testing laboratory was established at IARI in
   a] 1971 b] 1961
   c] 1981 d] 1991
43. International Seed Testing Association was organized on :
   a] 10 July, 1924 b] 12 July, 1925
   c] 12 Feb, 1942 d] 10 Feb 1924
44. First private seed company was
   a] Monsanto b] Namdhari
   c] Sutton & Sons d] Takii
45. Seedcoat is derived from
   a] Testa b] Embryo
   c] Endosperm d] Nucellus

Matching

1. **Match the following :**

a]	Maize	i]	Hypogeal
b]	Barley	ii]	Epigeael
c]	Pea		
d]	Castor		

2. **Match the following :**

a]	Onion	i]	Hypogeal
b]	Groundnut	ii]	Epigeal
c]	Cucurbita		
d]	Pea		

3. **Match the crop with its germination :**

a]	Rice	i]	80
b]	Pigeon pea	ii]	75
c]	Cotton	iii]	65
d]	Maize	iv]	90

4. **Match the crop with isolation distance (m):**

a]	Brinjal	i]	100
b]	Potato	i]	100
c]	Bhindi	iii]	200
d]	Rice	iv]	3

5. **Match the information :**

a]	Tetrazolium test	i]	Test of genetic purity of seeds
b]	Grow our test	ii]	Test of seed viability
c]	Breeder seed	iii]	Initial seed of a variety
d]	Foundation seed	iv]	Progeny of nucleus seed
e]	Nucleus seed	v]	Progeny of breeder seed
(f)	Certified seed	vi]	Progeny of foundation seed.

6. **Match the physical purity with crop in which its permissible for foundation and certified seeds:**

a]	95%	i]	Ground nut and spinach
b]	96%	ii]	Carrot
c]	97%	iii]	Wheat, barley and rice
d]	98%	iv]	Soyabean

7. **Match the objectional weed seed with the crops.**

a]	Wheat	i]	*Convolulus arvensis*
b]	Mustard	ii]	*Melilotus spp.*
c]	Methi	iii]	*Cuscuta spp.*
d]	Lucern	iv]	*Agemone mexicana*

8. **Match the proper isolation distance with the crop given below:**

a]	Wheat	i]	20
b]	Chicken pea	ii]	3
c]	Pigeon pea	iii]	50
d]	Cotton	iv]	200
e]	Maize	v]	1000
(f)	Pearl millet	vi]	400

9. **Match the prescribed germination percentage of foundation seed with the crops given below :**

a]	Rice	i]	70
b]	Groundnut	ii]	80
c]	Maize	iii]	65
d]	Cotton	iv]	90
e]	Pigeon pea	v]	75

Glossary of Technical Terms

Achene A hard, dry, one-chambered, one-seeded indehiscent fruit. The fruit wall is not joined with the seed coat. Example: spinach and sunflower.

Abnormal Seedlings. Seedlings which are unable to develop into normal plants.

Adaptation Some kind of modification that makes an organism better suited to its environment, perhaps more fit. An adaptation is a divergence in characteristic(s) as a result of natural selection. Local adaptation means that the selection pressure is something specific to that local environment.

AFLP DNA Fingerprinting Amplified Fragment Length Polymorphism is a particular molecular marker technique.

Allele A form of a gene. If a gene sits at the same place (locus) on a chromosome, the chromosome is said to be homologous.

Allozymes A type of molecular marker technique.

Androdioecious A species where the male and hermaphrodite flowers are borne on separate plants.

Angiosperm A plant whose seeds are formed within a fruit. Angiosperms, or flowering plants, are the dominant plants in the world today. Virtually all crop plants are of this plant class.

Annual A plant that completes it life cycle, from seed to seed, in one growing season.

Anther The saclike structure in which pollen is formed in a flower.

Anthesis The period from when the flower opens to when seed is set.

Apomixis Asexual reproduction. Can also be used to describe the production of seeds without sex.

Asexual Reproduction The process of producing new individuals by vegetative propagation – for example, using cuttings. All offspring are clones or genetically identical to the parent plant.

Backcrossing A technique used in plant breeding where the aim is not to create new varieties but to modify existing ones.

Berry Fruit with soft flesh surrounding one or more seeds.

Biennial A plant that completes its life cycle in two years. Generally, in the first year, growth occurs followed by flowering, fruiting and seed production in the second year.

Binomial The two-part scientific name consisting of a genus name and a species, cultivar, group, series, or hybrid epithet, denoting an individual within a genus.

Biodiversity Biodiversity is a commonly used abbreviation of the two words, "biological" and "diversity". It simply refers to the diversity, or variety, of plants and other living things in a particular area, region, or the planet as a whole system.

Did You Know ?

Smallest duck

The Cotton Teal or Quacky Duck (*Nettapus corfomandelianus*) measures 33 cm. It is found over most of India.

Blanch From the Old French word *blanchir*, to whiten. As a gardening term, it is used to describe a process of excluding light so as to make the plant material lighter in color and more tender. For example, this is commonly done with endive.

Bolt An Old English word for arrow, to bolt is a verb used to describe a plant prematurely producing seed.

Breeder Seed Progeny of nucleus seed or breeder seed produced under the strict supervision of original or sponsored plant breeder.

Carpel A floral leaf bearing ovules along the margins.

Certified Naturally Grown The Certified Naturally Grown™ label is a non-profit alternative eco-labeling program for small farms that grow using USDA Organic methods but are NOT a part of the USDA Certified Organic program.

Certified Seed Progeny of either foundation or registered or certified seed. Certification is done by State Seed Certification Agency.

Chasmogamy Flowers that open to expose anthers allowing cross-pollination.

Chromosome The structure that carries genetic information.

Cleistogamous Flowers that remain closed so that anthers shed pollen inside, resulting in self-fertilisation.

Cline This is the gradual change in characteristics of a population over large geographic distances.

Clonal Seed Orchard A seed orchard established using selected clonal material.

Cold Stratification Many herbs require the extra "conditioning" or "stratification" step. Cold stratification or conditioning is simply a method of letting the seeds know that it is time to wake up. It can simply mean sowing the seeds into flats outdoors while the weather is still cold and allowing them to wake up naturally. You can also mix the seeds with damp, sterile peat in a plastic bag and place them into the refrigerator. The time required is greatly dependant on the seed type. Once this is done, sprinkle the mixture on top of pots that you have filled with sterile, pre-moistened potting mix and place in a warm location (top of the refrigerator). Keep moist until germination occur and treat as you would any other seed you are germinating in pots.

Cultivar A variety of plant that has been produced and maintained through cultivation. Not a wild variety. A

GS

Cricketer Lance Clusener is also known as 'Zulu'.

word from the combination of the two words, "cultivated variety".

Damping-off A single, collective term used to describe underground, soil line, or crown rots of seedlings due to unknown causes. Many soil borne diseases and fungi can cause these symptoms of sudden plant death

Dead Seeds Inviable seeds.

Defective Seeds Seeds which are broken disease infested, insect damaged, undeveloped and unfit for germination.

Diploid The plant has two sets of chromosomes.

Ecotype A subspecies or race that is presumed to have adaptation(s) to a particular set of environmental conditions.

Embryo The result of fertilisation, it is the white materials under the seed coat.

Endemism A species is said to be endemic, when it only grows in a particular area.

F1, F2 The 'F' stands for Filial. The F1 generation is the first generation of offspring, produced by crossing two plants homozygous for contrasting alleles. The F2 generation is produced by selfing the F1 generation. Thus, the F1 generation would have less variation than the F2 generation.

Fecundity The rate of reproduction. Low fecundity = low number of offspring.

Fitness Some times this can get a little confused with 'healthy'. Healthy implied a physical fitness, however, there is also genetic fitness. Thus, a plant is said to be fit, when it can reproduce successfully, or produce viable offspring that themselves successfully reproduce. In this way progeny significantly contribute to the gene pool of subsequent generations and less fit individuals will contribute less.

Foundation Seed Primary source of seed of a genetically identified variety from which all increases are made.

Fresh Ungerminated Seeds Viable seeds which can abort water but do not germinate and remain fresh in germination test.

Fruit A fruit is a ripened ovary of a plant along with any attached parts that developed with it from the flower.

Gene Flow The exchange of genetic material between populations (transfer of pollen). Its rate is a function of the mating system.

Gene In its strictest sense it is the unit of inheritance. It can be seen as a small bit of DNA that codes for a protein that subsequently produces some character in a plant.

Gene Pool The sum of all the genes of all the individuals in a population.

Genecology The study of the genetic structure of a population and its relationship with the environment.

Genetic Drift This is the random change in allele frequencies due to chance. It is important to note that differences in phenotype or genotype due to genetic drift do not reflect adaptive differences.

Genetic Engineering Genetic engineering involves laboratory techniques to change the DNA of living organisms

Genetic Erosion The loss of genetic variation, associated with a narrowing of the genetic base or a loss of heterozygosity. Can be used within the context of agricultural varieties where a narrow genetic base results following breeding or selection of favourable characteristics and the loss of other genes.

Genetic Provenance Sub-groups of a species or sub-species that are most genetically similar.

Genetic Purity Absence of seeds of other variety of the same crop as well as of other crops.

Genotype The genetic constitution of a plants as opposed to it physical appearance (phenotype).

Geographic Provenance The often arbitrary sub-groups of a species based on geographic source. Used to describe seed or propagule sources.

Germination Emergence of normal seedlings from the seeds under ideal conditions of light temperature, moisture, oxygen and nutrients. Based on germination, seed is divided into five components, *viz.*, normal seedlings, abnormal seedlings, hard seeds, fresh ungerminated seeds and dead seeds.

Glaucous Adjective to describe a plant or fruit that are covered in a grayish, whitish, or bluish waxy or powdery substance. From the Greek, *glaukos and* Latin, *glaucus* meaning "blue-gray, green."

Hard Seeds. Seeds which do not absorb water.

Harden Off (hardening) The straightforward process of gradually acclimating a tender seedling, germinated and nurtured under artificial conditions, for life in the wilds of the garden. For most tender plants (tomatoes, peppers, etc.), hardening off for two weeks is desirable. Examine your garden on a frosty morning searching for areas that are relatively free from frost. For the first week, place your plants outside during the warmest part of the day. Initially limit their exposure, bringing them back indoors and under cover, after a few hours. During the second week, gradually extend the amount of time you leave them outside, bringing them back in during the evening. By the end of the second week, assuming that there is no longer any threat of frost, your plants should be ready for transplanting into the garden.

Heirloom Seed The definition does vary from person to person, company to company. It has become very popular as a marketing term. The Victory Seed Company adheres to the purest form of the definition. That is, an heirloom plant variety is one that has been valued by a family, tenderly and carefully preserved, and handed along from generation to generation. We also concede that there are some "heirlooms" whose origins were as a commercial release. That is, they were

introduced by a seed company or seedsman. These old "commercial heirlooms" have value and are worthy of protection. At a minimum, an heirloom variety must be, open-pollinated seed, not an unstable hybrid, and certainly not genetically modified! The oldest reference we have been able to find using the word "heirloom" as an adjective for seeds, plants and bulbs is on page 30 of the book entitled, *Pioneer American Gardening*, compiled by Elvnia Slosson, copyright 1951, Published by Coward-McCann, Inc., New York.

Heterosis Hybrid vigour.

Heterozygosity A heterozygous individual will have different alleles.

Hexaploid The plant has five sets of chromosomes.

Homozygous A homozygous individual will have identical alleles.

Hybrid A plant resulting from the cross mating of distinctly different parental types. From the Latin word *hybrida* or offspring of a domesticated sow and a wild boar.

Hybridisation The process where seed results from the transfer of pollen from one plant to another of a different taxon.

Improved Seed Seed of a released and popular variety produced by scientific method. Improved seed is of five types. *viz*. nucleus seed, breeder seed, foundation seed, registered seed and certified seed.

Inbreeding Depression Self-fertilisation is the most extreme form of inbreeding. Inbreeding increased homozygosity and hence decreased genetic variability. When the frequency of homozygosity (genetically similar individuals) is associated with an increase in lethal genes, a decline in vigour in a species can occur and is termed inbreeding depression. It can be associated with predominantly out-crossing species forced to mate with close relatives, thus increasing the frequency of undesirable characteristics common to each.

Indehiscent Not splitting open when ripe.

Inert Matter Non-living materials such as sand, pebbles, soil particles, straw, etc.

Introgression This is where genes from one species or subspecies get incorporated into another related species or subspecies

Isolation Distance Separation of the field of a variety from that of another variety of same crop to prescribed standard distance to avoid contamination.

Keeper A storage variety.

Leggy The tendency of plants to grow tall, tin and straggly. Typically caused by inadequate light.

Loam Loam is soil that is rich, fertile and is whose moisture retention abilities are balanced. The roots are from the Middle English word *lam* which translated means clay.

Mixed Mating A combination of selfing and cross-pollination reproductive systems

Morphology The form of the external structure. Anatomy refers to the internal structure

Natural Selection Differential reproduction and survival that results in adaptive characteristics.

Normal Seedlings Seedlings with normal growth of root and shoot.

Nucleus Seed Initial seed of an improved variety, limited in quantity and produced by originating plant breeder or institution.

Open Pollination Pollination system where pollen can flow freely between individuals.

Open-pollinated Seed A seed which produces offspring just like the parent plants. Open-pollinated seed allows growers to harvest and save seed for the following year.

Organic, Certified Certified Organic refers to products grown under guidelines as mandated by the National Standards on Organic Agriculture. To become certified, growers must keep very detailed records, adhere to the standards, have soil tested, and pay certification fees and duties. Only approved fertilizers and pest controls are employed. Examples of organizations who perform these certifications are OCIA, Oregon Tilth, California Certified Organic Farmers, and the Northeast Organic Farming Association.

Organically Grown Many small growers are strict and responsible gardeners but either cannot afford the fees associated with the USDA's certification process or take issues with the standards. Some believe that the USDA, who has affectively taken legal ownership over the use of the word *organic* as applied to marketing, is not strict enough. Since the work of many small growers is a labor of love, there is an exemption set in the national standard allowing a product to be labeled as "Organically Grown" if the grower meets the same requirements and does not have total gross organic sales of over $5,000 annually. In order for us to sell seed labeled as organic, we would be required to become a certified organic handler. This of course comes at a cost and at this time - a cost which is prohibitive to us. Much like the original grassroot efforts of the organic movement of decades ago, there are alternatives cropping up. One such alternative is the **Certified Naturally Grown™** label. It is a non-profit alternative eco-labeling program for small farms that grow using USDA Organic methods but are NOT a part of the USDA Certified Organic program.

Outbreeding Depression This is reduction in fitness as a result from hybridisation (crossing) between genetically distinct individuals, the opposite of the above.

Overseed When a seed has quality problems such as suspected low germination rates or physical damage, a farmer of gardener may choose to overseed. That is, plant denser than the normal seed spacing. This is intended to compensate for bad seeds and still result in a good stand of plants.

Paclobutrazol A chemical used to retard growth and increase early flowering in some eucalyptus.

Perennial A plant that lives from year to year and typically involves many reproductive cycles. By definition, for three seasons or more. From the Latin word *perennis*, "through the year".

Physical Purity Freedom from inert matter and defective seeds.

Pistil The seed-bearing organ of the flower, composed of stigma, style, and ovary.

Plant Breeders' Rights Legal rights provided to an originating plant breeder or owner of a variety to regulate production and marketing of his variety.

Plant Regulators Synthetic compounds that induce growth responses in plants. Also known as "growth regulators."

Pollen Tiny (often microscopic) bodies that are borne in the anthers of flowers and contains the male generative cells.

Polyploid The plant has three or more completes sets of chromosomes.

Primary Dormancy The main dormancy mechanism which must be overcome before germination can occur.

Reciprocal Transplant Experiments Experiments where populations of plants are grown in their 'local' environments and other locations, and compared to other populations. Used to determine differences in adaptability between provenances.

Registered Seed Progeny of either foundation seed or registered seed. In India, this category is generally omitted and certified seed is produced directly from foundation seed.

Resistance The ability of plants to suppress or retard the activities of a specified pest or pathogen. Also, the ability of plants to withstand a specific environmental or chemical stress.

Rhizome Rhizomes are underground, horizontal stems that produce shoots and roots. From the Greek word *rhiza*, root.

Rugose Wrinkled, usually covered with wrinkles, the venation seeming impressed into the surface.

Roguing Process of removal of off types (phenotypically different) plants from the field of an improved variety to avoid contamination.

Scarify (scarification) In plant types that have hard seed coats, germination can take months to years to naturally occur. Texas Bluebonnets are an example of this. To promote fast germination, growers scarify the seed coat. Scarification is the manual scratching or nicking of the seed coat as a means of simulate the natural weathering process. Once scarified, most seed will germinate quickly and should be kept watered until germination occurs and plants become established. There are three common methods used for scarification: 1. Physically nick the seeds with a knife, 2. Rub the seeds with fine sandpaper & 3. Freeze the seeds overnight then then quickly pour boiling water over the seeds and allow to soak at room temperature for several hours prior to planting. The key is to experiment and find out what works best for you and for the seed variety you are working with.

Seed In broad sense, any plant part which is used for commercial multiplication of a crop. In strict sense, the product of fertilized ovule that consists of embryo, seed coat and cotyledon (s).

Seed Dormancy A period of inactivity when growth and developmental processes are deferred.

Seed Orchard A plantation of trees selected for a particular characteristic, that is intensively managed to produce seed.

Seed Production Area A stand of unselected trees where seed production is one, but not necessarily the main purpose.

Seed Certification A legal system which ensures production of high quality seed in terms of genetic purity and germination. There are four basic requirements for seed certification *viz.*, improved variety, genetic purity, physical purity and germination.

Seed Testing Process of evaluation of seeds in terms of purity and germination.

Self-incompatibility Self sterility, inability to produce viable seed on the same plant. Whilst the exact mechanisms are largely unknown this means that a plant is obliged to cross-fertilise with another plant.

Selfing Also called self-pollination it is the transfer of pollen from the anthers (male) to the stigma (female) of the same flower. That is mating with yourself. It is important to note the selfing does not always preclude cross pollination.

Soil Seedbank Seed that lies within the soil (surface or under).

Spatial Autocorrelation Analysis When we look at a plant population, individual plants will have a spatial relationship with each other, for example, plants close to each other may be more similar than those far away from each other. As such, a correlation between individual plants occurs and the data set is said to lack independence. Sometimes when a statistical analysis is undertaken you can get more significant results than the data truly have because the observations you make on the individual plant are not independent.

Sport A phenomenon whereby a spontaneous mutation occurs within an otherwise stable plant variety. It is caused by a genetic mutation and not as a result of cross pollination.

Sucker The unwanted shoots from the stem or roots that draw nutrients and sap the plants fruiting and flowering ability.

Susceptibility The inability of plants to restrict the activities of a specified pest or pathogen. Also, the inability to

Sales tax is the exclusive jurisdiction of the state governments.

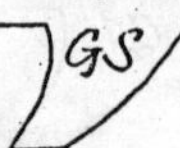

withstand a specific environmental or chemical stress.

Sustainable Agriculture "A sustainable agriculture must be economically viable, socially responsible, and ecologically sound. The economic, social, and ecological are interrelated, and all are essential to sustainability. An agriculture that uses up or degrades its natural resource base, or pollutes the natural environment, eventually will lost its ability to produce. It's not sustainable. An agriculture that isn't profitable, at least over time, will not allow its farmers to stay in business. It's not sustainable. An agriculture that fails to meet the needs of society, as producers and citizens as well as consumers, will not be sustained by society. It's not sustainable. A sustainable agriculture must be all three - ecologically sound, economically viable, and socially responsible. And the three must be in harmony"

Tetraploid The plant has four sets of chromosomes.

Tetrazolium This is a staining method that has been around for some 40 years and provides a reliable way for assessing seed quality (whether seed is viable). It does this by colouring the healthy embryo in the seed red, non-viable seed will not colour (Moore, 1973). It won't tell you when seed will germinate, just that it is capable of germinating.

Tolerance The ability of plants to endure a specified pest, pathogen, environmental pressure or chemical stress. A tolerant variety will sustain less damage than a susceptible variety when grown under the same conditions.

You Don't Know!

Polyembryology

The phenomenon of several larvae developing from a single egg is common among certain Chalcid wasps. In parasitic wasps of the family *Encytidae,* 1,500 young develop from a single egg.

Variegated Having streaks or spots of different colors; as variegated ivy.

Varietal Deterioration Permanent reduction either in the genetic or agronomic value of a released variety. Varietal purity can be maintained by keeping proper isolation distances and removal of off types.

Variety A genotype released for commercial cultivation either by State Variety Release Committee or Central Variety Release Committee.

Viability Whether or not a seed is alive. Usually expressed as a percentage of viable seeds in a seedlot.

Victory Garden "A wartime vegetable garden developed to increase food production esp. By home gardeners."

Zygote Results from the fusion of two gametes before mitosis or meiosis.

GS

Nitrogen element is generally found in Indian soils.

You Don't Know!

First satyagraha

In 1917, Mahatma Gandhi offered the first satyagraha in the Champarana district of Bihar for the indigo cultivators.

Each Word Counts

The science of relationship between organism and their environemnt.

Agro-Ecosystem

Fig. Agro ecosystem deteminants

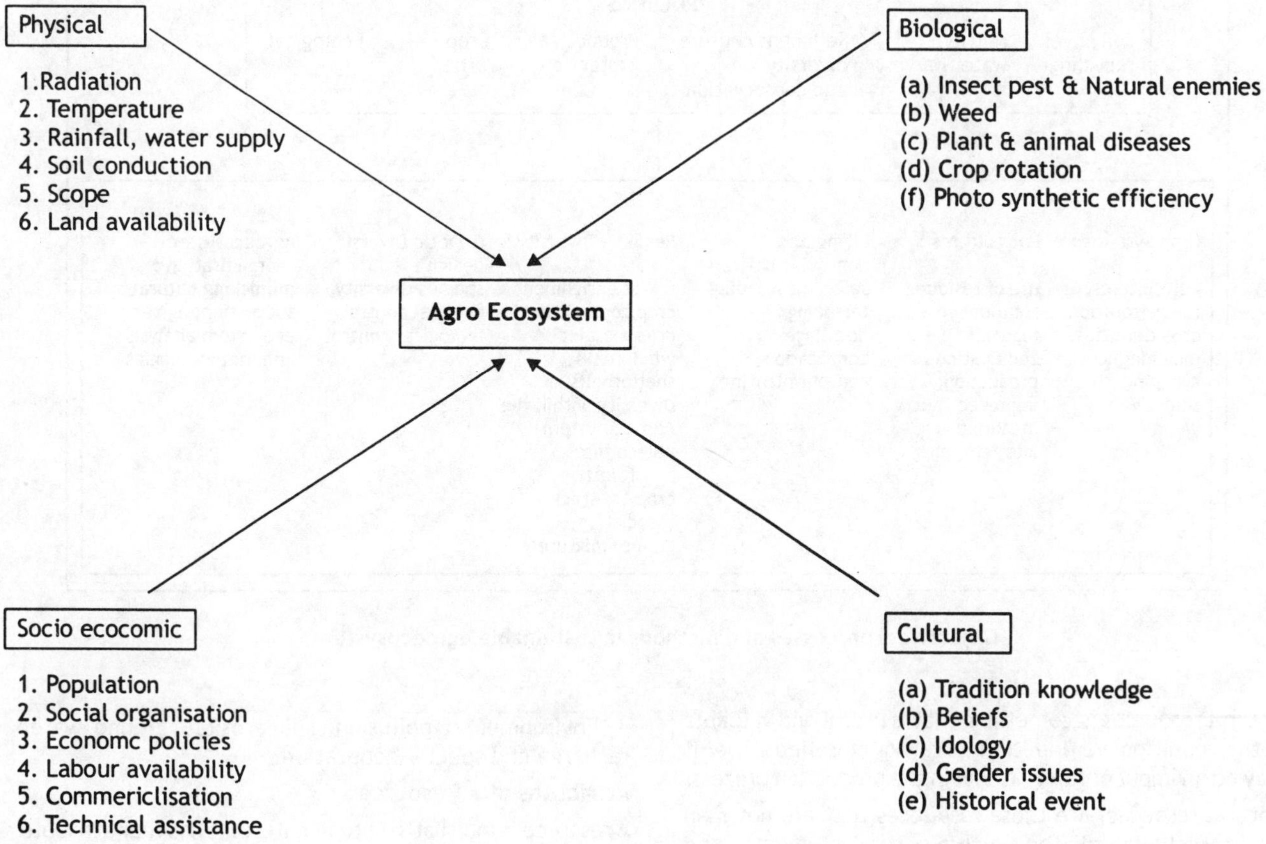

Physical

1.Radiation
2. Temperature
3. Rainfall, water supply
4. Soil conduction
5. Scope
6. Land availability

Biological

(a) Insect pest & Natural enemies
(b) Weed
(c) Plant & animal diseases
(d) Crop rotation
(f) Photo synthetic efficiency

Agro Ecosystem

Socio ecocomic

1. Population
2. Social organisation
3. Economc policies
4. Labour availability
5. Commericlisation
6. Technical assistance

Cultural

(a) Tradition knowledge
(b) Beliefs
(c) Idology
(d) Gender issues
(e) Historical event

Sustainable Agro-Ecosystem

Article 136 talks about special leave petition.

Natural resources

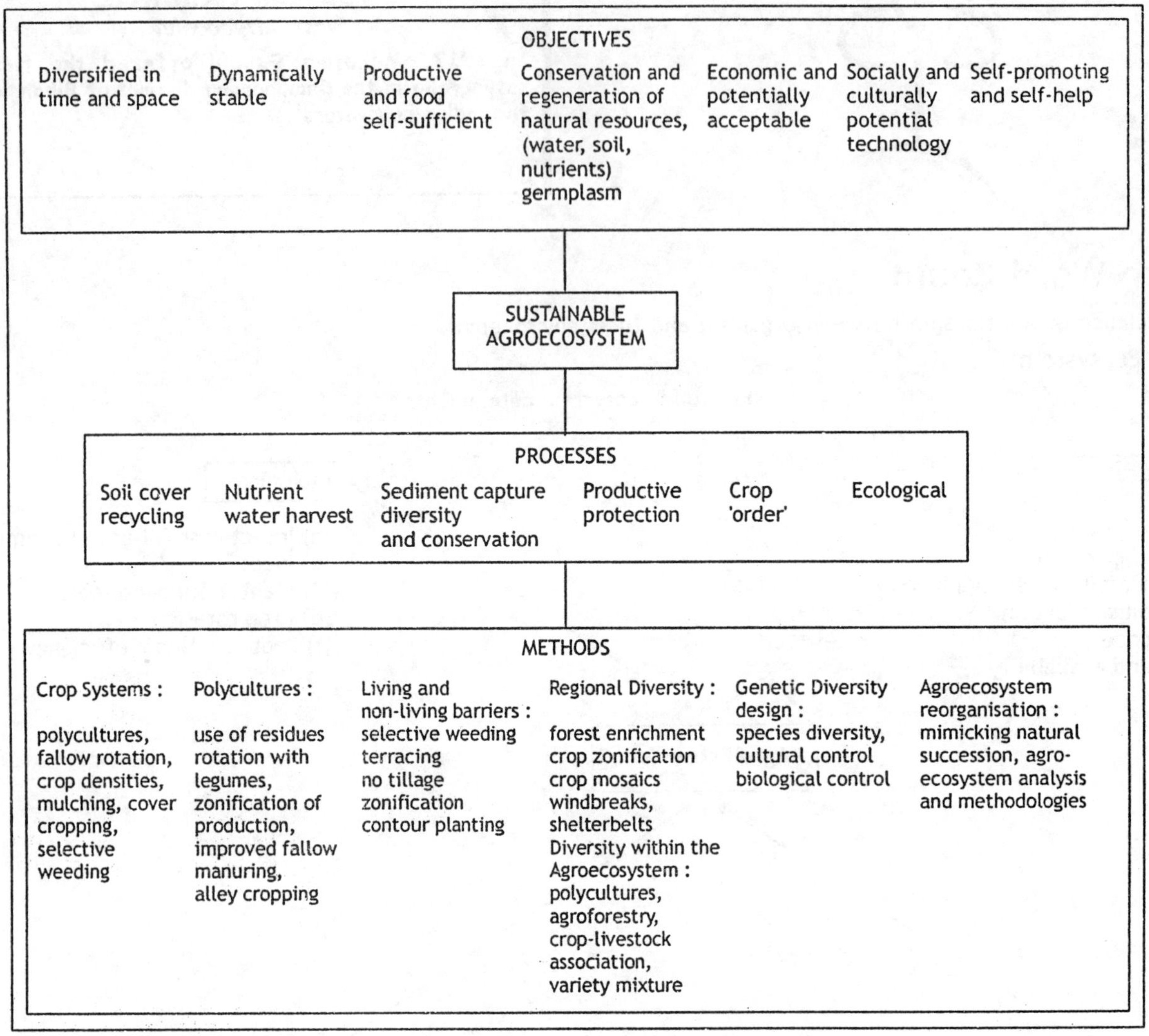

OBJECTIVES

Diversified in time and space	Dynamically stable	Productive and food self-sufficient	Conservation and regeneration of natural resources, (water, soil, nutrients) germplasm	Economic and potentially acceptable	Socially and culturally potential technology	Self-promoting and self-help

SUSTAINABLE AGROECOSYSTEM

PROCESSES

Soil cover recycling	Nutrient water harvest	Sediment capture diversity and conservation	Productive protection	Crop 'order'	Ecological

METHODS

Crop Systems :	Polycultures :	Living and non-living barriers :	Regional Diversity :	Genetic Diversity design :	Agroecosystem reorganisation :
polycultures, fallow rotation, crop densities, mulching, cover cropping, selective weeding	use of residues rotation with legumes, zonification of production, improved fallow manuring, alley cropping	selective weeding terracing no tillage zonification contour planting	forest enrichment crop zonification crop mosaics windbreaks, shelterbelts Diversity within the Agroecosystem : polycultures, agroforestry, crop-livestock association, variety mixture	species diversity, cultural control biological control	mimicking natural succession, agro-ecosystem analysis and methodologies

Objectives, processes and methods in sustainable agroecosystem

Natural resources is something that is useful and valuable in the condition we find it. In its raw or modified state, it may be an imput or raw material in the production process.

Natural resouces, are those resoueces that are not man made, including all the earth's natural elements, and environmental forces.

Natural resouces fall into four categories as follows

1. Basic natural resources such as land, water and air.
2. Natural resource commodities such as timber and fish.
3. Environmental amenities such as clean air and scenic views.
4. Environmental pollution, soil erosion, ground water recharge and species generation.

Attributes of a Resource

A resource is multi attribute in nature. It has four attribute as follows

1. **Quantity** : A resource available in nature may be infinite or finite in quantity- (*e.g.*) Coal, iron ore and petroleum products.

2. **Quality** : A resource will be different inquality also. (e.g.) Recovery of iron from iron ore will be high or low depending upon the quality of ore excaved from different

Red soil is found at Chhotanagpur plateau.
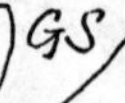

parts of the world.

3. **Time:** It refers to relative abundance and scarcity of resources at different point of time. (e.g.) A country ecomes rich in resources by discovering new resources or by finding new uses for already known resources.

4. **Space:** It refers to availability of different resources in different parts of the world. While availability is not uniform in all the regions, demand for resources is more or less same for all the people.

Importance of natural resources

Natural resource includes land, water, fisheries, mineral resources, forest, marine resources, climate, and topography. Some of resources are known to man. But man utilises small portion of land resources and other surface resources. There are large amount of undiscovered resources under the surface. Importance should be given to explore natural resources in surface and undersurface.

Classification of Resources

Based on the attributes, resources can be classified as

1. ***Exhaustible Resource or Stock Resource***

These resourceS—are finite in quantity and hence its exploitation and use should be in an optimum way. Spacio-temporal (space and time) dimensions of stock resource are important for its optimum use.

Examples

A. Non renewable resource Coal, oil, iron, other metal, old-growth timber and soils.

B. Renewable resources Water stored in tanks, solar energy stored in cells.

2. ***Inexhaustible Resource or Flow resource***

These resources can be exploited continuously without diminishing the stock. It must be utilised when it is available.

A. Non renewable resource
Wind, flowing river water.
B. Renewable resources
Young-growth timber.

3. ***BiologicalResources***

These resources represent complex category of resources. It involves solar energy, hydrological resources and soil nutrients.

(*e.g.*) Crops, Forest and Animals.

Man derives value from a complex system of solar, atmospheric, geological, hydrological and biological resources. There are several factors that influence the scarcity of natural resources as follows

1. ***India's Population Growth and Demography***: It is a fundamental determinant of raising demand for natural resources. Population growth increases pressure on the resources required for subsistence land for producing crops and grazing animals, forests required for fuelwood and grazing animals, forests required for fuelwood and other products.

Knowledge World!

Only pitcher plant

Nepenthes khasiana, **which captures and digests insects, is found in the Khasi and Jaintia Hills, Meghalaya, where there is a sanctuary for this rare plant.**

2. ***Income Growth***: As people become wealthier, their consumption increases and its form changes. Rich people consume more energy in the form of transportation, heating, and air conditioning. They seek improved housing that uses more space and they buy more manufactured goods. Raising income mean fewer people depend directly on natural resources for their subsistence needs.

3. ***Market Forces***: Pressure on natural resources also increases demands in national and international markets.

4. ***Technology*** : Improved technologies to harvest and harness natural resources have increased demand for them over the last century.

(*e.g.*) Better transportation infrastructure made forest areas more accessible to distinct markets; improvements in pumps and power sources made it economical to tap scarce ground water supplies. Improved agricultural technologies and production practices have stimulated vast increases in food production on India's limited area of agricultural land.

Mitigating Scarcity of Resources

All nations and regions are not uniformly endowed with natural resources have to mitigate resource scarcity for economic development. The various possibilities :

1. ***Substitution***: Scarce resource can be substituted with resource available in plenty. (*e.g.*) Aluminium can be used for copper.

2. ***Technological change***: Technological improvements mitigate resource scarcity in a variety of ways.

(a) Better efficiency in resource use. (*e.g.*) Better technology helps to recover more metal from its ore.

3. ***Trade***: Import of resources mitigates resource scarcity. It depends on infrastructural facilities like ports, roads.

4. ***Discovery*** : Discovery of new resources by better exploitation techniques. (*e.g.*) Discovery of new fields or mines.

5. ***Recycling***: Recycling of once used resources is a very important method of mitigating resource scarcity.

(*e.g.*) Waste paper, metal.

The Constitution created the Supreme Court of India.

Guiding Principles of Resource Development

The principle objective of resource development is to maximise national output. For this purpose, there should be optimum utilisation of resources not only for short period but also for sustained long period. The following principles have been evolved for resource development:

Economic use of resources to achieve minimum waste.

Sustained use of economic resources through conservation of renewable resources and economic use of exhaustible resources.

To increase multipurpose use of resource.

Integrated planning in use of natural resources. Minimising transport cost by locating industries nearer to source of raw material.

Optimum utilization of electrical power and other resources.

To employ modern technology to explore possibility of bringing more natural resources.

ENVIRONMENT

Environmental Problems

There are three types of environmental problems.

(I) Water related problem (II) Soil related problem and (III) Technology related problem.

(1) Water Related Problem

The problems are as follows :

(i) Increasing Water Table: In some of the areas, ground water level increases to ground surface. Water table less than 1.5 m below the ground level is considered to be harmful for agricultural operation.

(ii) Loss of Irrigation Water: The improper water management techniques result in large extent of loss of irrigation water. This is mainly because canal system and its branches are not properly lined. The seepage through improper lines is estimated the heavy loss of irrigation water.

(iii) Water Logging: Productivity in water logged areas decreases considerably.

(iv) Decreasing Water Quality : Quality of water to the agricultural sector decreases. This factor should be taken care of while programming major and minor irrigation systems.

(II) Soil Related Problem

Soil degradation is another problem in agriculture.

(i) *Salinization* : Seepage through unlined irrigation stream, improper land planning is main cause leading to soil salinity.

(ii) *Wind and water erosion of soil*: Erosion of soil due to wind and water are also given importance.

(iii) *Increasing toxicity in soil*

(III) Technology Related Problem

This problem are more due to application of chemical fertilizer and pesticides.

(i) High quality application of fertilizer and pesticide.

(ii) Increase in chemical content of soil and related degradation.

(iii) Higher toxicity affecting root zone of plants. Overdoses of chemicals and pesticides affect the soil structure and make it difficult for further cultivation.

The knowledge on proper methods of application should be given to the farmers through better extension methods.

Remedial Measures

To increase the agricultural productivity, proper remedial measures are to be taken for the above said three problems. Since the area under cultivation is a serious limiting factor in India, proper technology measures should be taken to increase the productivity.

Integrated Approach for Sustainability in Agriculture

1. ***Resources***

Problem

(1) Salinity (2) Cultivable waste (3) Flood (4) Drainage (5) Improper distribution of irrigation water.

Remedy

(1) Soil reclamation (2) Quality improvement (3) Land management (4) Ground water management and (5) Policy measures.

2. ***Cropping System***

Problem

Monocropping affect the fertility of the soil.

Remedy

Crop substitution for higher valued crops, increasing cropping intensity, policy measures relating to adding the mixed farming concepts to improve the productivity and income level of farms.

3. ***Inputs***

Problem

Inefficient distribution system of seed, chemical and fertilizer.

Remedy

Privatization, bulk purchase and storage in proper places and policy measure to strengthen their infrastructure to facilitate proper distribution of inputs in time and correct place.

4. ***Technology***

Problem

Recent technologies are costly.

'Mydoom' is a computer virus that create havoc in the software and internet provider companies.

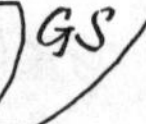

Remedy

Cost effective technology measure should be formulated to facilitate the poor farmer to adopt the technology. Policy for reorienting the measures regarding the cost-effective technologies are as follows:

Incentives for higher productivity through technical change.

Remunerative prices, subsidy on inputs.

Manpower improvement in training and management skills.

Infrastructure administration management.

Farmer responsible, farmer friendly administrators are needed.

Ecological Approach for Environmental Problems

The key ecological factors are land, water, vegetation, air, sea and wild life. The threats to them include overuse and pollution leading to soil erosion, salination, fertility decline, and degradation of plant quality, desertification, and poisoning via polluted water and airborne chemicals and radiation. There are limits to ecological carrying and self-regenerating capacity. They can be increased and but not yet surpassed.

To despair, ignore or reject an ecological message of doom is not likely to protect ecology. But the decision lies with thousands of farmers who must understand and act right way and in time.

Influence of Climate on Crops

Climate is the most dominating factor influencing the suitability of a crop to a particular region. The yield potential of a crop mainly depends on climate. More than 50% of variation in yield of a crop is due to climatic differences.

The type of vegetation or crops determined by climate.

Solar Radiation

(i) Thermal effect of Radiation: More than 70% of the solar radiation absorbed by plants is converted into heat. This heat energy is utilized for transpiration for convective heat exchange with the, ssurroundings. These changes determine the temperature of leaves and of other plant parts.

(ii) Photosynthetic effect of Radiation: A portion of solar radiation upto 28% of energy is used in photosynthesis.

(iii) Photoperiodic effect: It plays an important role as regulator and controller of growth and development.

Temperature

Cardinal Temperature

For each species of plants, there are maximum and minimum limits of temperature as at which growth is negligible optimum temperature at which growth is maximum. Most of the crop plants grow at 15°C to 30°C.

Quiz Kit!

First municipalities

Municipal corporations for civic amenities were set up for the first time between 1880 and 1884 when Lord Rippon was Viceroy of India.

Influence on Growth Substances

At optimum temperature, the activity of the growth promoters are high and activity of the growth regulator is low with the result that growth, rate is increased. At high or low temperature, the balance of growth substances change and affect growth.

Influence of Temperature on Development

Temperature has great influence on developmental rates like rate of germination, leaf initiation, tillering, flowering. Spikelet initiation and grain filling. All the development processes proceed at a faster rate at higher temperature.

Low Temperature Stress

Cell division and cell elongation.

Photosynthesis.

Water transport.

Vegetative growth.

Reproductive growth.

Rainfall

Rainfall analysis helps in taking decisions on time of sowing. Scheduling of irrigation, time of harvesting etc.

Amount of Rainfall

Rainfall may also be in excess of the optimum and thereby cause reduced yields which may appear paradescical to semi arid climates. When the rainfall is concentrated in 4-5 months of the year, there may be periods when the rate of precipitation exceeds the intake rate of soil.

Intensity of Rainfall

It influences erosion of soil.

Relative Humidity

Relative humidity has considerable influence on evapotranspiration and hence on the water requirement of crops.

Photosynthesis

When relative humidity is low, transpiration increases causing water deficit in plant. Water deficit cause partial or full closure of stomata and increase mesophyll resistance blocking entry of CO_2.

Pests

The incidence of insect pests and diseases is higher under

GK **Customs, excise duty and sales taxes are the main forms of indirect taxes.** GS high humidity conditions.

Grain Yield

Very high or very low relative humidity do not conducive for high grain yield. Impact of changing environment on cropping pattern as indicators of environment

Factors Responsible for Environment Change

- Human population
- Urbanisation and industrialisation
- Fossil fuel utilization -over exploitation
- Deforestation
- Mismanagement in agriculture
- Emission of green house gases

Impact in India

(i) Increasing temperature

- Tree line shifting towards the higher altitude (Apple in Shimla).
- Delaying of sowing of wheat crop.
- Rabi crop in north India production is under threatened (wheat, rapeseed, chickpea).

(ii) Precipitation

- All mathematical models projection is increase in precipitation in India.
- But the intensity will be more i.e. less rainy days which cause floods and drought.
- Which cause production decline.

(iii) CO_2 level increase

- Favourable for C_3 plants upto some extend.
- But increase the C_3 weed competition to C_4 plants.
- Increase the production and reduce the quality of the produce.

Finally due to climate change the wheat production in north India will be severely affected and the people should change their food habit towards rice.

POLLUTION

Effect of Industrial Effluents on Soil and Water Pollution

***(i)* Major industrial effluents**

(a) Sugar factory

(b) Tannery

(c) Distillery

(d) Textile and dye

(e) Paper mill

***(ii)* Effect on soil health**

(i) Chemical Properties: CEC, EC, pH, Heavy metal accumulation increase alkalinity, salinity and acidity of the soil.

(ii) Physical properties: Decrease the porosity and aeration, affect the structure and infiltration.

(iii) Biogical properties: Reduce the population of microorganism and fauna of the soil.

(iv) Effect on water: Surface water pollution

Ground Water Pollution

A. *Surface water pollution by direct discharge of effluents.*

- Heavy metal accumulation.
- Eutrophication.
- Reduce the productivity of the system.
- Biomagnification.

B. *Ground water pollution by infiltration*

- Reduce the water quality by increasing salinity alkalinity and heavy metal.
- Reduce the plant productivity by affecting its metabolism.
- Not fit for drinking and irrigation.
- Major pollution fluoride, arsenic and nitrate.

Air Pollutans on Vegetation

Pollutant	*Injury symptoms*	*Age of leaf affected the most*	*Leaf part affected*
SO_2	Chlorosis, inter-venial bleaching, tip and marginal necrosis; similar to drought, insect and chilling injury.	Middle aged most sensitive; older least sensitive	Mesophyll cells.
O_3	Fleck, stipple, bleached and necrotic spotting, pigmentatin; browning of conifer needle tips	Oldest most sensitive;	Palisade or spongy parenchyma in leaves
PAN	Glazing, silvering or bronzing of lower leaf surface	youngest most sensitive	Spongy parenchyma tissue
NO_2	Irrigular, white or brown collapsed lesions on near leaf margin	Middle-aged leaves most sensitive	Mesophyll cells
HF	Tip and marginal leaf burning, dwarfing, leaf abscission, narrow brown-red band, separate necrotic tissues from green tissues; fungal, cold an dhigh thermal injuries make similar marking	Youngest leaves most sensitive cells	Epidermis and nesophyll

Bangar and Khadar are the names given to the old and new alluvial soils respectively.

GS

Pollutant	*Injury symptoms*	*Age of leaf affected the most*	*Leaf part affected*
C_2H_4	Sepal withering, leaf abnormality; flower shedding, failure or proper leaf opening; excess water stress shows similar symptoms	Young leaves recover but older do not recover fully	All leaf part
Cl_2	Intervenial bleaching, marginal and tip burning, leaf abscission	Matutre leaves more sensitive	Edpidermis and mesophyll
NH_3	Cooked green appearance, becoming brown after drying	Mature leaves most sensitive	Complete leaf tissue
HCL	leaf margin necrossi on broad leaves, acid type necrotic leasion, tip burning in fir	Oldest leaves most sensitive cells	Epidermis and mesophyll
H_2S	Basal and marginal scorching	Youngest leaves most sensitive	All leaves
H_2SO_4	Necrotic spots on upper (Acid rain) conducive for leaf damage	All plant leaf surface, high humidity more	All leaf parts tissued

Name of sensitive and tolerant plant species to various air pollutions

Pollutant	*Sensitive Plants*	*Tolerant plants*
Sulphur dioxide	Alfalfa, Barley, cotton, wheat, soybean, clover, rye, pine, safflower, apple, squash, tomato	Cantaloupe, celery, corn, oak
Hydrogen sulphide	Chinese apricot, gladiolus, grape, pine, sweet corn, peach, plum.	Aspergus, cherry, cotton, pear, squash strawberry, tomato, wheat
Peroxy acetyl nitrate	Annual bluegrass, lettuce spinach, beans, beet, pepper, tobacco, clover, mint.	Broccoli, chrystanthemum, corn, cotton, sorghum, cabbage, onion, cauliflower, carrot, cucumber, strawberry
Ozone	Oak, potato, radish, soybean, tobaco, tomato, white pine	Beet, geranium, gladiolus, maple, mint, pepper, rice
Ethylene	Cotton, cowpea, orchid, tomato	Grasses, lettuce

GK GS

Chief Justice and twenty five other judges are there in the Supreme Court.

Points to Ponder

- Term ecology was proposed by Reiter
- The study of interaction between living organisms and their environment is called Ecology
- Synecology is the study of Communities
- Branch of Botany dealing with distribution of plants on earth's surface is called Phytogeography
- Pedology is the study of Soils
- Plants which are attached to the stones and rocks are Lithophytes
- Plants which grow in cold areas are Psychrophytes
- Plants growing in soils having sand and gravel are Psammophytes
- All the living organisms of the earth constitute Biosphere
- Biosphere is in which individuals interact with each other
- Endemic plants are those which are restricted to a certain area
- Scientists celebrate 5th June as World Environment Day
- Plants are killed in winter by frost because of desiccation and mechanical damage to the tissues
- Biomes are major ecological groupings of animals and plants
- A Community is organisms living in habitat
- Autecology refers to the ecological study of individual organism
- Transition zone between two vegetational types or vegetational regions is known as Ecotone
- Oxalophytes grow on acidic soil
- The most favoured habitat of plant is called ecological niche
- Virapary is characteristic feature of Halophytes
- A climax community is recognizable because it is Composed of trees
- Plant growing well in full sunlight are called Heliophytes
- Plants growing in saline soil water are called Halophytes
- Commensalism is when one of the partners is benefited
- All living organisms on or around the earth constitute a Biosphere
- Niches is our index of environmental conditions and environmental changes.
- Plants grow sparsely in arid regions of the world because of several factors combining together.
- The world ecosystem was first coined by A. G. Tansley
- An ecosystem may be defined as different communities of plants, animals and microbes together with their physico-chemical environments
- The cycling of elements of an ecosystem is called Bio-geochemical cycle
- An ecosystem resists changes because it is in a state of Homeostatis
- A food chain consists of producers, consumers and decomposers
- The pyramid of energy is always upright
- Green plants are very important because they are producers
- In a food chain initial organisms are Photosynthetes (primary producers)
- A food chain starts with Photosynthesis
- Equilibrium in atmosphere is maintained by producers, consumers and decomposers
- Primary source of energy in an ecosystem is sunlight (Solar energy)
- In a forest ecosystem green plants are Primary producers
- 10% energy is utilised by Herbivores from the food they ingest?
- In a food chain of grassland ecosystem the top consumers are Carnivores
- In an ecosystem the population of Primary producers is the largest
- The first link in a food chain is always a green plant because it has the capacity to synthesize organic food
- In a tree ecosystem the pyramid of number is Inverted
- Which of the following is the logical sequence of carbon cycle? Photosynthesis-consumer-decomposer
- A depleting source of energy is Fossil fuels
- Which of the following is a renewable source of energy? Trees
- Maximum solar energy can be trapped by more cultivation of crop plants
- When one organism is benefited without affecting the other is called Symbiosis
- B.O.D. stands for Biochemical oxygen demand
- Environmental Planning Organisation is NEERI
- Maintenance of soil fertility without addition of nutrients is due to activity of micro-organisms
- Deforestation has an alarming effect on Soil erosion
- Soil erosion can be prevented by Afforestation
- Soil erosion can be prevented by planting trees
- Eroded soils are devoid of plant nutrients
- Ecological factors are the factors which work In complex combinations interacting and modifying the effects of one another
- Mulching helps in moisture conservation
- Mangrove vegetation is found in Sunderbans.
- Grassland with scattered trees is called Savannah

The wheels on the Indian Flag symbolize : Justice.

GS

- The earliest settlers on barren lands are generally Lichens.
- The extent of rainfall characteristic of tropical deciduous forests lies in the range of 500-750 mm
- Conifers are found in eastern and western Himalayas
- The plant that is characteristic component of mangrove vegetation is *Rhizophora mangel*
- Desert plants are mostly Succulents
- Alpine vegetation is seen at the top of mountains
- Which of the following is a biome? Deciduous forest
- The typical biome of an area having 20 to 40 inches of rainfall is Grassland
- Lichens is pioneer of vegetation.
- Tropical evergreen forests are found in Assam
- The largest vegetational group in known as Formation
- Plant succession taking place in sandy area is called Psammosere
- Rain forests occur in Western Ghat and Eastern Himalayas
- Sea shore vegetation is called Halophytic vegetation
- *Rhizophora* is a characteristic component of Mangrove vegetation
- Water plants usually have a reduced vascular system
- An aquatic plant with floating leaves has stomata only on upper surface
- In submerged hydrophytes which of the following is present? Air-spaces
- Aerenchyma is present in which of the following plants? *Potamogeton*
- Heterophyllous plants are found as Rooted and floating hydrophytes
- *Typha* is an example of an emerged hydrophyte
- An American plant which had become a troublesome aquatic weed in India is *Eichhornia crassipes*
- which plant among the following is completely devoid of roots? *Ceratophyllum* (or) *Wolffia*
- Proximity of water level at low depths of soils results in development of Shallow roots
- Xerophytes have longer root system because Roots grow deeper to absorb water from deep water tables
- Plants growing in xerophytic conditions have Extensively developed root system small and thick leaves
- Sunken stomata and multiple epidermis are found in the leaves of *Nerium*
- Which group of plants has to face physiological dryness? Xerophytes
- Succulents (xerophytes) are likely to be found in Deserts (dry regions)
- Which of the following shows xerophytic characters? All

Knowledge World!

First scientific institute

The Indian Association for the Cultivation of Science was established in Kolkata in 1876 by Dr. Mahendra Lal Sarkar for conducting research in science. C.V. Raman later discovered the Raman Effect here.

Lithophytes

- Vegetation of any place is primarily determined by Rainfall
- The equilibrium of the species is disturbed by Selective grazing
- Grazing is an example of Negative pollution
- Silent valley is situated in Kerala
- If all the plants of the world die, all the animals will also die due to shortage of Oxygen
- Green plants form First trophic level
- One of the effects of dust is to cause Closure of stomata
- Increasing accumulation of DDT in organisms of a food chain in higher trophic levels is known as Biological amplification
- MAB stands for Man and biosphere
- Which of the following functions performed by a forest help most in controlling drought? Forests act as water sheds
- Autotrophs has the greatest amount of energy.
- The instrument which measures wind velocity is Anemometer
- Myrmecophily is found in *Acacia*
- As the number of organisms increases in a population Biotic potential increases
- Largest ecosystem of the world are Oceans
- Plants in a particular locality constitute Flora
- Study of interrelationship between species and its environment of a forest is called Forest ecology
- The *term Oekologie* (*Ecology*) was coined by Haeckel
- Thick cuticles on leaves are typical of plants growing in Dry habitats.
- The effect of pollution is first and most marked on Natural balance of our environment
- NEERI is situated at Nagpur
- Maximum contribution of O_2 is from Phytoplankton
- The succession initiated in a pond of about 6 meters depth exhibits one of the following stages Submerged plants ® floating plants ® Reeds ® herbs shrubs ® trees
- Indicator plants which can be used to indicate atmospheric pollution by SO_2 are *Lichens like Usnea*
- The term biosphere is used for the zone of earth where life exists in the hydrosphere

GK **The Central Government is empowered to levy customs and excise duty.** GS

- Ephemerals are xerophytes that are Drought escaping
- Recurrence of daily and seasonal changes in a community is known as Species dominance
- Greatest problem for water conservation is to reduce the amount of Evaporation
- Renewable source of energy is Biomass
- The total energy fixed by a gram plant (*Cicer arietinum*) in a ecosystem on the whole is called Gross production
- The region consisting of long and severe winters and growing season consisting of few months of summers is Taiga ecosystem
- Flow of energy declines from low to higher trophic level in ecosystem is mainly explained by Second law of thermodynamics
- The pyramid that can not be inverted in a stable ecosystem is pyramid of Energy
- Relative to roots, shoot can be very massive in plants of Moist tropical forests
- Soil erosion can be prevented by Over grazing
- Semi aquatic plant is Rice.
- The term biocoenosis was coined by Carl Mobius
- A natural food web contains several trophic levels
- Earth summit at Rio de Janerio was related to Conservation of genetic resources of plants and animals.
- Which of the following enhances B.O.D. of water? Sugar mill effluents
- Major pollution causing agent is Man
- Water pollution is caused by Industrial effluents
- The presence of Ozone (O_3) in the atmosphere of earth helps in checking the penetration of ultraviolet rays to earth.
- In Northern Indian plains there is an increase in floods in recent times due to Deforestation in the catchment area
- The molecular action of ultraviolet lights is mainly reflected through destruction of hydrogen bonds between DNA
- Radiation is a health hazard because it causes Leukemia.
- It is said the Taj may be destroyed due to Air pollutants released from oil refinery of Mathura
- SO_2 and NO_2 cause pollution by increasing Acidity
- Acidic rains are caused by SO_2, NO_2
- Biological treatment of water pollution is done with the help of Phytoplanktons
- CO_2 is usually not an air pollutant.
- Non-ionising radiations are U.V. rays
- Carbon mono-oxide is a pollutant because it combines with haemoglobin
- Spraying of D.D.T. on crops produces pollution of air, soil and water
- The ultimate environmental hazard to mankind is Nuclear pollution
- Thermal pollution of water bodies is due to Discharge of heat (hot water) from power plants
- Ozone layer is disturbed by Supersonic jets
- Smog is a common pollutant in places having Low temperature
- The component of a living cell affected by the pollutant SO_2 is All cell membrane system
- Most outstanding danger at present for survival of living beings on earth is Radiation hazards
- Sr ^{90} is responsible for atmospheric pollution?
- Minamata disease in Japan was caused by pollution of water by Mercury
- Main air pollutant is CO
- Ozone layer is destroyed due to Chlorofluorocarbon (CFC)
- Greenhouse effect is due to the presence of CO_2 layer in the atmosphere
- Deforestation will decrease Rainfall.
- Sheet erosion is caused by Heavy rains.
- Major aerosol pollutant present in the jet plane emission is Flourocarbon
- Sewage water can be purified for recycling with the action of Micro-organisms
- Increase in the percentage of fauna and decrease in flora may be dangerous because it enhances Percentage of CO_2
- Gases referred to as "Greenhouse gases" are Chlorofluorocarbon CO_2, CH_4, NO_2
- The substance which produces resistance against carcinogen is Tocofarol
- Which is not a greenhouse effect gas? H2
- The country which hosted the first World Earth Summit on conservation of environment is Brazil
- Chipko movement of connected with Plant conservation
- A pollutant is an agent which causes change in Local flora
- Gas responsible for Bhopal tragedy is Methyl isocyanate
- Polluted water can be purified by using Microorganisms
- Which of the following damages WBC, bone-marrows and lymph nodes? I^{131}
- Which of the following is extreme example of noise pollution? Sonic booms produced by aircraft
- Photocopying machines and other electrical equipment produce one of the following Ozone
- Harmful ultraviolet radiations coming from the sun cause Skin cancer
- Excessive inhalation of manganese causes Pneumonia
- Which disease in children is caused by the intensive use of nitrate fertilizers? Methemoglobinemia
- Greenhouse effect is related to Global warming

GK **'Creeping Death' is the name given to the soil erosion.** GS

Multiple Choice

1. Among the following which gas is found in maximum concentration in the atmosphere
   a] Rn b] Xe
   c] Kr d] Ar
2. Ventilation coefficient is
   a] Maximum mixing depth divided by average wind speed.
   b] Product of maximum mixing depth and maximum wind speed.
   c] Maximum mixing height divided by maximum wind speed.
   d] Product of maximum mixing depth and average wind speed.
3. Mosquito repellent coils/mats contain
   a] Paraquat b] BHC
   c] Toxaphene d] Derivatives of Allethrin
4. The concentration of CO_2 in the atmosphere is increasing at the rate of about
   a] 4 % b] 2 %
   c] 0.4 % d] 0.2 %
5. 1 Dobson unit at standard temperature and pressure is equal to ozone column thickness of
   a] 1 mm b] 10 mm
   c] 100 mm d] 0.01 mm
6. Who is the director of the Centre for Science and Environment?
   a] Anil Agarwal b] Sunita Narayan
   c] Menka Gandhi d] Medha Patekar
7. Which CFC was discovered first?
   a] CFC- 11 b] CFC- 12
   c] CFC- 114 d] CFC- 115
8. The concentration of ozone is found maximum in
   a] Troposphere b] Upper Stratosphere
   c] Lower Stratosphere d] Mesosphere
9. The National Institute of Oceanography is presently situated at
   a] Kerala b] Calicut
   c] Goa d] Cochin
10. Mycorrhiza help in the uptake of
    a] Potassium b] Phosphorus
    c] Nitrate d] Boron
11. In which area is Chaparral vegetation found ?
    a] Deccan plateau
    b] Coastal India
    c] Mediterranean areas
    d] Between equator & temperate
12. The residence time of N_2O in the atmosphere is approximately
    a] 15 years b] 150 years
    c] 15 weeks d] 15 days

Quiz Kit!

First ticket reservation and travel

Baroner Sir Jamsetjee Jeejeebhoy reserved the whole train and traveled from Bombay to Thane and back, along with some members of his family, on April 17, 1853.

13. Which elemental cycle has no atmospheric reservoir?
    a] Sulphur b] Carbon
    c] Nitrogen d] Phosphorus
14. Muscovite is the example of which group of minerals?
    a] Feldspar b] Ferromagnesium
    c] Quartz d] Mica
15. Highest porosity is found in
    a] Sand b] Clay
    c] Gravel d] Silt
16. Solubility of gases in water can be calculated using
    a] Darcy's law b] Henry's law
    c] Avagadro's law d] Stoke's law
17. The main constituent of biogas is
    a] H_2 b] H_2S
    c] N_2 d] CH_4
18. The actual global average surface air temperature is about
    a] 331 K b] 288 K
    c] 18°C d] 21°K
19. The Ministry of Environment was set up in
    a] 1970 b] 1975
    c] 1980 d] 1985
20. Age of earth is approximately
    a] 3.5 billion b] 4.5 billion
    c] 45 million d] 3.5 million
21. After methane and carbon dioxide which gas is found in highest concentration in biogas
    a] CO b] H_2
    c] N_2 d] H_2S
22. The target organ of cadmium toxicity
    a] Lung b] Liver
    c] Kidney d] Bones
23. Nalgonda technique is used for
    a] Chloride b] Fluoride
    c] Bromide d] Cadmium
24. The process used for removal of water hardness is
    a] Zeolite process b] Haber's process
    c] Ostwald process d] None of the above
25. Who gave the term ecosystem and when?
    a] A.G. Tansley (1935) b] A.G. Tansley (1925)
    c] E.P. Odum (1935) d] E.P. Odum (1925)

GK **Interpretation of Constitution comes under Appellate Jurisdiction of Supreme Court.** GS

26. IPCC came into existence in
a] 1978 b] 1968
c] 1988 d] 1998

27. The first international conference on Environment Education was held in
a] Tbilisi b] New Delhi
c] Bombay d] Turkey

28. Kyoto Protocol came into force on
a] 16 Feb 2004 b] 16 Feb 2005
c] 16 Feb 2002 d] 16 Feb 2000

29. Indian Institute of Petroleum is situated at
a] New Delhi b] Goa
c] Deharadun d] Bombay

30. Which of the following is mainly responsible for eutrophication?
a] Phosphateb] Nitrate
c] Carbonate d] Sulphate

31. The normal lapse rate of temperature per kilometer is
a] 6.4°C b] 4.6°C
c] 10° C d] 9°C

32. The albedo of the earth as a whole is
a] 25 % b] 50 %
c] 30 % d] 10 %

33. Atmospheric humidity is measured by
a] Radiometer b] Hygrometer
c] Hydrometer d] Micrometer

34. The origin of simplest life is attributed to
a] Proteozoic Era b] Cambrian Era
c] Archaezoic Era d] None of the above

35. The maximum permissible limit of free residual Chlorine in water is
a] 2 ppm b] 0.02 ppm
c] 0.2 ppm d] 20 ppm

36. Radiosonde is used to study
a] Earth's albedo at surface
b] Estimate pollutants in air
c] Atmospheric moisture content
d] Upper atmosphere's conditions

37. Which of the following enzymes is involved in the primary carboxylation in C_4 plants?
a] RUBP Carboxylase b] PEP
c] Oxygenase d] None of the above

38. The compound used for artificial rain-making/ cloud seeding
a] Dry ice b] AgI
c] $HgCl_2$ d] Both a & b

39. How many agro-climatic zones are found in India
a] 15 b] 16
c] 17 d] 18

40. About 50 % of the atmosphere lies below
a] 5.6 km b] 10 km
c] 15 km d] 30 km

41. The atmospheric layer reflecting radio waves is called
a] Homosphere b] Ionosphere
c] Ozonosphere d] None of the above

42. The specific heat of water
a] 1 cal/ gm/°C b] 4.18 cal/ gm/°C
c] 1 Joule/ gm/°C d] 10 cal/gm/°C

43. The first Biosphere Reserve in India was
a] Sundarban b] Nanda Devi
c] Nilgiri d] Nokrek

44. In which year Biodiversity Act was proposed?
a] 2000 b] 2002
c] 2004 d] 1998

45. Headquarters of UNEP is situated at
a] Kenya b] Switzerland
c] U.S.A. d] Britain

46. At which temp. density of water is found maximum
a] 0°C b] 100° C
c] 15°C d] 4°C

47. The important GHG mainly released from paddy field
a] CO_2 b] N_2O
c] CH_4 d] All of the above.

48. What is the study of trees as individuals in relation to their environment known as?
a] Forest ecology b] Forest synecology
c] Forest Autecology d] All of the above.

49. Demography is the statistical study of
a] Human society b] Human population
c] Human settlement d] Human life

50. The pH of normal rain is
a] 6.5 b] 5.6
c] 4.6 d] 3.6

51. Which term represents the sum total of life on earth
a] Biomass b] Biosphere
c] Biosphere d] Biome

52. Any living thing that successfully competes with people for food space, or other essential needs is called
a] Virus b] Bug
c] Parasite d] Pest

53. Of the following environmental assessment terms, which one deals exclusively with the carbon content of the environment
a] BOD b] COD
c] TOC d] POC

54. The dominance of a new genetic form as a result of environment change is called
a] Adaptation b] Natural selection
c] Succession d] Synergism

55. What does the term overkill deal with
a] Pesticidal poisoning
b] Soil erosion

GK GS

During the reign of emperor Chandragupta Vikramaditya the Chinese traveller Fa-Hein visited India.

c] Nuclear holocaust d] Global warming

56. **What is a chemical substance or physical agent capable of inducing inheritable genetic change called**
a] Carcinogen b] Mutagen
c] Teratogen d] Tumorogen

57. **When lakes become acidic due to acid rain, this is added to counteract the acidity**
a] Soil b] Sand
c] Lime d] None of the above

58. **Which technique can map the concentration of SO_2 over a whole town by operating a Gadget from one location**
a] LIDAR
b] Spectrophotometer
c] Gas Chromatography
d] Mass Spectroscopy

59. **Air pollution can be controlled & reduced considerably, but one of the following factors comes in its way.**
a] Politics b] Economics
c] Manpower d] Geography

60. **Photocopying & other electrical equipment produce one of the following pollutants**
a] Methane b] Ozone
c] Hydrogen d] Nitrogen oxides

61. **Which harmful gas is emitted by masonry building materials, even ground water ?**
a] H_2S b] Radon
c] Ammonia d] CO_2

62. **Which is the most abundant of all the hydrocarbon pollutants in the atmosphere ?**
a] Propane b] Methane
c] Butane d] Benzene

63. **Which oil -tanker accident first alerted the public to the grave problem of oil spills in oceans ?**
a] Agro merchant b] Ocean Eagle
c] Exxon Valdez d] Torrey Canyon

64. **The process of preparing compost with the help of earth worm is known as**
a] Composting b] Bioslury
c] Vermi composting d] Maturing

65. **Mycorrhizae is the association of**
a] Higher plants & fungi
b] Algae & fungi
c] Lower plants & fungi
d] Both A & C

66. **When was Ganga action plan launched ?**
a] June 1985 b] December 1985
c] May 1984 d] July 1984

67. **Largest salt water lake in India**
a] Lonar b] Chilka
c] Sambhar d] Wullar

68. **The main pollutant of London smog was**
a] SO_2 b] NO_2
c] PAN d] Ozone

69. **Which detector for DDT should be used in Gas chromatography ?**
a] Electron capture detector
b] Flame ionised detector
c] Thermal conductivity detector
d] None of the above

70. **In which of the following, inverted pyramid of biomass is found ?**
a] Grassland ecosystem
b] Pond ecosystem
c] Desert ecosystem
d] Forest ecosystem

71. **In denitrification process the nitrogen is released in form of**
a] NH_3 b] N_2
c] N_2O d] Both b & c

72. **Who coined the term symbiosis ?**
a] A.G. Tansley b] De Bary
c] Clement d] McDougall

73. **Which type of symbiosis is obligatory ?**
a] Mutualism b] Protocooperation
c] Commensalism d] Amensalism

74. **Concept of Hypervolume Niche was given by**
a] Grinnel b] Clement
c] Hutchinson d] Odum

75. **Vehicular pollution emits mainly**
a] SO_2 b] CO
c] NO d] Ozone

76. **Of the following classes of beings, which one is least sensitive to nuclear radiations?**
a] Single celled organism
b] Amphibians
c] Reptiles
d] Birds

77. **Which region of the seas & oceans are the most polluted?**
a] Estuarine b] Coastal
c] Sea depths d] Coral

78. **Of the following types of nuclear bombs, which one was tested on a Pacific atoll?**
a] Atom bomb b] Hydrogen bomb
c] Neutron bomb d] None of the above

79. **Which were the first living beings to establish themselves on rocky slopes?**
a] Toads d] Grasses
c] Lichens d] Frogs

80. **What is lost when one organism consume another ?**
a] Food b] Water
c] Energy d] Chemicals

The Indian income tax is direct and progressive.

81. Who formulated the ecological concept of the pyramid of numbers?
a] Charles Elton b] Paul R. Ehrlich
c] Paul Colivaux d] All of the above

82. Where was the Mitti bachao (Save the soil) movement launched in India ?
a] Thane, Maharashtra
b] Mysore, Karnataka
c] Darbhanga, Bihar
d] Hoshangabad, M.P.

83. When was the use of DDT banned for agricultural purposes in India ?
a] 1962 b] 1985
c] 1974 d] 1971

84. For the production of biogas the Indian biogas plant need
a] Fire wood b] Cattle dung
c] Agricultural waste d] Kerosene

85. Which forest area in India was first brought under control & protection ?
a] Malabar b] Konkan
c] Garhwal d] Sunderbans

86. When was the first National Forest Policy formulated?
a] 1948 b] 1980
c] 1964 d] 1952

87. Which plant is known as the gasoline plant ?
a] *Salvadora persica*
b] *Sterculia foetida*
c] *Thevetia peruviana*
d] *Euphoribia lathyris*

88. When did the Three Mile Island disaster occur ?
a] 1972 b] 1979
c] 1980 d] 1976

89. Flyash is the environmental pollutant generated by
a] Thermal power plant
b] Oil refinery
c] Fertilizer plant
d] Strip mining

90. Which of the following is known as liquid gold ?
a] Water b] Petroleum
c] Mercury d] Mustard oil

91. Which category of wastewater doesn't require seeding during a BOD test?
a] Distillery spent wash
b] Dyeing unit effluent
c] Domestic sewage
d] Pulp & paper mill effluent

92. Permanent hardness of water is caused by
a] Carbonates & chlorides
b] Bicarbonates & sulphates
c] Carbonates & bicarbonates
d] Chlorides & sulphates

93. The most commonly used method for desalination of water is
a] Distillation b] Reverse osmosis
c] Electrodialysis d] Flash evaporation

94. Elemental chlorine is widely used in
a] Metallurgy b] Water purification
c] Process industry d] Deodrants

95. The single largest class of insecticides of total registered pesticides in the world is
a] Organochlorine b] Organophosphate
c] Carbamate d] Pyrethroids

96. Cup anemometer is used for measuring
a] Water evaporation b] Wind speed
c] Wind direction d] Water flow

97. Which type of humus is acidic in nature ?
a] Mor b] Mull
c] Moder d] All of the above

98. Lambert's & Beer's law is used in
a] Spectrophotometer b] Chromatography
c] Potentiometer d] pH meter

99. Law of minimum was given by
a] Shelford b] Liebig
c] Blackman d] Clement

100. When was project tiger launched in India ?
a] 1972 b] 1973
c] 1978 d] 1974

101. When was the Jawaharlal Nehru Community Biodiversity movement launched in India?
a] 1988 b] 1989
c] 1969 d] 1985

Tank irrigation is the general mode of irrigation in peninsular India.

Matching

1. Match important environmental conventions

a]	Montreal protocol	i]	1992
b]	Helsinki declaration	ii]	1990
c]	London conference	iii]	1989
d]	Copenhagen conference	iv]	1987

2. Match statewise mangrove cover in India.

a]	West Bengal	i]	4.41 %
b]	Orissa	ii]	43.63%
c]	Andra Pradesh	iii]	0.43%
d]	Tamil Nadu	iv]	8.15%

3. Match the following

a]	O_1	i]	Ozone
b]	O_2	ii]	Molecular oxygen
c].	O_3	iii]	Nascent or atmospheric oxygen
d]	O_1+O_2O_3 iv]	iv]	UVR

4. Match different atmospheric layers.

a]	Upto 17 km in altitude	i]	Thermosphere
b]	Upto 50 km in altitude	ii]	Mesosphere
c]	From 50 km to 85 km in altitude	iii]	Toposphere
d]	Above 50 km in altitude	iv]	Strotosphere

5. Match air pollution episodes

a]	Meuse valley, Belgium	i]	1948
b]	Donora, Pennsylvania	ii]	1984
c]	London smog	iii]	1930
d]	Bhopal MIC gas tragedy	iv]	1952

6. Match commonly-used radiation material with their half-life

a]	Carbon-14	i]	5, 568 yrs.
b]	Cobalt-60	ii]	5.24 yrs.
c]	Iodine-131	iii]	8.08 yrs.
d]	Phosphorus-32	iv]	14.3 yrs.

7. Match commonly used radiation with material.

a]	Beta	i]	Strornium-90
b]	Beta, gamma	ii]	Uranium-238
c]	Alpha, gamma	iii]	Uranium-235
d]	Alpha, beta, gamma	iv]	Iodine-131

8. Match wetland of India.

a]	Bhoj	i]	Madhya Pradesh
b]	Harikeii]	ii]	Punjab
c]	Tso Morasi	iii]	J & K
d]	Pichol	iv]	Rajasthan

9. Match the following

a]	Hilaire	i]	First proposed the term 'Ecology'.
b]	Reiter	ii]	Introduced the Term 'Oekologie'.
c]	E. Haeckel	iii]	Used the term 'Ehtology'.
d]	Warming	iv]	First employed ecology for the study of plants.

10. Match ecological factors

a]	Climatic factors	i]	Altitude, mountain, direction, steepness.
b]	Topographic factors	ii]	Light, temperature, rainfall, humidity.
c]	Edaphic factors	iii]	Includes all kinds of interactions between different forms of life.
d]	Biotic factors	iv]	Soil formation, soil profile, minerals and organic matter in soil, soil water and air.

The President is not bound by the advice given by Supreme Court on the matters referred to it.

11. **Match the following**

a]	Megatherms	i]	30-45°C
b]	Mesotherms	ii]	20-30 °C
c]	Microtherms	iii]	10-20 °C
d]	Hekistotherms	iv]	0-10 °C

12. **Match the precipitation forms.**

a]	Fog	i]	*Ola.*
b]	Mist	ii]	*Kuhra*
c]	Den	iii]	*Dhundh*
d]	Hail	iv]	*Oas*

13. **Match the following**

a]	Sendtner and Derner	i]	Introduced community aspect in ecology
b]	Schroeter and Dirchner	ii]	Term 'Ecosystem'
c]	Tansely	iii]	Term 'Synecology'
d]	K. Mobius	iv]	Term 'Bioecoenogenesis'

14. **Match types of plants based on soil types**

a]	Lithophytes	i]	On sand
b]	Psammophytes	ii]	On acidic soil
c]	Halophytes	iii]	On saline soil
d]	Oxylophytes	iv]	On rocks

15. **Match intensity of sound**

a]	10 dB	i]	Average living room
b]	20 dB	ii]	Libraries with soft whisper
c]	30 dB	iii]	Very quiet place
d]	40 dB	iv]	Rustling of leaves

16. **Match intensity of sound**

a]	Light traffic noise	i]	50 dB
b]	Normal conversations	ii]	60 dB
c]	Cars, motorcycles, trucks	iii]	70-80 dB
d]	Jet plane upto 300 heights	iv]	100-110 dB

17 **Match sound intensities**

a]	110-120 dB	i]	Ultrasonics
b]	150 dB	ii]	Rocket engine
c]	180-195 dB	iii]	Jet plane at take off
d]	above 20,000 htz/sec.	iv]	Truck horns

18 **Match the following**

a]	CEP	i]	Lucknow
b]	CWPB	ii]	New Delhi
c]	EPA	iii]	U.S.A.
d]	ITRC	iv]	Switzerland

19 **Match the following**

a]	IUCN	i]	Dehradun
b]	UNEP	ii]	Geneva
c]	WHO	iii]	Switzerland
d]	FRI	iv]	Nairobi

20. **Match the following**

a]	Mercury	i]	Minamata disease
b]	Lead	ii]	Affects bone marrow
c]	Cadmium	iii]	"Ouch-ouch" of bone
d]	Arsenic	iv]	Black foot disease

GS

Kathakali is a form of dance drama from Kerala of India.

21 Match annual rainfall with vegetation

a]	0.13.24 cm	i]	Semi-arid grassland
b]	13.25-35.1 cm	ii]	Desert
c]	35.2-63.5 cm	iii]	Humid subtropical forest.
d]	63.6-114.3 cm	iv]	Savanna

22. Match the constituent with their percentage (by volume)

a]	Ozone	i]	0.000114
b]	Xenon	ii]	0.000001
c]	Hydrogen	iii]	0.0000087
d]	Krypton	iv]	0.00005

23. Match the following

a]	L_1	i]	Top layer or canopy layer of trees
b]	L_2	ii]	Middle layer or shrubby layer.
c]	L_3	iii]	Herbaceous or ground flora
d]	L_4	iv]	Ground stratum like mosses, thallophytes, lichens

24. Match types of plants based on soil types

a]	Chersophytes	i]	On wastelands
b]	Eremophytes	ii]	On deserts
c]	Psilophytes	iii]	In Savanna
d]	Chasmophytes	iv]	In crevices of rocks

25. Match habitat with salt content.

a]	Miohalophytes	i]	5% NaCl and above
b]	Mesohalophytes	ii]	Between 0.5-1% NaCl
c]	Mesoeuhalophytes	iii]	Below 0.5% NaCl
d]	Eneuhalophytes	iv]	1% NaCl and up

26. Match plant indicators with soil characteristics

a]	*Zizyphus nummularia*	i]	Good soil for agriculture
b]	*Butea monosperma*	ii]	Heavy alkaline
c]	*Rumex acetosella*	iii]	Acid grassland soil
d]	*Pinus*	iv]	Uranium rich soil

27. Match soil characteristic with plant indicators

a]	Saline water condition	i]	*Sueda spp.*
b]	Sandy loam type soil	ii]	*Andropogon scoparium*
c]	Recently distributed	iii]	*Argemone mexicana*
d]	Nitrate rich soil	iv]	*Rumex species*

28. Match plant indicators with minerals.

a]	Diamond	i]	Vallozoacandida
b]	Gold	ii]	Equisetum arvense
c]	Silver	iii]	Eriogonim ovalifolium
d]	Mercury	iv]	Stellaria setacea

29. Match pollutant with plant indicator.

a]	Ozone	i]	Salvia, Dahlia
b]	SO_2	ii]	Ficus, Xenia
c]	Hydrogen fluoride	iii]	Gladiolus, Pinus
d]	Feroxi-acetyl nitrate	iv]	Pitunia, Primrose

30. Match the following

a]	Combustion	i]	Waste recovery
b]	Manufacturing	ii]	Transportation
c]	Agriculture	iii]	Spray painting
d]	Solvent usage	iv]	Field burning

AGMARK is a quality guarantee stamp for agro-products like eggs, ghee, honey, spices, etc.

Glossary of Technical Terms

Abiotic Factor A non-living component of the environment, such as soil, nutrients, light, fire, or moisture.

Adaptation (1) Any aspect of an organism or its parts that is of value in allowing the organism to withstand the conditions of the environment. (2) The evolutionary process by which a species' genome and phenotypic characteristics change over time in response to changes in the environment.

Aerobic Aerobic organisms require oxygen for their life processes.

Agroecology The science of applying ecological concepts and principles to the design and management of sustainable agroecosystems.

Agroecosystem An agricultural system understood as an ecosystem.

Agroforestry The practice of including trees in crop- or animal-production agroecosystems.

Allelopathy An interference interaction in which a plant releases into the environment a compound that inhibits or stimulates the growth or development of other plants.

Alluvium Soil that has been transported to its present location by water flow (alluvial soil).

Alpha Diversity The variety of species in a particular location in one community or agroecosystem.

Amensalism An inter-organism interaction in which one organism negatively impacts another organism without receiving any direct benefit itself.

Anaerobic Anaerobic organisms do not require oxygen for their life processes, in fact oxygen is toxic to many of them. Most anaerobic organisms are bacteria or archaeans.

Aquaculture the production of food and feed using aquatic agroecosystems.

Autotroph An organism that satisfies its need for organic food molecules by using the energy of the sun, or of the oxidation of inorganic substances, to convert inorganic molecules into organic molecules. Green plants are autotrophs.

Beneficial Insects Beneficial insects are predators, parasites, or competitors of insect pests, helping to regulate pest populations without harm to crops.

Beta Diversity The difference in the assemblage of species from one location or habitat to another nearby location or habitat, or from one part of an agroecosystem to another.

Biogeochemical Cycle The manner in which the atoms of an element critical to life (such as carbon, nitrogen, or phosphorus) move from the bodies of living organisms to the physical environment and back again.

Biogeochemistry The study of biogeochemical cycling.

Biological Nitrogen Fixation The fixation, by bacterial cells, of atmospheric Nitrogen gas into organic compounds useful for life. Nitrogen-fixing bacteria exist in the soil and in association and symbiosis with plants or fungi.

Biomass The mass of all the organic matter in a given system at a given point in time.

Biotic Factor An aspect of the environment related to organisms or their interactions.

Boundary Layer A layer of air saturated with water vapor (from transpiration) that forms next to a leaf surface when there is no air movement.

Buffer Zone A less-intensively-managed and less-disturbed area at the margins of an agroecosystem that protects the adjacent natural system from the potential negative impacts of agricultural activities and management.

Bulk Density The mass of soil per unit of volume.

C:N Ratio The ratio of carbon to nitrogen in a material. The decomposition of materials is regulated in part by this ratio, so materials with different **C:N ratios** are usually mixed to improve decomposition rates in composting. The **C:N ratio** for optimal biological activity is about 25:1, with higher values being nitrogen limited and lower values being carbon limited. The average C:N ratio for soils is about 10:1.

C3 Pathway The most common metabolic pathway for photosynthetic *carbon fixation* in plants.

C4 Pathway A metabolic pathway for photosynthetic *carbon fixation* that is common in tropical plants that have high rates of growth and photsynthesis and are adapted to high temperatures, high light intensity, low CO_2 concentrations and low water availability. C4 metabolism is especially common in grass species.

CAM (Crassulacean Acid Metabolism), a metabolic pathway adapation of certain plants that allows them to take up CO_2 at night instead of during the day, greatly reducing transpirational water loss during photosynthesis. This type of metabolism is common in *xerophytes*.

Capillary Water The water that fills the micropores of the soil and is held to soil particles with a force between 0.3 and 31 bars of suction. Much of this water (that portion held to particles with less than 15 bars of suction) is readily available to plant roots.

Carbon Dioxide Compensation Point The concentration of carbon dioxide in a plant's chloroplasts below which the amount of photosynthate produced fails to compensate for the amount of amount of photosynthate used in respiration.

Carbon Fixation The part of the photosynthetic process in which carbon atoms are extracted from atmospheric carbon dioxide and used to make simple organic compounds that eventually become glucose.

Carbon Partitioning The manner in which a plant allocates to different plant parts the photosynthate it produces.

Carbonaceous Rich in carbon.

Carnivore Literally, an organism that eats meat. Most carnivores are animals, but a few fungi, plants, and protists are as well.

Catabatic Warming The process that occurs when a large air mass expands after having been forced over a mountain range and becomes warmer and dryer as a result of the expansion.

Cation Exchange Capacity (CEC) A measurement of a soil's

Tamil Nadu has the largest area under tank irrigation.

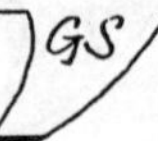

ability to bind positively charged ions (cations), which include many important nutrients. Depends on the amount and type of clay and the amount and humification of organic matter in soil. Most of the major cation nutrients are held in soil by CEC (calcium, magnesium, potassium).

Cellulose The chief component of the cell walls of plants, composed of a long chain of repeating glucose units. Cellulose can only be digested by bacteria, which often grow symbiotically in the guts of animals like ruminants and termites.

Chelation A reaction between a metallic ion and an organic compound that removes the metallic ion from solution. Chelation is a natural reaction in most soils and is enhanced by organic matter, especially humus. Chelation is similar to cation exchange, except that it usually is more stable in holding ions. Most of the trace cation nutrients are held in soil by chelation (copper, iron, manganese).

Chromosome Genetic material, composed largely of DNA and protein in the nucleus of cells.

Climax In classical ecological theory, the end point of the successional process; today, we refer instead to the stage of maturity reached when successional development shifts to dynamic change around an equilibrium point.

Clone An individual produced asexually from the tissues, cells, or genome of another individual. A clone is genetically identical to the individual from which it was derived.

Cold Air Drainage The flow of cold air down a slope at night, when re-radiation of heat (and therefore cooling of air) occurs more rapidly at higher elevations.

Colluvium Soil that has been transported to its present location by the actions of gravity.

Commensalism An inter-organism interaction in which one organism is aided by the interaction and the other is neither benefited nor harmed.

Community All the organisms living together in a particular location.

Community Supported Agriculture (CSA) A nationwide movement linking local consumers and farmers into communities. Typical CSA's consist of a group of consumer shareholders that pay a sum in advance in exchange for a regular selection (weekly, biweekly) of in-season crops produced by a farm.

Compensating Factor A factor of the environment that overcomes, eliminates, or modifies the impact of another factor.

Competition An interaction in which two organisms remove from the environment a limited resource that both require, and both organisms are harmed in the process. Competition can occur between members of the same species and between members of different species.

Composite A plant in the Compositae (Asteraceae) family.

Compost mixed decayed and decaying organic matter with available nutrients useful for fertilizer.

Composting the management of organic materials to produce *compost*.

Consumer An organism that ingests other organisms (or their parts or products) to obtain its food energy.

Did You Know ?

First colour film

Can-Can Dance **was shot first to be shown in colour (hand-painted) at the Tivoli Theatre, Mumbai, in 1897.**

Continental Influence The climatic effect of being distant from the moderating effects of a large body of water.

Coriolis Effect The deflection of air currents in atmospheric circulation cells due to the rotation of the earth.

Cross-pollination The fertilization of a flower by pollen from the flower of another individual of the same species.

Cultural Energy Inputs Forms of energy used in agricultural production that come from sources controlled or provided by humans.

Cytosterility A genetically-controlled condition of male sterility in the breeding line of a self-pollinating crop variety. A breeding line with cytosterility is used as the seed-producing parental line in the production of hybrid seed.

Dark Reactions The processes of photosynthesis that do not require light; specifically, the carbon-fixing and sugar-synthesizing processes of the Calvin cycle.

Decomposer A fungal or bacterial organism that obtains its nutrients and food energy by breaking down dead organic and fecal matter and absorbing some of its nutrient content.

Decomposition The process by which materials are broken down into simpler compounds by decomposers.

Density-dependent Directly linked to population density. This term is usually used to describe growth-limiting feedback mechanisms in a population of organisms.

Density-independent Not directly linked to population density. This term is usually used to describe growth-limiting feedback mechanisms in a population of organisms.

Detritivore An organism that feeds on dead organic and fecal matter.

Detritus Accumulated organic debris from dead organisms, often an important source of nutrients in a food web.

Detrivore Any organism which obtains most of its nutrients from the detritus in an ecosystem.

Dew Point The temperature at which relative humidity reaches 100% and water vapor is able to condense into water droplets. The dew point varies depending on the absolute water vapor content of the air.

Directed Selection The process of controlling genetic change in domesticated plants through manipulation of the plants' environment and their breeding process.

Disturbance An event or short-term process that alters a community or ecosystem by changing the relative population levels of at least some of the component species.

Diversity (1) The number or variety of species in a location, community, ecosystem, or agroecosystem. (2) The degree of heterogeneity of the biotic components of an ecosystem or agroecosystem (see *ecological diversity*).

Parliament has the authority to set common High Courts for States.

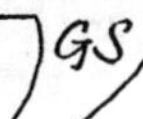

DNA (Deoxyribonucleic acid) The primary genetic material of life, containing two nucleotide chains in a double-helix.

Domestication The process of altering, through directed selection, the genetic makeup of a species so as to increase the species' usefulness to humans.

Dominant Species The species with the greatest impact on both the biotic and abiotic components of its community.

Dynamic Equilibrium A condition characterized by an overall balance in the processes of change in an ecosystem, made possible by the system's resiliency, and resulting in relative stability of structure and function despite constant change and small-scale disturbance.

Easily available water That portion of water held in the soil that can be readily absorbed by plant roots—usually capillary water between 0.3 and 15 bars of suction.

Ecological Diversity The degree of heterogeneity of an ecosystem's or agroecosystem's species makeup, genetic potential, vertical spatial structure, horizontal spatial structure, trophic structure, ecological functioning, and change over time.

Ecological Energy Inputs Forms of energy used in agricultural production that come directly from the sun.

Ecological Niche An organism's place and function in the environment, defined by its utilization of resources.

Ecosystem A functional system of complementary relations between living organisms and their environment within a certain physical area.

Ecotone A zone of gradual transition between two distinct ecosystems, communities, or habitats.

Ecotype A population of a species that differs genetically from other populations of the same species because local conditions have selected for certain unique physiological or morphological characteristics.

Edaphic of the soil, or influenced by the soil. For example, some edaphic factors that influence soil organisms are pH, organic matter content, and hydrology.

Edge Effect The phenomenon of an edge community, or ecotone, having greater ecological diversity than the neighboring communities.

Emergent Property A characteristic of a system that derives from the interaction of its parts and is not observable or inherent in the parts considered separately.

Environmental Complex The composite of all the individual factors of the environment acting and interacting in concert.

Environmental Resistance The genetically-based ability of an organism to withstand stresses, threats, or limiting factors in the environment.

Eolian Soil Soil that has been transported to its current location by the actions of wind (*aeolian* is an acceptable alternative spelling).

Epiphyll A plant that uses the leaf of another plant for support, but that draws no nutrients from the host plant.

Epiphyte A plant that uses the trunk or stem of another plant for support, but that draws no nutrients from the host plant.

Evapotranspiration All forms of evaporation of liquid water from the earth's surface, including the evaporation of bodies of water and soil moisture and the evaporation from leaf surfaces that occurs as part of transpiration.

Field Capacity The amount of water the soil can hold once gravitational water has drained away; this water is mostly capillary water held to soil particles with at least 0.3 bars of suction.

Food and Agriculture Organization The United Nations organization dedicated to agriculture.

Food Chain /Food Web All the interactions of predator and prey, included along with the exchange of nutrients into and out of the soil. These interactions connect the various members of an ecosystem, and describe how energy passes from one organism to another.

Food Demand The amount of food needed to support a given population or individual for a given amount of time.

Food Security Can be defined as the "state in which all persons obtain a nutritionally adequate, culturally acceptable diet at all times through local non-emergency sources."

Food System The interconnected meta-system of agroecosystems, their economic, social, cultural, and technological support systems, and systems of food distribution and consumption.

Forb A non-grass, herbaceous species (generally legumes and composites)

Gene Flow The exchange of genetic factors within and between populations by interbreeding or migration.

Gene The basic unit of inheritance comprising a specific sequence of nucleotides on a DNA chain that has a specific function and occupies a specific locus on a chromosome.

Generalist A species that tolerates a broad range of environmental conditions; a generalist has a broad ecological niche.

Genetic Engineering Transfer, by biotechnological methods, of genetic material from one organism to another.

Genetic Erosion The loss of genetic diversity in domesticated organisms that has resulted from human reliance on a few genetically uniform varieties of food crop plants and animals.

Genetic Vulnerability The susceptibility of genetically uniform crops to damage or destruction caused by outbreaks of a disease or pest or unusually poor weather conditions or climatic change.

Genotype An organism's genetic information, considered as a whole.

Geographic Information System (GIS) A computer-based system for managing and analyzing spatially-referenced data. Examples include land-use, topographic and soils data.

Glacial Soil Soil that has been transported to its current location by the movement of glaciers.

Grassland Region in which the climate is dry for long periods of the summer, and freezes in the winter. Grasslands are characterized by grasses and other erect herbs, usually without trees or shrubs. Grasslands occur in the dry temperate interiors of continents, and first appeared in the Miocene.

Gravitational Water That portion of water in the soil not held strongly enough by adhesion to soil particles to resist the downward pull of gravity.

Green Manure Organic matter added to the soil when a cover crop (often leguminous) is tilled in.

Gross Primary Productivity The rate of conversion of solar energy into biomass in an ecosystem.

Habitat The particular environment, characterized by a specific set of environmental conditions, in which a given species occurs.

Hardening Subjecting a seedling or plant to cooler temperatures in order to increase its resistance to more extreme cold.

Herbaceous Non-woody.

Herbivore An animal that feeds exclusively or mainly on plants. Herbivores convert plant biomass into animal biomass.

Heterosis The production of an exceptionally vigorous and/or productive hybrid progeny from a directed cross between two pure-breeding plant lines.

Heterotroph An organism that consumes other organisms to meet its energy needs.

Horizons Visually distinguishable layers in the soil profile.

Host Organism which serves as the habitat for a parasite, or possibly for a symbiont. A host may provide nutrition to the parasite or symbiont, or simply a place in which to live.

Host-specificity The extent to which a parasite is restricted in the host species it can use.

Humification The decomposition or metabolization of organic material in the soil into humus.

Humus The fraction of organic matter in the soil resulting from decomposition and mineralization of organic material.

Hybrid Vigor The production of an exceptionally vigorous and/or productive hybrid progeny from a directed cross between two pure-breeding plant lines. A synonym for *heterosis*.

Hydration The addition of water molecules to a mineral's chemical structure.

Hydrological Cycle The process encompassing the evaporation of water from the earth's surface, its condensation in the atmosphere, and its return to the surface through precipitation.

Hydrology The properties, distribution, and circulation of water.

Hydrolysis Replacement of cations in the structure of a silicate mineral with hydrogen ions, resulting in the decomposition of the mineral.

Hydrophyte A plant that is adapted to grow in water or very wet environments.

Hydroxide Clay A mineral component of the soil without definite crystalline structure composed of hydrated iron and aluminum oxides.

Hygroscopic Water The moisture that is held the most tightly to soil particles, usually with more than 31 bars of suction; it can remain in soil after oven drying.

Importance Value A measure of a species' presence in an ecosystem or community—such as number of individuals, biomass, or productivity—that can be used to determine the species' contribution to the diversity of the system.

Inoculum The initial organism or organisms that establish a new colony or population; breeding stock. The term is commonly applied to the application of small amounts of microrganisms to establish new populations in soils or composts, for example, Rhizobium bacteria are commonly applied as an inoculum to legume seeds at planting time.

Insolation The conversion, at the earth's surface, of short wave solar energy into long wave heat energy.

Integrated Pest Management Pest control using an array of complementary approaches including natural predators, parasites, pest-resistant varieties, pesticides, and other biological and environmental control practices.

Intercropping Planting more than one crop in a field using a regular pattern that interleaves each crop in some pattern. A form of polyculture.

Intermediate Disturbance Hypothesis The theory that diversity and productivity in natural ecosystems are highest when moderate disturbance occurs periodically but not too frequently.

Interspecific Competition Competition for resources among individuals of different species.

Intraspecific Competition Competition for resources among individuals of the same species.

Inversion The sandwiching of a layer of warm air between two layers of cold air in a valley.

***K*-strategist** A species that lives in conditions where mortality is density-dependent; a typical *K*-strategist has a relatively long lifespan and invests a relatively large amount of energy in each of the few offspring it produces.

Land Utilization Type (LUT): A kind of land use described or defined in a degree of detail greater than that of a major kind of land use. In the context of irrigated agriculture, a land utilization type refers to a crop, crop combination or cropping system with specified irrigation and management methods in a defined technical and socio-economic setting. In the context of rainfed agriculture, a land utilization type refers to a crop, crop combination or cropping system with a specified technical and socio-economic setting. A land utilization type in forestry consists of a technical specifications in a given physical, economic and social setting. Defined by FAO.

Landrace A locally-adapted strain of a species bred through traditional methods of directed selection. Also called **farmer varieties.**

Landscape Ecology The study of environmental factors and interactions at a scale that encompasses more than one ecosystem at a time.

Leaf-area Index A measure of leaf cover above a certain area of ground, given by the ratio of total leaf surface area to ground surface area.

Legume A plant in the Leguminosae (Fabaceae) family. Most species in this family can fix nitrogen.

Light Compensation Point The level of light intensity needed for a plant to produce an amount of photosynthate equal to the amount it uses for respiration.

Light Reactions The components of photosynthesis in which light energy is converted into chemical energy in the form of ATP and NADPH.

Lignin an amorphous polymer related to cellulose that cements

Rajasthan has the largest area under well irrigation.

cell walls, helping them stay rigid. Lignin is highly resistant to decomposition.

Limiting Nutrient A nutrient not present in the soil in sufficient quantity to support optimal plant growth.

Living Mulch A cover crop that is interplanted with the primary crop(s) during the growing season.

Lodging The flattening of a crop plant or crop stand by strong wind, usually involving uprooting or stem breakage.

Macronutrient A nutrient plants need in large quantities; the macronutrients include carbon, nitrogen, oxygen, phosphorus, sulfur, and water.

Maritime Influence The moderating effect of a nearby large body of water, such as an ocean, on the weather and climate of an area.

Mass Selection The traditional method of directed selection, in which seed is collected from those individuals in a population that show one or more desirable traits and then used for planting the next crop.

Mesophyte A plant that is adapted to environments that are neither very dry nor very wet. Compare with xerophyte and hydrophyte.

Microclimate The environmental conditions in the immediate vicinity of an organism.

Micronutrient A nutrient necessary for plant survival but needed in relatively small quantities.

Mineralization The process by which organic residues in the soil are broken down to release mineral nutrients that can be utilized by plants.

Mountain Wind The downslope movement of air at night that occurs as the upper slopes of a mountain cool more rapidly than those below.

Multiple Cropping The cultivation of two or more crops in succession or with some overlap in the same field within one year. Double-cropping of rice after wheat is an example. When crops overlap in time, multiple cropping is a form of polyculture.

Mutualism An interaction in which two organisms impact each other positively; neither is as successful in the absence of the interaction.

Mycorrhizae Symbiotic fungal connections with plant roots through which a fungal organism provides water and nutrients to a plant and the plant provides sugars to the fungi.

Natural Selection The process by which adaptive traits increase in frequency in a population due to the differential reproductive success of the individuals that possess the traits.

Net Primary Productivity (NPP) The difference between the rate of conversion of solar energy into biomass in an ecosystem and the rate at which energy is used to maintain the producers of the system.

Niche Amplitude The size or range of one or more of the dimensions of the multidimensional space encompassed by a particular species' niche. The niche amplitude of a generalist species is larger than that of a specialist species.

Niche Breadth Essentially a synonym for *niche amplitude*.

Niche Diversity Differences in the resource-use patterns of similar species that allow them to coexist successfully in the same environment.

Nitrogen Fixation The conversion of gaseous nitrogen into a form usable by plants. Ususally by bacteria.

Nocturnal Active only at night.

Non-governmental Organization (NGO) An organization that is not part of a government. Private foundations and development organizations are examples.

Nucleotide A subunit of DNA or RNA molecules, part of the genetic code.

Nutrient Any element or simple compound necessary for the health and survival of an organism. This includes air and water, as well as food.

Nutrient Cycling All the processes by which nutrients are transferred from one organism to another. For instance, the carbon cycle includes uptake of carbon dioxide by plants, ingestion by animals, and respiration and decay of the animal.

Omnivore Literally, an organism that will eat anything. Refers to animals who do not restrict their diet to just plants or other animals.

Open Pollination The natural dispersal of pollen among all the members of a cross-pollinating crop population, resulting in the maximum degree of genetic mixing and diversity.

Organic Agriculture Systems of agricultural production that do not use any synthetic chemical inputs.

Organic Generally refers to those substances produced by the metabolism of a living organism, especially carbon-containing compounds.

Organic Matter Material containing molecules based on Carbon, usually referring to soil organic matter.

Organism An individual of a species.

Overyielding The production of a yield by an intercrop that is larger than the yield produced by planting the component crops in monoculture on an equivalent area of land.

Oxidation The loss of electrons from an atom that accompanies the change from a reduced to an oxidized state.

Parasite An organism that uses another organism for food and thus harms the other organism.

Parasitism An interaction in which one organism feeds on another organism, harming (but generally not killing) it.

Parasitoid A parasite that feeds on predators or other parasites.

Patchiness A measurement of the diversity of successional stages present in a specific area.

Patchy Landscape A landscape with a diversity of successional stages or habitat types.

Percolation Water movement through the soil due to the pull of gravity.

Permanent Wilting Point The level of soil moisture below which a plant wilts and is unable to recover.

pH The logarithm of the reciprocal of the concentration of hydrogen ions in a medium, like water or soil (log10{1/[H+]}). pH values range from 0 to 14, giving the relative acidity or alkalinity of a medium, with a pH of 7 being neutral, and lower values being acidic, higher values, alkaline

Phenotype The physical expression of the genotype; an organism's physical characteristics.

Photoperiod The total number of hours of daylight.

Photorespiration The energetically-wasteful substitution of oxygen for carbon dioxide in the dark reactions of

photosynthesis, which occurs when plant stomata close and carbon dioxide concentration declines.

Photosynthate The simple-sugar end products of photosynthesis.

Phytoplankton Tiny, free-floating, photosynthetic organisms in aquatic systems. They include diatoms, desmids, and dinoflagellates.

Plankton Tiny, free-floating organisms of the ocean or other aquatic systems. They may be phytoplankton or zooplankton.

Pollinator Animal which carries pollen from one seed plant to another, unwittingly aiding the plant in its reproduction. Common pollinators include insects, especially bees, butterflies, and moths, birds, and bats.

Polyculture Cropping systems in which different crop species are grown in mixtures in the same field at the same time.

Polyploid Having three or more times the haploid number of chromosomes.

Population A group of individuals of the same species that live in the same geographic region.

Potential Niche The maximum possible distribution of a species in the environment.

Predation An interaction in which one organism kills and consumes another.

Predator Organism which hunts and eats other organisms. This includes both carnivores, which eat animals, and herbivores, which eat plants.

Prescribed Burn A fire set and controlled by humans to achieve some management objective, such as improving pasture in grazing systems.

Prevailing Winds The general wind patterns characteristic of broad latitudinal belts on the earth's surface.

Prey Organism hunted and eaten by a predator.

Primary Production The amount of light energy converted into plant biomass in a system.

Primary Succession Ecological succession on a site that was not previously occupied by living organisms.

Producer An organism that converts solar energy into biomass.

Production Harvest output or yield.

Productivity Index A measure of the amount of biomass invested in the harvested product in relation to the total amount of standing biomass present in the rest of the system.

Productivity The ecological processes and structures in an agroecosystem that enable production.

Protein A compound formed from a chain of amino acids. Proteins are present in all living things, and are used for enzymes, hormones and other essential molecules.

Protocooperation An interaction in which both organisms are benefited if the interaction occurs, but neither are harmed if it does not occur.

Rainfed Agroecosystem A farming system in which crop water needs are met by natural precipitation.

Realized Niche The actual distribution of an organism in the environment (compare with *potential niche*).

Reference Man A standard "human unit" for populations, based on the food consumption requirements of a "standard human", most often a male of a certain age and occupation (a holdover, perhaps, but useful for some purposes). Reference man units, also called "refman"

Quiz Kit!

First amendment to the Constitution

Bill No. 48 of 1951 was introduced in the Parliament on May 12, 1951. The President gave his assent to the Bill on June 18, 1951.

units, are used by nutritionists to estimate the food demand of populations, to account for differing food demand between individuals (babies versus adults, etc.). Refman units should be used with caution, because some segments of the population, such as pregnant women and developing children, can have specific nutritional demands that cannot be standardized properly by this method.

Regolith The layer or mantle of unconsolidated material (soil and mineral subsoil) between the soil surface and the solid bedrock of the earth below.

Relative Humidity The ratio of the actual water content of the air to the amount of water the air is capable of holding at a particular temperature.

Relative Rate of Light Transmission The percentage of the total incident light at the canopy of a system that reaches the ground.

Relay Cropping Cropping systems in which two or more crops are grown in sequence in the same field in the same year, with little or no overlap in time. Not a true form of polyculture because very little interspecies interaction usually occurs in these systems.

Renewable Resources Resources that are capable of being regenerated or replaced by ecological processes on a time scale relevant to their use. These resources, such as biomass or energy from animal traction, are contrasted with non-renewable resources, such as fossil fuels and mined products.

Reproduction Process by which new organisms are generated. Reproduction may be sexual, involving the fusion of gametes, or asexual.

Residual Soil Soil formed at its current location.

Response A physiological change in a plant that is induced by an outside, usually environmental, condition.

Rhizobium A genus of heterotrophic soil bacteria capable of forming symbiotic nodules on the roots of legume plants to fix atmospheric nitrogen.

***r*-strategist** A species that exists in relatively harsh environmental conditions and whose mortality is generally determined by density-independent factors; an *r*-strategist allocates more energy to reproduction than to growth.

Ruminant An animal with a multiple stomach (polygastric) system of digestion capable of digesting cellulose.

Safe Site A specific location that provides the environmental conditions necessary for seed germination and initial growth of the seedling.

Salinity A measure of the salt concentration of water. Higher salinity means more dissolved salts.

Saltation The transport of small soil particles just above the

GK GS

Warren Hastings was the first British Governor General of India.

soil surface by wind.

Saprophyte Organism which feeds on dead and decaying organisms, allowing the nutrients to be recycled into the ecosystem. Fungi and bacteria are two groups with many important saprophytes.

Saturation Point The level of light intensity at which photosynthetic pigments are completely stimulated and unable to make use of additional light.

Seaweed Any large photosynthetic protist, including rhodophytes and kelps. Seaweeds are not true plants, but like plants they can make their own food.

Secondary Succession Succession on a site that was previously occupied by living organisms but that has undergone severe disturbance.

Seed Bank The total seed presence in the soil.

Self-pollination The fertilization of the egg of a plant by its own pollen.

Shannon Index A measure of the species diversity of an ecosystem based on information theory.

Shifting Cultivation Farming systems that alternate periods of annual cropping with extended fallow periods. "Slash and burn" systems of shifting cultivation use fire to clear fallow areas for cropping.

Silicate Clay A soil component made up primarily of microscopic aluminum silicate plates.

Simpson Index A measure of the species diversity of an ecosystem based on the concept of dominance.

Slash and Burn A type of shifting cultivation that uses fire to clear fallow areas for cropping.

Slope Wind Air movement caused by the different heating and cooling rates of mountain slopes and valleys.

Soil Creep The movement of large soil particles along the soil surface by wind.

Soil Health The capacity of a soil to function within ecosystem boundaries to sustain biological productivity, maintain environmental quality, and promote plant and animal health. Often used synonymously with soil quality.

Soil Profile The set of observable horizontal layers in a vertical cross section of soil.

Soil Quality Used synonymously with soil health.

Soil Solution The liquid phase of the soil, made up of water and its dissolved solutes.

Solution The process by which soluble minerals in the regolith are dissolved into water.

Specialist Organism which has adopted a lifestyle specific to a particular set of conditions. Contrast with generalist.

Species Evenness The degree of heterogeneity in the spatial distribution of species in a community or ecosystem.

Species Richness The number of different species in a community or ecosystem.

Stakeholders All parties who will be affected by the results of planning decisions.

Standing Crop The total biomass of plants in an ecosystem at a specific point in time.

Stomata The openings on a leaf surface through which gases enter and leave the internal leaf environment.

Substrate "Supporting surface" on which an organism grows. The substrate may simply provide structural support, or may provide water and nutrients. A substrate may be inorganic, such as rock or soil, or it may be organic, such as wood.

Succession The process by which one community gives way to another.

Successional Mosaic A patchwork of habitats or areas in different stages of succession.

Symbiosis A relationship between different organisms that live in direct contact.

Synthetic variety A crop or horticultural variety produced through the cross-pollination of a limited number of parents that cross well and have certain desirable traits.

Temperate Region in which the climate undergoes seasonal change in temperature and moisture. Temperate regions of the earth lie primarily between 30 and 60 degrees latitude in both hemispheres.

Terrestrial Living on land, as opposed to marine or aquatic.

Thermophilic Thermophilic ("heat loving") organisms grow well under high temperature conditions. Thermophilic composting is the management of the composting process to maintain high temperature conditions, favoring the destruction of disease-causing organisms.

Tilth The combination of the characteristics of soil crumb structure, porosity, and ease of tillage.

Transpiration The evaporation of water through the stomata of a plant, which causes a flow of water from the soil through the plant and into the atmosphere.

Transported Soil Soil that has been moved to its current location by environmental forces.

Tree Tall plant with a central trunk. The term does not imply anything about relationships, but is a growth pattern that has evolved several times in plants.

Trophic Level A location in the hierarchy of feeding relationships within an ecosystem.

Trophic Structure The organization of feeding and energy-transfer relationships that determine the path of energy flow through a community or ecosystem.

Tropical Region in which the climate undergoes little seasonal change in either temperature or rainfall. Tropical regions of the earth lie primarily between 30 degrees north and south of the equator.

Valley Wind Air movement that occurs when the heating of a valley causes warm air to rise up adjacent mountain slopes.

Vavilov Centers of Domestication Areas of the world where crop plants are believed to be originally domesticated from their wild relatives. Named after the Russian scientist: N.I. Vavilov.

Vermicomposting Composting systems based on intensive management of worms, usually in specialized containers.

Vernalization The process in which a seed is subjected to a period of cold, causing changes that allow germination to occur.

Water of Hydration Water that is chemically bound to soil particles.

Xerophyte A plant that is adapted to growing under very dry conditions and extended droughts.

Zooplankton Tiny, free-floating organisms in aquatic systems. Unlike phytoplankton, zooplankton cannot produce their own food, and so are consumers.

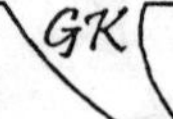

The percentage of tube wells is maximum in Uttar Pradesh.

GS

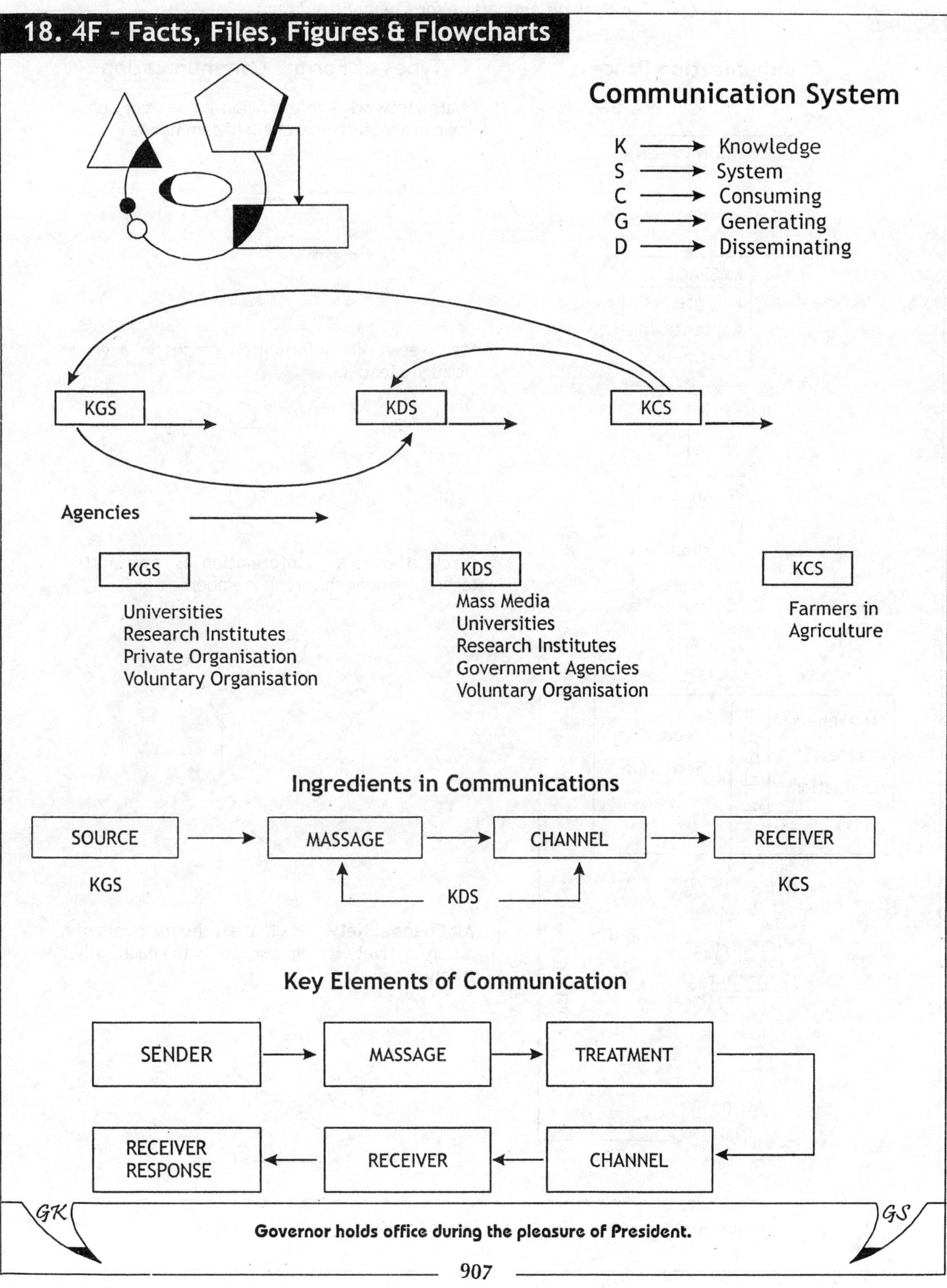

Communication System
K → Knowledge
S → System
C → Consuming
G → Generating
D → Disseminating
KGS
KDS
KCS
Agencies
KGS
Universities
Research Institutes
Private Organisation
Voluntary Organisation
KDS
Mass Media
Universities
Research Institutes
Government Agencies
Voluntary Organisation
KCS
Farmers in Agriculture
Ingredients in Communications
SOURCE
MASSAGE
CHANNEL
RECEIVER
KGS
KDS
KCS
Key Elements of Communication
SENDER
MASSAGE
TREATMENT
CHANNEL
RECEIVER
RECEIVER RESPONSE

Communication Process

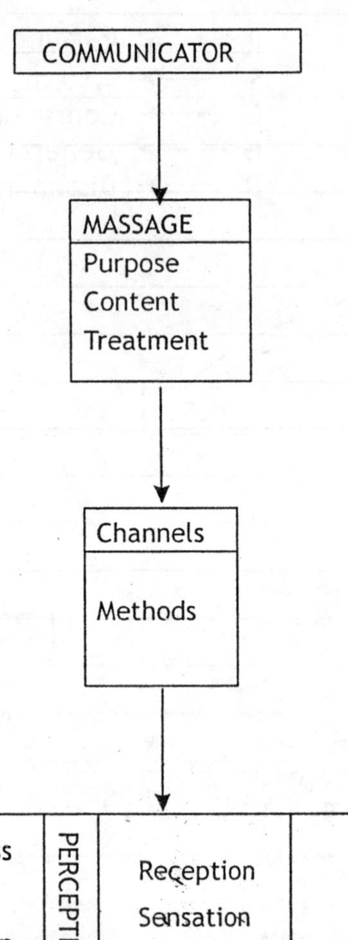

Types of Formal Communication

Chain Network - Information passes only up or down in a hierarchical chain of command.

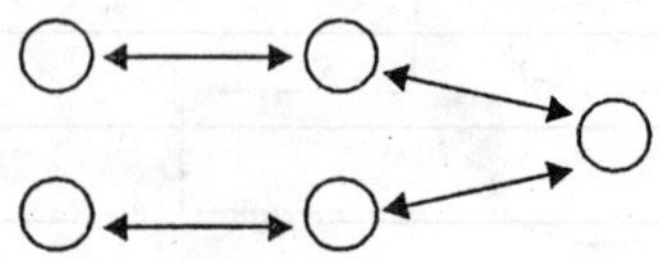

Star Network - Information passes in a group through a leader.

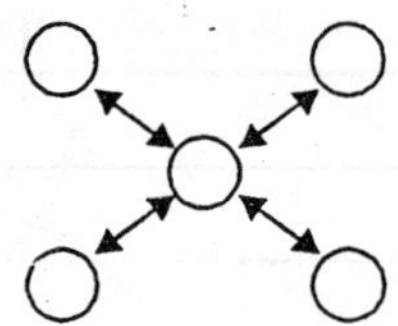

Circle Network - Information is transmitted laterally among the group members.

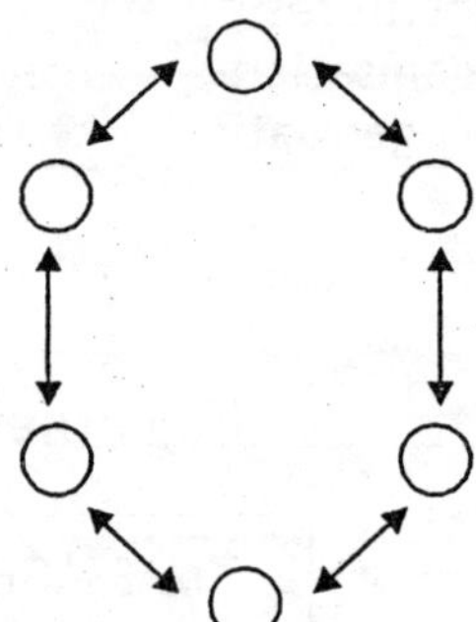

All Channel Network - In it all the members of a group actively communicate with each other freely.

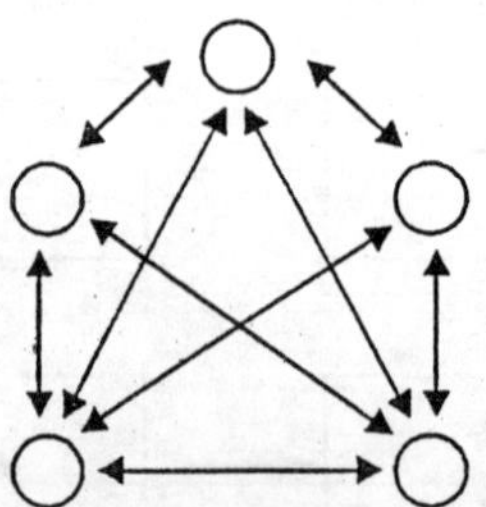

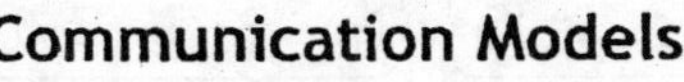
Communication Models

Speaker
Speech
Audience
Aristotel Model

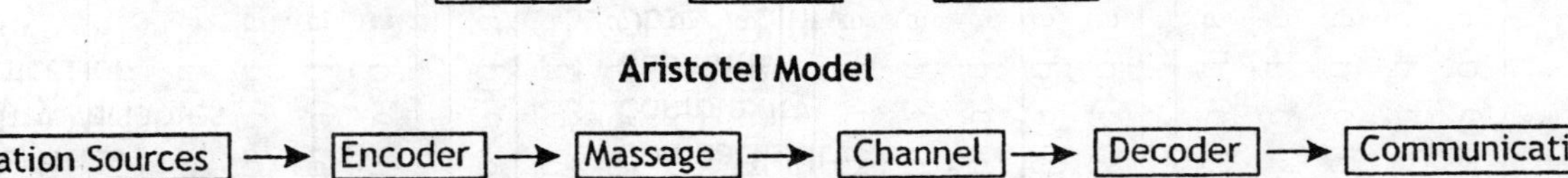
Communication Sources
Encoder
Massage
Channel
Decoder
Communication Receiver
Berlo's Model

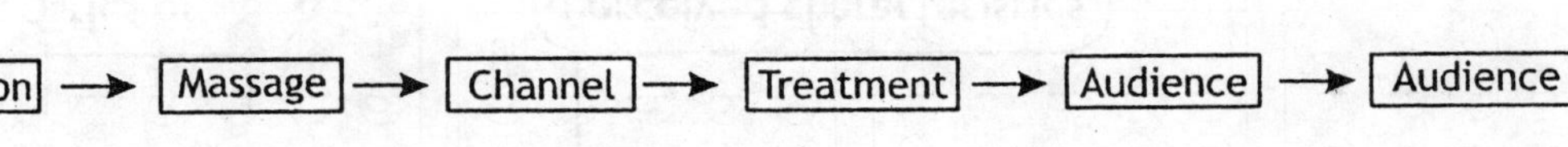
Communication
Massage
Channel
Treatment
Audience
Audience Response
Leagan's Model

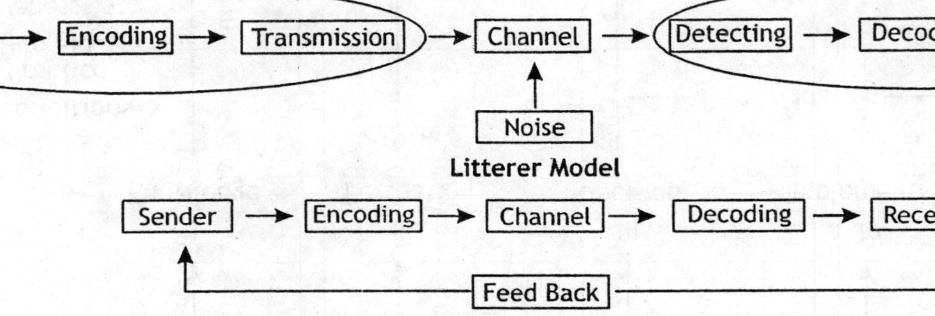
Selection
Encoding
Transmission
Channel
Noise
Detecting
Decoding
Selection
Litterer Model
Sender
Encoding
Channel
Decoding
Receiver
Feed Back
Westley & Mcleans Model

Source
Transmitter
Signal
Receiver
Destination
Shammon-Weaver Model

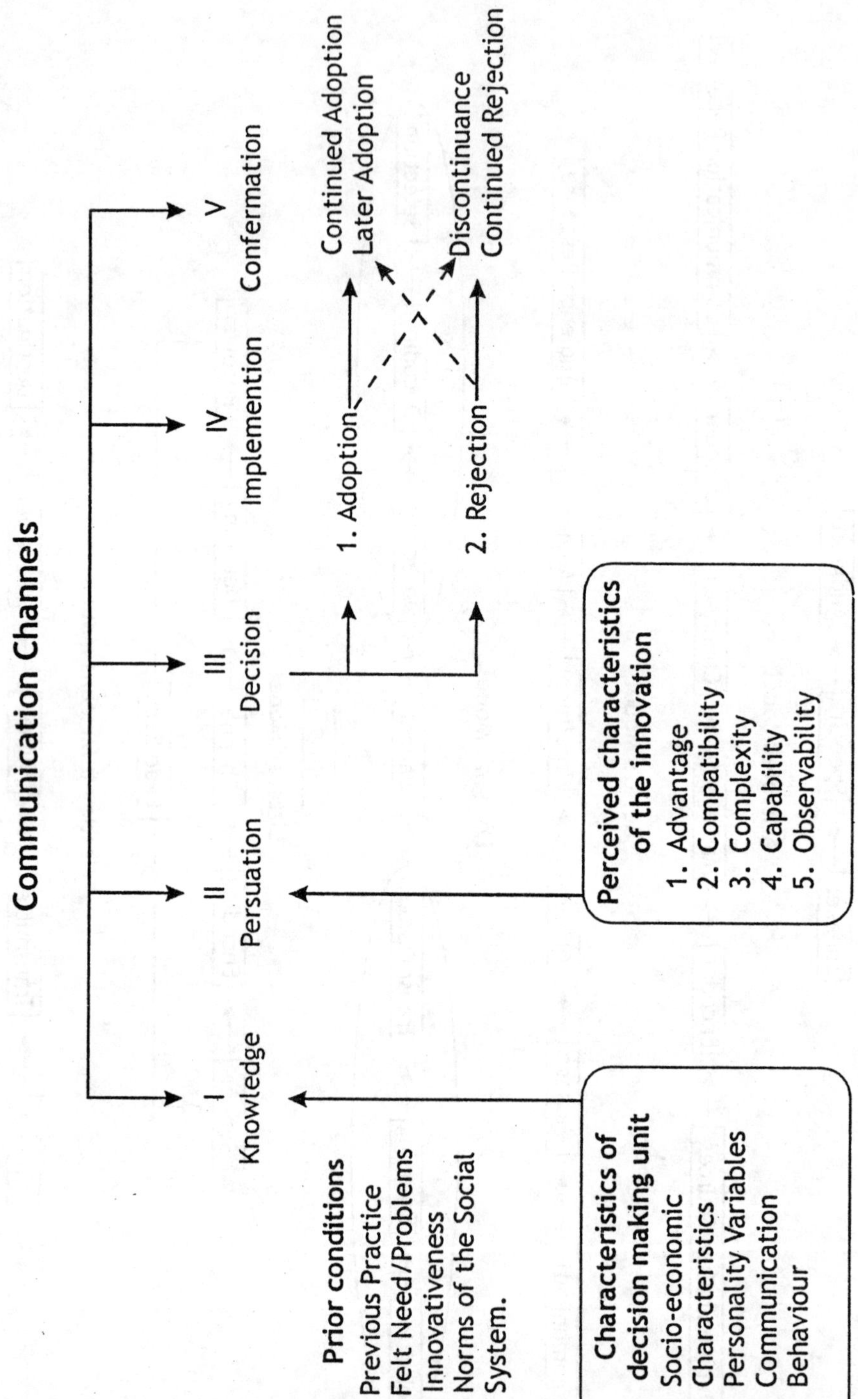

Communication Channels

I Knowledge → II Persuation → III Decision → IV Implemention → V Confermation

Prior conditions

1. Previous Practice
2. Felt Need/Problems
3. Innovativeness
4. Norms of the Social System.

Characteristics of decision making unit

1. Socio-economic Characteristics
2. Personality Variables
3. Communication Behaviour

Perceived characteristics of the innovation

1. Advantage
2. Compatibility
3. Complexity
4. Capability
5. Observability

Decision:

1. Adoption → Continued Adoption / Discontinuance
2. Rejection → Continued Rejection / Later Adoption

GS

Article 163 states that the Governor acts on the advice of Councils of Ministers.

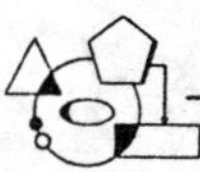

Methods of Training

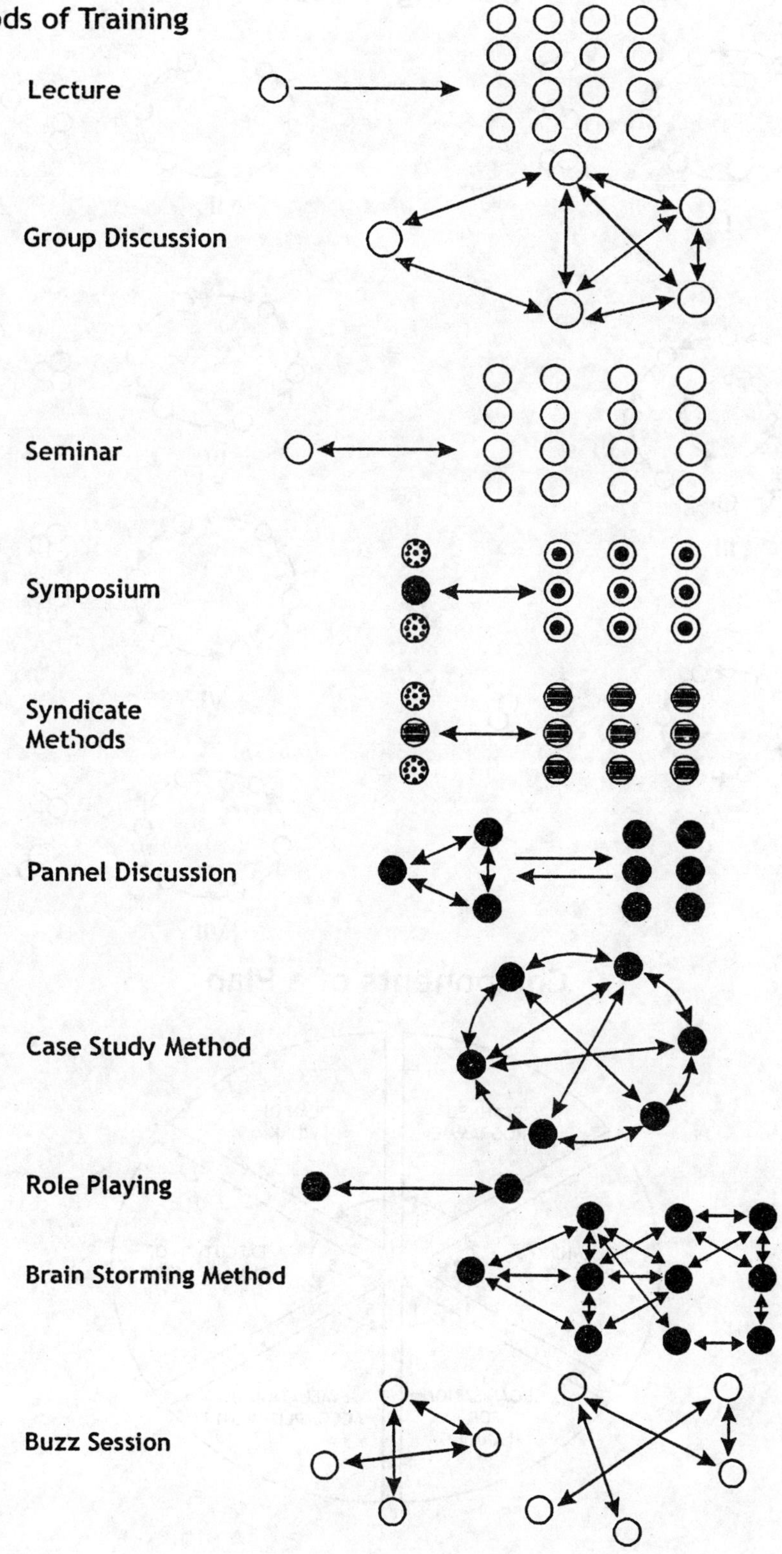

Lecture
Group Discussion
Seminar
Symposium
Syndicate Methods
Pannel Discussion
Case Study Method
Role Playing
Brain Storming Method
Buzz Session

GK

'Versailles' was the peace treaty with Germany signed after World War I in 1919.

GS

Training Process

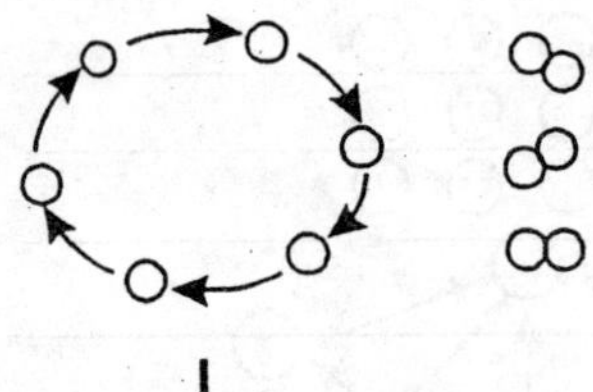

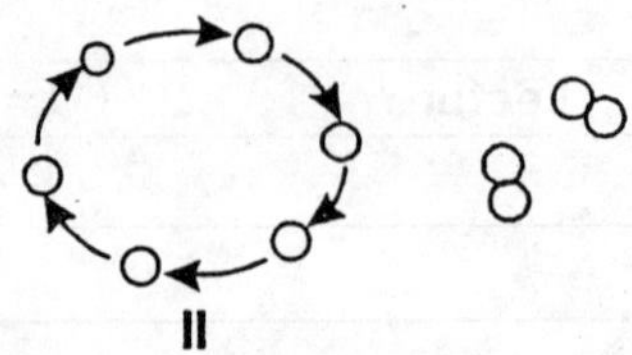

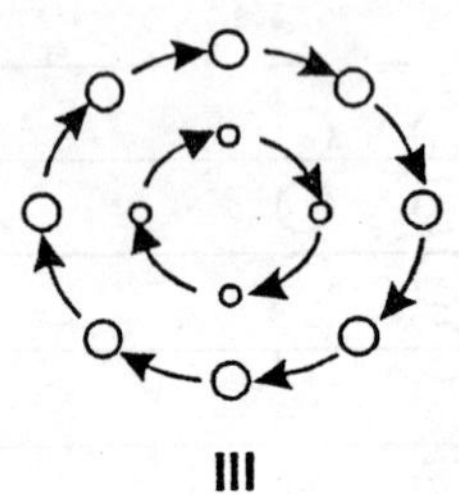

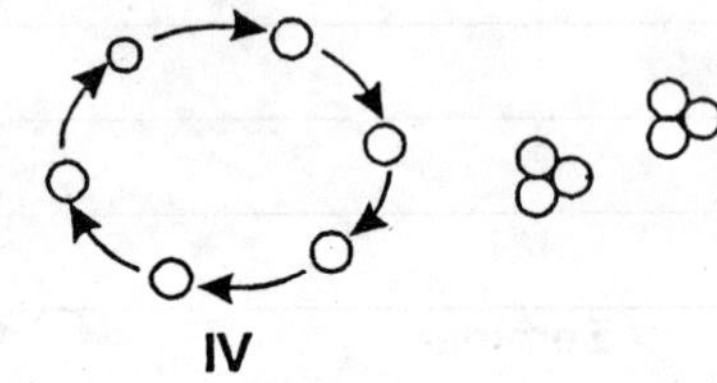

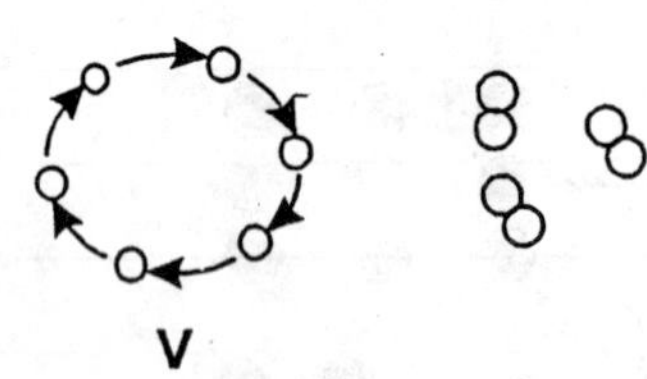

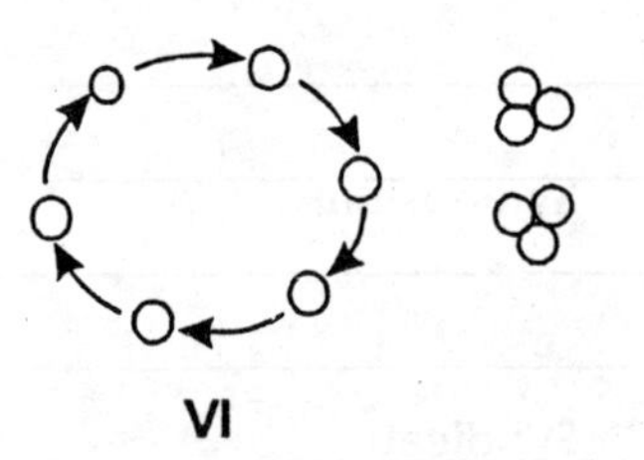

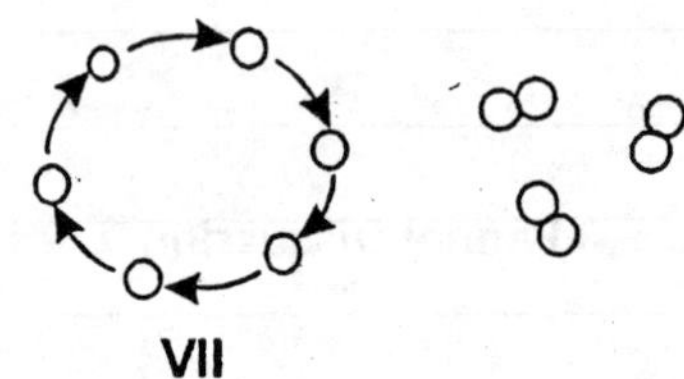

Components of a Plan

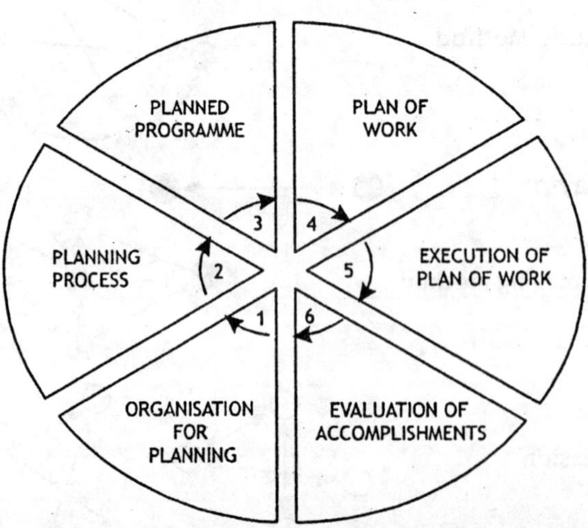

Extension Programme Planning Model

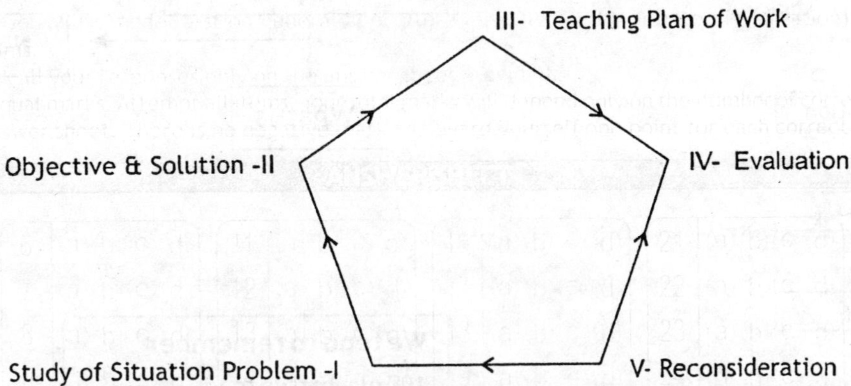

Extension Education Process

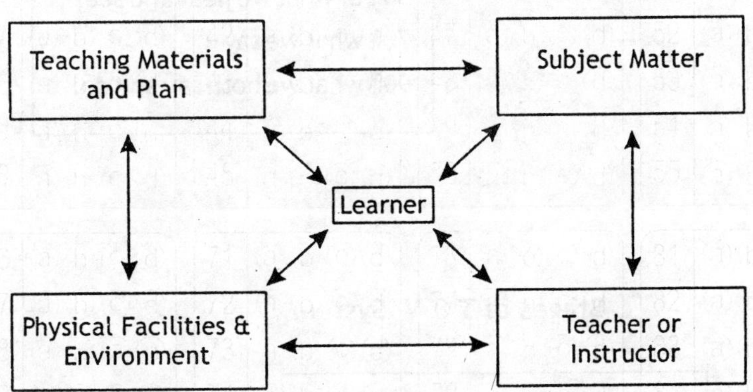

Elements of Effective Learning

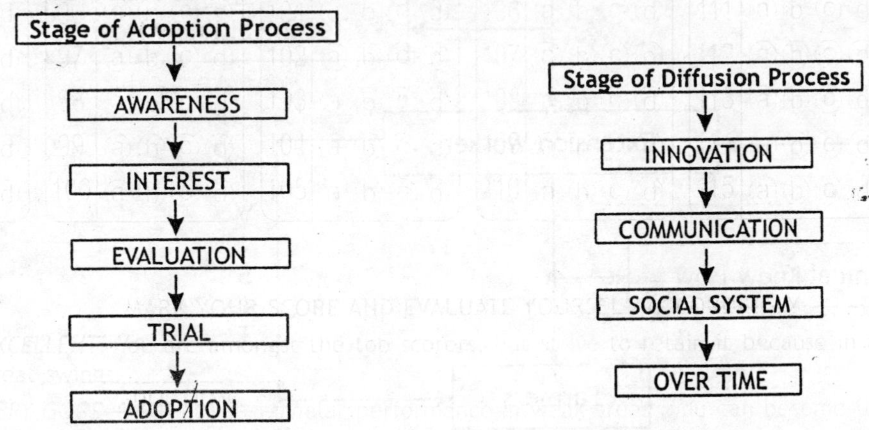

Yes! Advocate General has the right to speak in State Legislature.

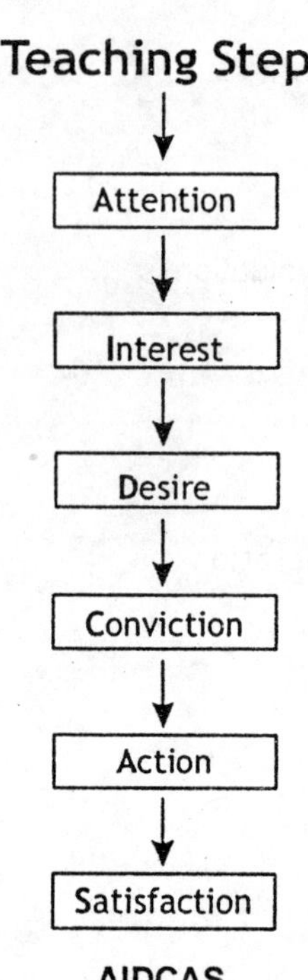

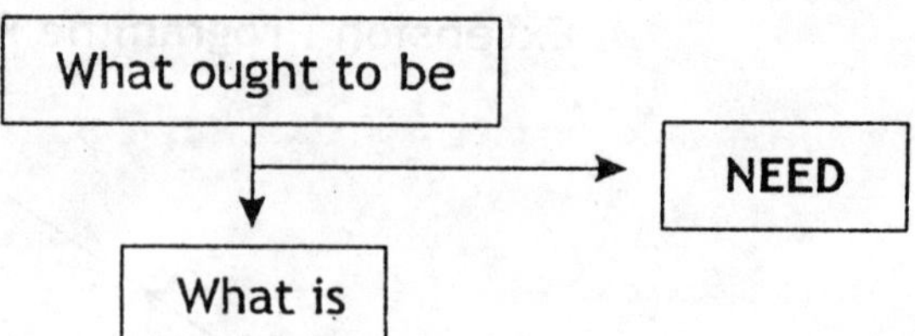

We tend to remember

10% of what we read

20% of what we hear

30% of what we see

50% of what we hear and see

70% what we say

90% what we both say and do.

Stages of T & V System

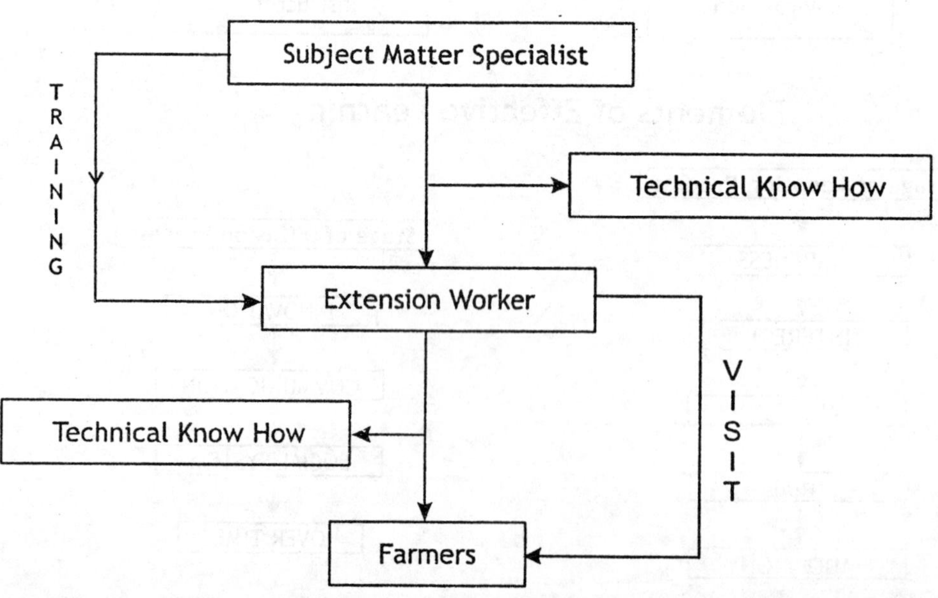

Obstacles to Agricultural Economic Development

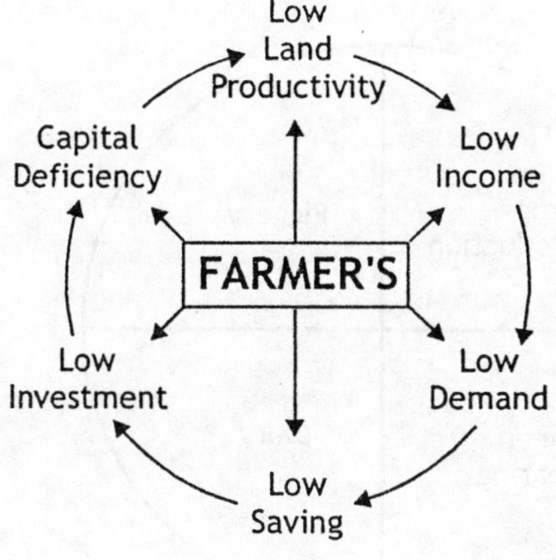

Rewards

Three C's of credit

Ⓒ Character : Strongly support risk bearing ability

Ⓒ Capacity : Direct bearing on repayment capacity

Ⓒ Capital : Mechanism strongly back bearing ability of borrower

Three R's of credit

Ⓡ Returns

Ⓡ Repayment capacity

Ⓡ Risk bearing ability

Types / Systems of Farming

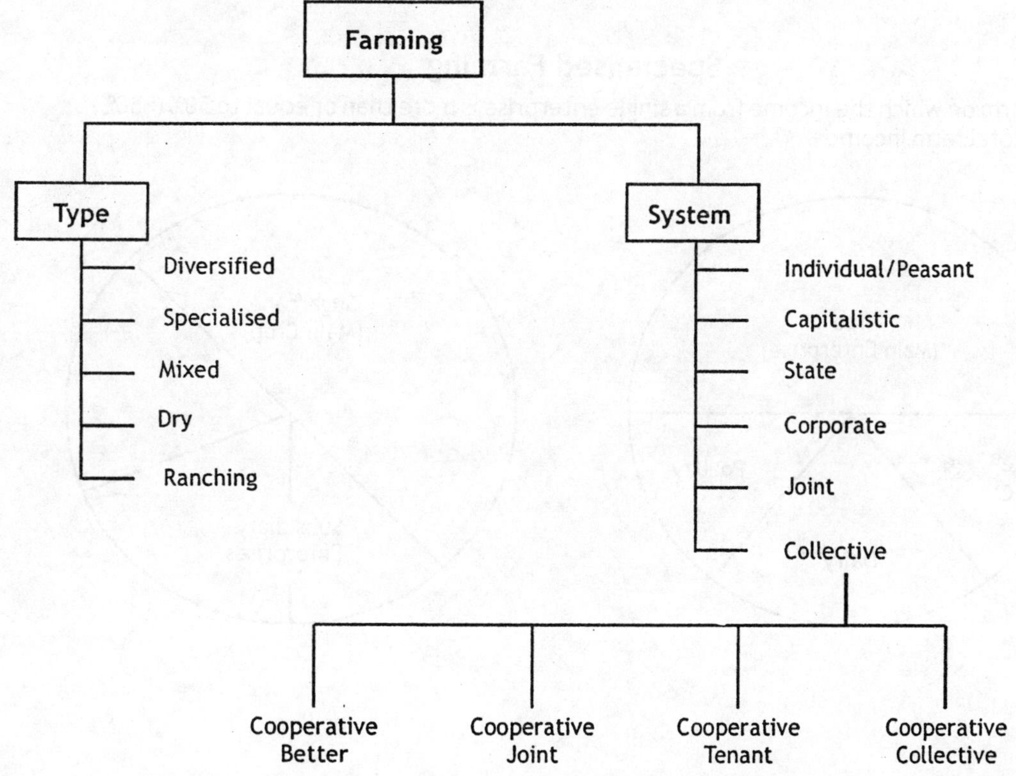

Diversified Farming

A farm on which the income from a single product is less than 50% (< 50%) of the total farm income.

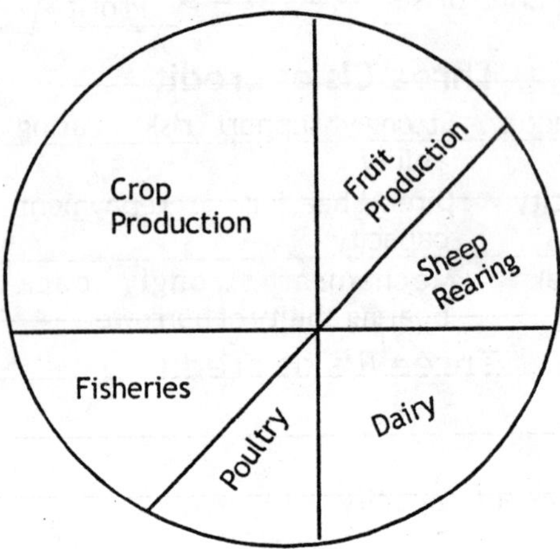

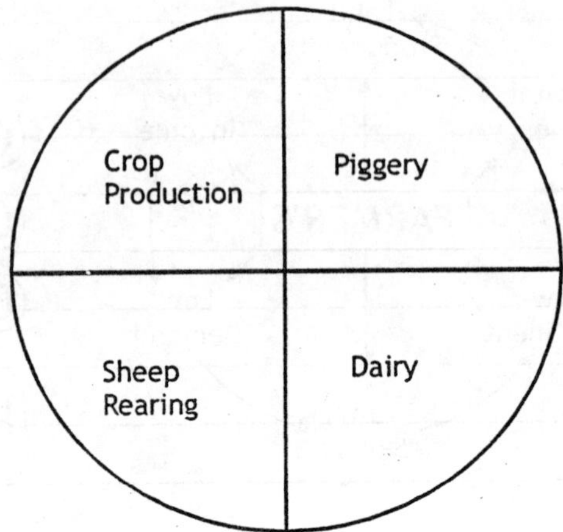

Specialised Farming

A farm on which the income from a single enterprise is more than or equal to 50% (≥50%) of total farm income.

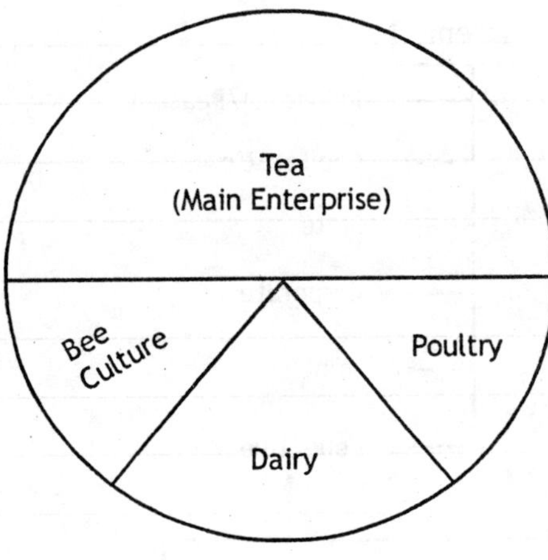

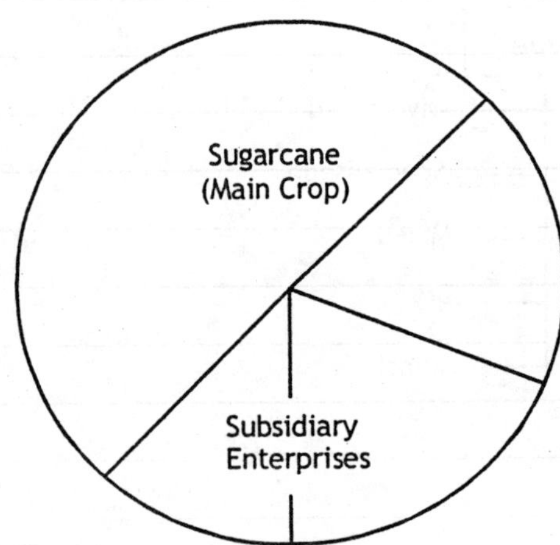

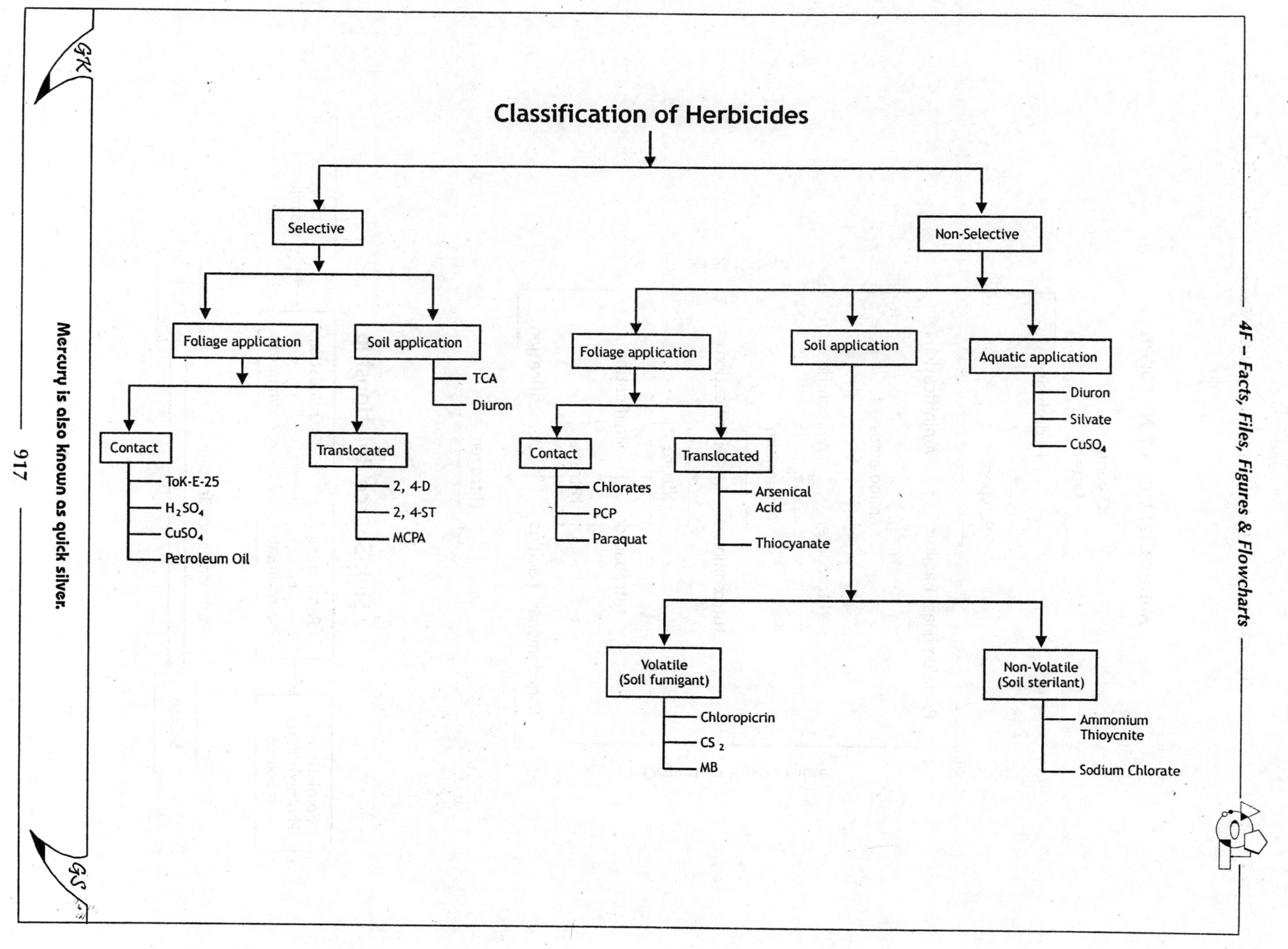
Classification of Herbicides
Selective
Non-Selective
Foliage application
Soil application
TCA
Diuron
Contact
Translocated
ToK-E-25
H_2SO_4
$CuSO_4$
Petroleum Oil
2, 4-D
2, 4-ST
MCPA
Foliage application
Soil application
Aquatic application
Diuron
Silvate
$CuSO_4$
Contact
Translocated
Chlorates
PCP
Paraquat
Arsenical Acid
Thiocyanate
Volatile (Soil fumigant)
Non-Volatile (Soil sterilant)
Chloropicrin
CS_2
MB
Ammonium Thioycnite
Sodium Chlorate

Mineralization of Nitrogen

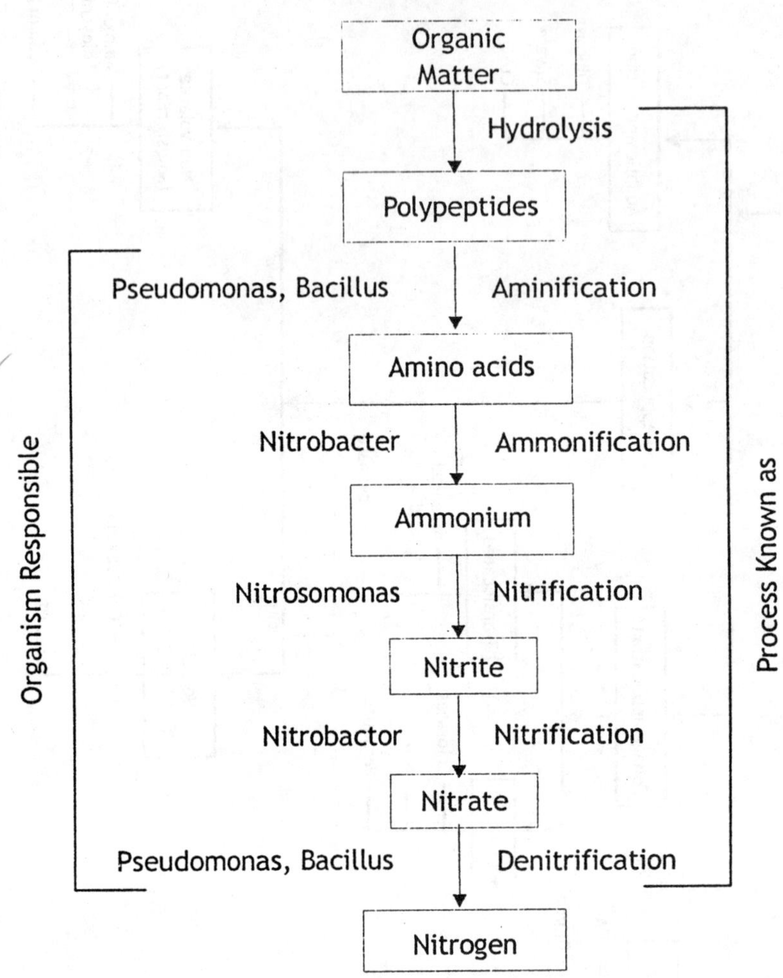

Sink-Source Relationship

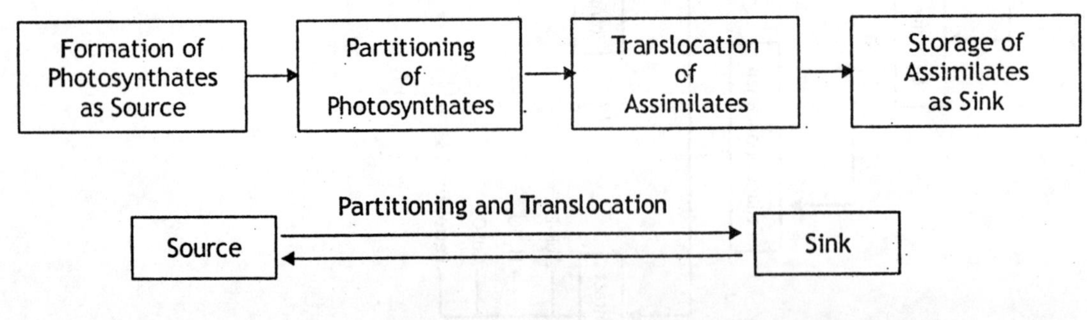

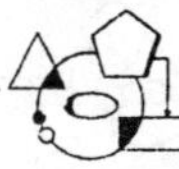

Soil Profile

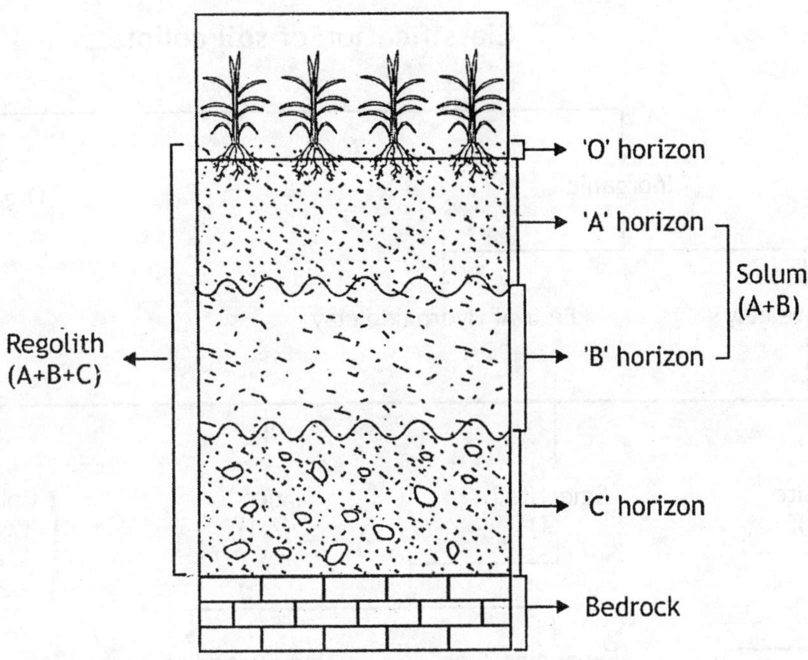

Factors of Soil formation

S = f (cl, o, r, p, t ....)

S = Soil formation
f = factors
cl = climate
o = organism
(cl, o) Active factor
r = relief/topography
p = parent material
(r, p) Passive factor
t = time or age of land ← Neutral factor

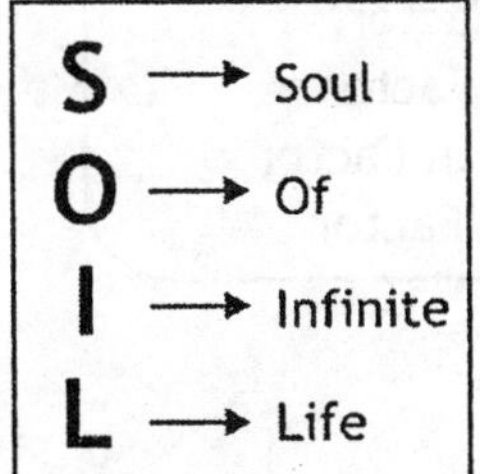

Components of Soil

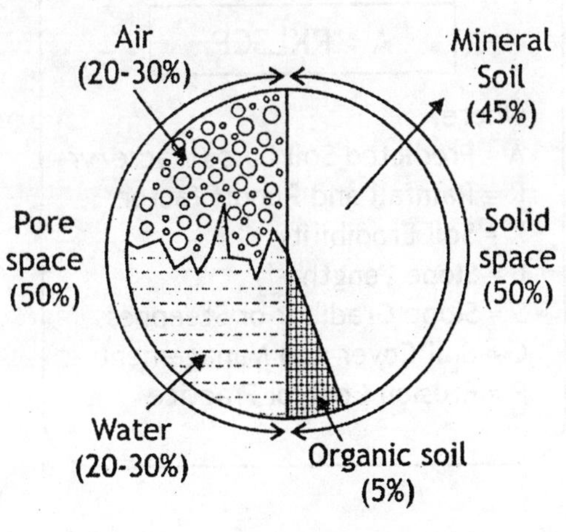

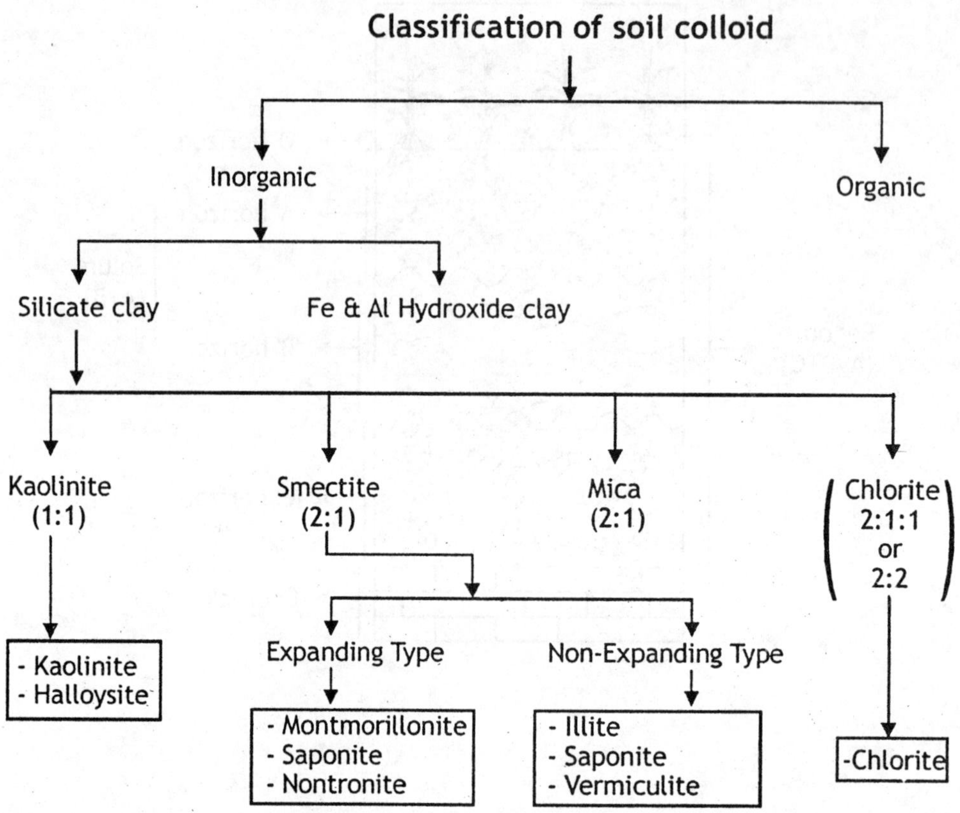

Soil Loss Equation due to Water Erosion

A = RKLSCP

where,

A = Predicted Soil Loss (t/acre/yr)
R = Rainfall and Runoff factor
K = Soil Erodibility
L = Slope Length
S = Slope Gradient or Steepness
C = Soil Cover and Management
P = Erosion Control Practice

Soil Loss Equation due to Wind Erosion

E = IRKFCWDB

where,

E = Soil Loss by Wind Erosion
I = Soil Cloddiness Factor
K = Surface Roughness Factor
F = Soil Textural Class Factor
C = Local Wind Factor
W = Field Width Factor
D = Wind Direction Factor
B = Wind Barrier Factor

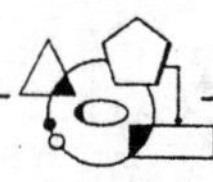

Different layers of soil water

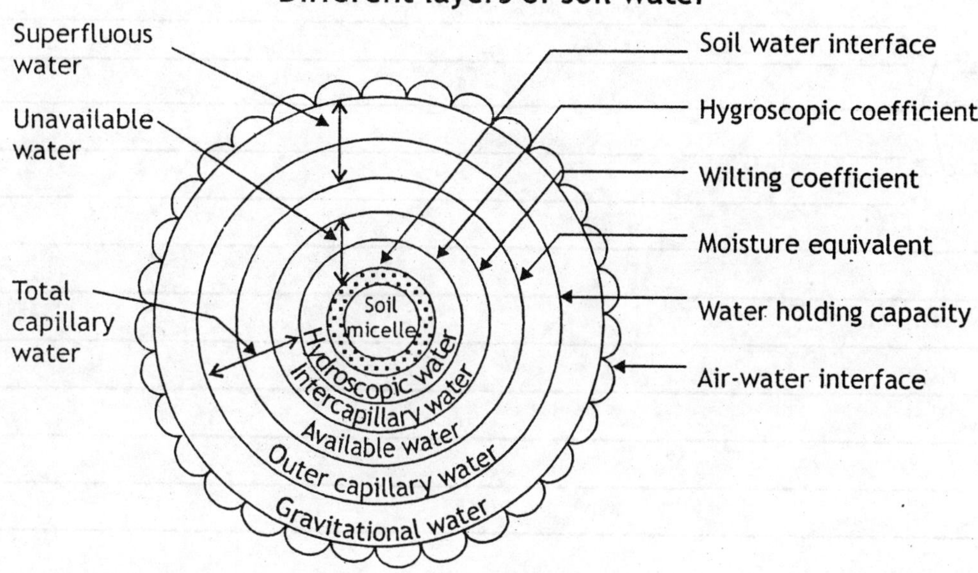

Soil Texture

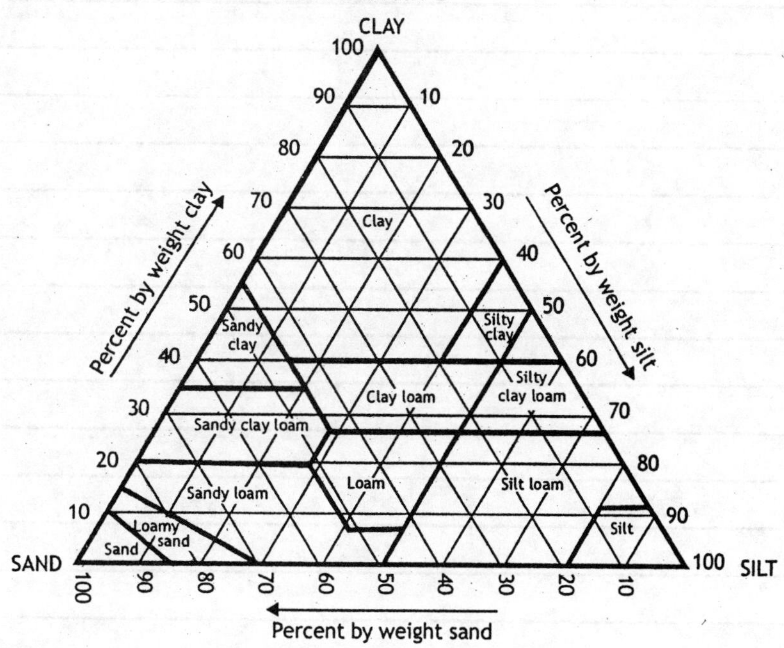

Triangular textural diagram (sand, silt and clay particle sizes of 2-0.02 mm, 0.02-0.002 mm and < 0.002 mm, respectively)

J&K have Salal project.

Notes

GK GS

Article 246 and 245 describe the legislative jurisdiction of the Union and the States.

19. Agri-sets for Zero Error in Agri-tests

The questions given in this chapter are MCQs model test papers on General Agriculture asked in various Agricultural Competitive Exams.

Take the Questions as you would, under the actual examination for TOP (Target Oriented Preparation).

Make your answers on the answer sheet provided along with these sets. Tally your score and tie up the loose end for the Agricultural Competitive Exams.

Remember

Winners make it happen;
Losers let it happen.

GK **Indian agriculture is intensive subsistence type.** GS

Ask Yourself

- Do you have a plan of action
- Do you have a clearly defined purpose
- What effort are you putting into preparation
- What price are you willing to pay? How far are you willing to go
- Do you have the patience to withstand the gestation period
- Are you willing to practice toward excellence
- Do you have any firm principles to stand on
- Do you have the "can do" attitude

PREPARATION = Purpose + Principle + Planning + Practice + Perseverance + Patience + Pride

Set-0

1. **Consider the following statements:**
   1. India is the largest producer of safflower oil in the world.
   2. Areawise, India is the largest cultivator of tea in the world.
   Which of the statements given above is/are correct?
   a] 1 only b] 2 only
   c] Both 1 and 2 d] Neither 1 nor 2
2. **Which one among the following components of soil organic matter is most resistant to decomposition?**
   a] Cellulose b] Hemicellulose
   c] Starch d] Lignin
3. **Which one of the following is a measure of the net positive charge held on soil exchange complex?**
   a] Anion exchange capacity
   b] Cation exchange capacity
   c] Base exchange capacity
   d] Electricaly conductivity
4. **Volatilisation loss of ammonia in soil increases with the increase in :**
   a] Calcium carbonate content of soil
   b] Moisture content of soil
   c] Clay content of soil
   d] Cation exchange capacity of soil
5. **The deficiency of which one of the following causes Khaira disease?**
   a] Copper
   b] Iron
   c] Phosphorus
   d] Zinc
6. **Which one of the following factors is responsible for poor fruit setting in custard apple ?**
   a] Dioecious nature b] Heterostyly
   c] Dichogamy d] Self-incompatibility
7. **Which of the following scientists proposed the Critical Level Concept for nutrients in soil ?**
   a] Jackson b] Cate and Nelson
   c] Stem d] Schofield
8. **Match List I (Nutrient Element) with List II (Source in Soil) and select the correct answer using the codes given below the lists:**

List I	List II
a] Calcium	1. Apatite
b] Nitrogen	2. Dolomite
c] Phosphorus	3. Feldspar
d] Potassium4.	Organics

	A	B	C	D
a]	2	1	4	3
b]	3	4	1	2
c]	3	1	4	2
d]	2	4	1	3

9. **Which one of the following organisms is involved in the formation of N_2O?**
   a] *Thiobacillus ferroxidans*
   b] *Paracoccus denitrificans*
   c] *Nitrosomonas europaea*
   d] *Desulfovibrio desulfuricans*
10. **Which one of the following is the correct order regarding thermal conductivity of soil?**
   a] Peat > Clay > Loam > Sand
   b] Clay > Loam > Sand > Peat
   c] Loam > Sand > Peat > Clay
   d] Sand > Loam > Clay > Peat
11. **Ammonium fixation in the soil is due to**
   a] Microbial assimilation
   b] Microbial fixation
   c] Microbial immobilization
   d] Soil exchange reaction
12. **Which one of the following soil properties does not change by cropping and cultivation?**
   a] Bulk density b] Particle density
   c] Porosity d] Permeability
13. **Size of the spray drop from a mist sprayer varies from?**
   a] 1-50 μm b] 50-100 μm
   c] 100-200 μm d] 200-300 μm
14. **Which of the following is a pest of apple?**
   a] *Sylepta lunalis* b] *Nodostoma subscotalum*
   c] *Cydia hemidoxa* d] *Eriosoma lanigerum*
15. **Cartap hydrochloride, a contact & stomach poison insecticide is isolated from**
   a] *Streptomyces grisea*
   b] *Streptomyces avermitilis*
   c] *Lumbriconeseis heteropodo*
   d] *Bacillus subtilis*
16. **The use of flood jet nozzle is required for**
   a] Minimizing the drift of chemicals
   b] Increasing the droplet size
   c] Quick delivery of spray fluid
   d] Better adhesion of spray fluid

17. Consider the following statements :
1. The sequence of nucleotides recognized by a restriction endonuclease is present in only one strand of DNA, running in 3' - 5' direction.
2. At present, more than a hundred restriction endonucleases have been discovered
3. Reverse transcriptase synthesizes DNA using RNA as a template.

Which of the statements given above are correct ?

a] 1 and 2 b] 2 and 3
c] 1 and 3 d] 1, 2 and 3

18. Where is National Institute of Agricultural Marketing located?

a] Jaipur b] Karnal
c] Lucknow d] Vijayawada

19. With reference to photosynthesis, consider the following statements :

(1) Ferridoxin is located in the thylakoid membrane.

(2) Phaeophytin is actually a chlorophyll b molecule that does not contain a magnesium ion.

(3) The reactions that break down water and produce oxygen and protons are located in stroma.

Which of the statements given above is/are correct?

a] 1 only b] 1 and 2
c] 1 and 3 d] 2 and 3

20. Which one of the following pairs is *not* correctly matched?

Substance		Part of the plant yielding substance
a] Cocaine	:	Leaves
b] Colchicin	:	Bulb
c] Digitalin	:	Root
d] Opium	:	Fruit

21. With reference to flowering plants, consider the following statements :

(1) Pollen grains contain auxin.

(2) Pollination prevents ovary abscission.

(3) Normal seeds synthesize auxins, gibberellins & cytokinins.

Which of the statements given above are correct ?

a] 1 and 2 b] 2 and 3
c] 1 and 3 d] 1, 2 and 3

Directions : The following 9 (Nine) items consist of two statements: one labelled as the 'Assertion (A)' and the other as 'Reason (R)'. You are to examine these two statements carefully and select the answer to these items using the codes given below:

a] Both A and R are individually true and R is the correct explanation of A.
b] Both A and R are individually true but R is not the correct explanation of A.
c] A is true but R is false.
d] A is false but R is true.

22/ Assertion (A) : In order to overcome the problems facing Indian agriculture, the Government of India evolved the Macro Management of Agriculture by integrating 27 identified schemes.

Reason (R) : As Agriculture is a subject under Concurrent List in the Seventh Schedule of the Constitution of India, the primary responsibility for increasing agricultural production rests with the Union Govt.

23. Assertion (A) : Without NAD^+, ATP cannot be formed in glycolysis.

Reason (R) : Witnout NAD^+, 3-phosphoglyceraldehyde cannot be oxidized to 1, 3-diphosphoglycerate.

24. Assertion (A) : Ethephon is used to hasten ripening in banana.

Reason (R) : Ethephon releases auxin.

25. Assertion (A) : A change in the ratio of input-prices correspondingly changes the slope of the iso-cost line.

Reason (R) : The slope of the iso-cost line indicates the ratio of factor-prices.

26. Assertion (A) : Dwarf wheat is always sown shallow compared to tall wheat.

Reason (R) : The coleoptile length is longer in case of Mexican wheat as compared to tall wheat.

27. Assertion (A) : The rosette habit of cabbage can be changed drastically by the application of Gibberellic Acid (GA).

Reason (R) : Since Gibberellic Acid helps in cell division and differentiation, the rosette habit of cabbage is changed.

28. Assertion (A) : With reference to soils, isomorphous substitution is an important mechanism through which negative charges are developed.

Reason (R) : Isomorphous substitution of lower valent ions occurs by higher valent ions.

29. Assertion (A) : The rotating disc aerosol generator is not useful for generating aerosol of pesticides like pyrethrum.

Reason (R) : Pyrethrum is a heat sensitive pesticide of plant origin.

30. Assertion (A) : All young soils are deficient in nitrogen.

Reason (R) : Nitrogen is not a significant component of any type of rock.

31. Which one of the following pairs is not correctly matched?

Crop		Important moisture sensitive stage
a] Rice	:	Panicle initiation, flowering
b] Sugarcane	:	Germination
c] Cotton	:	Flowering, boll development
d] Chillies	:	Flowering

Elements such as Fluorine, Chlorine and Bromine are Halogens.

GS

32. Consider the following statements :
1. A gram of guava fruit has more vitamin C than a gram of orange fruit.
2. Papaya does not contain vitamin C.
3. Jackfruit has a digestive enzyme bromelin.
Which of the statements give above is/are correct ?
a] 1 only b] 1 and 2
c] 1 and 3 c] 2 and 3

33. Anti-cancer drug vincristin is obtained from.
a] *Piper nigrum* b] *Rauwolfia serpentina*
c] *Hemidesmus indicus* d] *Catharanthus roseus*

34. Consider the following statements :
1. All the proteins of mitochondria are encoded by mitochondrial genome.
2. The chloroplast DNA does not code for all chloroplast proteins,
Which of the statements given above is/are correct ?
a] 1 only b] 2 only
c] Both 1 and 2 d] Neither 1 nor 2

35. Match List I (Plant) with List II (Propagation) and select the correct answer using the codes given below the lists :

List I	List II
A. Pineapple	1. Soft wood cutting
B. Citrus	2. Root cutting
C. Peach	3. Budding
D. Sapota	4. Sucker
	5. Stone grafting

	A	B	C	D
a]	4	3	2	1
b]	5	2	3	4
c]	5	3	2	4
d]	4	2	3	1

36. Consider the following statements :
1. The strawberry propagates by runners.
2. The chrysanthemum propagates by suckers
Which of the statements given above is/are correct?
a] 1 only b] 2 only
c] Both 1 and 2 d] Neither 1 nor 2

37. Which one of the following fruit crops is highly susceptible to salinity?
a] Date palm (*Phoenix dactylifera*)
b] Ber (*Zizyphus jujuba*)
c] Pomegranate (*Punica granatum*)
d] Almond (*Amygdalus communis*)

38. Which one of the following is the best test for ripeness of table grapes ?
a] Colour of the fruit
b] Sweetness and change in seed colour
c] Softness of the fruit
d] Ease of separation of grapes from the vines

39. Match List I (Crop) with List II (Cultivar) and select the correct answer using the codes given below the lists :

List I	List II
A. Banana	1. Sonaka
B. Grapes	2. Poovan
C. Carrot	3. Super Star
D. Rose	4 Pusa Kesar

	A	B	C	D
a]	3	1	4	2
b]	3	4	1	2
c]	2	4	1	3
d]	2	1	4	3

40. Consider the following statements:
1. In Indian agriculture, oil seeds are next to sugarcane in area coverage, production and value.
2. India's share in the world production of mango is more than 45%.
Which of the statements given above is/are correct?
a] 1 only b] 2 only
c] Both 1 and 2 d] Neither 1 nor 2

41. Which one of the following fruits is good in curing diabetes ?
a] Ber (*Zizyphus jujuba*)
b] Date Palm (*Phoenix dactylifera*)
c] Apple (*Pyrus malus*)
d] Jamun (*Eugenia jambolana or Syzygium cuminii*)

42. Consider the following statements :
1. The medicinal plant *Glycyrrhiza glabra* is a legume.
2. The leaves of *Cassia angustifolia* are purgative.
Which of the statements given above is/are correct ?
a] 1 only b] 2 only
c] Both 1 and 2 d] Neither 1 nor 2

43. Which one of the following chemicals is used to keep the fruits and vegetables firm ?
a] Calcium chloride
b] Sodium benzoate
c] Ascorbic acid
d] Sodium chloride

44. Consider the following statements :
1. Sunn hemp is obtained from *Crotalaria juncea* of family Papilionaceae.
2. Jute (*Crochorus capsularis*) fibres are actualy the scelerenchymatous tissue of secondary phloem.
3. The stem of narcotic plant *Cannabis sativa* yields fibres used in making ropes and sacks.
Which of the statements given above are correct ?
a] 1 and 2
b] 2 and 3
c] 1 and 3
d] 1, 2 and 3

GK **Aus, Aman and Boro are the 3 types of crops of rice in India.** GS

45. The heat treatment given before canning of vegetables with boiling water or steam followed by cooling is termed as
a] Blanching b] Brining
c] Clinching d] Exhausting

46. Consider the following statements :
The causes of deterioration in canned fruit products are
a] Bacteria b] Iron salts
c] Copper salts
Which of the statements given above is/are correct ?
a] 1 only b] 1 and 2
c] 2 and 3 d] 1, 2 and 3

47. When should the fertilizer be applied to lawn grass ?
a] When it is available at the lowest price
b] In summer time for cool season grasses
c] When the grass begins the dormant rest cycle
d] Just before the beginning of the active growth cycle of the grass.

48. With reference to mutagenic agents, consider the following statements :
1. Acridine dyes cause frame shift mutations.
2. Ultraviolet light causes point mutations.
3. Five-bromouracil causes transition mutation.
Which of the statements given above are correct?
a] 1 and 2 b] 2 and 3
c] 1 and 3 d] 1, 2 and 3

49. What is the amount of ammonium nitrogen and nitrate nitrogen presence in 25 kg of calcium ammonium nitrate (20% N)?
a] 1.5. kg ammonium nitrogen and 3.5 kg nitrate nitrogen
b] 2.5 kg ammonium nitrogen and 2.5 kg nitrate nitrogen.
c] 3.5 kg ammonium nitrogen and 1.5 kg nitrate nitrogen.
d] 4.5 kg ammonium nitrogen and 0.5 kg nitrate nitrogen.

50. Match List I (Disease) with List II (Crop) and select the correct answer using the codes given below the lists :

	List I		List II
A.	Late blight	1.	Wheat
B.	Loose smut	2.	Potato
C.	Blast	3.	Pear
D.	Fire blight	4.	Rice

	A	B	C	D
a]	3	4	1	2
b]	2	4	1	3
c]	2	1	4	3
d]	3	1	4	2

51. The interspecific semiochemicals which provide adaptive advantage to the emitter (producer) species are known as.
a] Allomones b] Hormones
c] Kairomones d] Pheromones

52. In which one of the following crops has the outbreak of woolly aphid been noticed recently ?
a] Maize b] Barley
c] Sugarcane d] Cotton

53. Agronomic research carried out on farmer's field with active participation of the farmer is called
a] On-farm research b] Adaptive research
c] Operational research d] Diversified research

54. Match list I (Weed) with list II (Family) and select the correct answer using the codes given below the lists:

List I	List II
A. Parthenium hysterophorus	1. Aizoaceae
B. Trianthema portulacastrum	2. Compositae
C. Eichhomia crassipes	3. Pontederiaceae

	A	B	C
a]	1	2	3
b]	3	1	2
c]	2	3	1
d]	2	1	3

55. With reference to root hairs of plants, consider the following statements :
1. Root hairs are the extensions of the cortex cells of the root.
2. The cell wall of root hairs acts as a semi-permeable membrane.
Which of these statements is/are correct ?
a] 1 only b] 2 only
c] Both 1 and 2 d] Neither 1 nor 2

56. The nozzle used to produce fog is
a] Gaseous energy nozzle
b] Thermal energy nozzle
c] Centrifugal energy nozzle
d] Kinetic energy nozzle

57. The critical stages of irrigation in cotton are
a] Vegetation and flowering
b] Squaring and flowering
c] Flowering and bolling
d] Bolling and boll opening

58. Which one of the following is most remunerative in low and farming system ?
a] Rice -Fish culture b] Rice -Poultry -Fish culture
c] Rice -Dairy d] Rice -Silviculture

59. Which one of the following companion cropping systems is most prevalent in Andaman and Nicobar Islands ?
a] Coconut + Sugarcane b] Coconut + Maize
c] Coconut + Black pepper d] Coconut + Sunflower

60. Which one of the following aphids feeds on pulses ?
a] Rhopalosiphum maidis b] Myzus persicae
c] Lipaphis erysimi d] Aphis cracivora

61. Which of the following is the characteristic of clay particles in the soil ?
a] Less than 0.002 mm in diameter and consist of primary minerals
b] Less than 0.002 mm in diameter and consist of primary and secondary minerals
c] 0.002 to 0.05. mm in diameter and consist of secondary minerals
d] Less than 0.002 mm in diameter and consist of secondary minerals

62. Consider the following :
1. *Buddlea asiatica* 2. *Duranta plumerii*
3. Carnation (*Dianthus caryophyllus*)
Which of the above is/are suitable for cut-flower industry ?
a] 1 and 2 b] 2 only
c] 1 and 3 d] 3 only

63. Kewada or screwpine (*Pandanus*), an important aromatic plant, is found in plenty in
a] Coastal tracts of Orissa
b] Sand dunes of Rajasthan
c] Foothills of Himachal Pradesh
d] Cold desert of Ladakh

64. The fruit pulp of which one of the following medicinal plants is used as laxative ?
a] *Atropa belladonna*
b] *Anethum graveolens*
c] *Cassia fistula*
d] *Rauwolfia serpentina*

65. Which one of the following pairs is responsible for tundu disease of wheat ?
a] Nematode and Bacterium
b] Bacterium and Fungus
c] Nematode and Fungus
d] Virus and Nematode

66. In which one of the following characters do mycoplasmas differ from bacteria ?
a] Reproduction of fission
b] Cell wall
c] Culture in cell free medium
d] Visibility under optical microscope

67. Which of the following is correct regarding viroid ?
a] RNA containing 50 -100 base pairs with protein coat
b] DNA containing 50 - 100 base pairs with protein coat
c] RNA containing 250 - 400 nucleotides without protein coat
d] DNA containing 250 - 400 nucleotides without protein coat

68. Match List I (Common Name) with List II (Scientific Name) and select the correct answer using the codes given below the lists :

List I	List II
A. Ring nematode	1. *Hoplolaimus* sp.
B. Spiral nematode	2. *Criconemoides* sp.
C. Lesion nematode	3. *Pratylenchus* sp.
D. Lance nematode	4. *Helicotylenchus* sp.

	A	B	C	D
a]	1	3	4	2
b]	2	4	3	1
c]	1	4	3	2
d]	2	3	4	1

69. The amount of carbendazim 50% wettable powder required to prepare 1000 ppm solution is
a] 1g in 1 litre of water
b] 0.1g in 1 litre of water
c] 2g in 1 litre of water
d] 0.2g in 1 litre of water

70. After initiation of crown gall disease caused by *Agrobacterium tumefaciens*, its spread and development
a] Depend on growth and reproduction of bacteria
b] Depend on auxin production by bacteria
c] Depend on non-plasmid genome of bacteria
d] Is independent of bacteria

71. Consider the following statements:
Modern soil classification (taxonomy) is based on
1. Soil morphology 2. Soil genesis
3. Climate of soil formation 4. Age of soil
Which of the statements given above are correct ?
a] 1, 2 and 3
b] 2 and 3
c] 1, 2 and 4
d] 1 and 4

72. Which one of the following pairs is *not* correctly matched ?
a] Acid soils : Al toxicity
b] Saline alkali soils : K toxicity
c] Waterlogged soils : Methane emission
d] Sandy soils : Nutrient leaching

73. The bushy appearance with dead heart in sugarcane at 6th internode is due to
a] Excess irrigation b] Excess nitrogen
c] Top borer d] Mealy bug

74. Which one among the following is the pneumatic sprayer ?
a] Foot sprayer
b] Stirrup sprayer
c] Hand compression sprayer
d] Rocker sprayer

GK

Common name of 'Patella bone' is 'Knee cap'.

GS

75. Match List I (Agroclimatic Zones) with List II (Main Strategy of Development) and select the correct answer using the codes given below the lists :

	List I		List II
A.	Lower Gangetic plains	1.	Reclamation of ravinous area
B.	Central plateau and hills	2.	Emphasis on increasing the tree cover
C.	Western dry region	3.	Crop diversification
D.	Trans-gangetic plains	4.	Minor irrigation programmes

	A	B	C	D
a]	3	1	2	4
b]	4	1	2	3
c]	3	2	1	4
d]	4	2	1	3

76. Consider the following statements :

1. For increasing the production of oilseeds/edible oils and to attain self-sufficiency in their production, a centrally-sponsored Oilseeds Production Programme is being implemented in 28 states.

2. Convergence of Oilseeds Production Programme with Watershed Development Programme is one of the thrust areas identified for increasing the production of oilseeds during the Tenth Five-Year-Plan.

Which of the statements given above is/are correct?

a] 1 only
b] 2 only
c] Both 1 and 2
d] Neither 1 nor 2

77. The market equilibrium for an agricultural commodity is determined by

a] The market demand for the commodity
b] The market supply of the commodity
c] The balancing of the forces of demand and supply for the commodity
d] Imports and exports

78. Consider the following crops :

1. Cereal crops
2. Plantation crops
3. Pulses and oilseeds

Which of the above crops are exempted from enforcement of ceiling on land holdings ?

a] 1 and 2 b] 2 only
c] 2 and 3 d] 3 only

79. Which one of the following structural changes is becoming increasingly popular in the agricultural marketing system in India ?

a] Integration
b] Specialization
c] Diversification
d] Decentralisation

80. Consider the following statements :

1. Vacuoles in plant cells have single membrane.
2. Mitochondrial DNA lacks histones.
3. Smooth Endoplasmic Reticulum is involved in lipid synthesis and membrane assembly.

Which of the statements given above is/are correct?

a] 1 only
b] 1 and 2
c] 2 and 3
d] 1, 2 and 3

81. Which one of the following refers to inter-regional allocation of agricultural production credit ?

a] Proportionate number of farmers
b] Variation in the cost of production
c] Proportionate cultivated area
d] Differential productivity of capital

82. Consider the following statements :

The presence of very large number of middlemen in agricultural marketing could be attributed to

1. The demand for farm products over a large area.
2. Highly scattered agricultural production.
3. Difficult transportation of agricultural products which also involves special care.

Which of the statements given above is/are correct?

a] 1 and 2
b] 2 and 3
c] 3 only
d] 1, 2 and 3

83. To produce more output per unit of land, a cultivator will

a] Resort to diversification
b] Use land intensively
c] Use land extensively
d] Use land intensively as well as extensively

84. Which of the following characterizes co-operative joint farming ?

a] Ownership and cultivation are individual practices
b] Ownership is individual and the reward is based on the contribution of labour
c] Cultivation is joint and the reward is based on the contribution of land
d] Cultivation is joint but the reward is not based on the contribution of land

85. A farm is said to be diversified when

a] Crop enterprise is mixed with livestock enterprise
b] Any one of the enterprises contributes more than 50% of the total income
c] None of the enterprises contributes more than 50% of the total income
d] Crop enterprise is combined with forestry

GK **Jhoom is the name given to shifting cultivation in India.** GS

86. Consider the following statements :

1. The smallest unit of DNA which is capable of undergoing crossing over and recombination is called cistron.

2. In protein synthesis, the enzyme that catalyzes peptide bonding is located in the smaller subunit of ribosomes.

3. In cell cycle, replication of DNA occurs during G_1 phase.

Which of the statements given above is/are correct?

a] 1 and 2 b] 2 only
c] 2 and 3 d] 3 only

87. Which one of the following types of soils is most suitable for groundnut cultivation ?

a] Sandy and sandy-loam soil
b] Loam and clay soil
c] Clayey soil
d] Silty soil

88. The concave Production Possibility Curve (PPC) is very common in agriculture because of

a] Constant rates of substitution between two products
b] Increasing rates of substitution between two products
c] Increasing rates of substitution between two factors
d] Increasing rates of substitution between two factors and between two products

89. Scientific name of macaroni wheat is

a] *Triticum aestivum* b] *Triticum dicoccum*
c] *Triticum duram* d] *Triticum vulgare*

90. Match List I with List II and select the correct answer using the codes given below the lists :

List I (Crop)	List II (Variety)
A. Castor	1. BR 13
B. Pigeonpea	2. Lal Bahadur
C. Maize	3. Kissan
D. Wheat	4. NPH 1

	A	B	C	D
a]	4	3	1	2
b]	4	1	3	2
c]	2	1	3	4
d]	2	3	1	4

91. Consider the following statements :

1. Muriate of potash is not suitable for sugarcane.

2. Saltpetre contains 65% K_2O.

3. Single superphosphate contains gypsum,

4. Ammonium phosphates are completely water soluble.

Which of the statements given above are correct ?

a] 1 and 2 b] 1, 3 and 4
c] 3 and 4 d] 1, 2, 3 and 4

92. With reference to the food grain cultivation next to Uttar Pradesh, which one among the following states has the highest irrigated area ?

a] Andhra Pradesh b] Maharashtra
c] Punjab d] West Bengal

93. The Chromosome theory of inheritance was first postulated by

a] Avery, McCarty and MacLeod
b] Frederick Griffith
c] Morgan and Sturtevant
d] Sutton and Boveri

94. In meiosis, reduction in the number of chromosomes occurs in

a] Diakinesis b] Anaphase I
c] Metaphase I d] Anaphase II

95. Which one of the following is the correct sequence in the biosynthetic pathway of Indole Acitic Acid (IAA)?

a] Tryptophan -Indole pyruvic acid - Indoleacetaldehyde - IAA
b] Indole pyruvic acid -Indoleacetaldehyde - Tryptophan - IAA
c] Tryptophan -Indoleacetaldehyde -Indole pyruvic acid - IAA
d] Indoleacetaldehyde -Indole pyruvic acid - Tryptophan - IAA

96. Which one of the following pairs is *not* correctly matched ?

a] Cucurbit : Self pollinated
b] French bean : Self pollinated
c] Brinjal : Often cross pollinated
d] Pea : Self pollinated

97. What is RNA splicing ?

a] Termination of RNA synthesis at specific base sequences within the DNA molecule
b] The excision of the introns and the formation of final mRNA molecule by joining the exons
c] The beginning of transcription as soon as the RNA polymerase-promoter complex is formed and an appropriate nucleotide binds to the enzyme
d] Release of newly formed RNA after the termination of transcription

98. Which one of the following phytohormones controls the apical dominance in plants ?

a] Auxin b] Cytokinin
c] Ethylene d] Giberellin

99. Which one of the following processes is most adversely affected by the deficiency of magnesium in plants ?

a] Defoliation
b] Upward translocation of nutrients
c] Downward movement of nutrients
d] Photosynthesis and carbohydrate metabolism

GK GS

Dhabar Commission was appointed on tribal area.

100. Consider the following amino acids :
1. Cysteine
2. Methionine
3. Glutamic acid
Which of the above contain sulphur?
a] 1 and 2 b] 2 and 3
c] 1 and 3 d] 1, 2 and 3

101. Which one of the following pairs is *not* correctly matched?
a] Copper : Plastocyanin
b] Iron : Phytochrome
c] Magnesium : Chlorophyll
d] Cobalt : Vitamin B_{12}

102. With reference to photorespiration in plants, consider the following statements :
1. With the increase in temperature and oxygen concentration, the affinity of RuBP carboxylase decreases for CO_2 and increase for O_2.
2. The peroxisomes present in the cells metabolise glycolate into glycine; and glycine into serine and CO_2.
3. In plants adapted to C4 pathway of photosynthesis to overcome photorespiratory losses, the CO_2 is fixed in mesophyll cells.
Which of the statements given above are correct?
a] 1 and 2 b] 2 and 3
c] 1 and 3 d] 1, 2 and 3

103. In marigold and sweet pea plants, which one of the following conditions favours cross pollination ?
a] Dichogamy b] Herkogamy
c] Self-sterility d] Unisexuality

104. Consider the following statements :
1. India imports the entire potassic fertilizers as there is no indigenous source available.
2. The all-India average fertilizer consumption is 140 kg/ha.
Which of the statements given above is/are correct?
a] 1 only b] 2 only
c] Both 1 and 2 d] Neither 1 nor 2

105. Which one among the following elements is a component of cell membrane ?
a] Sulphur b] Zinc
c] Molybdenum d] Phosphorus

106. Match List I with List II and select the correct answer using the codes given below the lists :

List I (Pest)	List II (Crop)
A. San Jose Scale	1. Coconut
B. Stem fly	2. Apple
C. Blackheaded caterpillar	3. Soyabean
D. Fruit sucking moths	4. Citrus

	A	B	C	D
a]	4	1	3	2
b]	4	3	1	2
c]	2	3	1	4
d]	2	1	3	4

107. Conversion of NO_3 -N to N_2, NO and N_2O is known as
a] Oxidation b] Denitrification
c] Mineralisation d] Nitrification

108. Consider the following statements :
1. Ethyl methane sulphonate is widely employed in the artificial induction of polyploidy.
2. Colchicin is used for inducing gene mutations.
Which of the statements given above is/are correct ?
a] 1 only b] 2 only
c] Both 1 and 2 d] Neither 1 nor 2

109. Which one of the following causes a disease in wheat due to which small, linear or oblong dark brown blotches, studded with minute black dots, appear on floral bracts and nodal tissues of culms ?
a] *Alternaria triticina*
b] *Dilophosopora rolfsii*
c] *Helminthosporium sativum*
d] *Septoria nodorum*

110. Consider the following statements :
Tomato hybrids are becoming popular due to
1. High yield
2. Resistance to insects
3. Uniform fruit size
Which of the statements given above is/are correct ?
a] 1 only b] 2 and 3
c] 1 and 3 d] 1, 2 and 3

111. Synthetic variety is developed by
a] Crossing inbred lines tested for GCA
b] Mixing seeds of inbred lines
c] Crossing inbred lines tested for SCA
d] Mixing seeds of open-pollinated cultivars

112. Consider the following statements :
1. In autopolyploids, the time of blooming is earlier and also shortened due to fast growth rate
2. Fertility level and seed set are very high in induced polyploids.
3. *Gossypium hirsutum* is an example of amphidiploidy.
Which of the statements given above is/are correct?
a] 1 and 2 b] 2 only
c] 3 only d] 1 and 3

113. Tift-23A, a cytoplasmic male sterile line, is used for development of hybrids in
a] Sorghum b] Maize
c] Rice d] Pearl millet

114. Consider the following statements :
Haploids can be artificially produced by
1. X-ray treatment
2. Delayed pollination
3. Anther culture
Which of the statements given above is/are correct ?
a] 1 and 2 b] 1 and 3
c] 3 only d] 1, 2 and 3

GK

First case of SARS "Severe Acute Respiratory Syndrome) have been reported from the Guangdong province of Southern China.

GS

115. The male sterility used in the development of hybrid wheat has been derived from

a] *Triticum dicoccum*

b] *Triticum durum*

c] *Triticum timopheevi*

d] *Triticum monococcum*

116. Consider the following statements :

1. All tRNA molecules have guanine residue at 5' end.

2. The amino acid is accepted by tRNA at 3' end only,

3. All tRNA molecules have CCA sequence at 3' end,

Which of the statements given above are correct ?

a] 1 and 2

b] 2 and 3

c] 1 and 3

d] 1, 2 and 3

117. The Wobble hypothesis regarding genetic code (codons and anticodons) was proposed by

a] Nirenberg

b] Leder

c] Khorana

d] Crick

118. Consider the following characters of *Pisum sativum*:

1. Colour of cotyledon 2. Colour of seed coat

3. Shape of seed 4. Length of pod

Which of the above were taken into account in Mendel's experiments on hybridization ?

a] 1, 2 and 3 b] 2, 3 and 4

c] 1 and 4 d] 1, 2, 3 and 4

119. Which one of the following statements is *not* correct?

a] Rough Endoplasmic Reticulum is particularly well developed in cells actively engaged in protein synthesis

b] Ribosomes have large contents of lipids in their compositions

c] Golgi complex is involved in the formation of cell plate during cell division

d] In germinating seeds, glyoxysomes convert fatty acids into sugars

120. The development of seed without sexual fusion of male and female gametes is known as

a] Vegetative reproduction

b] Apomixis

c] Self-incompatibility

d] Apospory

Set-1

1. **Nematodes come under the category of :**
   a] Symbiotic nitrogen fixtures
   b] Blue green algae
   c] Plant parasties
   d] Insect pests
2. **Out of the total cropped area of India, forest occupy only:**
   a] 26% b] 33%
   c] 35% d] 19.5%
3. **The first Director General of Indian Council of Agricultural Research was:**
   a] Dr. M.S. Swaminathan
   b] Dr. N.S. Randhawa
   c] Dr. G.S. Randhawa
   d] Dr. B.P. Pal
4. **Indian Institute of Horticulture Research for northern plain is situated at :**
   a] Hessargatha, Bangalore
   b] Lucknow
   c] Patna
   d] Kolkata
5. **Apple can successfully be grown at the following elevation:**
   a] 500 meter above sea level
   b] 1000 meter above sea level
   c] 2000 meter above sea level
   d] 200 meter above sea level
6. **The successful sub-tropical variety of peach is :**
   a] Alexander b] Alberta
   c] Florda Sun d] Florida Sun
7. **The fertilizer containing nitrogen in amide form is:**
   a] Urea
   b] C.A.N.
   c] Ammonium Sulphate
   d] Ammonium Phosphate
8. **Near the canal banks soils become:**
   a] Alkaline
   b] Acidic
   c] Neutral
   d] Well drained
9. **The functions of leaves are:**
   a] To synthesize food in presence of sunlight
   b] Intake of oxygen and release of carbon dioxide for respiration
   c] Both of the above
   d] None of these
10. **Azatobactor is useful in fixing of nitrogen from air in:**
    a] Non-leguminous crops like rice and wheat
    b] Leguminous crops
    c] None of the above
    d] Soyabean
11. **The cost of production of the crop can be minimized by using economic principle:**
    a] Law of diminishing return
    b] Law of opportunity cost
    c] Cost principle
    d] Principle of least cost combination
12. **Most of the plants obtain nitrogen from the soil in the form of:**
    a] Nitrite b] Nitrate
    c] Free nitrogen gas d] Nitric acid
13. **Which of the following is a micronutrient?**
    a] Calcium b] Zinc
    c] Nitrogen d] Phosphorus
14. **Which of the plant growth regulator is responsible for breaking dormancy?**
    a] Auxin b] Cytokinin
    c] Gibberellic acid d] 2, 4-D
15. **Tangelo is a cross between:**
    a] Tangerin and Pummelo
    b] Tangerin and Grape fruit
    c] Tangerin and Sweet orange
    d] Tangerin and Mandarin
16. **Which is the variety of brinjal?**
    a] Pusa Kesar b] Pusa Kranti
    c] Pusa Ruby d] Pusa Chetaki
17. **During rainy season wooden doors made of ply or soft wood generally swell up due to:**
    a] Osmosis
    b] Imbibition
    c] Diffusion
    d] Bad workmanship
18. **Deficiency of which of the following element causes general yellowing and poor growth of the plant?**
    a] Phosphorus b] Nitrogen
    c] Calcium d] Boron
19. **The plants growing in shady places adopt themselves by having following modification for more efficient carbon assimilation:**
    a] Leaves become broader
    b] Leaves becomes narrower
    c] Stem is modified into thorn
    d] Leaves become leathery

20. How many elements have so far been found essential for plant growth?
a] 20 b] 16
c] 30 d] 18

21. Which of the following statements is correct?
a] Kufri Badshah is late blight resistant variety of potato
b] Pant Bahar is a variety of brinjal
c] Pant C1 is a variety of tomato
d] Pusa Sawani is a variety of onion

22. Which one of the following Agricultural Universities was started in 1960?
a] C.S. Azad University of Agriculture and Tech., Kanpur
b] Punjab Agricultural University, Ludhiana
c] U.P. Agricultural University, Pantnagar
d] Rajendra Agri. University, Pusa, Bihar

23. Papaya fruits are rich in the following:
a] Vitamin A b] Vitamin D
c] Vitamin E d] Vitamin C

24. 'Sugar Baby' is the variety of :
a] Muskmelon b] Watermelon
c] Pumpkin d] Bittergourd

25. The cultivated muskmelons are mostly:
a] Monoecious b] Gynoecious
c] Androecious d] Andromonoecious

26. Which of the following statements is correct?
a] Rose, apple, peach, loquat belong to the family Rosaceae
b] Mango, guava and papaya belong to family Anacardiaceae
c] Potato, tomato, brinjal and peas belong to family Solanaceae
d] Peach belongs to family Leguminosae

27. Principle of equi-marginal returns is applied when:
a] The resources are unlimited
b] The resource prices are high
c] The prices of resources are varying
d] The resources are limited

28. Do the T.C. and the V.C. increase with increase in the output?
a] Never b] Yes
c] Partially d] With decrease in output

29. When a new variety of crop is introduced:
a] Partial budgeting is required
b] Full budgeting is a must
c] No budgeting is needed
d] Only planning is needed

30. The objective of preparation of calendar of operations is:
a] To prepare a cropping scheme
b] To calculate the cost of production
c] To assess the need of human and bullock power
d] To adjust farm enterprises

31. The farm plan would be based on the availability of efficiency of various factors of production:
a] True b] False
c] Partially true d] Incomplete statement

32. Farm budget is a tool of farm management analysis:
a] True
b] False
c] Never includes cost
d] It is a futile exercise

33. Extension workers should be well versed with economic principles:
a] Never
b] Partially
c] Without it they cannot give effective advice from business point of view
d] This is not necessary

34. Subsistence farming is characterized by:
a] Enough surplus for sale
b] Distress sale
c] Can hire large number of labourers
d] The resource structure is rich

35. Mixed farming is when:
a] Crop production is combined with livestock enterprise and 50-50% income is from both
b] Entire farm land is sown under mixed crops
c] Only milch cattle are raised on the farm
d] Income from milch animals is 95%

36. The agro-climatic zones in which India is divided are:
a] Ten b] Fifteen
c] Twelve d] Six

37. When a farm is managed by Government officials it is called a:
a] Peasant farm
b] Capitalistic farm
c] Collective farm
d] State farm

38. Maximum rose oil is produced in which of following country.
a] Bulgaria b] Italy
c] India d] China

39. When $\frac{\Delta X_1}{\Delta X_2} = \frac{Px_2}{Px_1}$ then we can:
a] Employ any combination of X_1 and X_2
b] Do nothing
c] Go on substituting X_1 for X_2
d] Employ X_2 only

40. The thrust of the agricultural programmes during the Eighth Plan is on:
a] Increasing production of rice
b] Increasing production of wheat
c] Increasing production of oilseeds, fruits, vegetables and pulses
d] Increasing production of cotton

GK GS

Bacteria are cosmopolitan.

41. The intensity of following one year crop rotation 'maize-potato-onion' is:
a] 100% b] 150%
c] 200% d] 300%

42. U.P. 319 is a variety of:
a] Paddy b] Wheat
c] Jowar d] Bajra

43. Which variety of the following relates to oilseeds?
a] CSH-4 b] Janak
c] Ganga-5 d] Varuna

44. Which of the following is commonly used for green manuring in India?
a] Dhaincha b] Lentil
c] Pea d] Potato

45. Most effective phosphatic fertilizer for crops in acidic soil is:
a] DAP
b] Rock phosphate
c] SSP
d] None of these

46. (i) Washington (ii) Honey dew (iii) Singapore and (iv) Ceylon varieties relate to:
a] Mango b] Banana
c] Apple d] Papaya

47. Give the name of recommended variety of 'loquat' for Uttar Pradesh:
a] Amrapali b] Queen
c] Lucknow 49 d] Tanaka

48. Under T&V System who is mainly responsible to make regular contact with farmers?
a] Dy. Director (Ext.)
b] Kisan Sahayak/Sevak
c] SDO
d] BDO

49. Which weedicide is used to kill weeds of broad leaves in legumes?
a] 2, 4-D
b] 2, 4, 5-T
c] 2, 4-DB
d] None of these

50. Tick, which disease is commonly found in Jowar fields?
a] Wilt b] Blast
c] Leaf spot d] Grain smut

51. The headquarter of Asian Vegetable Research Development Centre is located in:
a] Taiwan b] Thailand
c] Indonesia d] Malaysia

52. *Annona muriacta* L. is the botanical name of:
a] Sitaphal b] Ramphal
c] Cherimoya d] Sour-sop

53. Pusa Komal is a variety of:
a] Cowpea b] French bean
c] Guar d] Field pea

54. In a balanced diet the amount of vegetables for an adult is 300 gm. The ideal rates in which they should be used is:

	Green leafy vegetables	Root & tuber crop	Others
a]	125 gm.	75 gm.	100 gm
b]	100 gm.	75 gm.	125 gm.
c]	75 gm.	100 gm.	125 gm.
d]	100 gm.	125 gm.	75 gm.

55. The term 'Pomology' has been derived from the language?
a] Greek b] Latin
c] Arabic d] French

56. Banana is commonly propagted by:
a] Suckers b] Bulb
c] Corm d] Tubers

57. Gladiolus is propagated by.....
a] Corm b] Bulb
c] Rhizome d] Tuber

58. The "little leaf" disease of brinjal is caused by :
a] Fungus b] Bacteria
c] Nutrition d] None of the above

59. Maximum area under cashew-nut is in the state of :
a] West Bengal b] Orissa
c] Andhra Pradesh d] Kerala

60. 'Indian Horticulture' is being published by :
a] Indian Academy of Horticulture Sciences
b] Horticultural Society of India
c] Indian Council of Agricultural Research, New Delhi
d] Indian Agricultural Research Institute, New Delhi

61. Kinnow mandarin was developed by :
a] L. Moss b] W.B. Hayes
c] T. Tanaka d] H.B. Frost

62. Pollination in mango takes place by:
a] Housefly b] Wind
c] Whitefly d] None of these

63. Phalsa is commonly propagated by:
a] Seed b] Cutting
c] Budding d] Grafting

64. Polyembryonic variety of mango is:
a] Chandrakaran b] Ratna
c] Neelam d] Mallika

65. Wind breaks are planted in orchards towards :
a] South-west b] North-west
c] North-east d] East-south

66. Botanically 'pineapple' fruit is :
a] Sorosis b] Balausta
c] Berry d] Pome

67. Mycorrhizal association is found in:
a] Papaya b] Litchi
c] Bael d] Ber

GK GS
Shali is the name of a type of summer rice in Kashmir valley.

68. Caprification is related to :
a] Mango b] Guava
c] Banana d] Fig

69. The system of planting fruit plants followed in hills is:
a] Square b] Rectangular
c] Hexagonald] Contour

70. The most commonly used plant growth regulator for rooting of cuttings is :
a] IBA b] Cycocel
c] Ethephon d] GA_3

71. The term 'gene' was coined for the first time by :
a] Darwin b] Mandel
c] Bateson d] Johannson

72. Condition when a gene governs more than one trait is :
a] Multiple allele b] Polygene
c] Pleiotropy d] None of these

73. Chromosome number is reduced to half in :
a] Meiosis I b] Meiosis II
c] Mitosis d] Cytokinesis

74. International Crop Research Institute for semi-arid-tropics is situated at :
a] Jodhpur b] Bharatpur
c] Hyderabad d] Colombo

75. *Brassica juncea* is :
a] Aneupolyploid b] Haploid
c] Autotetraploid d] Allotetraploid

76. Horizontal resistance is :
a] Resistance against one race of pathogen
b] Resistance against all the races of pathogen
c] Resistance against lodging
d] None of the above

77. Clonal selection can be used in :
a] Wheat b] Groundnut
c] Potato d] Mustard

78. Single cross is a cross between :
a] Two open pollinated varieties
b] Two inbreds
c] Inbred and variety
d] Hybrid and variety

79. Foundation seed is produced from:
a] Nucleus seed b] Breeder seed
c] Quality seed d] Certified seed

80. Pulse crop which does not fix nitrogen from atmosphere easily is:
a] Arhar b] Pea
c] Soyabean d] Rajma

81. If only three irrigations are available indicate critical stages of wheat when these will be applied :
a] Crown root initiation, flowering, doughing stage
b] Critical root initiation, tillering, doughing stage
c] Tillering, jointing, milking
d] Critical root initiation, late jointing and dough stage

82. What is the content of organic matter in alluvial soil:
a] 0.86 - 6.53 b] 0.34 - 0.77
c] 0.28 - 1.10 d] 5.9 - 8.7

83. The group of organisms which convert ammonia into nitrites and nitrites to nitrates are respectively:
a] Nitrosomonas, Nitrobactor
b] Pseudomonas, Azotobactor
c] Streptomyces, Bacillus anthracis
d] Micromonospora, Nocardia

84. Which of the following elements is not essential element for plants but proves to be beneficial for some ?
a] Copper b] Boron
c] Sodium d] Iodine

85. Root knot is caused by which of the following nematodes ?
a] Aphlencoides
b] Melidogyne species
c] Trichodorous
d] Both (b) and (c)

86. The most efficient use of phosphorus is achieved by:
a] Broadcasting at the sowing time
b] Top dressing 21 days after sowing
c] Basal placement at the time of sowing
d] Foliar spray

87. Highest area of wheat is in :
a] Uttar Pradesh b] Punjab
c] Bihar d] Madhya Pradesh

88. Which of the following nutrient is cycled in the soil through atmosphere ?
a] Phosphorus b] Potash
c] Zinc d] Nitrogen

89. Which of the following is shortest duration variety of vegetable pea?
a] Arkel b] Bonneville
c] Azad Pea d] Asauji

90. What is the chemical control of potato blights ?
a] Spraying of endosulfan
b] Soil and seed treatment
c] Spraying of Gamma BHC
d] Spraying of Dithane M-45

91. Asepsis is :
a] Cleaning and grading of fruits and vegetables
b] Blanching of fruits and vegetables
c] Peeling of fruits and vegetables
d] None of these

92. Browning of cut apples is due to :
a] Bacteria b] Yeasts
c] Moulds d] Enzymes

93. Refractometer is used to determine:
a] Minerals b] TSS
c] Vitamins d] None of these

GK **Article 248 gives residuary powers to Union.** GS

94. Preservative for squash is :
a] Ptoassium metabisulphite
b] Vinegar
c] Salt
d] Citric acid

95. C.F.T.R.I. is situated at :
a] Mysore b] New Delhi
c] Lucknow d] None of these

96. Bulbil is a common feature in ____
a] Tiger lily b] Tulip
c] Dahlia d] Day lily

97. Fruit are balanced before canning at :
a] 80°C b] 90°C
c] 100°C d] None of these

98. End point of jelly at sea level is :
a] 95.5°C b] 100.5°C
c] 105.5°C d] None of these

99. The causal organism of 'red rot' a fungal disease of sugarcane, is present in :
a] Soil
b] Climate
c] Seed (planting material)
d] Virus

100. The most suitable chemical for seed treatment of wheat against 'Karnal Bunt' is :
a] B.H.C. b] Agrosan G.N.
c] Vitavax d] Captan

101. Origin of groundnut is :
a] India b] Brazil
c] Africa d] Japan

102. Cormel is a swollen end of ____
a] Primary root b] Fabrous root
c] Stolon d] Axillary bud

103. The fruit-fly belongs to :
a] Phycitidae b] Tephritidae
c] Tachinidae d] Coccinellidae

104. The insecticide act was passed by Parliament in the year :
a] 1973 b] 1965
c] 1968 d] 1972

105. The locust deposits its eggs :
a] On lower surface of leaves
b] In the soil
c] In the tender parts of a twig
d] In dry grasses

106. The rate of fumigation of EDB per quintal grain is :
a] 1 ml. b] 3 ml.
c] 6 ml. d] 10 ml.

107. Hopper burn is the result of attack of the following insect pest :
a] Grasshopper b] Sugarcane leafhopper
c] Brown planthopper d] Green leafhopper

108. Which of the following insecticide is safest for use ?
a] Methyl parathion b] Carbaryl
c] Phosphomidan d] DDT

109. Shobha is mutant cultivar of ____ developed at IIHR, Banglore.
a] Tulip b] Narcissus
c] Hibiscus d] Gladiolus

110. White grub is the pest of :
a] Rice b] Wheat
c] Groundnut d] Mustard

111. Integrated pest management refers to :
a] It is the use of zero pesticides
b] Use of zero tillage
c] Use of zero fertilizers
d] Management of all above judiciously

112. The deficiency of thiamine (B_1) in human body causes:
a] Night-blindness b] Beri-Beri
c] Pellagra d] Scurvy

113. Dissemination of plant pathogens take place only by:
a] Wind
b] Wind and water
c] Wind, water and soil
d] Wind, water, soil and insects

114. One of the factors causing malformation of mango is:
a] Fungus b] Nematodes
c] Virus d] Bacteria

115. The zinc deficiency in paddy plants causes :
a] Blight disease b] Ergot disease
c] Smut disease d] Khaira disease

116. The flower farming is scientifically called :
a] Pisiculture b] Arboriculture
c] Olericulture d] Floriculture

117. Developing mango fruits showing water soaked greyish patches on the fruit surfaces, finally turning brown necrotic area extending into the mesocarp is the typical symptom of :
a] Internal necrosis disease
b] Black tip of mango
c] Spongy tissue
d] Something else

118. Which part of litchi is edible ?
a] Caruncle
b] Fleshy thalamus
c] Juicy placental hair
d] Aril

119. The combination of sex type flowers set in mango are :
a] Male and female
b] Male and hermaphrodite
c] Female and hermaphrodite
d] Male, female and hermaphrodite

120. Panama wilt is a disease of :
a] Apple b] Banana
c] Pear d] Peach

GK GS
Ranikhet disease of fowl is also known as Doyle's disease.

Set-2

1. **The unconsolidated material on the underlaying rock is called**
   a] regolith b] soil
   c] solum d] earth
2. **If a soil has a bulk density of 1.50 mg/m³ and a particle density of 2.65 mg/m³, then the percentage pore space of that soil will be**
   a] 35.0 b] 38.5
   c] 43.4 d] 48.5
3. **Active soil forming factors would include**
   a] Vegetation and topography
   b] Time, topography and climate
   c] Time and vegetation
   d] Vegetation and climate
4. **In Asia which of following is largest exporter of flowers.**
   a] India b] China
   c] Korea d] Japan
5. **Which one of the following minerals is least weatherable in soil?**
   a] Olivine b] Quartz
   c] Calcite d] Biotite
6. **The given diagram is that of a soil textural triangle. A soil having 15% clay, 65% sand and 20% silt will be classified, on the basis of the given diagram, as**

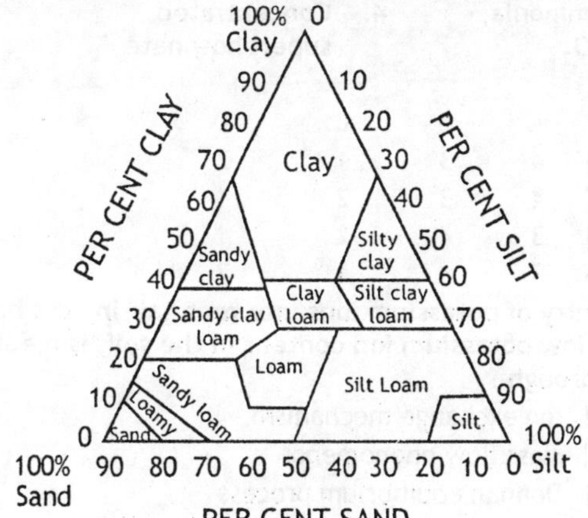

   a] Sandy clay loam
   b] Loam
   c] Loamy sand
   d] Sandy loam
7. **Match List-I (Microbial reactions) with List-II (Soil organisms) and select the correct answer using the codes given below the lists:**

	List-I		List-II
A.	Sulphate reduction	1.	*Thiobacillus*
B.	Sulphite oxidation	2.	*Desulfotomaculum*
C.	Nitrite oxidation	3.	*Rhodospirillum*
D.	Nitrogen fixation	4.	*Nitrobacter*

CODES:

	A	B	C	D
a]	1	2	3	4
b]	2	1	4	3
c]	2	1	3	4
d]	1	2	4	3

8. **American florist trade has reached around ____ billion US $.**
   a] 3 b] 4
   c] 8 d] 15
9. **The given chart shows the classification of irrigation waters:**

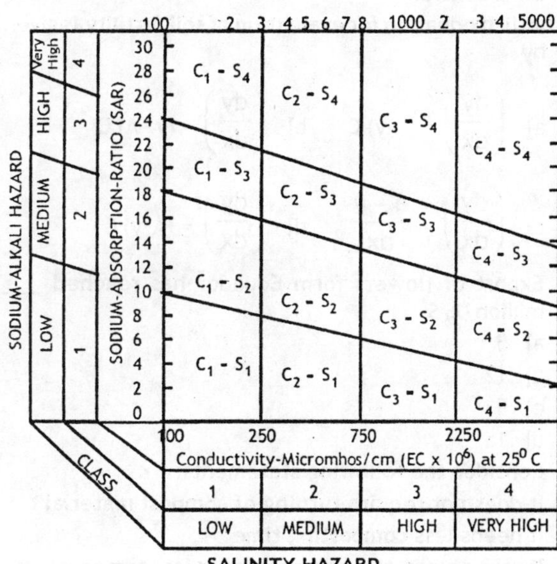

Irrigation water having Na+ = 15 me/litre, (Ca^{2+} + Mg^{2+}) = 8 me/litre and conductivity = EC x 10^6 = 2300 μmho/cm can be classified as
   a] C_4-S_2 b] C_3-S_3
   c] C_4-S_3 d] C_1-S_1

GK GS
Uttar Pradesh is the largest producer of wheat.

10. The following figure shows the relationship between nutrient concentration in the plants and yield of crops:

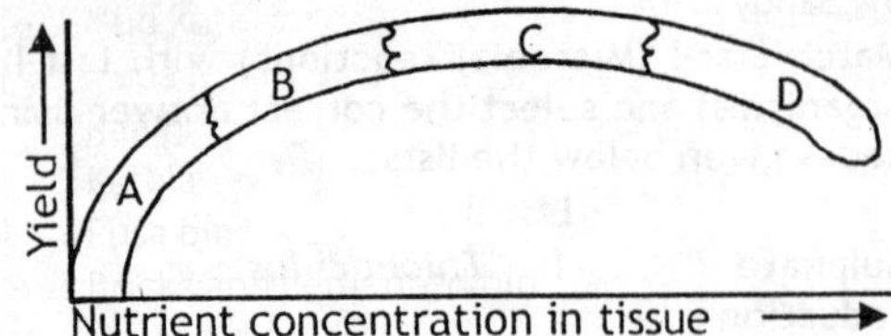

which portion(s) of the curve falls(s) in hidden hunger zone?
a] A and B b] B and C
c] B alone d] A alone

11. The best estimate of the available P in phosphatic fertilizers is obtained from the
a] Water soluble fraction
b] Citrate-insoluble fraction
c] Citrate soluble fraction
d] Water soluble and citrate soluble fractions

12. Given that

$\left(\frac{y}{x}\right)$ = increase of yield per unit of growth factor

A = maximum yield
y = yield without growth factor
C = a constant

Mitscherlich equation for evaluation of soil fertility is given by

a] $\left(\frac{dy}{dx}\right) = (A-y)\,C$ b] $\left(\frac{dy}{dx}\right) = (y - A)\,C$

c] $\left(\frac{dy}{dx}\right) = \frac{a-y}{dx}$ d] $\left(\frac{dy}{dx}\right) = (A+y)\,C$

13. Export of flowers form Ecuador has reached ____ million US $
a] 5
b] 12
c] 9
d] 15

14. Consider the following statements:
1. It does not require turning of compost material
2. It needs less composting time
3. It takes place at a comparatively low temperature
4. It has higher enzymatic activity

Of these statements
a] 1, 2, 3 and 4 are correct
b] 1 and 2 are correct
c] 2, 3 and 4 are correct
d] 1, 3 and 4 are correct

15. Match List-I with List-II and select the correct answer using the codes given below the lists:

List-I	List-II
A. Gypsum	1. Mangnesium fertilizer
B. 17 : 17 : 17	2. Micronutrient fertilizer
C. Serpentine	3. Calcium fertilizer
D. Chelates	4. Compound/complex fertilizer

CODES:

	A	B	C	D
a]	3	2	1	4
b]	3	4	1	2
c]	1	2	3	4
d]	1	4	3	2

16. Sulphate of Potash is preferable to Muriate of Potash (MOP) for tobacco because
a] The chlorine in the MOP affects the burning quality of tobacco
b] The chlorine in the MOP increases the nicotine content of tobacco
c] The application of MOP results in the luxury consumption of potassium
d] The sulphate ion helps to increase the availability of other nutrients

17. Match List-I (Reactants used for the preparation of fertilizers) with List-II (Fertilizers produced) and select the correct answer using the codes given below the lists:

List-I	List-II
A. Rock phosphate, phosphoric acid	1. Urea
B. Rock phosphate, phosphoric acid, nitric acid, ammonia	2. Ammonium sulphate
C. Ammonia, gypsum, CO_2	3. Nitro-phosphate
D. Ammonia, CO_2	4. Concentrated superphosphate

CODES:

	A	B	C	D
a]	4	3	1	2
b]	4	3	2	1
c]	3	4	2	1
d]	3	4	1	2

18. Entry of potassium ions into root hair in soils having a low potassium ion content in the soil, is mediated through
a] Ion exchange mechanism
b] Mass flow phenomenon
c] Donnan equilibrium process
d] Utilisation of metabolic energy

GK GS

Taxes on stamp duty belong to states.

19. Consider the following statements:
According to cohesion theory for the ascent of sap, water moves from the roots through the stems to the leaves in tall trees because of
1. Forces of root pressure
2. The gradient in decreasing water potentials from the soil, through the plant to the leaves
3. Forces of adhesion of water to cell walls, especially in the leaves
4. Forces of cohesion between water molecules

Of these statements
a] 1, 2 and 3 are correct
b] 2, 3 and 4 are correct
c] 1, 3 and 4 are correct
d] 1, 2 and 4 are correct

20. Which one of the following physiological processes requires close coordination of three different organelles such as mitochondria, chloroplast and peroxisome?
a] Photosynthesis b] Photorespiration
c] Respiration d] Protein synthesis

21. Which of the following are characteristic of C_4 plants?
1. There are two CO_2 acceptors, namely PEP and RuBP.
2. RuBP carboxylase is present in the mesophyll cells.
3. The first stable product is a 4-carbon compound.
4. Oxygen does not have any inhibitory effect on the process.

Select the correct answer using the codes given below:

CODES:
a] 1, 2 and 4 b] 1, 3 and 4
c] 1, 2 and 3 d] 2, 3 and 4

22. Which one of the following is a set of high energy products of the light reaction of photosynthesis that are used in dark reaction?
a] ATP and plastocynine
b] ATP and NADP
c] ATP and (NADPH + H^+)
d] Plastoquinone and ferridoxine

23. Which of the following pairs of hormones and bioassay techniques are correctly matched?
1. Cytokinin Radish cotyledon test
2. Gibberellic acid Rice second leaf test
3. Indole-3-acetic acid Coleoptile curvature test

Select the correct answer using the codes given below:

CODES:
a] 2 and 3 b] 1 and 2
c] 1 and 3 d] 1, 2 and 3

24. Maximum consumption of cut roses is during which of following event.
a] Valentine's day b] Mother's day
c] Easter d] X-mas day

25. Match List-I (Source of character) with List-II (Character exploited) and select the correct answer using the codes given below the lists:

List-I		List-II	
A.	Norin-10	1.	Resistance to *Puccinia recondita* causing leaf rust in wheat
B.	Opaque-2	2.	Semidwarfing habit in rice
C.	Dee-gee-woogen	3.	Protein content in maize
D.	Ir genes	4.	Short stature in wheat

CODES:

	A	B	C	D
a]	4	3	1	2
b]	4	3	2	1
c]	3	4	1	2
d]	3	4	2	1

26. Given that the somatic chromosome number of *Triticum aestivum* is 2n = 6x = 42, which one of the following pairs is correctly matched?
a] Monosome2n = 40
b] Trisome 2n = 42
c] Tetrasome 2n = 44
d] Nullisome 2n = 41

27. When breaks occur in two chromosomes simultaneously in a nucleus and the broken chromosomes rejoin in a new manner, it results in
a] Deletion b] Duplication
c] Translocation d] Inversion

28. Consider the following statements:
1. Heterochromatin is concentrated near the nuclear envelope and near the nucleolus.
2. Euchromatin is visible during interphase and is distributed throughout the nucleoplasm.
3. Euchromatin is concentrated near the nucleolus and near the nuclear envelope.

Of these statements
a] 1, 2 and 3 are correct
b] 1 and 2 are correct
c] 2 and 3 are correct
d] 1 and 3 are correct

29. The segregation of individuals in the F_2 or in a later generation of a cross, which shows a more extreme development of a character than either parent is termed as
a] Hybridisation b] Heterosis
c] Linkage d] Transgressive segregates

30. Heritability may be defined as the
a] Interaction product of genotype with environment
b] Sum total of hereditary material present in a species
c] Degree of resemblance between the original and the selected plants
d] Proportion of phenotypes variability which is due to heredity

31. Evaluation of inbreds in maize is carried out in four stages. Open-pollinated variety is involved in the
a] Phenotypic evaluation
b] Top cross evaluation
c] Single cross evaluation
d] Double cross evaluation

32. Multiline varieties are mixture of
a] Several morphologically different plants having identical diseases resistance genes
b] Bulked progenies of different inbred lines
c] Several similar pure lines having different genes for disease resistance
d] Morphologically identical plants with different genetic make-up

33. Indian floriculture industry is growing at ____ per cent per annum.
a] 15-20 b] 10-15
c] 5-10 d] 30-35

34. The mechanims of male sterility in self-pollinated crops can be successfully utilised to obtain
a] Selfed seed
b] Mutated seed without radiation treatment
c] Hybrid seed without emasculation
d] Sterile seed

35. The basic difference between a synthetic and composite variety lines in
a] The difference in synthesis
b] Testing for nicking ability
c] The mode of pollination
d] The number of components

36. Which of the following steps are involved in the utilization of hybrid vigour?
1. Production of uniform homozygous inbred lines.
2. Crossing inbreeds in combination that give uniform and productive single cross F_1 hybrids.
3. Crossing single crosses in combination to give productive double cross hybrids.
4. Mixing single cross F_1 hybrids to give productive double cross hybrids.

Select the correct answer using the codes given below:
CODES:
a] 1, 2 and 4 b] 1, 3, and 4
c] 1, 2 and 3 d] 2, 3 and 4

37. A high yielding semi-dwarf variety of rice was first developed at
a] CRRI, Cuttack b] IARI, New Delhi
c] ICRISAT, Hyderabad d] IRRI, Manila

38. The most popular extra-early maturing variety in pigeon pea is
a] UPAS-120
b] Gulbarga
c] Type 7
d] Type 17

39. 'Varalaxmi' is a hybrid developed in
a] Rice b] Cotton
c] Brinjal d] Brassica

40. Which of the following are the correct sets of crop, variety and breeding method employed to evolve the given variety?
Crop – Variety – Breeding method
1. Wheat – Sharbati Sonara – Mutation breeding
2. Sunflower – APSH-11 – Heterosis breeding
3. Maize – Amber – Population improvement
4. Groundnut – TMV-2 – Pedigree selection

Select the correct answer using the codes given below
CODES:
a] 2, 3 and 4 b] 1, 3 and 4
c] 1, 2 and 4 d] 1, 2 and 3

41. Transfer of resistance gene to high yielding self-pollinated varieties is achieved by
a] hybridization and pedigree selection
b] recurrent backcrossing and selection
c] hybridization and bulk selection
d] hybridization and mass selection

42. To which one of the following species does India's indigenous cotton belong?
a] *Gossypium barbadense*
b] *Gossypium hirsutum*
c] *Gossypium arboreum*
d] *Gossypium herbacium*

43. Total are under flower crops in our during 2000-01 was ____ thousand hectares.
a] 100 b] 35
c] 24 d] 65

44. Which of the following statements can be associated with the cultivation of mango as it is being practised in India?
1. Varieties of fruit trees like mango being grown are inferior and they are often propagated by seedlings.
2. Socio-economic status of most of the mango growers is low and they have small-sized holdings.
3. Marketing of mango is chaotic in most of the mango growing parts of India.
4. Knowledge about right type of mango varieties and their culture practices is not available.

Select the correct answer using the codes given below:
CODES:
a] 1, 2 and 4 b] 2, 3 and 4
c] 1, 3 and 4 d] 1, 2 and 3

45. Which of the following pairs are correctly matched?
1. Date palm Monoecious
2. Papaya Polygamous
3. Walnut Monoecious
4. Pistachio nut Dioecious

Select the correct answer using the codes given below:
CODES:
a] 2, 3 and 1 b] 1, 2 and 3
c] 1, 2 and 4 d] 1, 3 and 4

GK **India exports its 60% of tea production to Great Britain.** GS

46. 'Ratna' variety of mango is a cross between
    a] Alphonso and Banganapalli
    b] Neelum and Langra
    c] Neelum and Alphonso
    d] Dasehari and Neelum

47. Which of the following pairs are correctly matched?

1.	Ikebana	Paved path in a garden
2.	Pinjore garden	Mughal garden
3.	Topiary shrubs	Clipping and shearning
4.	Carnation	Cut flower

Select the correct answer using the codes given below:
CODES:
    a] 2, 3 and 4   b] 1, 2 and 4
    c] 1, 3 and 4   d] 1, 2 and 3

48. Consider the following species:
1. *Polyalthia longifolia*
2. *Acalypha hispida*
3. *Pentas lanceolata*
4. *Ipomoea palmata*

Among these, the ones which can be planted in the shrubbery would include
    a] 1 and 2
    b] 2 and 3
    c] 3 and 4
    d] 1, 2, 3 and 4

49. Which one of the following chemicals is used for the preservation of fruit products?
    a] Potassium sulphate
    b] Potassium manganate
    c] Potassium metabisulphide
    d] Potassium metabisulphite

50. Pollination of the fig flower by the blastophaga wasp takes place through
    a] Fasciation
    b] Parthenogenesis
    c] Caprification
    d] Apomixis

51. Fruit stored in a cold chamber exhibit longer storage life because
    a] Exposure to sunlight is prevented
    b] The concentration of carbon dioxide in the environment is increased
    c] The rate of respiration is decreased
    d] There is an increase in humidity

52. Multistorey cropping system is practised in
    a] Jammu and Kashmir
    b] Karnataka and Kerala
    c] Bihar and Madhya Pradesh
    d] Rajasthan and Haryana

53. Which of the following statements regarding crop rotation are correct?
1. Susceptible vegetable to a particular pest should be followed by a resistant one.
2. Heavy feeder should be followed by low feeder.
3. Deep rooted crops should be followed by the same type of crop.
4. Legume must be followed by a non-legume.

Select the correct answer using the codes given below:
CODES:
    a] 2, 3 and 4
    b] 1, 2 and 3
    c] 1, 2 and 4
    d] 1, 3 and 4

54. Which one of the following vegetables gives maximum calories per 100 gram of edible matter?
    a] Spinach
    b] Water melon
    c] Sweet potato
    d] Tomato

55. Which of the following structures help the fungi to survive under adverse conditions?
1. Zoospores
2. Chalmydospores
3. Sclerotia
4. Conidiospores

Select the correct answer using the codes given below:
CODES:
    a] 3 and 4   b] 2 and 4
    c] 1, 2 and 3   d] 2 and 3

56. Modification of floral parts into leafy structures by pathogenic infection is called
    a] hyperplasia   b] hypertrophy
    c] phyllody   d] witch's broom

57. Match List-I (Diseases) with List-II (Disease incitants) select the correct answer using the codes given below the lists:

List-I		List-II	
A.	Karnal bunt of wheat	1.	*Sclerospora graminicola*
B.	Late blight of potato	2.	*Tilletia indica*
C.	Flax rust	3.	*Phytophthora infestans*
D.	Downy mildew of bajra	4.	*Melamspora lini*

CODES:

	A	B	C	D
a]	3	2	4	1
b]	2	3	4	1
c]	2	3	1	4
d]	3	2	1	4

GK There are five members in Finance Commission. GS

58. Consider the following statements:
"Gundhi bug" causes damage to paddy crop by
1. Sucking juice from the developing grains
2. Sucking juice from the leaves
3. Sucking juice from the culm
4. Affecting the quality and quantity of grain

Of these statements
a] 2 and 4 are correct
b] 1, 3 and 4 are correct
c] 2 alone is correct
d] 1 alone is correct

59. Two of the more important diseases of potato whose spread in India has been successfully restricted through plant quarantine measures include
a] Brown rot and late blight
b] Wart and golden nematode
c] Ring rot and mosaic
d] Bacterial wilt and charcoal rot

60. Severe incidence of red hairy caterpillar occurs under which one of the following sets of conditions?
a] High temperature and high soil moisture
b] High temperature and low soil moisture
c] Low temperature and high soil moisture
d] Low temperature and low soil moisture

61. Match List-I with List-II and select the correct answer using the codes given below the lists:

List-I (*Pathogenic genera of fungi*)	List-II (*Symptoms and effects*)
A. Cystopus	1. Red rust
B. Protomyces	2. Green ear
C. Sclerospora	3. White rust
D. Taphrina	4. Leaf curl
	5. Stem gall

CODES:

	A	B	C	D
a]	1	5	2	4
b]	3	5	4	1
c]	4	1	2	3
d]	3	5	2	4

62. Consider the following stages:
1. Spore germination
2. Embryo infection
3. Spore on flower stigma
4. Infected grains

The correct sequence of these stages in the development of loose smut disease in wheat is
a] 3, 1, 4, 2 b] 1, 3, 4, 2
c] 1, 3, 2, 4 d] 3, 1, 2, 4

63. Which one of the following is the vector for yellow vein mosaic of bhindi (okra)?
a] Brown plant hopper
b] Green plant hopper
c] Aphids
d] White flies

64. Consider the following steps:
1. Sampling
2. Surveillance
3. Control strategy

The correct sequence of these steps in the management of pests and diseases is
a] 1, 2, 3
b] 2, 1, 3
c] 2, 3, 1
d] 1, 3, 2

65. Which one of the following statements correctly defines systemic insecticides?
a] An insecticide which enters into the system of insects
b] An insecticide which is absorbed into the plant system
c] An insecticide which is absorbed and translocated in the system of the plant
d] An insecticide that possesses translaminar activity

66. Which one of the following groups consists of systemic fungicides?
a] Bavistin, Topsin-M, Captan
b] Plantvax, Emisan, Blitox
c] Vitavax, Bavistin, Calixin
d] Ziram, Plantvax, Blitox

67. Bordeaux Mixture was discovered by P.A. Millardet of France during the year, 1882 following his chance observation of a farmer's practice for protection against
a] *Plasmopara viticola* on grapevine
b] *Uncinula necator* on grapevine
c] *Podosphaera leucotricha* on apple
d] *Venturia inequalis* on apple

68. Which one of the following sprayers will cause more loss of spray droplets by drift?
a] Low volume sprayer
b] High volume sprayer
c] Ultra low volume sprayer
d] Duster

69. Which of the following would determine the quantity of spray needed per hectare while using manually operated knapsack sprayers?
1. Swath of spray
2. Walking speed of the person doing the spraying
3. Calibration of the sprayer
4. Type of nozzle

Select the correct answer using the codes given below:
CODES:
a] 1, 2 and 4
b] 1, 2 and 3
c] 1, 3 and 4
d] 2, 3 and 4

GK GS
Lockjaw is bacterial disease in animals.

70. Which of the following safeguards are necessary for a proper operation and maintenance of plant protection equipment?
1. Lubrication should be followed faithfully using the specified lubricant.
2. A multigrade motor oil should not be mixed with petrol as a fuel for a 2-stroke engine.
3. The oil and fuel should be poured separately into the tank.

Select the correct answer using the codes given below:
CODES:
a] 2 and 3 b] 1 and 2
c] 1 and 3 d] 1, 2 and 3

71. Which one of the following remedies is to be suggested in case of failure of a manually operated sprayer to retain pressure?
a] Tightening of loose nuts and clamps and replacement of gaskets
b] Tightening of lid of the tank, replacement of the gasket, if necessary
c] Opening the nozzle and cleaning its various parts especially the orifice
d] Straightening the plunger rod if found bent

72. While adopting a new innovation in agriculture, farmers go through the following four stages in the mental process:
a] Awareness b] Interest
c] Action d] Satisfaction

The correct sequence of these four stage is
a] 1, 2, 4, 3 b] 2, 1, 3, 4
c] 1, 2, 3, 4 d] 2, 1, 4, 3

73. The fruits of research and development findings could reach the poorest of the poor through
a] Research personnel themselves
b] NGO's
c] A strong and devoted extension agency
d] Agricultural universities and colleges

74. In a communication process, information flows through the following four elements:
1. Message
2. Communicator
3. Channel
4. Audience

The correct sequence of flow of information is
a] 2, 3, 1, 4 b] 3, 2, 1, 4
c] 1, 3, 2, 4 d] 1, 2, 3, 4

75. For providing inputs like quality seeds, fertilizers and pesticides, the agency that is present at the Village Panchayat Samiti level is
a] NABARD
b] Nationalised banks
c] Cooperative society
d] Insurance companies

The following thirteen items consist of two statements, one labelled the 'Assertion A' and the other labelled the 'Reason R'. You are to examine these two statement carefully and decide if the Assertion A and Reasons R are individually true and if so, whether the Reason is a correct explanation of the Assertion. Select your answer to these items using the codes given below and mark your answer sheet accordingly.

CODES:
a] Both A and R are true and R is the correct explanation of A
b] Both A and R are true but is not a correct explanation of A
c] A is true but R is false
d] A is false but R is true

76. Assertion (A) : In India, area under wheat is more than what is necessary in a normal year, while area under oil seeds and pulses is still below the required levels.
Reason (R) : Area under pulses and oil seeds have been shifted over to the production of wheat in recent years on account of higher prices and improved productivity potential of high yielding variety wheat seeds.

77. Assertion (A) : Alkali soils need more puddling for the cultivation of rice when compared to normal soils.
Reason (R) : Alkali soils have more exchangeable sodium on the exchange complex than the normal soil.

78. Assertion (A) : Organic manures with wider C/N ratio when added to the soil decompose rapidly in soil and increase the plant available nitrogen.
Reason (R) : Organic acids are produced during the decomposi-tion of organic manures.

79. Assertion (A) : The complex and mixed fertilizers with the same nutrient contents are equally effective in crop production.
Reason (R) : The complex fertilizers are granular, free flowing and easy to apply whereas fertilizer mixtures are physical blends of two or more straight fertilizers mechanically mixed.

80. Assertion (A) : When light falls on guard cells of stomata, their osmotic potentials become more negative and they open.
Reason (R) : Light causes an increase in the transport of K^+ ions from the accessory cells to the guard cells.

81. Assertion (A) : Respiratory quotient for respiration of materials rich in carbohydrates is usually about 1.
Reason (R) : Carbohydrate molecules have the same number of atoms as the oxygen atoms.

82. Assertion (A) : Doubling the chromosomes of F_1 of an interspecific cross leads to fertility.
Reason (R) : By doubling the chromosomes of F_1 of an interspecific hybrid, each and every chromosome gets its homologue to form bivalents.

GK **Portuguese brought plants of tobacco to India.** GS

83. Assertion (A) : Self-incompatibility is a useful tool for the development of hybrid varieties in several crop species.
Reason (R) : Production and maintenance of inbred lines is tedious and costly in the case of self-incompatible species.

84. Assertion (A) : Cultivation of maize is not recommended as inter-crop in a mango orchard.
Reason (R) : Growth of maize remains poor in a mango orchard.

85. Assertion (A) : Vitamin C requirement of human body has to be met by exogenous sources such as fruit and vegetables.
Reason (R) : Vitamin C can neither be synthesised nor stored in the human body.

86. Assertion (A) : Seed treatment with chlorpyriphos is done in groundnut.
Reason (R) : It helps in controlling white grub.

87. Assertion (A) : Integrated pest management is the best way of controlling diseases and pests of crop plants.
Reason (R) : This ensures that all the insects and pathogens are killed effectively.

88. Assertion (A) : Survey and surveillance is the backbone of integrated pest control.
Reason (R) : Integrated control can be achieved by integrating chemical, biological, cultural, regulatory and mechanical methods of pest control.

89. Flower products export form India during 2000-01 was Rs _____ Crores.
a] 123.12 b] 65.3
c] 30.6 d] 78.5

90. Flower production in our country is about _____ MT.
a] 330 b] 400
c] 250 d] 500

91. "Mechanisation in agriculture causes displacement of agricultural labour. "Which one of the following statements does not fully support this viewpoint?
a] There is a greater outflow of agricultural labourers after mechanisation
b] There is a tendency among agricultural increases after mechanisation
c] There is a tendency among agricultural labour to improve their skills
d] The marginal productivity of labour is declining

92. Consider the following statements:
An 'operational holding' is one which includes all lands used wholly or partially for agricultural production. The improvement in the size of operational holdings has largely been due to
1. Extension of irrigation facilities
2. Land reclamation schemes
3. Consolidation of holdings
Of these statements
a] 1, 2 and 3 are correct
b] 1 and 3 are correct
c] 2 and 3 correct
d] 1 and 2 are correct

93. In terms of the area sown, which one of the following crops occupies the most important position in India?
a] Jute b] Cotton
c] Mustard d] Groundnut

94. Among the states growing cotton as rainfed crop, the one which obtains the highest yield in the country is
a] Maharashtra b] Haryana
c] Gujarat d] Punjab

95. North-east monsoon causes rainfall in the south-eastern part of the peninsular India during
a] November-December
b] February-March
c] September-October
d] May-June

96. Which one of the following pairs of states dominates at present in pigeon pea cultivation in India, on account of favourable agro-climatic conditions?
a] Bihar and Haryana
b] Uttar Pradesh and Maharashtra
c] Madhya Pradesh and Rajasthan
d] Kanataka and Andhra Pradesh

97. When stability of prices of a commodity is explained by the Cob-Webb model, which of the following conditions pertaining to the demand and supply functions should hold true?
1. The intercept of the demand function should be greater than the intercept of the supply function.
2. The intercept of the supply function should be greater than the intercept of the demand function.
3. The slope of the demand function should be greater than the slope of the supply function.
4. The slope of the supply function should be greater than the slope of the demand function.
Select the correct answer using the codes given below:
CODES:
a] 2 and 4 b] 2 and 3
c] 1 and 4 d] 1 and 2

98. In our country how many private units are engage in export of flowers.
a] 500 b] 400
c] 100 d] 300

99. Match List-I with List-II and select the correct answer using the codes given below the lists:

List-I		List-II	
A.	Minimum support price	1.	Farmer
B.	Re-finance	2.	Govt. of India
C.	Price taker	3.	NABARD

GK **Neogi was the first Chairman of Finance Commission.** GS

D. Storage 4. Time utility
E. Transportation 5. Space utility

CODES:

	A	B	C	D	E
a]	2	3	1	5	4
b]	2	3	5	1	4
c]	3	2	1	4	5
d]	3	2	5	1	4

100. Which one of the following principles will be applicable when a farmer is faced with problem of purchase of heifers versus milch cows on his dairy farm?

a] Cost principle
b] Opportunity cost principle
c] Comparative advantage principle
d] Time comparison principle

101. Match List -I (Decisions) with List-II (Techniques) and select the correct answer using the codes given below the lists:

List-I		List-II	
A.	Capital investment	1.	Least Cost Combination
B.	Economically optimum level of resource use	2.	Pay-back Period
C.	Economies of farm size	3.	Linear Programming
D.	Optimum enterprise combination under constraining resources	4.	Average Cost Curve

CODES:

	A	B	C	D
a]	2	1	3	4
b]	2	1	4	3
c]	1	2	4	3
d]	1	2	3	4

102. Cropping intensity and gestation period of the crops grown are

a] inversely related
b] directly related
c] exponentially related
d] not interrelated

103. Which one of the following production functions best explains the returns to scale effect ?

a] $Y = \alpha x_1 x_2 \ldots\ldots\ldots\ldots\ldots x_n$
b] $Y = \alpha_0 + \alpha_1 x_1 + \alpha_n x_n$
c] $Y/x_1 = \alpha_0 + \alpha_1 x_1 + \alpha_2 (x_2/x_1) \ldots\ldots \alpha_n (x_n/x_1)$
d] $Y = \alpha_0 + \alpha_1 x + \alpha_2 x^2 + \ldots\ldots\ldots + \alpha_n x^n$

104. Keeping in view the available technology for Kharif and Rabi crops, which one of the following patterns would be more in demand for crop planning?

a] Multiple cropping
b] Monoculture
c] Location specific cropping
d] Mixed cropping

105. At present, the farming system of India has become

a] Export oriented
b] Technology oriented
c] Market oriented
d] Domestic need oriented

106. The anti-quality constituent in young seedlings of forage sorghum is

a] Coumarins b] Terpenes
c] Acetic acid d] Hydrocyanic acid

107. Which one of the following species belongs to pop maize?

a] Everta b] Amylacea
c] Indurata d] Ceratina

108. High yield of maize is obtained in regions where the atmospheric temperature during the growing season is

a] High during the day and low during the night
b] High during the day and high during the night
c] Low during the day and high during the night
d] Low during the day and low during the night

109. Seed potato from the hilly regions of India is preferred because

a] It is cheaper
b] Good varieties are grown in those regions
c] It has better germination percentage
d] It is disease-free

110. Which of the following pairs of common name and scientific name are correctly matched?

1. Proso millet – *Panicum miliaceum*
2. Kodo millet – *Paspalum scorbiculatum*
3. Italian millet – *Panicum antidotale*
4. Finger millet – *Elucine coracana*

Select the correct answer using the codes given below :

CODES:

a] 1, 2 and 3 b] 1, 3 and 4
c] 1, 2 and 4 d] 2, 3 and 4

111. The dwarf wheat varieties should be

a] Sown at the same depth at which the tall wheat varieties are sown
b] Sown at a depth deeper than that for tall wheat varieties
c] Sown at a depth shallower than that for tall wheat varieties
d] Broadcast and incorporated in the top soil

112. Which one of the following chemical is used to treat seed potato to break its dormancy?

a] Ethral
b] Thiourea
c] Indole acetic acid
d] Naphthalene acetic acid

GK **Causative pathogen of AIDS is HTLV-III.** GS

113. The interculture operation in standing broadcast rice crop is commonly known as
a] Thinning b] Spudding
c] Rotation d] Beushening

114. Consider the following statements:
In paddy, fertilizer application should be directed to the
1. oxidised zone
2. reduced zone
3. surface
Of these statements
a] 1 alone is correct
b] 2 alone is correct
c] 2 and 3 are correct
d] 1 and 3 are correct

115. The following operations are practised in the cultivation of tobacco:
1. Priming
2. Topping
3. Desuckering
The correct sequence of these operations in the field is
a] 3, 2, 1 b] 3, 1, 2
c] 2, 3, 1 d] 2, 1, 3

116. Consider the following statements:
A mulching material with high albedo value is preferred because
1. It reflects back more of the incident radiation
2. It reduces considerably the downward transmission of heat
3. It has an evaporative cooling effect
a] 1 and 3 are correct
b] 1 and 2 are correct
c] 2 and 3 are correct
d] 1, 2 and 3 are correct

117. Maximum area in our country is in which flower crop.
a] Jasmine b] Tuberose
c] Marigold d] Crossandra

118. Match List-I (Common names of weeds) with List-II (Their scientific names) and select the correct using the codes given below the lists:

List-I	List-II
A. Nut grass	1. *Echinochola crusgalli*
B. Bermuda grass	2. *Cyperus rotandus*
C. Yellow nut sedge	3. *Cyndon dactylon*
D. Barynard grass	4. *Cyperus esculentus*

CODES:

	A	B	C	D
a]	1	3	4	2
b]	2	3	4	1
c]	2	4	3	1
d]	1	4	3	2

119. If the Land Utilisation Index (LUI) of a crop is 0.50, then the duration of that crop will be
a] 183 days
b] 172 days
c] 200 days
d] 135 days

120. If a farmer practices 'sugarcane-cotton-wheat', the cropping intensity at his farm will be
a] 100% b] 150%
c] 200% d] 300%

GK Punjab has the highest productivity of cotton. GS

Set-3

1. Organic matter content of "histosols" and "hydromorphic" soil is
   a] 30-40% b] 50-57%
   c] 60-70% d] 12-18%
2. Consider the following statements regarding the causes of low crop productivity of sandy soils:
1. Sandy soils have very low water retention capacity.
2. Sandy soils easily fix applied phosphatic fertilizer.
3. Applied plant nutrients easily leach out through drainage.
4. Sandy soils have very low organic matter content.

Of these statements
   a] 1, 2 and 3 are correct
   b] 1, 2 and 4 are correct
   c] 1, 3 and 4 are correct
   d] 2, 3 and 4 are correct

The cation exchange capacity (in units of milli equivalents per 100 gram of soil) of kaolinite clay varies between
   a] 3 and 15 b] 16 and 40
   c] 41 and 80 d] 81 and 100

4. The phosphorus content as P in the plough layer of one hectare of soil containing 0.1% P_2O_5 is
   a] 430 kg. b] 860 kg.
   c] 1290 kg. d] 1500 kg.
5. Match List-I (Principal soils of India) with List-II (Taxonomic soil order) and select the correct answer using the codes given below the lists:

	List-I		List-II
A.	Black soils of Maharashtra	1.	Alfisol
B.	Laterite soils of Kerala	2.	Entisol
C.	Red soils of Chotanagpur	3.	Ultisol
D.	Ganagetic alluvial soils	4.	Vertisol

CODES:

	A	B	C	D
a]	3	4	1	2
b]	3	4	2	1
c]	4	3	2	1
d]	4	3	1	2

6. Match List-I (Soil bacteria) with List-II (Microbiological reactions) and select the correct answer using the codes given below the lists:

	List -I		List-II
A.	Nitrobacter	1.	Symbiotic fixation of atmospheric nitrogen
B.	Azotobacter	2.	Non-symbiotic fixation of atmospheric nitrogen
C.	Nitrosomonas	3.	Conversion of ammonia into nitrite in soil
D.	Rhizobium	4.	Oxidation of nitrite to nitrate

CODES:

	A	B	C	D
a]	4	2	3	1
b]	2	4	1	3
c]	4	2	1	3
d]	2	4	3	1

7. "Whip-tail" symptom in cauliflower is due to the deficiency of
   a] Copper b] Molybdenum
   c] Iron d] Zinc
8. Match List-I with List-II and select the correct answer using the codes given below the list:

	List-I (*Nutient element*)		List-II (*Function in the plant*)
A.	Nitrogen	1.	Reduction of nitrate
B.	Phosphorus	2.	Root growth
C.	Potassium	3.	Acceleration of vegetative growth
D.	Molybdenum	4.	Translocation of photosynthates

CODES:

	A	B	C	D
a]	3	2	4	1
b]	4	2	1	3
c]	4	3	2	1
d]	3	1	4	2

9. 'Crop logging' is a method of
   a] Soil fertility evaluation
   b] Plant analysis for assessing requirement of nutrients for crop production
   c] Assessing crop damage
   d] Testing suitability of fertilizers
10. In a compost pit, the heat of combustion of organic matter raises the temperature when decay is occurring rapidly. During decay, the temperature rises by
   a] 75 to 90^0 C b] 60 to 80^0 C
   c] 50 to 72^0 C d] 40 to 60^0 C

Under Article 293 Government of India gives loans to States.

11. Which of the following pairs of *Rhizobium* spp. and the host genera on which the *Rhizobium* spp. subsists, are correctly matched?

1. *R. leguminosorum* *Pisum*
2. *R. japonicum* *Glycine*
3. *R. meliloti* *Medicago*

Select the correct answer using the codes given below:

CODES:

a] 1, 2 and 3 b] 1 and 2
c] 1 and 3 d] 2 and 3

12. Poor nitrification rate of non-edible mahua (*Madhuca latifolia*) cake is due to
a] The presence of oil
b] Wide C : N ratio
c] Low nitrogen content
d] The presence of certain alkaloids

13. Which one of the following pairs is correctly matched?

	Fertilizers	Nutrient content
a]	Urea	42.0. (% N)
b]	Diammonium phosphate	20.0 (% P)
c]	Ammonium sulphate	21.5 (% N)
d]	Muriate of potash	47.5 (% K)

14. The neutralising values of the following liming materials are such that
a] $CaO > Ca(OH)_2 > CaCO_3 > CaMg(CO_3)_2$
b] $CaO > Ca(OH)_2 > CaMg(CO_3)_2 > CaCO_3$
c] $Ca(OH)_2 > CaO > CaMg(CO_3)_2 > CaCO_3$
d] $CaCO_3 > CaO > Ca(OH)_2 > CaMg(CO_3)_2$

15. The nitrogenous fertilizer of choice for tea is
a] Calcium ammonium nitrate
b] Urea
c] Ammonium sulphate
d] Ammonium nitrate

16. Transport of phosphate from the cytosol (pH about 7.0) into vacuoles (pH nearly 5.5) involves
a] $[H_2PO_4]^-$ alone
b] $H_2PO_4]^{2-}$ alone
c] $[PO_4]^{3-}$ alone
d] $[H_2PO_4]$- and $[HPO_4]^{2-}$

17. The number of essential elements required for the growth of most higher plants is
a] 3 b] 6
c] 8 d] 16

18. Consider the following symptoms:
1. Younger leaves show signs of deficiency first.
2. High amide and nitrate nitrogen.
3. Effect more pronounced on shots when compared to roots and cruciferous plants develop narrow leaf lamina.
4. Blockage of the synthesis of cystine, methionine etc.

Development of the above-mentioned symptoms is attributable to the deficiency of
a] Calcium
b] Magnesium
c] Nitrogen
d] Sulphur

19. Diminished carbohydrate status of the plant placed under high nitrogen supply is due to
a] Lower photosynthetic activity
b] Vigorous protein synthesis
c] Increased transpiration rate
d] Increased vegetative growth

20. Match List-I (Symptoms) with List-II (Causes) and select the correct answer using the codes given below the lists:

	List-I		List-II
A.	Khaira disease of rice	1.	Phosphorus deficiency
B.	Browning of cauliflower	2.	Molybdenum deficiency
C.	Purple colouration of maize leaves	3.	Potassium deficiency
D.	Firing of tobacco leaves	4.	Zinc deficiency

CODES:

	A	B	C	D
a]	2	4	3	1
b]	2	4	1	3
c]	4	2	3	1
d]	4	2	1	3

21. The uptake of oxygen and production of carbon dioxide in light by photosynthesising tissue is called
a] Respiration
b] Photorespiration
c] Ground respiration
d] Salt respiration

22. The link between glycolysis and citric acid cycle is the oxidative decarboxylation reaction to form
a] Pyruvate b] Acetyl CoA
c] Oxaloacetate d] Citrate

23. Indole acetic acid belongs to which one of the following groups of plant growth hormones?
a] Gibberellins b] Auxins
c] Kinins d] Vitamins

24. In plants, growth rate, protein synthesis and potassium uptake are stimulated by
a] Cytokinins b] Enzyme
c] Auxin d] Ethylene

25. The growth hormone which overcomes genetic dwarfism in certain plants and causes elongation of intact plants is
a] Gibberellin b] Auxin
c] Abscisic acid d] Cytokinin

Louping ill is Viral disease of sheep.

GS

26. In maize, 2n chromosome number = 20. The number of chromosomes in
    1. endosperm cells
    2. pollen mother cell
    3. pollen tube nucleus and
    4. root-tip cells

would be respectively

	1	2	3	4
a]	30	20	20	10
b]	20	10	20	20
c]	20	20	10	20
d]	30	20	10	20

27. Consider the following statements:
    The formation of a loop during pachytene indicate

1. Translocation
2. Deficiency
3. Inversion
4. Duplication

Of these statements

a] 1 and 3 are correct b] 2 and 4 are correct
c] 1, 2 and 3 are correct d] 2, 3 and 4 are correct

28. Genotype x Environment interaction would take place when there is/are
    a] One genotype and two environments
    b] Two genotypes and one environment
    c] Three genotypes and one environment
    d] Two or more genotypes and two or more environments
29. The situation where an egg cell is developed into an embryo without fertilization, is described as
    a] Apomixis
    b] Parthenocarpy
    c] Sexual reproduction
    d] Parthenogenesis
30. An anticodon is a sequence of three nitrogenous bases found on
    a] DNA b] messenger RNA
    c] ribosomal RNA d] transfer RNA
31. What is the correct sequence of the following during the process of mitosis?

1. The chromosomes are in their most condensed form having moved to the equatorial plane of the spindle.
2. The centromere interacts with the spindle apparatus.
3. The chromosomes consist of long threads paired along their entire length.
4. Division of the cytoplams.
5. Sister chromatids move apart to opposite poles.
6. The nuclear membrane breaks down and spindle shaped structure of microtubules is organised.

Select the correct answer using the codes given below:

CODES:

a] 3, 6, 2, 1, 5, 4 b] 3, 6, 1, 2, 5, 4
c] 3, 2, 6, 5, 1, 4 d] 3, 2, 6, 1, 5, 4

32. Epistasis should be carefully distinguished from dominance, which referes to non-additivity of alleles at
    a] The same locus of a homologous chromosome
    b] Different loci of a homologous chromosome
    c] The same locus of non-homologous chromosomes
    d] Different loci of non-homologous chromosomes
33. Total area under protected cultivation in the world is around ____ thousand hectares.
    a] 20 b] 50
    c] 60 d] 100
34. Composites are developed by using
    a] Advanced generation seed mixtures obtained from high yielding intervarietal or interracial crosses
    b] Hybrid directly from intervarietal crosses
    c] Recurrent selections for specific combining ability
    d] The hybrid of an inbred with an open-pollinated variety
35. Plants grown from the seed of an open-pollinated variety of maize
    a] Vary in plant height as well as seed characters
    b] Are uniform in plant height and seed characters
    c] Vary in plant height but are uniform in seed characters
    d] Are uniform in plant height but vary in seed characters
36. Superiority of heterozygote leads to
    a] Fixation of recessive allele
    b] Fixation of dominant allele
    c] Maintenance of both the alleles
    d] Depression of inbreeding
37. Consider the following statements:
    Composites are more popular than the hybrid varieties because the seeds produced by a composite variety

1. Can be used again as seed for 3 to 4 years.
2. Have negligible inbreeding effect.
3. Make varietal maintenance easy.

Of these statements

a] 1, 2 and 3 are correct
b] 1 and 2 are correct
c] 2 and 3 are correct
d] 1 and 3 are correct

38. Consider the following international crop improvement institutes:

1. International Centre for Maize and Wheat Improvement, Mexico
2. International Crops Research Institute for the Semi-Arid Tropics, India
3. International Rice Research Institute, Philippines
4. International Centre for Potato, Peru

The correct chronological order in which these institutes were established is:

a] 1, 3, 2, 4 b] 3, 1, 2, 4
c] 3, 1, 4, 2 d] 1, 3, 4, 2

GK

Mangari cotton is sown in Andhra Pradesh.

GS

39. The initial gene pool of a composite population is composed of
a] Inbred lines
b] Isolines
c] Pure lines
d] Single plants derived from crosses and /or germplams lines

40. Global Floriculture industry is growing at a modest rate of ... per cent per annum.
a] 10-15 b] 5-10
c] 30 d] 20-25

41. In cotton, the popularly grown commercial hybrid 'Varalxami' is the product of
a] Intraspecific hybridization in *Gossypium hirsutum*
b] Intraspecific hybridization in *Gossypium barbadense*
c] Interspecific hybridization between *Gossypium hirsutum* and *Gossypium barbadense*
d] Interspecific hybridization between *Gossypium arboreum* and *Gossypium herbaceum*

42. Jawahar, Vikram, Kisan and Sona, the important varieties of maize are
a] Double-cross hybrids
b] Synthetic varieties
c] Composite varieties
d] Single-cross hybrids

43. The initial gene pool of a synthetic population is composed of
a] Plants from an open-pollinated variety
b] Pure lines
c] Inbred lines
d] Isolines

44. Consider the following sets of varieties of wheat :
1. Sonalika and Safed Lerma
2. Lerma Rojo and Chhoti Lerma
3. Heera and Moti
The 3-gene dwarf varieties of wheat would include
a] 1, 2, and 3 b] 1 alone
c] 2 and 3 d] 3 alone

45. In maize, 'Ganga-Safed-2' (GS-2) is a
a] Single-cross hybrid
b] Double-cross hybrid
c] Three-way cross hybrid
d] Double topcross hybrid

46. Litchi does not flower in South Indian plateau in winter months
a] For want of sufficient low temperature in winter months
b] Because of the high temperature prevalent at the time of flower formation
c] Because of high relative humidity present at the time of flower formation
d] Because of the excessive light stimulus present at the time of flower formation

47. In the case of potato, the seed rate per hectare is
a] 1 quintal
b] 5 quintals
c] 10 quintals
d] 15 quintals

48. Grapes are pruned only once in northern India while they are pruned twice in southern India. This difference is due to the fact that grapes
a] Shed leaves in winter in northern India while they remain evergreen in southern India
b] Are early in bearing in northern India while they bear late in southern India
c] Are less prone to diseases in northern India while they heavily in southern India
d] Are less prone to diseases in northern India while they are more prone to diseases in southern India

49. Which one of the following rose cultivars has striped flowers?
a] Abhisarcia b] Bhim
c] Chitchore d] Mrinalini

50. Consider the following species:
1. Bougainvillea
2. Hibiscus
3. Lantana
4. Pointsettia
The species where the foliage beauty is due to modification of leaves as bracts would include
a] 1 and 2 b] 2 and 3
c] 3 and 4 d] 1 and 4

51. Which of the following in No. one foliage plant at global level (Trade value)
a] Aglaonema
b] Dieffenbachia
c] Ficus elastica
d] Crodyline

52. Consider the following statements:
Mulching of banana crop with black polythene reduces the population of weed because it
1. Restricts the supply of oxygen.
2. Results in excessive moisture in the root zone of weeds.
3. Obstructs light and hence the weed starves.
Of these statements
a] 1 and 2 are correct
b] 2 and 3 are correct
c] 3 alone is correct
d] 1, 2 and 3 are correct

53. Which one of the following fruits contains more fat than the rest?
a] Apple
b] Guava
c] Avocado
d] Jack fruit

54. Match List-I (Fruits) with List-II (Carbohydrate content) and select the correct answer using the codes given below the lists:

List-I		List-II	
A.	Jack fruit	1.	11%
B.	Banana	2.	14%
C.	Guava (plains)	3.	36%
D.	Mango (ripe)	4.	19%

	A	B	C	D
a]	4	3	2	1
b]	4	3	1	2
c]	3	4	2	1
d]	3	4	1	2

55. Which one of the following fruits is cured in smoke for ripening?
 a] Mango b] Orange
 c] Jackfruit d] Banana

56. Lemon squash is preserved for a longer period without discolouration by the use of
 a] Malic acid and potassium metabisulphite
 b] Citric acid and potassium metabisulphite
 c] Benzoic acid and citric acid
 d] Potassium metabisulphite

57. The flat sour spoilage of canned vegetable is due to
 a] Development of sour odour without gas production by microorganisms
 b] Gas formation by the microorganisms
 c] Corrosion of the tin plate
 d] A decrease in pH

58. Match List-I (Diseases) with List-I (Causal organisms) and select the correct answer using the codes given below the lists:

List-I		List-II	
A.	Citrus canker	1.	*Pseudomonas solanacearum*
B.	Black rot	2.	*Agrobacterium tumefaciens*
C.	Brown rot	3.	Xanthomonas compestris
D.	Crown gall	4.	*Xanthomonas campestris* pv. *citri*

CODES:

	A	B	C	D
a]	4	3	1	2
b]	3	4	2	1
c]	3	4	1	2
d]	4	3	2	1

59. Which of the following pairs are correctly matched?
1. Brown spot of rice —*Helminthosporium oryzae*
2. Karnal bunt of wheat — *Neovossia indica*
3. Bunt of rice — *Neovossia horrida*
4. Hill bunt of wheat — *Ustilago tritici*

Select the correct answer using the codes given below:

CODES:
 a] 2, 3 and 4 b] 1, 2 and 4
 c] 1, 3 and 4 d] 1, 2 and 3

60. Loose smut of wheat can be effectively controlled by
 a] Crop rotation
 b] Soil treatment with non-systemic fungicides
 c] Seed treatment with systemic fungicides
 d] Spraying the crop with fungicides

61. Which one of the following correctly defines the period between inoculation and appearance of the first disease symptom?
 a] Penetration period b] Latent period
 c] Incubation period d] Infection period

62. Match List-I (Causal agent) with List-II (Diseases caused) and select the correct answer using the codes given below the lists:

List-I		List-II	
A.	Virus	1.	Sesamum phyllody
B.	Mycoplasma	2.	Potato spindle tuber
C.	Fungus	3.	Yellow vein mosaic of okra
D.	Viroid	4.	Downy mildew of bajra

CODES:

	A	B	C	D
a]	1	3	2	4
b]	3	1	4	2
c]	3	1	2	4
d]	1	3	4	2

63. Match List-I (Parasite) with List-II (Nature of the parasite) and select the correct answer using the codes given below the lists:

List-I		List-II	
A.	Striga	1.	Semi-parasite on stem
B.	Cascuta	2.	Semi-parasite on root
C.	Loranthus	3.	Total parasite on root
D.	Orobanche	4.	Total parasite on stem

CODES:

	A	B	C	D
a]	1	2	3	4
b]	2	4	3	1
c]	2	4	1	3
d]	1	2	4	3

64. The two important diseases of sugarcane would include
 a] Little leaf and bacterial wilt
 b] Red rot and grassy shoot
 c] Red rot and bunt
 d] Bakanae diseases and stem canker

65. Which one of the following statements correctly defines economic threshold level?
 a] The pest population level at which control measures should be taken to prevent the pest population from reaching economic injury level
 b] The lowest pest population density that will cause economic damage
 c] The optimum pest population density that will cause economic damage
 d] The pest population level at which application of pest control measures will be uneconomic

GS

Gumboro disease in fowl is caused due to virus.

66. Which one of the following statements correctly defines the "alternate host" of a pathogen/parasite?
   a] A host on which a parasite attacks simultaneously with the main host
   b] A host which grow in alternate season or year
   c] One of the two kinds of hosts on which a parasitiic fungus develops to complete its life cycle
   d] A host on which, the parasite does not attack in normal conditions.
67. Which one of the following is the best method of controlling ear cockle diseases of wheat?
   a] Application of solar heat
   b] Mechanical separation of infected seed by floatation
   c] Application of balanced dose of fertilizer
   d] Treatment of seed with pesticide
68. Maximum growth of floriculture has been noticed in which country in the last decade.
   a] Ecuador b] Zimbabwe
   c] Spain d] Belgium
69. Mycoplasma is sensitive to
   a] Pencillin b] Bavistin
   c] Tetracycline d] Calixin
70. Which of the following come under the category of cultural control of insects?
1. Change in sowing time
2. Destruction of collateral hosts
3. Killing effect of Sun's rays
4. Mixed cropping

CODES:
   a] 1, 2,3 and 4 b] 2 and 4
   c] 1, 2 and 4 d] 1 and 2
71. Which one of the following can be classified as autocidal technique of insect control?
   a] Introduction of bio-agents
   b] Evolving resistant cultivars
   c] Use of insect growth regulators
   d] Release of sterile males
72. Which one of the following sprayers requires only one-third of the quantity of water while keeping the same quantity of active ingredient of the pesticide per hectare?
   a] Foot/pedal sprayer
   b] Mist blower
   c] Rocking sprayer
   d] Knapsack/backpack sprayer
73. Which one of the following sprayers would require the lowest volume of spray solution to cover a hectare of rice crop?
   a] Power sprayer b] Foot sprayer
   c] Knapsack sprayer d] Hand sprayer
74. The quantity of liquid insecticide with 25% active ingredient required for preparing 500 litres of the spray fluid of 0.25% strength is
   a] 1.25 litre b] 2.5 litre
   c] 5.0 litre d] 10 litre
75. The first step in Extension Programme Planning at local level is
   a] Determining the objectives
   b] Analysis of the existing situation
   c] Deciding the extension teaching methods to be used
   d] Determining the targets
76. While working as an extension specialist in a village, one finds that the package of practices are not being followed despite the best communication methods one had employed for the purpose. One finds the following reasons for non-adaptivity:
1. 60% are laggards
2. 20% are innovative but orthodox
3. 20% are old and orthodox

The best remedial measure would be to
   a] adopt 1 and 3 b] adopt 2 alone
   c] avoid 3 d] adopt 2 and make them team leaders of 1
77. Consider the following institutions:
1. Zila parishad
2. Gram panchayat
3. Village school
4. Village co-operative

The basic rural institutions would include
   a] 1, 2 and 3 b] 2 and 4
   c] 2, 3 and 4 d] 1, 3 and 4

The following sixteen (15) items consists of two statements, one labelled the 'Assertion A' and the other labelled the 'Reasons R'. You are to examine these two statements carefully and decide if the Assertion A and the Reason R are individually true and if so, whether the Reason is a correct explanation of the Assertion. Select your answers to these items using the codes given below and mark your answer sheet accordingly.

CODES:
a] Both A and R are true and R is the correct explanation of A
b] Both A and R are true but R is not a correct explanation of A
c] A is true but R is false
d] A is false but R is true

78. Assertion (A) : Divisibility of input and output factors is an essential requirement of linear programming model.
   Reason (R) : Nature of input and output factors has to be continuous in a linear programming model.
79. Assertion (A) : Region II is a rational region of production.
   Reason (R) : Region II of a production function has a point of maximum profit.

Bailadila iron ore is exported to Japan.

80. Assertion (A) : Late-sown wheat crop requires less amount of fertilizer than does the crop sown at the optimal date.
    Reason (R) : Growth duration of late-sown wheat is less than that of the crop sown at the optimal date.
81. Assertion (A) : Potato seeds are treated with one per cent aqueous solution of thiourea for an hour before sowing.
    Reason (R) : Potato seeds have a dormancy period.
82. Assertion (A) : A deep soil profile regulates the free movement of nutrients for a good crop growth.
    Reason (R) : Eluviation is the result of illuviation process favouring accumulation of nutrient elements and colloidal content in B horizon.
83. Assertion (A) : Shale is an igneous rock.
    Reason (R) : Shales were formed by sedimentation of fine clay in sea-bed in remote past.
84. Assertion (A) : Granular structure is a characteristic of grassland soils.
    Reason (R) : Granular soils are rich in organic matter content.
85. Assertion (A) : Low availability of phosphorus is a problem in both acidic and alkaline soils.
    Reasons (R) : Soil pH regulates the availability of different soil nutrients.
86. Assertion (A) : Farmyard manure is beneficial in sandy as well as clayey soils.
    Reason (R) : Farmyard manure contains all the three major nutrients, namely on an average 0.5, 0.25 and 0.5 per cent of N_1 P_2O_5 and K_2O respectively.
87. Assertion (A) : Self-pollination is common in tomato.
    Reason (R) : The anthers are split longitudinally in a turned inward manner so that the pollen falls directly on the stigma.
88. Assertion (A) : Late cauliflower can be grown in rotation with *kharif* crop and lady's finger.
    Reason (R) : Cauliflower can be grown in between the rows of lady's finger crop.
89. Assertion (A) : Both nymphs and adults of an insect species have the same type of mouth parts.
    Reason (R) : Such species lack completely metamorphosis.
90. Assertion (A) : For successful farming, pest must be controlled.
    Reason (R) : All insects are crop pests.
91. Assertion (A) : The pump of a pneumatic or compression sprayer is required to be operated continuously while sprayings.
    Reason (R) : Initial operation of the pump develop certain amount of pressure on the spray fluid within the tank.
92. Assertion (A) : An extension worker may reached a large number of people through mass-media at low cost per unit of coverage.
    Reason (R) : The mass-media enable extension workers to greatly increase their teaching intensity.
93. In cut foliage trade, which of following is leading exporter.
    a] USA b] Italy
    c] Costa Rica d] Netherlands
94. Which of the following statements are correct?
1. Rainfed agriculture contributes about 44% to the total food production in India.
2. In India, the area under wheat lies somewhere between 23 and 24 million hectares.
3. India's productivity in rice is about 1.75 tonnes/ha.
4. According to the National Wastelands Development Board, an area of about 75 million hectares is under wastelands.

Select the correct answer using the codes given below:

CODES:
    a] 2, 3 and 4 b] 1, 3 and 4
    c] 1, 2 and 4 d] 1, 2 and 3
95. Consider the following statements:
    Agro-climatic zoning of cultivable land is based on
1. Geographic location of the region.
2. The water surplus condition
3. The water deficit condition

Of these statements
    a] 1 alone is correct
    b] 1 and 2 are correct
    c] 3 alone is correct
    d] 1 and 3 are correct
96. The given diagrams correspond to

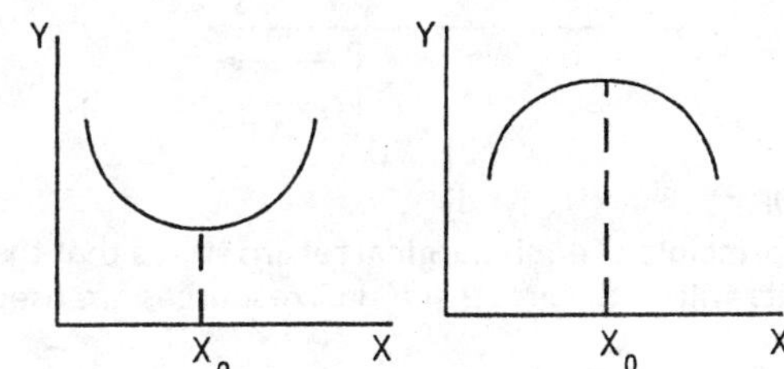

    a] Quadratic equation
    b] Cobb-Douglas equation
    c] Monotonic equation
    d] Non-monotonic equation
97. At a particular point on a Neo-classic Production Function graph, the magnitude of TPP, APP and MPP are 18, 6 and 2 respectively. The E_p (elasticity of production at this point would be
    a] 3 b] 2.57
    c] 1.16 d] 0.85

98. Consider the following situations:
1. Output remaining constant from smaller labour force.
2. An increase in output resulting from an increase in labour, other thing remaining constant.
3. Output remaining constant following an increase in the working hours per day.
Example(s) of increased productivity of labour would include
a] 1 alone b] 1 and 3
c] 2 and 3 d] 1, 2 and 3

99. Which one of the following economic principles helps a farmer to allocate his limited resources to two different enterprises?
a] The law of variable proportions
b] The law of comparative advantages
c] The law of equimarginal returns
d] The principle of fixed and variable costs

100. In the given figure, product Y_1 is denoted on the Y-axis and product Y_2 is denoted on the X-axis. The price of a unit of Y_1 and Y_2 are P_{Y1} and P_{Y2} respectively. Products Y_1 and Y_2 are substitutes in the range of

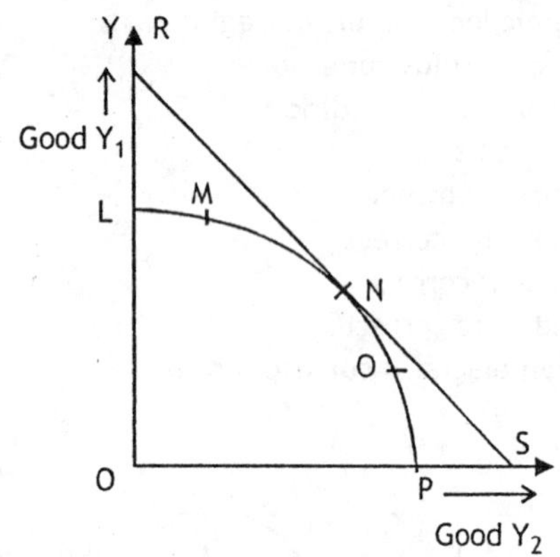

a] LM b] MO
c] OP d] LP

101. The principle of equi-marginal return states that the profits will be the greatest if unit resources are used where
a] Marginal returns are the highest
b] Average returns are the highest
c] Total returns are the highest
d] Net returns are the highest

102. Consider the following statements:
1. Iso-cost lines are always straight lines.
2. Iso-cost lines farther to the right portray higher cost and those nearer to the left portray lower cost.
3. The slope of an Iso-cost line cannot be expressed as the ratio of the price of one input to the price of another input.
The properties on an Iso-cost line would include
a] 1 and 3 b] 2 and 3
c] 1 and 2 d] 1, 2 and 3

103. Match List-I with List-II and select the correct answer using the codes given below the lists:

List-I		List-II	
A.	Factors	1.	Technical complements
B.	Resources	2.	Technical substitutes
C.	Irrational resource combinations	3.	Technical complimentarity
D.	Farm planning tools	4.	Budgets & programming

CODES:

	A	B	C	D
a]	2	1	3	4
b]	1	2	3	4
c]	1	2	4	3
d]	2	1	4	3

104. If the production is homogeneous degree one, then the scale elasticity of output will be
a] Equal to unity
b] Less than unity
c] Greater than unity
d] Equal to zero

105. If technical progress leads to a rise in labour productivity of 5%, then an increase of 5% in the wage rate would imply that
a] Workers will derive all of the benefits
b] There will be cost-push inflation
c] Unit labour costs will remain unchanged
d] Profit per unit will invariably fall

106. The inflection point on total product curve corresponds to the
a] Maximum of marginal product curve
b] Maximum of average product curve
c] Maximum of total product curve
d] Minimum of marginal product curve

107. Which one of the following types of crop cultivations will be most appropriate and effective in forests?
a] Extensive cropping b] Intensive cropping
c] Mixed cropping d] Inter-cropping

108. In wheat, seminal roots
a] Induce extra strength against lodging
b] Are responsible for nourishing the plants during early stages of the crop
c] Take care of the plants during the later stages of the crop
d] Appear during the vegetative phase of the crop

109. The latest botanical name of green gram is
a] *Vigna radiata* (L.) Wilezek
b] *Vigna mungo* (L.) Hepper
c] *Phaseolus mungo* (L.)
d] *Phaseolus vulgaris* (L.)

GK **CRD (Chronic Respiratory Disease) is mycoplasma disease of fowl.** GS

110. Consider the following statements:
Pudding of soil for transplanting paddy is done to
1. Control weeds
2. Reduce rate of infiltration
3. Improve the soil structure
4. Increase aeration
Of these statements
a] 1, 2, 3 and 4 are correct
b] 2 and 4 are correct
c] 1 and 2 are correct
d] 1, 3 and 4 are correct

111. Match List-I (Crop) with List-II (Optimum time of sowing in North India) and select the correct answer using the codes given below the lists:

List-I	List-II
A. Cotton	1. First fortnight of November
B. Barley	2. Second fortnight of October
C. Gram	3. Whole of the month of May
D. Groundnut	4. First fortnight of July

CODES:

	A	B	C	D
a]	3	1	4	2
b]	3	1	2	4
c]	1	3	2	4
d]	1	3	4	2

112. Match List-I (Crop) with List-II (Important diseases) and select the correct answer using the codes given below the list:

List-I	List-II
A. Groundnut	1. Blast
B. Rice	2. Tikka diseases
C. Wheat	3. Late blight
D. Potato	4. Black rust

CODES:

	A	B	C	D
a]	1	2	3	4
b]	1	2	4	3
c]	2	1	4	3
d]	2	1	3	4

113. Irrigation efficiency is very low in irrigated farming in India mainly because
a] Large irrigated area is under canal irrigation system
b] A major part of the soil types is not suitable for irrigation
c] Of the poor quality of irrigation water
d] Of the adverse climatic conditions

114. Which one of the following irrigation methods is best situated for an undulating topography?
a] Flow irrigation b] Check basin method
c] Sprinkler irrigation d] Furrow irrigation

115. Which one of the following is not a component of the moisture potential in soils?
a] Matric potential b] Turgor potential
c] Gravity potential d] Osmotic potential

116. The rice crop tolerates submergence because
a] The plant does not need oxygen
b] The plant has mechanism to transport oxygen through leaves
c] Bacteria keep the rhizosphere oxidised
d] Oxygen penetrating to root zone through air movement is sufficient

117. The crop suitable for cultivation in saline sodic soil would include
a] soyabean and rice
b] soyabean and sugarbeet
c] sugarbeet and rice
d] rice and maize

118. Which one of the following is a pair preemergence herbicides?
a] Paraquat and butachlor
b] 2,4-D and thiobencarb
c] Dalapon and triazine
d] Fluchloralin and pendimethalin

119. Match List-I (Crop) with List-II (Weed specific to the crop) and select the correct answer using the codes given below the lists:

List-I	List-II
A. Rice (lowland)	1. *Phalaris minor*
B. Maize	2. *Echinochloa crusgalli*
C. Wheat	3. *Lantana camara*
D. Uncultivated land	4. *Euphorbia hirta*

CODES:

	A	B	C	D
a]	2	3	1	4
b]	1	2	4	3
c]	1	3	2	4
d]	2	4	1	3

120. Consider the following crops:
1. Wheat 2. Potato
3. Maize 4. Moong
The correct sequence of sowing these crops in a cropping sequence planned under irrigated conditions for north-western plains of India would be
a] 4, 3, 1, 2 b] 4, 3, 2, 1
c] 3, 4, 1, 2 d] 3, 4, 2, 1

GK GS

Bababudan hills of Karnataka are famous for iron ore.

Set-4

1. **The volume (in litres) of one cumec flow of water in one hour is**
   a] 2.6×10^6 b] 3.6×10^6
   c] 4.6×10^6 d] 5.6×10^6

2. **A farmer is to select suitable crops for a drought-prone unirrigated area. Which one of the following groups of crops should he choose for mixed cropping?**
   a] Sesame, maize, oats
   b] Sorghum, wheat
   c] Sesame, pearl millet, niger
   d] Maize, upland rice, berseem

3. **Which one of the following can be suitably grown as a mixed crop with wheat ?**
   a] Cabbage b] Cotton
   c] Jowar d] Mustard

4. **Consider the following statements:**

Relay cropping implies

1. **Sowing of the second crop when the first crop has attained maturity and is nearing harvest.**
2. **A situation where two or more crops occupy a habitat with spatial annidation.**
3. **Overlapping of a part of the life span, of the two crops without any competition.**

Of the above statements :
   a] 1, 2 and 3 are correct
   b] 1 and 2 are correct
   c] 2 and 3 are correct
   d] 1 and 3 are correct

5. ***Imperata cylindrica* is associated with**
   a] Sugarbeetb] Tea
   c] Lentil d] Cereals

6. **Match List-I (crops) with List-II (weedicides) and select the correct answer using the codes given below the lists :**

List-I		List-II	
A.	Wheat	1.	Simazine
B.	Sugarcane	2.	Isoproturon
C.	Chickpea	3.	Thiobencarb
D.	Rice-paddy	4.	Fluchloralin

CODES :

	A	B	C	D
a]	1	2	4	3
b]	2	1	4	3
c]	1	2	3	4
d]	2	1	3	4

7. **According to the modern concepts, the major function of inter-row tillage is**
   a] moisture conservation through soil mulching
   b] improved soil aeration
   c] improved soil granulation
   d] moisture conservation through weed destruction

8. **"Ratooning" is practised as a matter of routine in**
   a] Sugarbeet and sugarcane
   b] Sugarcane and maize
   c] Sugarcane and Napier grass
   d] Sugarcane only

9. **DAPOG method of crop raising is associated with**
   a] Interculture operation
   b] Nursery raising
   c] Seed-bed preparation
   d] Intercultivation

10. **Where adequate irrigation water is available, four to six irrigations are applied to the wheat crop. At which one of the following stages of growth is the first irrigation recommended ?**
    a] Crown-root initiation stage
    b] Tillering stage
    c] Critical stage of growth
    d] Flowering stage

11. **In high P-fixing soils, water soluble phosphatic fertilizers are applied to the crops essentially as**
    a] Broadcast
    b] Placement below the seed
    c] Placement above the seed
    d] Top dressing

12. **Spreading varieties of groundnut are preferred in areas receiving late rains in the kharif season, because**
    a] They are of longer duration
    b] Their water requirements are high
    c] Their seed is dormant
    d] Their seed is bold

13. **Which one of the following parts of the tobacco plant synthesises nicotine ?**
    a] Root b] Stem
    c] Branch d] Leaf

14. **In the wet season, the initiation of panicle primordia in rice variety of 120 days' duration takes place in:**
    a] 60 to 65 days after sowing
    b] 65 to 70 days after sowing
    c] 70 to 75 days after sowing
    d] 75 to 80 days after sowing

GK

The state reorganization Act 1956 divides the country into 5 zones.

GS

15. Match List-I (fertilizer) with List-II (equivalent acidity) and select the correct answer using the codes given below the lists :

List-I		List-II
A.	Urea	1. 148
B.	Ammonium sulphate	2. 128
C.	Ammonium chloride	3. 110
D.	Anhydrous ammonia	4. 84
		5. 64

CODES

	A	B	C	D
a]	5	2	3	4
b]	4	3	2	1
c]	2	5	4	3
d]	4	1	2	3

16. Consider the following States/UT
1. Pondicherry
2. Punjab
3. Andhra Pradesh
4. Haryana

As per the 1991-92 data, the correct sequence of the decreasing order in respect of the consumption of N + P_2O_5 + K_2O fertilizer nutrients (in kg/ha/year) is
a] 1, 2, 4, 3 b] 2, 1, 3, 4
c] 1, 2, 3, 4 d] 2, 1, 4, 3

17. Which one of the following oil-cakes, in addition to its manurial value, acts also as a nitrification inhibitor?
a] Mustard b] Groundnut
c] Neem d] Sesamum (*til*)

18. Which one of the following organic manures has the narrowest C/N ratio ?
a] Biogas slurry b] Farm-yard manure
c] Compost d] Groundnut cake

19. Biofertilizer is a :
a] Mixture of organic manure and micro organisms
b] Mixture of inorganic fertilizer and micro-organisms
c] Culture having the desired strain of micro-organisms
d] Decomposed compost enriched with mineral solubilising bacteria

20. Maximum share of foliage plants in the world market is again form which of the following country.
a] Germany b] China
c] Netherlands d] Denmark

21. Match List-I with List-II and select the correct answer using the codes given below the lists :

List-I (Name of fertilizer)		List-II (Type of fertilizer)
A.	Ammonium sulphate	1. Biofertilizer
B.	Diammonium phosphate	2. Straight fertilizer
C.	Algae manure	3. Organic
D.	F.Y.M.	4. Complex fertilizer

CODES

	A	B	C	D
a]	2	4	3	1
b]	3	2	4	1
c]	2	4	1	3
d]	4	2	3	1

22. Bray suggested certain modification in the Mitscherlich equation. The modified equation is written as
a] (dy/dx) = (A - y) C
b] A = B (1 - y) / y
c] log (A - y) = log A - Cx
d] log (A - y) = log A - C_1B - Cx

23. Match List-I (micronutrients) with List-II (typical deficiency symptoms) and select the correct answer using the codes given below the lists :

List-I		List-II
A.	Manganese	1. Speckled yellow of sugarbeet
B.	Molybdenum	2. White bud of maize
C.	Zinc	3. Whiptail in cauliflower
D.	Copper	4. Reclamation disease of cereals

CODES :

	A	B	C	D
a]	1	2	3	4
b]	1	3	2	4
c]	2	4	3	1
d]	2	3	4	1

24. Which one of the following minerals is the main source of boron in soils ?
a] Pyrolsite b] Tourmaline
c] Haematite d] Hemimorphite

25. Criteria of essentiality of nutrients for plants were given by
a] Arnon b] Stout
c] Russel d] Liebig

26. The mineral chlorite has an extra layer of
a] Montmorillonite b] Illite
c] Gibbsite d] Brucite

27. The concentration of H^+ and OH^- and the pH of a 0.1 N HCl solution is

	H^+	OH^-	pH
a]	10^{-13} g ion/l	10^{-1} g ion/l	1
b]	10^{-1} g ion/l	10^{-13} g ion/l	13
c]	10^{-1} g ion/l	10^{-13} g ion/l	1
d]	10^{-13} g ion/l	10^{-1} g ion /l	13

GK GS

First regular Indian Census was conducted in 1881.

28. Apatite is a group of minerals which mainly contains
 a] Sulphur b] phosphorus
 c] Nitrogen d] Titanum

29. Chlorite belongs to the
 a] 2 : 2 type b] 1 : 1 type
 c] 2 : 1 type d] Amorphous type

30. Which of the following pairs are correctly matched?

1. Humus - Ligno-protein complex
2. Humin - Insoluble in dilute alkali
3. Fulvic acid - High molecular weight
4. Humic acid - Insoluble in dilute acid

Select the correct answer using the codes given below :
CODES :
 a] 2, 3 and 4 b] 1, 2 and 3
 c] 1, 2 and 4 d] 1, 3 and 4

31. Consider the following statements:
Soil is a medium for plant growth because
1. It provides mechanical support and nutrients to the plants.
2. It contains beneficial microorganisms.
3. Plants cannot be grown without soil

Of these statements :
 a] 1 alone is correct
 b] 2 and 3 are correct
 c] 1 and 3 are correct
 d] 1, 2 and 3 are correct

32. The horizons which make up the profile of a forest soil would include :
 a] A, B and C b] A, C and O
 c] B and O d] A, B, C and O

33. The soil layer of maximum leaching indicated by the symbol
 a] A_1 b] A_2
 c] B_1 d] B_2

34. When a group of phenotypically similar appearing plants is selected and harvested and their seeds are bulked, the process is known as :
 a] Pedigree breeding
 b] Mass selection
 c] Bulk method of breeding
 d] Pure line selection

35. The effect of the genotype of the pollen grain on the phenotype of the seed is termed as :
 a] Apospory b] Pollenia
 c] Endopolyploidy d] Xenia

36. Clistogamy encourages :
 a] Self-pollination b] Cross-pollination
 c] Apomixis d] Inbreeding depression

37. Exchange of genetic material takes place between non-sister chromatids of homologous chromosomes during the :
 a] Four strand stage in prophase I
 b] Two strand stage in prophase I
 c] Metaphase I stage in meiosis
 d] Two strand stage in prophase II

38. Heterosis over the better parent is called :
 a] Relative heterosis
 b] Standard heterosis
 c] Pseudoheterosis
 d] Heterobeltiosis

39. Which of the following are the base pairs of a DNA molecule ?

1. Adenine - Thymine
2. Cystine - Guanine
3. Adenine - Uracil
4. Thymine - Guanine

Select the correct answer using the codes given below :
CODES :
 a] 1 and 4 b] 3 and 4
 c] 1 and 2 d] 2 and 3

40. DNA exists as a double helix as a result of hydrogen bonding between :
 a] Sugar molecules b] Phosphate groups
 c] Nucleohistones d] Nucleosides

41. When a plant heterozygous for two given characters Tt Rr, both of them are completely dominant, is self-pollinated, the resulting progeny will consists of :
 a] Six phenotypes
 b] Four phenotypes
 c] Three phenotypes
 d] Two phenotypes

42. Sterility is most pronounced in the F_1 generation following :
 a] Intervarietal hybridization
 b] Intravarietal hybridization
 c] Intraspecific hybridization
 d] Interspecific hybridization

43. Which one of the following categories of seeds is the farmer expected to buy afresh everytime?
 a] Variety b] Hybrid
 c] Synthetic d] Composite

44. Hybrid seeds can be produced through open pollination with the use of :
 a] Inbreds b] Clones
 c] Pure lines d] Male sterile lines

45. Sterility/incompatibility of distant crosses can be overcome to produce hybrids, by :
 a] Pollen culture b] Anther culture
 c] Explant culture d] Embryo culture

46. A three way cross in maize is defined as a cross between a single cross and :
 a] An open pollinated variety
 b] Synthetic
 c] An inbred line
 d] Another single cross

GK **Rajasthan is the largest producer of gypsum.** GS
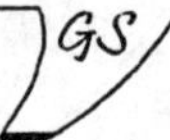

47. The production of double cross hybrid in maize was first suggested by :
a] Shull b] East
c] Jones d] Hull

48. Which one of the following improved crop varieties has been evolved through induced mutagenesis ?
a] IR-36 variety of rice
b] Aruna variety of castor
c] UP-262 variety of wheat
d] Varuna variety of mustard

49. The sugarcane (*Saccharum officinarum*, 2n = 80) variety are cytologically
a] Autopolyploids
b] Octoploids
c] Segmental polyploids
d] Complex polyploids

50. Biggest market of flower products of about 2 billion US $ is in ...
a] USA b] China
c] Germany d] Japan

51. The first sunflower hybrid developed in India is :
a] MSFH-8 b] KBSH-1
c] APSH-11 d] BSH-1

52. The 'Sonalika' variety of wheat has dwarfing genes derived from :
a] Rht 1-10 b] Stu-1
c] Norin-1 d] Norin-10

53. Match List-I (crops) with List-II (varieties) and select the correct answer using the codes given below the lists :

List-I	List-II
A. Cotton	1. Ratna
B. Potato	2. Sujata
C. Rice	3. Girija
D. Wheat	4. Kufri Sinduri

CODES :

	A	B	C	D
a]	4	2	1	3
b]	2	4	1	3
c]	4	2	3	1
d]	2	4	3	1

54. Essential elements are those
a] Which could be replaced by other elements with similar properties
b] Which are required in large quantities
c] Whose deficiency would prevent the completion of plant's life cycle
d] Which would promote plant growth

55. Molybdenum is required by plants because it is a cofactor for
a] Cytochrome oxidase b] Nitrate reductase
c] Nitrite reductase d] Phosphatase

56. Which of the following pairs are correct matched ?
1. Copper 'Die back' disease of citrus
2. Manganese 'Water core' in turnip
3. Boron 'Heart rot' of sugarbeet

CODES :
a] 1, 2 and 3 b] 1 and 3
c] 1 and 2 d] 2 and 3

57. *Khaira* disease of rice can be controlled by spraying
a] Copper sulphate b] Manganese sulphate
c] Borax d] Zinc sulphate

58. Which of the following are features of CO_2 metabolism of succulent plants with CAM?
a] Nocturnal stomatal opening
b] Night-time transpiration
c] Night-time CO_2 uptake
d] Decrease in acidity during the night

Select the correct answer using the codes given below :
CODES:
a] 2, 3 and 4 b] 1, 2 and 4
c] 1, 3 and 4 d] 1, 2 and 3

59. There is a definite relationship between the extent of accumulation of a particular metal ion and the extent of stomatal opening. The metal ion in question is :
a] Calcium b] Magnesium
c] Potassium d] Sodium

60. The photosynthetic pigment which occurs in red and blue green algae is :
a] Chlorophyll b] Carotenoid
c] Cytochrome d] Phycobilin

61. Evidence for universal nature of flowering hormone has emerged from experiments involving
a] Defoliation
b] Reciprocal grafting
c] Night interruption
d] Application of hormones

62. Substituted adenine compounds which promote cell division in plant tissues would include :
a] Cytokinins b] Gibberellins
c] Morphactins d] Triacontanols

63. When meristems of plants become dormant, there is:
a] Inhibition of auxins
b] Inhibition of cytokinins
c] Inhibition of gibberellins
d] Accumulation of abscissic acid

64. Match List-I with List-II and select the correct answer, using the codes given below the lists :

List-I (Purpose)	List-II (Most appropriate temperature)
A. For pasteurization of less acidic vegetable products	1. 60^0C
B. For pasteurization of fruit juices	2. $87^0 - 88^0$ C
C. For killing yeasts, causing non-poisonous type of decay in fruit & vegetable products	3. $116^0 - 121^0$ C
D. For blanching okra before canning	4. 10^0C

GK **52nd amendment is related to political defection.** GS

CODES :

	A	B	C	D
a]	3	5	4	1
b]	5	1	4	2
c]	5	2	1	4
d]	3	5	1	2

65. Survey is. caused by the deficiency of
a] Carotene b] Riboflavin
c] Ascorbic acid d] Alpha tocopherol

66. Which one of the following pairs is a rich source of iron ?
a] Sapota (chiku) and ash-gourd
b] Papaya and cabbage
c] Amaranthus and banana
d] Brinjal and citrus

67. Which one of the following crops can be grown successfully as an intercrop in mango orchard which has 10-year old trees ?
a] Wheat b] Onion
c] Maize d] Turmeric

68. Which one of the following crops is best suited as inter-crop in banana ?
a] Papaya b] Persimmon
c] Ginger d] Plum

69. 'Baradari' is one of the main features of
a] Italian garden b] English garden
c] Mughal garden d] Japanese garden

70. Match List-I (flowering trees) with List-II (colour of the flower) and select the correct answer using the codes given below the lists :

List-I	List-II
A. *Amherstina nobilis*	1. Bright flaming scarlet orange
B. *Butea monosperma*	2. Vermilion-colour-tipped yellow
C. *Callistemon lanceolatus*	3. Scarlet
D. *Cassia siamia*	4. Yellow

	A	B	C	D
a]	2	1	3	4
b]	2	3	1	4
c]	4	3	2	1
d]	4	1	3	2

71. Which one of the following species of climbers is valued for the beauty of its foliage?
a] *Antigonan leptopus* b] *Clerodendron splendens*
c] *Monstera deliciosa* d] *Tecoma grandiflora*

72. The main insect reported to be responsible for affecting pollination in mango is
a] Honey bee (*Apis indica*)
b] Common house fly (*Musca domestica*)
c] A dipterous fly (*Melipona* spp.)
d] A hover fly (*Syrphidae* spp.)

73. Match List-I (fruits) with List-I (commercial methods of propagation) and select the correct answer using the codes given below the lists:

List-I	List-II
A. Mango	1. Seed
B. Aonla	2. Air layering
C. Papaya	3. Budding
D. Jack fruit	4. Veneer grafting

	A	B	C	D
a]	4	3	1	2
b]	4	3	2	1
c]	1	4	3	2
d]	1	4	2	3

74. Brinjal is considered to have had its origin in
a] China b] Africa
c] India d] Japan

75. Which of foilowing is No. one flower in international cut flower market.
a] Rose b] Tulip
c] Chrysanthemum d] Orchid

76. Consider the following fruits:
1. Mandarin 2. Peach
3. Mango 4. Banana

The correct sequence in relation to the cold tolerance (low to high) of these fruits is
a] 4, 3, 1, 2 b] 3, 4, 2, 1
c] 3, 4, 1, 2 d] 4, 3, 2, 1

77. Coffee rust is caused by
a] *Uromyces hobsoni* b] *Hemileia vastatrix*
c] *Cronartium ribicola* d] *Ravenelia emblicae*

78. The plant disease primarily responsible for the "Great Bengal Famine" of 1943 was
a] Blast disease of rice
b] Bacterial blight of rice
c] Udbatta disease of rice
d] Brown leaf spot of rice

79. Pectinolytic enzymes of plant pathogens pay a major role during pathogenesis in diseases like
a] Rusts
b] Downy mildew
c] Wilts, damping-off and root/footrots
d] Powdery mildew

80. Which one of the following insects is a gregarious pest?
a] Sorghum shoot fly b] Spotted bollworm
c] Migratory locust d] Rice caseworm

81. Match List-I with List-II and select the correct answer using the codes given below the lists:

List -I (Crops)	List-II (Smut diseases)
A. Barley	1. Grain smut
B. Paddy	2. Whip smut
C. Sorghum	3. False smut
D. Sugarcane	4. Flag smut
	5. Covered smut
	6. Stinking smut

CODES:

	A	B	C	D
a]	5	3	2	6
b]	5	3	1	2
c]	1	4	2	6
d]	1	5	4	6

82. Match List-I with List-II and select the correct answer using the codes given below the lists:

List-I (Disease)	List-II (Causal organism)
A. Wart disease of potato	1. *Phytophthora parasiticia*
B. White rust of crucifers	2. *Albugo candida*
C. Blister blight of tea	3. *Uncinula necator*
D. Panama disease of banana	4. *Synchytrium endobioticum*
	5. *Edobasidium vexans*
	6. *Fusarium oxysporum* f.sp. *cubense*

CODES:

	A	B	C	D
a]	2	4	3	1
b]	4	2	5	6
c]	1	2	3	6
d]	2	4	5	3

83. Which one of the following is the correct mode of infection in Ergot of bajra (pearl millet)?
   a] Shoot infection
   b] Local infection of blossom
   c] Seedling infection
   d] Systemic infection

84. Which one of the following truly diagnoses the infestation by the sorghum shootfly?
   a] Formation of 'dead heart' in the early seedling stage of sorghum which can be easily pulled out
   b] Formation of 'dead heart' and a bunchy top during the later seedling stage of sorghum
   c] Formation of 'dead heart' which cannot be easily pulled out during the later seedling stage of sorghum
   d] Formation of silvery shoot during the early seedling stage of sorghum

85. Tomato leaf curl virus is transmitted by
   a] Aphid b] Whitefly
   c] Grasshopper d] Butterfly

86. Consider the following statements:
   1. Crop rotation 2. Mixed cropping
   3. Resistant variety 4. Soil amendment

Of these statements
   a] 1 and 4 are correct
   b] 1 and 2 are correct
   c] 2, 3 and 4 are correct
   d] 1, 2, 3 and 4 correct

87. Which one of the following pests is controlled through the application of a grease band around the tree trunk?
   a] Mango mealy bug
   b] Trunk borer of mango
   c] Codling moth on apple
   d] Woolly aphis on apple

88. Identify the correct sequence in which fungitoxic activity of the following were reported:
1. Dithiocarbametes
2. Bordeaux mixture
3. Captan
4. Oxathin derivatives

CODES:
   a] 1, 2, 3, 4 b] 2, 1, 4, 3
   c] 2, 1, 3, 4 d] 1, 2, 4, 3

89. The quantity of endosulfan 35% EC required for treating an area which needs 1000 litres of spray fluid at 0.05% (a.i.) strength is
   a] 35 x 0.05/1000 litres
   b] 2.5 x 1000/1000 litres
   c] 1000 x 0.05 / 35 litres
   d] 35 x 1000 /0.05 litres

90. Which one of the following tissues is most suitable for tissue culture to produce virus-free plants of potato?
   a] Leaf tissue b] Pollen
   c] Embryo tissue d] Meristem tip tissue

91. The size (in micrometers in diameter) of droplets produced by high volume sprayers should most appropriately be
   a] 200 to 500 b] 100 to 150
   c] 81 to 100 d] 30 to 4

92. Consider the following statements:

The rate of application of spray fluid per unit area is regulated by the
1. Operating pressure
2. Size of nozzle apperture
3. Swath width
4. Speed of the operator

Of these statements
   a] 1, 2 and 3 are correct
   b] 3 and 4 are correct
   c] 1, 2 and 4 are correct
   d] 1, 2,3 and 4 are correct

93. Spraying of one hectare of cotton crop is done using 500 litres of spray fluid prepared from 2 kg. of carbaryl 50% water dispersible powder. The concentration of active ingredient of carbaryl in the spray fluid is
   a] 1% b] 0.2%
   c] 0.1% d] 0.5%

94. Consider the following categories of adopters of an improved practice:
1. Early adopters 2. Early majority
3. Late majority 4. Laggards
5. Innovators

The correct sequence of the adopters' category is
   a] 1, 2, 3, 4, 5 b] 5, 1, 2, 3, 4
   c] 2, 1, 3, 5, 4 d] 5, 2, 1, 3, 4

95. The following are the four important reports on extension development:
1. The Royal Commission's Report
2. Agricultural Administration Team Report
3. Grow More Food Enquire Committee Report
4. Nalagarh Committee Report

The correct chronological order of these report is
 a] 1, 3, 2, 4 b] 3, 1, 2, 4
 c] 3, 1, 4, 2 d] 1, 3, 4, 2

96. The basic unit of development under the Integrated Rural Development Programme is a
 a] District
 b] Community development block
 c] Village
 d] Family

97. The sources of information are classified into the following three groups on the basis of their origin within and outside the social system:
1. Personal localite
2. Personal cosmopolite
3. Cosmopolite
4. The correct sequence of these groups in terms of their use is
 a] 1, 2, 3 b] 1, 3, 2
 c] 2, 3, 1 d] 2, 1, 3

The following eight items consist of two statements, one labelled the 'Assertion A' and the other labelled the 'Reason R'. You are to examine these two statements carefully and decide if the Assertion A and Reason R are individually true and if so, whether the Reason is a correct explanation of the Assertion. Select your answers to these items using the codes given below and mark your answer sheet accordingly.

CODES:
 a] Both A and R are true and R is the correct explanation of A
 b] Both A and R are true but R is not a correct explanation of A
 c] A is true but R is false
 d] A is false but R is true

98. Assertion (A) : Soil moisture moves from wet to dry zone.
 Reason (R) : Soil moisture moves in response to a potential gradient.

99. Assertion (A) : In sandy soil, the nutrient availability is low.
 Reason (R) : The water holding capacity of a sandy soil is poor.

100. Assertion : Berseem is the best intercrop for citrus orchards.
 Reason (R) Berseem enriches the orchard soil.

101. Assertion (A) : The cabbage seed production is done in the plains of North India.
 Reason (R) : The soils of the plains of North India are fertile.

102. Assertion (A) : Low volume (LV) sprays are more economic than high volume (HV) sprays.
 Reason (R) : Less quantity of pesticide is required per unit area for LV sprays when compared to the HV sprays.

103. Assertion (A) : Integrated pest management is a part of sustained agriculture.
 Reason (R) : Integrated pest management protects the crop against harmful pests with minimum pollution hazard to the environment and leads to increased food production.

104. Assertion (A) : Epiphytotic of late blight of potato in 1845 forced man to realise the importance of plant diseases.
 Reason (R) : In 1845 there was famine in Ireland and many people died of hunger and disease.

105. Assertion (A) : The basic thrust of the lab to land programme is increased productivity, additional income availability and greater employment opportunity.
 Reason (R) : The lab to land programme is has now been integrated with Krishi Vigyan Kendras (farm science centres) in the country.

106. A high percentage of cropped area in India is covered by rice because
 a] There is adequate rainfall
 b] Rice can be grown in different seasons
 c] Rice is adaptable to most types of land and agroclimate
 d] A large part of the population of India is rice-eating

107. Consider the following statements:

Agro-climatic zoning of cultivable land is based on
1. Water deficit condition
2. Water surplus condition
3. Geographical location of the region

Of the above statements
 a] 1, 2 and 3 are correct
 b] 1 and 3 are correct
 c] 2 and 3 are correct
 d] 1 and 2 are correct

108. Which one of the following approaches has been adopted by the Planning Commission in delinearing India into agro-climatic zones?
 a] Project approach
 b] Sectoral approach
 c] Crop production approach
 d] Holistic approach

109. Consider the following four steps involved in the study of agricultural production economics:
1. Determine and define the conditions which provide for optimum use of resources.
2. Delineate means and methods for changing the

existing use of resources to the optimum level.
3. Determine the extent to which the existing use of resources deviates from the optimum.
4. Analyze the factors or forces which are responsible for the existing production patterns and resource use.

The proper sequence of the steps in such study should be
a] 1, 2, 3, 4
b] 1, 3, 4, 2
c] 4, 1, 3, 2
d] 2, 3, 1, 4

110. Though all categories of farmers are interested in higher returns from their enterprises, what they-really try to maximise is
a] Net profit
b] Profit as percentage of investment
c] Farm business income
d] Gross profit

111. Given,
$Y = 25\ X_1^{0.35}\ X_2^{0.49}$, where
Y = output,
X_1 = input factor 1,
X_2 = input factor 2,
the nature of returns to scale will be
a] Increasing
b] Decreasing
c] Constant
d] Fluctuating

112. The given diagram shows break-even point of bullock power and tractor power. In this context, which of the following decisions would be termed rational? (Y-axis in the diagram refers to expenditure on bullock or tractor power in thousand of rupees and the X-axis referes to the size of the holding in hectares).

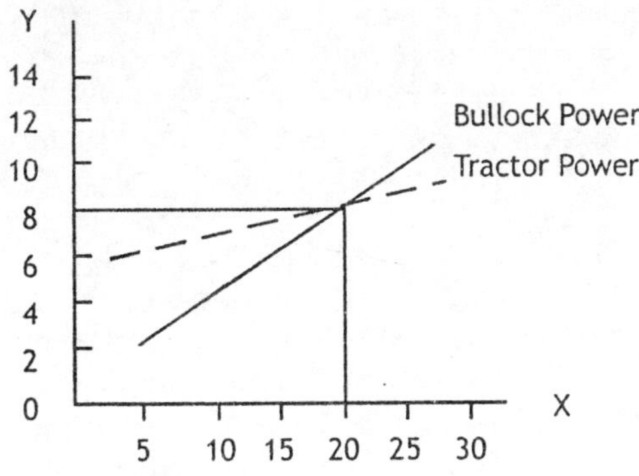

1. Bullock power should be used in place of tractor power.
2. If the size of the holding is 20 hectares, either bullock or tractor power could be used.
3. Tractor power is more economical than bullock power for farm sizes over 20 hectares.

Select the correct answer using the codes given below :
CODES:

a] 1, 2 and 3	b] 1 and 2
c] 2 and 3	d] 1 and 3

113. Consider the following data relating to a farm:

Total assets	:	Rs. 3000/-
Working assets	:	Rs. 800/-
Current assets	:	Rs. 1500/-
Current liabilities	:	Rs. 200/-
Total liabilities	:	Rs. 900/-

The net capital ratio in this case would be
a] 3.33
b] 4.0
c] 7.5
d] 11.5

114. In the given diagram which one of the following segments illustrates feasible zone in a linear programming problem?

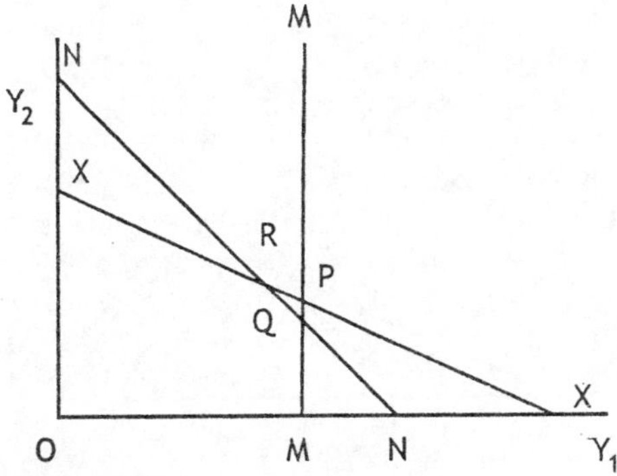

a] OXRQM
b] RPQ
c] QMN
d] XNR

115. The term 'operational holding' referes to the
a] Land area owned by a farmer
b] Cropped area cultivated by a farmer.
c] Net land area cultivated by a farmer
d] Land area leased in by a farmer

116. Which one of the following formulae gives the correct measure of 'Rate of Capital Turnover?
a] (Gross Income/Total Farm Assets) x 100
b] (Net Income /Total Farm Assets) x 100
c] (Farm Business Income / Total Farm Assets)x 100
d] (Capital Investment/Total Farm Assets) x 100

GK **State with lowest density of population : Arunachal Pradesh.** GS

117. Sufficient condition of farm efficiency is generally referred to as choice indicator because it

a] Is a valid criterion of the assumption of certainty

b] Is consistent with the choices made under perfect competition

c] Is consistent with the welfare increase to the farming population

d] Helps to make allocation compatible with the given objectives

118. Farm budgets are used to

a] Compare economic profitability of different production activities within or among farms

b] Indicate whether a proposed change will be profitable under a given set of factors

c] Explore conditions under which certain farm practices become profitable or unprofitable

d] To all the above

119. In mixed farming, the contribution of livestock to gross farm income is

a] At least 10 per cent

b] At least 30 per cent

c] At least 49 per cent

d] More than 50 per cent

120. The object of a subsistence farm is to produce

a] Essential food crops for family and market

b] Essential food crops for family only

c] Food and fibre crops for family and market

d] Vegetable crops for Family only.

Most of the north Indian Scheduled Caste people belong to Mediterranean race.

GS

Set-5

1. The normal eye colour in *Drosophila* is red. But mutants occur having white eye and also different shades ranging between white and red which are all recessive to red, while white colour is recessive to all others. This phenomenon is due to :
   a] Pseudodominance b] Pseudoallelism
   c] Co-dominance d] Polygenesis
2. Albinism in plants is associated with :
   a] Epistasis b] Recessive lethal
   c] Dominant lethal d] Chromosome duplication
3. The largest of cut flowers in the world is
   a] USA b] Colombia
   c] Israel d] Netherlands
4. In sweetpeas (*Lathyrus odoratus*) gene C or P alone produces white flower, whereas both C and P are needed to produce purple-coloured-flower. In a particular cross between two white-flowered plants, the offsprings were one-half purple and one-half white. The possible genotypes of the parents is :
   a] CcPp x ccPp b] CCpp x ccPP
   c] ccPp x CCpp d] Ccpp x ccPp
5. Match List-I (Crop plants) with List-II (Ploidy levels of the plants) and select the correct answer using the codes given below the lists :

	List I		List-II
A.	Bread wheat	1.	Diploid
B.	Pea	2.	Triploid
c.	Sea-island cotton	3.	Allo-hexaploid
D.	Banana	4.	Allo-tetraploid

CODES

	A	B	C	D
a]	2	1	4	3
b]	3	1	4	2
c]	2	4	3	1
d]	3	4	1	2

6. Which of the following pairs of definitions and terms are correctly matched ?

Select the correct answer using the codes given below :

1. The immediate effect of pollen on the character of endosperm – Xenia
2. A cytoplasmic-borne unit of heredity – Plasmogene
3. A haploid chromosome in an otherwise normal diploid individual – Monosome

CODES
   a] 1, 2 and 3 b] 1 and 2
   c] 1 and 3 d] 2 and 3
7. The condition in which a single gene influences more than one trait is known as :
   a] Epistasis b] Pleiotropy
   c] Polarity mutation d] Phenocopy
8. Which of the following type(s) of male sterility is/are used in commercial seed production of double cross hybrid maize ?
   a] Cytoplasmic alone
   b] Genetic alone
   c] Cytoplasmic and genetic
   d] Cytoplasmic and genetic with restorers
9. The composite variety has an advantage over the hybrid variety in respect of :
   a] Yield b] Synthesis
   c] Homogeneity d] Seed production
10. Consider the following statements:

1. Wheat is self-pollinated
2. Inbreeding depression is expected to be more pronounced in rapeseed than in mustard.
3. F_1 seeds alone are used for cultivation in both hybrid and composite varieties.
4. Hybrid vigour is expected to be more pronounced in bajra than in Bengal gram.

Of these statements
   a] 1, 2, 3 and 4 are correct
   b] 1 and 2 are correct
   c] 2, 3 and 4 are correct
   d] 1, 3 and 4 are correct
11. Consider the following statements:

Triticale

1. Is a cross between species of two different genera *Secale cereale* (2n = 14) and *Triticum vulgare* (2n = 42).
2. Is amphidiploid with 56 chromosomes.
3. Has high degree of fertility

Of these statements :
   a] 1 and 2 are correct b] 1 and 3 are correct
   c] 2 and 3 are correct d] 1, 2 and 3 are correct

GK GS

Under Article 258A the states can give their power to legislate to the Parliament.

12. Efforts are on in the Mexican wheat breeding programme to develop multiline varieties for greater genetic diversification to reduce rust infestation. What is the correct sequence of the following steps in the procedure adopted to develop a multiline variety ?

1. Choosing a commercially acceptable variety for the recurrent parent.
2. Introducing concurrently, different genes for rust resistance into the recurrent variety by separate backcross programme.
3. Compositing 5 to 10 different backcross derived lines of the recurrent variety each with a different gene for rust resistance.
4. Increasing the composite and growing as a commercial variety.

Select the correct answer using the codes given below :

CODES

a] 3, 1, 2, 4 b] 1, 3, 2, 4
c] 1, 2, 3, 4 d] 3, 2, 1, 4

13. Match List-I (Crop) with List-II (Variety) and select the correct answer using the codes given below the lists :

	List-I		List-II
A.	Cauliflower	1.	White Vienna
B.	Cabbage	2.	Golden acre
C.	Knolkhol	3.	Japanese white
D.	Radish	4.	Dania

CODES

	A	B	C	D
a]	4	2	3	1
b]	4	2	1	3
c]	2	4	3	1
d]	2	4	1	3

14. Wheat is of
a] Mediterranean origin
b] Indian origin
c] Mexican origin
d] Chinese origin

15. The components of synthetic population would have already been tested for
a] Specific combining ability
b] Genetic advance
c] General combining ability
d] Both general and specific combining ability

16. The popular variety, Ganga Safed-2 of maize is a
a] Single cross hybrid
b] Double cross hybrid
c] Three-way cross hybrid
d] Double topcross hybrid

17. The mutation induced nullisomic variety of sugarcane is
a] Co 8504 b] Co 8153
c] Co 527 d] Co 8152

18. In pedigree method of breeding, single plants are selected in which one of the following generations ?
a] F_5 b] F_3
c] F_2 d] F_1

19. Pointed gourd (*Tricosanthes dioica*) is a
a] Monoecious plant
b] Hermaphrodite plant
c] Dioecious plant
d] Polygamous plant

20. The family Anacardiaceae to which mango belongs, contains a number of plants of horticultural interest. List I contains the name of the plants and List II contains their botanical names. Match List I with List II and select the correct answer using the codes given below the list :

	List I		List II
A.	Cashew	1.	*Schinus molle*
B.	Hog plum	2.	*Buchanania latifolia*
C.	Small nut (Chiraunji)	3.	*Spondias mangifera*
D.	Pepper-tree	4.	*Anacardium occidentale*

CODES

	A	B	C	D
a]	4	3	2	1
b]	3	2	1	4
c]	2	3	1	4
d]	3	4	1	2

21. Pusa Delicious a gynodioecious variety of papaya has which of the following sex forms ?
(1) Pistillate
(2) Staminate
(3) Hermaphrodite

CODES

a] 1 and 2 b] 1 and 3
c] 2 and 3 d] 1, 2 and 3

22. Which of the following reason/reasons is/are responsible for seedlessness in Pusa Bedana cultivar of watermelon?

1. Vines do not have male flowers.
2. Male and female flowers do not open at one time.
3. Vines suffer from deficiency of Boron.
4. Cultivar is triploid in genetic constitution.

Select the correct answer using the codes given below :

CODES :

a] 1 and 2 b] 2 and 3
c] 3 and 4 d] 4 alone

23. Which one of the following genetic factors is responsible for poor fruit set in custard apple ?
a] Heterostyly
b] Dioecious nature
c] Dichogamy
d] Self-incompatibility

GK **State with highest density of population : West Bengal.** GS

24. Which of the following species are climbers ?
1. *Beaumontia grandiflora*
2. *Mussaenda erythrophylla*
3. *Lonicera japonica*
4. *Peltophorum ferrugineum*

CODES

a] 1 and 2 b] 2 and 3
c] 1 and 3 d] 3 and 4

25. Match List-I with List-II and select the correct answer using the codes given below the lists :

	List-I (Name of ornamental plants)		List-II (Colour of flowers)
A.	*Tabebuia chrysantha*	1.	Red
B.	*Amherstia nobilis*	2.	Yellow
C.	*Monodora grandiflora*	3.	Bluish purple
D.	*Jacaranda acutifolia*	4.	Greenish-yellow

CODES

	A	B	C	D
a]	1	2	4	3
b]	2	1	3	4
c]	2	1	4	3
d]	1	2	3	4

26. Which of the following pairs are sets of an orchard crop and a good companion crop?
1. Mango and pipalamool
2. Ber and brinjal
3. Sapota and sugarcane
4. Banana and turmeric

Select the correct answer using the codes given below :

CODES

a] 1 and 2 b] 1 and 4
c] 3 and 4 d] 2 and 4

27. Which one of the following types of vegetable gardens normally has the highest crop intensity?

a] Truck garden
b] Kitchen garden
c] Vegetable garden for processing
d] Market garden

28. Match List-I with List-II and select the correct answer using the codes given below the lists :

	List-I (Fruit)		List-II (Vit. A/Carotene per 100 g edible part)
A.	Ripe mango	1.	1710 I.U.
B.	Ripe papaya	2.	2020 I.U.
C.	Persimmon	3.	600 I.U.
D.	Dates	4.	4800 I.U.

CODES

	A	B	C	D
a]	4	2	1	3
b]	1	2	3	4
c]	4	1	2	3
d]	3	4	1	2

29. Match List-I with List-II and select the correct answer using the codes given below the lists :

	List-I (Crop)		List-II (Characteristic)
A.	Tomato	1.	Papain
B.	Persimmon	2.	Lycopine
C.	Chillies	3.	Soluble tannin
D.	Papaya	4.	Capsicin

CODES

	A	B	C	D
a]	2	3	4	1
b]	3	4	1	2
c]	2	1	4	3
d]	4	3	2	1

30. Match List-I (Fruit crop) with List-II (Physiological disorders during storage) and select the correct answer using the codes given below the lists :

	List-I		List-II
A.	Apple	1.	Cold scald
B.	Peach	2.	Bitter pit
C.	Citrus	3.	Chilling injury
D.	Mango	4.	Woolliness

CODES

	A	B	C	D
a]	3	2	4	1
b]	2	4	1	3
c]	3	4	1	2
d]	2	1	4	3

31. What is the correct sequence of the following steps in the preparation of pineapple jam?
1. Cooking the fruit
2. Addition of pectin
3. Addition of sugar
4. Boiling and removal of scum

CODES

a] 1, 3, 4, 2 b] 3, 1, 4, 2
c] 1, 3, 2, 4 d] 3, 1, 2, 4

32. Match List-I (Crop species) with List-II (Serious disease) and select the correct answer using the codes given below the lists :

	List-I		List-II
A.	Rice	1.	Green ear
B.	Wheat	2.	Bacterial leaf blight
C.	Maize	3.	Stalk rot
D.	Bajra	4.	Alternaria leaf blight

CODES

	A	B	C	D
a]	1	2	3	4
b]	2	3	1	4
c]	2	4	3	1
d]	4	3	2	1

33. Which of the following pairs of wheat rusts and alternate hosts are correctly matched ?

Select the correct answer using the codes given below.

1. *Puccinia graminis tritici* - *Berberis vulgaris*
2. *Puccinia recondita tritci* - *Thalictrum* spp.
3. *Puccinia striiformis* - *Phalaris minor*

CODES

a] 1, 2 and 3 b] 1 and 2
c] 1 and 3 d] 2 and 3

34. Match List-I (Crop disease) with List-II (Pathogen) and select the correct answer using the codes given below the lists :

	List-I		List-II
A.	Tikka disease of groundnut	1.	*Erysiphe graminis* f. sp. *hordei*
B.	Powdery mildew of barley	2.	*Puccinia striiformis*
C.	Yellow rust of wheat	3.	*Plasmopara viticola*
D.	Downy mildew of grapes	4.	*Cercospora personata*

CODES

	A	B	C	D
a]	2	1	4	3
b]	2	3	1	4
c]	4	1	2	3
d]	4	3	2	1

35. Which one of the following pairs of crop and disease is correct matched ?
    a] Cabbage — Early blight
    b] Cucumber — Club root
    c] Potato — Black wart
    d] Tomato — White rust

36. Which one of the following pairs is not correctly matched?
    a] Karnal bunt of wheat —*Neovossia horrida*
    b] False smut of rice – *Ustilaginoidea virens*
    c] Hill bunt of wheat – *Tilletia foetida*
    d] Ear cockle of wheat – *Anguina tritici*

37. Plant pathogenic viroids contain
    a] RNA only
    b] RNA and protein coat
    c] DNA and protein coat
    d] DNA only

38. Loose smut of wheat is a/an
    a] Externally seed-borne disease
    b] Externally seed-borne and soil-borne disease
    c] Air-borne disease
    d] Internally seed-borne disease

39. Which of the following are the sources of survival of the wilt pathogen of sugarcane in India ?
1. Ratoon of infected crop
2. Diseased plant debris
3. Setts of diseased crop
4. Collateral hosts

Select the correct answer using the codes given below :
CODES
    a] 1, 2, 3 and 4 b] 2 and 3
    c] 3 and 4 d] 1, 2 and 3

40. Which one of the following is the correct order in which they appear during biocontrol operations ?
    a] Parasite, Release, Mass rearing, Parasitism
    b] Mass rearing, Parasite, Release, Parasitism
    c] Release, Mass rearing, Parasite, Parasitism
    d] Parasite, Mass rearing, Release, Parasitism

41. Indian share in the global flower market is about____
    a] 5% b] 2%
    c] 10% d] 0.6%

42. Mycoplasma is sensitive to
    a] Penicillin b] Tetracycline
    c] Calixin d] Streptomycin

43. Whorl application of granular insecticide provides effective control measure against :
    a] Spotted bollworm
    b] Maize stem borer
    c] Stem borer of paddy
    d] Mango stem borer

44. Match List-I (Principles of disease control) with List-II (Practices) and select the correct answer using the codes given below the lists :

	List-I		List-II
A.	Exclusion	1.	Spraying a common fungicide
B.	Avoidance	2.	Chemical seed treatment
C.	Eradication	3.	Isolation in time and space
D.	Protection	4.	Plant quarantine

CODES

	A	B	C	D
a]	3	4	2	1
b]	3	4	1	2
c]	4	3	1	2
d]	4	3	2	1

45. Which one of the following insecticides is capable of controlling mites as well ?
    a] Endosulfan
    b] Toxaphene
    c] Cypermethrin
    d] Monocrotophos

46. The safest method of disposing off the left over pesticide is, disposal by :
    a] Pouring in the drain
    b] Pouring in a pond
    c] Pouring in a moving stream
    d] Burying in the soil

47. Which of the following types of manually operated sprayers are suitable for spraying 4 to 5 metre tall plantations ?
1. Hand compression sprayer
2. Knapsack sprayer
3. Foot sprayer
4. Rocking sprayer

Select the correct answer using the codes given below :
CODES
    a] 1, 2, 3 and 4 b] 2 and 4
    c] 1, 2 and 3 d] 3 and 4

48. Which of the following protective measures are necessary for a person manually dusting pesticide on crop in uncertain wind conditions ?
a] Wearing plastic aprons
b] Use of plastic gloves
c] Use of plastic shoes
d] Use of nose filters

Select the correct answer using the codes given below :
CODES
a] 1, 2, 3, 4 and 5 b] 1, 2, 4 and 5
c] 2, 3, and 4 d] 1, 3, and 5

49. If one gram of a pesticide formulation containing 50% active ingredient is mixed with one litre of water, what will be the concentration of the active ingredient in the spray fluid ?
a] 0.5 ppm b] 5.0 ppm
c] 50 ppm d] 500 ppm

50. The main purpose of T and V programme is to :
a] Identify research problems
b] Bring about coordination in the administration of agricultural extension
c] Update the technical knowledge and skills of agricultural extension workers
d] Help farmers in the adoption of latest innovations in agriculture

51. An extension worker practically explains Japanese method of paddy cultivation to a group of farmers and also shows its superiority over the traditional method. In the process, which of the following communication methods have been used by him?
1. Method demonstration
2. Field visit
3. Result demonstration
4. Study tour

Select the correct answer using the codes given below :
CODES
a] 1, 2, 3 and 4 b] 1 and 3
c] 2 and 3 d] 1, 2 and 4

52. Consider the following statements: Extension in the long run means change in
1. Knowledge
2. Skill
3. Work
4. Attitude

Of these statements
a] 1, 2 and 3 are correct
b] 1, 3 and 4 are correct
c] 1, 2 and 4 are correct
d] 2, 3 and 4 are correct

53. Total floriculture trade in the worlds is billion US $
a] 7 b] 20
c] 10 d] 50

Directions : The following Ten items consist of two statements, one labelled the 'Assertion A' and the other labelled the 'Reason R'. You are to examine these two statements carefully and decide if the Assertion A and Reason R are individually true and if so, whether the Reason is a correct explanation of the Assertion. Select your answers to these items using the codes given below and mark your answer sheet accordingly.

CODES
a] Both A and R are true and R is the correct explanation of A.
b] Both A and R are true but R is not a correct explanation of A.
c] A is true but R is false
d] A is false but R is true

54. Assertion (A) : Soil organic matter plays a pivotal role in sustaining soil productivity and fertility.
Reason (R) : Plant nutrients are slowly released from soil organic matter.

55. Assertion (A) : Cultivation of rice in the non-traditional areas in India has increased.
Reason (R) : Food habits of the people of non-rice growing areas have changed.

56. Assertion (A) : In plant cells plasmalemma controls the flow of solutes into the cytoplasm.
Reason (R): In plant cells, tonoplast regulates the transport of solutes into and out of the vacuoles.

57. Assertion (A) : The expected genetic gain as calculated by the formula
$Gs = k. \sigma_p.H$
is applications to a mixture of purelines or clones, but it is not applicable to segregating generations.
Reason (R) : Purelines or clones give rise to progenies which are identical in genotype with the parent family and the plants in segregating generations are likely to be heterozygous for a few or several genes.

58. Assertion (A) : The F_1 hybrids of *Solanum melongena* Muktakeshi x Banaras giant and Muktakeshi x Pusa Kranti exhibited increase over better parent as regards yield.
Reason (R) : Exploitation of hybrid vigour has attained considerable importance in most of the vegetables.

59. Assertion (A) : Fruit quality is improved on pruning fruit trees.
Reason (R) : Pruning improves fruit : leaf ratio.

60. Assertion (A) : Precooling of fruit and vegetables is done to reduce the rate of ripening process and extending shelf life.
Reason (R) : Precooling is done to immediately remove the field heat.

61. Assertion (A) : Philosophy of pest control based on eradication of pest species is the antithesis of integrated pest control.
Reason (R) : Eradication of pest species leads to an unstable biotic community.

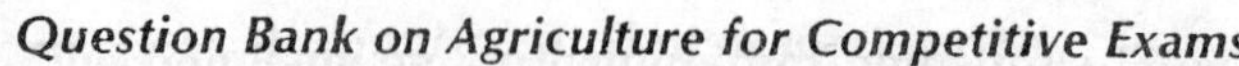

62. Assertion (A) : The broad spectrum insecticide needs to be used for all agricultural crops.
Reason (R) : Broad spectrum insecticide kills a large variety of insects.

63. Assertion (A) : Diffusion is the sub-set of communication as it involves specialised message (innovation) to be communicated.
Reason (R) : Communication of innovation is followed by adoption.

64. Biggest international flower market is located at
a] Washington b] Singapore
c] London d] Alsmeer

65. Which one of the following pairs is correctly matched?
a] Common cost - Cost associated with variable resources.
b] Average cost - A study of returns and costs.
c] Marginal cost - Additional cost to produce an additional unit of output.
d] Prime cost - Cost related to fixed resources.

66. Which one of the following pairs is not correctly matched ?
a] Elasticity of production greater than unity - Stage II of classical production function
b] Negative marginal product - Stage III of classical production function
c] Negative elasticity of production - Stage III of classical production function
d] Average product decreases but is greater than marginal product - Stage II of classical production function.

67. Colombia is a largest exporter of flower and its share in international flower trade is about %
a] 5 b] 15
c] 40 d] 1

68. Consider the following factors :
1. Soil characteristics
2. Mean annual rainfall
3. Potential evapotranspiration
4. Number of drought days in a year
The Planning Commission in India has divided the country into different agroclimatic regions based on factors
a] 1, 2 and 3
b] 1, 2 and 4
c] 1, 3 and 4
d] 2, 3 and 4

69. Which one of the following relationships will provide least cost combination of input use ?
a] $\frac{\Delta X_1}{\Delta X_2} = \frac{PX_2}{PX_1}$ b] $\frac{\Delta X_1}{\Delta X_2} > \frac{PX_2}{PX_1}$
c] $\frac{\Delta X_1}{\Delta X_2} < \frac{PX_2}{PX_1}$ d] $\frac{\Delta X_1}{\Delta X_2} = \frac{PX_1}{PX_2}$

70. Which one of the following principles is employed in production when resources are scarce ?
a] Added cost is equal to added returns
b] Equi-marginal principle
c] Comparative advantage
d] None of the above

71. Per capita consumption of cut flowers is highest in which country.
a] India b] USA
c] Switzerland d] Germany

72. Suppose that cost A_2 on a farm is Rs. 26,450 and the rent of the leased-in land is Rs. 2,250. Cost A_1 on this farm will be :
a] Rs. 24,000 b] Rs. 24,200
c] Rs. 28,000 d] Rs. 28,700

73. Consider the following statements:
Diversification of farm production helps farmers through stabilising:
1. Agricultural prices
2. Farm incomes
3. Agricultural production
Of these statements
a] 1 and 2 are correct
b] 1 and 3 are correct
c] 1, 2 and 3 are correct
d] 2 and 3 are correct

74. A normative farm plan is one which the farmer :
a] Has already put into operation on his farm
b] Should adopt during the current Five Year Plan period
c] Had actually adopted in the earlier Five Year Plan periods
d] Should adopt

75. In mixed farming, the contribution of livestock to gross farm income is
a] At least 10% b] At least 30%
c] At least 40% d] More than 50%

76. During the period 1990 to 1995, the contribution of agriculture sector to the gross domestic product of India has been in the range of :
a] 20 to 30 per cent
b] 30 to 40 per cent
c] 40 to 50 per cent
d] 50 to 60 per cent

77. Given that
$Y = 30 X_1^{0.25} X_2^{0.06}$
where Y = output
X_1 = input factor 1, and
X_2 = input factor 2,
the nature of return to scale will be
a] Constant b] Increasing
c] Decreasing d] Fluctuating

GK GS

Scattered type of settlement is found at Malabar coast.

78. The given diagram shows isoquants for different combinations of two input factors X_1 and X_2. Consider the following statements in this regard :

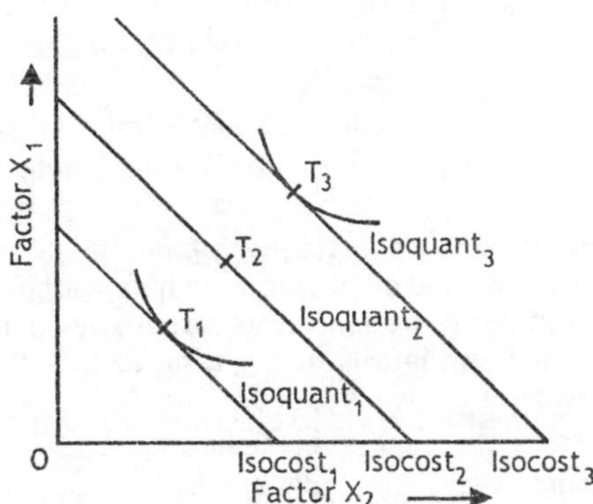

Higher production can be achieved by

1. Moving from isoquant$_1$ to isoquant$_3$
2. Moving from isoquant$_3$ to isoquant$_1$
3. Increasing the budget outlay

Of these statements :

a] 2 alone is correct
b] 1 alone is correct
c] 2 and 3 are correct
d] 1 and 3 are correct

79. In the given diagram, which one of the following segments represents feasible zone in a linear programming problem ?

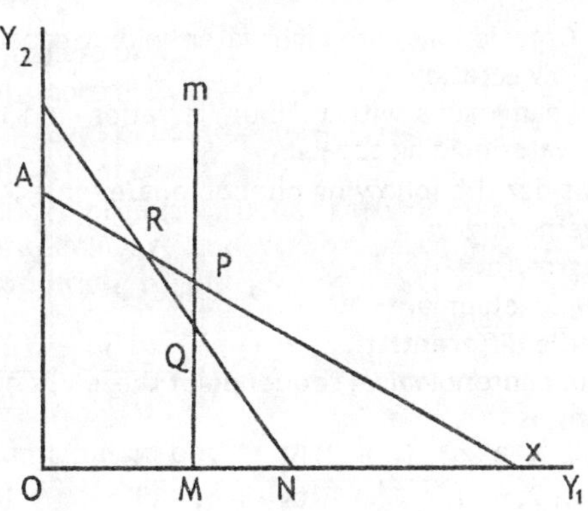

a] RPQ b] QMN
c] xNR d] OARQM

80. Match List-I with List-II and select the correct answer using the codes given below the lists :

	List-I (Crop)		List-II (Event)
A.	Sunflower	1.	Gypsum application
B.	Groundnut	2.	Potash application
C.	Coconut	3.	Poor seed setting
D.	Soyabean	4.	Phosphate application

CODES

	A	B	C	D
a]	1	3	2	4
b]	3	1	4	2
c]	3	1	2	4
d]	1	3	4	2

81. Match List-I with List-II and select the correct answer using the codes given below the lists :

	List-I (Crop)		List-II (Term used in relation to the crop)
A.	Corn	1.	Retting
B.	Cotton	2.	Tasselling
C.	Peanut	3.	Ginning
D.	Jute	4.	Pegging

	A	B	C	D
a]	1	2	3	4
b]	2	3	4	1
c]	2	3	1	4
d]	1	3	4	2

82. Match List-I (Host legume) with List-II (Rhizobium sps.) and select the correct answer using the codes given below the lists :

	List-I		List-II
A.	Fenugreek	1.	*R. meliloti*
B.	Clovers	2.	*R. phaseoli*
C.	Pease	3.	*R. trifolii*
D.	Beans	4.	*R. leguminosarum*

CODES

	A	B	C	D
a]	1	2	4	3
b]	1	3	2	4
c]	3	1	2	4
d]	1	3	4	2

83. Crop rotation is defined as a :

a] System of growing different kinds of crops on the same land.
b] System of growing different crops in succession on the same land.
c] Method of growing different crops
d] System of growing the same crop on different fields

84. Cropping pattern is defined as :

a] The proportion of area under different crops
b] Growing different crops in a recurrent succession
c] The proportion of area under different crops at a particular point of time.
d] A method of growing different crops

85. 'Mauritius', 'Mozambique' and 'Coromandel' are the varieties of

a] Chillies b] Mango
c] Pineapple d] Groundnut

GK GS

National Development Council was created in 1952.

86. Match List-I (Crop) with List-II (Variety) and select the correct answer using the codes given below the lists :

	List-I		List-II
A.	Pigeonpea	1.	Pusa 2-21
B.	Rice	2.	Karan 521
C.	Maize	3.	Bahar
D.	Barley	4.	Prabhat

CODES

	A	B	C	D
a]	3	1	4	2
b]	1	3	2	4
c]	3	1	2	4
d]	1	3	4	2

87. When irrigation water is available for two irrigations only, the wheat crop should receive these at :
a] CRI and late tillering stages
b] CRI and boot stages
c] CRI and flowering stages
d] CRI and milk stages

88. Dehaulming in potato is done to obtain :
a] Higher yield
b] Quality seed tuber
c] Higher starch content
d] Higher protein content

89. Consider the data presented in the following table :

Grain yield (in kg/ha) of Sorghum

Nitrogen levels	Plant density		
	S_1	S_2	S_3
N_1	4613	5452	5351
N_2	4816	5568	5423
N_3	5076	5905	5857
N_4	5466	6291	6219

CD (0.05)
N levels : 403
Plant Density : 435

Based on the given data, the nitrogen level(s) that recorded significantly higher grain yield than N_1 would be :
a] N_4 alone
b] N_4 and N_3
c] N_2, N_3 and N_4
d] N_3 alone

90. Match List-I (Herbicides) with List-II (Chemical group) and select the correct answer using the codes given below the lists :

	List-I		List-II
A.	Dalapon	1.	Quaternary ammonium compound
B.	2, 4, 5-T	2.	Symmetrical triazenes
C.	Simazine	3.	Phenoxy compounds
D.	Paraquat	4.	Aliphatic acids

CODES

	A	B	C	D
a]	3	4	1	2
b]	4	3	1	2
c]	4	3	2	1
d]	3	4	2	1

91. A farmer is to select a herbicide for controlling *Cyperus* spp. and dicot weeds on his sugarcane crop. The crop is followed, after its ratoon, by dhaincha in the kharif and lentils in the rabi. In this case, the herbicide of choice would be
a] Atrazine b] Dalapon
c] Diuron d] 2, 4-D

92. Which one of the following system needs irrigation throughout the year ?.
a] Diara land cropping system
b] Intercropping system
c] Mixed cropping system
d] Multiple cropping system

93. On the basis of total mass, the relative abundance of the given elements in the earth's crust is such that
a] Si > O_2 > Fe > Al b] O_2 > Al > Si > Fe
c] O_2 > Si > Fe > Al d] O_2 > Si > Al > Fe

94. Which one of the following sets of soil conditions is more suitable for plant growth ?
a] Sandy soils with high aeration and low water holding capacity
b] Clay soils with high water holding capacity and low aeration
c] Organic soils with high water holding capacity and low aeration
d] Loamy soils with medium aeration and medium water holding capacity

95. Consider the following phenomena/events :
1. Weathering
2. Biogrowth
3. Soil development
4. Profile differentiation

The correct chronological sequence of these phenomena/event is
a] 2, 1, 3, 4 b] 1, 2, 4, 3
c] 1, 2, 3, 4 d] 2, 1, 4, 3

96. Consider the following materials :
1. Montmorillonite
2. Humus
3. Kaolinite
4. Illite

The correct sequence of these materials in decreasing order of their CEC is
a] 1, 2, 3, 4 b] 2, 1, 3, 4
c] 2, 1, 4, 3 d] 1, 2, 4, 3

GK 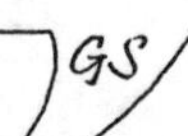 GS

State with lowest population : Sikkim.

97. **Phosphorus fixation refers to the phenomenon of**
    a] Absorption of phosphorus by the plant and synthesis of biological compounds
    b] Leaching of phosphorus from soil by percolating waters
    c] Conversion of soluble phosphorus into plant unavailable forms
    d] Anion exchange with colloidal constituents of soil

98. **Which one of the following pairs is not correctly matched ?**
    a] Expanding type of clay minerals – Higher water holding capacity
    b] High calcium content in soil – Iron chlorosis
    c] Water inundated soils – High nitrogen mineralisation
    d] Sandy soils – Split application of fertilizers

99. **Tolerance to soil salinity of the given set of crops is such that**
    a] Barley > wheat > upland paddy > beans
    b] Barley > wheat > beans > upland paddy
    c] Wheat > barley > beans > upland paddy
    d] Wheat > barley > upland paddy > beans

100. **Hydraulic conductivity of soils varies directly with**
    a] Capillary porosity
    b] Non-capillary porosity
    c] Total porosity
    d] Water holding capacity

101. **Total area under flower crops in the world is about _____ million hectares.**
    a] 0.25 b] 2.5
    c] 10 d] 50

102. **There is a net increase in the mineralisation of organic nitrogen in soil when**
    a] C : N ratio is 20 to 30 : 1
    b] C : N ratio is below 20 : 1
    c] C : N ratio is above 30 : 1
    d] Nitrogen is taken up by the crops

103. **World trade in foliage plants is about.... billion US $**
    a] 5 b] 3.1
    c] 10 d] 15

104. **Match List-I (Nutrients) with List-II (Soil test technique) and select the correct answer using the codes given below the lists :**

	List-I		List-II
A.	Available N	1.	Morgan's method
B.	Available P	2.	Chromic acid digestion method
C.	Available K	3.	Olsen's method
D.	Organic matter	4.	Alkaline permanganate method

CODES

	A	B	C	D
a]	4	3	2	1
b]	3	4	1	2
c]	4	3	1	2
d]	3	4	2	1

105. **What would be the concentration of nutrients in plants when they are suffering from hidden hunger?**
1. **Above deficiency symptom zone**
2. **Below profitable yield zone**
3. **Below deficiency symptom zone**
4. **Above profitable yield zone**

Select the correct answer using the codes given below :
CODES
    a] 1 and 4
    b] 2 and 3
    c] 1 and 2
    d] 3 and 4

106. **A soil with 20 me/100 g. CEC and 5 me/100 g exchangeable H will have a base saturation of**
    a] 25.0%
    b] 75.0%
    c] 80.0%
    d] 133.3%

107. **Match List-I (Biofertilizers) with List-II (Properties) and select the correct answer using the codes given below the lists :**

	List-I		List-II
A.	*Azospirillum*	1.	Helps in phosphorus uptake
B.	*Azotobacter*	2.	Floating aquatic fern
C.	*VAM*	3.	Grows on and inside the roots of graminaceous crops
D.	*Azolla*	4.	Free-living nitrogen fixing bacterium

CODES

	A	B	C	D
a]	4	3	2	1
b]	3	4	2	1
c]	3	4	1	2
d]	4	3	1	2

108. **Maximum share of foliage plants in the world market is again form which of the following country.**
    a. Germany b. China
    c. Netherlands d. Denmark

109. **In respect of a sugarcane crop to be grown in an acidic soil having high P-fixing capacity, the fertilizer of choice would be**
    a] Triple superphosphate
    b] Finely ground rock phosphate
    c] Coarse rock phosphate
    d] Superphosphate

110. **For preparing a mixed fertilizer, 100 kg. of ammonium sulphate and 125 kg. of single superphosphate were mixed. 25 kg. of filler was added to this mixture. In terms of N - P_2O_5 - K_2O, the fertilizer grade of this mixture will be :**
    a] 5 - 5 - 0 b] 5 - 10 - 0
    c] 8 - 8 - 0 d] 10 - 10 - 0

GK

Mumbai is the biggest port of India.

GS

111. Use of materials like sulphur, polyform UF, isobutylidene diurea etc. improves fertilizer use efficiency. When such additives are used, the N-fertilizers are termed as
a] Prilled urea
b] Controlled release nitrogenous fertilizers
c] Slow nitrifying N-fertilizers
d] Complex fertilizers

112. Consider the following statements regarding enzymes in living plants:
1. They are specific.
2. They are sensitive to heat.
3. The reactions caused by enzymes are reversible.
4. They exist in a colloidal state.

Of these statements :
a] 1 and 2 are correct
b] 1, 3 and 4 are correct
c] 2, 3 and 4 are correct
d] 1, 2, 3 and 4 are correct

113. Which one of the following plant nutrients is useful in increasing resistance to diseases and insect pests?
a] Calcium b] Phosphorus
c] Nitrogen d] Potassium

114. Parts of cell walls in plants are given in List I and their distinctive components are given in List II. Match the two lists and select the correct answer using the codes given below the lists :

	List-I	List-II
A.	Primary wall	1. Hemicellulose
B.	Secondary wall	2. Lignin
C.	Middle lamella	3. Lipids and proteins
D.	Plasmalemma	4. Pectin

CODES

	A	B	C	D
a]	1	2	3	4
b]	2	1	4	3
c]	2	1	3	4
d]	1	2	4	3

115. Soils with high pH are generally deficient in :
a] Zn and Mn b] B and Fe
c] Cu and Mod] Ca and Mg

116. During the conversion of each molecule of glucose into pyruvic acid through the glycolytic cycle :
a] One molecule of ATP is consumed and two molecules of ATP are generated
b] Two molecules of ATP are consumed and two molecules of ATP are generated
c] One molecule of ATP is consumed and four molecules of ATP are generated
d] Two molecules of ATP are consumed and four molecules of ATP are generated

117. Which is the correct chronological sequence of the following discoveries related to mechanism of photosynthesis ?
1. Calvin's cycle
2. Hatch and Slack cycle
3. Hill reaction
4. Red drop

Choose the correct answer using the codes given below :

CODES
a] 3, 4, 2, 1 b] 4, 3, 2, 1
c] 3, 4, 1, 2 d] 4, 3, 1, 2

118. Match List I (Plant growth regulators) with List II (Possible precursors) and select the correct answer using the codes given below the lists :

	List-I	List-II
A.	ABA	1. Tryptophan
B.	Ethylene	2. Terpenoids
C.	GA_3	3. Fatty acids
D.	IAA	4. Carotenoids

CODES

	A	B	C	D
a]	4	2	3	1
b]	2	3	1	4
c]	4	3	2	1
d]	3	2	1	4

119. Which of the following chemicals are responsible for the induction of female flower ?
1. Ethephone
2. GA_3
3. NAA
4. Morphactine

Select the correct answer using the codes given below :
a] 1 and 4 b] 2 and 4
c] 1 and 4 d] 3 and 4

120. The main function of maleic hydrazide is considered to be that of a/an
a] Antiauxin
b] Antagonist to gibberellin
c] Inhibitor
d] Growth retardant

GK GS

Three types of emergencies can be imposed under the Constitution.

Set-6

1. **Male sterility is common in**
   a] Petunia b] Marigold
   c] Antirrihunum d] All of these
2. **Self-incompatibility is not common in______**
   a] Ageratum b] Petunia
   c] English daisy d] Rose
3. **The Central Agmark Lab is located at:**
   a] Mumbai b] Nagpur
   c] Bangalore d] Kolkata
4. **Read the following statements and mark your answer according to code given below:**

I. For variable input: APP= MPP in the Ist stage
II. For variable input: APP= MPP at the end of Ist stage
III. For variable input: APP = MPP at the start of IInd stage
IV. For variable input: APP=MPP in the IInd stage

CODES:
   a] I and II are correct
   b] II and III are correct
   c] I and IV are correct
   d] II and IV are correct
5. **MPP (Marginal Physical Product) of a variable input is negative:**
   a] Only in I stage
   b] Only in II stage
   c] Only in III stage
   d] In II and III stage
6. **I.B.R.D. is synonymous to:**
   a] World Bank
   b] Asian Development Bank
   c] International Monetary Fund
   d] World Trade Organisation
7. **Match List -I with List-II and select the correct answer using the codes given below the lists:**

List-I (Farm Incomes)	List-II (Measure)
A. Family Labour Income	1. G.I. - cost A1
B. Farm Business Income (owner farmer)	2. G.I. - cost A2
C. Net Income	3. G.I. - cost B
D. Farm Business Income (tenant farmer)	4. G.I. - cost C

CODES:

	A	B	C	D
a]	2	1	3	4
b]	1	2	4	3
c]	1	2	3	4
d]	3	1	4	2

8. **Principle of Extension Education is:**
   a] Learning by hearing
   b] Learning by seeding
   c] Learning by reading
   d] Learning by doing
9. **Goal of Extension Education is:**
   a] To promote income of the farmers
   b] To promote production of the crops
   c] To promote new crops
   d] To promote scientific outlook
10. **Which one of the following is correct match:**
   a] Etawah Pilot Project Albert Mayer
   b] Grow More Food J.L. Nehru
   c] Cooperative Movement M.K. Gandhi
   d] Sewagram Project Vinoba Bhave
11. **What is full form of K.G.K. :**
   a] Kisan Gyan Kendra
   b] Kisan Guar Kendra
   c] Krishi Gyan Kendra
   d] Kisan Gossip Kendra
12. **What is full form of K.V.K.:**
   a] Krishi Vigyan Kranti
   b] Krishi Vigyan Kendra
   c] Kisan Vividha Kendra
   d] Krishi Vivad Kendra
13. **The first K.V.K. was established in 1974 at :**
   a] Nagpur b] Nilokheri
   c] Ludhiana d] Pondichery
14. **Which of the following rural development project was launched before independence:**
   a] Etawah Pilot Project
   b] Cooperative Movement
   c] Community Development
   d] National Extension Service
15. **Television broadcast for rural development in India, started in:**
   a] 1947 b] 1957
   c] 1967 d] 1977

GK GS
The first food park of Asia is being established at Dankuni near Kolkata.

16. Cooperative society is basic institution for:
    a] Political growth of the villagers
    b] Social growth of the villagers
    c] Cultural growth of the villagers
    d] Socio-economic growth of the villagers
17. Radio mass medium is characterised by:
    a] One way with immediate feedback
    b] One way without instant audience-response
    c] One way and colourful
    d] One way and timeless
18. Which one of the following is not correctly matched:

	Approach	Technique
a]	Mass	Exhibition
b]	Individual	Farm visit
c]	Group	Mela
d]	Mass	Newspaper

19. Purpose of Extension Evaulation is:
    a] To identify the weak points
    b] To identify the strong points
    c] To identify the gaps and errors
    d] To identify all the above
20. Read the following statements and mark your answer according to the code given below:

I C.D. is a programme
II C.D. is a cultural event
III C.D. is a social event
IV C.D. is a project

CODES:
    a] I and III are correct
    b] I and IV are correct
    c] II and III are correct
    d] II and IV are correct
21. Read the following statements and mark your answer according to the code given below:

I Extension evaluation is a planned process
II Extension evaluation is spontaneous
III Extension evaluation is indivisible portion of teaching.
IV Extension evaluation is beginning process.

CODES:
    a] I and III are correct
    b] II and III are correct
    c] I and IV are correct
    d] II and IV are correct
22. Read the following statements and mark your answer according to the code given below:

I Newspaper is popular medium of communication in rural areas
II Newspaper is written form of communication
III Film show is audio-visual form of communication
IV Film show is popular medium of communication in rural areas.

CODES:
    a] I and II are correct
    b] II and III are correct
    c] III and IV are correct
    d] I and IV are correct
23. Consider the following statements:
    Assertion (A) : Film is not effective medium in rural areas
    Reason (R) : Film cannot be projected with low voltage power
    Select the answer from the code given below:
    a] A is true but R is false
    b] R is true but A is false
    c] Both A and R are correct and R is the correct explanation of A
    d] Both A and R are correct but R is not a correct explanation of A
24. Consider the following statements:
    Assertion (A) : Farmers posses keen interest in training programs at block headquarter.
    Reason (R) : Farmers enjoy free meal during training period
    Select the correct answer from the code given below:
    a] A is the true but R is false
    b] R is true but A is false
    c] Both A and R are correct and R is the correct explanation of A
    d] Both A and R are true but R is not a correct explanation of A
25. Match List-I with List-II and select the correct answer using the code given below:

List-I (Form)	List-II (Technique)
A. Spoken	1. Folder
B. Written	2. Tape recorder
C. Projected	3. Film
D. Non-projected	4. Puppet

CODES:

	A	B	C	D
a]	2	1	3	4
b]	1	2	3	4
c]	2	3	1	4
d]	4	3	2	1

26. Match List-I with List-II and select the correct answer using the code given below:

List-I (Aids)	List-II (Method of contact)
A. Model	1. Group contact
B. Slide	2. Mass contact
C. Telephone call	3. Personal contact
D. Poster	4. Mass contact

CODES:

	A	B	C	D
a]	2	3	4	1
b]	1	2	3	4
c]	3	4	2	1
d]	3	2	4	1

27. Which of the following statements represent true concept of Extension Evaluation?
a] Extension evaluation begins before the start of an extension programme
b] Extension evaluation begins at the mid step of the programme
c] Extension evaluation begins at the last step of the programme
d] Extension evaluation is inbuilt in each step of the programme

28. People's participation in an extension programme is significant when:
a] Literate section of villagers participate
b] Gram Panchayat members participate
c] Majority of villagers participate
d] Local leaders participate

29. Which of the following is not a method of group communication?
a] Demonstration b] Symposium
c] Flannel graph d] Circular letter

30. Lab to Land Programme was started by:
a] Govt. of India b] Govt. of U.P.
c] Smt. Indira Gandhi d] ICAR

31. Cultivation of durum wheat is primarily confined:
a] Central India
b] Southern India
c] Central and Southern India
d] U.P. Hills

32. Tick the correct agro-climatic zone of U.P. which has highest average population density per km^2.
a] Bundelkhand zone
b] Central plains zone
c] Hills zone
d] North-eastern plains zone

33. Which country is the place of origin of potato?
a] South America b] South Canada
c] South Africa d] South China

34. Which one of the following is not correctly matched?

Crop	Test weight in gm.
a] Pearl-millet	5-7
b] Wheat	20-22
c] Mungbean	34-36
d] Sunflower	60-62

35. Which of the following intercropping system has synergistic effect on component crop?
a] Autumn planted Sugarcane + Mustard
b] Autumn planted Sugarcane + Chickpea
c] Autumn planted Sugarcane + Wheat
d] Autumn planted Sugarcane + Potato

36. The most concentrated fertilizer used for nutrient supply is:
a] Urea
b] DAP
c] Anhydrous Ammonia
d] SSP

37. Which one is the recommended optimum plant population for hybrid maize?
a] 40-50 thousands per hectare
b] 50-60 thousands per hectare
c] 60-70 thousands per hectare
d] 70-80 thousands per hectare

38. Which of the following seed rate of TPS is recommended for planting one hectare area?
a] 150 g. b] 250 g.
c] 350 g. d] 450 g.

39. Which of the following is more suitable for control of *Phalaris minor* in wheat field?
a] Sulpho-sulphuran
b] Pendimethalin
c] Isoproturon
d] Methabenzthiazuron

40. Maize (for green cobs)- Autumn Cane-Ratoon-Wheat will have cropping intensity of:
a] 133% b] 233%
c] 333% d] 433%

41. Which one of the following can be assessed by using equation?

$$\frac{\text{Harvest Diversity Index x 10000}}{\text{Multiple Cropping Index}}$$

a] Diversity Index
b] Multiple Cropping Index
c] Simultaneous Cropping Index
d] Cropping Intensity Index

42. Chemical which regulates opening and closing of stomata is:
a] Kaoline b] 2, 4-D
c] PMA d] Paraquat

43. Which one of the following is nonselective herbicide?
a] Alachlor b] Butachlor
c] Paraquat d] Atrazine

44. Which one of the following is not narrow leaved weeds?
a] *Cynodon dactylon* b] *Cyprus rotundus*
c] *Setaria glouca* d] *Melilotus indica*

45. Which one of the following statement is correct:
a] Blue green alage fix 20-30 k.g N/ha. in a rice field.
b] Blue green alage fix 40-50 k.g N/ha. in a rice field.
c] Blue green alage fix 8-10 k.g N/ha. in a rice field.
d] Blue green alage fix 60-70 k.g N/ha. in a rice field.

46. Which of the following is not Kharif seasons weed:
a] *Chenopodium album*
b] *Amaranthus virdis*
c] *Echynochloa colonum*
d] *Comelina benghalensis*

47. Thimma is an important cultivar of _______
a] Rose b] Bougainvillea
c] Tuberose d] Tulip

48. Which is the correct row ratio for intercropping of mustard with wheat:
a] 5:1 b] 9:1
c] 6:1 d] 10:1

49. The correct group of plant nutrients which act as energy exchanges?
a] K, Na, Ca and Mg
b] C, N, P and S
c] H and O
d] Fe, Mn, Mo, Cu, B and Zn

50. Which gas is released from paddy fields?
a] CH_4 b] H_2S
c] CO_2 d] NH_3

51. Short duration crop grown between two main crops is known as:
a] Companion crop b] Catch crop
c] Inter crop d] Relay crop

52. 'Peat' soil are generally characterised as:
a] Black, heavy and acidic soil
b] Red, light and acidic soil
c] Black, light and alkaline soil
d] Black, heavy and neutral soil

53. Solubility of rock phosphate can be improved by:
a] Nitrosomonas b] Nitrobacter
c] Azotobacter d] Bacillus polymexa

54. The principle way in which P and K ions move from the soil to the root of field crops:
a] Root interception b] Mass flow
c] Diffusion d] None

55. Biuret content in urea should not exceed in foliar spray on wheat more than the following one:
a] 0.55 b] 1.0%
c] 1.5% d] 2.0%

56. P content in DAP is:
a] 20% b] 36%
c] 46% d] 50%

57. Which element causes dispersion of soil particles:
a] Na b] K
c] Ca d] Mg

58. Which element mainly the plant uptake of :
a] Nitrogen b] Phsophorus
c] Potassium d] Molybdenum

59. Immobilization of sulphur takes place when the 'S' content of organic matter is less than:
a] 0.60% b] 0.45%
c] 0.30% d] 0.15%

60. Read the following statements and mark your answer according to the code given below:
I Soil acidity increases with exchangeable aluminium
II Soil acidity increases with exchangeable hydrogen
III Soil alkalinity increases with exchangeable calcium
IV Soil alkalinity increases with exchangeable aluminium

CODES:
a] I and II are correct
b] I and II are correct
c] II and III are correct
d] III and IV are correct

Consider the following statements:

61. Assertion (A) : Alkali soils are poor in nitrogen
Reason (R) : Nitrogen is leached from alkali soil

Select the correct answer from the code given below:
a] A is true but R is wrong
b] R is true but A is false
c] Both A and R are correct
d] Both A and R are false

62. Neem cake is used to increase the efficiency of urea by:
a] Reducing leaching and denitrification losses
b] Reducing leaching losses only
c] Reducing denitrification losses only
d] Reducing ammonia volatilization

63. The upper limit of available water in soil is determined at:
a] 1/3 bar
b] 1/2 bar
c] 2/3 bar
d] 1 bar

64. Carbon dioxide content in soil air is:
a] 0.03% b] 0.003%
c] 0.25% d] 0.50%

65. How much is the porosity percentage if the particle density of soil is 2.6 and bulk density is 1.3:
a] 35%
b] 40%
c] 50%
d] 55%

66. Read the following statements and mark your answer according to the code given below:
I Organic matter improves the structure of clayey soil
II Organic matter improves the structure of sandy soil
III Sandy soils are heavy in texture
IV Clayey soils are lighter in texture

CODES:
a] I and II are correct
b] I and III are correct
c] I and IV are correct
d] III and IV are correct

67. Read the following statements and mark your answer according to the code given below :
I Alfisols are lighter in texture
II Vertisols are heavier in texture
III Entisols have low potassium content
IV Ultisols have low phosphate fixation
CODES:
a] I and II are correct
b] I and III are correct
c] I and IV are correct
d] III and IV are correct

68. Potassium content is maximum in:
a] Feldspar b] Pyroxine
c] Amphibole d] Olivine

69. Urea supergranule is efficient source of nitrogen for waterlogged paddy because:
a] Its leaching losses are minimum
b] Its ammonia volatilization losses are low
c] Its fixation in soil is low
d] Denitrification losses from it are minimum

70. Which element has greater significance in biological nitrogen fixation:
a] Nickel b] Sulphur
c] Potassium d] Molybdenum

71. Chemical formula of single superphosphate is:
a] $Ca(H_2PO_4)_2$
b] $CaHPO_4$
c] $Ca_3(PO_4)_2$
d] $Ca(H_2PO_4)_2\ CaSO_4.2H_2O$

72. In which of the following, fruits are botanically called as 'Hesperidium'.
a] Jack fruit b] Orange
c] Litchi d] Apple

73. 'Olour' is polyembryonic variety of which fruit:
a] Guava b] Citrus
c] Mango d] Strawberry

74. Which of the following is a stem vegetable:
a] Carrot b] Sweet Potato
c] Knol-khol d] Radish

75. Edible banana fruit is seedless because of:
a] Embryo abortion
b] Absence of ovule
c] Vegetative parthenocarpy
d] Stimulated parthenocarpy

76. 'Ratna' variety of mango is a cross of:
a] Neelam x Dashehari b] Dashehari x Neelam
c] Neelam x Alphonso d] None of these

77. Sodium benzoate is added for the preservation of fruit juice in the concentration of :
a] 0.01 - 0.5% b] 0.06 - 0.10%
c] 1.1 - 1.5% d] 1.6 - 2.0%

78. Which of the following is ethylene absorbent?
a] $KMnO_4$ b] KNO_3
c] K_2SO_4 d] KCl

79. Which is dwarfing rootstock for mango?
a] Olour b] Totapari Red Small
c] Alphonso d] Dashehari

80. Which one of the following is dioecious plant?
a] Peach b] Pomegranate
c] Pea d] Pointed gourd

81. Which one of the following is evergreen temperate plant?
a] Apricot b] Chilgoza
c] Plum d] Almond

82. Which one of the following factor is responsible for higher yield under protected cultivation?
a] High photosynthetic efficiency
b] High O_2 availability
c] High temperature
d] All the above

83. Which one of the following is not a variety of mango?
a] Arka Aruna b] Arka Puneet
c] Arka Anmol d] Arka Hans

84. IAA (Indole-3-Acetic Acid) belongs to which group of plant hormone:
a] Auxin b] Gibberellin
c] Cytokinin d] None of these

85. Which of the following fruits are indigenous to India?
a] Mango, Banana b] Bael, Jack fruit
c] Aonla, Phasla d] All of the above

86. Wax coating treatment enhances the shelf life of fruits because it blocks:
a] Transpiration b] Respiration
c] Ripening process d] None of the above

87. Yellow coloured fruits and vegetables are rich source of:
a] Vit. 'E' b] Vit. 'C;
c] Vit. 'A' d] Vit. 'B'

88. Pusa giant variety of papaya is:
a] Dioecious b] Gynodioecious
c] Monoecious d] Andromonoecious

89. Which onion variety is suitable for Kharif:
a] Pusa Madhvi b] Agrifound Dark Red
c] Punjab Red Round d] None of the above

90. Which of the following is a ripening hormone?
a] Ethylene b] Cytokinin
c] Auxin d] Gibberellin

91. The isolation distance for producing certified seed of cole crops is maintained at:
a] 0.5 km. b] 1.0 km.
c] 1.5 km. d] 2.0 km.

92. Which of the following fruit is not suitable for jam making?
a] Mango b] Banana
c] Aonla d] Lemon

GK

The surface temperature of the Sun is 6000°C.

GS

93. The bitterness in Colocasia corns is due to the presence of:

a] Calcium oxalate
b] Calcium carbonate
c] Calcium chloride
d] Potassium oxalate

94. Pruning in ber is done:

a] During rainy season
b] During winter
c] During spring
d] During summer

95. The primary objective of green house cultivation of vegetables is.:

a] To get early yield
b] To get off season yield
c] To protect the crop from diseases
d] To protect the crop from excessive rains

96. Which one of the following fruits is a rich source of iron:

a] Mango b] Dateplam
c] Apple d] Karonda

97. Potato is planted:

a] In April in high hills of U.P.
b] In July in high hills of U.P.
c] In April in mid hills of U.P.
d] In Feb. in high hills of U.P.

98. Rose oil is primarily extrated form which of following Rosa species.

a] Odorata b] Grandiflorum
c] Indica d] Gallica

99. Spanish jasmine is common name of which Jasminum species.

a] Sambac b] Grandiflorum
c] Dahlia d] Flexible

100. Powdery mildew disease is a serious problem in which of the following crops?

a] Capsicum b] Okra
c] Cucurbits d] Pea

101. Bonsai culture has originated in which country.

a] China b] Korea
c] India d] Japan

102. Which of the following is controlled by Gibberellic acid?

a] Fruit fall
b] Ripening of fruits
c] Vegetative growth
d] Prevention of the loss of flowers

103. Auxin is :

a] An Enzyme b] A Vitamin
c] A Hormone d] A Protein

104. In case of mungbean seed:

a] The amount of carbohydrate is higher than that of protein.
b] The amount of protein is higher than that of carbohydrate.
c] Protein and carbohydrate are present in equal proportion.
d] There is no carbohydrate

105. Identify plant micro-nutrient:

a] Hydrogen b] Sulphur
c] Boron d] Magnesium

106. Anthesis is a phenomenon which occurs:

a] When the flower opens first
b] After pollination
c] When anthers are formed in the flower
d] When fruit drops due to low soil moisture

107. Whip-tail in cauliflower is caused by:

a] Excess of Boron
b] Deficiency of Boron
c] Excess of Molybdenum
d] Deficiency of Molybdenum

108. Rancidity in sunflower oil is caused by:

a] Reduction b] Oxidation
c] Esterification d] Nitrification

109. Which one rice variety has been developed through mutation breeding:

a] Jaya b] Padma
c] Jagannath d] Rasi

110. Opaque-2 maize composites are richer as compared to normal maize in:

a] Tryptophan
b] Lysine
c] Tryptophan and Lysine both
d] Protein

111. Triticale is:

a] Interspecific cross b] Intraspecific cross
c] Integreneric cross d] Intrageneric cross

112. Cabbage is cross pollinated owing to:

a] Floral morphology
b] Protandry
c] Protogyny
d] Self incompatibility

113. Haploids are having:

a] Somatic chromosome number
b] Small chromosomes
c] Very long chromosomes
d] Gametic chromosome number

114. Bulk population breeding is suitable for:

a] Fruit crops b] Vegetable crops
c] Small grain crops d] Flower crops

115. In sugarcane breeding the initial selection after hybridisation is done in the generation :

a] F_0 b] F_1
c] F_2 d] F_6

GK GS

Article 356 talks about State emergency.

116. **The most desirabie principle of insect control is:**
   a] Chemical control
   b] Physical control
   c] Cultural and mechanical control
   d] Integrated control

117. **Which of following is most common rooting hormone**
   a] IBA   b] NAA
   c] 2, 4-D   d] GA

118. **The Food Corporation of India was established in :**
   a] 1960   b] 1962
   c] 1965   d] 1967

119. **Marketing surplus is:**
   a] Total Production : Total Family Requirement
   b] Total Consumption : Marginal Consumption
   c] Total Cost : Marginal Cost
   d] Total Production : Total Cost

120. **Line passing through the least cost points in the isoquants map is called:**
   a] Isocline   b] Ridge line
   c] Expansion path   d] Isoquant curve

Money order system was started in 1880 in India.

Set-7

1. Genes which suppress or enhance the expression of other genes are called
   a] Dominant genes
   b] Recessive genes
   c] Modifier genes
   d] Duplicate genes
2. Which of the following conditions favour natural cross pollination?
1. Monoecious
2. Cleistogamy
3. Protogyny
4. Protandry

Select the correct answer using the codes given below:
CODES:
   a] 1, 2 and 3 b] 2, 3 and 4
   c] 1, 3 and 4 d] 1, 2 and 4
3. A character under study is governed by four allelic pairs. In such a situation, the number of possible genotypes in F_2 generation will be:
   a] 64 b] 84
   c] 87 d] 256
4. Crocus is propagated by ____
   a] Tuber b] Bulb
   c] Corm d] Rhizome

Select the correct answer using the codes given below:
CODES:
   a] 1, 2, 3 and 4 b] 1 and 4
   c] 2 and 3 d] 1, 2 and 3
5. Match List-I (Situation) with List-II (Effect) and select the correct answer using the codes given below the lists:

List-I		List-II	
A.	Cytoplasmic male sterility	1.	Induces dwarfism in wheat
B.	Norin-10 gene	2.	Hybrid seed production
C.	Ear to row breeding	3.	Selection procedure in a highly heritable crop
D.	Nullisomic	4.	Plant with one chromosome less than normal

CODES:

	A	B	C	D
a]	2	1	4	3
b]	1	2	4	3
c]	2	1	3	4
d]	1	2	3	4

6.

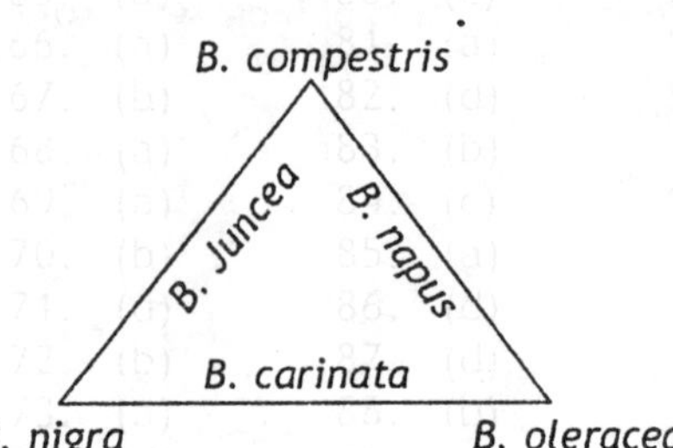

On the basis of the above Brassic triangle, the chromosome number of *Brassica carinata* will be
   a] 4n = 32 b] 4n = 34
   c] 4n = 36 d] 4n = 38
7. The phenotypic ratios of progenies in a cross between Aa Bb x aa bb are as follows:

Phenotypes	Ratio
AB	0.31%
Ab	47.0%
aB	48.6%
ab	0.33%

This may be due to the fact that the genes
   a] A and B are in repulsion phase
   b] A and B are in coupling phase
   c] A and B are linked but situated quite far apart on the chromosome
   d] A and B do not belong to the same linkage group
8. Consider the following steps:
1. Testing the combining ability
2. Production of inbred lines
3. Selection of inbred lines
4. Production of F_1 seed.

The correct sequence of these steps in the process of hybrid seed production is
   a] 1, 2, 3, 4 b] 1, 2, 4, 3
   c] 2, 1, 4, 3 d] 2, 1, 3, 4
9. The general combining ability of a maize inbred may be defined as the average performance of the inbred
   a] In a series of crosses
   b] At least for three years in a particular location
   c] Across location representing different soil and climatic conditions
   d] At several locations over a number of years

10. Birbal Sahani a cultivar of ____ is developed at NBRI Lucknow.
a] Rose b] Marigold
c] Bougainvillea d] Chrysanthemum

11. Consider the following steps:
1. Evaluation of parents
2. Emasculation
3. Pollination
4. Bagging and tagging

The correct sequence of these steps in a breeding programme is
a] 1, 2, 3, 4 b] 1, 2, 4, 3
c] 2, 1, 3, 4 d] 2, 1, 4, 3

12. What would be consequences if the two individuals, female (S_1S_2) is crossed with male (S_3S_4) carrying S alleles (self-incompatible gene)?
a] 100% self compatible
b] 100% self incompatible
c] 50% self compatible
d] 50% self incompatible

13. The principal male sterile line introduced for hybrid pearl millet production in India was
a] Tift 33 A b] BJ -104
c] MBH-118 d] CO-2

14. In pigeonpea, the hybrid seed production is made possible due to
a] Genetic MS lines b] CMS lines
c] CGMS lines d] Embryo rescue

15. Which one of the following genes was utilised in the production of high protein composites in maize?
a] Rht-1 b] Dee Zee Woo Zen
c] Opaque-2 d] Norin

16. Match List-1 (Crop) with List-II (Variety) and select the correct answer using the codes given below the lists:

List-I		List-II	
A.	Rice	1.	UP 262
B.	Wheat	2.	Hybrid Ganga
C.	Maize	3.	Red Kafir
D.	Sorghum	4.	IET 4094

CODES:

	A	B	C	D
a]	4	1	3	2
b]	1	4	2	3
c]	4	1	2	3
d]	1	4	3	2

17. Reselection generation after generation within the intermating of line selects is called
a] Pure line selection
b] Mass selection
c] Transgressive selection
d] Recurrent selection

18. Consider the following fruit crops:
1. Banana
2. Citrus
3. Mango
4. Guava

The correct sequence of these crops in decreasing order of their cultivated acreage in India is
a] 3, 1, 4, 2 b] 3, 1, 2, 4
c] 1, 3, 2, 4 d] 1, 3, 4, 2

19. Which one of the following is a monoembryonic citrus?
a] Sweet orange b] Acid lime
c] Mandarin d] Pummelo

20. Match List-I (Crop) with List-II (Variety) and select the correct answer using the codes given below the lists:

List-I		List-II	
A.	Banana	1.	Arka Puneet
B.	Cabbage	2.	Gros Michel
C.	Mango	3.	Pride of India
D.	Okra	4.	Pusa Sawani

CODES:

	A	B	C	D
a]	3	2	4	1
b]	2	3	4	1
c]	2	3	1	4
d]	3	2	1	4

21. Tangors are the hybrids of
a] Mandarin x Sweet orange
b] Mandarin x Grape fruit
c] Lemon x Lime
d] Tangelos x Sweet orange

22. In Quincunx system of orchard planting, the number of plants required is
a] almost double the number of plants required in square system
b] 50% more than that required in the square system
c] 15% more than required in the square system
d] 5% more than that required in the square system

23. Nucellar seedlings are
a] inferior to their parents
b] weaker and resistant to diseases
c] vigorous and true-to-type
d] superior to their parents

24. Match List-I (Crop) with List-II (Recommended plant density/ha) and select the correct answer using the codes given below the lists:

List-I		List-II	
A.	Banana	1.	70
B.	Guava	2.	660
C.	Fig	3.	3000
D.	Mango	4.	270

GK **Part XVIII of Constitution contains provisions for emergency.** GS

CODES:

	A	B	C	D
a]	4	3	2	1
b]	3	4	1	2
c]	4	3	1	2
d]	3	4	2	1

25. Match List-I (Crop) with List-II (Problems associated with the crop) and select the correct answer using the codes given below the lists:

List-I	List-II
A. Mango	1. Poor fruit set
B. Grape	2. Malformation
C. Pineapple	3. Powdery mildew
D. Custard apple	4. Fasciation

CODES:

	A	B	C	D
a]	3	2	4	1
b]	2	3	4	1
c]	3	2	1	4
d]	2	3	1	4

26. Which of the following sets of crops would be most suitable for intercropping in a bearing young mango orchard?
 a] Soyabean and cowpea
 b] Coriander and cowpea
 c] Soyabean and coriander
 d] Soyabean, coriander and cowpea
27. For pot culture which one of following is important cultivar of bougainvillea.
 a] Begam Sikander b] Surehka
 c] Archana d] Shweta
28. Consider the following statements:
 Regular consumption of fruit and vegetables is necessary to meet the vit. C needs of the human body because
1. All fruits and vegetables contain vit. C
2. Vit. C can neither be synthesized nor stored in human body
3. Vit. C controls rikettsiasis.

Of these statements
 a] 2 alone is correct
 b] 1 and 3 are correct
 c] 1 alone is correct
 d] 1, 2 and 3 are correct
29. Match List-I (Grape cultivars) with List-I (Uses) and select the correct answer using the codes given below the lists:

List-I	List-II
A. Thompson seedless	1. Wine
B. Chemma Sahibi	2. Raisin
C. Peral-of-Casaba	3. Canning
D. Perlette	4. Juice

CODES:

	A	B	C	D
a]	1	2	3	4
b]	2	1	3	4
c]	2	1	4	3
d]	1	2	4	3

30. Amar Shola is a cultivar of
 a] Marigold b] China aster
 c] Verbena d] Amaranthus
31. *Rhizoctonia solani* causes a tuber borne disease of potato, called
 a] Black scurf b] Charcoal rot
 c] Common scab d] Powdery scab
32. What is the correct sequence of the following historical plant diseases epiphytotics?
1. Coffee rust epiphytotics in Ceylon
2. Potato late blight epiphytotics in Ireland
3. Rice leaf spot (*Helminthosporium*) epiphytotics in Bengal.

Select the correct answer using the codes given below:

Cdoes:
 a] 1, 3, 2 b] 1, 2, 3
 c] 3, 1, 2 d] 2, 1, 3
33. Which of the following pairs is correctly matched?
 a] White fly — Yellow mosaic of green gram
 b] Aphid — Mango malformation
 c] Thrip — Groundnut tikka disease
 d] Brown plant hopper — Blast of rice
34. Consider the following stages in the life cycle of the fungus that causes black or stem rust of wheat:
1. Telia
2. Uredia
3. Spermogonia
4. Aecia

The correct sequence of these stages in the life cycle of the fungus is
 a] 1, 2, 3, 4 b] 2, 1, 4, 3
 c] 2, 1, 3, 4 d] 1, 2, 4, 3
35. Which one of the following is the cause of black heart of potato?
 a] Copper deficiency
 b] Boron deficiency
 c] Potassium deficiency
 d] Oxygen deficiency
36. Match List-I with List-II and select the correct answer using the codes given below the lists:

List-I (*Disease*)	List-II (*Genera of fungi*)
A. Bunts	1. Plasmopara, Bremia
B. Downy	2. Neovossia,

	Mildews		Tilletia
C.	Rusts	3.	Sphacelotheca, Tolyposporium
D.	Wilts	4.	Fusarium, Verticillium
		5.	Melampspora, Uromyces

CODES:

	A	B	C	D
a]	2	1	5	4
b]	2	4	3	1
c]	3	2	1	4
d]	4	1	5	3

37. French marigold is ____
   a] Diploid   b] Tetraploid
   c] Triploid   d] Aneuploid
38. Which one of the following fungicides is ineffective against loose smut of wheat?
   a] Baytan   b] Bavistin 50W
   c] Thiram   d] Vitavax 75W
39. Which one of the following types of fungicides is of greatest importance in protecting plants from fungal pathogenes?
   a] Therapeutant fungicides
   b] Eradicant fungicides
   c] Protectant fungicides
   d] Systemic fungicides
40. Which one of the following sets of abbreviations correctly designates pesticide formulation in the solid state?
   a] SD, EC and G   b] SD, EC and WP
   c] SD, G and WP   d] EC, G and WP
41. Which one of the following pesticides in banned in India?
   a] Endrin   b] BHC
   c] Malathion   d] Endosulfan
42. Which of the following would cause development of insect resistance to insecticides?
1. Sub-lethal dose of the insecticide applied
2. Repetitive use of the same insecticide
3. Host plant factors
4. Rigours of abiotic factors

Select the correct answer using the codes given below:

CODES:
   a] 2 and 4   b] 1 and 2
   c] 1, 2 and 3   d] 1, 2, 3 and 4
43. Match List-I with List-II and select the correct answer using the codes given below the lists:

List-I (Pesticides of plant origin)		List-II (Sources)	
A.	Neem Products	1.	*Azadirachta indica*
B.	Nicotine	2.	*Chrysanthemum cinerariaefolium*
C.	Pyrethrum	3.	*Derris elliptica*
D.	Rotenone	4.	*Nicotiana rustica*

CODES:

	A	B	C	D
a]	4	1	2	3
b]	1	4	2	3
c]	1	4	3	2
d]	4	1	3	2

44. Consider the following statements:
Microbial pesticides like *Baccilus thuringiensis* (Bt) and Nuclear Polyhedrosis Viruses (NPV) are well-suited to the integrated pest management programme because
1. The development of resistance in insects against pesticides is nil
2. There is much less possibility of resistance development against such pesticides
3. They are safe and ecofriendly
4. They are inexpensive

Of these statements
   a] 1, 2, 3 and 4 are correct
   b] 2, 3 and 4 are correct
   c] 3 and 4 correct
   d] 2 and 3 are correct
45.

EIL
ETL
EQ
Insect Population
Time

Equilibrum position (EQ), Economics Threshold Level (ETL) and Economic Injury Level (EIL) of an insect pest attacking a particular crop is shown in the above figure. The correct control strategy for this situation would be
   a] Spray insecticide at EIL
   b] Spray insecticide at ETL
   c] Spray insecticide below ETL
   d] None of the above as insecticide spray is not required.
46. Which one of the following is the most suitable substance for calibrating hand sprayer before taking up spraying with pesticides?
   a] Linseed oil   b] Mustard oil
   c] Water   d] Kerosene oil
47. Match List-I (Name of equipment) with List-II (Types) and select the correct answer using the codes given below the lists:

	List-I		List-II
A.	Battery operated CDA sprayer	1.	High volume sprayer
B.	Knapsack power sprayer	2.	Fumigation
C.	Rocking sprayer	3.	Low volume sprayer
D.	Soil injector	4.	ULV sprayer

CODES:

	A	B	C	D
a]	3	4	2	1
b]	4	3	1	2
c]	3	4	1	2
d]	4	3	2	1

48. What is the amount of Bavistin 50 W reqiured to be added to 1000 litres of water to make a spray mixture of 0.05% a.i. concentration?
a] 0.5 kg. b] 1.0 kg.
c] 1.5 kg. d] 2.0 kg.

49. The objective of TRYSEM (Training of Rural Youth for Self-Employment) is to provide technical skills to the rural youth from families below the poverty line. The target group comprises rural youth between the ages of
a] 20 and 25 b] 18 and 35
c] 14 and 35 d] 14 and 30

50. Consider the following statements:
In interpersonal channels of communication
1. Message flow tends to be one-way
2. Quantum of feedback available is very high
3. Communication is face to face
4. Speed of reaching a large audience is relatively slow
Of these statements
a] 1, 2, 3 and 4 are correct
b] 1 and 4 are correct
c] 2, 3 and 4 are correct
d] 2 and 3 are correct

51. Galdioulus has diploid chromosome number ____
a] 60 b] 30
c] 20 d] 40

52. Which one of the following is the special scheme launched by the ICAR for transfer of technology to the farmers?
a] All India Co-ordinated Research Projects
b] Operatioanl Research Project
c] Lab to Land Programme
d] National Demonstration Programme

Directions : The following 9 (Nine) items consist of two statements, one labelled the 'Assertion A' and the other labelled the 'Reason R'. You are to examine these two statements carefully and decide if the Assertion A and Reason R are individually true and if so, whether the Reason is a correct explanation of the Assertion. Select your answers to these items using the codes given below and mark your answer sheet accordingly.

CODES:
a] Both A and R are true and R is the correct explanation of A
b] Both A and R are true but R is not a correct explanation of A
c] A is true but R is false
d] A is false but R is true

53. Assertion (A) : Green manuring in soil results in better utilization of phosphorus by subsequent crops.
Reason (R) : Decomposition of green manure releases carbon dioxide which dissolves in water to form carbonic acid which in turn helps dissolution of phosphorus.

54. Assertion (A) : CCC leads to stunting of plant growth.
Reason (R) : CCC inhibits endogenous synthesis of gibberellins.

55. Assertion (A) : Linkage may be a boon to a plant breeder if desirable genes are linked.
Reason (R) : Linkage increases the frequency of parental types and reduces the frequency of recombinant types when compared to that expected with independent segregation.

56. Assertion (A) : Some rice varieties are photo-sensitive and some are photo-insensitive. Hybridization between these two categories of varieties is very difficult.
Reason (R) : Synchronization of flowering of these types normally does not occur.

57. Assertion (A) : The botanical name of noble cane is *Saccharum officinarum.*
Reason (R) : There is no difference between *Saccharum officinarum* and *S. spontaneum.*

58. Assertion (A) : Frequent picking of tender young fruits increases the yield of Okra.
Reason (R) : Frequent picking promotes successive fruit development.

59. Assertion (A) : Calcium carbide is used to hasten the ripening of mango.
Reason (R) : Calcium carbide releases acetylene which promotes fruit ripening.

60. Assertion (A) : Tobacco mosaic virus (TMV) has a very wide host range.
Reason (R) : TMV has an efficient aphid vector.

61. Assertion (A) : Aircraft spraying may be done from greater heights than dusting.
Reason (R) : Spray droplets fall faster than dust particles.

62. Samrat a tetraploid cultivar of ____ developed at NBRI Lucknow.
a] Amaryllis b] Tuberose
c] Rose d] Verbena

63. Rajat Rekha is a mutant cultivar of ____
a] Dahlia b] Tuberose
c] Rose d] Portulaca

64. The rose in India book is written by ___
a] B.P. Pal b] Vishnu Swarup
c] M.S. Randhawa d] S.L. Jindal

65. In the field of Indian agriculture, the following organizations play an important role:
1. NABARD
2. Agricultural Prices Commission
3. Regional Rural Banks
4. NAFED
The correct chronological order in which these were established is
a] 4, 2, 1, 3 b] 2, 4, 1, 3
c] 4, 2, 3, 1 d] 2, 4, 3, 1

66. Most of the cultivars of chrysanthemum are ____
a] Short day b] Long day
c] Day neutral d] None of above

67. All categories of Indian farmers are becoming increasingly conscious of the geographical development of crop husbandry. Which one of the following is the main reason behind this?
a] Higher labour efficiency
b] Increased irrigation efficiency
c] Suitable ecological efficiency
d] Acceleration in production efficiency

68. Which marginal rate of factor substitution

$\frac{\Delta X_1}{\Delta X_2}$ is more than the inverse price ratio

$\frac{PX_2}{PX_1}$ then for better returns one should use
a] more of X_1
b] more of X_2
c] more of both X_1 and X_2
d] less of both X_1 and X_2

69. Under product-product relationship, if a farmer is producing Y_1 and Y_2 products under non-optimal condition, that is,

$\frac{\Delta Y_1}{\Delta Y_2} > \frac{PY_2}{PY_1}$ then for the most profitable combination of Y_1 and Y_2 he should
a] increase the output of Y_1
b] increase the output of Y_2
c] increase the output of Y_1 and Y_2
d] decrease the output of Y_1 and Y_2

70. For getting large sized flowers which of following operation is important.
a] Pinching b] Disbudding
c] Dis-shooting d] Pruning

71. Which of the following are the characteristics of market structures?
1. Large number of producers are in the industry.
2. Producers tend to make homogeneous products.
3. Non-price competition is absent.
4. Producers can sell any quantity at the going price.
5. Producers are generally willing to produce even at unremunerative prices.
Select the correct answer using the codes given below:
CODES:
a] 1, 2, 3, 4 and 5
b] 2, 3 and 5
c] 1, 4 and 5
d] 1, 2, 3 and 4

72. Which one of following is growth retardant.
a] IAA b] NAA
c] B-Nine d] IBA

73.

I.
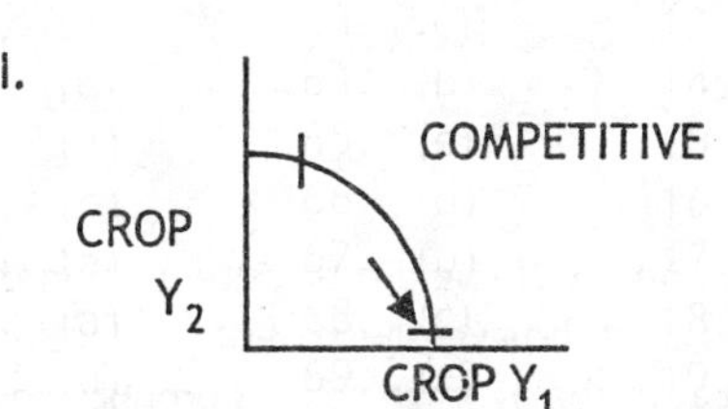

II.
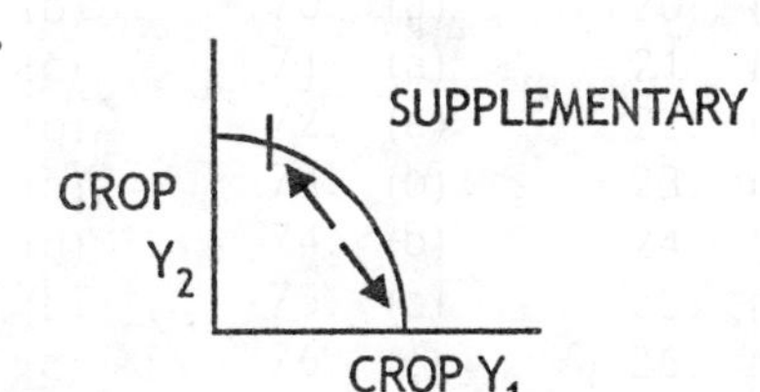

III.
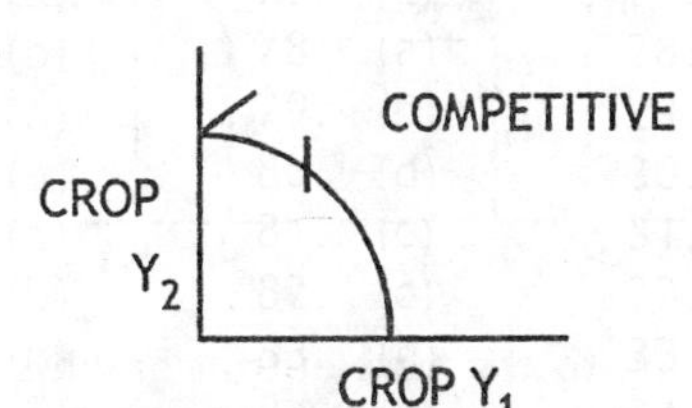

IV.
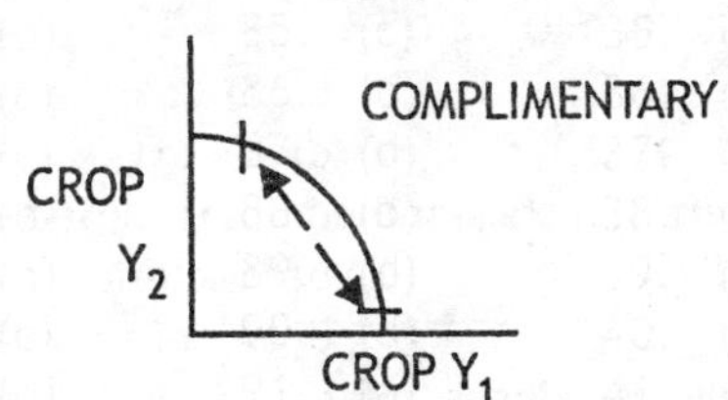

If the data obtained from a particular dose of input in the production of two partially competitive crops are plotted as shown in the above diagrams then the diagram relating to the most profitable crop combination would be the one shown in
a] I b] II
c] III d] IV

Basalt is extrusive igneous rock.

74. Match List-I with List-II and select the correct answer using the codes given below the lists:

List-I		List-II
A.	VMP = MFC	1. Complementary goods
B.	$\frac{\Delta X_1}{\Delta X_2} = \frac{Px2}{Px1}$	2. Product-Product relationship
C.	$\frac{\Delta Y_1}{\Delta Y_2} = \frac{Py2}{Py1}$	3. Factor-Factor relationship
D.	$\frac{MPPx1}{Px1} = \frac{MPPx2}{Px2}$	4. Factor-Product relationship
		5. Equi-marginal returns

CODES:

	A	B	C	D
a]	4	3	2	1
b]	4	3	2	5
c]	4	3	5	2
d]	2	4	3	5

75. Production should be discontinued when the price is
   a] Less than the average cost of production
   b] Equal to the average cost of production
   c] Equal to the marginal cost of production
   d] Less than the average variable cost of production

76.

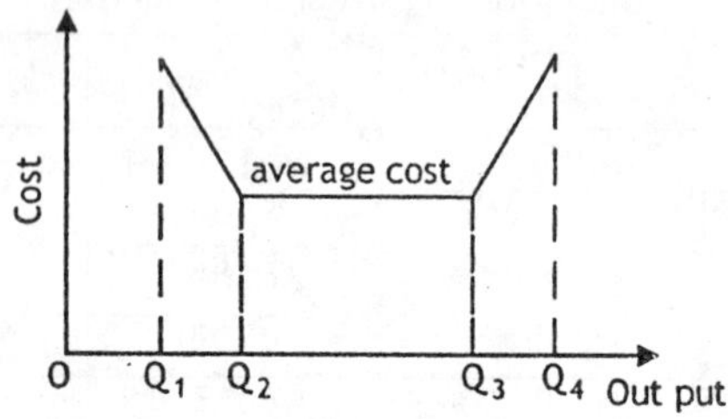

According to the cost-output graph shown in the above figure, increasing returns to scale will be available in Q_1 to Q_2
   a] Q_1 to Q_2 b] Q_1 to Q_3
   c] Q_2 to Q_4 d] Q_3 to Q_4

77. Aspermy in chrysanthemum is caused by ____
   a] Fungus b] Bacteria
   c] Virus d] All of above

78. Gladiolus belongs to family ____
   a] Liliaceae b] Iridaceae
   c] Begoniaceae d] Agavaceae

79. Groundnut is not an economical crop for deep black soils because
   a] Phosphorus requirement is high
   b] Swelling and shrinkage affect root system
   c] Peg penetration is very poor
   d] Drainage conditions are poor

80. Which of the following are used as fibre crops?
1. *Corchorus capsularis*
2. *Helianthus annus*
3. *Linum usitatissimum*
4. *Sesamum indicum*

Select the correct answer using the codes given below:
CODES:
   a] 1 and 2 b] 1 and 3
   c] 2 and 4 d] 3 and 4

81. In gladiolus flowering size corm having ____ cm diameter is optimum.
   a] 1-2 b] 2-3
   c] 3-5 d] 5 or above

82. Which of the following agro-climatic factors lead to successful sugar beet production?
1. Loam or sandy loam soils.
2. Soil salinity between 1.0 and 1.5%.
3. Average temperature of 30°C and above throughout the growing period.
4. Average temperature between 19.5°C and 22°C throughout the growing period.

Select the correct answer using the codes given below:
CODES:
   a] 2 and 4 b] 1, 2 and 4
   c] 1, 2 and 3 d] 1 and 3

83. Which of the following would one find in a submerged rice field?
1. Ammoniacal nitrogen
2. Ferrous sulphide
3. Reduced state

Select the correct answer using the codes given below:
CODES:
   a] 3 alone b] 1 alone
   c] 1 and 2 d] 1, 2 and 3

84. Water use efficiency is defined as
   a] dry matter produced per unit of water applied crop
   b] economic yield per unit of irrigation water applied
   c] ratio of water used by the crop to the irrigation water applied
   d] ratio of net irrigation to gross irrigation including other losses

85. Gross water requirement of a crop is the
   a] sum of net irrigation requirement, effective rainfall and other unavoidable losses during irrigation.
   b] total available water which is present in the soil profile during the crop period
   c] same as total rainfall in the crop period under heavy rainfall areas
   d] evapotranspiration and water used by the crop for its metabolic activities

86. Which of the following statements explain the true relationship between soil water potential and moisture content of loam sandy and clay soils?
1. Clay soils hold much more water at a given potential than loamy soils.
2. Loamy soils hold much more water at a given potential than clayey soils.
3. Loamy soils hold much more water at a given potential than sandy soils.
4. Clay soils hold much less water at a given potential than loamy soils.

Select the correct answer using the codes given below:
CODES:
a] 1 and 2 b] 2 and 3
c] 3 and 4 d] 1 and 3

87. Match List-I (Herbicides) with List-II (Chemical group) and select the correct answer using the codes given below the lists:

List-I	List-II
A. Paraquat	1. Phenoxy alkanoic acid
B. Dalapon	2. Aliphatic acids
C. 2-4-D	3. Bipyridillium compounds
D. Alachlor	4. Acetanilides

CODES:

	A	B	C	D
a]	3	2	4	1
b]	3	2	1	4
c]	2	3	1	4
d]	2	3	4	1

88. Match List-I (Crop) with List-II (Associated weed) and select the correct answer using the codes given below the lists:

List-I	List-II
A. Tobacco	1. *Cichorium intybus*
B. Lucern	2. *Cuscuta* sp.
C. Berseem	3. *Sorghum halepense*
D. Sugarcane	4. *Orobanche* sp.

CODES:

	A	B	C	D
a]	4	2	3	1
b]	2	4	1	3
c]	2	4	3	1
d]	4	2	1	3

89. Match List-I (Common name of herbicide) with List-II (Trade name of the herbicide) and select the correct answer using the codes given below the lists:

List-I	List-II
A. Alachlor	1. Basalin
B. Oxadiazon	2. Stomp
C. Pendimethalin	3. Ronstar
D. Fluchloralin	4. Lasso

CODES:

	A	B	C	D
a]	3	4	1	2
b]	4	3	1	2
c]	4	3	2	1
d]	3	4	2	1

90. Corm dormancy in gladiolus is broken by, which of following chemicals.
a] IAA b] GA
c] SADH d] 2, 4-D

91. The primary aim of multiple cropping is to
a] Have higher gross return
b] Increase food production
c] Maintain soil fertility
d] Provide technology for achieving the goal of higher level of land use intensity

92. The main objective of growing the "catch" crop is to
a] Add more residue to the soil
b] Suppress weeds
c] Prevent cracking of soil
d] Get an additional income without further investment

93. Corm dormancy in gladiolus is due to ______ acids.
a] Linolenic b] Linoleic
c] Stearic d] All of these

94. Silica-sesquioxide ratio is useful in the evaluation of soil in respect of
a] Nutrient availability
b] Silicon content
c] Iron and aluminum content
d] Its age

95. Consider the following statements:
Cation exchange of soil colloidal clay is due to
1. silicon tetrahedral layer
2. isomorphous substitution
3. hydration of clays
4. broken edges of clay lattices

Of these statements
a] 1 and 2 are correct
b] 2 and 4 are correct
c] 3 and 4 are correct
d] 1 and 3 are correct

96. The ratio of 'organic carbon' to 'organic matter' in soils is
a] 1.0 : 1.7 b] 1.0 : 2.0
c] 1.7 : 1.0 d] 20.0 : 1.0

97. A given soil sample contains
organic carbon = 0.6%
phosphorus = 0.1%
potassium = 1%
The per cent organic matter, P_2O_5 and K_2O contents of the soil is

	Organic matter	P_2O_5	K_2O
a]	1.02	0.29	1.2
b]	1.02	0.229	1.2

GK Sandstone is the most familiar sedimentary rock. GS

c] 1.2 2.29 0.83

d] 1.7 0.86 0.43

98. The molarity of 100 ml. of a solution containing 0.36 g of NaOH will be

a] 0.009 b] 0.090

c] 4.100 d] 1.440

99. The bulk density of a soil with particle density of 2.65 g/cc and pore space 40% would be

a] 1.29 g/cc b] 1.45 g/cc

c] 1.59 g/cc d] 1.72 g/cc

100. Consider the following types of erosion:

1. Gully erosion
2. Splash erosion
3. Normal erosion
4. Rill erosion
5. Sheet erosion

The correct sequence in terms of the increasing order of soil loss from the field due to these types of erosions is

a] 2, 3, 4, 5, 1 b] 2, 3, 5, 4, 1

c] 3, 2, 4, 1, 5 d] 3, 2, 5, 4, 1

101.

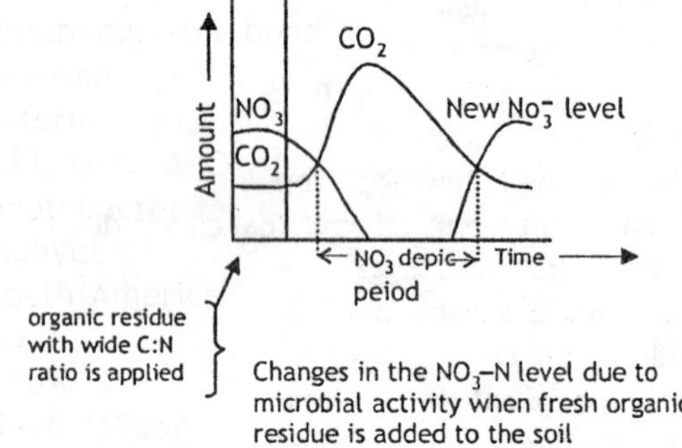

Consider the following statements regarding the diagram given above:

1. Nitrate is retained by the soil and hence the nitrate-nitrogen is temporarily immobilised during decomposition.
2. Higher nitrate level observed at the end of decomposition is due to the release of this nutrient by the soil.
3. The peak level of CO_2 and NO_3 observed in the diagram indicates the maximum activity of microbes developed during decomposition

Of these statements

a] 1, 2 and 3 are correct

b] 1 and 2 are correct

c] 2 and 3 are correct

d] 3 alone is correct

102. Non-essential but useful nutrients for plants would include

a] carbon and vanadium

b] vanadium and chlorine

c] vanadium and sodium

d] sodium and chlorine

103. Which of the following are the criteria for determining the essentiality of nutrient elements?

1. A deficiency of the particular element makes it impossible for the plant to grow normally or complete its life cycle.
2. The deficiency symptoms can be corrected only by supplying the nutrient in question.
3. The nutrient is abundantly present in the soil for rapid plant absorption.
4. The element is directly involved in the plant metabolism.

Select the correct answer using the codes given below:

a] 1 and 3 b] 1, 2 and 4

c] 2, 3 and 4 d] 1, 2, 3 and 4

104. Which of following is most serious fungus in gladiolus.

a] Fusarium b] Alternation

c] Botrytis d] Curvularia

105. "A-value" technique is used to estimate the available phosphorus in the soil. It is calculated using the formula

$$A = \frac{B(1-7)}{y}$$

where, 'A' is the amount of available soil phosphorus in kg/ha, 'B' is the amount of added fertilizer phosphorus in kg/ha and 'Y' is the

a] yield of the plant grown in the soil

b] total quantity of phosphorus present in the soil

c] total quantity of phosphorus present in the plant

d] fraction of the phosphorus in the plant which is derived from the phosphatic fertilizer added to the soil

106. Match List-I (Technique) with List-II (Organism) and select the correct answer using the codes given below the lists:

List-I	List-II
A. Colwell technique for B deficiency	1. *Aspergillus niger*
B. Sackett and Stewart method for P deficiency	2. *Cunninghamela* sp.
C. Mulder method for Cu and Mg	3. Sunflower
D. Mehlich plaque method for P deficiency	4. *Azotobacter*

GK Acancagua is the highest peak of Andes mountains. GS

CODES:

	A	B	C	D
a]	4	3	2	1
b]	3	4	2	1
c]	3	4	1	2
d]	4	3	1	2

107. Which one of the following nitrogen factors would bring down wide C : N ratio material of 100 : 1 to the desirable C : N ratio of 40 : 1 for compost making ?

a] 1.0 b] 1.5
c] 2.0 d] 2.5

108. Consider the following characteristics of legume root nodule:

1. Large and elongated
2. Often clustered on primary root
3. Pink to red centre.

Among these, the effective root nodules would include

a] 2 and 3 b] 1 and 2
c] 1 and 3 d] 1, 2 and 3

109. Consider the following statements regarding the dung slurry coming out of a biogas plant:

1. The slurry contains nitrogen both in the ammoniacal and non-ammoniacal form.
2. The nitrogen percentage of the slurry is much higher than that of the dung used in the gas plant.
3. The carbon-nitrogen ratio of the slurry from the gas plant is 10:1

Of these statements

a] 1 and 2 are correct
b] 1 and 3 are correct
c] 2 and 3 are correct
d] 1, 2 and 3 are correct

110. Nitrogen fertilizer use efficiency in rice can be increased by using

a] sulphur-coated urea
b] blue green algae
c] urea super granules
d] sulphur-coated urea and urea super granules

111. Following fertilizers are available for preparing a fertilizer mixture:

1. Ammonium sulphate
2. Urea
3. Single superphosphate
4. Basic slag
5. Muriate of potash

A combination of the above- mentioned fertilizers which can be mixed and kept for three months or more is

a] 1, 4 and 5 b] 2, 3 and 4
c] 2, 4 and 5 d] 1, 3 and 5

113. Which one of the following mechanisms is valid for non-competitive enzyme inhibition?

The inhibitor

a] does not combine with the enzyme
b] has close structural similarly to the substrate
c] combines with the enzyme at a location different from where the substrate attaches itself
d] forms a complex with the substrate at a specific point to improve efficiency

113. Plant roots are

a] positively phototropic
b] negatively hydrotropic
c] negatively geotropic
d] positively geotropic

114. Which one of the following nutrients plays the most effective role in controlling the rate of transpiration?

a] Phosphorus b] Potasssium
c] Zinc d] Magnesium

115. Match List I (Diagnostic symtpoms) with List II (Deficient nutrient) and select the correct answer using the codes given below the lists:

List-I		List-II	
A.	Death of apical buds and leathery leaves	1.	Zinc
B.	Uniform paling of plants without much of stunting	2.	Sulphur
C.	Uniform paling of top leaves & lower leaves remain green	3.	Molybdenum
D.	Interveinal chlorosis of newly formed leaves and reduction in leaf size	4.	Boron

CODES:

	A	B	C	D
a]	4	3	1	2
b]	3	4	1	2
c]	3	4	2	1
d]	4	3	2	1

116. Consider the following processes:

1. Oxidation of carbohydrates
2. Oxidation of pyruvic acid
3. Release of CO_2 water and energy

The correct sequence of these processes in respiration of upland paddy grown at an altitude of 725 metres is

a] 1, 3, 2 b] 3, 1, 2
c] 2, 1, 3 d] 1, 2, 3

117. Match List I and List II and selected the correct answer using the codes given below the lists:

List-I		List-II	
A.	Transport of electrons from Ferredoxin to P 700	1.	Oxidative phosphorylation
B.	Transport of electrons from PS II to PS I	2.	Substrate level phosphorylation

GK

CAT was set up on October 2, 1985.

GS

C.	Generation of ATP by mitochondrial oxidation of NADPH	3.	Cyclic phosphorylation
D.	Generation of ATP by the conversion of PEP to pyruvate	4.	Non-cyclic phosphorylation

CODES:

	A	B	C	D
a]	3	4	1	2
b]	2	3	4	1
c]	1	2	3	4
d]	4	1	2	3

118. Almost 90% of the total output of photosynthesis in the form of glucose is credited to

a] Green planktonic microflora of the oceans and seas
b] Grasses and shrubs
c] Gymnosperms
d] Angiosperms

119. Which of the following is/are the reason(s) for poor fruit set of tomato?

1. Average atmospheric temperature higher than the optimum.
2. Average atmospheric temperature lower than the optimum.
3. Severe attack by the leaf curl virus.
4. Wide fluctuations in the average atmospheric temperature.

Select the correct answer using the codes given below:

CODES:

a] 1 and 4
b] 3 alone
c] 2, 3 and 4
d] 1, 2, 3 and 4

120. Which one of the following plant hormones shows energy dependent polar transport?

a] Cytokinin
b] Ethylene
c] Indole-3-acetic acid
d] Gibberellic acid

Vosges of France and Black forest of Germany are Block mountain. GS

Set-8

1. Developments of farm sector gives a boost to secondary sector and development of primary and secondary sectors give a boost to the tertiary sector. This is called the "Product Contribution of Agriculture". This concept was coined by
   a] R.H. Tuck b] Ragner Frisch
   c] Simon Kuznet d] F.H. Knight
2. Agro-climatic Regional Planning was initiated in India with the inception of
   a] The First Five Year Plan
   b] Grow More Food Campaign
   c] In late eighties
   d] The Seventh Five Year Plan
3. Rice production is highest in areas having
1. High rainfall
2. Canal irrigated lands
3. Bright sunny days during and after anthesis

Choose the correct combination using the codes given below:
   a] 1, 2 and 3 b] 1 and 2
   c] 2 and 3 d] 1 and 3
4. The Ecosymstem determining agroecological zones has two components viz. biotic and abiotic. Abiotic components comprise of
   a] Light, water and animals
   b] Minerals, plants and soil organisms
   c] Materials like water, minerals, salts, atmospheric gases; energy like light, heat and stored energy
   d] Water, wind and plants
5. Farm Management as a resource allocating proposition is related to
   a] Inter-farm management
   b] Intra-farm management
   c] Both inter and intra farm management
   d] Inter-regional management.
6. In the figure given below product Y_1 is denoted on Y axis and product Y| on the X axis. The price of a unit of Y_1 and Y_2 are PY_1 and PY_2 respectively.

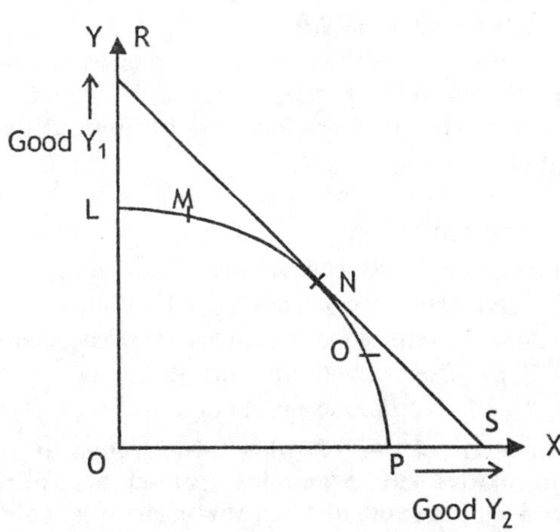

Point N in the figure corresponds to the point where

   a] $\frac{\Delta Y_1}{\Delta Y_2} = \frac{PY_1}{PY_2}$ b] $\frac{\Delta Y_1}{\Delta Y_2} = \frac{PY_2}{PY_1}$

   c] $\frac{Y_2}{Y_1} = \frac{PY_1}{PY_2}$ d] $\frac{\Delta PY_1}{\Delta PY_2} = \frac{Y_2}{Y_1}$

7. The paid-out costs of a farmer is Rs. 500 and fixed costs is Rs. 250. If the gross returns from farm business is Rs. 3,700, the capital turnover would be
   a] 17.42 b] 9.80
   c] 4.93 d] 3.33
8. Queen Elizabeth is a common cultivar of _____.
   a] Rose b] Marigold
   c] Carnation d] Dahlia
9. In the short run, Average Cost, Average Variable Cost and Marginal Cost curve are 'U' shaped due to the operation of
   a] Law of equimarginal returns
   b] Substitution between proportion
   c] Law of variable proportion
   d] Time comparison principle
10. Farm system of the Western Himalayan Region of India is mostly dominated by
    a] Arable farming b] Plantation crops
    c] Pasture d] Horticulture
11. The size of holding which provides full employment to the farm family provides a reasonable standard of living could be termed as
    a] Economic holding b] Optimum holding
    c] Basic holding d] Family holding
12. Ability of a system (say agriculture) to maintain inter-regional equity and enhance, if required, the flow of its products and services by way of link ,es with other systems without damaging its own long-term potential is called

Political equality is stressed under Article 325.

GS

a] An efficient system
b] A sustainable system
c] A productive system
d] None of the above

13. **Crop mixtures have proved biologically most dynamic under**
a] Water stress conditions
b] Irrigated conditions
c] High soil fertility conditions
d] Large holding conditions

14. **In Jute growing areas the usual alternate crop is**
a] Sugarcane b] Cotton
c] Rice d] Wheat

15. **Which one of the following terms issued to denote the relative land area under sole crop required to produce the same yield as obtained under sole crop required to produce the same yield as obtained under a mixed or an intercropping system at the same management level?**
a] Land utilization index
b] Cropping intensity
c] Land equivalent ratio
d] Relay cropping

16. **Pulses fit well in cropping systems as they are**
a] Short duration crops
b] Long duration crops
c] Diseases-resistant crops
d] Moisture stress-resistant crops

17. **As per principles of crop rotation pulses should be included in rotation because**
a] They increase total pulse production
b] They release land quickly for the second crop
c] They help in maintenance of soil fertility
d] They increase the field of succeeding cereal crop.

18. **Which one of the following pairs is not correctly matched?**

a] Relay cropping	:	Sowing pulse in rice crop prior to harvest
b] Inter cropping	:	Wheat after rice
c] Multi-tier cropping	:	Coconut-Pepper - Cacao-Pineapple
d] Sole cropping	:	Potato

19. **Which one of the following is the principal 'N' fixing algal community of blue-green algae?**
a] Rhizobia b] Anabaena
c] Azotobactor d] Azospirillium

20. **Groundnut pegs when developed in the soil form**
a] Roots b] Stems
c] Tubers d] Fruits

21. **Fire curing is followed in**
a] Chewing type of tobacco
b] Bidi tobacco
c] Hookah tobacco
d] Cheroot tobacco

22. **Consider the following physiological stages of wheat**
1. Crown root initiation 2. Seedling
3. Tillering 4. Ear emergence
The correct sequence of the above physiological stage is
a] 2, 3, 1 and 4 b] 2, 3, 4 and 1
c] 2, 1, 3 and 4 d] 1, 2, 3 and 4

23. **Low quality of Jute fibre is attributed to**
1. Discolouration of fibre
2. Very short fibre strands
3. Reduction of cellulose in fibre
4. Reduction of cell wall strength
Over-retting decreases the fibre quality due to
a] 3 and 4 b] 1 and 3
c] 1 and 4 d] 1 and 2

24. **Which of the following two crops are responsible for almost 75% of pulse production in India?**
a] Gram and Moongbean
b] Pigeonpea and Moongbean
c] Moongbean and Lentil
d] Gram and Pigeonpea

25. **Triticale is a cross between**
a] Wheat and Barley b] Barley and Rye
c] Wheat and Rye d] Wheat and Oat.

26. **An International Crop Research Institute called CIMMYT in Mexico deals with two crops grown widely in India. These crops are**
a] Rice and jute b] Rice and wheat
c] Maize and wheat d] Maize and millet

27. **Which one of the following occupies the highest volume per cent in a fine sandy loam surface soil in good condition for plant growth?**
a] Organic matter
b] Pore space which includes air and/or water
c] Mineral matter
d] Others including macro and micro flora and fauna

28. **Pedology considers soil as a**
a] Natural body
b] Purely a synthesised one
c] Medium for plant growth
d] Powdered rock mass of the earth's crust

29. **Chlorites are basically silicates of**
a] Magnesium with some iron and aluminum
b] Iron with some magnesium and aluminium
c] Aluminium with some iron and magnesium
d] Iron and aluminium.

30. **The cation exchange capacity of a mineral soil containing predominantly kaolinitic clay minerals generally ranges between**
a] 3-15 m.e. per 100 gm. soil
b] 15-40 m.e. per 100 gm. soil
c] 80-100 m.e. per 100 gm. soil
d] more than 100 m.e. per 100 gm. soil

31. **In soil science micelle stands for**
    a] Colloidal particle carrying positive charge
    b] Colloidal particle carrying negative charge
    c] Micropores in between textural particles
    d] Negatively charged silt particles with high CEC
32. **Soils high in swelling clays which crack widely upon drying resulting in shrinking, shearing and soil mass movement are characteristics of**
    a] Ultisols b] Entisols
    c] Podzols d] Vertisols
33. **Total porosity of a soil with 1.325 g/cc B.D. and 2.65 g/cc P.D. is (BD: Bulk density, PD: Particle density)**
    a] 30% b] 40%
    c] 50% d] 60%
34. **Which one of the following micro-organisms is responsible for maximum nutrient cycling in the soil?**
    a] Bacteria b] Actinomycetes
    c] Fungi d] Algae
35. **Which one of the following statements is correct?**
    a] Rhizobia fix nitrogen in soil non-symbiotically
    b] Azotobactors fix nitrogen symbiotically in association with legumes
    c] Blue green algae fix nitrogen in paddy fields under submerged conditions
    d] Clostridia fix nitrogen symbiotically in association with legumes
36. **When organic matter with wide C: N ratio (70:1) is added to the soil, the net effect is the disappearance of mineral nitrogen due to**
    a] Mineralization
    b] Immobilization
    c] Nitrogen transformations
    d] Nitrogen reduction
37. **The addition of excess soluble phosphatic fertilizers to soils may affect plant growth by**
    a] inhibting the growth of microorganisms
    b] precipitating iron and aluminium
    c] reducing the availability of zinc
    d] reducing the availability of calcium
38. **As plants grow to maturity, the concentration of plant nutrients changes. What kind of change occurs in respect of N, P, K, Ca and Mg?**
    a] Concentration of all the five elements decreases
    b] N and P decrease, K, Ca and Mg increase
    c] N, P and K decrease, Ca and Mg increase
    d] N, P and K increase, Ca and Mg decrease
39. **In order to assess the immediate fertility level of a given soil for potassium, which one of the following forms of potassium present in soil would one test for?**
    a] Total K b] Exchangeable K
    c] Water soluble K d] K held in the silt
40. **What culture should be given priority in groundnut cultivation?**
    a] Azospirilla b] Mycorrhiza
    c] Phosphobacteria d] Rhizobia
41. **Which one of the following concentrated organic manures has maximum phosphorus content?**
    a] Fish meal b] Stera meal
    c] Poultry manure d] Guano
42. **Match List-I with List-II and select the correct answer using the codes given below the lists:**

	List-I (Classes of fertiliser)		List-II (Example of different classes)
A.	Organic fertiliser	1.	Urea
B.	Inorganic fertiliser	2.	Isobutylidene diurea
C.	Biofertiliser	3.	Ammonium sulphate
D.	Slow release fertiliser	4.	Azolla
		5.	2-chloro-6-pyridene

CODES:

	A	B	C	D
a]	1	2	3	5
b]	1	3	4	2
c]	2	1	5	4
d]	5	1	4	3

43. **What is the effect of application of phosphatic fertilizer to a green manure leguminous crop on the availability of phosphorus to the succeeding crop?**
    a] Greater amount of phosphorus is taken up and converted into organic form, which is released on decomposition of green matter
    b] Green manure increases the availability in inorganic form to the next crop
    c] It does not affect the availability of phosphorus to succeeding crop
    d] It fixes the phosphorus through humus and reduces the availability
44. **The nitrogenous fertilizer manufactured in India which occupies the first place in production today is**
    a] Ammonium sulphate
    b] Ammonium phosphate
    c] Urea
    d] Calcium ammonium nitrate
45. **Which one of the following fertilisers is least hygroscopic?**
    a] Urea
    b] Ammonium sulphate
    c] Diammonium phosphate
    d] Calcium ammonium nitrate
46. **Diamonium phosphate requirement for supplying 45 kg. N and 115 kg. P_2O_5 to one hectare Peanut crop is**
    a] 100 kg. b] 150 kg.
    c] 200 kg. d] 250 kg.

47. **Maximum absorption of water by roots takes place through the**
a] Zone of elongation b] Zone of maturation
c] Root hair d] Root cap

48. **Long distance translocation of nitrogenous assimilations in plant occurs fats and carbohydrate**
a] Through cambium only
b] Mainly through the phloem
c] Mainly through the xylem-inorganic salts
d] Equally through xylem, phloem and cambium

49. **Zinc, iron, copper, boron, manganese and molybdenum are called micronutrients because they are**
a] nutrients of microscopic significance
b] nutrients which produce microscopic effects
c] essential nutrients but are required in minute quantities
d] nutrients required by micro-organisms

50. **Which one of the following often acts as the 'limiting factor' synthesis under Indian conditions?**
a] Sunlight b] Carbon dioxide
c] Water d] Chlorophyll

51. **The strongest chemical agent for breaking dormancy is**
a] Potassium nitrate b] Gibberellin
c] Ethylene d] Cumerin

52. **The optimum cardinal temperature points for germination of rice seeds are**
a] 18-20°C b] 20-22°C
c] 30-32°C d] 37-39°C

53. **All cytokinins are derivatives of**
a] Cytosine b] Adenine
c] Uracil d] Guanine

54. **Under normal aerobic conditions, pyruvic acid is oxidised to carbon dioxide and water in cellular organelles known as**
a] Golgi apparatus b] Ribosome
c] Chloroplast d] Mitochondria

55. **Interveinal chlorosis of the younger leaves of the plants is the characteristic symptom of**
a] Boron deficiency b] Zinc deficiency
c] Iron deficiency d] Phosphorus deficiency

56. **The high yielding rust resistant wheat variety 'Kalyan Sona' become susceptible to rust some years after its release and cultivation because of**
a] Variation in the environment
b] Mutation in the host plant
c] Resurgence on the part of the pathogen
d] Emergence of new races in the pathogen

57. **The first maize hybrid developed in India was**
a] Vikram b] Ganga-2
c] Vijay d] Kissan

58. **What is the correct sequence of the following for the production of hybrid maize?**
1. **Selfing selected open-pollinated populations through several generations to produce homozygous inbred lines**
2. **Selecting desirable plants in open-pollinated populations**
3. **Crossing chosen lines**

Select the correct answer using the codes given below:
a] 1, 2 and 3 b] 2, 3 and 1
c] 2, 1 and 3 d] 3, 2 and 1

59. **Genetics is the study of**
a] Genes
b] Gene Interaction
c] Heredity and Variation
d] DNA

60. **At the molecular level point mutations are due to**
a] Destruction of the bases
b] Destruction of the double helix
c] Shifting of a portion of the helix and bases
d] Alterations of sequences of base in DNA

61. **The rice variety containing "Dee-geo-woo-gen" gene is**
a] Indrasan b] Basmati
c] Tilak d] I.R.-8

62. **Specific combining ability of the parents should be high in order to develop**
a] Hybrids b] Composites
c] Varieties d] Synthetics

63. **A population at equilibrium developed from intermating a number of inbred lines is called a**
a] Composite variety b] Synthetic variety
c] Hybrid variety d] Pureline variety

64. **In sorghum somatic chromosome number is 20. What would be its chromosome number in the endosperm?**
a] 10 b] 20
c] 30 d] 40

65. **Heterosis is commercially exploited in:**
a] cross-pollinated species
b] often cross-pollinated species
c] self-pollinated species
d] vegetatively propagated species

66. **In which one of the following crops maximum exploitation of hybrid vigour during last three decades has been done?**
a] Wheat b] Rice
c] Maize d] Cotton

67. **A homozygous line developed by self fertilization in a cross pollinated crop is termed as**
a] Pureline b] Inbred
c] Clone d] Mutant

Cotopaxi in Equator is the highest active volcanic mountain of world. GS

68. There are four stages of development of Genetics up to 20th Century. The correct order of their occurrence is
    a] Darwinism, Lamarckism, Mendelism, Weismannism
    b] Lamarckism, Darwinism, Weismannism, Mendelism
    c] Lamarckism, Weismannism, Darwinism, Mendelism
    d] Weismannism, Lamarckism, Darwinism, Mendelism
69. A method of breeding most commonly used when the desired variation is required to be induced in a vegetatively propagated crop is
    a] Polyploidy breeding
    b] Pedigree method of breeding
    c] Mutation breeding
    d] Backcross method of breeding
70. With four bases A, U, G and C the possible number of triplet codons are 64. If there is no uracil, the number of codons will be
    a] 3/64 b] 9/64
    c] 27/64 d] None of the above
71. A mendelian population means
    a] Self pollinated population
    b] Inbred population
    c] Random-mating population
    d] Hybrid population
72. Which one of the following is the site of protein synthesis?
    a] Ribosomes b] t-RNA
    c] m-RNA d] r-RNA
73. When homologous chromosomes fail to pair in Prophase-I of meiosis, the phenomenon is known as
    a] Desynapsis b] Non-disjunction
    c] Asynapsis d] Disjunction
74. Which one of the following chemicals is used to induce polyploidy?
    a] Ethyl methane sulphonate
    b] Methyl methane sulphonate
    c] Colchicine
    d] Nitrous acid
75. Which one of the following will be most suitable for growing under arid zone of India?
    a] Jack fruit b] Papaya
    c] Sweet Orange d] Ber (Jujube)
76. Which of the following stone fruits are grown in the plains of North India?
1. Peach
2. Plum
3. Pear

Select the correct answer from the codes given below:
    a] 1 and 2 b] 2 and 3
    c] 1 only d] All 1, 2 and 3
77. In North Indian plains which is a superior rootstock of roses.
    a] Rosa indica b] Rosa multiflora
    c] Rosa bourboniana d] Rosa gallica
78. Which one of the following soils is best for vegetable cultivation?
    a] Sandy b] Sandy loam
    c] Clay loam d] Clay
79. The most widely used plant growth substance in grape culture for reducing compactness of bunches and improving berry size is
    a] N.A.A. b] GA_3
    c] I.A.A. d] 2, 4-D
80. The ill effects of submergence of roots of deciduous fruit plants in water for very long time, is due primarily to
    a] Lack of nutrients
    b] Lack of aeration
    c] Excess of moisture
    d] Poor absorption of moisture
81. Which one of the following groups of crops is most suitable and recommended for intercropping in the young orchards?
    a] Short duration fodders
    b] Short duration fruit crops
    c] Poplars and Eucalyptus.
    d] Short duration legume vegetables
82. The main objective of growing intercrops with the main fruit crop is to
    a] Help the fruit crop to grow better
    b] Improve the soil fertility
    c] Check soil erosion
    d] None of the above for addition and income
83. In relation to crop rotation which one of the following statements is incorrect?
    a] Legume may be followed by non-legume
    b] Vegetable susceptible to a particular pest should be followed by a resistant one
    c] Heavy feeders should be followed by low feeders
    d] Deep rooted crop should be followed by same type of crop
84. Which one of the following vegetables is the richest source of protein?
    a] Pea b] Fenugreek
    c] Pointed Gourd d] Cucumber
85. Which one of the following is the main cause for development of colour pigment in the flavedo of mature ripe orange fruit?
    a] Hesperidin b] Naringin
    c] Carotene d] Chlorophyll
86. Which one of the following fruits will be most useful in human diet for avoiding scurvy disease?
    a] Mandarin Orange b] Guava
    c] Aonla d] Apple
87. Citrus fruits can be preserved for a few weeks at room temperature by
    a] submerging them in cold water and then drying
    b] keeping them in well ventilated room
    c] putting them in an atmosphere rich in CO_2
    d] by wrapping individual fruit in HDP envelopes

The Election Commission was set up under Article 324.

88. Sodium benzoate is added for preservation of most of fruit juices in the concentration of
a] 0.01-.05%
b] 0.06-.10%
c] 1.1-1.5%
d] 1.6-2.0%

89. What is the chemical formula of potassium metabisulphite, the most common preservative?
a] KMS b] K_2SO_2
c] KSO_2 d] $K_2SO_2O_3$

90. What is the correct sequence in which the following events occur during pathogenesis or disease cycle?
1. Penetration
2. Colonization
3. Infection
4. Exit of pathogen
Select the correct answer from the codes given below:
a] 1, 3, 2, 4 b] 1, 2, 3, 4
c] 3, 1, 2, 4 d] 4, 2, 3, 1

91. Which one of the following is the cause of "spreading decline" of citrus?
a] Citrus nematode
b] Burrowing nematode
c] Fungal diseases
d] Nutritional deficiency

92. The incidence of "White grubs" is heavy in
a] Heavily water-logged fields
b] Dry fields having sandy loam soil
c] Dry fields of heavy clay soil
d] Desert land having little vegetation

93. Which one of the following helps in maintaining the shape of nematodes?
a] Pseudocoelomic fluid
b] Cuticle and muscles
c] Pseudocoelome
d] Annulation of cuticle

94. Green leaf hopper of paddy is the primary vector of
a] Tungro disease b] Sheath rot disease
c] Kresek disease d] Ufra disease

95. Bacterial leaf blight of rice caused by *Xanthomonas oryzae* can be identified by
a] Wilting of the plant b] Yellowing of leaves
c] Ooze test d] Defoliation

96. White ear-head in paddy is caused by
a] Gall fly b] Green leaf hopper
c] Yellow stem borer d] Stink bug

97. Match List-I with List-II and select the correct answer using the codes given below the lists:

List-I (*Contribution*)	List-II (*Scientist*)
A. A fungus is the cause of wheat-bunt disease	1. K.C. Mehta
B. Modern techniques of growing microorganisms in pure culture	2. T.O. Diener
C. Discovery of diseases cycle of wheat rust in India before 1947	3. Prevost
D. Introduced the term 'viroid'	4. Brefeld

	A	B	C	D
a]	1	2	3	4
b]	2	3	1	4
c]	3	4	1	2
d]	4	1	2	3

98. Which one of the following types of spraying is best done by manually operated sprayer having flat-fan nozzle?
a] High volume b] Low volume
c] Ultra low volume d] Semi-low volume

99. Domestic quarantine exists in India against
1. Panama disease of banana
2. Potato root eelworm
3. San Jose scale
4. Sugarbeet eelworm
Select the correct answer from the codes given below
CODES:
a] 1 and 4 b] 1 and 2
c] 2 and 3 d] 2, 3 and 4

100. The spray is classified as aerosol when average droplet size (volume median diameter) is
a] 50 μ b] 50-100μ
c] 100-200μ d] 200-400μ

101. Match List-I with List-II and select the correct answer using the codes given below the lists:

List-I (Equipments)	List-II (Insecticides/ pesticides)
A. High volume sprayer	1. Dusts
B. Blowers	2. Wettable powder
C. Back pack sprayer	3. Liquids
D. Rocker sprayer	4. Emulsions

CODES:

	A	B	C	D
a]	1	2	4	3
b]	2	1	3	4
c]	3	4	1	2
d]	4	3	2	1

GK — Nagoya city of Japan is called 'Detroit of Japan' because of the excellent automobile industry. — GS

102. Pneumatic sprayers have been used in pest control for many years. A pneumatic sprayer
   a] Operates using wind or air
   b] Contains poisonous pesticide
   c] Is a knapsack type sprayer
   d] Operates automatically

103. Solar energy treatment of wheat seed against loose smut was given by
   a] Jensen in 1889
   b] Swingle in 1892
   c] Luthra and Sattar in 1934
   d] Freeman and Johnson in 1909

104. Which one of the following equipments can be effectively used for spraying an orchard crop?
   a] Barrel pump b] Sikar pump
   c] Foot sprayer d] Heli sprayer

105. Match List-I with List-II and select the correct answer using the codes below the lists:

List-I (Equipments)		List-II (Mode of Use)	
A.	Rotary duster	1.	Ultra low volume (ULV) formation
B.	Package duster	2.	Small scale use
C.	Heli sprayer	3.	Feeding brush
D.	Pedel pump	4.	High volume (HV) spray

CODES:

	A	B	C	D
a]	2	1	4	3
b]	3	2	1	4
c]	1	3	4	2
d]	4	3	2	1

106. In case of rotary type of duster the air current is developed by
   a] Bellows
   b] Fans
   c] Self-propelling jacket
   d] Self-propelling blades

107. Which one of the following can be classified as hydraulic energy sprayer?
   a] Knapsack mist blower
   b] Controlled drip applicator sprayer
   c] Knapsack sprayer
   d] Wheel barrow sprayer

108. The two important principles of extension education are
   a] participation and leadership
   b] participation and classroom teaching
   c] leadership and classroom teaching
   d] leadership and subject-matter knowledge

109. Two which one of the following categories the discipline of extension education belong?
   a] Pure science
   b] Applied biological science
   c] Applied behavioural science
   d] Art

110. Which one of the following extension programmes linked farm plan with credit for farmers?
   a] Integrated Rural Development Programme
   b] Community Development Programme
   c] Intensive Agricultural Development Programme
   d] Etawah Pilot Project

111. Puppet show in villages as a method for transfer of technology is an example of
   a] Mass media
   b] Group media
   c] Individualized media
   d] Tribal media

The following 9 items consist of two statements, one labelled the 'Assertion A' and the other labelled the 'Reason R'. You are to examine these two statements carefully and decide if the Assertion A and the Reason R are individually true and if so, whether the Reason is a correct explanation of the Assertion. Select your answer to these items using the codes given below and mark your answer sheet accordingly:

CODES:
   a] Both A and R are true and R is the correct explanation of A
   b] Both A and R are true but R is not a correct explanation of A
   c] A is true but R is false
   d] A is false but R is true

112. Assertion (A) : Since independence the share of agriculture in national income has been decreasing steadily.
   Reason (R) : The percentage of population dependent on agriculture has by and large remained constant since independence.

113. Assertion (A) : During the green revolution both the yield per hectare and area under wheat increased sharply.
   Reason (R) : India is following the former Soviet pattern of agricultural development.

114. Assertion (A) : The wage rate of agricultural labourers is not equal at all times in the year.
   Reason (R) : The seasonality of farm work causes seasonal variations in the opportunity cost of labour.

115. Assertion (A) : Extension is taken as essential and major part of technology transfer.
   Reason (R) : It is incorrect to equate 'Agril. Extension' with the term 'Technology Transfer' that includes

the additional functions of input supply and agri.-services.

116. Assertion (A) : Pulses with tap root system do not complete for nutrients and moisture with cereals having fibrous root system.

Reason (R) : Pulses when included in inter-cropping with cereals act as complimentary with cereals having fibrous root system.

117. Assertion (A) : Symbiotic nitrogen fixation occurs in the nodules of leguminous plants.

Reason (R) : Nitrate reductase activity occurs in Jules.

118. Assertion (A) : All pests are not insects.

Reason (R) : There are a number of helpful and productive insects.

119. Assertion (A) : A multigrade motor oil should be mixed with petrol used as fuel for a 2-strokee engine for motorized knapsack sprayer.

Reason (R) : Multigrade motor oils are blended from different grades of lubricating oils and are suitable for engines subjected to great temperature variations.

120. Assertion (A) : Smut diseases (loose and covered smut) of wheat are seed-borne and can be effectively controlled by treating the seeds with a systemic fungicide.

Reason (R) : Systemic fungicide used (at the recommended dose) have no adverse effect on seed germination.

Set-9

1. Electrical conductivity of saturation extract of saline-sodic soils in m mhos per cm is:
   a] 2.0 b] between 2.0 and 3.0
   c] 3.5 d] More than 4.0
2. Deficiency symptoms of calcium on plants are first noticed at:
   a] Lower leaves b] Middle leaves
   c] Terminal leaves d] All leaves
3. Kinnow is a hybrid variety of:
   a] Sweet orange b] Mandarin
   c] Grape fruit d] Lemon
4. Onion variety suitable for growing in plains during the rainy season is:
   a] Hisar-2 b] Pusa-red
   c] Patna-red d] Niphad-53
5. Which variety is resistant to mango malflormation?
   a] Langra b] Mallika
   c] Deshehari d] None
6. When organic matter is added to the soil:
   a] Bulk density is increased
   b] Bulk density is decreased
   c] Bulk density does not change
   d] None of these
7. The availability of molybdenum is more at:
   a] pH-5.0 b] pH-7.0
   c] pH-7.5 d] pH-9.0
8. Which one of the following pairs is correctly matched?
   a] Pineapple - Berry
   b] Apple - Pepo
   c] Strawberry - Sorosis
   d] Mango - Drupe
9. "Seed plot technique" is followed in :
   a] Wheat b] Paddy
   c] Potato d] Bajra
11. Azotobacter fixes atmospheric nitrogen:
   a] Symbiotically
   b] Non-symbiotically
   c] Both symbiotically and non-symbiotically
   d] None of the above is correct
12. The soil is well flocculated if it has more of:
   a] Sodium b] Zinc
   c] Potassium d] Calcium
13. Rock phosphate is used in :
   a] Neutral soil b] Acidic soil
   c] Alkaline soil d] Saline soil
14. Improved Golden Yellow and California Advance are varieties of:
   a] Mandarin b] Loquat
   c] Apple d] Papaya
15. Papain is extracted from:
   a] Poppy b] Papaya
   c] Pine d] Rubber
16. What will be the moisture content of soil at 50% A.S.M., if it has field capacity and permanent wilting point values of 20% and 8% respectively?
   a] 18% b] 14%
   c] 12% d] 10%
17. For control of weeds in peas, herbicide used is :
   a] 2, 4-D
   b] Dosanax
   c] M C P B
   d] Atrazine
18. Montmorillonite clay has silica and alumina layers respectively in the following proportion:
   a] 2 : 1 b] 1 : 2
   c] 3 : 2 d] 2 : 3
19. Which one of the following micro-organisms is involved in nitrification in the soil?
   a] Pseudomonas b] Micro-coccus
   c] Thio-bacillus d] Nitrosomonas
20. White grub beetles lay eggs on /in:
   a] Twigs b] Underside of leaves
   c] Soil d] Muddy water
21. Pyrethrum is extracted from the :
   a] Leaves of *Nicotiana tobacum*
   b] Flowers of *Nicotiana tobacum*
   c] Stems of *Chrysanthemum* Sp.
   d] Flower petal of *Chrysanthemum* sp.
22. Which one of the following is a growth retardant:
   a] 2, 4, 5-T b] Cycocel
   c] CEPA d] TIBA
23. The correct pair is :
   a] Rice-white grub b] Potato-white ants
   c] Sugarcane-weevil d] Gram-pod borer
24. Pollen basket is found in the hind legs of:
   a] Termite b] Honeybee
   c] Ant d] Butterfly
25. Why insecticidal spraying should be avoided when crops are in flowering stage?
   a] To prevent flower dropping
   b] To facilitate proper fruit setting
   c] To protect bees
   d] To prevent insecticidal residues

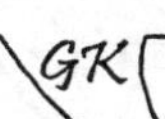

GK GS

K.S. Kelelker was the first chairman of the committee appointed for welfare of OBC.

26. **Which one is the anticoagulant?**
a] Zinc phosphide
b] Aluminium phosphide
c] Magnesium oxide
d] Bromodiolone

27. **Under integrated system of insect control we integrate:**
a] Cultural and biological methods
b] Biological and legal methods
c] Physical and chemical methods
d] All types of methods

28. **The least toxic insecticides to human beings belong to the following group:**
a] Organochlorines b] Organophosphates
c] Carbamates d] Synthetic pyrethroids

29. **Yellow mosaic in soyabean is spread by :**
a] Seeds b] Soils
c] Air d] White fly

30. **Loose smut of wheat is controlled by:**
a] Seed treatment b] Soil sterilization
c] Crop rotation d] Thriam spray on crop

31. **Systemic insecticide used to control pests is:**
a] Endosulphan b] Chlorpyriphos
c] Phosphamidon d] Quinophos

32. **Orobanche is a :**
a] Total root parasite
b] Total stem parasite
c] Partial root parasite
d] Partial stem parasite

33. **Canker diseases of citrus is caused by :**
a] Bacteria b] Fungus
c] Virus d] Nematodes

34. **'Tangru' is a viral disease of:**
a] Rice b] Citrus
c] Wheat d] Tobacco

35. **When marginal product is increasing at increasing rate then:**
a] Total product increases at a decreasing rate
b] Total product increases at a increasing rate
c] Average product decreases
d] None of the above

36. **Bharat Sundari is a famous cultivar of ____ developed at IIHR, Banglore.**
a] Rose b] Hibiscus
c] Gladiolus d] Bougainvillea

37. **Profit is maximised when:**
a] Marginal Cost < Marginal return
b] Marginal Cost = Marginal return
c] Marginal Cost > Marginal return
d] None of the above

38. **Television, radio, fairs and exhibitions are the media of :**
a] Group Communication
b] Individual Communication
c] Community Communication
d] Mass Communication

39. **Central Soil Salinity Research Institute is located at:**
a] Jodhpur b] Karnal
c] Dehradoon d] Hyderabad

40. **Highest number of sugar mills in India is located in :**
a] Maharastra b] Tamil Nadu
c] Uttar Pradesh d] Bihar

41. **Thornless rose rootstock was developed at ____**
a] IARI b] NBRI
c] IIHR d] BARC

42. **The term extension was first used by:**
a] America b] England
c] India d] Japan

43. **Serialise the following stages of adoption process:**
A. Interest
B. Evaluation
C. Trial
D. Awareness
Pick out the answer from the following:
a] A B C D b] C A B D
c] D A C B d] C B A D

44. **Sustainable agriculture can be achieved by :**
a] More resources
b] More subsidies
c] More technological support
d] Better prices of the produce

45. **Which one is not a system of farming?**
a] State farming
b] Collective farming
c] Cooperative farming
d] Dry farming

46. **Which of following Floridunda rose cultivar has been patented in USA.**
a] Chitchor b] China
c] Mohini d] Sindhoor

47. **Mendelian factors segregate in the process of:**
a] Hybridization b] Mitosis
c] Breeding d] Meiosis

48. **Plants bend towards light because:**
a] They need light for respiration
b] Light attracts them
c] They need light for photosynthesis
d] Cells on shaded side elongate more

49. **What is a:**
a] C_3 plant b] C_4 plant
c] Both d] None of the above

GK **Krivoi Rog is the famous iron ore field in Ukrain.** GS

50. Paddy has :
a] 3 stamens b] 4 stamens
c] 5 stamens d] 6 stamens

51. Cultivation of hybrid rice is most popular in:
a] Japan b] Korea
c] China d] India

52. International Crops Research Institute for Semi-Arid Tropics is situated at:
a] Jodhpur b] Bharatpur
c] Hyderabad d] Colombo

53. Self-Compatibility is helpful in the development of :
a] Synthetic variety
b] Hybrid variety
c] Pure line variety
d] Composite variety

54. Seedless variety of Guava is:
a] Haploid b] Diploid
c] Anueploid d] Tetraploid

55. The translocation of- sugars from source to sink takes place through:
a] Sieve tubes of the phloem
b] Companion cells
c] Xylem vessels
d] Medullary rays

56. Which one of the following amino acids is synthesized during photorespiratory metabolism?
a] Alanine b] Glutamic acid
c] Serine d] Tryptophan

57. Gibberellins produces:
a] Internodal elongation
b] Curvature of coleoptile
c] Root initiation
d] Cell division

58. Multilines are produced by mixing equal amounts of seeds of:
a] Genetically dissimilar lines
b] Isogenic lines
c] Inbred lines
d] Synthetic lines

59. The rate of- water absorption by plants is decreased due to deficiency of :
a] Boron b] Potassium
c] Iron d] Manganese

60. The secondary plant element is:
a] Nitrogen b] Calcium
c] Zinc d] Boron

61. The chromosome number of green gram is:
a] 12 b] 24
c] 36 d] 48

62. Which one of the following is an auxin?
a] BA b] SADH
c] PCPA d] MH

63. Segregation of gene occurs in :
a] F_1 generation
b] F_2 generation
c] Genetic formation of F_1
d] F_3 generation

64. Pungency in mustard oil is caused by :
a] Amino b] acid
c] Glucosilates d] None of the above

65. First cultivar of rose released by Dr. B.P. Pal was ____ in 1962.
a] Rose Sherbet b] Dulhan
c] Aruna d] Mehak

66. In a cell, energy in the form of ATP is formed in :
a] Ribosomes b] Mitochondria
c] Nucleus d] Golgi complex

67. Seed viability is tested by use of:
a] Bromide b] Iodide
c] Chloride d] None of the above

68. The weight of a half mole of water is:
a] 0.5 gram b] 1.0 gram
c] 9.0 gram d] 18.0 gram

69. Ratio obtained in Inhibitory Gene Action is:
a] 9 : 7 b] 9 : 3 : 4
c] 15 : 1 d] 13 : 3

70. 'Azolla' (a nitrogen fixing organism) is in symbiotic associated with :
a] Fern and blue green algae
b] Bryophyta and blue green algae
c] Fungi and blue green algae
d] Nostoc and blue green algae

71. What is the hereditary material in chromosomes?
a] RNA b] ATP
c] DNA d] ADP

72. Dark reaction of photosynthesis occurs in :
a] Grana b] Stroma
c] Thylakoid d] Osmium particles

73. The core metal in chlorophyll is :
a] Fe b] Mg
c] Mn d] Cu

74. Which of the following crops has protogyny nature?
a] Jowar b] Bajra
c] Maize d] Sugarbeet

75. Pedigree selection is used in :
a] Cross-pollinated plants
b] For crosses at intergeneric level
c] Self-pollinated crops
d] Plants difficult to be hybridized

76. The stomata in succulents open in:
a] Day time b] Night time
c] Early morning d] Do not open at all

GK

Article 366 defines an Anglo-Indian.

GS

77. **Phosphorus is taken by plants in the form of :**
a] Phosphorus chloride
b] Phosphoric acid
c] Phosphorus sulphite
d] Phosphorus carbonates

78. **Which of the cell-organ functions are suicide bag?**
a] Vacuole b] Mitochondria
c] Lysozome d] Golgi complex

79. **"Ear cockel" disease of wheat is caused by:**
a] Fungi b] Bacteria
c] Virus d] Nematode

80. **'Saw fly' damages toria crop at :**
a] Early stage of crop growth
b] Filling of siliqua
c] Maturity stage
d] None of the above

81. **'Winter Banana' is a variety of :**
a] Grape b] Pineapple
c] Banana d] Apple

82. **Which one is not correctly matched?**

Crop	**Seed rate (kg/ha)**
a] Gram	80
b] Til	4
c] Mustard	10
d] Sunflower	12

83. **Origin of soyabean is :**
a] India b] U.S.A.
c] Japan d] China

84. **Central Upland Rice Research Station is located at :**
a] Raipur b] Masodha
c] Hazaribagh d] Cuttack

85. **International Bougainvillea registration center is located at____**
a] IIHR b] NBRI
c] IARI d] TNAU

86. **Which of following is growing as a long stem cut rose in India.**
a] Rakagandha b] Anurag
c] Dr. Homi Bhabha d] Mother Teressa

87. **Green revolution in wheat is basically attributed to :**
a] Greater use of seeds of high yield varieties
b] Higher use of N, P, K fertilizers
c] Greater availability of irrigation
d] Greater use of plant protection chemicals

88. **Which one of the following vegetables contains maximum vitamin A?**
a] Radish b] Turnip
c] Brinjal d] Carrot

89. **Floribunda cultivar ____ of rose is also used as cut flower.**
a] Shola b] Rose Sherbet
c] Sindhoor d] Mohani

90. **Chromosome number in bread wheat is :**
a] 22 b] 32
c] 42 d] 62

91. **Which one of the following is a false fruit?**
a] Mango b] Citrus
c] Guava d] Tomato

92. **Choose the right pair:**
a] Arkel-Pea
b] Dania-Coriander
c] Arka Chandan - Watermelon
d] Ratna-Radish

93. **Whip tail in cauliflower is caused due to :**
a] Deficiency of iron
b] Deficiency of boron
c] Deficiency of molybdenum
d] Deficiency of sulphur

94. **The edible portion of sweet potato is :**
a] Modified root b] Modified stem
c] Leaf d] Fruit

95. **SAR of alkali soil is :**
a] Less than 15 b] More than 15
c] Less than 13 d] More than 13

96. **Urea hydrolysis in the soil produces:**
a] Uric acid
b] Ammonium carbomate
c] Ammonium carbonate
d] Amide

97. **Paddy crop can be successfully grown under waterlogged conditions because:**
a] Paddy roots do not require O_2
b] Paddy roots have capacity to absorb atmospheric O_2 directly
c] Needed O_2 is supplied from the foliar parts of paddy to the roots
d] Needed O_2 is supplied by the micro-organisms present in the soil

98. **'Spongy tissue' is a physiological disorder in the fruit of :**
a] Mango b] Litchi
c] Guava d] Pomegranate

99. **Sunflower oil is rated as a good quality oil because it contains:**
a] Low quantity of saturated fatty acids
b] Low quantity of unsaturated fatty acids
c] High quantity of saturated fatty acids
d] High quantity of unsaturated fatty acids

100. **Banjaran cultivar of rose belongs to ____ category.**
a] Hybrid Tea b] Polyantha
c] Floribunda d] Tea

101. **What is the form in which zinc is absorbed by the plants?**
a] Zn+ b] ZN++
c] ZN+++ d] None of these

102. **Vegetative parthenocarpy occurs in:**
a] Peach and Plum
b] Bael and Ber
c] Banana and Pineapple
d] Mango and Cashewnut

Lorraine is the famous iron ore field of France.

103. Which of following is very good rootstock of bougainvillea.
a] Dr. R. R Pal b] Sonnet
c] Thimma d] Stanz(a)

104. Which group of microbes increases phosphate solubility?
a] Azatobactor
b] Pseudomonas, Thiobacillus
c] Clastridium
d] Rhizobium

105. Maturity in sugarcane can easily be determined by hand refractometer when its reading is:
a] 10 - 15 b] 17 - 18
c] 20 - 22 d] More than 22

106. Pusa Basanti is a open pollinated cultivar of ____
a] Zinnia
b] Chrysanthemum
c] Rose
d] Marigold

107. Main objective of puddling in paddy is to :
a] Destroy insects-pests
b] Create impermeable layer in the soil
c] Mix the fertilizer in the soil
d] None of the above is true

108. Sulphur coated urea contains nitrogen:
a] 36% b] 25%
c] 40% d] 46%

109. Mayur a cultivar of gladiolus was developed at ____
a] NBRI b] IARI
c] IIHR d] PAU

110. Greening of potato results in :
a] Increase in nutritional quality
b] Decrease in nutritional quality
c] Increase in disease resistance
d] Decrease in disease resistance

111. Hawaiin Hibiscus are propagated easily by ____ budding.
a] T b] Ring
c] Patch d] Chip

112. The size of the clay particle in soil falls in one of the range:
a] 0.25 - 0.10 mm. b] 0.10 - 0.05 mm.
c] 0.05 - 0.002 mm. d] < 0.002 mm.

113. Which one of the following is not correctly matched:

	Crop	Disease
a]	Mustard	Alternaria blight
b]	Urd (Black gram)	Sterility mosaic
c]	Soyabean	Yellow mosaic
d]	Groundnut	Bud necrosis

114. Variety of arhar suitable for double cropping system is :
a] Bahar b] Type 17
c] Type 7 d] UPAS 120

115. Potato is propagated through :
a] Tuber b] Micro-tuber
c] T P S d] All of these

116. Loss of electron is known as :
a] Oxidation b] Reduction
c] Catalysis d] Hydrolysis

117. Bottom heating in ___ cuttings enhances rooting percentage.
a] Hibiscus b] Rose
c] Bougainvillea d] Dahlia

118. Internal necrosis of amla is due to:
a] Zinc deficiency
b] Sulphur deficiency
c] Boron deficiency
d] Iron deficiency

119. Suitable temperature for tuberization in potato is :
a] 12^0 -14^0C
b] 18^0 - 20^0C
c] 22^0 - 24^0C
d] 28^0 - 30^0C

120. Ammonia is converted into nitrate by:
a] Fungus b] Algae
c] Earthworms d] Bacteria

Set-10

1. The most convenient and safer chemical to control stored grain insects in rural areas is :
   a] E D B b] Celphos
   c] D D T d] Naphthalene ball
2. Insects with sucking type of mouth parts need following group of insecticides for effective control :
   a] Stomach poisons b] Systemic poisons
   c] Contact poisons d] Fumigants
3. Insecticides under Integrated Control System are applied at which one of the following stage :
   a] General equilibrium level
   b] Economic threshold level
   c] Economic injury level
   d] Total loss level
4. Cynogas pump is a :
   a] Sprayer b] Duster
   c] Flame thrower d] Fumigator
5. Which one of the following is the richest source of Vitamin 'A'?
   a] Banana b] Wheat
   c] Rice d] Mango
6. 'Aphids' in mustard can be controlled by the use of :
   a] Malathion b] Vitavex
   c] thiram d] Dithane-Z
7. Dahila is propagated by ____
   a] Tuber b] Tuberous root
   c] Rhizome d] Root
8. Maize-toria-wheat-moong rotation is termed as a system of :
   a] Mixed cropping b] Relay cropping
   c] Multiple cropping d] Intercropping
9. Three varieties with two seed rate were replicated thrice in an experiment. The number of degrees of freedom will be :
   a] (3 x 2 x 3) - 1 = 17
   b] (3 x 2 x 3) - 2 = 16
   c] (3 x 2 x 3) - 3 = 15
   d] (3 x 2 x 3) + 1 = 19
10. Origin of brinjal is :
    a] Africa b] India
    c] China d] South America
11. Coarse grained crops are grown on:
    a] High rainfall areas
    b] Water-logged areas
    c] Lowland areas
    d] Upland areas
12. Swami Vinayannanda of R.K. Math has done pioneer work in ____
    a] Dahila b] Gladious
    c] Rose d] Marigold
13. Which one of the following crops is grown in all seasons :
    a] Sugarcane b] Groundnut
    c] Lentil d] Maize
14. Growing plants of different heights in the same field at the same time is known as :
    a] Multiple-cropping
    b] Multi-storeyed cropping
    c] Relay cropping
    d] Mixed cropping
15. Water use efficiency is the highest in case of :
    a] Flood irrigation
    b] Border irrigation
    c] Sprinkler irrigation
    d] Drip irrigation
16. Potato tuber is modified form of :
    a] Root b] Stem
    c] Leaf d] Flower
17. Berseem is a fodder crop and is recommended for sowing in the months of :
    a] March-April
    b] June-July
    c] October-November
    d] January-February
18. Dahlia was first introduced by Agri- Horticultural Society of India at Calcutta in ____
    a] 1935 b] 1857
    c] 1982 d] 1905
19. The late cutting method of dahlia propagation was developed in ____ at Calcutta.
    a] 1935 b] 1857
    c] 1960 d] 1981
20. Richest source of Nitrogen is :
    a] Dried blood
    b] Raw bone meal
    c] Fish manure
    d] Dry activated sludge
21. Chrysanthemum in Hindi is known as ____.
    a] Gyaneshwari b] Guldaudi
    c] Daud d] Mum

GK **Masai people are found in East Africa.** GS

22. Geological erosion is also known as:
 a] Accelerated erosion
 b] Natural erosion
 c] Soil erosion
 d] Water erosion

23. Production of a new variety or a hybrid by crossing together two genotypically different individuals is referred to as :
 a] Mass selection b] Recombination
 c] Hybridization d] None of these

24. Which is the hereditary material in chromosomes :
 a] RNA b] ATP
 c] DNA d] ADP

25. Clonal selection is a common practice in which of the following:
 a] Mango b] Rice
 c] Sugarcane d] Banana

26. Which one of the following is not an element of communication :
 a] Message b] Feed-back
 c] Channel d] Fidelity

27. The profit can be maximised by the use of Law of Diminishing Return when :
 a] TP is maximum b] MP = AP
 c] MP > AP d] MC = MP

28. Which system of planting in an orchard would facilitate a farmer to plant maximum number of guava plants :
 a] Hexagonal b] Rectangular
 c] Contour d] Square

29. Browning in cauliflower can be corrected by use of :
 a] Ammonium molybdate
 b] Zinc sulphate
 c] Boric acid
 d] Urea

30. Vijay, Amber, Sona and Kisan are composite varieties of :
 a] Wheat b] Rice
 c] Maize d] Jowar

31. The seed rate of mustard per hectare is :
 a] 1 - 2 kg. b] 3 - 4 kg.
 c] 5 - 6 kg. d] 7 - 8 kg.

32. Planting of succeeding crop before harvest of standing crop is known as :
 a] Mixed cropping
 b] Relay cropping
 c] Intercropping
 d] Multiple cropping

33. Fruit and shoot borer is a serious pest of :
 a] Bhindi b] Brinjal
 c] Tomato d] Cabbage

34. Which one of the following chemicals is used for treatment of sugarcane setts :
 a] Dithane M-45 b] Agrosan G.N.
 c] Dimecron d] Aretan

35. Soil fertility is reduced due to :
 a] Continuous cropping
 b] Imbalanced use of fertilizers
 c] Over irrigation
 d] Poor drainage

36. In sugarcane wheat intercropping, sugarcane is planted in the month of :
 a] October b] February
 c] March d] April

37. Berseem seeds are inoculated with:
 a] Azolla culture
 b] Azotobacter culture
 c] Rhizobium culture
 d] None of these

38. Scorching or burning on margins of bottom leaves manifests the deficiency of :
 a] Nitrogen b] Zinc
 c] Copper d] Potassium

39. Which one is barley variety :
 a] Sonalika b] Naveen
 c] Azad d] Sona

40. Parthenocarpy occurs in :
 a] Mango b] Peach
 c] Jackfruit d] Banana

41. 'Diara' lands are :
 a] Areas flooded by sea
 b] Areas created by sand storms
 c] Areas on both sides of canal
 d] Areas located on either side of river which get flooded every year

42. In china and Japan Chrysanthemum is known as ______.
 a] Hill queen
 b] Autumn queen
 c] Japanese queen
 d] Chinese queen

43. Which one is a monocot vegetable?
 a] Carrot b] Parwal
 c] Fenugreek d] Colocasia

44. Which one is an insect trap crop ?
 a] Arhar b] Paddy
 c] Bhindi d] Potato

45. Non-essential but beneficial element of plant is :
 a] Calcium b] Sodium
 c] Potassium d] Magnesium

46. Carbon is an essential element for plant growth. Plants take it from :
 a] Soil b] Water
 c] Air d] Light

47. Winter maize is recommended to be sown on slopes of ridges :
 a] Southern b] Northern

GK **Article 343 describes the official languages.** GS

c] Eastern d] Western

48. Motha belongs to the family :
a] Gramineae b] Malvaceae
c] Cyperaceae d] Liliaceae

49. The ideal transplanting age of onion seedlings is between :
a] 2 - 4 wk b] 5 - 6 wk
c] 8 - 10 wk d] 12 - 14 wk

50. The most abundant acid in grape is :
a] Tartaric acid b] Citric acid
c] Ascorbic acid d] Mallic acid

51. During snowfall, the trees :
a] Do not respire
b] Do not photosynthesise
c] Show minimum life function
d] Show maximum transpiration

52. The Elasticity of Production (EP) at the starting point of Stage II, in Law of Diminishing Return would be :
a] > 1.0 b] 1.0
c] < 1.0 d] 0.0

53. Extension education focuses upon development of :
a] An individual b] A family
c] A social group d] A community

54. The category of seed being the costliest one is :
a] Foundation b] Truthful
c] Certified d] Breeder

55. Repaying capacity of loan of a farmer is decided on the basis of :
a] Total cultivated land
b] Crops and livestock raised
c] Gross income of the farmer
d] Net income of the farmer

56. The Block Development Council is chaired by :
a] Village Development Officer
b] Block Pramukh
c] Block Development Officer
d] District Agricultural Officer

57. The Directorate of Extension at the National level is located in the Ministry of :
a] Agriculture and Co-operation
b] Communication
c] Human Resource Development
d] Social Welfare

58. Which Plan ended before the completion of five years:
a] Fourth Plan b] Fifth Plan
c] Sixth Plan d] Third Plan

59. Appropriate time for planting of grapes in U.P. is :
a] June-July
b] March-April
c] September-October
d] December-January

60. Pusa Naveen is a variety of :
a] Bottle-gourd b] Bitter-gourd
c] Cauliflower d] Radish

61. The temperate fruit is :
a] Apple b] Guava
c] Kagzi lime d] Mango

62. Seedless fruits of banana are due to :
a] Seedlessness
b] Absence of ovules
c] Vegetative parthenocarpy
d] Stimulative parthenocarpy

63. For raising seedlings for planting an acre of papaya garden, quantity of seed required would be :
a] 100 - 200 gm. b] 50 - 100 gm.
c] 250 - 300 gm. d] 350 gm.

64. 'First Prize' is a variety of :
a] Bougainvillea b] Rose
c] Dahalia d] Chrysanthemum

65. The fruit richest in ascorbic acid is :
a] Lime b] Lemon
c] Aonla d] Jamun

66. Potassic fertilizer containing maximum quantity of potash is:
a] Muriate of potash
b] Sulphate of potash
c] Potassium nitrate
d] Potassium ammonium nitrate

68. Dr. M (A) Kher eminent scientist for NBRI Lucknow was associated with ______
a] Chrysanthemum
b] Dry flower
c] Rose
d] China aster

69. The farm-business income is:
a] Farm net income
b] The farm expenditure
c] The family labour income
d] The farm net income + unpaid charges of family labour + interest on owned capital

70. The extension work must be based on needs and interests of :
a] People b] Leader
c] Official d] Government

71. Which is the most effective light in photosynthesis:
a] Blue b] Green
c] Yellow d] Red

72. In glycolysis, conversion of a molecule of glucose to two molecules of Pyruvic acid results in a net gain of:
a] 8 ATP molecules
b] 38 ATP molecules
c] 2 ATP molecules
d] 36 ATP molecules

73. The number of ATP that forms during the Kreb's cycle of respiration will be:
a] 8 b] 15
c] 22 d] 30

GK GS

Blind valley is Karst Landform.

74. **Genes are present:**
a] Near the chromosomes
b] On the chromosomes
c] In the chromosomes
d] In between the chromosomes

75. **Traslocation of water takes place through which one of the following tissue system:**
a] Xylem b] Phloem
c] Cambium d] Cortical tissue

76. **The plants meet their carbon requirements by absorbing:**
a] Soil minerals
b] Soil organic matter
c] Carbon monoxide from atmosphere
d] Carbon dioxide from atmosphere

77. **Bending of plants towards light is called:**
a] Hydrotropism b] Geotropism
c] Phototropism d] Chemotropism

78. **Pink ballworm, grey weevil, red bug and dusty bug are the pests of:**
a] Cotton b] Jute
c] Sunflower d] Sugarcane

79. **Which one of the following is the scientific name of silkworm?**
a] *Bombyx mori* b] *Apis indica*
c] *Triplusia niva* d] *Oxya velox*

80. **Citrus orchards can successfully be intercropped with:**
a] Aonla b] Karaunda
c] Papaya d] Guava

81. **Removal of undesirable auxillary buds at initial stage from flowering plants is known as:**
a] Clipping b] Notching
c] Budding d] Topping

82. **Salt tolerant crop is:**
a] Cowpea b] Pea
c] Sugarbeetd] Muskmelon

83. **'Yellow Vein Mosaic' is a disease of:**
a] Citrus b] Okra
c] Litchi d] Loquat

84. **The cheapest method of preservation at the household stage is:**
a] Canning b] Freezing
c] Jellying d] Sun drying

85. **Plants having male and female flowers on the same plant are known as:**
a] Monophrodite b] Hermaphrodite
c] Monoecious d] Apomixis

86. **Avocado is also known as butterfruit because it is rich in:**
a] Sugar b] Fat
c] Protein d] Fat and protein both

87. **Which one of the following pairs is correctly matched?**
a] Red soil Calcareous
b] Alkali soil pH-9
c] Black soil Rich in Kaolinite
d] Laterite soil pH-8

88. **Nitrogen is a major nutrient because:**
a] It is available in soil more
b] It is constituent of protein
c] It is available in atmosphere more
d] It is required by plants more

89. **'Peat' soils are generally:**
a] Black, heavy and acidic
b] Red, light and acidic
c] Black, light and basic
d] Black, heavy and neutral

90. **A branch of science that deals with the study of origin of the soil, its classification and description is known as:**
a] Edaphology b] Ecology
c] Pedology d] Geology

91. **The number of essential elements for plant growth is :**
a] 8 b] 12
c] 16 d] 20

92. **Soil-amendment used to reclaim usar (barren) land is:**
a] Calcium sulphate
b] Sodium sulphate
c] Potassium sulphate
d] Calcium carbonate

93. **The salt affected soils are divided in how many classes depending upon electrical conductivity and sodium contents of their solution extract?**
a] Two b] Three
c] Four d] Five

94. **Which micro-organism is present in large quantities in soil?**
a] Algae b] Bacteria
c] Fungi d] Protozoa

95. **Sugar content shall be highest if the refractometer reading is:**
a] 20 b] 25
c] 11-12 d] 17-18

96. **Origin of maize crop is:**
a] India b] America
c] Africa d] England

97. **Concept of plant ideotype was first propounded by:**
a] C.M. Donald
b] J.D. Easin
c] Meyer and Anderson
d] M.S. Swaminathan

98. **Ratio of grain to stalk in Pigeonpea is:**
a] 1 : 2 b] 1 : 3
c] 1 : 10 d] 1 : 6-8

99. **Brix per cent in sugarcane juice is the indicator of :**
a] Salt content in juice
b] Sucrose per cent in juice
c] Total solids in juice
d] None of these

GK GS

First Language Commission was set up in 1956.

100. Which one of the following is the example of parallel intercropping?
   a] maize + Soyabean
   b] Sugarcane + Mustard
   c] Potato + Mustard
   d] Sugarcane + Potato

101. Need based and skill oriented vocational training to practicing farmers and employed extension personnel is provided through:
   a] Training and Visit
   b] Krishi Vigyan Kendra
   c] National Demonstration
   d] Lab to Land Programme

102. The new Panchayat Raj Act (73rd Constitution Amendment) was enacted all over the country from:
   a] 24th April 1993
   b] 24th April 1994
   c] 24th April 1995
   d] None of the above

103. Which of the following schemes have been merged under Jawahar Rozgar Yojna?
   1. NREP
   2. RLREGP
   3. IRDP
   4. TRYSEM

CODES
   a] 1 and 2 b] 1 and 3
   c] 3 and 4 d] All the above

104. Albert Mayer is the name associated with:
   a] Etawah Pilot Project
   b] Nilokheri Development Project
   c] Farraka Development Project
   d] Shantiniketan

105. Plant evolution is caused by natural selection among communities. The theory of Natural Selection was proposed by:
   a] Hooker b] Gamble
   c] Haldane d] Darwin

106. Chromosome number is reduced to half in:
   a] Meiosis I b] Meiosis II
   c] Mitosis d] Cytokinesis

107. Self-compatibility is helpful in the development of:
   a] Synthetic variety
   b] Hybrid variety
   c] Pure line variety
   d] Composite variety

108. Appropriate breeding methoa for transferring disease resistance from wild donor source to an established variety is:
   a] Pure line method
   b] Pedigree method
   c] Back cross method
   d] Bulk method

109. Foundation seed is produced from:
   a] Nucleus seed b] Breeder seed
   c] Quality seed d] Certified seed

110. Specific combining ability of the parents should be high in order to develop:
   a] Pure line b] Synthetics
   c] Composites d] Hybrids

111. Name a cereal variety developed by mutation breeding in India:
   a] Sonalika b] Jagannath
   c] Kiran d] Savita

112. The cornerstone of all plant breeding practices is:
   a] Introduction b] Selection
   c] Hybridization d] Selfing

113. Before transplanting of paddy, puddling is done:
   a] To increase the availability of nutrients
   b] To increase the activity of micro-organisms
   c] To reduce the percolation of water in the soil
   d] To increase the percolation of water in the soil

114. Name the first phase of water erosion:
   a] Sheet erosion b] Rill erosion
   c] Splash erosion d] None of these

115. 'Operation Flood Programme' relates to:
   a] Conserving soil
   b] Controlling flood
   c] Boosting milk supply
   d] Funds for flood victims

116. Nematodes are:
   a] Symbiotic nitrogen fixers
   b] Blue green algae
   c] Insects
   d] Plant pathogens

117. Level of Optimum Production is available in:
   a] I stage of Production
   b] II stage of Production
   c] III stage of Production
   d] None of the above

118. Which one of the following kind of budgeting would you advise to a new farm planner?
   a] Enterprise budgeting
   b] Complete budgeting
   c] Partial budgeting
   d] None of these

119. Soils brought by Ganga-Yamuna and Indus rivers and their tributaries are deep and are known as:
   a] Bhabar soils b] Vindhyan soils
   c] Desert soils d] Alluvial soils

120. In grafting, the lower part of the stem containing roots is called:
   a] Stock b] Scion
   c] Both d] None of these

Set-11

1. Name the crop having maximum area in India
   a] wheat b] sugarcane
   c] rice d] gram
2. Which hormone causes milk injection
   a] estrogen b] oxytocin
   c] thyroxine d] parathormone
3. Citric acid and fructose in bovine seminal plasma is contributed by
   a] prostate gland b] cowper's gland
   c] seminal vesicles d] epididymus
4. Name the most common soil type available in India
   a] red soil b] alluvial soil
   c] black soil d] lateritic soil
5. Entomology deals with
   a] weed management b] birds management
   c] plant insects d] water management
6. Colustrum coagulates during heating due to the higher concentration of
   a] albumin & globulin b] fat
   c] casein d] carotene
7. Which one of the following grains is most nutritive?
   a] wheat b] rice
   c] gram d] soyabean
8. What is the sugar content in sugarcane?
   a] 12% b] 20%
   c] 30% d] 8%
9. The percentage of irrigated area in India is
   a] 30% b] 40%
   c] 60% d] 80%
10. Crop lodging refers to
    a] cutting trees to logs
    b] record of crop from sowing to harvesting
    c] falling of plants horizontally on the ground
    d] determination of nutrient status in plants
11. Crop logging refers to
    a] falling of plants on the ground
    b] determination of nutrient status in plants
    c] cutting trees to logs
    d] progress record of the crop
12. The fat content of egg yolk is
    a] 11.5 per cent b] 16.5 per cent
    c] 32.5 per cent d] 22.5 per cent
13. Green revolution in India was brought about by
    a] development of irrigation facilities
    b] fertilizer availability
    c] improved production technology
    d] development of dwarf varieties of wheat and rice
14. What is the main reason of higher production of rice crop in summer season than in rainy season
    a] more sunshine hoursb] less diseases & insects
    c] less loss of fertilizer d] all the above
15. Which green fodder is more suitable for getting more milk from cows?
    a] oats b] maize
    c] bajra d] cowpea
16. What is the main reason that wheat is mainly grown in northern states?
    a] Irrigation facilities b] temperature
    c] soil type d] photoperiod
17. Which one of the following crops is not photosensitive?
    a] dwarf wheat b] sugarcane
    c] potato d] soyabean
18. Which one of the following is not a micro-nutrient?
    a] zinc b] sulphur
    c] iron d] boron
19. What do you mean by 'Water harvesting'?
    a] collection of rain water for use of irrigation later
    b] water conservation in soil
    c] better utilization of ground water
    d] to check water percolation in soil
20. Which one of the following is not edible oil
    a] soyabean b] sesamum
    c] sunflower d] linseed
21. Stem canker and black scurf of potato is caused by
    a] one fungus b] two fungi
    c] one fungus and one bacterium d] one bacterium
22. Anemometer is used
    a] soil moisture b] wind direction
    c] relative humidity d] wind speed
23. The characters expressed by a hybrid are termed as
    a] contrasting characters b] dominant characters
    c] recessive characters d] parental characters
24. The capacity of being inherited in respect to a character is called
    a] hereditary b] heritability
    c] competence d] heredity

B.G. Kher was the chairman of first Language Commission.

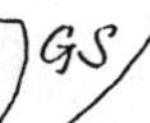

25. The increase in vigour of an offspring in relation to its parents is termed as
a] hybrid vigour b] heredity
c] hybridization d] compatibility
26. Failure to obtain fertilization and seed production after cross pollination is known as
a] compatibility b] incomplete fertilization
c] incompatibility d] cross over
27. What is 'crossing over'?
a] interchange of nonsister chromatids in homologous chromosome
b] Interchange of nonsister chromatids in non homologous chromosome
c] interchange of segments of nonhomologous chromosomes
d] none of the above
28. Adjustment of species of population to a changed environment over a period of time is known as
a] adaptation b] adoption
c] acclimatization d] adjustment
29. The most common method of crop improvement is
a] introduction b] hybridization
c] mutation breeding d] selection
30. Hybridization causes
a] production of genes
b] production of new chromosomes
c] reshuffling of genes
d] none of the above
31. Hybridization between species of the same genus is called
a] Intrageneric hybridization
b] intergenic hybridization
c] introgressive hybridization
d] interspecific hybridization
32. Removal of stamens from female parent plants is known as
a] pollinationb] sterilization
c] fertilization d] emasculation
33. Pure line is a progeny descending solely by self pollination from a single
a] heterogenous plant b] homozygous plant
c] heterozygous plant d] homogenous plant
34. The family of potato plant
a] solanaceae b] graminae
c] malvaceae d] tiliaceae
35. When pollen of a plant falls on the stigma of the flower of same plant, it is called
a] self pollination b] cross pollination
c] fertilization d] propagation
36. When pollen of a plant falls on the stigma of another plant, it is called
a] self pollination b] cross pollination
c] introgression d] emasculation
37. When male and female reproductive parts are in separate flowers on the same plant, it is termed as
a] monoecious b] dioecious
c] monoadalphous d] none of the above
38. The thermophillic bacteria in raw milk are
a] bacillus b] lactobacillus
c] staphylococcus d] none of these
39. What are antibodies?
a] modified antigen b] globulin molecule
c] lymphocytes d] plasma cells
40. When blood comes out with urine it is known as
a] haemoglobinura b] haemalurea
c] blood urea d] hematemesis
41. The haploid phase in plants is called
a] spermatophyte b] gametophyte
c] zygote d] sporophyte
42. The 'National flower of India' is
a] rose b] lotus
c] marigold d] lily
43. Cells with more than two sets of chromosomes are known as
a] haploid b] diploid
c] monoploid d] polyploid
44. Maximum number of genetic studies have been conducted on
a] rabbit b] monkey
c] drosophilad] pig
45. Laws of heredity was rediscovered in the year
a] 1807 b] 1887
c] 1900 d] 1910
46. Meiotic cell division is also called
a] horizontal division b] vertical division
c] equatorial division d] reduced division
47. The stage of cell division when chromatids are arranged parallel to equatorial plate is known as
a] metaphase b] prophase
c] telophase d] anaphase
48. The process of formation of male gametes in plants is known as
a] pollination b] fertilization
c] sporogenesis d] microsporogenesis
49. Gametes in plants as well as in animals are
a] polyploid b] diploid
c] hexaploid d] haploid
50. In higher plants seed formation is a result of
a] fertilization b] cross pollination
c] self pollination d] double fertilization
51. If R is dominant gene of red colour of flower, then which of the plants with the following genotype will not produce red flower?
a] RR b] rr
c] Rr d] rR

GK **Inselberg is the name given to isolated residual hills of deserts.** GS

52. Spices and condiments are grouped under
a] industrial crops b] commercial crops
c] food adjuncts d] field crops

53. If gene 'W' is lethal in 'w' form, which of the following plants will not survive ?
a] ww b] W W
c] Ww d] wW

54. Artificial insemination had its origin in
a] 1922 b] 1720
c] 1888 d] 1920

55. Genes, which suppress or enhance the expression of other genes, are called
a] dominant genes b] modifier genes
c] recessive genes d] none of the above

56. The element, which is a major component of many enzymes is
a] zinc b] manganese
c] sulphur d] copper

57. The process of marking the expression of a non-allelic gene by another gene is known as
a] expressivity b] interaction
c] epistasis d] hypostasis

58. Genes are composed of
a] protein b] RNA
c] lipids d] DNA

59. Who received the Nobel prize for DNA model?
a] H.G. Khorana b] G. Mendel
c] Watson and Crick d] Linus Pauling

60. Who was awarded Nobel prize for DNA coding?
a] Watson and Crick b] H.G. Khorana
c] Linus Pauling d] Warburg

61. Which one of the following base is not found in RNA?
a] adenine b] guanine
c] uracil d] thymine

62. Which one of the following base is not found in DNA?
a] guanine b] cytosine
c] uracil d] thymine

63. In which part of the cell does protein synthesis take place?
a] mitochondria b] chloroplast
c] periosomes d] ribosomes

64. Name the cell organelle where photosynthesis take place
a] ribosome b] endoplasmic reticulum
c] mitochondria d] chloroplast

65. The physical basis of heredity is known as
a] gene pool b] filial generation
c] gene frequency d] none of the above

66. The non-protein part of enzyme is called
a] coenzyme b] cofactor
c] prosthetic group d] all of these

67. Vaccination is essential for the prevention of which of the following diseases
a] FMD b] rinderpest
c] black quarter d] all of these

68. Any agent, which is capable of inducing mutation is called
a] mutant b] mutagen
c] mutan d] none of the above

69. *Parturient paresis* can be prevented by feeding a diet preparation
a] high in calcium
b] high in phosphorus and low in calcium
c] high in calcium and low in phosphorus
d] both a] and c]

70. Smallest unit of DNA that can undergo change resulting in a mutation is called
a] mutable gene b] mutant
c] mutan d] mutator gene

71. A unit of DNA molecule containing phosphate, a sugar and an organic base is called
a] nucleotide b] nucleoside
c] nucleoprotein d] codon

72. Chromosomes are made up of
a] fatty acids b] nucleic acids
c] nucleo proteins d] lipids

73. Formation of seed without fertilization is called as
a] parthenocarpy b] apogamy
c] vegetative propagation d] none of the above

74. A cow can best be judged when
a] she is in full production b] she is dry
c] 3 years old d] she has just calved

75. The productivity of cereals is usually higher than pulses and oil seeds, Why?
a] cereals accumulate protein
b] cereals accumulate lipids
c] cereals accumulate carbohydrates
d] cereals take longer duration

76. The function of the pericycle in root hair zone is formation of
a] leaves b] stem
c] starch d] lateral roots

77. Pulse flowers are fed by
a] ash weevilb] spiny borer
c] blister beetle d] pod fly

78. The function of xylem in plants is to carry
a] food b] water
c] nutrients d] photosynthates

79. Which one of the following crops thrives well at 40-50°F?
a] peach b] guava
c] apple d] mango

GK **First Minority Commission was set up in 1979.** GS

80. **Calf hood vaccination is practised in brucellosis because**
    a] calves are antibody producers
    b] calves are easy to vaccinate
    c] calves do not show vaccination reaction
    d] none of these
81. **The specific cause of Rinderpest**
    a] fungus b] bacteria
    c] virus d] protozoa
82. **Fertilization of ova occur in**
    a] ovary b] uterus
    c] oviduct d] infundibulum
83. **Banana is a**
    a] monocot herb b] monocot shrub
    c] monocot tree d] dicot herb
84. **Girdling in grapevine is caused by**
    a] thrips b] aphid
    c] stem girdler d] sphingid moth
85. **Which one of the following is a cool season crop ?**
    a] potato b] French bean
    c] cucumber d] bottle gourd
86. **Which one of the following plants is hardy in nature?**
    a] apple b] date
    c] citrus d] guava
87. **Ranikhet disease is caused by**
    a] fowl herpes virus b] corona virus
    c] entero virus d] paramyxo virus
88. **Which one of the following is a non-hardy crop?**
    a] plum b] citrus
    c] rose d] cherry
89. **One of the organs where microbial digestion in non ruminant animals occurs**
    a] stomach b] small intestine
    c] large intestine d] none of the above
90. **Pink colour of onion is due to**
    a] quercetin b] anthocyanin
    c] carotene d] none of the above
91. **Cereal grain byproducts are**
    a] rich in Ca
    b] rich in P
    c] low in P and rich in Ca
    d] rich in P and low in Ca
92. **Homeothermy refers to**
    a] maintenance of body weight
    b] maintenance of body temperature
    c] maintenance of feed intake
    d] maintenance of energy
93. **The initial food substances are made in**
    a] leaves b] stem
    c] root d] flower
94. **Father of soil Microbiology is**
    a] S.N. Winogradsky b] Beijerinck
    c] A. Waksman d] Leibig
95. **Rhizosphere refers to the small area**
    a] around the root b] on the root
    c] near the primary root d] all of the above
96. **The microorganisms using inorganic carbon source and sunlight as energy source are called**
    a] photorganotrophs b] chemo heterotrophs
    c] photo autotrophs d] chemo autotrophs
97. **Which fruit is known as 'King of temperate fruits'**
    a] almond b] apple
    c] peach d] walnut
98. **Which of the following animal is a ruminant**
    a] horse b] camel
    c] dog d] pig
99. **The obligate association between two organisms where both are benefited by each other is called**
    a] protocooperation b] antagonism
    c] symbiosis d] commensalism
100. **The association of two organisms where one of them get the benefit from the other but the other remained unaffected is called**
    a] protocooperation b] antagonism
    c] symbiosis d] commensalisms

GS

Drumlins glacial landform is called 'Basket of eggs topography'.

Set-12

1. To detect incidence of brucellosis in a herd, we prefer
   a] milk ring test b] plate agglutination test
   c] tube agglutination test d] all of these
2. The micro-organisms which are unable to survive in the absence of O_2 are called
   a] aerobe b] obligate aerobe
   c] facultative aerobe d] none of the above
3. The capacity of nitrogen fixation is more with
   a] free living microbes b] symbiotic association
   c] blue green algae d] symbiotic dissociation
4. Which one of the following is the toughest mineral?
   a] zirconium b] graphite
   c] mica d] quartz
5. Which is the most stable mineral from the following
   a] graphite b] quartz
   c] montmorillonite d] kaolinite
6. Which one of the soil will have more available water?
   a] sandy loam b] loam
   c] clay loam d] silty loam
7. India ranks ----- in milk production in the world
   a] first b] third
   c] second d] fourth
8. An element is essential for plant if
   a] the plant is not able to complete its life-cycle without it
   b] it is present in the soil
   c] it is present in the plant
   d] it catalyses the reaction of other elements in the plant
9. pF is defined as
   a] the logarithm of soil moisture potential
   b] the logarithm of soil moisture tension expressed in cm height of a column of water
   c] the logarithm of soil water tension
   d] soil moisture tension expressed in the height of Hg column
10. Apple scab disease, which earlier was confined only to Kashmir in India, spread to Himachal Pradesh after
    a] 1943 b] 1965
    c] 1973 d] 1985
11. The most valuable fertilizing constituents of cow manure are in
    a] the urine b] the dung
    c] the bedding d] none of these
12. The cost of various operations from land preparation to threshing of the crop is known as
    a] fixed cost b] cost of cultivation
    c] cost of farming d] none of the above
13. Dead heart in sugarcane is caused by
    a] Pyrilla b] termite
    c] early shoot borer d] white grub
14. The person who is hired by the farmer on yearly basis on some agreed wages is known as
    a] permanent labour b] temporary labour
    c] family labour d] worker
15. The inflorescence of banana is a
    a] spike b] spadix
    c] raceme d] umbel
16. If we want to impress the people about a particular recommended practice suitable for a specific area, then effective extension method is
    a] demonstration b] T.V.
    c] pamphlets d] radio
17. In animal nutrition, the total energy is expressed as
    a] creative phosphate b] ATP
    c] AMP d] Kilo calories
18. The specific drug for the treatment of coccidiosis in young calves is
    a] amproleum b] antrycide prosalt
    c] gentamycin d] kanamycin
19. What is the most important source of irrigation in India
    a] canal b] pond
    c] charas d] tubewell
20. Sericulture is not common in India due to
    a] unavailability of insect
    b] high cost involved
    c] unavailability of local market
    d] unfavourable climatic conditions
21. Which one of the following crop requires maximum water to grow
    a] wheat b] mango
    c] mustard d] banana
22. Which one of the following crop requires maximum amount of nitrogen
    a] wheat b] rice
    c] banana d] sugarcane

At present, 22 regional languages have been listed in the 8th Schedule.

GS

23. Which one of the following supplies more quantity of plant nutrients?
a] castor cake b] farmyard manure
c] compost d] night soil

24. Mango malformation results in
a] enhanced ripening b] delayed ripening
c] regular fruiting bearing d] irregular/alternate

25. The earliest pregnancy diagnosis test for bovines is
a] rectal examination b] progesterone assay
c] ultra sound d] radiography

26. What is the function of milk veins?
a] to carry milk from different quarters
b] to supply blood from heart to udder
c] to carry blood from udder towards heart
d] none of the above

27. Control of rinderpest in India has been possible due to
a] quarantine
b] passive immunization
c] immuno modulation
d] vaccination

28. 'Tikka' disease is observed in
a] soyabean b] groundnut
(9) mungbean d] cotton

29. Which one of the following political leaders contributed maximum to the green revolution?
a] C. Subramanyam b] Sardar Swaran Singh
c] Jagjivan Ram d] Rao Birendra Singh

30. The tundu disease of wheat is usually associated with a nematode known as
a] *Anguina tritici* b] *Meloidogyne incognitai*
c] *Meloidogyne javanica* d] *Heterodera avevae*

31. Name the fruit crop having maximum production in India
a] Mango b] Banana
c] Guava d] Orange

32. Honeybee is a
a] pest on human b] beneficial insect
c] parasite d] scavenger

33. Butterflies and moths are
a] Insects b] Mites
c] Ticks d] Snails

34. The nitrogenous wastes of earthworm are known as
a] vermicastings b] vermicompost
c] vermifaeces d] excreta

35. Smoking in orchard during night is useful to ward off
a] fruit fly b] fruit piercing moth
c] stem borer d] fruit borer

36. Which one of the following poultry disease is caused by an RNA virus
a] Marek's disease
b] fowl pox
c] infectious bursal disease
d] infectious laryngotrachitis

37. What is the percentage of proteins of feed?
a] 16% b] 6.25%
c] 20% d] none of these

38. Soyabean is highly rich in
a] protein b] carbohydrate
c] lipids d] vitamins

39. The food, which is commonly given to chicks for rearing in farms, is
a] rice b] groundnut
c] maize d] bajra

40. The vegetable consumed to overcome the deficiency of night blindness is
a] radish b] tomato
c] carrot d] cabbage

41. The plants that grow in desert are known as
a] Halophytes b] Xerophytes
c] Bryophytes d] Pteridophytes

42. Pick the odd one out
a] rice b] wheat
c] maize d] chickpea

43. The flour from which crop is used to prepare bread
a] wheat b] maize
c] jowar d] bajra

44. The staple food crop of India is
a] wheat b] maize
c] rice d] jowar

45. Mosquito is a
a] Parasite b] predator
c] pest d] mite

46. Early shoot borer of sugarcane is
a] *Chilo partellus*
b] *Chilo infuscatellus*
c] *Tryporyza excerptalis*
d] *Sesamia inferens*

47. DDT is commonly sprayed in evening hours in the colony of residential areas to control
a] Housefly b] cockroach
c] ants d] mosquito

48. Haemoglobinuria in buffaloes can be treated more effectively by
a] intravenous calcium b] sodium acid phosphate
c] antifibrinolytic drugs d] dextro saline

49. Rice is a
a] cereal b] pulses
c] millet d] none of these

GK Mt. Stromboli volcano is named 'Lighthouse of the Mediterranean'. GS

50. **Which one of the following diseases is more fatal to animals?**
a] black quarter b] foot and mouth disease
c] cow pox d] Johne's disease

51. **Ulcers on the mucosa of lips and gums is the symptoms of**
a] haemorrhagic septicaemia
b] piroplasmosii
c] anthrax
d] rinderpest

52. **Element, which is necessary for synthesis and maintenance of chlorophyll in plants**
a] Nitrogen b] Sulphur
c] Iron d] Zinc

53. **Hopperburn symptom in cotton is caused by**
a] aphid b] whitefly
c] thrips d] leaf hopper

54. **For economic production of grape the only single cultural practice is**
a] pruning
b] irrigation
c] fertilization
d] control of insects and diseases

55. **Common scab of potato is caused by**
a] *Spongospora subterranean*
b] *Rhizoctonic solani*
c] *Streptomyces scabies*
d] *Pseudomonas solanaceanum*

56. **Gregor John Mendel is related to which field**
a] Genetics b] Physics
c] Chemistryd] Taxonomy

57. **Late blight of potato was introduced in India for the first time in**
a] Shimla hillsb] Darjeeling
c] Nilgiri hills d] None of these

58. **Rinderpest virus belongs to the genus**
a] paramyxo virus b] morbilli virus
c] pneumo virus d] myxo virus

59. **Plant regulator which improves fruit set in tomato is**
a] gibbrellic acid
b] 2, 4-dichlorophenoxy acetic acid
c] Malic hydrazide
d] Ethrel

60. **The total number of State Agricultural Universities (SAUs) in India is**
a] 28 b] 25
c] 30 d] 32

61. **Indian Council of Agricultural Research (ICAR) is headed by**
a] Union Minister of Agriculture
b] IAS officer
c] Director-General
d] Secretary-General

62. **Stalk rot of maize is caused by**
a] *Erwinia chrysanthemum Pv zea*
b] *Xanthomoanas citrii*
c] *Clavibacter xyli*
d] *Pseudomonas solanacearum*

63. **Bacteria cannot survive in highly salted pickle because**
a] salts inhibit reproduction
b] bacteria do not get enough light for photosynthesis
c] they become plasmolysed and consequently killed
d] pickle does not contain nutrients necessary for bacteria to survive

64. **When ATP is converted into ADP, it releases**
a] Energy b] Hormones
c] Enzymes d] All of these

65. **Fertilisation is a**
a] source of variation
b] occurs in all plants and animals
c] is fusion of the gametes
d] all of these

66. **In electron transport system, electron combines to**
a] O_2 b] H_2O
c] H_2 d] Cytochrome

67. **Time of flowering of mango in north India is**
a] February-March b] April-May
c] July-August d] September-October

68. **Plants growing in xerophytic condition have**
a] Large flower
b] Extensive air spaces
c] Extensively developed root system, small and thick leaves
d] All of these

69. **Plants that have capacity to withstand wide range of temperature (heat) tolerance are said to be**
a] Mesothermic b] Stenothermic
c] Monothermic d] Eurythermic

70. **Spice crop is largely exported from India**
a] black pepper b] cardamom
c] canine d] turmeric

71. **Powdery mildew of apple is caused by**
a] *Erysiphe polygoni*
b] *Podosphaera leucotricha*
c] *Sphaerotheca fuliginea*
d] *Uncinula necator*

72. **The science, which belongs to vegetable culture, is known as**
a] Olericulture b] Vermiculture
c] Sericulture d] Horticulture

73. **The genetic material found in viruses is**
a] Plasmid b] RNA
c] DNA d] Both b & c

GK GS
The Constitution of J&K came into force on January 26, 1952.

74. Most dairy calves on the farm are raised
a] by hand b] on the dam
c] on foster cow d] on butter milk

75. The most destructive phase of the bacterial leaf blight disease of rice is known as
a] Kresek b] Necrosis
c] Blight d] Die Back

76. Monophagous pest of mango is
a] stem borer b] fruit sucking moth
c] nut weevild] blossom midge

77. Richest source of vitamin A is
a] papaya b] bael
c] ripe banana d] ripe mango

78. Excessive elongation of plant stem is due to
a] Cytokinin b] Indole Acetic Acid (IAA)
c] Gibberelin d] Auxins

79. Spraying of DDT produces pollution of
a] only air b] air, water and soil
c] only water d] only air and soil

80. In Bacteria, respiration takes place in
a] nucleus b] mitochondria
c] mesosomes d] cellwall

81. Mango malformation is due to
a] mango hopper b] mango aphid
c] mango bud mite d] mango mealy bug

82. Which breed of buffaloes is highest fat producer in milk?
a] Murrah b] Bhadawri
c] Mehsana d] Jafrabadi

83. In coconut the inflorescence is a
a] Panicle b] compound spike
c] spadix d] spike

84. Enzymes of citric acid or kreb's cycle occur in
a] inner membrane space
b] mitochondrial matrix
c] outer mitochondrial matrix
d] inner mitochondrial membrane

85. Best method of budding guava
a] T-budding
b] ring budding
c] patch forkert budding
d] improved patch/forkert budding

86. During seed germination stored food in seed is mobilized by
a] chloroplast b] gibberelin
c] golgi body d] nucleus

87. Cotton is formed by
a] outgrowths of seed coat
b] xylem fibres
c] outgrowth of the pericarp d] none of these

88. DDT is a
a] Insecticide b] Acaricide
c] Rodenticide d] Fungicide

89. In plants, during respiration
a] CO_2 is utilised and O_2 is released
b] O_2 is utilised and CO_2 is released
c] both are utilized
d] none of these

90. Potometer is used for measuring
a] Photosynthesis b] transpiration
c] respiration d] stomatal opening

91. Proteins are broken down to polypeptides and amino acids due to the action of
a] Zymase b] Amylase
c] Trypsin d] Lipase

92. Sorghum earhead bug lays eggs
a] in soil
b] in the stem
c] in between anthers or glumes
d] in flower buds

93. Rice thrips occurs as a pest in
a] nursery
b] main field
c] nursery and newly-transplanted rice
d] crop / nearing maturity

94. Who proposed the double helix model of DNA structure ?
a] Robert Brown b] Zinder and Joshua Lederberg
c] Frederick Griffith d] Watson and Crick

95. Maize seed is a
a] Dicot seed b] Cotyledon seed
c] Endospermic d] Non endospermic

96. Colostrum coagulates on heating due to the presence of
a] glyceridesb] albumen & globulin
c] high fat d] high lactose

97. Plant pathology deals with
a] plant insects b] weed science
c] plant diseases d] seed production

98. Used slurry from gobar gas plant contains nitrogen
a] 2% b] 10-20%
c] 30% d] 10%

99. Photosynthetic bacteria have
a] Leucoplasts b] Chromoplasts
c] Chromatophore d] Chloroplasts

100. For canning, tomato fruits are picked at
a] immature green stage
b] mature green stage
c] half-ripe or pink stage
d] red ripe stage

Falkland current is the cold current touching southern South American tip. GS

Set-13

1. To sterilize glasswares and culture media, equipment used is
   a] Dry oven b] Autoclave
   c] Dehumidifier d] Incubator
2. A common but effective disinfectant for water trough of cattle farm is
   a] phenol b] potassium permanganate
   c] lime d] washing soda
3. The edible part of cauliflower is known as
   a] head b] curd
   c] pods d] cob
4. Tuber production stops totally at a temperature of
   a] 25°C b] 30°C
   c] 35°C d] none of the above
5. The area where fruit crops are grown is called
   a] Garden b] Greenhouse
   c] Polyhouse d] Orchard
6. Which one of the following is a subtropical fruit
   a] apple b] banana
   c] coconut d] jujuba
7. Which one of the following is not a true fruit
   a] mango b] apple
   c] guava d] ber
8. What percentage of total holdings are occupied by small and marginal farmers
   a] 70% b] 50%
   c] 35% d] 25%
9. The most commonly used pesticide is
   a] phorate b] BHC
   c] DDT d] endosulfan
10. Cereals account for what percentage of total cropped area
   a] 30% b] 50%
   c] 70% d] 85%
11. Which of the following has maximum TDN (Total digestible nutrients)
   a] lucerne b] berseem
   c] oat d] maize
12. India ranks- in vegetable production
   a] first b] second
   c] third d] fourth
13. How much total N, P and K fertilizer is consumed per hectare per year in India?
   a] 32 kg b] 56 kg
   c] 75 kg d] 105 kg
14. Which of the following category of fertilizer is totally imported in India?
   a] Phosphorus b] Potassium
   c] Sulphur d] Zinc
15. Who conclusively proved that tuberculosis is a bacterial disease ?
   a] Robert Koch b] Charles Best
   c] Banting d] Claire
16. Tea mosquito bug infects
   a] tea b] guava
   c] cashew d] all the above
17. Earthing of potato crop is done
   a] 20 days after sowing b] 40 days after sowing
   c] 50 days after sowing d] none of the above
18. The most appropriate method of dehorning (disbudding) calves is
   a] caustic potash stickb] electrical dehorner
   c] surgical removal d] none of these
19. Rose is a _________ flowering plant
   a] woody b] herbaceous
   c] semiherbaceous d] none
20. What is the unit for measuring the viscosity of milk
   a] dyne b] poise
   c] ohm d] mho
21. Approximate feed consumption per week in 2 weeks old 100 birds is
   a] 7 kg b] 13 kg
   c] 10 kg d] 8 kg
22. Normal respiration rate in cattle is
   a] 30-32 b] 25-30
   c] 20-28 d] 21-27
23. Citrus greening disease is transmitted by
   a] *Diaphorina citri* b] *Typoriza erythreae*
   c] Both a and b d] None of the above
24. Infectious disease of cattle characterized by inflammation of sensitive tissues of the feet is called
   a] necrobacillotis b] pododermatitis
   c] foot abscess d] rhimitis
25. The chemical agents which kill bacteria are known as
   a] antisepticb] disinfectant
   c] germicide d] deodrant
26. Poultry infection which passes through egg is
   a] toxoplasmosis b] ascariasis
   c] bacillary white diarrhoea d] Ranikhet disease

GK

The strength of Panchayat varies between 5 and 31.

GS

27. Cauliflower curds can be stored for 30 days at
    a] 0°C with 85-90 per cent R.H.
    b] 10°C with 70-80 per cent R.H.
    c] 15°C with 60-70per cent R.H.
    d] none of the above

28. Dithane M-45 is available now as
    a] Emisan b] Indophil M-45
    c] Thiram d] Bavistin

29. First artificial insemination was conducted by
    a] Lazaro Spallanzani b] Demetri Vekiff
    c] Demertzia P.N. d] Schimitz J.A.

30. The cow normally produces
    a] one ovum b] two ovum
    c] three ovum d] four ovum

31. Banana is essentially a
    a] temperate crop b] subtropical crop
    c] tropical crop d] subtemperate

32. Which pathogen is associated with Irish famine
    a] *Bremia lactucae* b] *Sclerosporea gaminicola*
    c] *Phytophthora infestans* d] none of these

33. The most common outside source of FSH is
    a] PMS b] placenta extract
    c] ovary extract d] none of the above

34. Which one is fundamental unit of genetics ?
    a] chromosome b] D.N.A.
    c] R.N.A. d] gene

35. The blood of cattle contain the following percentage of sodium
    a] 0.22 b] 0.25
    c] 0.20 d] 0.18

36. Wobbly gait in hens is produced by the deficiency of
    a] folic acid b] thiamine
    c] vitamin A d] vitamin C

37. What of the following diagnostic test is used for the diagnosis of Mastitis ?
    a] white side test b] urease test
    c] agghitiration test d] complement fixation test

38. Name the crop having maximum production in India
    a] maize b] wheat
    c] groundnut d] rice

39. Mass selection is rarely used in
    a] allogamous species
    b] autogamous species
    c] asexually propagated species
    d] seed propagated species

40. Binomial nomenclature of plants is written in which language?
    a] English b] Roman
    c] Latin d] French

41. ___________is called as 'food of God'
    a] tea b] coffee
    c] cashew d] cocoa

42. Avian contagious epithelioma is also known as
    a] fowl plague b] fowl cholera
    c] fowl pox d] fowl leukaemia

43. Indicate the nitrogenous substance in milk
    a] lactose b] cholesterol
    c] uric acid d] carotene

44. Coffee, Cocoa, Rubber and Cardamom are grouped under
    a] commercial crops b] ratoon crops
    c] plantation crops d] spices & condiments

45. 'Night blindness', a disease of eye is caused due to deficiency of
    a] ascorbic acid b] vitamin A
    c] riboflavin d] folic acid

46. Dioecious form of sex is found in
    a] citrus b] aonla
    c] papaya d] litchi

47. When petals are arranged in two or more than two whorls, the flower is known as
    a] apetalous b] single flower
    c] double flower d] bisexual flower

48. The national fruit of India is
    a] aonla b] jackfruit
    c] mango d] apple

49. Ratio of protein and carbohydrate in diet should be
    a] 1 : 1 b] 2 : 5
    c] 2 : 1 d] 1 : 6

50. The following vaccines are used for immunisation against Ranikhet disease of poultry
    a] R_2b b] F
    c] MD d] F_2

51. False smut of rice is caused by
    a] *Sclerotinia sclerotiorum*
    b] *Claviceps purpurea*
    c] *Claviceps oryzae*
    d] *Ustilago oryzae*

52. Vitamin A requirement for an adult man is
    a] 1000 IU/ day b] 2000 IU/day
    c] 1500 IU/day d] 2500 IU/day

53. First 'State Agricultural University' of India is situated at
    a] Kanpur b] Pantnagar
    c] Solan d] Bangalore

54. Brown rot or wilt disease of potato is caused by
    a] fungus b] bacterium
    c] nematode d] virus

55. The most suitable time for vaccination against H.S. is
    a] October b] December
    c] Just before monsoon d] February

GK GS

Troposphere of atmosphere has all the weather phenomenon.

56. 'Tissue culture' is a method of propagation
a] sexual b] asexual
c] natural d] apomixis

57. 'Ikebana' is a method of
a] growing of roses
b] growing of vegetables
c] planting of chrysanthemum
d] flower arrangement

58. The ability of germination of a seed is known as
a] purity b] initiation
c] immunity d] viability

59. A high milk yield in cows may be accompanied by a fat test that is
a] high b] low
c] medium d] erratic

60. 'Onion' contains
a] Na b] S
c] P d] Mo

61. The edible part of litchi is
a] aril b] endosperm
c] placenta d] mesocarp

62. Immediately after harvest, the tubers have a rest or a dormant period of
a] 1 to 2 months b] 2 to 3 months
c] 3 to 4 months d] none of the above

63. Most of the glycogen in the animal body is present in
a] lung b] liver
c] spleen d] heart

64. Iron deficiency is more common in
a] flower crops b] vegetables
c] fruit crops d] pulses

65. Growing of vegetables in water is known as
a] water harvesting b] hydrology
c] hydroponics d] floating garden

66. 'Tree culture' is a synonym of
a] arboriculture b] silviculture
c] horticulture d] floriculture

67. An iodine rich feed ingredient for poultry is
a] fish meal b] groundnut meal
c] coconut meal d] meat meal

68. The outstanding value of the cow as a farm animal is her ability to
a] use roughage b] use grain
c] produce urine d] produce manure

69. The cheap and richest source of pectin is
a] apple b] bael
c] guava d] mango

70. Bacterial leaf blight of rice is caused by
a] *Xanthomonas malvacerum*
b] *Xanthomonas campestris Pv oryzae*
c] *Pseudomonas rubrileans*
d] *Corynebacterium tritici*

71. *In vitro* culture consists of
a] pollen culture b] tissue culture
c] embryo culture d] all of these

72. In a garden which colour dominates round the year?
a] green b] red
c] yellow d] blue

73. An asexual progeny of a single homozygous plant is known as
a] variety b] race
c] strain d] clone

74. Vegetables are called as protective food because they are rich in
a] fats b] vitamins & minerals
c] proteins d] sugar

75. 'Yellow colour' in turmeric is due to
a] curcumin b] carotene
c] turmarin d] lycopene

76. Damping-off of seedlings is caused by
a] only *Pythium aphanidermatium*
b] *Pythium and Phytophthora* spp.
c] many fungi including *Pythium* spp.
d] all of the above

77. Marigold produce flowers after__________ days after planting
a] 55-60 b] 100-120
c] 180-200 d] 200-250

78. Amongst the exotic cattle breeds, which one has lowest fat content in milk
a] Jersey b] Holstein
c] Brown Swiss d] Guernsey

79. 'Cobalt is the part of
a] vitamin A b] vitamin B
c] vitamin C d] vitamin D

80. A mature and ripened ovary contain seed is termed as
a] embryo b] stone
c] fruit d] grain

81. Reciprocal recurrent selection is used when a character is governed by
a] additive genes
b] non-additive genes
c] both additive and non-additive genes
d] all of the above

82. The caloric value of 100 gm of cow's milk is
a] 50 b] 75
c] 100 d] 150

83. Which is the largest producer of tea in the world?
a] Brazil
b] China
c] India
d] Mexico

84. Cauliflower requires an optimum monthly average temperature of
a] 5-10°C b] 10-15°C -
c] 15-20°C d] none of the above

85. Apple is a type of
a] achene b] samara
c] caryopsis d] false fruit

86. Which is an aesthetic branch of horticulture?
a] pomology b] olericulture
c] floriculture d] hydroponics

87. Continuous salivation is a symptom of
a] paralytic form of rabies
b] foot and mouth disease
c] haemorrhagic septecimia
d] all of these

88. Anaemia is caused due to deficiency of
a] citric acid b] ascorbic acid
c] folic acid d] iron

89. Which is the largest fruit of the world?
a] bottle gourd b] watermelon
c] jack fruit d] snake gourd

90. Which is the richest source of vitamin C?
a] wood apple b] bael
c] citrus d] aonla

91. Which fruit is recommended for sugar patient?
a] aonla b] jamun
c] bael d] banana

92. Which of following is an economic factor that affects the type or system of farming?
a] availability of capital b] water
c] land d] all of them

93. Fowl pox vaccination in poultry should be done at what age
a] 5 -7 weeks b] 2 -4 weeks
c] 9-11 weeks d] 16-18 weeks

94. Vegetables are rich source of
a] vitamins b] proteins
c] minerals d] carbohydrates

95. Pruning is an essential practice for fruiting of
a] jackfruit b] grape
c] citrus d] guava

96. Which is a new addition under plantation crops?
a] rubber b] oil palm
c] tea d] coconut

97. Scurvy is caused due to deficiency of
a] Vitamin A b] Vitamin B
c] Vitamin C d] Vitamin K

98. The most suitable method of identification of lack of certain element in a plant is
a] sand culture b] solution culture
c] rock culture d] tensiometer

99. The incubation period of poultry egg is
a] 15 days b] 18 days
c] 21 days d] 24 days

100. A refrigerant commonly used in Indian Cold Storage is
a] ammonia b] carbide
c] ethylene d] sodium benzoate

GK GS

Equatorial Low Pressure Belt is called Doldrums.

Set-14

1. Which of the following correctly represent the thumb rule for feeding of buffaloes?
   a] one kg of concentrates for each half litre of milk
   b] one kg of concentrates for each litre of milk
   c] one kg of concentrates for two litres of milk
   d] one kg of concentrates for two and half litres of milk
2. Treatment of stringhalt in bovines involves
   a] severing of middle patellar ligament
   b] severing of medial patellar ligament
   c] severing of lateral patellar ligament
   d] none of these
3. Commercial life of rose is _______ years
   a] eight b] two
   c] fifteen d] twenty five
4. Seed rate required for raising onion seedlings for one acre is
   a] 2-3 kg b] 4-5 kg
   c] 8-10 kg d] none of the above
5. Greenhouse plants are also known as
   a] sun-loving plants b] shade-loving plants
   c] xerophytic plants d] climbers
6. Term 'Two leaf and one bud' is related with
   a] coffee b] tea
   c] cocoa d] tobacco
7. A corm is differentiated from a bulb, because it is
   a] hollow b] semi-hollow
   c] solid d] root
8. Respiratory quotient (RQ) value of protein is
   a] 1.0 b] 0.8
   c] 0.7 d] 0.6
9. Edible portion of apple is
   a] embryo b] endosperm
   c] ovary d] thalamus
10. A popular delicious fruit of India is
    a] banana b] litchi
    c] apple d] pineapple
11. The leading producer of spices in the world is
    a] India b] Brazil
    c] Mexico d] Indonesia
12. Which crop is used as an emblem of the United Nations?
    a] apple b] mango
    c] olive d] walnut
13. Sour lime and lemon require climate
    a] tropical b] subtropical
    c] temperate d] tropics and subtropics both
14. 'Rickets' disease is caused due to deficiency of
    a] vitamin A b] vitamin B
    c] vitamin C d] vitamin D
15. Some fruits can be ripened well after their harvesting as well as on the tree. These fruits are known as
    a] climacteric b] non climacteric
    c] false fruits d] delicious
16. Tomato is rich in
    a] vitamin A b] riboflavin
    c] vitamin E d] vitamin C
17. The test to determine amount of antitoxin in blood is known as
    a] Eherlich test b] Schick test
    c] Ramon's test d] Flocculation test
18. Ampilography is the cultivation of
    a] cole-crops b] vine-crops
    c] bulb-crops d] tuber-crops
19. Which is referred to as plantation crop
    a] grape b] potato
    c] rubber d] pineapple
20. Who discovered the vitamin?
    a] Sumner b] Funk
    c] Lunin d] Sangour
21. Name the most common disease transmitted to humans through cows rule
    a] smallpox b] tuberculosis
    c] malaria d] milk fever
22. Lack in reproduction capacity occurs due to deficiency of
    a] vitamin A b] vitamin C
    c] vitamin K d] vitamin E
23. The beauty of trees planted along the water canals get enhanced due to
    a] refraction b] reflection
    c] dispersion d] colourful stem
24. The normal body temperature of poultry birds
    a] 98.4-98.6°C b] 101.5-103.5°C
    c] 105.8-107.8°C d] none of the above
25. The major factors considered at the time of planting of trees on road side are
    a] shade & timber b] flowering only.
    c] shape & height d] oil content in leaves

GK **Pradhan is chairman of Panchayat Samiti.** GS

26. **Addition of citric acid in vase solution is helpful in**
a] causing stem closing
b] increasing water uptake
c] preventing water uptake
d] reducing bending of bud

27. **Clove, a spice is taken from which part of plant**
a] root b] leaf
c] stem d] flower

28. **Hopper burn in cotton is caused due to**
a] phytotoxemia b] defoliation
c] oviposition d] desapping

29. **Which is known as Garden city of India?**
a] Bangalore b] Chandigarh
c] New Delhi d] Mysore

30. **Which soil is best for vegetable production?**
a] Sandy b] sandy loam
c] clay loam d] clay

31. **Red Sindhi is a**
a] draft breed b] milk breed
c] dual purpose breed d] none of the above

32. **What should be the TDN content in a balanced ration for cattle?**
a] 70- 75% b] 50-55%
c] 45-50% d] 55-60%

33. **The richest source of protein is**
a] pea b] fenugreek
c] pointed gourd d] potato

34. **Which is most suitable for avoiding 'scurvy disease'**
a] lemon b] guava
c] aonla d] apple

35. **Which are water soluble vitamins?**
a] Vitamins A and C b] Vitamins A and D
c] Vitamins B and C d] Vitamins B and E

36. **Ikebana is a Japanese art of**
a] modern painting b] tree farming
c] freestyle wrestling d] flower arrangement

37. **The main function of hormones is**
a] stimulation b] to start
c] to inhibit d] all of above

38. **Pellagra is caused due to deficiency of**
a] thiamine b] riboflavin
c] niacin d] thyroxin

39. **Which vegetable crop occupies maximum acreage in India?**
a] tomato b] onion
c] cole-crops d] potato

40. **Underground stem of banana is a**
a] runner b] corm
c] rhizome d] pseudostem

41. **In cross-pollinated species, mass selected variety is a mixture of several**
a] inbred b] heterozygotes
c] pure lines d] homo and heterozygotes

42. **Which vegetable has high calorie value?**
a] amaranthus b] cow pea
c] potato d] spinach

43. **Plantation and spices are main crops of**
a] eastern plateau and hill zone
b] western plateau and hill zone
c] southern plateau and hill zone
d] central plateau and hill zone

44. **Chromosome number reduced to half in**
a] meiosis-I b] meiosis-II
c] mitosis d] cytokinesis

45. **Remedy for scurvy disease, which fruit is the most suitable?**
a] guava b] aonla
c] chilli d] Barbados cherry

46. **Rose belongs to family**
a] solanaceae b] rosaceae
c] malvaceae d] cruciferae

47. **Sweet potato is a modified form of**
a] potato b] root
c] stem d] rhizome

48. **The hereditary material in chromosomes is**
a] RNA b] DNA
c] ATP d] ADP

49. **Which constituent affects freezing point of milk?**
a] fat b] protein
c] lactose d] none of the above

50. **Grape is rich source of**
a] ascorbic acid b] tartaric acid
c] mallic acidd] citric acid

51. **During snowfall, plants**
a] show minimum life function
b] show maximum transpiration
c] become more active
d] do not respire

52. **The richest source of ascorbic acid is**
a] anola b] Bardados cherry
c] lime d] guava

53. **Rose is planted in the months of**
a] May-June
b] June-July
c] End of September -first fortnight of October
d] April-May

54. **Bending of plants towards light is called**
a] hydrotropism b] geotropism
c] chemotropism d] phototropism

55. **Plant bearing pistillate and staminate flowers separately on same plant are known as**
a] bisexual b] monoecious
c] dioecious d] hermaphrodite

GK **Willy-willies is the name of the tropical cyclones in north-western Australia.** GS

56. The richest source of fat is
a] banana b] cashew nut
c] coconut d] almond

57. Anti-sterility vitamin is
a] vitamin K b] vitamin E
c] vitamin D d] vitamin B_2

58. Greenhouse cultivation is mostly applied for growing of
a] vegetables b] cut-flowers
c] vine-crops d] shade-loving trees

59. Which one is the softest mineral ?
a] talc b] quartz
c] montmorillonite d] beidelite

60. An equipment used for removing moisture from fruit & vegetables juices by evaporation is
a] evaporator b] transpirator
c] autoclave d] dehydrator

61. Seeds without endosperm are known as
a] monoembryonic seeds b] polyembryonic seeds
c] exalbuminous seeds d] apomictic seeds

62. Dual purpose breed of cows is
a] Sahiwal b] Jersey
c] Hariana d] Karan Swiss

63. A rich source of vitamin B_1 (Thiamin) is
a] mango b] guava
c] cashew-nut d] karonda

64. In fruit canning and pickling, salt acts as a/an
a] oxidant b] antioxidant
c] reductantd] enzyme

65. Aroma in fruits is due to
a] acids b] hydroxy compounds
c] esters d] xanthophylls

66. Thieleriasis is caused by
a] virus b] bacteria
c] protozoa d] none of these

67. Which preservative is the most suitable for coloured juices?
a] KMS b] sodium benzoate
c] hydrogen peroxide d] sorbic acid

68. Vegetables require a sterilization temperature of
a] 100°C b] 105°C
c] 116°C d] 121°C

69. A cross between a F_1 hybrid with any of its parent is known as
a] back cross b] hybridization
c] crossing over d] multiple cross

70. A plant which provides pollen grain for pollination in another plant is known as
a] pollinizer b] pollinator
c] genotype d] donor

71. Which of the following parameters are used in purchase of dairy cows for establishing a dairy herd
a] milk yield in pail b] parity
c] stage of lactation d] feed consumption
Codes:
a] 1, 2 and 3 b] 1, 2 and 4
c] 3 and 4 d] 1, 2, 3 & 4

72. Dehydrated grapes are known as
a] raisin b] resin
c] nut d] dried berry

73. The largest size fruit grown in India is
a] custard apple b] jackfruit
c] pineapple d] bael

74. Which fruit is useful for curing diabetes?
a] aonla b] jamun
c] bael d] guava

75. Which is the leading fruit crop of India?
a] mango b] banana
c] citrus d] guava

76. The plants in which staminate, pistillate and perfect flowers are borne on the same plant are known as
a] andromonoecious b] androecious
c] trimonoecious d] dioecious

77. Refractometer is used for measuring
a] salt per cent b] acid per cent
c] sugar per cent d] fibre content

78. H.S. occurs in the severe form of in
a] cattle and buffaloes b] horses and mules
c] dogs and cats d] pigs and rodents

79. Mango malformation a serious problem in India is in the fruit zone
a] southern b] northern
c] eastern d] western

80. Best intercrops selected for growing in the orchards should be
a] cereal crops b] cucurbit
c] brassica d] legumes

81. Fruit crop exported on large scale from India
a] grape b] banana
c] mango d] citrus

82. Aroma in coffee is due to..
a] fermentation b] drying
c] roasting d] none

83. The economic part of sandalwood tree is
a] leaves b] heartwood & roots
c] inflorescence d] bark

84. The state which leads in fruit production is
a] Maharashtra b] Tamil Nadu
c] Andhra Pradesh d] Orissa

85. RNA is synthesized in
a] cytoplasm b] nucleus
c] ribosome d] all

Panchayati Raj was given Constitutional status under 73rd Amendment.

GS

86. The backbone of Indian economy is
a] Industry
b] agriculture
c] manufacturing sector
d] information technology

87. Chillis are good source of
a] protein b] fat
c] carbohydrates d] vitamin

88. Back cross is crossing with
a] homozygous recessive female parents
b] either of parents
c] recurrent parents
d] none

89. In areas having sufficient availability of green fodder, the preferred exotic breed for crossing Indian cattle is
a] Brown Swiss b] Ayreshire
c] Holstein -Friesian d] Jersey

90. The state which leads in vegetable production is
a] Bihar b] Punjab
c] U.P. d] West Bengal

91. Rabi pulse cultivated on large scale is
a] chick pea b] pigeon pea
c] cow pea d] none

92 Nitrogenous fertilizer with highest nitrogen content
a] urea b] anhydrous ammonia
c] ammonium chloride d] none

93. Crossing over takes place in
a] zygotene b] pachytene
c] diplotene d] leptotene

94. Rate of photosynthesis is maximum in
a] red light b] black light
c] green light d] yellow light

95. Which one of the following is a warm season crop
a] cabbage b] tomato
c] cucumber d] potato

96. Abortions in the last trimester of pregnancy in cows signifies infection with
a] Campylobacter b] Brucella abortus
c] Johne's disease d] Bovine coital exanthema

97. The micro organisms which uses the organic compounds for their energy and carbon source are called
a] heterotrophs b] autotrophs
c] phototrophs d] chemotrophs

98. Tapioca meal can safely be included in the poultry diet up to 10% because
a] it contains linamarin
b] it is low in protein
c] it is always supplemented with methionine
d] it is a good source of energy

99. The 'animal protein factor' (APF) is also designated to
a] the feeds of animal origin
b] the ratio of animal flesh to animal weight
c] the ratio of animal protein to animal flesh
d] vitamin B_{12}

100. Which is not a greenhouse gas?
a] CO_2 b] CH_4
c] CO d] O_2

GK **Tornado is tropical cyclone.** GS

Set-15

1. **Glycolysis is also called as**
   a] EMP pathway b] krebs cycle
   c] both d] none
2. **Syngamy refer to**
   a] fusion to sperm with nucleus
   b] fusion of one sperm with egg
   c] fusion of sperm with synergids
   d] none
3. **The total solids content of cow's milk is approx.**
   a] 12% b] 14.5%
   c] 15.3% d] 10.0%
4. **Fowl pox in poultry is caused by**
   a] virus b] bacteria
   c] parasite d] insect
5. **Which constituent of milk gets mainly changed during its conversion into curd?**
   a] milk casein b] milk fat
   c] milk carbohydrate d] milk vitamins
6. **Rapid and convenient method of grafting mango**
   a] side b] veener
   c] tongue d] cleft
7. **National flower of India is**
   a] rose b] lotus
   c] carnation d] marigold
8. **Dry fruits like cashew & almond are richest source of**
   a] vitamin B b] fat
   c] vitamin K d] carbohydrates
9. **Vitamin C is maximum in**
   a] guava b] aonla
   c] papaya d] mango
10. **Urea can be classified as feeding stuffs as**
    a] high energy concentrate
    b] low energy roughage
    c] true protein source
    d] non protein nitrogen
11. **Carrot is a rich source**
    a] vitamin A b] vitamin B
    c] vitamin C d] vitamin D
12. **Which fatty acid chiefly influences production of hardness or softness of milk fat?**
    a] palmitic b] oleic
    c] stearic d] caproic
13. **In tomato red colour is due to**
    a] anthocyanin b] lycopene
    c] xanthophylls d] none
14. **Language, the extension system used is**
    a] demonstration b] leaflet
    c] charts d] none of these
15. **Low temperature preservation prevent growth of micro-organism due to**
    a] reduction in growth of micro organism
    b] increase in growth of micro organism
    c] killing of micro organism
    d] all of the above
16. **Treatment of fruits and vegetables by slight temperature for small time is known as**
    a] sterilization b] blanching
    c] pasteurization d] none of these
17. **Which is the ripening hormone?**
    a] auxin b] NAA
    c] IBA d] ethylene
18. **Good source of carbohydrates, vitamins, fats, minerals & protein is**
    a] vegetables b] egg
    c] milk d] none of these
19. **Pungency in onion is due to**
    a] allyl propyl disulphide b] diallyl disulphide -
    c] capsacin d] none of these
20. **Most important cereal crop in India is**
    a] rice b] wheat
    c] sorghum d] barley
21. **Sugarcane photoperiodically is a plant of**
    a] short day b] long day
    c] day neutral d] none
22. **Which part of the country contributes maximum apple production?**
    a] Jammu & Kashmir b] Himachal Pradesh
    c] Madhya Pradesh d] Uttar Pradesh
23. **Which is non mineral nutrient essential for plant?**
    a] carbon b] nitrogen
    c] phosphorus d] potassium
24. **Selfing of a plant increases**
    a] homogeneity b] homozygosity
    c] heterozygosity d] none of the above
25. **India is second largest producer of vegetables after**
    a] Brazil b] China
    c] Africa d] Japan

GK | GS
The Panchayati Raj has been placed under 11th Schedule.

26. Thailand ranks first in production of the fruit
a] banana b] mango
c] pineapple d] orange

27. Non climacteric fruit is
a] which ripe after the maturity
b] which ripe before the maturity
c] which ripe at the time of maturity
d] none of these

28. Thick yellowish discharge from the vagina of cow indicates
a] pyometra b] heat cycle
c] mastitis d] vaginitis

29. *Phyllocnistis citrella* is pest of citrus and it
a] webs the inflorescence b] bores the fruit
c] bores the stem d] mines the leaves

30. Which CHOs is used to enhance vase life of flowers
a] glucose b] maltose
c] sucrose d] none of these

31. Isotope have
a] same number of protons
b] same number of neutrons
c] same number of electrons
d] none of these

32. CO_2 accepter in C_3 plants is
a] PEP b] RUBP
c] both a and b d] none of these

33. Ripening fruits emit which type of gas
a] ammonia b] ethylene
c] methane d] all of the above

34. For vitamin C rich source is
a] banana b] mango
c] guava d] all of the above

35. When two heterozygous plants are crossed, segregation is earliest seen in
a] F1 generation b] F2 generation
c] test crossd] none

36. Which crop needs maximum water?
a] barley b] maize
c] sorghum d] sugarcane

37. Most popular broiler breed (strain) in India is
a] Babcock b] astrolop
c] minorca d] none of these

38. National flower of Japan is
a] rose b] chrysanthemum
c] tulip d] carnation

39. Beta carotene is the source of
a] vitamin A b] riboflavin
c] vitamin C d] niacin

40. Vaccination against rinderpest in crossbreed animals (cattle) should be carried out
a] every 6 months b] every year
c] every two years d] every five years

41. A disease having regular occurrence at a particular area is called
a] epidemic b] endemic
c] sporadic d] none of these

42. The movement of water from root to leaves is due to
a] high potential of water in roots than leaves
b] low potential of water in roots than leaves
c] equal potential of water
d] no relation of water potential

43. Which vegetable is underground stem?
a] potato b] radish
c] carrot d] spinach

44. Photophosphorylation is
a] hill reaction
b] formation of ATP in presence of light
c] formation of NADP in presence of light
d] none of these

45. Which is non climacteric in nature?
a] pineapple b] mango
c] mango d] avacado

46. The term viviparous means
a] germination of seed on mother plant
b] germination of seed in field
c] germination of seed on monecious plant
d] germination of seed on undesirable place

47. Which of the following pigment is responsible for red colour in rose
a] globuline b] anthocyanin
c] xanthophylls d] all of these

48. The term pruning refers to
a] providing shape to plant b] providing space for growth
c] for better fruiting d] all of these

49. Where does DNA synthesis occur
a] interphase b] prophase
c] sphase d] G1 phase

50. In tea, which of the following beverage substance is present?
a] caffeine b] thiobromin
c] piperine d] none of these

51. Which is non bulb forming member of onion family
a] leek b] garlic
c] onion d] none of these

52. The best quality tea grown in the world is at
a] Nilgiri (India) b] China
c] Darjeeling (India) d] Nainital (India)

53. Which of the following is the narcotic drug
a] poppy b] tea
c] coffee d] all of these

54. The Rashtrapathi Garden is example of which type of garden?
a] Mughal b] Italian
c] English d] Japanese

GK **Tropical rain forest in the Amazon low lands is Selves.** GS

55. The cow should not be milked before calving because
(i) the birth of the calf may be delayed
(ii) even though the udder becomes swollen and hardened no injury is likely to result
(iii) the cow is in a very nervous condition at this time
a] i + ii b] ii + iii
c] iii + i d] all of these

56. In freezing which principle is followed
a] condensation b] crystallization
c] drying d] all of the above

57. Banana prefers which type of climate?
a] hot b] dry
c] cold d] humid

58. Cryo-preservation is associated with
a] liquid nitrogen b] liquid oxygen
c] liquid potassium d] liquid carbon dioxide

59. The fruits & vegetables are known as
a] nutritive food b] energy food
c] protective food d] none of these

60. Vaccine contains
a] antigen b] antibody
c] agglutinin d] all of these

61. The inflorescence of sugarcane is known as
a] spike b] arrow
c] cymose d] none of these

62. Extension method is
a] learning process
b] teaching process
c] learning & teaching process
d] none of these

63. Which is the rich source of vitamin C?
a] cabbage b] tomato
c] chilli d] all of these

64. The pungency of chilli is due to
a] capsanthin b] capsicin
c] isothiocynate d] none of these

65. Which of the following is the flower preservative?
a] $KMnO_3$ b] KNO_3
c] $HGCl_2$ d] none of these

66. Antibodies are found in
a] blood b] serum
c] plasma d] none of the above

67. The natural rubber is mostly used in which industry?
a] tyre b] sheet
c] box d] none of these

68. The 20th century vegetable
a] cluster bean b] amaranthus
c] brinjal d] winged bean

69. Which of the following is used as a cut flower?
a] marigold b] jasmine
c] gladiolus d] all of these

70. Which one of the following systems should be recommended for successful management of pasture?
a] rotational grazing b] annual pasture
c] short rotation d] permanent

71. Which cucurbit have highest vitamin A?
a] bitter gourd b] cucumber
c] pumpkin d] all of these

72. Which part of tea plant is harvested
a] a bud & young leaves b] mature leaves
c] bark d] all of these

73. Most limiting amino acid in soyabean-based poultry diet is
a] lysine b] methionine
c] arginine d] tryptophan

74. Protein content in groundnut is
a] 22% b] 25%
c] 40% d] 43%

75. Maximum oil content is in
a] oil palm b] soyabean
c] coconut d] none of these

76. Photosynthesis occur in
a] chloroplast b] mitochondria
c] golgi bodies d] none of these

77. The ability of cell to generate into a whole plant is known as
a] regeneration capacity
b] totipotency
c] autolysis
d] none of these

78. Rice has been classified as
a] Short day plant b] long day plant
c] day neutral plant d] none of these

79. The fruit of wheat is known as
a] pod b] capsule
c] siliqua d] caryopsis

80. Isotone have
a] same number of protons
b] same number of neutrons
c] same number of electrons
d] none of these

81. Carnation is the national flower of
a] Japan b] China
c] Spain d] Holland

82. Tissue culture is common in
a] rose b] orchid
c] gladiolus d] palms

83. Element which is an important constituent of organic compound is
a] carbon b] phosphorus
c] potassium d] nitrogen

84. **The metal nutrient present in the structure of chlorophyll is**
a] Fe b] Zn
c] Ca d] Mg

85. **A cross in which order of male & female parent in reversed is called**
a] reciprocal cross b] back cross
c] test crossd] top cross

86. **The characteristic symptom of pink bollworm damage in cotton is**
a] Flared squares
b] Rosetted flowers
c] Terminal shoot drying
d] Small circular bore holes on small bolls

87. **A unit of DNA molecule containing only a sugar and an organic base is called**
a] nucleotide b] nucleoside
c] codon d] nucleus

88. **For measuring leaf area the following instrument is used**
a] leaf gaugeb] anemometer
c] potometer d] barometer

89. **Hormone promoting the root tuber growth is**
a] auxin b] GA
c] cytokinin d] ethylene

90. **Fruit which is rich in pectin & acid is**
a] jamun b] apple
c] pear d] pomegranate

91. **Vitamin C is commonly known as**
a] citric acid b] acetic acid
c] lactic acid d] ascorbic acid

92. **Acetic acid is present in**
a] nectar b] sweet chutney
c] vinegar d] apple cider

93. **The pigment present in watermelon is**
a] carotenoid b] anthocyanin
c] lycopene d] both b & c

94. **Which is not a legume crop**
a] pea b] beans
c] wheat d] fenugreek

95. **Which of the following is a draft breed?**
a] Malwi b] Sahiwal
c] Gir d] Kankry

96. **Bone meal is a good source of**
a] carbohydrates b] fat
c] phosphorus d] calcium & phosphorus

97. **Sequence of growing crop on a given piece of land is referred to as**
a] crop management b] crop production
c] crop science d] crop rotation

98. **Alternate form of gene at the same locus are referred to as**
a] chromosome b] plastic
c] allele d] dominant

99. **Animal cells & plant cells can be differentiated by**
a] conductivity b] shape
c] size d] presence or absence of cell-wall

100. **In presence of sunlight, CO_2 & water (plus chlorophyll) are converted into carbohydrates is called**
a] respiration b] metabolism
c] photosynthesis d] solar radiation

GK GS
Deciduous type of trees are generally found in Tropical monsoonal climate.

Set-16

1. NPK are regarded in plant nutrition as
   a] micro nutrient b] macro nutrient
   c] trace element d] non essential,
2. Yellow colour of cow milk is due to
   a] fat b] lactose
   c] protein d] carotene
3. Respiration in plants is essentially a process related to the following
   a] oxidation b] evaporation
   c] transpiration d] none of these
4. Scientific name of mango nut weevil is
   a] *Rhynchaenus mangiferae*
   b] *Sternochetus mangiferae*
   c] *Aceria mangifera*
   d] *Myllocerus discolor*
5. DNA contain following number of nitrogenous base repeated in various sequences
   a] 10 b] 2
   c] 4 d] 1
6. Cryopreservation is done by liquid nitrogen at temperature of
   a] 200°C b] 0°C
   c] -196°C d] 4°C
7. Most important temperate vegetable is
   a] Indian cauliflower b] cabbage
   c] onion d] radish
8. Which protein is most abundant in milk
   a] albumin b] globule
   c] casein d] α-globulin
9. Sweet potato is primarily a crop of
   a] tropical region b] temperate region
   c] arid region d] none
10. Colostrum must be fed to the newborn calf because it is rich in
    (I) lactose (II) lactoglobulin (III) iron
    Select the correct answer
    a] only I and II are correct
    b] only I and III are correct
    c] only II and III are correct
    d] only II is correct
11. Indicate the portion of milk in a milking which has highest percentage of fat
    a] fore milk b] last milk
    c] mid milking d] strippings
12. Haploid refers to
    a] two sets of chromosomes
    b] half set of chromosomes
    c] three sets of chromosomes
    d] single set of chromosomes
13. The first scientific-research in artificial insemination of domestic animals was conducted with
    a] cow b] monkey
    c] mare d] dog
14. In general fruit & vegetables are not rich source of
    a] vitamin A b] vitamin B_1
    c] vitamin C d] mineral
15. Unit of heredity of
    a] chromosome b] cell
    c] chromatidd] gene
16. Mango & banana are
    a] climacteric
    b] non climacteric
    c] nature of ripening depend upon var
    d] none of these
17. Fruit & vegetables are rich in dietary fibres which (protect) from
    a] colon cancer b] protein deficiency
    c] fat deficiency d] sugar deficiency
18. Crossing over during meiosis results in
    a] promoting linkage b] breaking linkage
    c] help in mutation d] none of these
19. Rich source of protein (>12%) is
    a] walnut b] cashewnut
    c] peanut d] mango
20. Which fruit crop can be grown under extreme drought condition
    a] date palm b] banana
    c] mango d] guava
21. Which breed of cows is highest fat producer in milk
    a] Red sindhi b] Holstein
    c] Karan-fris d] Karan Swiss
22. Which vegetable is rich source of calcium?
    a] fenugreek
    b] amaranthus
    c] potato
    d] tomato
23. Daily requirement of vegetable per capita/day is
    a] 150 gram b] 200 gram
    c] 285 gram d] 400 gram

GK GS

President appoints special officer for SC & ST.

24. **Biggest rose garden in the world is situated at**
a] New Delhi b] France
c] New York d] Frankfurt
25. **Ikebana is**
a] flower arrangement b] bonsai culture
c] soil culture d] all
26. **Which crop is richest source of oil?**
a] groundnut b] castor
c] sesamum d] oil palm
27. **Calving interval is the sum total of**
a] lactation length and gestation length
b] gestation length and dry days
c] service period and gestation days
d] open days and dry days
28. **In one turn of genetic model of DNA, how much distance is covered**
a] 3.4 Å b] 34 Å
c] 36 Å d] 38 Å
29. **In the term horticulture, the word Hortus is**
a] Greek b] Latin
c] English d] Italian
30. **Sugarcane mealy bug is seen**
a] on the leaf b] stem
c] roots d] under leaf sheath
31. **Transplastomics is**
a] transfer of mitochondria
b] transfer of chloroplast
c] transfer of cytoplasm
d] all of these
32. **Pick the monocot crop**
a] pineapple b] date palm
c] banana d] all of these
33. **A good way to improve a herd at low cost is by**
a] buying better cows
b] use of a pure bred bull
c] buying good bred heifers
d] giving better treatment to cows
34. **Growing of crops in between kharif and rabi seasons is**
a] relay cropping b] catch crop
c] zaid cropping d] none of these
35. **Which of the following poultry diseases is a self-limiting disease?**
a] fowl pox b] coccidiosis
c] ascariasis d] none of these
36. **Feed stuff containing more than 8% crude fibre is called**
a] concentrates b] feed supplements
c] roughness d] none of these
37. **Bitterness in groundnut is due to ________**
a] alkaloid b] tannins
c] chemical spraying d] aflatoxin
38. **Yellow revolution is related to**
a] fishery b] agroindustry
c] onion d] oilseed
39. **The ________ is done to obtain virus free plants**
a] pollen culture b] cell culture
c] embryo culture d] meristem culture
40. **________ is the site of β-oxidation in cell**
a] chloroplast b] vacuole
c] mitochondria d] ribosomes
41. **White ear in rice is caused due to**
a] stem borer b] leaf foler
c] earhead bug d] white tip nematode
42. **India stands 1st in the production of__________**
a] tomato b] cabbage
c] onion d] okra
43. **Coconut water is rich source of____________**
a] Vitamin B_1 b] Vitamin B_2
c] Vitamin B_6 d] Vitamin C + cyanocobalamin
44. **________ gardens are also called as nature in miniature**
a] Japanese b] Mughal
c] Italian d] English
45. **reagent is used for determination of chlorophyll**
a] acetic acid b] ethanol
c] petroleum ether d] acetone
46. **The micro organism responsible for ropiness of milk is**
a] *Streptococcus liquifaciens*
b] *Pseudomonas flourscens*
c] *Bacillus caset*
d] *Torula amaia*
47. **Whisky is a fermented product of ________**
a] fruit b] malt
c] grain mash d] cocoa
48. **__________ is a thermosensitive vitamin**
a] Vitamin A b] Vitamin E
c] Vitamin C d] Vitamin D
49. **__________ is the richest source of sugar**
a] sapota b] karonda
c] dates d] mango
50. **India ranks __________ in rice production**
a] first b] second
c] fourth d] fifth
51. **Mushrooms are an excellent source of______ mineral that the body needs to produce red blood cells**
a] Mn b] Cu
c] Fe d] Cl
52. **Per capita net availability of food grains was ________gm/day in 2004-2005**
a] 467 b] 480
c] 520 d] 580

Petroleum is liquid gold.

53. Pairing of homologous chromosomes during zygotene is called
a] chiasma b] fertilization
c] synapsis d] double fertilization

54. The greenhouse gas that is released from paddy fields is__________
a] CO_2 b] CH_4
c] N_2O d] NH_3

55. Pellagra is due to deficiency of _______
a] niacin b] folic acid
c] biotin d] pyridoxine

56. The law most valid in agricultural production ______
a] law of equi-marginal returns
b] law of diminishing returns
c] law of MPP
d] both a and b

57. Milk is deficient in
a] Ca b] P_2O_5
c] Fe d] Lactose

58. ______instrument is used to measure growth of plant
a] fathometer b] galvanometer
c] crescograph d] anemometer

59. India is the largest producer, consumer as well as exporter of
a] spices b] fruits
c] milk d] rice

60. Which of the following poultry diseases cannot be prevented by vaccination?
a] Chronic respiratory disease
b] Marek's disease
c] Ranikhet disease
d] None of the above

61. Name the state which is known as "Fruit bowl of India"
a] Andhra Pradesh b] Uttar Pradesh
c] Gujarat d] Maharashtra

62. Operation flood is related to ___________
a] drought-prone areas
b] flood control
c] dairy development
d] multipurposes dams

63. The disadvantage of dehorning a bull is that it
a] reduces his prepotency
b] reduces his good luck
c] disqualifies him in the show ring
d] makes him more vicious

64. Black tip or Mango necrosis is caused by
a] *Glomerella cingulata*
b] Zinc deficiency
c] Smoke and sulphur dioxide emitted from the brick kilns
d] None of these

65. Type of soil predominant in India
a] red soil b] alluvial soil
c] humous soil d] preaty soil

66. Vitamin containing Cobalt as a constituent is______
a] B_2 b] B_6
c] B_1 d] B_{12}

67. The milk fat percentage is highest in which of the following animals?
a] mare b] buffalo
c] goat d] cow

68. The agricultural universities in India were established on pattern of__________
a] Land Grant College of USA
b] Agricultural Universities of Great Britain
c] Cornell University
d] Tennessee Valley Authority

69. Vitamin name was suggested by__________
a] Strasburger b] Robert Hooke
c] Robert Brown d] Casimir Funk

70. Antibodies are produced by those cells of the animal body which synthesize
a] haemoglobins b] albumins
c] globulins d] none of these

71. Premier oilseed crop of india is________
a] groundnut b] tobacco
c] sugarcane d] rice

72. The unit of rural society is________
a] town b] village panchayat
c] village d] co-operative society

73. Which of the following is a milk protein
a] Lactin b] Galactose
c] Caesin d] Collagen

74. The most abundant element in the living system is
a] oxygen b] nitrogen
c] hydrogen d] carbon

75. Power house of cell is
a] chloroplast b] nucleus
c] mitochondria d] ER

76. _____fruit is called as poor man's fruit
a] ber b] custard apple
c] guava d] aonla

77. The vegetable which is richest source of vitamin A is
a] turnip leaves b] carrot
c] pumpkin yellow d] drumstick leaves

78. The vegetable crop that tops in the share of total vegetable production
a] brinjal b] tomato
c] potato d] cauliflower

79. ELISA test is done to find out_______ diseases
a] fungal b] bacterial
c] viral d] all of these

GK **The term of office of CEC is 5 years.** GS

80. **The pH of nearly all food is below**
a] 6 b] 6.5
c] 7.0 d] 8.0

81. **Introductions are used as**
a] a direct variety b] a variety after selection
c] a parent of hybrid d] all of the above

82. **In crop plants, clonal selection is used in**
a] non-flowering species
b] low seed setting species
c] apomictic species
d] all of the above

83. **Synthetic variety is a mixture of**
a] all possible crosses among inbreds of good general combining ability
b] several pure lines
c] several open pollinated genotypes
d] several inbred lines

84. **The yield of a synthetic variety is always higher than**
a] open pollinated variety
b] single cross hybrid
c] three way cross hybrid
d] double cross hybrid

85. **Heterosis results due to**
a] dominance b] over dominance
c] epistasis d] all of the above

86. **Late blight of potato is caused by**
a] *Phytophthora infestans*
b] *Pythium aphanidematum*
c] *Plasmopara viticola*
d] *Alternaria alternata*

87. **Bordeaux mixture was developed by**
a] Anton de Bary b] P.M.A. Millardet
c] E.J. Butler d] J.C. Walker

88. **Fungicides which are absorbed into the system of the plant and move to the remote site of infection are known as**
a] systematic b] systemic
c] dressers d] fungistatic

89. **Wart disease of potato is caused by**
a] *Olpidium brassicae*
b] *Plasmodiophora brassicae*
c] *Synchytrium endobioticum*
d] *Physoderma zea-maydis*

90. **The intermediate host of eye worm of poultry is**
a] Stomoxys b] Pycnoscellis
c] *Musca domestica* d] Tabanus

91. **Seed potato is not transported from Darjelling area to other parts of the country because of**
a] domestic quarantine
b] high transport cost involved
c] it does not perform well In the plains
d] All of the above

92. **Viruses which can kill the bacteria are known as**
a] virion b] antibodies
c] bacteriophages d] none of the above

93. **Pathogen responsible for Bengal famine was**
a] *Helminthosporium oryzae*
b] *Helminthosporium graminearum*
c] *Alternaria alternata*
d] *Curvularia lunata*

94. **Thiram is used as**
a] seed dresser b] foliar spray
c] growth promoter d] soil applicant

95. **Less degraded protein supplement for ruminant is**
a] urea b] barley
c] soyabean meal d] fish meal

96. **Karathane is**
a] fungicide b] insecticide
c] nematicide d] bactericide

97. **When a disease is more or less constantly present from year to year in a moderate to severe form, in a particular country or part of the earth then it is known as**
a] epiphytotic b] sporadic
c] pandemic d] endemic

98. **Microscope was developed for the first time by**
a] Theophrastus b] Shakespeare
c] Leeuwenhoek d] Louis Pasteur

99. **Gene to gene hypothesis of disease resistance and susceptibility was given by**
a] Erikson b] Biffen
c] Flor d] Blakslee

100. **The abnormal increase in the size of a plant organ is known as**
a] hyperplasia b] hypertrophy
c] both a and b d] none of the above

Marie Curie died in 1934.

Set-17

1. **Murrah breed of buffaloes has originated from**
   a] China b] Egypt
   c] Thailand d] India
2. **Bacteria having one flagellum at each end of the cell are known as**
   a] atrichous b] lophotrichous
   c] amphitrichous d] monotrichous
3. **Hyphal structures sent into the host cell by different fungi to absorb the nutrients are known as**
   a] appressoria b] haustoria
   c] sclerotia d] rhizobia
4. **Citrus canker disease is caused by**
   a] *Xanthomonas campestris* Pv. *citri*
   b] *Phytophthora citrophthora*
   c] Citrus greening virus
   d] Citrus exocortis virus
5. **Artificial insemination in cattle was done first time in India at**
   a] Lucknow b] Mathura
   c] Izatnagar d] Rishikesh
6. **Raising the pH of the field soil to 7 or above by adding line (calcium carbonate) gives good control of**
   a] wart disease of potato
   b] late blight of potato
   c] soft rot of potato
   d] club root of crucifers
7. **Which one of the following causes fruit rot of crucifers?**
   a] *Pythium aphanidermatum*
   b] *Sclerotinia sclerotiorum*
   c] *Phytophthora parasitica*
   d] *Olpidium brassicae*
8. **Pure bred bulls should be used because**
   a] they are easier to handle
   b] they cost more than grades
   c] the sire is half the herd
   d] they are prepotent
9. **What is the proper age for vaccination against Brucellosis?**
   a] 6 months b] 1.5 years
   c] 2.5 years d] 3.5 years
10. **Downy mildew fungi, are**
    a] only seed-borne
    b] only soil-borne
    c] both seed and soil-borne
    d] none of the above
11. **Which one of the following is the correct method to get more milk from a crossbred cow?**
    a] feeding nutritious forages
    b] feeding balanced ration
    c] feeding thyroprotein
    d] feeding low quality forage with high amount of ration
12. **Downy mildew of crucifers caused by**
    a] *Albugo candida* b] *Peronospora parasitica*
    c] Both a & b d] None of the above
13. **Characteristic symptoms of soft rot of apple appear on**
    a] mature apples b] green apples
    c] both a and b d] none of the above
14. **Ring spot disease of sugarcane is caused by**
    a] *Ascochyta* sp. b] *Colletotrichum falcatum*
    c] *Clavibacter xyli* d] *Leptosphaeria sacchari*
15. **Solar energy treatment of wheat seed is recommended for the control of**
    a] loose smut b] karnal bunt
    c] hill bunt d] black rust
16. **The best way to control the disease spread of smut of sugarcane is to**
    a] remove the smutted whips from the field
    b] discourage the practice of ratooning
    c] avoid planting of setts from smutted canes
    d] all of them
17. **The presence of foul smell (stinking smell) in hill bunt or karnal bunt infected fields is due to the production of**
    a] formaline b] trimethylamine
    c] phenylalanine d] lysine
18. **Karnal bunt of wheat is caused by**
    a] *Neovossia horrida* b] *Neovossia indica*
    c] *Tilletia carries* d] *Tilletia foetida*
19. **Which of the following fungi causes black rust of wheat**
    a] *Puccinia graminis tritici*
    b] *Puccinia recondita*
    c] *Puccinia striformis*
    d] none of the above
20. **Stripe rust of wheat is also known as**
    a] brown rust b] black rust
    c] yellow rust d] stem rust

GK **Kacedonia is associated with Alexander** GS

21. **Which of the following rusts is the earliest to appear on wheat in India**
    a] black rust b] yellow rust
    c] brown rust d] none of the above
22. **Rice blast disease is caused by**
    a] *Xanthomonas oryzae*
    b] *Drechslera oryzae*
    c] *Pyricularia oryzae*
    d] *Rhizopus oryzae*
23. **Kitazin and Hinosan are fungicides used for control of**
    a] paddy blast
    b] early blight of potato
    c] sheath blight of paddy
    d] brown spot of rice
24. **Fusarium wilt of banana is also known as**
    a] bunchy top b] panama disease
    c] both a and b d] none of the above
25. **The activity of *Rhizoctonia solani* causing stem canker and black scurf of potato can be suppressed by a biocontrol agent**
    a] *Gliocladium* b] *Aspergillus*
    c] *Trichoderma* d] *Entomophthora*
26. **Black heart of potato is caused by group of organism**
    a] bacteria
    b] fungi
    c] poor ventilation in the storage (lack of oxygen)
    d] all of the above
27. **Thiram is mainly used as**
    a] foliage spray b] fumigant
    c] seed dresser d] all of these
28. **Potato is native of**
    a] Europe b] South America
    c] Africa d] None of the above
29. **Tuber production in potato is maximum at a temperature of**
    a] 10°C b] 15°C
    c] 20°C d] None of the above
30. **What type of milking pail is best for clean milk production**
    a] earthen pot b] domed bucket
    c] ordinary tin bucket d] plastic bucket
31. **The ideal soil for potato is one which is having a pH range of**
    a] 4.5 to 5.0 b] 5.2 to 6.4
    c] 6.5 to 7.5 d] none of the above
32. **Which fruit is rich source of fat?**
    a] amla b] bael
    c] avocado d] all
33. **Seed rate per acre in potato varies from**
    a] 6 to 10 quintals b] 15 to quintals
    c] 20 to 25 quintals d] none of the above
34. **Corky outgrowths on guava is caused by the feeding of the insect**
    a] Guava fruit borer b] Guava mealy bug
    c] Tea mosquito bug d] Green bug
35. **Viral diseases of potato are spread by**
    a] white fliesb] jassids
    c] aphids d] none of the above
36. **Haulms in potato seed crop are cut during**
    a] End of December or 1st week of January
    b] second fortnight of January
    c] 1st fortnight of February
    d] none of the above
37. **For long storage, potato tubers should be kept at**
    a] 5-10°C b] 10-15°C
    c] 15-20°C d] none of the above
38. **For keeping the sugar content low, potato tubers should be stored at a temperature**
    a] 5°C b] 10°C
    c] 15°C d] none of the above
39. **Very Important vegetable included in solanaceous fruit group is**
    a] tomato b] potato
    c] okra d] sweet potato
40. **Cultivated tomato belongs to the species**
    a] hirsutum b] peruvianum
    c] pimpinellifolium d] esculentum
41. **Average monthly temperature required for better growth and fruit-set in tomato is**
    a] 21°C to 23°C b] 30°C to 35°C
    c] 35° C to 40°C d] 5°C to 10°C
42. **The best soil for tomato growing is**
    a] sandy loam
    b] loam
    c] clay
    d] loamy sand
43. **Seed rate of tomato crop for an area of one acre is**
    a] 400- 500 g b] 300 g
    c] 100 g d] 250 g
44. **Cracking of tomato fruit is brought about by**
    a] restricted watering
    b] a period of drought followed by sudden heavy watering
    c] overwatering
    d] insufficient watering
45. **Pruning and training of tomato plants to a single stem helps in getting**
    a] high total yield b] high marketable yield
    c] high early yield d] increased fruit size
46. **Pollination and fertilization processes in tomato are impaired due to night temperature below**
    a] 13°C b] 5°C
    c] 10°C d] 20°C

GK **Population of Seychelles is around 80,000.** GS

47. For distant transportation, tomato fruits are picked at
a] immature green stage b] mature green stage
c] turning stage d] red ripe stage

48. Best temperature for storing tomato fruits is
a] 12°C -15°C b] 4.5°C
c] 5°C -10°Cd] 15°C -20°C

49. The most serious virus disease of tomato in the rainy and autumn season is
a] TMV b] CMV
c] leaf curl d] tomato spotted wilt virus

50. For getting an early crop of brinjal, the sowing may be preferably done in
a] silt loam b] sandy loam
c] clayey soil d] none of the above

51. Little leaf disease of brinjal is caused by
a] mycoplasma b] bacteria
c] fungi d] none of the above

52. The most important state growing chilli is
a] Karnataka b] Uttar Pradesh
c] Andhra Pradesh d] None of the above

53. Bud, blossom and fruit drop in chilli, in general, is due to
a] deficiency of nitrogen
b] humid climate
c] unfavourable temperature and water supply
d] none of the above

54. Number of irrigations generally applied to chilli are
a] 3-4 b] 6-8
c] 15-16 d] none of the above

55. Blossom end rot in chilli is due to
a] fungal disease
b] viral disease
c] non-parasitic trouble
d] none of the above

56. Pollinating agent in chilli is
a] house fly b] bee
c] wind d] none of the above

57. All cole crops belong to the family
a] cruciferae b] umbelliferae
c] brassicae d] none of the above

58. By which of the following method, the benefit of first hand knowledge, goodwill and confidence can be developed
a] telephone
b] demonstration
c] visiting homes & farms
d] all of these

59. Which is a seedless fruit
a] pineapple b] apple
c] sapota d] strawberry

60. The smallest seeds are found in
a] onion b] candy tuft
c] orchid d] mustard

61. Whiptail in cauliflower is caused due to the deficiency of
a] boron b] molybdenum
c] potassium d] none of the above

62. Black rot in cauliflower is caused by
a], fungal disease b] bacterial disease
c] viral disease d] none of the above

63. Radish belongs to the genus
a] brassica b] raphanus
c] daucus d] none of the above

64. The system of pollination in radish is
a] self pollination b] cross pollination
c] often cross pollination d] none of the above

65. The most important insect attacking radish is
a] red pumpkin beetle b] aphid
c] thrip d] none of the above

66. Orange-coloured varieties of carrot are rich source of
a] carotene b] lycopene
c] anthocyanin d] none of the above

67. Carrot belongs to the family
a] cruciferae b] malveceae
c] umbelliferae d] none of the above

68. Carrot yellows disease is caused by
a] fungus b] bacterium
c] virus d] none of the above

69. Major component of sugar in radish Is
a] fructose b] glucose
c] sucrose d] none of the above

70. Characteristic pungent flavour of radish is due to the presence of
a] isothiocyanates b] sinigrin
c] glycoproteins d] none of the above

71. The edible portion of radish develops from
a] primary root only
b] hypocotyls only
c] both primary root and hypocotyls
d] none of the above

72. The inflorescence of radish is a typical
a] raceme b] umbel
c] cymose d] none of the above

73. Onion belongs to the family
a] liliaceae b] araceae
c] amrylllidaceae d] none of the above

74. Rich source of fat is
a] cashew b] walnut
c] almond d] none of these

75. The insect which favours the development of purple blotch disease of onion is
a] aphid b] whitefly
c] thrip d] none of the above

76. Papaya fruit contains a proteolytic enzyme
a] amylase b] bromeliase
c] papain d] PME

77. Botanically banana is fruit
a] berry b] pome
c] stone d] hesperidium

78. Per capita fruit requirement in our diet is
a] 85 g b] 185 g
c] 70 g d] 170 g

79. Normal pulse rate in buffalo is
a] 30-40 b] 20-30
c] 50-70 d] 70-90

80. Beriberi disease is due to the deficiency of a
a] Vit. C b] Vit. D
c] Vit. E d] Vit. B_1

81. The soil fit for orchard planting whose lime content is
a] below 5% b] 5-10%
c] 10-20% d] above 20%

82. Best time of planting evergreen fruits is
a] December-January b] February -March
c] July -August d] September- October

83. Best time for planting deciduous plants
a] December -January b] February- March
c] July -August d] September- October

84. Pit size used for planting most of the fruit plants
a] 0.5 x 0.5 m b] 1 x 1 m
c] 1.5 x 1.5 m d] 2 x 2 m

85. Planting distance of grape vines in general
a] 1.5 x 2 m b] 2.5 x 5 m
c] 3 x 3 m d] 3.5 x 3.5 m

86. Mango is planted at a distance of
a] 6 x 6 m b] 8 x 8 m
c] 9 x 9 m d] 9-10m

87. Banana is planted in
a] July-August
b] September-October
c] November-December
d] January-February

88. Type of banana suckers commonly used for planting
a] knife suckers b] razor suckers
c] sword suckers d] fork suckers

89. Soil best suited for planting papaya is
a] well drained loam b] well-drained clay
c] sandy loam d] any well-drained soil

90. The ideal soil depth for planting fruit trees is
a] 1 m b] 2 m
c] 3 m d] 4 m

91. An anti-haemorrhagic vitamin is
a] vitamin A b] vitamin E
c] vitamin K d] vitamin C

92. Top working in temperate fruit plants in dormancy is best done by the technique
a] tongue grafting b] T -budding
c] veneer grafting d] chip budding

93. Dying of plant organs, especially stems or branches, from tip to backward is known as
a] sun scald b] die-back
c] wilting d] necrosis

94. Plants produced from asexual methods are
a] vigorous
b] true-to-type
c] susceptible to diseases
d] having long juvenile period

95. The layer of plant tissues lies between system and phloem of a stem/shoot
a] cambium b] sclerenchyma
c] collenchyma d] mucilaginous membrane

96. The process of regeneration of roots in the absence of sunlight.
a] etiolation b] photosynthesis
c] humidification d] rhizogenesis

97. CAN, ammonium sulphate and urea are the major fertilizers which supply the nutrient called
a] nitrogen b] phosphorus
c] potassium d] calcium

98. Soil management practice wherein the fruit trees are grown in permanent grasses without any hoeing and adding of litter
a] clean culture b] grass culture
c] sod culture d] mulching

99. Crops which are grown in the orchard to add organic matter and nitrogen to improve soil structure
a] intercrops b] cover crops
c] green manuring d] leguminous crops

100. The chief growth stimulating nutrient
a] nitrogen b] phosphorus
c] potassium d] calcium

Set-18

1. System of irrigation best suited under water scarce arid regions is
   a] sprinkling b] drip/trickle
   c] furrow d] basin
2. System of management which helps to conserve soil moisture, raises soil temperature in winter and suppresses weed growth is
   a] sod b] mulching
   c] cover crops d] clean culture
3. The nutrient, which improves fruit quality and giving rigidity to twigs, shoots, spurs etc. is
   a] calcium b] potassium
   c] nitrogen d] phosphorus
4. The period during which water scarcity affects fruit size, quality, maturity and causes fruit drop is
   a] growth period
   b] dormant period
   c] fruit development period
   d] post-harvesting period
5. Sod culture is beneficial for the fruit culture in areas
   a] plains b] valley
   c] hills d] coast
6. Loose smut of wheat can be controlled by seed treatment with
   a] vitavax b] thiram
   c] agallol d] captan
7. Green manuring crops improve the soil
   a] texture b] structure
   c] irrigation requirement d] sun burning
8. Training of fruit trees influence their
   a] shape b] framework
   c] fruit production d] both shape and framework
9. In citrus interveinal chlorosis, scaling and rosette formation of new leaves is due to the deficiency of a nutrient
   a] nitrogen b] potassium
   c] zinc d] iron
10. National fruit of India is
    a] banana b] apple
    c] mango d] citrus
11. Best method for mass multiplication of mango is
    a] T-budding b] air layering
    c] march grafting d] side grafting
12. A serious physiological disorder of mango fruit near the brick kilns is
    a] spongy tissue b] black spot
    c] black tip d] malformation
13. Mature mango fruits can be artificially ripened by the treatment
    a] carbaryl @1.5 g/kg b] 2,4- D @1.5 9/kg
    c] GA_3 @ 1.5 g/kg d] Ethephon @1.5 g/kg
14. Powdery mildew a serious white fungal growth on panicles during February-March can be checked by spraying
    a] Karathane 0.1% b] Sevin 1.0%
    c] Aerosal0.5% d] Captaf 0.3%
15. Mango belongs to the family
    a] rutaceae b] myrtaceae
    c] sapindaceae d] anacardiaceae
16. Banana belongs to the family
    a] rutaceae b] myrtaceae
    c] rosaceae d] musaceae
17. Banana is the richest source of
    a] carbohydrate b] vitamin A
    c] vitamin B d] iron
18. Commercial propagation method of banana is
    a] seed b] cuttings
    c] suckers d] budding
19. Best season of planting for banana is
    a] spring b] autumn
    c] rainy d] winter
20. Banana needs heavy manuring because it is
    a] herb b] shrub
    c] deep rooted d] shallow-rooted
21. Moisture requirement of banana is
    a] nil b] low
    c] high d] medium
22. Pseudostem of banana is the rich source of
    a] protein b] fat
    c] starch d] minerals
23. Sweet orange is harvested at full ripening on the trees because it is
    a] climacteric
    b] non-climacteric
    c] climatic
    d] non-climatic
24. Sweet orange requires the climate of
    a] dry b] semi- arid
    c] humid d] dry and semi -arid

GK **Geoffery Chaucer is known as Father of English Poetry** GS

25. **Sweet orange is generally a fruit with**
a] tight skinned b] loose skinned
c] partially loose d] partially tight

26. **Mosambi variety of sweet orange is**
a] smooth skinned
b] smooth with longitudinal stripes
c] rough
d] smooth with longitudinal stripes with a circular ring around the apex

27. **Sweet oranges are generally consumed as**
a] table fruitb] squash
c] pickle d] juice and squash

28. **Seed of sweet orange is**
a] mono embryony d] diembryony
c] triembryony d] polyembryony

29. **Sour lime and lemons are good source of**
a] A b] B_1
c] B_2 d] C

30. **Sour lime is commonly called**
a] Lime b] Kaghzi lime
c] Neebu d] Khatta nimbu

31. **Sour lime is usually propagated by**
a] seeds b] cuttings
c] air layering d] T-budding

32. **Sour lime is**
a] tree b] shrub
c] bush d] semitree

33. **In India sour lime and lemon are generally used as**
a] jam b] pickle
c] juice d] pickle and juice

34. **Sour lime prefers to grow in a temperature**
a] below 4°C b] 10-30°C
c] 30-35°C d] above 35°C

35. **Name of the grapefruit is because of**
a] taste like grape
b] size of grape
c] bearing in clusters, alkaloid characteristics
d] none of the above

36. **The grapefruit is mainly used as**
a] dessert fruit b] pickles
c] squash d] wine

37. **Grapefruit and pummelo thrive well in a climate**
a] hot and humid b] hot and dry
c] cold and humid d] cold and dry

38. **Grapefruit is commercially propagated by**
a] seeds b] cutting
c] air layering d] T-budding

39. **Papaya belongs to the family**
a] sapotaceae b] moraceae
c] myrtaceae d] caricaceae

40. **Papaya bears flowers of the kind**
a] staminate b] pistillate
c] hermaphrodite d] any kind

41. **In papaya pollination is by pollinator**
a] insect b] wind
c] water d] animal / man

42. **The yellow pigment in papaya fruit is**
a] anthocyanin b] xanthophylls
c] caricaxanthin d] carotene

43. **Papaya is the richest source of proteolytic enzyme**
a] protease b] papain
c] xanthain d] PME

44. **Papaya climatically is**
a] temperate b] tropical
c] subtropical d] sub-temperate

45. **Papaya is commercially propagated by**
a] budding b] layering
c] cutting d] seed

46. **In papaya serious disease of papaya is**
a] fruit rot b] damping off
c] leaf curl d] virus

47. **Best soil texture for papaya cultivation**
a] heavy b] medium heavy
c] loam d] well drained loam

48. **Guava is the rich source of vitamin**
a] vitamin A b] vitamin B_1
c] vitamin 82 d] vitamin C

49. **Major product from ripe guava is**
a] jam b] nectar
c] jelly d] squash

50. **Botanically guava is**
a] drupe b] pome
c] berry d] nut

51. **Guava is commercially propagated by**
a] cutting b] T-budding
c] patch-budding d] layering

52. **Guava decline is due to**
a] wilt b] anthracnose
c] bacterial canker d] virus

53. **To avoid rainy season crop and flower shedding at full bloom is done by spraying growth regulator**
a] NAA @ 600 ppm
b] 1BA @ 600 ppm
c] 2,4-D @ 600 ppm
d] 2,4,5- T @ 600 ppm

54. **Serious pest in the neglected guava orchard is**
a] stem borer b] fruit fly
c] aphid d] case worm ,

55. **Family of guava is**
a] rosaceae b] carciaceae
c] myrtaceae d] vitaceae

GK GS

Adolt Hitler is known as Fuhrer

56. Bronzing in guava is due to deficiency of nutrient
a] nitrogen
b] phosphorus
c] potassium
d] phosphorus, potassium, zinc complex

57. Bronzing is caused due to a stress
a] water b] heat
c] nutrient d] frost

58. Rainy season crop of guava is less economical because of
a] high rainfall b] high invasion of fruit fly
c] fruit size small d] no proper ripening

59. Best method of guava propagation is
a] cutting b] stooling
c] inarching d] forkert-budding

60. Grape belongs to the family
a] vitaceae b] turaceae
c] spaindaceae d] rosaceae

61. In North India climatic conditions are not favourable for grape cultivation because of
a] short growth duration
b] extremes of low temperature
c] unsuitable varieties
d] prolonged rainy weather

62. In southern and western India the grape cultivation is highly successful because of
a] prolonged moderate climate
b] prolonged rainy season
c] absence of low temperature in winter
d] hot and dry climate

63. The grapes are commercially propagated through
a] seeds b] hardwood cuttings
c] layering d] grafting

64. Only economical system of training grapes vines in India is
a] head b] telephone
c] arbour/bower d] kniffin

65. Only disease which devitalizes the growth of vines is
a] powdery mildew b] anthracnose
c] downy mildew d] foot rot

66. Serious pest of grape vine is
a] leaf folder b] thrips
c] beetle d] termites

67. Improvement of berry quality in shape has been enhanced by the use of growth regulators
a] Indole butyric acid (IBA)
b] gibberellic acid (GA_3)
c] naphthalene acetic acid (NM) and
d] 2,4-D

68. Grapes are the richest source of sugars like
a] sucrose b] fructose
c] glucose d] glucose and fructose

69. Floriculture deals with growing and marketing of
a] flowers b] fruits
c] vegetable d] field crops

70. Landscaping deals with the of tract of land
a] beautification, with ornamental plants
b] field crops
c] fruit plants
d] vegetables

71. Marigold flowers produces
a] odour b] essential oil
c] fragrant d] none

72. Chrysanthemum is propagated through
a] terminal cuttings b] hardwood cuttings
c] semihardwood cuttings d] layering

73. Chrysanthemum is a________ plant
a] biennial b] perennial
c] annual d] none

74. Planting time of chrysanthemum is ________
a] July-August b] November-December
c] May-June d] February-March

75. Roses are wintered in month of
a] April-May
b] December-January
c] End of September to 1st fortnight of October
d] June-July

76. Roses are pruned in the month of
a] June-July
b] April-May
c] May-June
d] End of September to 1st fortnight of October

77. Rose flowers are harvested for long distance market________
a] Ist petal unfurled b] 4-5 petals unfurled
c] 5-10 petals unfurled d] 8-10 petals unfurled

78. Marigold is propagated through
a] layering b] grafting
c] division d] seed

79. The seed rate of marigold for one acre is
a] 100 b] 500
c] 1000 d] 2000 g

80. Feeding of leguminous leaves causes frothy bloat in cattle because of its
a] carbohydrate content
b] chlorophyll
c] soluble protein content
d] cellulose content in excess

81. What is the cause of Haemorrhagic septicemia (H.S.) in cattle?
a] *Mycobacterium*
b] *Bacillus anthraicis*
c] *Pasteurella multocida*
d] *Actinobacillus liginietresi*

GK GS

Napolean Bonaparte is known as Man of Destiny

82. The most hygienic and efficient method of milking is
a] wet full hand
b] dry full hand
c] wet fisting with thumb in
d] dry fisting with thumb in

83. Latex is extracted from
a] rubber b] bajra
c] teak d] cotton

84. The phenomenon in which certain pollutants gets accumulated in the tissues are in increasing concentrations along a food chain is known as
a] Eutrophication b] Bioconcentration
c] Biomagnification d] Bioliquification

85. Pick the odd one out-
a] radish b] carrot
c] beetroot d] bhendi

86. The crop, which is used for preparing sweet corn, is
a] jowar b] ragi
c] red gram d] maize

87. Pigeonpea is a
a] millet b] pulse
c] vegetable d] cereal

88. The major structural element of cell and cell organelle in plants is
a] potassium b] phosphorus
c] sulphur d] nitrogen

89. The element which is an important constituent of plant cell walls is
a] sulphur
b] calcium (calcium pectate)
c] phosphorus
d] zinc

90. The food crop, which requires cool winter to grow is
a] Wheat b] Rice
c] Jowar d] Sunflower

91. Dead heart in sugarcane is caused by
a] *Pyrilla* b] Termite
c] *Chilo infuscatefllus* d] White grub

92. Cabbage is a
a] fruit b] inflorescence
c] flower d] bud

93. The oilseed crop where the kernels are produced beneath soil is
a] safflower b] castor
c] groundnut d] mustard

94. The ingredients like caffeine is present in
a] cashew b] clove
c] pepper d] coffee

95. Coffee powder is extracted from which plant part of coffee?
a] leaves b] stem
c] buds d] berry

96. Outbreak of Ranikhet disease in poultry in India is caused by
a] mesogenic strain of the virus
b] lentogenic strain of virus
c] velogenic strain of the virus
d] both mesogenic and lentogenic strains ofvirus

97. The mutations of micro-organisms most commonly occur during
a] synthesis of RNA b] utilization of ATP
c] DNA replication d] oxidation of ADP

98. Best soil for planting an orchard is
a] sandy b] silt
c] loam d] clay

99. The fruits of coffee are commonly known as
a] kernals b] berries
c] nuts d] capsules

100. The fruit of cotton is called
a] capsules b] boll
c] berries d] kernal

20. Evaluate Yourself

Self Evaluation must be SMART

S - Systematic
M - Meaningful
A - Accurate
R - Realistic
T - Time Bound

1 First Indian director of IARI was. (J.R.F. 1994)

____

2 Father of modern plant pathology. (Ph.D. 1997)

____

3 Bordeaux mixture was developed by. (J.R.F. 1995)

____

4 Wart disease of potato is caused by. (J.R.F. 1994)

____

5 Nitrogen fixation in rice field is carried out by which blue green algae? (Ph.D. 1997)

____

6 Total area of India is. (J.R.F. 1994)

____

7 Forest area in India. (Ph.D. 1998) (J.R.F. 2003)

____

8 Which country tops in the sugar production? (J.R.F. 1994)

____

9 Which state is highest producer of rice? (Ph.D. 1997)

____

10 In India which state is leading sugar producer? (J R.F. 1995)

____

11 The most critical stage during the growth of wheat is. (J.R.F. 1994) (J.R.F. 2003)

____

12 Urea contains % of nitrogen. (J.R.F. 1995) (J.R.F. 2003)

____

13 Pusa Ruby is a variety of crop. (J.R.F. 2001)

____

14 Net sown area of India is. (Ph.D. 1998)

____

15 Gross cropped area of India is. (J.R.F. 1995)

____

16 KVK was recommended by which committee? (Ph.D. 1997)

____

17 Inflorescence of sugarcane is known as. (J.R.F. 1994)

____

18 Red colour of tomato is due to. (J.R.F. 1994)

____

19 Richest source of Vitamin C is. (Ph.D. 1997)

____

20 Which Indian scientist shared world food prize for miracle maize? (Ph.D. 1999)

____

21 Power tiller is most suitable for the cultivation of. (J.R.F. 1994)

____

22 Harvesting of paddy is done at moisture content. (Ph.D. 1997)

____

23 National Seed Corporation was established in. (J.R.F. 1995)

____

24 Seed Act was passed in. (J.R.F. 1994)

____

25 Loose smut of wheat is. (J.R.F. 1995)

____

26 Tz test is done for. (Ph.D. 1997)

____

GK **Damodar river is called Bengal's sorrow.** GS

27 Establishment of NABARD. (J.R.F. 2001)

28 Cereals are deficient in which amino acid? (J.R.F. 1994)

29 Harvest index is. (J.R.F. 1998)

30 Kresek in rice is caused by. (Ph.D. 1997)

31 Grassy shoot disease of sugarcane is caused by. (J.R.F. 1994)

32 Bunchy top of banana is caused by. (J.R.F. 1995)

33 Which oil seed crop leads in the production? (J.R.F.2091)

34 Which state has highest production of soyabean in India? (J.R.F. 1998)

35 Which state in India is the leading producer of coffee? (J.R.F. 1994)

36 The name of the macaroni wheat is. (Ph.D. 1997)

37 Hand refrectometer reading for sugarcane maturity is. (J.R.F. 1995)

38 Which portion of cane is best suited to be used as seed/set? (J.R.F. 2001) (J.R.F. 2003)

39 In mango edible part is. (Ph.D. 1997)

40 Amino acid which is deficient in legumes? (J.R.F. 1994)

41 The plants growing in saltwater are known as. (J.R.F. 1995)

42 Granite is rock. (J.R.F. 2001)

43 Sugarcane is......... plant. (J.R.F. 1994), (Ph.D. 1997)

44 Kresek symptom is present in. (J.R.F. 1995)

45 Pahala blight of sugarcane is caused due to deficiency of. (J.R.F. 1995)

46 Reclamation disease due to deficiency of. (Ph.D. 1997)

47 Contribution of live stock to agriculture GDP. (J.R.F. 1994)

48 What is the amount of fertilizers consumed in India annually? (J.R.F. 2001)

49 The term 'Green Revolution' was coined by. (J.R.F. 1998)

50 R.B.I. was established in which year? (Ph.D. 1997)

51 Central Soil Salinity Research Institute is situated in. (J.R.F.1994), J.R.F. 2003)

52 Late blight of potato is caused by. (Ph.D. 1999)

53 Hormone related to drought tolerance is. (Ph.D. 1997)

54 Apical bud dominance is caused by which hormone? (J.R.F. 1994)

55 A hormone used as a herbicide. (J.R.F. 2001)

56 Proteins are made up of. (Ph.D. 1997)

57 How many nutrients are termed as essential elements for plants? (Ph.D. 2002)

58 Translocation of water and nutrients from roots to above ground parts of plants takes place through. (J.R.F. 1994)

59 Causal agent of mad cow disease is. (Ph.D. 1999)

60 Dormancy breaking hormone is. (J.R.F. 2001)

61 Total registered pesticide in India is. (J.R.F 1994)

62 ICAR day is celebrated on. (Ph.D. 1999)

63 ICRISAT is situated at. (J.R.F. 1996)

64 Double helical structure of DNA was given by. (J.R.F. 1994)

65 Which wind cause rainfall in Tamil Nadu? (J.R.F. 2001)

66 Ring worm disease is caused by. (J.R.F. 1994)

67 Total number of KVK in India. (Ph.D. 2002)

68 National Agriculture Insurance Scheme was introduced in the year. (Ph.D. 2002)

69 Who discovered vitamin? (J.R.F. 1994)

70 Theory of evolution was given by. (Ph.D. 1999)

71 Laws of heredity was given by. (B.H.U. Pre PG. 2002)

72 Instrument used to measure atmospheric pressure. (J.R.F. 1994) (J.R.F. 2003)

73 Mycorrhiza is a symbiotic association between. (J.R.F. 1996)

74 Growing of plants under soil less condition is called. (J.R.F. 1994)

75 Stress hardening in plants can be active by. (J.R.F. 2001)

76 Flowering hormone used in pineapple. (J.R.F. 1994)

77 Photorespiration is process. (Ph.D. 1999)

78 The major form of transport carbohydrate in higher plant is. (Ph.D. 2002)

79 Element contributing to the disease & drought resistance. (J.R.F. 1998)

80 Khaira disease is caused by the deficiency of. (J.R.F. 2001) (J.R.F. 2003)

81 Little leaf of citrus is caused by. (Ph.D. 2002)

82 Complex disease in association with nematode. (B.H.U. Pre PG. 2002)

83 Solar heat treatment is recommended against. (J.R.F. 1996)

84 The element involved in energy transfer and storage in plants is. (J.R.F. 2001)

85 Muriate of Potash (MOP) is chemically. (B.H.U. Pre PG. 2002)

86 Maximum allowable biurate content of urea is. (J.R.F. 2003)

87 For the maximization of net revenue the condition is. (Ph.D. 1999)

88 Over the years the contribution of agriculture to GDP of Indian economy has been. (Ph.D. 2002)

89 On the indifference curve, the utility is. (Ph.D. 1999)

90 The 'U' shape of cost curves could be best explained by the law of. (Ph.D. 2002)

GK **Walt Disney died in 1966.** GS

91 For the consumer to be in equilibrium the necessary condition is that the ratio of. (Ph.D. 1999)

92 Indifference curve approach could be attributed to. (Ph.D. 2002)

93 Pineapple is a......................plant. (J.R.F. 1995)

94 In the tropical climate.............plants are more produce. (Ph.D. 1998)

95 Mg^{+2} is component of. (J.R.F. 2003) (Ph.D. 1999)

96 Heart rot of sugarbeet is caused by. (J.R.F. 1995)

97 2,4- D is used for...............type weeds control. (B.H.U. Pre PG. 2002)

98 The berry size of Thomson Seedless grapes increased by (Ph.D. 1998) the hormone.

99 The optimum spacing for wheat is (line to line). (J.R.F. 1995)

100 The non-traditional area for cultivating wheat is. (B.H.U. Pre PG. 2002)

101 The all India average for wheat yield is. (Ph.D. 2002)

102 The haploid number of chromosome in rice is. (Ph.D. 1998)

103 Rice is considered as...................a plant. (J.R.F. 1996)

104 The optimum depth of puddling in rice is. (Ph.D. 2002)

105 In wet nursery for rice, the level of water is maintained at. (B.H.U. Pre PG. 2002)

106 In field paddy grains is harvested when per cent moisture is. (J.R.F. 1996)

107 Total water requirement of the sugarcane crop is. (J.R.F. 1995)

108 Sugarcane is irrigated at every......days during its growing period. (J.R.F. 1996)

109 In India, the predominant species of cultivated cotton is. (B.H.U. Pre PG. 2002)

110 Regurs refers to. (Ph.D. 1998)

111 An element not essential for plant growth is. (B.H.U. Pre PG 1994)

112 N content of F.Y.M. and urban compost respectively is. (B.H.U. Pre PG 2002)

113 Major P fertilizer in India is. (J.R.F. 1995)

114 Saltation is a type of. (J.R.F. 1996)

115 Very few poor (below poverty line) people is in. (J.R.F. 1991)

116 Biologically active form of glucose is. (Ph.D. 1998)

117 The bond present in sugars. (Ph.D. 1999)

118 The most abundant protein present in the world is. (J.R.F. 1995)

119 The form of amino acids present in living organisms is. (B.H.U. Pre P G. 2002)

120 The pairs are strongly bound with each other. (Ph.D. 1999)

GK Tenzing Norgay is known as Tiger of Snows GS

121 Biofertilizer more suited for sugarcane is. (Ph.D. 1998)

122 Lines joining equal rainfall called. (J.R.F. 1995)

123 Origin of maize. (B.H.U. Pre PG. 2002)

124 Depth of sowing of soyabean seed. (J.R.F. 2003)

125 Cropping intensity of India. (Ph. D. 2002)

126 First rice variety introduced in India. (J.R.F. 2001)

127 Nicotine content in tobacco is related with. (Ph.D.1999)

128 Boro rice is transplanted in. (Ph.D. 1998)

129 Glyphosate is a.............herbicide. (J.R.F. 1995)

130 Indian mustard is. (J.R.F. 2001)

131 Mass flow is affected by. (Ph.D. 2002)

132 Carbon content in organic matter. (Ph.D. 1998)

133 Desi cotton known as. (J.R.F. 1995)

134 Metric suctions is measured by. (B.H.U. Pre PG. 2002)

135 Nitrogen biofertilizer for wheat. (J.R.F. 1996)

136 Brix measures. (J.R.F. 1995)

137 Verticle mulch is used in soils. (J.R.F. 2001)

138 Design used when fertility gradient is in two direction. (Ph.D. 1998)

139 Criteria of essentiality was given by. (B.H.U. Pre PG. 2002)

140 Nucleus was discovered by. (Ph.D. 1999)

141 Segregation occurs during. (J.R.F. 1996)

142 The longest mitotic phase is. (J.R.F. 1995)

143 Tetrad is seen in. (Ph.D. 1998)

144 Chaisma is seen during. (J.R.F. 2003)

145 Multilines in wheat are produced by. (Ph.D. 2002)

146 The selection procedure which provides the maximum improvement ever the base population is? (Ph.D. 2002)

147 Most dangerous disease of potato. (J.R.F. 1995)

148 Phyllody disease in plants is caused by. (Ph.D. 1998)

149 Father of plant nematology. (J.R.F.1996)

150 Rice production is highest in country. (B.H.U. Pre PG. 2002)

151 Single super phosphate contains sulphur. (J.R.F. 1995)

152 Most prominent soil group of India. (J.R.F. 2003)

153 Murate of potash is chemically known as. (J.R.F. 1996)

154 Highest contribution to vegetable oil. (Ph.D. 2000)

155 Explosive fertilizer is. (B.H.U. Pre PG. 2001)

156 Most mutation tolerant. (Ph.D. 1998)

GK

Automatic rifle was invented by John Browning in 1918.

GS

157 The person acting as a connecting link between higher department official & farming community in T & V system, known as. (Ph.D. 1999)

158 The pest which attacks all the parts of the plant is? (J.R.F. 1995)

159 White grubs prefer to lay eggs on. (Ph.D. 2002)

160 Pesticide consumption in India is. (B.H.U. Pre P.G. 2002)

161 Micro nutrient deficient in India. (J.R.F. 1991)

162 Mantek disease of rice is caused by. (J.R.F. 1996)

163 International pest is. (Ph.D. 1998)

164 The first Fisheries University was. (Ph.D.1995)

165 Indian Institute of Tropical Meteorology is. (J.R.F. 1996)

166 The second Indian recipient of world food prize was. (Ph.D.1996)

167 Spike tooth harrow is a. (Ph.D. 1999)

68 India ranks .... rank in fruit production. (J.R.F. 1996)

169 Club root of cauliflower is caused by. (Ph.D. 1995)

170 Sheath blight of rice is caused by. (Ph.D. 1995)

171 Father of Microbiology. (B.H.U. Pre PG. 2001)

172 Bacterium was discovered by. (B.H.U. Pre PG. 2001)

173 The variety of wheat which is resistant to all the three rusts is? (Ph.D. 1995)

174 In sugarcane, taking of ratoon crop is advisable only. (B.H.U. Pre PG. 2002)

175 Wilt disease is.................disease. (Ph.D. 2002)

176 Vector of leaf curl disease in cotton. (Ph.D. 1995)

177 Flared square symptoms seen in cotton due to the. (Ph.D. 1998)

178 The number of electrons required for conversion of NO_3 to NH (Ph.D. 2002)

179 Conversion of fat to sugar occurs in. (Ph.D. 1995)

180 The net requirement for assimilation of CO_2 in C_4 plants is. (Ph.D. 1997)

181 The primary acceptor of electron in PS is. (J.R.F. 2001)

182 Gneiss is a................rock. (Ph.D. 1995)

183 Under the USDA system, silt is classified as having a particle size of. (J.R.F. 1997)

184 Hygroscopic water is held at a tension of atmosphere. (J.R.F. 1997)

185 Kaolinite is a layer silicate type. (Ph.D. 1998)

186 Clays are.................minerals. (J.R.F. 1998)

187 C : N ratio of organic matter is. (Ph.D. 1995)

188 CAN is.................fertilizer. (J.R.F. 1998)

GK

Adam Smith is known as the Father of Economics

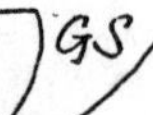

189 Most widely used for correcting soil acidity. (J.R.F. 2003)

190 Largest importer of cut flowers in the world. (J.R.F. 1997)

191 Density of water is maximum at.............. ^0C. (Ph.D. 1995)

192 Diara cultivation method is followed in. (J.R.F. 1997)

193 The irrigation method which is suitable for saline soils? (J.R.F. 1997)

194 The role of extension education in India is performed by. (J.R.F. 1997)

195 Extension is a. (Ph.D. 1995)

196 Oldest method of selection is. (Ph.D. 1998)

197 The limit of the regression coefficient is. (J.R.F. 1998)

198 Highly salt tolerant crop. (J.R.F. 2001)

199 Formation of male flowers induced by. (J.R.F. 2003)

200 Cytokinin is mostly synthesized in. (Ph.D. 2002)

201 Recent method of control of boll worms is. (J.R.F. 2003)

202 Orobanche is a parasite associated with. (Ph.D. 1998)

203 Dr. B.P. Pal is associated with. (J.R.F. 1998)

204 Art of giving shape to shrubs resembling to figure is called as. (B.H.U. Pre PG. 2002)

205 The dwarf variety of mango is. (J.R.F. 1997)

206 Mango malformation can be checked by. (Ph.D. 1995)

207 Photorespiration occurs in. (J.R.F. 2003)

208 Photosynthesis is an. (J.R.F. 1997)

209 Root promoting hormone is. (J.R.F. 1998)

210 Fruit of rose is known as. (Ph.D. 2002)

211 What concentration of sugar is used for preservation? (J.R.F. 2001)

212 Fruit crop which requires the highest number of irrigations? (Ph.D. 1995)

213 Photosystem II is absent in. (Ph.D. 1998)

214 Conversion of fat into carbohydrate is. (J.R.F. 1998)

215 Peroxisome is a. (J.R.F. 1997)

216 The main site for the dark reaction of photosynthesis is. (J.R.F. 2001)

217 The F_2 ratio for complementary interaction is. (Ph.D. 1998)

218 The shortest phase of all the mitosis phases is. (J.R.F. 1997)

219 Longest phase of all the mitosis phases is. (J.R.F. 2003)

220 The site of protein synthesis is. (J.R.F. 1997)

221 Operation flood is related to. (J.R.F. 1997)

GK GS

Animal Farm is written by George Orwell.

222 Bench terracing is done when the slope is more than. (J.R.F. 1998)

223 Mass per unit volume is called as. (J.R.F.1996)

224 Disk plough are used when the soil is. (B.H.U. Pre PG. 2002)

225 Post-harvest losses for cereals account for.....of total production. (Ph.D. 1999)

226 National Seed Act passed. (J.R.F. 1998)

227 ICAR was initiated as per recommendation of. (J.R.F. 1997)

228 Mango variety which is suitable for high density planting is? (Ph.D. 1998)

229 Mallika is a cross of. (J.R.F. 2000)

230 T & V was first started in. (J.R.F. 2001)

231 Indian Journal of Agriculture Sciences is published by. (J.R.F. 1997)

232 The design to be followed for one directional fertility gradient is. (J.R.F. 1998)

233 The distribution where the sp. is equal to the root of mean is. (J.R.F. 2000)

234 Maximum contribution of the crop to the production of cereals in the country. (J.R.F. 2001)

235 Highest sugarcane yield per hectare on the state. (J.R.F. 2000)

236 Distribution of mean, median, mode is a. (Ph.D. 1995)

237 Nursery area required for seedling of rice for one hectare field is. (J.R.F. 1998)

238 Marble is a. (Ph.D. 1995)

239 Time duration for adsali crop of sugarcane. (J.R.F. 2000)

240 Crossing over occurs in. (J.R.F. 2003)

241 First man made cereal is. (J.R.F.2000) (J.R.F.1998)

242 Triticale is a crossing of. (Ph.D. 1994)

243 Interveinal chlorosis occurs due to. (J.R.F.1996)

244 Saline tolerant fruit crop is. (Ph.D. 1995)

245 Optimum pH for rice. (J.R.F. 1998)

246 Dead heart & white ear associated with the crop. (J.R.F. 2000)

247 Formula of urea. (J.R.F.1996)

248 M.O.P. contains K_2O. (Ph.D. 1994)

249 Fruit ripening hormone is. (J.R.F. 2001)

250 Water use efficiency higher in plants. (J.R.F. 2003)

251 Ooze test is done for detecting. (J.R.F. 1998)

252 Maximum arable land in country. (J.R.F. 1996)

253 Malathion is a. (Ph.D. 1994)

254 Family of sugarcane. (J.R.F. 2000)

GK **Bethlehem is associated with Christ** GS

255 Most widely grown rabi pulse crop. (J.R.F. 2001)

256 Unit of rural society. (Ph.D. 1994)

257 Fruit type of guava. (Ph.D. 1994)

258 The term genetics was coined by. (J.R.F. 2000)

259 Sodic soils reclaimed by. (J.R.F. 1998)

260 I.I.S.R. is situated at. (Ph.D. 1994)

261 IBPGR is situated at. (J.R.F. 2000)

262 Proportion of sand, silt, clay known as. (Ph.D. 1995)

263 Cheapest N contains fertilizer is. (Ph.D. 1994)

264 Total land capability classes are. (Ph.D. 1994)

265 Bacteria change nitrite to nitrate are. (B.H.U. Pre PG. 2002)

266 Akiochi disease is due to. (J.R.F. 2000)

267 First product of urea hydrolysis. (J.R.F. 1998)

268 Azolla is a. (Ph.D. 1994)

269 Functional nutrients given by. (J.R.F. 2000)

270 State leading in wheat production is. (Ph.D. 2002)

271 NABARD was setup on the recommendation of. (B.H.U. Pre PG. 2001)

272 White rust of crucifers is caused by. (J.R.F. 2000)

273 The hard fruits of citrus is due to the. (Ph.D. 1999)

274 Which is the antidote of insect poisoning? (B.H.U. Pre PG. 2001)

275 Tetrasomic is. (J.R.F. 2003)

276 Precursor of IAA is. (Ph.D. 1999)

277 Fruit of okra is. (J.R.F. 2000)

278 Vitamin, containing cobalt as a constituent is. (J.R.F. 2000)

279 Free living nitrogen fixing organism is. (Ph.D. 1994)

280 Dwarfing gene in rice (Ph.D. 1994)

281 Centre of origin of wheat is. (J.R.F. 1996)

282 Total deemed universities under ICAR. (Ph.D. 1994)

283 Total National Bureau are. (J.R.F. 2002)

284 Lab to land programme started in the year. (J.R.F. 2000)

285 Disease which was discovered in Haryana? (J.R.F. 1993)

286 Chromosomal theory of inheritance was proposed by. (J.R.F. 2003)

287 Ufra disease in rice is caused by. (Ph.D. 1994)

288 The fruit of mustard is known as. (J.R.F. 2000) (J.R.F. 1998)

289 N content in ammonium sulphate is. (J.R.F. 2003)

GK GS

Fastest land mammal is Chetah (100 km per hour)

290 Herbicides are not used in the dust formulation because of. (Ph.D. 1994)

291 The fungicide which is used for smut? (Ph.D. 1994)

292 First stable product formed in C_3 plants is. (Ph.D. 1995)

293 The greenhouse gas that's released from paddy field is. (J.R.F. 1998)

294 B.P. Pal is a variety of. (J.R.F. 2000)

295 Over the years the per cent contribution of agriculture to G.D.P. is. (J.R.F.2003)

296 I.A.D.P. programme started in the year. (Ph.D. 2002)

297 Person associated with Gurgoan pilot project is. (J.R.F. 1993)

298 Net irrigated area of India. (Ph.D. 1994)

299 What is the price called which is fixed by government? (J.R.F.2003)

300 Powerhouse of cell is. (B.H.U. Pre PG. 2002)

301 The per cent land resource of India in world is. (J.R.F. 1993)

302 Headquarter of W.T.O. is located at. (J.R.F. 2000)

303 The highest award presented to an agricultural scientist in the country is. (J.R.F. 1993)

304 The growth rate projected in the 10th five year plan is. (J.R.F. 2003)

305 I.A.R.I. is a. (J.R.F. 2001)

306 Highest C.E.C. found in. (Ph.D. 1995)

307 E.S.P. of normal soils is. (J.R.F. 1998)

308 Sugar turn out from sugarcane in India is. (J.R.F. 1993)

309 Permanent wilting point is observed at. (J.R.F.2000)

310 Mango is mostly propagated through. (Ph.D. 1994)

311 Microorganism associated with the symbiotic N_2 fixation in non legumes is. (Ph.D. 1994)

312 Emmer wheat is. (Ph.D.1995)

313 Net capital ratio is. (Ph.D. 1999)

314 Catkin is a inflorescence of. (J.R.F. 1998)

315 Black heart of potato is due to. (Ph.D. 2002)

316 The Arka series of varieties are released from. (J.R.F. 2000)

317 Major agricultural import in India is. (J.R.F. 1993)

318 Which programme also known as package programme? (B.H.U. Pre PG 2002)

319 Whip tail of cauliflower is due to the deficiency. (J.R.F. 1993)

320 Gross cropped area in India is. (J.R.F. 2000)

321 Yellow rust in wheat is caused by. (J.R.F. 1993)

322 Seedless mango variety is. (J.R.F. 1993)

GK

Largest insect is Goliath beetle (110 mm long, 100 gm)

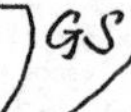

GS

323 Pink revolution refers to. (Ph.D. 2001)

324 Correlation coefficient ranges between. (J.R.F. 2000) (J.R.F. 1998)

325 Pusa Snow Ball is a variety of. (Ph.D. 1995)

326 CRIDA is located at. (J.R.F. 1998)

327 India's share in the fruit production in the world is. (J.R.F. 2000)

328 Fertilizers which are completely imported? (J.R.F. 1993)

329 The contribution of agriculture to GDP of India is. (Ph.D. 2003)

330 Present D.G. of ICAR. (B.H.U. Pre PG. 2002)

331 Moisture content for safe storage of cereals is. (Ph.D. 2001)

332 Lunishree is a variety of. (Ph.D. 2000)

333 Highest cotton production in India. (J.R.F. 1993)

334 Supporting price of wheat in 2002. (J.R.F. 2003)

335 Minimum support price is formulated by. (Ph.D. 2001)

336 Central Agricultural University is located at. (J.R.F. 1993)

337 Constituent of wheat affecting its backing quality is. (Ph.D. 2001)

338 I.G.F.R.I. is situated at. (J.R.F. 1993)

339 CRI stage occurs in wheat in. (J.R.F. 1998)

340 Bulk density is high in. (Ph.D. 1999)

341 Suicidal bags of cell are. (Ph.D. 1995)

342 The fertilizer also called nitro chalk. (B.H.U. Pre PG. 2002)

343 Regression coefficient varies between. (Ph.D. 2001) (J.R.F. 1998)

344 Soil transported through wind is. (J.R.F. 1993)

345 Formation of mRNA from DNA is called. (J.R.F. 1993)

346 The ion generally dominant in soil solution is. (Ph.D. 2001)

347 Fixation in soil is a problem in case of. (Ph.D. 2002)

348 Luxury consumption is generally associated with. (Ph.D. 2001)

349 The journal 'Indian Farming' is published by. (Ph.D. 1999)

350 Pride fruit of India is. (J.R.F. 1998)

351 Solar constant is. (Ph.D. 1999)

352 C.V. of rainfall is more in. (Ph.D. 1995)

353 Striga is type parasite. (B.H.U. Pre PG. 2001)

354 Lady bird beetle is a. (Ph.D. 2001)

355 Monophagus pest is. (B.H.U. Pre PG. 2001)

GK Crossword puzzle ws invented by Arthur Wynne in 1913. GS

356 Maximum production of potato in. (B.H.U. Pre P.G. 2001)

357 CIPHET is situated at. (Ph.D. 1995)

358 First project launched in India. (Ph.D. 2001)

359 Maximum oil seed producer crop in India. (J.R.F. 1998)

360 Maximum nutrient uptake by plants. (Ph.D. 2001)

361 Nutrient available in combined form. (B.H.U. Pre P.G. 2002)

362 Sulphur is available in which form? (J.R.F. 1993)

363 Sugarcane sets required per ha. (Ph.D. 2001)

364 Most efficient irrigation system. (B.H.U. Pre P. G. 2001)

365 How many Deemed Universities under ICAR? (J.R.F. 1993)

366 Insects which attack rice at night. (Ph.D. 2001)

367 Formula of gypsum. (J.R.F.1993)

368 Maximum productivity among cereals in world. (J.R.F.1998)

369 Queen of cereals known as. (J.R.F. 1993)

370 Gold of America known as. (Ph.D. 1991)

371 Back bone of America is the crop. (B.H.U. Pre P. G. 2002)

372 Hulling % in rice is. (Ph.D. 1999)

373 Weight of cotton bale is. (J.R.F. 1998)

374 Which is not a measure of central tendency? (J.R.F. 1999)

375 Pungency in onion is due to presence of. (J.R.F. 1992)

376 CEC of humus is. (Ph.D. 1999)

377 Movement from higher concentration to lower concentration knwon as. (J.R.F. 1999)

378 'V' shape pattern of yellowing shows. (Ph.D. 2001)

379 Fiber of cotton contains. (J.R.F. 1999)

380 Available water lies between. (J.R.F. 1992)

381 Growing subsidiary crop between widely rowed space of main crop is called. (Ph.D. 1992)

382 Least degree of freedom in ANOVA is. (J.R.F. 1999)

383 Sugarcane is a. (J.R.F. 1992)

384 Isotopes have same number of. (J.R.F. 1999)

385 N_2 deficiency occurs in plants on. (J.R.F. 1992)

386 Highest K_2O containing fertilizer. (B.H.U. Pre PG. 2002)

387 Boron is harmful for plants when more than. (Ph.D. 2001)

388 Which is not a measure of dispersion? (J.R.F. 1992)

389 Irish famine (1845) was caused by. (J.R.F. 1992)

390 Production of ATP in the presence of light is known as. (J.R.F. 1999)

391 For seed purpose carrot is grown as. (Ph.D. 2002)

GK GS
Invisible Man is written by H.G. Wells.

392 Soil moisture is measured by. (J.R.F. 1999)

393 Bacteria which convert NH_4^+ to NO_2^- (J.R.F. 1992)

394 Physical condition of soil is known as. (J.R.F. 1992)

395 Infrared thermometer used for. (Ph.D. 2001)

396 Most serious disease in sugarcane. (J.R.F. 1999)

397 Ribosome found in mitochondria. (Ph.D. 1999)

398 Black soils found in. (J.R.F. 1992)

399 World staple food is. (Ph.D. 1999)

400 3m deep, more than 18m wide gully known as. (J.R.F. 1992)

401 In which fertilizer N & P present in highest amount. (B.H.U. Pre PG. 2001)

402 Leading state in acreage of rice. (J.R.F. 1999)

404 Which crop having highest percentage of irrigation? (J.R.F. 1991)

404 Sugarcane sowing in trench method to. (Ph.D. 2001)

405 Dough stage means. (J.R.F. 2003)

406 C : N ratio of most arable soils. (J.R.F. 1999)

407 Noble cane is. (Ph.D. 1999)

408 Rediscovery of Mendel laws by. (J.R.F. 1992)

409 Pattern used in dry land. (J.R.F. 1992)

410 Parboiling in rice conserves. (Ph.D. 2001)

411 Maize is a. (J.R.F. 2001)

412 In 1943 Bengal famine was due to. (J.R.F. 1992)

413 Maximum total porosity found in. (J.R.F. 1999)

414 Most frost effected crop. (J.R.F. 1992)

415 H.D. 2329 is a variety of. (Ph.D. 2001)

416 Drip irrigation introduced in India from. (B.H.U. Pre PG. 2002)

417 N% in CAN is. (J.R.F. 1999)

418 Highest amount of rainfall received in India by. (J.R.F. 1992)

419 The river basins have more utilizable flow is. (B.H.U. Pre PG. 2001)

420 First agricultural chemist of Imperial Agriculture Research Station was. (Ph.D. 2001)

421 Pseudo cereal is. (J.R.F. 1999)

422 King of fruits. (J.R.F. 1992)

423 Queen of spices. (J.R.F. 1999)

424 How water rises in plant? (J R.F. 1991)

425 The first fertilizer produced in India was. (Ph.D. 2001)

426 Oil content in soyabean. (J.R.F. 1999)

427 Hand book of Agriculture published by. (B.H.U. Pre PG. 2001)

428 Growth of plants towards lights known as. (J.R.F. 2003)

GK GS

Louis Pasteur, French chemist is known as founder of Microbiology

429 AGMARK is. (J.R.F. 1991)

430 Diameter of clay particles. (Ph.D. 2001)

431 Available form of N in soil. (Ph.D. 1999)

432 E.S.P. of alkali soils. (J.R.F. 1999)

433 The normal ratio of rice/paddy is. (B.H.U. Pre PG 2001)

434 Crop for attracting insects. (J.R.F. 1999)

435 Wind velocity can be measured by. (J.R.F. 1999)

436 Deflocculation of soil particles occurs by. (J.R.F. 1992)

437 Planting before harvest of main crop known as. (J.R.F. 1992)

438 Which is not a greenhouse gas? (J.R.F. 1999)

439 Urea is maximum marketed as. (J.R.F. 2003)

440 Maximum metabolic activity of plant is at. (Ph.D. 2001)

441 ICMR recommendation for pulse consumption/day/capita. (Ph.D. 2003)

442 1 gm. glucose has. (Ph.D. 2001)

443 Harvest Index in wheat is. (Ph.D. 1999)

444 Leaf curl of tomato is due to. (J.R.F. 1991)

445 The crop which prefer ammonical form of N. (J.R.F. 1999)

446 PUFA content is highest in. (J.R.F. 1992)

447 Interospecific hybrid are more common in. (B.H.U. Pre PG. 2001)

448 Shedding of plant parts is due to. (Ph.D. 2001)

449 Soil is clay when clay separates are minimum of. (J.R.F. 1992)

450 Maximum N loss in rice in form of. (J.R.F. 1999)

451 Head of Planning Commission is. (Ph.D. 1994)

452 Climate for arhar is. (J.R.F. 1997)

453 Which fertilizer is supplied to the fields poor in phosphorous ? (Ph.D. 1996)

454 High yielding variety of wheat produced by in world. (Ph.D. 1994)

455 Maximum P available at pH. (J.R.F. 2003)

456 Chemical used for delinting of cotton. (Ph.D. 1994)

457 Highest nutrient content in which fertilizer? (Ph.D. 1999)

458 Sowing time of cotton in north India. (Ph.D. 1994)

459 Soil erosion in India is. (B.H.U. Pre P.G. 2001)

460 No. of eggs/sec. laid by termite. (Ph.D. 1994)

461 Which is a stable element? (Ph.D. 1999)

462 Nature & Properties of Soil is written by. (Ph.D. 1994)

463 Which gas is dominant in waterlogged soils? (Ph.D. 1994)

Butterfield 8 is written by John O' Hara.

464 Rothemsted Agricultural Research Station was founded by. (Ph.D. 1999)

465 Protein content in wheat. (Ph.D. 1994)

466 Curing is related to the crop. (J.R.F. 1999)

467 Which microorganism is responsible for maximum nutrient cycling in the soil? (J.R.F. 2003)

468 The extension programme linked form planned with credit for farmers. (Ph.D. 1999)

469 Ratooning is practiced in crop. (Ph.D. 1994)

470 Molybdenum is required by plants because it is a cofactor for. (Ph.D. 1994)

471 The basic unit of development under the Integrated Rural Development Programme is a. (Ph.D. 1994)

472 Botanical name of green gram is. (Ph.D. 1994)

473 Irrigation method best for undulating topography. (Ph.D. 1994)

474 Present farming system of India has become. (J.R.F. 2003)

475 The chemical widely used to treat seed of potato to break its dormancy. (J.R.F. 2003)

476 Wheat is a origin of. (Ph.D. 1994)

477 Azotobacter fixes atmospheric nitrogen. (Ph.D. 1994)

478 Rock phosphate used in. (Ph.D. 1999)

479 Cycocel is a. (J.R.F. 1999)

480 Wheat is a. (J.R.F. 2003)

481 S.A.R. of alkali soils is. (Ph.D. 1994)

482 Most of the plant obtain nitrogen from the soil in the form of. (Ph.D. 1994)

483 Total agroclimatic zones are in India. (J.R.F. 1999)

484 Foundation seed is produced by. (Ph.D. 1994)

485 Pulse crop doesn't fix N from atmosphere. (Ph.D. 1994)

486 If only three irrigations are available indicate critical stages of wheat when these will be applied. (B.H.U. Pre P.G. 2001)

487 Potato tuber is a modified form of. (B.H.U. Pre P.G. 2001)

488 The most efficient use of phosphorous is achieve by. (J.R.F. 1998)

489 'Goodness of Fit' is used for distribution of. (Ph.D. 1999)

490 Growing plants of different heights in the same field at the same time is known as. (B.H.U. Pre PG. 2001)

491 Water use efficiency is the highest in case of. (J.R.F. 2003)

492 Type of soil water which is most useful for plants. (J.R.F. 1991)

493 Scorching of burning on margins of bottom leaves manifests the deficiency of. (Ph.D. 1999)

494 Plants take carbon from. (J.R.F. 1991)

495 Concept of plant ideotype was first propounded by. (J.R.F. 1991)

496 The nutrient plays important role in controlling the rate of transpiration. (J.R.F. 1991)

GK GS

Robert Browning is a English poet of the Victorian Era.

497 Highest contribution in G.D.P. (B.H.U. Pre PG. 2001)

498 IVLP is. (Ph.D. 1998)

499 Very few poor (below poverty line) people is in. (J.R.F. 1991)

500 Micro nutrient deficient in Indian soils. (J.R.F. 1991)

501 In waterlogged area which gas is found abundantly? (J.R.F. 1999)

502 What means photo-respiration? (J.R.F. 1999)

503 Scientific name of berseem is. (J.R.F. 1999)

504 The insect boll worm commonly found on. (J.R.F. 1994)

505 Most destructive disease of sugarcane is. (Ph.D. 1998)

506 Global warming is caused due to. (J.R.F. 2001)

507 Primary consumers are. (J.R.F. 1993)

508 The edible part of sweet potato (Ipomoea batata) is. (Ph.D. 2001)

509 Potatoes are borne on. (J.R.F. 1998)

510 An example of albuminous seed showing hypogeal germination is. (Ph.D. 2000)

511 The part that develops into seed is. (J.R.F. 1998)

512 Seeds with tigers claw like structures are found in. (B.H.U. Pre PG. 2001)

513 Study on seasonal flowering in shrubs and trees is known as. (Ph.D. 2001)

514 Fruit Research Institute in India is located in. (J.R.F. 1993)

515 Die back of shoots occur due to deficiency of. (Ph.D. 1994)

516 Grey spots on leaves occur due to deficiency of. (B.H.U. Pre PG. 1994)

517 Yellowing of tea leaves occur due to the deficiency of. (J.R.F. 1996)

518 Indian soil is mainly deficient of. (J.R.F. 1994)

519 Lowest rainfall in India. (B.H.U. Pre PG. 1999)

520 Highest water erosion in the state. (J.R.F. 1994)

521 The state which have oldest Agricultural University. (Ph.D. 1997)

522 State which have highest area of saline soil? (Ph.D. 2001)

523 Highest area of acidic soils in the state is. (J.R.F. 2002)

524 Total State Agricultural Universities (SAUs) are. (B.H.U. Pre PG. 1998)

GK GS

Safety Razor was invented by King C. Gillette in 1895.

3. General Agriculture

Multiple Choice Questions

1. (a)
2. (c)
3. (b)
4. (a)
5. (b)
6. (d)
7. (b)
8. (c)
9. (b)
10. (a)
11. (a)
12. (c)
13. (b)
14. (c)
15. (a)
16. (c)
17. (d)
18. (c)
19. (c)
20. (b)
21. (a)
22. (b)
23. (b)
24. (d)
25. (b)
26. (b)
27. (a)
28. (b)
29. (a)
30. (b)
31. (d)
32. (c)
33. (b)
34. (d)
35. (b)
36. (a)
37. (c)
38. (b)
39. (a)
40. (d)
41. (c)
42. (a)
43. (a)
44. (b)
45. (b)
46. (d)
47. (b)
48. (b)
49. (c)
50. (c)
51. (c)
52. (b)
53. (c)
54. (d)
55. (d)
56. (c)
57. (a)
58. (d)
59. (a)
60. (c)
61. (c)
62. (b)
63. (c)
64. (a)
65. (c)
66. (d)
67. (a)
68. (c)
69. (d)
70. (b)
71. (c)
72. (a)
73. (d)
74. (b)
75. (c)
76. (d)
77. (c)
78. (a)
79. (a)
80. (c)
81. (d)
82. (a)
83. (d)
84. (c)
85. (a)
86. (b)
87. (c)
88. (b)
89. (c)
90. (d)
91. (c)
92. (a)
93. (c)
94. (a)
95. (b)
96. (a)
97. (c)
98. (d)
99. (a)
100. (a)
101. (b)
102. (c)
103. (c)
104. (c)
105. (d)
106. (b)
107. (a)
108. (a)
109. (c)
110. (b)
111. (c)
112. (b)
113. (d)
114. (a)
115. (c)
116. (b)
117. (c)
118. (c)
119. (b)
120. (b)
121. (c)
122. (a)
123. (c)
124. (a)
125. (a)
126. (b)
127. (a)
128. (a)
129. (c)
130. (a)
131. (c)
132. (d)
133. (b)
134. (a)
135. (d)
136. (c)
137. (c)
138. (b)
139. (d)
140. (c)
141. (c)
142. (c)
143. (d)
144. (d)
145. (c)
146. (c)
147. (d)
148. (d)
149. (d)
150. (d)
151. (d)
152. (c)
153. (c)
154. (a)
155. (d)
156. (c)
157. (d)
158. (b)
159. (a)
160. (d)
161. (a)
162. (a)
163. (c)
164. (c)
165. (b)
166. (b)
167. (c)
168. (c)
169. (a)
170. (a)
171. (d)
172. (c)
173. (a)
174. (c)
175. (a)
176. (d)
177. (c)
178. (b)
179. (c)
180. (a)
181. (c)
182. (b)
183. (d)
184. (c)
185. (c)
186. (b)
187. (b)
188. (d)
189. (b)
190. (c)
191. (b)
192. (d)
193. (a)
194. (a)
195. (c)
196. (d)
197. (b)
198. (a)
199. (d)
200. (b)
201. (c)
202. (b)
203. (a)
204. (b)
205. (d)
206. (a)
207. (c)
208. (b)
209. (b)
210. (b)
211. (a)
212. (b)
213. (b)
214. (b)
215. (b)
216. (b)
217. (b)
218. (a)
219. (d)
220. (a)
221. (d)
222. (a)
223. (d)
224. (c)
225. (b)
226. (b)
227. (a)
228. (d)
229. (a)
230. (a)
231. (b)
232. (d)
233. (a)
234. (c)
235. (c)
236. (d)
237. (b)
238. (b)
239. (b)
240. (b)
241. (c)
242. (a)
243. (d)
244. (a)
245. (d)
246. (c)
247. (c)
248. (b)
249. (a)
250. (d)
251. (a)
252. (c)
253. (c)
254. (c)
255. (c)
256. (d)
257. (b)
258. (c)
259. (a)
260. (a)
261. (d)
262. (d)
263. (b)
264. (d)
265. (b)
266. (c)
267. (a)
268. (b)
269. (b)
270. (c)
271. (a)
272. (d)
273. (d)
274. (c)
275. (c)
276. (d)
277. (b)
278. (b)
279. (c)
280. (b)
281. (b)
282. (b)
283. (d)
284. (a)
285. (c)
286. (b)
287. (a)
288. (b)
289. (d)
290. (d)
291. (b)
292. (a)
293. (a)
294. (d)
295. (c)
296. (d)
297. (a)
298. (b)
299. (a)
300. (c)
301. (c)
302. (b)
303. (d)
304. (d)
305. (d)
306. (b)
307. (b)
308. (b)
309. (a)
310. (a)
311. (b)
312. (d)
313. (a)
314. (b)
315. (b)
316. (b)
317. (a)
318. (c)
319. (b)
320. (c)
321. (b)
322. (d)
323. (a)
324. (a)
325. (d)
326. (c)
327. (c)
328. (b)
329. (c)
330. (b)
331. (b)
332. (a)
333. (b)
334. (d)
335. (b)
336. (c)
337. (d)
338. (c)
339. (b)
340. (c)
341. (c)
342. (c)
343. (c)
344. (c)
345. (a)
346. (c)
347. (d)
348. (d)
349. (d)
350. (d)
351. (b)
352. (d)

GK **Mahatma Gandhi is written by Romain Rolland.** GS

353. (d)
354. (a)
355. (a)
356. (b)
357. (b)
358. (c)
359. (a)
360. (c)
361. (c)
362. (c)
363. (d)
364. (c)
365. (d)
366. (b)
367. (c)
368. (b)
369. (a)
370. (b)
371. (a)
372. (a)
373. (c)
374. (d)
375. (b)
376. (a)
377. (c)
378. (b)
379. (a)
380. (c)
381. (d)
382. (a)
383. (d)
384. (d)
385. (c)
386. (a)
387. (a)
388. (b)
389. (c)
390. (b)
391. (a)
392. (a)
393. (b)
394. (a)
395. (b)
396. (b)
397. (d)
398. (a)
399. (b)
400. (a)
401. (d)
402. (d)
403. (c)
404. (a)
405. (d)
406. (a)
407. (b)
408. (b)
409. (a)
410. (b)
411. (b)
412. (b)
413. (c)
414. (d)
415. (a)
416. (c)
417. (c)
418. (b)
419. (a)
420. (a)
421. (a)
422. (a)
423. (b)
424. (d)
425. (c)
426. (d)
427. (b)
428. (c)
429. (a)
430. (d)
431. (c)
432. (d)
433. (d)
434. (d)
435. (a)
436. (c)
437. (d)
438. (d)
439. (b)
440. (b)
441. (b)
442. (b)
443. (a)
444. (b)
445. (d)
446. (d)
447. (c)
448. (b)
449. (d)
450. (b)
451. (d)
452. (a)
453. (b)
454. (b)
455. (a)
456. (c)
457. (c)
458. (c)
459. (c)
460. (b)
461. (c)
462. (d)
463. (a)
464. (c)
465. (c)
466. (c)
467. (c)
468. (b)
469. (c)
470. (d)
471. (c)
472. (b)
473. (a)
474. (c)
475. (a)
476. (c)
477. (b)
478. (a)
479. (c)
480. (b)
481. (a)
482. (b)
483. (c)
484. (c)
485. (d)
486. (c)
487. (a)
488. (a)
489. (d)
490. (c)
491. (a)
492. (a)
493. (a)
494. (d)
495. (c)
496. (d)
497. (a)
498. (a)
499. (d)
500. (a)
501. (a)
502. (b)
503. (c)
504. (d)
505. (d)
506. (b)
507. (b)
508. (b)
509. (b)
510. (d)
511. (b)
512. (b)
513. (a)
514. (d)
515. (d)
516. (d)
517. (a)
518. (b)
519. (c)
520. (a)
521. (b)
522. (a)
523. (a)
524. (b)
525. (b)
526. (c)
527. (b)
528. (a)
529. (b)
530. (a)
531. (a)
532. (b)
533. (c)
534. (c)
535. (a)
536. (a)
537. (b)
538. (c)
539. (d)
540. (a)
541. (a)
542. (a)
543. (c)
544. (c)
545. (a)
546. (b)
547. (c)
548. (a)
549. (c)
550. (a)
551. (d)
552. (c)
553. (d)
554. (c)
555. (d)
556. (c)
557. (b)
558. (c)
559. (c)
560. (a)
561. (c)
562. (c)
563. (b)
564. (d)
565. (c)
566. (b)
567. (d)
568. (b)
569. (b)
570. (d)
571. (a)
572. (c)
573. (b)
574. (b)
575. (d)
576. (d)
577. (c)
578. (c)
579. (d)
580. (a)
581. (a)
582. (b)
583. (b)
584. (a)
585. (d)
586. (a)
587. (b)
588. (b)
589. (a)
590. (d)
591. (a)
592. (d)
593. (a)
594. (d)
595. (d)
596. (a)
597. (c)
598. (d)
599. (a)
600. (b)
601. (c)
602. (a)
603. (b)
604. (c)
605. (c)
606. (c)
607. (b)
608. (d)
609. (c)
610. (c)
611. (c)
612. (b)
613. (b)
614. (b)
615. (c)
616. (b)
617. (b)
618. (c)
619. (b)
620. (c)
621. (d)
622. (c)
623. (b)
624. (d)
625. (a)
626. (b)
627. (d)
628. (b)
629. (c)
630. (a)
631. (a)
632. (a)
633. (a)
634. (b)
635. (a)
636. (b)
637. (b)
638. (c)
639. (b)
640. (a)
641. (d)
642. (a)
643. (b)
644. (b)
645. (a)
646. (b)
647. (a)
648. (d)
649. (d)
650. (c)
651. (a)
652. (c)
653. (b)
654. (b)
655. (a)
656. (d)
657. (d)
658. (a)
659. (c)
660. (a)
661. (b)
662. (c)
663. (a)
664. (a)
665. (d)
666. (b)
667. (b)
668. (b)
669. (b)
670. (c)
671. (b)
672. (d)
673. (c)
674. (b)
675. (b)
676. (a)
677. (b)
678. (b)
679. (d)
680. (d)
681. (a)
682. (b)
683. (d)
684. (d)
685. (b)
686. (d)
687. (a)
688. (a)
689. (c)
690. (a)
691. (c)
692. (a)
693. (b)
694. (a)
695. (b)
696. (b)
697. (a)
698. (c)
699. (b)
700. (c)
701. (b)
702. (b)
703. (a)
704. (d)
705. (a)
706. (d)
707. (c)
708. (b)
709. (d)
710. (b)
711. (c)
712. (b)
713. (d)
714. (c)
715. (a)
716. (c)
717. (a)
718. (b)
719. (d)
720. (b)
721. (c)
722. (b)
723. (a)
724. (a)
725. (d)
726. (d)
727. (c)
728. (c)
729. (a)
730. (c)

731. (b)
732. (d)
733. (d)
734. (b)
735. (a)
736. (b)
737. (d)
738. (d)
739. (a)
740. (c)
741. (a)
742. (d)
743. (c)
744. (c)
745. (b)
746. (a)
747. (b)
748. (b)
749. (b)
750. (b)
751. (c)
752. (a)
753. (b)
754. (d)
755. (c)
756. (b)
757. (b)
758. (c)
759. (a)
760. (b)
761. (c)
762. (b)
763. (c)
764. (d)
765. (b)
766. (d)
767. (c)
768. (b)
769. (a)
770. (b)
771. (c)
772. (b)
773. (c)
774. (c)
775. (b)
776. (a)
777. (a)
778. (a)
779. (a)
780. (b)
781. (a)
782. (d)
783. (a)
784. (c)
785. (c)
786. (a)
787. (b)
788. (a)
789. (d)
790. (a)
791. (c)
792. (a)
793. (b)
794. (d)
795. (b)
796. (c)
797. (c)
798. (a)
799. (c)
800. (b)
801. (a)
802. (c)
803. (c)
804. (b)
805. (a)
806. (b)
807. (a)
808. (d)
809. (b)
810. (a)
811. (c)
812. (b)
813. (b)
814. (a)
815. (a)
816. (a)
817. (c)
818. (a)
819. (b)
820. (a)
821. (b)
822. (a)
823. (a)
824. (a)
825. (a)
826. (c)
827. (a)
828. (d)
829. (b)
830. (b)
831. (a)
832. (a)
833. (d)
834. (c)
835. (c)
836. (b)
837. (b)
838. (a)
839. (a)
840. (c)
841. (b)
842. (a)
843. (d)
844. (a)
845. (b)
846. (a)
847. (a)
848. (a)
849. (d)
850. (a)
851. (a)
852. (a)
853. (b)
854. (c)
855. (b)
856. (d)
857. (a)
858. (c)
859. (c)
860. (b)
861. (d)
862. (a)
863. (b)
864. (a)
865. (c)
866. (a)
867. (c)
868. (b)
869. (a)
870. (a)
871. (b)
872. (b)
873. (a)
874. (a)
875. (b)
876. (b)
877. (b)
878. (c)
879. (b)
880. (c)
881. (c)
882. (c)
883. (b)
884. (a)
885. (b)
886. (c)
887. (c)
888. (a)
889. (a)
890. (d)
891. (b)
892. (b)
893. (a)
894. (b)
895. (b)
896. (d)
897. (c)
898. (b)
899. (b)
900. (b)
901. (c)
902. (c)
903. (b)
904. (c)
905. (d)
906. (d)
907. (d)
908. (d)
909. (d)
910. (d)
911. (d)
912. (b)
913. (a)
914. (b)
915. (b)
916. (c)
917. (a)
918. (d)
919. (b)
920. (c)
921. (a)
922. (b)
923. (a)
924. (d)
925. (b)
926. (a)
927. (c)
928. (c)
929. (d)
930. (a)
931. (a)
932. (b)
933. (a)
934. (a)
935. (a)
936. (a)
937. (d)
938. (b)
939. (b)
940. (b)
941. (b)
942. (a)
943. (d)
944. (d)
945. (a)
946. (d)
947. (d)
948. (c)
949. (d)
950. (d)
951. (a)
952. (c)
953. (a)
954. (c)
955. (c)
956. (c)
957. (b)
958. (a)
959. (d)
960. (b)
961. (d)
962. (d)
963. (c)
964. (a)
965. (c)
966. (c)
967. (b)
968. (a)
969. (d)
970. (c)
971. (a)
972. (d)
973. (b)
974. (c)
975. (c)
976. (c)
977. (d)
978. (d)
979. (d)
980. (d)
981. (b)
982. (d)
983. (a)
984. (c)
985. (b)
986. (d)
987. (d)
988. (a)

4. Agronomy

Multiple Choice Questions

1. (c)
2. (d)
3. (a)
4. (d)
5. (c)
6. (a)
7. (c)
8. (d)
9. (b)
10. (a)
11. (b)
12. (a)
13. (c)
14. (c)
15. (b)
16. (b)
17. (d)
18. (b)
19. (b)
20. (a)
21. (a)
22. (a)
23. (d)
24. (a)
25. (c)
26. (a)
27. (d)
28. (b)
29. (a)
30. (b)
31. (a)
32. (b)
33. (b)
34. (d)
35. (a)
36. (b)
37. (d)
38. (d)
39. (b)
40. (b)
41. (b)
42. (d)
43. (a)
44. (c)
45. (d)
46. (a)
47. (c)
48. (c)
49. (c)
50. (a)
51. (d)
52. (c)
53. (a)
54. (c)
55. (c)
56. (c)
57. (b)
58. (a)
59. (c)
60. (a)
61. (a)
62. (b)
63. (a)
64. (a)
65. (d)
66. (c)
67. (d)
68. (a)
69. (d)
70. (b)
71. (c)
72. (c)
73. (c)
74. (a)
75. (c)
76. (a)
77. (b)
78. (d)
79. (a)
80. (a)
81. (c)
82. (a)
83. (c)
84. (a)
85. (a)
86. (a)
87. (b)
88. (a)
89. (b)
90. (c)
91. (a)
92. (b)
93. (c)
94. (b)
95. (b)
96. (a)
97. (b)
98. (a)
99. (b)
100. (c)
101. (d)
102. (a)
103. (a)
104. (c)
105. (b)
106. (c)
107. (a)
108. (b)
109. (c)
110. (b)
111. (b)
112. (b)
113. (c)
114. (d)
115. (a)
116. (c)
117. (c)
118. (b)
119. (b)
120. (a)
121. (a)
122. (a)
123. (a)
124. (a)
125. (a)
126. (a)
127. (a)
128. (b)
129. (a)
130. (d)
131. (d)
132. (b)
133. (c)
134. (d)
135. (d)
136. (b)
137. (d)
138. (b)
139. (b)
140. (c)
141. (a)
142. (a)
143. (c)
144. (a)
145. (a)
146. (b)
147. (b)
148. (b)
149. (d)
150. (a)
151. (b)
152. (c)
153. (a)
154. (c)
155. (b)
156. (a)
157. (a)
158. (c)
159. (c)
160. (b)
161. (b)
162. (a)
163. (b)
164. (a)
165. (b)
166. (c)
167. (a)
168. (c)
169. (b)
170. (a)
171. (c)
172. (d)
173. (c)
174. (b)
175. (d)
176. (a)
177. (b)
178. (c)
179. (b)
180. (c)
181. (b)
182. (c)
183. (d)
184. (c)
185. (a)
186. (c)
187. (b)
188. (d)
189. (a)
190. (a)
191. (c)
192. (c)
193. (b)
194. (d)
195. (b)
196. (b)
197. (d)
198. (a)
199. (a)
200. (d)
201. (c)
202. (b)
203. (b)
204. (a)
205. (d)
206. (b)
207. (a)
208. (d)
209. (a)
210. (b)
211. (d)
212. (b)
213. (a)
214. (a)
215. (c)
216. (c)
217. (a)
218. (c)
219. (d)
220. (d)
221. (b)
222. (d)
223. (c)
224. (b)
225. (b)
226. (c)
227. (a)
228. (a)
229. (c)
230. (c)
231. (b)
232. (d)
233. (a)
234. (b)
235. (c)
236. (c)
237. (a)
238. (c)
239. (b)
240. (a)
241. (b)
242. (b)
243. (b)
244. (c)
245. (c)
246. (d)
247. (a)
248. (a)
249. (b)
250. (c)
251. (d)
252. (d)
253. (c)
254. (c)
255. (c)
256. (a)
257. (c)
258. (c)
259. (a)
260. (a)
261. (d)
262. (d)
263. (b)
264. (d)
265. (c)
266. (b)
267. (c)
268. (c)
269. (a)
270. (c)
271. (d)
272. (a)
273. (a)
274. (a)
275. (a)
276. (a)
277. (c)
278. (a)
279. (c)
280. (a)
281. (b)
282. (a)
283. (c)
284. (c)
285. (b)
286. (c)
287. (c)
288. (c)
289. (d)
290. (a)
291. (c)
292. (d)
293. (d)
294. (a)
295. (a)
296. (a)
297. (a)
298. (a)
299. (a)
300. (b)
301. (d)
302. (b)
303. (d)
304. (c)
305. (b)
306. (a)
307. (d)
308. (d)
309. (a)
310. (c)
311. (c)
312. (d)
313. (b)
314. (b)
315. (a)
316. (a)
317. (c)
318. (a)
319. (d)
320. (b)
321. (a)
322. (a)
323. (b)
324. (d)
325. (b)
326. (b)
327. (c)
328. (c)
329. (a)
330. (c)
331. (b)
332. (b)
333. (c)
334. (c)
335. (a)
336. (d)
337. (a)
338. (b)
339. (c)
340. (a)
341. (d)
342. (a)
343. (b)
344. (a)
345. (c)

GK **UCI stands for Universal Childhood Immunisation** GS

346. (b)
347. (a)
348. (b)
349. (b)
350. (d)
351. (b)
352. (d)
353. (d)
354. (b)
355. (d)
356. (d)
357. (a)
358. (c)
359. (b)
360. (d)
361. (d)
362. (d)
363. (b)
364. (b)
365. (b)
366. (a)
367. (b)
368. (b)
369. (a)
370. (a)
371. (c)
372. (b)
373. (b)
374. (a)
375. (a)
376. (a)
377. (a)
378. (b)
379. (c)
380. (c)
381. (a)
382. (a)
383. (c)
384. (a)
385. (a)
386. (c)
387. (b)
388. (c)
389. (d)
390. (c)
391. (b)
392. (a)
393. (c)
394. (a)
395. (d)
396. (a)
397. (c)
398. (b)
399. (a)
400. (a)
401. (a)
402. (a)
403. (b)
404. (c)
405. (c)
406. (a)
407. (d)
408. (c)
409. (d)
410. (a)
411. (b)
412. (a)
413. (c)
414. (b)
415. (d)
416. (b)
417. (d)
418. (d)
419. (c)
420. (c)
421. (c)
422. (c)
423. (c)
424. (d)
425. (b)
426. (a)
427. (b)
428. (c)
429. (d)
430. (a)
431. (c)
432. (a)
433. (b)
434. (b)
435. (b)
436. (c)
437. (c)
438. (b)
439. (c)
440. (d)
441. (c)
442. (c)
443. (a)
444. (c)
445. (b)
446. (a)
447. (b)
448. (a)
449. (a)
450. (a)
451. (b)
452. (a)
453. (b)
454. (c)
455. (c)
456. (d)
457. (a)
458. (c)
459. (b)
460. (b)
461. (c)
462. (a)
463. (b)
464. (c)
465. (b)
466. (c)
467. (c)
468. (a)
469. (c)
471. (a)
472. (a)
473. (b)
474. (d)
475. (a)
476. (d)
477. (c)
478. (b)
479. (c)
480. (b)
481. (c)
482. (d)
483. (c)
484. (d)
485. (d)
486. (d)
487. (b)
488. (b)
489. (b)
490. (a)
491. (b)
492. (d)
493. (c)
494. (a)
495. (a)
496. (a)
497. (b)
498. (c)
499. (c)
500. (b)
501. (c)
502. (c)
503. (a)
504. (a)
505. (a)
506. (c)
507. (c)
508. (a)
509. (c)
510. (d)
511. (d)
512. (a)
513. (c)
514. (b)
515. (b)
516. (b)
517. (d)
518. (b)
519. (b)
520. (b)
521. (c)
522. (a)
523. (d)
524. (b)
525. (a)
526. (d)
527. (b)
528. (b)
529. (b)
530. (c)
531. (d)
532. (a)
533. (a)
534. (a)
535. (a)
536. (c)
537. (c)
538. (d)
539. (d)
540. (b)
541. (c)
542. (c)
543. (c)
544. (a)
545. (c)
546. (c)
547. (b)
548. (a)
549. (c)
550. (b)
551. (b)
552. (a)
553. (d)
554. (d)
555. (a)
556. (a)
557. (b)
558. (b)
559. (b)
560. (b)
561. (a)
562. (a)
563. (b)
564. (a)
565. (b)
566. (a)
567. (c)
568. (b)
569. (b)
570. (a)
571. (b)
572. (b)
573. (c)
574. (b)
575. (c)
576. (c)
577. (c)
578. (b)
579. (c)
580. (a)
581. (b)
582. (a)
583. (a)
584. (a)
585. (d)
586. (a)
587. (c)
588. (d)
589. (d)
590. (b)
591. (a)
592. (c)
593. (c)
594. (c)
595. (b)
596. (c)
597. (a)
598. (d)
599. (d)
600. (a)
601. (d)
602. (a)
603. (b)
604. (a)
605. (b)
606. (a)
607. (c)
608. (a)
609. (a)
610. (a)
611. (c)
612. (b)
613. (d)
614. (c)
615. (c)
616. (a)
617. (d)
618. (c)
619. (b)
620. (a)
621. (b)
622. (c)
623. (b)
624. (c)
625. (d)
626. (a)
627. (a)
628. (b)
629. (b)
630. (c)
631. (d)
632. (c)
633. (d)
634. (a)
635. (b)
636. (a)
637. (b)
638. (c)
639. (a)
640. (b)
641. (b)
642. (b)
643. (b)
644. (c)
645. (a)
646. (a)
647. (b)
648. (b)
649. (c)
650. (c)
651. (b)
652. (d)
653. (c)
654. (b)
655. (d)
656. (b)
657. (c)
658. (d)
659. (b)
660. (d)
661. (c)
662. (a)
663. (d)
664. (d)
665. (b)
666. (a)
667. (a)
668. (a)
669. (a)
670. (a)
671. (d)
672. (a)
673. (a)
674. (b)
675. (b)
676. (a)
677. (d)
678. (c)
679. (d)
680. (a)
681. (b)
682. (c)
683. (c)
684. (b)
685. (d)
686. (a)
687. (b)
688. (d)
689. (a)
690. (d)
691. (d)
692. (b)
693. (d)
694. (c)
695. (c)
696. (d)
697. (c)
698. (a)
699. (a)
700. (c)
701. (a)
702. (d)
703. (a)
704. (d)
705 (d)
706. (a)
707. (b)
708. (d)
709. (d)

GK GS

Hovercraft was invented by CHristopher Cockerell in 1955.

Agronomy : Matching

S.No.	a	b	c	d
1.	iii	i	ii	iv
2.	ii	iii	iv	i
3.	iii	iv	i	ii
4.	iv	i	ii	iii
5.	iii	ii	iv	i
6.	iii	iv	i	ii
7.	ii	iv	i	iii
8.	iii	i	iv	ii
9.	iv	iii	i	ii
10.	ii	iii	iv	i
11.	iv	ii	i	iii
12.	iii	i	iv	ii
13.	iv	i	ii	iii
14.	ii	iii	i	iv
15.	iii	iv	i	ii
16.	iii	ii	iv	i
17.	ii	i	iv	iii
18.	iv	ii	i	iii
19.	ii	iii	iv	i
20.	ii	i	iv	iii
21.	iv	i	ii	iii
22.	ii	iii	iv	i
23.	iv	i	ii	iii
24.	iii	i	iv	ii
25.	iv	i	ii	iii
26.	iii	i	iv	ii
27.	ii	iii	iv	i
28.	iv	i	ii	iii
29.	iv	iii	i	ii
30.	iv	i	iii	ii
31.	iii	iv	ii	i
32.	ii	iii	iv	i
33.	iii	i	iv	ii
34.	iv	i	ii	iii
35.	ii	iv	i	iii
36.	iv	i	ii	iii
37.	iv	iii	i	ii
38.	iv	ii	i	iii
39.	iii	i	iv	ii
40.	ii	iii	iv	i
41.	iv	i	iii	ii
42.	ii	iii	iv	i
43.	iii	i	ii	iv
44.	ii	iv	i	iii
45.	i	iii	iv	ii
46.	ii	i	iv	iii
47.	iii	i	iv	ii
48.	iv	iii	i	ii
49.	iii	ii	iv	i
50.	iii	i	iv	ii
51.	ii	iv	i	iii

S.No.	a	b	c	d
52.	ii	iv	i	iii
53.	iv	i	ii	iii
54.	ii	iv	i	iii
55.	iv	iii	i	ii
56.	iv	i	ii	iii
57.	ii	iv	i	iii
58.	iii	i	iv	ii
59.	ii	iv	iii	i
60.	iv	i	ii	iii
61.	ii	iii	iv	i
62.	iii	i	iv	ii
63.	ii	iv	i	iii
64.	iii	ii	i	iv
65.	iii	i	iv	ii
66.	iv	ii	i	iii
67.	iii	i	iv	ii
68.	iii	i	ii	iv
69.	iv	i	ii	iii
70.	ii	iv	i	iii
71.	ii	iii	i	iv
72.	iii	i	iv	ii
73.	ii	iv	iii	i
74.	ii	iii	iv	i
75.	iii	iv	ii	i
76.	ii	i	iv	iii
77.	iv	i	ii	iii
78.	iv	i	ii	iii
79.	iv	i	iii	ii
80.	iii	iv	i	ii
81.	iv	i	iii	ii
82.	ii	iii	iv	i
83.	iii	i	iv	ii
84.	ii	i	iv	iii
85.	iv	i	ii	iii
86.	iii	iv	ii	i
87.	iii	i	iv	ii
88.	ii	iii	iv	i
89.	iv	iii	i	ii
90.	iv	iii	i	ii
91.	iii	ii	i	iv
92.	ii	iii	iv	i
93.	iii	i	iv	ii
94.	i	iv	iii	ii
95.	iii	i	iv	ii
96.	iii	iv	i	ii
97.	iii	i	iv	ii
98.	iv	i	ii	iii
99.	ii	iii	iv	ii
100.	iv	i	ii	iii
101.	ii	i	iv	iii
102.	iv	iii	ii	i
103.	iii	i	iv	ii

5. Horticulture

Multiple Choice Questions

1. (a)
2. (c)
3. (a)
4. (d)
5. (a)
6. (d)
7. (d)
8. (b)
9. (a)
10. (b)
11. (d)
12. (b)
13. (d)
14. (a)
15. (d)
16. (b)
17. (a)
18. (b)
19. (b)
20. (b)
21. (c)
22. (c)
23. (a)
24. (a)
25. (d)
26. (c)
27. (b)
28. (d)
29. (c)
30. (d)
31. (a)
32. (a)
33. (a)
34. (c)
35. (c)
36. (c)
37. (d)
38. (c)
39. (a)
40. (b)
41. (c)
42. (d)
43. (b)
44. (a)
45. (b)
46. (d)
47. (c)
48. (d)
49. (d)
50. (a)
51. (a)
52. (a)
53. (b)
54. (a)
55. (d)
56. (a)
57. (d)
58. (c)
59. (d)
60. (c)
61. (c)
62. (c)
63. (a)
64. (d)
65. (a)
66. (a)
67. (a)
68. (b)
69. (c)
70. (a)
71. (a)
72. (d)
73. (b)
74. (c)
75. (d)
76. (c)
77. (d)
78. (c)
79. (a)
80. (a)
81. (a)
82. (d)
83. (c)
84. (c)
85. (d)
86. (a)
87. (a)
88. (b)
89. (c)
90. (d)
91. (d)
92. (b)
93. (a)
94. (c)
95. (a)
96. (a)
97. (b)
98. (b)
99. (a)
100. (c)
101. (a)
102. (a)
103. (b)
104. (b)
105. (a)
106. (a)
107. (c)
108. (c)
109. (c)
110. (a)
111. (b)
112. (a)
113. (a)
114. (a)
115. (a)
116. (c)
117. (c)
118. (b)
119. (a)
120. (c)
121. (a)
122. (b)
123. (d)
124. (a)
125. (a)
126. (b)
127. (b)
128. (a)
129. (c)
130. (d)
131. (d)
132. (c)
133. (b)
134. (b)
135. (d)
136. (c)
137. (c)
138. (b)
139. (a)
140. (c)
141. (a)
142. (b)
143. (d)
144. (b)
145. (c)
146. (b)
147. (a)
148. (d)
149. (d)
150. (b)
151. (b)
152. (a)
153. (d)
154. (a)
155. (c)
156. (a)
157. (d)
158. (d)
159. (c)
160. (a)
161. (d)
162. (c)
163. (c)
164. (b)
165. (b)
166. (a)
167. (c)
168. (b)
169. (d)
170. (d)
171. (d)
172. (b)
173. (a)
174. (a)
175. (a)
176. (a)
177. (b)
178. (d)
179. (b)
180. (c)
181. (d)
182. (b)
183. (b)
184. (c)
185. (c)
186. (a)
187. (b)
188. (d)
189. (b)
190. (a)
191. (a)
192. (a)
193. (a)
194. (c)
195. (c)
196. (a)
197. (d)
198. (b)
199. (a)
200. (b)
201. (d)
202. (d)
203. (d)
204. (a)
205. (c)
206. (c)
207. (d)
208. (c)
209. (a)
210. (a)
211. (d)
212. (c)
213. (b)
214. (a)
215. (a)
216. (b)
217. (a)
218. (a)
219. (b)
220. (c)
221. (d)
222. (d)
223. (c)
224. (a)
225. (d)
226. (b)
227. (a)
228. (d)
229. (c)
230. (a)
231. (d)
232. (c)
233. (d)
234. (a)
235. (a)
236. (c)
237. (b)
238. (a)
239. (c)
240. (a)
241. (a)
242. (c)
243. (a)
244. (b)
245. (b)
246. (d)
247. (a)
248. (a)
249. (a)
250. (b)
251. (c)
252. (b)
253. (a)
254. (c)
255. (d)
256. (d)
257. (b)
258. (a)
259. (d)
260. (b)
261. (a)
262. (c)
263. (a)
264. (a)
265. (b)
266. (a)
267. (a)
268. (c)
269. (c)
270. (d)
271. (c)
272. (b)
273. (c)
274. (a)
275. (d)
276. (a)
277. (a)
278. (a)
279. (a)
280. (a)
281. (b)
282. (c)
283. (a)
284. (d)
285. (d)
286. (a)
287. (a)
288. (b)
289. (d)
290. (c)
291. (a)
292. (a)
293. (a)
294. (a)
295. (a)
296. (a)
297. (c)
298. (b)
299. (d)
300. (d)
301. (d)
302. (d)
303. (d)
304. (d)
305. (d)
306. (d)
307. (a)
308. (c)
309. (a)
310. (b)
311. (a)
312. (a)
313. (a)
314. (d)
315. (d)
316. (b)
317. (a)
318. (b)
319. (a)
320. (d)
321. (d)
322. (c)
323. (d)
324. (d)
325. (d)
326. (b)
327. (d)
328. (c)
329. (b)
330. (b)
331. (d)
332. (b)
333. (b)
334. (d)
335. (b)
336. (a)
337. (d)
338. (d)
339. (c)
340. (b)
341. (a)
342. (d)
343. (b)
344. (c)
345. (c)
346. (d)
347. (d)
348. (b)
349. (c)
350. (c)
351. (d)
352. (b)
353. (b)
354. (d)
355. (c)
356. (c)
357. (c)
358. (b)
359. (c)
360. (c)
361. (a)
362. (d)
363. (d)
364. (b)
365. (a)
366. (c)
367. (a)
368. (d)
369. (a)
370. (c)
371. (b)
372. (a)
373. (c)

GK GS

Linoleum was invented by Frederick Walton in 1860.

Horticulture : Matching

S.No.	a	b	c	d
1.	ii	iii	iv	i
2.	iii	iv	i	ii
3.	iv	i	ii	iii
4.	ii	iii	iv	i
5.	iii	i	iv	ii
6.	iii	i	iv	ii
7.	ii	iii	iv	i
8.	iii	i	iv	ii
9.	ii	iv	i	iii
10.	iv	i	iii	ii
11.	ii	iv	i	iii
12.	iii	i	iv	ii
13.	iv	i	ii	iii
14.	iii	i	iv	ii
15.	ii	iv	i	iii
16.	iv	i	ii	iii
17.	iii	iv	ii	i
18.	iv	i	ii	iii
19.	iii	i	iv	ii
20.	iv	i	iii	ii
21.	ii	iv	i	iii
22.	iv	iii	i	ii
23.	ii	iv	i	iii
24.	iii	i	iv	ii
25.	iv	iii	i	ii
26.	ii	iv	i	iii
27.	iii	i	iv	ii
28.	iv	iii	ii	i
29.	ii	iv	i	iii
30.	iii	i	iv	ii
31.	ii	iv	i	iii
32.	iii	i	iv	ii
33.	iii	i	iv	ii
34.	iv	i	ii	iii
35.	ii	iv	i	iii
36.	iii	iv	ii	i
37.	ii	iii	i	iv

S.No.	a	b	c	d
38.	iv	i	ii	iii
39.	ii	iv	iii	i
40.	iii	i	iv	ii
41.	iv	iii	i	ii
42.	iii	i	iv	ii
43.	iv	iii	i	ii
44.	iii	iv	i	ii
45.	ii	iv	i	iii
46.	ii	iv	i	iii
47.	iii	i	iv	ii
48.	iv	i	ii	iii
49.	ii	iv	i	iii
50.	iii	i	iv	ii
51.	iv	iii	i	ii
52.	ii	iv	i	iii
53.	iii	i	iv	ii
54.	iv	iii	i	ii
55.	ii	i	iv	iii
56.	i	iii	iv	ii
57.	iii	i	iv	ii
58.	ii	iv	i	iii
59.	iv	i	ii	iii
60.	ii	iv	i	iii
61.	iv	iii	ii	i
62.	ii	iv	i	iii
63.	iii	i	iv	ii
64.	ii	iv	i	iii
65.	iii	i	iv	ii
66.	iii	i	iv	ii
67.	ii	iv	i	iii
68.	iv	i	ii	iii
69.	iii	iv	i	ii
70.	ii	iii	iv	i
71.	iv	ii	i	iii
72.	i	ii	iii	iv
73.	i	iv	iii	ii
74.	iii	iv	ii	i

GK GS

M.S. Golwalker is known as Guruji

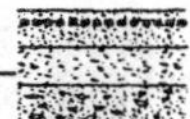

6. Soil Science

Multiple Choice Questions

1. (a)
2. (c)
3. (a)
4. (c)
5. (b)
6. (a)
7. (c)
8. (a)
9. (b)
10. (b)
11. (d)
12. (b)
13. (c)
14. (a)
15. (c)
16. (b)
17. (b)
18. (a)
19. (b)
20. (a)
21. (b)
22. (c)
23. (a)
24. (b)
25. (c)
26. (b)
27. (c)
28. (d)
29. (b)
30. (a)
31. (a)
32. (a)
33. (b)
34. (b)
35. (d)
36. (a)
37. (b)
38. (a)
39. (b)
40. (c)
41. (a)
42. (b)
43. (a)
44. (c)
45. (b)
46. (c)
47. (c)
48. (a)
49. (b)
50. (c)
51. (d)
52. (d)
53. (b)
54. (a)
55. (a)
56. (d)
57. (b)
58. (b)
59. (b)
60. (a)
61. (c)
62. (a)
63. (d)
64. (b)
65. (b)
66. (a)
67. (a)
68. (a)
69. (c)
70. (a)
71. (a)
72. (a)
73. (c)
74. (c)
75. (c)
76. (b)
77. (b)
78. (b)
79. (d)
80. (a)
81. (c)
82. (a)
83. (c)
84. (c)
85. (c)
86. (b)
87. (c)
88. (a)
89. (b)
90. (a)
91. (a)
92. (b)
93. (c)
94. (a)
95. (c)
96. (a)
97. (b)
98. (a)
99. (a)
100. (a)
101. (b)
102. (c)
103. (c)
104. (b)
105. (d)
106. (b)
107. (c)
108. (b)
109. (d)
110. (a)
111. (b)
112. (a)
113. (d)
114. (c)
115. (a)
116. (c)
117. (c)
118. (a)
119. (c)
120. (a)
121. (b)
122. (d)
123. (a)
124. (a)
125. (b)
126. (a)
127. (a)
128. (a)
129. (a)
130. (c)
131. (b)
132. (a)
133. (b)
134. (b)
135. (b)
136. (d)
137. (c)
138. (a)
139. (b)
140. (a)
141. (b)
142. (c)
143. (d)
144. (b)
145. (a)
146. (b)
147. (a)
148. (c)
149. (a)
150. (b)
151. (c)
152. (b)
153. (c)
154. (c)
155. (a)
156. (a)
157. (d)
158. (b)
159. (a)
160. (c)
161. (a)
162. (c)
163. (b)
164. (b)
165. (c)
166. (c)
167. (a)
168. (b)
169. (c)
170. (d)
171. (a)
172. (a)
173. (c)
174. (a)
175. (a)
176. (c)
177. (d)
178. (b)
179. (b)
180. (a)
181. (a)
182. (a)
183. (a)
184. (a)
185. (b)
186. (d)
187. (a)
188. (a)
189. (c)
190. (b)
191. (a)
192. (b)
193. (d)
194. (b)
195. (a)
196. (a)
197. (d)
198. (b)
199. (b)
200. (a)
201. (a)
202. (a)
203. (a)
204. (c)
205. (a)
206. (a)
207. (a)
208. (c)
209. (a)
210. (a)
211. (b)
212. (d)
213. (a)
214. (a)
215. (b)
216. (a)
217. (b)
218. (d)
219. (b)
220. (b)
221. (a)
222. (c)
223. (a)
224. (d)
225. (b)
226. (a)
227. (a)
228. (c)
229. (b)
230. (d)
231. (b)
232. (b)
233. (a)
234. (c)
235. (d)
236. (d)
237. (a)
238. (c)
239. (b)
240. (b)
241. (a)
242. (b)
243. (c)
244. (c)
245. (c)
246. (a)
247. (b)
248. (a)
249. (c)
250. (b)
251. (a)
252. (b)
253. (a)
254. (d)
255. (b)
256. (b)
257. (a)
258. (c)
259. (d)
260. (c)
261. (a)
262. (d)
263. (a)
264. (c)
265. (b)
266. (b)
267. (b)
268. (a)
269. (c)
270. (c)
271. (b)
272. (c)
273. (c)
274. (c)
275. (b)
276. (d)
277. (a)
278. (a)
279. (d)
280. (a)
281. (c)
282. (a)
283. (a)
284. (d)
285. (b)
286. (d)
287. (a)
288. (c)
289. (c)
290. (c)
291. (a)
292. (a)
293. (c)
294. (d)
295. (c)
296. (b)
297. (a)
298. (b)
299. (a)
300. (d)
301. (a)
302. (c)
303. (b)
304. (c)
305. (c)
306. (d)
307. (c)
308. (b)
309. (c)
310. (b)
311. (b)
312. (a)
313. (c)
314. (a)
315. (b)
316. (c)
317. (a)
318. (c)
319. (a)
320. (c)
321. (a)
322. (a)
323. (d)
324. (c)
325. (a)
326. (b)
327. (c)
328. (b)
329. (a)
330. (a)
331. (d)
332. (d)
333. (c)
334. (b)
335. (c)

336. (c)
337. (a)
338. (c)
339. (a)
340. (a)
341. (c)
342. (d)
343. (a)
344. (a)
345. (b)
346. (d)
347. (a)
348. (b)
349. (b)
350. (b)
351. (a)
352. (a)
353. (a)
354. (a)
355. (a)
356. (a)
357. (a)
358. (b)
359. (a)
360. (d)
361. (d)
362. (c)
363. (c)
364. (d)
365. (a)
366. (c)
367. (c)
368. (c)
369. (a)
370. (d)
371. (a)
372. (d)
373. (a)
374. (a)
375. (b)
376. (d)
377. (a)
378. (d)
379. (a)
380. (c)l
381. (c)
382. (b)
383. (c)
384. (b)
385. (a)
386. (a)
387. (b)
388. (a)
389. (a)
390. (a)
391. (a)
392. (b)
393. (d)
394. (b)
395. (b)
396. (b)
397. (c)
398. (b)
399. (b)
400. (b)
401. (a)
402. (a)
403. (c)
404. (b)
405. (d)
406. (a)
407. (c)
408. (c)
409. (b)
410. (b)
411. (b)
412. (a)
413. (d)
414. (b)
415. (a)
416. (a)
417. (c)
418. (c)
419. (a)
420. (b)
421. (a)
422. (a)
423. (b)
424. (c)
425. (b)
426. (b)
427. (b)
428. (d)
429. (a)
430. (a)
431. (b)
432. (b)
433. (c)
434. (b)
435. (a)
436. (a)
437. (c)
438. (b)
439. (c)
440. (c)
441. (d)
442. (c)
443. (b)
444. (d)
445. (b)
446. (a)
447. (b)
448. (c)
449. (a)
450. (d)
451. (a)
452. (b)
453. (b)
454. (a)
455. (c)
456. (d)
457. (b)
458. (c)
459. (c)
460. (c)
461. (b)
462. (b)
463. (c)
464. (d)
465. (b)
466. (d)
467. (a)
468. (b)
469. (c)
470. (a)
471. (c)
472. (b)
473. (c)
474. (c)
475. (c)
476. (c)
477. (d)
478. (b)
479. (a)
480. (c)
481. (d)
482. (c)
483. (a)
484. (c)
485. (b)
486. (a)
487. (d)
488. (b)
489. (b)
490. (b)
491. (a)
492. (c)
493. (c)
494. (b)
495. (a)
496. (c)
497. (a)
498. (b)
499. (b)
500. (a)
501. (a)
502. (a)
503. (b)
504. (a)
505. (d)
506. (c)
507. (c)
508. (a)
509. (c)
510. (a)
511. (a)
512. (d)
513. (c)
514. (c)
515. (b)
516. (a)
517. (c)
518. (c)
519. (c)
520. (a)
521. (d)
522. (c)
523. (a)
524. (b)
525. (a)
526. (a)
527. (c)
528. (c)
529. (a)
530. (a)
531. (d)
532. (d)
533. (d)
534. (a)
535. (a)
536. (a)
537. (a)
538. (d)
539. (b)
540. (c)
541. (d)
542. (c)
543. (d)
544. (d)
545. (d)
546. (d)
547. (d)
548. (d)
549. (b)
550. (d)
551. (d)
552. (b)
553. (a)
554. (b)
555. (a)
556. (d)
557. (b)
558. (b)
559. (d)
560. (d)
561. (b)
562. (b)
563. (b)
564. (b)
565. (d)
566. (b)
567. (b)
568. (d)
569. (b)
570. (d)
571. (c)
572. (b)
573. (b)
574. (b)
575. (d)
576. (b)
577. (a)
578. (d)
579. (c)
580. (a)
581. (b)

Soil Science : Matching

S.No.	a	b	c	d	S.No.	a	b	c	d
1.	iii	iv	i	ii	12.	ii	iii	i	iv
2.	iii	iv	i	ii	13.	iv	iii	i	ii
3.	ii	iv	iii	i	14.	iv	i	ii	iii
4.	i	iii	iv	ii	15.	iii	iv	i	ii
5.	ii	i	iv	iii	16.	iii	iv	i	ii
6.	iii	iv	i	ii	17.	iv	i	ii	iii
7.	iv	iii	ii	i	18.	iii	iv	i	ii
8.	iv	iii	i	ii	19.	iv	i	ii	iii
9.	iv	ii	iii	i	20.	iv	iii	ii	i
10.	iv	i	ii	iii	21.	iii	i	iv	ii
11.	iv	ii	i	iii	22.	ii	iv	iii	i

S.No.	a	b	c	d
23.	iv	i	ii	iii
24.	iii	iv	i	ii
25.	iii	ii	i	iv
26.	iii	iv	ii	i
27.	iv	i	ii	iii
28.	iii	ii	iv	i
29.	ii	iv	i	iii
30.	iv	i	ii	iii
31.	iii	i	iv	ii
32.	ii	iii	iv	i
33.	iii	i	iv	ii
34.	iii	iv	ii	i
35.	iv	ii	i	iii
36.	iii	iv	i	ii
37.	iv	iii	i	ii
38.	ii	iv	i	iii
39.	ii	iii	i	iv
40.	iv	ii	i	iii
41.	iii	iv	i	ii
42.	iii	i	ii	iv
43.	ii	iv	i	iii
44.	iii	i	iv	ii
45.	ii	iii	iv	i
46.	iii	ii	iv	i
47.	iv	i	ii	iii
48.	iii	i	iv	ii

S.No.	a	b	c	d
49.	iv	iii	ii	i
50.	iii	iv	ii	i
51.	iii	i	iv	ii
52.	iv	i	ii	iii
53.	iv	ii	i	iii
54.	iii	i	iv	ii
55.	iii	iv	ii	i
56.	iii	i	iv	ii
57.	ii	i	iv	iii
58.	iv	i	ii	iii
59.	iii	i	iv	ii
60.	iv	iii	ii	i
61.	ii	iv	i	iii
62.	iii	ii	iv	i
63.	iv	i	ii	iii
64.	iv	i	ii	iii
65.	iii	iv	i	ii
66.	ii	iv	i	iii
67.	iii	iv	i	ii
68.	iii	iv	i	ii
69.	iv	i	ii	iii
70.	iii	iv	i	ii
71	ii	iv	i	iii
72.	iv	i	ii	iii
73.	iii	i	iv	ii
74.	i	ii	iii	iv

GK **Electric Flat Iron was invented by H.W. Seeley in 1882.** GS

7. Genetics, Plant Breeding and Biotechnology

Multiple Choice Questions

1. (d)	51. (c)	101. (a)	151. (d)	201. (b)	251. (d)	301. (b)
2. (c)	52. (d)	102. (c)	152. (a)	202. (a)	252. (a)	302. (d)
3. (a)	53. (d)	103. (a)	153. (d)	203. (c)	253. (c)	303. (b)
4. (c)	54. (b)	104. (b)	154. (d)	204. (b)	254. (b)	304. (d)
5. (c)	55. (d)	105. (a)	155. (a)	205. (c)	255. (c)	305. (b)
6. (c)	56. (a)	106. (d)	156. (a)	206. (b)	256. (c)	306. (d)
7. (a)	57. (c)	107. (b)	157. (a)	207. (d)	257. (d)	307. (c)
8. (c)	58. (d)	108. (b)	158. (c)	208. (d)	258. (c)	308. (c)
9. (d)	59. (c)	109. (c)	159. (a)	209. (b)	259. (c)	309. (b)
10. (a)	60. (d)	110. (a)	160. (c)	210. (d)	260. (a)	310. (c)
11. (b)	61. (d)	111. (a)	161. (b)	211. (a)	261. (b)	311. (b)
12. (c)	62. (a)	112. (a)	162. (d)	212. (a)	262. (d)	312. (c)
13. (a)	63. (b)	113. (d)	163. (c)	213. (a)	263. (c)	313. (a)
14. (a)	64. (b)	114. (c)	164. (c)	214. (d)	264. (b)	314. (d)
15. (d)	65. (c)	115. (d)	165. (c)	215. (d)	265. (b)	315. (d)
16. (c)	66. (c)	116. (c)	166. (b)	216. (b)	266. (d)	316. (b)
17. (b)	67. (d)	117. (b)	167. (c)	217. (a)	267. (d)	317. (c)
18. (c)	68. (b)	118. (d)	168. (a)	218. (c)	268. (d)	318. (d)
19. (c)	69. (d)	119. (c)	169. (c)	219. (a)	269. (b)	319. (d)
20. (b)	70. (d)	120. (a)	170. (a)	220. (a)	270. (b)	320. (d)
21. (a)	71. (c)	121. (d)	171. (a)	221. (d)	271. (c)	321. (c)
22. (a)	72. (b)	122. (d)	172. (c)	222. (b)	272. (c)	322. (a)
23. (d)	73. (a)	123. (c)	173. (c)	223. (c)	273. (a)	323. (b)
24. (b)	74. (c)	124. (b)	174. (a)	224. (b)	274. (b)	324. (d)
25. (a)	75. (c)	125. (b)	175. (c)	225. (c)	275. (b)	325. (d)
26. (a)	76. (a)	126. (a)	176. (c)	226. (c)	276. (a)	326. (a)
27. (d)	77. (b)	127. (d)	177. (c)	227. (a)	277. (a)	327. (a)
28. (c)	78. (d)	128. (a)	178. (d)	228. (a)	278. (a)	328. (d)
29. (c)	79. (d)	129. (b)	179. (c)	229. (c)	279. (a)	329. (b)
30. (c)	80. (d)	130. (a)	180. (d)	230. (d)	280. (b)	330. (d)
31. (a)	81. (b)	131. (d)	181. (c)	231. (c)	281. (d)	331. (b)
32. (c)	82. (d)	132. (d)	182. (a)	232. (a)	282. (a)	332. (b)
33. (b)	83. (a)	133. (c)	183. (c)	233. (a)	283. (d)	333. (a)
34. (d)	84. (a)	134. (a)	184. (a)	234. (a)	284. (b)	334. (c)
35. (b)	85. (a)	135. (c)	185. (c)	235. (b)	285. (a)	335. (c)
36. (b)	86. (c)	136. (d)	186. (a)	236. (d)	286. (b)	336. (a)
37. (a)	87. (c)	137. (a)	187. (c)	237. (c)	287. (b)	337. (c)
38. (c)	88. (d)	138. (a)	188. (c)	238. (b)	288. (d)	338. (d)
39. (a)	89. (a)	139. (b)	189. (c)	239. (c)	289. (b)	339. (b)
40. (b)	90. (c)	140. (b)	190. (d)	240. (c)	290. (a)	340. (b)
41. (b)	91. (c)	141. (c)	191. (c)	241. (d)	291. (b)	341. (a)
42. (c)	92. (b)	142. (b)	192. (a)	242. (b)	292. (d)	342. (b)
43. (d)	93. (a)	143. (b)	193. (b)	243. (a)	293. (b)	343. (a)
44. (a)	94. (a)	144. (b)	194. (b)	244. (b)	294. (d)	344. (c)
45. (d)	95. (b)	145. (a)	195. (c)	245. (a)	295. (d)	345. (a)
46. (a)	96. (a)	146. (c)	196. (d)	246. (d)	296. (c)	346. (a)
47. (d)	97. (b)	147. (a)	197. (b)	247. (c)	297. (a)	347. (b)
48. (d)	98. (b)	148. (a)	198. (a)	248. (b)	298. (a)	348. (b)
49. (a)	99. (b)	149. (d)	199. (c)	249. (c)	299. (a)	349. (a)
50. (b)	100. (c)	150. (a)	200. (d)	250. (a)	300. (c)	350. (d)

351. (a)	377. (d)	403. (a)	429. (d)	454. (c)	479. (c)	504. (c)
352. (c)	378. (a)	404. (b)	430. (c)	455. (b)	480. (d)	505. (d)
353. (b)	379. (c)	405. (c)	431. (b)	456. (b)	481. (a)	506. (b)
354. (c)	380. (c)	406. (b)	432. (b)	457. (b)	482. (c)	507. (b)
355. (c)	381. (a)	407. (c)	433. (b)	458. (b)	483. (a)	508. (d)
356. (c)	382. (b)	408. (d)	434. (d)	459. (c)	484. (a)	509. (d)
357. (c)	383. (b)	409. (c)	435. (a)	460. (a)	485. (b)	510. (c)
358. (d)	384. (d)	410. (c)	436. (a)	461. (d)	486. (d)	511. (d)
359. (a)	385. (d)	411. (c)	437. (d)	462. (c)	487. (c)	512. (c)
360. (c)	386. (a)	412. (c)	438. (b)	463. (b)	488. (a)	513. (b)
361. (c)	387. (c)	413. (d)	439. (a)	464. (d)	489. (a)	514. (c)
362. (a)	388. (a)	414. (a)	440. (c)	465. (b)	490. (b)	515. (b)
363. (c)	389. (b)	415. (b)	441. (a)	466. (a)	491. (b)	516. (c)
364. (d)	390. (c)	416. (d)	442. (a)	467. (c)	492. (c)	517. (a)
365. (d)	391. (d)	417. (a)	443. (a)	468. (c)	493. (b)	518. (c)
366. (c)	392. (b)	418. (b)	444. (b)	469. (b)	494. (c)	519. (d)
367. (d)	393. (d)	419. (a)	445. (b)	470. (b)	495. (d)	520. (c)
368. (b)	394. (a)	420. (b)	446. (d)	471. (a)	496. (a)	521. (b)
369. (a)	395. (b)	421. (c)	447. (c)	472. (b)	497. (b)	522. (c)
370. (a)	396. (c)	422. (a)	448. (c)	473. (b)	498. (c)	523. (a)
371. (a)	397. (a)	423. (d)	449. (d)	474. (b)	499. (c)	524. (b)
372. (d)	398. (d)	424. (a)	450. (c)	475. (b)	500. (d)	525. (a)
373. (d)	399. (a)	425. (d)	451. (b)	476. (c)	501. (c)	526. (c)
374. (d)	400. (c)	426. (b)	452. (c)	477. (d)	502. (a)	527. (a)
375. (b)	401. (d)	427. (b)	453. (a)	478. (c)	503. (b)	528. (d)
376. (d)	402. (c)	428. (b)				

Genetics, Plant Breeding and Biotechnology : Matching

S.No.	a	b	c	d	S.No.	a	b	c	d
1.	iv	i	iii	ii	17.	iv	iii	i	ii
2.	iii	i	iv	ii	18.	ii	iii	iv	i
3.	iii	i	iv	ii	19.	iv	i	iii	ii
4.	iv	i	ii	iii	20.	iii	iv	ii	i
5.	ii	iv	i	iii	21.	i	iii	iv	ii
6.	iii	i	iv	ii	22.	iv	i	ii	iii
7.	ii	iv	iii	i	23.	ii	i	iii	iv
8.	iii	iv	i	ii	24.	iv	i	ii	iii
9.	iii	i	iv	ii	25.	iii	i	iv	ii
10.	iv	ii	i	iii	26.	ii	iii	iv	i
11.	iii	i	iv	ii	27.	iii	i	iv	ii
12.	iii	iv	ii	i	28.	ii	iv	i	iii
13.	ii	iv	iii	i	29.	ii	i	iv	iii
14.	iii	iv	i	ii	30.	iv	i	ii	iii
15.	iv	i	ii	iii	31.	ii	iv	i	iii
16.	ii	iii	iv	i					

8. Entomology

Multiple Choice Questions

1. (a)	49. (c)	97. (d)	145. (b)	193. (b)	241. (a)	289. (b)
2. (d)	50. (b)	98. (c)	146. (d)	194. (a)	242. (c)	290. (b)
3. (d)	51. (c)	99. (a)	147. (c)	195. (d)	243. (b)	291. (b)
4. (b)	52. (a)	100. (c)	148. (b)	196. (c)	244. (b)	292. (c)
5. (b)	53. (d)	101. (a)	149. (b)	197. (a)	245. (c)	293. (b)
6. (a)	54. (a)	102. (a)	150. (b)	198. (c)	246. (b)	294. (c)
7. (c)	55. (c)	103. (c)	151. (a)	199. (d)	247. (a)	295. (c)
8. (a)	56. (b)	104. (a)	152. (c)	200. (b)	248. (b)	296. (a)
9. (d)	57. (a)	105. (b)	153. (d)	201. (a)	249. (b)	297. (c)
10. (c)	58. (b)	106. (d)	154. (a)	202. (c)	250. (a)	298. (a)
11. (b)	59. (b)	107. (b)	155. (a)	203. (b)	251. (b)	299. (b)
12. (b)	60. (b)	108. (d)	156. (b)	204. (d)	252. (d)	300. (a)
13. (c)	61. (d)	109. (c)	157. (a)	205. (a)	253. (b)	301. (c)
14. (c)	62. (b)	110. (a)	158. (b)	206. (b)	254. (c)	302. (b)
15. (b)	63. (a)	111. (c)	159. (a)	207. (b)	255. (b)	303. (b)
16. (a)	64. (b)	112. (b)	160. (c)	208. (a)	256. (a)	304. (b)
17. (a)	65. (d)	113. (c)	161. (a)	209. (c)	257. (c)	305. (b)
18. (c)	66. (a)	114. (d)	162. (c)	210. (c)	258. (b)	306. (a)
19. (d)	67. (d)	115. (c)	163. (b)	211. (c)	259. (c)	307. (c)
20. (a)	68. (b)	116. (c)	164. (c)	212. (c)	260. (b)	308. (d)
21. (d)	69. (d)	117. (b)	165. (b)	213. (a)	261. (b)	309. (c)
22. (b)	70. (b)	118. (b)	166. (b)	214. (a)	262. (c)	310. (b)
23. (b)	71. (b)	119. (b)	167. (b)	215. (c)	263. (a)	311. (d)
24. (d)	72. (d)	120. (d)	168. (d)	216. d.	264. (b)	312. (a)
25. (c)	73. (c)	121. (c)	169. (c)	217. (c)	265. (b)	313. (c)
26. (a)	74. (b)	122. (b)	170. (c)	218. (a)	266. (b)	314. (b)
27. (a)	75. (d)	123. (c)	171. (a)	219. (b)	267. (b)	315. (d)
28. (a)	76. (a)	124. (b)	172. (c)	220. (a)	268. (c)	316. (b)
29. (b)	77. (d)	125. (c)	173. (c)	221. (a)	269. (c)	317. (b)
30. (d)	78. (d)	126. (a)	174. (b)	222. (b)	270. (b)	318. (b)
31. (c)	79. (a)	127. (d)	175. (d)	223. (d)	271. (a)	319. (c)
32. (a)	80. (c)	128. (c)	176. (d)	224. (b)	272. (c)	320. (d)
33. (d)	81. (d)	129. (b)	177. (b)	225. (c)	273. (d)	321. (b)
34. (b)	82. (a)	130. (d)	178. (c)	226. (c)	274. (a)	322. (d)
35. (c)	83. (b)	131. (d)	179. (b)	227. (c)	275. (d)	323. (a)
36. (a)	84. (b)	132. (c)	180. (a)	228. (a)	276. (b)	324. (c)
37. (c)	85. (a)	133. (b)	181. (d)	229. (d)	277. (c)	325. (b)
38. (c)	86. (c)	134. (b)	182. (c)	230. (a)	278. (c)	326. (d)
39. (a)	87. (b)	135. (c)	183. (d)	231. (b)	279. (c)	327. (d)
40. (c)	88. (a)	136. (b)	184. (c)	232. (b)	280. (a)	328. (d)
41. (a)	89. (d)	137. (a)	185. (c)	233. (b)	281. (a)	329. (d)
42. (a)	90. (b)	138. (c)	186. (a)	234. (c)	282. (a)	330. (d)
43. (c)	91. (b)	139. (c)	187. (d)	235. (b)	283. (d)	331. (b)
44. (b)	92. (b)	140. (b)	188. (b)	236. (c)	284. (a)	332. (d)
45. (c)	93. (b)	141. (c)	189. (a)	237. (c)	285. (b)	333. (d)
46. (a)	94. (b)	142. (a)	190. (a)	238. (c)	286. (b)	334. (b)
47. (d)	95. (d)	143. (d)	191. (d)	239. (c)	287. (b)	335. (a)
48. (c)	96. (d)	144. (a)	192. (a)	240. (b)	288. (c)	336. (c)

GK GS

Australia is called the Land of Kangaroo.

337. (b)	344. (b)	351. (a)	358. (d)	365. (c)	372. (b)	379. (a)
338. (b)	345. (a)	352. (b)	359. (d)	366. (a)	373. (d)	380. (b)
339. (d)	346. (d)	353. (c)	360. (c)	367. (b)	374. (b)	381. (b)
340. (a)	347. (b)	354. (a)	361. (c)	368. (b)	375. (d)	382. (c)
341. (a)	348. (a)	355. (d)	362. (a)	369. (b)	376. (c)	383. (c)
342. (b)	349. (c)	356. (c)	363. (c)	370. (d)	377. (a)	
343. (a)	350. (a)	357. (b)	364. (c)	371. (a)	378. (a)	

Entomology : Matching

S.No.	a	b	c	d	S.No.	a	b	c	d
1.	ii	i	iv	iii	41.	iii	iv	ii	i
2.	iv	iii	i	ii	42.	iv	iii	ii	i
3.	ii	i	iv	iii	43.	iv	ii	iii	i
4.	iv	ii	i	iii	44.	iv	i	ii	iii
5.	iii	iv	ii	i	45.	ii	iii	i	iv
6.	ii	i	iv	iii	46.	iv	i	ii	iii
7.	ii	i	iv	iii	47.	iv	i	iii	ii
8.	iii	i	ii	iv	48.	iii	iv	i	ii
9.	ii	i	iv	iii	49.	ii	iii	iv	i
10.	ii	iii	iv	i	50.	iv	iii	ii	i
11.	ii	iii	iv	i	51.	iii	iv	ii	i
12.	iii	iv	i	ii	52.	iii	iv	ii	i
13.	ii	i	iv	iii	53.	iii	iv	ii	i
14.	ii	i	iv	iii	54.	iv	iii	i	ii
15.	i	iii	ii	iv	55.	ii	iv	i	iii
16.	ii	i	iv	iii	56.	iv	iii	i	ii
17.	ii	iii	iv	i	57.	iv	iii	ii	i
18.	iv	iii	ii	i	58.	ii	iv	i	iii
19.	iv	iii	i	ii	59.	iv	iii	i	ii
20.	ii	iv	i	iii	60.	iii	iv	i	ii
21.	iv	iii	ii	i	61.	i	iv	ii	iii
22.	iv	iii	ii	i	62.	iv	ii	iii	i
23.	iv	i	ii	iii	63.	i	iii	iv	ii
24.	iii	i	iv	ii	64.	ii	i	iv	iii
25.	iii	iv	ii	i	65.	iii	iv	ii	i
26.	iv	iii	ii	i	66.	iii	iv	ii	i
27.	iii	i	iv	ii	67.	iv	ii	i	iii
28.	iv	iii	i	ii	68.	ii	i	iii	iv
29.	iv	iii	ii	i	69.	ii	i	iv	iii
30.	iv	iii	i	ii	70.	iii	i	iv	ii
31.	ii	iii	iv	i	71.	iv	iii	ii	i
32.	iv	iii	ii	i	72.	ii	iv	i	iii
33.	iv	iii	i	ii	73.	ii	iv	iii	i
34.	iii	iv	i	ii	74.	iii	i	iv	ii
35.	ii	i	iii	iv	75.	ii	iv	i	iii
36.	ii	iii	i	iv	76.	iv	i	ii	iii
37.	ii	iv	i	iii	77.	ii	iii	iv	i
38.	i	iv	ii	iii	78.	iv	ii	i	iii
39.	ii	iii	iv	i	79.	iv	i	ii	iii
40.	ii	i	iv	iii	80.	iii	iv	i	ii

S.No.	a	b	c	d
81.	ii	iii	iv	i
82.	iv	i	ii	iii
83.	iii	ii	iv	i
84.	ii	iii	iv	i
85.	iv	i	ii	iii
86.	iii	iv	ii	i
87.	iv	i	ii	iii
88.	ii	iv	i	iii
89.	iii	i	iv	ii
90.	iii	i	ii	iv
91.	ii	iii	iv	i
92.	iii	i	iv	ii
93.	ii	iv	i	iii
94.	ii	iv	i	iii
95.	iii	i	iv	ii
96.	ii	iv	i	iii
97.	iv	i	ii	iii
98.	iii	i	iv	ii
99.	iv	i	ii	iii
100.	ii	iv	i	iii
101.	ii	iv	i	iii
102.	iii	i	iv	ii
103.	iv	iii	ii	i
104.	i	ii	iii	iv
105.	i	iii	ii	iv
106.	iv	iii	i	ii
107.	ii	iii	iv	i
108.	i	iii	ii	iv
109.	iii	iv	i	ii
110.	iii	i	ii	iv
111.	iv	iii	i	ii

9. Plant Pathology

Multiple Choice Questions

1. (a)	32. (c)	63. (c)	94. (b)	125. (b)	156. (c)	187. (b)
2. (b)	33. (b)	64. (c)	95. (b)	126. (a)	157. (d)	188. (d)
3. (b)	34. (a)	65. (c)	96. (a)	127. (a)	158. (c)	189. (a)
4. (b)	35. (b)	66. (c)	97. (a)	128. (c)	159. (b)	190. (b)
5. (c)	36. (c)	67. (c)	98. (b)	129. (c)	160. (a)	191. (b)
6. (a)	37. (c)	68. (d)	99. (b)	130. (a)	161. (c)	192. (d)
7. (c)	38. (c)	69. (b)	100. (a)	131. (c)	162. (b)	193. (a)
8. (a)	39. (a)	70. (a)	101. (a)	132. (a)	163. (b)	194. (c)
9. (a)	40. (b)	71. (b)	102. (c)	133. (c)	164. (b)	195. (a)
10. (b)	41. (b)	72. (b)	103. (d)	134. (c)	165. (b)	196. (a)
11. (c)	42. (c)	73. (c)	104. (c)	135. (c)	166. (b)	197. (c)
12. (a)	43. (a)	74. (a)	105. (b)	136. (b)	167. (d)	198. (d)
13. (b)	44. (c)	75. (b)	106. (a)	137. (c)	168. (c)	199. (b)
14. (b)	45. (a)	76. (c)	107. (d)	138. (c)	169. (c)	200. (a)
15. (c)	46. (a)	77. (d)	108. (b)	139. (b)	170. (a)	201. (b)
16. (d)	47. (b)	78. (c)	109. (b)	140. (b)	171. (b)	202. (a)
17. (d)	48. (b)	79. (c)	110. (a)	141. (b)	172. (d)	203. (b)
18. (b)	49. (b)	80. (c)	111. (b)	142. (b)	173. (d)	204. (b)
19. (b)	50. (b)	81. (a)	112. (a)	143. (a)	174. (b)	205. (a)
20. (d)	51. (b)	82. (d)	113. (a)	144. (b)	175. (c)	206. (c)
21. (a)	52. (c)	83. (d)	114. (c)	145. (a)	176. (d)	207. (b)
22. (c)	53. (c)	84. (d)	115. (d)	146. (b)	177. (b)	208. (b)
23. (c)	54. (a)	85. (b)	116. (a)	147. (b)	178. (b)	209. (d)
24. (c)	55. (d)	86. (b)	117. (b)	148. (d)	179. (d)	210. (d)
25. (a)	56. (c)	87. (a)	118. (d)	149. (b)	180. (d)	211. (b)
26. (d)	57. (b)	88. (a)	119. (b)	150. (b)	181. (b)	212. (c)
27. (b)	58. (c)	89. (b)	120. (b)	151. (c)	182. (c)	213 (b)
28. (d)	59. (c)	90. (b)	121. (b)	152. (b)	183. (a)	214. (d)
29. (a)	60. (b)	91. (c)	122. (c)	153. (c)	184. (b)	215. (a)
30. (b)	61. (c)	92. (a)	123. (d)	154. (c)	185. (b)	
31. (b)	62. (a)	93. (a)	124. (a)	155. (a)	186. (d)	

Plant Pathology : Matching

S.No.	a	b	c	d
1.	ii	iv	iii	i
2.	iv	i	ii	iii
3.	ii	iii	iv	i
4.	i	iv	ii	iv
5.	iv	iii	i	ii
6.	iii	i	iv	ii
7.	iv	iii	i	ii
8.	ii	iv	i	iii
9.	ii	iii	iv	i
10.	iv	i	ii	iii
11.	iii	i	iv	ii
12.	ii	iii	i	iv
13.	iv	i	ii	iii
14.	iii	i	iv	ii
15.	ii	iv	i	iii
16.	ii	iii	iv	i
17.	iii	i	iv	ii
18.	ii	iii	iv	i
19.	iv	i	ii	iii
20.	ii	i	iv	iii
21.	iii	iv	i	ii
22.	ii	iii	iv	i
23.	ii	iii	iv	i
24.	iv	i	ii	iii

Telegraph was invented by M. Lammond in 1787.

S.No.	a	b	c	d
25.	ii	iii	iv	i
26.	iii	i	iv	ii
27.	ii	i	iv	iii
28.	iv	i	ii	iii
29.	iii	i	iv	ii
30.	ii	iv	i	iii
31.	i	iv	ii	iii
32.	iii	iv	i	ii
33.	ii	iii	i	iv
34.	iv	i	ii	iii
35.	iii	i	ii	iv
36.	ii	iv	i	iii
37.	iv	i	ii	iii
38.	iii	iv	i	ii
39.	iv	i	ii	iii
40.	ii	iv	i	iii
41.	iv	i	ii	iii
42.	iii	i	iv	ii
43.	iii	i	iv	ii
44.	iv	iii	ii	i
45.	ii	iv	i	iii
46.	iv	i	ii	iii
47.	ii	iii	iv	i
48.	ii	iv	i	iii

10. Agricultural Economics and Farm Management

Multiple Choice Questions

1. (c)
2. (b)
3. (b)
4. (a)
5. (b)
6. (b)
7. (d)
8. (c)
9. (b)
10. (b)
11. (b)
12. (c)
13. (b)
14. (d)
15. (a)
16. (c)
17. (b)
18. (d)
19. (b)
20. (a)
21. (a)
22. (c)
23. (c)
24. (a)
25. (b)
26. (b)
27. (d)
28. (c)
29. (a)
30. (b)
31. (c)
32. (a)
33. (b)
34. (c)
35. (a)
36. (c)
37. (a)
38. (d)
39. (c)
40. (a)
41. (a)
42. (c)
43. (d)
44. (b)
45. (b)
46. (a)
47. (d)
48. (a)
49. (c)
50. (a)
51. (a)
52. (c)
53. (b)
54. (a)
55. (b)
56. (d)
57. (b)
58. (d)
59. (a)
60. (c)
61. (c)
62. (d)
63. (d)
64. (b)
65. (b)
66. (b)
67. (d)
68. (c)
69. (a)
70. (a)
71. (c)
72. (a)
73. (a)
74. (b)
75. (b)
76. (b)
77. (a)
78. (a)
79. (c)
80. (d)
81. (c)
82. (b)
83. (c)
84. (d)
85. (c)
86. (b)
87. (a)
88. (a)
89. (a)
90. (a)
91. (d)
92. (a)
93. (c)
94. (a)
95. (b)
96. (b)
97. (a)
98. (a)
99. (d)
100. (c)
101. (d)
102. (a)
103. (c)
104. (b)
105. (a)
106. (a)
107. (b)
108. (c)
109. (a)
110. (c)
111. (a)
112. (b)
113. (c)
114. (b)
115. (b)
116. (a)
117. (d)
118. (c)
119. (d)
120. (c)
121. (a)
122. (a)
123. (d)
124. (a)
125. (c)
126. (c)
127. (c)
128. (b)
129. (d)
130. (c)
131. (a)
132. (a)
133. (b)
134. (d)
135. (c)
136. (d)
137. (a)
138. (a)
139. (c)
140. (b)
141. (a)
142. (b)
143. (d)
144. (c)
145. (a)
146. (d)
147. (a)
148. (a)
149. (a)
150. (a)
151. (c)
152. (c)
153. (c)
154. (b)
155. (a)
156. (b)
157. (a)
158. (d)
159. (b)
160. (c)
161. (b)
162. (a)
163. (a)
164. (b)
165. (b)
166. (b)
167. (b)
168. (a)
169. (d)
170. (b)
171. (b)
172. (c)
173. (c)
174. (b)
175. (a)
176. (b)
177. (a)
178. (a)
179. (b)
180. (b)
181. (a)
182. (c)
183. (c)
184. (c)
185. (a)
186. (a)
187. (b)
188. (b)
189. (a)
190. (c)
191. (b)
192. (b)
193. (a)
194. (b)
195. (a)
196. (b)
197. (a)
198. (b)
199. (a)
200. (a)
201. (d)
202. (d)
203. (b)
204. (a)
205. (d)
206. (b)
207. (a)
208. (d)
209. (c)
210. (b)
211. (d)
212. (c)
213. (c)
214. (a)
215. (b)
216. (c)
217. (c)
218. (a)
219. (c)
220. (a)
221. (a)
222. (c)
223. (d)
224. (d)
225. (a)
226. (c)
227. (c)
228. (b)
229. (c)
230. (a)
231. (b)
232. (a)
233. (c)
234. (b)
235. (c)
236. (d)
237. (a)
238. (d)
239. (c)
240. (a)
241. (a)
242. (b)
243. (a)
244. (b)
245. (c)
246. (b)
247. (c)
248. (d)
249. (b)
250. (d)
251. (b)
252. (b)
253. (b)
254. (a)
255. (d)
256. (b)
257. (a)
258. (d)
259. (b)
260. (c)
261. (a)
262. (b)
263. (a)
264. (b)
265. (a)
266. (c)
267. (c)
268. (d)
269. (a)
270. (b)
271. (b)
272. (b)
273. (b)
274. (a)
275. (c)
276. (d)
277. (c)
278. (a)
279. (b)
280. (b)
281. (a)
282. (c)
283. (c)
284. (d)
285. (a)
286. (b)
287. (c)
288. (c)
289. (b)
290. (a)
291. (b)
292. (b)
293. (d)
294. (b)
295. (d)
296. (c)
297. (d)
298. (a)
299. (a)
300. (c)
301. (c)
302. (c)
303. (b)
304. (a)
305. (a)
306. (c)
307. (b)
308. (b)
309. (b)
310. (c)
311. (c)
312. (a)
313. (c)
314. (b)
315. (c)
316. (a)
317. (c)
318. (a)
319. (a)
320. (d)
321. (a)
322. (c)
323. (c)
324. (c)
325. (c)
326. (b)
327. (a)
328. (c)
329. (d)
330. (a)
331. (a)
332. (a)
333. (b)
334. (c)
335. (b)
336. (b)

Genghis Khan, the Mongoal conqurer died in 1227.

337. (b)
338. (d)
339. (d)
340. (d)
341. (a)
342. (a)
343. (d)
344. (a)
345. (b)
346. (d)
347. (d)
348. (c)
349. (a)
350. (b)
351. (c)
352. (c)
353. (a)
354. (c)
355. (b)
356. (d)
357. (a)
358. (c)
359. (a)
360. (a)
361. (d)
362. (a)
363. (d)
364. (a)
365. (c)
366. (c)
367. (c)
368. (c)
369. (b)
370. (c)
371. (d)
372. (a)
373. (d)
374. (b)
375. (d)
376. (b)
377. (d)
378. (b)
379. (a)
380. (a)
381. (a)
382. (a)
383. (b)
384. (a)
385. (c)
386. (c)

Agricultural Economics and Farm Management : Matching

S.No.	a	b	c	d
1.	ii	iii	i	iv
2.	iii	i	iv	ii
3.	ii	iv	iii	i
4.	ii	iii	iv	i
5.	iv	i	ii	iii
6.	ii	i	iv	iii
7.	iii	i	iv	ii
8.	iv	iii	i	ii
9.	iv	iii	ii	i
10.	iii	i	ii	iv
11.	iii	i	iv	ii
12.	iv	ii	iii	i

11. Extension Education

Multiple Choice Questions

1. (b)
2. (a)
3. (c)
4. (b)
5. (c)
6. (a)
7. (b)
8. (d)
9. (d)
10. (b)
11. (c)
12. (c)
13. (a)
14. (b)
15. (a)
16. (a)
17. (a)
18. (a)
19. (c)
20. (a)
21. (c)
22. (a)
23. (b)
24. (b)
25. (a)
26. (d)
27. (d)
28. (d)
29. (d)
30. (a)
31. (b)
32. (b)
33. (d)
34. (b)
35. (a)
36. (d)
37. (b)
38. (b)
39. (c)
40. (c)
41. (a)
42. (c)
43. (a)
44. (b)
45. (a)
46. (a)
47. (a)
48. (d)
49. (b)
50. (a)
51. (a)
52. (c)
53. (d)
54. (b)
55. (c)
56. (c)
57. (c)
58. (a)
59. (a)
60. (d)
61. (a)
62. (b)
63. (d)
64. (a)
65. (c)
66. (b)
67. (b)
68. (d)
69. (a)
70. (b)
71. (a)
72. (b)
73. (a)
74. (a)
75. (d)
76. (a)
77. (b)
78. (b)
79. (a)
80. (b)
81. (b)
82. (b)
83. (d)
84. (c)
85. (a)
86. (b)
87. (b)
88. (d)
89. (c)
90. (d)
91. (c)
92. (b)
93. (a)
94. (c)
95. (a)
96. (d)
97. (d)
98. (b)
99. (d)
100. (c)
101. (b)
102. (c)
103. (a)
104. (d)
105. (c)
106. (a)
107. (b)
108. (c)
109. (d)
110. (a)
111. (a)
112. (c)
113. (b)
114. (a)
115. (c)
116. (b)
117. (b)
118. (d)
119. (b)
120. (d)
121. (b)
122. (c)
123. (c)
124. (d)
125. (c)
126. (a)
127. (b)
128. (a)
129. (b)
130. (d)
131. (a)
132. (b)
133. (d)
134. (c)
135. (b)
136. (c)
137. (b)
138. (c)
139. (b)
140. (b)
141. (b)
142. (d)
143. (c)
144. (d)
145. (c)
146. (a)
147. (c)
148. (b)
149. (d)
150. (a)
151. (d)
152. (b)
153. (a)
154. (a)
155. (b)
156. (c)
157. (c)
158. (a)
159. (b)
160. (d)
161. (c)
162. (c)
163. (c)
164. (c)
165. (d)
166. (a)
167. (b)
168. (d)
169. (d)
170. (c)
171. (a)
172. (c)
173. (b)
174. (b)
175. (d)
176. (d)
177. (d)
178. (a)
179. (a)
180. (d)
181. (c)
182. (a)
183. (d)
184. (a)
185. (a)
186. (c)
187. (a)
188. (d)
189. (c)
190. (c)
191. (b)
192. (b)
193. (a)
194. (d)
195. (d)
196. (a)
197. (d)
198. (d)
199. (a)
200. (a)
201. (d)
202. (a)
203. (c)
204. (d)
205. (b)
206. (c)
207. (d)
208. (c)
209. (b)
210. (c)
211. (a)

Extension Education : Matching

S.No.	a	b	c	d	e
1.	iii	i	iv	ii	-
2.	iii	iv	i	ii	-
3.	iv	i	ii	v	iii
4.	iii	iv	ii	i	iv
5.	iv	iii	i	v	ii
6.	iv	i	v	iii	ii
7.	iii	v	ii	i	iv
8.	iv	iii	ii	v	i
9.	iii	i	iv	ii	-
10.	ii	iv	i	iii	-

S.No.	a	b	c	d	e
11.	iv	i	ii	iii	-
12.	iii	i	v	ii	iv
13.	iv	iii	ii	i	-
14.	i	iii	ii	iv	-
15.	iii	iv	ii	i	-
16.	i	ii	iv	iii	-
17.	ii	iii	i	iv	-
18.	i	ii	iii	iv	-
19.	iii	i	iv	ii	

GK GS

Caustic lotion is Silver nitrate.

12. Animal Husbandry

Multiple Choice Questions

1. (c)	47. (b)	93. (a)	139. (b)	185. (b)	231. (c)	277. (c)
2. (b)	48. (a)	94. (d)	140. (b)	186. (c)	232. (b)	278. (c)
3. (a)	49. (c)	95. (b)	141. (a)	187. (c)	233. (c)	279. (b)
4. (d)	50. (d)	96. (b)	142. (d)	188. (b)	234. (d)	280. (b)
5. (a)	51. (b)	97. (b)	143. (a)	189. (a)	235. (a)	281. (c)
6. (d)	52. (c)	98. (a)	144. (a)	190. (c)	236. (c)	282. (d)
7. (c)	53. (b)	99. (a)	145. (a)	191. (c)	237. (a)	283. (b)
8. (d)	54. (d)	100. (c)	146. (a)	192. (d)	238. (b)	284. (a)
9. (d)	55. (b)	101. (b)	147. (a)	193. (a)	239. (d)	285. (b)
10. (b)	56. (b)	102. (c)	148. (b)	194. (b)	240. (d)	286. (b)
11. (d)	57. (b)	103. (c)	149. (b)	195. (d)	241. (c)	287. (b)
12. (a)	58. (c)	104. (a)	150. (b)	196. (d)	242. (d)	288. (c)
13. (b)	59. (a)	105. (b)	151. (a)	197. (d)	243. (a)	289. (b)
14. (c)	60. (d)	106. (a)	152. (c)	198. (a)	244. (c)	290. (a)
15. (c)	61. (d)	107. (c)	153. (c)	199. (a)	245. (b)	291. (b)
16. (b)	62. (c)	108. (b)	154. (b)	200. (c)	246. (c)	292. (b)
17. (b)	63. (c)	109. (a)	155. (c)	201. (c)	247. (d)	293. (c)
18. (b)	64. (a)	110. (d)	156. (b)	202. (a)	248. (b)	294. (a)
19. (a)	65. (b)	111. (a)	157. (c)	203. (c)	249. (c)	295. (b)
20. (d)	66. (b)	112. (d)	158. (a)	204. (a)	250. (b)	296. (b)
21. (c)	67. (b)	113. (a)	159. (a)	205. (a)	251. (b)	297. (c)
22. (b)	68. (d)	114. (c)	160. (c)	206. (c)	252. (d)	298. (c)
23. (c)	69. (d)	115. (c)	161. (a)	207. (d)	253. (c)	299. (a)
24. (a)	70. (c)	116. (b)	162. (d)	208. (b)	254. (c)	300. (d)
25. (a)	71. (d)	117. (c)	163. (a)	209. (a)	255. (c)	301. (a)
26. (d)	72. (a)	118. (d)	164. (c)	210. (b)	256. (c)	302. (a)
27. (c)	73. (c)	119. (c)	165. (a)	211. (a)	257. (c)	303. (a)
28. (b)	74. (c)	120. (c)	166. (b)	212. (d)	258. (a)	304. (c)
29. (c)	75. (a)	121. (a)	167. (c)	213. (b)	259. (d)	305. (c)
30. (a)	76. (a)	122. (a)	168. (d)	214. (c)	260. (c)	306. (a)
31. (a)	77. (b)	123. (b)	169. (b)	215. (a)	261. (d)	307. (c)
32. (d)	78. (b)	124. (b)	170. (d)	216. (a)	262. (c)	308. (c)
33. (b)	79. (c)	125. (c)	171. (a)	217. (a)	263. (b)	309. (a)
34. (a)	80. (c)	126. (d)	172. (a)	218. (c)	264. (b)	310. (d)
35. (c)	81. (d)	127. (c)	173. (d)	219. (b)	265. (d)	311. (a)
36. (b)	82. (d)	128. (b)	174. (a)	220. (d)	266. (d)	312. (a)
37. (d)	83. (d)	129. (b)	175. (c)	221. (d)	267. (b)	313. (c)
38. (a)	84. (b)	130. (d)	176. (c)	222. (d)	268. (c)	314. (b)
39. (b)	85. (c)	131. (b)	177. (a)	223. (a)	269. (b)	315. (a)
40. (c)	86. (d)	132. (b)	178. (b)	224. (d)	270. (a)	316. (d)
41. (d)	87. (c)	133. (b)	179. (b)	225. (b)	271. (c)	317. (c)
42. (b)	88. (b)	134. (b)	180. (b)	226. (b)	272. (d)	318. (a)
43. (a)	89. (b)	135. (c)	181. (d)	227. (c)	273. (a)	319. (b)
44. (b)	90. (c)	136. (b)	182. (d)	228. (c)	274. (b)	320. (c)
45. (d)	91. (c)	137. (a)	183. (b)	229. (c)	275. (a)	321. (b)
46. (d)	92. (a)	138. (c)	184. (c)	230. (d)	276. (a)	322. (a)

GK GS

Sir Winsten Churcill won Nobel prize for Literature.

323. (b)	360. (a)	397. (b)	434. (c)	471. (c)	508. (b)	545. (a)
324. (c)	361. (b)	398. (a)	435. (a)	472. (c)	509. (d)	546. (d)
325. (d)	362. (c)	399. (d)	436. (c)	473. (b)	510. (c)	547. (d)
326. (a)	363. (a)	400. (a)	437. (d)	474. (c)	511. (a)	548. (c)
327. (c)	364. (b)	401. (b)	438. (c)	475. (b)	512. (b)	549. (b)
328. (c)	365. (c)	402. (d)	439. (c)	476. (d)	513. (b)	550. (c)
329. (d)	366. (a)	403. (b)	440. (c)	477. (c)	514. (b)	551. (c)
330. (d)	367. (d)	404. (d)	441. (c)	478. (b)	515. (c)	552. (b)
331. (c)	368. (c)	405. (a)	442. (c)	479. (d)	516. (d)	553. (d)
332. (d)	369. (d)	406. (c)	443. (c)	480. (c)	517. (a)	554. (d)
333. (a)	370. (d)	407. (b)	444. (a)	481. (a)	518. (a)	555. (c)
334. (c)	371. (a)	408. (c)	445. (d)	482. (a)	519. (b)	556. (c)
335. (a)	372. (d)	409. (b)	446. (b)	483. (d)	520. (a)	557. (c)
336. (b)	373. (a)	410. (a)	447. (b)	484. (d)	521. (a)	558. (b)
337. (c)	374. (b)	411. (c)	448. (d)	485. (b)	522. (b)	559. (d)
338. (a)	375. (d)	412. (b)	449. (c)	486. (c)	523. (a)	560. (b)
339. (a)	376. (c)	413. (c)	450. (c)	487. (c)	524. (c)	561. (b)
340. (c)	377. (b)	414. (a)	451. (b)	488. (a)	525. (c)	562. (d)
341. (b)	378. (b)	415. (d)	452. (c)	489. (b)	526. (d)	563. (c)
342. (b)	379. (a)	416. (d)	453. (b)	490. (d)	527. (c)	564. (c)
343. (b)	380. (c)	417. (b)	454. (a)	491. (b)	528. (d)	565. (d)
344. (a)	381. (b)	418. (b)	455. (c)	492. (d)	529. (b)	566. (a)
345. (a)	382. (d)	419. (c)	456. (a)	493. (c)	530. (b)	567. (b)
346. (c)	383. (c)	420. (b)	457. (d)	494. (b)	531. (d)	568. (c)
347. (b)	384. (a)	421. (c)	458. (a)	495. (b)	532. (c)	569. (d)
348. (b)	385. (d)	422. (d)	459. (d)	496. (d)	533. (c)	570. (a)
349. (c)	386. (d)	423. (c)	460. (a)	497. (a)	534. (c)	571. (b)
350. (a)	387. (a)	424. (c)	461. (a)	498. (d)	535. (c)	572. (b)
351. (c)	388. (a)	425. (a)	462. (a)	499. (b)	536. (a)	573. (c)
352. (c)	389. (c)	426. (c)	463. (c)	500. (c)	537. (a)	574. (d)
353. (a)	390. (a)	427. (c)	464. (a)	501. (b)	538. (c)	575. (c)
354. (d)	391. (d)	428. (c)	465. (a)	502. (a)	539. (b)	576. (c)
355. (a)	392. (a)	429. (d)	466. (b)	503. (a)	540. (d)	577. (a)
356. (b)	393. (d)	430. (d)	467. (c)	504. (a)	541. (a)	578 (a)
357. (a)	394. (d)	431. (a)	468. (b)	505. (a)	542. (a)	579. (c)
358. (a)	395. (a)	432. (d)	469. (b)	506. (c)	543. (b)	580. (a)
359. (b)	396. (b)	433. (a)	470. (a)	507. (b)	544. (d)	

Animal Husbandry : Matching

S.No.	a	b	c	d	e
1.	iv	i	ii	iii	-
2.	iv	i	ii	iii	-
3.	iv	i	ii	iii	-
4.	iv	i	ii	iii	-
5.	iii	iv	i	ii	-
6.	iv	i	ii	iii	-
7.	iv	i	ii	iii	-
8.	iii	i	iv	ii	-
9.	iv	i	ii	iii	-

S.No.	a	b	c	d	e
10.	iii	v	i	ii	iv
11.	iii	i	v	ii	iv
12.	ii	v	i	iv	iii
13.	iii	i	iv	ii	-
14.	v	iii	ii	iv	i
15.	iii	i	iv	ii	-
16.	iv	i	ii	iii	v
17.	v	iii	i	ii	iv
18.	iii	v	i	ii	iv

GK GS

New Orleans is situated on the banks of river Mississippi

S.No.	a	b	c	d	e	S.No.	a	b	c	d	e
19.	v	iii	i	ii	iv	42.	iv	i	ii	v	iii
20.	iv	i	ii	v	iii	43.	iv	i	v	ii	iii
21.	iii	v	iv	i	ii	44.	ii	v	iv	iii	i
22.	iv	iii	i	v	ii	45.	v	i	iv	ii	iii
23.	iii	v	i	ii	iv	46.	iii	iv	i	ii	v
24.	v	iii	i	ii	iv	47.	ii	v	i	iii	iv
25.	ii	v	i	iii	iv	48.	v	i	iv	ii	iii
26.	iv	i	v	ii	iii	49.	iii	v	i	ii	iv
27.	v	i	iv	ii	iii	50.	iv	i	v	ii	iv
28.	iii	iv	i	v	ii	51.	v	i	iv	ii	iii
29.	ii	v	i	iii	iv	52.	iii	v	i	iv	ii
30.	iv	i	v	ii	iii	53.	iii	i	v	ii	iv
31.	ii	iv	i	v	iii	54.	iv	v	i	iii	ii
32.	iii	i	iv	v	ii	55.	iii	i	iv	ii	v
33.	ii	iv	i	v	iii	56.	iii	i	iv	ii	-
34.	iv	i	v	ii	iii	57.	iii	ii	i	iv	-
35.	v	ii	i	iv	iii	58.	iv	i	ii	v	iii
36.	iii	i	iv	ii	-	59.	iii	i	iv	ii	-
37.	iv	v	ii	i	iii	60.	iv	iii	i	ii	-
38.	ii	iv	i	iii	v	61.	ii	i	iv	iii	-
39.	ii	iv	i	v	iii	62.	iii	ii	iv	i	-
40.	iv	v	i	iii	ii	63.	iv	iii	ii	i	-
41.	iii	v	iv	i	ii	64.	i	iii	ii	iv	-

13. Plant Physiology

Multiple Choice Questions

1. (c)
2. (b)
3. (c)
4. (b)
5. (d)
6. (b)
7. (c)
8. (d)
9. (d)
10. (c)
11. (c)
12. (c)
13. (b)
14. (c)
15. (d)
16. (c)
17. (d)
18. (c)
19. (a)
20. (d)
21. (b)
22. (d)
23. (c)
24. (b)
25. (b)
26. (a)
27. (c)
28. (b)
29. (d)
30. (d)
31. (c)
32. (a)
33. (a)
34. (a)
35. (b)
36. (c)
37. (b)
38. (b)
39. (d)
40. (a)
41. (d)
42. (d)
43. (b)
44. (a)
45. (c)
46. (c)
47. (b)
48. (c)
49. (d)
50. (d)
51. (d)
52. (a)
53. (b)
54. (c)
55. (b)
56. (d)
57. (b)
58. (d)
59. (b)
60. (a)
61. (b)
62. (a)
63. (a)
64. (c)
65. (b)
66. (b)
67. (a)
68. (a)
69. (c)
70. (c)
71. (b)
72. (b)
73. (b)
74. (b)
75. (a)
76. (a)
77. (b)
78. (c)
79. (d)
80. (d)
81. (a)
82. (c)
83. (a)
84. (b)
85. (a)
86. (c)
87. (b)
88. (c)
89. (b)
90. (b)
91. (a)
92. (b)
93. (a)
94. (d)
95. (c)
96. (b)
97. (d)
98. (b)
99. (c)
100. (c)
101. (a)
102. (c)
103. (b)
104. (b)
105. (a)
106. (d)
107. (b)
108. (a)
109. (b)
110. (b)
111. (c)
112. (c)
113. (c)
114. (b)
115. (b)
116. (c)
117. (b)
118. (a)
119. (d)
120. (b)
121. (c)
122. (b)
123. (d)
124. (a)
125. (b)
126. (b)
127. (a)
128. (d)
129. (c)
130. (a)
131. (a)
132. (b)
133. (d)
134. (a)
135. (b)
136. (d)
137. (c)
138. (b)
139. (b)
140. (a)
141. (a)
142. (a)
143. (d)
144. (b)
145. (d)
146. (a)
147. (b)
148. (c)
149. (c)
150. (d)
151. (c)
152. (b)
153. (a)
154. (d)
155. (a)
156. (b)
157. (c)
158. (b)
159. (c)
160. (b)
161. (b)
162. (c)
163. (c)
164. (b)
165. (b)
166. (b)
167. (a)
168. (b)
169. (a)
170. (d)
171. (c)
172. (d)
173. (d)
174. (d)
175. (b)
176. (c)
177. (c)
178. (c)
179. (a)
180. (c)
181. (a)
182. (a)
183. (a)
184. (b)
185. (c)
186. (d)
187. (b)
188. (c)
189. (a)
190. (c)
191. (c)
192. (c)
193. (c)
194. (b)
195. (b)
196. (d)
197. (c)
198. (d)
199. (b)
200. (c)
201. (b)
202. (c)
203. (a)
204. (c)
205. (c)
206. (d)
207. (c)
208. (b)
209. (b)
210. (b)
211. (c)
212. (b)
213. (c)
214. (a)
215. (c)
216. (a)
217. (b)
218. (a)
219. (b)
220. (c)
221. (a)
222. (b)
223. (b)
224. (c)
225. (b)
226. (c)
227. (a)
228. (b)
229. (c)
230. (b)
231. (a)
232. (b)
233. (b)
234. (d)
235. (c)
236. (c)
237. (d)
238. (a)
239. (c)
240. (c)
241. (b)
242. (d)
243. (b)
244. (b)
245. (a)
246. (c)
247. (c)
248. (c)
249. (a)
250. (c)
251. (d)
252. (c)
253. (a)
254. (d)
255. (c)
256. (b)
257. (c)
258. (c)
259. (a)
260. (c)
261. (a)
262. (d)
263. (c)
264. (d)
265. (c)
266. (a)
267. (a)
268. (b)
269. (b)
270. (b)
271. (a)
272. (b)
273. (a)
274. (c)
275. (b)
276. (a)
277. (c)
278. (d)
279. (a)
280. (d)
281. (b)
282. (d)
283. (c)
284. (b)
285. (c)
286. (c)
287. (c)
288. (a)
289. (b)
290. (c)
291. (a)
292. (c)
293. (b)
294. (d)
295. (a)
296. (b)
297. (b)
298. (b)
299. (b)
300. (b)
301. (b)
302. (b)
303. (a)
304. (b)
305. (a)
306. (b)
307. (a)
308. (a)
309. (c)

GK GS

Red Earth and Pouring Rain is written by Vikram Chandra.

Plant Physiology : Matching

S.No.	a	b	c	d
1.	iii	iv	i	ii
2.	iii	iv	i	ii
3.	iii	i	ii	iv
4.	iv	i	iii	ii
5.	iii	i	iv	ii
6.	iii	iv	i	ii
7.	iv	i	ii	iii
8.	ii	iv	i	iii
9.	ii	i	iv	iii
10.	iii	ii	i	iv
11.	ii	iii	iv	i
12.	iii	i	iv	ii
13.	iv	i	ii	iii
14.	iii	ii	iv	i
15.	iv	i	iii	ii

S.No.	a	b	c	d
16.	ii	iv	i	iii
17.	iii	iv	ii	i
18.	iv	i	ii	iii
19.	iii	iv	i	ii
20.	iv	i	ii	iii
21.	iii	i	iv	ii
22.	iv	ii	i	iii
23.	iii	ii	iv	i
24.	iv	i	ii	iii
25.	iii	ii	iv	i
26.	iii	ii	iv	i
27.	ii	iv	i	iii
28.	ii	iv	i	iii
29.	i	iv	iii	ii
30.	iii	iv	ii	i

14. Agricultural Engineering

Multiple Choice Questions

1. (a)
2. (b)
3. (a)
4. (b)
5. (a)
6. (c)
7. (b)
8. (b)
9. (a)
10. (b)
11. (c)
12. (b)
13. (a)
14. (a)
15. (b)
16. (d)
17. (b)
18. (a)
19. (a)
20. (b)
21. (b)
22. (d)
23. (b)
24. (a)
25. (c)
26. (c)
27. (c)
28. (a)
29. (a)
30. (b)
31. (c)
32. (a)
33. (b)
34. (a)
35. (b)
36. (b)
37. (b)
38. (a)
39. (a)
40. (b)
41. (a)
42. (b)
43. (a)
44. (c)
45. (c)
46. (c)
47. (a)
48. (c)
49. (d)
50. (a)
51. (b)
52. (d)
53. (b)
54. (a)
55. (c)
56. (b)
57. (d)
58. (d)
59. (a)
60. (b)
61. (a)
62. (b)
63. (d)
64. (b)
65. (b)
66. (a)
67. (b)
68. (d)
69. (a)
70. (a)
71. (b)
72. (d)
73. (a)
74. (d)
75. (a)
76. (a)
77. (b)
78. (a)
79. (b)
80. (b)
81. (b)
82. (a)
83. (b)
84. (b)
85. (b)
86. (a)
87. (c)
88. (d)
89. (d)
90. (c)
91. (a)
92. (c)
93. (a)
94. (d)
95. (a)
96. (b)
97. (a)
98. (d)
99. (d)
100. (d)
101. (d)
102. (b)
103. (b)
104. (a)
105. (d)
106. (a)
107. (a)
108. (a)
109. (a)
110. (d)
111. (a)
112. (a)
113. (a)
114. (b)
115. (a)
116. (a)
117. (b)
118. (d)
119. (b)
120. (b)
121. (a)
122. (d)
123. (b)
124. (c)
125. (a)
126. (d)
127. (a)
128. (c)
129. (a)
130. (d)
131. (a)
132. (a)
133. (a)
134. (b)
135. (a)
136. (d)
137. (d)
138. (b)
139. (a)
140. (c)
141. (b)
142. (a)
143. (c)
144. (d)
145. (c)
146. (a)
147. (a)
148. (b)
149. (c)
150. (d)
151. (a)
152. (a)
153. (b)
154. (c)
155. (c)
156. (d)
157. (a)
158. (b)
159. (a)
160. (c)
161. (a)
162. (a)
163. (c)
164. (b)
165. (a)
166. (a)
167. (b)
168. (a)
169. (b)
170. (b)
171. (b)
172. (a)
173. (a)
174. (b)
175. (b)
176. (c)
177. (a)
178. (b)
179. (a)
180. (c)
181. (b)
182. (a)
183. (a)
184. (b)
185. (a)
186. (a)
187. (a)
188. (c)
189. (a)
190. (a)
191. (b)
192. (d)
193. (c)
194. (c)
195. (c)
196. (d)
197. (b)
198. (c)
199. (b)
200. (a)
201. (b)
202. (a)
203. (d)
204. (a)
205. (c)
206. (c)
207. (d)
208. (b)
209. (a)
210. (a)
211. (c)
212. (d)
213. (b)
214. (a)
215. (a)
216. (a)
217. (b)
218. (c)
219. (c)
220. (a)
221. (c)
222. (d)
223. (d)
224. (c)
225. (a)
226. (b)
227. (b)
228. (d)
229. (d)
230. (b)
231. (b)
232. (c)
233. (b)
234. (a)
235. (d)
236. (a)
237. (a)
238. (a)
239. (d)
240. (c)
241. (c)
242. (c)
243. (d)
244. (d)
245. (b)
246. (c)
247. (a)
248. (c)
249. (b)
250. (d)
251. (c)
252. (b)
253. (b)
254. (d)
255. (b)
256. (c)
257. (c)
258. (c)
259. (d)
260. (c)
261. (c)
262. (a)
263. (d)
264. (b)
265. (a)
266. (d)
267. (a)
268. (a)
269. (a)
270. (d)
271. (a)
272. (b)
273. (c)
274. (c)
275. (d)
276. (d)
277. (a)
278. (b)
279. (a)
280. (d)
281. (d)
282. (c)
283. (a)
284. (a)
285. (c)
286. (a)
287. (a)
288. (a)
289. (a)
290. (a)
291. (c)
292. (a)
293. (b)
294. (a)
295. (c)
296. (a)
297. (d)
298. (b)
299. (b)
300. (b)
301. (a)
302. (c)
303. (c)
304. (c)
305. (b)
306. (b)
307. (b)
308. (a)
309. (b)
310. (c)
311. (b)
312. (d)
313. (a)
314. (d)
315. (a)
316. (b)
317. (b)
318. (b)
319. (b)
320. (b)
321. (c)
322. (a)
323. (a)
324. (b)
325. (d)
326. (c)
327. (d)
328. (a)
329. (c)
330. (b)
331. (d)
332. (c)
333. (a)
334. (a)
335. (b)
336. (c)

GK GS

Radha Mohan Cup is related to Polo.

337. (c)
338. (a)
339. (c)
340. (c)
341. (b)
342. (a)
343. (a)
344. (c)
345. (a)
346. (b)
347. (a)
348. (b)
349. (b)
350. (a)
351. (c)
352. (a)
353. (c)
354. (a)
355. (b)
356. (a)
357. (c)
358. (a)
359. (b)
360. (a)
361. (b)
362. (a)
363. (b)
364. (b)
365. (b)
366. (c)
367. (b)
368. (c)
369. (c)
370. (c)
371. (d)
372. (c)
373. (d)
374. (a)
375. (b)
376. (c)
377. (c)
378. (b)
379. (a)
380. (c)
381. (c)
382. (c)
383. (b)
384. (a)
385. (c)
386. (a)
387. (c)
388. (d)
389. (c)
390. (b)
391. (b)
392. (c)
393. (d)
394. (b)
395. (c)
396. (d)
397. (c)
398. (b)
399. (a)
400. (c)
401. (a)
402. (d)
403. (b)
404. (d)
405. (a)
406. (c)
407. (c)
408. (b)
409. (b)
410. (d)
411. (c)
412. (a)
413. (c)
414. (b)
415. (b)
416. (d)
417. (c)
418. (c)
419. (b)
420. (d)
421. (b)
422. (c)
423. (c)
424. (c)
425. (a)
426. (a)
427. (b)
428. (a)
429. (a)
430. (d)
431. (b)
432. (c)
433. (a)
434. (c)
435. (c)
436. (b)
437. (a)
438. (b)
439. (c)
440. (d)
441. (d)
442. (c)
443. (b)
444. (b)
445. (d)
446. (a)
447. (d)
448. (b)
449. (a)
450. (b)
451. (b)
452. (c)
453. (d)
454. (c)
455. (c)
456. (c)
457. (b)
458. (c)
459. (a)
460. (a)
461. (c)
462. (b)
463. (d)
464. (c)
465. (a)
466. (a)
467. (c)
468. (d)
469. (a)
470. (a)
471. (d)
472. (a)
473. (b)
474. (a)
475. (b)
476. (a)
477. (d)
478. (d)
479. (a)
480. (b)
481. (c)
482. (b)
483. (a)
484. (a)
485. (b)
486. (a)
487. (a)
488. (d)
489. (d)
490. (a)
491. (a)
492. (d)
493. (c)
494. (a)
495. (c)
496. (a)
497. (a)
498. (c)
499. (c)
500. (b)
501. (c)
502. (b)
503. (c)
504. (b)
505. (b)
506. (b)
507. (b)
508. (b)
509. (d)
510. (b)
511. (b)
512. (a)
513. (b)
514. (b)
515. (b)
516. (b)
517. (a)
518. (a)
519. (a)
520. (b)
521. (b)
522. (d)
523. (c)
524. (a)
525. (a)
526. (a)
527. (c)
528. (a)
529. (c)
530. (a)
531. (a)
532. (b)
533. (d)
534. (c)
535. (a)
536. (c)
537. (c)
538. (c)
539. (c)
540. (c)
541. (b)
542. (b)
543. (a)
544. (b)
545. (d)
546. (c)
547. (c)
548. (c)
549. (b)
550. (a)
551. (c)
552. (c)
553. (a)
554. (d)
555. (d)
556. (b)
557. (c)
558. (d)
559. (a)
560. (b)
561. (b)
562. (a)
563. (d)
564. (b)
565. (c)
566. (d)
567. (b)
568. (c)
569. (c)
570. (c)
571. (c)
572. (d)
573. (d)
574. (a)
575. (b)
576. (c)
577. (c)
578. (b)
579. (c)
580. (d)
581. (c)
582. (a)
583. (a)
584. (b)
585. (c)
586. (a)
587. (a)
588. (c)
589. (a)
590. (b)
591. (c)
592. (b)
593. (c)
594. (b)
595. (b)
596. (d)
597. (a)
598. (c)
599. (c)
600. (a)
601. (c)
602. (b)
603. (a)
604. (d)
605. (d)
606. (c)
607. (a)
608. (c)
609. (c)
610. (d)
611. (d)
612. (b)
613. (a)
614. (c)
615. (b)
616. (c)
617. (d)
618. (b)
619. (b)
620. (b)
621. (b)
622. (a)
623. (c)
624. (c)
625. (a)
626. (b)
627. (a)
628. (d)
629. (a)
630. (b)
631. (b)
632. (b)
633. (a)
634. (a)
635. (d)
636. (d)
637. (a)
638. (b)
639. (a)
640. (c)
641. (d)
642. (c)
643. (a)
644. (a)
645. (a)
646. (b)
647. (c)
648. (b)
649. (b)
650. (a)
651. (a)
652. (c)
653. (d)
654. (a)
655. (c)
656. (b)
657. (c)
658. (d)
659. (c)
660. (a)
661. (c)
662. (b)
663. (d).

664. (a)
665. (b)
666. (c)
667. (b)
668. (b)
669. (b)
670. (a)
671. (a)
672. (c)
673. (b)
674. (d)
675. (a)
676. (d)
677. (a)
678. (a)
679. (b)
680. (c)
681. (c)
682. (a)
683. (d)
684. (b)
685. (c)
686. (a)
687. (c)
688. (b)
689. (c)
690. (a)
691. (b)
692. (b)
693. (d)
694. (c)
695. (b)
696. (a)
697. (b)
698. (d)
699. (b)

GK GS

Zend Avesta is the religious book of the Parsis.

700. (a)
701. (b)
702. (b)
703. (b)
704. (b)
705. (a)
706. (c)
707. (a)
708. (a)
709. (a)
710. (a)
711. (b)
712. (a)
713. (b)
714. (c)
715. (c)
716. (b)
717. (c)
718. (d)
719. (b)
720. (a)
721. (b)
722. (d)
723. (a)
724. (b)
725. (b)
726. (b)
727. (a)
728. (d)
729. (b)
730. (b)
731. (c)
732. (b)
733. (a)
734. (a)
735. (b)
736. (c)
737. (b)
738. (a)
739. (b)
740. (b)
741. (c)
742. (b)
743. (c)
744. (b)
745. (c)
746. (a)
747. (b)
748. (d)
749. (c)
750. (c)
751. (b)
752. (b)
753. (c)
754. (d)
755. (a)
756. (a)
757. (b)
758. (d)
759. (c)
760. (b)
761. (b)
762. (d)
763. (a)
764. (c)
765. (c)
766. (d)
767. (d)
768. (a)
769. (d)
770. (a)
771. (d)
772. (d)
773. (c)
774. (d)
775. (a)
776. (b)
777. (c)
778. (b)
779. (c)
780. (a)
781. (b)
782. (d)
783. (b)
784. (a)
785. (c)
786. (b)
787. (c)
788. (a)
789. (c)
790. (b)
791. (b)
792. (d)
793. (d)
794. (c)
795. (c)
796. (b)
797. (d)
798. (b)
799. (d)
800. (c)
801. (a)
802. (b)
803. (c)
804. (a)
805. (b)
806. (a)
807. (b)
808. (c)
809. (b)
810. (c)
811. (c)
812. (a)
813. (b)
814. (c)
815. (a)
816. (d)
817. (b)
818. (b)
819. (a)
820. (d)
821. (a)
822. (c)
823. (c)
824. (a)
825. (c)
826. (b)
827. (a)
828. (b)
829. (b)
830. (c)
831. (c)
832. (b)
833. (d)
834. (a)
835. (b)
836. (c)
837. (c)
838. (a)
839. (d)
840. (c)
841. (b)
842. (b)
843. (c)
844. (c)
845. (a)
846. (b)
847. (c)
848. (b)
849. (b)
850. (b)
851. (a)
852. (b)
853. (b)
854. (a)
855. (a)
856. (c)
857. (a)
858. (c)
859. (b)
860. (c)
861. (b)
862. (c)
863. (b)
864. (b)
865. (a)
866. (c)
867. (a)
868. (b)
869. (d)
870. (b)
871. (b)
872. (c)
873. (c)
874. (a)
875. (d)
876. (b)
877. (a)
878. (a)
879. (b)
880. (b)
881. (c)
882. (b)
883. (d)
884. (d)
885. (c)
886. (d)
887. (b)
888. (b)
889. (c)
890. (b)
891. (d)
892. (c)
893. (c)
894. (b)
895. (d)
896. (c)
897. (c)
898. (d)
899. (c)
900. (b)
901. (d)
902. (b)
903. (d)
904. (c)
905. (b)
906. (a)
907. (b)
908. (a)
909. (b)
910. (a)
911. (b)
912. (d)
913. (d)
914. (a)
915. (c)
916. (b)
917. (d)
918. (c)
919. (d)
920. (c)
921. (a)
922. (c)
923. (c)
924. (d)
925. (d)
926. (c)
927. (a)
928. (c)
929. (a)
930. (b)
931. (d)
932. (a)
933. (b)
934. (b)
935. (c)
936. (d)
937. (d)
938. (c)
939. (b)
940. (d)
941. (d)
942. (a)
943. (d)
944. (d)
945. (a)
946. (d)
947. (d)
948. (d)
949. (b)
950. (b)
951. (b)
952. (c)
953. (d)
954. (c)
955. (c)
956. (d)
957. (c)
958. (d)
959. (a)
960. (b)
961. (b)
962. (c)
963. (c)
964. (d)
965. (c)
966. (d)
967. (c)
968. (d)
969. (a)
970. (b)
971. (b)
972. (c)
973. (a)
974. (d)
975. (c)
976. (b)
977. (d)
978. (c)
979. (b)
980. (d)
981. (b)
982. (d)
983. (c)
984. (a)
985. (a)
986. (d)
987. (a)
988. (a)
989. (b)
990. (a)
991. (c)
992. (a)
993. (d)
994. (a)
995. (a)
996. (c)
997. (a)
998. (c)
999. (a)
1000. (c)
1001. (c)
1002. (b)
1003. (a)
1004. (b)
1005. (c)
1006. (b)
1007. (c)
1008. (b)
1009. (b)
1010. (b)
1011. (b)
1012. (c)
1013. (b)
1014. (a)
1015. (b)
1016. (c)
1017. (c)
1018. (d)
1019. (b)
1020. (c)
1021. (d)
1022. (b)
1023. (c)
1024. (a)
1025. (a)
1026. (b)
1027. (a)
1028. (a)
1029. (b)
1030. (b)
1031. (c)
1032. (c)
1033. (d)
1034. (b)
1035. (d)
1036. (b)
1037. (c)
1038. (b)
1039. (c)
1040. (b)
1041. (c)
1042. (b)
1043. (b)
1044. (b)
1045. (b)
1046. (c)
1047. (d)
1048. (c)
1049. (b)
1050. (c)
1051. (b)
1052. (d)
1053. (d)
1054. (a)
1055. (a)
1056. (b)
1057. (b)
1058. (b)
1059. (a)
1060. (c)
1061. (d)
1062. (b)
1063. (c)

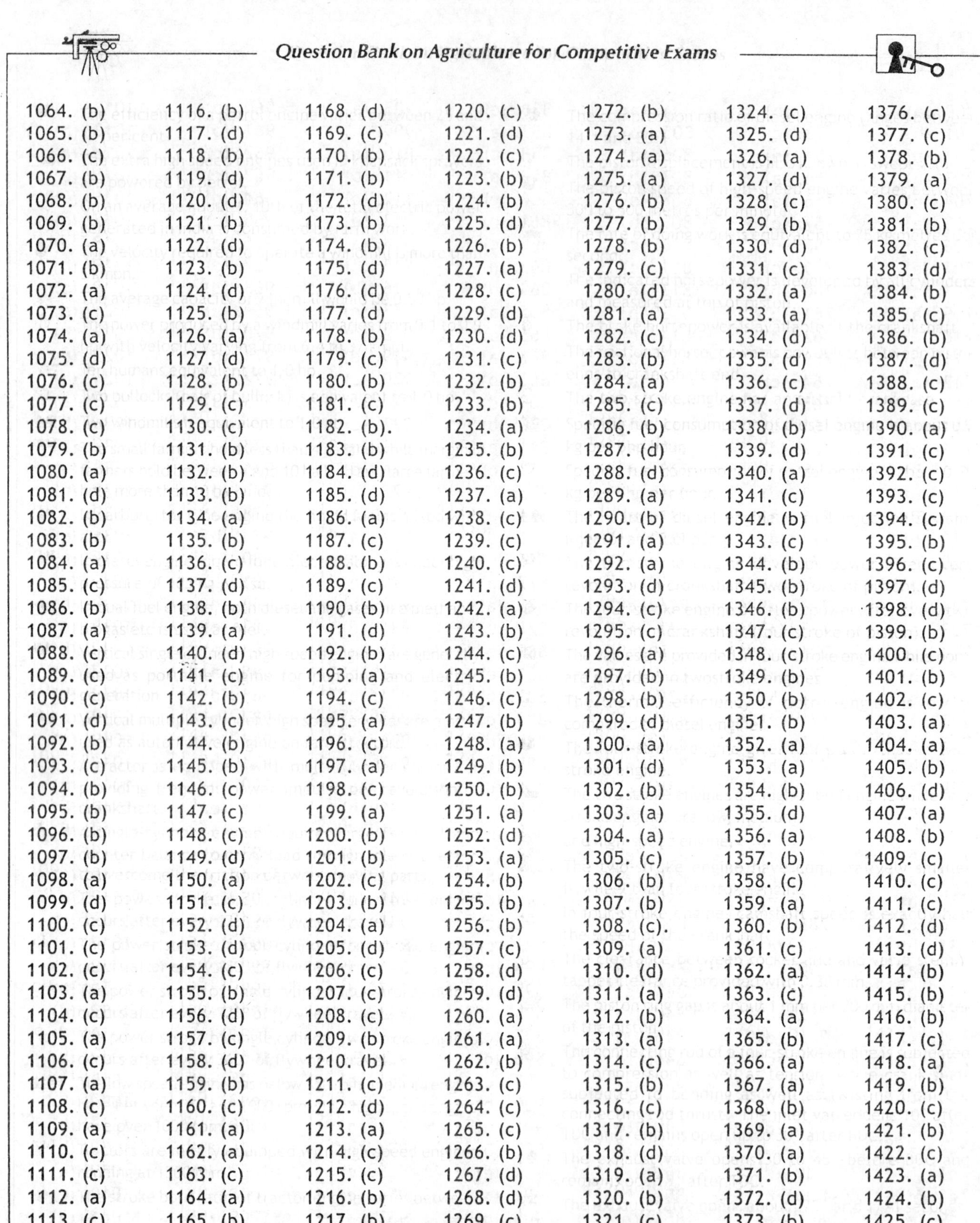

1064. (b)	1116. (b)	1168. (d)	1220. (c)	1272. (b)	1324. (c)	1376. (c)
1065. (b)	1117. (d)	1169. (c)	1221. (d)	1273. (a)	1325. (d)	1377. (c)
1066. (c)	1118. (b)	1170. (b)	1222. (c)	1274. (a)	1326. (a)	1378. (b)
1067. (b)	1119. (d)	1171. (b)	1223. (b)	1275. (a)	1327. (d)	1379. (a)
1068. (b)	1120. (d)	1172. (d)	1224. (b)	1276. (b)	1328. (c)	1380. (c)
1069. (b)	1121. (b)	1173. (b)	1225. (d)	1277. (b)	1329. (b)	1381. (b)
1070. (a)	1122. (d)	1174. (b)	1226. (b)	1278. (b)	1330. (d)	1382. (c)
1071. (b)	1123. (b)	1175. (d)	1227. (a)	1279. (c)	1331. (c)	1383. (d)
1072. (a)	1124. (d)	1176. (d)	1228. (c)	1280. (d)	1332. (a)	1384. (b)
1073. (c)	1125. (b)	1177. (d)	1229. (d)	1281. (a)	1333. (a)	1385. (c)
1074. (a)	1126. (a)	1178. (b)	1230. (d)	1282. (c)	1334. (d)	1386. (b)
1075. (d)	1127. (d)	1179. (c)	1231. (c)	1283. (b)	1335. (d)	1387. (d)
1076. (c)	1128. (b)	1180. (b)	1232. (b)	1284. (a)	1336. (c)	1388. (c)
1077. (c)	1129. (c)	1181. (c)	1233. (b)	1285. (c)	1337. (d)	1389. (c)
1078. (c)	1130. (c)	1182. (b).	1234. (a)	1286. (b)	1338. (b)	1390. (a)
1079. (b)	1131. (b)	1183. (b)	1235. (b)	1287. (d)	1339. (d)	1391. (c)
1080. (a)	1132. (b)	1184. (d)	1236. (c)	1288. (d)	1340. (a)	1392. (c)
1081. (d)	1133. (b)	1185. (d)	1237. (a)	1289. (a)	1341. (c)	1393. (c)
1082. (b)	1134. (a)	1186. (a)	1238. (c)	1290. (b)	1342. (b)	1394. (c)
1083. (b)	1135. (b)	1187. (c)	1239. (c)	1291. (a)	1343. (c)	1395. (b)
1084. (a)	1136. (c)	1188. (b)	1240. (c)	1292. (a)	1344. (b)	1396. (c)
1085. (c)	1137. (d)	1189. (c)	1241. (d)	1293. (d)	1345. (b)	1397. (d)
1086. (b)	1138. (b)	1190. (b)	1242. (a)	1294. (c)	1346. (b)	1398. (d)
1087. (a)	1139. (a)	1191. (d)	1243. (b)	1295. (c)	1347. (c)	1399. (b)
1088. (c)	1140. (d)	1192. (b)	1244. (c)	1296. (a)	1348. (c)	1400. (c)
1089. (c)	1141. (c)	1193. (a)	1245. (b)	1297. (b)	1349. (b)	1401. (b)
1090. (c)	1142. (b)	1194. (c)	1246. (c)	1298. (b)	1350. (b)	1402. (c)
1091. (d)	1143. (c)	1195. (a)	1247. (b)	1299. (d)	1351. (b)	1403. (a)
1092. (b)	1144. (b)	1196. (c)	1248. (a)	1300. (a)	1352. (a)	1404. (a)
1093. (c)	1145. (b)	1197. (a)	1249. (b)	1301. (d)	1353. (a)	1405. (b)
1094. (b)	1146. (c)	1198. (c)	1250. (b)	1302. (b)	1354. (b)	1406. (d)
1095. (b)	1147. (c)	1199. (a)	1251. (a)	1303. (a)	1355. (d)	1407. (a)
1096. (b)	1148. (c)	1200. (b)	1252. (d)	1304. (a)	1356. (a)	1408. (b)
1097. (b)	1149. (d)	1201. (b)	1253. (a)	1305. (c)	1357. (b)	1409. (c)
1098. (a)	1150. (a)	1202. (c)	1254. (b)	1306. (b)	1358. (c)	1410. (c)
1099. (d)	1151. (c)	1203. (b)	1255. (b)	1307. (b)	1359. (a)	1411. (c)
1100. (c)	1152. (d)	1204. (a)	1256. (b)	1308. (c).	1360. (b)	1412. (d)
1101. (c)	1153. (d)	1205. (d)	1257. (c)	1309. (a)	1361. (c)	1413. (d)
1102. (c)	1154. (c)	1206. (c)	1258. (d)	1310. (d)	1362. (a)	1414. (b)
1103. (a)	1155. (b)	1207. (c)	1259. (d)	1311. (a)	1363. (c)	1415. (b)
1104. (c)	1156. (d)	1208. (c)	1260. (a)	1312. (b)	1364. (a)	1416. (b)
1105. (a)	1157. (c)	1209. (b)	1261. (a)	1313. (a)	1365. (b)	1417. (c)
1106. (c)	1158. (d)	1210. (b)	1262. (b)	1314. (c)	1366. (a)	1418. (a)
1107. (a)	1159. (b)	1211. (c)	1263. (c)	1315. (b)	1367. (a)	1419. (b)
1108. (c)	1160. (c)	1212. (d)	1264. (c)	1316. (c)	1368. (b)	1420. (c)
1109. (a)	1161. (a)	1213. (a)	1265. (c)	1317. (b)	1369. (a)	1421. (b)
1110. (c)	1162. (a)	1214. (c)	1266. (b)	1318. (d)	1370. (a)	1422. (c)
1111. (c)	1163. (c)	1215. (c)	1267. (d)	1319. (a)	1371. (b)	1423. (a)
1112. (a)	1164. (a)	1216. (b)	1268. (d)	1320. (b)	1372. (d)	1424. (b)
1113. (c)	1165. (b)	1217. (b)	1269. (c)	1321. (c)	1373. (b)	1425. (c)
1114. (d)	1166. (c)	1218. (b)	1270. (c)	1322. (b)	1374. (d)	1426. (b)
1115. (c)	1167. (d)	1219. (b)	1271. (d)	1323. (c)	1375. (a)	1427. (d)

GK

Pablo Picasso was a Spanish painter, sculptor and ceremacist.

GS

Agricultural Engineering : Matching

	a	b	c	d	e	f	g	h	i	j	k	l	m	n	o
1.	iv	vii	v	i	ix	x	ii	iii	vi	viii					
2.	iii	x	ix	iv	v	i	ii	viii	vii	vi					
3.	x	ix	vi	viii	ii	vii	v	iv	iii	i					
4.	vi	iv	i	vii	viii	x	v	ii	iii	ix					
5.	vii	viii	iv	v	iii	ii	i	x	vi	ix					
6.	iv	ix	v	viii	vi	vii	x	i	iii	ii					
7.	v	iv	iii	i	ii	vii	viii	x	vi	ix					
8.	vii	ix	viii	iii	v	iv	x	vi	i	ii					
9.	ii	vi	iii	v	iv	viii	vii	x	ix	i					
10.	ix	vii	v	iv	vi	viii	iii	x	i	ii					
11.	ii	v	iii	vii	iv	vi	viii	x	i	ix					
12.	viii	ix	x	vii	vi	v	iv	iii	ii	i					
13.	vii	i	viii	vi	ix	iv	v	iii	x	ii					
14.	ii	iv	iii	vii	vi	v	viii	ix	x	i					
15.	v	ix	i	viii	ii	x	iv	vi	vii	iii					
16.	x	vi	i	vii	viii	ii	iii	iv	ix	v					
17.	viii	vi	x	i	vii	ix	iii	iv	ii	v					
18.	vi	viii	ix	ii	vii	iii	x	i	v	iv					
19.	v	vii	vi	x	viii	ii	xii	iii	xiii	i	iv	xi	ix		
20.	iv	viii	v	i	vi	iii	ii	ix	x	vii					
21.	v	iii	viii	x	iv	ii	ix	vii	i	vi					
22.	vi	v	i	ix	iii	viii	ii	iv	x	vii					
23.	iv	v	vii	ix	x	i	iii	ii	vi	viii					
24.	iii	iv	viii	i	x	x	vi	ii	vii	v					
25.	iv	viii	ix	x	vii	ii	i	iii	v	vi					
26.	v	iv	vii	i	ii	viii	iii	x	vi	ix					
27.	vi	iv	ix	i	ii	x	iii	viii	v	vii					
28.	ix	vi	x	v	vii	ii	iv	iii	i	viii					
29.	iv	vi	x	i	vii	ix	viii	iii	v	ii					
30.	viii	vi	x	ii	iv	i	iii	v							
31.	ii	v	iii	vii	i	iv	vi								
32.	iv	i	v	vi	ii	iii	viii	ix	x	vii					
33.	ii	v	viii	ix	vi	vii	i	x	iii	iv					
34.	iv	vii	viii	ix	vi	xii	v	i	xi	iii	x	ii			
35.	iv	v	vi	vii	viii	ix	ii	iii	x	i					
36.	iv	v	i	ii	iii	ix	vi	x	vii	viii					
37.	vi	vii	iv	v	ii	xi	x	iii	i	viii	ix				
38.	ix	x	vii	vi	viii	v	iv	iii	i	ii					
39.	v	vi	i	ii	iv	iii									
40.	iii	vii	vi	viii	vii	ii	i	iv							
41.	vi	i	iv	vii	iii	v	v								
42.	x	vi	vii	xii	xi	xiv	xiii	iii	i	viii	ii	iv	ix	xv	v
43.	iii	iv	vii	vi	ii	i	v	x	viii	ix					

Aristotle was a Greek teacher and philosopher.

1428. (a)
1429. (b)
1430. (b)
1431. (b)
1432. (b)
1433. (c)
1434. (c)
1435. (d)
1436. (c)
1437. (b)
1438. (a)
1439. (c)
1440. (d)
1441. (c)
1442. (a)
1443. (a)
1444. (b)
1445. (c)
1446. (c)
1447. (a)
1448. (b)
1449. (d)
1450. (a)
1451. (b)
1452. (d)
1453. (b)
1454. (a)
1455. (a)
1456. (c)
1457. (b)
1458. (c)
1459. (a)
1460. (b)
1461. (b)
1462. (b)
1463. (b)
1464. (d)
1465. (c)
1466. (b)
1467. (c)
1468. (b)
1469. (c)
1470. (b)
1471. (a)
1472. (a)
1473. (b)
1474. (c)
1475. (b)
1476. (a)
1477. (b)
1478. (c)
1479. (d)
1480. (b)
1481. (d)
1482. (b)
1483. (a)
1484. (d)
1485. (b)
1486. (b)
1487. (c)
1488. (c)
1489. (c)
1490. (a)
1491. (c)
1492. (c)
1493. (b)
1494. (c)
1495. (b)
1496. (b)
1497. (a)
1498. (b)
1499. (d)
1500. (c)
1501. (b)
1502. (a)
1503. (b)
1504. (b)
1505. (d)
1506. (c)
1507. (d)
1508. (c)
1509. (d)
1510. (c)
1511. (b)
1512. (b)
1513. (c)
1514. (c)
1515. (c)
1516. (b)
1517. (c)
1518. (b)
1519. (b)
1520. (c)
1521. (c)
1522. (b)
1523. (c)
1524. (d)
1525. (b)
1526. (b)
1527. (c)
1528. (b)
1529. (b)
1530. (d)
1531. (c)
1532. (a)
1533. (b)
1534. (a)
1535. (d)
1536. (c)
1537. (d)
1538. (b)
1539. (c)
1540. (d)
1541. (a)
1542. (b)
1543. (c)
1544. (d)
1545. (a)
1546. (a)
1547. (b)
1548. (a)
1549. (d)
1550. (c)
1551. (a)
1552. (d)
1553. (b)
1554. (b)
1555. (b)
1556. (c)
1557. (a)
1558. (b)
1559. (a)
1560. (a)
1561. (c)
1562. (a)
1563. (b)
1564. (b)
1565. (c)
1566. (d)
1567. (c)
1568. (c)
1569. (b)
1570. (c)
1571. (b)
1572. (b)
1573. (b)
1574. (a)
1575. (b)
1576. (c)
1577. (a)
1578. (c)
1579. (b)
1580. (b)
1581. (a)
1582. (c)
1583. (d)
1584. (b)
1585. (a)
1586. (c)
1587. (b)
1588. (b)
1589. (c)
1590. (c)
1591. (b)
1592. (c)
1593. (c)
1594. (b)
1595. (a)
1596. (b)
1597. (d)
1598. (d)
1599. (a)
1600. (b)
1601. (d)
1602. (b)
1603. (d)
1604. (c)
1605. (c)
1606. (a)
1607. (a)
1608. (a)
1609. (b)
1610. (b)
1611. (b)
1612. (d)
1613. (d)
1614. (b)
1615. (a)
1616. (a)
1617. (d)
1618. (c)
1619. (d)
1620. (d)
1621. (c)
1622. (c)
1623. (a)
1624. (d)
1625. (b)
1626. (c)
1627. (b)
1628. (a)
1629. (a)
1630. (b)
1631. (b)
1632. (d)
1633. (d)
1634. (a)
1635. (b)
1636. (b)
1637. (b)
1638. (d)
1639. (b)
1640. (c)
1641. (d)
1642. (b)
1643. (a)
1644. (a)
1645. (a)
1646. (c)
1647. (c)
1648. (d)
1649. (b)
1650. (c)
1651. (b)
1652. (c)
1653. (c)
1654. (a)
1655. (c)
1656. (c)
1657. (a)
1658. (a)
1659. (d)
1660. (b)
1661. (c)
1662. (a)
1663. (d)
1664. (c)
1665. (c)
1666. (a)
1667. (a)
1668. (d)
1669. (a)
1670. (b)
1671. (a)
1672. (a)
1673. (d)
1674. (b)
1675. (c)
1676. (a)
1677. (c)
1678. (b)
1679. (a)
1680. (b)
1681. (a)
1682. (d)
1683. (d)
1684. (a)
1685. (c)
1686. (b)
1687. (a)
1688. (a)
1689. (a)
1690. (b)
1691. (b)
1692. (b)
1693. (a)
1694. (b)
1695. (a)
1696. (c)
1697. (c)
1698. (b)
1699. (a)
1700. (c)
1701. (b)
1702. (c)
1703. (d)
1704. (a)
1705. (c)
1706. (a)
1707. (a)
1708. (d)
1709. (c)
1710. (a)
1711. (d)
1712. (c)
1713. (a)
1714. (a)
1715. (a)
1716. (a)
1717. (a)
1718. (a)
1719. (a)
1720. (c)
1721. (c)
1722. (a)
1723. (b)
1724. (c)
1725. (a)
1726. (b)
1727. (d)
1728. (a)
1729. (b)
1730. (a)
1731. (b)
1732. (b)
1733. (b)
1734. (b)
1735. (a)
1736. (a)
1737. (d)
1738. (c)
1739. (a)
1740. (c)
1741. (b)
1742. (b)
1743. (c)

	a	b	c	d	e	f	g	h	i	j	k	l	m	n	o
44.	iii	iv	v	ii	i	vii	vi								
45.	vi	iii	ix	viii	i	iv	v	vii	ii						
46.	v	viii	x	i	ii	ix	iii	vi	iv	vii					
47.	vi	x	i	ix	ii	v	viii	iii	iv	vii					
48.	xi	iv	ix	i	ii	xii	v	vii	iii	x	vi	viii			
49.	v	viii	ii	i	vii	iii	iv	ix	vi						
50.	xi	viii	i	ix	ii	x	iii	iv	vii	xiv	xiii	v	xii	xv	vi
51.	iv	i	v	ii	vi	iii									
52.	iv	ii	vii	vi	i	viii	iii	v							
53.	ii	v	vi	viii	iii	i	iv	ix	x	vii					
54.	iii	vi	i	ii	vii	iv	v								
55.	ix	viii	x	iii	i	ii	iv	vii	v	vi					
56.	iv	ix	v	vi	xi	i	x	iii	ii	xii	viii	vii			

15. Agricultural Statistics

Multiple Choice Questions

1. (a)
2. (a)
3. (a)
4. (b)
5. (a)
6. (c)
7. (b)
8. (c)
9. (b)
10. (c)
11. (c)
12. (d)
13. (c)
14. (a)
15. (a)
16. (c)
17. (b)
18. (a)
19. (c)
20. (b)
21. (b)
22. (c)
23. (a)
24. (c)
25. (a)
26. (a)
27. (c)
28. (b)
29. (c)
30. (b)
31. (c)
32. (c)
33. (a)
34. (c)
35. (c)
36. (c)
37. (d)
38. (a)
39. (a)
40. (b)
41. (b)
42. (b)
43. (b)
44. (d)
45. (a)
46. (a)
47. (a)
48. (a)
49. (a)
50. (d)
51. (b)
52. (d)
53. (c)
54. (b)

16. Seed Science

Multiple Choice Questions

1.	(c)	8.	(c)	15.	(c)	22.	(b)	29.	(c)	35.	(a)	41.	(b)
2.	(c)	9.	(a)	16.	(d)	23.	(b)	30.	(a)	36.	(c)	42.	(b)
3.	(b)	10.	(a)	17.	(d)	24.	(c)	31.	(b)	37.	(b)	43.	(a)
4.	(d)	11.	(a)	18.	(c)	25.	(a)	32.	(d)	38.	(c)	44.	(c)
5.	(d)	12.	(d)	19.	(a)	26.	(d)	33.	(a)	39.	(d)	45.	(a)
6.	(d)	13.	(b)	20.	(b)	27.	(b)	34.	(c)	40.	(b)		
7.	(c)	14.	(d)	21.	(b)	28.	(d)						

Seed Science : Matching

	a	b	c	d	e	f
1.	i	i	i	ii		
2.	ii	ii	ii	i		
3.	i	ii	iii	iv		
4.	i	ii	iii	iv		
5.	ii	i	iv	v	iii	vi
6.	ii	i	iv	iii		
7.	i	iv	ii	iii		
8.	ii	i	iv	iii	vi	v
9.	ii	i	iv	iii	v	

17. Ecology and Environment

Multiple Choice Questions

1. (d)	16. (b)	31. (a)	46. (d)	61. (b)	76. (a)	91. (c)
2. (d)	17. (d)	32. (c)	47. (c)	62. (b)	77. (b)	92. (d)
3. (d)	18. (b)	33. (b)	48. (c)	63. (d)	78. (b)	93. (b)
4. (c)	19. (c)	34. (c)	49. (b)	64. (c)	79. (c)	94. (c)
5. (d)	20. (b)	35. (c)	50. (b)	65. (a)	80. (c)	95. (b)
6. (b)	21. (b)	36. (d)	51. (c)	66. (a)	81. (a)	96. (b)
7. (b)	22. (c)	37. (b)	52. (d)	67. (b)	82. (d)	97. (a)
8. (c)	23. (b)	38. (d)	53. (c)	68. (a)	83. (b)	98. (a)
9. (c)	24. (a)	39. (a)	54. (b)	69. (a)	84. (c)	99. (b)
10. (b)	25. (a)	40. (a)	55. (c)	70. (b)	85. (a)	100. (b)
11. (c)	26. (c)	41. (b)	56. (b)	71. (d)	86. (d)	101. (b)
12. (b)	27. (b)	42. (a)	57. (c)	72. (b)	87. (d)	
13. (d)	28. (b)	43. (c)	58. (a)	73. (a)	88. (b)	
14. (d)	29. (c)	44. (a)	59. (b)	74. (c)	89. (a)	
15. (b)	30. (a)	45. (a)	60. (b)	75. (b)	90. (b)	

Ecology and Environment : Matching

	a	b	c	d
1.	iv	iii	ii	i
2.	ii	i	iv	iii
3.	iii	ii	i	iv
4.	iii	iv	ii	i
5.	iii	i	iv	ii
6.	i	ii	iii	iv
7.	i	iv	iii	ii
8.	i	ii	iii	iv
9.	iii	ii	i	iv
10.	ii	i	iv	iii
11.	i	ii	iii	iv
12.	iii	ii	i	iv
13.	i	iii	ii	iv
14.	iv	i	iii	ii
15.	iv	iii	ii	i

	a	b	c	d
16.	i	ii	iii	iv
17.	iv	iii	ii	i
18.	iv	ii	iii	i
19.	iii	iv	ii	i
20.	i	ii	iii	iv
21.	ii	i	iv	iii
22.	ii	iii	iv	i
23.	iv	iii	ii	i
24.	i	ii	iii	iv
25.	iii	ii	i	iv
26.	i	ii	iii	iv
27.	i	ii	iii	iv
28.	i	ii	iii	iv
29.	i	ii	iii	iv
30.	ii	i	iv	iii

19. Agri-sets for Zero Error in Agri-tests

SET-0

1. (c)
2. (d)
3. (b)
4. (a)
5. (d)
6. (c)
7. (b)
8. (d)
9. (b)
10. (d)
11. (d)
12. (b)
13. (b)
14. (d)
15. (c)
16. (a)
17. (b)
18. (a)
19. (a)
20. (c)
21. (d)
22. (c)
23. (a)
24. (c)
25. (a)
26. (c)
27. (c)
28. (b)
29. (a)
30. (a)
31. (b)
32. (c)
33. (d)
34. (b)
35. (a)
36. (c)
37. (d)
38. (d)
39. (d)
40. (b)
41. (d)
42. (c)
43. (a)
44. (d)
45. (a)
46. (d)
47. (d)
48. (d)
49. (b)
50. (c)
51. (a)
52. (c)
53. (a)
54. (d)
55. (d)
56. (b)
57. (c)
58. (b)
59. (c)
60. (d)
61. (d)
62. (d)
63. (a)
64. (c)
65. (a)
66. (b)
67. (c)
68. (b)
69. (c)
70. (c)
71. (c)
72. (b)
73. (c)
74. (c)
75. (b)
76. (c)
77. (c)
78. (b)
79. (a)
80. (d)
81. (b)
82. (d)
83. (b)
84. (c)
85. (c)
86. (a)
87. (a)
88. (b)
89. (c)
90. (b)
91. (b)
92. (c)
93. (d)
94. (b)
95. (a)
96. (a)
97. (b)
98. (a)
99. (d)
100. (a)
101. (b)
102. (c)
103. (a)
104. (a)
105. (d)
106. (c)
107. (b)
108. (d)
109. (a)
110. (d)
111. (a)
112. (c)
113. (d)
114. (d)
115. (c)
116. (d)
117. (d)
118. (a)
119. (b)
120. (b)

SET-1

1. (c)
2. (d)
3. (d)
4. (b)
5. (c)
6. (c)
7. (a)
8. (a)
9. (c)
10. (a)
11. (c)
12. (b)
13. (b)
14. (c)
15. (a)
16. (b)
17. (b)
18. (b)
19. (a)
20. (b)
21. (a)
22. (c)
23. (a)
24. (b)
25. (d)
26. (a)
27. (a)
28. (b)
29. (a)
30. (a)
31. (a)
32. (a)
33. (c)
34. (b)
35. (a)
36. (b)
37. (d)
38. (b)
39. (a)
40. (c)
41. (d)
42. (b)
43. (d)
44. (a)
45. (b)
46. (d)
47. (d)
48. (b)
49. (a)
50. (d)
51. (a)
52. (d)
53. (a)
54. (a)
55. (b)
56. (a)
57. (a)
58. (d)
59. (d)
60. (c)
61. (d)
62. (a)
63. (a)
64. (a)
65. (b)
66. (a)
67. (b)
68. (d)
69. (d)
70. (d)
71. (d)
72. (c)
73. (a)
74. (c)
75. (d)
76. (b)
77. (c)
78. (b)
79. (b)
80. (d)
81. (b)
82. (b)
83. (a)
84. (c)
85. (b)
86. (c)
87. (a)
88. (d)
89. (a)
90. (d)
91. (a)
92. (d)
93. (b)
94. (a)
95. (a)
96. (a)
97. (c)
98. (c)
99. (c)
100. (b)
101. (b)
102. (c)
103. (b)
104. (c)
105. (b)
106. (b)
107. (c)
108. (b)
109. (d)
110. (c)
111. (a)
112. (b)
113. (d)
114. (a)
115. (d)
116. (d)
117. (a)
118. (d)
119. (b)
120. (b)

SET-2

1. (a)
2. (c)
3. (d)
4. (d)
5. (b)
6. (d)
7. (c)
8. (b)
9. (a)
10. (c)
11. (d)
12. (a)
13. (c)
14. (a)
15. (b)
10. (a)
17. (b)
18. (d)
19. (b)
20. (b)
21. (b)
22. (c)
23. (d)
24. (a)
25. (b)
26. (c)
27. (c)
28. (c)
29. (d)
30. (d)
31. (b)
32. (c)
33. (a)
34. (c)
35. (b)
36. (c)
37. (d)
38. (a)
39. (b)
40. (c)
41. (b)
42. (c)
43. (a)
44. (d)
45. (d)
46. (c)
47. (a)
48. (b)
49. (d)
50. (c)
51. (c)
52. (b)
53. (c)
54. (c)
55. (d)
56. (c)
57. (b)
58. (d)
59. (b)
60. (a)
61. (d)
62. (d)
63. (d)
64. (b)
65. (c)
66. (c)
67. (a)
68. (c)
69. (a)
70. (d)
71. (b)
72. (c)
73. (c)
74. (a)
75. (c)
76. (a)
77. (a)
78. (d)
79. (b)
80. (a)
81. (a)
82. (a)
83. (b)
84. (c)
85. (a)
86. (a)
87. (a)
88. (b)
89. (a)
90. (b)
91. (b)
92. (a)
93. (d)
94. (c)
95. (a)
96. (b)
97. (b)

GK GS

Shakti Sthala is associated with Indira Gandhi

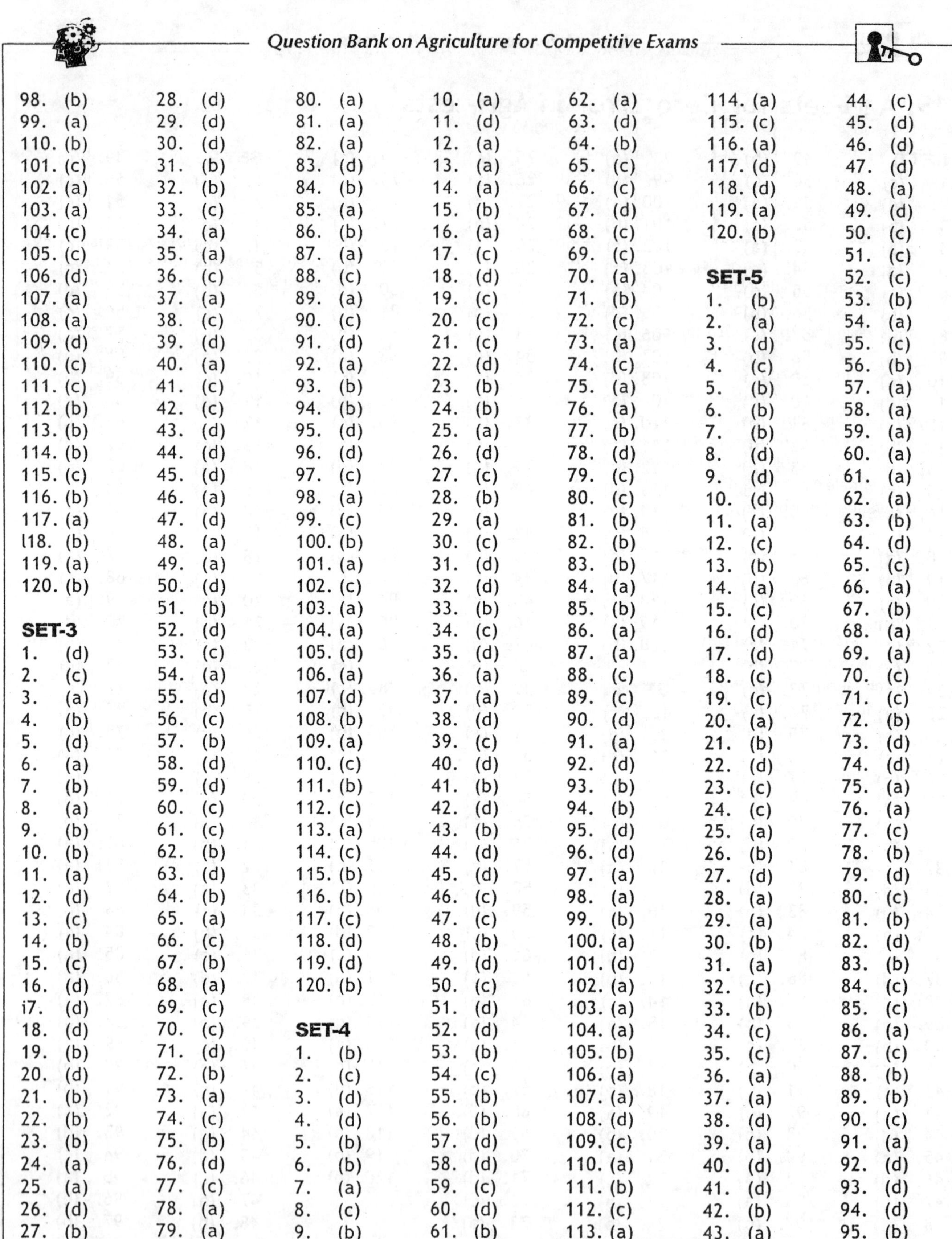

98. (b)
99. (a)
110. (b)
101. (c)
102. (a)
103. (a)
104. (c)
105. (c)
106. (d)
107. (a)
108. (a)
109. (d)
110. (c)
111. (c)
112. (b)
113. (b)
114. (b)
115. (c)
116. (b)
117. (a)
118. (b)
119. (a)
120. (b)

SET-3

1. (d)
2. (c)
3. (a)
4. (b)
5. (d)
6. (a)
7. (b)
8. (a)
9. (b)
10. (b)
11. (a)
12. (d)
13. (c)
14. (b)
15. (c)
16. (d)
17. (d)
18. (d)
19. (b)
20. (d)
21. (b)
22. (b)
23. (b)
24. (a)
25. (a)
26. (d)
27. (b)
28. (d)
29. (d)
30. (d)
31. (b)
32. (b)
33. (c)
34. (a)
35. (a)
36. (c)
37. (a)
38. (c)
39. (d)
40. (a)
41. (c)
42. (c)
43. (d)
44. (d)
45. (d)
46. (a)
47. (d)
48. (a)
49. (a)
50. (d)
51. (b)
52. (d)
53. (c)
54. (a)
55. (d)
56. (c)
57. (b)
58. (d)
59. (d)
60. (c)
61. (c)
62. (b)
63. (d)
64. (b)
65. (a)
66. (c)
67. (b)
68. (a)
69. (c)
70. (c)
71. (d)
72. (b)
73. (a)
74. (c)
75. (b)
76. (d)
77. (c)
78. (a)
79. (a)
80. (a)
81. (a)
82. (a)
83. (d)
84. (b)
85. (a)
86. (b)
87. (a)
88. (c)
89. (a)
90. (c)
91. (d)
92. (a)
93. (b)
94. (b)
95. (d)
96. (d)
97. (c)
98. (a)
99. (c)
100. (b)
101. (a)
102. (c)
103. (a)
104. (a)
105. (d)
106. (a)
107. (d)
108. (b)
109. (a)
110. (c)
111. (b)
112. (c)
113. (a)
114. (c)
115. (b)
116. (b)
117. (c)
118. (d)
119. (d)
120. (b)

SET-4

1. (b)
2. (c)
3. (d)
4. (d)
5. (b)
6. (b)
7. (a)
8. (c)
9. (b)
10. (a)
11. (d)
12. (a)
13. (a)
14. (a)
15. (b)
16. (a)
17. (c)
18. (d)
19. (c)
20. (c)
21. (c)
22. (d)
23. (b)
24. (b)
25. (a)
26. (d)
27. (c)
28. (b)
29. (a)
30. (c)
31. (d)
32. (d)
33. (b)
34. (c)
35. (d)
36. (a)
37. (a)
38. (d)
39. (c)
40. (d)
41. (b)
42. (d)
43. (b)
44. (d)
45. (d)
46. (c)
47. (c)
48. (b)
49. (d)
50. (c)
51. (d)
52. (d)
53. (b)
54. (c)
55. (b)
56. (b)
57. (d)
58. (d)
59. (c)
60. (d)
61. (b)
62. (a)
63. (d)
64. (b)
65. (c)
66. (c)
67. (d)
68. (c)
69. (c)
70. (a)
71. (b)
72. (c)
73. (a)
74. (c)
75. (a)
76. (a)
77. (b)
78. (d)
79. (c)
80. (c)
81. (b)
82. (b)
83. (b)
84. (a)
85. (b)
86. (a)
87. (a)
88. (c)
89. (c)
90. (d)
91. (a)
92. (d)
93. (b)
94. (b)
95. (d)
96. (d)
97. (a)
98. (a)
99. (b)
100. (a)
101. (d)
102. (a)
103. (a)
104. (a)
105. (b)
106. (a)
107. (a)
108. (d)
109. (c)
110. (a)
111. (b)
112. (c)
113. (a)
114. (a)
115. (c)
116. (a)
117. (d)
118. (d)
119. (a)
120. (b)

SET-5

1. (b)
2. (a)
3. (d)
4. (c)
5. (b)
6. (b)
7. (b)
8. (d)
9. (d)
10. (d)
11. (a)
12. (c)
13. (b)
14. (a)
15. (c)
16. (d)
17. (d)
18. (c)
19. (c)
20. (a)
21. (b)
22. (d)
23. (c)
24. (c)
25. (a)
26. (b)
27. (d)
28. (a)
29. (a)
30. (b)
31. (a)
32. (c)
33. (b)
34. (c)
35. (c)
36. (a)
37. (a)
38. (d)
39. (a)
40. (d)
41. (d)
42. (b)
43. (a)
44. (c)
45. (d)
46. (d)
47. (d)
48. (a)
49. (d)
50. (c)
51. (c)
52. (c)
53. (b)
54. (a)
55. (c)
56. (b)
57. (a)
58. (a)
59. (a)
60. (a)
61. (a)
62. (a)
63. (b)
64. (d)
65. (c)
66. (a)
67. (b)
68. (a)
69. (a)
70. (c)
71. (c)
72. (b)
73. (d)
74. (d)
75. (a)
76. (a)
77. (c)
78. (b)
79. (d)
80. (c)
81. (b)
82. (d)
83. (b)
84. (c)
85. (c)
86. (a)
87. (c)
88. (b)
89. (b)
90. (c)
91. (a)
92. (d)
93. (d)
94. (d)
95. (b)

96. (c)
97. (c)
98. (c)
99. (a)
100. (b)
101. (b)
102. (b)
103. (b)
104. (a)
105. (c)
106. (a)
107. (c)
108. (c)
109. (b)
110. (c)
111. (b)
112. (d)
113. (d)
114. (d)
115. (a)
116. (d)
1i7. (d)
118. (c)
119. (c)
120. (a)

SET-6

1. (d)
2. (d)
3. (b)
4. (b)
5. (c)
6. (a)
7. (d)
8. (d)
9. (d)
10. (a)
11. (c)
12. (b)
13. (d)
14. (b)
15. (d)
16. (d)
17. (b)
18. (c)
19. (d)
20. (b)
21. (a)
22. (b)
23. (b)
24. (d)
25. (a)
26. (b)
27. (c)
28. (c)
29. (d)
30. (d)
31. (c)
32. (b)
33. (a)
34. (b)
35. (d)
36. (c)
37. (c)
38. (a)
39. (c)
40. (a)
41. (a)
42. (c)
43. (c)
44. (d)
45. (a)
46. (a)
47. (a)
48. (b)
49. (c)
50. (a)
51. (b)
52. (a)
53. (d)
54. (c)
55. (c)
56. (c)
57. (a)
58. (b)
59. (b)
60. (b)
61. (a)
62. (a)
63. (a)
64. (c)
65. (c)
66. (a)
67. (a)
68. (a)
69. (a)
70. (d)
71. (d)
72. (b)
73. (c)
74. (c)
75. (c)
76. (c)
77. (b)
78. (a)
79. (a)
80. (d)
81. (b)
82. (d)
83. (d)
84. (a)
85. (d)
86. (b)
87. (c)
88. (a)
89. (b)
90. (a)
91. (b)
92. (d)
93. (a)
94. (d)
95. (b)
96. (d)
97. (a)
98. (b)
99. (b)
100. (d)
101. (a)
102. (a)
103. (c)
104. (a)
105. (c)
106. (a)
107. (b)
108. (c)
109. (c)
110. (b)
111. (c)
112. (d)
113. (d)
114. (c)
115. (c)
116. (d)
117. (a)
118. (c)
119. (a)
120. (c)

SET-7

1. (c)
2. (c)
3. (d)
4. (c)
5. (c)
6. (d)
7. (a)
8. (d)
9. (a)
10. (d)
11. (b)
12. (a)
13. (a)
14. (a)
15. (c)
16. (c)
17. (d)
18. (b)
19. (d)
20. (c)
21. (a)
22. (a)
23. (c)
24. (d)
25. (b)
26. (d)
27. (a)
28. (a)
29. (c)
30. (d)
31. (a)
32. (d)
33. (a)
34. (a)
35. (d)
36. (a)
37. (b)
38. (c)
39. (c)
40. (c)
41. (a)
42. (d)
43. (b)
44. (d)
45. (b)
46. (a)
47. (b)
48. (b)
49. (b)
50. (c)
51. (a)
52. (c)
53. (c)
54. (a)
55. (a)
56. (b)
57. (c)
58. (a)
59. (a)
60. (b)
61. (c)
62. (a)
63. (b)
64. (a)
65. (d)
66. (a)
67. (c)
68. (b)
69. (a)
70. (b)
71. (d)
72. (c)
73. (a)
74. (b)
75. (c)
76. (b)
77. (c)
78. (b)
79. (c)
80. (b)
81. (c)
82. (b)
83. (d)
84. (a)
85. (a)
86. (d)
87. (b)
88. (d)
89. (c)
90. (b)
91. (d)
92. (d)
93. (d)
94. (c)
95. (b)
96. (a)
97. (b)
98. (b)
99. (c)
100. (d)
101. (a)
102. (c)
103. (b)
104. (a)
105. (d)
106. (c)
107. (d)
108. (c)
109. (d)
110. (d)
111. (d)
112. (c)
113. (d)
114. (b)
115. (d)
116. (d)
117. (a)
118. (a)
119. (d)
120. (c)

SET-8

1. (b)
2. (d)
3. (a)
4. (c)
5. (c)
6. (b)
7. (c)
8. (a)
9. (c)
10. (d)
11. (a)
12. (b)
13. (a)
14. (c)
15. (c)
16. (a)
17. (c)
18. (b)
19. (b)
20. (d)
21. (a)
22. (c)
23. (d)
24. (d)
25. (c)
26. (c)
27. (b)
28. (a)
29. (a)
30. (a)
31. (b)
32. (d)
33. (c)
34. (a)
35. (c)
36. (b)
37. (c)
38. (c)
39. (b)
40. (d)
41. (d)
42. (b)
43. (a)
44. (c)
45. (c)
46. (d)
47. (c)
48. (b)
49. (c)
50. (b)
51. (a)
52. (c)
53. (b)
54. (d)
55. (c)
56. (d)
57. (b)
58. (c)
59. (c)
60. (d)
61. (d)
62. (a)
63. (b)
64. (c)
65. (a)
66. (c)
67. (b)
68. (c)
69. (c)
70. (c)
71. (c)
72. (d)
73. (c)
74. (c)
75. (d)
76. (c)
77. (a)
78. (b)
79. (b)
80. (b)
81. (d)
82. (d)
83. (d)
84. (a)
85. (a)
86. (c)
87. (b)
88. (b)
90. (a)
91. (b)
92. (b)
93. (a)
94. (a)

GK Newyork is situated on the banks of river Hudson GS

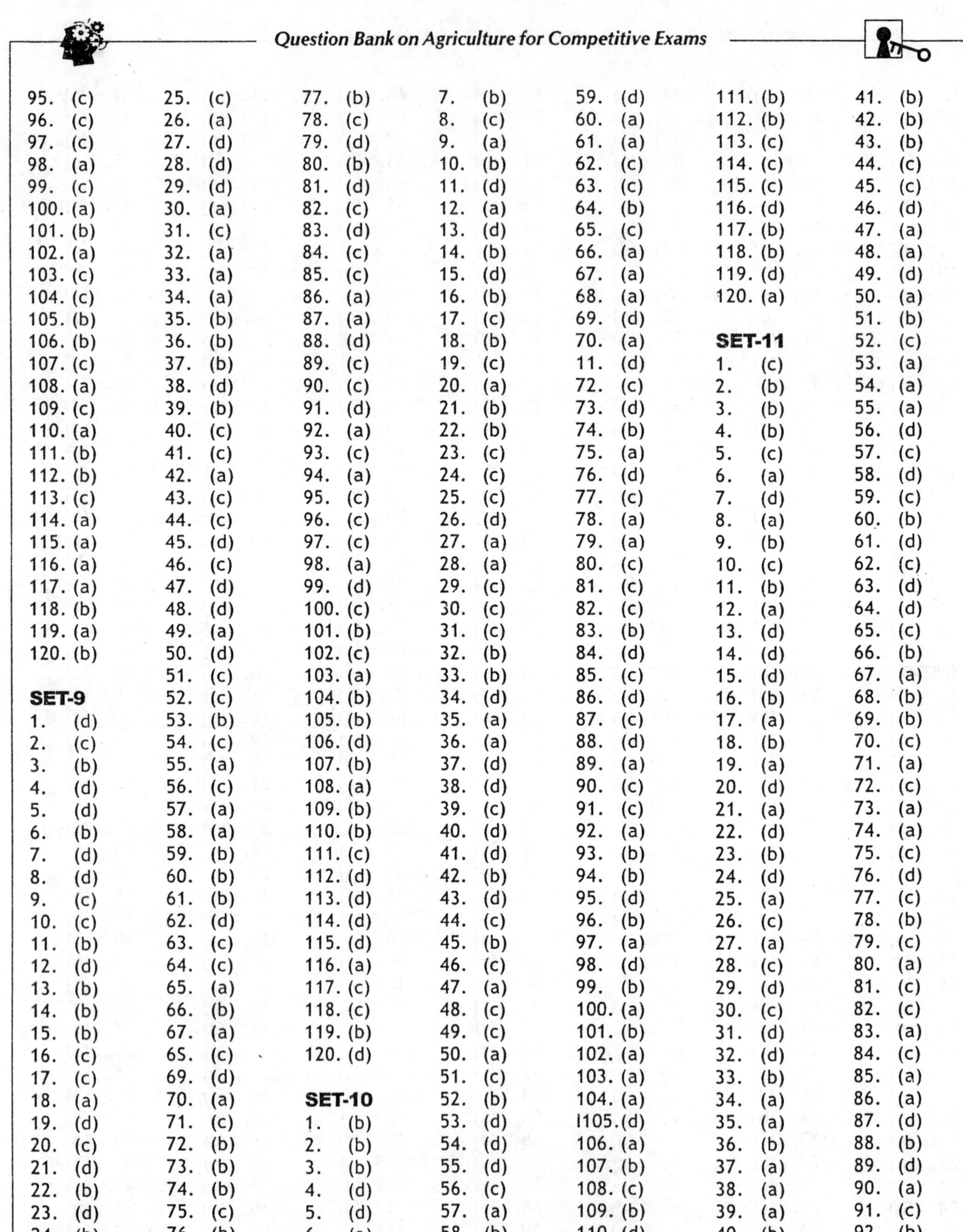

95. (c)
96. (c)
97. (c)
98. (a)
99. (c)
100. (a)
101. (b)
102. (a)
103. (c)
104. (c)
105. (b)
106. (b)
107. (c)
108. (a)
109. (c)
110. (a)
111. (b)
112. (b)
113. (c)
114. (a)
115. (a)
116. (a)
117. (a)
118. (b)
119. (a)
120. (b)

SET-9

1. (d)
2. (c)
3. (b)
4. (d)
5. (d)
6. (b)
7. (d)
8. (d)
9. (c)
10. (c)
11. (b)
12. (d)
13. (b)
14. (b)
15. (b)
16. (c)
17. (c)
18. (a)
19. (d)
20. (c)
21. (d)
22. (b)
23. (d)
24. (b)
25. (c)
26. (a)
27. (d)
28. (d)
29. (d)
30. (a)
31. (c)
32. (a)
33. (a)
34. (a)
35. (b)
36. (b)
37. (b)
38. (d)
39. (b)
40. (c)
41. (c)
42. (a)
43. (c)
44. (c)
45. (d)
46. (c)
47. (d)
48. (d)
49. (a)
50. (d)
51. (c)
52. (c)
53. (b)
54. (c)
55. (a)
56. (c)
57. (a)
58. (a)
59. (b)
60. (b)
61. (b)
62. (d)
63. (c)
64. (c)
65. (a)
66. (b)
67. (a)
6S. (c)
69. (d)
70. (a)
71. (c)
72. (b)
73. (b)
74. (b)
75. (c)
76. (b)
77. (b)
78. (c)
79. (d)
80. (b)
81. (d)
82. (c)
83. (d)
84. (c)
85. (c)
86. (a)
87. (a)
88. (d)
89. (c)
90. (c)
91. (d)
92. (a)
93. (c)
94. (a)
95. (c)
96. (c)
97. (c)
98. (a)
99. (d)
100. (c)
101. (b)
102. (c)
103. (a)
104. (b)
105. (b)
106. (d)
107. (b)
108. (a)
109. (b)
110. (b)
111. (c)
112. (d)
113. (d)
114. (d)
115. (d)
116. (a)
117. (c)
118. (c)
119. (b)
120. (d)

SET-10

1. (b)
2. (b)
3. (b)
4. (d)
5. (d)
6. (a)
7. (b)
8. (c)
9. (a)
10. (b)
11. (d)
12. (a)
13. (d)
14. (b)
15. (d)
16. (b)
17. (c)
18. (b)
19. (c)
20. (a)
21. (b)
22. (b)
23. (c)
24. (c)
25. (c)
26. (d)
27. (a)
28. (a)
29. (c)
30. (c)
31. (c)
32. (b)
33. (b)
34. (d)
35. (a)
36. (a)
37. (d)
38. (d)
39. (c)
40. (d)
41. (d)
42. (b)
43. (d)
44. (c)
45. (b)
46. (c)
47. (a)
48. (c)
49. (c)
50. (a)
51. (c)
52. (b)
53. (d)
54. (d)
55. (d)
56. (c)
57. (a)
58. (b)
59. (d)
60. (a)
61. (a)
62. (c)
63. (c)
64. (b)
65. (c)
66. (a)
67. (a)
68. (a)
69. (d)
70. (a)
11. (d)
72. (c)
73. (d)
74. (b)
75. (a)
76. (d)
77. (c)
78. (a)
79. (a)
80. (c)
81. (c)
82. (c)
83. (b)
84. (d)
85. (c)
86. (d)
87. (c)
88. (d)
89. (a)
90. (c)
91. (c)
92. (a)
93. (b)
94. (b)
95. (d)
96. (b)
97. (a)
98. (d)
99. (b)
100. (a)
101. (b)
102. (a)
103. (a)
104. (a)
I105. (d)
106. (a)
107. (b)
108. (c)
109. (b)
110. (d)
111. (b)
112. (b)
113. (c)
114. (c)
115. (c)
116. (d)
117. (b)
118. (b)
119. (d)
120. (a)

SET-11

1. (c)
2. (b)
3. (b)
4. (b)
5. (c)
6. (a)
7. (d)
8. (a)
9. (b)
10. (c)
11. (b)
12. (a)
13. (d)
14. (d)
15. (d)
16. (b)
17. (a)
18. (b)
19. (a)
20. (d)
21. (a)
22. (d)
23. (b)
24. (d)
25. (a)
26. (c)
27. (a)
28. (c)
29. (d)
30. (c)
31. (d)
32. (d)
33. (b)
34. (a)
35. (a)
36. (b)
37. (a)
38. (a)
39. (a)
40. (b)
41. (b)
42. (b)
43. (b)
44. (c)
45. (c)
46. (d)
47. (a)
48. (a)
49. (d)
50. (a)
51. (b)
52. (c)
53. (a)
54. (a)
55. (a)
56. (d)
57. (c)
58. (d)
59. (c)
60. (b)
61. (d)
62. (c)
63. (d)
64. (d)
65. (c)
66. (b)
67. (a)
68. (b)
69. (b)
70. (c)
71. (a)
72. (c)
73. (a)
74. (a)
75. (c)
76. (d)
77. (c)
78. (b)
79. (c)
80. (a)
81. (c)
82. (c)
83. (a)
84. (c)
85. (a)
86. (a)
87. (d)
88. (b)
89. (d)
90. (a)
91. (c)
92. (b)

GK Vijay Ghat is associated with Lal Bahadur Shastri GS

93. (a)
94. (a)
95. (a)
96. (c)
97. (b)
98. (b)
99. (c)
100. (d)

SET12

1. (c)
2. (b)
3. (b)
4. (b)
5. (c)
6. (a)
7. (d)
8. (a)
9. (b)
10. (c)
11. (b)
12. (a)
13. (d)
14. (d)
15. (d)
16. (b)
17. (a)
18. (b)
19. (a)
20. (d)
21. (a)
22. (d)
23. (b)
24. (d)
25. (a)
26. (c)
27. (a)
28. (c)
29. (d)
30. (c)
31. . (d)
32. (d)
33. (b)
34. (a)
35. (a)
36. (b)
37. (a)
38. (a)
39. (a)
40. (b)
41. (b)
42. (b)
43. (b)
44. (c)
45. (c)
46. (d)
47. (a)
48. (a)
49. (d)
50. (a)
51. (b)
52. (c)
53. (a)
54. (a)
55. (a)
56. (d)
57. (c)
58. (d)
59. (c)
60. (b)
61. (d)
62. (c)
63. (d)
64. (d)
65. (c)
66. (b)
67. (a)
68. (b)
69. (b)
70. (c)
71. (a)
72. (c)
73. (a)
74. (a)
75. (c)
76. (d)
77. (c)
78. (b)
79. (c)
80. (a)
81. (c)
82. (c)
83. (a)
84. (c)
85. (a)
86. (a)
87. (d)
88. (b)
89. (d)
90. (a)
91. (c)
92. (b)
93. (a)
94. (a)
95. (a)
96. (c)
97. (b)
98. (b)
99. (c)
100. (d)

SET13

1. (b)
2. (c)
3. (b)
4. (b)
5. (d)
6. (d)
7. (b)
8. (a)
9. (c)
10. (c)
11. (c)
12. (b)
13. (a)
14. (a)
15. (a)
16. (d)
17. (b)
18. (a)
19. (a)
20. (b)
21. (b)
22. (a)
23. (c)
24. (b)
25. (c)
26. (c)
27. (a)
28. (b)
29. (a)
30. (a)
31. (c)
32:(c)
33. (a)
34. (d)
35. (a)
36. (c)
37. (b)
38. (d)
39. (c)
40. (c)
41. (c)
42. (c)
43. (d)
44. (c)
45. (b)
46. (c)
47. (c)
48. (c)
49. (d)
50. (a)
51. (c)
52. (b)
53. (b)
54. (b)
55 (c)
56. (b)
57. (d)
58. (d)
59. (b)
60. (b)
61. (a)
62. (b)
63. (b)
64. (c)
65. (c)
66. (a)
67. (a)
68. (a)
69. (c)
70. (b)
71. (d)
72. (a)
73. (d)
74. (b)
75. (a)
76. (c)
77. (a)
78. (b)
79. (b)
80. (c)
81. (c)
82. (b)
83. (c)
84. (c)
85. (d)
86. (c)
87. (a)
88. (d)
89. (c)
90. (d)
91. (b)
92. (a)
93. (a)
94. (c)
95. (b)
96. (b)
97. (c)
98. (b)
99. (c)
100. (a)

SET14

1. (d)
2. (a)
3. (a)
4. (b)
5. (b)
6. (b)
7. (c)
8. (b)
9. (d)
10. (a)
11. (a)
12. (c)
13. (a)
14. (d)
15. (a)
16. (a)
17. (a)
18. (b)
19. (c)
20. (b)
21. (b)
22. (d)
23. (b)
24. (c)
25. (a)
26. (b)
27. (d)
28. (a)
29. (a)
30. (b)
31. (b)
32. (a)
33. (b)
34. (c)
35. (c)
36. (d)
37. (a)
38. (c)
39. (d)
40. (c)
41. (d)
42. (c)
43. (c)
44. (a)
45. (d)
46. (b)
47. (b)
48. (b)
49. (d)
50. (b)
51. (a)
52. (b)
53. (c)
54. (d)
55. (b)
56. (d)
57. (b)
58. (b)
59. (a)
60. (a)
61. (c)
62. (c)
63. (c)
64. (b)
65. (c)
66. (d)
67. (b)
68. (c)
69. (a)
70. (a)
71. (a)
72. (a)
73. (b)
74. (b)
75. (a)
76. (c)
77. (c)
78. (a)
79. (b)
80. (d)
81. (c)
82. (c)
83. (d)
84. (a)
85. (c)
86. (b)
87. (d)
88. (b)
89. (d)
90. (d)
91. (a)
92. (b)
93. (b)
94. (a)
95. (a)
96. (b)
97. (a)
98. (a)
99. (c)
100. (d)

SET15

1. (a)
2. (b)
3. (a)
4. (a)
5. (c)
6. (a)
7. (b)
8. (b)
9. (b)
10. (d)
11. (a)
12. (a)
13. (b)
14. (b)
15. (a)
16. (b)
17. (d)
18. (a)
19. (b)
20. (a)
21. (a)
22. (a)
23. (a)
24. (b)
25. (b)
26. (c)
27. (c)
28. (a)
29. (d)
30. (c)
31. (a)
32. (b)
33. (b)
34. (c)
35. (a)
36. (d)
37. (b)
38. (b)
39. (a)
40. (c)
41. (b)
42. (b)
43. (a)
44. (b)
45. (a)
46. (a)
47. (b)
48. (c)

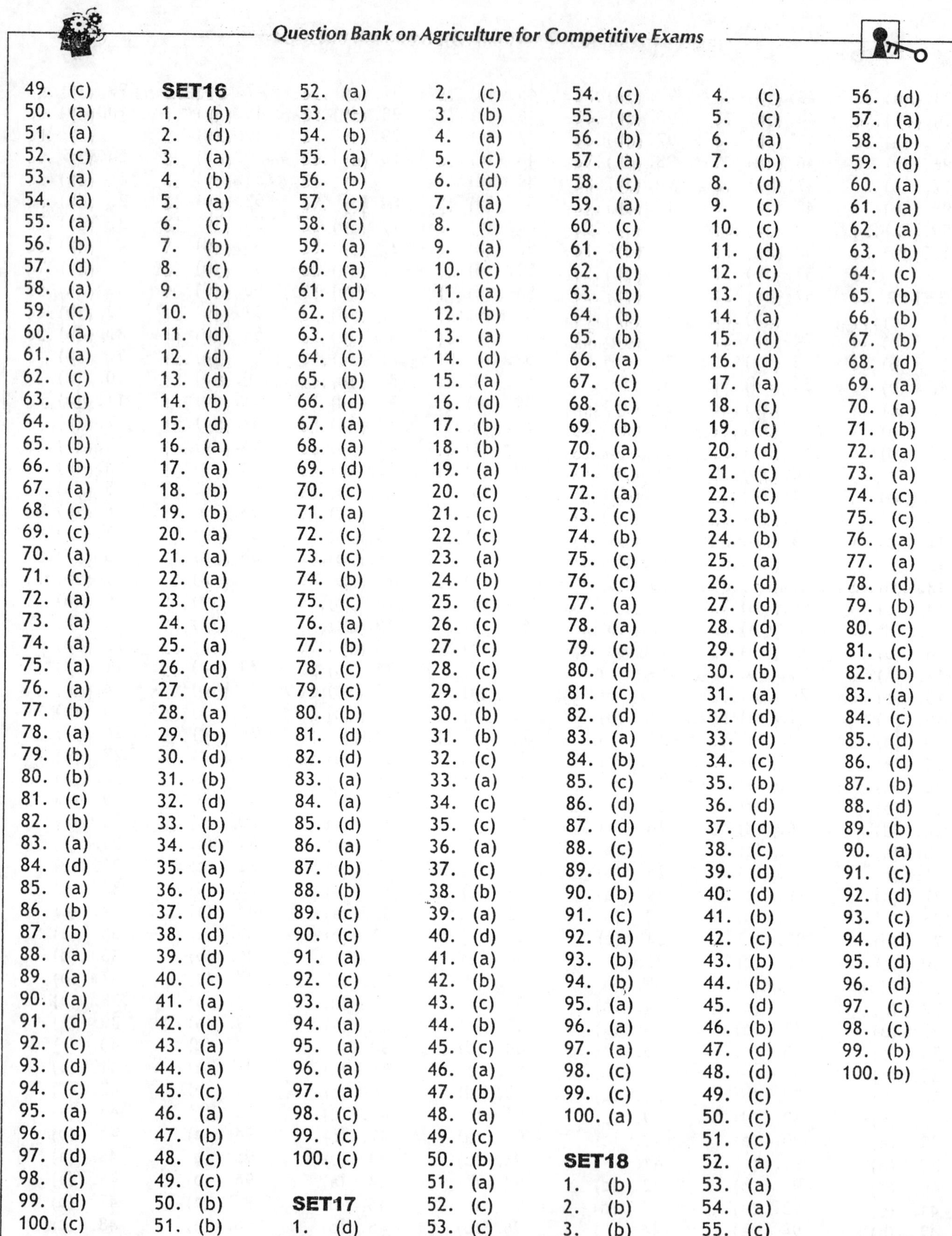

49. (c)
50. (a)
51. (a)
52. (c)
53. (a)
54. (a)
55. (a)
56. (b)
57. (d)
58. (a)
59. (c)
60. (a)
61. (a)
62. (c)
63. (c)
64. (b)
65. (b)
66. (b)
67. (a)
68. (c)
69. (c)
70. (a)
71. (c)
72. (a)
73. (a)
74. (a)
75. (a)
76. (a)
77. (b)
78. (a)
79. (b)
80. (b)
81. (c)
82. (b)
83. (a)
84. (d)
85. (a)
86. (b)
87. (b)
88. (a)
89. (a)
90. (a)
91. (d)
92. (c)
93. (d)
94. (c)
95. (a)
96. (d)
97. (d)
98. (c)
99. (d)
100. (c)

SET16

1. (b)
2. (d)
3. (a)
4. (b)
5. (a)
6. (c)
7. (b)
8. (c)
9. (b)
10. (b)
11. (d)
12. (d)
13. (d)
14. (b)
15. (d)
16. (a)
17. (a)
18. (b)
19. (b)
20. (a)
21. (a)
22. (a)
23. (c)
24. (c)
25. (a)
26. (d)
27. (c)
28. (a)
29. (b)
30. (d)
31. (b)
32. (d)
33. (b)
34. (c)
35. (a)
36. (b)
37. (d)
38. (d)
39. (d)
40. (c)
41. (a)
42. (d)
43. (a)
44. (a)
45. (c)
46. (a)
47. (b)
48. (c)
49. (c)
50. (b)
51. (b)
52. (a)
53. (c)
54. (b)
55. (a)
56. (b)
57. (c)
58. (c)
59. (a)
60. (a)
61. (d)
62. (c)
63. (c)
64. (c)
65. (b)
66. (d)
67. (a)
68. (a)
69. (d)
70. (c)
71. (a)
72. (c)
73. (c)
74. (b)
75. (c)
76. (a)
77. (b)
78. (c)
79. (c)
80. (b)
81. (d)
82. (d)
83. (a)
84. (a)
85. (d)
86. (a)
87. (b)
88. (b)
89. (c)
90. (c)
91. (a)
92. (c)
93. (a)
94. (a)
95. (a)
96. (a)
97. (a)
98. (c)
99. (c)
100. (c)

SET17

1. (d)
2. (c)
3. (b)
4. (a)
5. (c)
6. (d)
7. (a)
8. (c)
9. (a)
10. (c)
11. (a)
12. (b)
13. (a)
14. (d)
15. (a)
16. (d)
17. (b)
18. (b)
19. (a)
20. (c)
21. (c)
22. (c)
23. (a)
24. (b)
25. (c)
26. (c)
27. (c)
28. (c)
29. (c)
30. (b)
31. (b)
32. (c)
33. (a)
34. (c)
35. (c)
36. (a)
37. (c)
38. (b)
39. (a)
40. (d)
41. (a)
42. (b)
43. (c)
44. (b)
45. (c)
46. (a)
47. (b)
48. (a)
49. (c)
50. (b)
51. (a)
52. (c)
53. (c)
54. (c)
55. (c)
56. (b)
57. (a)
58. (c)
59. (a)
60. (c)
61. (b)
62. (b)
63. (b)
64. (b)
65. (b)
66. (a)
67. (c)
68. (c)
69. (b)
70. (a)
71. (c)
72. (a)
73. (c)
74. (b)
75. (c)
76. (c)
77. (a)
78. (a)
79. (c)
80. (d)
81. (c)
82. (d)
83. (a)
84. (b)
85. (c)
86. (d)
87. (d)
88. (c)
89. (d)
90. (b)
91. (c)
92. (a)
93. (b)
94. (b)
95. (a)
96. (a)
97. (a)
98. (c)
99. (c)
100. (a)

SET18

1. (b)
2. (b)
3. (b)
4. (c)
5. (c)
6. (a)
7. (b)
8. (d)
9. (c)
10. (c)
11. (d)
12. (c)
13. (d)
14. (a)
15. (d)
16. (d)
17. (a)
18. (c)
19. (c)
20. (d)
21. (c)
22. (c)
23. (b)
24. (b)
25. (a)
26. (d)
27. (d)
28. (d)
29. (d)
30. (b)
31. (a)
32. (d)
33. (d)
34. (c)
35. (b)
36. (d)
37. (d)
38. (c)
39. (d)
40. (d)
41. (b)
42. (c)
43. (b)
44. (b)
45. (d)
46. (b)
47. (d)
48. (d)
49. (c)
50. (c)
51. (c)
52. (a)
53. (a)
54. (a)
55. (c)
56. (d)
57. (a)
58. (b)
59. (d)
60. (a)
61. (a)
62. (a)
63. (b)
64. (c)
65. (b)
66. (b)
67. (b)
68. (d)
69. (a)
70. (a)
71. (b)
72. (a)
73. (a)
74. (c)
75. (c)
76. (a)
77. (a)
78. (d)
79. (b)
80. (c)
81. (c)
82. (b)
83. (a)
84. (c)
85. (d)
86. (d)
87. (b)
88. (d)
89. (b)
90. (a)
91. (c)
92. (d)
93. (c)
94. (d)
95. (d)
96. (d)
97. (c)
98. (c)
99. (b)
100. (b)

20. Evaluate Yourself

1. Dr.B. Vishvanath
2. Anton de Bary
3. P.M.A. Millardet
4. Synchitriem endobioticum
5. Azolla
6. 32,87263 sq. km.
7. 19.39%
8. India
9. West Bengal
10. Uttar Pradesh
11. CRI stage
12. 46%
13. Tomato
14. 143 million hectare
15. 193 million hectare
16. Mohan Singh Mehta Committee
17. Arrow
18. Lycopene
19. Barbadose cherry
20. Dr. Surinder K. Vasal
21. Paddy
22. 21-23%
23. 1963
24. 1966
25. Internally seed borne
26. Viability & Vigour
27. 12 July, 1982
28. Lysine
29. Economic yield/Biological yield x 100
30. Xanthomonas oryzae
31. Mycoplasma
32. Virus
33. Groundnut (33%)
34. Madhya Pradesh
35. Karnataka
36. Triticum durum
37. 20
38. Top 1/3rd to 1/2
39. Mesocarp
40. Methionine
41. Halophites
42. Ignite
43. C_4
44. Bacterial Leaf Blight
45. Mn
46. Cu
47. 25%
48. 19 million tones
49. William Gaude
50. 1935
51. Karnal
52. Phytophthora infestans
53. Abscisic acid
54. Auxin
55. 2.4-D
56. Amino acids
57. 16
58. Xylem
59. Prions
60. Cytokinin
61. 164
62. 16 July
63. Hyderabad
64. Watson & Crick
65. North East monsoon
66. Fungus
67. 314 (March, 2001)
68. 1999-2000 (Rabi)
69. Funk
70. Charles Darwin
71. Gregor Mendal
72. Barometer
73. Fungi & roots of higher plants
74. Hydroponics
75. ABA
76. $H_2C = CH_2$
77. Energy Spending
78. Sucrose
79. Potassium
80. Zn
81. MLO's
82. Yellow ear rot
83. Ustilago segetum tritici
84. Phosphorous
85. KCl
86. 1.5%
87. $\frac{\Delta Y_1}{\Delta X_1} = \frac{PX_1}{PY_1}$
88. Decreasing
89. Constant
90. Variable properties
91. Marginal utilities is equal to price ratio
92. Hicks
93. CAM

94. C_4 plants
95. Chlorophyll
96. Boron deficiency
97. Broad leaf
98. GA_3
99. 22.5 cm
100. Eastern India
101. 12-14 qt/ha
102. 12
103. Short day plant
104. 5 cm
105. 5 cm
106. 20-25%
107. 200-300 cm
108. 8-12 days
109. Gossipium hersutum
110. Black soil
111. Iodine
112. 0.5 add 1.4%
113. DAP
114. Wind erosion
115. Punjab
116. D-form
117. Phosphodiester bond
118. Rubisco
119. L-form
120. A+T, G+C, A+C
121. Azotobacter
122. Isohyet
123. South America
124. 3 cm.
125. 140%
126. IR -8 (1966)
127. Nitrogen
128. Nov. - Dec.
129. Non selective
130. Brassica juncea
131. Transpiration
132. 58%
133. Gossipium arboreum
134. Tensiometer
135. Azotobacter
136. TSS
137. Black cotton soils
138. L.S.D.
139. Arnon & Stout
140. Robert Brown
141. Meiosis only
142. Prophase
143. Pachytene
144. Diplotene
145. Back cross breeding
146. Pure line selection
147. Late blight of potato
148. Mycoplasma
149. Bastian
150. China
151. 12%
152. Alluvial soil
153. KCI
154. Groundnut
155. Ammonium nitrate
156. Polyploids
157. Village Extension Worker
158. Termite
159. Sandy soil
160. 450 g/ha.
161. Zn
162. Rice root nematode
163. Schistocerca gregaria
164. CIFE, Mumbai
165. Pune
166. Dr. Varghese Kurien
167. Secondary tillage implement
168. First
169. Plasmodiophara brassicae
170. Rhizoctonia solani
171. Louis Pasteur
172. Leeuwenhoek
173. Choti Lerma
174. Once time
175. Soil borne
176. White flies
177. Spotted boll worm
178. 8
179. Glyoxysomes
180. 5
181. Pheophytin
182. Metamorphic
183. 0.02 to 0.002 mm.
184. 31 or more
185. 1:1 type
186. Secondary
187. 10 : 1
188. Neutral
189. Lime
190. Germany
191. 4°C
192. Cucurbits
193. Flood method

GK

Akyab is situated on the banks of river Irrawady

GS

194. State Agricultural University (SAUs)
195. Two way flow of message
196. Mass selection
197. 0 to 1
198. Barley
199. GA_3
200. Root tips
201. Bt. transgenic plants
202. Tobacco
203. Breeding
204. Topiary
205. Amrapali
206. NAA Spray
207. Chloroplast
208. Oxidation-reduction process
209. IBA
210. Hips
211. 60-70%
212. Banana
213. C4 plants
214. Glyoxylate cycle
215. Single membrane organelle
216. Stroma
217. 9: 7
218. Anaphase
219. Prophase
220. Ribosomes
221. Dairy development
222. 15%
223. Bulk density
224. Tough
225. 10%
226. 1966
227. Royal Commission on Agriculture (1925)
228. Amrapali
229. Neelam x Dashari
230. Rajasthan
231. ICAR
232. R.B.D.
233. Poisson
234. Rice
235. Tamilnadu
236. Normal distribution
237. 0.10 ha.
238. Metamorphic rock
239. 18 months
240. Pachytene
241. Triticale
242. Wheat x Rye
243. Fe deficiency
244. Date palm
245. 4-6 pH
246. Rice
247. $CO(NH_2)_2$
248. 58-60%
249. Ethylene
250. CAM plants
251. Bacteria
252. Australia
253. Systemic organophosphate
254. Graminae
255. Bengal Gram
256. Village
257. Berry
258. Bateson
259. Gypsum
260. Lucknow
261. Italy, Rome
262. Soil texture
263. Urea
264. 8
265. Nitrobacter
266. Sulphur toxicity
267. Ammonium carbamate
268. Blue green algae
269. Nicholas
270. Uttar Pradesh
271. Siva Raman Committee, Narasimhan Committee
272. Albugo candida
273. B deficiency (Ph.D. 1999)
274. Atropin
275. 2n + 2
276. Tryptophan
277. Capsule
278. Vit. B_{12}
279. Azotobacter
280. Dee gee woo gene
281. Mexico
282. 4
283. 283
284. 1979
285. Karnal bunt
286. Sutton & Bovery
287. Ditylenchus sp. (Nematode)
288. Siliqua
289. 21%
290. Drift hazard
291. Vitavax
292. PGA
293. CH_4

294. Rose
295. Decreasing
296. 1960
297. F.L. Brayne
298. 57 m ha
299. Minimum Support Price
300. Mitochondria
301. 2.4%
302. Geneva
303. Rafi Ahmad Kidwai Award
304. 8.0%
305. Deemed University
306. Vermiculite
307. Less than 15%
308. 8-10%
309. 15 bar
310. Veneer grafting
311. Azolla
312. Triticum dicoccum
313. $\frac{\textbf{Total assets}}{\textbf{Total liabilities}}$
314. Cauliflower
315. Poorly drained soil
316. IIHR Bangalore
317. Edible oils
318. IADP
319. Mo
320. 193 m ha
321. Puccinia striformis
322. Sindhu
323. Onion
324. -1 to +1
325. Cauliflower
326. Hyderabad
327. 10%
328. Potassic
329. 22.2%
330. Mangala Rai
331. 12-14%
332. Super Rice
333. Maharashtra
334. 610
335. CACP
336. Imphal
337. Glutein
338. Jhansi
339. 21 days
340. Sandy soil
341. Lysosomes
342. CAN
343. -OC to +OC
344. Aeolian soil
345. Transcription
346. Ca^{2+}
347. Phosphatic fertilizers
348. Potassium
349. ICAR
350. Mango
351. 1.94 cal/cm^2/sec.
352. Thar desert
353. Semi root
354. Predator
355. Yellows stem borer
356. Uttar Pradesh
357. Ludhiana
358. Etawah
359. Groundnut
360. K^+
361. Nitrogen
362. SO_4^{-2}
363. 35,000-40,000
364. Drip irrigation
365. 4
366. Army worm
367. $CaSO_4.2H_2O$
368. Maize
369. Maize
370. Soyabean
371. Maize
372. 65%
373. 170 kg.
374. Range
375. Sulphur compound
376. 200- 300
377. Diffusion
378. N deficiency
379. Cellulose
380. 1/3 to 15 bar
381. Intercropping
382. 12
383. Quantitative short day plant
384. Proton
385. Lower leaves
386. KCl (60%)
387. 3 PPM conc.
388. Coefficient of variation
389. Phytophthora infestans
390. Photo-phosphorylation
391. Binneal
392. Tensiometer

GK **R.S. V.P. stands for Respondez S'il vous Plait (Reply if you please).** GS

393. Nitrosomonas
394. Soil tilth
395. Crop canopy temperature
396. Red rot
397. 70s type
398. Maharashtra
399. Wheat
400. Small gully
401. D.A.P.
402. West Bengal
403. Wheat
404. Prevent lodging
405. Milking to just maturing
406. 10 : 1
407. Saccahrum officinerum (Ph.D. 1999)
408. Devries, Correns, Tschermak
409. Broadcasting
410. Vitamin B
411. C4 plant
412. Blast of rice (Pyricularia Oryzae)
413. Clay soils
414. Gram
415. Wheat
416. Israel
417. 25%
418. South West Monsoon
419. Ganga
420. J. W. Leather
421. Buck Wheat
422. Mango
423. Cardamom
424. By Transpiration Pull
425. S.S.P.
426. 20%
427. ICAR
428. Photoperiodism
429. Quality of food product
430. <.002
431. NH_4 & NO_3
432. >15%
433. 2/3
434. Trap crop
435. Anemometer
436. Na
437. Relay cropping
438. O_2
439. Prills
440. 22-30°C
441. 85 gram/day
442. 45 gram fat
443. 35 -45
444. Virus
445. Rice
446. Sunflower
447. Cotton
448. ABA
449. 30%
450. Denitrification
451. Prime Minister
452. Warm tropical
453. Bone meal
454. Borlog
455. 6.5 -7.0 pH
456. H_2SO_4
457. S.S.P.
458. 5 May
459. 16 tonnes/ha/year
460. 60
461. Zr
462. N.C. Brady
463. CO_2
464. Larves & Gilbert
465. 12-14%
466. Tobacco
467. Bacteria
468. IRDP
469. Sugarcane
470. Nitrate reductase
471. Family
472. Vigna radiata
473. Sprinkler irrigation
474. Market Oriented
475. Thiourea
476. Mediterranean
477. Non symbiotically
478. Acidic soil
479. Growth retardant
480. C_3 plant
481. >13
482. Nitrate
483. 15
484. Breeder seed
485. Rajma
486. CRI, tillering, doughing stage
487. Stem
488. Basal placement at the time of sowing
489. Chi-square test
490. Multi storied cropping
491. Drip irrigation
492. Capillary water

493. Potassium
494. Air
495. C.M. Donald
496. Potassium
497. Milk
498. Institute Village Linkage Programme
499. Punjab
500. Zn
501. CH_4
502. Production of ATP
503. Trifolium alexandrium
504. Cotton
505. Red rot of sugarcane
506. Greenhouse effect
507. Zooplankatons
508. Adventitious root
509. Stolons
510. Maize
511. Ovule
512. Martynia
513. Floriculture
514. Sabour, Bhagalpur (Bihar)
515. Copper
516. Manganese
517. Sulphur
518. Nitrogen
519. Jaiselmer
520. West Bengal
521. Uttaranchal
522. Gujarat
523. West Bengal
524. 34

AGRISET - 0

ANSWER SHEET FOR GENERAL AGRICULTURE MODEL PAPER

- I have provided you MCQs model test on general agriculture for TOP (Target Oriented Preparation) for the Agricultural Competitive Exams.
- You have to mark all your responses only on the answer sheet provided.
- All items carry equal marks. Attempt all items. Your total marks will depend only on the number of correct responses marked by you in the Answer sheet. There is no negative marking. Award yourself one point for each correct answer.

ANSWER SHEET

1	ⓐⓑⓒⓓ	6	ⓐⓑⓒⓓ	11	ⓐⓑⓒⓓ	16	ⓐⓑⓒⓓ	21	ⓐⓑⓒⓓ	26	ⓐⓑⓒⓓ
2	ⓐⓑⓒⓓ	7	ⓐⓑⓒⓓ	12	ⓐⓑⓒⓓ	17	ⓐⓑⓒⓓ	22	ⓐⓑⓒⓓ	27	ⓐⓑⓒⓓ
3	ⓐⓑⓒⓓ	8	ⓐⓑⓒⓓ	13	ⓐⓑⓒⓓ	18	ⓐⓑⓒⓓ	23	ⓐⓑⓒⓓ	28	ⓐⓑⓒⓓ
4	ⓐⓑⓒⓓ	9	ⓐⓑⓒⓓ	14	ⓐⓑⓒⓓ	19	ⓐⓑⓒⓓ	24	ⓐⓑⓒⓓ	29	ⓐⓑⓒⓓ
5	ⓐⓑⓒⓓ	10	ⓐⓑⓒⓓ	15	ⓐⓑⓒⓓ	20	ⓐⓑⓒⓓ	25	ⓐⓑⓒⓓ	30	ⓐⓑⓒⓓ
31	ⓐⓑⓒⓓ	36	ⓐⓑⓒⓓ	41	ⓐⓑⓒⓓ	46	ⓐⓑⓒⓓ	51	ⓐⓑⓒⓓ	56	ⓐⓑⓒⓓ
32	ⓐⓑⓒⓓ	37	ⓐⓑⓒⓓ	42	ⓐⓑⓒⓓ	47	ⓐⓑⓒⓓ	52	ⓐⓑⓒⓓ	57	ⓐⓑⓒⓓ
33	ⓐⓑⓒⓓ	38	ⓐⓑⓒⓓ	43	ⓐⓑⓒⓓ	48	ⓐⓑⓒⓓ	53	ⓐⓑⓒⓓ	58	ⓐⓑⓒⓓ
34	ⓐⓑⓒⓓ	39	ⓐⓑⓒⓓ	44	ⓐⓑⓒⓓ	49	ⓐⓑⓒⓓ	54	ⓐⓑⓒⓓ	59	ⓐⓑⓒⓓ
35	ⓐⓑⓒⓓ	40	ⓐⓑⓒⓓ	45	ⓐⓑⓒⓓ	50	ⓐⓑⓒⓓ	55	ⓐⓑⓒⓓ	60	ⓐⓑⓒⓓ
61	ⓐⓑⓒⓓ	66	ⓐⓑⓒⓓ	71	ⓐⓑⓒⓓ	76	ⓐⓑⓒⓓ	81	ⓐⓑⓒⓓ	86	ⓐⓑⓒⓓ
62	ⓐⓑⓒⓓ	67	ⓐⓑⓒⓓ	72	ⓐⓑⓒⓓ	77	ⓐⓑⓒⓓ	82	ⓐⓑⓒⓓ	87	ⓐⓑⓒⓓ
63	ⓐⓑⓒⓓ	68	ⓐⓑⓒⓓ	73	ⓐⓑⓒⓓ	78	ⓐⓑⓒⓓ	83	ⓐⓑⓒⓓ	88	ⓐⓑⓒⓓ
64	ⓐⓑⓒⓓ	69	ⓐⓑⓒⓓ	74	ⓐⓑⓒⓓ	79	ⓐⓑⓒⓓ	84	ⓐⓑⓒⓓ	89	ⓐⓑⓒⓓ
65	ⓐⓑⓒⓓ	70	ⓐⓑⓒⓓ	75	ⓐⓑⓒⓓ	80	ⓐⓑⓒⓓ	85	ⓐⓑⓒⓓ	90	ⓐⓑⓒⓓ
91	ⓐⓑⓒⓓ	96	ⓐⓑⓒⓓ	101	ⓐⓑⓒⓓ	106	ⓐⓑⓒⓓ	111	ⓐⓑⓒⓓ	116	ⓐⓑⓒⓓ
92	ⓐⓑⓒⓓ	97	ⓐⓑⓒⓓ	102	ⓐⓑⓒⓓ	107	ⓐⓑⓒⓓ	112	ⓐⓑⓒⓓ	117	ⓐⓑⓒⓓ
93	ⓐⓑⓒⓓ	98	ⓐⓑⓒⓓ	103	ⓐⓑⓒⓓ	108	ⓐⓑⓒⓓ	113	ⓐⓑⓒⓓ	118	ⓐⓑⓒⓓ
94	ⓐⓑⓒⓓ	99	ⓐⓑⓒⓓ	104	ⓐⓑⓒⓓ	109	ⓐⓑⓒⓓ	114	ⓐⓑⓒⓓ	119	ⓐⓑⓒⓓ
95	ⓐⓑⓒⓓ	100	ⓐⓑⓒⓓ	105	ⓐⓑⓒⓓ	110	ⓐⓑⓒⓓ	115	ⓐⓑⓒⓓ	120	ⓐⓑⓒⓓ

MARK YOUR SCORE AND EVALUATE YOURSELF ACCORDINGLY

Above 100: *EXCELLENT.* You are amongst the top scorers, but strive to retain it because in agriculture there are great swings.

80 to 100: *VERY GOOD.* With about a similar performance in weak areas, you can become the top scorer. But be aware! there are many a slip between cup and lips.

70 to 80: *FAIR.* You still have a long way to go. Success is merely a chance factor.

Less than 70: *POOR*: Why are you even thinking of Agricultural Competitive Exam now. Work, Work and Work more. Who knows, your labour will pay off, as it does as a rule.

Notes

AGRISET - 1

ANSWER SHEET FOR GENERAL AGRICULTURE MODEL PAPER

- I have provided you MCQs model test on general agriculture for TOP (Target Oriented Preparation) for the Agricultural Competitive Exams.
- You have to mark all your responses only on the answer sheet provided.
- All items carry equal marks. Attempt all items. Your total marks will depend only on the number of correct responses marked by you in the Answer sheet. There is no negative marking. Award yourself one point for each correct answer.

ANSWER SHEET

1	ⓐⓑⓒⓓ	6	ⓐⓑⓒⓓ	11	ⓐⓑⓒⓓ	16	ⓐⓑⓒⓓ	21	ⓐⓑⓒⓓ	26	ⓐⓑⓒⓓ
2	ⓐⓑⓒⓓ	7	ⓐⓑⓒⓓ	12	ⓐⓑⓒⓓ	17	ⓐⓑⓒⓓ	22	ⓐⓑⓒⓓ	27	ⓐⓑⓒⓓ
3	ⓐⓑⓒⓓ	8	ⓐⓑⓒⓓ	13	ⓐⓑⓒⓓ	18	ⓐⓑⓒⓓ	23	ⓐⓑⓒⓓ	28	ⓐⓑⓒⓓ
4	ⓐⓑⓒⓓ	9	ⓐⓑⓒⓓ	14	ⓐⓑⓒⓓ	19	ⓐⓑⓒⓓ	24	ⓐⓑⓒⓓ	29	ⓐⓑⓒⓓ
5	ⓐⓑⓒⓓ	10	ⓐⓑⓒⓓ	15	ⓐⓑⓒⓓ	20	ⓐⓑⓒⓓ	25	ⓐⓑⓒⓓ	30	ⓐⓑⓒⓓ
31	ⓐⓑⓒⓓ	36	ⓐⓑⓒⓓ	41	ⓐⓑⓒⓓ	46	ⓐⓑⓒⓓ	51	ⓐⓑⓒⓓ	56	ⓐⓑⓒⓓ
32	ⓐⓑⓒⓓ	37	ⓐⓑⓒⓓ	42	ⓐⓑⓒⓓ	47	ⓐⓑⓒⓓ	52	ⓐⓑⓒⓓ	57	ⓐⓑⓒⓓ
33	ⓐⓑⓒⓓ	38	ⓐⓑⓒⓓ	43	ⓐⓑⓒⓓ	48	ⓐⓑⓒⓓ	53	ⓐⓑⓒⓓ	58	ⓐⓑⓒⓓ
34	ⓐⓑⓒⓓ	39	ⓐⓑⓒⓓ	44	ⓐⓑⓒⓓ	49	ⓐⓑⓒⓓ	54	ⓐⓑⓒⓓ	59	ⓐⓑⓒⓓ
35	ⓐⓑⓒⓓ	40	ⓐⓑⓒⓓ	45	ⓐⓑⓒⓓ	50	ⓐⓑⓒⓓ	55	ⓐⓑⓒⓓ	60	ⓐⓑⓒⓓ
61	ⓐⓑⓒⓓ	66	ⓐⓑⓒⓓ	71	ⓐⓑⓒⓓ	76	ⓐⓑⓒⓓ	81	ⓐⓑⓒⓓ	86	ⓐⓑⓒⓓ
62	ⓐⓑⓒⓓ	67	ⓐⓑⓒⓓ	72	ⓐⓑⓒⓓ	77	ⓐⓑⓒⓓ	82	ⓐⓑⓒⓓ	87	ⓐⓑⓒⓓ
63	ⓐⓑⓒⓓ	68	ⓐⓑⓒⓓ	73	ⓐⓑⓒⓓ	78	ⓐⓑⓒⓓ	83	ⓐⓑⓒⓓ	88	ⓐⓑⓒⓓ
64	ⓐⓑⓒⓓ	69	ⓐⓑⓒⓓ	74	ⓐⓑⓒⓓ	79	ⓐⓑⓒⓓ	84	ⓐⓑⓒⓓ	89	ⓐⓑⓒⓓ
65	ⓐⓑⓒⓓ	70	ⓐⓑⓒⓓ	75	ⓐⓑⓒⓓ	80	ⓐⓑⓒⓓ	85	ⓐⓑⓒⓓ	90	ⓐⓑⓒⓓ
91	ⓐⓑⓒⓓ	96	ⓐⓑⓒⓓ	101	ⓐⓑⓒⓓ	106	ⓐⓑⓒⓓ	111	ⓐⓑⓒⓓ	116	ⓐⓑⓒⓓ
92	ⓐⓑⓒⓓ	97	ⓐⓑⓒⓓ	102	ⓐⓑⓒⓓ	107	ⓐⓑⓒⓓ	112	ⓐⓑⓒⓓ	117	ⓐⓑⓒⓓ
93	ⓐⓑⓒⓓ	98	ⓐⓑⓒⓓ	103	ⓐⓑⓒⓓ	108	ⓐⓑⓒⓓ	113	ⓐⓑⓒⓓ	118	ⓐⓑⓒⓓ
94	ⓐⓑⓒⓓ	99	ⓐⓑⓒⓓ	104	ⓐⓑⓒⓓ	109	ⓐⓑⓒⓓ	114	ⓐⓑⓒⓓ	119	ⓐⓑⓒⓓ
95	ⓐⓑⓒⓓ	100	ⓐⓑⓒⓓ	105	ⓐⓑⓒⓓ	110	ⓐⓑⓒⓓ	115	ⓐⓑⓒⓓ	120	ⓐⓑⓒⓓ

MARK YOUR SCORE AND EVALUATE YOURSELF ACCORDINGLY

Above 100: *EXCELLENT*. You are amongst the top scorers, but strive to retain it because in agriculture there are great swings.

80 to 100: *VERY GOOD*. With about a similar performance in weak areas, you can become the top scorer. But be aware! there are many a slip between cup and lips.

70 to 80: *FAIR*. You still have a long way to go. Success is merely a chance factor.

Less than 70: *POOR*: Why are you even thinking of Agricultural Competitive Exam now. Work, Work and Work more. Who knows, your labour will pay off, as it does as a rule.

Notes

AGRISET - 2

ANSWER SHEET FOR GENERAL AGRICULTURE MODEL PAPER

- I have provided you MCQs model test on general agriculture for TOP (Target Oriented Preparation) for the Agricultural Competitive Exams.
- You have to mark all your responses only on the answer sheet provided.
- All items carry equal marks. Attempt all items. Your total marks will depend only on the number of correct responses marked by you in the Answer sheet. There is no negative marking. Award yourself one point for each correct answer.

ANSWER SHEET

1	ⓐⓑⓒⓓ	6	ⓐⓑⓒⓓ	11	ⓐⓑⓒⓓ	16	ⓐⓑⓒⓓ	21	ⓐⓑⓒⓓ	26	ⓐⓑⓒⓓ
2	ⓐⓑⓒⓓ	7	ⓐⓑⓒⓓ	12	ⓐⓑⓒⓓ	17	ⓐⓑⓒⓓ	22	ⓐⓑⓒⓓ	27	ⓐⓑⓒⓓ
3	ⓐⓑⓒⓓ	8	ⓐⓑⓒⓓ	13	ⓐⓑⓒⓓ	18	ⓐⓑⓒⓓ	23	ⓐⓑⓒⓓ	28	ⓐⓑⓒⓓ
4	ⓐⓑⓒⓓ	9	ⓐⓑⓒⓓ	14	ⓐⓑⓒⓓ	19	ⓐⓑⓒⓓ	24	ⓐⓑⓒⓓ	29	ⓐⓑⓒⓓ
5	ⓐⓑⓒⓓ	10	ⓐⓑⓒⓓ	15	ⓐⓑⓒⓓ	20	ⓐⓑⓒⓓ	25	ⓐⓑⓒⓓ	30	ⓐⓑⓒⓓ
31	ⓐⓑⓒⓓ	36	ⓐⓑⓒⓓ	41	ⓐⓑⓒⓓ	46	ⓐⓑⓒⓓ	51	ⓐⓑⓒⓓ	56	ⓐⓑⓒⓓ
32	ⓐⓑⓒⓓ	37	ⓐⓑⓒⓓ	42	ⓐⓑⓒⓓ	47	ⓐⓑⓒⓓ	52	ⓐⓑⓒⓓ	57	ⓐⓑⓒⓓ
33	ⓐⓑⓒⓓ	38	ⓐⓑⓒⓓ	43	ⓐⓑⓒⓓ	48	ⓐⓑⓒⓓ	53	ⓐⓑⓒⓓ	58	ⓐⓑⓒⓓ
34	ⓐⓑⓒⓓ	39	ⓐⓑⓒⓓ	44	ⓐⓑⓒⓓ	49	ⓐⓑⓒⓓ	54	ⓐⓑⓒⓓ	59	ⓐⓑⓒⓓ
35	ⓐⓑⓒⓓ	40	ⓐⓑⓒⓓ	45	ⓐⓑⓒⓓ	50	ⓐⓑⓒⓓ	55	ⓐⓑⓒⓓ	60	ⓐⓑⓒⓓ
61	ⓐⓑⓒⓓ	66	ⓐⓑⓒⓓ	71	ⓐⓑⓒⓓ	76	ⓐⓑⓒⓓ	81	ⓐⓑⓒⓓ	86	ⓐⓑⓒⓓ
62	ⓐⓑⓒⓓ	67	ⓐⓑⓒⓓ	72	ⓐⓑⓒⓓ	77	ⓐⓑⓒⓓ	82	ⓐⓑⓒⓓ	87	ⓐⓑⓒⓓ
63	ⓐⓑⓒⓓ	68	ⓐⓑⓒⓓ	73	ⓐⓑⓒⓓ	78	ⓐⓑⓒⓓ	83	ⓐⓑⓒⓓ	88	ⓐⓑⓒⓓ
64	ⓐⓑⓒⓓ	69	ⓐⓑⓒⓓ	74	ⓐⓑⓒⓓ	79	ⓐⓑⓒⓓ	84	ⓐⓑⓒⓓ	89	ⓐⓑⓒⓓ
65	ⓐⓑⓒⓓ	70	ⓐⓑⓒⓓ	75	ⓐⓑⓒⓓ	80	ⓐⓑⓒⓓ	85	ⓐⓑⓒⓓ	90	ⓐⓑⓒⓓ
91	ⓐⓑⓒⓓ	96	ⓐⓑⓒⓓ	101	ⓐⓑⓒⓓ	106	ⓐⓑⓒⓓ	111	ⓐⓑⓒⓓ	116	ⓐⓑⓒⓓ
92	ⓐⓑⓒⓓ	97	ⓐⓑⓒⓓ	102	ⓐⓑⓒⓓ	107	ⓐⓑⓒⓓ	112	ⓐⓑⓒⓓ	117	ⓐⓑⓒⓓ
93	ⓐⓑⓒⓓ	98	ⓐⓑⓒⓓ	103	ⓐⓑⓒⓓ	108	ⓐⓑⓒⓓ	113	ⓐⓑⓒⓓ	118	ⓐⓑⓒⓓ
94	ⓐⓑⓒⓓ	99	ⓐⓑⓒⓓ	104	ⓐⓑⓒⓓ	109	ⓐⓑⓒⓓ	114	ⓐⓑⓒⓓ	119	ⓐⓑⓒⓓ
95	ⓐⓑⓒⓓ	100	ⓐⓑⓒⓓ	105	ⓐⓑⓒⓓ	110	ⓐⓑⓒⓓ	115	ⓐⓑⓒⓓ	120	ⓐⓑⓒⓓ

MARK YOUR SCORE AND EVALUATE YOURSELF ACCORDINGLY

Above 100: *EXCELLENT*. You are amongst the top scorers, but strive to retain it because in agriculture there are great swings.

80 to 100: *VERY GOOD*. With about a similar performance in weak areas, you can become the top scorer. But be aware! there are many a slip between cup and lips.

70 to 80: *FAIR*. You still have a long way to go. Success is merely a chance factor.

Less than 70: *POOR*: Why are you even thinking of Agricultural Competitive Exam now. Work, Work and Work more. Who knows, your labour will pay off, as it does as a rule.

Notes

AGRISET - 3

ANSWER SHEET FOR GENERAL AGRICULTURE MODEL PAPER

- I have provided you MCQs model test on general agriculture for TOP (Target Oriented Preparation) for the Agricultural Competitive Exams.
- You have to mark all your responses only on the answer sheet provided.
- All items carry equal marks. Attempt all items. Your total marks will depend only on the number of correct responses marked by you in the Answer sheet. There is no negative marking. Award yourself one point for each correct answer.

ANSWER SHEET

1 ⓐⓑⓒⓓ	6 ⓐⓑⓒⓓ	11 ⓐⓑⓒⓓ	16 ⓐⓑⓒⓓ	21 ⓐⓑⓒⓓ	26 ⓐⓑⓒⓓ
2 ⓐⓑⓒⓓ	7 ⓐⓑⓒⓓ	12 ⓐⓑⓒⓓ	17 ⓐⓑⓒⓓ	22 ⓐⓑⓒⓓ	27 ⓐⓑⓒⓓ
3 ⓐⓑⓒⓓ	8 ⓐⓑⓒⓓ	13 ⓐⓑⓒⓓ	18 ⓐⓑⓒⓓ	23 ⓐⓑⓒⓓ	28 ⓐⓑⓒⓓ
4 ⓐⓑⓒⓓ	9 ⓐⓑⓒⓓ	14 ⓐⓑⓒⓓ	19 ⓐⓑⓒⓓ	24 ⓐⓑⓒⓓ	29 ⓐⓑⓒⓓ
5 ⓐⓑⓒⓓ	10 ⓐⓑⓒⓓ	15 ⓐⓑⓒⓓ	20 ⓐⓑⓒⓓ	25 ⓐⓑⓒⓓ	30 ⓐⓑⓒⓓ
31 ⓐⓑⓒⓓ	36 ⓐⓑⓒⓓ	41 ⓐⓑⓒⓓ	46 ⓐⓑⓒⓓ	51 ⓐⓑⓒⓓ	56 ⓐⓑⓒⓓ
32 ⓐⓑⓒⓓ	37 ⓐⓑⓒⓓ	42 ⓐⓑⓒⓓ	47 ⓐⓑⓒⓓ	52 ⓐⓑⓒⓓ	57 ⓐⓑⓒⓓ
33 ⓐⓑⓒⓓ	38 ⓐⓑⓒⓓ	43 ⓐⓑⓒⓓ	48 ⓐⓑⓒⓓ	53 ⓐⓑⓒⓓ	58 ⓐⓑⓒⓓ
34 ⓐⓑⓒⓓ	39 ⓐⓑⓒⓓ	44 ⓐⓑⓒⓓ	49 ⓐⓑⓒⓓ	54 ⓐⓑⓒⓓ	59 ⓐⓑⓒⓓ
35 ⓐⓑⓒⓓ	40 ⓐⓑⓒⓓ	45 ⓐⓑⓒⓓ	50 ⓐⓑⓒⓓ	55 ⓐⓑⓒⓓ	60 ⓐⓑⓒⓓ
61 ⓐⓑⓒⓓ	66 ⓐⓑⓒⓓ	71 ⓐⓑⓒⓓ	76 ⓐⓑⓒⓓ	81 ⓐⓑⓒⓓ	86 ⓐⓑⓒⓓ
62 ⓐⓑⓒⓓ	67 ⓐⓑⓒⓓ	72 ⓐⓑⓒⓓ	77 ⓐⓑⓒⓓ	82 ⓐⓑⓒⓓ	87 ⓐⓑⓒⓓ
63 ⓐⓑⓒⓓ	68 ⓐⓑⓒⓓ	73 ⓐⓑⓒⓓ	78 ⓐⓑⓒⓓ	83 ⓐⓑⓒⓓ	88 ⓐⓑⓒⓓ
64 ⓐⓑⓒⓓ	69 ⓐⓑⓒⓓ	74 ⓐⓑⓒⓓ	79 ⓐⓑⓒⓓ	84 ⓐⓑⓒⓓ	89 ⓐⓑⓒⓓ
65 ⓐⓑⓒⓓ	70 ⓐⓑⓒⓓ	75 ⓐⓑⓒⓓ	80 ⓐⓑⓒⓓ	85 ⓐⓑⓒⓓ	90 ⓐⓑⓒⓓ
91 ⓐⓑⓒⓓ	96 ⓐⓑⓒⓓ	101 ⓐⓑⓒⓓ	106 ⓐⓑⓒⓓ	111 ⓐⓑⓒⓓ	116 ⓐⓑⓒⓓ
92 ⓐⓑⓒⓓ	97 ⓐⓑⓒⓓ	102 ⓐⓑⓒⓓ	107 ⓐⓑⓒⓓ	112 ⓐⓑⓒⓓ	117 ⓐⓑⓒⓓ
93 ⓐⓑⓒⓓ	98 ⓐⓑⓒⓓ	103 ⓐⓑⓒⓓ	108 ⓐⓑⓒⓓ	113 ⓐⓑⓒⓓ	118 ⓐⓑⓒⓓ
94 ⓐⓑⓒⓓ	99 ⓐⓑⓒⓓ	104 ⓐⓑⓒⓓ	109 ⓐⓑⓒⓓ	114 ⓐⓑⓒⓓ	119 ⓐⓑⓒⓓ
95 ⓐⓑⓒⓓ	100 ⓐⓑⓒⓓ	105 ⓐⓑⓒⓓ	110 ⓐⓑⓒⓓ	115 ⓐⓑⓒⓓ	120 ⓐⓑⓒⓓ

MARK YOUR SCORE AND EVALUATE YOURSELF ACCORDINGLY

Above 100: *EXCELLENT.* You are amongst the top scorers, but strive to retain it because in agriculture there are great swings.

80 to 100: *VERY GOOD.* With about a similar performance in weak areas, you can become the top scorer. But be aware! there are many a slip between cup and lips.

70 to 80: *FAIR.* You still have a long way to go. Success is merely a chance factor.

Less than 70: *POOR*: Why are you even thinking of Agricultural Competitive Exam now. Work, Work and Work more. Who knows, your labour will pay off, as it does as a rule.

Notes

AGRISET - 4

ANSWER SHEET FOR GENERAL AGRICULTURE MODEL PAPER

- I have provided you MCQs model test on general agriculture for TOP (Target Oriented Preparation) for the Agricultural Competitive Exams.
- You have to mark all your responses only on the answer sheet provided.
- All items carry equal marks. Attempt all items. Your total marks will depend only on the number of correct responses marked by you in the Answer sheet. There is no negative marking. Award yourself one point for each correct answer.

ANSWER SHEET

1	ⓐⓑⓒⓓ	6	ⓐⓑⓒⓓ	11	ⓐⓑⓒⓓ	16	ⓐⓑⓒⓓ	21	ⓐⓑⓒⓓ	26	ⓐⓑⓒⓓ
2	ⓐⓑⓒⓓ	7	ⓐⓑⓒⓓ	12	ⓐⓑⓒⓓ	17	ⓐⓑⓒⓓ	22	ⓐⓑⓒⓓ	27	ⓐⓑⓒⓓ
3	ⓐⓑⓒⓓ	8	ⓐⓑⓒⓓ	13	ⓐⓑⓒⓓ	18	ⓐⓑⓒⓓ	23	ⓐⓑⓒⓓ	28	ⓐⓑⓒⓓ
4	ⓐⓑⓒⓓ	9	ⓐⓑⓒⓓ	14	ⓐⓑⓒⓓ	19	ⓐⓑⓒⓓ	24	ⓐⓑⓒⓓ	29	ⓐⓑⓒⓓ
5	ⓐⓑⓒⓓ	10	ⓐⓑⓒⓓ	15	ⓐⓑⓒⓓ	20	ⓐⓑⓒⓓ	25	ⓐⓑⓒⓓ	30	ⓐⓑⓒⓓ
31	ⓐⓑⓒⓓ	36	ⓐⓑⓒⓓ	41	ⓐⓑⓒⓓ	46	ⓐⓑⓒⓓ	51	ⓐⓑⓒⓓ	56	ⓐⓑⓒⓓ
32	ⓐⓑⓒⓓ	37	ⓐⓑⓒⓓ	42	ⓐⓑⓒⓓ	47	ⓐⓑⓒⓓ	52	ⓐⓑⓒⓓ	57	ⓐⓑⓒⓓ
33	ⓐⓑⓒⓓ	38	ⓐⓑⓒⓓ	43	ⓐⓑⓒⓓ	48	ⓐⓑⓒⓓ	53	ⓐⓑⓒⓓ	58	ⓐⓑⓒⓓ
34	ⓐⓑⓒⓓ	39	ⓐⓑⓒⓓ	44	ⓐⓑⓒⓓ	49	ⓐⓑⓒⓓ	54	ⓐⓑⓒⓓ	59	ⓐⓑⓒⓓ
35	ⓐⓑⓒⓓ	40	ⓐⓑⓒⓓ	45	ⓐⓑⓒⓓ	50	ⓐⓑⓒⓓ	55	ⓐⓑⓒⓓ	60	ⓐⓑⓒⓓ
61	ⓐⓑⓒⓓ	66	ⓐⓑⓒⓓ	71	ⓐⓑⓒⓓ	76	ⓐⓑⓒⓓ	81	ⓐⓑⓒⓓ	86	ⓐⓑⓒⓓ
62	ⓐⓑⓒⓓ	67	ⓐⓑⓒⓓ	72	ⓐⓑⓒⓓ	77	ⓐⓑⓒⓓ	82	ⓐⓑⓒⓓ	87	ⓐⓑⓒⓓ
63	ⓐⓑⓒⓓ	68	ⓐⓑⓒⓓ	73	ⓐⓑⓒⓓ	78	ⓐⓑⓒⓓ	83	ⓐⓑⓒⓓ	88	ⓐⓑⓒⓓ
64	ⓐⓑⓒⓓ	69	ⓐⓑⓒⓓ	74	ⓐⓑⓒⓓ	79	ⓐⓑⓒⓓ	84	ⓐⓑⓒⓓ	89	ⓐⓑⓒⓓ
65	ⓐⓑⓒⓓ	70	ⓐⓑⓒⓓ	75	ⓐⓑⓒⓓ	80	ⓐⓑⓒⓓ	85	ⓐⓑⓒⓓ	90	ⓐⓑⓒⓓ
91	ⓐⓑⓒⓓ	96	ⓐⓑⓒⓓ	101	ⓐⓑⓒⓓ	106	ⓐⓑⓒⓓ	111	ⓐⓑⓒⓓ	116	ⓐⓑⓒⓓ
92	ⓐⓑⓒⓓ	97	ⓐⓑⓒⓓ	102	ⓐⓑⓒⓓ	107	ⓐⓑⓒⓓ	112	ⓐⓑⓒⓓ	117	ⓐⓑⓒⓓ
93	ⓐⓑⓒⓓ	98	ⓐⓑⓒⓓ	103	ⓐⓑⓒⓓ	108	ⓐⓑⓒⓓ	113	ⓐⓑⓒⓓ	118	ⓐⓑⓒⓓ
94	ⓐⓑⓒⓓ	99	ⓐⓑⓒⓓ	104	ⓐⓑⓒⓓ	109	ⓐⓑⓒⓓ	114	ⓐⓑⓒⓓ	119	ⓐⓑⓒⓓ
95	ⓐⓑⓒⓓ	100	ⓐⓑⓒⓓ	105	ⓐⓑⓒⓓ	110	ⓐⓑⓒⓓ	115	ⓐⓑⓒⓓ	120	ⓐⓑⓒⓓ

MARK YOUR SCORE AND EVALUATE YOURSELF ACCORDINGLY

Above 100: *EXCELLENT*. You are amongst the top scorers, but strive to retain it because in agriculture there are great swings.

80 to 100: *VERY GOOD*. With about a similar performance in weak areas, you can become the top scorer. But be aware! there are many a slip between cup and lips.

70 to 80: *FAIR*. You still have a long way to go. Success is merely a chance factor.

Less than 70: *POOR*: Why are you even thinking of Agricultural Competitive Exam now. Work, Work and Work more. Who knows, your labour will pay off, as it does as a rule.

Notes

AGRISET - 5

ANSWER SHEET FOR GENERAL AGRICULTURE MODEL PAPER

- I have provided you MCQs model test on general agriculture for TOP (Target Oriented Preparation) for the Agricultural Competitive Exams.
- You have to mark all your responses only on the answer sheet provided.
- All items carry equal marks. Attempt all items. Your total marks will depend only on the number of correct responses marked by you in the Answer sheet. There is no negative marking. Award yourself one point for each correct answer.

ANSWER SHEET

1	ⓐⓑⓒⓓ	6	ⓐⓑⓒⓓ	11	ⓐⓑⓒⓓ	16	ⓐⓑⓒⓓ	21	ⓐⓑⓒⓓ	26	ⓐⓑⓒⓓ
2	ⓐⓑⓒⓓ	7	ⓐⓑⓒⓓ	12	ⓐⓑⓒⓓ	17	ⓐⓑⓒⓓ	22	ⓐⓑⓒⓓ	27	ⓐⓑⓒⓓ
3	ⓐⓑⓒⓓ	8	ⓐⓑⓒⓓ	13	ⓐⓑⓒⓓ	18	ⓐⓑⓒⓓ	23	ⓐⓑⓒⓓ	28	ⓐⓑⓒⓓ
4	ⓐⓑⓒⓓ	9	ⓐⓑⓒⓓ	14	ⓐⓑⓒⓓ	19	ⓐⓑⓒⓓ	24	ⓐⓑⓒⓓ	29	ⓐⓑⓒⓓ
5	ⓐⓑⓒⓓ	10	ⓐⓑⓒⓓ	15	ⓐⓑⓒⓓ	20	ⓐⓑⓒⓓ	25	ⓐⓑⓒⓓ	30	ⓐⓑⓒⓓ
31	ⓐⓑⓒⓓ	36	ⓐⓑⓒⓓ	41	ⓐⓑⓒⓓ	46	ⓐⓑⓒⓓ	51	ⓐⓑⓒⓓ	56	ⓐⓑⓒⓓ
32	ⓐⓑⓒⓓ	37	ⓐⓑⓒⓓ	42	ⓐⓑⓒⓓ	47	ⓐⓑⓒⓓ	52	ⓐⓑⓒⓓ	57	ⓐⓑⓒⓓ
33	ⓐⓑⓒⓓ	38	ⓐⓑⓒⓓ	43	ⓐⓑⓒⓓ	48	ⓐⓑⓒⓓ	53	ⓐⓑⓒⓓ	58	ⓐⓑⓒⓓ
34	ⓐⓑⓒⓓ	39	ⓐⓑⓒⓓ	44	ⓐⓑⓒⓓ	49	ⓐⓑⓒⓓ	54	ⓐⓑⓒⓓ	59	ⓐⓑⓒⓓ
35	ⓐⓑⓒⓓ	40	ⓐⓑⓒⓓ	45	ⓐⓑⓒⓓ	50	ⓐⓑⓒⓓ	55	ⓐⓑⓒⓓ	60	ⓐⓑⓒⓓ
61	ⓐⓑⓒⓓ	66	ⓐⓑⓒⓓ	71	ⓐⓑⓒⓓ	76	ⓐⓑⓒⓓ	81	ⓐⓑⓒⓓ	86	ⓐⓑⓒⓓ
62	ⓐⓑⓒⓓ	67	ⓐⓑⓒⓓ	72	ⓐⓑⓒⓓ	77	ⓐⓑⓒⓓ	82	ⓐⓑⓒⓓ	87	ⓐⓑⓒⓓ
63	ⓐⓑⓒⓓ	68	ⓐⓑⓒⓓ	73	ⓐⓑⓒⓓ	78	ⓐⓑⓒⓓ	83	ⓐⓑⓒⓓ	88	ⓐⓑⓒⓓ
64	ⓐⓑⓒⓓ	69	ⓐⓑⓒⓓ	74	ⓐⓑⓒⓓ	79	ⓐⓑⓒⓓ	84	ⓐⓑⓒⓓ	89	ⓐⓑⓒⓓ
65	ⓐⓑⓒⓓ	70	ⓐⓑⓒⓓ	75	ⓐⓑⓒⓓ	80	ⓐⓑⓒⓓ	85	ⓐⓑⓒⓓ	90	ⓐⓑⓒⓓ
91	ⓐⓑⓒⓓ	96	ⓐⓑⓒⓓ	101	ⓐⓑⓒⓓ	106	ⓐⓑⓒⓓ	111	ⓐⓑⓒⓓ	116	ⓐⓑⓒⓓ
92	ⓐⓑⓒⓓ	97	ⓐⓑⓒⓓ	102	ⓐⓑⓒⓓ	107	ⓐⓑⓒⓓ	112	ⓐⓑⓒⓓ	117	ⓐⓑⓒⓓ
93	ⓐⓑⓒⓓ	98	ⓐⓑⓒⓓ	103	ⓐⓑⓒⓓ	108	ⓐⓑⓒⓓ	113	ⓐⓑⓒⓓ	118	ⓐⓑⓒⓓ
94	ⓐⓑⓒⓓ	99	ⓐⓑⓒⓓ	104	ⓐⓑⓒⓓ	109	ⓐⓑⓒⓓ	114	ⓐⓑⓒⓓ	119	ⓐⓑⓒⓓ
95	ⓐⓑⓒⓓ	100	ⓐⓑⓒⓓ	105	ⓐⓑⓒⓓ	110	ⓐⓑⓒⓓ	115	ⓐⓑⓒⓓ	120	ⓐⓑⓒⓓ

MARK YOUR SCORE AND EVALUATE YOURSELF ACCORDINGLY

Above 100: *EXCELLENT.* You are amongst the top scorers, but strive to retain it because in agriculture there are great swings.

80 to 100: *VERY GOOD.* With about a similar performance in weak areas, you can become the top scorer. But be aware! there are many a slip between cup and lips.

70 to 80: *FAIR.* You still have a long way to go. Success is merely a chance factor.

Less than 70: *POOR:* Why are you even thinking of Agricultural Competitive Exam now. Work, Work and Work more. Who knows, your labour will pay off, as it does as a rule.

Notes

AGRISET - 6

ANSWER SHEET FOR GENERAL AGRICULTURE MODEL PAPER

- I have provided you MCQs model test on general agriculture for TOP (Target Oriented Preparation) for the Agricultural Competitive Exams.
- You have to mark all your responses only on the answer sheet provided.
- All items carry equal marks. Attempt all items. Your total marks will depend only on the number of correct responses marked by you in the Answer sheet. There is no negative marking. Award yourself one point for each correct answer.

ANSWER SHEET

1 ⓐⓑⓒⓓ	6 ⓐⓑⓒⓓ	11 ⓐⓑⓒⓓ	16 ⓐⓑⓒⓓ	21 ⓐⓑⓒⓓ	26 ⓐⓑⓒⓓ
2 ⓐⓑⓒⓓ	7 ⓐⓑⓒⓓ	12 ⓐⓑⓒⓓ	17 ⓐⓑⓒⓓ	22 ⓐⓑⓒⓓ	27 ⓐⓑⓒⓓ
3 ⓐⓑⓒⓓ	8 ⓐⓑⓒⓓ	13 ⓐⓑⓒⓓ	18 ⓐⓑⓒⓓ	23 ⓐⓑⓒⓓ	28 ⓐⓑⓒⓓ
4 ⓐⓑⓒⓓ	9 ⓐⓑⓒⓓ	14 ⓐⓑⓒⓓ	19 ⓐⓑⓒⓓ	24 ⓐⓑⓒⓓ	29 ⓐⓑⓒⓓ
5 ⓐⓑⓒⓓ	10 ⓐⓑⓒⓓ	15 ⓐⓑⓒⓓ	20 ⓐⓑⓒⓓ	25 ⓐⓑⓒⓓ	30 ⓐⓑⓒⓓ
31 ⓐⓑⓒⓓ	36 ⓐⓑⓒⓓ	41 ⓐⓑⓒⓓ	46 ⓐⓑⓒⓓ	51 ⓐⓑⓒⓓ	56 ⓐⓑⓒⓓ
32 ⓐⓑⓒⓓ	37 ⓐⓑⓒⓓ	42 ⓐⓑⓒⓓ	47 ⓐⓑⓒⓓ	52 ⓐⓑⓒⓓ	57 ⓐⓑⓒⓓ
33 ⓐⓑⓒⓓ	38 ⓐⓑⓒⓓ	43 ⓐⓑⓒⓓ	48 ⓐⓑⓒⓓ	53 ⓐⓑⓒⓓ	58 ⓐⓑⓒⓓ
34 ⓐⓑⓒⓓ	39 ⓐⓑⓒⓓ	44 ⓐⓑⓒⓓ	49 ⓐⓑⓒⓓ	54 ⓐⓑⓒⓓ	59 ⓐⓑⓒⓓ
35 ⓐⓑⓒⓓ	40 ⓐⓑⓒⓓ	45 ⓐⓑⓒⓓ	50 ⓐⓑⓒⓓ	55 ⓐⓑⓒⓓ	60 ⓐⓑⓒⓓ
61 ⓐⓑⓒⓓ	66 ⓐⓑⓒⓓ	71 ⓐⓑⓒⓓ	76 ⓐⓑⓒⓓ	81 ⓐⓑⓒⓓ	86 ⓐⓑⓒⓓ
62 ⓐⓑⓒⓓ	67 ⓐⓑⓒⓓ	72 ⓐⓑⓒⓓ	77 ⓐⓑⓒⓓ	82 ⓐⓑⓒⓓ	87 ⓐⓑⓒⓓ
63 ⓐⓑⓒⓓ	68 ⓐⓑⓒⓓ	73 ⓐⓑⓒⓓ	78 ⓐⓑⓒⓓ	83 ⓐⓑⓒⓓ	88 ⓐⓑⓒⓓ
64 ⓐⓑⓒⓓ	69 ⓐⓑⓒⓓ	74 ⓐⓑⓒⓓ	79 ⓐⓑⓒⓓ	84 ⓐⓑⓒⓓ	89 ⓐⓑⓒⓓ
65 ⓐⓑⓒⓓ	70 ⓐⓑⓒⓓ	75 ⓐⓑⓒⓓ	80 ⓐⓑⓒⓓ	85 ⓐⓑⓒⓓ	90 ⓐⓑⓒⓓ
91 ⓐⓑⓒⓓ	96 ⓐⓑⓒⓓ	101 ⓐⓑⓒⓓ	106 ⓐⓑⓒⓓ	111 ⓐⓑⓒⓓ	116 ⓐⓑⓒⓓ
92 ⓐⓑⓒⓓ	97 ⓐⓑⓒⓓ	102 ⓐⓑⓒⓓ	107 ⓐⓑⓒⓓ	112 ⓐⓑⓒⓓ	117 ⓐⓑⓒⓓ
93 ⓐⓑⓒⓓ	98 ⓐⓑⓒⓓ	103 ⓐⓑⓒⓓ	108 ⓐⓑⓒⓓ	113 ⓐⓑⓒⓓ	118 ⓐⓑⓒⓓ
94 ⓐⓑⓒⓓ	99 ⓐⓑⓒⓓ	104 ⓐⓑⓒⓓ	109 ⓐⓑⓒⓓ	114 ⓐⓑⓒⓓ	119 ⓐⓑⓒⓓ
95 ⓐⓑⓒⓓ	100 ⓐⓑⓒⓓ	105 ⓐⓑⓒⓓ	110 ⓐⓑⓒⓓ	115 ⓐⓑⓒⓓ	120 ⓐⓑⓒⓓ

MARK YOUR SCORE AND EVALUATE YOURSELF ACCORDINGLY

Above 100: *EXCELLENT*. You are amongst the top scorers, but strive to retain it because in agriculture there are great swings.

80 to 100: *VERY GOOD*. With about a similar performance in weak areas, you can become the top scorer. But be aware! there are many a slip between cup and lips.

70 to 80: *FAIR*. You still have a long way to go. Success is merely a chance factor.

Less than 70: *POOR*: Why are you even thinking of Agricultural Competitive Exam now. Work, Work and Work more. Who knows, your labour will pay off, as it does as a rule.

Notes

AGRISET - 7

ANSWER SHEET FOR GENERAL AGRICULTURE MODEL PAPER

- I have provided you MCQs model test on general agriculture for TOP (Target Oriented Preparation) for the Agricultural Competitive Exams.
- You have to mark all your responses only on the answer sheet provided.
- All items carry equal marks. Attempt all items. Your total marks will depend only on the number of correct responses marked by you in the Answer sheet. There is no negative marking. Award yourself one point for each correct answer.

ANSWER SHEET

1 ⓐⓑⓒⓓ	6 ⓐⓑⓒⓓ	11 ⓐⓑⓒⓓ	16 ⓐⓑⓒⓓ	21 ⓐⓑⓒⓓ	26 ⓐⓑⓒⓓ
2 ⓐⓑⓒⓓ	7 ⓐⓑⓒⓓ	12 ⓐⓑⓒⓓ	17 ⓐⓑⓒⓓ	22 ⓐⓑⓒⓓ	27 ⓐⓑⓒⓓ
3 ⓐⓑⓒⓓ	8 ⓐⓑⓒⓓ	13 ⓐⓑⓒⓓ	18 ⓐⓑⓒⓓ	23 ⓐⓑⓒⓓ	28 ⓐⓑⓒⓓ
4 ⓐⓑⓒⓓ	9 ⓐⓑⓒⓓ	14 ⓐⓑⓒⓓ	19 ⓐⓑⓒⓓ	24 ⓐⓑⓒⓓ	29 ⓐⓑⓒⓓ
5 ⓐⓑⓒⓓ	10 ⓐⓑⓒⓓ	15 ⓐⓑⓒⓓ	20 ⓐⓑⓒⓓ	25 ⓐⓑⓒⓓ	30 ⓐⓑⓒⓓ
31 ⓐⓑⓒⓓ	36 ⓐⓑⓒⓓ	41 ⓐⓑⓒⓓ	46 ⓐⓑⓒⓓ	51 ⓐⓑⓒⓓ	56 ⓐⓑⓒⓓ
32 ⓐⓑⓒⓓ	37 ⓐⓑⓒⓓ	42 ⓐⓑⓒⓓ	47 ⓐⓑⓒⓓ	52 ⓐⓑⓒⓓ	57 ⓐⓑⓒⓓ
33 ⓐⓑⓒⓓ	38 ⓐⓑⓒⓓ	43 ⓐⓑⓒⓓ	48 ⓐⓑⓒⓓ	53 ⓐⓑⓒⓓ	58 ⓐⓑⓒⓓ
34 ⓐⓑⓒⓓ	39 ⓐⓑⓒⓓ	44 ⓐⓑⓒⓓ	49 ⓐⓑⓒⓓ	54 ⓐⓑⓒⓓ	59 ⓐⓑⓒⓓ
35 ⓐⓑⓒⓓ	40 ⓐⓑⓒⓓ	45 ⓐⓑⓒⓓ	50 ⓐⓑⓒⓓ	55 ⓐⓑⓒⓓ	60 ⓐⓑⓒⓓ
61 ⓐⓑⓒⓓ	66 ⓐⓑⓒⓓ	71 ⓐⓑⓒⓓ	76 ⓐⓑⓒⓓ	81 ⓐⓑⓒⓓ	86 ⓐⓑⓒⓓ
62 ⓐⓑⓒⓓ	67 ⓐⓑⓒⓓ	72 ⓐⓑⓒⓓ	77 ⓐⓑⓒⓓ	82 ⓐⓑⓒⓓ	87 ⓐⓑⓒⓓ
63 ⓐⓑⓒⓓ	68 ⓐⓑⓒⓓ	73 ⓐⓑⓒⓓ	78 ⓐⓑⓒⓓ	83 ⓐⓑⓒⓓ	88 ⓐⓑⓒⓓ
64 ⓐⓑⓒⓓ	69 ⓐⓑⓒⓓ	74 ⓐⓑⓒⓓ	79 ⓐⓑⓒⓓ	84 ⓐⓑⓒⓓ	89 ⓐⓑⓒⓓ
65 ⓐⓑⓒⓓ	70 ⓐⓑⓒⓓ	75 ⓐⓑⓒⓓ	80 ⓐⓑⓒⓓ	85 ⓐⓑⓒⓓ	90 ⓐⓑⓒⓓ
91 ⓐⓑⓒⓓ	96 ⓐⓑⓒⓓ	101 ⓐⓑⓒⓓ	106 ⓐⓑⓒⓓ	111 ⓐⓑⓒⓓ	116 ⓐⓑⓒⓓ
92 ⓐⓑⓒⓓ	97 ⓐⓑⓒⓓ	102 ⓐⓑⓒⓓ	107 ⓐⓑⓒⓓ	112 ⓐⓑⓒⓓ	117 ⓐⓑⓒⓓ
93 ⓐⓑⓒⓓ	98 ⓐⓑⓒⓓ	103 ⓐⓑⓒⓓ	108 ⓐⓑⓒⓓ	113 ⓐⓑⓒⓓ	118 ⓐⓑⓒⓓ
94 ⓐⓑⓒⓓ	99 ⓐⓑⓒⓓ	104 ⓐⓑⓒⓓ	109 ⓐⓑⓒⓓ	114 ⓐⓑⓒⓓ	119 ⓐⓑⓒⓓ
95 ⓐⓑⓒⓓ	100 ⓐⓑⓒⓓ	105 ⓐⓑⓒⓓ	110 ⓐⓑⓒⓓ	115 ⓐⓑⓒⓓ	120 ⓐⓑⓒⓓ

MARK YOUR SCORE AND EVALUATE YOURSELF ACCORDINGLY

Above 100: *EXCELLENT*. You are amongst the top scorers, but strive to retain it because in agriculture there are great swings.

80 to 100: *VERY GOOD*. With about a similar performance in weak areas, you can become the top scorer. But be aware! there are many a slip between cup and lips.

70 to 80: *FAIR*. You still have a long way to go. Success is merely a chance factor.

Less than 70: *POOR*: Why are you even thinking of Agricultural Competitive Exam now. Work, Work and Work more. Who knows, your labour will pay off, as it does as a rule.

Notes

AGRISET - 8

ANSWER SHEET FOR GENERAL AGRICULTURE MODEL PAPER

- I have provided you MCQs model test on general agriculture for TOP (Target Oriented Preparation) for the Agricultural Competitive Exams.
- You have to mark all your responses only on the answer sheet provided.
- All items carry equal marks. Attempt all items. Your total marks will depend only on the number of correct responses marked by you in the Answer sheet. There is no negative marking. Award yourself one point for each correct answer.

ANSWER SHEET

1	ⓐⓑⓒⓓ	6	ⓐⓑⓒⓓ	11	ⓐⓑⓒⓓ	16	ⓐⓑⓒⓓ	21	ⓐⓑⓒⓓ	26	ⓐⓑⓒⓓ
2	ⓐⓑⓒⓓ	7	ⓐⓑⓒⓓ	12	ⓐⓑⓒⓓ	17	ⓐⓑⓒⓓ	22	ⓐⓑⓒⓓ	27	ⓐⓑⓒⓓ
3	ⓐⓑⓒⓓ	8	ⓐⓑⓒⓓ	13	ⓐⓑⓒⓓ	18	ⓐⓑⓒⓓ	23	ⓐⓑⓒⓓ	28	ⓐⓑⓒⓓ
4	ⓐⓑⓒⓓ	9	ⓐⓑⓒⓓ	14	ⓐⓑⓒⓓ	19	ⓐⓑⓒⓓ	24	ⓐⓑⓒⓓ	29	ⓐⓑⓒⓓ
5	ⓐⓑⓒⓓ	10	ⓐⓑⓒⓓ	15	ⓐⓑⓒⓓ	20	ⓐⓑⓒⓓ	25	ⓐⓑⓒⓓ	30	ⓐⓑⓒⓓ
31	ⓐⓑⓒⓓ	36	ⓐⓑⓒⓓ	41	ⓐⓑⓒⓓ	46	ⓐⓑⓒⓓ	51	ⓐⓑⓒⓓ	56	ⓐⓑⓒⓓ
32	ⓐⓑⓒⓓ	37	ⓐⓑⓒⓓ	42	ⓐⓑⓒⓓ	47	ⓐⓑⓒⓓ	52	ⓐⓑⓒⓓ	57	ⓐⓑⓒⓓ
33	ⓐⓑⓒⓓ	38	ⓐⓑⓒⓓ	43	ⓐⓑⓒⓓ	48	ⓐⓑⓒⓓ	53	ⓐⓑⓒⓓ	58	ⓐⓑⓒⓓ
34	ⓐⓑⓒⓓ	39	ⓐⓑⓒⓓ	44	ⓐⓑⓒⓓ	49	ⓐⓑⓒⓓ	54	ⓐⓑⓒⓓ	59	ⓐⓑⓒⓓ
35	ⓐⓑⓒⓓ	40	ⓐⓑⓒⓓ	45	ⓐⓑⓒⓓ	50	ⓐⓑⓒⓓ	55	ⓐⓑⓒⓓ	60	ⓐⓑⓒⓓ
61	ⓐⓑⓒⓓ	66	ⓐⓑⓒⓓ	71	ⓐⓑⓒⓓ	76	ⓐⓑⓒⓓ	81	ⓐⓑⓒⓓ	86	ⓐⓑⓒⓓ
62	ⓐⓑⓒⓓ	67	ⓐⓑⓒⓓ	72	ⓐⓑⓒⓓ	77	ⓐⓑⓒⓓ	82	ⓐⓑⓒⓓ	87	ⓐⓑⓒⓓ
63	ⓐⓑⓒⓓ	68	ⓐⓑⓒⓓ	73	ⓐⓑⓒⓓ	78	ⓐⓑⓒⓓ	83	ⓐⓑⓒⓓ	88	ⓐⓑⓒⓓ
64	ⓐⓑⓒⓓ	69	ⓐⓑⓒⓓ	74	ⓐⓑⓒⓓ	79	ⓐⓑⓒⓓ	84	ⓐⓑⓒⓓ	89	ⓐⓑⓒⓓ
65	ⓐⓑⓒⓓ	70	ⓐⓑⓒⓓ	75	ⓐⓑⓒⓓ	80	ⓐⓑⓒⓓ	85	ⓐⓑⓒⓓ	90	ⓐⓑⓒⓓ
91	ⓐⓑⓒⓓ	96	ⓐⓑⓒⓓ	101	ⓐⓑⓒⓓ	106	ⓐⓑⓒⓓ	111	ⓐⓑⓒⓓ	116	ⓐⓑⓒⓓ
92	ⓐⓑⓒⓓ	97	ⓐⓑⓒⓓ	102	ⓐⓑⓒⓓ	107	ⓐⓑⓒⓓ	112	ⓐⓑⓒⓓ	117	ⓐⓑⓒⓓ
93	ⓐⓑⓒⓓ	98	ⓐⓑⓒⓓ	103	ⓐⓑⓒⓓ	108	ⓐⓑⓒⓓ	113	ⓐⓑⓒⓓ	118	ⓐⓑⓒⓓ
94	ⓐⓑⓒⓓ	99	ⓐⓑⓒⓓ	104	ⓐⓑⓒⓓ	109	ⓐⓑⓒⓓ	114	ⓐⓑⓒⓓ	119	ⓐⓑⓒⓓ
95	ⓐⓑⓒⓓ	100	ⓐⓑⓒⓓ	105	ⓐⓑⓒⓓ	110	ⓐⓑⓒⓓ	115	ⓐⓑⓒⓓ	120	ⓐⓑⓒⓓ

MARK YOUR SCORE AND EVALUATE YOURSELF ACCORDINGLY

Above 100: *EXCELLENT*. You are amongst the top scorers, but strive to retain it because in agriculture there are great swings.

80 to 100: *VERY GOOD*. With about a similar performance in weak areas, you can become the top scorer. But be aware! there are many a slip between cup and lips.

70 to 80: *FAIR*. You still have a long way to go. Success is merely a chance factor.

Less than 70: *POOR*: Why are you even thinking of Agricultural Competitive Exam now. Work, Work and Work more. Who knows, your labour will pay off, as it does as a rule.

Notes

AGRISET - 9

ANSWER SHEET FOR GENERAL AGRICULTURE MODEL PAPER

- I have provided you MCQs model test on general agriculture for TOP (Target Oriented Preparation) for the Agricultural Competitive Exams.
- You have to mark all your responses only on the answer sheet provided.
- All items carry equal marks. Attempt all items. Your total marks will depend only on the number of correct responses marked by you in the Answer sheet. There is no negative marking. Award yourself one point for each correct answer.

ANSWER SHEET

1 ⓐⓑⓒⓓ	6 ⓐⓑⓒⓓ	11 ⓐⓑⓒⓓ	16 ⓐⓑⓒⓓ	21 ⓐⓑⓒⓓ	26 ⓐⓑⓒⓓ
2 ⓐⓑⓒⓓ	7 ⓐⓑⓒⓓ	12 ⓐⓑⓒⓓ	17 ⓐⓑⓒⓓ	22 ⓐⓑⓒⓓ	27 ⓐⓑⓒⓓ
3 ⓐⓑⓒⓓ	8 ⓐⓑⓒⓓ	13 ⓐⓑⓒⓓ	18 ⓐⓑⓒⓓ	23 ⓐⓑⓒⓓ	28 ⓐⓑⓒⓓ
4 ⓐⓑⓒⓓ	9 ⓐⓑⓒⓓ	14 ⓐⓑⓒⓓ	19 ⓐⓑⓒⓓ	24 ⓐⓑⓒⓓ	29 ⓐⓑⓒⓓ
5 ⓐⓑⓒⓓ	10 ⓐⓑⓒⓓ	15 ⓐⓑⓒⓓ	20 ⓐⓑⓒⓓ	25 ⓐⓑⓒⓓ	30 ⓐⓑⓒⓓ
31 ⓐⓑⓒⓓ	36 ⓐⓑⓒⓓ	41 ⓐⓑⓒⓓ	46 ⓐⓑⓒⓓ	51 ⓐⓑⓒⓓ	56 ⓐⓑⓒⓓ
32 ⓐⓑⓒⓓ	37 ⓐⓑⓒⓓ	42 ⓐⓑⓒⓓ	47 ⓐⓑⓒⓓ	52 ⓐⓑⓒⓓ	57 ⓐⓑⓒⓓ
33 ⓐⓑⓒⓓ	38 ⓐⓑⓒⓓ	43 ⓐⓑⓒⓓ	48 ⓐⓑⓒⓓ	53 ⓐⓑⓒⓓ	58 ⓐⓑⓒⓓ
34 ⓐⓑⓒⓓ	39 ⓐⓑⓒⓓ	44 ⓐⓑⓒⓓ	49 ⓐⓑⓒⓓ	54 ⓐⓑⓒⓓ	59 ⓐⓑⓒⓓ
35 ⓐⓑⓒⓓ	40 ⓐⓑⓒⓓ	45 ⓐⓑⓒⓓ	50 ⓐⓑⓒⓓ	55 ⓐⓑⓒⓓ	60 ⓐⓑⓒⓓ
61 ⓐⓑⓒⓓ	66 ⓐⓑⓒⓓ	71 ⓐⓑⓒⓓ	76 ⓐⓑⓒⓓ	81 ⓐⓑⓒⓓ	86 ⓐⓑⓒⓓ
62 ⓐⓑⓒⓓ	67 ⓐⓑⓒⓓ	72 ⓐⓑⓒⓓ	77 ⓐⓑⓒⓓ	82 ⓐⓑⓒⓓ	87 ⓐⓑⓒⓓ
63 ⓐⓑⓒⓓ	68 ⓐⓑⓒⓓ	73 ⓐⓑⓒⓓ	78 ⓐⓑⓒⓓ	83 ⓐⓑⓒⓓ	88 ⓐⓑⓒⓓ
64 ⓐⓑⓒⓓ	69 ⓐⓑⓒⓓ	74 ⓐⓑⓒⓓ	79 ⓐⓑⓒⓓ	84 ⓐⓑⓒⓓ	89 ⓐⓑⓒⓓ
65 ⓐⓑⓒⓓ	70 ⓐⓑⓒⓓ	75 ⓐⓑⓒⓓ	80 ⓐⓑⓒⓓ	85 ⓐⓑⓒⓓ	90 ⓐⓑⓒⓓ
91 ⓐⓑⓒⓓ	96 ⓐⓑⓒⓓ	101 ⓐⓑⓒⓓ	106 ⓐⓑⓒⓓ	111 ⓐⓑⓒⓓ	116 ⓐⓑⓒⓓ
92 ⓐⓑⓒⓓ	97 ⓐⓑⓒⓓ	102 ⓐⓑⓒⓓ	107 ⓐⓑⓒⓓ	112 ⓐⓑⓒⓓ	117 ⓐⓑⓒⓓ
93 ⓐⓑⓒⓓ	98 ⓐⓑⓒⓓ	103 ⓐⓑⓒⓓ	108 ⓐⓑⓒⓓ	113 ⓐⓑⓒⓓ	118 ⓐⓑⓒⓓ
94 ⓐⓑⓒⓓ	99 ⓐⓑⓒⓓ	104 ⓐⓑⓒⓓ	109 ⓐⓑⓒⓓ	114 ⓐⓑⓒⓓ	119 ⓐⓑⓒⓓ
95 ⓐⓑⓒⓓ	100 ⓐⓑⓒⓓ	105 ⓐⓑⓒⓓ	110 ⓐⓑⓒⓓ	115 ⓐⓑⓒⓓ	120 ⓐⓑⓒⓓ

MARK YOUR SCORE AND EVALUATE YOURSELF ACCORDINGLY

Above 100: *EXCELLENT*. You are amongst the top scorers, but strive to retain it because in agriculture there are great swings.

80 to 100: *VERY GOOD*. With about a similar performance in weak areas, you can become the top scorer. But be aware! there are many a slip between cup and lips.

70 to 80: *FAIR*. You still have a long way to go. Success is merely a chance factor.

Less than 70: *POOR*: Why are you even thinking of Agricultural Competitive Exam now. Work, Work and Work more. Who knows, your labour will pay off, as it does as a rule.

Notes

AGRISET - 10

ANSWER SHEET FOR GENERAL AGRICULTURE MODEL PAPER

- I have provided you MCQs model test on general agriculture for TOP (Target Oriented Preparation) for the Agricultural Competitive Exams.
- You have to mark all your responses only on the answer sheet provided.
- All items carry equal marks. Attempt all items. Your total marks will depend only on the number of correct responses marked by you in the Answer sheet. There is no negative marking. Award yourself one point for each correct answer.

ANSWER SHEET

No.	Options	No.	Options	No.	Options	No.	Options	No.	Options	No.	Options
1	ⓐⓑⓒⓓ	6	ⓐⓑⓒⓓ	11	ⓐⓑⓒⓓ	16	ⓐⓑⓒⓓ	21	ⓐⓑⓒⓓ	26	ⓐⓑⓒⓓ
2	ⓐⓑⓒⓓ	7	ⓐⓑⓒⓓ	12	ⓐⓑⓒⓓ	17	ⓐⓑⓒⓓ	22	ⓐⓑⓒⓓ	27	ⓐⓑⓒⓓ
3	ⓐⓑⓒⓓ	8	ⓐⓑⓒⓓ	13	ⓐⓑⓒⓓ	18	ⓐⓑⓒⓓ	23	ⓐⓑⓒⓓ	28	ⓐⓑⓒⓓ
4	ⓐⓑⓒⓓ	9	ⓐⓑⓒⓓ	14	ⓐⓑⓒⓓ	19	ⓐⓑⓒⓓ	24	ⓐⓑⓒⓓ	29	ⓐⓑⓒⓓ
5	ⓐⓑⓒⓓ	10	ⓐⓑⓒⓓ	15	ⓐⓑⓒⓓ	20	ⓐⓑⓒⓓ	25	ⓐⓑⓒⓓ	30	ⓐⓑⓒⓓ
31	ⓐⓑⓒⓓ	36	ⓐⓑⓒⓓ	41	ⓐⓑⓒⓓ	46	ⓐⓑⓒⓓ	51	ⓐⓑⓒⓓ	56	ⓐⓑⓒⓓ
32	ⓐⓑⓒⓓ	37	ⓐⓑⓒⓓ	42	ⓐⓑⓒⓓ	47	ⓐⓑⓒⓓ	52	ⓐⓑⓒⓓ	57	ⓐⓑⓒⓓ
33	ⓐⓑⓒⓓ	38	ⓐⓑⓒⓓ	43	ⓐⓑⓒⓓ	48	ⓐⓑⓒⓓ	53	ⓐⓑⓒⓓ	58	ⓐⓑⓒⓓ
34	ⓐⓑⓒⓓ	39	ⓐⓑⓒⓓ	44	ⓐⓑⓒⓓ	49	ⓐⓑⓒⓓ	54	ⓐⓑⓒⓓ	59	ⓐⓑⓒⓓ
35	ⓐⓑⓒⓓ	40	ⓐⓑⓒⓓ	45	ⓐⓑⓒⓓ	50	ⓐⓑⓒⓓ	55	ⓐⓑⓒⓓ	60	ⓐⓑⓒⓓ
61	ⓐⓑⓒⓓ	66	ⓐⓑⓒⓓ	71	ⓐⓑⓒⓓ	76	ⓐⓑⓒⓓ	81	ⓐⓑⓒⓓ	86	ⓐⓑⓒⓓ
62	ⓐⓑⓒⓓ	67	ⓐⓑⓒⓓ	72	ⓐⓑⓒⓓ	77	ⓐⓑⓒⓓ	82	ⓐⓑⓒⓓ	87	ⓐⓑⓒⓓ
63	ⓐⓑⓒⓓ	68	ⓐⓑⓒⓓ	73	ⓐⓑⓒⓓ	78	ⓐⓑⓒⓓ	83	ⓐⓑⓒⓓ	88	ⓐⓑⓒⓓ
64	ⓐⓑⓒⓓ	69	ⓐⓑⓒⓓ	74	ⓐⓑⓒⓓ	79	ⓐⓑⓒⓓ	84	ⓐⓑⓒⓓ	89	ⓐⓑⓒⓓ
65	ⓐⓑⓒⓓ	70	ⓐⓑⓒⓓ	75	ⓐⓑⓒⓓ	80	ⓐⓑⓒⓓ	85	ⓐⓑⓒⓓ	90	ⓐⓑⓒⓓ
91	ⓐⓑⓒⓓ	96	ⓐⓑⓒⓓ	101	ⓐⓑⓒⓓ	106	ⓐⓑⓒⓓ	111	ⓐⓑⓒⓓ	116	ⓐⓑⓒⓓ
92	ⓐⓑⓒⓓ	97	ⓐⓑⓒⓓ	102	ⓐⓑⓒⓓ	107	ⓐⓑⓒⓓ	112	ⓐⓑⓒⓓ	117	ⓐⓑⓒⓓ
93	ⓐⓑⓒⓓ	98	ⓐⓑⓒⓓ	103	ⓐⓑⓒⓓ	108	ⓐⓑⓒⓓ	113	ⓐⓑⓒⓓ	118	ⓐⓑⓒⓓ
94	ⓐⓑⓒⓓ	99	ⓐⓑⓒⓓ	104	ⓐⓑⓒⓓ	109	ⓐⓑⓒⓓ	114	ⓐⓑⓒⓓ	119	ⓐⓑⓒⓓ
95	ⓐⓑⓒⓓ	100	ⓐⓑⓒⓓ	105	ⓐⓑⓒⓓ	110	ⓐⓑⓒⓓ	115	ⓐⓑⓒⓓ	120	ⓐⓑⓒⓓ

MARK YOUR SCORE AND EVALUATE YOURSELF ACCORDINGLY

Above 100: *EXCELLENT*. You are amongst the top scorers, but strive to retain it because in agriculture there are great swings.

80 to 100: *VERY GOOD*. With about a similar performance in weak areas, you can become the top scorer. But be aware! there are many a slip between cup and lips.

70 to 80: *FAIR*. You still have a long way to go. Success is merely a chance factor.

Less than 70: *POOR*: Why are you even thinking of Agricultural Competitive Exam now. Work, Work and Work more. Who knows, your labour will pay off, as it does as a rule.

Notes

AGRISET - 11

ANSWER SHEET FOR GENERAL AGRICULTURE MODEL PAPER

- I have provided you MCQs model test on general agriculture for TOP (Target Oriented Preparation) for the Agricultural Competitive Exams.
- You have to mark all your responses only on the answer sheet provided.
- All items carry equal marks. Attempt all items. Your total marks will depend only on the number of correct responses marked by you in the Answer sheet. There is no negative marking. Award yourself one point for each correct answer.

ANSWER SHEET

1	ⓐⓑⓒⓓ	6	ⓐⓑⓒⓓ	11	ⓐⓑⓒⓓ	16	ⓐⓑⓒⓓ	21	ⓐⓑⓒⓓ
2	ⓐⓑⓒⓓ	7	ⓐⓑⓒⓓ	12	ⓐⓑⓒⓓ	17	ⓐⓑⓒⓓ	22	ⓐⓑⓒⓓ
3	ⓐⓑⓒⓓ	8	ⓐⓑⓒⓓ	13	ⓐⓑⓒⓓ	18	ⓐⓑⓒⓓ	23	ⓐⓑⓒⓓ
4	ⓐⓑⓒⓓ	9	ⓐⓑⓒⓓ	14	ⓐⓑⓒⓓ	19	ⓐⓑⓒⓓ	24	ⓐⓑⓒⓓ
5	ⓐⓑⓒⓓ	10	ⓐⓑⓒⓓ	15	ⓐⓑⓒⓓ	20	ⓐⓑⓒⓓ	25	ⓐⓑⓒⓓ
26	ⓐⓑⓒⓓ	31	ⓐⓑⓒⓓ	36	ⓐⓑⓒⓓ	41	ⓐⓑⓒⓓ	46	ⓐⓑⓒⓓ
27	ⓐⓑⓒⓓ	32	ⓐⓑⓒⓓ	37	ⓐⓑⓒⓓ	42	ⓐⓑⓒⓓ	47	ⓐⓑⓒⓓ
28	ⓐⓑⓒⓓ	33	ⓐⓑⓒⓓ	38	ⓐⓑⓒⓓ	43	ⓐⓑⓒⓓ	48	ⓐⓑⓒⓓ
29	ⓐⓑⓒⓓ	34	ⓐⓑⓒⓓ	39	ⓐⓑⓒⓓ	44	ⓐⓑⓒⓓ	49	ⓐⓑⓒⓓ
30	ⓐⓑⓒⓓ	35	ⓐⓑⓒⓓ	40	ⓐⓑⓒⓓ	45	ⓐⓑⓒⓓ	50	ⓐⓑⓒⓓ
51	ⓐⓑⓒⓓ	56	ⓐⓑⓒⓓ	61	ⓐⓑⓒⓓ	66	ⓐⓑⓒⓓ	71	ⓐⓑⓒⓓ
52	ⓐⓑⓒⓓ	57	ⓐⓑⓒⓓ	62	ⓐⓑⓒⓓ	67	ⓐⓑⓒⓓ	72	ⓐⓑⓒⓓ
53	ⓐⓑⓒⓓ	58	ⓐⓑⓒⓓ	63	ⓐⓑⓒⓓ	68	ⓐⓑⓒⓓ	73	ⓐⓑⓒⓓ
54	ⓐⓑⓒⓓ	59	ⓐⓑⓒⓓ	64	ⓐⓑⓒⓓ	69	ⓐⓑⓒⓓ	74	ⓐⓑⓒⓓ
55	ⓐⓑⓒⓓ	60	ⓐⓑⓒⓓ	65	ⓐⓑⓒⓓ	70	ⓐⓑⓒⓓ	75	ⓐⓑⓒⓓ
76	ⓐⓑⓒⓓ	81	ⓐⓑⓒⓓ	86	ⓐⓑⓒⓓ	91	ⓐⓑⓒⓓ	96	ⓐⓑⓒⓓ
77	ⓐⓑⓒⓓ	82	ⓐⓑⓒⓓ	87	ⓐⓑⓒⓓ	92	ⓐⓑⓒⓓ	97	ⓐⓑⓒⓓ
78	ⓐⓑⓒⓓ	83	ⓐⓑⓒⓓ	88	ⓐⓑⓒⓓ	93	ⓐⓑⓒⓓ	98	ⓐⓑⓒⓓ
79	ⓐⓑⓒⓓ	84	ⓐⓑⓒⓓ	89	ⓐⓑⓒⓓ	94	ⓐⓑⓒⓓ	99	ⓐⓑⓒⓓ
80	ⓐⓑⓒⓓ	85	ⓐⓑⓒⓓ	90	ⓐⓑⓒⓓ	95	ⓐⓑⓒⓓ	100	ⓐⓑⓒⓓ

MARK YOUR SCORE AND EVALUATE YOURSELF ACCORDINGLY

Above 85: *EXCELLENT*. You are amongst the top scorers, but strive to retain it because in agriculture there are great swings.

65 to 85: *VERY GOOD*. With about a similar performance in weak areas, you can become the top scorer. But be aware! there are many a slip between cup and lips.

55 to 65: *FAIR*. You still have a long way to go. Success is merely a chance factor.

Less than 55: *POOR*: Why are you even thinking of Agricultural Competitive Exam now. Work, Work and Work more. Who knows, your labour will pay off, as it does as a rule.

Notes

AGRISET - 12

ANSWER SHEET FOR GENERAL AGRICULTURE MODEL PAPER

- I have provided you MCQs model test on general agriculture for TOP (Target Oriented Preparation) for the Agricultural Competitive Exams.
- You have to mark all your responses only on the answer sheet provided.
- All items carry equal marks. Attempt all items. Your total marks will depend only on the number of correct responses marked by you in the Answer sheet. There is no negative marking. Award yourself one point for each correct answer.

ANSWER SHEET

1 ⓐⓑⓒⓓ	6 ⓐⓑⓒⓓ	11 ⓐⓑⓒⓓ	16 ⓐⓑⓒⓓ	21 ⓐⓑⓒⓓ
2 ⓐⓑⓒⓓ	7 ⓐⓑⓒⓓ	12 ⓐⓑⓒⓓ	17 ⓐⓑⓒⓓ	22 ⓐⓑⓒⓓ
3 ⓐⓑⓒⓓ	8 ⓐⓑⓒⓓ	13 ⓐⓑⓒⓓ	18 ⓐⓑⓒⓓ	23 ⓐⓑⓒⓓ
4 ⓐⓑⓒⓓ	9 ⓐⓑⓒⓓ	14 ⓐⓑⓒⓓ	19 ⓐⓑⓒⓓ	24 ⓐⓑⓒⓓ
5 ⓐⓑⓒⓓ	10 ⓐⓑⓒⓓ	15 ⓐⓑⓒⓓ	20 ⓐⓑⓒⓓ	25 ⓐⓑⓒⓓ
26 ⓐⓑⓒⓓ	31 ⓐⓑⓒⓓ	36 ⓐⓑⓒⓓ	41 ⓐⓑⓒⓓ	46 ⓐⓑⓒⓓ
27 ⓐⓑⓒⓓ	32 ⓐⓑⓒⓓ	37 ⓐⓑⓒⓓ	42 ⓐⓑⓒⓓ	47 ⓐⓑⓒⓓ
28 ⓐⓑⓒⓓ	33 ⓐⓑⓒⓓ	38 ⓐⓑⓒⓓ	43 ⓐⓑⓒⓓ	48 ⓐⓑⓒⓓ
29 ⓐⓑⓒⓓ	34 ⓐⓑⓒⓓ	39 ⓐⓑⓒⓓ	44 ⓐⓑⓒⓓ	49 ⓐⓑⓒⓓ
30 ⓐⓑⓒⓓ	35 ⓐⓑⓒⓓ	40 ⓐⓑⓒⓓ	45 ⓐⓑⓒⓓ	50 ⓐⓑⓒⓓ
51 ⓐⓑⓒⓓ	56 ⓐⓑⓒⓓ	61 ⓐⓑⓒⓓ	66 ⓐⓑⓒⓓ	71 ⓐⓑⓒⓓ
52 ⓐⓑⓒⓓ	57 ⓐⓑⓒⓓ	62 ⓐⓑⓒⓓ	67 ⓐⓑⓒⓓ	72 ⓐⓑⓒⓓ
53 ⓐⓑⓒⓓ	58 ⓐⓑⓒⓓ	63 ⓐⓑⓒⓓ	68 ⓐⓑⓒⓓ	73 ⓐⓑⓒⓓ
54 ⓐⓑⓒⓓ	59 ⓐⓑⓒⓓ	64 ⓐⓑⓒⓓ	69 ⓐⓑⓒⓓ	74 ⓐⓑⓒⓓ
55 ⓐⓑⓒⓓ	60 ⓐⓑⓒⓓ	65 ⓐⓑⓒⓓ	70 ⓐⓑⓒⓓ	75 ⓐⓑⓒⓓ
76 ⓐⓑⓒⓓ	81 ⓐⓑⓒⓓ	86 ⓐⓑⓒⓓ	91 ⓐⓑⓒⓓ	96 ⓐⓑⓒⓓ
77 ⓐⓑⓒⓓ	82 ⓐⓑⓒⓓ	87 ⓐⓑⓒⓓ	92 ⓐⓑⓒⓓ	97 ⓐⓑⓒⓓ
78 ⓐⓑⓒⓓ	83 ⓐⓑⓒⓓ	88 ⓐⓑⓒⓓ	93 ⓐⓑⓒⓓ	98 ⓐⓑⓒⓓ
79 ⓐⓑⓒⓓ	84 ⓐⓑⓒⓓ	89 ⓐⓑⓒⓓ	94 ⓐⓑⓒⓓ	99 ⓐⓑⓒⓓ
80 ⓐⓑⓒⓓ	85 ⓐⓑⓒⓓ	90 ⓐⓑⓒⓓ	95 ⓐⓑⓒⓓ	100 ⓐⓑⓒⓓ

MARK YOUR SCORE AND EVALUATE YOURSELF ACCORDINGLY

Above 85: *EXCELLENT*. You are amongst the top scorers, but strive to retain it because in agriculture there are great swings.

65 to 85: *VERY GOOD*. With about a similar performance in weak areas, you can become the top scorer. But be aware! there are many a slip between cup and lips.

55 to 65: *FAIR*. You still have a long way to go. Success is merely a chance factor.

Less than 55: *POOR*: Why are you even thinking of Agricultural Competitive Exam now. Work, Work and Work more. Who knows, your labour will pay off, as it does as a rule.

Notes

AGRISET - 13

ANSWER SHEET FOR GENERAL AGRICULTURE MODEL PAPER

- I have provided you MCQs model test on general agriculture for TOP (Target Oriented Preparation) for the Agricultural Competitive Exams.
- You have to mark all your responses only on the answer sheet provided.
- All items carry equal marks. Attempt all items. Your total marks will depend only on the number of correct responses marked by you in the Answer sheet. There is no negative marking. Award yourself one point for each correct answer.

ANSWER SHEET

1	ⓐⓑⓒⓓ	6	ⓐⓑⓒⓓ	11	ⓐⓑⓒⓓ	16	ⓐⓑⓒⓓ	21	ⓐⓑⓒⓓ
2	ⓐⓑⓒⓓ	7	ⓐⓑⓒⓓ	12	ⓐⓑⓒⓓ	17	ⓐⓑⓒⓓ	22	ⓐⓑⓒⓓ
3	ⓐⓑⓒⓓ	8	ⓐⓑⓒⓓ	13	ⓐⓑⓒⓓ	18	ⓐⓑⓒⓓ	23	ⓐⓑⓒⓓ
4	ⓐⓑⓒⓓ	9	ⓐⓑⓒⓓ	14	ⓐⓑⓒⓓ	19	ⓐⓑⓒⓓ	24	ⓐⓑⓒⓓ
5	ⓐⓑⓒⓓ	10	ⓐⓑⓒⓓ	15	ⓐⓑⓒⓓ	20	ⓐⓑⓒⓓ	25	ⓐⓑⓒⓓ
26	ⓐⓑⓒⓓ	31	ⓐⓑⓒⓓ	36	ⓐⓑⓒⓓ	41	ⓐⓑⓒⓓ	46	ⓐⓑⓒⓓ
27	ⓐⓑⓒⓓ	32	ⓐⓑⓒⓓ	37	ⓐⓑⓒⓓ	42	ⓐⓑⓒⓓ	47	ⓐⓑⓒⓓ
28	ⓐⓑⓒⓓ	33	ⓐⓑⓒⓓ	38	ⓐⓑⓒⓓ	43	ⓐⓑⓒⓓ	48	ⓐⓑⓒⓓ
29	ⓐⓑⓒⓓ	34	ⓐⓑⓒⓓ	39	ⓐⓑⓒⓓ	44	ⓐⓑⓒⓓ	49	ⓐⓑⓒⓓ
30	ⓐⓑⓒⓓ	35	ⓐⓑⓒⓓ	40	ⓐⓑⓒⓓ	45	ⓐⓑⓒⓓ	50	ⓐⓑⓒⓓ
51	ⓐⓑⓒⓓ	56	ⓐⓑⓒⓓ	61	ⓐⓑⓒⓓ	66	ⓐⓑⓒⓓ	71	ⓐⓑⓒⓓ
52	ⓐⓑⓒⓓ	57	ⓐⓑⓒⓓ	62	ⓐⓑⓒⓓ	67	ⓐⓑⓒⓓ	72	ⓐⓑⓒⓓ
53	ⓐⓑⓒⓓ	58	ⓐⓑⓒⓓ	63	ⓐⓑⓒⓓ	68	ⓐⓑⓒⓓ	73	ⓐⓑⓒⓓ
54	ⓐⓑⓒⓓ	59	ⓐⓑⓒⓓ	64	ⓐⓑⓒⓓ	69	ⓐⓑⓒⓓ	74	ⓐⓑⓒⓓ
55	ⓐⓑⓒⓓ	60	ⓐⓑⓒⓓ	65	ⓐⓑⓒⓓ	70	ⓐⓑⓒⓓ	75	ⓐⓑⓒⓓ
76	ⓐⓑⓒⓓ	81	ⓐⓑⓒⓓ	86	ⓐⓑⓒⓓ	91	ⓐⓑⓒⓓ	96	ⓐⓑⓒⓓ
77	ⓐⓑⓒⓓ	82	ⓐⓑⓒⓓ	87	ⓐⓑⓒⓓ	92	ⓐⓑⓒⓓ	97	ⓐⓑⓒⓓ
78	ⓐⓑⓒⓓ	83	ⓐⓑⓒⓓ	88	ⓐⓑⓒⓓ	93	ⓐⓑⓒⓓ	98	ⓐⓑⓒⓓ
79	ⓐⓑⓒⓓ	84	ⓐⓑⓒⓓ	89	ⓐⓑⓒⓓ	94	ⓐⓑⓒⓓ	99	ⓐⓑⓒⓓ
80	ⓐⓑⓒⓓ	85	ⓐⓑⓒⓓ	90	ⓐⓑⓒⓓ	95	ⓐⓑⓒⓓ	100	ⓐⓑⓒⓓ

MARK YOUR SCORE AND EVALUATE YOURSELF ACCORDINGLY

Above 85: *EXCELLENT*. You are amongst the top scorers, but strive to retain it because in agriculture there are great swings.

65 to 85: *VERY GOOD*. With about a similar performance in weak areas, you can become the top scorer. But be aware! there are many a slip between cup and lips.

55 to 65: *FAIR*. You still have a long way to go. Success is merely a chance factor.

Less than 55: *POOR*: Why are you even thinking of Agricultural Competitive Exam now. Work, Work and Work more. Who knows, your labour will pay off, as it does as a rule.

Notes

AGRISET - 14

ANSWER SHEET FOR GENERAL AGRICULTURE MODEL PAPER

- I have provided you MCQs model test on general agriculture for TOP (Target Oriented Preparation) for the Agricultural Competitive Exams.
- You have to mark all your responses only on the answer sheet provided.
- All items carry equal marks. Attempt all items. Your total marks will depend only on the number of correct responses marked by you in the Answer sheet. There is no negative marking. Award yourself one point for each correct answer.

ANSWER SHEET

1	ⓐⓑⓒⓓ	6	ⓐⓑⓒⓓ	11	ⓐⓑⓒⓓ	16	ⓐⓑⓒⓓ	21	ⓐⓑⓒⓓ
2	ⓐⓑⓒⓓ	7	ⓐⓑⓒⓓ	12	ⓐⓑⓒⓓ	17	ⓐⓑⓒⓓ	22	ⓐⓑⓒⓓ
3	ⓐⓑⓒⓓ	8	ⓐⓑⓒⓓ	13	ⓐⓑⓒⓓ	18	ⓐⓑⓒⓓ	23	ⓐⓑⓒⓓ
4	ⓐⓑⓒⓓ	9	ⓐⓑⓒⓓ	14	ⓐⓑⓒⓓ	19	ⓐⓑⓒⓓ	24	ⓐⓑⓒⓓ
5	ⓐⓑⓒⓓ	10	ⓐⓑⓒⓓ	15	ⓐⓑⓒⓓ	20	ⓐⓑⓒⓓ	25	ⓐⓑⓒⓓ
26	ⓐⓑⓒⓓ	31	ⓐⓑⓒⓓ	36	ⓐⓑⓒⓓ	41	ⓐⓑⓒⓓ	46	ⓐⓑⓒⓓ
27	ⓐⓑⓒⓓ	32	ⓐⓑⓒⓓ	37	ⓐⓑⓒⓓ	42	ⓐⓑⓒⓓ	47	ⓐⓑⓒⓓ
28	ⓐⓑⓒⓓ	33	ⓐⓑⓒⓓ	38	ⓐⓑⓒⓓ	43	ⓐⓑⓒⓓ	48	ⓐⓑⓒⓓ
29	ⓐⓑⓒⓓ	34	ⓐⓑⓒⓓ	39	ⓐⓑⓒⓓ	44	ⓐⓑⓒⓓ	49	ⓐⓑⓒⓓ
30	ⓐⓑⓒⓓ	35	ⓐⓑⓒⓓ	40	ⓐⓑⓒⓓ	45	ⓐⓑⓒⓓ	50	ⓐⓑⓒⓓ
51	ⓐⓑⓒⓓ	56	ⓐⓑⓒⓓ	61	ⓐⓑⓒⓓ	66	ⓐⓑⓒⓓ	71	ⓐⓑⓒⓓ
52	ⓐⓑⓒⓓ	57	ⓐⓑⓒⓓ	62	ⓐⓑⓒⓓ	67	ⓐⓑⓒⓓ	72	ⓐⓑⓒⓓ
53	ⓐⓑⓒⓓ	58	ⓐⓑⓒⓓ	63	ⓐⓑⓒⓓ	68	ⓐⓑⓒⓓ	73	ⓐⓑⓒⓓ
54	ⓐⓑⓒⓓ	59	ⓐⓑⓒⓓ	64	ⓐⓑⓒⓓ	69	ⓐⓑⓒⓓ	74	ⓐⓑⓒⓓ
55	ⓐⓑⓒⓓ	60	ⓐⓑⓒⓓ	65	ⓐⓑⓒⓓ	70	ⓐⓑⓒⓓ	75	ⓐⓑⓒⓓ
76	ⓐⓑⓒⓓ	81	ⓐⓑⓒⓓ	86	ⓐⓑⓒⓓ	91	ⓐⓑⓒⓓ	96	ⓐⓑⓒⓓ
77	ⓐⓑⓒⓓ	82	ⓐⓑⓒⓓ	87	ⓐⓑⓒⓓ	92	ⓐⓑⓒⓓ	97	ⓐⓑⓒⓓ
78	ⓐⓑⓒⓓ	83	ⓐⓑⓒⓓ	88	ⓐⓑⓒⓓ	93	ⓐⓑⓒⓓ	98	ⓐⓑⓒⓓ
79	ⓐⓑⓒⓓ	84	ⓐⓑⓒⓓ	89	ⓐⓑⓒⓓ	94	ⓐⓑⓒⓓ	99	ⓐⓑⓒⓓ
80	ⓐⓑⓒⓓ	85	ⓐⓑⓒⓓ	90	ⓐⓑⓒⓓ	95	ⓐⓑⓒⓓ	100	ⓐⓑⓒⓓ

MARK YOUR SCORE AND EVALUATE YOURSELF ACCORDINGLY

Above 85: *EXCELLENT*. You are amongst the top scorers, but strive to retain it because in agriculture there are great swings.

65 to 85: *VERY GOOD*. With about a similar performance in weak areas, you can become the top scorer. But be aware! there are many a slip between cup and lips.

55 to 65: *FAIR*. You still have a long way to go. Success is merely a chance factor.

Less than 55: *POOR*: Why are you even thinking of Agricultural Competitive Exam now. Work, Work and Work more. Who knows, your labour will pay off, as it does as a rule.

Notes

AGRISET - 15

ANSWER SHEET FOR GENERAL AGRICULTURE MODEL PAPER

- I have provided you MCQs model test on general agriculture for TOP (Target Oriented Preparation) for the Agricultural Competitive Exams.
- You have to mark all your responses only on the answer sheet provided.
- All items carry equal marks. Attempt all items. Your total marks will depend only on the number of correct responses marked by you in the Answer sheet. There is no negative marking. Award yourself one point for each correct answer.

ANSWER SHEET

No.	Options	No.	Options	No.	Options	No.	Options	No.	Options
1	ⓐⓑⓒⓓ	6	ⓐⓑⓒⓓ	11	ⓐⓑⓒⓓ	16	ⓐⓑⓒⓓ	21	ⓐⓑⓒⓓ
2	ⓐⓑⓒⓓ	7	ⓐⓑⓒⓓ	12	ⓐⓑⓒⓓ	17	ⓐⓑⓒⓓ	22	ⓐⓑⓒⓓ
3	ⓐⓑⓒⓓ	8	ⓐⓑⓒⓓ	13	ⓐⓑⓒⓓ	18	ⓐⓑⓒⓓ	23	ⓐⓑⓒⓓ
4	ⓐⓑⓒⓓ	9	ⓐⓑⓒⓓ	14	ⓐⓑⓒⓓ	19	ⓐⓑⓒⓓ	24	ⓐⓑⓒⓓ
5	ⓐⓑⓒⓓ	10	ⓐⓑⓒⓓ	15	ⓐⓑⓒⓓ	20	ⓐⓑⓒⓓ	25	ⓐⓑⓒⓓ
26	ⓐⓑⓒⓓ	31	ⓐⓑⓒⓓ	36	ⓐⓑⓒⓓ	41	ⓐⓑⓒⓓ	46	ⓐⓑⓒⓓ
27	ⓐⓑⓒⓓ	32	ⓐⓑⓒⓓ	37	ⓐⓑⓒⓓ	42	ⓐⓑⓒⓓ	47	ⓐⓑⓒⓓ
28	ⓐⓑⓒⓓ	33	ⓐⓑⓒⓓ	38	ⓐⓑⓒⓓ	43	ⓐⓑⓒⓓ	48	ⓐⓑⓒⓓ
29	ⓐⓑⓒⓓ	34	ⓐⓑⓒⓓ	39	ⓐⓑⓒⓓ	44	ⓐⓑⓒⓓ	49	ⓐⓑⓒⓓ
30	ⓐⓑⓒⓓ	35	ⓐⓑⓒⓓ	40	ⓐⓑⓒⓓ	45	ⓐⓑⓒⓓ	50	ⓐⓑⓒⓓ
51	ⓐⓑⓒⓓ	56	ⓐⓑⓒⓓ	61	ⓐⓑⓒⓓ	66	ⓐⓑⓒⓓ	71	ⓐⓑⓒⓓ
52	ⓐⓑⓒⓓ	57	ⓐⓑⓒⓓ	62	ⓐⓑⓒⓓ	67	ⓐⓑⓒⓓ	72	ⓐⓑⓒⓓ
53	ⓐⓑⓒⓓ	58	ⓐⓑⓒⓓ	63	ⓐⓑⓒⓓ	68	ⓐⓑⓒⓓ	73	ⓐⓑⓒⓓ
54	ⓐⓑⓒⓓ	59	ⓐⓑⓒⓓ	64	ⓐⓑⓒⓓ	69	ⓐⓑⓒⓓ	74	ⓐⓑⓒⓓ
55	ⓐⓑⓒⓓ	60	ⓐⓑⓒⓓ	65	ⓐⓑⓒⓓ	70	ⓐⓑⓒⓓ	75	ⓐⓑⓒⓓ
76	ⓐⓑⓒⓓ	81	ⓐⓑⓒⓓ	86	ⓐⓑⓒⓓ	91	ⓐⓑⓒⓓ	96	ⓐⓑⓒⓓ
77	ⓐⓑⓒⓓ	82	ⓐⓑⓒⓓ	87	ⓐⓑⓒⓓ	92	ⓐⓑⓒⓓ	97	ⓐⓑⓒⓓ
78	ⓐⓑⓒⓓ	83	ⓐⓑⓒⓓ	88	ⓐⓑⓒⓓ	93	ⓐⓑⓒⓓ	98	ⓐⓑⓒⓓ
79	ⓐⓑⓒⓓ	84	ⓐⓑⓒⓓ	89	ⓐⓑⓒⓓ	94	ⓐⓑⓒⓓ	99	ⓐⓑⓒⓓ
80	ⓐⓑⓒⓓ	85	ⓐⓑⓒⓓ	90	ⓐⓑⓒⓓ	95	ⓐⓑⓒⓓ	100	ⓐⓑⓒⓓ

MARK YOUR SCORE AND EVALUATE YOURSELF ACCORDINGLY

Above 85: *EXCELLENT*. You are amongst the top scorers, but strive to retain it because in agriculture there are great swings.

65 to 85: *VERY GOOD*. With about a similar performance in weak areas, you can become the top scorer. But be aware! there are many a slip between cup and lips.

55 to 65: *FAIR*. You still have a long way to go. Success is merely a chance factor.

Less than 55: *POOR*: Why are you even thinking of Agricultural Competitive Exam now. Work, Work and Work more. Who knows, your labour will pay off, as it does as a rule.

Notes

AGRISET - 16

ANSWER SHEET FOR GENERAL AGRICULTURE MODEL PAPER

- I have provided you MCQs model test on general agriculture for TOP (Target Oriented Preparation) for the Agricultural Competitive Exams.
- You have to mark all your responses only on the answer sheet provided.
- All items carry equal marks. Attempt all items. Your total marks will depend only on the number of correct responses marked by you in the Answer sheet. There is no negative marking. Award yourself one point for each correct answer.

ANSWER SHEET

1	ⓐⓑⓒⓓ	6	ⓐⓑⓒⓓ	11	ⓐⓑⓒⓓ	16	ⓐⓑⓒⓓ	21	ⓐⓑⓒⓓ
2	ⓐⓑⓒⓓ	7	ⓐⓑⓒⓓ	12	ⓐⓑⓒⓓ	17	ⓐⓑⓒⓓ	22	ⓐⓑⓒⓓ
3	ⓐⓑⓒⓓ	8	ⓐⓑⓒⓓ	13	ⓐⓑⓒⓓ	18	ⓐⓑⓒⓓ	23	ⓐⓑⓒⓓ
4	ⓐⓑⓒⓓ	9	ⓐⓑⓒⓓ	14	ⓐⓑⓒⓓ	19	ⓐⓑⓒⓓ	24	ⓐⓑⓒⓓ
5	ⓐⓑⓒⓓ	10	ⓐⓑⓒⓓ	15	ⓐⓑⓒⓓ	20	ⓐⓑⓒⓓ	25	ⓐⓑⓒⓓ
26	ⓐⓑⓒⓓ	31	ⓐⓑⓒⓓ	36	ⓐⓑⓒⓓ	41	ⓐⓑⓒⓓ	46	ⓐⓑⓒⓓ
27	ⓐⓑⓒⓓ	32	ⓐⓑⓒⓓ	37	ⓐⓑⓒⓓ	42	ⓐⓑⓒⓓ	47	ⓐⓑⓒⓓ
28	ⓐⓑⓒⓓ	33	ⓐⓑⓒⓓ	38	ⓐⓑⓒⓓ	43	ⓐⓑⓒⓓ	48	ⓐⓑⓒⓓ
29	ⓐⓑⓒⓓ	34	ⓐⓑⓒⓓ	39	ⓐⓑⓒⓓ	44	ⓐⓑⓒⓓ	49	ⓐⓑⓒⓓ
30	ⓐⓑⓒⓓ	35	ⓐⓑⓒⓓ	40	ⓐⓑⓒⓓ	45	ⓐⓑⓒⓓ	50	ⓐⓑⓒⓓ
51	ⓐⓑⓒⓓ	56	ⓐⓑⓒⓓ	61	ⓐⓑⓒⓓ	66	ⓐⓑⓒⓓ	71	ⓐⓑⓒⓓ
52	ⓐⓑⓒⓓ	57	ⓐⓑⓒⓓ	62	ⓐⓑⓒⓓ	67	ⓐⓑⓒⓓ	72	ⓐⓑⓒⓓ
53	ⓐⓑⓒⓓ	58	ⓐⓑⓒⓓ	63	ⓐⓑⓒⓓ	68	ⓐⓑⓒⓓ	73	ⓐⓑⓒⓓ
54	ⓐⓑⓒⓓ	59	ⓐⓑⓒⓓ	64	ⓐⓑⓒⓓ	69	ⓐⓑⓒⓓ	74	ⓐⓑⓒⓓ
55	ⓐⓑⓒⓓ	60	ⓐⓑⓒⓓ	65	ⓐⓑⓒⓓ	70	ⓐⓑⓒⓓ	75	ⓐⓑⓒⓓ
76	ⓐⓑⓒⓓ	81	ⓐⓑⓒⓓ	86	ⓐⓑⓒⓓ	91	ⓐⓑⓒⓓ	96	ⓐⓑⓒⓓ
77	ⓐⓑⓒⓓ	82	ⓐⓑⓒⓓ	87	ⓐⓑⓒⓓ	92	ⓐⓑⓒⓓ	97	ⓐⓑⓒⓓ
78	ⓐⓑⓒⓓ	83	ⓐⓑⓒⓓ	88	ⓐⓑⓒⓓ	93	ⓐⓑⓒⓓ	98	ⓐⓑⓒⓓ
79	ⓐⓑⓒⓓ	84	ⓐⓑⓒⓓ	89	ⓐⓑⓒⓓ	94	ⓐⓑⓒⓓ	99	ⓐⓑⓒⓓ
80	ⓐⓑⓒⓓ	85	ⓐⓑⓒⓓ	90	ⓐⓑⓒⓓ	95	ⓐⓑⓒⓓ	100	ⓐⓑⓒⓓ

MARK YOUR SCORE AND EVALUATE YOURSELF ACCORDINGLY

Above 85: *EXCELLENT*. You are amongst the top scorers, but strive to retain it because in agriculture there are great swings.

65 to 85: *VERY GOOD*. With about a similar performance in weak areas, you can become the top scorer. But be aware! there are many a slip between cup and lips.

55 to 65: *FAIR*. You still have a long way to go. Success is merely a chance factor.

Less than 55: *POOR*: Why are you even thinking of Agricultural Competitive Exam now. Work, Work and Work more. Who knows, your labour will pay off, as it does as a rule.

Notes

AGRISET - 17

ANSWER SHEET FOR GENERAL AGRICULTURE MODEL PAPER

- I have provided you MCQs model test on general agriculture for TOP (Target Oriented Preparation) for the Agricultural Competitive Exams.
- You have to mark all your responses only on the answer sheet provided.
- All items carry equal marks. Attempt all items. Your total marks will depend only on the number of correct responses marked by you in the Answer sheet. There is no negative marking. Award yourself one point for each correct answer.

ANSWER SHEET

1	ⓐⓑⓒⓓ	6	ⓐⓑⓒⓓ	11	ⓐⓑⓒⓓ	16	ⓐⓑⓒⓓ	21	ⓐⓑⓒⓓ
2	ⓐⓑⓒⓓ	7	ⓐⓑⓒⓓ	12	ⓐⓑⓒⓓ	17	ⓐⓑⓒⓓ	22	ⓐⓑⓒⓓ
3	ⓐⓑⓒⓓ	8	ⓐⓑⓒⓓ	13	ⓐⓑⓒⓓ	18	ⓐⓑⓒⓓ	23	ⓐⓑⓒⓓ
4	ⓐⓑⓒⓓ	9	ⓐⓑⓒⓓ	14	ⓐⓑⓒⓓ	19	ⓐⓑⓒⓓ	24	ⓐⓑⓒⓓ
5	ⓐⓑⓒⓓ	10	ⓐⓑⓒⓓ	15	ⓐⓑⓒⓓ	20	ⓐⓑⓒⓓ	25	ⓐⓑⓒⓓ
26	ⓐⓑⓒⓓ	31	ⓐⓑⓒⓓ	36	ⓐⓑⓒⓓ	41	ⓐⓑⓒⓓ	46	ⓐⓑⓒⓓ
27	ⓐⓑⓒⓓ	32	ⓐⓑⓒⓓ	37	ⓐⓑⓒⓓ	42	ⓐⓑⓒⓓ	47	ⓐⓑⓒⓓ
28	ⓐⓑⓒⓓ	33	ⓐⓑⓒⓓ	38	ⓐⓑⓒⓓ	43	ⓐⓑⓒⓓ	48	ⓐⓑⓒⓓ
29	ⓐⓑⓒⓓ	34	ⓐⓑⓒⓓ	39	ⓐⓑⓒⓓ	44	ⓐⓑⓒⓓ	49	ⓐⓑⓒⓓ
30	ⓐⓑⓒⓓ	35	ⓐⓑⓒⓓ	40	ⓐⓑⓒⓓ	45	ⓐⓑⓒⓓ	50	ⓐⓑⓒⓓ
51	ⓐⓑⓒⓓ	56	ⓐⓑⓒⓓ	61	ⓐⓑⓒⓓ	66	ⓐⓑⓒⓓ	71	ⓐⓑⓒⓓ
52	ⓐⓑⓒⓓ	57	ⓐⓑⓒⓓ	62	ⓐⓑⓒⓓ	67	ⓐⓑⓒⓓ	72	ⓐⓑⓒⓓ
53	ⓐⓑⓒⓓ	58	ⓐⓑⓒⓓ	63	ⓐⓑⓒⓓ	68	ⓐⓑⓒⓓ	73	ⓐⓑⓒⓓ
54	ⓐⓑⓒⓓ	59	ⓐⓑⓒⓓ	64	ⓐⓑⓒⓓ	69	ⓐⓑⓒⓓ	74	ⓐⓑⓒⓓ
55	ⓐⓑⓒⓓ	60	ⓐⓑⓒⓓ	65	ⓐⓑⓒⓓ	70	ⓐⓑⓒⓓ	75	ⓐⓑⓒⓓ
76	ⓐⓑⓒⓓ	81	ⓐⓑⓒⓓ	86	ⓐⓑⓒⓓ	91	ⓐⓑⓒⓓ	96	ⓐⓑⓒⓓ
77	ⓐⓑⓒⓓ	82	ⓐⓑⓒⓓ	87	ⓐⓑⓒⓓ	92	ⓐⓑⓒⓓ	97	ⓐⓑⓒⓓ
78	ⓐⓑⓒⓓ	83	ⓐⓑⓒⓓ	88	ⓐⓑⓒⓓ	93	ⓐⓑⓒⓓ	98	ⓐⓑⓒⓓ
79	ⓐⓑⓒⓓ	84	ⓐⓑⓒⓓ	89	ⓐⓑⓒⓓ	94	ⓐⓑⓒⓓ	99	ⓐⓑⓒⓓ
80	ⓐⓑⓒⓓ	85	ⓐⓑⓒⓓ	90	ⓐⓑⓒⓓ	95	ⓐⓑⓒⓓ	100	ⓐⓑⓒⓓ

MARK YOUR SCORE AND EVALUATE YOURSELF ACCORDINGLY

Above 85: *EXCELLENT*. You are amongst the top scorers, but strive to retain it because in agriculture there are great swings.

65 to 85: *VERY GOOD*. With about a similar performance in weak areas, you can become the top scorer. But be aware! there are many a slip between cup and lips.

55 to 65: *FAIR*. You still have a long way to go. Success is merely a chance factor.

Less than 55: *POOR*: Why are you even thinking of Agricultural Competitive Exam now. Work, Work and Work more. Who knows, your labour will pay off, as it does as a rule.

Notes

AGRISET - 18

ANSWER SHEET FOR GENERAL AGRICULTURE MODEL PAPER

- I have provided you MCQs model test on general agriculture for TOP (Target Oriented Preparation) for the Agricultural Competitive Exams.
- You have to mark all your responses only on the answer sheet provided.
- All items carry equal marks. Attempt all items. Your total marks will depend only on the number of correct responses marked by you in the Answer sheet. There is no negative marking. Award yourself one point for each correct answer.

ANSWER SHEET

1	ⓐⓑⓒⓓ	6	ⓐⓑⓒⓓ	11	ⓐⓑⓒⓓ	16	ⓐⓑⓒⓓ	21	ⓐⓑⓒⓓ
2	ⓐⓑⓒⓓ	7	ⓐⓑⓒⓓ	12	ⓐⓑⓒⓓ	17	ⓐⓑⓒⓓ	22	ⓐⓑⓒⓓ
3	ⓐⓑⓒⓓ	8	ⓐⓑⓒⓓ	13	ⓐⓑⓒⓓ	18	ⓐⓑⓒⓓ	23	ⓐⓑⓒⓓ
4	ⓐⓑⓒⓓ	9	ⓐⓑⓒⓓ	14	ⓐⓑⓒⓓ	19	ⓐⓑⓒⓓ	24	ⓐⓑⓒⓓ
5	ⓐⓑⓒⓓ	10	ⓐⓑⓒⓓ	15	ⓐⓑⓒⓓ	20	ⓐⓑⓒⓓ	25	ⓐⓑⓒⓓ
26	ⓐⓑⓒⓓ	31	ⓐⓑⓒⓓ	36	ⓐⓑⓒⓓ	41	ⓐⓑⓒⓓ	46	ⓐⓑⓒⓓ
27	ⓐⓑⓒⓓ	32	ⓐⓑⓒⓓ	37	ⓐⓑⓒⓓ	42	ⓐⓑⓒⓓ	47	ⓐⓑⓒⓓ
28	ⓐⓑⓒⓓ	33	ⓐⓑⓒⓓ	38	ⓐⓑⓒⓓ	43	ⓐⓑⓒⓓ	48	ⓐⓑⓒⓓ
29	ⓐⓑⓒⓓ	34	ⓐⓑⓒⓓ	39	ⓐⓑⓒⓓ	44	ⓐⓑⓒⓓ	49	ⓐⓑⓒⓓ
30	ⓐⓑⓒⓓ	35	ⓐⓑⓒⓓ	40	ⓐⓑⓒⓓ	45	ⓐⓑⓒⓓ	50	ⓐⓑⓒⓓ
51	ⓐⓑⓒⓓ	56	ⓐⓑⓒⓓ	61	ⓐⓑⓒⓓ	66	ⓐⓑⓒⓓ	71	ⓐⓑⓒⓓ
52	ⓐⓑⓒⓓ	57	ⓐⓑⓒⓓ	62	ⓐⓑⓒⓓ	67	ⓐⓑⓒⓓ	72	ⓐⓑⓒⓓ
53	ⓐⓑⓒⓓ	58	ⓐⓑⓒⓓ	63	ⓐⓑⓒⓓ	68	ⓐⓑⓒⓓ	73	ⓐⓑⓒⓓ
54	ⓐⓑⓒⓓ	59	ⓐⓑⓒⓓ	64	ⓐⓑⓒⓓ	69	ⓐⓑⓒⓓ	74	ⓐⓑⓒⓓ
55	ⓐⓑⓒⓓ	60	ⓐⓑⓒⓓ	65	ⓐⓑⓒⓓ	70	ⓐⓑⓒⓓ	75	ⓐⓑⓒⓓ
76	ⓐⓑⓒⓓ	81	ⓐⓑⓒⓓ	86	ⓐⓑⓒⓓ	91	ⓐⓑⓒⓓ	96	ⓐⓑⓒⓓ
77	ⓐⓑⓒⓓ	82	ⓐⓑⓒⓓ	87	ⓐⓑⓒⓓ	92	ⓐⓑⓒⓓ	97	ⓐⓑⓒⓓ
78	ⓐⓑⓒⓓ	83	ⓐⓑⓒⓓ	88	ⓐⓑⓒⓓ	93	ⓐⓑⓒⓓ	98	ⓐⓑⓒⓓ
79	ⓐⓑⓒⓓ	84	ⓐⓑⓒⓓ	89	ⓐⓑⓒⓓ	94	ⓐⓑⓒⓓ	99	ⓐⓑⓒⓓ
80	ⓐⓑⓒⓓ	85	ⓐⓑⓒⓓ	90	ⓐⓑⓒⓓ	95	ⓐⓑⓒⓓ	100	ⓐⓑⓒⓓ

MARK YOUR SCORE AND EVALUATE YOURSELF ACCORDINGLY

Above 85: *EXCELLENT*. You are amongst the top scorers, but strive to retain it because in agriculture there are great swings.

65 to 85: *VERY GOOD*. With about a similar performance in weak areas, you can become the top scorer. But be aware! there are many a slip between cup and lips.

55 to 65: *FAIR*. You still have a long way to go. Success is merely a chance factor.

Less than 55: *POOR*: Why are you even thinking of Agricultural Competitive Exam now. Work, Work and Work more. Who knows, your labour will pay off, as it does as a rule.

Notes

Notes

Notes